SINGLE VARIABLE CALCULUS

Early Transcendentals

Sixth Edition

C. HENRY EDWARDS
The University of Georgia, Athens

DAVID E. PENNEY
The University of Georgia, Athens

Prentice Hall

Prentice Hall, Upper Saddle River, New Jersey 07458

Library of Congress Cataloging-in-Publication Data

Edwards, C. H. (Charles Henry)
[Calculus: early transcendentals. Selections]
Single variable calculus: early transcendentals / C. Henry Edwards, David E. Penney—6th ed.
p. cm.
Selections from Calculus: early transcendentals. 6th ed. 2002.
Includes bibliographical references and index.
ISBN 0-13-041407-7
1. Calculus. I. Title.

QA303.2 .E38 2002
515—dc21 2001052355

Acquisition Editor: George Lobell
Editor in Chief: Sally Yagan
Vice President/Director of Production and Manufacturing: David W. Riccardi
Executive Managing Editor: Kathleen Schiaparelli
Senior Managing Editor: Linda Mihatov Behrens
Production Editor: Barbara Mack
Assistant Managing Editor, Math Media Production: John Matthews
Media Production Editors: Donna Crilly, Wendy Perez
Manufacturing Buyer: Alan Fischer
Manufacturing Manager: Trudy Pisciotti
Marketing Manager: Angela Battle
Marketing Assistant: Rachel Beckman
Assistant Editor of Media: Vince Jansen
Editorial Assistant/Supplements Editor: Melanie Van Benthuysen
Art Director: Jonathan Boylan
Assistant to the Art Director: John Christiana
Interior Designer: Donna Young
Cover Designer: Bruce Kenselaar
Art Editor: Thomas Benfatti
Managing Editor, Audio/Video Assets: Grace Hazeldine
Creative Director: Carole Anson
Director of Creative Services: Paul Belfanti
Photo Researcher: Karen Pugliano
Photo Editor: Beth Boyd
Cover Photo: Antonio Martinelli/Arche De La Defense, Sunset View–Paris, Arch. J. O. Spreckelsen-P. Andreu
Art Studio: Network Graphics

Prentice Hall

Printed in the United States of America
10 9 8 7 6 5 4 3 2 1

ISBN 0-13-041407-7

Pearson Education LTD., *London*
Pearson Education Australia PTY, Limited, *Sydney*
Pearson Education Singapore, Pte. Ltd
Pearson Education North Asia Ltd, *Hong Kong*
Pearson Education Canada, Ltd., *Toronto*
Pearson Educacion de Mexico, S.A. de C.V.
Pearson Education–Japan, *Tokyo*
Pearson Education Malaysia, Pte. Ltd

CONTENTS

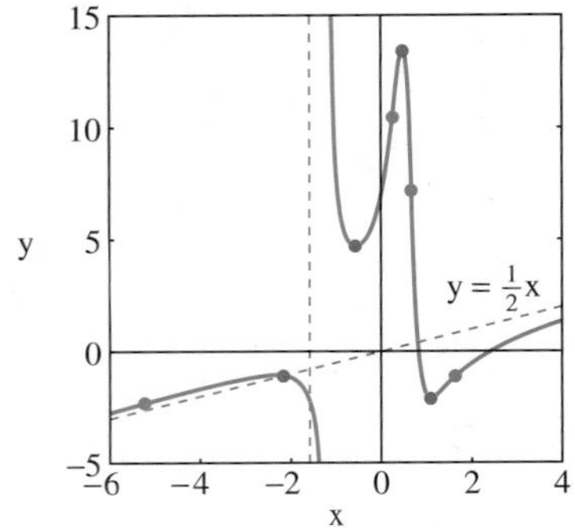

15
10
5
0
−5
y
−6
−4
−2
0
2
4
x
$y = \frac{1}{2}x$

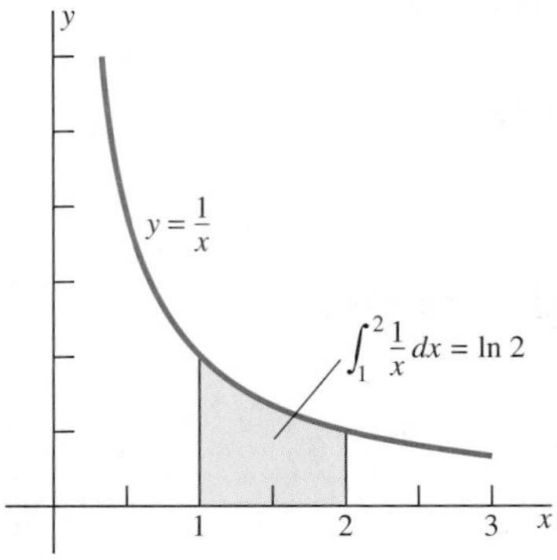

y
$y = \frac{1}{x}$
$\int_1^2 \frac{1}{x}\,dx = \ln 2$
1
2
3
x

CHAPTER 6 APPLICATIONS OF THE INTEGRAL 397

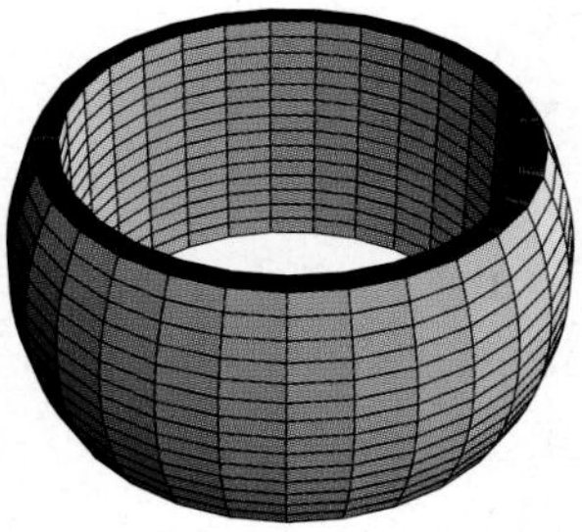

CHAPTER 7 TECHNIQUES OF INTEGRATION 491

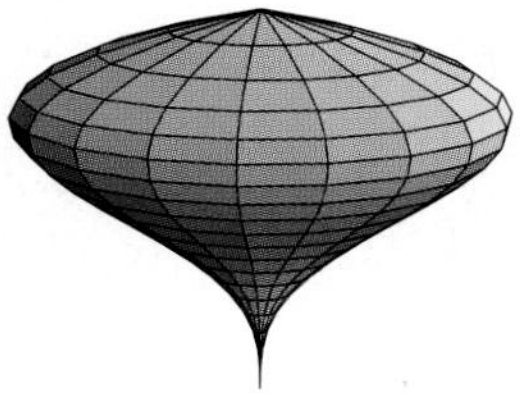

CHAPTER 8 DIFFERENTIAL EQUATIONS 547

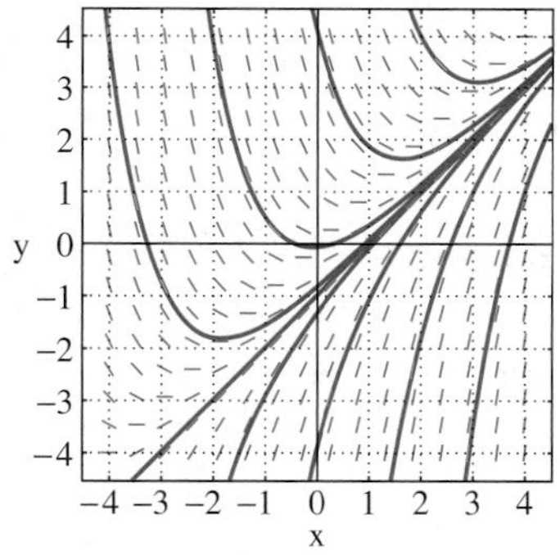

CHAPTER 9 POLAR COORDINATES AND PARAMETRIC CURVES 625

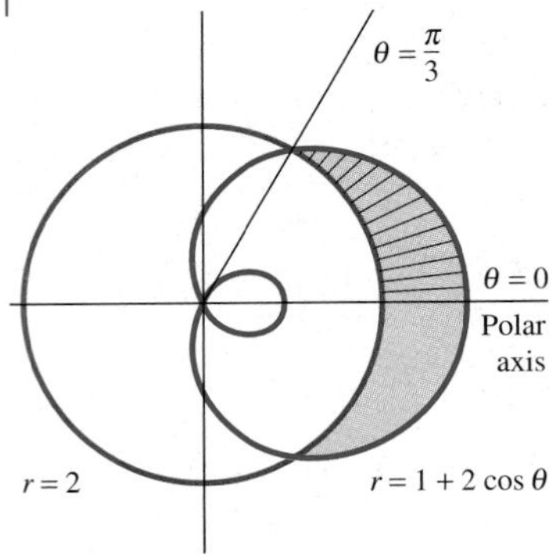

CHAPTER 10 INFINITE SERIES 683

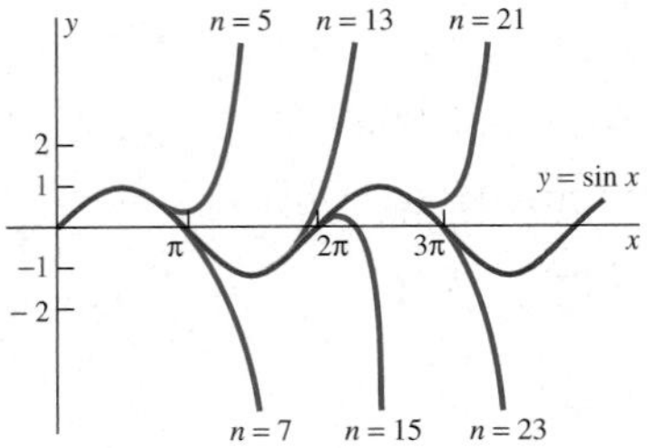

CHAPTER 11 VECTORS AND MATRICES 773

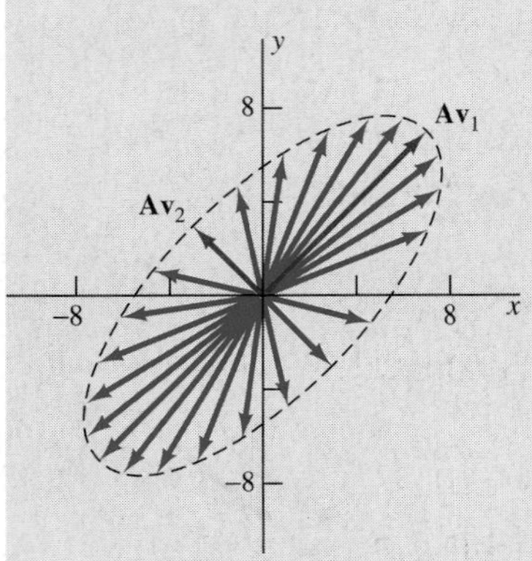

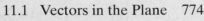

The shaded contents do not appear in Single Variable Calculus.

SINGLE VARIABLE CALCULUS

Early Transcendentals

CHAPTER 12 CURVES AND SURFACES IN SPACE 845

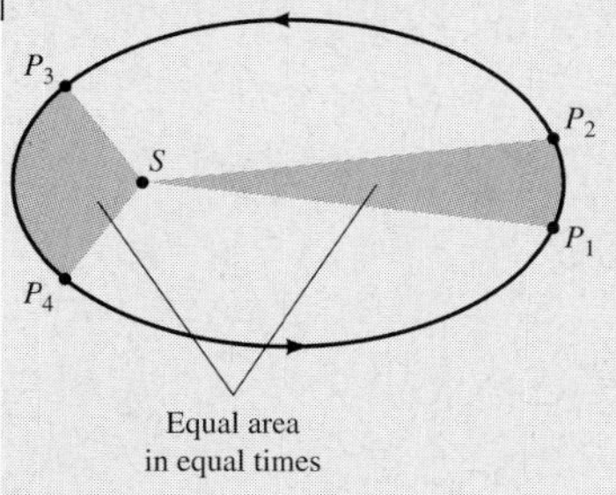

CHAPTER 13 PARTIAL DIFFERENTIATION 893

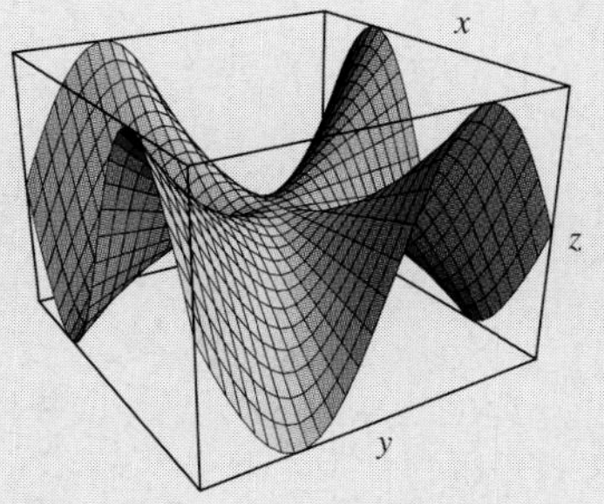

CHAPTER 14 MULTIPLE INTEGRALS 993

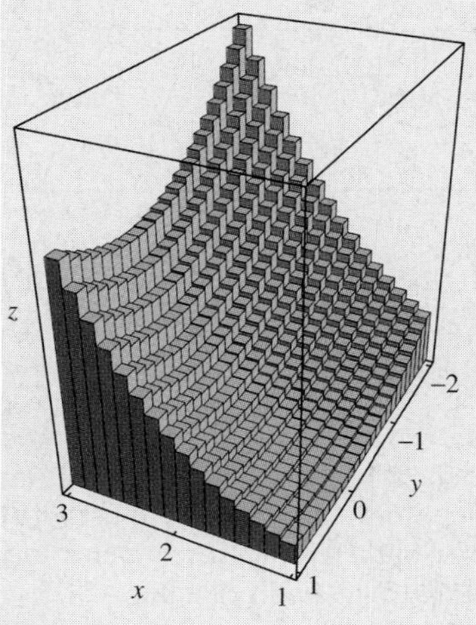

CHAPTER 15 VECTOR CALCULUS 1067

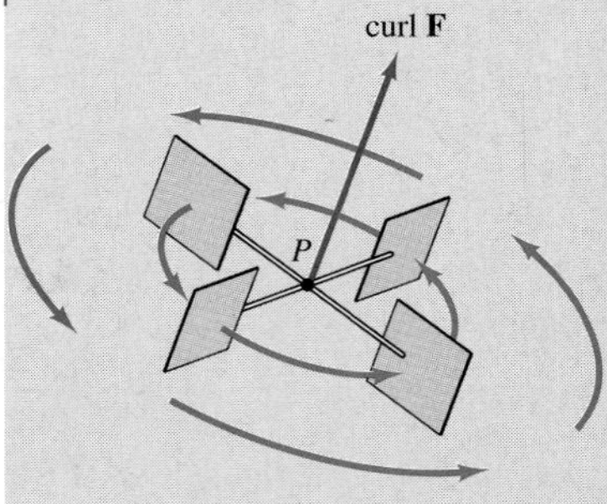

PHOTO CREDITS p. 1. CORBIS p. 53 (top left) Hulton/Archive; (bottom left) IBM Corporation; (right) Navy Visual News Service p. 101 Wards Science/Science Source p. 162 Richard Megna/Fundamental Photographs p. 217 (top left) The Royal Society of London; (bottom right) David E. Penney p. 301 Art Resource, N.Y. p. 312 Richard Megna/Fundamental Photographs p. 397 The Granger Collection p. 439 Lucas Bruno/AP/Wide World Photos p. 442 Omnia/Hulton/Archive p. 491 (top left) Library of Congress; (bottom right) David E. Penney p. 547 CORBIS p. 625 (top left) Stock Montage, Inc./Historical Pictures Collection; (bottom right) Stephen Gerard/Science Source/Photo Researchers, Inc. p. 683 (top left) University of Cambridge; (bottom right) Image by David E. Penney

ABOUT THE AUTHORS

C. Henry Edwards is emeritus professor of mathematics at the University of Georgia. He earned his Ph.D. at the University of Tennessee in 1960, and recently retired after 40 years of classroom teaching (including calculus or differential equations almost every term) at the universities of Tennessee, Wisconsin, and Georgia, with a brief interlude at the Institute for Advanced Study (Princeton) as an Alfred P. Sloan Research Fellow. He has received numerous teaching awards, including the University of Georgia's *honoratus* medal in 1983 (for sustained excellence in honors teaching), its Josiah Meigs award in 1991 (the institution's highest award for teaching), and the 1997 state-wide Georgia Regents award for research university faculty teaching excellence. His scholarly career has ranged from research and dissertation direction in topology to the history of mathematics to computing and technology in the teaching and applications of mathematics. In addition to being author or co-author of calculus, advanced calculus, linear algebra, and differential equations textbooks, he is well-known to calculus instructors as author of *The Historical Development of the Calculus* (Springer-Verlag, 1979). During the 1990s he served as a principal investigator on three NSF-supported projects: (1) A school mathematics project including Maple for beginning algebra students, (2) A Calculus-with-*Mathematica* program, and (3) A MATLAB-based computer lab project for numerical analysis and differential equations students.

David E. Penney, University of Georgia, completed his Ph.D. at Tulane University in 1965 (under the direction of Prof. L. Bruce Treybig) while teaching at the University of New Orleans. Earlier he had worked in experimental biophysics at Tulane University and the Veteran's Administration Hospital in New Orleans under the direction of Robert Dixon McAfee, where Dr. McAfee's research team's primary focus was on the active transport of sodium ions by biological membranes. Penney's primary contribution here was the development of a mathematical model (using simultaneous ordinary differential equations) for the metabolic phenomena regulating such transport, with potential future applications in kidney physiology, management of hypertension, and treatment of congestive heart failure. He also designed and constructed servomechanisms for the accurate monitoring of ion transport, a phenomenon involving the measurement of potentials in microvolts at impedances of millions of megohms. Penney began teaching calculus at Tulane in 1957 and taught that course almost every term with enthusiasm and distinction until his retirement at the end of the last millennium. During his tenure at the University of Georgia he received numerous University-wide teaching awards as well as directing several doctoral dissertations and seven undergraduate research projects. He is the author of research papers in number theory and topology and is the author or co-author of textbooks on calculus, computer programming, differential equations, linear algebra, and liberal arts mathematics.

PREFACE

Contemporary calculus instructors and students face traditional challenges as well as new ones that result from changes in the role and practice of mathematics by scientists and engineers in the world at large. As a consequence, this sixth edition of our calculus textbook is its most extensive revision since the first edition appeared in 1982.

Two entire chapters of the fifth edition have been replaced in the table of contents by two new ones; most of the remaining chapters have been extensively rewritten. Nearly 160 of the book's over 800 worked examples are new for this edition and the 1850 figures in the text include 250 new computer-generated graphics. Almost 800 of its 7250 problems are new, and these are augmented by over 330 new conceptual discussion questions that now precede the problem sets. Moreover, almost 1100 new true/false questions are included in the Study Guides on the new CD-ROM that accompanies this edition. In summary, almost 2200 of these 8650-plus problems and questions are new, and the text discussion and explanations have undergone corresponding alteration and improvement.

PRINCIPAL NEW FEATURES

The current revision of the text features

- **Early transcendentals** fully integrated in Semester I.
- **Differential equations** and applications in Semester II.
- **Linear systems and matrices** in Semester III.

Complete coverage of the calculus of transcendental functions is now fully integrated in Chapters 1 through 6—with the result that the Chapter 7 and 8 titles in the 5th edition table of contents do not appear in this 6th edition.

A new chapter on differential equations (Chapter 8) now appears immediately after Chapter 7 on techniques of integration. It includes both direction fields and Euler's method together with the more elementary symbolic methods (which exploit techniques from Chapter 7) and interesting applications of both first- and second-order equations. Chapter 10 (Infinite Series) now ends with a new section on power series solutions of differential equations, thus bringing full circle a unifying focus of second-semester calculus on elementary differential equations.

Linear systems and matrices, ending with an elementary treatment of eigenvalues and eigenvectors, are now introduced in Chapter 11. The subsequent coverage of multivariable calculus now integrates matrix methods and terminology with the traditional notation and approach—including (for instance) introduction and extensive application of the chain rule in matrix-product form.

NEW LEARNING RESOURCES

Conceptual Discussion Questions

The set of problems that concludes each section is now preceded by a brief **Concepts: Questions and Discussion** set consisting of several open-ended conceptual questions that can be used for either individual study or classroom discussion.

The Text CD-ROM

The content of the new CD-ROM that accompanies this text is fully integrated with the textbook material, and is designed specifically for use hand-in-hand with study of the book itself. This CD-ROM features the following resources to support learning and teaching:

- **Interactive True/False Study Guides** that reinforce and encourage student reading of the text. Ten author-written questions for each section carefully guide students through the section, and students can request individual hints suggesting where in the section to look for needed information.
- **Live Examples** feature dynamic multimedia and computer algebra presentations—many accompanied by audio explanations—which enhance student intuition and understanding. These interactive examples expand upon many of the textbook's principal examples; students can change input data and conditions and then observe the resulting changes in step-by-step solutions and accompanying graphs and figures. **Walkthrough videos** demonstrate how students can interact with these live examples.
- **Homework Starters** for the principal types of computational problems in each textbook section, featuring both interactive presentations similar to the live examples and (web-linked) voice-narrated videos of pencil-and-paper investigations illustrating typical initial steps in the solution of selected textbook problems.
- **Computing Project Resources** support most of the over three dozen projects that follow key sections in the text. For each such project marked in the text by a CD-ROM icon, more extended discussions illustrating Maple, *Mathematica,* MATLAB, and graphing calculator investigations are provided. Computer algebra system commands can be copied and pasted for interactive execution.
- **Hyperlinked Maple Worksheets** contributed by Harald Pleym of Telemark University College (Norway) constitute an interactive version of essentially the whole textbook. Students and faculty using Maple can change input data and conditions in most of the text examples to investigate the resulting changes in step-by-step solutions and accompanying graphs and figures.
- **PowerPoint Presentations** provide classroom projection versions of about 350 of the figures in the text that would be least convenient to reproduce on a blackboard.
- **Website** The contents of the CD-ROM together with additional learning and teaching resources are maintained and updated at the textbook website **www.prenhall.com/edwards,** which includes a Comments and Suggestions center where we invite response from both students and instructors.

Computerized Homework Grading System

About 2000 of the textbook problems are incorporated in an automated grading system that is now available. Each problem solution in the system is structured algorithmically so that students can work in a computer lab setting to submit homework assignments for automatic grading. (There is a small annual fee per participating student.)

New Solutions Manuals

The entirely new 1900-page **Instructor's Solutions Manual** (available in two volumes) includes a detailed solution for every problem in the book. These solutions were written exclusively by the authors and have been checked independently by others.

The entirely new 950-page **Student Solutions Manual** (available in two volumes) includes a detailed solution for every odd-numbered problem in the text. The answers (alone) to most of these odd-numbered problems are included in the answers section at the back of this book.

New Technology Manuals Each of the following manuals is available shrink-wrapped with any version of the text for half the normal price of the manual (all of which are inexpensive):

- Jensen, *Using MATLAB in Calculus* (0-13-027268-X)
- Freese/Stegenga, *Calculus Concepts Using Derive* (0-13-085152-3)
- Gresser, *TI Graphing Calculator Approach, 2e* (0-13-092017-7)
- Gresser, *A Mathematica Approach, 2e* (0-13-092015-0)
- Gresser, *A Maple Approach, 2e* (0-13-092014-2)

THE TEXT IN MORE DETAIL . . .

In preparing this edition, we have taken advantage of many valuable comments and suggestions from users of the first five editions. This revision was so pervasive that the individual changes are too numerous to be detailed in a preface, but the following paragraphs summarize those that may be of widest interest.

▼ **New Problems** Most of the almost 800 new problems lie in the intermediate range of difficulty, neither highly theoretical nor computationally routine. Many of them have a new technology flavor, suggesting (if not requiring) the use of technology ranging from a graphing calculator to a computer algebra system.

▼ **Discussion Questions and Study Guides** We hope the 330 conceptual discussion questions and 1080 true/false study-guide questions constitute a useful addition to the traditional fare of student exercises and problems. The True/False Study Guide for each section provides a focus on the key ideas of the section, with the single goal of motivating guided student reading of the section.

▼ **Examples and Explanations** About 20% of the book's worked examples are either new or significantly revised, together with a similar percentage of the text discussion and explanations. Additional computational detail has been inserted in worked examples where students have experienced difficulty, together with additional sentences and paragraphs in similar spots in text discussions.

▼ **Project Material** Many of the text's almost 40 projects are new for this edition. These appear following the problem sets at the ends of key sections throughout the text. Most (but not all) of these projects employ some aspect of modern computational technology to illustrate the principal ideas of the preceding section, and many contain additional problems intended for solution with the use of a graphing calculator or computer algebra system. Where appropriate, project discussions are significantly expanded in the CD-ROM versions of the projects.

▼ **Historical Material** Historical and biographical chapter openings offer students a sense of the development of our subject by real human beings. Indeed, our exposition of calculus frequently reflects the historical development of the subject—from ancient times to the ages of Newton and Leibniz and Euler to our own era of new computational power and technology.

TEXT ORGANIZATION

▼ **Introductory Chapters** Instead of a routine review of precalculus topics, Chapter 1 concentrates specifically on functions and graphs for use in mathematical modeling. It includes a section cataloging informally the elementary transcendental functions of calculus, as background to their more formal treatment using calculus itself. Chapter 1 concludes with a section addressing the question "What *is* calculus?" Chapter 2 on limits begins with a section on tangent lines to motivate the official introduction of limits in Section 2.2. Trigonometric limits are treated throughout Chapter 2 in order to encourage a richer and more visual introduction to the limit concept.

▼ **Differentiation Chapters** The sequence of topics in Chapters 3 and 4 differs a bit from the most traditional order. We attempt to build student confidence by introducing topics more nearly in order of increasing difficulty. The chain rule appears quite early (in Section 3.3) and we cover the basic techniques for differentiating algebraic functions before discussing maxima and minima in Sections 3.5 and 3.6. Section 3.7 treats the derivatives of all six trigonometric functions, and Section 3.8 (much strengthened for this edition) introduces the exponential and logarithmic functions. Implicit differentiation and related rates are combined in a single section (Section 3.9). The authors' fondness for Newton's method (Section 3.10) will be apparent.

The mean value theorem and its applications are deferred to Chapter 4. In addition, a dominant theme of Chapter 4 is the use of calculus both to construct graphs of functions and to explain and interpret graphs that have been constructed by a calculator or computer. This theme is developed in Sections 4.4 on the first derivative test and 4.6 on higher derivatives and concavity. But it may also be apparent in Sections 4.8 and 4.9 on l'Hôpital's rule, which now appears squarely in the context of differential calculus and is applied here to round out the calculus of exponential and logarithmic functions.

▼ **Integration Chapters** Chapter 5 begins with a section on antiderivatives—which could logically be included in the preceding chapter, but benefits from the use of integral notation. When the definite integral is introduced in Sections 5.3 and 5.4, we emphasize endpoint and midpoint sums rather than upper and lower and more general Riemann sums. This concrete emphasis carries through the chapter to its final section on numerical integration.

Chapter 6 begins with a largely new section on Riemann sum approximations, with new examples centering on fluid flow and medical applications. Section 6.6 is a new treatment of centroids of plane regions and curves. Section 6.7 gives the integral approach to logarithms, and Sections 6.8 and 6.9 cover both the differential and the integral calculus of inverse trigonometric functions and of hyperbolic functions.

Chapter 7 (Techniques of Integration) is organized to accommodate those instructors who feel that methods of formal integration now require less emphasis, in view of modern techniques for both numerical and symbolic integration. Integration by parts (Section 7.3) precedes trigonometric integrals (Section 7.4). The method of partial fractions appears in Section 7.5, and trigonometric substitutions and integrals involving quadratic polynomials follow in Sections 7.6 and 7.7. Improper integrals appear in Section 7.8, with new and substantial subsections on special functions and probability and random sampling. This rearrangement of Chapter 7 makes it more convenient to stop wherever the instructor desires.

▼ **Differential Equations** This entirely new chapter begins with the most elementary differential equations and applications (Section 8.1) and then proceeds to introduce both graphical (slope field) and numerical (Euler) methods in Section 8.2. Subsequent sections of the chapter treat separable and linear first-order differential equations and (in more depth than usual in a calculus course) applications such as population growth (including logistic and predator-prey populations) and motion with resistance. The final two sections of Chapter 8 treat second-order linear equations and applications to mechanical vibrations. Instructors desiring still more coverage of differential equations can arrange with the publisher to bundle and use appropriate sections of Edwards and Penney, **Differential Equations: Computing and Modeling 2/e** (Prentice Hall, 2000).

▼ **Parametric Curves and Polar Coordinates** The principal change in Chapter 9 is the replacement of three separate sections in the 5th edition on parabolas, ellipses, and hyperbolas with a single Section 9.6 that provides a unified treatment of all the conic sections.

▼ **Infinite Series** After the usual introduction to convergence of infinite sequences and series in Sections 10.2 and 10.3, a combined treatment of Taylor polynomials and Taylor series appears in Section 10.4. This makes it possible for the instructor to experiment with a briefer treatment of infinite series, but still offer exposure to the Taylor series that are so important for applications. The principal change in Chapter 10 is the addition of a new final section on power series methods and their use to introduce new transcendental functions, thereby concluding the middle third of the book with a return to differential equations.

▼ **Vectors and Matrices** After covering vectors in its first four sections, Chapter 11 continues with three sections on solution of linear systems (through elementary Gaussian elimination), matrices (through calculation of inverse matrices), and eigenvalues and eigenvectors and their use in classification of rotated conics. This introduction of linear systems and matrices provides the preparation required for the matrix-oriented multivariable calculus of Chapter 13. The intervening Chapter 12 (Curves and Surfaces in Space) includes discussion of Kepler-Newton motion of planets and satellites. The chapter includes also a brief application of eigenvalues to the discussion of rotated quadric surfaces in space.

▼ **Multivariable Calculus** Appropriately enough, the introduction and initial application of partial derivatives is traditional. But, beginning with the introduction of the multivariable chain rule in matrix-product form, matrix notation and terminology is used consistently in the remainder of Chapter 13. This approach affords a more clear-cut treatment of differentiability and linear approximation of multivariable functions, as well as of directional derivatives and Lagrange multipliers. We conclude Chapter 13 with a classification of critical points based on eigenvalues of the (Hessian) matrix of second derivatives (thereby generalizing directly the second derivative test of single-variable calculus). As a final illustration of the utility of matrix methods, this approach unifies the standard discriminant-based classification of two-variable critical points with the analogous classification of critical points for functions of three or more variables. Matrix methods (naturally) are needed less frequently in Chapters 14 (Multiple Integrals) and 15 (Vector Calculus), but appear whenever a change of variables in a multiple integral is involved.

OPTIONS IN TEACHING CALCULUS

The present version of the text is accompanied by a more traditional version that treats transcendental functions later in single variable calculus and omits matrices in multivariable calculus. Both versions of the complete text are also available in two-volume split editions. By appropriate selection of first and second volumes, the instructor can therefore construct a complete text for a calculus sequence with

- Early transcendentals in single variable calculus and matrices in multivariable calculus;
- Early transcendentals in single variable calculus but traditional coverage of multivariable calculus;
- Transcendental functions delayed until after the integral in single variable calculus, but matrices used in multivariable calculus;
- Neither early transcendentals in single variable calculus nor matrices in multivariable calculus.

ACKNOWLEDGMENTS

All experienced textbook authors know the value of critical reviewing during the preparation and revision of a manuscript. In our work on this edition we have profited greatly from the unusually detailed and constructive advice of the following very able reviewers:

- Kenzu Abdella—Trent University
- Martina Bode—Northwestern University
- David Caraballo—Georgetown University
- Tom Cassidy—Bucknell University
- Lucille Croom—Hunter College
- Yuanan Diao—University of North Carolina at Charlotte
- Victor Elias—University of Western Ontario
- Haitao Fan—Georgetown University
- James J. Faran, V—The State University of New York at Buffalo
- K. N. Gowrisankaran—McGill University
- Qing Han—University of Notre Dame
- Melvin D. Lax—California State University, Long Beach
- Robert H. Lewis—Fordham University
- Allan B. MacIsaac—University of Western Ontario
- Rudolph M. Najar—California State University, Fresno
- Bill Pletsch—Albuquerque Technical and Vocational Institute
- Nancy Rallis—Boston College
- Robert C. Reilly—University of California, Irvine
- James A. Reneke—Clemson University
- Alexander Retakh—Yale University
- Carl Riehm—McMaster University
- Ira Sharenow—University of Wisconsin, Madison
- Kay Strangman—University of Wisconsin, Madison
- Sophie Tryphonas—University of Toronto at Scarborough
- Kamran Vakili—Princeton University
- Cathleen M. Zucco-Teveloff—Trinity College

Many of the best improvements that have been made must be credited to colleagues and users of the previous five editions throughout the United States, Canada, and abroad. We are grateful to all those, especially students, who have written to us, and hope they will continue to do so. We thank the accuracy checkers of M. and N. Toscano, who verified the solution of every worked-out example and odd-numbered answer, as well as all of the solutions in the Instructor's and Student Solutions Manuals. We believe that the appearance and quality of the finished book is clear testimony to the skill, diligence, and talent of an exceptional staff at Prentice Hall. We owe special thanks to George Lobell, our mathematics editor, whose advice and criticism guided and shaped this revision in many significant and tangible ways, as did the constructive comments and suggestions of Ed Millman, our developmental editor. We also thank Gale Epps and Melanie Van Benthuysen for their highly varied and detailed services in aid of editors and authors throughout the work of revision. The visual graphics of this text have been widely praised in previous editions, and it is time for us to thank Ron Weickart of Network Graphics, who has worked with us through the past three editions. Barbara Mack, our production editor, expertly combined varied sources of graphic and text material and smoothly managed the whole process of book production. Our art director, Jonathan Boylan, supervised and coordinated the attractive design and layout of the text and cover for this edition. Vince Jansen coordinated the production of the CD-ROM, for which we thank especially Robert Curtis and Lee Wayand for their interactive examples and Harald Pleym for his Maple worksheets. Finally, we again are unable to thank Alice Fitzgerald Edwards and Carol Wilson Penney adequately for their unrelenting assistance, encouragement, support, and patience extending through six editions and over two decades of work on this textbook.

C. Henry Edwards
hedwards@math.uga.edu

David E. Penney
dpenney@math.uga.edu

SINGLE VARIABLE CALCULUS

Early Transcendentals

FUNCTIONS, GRAPHS, AND MODELS

1

René Descartes (1596–1650)

The seventeenth-century French scholar René Descartes is perhaps better remembered today as a philosopher than as a mathematician. But most of us are familiar with the "Cartesian plane" in which the location of a point P is specified by its coordinates (x, y).

As a schoolboy Descartes was often permitted to sleep late because of allegedly poor health. He claimed that he always thought most clearly about philosophy, science, and mathematics while he was lying comfortably in bed on cold mornings. After graduating from college, where he studied law (apparently with little enthusiasm), Descartes traveled with various armies for a number of years, but more as a gentleman soldier than as a professional military man.

In 1637, after finally settling down (in Holland), Descartes published his famous philosophical treatise *Discourse on the Method* (of Reasoning Well and Seeking Truth in the Sciences). One of three appendices to this work sets forth his new "analytic" approach to geometry. His principal idea (published almost simultaneously by his countryman Pierre de Fermat) was the correspondence between an *equation* and its *graph,* generally a curve in the plane. The equation could be used to study the curve and vice versa.

Suppose that we want to solve the equation $f(x) = 0$. Its solutions are the intersection points of the graph of $y = f(x)$ with the x-axis, so an accurate picture of the curve shows the number and approximate locations of the solutions of the equation. For instance, the graph

$$y = x^3 - 3x^2 + 1$$

has three x-intercepts, showing that the equation

$$x^3 - 3x^2 + 1 = 0$$

has three real solutions—one between -1 and 0, one between 0 and 1, and one between 2 and 3. A modern graphing calculator or computer program can approximate these solutions more accurately by magnifying the regions in which they are located. For instance, the magnified center region shows that the corresponding solution is $x \approx 0.65$.

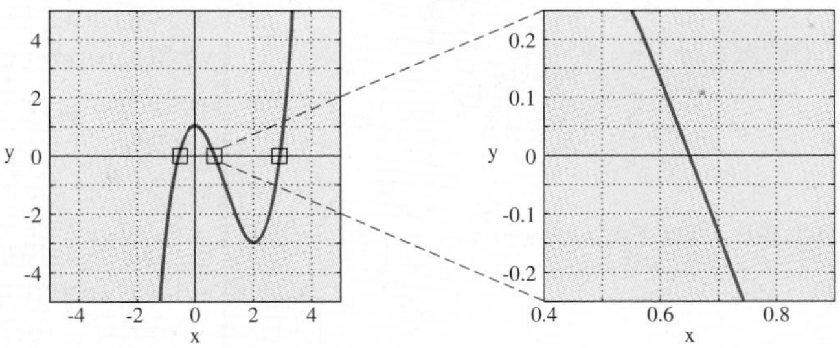

The graph $y = x^3 - 3x^2 + 1$

1.1 FUNCTIONS AND MATHEMATICAL MODELING

Calculus is one of the supreme accomplishments of the human intellect. This mathematical discipline stems largely from the seventeenth-century investigations of Isaac Newton (1642–1727) and Gottfried Wilhelm Leibniz (1646–1716). Yet some of its ideas date back to the time of Archimedes (287–212 B.C.) and originated in cultures as diverse as those of Greece, Egypt, Babylonia, India, China, and Japan. Many of the scientific discoveries that have shaped our civilization during the past three centuries would have been impossible without the use of calculus.

The principal objective of calculus is the analysis of problems of change (of motion, for example) and of content (the computation of area and volume, for instance). These problems are fundamental because we live in a world of ceaseless change, filled with bodies in motion and phenomena of ebb and flow. Consequently, calculus remains a vibrant subject, and today this body of conceptual understanding and computational technique continues to serve as the principal quantitative language of science and technology.

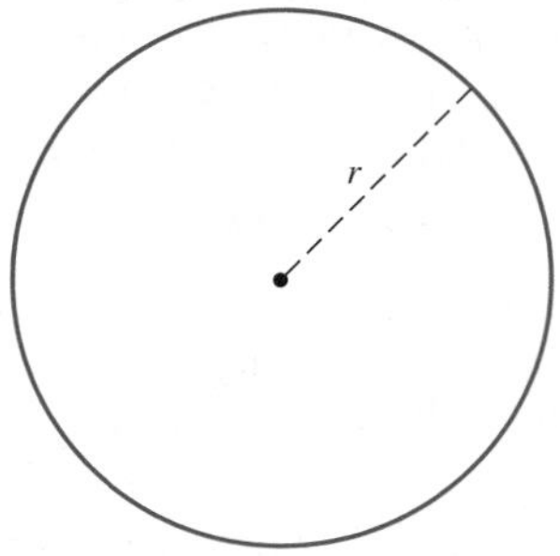

FIGURE 1.1.1 Circle: area $A = \pi r^2$, circumference $C = 2\pi r$.

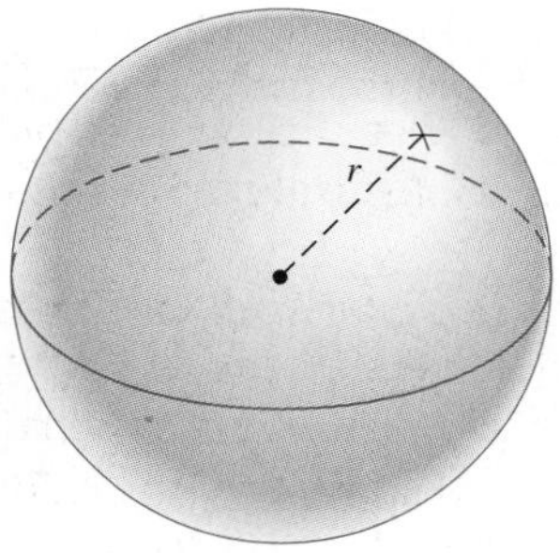

FIGURE 1.1.2 Sphere: volume $V = \frac{4}{3}\pi r^3$, surface area $S = 4\pi r^2$.

Functions

Most applications of calculus involve the use of real numbers or *variables* to describe changing quantities. The key to the mathematical analysis of a geometric or scientific situation is typically the recognition of relationships among the variables that describe the situation. Such a relationship may be a formula that expresses one variable as a *function* of another. For example:

- The area A of a circle of radius r is given by $A = \pi r^2$ (Fig. 1.1.1). The volume V and surface area S of a sphere of radius r are given by

$$V = \tfrac{4}{3}\pi r^3 \quad \text{and} \quad S = 4\pi r^2,$$

respectively (Fig. 1.1.2).
- After t seconds (s) a body that has been dropped from rest has fallen a distance

$$s = \tfrac{1}{2}gt^2$$

feet (ft) and has speed $v = gt$ feet per second (ft/s), where $g \approx 32$ ft/s^2 is gravitational acceleration.
- The volume V (in liters, L) of 3 grams (g) of carbon dioxide at 27°C is given in terms of its pressure p in atmospheres (atm) by $V = 1.68/p$.

DEFINITION Function
A real-valued **function** f defined on a set D of real numbers is a rule that assigns to each number x in D exactly one real number, denoted by $f(x)$.

The set D of all numbers for which $f(x)$ is defined is called the **domain** (or **domain of definition**) of the function f. The number $f(x)$, read "f of x," is called the **value** of the function f at the number (or point) x. The set of all values $y = f(x)$ is called the **range** of f. That is, the range of f is the set

$$\{y : y = f(x) \quad \text{for some } x \text{ in } D\}.$$

In this section we will be concerned more with the domain of a function than with its range.

EXAMPLE 1 The squaring function defined by

$$f(x) = x^2$$

assigns to each real number x its square x^2. Because every real number *can* be squared, the domain of f is the set $\boldsymbol{R}$ of all real numbers. But only nonnegative numbers are squares. Moreover, if $a \geqq 0$, then $a = (\sqrt{a})^2 = f(\sqrt{a})$, so a is a square. Hence the range of the squaring function f is the set $\{y : y \geqq 0\}$ of all nonnegative real numbers. ◆

Functions can be described in various ways. A *symbolic* description of the function f is provided by a formula that specifies how to compute the number $f(x)$ in terms of the number x. Thus the symbol $f(\ )$ may be regarded as an operation that is to be performed whenever a number or expression is inserted between the parentheses.

EXAMPLE 2 The formula

$$f(x) = x^2 + x - 3 \qquad \textbf{(1)}$$

defines a function f whose domain is the entire real line $\boldsymbol{R}$. Some typical values of f are $f(-2) = -1$, $f(0) = -3$, and $f(3) = 9$. Some other values of the function f are

$$\begin{aligned} f(4) &= 4^2 + 4 - 3 = 17, \\ f(c) &= c^2 + c - 3, \\ f(2+h) &= (2+h)^2 + (2+h) - 3 \\ &= (4 + 4h + h^2) + (2+h) - 3 = h^2 + 5h + 3, \quad \text{and} \\ f(-t^2) &= (-t^2)^2 + (-t^2) - 3 = t^4 - t^2 - 3. \end{aligned}$$ ◆

When we describe the function f by writing a formula $y = f(x)$, we call x the **independent variable** and y the **dependent variable** because the value of y depends—through f—upon the choice of x. As the independent variable x changes, or varies, then so does the dependent variable y. The way that y varies is determined by the rule of the function f. For example, if f is the function of Eq. (1), then $y = -1$ when $x = -2$, $y = -3$ when $x = 0$, and $y = 9$ when $x = 3$.

You may find it useful to visualize the dependence of the value $y = f(x)$ on x by thinking of the function f as a kind of machine that accepts as input a number x and then produces as output the number $f(x)$, perhaps displayed or printed (Fig. 1.1.3).

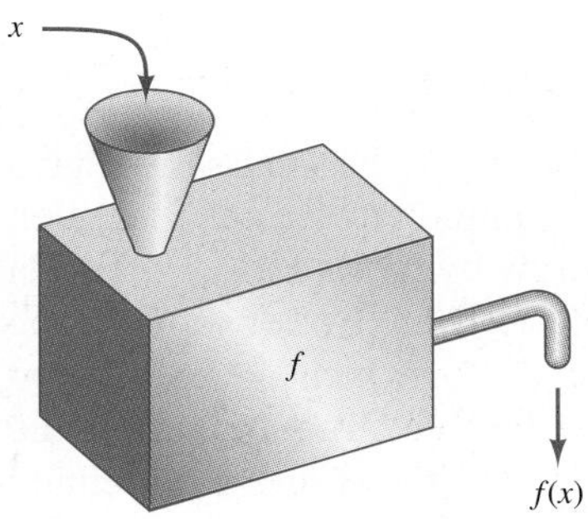

FIGURE 1.1.3 A "function machine."

One such machine is the square root key of a simple pocket calculator. When a nonnegative number x is entered and this key is pressed, the calculator displays (an approximation to) the number $\sqrt{x}$. Note that the domain of this *square root function* $f(x) = \sqrt{x}$ is the set $[0, +\infty)$ of all nonnegative real numbers, because no negative number has a real square root. The range of f is also the set of all nonnegative real numbers, because the symbol $\sqrt{x}$ always denotes the *nonnegative* square root of x. The calculator illustrates its "knowledge" of the domain by displaying an error message if we ask it to calculate the square root of a negative number (or perhaps a complex number, if it's a more sophisticated calculator).

EXAMPLE 3 Not every function has a rule expressible as a simple one-part formula such as $f(x) = \sqrt{x}$. For instance, if we write

$$h(x) = \begin{cases} x^2 & \text{if } x \geqq 0, \\ \sqrt{-x} & \text{if } x < 0, \end{cases}$$

then we have defined a perfectly good function with domain $\boldsymbol{R}$. Some of its values are $h(-4) = 2$, $h(0) = 0$, and $h(2) = 4$. By contrast, the function g in Example 4 is defined initially by means of a verbal description rather than by means of formulas. ◆

EXAMPLE 4 For each real number x, let $g(x)$ denote the greatest integer that is less than or equal to x. For instance, $g(2.5) = 2$, $g(0) = 0$, $g(-3.5) = -4$, and $g(\pi) = 3$. If n is an integer, then $g(x) = n$ for every number x such that $n \leqq x < n + 1$. This function g is called the **greatest integer function** and is often denoted by

$$g(x) = [\![x]\!].$$

Thus $[\![2.5]\!] = 2$, $[\![-3.5]\!] = -4$, and $[\![\pi]\!] = 3$. Note that although $[\![x]\!]$ is defined for all x, the range of the greatest integer function is not all of $\boldsymbol{R}$, but the set $\boldsymbol{Z}$ of all *integers*. ◆

The name of a function need not be a single letter such as f or g. For instance, think of the trigonometric functions $\sin(x)$ and $\cos(x)$ with the names sin and cos.

EXAMPLE 5 Another descriptive name for the greatest integer function of Example 4 is

$$\text{FLOOR}(x) = [\![x]\!]. \tag{2}$$

(We think of the integer n as the "floor" beneath the real numbers lying between n and $n + 1$.) Similarly, we may use $\text{ROUND}(x)$ to name the familiar function that "rounds off" the real number x to the nearest integer n, except that $\text{ROUND}(x) = n + 1$ if $x = n + \frac{1}{2}$ (so we "round upward" in case of ambiguity). Round off enough different numbers to convince yourself that

$$\text{ROUND}(x) = \text{FLOOR}\left(x + \tfrac{1}{2}\right) \tag{3}$$

for all x.

Closely related to the FLOOR and ROUND functions is the "ceiling function" used by the U.S. Postal Service; $\text{CEILING}(x)$ denotes the least integer that is not less than the number x. In 2001 the postage rate for a first-class letter was 34¢ for the first ounce and 21¢ for each additional ounce or fraction thereof. For a letter weighing $w > 0$ ounces, the number of "additional ounces" involved is $\text{CEILING}(w) - 1$. Therefore the postage $s(w)$ due on this letter is given by

$$s(w) = 34 + 21 \cdot [\text{CEILING}(w) - 1] = 13 + 21 \cdot \text{CEILING}(w).$$ ◆

Domains and Intervals

The function f and the value or expression $f(x)$ are different in the same sense that a machine and its output are different. Nevertheless, it is common to use an expression like "the function $f(x) = x^2$" to define a function merely by writing its formula. In this situation the domain of the function is not specified. Then, by convention, the **domain of the function** f is the set of all real numbers x for which the expression $f(x)$ makes sense and produces a real number y. For instance, the domain of the function $h(x) = 1/x$ is the set of all nonzero real numbers (because $1/x$ is defined precisely when $x \neq 0$).

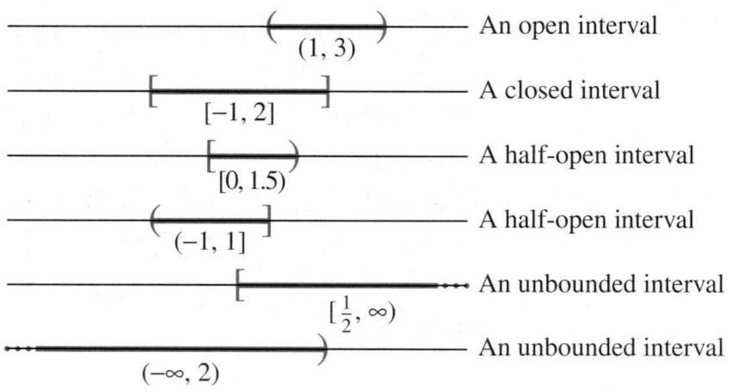

FIGURE 1.1.4 Some examples of intervals of real numbers.

Domains of functions frequently are described in terms of *intervals* of real numbers (Fig. 1.1.4). (Interval notation is reviewed in Appendix A.) Recall that a **closed interval** $[a, b]$ contains both its endpoints $x = a$ and $x = b$, whereas the **open interval** (a, b) contains neither endpoint. Each of the **half-open intervals** $[a, b)$ and $(a, b]$ contains exactly one of its two endpoints. The **unbounded interval** $[a, \infty)$

contains its endpoint $x = a$, whereas $(-\infty, a)$ does not. The previously mentioned domain of $h(x) = 1/x$ is the *union* of the unbounded intervals $(-\infty, 0)$ and $(0, \infty)$.

EXAMPLE 6 Find the domain of the function $g(x) = \dfrac{1}{2x+4}$.

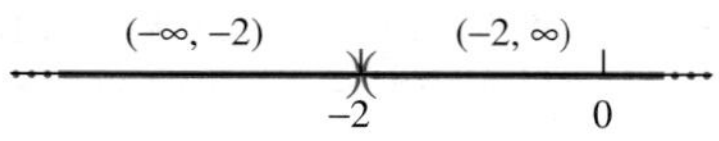

FIGURE 1.1.5 The domain of $g(x) = 1/(2x+4)$ is the union of two unbounded open intervals.

Solution Division by zero is not allowed, so the value $g(x)$ is defined precisely when $2x + 4 \neq 0$. This is true when $2x \neq -4$, and thus when $x \neq -2$. Hence the domain of g is the set $\{x : x \neq 2\}$, which is the union of the two unbounded open intervals $(-\infty, -2)$ and $(-2, \infty)$, shown in Fig. 1.1.5. ◆

EXAMPLE 7 Find the domain of $h(x) = \dfrac{1}{\sqrt{2x+4}}$.

Solution Now it is necessary not only that the quantity $2x + 4$ be nonzero, but also that it be positive, in order that the square root $\sqrt{2x+4}$ is defined. But $2x + 4 > 0$ when $2x > -4$, and thus when $x > -2$. Hence the domain of h is the single unbounded open interval $(-2, \infty)$. ◆

Mathematical Modeling

The investigation of an applied problem often hinges on defining a function that captures the essence of a geometrical or physical situation. Examples 8 and 9 illustrate this process.

EXAMPLE 8 A rectangular box with a square base has volume 125. Express its total surface area A as a function of the edge length x of its base.

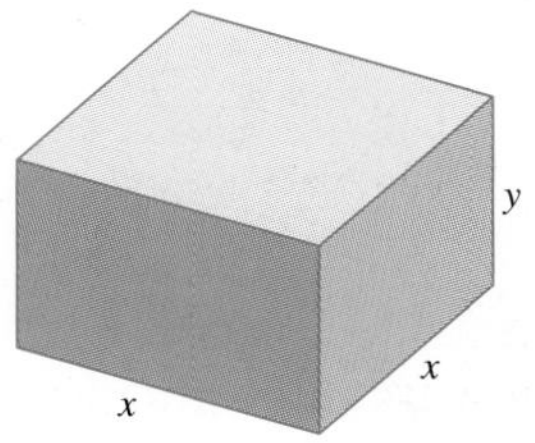

FIGURE 1.1.6 The box of Example 8.

Solution The first step is to draw a sketch and to label the relevant dimensions. Figure 1.1.6 shows a rectangular box with square base of edge length x and with height y. We are given that the volume of the box is

$$V = x^2 y = 125. \tag{4}$$

Both the top and the bottom of the box have area x^2 and each of its four vertical sides has area xy, so its total surface area is

$$A = 2x^2 + 4xy. \tag{5}$$

But this is a formula for A in terms of the *two* variables x and y rather than a function of the *single* variable x. To eliminate y and thereby obtain A in terms of x alone, we solve Eq. (4) for $y = 125/x^2$ and then substitute this result in Eq. (5) to obtain

$$A = 2x^2 + 4x \cdot \frac{125}{x^2} = 2x^2 + \frac{500}{x}.$$

Thus the surface area is given as a function of the edge length x by

$$A(x) = 2x^2 + \frac{500}{x}, \quad 0 < x < +\infty. \tag{6}$$

It is necessary to specify the domain because negative values of x make sense in the *formula* in (5) but do not belong in the domain of the *function* A. Because every $x > 0$ determines such a box, the domain does, in fact, include all positive real numbers. ◆

COMMENT In Example 8 our goal was to express the dependent variable A as a *function* of the independent variable x. Initially, the geometric situation provided us instead with

1. The *formula* in Eq. (5) expressing A in terms of both x and the additional variable y, and
2. The *relation* in Eq. (4) between x and y, which we used to eliminate y and thereby express A as a function of x alone.

We will see that this is a common pattern in many different applied problems, such as the one that follows.

The Animal Pen Problem You must build a rectangular holding pen for animals. To save material, you will use an existing wall as one of its four sides. The fence for the other three sides costs \$5/ft, and you must spend \$1/ft to paint the portion of the wall that forms the fourth side of the pen. If you have a total of \$180 to spend, what dimensions will maximize the area of the pen you can build?

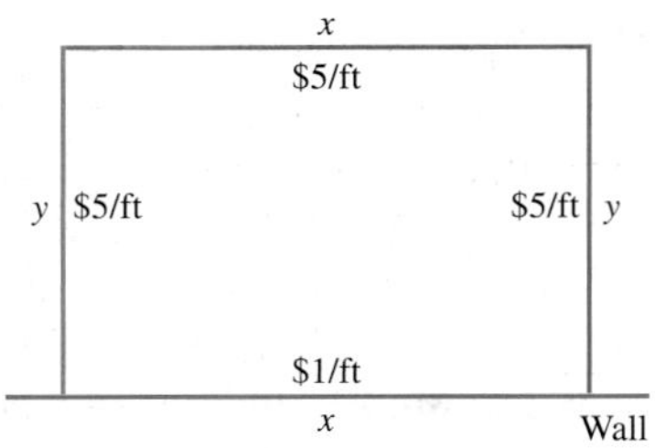

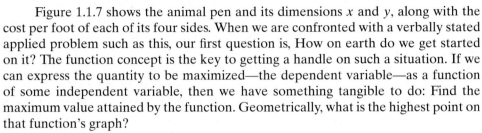

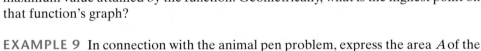

FIGURE 1.1.7 The animal pen.

Figure 1.1.7 shows the animal pen and its dimensions x and y, along with the cost per foot of each of its four sides. When we are confronted with a verbally stated applied problem such as this, our first question is, How on earth do we get started on it? The function concept is the key to getting a handle on such a situation. If we can express the quantity to be maximized—the dependent variable—as a function of some independent variable, then we have something tangible to do: Find the maximum value attained by the function. Geometrically, what is the highest point on that function's graph?

EXAMPLE 9 In connection with the animal pen problem, express the area A of the pen as a function of the length x of its wall side.

Solution The area A of the rectangular pen of length x and width y is

$$A = xy. \tag{7}$$

When we multiply the length of each side in Fig. 1.1.7 by its cost per foot and then add the results, we find that the total cost C of the pen is

$$C = x + 5y + 5x + 5y = 6x + 10y.$$

So

$$6x + 10y = 180, \tag{8}$$

because we are given $C = 180$. Choosing x to be the independent variable, we use the relation in Eq. (8) to eliminate the additional variable y from the area formula in Eq. (7). We solve Eq. (8) for y and substitute the result

$$y = \tfrac{1}{10}(180 - 6x) = \tfrac{3}{5}(30 - x) \tag{9}$$

in Eq. (7). Thus we obtain the desired function

$$A(x) = \tfrac{3}{5}(30x - x^2)$$

that expresses the area A as a function of the length x.

In addition to this formula for the function A, we must also specify its domain. Only if $x > 0$ will actual rectangles be produced, but we find it convenient to include the value $x = 0$ as well. This value of x corresponds to a "degenerate rectangle" of base length zero and height

$$y = \tfrac{3}{5} \cdot 30 = 18,$$

a consequence of Eq. (9). For similar reasons, we have the restriction $y \geqq 0$. Because

$$y = \tfrac{3}{5}(30 - x),$$

it follows that $x \leqq 30$. Thus the complete definition of the area function is

$$A(x) = \tfrac{3}{5}(30x - x^2), \quad 0 \leqq x \leqq 30. \tag{10}$$

◆

COMMENT The domain of a function is a necessary part of its definition, and for each function we must specify the domain of values of the independent variable. In applications, we use the values of the independent variable that are relevant to the problem at hand.

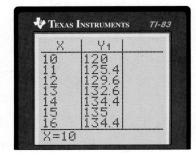

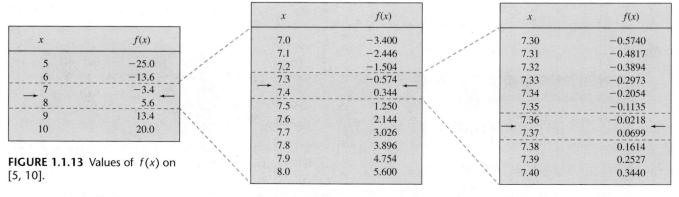

Example 9 illustrates an important part of the solution of a typical applied problem—the formulation of a **mathematical model** of the physical situation under study. The area function $A(x)$ defined in (10) provides a mathematical model of the animal pen problem. The shape of the optimal animal pen can be determined by finding the maximum value attained by the function A on its domain of definition.

x	$A(x)$
0	0
5	75
10	120
15	135 ←
20	120
25	75
30	0

FIGURE 1.1.8 Area $A(x)$ of a pen with side of length x.

Numerical Investigation

Armed with the result of Example 9, we might attack the animal pen problem by calculating a table of values of the area function $A(x)$ in Eq. (10). Such a table is shown in Fig. 1.1.8. The data in this table suggest strongly that the maximum area is $A = 135\ \text{ft}^2$, attained with side length $x = 15$ ft, in which case Eq. (9) yields $y = 9$ ft. This conjecture appears to be corroborated by the more refined data shown in Fig. 1.1.9.

Thus it seems that the animal pen with maximal area (costing \$180) is $x = 15$ ft long and $y = 9$ ft wide. The tables in Figs. 1.1.8 and 1.1.9 show only *integral* values of x, however, and it is quite possible that the length x of the pen of maximal area is *not* an integer. Consequently, numerical tables alone do not settle the matter. A new mathematical idea is needed in order to *prove* that $A(15) = 135$ is the maximum value of

$$A(x) = \tfrac{3}{5}(30x - x^2), \qquad 0 \leqq x \leqq 30$$

for *all* x in its domain. We attack this problem again in Section 1.2.

x	$A(x)$
10	120
11	125.4
12	129.6
13	132.6
14	134.4
15	135 ←
16	134.4
17	132.6
18	129.6
19	125.4
20	120

FIGURE 1.1.9 Further indication that $x = 15$ yields maximal area $A = 135$.

Tabulation of Functions

Many scientific and graphing calculators allow the user to program a given function for repeated evaluation, and thereby to painlessly compute tables like those in Figs. 1.1.8 and 1.1.9. For instance, Figs. 1.1.10 and 1.1.11 show displays of a calculator prepared to calculate values of the dependent variable

$$y_1 = A(x) = (3/5)(30x - x^2),$$

and Fig. 1.1.12 shows the calculator's resulting version of the table in Fig. 1.1.9.

The use of a calculator or computer to tabulate values of a function is a simple technique with surprisingly many applications. Here we illustrate a method of solving approximately an equation of the form $f(x) = 0$ by *repeated tabulation* of values $f(x)$ of the function f.

As a specific example, suppose that we ask what value of x in Eq. (10) yields an animal pen of area $A = 100$. Then we need to solve the equation

$$A(x) = \tfrac{3}{5}(30x - x^2) = 100,$$

which is equivalent to the equation

$$f(x) = \tfrac{3}{5}(30x - x^2) - 100 = 0. \qquad \textbf{(11)}$$

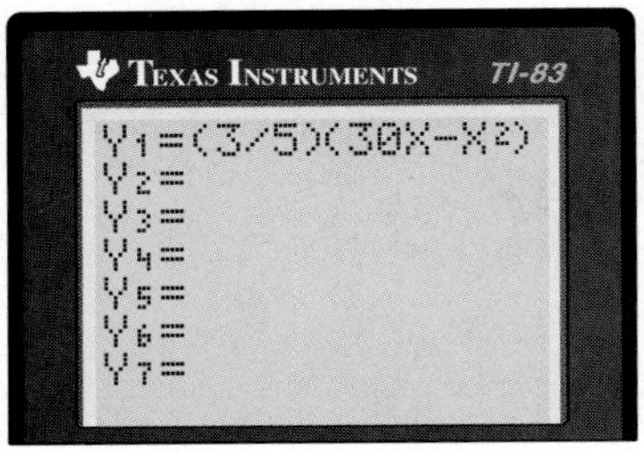

FIGURE 1.1.10 A calculator programmed to evaluate $A(x) = (3/5)(30x - x^2)$.

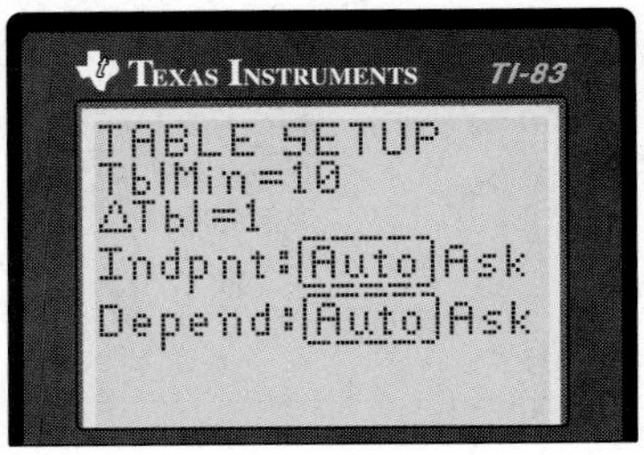

FIGURE 1.1.11 The table setup.

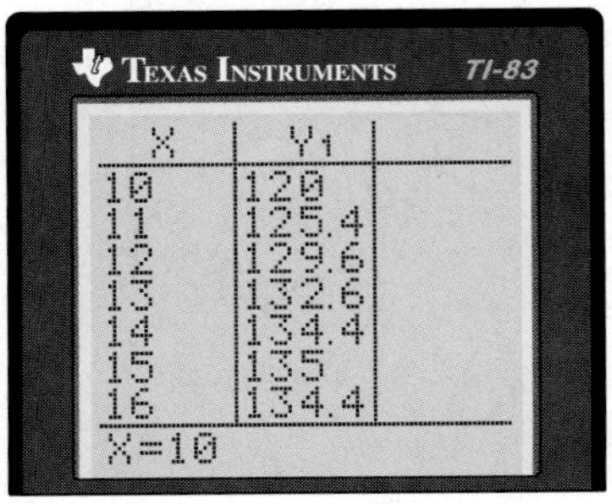

FIGURE 1.1.12 The resulting table.

This is a quadratic equation that could be solved using the quadratic formula of basic algebra, but we want to take a more direct, numerical approach. The reason is that the numerical approach is applicable even when no simple formula (such as the quadratic formula) is available.

The data in Fig. 1.1.8 suggest that one value of x for which $A(x) = 100$ lies somewhere between $x = 5$ and $x = 10$ and that a second such value lies between $x = 20$ and $x = 25$. Indeed, substitution in Eq. (11) yields

$$f(5) = -25 < 0 \quad \text{and} \quad f(10) = 20 > 0.$$

The fact that $f(x)$ is *negative* at one endpoint of the interval $[5, 10]$ but *positive* at the other endpoint suggests that $f(x)$ is *zero* somewhere between $x = 5$ and $x = 10$.

To see *where*, we tabulate values of $f(x)$ on $[5, 10]$. In the table of Fig. 1.1.13 we see that $f(7) < 0$ and $f(8) > 0$, so we focus next on the interval $[7, 8]$. Tabulating $f(x)$ on $[7, 8]$ gives the table of Fig. 1.1.14, where we see that $f(7.3) < 0$ and $f(7.4) > 0$.

We therefore tabulate $f(x)$ once more, this time on the interval $[7.3, 7.4]$. In Fig. 1.1.15 we see that

$$f(7.36) \approx -0.02 \quad \text{and} \quad f(7.37) \approx 0.07.$$

Because $f(7.36)$ is considerably closer to zero than is $f(7.37)$, we conclude that the desired solution of Eq. (11) is given approximately by $x \approx 7.36$, accurate to two decimal places. If greater accuracy were needed, we could continue to tabulate $f(x)$ on smaller and smaller intervals.

If we were to begin with the interval $[20, 25]$ and proceed similarly, we would find the second value $x \approx 22.64$ such that $f(x) = 0$. (You should do this for practice.)

Finally, let's calculate the corresponding values of the width y of the animal pen such that $A = xy = 100$:

- If $x \approx 7.36$, then $y \approx 13.59$.
- If $x \approx 22.64$, then $y \approx 4.42$.

Thus, under the cost constraint of the animal pen problem, we can construct either a 7.36-ft by 13.59-ft or a 22.64-ft by 4.42-ft rectangle, both of area 100 ft^2.

The layout of Figs. 1.1.13 through 1.1.15 suggests the idea of repeated tabulation as a process of successive numerical magnification. This method of repeated tabulation can be applied to a wide range of equations of the form $f(x) = 0$. If the interval $[a, b]$ contains a solution and the endpoint values $f(a)$ and $f(b)$ differ in sign, then we can approximate this solution by tabulating values on successively

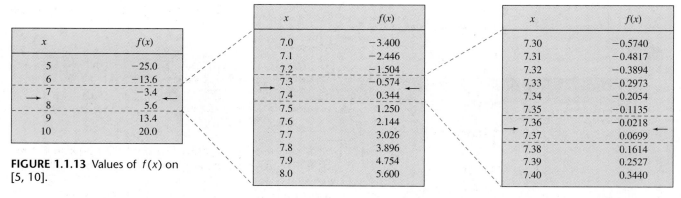

x	$f(x)$
5	−25.0
6	−13.6
7	−3.4
8	5.6
9	13.4
10	20.0

FIGURE 1.1.13 Values of $f(x)$ on [5, 10].

x	$f(x)$
7.0	−3.400
7.1	−2.446
7.2	−1.504
7.3	−0.574
7.4	0.344
7.5	1.250
7.6	2.144
7.7	3.026
7.8	3.896
7.9	4.754
8.0	5.600

FIGURE 1.1.14 Values of $f(x)$ on [7, 8].

x	$f(x)$
7.30	−0.5740
7.31	−0.4817
7.32	−0.3894
7.33	−0.2973
7.34	−0.2054
7.35	−0.1135
7.36	−0.0218
7.37	0.0699
7.38	0.1614
7.39	0.2527
7.40	0.3440

FIGURE 1.1.15 Values of $f(x)$ on [7.3, 7.4].

smaller subintervals. Problems 57 through 66 and the project at the end of this section are applications of this concrete numerical method for the approximate solution of equations.

1.1 TRUE/FALSE STUDY GUIDE

1.1 CONCEPTS: QUESTIONS AND DISCUSSION

1. Can a function have the same value at two different points? Can it have two different values at the same point x?
2. Explain the difference between a dependent variable and an independent variable. A change in one both causes and determines a change in the other. Which one is the "controlling variable"?
3. What is the difference between an open interval and a closed interval? Is every interval on the real line either open or closed? Justify your answer.
4. Suppose that S is a set of real numbers. Is there a function whose domain of definition is precisely the set S? Is there a function defined on the whole real line whose range is precisely the set S? Is there a function that has the value 1 at each point of S and the value 0 at each point of the real line $\boldsymbol{R}$ not in S?
5. Figure 1.1.6 shows a box with square base and height y. Which of the following two formulas would suffice to define the volume V of this box as a function of y?

   (a) $V = x^2 y$; (b) $V = y(10 - 2y)^2$.

   Discuss the difference between a formula and a function.
6. In the following table, y is a function of x. Determine whether or not x is a function of y.

x	0	2	4	6	8	10
y	−1	3	8	7	3	−2

1.1 PROBLEMS

In Problems 1 through 4, find and simplify each of the following values: (a) $f(-a)$; (b) $f(a^{-1})$; (c) $f(\sqrt{a})$; (d) $f(a^2)$.

1. $f(x) = \dfrac{1}{x}$

2. $f(x) = x^2 + 5$

3. $f(x) = \dfrac{1}{x^2+5}$

4. $f(x) = \sqrt{1 + x^2 + x^4}$

In Problems 5 through 10, find all values of a such that $g(a) = 5$.

5. $g(x) = 3x + 4$

6. $g(x) = \dfrac{1}{2x - 1}$

7. $g(x) = \sqrt{x^2 + 16}$

8. $g(x) = x^3 - 3$

9. $g(x) = \sqrt[3]{x + 25}$

10. $g(x) = 2x^2 - x + 4$

In Problems 11 through 16, compute and then simplify the quantity $f(a + h) - f(a)$.

11. $f(x) = 3x - 2$

12. $f(x) = 1 - 2x$

13. $f(x) = x^2$

14. $f(x) = x^2 + 2x$

15. $f(x) = \dfrac{1}{x}$

16. $f(x) = \dfrac{2}{x + 1}$

In Problems 17 through 20, find the range of values of the given function.

17. $f(x) = \begin{cases} \dfrac{x}{|x|} & \text{if } x \neq 0; \\ 0 & \text{if } x = 0 \end{cases}$

18. $f(x) = [\![3x]\!]$ (Recall that $[\![x]\!]$ is the largest integer not exceeding x.)

19. $f(x) = (-1)^{[\![x]\!]}$

20. $f(x)$ is the first-class postage (in cents) for a letter mailed in the United States and weighing x ounces, $0 < x < 12$. As of January 7, 2001 the postage rate for such a letter was 34¢ for the first ounce plus 21¢ for each additional ounce or fraction thereof.

In Problems 21 through 35, find the largest domain (of real numbers) on which the given formula determines a (real-valued) function.

21. $f(x) = 10 - x^2$

22. $f(x) = x^3 + 5$

23. $f(t) = \sqrt{t^2}$

24. $g(t) = \left(\sqrt{t}\right)^2$

25. $f(x) = \sqrt{3x - 5}$

26. $g(t) = \sqrt[3]{t + 4}$

27. $f(t) = \sqrt{1 - 2t}$

28. $g(x) = \dfrac{1}{(x + 2)^2}$

29. $f(x) = \dfrac{2}{3 - x}$

30. $g(t) = \sqrt{\dfrac{2}{3 - t}}$

31. $f(x) = \sqrt{x^2 + 9}$

32. $h(z) = \dfrac{1}{\sqrt{4 - z^2}}$

33. $f(x) = \sqrt{4 - \sqrt{x}}$

34. $f(x) = \sqrt{\dfrac{x + 1}{x - 1}}$

35. $g(t) = \dfrac{t}{|t|}$

36. Express the area A of a square as a function of its perimeter P.

37. Express the circumference C of a circle as a function of its area A.

38. Express the volume V of a sphere as a function of its surface area S.

39. Given: 0°C is the same as 32°F, and a temperature change of 1°C is the same as a change of 1.8°F. Express the Celsius temperature C as a function of the Fahrenheit temperature F.

40. Show that if a rectangle has base x and perimeter 100 (Fig. 1.1.16), then its area A is given by the function

$$A(x) = x(50 - x), \quad 0 \leqq x \leqq 50.$$

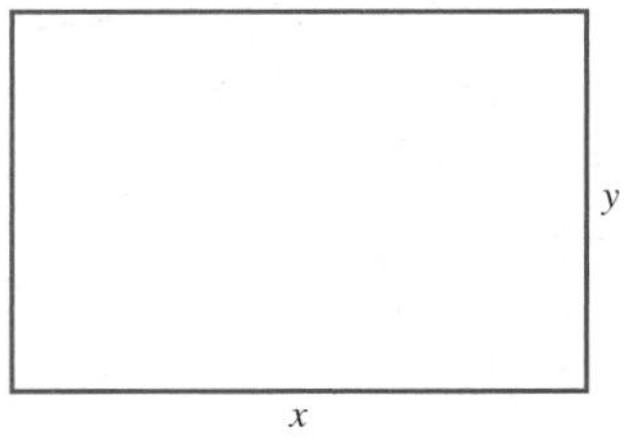

FIGURE 1.1.16 $A = xy$ (Problem 40).

41. A rectangle with base of length x is inscribed in a circle of radius 2 (Fig 1.1.17). Express the area A of the rectangle as a function of x.

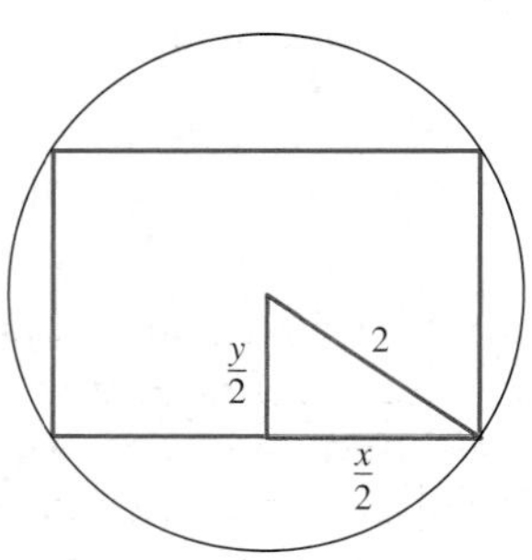

FIGURE 1.1.17 $A = xy$ (Problem 41).

42. An oil field containing 20 wells has been producing 4000 barrels of oil daily. For each new well that is drilled, the daily production of each well decreases by 5 barrels per day. Write the total daily production of the oil field as a function of the number x of new wells drilled.

43. Suppose that a rectangular box has volume 324 cm^3 and a square base of edge length x centimeters. The material for the base of the box costs 2¢/cm^2, and the material for its top and four sides costs 1¢/cm^2. Express the total cost of the box as a function of x. See Fig. 1.1.18.

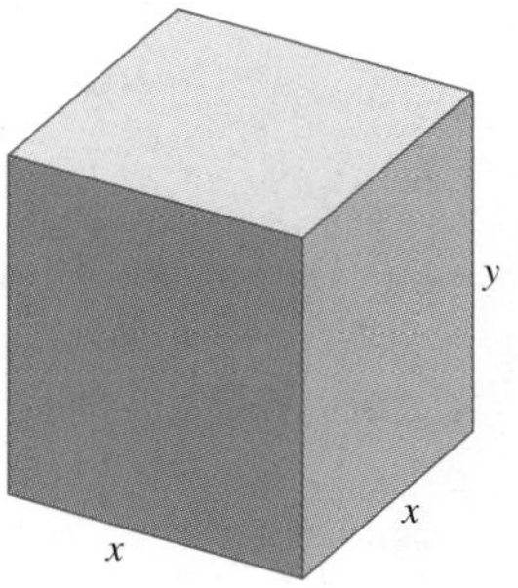

FIGURE 1.1.18 $V = x^2y$ (Problem 43).

44. A rectangle of fixed perimeter 36 is rotated around one of its sides S to generate a right circular cylinder. Express the volume V of this cylinder as a function of the length x of the side S. See Fig. 1.1.19.

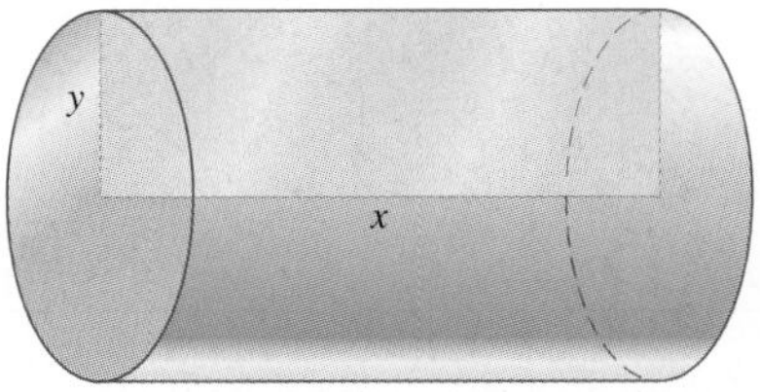

FIGURE 1.1.19 $V = \pi xy^2$ (Problem 44).

45. A right circular cylinder has volume 1000 $in.^3$ and the radius of its base is r inches. Express the total surface area A of the cylinder as a function of r. See Fig. 1.1.20.

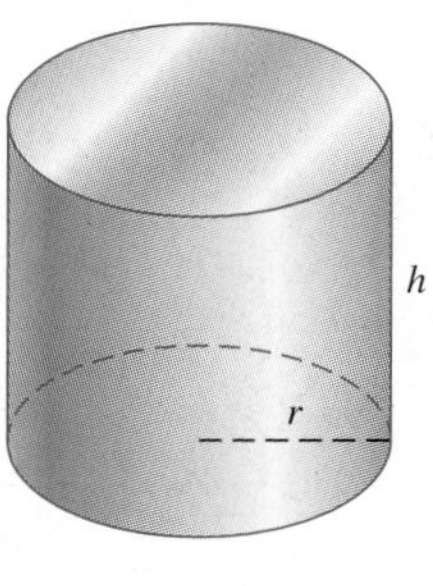

FIGURE 1.1.20 $V = \pi r^2 h$ (Problem 45).

46. A rectangular box has total surface area 600 cm^2 and a square base with edge length x centimeters. Express the volume V of the box as a function of x.

47. An open-topped box is to be made from a square piece of cardboard of edge length 50 in. First, four small squares, each of edge length x inches, are cut from the corners of the cardboard (Fig. 1.1.21). Then the four resulting flaps are turned up—folded along the dotted lines—to form the four sides of the box, which will thus have a square base and a depth of x inches (Fig. 1.1.22). Express its volume V as a function of x.

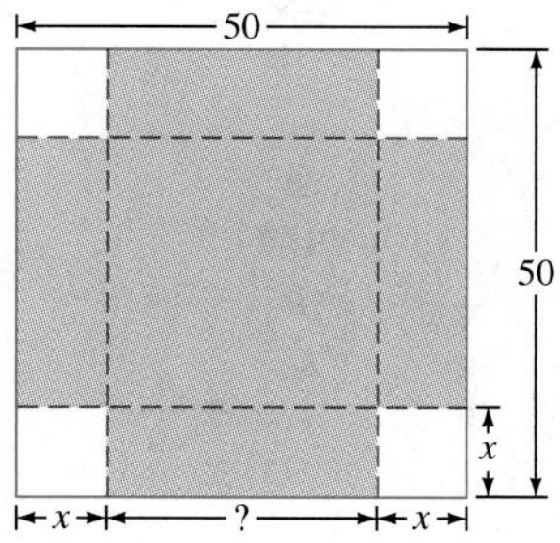

FIGURE 1.1.21 Fold the edges up to make a box (Problem 47).

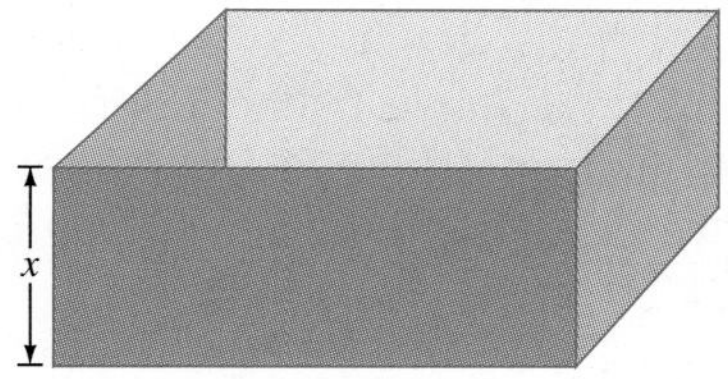

FIGURE 1.1.22 The box of Problem 47.

48. Continue Problem 40 by numerically investigating the area of a rectangle of perimeter 100. What dimensions (length and width) would appear to maximize the area of such a rectangle?

49. Determine numerically the number of new oil wells that should be drilled to maximize the total daily production of the oil field of Problem 42.

50. Investigate numerically the total surface area A of the rectangular box of Example 8. Assuming that both $x \geqq 1$ and $y \geqq 1$, what dimensions x and y would appear to minimize A?

Problems 51 through 56 deal with the functions CEILING, FLOOR, *and* ROUND *of Example 5.*

51. Show that $\text{CEILING}(x) = -\text{FLOOR}(-x)$ for all x.

52. Suppose that k is a constant. What is the range of the function $g(x) = \text{ROUND}(kx)$?

53. What is the range of the function $g(x) = \frac{1}{10}\text{ROUND}(10x)$?

54. Recalling that $\pi \approx 3.14159$, note that $\frac{1}{100}\text{ROUND}(100\pi) = 3.14$. Hence define (in terms of ROUND) a function $\text{ROUND2}(x)$ that gives the value of x rounded accurate to two decimal places.

55. Define a function $\text{ROUND4}(x)$ that gives the value of x rounded accurate to four decimal places, so that $\text{ROUND4}(\pi) = 3.1416$.

56. Define a function $\text{CHOP4}(x)$ that "chops off" (or discards) all decimal places of x beyond the fourth one, so that $\text{CHOP4}(\pi) = 3.1415$.

In Problems 57 through 66, a quadratic equation $ax^2+bx+c = 0$ and an interval $[p, q]$ containing one of its solutions are given. Use the method of repeated tabulation to approximate this solution with two digits correct or correctly rounded to the right of the decimal. Check that your result agrees with one of the two solutions given by the quadratic formula,

$$x = \frac{-b \pm \sqrt{b^2 - 4ac}}{2a}.$$

57. $x^2 - 3x + 1 = 0, \quad [0, 1]$

58. $x^2 - 3x + 1 = 0, \quad [2, 3]$

59. $x^2 + 2x - 4 = 0, \quad [1, 2]$

60. $x^2 + 2x - 4 = 0, \quad [-4, -3]$

61. $2x^2 - 7x + 4 = 0, \quad [0, 1]$

62. $2x^2 - 7x + 4 = 0, \quad [2, 3]$

63. $x^2 - 11x + 25 = 0, \quad [3, 4]$

64. $x^2 - 11x + 25 = 0, \quad [7, 8]$

65. $3x^2 + 23x - 45 = 0, \quad [1, 2]$

66. $3x^2 + 23x - 45 = 0, \quad [-10, -9]$

1.1 Project: A Square Wading Pool

Beginning with a given square piece of tin, you want to design a wading pool in the manner indicated in Figs. 1.1.21 and 1.1.22. Your task is to investigate the maximum volume of the pool that can be constructed and how to design a wading pool of specified volume. The key is to express the volume of the wading pool as a function $V(x)$ of the edge length x of each corner square (as in Problem 47). You can then use a calculator or computer for appropriate repeated tabulation of this function.

1.2 GRAPHS OF EQUATIONS AND FUNCTIONS

Graphs and equations of *straight lines* in the xy-coordinate plane are reviewed in Appendix B. Recall the **slope-intercept equation**

$$y = mx + b \tag{1}$$

of the straight line with **slope** $m = \tan\phi$, **angle of inclination** ϕ, and y-**intercept** b (Fig. 1.2.1). The "rise over run" definition

$$m = \frac{\text{rise}}{\text{run}} = \frac{\Delta y}{\Delta x} = \frac{y_2 - y_1}{x_2 - x_1} \tag{2}$$

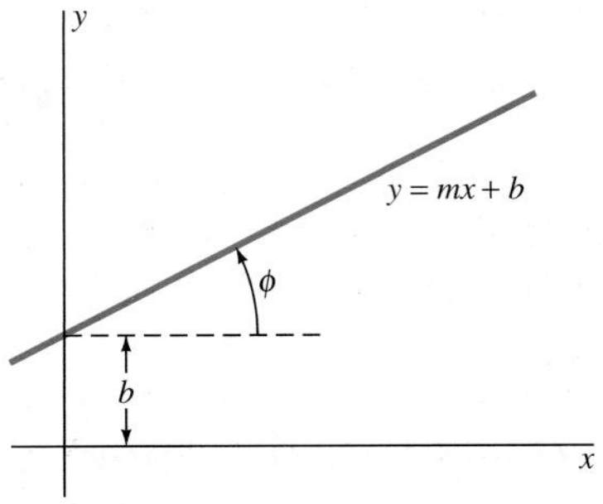

FIGURE 1.2.1 A line with y-intercept b and inclination angle ϕ.

of the slope (Fig. 1.2.2) leads to the **point-slope equation**

$$y - y_0 = m(x - x_0) \tag{3}$$

of the straight line with slope m that passes through the point (x_0, y_0)—see Fig. 1.2.3. In either case a point (x, y) in the xy-plane lies on the line if and only if its coordinates x and y satisfy the indicated equation.

If $\Delta y = 0$ in Eq. (2), then $m = 0$ and the line is *horizontal*. If $\Delta x = 0$, then the line is *vertical* and (because we cannot divide by zero) the slope of the line is not defined. Thus:

- Horizontal lines have slope zero.
- Vertical lines have no defined slope at all.

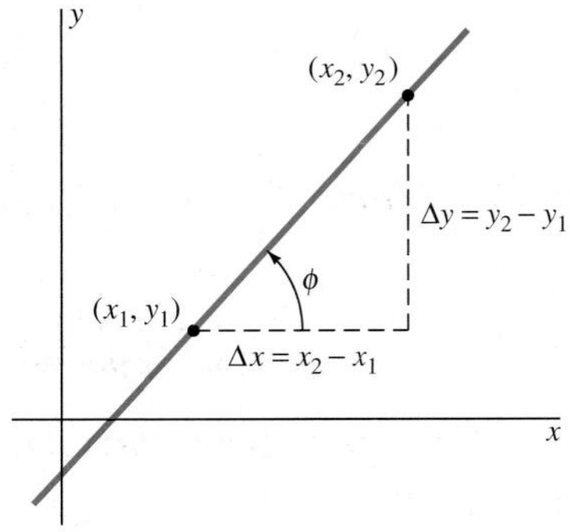

FIGURE 1.2.2 Slope $m = \tan\phi = \dfrac{\Delta y}{\Delta x}$.

EXAMPLE 1 Write an equation of the line L that passes through the point $P(3, 5)$ and is parallel to the line having equation $y = 2x - 4$.

Solution The two parallel lines have the same angle of inclination ϕ (Fig. 1.2.4) and therefore have the same slope m. Comparing the given equation $y = 2x - 4$ with the slope-intercept equation in (1), we see that $m = 2$. The point-slope equation therefore gives

$$y - 5 = 2(x - 3)$$

—alternatively, $y = 2x - 1$, for an equation of the line L. ◆

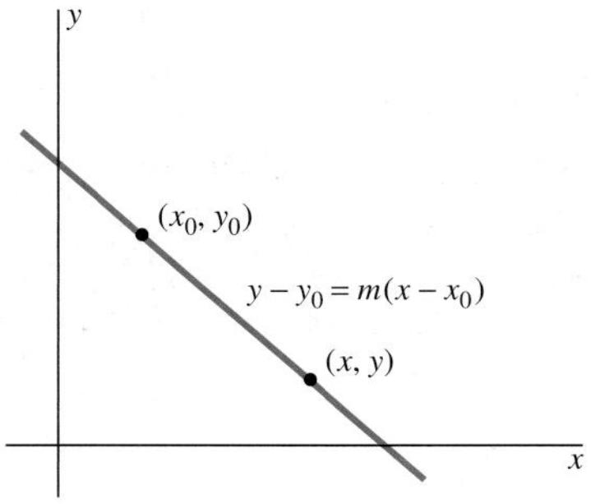

FIGURE 1.2.3 The line through (x_0, y_0) with slope m.

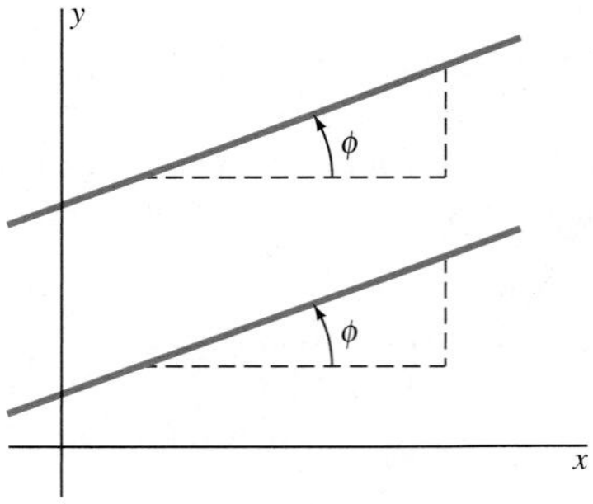

FIGURE 1.2.4 Parallel lines have the same slope $m = \tan\phi$.

Equations (1) and (3) can both be put into the form of the general linear equation

$$Ax + By = C. \tag{4}$$

Conversely, if $B \neq 0$, then we can divide the terms in Eq. (4) by B and solve for y, thereby obtaining the point-slope equation of a straight line. If $A = 0$, then the resulting equation has the form $y = H$, the equation of a *horizontal line* with slope zero. If $B = 0$ but $A \neq 0$, then Eq. (4) can be solved for $x = K$, the equation of a *vertical line* (having no slope at all). In summary, we see that if the coefficients A and B are not both zero, then Eq. (4) is the equation of some straight line in the plane.

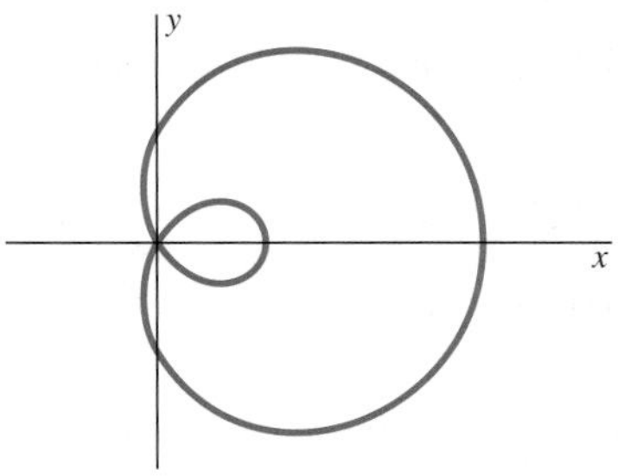

FIGURE 1.2.5 The graph of the equation $x^2 + y^2 = (x^2 + y^2 - 2x)^2$.

Graphs of More General Equations

A straight line is a simple example of the graph of an equation. By contrast, a computer-graphing program produced the exotic curve shown in Fig. 1.2.5 when asked to picture the set of all points (x, y) satisfying the equation

$$x^2 + y^2 = (x^2 + y^2 - 2x)^2.$$

Both a straight line and this complicated curve are examples of *graphs* of equations.

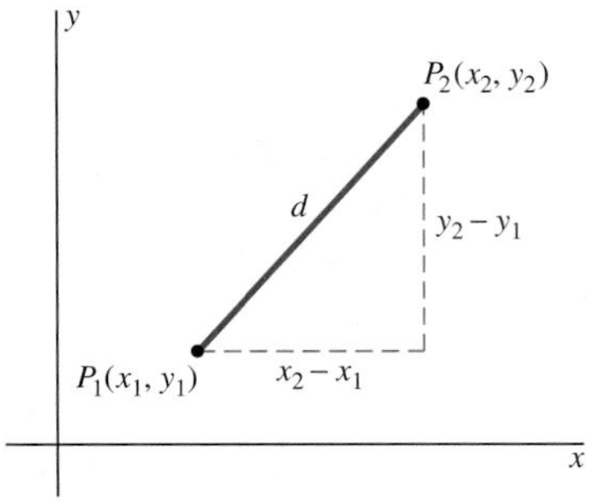

FIGURE 1.2.6 The Pythagorean theorem implies the **distance formula** $d = \sqrt{(x_2 - x_1)^2 + (y_2 - y_1)^2}$.

DEFINITION Graph of an Equation

The **graph** of an equation in two variables x and y is the set of all points (x, y) in the plane that satisfy the equation.

For example, the distance formula of Fig. 1.2.6 tells us that the graph of the equation

$$x^2 + y^2 = r^2 \tag{5}$$

is the circle of radius r centered at the origin $(0, 0)$. More generally, the graph of the equation

$$(x - h)^2 + (y - k)^2 = r^2 \tag{6}$$

is the circle of radius r with center (h, k). This also follows from the distance formula, because the distance between the points (x, y) and (h, k) in Fig. 1.2.7 is r.

EXAMPLE 2 The equation of the circle with center (3, 4) and radius 10 is

$$(x - 3)^2 + (y - 4)^2 = 100,$$

which may also be written in the form

$$x^2 + y^2 - 6x - 8y - 75 = 0.$$ ◆

Translates of Graphs

Suppose that the xy-plane is shifted rigidly (or *translated*) by moving each point h units to the right and k units upward. (A negative value of h or k corresponds to a leftward or downward movement.) That is, each point (x, y) of the plane is moved to the point $(x + h, y + k)$; see Fig. 1.2.8. Then the circle with radius r and center $(0, 0)$

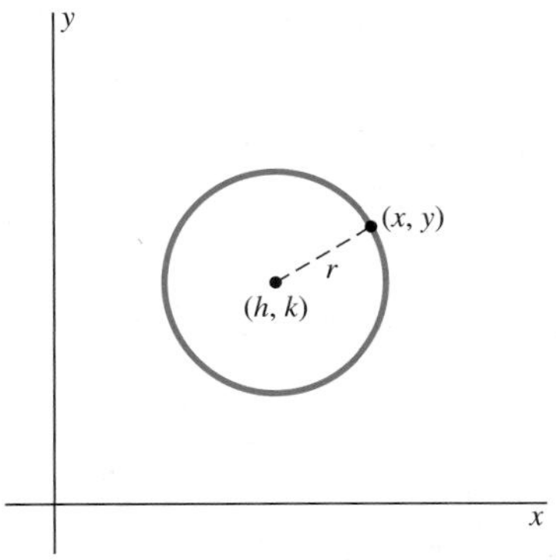

FIGURE 1.2.7 A translated circle.

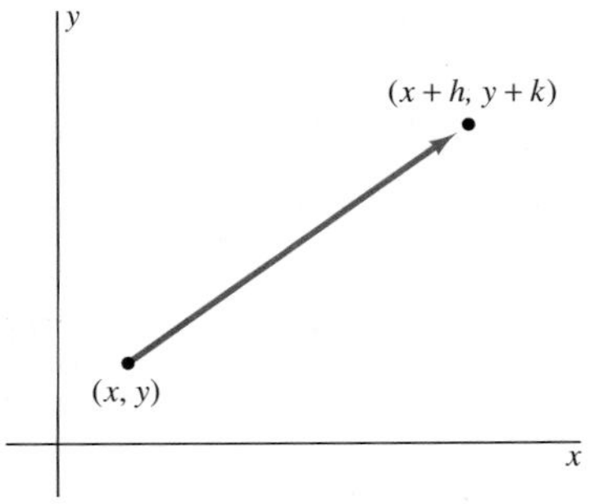

FIGURE 1.2.8 Translating a point.

is translated to the circle with radius r and center (h, k). Thus the general circle described by Eq. (6) is a *translate* of the origin-centered circle. Note that the equation of the translated circle is obtained from the original equation by replacing x with $x - h$ and y with $y - k$. This observation illustrates a general principle that describes equations of translated (or "shifted") graphs.

Translation Principle

When the graph of an equation is translated h units to the right and k units upward, the equation of the translated curve is obtained from the original equation by replacing x with $x - h$ and y with $y - k$.

Observe that we can write the equation of a translated circle in Eq. (6) in the general form

$$x^2 + y^2 + ax + by = c. \tag{7}$$

What, then, can we do when we encounter an equation already of the form in Eq. (7)? We first recognize that it is an equation of a circle. Next, we can discover its center and radius by the technique of *completing the square.* To do so, we note that

$$x^2 + ax = \left(x + \frac{a}{2}\right)^2 - \frac{a^2}{4},$$

which shows that $x^2 + ax$ can be made into the perfect square $(x + \frac{1}{2}a)^2$ by adding to it the square of *half* the coefficient of x.

EXAMPLE 3 Find the center and radius of the circle that has the equation

$$x^2 + y^2 - 4x + 6y = 12.$$

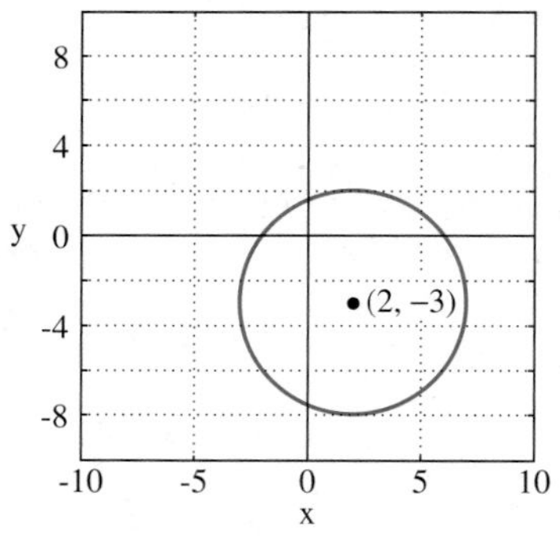

FIGURE 1.2.9 The circle of Example 3.

Solution We complete the square separately for each of the variables x and y. This gives

$$(x^2 - 4x + 4) + (y^2 + 6y + 9) = 12 + 4 + 9;$$

$$(x - 2)^2 + (y + 3)^2 = 25.$$

Hence the circle—shown in Fig. 1.2.9—has center $(2, -3)$ and radius 5. Solving the last equation for y gives

$$y = -3 \pm \sqrt{25 - (x - 2)^2}.$$

Thus the whole circle consists of the graphs of the *two* equations

$$y = -3 + \sqrt{25 - (x - 2)^2}$$

and

$$y = -3 - \sqrt{25 - (x - 2)^2}$$

that describe its upper and lower semicircles. ◆

Graphs of Functions

The graph of a function is a special case of the graph of an equation.

DEFINITION Graph of a Function

The **graph** of the function f is the graph of the equation $y = f(x)$.

Thus the graph of the function f is the set of all points in the plane that have the form $(x, f(x))$, where x is in the domain of f. (See Fig. 1.2.10.) Because the second

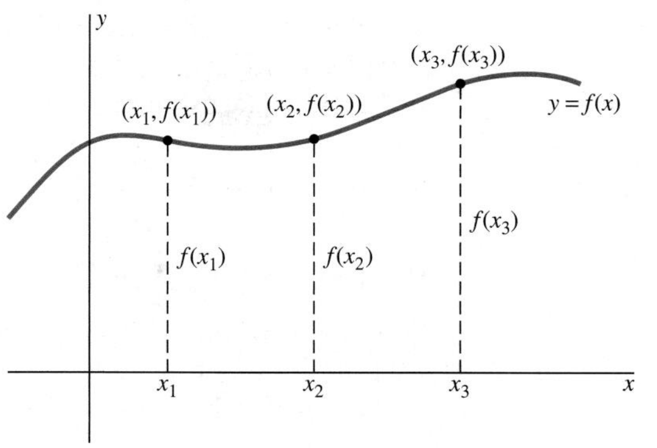

FIGURE 1.2.10 The graph of the function f.

coordinate of such a point is uniquely determined by its first coordinate, we obtain the following useful principle:

The Vertical Line Test

Each vertical line through a point in the domain of a function meets its graph in exactly one point.

Thus no vertical line can intersect the graph of a function in more than one point. For instance, it follows that the curve in Fig. 1.2.5 cannot be the graph of a *function,* although it *is* the graph of an equation. Similarly, a circle cannot be the graph of a function.

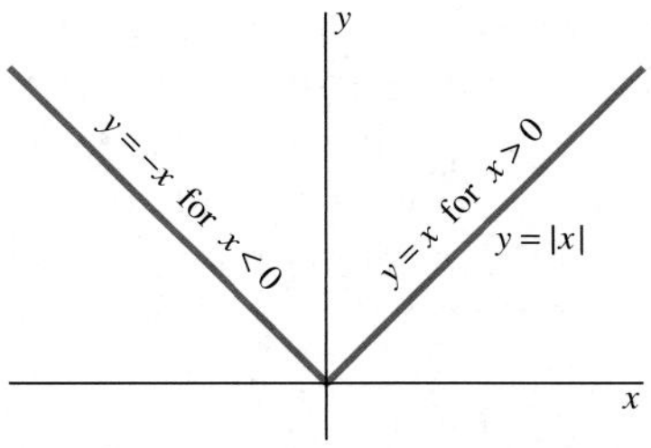

FIGURE 1.2.11 The graph of the absolute value function $y = |x|$ of Example 4.

EXAMPLE 4 Construct the graph of the absolute value function $f(x) = |x|$.

Solution Recall that

$$|x| = \begin{cases} x & \text{if } x \geqq 0, \\ -x & \text{if } x < 0. \end{cases}$$

So the graph of $y = |x|$ consists of the right half of the line $y = x$ together with the left half of the line $y = -x$, as shown in Fig. 1.2.11. ◆

EXAMPLE 5 Sketch the graph of the reciprocal function

$$f(x) = \frac{1}{x}.$$

Solution Let's examine four natural cases.

1. When x is positive and numerically large, $f(x)$ is small and positive.
2. When x is positive and near zero, $f(x)$ is large and positive.
3. When x is negative and numerically small (negative and close to zero), $f(x)$ is large and negative.
4. When x is large and negative (x is negative but $|x|$ is large), $f(x)$ is small and negative (negative and close to zero).

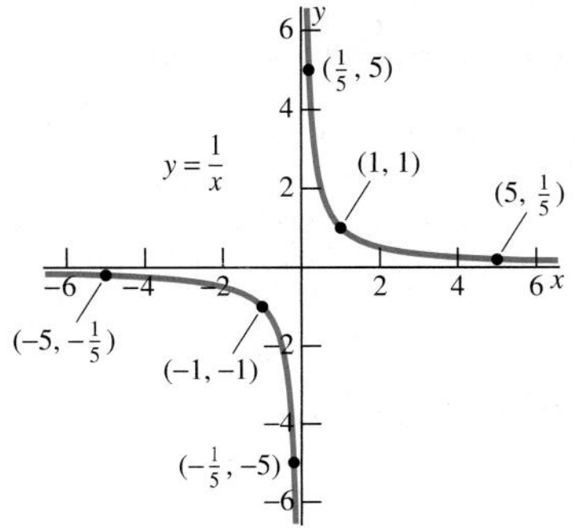

FIGURE 1.2.12 The graph of the reciprocal function $y = 1/x$ of Example 5.

To get started with the graph, we can plot a few points, such as

$$(1, 1), (-1, -1), \left(5, \tfrac{1}{5}\right), \left(\tfrac{1}{5}, 5\right), \left(-5, -\tfrac{1}{5}\right), \text{ and } \left(-\tfrac{1}{5}, -5\right).$$

The locations of these points, together with the four cases just discussed, suggest that the actual graph resembles the one shown in Fig. 1.2.12. ◆

Figure 1.2.12 exhibits a "gap," or "discontinuity," in the graph of $y = 1/x$ at $x = 0$. Indeed, the gap is called an *infinite discontinuity* because y increases without

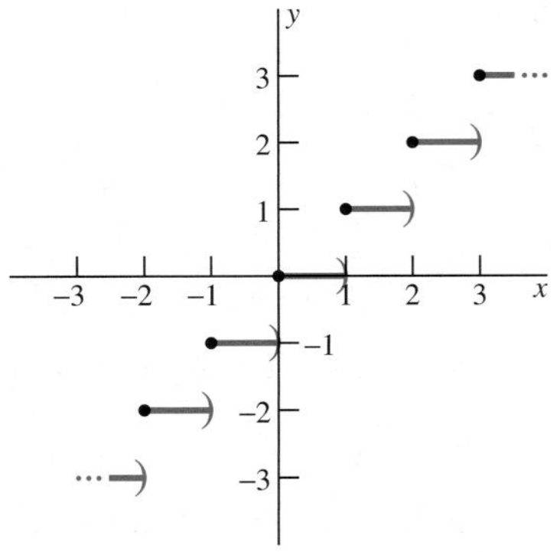

FIGURE 1.2.13 The graph of the greatest integer function $f(x) = [\![x]\!]$ of Example 6.

bound as x approaches zero from the right, whereas y decreases without bound as x approaches zero from the left. This phenomenon generally is signaled by the presence of denominators that are zero at certain values of x, as in the case of the functions

$$f(x) = \frac{1}{1-x} \quad \text{and} \quad f(x) = \frac{1}{x^2},$$

which we ask you to graph in the problems.

EXAMPLE 6 Figure 1.2.13 shows the graph of the greatest integer function $f(x) = [\![x]\!]$ in Example 4 in Section 1.1. Note the "jumps" that occur at integral values of x. On calculators, the greatest integer function is sometimes denoted by **INT**; in some programming languages, it is called `FLOOR`. ◆

EXAMPLE 7 Graph the function with the formula

$$f(x) = x - [\![x]\!] - \tfrac{1}{2}.$$

Solution Recall that $[\![x]\!] = n$, where n is the greatest integer not exceeding x—thus $n \leqq x < n + 1$. Hence if n is an integer, then

$$f(n) = n - n - \tfrac{1}{2} = -\tfrac{1}{2}.$$

This implies that the point $(n, -\frac{1}{2})$ lies on the graph of f for each integer n. Next, if $n \leqq x < n + 1$ (where, again, n is an integer), then

$$f(x) = x - n - \tfrac{1}{2}.$$

Because $y = x - n - \frac{1}{2}$ has as its graph a straight line of slope 1, it follows that the graph of f takes the form shown in Fig. 1.2.14. This *sawtooth function* is another example of a discontinuous function. The values of x where the value of $f(x)$ makes a jump are called **points of discontinuity** of the function f. Thus the points of discontinuity of the sawtooth function are the integers. As x approaches the integer n from the left, the value of $f(x)$ approaches $+\frac{1}{2}$, but $f(x)$ abruptly jumps to the value $-\frac{1}{2}$ when $x = n$. A precise definition of continuity and discontinuity for functions appears in Section 2.4. Figure 1.2.15 shows a graphing calculator prepared to graph the sawtooth function. ◆

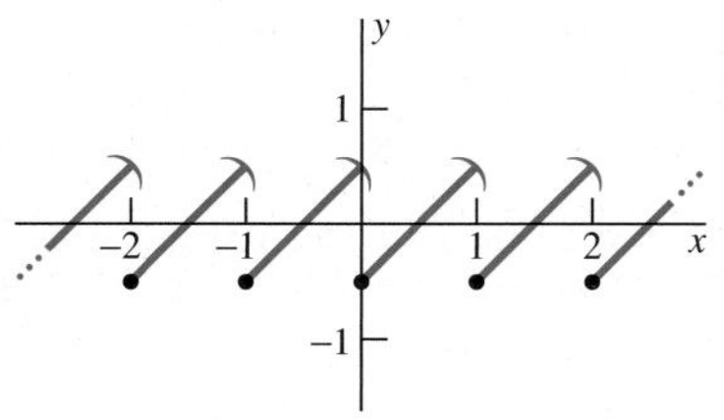

FIGURE 1.2.14 The graph of the sawtooth function $f(x) = x - [\![x]\!] - \frac{1}{2}$ of Example 7.

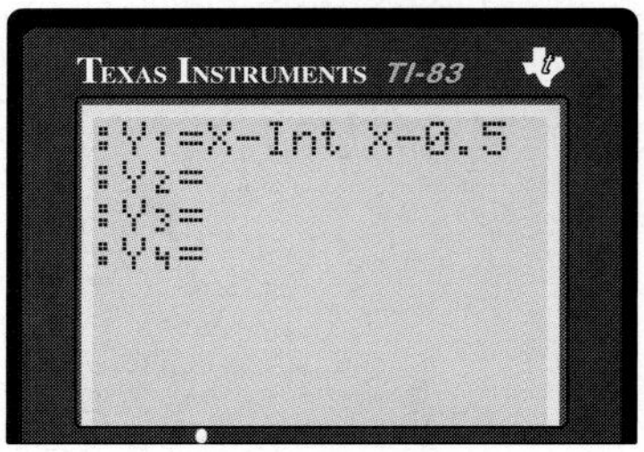

FIGURE 1.2.15 A graphing calculator prepared to graph the sawtooth function of Example 7.

Parabolas

The graph of a *quadratic* function of the form

$$f(x) = ax^2 + bx + c \quad (a \neq 0) \tag{8}$$

is a *parabola* whose shape resembles that of the particular parabola in Example 8.

EXAMPLE 8 Construct the graph of the parabola $y = x^2$.

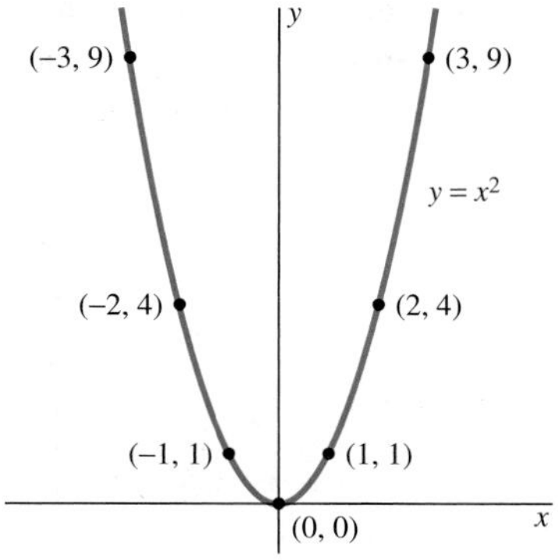

FIGURE 1.2.16 The graph of the parabola $y = x^2$ of Example 8.

Solution We plot some points in a short table of values.

x	-3	-2	-1	0	1	2	3
$y = x^2$	9	4	1	0	1	4	9

When we draw a smooth curve through these points, we obtain the curve shown in Fig. 1.2.16. ◆

The parabola $y = -x^2$ would look similar to the one in Fig. 1.2.16 but would open downward instead of upward. More generally, the graph of the equation

$$y = ax^2 \tag{9}$$

is a parabola with its *vertex* at the origin, provided that $a \neq 0$. This parabola opens upward if $a > 0$ and downward if $a < 0$. [For the time being, we may regard the vertex of a parabola as the point at which it "changes direction." The vertex of a parabola of the form $y = ax^2$ ($a \neq 0$) is always at the origin. A precise definition of the *vertex* of a parabola appears in Chapter 9.]

EXAMPLE 9 Construct the graphs of the functions $f(x) = \sqrt{x}$ and $g(x) = -\sqrt{x}$.

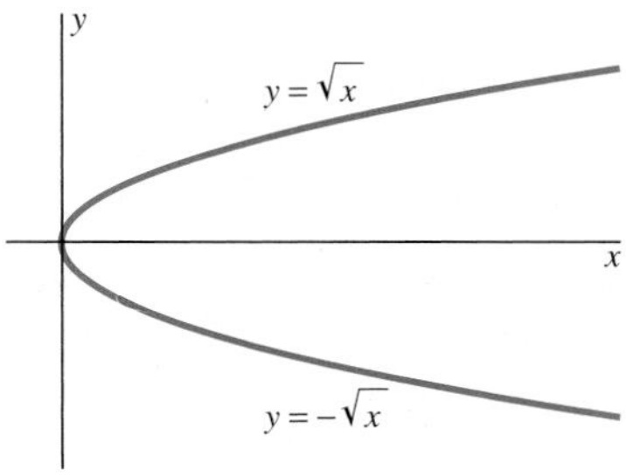

FIGURE 1.2.17 The graph of the parabola $x = y^2$ of Example 9.

Solution After plotting and connecting points satisfying $y = \pm\sqrt{x}$, we obtain the parabola $y^2 = x$ shown in Fig. 1.2.17. This parabola opens to the right. The upper half is the graph of $f(x) = \sqrt{x}$, the lower half is the graph of $g(x) = -\sqrt{x}$. Thus the union of the graphs of these two functions is the graph of the *single* equation $y^2 = x$. (Compare this with the circle of Example 3.) More generally, the graph of the equation

$$x = by^2 \tag{10}$$

is a parabola with its vertex at the origin, provided that $b \neq 0$. This parabola opens to the right if $b > 0$ (as in Fig. 1.2.17), but to the left if $b < 0$. ◆

The *size* of the coefficient a in Eq. (9) [or of b in Eq. (10)] determines the "width" of the parabola; its *sign* determines the direction in which the parabola opens. Specifically, the larger $a > 0$ is, the steeper the curve rises and hence the narrower the parabola is. (See Fig. 1.2.18.)

The parabola in Fig. 1.2.19 has the shape of the "standard parabola" in Example 8, but its vertex is located at the point (h, k). In the indicated uv-coordinate

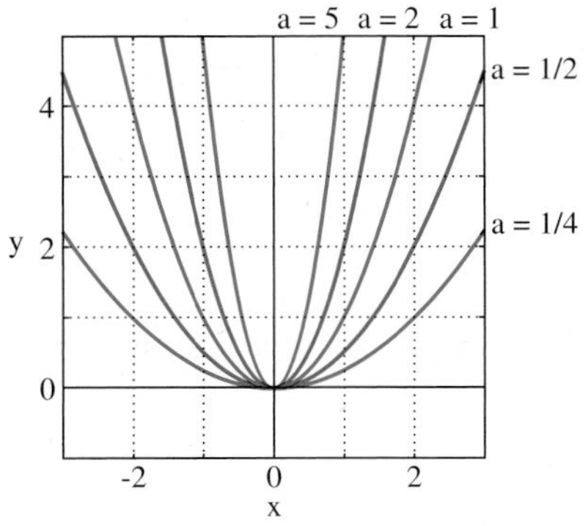

FIGURE 1.2.18 Parabolas with different widths.

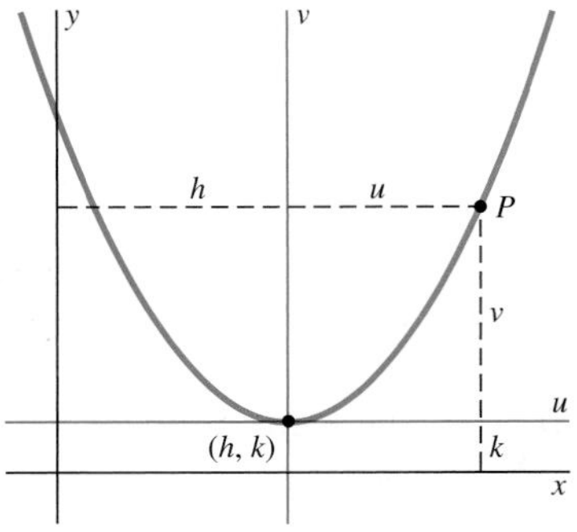

FIGURE 1.2.19 A translated parabola.

system, the equation of this parabola is $v = u^2$, in analogy with Eq. (9) with $a = 1$. But the uv-coordinates and xy-coordinates are related as follows:

$$u = x - h, \quad v = y - k.$$

Hence the xy-coordinate equation of this parabola is

$$y - k = (x - h)^2. \tag{11}$$

Thus when the parabola $y = x^2$ is translated h units to the right and k units upward, the equation in (11) of the translated parabola is obtained by replacing x with $x - h$ and y with $y - k$. This is another instance of the *translation principle* that we observed in connection with circles.

More generally, the graph of any equation of the form

$$y = ax^2 + bx + c \quad (a \neq 0) \tag{12}$$

can be recognized as a translated parabola by first completing the square in x to obtain an equation of the form

$$y - k = a(x - h)^2. \tag{13}$$

The graph of this equation is a parabola with its vertex at (h, k).

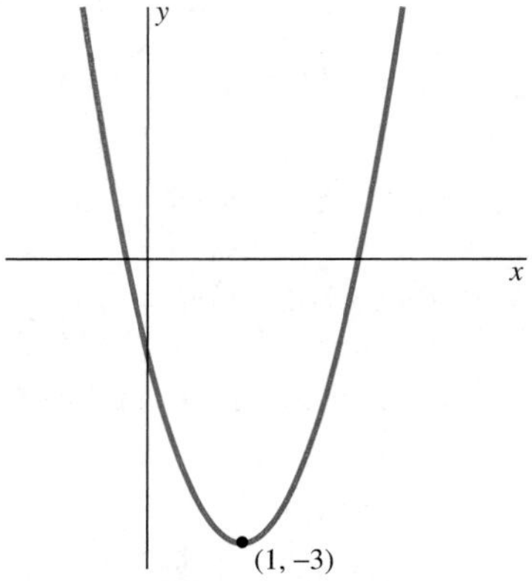

FIGURE 1.2.20 The parabola $y = 2x^2 - 4x - 1$ of Example 10.

EXAMPLE 10 Determine the shape of the graph of the equation

$$y = 2x^2 - 4x - 1. \tag{14}$$

Solution If we complete the square in x, Eq. (14) takes the form

$$y = 2(x^2 - 2x + 1) - 3;$$
$$y + 3 = 2(x - 1)^2.$$

Hence the graph of Eq. (14) is the parabola shown in Fig. 1.2.20. It opens upward and its vertex is at (1, –3). ◆

Applications of Quadratic Functions

In Section 1.1 we saw that a certain type of applied problem may call for us to find the maximum or minimum attained by a certain function f. If the function f is a quadratic function as in Eq. (8), then the graph of $y = f(x)$ is a parabola. In this case the maximum (or minimum) value of $f(x)$ corresponds to the highest (or lowest) point of the parabola. We can therefore find this maximum (or minimum) value graphically—at least approximately—by zooming in on the vertex of the parabola.

For instance, recall the animal pen problem of Section 1.1. In Example 9 there we saw that the area A of the pen (see Fig. 1.2.21) is given as a function of its base length x by

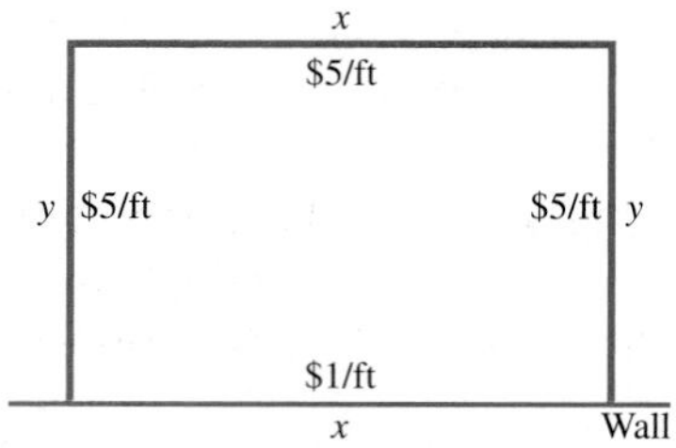

FIGURE 1.2.21 The animal pen.

$$A(x) = \tfrac{3}{5}(30x - x^2), \quad 0 \leqq x \leqq 30. \tag{15}$$

Figure 1.2.22 shows the graph $y = A(x)$, and Figs. 1.2.23, 1.2.24, and 1.2.25 show successive magnifications of the region near the high point (vertex) of the parabola. The dashed rectangle in each figure is the viewing window for the next. Figure 1.2.25 makes it *seem* that the maximum area of the pen is $A(15) = 135$. It is clear from the figure that the maximum value of $A(x)$ is within 0.001 of $A = 135$.

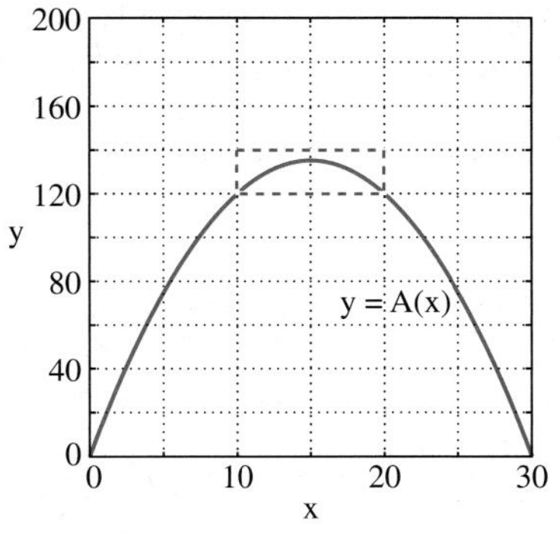

FIGURE 1.2.22 The graph $y = A(x)$.

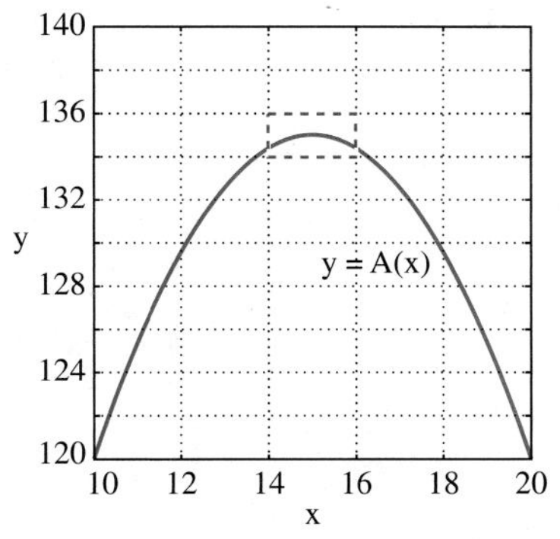

FIGURE 1.2.23 The first zoom.

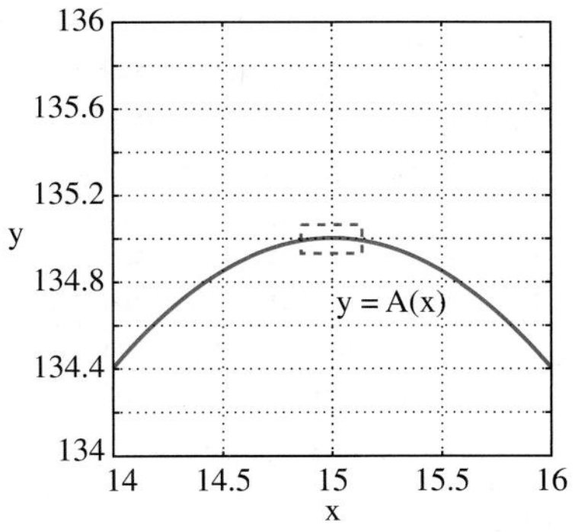

FIGURE 1.2.24 The second zoom.

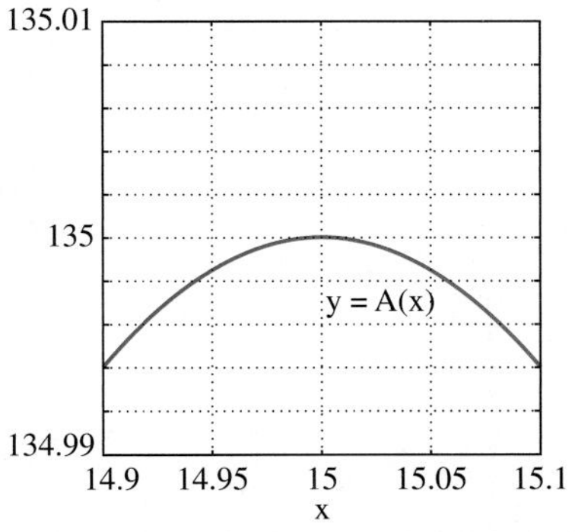

FIGURE 1.2.25 The third zoom.

We can verify by completing the square as in Example 10 that the maximum value is *precisely* $A(15) = 135$:

$$A = -\tfrac{3}{5}(x^2 - 30x) = -\tfrac{3}{5}(x^2 - 30x + 225 - 225)$$
$$= -\tfrac{3}{5}(x^2 - 30x + 225) + 135;$$

that is,

$$A - 135 = -\tfrac{3}{5}(x - 15)^2. \qquad \textbf{(16)}$$

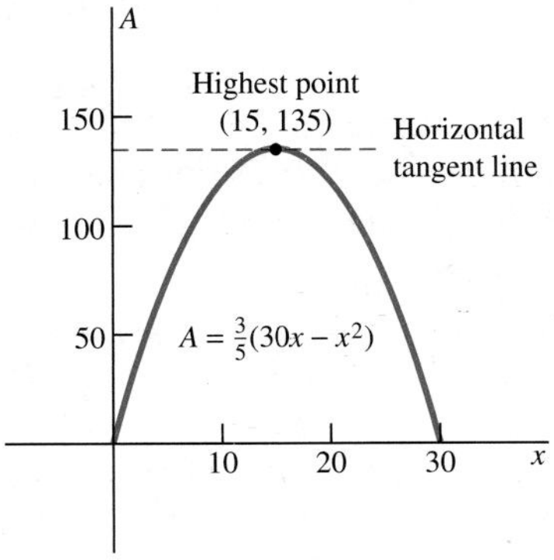

FIGURE 1.2.26 The graph of $A(x) = \frac{3}{5}(30x - x^2)$ for $0 \leqq x \leqq 30$.

It follows from Eq. (16) that the graph of Eq. (15) is the parabola shown in Fig. 1.2.26, which opens downward from its vertex (15, 135). This *proves* that the maximum value of $A(x)$ on the interval [0, 30] is the value $A(15) = 135$, as both our numerical investigations in Section 1.1 and our graphical investigations here suggest. And when we glance at Eq. (16) in the form

$$A(x) = 135 - \tfrac{3}{5}(x - 15)^2,$$

it's clear and unarguable that the maximum possible value of $135 - \frac{3}{5}u^2$ is 135 when $u = x - 15 = 0$—that is, when $x = 15$.

The technique of completing the square is quite limited: It can be used to find maximum or minimum values only of *quadratic* functions. One of the goals in calculus is to develop a more general technique that can be applied to a far wider variety of functions.

The basis of this more general technique lies in the following observation. Visual inspection of the graph of

$$A(x) = \tfrac{3}{5}(30x - x^2)$$

in Fig. 1.2.26 suggests that the line tangent to the curve at its highest point is horizontal. If we *knew* that the tangent line to a graph at its highest point must be horizontal, then our problem would reduce to showing that (15, 135) is the only point of the graph of $y = A(x)$ at which the tangent line is horizontal.

But what do we mean by the *tangent line* to an arbitrary curve? We pursue this question in Section 2.1. The answer will open the door to the possibility of finding the maximum and minimum values of a wide variety of functions.

Graphic, Numeric, and Symbolic Viewpoints

An equation $y = f(x)$ provides a *symbolic* description of the function f. A table of values of f (like those in Section 1.1) is a *numeric* representation of the function, whereas this section deals largely with *graphic* representations of functions. Interesting applications often involve looking at the same function from at least two of these three viewpoints.

EXAMPLE 11 Suppose that a car begins (at time $t = 0$ hours) in Athens, Georgia (position $x = 0$ miles) and travels to Atlanta (position $x = 60$) with a constant speed of 60 mi/h. The car stays in Atlanta for exactly one hour, then returns to Athens, again with a constant speed of 60 mi/h. Describe the car's "position function" both graphically and symbolically.

Solution It's fairly clear that $x = 60t$ during the 1-hour trip from Athens to Atlanta; for instance, after $t = \frac{1}{2}$ hour the car has traveled halfway, so $x = 30 = \frac{1}{2} \cdot 60$. During the next hour, $1 \leqq t \leqq 2$, the car's position is constant, $x \equiv 60$. And perhaps you can see that during the return trip of the third hour, $2 \leqq t \leqq 3$, the car's position is given by

$$x = 60 - 60(t - 2) = 180 - 60t$$

(so that $x(2) = 60$ and $x(3) = 0$). Thus the position function $x(t)$ is defined symbolically by

$$x(t) = \begin{cases} 60t & \text{if } 0 \leqq t \leqq 1, \\ 60 & \text{if } 1 < t \leqq 2, \\ 180 - 60t & \text{if } 2 < t \leqq 3. \end{cases}$$

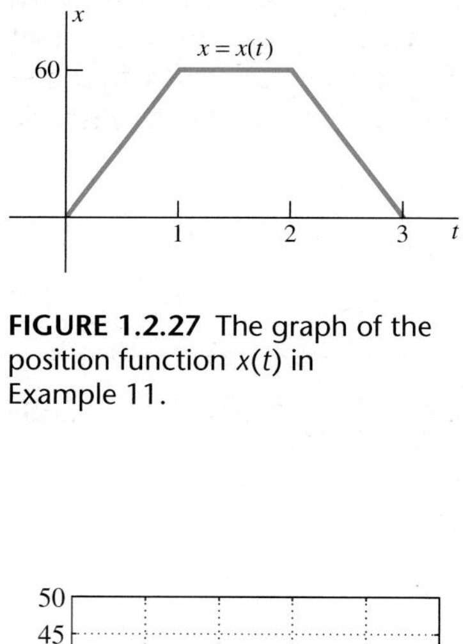

FIGURE 1.2.27 The graph of the position function $x(t)$ in Example 11.

The domain of this function is the t-interval [0, 3] and its graph is shown in Fig. 1.2.27, where we denote both the function and the dependent variable by the same symbol x (an abuse of notation that's not uncommon in applications). ◆

EXAMPLE 12 During the decade of the 1980s the population P (in thousands) of a small but rapidly growing city was recorded in the following table.

Year	1980	1982	1984	1986	1988	1990
t	0	2	4	6	8	10
P	27.00	29.61	32.48	35.62	39.07	42.85

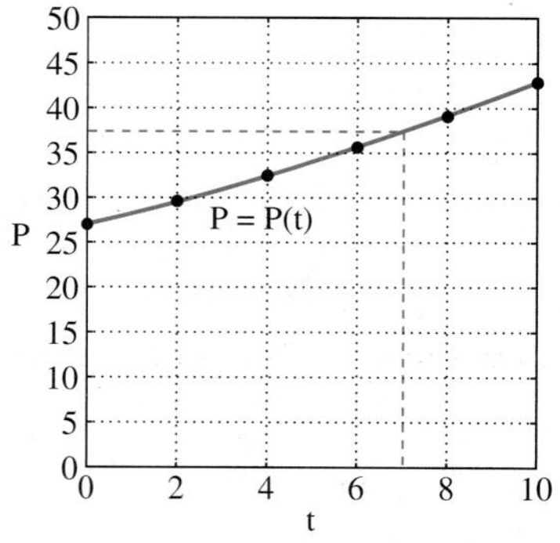

FIGURE 1.2.28 The population function of Example 12.

Estimate the population of this city in the year 1987.

Solution Figure 1.2.28 shows a graph of the population function $P(t)$ obtained by connecting the six given data points $(t, P(t))$ with a smooth curve. A careful measurement of the height of the point on this curve at which $t = 7$ yields the approximate population $P(7) \approx 37.4$ (thousand) of the city in 1987. ◆

1.2 TRUE/FALSE STUDY GUIDE

1.2 CONCEPTS: QUESTIONS AND DISCUSSION

1. Two general forms of equations of straight lines are reviewed at the beginning of this section. Describe a straight line for which the *slope-intercept equation* would be the one more convenient to use in writing an equation of the line. Then describe a line for which the *point-slope equation* would be more convenient.
2. (a) What is the difference between a line that has slope zero and a line that has no slope? If two lines are perpendicular and one of them has slope zero, what is the slope of the other line? (b) Let L_1 and L_2 be two perpendicular lines having slopes m_1 and m_2, respectively. Theorem 2 in Appendix B asserts that L_1 and L_2 are perpendicular if and only if $m_1m_2 = -1$. Is this assertion true in case L_1 is the x-axis and L_2 is the y-axis? Or is there an oversight in the statement of Theorem 2 in Appendix B?
3. (a) Sketch the graph of the equation $|x| + |y| = 1$. Is this graph the graph of some function? Justify your answer. (b) Repeat part (a), but with the equation $|x + y| = 1$.
4. (a) Suppose that f is a function such that $f(x) > 0$ for all real x. Discuss the question of whether the graph of the given equation is the graph of *some* function.

   (i) $y^2 = f(x)$; (ii) $|y| = f(x)$; (iii) $y = |f(x)|$.

   (b) Repeat part (a), but assume that $f(x) < 0$ for all x. (c) Repeat part (a), but assume that f has both positive and negative values. For instance, sketch the graphs of the equations in (i), (ii), and (iii) if $f(x) = x^2 - 1$.
5. Newspaper articles often describe or refer to functions (either explicitly or implicitly) but rarely contain equations. Find and discuss examples of numeric and graphic representations of functions in a typical issue of your local newspaper. Also see if you can find a reference to a function that is described verbally but without either a graphic or a numeric representation.

1.2 PROBLEMS

In Problems 1 through 10, write an equation of the line L described and sketch its graph.

1. L passes through the origin and the point $(2, 3)$.
2. L is vertical and has x-intercept 7.
3. L is horizontal and passes through $(3, -5)$.
4. L has x-intercept 2 and y-intercept -3.
5. L passes through $(2, -3)$ and $(5, 3)$.
6. L passes through $(-1, -4)$ and has slope $\frac{1}{2}$.
7. L passes through $(4, 2)$ and has angle of inclination $135°$.
8. L has slope 6 and y-intercept 7.
9. L passes through $(1, 5)$ and is parallel to the line with equation $2x + y = 10$.
10. L passes through $(-2, 4)$ and is perpendicular to the line with equation $x + 2y = 17$.

Sketch the translated circles in Problems 11 through 16. Indicate the center and radius of each.

11. $x^2 + y^2 = 4x$
12. $x^2 + y^2 + 6y = 0$
13. $x^2 + y^2 + 2x + 2y = 2$
14. $x^2 + y^2 + 10x - 20y + 100 = 0$
15. $2x^2 + 2y^2 + 2x - 2y = 1$
16. $9x^2 + 9y^2 - 6x - 12y = 11$

Sketch the translated parabolas in Problems 17 through 22. Indicate the vertex of each.

17. $y = x^2 - 6x + 9$
18. $y = 16 - x^2$
19. $y = x^2 + 2x + 4$
20. $2y = x^2 - 4x + 8$
21. $y = 5x^2 + 20x + 23$
22. $y = x - x^2$

The graph of the equation $(x - h)^2 + (y - k)^2 = C$ is a circle if $C > 0$, is the single point (h, k) if $C = 0$, and contains no points if $C < 0$. (Why?) Identify the graphs of the equations in Problems 23 through 26. If the graph is a circle, give its center and radius.

23. $x^2 + y^2 - 6x + 8y = 0$

24. $x^2 + y^2 - 2x + 2y + 2 = 0$

25. $x^2 + y^2 + 2x + 6y + 20 = 0$

26. $2x^2 + 2y^2 - 2x + 6y + 5 = 0$

Sketch the graphs of the functions in Problems 27 through 50. Take into account the domain of definition of each function, and plot points as necessary.

27. $f(x) = 2 - 5x, \quad -1 \leqq x \leqq 1$

28. $f(x) = 2 - 5x, \quad 0 \leqq x < 2$

29. $f(x) = 10 - x^2$

30. $f(x) = 1 + 2x^2$

31. $f(x) = x^3$

32. $f(x) = x^4$

33. $f(x) = \sqrt{4 - x^2}$

34. $f(x) = -\sqrt{9 - x^2}$

35. $f(x) = \sqrt{x^2 - 9}$

36. $f(x) = \dfrac{1}{1 - x}$

37. $f(x) = \dfrac{1}{x + 2}$

38. $f(x) = \dfrac{1}{x^2}$

39. $f(x) = \dfrac{1}{(x - 1)^2}$

40. $f(x) = \dfrac{|x|}{x}$

41. $f(x) = \dfrac{1}{2x + 3}$

42. $f(x) = \dfrac{1}{(2x + 3)^2}$

43. $f(x) = \sqrt{1 - x}$

44. $f(x) = \dfrac{1}{\sqrt{1 - x}}$

45. $f(x) = \dfrac{1}{\sqrt{2x + 3}}$

46. $f(x) = |2x - 2|$

47. $f(x) = |x| + x$

48. $f(x) = |x - 3|$

49. $f(x) = |2x + 5|$

50. $f(x) = \begin{cases} |x| & \text{if } x < 0, \\ x^2 & \text{if } x \geqq 0 \end{cases}$

Sketch graphs of the functions given in Problems 51 through 56. Indicate any points of discontinuity.

51. $f(x) = \begin{cases} 0 & \text{if } x < 0, \\ 1 & \text{if } x \geqq 0 \end{cases}$

52. $f(x) = \begin{cases} 1 & \text{if } x \text{ is an integer,} \\ 0 & \text{otherwise} \end{cases}$

53. $f(x) = [\![2x]\!]$

54. $f(x) = \dfrac{x - 1}{|x - 1|}$

55. $f(x) = [\![x]\!] - x$

56. $f(x) = [\![x]\!] + [\![-x]\!] + 1$

In Problems 57 through 64, use a graphing calculator or computer to find (by zooming) the highest or lowest (as appropriate) point P on the given parabola. Determine the coordinates of P with two digits to the right of the decimal correct or correctly rounded. Then verify your result by completing the square to find the actual vertex of the parabola.

57. $y = 2x^2 - 6x + 7$

58. $y = 2x^2 - 10x + 11$

59. $y = 4x^2 - 18x + 22$

60. $y = 5x^2 - 32x + 49$

61. $y = -32 + 36x - 8x^2$

62. $y = -53 - 34x - 5x^2$

63. $y = 3 - 8x - 3x^2$

64. $y = -28 + 34x - 9x^2$

In Problems 65 through 68, use the method of completing the square to graph the appropriate function and thereby determine the maximum or minimum value requested.

65. If a ball is thrown straight upward with initial velocity 96 ft/s, then its height t seconds later is $y = 96t - 16t^2$ (ft). Determine the maximum height that the ball attains.

66. Find the maximum possible area of the rectangle described in Problem 40 of Section 1.1.

67. Find the maximum possible value of the product of two positive numbers whose sum is 50.

68. In Problem 42 of Section 1.1, you were asked to express the daily production of a specific oil field as a function $P = f(x)$ of the number x of new oil wells drilled. Construct the graph of f and use it to find the value of x that maximizes P.

In Problems 69 through 72 write a symbolic description of the function whose graph is pictured. You may use the greatest integer function of Examples 6 and 7.

69. Figure 1.2.29

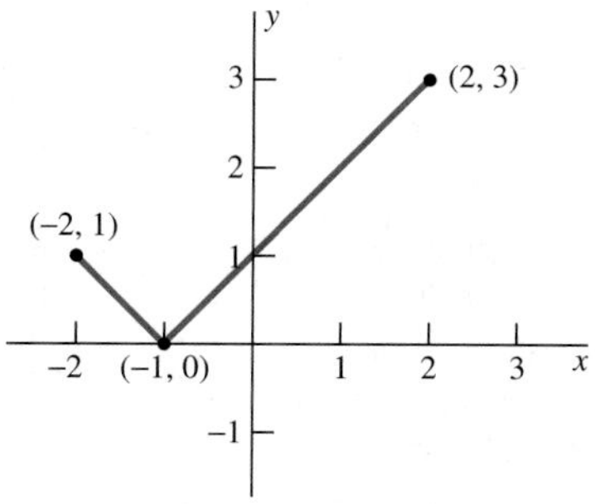

FIGURE 1.2.29 Problem 69.

70. Figure 1.2.30

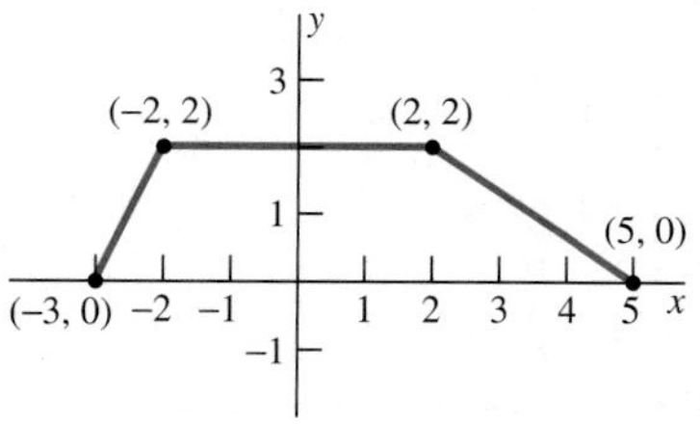

FIGURE 1.2.30 Problem 70.

71. Figure 1.2.31

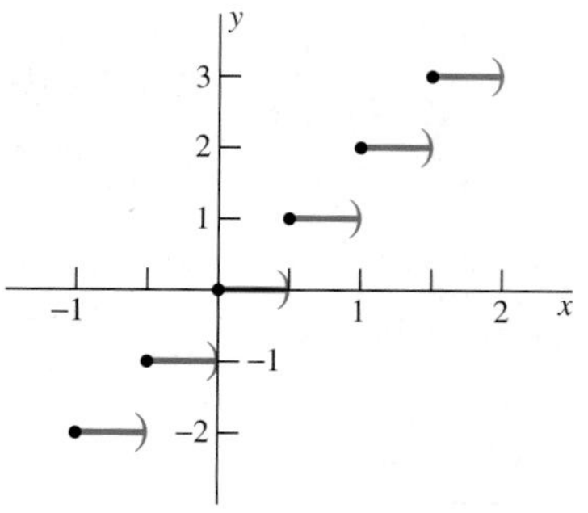

FIGURE 1.2.31 Problem 71.

72. Figure 1.2.32

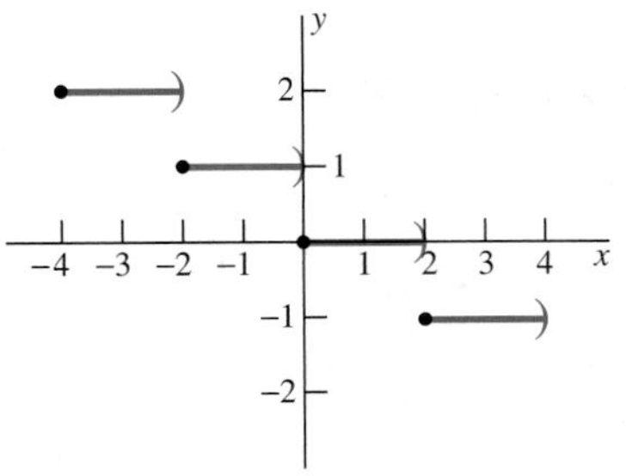

FIGURE 1.2.32 Problem 72.

Each of Problems 73 through 76 describes a trip you made along a straight road connecting two cities 120 miles apart. Sketch the graph of the distance x from your starting point (in miles) as a function of the time t elapsed (in hours). Also describe the function $x(t)$ symbolically.

73. You traveled for one hour at 45 mi/h, then realized you were going to be late, and therefore traveled at 75 mi/h for the next hour.

74. You traveled for one hour at 60 mi/h, stopped for a half hour while a herd of caribou crossed the road, then drove on toward your destination for the next hour at 60 mi/h.

75. You traveled for one hour at 60 mi/h, were suddenly engulfed in a dense fog, and drove back home at 30 mi/h.

76. You traveled for a half hour at 60 mi/h, suddenly remembered you had left your wallet at home, drove back at 60 mi/h to get it, and finally drove for two hours at 60 mi/h toward your original destination.

77. Suppose that the cost C of printing a pamphlet of at most 100 pages is a linear function of the number p of pages it contains. It costs \$1.70 to print a pamphlet with 34 pages, whereas a pamphlet with 79 pages costs \$3.05. (a) Express C as a function of p. Use this function to find the cost of printing a pamphlet with 50 pages. (b) Sketch the straight line graph of the function $C(p)$. Tell what the slope and the C-intercept of this line mean—perhaps in terms of the "fixed cost" to set up the press for printing and the "marginal cost" of each additional page printed.

78. Suppose that the cost C of renting a car for a day is a linear function of the number x of miles you drive that day. On day 1 you drove 207 miles and the cost was \$99.45. On day 2 you drove 149 miles and the cost was \$79.15. (a) Express C as a function of x. Use this function to find the cost for day 3 if you drove 175 miles. (b) Sketch the straight line graph of the function $C(x)$. Tell what the slope and the C-intercept of this line mean—perhaps in terms of fixed and marginal costs as in Problem 77.

79. For a Federal Express letter weighing at most one pound sent to a certain destination, the charge C is \$8.00 for the first 8 ounces plus 80¢ for each additional ounce or fraction thereof. Sketch the graph of this function C of the total number x of ounces, and describe it symbolically in terms of the greatest integer function of Examples 6 and 7.

80. In a certain city, the charge C for a taxi trip of at most 20 miles is \$3.00 for the first 2 miles (or fraction thereof), plus 50¢ for each half-mile (or part thereof) up to a total of 10 miles, plus 50¢ for each mile (or part thereof) over 10 miles. Sketch the graph of this function C of the number x of miles and describe it symbolically in terms of the greatest integer function of Examples 6 and 7.

81. The volume V (in liters) of a sample of 3 g of carbon dioxide at 27°C was measured as a function of its pressure p (in atmospheres) with the results shown in the following table:

p	0.25	1.00	2.50	4.00	6.00
V	6.72	1.68	0.67	0.42	0.27

Sketch the graph of the function $V(p)$ and use the graph to estimate the volumes of the gas sample at pressures of 0.5 and 5 atmospheres.

82. The average temperature T (in °F) in Athens, Georgia was measured at two-month intervals, with the results shown in the following table:

Date	Jul 15	Sep 15	Nov 15	Jan 15	Mar 15	May 15
T	79.1	70.2	52.3	43.4	52.2	70.1

Sketch the graph of T as a function of the number of days after July 15. Then use your graph to estimate the average temperature on October 15 and on April 15.

1.2 Project: A Broken Tree

A 50-ft tree stands 10 ft from a fence 10 feet high. The tree is suddenly "broken" part of the way up. You are to determine the height of the break so that the tree falls with its trunk barely touching the top of the fence when the tip of the tree strikes the ground on the other side of the fence. The key is the use of simple geometry to derive the equations

$$y = \frac{100}{x - 10}, \tag{1}$$

$$(y + 10)^2 = 2500 - 100x \tag{2}$$

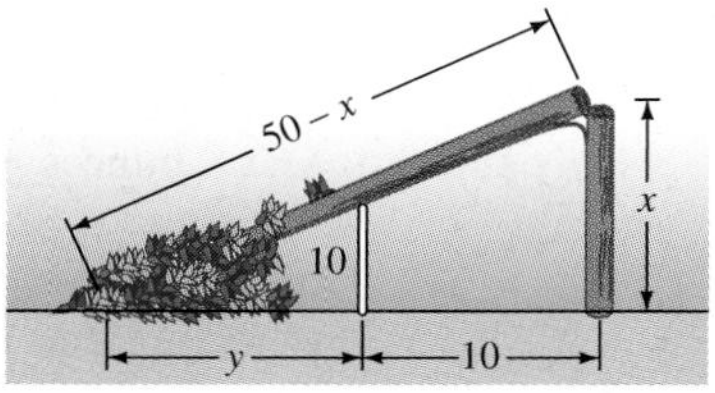

FIGURE 1.2.33 The broken tree.

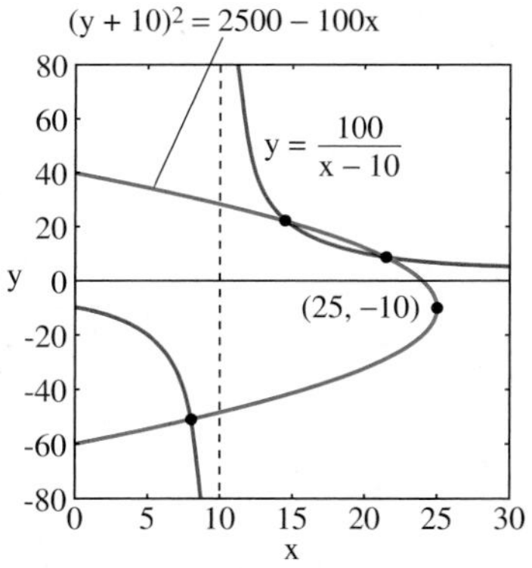

FIGURE 1.2.34 The hyperbola and parabola in the broken tree investigation.

relating the lengths x and y indicated in Fig. 1.2.33. The graph of Eq. (1) is a translated *rectangular hyperbola*, while the graph of Eq. (2) is a translated parabola (Fig. 1.2.34). You can use a graphing calculator or computer to locate the pertinent point(s) of intersection of these two graphs.

1.3 POLYNOMIALS AND ALGEBRAIC FUNCTIONS

In this section and the next we briefly survey a variety of functions that are used in applications of calculus to describe and model changing phenomena in the world around us. Our viewpoint here is largely graphical. The objective is for you to attain a general understanding of major differences between different types of functions. In later chapters we use calculus to investigate further the graphs presented here.

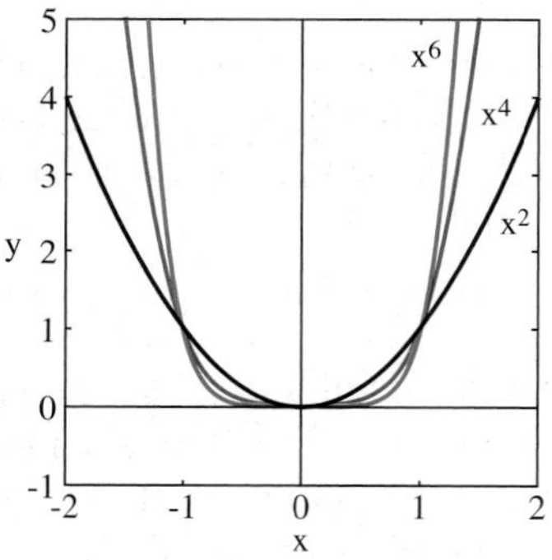

FIGURE 1.3.1 Graphs of power functions of even degree (Example 1).

Power Functions

A function of the form $f(x) = x^k$ (where k is a constant) is called a **power function**. If $k = 0$ then we have the constant function $f(x) \equiv 1$. The shape of the graph of a power function with exponent $k = n$, a positive integer, depends on whether n is even or odd.

EXAMPLE 1 The graphs of the *even*-degree power functions $x^2, x^4, x^6, \ldots$ all "cup upward," as indicated in Fig. 1.3.1. If $n > 2$ is an *even* integer then the graph $y = x^n$ resembles the parabola $y = x^2$, but is flatter near the origin and steeper when $|x| > 1$.

The graphs of the *odd*-degree power functions $x^1, x^3, x^5, \ldots$ all go "from southwest to northeast," as indicated in Fig. 1.3.2. If $n > 1$ is an *odd* integer then the graph $y = x^n$ resembles that of $y = x^3$, but again is flatter near the origin and steeper when $|x| > 1$. ◆

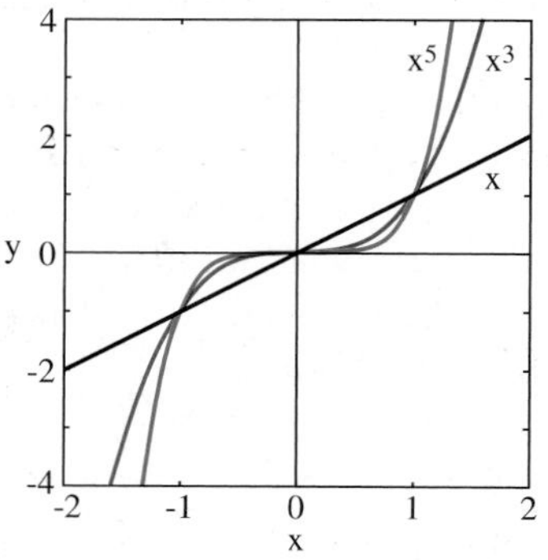

FIGURE 1.3.2 Graphs of power functions of odd degree (Example 1).

Note that all the power function graphs in Figs. 1.3.1 and 1.3.2 pass through the origin, through the point (1, 1), and either through (−1, 1) or (−1, −1), depending on whether n is even or odd. In either case, x^n increases numerically (either positively or negatively) as x does. Would you agree that the notation

$$x^n \to +\infty \quad \text{as} \quad x \to +\infty, \qquad x^n \to \begin{cases} +\infty & \text{as } x \to -\infty & \text{if } n \text{ is even,} \\ -\infty & \text{as } x \to -\infty & \text{if } n \text{ is odd} \end{cases}$$

(with the arrow signifying "goes to") provides a convenient and suggestive description of the general features, when $|x|$ becomes large, of the graphs in Figs. 1.3.1 and 1.3.2?

The graph $y = x^k$ may have a quite different appearance if the exponent k is not a positive integer. If k is a negative integer—say, $k = -m$ where m is a positive integer—then

$$f(x) = x^k = x^{-m} = \frac{1}{x^m},$$

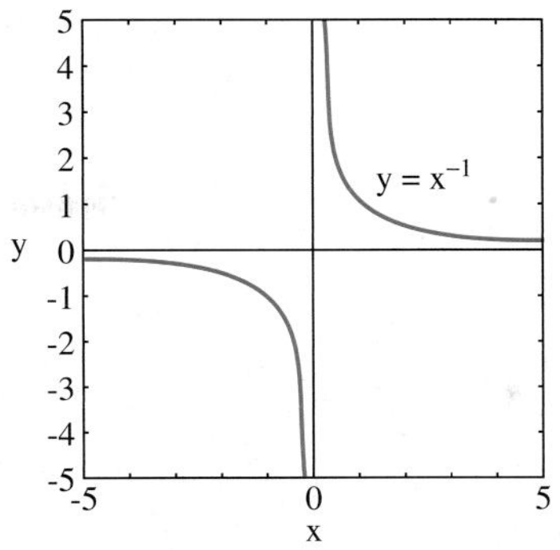

FIGURE 1.3.3 $y = \dfrac{1}{x}$.

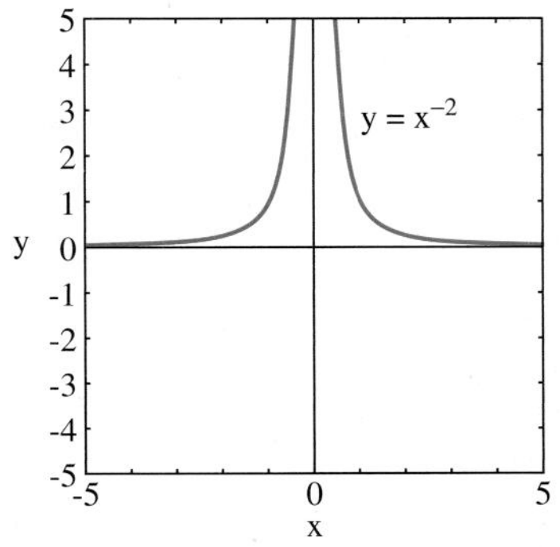

FIGURE 1.3.4 $y = \dfrac{1}{x^2}$.

so in this case the power function is the reciprocal of a function like those in Example 1. Figures 1.3.3 and 1.3.4 show the graphs of

$$y = x^{-1} = \frac{1}{x} \quad \text{and} \quad y = x^{-2} = \frac{1}{x^2},$$

respectively. Observe that 0 is not in the domain of such a function. Moreover, the reciprocal of a number close to zero is very large in magnitude, which explains the behavior of these graphs near zero: In both graphs, $|y|$ is very large—so the point (x, y) is either very high or very low—when x is close to zero.

The graph $y = x^k$ may be undefined if $x \leqq 0$ and k is not an integer. In the simplest such case, when k is *irrational,* we do not attempt to define x^k if $x < 0$, so the graph of x^k exists only for $x \geqq 0$.

The situation is still more complicated if the exponent k is not an integer. We do not (at present) attempt to define the expression x^k if k is irrational—that is, not a quotient of integers. But if $k = m/n$ is rational, with the integers m and n having no common integral factor larger than 1, then we can write

$$x^k = x^{m/n} = \sqrt[n]{x^m},$$

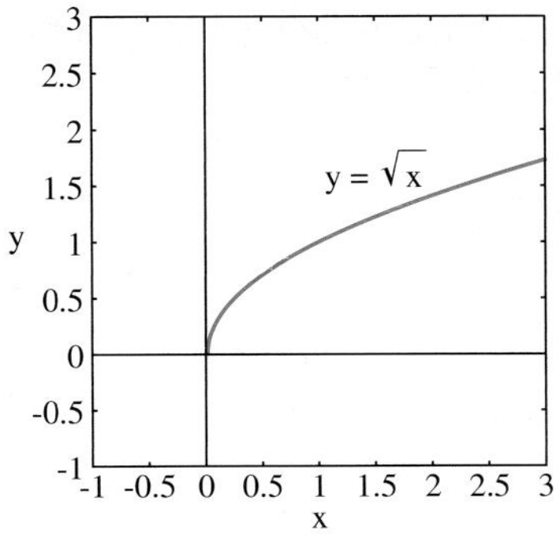

FIGURE 1.3.5 $y = x^{1/2}$.

and thereby interpret $f(x) = x^k$ as a "root function." If n is odd then $\sqrt[n]{x^m}$ is defined for all real x if m is positive and for all nonzero values of x if m is negative. But if n is even and m is odd, then the root $\sqrt[n]{x^m}$ is not defined for negative x values.

The typical behavior of such root functions is illustrated by the graphs of $y = x^{1/2} = \sqrt{x}$ and $y = x^{1/3} = \sqrt[3]{x}$ shown in Figs. 1.3.5 and 1.3.6. The square root $\sqrt{x}$ is defined only for $x \geqq 0$. The cube root $\sqrt[3]{x}$ is defined for all x, but observe that its graph appears to be tangent to the y-axis at the origin.

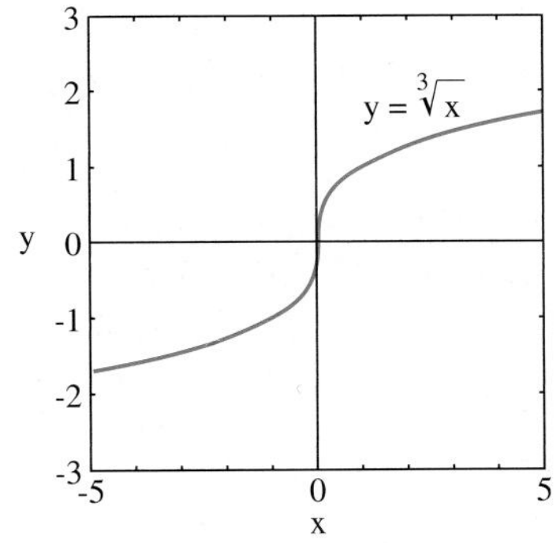

FIGURE 1.3.6 $y = x^{1/3}$.

Combinations of Functions

Many varied and complicated functions can be assembled out of simple "building-block functions." Here we discuss some of the ways of combining functions to obtain new ones.

Suppose that f and g are functions and that c is a fixed real number. The **(scalar) multiple** cf, the **sum** $f + g$, the **difference** $f - g$, the **product** $f \cdot g$, and the **quotient** f/g are the new functions with the following formulas:

$$(cf)(x) = c \cdot f(x), \tag{1}$$

$$(f + g)(x) = f(x) + g(x), \tag{2}$$

$$(f - g)(x) = f(x) - g(x), \tag{3}$$

$$(f \cdot g)(x) = f(x) \cdot g(x), \quad \text{and} \tag{4}$$

$$\left(\frac{f}{g}\right)(x) = \frac{f(x)}{g(x)}. \tag{5}$$

The combinations in Eqs. (2) through (4) are defined for every number x that lies both in the domain of f *and* in the domain of g. In Eq. (5) we must also require that $g(x) \neq 0$.

EXAMPLE 2 Let $f(x) = x^2 + 1$ and $g(x) = x - 1$. Then:

$$\begin{aligned}(3f)(x) &= 3(x^2 + 1),\\ (f+g)(x) &= (x^2+1) + (x-1) = x^2 + x,\\ (f-g)(x) &= (x^2+1) - (x-1) = x^2 - x + 2,\\ (f\cdot g)(x) &= (x^2+1)(x-1) = x^3 - x^2 + x - 1, \quad \text{and}\\ \left(\frac{f}{g}\right)(x) &= \frac{x^2+1}{x-1} \quad (x \neq 1).\end{aligned}$$

◆

EXAMPLE 3 If $f(x) = \sqrt{1-x}$ for $x \leqq 1$ and $g(x) = \sqrt{1+x}$ for $x \geqq -1$, then the sum and product of f and g are defined where *both* f and g are defined. Thus the domain of both

$$f(x) + g(x) = \sqrt{1-x} + \sqrt{1+x}$$

and

$$f(x)\cdot g(x) = \sqrt{1-x}\,\sqrt{1+x} = \sqrt{1-x^2}$$

is the closed interval $[-1, 1]$. But the domain of the quotient

$$\frac{f(x)}{g(x)} = \frac{\sqrt{1-x}}{\sqrt{1+x}} = \sqrt{\frac{1-x}{1+x}}$$

is the half-open interval $(-1, 1]$, because $g(-1) = 0$. ◆

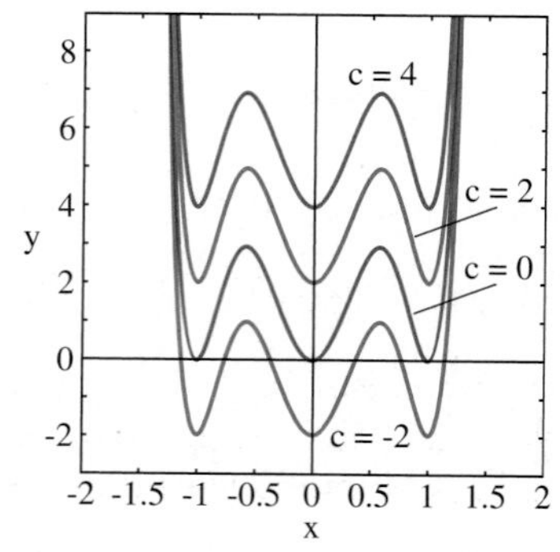

FIGURE 1.3.7 $y = 20x^2(x^2-1)^2 + c$ for $c = -2, 0, 2, 4$.

The results of algebraic operations can sometimes be visualized with the aid of geometric interpretations of the operations. Figures 1.3.7 through 1.3.10 show the results of various operations involving the function $f(x) = 20x^2(x^2-1)^2$. Adding a constant simply shifts the graph vertically, as in Fig. 1.3.7, which shows $y = f(x) + c$ for $c = -2$, 0, 2, and 4. Multiplication by a positive constant c expands (if $c > 1$) or contracts (if $0 < c < 1$) the graph in the vertical direction, as in Fig. 1.3.8, which shows $y = cf(x)$ for $c = 1$, 2, and 3. Figure 1.3.9 shows $y = f(x)$ and the parabola $y = 2x^2$, whereas Fig. 1.3.10 shows the graph $y = 2x^2 + f(x)$, obtained by adding the ordinates of the two curves.

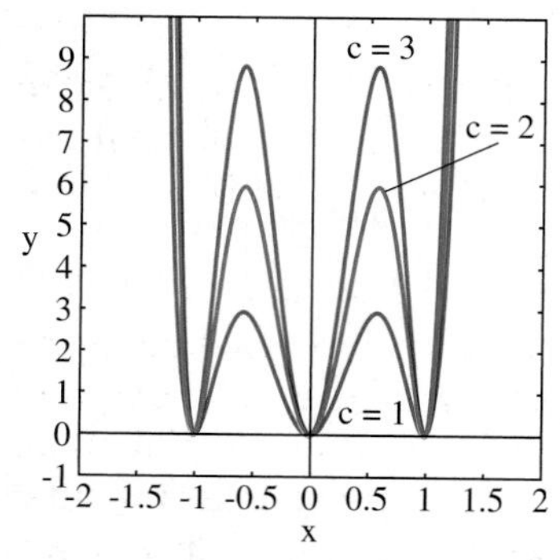

FIGURE 1.3.8 $y = c\cdot 20x^2(x^2-1)^2$ for $c = 1, 2, 3$.

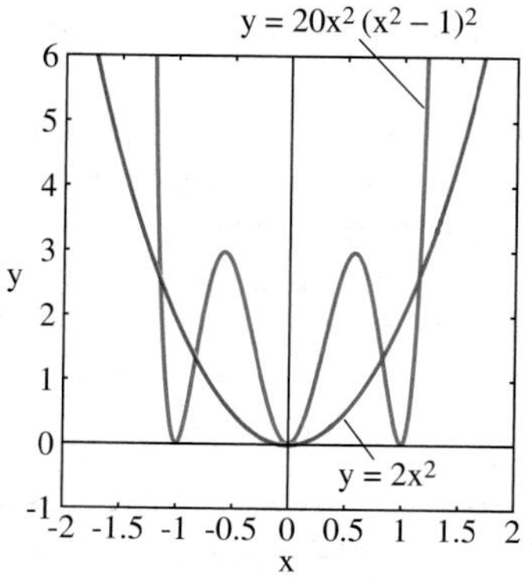

FIGURE 1.3.9 $y = 2x^2$ and $y = 20x^2(x^2-1)^2$.

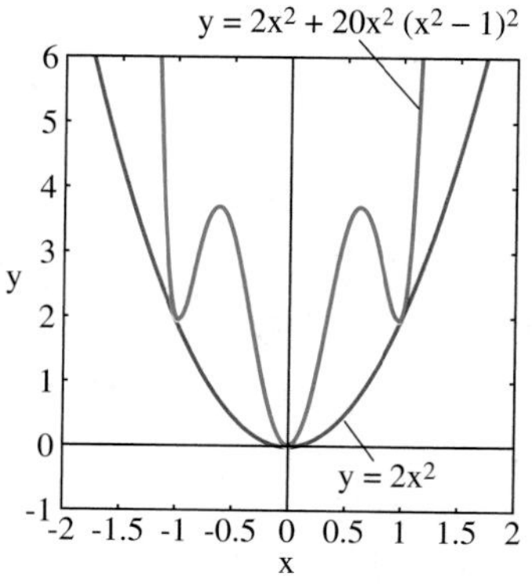

FIGURE 1.3.10 $y = 2x^2$ and $y = 2x^2 + 20x^2(x^2-1)^2$.

Polynomials

A **polynomial** of **degree** n is a function of the form

$$p(x) = a_n x^n + a_{n-1}x^{n-1} + \cdots + a_2 x^2 + a_1 x + a_0 \quad \textbf{(6)}$$

where the coefficients $a_0, a_1, \ldots, a_n$ are fixed real numbers and $a_n \neq 0$. Thus an nth-degree polynomial is a sum of constant multiples of the **power functions**

$$1, \quad x, \quad x^2, \quad \ldots, \quad x^{n-1}, \quad x^n.$$

A first-degree polynomial is simply a *linear function* $a_1 x + a_0$ whose graph is a straight line. A second-degree polynomial is a *quadratic function* whose graph $y = a_2 x^2 + a_1 x + a_0$ is a parabola (see Section 1.2).

Recall that a **zero** of the function f is a solution of the equation

$$f(x) = 0.$$

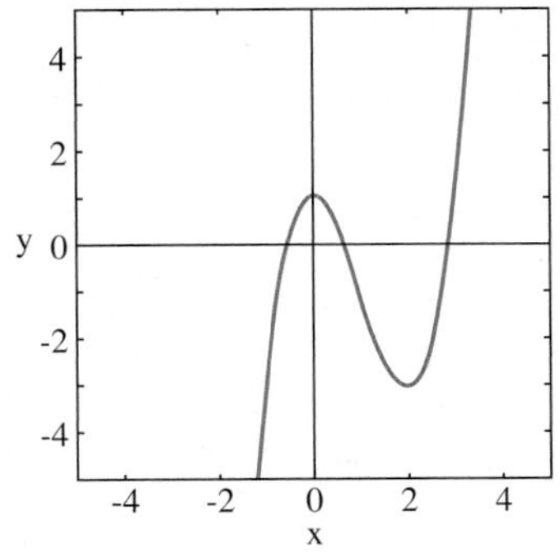

FIGURE 1.3.11 $f(x) = x^3 - 3x^2 + 1$ has three real zeros (Example 4).

Is it obvious to you that *the zeros of $f(x)$ are precisely the x-intercepts of the graph*

$$y = f(x)?$$

Indeed, a major reason for being interested in the graph of a function is to see the number and approximate locations of its zeros.

A key to understanding graphs of higher-degree polynomials is the *fundamental theorem of algebra*. It states that every nth-degree polynomial has n zeros (possibly complex, possibly repeated). It follows that an nth-degree polynomial has *no more than n* distinct real zeros.

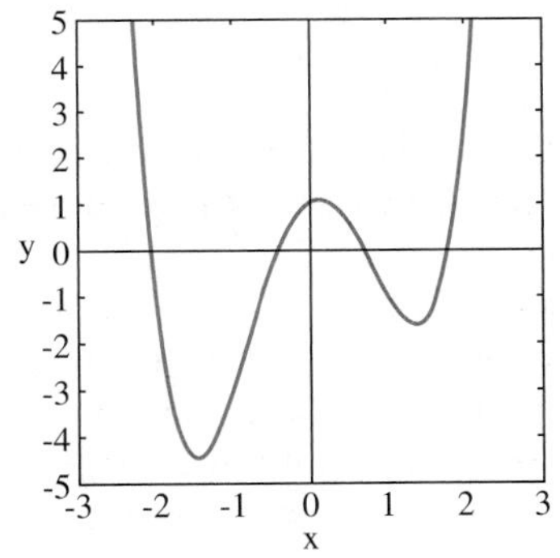

FIGURE 1.3.12 $f(x) = x^4 - 4x^2 + x + 1$ has four real zeros (Example 4).

EXAMPLE 4 Figures 1.3.11 and 1.3.12 exhibit polynomials that both have the maximum number of real zeros allowed by the fundamental theorem of algebra. But the graphs of power functions in Figs. 1.3.1 and 1.3.2 show that a high-degree polynomial may have only a single real zero. And the quadratic function

$$f(x) = x^2 + 4x + 13 = (x + 2)^2 + 9$$

has no real zeros at all. (Why not?) Figure 1.3.7 includes graphs of sixth-degree polynomials having six, three, or no zeros. Indeed, an nth-degree polynomial can have any number of zeros from 0 to n if n is even (from 1 to n if n is odd). ◆

A polynomial behaves "near infinity"—that is, outside an interval on the x-axis containing its real zeros—in much the same way as a power function of the same degree. If $p(x)$ is a polynomial of *odd* degree, then $y = p(x)$ goes in opposite (vertical) directions as x goes to $-\infty$ and to $+\infty$ (like the cubic polynomial graph in Fig. 1.3.11). But if $p(x)$ is a polynomial of *even* degree, then $y = p(x)$ goes in the same (vertical) direction as x goes to $-\infty$ and to $+\infty$ (like the 4th-degree polynomial graph in Fig. 1.3.12).

Between the extremes to the left and right, where $|x|$ is large, an nth-degree polynomial has at most $n - 1$ "bends"—like the 2 bends of the 3rd-degree polynomial graph in Fig. 1.3.11 and the 3 bends of the 4th-degree polynomial graph in Fig. 1.3.12. In Chapter 4 we will use calculus to see why this is so (and to make precise the notion of a "bend" in a curve).

Calculator/Computer Graphing

A typical calculator or computer graphing utility shows (on its graphics screen or monitor) only that portion of a graph $y = f(x)$ that lies within a selected rectangular **viewing window** of the form

$$\{(x, y) : a \leqq x \leqq b \quad \text{and} \quad c \leqq y \leqq d\}.$$

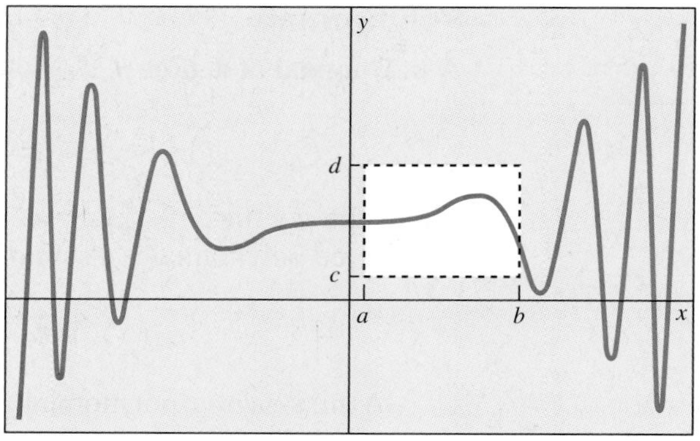

FIGURE 1.3.13 The viewing window $a \leqq x \leqq b$, $c \leqq y \leqq d$.

The parts of the graph that lie outside this viewing window remain unseen (Fig. 1.3.13). With a calculator the maximum and minimum x- and y-values may be entered explicitly in a form such as

$$\text{Xmin} = a \qquad \text{Ymin} = c$$

$$\text{Xmax} = b \qquad \text{Ymax} = d$$

Frequently the user must specify the ***x*-range** $[a, b]$ and the ***y*-range** $[c, d]$ carefully so that the viewing window will show the desired portion of the graph. The calculator or computer's "default window" may provide only a starting point.

EXAMPLE 5 Construct a graph that exhibits the principal features of the cubic polynomial

$$y = x^3 + 12x^2 + 5x - 66. \tag{7}$$

Solution We anticipate a graph that looks somewhat like the cubic graph in Fig. 1.3.11, one that goes "from southwest to northeast," perhaps with a couple of bends in between. But when we enter Eq. (7) in a typical graphing calculator with default viewing window $-10 \leqq x \leqq 10$, $-10 \leqq y \leqq 10$, we get the result shown in Fig. 1.3.14. Evidently our viewing window is not large enough to show the expected behavior.

Doubling each dimension of the viewing window, we get the result in Fig. 1.3.15. Now we see the three zeros that a cubic polynomial can have, as well as some possibility of two bends, but it appears that magnification in the y-direction is indicated. Perhaps we need a y-range measuring in the hundreds rather than the tens. With the viewing window $-20 \leqq x \leqq 20$, $-200 \leqq y \leqq 200$ we finally get the satisfying graph shown in Fig. 1.3.16.

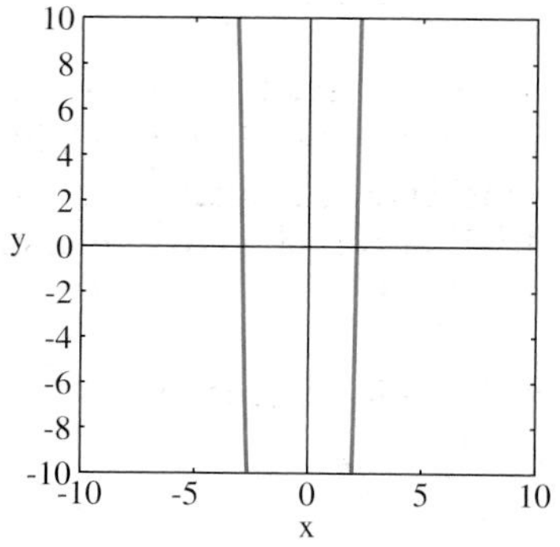

FIGURE 1.3.14 $y = x^3 + 12x^2 + 5x - 66$ with viewing window $-10 \leqq x \leqq 10$, $-10 \leqq y \leqq 10$.

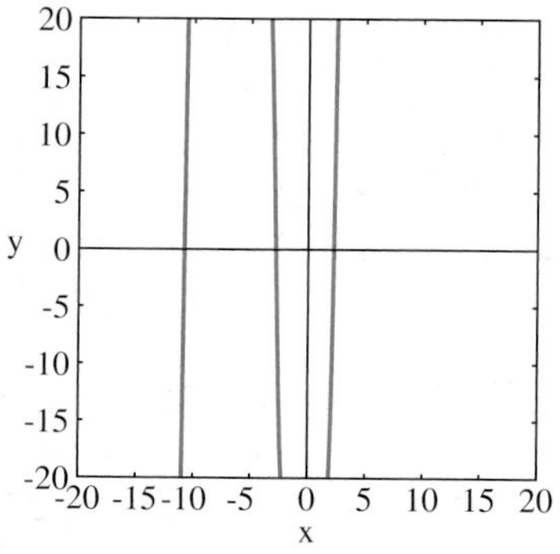

FIGURE 1.3.15 $y = x^3 + 12x^2 + 5x - 66$ with viewing window $-20 \leqq x \leqq 20$, $-20 \leqq y \leqq 20$.

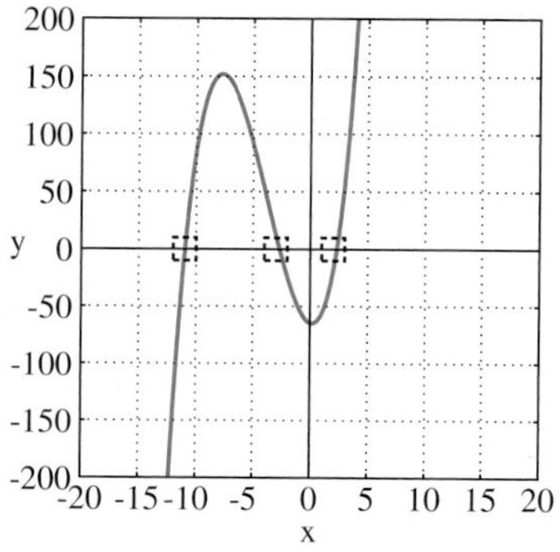

FIGURE 1.3.16 $y = x^3 + 12x^2 + 5x - 66$ with viewing window $-20 \leqq x \leqq 20$, $-200 \leqq y \leqq 200$.

Once we have zoomed out to see the "big picture," we can zoom in on points of interest. For instance, Fig. 1.3.16 indicates "zoom boxes" locating the three zeros of the polynomial in (7). Apparently these zeros are located at or near the points $x = -11$, $x = -3$, and $x = 2$. Each can be approximated graphically as closely as you please (subject to the limitations of your computer) by the method of successive magnifications. (See if you can convince yourself that these three zeros are *exactly* the indicated integers. How could you verify that this actually is true?) ◆

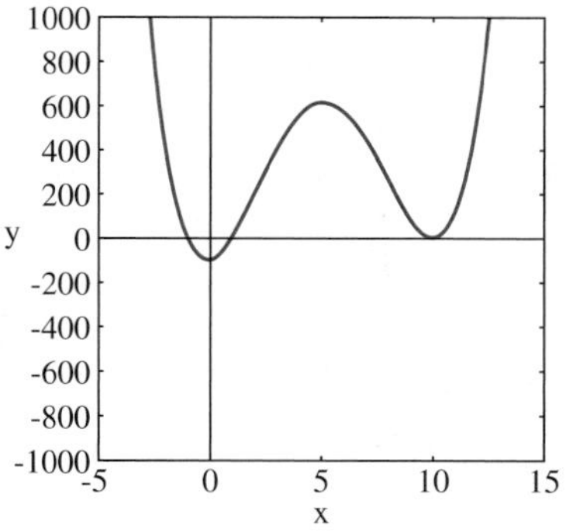

FIGURE 1.3.17 $y = (x^2 - 1)(x - 10)(x - 10.1)$ with viewing window $-5 \leqq x \leqq 15$, $-1000 \leqq y \leqq 1000$.

EXAMPLE 6 Investigate the graph of the quartic (fourth-degree) polynomial

$$f(x) = (x^2 - 1)(x - 10)(x - 10.1) = x^4 - (20.1)x^3 + 100x^2 + (20.1)x - 101. \quad \textbf{(8)}$$

Solution Here we know the zeros $x = -1$, 1, 10, and 10.1 in advance, so it makes sense to choose an x-range that includes all four. Noting that $f(0) = -101$, we suspect that a y-range measuring in the hundreds is indicated. Thus with the viewing window $-5 \leqq x \leqq 15$, $-1000 \leqq y \leqq 1000$, we get the attractive graph in Fig. 1.3.17. Observe that with its three bends it resembles the quartic graph in Fig. 1.3.12.

But now the behavior of the graph near the point $x = 10$ is unclear. Does it dip beneath the x-axis or not? We select the viewing window $9.5 \leqq x \leqq 10.5$, $-1 \leqq y \leqq 1$ to magnify this area and get the result in Fig. 1.3.18. This is a case where it appears that different plots on different scales are required to show all the behavior of the graph. ◆

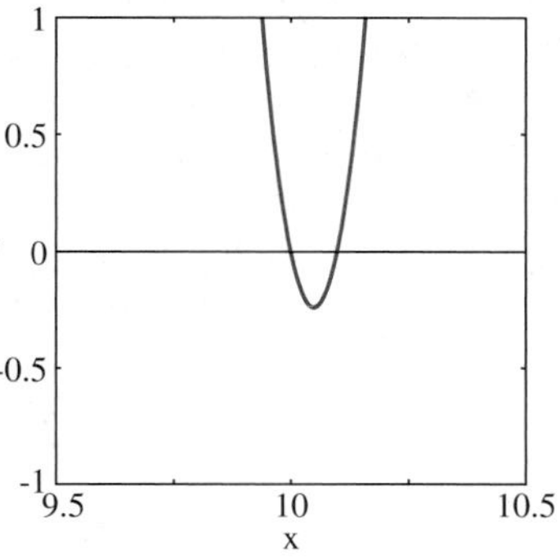

FIGURE 1.3.18 $y = (x^2 - 1)(x - 10)(x - 10.1)$ with viewing window $9.5 \leqq x \leqq 10.5$, $-1 \leqq y \leqq 1$.

Our graphs in Examples 5 and 6 exhibit the maximum possible number of zeros and bends for the polynomials in Eqs. (7) and (8), so we are fairly confident that our investigations reveal the main qualitative features of the graphs of these polynomials. But only with the calculus techniques of Chapter 4 can we be certain of the structure of a graph. For instance, a polynomial graph can exhibit fewer than the maximum possible number of bends, but at this stage we cannot be certain that more bends are not hidden somewhere, perhaps visible only on a scale different from that of the viewing window we have selected.

Rational Functions

Just as a rational number is a quotient of two integers, a **rational function** is a quotient

$$f(x) = \frac{p(x)}{q(x)} \quad \textbf{(9)}$$

of two polynomials $p(x)$ and $q(x)$. Graphs of rational functions and polynomials have several features in common. For instance, a rational function has only a finite number of zeros, because $f(x)$ in Eq. (9) can be zero only when the numerator polynomial $p(x)$ is zero. Similarly, the graph of a rational function has only a finite number of bends.

But the denominator polynomial $q(x)$ in Eq. (9) may have a zero at a point $x = a$ where the numerator is nonzero. In this case the value of $f(x)$ will be very large in magnitude when x is close to a. This observation implies that the graph of a rational function may have a feature that no polynomial graph can have—an *asymptote.*

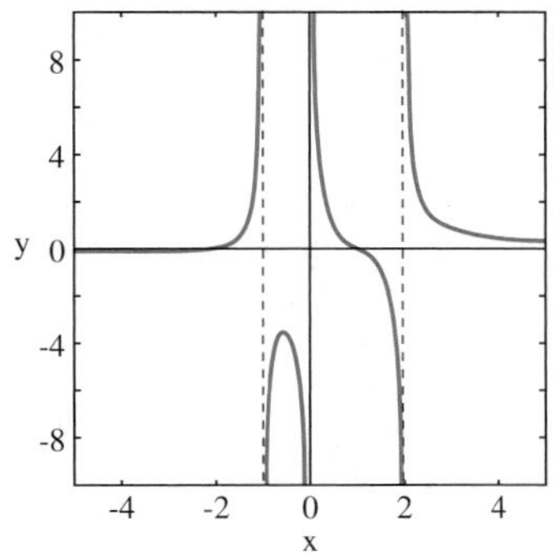

FIGURE 1.3.19 The graph of the rational function in Eq. (10) (Example 7).

EXAMPLE 7 Figure 1.3.19 shows the graph of the rational function

$$f(x) = \frac{(x+2)(x-1)}{x(x+1)(x-2)}. \quad \textbf{(10)}$$

Note the x-intercepts $x = -2$ and $x = 1$, corresponding to the zeros of the numerator $(x + 2)(x - 1)$. The vertical lines $x = -1$, $x = 0$, and $x = 2$ shown in the graph correspond to the zeros of the denominator $x(x+1)(x-2)$. These vertical lines are *asymptotes* of the graph of f. ◆

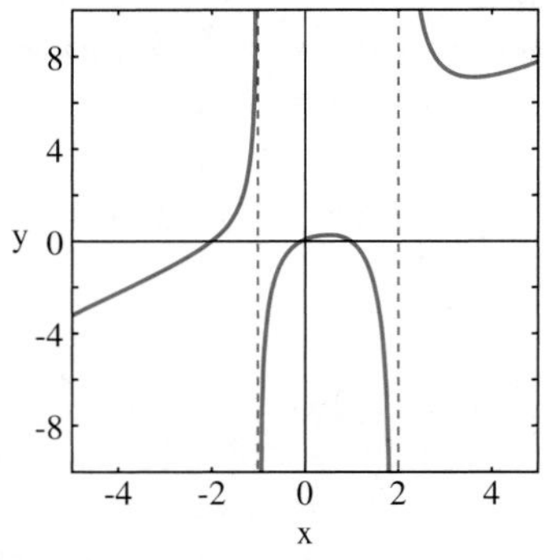

FIGURE 1.3.20 The graph of the rational function in Eq. (11) (Example 8).

EXAMPLE 8 Figure 1.3.20 shows the graph of the rational function

$$f(x) = \frac{x(x+2)(x-1)}{(x+1)(x-2)}. \tag{11}$$

The x-intercepts $x = -2$, $x = 0$, and $x = 1$ correspond to the zeros of the numerator, whereas the asymptotes $x = -1$ and $x = 2$ correspond to the zeros of the denominator. ◆

It should be clear that—by counting x-intercepts and asymptotes—you could match the rational functions in Eqs. (10) and (11) with their graphs in Figs. 1.3.19 and 1.3.20 without knowing in advance which was which.

Algebraic Functions

An **algebraic function** is one whose formula can be constructed beginning with power functions and applying the algebraic operations of addition, subtraction, multiplication by a real number, multiplication, division, and/or root-taking. Thus polynomials and rational functions are algebraic functions. But whereas every polynomial is defined everywhere on the real line, and every rational function is defined everywhere except at the (finitely many) real zeros of its denominator (which correspond to vertical asymptotes), the domain of definition of an algebraic function may be quite limited. For instance, Figs. 1.3.21 and 1.3.22 show the graphs of the algebraic functions

$$f(x) = \sqrt[4]{16 - x^4} \quad \text{and} \quad g(x) = \sqrt{x^2 - 16}$$

on the bounded and unbounded intervals (respectively) where they are defined.

The graph of every polynomial or rational function looks "smooth" at every point where it is defined, but the graph of an algebraic function may exhibit "corners" or sharp "cusps" where it does not look smooth. For instance, look at the graphs in Figs. 1.3.23 and 1.3.24 of the algebraic functions

$$f(x) = \sqrt{x^2} = |x| \quad \text{and} \quad g(x) = \sqrt[3]{x^2(x-2)^2}.$$

In Chapter 3 we will use concepts of calculus to say precisely what is meant by a smooth graph.

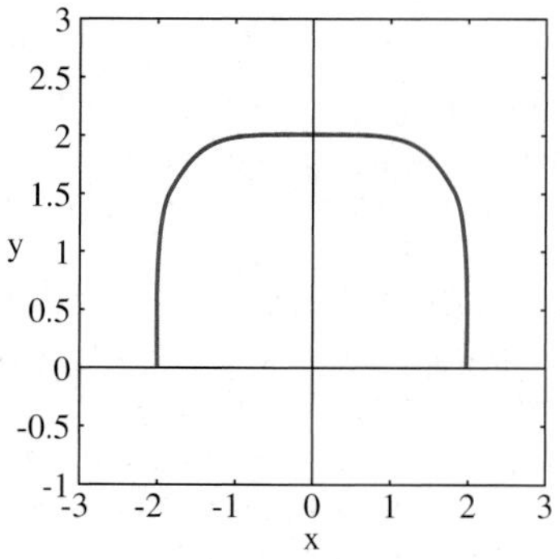

FIGURE 1.3.21 $y = \sqrt[4]{16 - x^4}$ on $[-2, 2]$.

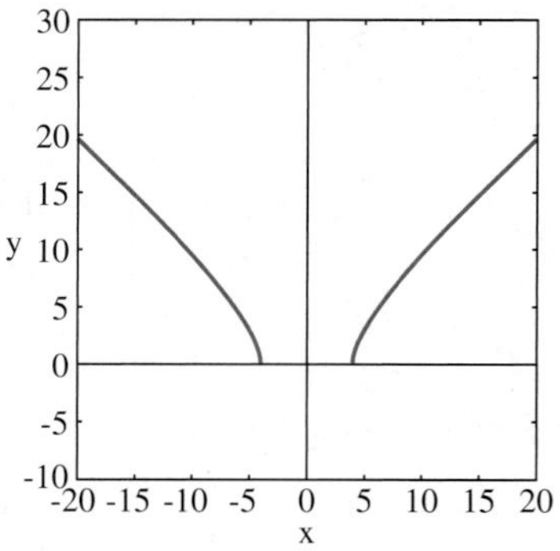

FIGURE 1.3.22 $y = \sqrt{x^2 - 16}$ on $(-\infty, -4) \cup (4, \infty)$.

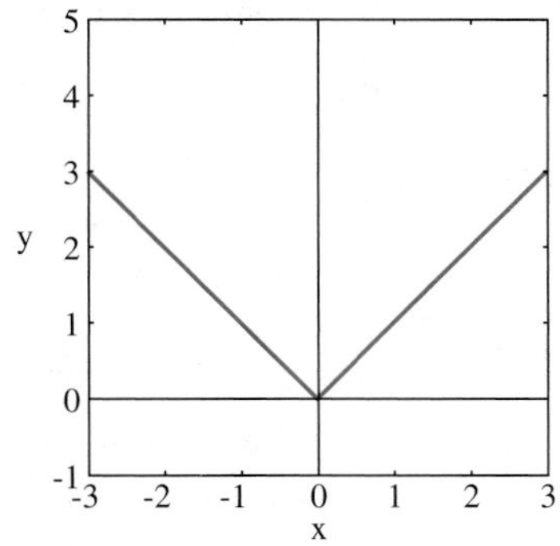

FIGURE 1.3.23 $y = |x|$ with a "corner" at the origin.

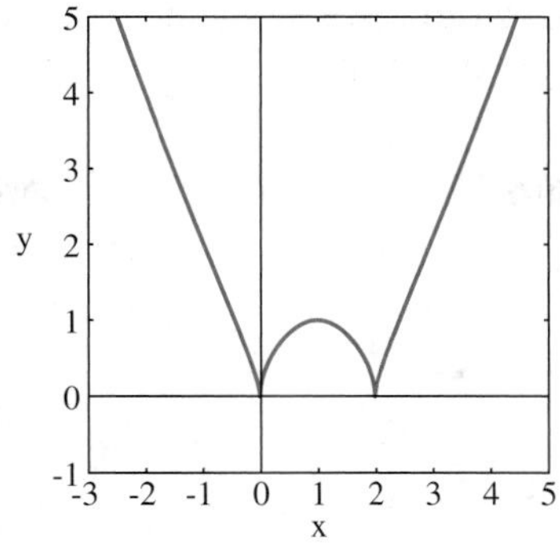

FIGURE 1.3.24 $y = \sqrt[3]{x^2(x-2)^2}$ with "cusps" at (0, 0) and (2, 0).

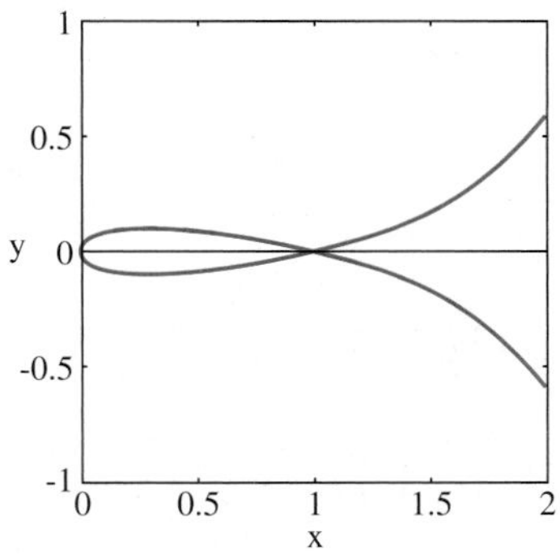

FIGURE 1.3.25 $y = \pm(0.2969\sqrt{x} - 0.126x - 0.3516x^2 + 0.2843x^3 - 0.10151x^4)$.

Figure 1.3.25 shows the graphs of the two algebraic functions defined by

$$y = \pm\left(0.2969\sqrt{x} - 0.126x - 0.3516x^2 + 0.2843x^3 - 0.10151x^4\right). \quad \textbf{(12)}$$

The loop describes the cross-sectional profile of the NASA 0012 airfoil as designed by aeronautical engineers.

1.3 TRUE/FALSE STUDY GUIDE

1.3 CONCEPTS: QUESTIONS AND DISCUSSION

1. In each of the following eight cases, give an example of a function as described or explain why no such function exists.
   (a) A polynomial function of degree less than 2 whose graph lies entirely above the x-axis.
   (b) A polynomial of positive degree whose graph lies entirely beneath the x-axis.
   (c) A polynomial of positive degree and with positive leading coefficient whose graph lies entirely below the x-axis (the *leading coefficient* of a polynomial is the coefficient of its term of highest degree).
   (d) A polynomial of odd degree with negative leading coefficient whose graph does not intersect the x-axis.
   (e) A polynomial whose graph lies entirely between the lines $y = -1$ and $y = 1$.
   (f) A polynomial whose graph contains points above the line $y = 1$ and below the line $y = -1$, but contains no points between those two lines.
   (g) A rational function that has both positive and negative values but is never zero.
   (h) A nonconstant rational function that is never zero and has no vertical asymptote.

2. In each of the following five cases write the formula of a specific function as described. Also sketch a typical graph of such a function (not necessarily the same one you defined symbolically).
   (a) A quadratic polynomial with no real zeros.
   (b) A cubic polynomial with exactly one real zero $x \neq 0$.
   (c) A cubic polynomial with exactly two distinct real zeros.
   (d) A quartic polynomial with exactly two distinct real zeros.
   (e) A quartic polynomial with exactly three distinct real zeros.

3. Which of the following algebraic functions agrees with some polynomial function?
(a) $f(x) = \sqrt{x^2 + 2x + 1}$ (b) $f(x) = \sqrt{x^4 + 4x + 4}$
(c) $f(x) = \sqrt[3]{(x-1)^3}$ (d) $f(x) = \sqrt[3]{(x-2)^2}$

1.3 PROBLEMS

In Problems 1 through 6, find $f + g$, $f \cdot g$, and f/g, and give the domain of definition of each of these new functions.

1. $f(x) = x + 1, \quad g(x) = x^2 + 2x - 3$

2. $f(x) = \dfrac{1}{x-1}, \quad g(x) = \dfrac{1}{2x+1}$

3. $f(x) = \sqrt{x}, \quad g(x) = \sqrt{x-2}$

4. $f(x) = \sqrt{x+1}, \quad g(x) = \sqrt{5-x}$

5. $f(x) = \sqrt{x^2+1}, \quad g(x) = \dfrac{1}{\sqrt{4-x^2}}$

6. $f(x) = \dfrac{x-1}{x-2}, \quad g(x) = \dfrac{x+1}{x+2}$

In Problems 7 through 12, match the given polynomial with its graph among those shown in Figs. 1.3.26 through 1.3.31. Do not use a graphing calculator or a computer. Instead, consider the degree of the polynomial, its indicated number of zeros, and its behavior for $|x|$ large.

7. $f(x) = x^3 - 3x + 1$

8. $f(x) = 1 + 4x - x^3$

9. $f(x) = x^4 - 5x^3 + 13x + 1$

10. $f(x) = 2x^5 - 10x^3 + 6x - 1$

11. $f(x) = 16 + 2x^2 - x^4$

12. $f(x) = x^5 + x$

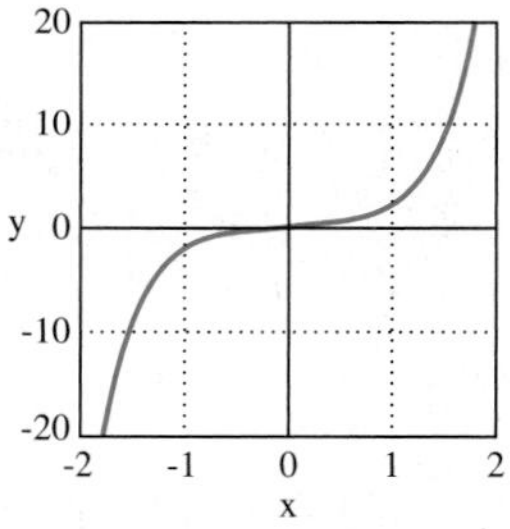

FIGURE 1.3.26

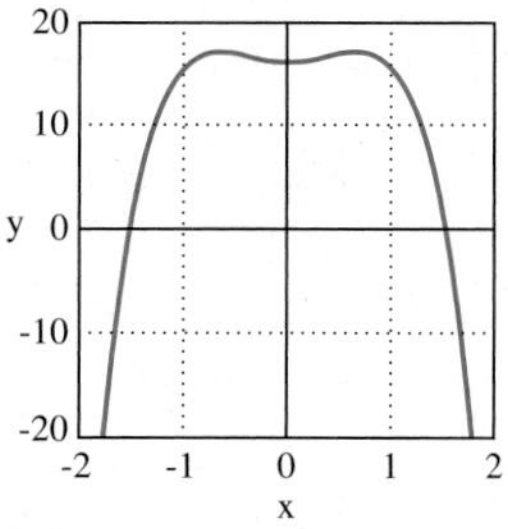

FIGURE 1.3.27

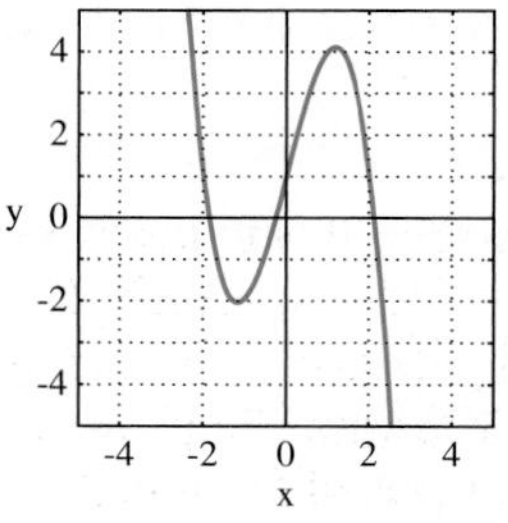

FIGURE 1.3.28

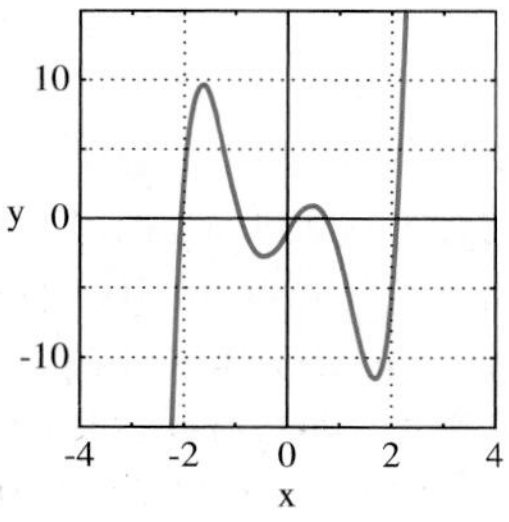

FIGURE 1.3.29

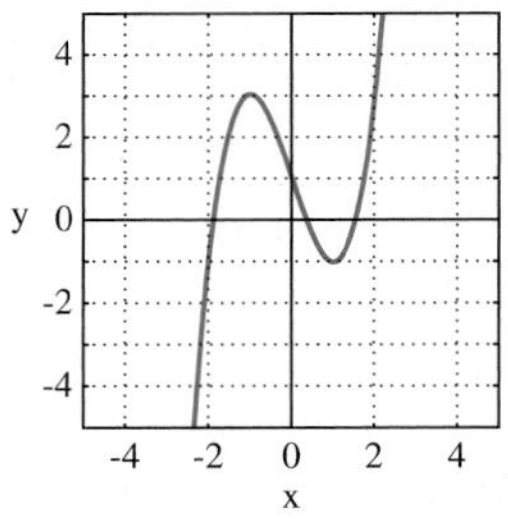

FIGURE 1.3.30

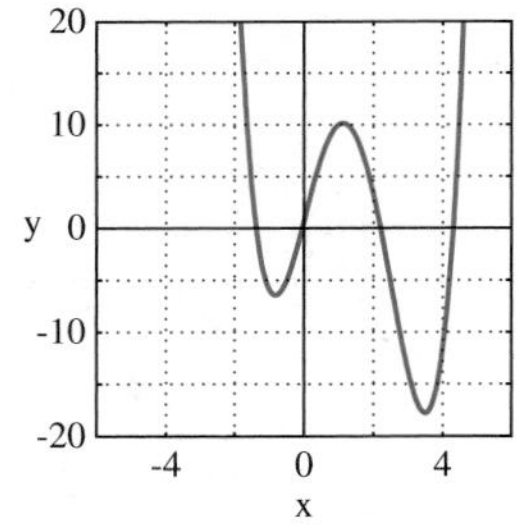

FIGURE 1.3.31

In Problems 13 through 16, use the vertical asymptotes of the given rational function (rather than a graphing calculator or computer) to match it with its graph among those shown in Figs. 1.3.32 through 1.3.35.

13. $f(x) = \dfrac{1}{(x+1)(x-2)}$

14. $f(x) = \dfrac{x}{x^2 - 9}$

15. $f(x) = \dfrac{3}{x^2+1}$

16. $f(x) = \dfrac{x^2+1}{x^3-1}$

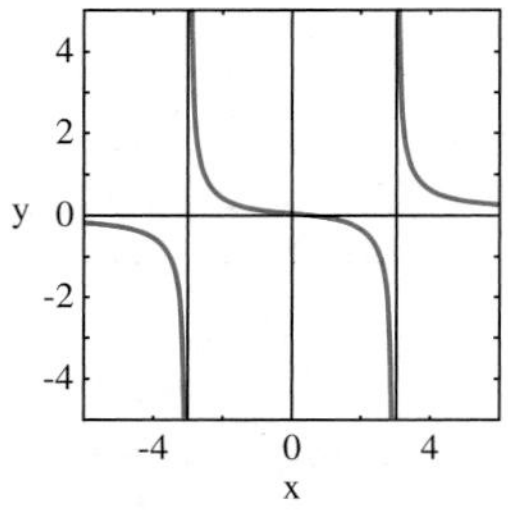

FIGURE 1.3.32

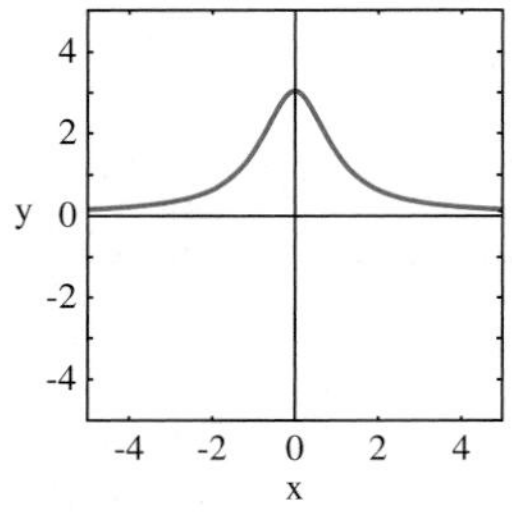

FIGURE 1.3.33

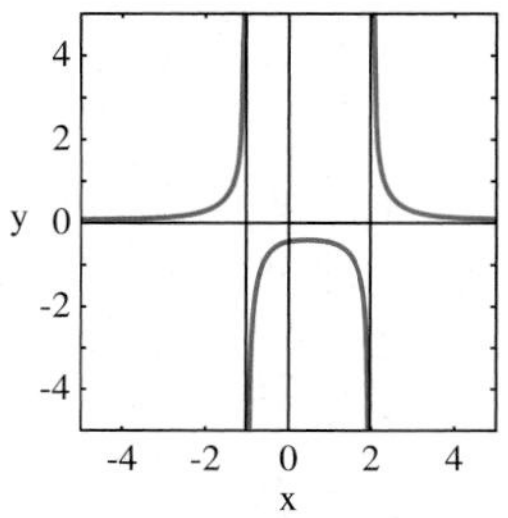

FIGURE 1.3.34

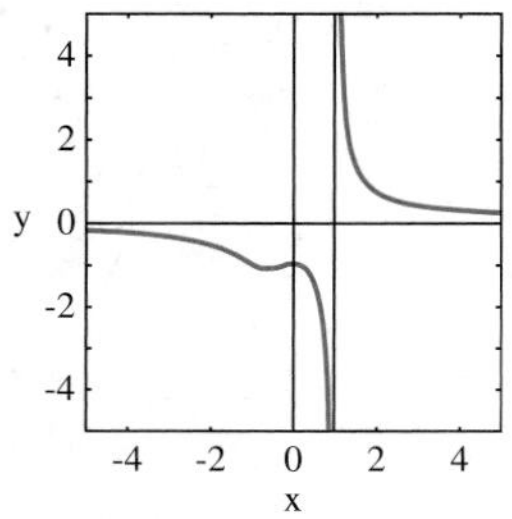

FIGURE 1.3.35

In Problems 17 through 20, use primarily the domain of definition of the given algebraic function (rather than a graphing calculator or computer) to match it with its graph among those in Figs. 1.3.36 through 1.3.39.

17. $f(x) = x\sqrt{x+2}$

18. $f(x) = \sqrt{2x - x^2}$

19. $f(x) = \sqrt{x^2 - 2x}$

20. $f(x) = 2\sqrt[3]{x^2 - 2x}$

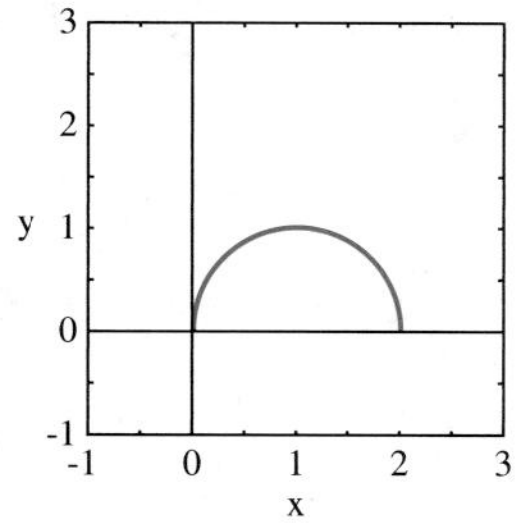

FIGURE 1.3.36

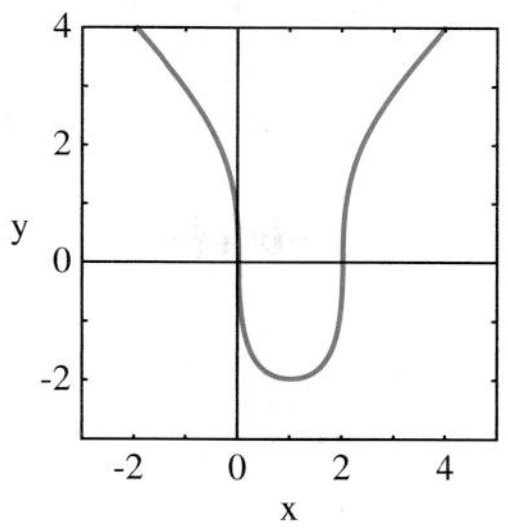

FIGURE 1.3.37

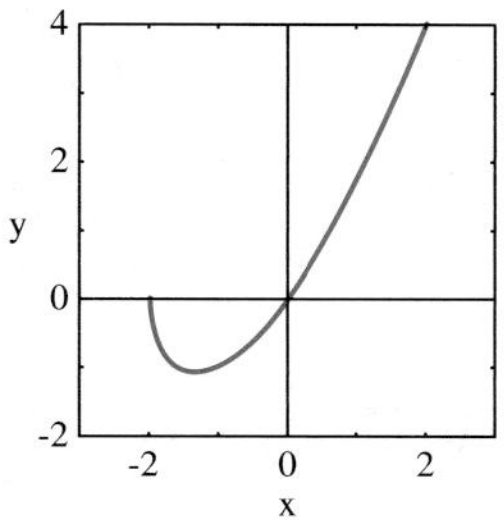

FIGURE 1.3.38

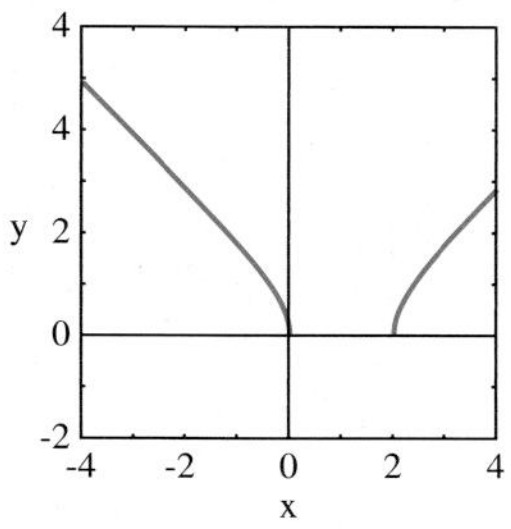

FIGURE 1.3.39

In Problems 21 through 30 use a graphing calculator or computer to determine one or more appropriate viewing windows to exhibit the principal features of the graph $y = f(x)$. In particular, determine thereby the number of real solutions of the equation $f(x) = 0$ and the approximate location (to the nearest integer) of each of these solutions.

21. $f(x) = x^3 - 3x + 1$

22. $f(x) = x^3 - 3x + 2$

23. $f(x) = x^3 - 3x + 3$

24. $f(x) = 2x^4 - 6x^3 + 10x - 5$

25. $f(x) = 2x^4 - 6x^3 + 10x - 6$

26. $f(x) = 2x^4 - 6x^3 + 10x - 7$

27. $f(x) = x^3 - 50x - 100$

28. $f(x) = x^4 + 20x^3 - 50x - 30$

29. $f(x) = x^5 + 5x^4 - 100x^3 - 200x^2 + 2500x - 3500$

30. $f(x) = x^6 - 250x^4 + 2500x^2 - 2500$

In Problems 31 through 37, determine how the graph $y = f(x)$ changes when the value of c is changed within the given interval. With a graphing calculator or computer you should be able to plot graphs with different values of c on the same screen.

31. $f(x) = x^3 - 3x + c,\ -5 \leqq c \leqq 5$

32. $f(x) = x^3 + cx,\ -5 \leqq c \leqq 5$

33. $f(x) = x^3 + cx^2,\ -5 \leqq c \leqq 5$

34. $f(x) = x^4 + cx^2,\ -5 \leqq c \leqq 5$

35. $f(x) = x^5 + cx^3 + x,\ -5 \leqq c \leqq 5$

36. $f(x) = \dfrac{1}{1 + cx^2},\ 1 \leqq c \leqq 10$

37. $f(x) = \sqrt{\dfrac{x^2}{c^2 - x^2}},\ 1 \leqq c \leqq 10,\ x \text{ in } (-c, c)$

38. Use the graphical method of repeated magnifications to find both the length and the maximum width of the airfoil shown in Fig. 1.3.25. Determine each accurate to three decimal places.

1.3 Project: A Leaning Ladder

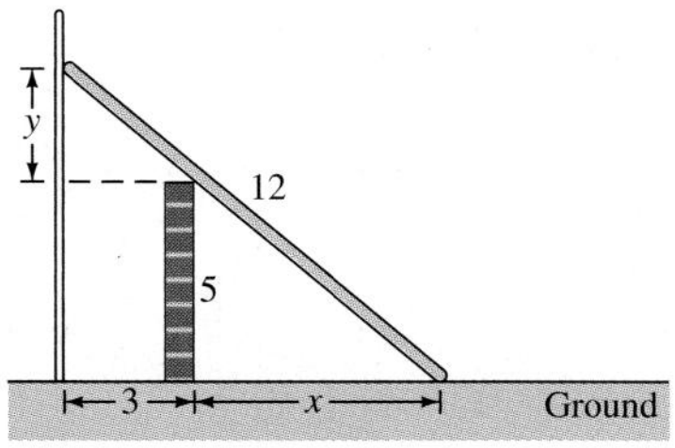

FIGURE 1.3.40 The leaning ladder.

A 12-ft ladder leans across a 5-ft fence and touches a high wall located 3 ft behind the fence. You are to find the distance from the foot of the ladder to the bottom of the fence. The key is the use of simple geometry to derive the equations

$$xy = 15 \quad \text{and} \quad (x + 3)^2 + (y + 5)^2 = 144$$

relating the lengths x and y indicated in Fig. 1.3.40. Can you eliminate y to find a quartic polynomial equation that x must satisfy? If so, then you can use a graphing calculator or computer to approximate the possible values of x by the method of repeated magnification.

1.4 TRANSCENDENTAL FUNCTIONS

Continuing the survey of elementary functions begun in Section 1.3, we now review briefly the most familiar nonalgebraic functions that are studied in calculus. These include the *trigonometric functions* that are used to model periodic phenomena—phenomena of ebb and flow, involving quantities that oscillate with the passage of time—and the *exponential* and *logarithmic functions* that are used to model phenomena of growth and decay—involving quantities that either increase steadily or decrease steadily as time passes. We also introduce *composition* of functions, a new way (in addition to the algebraic operations of Section 1.3) of combining familiar functions to form new ones.

Trigonometric Functions

A review of trigonometry is included in Appendix C. In elementary trigonometry a trigonometric function such as $\sin A$, $\cos A$, or $\tan A$ ordinarily is first defined as a function of an *angle* A in a right triangle. But here a trigonometric function of a *real number* x corresponds to that function of an angle measuring x radians. Thus

$$\sin\frac{\pi}{6}=\frac{1}{2},\qquad \cos\frac{\pi}{6}=\frac{\sqrt{3}}{2},\qquad\text{and}\qquad \tan\frac{\pi}{6}=\frac{\sin\frac{\pi}{6}}{\cos\frac{\pi}{6}}=\frac{1}{\sqrt{3}}$$

because $\pi/6$ is the radian measure of an angle of 30°. Recall that

$$\pi \text{ radians} = 180 \text{ degrees}, \tag{1}$$

so

$$1 \text{ rad} = \frac{180}{\pi} \text{ deg} \quad\text{and}\quad 1 \text{ deg} = \frac{\pi}{180} \text{ rad}.$$

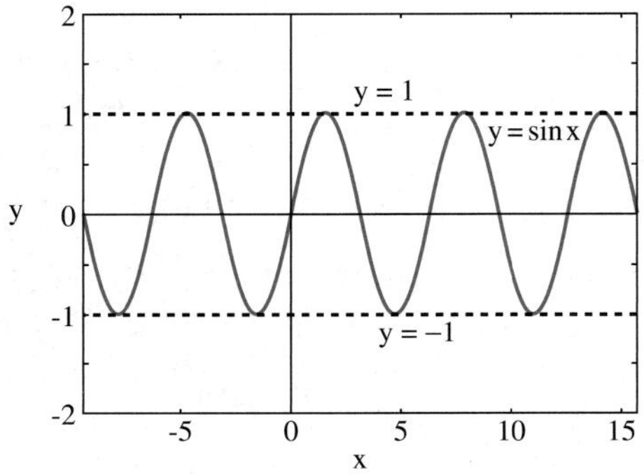

FIGURE 1.4.1 $y = \sin x$.

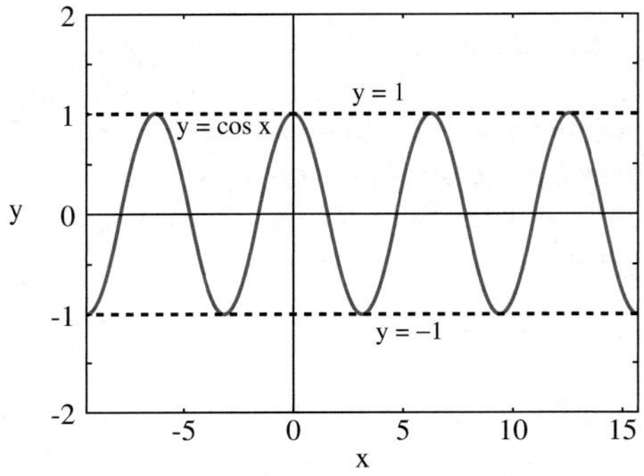

FIGURE 1.4.2 $y = \cos x$.

Figures 1.4.1 and 1.4.2 show the graphs $y = \sin x$ and $y = \cos x$ of the sine and cosine functions, respectively. The value of each oscillates between $+1$ and -1, exhibiting the characteristic *periodicity* of the trigonometric functions:

$$\sin(x + 2\pi) = \sin x \quad\text{and}\quad \cos(x + 2\pi) = \cos x \tag{2}$$

for all x.

If we translate the graph $y = \cos x$ by $\pi/2$ units to the right, we get the graph $y = \sin x$. This observation corresponds to the familiar relation

$$\cos\left(x - \frac{\pi}{2}\right) = \cos\left(\frac{\pi}{2} - x\right) = \sin x. \tag{3}$$

EXAMPLE 1 Figure 1.4.3 shows the translated sine curve obtained by translating the origin to the point $(1, 2)$. Its equation is obtained upon replacing x and y in $y = \sin x$ with $x - 1$ and $y - 2$, respectively:

$$y - 2 = \sin(x - 1); \qquad \text{that is,}$$
$$y = 2 + \sin(x - 1). \qquad ◆$$

The world around us is full of quantities that oscillate like the trigonometric functions. Think of the alternation of day and night, the endless repetition of the

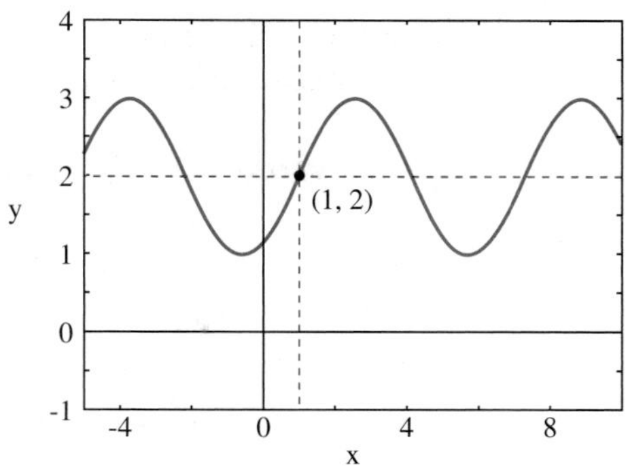

FIGURE 1.4.3 The translated sine curve $y - 2 = \sin(x - 1)$.

seasons, the monthly cycle of the moon, the rise and fall of the tides, the beat of your heart.

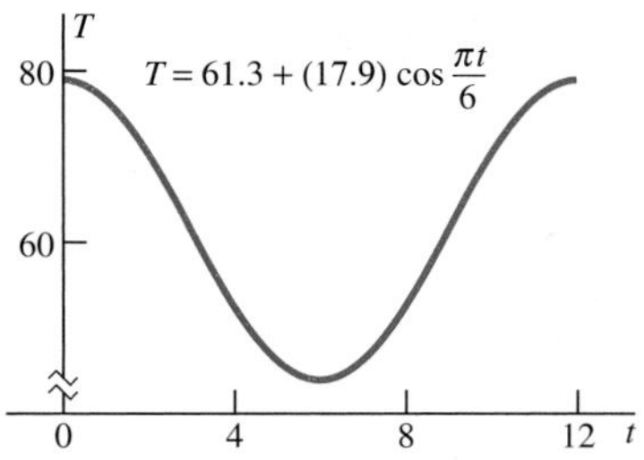

FIGURE 1.4.4 Average daily temperature in Athens, Georgia, t months after July 15 (Example 2).

EXAMPLE 2 Figure 1.4.4 shows the cosine-like behavior of temperatures in Athens, Georgia. The average temperature T (in °F) during a 24-hr day t months after July 15 is given approximately by

$$T = T(t) = 61.3 + 17.9\cos\frac{\pi t}{6}. \qquad \textbf{(4)}$$

For instance, on a typical October 15 (three months after July 15) the average temperature is

$$T(3) = 61.3 + 17.9\cos\frac{3\pi}{6} = 61.3 \quad (^\circ\text{F})$$

because $\cos(3\pi/6) = \cos(\pi/2) = 0$. Thus the "midpoint" of fall weather in Athens—when the average daily temperature is midway between summer's high and winter's low—occurs about three weeks after the official beginning of fall (on or about September 22). Note also that

$$T(t + 12) = 61.3 + 17.9\cos\left(\frac{\pi t}{6} + 2\pi\right) = 61.3 + 17.9\cos\left(\frac{\pi t}{6}\right) = T(t)$$

(why?), in agreement with the yearly 12-month cycle of average weather. ◆

The periodicity and oscillatory behavior of the trigonometric functions make them quite unlike polynomial functions. Because

$$\sin n\pi = 0 \quad \text{and} \quad \cos\left(\frac{2n+1}{2}\pi\right) = 0 \qquad \textbf{(5)}$$

for $n = 0, 1, 2, 3, \dots$, we see that the simple trigonometric equations

$$\sin x = 0 \quad \text{and} \quad \cos x = 0 \qquad \textbf{(6)}$$

have *infinitely many solutions.* In contrast, a polynomial equation can have only a finite number of solutions.

Figure 1.4.5 shows the graph of $y = \tan x$. The x-intercepts correspond to the zeros of the numerator $\sin x$ in the relation

$$\tan x = \frac{\sin x}{\cos x}, \qquad \textbf{(7)}$$

whereas the vertical asymptotes correspond to the zeros of the denominator $\cos x$. Observe the "infinite gaps" in the graph $y = \tan x$ at these odd-integral multiples of $\pi/2$. We call these gaps *discontinuities,* phenomena we discuss further in Chapter 2.

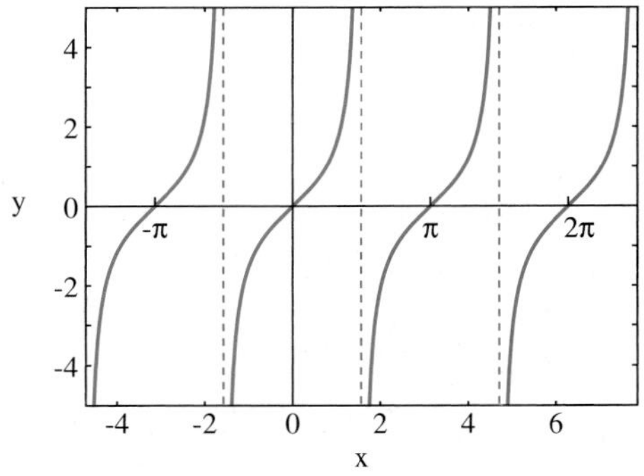

FIGURE 1.4.5 $y = \tan x$.

Composition of Functions

Many varied and complex functions can be "put together" by using quite simple "building-block" functions. In addition to adding, subtracting, multiplying, or dividing two given functions, we can also combine functions by letting one function act on the output of the other.

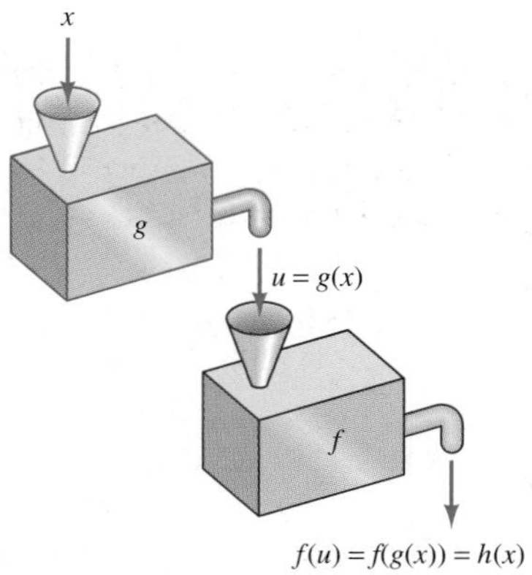

FIGURE 1.4.6 The composition of f and g.

DEFINITION Composition of Functions

The **composition** of the two functions f and g is the function $h = f \circ g$ defined by

$$h(x) = f(g(x)) \tag{8}$$

for all x in the domain of g such that $u = g(x)$ is in the domain of f. (The right-hand side in Eq. (8) is read "f of g of x.")

Thus the output $u = g(x)$ of the function g is used as the input to the function f (Fig. 1.4.6). We sometimes refer to g as the *inner function* and to f as the *outer function* in Eq. (8).

EXAMPLE 3 If $f(x) = \sqrt{x}$ and $g(x) = 1 - x^2$, then

$$f(g(x)) = \sqrt{1 - x^2} \quad \text{for } |x| \leqq 1,$$

whereas

$$g(f(x)) = 1 - \left(\sqrt{x}\right)^2 = 1 - x \quad \text{for } x \geqq 0.$$

◆

The $f(g(x))$ notation for compositions is most commonly used in ordinary computations, whereas the $f \circ g$ notation emphasizes that the composition may be regarded as a new kind of combination of the functions f and g. But Example 3 shows that $f \circ g$ is quite unlike the product fg of the two functions f and g, for

$$f \circ g \neq g \circ f,$$

whereas $fg = gf$ (because $f(x) \cdot g(x) = g(x) \cdot f(x)$ whenever $f(x)$ and $g(x)$ are defined). So remember that composition is quite different in character from ordinary multiplication of functions.

EXAMPLE 4 If

$$f(x) = x^2 \quad \text{and} \quad g(x) = \cos x,$$

then the functions

$$f(x)g(x) = x^2 \cos x,$$
$$f(g(x)) = \cos^2 x = (\cos x)^2, \quad \text{and}$$
$$g(f(x)) = \cos x^2 = \cos(x^2)$$

are defined for all x. Figures 1.4.7 through 1.4.9 illustrate vividly how different these three functions are.

◆

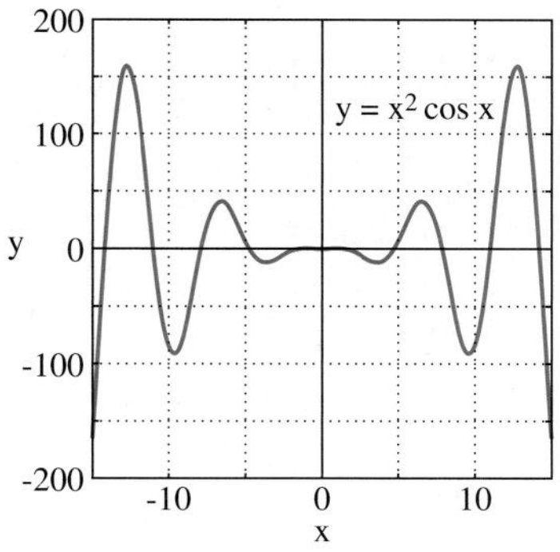

FIGURE 1.4.7 $y = x^2 \cos x$ (Example 4).

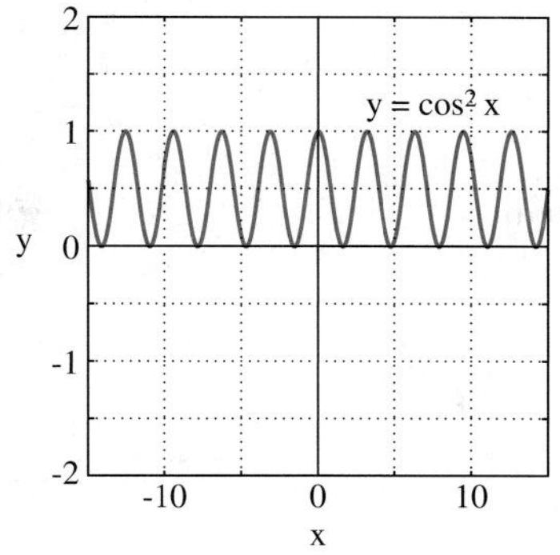

FIGURE 1.4.8 $y = \cos^2 x$ (Example 4).

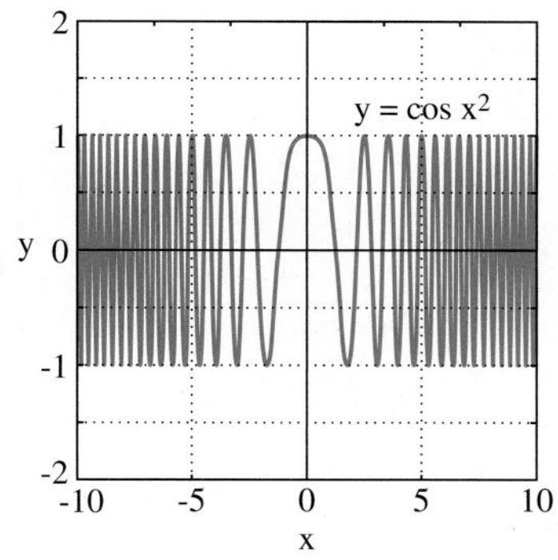

FIGURE 1.4.9 $y = \cos x^2$ (Example 4).

EXAMPLE 5 Given the function $h(x) = (x^2+4)^{3/2}$, find two functions f and g such that $h(x) = f(g(x))$.

Solution It is technically correct—but useless—simply to let $g(x) = x$ and $f(u) = (u^2+4)^{3/2}$. We seek a nontrivial answer here. To calculate $(x^2+4)^{3/2}$, we must first calculate x^2+4. So we choose $g(x) = x^2+4$ as the inner function. The last step is to raise $u = g(x)$ to the power $\frac{3}{2}$, so we take $f(u) = u^{3/2}$ as the outer function. Thus if

$$f(x) = x^{3/2} \quad \text{and} \quad g(x) = x^2+4,$$

then $f(g(x)) = f(x^2+4) = (x^2+4)^{3/2} = h(x)$. ◆

Exponential Functions

An **exponential function** is a function of the form

$$f(x) = a^x, \tag{9}$$

where the **base** a is a fixed positive real number—a constant. Note the difference between an exponential function and a power function. In the power function x^n, the *variable* x is raised to a *constant* power; in the exponential function a^x, a *constant* is raised to a *variable* power.

Many computers and programmable calculators use the notation a ^ x to denote the exponential a^x (a few use a ↑ x). If $a > 1$, then the graph $y = a^x$ looks much like those in Fig. 1.4.10, which shows $y = 2^x$ and $y = 10^x$. The graph of an exponential function with base a, $a > 1$, lies entirely above the x-axis and rises steadily from left to right. Therefore, such a graph is nothing like the graph of a polynomial or trigonometric function. The larger the base a, the more rapid the rate at which the curve $y = a^x$ rises (for $x > 0$). Thus $y = 10^x$ climbs more steeply than $y = 2^x$.

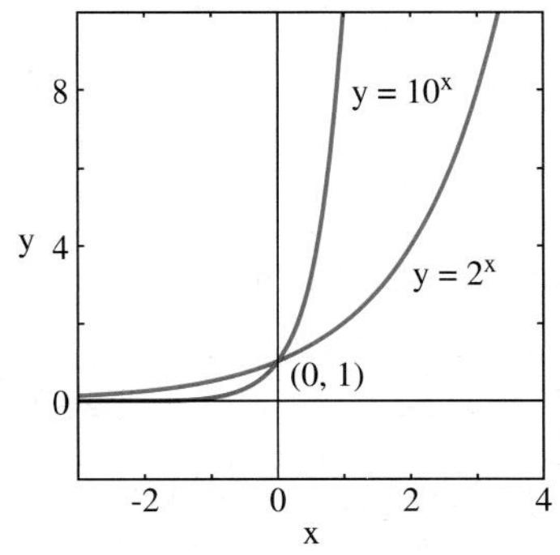

FIGURE 1.4.10 Increasing exponential functions $y = 2^x$ and $y = 10^x$.

EXAMPLE 6 Every exponential function (with base $a > 1$) increases *very rapidly* when x is large. The following table comparing values of x^2 with 2^x exhibits vividly the rapid rate of increase of the exponential function 2^x, even compared with the power function x^2, which increases at a more restrained rate as x increases.

x	x^2	2^x
10	100	1024
20	400	1048576
30	900	1073741824
40	1600	1099511627776
50	2500	1125899906842624
60	3600	1152921504606846976
70	4900	1180591620717411303424
80	6400	1208925819614629174706176
90	8100	1237940039285380274899124224
100	10000	1267650600228229401496703205376

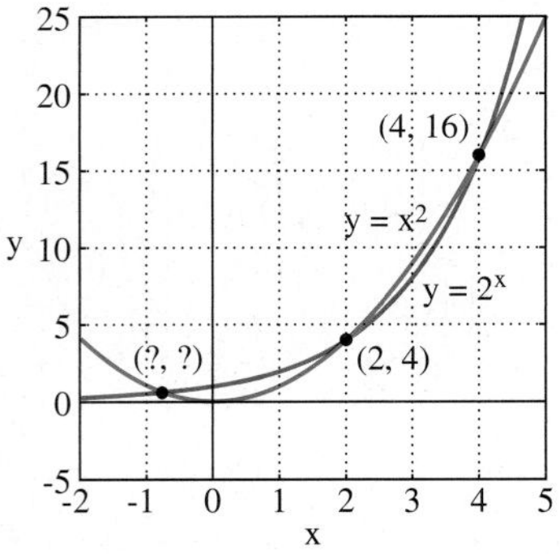

FIGURE 1.4.11 $y = x^2$ and $y = 2^x$.

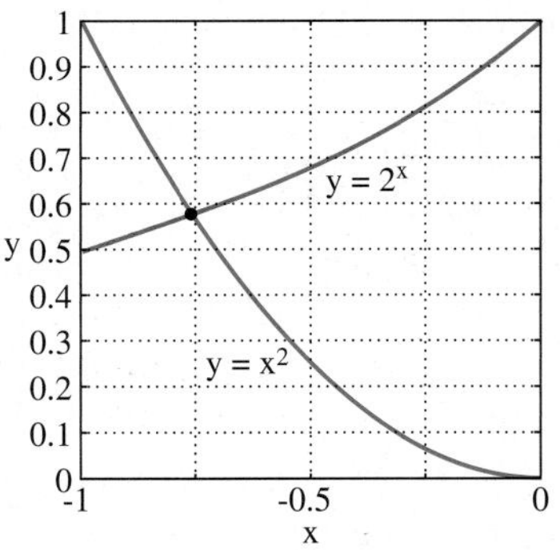

FIGURE 1.4.12 A magnification of 1.4.11 showing the negative solution.

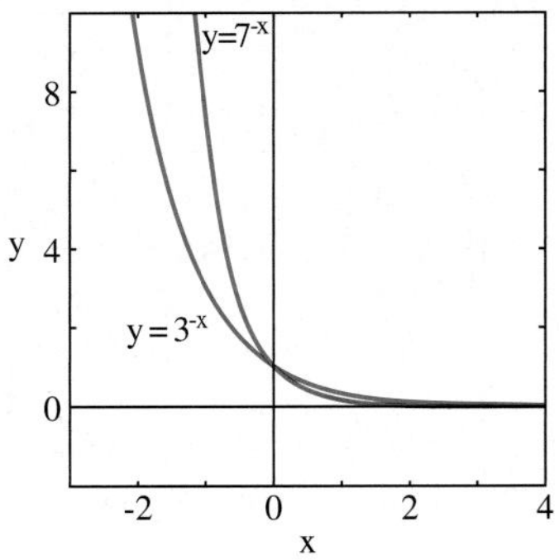

FIGURE 1.4.13 Decreasing exponential functions $y = 3^{-x}$ and $y = 7^{-x}$.

The comparison between x^2 and 2^x for smaller values of x is interesting in a different way. The graphs of $y = x^2$ and $y = 2^x$ in Fig. 1.4.11 indicate that the equation $x^2 = 2^x$ has three solutions between $x = -2$ and $x = 5$. Is it clear to you that $x = 2$ and $x = 4$ are *exact* solutions? The "zoom" shown in Fig. 1.4.12 indicates that the negative solution is a bit less than -0.75. Perhaps you can zoom once more and find the value of this negative solution accurate to at least two decimal places. ◆

If we replace x in Eq. (9) with $-x$, we get the function a^{-x}. Its graph $y = a^{-x}$ *falls* from left to right if $a > 1$. Figure 1.4.13 shows the graphs $y = 3^{-x}$ and $y = 7^{-x}$.

Whereas trigonometric functions are used to describe periodic phenomena of ebb and flow, exponential functions are used to describe natural processes of steady growth or steady decline.

EXAMPLE 7 Let $P(t)$ denote the number of rodents after t months in a certain prolific population that doubles every month. If there are $P(0) = 10$ rodents initially, then there are

- $P(1) = 10 \cdot 2^1 = 20$ rodents after 1 month,
- $P(2) = 10 \cdot 2^2 = 40$ rodents after 2 months,
- $P(3) = 10 \cdot 2^3 = 80$ rodents after 3 months,

and so forth. Thus the rodent population after t months is given by the exponential function

$$P(t) = 10 \cdot 2^t \qquad \textbf{(10)}$$

if t is a nonnegative integer. Under appropriate conditions, Eq. (10) gives an accurate approximation to the rodent population even when t is not an integer. For instance, this formula predicts that after $t = 4\frac{1}{2}$ months, there will be

$$P(4.5) = 10 \cdot 2^{4.5} \approx 226.27 \approx 226 \quad \text{rodents.}$$ ◆

EXAMPLE 8 Suppose that you invest \$5000 in a money-market account that pays 8% interest compounded annually. This means that the amount in the account is multiplied by 1.08 at the end of each year. Let $A(t)$ denote the amount in your account at the end of t years. Then,

- $A(1) = 5000 \cdot 1.08^1$ (\$5400.00) after 1 yr,
- $A(2) = 5000 \cdot 1.08^2$ (\$5832.00) after 2 yr,
- $A(3) = 5000 \cdot 1.08^3$ (\$6298.56) after 3 yr,

and so on. Thus after t years (t a nonnegative integer), the amount in your account is given by the exponential function

$$A(t) = 5000 \cdot 1.08^t. \tag{11}$$

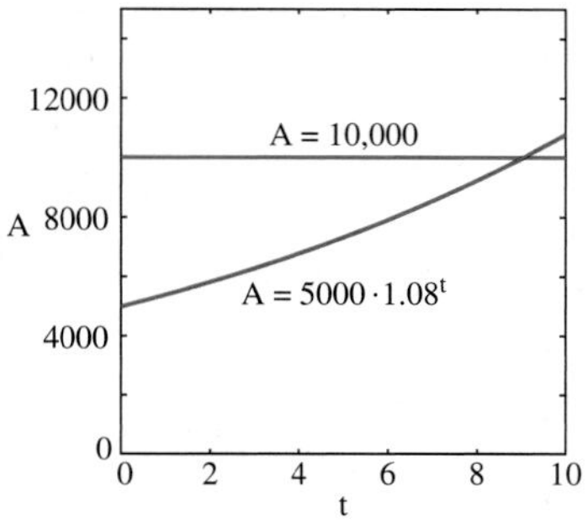

FIGURE 1.4.14 The graph for Example 8.

Figure 1.4.14 shows the graph $A(t) = 5000 \cdot 1.08^t$ as well as the horizontal line $A = 10{,}000$. From this graph we see, for instance, that the amount in the account has doubled (to \$10,000) after approximately $t = 9$ yr. We could approximate the "doubling time" more accurately by magnifying the graph near the intersection of the horizontal line and the rising curve. ◆

Example 9 exhibits a function that combines the steady decrease of an exponential function with negative exponent with the oscillation of a trigonometric function.

EXAMPLE 9 The function

$$y(t) = 3 \cdot 2^{-t} \cos 4\pi t \tag{12}$$

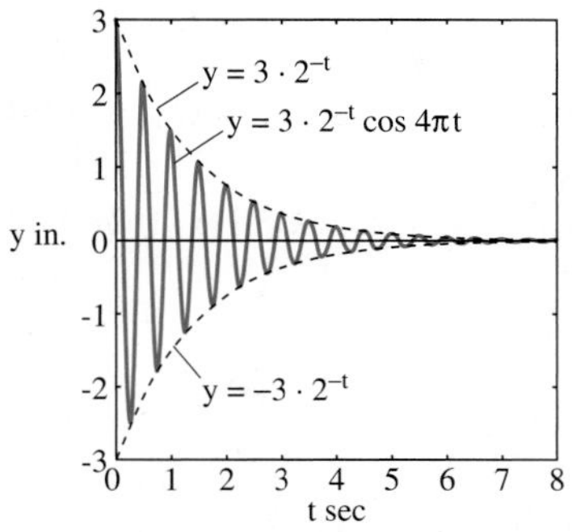

FIGURE 1.4.15 $y(t) = 3 \cdot 2^{-t} \cos 4\pi t$ (Example 9).

might describe the amplitude y, in inches, of the up-and-down vibrations of a car with very poor shock absorbers t seconds after it hits a deep pothole. Can you see that Eq. (12) describes an initial ($t = 0$) amplitude of 3 inches that halves every second, while two complete up-and-down oscillations occur every second? (The factor $3 \cdot 2^{-t}$ is the decreasing amplitude of the vibrations, while the function $\cos 4\pi t$ has period $\frac{1}{2}$ s.) Figure 1.4.15 shows the graph of $y(t)$. The curve described in Eq. (12) oscillates between the two curves $y(t) = \pm 3 \cdot 2^{-t}$. It appears that the car's vibrations subside and are negligible after 7 or 8 seconds. ◆

Logarithmic Functions

In analogy with the inverse trigonometric functions that you may have seen in trigonometry, logarithms are "inverse" to exponential functions. The **base** ***a*** **logarithm** of the positive number x is the power to which a must be raised to get x. That is,

$$y = \log_a x \quad \text{if} \quad a^y = x. \tag{13}$$

The **LOG** key on most calculators gives the base 10 (*common*) *logarithm* $\log_{10} x$. The **LN** key gives the *natural logarithm*

$$\ln x = \log_e x,$$

where e is a special irrational number:

$$e = 2.71828182845904523536\ldots.$$

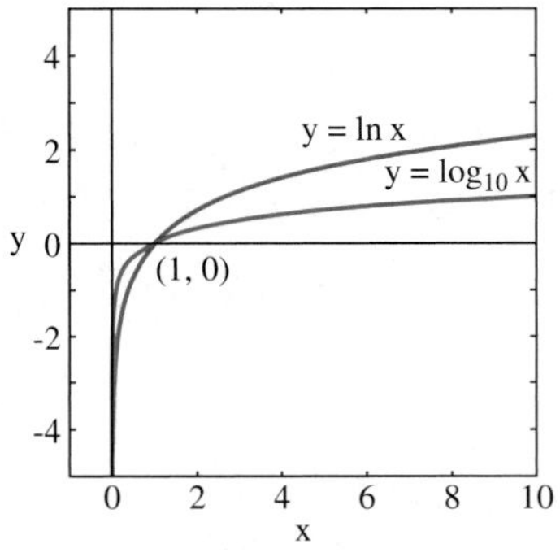

FIGURE 1.4.16 The common and natural logarithm functions.

You'll see the significance of this strange-looking base in Chapter 3.

Figure 1.4.16 shows the graphs $y = \ln x$ and $y = \log_{10} x$. Both graphs pass through the point $(1, 0)$ and rise steadily (though slowly) from left to right. Because exponential functions never take on zero or negative values, neither zero nor any negative number is in the domain of any logarithmic function.

The facts that $\log_{10} 100{,}000 = 5$ and $\log_{10} 1{,}000{,}000 = 6$ indicate that the function $\log x = \log_{10} x$ increases quite slowly as x increases. Whereas Example 6 above illustrates the fact that an exponential function a^x (with $a > 1$) increases more rapidly than any power function as $x \to \infty$, Example 10 illustrates the fact that a logarithmic function increases *more slowly* than any power function.

EXAMPLE 10 In the following table we compare the rate of growth of the power function $f(x) = \frac{1}{2}x^{1/5}$ with that of the logarithm function $g(x) = \log x$.

x	$f(x) = \frac{1}{2}x^{1/5}$	$g(x) = \log x$
20000	3.62390	4.30103
40000	4.16277	4.60206
60000	4.51440	4.77815
80000	4.78176	4.90309
100000	5	5
120000	5.18569	5.07918
140000	5.34805	5.14613
160000	5.49279	5.20412
180000	5.62373	5.25527
200000	5.74349	5.30103

It appears here and in Fig. 1.4.17 that $\log x$ is smaller than $\frac{1}{2}x^{1/5}$ when $x > 100{,}000$. Figure 1.4.18 shows that $\log x$ initially is smaller than $\frac{1}{2}x^{1/5}$, but "catches up and passes" $\frac{1}{2}x^{1/5}$ somewhere around (although a bit less than) $x = 5$. Then $\frac{1}{2}x^{1/5}$ in turn catches up and passes $\log x$ at $x = 100{,}000$. When $x = 10^{50}$, $\frac{1}{2}x^{1/5} = 5{,}000{,}000{,}000$, but the value of $\log x$ is only 50. ◆

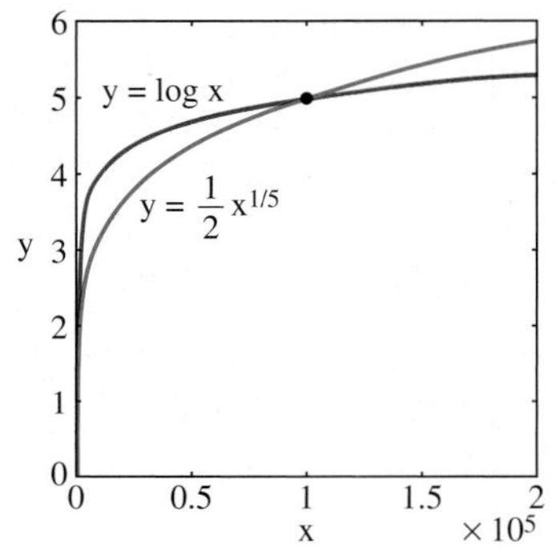

FIGURE 1.4.17 $\frac{1}{2}x^{1/5}$ passes $\log x$.

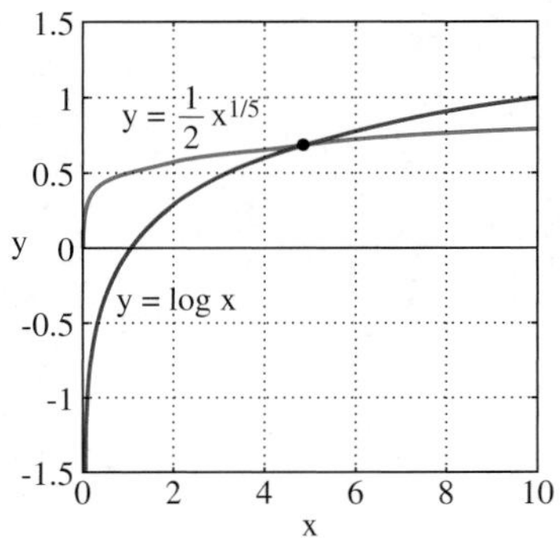

FIGURE 1.4.18 $\log x$ passes $\frac{1}{2}x^{1/5}$.

Transcendental Equations

The trigonometric, exponential, and logarithmic functions are called *transcendental* functions. As we saw in Eqs. (5) and (6), an equation that includes transcendental functions can have infinitely many solutions. But it also may have only a finite number of solutions. Determining whether the number of solutions is finite or infinite can be difficult. One approach is to write the given equation in the form

$$f(x) = g(x), \tag{14}$$

where both the functions f and g are readily graphed. Then the real solutions of Eq. (14) correspond to the intersections of the two graphs $y = f(x)$ and $y = g(x)$.

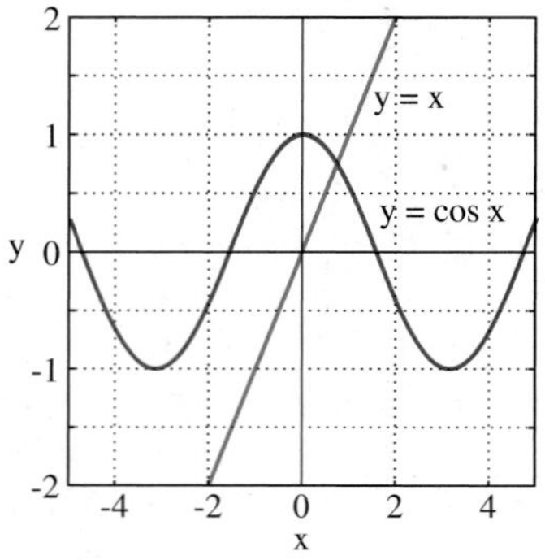

FIGURE 1.4.19 Solving the equation $x = \cos x$ of Example 11.

EXAMPLE 11 The single point of intersection of the graphs $y = x$ and $y = \cos x$, shown in Fig. 1.4.19, indicates that the equation

$$x = \cos x$$

has only a single solution. Moreover, from the graph you can glean the additional information that the solution lies in the interval $(0, 1)$. ◆

EXAMPLE 12 The graphs of $y = 1 - x$ and $y = 3\cos x$ are shown in Fig. 1.4.20. In contrast with Example 11, there are three points of intersection of the graphs. This

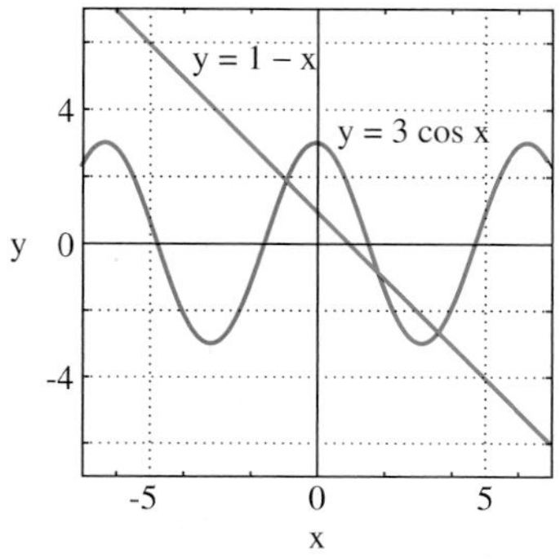

FIGURE 1.4.20 Solving the equation $1 - x = 3\cos x$ of Example 12.

makes it clear that the equation

$$1 - x = 3\cos x$$

has one negative solution and two positive solutions. They could be approximated by (separately) zooming in on the three intersection points. ◆

Can You Believe What You See on Your Calculator/Computer Screen?

The examples we give next show that the short answer to this question is "not always." One reason is that a typical graphing calculator or simple computer program plots only a finite number of equally spaced points on the curve $y = f(x), a \leqq x \leqq b$, joining the selected points with straight line segments. If the plotted points are sufficiently close, then the resulting graph may look to the unaided eye like a smooth curve, but it may miss some essential features that would be revealed if more points were plotted.

EXAMPLE 13 A 1-ampere alternating current with frequency 60 Hz (Hertz; cycles per second) is described by the function

$$I(t) = \sin 120\pi t. \tag{15}$$

The absolute value $|I(t)|$ gives the magnitude (in amperes) of the current at time t, which flows in one direction when $I > 0$ the opposite direction when $I < 0$. A simple computer program was used to plot the alleged graphs of $I(t)$ shown in Figs. 1.4.21 through 1.4.23. The graph in Fig. 1.4.21 is plotted on the interval $-1 \leqq t \leqq 1$, where we should see 120 complete oscillations because the period of $I(t)$ in Eq. (14) is $1/60$ s. But instead the figure shows exactly one oscillation, so something has gone badly wrong. The graph in Fig. 1.4.22 is plotted on the interval $-\frac{1}{2} \leqq t \leqq \frac{1}{2}$, and whatever it is has gone from merely wrong to outright bizarre. Finally, in Fig. 1.4.23 the graph is plotted on the interval $-\frac{1}{30} \leqq t \leqq \frac{1}{30}$ of length $\frac{4}{60}$, so we should see exactly 4 complete oscillations. And indeed we do, so we've finally got a correct graph of the current function in Eq. (14). ◆

Here's an explanation of what went wrong at first in Example 13. The computer was programmed to plot values at exactly 120 equally spaced points of the interval desired. So in Fig. 1.4.21 we're plotting only 1 point per cycle—not nearly enough to capture the actual shape of the curve—and only 2 points per cycle in Fig. 1.4.22. But in Fig. 1.4.23 we're plotting 30 points per cycle, and this gives an accurate representation of the actual graph.

The incorrect graph in Fig. 1.4.21—which seems to portray an oscillation with the incorrect period of 2 s, instead of the correct $\frac{1}{60}$ s—is an example of the phenomenon of *aliasing.* Another example of aliasing, occasionally seen in old Western movies, is the wagon wheel that appears to rotate slowly in the wrong direction.

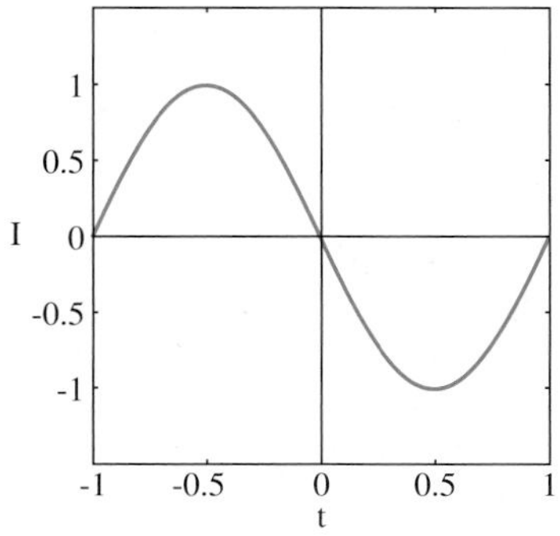

FIGURE 1.4.21 On the interval [−1, 1] it's wrong.

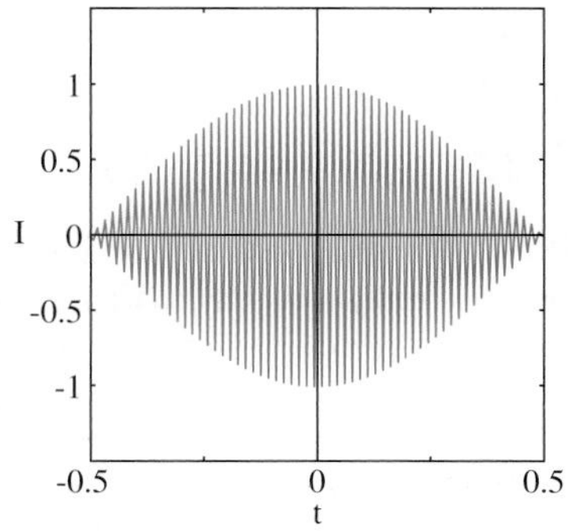

FIGURE 1.4.22 On the interval [−1/2, 1/2] it's bizarre.

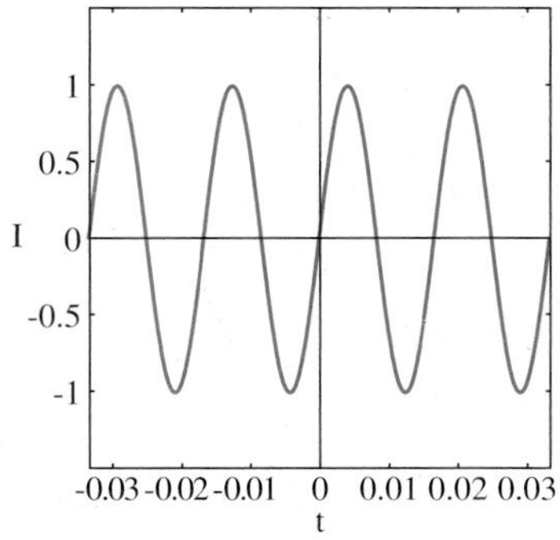

FIGURE 1.4.23 On the interval [−1/30, 1/30] it's correct!

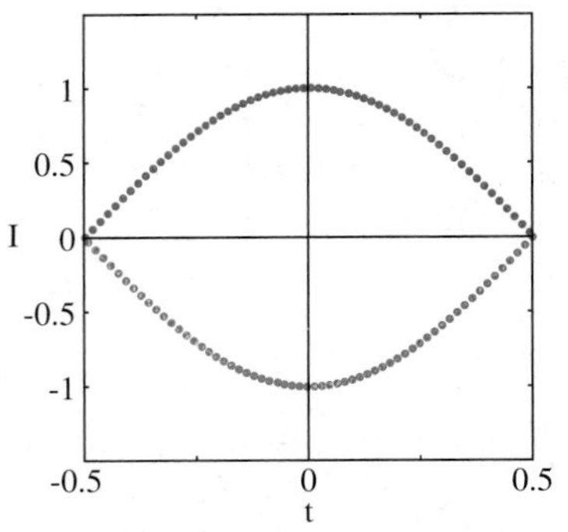

FIGURE 1.4.24 Individual plotted points that are joined by line segments in Fig. 1.4.22.

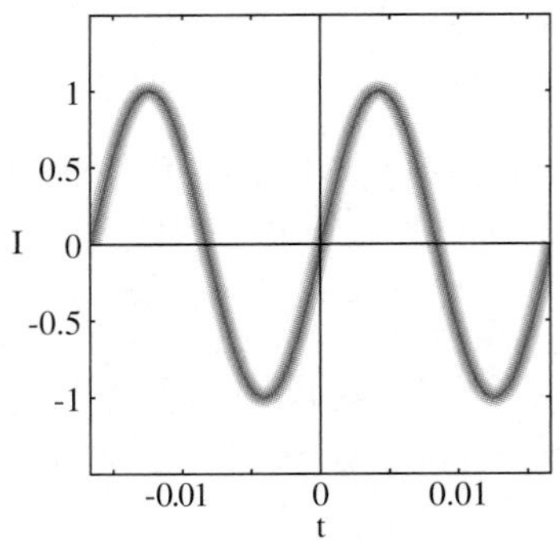

FIGURE 1.4.25 $I(t) = \sin 120\pi t + 0.01 \sin 12000\pi t$ on the interval $-1/60 \leqq t \leqq 1/60$.

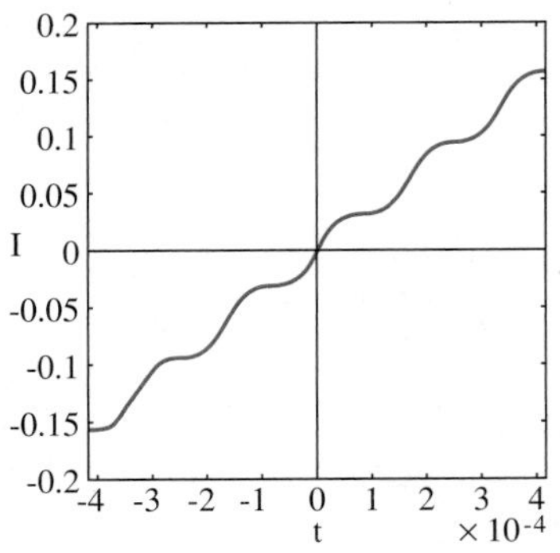

FIGURE 1.4.26 $I(t) = \sin 120\pi t + 0.01 \sin 12000\pi t$ on the interval $-1/2400 \leqq t \leqq 1/2400$.

REMARK The aliasing phenomenon exhibited in Figs. 1.4.21 and 1.4.22 is heavily dependent on the precise number of points being plotted. A plotting device (such as graphing calculator) that uses a fixed number of plotting points is susceptible to aliasing. More sophisticated graphing utilities may avoid aliasing by using a variable number of nonuniformly spaced plotting points.

Figure 1.4.22 consists largely of line segments joining consecutive points that are far apart. Figure 1.4.24 shows how that incorrect graph came about; points 1, 3, 5, 7, . . . , 117, 119 in the interval $[-0.5, 0.5]$ are plotted in red, whereas points 2, 4, 6, . . . , 118, 120 are plotted in blue. Now you can see what happened when the computer plotted line segments joining point 1 to point 2, point 2 to point 3, and so forth.

One moral of Example 13 is that it pays to know what you're looking for in a graph. If the graph looks markedly different in windows of different sizes, this is a clue that something's wrong.

Whereas in Example 13 we got anomalous results by plotting the graph in windows of different sizes, the next example illustrates a situation where we *must* plot graphs on different scales in order to see the whole picture.

EXAMPLE 14 Now suppose that a high-frequency (6000 Hz) current of 0.01 ampere is added to the current in Eq. (15), so the resulting current is described by

$$I(t) = \sin 120\pi t + (0.01) \sin 12000\pi t. \quad \textbf{(16)}$$

When we plot Eq. (16) on the interval $-\frac{1}{60} \leqq t \leqq \frac{1}{60}$, we get the graph shown in Fig. 1.4.25. It looks like two cycles of the original current in (15), although the plot is perhaps a bit "fuzzy." To see the effect of the added second term in Eq. (16) we must plot the graph on a much magnified scale, as in Fig. 1.4.26. The "fuzz" in Fig. 1.4.25 has now been magnified to show clearly the high-frequency oscillations with period $\frac{1}{6000}$ s. ◆

1.4 TRUE/FALSE STUDY GUIDE

1.4 CONCEPTS: QUESTIONS AND DISCUSSION

Each of the following items describes a particular population numbering $P(t)$ at time t. Tell whether you think the function $P(t)$ seems more likely to be a linear, quadratic, polynomial, root, rational, trigonometric, exponential, or logarithmic function of t. In each case write a specific function satisfying the given description.

1. The population triples every five years.
2. The population increases by the same amount each year.
3. The population oscillates every five years between a maximum of 120 and a minimum of 80.

4. The population decreases for a time, reaches a minimum value, then increases thereafter (getting larger and larger as time goes on).
5. The population increases for a time and reaches a maximum value, then decreases for a time and reaches a minimum value, and thereafter increases (becoming larger and larger).
6. The population increases each year, but by a smaller percentage than it increased in the preceding year.
7. The population decreases by the same percentage each year.
8. The population increases for a time, reaches a maximum value, and decreases thereafter (apparently dying out) with $P(t)$ approaching zero as t increases.

1.4 PROBLEMS

In Problems 1 through 10, match the given function with its graph among those shown in Figs. 1.4.27 through 1.4.36. Try to do this without using your graphing calculator or computer.

1. $f(x) = 2^x - 1$
2. $f(x) = 2 - 3^{-x}$
3. $f(x) = 1 + \cos x$
4. $f(x) = 2 - 2\sin x$
5. $f(x) = 1 + 2\cos x$
6. $f(x) = 2 - \sin x$

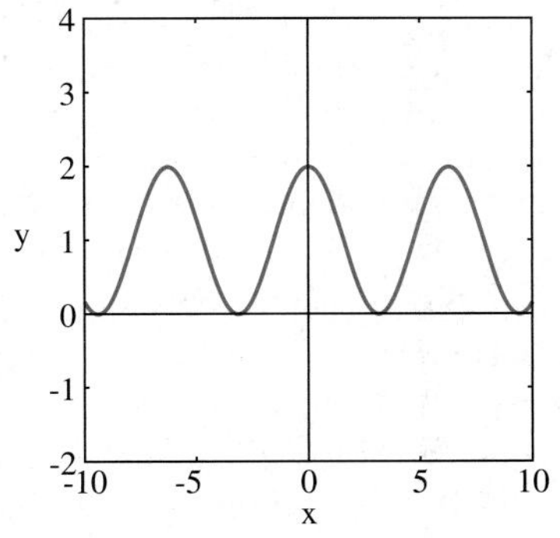

FIGURE 1.4.27

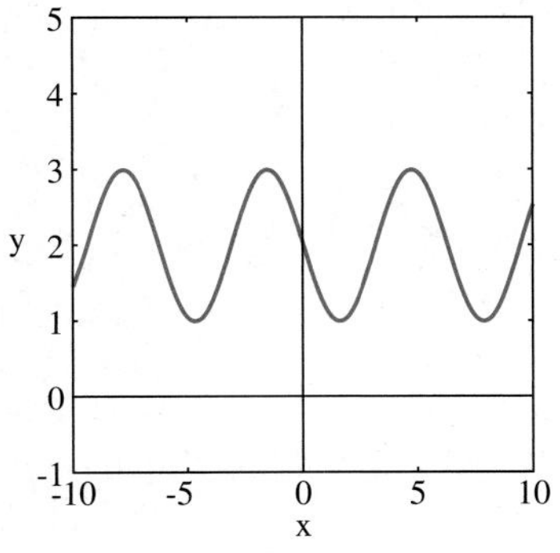

FIGURE 1.4.28

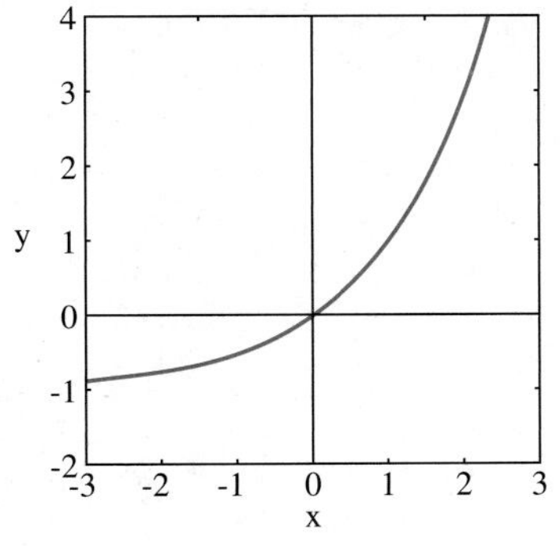

FIGURE 1.4.29

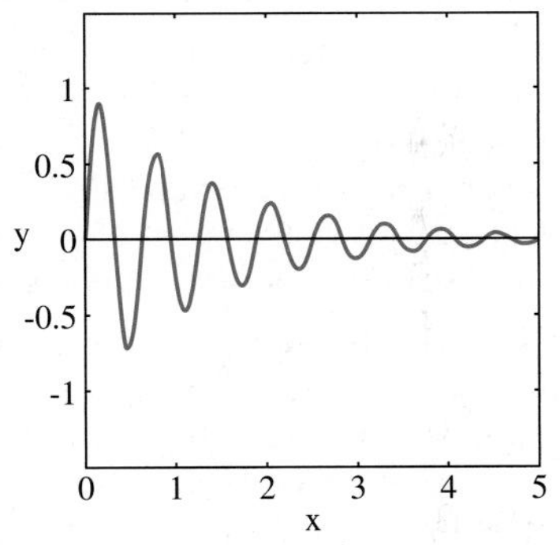

FIGURE 1.4.30

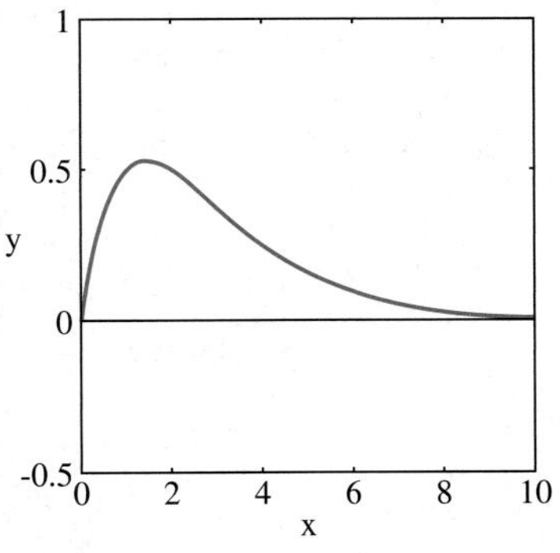

FIGURE 1.4.31

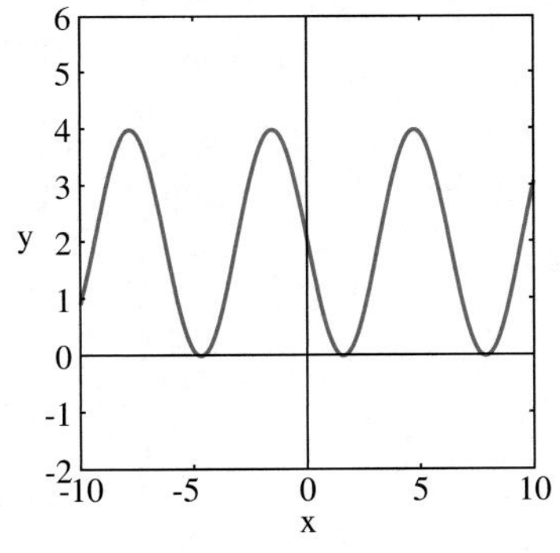

FIGURE 1.4.32

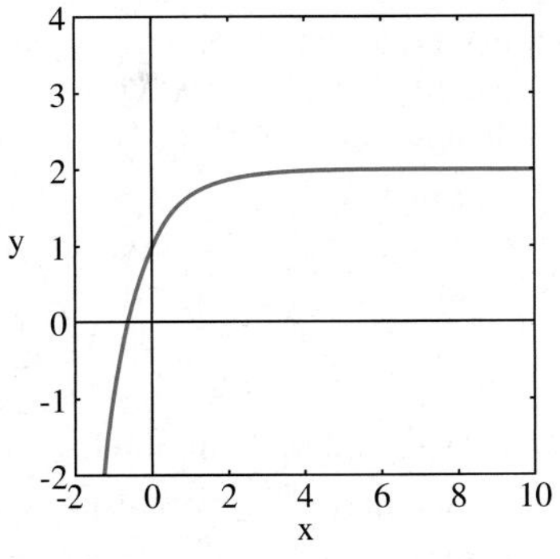

FIGURE 1.4.33

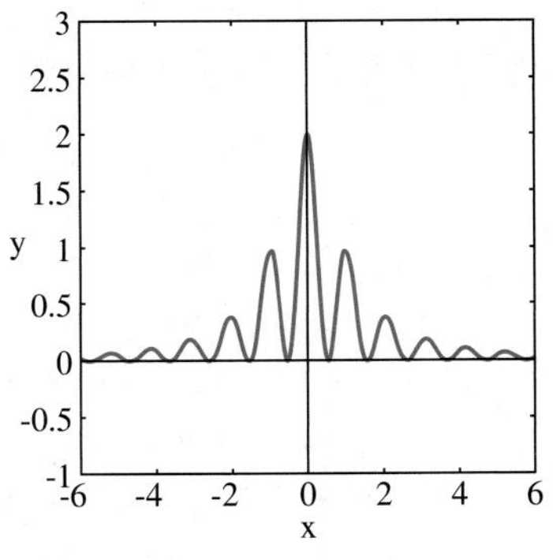

FIGURE 1.4.34

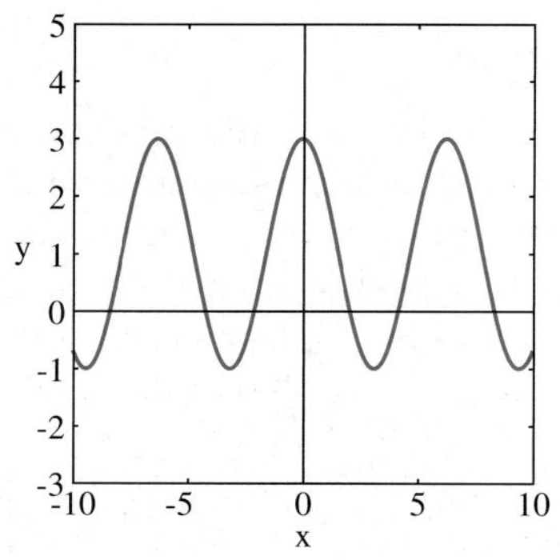

FIGURE 1.4.35

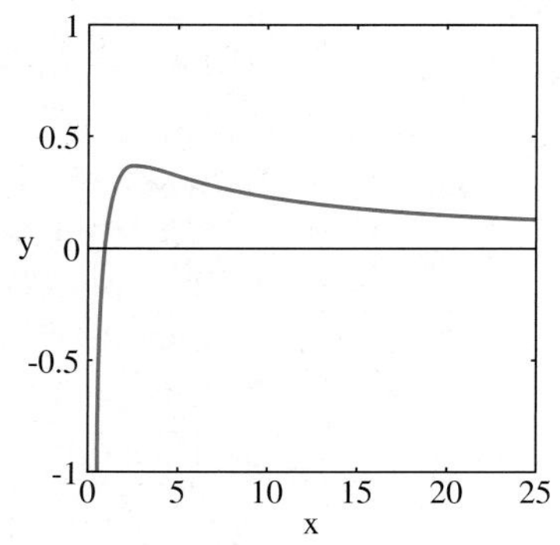

FIGURE 1.4.36

7. $f(x) = \dfrac{x}{2^x}$

8. $f(x) = \dfrac{\log x}{x}$

9. $f(x) = \dfrac{1 + \cos 6x}{1 + x^2}$

10. $f(x) = 2^{-x} \sin 10x$

In Problems 11 through 20, find $f(g(x))$ and $g(f(x))$.

11. $f(x) = 1 - x^2$, $g(x) = 2x + 3$

12. $f(x) = -17$, $g(x) = |x|$

13. $f(x) = \sqrt{x^2 - 3}$, $g(x) = x^2 + 3$

14. $f(x) = x^2 + 1$, $g(x) = \dfrac{1}{x^2 + 1}$

15. $f(x) = x^3 - 4$, $g(x) = \sqrt[3]{x + 4}$

16. $f(x) = \sqrt{x}$, $g(x) = \cos x$

17. $f(x) = \sin x$, $g(x) = x^3$

18. $f(x) = \sin x$, $g(x) = \cos x$

19. $f(x) = 1 + x^2$, $g(x) = \tan x$

20. $f(x) = 1 - x^2$, $g(x) = \sin x$

In Problems 21 through 30, find a function of the form $f(x) = x^k$ (you must specify k) and a function g such that $f(g(x)) = h(x)$.

21. $h(x) = (2 + 3x)^2$

22. $h(x) = (4 - x)^3$

23. $h(x) = \sqrt{2x - x^2}$

24. $h(x) = (1 + x^4)^{17}$

25. $h(x) = (5 - x^2)^{3/2}$

26. $h(x) = \sqrt[3]{(4x - 6)^4}$

27. $h(x) = \dfrac{1}{x + 1}$

28. $h(x) = \dfrac{1}{1 + x^2}$

29. $h(x) = \dfrac{1}{\sqrt{x + 10}}$

30. $h(x) = \dfrac{1}{(1 + x + x^2)^3}$

In Problems 31–40, use a graphing calculator or computer to determine the number of real solutions by inspecting the graph of the given equation.

31. $x = 2^{-x}$

32. $x + 1 = 3 \cos x$

33. $x - 1 = 3 \cos x$

34. $x = 5 \cos x$

35. $x = 7 \cos x$

36. $2 \log_{10} x = \cos x \quad (x > 0)$

37. $\log_{10} x = \cos x \quad (x > 0)$

38. $x^2 = 10 \cos x$

39. $x^2 = 100 \sin x$

40. $x = 5 \cos x + 10 \log_{10} x \quad (x > 0)$

41. Consider the population of Example 7 in this section, which starts with 10 rodents and doubles every month. Determine graphically (that is, by zooming) how long it will take this population to grow to 100 rodents. (Assume that each month is 30 days long and obtain an answer correct to the nearest day.)

42. Consider the money-market account of Example 8, which pays 8% annually. Determine graphically how long it will take the initial investment of \$5000 to triple.

43. In 1980 the population P of Mexico was 67.4 million and was growing at the rate of 2.6% per year. If the population continues to grow at this rate, then t years after 1980 it will be $P(t) = 67.4 \cdot (1.026)^t$ (millions). Determine graphically how long it will take the population of Mexico to double.

44. Suppose that the amount A of ozone in the atmosphere decreases at the rate of 0.25% per year, so that after t years the amount remaining is $A(t) = A_0(0.9975)^t$, where A_0 denotes the initial amount. Determine graphically how long it will take for only half the original amount of ozone to be left. Does the numerical value of A_0 affect this answer?

45. The nuclear accident at Chernobyl left the surrounding region contaminated with strontium-90, which initially was emitting radiation at approximately 12 times the level safe for human habitation. When an atom of strontium-90 emits radiation, it decays to a nonradioactive isotope. In this way, about 2.5% of the strontium-90 disappears each year. Then the amount of radiation left after t years will be $A(t) = 12 \cdot (0.975)^t$ (measured in "safe units" of radiation). Determine graphically how long (after the original accident) it will be until the region measures only 1 safe unit, and it is therefore safe for humans to return.

46. Refer to Example 6 of this section; determine graphically the value (accurate to three decimal places) of the negative solution of the equation $x^2 = 2^x$.

47. Refer to Example 10 of this section; determine graphically the value (accurate to three decimal places) of the solution near $x = 5$ of the equation $\log_{10} x = \frac{1}{2}x^{1/5}$.

48. The equation $x^{10} = 3^x$ has three real solutions. Graphically approximate each of them accurate to two decimal places.

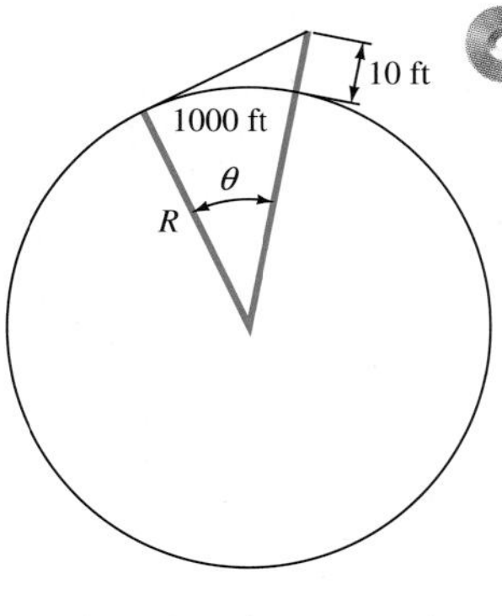

FIGURE 1.4.37 The asteroid problem.

1.4 Project: A Spherical Asteroid

You land your space ship on a spherical asteroid between Earth and Mars. Your copilot walks 1000 feet away along the asteroid's smooth surface carrying a 10-ft rod and thereby vanishes over the horizon. When she places one end of the rod on the ground and holds it straight up and down, you—lying flat on the ground—can just barely see the tip of the rod. Use this information to find the radius R of this asteroid (in miles). The key will be to derive a pair of equations relating R and the angle θ indicated in Fig. 1.4.37. (Think of the right triangle shown there and of the relationship between circular arc length and subtended central angle.) You can then attempt to solve these equations graphically. You should find plenty of solutions. But which of them gives the radius of the asteroid?

1.5 PREVIEW: WHAT *IS* CALCULUS?

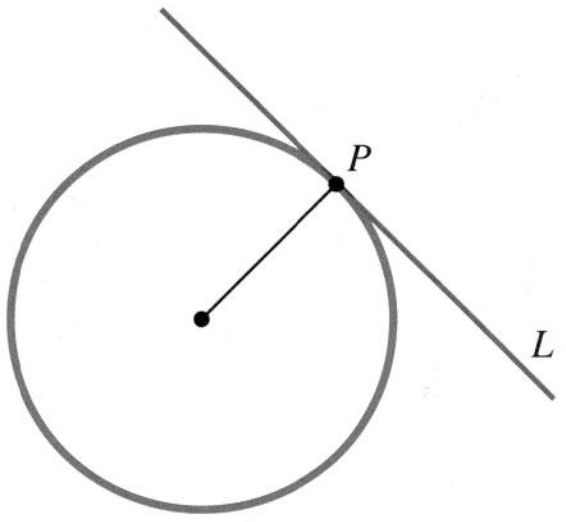

FIGURE 1.5.1 The tangent line L touches the circle at the point P.

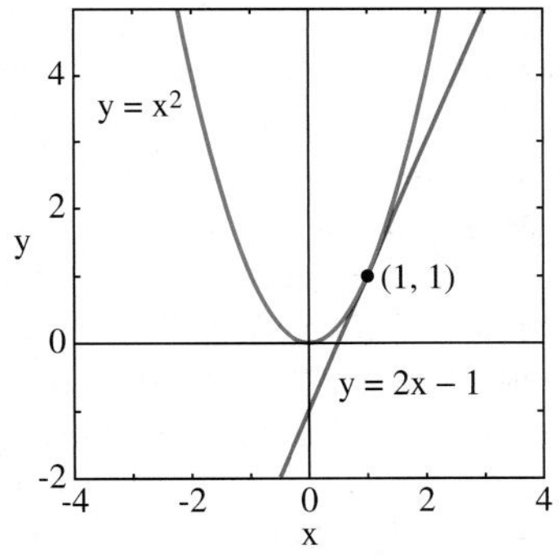

FIGURE 1.5.2 The line tangent to the parabola $y = x^2$ at the point (1, 1).

Surely this question is on your mind as you begin a study of calculus that may extend over two or three terms. Following our review of functions and graphs in Sections 1.1 through 1.4, we can preview here at least the next several chapters, where the central concepts of calculus are developed.

The Two Fundamental Problems

The body of computational technique that constitutes "the calculus" revolves around two fundamental geometric problems that people have been investigating for more than 2000 years. Each problem involves the graph $y = f(x)$ of a given function.

The first fundamental problem is this: What do we mean by the *line tangent* to the curve $y = f(x)$ at a given point? The word *tangent* stems from the Latin word *tangens,* for "touching." Thus a line tangent to a curve is one that "just touches" the curve. Lines tangent to circles (Fig. 1.5.1) are well known from elementary geometry. Figure 1.5.2 shows the line tangent to the parabola $y = x^2$ at the point (1, 1). We will see in Section 2.1 that this particular tangent line has slope 2, so its point-slope equation is

$$y - 1 = 2 \cdot (x - 1); \quad \text{that is,} \quad y = 2x - 1.$$

Our first problem is how to find tangent lines in more general cases.

> **The Tangent Problem**
>
> Given a point $P(x, f(x))$ on the curve $y = f(x)$, how do we calculate the slope of the tangent line at P (Fig. 1.5.3)?

We begin to explore the answer to this question in Chapter 2. If we denote by $m(x)$ the slope of the tangent line at $P(x, f(x))$, then m is a *new function.* It might informally be called a *slope-predictor* for the curve $y = f(x)$. In calculus this slope-predictor function is called the **derivative** of the function f. In Chapter 3 we learn to calculate derivatives of a variety of functions, and in both Chapter 3 and Chapter 4 we see numerous applications of derivatives in solving real-world problems. These three chapters introduce part of calculus called *differential calculus*.

The tangent problem is a geometric problem—a purely mathematical question. But its answer (in the form of derivatives) is the key to the solution of diverse applied problems in many scientific and technical areas. Examples 1 and 2 may suggest to you the *connections* that are the key to the pivotal role of calculus in science and technology.

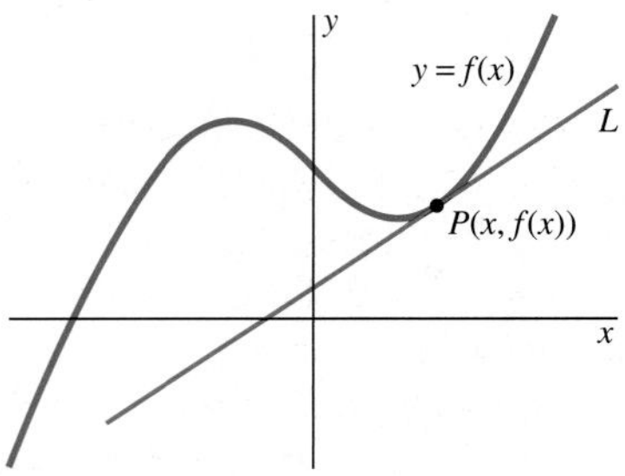

FIGURE 1.5.3 What is the slope of the line L tangent to the graph $y = f(x)$ at the point $P(x, f(x))$?

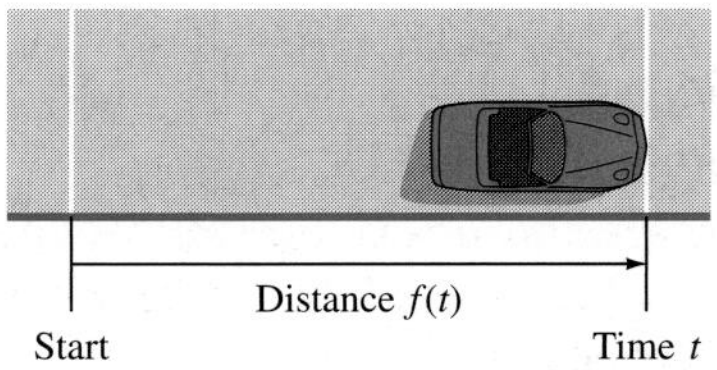

FIGURE 1.5.4 A car on a straight road (Example 1).

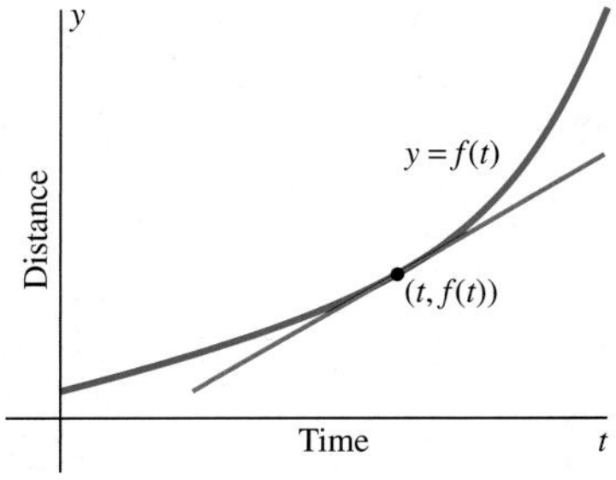

FIGURE 1.5.5 The slope of the tangent line at the point $(t, f(t))$ is the velocity at the time t (Example 1).

EXAMPLE 1 Suppose that you're driving a car along a long, straight road (Fig. 1.5.4). If $f(t)$ denotes the *distance* (in miles) the car has traveled at time t (in hours), then the slope of the line tangent to the curve $y = f(t)$ at the point $(t, f(t))$ (Fig. 1.5.5) is the *velocity* (in miles per hour) of the car at time t. ◆

EXAMPLE 2 Suppose that $f(t)$ denotes the number of people in the United States who have a certain serious disease at time t (measured in days from the beginning of the year). Then the slope of the line tangent to the curve $y = f(t)$ at the point $(t, f(t))$ (Fig. 1.5.6) is the *rate of growth* (the number of persons newly affected per day) of the diseased population at time t. ◆

Note The truth of the statements made in these two examples is *not* obvious. To understand such things is one reason you study calculus! We return to the concepts of velocity and rate of change at the beginning of Chapter 3.

Here we will be content with the observation that the slopes of the tangent lines in Examples 1 and 2 at least have the correct *units*. If in the time-distance plane of Example 1 we measure time t (on the horizontal axis) in seconds and distance y (on the vertical axis) in feet (or meters), then the slope (ratio of rise to run) of a straight line has the dimensions of feet (or meters) per second—the proper units for velocity (Fig. 1.5.7). Similarly, if in the ty-plane of Example 2 time t is measured in months and y is measured in persons, then the slope of a straight line has the proper units of persons per month for measuring the rate of growth of the afflicted population (Fig. 1.5.8).

The second fundamental problem of calculus is the problem of *area*. Given the graph $y = f(x)$, what is the area between this graph and the x-axis over the interval $[a, b]$?

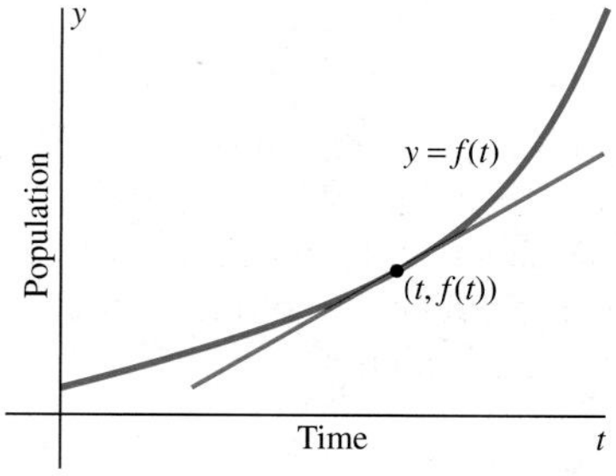

FIGURE 1.5.6 The rate of growth of $f(t)$ at the time t is the slope of the tangent line at the point $(t, f(t))$ (Example 2).

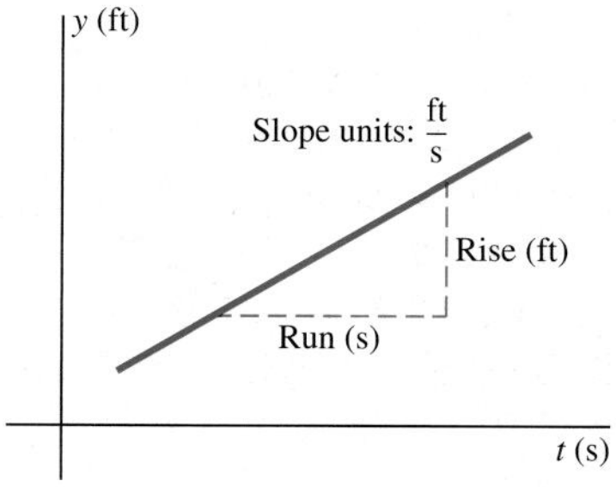

FIGURE 1.5.7 Here slope has the dimensions of velocity (ft/s).

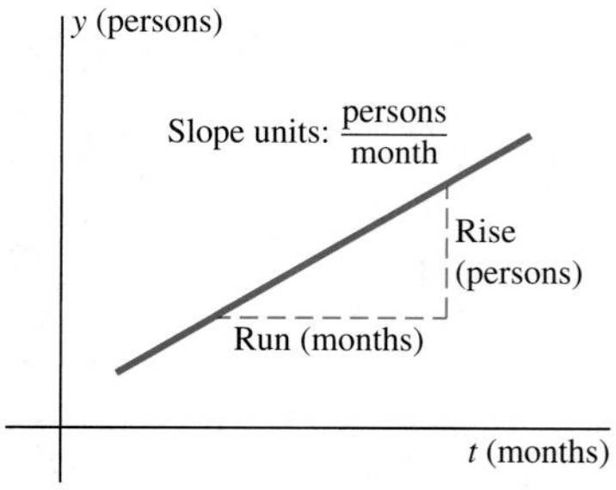

FIGURE 1.5.8 Here slope has the dimensions of rate of change of population.

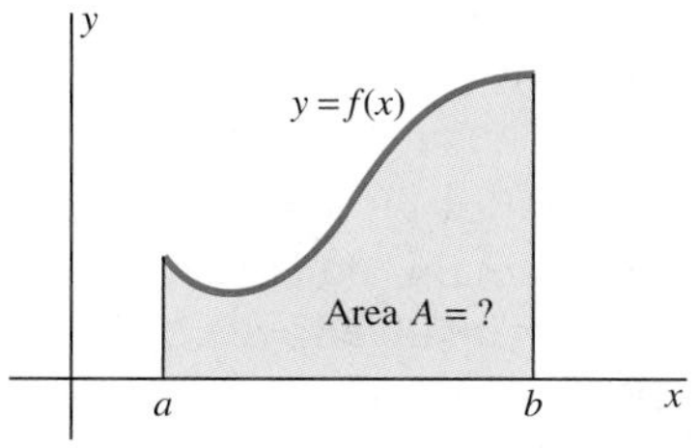

FIGURE 1.5.9 The area problem.

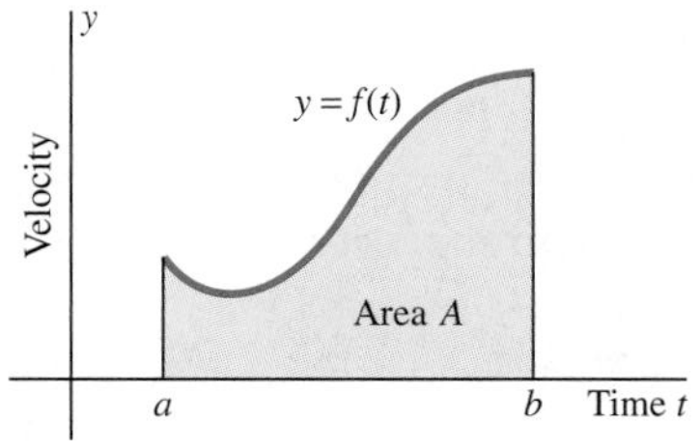

FIGURE 1.5.10 The area A under the velocity curve is equal to the distance traveled during the time interval $a \leqq t \leqq b$ (Example 3).

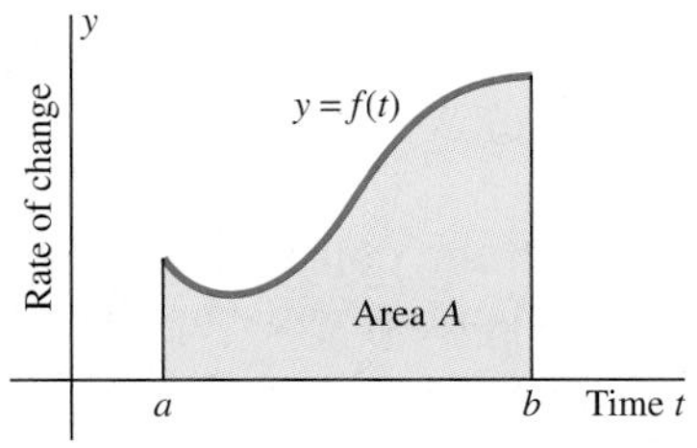

FIGURE 1.5.11 The area A under the rate-of-change curve is equal to the net change in the population from time $t = a$ to $t = b$ (Example 4).

The Area Problem

If $f(x) \geqq 0$ for x in the interval $[a, b]$, how do we calculate the area A of the plane region that lies between the curve $y = f(x)$ and the x-axis over the interval $[a, b]$ (Fig. 1.5.9)?

We begin to explore the answer to this second question in Chapter 5. In calculus the area A is called an *integral* of the function f. Chapters 5 and 6 are devoted to the calculation and application of integrals. These two chapters introduce the other part of calculus, which is called *integral calculus*.

Like the tangent problem, the area problem is a purely mathematical question, but its answer (in the form of integrals) has extensive ramifications of practical importance. Examples 3 and 4 have an obvious kinship with Examples 1 and 2.

EXAMPLE 3 If $f(t)$ denotes the *velocity* of a car at time t, then the area under the curve $y = f(t)$ over the time interval $[a, b]$ is equal to the *distance* traveled by the car between time $t = a$ and time $t = b$ (Fig. 1.5.10). ◆

EXAMPLE 4 If $f(t)$ denotes the *rate of growth* of a diseased population at time t, then the area under the curve $y = f(t)$ over the time interval $[a, b]$ is equal to the net change in the *size* of this population between time $t = a$ and time $t = b$ (Fig. 1.5.11). ◆

When we discuss integrals in Chapter 5, you will learn why the statements in Examples 3 and 4 are true.

The Fundamental Relationship

Examples 1 and 3 are two sides of a certain coin: There is an "inverse relationship" between the *distance* traveled and the *velocity* of a moving car. Examples 2 and 4 exhibit a similar relationship between the *size* of a population and its *rate of change*.

Both the distance/velocity relationship and the size/rate-of-change relationship illustrated by Examples 1 through 4 are consequences of a deep and fundamental relationship between the tangent problem and the area problem. This more general relationship is described by the *fundamental theorem of calculus,* which we discuss in Chapter 5. It was discovered in 1666 by Isaac Newton at the age of 23 while he was still a student at Cambridge University. A few years later it was discovered independently by Gottfried Wilhelm Leibniz, who was then a German diplomat in Paris who studied mathematics privately. Although the tangent problem and the area problem had, even then, been around for almost 2000 years, and much progress on separate solutions had been made by predecessors of Newton and Leibniz, their joint discovery of the fundamental relationship between the area and tangent problems made them famous as "the inventors of the calculus."

Applications of Calculus

So calculus centers around the computation *and application* of derivatives and integrals—that is, of tangent line slopes and areas under graphs. Throughout this textbook, you will see concrete applications of calculus to different areas of science and technology. The following list of a dozen such applications gives just a brief indication of the extraordinary range and real-world power of calculus.

- Suppose that you make and sell tents. How can you make the biggest tent from a given amount of cloth and thereby maximize your profit? (Section 3.6)
- You throw into a lake a cork ball that has one-fourth the density of water. How deep will it sink in the water? (Section 3.10)

- A driver involved in an accident claims he was going only 25 mi/h. Can you determine from his skid marks the actual speed of his car at the time of the accident? (Section 5.2)
- The great pyramid of Khufu at Gizeh, Egypt, was built well over 4000 years ago. No personnel records from the construction remain, but nevertheless we can calculate the approximate number of laborers involved. (Section 6.5)
- Suppose that you win the Florida lottery and decide to use part of your winning to purchase a "perpetual annuity" that will pay you and your heirs (and theirs, *ad infinitum*) $10,000 per year. What is a fair price for an insurance company to charge you for such an annuity? (Section 7.8)
- If the earth's population continues to grow at its present rate, when will there be "standing room only"? (Section 8.1)
- The factories polluting Lake Erie are forced to cease dumping wastes into the lake immediately. How long will it take for natural processes to restore the lake to an acceptable level of purity? (Section 8.4)
- In 1845 the Belgian demographer Verhulst used calculus to predict accurately the course of U.S. population growth (to within 1%) well into the twentieth century, long after his death. How? (Section 8.5)
- What explains the fact that a well-positioned reporter can eavesdrop on a quiet conversation between two diplomats 50 feet away in the Whispering Gallery of the U.S. Senate, even if this conversation is inaudible to others in the same room? (Section 9.6)
- Suppose that Paul and Mary alternately toss a fair six-sided die in turn until one wins the pot by getting the first "six." How advantageous is it to be the one who tosses first? (Section 10.3)
- How can a submarine traveling in darkness beneath the polar icecap keep accurate track of its position without being in radio contact with the rest of the world? (Section 12.1)
- Suppose that your club is designing an unpowered race car for the annual downhill derby. You have a choice of solid wheels, bicycle wheels with thin spokes, or even solid spherical wheels (like giant ball bearings). Can you determine (without time-consuming experimentation) which will make the race car go the fastest? (Section 14.5)
- Some bullets have flattened tips. Is it possible that an artillery shell with a flat-tipped "nose cone" may experience less air resistance—and therefore travel farther—than a shell with a smoothly rounded tip? (Section 15.5)

1.5 TRUE/FALSE STUDY GUIDE

CHAPTER 1 REVIEW: DEFINITIONS AND CONCEPTS

Use this list as a guide to key topics that you may need to review or discuss.

1. The definition of a function
2. The domain and range of a function
3. Dependent and independent variables
4. Open and closed interval notation
5. The idea of a mathematical model
6. Numerical investigation of values of functions by repeated tabulation
7. Slope-intercept and point-slope equations of straight lines
8. The graph of an equation
9. The graph of a function
10. The translation principle for graphs of equations
11. The vertical line test for graphs of functions
12. Equations and translates of circles
13. Parabolas and graphs of quadratic functions
14. Contrasting graphic, numeric, and symbolic views of functions
15. Graphical solution of equations by the method of repeated magnification
16. Algebraic combinations of functions

17. The maximum possible number of zeros of a polynomial

18. The behavior of the polynomial $p(x)$ for $|x|$ large

19. Location of the vertical asymptotes of a rational function

20. Qualitative comparison of graphs of power functions, polynomials, rational functions, and algebraic functions

21. Periods of trigonometric functions

22. Rates of growth of exponential and logarithmic functions

23. Qualitative comparison of the graphs of trigonometric, exponential, and logarithmic functions

24. Composition of functions

25. Determining a viewing window that exhibits the principal features of the graph of a function

26. Possible pitfalls of calculator and computer graphing

CHAPTER 1 MISCELLANEOUS PROBLEMS

In Problems 1 through 10, find the domain of definition of the function with the given formula.

1. $f(x) = \sqrt{x-4}$

2. $f(x) = \dfrac{1}{2-x}$

3. $f(x) = \dfrac{1}{x^2-9}$

4. $f(x) = \dfrac{x}{x^2+1}$

5. $f(x) = \left(1+\sqrt{x}\right)^3$

6. $f(x) = \dfrac{x+1}{x^2-2x}$

7. $f(x) = \sqrt{2-3x}$

8. $f(x) = \dfrac{1}{\sqrt{9-x^2}}$

9. $f(x) = (x-2)(4-x)$

10. $f(x) = \sqrt{(x-2)(4-x)}$

11. In accord with Boyle's law, the pressure p (lb/in.^{2}) and volume V (in.^{3}) of a certain gas satisfy the condition $pV = 800$. What is the range of possible values of p, given $100 \leqq V \leqq 200$?

12. The relationship between the Fahrenheit temperature F and the Celsius temperature C is given by

$$F = 32 + \tfrac{9}{5}C.$$

If the temperature on a given day ranges from a low of 70°F to a high of 90°F, what is the range of temperature in degrees Celsius?

13. An electric circuit contains a battery that supplies E volts in series with a resistance of R ohms (Fig. 1.MP.1). Then the current of I amperes that flows in the circuit satisfies Ohm's law, $E = IR$. If $E = 100$ and $25 < R < 50$, what is the range of possible values of I?

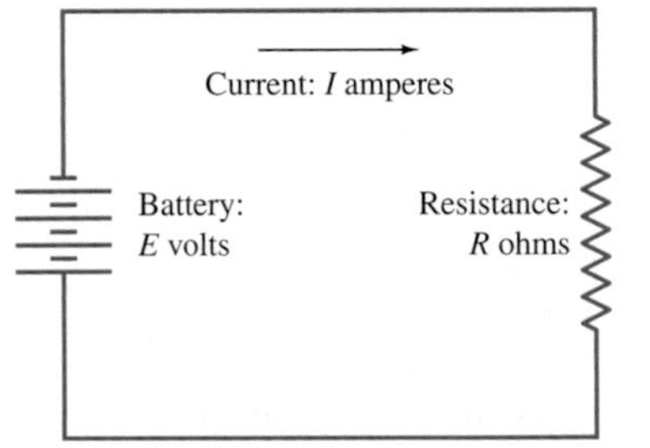

FIGURE 1.MP.1 The simple electric circuit of Problem 13.

14. The period T (in seconds) of a simple pendulum of length L (in feet) is given by $T = 2\pi\sqrt{L/32}$. If $3 < L < 4$, what is the range of possible values of T?

15. Express the volume V of a cube as a function of its total surface area S.

16. The height of a certain right circular cylinder is equal to its radius. Express its total surface area A (including both ends) as a function of its volume V.

17. Express the area A of an equilateral triangle as a function of its perimeter P.

18. A piece of wire 100 in. long is cut into two pieces of lengths x and $100 - x$. The first piece is bent into the shape of a square, the second into the shape of a circle. Express as a function of x the sum A of the areas of the square and circle.

In Problems 19 through 24, write an equation of the straight line L described.

19. L passes through $(-3, 5)$ and $(1, 13)$.

20. L passes through $(4, -1)$ and has slope -3.

21. L has slope $\frac{1}{2}$ and y-intercept -5.

22. L passes through $(2, -3)$ and is parallel to the line with equation $3x - 2y = 4$.

23. L passes through $(-3, 7)$ and is perpendicular to the line with equation $y - 2x = 10$. (Appendix B reviews slopes of perpendicular lines.)

24. L is the perpendicular bisector of the segment joining $(1, -5)$ and $(3, -1)$.

In Problems 25 through 34, match the given function with its graph among those shown in Figs. 1.MP.2 through 1.MP.11. Do this without using your graphing calculator or computer. Instead, rely on your knowledge of the general characteristics of polynomial, rational, algebraic, trigonometric, exponential, and logarithmic functions.

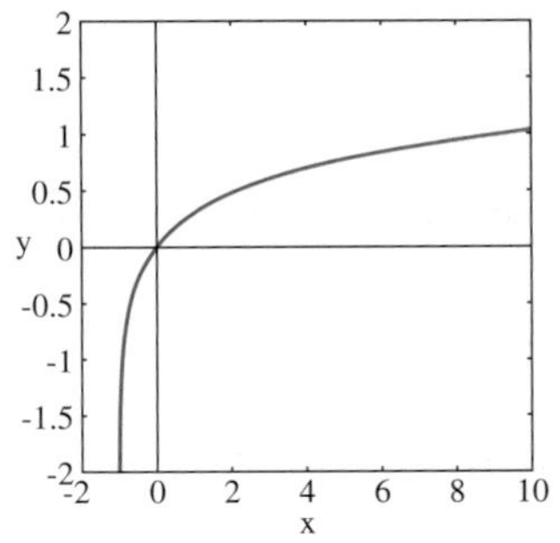

FIGURE 1.MP.2

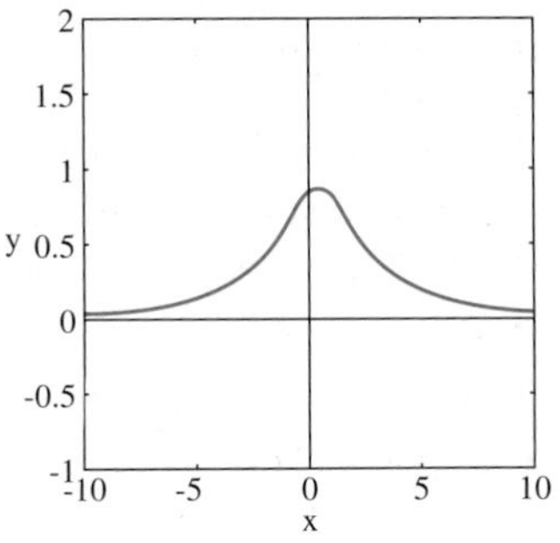

FIGURE 1.MP.3

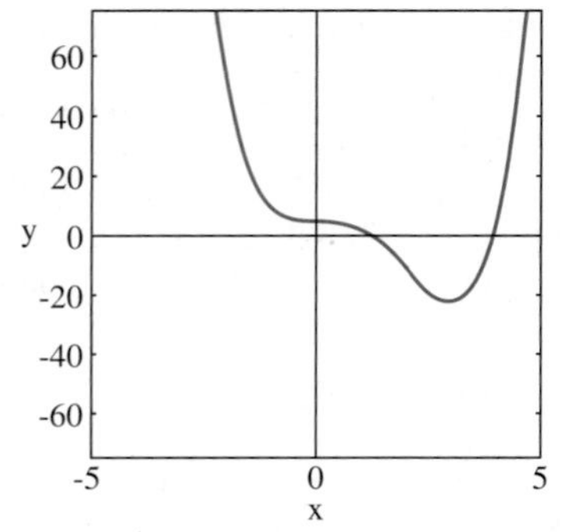

FIGURE 1.MP.4

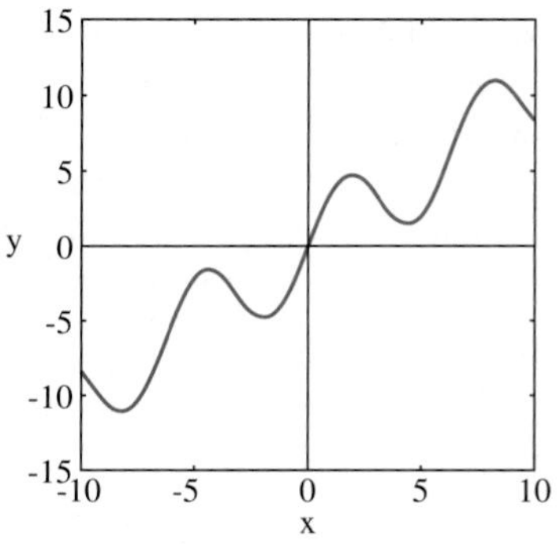

FIGURE 1.MP.5

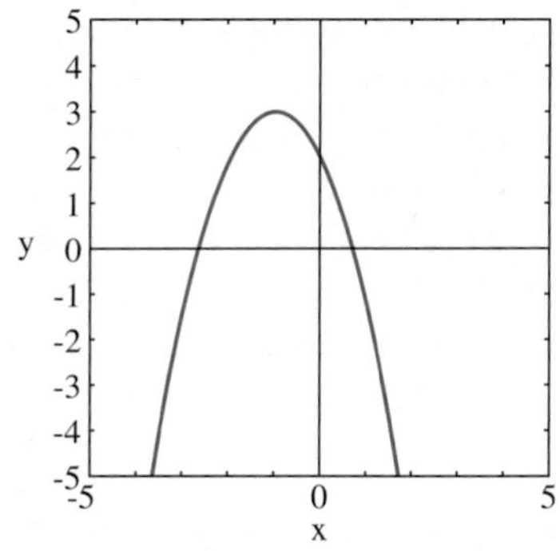

FIGURE 1.MP.6

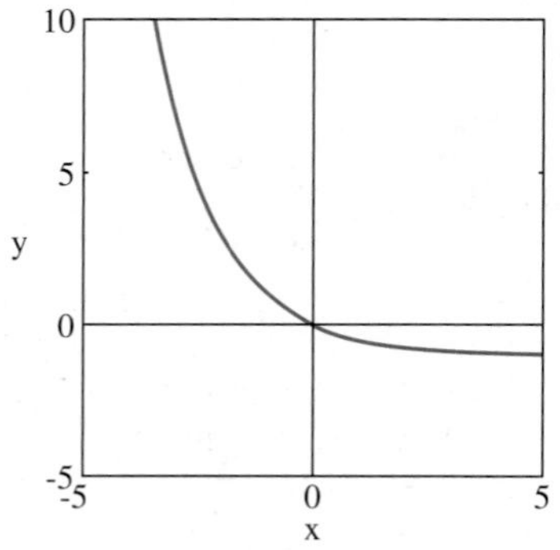

FIGURE 1.MP.7

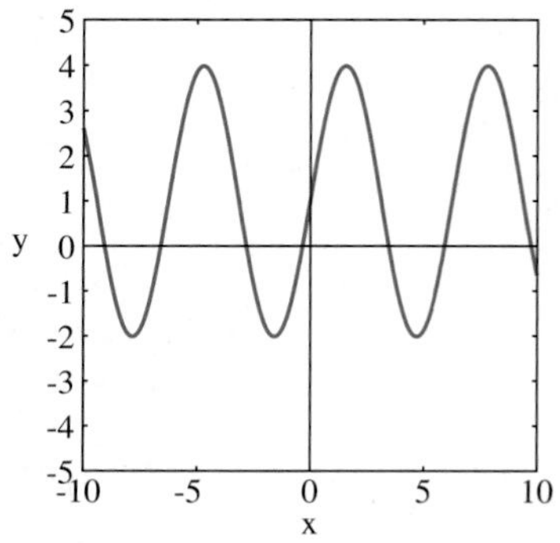

FIGURE 1.MP.8

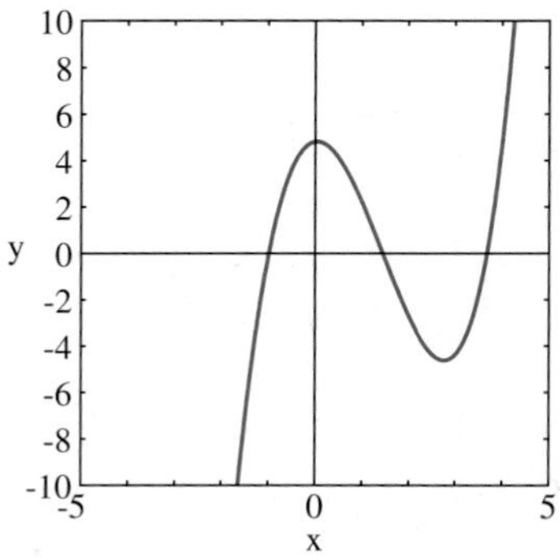

FIGURE 1.MP.9

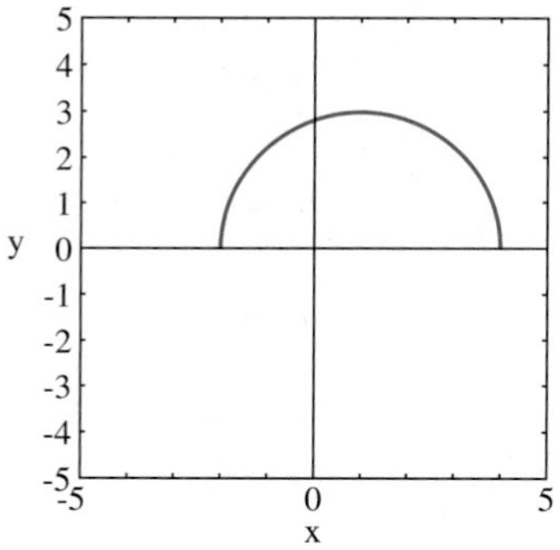

FIGURE 1.MP.10

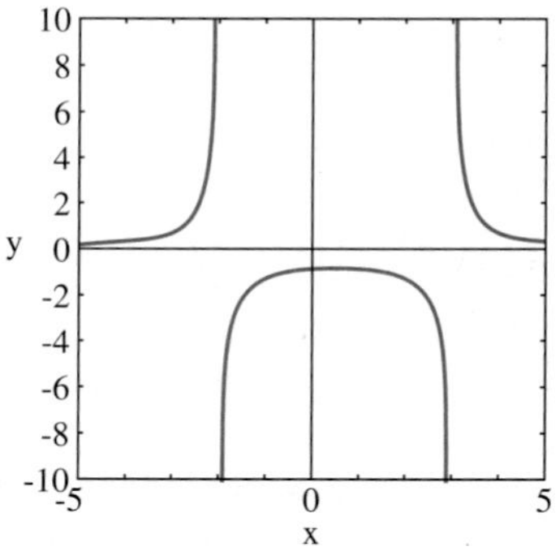

FIGURE 1.MP.11

25. $f(x) = 2 - 2x - x^2$

26. $f(x) = x^3 - 4x^2 + 5$

27. $f(x) = x^4 - 4x^3 + 5$

28. $f(x) = \dfrac{5}{x^2 - x - 6}$

29. $f(x) = \dfrac{5}{x^2 - x + 6}$

30. $f(x) = \sqrt{8 + 2x - x^2}$

31. $f(x) = 2^{-x} - 1$

32. $f(x) = \log_{10}(x + 1)$

33. $f(x) = 1 + 3\sin x$

34. $f(x) = x + 3\sin x$

Sketch the graphs of the equations and functions given in Problems 35 through 44.

35. $2x - 5y = 7$

36. $|x - y| = 1$

37. $x^2 + y^2 = 2x$

38. $x^2 + y^2 = 4y - 6x + 3$

39. $y = 2x^2 - 4x - 1$

40. $y = 4x - x^2$

41. $f(x) = \dfrac{1}{x + 5}$

42. $f(x) = \dfrac{1}{4 - x^2}$

43. $f(x) = |x - 3|$

44. $f(x) = |x - 3| + |x + 2|$

45. Apply the triangle inequality (of Appendix A) twice to show that

$$|a + b + c| \leqq |a| + |b| + |c|$$

for arbitrary real numbers a, b, and c.

46. Write $a = (a - b) + b$ to deduce from the triangle inequality (of Appendix A) that

$$|a| - |b| \leqq |a - b|$$

for arbitrary real numbers a and b.

47. Solve the inequality $x^2 - x - 6 > 0$. [*Suggestion:* Conclude from the factorization

$$x^2 - x - 6 = (x - 3)(x + 2)$$

that the quantities $x - 3$ and $x + 2$ must be either both positive or both negative for the inequality to hold. Consider the two cases separately to conclude that the solution set is $(-\infty, -2) \cup (3, +\infty)$.]

Use the method of Problem 47 to solve the inequalities in Problems 48 through 50.

48. $x^2 - 3x + 2 < 0$

49. $x^2 - 2x - 8 > 0$

50. $2x \geqq 15 - x^2$

The remaining problems require the use of an appropriate calculator or computer. In Problems 51 through 56, use either the method of repeated tabulation or the method of successive zooms (or both) to find the two roots (with three digits to the right of the decimal point correct or correctly rounded) of the given quadratic equation. You may check your work with the aid of the quadratic formula and an ordinary calculator.

51. $x^2 - 5x - 7 = 0$

52. $3x^2 - 10x - 11 = 0$

53. $4x^2 - 14x + 11 = 0$

54. $5x^2 + 24x - 35 = 0$

55. $8x^2 + 33x - 36 = 0$

56. $9x^2 + 74x - 156 = 0$

In Problems 57 through 62, apply either the method of repeated tabulation or the method of successive zooms (or both) to find the lowest point on the given parabola. You may check your work by completing the square.

57. $y = x^2 - 5x + 7$

58. $y = 3x^2 - 10x + 11$

59. $y = 4x^2 - 14x + 11$

60. $y = 5x^2 + 24x + 35$

61. $y = 8x^2 + 33x + 35$

62. $y = 9x^2 + 74x + 156$

63. Figure 1.MP.12 shows a 10-cm by 7-cm portrait that includes a border of width x on the top and bottom and of width $2x$ on either side. The area of the border is itself 20 cm^2. Use either repeated tabulation or successive zooms to find x.

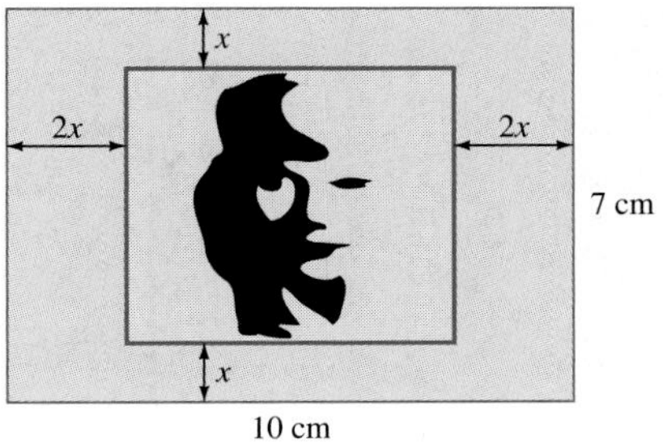

FIGURE 1.MP.12 The bordered portrait of Problem 63.

64. A mail-order catalog lists a 60-in. by 35-in. tablecloth that shrinks 7% in area when first washed. The catalog description also implies that the length and width will *both* decrease by the same amount x. Use numerical (tabulation) or graphical (zoom) methods to find x.

Determine graphically the number of real solutions of each equation in Problems 65 through 70.

65. $x^3 - 7x + 3 = 0$

66. $x^4 - 3x^2 + 4x - 5 = 0$

67. $\sin x = x^3 - 3x + 1$

68. $\cos x = x^4 - x$

69. $\cos x = \log_{10} x$

70. $10^{-x} = \log_{10} x$

PRELUDE TO CALCULUS 2

Ada Byron (1815–1852)

The modern computer programming language Ada is named in honor of Ada Byron, daughter of the English poet Lord Byron. Her interest in science and mathematics led her around 1840 to study the Difference Engine, a gear-based mechanical calculator that the mathematician Charles Babbage had built to compute tables of values of functions. By then he was designing his much more advanced Analytic Engine, an elaborate computing machine that would have been far ahead of its time if it had been completed. In 1843 Ada Byron wrote a series of brief essays explaining the planned operation of the Analytical Engine and its underlying mathematical principles. She included a prototype "computer program" to illustrate how its calculations were to be "programmed" in advance, using a deck of punched cards to specify its instructions.

The Difference Engine

Calculus has been called "the calculating engine par excellence." But in our own time the study and applications of calculus have been reshaped by electronic computers. Throughout this book we illustrate concepts of calculus by means of graphic, numeric, and symbolic results generated by computers. In Chapter 2 we exploit computational technology systematically in the investigation of limits.

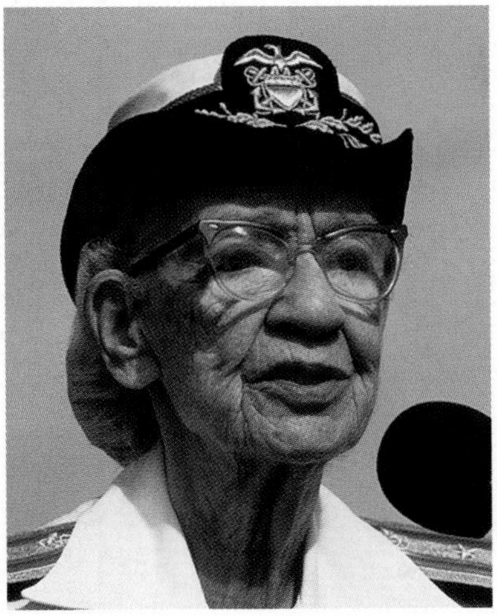

Grace Murray Hopper (1906–1992)

Almost exactly a century after the death of Ada Byron, the first modern computer compiler (for translation of human-language programs into machine-language instructions) was developed by Grace Murray Hopper. As a mathematician and U.S. Navy officer, Hopper had worked with the very first modern electronic computers developed during and immediately after World War II. In 1967 she was recalled to active duty to lead efforts to standardize the computer language COBOL for the Navy. In 1985 at the age of 79, she became Rear Admiral Grace Hopper. In 1986 she was retired—as the Navy's oldest commissioned officer on active duty—in a ceremony held aboard the U.S.S. Constitution, the Navy's oldest commissioned warship.

2.1 TANGENT LINES AND SLOPE PREDICTORS

In Sections 1.2 and 1.5 we saw that certain applied problems raise the question of what is meant by the *tangent line* at a specified point of a general curve $y = f(x)$. In this section we see that this "tangent-line problem" leads to the limit concept, which we pursue further in Section 2.2.

In elementary geometry the line tangent to a circle at a point P is defined as the straight line through P that is perpendicular to the radius (OP) to that point (Fig. 2.1.1). A general graph $y = f(x)$ has no radius for us to use, but the line tangent to the graph at the point P should be the straight line through P that has—in some sense—the same direction at P as the curve itself. Because a line's "direction" is determined by its slope, our plan for defining a line tangent to a curve amounts to finding an appropriate "slope-prediction formula" that will give the proper slope of the tangent line. Example 1 illustrates this approach in the case of one of the simplest of all nonstraight curves, the parabola with equation $y = x^2$.

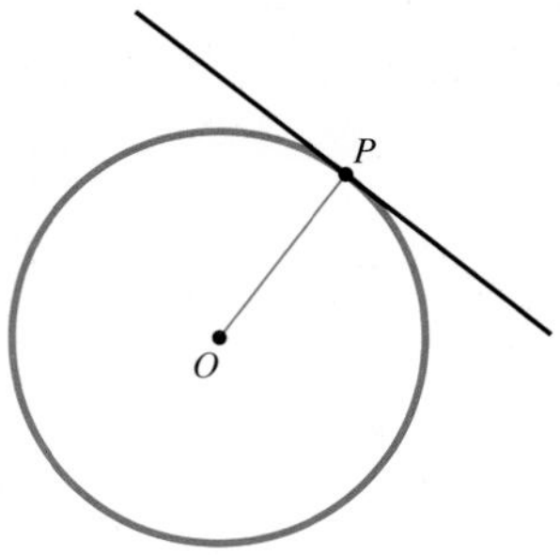

FIGURE 2.1.1 The line tangent to the circle at the point P is perpendicular to the radius OP.

EXAMPLE 1 Determine the slope of the line L tangent to the parabola $y = x^2$ at the point $P(a, a^2)$.

Solution Figure 2.1.2 shows the parabola $y = x^2$ and a typical point $P(a, a^2)$ on it. The figure also shows a visual guess of the direction of the desired tangent line L at P. We must find the slope of L.

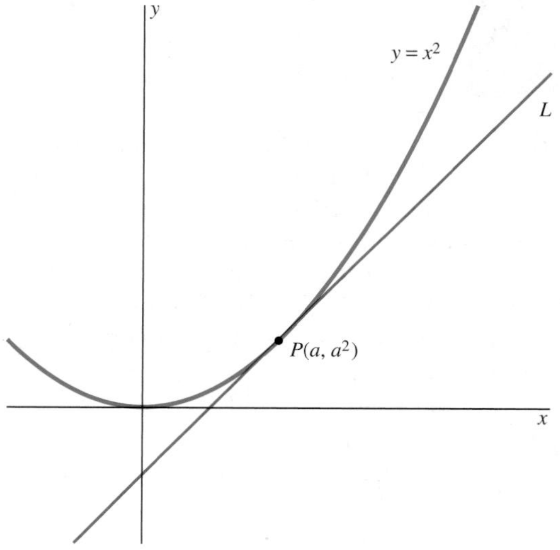

FIGURE 2.1.2 The tangent line at P should have the same direction as the curve does at P (Example 1).

We cannot immediately calculate the slope of L, because we know the coordinates of only *one* point $P(a, a^2)$ of L. Hence we begin with another line whose slope we *can* compute. Figure 2.1.3 shows the **secant line** K that passes through the point P and the nearby point $Q(b, b^2)$ of the parabola $y = x^2$. Let us write

$$h = \Delta x = b - a$$

for the difference of the x-coordinates of P and Q. (The notation Δx is as old as calculus itself, and it means now what it did 300 years ago: an **increment,** or *change,* in the value of x.) Then the coordinates of Q are given by the formulas

$$b = a + h \quad \text{and} \quad b^2 = (a + h)^2.$$

Hence the difference in the y-coordinates of P and Q is

$$\Delta y = b^2 - a^2 = (a + h)^2 - a^2.$$

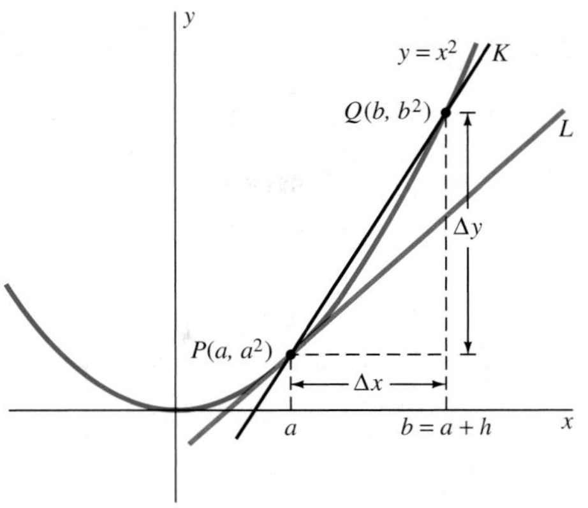

FIGURE 2.1.3 The secant line K passes through the two points P and Q, which we can use to determine its slope (Example 1).

Because P and Q are two different points, we can use the definition of slope to calculate the slope m_{PQ} of the secant line K through P and Q. If you change the value of $h = \Delta x$, you change the line K and thereby change its slope. Therefore, m_{PQ} depends on h:

$$\begin{aligned} m_{PQ} &= \frac{\Delta y}{\Delta x} = \frac{(a+h)^2 - a^2}{(a+h) - a} \\ &= \frac{(a^2 + 2ah + h^2) - a^2}{h} = \frac{2ah + h^2}{h} = \frac{h(2a+h)}{h}. \end{aligned} \tag{1}$$

Because h is nonzero, we may cancel it in the final fraction. Thus we find that the slope of the secant line K is given by

$$m_{PQ} = 2a + h. \tag{2}$$

Now imagine what happens as you move the point Q along the curve closer and closer to the point P. (This situation corresponds to h approaching zero.) The line K still passes through P and Q, but it pivots around the fixed point P. As h approaches zero, the secant line K comes closer to coinciding with the tangent line L. This phenomenon is suggested in Fig. 2.1.4, which shows the secant line K approaching the tangent line L.

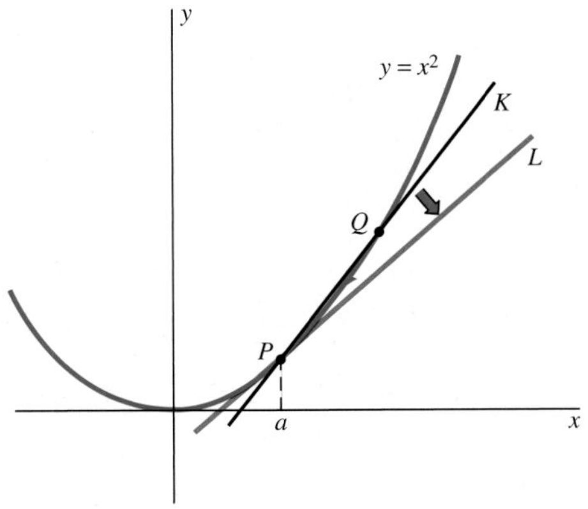

FIGURE 2.1.4 As $h \to 0$, Q approaches P, and K moves into coincidence with the tangent line L (Example 1).

Our idea is to *define* the tangent line L as the limiting position of the secant line K. To see precisely what this means, examine what happens to the slope of K as

K pivots into coincidence with L:

As h approaches zero,

Q approaches P, and so

K approaches L; meanwhile,

the slope of K approaches the slope of L.

Hence our question is this: As the number h approaches zero, what value does the slope $m_{PQ} = 2a + h$ approach? We can state this question of the "limiting value" of $2a + h$ by writing

$$\lim_{h\to 0} (2a + h) = ? \tag{3}$$

Here, "lim" is an abbreviation for the word "limit," and "$h \to 0$" is an abbreviation for the phrase "h approaches zero." Thus Eq. (3) asks, "What is the limit of $2a + h$ as h approaches zero?"

For any specific value of a we can investigate this question numerically by calculating values of $2a + h$ with values of h that become closer and closer to zero—such as the values $h = 0.1$, $h = -0.01$, $h = 0.001$, $h = -0.0001$, ..., or the values $h = 0.5, h = 0.1, h = 0.05, h = 0.01, \dots$. For instance, the tables of values in Figs. 2.1.5 and 2.1.6 indicate that with $a = 1$ and $a = -2$ we should conclude that

$$\lim_{h\to 0} (2 + h) = 2 \quad \text{and} \quad \lim_{h\to 0} (-4 + h) = -4.$$

More generally, it seems clear from the table in Fig. 2.1.7 that

$$\lim_{h\to 0} m_{PQ} = \lim_{h\to 0} (2a + h) = 2a. \tag{4}$$

h	$2+h$
0.1	2.1
0.01	2.01
0.001	2.001
0.0001	2.0002
↓	↓
0	2

FIGURE 2.1.5 As $h \to 0$ (first column), $2 + h$ approaches 2 (second column).

h	$-4+h$
0.5	−3.5
0.1	−3.9
0.05	−3.95
0.01	−3.99
0.005	−3.995
0.001	−3.999
↓	↓
0	−4

FIGURE 2.1.6 As $h \to 0$ (first column), $-4 + h$ approaches -4 (second column).

h	$2a+h$
0.01	$2a + 0.01$
0.001	$2a + 0.001$
⋮	⋮
↓	↓
0	$2a$

FIGURE 2.1.7 As $h \to 0$ (first column), $2a + h$ approaches $2a$ (second column) (Example 1).

This, finally, answers our original question: The slope $m = m(a)$ of the line tangent to the parabola $y = x^2$ at the point (a, a^2) is given by

$$m = 2a. \tag{5}$$

◆

The formula in Eq. (5) is a "slope predictor" for (lines tangent to) the parabola $y = x^2$. Once we know the slope of the line tangent to the curve at a given point of the curve, we can then use the point-slope formula to write an equation of this tangent line.

EXAMPLE 2 With $a = 1$, the slope predictor in Eq. (5) gives $m = 2$ for the slope of the line tangent to $y = x^2$ at the point $(1, 1)$. Hence an equation of this line is

$$y - 1 = 2(x - 1); \qquad \text{that is,} \quad y = 2x - 1.$$

With $a = -3$, Eq. (5) gives $m = -6$ as the slope of the line tangent at $(-3, 9)$, so an equation of the line tangent to the curve at this point is

$$y - 9 = -6(x + 3); \qquad \text{that is,} \quad y = -6x - 9. \qquad \blacklozenge$$

In Fig. 2.1.8 the parabola $y = x^2$ and its tangent line $y = 2x - 1$ passing through $(1, 1)$ are both graphed. The relationship between the curve and its tangent line is such that as we "zoom in" on the point of tangency, successive magnifications show less and less of a difference between the curve and the tangent line. This phenomenon is illustrated in Figs. 2.1.9 through 2.1.11.

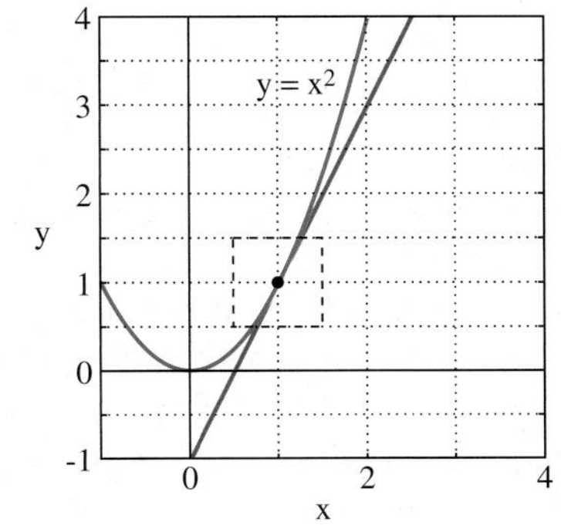

FIGURE 2.1.8 The parabola $y = x^2$ and its tangent line at $P(1, 1)$.

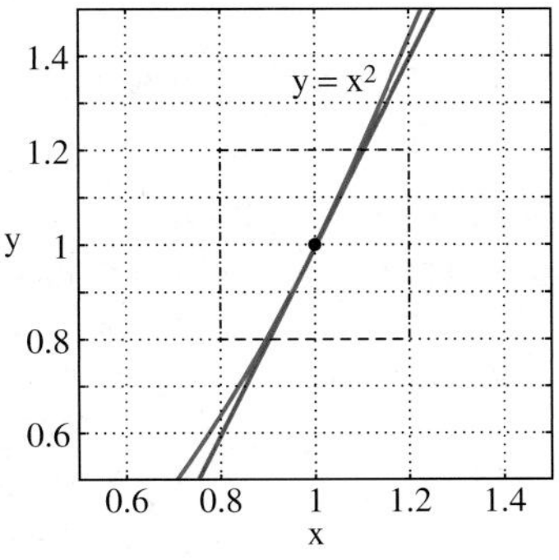

FIGURE 2.1.9 First magnification.

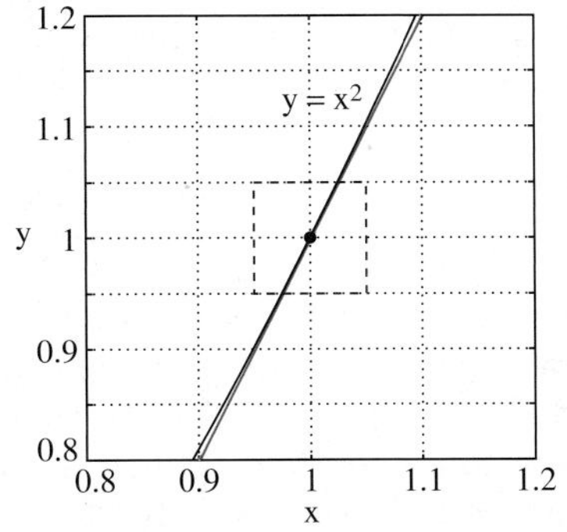

FIGURE 2.1.10 Second magnification.

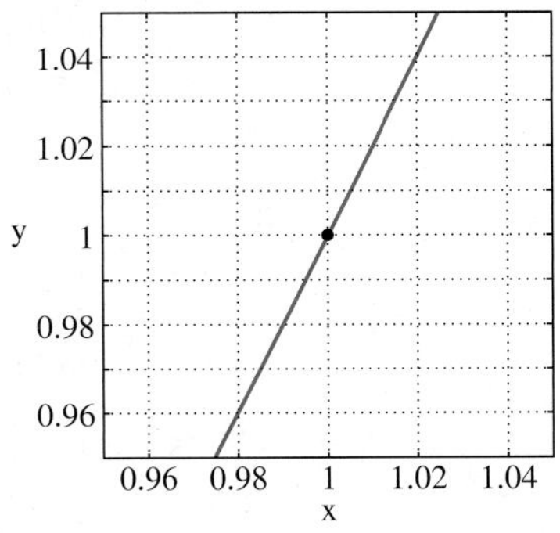

FIGURE 2.1.11 Can you see the difference?

REMARK In Example 1 we proceeded as though the concept of a tangent line to a curve were self-evident. The actual meaning of the slope-predictor result $m = 2a$ in Eq. (5) is this: Whatever is meant by the line tangent to the parabola $y = x^2$ at the point $P(a, a^2)$, it can only be the unique straight line through P with slope $m = 2a$. Thus we must *define* the line tangent to $y = x^2$ at P to be the line whose point-slope equation is $y - a^2 = 2a(x - a)$. Pictures like those in Figs. 2.1.8 through 2.1.11 certainly support our conviction that this definition is the correct one.

More General Slope Predictors

The general case of the line tangent to a curve $y = f(x)$ is scarcely more complicated than the special case $y = x^2$ of Example 1. Given the function f, suppose that we want to find the slope of the line L tangent to $y = f(x)$ at the point $P(a, f(a))$. As indicated in Fig. 2.1.12, let K be the secant line passing through the point P and the nearby point $Q(a + h, f(a + h))$ on the graph. The slope of this secant line is the **difference quotient**

$$m_{PQ} = \frac{\Delta y}{\Delta x} = \frac{f(a + h) - f(a)}{h} \quad (\text{with } h \neq 0). \qquad \textbf{(6)}$$

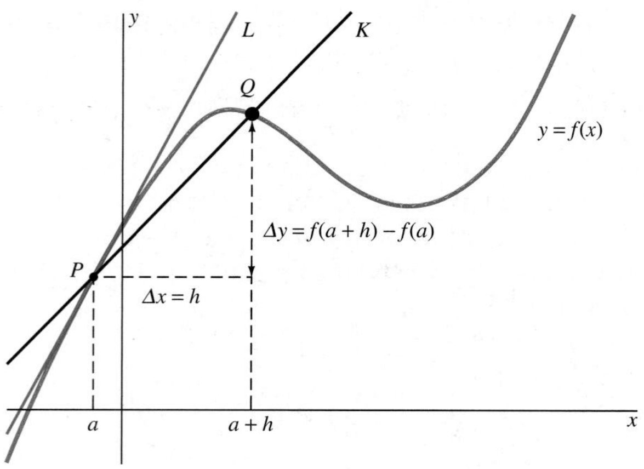

FIGURE 2.1.12 As $h \to 0$, $Q \to P$, and the slope of K approaches the slope of the tangent line L.

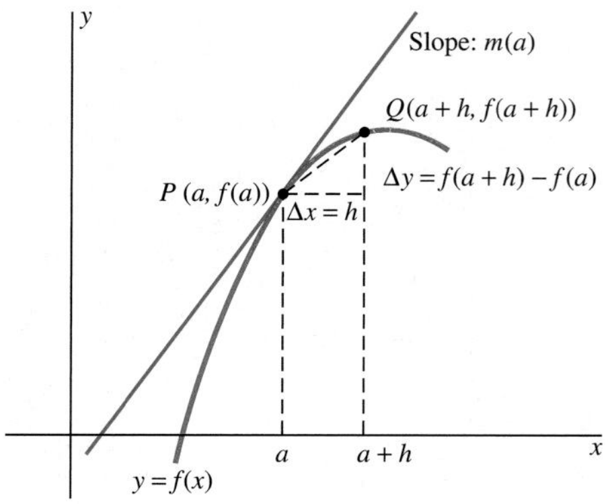

FIGURE 2.1.13 The slope of the tangent line at $(a,\ f(a))$ is
$$m(a) = \lim_{h \to 0} \frac{f(a+h) - f(a)}{h}.$$

We now force Q to approach the fixed point P along the curve $y = f(x)$ by making h approach zero. We ask whether m_{PQ} approaches some limiting value m as $h \to 0$. If so, we write

$$m = \lim_{h \to 0} \frac{f(a+h) - f(a)}{h}$$

and conclude that this number m is the slope of the line tangent to the graph $y = f(x)$ at the point $(a,\ f(a))$. Actually, this slope depends on a and we can indicate this by writing

$$m_a = \lim_{h \to 0} \frac{f(a+h) - f(a)}{h}. \tag{7}$$

If we can express the limiting value on the right explicitly in terms of a, then Eq. (7) yields a *slope predictor* for lines tangent to the curve $y = f(x)$. In this case the line tangent to the curve at the point $P(a,\ f(a))$ is *defined* to be the straight line through P that has slope m_a. This tangent line is indicated in Fig. 2.1.13.

In Chapter 3 we will acknowledge the fact that the slope m_a is somehow "derived" from the function f by calling this number the **derivative** of the function f at the point a. Indeed, much of Chapter 3 will be devoted to methods of calculating derivatives of various familiar functions. Most of these methods are based on the limit techniques of Sections 2.2 and 2.3, but the case of quadratic functions is sufficiently simple for inclusion here. Recall from Section 1.2 that the graph of any quadratic function is a parabola that opens either upward or downward.

THEOREM Parabolas and Tangent Lines

Consider the parabola $y = f(x)$ where

$$f(x) = px^2 + qx + r \tag{8}$$

(with $p \neq 0$). Then the line tangent to this parabola at the point $P(a,\ f(a))$ has slope

$$m_a = 2pa + q. \tag{9}$$

PROOF The slope of the secant line given in (6) may be simplified as follows:

$$m_{PQ} = \frac{f(a+h) - f(a)}{h} = \frac{[p(a+h)^2 + q(a+h) + r] - [pa^2 + qa + r]}{h}$$
$$= \frac{[p(a^2 + 2ah + h^2) + q(a+h) + r] - [pa^2 + qa + r]}{h} = \frac{2pah + ph^2 + qh}{h},$$

and therefore

$$m_{PQ} = 2pa + q + ph.$$

The numbers p, q, and a are fixed, so as $h \to 0$ the product ph approaches zero, much as in our computations in Example 1. Thus

$$m_a = \lim_{h\to 0} m_{PQ} = \lim_{h\to 0}(2pa + q + ph) = 2pa + q,$$

as claimed in Eq. (9). ◄

REMARK 1 Thus the formula $m_a = 2pa + q$ provides a ready slope predictor for lines tangent to the parabola with equation

$$y = px^2 + qx + r.$$

Given the coefficients p, q, r, and the number a, we need only substitute in this slope-predictor formula to obtain the slope m_a of the line tangent to the parabola at the point where $x = a$. We need *not* repeat the computational steps that were carried out in the derivation of the slope-predictor formula.

REMARK 2 If we replace a with x we get the **slope-predictor function**

$$m(x) = 2px + q. \tag{10}$$

Here m is a *function* whose value $m(x)$ at x is the slope of the line tangent to the parabola $y = f(x)$ at the point $P(x, f(x))$. Perhaps the visual scheme

$$\begin{array}{cccc} f(x) = px^2 & + qx & + & r \\ \downarrow & \downarrow & \downarrow & \downarrow \\ m(x) = 2px & + & q & + 0 \end{array}$$

makes this slope predictor easy for you to remember.

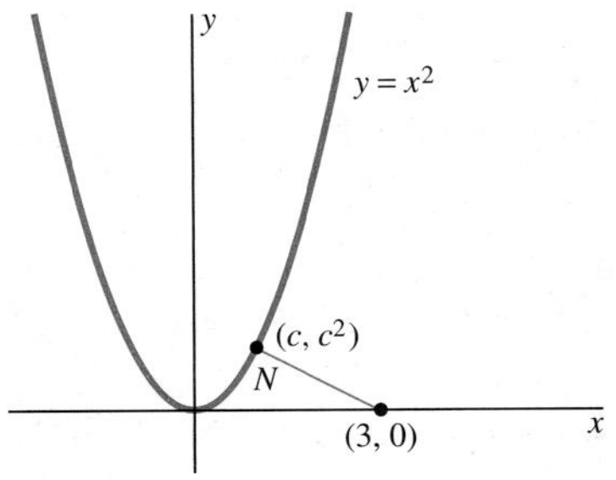

FIGURE 2.1.14 The normal line N from the point (3, 0) to the point (c, c^2) on the parabola $y = x^2$.

EXAMPLE 3 Find an equation of the line tangent to the parabola $y = 2x^2 - 3x + 5$ at the point where $x = -1$.

Solution Here we have $p = 2$, $q = -3$, $r = 5$, and the y-coordinate of our point is $2 \cdot (-1)^2 - 3 \cdot (-1) + 5 = 10$. Then Eq. (10) gives the slope predictor

$$m(x) = 2 \cdot 2x + (-3) = 4x - 3,$$

so the slope of the line tangent to the parabola at the point $(-1, 10)$ is $m(-1) = 4 \cdot (-1) - 3 = -7$. The point-slope equation of this tangent line is therefore

$$y - 10 = (-7)(x + 1); \qquad \text{that is,} \quad y = -7x + 3.$$ ◆

Normal Lines

How would you find the point $P(c, c^2)$ that lies on the parabola $y = x^2$ and is closest to the point (3, 0)? Intuitively, the line segment N with endpoints (3, 0) and P should be perpendicular, or *normal,* to the parabola's tangent line at P (Fig. 2.1.14). But if the slope of the tangent line is m, then—by Theorem 2 in Appendix B—the slope of the normal line is

$$m_N = -\frac{1}{m}. \tag{11}$$

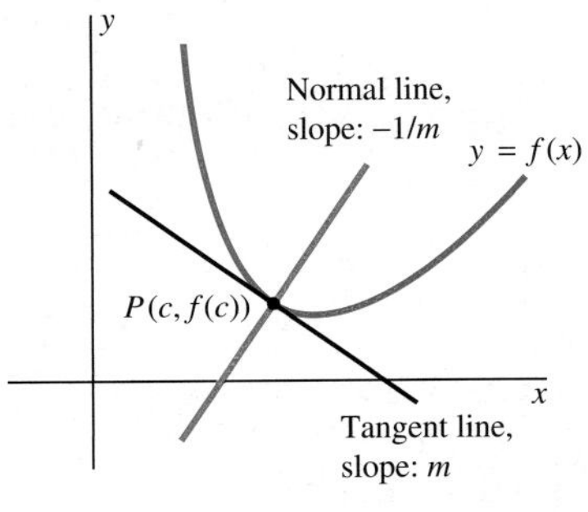

FIGURE 2.1.15 The tangent line and normal line through the point P on a curve.

(Theorem 2 tells us that if two perpendicular lines have nonzero slopes m_1 and m_2, then $m_1 m_2 = -1$.) More precisely, the **normal line** at a point P of a curve where the tangent line has slope m is *defined* to be the line through P with slope $m_N = -1/m$ (Fig. 2.1.15). Consequently, the parabolic slope predictor in (9) enables us to write equations of lines normal to parabolas as easily as equations of tangent lines.

EXAMPLE 4 In Example 3 we found that the line tangent to the parabola $y = 2x^2 - 3x + 5$ at the point $P(-1, 10)$ has slope -7. Therefore the slope of the line normal to that parabola at P is $m_N = -1/(-7) = \frac{1}{7}$. So the point-slope equation of the normal line is

$$y - 10 = \tfrac{1}{7}(x+1); \qquad \text{that is,} \quad y = \tfrac{1}{7}x + \tfrac{71}{7}. \qquad \blacklozenge$$

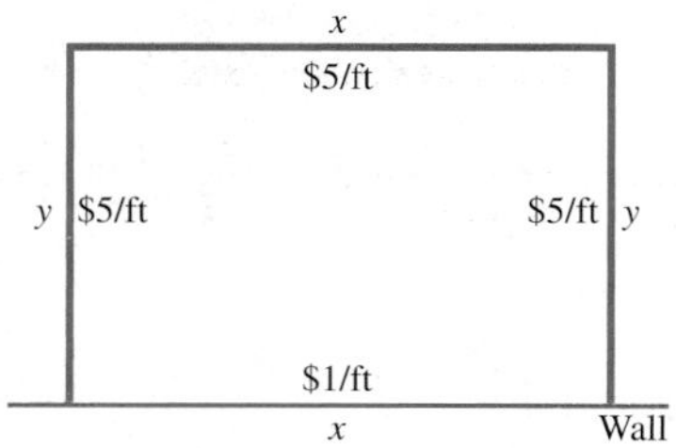

FIGURE 2.1.16 The animal pen.

The Animal Pen Problem Completed

Now we can apply our newfound knowledge of slope-predictor formulas to wrap up our continuing discussion of the animal pen problem of Section 1.1. In Example 9 there we found that the area A of the pen (see Fig. 2.1.16) is given as a function of its base length x by

$$A(x) = \tfrac{3}{5}(30x - x^2) = -\tfrac{3}{5}x^2 + 18x \tag{12}$$

for $0 \leqq x \leqq 30$. Therefore our problem is to find the maximum value of $A(x)$ for x in the closed interval $[0, 30]$.

Let us accept as intuitively obvious—we will see a proof in Chapter 3—the fact that the maximum value of $A(x)$ occurs at the high point where the line tangent to the parabola $y = A(x)$ is *horizontal,* as indicated in Fig. 2.1.17. But the function $A(x)$ in Eq. (12) is quadratic with $p = -\frac{3}{5}$ and $q = 18$ (compare (12) with (8)). Therefore the slope predictor in (10) implies that the slope of the tangent line at an arbitrary point $(x, A(x))$ of the parabola is given by

$$m = m(x) = 2px + q = -\tfrac{6}{5}x + 18.$$

We ask when $m = 0$ and find that this happens when

$$-\tfrac{6}{5}x + 18 = 0,$$

and thus when $x = 15$. In agreement with the result found by algebraic methods in Section 1.2, we find that the maximum possible area of the pen is

$$A(15) = \tfrac{3}{5}(30 \cdot 15 - 15^2) = 135 \quad (\text{ft}^2).$$

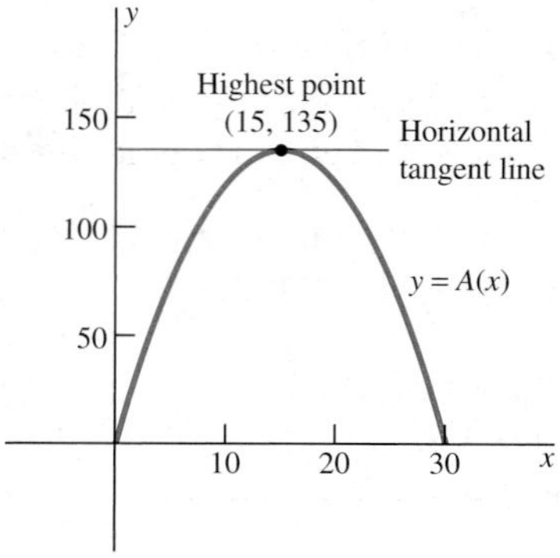

FIGURE 2.1.17 The graph of $y = A(x)$, $0 \leqq x \leqq 30$.

Numerical Investigation of Slopes

Suppose that you are given the function f and a specific numerical value of a. You can then use a calculator to investigate the value

$$m = \lim_{h \to 0} \frac{f(a+h) - f(a)}{h} \tag{13}$$

of the slope of the line tangent to the curve $y = f(x)$ at the point $(a, f(a))$. Simply calculate the values of the difference quotient

$$\frac{f(a+h) - f(a)}{h} \tag{14}$$

with successively smaller nonzero values of h to see whether a limiting numerical value is apparent.

EXAMPLE 5 Find by numerical investigation (an approximation to) the line tangent to the graph of

$$f(x) = x + \frac{1}{x} \tag{15}$$

at the point $(2, \frac{5}{2})$.

Solution Figure 2.1.18 shows a TI calculator prepared to calculate the difference quotient in Eq. (14) with the function f in Eq. (15). As indicated in Fig. 2.1.19, successive values of this quotient can then be calculated by brief "one-liners." Figure 2.1.20 shows an HP calculator prepared to define the same quotient; then evaluation of the expression `'M(2,0.0001)'` yields the approximate value $m \approx 0.75001$. In

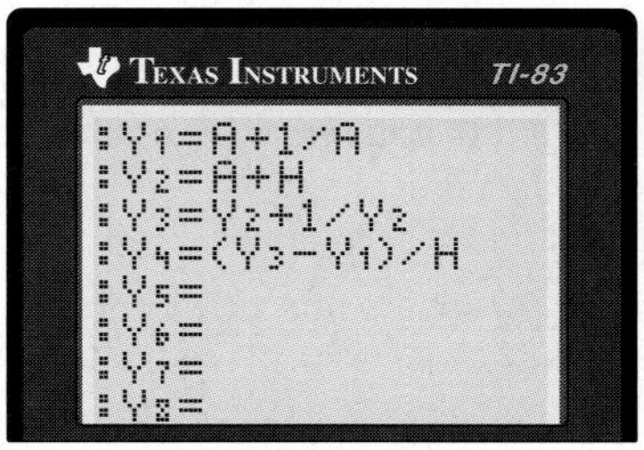

FIGURE 2.1.18 A calculator prepared to calculate $\dfrac{f(a+h)-f(a)}{h}$ with $f(x)=x+\dfrac{1}{x}$.

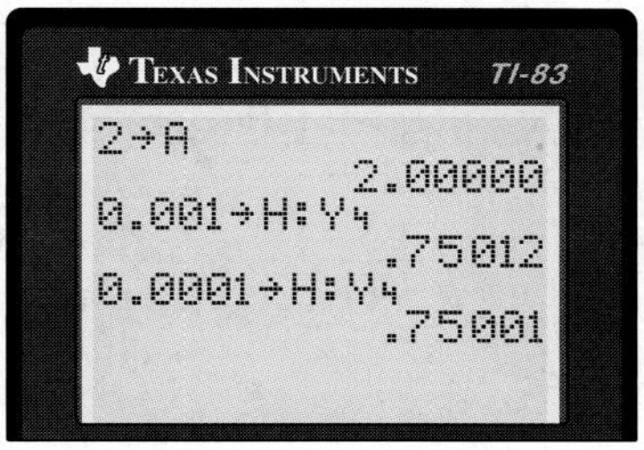

FIGURE 2.1.19 Approximating $\displaystyle\lim_{h\to 0}\frac{f(a+h)-f(a)}{h}$.

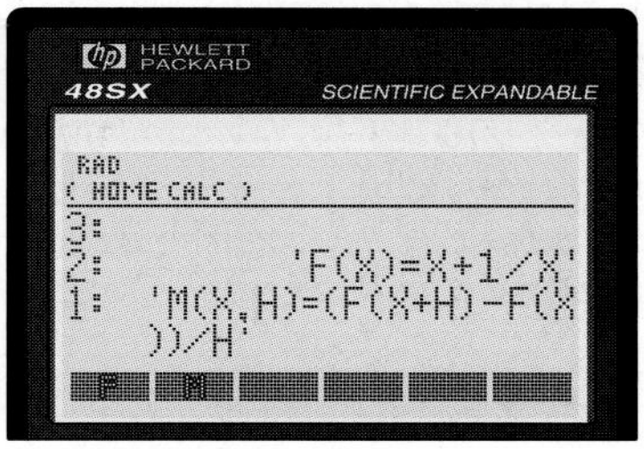

FIGURE 2.1.20 A calculator prepared to compute $\dfrac{f(x+h)-f(x)}{h}$.

h	$\dfrac{f(2+h)-f(2)}{h}$
0.1	0.76190
0.01	0.75124
0.001	0.75012
0.0001	0.75001
0.00001	0.75000
↓	↓
0	$\frac{3}{4}$

FIGURE 2.1.21 Numerical investigation of the limit in (13) with $f(x)=x+\dfrac{1}{x}$, $a=2$.

this way we get the table shown in Fig. 2.1.21, which suggests that the slope of the line tangent to the graph of $f(x)$ at the point $(2, \frac{5}{2})$ is $m=\frac{3}{4}$. If so, then the tangent line at this point has the point-slope equation

$$y-\tfrac{5}{2}=\tfrac{3}{4}(x-2); \qquad \text{that is,} \quad y=\tfrac{3}{4}x+1.$$

Our numerical investigation does not constitute a rigorous proof that this actually is the desired tangent line, but Figs. 2.1.22 and 2.1.23 showing the computer-generated graphs

$$y=x+\frac{1}{x} \quad \text{and} \quad y=\frac{3}{4}x+1$$

are strong evidence that we've got it right. (Do you agree?) ◆

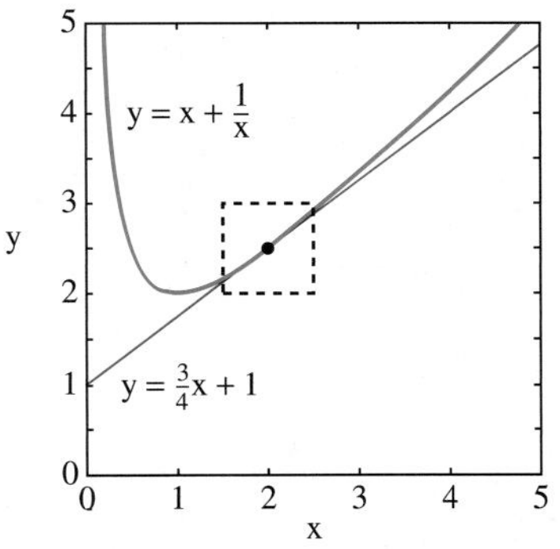

FIGURE 2.1.22 The curve and its tangent line (Example 5).

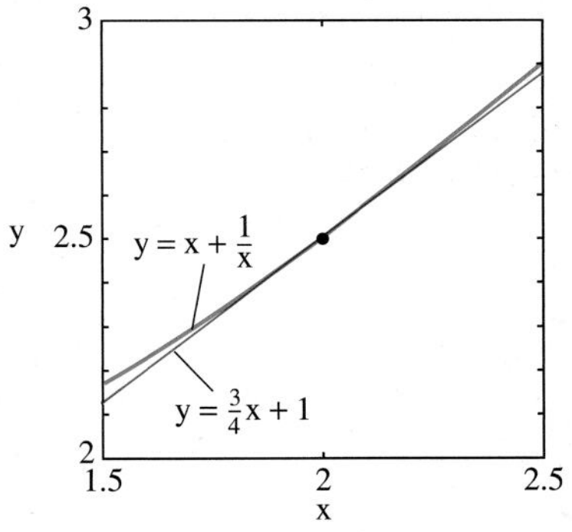

FIGURE 2.1.23 The curve and its tangent line magnified near $(2, \frac{5}{2})$.

2.1 TRUE/FALSE STUDY GUIDE

2.1 CONCEPTS: QUESTIONS AND DISCUSSION

1. What is the slope-predictor function for the straight line with equation $y=17x-21$?
2. Can two different parabolas with equations of the form $y=px^2+qx+r$ have the same slope-predictor function?
3. The vertex of the parabola with equation $y=px^2+qx+r$ is its highest point (if $p<0$) or its lowest point (if $p>0$). As indicated in Fig. 2.1.17, it is apparent that this vertex is the single point of the parabola at which the tangent line is horizontal. Is it true—for *any* given curve $y=f(x)$—that a point on the graph at which the tangent line is horizontal is either the highest or the lowest point on the graph?

2.1 PROBLEMS

In Problems 1 through 14, first apply the slope-predictor formula in (10) for quadratic functions to write the slope $m(a)$ of the line tangent to $y = f(x)$ at the point where $x = a$. Then write an equation of the line tangent to the graph of f at the point $(2, f(2))$.

1. $f(x) \equiv 5$

2. $f(x) = x$

3. $f(x) = x^2$

4. $f(x) = 1 - 2x^2$

5. $f(x) = 4x - 5$

6. $f(x) = 7 - 3x$

7. $f(x) = 2x^2 - 3x + 4$

8. $f(x) = 5 - 3x - x^2$

9. $f(x) = 2x(x + 3)$

10. $f(x) = 3x(5 - x)$

11. $f(x) = 2x - \left(\frac{x}{10}\right)^2$

12. $f(x) = 4 - (3x + 2)^2$

13. $f(x) = (2x + 1)^2 - 4x$

14. $f(x) = (2x+3)^2 - (2x - 3)^2$

In Problems 15 through 24, find all points of the curve $y = f(x)$ at which the tangent line is horizontal.

15. $y = 10 - x^2$

16. $y = 10x - x^2$

17. $y = x^2 - 2x + 1$

18. $y = x^2 + x - 2$

19. $y = x - \left(\frac{x}{10}\right)^2$

20. $y = x(100 - x)$

21. $y = (x + 3)(x - 5)$

22. $y = (x - 5)^2$

23. $y = 70x - x^2$

24. $y = 100\left(1 - \frac{x}{10}\right)^2$

In Problems 25 through 35, use the slope-predictor formula for quadratic functions as necessary. In Problems 25 through 27, write equations for both the line tangent to, and the line normal to, the curve $y = f(x)$ at the given point P.

25. $y = x^2$; $P(-2, 4)$

26. $y = 5 - x - 2x^2$; $P(-1, 4)$

27. $y = 2x^2 + 3x - 5$; $P(2, 9)$

28. Prove that the line tangent to the parabola $y = x^2$ at the point (x_0, y_0) intersects the x-axis at the point $(x_0/2, 0)$. See Fig. 2.1.24.

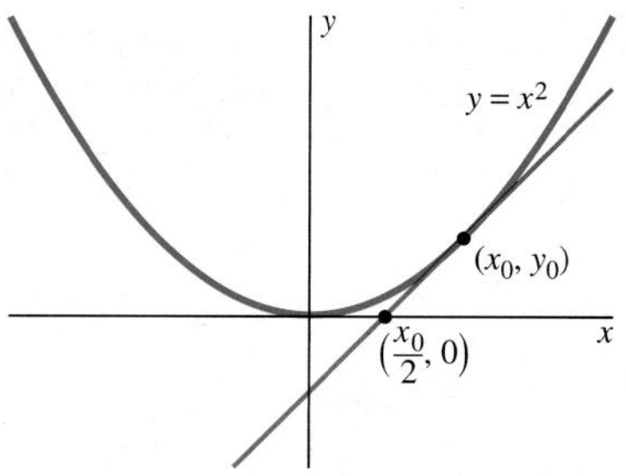

FIGURE 2.1.24 The parabola and tangent line of Problem 28.

29. If a ball is thrown straight upward with initial velocity 96 ft/s, then its height t seconds later is $y(t) = 96t - 16t^2$ feet. Determine the maximum height the ball attains by finding the point in the parabola $y(t) = 96t - 16t^2$ where the tangent line is horizontal.

30. According to Problem 40 of Section 1.1, the area of a rectangle with base of length x and perimeter 100 is $A(x) = x(50 - x)$. Find the maximum possible area of this rectangle by finding the point on the parabola $A = x(50 - x)$ at which the tangent line is horizontal.

31. Find the maximum possible value of the product of two positive numbers whose sum is 50.

32. Suppose that a projectile is fired at an angle of 45° from the horizontal. Its initial position is the origin in the xy-plane, and its initial velocity is $100\sqrt{2}$ ft/s (Fig. 2.1.25). Then its trajectory will be the part of the parabola $y = x - (x/25)^2$ for which $y \geqq 0$. (a) How far does the projectile travel (horizontally) before it hits the ground? (b) What is the maximum height above the ground that the projectile attains?

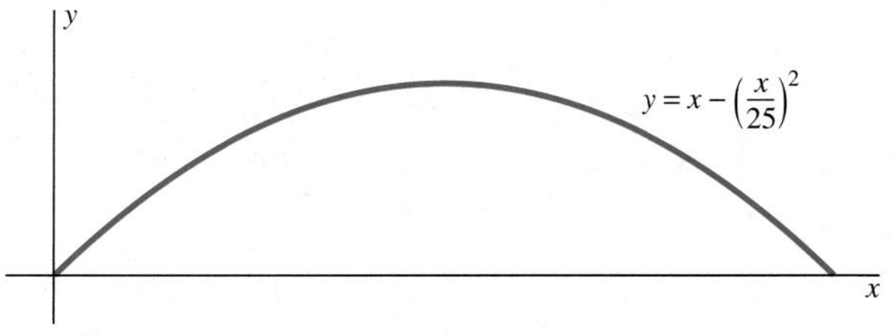

FIGURE 2.1.25 The trajectory of the projectile of Problem 32.

33. One of the two lines that pass through the point (3, 0) and are tangent to the parabola $y = x^2$ is the x-axis. Find an equation for the *other* line. (*Suggestion:* First find the value of the number a shown in Fig. 2.1.26.)

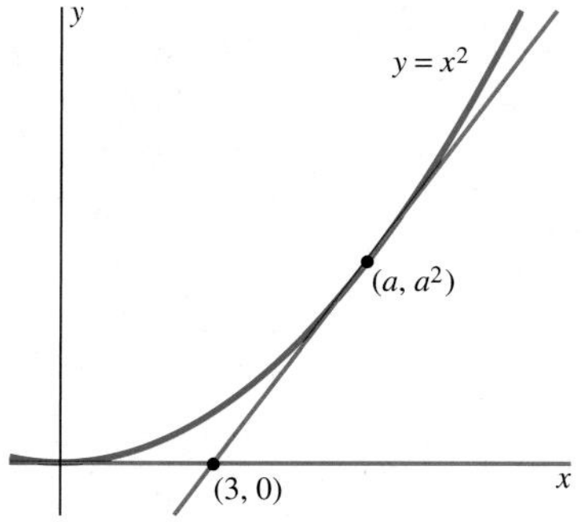

FIGURE 2.1.26 Two lines tangent to the parabola of Problem 33.

34. Write equations for the two straight lines that pass through the point (2, 5) and are tangent to the parabola $y = 4x - x^2$. (*Suggestion:* Draw a figure like Fig. 2.1.26.)

35. Between Examples 3 and 4 we raised—but did not answer—the question of how to locate the point on the graph of $y = x^2$ closest to the point (3, 0). It's now time for you to find that point. (*Suggestion:* Draw a figure like Fig. 2.1.26. The cubic equation you should obtain has one solution that is apparent by inspection.)

Let $P(a, f(a))$ be a fixed point on the graph of $y = f(x)$. If $h > 0$, then $Q(a + h, f(a + h))$ lies to the right, and $R(a - h, f(a - h))$ lies to the left, of P. Does Fig. 2.1.27 make it appear plausible—for $h > 0$ and h very small—that the slope

$$m_{RQ} = \frac{f(a + h) - f(a - h)}{2h} = \frac{1}{2}(m_{PQ} + m_{RP})$$

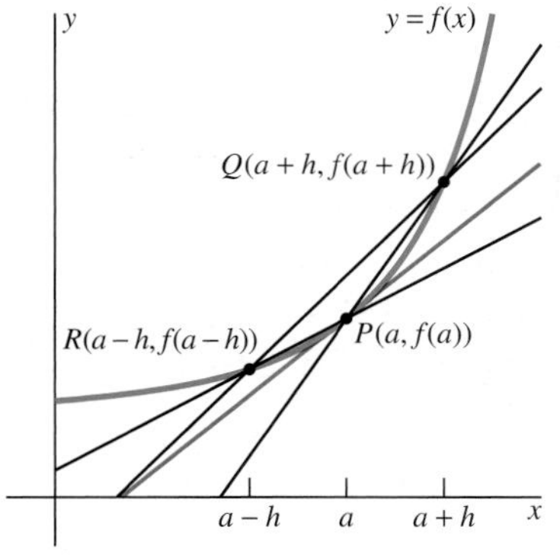

FIGURE 2.1.27 Three different approximations to the slope of a tangent line.

is generally an especially good approximation to the slope m of the line tangent to the graph at P? In particular, the "symmetric difference quotient" m_{RQ} is generally a better approximation to m than either the standard right-hand difference quotient

$$m_{PQ} = \frac{f(a+h) - f(a)}{h}$$

or the left-hand difference quotient

$$m_{RP} = \frac{f(a) - f(a-h)}{h}.$$

In Problems 36 through 48, use a calculator or computer to investigate numerically the slope m of the line tangent to the given graph at $P(a, f(a))$ by calculating both m_{PQ} and m_{RQ} for $h = 0.1, 0.01, 0.001, \ldots$. Check the resulting value of m by plotting both the graph of $y = f(x)$ and the alleged tangent line.

36. $f(x) = x^2$; $a = -1$

37. $f(x) = x^3$; $a = 2$

38. $f(x) = x^3$; $a = -1$

39. $f(x) = \sqrt{x}$; $a = 1$

40. $f(x) = \sqrt{x}$; $a = 4$

41. $f(x) = \dfrac{1}{x}$; $a = 1$

42. $f(x) = \dfrac{1}{x}$; $a = -\frac{1}{2}$

43. $f(x) = \cos x$; $a = 0$

44. $f(x) = \sin 10\pi x$; $a = 0$

45. $f(x) = \cos x$; $a = \frac{1}{4}\pi$

46. $f(x) = \sin 10\pi x$; $a = \frac{1}{20}$

47. $f(x) = \sqrt{25 - x^2}$; $a = 0$

48. $f(x) = \sqrt{25 - x^2}$; $a = 3$

2.1 Project: Numerical Slope Investigations

This project consists of numerical investigations of slopes based on the fact that a curve $y = f(x)$ and its tangent line at a fixed point $P(a, f(a))$ are virtually indistinguishable at a sufficiently high level of magnification (as illustrated in Figs. 2.1.8 through 2.1.11). Suppose that we plot the graph on a calculator or computer having a facility that lets us "grab" the xy-coordinates of selected points on the graph. For example, with some graphing calculators, you can simply press the **TRACE** key and then move the cursor along the graph, pausing to read the coordinates of any desired point. Suppose that you "zoom in" on the point P, and at the kth zoom record the coordinates (x_1, y_1) and (x_2, y_2) of two points located on opposite sides of P (as indicated in Fig. 2.1.28). Then you can approximate the slope m of the line tangent to the graph at P by calculating the value of the difference quotient

$$m_k = \frac{\Delta y}{\Delta x} = \frac{y_2 - y_1}{x_2 - x_1}.$$

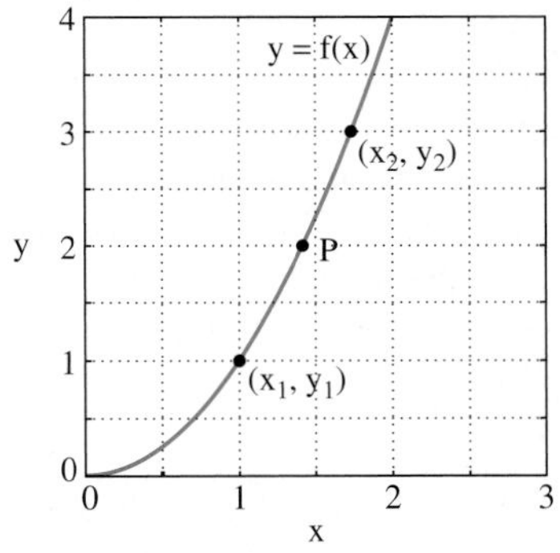

FIGURE 2.1.28 Points on either side of P.

After several successive zooms you may have a very accurate approximation to m. The CD-ROM material for this project lists several tangent line problems for you to investigate in this manner.

2.2 THE LIMIT CONCEPT

In Section 2.1 we defined the slope m of the line tangent to the graph $y = f(x)$ at the point $P(a, f(a))$ to be

$$m = \lim_{h \to 0} \frac{f(a+h) - f(a)}{h}. \tag{1}$$

The graph that motivated this definition is repeated in Fig. 2.2.1, with $a + h$ relabeled as x (so that $h = x - a$). We see that x approaches a as h approaches zero, so Eq. (1) can be written in the form

$$m = \lim_{x \to a} \frac{f(x) - f(a)}{x - a}. \tag{2}$$

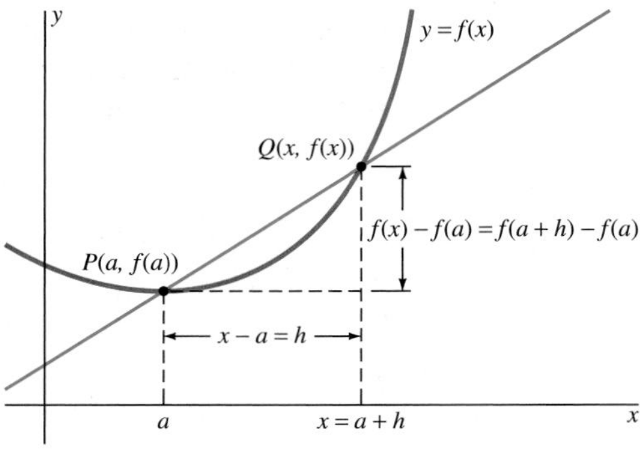

FIGURE 2.2.1 The slope m at $P(a, f(a))$ can be defined in this way: $m = \lim_{x \to a} \dfrac{f(x) - f(a)}{x - a}$.

Thus the computation of m amounts to the determination of the limit, as x approaches a, of the function

$$g(x) = \frac{f(x) - f(a)}{x - a}. \tag{3}$$

In order to develop general methods for calculating such limits, we need to investigate more fully the meaning of the statement

$$\lim_{x \to a} f(x) = L. \tag{4}$$

This is read "the limit of $f(x)$ as x approaches a is L." We sometimes write Eq. (4) in the concise form

$$f(x) \to L \quad \text{as} \quad x \to a.$$

The function f need not be defined at the point $x = a$ in order for us to discuss the limit of f at a. The actual value of $f(a)$—if any—actually is immaterial. It suffices for $f(x)$ to be defined for all points *other than* a in some **neighborhood** of a—that is, for all $x \neq a$ is some open interval containing a. This is exactly the situation for the function in Eq. (3), which is defined *except* at a (where the denominator is zero). The following statement presents the meaning of Eq. (4) in intuitive language.

Idea of the Limit

We say that the number L is the *limit* of $f(x)$ as x approaches a provided that we can make the number $f(x)$ as close to L as we please merely by choosing x sufficiently near, though not equal to, the number a.

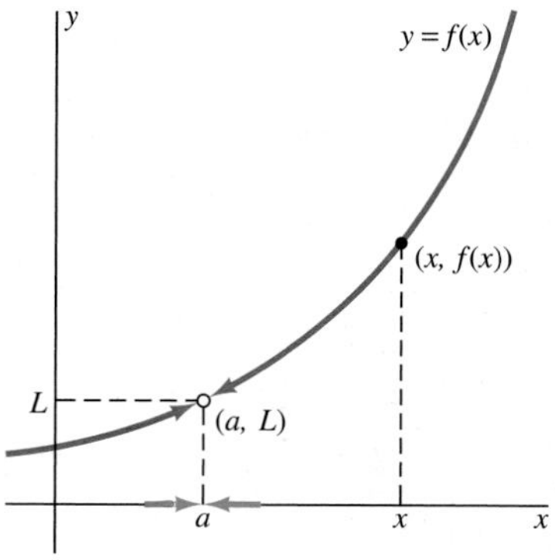

FIGURE 2.2.2 Graphical interpretation of the limit concept.

What this means, roughly, is that $f(x)$ tends to get closer and closer to L as x gets closer and closer to a. Once we decide how close to L we want $f(x)$ to be, it is necessary that $f(x)$ be that close to L for *all* x sufficiently close to (but not equal to) a.

Figure 2.2.2 shows a graphical interpretation of the limit concept. As x approaches a (from either side), the point $(x, f(x))$ on the graph $y = f(x)$ must approach the point (a, L).

In this section we explore the idea of the limit, mainly through the investigation of specific examples. A precise statement of the definition of the limit appears in Section 2.3.

EXAMPLE 1 Investigate the value of $\displaystyle\lim_{x \to 3} \frac{x - 1}{x + 2}$.

Investigation This is an investigation (rather than a solution) because numerical calculations may strongly suggest the value of a limit but cannot establish its value with certainty. The table in Fig. 2.2.3 gives values of

$$f(x) = \frac{x - 1}{x + 2},$$

correct to six rounded decimal places, for values of x that approach 3 (but are not equal to 3). The first and third columns of the table show values of x that approach 3 both from the left and from the right.

x	$\dfrac{x-1}{x+2}$	x	$\dfrac{x-1}{x+2}$
2	0.250000	4	0.500000
2.9	0.387755	3.1	0.411765
2.99	0.398798	3.01	0.401198
2.999	0.399880	3.001	0.400120
2.9999	0.399988	3.0001	0.400012
↓	↓	↓	↓
3	0.4	3	0.4

FIGURE 2.2.3 Investigating the limit in Example 1.

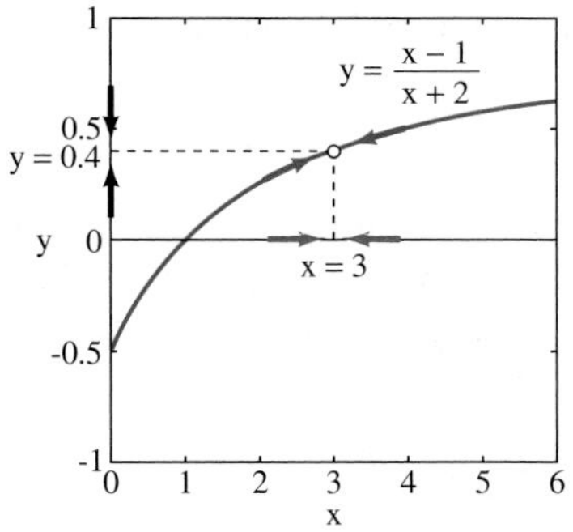

FIGURE 2.2.4 The limit in Example 1.

Now examine the table—read down the columns for x, because *down* is the table's direction for "approaches"—to see what happens to the corresponding values of $f(x)$. The data clearly suggest that

$$\lim_{x\to 3}\frac{x-1}{x+2}=\frac{2}{5}.$$ ◆

REMARK 1 The graph of $f(x)=(x-1)/(x+2)$ in Fig. 2.2.4 reinforces our guess that $f(x)$ is near $\frac{2}{5}$ when x is near 3. For still more reinforcement you can use a graphing calculator or computer to zoom in on the point on the graph where $x=3$.

REMARK 2 *Note that we did not simply substitute the value $x=3$ into the function $f(x)=(x-1)/(x+2)$ to obtain the apparent value $\frac{2}{5}=0.4$ of the limit.* Although such substitution would produce the correct answer in this particular case, in many limits it produces either an incorrect answer or no answer at all. (See Examples 2 and 3 and Problems 19 through 36 and 47 through 56.)

EXAMPLE 2 Investigate the value of $\displaystyle\lim_{x\to 2}\frac{x^2-4}{x^2+x-6}$.

Investigation The numerical data shown in Fig. 2.2.5 certainly suggest that

$$\lim_{x\to 2}\frac{x^2-4}{x^2+x-6}=\frac{4}{5}.$$ ◆

x	$\dfrac{x^2-4}{x^2+x-6}$	x	$\dfrac{x^2-4}{x^2+x-6}$
1	0.750000	3	0.833333
1.5	0.777778	2.5	0.818182
1.9	0.795918	2.1	0.803922
1.99	0.799599	2.01	0.800399
1.999	0.799960	2.001	0.800040
1.9999	0.799996	2.0001	0.800004
↓	↓	↓	↓
2	0.8	2	0.8

FIGURE 2.2.5 Investigating the limit in Example 2.

REMARK The function

$$f(x)=\frac{x^2-4}{x^2+x-6}$$

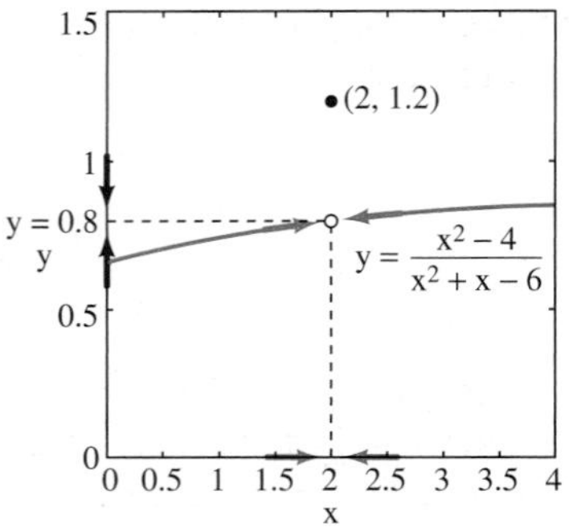

FIGURE 2.2.6 The limit in Example 2.

is not defined at $x = 2$, so we cannot merely substitute 2 for x. But if we let

$$g(x) = \begin{cases} \dfrac{x^2 - 4}{x^2 + x - 6} & \text{if } x \neq 2, \\ \dfrac{6}{5} & \text{if } x = 2, \end{cases}$$

then $g(x)$ *is* defined at $x = 2$ (and agrees with $f(x)$ elsewhere). Is it clear to you that f and g must have the same limit at $x = 2$? Figure 2.2.6 shows the graph $y = g(x)$, including the isolated point (2, 1.2) on its graph.

EXAMPLE 3 Investigate the value of $\displaystyle\lim_{t\to 0} \frac{\sqrt{t+25} - 5}{t}$.

Investigation Here we cannot make a guess by substituting $t = 0$ because the fraction

$$g(t) = \frac{\sqrt{t+25} - 5}{t}$$

is not defined when $t = 0$. But the numerical data shown in Fig. 2.2.7 indicate that

$$\lim_{t\to 0} \frac{\sqrt{t+25} - 5}{t} = \frac{1}{10}.$$

t	$\dfrac{\sqrt{t+25} - 5}{t}$
1.0	0.099020
0.5	0.099505
0.1	0.099900
0.05	0.099950
0.01	0.099990
0.005	0.099995
↓	↓
0	0.1

FIGURE 2.2.7 Investigating the limit in Example 3.

We can attempt to corroborate this result graphically by zooming in on the point $(1, \frac{1}{10})$. The plot shown in Fig. 2.2.8 does not contradict the indicated limit, but somehow is unconvincing because it "goes too far" and suggests (incorrectly!) that $g(t) = \frac{1}{10}$ for $t \neq 0$. The problem is that the scale on the y-axis is too coarse. The magnification shown in Fig. 2.2.9 *does* appear to substantiate the limiting value of $\frac{1}{10}$. ◆

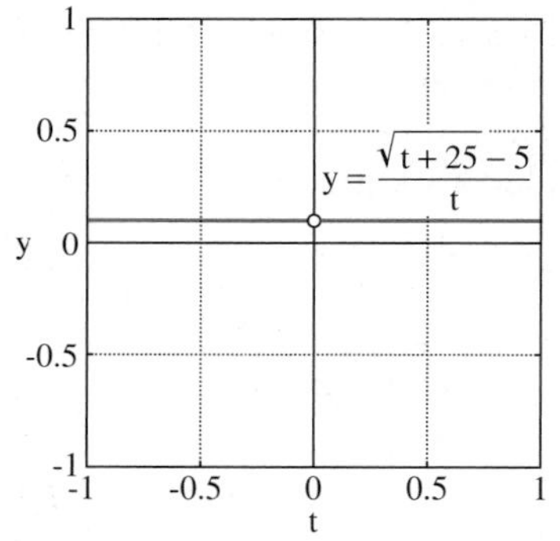

FIGURE 2.2.8 Graph of $g(t) = \dfrac{\sqrt{t+25} - 5}{t}$ for $-1 \leq t \leq +1$, $-1 \leq y \leq +1$.

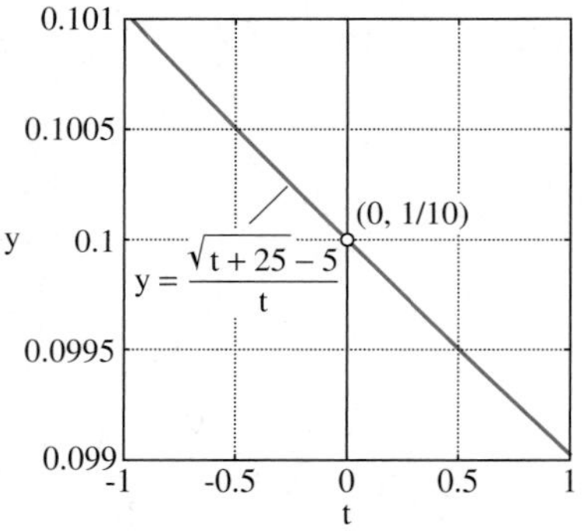

FIGURE 2.2.9 Graph of $g(t) = \dfrac{\sqrt{t+25} - 5}{t}$ for $-1 \leq t \leq +1$, $0.099 \leq y \leq 0.101$.

REMARK Can you see that, upon dividing each number in the second column of Fig. 2.2.7 by 10000, one might well suspect that

$$\lim_{t\to 0} \frac{\sqrt{t+25} - 5}{10000t} = 0? \qquad \textbf{(Wrong!)}$$

In fact, the value of this limit (as we will see in Example 13) is exactly $10^{-5} = 0.00001$, not zero. This fact constitutes a warning that numerical investigations of limits are not conclusive.

The numerical investigation in Example 3 is incomplete because the table in Fig. 2.2.7 shows values of the function $g(t)$ on only one side of the point $t = 0$. But in order that $\lim_{x\to a} f(x) = L$, it is necessary for $f(x)$ to approach L *both* as x approaches a from the left *and* as x approaches a from the right. If $f(x)$ approaches different values as x approaches a from different sides, then $\lim_{x\to a} f(x)$ does not exist. In Section 2.3 we discuss such *one-sided* limits in more detail.

EXAMPLE 4 Investigate $\lim_{x\to 0} f(x)$, given

$$f(x) = \frac{x}{|x|} = \begin{cases} 1 & \text{if } x > 0, \\ -1 & \text{if } x < 0. \end{cases}$$

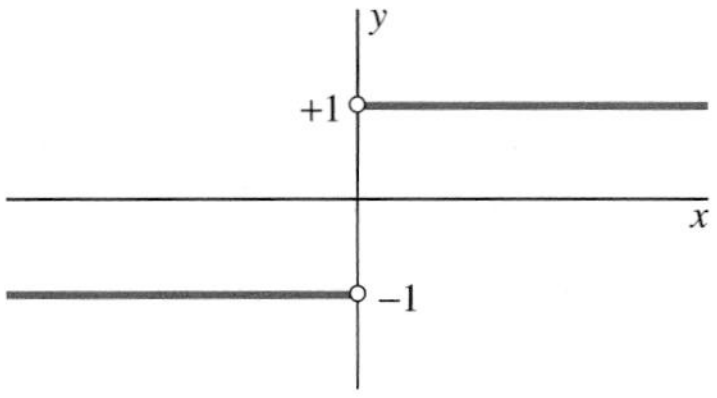

FIGURE 2.2.10 The graph of $f(x) = \dfrac{x}{|x|}$ (Example 4).

Solution From the graph of f shown in Fig. 2.2.10, it is apparent that $f(x) \to 1$ as $x \to 0$ from the right and that $f(x) \to -1$ as $x \to 0$ from the left. In particular, there are positive values of x as close to zero as we please such that $f(x) = 1$ and negative values of x equally close to zero such that $f(x) = -1$. Hence we cannot make $f(x)$ as close as we please to any *single* value of L *merely* by choosing x sufficiently close to zero. Therefore,

$$\lim_{x\to 0} \frac{x}{|x|} \quad \text{does not exist.}$$ ◆

In Example 5 the value obtained by substituting $x = a$ in $F(x)$ to find $\lim_{x\to a} F(x)$ is incorrect.

EXAMPLE 5 Evaluate $\lim_{x\to 0} F(x)$ where

$$F(x) = \begin{cases} 1 & \text{if } x \neq 0, \\ 0 & \text{if } x = 0. \end{cases}$$

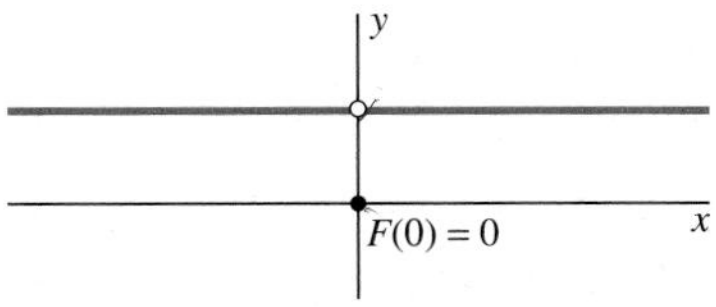

FIGURE 2.2.11 The graph of the function F of Example 5.

The graph of F is shown in Fig. 2.2.11.

Solution The fact that $F(x) = 1$ for *every* value of $x \neq 0$ in any neighborhood of zero implies that

$$\lim_{x\to 0} F(x) = 1.$$

But note that the value of the limit at $x = 0$ is *not* equal to the functional value $F(0) = 0$ there. ◆

The Limit Laws

Numerical investigations such as those in Examples 1 through 3 provide us with an intuitive feeling for limits and typically suggest the correct value of a limit. But most limit computations are based neither on merely suggestive (and imprecise) numerical estimates nor on direct (but difficult) applications of the definition of limit. Instead, such computations are performed most easily and naturally with the aid of the *limit laws* that we give next. These "laws" actually are *theorems,* whose proofs (based on the precise definition of the limit) are included in Appendix D.

Constant Law

If $f(x) \equiv C$, where C is a constant [so $f(x)$ is a **constant function**], then

$$\lim_{x\to a} f(x) = \lim_{x\to a} C = C. \qquad \textbf{(5)}$$

Sum Law

If both of the limits

$$\lim_{x\to a} f(x) = L \quad \text{and} \quad \lim_{x\to a} g(x) = M$$

exist, then

$$\lim_{x\to a}[f(x) \pm g(x)] = \left[\lim_{x\to a} f(x)\right] \pm \left[\lim_{x\to a} g(x)\right] = L \pm M. \qquad \textbf{(6)}$$

(The limit of a sum is the sum of the limits; the limit of a difference is the difference of the limits.)

Product Law

If both of the limits

$$\lim_{x\to a} f(x) = L \quad\text{and}\quad \lim_{x\to a} g(x) = M$$

exist, then

$$\lim_{x\to a}[f(x)g(x)] = \left[\lim_{x\to a} f(x)\right]\left[\lim_{x\to a} g(x)\right] = LM. \qquad \textbf{(7)}$$

(The limit of a product is the product of the limits.)

Quotient Law

If both of the limits

$$\lim_{x\to a} f(x) = L \quad\text{and}\quad \lim_{x\to a} g(x) = M$$

exist *and* if $M \neq 0$, then

$$\lim_{x\to a} \frac{f(x)}{g(x)} = \frac{\lim_{x\to a} f(x)}{\lim_{x\to a} g(x)} = \frac{L}{M}. \qquad \textbf{(8)}$$

(The limit of a quotient is the quotient of the limits, provided that the limit of the denominator is not zero.)

Root Law

If n is a positive integer and if $a > 0$ for even values of n, then

$$\lim_{x\to a} \sqrt[n]{x} = \sqrt[n]{a}. \qquad \textbf{(9)}$$

The case $n = 1$ of the root law is obvious:

$$\lim_{x\to a} x = a. \qquad \textbf{(10)}$$

Examples 6 and 7 show how the limit laws can be used to evaluate limits of polynomials and rational functions.

EXAMPLE 6

$$\lim_{x\to 3} (x^2 + 2x + 4) = \left(\lim_{x\to 3} x^2\right) + \left(\lim_{x\to 3} 2x\right) + \left(\lim_{x\to 3} 4\right)$$

$$= \left(\lim_{x\to 3} x\right)^2 + 2\left(\lim_{x\to 3} x\right) + \left(\lim_{x\to 3} 4\right) = 3^2 + 2\cdot 3 + 4 = 19. \quad ◆$$

EXAMPLE 7

$$\lim_{x\to 3} \frac{2x+5}{x^2+2x+4} = \frac{\lim_{x\to 3}(2x+5)}{\lim_{x\to 3}(x^2+2x+4)}$$

$$= \frac{2\cdot 3+5}{3^2+2\cdot 3+4} = \frac{11}{19}. \quad ◆$$

Note In Examples 6 and 7, we systematically applied the limit laws until we could simply substitute 3 for $\lim_{x\to 3} x$ at the final step. To determine the limit of a quotient of polynomials, we must verify before this final step that the limit of the denominator is not zero. If the denominator limit is zero, then the limit *may* fail to exist.

EXAMPLE 8 Investigate $\displaystyle\lim_{x\to 1} \frac{1}{(x-1)^2}$.

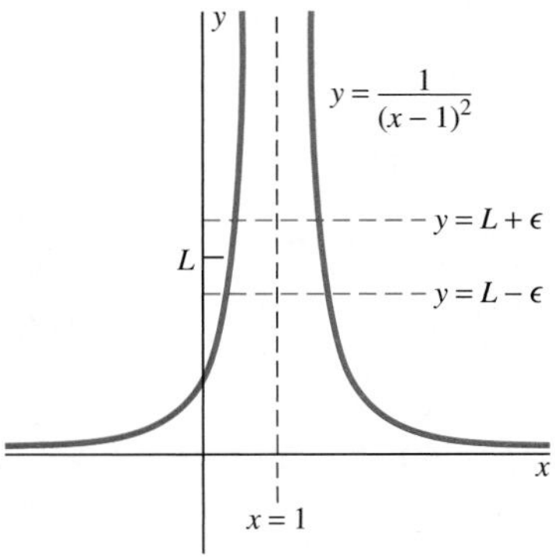

FIGURE 2.2.12 The graph of $y = \dfrac{1}{(x-1)^2}$ (Example 8).

Solution Because $\lim_{x \to 1}(x-1)^2 = 0$, we cannot apply the quotient law. Moreover, we can make $1/(x-1)^2$ arbitrarily large by choosing x sufficiently close to 1. Hence $1/(x-1)^2$ cannot approach any (finite) number L as x approaches 1. Therefore, the limit in this example does not exist. You can see the geometric reason if you examine the graph of $y = 1/(x-1)^2$ in Fig. 2.2.12. As $x \to 1$, the corresponding point (x, y) ascends the curve near the vertical line $x = 1$. It must therefore leave the indicated strip between the two horizontal lines $x = L - \epsilon$ and $x = L + \epsilon$ that bracket the proposed limit L. Thus, the point (x, y) cannot approach the point $(1, L)$ as $x \to 1$. ◆

EXAMPLE 9 Investigate $\displaystyle\lim_{x \to 2} \frac{x^2 - 4}{x^2 + x - 6}$.

Solution We cannot immediately apply the quotient law (as we did in Example 7) because the denominator approaches zero as x approaches 2. If the numerator were approaching some number *other* than zero as $x \to 2$, then the limit would fail to exist (as in Example 8). But here the numerator *does* approach zero, so there is a possibility that a factor of the numerator can be canceled with the same factor of the denominator, thus removing the zero-denominator problem. Indeed,

$$\begin{aligned}\lim_{x \to 2} \frac{x^2 - 4}{x^2 + x - 6} &= \lim_{x \to 2} \frac{(x-2)(x+2)}{(x-2)(x+3)} \\ &= \lim_{x \to 2} \frac{x+2}{x+3} = \frac{4}{5}.\end{aligned}$$

We can cancel the factor $x - 2$ because it is nonzero: $x \neq 2$ when we evaluate the limit as x approaches 2. Moreover, this verifies the numerical limit of 0.8 that we found in Example 2. ◆

Substitution of Limits

It is tempting to write

$$\begin{aligned}\lim_{x \to -4} \sqrt{x^2 + 9} &= \sqrt{\lim_{x \to -4} (x^2 + 9)} \\ &= \sqrt{(-4)^2 + 9} = \sqrt{25} = 5. \end{aligned} \qquad \textbf{(11)}$$

But can we simply "move the limit inside the radical" in Eq. (11)? To analyze this question, let us write

$$f(x) = \sqrt{x} \quad \text{and} \quad g(x) = x^2 + 9.$$

Then the function that appears in Eq. (11) is the composite function

$$f(g(x)) = \sqrt{g(x)} = \sqrt{x^2 + 9}.$$

(Recall that the left-hand expression in this equation is read "f of g of x.") Hence our question is whether or not

$$\lim_{x \to a} f(g(x)) = f\Big(\lim_{x \to a} g(x)\Big).$$

The next limit law answers this question in the affirmative, provided that the "outside function" f meets a certain condition; if so, then the limit of the composite function $f(g(x))$ as $x \to a$ may be found by substituting into the function f the limit of $g(x)$ as $x \to a$.

Substitution Law **Limits of Compositions**

Suppose that

$$\lim_{x\to a} g(x) = L \quad \text{and that} \quad \lim_{x\to L} f(x) = f(L).$$

Then

$$\lim_{x\to a} f(g(x)) = f\left(\lim_{x\to a} g(x)\right) = f(L). \tag{12}$$

Thus the condition under which Eq. (12) holds is that the limit of the *outer* function f not only exists at $x = L$, but also is equal to the "expected" value of f—namely, $f(L)$. In particular, because

$$\lim_{x\to -4}(x^2+9) = 25 \quad \text{and} \quad \lim_{x\to 25}\sqrt{x} = \sqrt{25} = 5,$$

this condition is satisfied in Eq. (11). Hence the computations shown there are valid.

In this section we use only the following special case of the substitution law. With $f(x) = x^{1/n}$, where n is a positive integer, Eq. (12) takes the form

$$\lim_{x\to a}\sqrt[n]{g(x)} = \sqrt[n]{\lim_{x\to a} g(x)}, \tag{13}$$

under the assumption that the limit of $g(x)$ exists as $x \to a$ (and is positive if n is even). With $g(x) = x^m$, where m is a positive integer, Eq. (13) in turn yields

$$\lim_{x\to a} x^{m/n} = a^{m/n}, \tag{14}$$

with the condition that $a > 0$ if n is even. Equations (13) and (14) may be regarded as generalized root laws. Example 10 illustrates the use of these special cases of the substitution law.

EXAMPLE 10

$$\begin{aligned}
\lim_{x\to 4}\sqrt[3]{3\sqrt{x^3}+20\sqrt{x}} &= \sqrt[3]{\lim_{x\to 4}(3x^{3/2}+20\sqrt{x}\,)} && \text{[using Eq. (13)]}\\
&= \left(\lim_{x\to 4} 3x^{3/2} + \lim_{x\to 4} 20\sqrt{x}\right)^{1/3} && \text{[using the sum law]}\\
&= \left(3\cdot 4^{3/2} + 20\sqrt{4}\right)^{1/3} && \text{[using Eq. (14)]}\\
&= (24+40)^{1/3} = \sqrt[3]{64} = 4.
\end{aligned}$$

◆

Slope-Predictor Functions

Our discussion of limits began with the slope

$$m_a = \lim_{h\to 0}\frac{f(a+h)-f(a)}{h} \tag{15}$$

of the line tangent to the graph $y = f(x)$ at the point $(a, f(a))$. The lines tangent to $y = f(x)$ at different points have different slopes. Thus if we replace a with x in Eq. (15), we get a *new function* defined by

$$m(x) = \lim_{h\to 0}\frac{f(x+h)-f(x)}{h}. \tag{16}$$

This function m may be regarded as a "slope predictor" for lines tangent to the graph $y = f(x)$. It is a new function *derived* from the original function $f(x)$, and in Chapter 3 we will call it the *derivative* of f.

EXAMPLE 11 In Section 2.1 we saw that the line tangent to the graph $y = px^2 + qx + r$ at the point where $x = a$ has slope $m_a = 2pa + q$. Hence the slope-predictor function for the quadratic function

$$f(x) = px^2 + qx + r \tag{17}$$

is the linear function

$$m(x) = 2px + q. \tag{18}$$

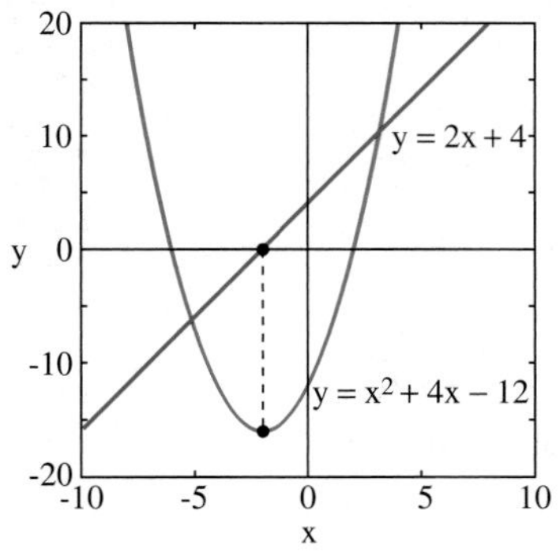

FIGURE 2.2.13 The parabola $y = x^2 + 4x - 12$ and its slope predictor $m(x) = 2x + 4$.

Figure 2.2.13 illustrates the case $p = 1$, $q = 4$, $r = -12$. It is worth noting that the x-intercept where $m(x) = 0$ corresponds to the point of the parabola $y = f(x)$ where the tangent line is horizontal. ◆

The slope-predictor definition in Eq. (16) calls for us to carry out the following four steps.

1. Write the definition of $m(x)$.
2. Substitute into this definition the formula of the given function f.
3. Make algebraic simplifications until Step 4 can be carried out.
4. Determine the value of the limit as $h \to 0$.

Note that x may be thought of as a *constant* throughout this computation—it is h that is the variable in this four-step process.

EXAMPLE 12 Find the slope-predictor function for the function

$$f(x) = x + \frac{1}{x}$$

that was investigated numerically in Example 5 of Section 2.1.

Solution The first two steps in the preceding list yield

$$m(x) = \lim_{h \to 0} \frac{f(x+h) - f(x)}{h} = \lim_{h \to 0} \frac{\left(x + h + \dfrac{1}{x+h}\right) - \left(x + \dfrac{1}{x}\right)}{h}.$$

We cancel the two copies of x in the numerator and proceed to simplify algebraically, first finding a common denominator in the numerator:

$$\begin{aligned} m(x) &= \lim_{h \to 0} \frac{h + \dfrac{1}{x+h} - \dfrac{1}{x}}{h} \\ &= \lim_{h \to 0} \frac{h(x+h)x + x - (x+h)}{h(x+h)x} \\ &= \lim_{h \to 0} \frac{h(x+h)x - h}{h(x+h)x}. \end{aligned}$$

Now we can divide numerator and denominator by h (because $h \neq 0$) and finally apply the sum, product, and quotient laws to evaluate the limit as $h \to 0$:

$$\begin{aligned} m(x) &= \lim_{h \to 0} \frac{h(x+h)x - h}{h(x+h)x} \\ &= \lim_{h \to 0} \frac{(x+h)x - 1}{(x+h)x} = \frac{x^2 - 1}{x^2} = 1 - \frac{1}{x^2}. \end{aligned}$$

For instance, the slope of the line tangent to

$$y = x + \frac{1}{x}$$

at the point $(2, \frac{5}{2})$ is $m(2) = \frac{3}{4}$ (thus verifying the result in Example 5 of Section 2.1). ◆

Example 13 illustrates an algebraic procedure often used in "preparing" functions before taking limits. This procedure can be applied when roots are present and resembles the simple computation

$$\begin{aligned}\frac{1}{\sqrt{5}-\sqrt{2}} &= \frac{1}{\sqrt{5}-\sqrt{2}} \cdot \frac{\sqrt{5}+\sqrt{2}}{\sqrt{5}+\sqrt{2}} \\ &= \frac{\sqrt{5}+\sqrt{2}}{5-2} = \frac{\sqrt{5}+\sqrt{2}}{3}.\end{aligned}$$

EXAMPLE 13 Find the slope-predictor function for the function $f(x) = \sqrt{x}$.

Solution

$$m(x) = \lim_{h\to 0} \frac{\sqrt{x+h}-\sqrt{x}}{h}. \tag{19}$$

To prepare the fraction for evaluation of the limit, we first multiply the numerator and denominator by the *conjugate* $\sqrt{x+h}+\sqrt{x}$ of the numerator:

$$\begin{aligned}m(x) &= \lim_{h\to 0} \frac{\sqrt{x+h}-\sqrt{x}}{h} \cdot \frac{\sqrt{x+h}+\sqrt{x}}{\sqrt{x+h}+\sqrt{x}} \\ &= \lim_{h\to 0} \frac{(x+h)-x}{h(\sqrt{x+h}+\sqrt{x})} \\ &= \lim_{h\to 0} \frac{1}{\sqrt{x+h}+\sqrt{x}}.\end{aligned}$$

Thus

$$m(x) = \frac{1}{2\sqrt{x}}. \tag{20}$$

(In the final step we used the sum, quotient, and root laws—we did not simply substitute 0 for h.) ◆

Note that if we equate the right-hand sides of Eqs. (19) and (20) and take $x = 25$, then we get the limit in Example 3:

$$\lim_{h\to 0} \frac{\sqrt{25+h}-5}{h} = \frac{1}{10}.$$

(The t in Example 3 has been replaced here with h.) And if we divide both sides by 10000 we find that

$$\lim_{h\to 0} \frac{\sqrt{25+h}-5}{10000h} = \frac{1}{100000} = 0.00001,$$

as claimed in the remark following Example 3.

2.2 TRUE/FALSE STUDY GUIDE

2.2 CONCEPTS: QUESTIONS AND DISCUSSION

1. The sum, product, and quotient laws imply that *if* the limits

$$\lim_{x\to a} f(x) \quad \text{and} \quad \lim_{x\to a} g(x) \tag{21}$$

both exist, *then* the limit

$$\lim_{x\to a} [f(x) \odot g(x)] \tag{22}$$

also exists—with the symbol $\odot$ denoting either $+$, $-$, $\times$, or $\div$ (and assuming in the case of division that $\lim_{x\to a} g(x) \neq 0$). Can you produce examples—in all four cases—of functions such that *neither* of the limits in (21) exists, but nevertheless the limit in (22) does exist? It may help to review the examples of nonexisting limits in this section.

2. Can you produce examples of functions f and g such that both

$$\lim_{x\to a} g(x) = b \quad \text{and} \quad \lim_{x\to b} f(x) = c$$

exist, but

$$\lim_{x\to a} f(g(x)) \neq f\left(\lim_{x\to a} g(x)\right)?$$

If so, why does this not contradict the substitution law of limits?

2.2 PROBLEMS

Apply the limit laws of this section to evaluate the limits in Problems 1 through 18. Justify each step by citing the appropriate limit law.

1. $\lim_{x\to 3} (3x^2 + 7x - 12)$

2. $\lim_{x\to -2} (x^3 - 3x^2 + 5)$

3. $\lim_{x\to 1} (x^2 - 1)(x^7 + 7x - 4)$

4. $\lim_{x\to -2} (x^3 - 3x + 3)(x^2 + 2x + 5)$

5. $\lim_{x\to 1} \dfrac{x+1}{x^2+x+1}$

6. $\lim_{t\to -2} \dfrac{t+2}{t^2+4}$

7. $\lim_{x\to 3} \dfrac{(x^2+1)^3}{(x^3-25)^3}$

8. $\lim_{z\to -1} \dfrac{(3z^2+2z+1)^{10}}{(z^3+5)^5}$

9. $\lim_{x\to 1} \sqrt{4x+5}$

10. $\lim_{y\to 4} \sqrt{27 - \sqrt{y}}$

11. $\lim_{x\to 3} (x^2 - 1)^{3/2}$

12. $\lim_{t\to -4} \sqrt{\dfrac{t+8}{25-t^2}}$

13. $\lim_{z\to 8} \dfrac{z^{2/3}}{z - \sqrt{2z}}$

14. $\lim_{t\to 2} \sqrt[3]{3t^3 + 4t - 5}$

15. $\lim_{w\to 0} \sqrt{(w-2)^4}$

16. $\lim_{t\to -4} \sqrt[3]{(t+1)^6}$

17. $\lim_{x\to -2} \sqrt[3]{\dfrac{x+2}{(x-2)^2}}$

18. $\lim_{y\to 5} \left(\dfrac{2y^2+2y+4}{6y-3}\right)^{1/3}$

In Problems 19 through 28, note first that the numerator and denominator have a common algebraic factor (as in Example 9). Use this fact to help evaluate the given limit.

19. $\lim_{x\to -1} \dfrac{x+1}{x^2-x-2}$

20. $\lim_{t\to 3} \dfrac{t^2-9}{t-3}$

21. $\lim_{x\to 1} \dfrac{x^2+x-2}{x^2-4x+3}$

22. $\lim_{y\to -1/2} \dfrac{4y^2-1}{4y^2+8y+3}$

23. $\lim_{t\to -3} \dfrac{t^2+6t+9}{t^2-9}$

24. $\lim_{x\to 2} \dfrac{x^2-4}{3x^2-2x-8}$

25. $\lim_{z\to -2} \dfrac{(z+2)^2}{z^4-16}$

26. $\lim_{t\to 3} \dfrac{t^3-9t}{t^2-9}$

27. $\lim_{x\to 1} \dfrac{x^3-1}{x^4-1}$

28. $\lim_{y\to -3} \dfrac{y^3+27}{y^2-9}$

In Problems 29 through 36, evaluate those limits that exist.

29. $\lim_{x\to 3} \dfrac{\frac{1}{x} - \frac{1}{3}}{x-3}$

30. $\lim_{t\to 0} \dfrac{\frac{1}{2+t} - \frac{1}{2}}{t}$

31. $\lim_{x\to 4} \dfrac{x-4}{\sqrt{x}-2}$

32. $\lim_{x\to 9} \dfrac{3-\sqrt{x}}{9-x}$

33. $\lim_{t\to 0} \dfrac{\sqrt{t+4}-2}{t}$

34. $\lim_{h\to 0} \dfrac{1}{h}\left(\dfrac{1}{\sqrt{9+h}} - \dfrac{1}{3}\right)$

35. $\lim_{x\to 4} \dfrac{x^2-16}{2-\sqrt{x}}$

36. $\lim_{x\to 0} \dfrac{\sqrt{1+x} - \sqrt{1-x}}{x}$

In Problems 37 through 46, use the four-step process illustrated in Examples 12 and 13 to find a slope-predictor function for the given function $f(x)$. Then write an equation for the line tangent to the curve $y = f(x)$ at the point where $x = 2$.

37. $f(x) = x^3$

38. $f(x) = \dfrac{1}{x}$

39. $f(x) = \dfrac{1}{x^2}$

40. $f(x) = \dfrac{1}{x+1}$

41. $f(x) = \dfrac{2}{x-1}$

42. $f(x) = \dfrac{x}{x-1}$

43. $f(x) = \dfrac{1}{\sqrt{x+2}}$

44. $f(x) = x^2 + \dfrac{3}{x}$

45. $f(x) = \sqrt{2x+5}$

46. $f(x) = \dfrac{x^2}{x+1}$

In Problems 47 through 56, the actual value of the given limit $\lim_{x\to a} f(x)$ is a rational number that is a ratio of two single-digit integers. Guess this limit on the basis of a numerical investigation in which you calculate $f(x)$ for $x = a \pm 0.1$, $x = a \pm 0.05$, $x = a \pm 0.01$, $x = a \pm 0.005$, and so on. Use other similar values of x near a as you wish.

47. $\lim_{x\to 0} \dfrac{(1+x)^2 - 1}{x}$

48. $\lim_{x\to 1} \dfrac{x^4-1}{x-1}$

49. $\lim_{x\to 0} \dfrac{\sqrt{x+9}-3}{x}$

50. $\lim_{x\to 4} \dfrac{x^{3/2}-8}{x-4}$

51. $\lim_{x\to 0} \dfrac{1}{x}\left[\dfrac{2}{(2+x)^3} - \dfrac{1}{4}\right]$

52. $\lim_{x\to 0} \dfrac{(3+x)^{-1} - (3-x)^{-1}}{x}$

53. $\lim_{x\to 0} \dfrac{\sin x}{x}$

54. $\lim_{x\to 0} \dfrac{1-\cos x}{x^2}$

55. $\lim_{x\to 0} \dfrac{x - \sin x}{x^3}$

56. $\lim_{x\to 0} \left(1 + \dfrac{1}{|x|}\right)^x$

57. In contrast with the rational-valued limits in Problems 47 through 56, the value of the limit

$$\lim_{x\to 0}(1+x)^{1/x}$$

is the famous irrational number e (of Chapter 3), whose three-place decimal approximation is $e \approx 2.718$. Numerically investigate this limit to approximate e accurate to five decimal places. Corroborate this value graphically by zooming in on the y-intercept of the curve $y=(1+x)^{1/x}$.

58. Verify graphically the limit

$$\lim_{x\to 0}\frac{\sin x}{x}$$

of Problem 53 by zooming in on the y-intercept of the curve $y=(\sin x)/x$.

59. Investigate the limit

$$\lim_{x\to 0}\frac{x-\tan x}{x^3}$$

both numerically and graphically. Determine its value accurate to four decimal places.

60. The value of

$$\lim_{x\to 0}\frac{\sin 2x}{\tan 5x}$$

is the ratio of two single-digit integers. Determine this value both numerically and graphically.

61. Calculate the value of

$$f(x)=\sin\frac{\pi}{x}$$

for $x=\frac{1}{2},\frac{1}{4},\frac{1}{8},\frac{1}{16},\dots$. What do you now conjecture to be the value of

$$\lim_{x\to 0}\sin\frac{\pi}{x}?$$

Next calculate $f(x)$ for $x=\frac{2}{3},\frac{2}{9},\frac{2}{27},\frac{2}{81},\dots$. Now what do you conclude?

62. To investigate the limit of $f(x)=\sin x+10^{-5}\cos x$ as $x\to 0$, set your graphing calculator or computer to display exactly four digits to the right of the decimal point. After calculating $f(x)$ with $x=0.1, 0.001, 0.00001, 0.0000001, \dots$, what do you conclude? (Your answer may depend on how your particular calculator works.) Now zoom in on the y-intercept of the curve $y=f(x)$ sufficiently to show that the value of the limit is nonzero. What is it?

63. Investigate numerically or graphically (or both) the value of the limit

$$\lim_{x\to 0}\left(\log_{10}\frac{1}{|x|}\right)^{-1/32}.$$

The actual value of this limit is zero, so you'll see that your calculator or computer cannot always be believed.

64. (a) Show that the slope of the line tangent to the graph of $y=10^x$ at the point $(0,1)$ is the number

$$L=\lim_{h\to 0}\frac{10^h-1}{h}.$$

Investigate this limit numerically and graphically. Do your results substantiate the fact that $L=\ln 10$, the value produced by the **LN** key on your calculator? (b) Show that the slope-predictor function for lines tangent to the graph $y=10^x$ is $m(x)=L\cdot 10^x$. Corroborate this fact by using a calculator or computer to plot the graph of $y=10^x$ and its predicted tangent lines at several different points.

2.2 Project: Limits, Slopes, and Logarithms

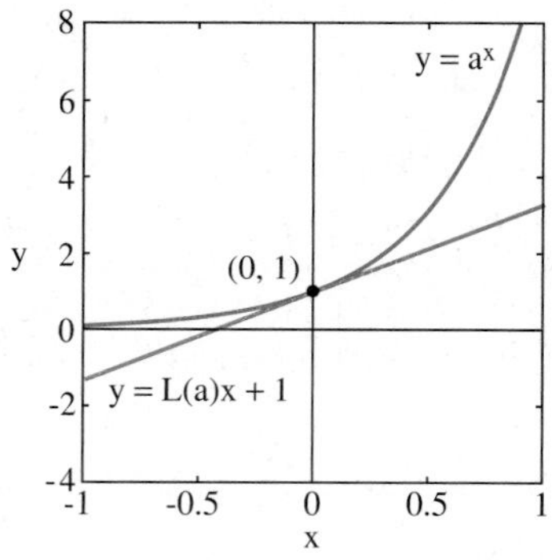

FIGURE 2.2.14 The graph of $y=a^x$ and its tangent line at the point (0,1).

Generalize the result in Problem 64 of this section. First refer to Fig. 2.2.14. Then suppose that a is a positive constant. Show that the slope of the line tangent to the graph of $y=a^x$ at the point $(0,1)$ is the number

$$L(a)=\lim_{h\to 0}\frac{a^h-1}{h}. \qquad (1)$$

(Note how the notation of functions is used in Eq. (1) to emphasize the dependence of the slope on the base constant a.) Next choose at random a pair of positive integers a and b and investigate the numerical values of $L(a)$, $L(b)$, and $L(ab)$. Are your results consistent with the fact that

$$L(ab)=L(a)+L(b), \qquad (2)$$

in analogy with the law of logarithms

$$\log ab=\log a+\log b? \qquad (3)$$

At this point the connection between Eqs. (2) and (3) is surely an enigma rather than an explanation. The mystery will be explained in Section 3.8, in which we study *natural logarithms*. For now, use the **LN** key on your calculator to find $\ln a$, $\ln b$, and $\ln ab$; compare these with your earlier values of $L(a)$, $L(b)$, and $L(ab)$. You can also follow up these investigations with a computer algebra system: Use it to attempt to evaluate the limit in Eq. (1) symbolically, and then compare the symbolic result with your numerical results.

2.3 MORE ABOUT LIMITS

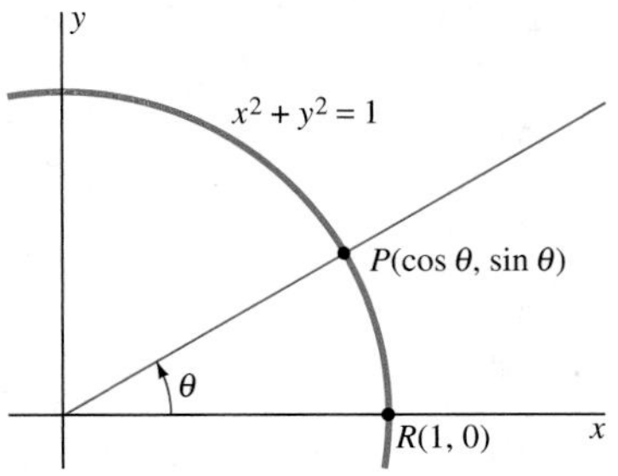

FIGURE 2.3.1 An angle θ.

To investigate limits of trigonometric functions, we begin with Fig. 2.3.1, which shows an angle θ with its vertex at the origin, its initial side along the positive x-axis, and its terminal side intersecting the unit circle at the point P. By the definition of the sine and cosine functions, the coordinates of P are $P(\cos\theta, \sin\theta)$. From geometry we see that, as $\theta \to 0$, the point $P(\cos\theta, \sin\theta)$ approaches the point $R(1, 0)$. Hence $\cos\theta \to 1$ and $\sin\theta \to 0$ as $\theta \to 0$ through positive values. A similar picture gives the same result for negative values of θ, so we see that

$$\lim_{\theta\to 0} \cos\theta = 1 \quad \text{and} \quad \lim_{\theta\to 0} \sin\theta = 0. \tag{1}$$

Equation (1) says simply that the *limits* of the functions $\cos\theta$ and $\sin\theta$ as $\theta \to 0$ are equal to their respective *values* at $\theta = 0$: $\cos 0 = 1$ and $\sin 0 = 0$.

The limit of the quotient $(\sin\theta)/\theta$ as $\theta \to 0$ plays a special role in the calculus of trigonometric functions. For instance, it is needed to find slopes of lines tangent to trigonometric graphs such as $y = \cos x$ and $y = \sin x$.

Note that the value of the quotient $(\sin\theta)/\theta$ is not defined when $\theta = 0$. (Why not?) But a calculator set in *radian mode* provides us with the numerical evidence shown in Fig. 2.3.2. This table strongly suggests that the limit of $(\sin\theta)/\theta$ is 1 as $\theta \to 0$. This conclusion is supported by the graph of $y = (\sin x)/x$ shown in Fig. 2.3.3, where it appears that the point (x, y) on the curve is near $(0, 1)$ when x is near zero. Later in this section we provide a proof of the following result.

θ	$\dfrac{\sin\theta}{\theta}$
±1.0	0.84147
±0.5	0.95885
±0.1	0.99833
±0.05	0.99958
±0.01	0.99998
±0.005	1.00000
±0.001	1.00000
⋮	⋮
↓	↓
0	1

FIGURE 2.3.2 The numerical data suggest that $\lim_{\theta\to 0} \dfrac{\sin\theta}{\theta} = 1$.

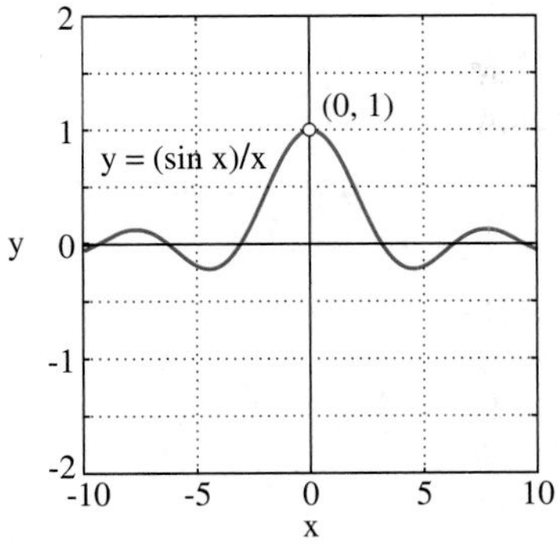

FIGURE 2.3.3 $y = \dfrac{\sin x}{x}$ for $x \neq 0$.

THEOREM 1 The Basic Trigonometric Limit

$$\lim_{x\to 0} \frac{\sin x}{x} = 1. \tag{2}$$

As in Examples 1 and 2, many other trigonometric limits can be reduced to the one in Theorem 1.

EXAMPLE 1 Show that

$$\lim_{x\to 0} \frac{1 - \cos x}{x} = 0. \tag{3}$$

Solution We multiply the numerator and denominator in Eq. (3) by the "conjugate" $1 + \cos x$ of the numerator $1 - \cos x$. Then we apply the identity $1 - \cos^2 x = \sin^2 x$. This gives

$$\lim_{x\to 0} \frac{1 - \cos x}{x} = \lim_{x\to 0} \frac{1 - \cos x}{x} \cdot \frac{1 + \cos x}{1 + \cos x} = \lim_{x\to 0} \frac{\sin^2 x}{x(1 + \cos x)}$$

$$= \left(\lim_{x\to 0} \frac{\sin x}{x}\right)\left(\lim_{x\to 0} \frac{\sin x}{1 + \cos x}\right) = 1 \cdot \frac{0}{1+1} = 0.$$

In the last step we used *all* the limits in Eqs. (1) and (2). ◆

EXAMPLE 2 Evaluate $\displaystyle\lim_{x\to 0} \frac{\tan 3x}{x}$.

Solution

$$\lim_{x\to 0} \frac{\tan 3x}{x} = 3\left(\lim_{x\to 0} \frac{\tan 3x}{3x}\right) = 3\left(\lim_{\theta\to 0} \frac{\tan\theta}{\theta}\right) \qquad (\theta = 3x)$$

$$= 3\left(\lim_{\theta\to 0} \frac{\sin\theta}{\theta\cos\theta}\right) \qquad \left(\text{because } \tan\theta = \frac{\sin\theta}{\cos\theta}\right)$$

$$= 3\left(\lim_{\theta\to 0} \frac{\sin\theta}{\theta}\right)\left(\lim_{\theta\to 0} \frac{1}{\cos\theta}\right) \qquad \text{(by the product law of limits)}$$

$$= 3 \cdot 1 \cdot \frac{1}{1} = 3.$$

We used the fact that $\tan\theta = (\sin\theta)/(\cos\theta)$ as well as some of the limits in Eqs. (1) and (2). ◆

x	$\sin\dfrac{\pi}{x}$
1	0
0.5	0
0.1	0
0.05	0
0.01	0
0.005	0
0.001	0

FIGURE 2.3.4 Do you think that $\lim_{x\to 0} \sin\dfrac{\pi}{x} = 0$ (Example 3)?

Example 3 constitutes a *warning:* The results of a numerical investigation can be misleading unless they are interpreted with care.

EXAMPLE 3 The numerical data shown in the table of Fig. 2.3.4 suggest that the limit

$$\lim_{x\to 0} \sin\frac{\pi}{x} \tag{4}$$

has the value zero. But it appears in the graph of $y = \sin(\pi/x)$ (for $x \neq 0$), shown in Fig. 2.3.5, that the value of $\sin(\pi/x)$ oscillates infinitely often between $+1$ and -1 as $x \to 0$. Indeed, this fact follows from the periodicity of the sine function, because π/x increases without bound as $x \to 0$. Hence $\sin(\pi/x)$ cannot approach zero (or any other number) as $x \to 0$. Therefore the limit in (4) *does not exist.*

We can explain the potentially misleading results tabulated in Fig. 2.3.4 as follows: Each value of x shown there just happens to be of the form $1/n$, the reciprocal of an integer. Therefore,

$$\sin\frac{\pi}{x} = \sin\frac{\pi}{1/n} = \sin n\pi = 0$$

for every nonzero integer n. But with a different selection of "trial values" of x, we might have obtained the results shown in Fig. 2.3.6, which immediately suggest the nonexistence of the limit in (4). ◆

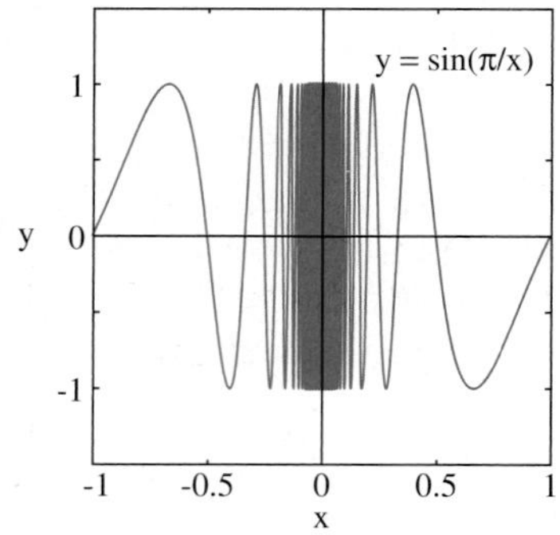

FIGURE 2.3.5 The graph of $y = \sin\dfrac{\pi}{x}$ shows infinite oscillation as $x \to 0$ (Example 3).

x	$\sin\dfrac{\pi}{x}$
$\frac{2}{9}$	$+1$
$\frac{2}{11}$	-1
$\frac{2}{101}$	$+1$
$\frac{2}{103}$	-1
$\frac{2}{1001}$	$+1$
$\frac{2}{1003}$	-1

FIGURE 2.3.6 Verify the entries in the second column (Example 3).

The Squeeze Law of Limits

A final property of limits that we will need is the *squeeze law* (also known as the "sandwich theorem"). It is related to the fact that taking limits preserves inequalities among functions.

Figure 2.3.7 illustrates how and why the squeeze law works and how it got its name. The idea is that $g(x)$ is trapped between $f(x)$ and $h(x)$ near a; both $f(x)$ and $h(x)$ approach the same limit L, so $g(x)$ must approach L as well. A formal proof of the squeeze law can be found in Appendix D.

Squeeze Law

Suppose that $f(x) \leqq g(x) \leqq h(x)$ for all $x \neq a$ in some neighborhood of a and also that

$$\lim_{x\to a} f(x) = L = \lim_{x\to a} h(x).$$

Then

$$\lim_{x\to a} g(x) = L$$

as well.

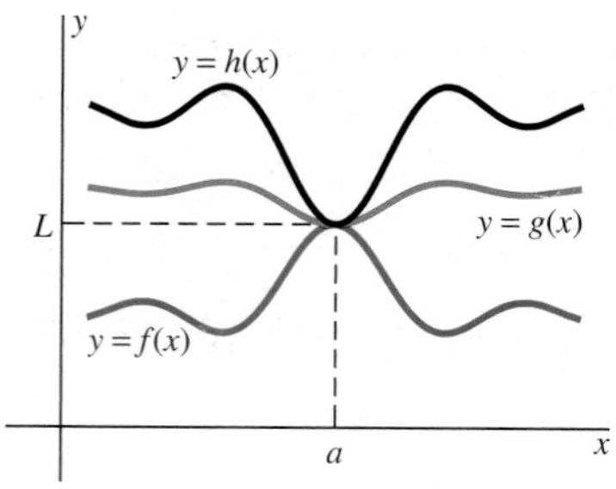

FIGURE 2.3.7 How the squeeze law works.

EXAMPLE 4 Figures 2.3.8 and 2.3.9 show two views of the graph of the function g defined for $x \neq 0$ by

$$g(x) = x \sin\frac{1}{x}.$$

As in Example 3, $\sin(1/x)$ oscillates infinitely often between $+1$ and -1 as $x \to 0$. Therefore the graph $y = g(x)$ bounces back and forth between the lines $y = +x$

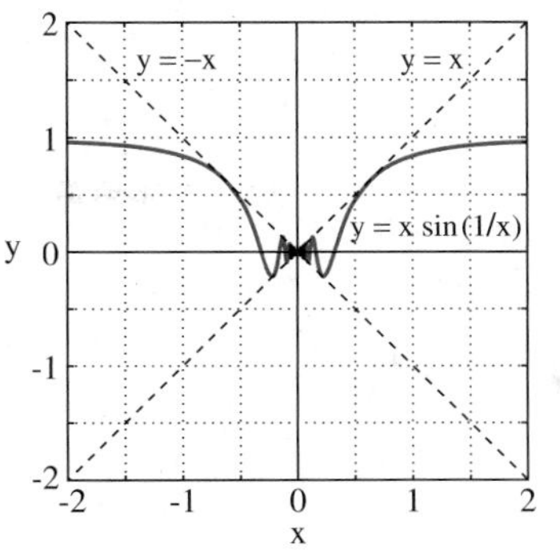

FIGURE 2.3.8 The graph of $g(x) = x \sin \frac{1}{x}$ for $x \neq 0$ (Example 4).

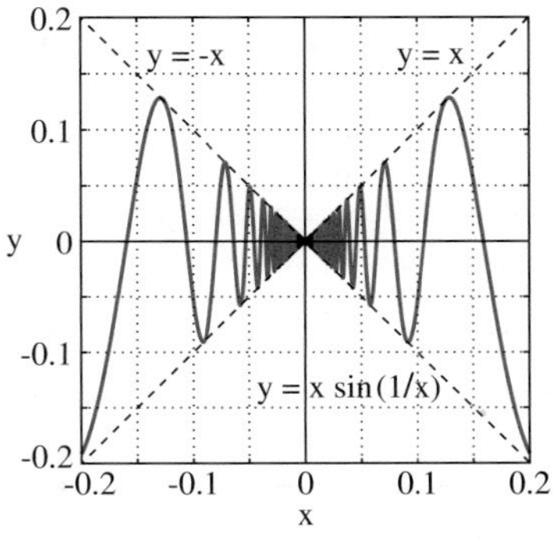

FIGURE 2.3.9 The graph magnified near the origin (Example 4).

and $y = -x$. Because $|\sin(1/x)| \leqq 1$ for all $x \neq 0$,

$$-|x| \leqq x \sin \frac{1}{x} \leqq +|x|$$

for all $x \neq 0$. Moreover, $\pm|x| \to 0$ as $x \to 0$, so with $f(x) = -|x|$ and $h(x) = +|x|$, it follows from the squeeze law of limits that

$$\lim_{x \to 0} x \sin \frac{1}{x} = 0. \tag{5}$$

◆

QUESTION Why *doesn't* the limit in Eq. (5) follow from the product law of limits with $f(x) = x$ and $g(x) = \sin(1/x)$?

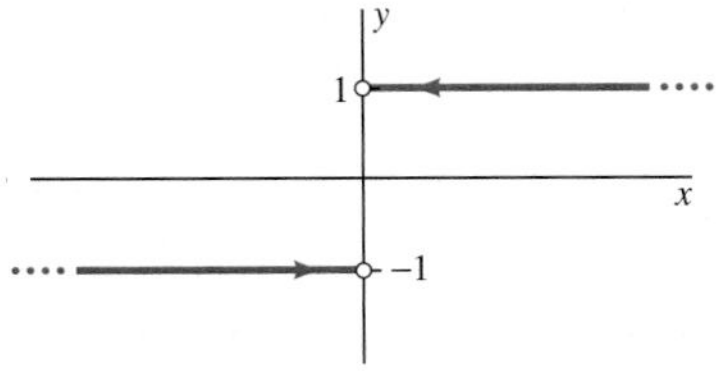

FIGURE 2.3.10 The graph of $f(x) = \dfrac{x}{|x|}$ again.

One-Sided Limits

In Example 4 of Section 2.2 we examined the function

$$f(x) = \frac{x}{|x|} = \begin{cases} 1 & \text{if } x > 0; \\ -1 & \text{if } x < 0. \end{cases}$$

The graph of $y = f(x)$ is shown in Fig. 2.3.10. We argued that the limit of $f(x)$ as $x \to 0$ does not exist because $f(x)$ approaches $+1$ as x approaches zero from the right, whereas $f(x) \to -1$ as x approaches zero from the left. A natural way of describing this situation is to say that at $x = 0$ the *right-hand limit* of $f(x)$ is $+1$ and the *left-hand limit* of $f(x)$ is -1.

Here we define and investigate such one-sided limits. Their definitions will be stated initially in the informal language we used in Section 2.2 to describe the "idea of the limit." To define the right-hand limit of $f(x)$ at $x = a$, we must assume that f is defined on an open interval immediately to the right of a. To define the left-hand limit, we must assume that f is defined on an open interval immediately to the left of a.

The Right-Hand Limit of a Function

Suppose that f is defined on the interval (a, c) immediately to the right of a. Then we say that the number L is the **right-hand limit** of $f(x)$ as x approaches a (from the right), and we write

$$\lim_{x \to a^+} f(x) = L, \tag{6}$$

provided that we can make the number $f(x)$ as close to L as we please merely by choosing the point x in (a, c) sufficiently close to a.

We may describe the right-hand limit in Eq. (6) by saying that $f(x) \to L$ as $x \to a^+$; that is, as x approaches a from the right. The symbol a^+ denotes the right-hand, or "positive," side of the number a (which may be positive, negative, or zero).

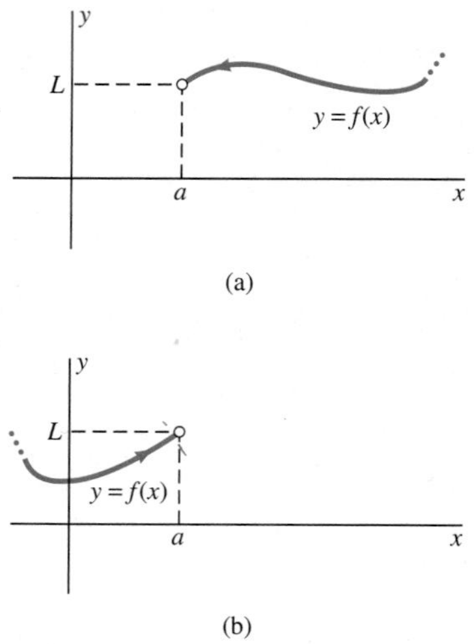

FIGURE 2.3.11 (a) The right-hand limit of $f(x)$ is L. (b) The left-hand limit of $f(x)$ is L.

For instance, we see in Fig. 2.3.10 that

$$\lim_{x \to 0^+} \frac{|x|}{x} = +1 \tag{7}$$

because $|x|/x$ is equal to $+1$ for all x to the right of zero. See Fig. 2.3.11(a) for a more general geometric interpretation of right-hand limits.

The Left-Hand Limit of a Function

Suppose that f is defined on the interval (c, a) immediately to the left of a. Then we say that the number L is the **left-hand limit** of $f(x)$ as x approaches a (from the left), and we write

$$\lim_{x \to a^-} f(x) = L, \tag{8}$$

provided that we can make the number $f(x)$ as close to L as we please merely by choosing the point x in (c, a) sufficiently close to a.

We may describe the left-hand limit in Eq. (8) by saying that $f(x) \to L$ as $x \to a^-$; that is, as x approaches a from the left. The symbol a^- denotes the left-hand or "negative" side of a.

For instance, we see in Fig. 2.3.10 that

$$\lim_{x \to 0^-} \frac{|x|}{x} = -1 \tag{9}$$

because $|x|/x$ is equal to -1 for all x to the left of zero. See Fig. 2.3.11(b) for a more general geometric interpretation of left-hand limits.

In Example 4 of Section 2.2 we argued (in essence) that, because the limits in Eqs. (7) and (9) are not equal, the corresponding two-sided limit

$$\lim_{x \to 0} \frac{|x|}{x}$$

does not exist. More generally, Theorem 2 (next) follows from careful consideration of the definitions of all the limits involved.

THEOREM 2 One-Sided Limits and Two-Sided Limits

Suppose that the function f is defined for $x \neq a$ in a neighborhood of the point a. Then the two-sided limit

$$\lim_{x \to a} f(x)$$

exists and is equal to the number L if and only if the one-sided limits

$$\lim_{x \to a^+} f(x) \quad \text{and} \quad \lim_{x \to a^-} f(x)$$

both exist and are equal to L.

Theorem 2 is particularly useful in showing that certain (two-sided) limits do *not* exist, frequently by showing that the left-hand and right-hand limits are not equal to each other.

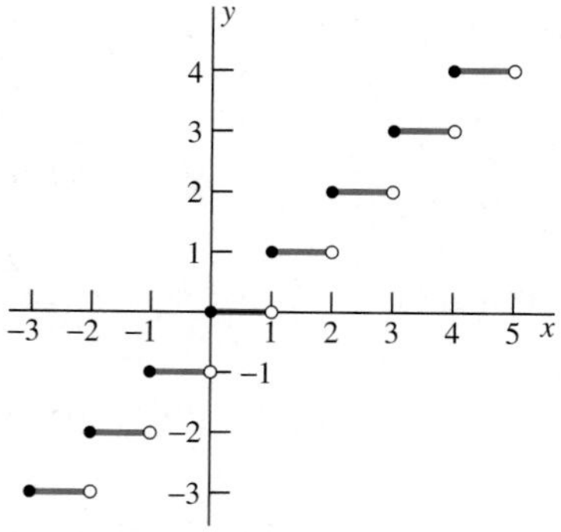

FIGURE 2.3.12 The graph of the greatest integer function $f(x) = [\![x]\!]$ (Example 5).

EXAMPLE 5 The graph of the greatest integer function $f(x) = [\![x]\!]$ is shown in Fig. 2.3.12. It should be apparent that if a is not an integer, then

$$\lim_{x \to a^+} [\![x]\!] = \lim_{x \to a^-} [\![x]\!] = \lim_{x \to a} [\![x]\!] = [\![a]\!].$$

But if $a = n$, an integer, then

$$\lim_{x \to n^-} [\![x]\!] = n - 1 \quad \text{and} \quad \lim_{x \to n^+} [\![x]\!] = n.$$

Because these left-hand and right-hand limits are not equal, it follows from Theorem 2 that the limit of $f(x) = [\![x]\!]$ does not exist as x approaches an integer n. ◆

EXAMPLE 6 According to the root law in Section 2.2,

$$\lim_{x\to a} \sqrt{x} = \sqrt{a} \quad \text{if } a > 0.$$

But the limit of $f(x) = \sqrt{x}$ as $x \to 0^-$ is not defined because the square root of a negative number is undefined. Hence f is undefined on every open interval containing zero. What we can say in the case $a = 0$ is that

$$\lim_{x\to 0^+} \sqrt{x} = 0,$$

and that the left-hand limit

$$\lim_{x\to 0^-} \sqrt{x}$$

does not exist. ◆

To each of the limit laws in Section 2.2 there correspond two *one-sided limit laws*—a right-hand version and a left-hand version. You may apply these one-sided limit laws in the same way you apply the two-sided limit laws in the evaluation of limits.

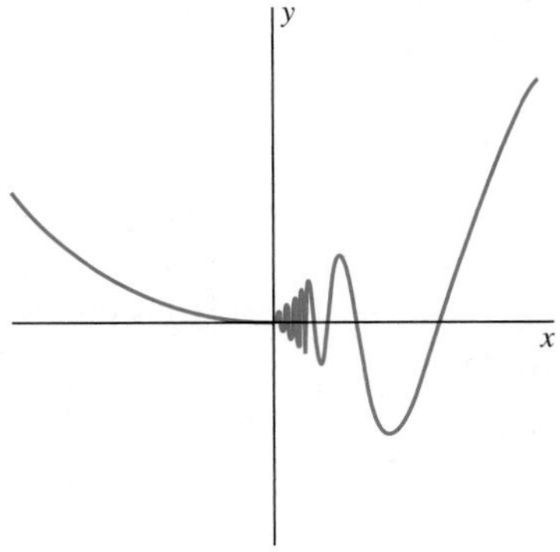

FIGURE 2.3.13 $y = f(x)$ (Example 7).

EXAMPLE 7 Figure 2.3.13 shows the graph of the function f defined by

$$f(x) = \begin{cases} x^2 & \text{if } x \leqq 0; \\ x \sin \dfrac{1}{x} & \text{if } x > 0. \end{cases}$$

Clearly

$$\lim_{x\to 0^-} f(x) = 0 \quad \text{and} \quad \lim_{x\to 0^+} f(x) = 0$$

by a one-sided version of the squeeze law (as in Example 4). It therefore follows from Theorem 2 that

$$\lim_{x\to 0} f(x) = 0.$$ ◆

EXAMPLE 8 Upon applying the appropriate one-sided limit laws, we find that

$$\lim_{x\to 3^-} \left(\frac{x^2}{x^2+1} + \sqrt{9-x^2} \right) = \frac{\lim_{x\to 3^-} x^2}{\lim_{x\to 3^-} (x^2+1)} + \sqrt{\lim_{x\to 3^-} (9-x^2)}$$
$$= \frac{9}{9+1} + \sqrt{0} = \frac{9}{10}.$$

Note that the two-sided limit at 3 is not defined because $\sqrt{9-x^2}$ is not defined when $x > 3$. ◆

Existence of Tangent Lines

Recall that the slope of the line tangent to the graph $y = f(x)$ at the point $P(a, f(a))$ is defined to be

$$m = \lim_{x\to a} \frac{f(x) - f(a)}{x - a} \tag{10}$$

provided that this (two-sided) limit exists. In this case an equation of the line tangent to the graph $y = f(x)$ at $P(a, f(a))$ is

$$y - f(a) = m(x - a).$$

If the limit in (10) does not exist, then we say that the curve $y = f(x)$ does not have a tangent line at the point P. The following example gives perhaps the simplest example of a function whose graph has a tangent line everywhere except at a single isolated point.

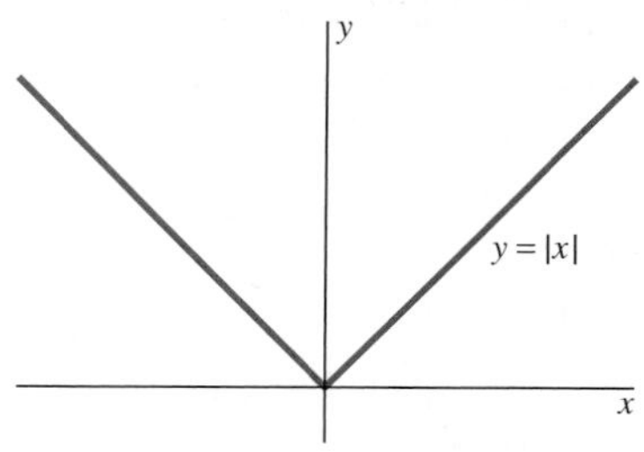

FIGURE 2.3.14 The graph of $f(x) = |x|$ has a corner point at (0, 0).

EXAMPLE 9 Show that the graph $y = |x|$ has no tangent line at the origin.

Solution Figure 2.3.14 shows the graph of the function $f(x) = |x|$. The sharp corner at the point (0, 0) makes it intuitively clear that there can be no tangent line there—surely no single straight line can be a good approximation to the shape of the graph at the origin. To verify this intuitive observation, note that when $a = 0$ we have

$$\frac{f(a+h) - f(a)}{h} = \frac{|h|}{h} = \begin{cases} -1 & \text{if } h < 0, \\ +1 & \text{if } h > 0. \end{cases}$$

Hence the left-hand limit of the quotient is -1, whereas the right-hand limit is $+1$. Therefore the two-sided limit in (10) does not exist, so the graph $y = |x|$ has no tangent line at the origin, where $a = 0$. ◆

QUESTION Does Fig. 2.3.14 make it clear to you that for $f(x) = |x|$ and $a \neq 0$, the value of the "slope limit" in (10) is given by

$$m = \begin{cases} -1 & \text{if } a < 0; \\ +1 & \text{if } a > 0? \end{cases}$$

It follows (as is apparent from Fig. 2.3.14) that the line $y = x$ is tangent to the graph $y = |x|$ at any point of the graph to the right of the origin, and that the line $y = -x$ is the tangent line at any point of the graph to the left of the origin.

Infinite Limits

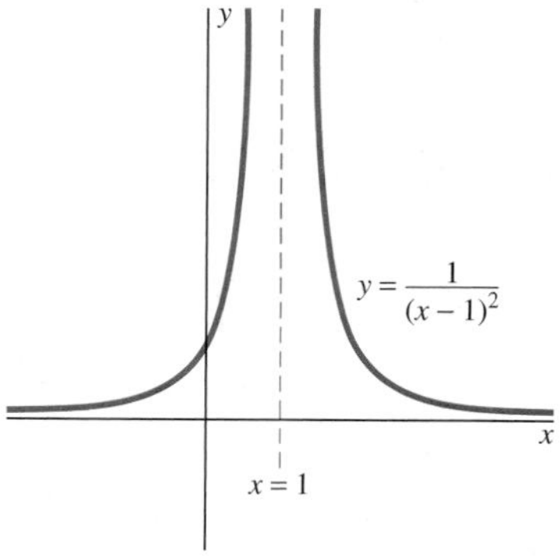

FIGURE 2.3.15 The graph of the function $f(x) = \dfrac{1}{(x-1)^2}$.

In Example 8 of Section 2.2, we investigated the function $f(x) = 1/(x-1)^2$; the graph of f is shown in Fig. 2.3.15. The value of $f(x)$ *increases without bound* (that is, eventually exceeds any preassigned number) as x approaches 1 either from the right or from the left. This situation can be described by writing

$$\lim_{x\to 1^-} \frac{1}{(x-1)^2} = +\infty = \lim_{x\to 1^+} \frac{1}{(x-1)^2}, \tag{11}$$

and we say that each of these one-sided limits is equal to "plus infinity."

CAUTION The expression

$$\lim_{x\to 1^+} \frac{1}{(x-1)^2} = +\infty \tag{12}$$

does not mean that there exists an "infinite real number" denoted by $+\infty$—there does not! Neither does it mean that the limit on the left-hand side in Eq. (12) exists—it does not! Instead, Eq. (12) is just a convenient way of saying *why* the right-hand limit in Eq. (12) does not exist: because the quantity $1/(x-2)^2$ increases without bound as $x \to 1^+$.

With similar provisos we may write

$$\lim_{x\to 1} \frac{1}{(x-1)^2} = +\infty \tag{13}$$

despite the fact that the (two-sided) limit in Eq. (13) does not exist. The expression in Eq. (13) is merely a convenient way of saying that the limit in Eq. (13) does not exist because $1/(x-1)^2$ increases without bound as $x \to 1$ from either side.

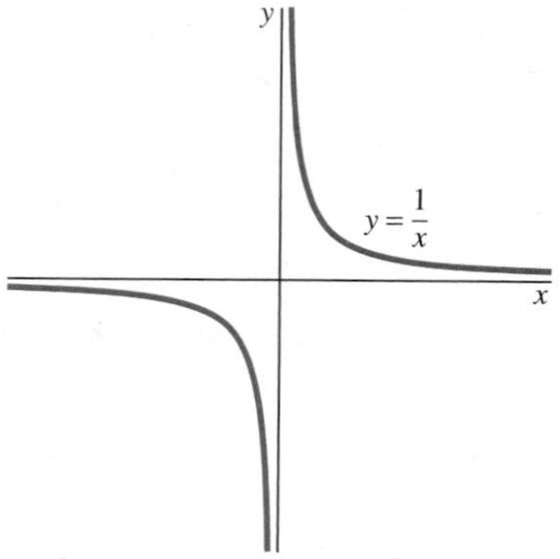

FIGURE 2.3.16 The graph of the function $f(x) = \dfrac{1}{x}$.

Now consider the function $f(x) = 1/x$; its graph is shown in Fig. 2.3.16. This function increases without bound as x approaches zero from the right but decreases without bound—it becomes less than any preassigned negative number—as x approaches zero from the left. We therefore write

$$\lim_{x\to 0^-} \frac{1}{x} = -\infty \quad \text{and} \quad \lim_{x\to 0^+} \frac{1}{x} = +\infty. \tag{14}$$

There is no shorthand for the two-sided limit in this case. We may say only that

$$\lim_{x\to 0} \frac{1}{x} \quad \text{does not exist.}$$

EXAMPLE 10 Investigate the behavior of the function

$$f(x) = \frac{2x+1}{x-1}$$

near the point $x = 1$, where the limit of $f(x)$ does not exist.

Solution First we look at the behavior of $f(x)$ just to the right of the number 1. If x is greater than 1 but close to 1, then $2x + 1$ is close to 3 and $x - 1$ is a small *positive* number. In this case the quotient $(2x + 1)/(x - 1)$ is a large positive number, and the closer x is to 1, the larger this positive quotient will be. For such x, $f(x)$ increases without bound as x approaches 1 from the right. That is,

$$\lim_{x\to 1^+} \frac{2x+1}{x-1} = +\infty, \tag{15}$$

as the data in Fig. 2.3.17 suggest.

x	$\frac{2x+1}{x-1}$	x	$\frac{2x+1}{x-1}$
1.1	32	0.9	−28
1.01	302	0.99	−298
1.001	3002	0.999	−2998
1.0001	30002	0.9999	−29998
⋮	⋮	⋮	⋮
↓	↓	↓	↓
1	$+\infty$	1	$-\infty$

FIGURE 2.3.17 The behavior of $f(x) = \dfrac{2x+1}{x-1}$ for x near 1 (Example 10).

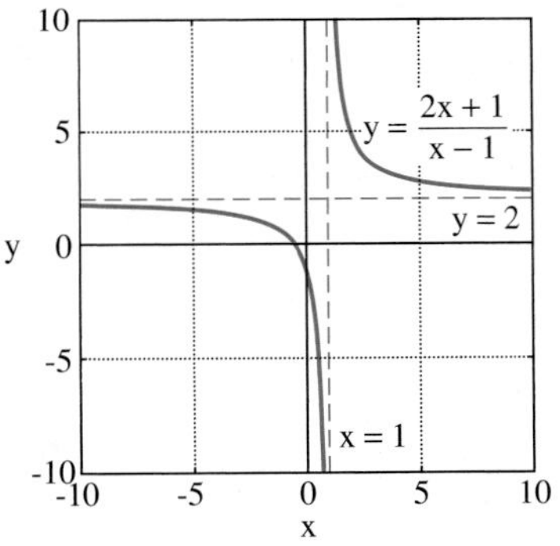

FIGURE 2.3.18 Graph of $f(x) = \dfrac{2x+1}{x-1}$.

If instead x is less than 1 but still close to 1, then $2x + 1$ is still close to 3, but now $x - 1$ is a *negative* number close to zero. In this case the quotient $(2x+1)/(x-1)$ is a (numerically) large negative number and decreases without bound as $x \to 1^-$. Hence we conclude that

$$\lim_{x\to 1^-} \frac{2x+1}{x-1} = -\infty. \tag{16}$$

The results in Eqs. (15) and (16) provide a concise description of the behavior of $f(x) = (2x + 1)/(x - 1)$ near the point $x = 1$. (See Fig. 2.3.18.) Finally, to remain consistent with Theorem 2 on one-sided and two-sided limits, we do *not* write

$$\lim_{x\to 1} \frac{2x+1}{x-1} = \infty. \qquad \textbf{(Wrong!)}$$

Do you see, however, that it would be correct to write

$$\lim_{x\to 1} \left| \frac{2x+1}{x-1} \right| = +\infty?$$

◆

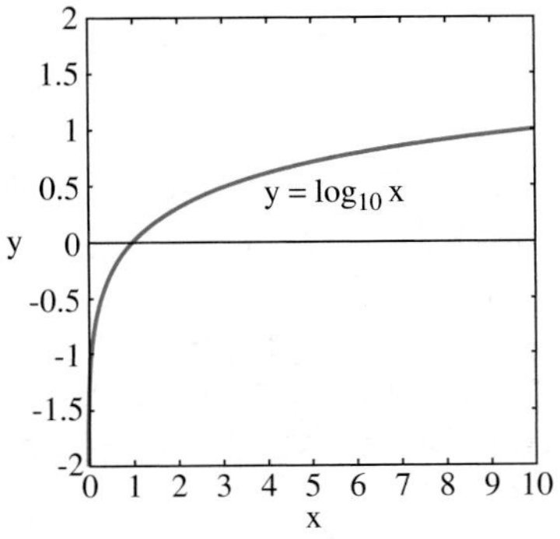

FIGURE 2.3.19 Graph of $f(x) = \log_{10} x$.

EXAMPLE 11 The graph of $f(x) = \log_{10} x$ is shown in Fig. 2.3.19. The graph makes it clear that

$$\lim_{x\to 0^+} \log_{10} x = -\infty.$$

But the left-hand limit of $f(x)$ at $x = 0$ does not exist because $\log_{10} x$ is not defined if $x \leq 0$. ◆

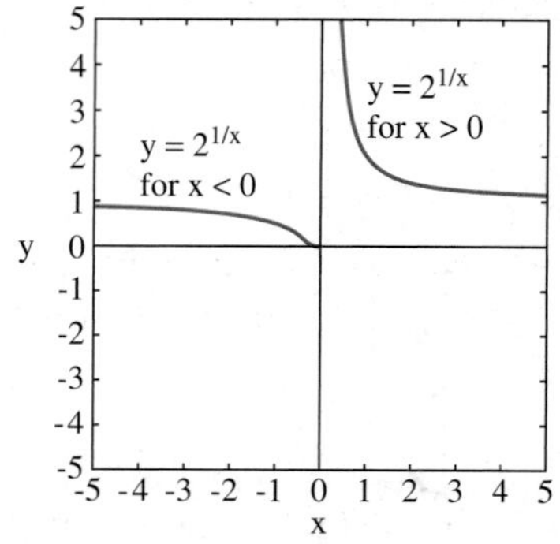

FIGURE 2.3.20 Graph of $f(x) = 2^{1/x}$.

EXAMPLE 12 Look at the graph of $y = 2^x$ in Fig. 1.4.10 to see that

$$\lim_{x \to 0^-} \frac{1}{x} = -\infty \quad \text{implies that} \quad \lim_{x \to 0^-} 2^{1/x} = 0$$

(because $2^t \to 0$ as $t \to -\infty$), whereas

$$\lim_{x \to 0^+} \frac{1}{x} = \infty \quad \text{implies that} \quad \lim_{x \to 0^+} 2^{1/x} = \infty$$

(because $2^t \to +\infty$ as $t \to +\infty$). These one-sided limits of $2^{1/x}$ at $x = 0$ are illustrated in Fig. 2.3.20. ◆

The Basic Trigonometric Limit

We now provide a geometric proof that

$$\lim_{\theta \to 0} \frac{\sin \theta}{\theta} = 1. \tag{17}$$

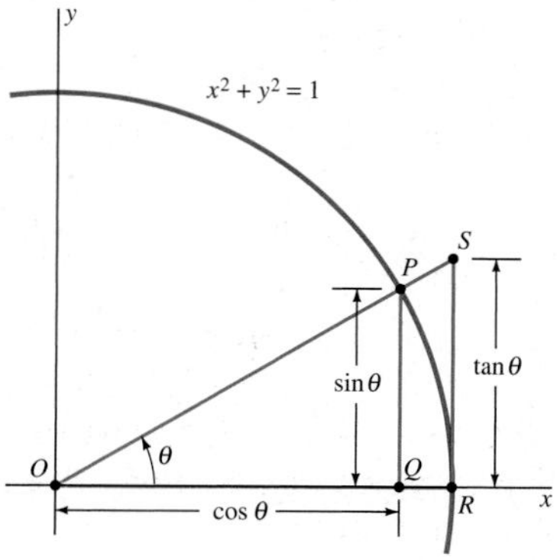

FIGURE 2.3.21 Aid to the proof of the basic trigonometric limit.

PROOF Figure 2.3.21 shows the angle θ, the triangles OPQ and ORS, and the circular sector OPR that contains the triangle OPQ and is contained in the triangle ORS. Hence

$$\text{area}(\triangle OPQ) < \text{area}(\text{sector } OPR) < \text{area}(\triangle ORS).$$

In terms of θ, this means that

$$\frac{1}{2}\sin\theta\cos\theta < \frac{1}{2}\theta < \frac{1}{2}\tan\theta = \frac{\sin\theta}{2\cos\theta}.$$

Here we use the standard formula for the area of a triangle to obtain the area of $\triangle OPQ$ and $\triangle ORS$. We also use the fact that the area of a circular sector in a circle of radius r is $A = \frac{1}{2}r^2\theta$ if the sector is subtended by a central angle of θ radians; here, $r = 1$. If $0 < \theta < \pi/2$, then we can divide each member of the last inequality by $\frac{1}{2}\sin\theta$ to obtain

$$\cos\theta < \frac{\theta}{\sin\theta} < \frac{1}{\cos\theta}.$$

We take reciprocals, which reverses the inequalities:

$$\cos\theta < \frac{\sin\theta}{\theta} < \frac{1}{\cos\theta}.$$

Now we apply the squeeze law of limits with

$$f(\theta) = \cos\theta, \qquad g(\theta) = \frac{\sin\theta}{\theta}, \qquad \text{and} \quad h(\theta) = \frac{1}{\cos\theta}.$$

Because it is clear from Eq. (1) (at the beginning of this section) that $f(\theta)$ and $h(\theta)$ both approach 1 as $\theta \to 0^+$, so does $g(\theta) = (\sin\theta)/\theta$. This geometric argument shows that $(\sin\theta)/\theta \to 1$ for *positive* values of θ that approach zero. But the same result follows for negative values of θ, because $\sin(-\theta) = -\sin\theta$. So we have proved Eq. (17). ◄

The Precise Definition of the Limit

When we say that $f(x)$ approaches the limiting value L as x approaches a, we imply that the behavior of the variable x *controls* the behavior of the value $f(x)$. As x approaches a, this *forces* the value of $f(x)$ to approach L. In Section 2.2 we said that $\lim_{x \to a} f(x) = L$ provided that we can make $f(x)$ as close to L as we please merely by choosing x sufficiently close to a (though not equal to a).

But how close is "sufficiently close"? We can say how close to L we want $f(x)$ to be by prescribing an *error tolerance.* Then the question is this: How close to a must x be in order to force the numerical difference $|f(x) - L|$—the "discrepancy" between $f(x)$ and L—to be less than the prescribed error tolerance. For instance:

- How close to a must x be to guarantee that $|f(x) - L| < 0.1$?
- How close to a must x be to guarantee that $|f(x) - L| < 0.01$?
- How close to a must x be to guarantee that $|f(x) - L| < 0.001$?

For any given error tolerance—however small it may be—we need to determine how close to a (but not equal to a) the variable x must be in order to satisfy that error tolerance.

EXAMPLE 13 Suppose that $a = 2$ and $f(x) = 5x - 3$. We could easily use the limit laws to show that $\lim_{x\to 2}(5x - 3) = 7$, so that $L = 7$. But let's instead begin afresh. We note first that

$$|f(x) - L| = |(5x - 3) - 7| = |5x - 10| = 5 \cdot |x - 2|.$$

Thus $|(5x - 3) - 7|$ is always 5 times $|x - 2|$. It follows that

- If $|x - 2| < 0.02$ then $|5x - 10| = 5 \cdot |x - 2| < 5 \cdot (0.02) = 0.1$.
- If $|x - 2| < 0.002$ then $|5x - 10| = 5 \cdot |x - 2| < 5 \cdot (0.002) = 0.01$.
- If $|x - 2| < 0.0002$ then $|5x - 10| = 5 \cdot |x - 2| < 5 \cdot (0.0002) = 0.001$.

More generally, we need only divide any given error tolerance $\epsilon > 0$ by 5 to get the "variance" in x that works:

$$\text{If} \quad |x - 2| < \frac{\epsilon}{5} \quad \text{then} \quad |(5x - 3) - 7| = 5 \cdot |x - 2| < 5 \cdot \frac{\epsilon}{5} = \epsilon. \tag{18}$$

Thus we can force $f(x) = 5x - 3$ to be within ϵ of $L = 7$ merely by requiring that x be within $\epsilon/5$ of $a = 2$. In this example it is also harmless if $x = 2$ as well—in which case $|(5x - 3) - 7| = 0$—but we include the requirement that $x \neq 2$ by writing $0 < |x - 2| < \epsilon/5$. Finally, if we write $\delta = \epsilon/5$ for this variance in x that forces an acceptable discrepancy in $f(x) = 5x - 3$, we conclude from (18) that

$$|(5x - 3) - 7| < \epsilon \quad \text{for all } x \text{ such that} \quad 0 < |x - 2| < \delta. \tag{19}$$

◆

The exact meaning of limits was debated vigorously—sometimes acrimoniously—during the 17th and 18th centuries. The condition in (19) illustrates the precise definition of the limit that was finally formulated by the German mathematician Karl Weierstrass (1815–1897) and is the definition accepted to this day.

DEFINITION The Limit

Suppose that $f(x)$ is defined in an open interval containing the point a (except possibly not at a itself). Then we say that the number L is the **limit of** $f(x)$ **as** x **approaches** a—and we write

$$\lim_{x\to a} f(x) = L$$

—provided that the following criterion is satisfied: Given any number $\epsilon > 0$, there exists a corresponding number $\delta > 0$ such that

$$|f(x) - L| < \epsilon \quad \text{for all } x \text{ such that} \quad 0 < |x - a| < \delta. \tag{20}$$

The condition in (20) can be rewritten in the form

$$\text{If} \quad 0 < |x - a| < \delta \quad \text{then} \quad |f(x) - L| < \epsilon,$$

or even more simply in the form

$$0 < |x - a| < \delta \quad \text{implies that} \quad |f(x) - L| < \epsilon. \tag{21}$$

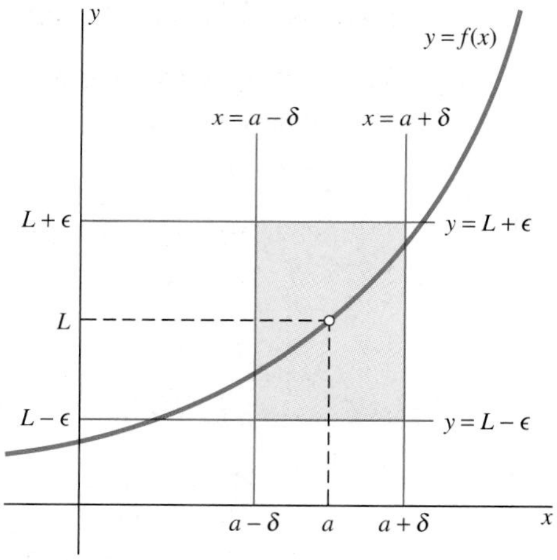

FIGURE 2.3.22 Geometric illustration of the limit definition.

Figure 2.3.22 illustrates this definition, which for obvious reasons is often called the "epsilon-delta" definition of limits. The points on the graph of $y = f(x)$ that satisfy the inequality $|f(x) - L| < \epsilon$ are those that lie between the horizontal lines $y = L - \epsilon$ and $y = L + \epsilon$. The points on this graph that satisfy the inequality $|x - a| < \delta$ are those that lie between the vertical lines $x = a - \delta$ and $x = a + \delta$. Consequently, the definition of the limit implies that $\lim_{x \to a} f(x) = L$ if and only if the following statement is true:

> Suppose that the two horizontal lines $y = L - \epsilon$ and $y = L + \epsilon$ (with $\epsilon > 0$) are given. Then it is possible to choose two vertical lines $x = a - \delta$ and $x = a + \delta$ (with $\delta > 0$) so that every point (with $x \neq a$) on the graph of $y = f(x)$ that lies between the two vertical lines must also lie between the two horizontal lines.

Figure 2.3.22 suggests that the closer together are the two horizontal lines, the closer together the two vertical lines will need to be. This is precisely what we mean by "making $f(x)$ closer to L by making x closer to a."

Application of the epsilon-delta definition of limits to establish a limit is usually a two-step process:

- Given $\epsilon > 0$, we first analyze the first inequality $|f(x) - L| < \epsilon$ in (20) to *estimate* or *deduce* a value of $\delta > 0$ that works.
- Then we attempt to *prove* that this value of δ works—that is, prove that $0 < |x - a| < \delta$ implies that $|f(x) - L| < \epsilon$.

EXAMPLE 14 Use the epsilon-delta definition of limits to prove that

$$\lim_{x \to 3} (13x - 29) = 10.$$

Solution Our analysis of the first inequality in (20) consists of noting that it takes the form

$$|(13x - 29) - 10| = |13x - 39| = 13 \cdot |x - 3| < \epsilon,$$

which boils down to $|x - 3| < \epsilon/13$. This leads us to *guess*—on the basis of rather strong circumstantial evidence—that the value $\delta = \epsilon/13$ will work. To *prove* this, we need only note that if $\delta = \epsilon/13$, then

$$0 < |x - 3| < \delta \quad \text{implies that} \quad |(13x - 29) - 10| = 13 \cdot |x - 3| < 13 \cdot \frac{\epsilon}{13} = \epsilon.$$

Thus $0 < |x - 3| < \delta$ implies that $|(13x - 29) - 10| < \epsilon$, as desired. ◆

EXAMPLE 15 Use the epsilon-delta definition of limits to prove that $\lim_{x\to 0} \sqrt[3]{x} = 0$.

Solution Our analysis of the first inequality in (20) consists of noting that it takes the form

$$\left|\sqrt[3]{x} - 0\right| = \left|\sqrt[3]{x}\right| = \sqrt[3]{|x|} < \epsilon,$$

which can be simplified to $|x| < \epsilon^3$. This leads us to *guess* that the value $\delta = \epsilon^3$ will work. To *prove* this, we need only note that if $\delta = \epsilon^3$, then

$$0 < |x - 0| < \delta \quad \text{implies that} \quad \left|\sqrt[3]{x} - 0\right| = \sqrt[3]{|x|} < \sqrt[3]{\epsilon^3} = \epsilon.$$

Thus $0 < |x - 0| < \delta$ implies that $|\sqrt[3]{x} - 0| < \epsilon$, as desired. ◆

Given a value of $\epsilon > 0$, it is frequently more difficult to *guess* a value of δ that works than to *prove* that it does; see Problems 75–84 and this section's project for additional practice. In Appendix D we use the epsilon-delta definition of limits to establish rigorously the laws of limits.

2.3 TRUE/FALSE STUDY GUIDE

2.3 CONCEPTS: QUESTIONS AND DISCUSSION

1. We have interpreted the statement $\lim_{x\to a} f(x) = L$ to mean (roughly) that "$f(x)$ tends to get closer and closer to L as x gets closer and closer to a." What would be meant by the statement that "$f(x)$ gets *steadily* closer to L as x gets steadily closer to a"? State it precisely, something along the lines that "$f(x)$ is still closer to L whenever x is still closer to a" (which is still not sufficiently precise). Does this follow from the statement that $\lim_{x\to a} f(x) = L$? It may help to think about the oscillatory function of Example 4.
2. Formulate precise epsilon-delta definitions of one-sided limits, as well as an M-delta definition of the infinite limit $\lim_{x\to a} f(x) = +\infty$. The latter definition should involve the inequality $f(x) > M$; illustrate it with a figure that is similar to Fig. 2.3.22, but involves only a single horizontal line.

2.3 PROBLEMS

Find the trigonometric limits in Problems 1 through 24. If you have a graphing calculator or a computer with graphing facility, verify that graphical evidence supports your answer.

1. $\displaystyle\lim_{\theta\to 0} \frac{\theta^2}{\sin\theta}$

2. $\displaystyle\lim_{\theta\to 0} \frac{\sin^2\theta}{\theta^2}$

3. $\displaystyle\lim_{\theta\to 0} \frac{1-\cos\theta}{\theta^2}$

4. $\displaystyle\lim_{\theta\to 0} \frac{\tan\theta}{\theta}$

5. $\displaystyle\lim_{t\to 0} \frac{2t}{(\sin t) - t}$

6. $\displaystyle\lim_{\theta\to 0} \frac{\sin(2\theta^2)}{\theta^2}$

7. $\displaystyle\lim_{x\to 0} \frac{\sin 5x}{x}$

8. $\displaystyle\lim_{z\to 0} \frac{\sin 2z}{z\cos 3z}$

9. $\displaystyle\lim_{x\to 0} \frac{\sin x}{\sqrt{x}}$

10. $\displaystyle\lim_{x\to 0} \frac{1-\cos 2x}{x}$

11. $\displaystyle\lim_{x\to 0} \frac{1}{x}\sin\frac{x}{3}$

12. $\displaystyle\lim_{\theta\to 0} \frac{(\sin 3\theta)^2}{\theta^2\cos\theta}$

13. $\displaystyle\lim_{x\to 0} \frac{1-\cos x}{\sin x}$

14. $\displaystyle\lim_{x\to 0} \frac{\tan 3x}{\tan 5x}$

15. $\displaystyle\lim_{x\to 0} x\sec x\csc x$

16. $\displaystyle\lim_{\theta\to 0} \frac{\sin 2\theta}{\theta}$

17. $\displaystyle\lim_{\theta\to 0} \frac{1-\cos\theta}{\theta\sin\theta}$

18. $\displaystyle\lim_{\theta\to 0} \frac{\sin^2\theta}{\theta}$

19. $\displaystyle\lim_{z\to 0} \frac{\tan z}{\sin 2z}$

20. $\displaystyle\lim_{x\to 0} \frac{\tan 2x}{3x}$

21. $\displaystyle\lim_{x\to 0} x\cot 3x$

22. $\displaystyle\lim_{x\to 0} \frac{x-\tan x}{\sin x}$

23. $\displaystyle\lim_{t\to 0} \frac{1}{t^2}\sin^2\left(\frac{t}{2}\right)$

24. $\displaystyle\lim_{x\to 0} \frac{\sin 2x}{\sin 5x}$

Use the squeeze law of limits to find the limits in Problems 25 through 28. Also illustrate each of these limits by graphing the functions f, g, and h (in the notation of the squeeze law) on the same screen.

25. $\displaystyle\lim_{x\to 0} x^2\cos 10x$

26. $\displaystyle\lim_{x\to 0} x^2\sin\frac{1}{x^2}$

27. $\displaystyle\lim_{x\to 0} x^2\cos\frac{1}{\sqrt[3]{x}}$

28. $\displaystyle\lim_{x\to 0} \sqrt[3]{x}\sin\frac{1}{x}$

Use one-sided limit laws to find the limits in Problems 29 through 48 or to determine that they do not exist.

29. $\lim_{x\to 0^+} (3 - \sqrt{x})$

30. $\lim_{x\to 0^+} (4 + 3x^{3/2})$

31. $\lim_{x\to 1^-} \sqrt{x-1}$

32. $\lim_{x\to 4^-} \sqrt{4-x}$

33. $\lim_{x\to 2^+} \sqrt{x^2-4}$

34. $\lim_{x\to 3^+} \sqrt{9-x^2}$

35. $\lim_{x\to 5^-} \sqrt{x(5-x)}$

36. $\lim_{x\to 2^-} x\sqrt{4-x^2}$

37. $\lim_{x\to 4^+} \sqrt{\dfrac{4x}{x-4}}$

38. $\lim_{x\to -3^+} \sqrt{6-x-x^2}$

39. $\lim_{x\to 5^-} \dfrac{x-5}{|x-5|}$

40. $\lim_{x\to -4^+} \dfrac{16-x^2}{\sqrt{16-x^2}}$

41. $\lim_{x\to 3^+} \dfrac{\sqrt{x^2-6x+9}}{x-3}$

42. $\lim_{x\to 2^+} \dfrac{x-2}{x^2-5x+6}$

43. $\lim_{x\to 2^+} \dfrac{2-x}{|x-2|}$

44. $\lim_{x\to 7^-} \dfrac{7-x}{|x-7|}$

45. $\lim_{x\to 1^+} \dfrac{1-x^2}{1-x}$

46. $\lim_{x\to 0^-} \dfrac{x}{x-|x|}$

47. $\lim_{x\to 5^+} \dfrac{\sqrt{(5-x)^2}}{5-x}$

48. $\lim_{x\to -4^-} \dfrac{4+x}{\sqrt{(4+x)^2}}$

For each of the functions in Problems 49 through 58, there is exactly one point a where both the right-hand and left-hand limits of $f(x)$ fail to exist. Describe (as in Example 10) the behavior of $f(x)$ for x near a.

49. $f(x) = \dfrac{1}{x-1}$

50. $f(x) = \dfrac{2}{3-x}$

51. $f(x) = \dfrac{x-1}{x+1}$

52. $f(x) = \dfrac{2x-5}{5-x}$

53. $f(x) = \dfrac{1-x^2}{x+2}$

54. $f(x) = \dfrac{1}{(x-5)^2}$

55. $f(x) = \dfrac{|1-x|}{(1-x)^2}$

56. $f(x) = \dfrac{x+1}{x^2+6x+9}$

57. $f(x) = \dfrac{x-2}{4-x^2}$

58. $f(x) = \dfrac{x-1}{x^2-3x+2}$

In Problems 59 and 60, find the left-hand and right-hand limits of $f(x)$ at $a = 2$. Does the two-sided limit of f exist there? Sketch the graph of $y = f(x)$.

59. $f(x) = \dfrac{x^2-4}{|x-2|}$

60. $f(x) = \dfrac{x^4-8x+16}{|x-2|}$

In Problems 61 through 68, do the following:

(a) Sketch the graph of the given function f.
(b) For each integer n, evaluate the one-sided limits

$$\lim_{x\to n^-} f(x) \quad \text{and} \quad \lim_{x\to n^+} f(x)$$

in terms of n.
(c) Determine those values of a for which $\lim_{x\to a} f(x)$ exists.

Recall that $[\![x]\!]$ denote the greatest integer that does not exceed x.

61. $f(x) = \begin{cases} 2 & \text{if } x \text{ is not an integer;} \\ 2+(-1)^x & \text{if } x \text{ is an integer.} \end{cases}$

62. $f(x) = \begin{cases} x & \text{if } x \text{ is not an integer;} \\ 0 & \text{if } x \text{ is an integer.} \end{cases}$

63. $f(x) = [\![10x]\!]$

64. $f(x) = (-1)^{[\![x]\!]}$

65. $f(x) = x - [\![x]\!] - \frac{1}{2}$

66. $f(x) = \left[\!\!\left[\dfrac{x}{2}\right]\!\!\right]$

67. $f(x) = [\![x]\!] + [\![-x]\!]$

68. $f(x) = \begin{cases} \dfrac{[\![x]\!]}{x} & \text{if } x \neq 0; \\ 0 & \text{if } x = 0. \end{cases}$

69. If $g(x) = \frac{1}{10}[\![10x]\!]$, the value of x to one decimal place *rounded down,* sketch the graph of g and determine the values of a such that $\lim_{x\to a} g(x)$ exists.

70. The **sign function** $\operatorname{sgn}(x)$ is defined as follows:

$$\operatorname{sgn}(x) = \begin{cases} \dfrac{x}{|x|} & \text{if } x \neq 0; \\ 0 & \text{if } x = 0. \end{cases}$$

Use the sign function to define two functions f and g whose limits as $x \to 0$ do not exist, but such that

(a) $\lim_{x\to 0} [f(x) + g(x)]$ does exist;
(b) $\lim_{x\to 0} f(x) \cdot g(x)$ does exist.

71. Let

$$f(x) = \begin{cases} x^2 & \text{if } x \text{ is rational;} \\ 0 & \text{if } x \text{ is irrational.} \end{cases}$$

Use the squeeze law of limits to show that $\lim_{x\to 0} f(x) = f(0) = 0$.

72. Sketch the graph of the function

$$f(x) = \frac{1}{1+2^{1/x}}$$

for $x \neq 0$. Then determine whether or not $\lim_{x\to 0} f(x)$ exists.

In Problems 73 and 74, first examine the value of $f(x)$ on intervals of the form

$$\frac{1}{n+1} < x < \frac{1}{n}$$

where n is an integer. Then determine whether or not $\lim_{x\to 0} f(x)$ exists. If your graphing calculator or computer has a greatest integer (or "floor") function, graph f to corroborate your answer.

73. $f(x) = x \cdot \left[\!\!\left[\dfrac{1}{x}\right]\!\!\right]$

74. $f(x) = x^2 \cdot \left[\!\!\left[\dfrac{1}{x}\right]\!\!\right]$

In Problems 75 through 84, use the epsilon-delta definition of limits to prove the given equation.

75. $\lim_{x\to -3} (7x - 9) = -30$

76. $\lim_{x\to 5} (17x - 35) = 50$

77. $\lim_{x\to 0^+} \sqrt{x} = 0$ *Suggestion:* First formulate a precise epsilon-delta definition of right-hand limits.

78. $\lim_{x\to 0} x^2 = 0$

79. $\lim_{x\to 2} x^2 = 4$ *Suggestion:* Note that

$$|x^2 - 4| = |x+2| \cdot |x-2|.$$

Then argue that if we agree to choose $\delta < 1$, then $|x-2| < \delta$ will imply that $|x+2| < 3$. (Why?) Then show that it works to choose δ to be the smaller of the two numbers 1 and $\epsilon/3$.

80. $\lim_{x\to 7} (x^2 - 5x - 4) = 10$ *Suggestion:* Note that

$$|(x^2 - 5x - 4) - 10| = |x+2| \cdot |x-7|.$$

Then argue that if we agree to choose $\delta < 1$, then $|x-7| < \delta$ will imply that $|x+2| < 9$. (Why?)

81. $\lim_{x\to 10} (2x^2 - 13x - 25) = 45$ *Suggestion:* Write

$$|(2x^2 - 13x - 25) - 45| = |2x+7| \cdot |x-10|.$$

Then argue that if we agree to choose $\delta < 1$, then $|x-10| < \delta$ will imply that $|2x+7| < 29$. (Why?)

82. $\lim_{x\to 2} x^3 = 8$ *Suggestion:* First verify that

$$|x^3 - 8| = |x^2 + 2x + 4| \cdot |x-2|.$$

Then argue that if we agree to choose $\delta < 1$, then $|x-2| < \delta$ will imply that $|x^2 + 2x + 4| < 19$. (Why?)

83. Generalize the approach of Problem 79 to prove that $\lim_{x\to a} x^2 = a^2$.

84. Generalize the approach of Problem 82 to prove that $\lim_{x\to a} x^3 = a^3$.

2.3 Project: Numerical Epsilon-Delta Limit Investigations

Figure 2.3.23 shows a steadily rising graph $y = f(x)$ that passes through the point (a, L). Given a single numerical value of $\epsilon > 0$, we can illustrate the limit $\lim_{x\to a} f(x) = L$ by solving the equations $f(x) = L \pm \epsilon$ graphically or numerically for the indicated values x_1 to the left of a such that $f(x_1) = L - \epsilon$ and x_2 to the right of a such that $f(x_2) = L + \epsilon$. If $\delta > 0$ is chosen smaller than either of the two indicated distances $\delta_1 = a - x_1$ and $\delta_2 = x_2 - a$, then the figure suggests that

$$0 < |x - a| < \delta \quad \text{implies that} \quad |f(x) - L| < \epsilon. \tag{21}$$

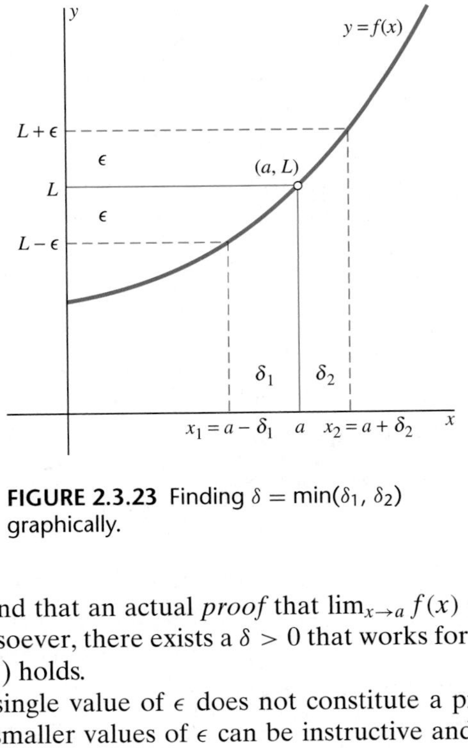

FIGURE 2.3.23 Finding $\delta = \min(\delta_1, \delta_2)$ graphically.

You should understand that an actual *proof* that $\lim_{x\to a} f(x) = L$ must show that, given *any* $\epsilon > 0$ whatsoever, there exists a $\delta > 0$ that works for *this* ϵ—meaning that the implication in (21) holds.

Doing it for a single value of ϵ does not constitute a proof, but doing it for several successively smaller values of ϵ can be instructive and perhaps convincing. Suppose, for example, that

$$f(x) = x^3 + 5x^2 + 10x + 98, \qquad a = 3, \quad \text{and} \quad L = 200.$$

Then, for a particular fixed value of $\epsilon > 0$, you can use a calculator or computer algebra system to solve the equations

$$x^3 + 5x^2 + 10x + 98 = 200 - \epsilon \quad \text{and} \quad x^3 + 5x^2 + 10x + 98 = 200 + \epsilon$$

numerically for the solutions x_1 and x_2 near 3. With $\epsilon = 1$, $\epsilon = 0.2$, and $\epsilon = 0.04$ you should obtain the following results.

ϵ	x_1	x_2	δ_1	δ_2	δ
1	2.98503	3.01488	0.01497	0.01488	0.01
0.2	2.99701	3.00298	0.00299	0.00298	0.002
0.04	2.99940	3.00060	0.00060	0.00060	0.0005

In the final column, each value of δ is (for safety) chosen a bit smaller than either δ_1 or δ_2, to be sure that it works with the corresponding value of ϵ. You might try a still smaller value such as $\epsilon = 0.001$ to find a corresponding value of δ that works. Then carry out a similar investigation to "verify" numerically a polynomial limit of your own selection.

2.4 THE CONCEPT OF CONTINUITY

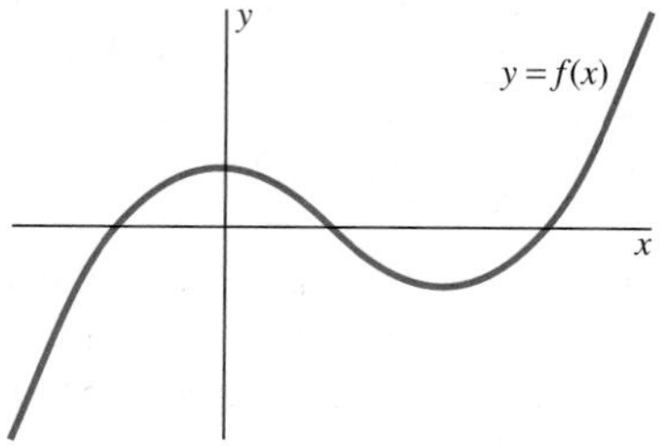

FIGURE 2.4.1 A continuous graph.

Anyone can see a drastic difference between the graphs in Figs. 2.4.1 and 2.4.2. Figure 2.4.1 is intended to suggest that the graph $y = f(x)$ can be traced with a continuous motion—without any jumps—of the pen from left to right. But in Fig. 2.4.2 the pen must make a sudden jump at $x = a$.

The concept of continuity isolates the property that the function f of Fig. 2.4.1 possesses but that the function g of Fig. 2.4.2 lacks. We first define *continuity* of a function at a single point.

DEFINITION Continuity at a Point

Suppose that the function f is defined in a neighborhood of a. We say that f is **continuous at** a provided that $\lim_{x \to a} f(x)$ exists and, moreover, that the value of this limit is $f(a)$. In other words, f is continuous at a provided that

$$\lim_{x \to a} f(x) = f(a). \qquad \textbf{(1)}$$

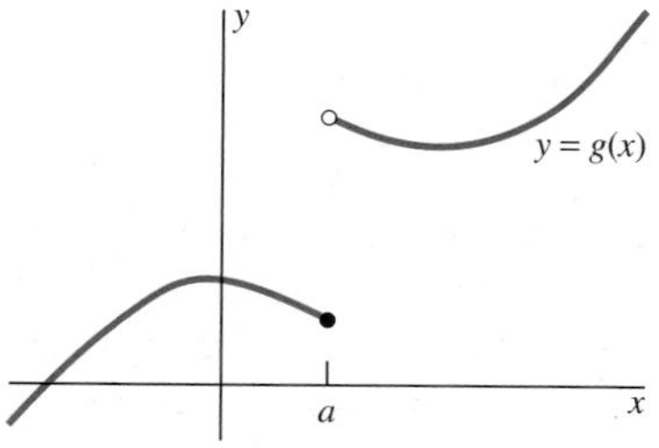

FIGURE 2.4.2 A graph that is not continuous.

Briefly, continuity of f at a means this:

The limit of f at a is equal to the value of f there.

Another way to put it is this: The limit of f at a is the "expected" value—the value that you would assign if you knew the values of f for $x \neq a$ in a neighborhood of a and you knew f to be "predictable." Alternatively, continuity of f at a means this: When x is close to a, $f(x)$ is close to $f(a)$.

Analysis of the definition of continuity shows us that to be continuous at the point a, the function f must satisfy the following three conditions:

1. The function f must be defined at a [so that $f(a)$ exists].
2. The limit of $f(x)$ as x approaches a must exist.
3. The numbers in conditions 1 and 2 must be equal:

$$\lim_{x \to a} f(x) = f(a).$$

If any one of these conditions is not satisfied, then f is not continuous at a. Examples 1 through 3 illustrate these three possibilities for *discontinuity* at a point. If the function f is *not* continuous at a, then we say that it is **discontinuous** there, or that a is a **discontinuity** of f. Intuitively, a discontinuity of f is a point where the graph of f has a "gap," or "jump," of some sort.

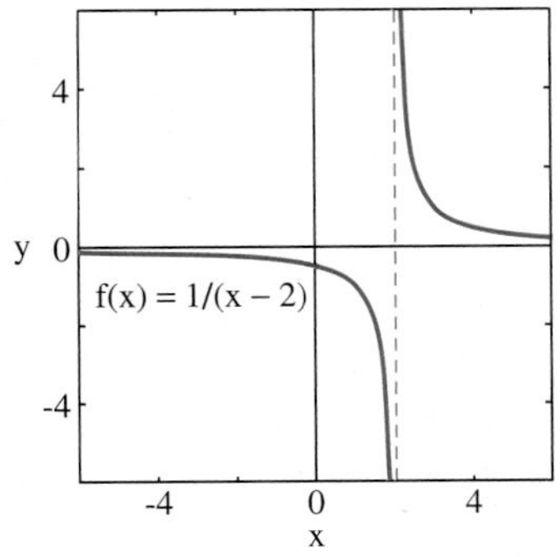

FIGURE 2.4.3 The function $f(x) = 1/(x-2)$ has an infinite discontinuity at $x = 2$ (Example 1).

EXAMPLE 1 Figure 2.4.3 shows the graph of the function f defined by

$$f(x) = \frac{1}{x-2} \quad \text{for } x \neq 2.$$

Because f is not defined at the point $x = 2$, it is not continuous there. Moreover, f has what might be called an *infinite discontinuity* at $x = 2$. ◆

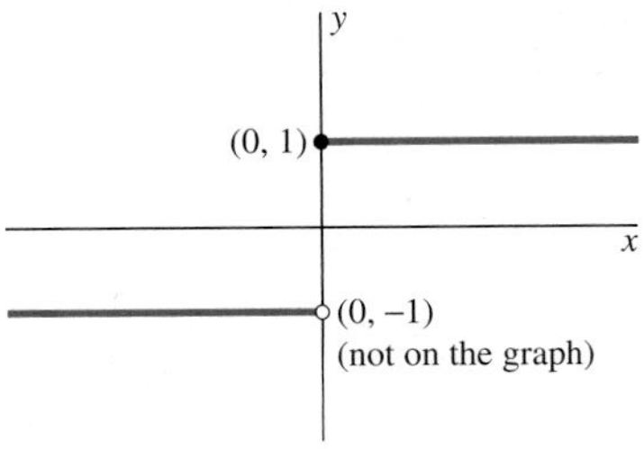

FIGURE 2.4.4 The function g has a finite jump discontinuity at $x = 0$ (Example 2).

EXAMPLE 2 Figure 2.4.4 shows the graph of the function g defined by

$$g(x) = \text{sgn}(x) = \begin{cases} +1 & \text{if } x \geqq 0; \\ -1 & \text{if } x < 0. \end{cases}$$

Its left-hand and right-hand limits at $x = 0$ are unequal, so $g(x)$ has no limit as $x \to 0$. Consequently, the function g is not continuous at $x = 0$; it has what might be called a *finite jump discontinuity* there. ◆

EXAMPLE 3 Figure 2.4.5 shows the graph of the function h defined by

$$h(x) = \begin{cases} \dfrac{\sin x}{x} & \text{if } x \neq 0; \\ 0 & \text{if } x = 0. \end{cases}$$

Because we saw in Section 2.3 that

$$\lim_{x\to 0} h(x) = \lim_{x\to 0} \frac{\sin x}{x} = 1,$$

whereas $h(0) = 0$, we see that the limit and the value of h at $x = 0$ are not equal. Thus the function h is not continuous there. As x moves from negative values through $x = 0$ to positive values, the value of $h(x)$ jumps from "near 1" to zero and back again. ◆

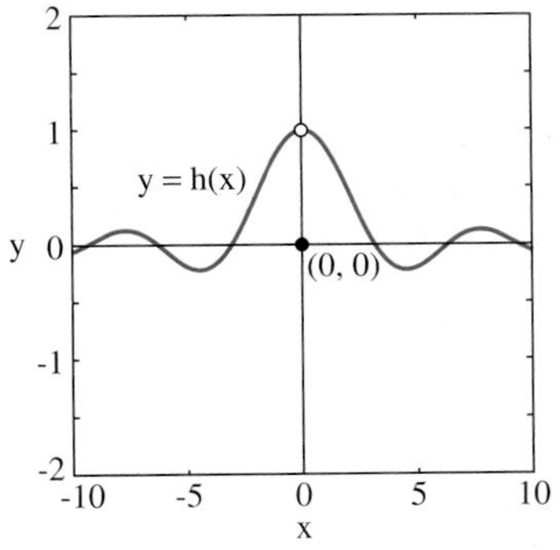

FIGURE 2.4.5 The point (0, 0) is on the graph; the point (0, 1) is not (Example 3).

The discontinuity at the origin in Example 3 is an example of a *removable* discontinuity. The point a where the function f is discontinuous is called a **removable discontinuity** provided that there exists a function F such that

- $F(x) = f(x)$ for all $x \neq a$ in the domain of definition of f, and
- This new function F is continuous at a.

The original function f may or may not be defined at a, but in any event the graphs of f and F differ only at $x = a$. Sometimes it is simpler to speak of "old" and "new" versions of the same function f. Thus we might say that a removable discontinuity of a function is one that can be removed by suitable definition—or, if necessary, redefinition—of the function at that single point.

REMARK The discontinuity at the origin of the function h in Example 3 is removable. The reason is that if we change the original value $h(0) = 0$ to $h(0) = 1$, then

$$\lim_{x\to 0} h(x) = \lim_{h\to 0} \frac{\sin x}{x} = 1 = h(0),$$

so h is now continuous at $x = 0$. By contrast, the discontinuities in the sawtooth function f of the next example are not removable, because we see genuine jumps or gaps in the graph that obviously cannot be removed simply by changing the values of f at these discontinuities.

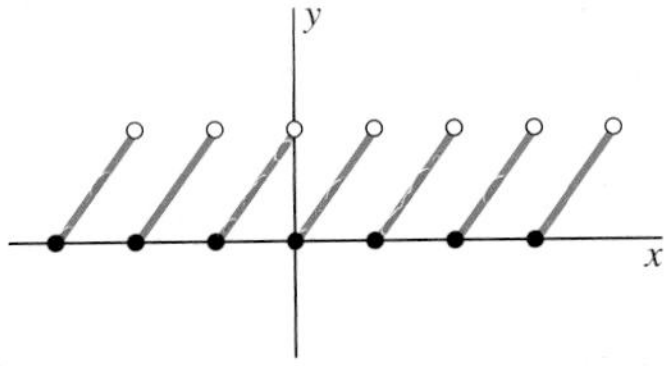

FIGURE 2.4.6 The "sawtooth function" of Example 4.

EXAMPLE 4 Figure 2.4.6 shows the graph of the function f defined by

$$f(x) = x - [\![x]\!].$$

As before, $[\![x]\!]$ denotes the largest integer no greater than x. If $x = n$, an integer, then $[\![n]\!] = n$, so $f(n) = 0$. On the open interval $(n, n+1)$, the graph of f is linear and has slope 1. It should be clear that f is

- Continuous at x if x is *not* an integer;
- Discontinuous at each integer point on the x-axis. ◆

Combinations of Continuous Functions

Frequently we are most interested in functions that *are* continuous. Suppose that the function f is defined on an open interval or a union of open intervals. Then we say simply that f is **continuous** if it is continuous at each point of its domain of definition.

It follows readily from the limit laws in Section 2.2 that *any constant multiple, sum, difference, or product of continuous functions is continuous.* That is, if c is a constant and the functions f and g are continuous at a, then so are the functions

$$cf, \qquad f+g, \qquad f-g, \qquad \text{and} \quad f \cdot g.$$

For instance, if f and g are continuous at a, then

$$\lim_{x\to a}[f(x)+g(x)] = \left(\lim_{x\to a} f(x)\right) + \left(\lim_{x\to a} g(x)\right) = f(a)+g(a),$$

so it follows that the sum $f+g$ is also continuous at a.

EXAMPLE 5 Because $f(x) = x$ and constant-valued functions are clearly continuous everywhere, it follows that the cubic polynomial function

$$f(x) = x^3 - 3x^2 + 1 = x \cdot x \cdot x + (-3) \cdot x \cdot x + 1$$

is continuous everywhere. ◆

More generally, it follows in a similar way that *every* **polynomial function**

$$p(x) = b_n x^n + b_{n-1}x^{n-1} + \cdots + b_1 x + b_0$$

is continuous at each point of the real line. In short, every polynomial is continuous everywhere.

If $p(x)$ and $q(x)$ are polynomials, then the quotient law for limits and the continuity of polynomials imply that

$$\lim_{x\to a} \frac{p(x)}{q(x)} = \frac{\lim_{x\to a} p(x)}{\lim_{x\to a} q(x)} = \frac{p(a)}{q(a)}$$

provided that $q(a) \neq 0$. Thus every **rational function**

$$f(x) = \frac{p(x)}{q(x)} \tag{2}$$

is continuous wherever it is defined—that is, wherever the denominator polynomial is nonzero. More generally, *the quotient of any two continuous functions is continuous at every point where the denominator is nonzero.*

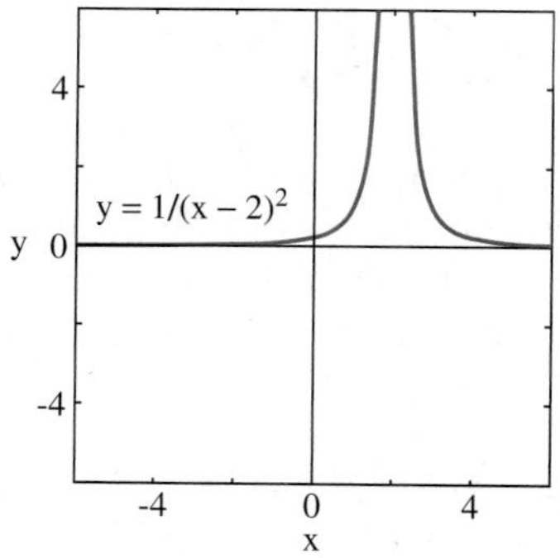

FIGURE 2.4.7 The function $f(x) = 1/(x-2)^2$ has an infinite discontinuity at $x = 2$.

At a point $x = a$ where the denominator in Eq. (2) is zero, $q(a) = 0$, there are two possibilities:

- If $p(a) \neq 0$, then f has an infinite discontinuity (as in Figs. 2.4.3 and 2.4.7) at $x = a$.
- Otherwise, f *may* have a removable discontinuity at $x = a$.

EXAMPLE 6 Suppose that

$$f(x) = \frac{x-2}{x^2 - 3x + 2}. \tag{3}$$

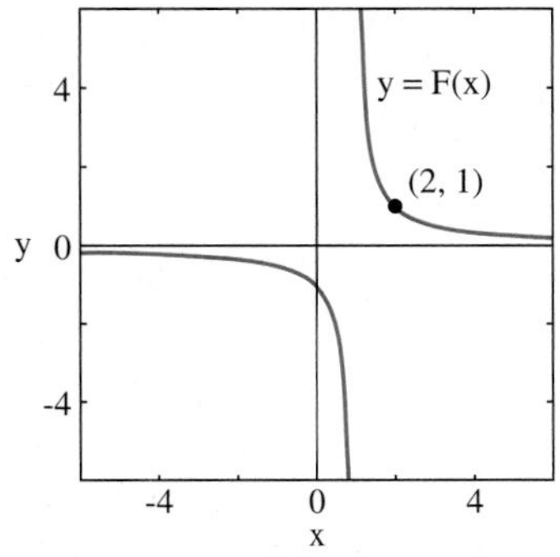

FIGURE 2.4.8 In Example 6, the graph $y = F(x)$ consists of the graph $y = f(x)$ with the single point (2,1) adjoined.

We factor the denominator: $x^2 - 3x + 2 = (x - 1)(x - 2)$. This shows that f is not defined at $x = 1$ and at $x = 2$. Thus the rational function defined in Eq. (3) is continuous except at these two points. Because cancellation gives

$$f(x) = \frac{x-2}{x^2 - 3x + 2} = \frac{1}{x-1}$$

except at the single point $x = 2$, the new function

$$F(x) = \frac{1}{x-1} \tag{4}$$

agrees with $f(x)$ if $x \neq 2$ but is continuous at $x = 2$ also, where $F(2) = 1$. Thus f has a removable discontinuity at $x = 2$; the discontinuity at $x = 1$ is not removable. (See Fig. 2.4.8.) ◆

Continuity of Trigonometric Functions

At the beginning of Section 2.3 we noted that

$$\lim_{x \to 0} \cos x = 1 \quad \text{and} \quad \lim_{x \to 0} \sin x = 0. \tag{5}$$

Because $\cos 0 = 1$ and $\sin 0 = 0$, the sine and cosine functions are continuous at $x = 0$ by definition. But this fact implies that they are continuous everywhere.

THEOREM 1 Continuity of Sine and Cosine
The functions $f(x) = \sin x$ and $g(x) = \cos x$ are continuous functions of x on the whole real line.

PROOF We give the proof only for $\sin x$; the proof for $\cos x$ is similar. (See Problem 67.) We want to show that $\lim_{x \to a} \sin x = \sin a$ for every real number a. If we write $x = a + h$, so that $h = x - a$, then $h \to 0$ as $x \to a$. Thus we need only show that

$$\lim_{h \to 0} \sin(a + h) = \sin a.$$

But the addition formula for the sine function yields

$$\begin{aligned} \lim_{h \to 0} \sin(a + h) &= \lim_{h \to 0} (\sin a \cos h + \cos a \sin h) \\ &= (\sin a)\left(\lim_{h \to 0} \cos h\right) + (\cos a)\left(\lim_{h \to 0} \sin h\right) = \sin a \end{aligned}$$

as desired; we used the limits in Eq. (5) in the last step. ◀

REMARK It now follows that the function

$$\tan x = \frac{\sin x}{\cos x} \tag{6}$$

is continuous except where $\cos x = 0$—that is, except when x is an *odd* integral multiple of $\pi/2$. As illustrated in Fig. 2.4.9, $\tan x$ has an infinite discontinuity at each such point.

Composition of Continuous Functions

Recall from Section 1.4 that the **composition** of the two functions f and g is the function $h = f \circ g$ defined by

$$h(x) = f(g(x))$$

for all x in the domain of g such that $u = g(x)$ is in the domain of f. Theorem 2 implies that functions built by forming compositions of *continuous* functions are themselves continuous.

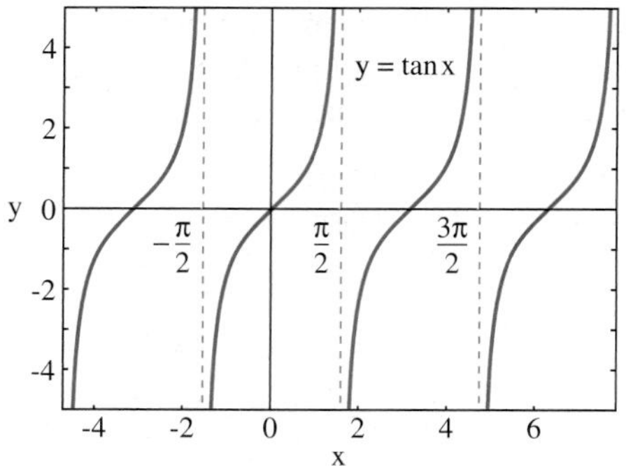

FIGURE 2.4.9 The function tan x has infinite discontinuities at $x = \pm\pi/2$, $\pm 3\pi/2, \ldots$.

THEOREM 2 Continuity of Compositions
The composition of two continuous functions is continuous. More precisely, if g is continuous at a and f is continuous at $g(a)$, then $f \circ g$ is continuous at a.

PROOF The continuity of g at a means that $g(x) \to g(a)$ as $x \to a$, and the continuity of f at $g(a)$ implies that $f(x) \to f(g(a))$ as $x \to g(a)$. Hence the substitution law for limits (Section 2.2) yields

$$\lim_{x \to a} f(g(x)) = f\left(\lim_{x \to a} g(x)\right) = f(g(a)),$$

as desired. ◄

Recall from the root law in Section 2.2 that

$$\lim_{x \to a} \sqrt[n]{x} = \sqrt[n]{a}$$

under the conditions that n is an integer and that $a > 0$ if n is even.Thus the nth-root function $f(x) = \sqrt[n]{x}$ is continuous everywhere if n is odd; f is continuous for $x > 0$ if n is even.

We may combine this result with Theorem 2. Then we see that *a root of a continuous function is continuous wherever it is defined.* That is, the composition

$$h(x) = \sqrt[n]{g(x)} = [g(x)]^{1/n}$$

of $f(x) = \sqrt[n]{x}$ and the function $g(x)$ is continuous at a if g is, assuming that $g(a) > 0$ if n is even (so that $\sqrt[n]{g(a)}$ is defined).

EXAMPLE 7 Show that the function

$$f(x) = \left(\frac{x - 7}{x^2 + 2x + 2}\right)^{2/3}$$

is continuous on the whole real line.

Solution Note first that the denominator

$$x^2 + 2x + 2 = (x + 1)^2 + 1$$

is never zero, because its smallest value (when $x = -1$) is $0^2 + 1 = 1$. Hence the rational function

$$r(x) = \frac{x - 7}{x^2 + 2x + 2}$$

is defined and continuous everywhere. It then follows from Theorem 2 and the continuity of the cube root function that

$$f(x) = [r(x)]^{2/3} = \sqrt[3]{[r(x)]^2}$$

is continuous everywhere—as suggested by its graph in Fig. 2.4.10, where we see the high point $(-1, 4)$ corresponding to the minimum value (at $x = -1$) of the denominator of $r(x)$. ◆

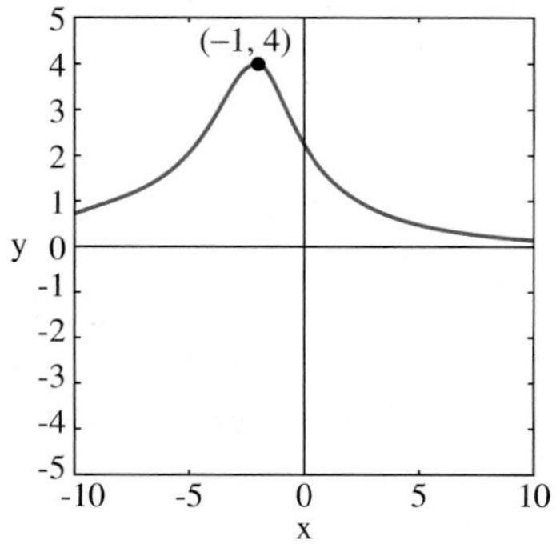

FIGURE 2.4.10 The graph $y = \left(\dfrac{x-7}{x^2+2x+2}\right)^{2/3}$.

EXAMPLE 8 (a) The exponential function $f(x) = 2^x$ is continuous everywhere, and therefore so is the composition $h(x) = 2^{\sin x}$ of f and the sine function. Refer to Fig. 2.4.11, where we see high and low points on the graph of $y = 2^{\sin x}$ corresponding to the high and low points on the graph of $y = \sin x$. (b) By contrast, the tangent function $\tan x$ has infinite discontinuities at odd integral multiples of $\pi/2$ (as shown in Fig. 2.4.9), and we see corresponding discontinuities in the composition $h(x) = 2^{\tan x}$ when we look at the graph in Fig. 2.4.12. These discontinuities are interesting in that, if a is an odd integral multiple of $\pi/2$, then

$$\lim_{x \to a^-} h(x) = \lim_{x \to a^-} 2^{\tan x} = +\infty,$$

whereas

$$\lim_{x \to a^+} h(x) = \lim_{x \to a^+} 2^{\tan x} = 0 = h(a).$$ ◆

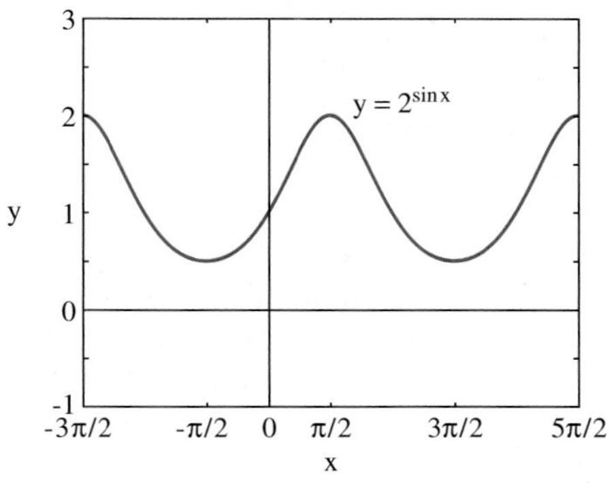

FIGURE 2.4.11 The function $h(x) = 2^{\sin x}$ is continuous everywhere.

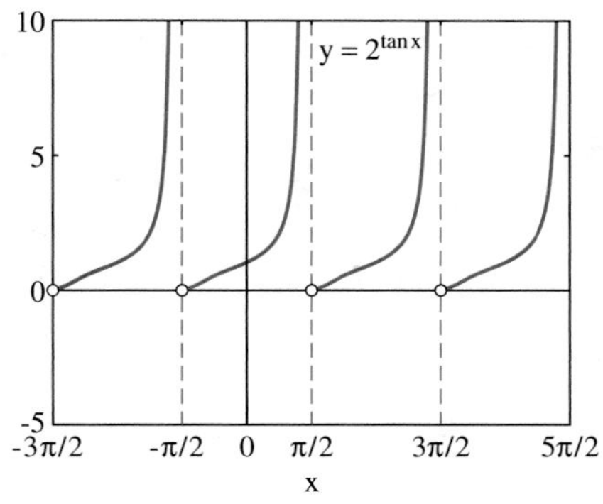

FIGURE 2.4.12 The function $h(x) = 2^{\tan x}$ has infinite discontinuities.

The function $h(x) = 2^{\tan x}$ of Example 8(b) illustrates the concept of *one-sided continuity.* It is convenient to say that the function f is

- **continuous from the left at** a if $\lim_{x \to a^-} = f(a)$, and is
- **continuous from the right at** a if $\lim_{x \to a^+} = f(a)$.

Thus $h(x) = 2^{\tan x}$ is continuous from the right at a if a is an odd integral multiple of $\pi/2$, but is *not* continuous from the left at a. Of course, a function that is continuous at a is automatically continuous at a both from the left and from the right.

Continuous Functions on Closed Intervals

An applied problem typically involves a function whose domain is a *closed interval.* For example, in the animal pen problem of Section 1.1, we found that the area A of the rectangular pen in Fig. 2.4.13 was expressed as a function of its base length x by

$$A = f(x) = \frac{3}{5}x(30 - x).$$

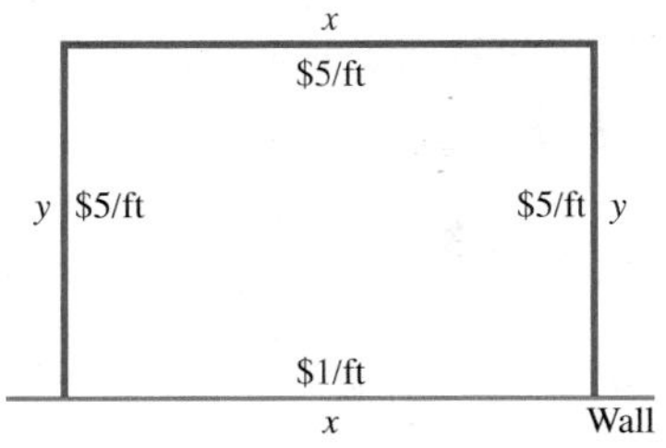

FIGURE 2.4.13 The animal pen.

Although this formula for f is meaningful for all x, only values in the closed interval $[0, 30]$ correspond to actual rectangles, so only such values are pertinent to the animal pen problem.

The function f defined on the closed interval $[a, b]$ is said to be **continuous on** $[a, b]$ provided that

- f is continuous at each point of the open integral (a, b),
- f is continuous from the right at the left-hand endpoint a, and
- f is continuous from the left at the right-hand endpoint b.

The last two conditions imply that, at each endpoint, the value of the function is equal to its limit from *within* the interval. For instance, every polynomial is continuous on every closed interval. The square root function $f(x) = \sqrt{x}$ is continuous from the right at 0 because $\lim_{x\to 0^+} \sqrt{x} = 0 = \sqrt{0}$. Therefore f is continuous on the closed interval $[0, 1]$ even though f is not defined for $x < 0$.

Continuous functions defined on closed intervals have very special properties. For example, every such function has the *intermediate value property* of Theorem 3. (A proof of this theorem is given in Appendix E.) We suggested earlier that continuity of a function is related to the possibility of tracing its graph without lifting the pen from the paper. Theorem 3, the *intermediate value theorem,* expresses this fact with precision.

THEOREM 3 Intermediate Value Property
Suppose that the function f is continuous on the closed interval $[a, b]$. Then $f(x)$ assumes every intermediate value between $f(a)$ and $f(b)$. That is, if K is any number between $f(a)$ and $f(b)$, then there exists at least one number c in (a, b) such that $f(c) = K$.

Figure 2.4.14 shows the graph of a typical continuous function f whose domain is the closed interval $[a, b]$. The number K is located on the y-axis, somewhere between $f(a)$ and $f(b)$. In the figure $f(a) < f(b)$, but this is not important. The horizontal line through K must cross the graph of f somewhere, and the x-coordinate of the point where graph and line meet yields the value of c. The number c is the one whose existence is guaranteed by the intermediate value property of the continuous function f.

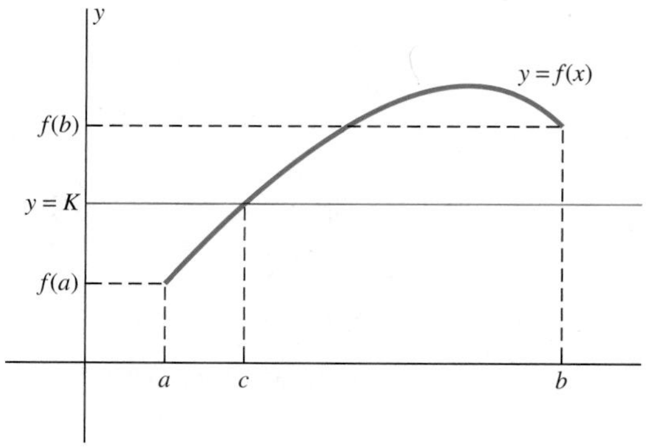

FIGURE 2.4.14 The continuous function f attains the intermediate value K at $x = c$.

Thus the intermediate value theorem implies that each horizontal line meeting the y-axis between $f(a)$ and $f(b)$ must cross the graph of the continuous function f somewhere. This is a way of saying that the graph has no gaps or jumps, suggesting that the idea of being able to trace such a graph without lifting the pen from the paper is accurate.

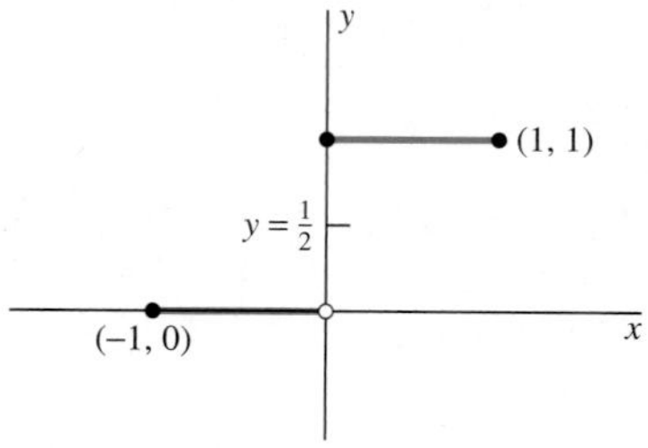

FIGURE 2.4.15 This discontinuous function does not have the intermediate value property (Example 9).

EXAMPLE 9 The discontinuous function defined on $[-1, 1]$ as

$$f(x) = \begin{cases} 0 & \text{if } x < 0, \\ 1 & \text{if } x \geqq 0 \end{cases}$$

does *not* attain the intermediate value $\frac{1}{2}$. See Fig. 2.4.15. ◆

Existence of Solutions of Equations

An important application of the intermediate value theorem is the verification of the existence of solutions of equations written in the form

$$f(x) = 0. \tag{7}$$

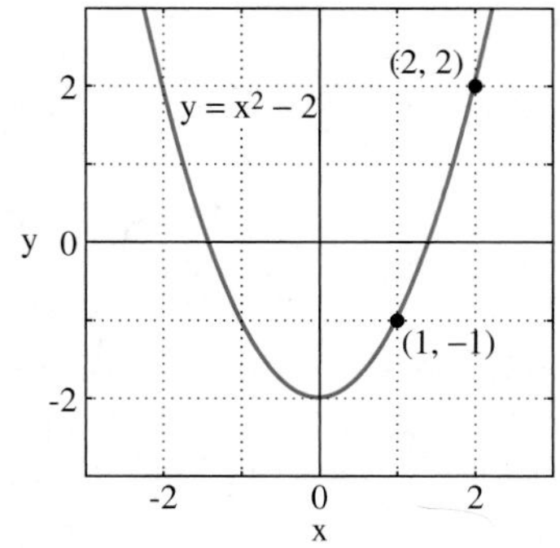

FIGURE 2.4.16 The graph of $f(x) = x^2 - 2$ (Example 10).

EXAMPLE 10 You could attempt to approximate the number $\sqrt{2}$ graphically by zooming in on the intersection of the parabola $y = x^2 - 2$ with the positive x-axis (Fig. 2.4.16). The x-coordinate of the intersection yields the positive solution of the equation

$$f(x) = x^2 - 2 = 0. \tag{8}$$

Perhaps it makes no sense to zoom in on this point unless we know that it's "really there." But we can see from Eq. (8) that

$$f(1) = -1 < 0, \qquad \text{whereas} \quad f(2) = 2 > 0.$$

We note that the function f is continuous on $[1, 2]$ (it is continuous everywhere) and that $K = 0$ is an intermediate value of f on the interval $[1, 2]$. Therefore, it follows from Theorem 3 that $f(c) = c^2 - 2 = 0$ for some number c in $(1, 2)$—that is, that

$$c^2 = 2.$$

This number c is the desired square root of 2. Thus it is the intermediate value property of continuous functions that guarantees the existence of the number $\sqrt{2}$: There *is* a real number whose square is 2. ◆

As indicated in Fig. 2.4.17, the solutions of Eq. (7) are simply the points where the graph $y = f(x)$ crosses the x-axis. Suppose that f is continuous and that we can find a closed interval $[a, b]$ (such as the interval $[1, 2]$ of Example 10) such that the value of f is positive at one endpoint of $[a, b]$ and negative at the other. That is, suppose that $f(x)$ *changes sign* on the closed interval $[a, b]$. Then the intermediate value property ensures that $f(x) = 0$ at some point of $[a, b]$.

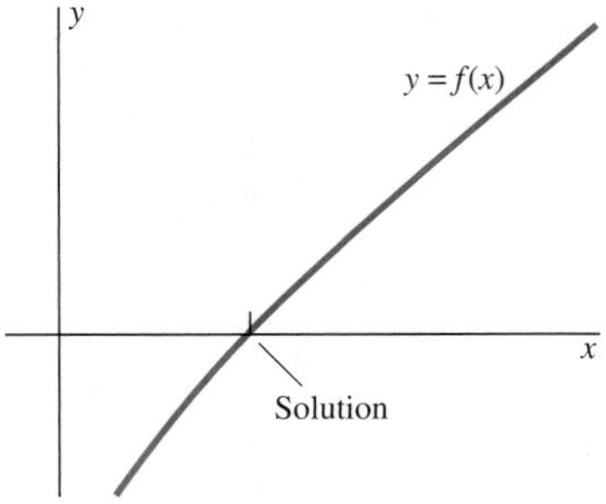

FIGURE 2.4.17 The solution of the equation $f(x) = 0$.

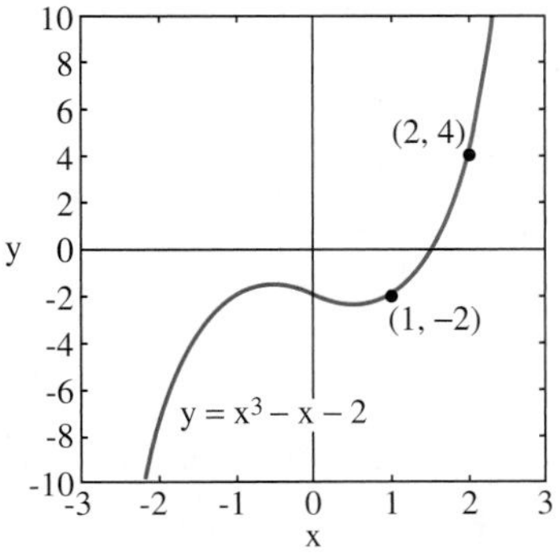

FIGURE 2.4.18 The equation $x^3 - x - 2 = 0$ of Example 11 appears to have a solution somewhere between $x = 1$ and $x = 2$.

EXAMPLE 11 The graph $y = x^3 - x - 2$ shown in Fig. 2.4.18 indicates that the equation

$$f(x) = x^3 - x - 2 = 0$$

has a solution somewhere between $x = 1$ and $x = 2$. Apply the intermediate value theorem to show that this actually is so.

Solution The function $f(x)$ is continuous on $[1, 2]$ because it is a polynomial and, therefore, is continuous everywhere. Because $f(1) = -2$ and $f(2) = +4$, the intermediate value theorem implies that every number between -2 and $+4$ is a value of $f(x)$ on $[1, 2]$. In particular,

$$-2 = f(1) < 0 < f(2) = +4,$$

so the intermediate value property of f implies that f attains the value 0 at some number c between $x = 1$ and $x = 2$. That is,

$$f(c) = c^3 - c - 2 = 0,$$

so $x = c$ is a solution in $(1, 2)$ of the equation $x^3 - x - 2 = 0$. ◆

The following example shows that not every suspected root of an equation $f(x) = 0$ that seems to be visible on a computer-plotted figure is actually there. Indeed, a graphing calculator or computer ordinarily is programmed to plot close but isolated points on the desired graph $y = f(x)$ and then join these points with line segments so short that the result looks like a smooth curve. In effect, the computer is *assuming* that the function f is continuous, whether or not it actually is continuous.

EXAMPLE 12 Figure 2.4.19 shows a computer plot of the graph of the function

$$f(x) = \frac{10 \cdot [\![1000x]\!] - 4995}{10000}.$$

The graph $y = f(x)$ appears indistinguishable from the line $y = x - \frac{1}{2}$, and in particular it appears that the equation $f(x) = 0$ has the solution $x = \frac{1}{2}$. But when we zoom in near this alleged solution we see the graph shown in Fig. 2.4.20. Now we see that the function f is discontinuous, and actually "jumps" across the x-axis without intersecting it. Thus the equation $f(x) = 0$ has no solution at all. ◆

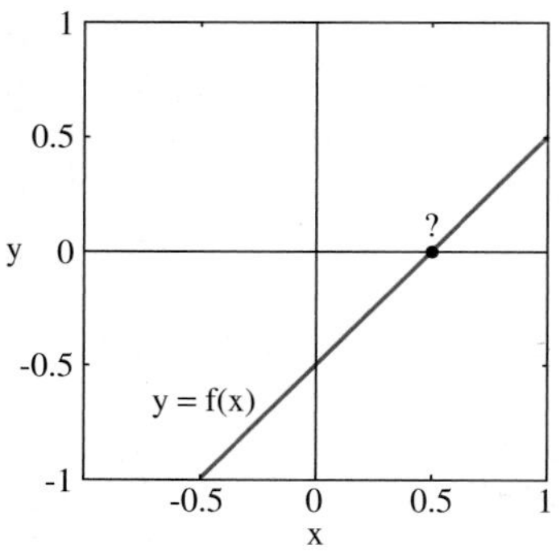

FIGURE 2.4.19 The graph $y = f(x)$ of Example 12 appears to have x-intercept $x = 0.5$.

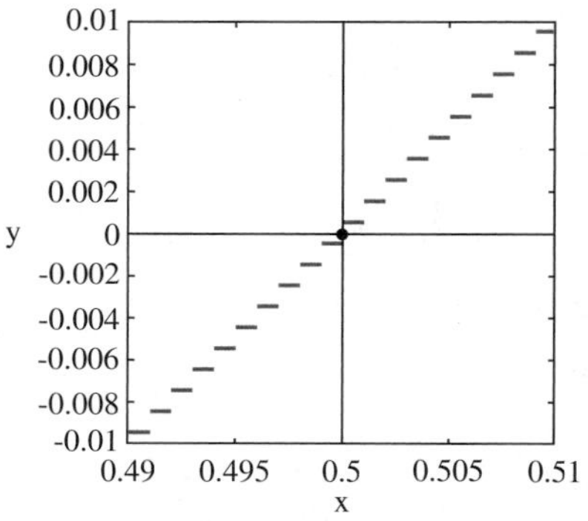

FIGURE 2.4.20 The graph in Example 12 jumps across the x-axis—there is no x-intercept.

2.4 TRUE/FALSE STUDY GUIDE

2.4 CONCEPTS: QUESTIONS AND DISCUSSION

1. Suppose that $a < b < c$. If the function f is continuous both on the closed interval $[a, b]$ and on the closed interval $[b, c]$, does it follow that f is continuous on $[a, c]$? If f is continuous on the closed interval $[n, n + 1]$ for every integer n, does it follow that f is continuous on the entire real line?
2. Suppose that the function f is continuous everywhere and that the composition $f(g(x))$ is continuous at $x = a$. Does it follow that $g(x)$ is continuous at a? *Suggestion:* Consider the possibility that $f(x) = |x|$.
3. Suppose that $p(x)$ is a polynomial of odd degree with positive leading coefficient. Then its graph $y = p(x)$ "heads southwest in the third quadrant" and "northeast in the first quadrant." Can you state more precisely what these

intuitive statements mean? Why does it follow that the equation $p(x) = 0$ always has at least one solution?

4. If $10^L = y$, then we call the number L the *base* 10 *logarithm* of y and write $L = \log y$. Assume that the exponential function $f(x) = 10^x$ is continuous everywhere (it is!) and—as suggested by its graph in Fig. 1.4.10—it has both arbitrarily large positive values and values arbitrarily close to zero. Then explain why the intermediate value theorem implies that every positive number has a base 10 logarithm.

2.4 PROBLEMS

In Problems 1 through 8, apply the limit laws and the theorems of this section to show that the given function is continuous for all x.

1. $f(x) = 2x^5 - 7x^2 + 13$

2. $f(x) = 7x^3 - (2x+1)^5$

3. $g(x) = \dfrac{2x-1}{4x^2+1}$

4. $g(x) = \dfrac{x^3}{x^2+2x+5}$

5. $h(x) = \sqrt{x^2+4x+5}$

6. $h(x) = \sqrt[3]{1-5x}$

7. $f(x) = \dfrac{1-\sin x}{1+\cos^2 x}$

8. $g(x) = \sqrt[4]{1-\sin^2 x}$

In Problems 9 through 14, apply the limit laws and the theorems of this section to show that the given function is continuous on the indicated interval.

9. $f(x) = \dfrac{1}{x+1}, \quad x > -1$

10. $f(x) = \dfrac{x-1}{x^2-4}, \quad -2 < x < 2$

11. $g(t) = \sqrt{9-4t^2}, \quad -\frac{3}{2} \leqq t \leqq \frac{3}{2}$

12. $h(z) = \sqrt{(z-1)(3-z)}, \quad 1 \leqq z \leqq 3$

13. $f(x) = \dfrac{x}{\cos x}, \quad -\frac{1}{2}\pi < x < \frac{1}{2}\pi$

14. $g(t) = \sqrt{1-2\sin t}, \quad -\frac{1}{6}\pi < t < \frac{1}{6}\pi$

In Problems 15 through 36, tell where the given function is continuous. Recall that when the domain of a function is not specified, it is the set of all real numbers for which the formula of the function is meaningful.

15. $f(x) = 2x + \sqrt[3]{x}$

16. $g(x) = x^2 + \dfrac{1}{x}$

17. $f(x) = \dfrac{1}{x+3}$

18. $f(t) = \dfrac{5}{5-t}$

19. $f(x) = \dfrac{1}{x^2+1}$

20. $g(z) = \dfrac{1}{z^2-1}$

21. $f(x) = \dfrac{x-5}{|x-5|}$

22. $h(x) = \dfrac{x^2+x+1}{x^2+1}$

23. $f(x) = \dfrac{x^2+4}{x-2}$

24. $f(t) = \sqrt[4]{4+t^4}$

25. $f(x) = \sqrt[3]{\dfrac{x+1}{x-1}}$

26. $F(u) = \sqrt[3]{3-u^3}$

27. $f(x) = \dfrac{3}{x^2-x}$

28. $f(z) = \sqrt{9-z^2}$

29. $f(x) = \dfrac{x}{\sqrt{4-x^2}}$

30. $f(x) = \sqrt{\dfrac{1-x^2}{4-x^2}}$

31. $f(x) = \dfrac{\sin x}{x^2}$

32. $g(\theta) = \dfrac{\theta}{\cos\theta}$

33. $f(x) = \dfrac{1}{\sin 2x}$

34. $f(x) = \sqrt{\sin x}$

35. $f(x) = \sin|x|$

36. $G(u) = \dfrac{1}{\sqrt{1+\cos u}}$

In Problems 37 through 48, find the points where the given function is not defined and is therefore not continuous. For each such point a, tell whether or not this discontinuity is removable.

37. $f(x) = \dfrac{x}{(x+3)^3}$

38. $f(t) = \dfrac{t}{t^2-1}$

39. $f(x) = \dfrac{x-2}{x^2-4}$

40. $G(u) = \dfrac{u+1}{u^2-u-6}$

41. $f(x) = \dfrac{1}{1-|x|}$

42. $h(x) = \dfrac{|x-1|}{(x-1)^3}$

43. $f(x) = \dfrac{x-17}{|x-17|}$

44. $g(x) = \dfrac{x^2+5x+6}{x+2}$

45. $f(x) = \begin{cases} -x & \text{if } x < 0; \\ x^2 & \text{if } x > 0 \end{cases}$

46. $f(x) = \begin{cases} x+1 & \text{if } x < 1; \\ 3-x & \text{if } x > 1 \end{cases}$

47. $f(x) = \begin{cases} 1+x^2 & \text{if } x < 0; \\ \dfrac{\sin x}{x} & \text{if } x > 0 \end{cases}$

48. $f(x) = \begin{cases} \dfrac{1-\cos x}{x} & \text{if } x < 0; \\ x^2 & \text{if } x > 0 \end{cases}$

In Problems 49 through 52, find a value of the constant c so that the function f(x) is continuous for all x.

49. $f(x) = \begin{cases} x+c & \text{if } x < 0, \\ 4-x^2 & \text{if } x \geqq 0 \end{cases}$

50. $f(x) = \begin{cases} 2x+c & \text{if } x \leqq 3, \\ 2c-x & \text{if } x > 3 \end{cases}$

51. $f(x) = \begin{cases} c^2 - x^2 & \text{if } x < 0, \\ 2(x-c)^2 & \text{if } x \geqq 0 \end{cases}$

52. $f(x) = \begin{cases} c^3 - x^3 & \text{if } x \leqq \pi, \\ c \sin x & \text{if } x > \pi \end{cases}$

In Problems 53 through 58, apply the intermediate value property of continuous functions to show that the given equation has a solution in the given interval.

53. $x^2 - 5 = 0$ on $[2, 3]$

54. $x^3 + x + 1 = 0$ on $[-1, 0]$

55. $x^3 - 3x^2 + 1 = 0$ on $[0, 1]$

56. $x^3 = 5$ on $[1, 2]$

57. $x^4 + 2x - 1 = 0$ on $[0, 1]$

58. $x^5 - 5x^3 + 3 = 0$ on $[-3, -2]$

In Problems 59 and 60, show that the given equation has three distinct roots by calculating the values of the left-hand side at $x = -3, -3, -1, 0, 1, 2,$ and 3 and then applying the intermediate value property of continuous functions on appropriate closed intervals.

59. $x^3 - 4x + 1 = 0$ **60.** $x^3 - 3x^2 + 1 = 0$

61. Suppose that you accept a job now (time $t = 0$) at an annual salary of \$25, 000 and are promised a 6% raise at the end of each year of employment. Explain why your salary in thousands of dollars after t years is given by the formula

$$S(t) = 25 \cdot (1.06)^{[\![t]\!]}.$$

Graph this function for the first 5 years and comment on its continuity.

62. Suppose that you accept the same job as in Problem 61, but now are promised a 1.5% raise at the end of each quarter (three months). (a) Write a formula for your salary (in thousands of dollars) after t years. (b) Graph this new salary function and comment on its continuity. (c) Which is the better deal, the promised salary of Problem 61 or the one of this problem?

63. Suppose that f and g are two functions both continuous on the interval $[a, b]$, and such that $f(a) = g(b) = p$ and $f(b) = g(a) = q$ where $p \neq q$. Sketch typical graphs of two such functions. Then apply the intermediate value theorem to the function $h(x) = f(x) - g(x)$ to show that $f(c) = g(c)$ at some point c of (a, b).

64. Suppose that today you leave your home in Estes Park, CO at 1 P.M. and drive to Grand Lake, arriving at 2 P.M. Tomorrow you leave your destination in Grand Lake at 1 P.M. and retrace the same route, arriving home at 2 P.M. Use Problem 63 as a suggestion to show that at some instant between 1 and 2 P.M. you are at precisely the same point on the road both days. What must you assume about the functions describing your location as a function of time each day?

65. Apply the intermediate value property of continuous functions to show that every positive number a has a square root. That is, given $a > 0$, prove that there exists a number r such that $r^2 = a$.

66. Apply the intermediate value property to prove that every real number has a cube root.

67. Show that the cosine function is continuous on the set of all real numbers. (*Suggestion:* Alter the proof of Theorem 1 of the continuity of the sine function.)

68. Determine where the function $f(x) = x + [\![x]\!]$ is continuous.

69. Suppose that $f(x) = 0$ if x is a rational number, whereas $f(x) = 1$ if x is irrational. Prove that f is discontinuous at every real number.

70. Suppose that $f(x) = 0$ if x is a rational number, whereas $f(x) = x^2$ if x is irrational. Prove that f is continuous only at the single point $x = 0$.

71. Figure 2.4.21 suggests that the equation $x = \cos x$ has a solution in the interval $(0, \pi/2)$. Use the intermediate value theorem to show that this is true. Then use your calculator to approximate this solution accurate to two decimal places.

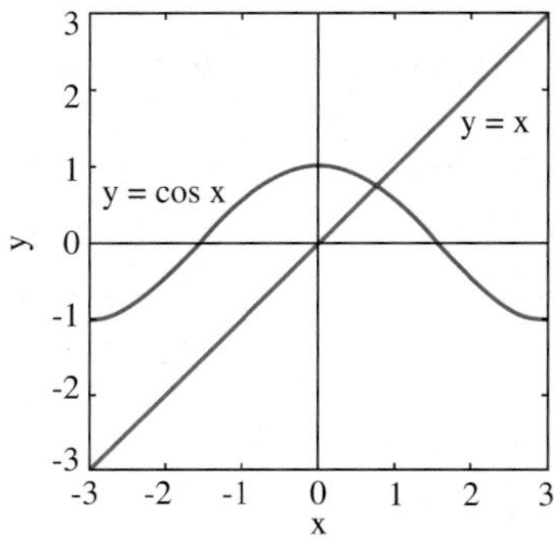

FIGURE 2.4.21 The graphs $y = x$ and $y = \cos x$ (Problem 71).

72. Figure 2.4.22 suggests that the equation $x = -5\cos x$ has at least three distinct solutions. Use the intermediate value theorem to show that this is true. Then use your calculator to approximate each of these solutions accurate to two decimal places.

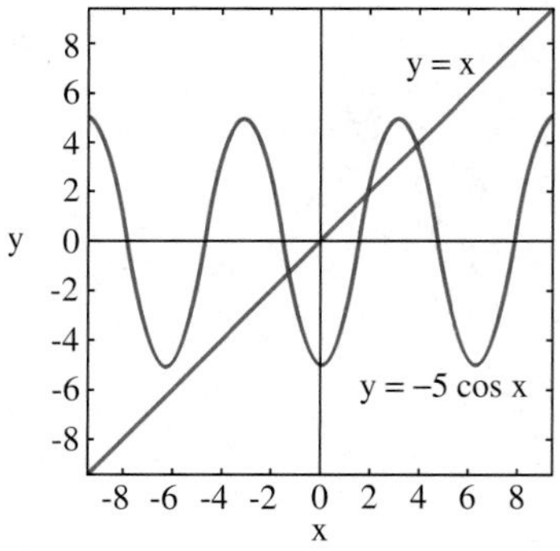

FIGURE 2.4.22 The graphs $y = x$ and $y = -5\cos x$ (Problem 72).

Investigate the continuity of each of the functions defined in Problems 73 through 78. For each discontinuity, determine whether the given function is continuous from the right and whether it is continuous from the left. Use a graphing calculator or computer if you find it helpful.

73. $f(x) = 2^{1/x}$ if $x \neq 0$; $f(0) = 0$

74. $f(x) = 2^{-1/x^2}$ if $x \neq 0$; $f(0) = 0$

75. $f(x) = \dfrac{1}{1 + 2^{1/x}}$ if $x \neq 0$; $f(0) = 1$

76. $f(x) = \dfrac{1}{1 + 2^{-1/x^2}}$ if $x \neq 0$; $f(0) = 1$

77. $f(x) = \dfrac{1}{1 + 2^{\tan x}}$ where meaningful;
$f(x) = 1$ otherwise

78. $f(x) = \dfrac{1}{1 + 2^{1/\sin x}}$ where meaningful;
$f(x) = 0$ otherwise

CHAPTER 2 REVIEW: DEFINITIONS, CONCEPTS, RESULTS

Use the following list as a guide to topics that you may need to review.

1. Slope predictors for tangent lines
2. Lines tangent to parabolas
3. Numerical investigation of slopes of lines tangent to more general curves
4. The idea of the limit of $f(x)$ as x approaches a
5. Numerical and graphical investigations of limits
6. The limit laws: constant, addition, product, quotient, root, substitution, and squeeze
7. Slope-predictor functions
8. The four-step process for finding slope-predictor functions
9. The basic trigonometric limit
10. Evaluation of limits of trigonometric functions
11. Right-hand and left-hand limits
12. The relation between one-sided and two-sided limits
13. The precise definition of the limit
14. Existence of tangent lines
15. Infinite limits
16. Continuity of a function at a point
17. Combinations of continuous functions
18. Continuity of polynomials and rational functions
19. Removable discontinuities
20. Continuity of trigonometric functions
21. Compositions of continuous functions
22. The intermediate value property of continuous functions
23. Existence of solutions of equations

CHAPTER 2 MISCELLANEOUS PROBLEMS

Apply the limit laws to evaluate the limits in Problems 1 through 40 or to show that the indicated limit does not exist, as appropriate.

1. $\displaystyle\lim_{x\to 0} (x^2 - 3x + 4)$

2. $\displaystyle\lim_{x\to -1} (3 - x + x^3)$

3. $\displaystyle\lim_{x\to 2} (4 - x^2)^{10}$

4. $\displaystyle\lim_{x\to 1} (x^2 + x - 1)^{17}$

5. $\displaystyle\lim_{x\to 2} \frac{1 + x^2}{1 - x^2}$

6. $\displaystyle\lim_{x\to 3} \frac{2x}{x^2 - x - 3}$

7. $\displaystyle\lim_{x\to 1} \frac{x^2 - 1}{1 - x}$

8. $\displaystyle\lim_{x\to -2} \frac{x + 2}{x^2 + x - 2}$

9. $\displaystyle\lim_{t\to -3} \frac{t^2 + 6t + 9}{9 - t^2}$

10. $\displaystyle\lim_{x\to 0} \frac{4x - x^3}{3x + x^2}$

11. $\displaystyle\lim_{x\to 3} (x^2 - 1)^{2/3}$

12. $\displaystyle\lim_{x\to 2} \sqrt{\frac{2x^2 + 1}{2x}}$

13. $\displaystyle\lim_{x\to 3} \left(\frac{5x + 1}{x^2 - 8}\right)^{3/4}$

14. $\displaystyle\lim_{x\to 1} \frac{x^4 - 1}{x^2 + 2x - 3}$

15. $\displaystyle\lim_{x\to 7} \frac{\sqrt{x + 2} - 3}{x - 7}$

16. $\displaystyle\lim_{x\to 1^+} \left(x - \sqrt{x^2 - 1}\right)$

17. $\displaystyle\lim_{x\to -4} \frac{\dfrac{1}{\sqrt{13 + x}} - \dfrac{1}{3}}{x + 4}$

18. $\displaystyle\lim_{x\to 1^+} \frac{1 - x}{|1 - x|}$

19. $\displaystyle\lim_{x\to 2^+} \frac{2 - x}{\sqrt{4 - 4x + x^2}}$

20. $\displaystyle\lim_{x\to -2^-} \frac{x + 2}{|x + 2|}$

21. $\displaystyle\lim_{x\to 4^+} \frac{x - 4}{|x - 4|}$

22. $\displaystyle\lim_{x\to 3^-} \sqrt{x^2 - 9}$

23. $\displaystyle\lim_{x\to 2^+} \sqrt{4 - x^2}$

24. $\displaystyle\lim_{x\to -3} \frac{x}{(x + 3)^2}$

25. $\displaystyle\lim_{x\to 2} \frac{x + 2}{(x - 2)^2}$

26. $\displaystyle\lim_{x\to 1^-} \frac{x}{x - 1}$

27. $\displaystyle\lim_{x\to 3^+} \frac{x}{x - 3}$

28. $\displaystyle\lim_{x\to 1^-} \frac{x - 2}{x^2 - 3x + 2}$

29. $\displaystyle\lim_{x\to 1^-} \frac{x + 1}{(x - 1)^3}$

30. $\displaystyle\lim_{x\to 5^+} \frac{25 - x^2}{x^2 - 10x + 25}$

31. $\displaystyle\lim_{x\to 0} \frac{\sin 3x}{x}$

32. $\displaystyle\lim_{x\to 0} \frac{\tan 5x}{x}$

33. $\displaystyle\lim_{x\to 0} \frac{\sin 3x}{\sin 2x}$

34. $\displaystyle\lim_{x\to 0} \frac{\tan 2x}{\tan 3x}$

35. $\displaystyle\lim_{x\to 0^+} \frac{x}{\sin \sqrt{x}}$

36. $\displaystyle\lim_{x\to 0} \frac{1 - \cos 3x}{2x}$

37. $\displaystyle\lim_{x\to 0} \frac{1 - \cos 3x}{2x^2}$

38. $\displaystyle\lim_{x\to 0} x^3 \cot x \csc x$

39. $\displaystyle\lim_{x\to 0} \frac{\sec 2x \tan 2x}{x}$

40. $\displaystyle\lim_{x\to 0} x^2 \cot^2 3x$

In Problems 41 through 46, apply your knowledge of lines tangent to parabolas (Section 2.1) to write a slope-predictor formula for the given curve $y = f(x)$. Then write an equation for the line tangent to $y = f(x)$ at the point $(1, f(1))$.

41. $f(x) = 3 + 2x^2$

42. $f(x) = x - 5x^2$

43. $f(x) = 3x^2 + 4x - 5$

44. $f(x) = 1 - 2x - 3x^2$

45. $f(x) = (x - 1)(2x - 1)$

46. $f(x) = \frac{x}{3} - \left(\frac{x}{4}\right)^2$

In Problems 47 through 53, use the "four-step process" of Section 2.3 to find a slope-predictor formula for the graph $y = f(x)$.

47. $f(x) = 2x^2 + 3x$

48. $f(x) = x - x^3$

49. $f(x) = \frac{1}{3 - x}$

50. $f(x) = \frac{1}{2x + 1}$

51. $f(x) = x - \frac{1}{x}$

52. $f(x) = \frac{x}{x + 1}$

53. $f(x) = \frac{x + 1}{x - 1}$

54. Find a slope-predictor formula for the graph

$$f(x) = 3x - x^2 + |2x + 3|$$

at the points where a tangent line exists. Find the point (or points) where no tangent line exists. Sketch the graph of f.

55. Write equations of the two lines through $(3, 4)$ that are tangent to the parabola $y = x^2$. (*Suggestion:* Let (a, a^2) denote either point of tangency; first solve for a.)

56. Write an equation for the circle with center $(2, 3)$ that is tangent to the line with equation $x + y + 3 = 0$.

In Problems 57 through 60, explain why each function is continuous wherever it is defined by the given formula. For each point a where f is not defined by the formula, tell whether a value can be assigned to $f(a)$ in such a way as to make f continuous at a.

57. $f(x) = \frac{1 - x}{1 - x^2}$

58. $f(x) = \frac{1 - x}{(2 - x)^2}$

59. $f(x) = \frac{x^2 + x - 2}{x^2 + 2x - 3}$

60. $f(x) = \frac{|x^2 - 1|}{x^2 - 1}$

61. Apply the intermediate value property of continuous functions to prove that the equation $x^5 + x = 1$ has a solution.

62. Apply the intermediate value property of continuous functions to prove that the equation $x^3 - 4x^2 + 1 = 0$ has three different solutions.

63. Show that there is a number x between 0 and $\pi/2$ such that $x = \cos x$.

64. Show that there is a number x between $\pi/2$ and π such that $\tan x = -x$. (*Suggestion:* First sketch the graphs of $y = \tan x$ and $y = -x$.)

65. Find how many straight lines through the point $(12, \frac{15}{2})$ are normal to the graph of $y = x^2$ and find the slope of each. (*Suggestion:* The cubic equation you should obtain has one root evident by inspection.)

66. A circle of radius r is dropped into the parabola $y = x^2$. If r is too large, the circle will not fall all the way to the bottom; if r is sufficiently small, the circle will touch the parabola at its vertex $(0, 0)$. (See Fig. 2.MP.1.) Find the largest value of r so that the circle will touch the vertex of the parabola.

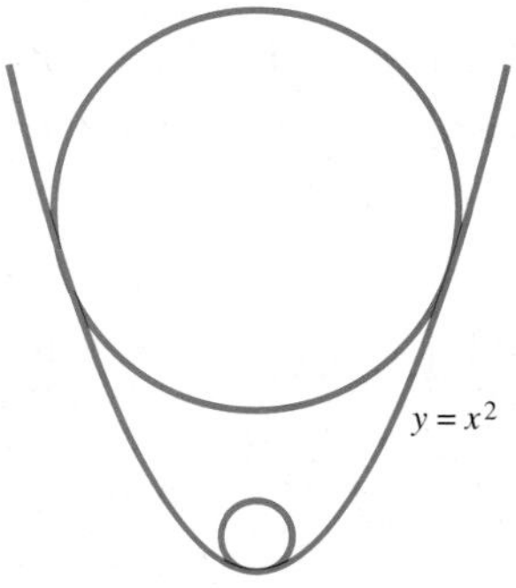

FIGURE 2.MP.1 If the circle is too large, it cannot touch the bottom of the parabola (Problem 66).

THE DERIVATIVE 3

Isaac Newton (1642–1727)

Isaac Newton was born in a rural English farming village on Christmas Day in 1642, three months after his father's death. When the boy was three, his mother remarried and left him with his grandmother. Nothing known about his childhood and early schooling hinted that his life and work would constitute a turning point in the history of humanity.

But due to the influence of an uncle who suspected hidden potential in young Isaac, Newton was able to enter Cambridge University in 1661. During the years 1665 and 1666, when Cambridge closed because of the bubonic plague then sweeping Europe, he returned to his country home and there laid the foundations for the three towering achievements of his scientific career—the invention of the calculus, the discovery of the spectrum of colors in light, and the theory of gravitation. Of these two years he later wrote that "in those days I was in the prime of my age of invention and minded mathematics and philosophy more than at any time since." Indeed, his thirties were devoted more to smoky chemical (and even alchemical) experiments than to serious mathematical investigations.

In his forties, while a mathematics professor at Cambridge, Newton wrote the *Principia Mathematica* (1687), perhaps the single most influential scientific treatise ever published. In it he applied the concepts of the calculus to explore the workings of the universe, including the motions of the earth, moon, and planets about the sun. A student is said to have remarked, "There goes the man that wrote a book that neither he nor anyone else understands." But it established for Newton such fame that upon his death in 1727 he was buried alongside his country's greats in Westminster Abbey with such pomp that the French philosopher Voltaire remarked, "I have seen a professor of mathematics . . . buried like a king who had done good to his subjects."

Shortly after his Cambridge graduation in 1665, Newton discovered a new method for solving an equation of the form $f(x) = 0$. Unlike special methods such as the quadratic formula that apply only to equations of special form, *Newton's method* can be used to approximate numerical solutions of virtually any equation. In Section 3.10 we present an iterative formulation of Newton's method that is especially adaptable to calculators and computers. There we describe how the combination of Newton's method with modern computer graphics has led to the generation of striking fractal images associated with the science of *chaos*. The pictures here result from the application of a complex-number version of Newton's method to the simple equation $x^3 + 1 = 0$.

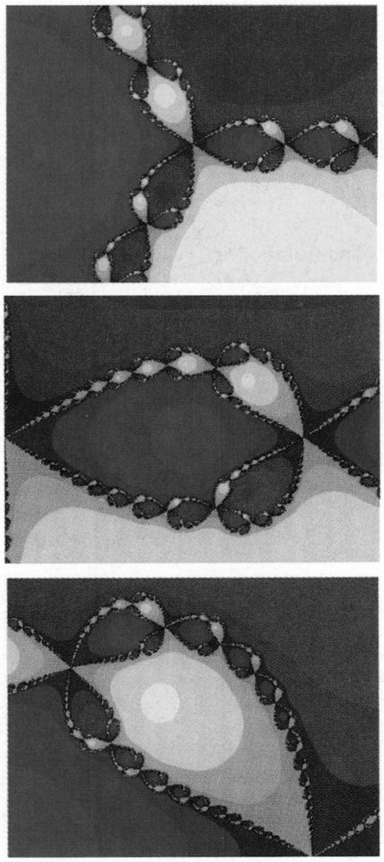

3.1 THE DERIVATIVE AND RATES OF CHANGE

In Section 2.1 we saw that the line tangent to the curve $y = f(x)$ (Fig. 3.1.1) at the point $P(a, f(a))$ has slope

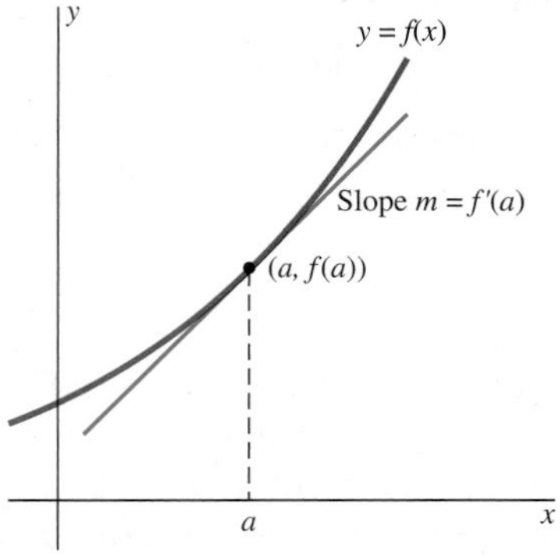

FIGURE 3.1.1 The geometric motivation for the definition of the derivative.

$$m = m(a) = \lim_{h \to 0} \frac{f(a+h) - f(a)}{h} \tag{1}$$

provided that this limit exists. As in the slope-prediction formulas of Section 2.2, we get a new function f'—the *derivative* of the original function f—when we replace the constant a in (1) with the independent variable x.

DEFINITION The Derivative

The **derivative** of the function f is the function f' defined by

$$f'(x) = \lim_{h \to 0} \frac{f(x+h) - f(x)}{h} \tag{2}$$

for all x for which this limit exists.

It is important to understand that when the limit in (2) is evaluated, we hold x fixed while h approaches zero. When we are specifically interested in the value $f'(a)$ of the derivative f' at the number $x = a$, we sometimes rewrite Eq. (2) in the form

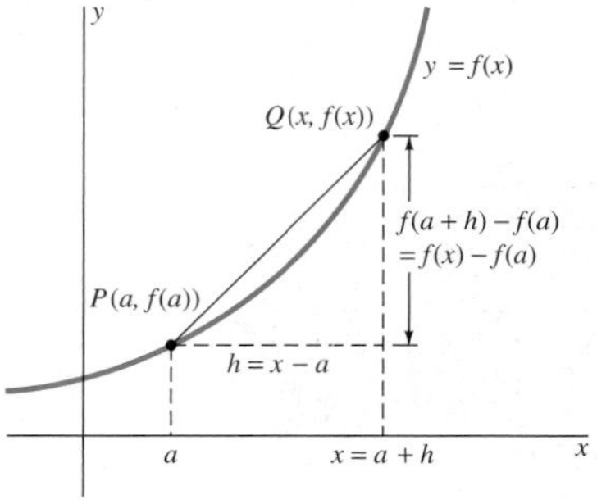

FIGURE 3.1.2 The notation in Eq. (3).

$$f'(a) = \lim_{h \to 0} \frac{f(a+h) - f(a)}{h} = \lim_{x \to a} \frac{f(x) - f(a)}{x - a}. \tag{3}$$

The second limit in Eq. (3) is obtained from the first by writing $x = a + h$, $h = x - a$, and by noting that $x \to a$ as $h \to 0$ (Fig. 3.1.2). The statement that these equivalent limits exist can be abbreviated as "$f'(a)$ exists." In this case we say that the function f is **differentiable** at $x = a$. The process of finding the derivative f' is called **differentiation** of f.

However it is found, the derivative f' is a slope predictor for lines tangent to the graph $y = f(x)$ of the original function f (Fig. 3.1.1).

The Derivative as Slope Predictor

The slope m of the line tangent to the graph $y = f(x)$ at the point $(a, f(a))$ where $x = a$ is

$$m = f'(a). \tag{4}$$

Application of the point-slope formula gives

$$y - f(a) = f'(a) \cdot (x - a) \tag{5}$$

as an equation of this tangent line.

Differentiating a given function f by direct evaluation of the limit in Eq. (3) involves carrying out four steps:

1. Write the definition in Eq. (2) of the derivative.
2. Substitute the expressions $f(x + h)$ and $f(x)$ as determined by the particular function f.
3. Simplify the result by algebraic methods until it is possible to . . .
4. Apply appropriate limit laws to finally evaluate the limit.

In Section 2.2 we used this same "four-step process" to calculate several slope-predictor functions—that is, derivatives. The limit calculations of Examples 12 and 13 in Section 2.2—where we found the derivatives of the functions

$$f(x) = x + \frac{1}{x} \quad \text{and} \quad f(x) = \sqrt{x}$$

—illustrate algebraic simplification techniques that frequently are useful in the evaluation of derivatives directly from the definition in Eq. (2).

EXAMPLE 1 First apply the definition of the derivative directly to differentiate the function

$$f(x) = \frac{x}{x+3}.$$

Then find the line tangent to the graph of f at the origin, where $f(0) = 0$.

Solution Steps 1 and 2 above give

$$f'(x) = \lim_{h\to 0} \frac{f(x+h) - f(x)}{h} = \lim_{h\to 0} \frac{\dfrac{x+h}{(x+h)+3} - \dfrac{x}{x+3}}{h}.$$

Then an algebraic simplification suggested by the common-denominator calculation

$$\frac{\dfrac{a}{b} - \dfrac{c}{d}}{h} = \frac{\dfrac{ad-bc}{bd}}{h} = \frac{ad-bc}{hbd}$$

yields

$$\begin{aligned} f'(x) &= \lim_{h\to 0} \frac{(x+h)(x+3) - x(x+h+3)}{h(x+h+3)(x+3)} \\ &= \lim_{h\to 0} \frac{3h}{h(x+h+3)(x+3)} = \lim_{h\to 0} \frac{3}{(x+h+3)(x+3)} \\ &= \frac{3}{\left(\lim\limits_{h\to 0}(x+h+3)\right)\left(\lim\limits_{h\to 0}(x+3)\right)}. \end{aligned}$$

We therefore find finally that

$$f'(x) = \frac{3}{(x+3)(x+3)} = \frac{3}{(x+3)^2}.$$

Substituting $a = 0$, $f(0) = 0$, and $f'(0) = \frac{1}{3}$ in Eq. (5) gives the equation $y = \frac{1}{3}x$ of the line tangent to the graph $y = x/(x+3)$ at the origin (0, 0) (Fig. 3.1.3). ◆

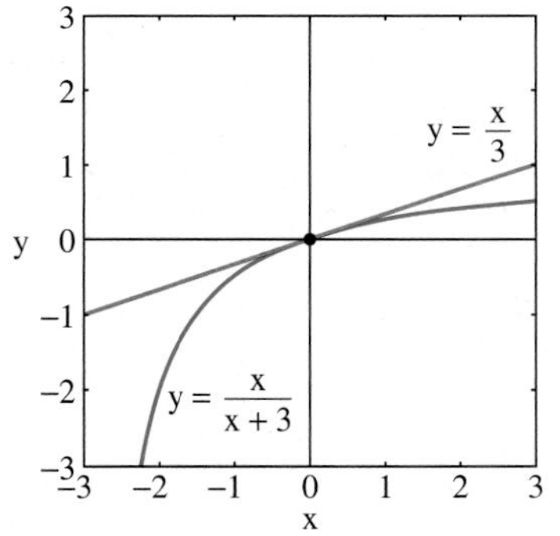

FIGURE 3.1.3 The tangent line $y = \frac{1}{3}x$ to the curve $y = x/(x+3)$ at the origin.

Even when the function f is rather simple, this four-step process for computing f' directly from the definition of the derivative can be time consuming. Also, Step 3 may require considerable ingenuity. Moreover, it would be very repetitious to continue to rely on this process. To avoid tedium, we want a fast, easy, short method for computing $f'(x)$.

That new method is one focus of this chapter: the development of systematic methods ("rules") for differentiating those functions that occur most frequently. Such functions include polynomials, rational functions, the trigonometric functions $\sin x$ and $\cos x$, and combinations of such functions. Once we establish these general differentiation rules, we can apply them formally, almost mechanically, to compute derivatives. Only rarely should we need to return to the definition of the derivative.

Figure 3.1.4 illustrates the slope-predictor function for a parabola that we exhibited in Eq. (10) of Section 2.1. Restated in the language of derivatives, this is an example of a "differentiation rule."

$$\begin{array}{rcccc} y & = (px^2) & + (qx) & + r \\ & \downarrow & \downarrow & \downarrow \\ m(x) & = 2(px) & + \quad q & + 0 \end{array}$$

FIGURE 3.1.4 Termwise construction of the slope-predictor function $m(x) = 2px + q$ for a parabola $y = px^2 + qx + r$. Note that the exponent 2 in the quadratic term px^2 comes "down out front"—yielding the linear term $2px$—while the linear term qx simply yields the constant q, and the constant term r just "disappears."

RULE Differentiation of Quadratic Functions
The derivative of the quadratic function

$$f(x) = ax^2 + bx + c \tag{6}$$

is the linear function

$$f'(x) = 2ax + b. \tag{7}$$

Note that this rule works in the same way no matter whether we denote the coefficients by a, b, and c as in Eqs. (6) and (7), or by p, q, and r as in Fig. 3.1.4.

It may be instructive to derive the differentiation formula in (7) directly from the definition of the derivative:

$$\begin{aligned} f'(x) &= \lim_{h \to 0} \frac{f(x+h) - f(x)}{h} \\ &= \lim_{h \to 0} \frac{[a(x+h)^2 + b(x+h) + c] - [ax^2 + bx + c]}{h} \\ &= \lim_{h \to 0} \frac{(ax^2 + 2ahx + ah^2 + bx + bh + c) - (ax^2 + bx + c)}{h} \\ &= \lim_{h \to 0} \frac{2ahx + ah^2 + bh}{h} \\ &= \lim_{h \to 0} (2ax + ah + b). \end{aligned}$$

Therefore

$$f'(x) = 2ax + b.$$

Once we know this rule, we need never again apply the definition of the derivative to differentiate a quadratic function.

EXAMPLE 2 (a) If $f(x) = 3x^2 - 4x + 5$, we can apply Eq. (7) to write the derivative immediately, without going through the four-step process:

$$f'(x) = 2 \cdot (3x) + (-4) = 6x - 4.$$

Figure 3.1.5 compares the graph of f with that of its derivative f'.

(b) Similarly, if $g(t) = 2t - 5t^2$, then

$$g'(t) = (2) + 2 \cdot (-5t) = 2 - 10t. \qquad ◆$$

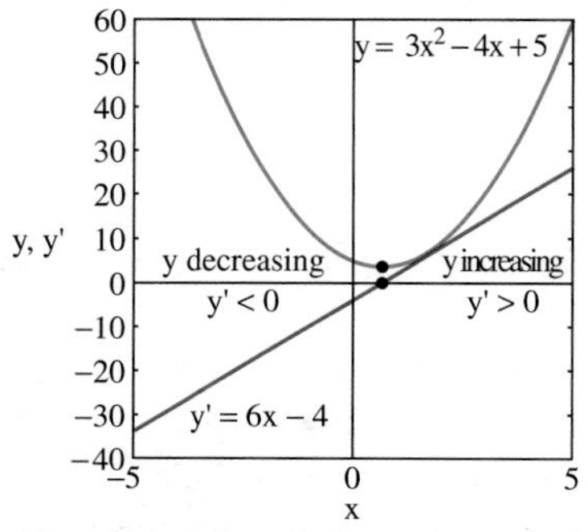

FIGURE 3.1.5 Note that the curve $y = f(x)$ is falling (from left to right) where the derivative $f'(x)$ is negative, and is rising where the derivative is positive.

It makes no difference what the name for the function is or whether we write x or t for the independent variable. This flexibility is valuable—in general, it is such adaptability that makes mathematics applicable to virtually every other branch of human knowledge. In any case, you should learn every differentiation rule in a form independent of the notation used to state it.

We develop additional differentiation rules in Sections 3.2 through 3.4. First, however, we must introduce new notation and a new interpretation of the derivative.

Differential Notation

An important alternative notation for the derivative originates from the early custom of writing Δx in place of h (because $h = \Delta x$ is an increment in x) and

$$\Delta y = f(x + \Delta x) - f(x)$$

for the resulting change (or increment) in y. The slope of the secant line K of Fig. 3.1.6

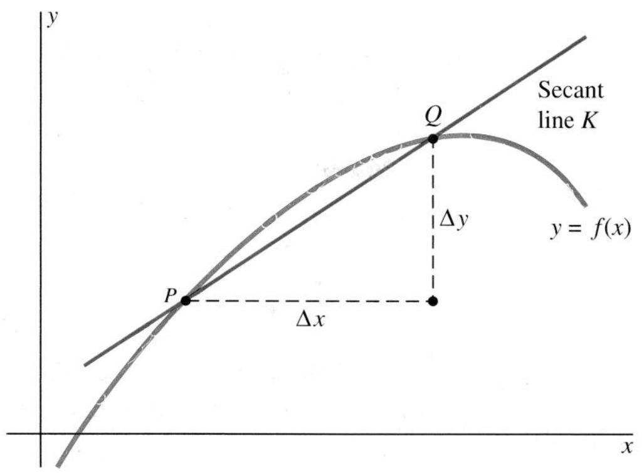

FIGURE 3.1.6 Origin of the dy/dx notation.

is then

$$m_{\text{sec}} = \frac{\Delta y}{\Delta x} = \frac{f(x+h) - f(x)}{h},$$

and the slope of the tangent line is

$$m = \frac{dy}{dx} = \lim_{\Delta x \to 0} \frac{\Delta y}{\Delta x}. \qquad \textbf{(8)}$$

Hence, if $y = f(x)$, we often write

$$\frac{dy}{dx} = f'(x). \qquad \textbf{(9)}$$

(The so-called *differentials* dy and dx are discussed carefully in Chapter 4.) The symbols $f'(x)$ and dy/dx for the derivative of the function $y = f(x)$ are used almost interchangeably in mathematics and its applications, so you need to be familiar with both versions of the notation. You also need to know that dy/dx is a single symbol representing the derivative; it is *not* the quotient of two separate quantities dy and dx.

EXAMPLE 2 (Continued) If $y = ax^2 + bx + c$, then the derivative in Eq. (7) in differential notation takes the form

$$\frac{dy}{dx} = 2ax + b.$$

Consequently,

$$\text{if} \quad y = 3x^2 - 4x + 5, \qquad \text{then} \quad \frac{dy}{dx} = 6x - 4;$$

$$\text{if} \quad z = 2t - 5t^2, \qquad \text{then} \quad \frac{dz}{dt} = 2 - 10t.$$ ◆

The letter d in the notation dy/dx stands for the word "differential." Whether we write dy/dx or dz/dt, the dependent variable appears "upstairs" and the independent variable "downstairs."

Rates of Change

The derivative of a function serves as a slope predictor for straight lines tangent to the graph of that function. Now we introduce the equally important interpretation of the derivative of a function as the rate of change of that function with respect to the independent variable.

We begin with the *instantaneous rate of change* of a function whose independent variable is time t. Suppose that Q is a quantity that varies with time t, and write $Q = f(t)$ for the value of Q at time t. For example, Q might be

- The size of a population (such as kangaroos, people, or bacteria);
- The number of dollars in a bank account;

- The volume of a balloon being inflated;
- The amount of water in a reservoir with variable inflow and outflow;
- The amount of a chemical product produced in a reaction; or
- The distance traveled t hours after the beginning of a journey.

The change in Q from time t to time $t + \Delta t$ is the **increment**

$$\Delta Q = f(t + \Delta t) - f(t).$$

The **average rate of change** of Q (per unit of time) is, by definition, the ratio of the change ΔQ in Q to the change Δt in t. Thus it is the quotient

$$\frac{\Delta Q}{\Delta t} = \frac{f(t + \Delta t) - f(t)}{\Delta t} \tag{10}$$

illustrated in Fig. 3.1.7.

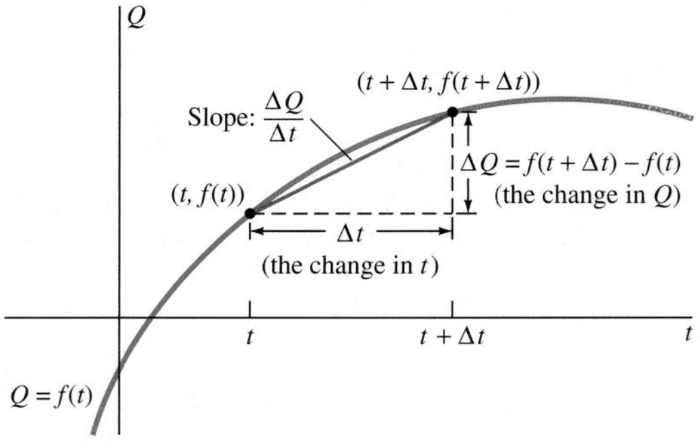

FIGURE 3.1.7 Average rate of change as a slope.

We define the **instantaneous rate of change** of Q (per unit of time) to be the limit of this average rate as $\Delta t \to 0$. That is, the instantaneous rate of change of Q is

$$\lim_{\Delta t \to 0} \frac{\Delta Q}{\Delta t} = \lim_{\Delta t \to 0} \frac{f(t + \Delta t) - f(t)}{\Delta t}. \tag{11}$$

But the right-hand limit in Eq. (11) is simply the derivative $f'(t)$. So we see that the instantaneous rate of change of $Q = f(t)$ is the derivative

$$\frac{dQ}{dt} = f'(t). \tag{12}$$

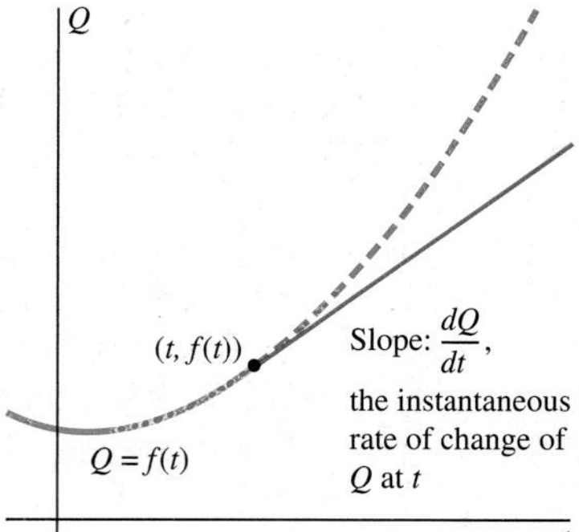

FIGURE 3.1.8 The relation between the tangent line at $(t, f(t))$ and the instantaneous rate of change of f at t.

To interpret intuitively the concept of instantaneous rate of change, think of the point $P(t, f(t))$ moving along the graph of the function $Q = f(t)$. As Q changes with time t, the point P moves along the curve. But suppose that suddenly, at the instant t, the point P begins to follow a straight-line path—like a whirling particle suddenly cut loose from its string. Then the new path of P would appear as in Fig. 3.1.8. The dashed curve in the figure corresponds to the "originally planned" behavior of Q (before P decided to fly off along the straight-line path). But the straight-line path of P (of constant slope) corresponds to the quantity Q "changing at a constant rate." Because the straight line is tangent to the graph $Q = f(t)$, we can interpret dQ/dt as the instantaneous rate of change of the quantity Q at the instant t:

The instantaneous rate of change of $Q = f(t)$ at time t is equal to the slope of the line tangent to the curve $Q = f(t)$ at the point $(t, f(t))$.

We can draw additional important conclusions. Because a positive slope corresponds to a rising tangent line and a negative slope corresponds to a falling tangent

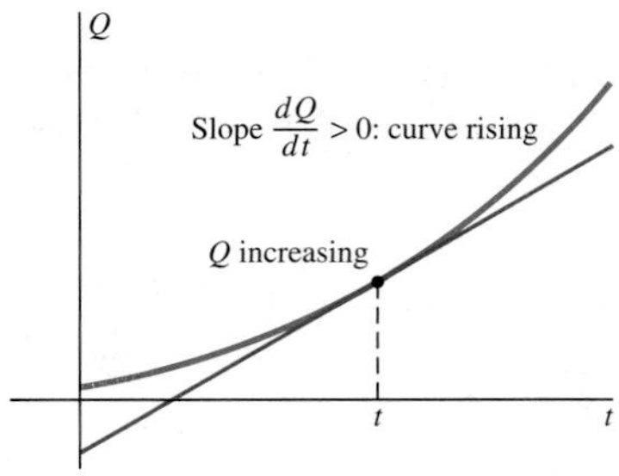

FIGURE 3.1.9 Quantity increasing—derivative positive.

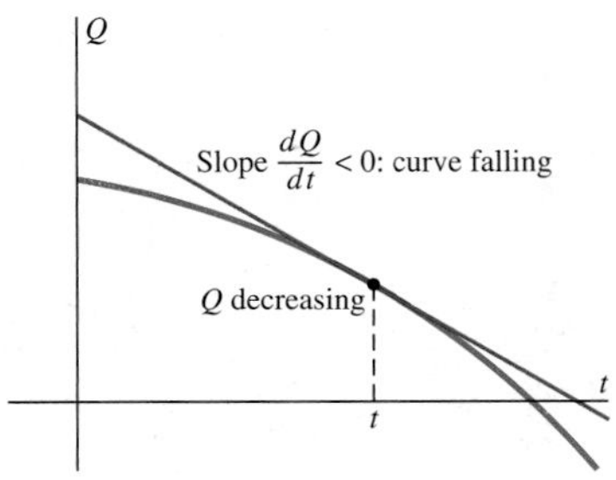

FIGURE 3.1.10 Quantity decreasing—derivative negative.

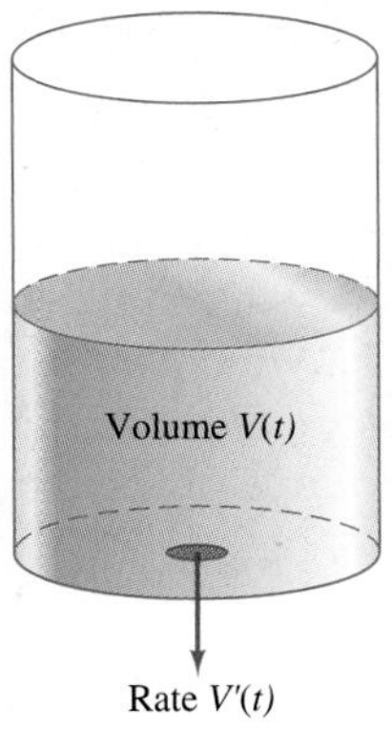

FIGURE 3.1.11 The draining tank of Example 3.

line (as in Figs. 3.1.9 and 3.1.10), we say that

$$Q \text{ is } \textit{increasing} \text{ at time } t \text{ if } \frac{dQ}{dt} > 0;$$
$$Q \text{ is } \textit{decreasing} \text{ at time } t \text{ if } \frac{dQ}{dt} < 0. \qquad \textbf{(13)}$$

Note The meaning of the phrase "$Q = f(t)$ is increasing *over* (or *during*) *the time interval from* $t = a$ *to* $t = b$" should be intuitively clear. The expressions in (13) give us a way to make precise what we mean by "$Q = f(t)$ is increasing *at time* t"—that is, at the instant t. Note also that the fact that a function is increasing at some instant does not necessarily imply that it continues to increase throughout some interval of time; this question is discussed in Section 4.3.

EXAMPLE 3 The cylindrical tank in Fig. 3.1.11 has a vertical axis and is initially filled with 600 gal of water. This tank takes 60 min to empty after a drain in its bottom is opened. Suppose that the drain is opened at time $t = 0$. Suppose also that the volume V of water remaining in the tank after t minutes is

$$V(t) = \tfrac{1}{6}(60 - t)^2 = 600 - 20t + \tfrac{1}{6}t^2$$

gallons. Find the instantaneous rate at which water is flowing out of the tank at time $t = 15$ (min) *and* at time $t = 45$ (min). Also find the average rate at which water flows out of the tank during the half hour from $t = 15$ to $t = 45$.

Solution The instantaneous rate of change of the volume $V(t)$ of water in the tank is given by the derivative

$$\frac{dV}{dt} = -20 + \tfrac{1}{3}t.$$

At the instants $t = 15$ and $t = 45$ we obtain

$$V'(15) = -20 + \tfrac{1}{3} \cdot 15 = -15$$

and

$$V'(45) = -20 + \tfrac{1}{3} \cdot 45 = -5.$$

The units here are gallons per minute (gal/min). The fact that $V'(15)$ and $V'(45)$ are negative is consistent with the observation that V is a decreasing function of t (as t increases, V decreases). One way to indicate this is to say that after 15 min, the water is flowing *out* of the tank at 15 gal/min; after 45 min, the water is flowing *out* at 5 gal/min. The instantaneous rate of change of V at $t = 15$ is -15 gal/min, and the instantaneous rate of change of V at $t = 45$ is -5 gal/min. We could have predicted the units, because $\Delta V/\Delta t$ is a ratio of gallons to minutes, and therefore its limit $V'(t) = dV/dt$ must be expressed in the same units.

During the time interval of length $\Delta t = 30$ min from time $t = 15$ to time $t = 45$, the *average* rate of change of the volume $V(t)$ is

$$\frac{\Delta V}{\Delta t} = \frac{V(45) - V(15)}{45 - 15} = \frac{\frac{1}{6}(60-45)^2 - \frac{1}{6}(60-15)^2}{45 - 15} = \frac{-300}{30}.$$

Each numerator in the last equation is measured in gallons—this is especially apparent when you examine the second numerator—and each denominator is measured in minutes. Hence the ratio in the last fraction is a ratio of gallons to minutes, so the average rate of change of the volume V of water *in* the tank is -10 gal/min. Thus the average rate of flow of water *out* of the tank during this half-hour interval is 10 gal/min. ◆

Our examples of functions up to this point have been restricted to those with formulas or verbal descriptions. Scientists and engineers often work with tables of values obtained from observations or experiments. Example 4 shows how the instantaneous rate of change of such a tabulated function can be estimated.

EXAMPLE 4 The table in Fig. 3.1.12 gives the U.S. population P (in millions) in the nineteenth century at 10-year intervals. Estimate the instantaneous rate of population growth in 1850.

Solution We take $t = 0$ (yr) in 1800, so $t = 50$ corresponds to the year 1850. In Fig. 3.1.13 we have plotted the given data and then added a freehand sketch of a smooth curve that fits these data.

t	Year	U.S. Population (Millions)
0	1800	5.3
10	1810	7.2
20	1820	9.6
30	1830	12.9
40	1840	17.1
50	1850	23.2
60	1860	31.4
70	1870	38.6
80	1880	50.2
90	1890	62.9
100	1900	76.0

FIGURE 3.1.12 Data for Example 4.

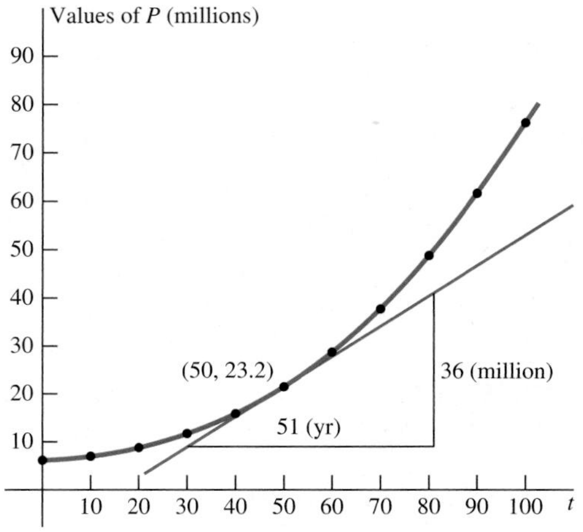

FIGURE 3.1.13 A smooth curve that fits the data of Fig. 3.1.12 well (Example 4).

We hope that this curve fitting the data is a good approximation to the true graph of the unknown function $P = f(t)$. The instantaneous rate of change dP/dt in 1850 is then the slope of the tangent line at the point (50, 23.2). We draw the tangent line as accurately as we can by visual inspection and then measure the base and height of the triangle in Fig. 3.1.13. In this way we approximate the slope of the tangent at $t = 50$ as

$$\frac{dP}{dt} \approx \frac{36}{51} \approx 0.71$$

millions of people per year (in 1850). Although there was no national census in 1851, we would expect the U.S. population then to have been approximately $23.2 + 0.7 = 23.9$ million. ◆

Velocity and Acceleration

Suppose that a particle moves along a horizontal straight line, with its location x at time t given by its **position function** $x = f(t)$. Thus we make the line of motion a coordinate axis with an origin and a positive direction; $f(t)$ is merely the x-coordinate of the moving particle at time t (Fig. 3.1.14).

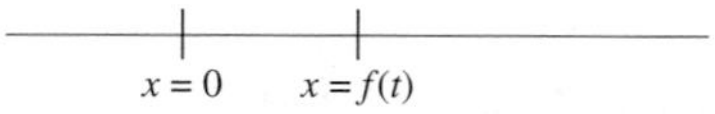

FIGURE 3.1.14 The particle in motion is at the point $x = f(t)$ at time t.

Think of the time interval from t to $t + \Delta t$. The particle moves from position $f(t)$ to position $f(t + \Delta t)$ during this interval. Its displacement is then the increment

$$\Delta x = f(t + \Delta t) - f(t).$$

We calculate the *average velocity* of the particle during this time interval exactly as we would calculate average speed on a long motor trip: We divide the distance by the time to obtain an average speed in miles per hour. In this case we divide the displacement of the particle by the elapsed time to obtain the **average velocity**

$$\overline{v} = \frac{\Delta x}{\Delta t} = \frac{f(t + \Delta t) - f(t)}{\Delta t}. \tag{14}$$

(The overbar is a standard symbol that usually connotes an average of some sort.) We define the **instantaneous velocity** v of the particle at the time t to be the limit of the average velocity $\overline{v}$ as $\Delta t \to 0$. That is,

$$v = \lim_{\Delta t \to 0} \frac{\Delta x}{\Delta t} = \lim_{\Delta t \to 0} \frac{f(t + \Delta t) - f(t)}{\Delta t}. \tag{15}$$

We recognize the limit on the right in Eq. (15)—it is the definition of the derivative of f at time t. Therefore, the velocity of the moving particle at time t is simply

$$v = \frac{dx}{dt} = f'(t). \tag{16}$$

Thus *velocity is instantaneous rate of change of position.* The velocity of a moving particle may be positive or negative, depending on whether the particle is moving in the positive or negative direction along the line of motion. We define the **speed** of the particle to be the *absolute value* $|v|$ of the velocity.

EXAMPLE 5 Figure 3.1.15 shows a car moving along the (horizontal) x-axis. Suppose that its position (in feet) at time t (in seconds) is given by

$$x(t) = 5t^2 + 100.$$

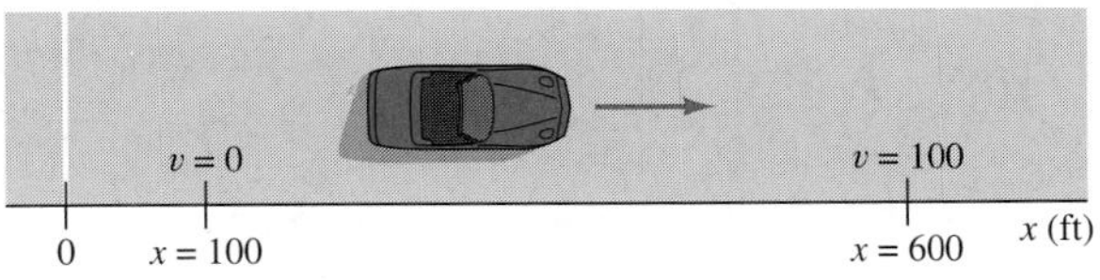

FIGURE 3.1.15 The car of Example 5.

Then its velocity at time t is

$$v(t) = x'(t) = 10t.$$

Because $x(0) = 100$ and $v(0) = 0$, the car starts at time $t = 0$ from rest—$v(0) = 0$—at the point $x = 100$. Substituting $t = 10$, we see that $x(10) = 600$ and $v(10) = 100$, so after 10 s the car has traveled 500 ft (from its starting point $x = 100$), and its speed then is 100 ft/s. ◆

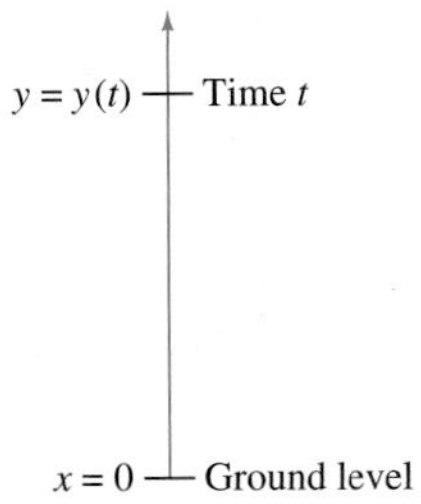

FIGURE 3.1.16 Vertical motion with position function $y(t)$.

Vertical Motion

In the case of vertical motion—such as that of a ball thrown straight upward—it is common to denote the position function by $y(t)$ rather than by $x(t)$. Typically, $y(t)$ denotes the height above the ground at time t, as in Fig. 3.1.16. But velocity is still the derivative of position:

$$v(t) = \frac{dy}{dt}.$$

Upward motion with y increasing corresponds to *positive velocity,* $v > 0$ (Fig. 3.1.17). *Downward motion* with y decreasing corresponds to *negative velocity,* $v < 0$.

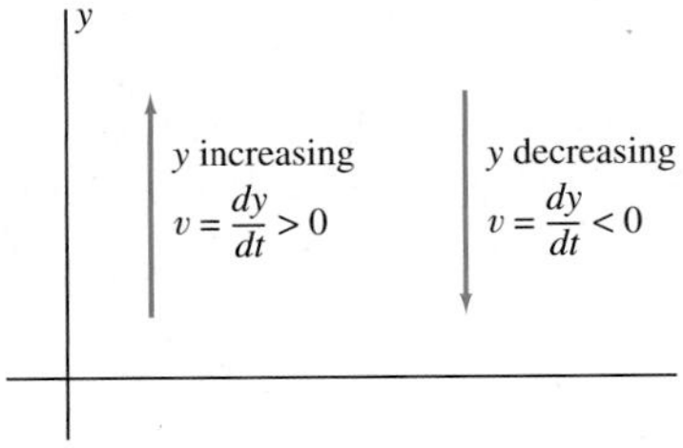

FIGURE 3.1.17 Upward motion and downward motion.

The case of vertical motion under the influence of constant gravity is of special interest. If a particle is projected straight upward from an initial height y_0 (ft) above the ground at time $t = 0$ (s) and with initial velocity v_0 (ft/s) *and* if air resistance is negligible, then its height y (in feet above the ground) at time t is given by a formula known from physics,

$$y(t) = -\tfrac{1}{2}gt^2 + v_0 t + y_0. \tag{17}$$

Here g denotes the *acceleration* due to the force of gravity. Near the surface of the earth, g is nearly constant, so we assume that it is exactly constant, and at the surface of the earth, $g \approx 32$ ft/s^2, or $g \approx 9.8$ m/s^2.

If we differentiate y with respect to time t, we obtain the velocity of the particle at time t:

$$v(t) = \frac{dy}{dt} = -gt + v_0. \tag{18}$$

The **acceleration** of the particle is defined to be the instantaneous time rate of change (derivative) of its velocity:

$$a = \frac{dv}{dt} = -g. \tag{19}$$

Your intuition should tell you that a body projected upward in this way will reach its maximum height at the instant that its velocity becomes zero—when $v(t) = 0$. (We shall see in Section 3.5 why this is true.)

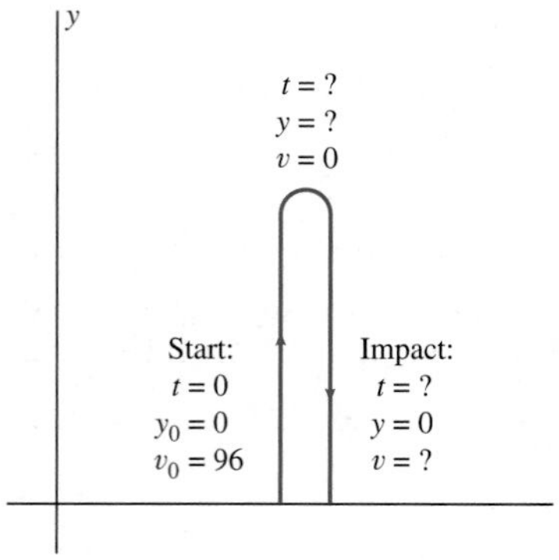

FIGURE 3.1.18 Data for the ball of Example 6.

EXAMPLE 6 Find the maximum height attained by a ball thrown straight upward from the ground with initial velocity $v_0 = +96$ ft/s. Also find the velocity with which it hits the ground upon its return.

Solution To begin the solution of a motion problem such as this, we sketch a diagram like Fig. 3.1.18, indicating both the given data and the data that are unknown at the time instants in question. Here we focus on the time $t = 0$ when the ball leaves the ground ($y = 0$), the unknown time when it reaches its maximum height with velocity $v = 0$, and the unknown time when it returns to the ground.

We begin by substituting $y_0 = 0$, $v_0 = 96$, and $g = 32$ in Eq. (17). Then the height of the ball at time t (so long as it remains aloft) is given by

$$y(t) = -16t^2 + 96t.$$

Then differentiation gives its velocity at time t,

$$v(t) = y'(t) = -32t + 96$$

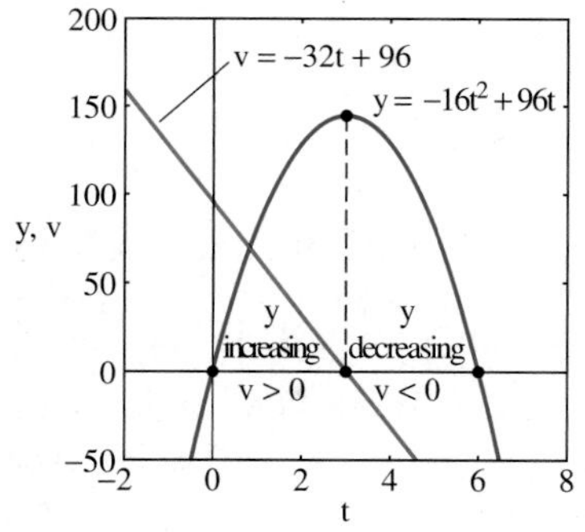

FIGURE 3.1.19 Note that the ball is rising when its velocity $v > 0$, falling when $v < 0$, and is at its apex when $v = 0$.

(see Fig. 3.1.19). The ball attains its maximum height when $v = 0$; that is, when

$$v(t) = -32t + 96 = 0.$$

This occurs when $t = 3$ (s). Substituting $t = 3$ in the height function $y(t)$ gives the maximum height of the ball,

$$y_{\max} = y(3) = -16 \cdot (3)^2 + 96 \cdot (3) = 144 \quad \text{(ft)}.$$

The ball returns to the ground when $y(t) = 0$. The equation

$$y(t) = -16t^2 + 96t = -16t(t - 6) = 0$$

has the two solutions $t = 0$ and $t = 6$. Thus the ball returns to the ground at time $t = 6$. The velocity with which it strikes the ground is

$$v(t) = (-32)(6) + 96 = -96 \quad \text{(ft/s)}.$$

◆

Other Rates of Change

The derivative of any function—not merely a function of time—may be interpreted as its instantaneous rate of change with respect to the independent variable. If $y = f(x)$, then the **average rate of change** of y (per unit change in x) on the interval $[x, x + \Delta x]$ is the quotient

$$\frac{\Delta y}{\Delta x} = \frac{f(x + \Delta x) - f(x)}{\Delta x}.$$

The **instantaneous rate of change** of y **with respect to** x is the limit, as $\Delta x \to 0$, of the average rate of change. Thus the instantaneous rate of change of y with respect to x is

$$\lim_{\Delta x \to 0} \frac{\Delta y}{\Delta x} = \frac{dy}{dx} = f'(x). \tag{20}$$

Example 7 illustrates the fact that a dependent variable may sometimes be expressed as two different functions of two different independent variables. The derivatives of these functions are then rates of change of the dependent variable with respect to the two different independent variables.

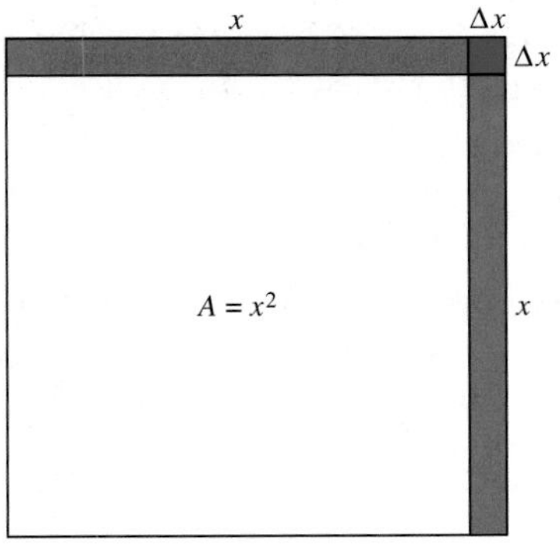

FIGURE 3.1.20 The square of Example 7:
$A + \Delta A = (x + \Delta x)^2;$
$\Delta A = 2x\Delta x + (\Delta x)^2;$
$\frac{\Delta A}{\Delta x} = 2x + \Delta x;$
$\frac{dA}{dx} = 2x.$

EXAMPLE 7 The area of a square with edge length x centimeters is $A = x^2$, so the derivative of A with respect to x,

$$\frac{dA}{dx} = 2x, \tag{21}$$

is the rate of change of its area A with respect to x. (See the computations in Fig. 3.1.20.) The units of dA/dx are square centimeters *per centimeter.* Now suppose that the edge length of the square is increasing with time: $x = 5t$, with time t in seconds. Then the area of the square at time t is

$$A = (5t)^2 = 25t^2.$$

The derivative of A with respect to t is

$$\frac{dA}{dt} = 2 \cdot 25t = 50t; \tag{22}$$

this is the rate of change of A with respect to time t, with units of square centimeters *per second.* For instance, when $t = 10$ (so $x = 50$), the values of the two derivatives of A in Eqs. (21) and (22) are

$$\left.\frac{dA}{dx}\right|_{x=50} = 2 \cdot 50 = 100 \quad (\text{cm}^2/\text{cm})$$

and

$$\left.\frac{dA}{dt}\right|_{t=10} = 50 \cdot 10 = 500 \quad (\text{cm}^2/\text{s}).$$

Thus A is increasing at the rate of 100 cm^2 per cm increase in x, and at the rate of 500 cm^2 per second increase in t. ◆

The notation dA/dt for the derivative suffers from the minor inconvenience of not providing a "place" to substitute a particular value of t, such as $t = 10$. The last lines of Example 7 illustrate one way around this difficulty.

Just as we can speak of whether the quantity $Q(t)$ is increasing or decreasing at time $t = a$—according as $Q'(a) > 0$ or $Q'(a) < 0$—we can ask whether the function $y = f(x)$ is an increasing or decreasing function of x. Thinking of rising tangent lines with positive slopes, and falling tangent lines with negative slopes, we say in analogy with (13) that

y is *increasing* at the point $x = a$ if $f'(a) > 0$;

y is *decreasing* at the point $x = a$ if $f'(a) < 0$.

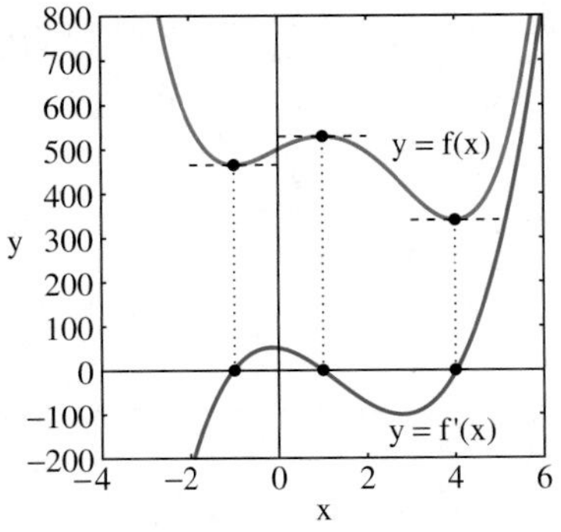

FIGURE 3.1.21 Correspondence between the function graph $y = f(x)$ and the derivative graph $y = f'(x)$.

EXAMPLE 8 Figure 3.1.21 shows the graphs $y = f(x)$ of a function and $y = f'(x)$ of its derivative. Observe that

- $y = f(x)$ has a horizontal tangent line at points where $f'(x) = 0$;
- $f(x)$ is increasing on open intervals where $f'(x) > 0$; and
- $f(x)$ is decreasing on open intervals where $f'(x) < 0$. ◆

3.1 TRUE/FALSE STUDY GUIDE

3.1 CONCEPTS: QUESTIONS AND DISCUSSION

1. The slope line in Fig. 3.1.5 looks as if it might be tangent to the parabola. Is it? If not, what's a simple way you could alter the equation of the parabola—without changing its slope line—in order to ensure that the line will be tangent to the altered parabola?
2. When a ball is tossed straight upward, it may appear to hover at the apex of its trajectory for a brief period of time. Does it?
3. You are pulled over by a policeman who claims that you did not stop properly at a stop sign. You argue that as you braked your car, its velocity was zero at a certain instant before you removed your foot from the brake pedal and proceeded through the intersection. The policeman replies that you nevertheless did not come to a full stop—that he is certain your velocity did not remain zero for even a hundredth of a second. What's the cause of this disagreement? Explain it with such convincing clarity that the judge will let you off without you paying a fine.
4. The ball of Example 6 took the same amount of time to rise from the ground to its highest point as to fall back to the ground. Is this always the case for a ball governed by Eqs. (17)and (18) of this section? *Suggestion:* In lieu of a blizzard of algebra, think about the symmetry of the parabola in Fig. 3.1.19.

3.1 PROBLEMS

In Problems 1 through 10, find the indicated derivative by using the differentiation rule in Eqs. (6) and (7):

$$\text{If} \quad f(x) = ax^2 + bx + c, \qquad \text{then} \quad f'(x) = 2ax + b.$$

1. $f(x) = 4x - 5$; find $f'(x)$.
2. $g(t) = 100 - 16t^2$; find $g'(t)$.
3. $h(z) = z(25 - z)$; find $h'(z)$.
4. $f(x) = 16 - 49x$; find $f'(x)$.
5. $y = 2x^2 + 3x - 17$; find dy/dx.
6. $x = 16t - 100t^2$; find dx/dt.
7. $z = 5u^2 - 3u$; find dz/du.
8. $v = 5y(100 - y)$; find dv/dy.
9. $x = -5y^2 + 17y + 300$; find dx/dy.
10. $u = 7t^2 + 13t$; find du/dt.

In Problems 11 through 20, apply the definition of the derivative (as in Example 1) to find $f'(x)$.

11. $f(x) = 2x - 1$
12. $f(x) = 2 - 3x$
13. $f(x) = x^2 + 5$
14. $f(x) = 3 - 2x^2$
15. $f(x) = \dfrac{1}{2x + 1}$
16. $f(x) = \dfrac{1}{3 - x}$
17. $f(x) = \sqrt{2x + 1}$
18. $f(x) = \dfrac{1}{\sqrt{x + 1}}$
19. $f(x) = \dfrac{x}{1 - 2x}$
20. $f(x) = \dfrac{x + 1}{x - 1}$

In Problems 21 through 25, the position function $x = f(t)$ of a particle moving in a horizontal straight line is given. Find its location x when its velocity v is zero.

21. $x = 100 - 16t^2$
22. $x = -16t^2 + 160t + 25$
23. $x = -16t^2 + 80t - 1$
24. $x = 100t^2 + 50$
25. $x = 100 - 20t - 5t^2$

In Problems 26 through 29, the height $y(t)$ (in feet at time t seconds) of a ball thrown vertically upward is given. Find the maximum height that the ball attains.

26. $y = -16t^2 + 160t$
27. $y = -16t^2 + 64t$
28. $y = -16t^2 + 128t + 25$
29. $y = -16t^2 + 96t + 50$

In Problems 30 through 35 (Figs. 3.1.22 through 3.1.27), match the given graph of the function f with that of its derivative, which appears among those in Fig. 3.1.28, parts (a) through (f).

30. Figure 3.1.22
31. Figure 3.1.23

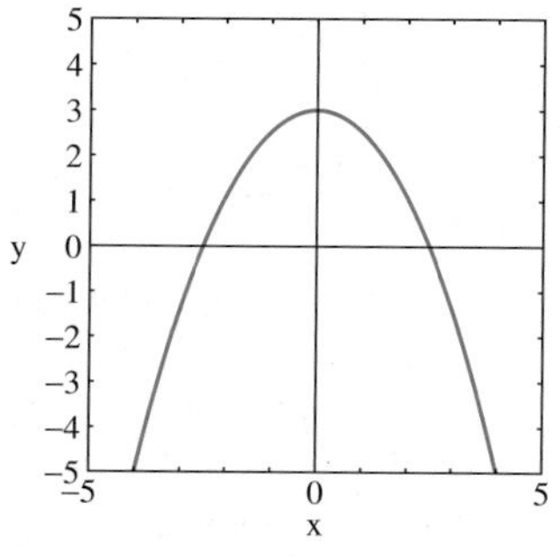

FIGURE 3.1.22

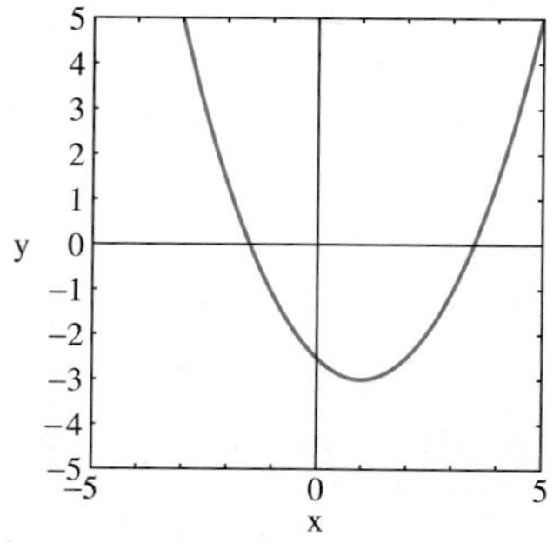

FIGURE 3.1.23

32. Figure 3.1.24

33. Figure 3.1.25

34. Figure 3.1.26

35. Figure 3.1.27

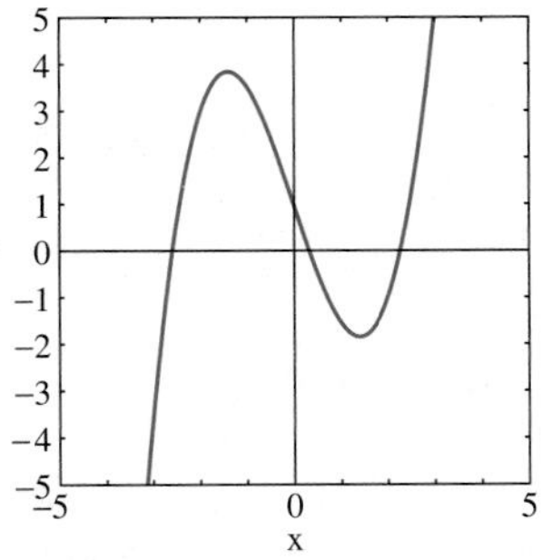

FIGURE 3.1.24

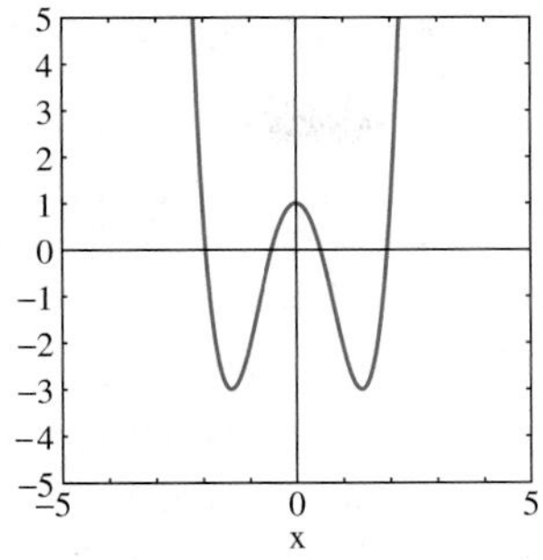

FIGURE 3.1.25

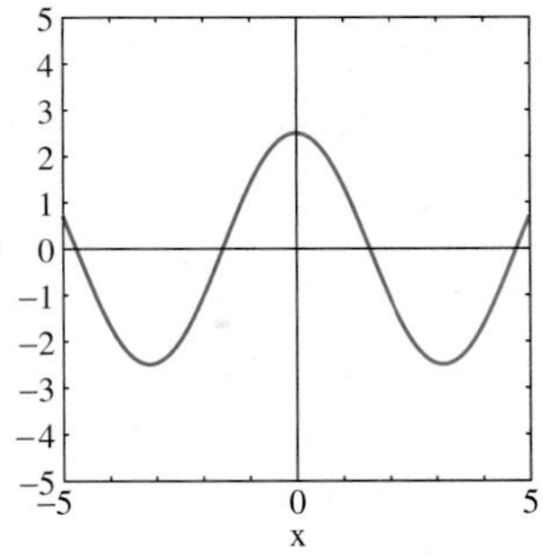

FIGURE 3.1.26

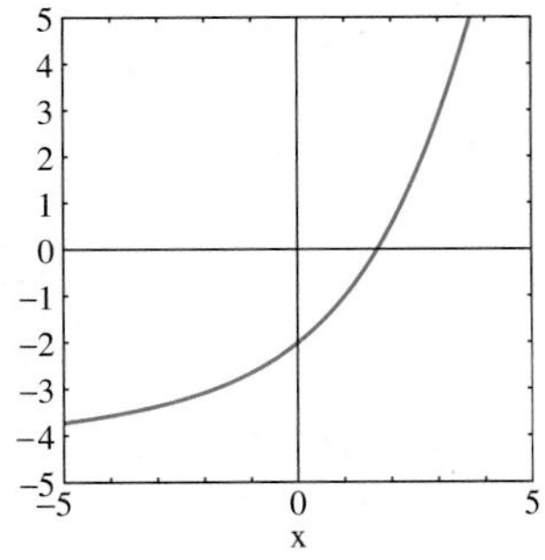

FIGURE 3.1.27

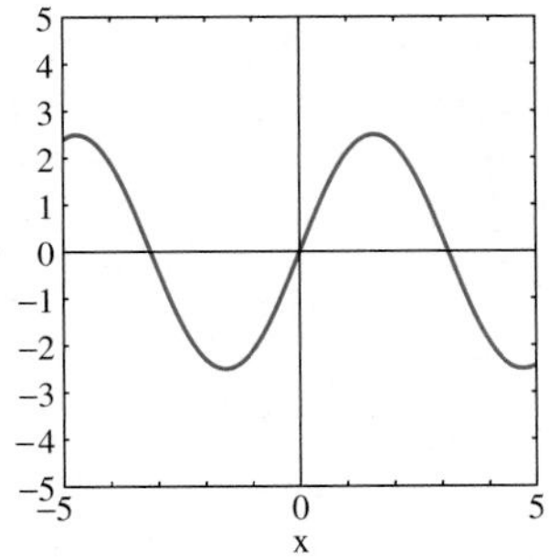

FIGURE 3.1.28(a)

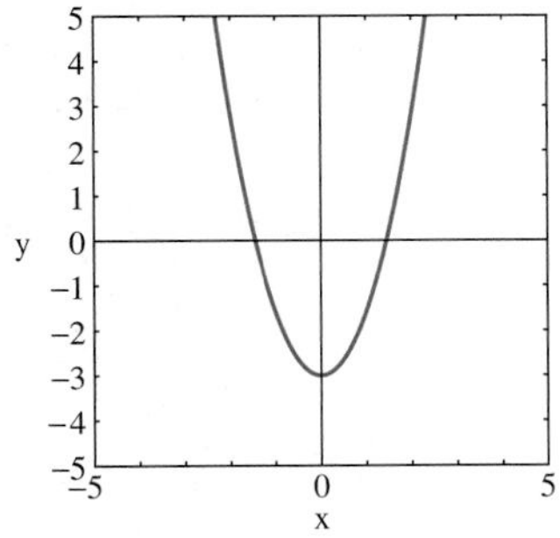

FIGURE 3.1.28(b)

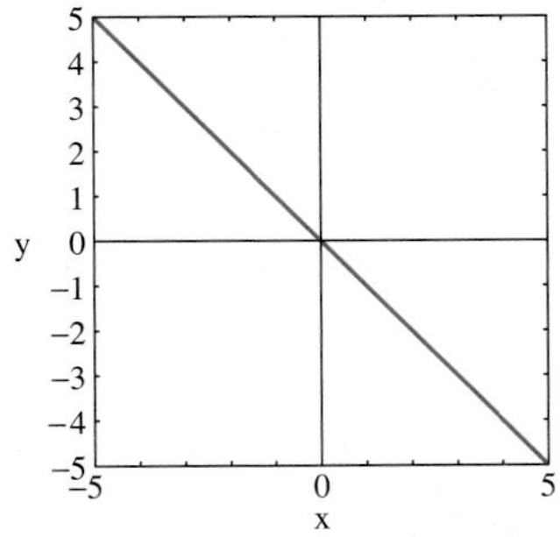

FIGURE 3.1.28(c)

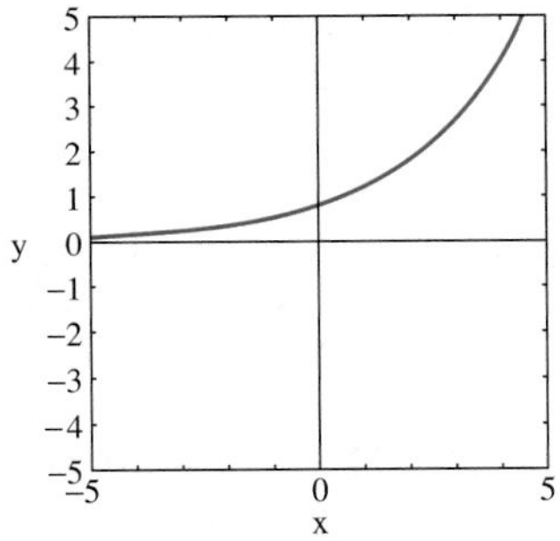

FIGURE 3.1.28(d)

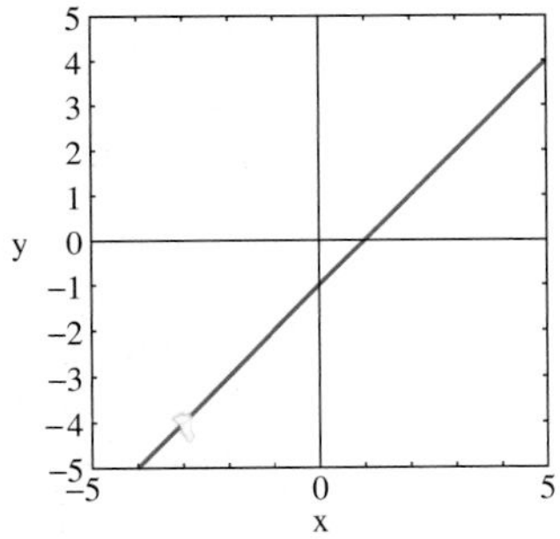

FIGURE 3.1.28(e)

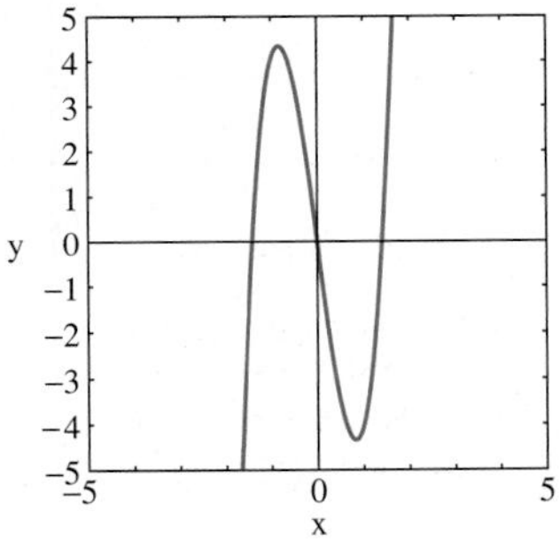

FIGURE 3.1.28(f)

36. The Celsius temperature C is given in terms of the Fahrenheit temperature F by $C = \frac{5}{9}(F - 32)$. Find the rate of change of C with respect to F and the rate of change of F with respect to C.

37. Find the rate of change of the area A of a circle with respect to its circumference C.

38. A stone dropped into a pond at time $t = 0$ s causes a circular ripple that travels out from the point of impact at 5 m/s. At what rate (in square meters per second) is the area within the circle increasing when $t = 10$?

39. A car is traveling at 100 ft/s when the driver suddenly applies the brakes ($x = 0, t = 0$). The position function of the skidding car is $x(t) = 100t - 5t^2$. How far and for how long does the car skid before it comes to a stop?

40. A water bucket containing 10 gal of water develops a leak at time $t = 0$, and the volume V of water in the bucket t seconds later is given by

$$V(t) = 10\left(1 - \frac{t}{100}\right)^2$$

until the bucket is empty at time $t = 100$. (a) At what rate is water leaking from the bucket after exactly 1 min has passed? (b) When is the instantaneous rate of change of V equal to the average rate of change of V from $t = 0$ to $t = 100$?

41. A population of chipmunks moves into a new region at time $t = 0$. At time t (in months), the population numbers

$$P(t) = 100[1 + (0.3)t + (0.04)t^2].$$

(a) How long does it take for this population to double its initial size $P(0)$? (b) What is the rate of growth of the population when $P = 200$?

42. The following data describe the growth of the population P (in thousands) of Gotham City during a 10-year period. Use the graphical method of Example 4 to estimate its rate of growth in 1989.

Year	1984	1986	1988	1990	1992	1994
P	265	293	324	358	395	427

43. The following data give the distance x in feet traveled by an accelerating car (that starts from rest at time $t = 0$) in the first t seconds. Use the graphical method of Example 4 to estimate its speed (in miles per hour) when $t = 20$ and again when $t = 40$.

t	0	10	20	30	40	50	60
x	0	224	810	1655	2686	3850	5109

In Problems 44 through 49, use the fact (proved in Section 3.2) that the derivative of $y = ax^3 + bx^2 + cx + d$ *is* $dy/dx = 3ax^2 + 2bx + c$.

44. Prove that the rate of change of the volume V of a cube with respect to its edge length x is equal to half the surface area A of the cube (Fig. 3.1.29).

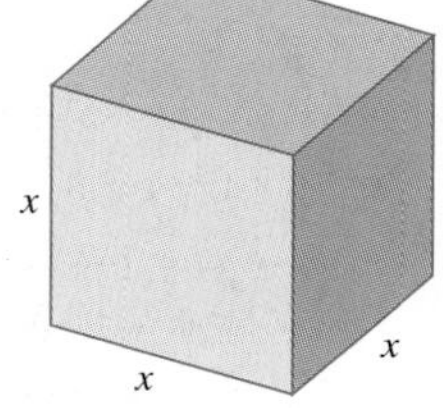

FIGURE 3.1.29 The cube of Problem 44—volume $V = x^3$, surface area $S = 6x^2$.

45. Show that the rate of change of the volume V of a sphere with respect to its radius R is equal to its surface area S (Fig. 3.1.30).

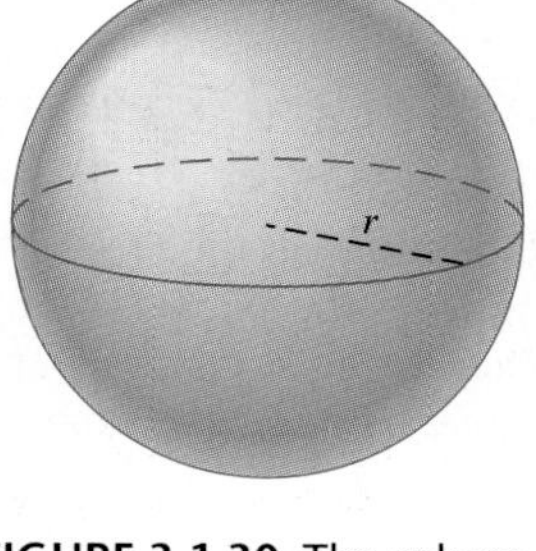

FIGURE 3.1.30 The sphere of Problem 45—volume $V = \frac{4}{3}\pi r^3$ surface area $S = 4\pi r^2$.

46. The height h of a certain cylinder whose height changes is always twice its radius r. Show that the rate of change of its volume V with respect to r is equal to its total surface area S (Fig. 3.1.31).

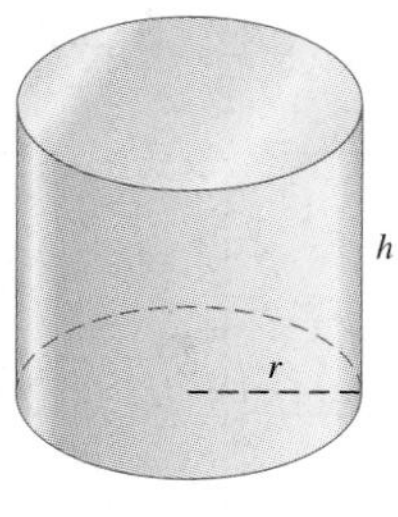

FIGURE 3.1.31 The cylinder of Problem 46—volume $V = \pi r^2 h$, surface area $S = 2\pi r^2 + 2\pi r h$.

47. A spherical balloon with an initial radius r of 5 in. begins to leak at time $t = 0$, and its radius t seconds later is $r = (60 - t)/12$ in. At what rate (in cubic inches per second) is air leaking from the balloon when $t = 30$?

48. The volume V (in liters) of 3 g of CO_2 at 27°C is given in terms of its pressure p (in atmospheres) by the formula $V = 1.68/p$. What is the rate of change of V with respect to p when $p = 2$ (atm)? (*Suggestion:* Use the fact that the derivative of $f(x) = c/x$ is $f'(x) = -c/x^2$ if c is a constant; you can establish this by using the definition of the derivative.)

49. As a snowball with an initial radius of 12 cm melts, its radius decreases at a constant rate. It begins to melt when $t = 0$ (h) and takes 12 h to disappear. (a) What is its rate of change of volume when $t = 6$? (b) What is its average rate of change of volume from $t = 3$ to $t = 9$?

50. A ball thrown vertically upward at time $t = 0$ (s) with initial velocity 96 ft/s and with initial height 112 ft has height function $y(t) = -16t^2 + 96t + 112$. (a) What is the maximum height attained by the ball? (b) When and with what impact speed does the ball hit the ground?

51. A spaceship approaching touchdown on the planet Gzyx has height y (meters) at time t (seconds) given by $y = 100 - 100t + 25t^2$. When and with what speed does it hit the ground?

52. The population (in thousands) of the city Metropolis is given by

$$P(t) = 100[1 + (0.04)t + (0.003)t^2],$$

with t in years and with $t = 0$ corresponding to 1980. (a) What was the rate of change of P in 1986? (b) What was the average rate of change of P from 1983 to 1988?

53. Suppose that during the 1990s the population P of a small city was given by

$$P(t) = 10 + t - 0.1t^2 + 0.006t^3$$

(with t in years and P is thousands). Taking $t = 0$ on January 1, 1990, find the time(s) during the 1990s at which the instantaneous rate of change of this population was equal to its average rate of change during the whole decade. (Use the differentiation formulas given in the instructions for Problem 44–49.)

Problems 54 through 60 involve the ***left-hand*** *and* ***right-hand derivatives*** *of f at a that are defined by*

$$f'_-(a) = \lim_{h \to 0^-} \frac{f(a+h) - f(a)}{h}$$

and

$$f'_+(a) = \lim_{h\to 0^+} \frac{f(a+h) - f(a)}{h},$$

(assuming these limits exist). Then $f'(a)$ exists if and only both the left-hand and right-hand derivatives exist and $f'_-(a) = f'_+(a)$.

54. (a) Find $f'_-(0)$ and $f'_+(0)$ given $f(x) = |x|$. (b) The function $f(x) = |2x - 10|$ is differentiable except at a single point. What is this point, and what are the values of its left-hand and right-hand derivatives of f there?

55. Sketch the graph of the given function f and determine if it is differentiable at $x = 0$:

(a) $f(x) = \begin{cases} x & \text{if } x < 0, \\ 2x & \text{if } x \geqq 0; \end{cases}$

(b) $f(x) = \begin{cases} x^2 & \text{if } x < 0, \\ 2x^2 & \text{if } x \geqq 0. \end{cases}$

56. Investigate the differentiability of the function f defined by

$$f(x) = \begin{cases} 2x+1 & \text{if } x < 1, \\ 4x - x^2 & \text{if } x \geqq 1. \end{cases}$$

57. Investigate the differentiability of the function f defined by

$$f(x) = \begin{cases} 11 + 6x - x^2 & \text{if } x < 3, \\ x^2 - 6x + 29 & \text{if } x \geqq 3. \end{cases}$$

58. Sketch the graph of the function $f(x) = x \cdot |x|$ and show that it is differentiable everywhere. Can you write a single (one-part) formula that gives the value of $f'(x)$ both for $x > 0$ and for $x < 0$?

59. Sketch the graph of the function $f(x) = x + |x|$. Then investigate its differentiability. Find the derivative $f'(x)$ where it exists; find the one-sided derivatives at the points where $f'(x)$ does not exist.

60. Repeat Problem 59, except with the function $f(x) = x \cdot (x + |x|)$.

3.2 BASIC DIFFERENTIATION RULES

Here we begin our systematic development of formal rules for finding the derivative f' of the function f:

$$f'(x) = \lim_{h\to 0} \frac{f(x+h) - f(x)}{h}. \qquad \textbf{(1)}$$

Some alternative notation for derivatives will be helpful.

When we interpreted the derivative in Section 3.1 as a rate of change, we found it useful to employ the dependent-independent variable notation

$$y = f(x), \quad \Delta x = h, \quad \Delta y = f(x + \Delta x) - f(x). \qquad \textbf{(2)}$$

This led to the "differential notation"

$$\frac{dy}{dx} = \lim_{\Delta x\to 0} \frac{\Delta y}{\Delta x} = \lim_{\Delta x\to 0} \frac{f(x+\Delta x) - f(x)}{\Delta x} \qquad \textbf{(3)}$$

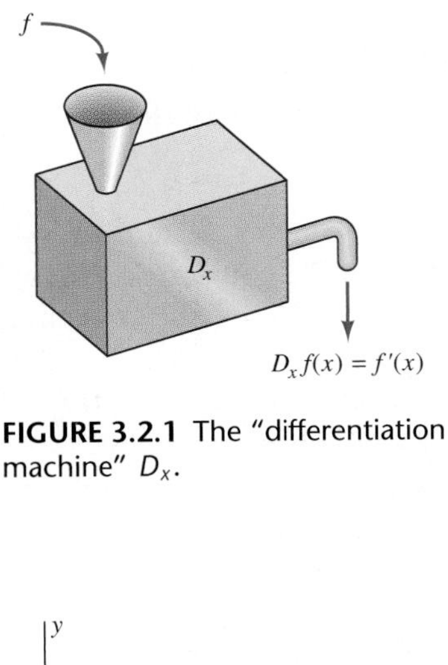

FIGURE 3.2.1 The "differentiation machine" D_x.

for the derivative. When you use this notation, remember that the symbol dy/dx is simply another notation for the derivative $f'(x)$; it is *not* the quotient of two separate entities dy and dx.

A third notation is sometimes used for the derivative $f'(x)$; it is $D_x f(x)$. Here, think of D_x as a "machine" that operates on the function f to produce its derivative $D_x f$ with respect to x (Fig. 3.2.1). Thus we can write the derivative of $y = f(x) = x^3$ in any of three ways:

$$f'(x) = \frac{dy}{dx} = D_x x^3 = 3x^2.$$

These three notations for the derivative—the function notation $f'(x)$, the differential notation dy/dx, and the operator notation $D_x f(x)$—are used almost interchangeably in mathematical and scientific writing, so you need to be familiar with each.

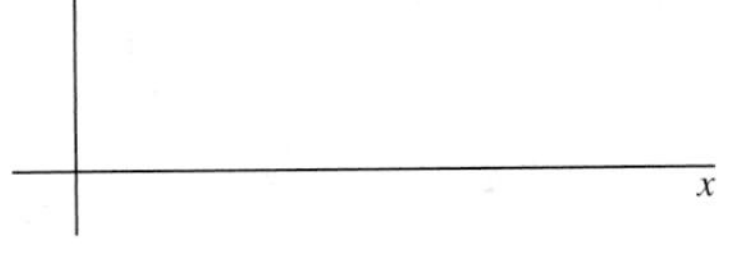

FIGURE 3.2.2 The derivative of a constant-valued function is zero (Theorem 1).

The Derivative of a Constant

Our first differentiation rule says that *the derivative of a constant function is identically zero.* Geometry makes this obvious, because the graph of a constant function is a horizontal straight line that is its own tangent line, with slope zero at every point (Fig. 3.2.2).

THEOREM 1 Derivative of a Constant

If $f(x) = c$ (a constant) for all x, then $f'(x) = 0$ for all x. That is,

$$\frac{dc}{dx} = D_x c = 0. \tag{4}$$

PROOF Because $f(x+h) = f(x) = c$, we see that

$$f'(x) = \lim_{h \to 0} \frac{f(x+h) - f(x)}{h} = \lim_{h \to 0} \frac{c - c}{h} = \lim_{h \to 0} \frac{0}{h} = 0.$$

The Power Rule

As motivation for the next rule, consider the following list of derivatives, all of which have already appeared in the text (or as problems). The first two are special cases of the formula $D_x(ax^2 + bx + c) = 2ax + b$.

$$D_x x = 1$$

$$D_x x^2 = 2x = 2 \cdot x^1$$

$$D_x x^3 = 3 \cdot x^2 \qquad \text{(Problem 37, Section 2.2)}$$

$$D_x \frac{1}{x} = D_x x^{-1} = -\frac{1}{x^2} = -1 \cdot x^{-2} \qquad \text{(Problem 38, Section 2.2)}$$

$$D_x \frac{1}{x^2} = D_x x^{-2} = -\frac{2}{x^3} = -2 \cdot x^{-3} \qquad \text{(Problem 39, Section 2.2)}$$

$$D_x \sqrt{x} = D_x x^{1/2} = \frac{1}{2\sqrt{x}} = \frac{1}{2} \cdot x^{-1/2} \qquad \text{(Example 13, Section 2.2)}$$

Each of these formulas fits the simple pattern

$$D_x x^n = n x^{n-1}, \tag{5}$$

in which the exponent n is simultaneously placed before the variable and, in the exponent, is decreased by 1. Thus it appears that the blanks in the pattern

$$D_x x^{\square} = \square\, x^{\square - 1}$$

can be filled with any (single) integer you please, or even the fraction $\frac{1}{2}$. But Eq. (5)—inferred from the preceding list of derivatives—is as yet only a conjecture. Nevertheless, many discoveries in mathematics are made by detecting such patterns, then proving that they hold universally.

Eventually we shall see that the formula in Eq. (5), called the **power rule,** is valid for all real numbers n. At this time we give a proof only for the case in which the exponent n is a *positive integer.*

THEOREM 2 Power Rule for a Positive Integer n

If n is a positive integer and $f(x) = x^n$, then

$$f'(x) = n x^{n-1}. \tag{6}$$

PROOF For a positive integer n, the identity

$$b^n - a^n = (b - a)(b^{n-1} + b^{n-2}a + b^{n-3}a^2 + \cdots + ba^{n-2} + a^{n-1})$$

is easy to verify by multiplication. Thus, if $b \neq a$, then

$$\frac{b^n - a^n}{b - a} = b^{n-1} + b^{n-2}a + b^{n-3}a^2 + \cdots + ba^{n-2} + a^{n-1}.$$

Because each of the n terms on the right-hand side approaches a^{n-1} as $b \to a$, this tells us that

$$\lim_{b \to a} \frac{b^n - a^n}{b - a} = na^{n-1}$$

by various limit laws. Now let $b = x + h$ and $a = x$, so that $h = b - a$. Then $h \to 0$ as $b \to a$, and hence

$$f'(x) = \lim_{h \to 0} \frac{(x+h)^n - x^n}{h} = nx^{n-1}. \tag{7}$$

This establishes Theorem 2. ◄

We need not always use the same symbols x and n for the independent variable and the constant exponent in the power rule. For instance,

$$D_t t^m = mt^{m-1} \quad \text{and} \quad D_z z^k = kz^{k-1}.$$

If it is perfectly clear what the independent variable is, the subscript may be dropped from D_x (or D_t, or D_z), as in Example 1.

EXAMPLE 1 $Dx^7 = 7x^6, \qquad Dt^{17} = 17t^{16}, \qquad Dz^{100} = 100z^{99}.$ ◆

The Derivative of a Linear Combination

To use the power rule to differentiate polynomials, we need to know how to differentiate *linear combinations.* A **linear combination** of the functions f and g is a function of the form $af + bg$ where a and b are constants. It follows from the sum and product laws for limits that

$$\lim_{x \to c} [af(x) + bg(x)] = a\left(\lim_{x \to c} f(x)\right) + b\left(\lim_{x \to c} g(x)\right) \tag{8}$$

provided that the two limits on the right in Eq. (8) both exist. The formula in Eq. (8) is called the **linearity property** of the limit operation. It implies an analogous linearity property of differentiation.

THEOREM 3 Derivative of a Linear Combination

If f and g are differentiable functions and a and b are fixed real numbers, then

$$D_x[af(x) + bg(x)] = a[D_x f(x)] + b[D_x g(x)]. \tag{9}$$

With $u = f(x)$ and $v = g(x)$, this takes the form

$$\frac{d(au + bv)}{dx} = a\frac{du}{dx} + b\frac{dv}{dx}. \tag{9$'$}$$

PROOF The linearity property of limits immediately gives

$$\begin{aligned} D_x[af(x) + bg(x)] &= \lim_{h \to 0} \frac{[af(x+h) + bg(x+h)] - [af(x) + bg(x)]}{h} \\ &= a\left(\lim_{h \to 0} \frac{f(x+h) - f(x)}{h}\right) + b\left(\lim_{h \to 0} \frac{g(x+h) - g(x)}{h}\right) \\ &= a[D_x f(x)] + b[D_x g(x)], \end{aligned}$$

as desired. ◄

Now take $a = c$ and $b = 0$ in Eq. (9). The result is

$$D_x[cf(x)] = cD_x f(x); \tag{10}$$

alternatively,

$$\frac{d(cu)}{dx} = c\frac{du}{dx}, \tag{10'}$$

Thus *the derivative of a constant multiple of a function is the same constant multiple of its derivative.*

EXAMPLE 2

(a) $D_x(16x^6) = 16 \cdot 6x^5 = 96x^5$.
(b) If $f(z) = 7z^3$, then $f'(z) = 21z^2$.
(c) $\dfrac{d}{du}(99u^{100}) = 9900u^{99}$. ◆

Next, take $a = b = 1$ in Eq. (9). We find that

$$D_x[f(x) + g(x)] = [D_x f(x)] + [D_x g(x)]. \tag{11}$$

In differential notation,

$$\frac{d(u+v)}{dx} = \frac{du}{dx} + \frac{dv}{dx}. \tag{11'}$$

Thus *the derivative of the sum of two functions is the sum of their derivatives.* Similarly, for differences we have

$$\frac{d(u-v)}{dx} = \frac{du}{dx} - \frac{dv}{dx}. \tag{12}$$

It's easy to see that these rules generalize to sums and differences of more than two functions. For example, repeated application of Eq. (11) to the sum of a finite number of differentiable functions gives

$$\frac{d(u_1 + u_2 + \cdots + u_n)}{dx} = \frac{du_1}{dx} + \frac{du_2}{dx} + \cdots + \frac{du_n}{dx}. \tag{13}$$

REMARK Equation (13) tells us that, when differentiating a sum of terms, we simply differentiate each term and then add the results.

EXAMPLE 3

$$\begin{aligned} D_x(36 + 26x + 7x^5 - 5x^9) &= 0 + 26 \cdot 1 + 7 \cdot 5x^4 - 5 \cdot 9x^8 \\ &= 26 + 35x^4 - 45x^8. \end{aligned}$$ ◆

The Derivative of a Polynomial

When we apply Eqs. (10) and (13) and the power rule to the polynomial

$$p(x) = a_n x^n + a_{n-1}x^{n-1} + \cdots + a_2x^2 + a_1x + a_0$$

(and thus differentiate termwise), we find the derivative *as fast as we can write it:*

$$p'(x) = na_nx^{n-1} + (n-1)a_{n-1}x^{n-2} + \cdots + 3a_3x^2 + 2a_2x + a_1. \tag{14}$$

With this result, it becomes a routine matter to write an equation for a line tangent to the graph of a polynomial.

EXAMPLE 4 Write an equation for the straight line that is tangent to the graph of $y = 2x^3 - 7x^2 + 3x + 4$ at the point $(1, 2)$.

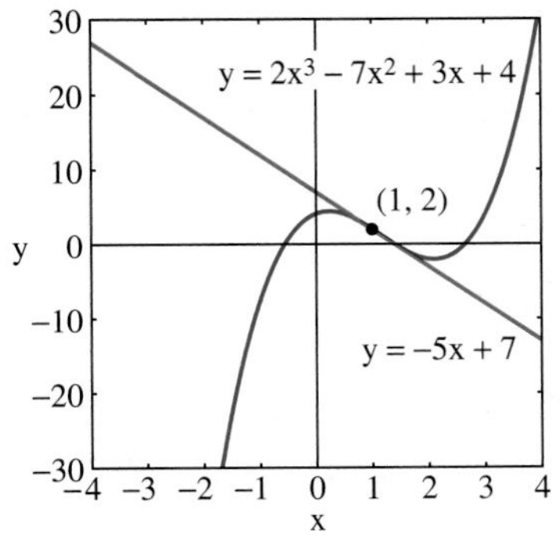

FIGURE 3.2.3 The graph $y = 2x^3 - 7x^2 + 3x + 4$ and its tangent line $y = -5x + 7$ at the point $(1, 2)$.

Solution We compute the derivative as in Eq. (14):

$$\frac{dy}{dx} = 2 \cdot 3x^2 - 7 \cdot 2x + 3 + 0 = 6x^2 - 14x + 3.$$

We substitute $x = 1$ in dy/dx and find that the slope of the tangent line at $(1, 2)$ is $m = -5$. So the point-slope equation of the tangent line is

$$y - 2 = -5(x - 1);$$

that is,

$$y = -5x + 7.$$

A calculator- or computer-generated picture like Fig. 3.2.3 provides suggestive visual evidence of the validity of this tangent line computation. ◆

EXAMPLE 5 The volume V (in cubic centimeters) of a given sample of water varies with changing temperature T. For T between 0°C and 30°C, the relation is given almost exactly by the formula

$$V = V_0[1 - (6.427 \times 10^{-5})T + (8.505 \times 10^{-6})T^2 - (6.790 \times 10^{-8})T^3],$$

where V_0 is the volume of the water (*not* ice) sample at 0°C. Suppose that $V_0 = 10^5$ cm^3. Find both the volume and the rate of change of volume with respect to temperature when $T = 20$°C.

Solution Substituting $V_0 = 10^5 = 100{,}000$ in the given volume formula yields

$$V(T) = 100{,}000 - (6.427)T + (0.8505)T^2 - (0.00679)T^3.$$

Then substituting $T = 20$ yields $V(20) \approx 100{,}157.34$, so the sample would expand by about 157 cm^3 if heated from 0°C to 20°C. The rate of change of volume V with respect to temperature T is given by

$$\frac{dV}{dT} = -6.427 + (1.7010)T - (0.02037)T^2,$$

and substituting $T = 20$ here yields

$$\left.\frac{dV}{dT}\right|_{T=20} \approx 19.45 \quad (\text{cm}^3/°\text{C}).$$

Thus we should expect the volume of the water sample to increase by slightly more than 19 cm^3 if it is heated by 1°C from 20°C to 21°C. In fact, direct substitution into the original volume formula gives

$$V(21) - V(20) \approx 19.88.$$

Finally, we note that the average rate of change of V with respect to T on the interval $19.5 \leqq T \leqq 20.5$ centered at $T = 20$ is

$$\frac{\Delta V}{\Delta T} = \frac{V(20.5) - V(19.5)}{20.5 - 19.5} \approx 19.44 \quad (\text{cm}^3/°\text{C}),$$

which is very close to the derivative dV/dT at $T = 20$. ◆

The Product Rule and the Quotient Rule

It might be natural to conjecture that the derivative of a product $f(x)g(x)$ is the product of the derivatives. This is *false!* For example, if $f(x) = g(x) = x$, then

$$D_x[f(x)g(x)] = D_x x^2 = 2x.$$

But

$$[D_x f(x)] \cdot [D_x g(x)] = (D_x x) \cdot (D_x x) = 1 \cdot 1 = 1.$$

In general, the derivative of a product is *not* merely the product of the derivatives. Theorem 4 tells us what it *is*.

THEOREM 4 The Product Rule

If f and g are differentiable at x, then fg is differentiable at x, and

$$D_x[f(x)g(x)] = f'(x)g(x) + f(x)g'(x). \tag{15}$$

With $u = f(x)$ and $v = g(x)$, this **product rule** takes the form

$$\frac{d(uv)}{dx} = u\frac{dv}{dx} + v\frac{du}{dx}. \tag{15'}$$

When it is clear what the independent variable is, we can write the product rule even more briefly:

$$(uv)' = u'v + uv'. \tag{15''}$$

PROOF We use an "add and subtract" device.

$$\begin{aligned}
D_x[f(x)g(x)] &= \lim_{h\to 0} \frac{f(x+h)g(x+h) - f(x)g(x)}{h} \\
&= \lim_{h\to 0} \frac{f(x+h)g(x+h) - f(x)g(x+h) + f(x)g(x+h) - f(x)g(x)}{h} \\
&= \lim_{h\to 0} \frac{f(x+h)g(x+h) - f(x)g(x+h)}{h} + \lim_{h\to 0} \frac{f(x)g(x+h) - f(x)g(x)}{h} \\
&= \left(\lim_{h\to 0} \frac{f(x+h) - f(x)}{h}\right)\left(\lim_{h\to 0} g(x+h)\right) + f(x)\left(\lim_{h\to 0} \frac{g(x+h) - g(x)}{h}\right) \\
&= f'(x)g(x) + f(x)g'(x).
\end{aligned}$$

In this proof we used the sum law and product law for limits, the definitions of $f'(x)$ and $g'(x)$, and the fact that

$$\lim_{h\to 0} g(x+h) = g(x).$$

This last equation holds because g is differentiable and therefore continuous at x (as we will see in Theorem 2 in Section 3.4).

In words, the product rule says that *the derivative of the product of two functions is formed by multiplying the derivative of each by the other and then adding the results.*

EXAMPLE 6 Find the derivative of

$$f(x) = (1 - 4x^3)(3x^2 - 5x + 2)$$

without first multiplying out the two factors.

Solution

$$\begin{aligned}
&D_x[(1 - 4x^3)(3x^2 - 5x + 2)] \\
&\quad = [D_x(1 - 4x^3)](3x^2 - 5x + 2) + (1 - 4x^3)[D_x(3x^2 - 5x + 2)] \\
&\quad = (-12x^2)(3x^2 - 5x + 2) + (1 - 4x^3)(6x - 5) \\
&\quad = -60x^4 + 80x^3 - 24x^2 + 6x - 5.
\end{aligned}$$

◆

We can apply the product rule repeatedly to find the derivative of a product of three or more differentiable functions $u_1, u_2, \dots, u_n$ of x. For example,

$$\begin{aligned} D[u_1u_2u_3] &= (u_1u_2)'u_3 + (u_1u_2)u_3' \\ &= (u_1'u_2 + u_1u_2')u_3 + u_1u_2u_3' \\ &= u_1'u_2u_3 + u_1u_2'u_3 + u_1u_2u_3'. \end{aligned}$$

Note that the derivative of each factor in the original product is multiplied by the other two factors and then the three resulting products are added. This is, indeed, the general result:

$$D(u_1u_2\cdots u_n) = u_1'u_2u_3\cdots u_{n-1}u_n + u_1u_2'u_3\cdots u_{n-1}u_n + \cdots + u_1u_2u_3\cdots u_{n-1}u_n', \tag{16}$$

where the sum in Eq. (16) has one term corresponding to each of the n factors in the product $u_1u_2\cdots u_n$. It is easy to establish this **extended product rule** (see Problem 62) one step at a time—next with $n = 4$, then with $n = 5$, and so forth.

Our next result tells us how to find the derivative of the reciprocal of a function if we know the derivative of the function itself.

THEOREM The Reciprocal Rule

If f is differentiable at x and $f(x) \neq 0$, then

$$D_x\left[\frac{1}{f(x)}\right] = -\frac{f'(x)}{[f(x)]^2}. \tag{17}$$

With $u = f(x)$, the reciprocal rule takes the form

$$\frac{d}{dx}\left(\frac{1}{u}\right) = -\frac{1}{u^2}\cdot\frac{du}{dx}. \tag{17$'$}$$

If there can be no doubt what the independent variable is, we can write

$$\left(\frac{1}{u}\right)' = -\frac{u'}{u^2}. \tag{17$''$}$$

PROOF As in the proof of Theorem 4, we use the limit laws, the definition of the derivative, and the fact that a function is continuous wherever it is differentiable (by Theorem 2 of Section 3.4). Moreover, note that $f(x+h) \neq 0$ for h near zero because $f(x) \neq 0$ and f is continuous at x. (See Problem 16 in Appendix D.) Therefore

$$D_x\left[\frac{1}{f(x)}\right] = \lim_{h\to 0}\frac{1}{h}\left(\frac{1}{f(x+h)} - \frac{1}{f(x)}\right) = \lim_{h\to 0}\frac{f(x) - f(x+h)}{hf(x+h)f(x)}$$

$$= -\left(\lim_{h\to 0}\frac{1}{f(x+h)f(x)}\right)\left(\lim_{h\to 0}\frac{f(x+h) - f(x)}{h}\right) = -\frac{f'(x)}{[f(x)]^2}.$$ ◄

EXAMPLE 7 With $f(x) = x^2 + 1$ in Eq. (17), we get

$$D_x\left(\frac{1}{x^2+1}\right) = -\frac{D_x(x^2+1)}{(x^2+1)^2} = -\frac{2x}{(x^2+1)^2}.$$ ◆

We now combine the reciprocal rule with the power rule for positive integral exponents to establish the power rule for negative integral exponents.

THEOREM 5 Power Rule for a Negative Integer n

If n is a negative integer, then $D_x x^n = nx^{n-1}$.

PROOF Let $m = -n$, so that m is a *positive* integer. If $x \neq 0$ then we can apply the reciprocal rule with $f(x) = x^m \neq 0$ and $f'(x) = mx^{m-1}$ (the latter by the power rule with positive integer exponent). This gives

$$D_x x^n = D_x\left(\frac{1}{x^m}\right) = -\frac{D_x(x^m)}{(x^m)^2} = -\frac{mx^{m-1}}{x^{2m}} = (-m)x^{-m-1} = nx^{n-1}.$$

Thus we have established that the rule in Theorem 5 holds precisely where the function being differentiated is defined—that is, where $x \neq 0$. ◀

EXAMPLE 8

$$D_x\left(\frac{5x^4 - 6x + 7}{2x^2}\right) = D_x\left(\tfrac{5}{2}x^2 - 3x^{-1} + \tfrac{7}{2}x^{-2}\right)$$

$$= \tfrac{5}{2}(2x) - 3(-x^{-2}) + \tfrac{7}{2}(-2x^{-3}) = 5x + \frac{3}{x^2} - \frac{7}{x^3}.$$

The key here was to "divide out" before differentiating. ◆

Now we apply the product rule and reciprocal rule to get a rule for differentiation of the quotient of two functions.

THEOREM 6 The Quotient Rule

If f and g are differentiable at x and $g(x) \neq 0$, then f/g is differentiable at x and

$$D_x\left[\frac{f(x)}{g(x)}\right] = \frac{f'(x)g(x) - f(x)g'(x)}{[g(x)]^2}. \tag{18}$$

With $u = f(x)$ and $v = g(x)$, this rule takes the form

$$\frac{d}{dx}\left(\frac{u}{v}\right) = \frac{v\dfrac{du}{dx} - u\dfrac{dv}{dx}}{v^2} \tag{18'}$$

If it is clear what the independent variable is, we can write the quotient rule in the form

$$\left(\frac{u}{v}\right)' = \frac{u'v - uv'}{v^2}. \tag{18''}$$

PROOF We apply the product rule to the factorization

$$\frac{f(x)}{g(x)} = f(x) \cdot \frac{1}{g(x)}.$$

This gives

$$D_x\left[\frac{f(x)}{g(x)}\right] = [D_x f(x)] \cdot \frac{1}{g(x)} + f(x) \cdot D_x\left[\frac{1}{g(x)}\right]$$

$$= \frac{f'(x)}{g(x)} + f(x) \cdot \left(-\frac{g'(x)}{[g(x)]^2}\right) = \frac{f'(x)g(x) - f(x)g'(x)}{[g(x)]^2}. \quad ◀$$

Note that the numerator in Eq. (18) is *not* the derivative of the product of f and g. And the minus sign means that the *order* of terms in the numerator is important.

EXAMPLE 9 Find $z'(t) = dz/dt$ if z is given by

$$z = \frac{1 - t^3}{1 + t^4}.$$

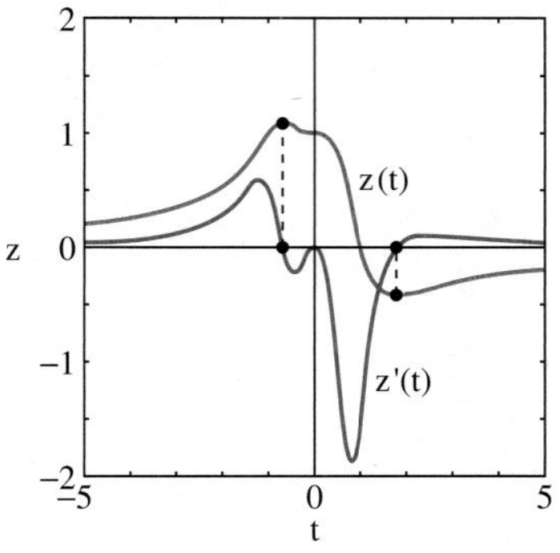

FIGURE 3.2.4 Graphs of the function $z(t)$ of Example 9 and its derivative $z'(t)$.

Solution Here, primes denote derivatives with respect to t. With t (rather than x) as the independent variable, the quotient rule gives

$$\frac{dz}{dt} = \frac{(1-t^3)'(1+t^4) - (1-t^3)(1+t^4)'}{(1+t^4)^2}$$

$$= \frac{(-3t^2)(1+t^4) - (1-t^3)(4t^3)}{(1+t^4)^2} = \frac{t^6 - 4t^3 - 3t^2}{(1+t^4)^2}.$$

Figure 3.2.4 shows computer-generated graphs of the function $z(t)$ and its derivative $z'(t)$. Observe that $z(t)$ is increasing on intervals where $z'(t)$ is positive and is decreasing on intervals where $z'(t)$ is negative (thus corroborating our computation of the derivative). A quick computer or calculator graph of a function and its alleged derivative will often reveal an error if one has been made. ◆

3.2 TRUE/FALSE STUDY GUIDE

3.2 CONCEPTS: QUESTIONS AND DISCUSSION

1. Theorems 2 and 5 in this section imply that the power rule $D_x x^n = nx^{n-1}$ holds provided that the integer n is nonzero. Does it also hold if $n = 0$? Can you think of a simple algebraic function whose derivative is a nonzero constant multiple of $1/x$?
2. Example 1 and the discussion preceding it may seem to imply that the power rule holds in the very general form $D[\text{whatever}]^n = n[\text{whatever}]^{n-1}$—more precisely, $D_x[f(x)]^n = n[f(x)]^{n-1}$. Is this true or false? When you're confronted with a question like this, don't just sit there. *Check it out!* Test the conjecture with specific choices for n and $f(x)$—perhaps $n = 7$ and $f(x) = x^{11}$. What are the simplest choices you can use to resolve the matter?

3.2 PROBLEMS

Apply the differentiation rules of this section to find the derivatives of the functions in Problems 1 through 40.

1. $f(x) = 3x^2 - x + 5$

2. $g(t) = 1 - 3t^2 - 2t^4$

3. $f(x) = (2x+3)(3x-2)$

4. $g(x) = (2x^2-1)(x^3+2)$

5. $h(x) = (x+1)^3$

6. $g(t) = (4t-7)^2$

7. $f(y) = y(2y-1)(2y+1)$

8. $f(x) = 4x^4 - \dfrac{1}{x^2}$

9. $g(x) = \dfrac{1}{x+1} - \dfrac{1}{x-1}$

10. $f(t) = \dfrac{1}{4-t^2}$

11. $h(x) = \dfrac{3}{x^2+x+1}$

12. $f(x) = \dfrac{1}{1-\dfrac{2}{x}}$

13. $g(t) = (t^2+1)(t^3+t^2+1)$

14. $f(x) = (2x^3-3)(17x^4-6x+2)$

15. $g(z) = \dfrac{1}{2z} - \dfrac{1}{3z^2}$

16. $f(x) = \dfrac{2x^3-3x^2+4x-5}{x^2}$

17. $g(y) = 2y(3y^2-1)(y^2+2y+3)$

18. $f(x) = \dfrac{x^2-4}{x^2+4}$

19. $g(t) = \dfrac{t-1}{t^2+2t+1}$

20. $u(x) = \dfrac{1}{(x+2)^2}$

21. $v(t) = \dfrac{1}{(t-1)^3}$

22. $h(x) = \dfrac{2x^3+x^2-3x+17}{2x-5}$

23. $g(x) = \dfrac{3x}{x^3+7x-5}$

24. $f(t) = \dfrac{1}{\left(t+\dfrac{1}{t}\right)^2}$

25. $g(x) = \dfrac{\dfrac{1}{x} - \dfrac{2}{x^2}}{\dfrac{2}{x^3} - \dfrac{3}{x^4}}$

26. $f(x) = \dfrac{x^3 - \dfrac{1}{x^2+1}}{x^4 + \dfrac{1}{x^2+1}}$

27. $y(x) = x^3 - 6x^5 + \frac{3}{2}x^{-4} + 12$

28. $x(t) = \dfrac{3}{t} - \dfrac{4}{t^2} - 5$

29. $y(x) = \dfrac{5-4x^2+x^5}{x^3}$

30. $u(x) = \dfrac{2x-3x^2+2x^4}{5x^2}$

31. $y(x) = 3x - \dfrac{1}{4x^2}$

32. $f(z) = \dfrac{1}{z(z^2+2z+2)}$

33. $y(x) = \dfrac{x}{x-1} + \dfrac{x+1}{3x}$

34. $u(t) = \dfrac{1}{1-4t^{-2}}$

35. $y(x) = \dfrac{x^3-4x+5}{x^2+9}$

36. $w(z) = z^2\left(2z^3 - \dfrac{3}{4z^4}\right)$

37. $y(x) = \dfrac{2x^2}{3x - \dfrac{4}{5x^4}}$

38. $z(t) = \dfrac{4}{(t^2 - 3)^2}$

39. $y(x) = \dfrac{x^2}{x + 1}$

40. $h(w) = \dfrac{w + 10}{w^2}$

In Problems 41 through 50, write an equation of the line tangent to the curve $y = f(x)$ at the given point P on the curve. Express the answer in the form $ax + by = c$.

41. $y = x^3$; $P(2, 8)$

42. $y = 3x^2 - 4$; $P(1, -1)$

43. $y = \dfrac{1}{x - 1}$; $P(2, 1)$

44. $y = 2x - \dfrac{1}{x}$; $P(0.5, -1)$

45. $y = x^3 + 3x^2 - 4x - 5$; $P(1, -5)$

46. $y = \left(\dfrac{1}{x} - \dfrac{1}{x^2}\right)^{-1}$; $P(2, 4)$

47. $y = \dfrac{3}{x^2} - \dfrac{4}{x^3}$; $P(-1, 7)$

48. $y = \dfrac{3x - 2}{3x + 2}$; $P(2, 0.5)$

49. $y = \dfrac{3x^2}{x^2 + x + 1}$; $P(-1, 3)$

50. $y = \dfrac{6}{1 - x^2}$; $P(2, -2)$

51. Apply the formula in Example 5 to answer the following two questions. (a) If 1000 cm^3 of water at 0°C is heated, does it initially expand or contract? (b) What is the rate (in cm^3/°C) at which it initially contracts or expands?

52. Susan's weight in pounds is given by the formula $W = (2 \times 10^9)/R^2$, where R is her distance in miles from the center of the earth. What is the rate of change of W with respect to R when $R = 3960$ mi? If Susan climbs a mountain, beginning at sea level, at what rate in ounces per (vertical) mile does her weight initially decrease?

53. The conical tank shown in Fig. 3.2.5 has radius 160 cm and height 800 cm. Water is running out of a small hole in the bottom of the tank. When the height h of water in the tank is 600 cm, what is the rate of change of its volume V with respect to h?

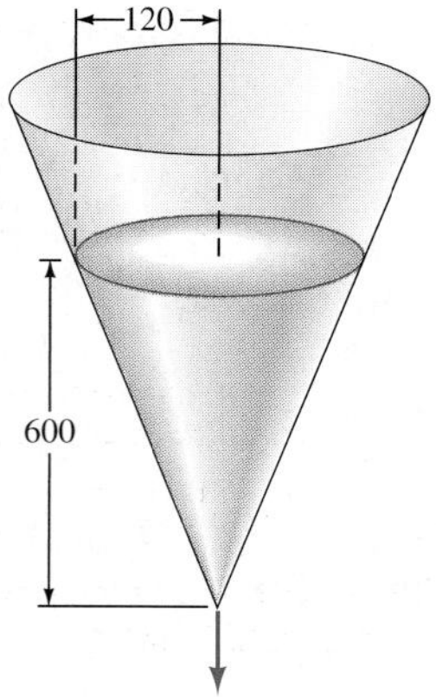

FIGURE 3.2.5 The leaky tank of Problem 53.

54. Find the x- and y-intercepts of the straight line that is tangent to the curve $y = x^3 + x^2 + x$ at the point (1, 3) (Fig. 3.2.6).

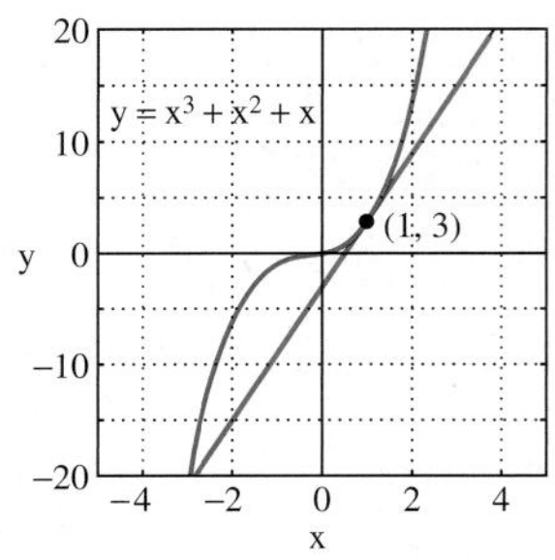

FIGURE 3.2.6 The tangent line of Problem 54.

55. Find an equation for the straight line that passes through the point (1, 5) and is tangent to the curve $y = x^3$. [*Suggestion:* Denote by (a, a^3) the point of tangency, as indicated in Fig 3.2.7. Find by inspection small integral solutions of the resulting cubic equation in a.]

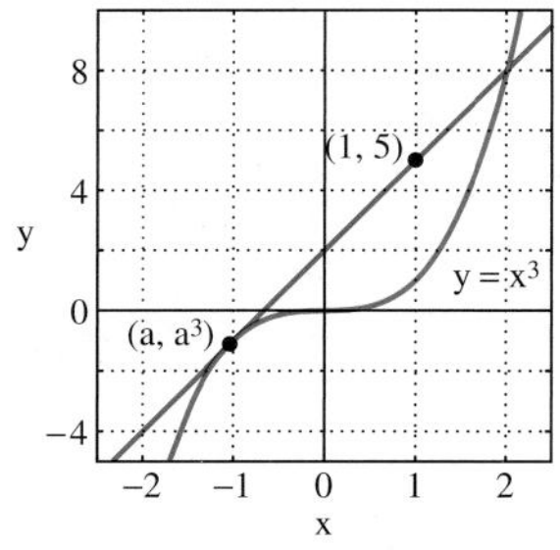

FIGURE 3.2.7 The tangent line of Problem 55.

56. Find *two* lines through the point (2, 8) that are tangent to the curve $y = x^3$. [See the suggestion for Problem 55.]

57. Prove that no straight line can be tangent to the curve $y = x^2$ at two different points.

58. Find the two straight lines of slope −2 that are tangent to the curve $y = 1/x$.

59. Let $n \geqq 2$ be a fixed but unspecified integer. Find the x-intercept of the line that is tangent to the curve $y = x^n$ at the point $P(x_0, y_0)$.

60. Prove that the curve $y = x^5 + 2x$ has no horizontal tangents. What is the smallest slope that a line tangent to this curve can have?

61. Apply Eq. (16) with $n = 3$ and $u_1 = u_2 = u_3 = f(x)$ to show that

$$D_x([f(x)]^3) = 3[f(x)]^2 \cdot f'(x).$$

62. (a) First write $u_1u_2u_3u_4 = (u_1u_2u_3)u_4$ to verify Eq. (16) for $n = 4$. (b) Then write $u_1u_2u_3u_4u_5 = (u_1u_2u_3u_4)u_5$ and apply the result in part (a) to verify Eq. (16) for $n = 5$.

63. Apply Eq. (16) to show that

$$D_x([f(x)]^n) = n[f(x)]^{n-1} \cdot f'(x)$$

if n is a positive integer and $f'(x)$ exists.

64. Use the result of Problem 63 to compute $D_x[(x^2+x+1)^{100}]$.

65. Use the result of Problem 63 to find $g'(x)$ given $g(x) = (x^3 - 17x + 35)^{17}$.

66. Find constants a, b, c, and d such that the graph of

$$f(x) = ax^3 + bx^2 + cx + d$$

has horizontal tangent lines at the points (0, 1) and (1, 0).

In connection with Problems 67 through 71, Figs. 3.2.8 through 3.2.11 show the curves

$$y = \frac{x^n}{1+x^2}$$

for $n = 0, 1, 2,$ *and* 3.

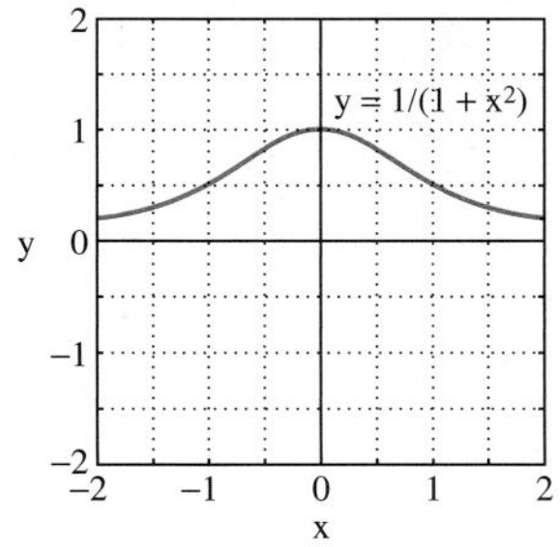

FIGURE 3.2.8 The graph of $y = \dfrac{1}{1+x^2}$.

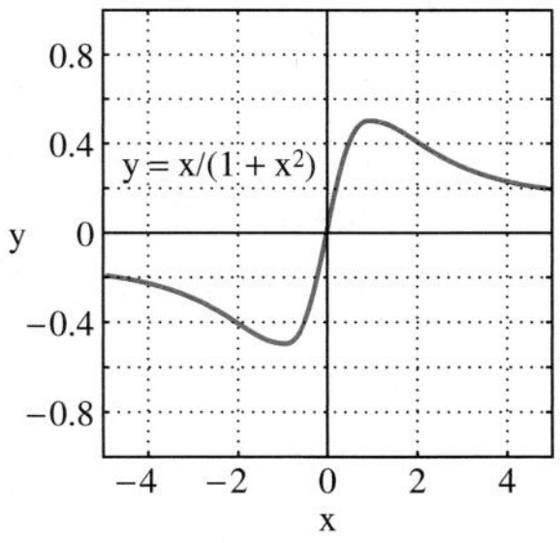

FIGURE 3.2.9 The graph of $y = \dfrac{x}{1+x^2}$.

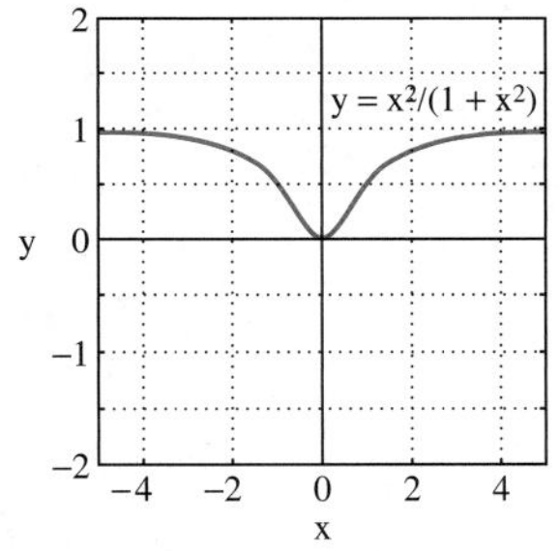

FIGURE 3.2.10 The graph of $y = \dfrac{x^2}{1+x^2}$.

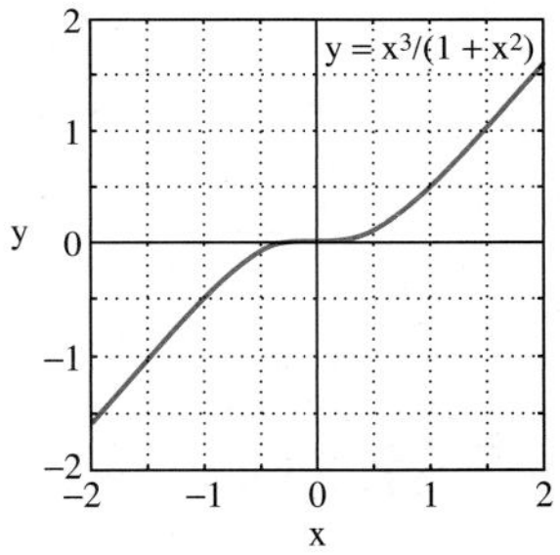

FIGURE 3.2.11 The graph of $y = \dfrac{x^3}{1+x^2}$.

67. Show that for $n = 0$ and $n = 2$, the curve has only a single point where the tangent line is horizontal (Figs. 3.2.8 and 3.2.10).

68. When $n = 1$, there are two points on the curve where the tangent line is horizontal (Fig. 3.2.9). Find them.

69. Show that for $n \geqq 3$, (0, 0) is the only point on the graph of

$$y = \frac{x^n}{1+x^2}$$

at which the tangent line is horizontal (Fig. 3.2.11).

70. Figure 3.2.12 shows the graph of the derivative $f'(x)$ of the function

$$f(x) = \frac{x^3}{1+x^2}.$$

There appear to be two points on the graph of $y = f(x)$ at which the tangent line has slope 1. Find them.

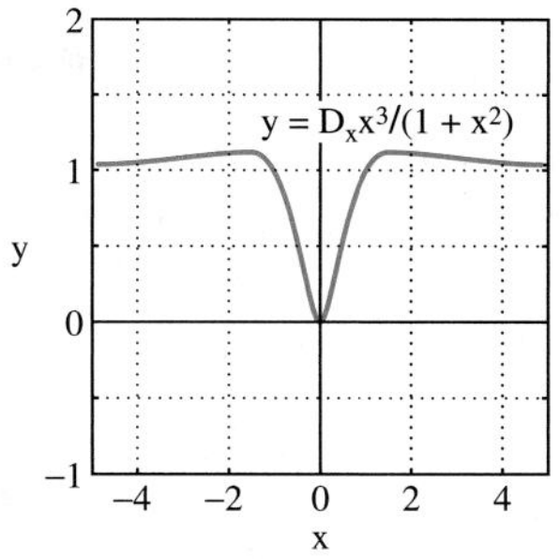

FIGURE 3.2.12 The graph of $y = D_x\left(\dfrac{x^3}{1+x^2}\right)$ of Problems 70 and 71.

71. It appears in Fig. 3.2.12 that there are three points on the curve $y = f'(x)$ at which the tangent line is horizontal. Find them.

72. Much of life on earth (as we know it) depends critically on the variation of water density with temperature. Consider a sample of water than has a volume of exactly 1000 cm^3 when measured at precisely 0°C. Figure 3.2.13 shows a graph of its volume function $V(T)$ as given by the formula in Example 5. The surprise is that, as the temperature is increased, the sample initially contracts rather than expands in volume. Evidently a minimal volume $V_m = V(T_m)$ occurs at a critical temperature $T_m \approx 4$ (°C). Given that the tangent line to the graph of V is horizontal at the point (T_m, V_m), find: (a) the numerical values of T_m and V_m, and (b) the temperature $T_1 \approx 8$ (°C) at which the volume of the sample is again exactly 1000 cm^3. *Comment:* Because water that's slightly warmer than the freezing point of 0°C is slightly denser than water at 0°C, the warmer water sinks to the bottom as a cooling lake freezes. But ice is less dense, so it floats on the surface. Consequently, ice at the surface traps somewhat warmer water at the bottom of the lake—which otherwise might freeze solid. This phenomenon is responsible for the survival and evolution of life forms that can withstand cold water but not freezing.

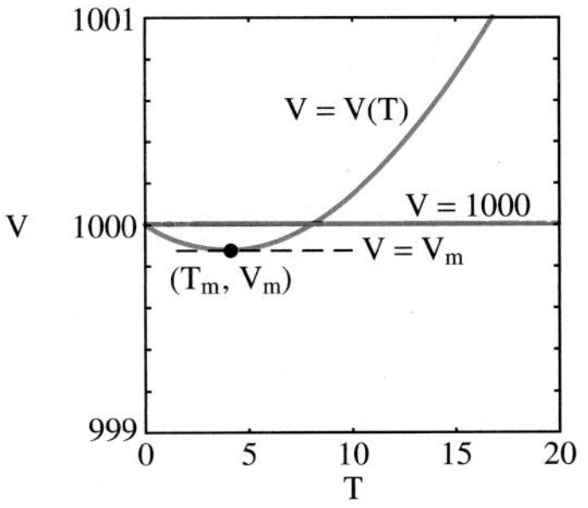

FIGURE 3.2.13 The temperature-volume graph of Problem 72.

In Problems 73 through 78, sketch the graph of the given function f and determine where it is differentiable. Recall the definition *of one-sided derivatives in Problem 54 of Section 3.1, as well as the fact that $f'(a)$ exists if and only if $f'_-(a) = f'_+(a)$.*

73. $f(x) = |x^3|$

74. $f(x) = x^3 + |x^3|$

75. $f(x) = \begin{cases} 2 + 3x^2 & \text{if } x < 1, \\ 3 + 2x^3 & \text{if } x \geqq 1 \end{cases}$

76. $f(x) = \begin{cases} x^4 & \text{if } x < 1, \\ 2 - \dfrac{1}{x^4} & \text{if } x \geqq 1 \end{cases}$

77. $f(x) = \begin{cases} \dfrac{1}{2 - x} & \text{if } x < 1, \\ x & \text{if } x \geqq 1 \end{cases}$

78. $f(x) = \begin{cases} \dfrac{12}{(5 - x)^2} & \text{if } x < 3, \\ x^2 - 3x + 3 & \text{if } x \geqq 3 \end{cases}$

3.3 THE CHAIN RULE

We saw in Section 3.2 how to differentiate powers of the independent variable, but we often need to differentiate powers of rather general (or even unknown) functions. For instance, suppose that

$$y = u^3 \tag{1}$$

where u is in turn a function of x. Then the extended product rule [Eq. (16) in Section 3.2] yields

$$\frac{dy}{dx} = D_x u^3 = D_x(u \cdot u \cdot u) = u' \cdot u \cdot u + u \cdot u' \cdot u + u \cdot u \cdot u'$$

where $u' = du/dx$. After we collect terms, we find that

$$\frac{dy}{dx} = 3u^2 u' = 3u^2 \frac{du}{dx}. \tag{2}$$

Is it a surprise that the derivative of u^3 is *not* simply $3u^2$, which you might expect in analogy with the correct formula $D_x x^3 = 3x^2$? There is an additional factor du/dx, whose presence may seem more natural if we differentiate y in Eq. (1) with respect to u, and write

$$\frac{dy}{du} = 3u^2.$$

Then the derivative formula in (2) takes the form

$$\frac{dy}{dx} = \frac{dy}{du} \cdot \frac{du}{dx}. \tag{3}$$

Equation (3), the **chain rule,** holds for *any* two differentiable functions $y = f(u)$ and $u = g(x)$. The formula in Eq. (2) is simply the special case of (3) with $f(u) = u^3$.

EXAMPLE 1 If

$$y = (3x^2 + 5)^{17},$$

it would be impractical to write the binomial expansion of the seventeenth power of $3x^2 + 5$ before differentiating. The `Expand` command in a typical computer algebra system yields a polynomial in x having 18 terms, some of which have 15-digit coefficients:

$$(3x^2 + 5)^{17} = 762939453125 + 7781982421857x^2 + \cdots$$
$$+ 186911613281250x^{18} + \cdots + 129140163x^{34}.$$

(Each ellipsis replaces seven omitted terms.) But if we simply write

$$y = u^{17} \quad \text{with} \quad u = 3x^2 + 5,$$

then

$$\frac{dy}{du} = 17u^{16} \quad \text{and} \quad \frac{du}{dx} = 6x.$$

Hence the chain rule yields

$$\begin{aligned}\frac{dy}{dx} &= \frac{dy}{du} \cdot \frac{du}{dx} = 17u^{16} \cdot 6x \\ &= 17(3x^2+5)^{16} \cdot 6x = 102x(3x^2+5)^{16}.\end{aligned}$$

◆

The formula in (3) is one that, once learned, is unlikely to be forgotten. Although dy/du and du/dx are *not* fractions—they are merely symbols representing the derivatives $f'(u)$ and $g'(x)$—it is much as though they *were* fractions, with the du in the first factor canceling the du in the second factor:

$$\frac{dy}{du} \cdot \frac{du}{dx} = \frac{dy}{\not{du}} \cdot \frac{\not{du}}{dx} = \frac{dy}{dx}. \qquad \textbf{[Invalid cancellation!]}$$

But you should realize that such "cancellation" no more proves the chain rule than canceling two copies of the symbol d proves that

$$\frac{dy}{dx} = \frac{\not{d}y}{\not{d}x} = \frac{y}{x} \qquad \textbf{[An absurdity!]}.$$

It is nevertheless an excellent way to *remember* the chain rule. Such manipulations with differentials are so suggestive (even when invalid) that they played a substantial role in the early development of calculus in the seventeenth and eighteenth centuries. Many formulas were thereby produced (and later proved valid), as were some formulas that were incorrect.

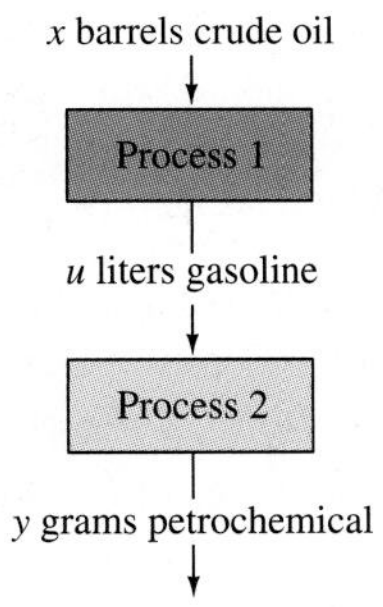

FIGURE 3.3.1 The two-process oil refinery (Example 2).

EXAMPLE 2 For a physical interpretation of the chain rule, imagine an oil refinery that first makes u liters of gasoline from x barrels of crude oil. Then, in a second process, the refinery makes y grams of a marketable petrochemical from the u liters of gasoline. (The two processes are illustrated in Fig. 3.3.1.) Then y is a function of u and u is a function of x, so the final output y is a function also of the input x. Consider the *units* in which the *derivatives* of these functions are measured.

$$\frac{dy}{du} : \frac{\text{g}}{\text{L}} \qquad \text{(grams of petrochemical per liter of gasoline)}$$

$$\frac{du}{dx} : \frac{\text{L}}{\text{barrel}} \qquad \text{(liters of gasoline per barrel of oil)}$$

$$\frac{dy}{dx} : \frac{\text{g}}{\text{barrel}} \qquad \text{(grams of petrochemical per barrel of oil)}$$

When we include the units in the chain rule equation

$$\frac{dy}{dx} = \frac{dy}{du}\frac{du}{dx},$$

we get

$$\frac{dy}{dx}\,\frac{\text{g}}{\text{barrel}} = \left(\frac{dy}{du}\,\frac{\text{g}}{\not{\text{L}}}\right) \cdot \left(\frac{du}{dx}\,\frac{\not{\text{L}}}{\text{barrel}}\right) = \left(\frac{dy}{du} \cdot \frac{du}{dx}\right) \frac{\text{g}}{\text{barrel}}.$$

The handy cancellation of units seems to confirm the validity of the chain rule (at least in this application). For example, if we get 3 g of petrochemical per liter of gasoline and 75 L of gasoline per barrel of oil, how could we fail to get $225 = 3 \cdot 75$ g of petrochemical per barrel of oil? ◆

The Chain Rule in Function Notation

Although Eq. (3) is a memorable statement of the chain rule in differential notation, it has the disadvantage of not specifying the values of the variables at which the derivatives are evaluated. This problem can be solved by the use of function notation for the derivatives. Let us write

$$y = f(u), \qquad u = g(x) \qquad y = h(x) = f(g(x)).$$

Then

$$\frac{du}{dx} = g'(x), \qquad \frac{dy}{dx} = h'(x),$$

and

$$\frac{dy}{du} = f'(u) = f'(g(x)).$$

Substituting these derivatives into the chain rule formula

$$\frac{dy}{dx} = \frac{dy}{du} \cdot \frac{du}{dx} \tag{3}$$

recasts it in the form

$$h'(x) = f'(g(x)) \cdot g'(x). \tag{4}$$

This version of the chain rule gives the derivative of the *composition* $h = f \circ g$ of two functions f and g in terms of *their* derivatives.

THEOREM 1 The Chain Rule

Suppose that g is differentiable at x and that f is differentiable at $g(x)$. Then the composition $h = f \circ g$ defined by $h(x) = f(g(x))$ is differentiable at x, and its derivative is

$$h'(x) = f'(g(x)) \cdot g'(x). \tag{4}$$

REMARK The chain rule in (4) shows that the derivative of the composition $h = f \circ g$ is a *product* of the derivatives of f and g. Note, however, that these two derivatives are evaluated at *different* points. The derivative g' of the *inner function* is evaluated at x, whereas the derivative f' of the *outer function* is evaluated at $g(x)$ (rather than at the same point x).

EXAMPLE 3 In Example 1 we applied the differential form of the chain rule in (3) to differentiate the function

$$h(x) = (3x^2 + 5)^{17}.$$

To apply the functional form of the chain rule in (4), we must first identify the outer function

$$f(x) = x^{17}, \qquad \text{for which} \quad f'(x) = 17x^{16},$$

and the inner function

$$g(x) = 3x^2 + 5, \qquad \text{for which} \quad g'(x) = 6x.$$

Then

$$\begin{aligned} h'(x) &= f'(g(x)) \cdot g'(x) \\ &= f'(3x^2 + 5) \cdot (3x^2 + 5)' \\ &= 17(3x^2 + 5)^{16} \cdot 6x = 102x(3x^2 + 5)^{16}. \end{aligned}$$

◆

The Proof of the Chain Rule

To *outline* a proof of the chain rule, suppose that we are given differentiable functions $y = f(u)$ and $u = g(x)$ and want to compute the derivative

$$\frac{dy}{dx} = \lim_{\Delta x \to 0} \frac{\Delta y}{\Delta x} = \lim_{\Delta x \to 0} \frac{f(g(x + \Delta x)) - f(g(x))}{\Delta x}. \tag{5}$$

The differential form of the chain rule suggests the factorization

$$\frac{\Delta y}{\Delta x} = \frac{\Delta y}{\Delta u} \frac{\Delta u}{\Delta x} \tag{6}$$

where

$$\Delta u = g(x + \Delta x) - g(x) \quad \text{and} \quad \Delta y = f(u + \Delta u) - f(u).$$

For x fixed, the factorization in Eq. (6) is valid if $g'(x) \neq 0$, because

$$g'(x) = \frac{du}{dx} = \lim_{\Delta x \to 0} \frac{\Delta u}{\Delta x} \neq 0$$

implies that $\Delta u \neq 0$ if $\Delta x \neq 0$ is sufficiently small—for if so, then $\Delta u = (\Delta u / \Delta x) \cdot \Delta x$ is the product of nonzero numbers. But the fact that g is differentiable, and therefore continuous, at the point x (see Theorem 2 in Section 3.4) implies that

$$\Delta u = g(x + \Delta x) - g(x) \to 0 \qquad \text{as} \quad \Delta x \to 0.$$

The product law of limits therefore gives

$$\frac{dy}{dx} = \lim_{\Delta x \to 0} \frac{\Delta y}{\Delta u} \cdot \frac{\Delta u}{\Delta x} = \left(\lim_{\Delta u \to 0} \frac{\Delta y}{\Delta u} \right) \cdot \left(\lim_{\Delta x \to 0} \frac{\Delta u}{\Delta x} \right) = \frac{dy}{du} \cdot \frac{du}{dx}.$$

Thus we have shown that $D_x[f(g(x))] = f'(g(x)) \cdot g'(x)$ at any point x at which $g'(x) \neq 0$. But if $g'(x) = 0$, then it is entirely possible that Δu *is* zero for some or all nonzero values of Δx approaching zero—in which case the factorization in (6) is invalid. Our proof of the chain rule is therefore incomplete. In Section 4.2 we give a proof that does not require the assumption that $g'(x) \neq 0$.

The Generalized Power Rule

If we substitute $g(x) = u$ and $g'(x) = du/dx$ into Eq. (4) with $h'(x) = D_x f(g(x)) = D_x f(u)$, we get the hybrid form

$$D_x[f(u)] = f'(u) \cdot \frac{du}{dx} \tag{7}$$

of the chain rule that frequently is the most useful form for purely computational purposes. Recall that the subscript x in D_x specifies that $f(u)$ is being differentiated with respect to x, *not* with respect to u.

Let us set $f(u) = u^n$ in Eq. (7), where n is an integer. Because $f'(u) = nu^{n-1}$, we thereby obtain

$$D_x u^n = nu^{n-1} \frac{du}{dx}, \tag{8}$$

the *chain rule version* of the power rule. If $u = g(x)$ is a differentiable function, then Eq. (8) implies that

$$D_x[g(x)]^n = n[g(x)]^{n-1} \cdot D_x[g(x)]. \tag{9}$$

[If $n - 1 < 0$, we must add the proviso that $g(x) \neq 0$ in order for the right-hand side in Eq. (9) to be meaningful.] We refer to this chain rule version of the power rule as the **generalized power rule.**

REMARK We may interpret the operator form in (9) as describing a chain rule procedure in which we work *from the outside to the inside*—differentiating first the outer function and then the inner function. This outside-inside procedure is illustrated in the next example.

EXAMPLE 4 To differentiate

$$y = \frac{1}{(2x^3 - x + 7)^2},$$

we first write

$$y = (2x^3 - x + 7)^{-2}$$

in order to apply the generalized power rule, Eq. (9), with $n = -2$. This gives

$$\frac{dy}{dx} = \underbrace{(-2)(2x^3 - x + 7)^{-3}}_{\text{derivative of outer function}} \cdot D_x(2x^3 - x + 7)$$

$$= (-2)(2x^3 - x + 7)^{-3} \cdot \underbrace{(6x^2 - 1)}_{\text{derivative of inner function}} = \frac{2(1 - 6x^2)}{(2x^3 - x + 7)^3}.$$ ◆

EXAMPLE 5 Find the derivative of the function

$$h(z) = \left(\frac{z-1}{z+1}\right)^5.$$

Solution The key to applying the generalized power rule is observing *what* the given function is a power *of*. Here,

$$h(z) = u^5, \qquad \text{where} \quad u = \frac{z-1}{z+1},$$

and z, not x, is the independent variable. Hence we apply first Eq. (8) and then the quotient rule to get

$$h'(z) = 5u^4\frac{du}{dz} = 5\left(\frac{z-1}{z+1}\right)^4 D_z\left(\frac{z-1}{z+1}\right)$$

$$= 5\left(\frac{z-1}{z+1}\right)^4 \cdot \frac{(1)(z+1) - (z-1)(1)}{(z+1)^2}$$

$$= 5\left(\frac{z-1}{z+1}\right)^4 \cdot \frac{2}{(z+1)^2} = \frac{10(z-1)^4}{(z+1)^6}.$$ ◆

The importance of the chain rule goes far beyond the power function differentiations illustrated in Examples 1, 4, and 5. We shall learn in later sections how to differentiate exponential, logarithmic, and trigonometric functions. Each time we learn a new differentiation formula—for the derivative $f'(x)$ of a new function $f(x)$—the formula in Eq. (7) immediately provides us with the chain rule version of that formula,

$$D_x f(u) = f'(u) D_x u.$$

The step from the power rule $D_x x^n = nx^{n-1}$ to the generalized power rule $D_x u^n = nu^{n-1} D_x u$ is our first instance of this general phenomenon.

Rate-Of-Change Applications

Suppose that the physical or geometric quantity p depends on the quantity q, which in turn depends on time t. Then the *dependent* variable p is a function both of the *intermediate* variable q and of the *independent* variable t. Hence the derivatives that appear in the chain rule formula

$$\frac{dp}{dt} = \frac{dp}{dq}\frac{dq}{dt}$$

are rates of change (as in Section 3.1) of these variables with respect to one another. For instance, suppose that a spherical balloon is being inflated or deflated. Then its volume V and its radius r are changing with time t, and

$$\frac{dV}{dt} = \frac{dV}{dr}\frac{dr}{dt}.$$

Remember that a positive derivative signals an increasing quantity and that a negative derivative signals a decreasing quantity.

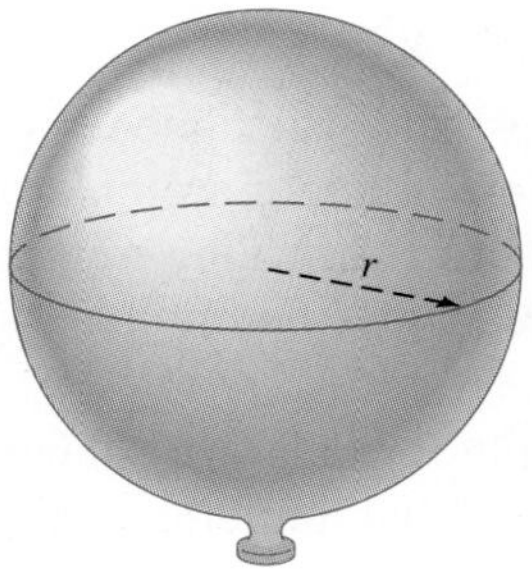

FIGURE 3.3.2 The spherical balloon with volume $V = \frac{4}{3}\pi r^3$.

EXAMPLE 6 A spherical balloon is being inflated (Fig. 3.3.2). The radius r of the balloon is increasing at the rate of 0.2 cm/s when $r = 5$ cm. At what rate is the volume V of the balloon increasing at that instant?

Solution Given $dr/dt = 0.2$ cm/s when $r = 5$ cm, we want to find dV/dt at that instant. Because the volume of the balloon is

$$V = \tfrac{4}{3}\pi r^3,$$

we see that $dV/dr = 4\pi r^2$. So the chain rule gives

$$\frac{dV}{dt} = \frac{dV}{dr}\cdot\frac{dr}{dt} = 4\pi r^2\frac{dr}{dt} = 4\pi(5)^2(0.2) \approx 62.83 \quad (\text{cm}^3/\text{s})$$

at the instant when $r = 5$ cm. ◆

In Example 6 we did not need to know r explicitly as a function of t. But suppose we are told that after t seconds the radius (in centimeters) of an inflating balloon is $r = 3 + (0.2)t$ (until the balloon bursts). Then the volume of this balloon is

$$V = \frac{4}{3}\pi r^3 = \frac{4}{3}\pi\left(3 + \frac{t}{5}\right)^3,$$

so dV/dt is given explicitly as a function of t by

$$\frac{dV}{dt} = \frac{4}{3}\pi(3)\left(3 + \frac{t}{5}\right)^2\left(\frac{1}{5}\right) = \frac{4}{5}\pi\left(3 + \frac{t}{5}\right)^2.$$

EXAMPLE 7 Imagine a spherical raindrop that is falling through water vapor in the air. Suppose that the vapor adheres to the surface of the raindrop in such a way that the time rate of increase of the mass M of the droplet is proportional to the surface area S of the droplet. If the initial radius of the droplet is, in effect, zero and the radius is $r = 1$ mm after 20 s, when is the radius 3 mm?

Solution We are given

$$\frac{dM}{dt} = kS,$$

where k is some constant that depends upon atmospheric conditions. Now

$$M = \tfrac{4}{3}\pi\rho r^3 \quad \text{and} \quad S = 4\pi r^2,$$

where ρ is the density of water. Hence the chain rule gives

$$4\pi kr^2 = kS = \frac{dM}{dt} = \frac{dM}{dr}\cdot\frac{dr}{dt};$$

that is,

$$4\pi kr^2 = 4\pi\rho r^2\frac{dr}{dt}.$$

This implies that

$$\frac{dr}{dt} = \frac{k}{\rho},$$

a constant. So the radius of the droplet grows at a *constant* rate. Thus if it takes 20 s for r to grow to 1 mm, it will take 1 min for r to grow to 3 mm. ◆

3.3 TRUE/FALSE STUDY GUIDE

3.3 CONCEPTS: QUESTIONS AND DISCUSSION

1. As a mathematical quiz show contestant, you are asked to calculate the value $F'(7)$ for the composition $F = f \circ g$. The functions f and g are unknown, but you are permitted to ask exactly three questions regarding numerical values of these functions and/or their derivatives at specified points. What three questions should you ask?
2. Write the function-notation form of the chain rule formula in Problem 63 for a composition $F = f \circ g \circ h$ of three functions. What numerical data are now needed to calculate the numerical value $F'(7)$?

3.3 PROBLEMS

Find dy/dx in Problems 1 through 12.

1. $y = (3x + 4)^5$ **2.** $y = (2 - 5x)^3$

3. $y = \dfrac{1}{3x - 2}$ **4.** $y = \dfrac{1}{(2x + 1)^3}$

5. $y = (x^2 + 3x + 4)^3$ **6.** $y = (7 - 2x^3)^{-4}$

7. $y = (2 - x)^4(3 + x)^7$ **8.** $y = (x + x^2)^5(1 + x^3)^2$

9. $y = \dfrac{x + 2}{(3x - 4)^3}$ **10.** $y = \dfrac{(1 - x^2)^3}{(4 + 5x + 6x^2)^2}$

11. $y = [1 + (1 + x)^3]^4$ **12.** $y = [x + (x + x^2)^{-3}]^{-5}$

In Problems 13 through 20, express the derivative dy/dx in terms of x.

13. $y = (u + 1)^3$ and $u = \dfrac{1}{x^2}$

14. $y = \dfrac{1}{2u} - \dfrac{1}{3u^2}$ and $u = 2x + 1$

15. $y = (1 + u^2)^3$ and $u = (4x - 1)^2$

16. $y = u^5$ and $u = \dfrac{1}{3x - 2}$

17. $y = u(1 - u)^3$ and $u = \dfrac{1}{x^4}$

18. $y = \dfrac{u}{u + 1}$ and $u = \dfrac{x}{x + 1}$

19. $y = u^2(u - u^4)^3$ and $u = \dfrac{1}{x^2}$

20. $y = \dfrac{u}{(2u + 1)^4}$ and $u = x - \dfrac{2}{x}$

In Problems 21 through 26, identify a function u of x and an integer $n \neq 1$ such that $f(x) = u^n$. Then compute $f'(x)$.

21. $f(x) = (2x - x^2)^3$ **22.** $f(x) = \dfrac{1}{2 + 5x^3}$

23. $f(x) = \dfrac{1}{(1 - x^2)^4}$ **24.** $f(x) = (x^2 - 4x + 1)^3$

25. $f(x) = \left(\dfrac{x + 1}{x - 1}\right)^7$ **26.** $f(x) = \dfrac{(x^2 + x + 1)^4}{(x + 1)^4}$

Differentiate the functions given in Problems 27 through 36.

27. $g(y) = y + (2y - 3)^5$ **28.** $h(z) = z^2(z^2 + 4)^3$

29. $F(s) = \left(s - \dfrac{1}{s^2}\right)^3$

30. $G(t) = \left(t^2 + 1 + \dfrac{1}{t}\right)^2$

31. $f(u) = (1 + u)^3(1 + u^2)^4$

32. $g(w) = (w^2 - 3w + 4)(w + 4)^5$

33. $h(v) = \left[v - \left(1 - \dfrac{1}{v}\right)^{-1}\right]^{-2}$

34. $p(t) = \left(\dfrac{1}{t} + \dfrac{1}{t^2} + \dfrac{1}{t^3}\right)^{-4}$

35. $F(z) = \dfrac{1}{(3 - 4z + 5z^5)^{10}}$

36. $G(x) = \{1 + [x + (x^2 + x^3)^4]^5\}^6$

In Problems 37 through 44, dy/dx can be found in two ways—one way using the chain rule, the other way without using it. Use both techniques to find dy/dx and then compare the answers. (They should agree!)

37. $y = (x^3)^4 = x^{12}$

38. $y = x = \left(\dfrac{1}{x}\right)^{-1}$

39. $y = (x^2 - 1)^2 = x^4 - 2x^2 + 1$

40. $y = (1 - x)^3 = 1 - 3x + 3x^2 - x^3$

41. $y = (x + 1)^4 = x^4 + 4x^3 + 6x^2 + 4x + 1$

42. $y = (x + 1)^{-2} = \dfrac{1}{x^2 + 2x + 1}$

43. $y = (x^2 + 1)^{-1} = \dfrac{1}{x^2 + 1}$

44. $y = (x^2 + 1)^2 = (x^2 + 1)(x^2 + 1)$

We shall see in Section 3.7 that $D_x[\sin x] = \cos x$ (provided that x is in radian measure). Use this fact and the chain rule to find the derivatives of the functions in Problems 45 through 48.

45. $f(x) = \sin(x^3)$ **46.** $g(t) = (\sin t)^3$

47. $g(z) = (\sin 2z)^3$ **48.** $k(u) = \sin(1 + \sin u)$

49. A pebble dropped into a lake creates an expanding circular ripple (Fig. 3.3.3). Suppose that the radius of the circle is increasing at the rate of 2 in./s. At what rate is its area increasing when its radius is 10 in.?

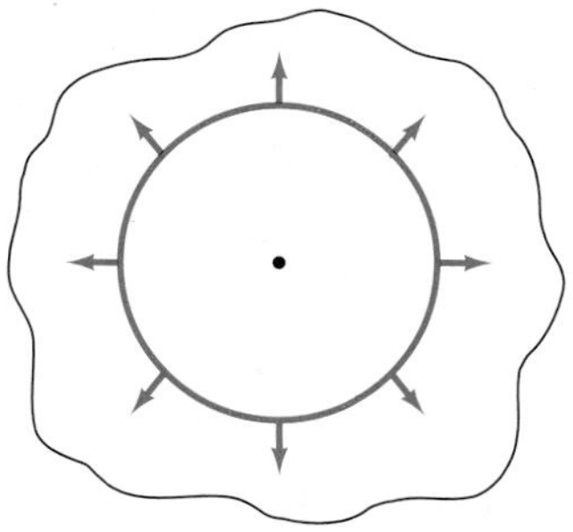

FIGURE 3.3.3 Expanding circular ripple in a lake (Problem 49).

50. The area of a circle is decreasing at the rate of 2π cm^2/s. At what rate is the radius of the circle decreasing when its area is 75π cm^2?

51. Each edge x of a square is increasing at the rate of 2 in./s. At what rate is the area A of the square increasing when each edge is 10 in.?

52. Each edge of an equilateral triangle is increasing at 2 cm/s (Fig. 3.3.4). At what rate is the area of the triangle increasing when each edge is 10 cm?

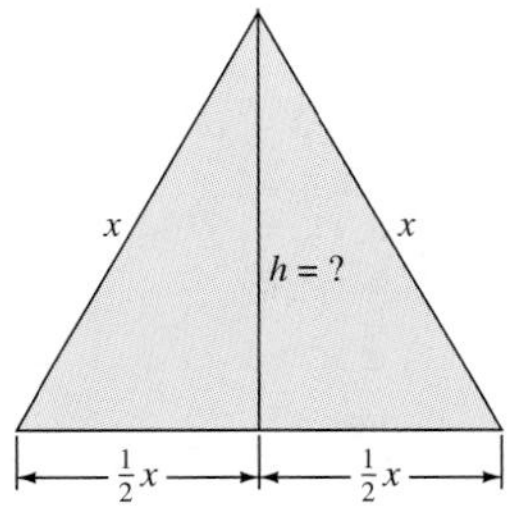

FIGURE 3.3.4 The triangle of Problem 52 with area $A = \frac{1}{2}xh$.

53. A cubical block of ice is melting in such a way that each edge decreases steadily by 2 in. every hour. At what rate is its volume decreasing when each edge is 10 in. long?

54. Find $f'(-1)$, given $f(y) = h(g(y))$, $h(2) = 55$, $g(-1) = 2$, $h'(2) = -1$, and $g'(-1) = 7$.

55. Given: $G(t) = f(h(t))$, $h(1) = 4$, $f'(4) = 3$, and $h'(1) = -6$. Find $G'(1)$.

56. Suppose that $f(0) = 0$ and that $f'(0) = 1$. Calculate the derivative of $f(f(f(x)))$ at $x = 0$.

57. Air is being pumped into a spherical balloon in such a way that its radius r is increasing at the rate of $dr/dt = 1$ cm/s. What is the time rate of increase, in cubic centimeters per second, of the balloon's volume when $r = 10$ cm?

58. Suppose that the air is being pumped into the balloon of Problem 57 at the constant rate of 200π cm^3/s. What is the time rate of increase of the radius r when $r = 5$ cm?

59. Air is escaping from a spherical balloon at the constant rate of 300π cm^3/s. What is the radius of the balloon when its radius is decreasing at the rate of 3 cm/s?

60. A spherical hailstone is losing mass by melting uniformly over its surface as it falls. At a certain time, its radius is 2 cm and its volume is decreasing at the rate of 0.1 cm^3/s. How fast is its radius decreasing at that time?

61. A spherical snowball is melting in such a way that the rate of decrease of its volume is proportional to its surface area. At 10 A.M. its volume is 500 in.^{3} and at 11 A.M. its volume is 250 in.^{3}. When does the snowball finish melting? (See Example 7.)

62. A cubical block of ice with edges 20 in. long begins to melt at 8 A.M. Each edge decreases at a constant rate thereafter and each is 8 in. long at 4 P.M. What was the rate of change of the block's volume at noon?

63. Suppose that u is a function of v, that v is a function of w, that w is a function of x, and that all these functions are differentiable. Explain why it follows from the chain rule that

$$\frac{du}{dx} = \frac{du}{dv} \cdot \frac{dv}{dw} \cdot \frac{dw}{dx}.$$

64. Let f be a differentiable function such that $f(1) = 1$. If $F(x) = f(x^n)$ and $G(x) = [f(x)]^n$ (where n is a fixed integer), show that $F(1) = G(1)$ and that $F'(1) = G'(1)$.

Recall from Example 13 in Section 2.2 that

$$D_x\left(\sqrt{x}\right) = \frac{1}{2\sqrt{x}}.$$

Use (only) this fact and the chain rule to calculate the derivative of each function given in Problems 65 through 68.

65. $h(x) = \sqrt{x+4}$

66. $h(x) = x^{3/2}$

67. $h(x) = (x^2+4)^{3/2}$

68. $h(x) = |x| = \sqrt{x^2}$

3.4 DERIVATIVES OF ALGEBRAIC FUNCTIONS

We saw in Section 3.3 that the chain rule yields the differentiation formula

$$D_x u^n = nu^{n-1}\frac{du}{dx} \tag{1}$$

if $u = f(x)$ is a differentiable function and the exponent n is an integer. We shall see in Theorem 1 of this section that this **generalized power rule** holds not only when the exponent is an integer, but also when it is a rational number $r = p/q$ (where p and q

are integers and $q \neq 0$). Recall that rational powers are defined in terms of integral roots and powers as follows:

$$u^{p/q} = \sqrt[q]{u^p} = \left(\sqrt[q]{u}\right)^p.$$

We consider first the case of a rational power of the independent variable x:

$$y = x^{p/q}, \tag{2}$$

where p and q are integers with q positive. In Problems 72 through 75 we illustrate the proof that the derivative of the root function $f(x) = \sqrt[q]{x}$ is given by

$$D_x\left(x^{1/q}\right) = \frac{1}{q}x^{(1/q)-1} = \frac{1}{q}x^{-(q-1)/q} \tag{3}$$

for $x > 0$; essentially the same proof works for $x < 0$ if q is odd (so that no even root of a negative number is involved). Thus the power rule—which we established in Section 3.2 only for integral exponents—also holds if the exponent of x is the reciprocal of a positive integer.

Consequently, we can apply Eq. (1) with $n = p$ and $u = x^{1/q}$ to differentiate the rational power of x in (2):

$$\begin{aligned} D_x\left[x^{p/q}\right] &= D_x\left[\left(x^{1/q}\right)^p\right] \\ &= p\left(x^{1/q}\right)^{p-1} \cdot D_x\left(x^{1/q}\right) \\ &= p\left(x^{1/q}\right)^{p-1} \cdot \frac{1}{q}x^{(1/q)-1} \\ &= \frac{p}{q}x^{(p/q)-(1/q)+(1/q)-1}; \end{aligned}$$

therefore

$$D_x\left[x^{p/q}\right] = \frac{p}{q}x^{(p/q)-1}.$$

Thus we have shown that the **power rule**

$$D_x x^r = r x^{r-1} \tag{4}$$

holds if the exponent $r = p/q$ is a rational number (subject to the conditions previously mentioned).

Using Eq. (4) we can differentiate a simple "radical" (or "root") function by first rewriting it as a power with a fractional exponent.

EXAMPLE 1

(a) $D_x\left[\sqrt{x}\right] = D_x\left[x^{1/2}\right] = \frac{1}{2}x^{-1/2} = \frac{1}{2\sqrt{x}}.$

(b) If $y = \sqrt{x^3}$, then $\frac{dy}{dx} = \frac{3}{2}x^{1/2} = \frac{3}{2}\sqrt{x}.$

(c) If $g(t) = \frac{1}{\sqrt[3]{t^2}} = t^{-2/3}$, then $g'(t) = -\frac{2}{3}t^{-5/3} = -\frac{2}{3\sqrt[3]{t^5}}.$ ◆

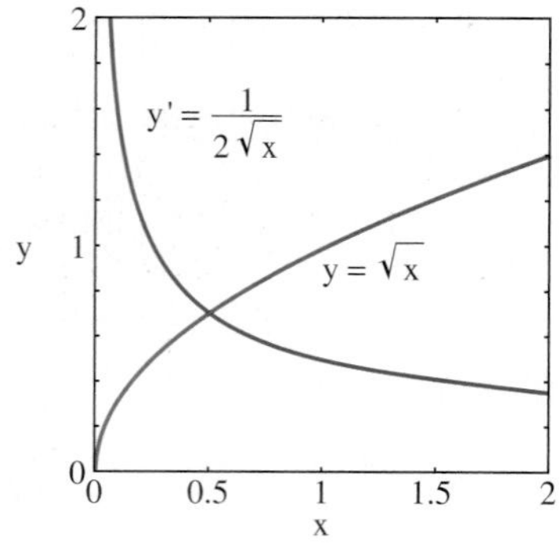

FIGURE 3.4.1 The graphs of $f(x) = \sqrt{x}$ and $f'(x) = \frac{1}{2\sqrt{x}}$.

REMARK In parts (a) and (b) of Example 1 it is necessary that $x \geqq 0$ in order that $\sqrt{x}$ be defined. In part (a) it is, moreover, necessary that $x \neq 0$; if $x = 0$ then the formula

$$D_x\left[\sqrt{x}\right] = \frac{1}{2\sqrt{x}}$$

would involve division by zero. Figure 3.4.1 shows the graphs of the function $f(x) = \sqrt{x}$ and its derivative $f'(x) = 1/(2\sqrt{x})$ for $x > 0$. Note that $f'(x) \to \infty$ as $x \to 0^+$, further emphasizing the fact that $f(x) = \sqrt{x}$ is not differentiable at $x = 0$.

The Generalized Power Rule

For the more general form of the power rule, let

$$y = u^r$$

where u is a differentiable function of x and $r = p/q$ is rational. Then

$$\frac{dy}{du} = ru^{r-1}$$

by Eq. (4), so the chain rule gives

$$\frac{dy}{dx} = \frac{dy}{du} \cdot \frac{du}{dx} = ru^{r-1}\frac{du}{dx}.$$

Thus

$$D_x u^r = ru^{r-1}\frac{du}{dx}, \tag{5}$$

which is the generalized power rule for rational exponents.

THEOREM 1 Generalized Power Rule
If r is a rational number, then

$$D_x[f(x)]^r = r[f(x)]^{r-1} \cdot f'(x) \tag{6}$$

wherever the function f is differentiable and the right-hand side is defined.

For the right-hand side in Eq. (6) to be "defined" means that $f'(x)$ exists, there is no division by zero, and no even root of a negative number appears.

EXAMPLE 2

$$D_x\left[\sqrt{4-x^2}\right] = D_x\left[(4-x^2)^{1/2}\right] = \tfrac{1}{2}(4-x^2)^{-1/2} \cdot D_x(4-x^2)$$
$$= \tfrac{1}{2}(4-x^2)^{-1/2} \cdot (-2x);$$
$$D_x\left[\sqrt{4-x^2}\right] = -\frac{x}{\sqrt{4-x^2}} \tag{7}$$

except where $x = \pm 2$ (division by zero) or where $|x| > 2$ (square root of a negative number). Thus Eq. (7) holds if $-2 < x < 2$. In writing derivatives of algebraic functions, we ordinarily omit such disclaimers unless they are pertinent to some specific purpose at hand. But note in Fig. 3.4.2 that if $f(x) = \sqrt{4-x^2}$ then $f'(x) \to +\infty$ as $x \to -2^+$ and $f'(x) \to -\infty$ as $x \to +2^-$. ◆

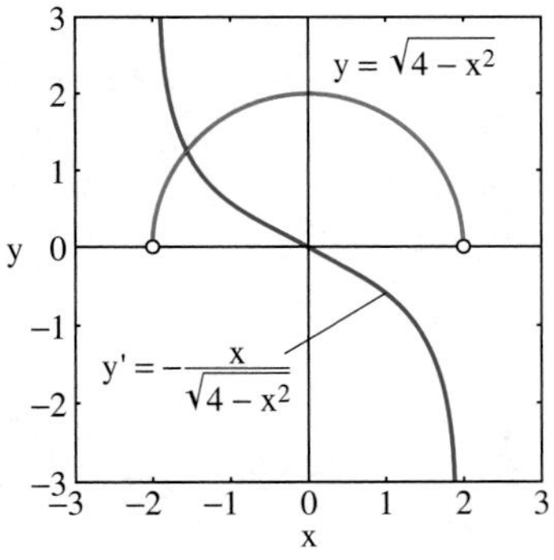

FIGURE 3.4.2 The graphs of $f(x) = \sqrt{4-x^2}$ and $f'(x) = \dfrac{-x}{\sqrt{4-x^2}}$.

A template for the application of the generalized power rule is

$$D_x([***]^n) = n[***]^{n-1}D_x[***],$$

where $***$ represents a function of x and (as we now know) n can be either an integer or a fraction (a quotient of integers).

But to differentiate a *power of a function,* we must first recognize *what function* it is a power *of*. So to differentiate a function involving roots (or radicals), we first "prepare" it for an application of the generalized power rule by rewriting it as a power function with fractional exponent. Examples 3 through 6 illustrate this technique.

EXAMPLE 3 If $y = 5\sqrt{x^3} - \dfrac{2}{\sqrt[3]{x}}$, then

$$y = 5x^{3/2} - 2x^{-1/3},$$

so

$$\frac{dy}{dx} = 5 \cdot \left(\frac{3}{2}x^{1/2}\right) - 2 \cdot \left(-\frac{1}{3}x^{-4/3}\right) = \frac{15}{2}x^{1/2} + \frac{2}{3}x^{-4/3} = \frac{15}{2}\sqrt{x} + \frac{2}{3\sqrt[3]{x^4}}.$$ ◆

EXAMPLE 4 With $f(x) = 3 - 5x$ and $r = 7$, the generalized power rule yields

$$\begin{aligned} D_x[(3-5x)^7] &= 7(3-5x)^6 D_x(3-5x) \\ &= 7(3-5x)^6(-5) = -35(3-5x)^6. \end{aligned}$$ ◆

EXAMPLE 5 With $f(x) = 2x^2 - 3x + 5$ and $r = \frac{1}{2}$, the generalized power rule yields

$$\begin{aligned} D_x\sqrt{2x^2-3x+5} &= D_x(2x^2-3x+5)^{1/2} \\ &= \frac{1}{2}(2x^2-3x+5)^{-1/2} D_x(2x^2-3x+5) \\ &= \frac{4x-3}{2\sqrt{2x^2-3x+5}}. \end{aligned}$$ ◆

EXAMPLE 6 If

$$x = \left[5t + \sqrt[3]{(3t-1)^4}\right]^{10}$$

then Eq. (5) with $u = 5t + (3t-1)^{4/3}$ and with independent variable t gives

$$\begin{aligned} \frac{dx}{dt} &= 10u^9 \cdot \frac{du}{dt} \\ &= 10\left[5t + (3t-1)^{4/3}\right]^9 \cdot D_t\left[5t + (3t-1)^{4/3}\right] \\ &= 10\left[5t + (3t-1)^{4/3}\right]^9 \cdot \left[D_t(5t) + D_t(3t-1)^{4/3}\right] \\ &= 10\left[5t + (3t-1)^{4/3}\right]^9 \cdot \left[5 + \tfrac{4}{3}(3t-1)^{1/3} \cdot 3\right]; \\ \frac{dx}{dt} &= 10\left[5t + (3t-1)^{4/3}\right]^9 \cdot \left[5 + 4(3t-1)^{1/3}\right]. \end{aligned}$$ ◆

Example 6 illustrates the fact that we apply the chain rule (or generalized power rule) by working from the *outside* to the *inside.* At each step the derivative of the outside function is multiplied by the derivative of the inside function. We continue until no "inside function" remains undifferentiated. Does the process remind you of peeling an onion, one layer at a time, until its core is reached?

Differentiability and Vertical Tangent Lines

Whereas polynomials and rational functions are both continuous and differentiable wherever they are defined, simple algebraic functions can be continuous at points where their derivatives do not exist.

EXAMPLE 7 If

$$f(x) = |x| = \sqrt{x^2}$$

denotes the absolute value function, then for $x \neq 0$ we find that

$$f'(x) = D_x\left[(x^2)^{1/2}\right] = \tfrac{1}{2}(x^2)^{-1/2}(2x) = \frac{x}{\sqrt{x^2}} = \frac{x}{|x|} = \begin{cases} -1 & \text{if } x < 0, \\ +1 & \text{if } x > 0. \end{cases}$$

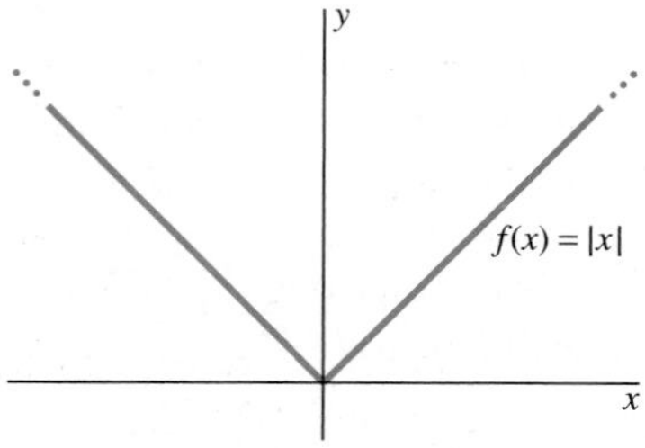

FIGURE 3.4.3 The graph of $f(x) = |x|$.

Thus f is differentiable at every point except possibly for the origin $x = 0$. In fact, the graph of $f(x) = |x|$ in Fig. 3.4.3 makes it clear that the difference quotient

$$\frac{f(x) - f(0)}{x - 0} = \frac{|x|}{x}$$

has left-hand limit -1 and right-hand limit $+1$ at $x = 0$. Thus the absolute value function is not differentiable at the isolated point $x = 0$, where the graph $y = |x|$ has a "corner point" rather than a tangent line. (Can you think of a continuous function whose graph has *infinitely many* such corner points?) ◆

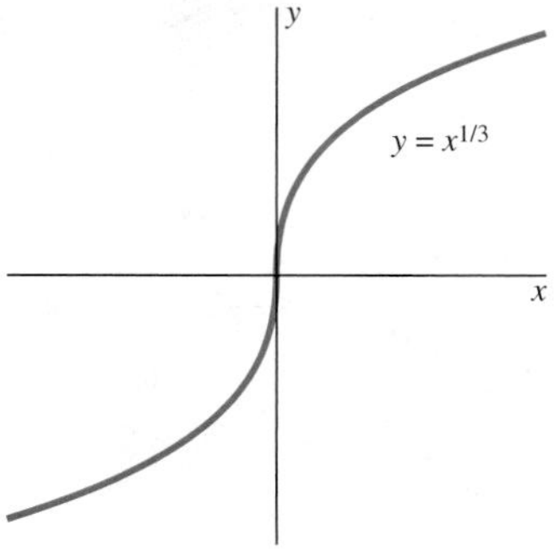

FIGURE 3.4.4 The graph of the cube root function.

EXAMPLE 8 Figure 3.4.4 shows the graph of the cube-root function

$$y = \sqrt[3]{x} = x^{1/3},$$

and illustrates another way in which a function can fail to be differentiable at an isolated point. Its derivative,

$$\frac{dy}{dx} = \frac{1}{3}x^{-2/3} = \frac{1}{3\sqrt[3]{x^2}},$$

increases without bound as $x \to 0$ but does not exist at $x = 0$. Therefore, the definition of tangent line does not apply to this graph at $(0, 0)$. Nevertheless, from the figure it seems appropriate to regard the vertical line $x = 0$ as the line tangent to the curve $y = x^{1/3}$ at the point $(0, 0)$. ◆

DEFINITION Vertical Tangent Line

The curve $y = f(x)$ has a **vertical tangent line** at the point $(a, f(a))$ provided that f is continuous at a and

$$|f'(x)| \to +\infty \quad \text{as} \quad x \to a. \qquad \textbf{(8)}$$

Thus the graph of the continuous function $f(x) = x^{1/3}$ of Example 8 has a vertical tangent line at the origin, even though f is not differentiable at $x = 0$. Note that the requirement that f be continuous at $x = a$ implies that $f(a)$ must be defined. Thus it would be pointless to ask about a line (vertical or not) tangent to the curve $y = 1/x$ where $x = 0$.

If f is defined (and differentiable) on only one side of $x = a$, we mean in Eq. (8) that $|f'(x)| \to +\infty$ as x approaches a from that side.

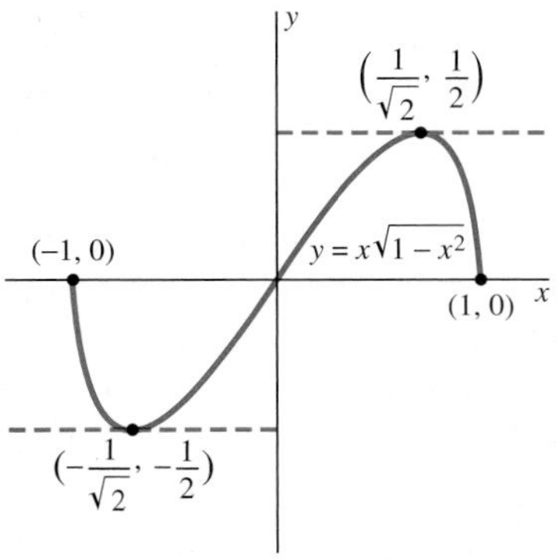

FIGURE 3.4.5 The graph of $f(x) = x\sqrt{1-x^2}$, $-1 \leqq x \leqq 1$ (Example 9).

EXAMPLE 9 Find the points on the curve

$$y = f(x) = x\sqrt{1-x^2}, \quad -1 \leqq x \leqq 1,$$

at which the tangent line is either horizontal or vertical.

Solution We differentiate using first the product rule and then the chain rule:

$$f'(x) = (1-x^2)^{1/2} + \frac{x}{2}(1-x^2)^{-1/2}(-2x)$$

$$= (1-x^2)^{-1/2}[(1-x^2) - x^2] = \frac{1-2x^2}{\sqrt{1-x^2}}.$$

Now $f'(x) = 0$ only when the numerator $1 - 2x^2$ is zero—that is, when $x = \pm 1/\sqrt{2}$. Because $f(\pm 1/\sqrt{2}) = \pm 1/2$, the curve has a horizontal tangent line at each of the two points $(1/\sqrt{2}, 1/2)$ and $(-1/\sqrt{2}, -1/2)$.

We also observe that the denominator $\sqrt{1-x^2}$ approaches zero as $x \to -1^+$ and as $x \to +1^-$. Because $f(\pm 1) = 0$, we see that the curve has a vertical tangent line at each of the two points $(1, 0)$ and $(-1, 0)$. The graph of f is shown in Fig. 3.4.5. ◆

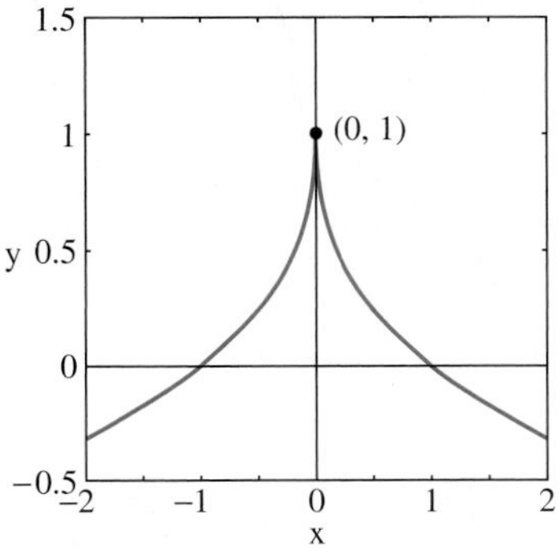

FIGURE 3.4.6 The graph of $y = 1 - \sqrt[5]{x^2}$ with a cusp at $(0, 1)$.

EXAMPLE 10 Figure 3.4.6 shows the graph of the function $f(x) = 1 - \sqrt[5]{x^2}$, which appears to have a sharp "cusp" (rather than a corner) at the point $(0, 1)$. Because the absolute value of the derivative $f'(x) = -\frac{2}{5}x^{-3/5}$ approaches $+\infty$ as $x \to 0$, the curve $y = f(x)$ has a vertical tangent at that point. ◆

Whereas the preceding examples show that a function can be continuous without being differentiable, the following theorem says that a function *is* continuous wherever it is differentiable. Thus differentiability of a function is a stronger condition than continuity alone.

THEOREM 2 Differentiability Implies Continuity
Suppose that the function f is defined in a neighborhood of a. If f is differentiable at a, then f is continuous at a.

PROOF Because $f'(a)$ exists, the product law for limits yields

$$\begin{aligned}\lim_{x\to a}[f(x)-f(a)] &= \lim_{x\to a}(x-a)\cdot\frac{f(x)-f(a)}{x-a}\\ &= \left(\lim_{x\to a}(x-a)\right)\left(\lim_{x\to a}\frac{f(x)-f(a)}{x-a}\right)\\ &= 0\cdot f'(a)=0.\end{aligned}$$

Thus $\lim_{x\to a} f(x) = f(a)$, so f is continuous at a. ◄

3.4 TRUE/FALSE STUDY GUIDE

3.4 CONCEPTS: QUESTIONS AND DISCUSSION

1. (a) Can you define a function that is continuous everywhere and has a "corner point" at each integer point $x = n$, but is differentiable at every other point of the real line? (b) Can you define a function that is continuous everywhere and has a vertical tangent line at each integer point $x = n$, but is differentiable at every other point of the real line?
2. Suppose that the function f has the following property: Every point x of the real line lies in a closed interval $[a, b]$ on which the graph of f is a semicircle having this interval as a diameter. Sketch a typical graph of such a function. Discuss the continuity and differentiability of f. *Remark:* The set of all endpoints of the closed intervals mentioned might (or might not) be the set of all integer points on the real line.
3. In Question 2, you may have assumed that each endpoint of the interval $[a, b]$ lies on exactly two such semicircles, one to the right and one to the left. Can you think of a function g whose graph consists entirely of semicircles, but does not satisfy this "two semicircles" condition? If so, discuss the differentiability of g. *Suggestion:* The construction of g might (or might not) involve the set $\{1, \frac{1}{2}, \frac{1}{3}, \frac{1}{4}, \frac{1}{5}, \dots\}$ of all reciprocals of positive integers.
4. Suppose that the function f is continuous everywhere. At how many points do you suspect that f can fail to be differentiable? What's the worst such function you can think of?

3.4 PROBLEMS

Differentiate the functions given in Problems 1 through 44.

1. $f(x) = 4\sqrt{x^5} + \dfrac{2}{\sqrt{x}}$

2. $g(t) = 9\sqrt[3]{t^4} - \dfrac{3}{\sqrt[3]{t}}$

3. $f(x) = \sqrt{2x+1}$

4. $h(z) = \dfrac{1}{\sqrt[3]{7-6z}}$

5. $f(x) = \dfrac{6-x^2}{\sqrt{x}}$

6. $\phi(u) = \dfrac{7+2u-3u^4}{\sqrt[3]{u^2}}$

7. $f(x) = (2x+3)^{3/2}$

8. $g(x) = (3x+4)^{4/3}$

9. $f(x) = (3-2x^2)^{-3/2}$

10. $f(y) = (4-3y^3)^{-2/3}$

11. $f(x) = \sqrt{x^3+1}$

12. $g(z) = \dfrac{1}{(z^4+3)^2}$

13. $f(x) = \sqrt{2x^2+1}$

14. $f(t) = \dfrac{t}{\sqrt{1+t^4}}$

15. $f(t) = \sqrt{2t^3}$

16. $g(t) = \sqrt{\dfrac{1}{3t^5}}$

17. $f(x) = (2x^2-x+7)^{3/2}$

18. $g(z) = (3z^2-4)^{97}$

19. $g(x) = \dfrac{1}{(x-2x^3)^{4/3}}$

20. $f(t) = [t^2+(1+t)^4]^5$

21. $f(x) = x\sqrt{1-x^2}$

22. $g(x) = \sqrt{\dfrac{2x+1}{x-1}}$

23. $f(t) = \sqrt{\dfrac{t^2+1}{t^2-1}}$

24. $h(y) = \left(\dfrac{y+1}{y-1}\right)^{17}$

25. $f(x) = \left(x-\dfrac{1}{x}\right)^3$

26. $g(z) = \dfrac{z^2}{\sqrt{1+z^2}}$

27. $f(v) = \dfrac{\sqrt{v+1}}{v}$

28. $h(x) = \left(\dfrac{x}{1+x^2}\right)^{5/3}$

29. $f(x) = \sqrt[3]{1 - x^2}$

30. $g(x) = \sqrt{x + \sqrt{x}}$

31. $f(x) = x(3 - 4x)^{1/2}$

32. $g(t) = \dfrac{t - (1 + t^2)^{1/2}}{t^2}$

33. $f(x) = (1 - x^2)(2x + 4)^{1/3}$

34. $f(x) = (1 - x)^{1/2}(2 - x)^{1/3}$

35. $g(t) = \left(1 + \dfrac{1}{t}\right)^2 (3t^2 + 1)^{1/2}$

36. $f(x) = x(1 + 2x + 3x^2)^{10}$

37. $f(x) = \dfrac{2x - 1}{(3x + 4)^5}$

38. $h(z) = (z - 1)^4(z + 1)^6$

39. $f(x) = \dfrac{(2x + 1)^{1/2}}{(3x + 4)^{1/3}}$

40. $f(x) = (1 - 3x^4)^5(4 - x)^{1/3}$

41. $h(y) = \dfrac{\sqrt{1 + y} + \sqrt{1 - y}}{\sqrt[3]{y^5}}$

42. $f(x) = \sqrt{1 - \sqrt[3]{x}}$

43. $g(t) = \sqrt{t + \sqrt{t + \sqrt{t}}}$

44. $f(x) = x^3\sqrt{1 - \dfrac{1}{x^2 + 1}}$

For each curve given in Problems 45 through 50, find all points on the graph where the tangent line is either horizontal or vertical.

45. $y = x^{2/3}$

46. $y = x\sqrt{4 - x^2}$

47. $y = x^{1/2} - x^{3/2}$

48. $y = \dfrac{1}{\sqrt{9 - x^2}}$

49. $y = \dfrac{x}{\sqrt{1 - x^2}}$

50. $y = \sqrt{(1 - x^2)(4 - x^2)}$

In Problems 51 through 56, first write an equation of the line tangent to the given curve $y = f(x)$ at the indicated point P. Then illustrate your result with a graphing calculator or computer by graphing both the curve and the tangent line on the same screen.

51. $y = 2\sqrt{x}$, at the point P where $x = 4$

52. $y = 3\sqrt[3]{x}$, at the point P where $x = 8$

53. $y = 3\sqrt[3]{x^2}$, at the point P where $x = -1$

54. $y = 2\sqrt{1 - x}$, at the point P where $x = \frac{3}{4}$

55. $y = x\sqrt{4 - x}$, at the point P where $x = 0$

56. $y = (1 - x)\sqrt{x}$, at the point P where $x = 4$

In Problems 57 through 62, match the given graph $y = f(x)$ of a function with the graph $y = f'(x)$ of its derivative among those shown in Figs. 3.4.13(a) through 3.4.13(f).

57. Figure 3.4.7

58. Figure 3.4.8

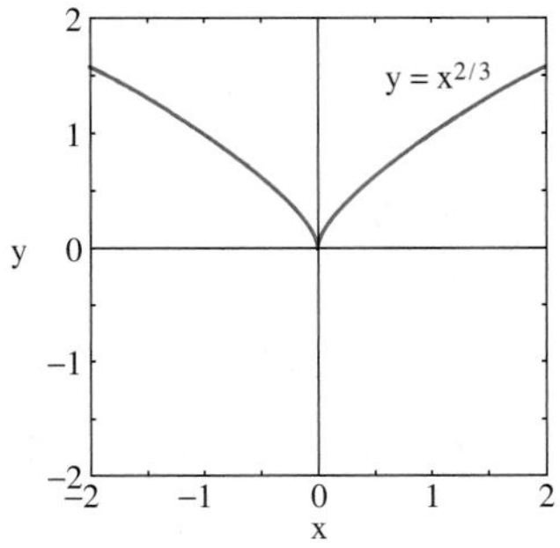

FIGURE 3.4.7 $y = x^{2/3}$ (Problem 57).

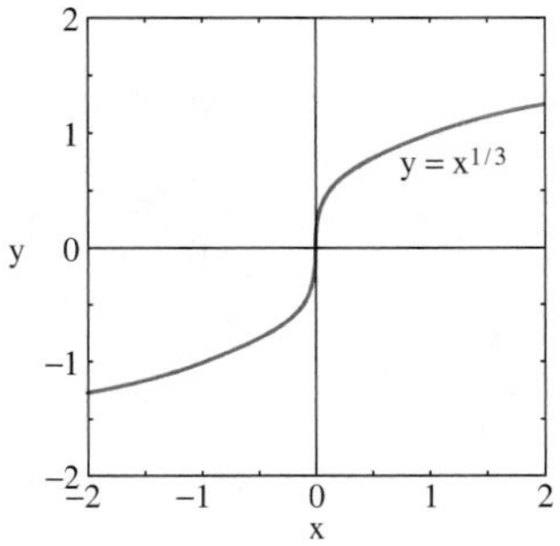

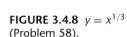

FIGURE 3.4.8 $y = x^{1/3}$ (Problem 58).

59. Figure 3.4.9

60. Figure 3.4.10

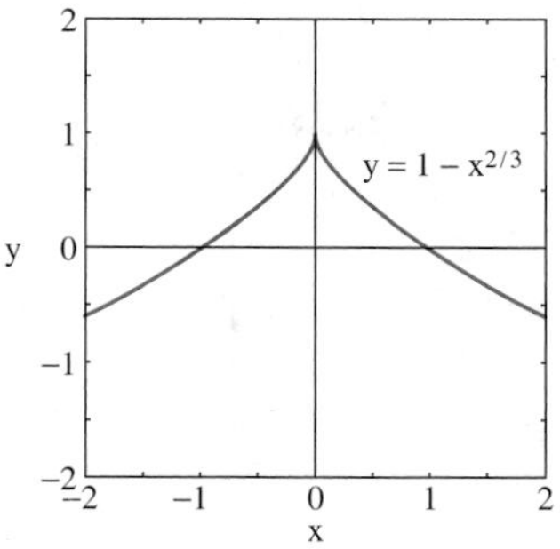

FIGURE 3.4.9 $y = 1 - x^{2/3}$ (Problem 59).

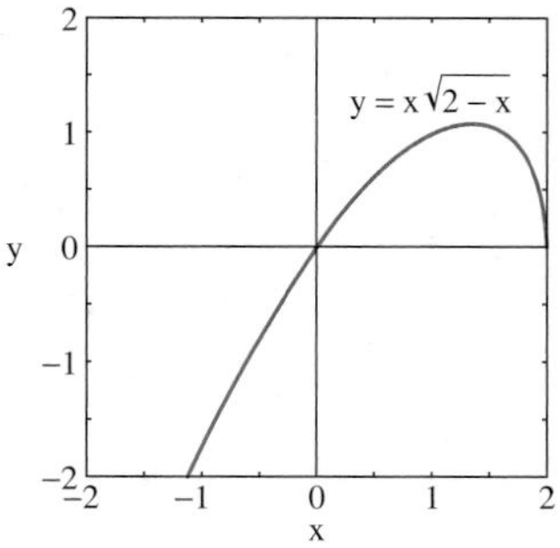

FIGURE 3.4.10 $y = x\sqrt{2 - x}$ (Problem 60).

61. Figure 3.4.11

62. Figure 3.4.12

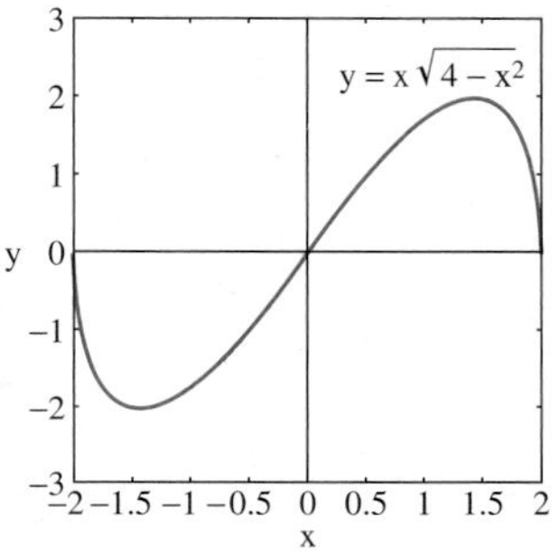

FIGURE 3.4.11 $y = x\sqrt{4 - x^2}$ (Problem 61).

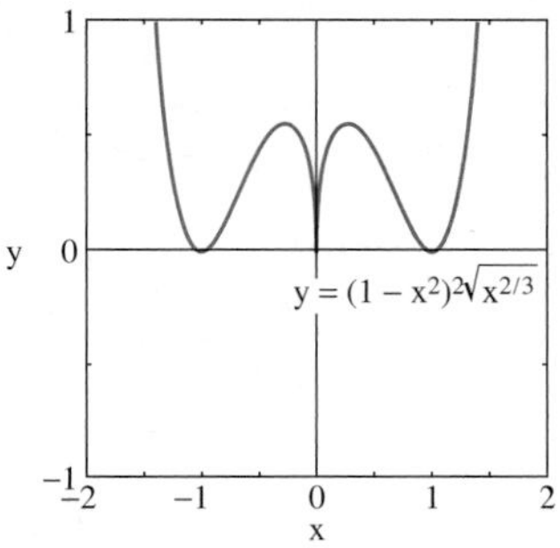

FIGURE 3.4.12 $y = (1 - x^2)^2\sqrt{x^{2/3}}$ (Problem 62).

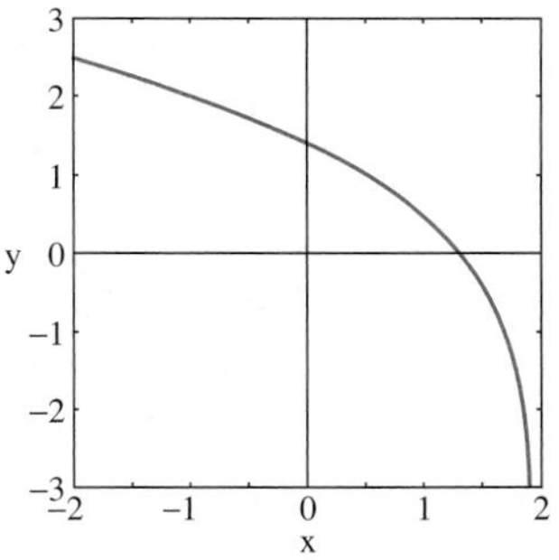

FIGURE 3.4.13(a)

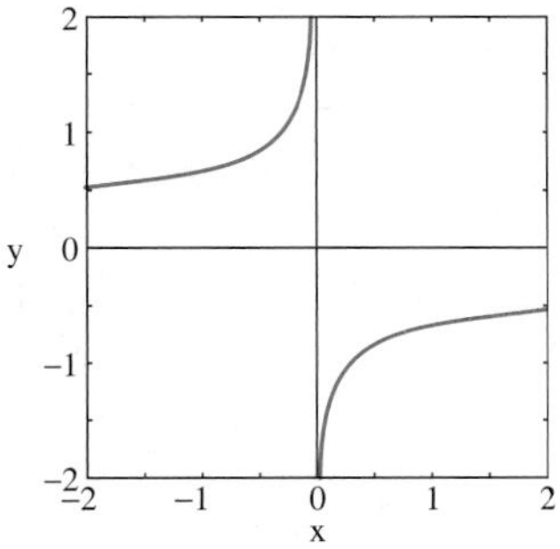

FIGURE 3.4.13(b)

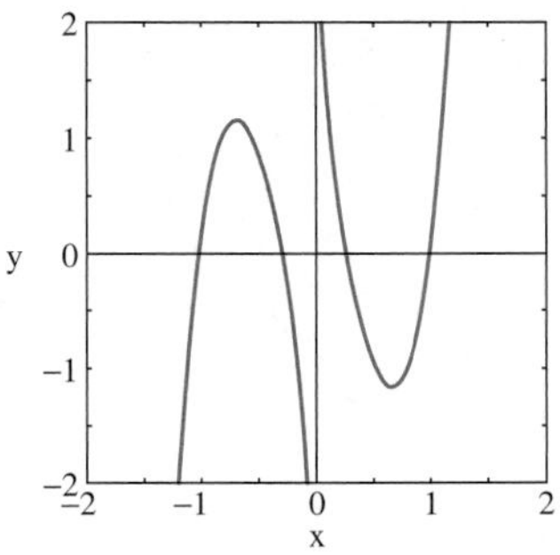

FIGURE 3.4.13(c)

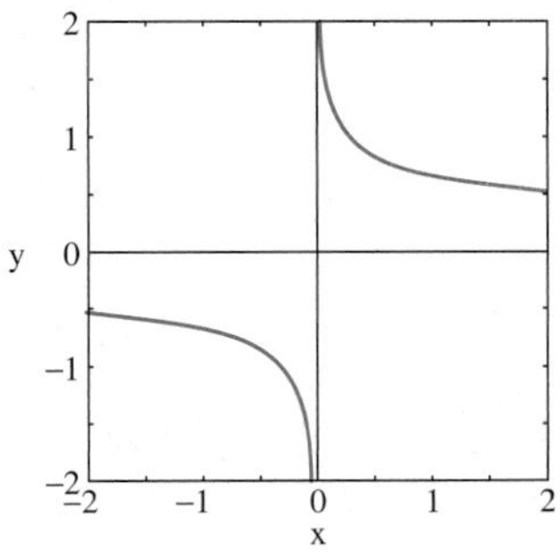

FIGURE 3.4.13(d)

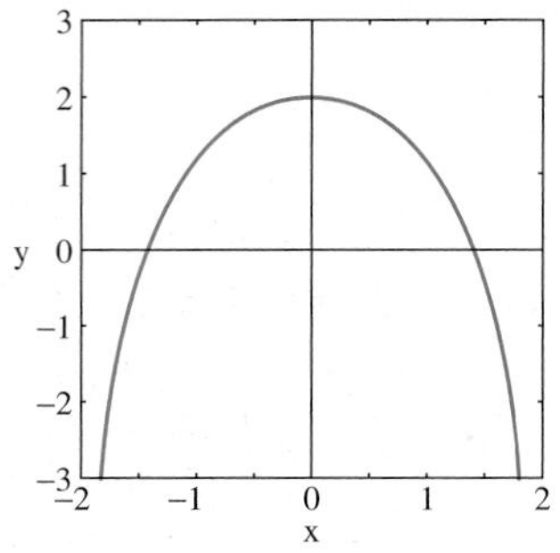

FIGURE 3.4.13(e)

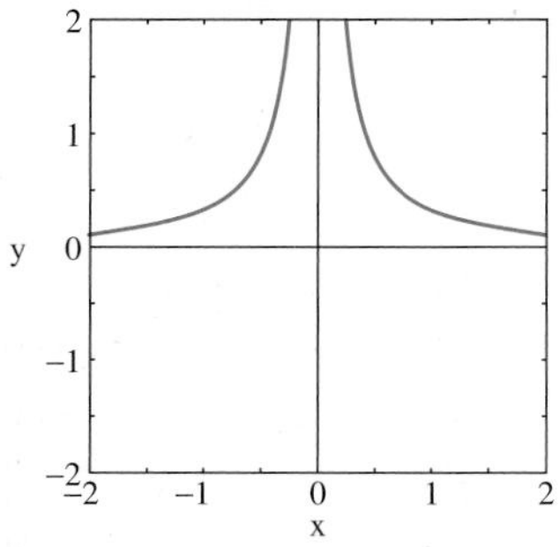

FIGURE 3.4.13(f)

63. The period of oscillation P (in seconds) of a simple pendulum of length L (in feet) is given by $P = 2\pi\sqrt{L/g}$, where $g = 32$ ft/s^2. Find the rate of change of P with respect to L when $P = 2$.

64. Find the rate of change of the volume $V = \frac{4}{3}\pi r^3$ of a sphere of radius r with respect to its surface area $A = 4\pi r^2$ when $r = 10$.

65. Find the two points on the circle $x^2 + y^2 = 1$ at which the slope of the tangent line is -2 (Fig. 3.4.14).

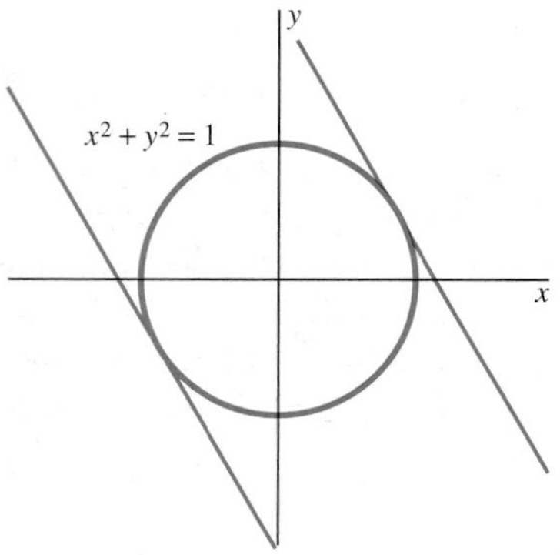

FIGURE 3.4.14 The two tangent lines of Problem 65.

66. Find the two points on the circle $x^2 + y^2 = 1$ at which the slope of the tangent line is 3.

67. Find a line through the point $P(18, 0)$ that is normal to the tangent line to the parabola $y = x^2$ at some point $Q(a, a^2)$ (see Fig. 3.4.15). (*Suggestion:* You will obtain a cubic equation in the unknown a. Find by inspection a small integral root r. The cubic polynomial is then the product of $a - r$ and a quadratic polynomial; you can find the latter by division of $a - r$ into the cubic.)

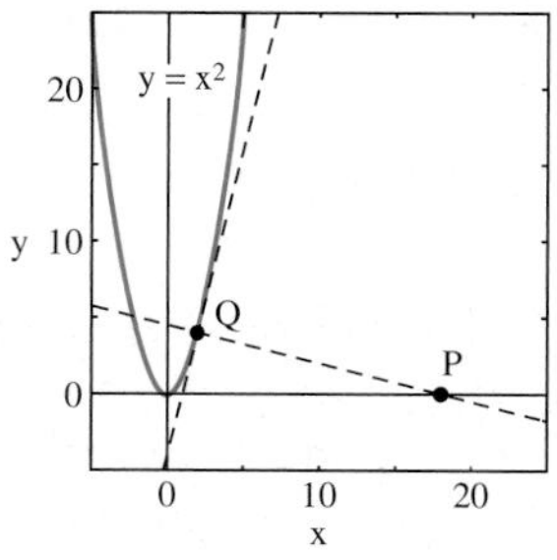

FIGURE 3.4.15 The tangent and normal of Problem 67.

68. Find three distinct lines through the point $P(3, 10)$ that are normal to the parabola $y = x^2$ (Fig. 3.4.16). (See the suggestion for Problem 67. This problem will require a certain amount of calculator-aided computation.)

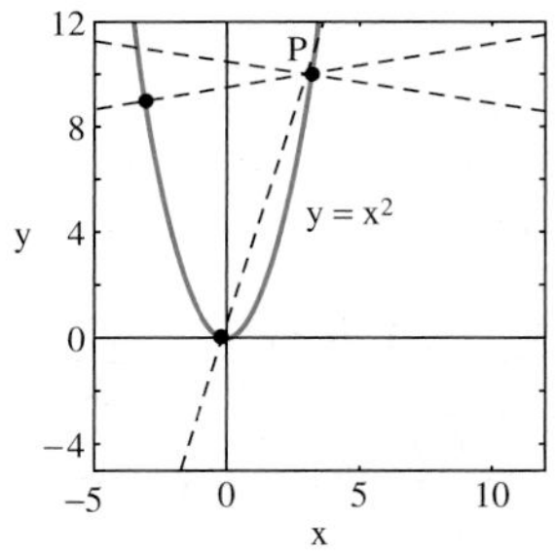

FIGURE 3.4.16 The three normal lines of Problem 68.

69. Find two distinct lines through the point $P(0, \frac{5}{2})$ that are normal to the curve $y = x^{2/3}$ (Fig. 3.4.17).

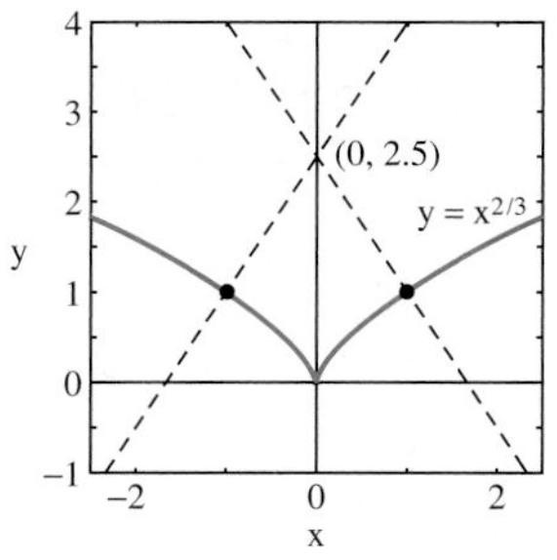

FIGURE 3.4.17 The two normal lines of Problem 69.

70. Verify that the line tangent to the circle $x^2 + y^2 = a^2$ at the point P is perpendicular to the radius OP (Fig. 3.4.18).

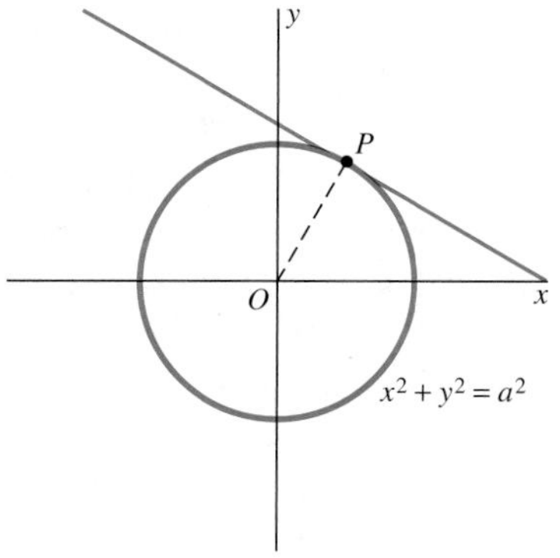

FIGURE 3.4.18 The circle, radius, and tangent line of Problem 70.

71. Consider the cubic equation $x^3 = 3x + 8$. If we differentiate each side with respect to x, we obtain $3x^2 = 3$, which has the two solutions $x = 1$ and $x = -1$. But neither of these is a solution of the original cubic equation. What went wrong? Why does differentiation of both sides of the cubic equation give an invalid result?

The derivation of the generalized power rule $D_x u^r = r u^{r-1} \cdot D_x u$ (for $r = p/q$, a rational number) provided in this section depends on the assumed differentiability of the qth root function $f(x) = x^{1/q}$. If $a > 0$ and q is a positive integer, then the derivative of f is given by

$$f'(a) = \lim_{x \to a} \frac{x^{1/q} - a^{1/q}}{x - a} \tag{9}$$

provided that this limit exists. Problems 72 through 75 illustrate the evaluation of this limit using the algebraic identity

$$s^q - t^q = (s - t)\underbrace{(s^{q-1} + s^{q-2}t + \cdots + st^{q-2} + t^{q-1})}_{q \text{ terms}}. \tag{10}$$

For instance, with $s = x^{1/q}$ and $t = a^{1/q}$ this identity yields (with $q = 2$, 3, and 5) the formulas

$$x - a = \left(x^{1/2} - a^{1/2}\right)\left(x^{1/2} + a^{1/2}\right), \tag{11}$$

$$x - a = \left(x^{1/3} - a^{1/3}\right)\left(x^{2/3} + x^{1/3}a^{1/3} + a^{2/3}\right), \tag{12}$$

and

$$x - a = \left(x^{1/5} - a^{1/5}\right)\left(x^{4/5} + x^{3/5}a^{1/5} + x^{2/5}a^{2/5} + x^{1/5}a^{3/5} + a^{4/5}\right). \tag{13}$$

72. Substitute (11) in the denominator in (9) to show that $D_x x^{1/2} = \frac{1}{2}x^{-1/2}$ for $x > 0$.

73. Substitute (12) in the denominator in (9) to show that $D_x x^{1/3} = \frac{1}{3}x^{-2/3}$ for $x > 0$.

74. Substitute (13) in the denominator in (9) to show that $D_x x^{1/5} = \frac{1}{5}x^{-4/5}$ for $x > 0$.

75. Finally, explain how Eq. (10) can be applied in the general case to prove that

$$D_x x^{1/q} = \frac{1}{q} x^{-(q-1)/q}$$

if $x > 0$ and q is a positive integer.

3.5 MAXIMA AND MINIMA OF FUNCTIONS ON CLOSED INTERVALS

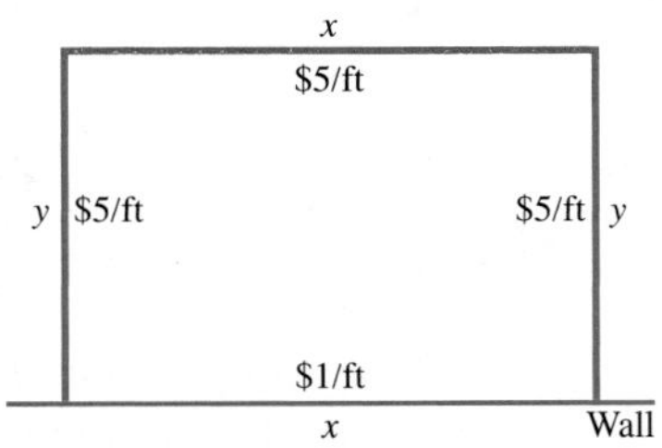

FIGURE 3.5.1 The animal pen.

In applications we often need to find the maximum (largest) or minimum (smallest) value that a specified quantity can attain. The animal pen problem posed in Section 1.1 is a simple yet typical example of an applied maximum-minimum problem. There we investigated the animal pen shown in Fig. 3.5.1, with the indicated dollar-per-foot cost figures for its four sides. We showed that if $180 is allocated for material to construct this pen, then its area $A = f(x)$ is given as a function of its base length x by

$$f(x) = \tfrac{3}{5}x(30 - x), \quad 0 \leqq x \leqq 30. \tag{1}$$

Hence the question of the largest possible area of the animal pen is equivalent to the purely mathematical problem of finding the maximum value attained by the function $f(x) = \frac{3}{5}x(30 - x)$ on the closed interval $[0, 30]$.

DEFINITION Maximum and Minimum Values
If c is in the closed interval $[a, b]$, then $f(c)$ is called the **minimum value** of $f(x)$ on $[a, b]$ if $f(c) \leqq f(x)$ for all x in $[a, b]$. Similarly, if d is in $[a, b]$, then $f(d)$ is called the **maximum value** of $f(x)$ on $[a, b]$ if $f(d) \geqq f(x)$ for all x in $[a, b]$.

Thus if $f(c)$ is the minimum value and $f(d)$ the maximum value of $f(x)$ on $[a, b]$, then

$$f(c) \leqq f(x) \leqq f(d) \tag{2}$$

for all x in $[a, b]$, and hence $f(x)$ attains no value smaller than $f(c)$ or larger than $f(d)$. In geometric terms, $(c, f(c))$ is a *low point* and $(d, f(d))$ is a *high point* on the curve $y = f(x)$, $a \leqq x \leqq b$, as illustrated in Figs. 3.5.2 and 3.5.3.

Theorem 1 (proved in Appendix E) says that a continuous function f on a closed interval $[a, b]$ attains a minimum value $f(c)$ and a maximum value $f(d)$, so the inequalities in (2) hold: The curve $y = f(x)$ over $[a, b]$ has both a lowest point and a highest point.

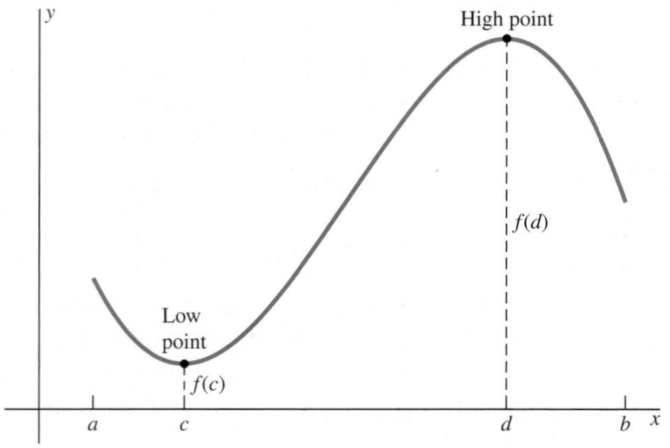

FIGURE 3.5.2 $f(c)$ is the minimum value and $f(d)$ is the maximum value of $f(x)$ on $[a, b]$.

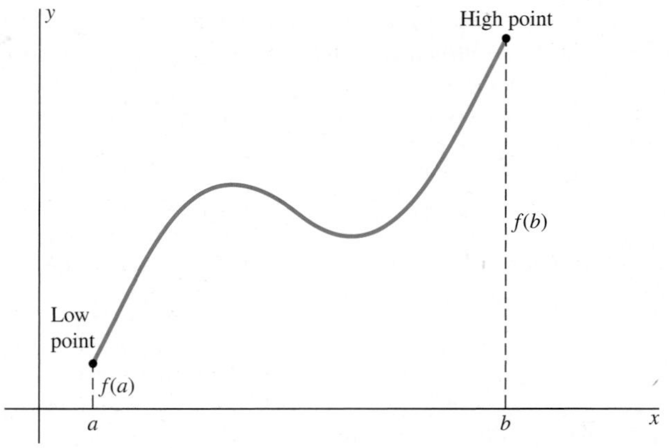

FIGURE 3.5.3 Maximum and minimum values can occur at the endpoints of an interval. Here $f(a)$ is the minimum value and $f(b)$ is the maximum value of $f(x)$ on $[a, b]$.

THEOREM 1 Maximum and Minimum Value Property
If the function f is continuous on the closed interval $[a, b]$, then there exist numbers c and d in $[a, b]$ such that $f(c)$ is the minimum value, and $f(d)$ the maximum value, of f on $[a, b]$.

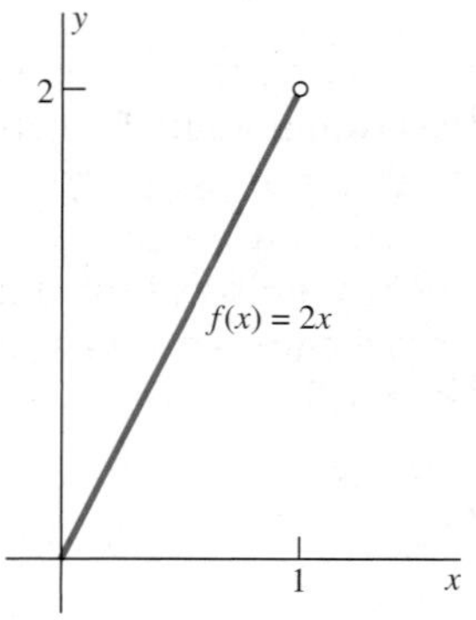

FIGURE 3.5.4 The graph of the function of Example 1.

In short, a continuous function defined on a closed and bounded interval attains both a minimum value and a maximum value at points of the interval. Hence we see it is the *continuity* of the function

$$f(x) = \frac{3}{5}x(30 - x)$$

on the *closed* interval $[0, 30]$ that guarantees that the maximum value of f exists and is attained at some point of the interval $[0, 30]$.

Suppose that the function f is defined on the interval I. Examples 1 and 2 show that if *either* f is not continuous *or* I is not closed, then f may fail to attain maximum and minimum values at points of I. Thus both hypotheses in Theorem 1 are necessary.

EXAMPLE 1 Let the continuous function $f(x) = 2x$ be defined only for $0 \leqq x < 1$, so that its domain of definition is not a closed interval. From the graph shown in Fig. 3.5.4, it is clear that f attains its minimum value 0 at $x = 0$. But $f(x) = 2x$ attains *no* maximum value at any point of $[0, 1)$. The only possible candidate for a maximum value would be the value 2 at $x = 1$, but $f(1)$ is not defined. ◆

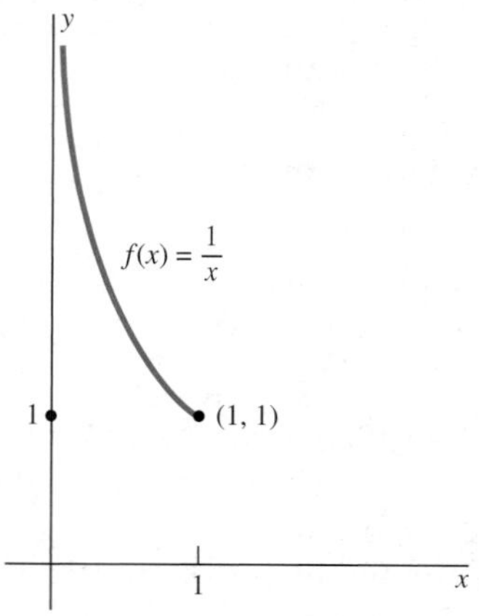

FIGURE 3.5.5 The graph of the function of Example 2.

EXAMPLE 2 The function f defined on the closed interval $[0, 1]$ with the formula

$$f(x) = \begin{cases} \dfrac{1}{x} & \text{if } 0 < x \leqq 1, \\ 1 & \text{if } x = 0 \end{cases}$$

is not continuous on $[0, 1]$ because $\lim_{x\to 0^+}(1/x)$ does not exist (Fig. 3.5.5). This function does attain its minimum value of 1 at $x = 0$ and also at $x = 1$. But it attains no maximum value on $[0, 1]$ because $1/x$ can be made arbitrarily large by choosing x positive and very close to zero. ◆

For a variation on Example 2, the function $g(x) = 1/x$ with domain the *open* interval $(0, 1)$ attains neither a maximum nor a minimum there.

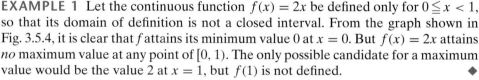

Local Maxima and Minima

Once we know that the continuous function f *does* attain minimum and maximum values on the closed interval $[a, b]$, the remaining question is this: Exactly *where* are these values located? We solved the animal pen problem in Section 2.1 on the basis of the following assumption, motivated by geometry: The function $f(x) = \frac{3}{5}x(30 - x)$ attains its maximum value on $[0, 30]$ at an interior point of that interval, a point at which the tangent line is horizontal. Theorems 2 and 3 of this section provide a rigorous basis for the method we used there.

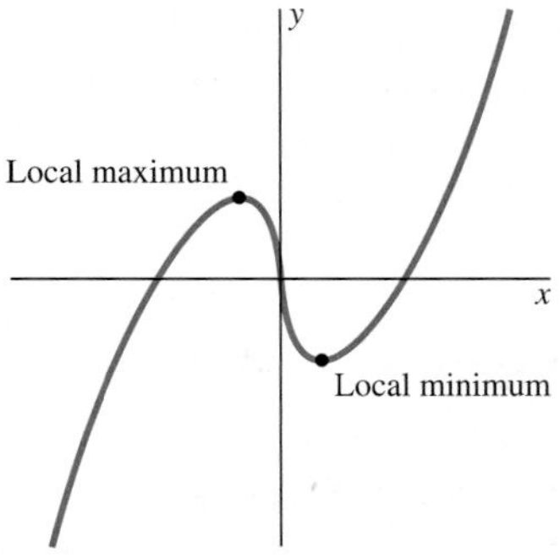

FIGURE 3.5.6 Local extrema.

We say that the value $f(c)$ is a **local maximum value** of the function f if $f(x) \leqq f(c)$ for all x sufficiently near c. More precisely, if this inequality holds for all x that are simultaneously in the domain of f and in some open interval containing c, then $f(c)$ is a local maximum of f. Similarly, we say that the value $f(c)$ is a **local minimum value** of f if $f(x) \geqq f(c)$ for all x sufficiently near c.

As Fig. 3.5.6 shows, a local maximum is a point such that no nearby points on the graph are higher, and a local minimum is one such that no nearby points on the graph are lower. A **local extremum** of f is a value of f that is either a local maximum or a local minimum.

THEOREM 2 Local Maxima and Minima

Suppose that f is differentiable at c and is defined on a open interval containing c. If $f(c)$ is either a local maximum value or a local minimum value of f, then $f'(c) = 0$.

Thus a local extremum of a *differentiable* function on an *open* interval can occur only at a point where the derivative is zero and, therefore, where the line tangent to the graph is horizontal.

PROOF OF THEOREM 2 Suppose, for instance, that $f(c)$ is a local maximum value of f. The assumption that $f'(c)$ exists means that the right-hand and left-hand limits

$$\lim_{h \to 0^+} \frac{f(c+h) - f(c)}{h} \quad \text{and} \quad \lim_{h \to 0^-} \frac{f(c+h) - f(c)}{h}$$

both exist and are equal to $f'(c)$.

If $h > 0$, then

$$\frac{f(c+h) - f(c)}{h} \leqq 0,$$

because $f(c) \geqq f(c + h)$ for all small positive values of h. Hence, by a one-sided version of the squeeze law for limits (in Section 2.2), this inequality is preserved when we take the limit as $h \to 0$. We thus find that

$$f'(c) = \lim_{h \to 0^+} \frac{f(c+h) - f(c)}{h} \leqq \lim_{h \to 0^+} 0 = 0.$$

Similarly, in the case $h < 0$, we find that

$$\frac{f(c+h) - f(c)}{h} \geqq 0.$$

Therefore,

$$f'(c) = \lim_{h \to 0^-} \frac{f(c+h) - f(c)}{h} \geqq \lim_{h \to 0^-} 0 = 0.$$

Because both $f'(c) \leqq 0$ and $f'(c) \geqq 0$, we conclude that $f'(c) = 0$. This establishes Theorem 2. ◄

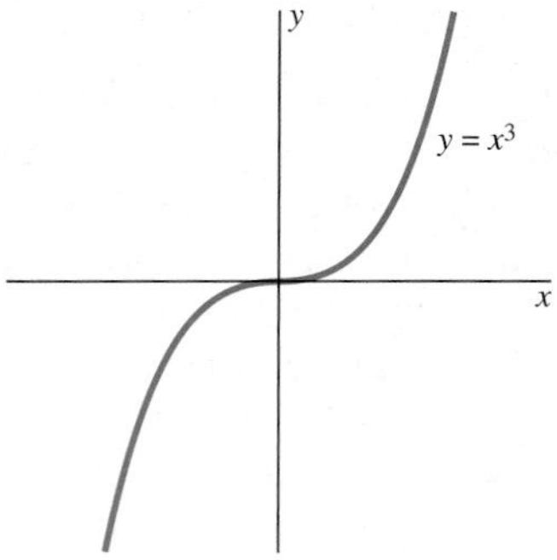

FIGURE 3.5.7 There is no extremum at $x = 0$ even though the derivative is zero there.

BEWARE The converse of Theorem 2 is false. That is, the fact that $f'(c) = 0$ is *not enough* to imply that $f(c)$ is a local extremum. For example, consider the function $f(x) = x^3$. Its derivative $f'(x) = 3x^2$ is zero at $x = 0$. But a glance at its graph (Fig. 3.5.7) shows us that $f(0)$ is *not* a local extremum of f.

Thus the equation $f'(c) = 0$ is a *necessary* condition for $f(c)$ to be a local maximum or minimum value for a function f that is differentiable on an open interval containing c. It is not a *sufficient* condition. The reason: $f'(x)$ may well be zero at points other than local maxima and minima. We give sufficient conditions for local maxima and minima in Chapter 4.

The Closed Interval Maximum-Minimum Method

In most types of optimization problems, we are less interested in the local extrema (as such) than in the *absolute,* or *global,* maximum and minimum values attained by a given continuous function. If f is a function with domain D, we call $f(c)$ the **absolute maximum value,** or **global maximum value,** of f on D provided that $f(c) \geqq f(x)$ for *all* x in D. Briefly, $f(c)$ is the largest value of f on D. It should be clear how the global minimum of f is to be defined. Figure 3.5.8 illustrates some local and global extrema. On the one hand, every global extremum is, of course, local as well. On the other hand, the graph shows local extrema that are not global.

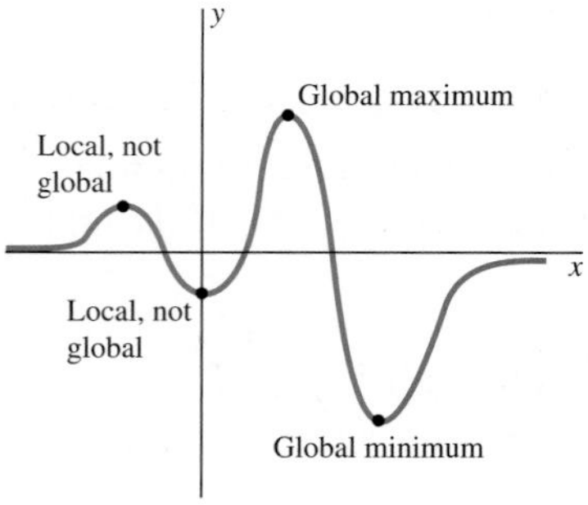

FIGURE 3.5.8 Some extrema are global; others are merely local.

Theorem 3 tells us that the absolute maximum and absolute minimum values of the continuous function f on the closed interval $[a, b]$ occur either at one of the endpoints a or b or at a *critical point* of f. The number c in the domain of f is called a **critical point** of f if either

- $f'(c) = 0$, or
- $f'(c)$ does not exist.

THEOREM 3 Absolute Maxima and Minima
Suppose that $f(c)$ is the absolute maximum (or absolute minimum) value of the continuous function f on the closed interval $[a, b]$. Then c is either a critical point of f or one of the endpoints a and b.

PROOF This result follows almost immediately from Theorem 2. If c is not an endpoint of $[a, b]$, then $f(c)$ is a local extremum of f on the open interval (a, b). In this case Theorem 2 implies that $f'(c) = 0$, provided that f is differentiable at c. ◀

As a consequence of Theorem 3, we can find the (absolute) maximum and minimum values of the function f on the closed interval $[a, b]$ as follows:

1. *Locate* the critical points of f: those points where $f'(x) = 0$ and those points where $f'(x)$ does not exist.
2. *List* the values of x that yield *possible* extrema of f: the two endpoints a and b and those critical points that lie in $[a, b]$.
3. *Evaluate* $f(x)$ at each point in this list of possible extrema.
4. *Inspect* these values of $f(x)$ to see which is the smallest and which is the largest.

The largest of the values in Step 4 is the absolute maximum value of f; the smallest, the absolute minimum. We call this procedure the **closed-interval maximum-minimum method.**

EXAMPLE 3 For our final discussion of the animal pen problem, let us apply the closed-interval maximum-minimum method to find the maximum and minimum values of the differentiable function

$$f(x) = \tfrac{3}{5}x(30 - x) = \tfrac{3}{5}(30x - x^2)$$

on the closed interval $[0, 30]$.

Solution The derivative of f is

$$f'(x) = \tfrac{3}{5}(30 - 2x),$$

which is zero only at the point $x = 15$ in $[0, 30]$. Including the two endpoints, our list of the only values of x that can yield extrema of f consists of 0, 15, and 30. We evaluate f at each:

$$f(0) = 0, \quad \longleftarrow \text{ absolute minimum}$$

$$f(15) = 135, \quad \longleftarrow \text{ absolute maximum}$$

$$f(30) = 0. \quad \longleftarrow \text{ absolute minimum}$$

Thus the maximum value of $f(x)$ on $[0, 30]$ is 135 (attained at $x = 15$), and the minimum value is 0 (attained both at $x = 0$ and at $x = 30$). ◆

EXAMPLE 4 Find the maximum and minimum values of

$$f(x) = 2x^3 - 3x^2 - 12x + 15$$

on the closed interval $[0, 3]$.

Solution The derivative of f is

$$f'(x) = 6x^2 - 6x - 12 = 6(x - 2)(x + 1).$$

So the critical points of f are the solutions of the equation

$$6(x - 2)(x + 1) = 0$$

and the numbers c for which $f'(c)$ does not exist. There are none of the latter, so the critical points of f occur at $x = -1$ and $x = 2$. The first of these is not in the domain of f; we discard it, and thus the only critical point of f in $[0, 3]$ is $x = 2$. Including the two endpoints, our list of all values of x that yield a possible maximum or minimum value of f consists of 0, 2, and 3. We evaluate the function f at each:

$$f(0) = 15, \quad \longleftarrow \text{ absolute maximum}$$

$$f(2) = -5, \quad \longleftarrow \text{ absolute minimum}$$

$$f(3) = 6.$$

Therefore the maximum value of f on $[0, 3]$ is $f(0) = 15$ and its minimum value is $f(2) = -5$. ◆

If in Example 4 we had asked for the maximum and minimum values of $f(x)$ on the interval $[-2, 3]$ (instead of the interval $[0, 3]$), then we would have included *both* critical points $x = -1$ and $x = 2$ in our list of possibilities. The resulting values of f would have been

$$f(-2) = 11,$$

$$f(-1) = 22, \quad \longleftarrow \text{ absolute maximum}$$

$$f(2) = -5, \quad \longleftarrow \text{ absolute minimum}$$

$$f(3) = 6.$$

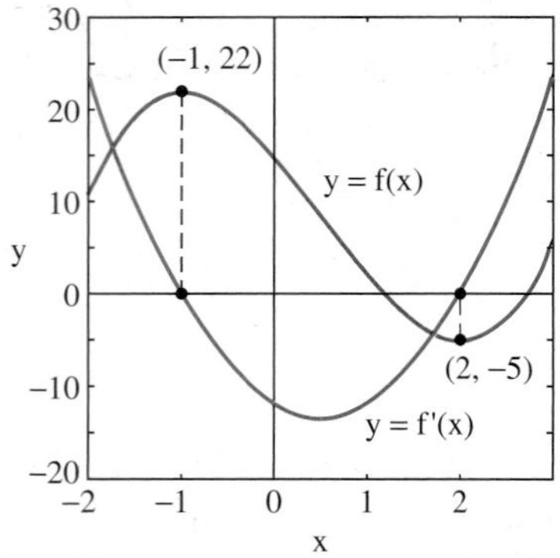

FIGURE 3.5.9 The critical points of the differentiable function $f(x)$ are the zeros of $f'(x)$.

Figure 3.5.9 shows both the curve $y = f(x)$ and the graph of its derivative. Note the vertical line segments joining high and low points on $y = f(x)$ with x-intercepts of $dy/dx = f'(x)$. Thus the figure illustrates the following fact:

> The critical points of a differentiable function $f(x)$ are the zeros of its derivative $f'(x)$.

On the basis of this principle, we can approximate a critical point of f graphically by "zooming in" on a zero of f'.

In Example 4 the function f was differentiable everywhere. Examples 5 and 6 illustrate the case of an extremum at a critical point where the function is not differentiable.

EXAMPLE 5 Find the maximum and minimum values of the function $f(x) = 3 - |x - 2|$ on the interval $[1, 4]$.

Solution If $x \leqq 2$, then $x - 2 \leqq 0$, so

$$f(x) = 3 - (2 - x) = x + 1.$$

If $x \geqq 2$, then $x - 2 \geqq 0$, so

$$f(x) = 3 - (x - 2) = 5 - x.$$

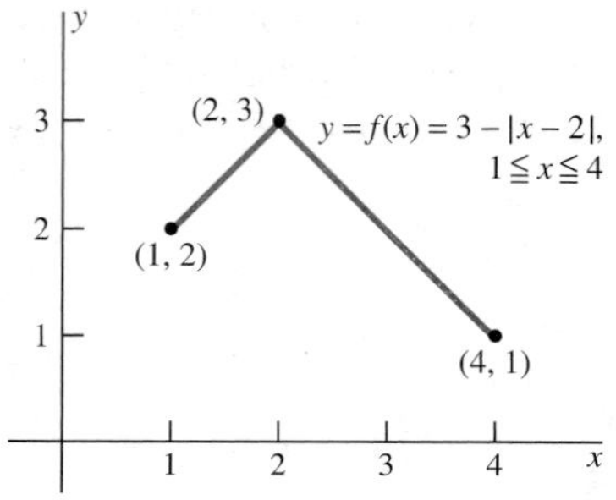

FIGURE 3.5.10 Graph of the function of Example 5.

Consequently, the graph of f looks like the one shown in Fig. 3.5.10. The only critical point of f in $[1, 4]$ is the point $x = 2$, because $f'(x)$ takes on only the two values $+1$ and -1 (and so is never zero), and $f'(2)$ does not exist. (Why not?) Evaluation of f at this critical point and at the two endpoints yields

$$f(1) = 2,$$
$$f(2) = 3, \qquad \longleftarrow \text{absolute maximum}$$
$$f(4) = 1. \qquad \longleftarrow \text{absolute minimum}$$

◆

EXAMPLE 6 Find the maximum and minimum values of

$$f(x) = 5x^{2/3} - x^{5/3}$$

on the closed interval $[-1, 4]$.

Solution Differentiating f yields

$$f'(x) = \frac{10}{3}x^{-1/3} - \frac{5}{3}x^{2/3} = \frac{5}{3}x^{-1/3}(2 - x) = \frac{5(2 - x)}{3x^{1/3}}.$$

Hence f has two critical points in the interval: $x = 2$, where $f'(x) = 0$, and $x = 0$, where $f'(x)$ does not exist (the graph of f has a vertical tangent at $(0, 0)$). When we evaluate f at these two critical points and at the two endpoints, we get

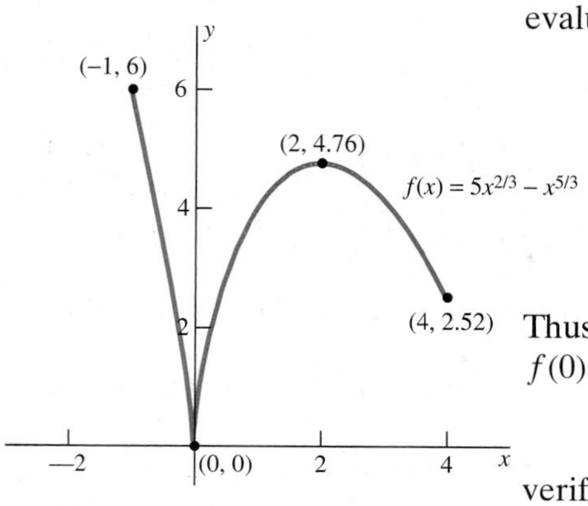

FIGURE 3.5.11 Graph of the function of Example 6.

$$f(-1) = 6, \qquad \longleftarrow \text{absolute maximum}$$
$$f(0) = 0 \qquad \longleftarrow \text{absolute minimum}$$
$$f(2) = 5 \cdot 2^{2/3} - 2^{5/3} \approx 4.76,$$
$$f(4) = 5 \cdot 4^{2/3} - 4^{5/3} \approx 2.52.$$

Thus the maximum value $f(-1) = 6$ occurs at an endpoint. The minimum value $f(0) = 0$ occurs at a point where f is not differentiable.

◆

By using a graphics calculator or computer with graphics capabilities, you can verify that the graph of the function f of Example 6 is that shown in Fig. 3.5.11. But in the usual case of a continuous function that has only finitely many critical points in a given closed interval, the closed-interval maximum-minimum method suffices to determine its maximum and minimum values without requiring any detailed knowledge of the graph of the function.

EXAMPLE 7 Figure 3.5.12 shows the graphs of the function

$$f(x) = 4x^4 - 11x^2 - 5x - 3$$

and its derivative

$$f'(x) = 16x^3 - 22x - 5$$

in the viewing window $-3 \leqq x \leqq 3$, $-30 \leqq y \leqq 30$. Evidently the maximum value of $f(x)$ on the closed interval $[-2, 2]$ is the left-endpoint value $f(-2) = 27$.

The lowest point on the graph of $y = f(x)$ and the corresponding zero of its derivative $dy/dx = f'(x)$ lie within the small boxes in the figure. To find this lowest point exactly we would need to solve the cubic equation $16x^3 - 22x - 5 = 0$. But the lowest point also can be located approximately by using a graphing calculator or computer to zoom in more closely.

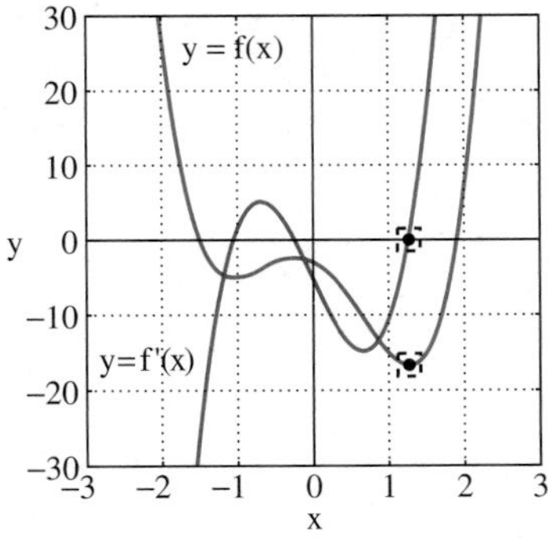

FIGURE 3.5.12 The graphs $y = f(x)$ and $y = f'(x)$.

If we attempt to zoom in on the lowest point without changing the "range factors" or "aspect ratios" of the viewing window, we get a picture like the one in Fig. 3.5.13. Here the magnified graph is indistinguishable from its horizontal tangent line at the low point, so it's impossible to gauge accurately the x-coordinate of the critical point.

Consequently, it is much more effective to zoom in on the corresponding zero of the derivative $f'(x)$. We can then locate the indicated critical point with much greater precision. Thus it is clear in Fig. 3.5.14 that the minimum value attained by $f(x)$ on $[-2, 2]$ is approximately $f(1.273) \approx -16.686$. ◆

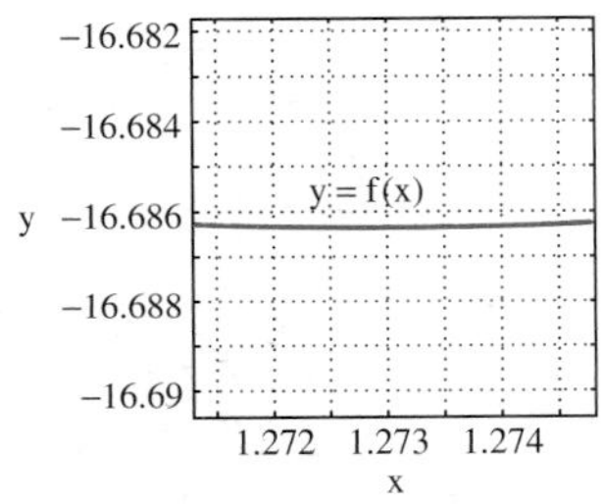

FIGURE 3.5.13 Zooming in on the minimum shown in Fig. 3.5.12.

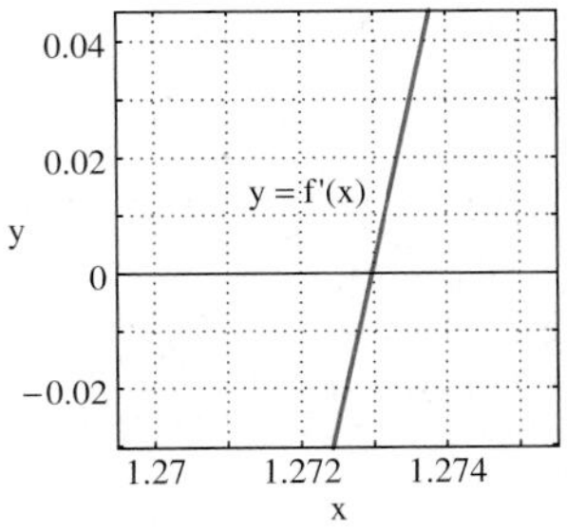

FIGURE 3.5.14 Zooming in instead on the zero of $f'(x)$ shown in Fig. 3.5.12.

3.5 TRUE/FALSE STUDY GUIDE

3.5 CONCEPTS: QUESTIONS AND DISCUSSION

1. Suppose that the function f is continuous on the closed interval $[a, b]$. In each of the five following cases, sketch a possible graph (if any) of f.
   (a) f has a single critical point c but neither a local minimum nor a local maximum in the open interval (a, b). Discuss the possibility that f is not differentiable at c and the possibility that f is differentiable there.
   (b) f has two critical points but only a single local extremum in (a, b).
   (c) f has both a local maximum and a local minimum, but only one critical point in (a, b).
   (d) f has exactly one local maximum, exactly one local minimum, and exactly three critical points in (a, b).
   (e) f has three local maxima but only a single local minimum in (a, b).
2. Can you give an example of a polynomial of odd degree that has neither a local minimum value nor a local maximum value? Can you give an example

of a polynomial of even degree that has neither an absolute minimum value nor an absolute maximum value?

3. Assume that you have located a point on the graph of a differentiable function where a local extremum occurs. Suppose that you zoom in on this point with a graphing calculator or computer, magnifying at each step by the same factor in the x-direction and the y-direction. Should the graph always look like a horizontal line (as in Fig. 3.5.13) after zooming in sufficiently closely in this manner?

3.5 PROBLEMS

In Problems 1 through 10, state whether the given function attains a maximum value or a minimum value (or both) on the given interval. [Suggestion: Begin by sketching a graph of the function.]

1. $f(x) = 1 - x$; $[-1, 1)$

2. $f(x) = 2x + 1$; $[-1, 1)$

3. $f(x) = |x|$; $(-1, 1)$

4. $f(x) = \dfrac{1}{\sqrt{x}}$; $(0, 1]$

5. $f(x) = |x - 2|$; $(1, 4]$

6. $f(x) = 5 - x^2$; $[-1, 2)$

7. $f(x) = x^3 + 1$; $[-1, 1]$

8. $f(x) = \dfrac{1}{x^2 + 1}$; $(-\infty, \infty)$

9. $f(x) = \dfrac{1}{x(1 - x)}$; $[2, 3]$

10. $f(x) = \dfrac{1}{x(1 - x)}$; $(0, 1)$

In Problems 11 through 40, find the maximum and minimum values attained by the given function on the indicated closed interval.

11. $f(x) = 3x - 2$; $[-2, 3]$

12. $f(x) = 4 - 3x$; $[-1, 5]$

13. $h(x) = 4 - x^2$; $[1, 3]$

14. $f(x) = x^2 + 3$; $[0, 5]$

15. $g(x) = (x - 1)^2$; $[-1, 4]$

16. $h(x) = x^2 + 4x + 7$; $[-3, 0]$

17. $f(x) = x^3 - 3x$; $[-2, 4]$

18. $g(x) = 2x^3 - 9x^2 + 12x$; $[0, 4]$

19. $h(x) = x + \dfrac{4}{x}$; $[1, 4]$

20. $f(x) = x^2 + \dfrac{16}{x}$; $[1, 3]$

21. $f(x) = 3 - 2x$; $[-1, 1]$

22. $f(x) = x^2 - 4x + 3$; $[0, 2]$

23. $f(x) = 5 - 12x - 9x^2$; $[-1, 1]$

24. $f(x) = 2x^2 - 4x + 7$; $[0, 2]$

25. $f(x) = x^3 - 3x^2 - 9x + 5$; $[-2, 4]$

26. $f(x) = x^3 + x$; $[-1, 2]$

27. $f(x) = 3x^5 - 5x^3$; $[-2, 2]$

28. $f(x) = |2x - 3|$; $[1, 2]$

29. $f(x) = 5 + |7 - 3x|$; $[1, 5]$

30. $f(x) = |x + 1| + |x - 1|$; $[-2, 2]$

31. $f(x) = 50x^3 - 105x^2 + 72x$; $[0, 1]$

32. $f(x) = 2x + \dfrac{1}{2x}$; $[1, 4]$

33. $f(x) = \dfrac{x}{x + 1}$; $[0, 3]$

34. $f(x) = \dfrac{x}{x^2 + 1}$; $[0, 3]$

35. $f(x) = \dfrac{1 - x}{x^2 + 3}$; $[-2, 5]$

36. $f(x) = 2 - \sqrt[3]{x}$; $[-1, 8]$

37. $f(x) = x\sqrt{1 - x^2}$; $[-1, 1]$

38. $f(x) = x\sqrt{4 - x^2}$; $[0, 2]$

39. $f(x) = x(2 - x)^{1/3}$; $[1, 3]$

40. $f(x) = x^{1/2} - x^{3/2}$; $[0, 4]$

41. Suppose that $f(x) = Ax + B$ is a linear function and that $A \neq 0$. Explain why the maximum and minimum values of f on a closed interval $[a, b]$ must occur at the endpoints of the interval.

42. Suppose that f is continuous on $[a, b]$ and differentiable on (a, b) and that $f'(x)$ is never zero at any point of (a, b). Explain why the maximum and minimum values of f must occur at the endpoints of the interval $[a, b]$.

43. Explain why every real number is a critical point of the greatest integer function $f(x) = [\![x]\!]$.

44. Prove that every quadratic function

$$f(x) = ax^2 + bx + c \quad (a \neq 0)$$

has exactly one critical point on the real line.

45. Explain why the cubic polynomial function

$$f(x) = ax^3 + bx^2 + cx + d \quad (a \neq 0)$$

can have either two, one, or no critical points on the real line. Produce examples that illustrate each of the three cases.

46. Define $f(x)$ to be the distance from x to the nearest integer. What are the critical points of f?

In Problems 47 through 52, match the given graph of the function with the graph of its derivative f' from those in Fig. 3.5.15, parts (a) through (f).

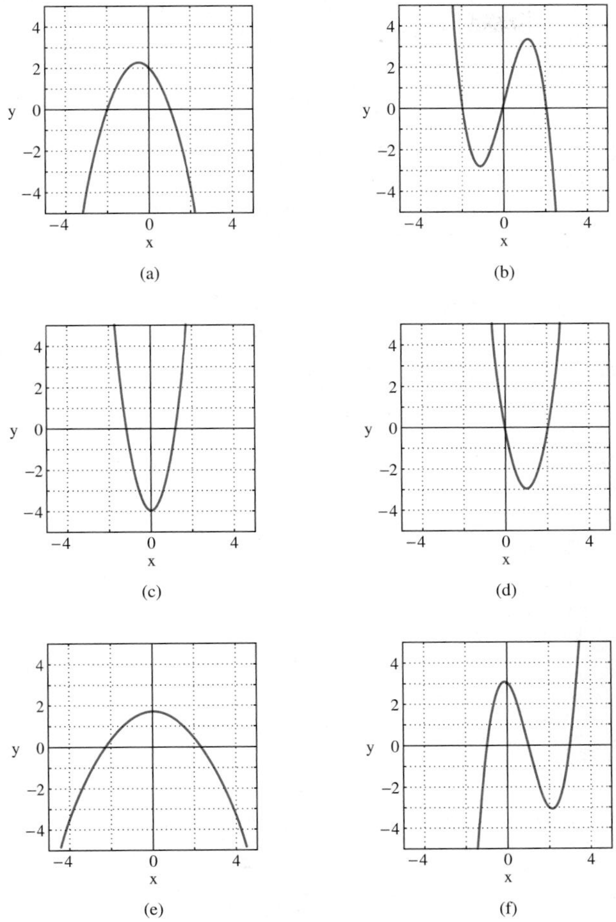

FIGURE 3.5.15

47. Figure 3.5.16

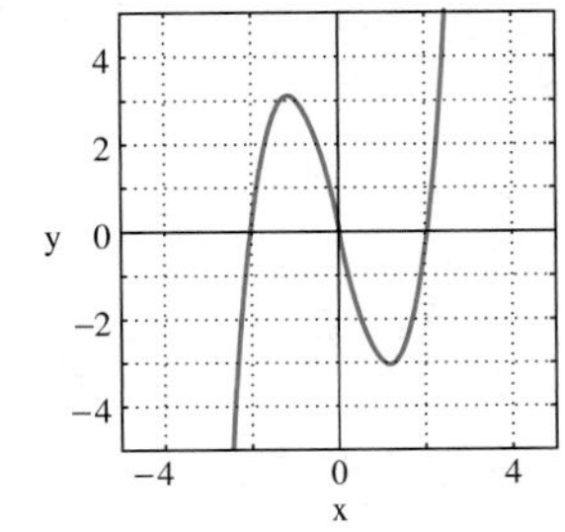

FIGURE 3.5.16

48. Figure 3.5.17

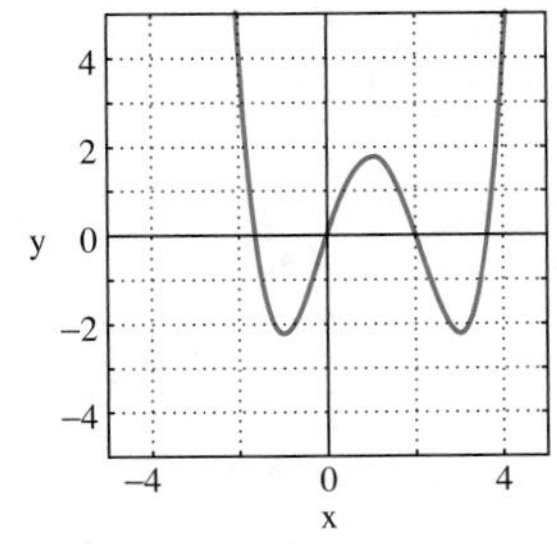

FIGURE 3.5.17

49. Fig. 3.5.18

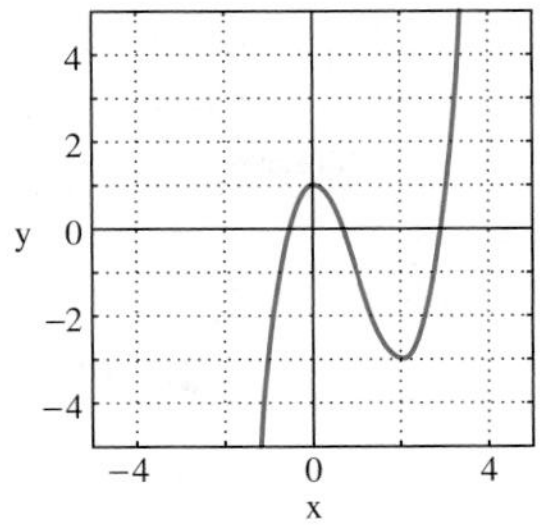

FIGURE 3.5.18

50. Fig. 3.5.19

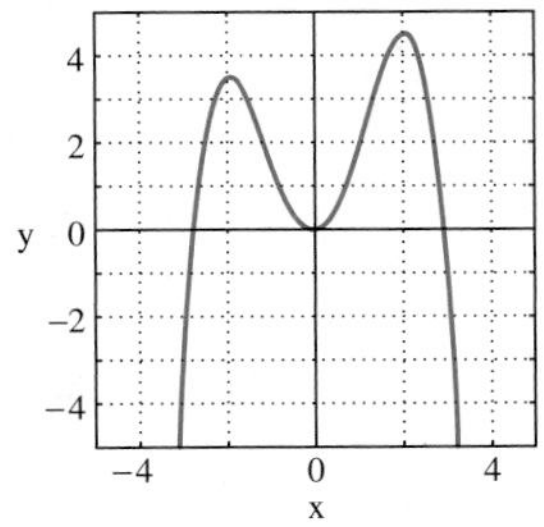

FIGURE 3.5.19

51. Fig. 3.5.20

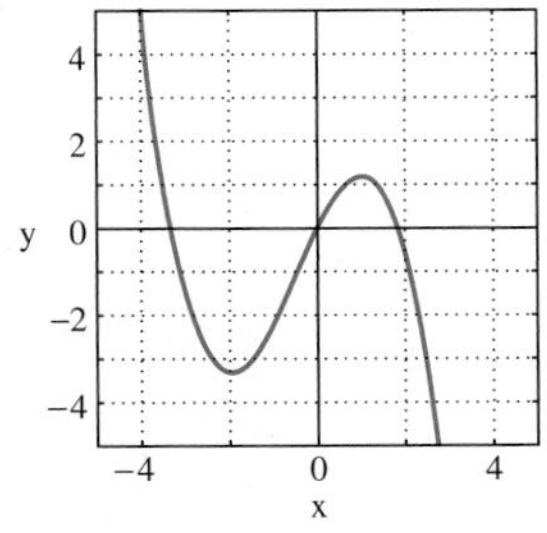

FIGURE 3.5.20

52. Fig. 3.5.21

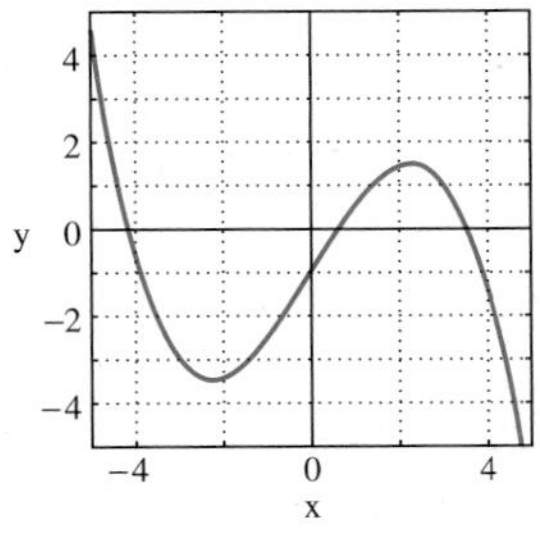

FIGURE 3.5.21

In Problems 53 through 60, find good approximations to the maximum and minimum values of the given function on the indicated closed interval by zooming in on the zeros of the derivative.

53. $f(x) = x^3 + 3x^2 - 7x + 10;\quad [-2, 2]$

54. $f(x) = x^3 + 3x^2 - 7x + 10;\quad [-4, 2]$

55. $f(x) = x^4 - 3x^3 + 7x - 5;\quad [-3, 3]$

56. $f(x) = x^4 - 5x^3 + 17x - 5;\quad [-3, 3]$

57. $f(x) = x^4 - 5x^3 + 17x - 5;\quad [0, 2]$

58. $f(x) = x^5 - 5x^4 - 15x^3 + 17x^2 + 23x;\quad [-1, 1]$

59. $f(x) = x^5 - 5x^4 - 15x^3 + 17x^2 + 23x;\quad [-3, 3]$

60. $f(x) = x^5 - 5x^4 - 15x^3 + 17x^2 + 23x;\quad [0, 10]$

3.5 Project: When Is Your Coffee Cup Stablest?

Your car has no cupholder, so you must place your filled coffee cup on the passenger seat beside you when you start out in the morning. Bitter experience has taught you that the cup is least stable—and most prone to spill—when it's completely full, but becomes more stable as you drink the coffee and thereby lower its level in the cup. Now you're ready to apply calculus to analyze this phenomenon.

Figure 3.5.22 shows a coffee cup partially filled with coffee. We will assume that it is stablest when the *centroid* of the cup-plus-coffee is lowest. The **centroid** of a solid cylinder or cylindrical shell is its geometric central point, and the y-coordinate $\overline{y}$ of the centroid of a composite body consisting of several pieces with masses m_1, m_2, and m_3 having centroids with respective y-coordinates y_1, y_2, and y_3 is given by

$$\overline{y} = \frac{m_1 y_1 + m_2 y_2 + m_3 y_3}{m_1 + m_2 + m_3}. \tag{1}$$

This formula means that $\overline{y}$ is an average of the y-coordinates y_1, y_2, and y_3 of the individual centroids, each weighted by the corresponding mass.

The simplified model of the coffee cup shown in Fig. 3.5.22 consists of the following:

- A side surface that is a cylindrical shell with height H, inner radius R, and thickness T, and
- A bottom that is a solid cylinder with radius $R + T$ and height B.

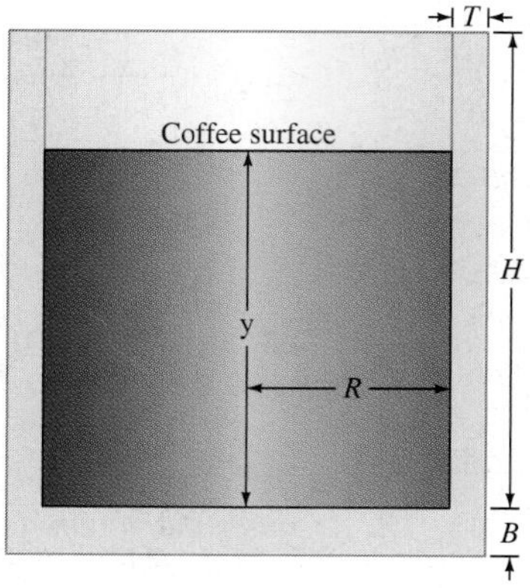

FIGURE 3.5.22 Coffee cup partially filled with coffee to depth y.

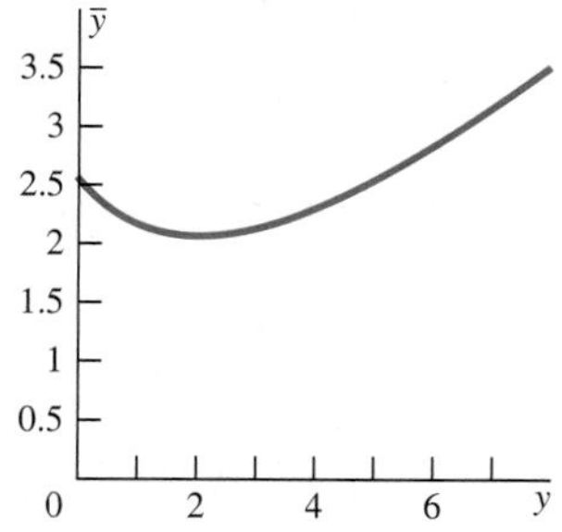

FIGURE 3.5.23 Centroid height $f(y)$ as a function of coffee depth y.

The cup is partially filled with coffee with depth y and density 1 g/cm^3. In the CD-ROM discussion of this project, we take $H = 8$, $R = 3$, $T = 0.5$, and $B = 1$ (all units are centimeters). Assuming also that the density of the material of the cup itself is $\delta = 1$ g/cm^3, we apply Eq. (1) to derive the function

$$f(y) = \frac{87 + 4y^2}{34 + 8y}, \quad 0 \leqq y \leqq 8 \tag{2}$$

giving the y-coordinate $\overline{y} = f(y)$ of the centroid of the cup-plus-coffee as a function of the depth y of the coffee in the cup.

Figure 3.5.23 shows the graph of the function f. It appears that the centroid is lowest when $y = 2$, and thus when the cup is about one-quarter filled with coffee. To find when $f'(y) = 0$, you can differentiate the function in (2) and simplify to obtain

$$f'(y) = \frac{2(4y^2 + 34y - 87)}{(4y + 17)^2}. \tag{3}$$

Thus you need only solve a quadratic equation to see where the numerator is zero: when $y = \frac{1}{4}(-17 \pm 7\sqrt{13})$. The positive solution gives the optimal depth $y \approx 2.0597$ cm of the coffee in your cup—just a bit more than a quarter of the height $H = 8$ cm of the cup.

Carry out this analysis with your own favorite coffee cup. Measure its physical dimensions H, R, T, and B. How can you determine the approximate density δ of its material?

3.6 APPLIED OPTIMIZATION PROBLEMS

This section is devoted to applied maximum-minimum problems (like the animal pen problem of Section 1.1) for which the closed-interval maximum-minimum method of Section 3.5 can be used. When we confront such a problem, there is an important first step: We must determine the quantity to be maximized or minimized. This quantity will be the dependent variable in our analysis of the problem.

This dependent variable must then be expressed as a function of an independent variable, one that "controls" the values of the dependent variable. If the domain of values of the independent variable—those that are pertinent to the applied problem—is a closed interval, then we may proceed with the closed-interval maximum-minimum method. This plan of attack can be summarized in the following steps:

1. *Find the quantity to be maximized or minimized.* This quantity, which you should describe with a word or short phrase and label with a descriptive letter, will be the dependent variable. Because it is a *dependent* variable, it depends on something else; that quantity will be the independent variable. Here we call the independent variable x.
2. *Express the dependent variable as a function of the independent variable.* Use the information in the problem to write the dependent variable as a function of x. Always draw a figure and *label the variables;* this is generally the best way to find the relationship between the dependent and independent variables. Use auxiliary variables if they help, but not too many, for you must eventually eliminate them. You *must* express the dependent variable as a function of the *single* independent variable x and various constants before you can compute any derivatives. Find the domain of this function as well as its formula. Force the domain to be a closed and bounded interval if possible—if the natural domain is an open interval, adjoin the endpoints if you can.
3. *Apply calculus to find the critical points.* Compute the derivative f' of the function f that you found in Step 2. Use the derivative to find the critical points—where $f'(x) = 0$ and where $f'(x)$ does not exist. If f is differentiable everywhere, then its only critical points occur where $f'(x) = 0$.
4. *Identify the extrema.* Evaluate f at each critical point in its domain *and* at the two endpoints. The values you obtain will tell you which is the absolute maximum and which is the absolute minimum. Of course, either or both of these may occur at more than one point.
5. *Answer the question posed in the original problem.* In other words, interpret your results. The answer to the original problem may be something other than merely the largest (or smallest) value of f. Give a precise answer to the specific question originally asked.

Observe how we follow this five-step process in Example 1.

EXAMPLE 1 A farmer has 200 yd of fence with which to construct three sides of a rectangular pen; an existing long, straight wall will form the fourth side. What dimensions will maximize the area of the pen?

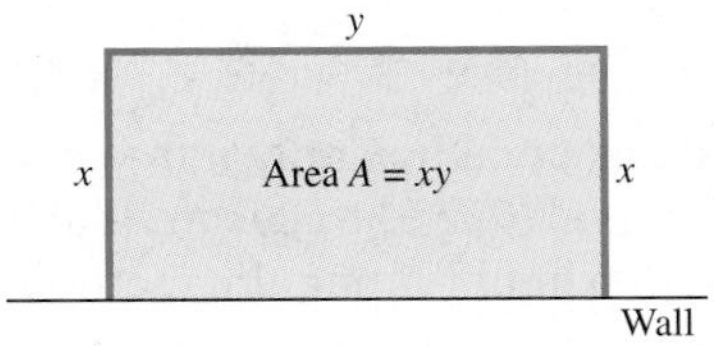

FIGURE 3.6.1 The rectangular pen of Example 1.

Solution We want to maximize the area A of the pen shown in Fig. 3.6.1. To get a formula for the *dependent* variable A, we observe that the area of a rectangle is the product of its base and its height. So we let x denote the length of each of the two sides of the pen perpendicular to the wall. We also let y denote the length of the side parallel to the wall. Then the area of the rectangle is given by the *formula*

$$A = xy.$$

Now we need to write A as a *function* of either x or y. Because all 200 yd of fence are to be used,

$$2x + y = 200, \qquad \text{so} \quad y = 200 - 2x \tag{1}$$

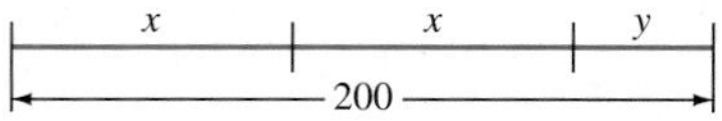

FIGURE 3.6.2 The relation in Eq. (1) between x and y (Example 1).

(We chose to express y in terms of x merely because the algebra is slightly simpler.) Next, we substitute this value of y into the formula $A = xy$ to obtain

$$A(x) = x(200 - 2x) = 200x - 2x^2. \tag{2}$$

This equation expresses the dependent variable A as a function of the independent variable x.

Before proceeding, we must find the domain of the function A. It is clear from Fig. 3.6.2 that $0 < x < 100$. But to apply the closed-interval maximum-minimum method, we need a closed interval. In this example, we may adjoin the endpoints to $(0, 100)$ to get the *closed* interval $[0, 100]$. The values $x = 0$ and $x = 100$ correspond to "degenerate" pens of area zero. Because zero cannot be the maximum value of A, there is no harm in thus enlarging the domain of the function A.

Now we compute the derivative of the function A in Eq. (2):

$$\frac{dA}{dx} = 200 - 4x.$$

Because A is differentiable, its only critical points occur when

$$\frac{dA}{dx} = 0;$$

that is, when

$$200 - 4x = 0.$$

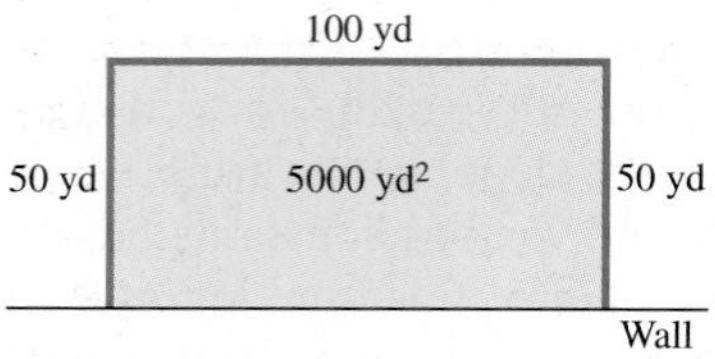

FIGURE 3.6.3 The pen with maximal area of Example 1.

So $x = 50$ is the only interior critical point. Including the endpoints, the extrema of A can occur only at $x = 0$, 50, and 100. We evaluate A at each:

$$A(0) = 0,$$
$$A(50) = 5000, \qquad \longleftarrow \text{ absolute maximum}$$
$$A(100) = 0.$$

Thus the maximal area is $A(50) = 5000$ (yd^2). From Eq. (1) we find that $y = 100$ when $x = 50$. Therefore, for the pen to have maximal area, each of the two sides perpendicular to the wall should be 50 yd long and the side parallel to the wall should be 100 yd long (Fig. 3.6.3). ◆

EXAMPLE 2 A piece of sheet metal is rectangular, 5 ft wide and 8 ft long. Congruent squares are to be cut from its four corners. The resulting piece of metal is to be folded and welded to form an open-topped box (Fig. 3.6.4). How should this be done to get a box of largest possible volume?

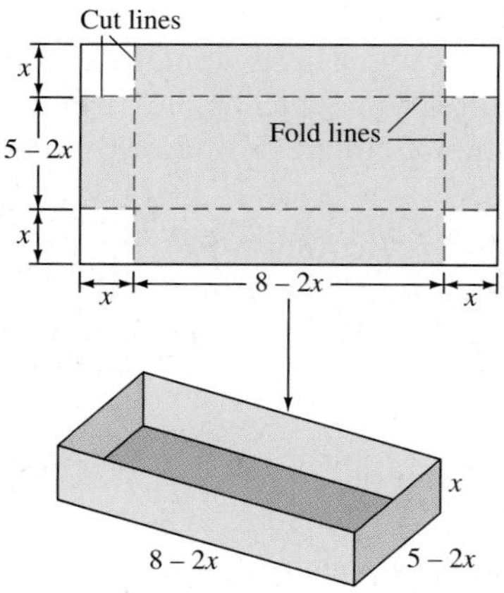

FIGURE 3.6.4 Making the box of Example 2.

Solution The quantity to be maximized—the dependent variable—is the volume V of the box to be constructed. The shape and thus the volume of the box are determined by the length x of the edge of each corner square removed. Hence x is a natural choice for the independent variable.

To write the volume V as a function of x, note that the finished box will have height x and its base will measure $8 - 2x$ ft by $5 - 2x$ ft. Hence its volume is given by

$$V(x) = x(5 - 2x)(8 - 2x) = 4x^3 - 26x^2 + 40x.$$

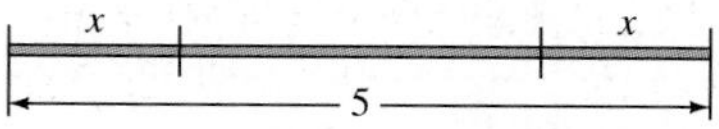

FIGURE 3.6.5 The 5-ft width of the metal sheet (Example 2).

The procedure described in this example will produce an actual box only if $0 < x < 2.5$ (Fig. 3.6.5). But we make the domain the *closed* interval $[0, 2.5]$ to ensure that a maximum of $V(x)$ exists and to use the closed-interval maximum-minimum method. The values $x = 0$ and $x = 2.5$ correspond to "degenerate" boxes of zero volume, so adjoining these points to $(0, 2.5)$ will affect neither the location of the absolute maximum nor its value.

Now we compute the derivative of V:

$$V'(x) = 12x^2 - 52x + 40 = 4(3x - 10)(x - 1).$$

The only critical points of the differentiable function V occur where

$$V'(x) = 0;$$

that is, where

$$4(3x - 10)(x - 1) = 0.$$

The solutions of this equation are $x = 1$ and $x = \frac{10}{3}$. We discard the latter because it does not lie in the domain $[0, 2.5]$ of V. So we examine these values of V:

$$V(0) = 0,$$
$$V(1) = 18, \quad \longleftarrow \text{absolute maximum}$$
$$V(2.5) = 0.$$

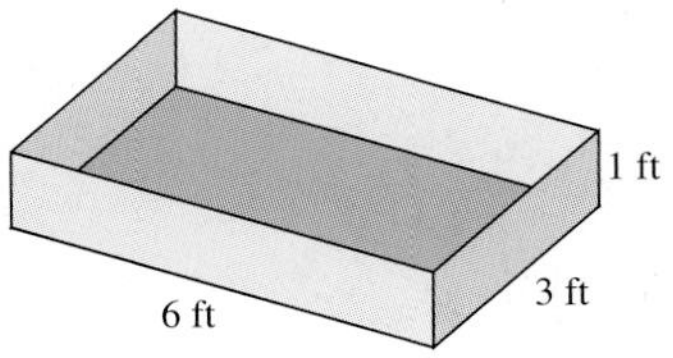

FIGURE 3.6.6 The box with maximal volume of Example 2.

Thus the maximum value of $V(x)$ on $[0, 2.5]$ is $V(1) = 18$. The answer to the question posed is this: The squares cut from the corners should be of edge length 1 ft each. The resulting box will measure 6 ft by 3 ft by 1 ft, and its volume will be 18 ft^3 (Fig. 3.6.6). ◆

For our next application of the closed-interval maximum-minimum method, let us consider a typical problem in business management. Suppose that x units of computer diskettes are to be manufactured at a *total cost* of $C(x)$ dollars. We make the simple (but not always valid) assumption that the cost function $C(x)$ is the sum of two terms:

- A constant term a representing the *fixed cost* of acquiring and maintaining production facilities (overhead), and
- A variable term representing the *additional cost* of making x units at, for example, b dollars each.

Then

the **total cost** is the **sum** of the **fixed cost** and the **additional cost,**

so the *cost function* $C(x)$ is given by

$$C(x) = a + bx. \tag{3}$$

We also assume that the number of units that can be sold (and hence will be manufactured) is a linear function of the selling price p, so that $x = m - np$ where m and n are positive constants. The minus sign indicates that an increase in selling price will result in a decrease in sales. If we solve this last equation for p, we get the *price function*

$$p(x) = A - Bx \tag{4}$$

(A and B are also constants).

The quantity to be maximized is profit, given here by the *profit function* $P(x)$, which is equal to the sales revenue minus the production costs. Thus

$$P(x) = xp(x) - C(x). \tag{5}$$

EXAMPLE 3 Suppose that the cost of publishing a small book is \$10,000 to set up the (annual) press run plus \$8 for each book printed. The publisher sold 7000 copies last year at \$13 each, but sales dropped to 5000 copies this year when the price was raised to \$15 per copy. Assume that up to 10,000 copies can be printed in a single press run. How many copies should be printed, and what should be the selling price of each copy, to maximize the year's profit on this book?

Solution The dependent variable to be maximized is the profit P. As independent variable we choose the number x of copies to be printed; also, $0 \leqq x \leqq 10{,}000$. The given cost information then implies that

$$C(x) = 10{,}000 + 8x.$$

Now we substitute into Eq. (4) the data $x = 7000$ when $p = 13$ as well as the data $x = 5000$ when $p = 15$. We obtain the equations

$$A - 7000B = 13, \qquad A - 5000B = 15.$$

When we solve these equations simultaneously, we find that $A = 20$ and $B = 0.001$. Hence the price function is

$$p(x) = 20 - \frac{x}{1000},$$

and thus the profit function is

$$P(x) = x\left(20 - \frac{x}{1000}\right) - (10{,}000 + 8x).$$

We expand and collect terms to obtain

$$P(x) = 12x - \frac{x^2}{1000} - 10{,}000, \quad 0 \leqq x \leqq 10{,}000.$$

Now

$$\frac{dP}{dx} = 12 - \frac{x}{500},$$

and the only critical points of the differentiable function P occur when

$$\frac{dP}{dx} = 0;$$

that is, when

$$12 - \frac{x}{500} = 0; \quad x = 12 \cdot 500 = 6000.$$

We check P at this value of x as well as the values of $P(x)$ at the endpoints to find the maximum profit:

$$\begin{aligned} P(0) &= -10{,}000, \\ P(6000) &= 26{,}000, \qquad \longleftarrow \text{ absolute maximum} \\ P(10{,}000) &= 10{,}000. \end{aligned}$$

Therefore, the maximum possible annual profit of \$26,000 results from printing 6000 copies of the book. Each copy should be sold for \$14, because

$$p(6000) = 20 - \frac{6000}{1000} = 14. \qquad \blacklozenge$$

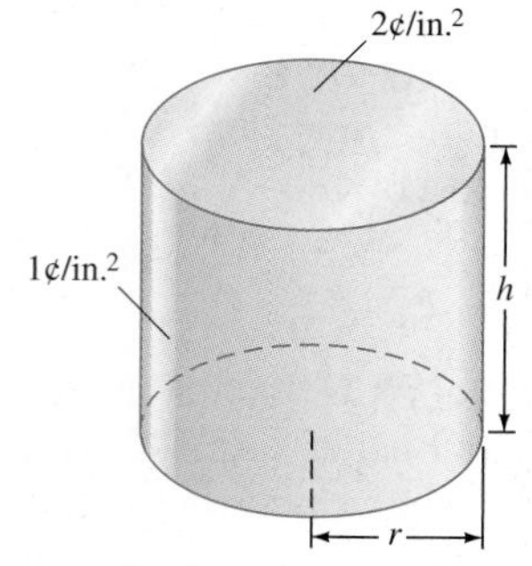

FIGURE 3.6.7 The cylindrical can of Example 4.

EXAMPLE 4 We need to design a cylindrical can with radius r and height h. The top and bottom must be made of copper, which will cost 2¢/in.² The curved side is to be made of aluminum, which will cost 1¢/in.² We seek the dimensions that will maximize the volume of the can. The only constraint is that the total cost of the can is to be 300π¢.

Solution We need to maximize the volume V of the can, which we can compute if we know its radius r and its height h (Fig. 3.6.7). With these dimensions, we find that

$$V = \pi r^2 h, \tag{6}$$

but we need to express V as a function of r alone (or as a function of h alone).

Both the circular top and bottom of the can have area πr^2 in.², so the area of copper to be used is $2\pi r^2$ and its cost is $4\pi r^2$ cents. The area of the curved side of the can is $2\pi rh$ in.^{2}, so the area of aluminum used is the same, and the aluminum costs $2\pi rh$ cents.

We obtain the total cost of the can by adding the cost of the copper to the cost of the aluminum. This sum must be 300π¢, and therefore

$$4\pi r^2 + 2\pi rh = 300\pi. \tag{7}$$

We eliminate h in Eq. (6) by solving Eq. (7) for h:

$$h = \frac{300\pi - 4\pi r^2}{2\pi r} = \frac{1}{r}(150 - 2r^2). \tag{8}$$

Hence

$$V = V(r) = (\pi r^2)\frac{1}{r}(150 - 2r^2) = 2\pi(75r - r^3). \tag{9}$$

To determine the domain of definition of V, we note from Eq. (7) that $4\pi r^2 < 300\pi$, so $r < \sqrt{75}$ for the desired can; with $r = \sqrt{75} = 5\sqrt{3}$, we get a degenerate can with height $h = 0$. With $r = 0$, we obtain *no* value of h in Eq. (8) and therefore no can, but $V(r)$ is nevertheless continuous at $r = 0$. Consequently, we can take the closed interval $[0, 5\sqrt{3}]$ to be the domain of V.

Calculating the derivative yields

$$V'(r) = 2\pi(75 - 3r^2) = 6\pi(25 - r^2).$$

Because $V(r)$ is a polynomial, $V'(r)$ exists for all values of r, so we obtain all critical points by solving the equation

$$V'(r) = 0;$$

that is,

$$6\pi(25 - r^2) = 0.$$

We discard the solution -5, as it does not lie in the domain of V. Thus we obtain only the single critical point $r = 5$ in $[0, 5\sqrt{3}]$. Now

$$V(0) = 0,$$
$$V(5) = 500\pi, \qquad \longleftarrow \text{ absolute maximum}$$
$$V(5\sqrt{3}) = 0.$$

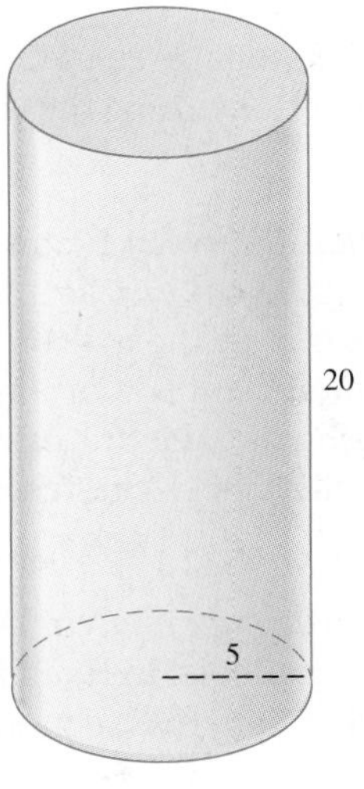

FIGURE 3.6.8 The can of maximal volume in Example 4.

Thus the can of maximum volume has radius $r = 5$ in., and Eq. (8) yields its height to be $h = 20$ in. Figure 3.6.8 shows such a can. ◆

EXAMPLE 5 (A Sawmill Problem) Suppose that you need to cut a beam with maximal rectangular cross section from a circular log of radius 1 ft. (This is the geometric problem of finding the rectangle of greatest area that can be inscribed in a circle of radius 1.) What are the shape and cross-sectional area of such a beam?

Solution Let x and y denote half the base and half the height, respectively, of the inscribed rectangle (Fig. 3.6.9). Apply the Pythagorean theorem to the small right triangle in the figure. This yields the equation

$$x^2 + y^2 = 1, \qquad \text{so} \quad y = \sqrt{1 - x^2}.$$

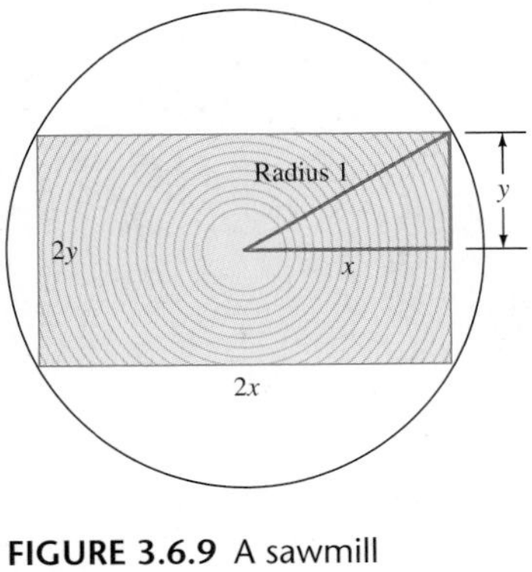

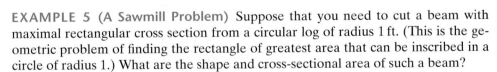

FIGURE 3.6.9 A sawmill problem—Example 5.

The area of the inscribed rectangle is $A = (2x)(2y) = 4xy$. You may now express A as a function of x alone:

$$A(x) = 4x\sqrt{1 - x^2}.$$

The practical domain of definition of A is $(0, 1)$, and there is no harm (and much advantage) in adjoining the endpoints, so you take $[0, 1]$ to be the domain. Next,

$$\frac{dA}{dx} = 4 \cdot (1 - x^2)^{1/2} + 2x(1 - x^2)^{-1/2}(-2x) = \frac{4 - 8x^2}{(1 - x^2)^{1/2}}.$$

You observe that $A'(1)$ does not exist, but this causes no trouble, because differentiability at the endpoints is not assumed in Theorem 3 of Section 3.5. Hence you need only solve the equation

$$A'(x) = 0;$$

that is,

$$\frac{4 - 8x^2}{\sqrt{1 - x^2}} = 0.$$

A fraction can be zero only when its numerator is zero and its denominator is *not,* so $A'(x) = 0$ when $4 - 8x^2 = 0$. Thus you find the only critical point of A in the open interval $(0, 1)$ to be $x = \sqrt{1/2} = \frac{1}{2}\sqrt{2}$ (and $2x = 2y = \sqrt{2}$). You evaluate A here and at the two endpoints to find that

$$A(0) = 0,$$
$$A(\tfrac{1}{2}\sqrt{2}) = 2, \qquad \longleftarrow \text{ absolute maximum}$$
$$A(1) = 0.$$

Therefore, the beam with rectangular cross section of maximal area is square, with edges $\sqrt{2}$ ft long and with cross-sectional area 2 ft^2. ◆

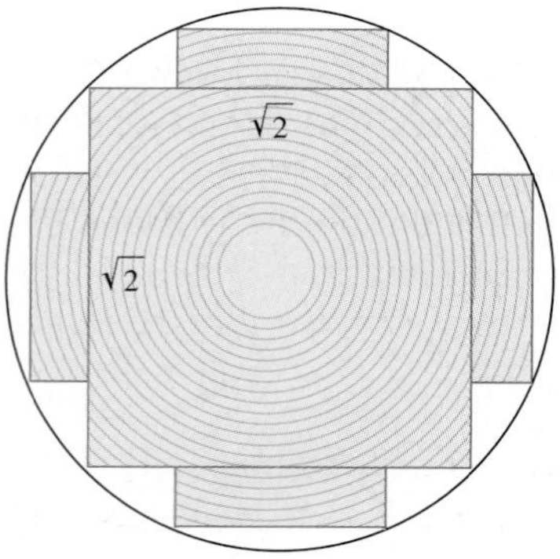

FIGURE 3.6.10 Cut four more beams after cutting one large beam.

In Problem 43 we ask you to maximize the total cross-sectional area of the four planks that can be cut from the four pieces of log that remain after cutting the square beam (Fig. 3.6.10).

Plausibility You should always check your answers for *plausibility*. In Example 5, the cross-sectional area of the log from which the beam is to be cut is $\pi \approx 3.14$ ft^2. The beam of maximal cross-section area 2 ft^2 thus uses a little less than 64% of the log. This *is* plausible. Had the fraction been an extremely inefficient 3% or a wildly optimistic 98%, you should have searched for an error in arithmetic, algebra, calculus, or logic (as you would had the fraction been −14% or 150%). Check the results of Examples 1 through 4 for plausibility.

Dimensions Another way to check answers is to use *dimensional analysis.* Work the problem with unspecified constants in place of the actual numbers. In Example 5, it would be good practice to find the beam of maximal rectangular cross section that can be cut from a circular log of radius R rather than radius 1 ft. You can always substitute the given value $R = 1$ at the conclusion of the solution. A brief solution to this problem might go as follows:

Dimensions of beam: base $2x$, height $2y$.

Area of beam: $A = 4xy$.

Draw a radius of the log from its center to one corner of the rectangular beam, as in Fig. 3.6.11. This radius has length R, so the Pythagorean theorem gives

$$x^2 + y^2 = R^2; \qquad y = \sqrt{R^2 - x^2}.$$

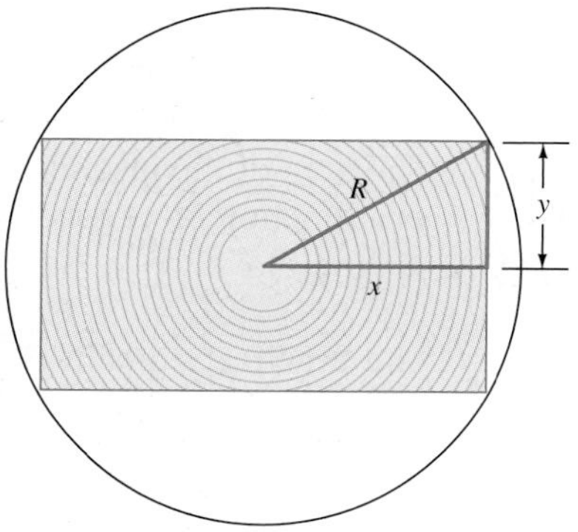

FIGURE 3.6.11 The log with radius R.

Area of beam:

$$A(x) = 4x\sqrt{R^2 - x^2}, \quad 0 \leqq x \leqq R.$$

$$A'(x) = 4(R^2 - x^2)^{1/2} + 2x(R^2 - x^2)^{-1/2}(-2x) = \frac{4R^2 - 8x^2}{\sqrt{R^2 - x^2}}.$$

$A'(x)$ does not exist when $x = R$, but that's an endpoint; we'll check it separately. $A'(x) = 0$ when $x = \frac{1}{2}R\sqrt{2}$ (ignore the negative root; it's not in the domain of A).

$$A(0) = 0,$$
$$A(\tfrac{1}{2}R\sqrt{2}) = 2R^2, \qquad \longleftarrow \text{ absolute maximum}$$
$$A(R) = 0.$$

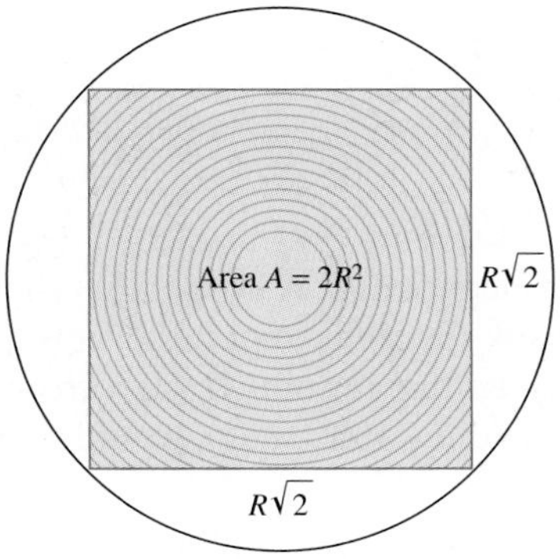

FIGURE 3.6.12 The inscribed square beam with maximal cross-sectional area.

Figure 3.6.12 shows the dimensions of the inscribed rectangle of maximal area.

Now you can check the results for dimensional accuracy. The value of x that maximizes A is a length (R) multiplied by a pure (dimensionless) numerical constant ($\frac{1}{2}\sqrt{2}$), so x has the dimensions of length—that's correct; had it been anything else, you would need to search for the error. Moreover, the maximum cross-sectional area of the beam is $2R^2$, the product of a pure number and the square of a length, thus having the dimensions of area. This, too, is correct.

EXAMPLE 6 We consider the reflection of a ray of light by a mirror M as in Fig. 3.6.13, which shows a ray traveling from point A to point B via reflection off M at the point P. We assume that the location of the point of reflection is such that the total distance $d_1 + d_2$ traveled by the light ray will be minimized. This is an application of *Fermat's principle of least time* for the propagation of light. The problem is to find P.

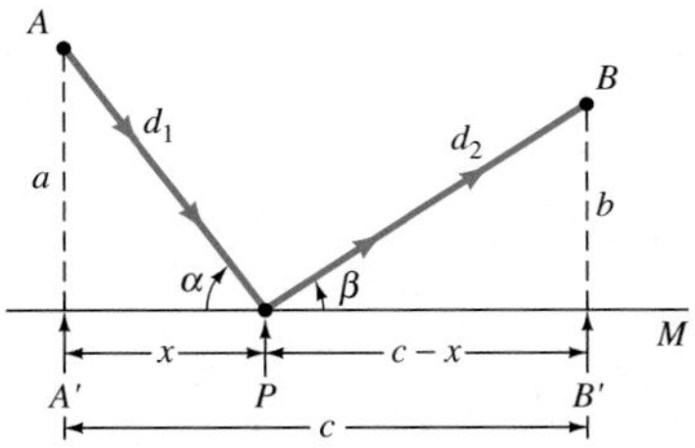

FIGURE 3.6.13 Reflection at P of a light ray by a mirror M (Example 6).

Solution Drop perpendiculars from A and B to the plane of the mirror M. Denote the feet of these perpendiculars by A' and B' (Fig. 3.6.13). Let a, b, c, and x denote the lengths of the segments AA', BB', $A'B'$, and $A'P$, respectively. Then $c - x$ is the length of the segment PB'. By the Pythagorean theorem, the distance to be minimized is then

$$d_1 + d_2 = f(x) = \sqrt{a^2 + x^2} + \sqrt{b^2 + (c - x)^2}. \tag{10}$$

We may choose as the domain of f the interval $[0, c]$, because the minimum of f must occur somewhere within that interval. (To see why, examine the picture you get if x is *not* in that interval.)

Then

$$f'(x) = \frac{x}{\sqrt{a^2 + x^2}} + \frac{(c - x)(-1)}{\sqrt{b^2 + (c - x)^2}}. \tag{11}$$

Recognizing the distances d_1 and d_2 in the denominators here, we see that

$$f'(x) = \frac{x}{d_1} - \frac{c - x}{d_2}. \tag{12}$$

Consequently, any horizontal tangent to the graph of f must occur over the point x determined by the equation

$$\frac{x}{d_1} = \frac{c - x}{d_2}. \tag{13}$$

At such a point, $\cos\alpha = \cos\beta$, where α is the angle of the incident light ray and β is the angle of the reflected ray (Fig. 3.6.13). Both α and β lie between 0 and $\pi/2$, and thus we find that $\alpha = \beta$. In short, the point P must be located so that the angle of incidence is equal to the angle of reflection, a familiar principle from physics. ◆

The computation in Example 6 has an alternative interpretation that is interesting, if somewhat whimsical. Figure 3.6.14 shows a feedlot 200 ft long with a water

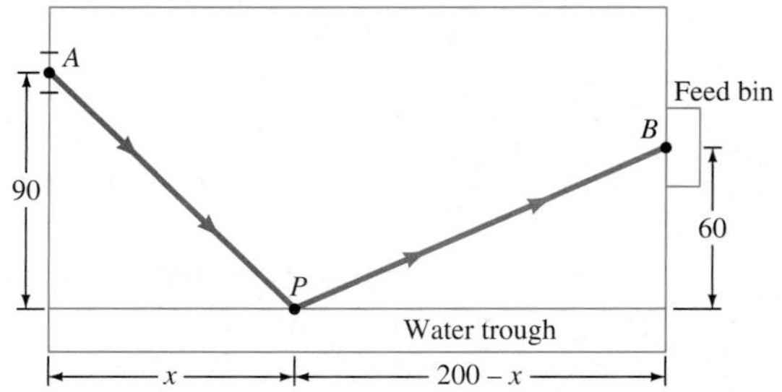

FIGURE 3.6.14 The feedlot.

trough along one edge and a feed bin located on an adjacent edge. A cow enters the gate at the point A, 90 ft from the water trough. She walks straight to point P, gets a drink from the trough, and then walks straight to the feed bin at point B, 60 ft from the trough. If the cow knew calculus, what point P along the water trough would she select to minimize the total distance she walks?

In comparing Figs. 3.6.13 and 3.6.14, we see that the cow's problem is to minimize the distance function f in Eq. (10) with the numerical values $a = 90$, $b = 60$, and $c = 200$. When we substitute these values and

$$d_1 = \sqrt{a^2 + x^2} \quad \text{and} \quad d_2 = \sqrt{b^2 + (c - x)^2}$$

in Eq. (13), we get

$$\frac{x}{\sqrt{8100 + x^2}} = \frac{200 - x}{\sqrt{3600 + (200 - x)^2}}.$$

We square both sides, clear the equation of fractions, and simplify. The result is

$$\begin{aligned} x^2[3600 + (200 - x)^2] &= (200 - x)^2(8100 + x^2); \\ 3600x^2 &= 8100(200 - x)^2; \qquad \text{(Why?)} \\ 60x &= 90(200 - x); \\ 150x &= 18{,}000; \\ x &= 120. \end{aligned}$$

Thus the cow should proceed directly to the point P located 120 ft along the water trough.

These examples indicate that the closed-interval maximum-minimum method is applicable to a wide range of problems. Indeed, applied optimization problems that seem as different as light rays and cows may have essentially identical mathematical models. This is only one illustration of the power of generality that calculus exploits so effectively.

3.6 TRUE/FALSE STUDY GUIDE

3.6 CONCEPTS: QUESTIONS AND DISCUSSION

1. How do you decide what is the dependent variable in an optimization problem? The independent variable? Discuss the differences in the roles played by dependent and independent variables in an optimization problem.
2. Discuss the differences among the following three items:
   - A *relation* among two or more variables describing an applied problem.
   - A *formula* giving the dependent variable in terms of other variables.
   - A *function* expressing the dependent variable in terms of an independent variable.

   Outline and contrast the roles played by relations, formulas, and functions in typical optimization problems.

3.6 PROBLEMS

1. Find two positive real numbers x and y such that their sum is 50 and their product is as large as possible.

2. Find the maximum possible area of a rectangle of perimeter 200 m.

3. A rectangle with sides parallel to the coordinate axes has one vertex at the origin, one on the positive x-axis, one on the positive y-axis, and its fourth vertex in the first quadrant on the line with equation $2x + y = 100$ (Fig. 3.6.15). What is the maximum possible area of such a rectangle?

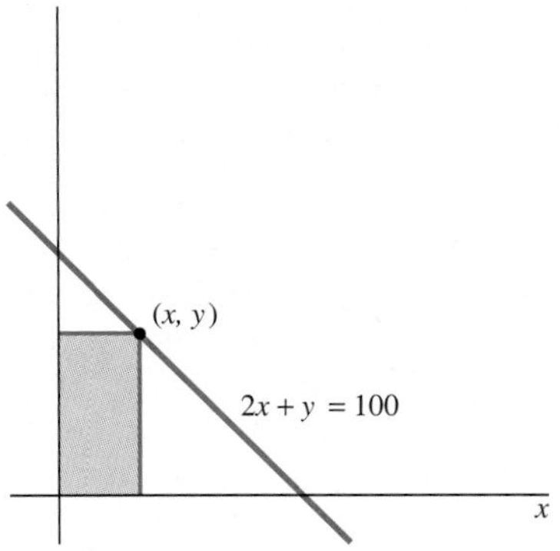

FIGURE 3.6.15 The rectangle of Problem 3.

4. A farmer has 600 m of fencing with which to enclose a rectangular pen adjacent to a long existing wall. He will use the wall for one side of the pen and the available fencing for the remaining three sides. What is the maximum area that can be enclosed in this way?

5. A rectangular box has a square base with edges at least 1 in. long. It has no top, and the total area of its five sides is 300 in.^{2} (Fig. 3.6.16). What is the maximum possible volume of such a box?

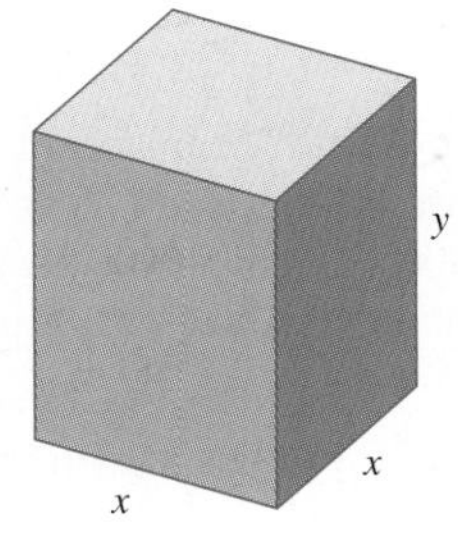

FIGURE 3.6.16 A box with square base and volume $V = x^2y$ (Problems 5, 17, and 20).

6. If x is in the interval $[0, 1]$, then $x - x^2$ is not negative. What is the maximum value that $x - x^2$ can have on that interval? In other words, what is the greatest amount by which a real number can exceed its square?

7. The sum of two positive numbers is 48. What is the smallest possible value of the sum of their squares?

8. A rectangle of fixed perimeter 36 is rotated around one of its sides, thus sweeping out a figure in the shape of a right circular cylinder (Fig. 3.6.17). What is the maximum possible volume of that cylinder?

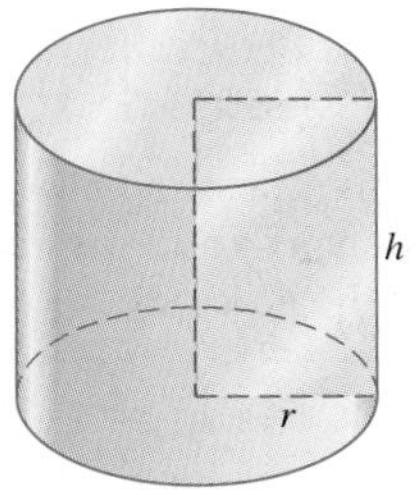

FIGURE 3.6.17 The rectangle and cylinder of Problem 8.

9. The sum of two nonnegative real numbers is 10. Find the minimum possible value of the sum of their cubes.

10. Suppose that the strength of a rectangular beam is proportional to the product of the width and the *square* of the height of its cross section. What shape beam should be cut from a cylindrical log of radius r to achieve the greatest possible strength?

11. A farmer has 600 yd of fencing with which to build a rectangular corral. Some of the fencing will be used to construct two interval divider fences, both parallel to the same two sides of the corral (Fig. 3.6.18). What is the maximum possible total area of such a corral?

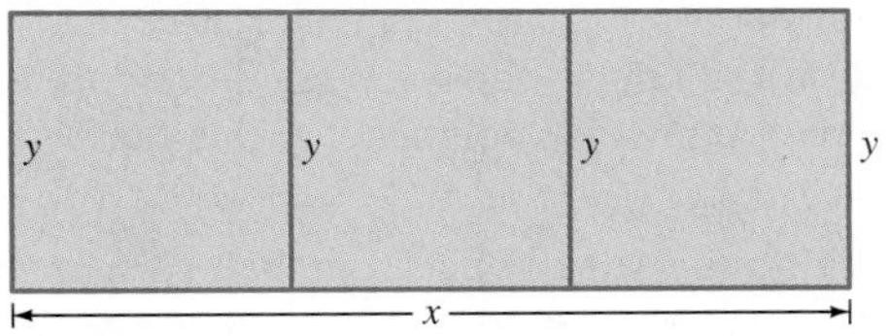

FIGURE 3.6.18 The divided corral of Problem 11.

12. Find the maximum possible volume of a right circular cylinder if its total surface area—including both circular ends—is 150π.

13. Find the maximum possible area of a rectangle with diagonals of length 16.

14. A rectangle has a line of fixed length L reaching from one vertex to the midpoint of one of the far sides (Fig. 3.6.19). What is the maximum possible area of such a rectangle?

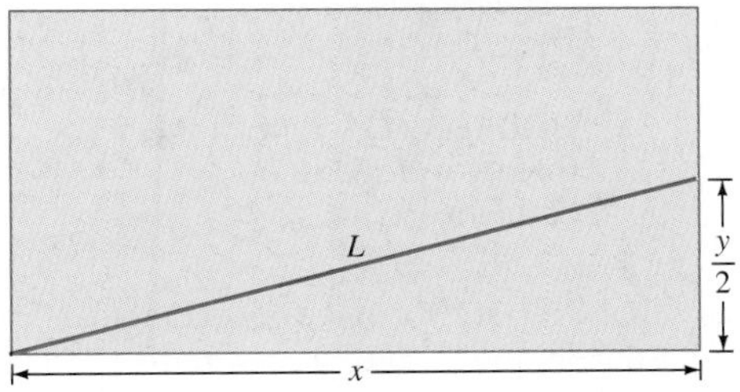

FIGURE 3.6.19 The rectangle of Problem 14.

15. The volume V (in cubic centimeters) of 1 kg of water at temperature T between 0°C and 30°C is very closely approximated by

$$V = 999.87 - (0.06426)T + (0.0085043)T^2 - (0.0000679)T^3.$$

At what temperature does water have its maximum density?

16. What is the maximum possible area of a rectangle with a base that lies on the x-axis and with two upper vertices that lie on the graph of the equation $y = 4 - x^2$ (Fig. 3.6.20)?

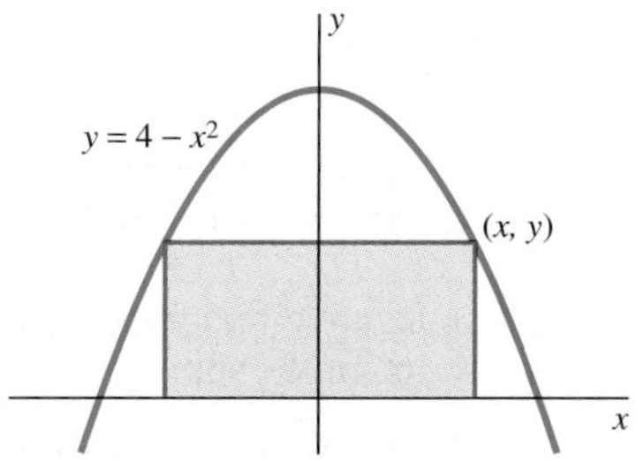

FIGURE 3.6.20 The rectangle of Problem 16.

17. A rectangular box has a square base with edges at least 1 cm long. Its total surface area is 600 cm^2. What is the largest possible volume that such a box can have?

18. You must make a cylindrical can with a bottom but no top from 300π in.^{2} of sheet metal. No sheet metal will be wasted; you are allowed to order a circular piece of any size for its base and any appropriate rectangular piece to make into its curved side so long as the given conditions are met. What is the greatest possible volume of such a can?

19. Three large squares of tin, each with edges 1 m long, have four small, equal squares cut from their corners. All twelve resulting small squares are to be the same size (Fig. 3.6.21). The three large cross-shaped pieces are then folded and welded to make boxes with no tops, and the twelve small squares are used to make two small cubes. How should this be done to maximize the total volume of all five boxes?

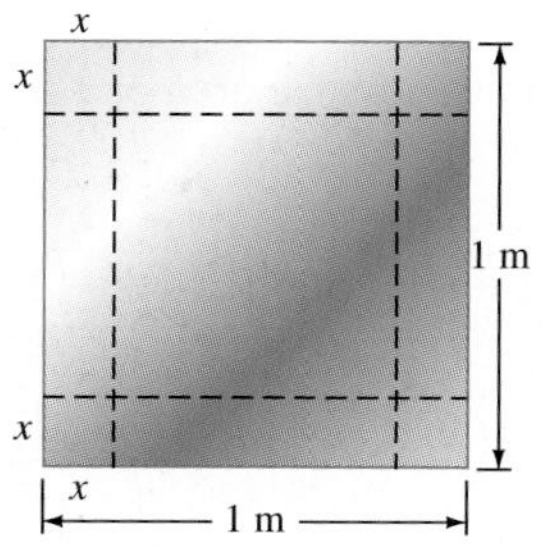

FIGURE 3.6.21 One of the three 1-m squares of Problem 19.

20. Suppose that you are to make a rectangular box with a square base from two different materials. The material for the top and four sides of the box costs \$1/ft^2; the material for the base costs \$2/ft^2. Find the dimensions of the box of greatest possible volume if you are allowed to spend \$144 for the material to make it.

21. A piece of wire 80 in. long is cut into at most two pieces. Each piece is bent into the shape of a square. How should this be done to minimize the sum of the area(s) of the square(s)? To maximize it?

22. A wire of length 100 cm is cut into two pieces. One piece is bent into a circle, the other into a square. Where should the cut be made to maximize the sum of the areas of the square and the circle? To minimize that sum?

23. A farmer has 600 m of fencing with which she plans to enclose a rectangular pasture adjacent to a long existing wall. She plans to build one fence parallel to the wall, two to form the ends of the enclosure, and a fourth (parallel to the ends of the enclosure) to divide it equally. What is the maximum area that can be enclosed?

24. A zookeeper needs to add a rectangular outdoor pen to an animal house with a corner notch, as shown in Fig. 3.6.22. If 85 m of new fence is available, what dimensions of the pen will maximize its area? No fence will be used along the walls of the animal house.

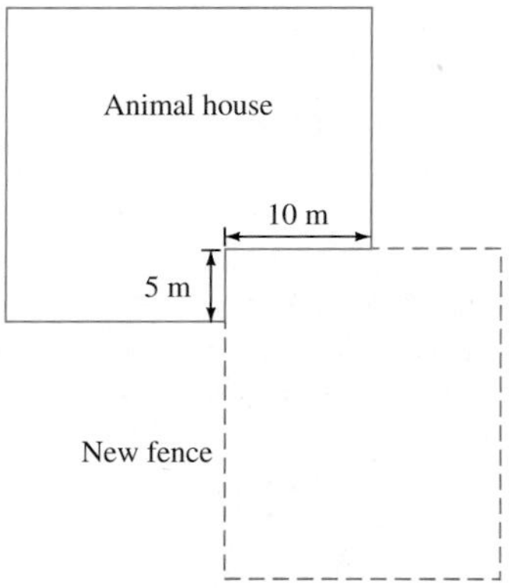

FIGURE 3.6.22 The rectangular pen of Problem 24.

25. Suppose that a post office can accept a package for mailing only if the sum of its length and its girth (the circumference of its cross section) is at most 100 in. What is the maximum volume of a rectangular box with square cross section that can be mailed?

26. Repeat Problem 25, but use a cylindrical package; its cross section is circular.

27. A printing company has eight presses, each of which can print 3600 copies per hour. It costs \$5.00 to set up each press for a run and $10 + 6n$ dollars to run n presses for 1 h. How many presses should be used to print 50,000 copies of a poster most profitably?

28. A farmer wants to hire workers to pick 900 bushels of beans. Each worker can pick 5 bushels per hour and is paid \$1.00 per bushel. The farmer must also pay a supervisor \$10 per hour while the picking is in progress, and he has additional miscellaneous expenses of \$8 per worker. How many workers should he hire to minimize the total cost? What will then be the cost per bushel picked?

29. The heating and cooling costs for a certain uninsulated house are \$500/yr, but with $x \leq 10$ in. of insulation, the costs are $1000/(2+x)$ dollars/yr. It costs \$150 for each inch (thickness) of insulation installed. How many inches of insulation should be installed to minimize the *total* (initial plus annual) costs over a 10-yr period? What will then be the annual savings resulting from this optimal insulation?

30. A concessionaire had been selling 5000 burritos each game night at 50¢ each. When she raised the price to 70¢ each, sales dropped to 4000 per night. Assume a linear relationship between price and sales. If she has fixed costs of \$1000 per night and each burrito costs her 25¢, what price will maximize her nightly profit?

31. A commuter train carries 600 passengers each day from a suburb to a city. It costs \$1.50 per person to ride the train. Market research reveals that 40 fewer people would ride the train for each 5¢ increase in the fare, 40 more for each 5¢ decrease. What fare should be charged to get the largest possible revenue?

32. Find the shape of the cylinder of maximal volume that can be inscribed in a sphere of radius R (Fig. 3.6.23). Show that the ratio of the height of the cylinder to its radius is $\sqrt{2}$ and that the ratio of the volume of the sphere to that of the maximal cylinder is $\sqrt{3}$.

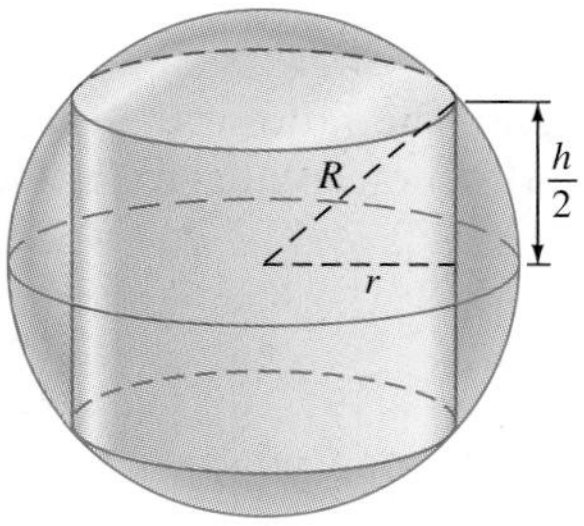

FIGURE 3.6.23 The sphere and cylinder of Problem 32.

33. Find the dimensions of the right circular cylinder of greatest volume that can be inscribed in a right circular cone of radius R and height H (Fig. 3.6.24).

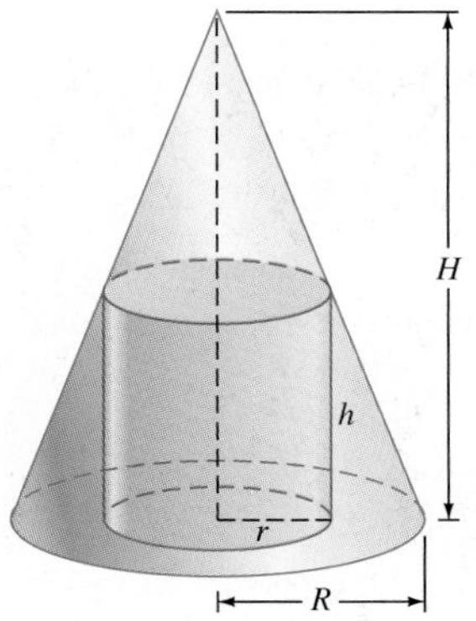

FIGURE 3.6.24 The cone and cylinder of Problem 33.

34. Figure 3.6.25 shows a circle of radius 1 in which a trapezoid is inscribed. The longer of the two parallel sides of the trapezoid coincides with a diameter of the circle. What is the maximum possible area of such a trapezoid. (*Suggestion:* A positive quantity is maximized when its square is maximized.)

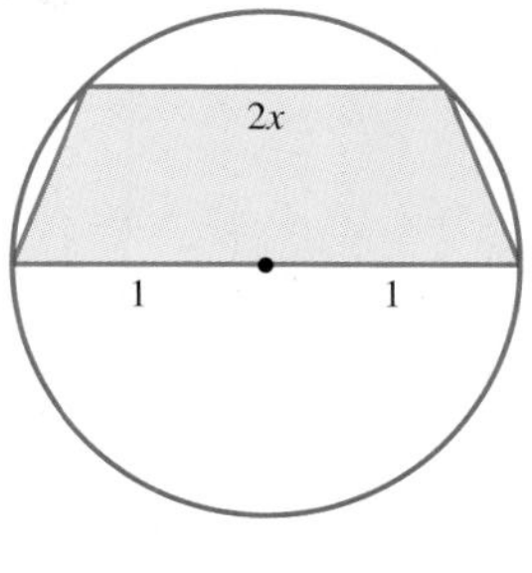

FIGURE 3.6.25 The circle and trapezoid of Problem 34.

35. Show that the rectangle of maximal perimeter that can be inscribed in a circle is a square.

36. Find the dimensions of the rectangle (with sides parallel to the coordinate axes) of maximal area that can be inscribed in the ellipse with equation

$$\frac{x^2}{25} + \frac{y^2}{9} = 1$$

(Fig. 3.6.26).

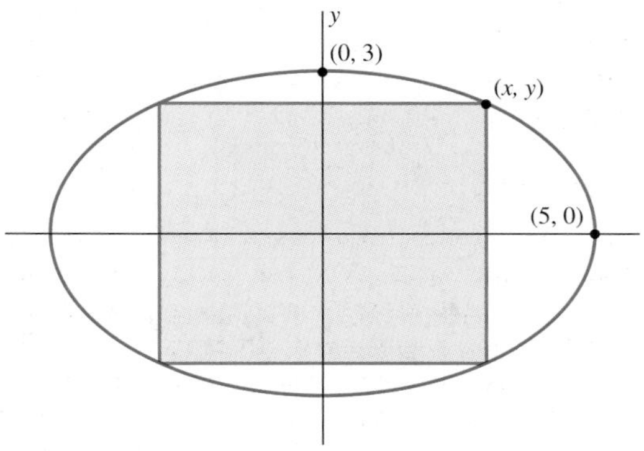

FIGURE 3.6.26 The ellipse and rectangle of Problem 36.

37. A right circular cone of radius r and height h has slant height $L = \sqrt{r^2 + h^2}$. What is the maximum possible volume of a cone with slant height 10?

38. Two vertical poles 10 ft apart are both 10 ft tall. Find the length of the shortest rope that can reach from the top of one pole to a point on the ground between them and then to the top of the other pole.

39. The sum of two nonnegative real numbers is 16. Find the maximum possible value and the minimum possible value of the sum of their cube roots.

40. A straight wire 60 cm long is bent into the shape of an L. What is the shortest possible distance between the two ends of the bent wire?

41. What is the shortest possible distance from a point on the parabola $y = x^2$ to the point $(0, 1)$?

42. Given: There is exactly one point on the graph of $y = \sqrt[3]{3x-4}$ that is closest to the origin. Find it. (*Suggestion:* See Fig. 3.6.27, and solve the equation you obtain by inspection.)

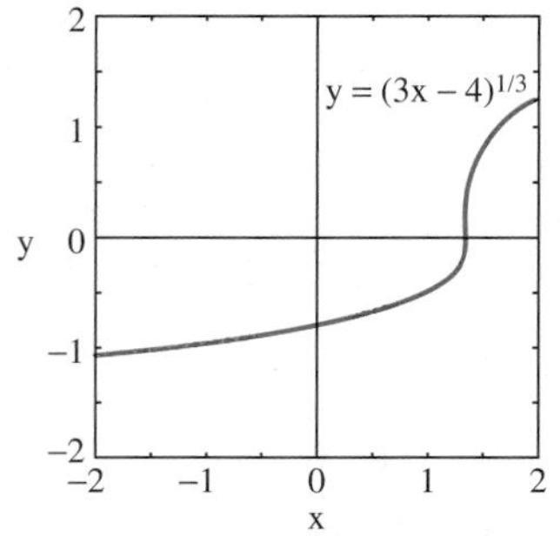

FIGURE 3.6.27 The curve of Problem 42.

43. Find the dimensions that maximize the cross-sectional area of the four planks that can be cut from the four pieces of the circular log of Example 5—the pieces that remain after a square beam has been cut from the log (Fig. 3.6.10).

44. Find the maximal area of a rectangle inscribed in an equilateral triangle with edges of length 1, as in Fig. 3.6.28.

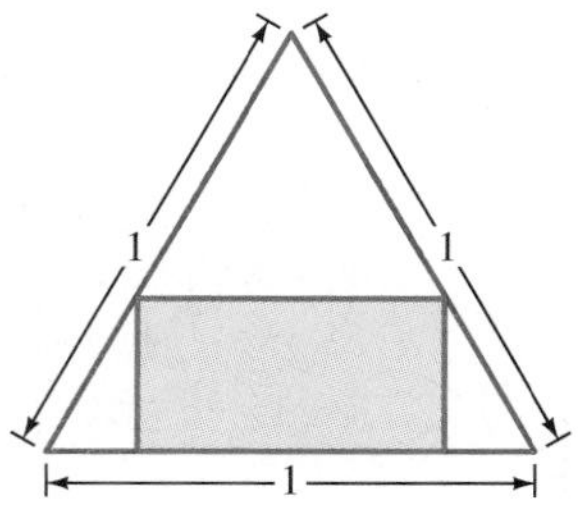

FIGURE 3.6.28 The rectangle and equilateral triangle of Problem 44.

45. A small island is 2 km off shore in a large lake. A woman on the island can row her boat 10 km/h and can run at a speed of 20 km/h. If she rows to the closest point of the straight shore, she will land 6 km from a village on the shore. Where should she land to reach the village most quickly by a combination of rowing and running?

46. A factory is located on one bank of a straight river that is 2000 m wide. On the opposite bank but 4500 m downstream is a power station from which the factory draws its electricity. Assume that it costs three times as much per meter to lay an underwater cable as to lay an aboveground cable. What path should a cable connecting the power station to the factory take to minimize the cost of laying the cable?

47. A company has plants that are located (in an appropriate coordinate system) at the points $A(0, 1)$, $B(0, -1)$, and $C(3, 0)$ (Fig. 3.6.29). The company plans to construct a distribution center at the point $P(x, 0)$. What value of x would minimize the sum of the distances from P to A, B, and C?

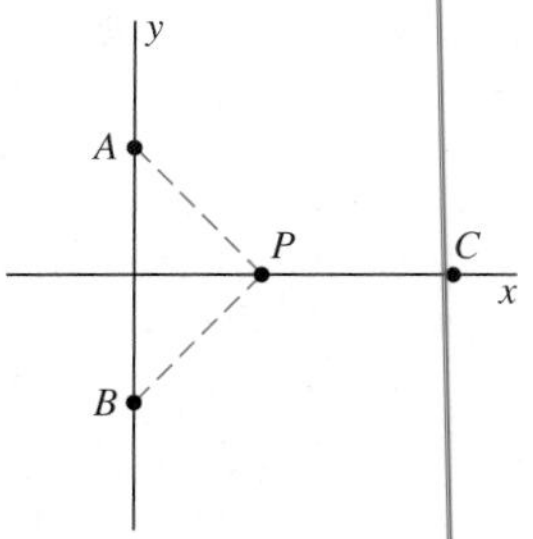

FIGURE 3.6.29 The locations in Problem 47.

48. Light travels at speed c in air and at a slower speed v in water. (The constant c is approximately 3×10^{10} cm/s; the ratio $n = c/v$, known as the **index of refraction,** depends on the color of the light but is approximate 1.33 for water.) Figure 3.6.30 shows the path of a light ray traveling from point A in air to point B in water, with what appears to be a sudden change in direction as the ray moves through the air-water interface. (a) Write the time T required for the ray to travel from A to B in terms of the variable x and the constants a, b, c, s, and v, all of which have been defined or are shown in the figure. (b) Show that the equation $T'(x) = 0$ for minimizing T is equivalent to the condition

$$\frac{\sin\alpha}{\sin\beta} = \frac{c}{v} = n.$$

This is **Snell's law:** The ratio of the sines of the angles of incidence and refraction is equal to the index of refraction.

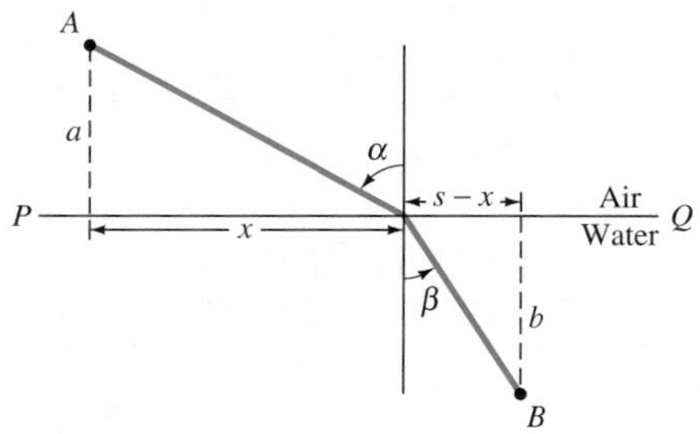

FIGURE 3.6.30 Snell's law gives the path of refracted light (Problem 48).

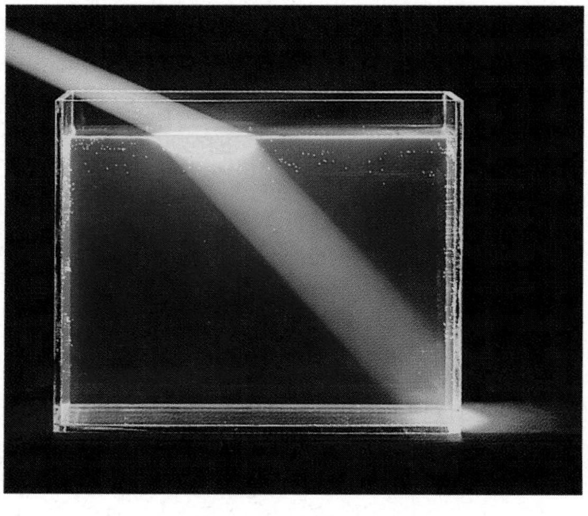

Refraction of light at an air-water interface

49. The mathematics of Snell's law (Problem 48) is applicable to situations other than the refraction of light. Figure 3.6.31 shows an east-west geologic fault that separates two towns at points A and B. Assume that A is a miles north of the fault, that B is b miles south of the fault, and that B is L miles east of A. We want to build a road from A to B. Because of differences in terrain, the cost of construction is C_1 (in millions of dollars per mile) north of the fault and C_2 south of it. Where should the point P be placed to minimize the total cost of road construction? (a) Using the notation in the figure, show that the cost is minimized when $C_1 \sin\theta_1 = C_2 \sin\theta_2$. (b) Take $a = b = C_1 = 1$, $C_2 = 2$, and $L = 4$. Show that the equation in part (a) is equivalent to

$$f(x) = 3x^4 - 24x^3 + 51x^2 - 32x + 64 = 0.$$

To approximate the desired solution of this equation, calculate $f(0)$, $f(1)$, $f(2)$, $f(3)$, and $f(4)$. You should find that $f(3) > 0 > f(4)$. Interpolate between $f(3)$ and $f(4)$ to approximate the desired root of this equation.

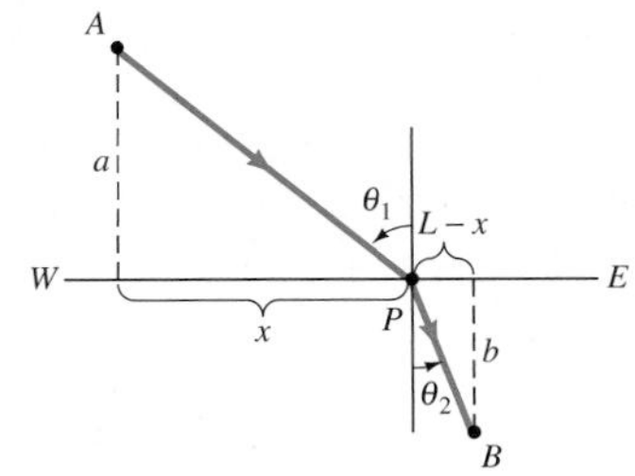

FIGURE 3.6.31 Building a road from A to B (Problem 49).

50. The sum of the volumes of two cubes is 2000 in.^{3} What should their edges x and y be to maximize the sum of their surface areas? To minimize it?

51. The sum of the surface areas of a cube and a sphere is 1000 in.^{2} What should their dimensions be to minimize the sum of their volumes? To maximize it?

52. Your brother has six pieces of wood with which to make the kite frame shown in Fig. 3.6.32. The four outer pieces with the indicated lengths have already been cut. How long should the lengths of the inner struts be to maximize the area of the kite?

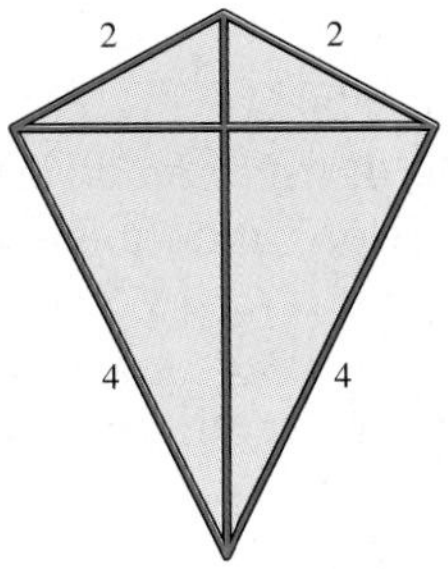

FIGURE 3.6.32 The kite frame (Problem 52).

Problems 53 through 55 deal with alternative methods of constructing a tent.

53. Figure 3.6.33 shows a 20-by-20-ft square of canvas tent material. Girl Scout Troop A must cut pieces from its four corners as indicated, so that the four remaining triangular flaps can be turned up to form a tent in the shape of a pyramid with a square base. How should this be done to maximize the volume of the tent?

Let A denote the area of the base of the tent and h its height. With x as indicated in the figure, show that the volume $V = \frac{1}{3}Ah$ of the tent is given by

$$V(x) = \tfrac{4}{3}x^2\sqrt{100 - 20x}, \quad 0 \leqq x \leqq 5.$$

Maximize V by graphing $V(x)$ and $V'(x)$ and zooming in on the zero of $V'(x)$.

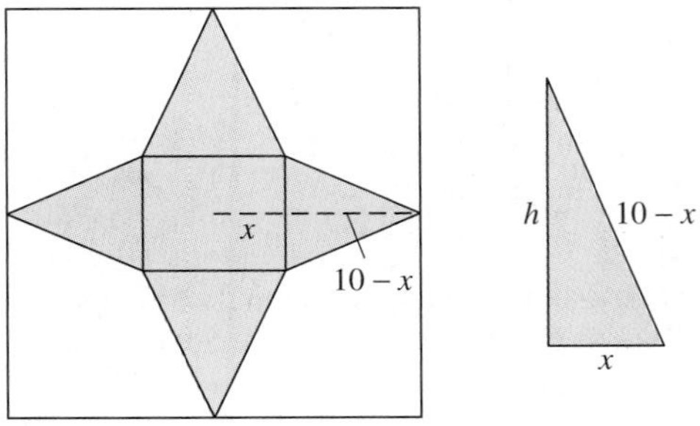

FIGURE 3.6.33 The canvas square—first attempt.

54. Girl Scout Troop B must make a tent in the shape of a pyramid with a square base from a similar 20-by-20-ft square of canvas but in the manner indicated in Fig. 3.6.34. With x as indicated in the figure, show that the volume of the tent is given by

$$V(x) = \tfrac{2}{3}x^2\sqrt{200 - 20x}, \quad 0 \leqq x \leqq 10.$$

Maximize V graphically as in Problem 53.

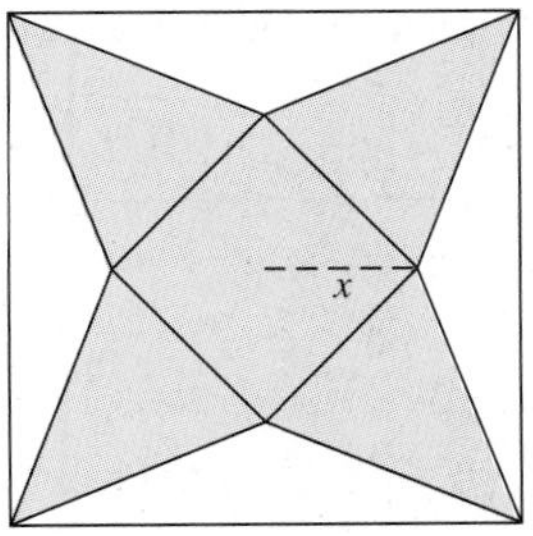

FIGURE 3.6.34 The canvas square—second attempt.

55. Solve Problems 53 and 54 analytically to verify that the maximal volume in Problem 54 is exactly $2\sqrt{2}$ times the maximal volume in Problem 53. It pays to think before making a tent!

Problems 56 and 57 deal with rectangular boxes with square base. Such a box is said to be closed *if it has both a (square) bottom and a top (as well as four vertical sides),* open *if it has a bottom but no top.*

56. Show that, among all closed square-based rectangular boxes with a given fixed total surface area, the one with maximal volume is a cube.

57. Show that, among all open square-based rectangular boxes with a given fixed total surface area, the one with maximal volume has height equal to half the length of the edge of its base.

Problems 58 through 60 deal with right circular cylinders. Such a "can" is said to be closed *if it has both a (circular) bottom and a top (as well as a curved side),* open *if it has a bottom but no top.*

58. Show that, among all closed cylindrical cans with a given fixed total surface area, the one with maximal volume has height equal to the diameter of its base.

59. Show that, among all open cylindrical cans with a given fixed total surface area, the one with maximal volume has height equal to the radius of its base.

60. Suppose that the bottom and curved side surface of a pop-top soft drink can have the same thickness. But, in order that the top not be ripped upon opening, it is three times as thick as the bottom. Show that, among all such soft drink cans made from a fixed total amount of material (including the triple-thick top), the one with maximal volume has height approximately twice its diameter. (Perhaps this is why soft drink cans look somewhat taller than soup or vegetable cans.) *Suggestion:* To simplify the computations, you may assume that the amount of material used to make a can of inner radius r, inner height h, and thickness t (except for the top, of thickness $3t$), is $\pi r^2 t + 2\pi rht + 3\pi r^2 t$. This will be quite accurate if t is very small in comparison with r and h.

61. Figure 3.6.35 shows a triangle bounded by the nonnegative coordinate axes and the line tangent to the curve $y = 1/(1 + x^2)$ at the first-quadrant point (x, y). Is it apparent that the area $A(x)$ of this triangle is very large when $x > 0$ is very close to zero? But your task is to find the maximum and minimum values of A for $\frac{1}{2} \leqq x \leqq 2$. It will be convenient to use a computer algebra system, both to find $A(x)$ and to solve the sixth-degree equation you should encounter.

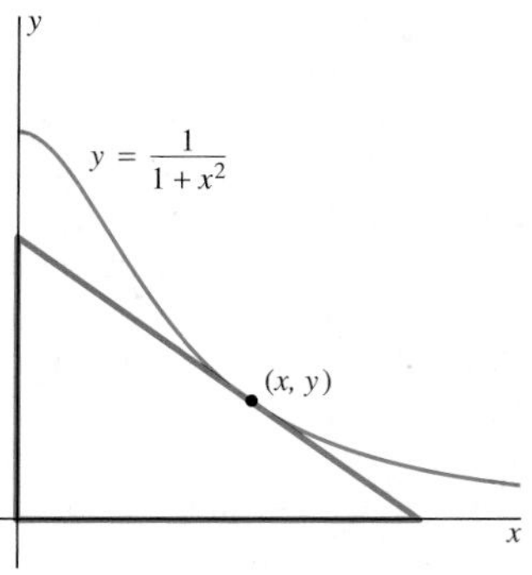

FIGURE 3.6.35 Triangle bounded by coordinate axes and a tangent line to the curve $y = \dfrac{1}{1 + x^2}$.

62. Figure 3.6.36 shows a one-mile-square city park. A local power company needs to run a power line from the northwest corner A of the park to the southeast corner B. To preserve the beauty of the park, only underground lines may be run through the park itself, but overhead lines are permissible along the boundary of the park. The power company plans to construct an overhead line a distance x along the west edge of the park, then from the southern end of this line continue with a straight power line to point B. If overhead lines cost \$40 thousand per mile and underground lines cost \$100 thousand per mile, how should the power company construct the line to minimize its total cost?

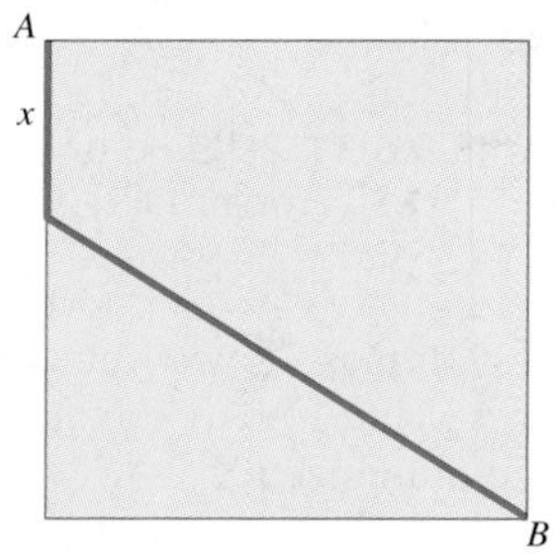

FIGURE 3.6.36 The one-mile-square park in central Villabuena.

3.7 DERIVATIVES OF TRIGONOMETRIC FUNCTIONS

In this section we begin our study of the calculus of trigonometric functions, focusing first on the sine and cosine functions. The definitions and the elementary properties of trigonometric functions are reviewed in Appendix C.

When we write $\sin\theta$ (or $\cos\theta$), we mean the sine (or cosine) of an angle of θ radians (rad). Recall the fundamental relation between radian measure and degree measure of angles:

$$\pi \text{ radians} = 180 \text{ degrees}. \qquad \textbf{(1)}$$

Upon division of both sides of this equation by π and 180, respectively, and abbreviating the units, we get the conversion relations

Radians	Degrees
0	0
$\pi/6$	30
$\pi/4$	45
$\pi/3$	60
$\pi/2$	90
$2\pi/3$	120
$3\pi/4$	135
$5\pi/6$	150
π	180
$3\pi/2$	270
2π	360
4π	720

FIGURE 3.7.1 Some radian-degree conversions.

$$1 \text{ rad} = \frac{180}{\pi} \text{ deg} \quad \text{and} \quad 1 \text{ deg} = \frac{\pi}{180} \text{ rad}.$$

Figure 3.7.1 shows radian-degree conversions for some frequently occurring angles.

The derivatives of the sine and cosine functions depend on the limits

$$\lim_{\theta \to 0} \frac{\sin \theta}{\theta} = 1, \qquad \lim_{\theta \to 0} \frac{1 - \cos \theta}{\theta} = 0 \tag{2}$$

that we established in Section 2.3. The addition formulas

$$\begin{aligned} \cos(x + y) &= \cos x \cos y - \sin x \sin y, \\ \sin(x + y) &= \sin x \cos y + \cos x \sin y \end{aligned} \tag{3}$$

are needed as well.

THEOREM 1 Derivatives of Sines and Cosines

The functions $f(x) = \sin x$ and $g(x) = \cos x$ are differentiable for all x, and

$$D_x \sin x = \cos x, \tag{4}$$

$$D_x \cos x = -\sin x. \tag{5}$$

PROOF To differentiate $f(x) = \sin x$, we begin with the definition of the derivative,

$$f'(x) = \lim_{h \to 0} \frac{f(x + h) - f(x)}{h} = \lim_{h \to 0} \frac{\sin(x + h) - \sin x}{h}.$$

Next we apply the addition formula for the sine and the limit laws to get

$$\begin{aligned} f'(x) &= \lim_{h \to 0} \frac{(\sin x \cos h + \sin h \cos x) - \sin x}{h} \\ &= \lim_{h \to 0} \left[(\cos x) \frac{\sin h}{h} - (\sin x) \frac{1 - \cos h}{h} \right] \\ &= (\cos x) \left(\lim_{h \to 0} \frac{\sin h}{h} \right) - (\sin x) \left(\lim_{h \to 0} \frac{1 - \cos h}{h} \right). \end{aligned}$$

The limits in Eq. (2) now yield

$$f'(x) = (\cos x)(1) - (\sin x)(0) = \cos x,$$

which proves Eq. (4). The proof of Eq. (5) is quite similar. (See Problem 72.) ◄

Examples 1 through 4 illustrate the application of Eqs. (4) and (5) in conjunction with the general differentiation formulas of Sections 3.2, 3.3, and 3.4 to differentiate various combinations of trigonometric and other functions.

EXAMPLE 1 The product rule yields

$$\begin{aligned} D_x(x^2 \sin x) &= (D_x x^2)(\sin x) + (x^2)(D_x \sin x) \\ &= 2x \sin x + x^2 \cos x. \end{aligned}$$

◆

EXAMPLE 2 If $y = \dfrac{\cos x}{1 - \sin x}$, then the quotient rule yields

$$\begin{aligned}\frac{dy}{dx} &= \frac{(D_x \cos x)(1 - \sin x) - (\cos x)[D_x(1 - \sin x)]}{(1 - \sin x)^2} \\ &= \frac{(-\sin x)(1 - \sin x) - (\cos x)(-\cos x)}{(1 - \sin x)^2} \\ &= \frac{-\sin x + \sin^2 x + \cos^2 x}{(1 - \sin x)^2} = \frac{-\sin x + 1}{(1 - \sin x)^2};\\ \frac{dy}{dx} &= \frac{1}{1 - \sin x}.\end{aligned}$$

◆

EXAMPLE 3 If $x = \cos^3 t$ and $u = \cos t$—so that $x = u^3$—then the chain rule yields

$$\frac{dx}{dt} = \frac{dx}{du}\frac{du}{dt} = (3u^2)(-\sin t) = (3\cos^2 t)(-\sin t) = -3\cos^2 t \sin t.$$

◆

EXAMPLE 4 If $g(t) = (2 - 3\cos t)^{3/2}$, then the chain rule yields

$$\begin{aligned} g'(t) &= \tfrac{3}{2}(2 - 3\cos t)^{1/2} D_t(2 - 3\cos t) \\ &= \tfrac{3}{2}(2 - 3\cos t)^{1/2}(3\sin t) = \tfrac{9}{2}(2 - 3\cos t)^{1/2}\sin t.\end{aligned}$$

◆

EXAMPLE 5 Write an equation of the line tangent to the curve $y = \cos^2 x$ at the point P on the graph where $x = 0.5$. Approximations are allowed.

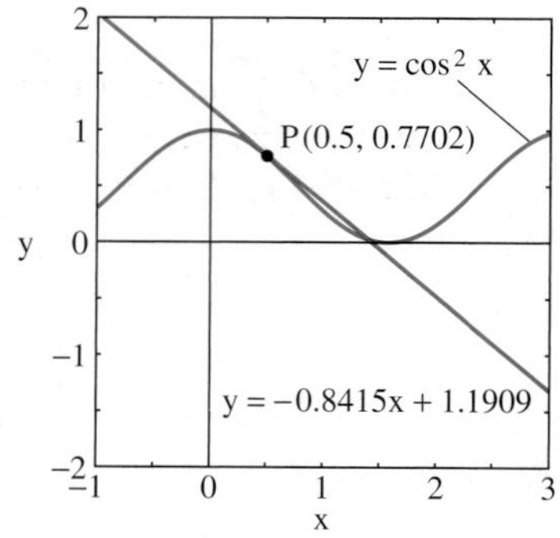

FIGURE 3.7.2 The curve $y = \cos^2 x$ and its tangent line at the point P where $x = 0.5$.

Solution The y-coordinate of P is $y(0.5) = (\cos 0.5)^2 \approx (0.8776)^2 \approx 0.7702$. Because

$$\frac{dy}{dx} = -2\cos x \sin x,$$

the slope of the tangent line at P is

$$m = \left.\frac{dy}{dx}\right|_{x=0.5} = -2(\cos 0.5)(\sin 0.5) \approx -0.8415.$$

Then the point-slope formula gives the (approximate) equation

$$y - 0.7702 = -(0.8415)(x - 0.5);$$

that is, $y = -(0.8415)x + 1.1909$, as the desired equation of the tangent line at P. Figure 3.7.2 shows the result of checking this computation by graphing both the curve $y = \cos^2 x$ and the line with this equation.

◆

The Remaining Trigonometric Functions

It is easy to differentiate the other four trigonometric functions, because they can be expressed in terms of the sine and cosine functions:

$$\begin{aligned} \tan x &= \frac{\sin x}{\cos x}, & \cot x &= \frac{\cos x}{\sin x}, \\ \sec x &= \frac{1}{\cos x}, & \csc x &= \frac{1}{\sin x}. \end{aligned} \qquad \textbf{(6)}$$

Each of these formulas is valid except where a zero denominator is encountered. Thus $\tan x$ and $\sec x$ are undefined when x is an odd integral multiple of $\pi/2$, and $\cot x$ and $\csc x$ are undefined when x is an integral multiple of π. The graphs of the six trigonometric functions appear in Fig. 3.7.3. There we show the sine and its reciprocal, the cosecant, in the same coordinate plane; we also pair the cosine with the secant but show the tangent and cotangent functions separately.

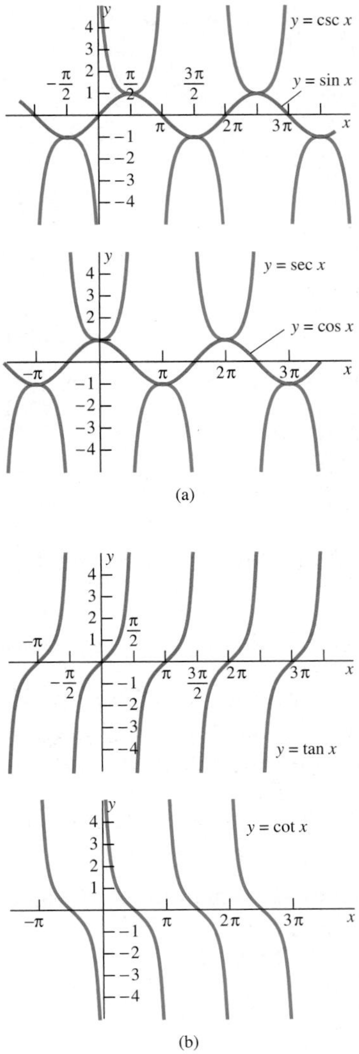

FIGURE 3.7.3 Graphs of the six trigonometric functions.

The functions in Eq. (6) can be differentiated by using the quotient rule and the derivatives of the sine and cosine functions. For example,

$$\tan x = \frac{\sin x}{\cos x},$$

so

$$D_x \tan x = \frac{(D_x \sin x)(\cos x) - (\sin x)(D_x \cos x)}{(\cos x)^2}$$

$$= \frac{(\cos x)(\cos x) - (\sin x)(-\sin x)}{\cos^2 x} = \frac{\cos^2 x + \sin^2 x}{\cos^2 x} = \frac{1}{\cos^2 x};$$

$$D_x \tan x = \sec^2 x.$$

As an exercise, you should derive in similar fashion the differentiation formulas in Eqs. (8) through (10) of Theorem 2.

THEOREM 2 Derivatives of Trigonometric Functions

The functions $f(x) = \tan x$, $g(x) = \cot x$, $p(x) = \sec x$, and $q(x) = \csc x$ are differentiable wherever they are defined, and

$$D_x \tan x = \sec^2 x, \qquad \textbf{(7)}$$

$$D_x \cot x = -\csc^2 x, \qquad \textbf{(8)}$$

$$D_x \sec x = \sec x \tan x, \qquad \textbf{(9)}$$

$$D_x \csc x = -\csc x \cot x. \qquad \textbf{(10)}$$

The patterns in the formulas of Theorem 2 and in Eqs. (4) and (5) make them easy to remember. The formulas in Eqs. (5), (8), and (10) are the "cofunction analogues" of those in Eqs. (4), (7), and (9), respectively. Note that the derivative formulas for the three cofunctions are those involving minus signs.

EXAMPLE 6

$$D_x(x \tan x) = (D_x x)(\tan x) + (x)(D_x \tan x)$$

$$= (1)(\tan x) + (x)(\sec^2 x) = \tan x + x \sec^2 x.$$

$$D_t(\cot^3 t) = D_t(\cot t)^3 = 3(\cot t)^2 D_t \cot t$$

$$= 3(\cot t)^2(-\csc^2 t) = -3\csc^2 t \cot^2 t.$$

$$D_z\left(\frac{\sec z}{\sqrt{z}}\right) = \frac{(D_z \sec z)\left(\sqrt{z}\right) - (\sec z)\left(D_z\sqrt{z}\right)}{\left(\sqrt{z}\right)^2}$$

$$= \frac{(\sec z)(\tan z)\left(\sqrt{z}\right) - (\sec z)\left(\frac{1}{2}z^{-1/2}\right)}{z}$$

$$= \tfrac{1}{2}z^{-3/2}(2z \tan z - 1)\sec z.$$ ◆

Chain Rule Formulas

Recall from Eq. (7) in Section 3.3 that the chain rule gives

$$D_x[g(u)] = g'(u)\frac{du}{dx} \qquad \textbf{(11)}$$

for the derivative of the composition $g(u(x))$ of two differentiable functions g and u. This formula yields a *chain rule version* of each new differentiation formula that we learn.

If we apply Eq. (11) first with $g(u) = \sin u$, then with $g(u) = \cos u$, and so on, we get the chain rule versions of the trigonometric differentiation formulas:

$$D_x \sin u = (\cos u)\frac{du}{dx}, \tag{12}$$

$$D_x \cos u = (-\sin u)\frac{du}{dx}, \tag{13}$$

$$D_x \tan u = (\sec^2 u)\frac{du}{dx}, \tag{14}$$

$$D_x \cot u = (-\csc^2 u)\frac{du}{dx}, \tag{15}$$

$$D_x \sec u = (\sec u \tan u)\frac{du}{dx}, \tag{16}$$

$$D_x \csc u = (-\csc u \cot u)\frac{du}{dx}. \tag{17}$$

The cases in which $u = kx$ (where k is a constant) are worth mentioning. For example,

$$D_x \sin kx = k\cos kx \quad \text{and} \quad D_x \cos kx = -k\sin kx. \tag{18}$$

The formulas in (18) provide an explanation of why radian measure is more appropriate than degree measure. Because it follows from Eq. (1) that an angle of degree measure x has radian measure $\pi x/180$, the "sine of an angle of x degrees" is a *new* and *different* function with the formula

$$\sin x^\circ = \sin\frac{\pi x}{180},$$

expressed on the right-hand side in terms of the standard (radian-measure) sine function. Hence the first formula in (18) yields

$$D_x \sin x^\circ = \frac{\pi}{180}\cos\frac{\pi x}{180},$$

so

$$D_x \sin x^\circ \approx (0.01745)\cos x^\circ.$$

The necessity of using the approximate value 0.01745 here—and indeed its very presence—is one reason why radians instead of degrees are used in the calculus of trigonometric functions: When we work with radians, we don't need such approximations.

EXAMPLE 7 If $y = 2\sin 10t + 3\cos \pi t$, then

$$\frac{dy}{dt} = 20\cos 10t - 3\pi \sin \pi t.$$ ◆

EXAMPLE 8

$$\begin{aligned}
&D_x(\sin^2 3x \cos^4 5x) \\
&\quad = [D_x(\sin 3x)^2](\cos^4 5x) + (\sin^2 3x)[D_x(\cos 5x)^4] \\
&\quad = 2(\sin 3x)(D_x \sin 3x)\cdot(\cos^4 5x) + (\sin^2 3x)\cdot 4(\cos 5x)^3\cdot(D_x \cos 5x) \\
&\quad = 2(\sin 3x)(3\cos 3x)(\cos^4 5x) + (\sin^2 3x)(4\cos^3 5x)(-5\sin 5x) \\
&\quad = 6\sin 3x\cos 3x\cos^4 5x - 20\sin^2 3x\sin 5x\cos^3 5x.
\end{aligned}$$ ◆

EXAMPLE 9 Differentiate $f(x) = \cos\sqrt{x}$.

Solution If $u = \sqrt{x}$, then $du/dx = 1/(2\sqrt{x})$, so Eq. (13) yields

$$D_x \cos\sqrt{x} = D_x \cos u = (-\sin u)\frac{du}{dx} = -\left(\sin\sqrt{x}\right)\frac{1}{2\sqrt{x}} = -\frac{\sin\sqrt{x}}{2\sqrt{x}}.$$

Alternatively, we can carry out this computation without introducing the auxiliary variable u:

$$D_x \cos\sqrt{x} = \left(-\sin\sqrt{x}\right)\cdot D_x\left(\sqrt{x}\right) = -\frac{\sin\sqrt{x}}{2\sqrt{x}}.$$

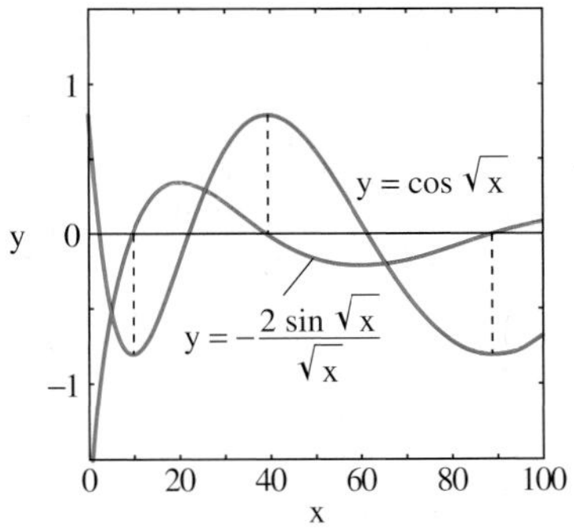

FIGURE 3.7.4 The curve $y = \cos\sqrt{x}$ and the constant multiple $y = -(2\sin\sqrt{x})/\sqrt{x}$ of its derivative.

In Fig. 3.7.4 we have plotted both the curve $y = y(x) = \cos\sqrt{x}$ and (to show the vertical scale more clearly) the constant multiple

$$y = 4y'(x) = -\frac{2\sin\sqrt{x}}{\sqrt{x}}$$

of its derivative. Note the correspondence in this figure between the local maxima and minima of the function $y(x) = \cos\sqrt{x}$ and the zeros of its derivative $y'(x)$ (which are the same as the zeros of $4y'(x)$). ◆

EXAMPLE 10 Differentiate

$$y = \sin^2(2x-1)^{3/2} = \left[\sin(2x-1)^{3/2}\right]^2.$$

Solution Here, $y = u^2$, where $u = \sin(2x-1)^{3/2}$, so

$$\begin{aligned}\frac{dy}{dx} = 2u\frac{du}{dx} &= 2\left[\sin(2x-1)^{3/2}\right]\cdot D_x\left[\sin(2x-1)^{3/2}\right]\\ &= 2\left[\sin(2x-1)^{3/2}\right]\left[\cos(2x-1)^{3/2}\right]\cdot D_x(2x-1)^{3/2}\\ &= 2\left[\sin(2x-1)^{3/2}\right]\left[\cos(2x-1)^{3/2}\right]\tfrac{3}{2}(2x-1)^{1/2}\cdot 2\\ &= 6(2x-1)^{1/2}\left[\sin(2x-1)^{3/2}\right]\left[\cos(2x-1)^{3/2}\right].\end{aligned}$$ ◆

EXAMPLE 11

$$D_x \tan 2x^3 = (\sec^2 2x^3)\cdot D_x(2x^3) = 6x^2\sec^2 2x^3.$$

$$\begin{aligned}D_t\cot^3 2t &= D_t(\cot 2t)^3 = 3(\cot 2t)^2\cdot D_t(\cot 2t)\\ &= (3\cot^2 2t)(-\csc^2 2t)\cdot D_t(2t)\\ &= -6\csc^2 2t\cot^2 2t.\end{aligned}$$

$$D_y\sec\sqrt{y} = \left(\sec\sqrt{y}\tan\sqrt{y}\right)\cdot D_y\sqrt{y} = \frac{\sec\sqrt{y}\tan\sqrt{y}}{2\sqrt{y}}.$$

$$\begin{aligned}D_z\sqrt{\csc z} &= D_z(\csc z)^{1/2} = \tfrac{1}{2}(\csc z)^{-1/2}\cdot D_z(\csc z)\\ &= \tfrac{1}{2}(\csc z)^{-1/2}(-\csc z\cot z) = -\tfrac{1}{2}(\cot z)\sqrt{\csc z}.\end{aligned}$$ ◆

Examples 12 and 13 illustrate the applications of trigonometric functions to rate-of-change and maximum-minimum problems.

EXAMPLE 12 A rocket is launched vertically and is tracked by a radar station located on the ground 5 mi from the launch pad. Suppose that the elevation angle θ of the line of sight to the rocket is increasing at 3° per second when $\theta = 60°$. What is the velocity of the rocket at this instant?

Solution First we convert the given data from degrees into radians. Because there are $\pi/180$ rad in 1°, the rate of increase of θ becomes

$$\frac{3\pi}{180} = \frac{\pi}{60}\quad\text{(rad/s)}$$

at the instant when

$$\theta = \frac{60\pi}{180} = \frac{\pi}{3}\quad\text{(rad).}$$

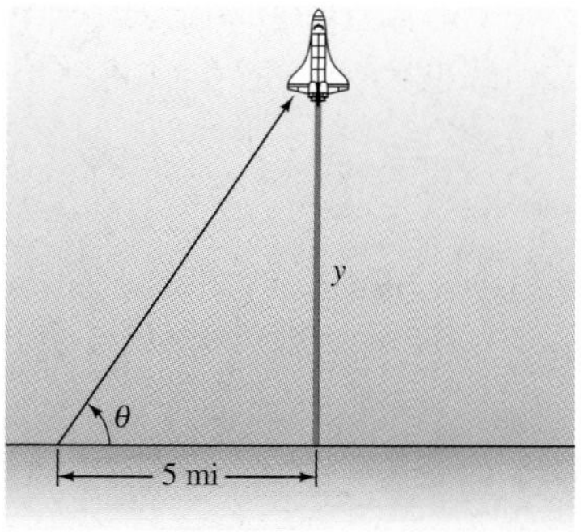

FIGURE 3.7.5 Tracking an ascending rocket (Example 12).

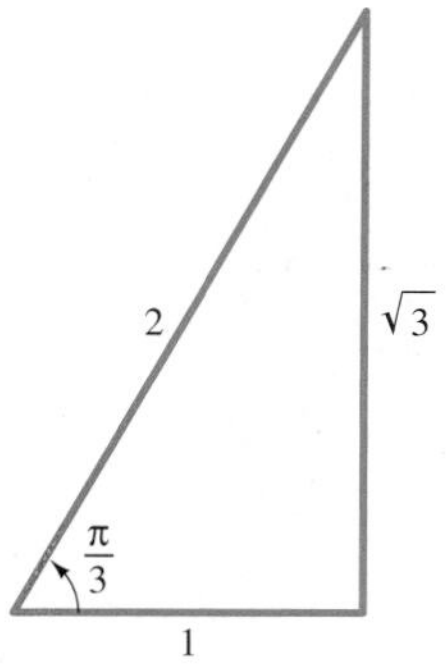

FIGURE 3.7.6 $\sec \frac{\pi}{3} = 2$ (Example 12).

From Fig. 3.7.5 we see that the height y (in miles) of the rocket is

$$y = 5 \tan \theta.$$

Hence its velocity is

$$\frac{dy}{dt} = \frac{dy}{d\theta} \cdot \frac{d\theta}{dt} = 5(\sec^2 \theta)\,\frac{d\theta}{dt}.$$

Because $\sec(\pi/3) = 2$ (Fig. 3.7.6), the velocity of the rocket is

$$\frac{dy}{dt} = 5 \cdot 2^2 \cdot \frac{\pi}{60} = \frac{\pi}{3} \quad \text{(mi/s)},$$

about 3770 mi/h, at the instant when $\theta = 60°$. ◆

EXAMPLE 13 A rectangle is inscribed in a semicircle of radius R (Fig. 3.7.7). What is the maximum possible area of such a rectangle?

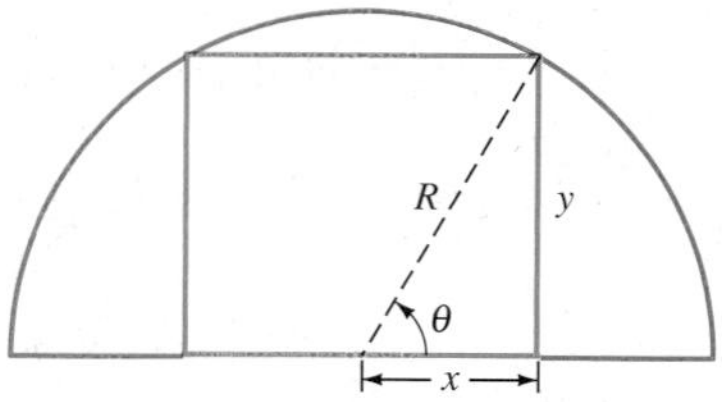

FIGURE 3.7.7 The rectangle of Example 13.

Solution If we denote the length of *half* the base of the rectangle by x and its height by y, then its area is $A = 2xy$. We see in Fig. 3.7.7 that the right triangle has hypotenuse R, the radius of the circle. So

$$x = R \cos \theta \quad \text{and} \quad y = R \sin \theta. \tag{19}$$

Each value of θ between 0 and $\pi/2$ corresponds to a possible inscribed rectangle. The values $\theta = 0$ and $\theta = \pi/2$ will yield degenerate rectangles.

We substitute the data in Eq. (19) into the formula $A = 2xy$ to obtain the area

$$\begin{aligned} A = A(\theta) &= 2(R \cos \theta)(R \sin \theta) \\ &= 2R^2 \cos \theta \sin \theta \end{aligned} \tag{20}$$

as a function of θ on the closed interval $[0, \pi/2]$. To find the critical points, we differentiate:

$$\frac{dA}{d\theta} = 2R^2(-\sin \theta \sin \theta + \cos \theta \cos \theta) = 2R^2(\cos^2 \theta - \sin^2 \theta).$$

Because $dA/d\theta$ always exists, we have critical points only if

$$\begin{aligned} \cos^2 \theta - \sin^2 \theta &= 0; \\ \sin^2 \theta &= \cos^2 \theta; \\ \tan^2 \theta &= 1; \\ \tan \theta &= \pm 1. \end{aligned}$$

The only value of θ in $[0, \pi/2]$ such that $\tan \theta = \pm 1$ is $\theta = \pi/4$.

Upon evaluation of $A(\theta)$ at each of the possible values $\theta = 0$, $\theta = \pi/4$, and $\theta = \pi/2$ (the endpoints and the critical point), we find that

$$A(0) = 0,$$

$$A\left(\frac{\pi}{4}\right) = 2R^2\left(\frac{1}{\sqrt{2}}\right)\left(\frac{1}{\sqrt{2}}\right) = R^2, \quad \longleftarrow \text{ absolute maximum}$$

$$A\left(\frac{\pi}{2}\right) = 0.$$

Thus the largest inscribed rectangle has area R^2, and its dimensions are $2x = R\sqrt{2}$ and $y = R/\sqrt{2}$. ◆

3.7 TRUE/FALSE STUDY GUIDE

3.7 CONCEPTS: QUESTIONS AND DISCUSSION

1. The function f is said to be **even** if $f(-x) = f(x)$ for all x, **odd** if $f(-x) = -f(x)$ for all x. For instance, the power function $f(x) = x^n$ is even if n is an even integer, but is odd if n is an odd integer. How can you determine if a function is even or odd by looking at its graph? Which of the six trigonometric functions are even and which are odd?
2. Give an example of a function (with domain the set of all real numbers) that is neither even nor odd. Find every function that is both even and odd.
3. The six trigonometric functions all have period 2π, meaning that $f(x + 2\pi) = f(x)$ for all x. Which of the trigonometric functions have period π? Determine the value of the constant k if the function $f(t) = A\cos kt + B\sin kt$ models:
   - The height of the tide at a certain beachfront location; with t in hours, the values of $f(t)$ repeat periodically every 12 h 25 min.
   - The average monthly rainfall in a certain locale; with t in months, the values of $f(t)$ repeat periodically every 12 months.
   - The average daily temperature in a certain locale; with t in days, the values of $f(t)$ repeat periodically every 365 days.
4. Considering the trigonometric functions $\sin x$, $\tan x$, $\sec x$, and their cofunctions, what is the pattern of *signs* of their derivatives? State a single short sentence telling which of the six derivative formulas include minus signs and which do not.

3.7 PROBLEMS

Differentiate the functions given in Problems 1 through 20.

1. $f(x) = 3\sin^2 x$
2. $f(x) = 2\cos^4 x$
3. $f(x) = x\cos x$
4. $f(x) = \sqrt{x}\sin x$
5. $f(x) = \dfrac{\sin x}{x}$
6. $f(x) = \dfrac{\cos x}{\sqrt{x}}$
7. $f(x) = \sin x\cos^2 x$
8. $f(x) = \cos^3 x\sin^2 x$
9. $g(t) = (1 + \sin t)^4$
10. $g(t) = (2 - \cos^2 t)^3$
11. $g(t) = \dfrac{1}{\sin t + \cos t}$
12. $g(t) = \dfrac{\sin t}{1 + \cos t}$
13. $f(x) = 2x\sin x - 3x^2\cos x$
14. $f(x) = x^{1/2}\cos x - x^{-1/2}\sin x$
15. $f(x) = \cos 2x\sin 3x$
16. $f(x) = \cos 5x\sin 7x$
17. $g(t) = t^3\sin^2 2t$
18. $g(t) = \sqrt{t}\cos^3 3t$
19. $g(t) = (\cos 3t + \cos 5t)^{5/2}$
20. $g(t) = \dfrac{1}{\sqrt{\sin^2 t + \sin^2 3t}}$

Find dy/dx in Problems 21 through 40.

21. $y = \sin^2\sqrt{x}$
22. $y = \dfrac{\cos 2x}{x}$
23. $y = x^2\cos(3x^2 - 1)$
24. $y = \sin^3 x^4$
25. $y = \sin 2x\cos 3x$
26. $y = \dfrac{x}{\sin 3x}$
27. $y = \dfrac{\cos 3x}{\sin 5x}$
28. $y = \sqrt{\cos\sqrt{x}}$

29. $y = \sin^2 x^2$

30. $y = \cos^3 x^3$

31. $y = \sin 2\sqrt{x}$

32. $y = \cos 3\sqrt[3]{x}$

33. $y = x \sin x^2$

34. $y = x^2 \cos\left(\dfrac{1}{x}\right)$

35. $y = \sqrt{x} \sin \sqrt{x}$

36. $y = (\sin x - \cos x)^2$

37. $y = \sqrt{x}(x - \cos x)^3$

38. $y = \sqrt{x} \sin \sqrt{x + \sqrt{x}}$

39. $y = \cos(\sin x^2)$

40. $y = \sin(1 + \sqrt{\sin x})$

Find dx/dt in Problems 41 through 60.

41. $x = \tan t^7$

42. $x = \sec t^7$

43. $x = (\tan t)^7$

44. $x = (\sec 2t)^7$

45. $x = t^7 \tan 5t$

46. $x = \dfrac{\sec t^5}{t}$

47. $x = \sqrt{t} \sec \sqrt{t}$

48. $x = \sec \sqrt{t} \tan \sqrt{t}$

49. $x = \csc\left(\dfrac{1}{t^2}\right)$

50. $x = \cot\left(\dfrac{1}{\sqrt{t}}\right)$

51. $x = \dfrac{\sec 5t}{\tan 3t}$

52. $x = \sec^2 t - \tan^2 t$

53. $x = t \sec t \csc t$

54. $x = t^3 \tan^3 t^3$

55. $x = \sec(\sin t)$

56. $x = \cot(\sec 7t)$

57. $x = \dfrac{\sin t}{\sec t}$

58. $x = \dfrac{\sec t}{1 + \tan t}$

59. $x = \sqrt{1 + \cot 5t}$

60. $x = \sqrt{\csc \sqrt{t}}$

In Problems 60 through 64, write an equation of the line that is tangent to the given curve $y = f(x)$ at the point P with the given x-coordinate. Then check the plausibility of your result by plotting both the curve and the line you found on the same screen.

61. $y = x \cos x;\quad x = \pi$

62. $y = \cos^2 x;\quad x = \pi/4$

63. $y = \dfrac{4}{\pi} \tan \dfrac{\pi x}{4};\quad x = 1$

64. $y = \dfrac{3}{\pi} \sin^2 \dfrac{\pi x}{3};\quad x = 5$

In Problems 65 through 68, find all points on the given curve $y = f(x)$ where the tangent line is horizontal.

65. $y = \cos 2x$

66. $y = x - 2 \sin x$

67. $y = \sin x \cos x$

68. $y = \dfrac{1}{3 \sin^2 x + 2 \cos^2 x}$

69. Figure 3.7.8 shows the graph $y = x - 2 \cos x$ and two lines of slope 1 both tangent to this graph. Write equations of these two lines.

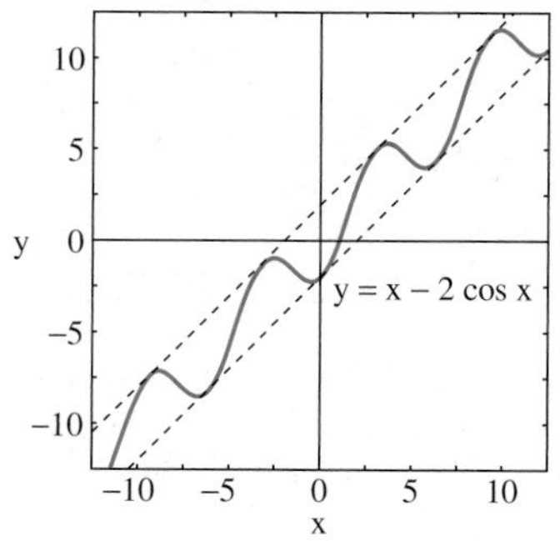

FIGURE 3.7.8 The curve $y = x - 2 \cos x$ and two tangent lines each having slope 1.

70. Figure 3.7.9 shows the graph

$$y = \frac{16 + \sin x}{3 + \sin x}$$

and its two horizontal tangent lines. Write equations of these two lines.

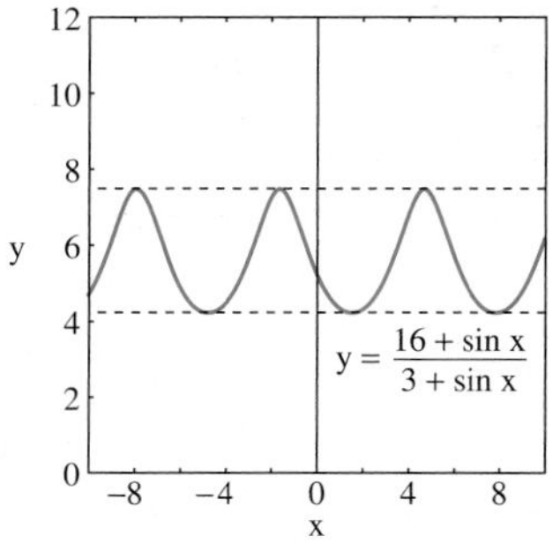

FIGURE 3.7.9 The curve $y = \dfrac{16 + \sin x}{3 + \sin x}$ and its two horizontal tangent lines.

71. Derive the differentiation formulas in Eqs. (8) through (10).

72. Use the definition of the derivative to show directly that $g'(x) = -\sin x$ if $g(x) = \cos x$.

73. If a projectile is fired from ground level with initial velocity v_0 and inclination angle α and if air resistance can be ignored, then its range—the horizontal distance it travels—is

$$R = \frac{1}{16} v_0^2 \sin \alpha \cos \alpha$$

(Fig. 3.7.10). What value of α maximizes R?

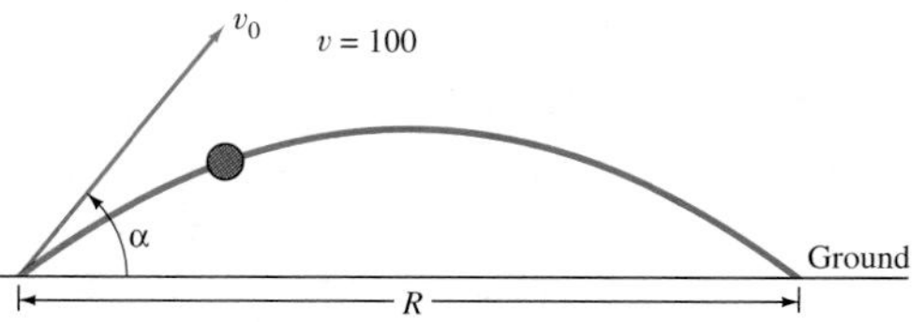

FIGURE 3.7.10 The projectile of Problem 73.

74. A weather balloon that is rising vertically is observed from a point on the ground 300 ft from the spot directly beneath the balloon (Fig. 3.7.11). At what rate is the balloon rising when the angle between the ground and the observer's line of sight is 45° and is increasing at 1° per second?

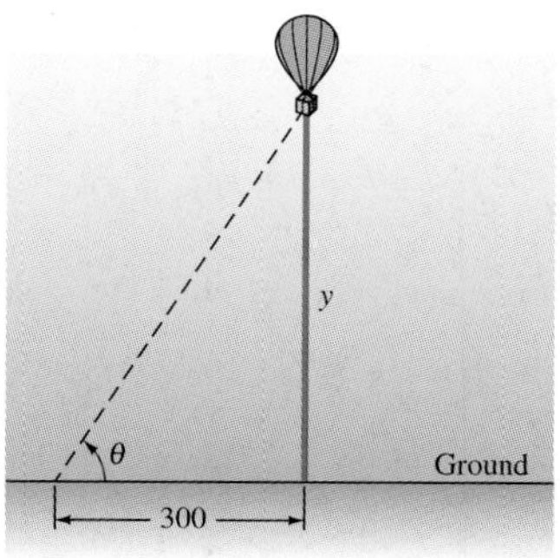

FIGURE 3.7.11 The weather balloon of Problem 74.

75. A rocket is launched vertically upward from a point 2 mi west of an observer on the ground. What is the speed of the rocket when the angle of elevation (from the horizontal) of the observer's line of sight to the rocket is 50° and is increasing at 5° per second?

76. A plane flying at an altitude of 25,000 ft has a defective airspeed indicator. To determine her speed, the pilot sights a fixed point on the ground. At the moment when the angle of depression (from the horizontal) of her line of sight is 65°, she notes that this angle is increasing at 1.5° per second (Fig. 3.7.12). What is the speed of the plane?

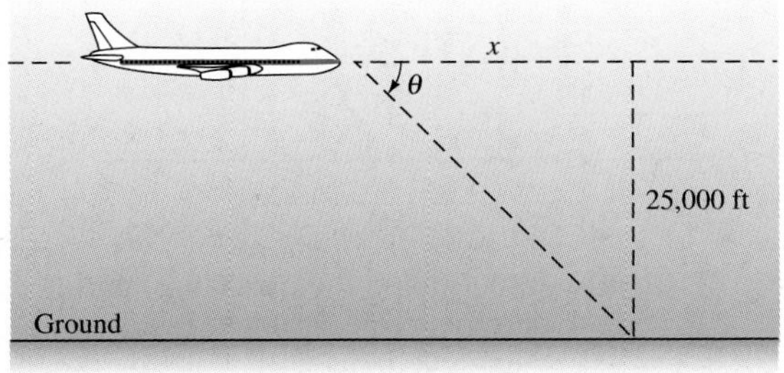

FIGURE 3.7.12 The airplane of Problem 76.

77. An observer on the ground sights an approaching plane flying at constant speed and at an altitude of 20,000 ft. From his point of view, the plane's angle of elevation is increasing at 0.5° per second when the angle is 60°. What is the speed of the plane?

78. Find the largest possible area A of a rectangle inscribed in the unit circle $x^2 + y^2 = 1$ by maximizing A as a function of the angle θ indicated in Fig. 3.7.13.

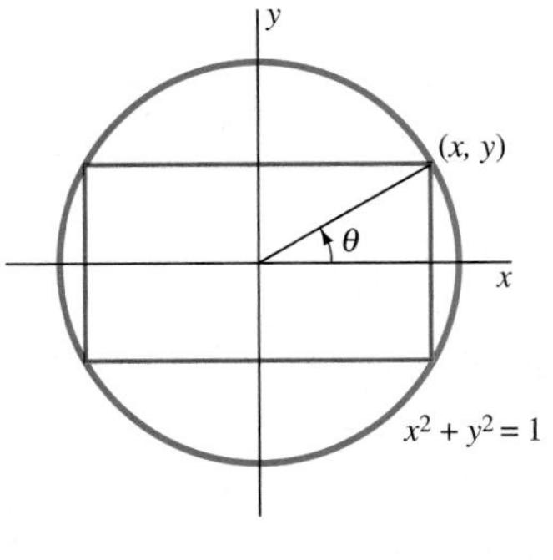

FIGURE 3.7.13 A rectangle inscribed in the unit circle (Problem 78).

79. A water trough is to be made from a long strip of tin 6 ft wide by bending up at an angle θ a 2-ft strip on each side (Fig. 3.7.14). What angle θ would maximize the cross-sectional area, and thus the volume, of the trough?

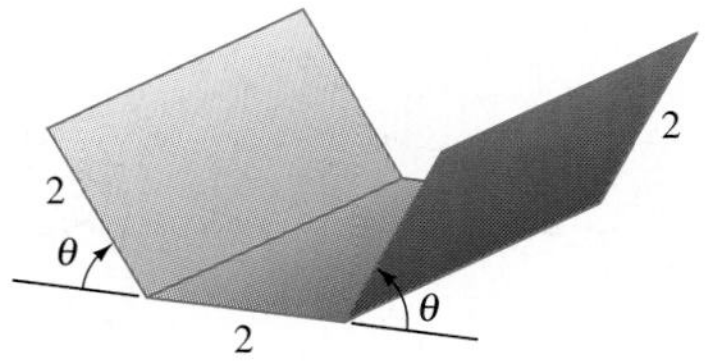

FIGURE 3.7.14 The water trough of Problem 79.

80. A circular patch of grass of radius 20 m is surrounded by a walkway, and a light is placed atop a lamppost at the circle's center. At what height should the light be placed to illuminate the walkway most strongly? The intensity of illumination I of a surface is given by $I = (k \sin\theta)/D^2$, where D is the distance from the light source to the surface, θ is the angle at which light strikes the surface, and k is a positive constant.

81. Find the minimum possible volume V of a cone in which a sphere of given radius R is inscribed. Minimize V as a function of the angle θ indicated in Fig. 3.7.15.

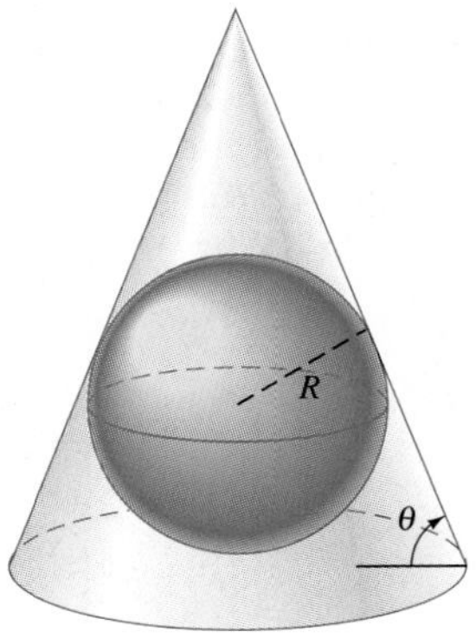

FIGURE 3.7.15 Finding the smallest cone containing a fixed sphere (Problem 81).

82. A very long rectangular piece of paper is 20 cm wide. The bottom right-hand corner is folded along the crease shown in Fig. 3.7.16, so that the corner just touches the left-hand side of the page. How should this be done so that the crease is as short as possible?

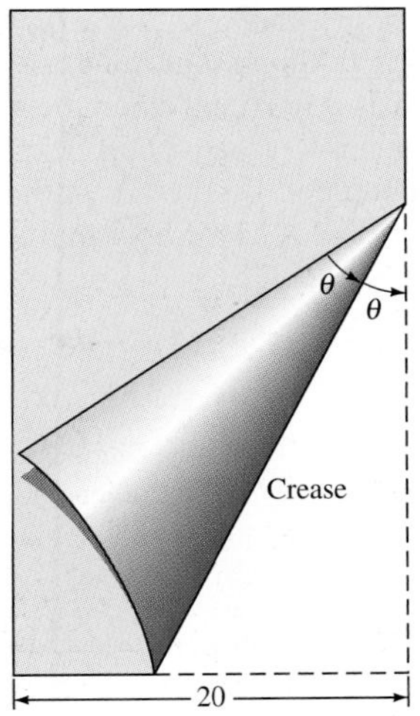

FIGURE 3.7.16 Fold a piece of paper; make the crease of minimal length (Problem 82).

83. Find the maximum possible area A of a trapezoid inscribed in a semicircle of radius 1, as shown in Fig. 3.7.17. Begin by expressing A as a function of the angle θ shown there.

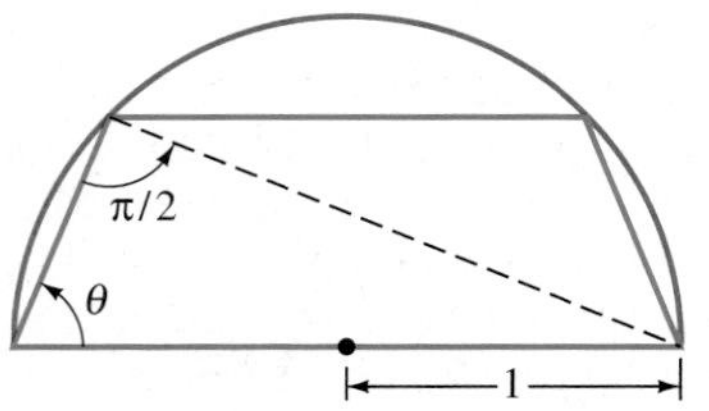

FIGURE 3.7.17 A trapezoid inscribed in a semicircle (Problem 83).

84. A logger must cut a six-sided beam from a circular log of diameter 30 cm so that its cross section is as shown in Fig. 3.7.18. The beam is symmetrical, with only two different internal angles α and β. Show that the cross section is maximal when the cross section is a regular hexagon, with equal sides and angles (corresponding to $\alpha = \beta = 2\pi/3$). Note that $\alpha + 2\beta = 2\pi$. (Why?)

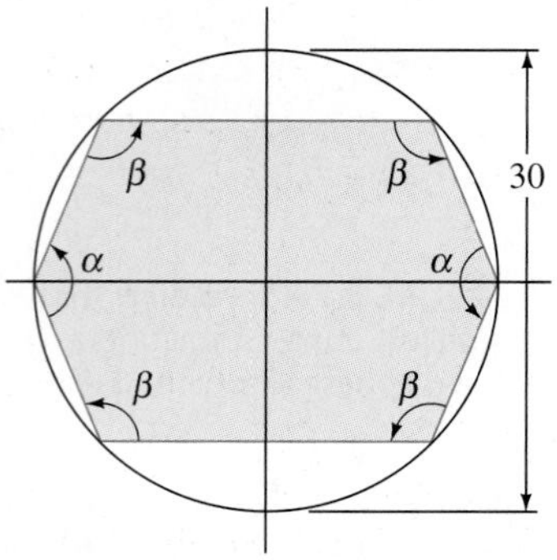

FIGURE 3.7.18 A hexagonal beam cut from a circular log (Problem 84).

85. Consider a circular arc of length s with its endpoints on the x-axis (Fig. 3.7.19). Show that the area A bounded by this arc and the x-axis is maximal when the circular arc is in the shape of a semicircle. [*Suggestion:* Express A in terms of the angle θ subtended by the arc at the center of the circle, as shown in Fig. 3.7.19. Show that A is maximal when $\theta = \pi$.]

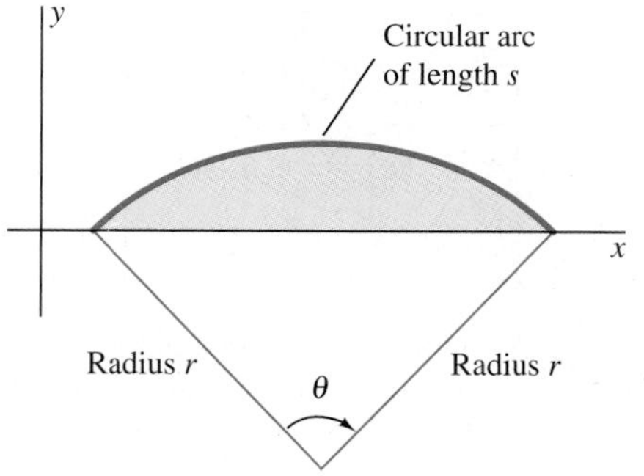

FIGURE 3.7.19 Finding the maximum area bounded by a circular arc and its chord (Problem 85).

86. A hiker starting at a point P on a straight road wants to reach a forest cabin that is 2 km from a point Q 3 km down the road from P (Fig. 3.7.20). She can walk 8 km/h along the road but only 3 km/h through the forest. She wants to minimize the time required to reach the cabin. How far down the road should she walk before setting off through the forest straight for the cabin? [*Suggestion:* Use the angle θ between the road and the path she takes through the forest as the independent variable.]

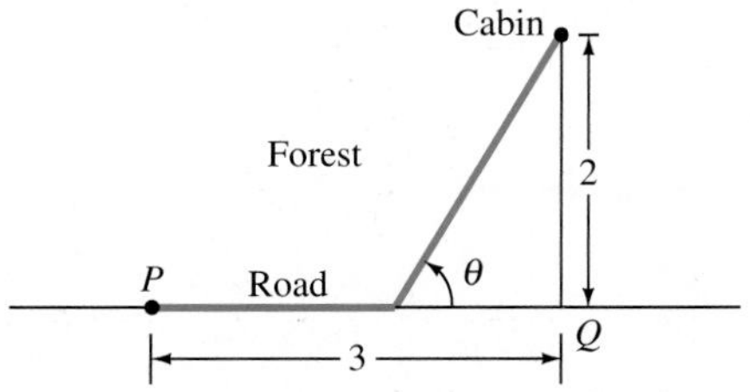

FIGURE 3.7.20 Finding the quickest path to the cabin in the forest (Problem 86).

87. Show that the function (graphed in Fig. 3.7.21)

$$f(x) = \begin{cases} x \sin \dfrac{1}{x} & \text{if } x \neq 0 \\ 0 & \text{if } x = 0 \end{cases}$$

(see Example 4 in Section 2.3) is *not* differentiable at $x = 0$. [*Suggestion:* Show that whether $z = 1$ or $z = -1$, there are arbitrarily small values of h such that $[f(h) - f(0)]/h = z$. Then use the *definition* of the derivative.]

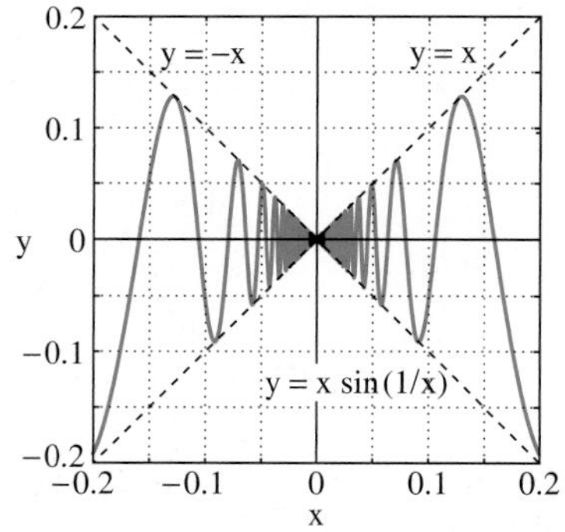

FIGURE 3.7.21 The graph of $y = x \sin \dfrac{1}{x}$ near $x = 0$.

88. Let

$$f(x) = \begin{cases} x^2 \sin \dfrac{1}{x} & \text{if } x \neq 0 \\ 0 & \text{if } x = 0 \end{cases}$$

(the graph of f appears in Figs. 3.7.22 and 3.7.23). Apply the definition of the derivative to show that f is differentiable at $x = 0$ and that $f'(0) = 0$.

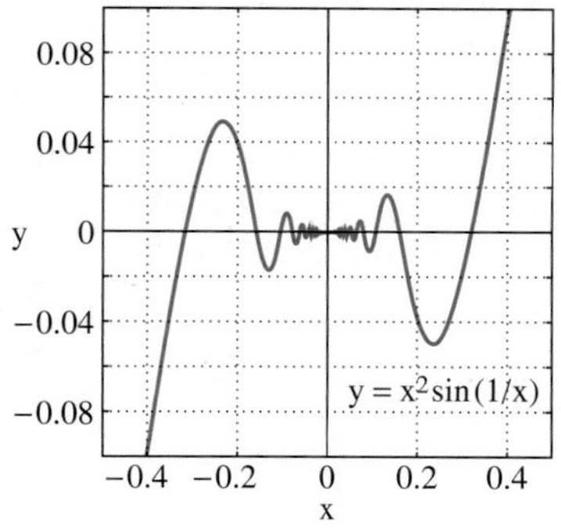

FIGURE 3.7.22 The graph of $y = x^2 \sin \dfrac{1}{x}$ (Problem 88).

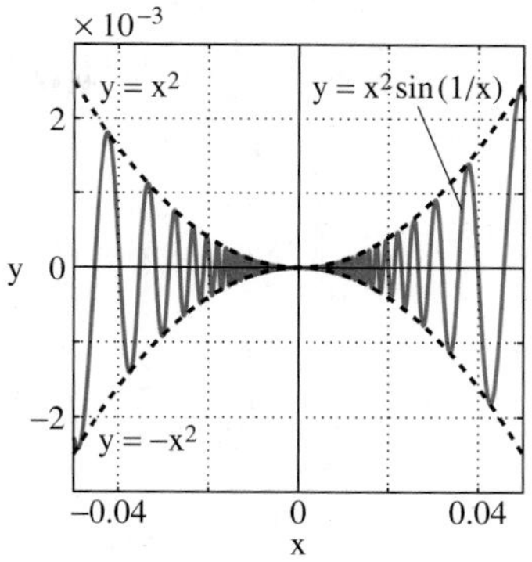

FIGURE 3.7.23 The graph in Fig. 3.7.22 magnified (Problem 88).

3.8 EXPONENTIAL AND LOGARITHMIC FUNCTIONS

Until now, we have concentrated on algebraic and trigonometric functions. Exponential and logarithmic functions complete the list of the so-called *elementary functions* that are most important in applications of calculus.

Exponential Functions

An *exponential function* is a function of the form

$$f(x) = a^x \tag{1}$$

where $a > 0$. Note that the exponent x is the variable here; the number a, called the *base,* is a constant. Thus

- An exponential function $f(x) = a^x$ is a constant raised to a variable power, whereas
- The power function $p(x) = x^k$ is a variable raised to a constant power.

In elementary algebra a *rational* power of the positive real number a is defined in terms of integral roots and powers. If n is a positive integer then

$$a^n = a \cdot a \cdot a \cdots a \qquad (n \text{ factors})$$

and

$$a^{-n} = \frac{1}{a^n}.$$

Next we learn that if $r = p/q$ where p and q are integers (with q positive), then the rational power a^r is defined by

$$a^{p/q} = \sqrt[q]{a^p} = \left(\sqrt[q]{a}\right)^p.$$

The following **laws of exponents** are then established for all *rational* exponents r and s:

$$\begin{aligned} a^{r+s} &= a^r \cdot a^s, & (a^r)^s &= a^{r \cdot s}, \\ a^{-r} &= \frac{1}{a^r}, & (ab)^r &= a^r \cdot b^r. \end{aligned} \tag{2}$$

Moreover, recall that

$$a^0 = 1 \tag{3}$$

for every positive real number a.

The following example illustrates the fact that applications often call for irrational exponents as well as rational exponents.

EXAMPLE 1 Consider a bacteria population $P(t)$ that begins (at time $t = 0$) with initial population $P(0) = 1$ (million) and doubles every hour thereafter. The growing population is given at 1-hour intervals as in the following table:

t	1	2	3	4	5	(hours)
P	2	4	8	16	32	(millions)

It is evident that $P(n) = 2^n$ if n is an integer. Now let's make the plausible assumption that the population increases by the same factor in any two time intervals of the same length—for example, if it grows by 10% in any one eight-minute interval, then it grows by 10% in any *other* eight-minute interval. If q is a positive integer and k denotes the factor by which the population increases during a time interval of length $\Delta t = 1/q$, then the population is given at successive time intervals of length $1/q$ as in the next table.

t	$\frac{1}{q}$	$\frac{2}{q}$	$\frac{3}{q}$	$\cdots$	$\frac{q}{q} = 1$	
P	k	k^2	k^3	$\cdots$	$k^q = 2$	(Why?)

We therefore see that $k = 2^{1/q}$. If p is another positive integer, then during p/q hours the population P will increase p times by the factor $k = 2^{1/q}$, so it follows that

$$P(p/q) = k^p = (2^{1/q})^p = 2^{p/q}.$$

Thus the bacteria population after t hours is given (in millions) by

$$P(t) = 2^t$$

if the exponent t is a rational number. But because time is not restricted to rational values alone, we surely ought to conclude that $P(t) = 2^t$ for *all* $t \geqq 0$. ◆

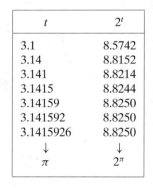

t	2^t
3.1	8.5742
3.14	8.8152
3.141	8.8214
3.1415	8.8244
3.14159	8.8250
3.141592	8.8250
3.1415926	8.8250
↓	↓
π	2^π

FIGURE 3.8.1 Investigating 2^π.

Investigation But what do we mean by an expression involving an irrational exponent, such as $2^{\sqrt{2}}$ or 2^π? To find the value of 2^π, we might work with (rational) finite decimal approximations to the irrational number $\pi = 3.1415926\cdots$. For example, a calculator gives

$$2^{3.1} = 2^{31/10} = \left(\sqrt[10]{2}\right)^{31} \approx 8.5742.$$

The approximate values shown in the table in Fig. 3.8.1 indicate that the bacteria population in Example 1 after π hours is

$$P(\pi) \approx 8.8250 \quad \text{(million)}.$$

Because any irrational number can be approximated arbitrarily closely by rational numbers, the preceding investigation suggests that the value of a^x—with irrational exponent x and a fixed base $a > 0$—can be regarded as a limit of the form

$$a^x = \lim_{r \to x} a^r \quad (r \text{ rational}). \tag{4}$$

Indeed, when the meaning of the limit in (4) is made precise, it provides one way of defining as well as calculating values of the exponential function $f(x) = a^x$ for all x.

On a calculator, the [∧] key (sometimes the [y^x] key) is ordinarily used to calculate values of exponential functions. For instance, Fig. 3.8.2 shows the result of graphing the function defined by `y = 2∧x`. We see the steadily rising graph (from left to right) of a function that is positive-valued for all x. Indeed, if r and s are positive rational numbers with $r < s$ and $a > 1$, then we note first that $a^{s-r} > 1$ (Why?) and

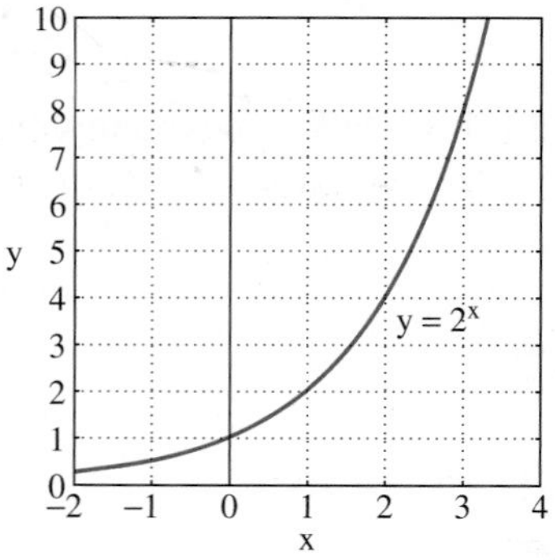

FIGURE 3.8.2 The graph $y = 2^x$.

then that

$$a^r < a^r \cdot a^{s-r} = a^{r+(s-r)} = a^s.$$

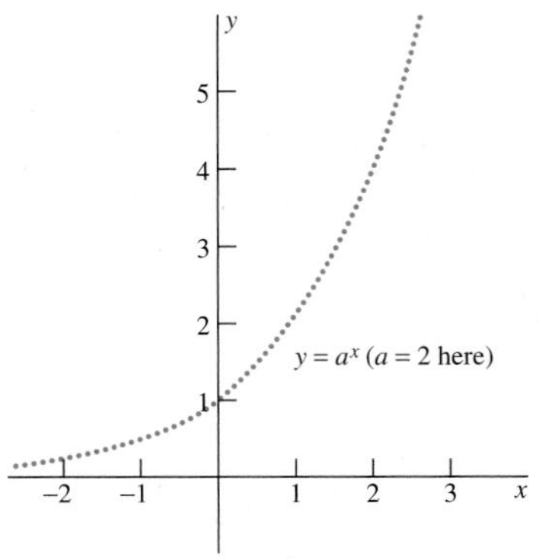

FIGURE 3.8.3 The graph of $y = a^x$ has "holes" if only rational values of x are used.

Thus $a^r < a^s$ whenever $0 < r < s$, so the exponential function $f(x) = a^x$ is certainly an increasing function if only positive rational values of the exponent are involved. A graphing calculator or computer actually plots only finitely many points (x, a^x), but the curve plotted in Fig. 3.8.2 looks connected because these points are plotted too close together for the eye to distinguish them.

By contrast, the graph in Fig. 3.8.3 is shown with a dotted curve to suggest that it is densely filled with tiny holes corresponding to the missing points (x, a^x) for which x is irrational. In Section 6.7 we will use calculus to show that these holes can be filled to obtain the graph of a continuous increasing function f with the following properties:

- $f(x)$ is defined for every real number x;
- $f(r) = a^r$ if r is rational; and
- the laws of exponents in (2) hold for irrational as well as rational exponents.

We therefore write $f(x) = a^x$ for all x and call f the **exponential function with base** a.

As illustrated in Fig. 3.8.4, the exponential function $f(x) = a^x$ with $a > 1$ increases rapidly as $x > 0$ increases, and the graphs of $y = a^x$ look qualitatively similar for different values of the base a so long as $a > 1$. The steep rate of increase of a^x for x positive and increasing is a characteristic feature of exponential functions. Figures 3.8.5 and 3.8.6 compare the graphs of the exponential function $y = 2^x$ and the quadratic function $y = x^2$.

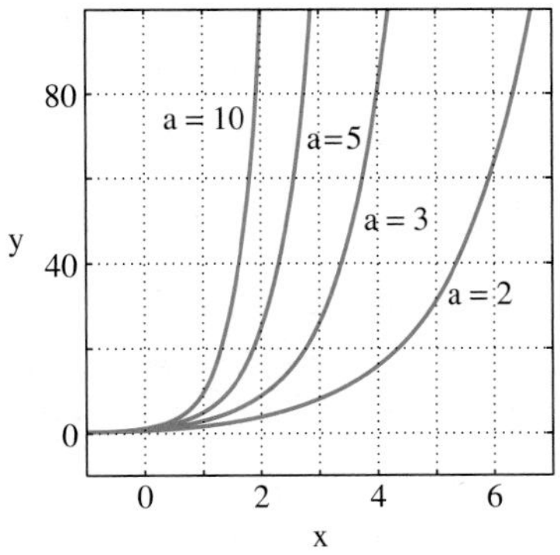

FIGURE 3.8.4 $y = a^x$ for $a = 2, 3, 5, 10$.

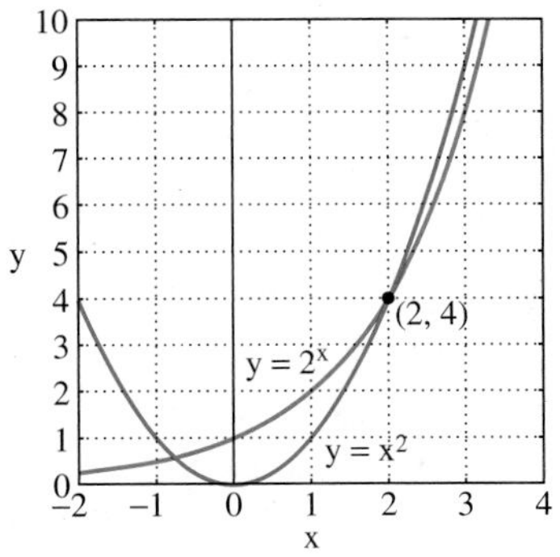

FIGURE 3.8.5 Here the graphs $y = 2^x$ and $y = x^2$ look similar for $x > 2$.

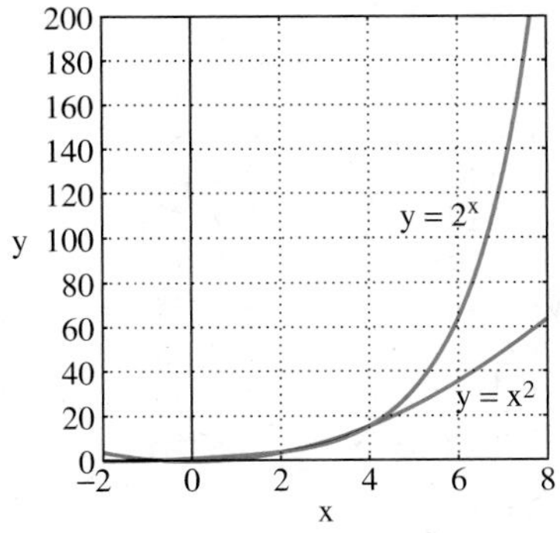

FIGURE 3.8.6 But here we see 2^x increasing much more rapidly than x^2.

Derivatives of Exponential Functions

To compute the derivative of the exponential function $f(x) = a^x$, we begin with the definition of the derivative and then use the first law of exponents in Eq. (2) to simplify. This gives

$$f'(x) = \lim_{h\to 0} \frac{f(x+h) - f(x)}{h} = \lim_{h\to 0} \frac{a^{x+h} - a^x}{h}$$

$$= \lim_{h\to 0} \frac{a^x a^h - a^x}{h} \qquad \text{(by the laws of exponents)}$$

$$= a^x \left(\lim_{h\to 0} \frac{a^h - 1}{h} \right) \qquad \text{(because } a^x \text{ is “constant” with respect to } h\text{).}$$

Under the assumption that $f(x) = a^x$ is differentiable, it follows that the limit

$$m(a) = \lim_{h \to 0} \frac{a^h - 1}{h} \tag{5}$$

exists. Although its value $m(a)$ depends on a, the limit is a constant as far as x is concerned. Thus we find that the derivative of a^x is a *constant multiple* of a^x itself:

$$D_x a^x = m(a) \cdot a^x. \tag{6}$$

h	$\frac{2^h - 1}{h}$	$\frac{3^h - 1}{h}$
0.1	0.718	1.161
0.01	0.696	1.105
0.001	0.693	1.099
0.0001	0.693	1.099

FIGURE 3.8.7 Investigating the values of $m(2)$ and $m(3)$.

Because $a^0 = 1$, we see from Eq. (6) that the constant $m(a)$ is the slope of the line tangent to the curve $y = a^x$ at the point (0, 1), where $x = 0$.

The numerical data shown in Fig. 3.8.7 suggest that $m(2) \approx 0.693$ and that $m(3) \approx 1.099$. The tangent lines with these slopes are shown in Fig. 3.8.8. Thus it appears that

$$D_x 2^x \approx (0.693) \cdot 2^x \quad \text{and} \quad D_x 3^x \approx (1.099) \cdot 3^x. \tag{7}$$

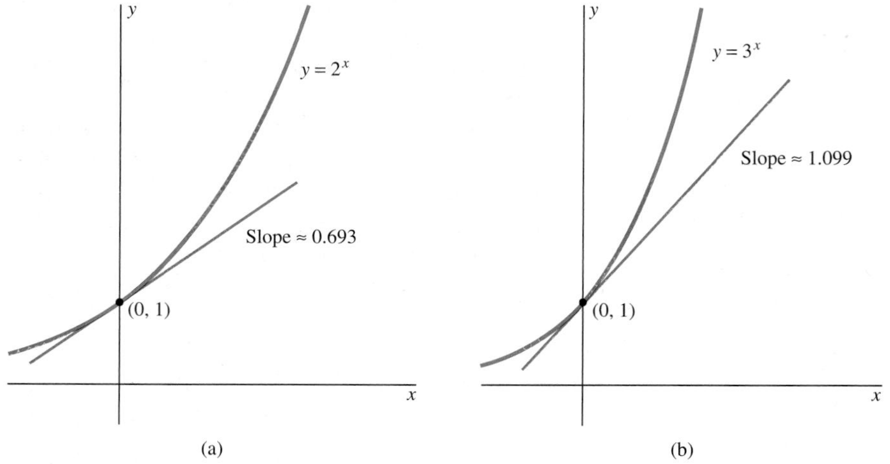

FIGURE 3.8.8 The graphs (a) $y = 2^x$ and (b) $y = 3^x$.

We would like somehow to avoid awkward numerical factors like those in Eq. (7). It seems plausible that the value $m(a)$ defined in Eq. (5) is a continuous function of a. If so, then because $m(2) < 1$ and $m(3) > 1$, the intermediate value theorem implies that $m(e) = 1$ (exactly) for some number e between 2 and 3. If we use this particular number e as the base, then it follows from Eq. (6) that the derivative of the resulting exponential function $f(x) = e^x$ is

$$D_x e^x = e^x. \tag{8}$$

Thus the function e^x is its own derivative. We call $f(x) = e^x$ the **natural exponential function.** Its graph is shown in Fig. 3.8.9.

We will see in Section 4.9 that the number e is given by the limit

$$e = \lim_{n \to \infty} \left(1 + \frac{1}{n}\right)^n.$$

Let us investigate this limit numerically. With a calculator we obtain the values in the table of Fig. 3.8.10. The evidence suggests (but does not prove) that $e \approx 2.718$ to three places. This number e is one of the most important special numbers in mathematics. It is known to be irrational; its value accurate to 15 places is

$$e \approx 2.71828\ 1828\ 459045.$$

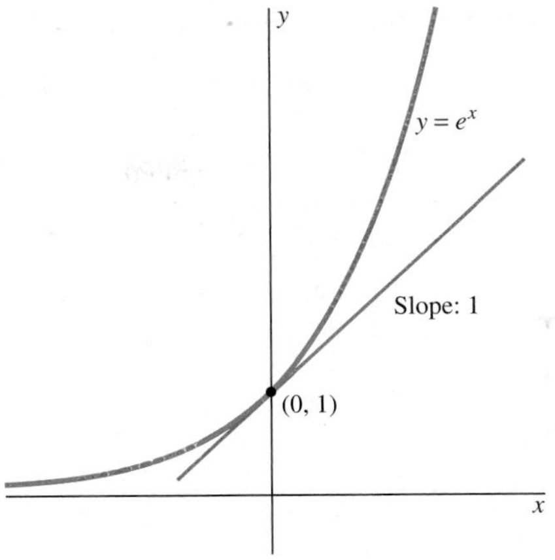

FIGURE 3.8.9 The graph $y = e^x$.

n	$\left(1+\frac{1}{n}\right)^n$
10	2.594
100	2.705
1,000	2.717
10,000	2.718
100,000	2.718

FIGURE 3.8.10 Numerical estimate of the number e.

The chain rule version of Eq. (8) is the differentiation formula

$$D_x e^u = e^u \frac{du}{dx}, \tag{9}$$

where u denotes a differentiable function of x. In particular,

$$D_x e^{kx} = ke^{kx}$$

if k is a constant. For instance, $D_x e^{-x} = -e^{-x}$ and $D_x e^{2x} = 2e^{2x}$.

EXAMPLE 2

(a) If $f(x) = x^2 e^{-x}$, then the product rule gives

$$\begin{aligned} f'(x) &= (D_x x^2)e^{-x} + x^2(D_x e^{-x}) \\ &= (2x)e^{-x} + x^2(-e^{-x}) \\ &= (2x - x^2)e^{-x}. \end{aligned}$$

(b) If $y = \dfrac{e^{2x}}{2x+1}$, then the quotient rule gives

$$\begin{aligned} \frac{dy}{dx} &= \frac{(D_x e^{2x})(2x+1) - (e^{2x})D_x(2x+1)}{(2x+1)^2} \\ &= \frac{(2e^{2x})(2x+1) - (e^{2x})(2)}{(2x+1)^2} = \frac{4xe^{2x}}{(2x+1)^2}. \end{aligned}$$

◆

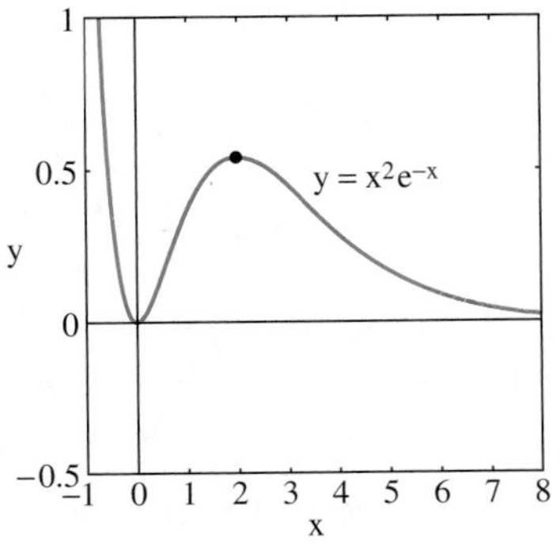

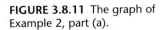

FIGURE 3.8.11 The graph of Example 2, part (a).

EXAMPLE 3 Figure 3.8.11 shows a computer plot of the graph of $f(x) = x^2 e^{-x}$. Find the coordinates of the indicated local maximum point on the curve in the first quadrant.

Solution The calculation in part (a) of Example 2 yields $f'(x) = 0$ when

$$(2x - x^2)e^{-x} = \frac{x(2-x)}{e^x} = 0,$$

so the only critical points of f are at $x = 0$ and $x = 2$. Thus the indicated first-quadrant critical point on the curve is $(2, f(2)) = (2, 4e^{-2}) \approx (2, 0.5413)$. ◆

Logarithms and Inverse Functions

In precalculus courses, the **base a logarithm function** $\log_a x$ is introduced as the "opposite" of the exponential function $f(x) = a^x$ with base $a > 1$. That is, $\log_a x$ is the power to which a must be raised to get x. Thus

$$y = \log_a x \quad \text{if and only if} \quad a^y = x. \tag{10}$$

With $a = 10$, this is the base 10 *common logarithm* $\log_{10} x$.

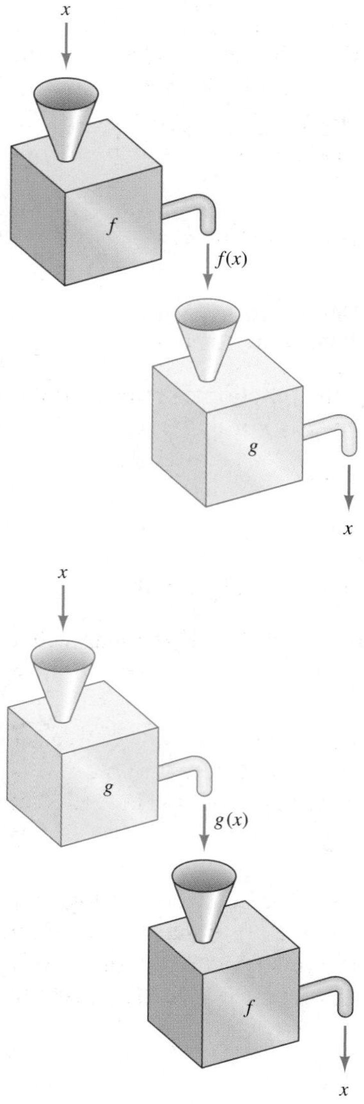

FIGURE 3.8.12 Inverse functions f and g. Each undoes the effect of the other.

EXAMPLE 4

$$\begin{aligned}
\log_{10} 1000 &= 3 && \text{because} & 1000 &= 10^3;\\
\log_{10}(0.1) &= -1 && \text{because} & 0.1 &= 10^{-1};\\
\log_2 16 &= 4 && \text{because} & 16 &= 2^4;\\
\log_3 9 &= 2 && \text{because} & 9 &= 3^2.
\end{aligned}$$ ◆

If $y = \log_a x$, then $a^y = x > 0$. Hence it follows that

$$a^{\log_a x} = x \tag{11a}$$

and

$$\log_a(a^y) = y. \tag{11b}$$

Thus the base a exponential and logarithmic functions are natural opposites, in the sense that each undoes the result of applying the other. Apply both in succession—in either order—and you're back where you started (Fig. 3.8.12). Example 5 gives other familiar pairs of functions that are *inverses* of each other.

EXAMPLE 5 The following are some pairs of *inverse functions:*

(a) $f(x) = x + 1$ and $g(x) = x - 1$.

Adding 1 and subtracting 1 are inverse operations; doing either undoes the other. Next, doubling and halving are inverse operations:

(b) $f(x) = 2x$ and $g(x) = \dfrac{x}{2}$.

A function can be its own inverse:

(c) $f(x) = \dfrac{1}{x}$ and $g(x) = \dfrac{1}{x}$. ◆

Like $f(x) = a^x$ and $g(x) = \log_a x$, each pair f and g of functions given in Example 5 has the property that

$$f(g(x)) = x \quad \text{and} \quad g(f(x)) = x \tag{12}$$

for all values of x in the domains of g and f, respectively. For instance, the functions $f(x) = x + 1$ and $g(x) = x - 1$ in part (a) of the example are defined for all x, and it is easy to check that

$$f(g(x)) = g(x) + 1 = (x - 1) + 1 = x$$

and

$$g(f(x)) = f(x) - 1 = (x + 1) - 1 = x$$

for every real number x.

DEFINITION Inverse Functions

The two functions f and g are **inverse functions,** or are **inverses** of each other, provided that

- The range of values of each function is the domain of definition of the other, and
- The relations in (12) hold for all x in the domains of g and f, respectively.

The following two examples illustrate the fact that care is required when we specify the domains of definition of the functions f and g to ensure that the condition in (12) is satisfied.

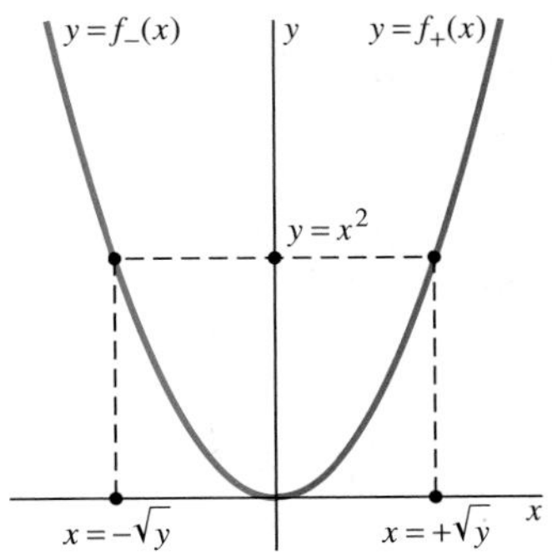

FIGURE 3.8.13 The function $f(x) = x^2$ and its restrictions f_- and f_+.

EXAMPLE 6 The function $f(x) = x^2$ is defined for all x and its range is the set of all nonnegative real numbers y; thus we write $f : (-\infty, +\infty) \longrightarrow [0, +\infty)$. As indicated in Fig. 3.8.13, it is a familiar fact that each positive number y has two different square roots, $g_+(y) = +\sqrt{y}$ and $g_-(y) = -\sqrt{y}$. (Recall that the symbol $\sqrt{y}$ unadorned with either sign *always* denotes the nonnegative square root of y.) The positive square root function $g_+(y) = +\sqrt{y}$ is defined for all $y \geqq 0$, as is $g_-(y)$, and they are the inverses of the two different squaring functions

$$f_+ : [0, +\infty) \longrightarrow [0, +\infty) \quad \text{defined by} \quad f_+(x) = x^2 \text{ for } x \geqq 0$$

and

$$f_- : (-\infty, 0] \longrightarrow [0, +\infty) \quad \text{defined by} \quad f_-(x) = x^2 \text{ for } x \leqq 0.$$

(The functions f_+ and f_- are obtained by "restricting" the function $f(x) = x^2$ to the nonnegative x-axis and the nonpositive x-axis, respectively.) For instance,

$$f_-(g_-(x)) = (-\sqrt{x})^2 = (\sqrt{x})^2 = x \quad \text{for all} \quad x \geqq 0$$

and

$$g_-(f_-(x)) = -\sqrt{x^2} = -\sqrt{(-x)^2} = -(-x) = x \quad \text{for all} \quad x \leqq 0.$$

Thus the functions f_- and g_- are inverse functions. You should verify similarly that the functions f_+ and g_+ are inverse functions. ◆

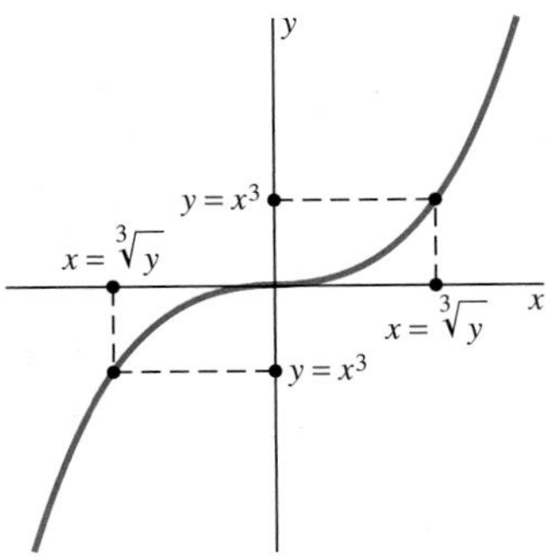

FIGURE 3.8.14 The function $f(x) = x^2$ has inverse $g(y) = \sqrt[3]{y}$ defined for all y.

EXAMPLE 7 In contrast with Example 6, the functions $f(x) = x^3$ and $g(x) = \sqrt[3]{x}$ are inverse functions defined for all x. The difference is that any real number x—whether positive, negative, or zero—has one and only one cube root (as indicated in Fig. 3.8.14). ◆

Because $a^x > 0$ for all x (as illustrated in Fig. 3.8.15), it follows that $\log_a x$ is defined *only* for $x > 0$. Because interchanging x and y in $a^y = x$ yields $y = a^x$, it follows from Eq. (10) that the graph of $y = \log_a x$ is the reflection in the line $y = x$ of the graph of $y = a^x$ and therefore has the shape shown in Fig. 3.8.15. Because

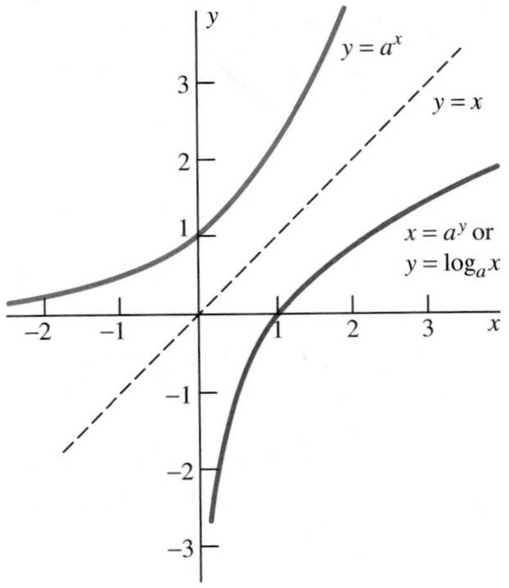

FIGURE 3.8.15 The graph of $x = a^y$ is the graph of the inverse function $\log_a x$ of the exponential function a^x. The case $a > 1$ is shown here.

$a^0 = 1$, it also follows that

$$\log_a 1 = 0,$$

so the intercepts in the figure are independent of the choice of a.

We can use the inverse-function relationship between $\log_a x$ and a^x to deduce, from the laws of exponents in Eq. (2), the following **laws of logarithms:**

$$\log_a xy = \log_a x + \log_a y, \qquad \log_a \frac{1}{x} = -\log_a x,$$
$$\log_a \frac{x}{y} = \log_a x - \log_a y, \qquad \log_a x^y = y \log_a x. \tag{13}$$

We will verify these laws of logarithms in Section 6.7.

Derivatives of Inverse Functions

Our interest in inverse-function pairs at this point stems from the following general principle: When we know the derivative of either of two inverse functions, then we can use the inverse-function relationship between them to *discover* the derivative of the other of the two functions. Theorem 1 is usually proved in an advanced calculus course.

THEOREM 1 Differentiation of an Inverse Function
Suppose that the differentiable function f is defined on the open interval I and that $f'(x) > 0$ for all x in I. Then f has an inverse function g, the function g is differentiable, and

$$g'(x) = \frac{1}{f'(g(x))} \tag{14}$$

for all x in the domain of g.

COMMENT 1 Theorem 1 is true also when the condition $f'(x) > 0$ is replaced with the condition $f'(x) < 0$. If we assume that g is differentiable, then we can derive the formula in Eq. (14) by differentiating with respect to x each side in the inverse-function relation

$$f(g(x)) = x.$$

When we differentiate each side, using the fact that this relation is actually an identity on some interval and using the chain rule on the left-hand side, the result is

$$f'(g(x)) \cdot g'(x) = 1.$$

When we solve this equation for $g'(x)$, the result is Eq. (14).

COMMENT 2 In order that the function f in Theorem 1 have an inverse function g, it is necessary (and sufficient) that, for each y in the range of f, there exists exactly one x in the domain of f such that $f(x) = y$. (We can then define $g(y) = x$.) Figure 3.8.14 indicates that this is so for the cubing function $f(x) = x^3$ of Example 7. In contrast, we see in Fig. 3.8.13 that each $y > 0$ in the range of the squaring function $f(x) = x^2$ of Example 6 corresponds to two different values of x—the positive and negative square roots of y. This is why the squaring function $f : (-\infty, +\infty) \longrightarrow [0, +\infty)$ has no (single) inverse function. The graph of f is the entire parabola in the figure. The right and left "halves" of the parabola are the graphs of the restrictions f_+ and f_- with inverse functions g_+ and g_-, respectively.

COMMENT 3 Equation (14) is easy to remember in differential notation. Let us write $x = f(y)$ and $y = g(x)$. Then $dy/dx = g'(x)$ and $dx/dy = f'(y)$. So Eq. (14) becomes the seemingly inevitable formula

$$\frac{dy}{dx} = \frac{1}{\dfrac{dx}{dy}}. \tag{15}$$

In using Eq. (15), it is important to remember that dy/dx is to be evaluated at x, but dx/dy is to be evaluated at the corresponding value of y; namely, $y = g(x)$.

EXAMPLE 8 In Section 3.4 we verified the power rule $D_x x^r = rx^{r-1}$ for rational values of the exponent r. But there we needed to know in advance that $D_x x^{1/q} = (1/q)x^{(1/q)-1}$ for every positive integer q. Now we observe that the power function $f(x) = x^q$, $x > 0$ certainly has a positive derivative: $f'(x) = qx^{q-1}$ for $x > 0$. Therefore Theorem 1 implies that its inverse function $g(x) = x^{1/q} = \sqrt[q]{x}$ exists and has derivative

$$D_x x^{1/q} = g'(x) = \frac{1}{f'(g(x))} = \frac{1}{q\left(x^{1/q}\right)^{q-1}} = \frac{1}{qx^{1-(1/q)}} = \frac{1}{q}x^{(1/q)-1},$$

as desired. Alternatively, we could use the approach of Comment 1 and simply write the identity $(x^{1/q})^q = x$. Then differentiation, using the chain rule on the left (and $D_x x \equiv 1$ on the right) gives the equation $q(x^{1/q})^{q-1} \cdot D_x x = 1$, so we can solve for $D_x x^{1/q}$. ◆

The Natural Logarithm

The natural exponential function $f(x) = e^x$ is defined for all x and $f'(x) = e^x > 0$. If f is the inverse function that consequently is guaranteed by Theorem 1 in this section, then $f(g(x)) = e^{g(x)} = x$. Thus $g(x)$ is "the power to which e must be raised to get x," and therefore is simply the logarithm function with base e: $g(x) = \log_e x$. The function g is therefore called the **natural logarithm** function. It is commonly denoted (on calculator keys, for instance) by the special symbol ln:

$$\ln x = \log_e x \quad (x > 0). \tag{16}$$

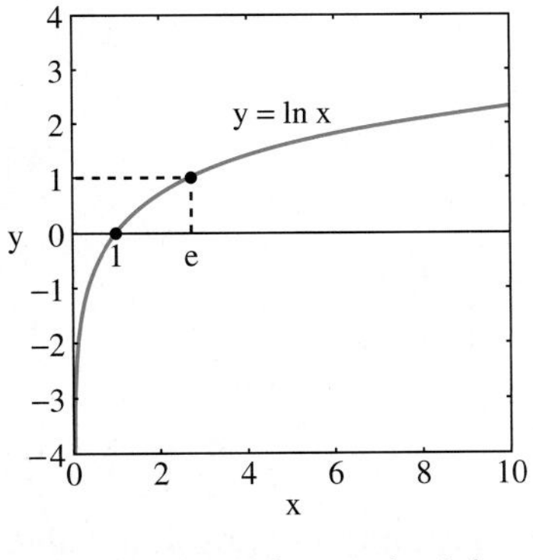

FIGURE 3.8.16 The graph of the natural logarithm function.

Because $e^x > 0$ for all x, it follows that $\ln x$ is defined only for $x > 0$. The graph of $y = \ln x$ is shown in Fig. 3.8.16, and appears to rise quite slowly when x is large. We note that $\ln 1 = 0$, so the graph has x-intercept $x = 1$, and that $\ln e = 1$ (because $\ln e = \log_e e = 1$).

The inverse function relations between $f(x) = e^x$ and $g(x) = \ln x$ are these:

$$e^{\ln x} = x \quad \text{for all } x > 0 \tag{17a}$$

and

$$\ln(e^x) = x \quad \text{for all } x. \tag{17b}$$

Derivatives of Logarithmic Functions

To differentiate the natural logarithm function, we can apply Eq. (14) in Theorem 1 (with $f(x) = e^x$ and $g(x) = \ln x$) and thereby write

$$g'(x) = \frac{1}{f'(g(x))} = \frac{1}{f'(\ln x)} = \frac{1}{e^{\ln x}} = \frac{1}{x}.$$

Alternatively, we could begin with Eq. (17a) and differentiate both sides with respect to x, as follows:

$$D_x e^{\ln x} = D_x x;$$

$$e^{\ln x} \cdot D_x \ln x = 1 \qquad \text{(by Eq. (9) with } u = \ln x\text{);}$$

$$x \cdot D_x \ln x = 1.$$

Thus we find either way that the derivative $g'(x) = D_x \ln x$ of the natural logarithm function is given by

$$D_x \ln x = \frac{1}{x} \tag{18}$$

for $x > 0$. Thus $\ln x$ is the hitherto missing function whose derivative is $x^{-1} = 1/x$.

Just as with exponentials, the derivative of a logarithm function with base other than e involves an inconvenient numerical factor. For instance, Problem 74 shows that

$$D_x \log_{10} x \approx \frac{0.4343}{x}. \tag{19}$$

The contrast between Eqs. (18) and (19) illustrates one way in which base e logarithms are "natural."

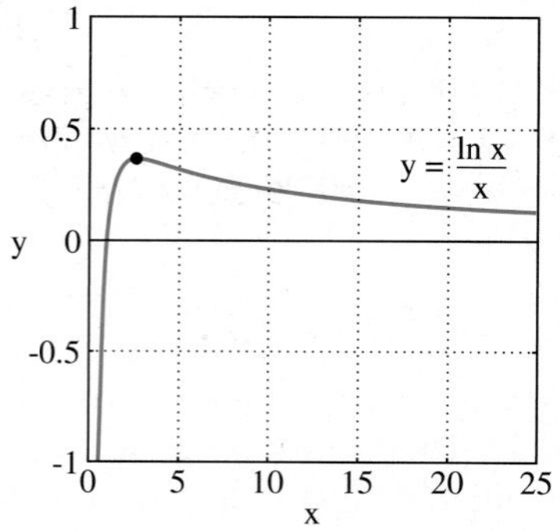

FIGURE 3.8.17 The graph of Example 9.

EXAMPLE 9 Figure 3.8.17 shows the graph of the function

$$f(x) = \frac{\ln x}{x}.$$

Find the coordinates of the indicated first-quadrant critical point on this curve.

Solution Equation (18) and the quotient rule yield

$$f'(x) = \frac{(D_x \ln x)(x) - (\ln x)(D_x x)}{x^2} = \frac{\frac{1}{x} \cdot x - (\ln x) \cdot 1}{x^2} = \frac{1 - \ln x}{x^2}.$$

Hence the only critical point of f occurs when $\ln x = 1$; that is, when $x = e$. Thus the critical point indicated in Fig. 3.8.17 is $(e, 1/e) \approx (2.718, 0.368)$. ◆

The chain-rule version of Eq. (18) is

$$D_x \ln u = \frac{1}{u} \cdot \frac{du}{dx} = \frac{u'}{u}, \tag{20}$$

where u is a positive-valued function of x and u' denotes $u'(x)$. If $u(x)$ has negative values, then the function $\ln |u|$ is defined wherever u is nonzero, so

$$D_x \ln |u| = \frac{1}{|u|} \cdot \frac{d|u|}{du} \cdot \frac{du}{dx}$$

by Eq. (20) with $|u|$ in place of u. But from the familiar graph of the absolute value function we see that

$$\frac{d|u|}{du} = \frac{|u|}{u} = \begin{cases} -1 & \text{if } u < 0, \\ +1 & \text{if } u > 0. \end{cases}$$

It therefore follows that

$$D_x \ln |u| = \frac{1}{u} \cdot \frac{du}{dx} \tag{21}$$

wherever the differentiable function $u(x)$ is nonzero. In particular,

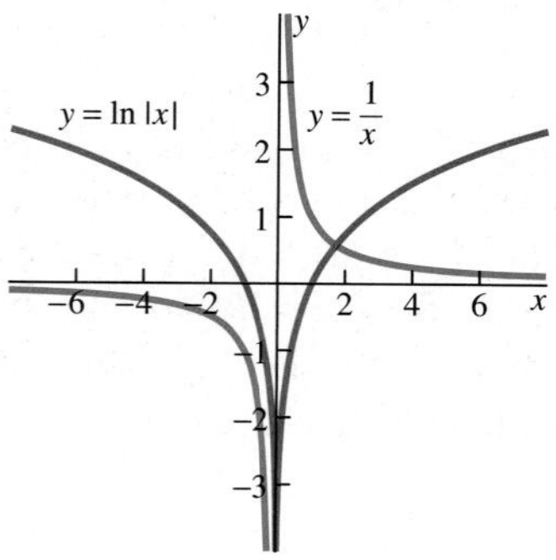

FIGURE 3.8.18 The function $f(x) = \ln |x|$ and its derivative $f'(x) = 1/x$.

$$D_x \ln |x| = \frac{1}{x} \tag{22}$$

if $x \neq 0$ (see Fig. 3.8.18).

EXAMPLE 10 With $u = 1 + x^2$ as the "inner function" in Eq. (20), we get

$$D_x \ln(1 + x^2) = \frac{u'}{u} = \frac{2x}{1 + x^2}. \quad ◆$$

EXAMPLE 11 Find the derivative of $y = \sqrt{1 + \ln x}$.

Solution Now $u = 1 + \ln x$ is the inner function, so

$$\begin{aligned}\frac{dy}{dx} &= \frac{1}{2}(1 + \ln x)^{-1/2} \cdot D_x(1 + \ln x)\\ &= \frac{1}{2}(1 + \ln x)^{-1/2} \cdot \frac{1}{x} = \frac{1}{2x\sqrt{1 + \ln x}}.\end{aligned} \quad ◆$$

EXAMPLE 12 Find the derivative of $y = \ln \sqrt{\dfrac{2x + 3}{4x + 5}}$.

Solution If we differentiated immediately, we'd find ourselves applying the quotient rule to differentiate the fraction within the radical. (Try it yourself!) It's simpler to apply laws of logarithms to simplify the given function *before* differentiating it:

$$y = \ln\left(\frac{2x + 3}{4x + 5}\right)^{1/2} = \frac{1}{2} \ln \frac{2x + 3}{4x + 5} = \frac{1}{2}[\ln(2x + 3) - \ln(4x + 5)].$$

Then

$$\frac{dy}{dx} = \frac{1}{2}\left(\frac{2}{2x + 3} - \frac{4}{4x + 5}\right) = \frac{1}{2x + 3} - \frac{2}{4x + 5} = -\frac{1}{8x^2 + 22x + 15}. \quad ◆$$

Logarithmic Differentiation

The derivatives of certain functions are most conveniently found by first differentiating their logarithms. This process—called **logarithmic differentiation**—involves the following steps for finding $f'(x)$.

1. Given $\qquad y = f(x).$
2. Take *natural* logarithms; then simplify, using laws of logarithms $\qquad \ln y = \ln f(x).$
3. Differentiate with respect to x $\qquad \dfrac{1}{y} \cdot \dfrac{dy}{dx} = D_x[\ln f(x)].$
4. Multiply both sides by $y = f(x)$ $\qquad \dfrac{dy}{dx} = f(x) D_x[\ln f(x)].$

REMARK If $f(x)$ is not positive-valued everywhere, then Steps 1 and 2 should be replaced with $y = |f(x)|$ and $\ln y = \ln |f(x)|$, respectively. The differentiation in Step 3 then leads to the result $dy/dx = f(x)D_x[\ln |f(x)|]$ in Step 4. In practice, we need not be overly concerned in advance with the sign of $f(x)$, because the appearance of what seems to be the logarithm of a negative quantity will signal the fact that absolute values should be used.

EXAMPLE 13 Find dy/dx, given

$$y = \frac{\sqrt{(x^2 + 1)^3}}{\sqrt[3]{(x^3 + 1)^4}}.$$

Solution The laws of logarithms give

$$\ln y = \ln \frac{(x^2+1)^{3/2}}{(x^3+1)^{4/3}} = \frac{3}{2}\ln(x^2+1) - \frac{4}{3}\ln(x^3+1).$$

Then differentiation with respect to x gives

$$\frac{1}{y}\cdot\frac{dy}{dx} = \frac{3}{2}\cdot\frac{2x}{x^2+1} - \frac{4}{3}\cdot\frac{3x^2}{x^3+1} = \frac{3x}{x^2+1} - \frac{4x^2}{x^3+1}.$$

Finally, to solve for dy/dx, we multiply both sides by

$$y = \frac{(x^2+1)^{3/2}}{(x^3+1)^{4/3}},$$

and we obtain

$$\frac{dy}{dx} = \left(\frac{3x}{x^2+1} - \frac{4x^2}{x^3+1}\right)\cdot\frac{(x^2+1)^{3/2}}{(x^3+1)^{4/3}}.$$ ◆

EXAMPLE 14 Find dy/dx, given $y = x^{x+1}$ for $x > 0$.

Solution If $y = x^{x+1}$, then

$$\ln y = \ln(x^{x+1}) = (x+1)\ln x;$$

$$\frac{1}{y}\cdot\frac{dy}{dx} = (1)(\ln x) + (x+1)\left(\frac{1}{x}\right) = 1 + \frac{1}{x} + \ln x.$$

Multiplying by $y = x^{x+1}$ gives

$$\frac{dy}{dx} = \left(1 + \frac{1}{x} + \ln x\right)x^{x+1}.$$ ◆

3.8 TRUE/FALSE STUDY GUIDE

3.8 CONCEPTS: QUESTIONS AND DISCUSSION

1. Example 5 lists three inverse function pairs. List several more inverse function pairs f and g of your own. In each case specify the domains of f and g and verify that f and g are indeed inverse functions.
2. Suppose that n is a positive integer. Discuss (as in Examples 6 and 7) the question of whether the power function $f(x) = x^n$ and the root function $g(x) = \sqrt[n]{x}$ are inverse functions defined for all x. How does the situation depend on whether n is even or odd? Discuss positive and negative nth roots if necessary. Specify the domain of definition of each function you mention and verify all claims you make.
3. Sketch the bell-shaped graph of the function
$$f(x) = \frac{1}{1+x^2}.$$
Explain why f (which is defined for all x) does not have an inverse function, but its restrictions f_+ and f_- to the positive and negative x-axes do have inverse functions g_+ and g_- (using notation similar to that in Example 6). Find formulas for $g_+(x)$ and $g_-(x)$ and specify the domain of definition of each of these two inverse functions.
4. Restrict the domain of each of the six trigonometric functions $\sin x$, $\cos x$, $\tan x$, $\cot x$, $\sec x$, and $\csc x$ to those points in the interval $0 < x < \pi$ at which they are defined. Consulting the graphs in Fig. 3.7.3 as necessary, determine which of these functions have inverse functions. Answer the same question if instead the domains are restricted to the interval $0 < x < \pi/2$.

3.8 PROBLEMS

Differentiate the functions in Problems 1 through 38.

1. $f(x) = e^{2x}$

2. $f(x) = e^{3x-1}$

3. $f(x) = e^{x^2}$

4. $f(x) = e^{4-x^3}$

5. $f(x) = e^{1/x^2}$

6. $f(x) = x^2 e^{x^3}$

7. $g(t) = te^{\sqrt{t}}$

8. $g(t) = (e^{2t} + e^{3t})^7$

9. $g(t) = (t^2 - 1)e^{-t}$

10. $g(t) = \sqrt{e^t - e^{-t}}$

11. $g(t) = e^{\cos t}$

12. $f(x) = xe^{\sin x}$

13. $g(t) = \dfrac{1 - e^{-t}}{t}$

14. $f(x) = e^{-1/x}$

15. $f(x) = \dfrac{1 - x}{e^x}$

16. $f(x) = e^{\sqrt{x}} + e^{-\sqrt{x}}$

17. $f(x) = e^{e^x}$

18. $f(x) = \sqrt{e^{2x} + e^{-2x}}$

19. $f(x) = \sin(2e^x)$

20. $f(x) = \cos(e^x + e^{-x})$

21. $f(x) = \ln(3x - 1)$

22. $f(x) = \ln(4 - x^2)$

23. $f(x) = \ln\sqrt{1 + 2x}$

24. $f(x) = \ln[(1 + x)^2]$

25. $f(x) = \ln\sqrt[3]{x^3 - x}$

26. $f(x) = \ln(\sin^2 x)$

27. $f(x) = \cos(\ln x)$

28. $f(x) = (\ln x)^3$

29. $f(x) = \dfrac{1}{\ln x}$

30. $f(x) = \ln(\ln x)$

31. $f(x) = \ln\left(x\sqrt{x^2 + 1}\right)$

32. $g(t) = t^{3/2}\ln(t + 1)$

33. $f(x) = \ln(\cos x)$

34. $f(x) = \ln(2\sin x)$

35. $f(t) = t^2\ln(\cos t)$

36. $f(x) = \sin(\ln 2x)$

37. $g(t) = t(\ln t)^2$

38. $g(t) = \sqrt{t}[\cos(\ln t)]^2$

In Problems 39 through 46, apply laws of logarithms to simplify the given function before finding its derivative.

39. $f(x) = \ln[(2x + 1)^3(x^2 - 4)^4]$

40. $f(x) = \ln\sqrt{\dfrac{1 - x}{1 + x}}$

41. $f(x) = \ln\sqrt{\dfrac{4 - x^2}{9 + x^2}}$

42. $f(x) = \ln\dfrac{\sqrt{4x - 7}}{(3x - 2)^3}$

43. $f(x) = \ln\dfrac{x + 1}{x - 1}$

44. $f(x) = x^2\ln\dfrac{1}{2x + 1}$

45. $g(t) = \ln\dfrac{t^2}{t^2 + 1}$

46. $f(x) = \ln\dfrac{\sqrt{x + 1}}{(x - 1)^3}$

In Problems 47 through 58, find dy/dx by logarithmic differentiation.

47. $y = 2^x$

48. $y = x^x$

49. $y = x^{\ln x}$

50. $y = (1 + x)^{1/x}$

51. $y = (\ln x)^{\sqrt{x}}$

52. $y = (3 + 2^x)^x$

53. $y = \dfrac{(1 + x^2)^{3/2}}{(1 + x^3)^{4/3}}$

54. $y = (x + 1)^x$

55. $y = (x^2 + 1)^{x^2}$

56. $y = \left(1 + \dfrac{1}{x}\right)^x$

57. $y = \left(\sqrt{x}\right)^{\sqrt{x}}$

58. $y = x^{\sin x}$

In Problems 59 through 62, write an equation of the line tangent to the graph of the given function at the indicated point.

59. $y = xe^{2x}$ at the point $(1, e^2)$

60. $y = e^{2x}\cos x$ at the point $(0, 1)$

61. $y = x^3\ln x$ at the point $(1, 0)$

62. $y = \dfrac{\ln x}{x^2}$ at the point (e, e^{-2})

In Problems 63 and 64, differentiate the given function $f(x)$ and its derivative in turn, several times in succession. Then give a likely formula for the result after n successive differentiations in this manner.

63. $f(x) = e^{2x}$

64. $f(x) = xe^x$

65. Figure 3.8.19 shows the graph of the function $f(x) = e^{-x/6}\sin x$, together with the graphs of its "envelope curves" $y = e^{-x/6}$ and $y = -e^{-x/6}$. Find the first local maximum point and the first local minimum point on the graph of f for $x > 0$.

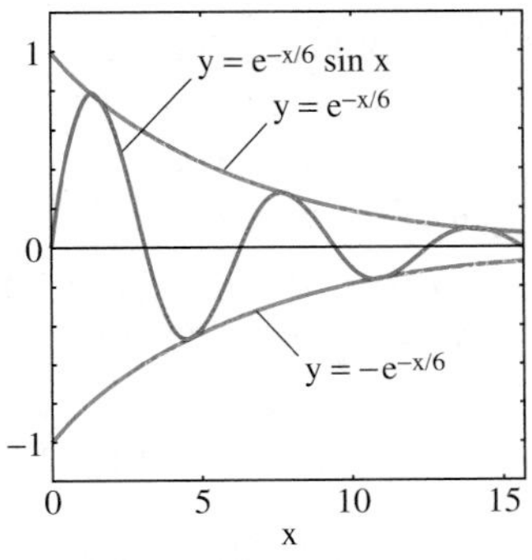

FIGURE 3.8.19 The graph for Problems 65 and 66.

66. Find the first two points of tangency of the curve $y = e^{-x/6}\sin x$ with the two envelope curves shown in Fig. 3.8.19. Are these the same as the two local extreme points found in Problem 65?

67. Find graphically the coordinates (accurate to three decimal places) of the intersection point of the graphs $y = e^x$ and $y = x^{10}$ indicated in Fig. 3.8.20.

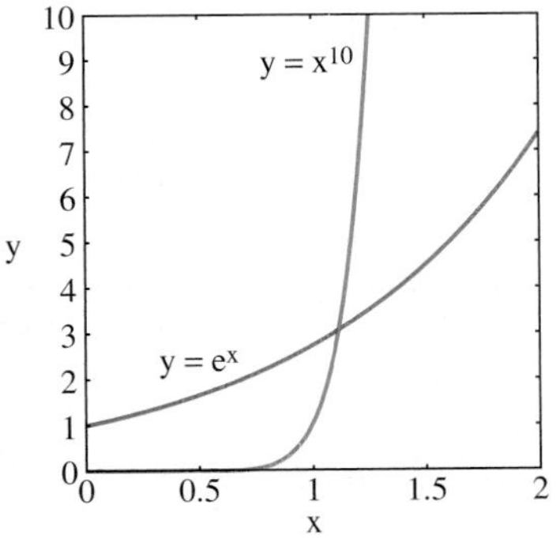

FIGURE 3.8.20 Comparing $y = e^x$ and $y = x^{10}$.

68. See Problem 67. Determine a viewing rectangle that reveals a second intersection point (with $x > 10$) of the graphs

$y = e^x$ and $y = x^{10}$. Then determine graphically the first three digits of the larger solution x of the equation $e^x = x^{10}$ (thus writing this solution in the form $p.qr \times 10^k$).

69. If we substitute $n = 10^k$ in $e = \lim_{n\to\infty}\left(1 + \frac{1}{n}\right)^n$, we get the limit

$$e = \lim_{k\to\infty}\left(1 + \frac{1}{10^k}\right)^{10^k}$$

that "converges" much more rapidly. Using a calculator or computer, substitute $k = 1, 2, 3, \dots, 8$ in turn to discover that $e \approx 2.71828$ accurate to five decimal places.

70. Suppose that u and v are differentiable functions of x. Show by logarithmic differentiation that

$$D_x(u^v) = v(u^{v-1})\frac{du}{dx} + (u^v \ln u)\frac{dv}{dx}.$$

Interpret the two terms on the right in relation to the special cases in which (a) u is a constant; (b) v is a constant.

71. Suppose that $y = uvw/pqr$, where u, v, w, p, q, and r are nonzero differentiable functions of x. Show by logarithmic differentiation that

$$\frac{dy}{dx} = y\cdot\left(\frac{1}{u}\frac{du}{dx} + \frac{1}{v}\frac{dv}{dx} + \frac{1}{w}\frac{dw}{dx} - \frac{1}{p}\frac{dp}{dx} - \frac{1}{q}\frac{dq}{dx} - \frac{1}{r}\frac{dr}{dx}\right).$$

Is the generalization—for an arbitrary finite number of factors in numerator and denominator—obvious?

72. Show that the number $\log_2 3$ is irrational. [*Suggestion:* Assume to the contrary that $\log_2 3 = p/q$ where p and q are positive integers; then express the consequence of this assumption in exponential form. Under what circumstances can an integral power of 2 equal an integral power of 3?]

73. The `log` key on the typical calculator denotes the base 10 logarithm $f(x) = \log_{10} x$.
(a) Use the definition of the derivative to show that

$$f'(1) = \lim_{h\to 0}\log_{10}(1+h)^{1/h}.$$

(b) Investigate the limit in (a) numerically to show that $f'(1) \approx 0.4343$.

74. The object of this problem is to differentiate the base 10 logarithm function of Problem 73.
(a) First use the known formula for $D_x e^u$ to show that $D_x 10^x = 10^x \ln 10$.
(b) Conclude from the chain rule that

$$D_x 10^u = 10^u(\ln 10)\frac{du}{dx}.$$

(c) Substitute $u = \log_{10} x$ in the inverse function identity $10^{\log_{10} x} = x$ and then differentiate using the result of part (b) to conclude that

$$D_x \log_{10} x = \frac{1}{x \ln 10} \approx \frac{0.4343}{x},$$

consistent with the result (for $x = 1$) of Problem 73.

3.8 Project: Discovering the Number e for Yourself

You can investigate the value of e by approximating the value of a such that

$$m(a) = \lim_{h\to 0}\frac{a^h - 1}{h} = 1.$$

You need use only available technology to calculate (with appropriate fixed values of a) values of the function $\phi(h) = (a^h - 1)/h$ with h sufficiently small that you can recognize (to appropriate accuracy) the value of the limit.

For instance, if you calculate $\phi(h)$ with $a = 2$ and with $a = 3$ for $h = 0.1, 0.01, 0.001, 0.0001, \dots,$ you should find that

$$m(2) \approx 0.6931 < 1 \quad \text{whereas} \quad m(3) \approx 1.0986 > 1.$$

It follows that the mysterious number e for which $m(e) = 1$ is somewhere between 2 and 3. Linear interpolation between the values of $m(2) \approx 0.6931$ and $m(3) \approx 1.0986$ suggests that $e \approx 2.7$ or $e \approx 2.8$ accurate to one decimal place.

Investigate the values of $m(2.7)$ and $m(2.8)$ to verify the entries shown in Fig. 3.8.21. Continue in this way to close in on the number e. Don't quit until you're convinced that $e \approx 2.718$ accurate to three decimal places.

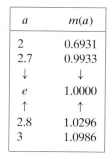

a	$m(a)$
2	0.6931
2.7	0.9933
↓	↓
e	1.0000
↑	↑
2.8	1.0296
3	1.0986

FIGURE 3.8.21 Closing in on the number e.

3.9 IMPLICIT DIFFERENTIATION AND RELATED RATES

A formula such as $y = x^3 \sin x$ defines y "explicitly" as a function of x. Most of the functions we have seen so far have been defined explicitly in this way. Nevertheless, a function can also be defined "implicitly" by an equation that can be solved for y in terms of x. Indeed, we will see that a single equation relating the two variables x and y can implicitly define two or more different functions of x.

EXAMPLE 1

(a) When we solve the equation

$$x - y^2 = 0$$

for $y = \pm\sqrt{x}$, we get the two *explicit* functions

$$f(x) = \sqrt{x} \quad \text{and} \quad g(x) = -\sqrt{x}$$

that we say are implicitly defined by the original equation. The graphs of these two functions—both defined for $x \geqq 0$—are the upper and lower branches of the parabola shown (in different colors) in Fig. 3.9.1. The whole parabola is the graph of the equation $x - y^2 = 0$ (or $x = y^2$) but is not the graph of any single function. (Why?)

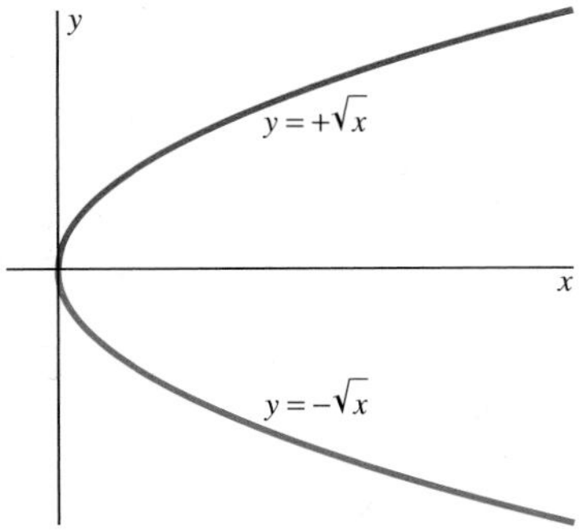

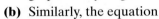

FIGURE 3.9.1 The equation $x - y^2 = 0$ implicitly defines the two functions $f(x) = \sqrt{x}$ and $g(x) = -\sqrt{x}$.

(b) Similarly, the equation

$$x^2 + y^2 = 100$$

implicitly defines the two continuous functions

$$f(x) = \sqrt{100 - x^2} \quad \text{and} \quad g(x) = -\sqrt{100 - x^2}$$

—both defined for $-10 \leqq x \leqq 10$—that correspond to the solutions $y = \pm\sqrt{100 - x^2}$ for y in terms of x. The graphs of f and g are the upper and lower semicircles of the whole circle $x^2 + y^2 = 100$ (shown in different colors in Fig. 3.9.2). ◆

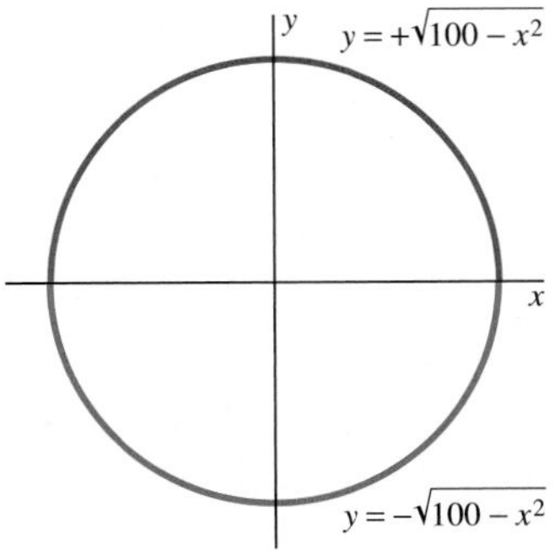

FIGURE 3.9.2 The equation $x^2 + y^2 = 100$ implicitly defines the two functions $f(x) = \sqrt{100 - x^2}$ and $g(x) = -\sqrt{100 - x^2}$.

Whereas the equations $x - y^2 = 0$ and $x^2 + y^2 = 100$ are readily solved for y in terms of x, an equation such as $x^3 + y^3 = 3xy$ or $\sin(x + 2y) = 2x \cos y$ may be difficult or impossible to solve for an implicitly defined function $y(x)$. And yet the derivative dy/dx can be calculated without first expressing y in terms of x. Here's how: We can use the chain rule and other basic differentiation rules to differentiate both sides of the given equation with respect to x (we think of x as the independent variable, although it is permissible to reverse the roles of x and y). We then solve the resulting equation for the derivative $y'(x) = dy/dx$ of the implicitly defined function $y(x)$. This process is called **implicit differentiation.** In the examples and problems of this section, we proceed on the assumption that our implicitly defined functions actually exist and are differentiable at almost all points in their domains. (The functions with the graphs shown in Fig. 3.9.2 are *not* differentiable at the endpoints of their domains.)

EXAMPLE 2 Use implicit differentiation to find the derivative of a differentiable function $y = f(x)$ implicitly defined by the equation

$$x^2 + y^2 = 100.$$

Solution The equation $x^2 + y^2 = 100$ is to be regarded as an *identity* that implicitly defines $y = y(x)$ as a function of x. Because $x^2 + [y(x)]^2$ is then a function of x, it has the same derivative as the constant function 100 on the right-hand side of the identity. Thus we may differentiate both sides of the *identity* $x^2 + y^2 = 100$ with respect to x and equate the results. We obtain

$$2x + 2y\frac{dy}{dx} = 0.$$

In this step, it is essential to remember that y is a function of x, so the chain rule yields $D_x(y^2) = 2y\,D_x y$.

Then we solve for

$$\frac{dy}{dx} = -\frac{x}{y}. \tag{1}$$

It may be surprising to see a formula for dy/dx containing both x and y, but such a formula can be just as useful as one containing only x. For example, the formula in Eq. (1) tells us that the slope of the line tangent to the circle $x^2 + y^2 = 100$ at the

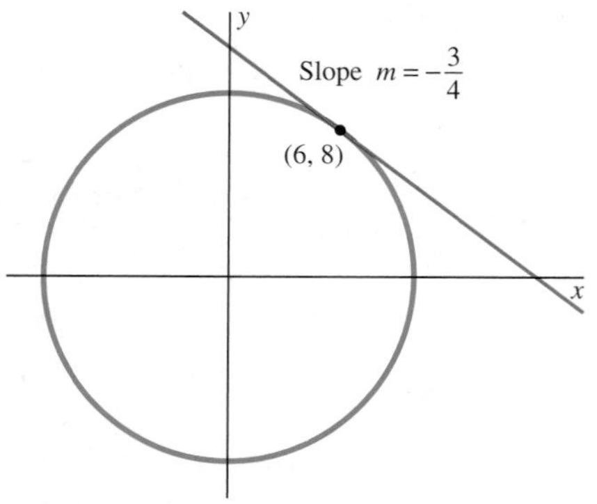

FIGURE 3.9.3 The circle $x^2 + y^2 = 100$ and the tangent line at the point (6, 8).

point (6, 8) is

$$\left.\frac{dy}{dx}\right|_{(6,8)} = -\frac{6}{8} = -\frac{3}{4}.$$

The circle and this line are shown in Fig. 3.9.3. ◆

Note If we solve for $y = \pm\sqrt{100 - x^2}$ in Example 1, then

$$\frac{dy}{dx} = \frac{-x}{\pm\sqrt{100 - x^2}} = -\frac{x}{y},$$

in agreement with Eq. (1). Thus Eq. (1) simultaneously gives us the derivatives of both the functions $y = +\sqrt{100 - x^2}$ and $y = -\sqrt{100 - x^2}$ implicitly defined by the equation $x^2 + y^2 = 100$.

EXAMPLE 3 The *folium of Descartes* is the graph of the equation

$$x^3 + y^3 = 3xy \qquad \textbf{(2)}$$

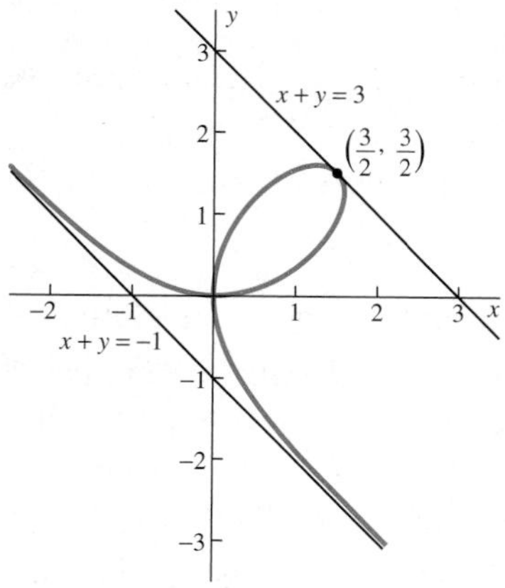

FIGURE 3.9.4 A tangent line and an apparent asymptote to the curve $x^3 + y^3 = 3xy$.

This curve was first proposed by René Descartes as a challenge to Pierre de Fermat (1601–1665) to find its tangent line at an arbitrary point. The project for this section tells how we constructed Fig. 3.9.4. It indicates that the second- and fourth-quadrant points on the graph for which $|x|$ and $|y|$ are both large lie very close to the straight line $x + y + 1 = 0$. In the first quadrant we see a loop shaped like a laurel leaf—hence the name *folium.* (Can you see directly from Eq. (2) that the third quadrant contains *no* points of the folium?) Here we want to find Fermat's answer as to the slope of a typical line tangent to the folium of Descartes.

Solution Equation (2) is a cubic equation in x, and we see in Fig. 3.9.4 three different branches of the graph over an interval to the right of the origin. When we ask a computer algebra system to solve the equation for these implicitly defined functions of x, it produces three different expressions, the simplest of which is

$$y = \frac{1}{2}\sqrt[3]{-4x^3 + 4\sqrt{x^6 - 4x^3}} + \frac{2x}{\sqrt[3]{-4x^3 + 4\sqrt{x^6 - 4x^3}}}.$$

(It turns out that this formula describes the upper part of the loop in Fig. 3.9.4.) Surely you would not relish explicit differentiation of this expression to find the slope of a line tangent to the folium. Fortunately, the alternative of implicit differentiation is available. We need only differentiate each side of Eq. (2) with respect to x, remembering that y is a function of x. Hence we use the chain rule to differentiate y^3 and the product rule to differentiate $3xy$. This yields

$$3x^2 + 3y^2\frac{dy}{dx} = 3y + 3x\frac{dy}{dx}.$$

We can now collect coefficients (those involving dy/dx and those not) and solve for the derivative:

$$(3y^2 - 3x)\frac{dy}{dx} = 3y - 3x^2;$$

$$\frac{dy}{dx} = \frac{y - x^2}{y^2 - x}. \quad \textbf{(3)}$$

For instance, at the point $P\left(\frac{3}{2}, \frac{3}{2}\right)$ of the folium, the slope of the tangent line is

$$\left.\frac{dy}{dx}\right|_{\left(\frac{3}{2}, \frac{3}{2}\right)} = \frac{\frac{3}{2} - \left(\frac{3}{2}\right)^2}{\left(\frac{3}{2}\right)^2 - \frac{3}{2}} = -1.$$

This result agrees with our intuition about the figure, because the evident symmetry of the folium around the line $y = x$ suggests that the tangent line at P should, indeed, have slope -1. The equation of this tangent line is

$$y - \tfrac{3}{2} = -\left(x - \tfrac{3}{2}\right);$$

that is, $x + y = 3$. ◆

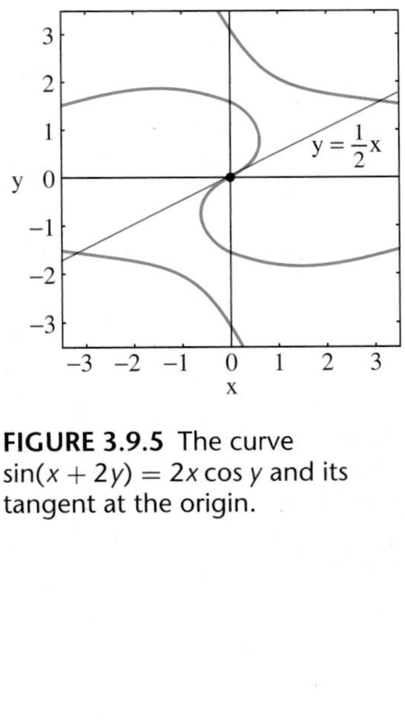

FIGURE 3.9.5 The curve $\sin(x + 2y) = 2x \cos y$ and its tangent at the origin.

EXAMPLE 4 Figure 3.9.5 shows a computer plot of the graph of the equation

$$\sin(x + 2y) = 2x \cos y. \quad \textbf{(4)}$$

Write the equation of the line tangent to this curve at the origin (0, 0).

Solution When we differentiate each side in (4) with respect to the independent variable x, regarding y as a function of x, we get

$$[\cos(x + 2y)] \cdot \left(1 + 2\frac{dy}{dx}\right) = 2\cos y - (2x \sin y)\frac{dy}{dx}. \quad \textbf{(5)}$$

We could collect coefficients and solve for the derivative dy/dx. But because we need only the slope $y'(0)$ at the origin, let us instead substitute $x = y = 0$ in Eq. (5). Noting that $\cos(0) = 1$ and $\sin(0) = 0$, we get the equation

$$1 + 2y'(0) = 2,$$

from which we see that $y'(0) = \frac{1}{2}$. The resulting tangent line $y = \frac{1}{2}x$ plotted in Fig. 3.9.5 "looks right," and thus corroborates the results of our calculations. ◆

Related Rates

A **related-rates** problem involves two or more quantities that vary with time and an equation that expresses some relationship between these quantities. Typically, the values of these quantities at some instant are given, together with all their time rates of change but one. The problem is usually to find the time rate of change that is *not* given, at some instant specified in the problem. Implicit differentiation, with respect to time t, of the equation that relates the given quantities will produce an equation that relates the *rates of change* of the given quantities. This is the key to solving a related-rates problem.

EXAMPLE 5 Suppose that $x(t)$ and $y(t)$ are the x- and y-coordinates at time t of a point moving around the circle with equation

$$x^2 + y^2 = 25. \quad \textbf{(6)}$$

Let us use the chain rule to differentiate both sides of this equation *with respect to time t*. This produces the equation

$$2x\frac{dx}{dt} + 2y\frac{dy}{dt} = 0. \tag{7}$$

If the values of x, y, and dx/dt are known at a certain instant t, then Eq. (7) can be solved for the value of dy/dt. It is *not* necessary to know x and y as functions of t. Indeed, it is common for a related-rates problem to contain insufficient information to express x and y as functions of t.

For instance, suppose that we are given $x = 3$, $y = 4$, and $dx/dt = 12$ at a certain instant. Substituting these values into Eq. (7) yields

$$2 \cdot 3 \cdot 12 + 2 \cdot 4 \cdot \frac{dy}{dt} = 0,$$

so we find that $dy/dt = -9$ at the same instant. ◆

EXAMPLE 6 A rocket that is launched vertically is tracked by a radar station located on the ground 3 mi from the launch site. What is the vertical speed of the rocket at the instant that its distance from the radar station is 5 mi and this distance is increasing at the rate of 5000 mi/h?

Solution Figure 3.9.6 illustrates this situation. We denote the altitude of the rocket (in miles) by y and its distance from the radar station by z. We are given

$$\frac{dz}{dt} = 5000 \qquad \text{when} \qquad z = 5.$$

We want to find dy/dt (in miles per hour) at this instant.

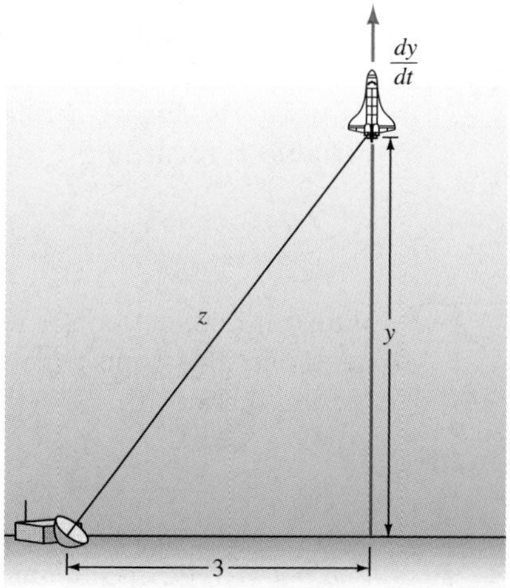

FIGURE 3.9.6 The rocket of Example 6.

We apply the Pythagorean theorem to the right triangle in the figure and obtain

$$y^2 + 9 = z^2$$

as a relation between y and z. From this we see that $y = 4$ when $z = 5$. Implicit differentiation then gives

$$2y\frac{dy}{dt} = 2z\frac{dz}{dt}.$$

We substitute the data $y = 4$, $z = 5$, and $dz/dt = 5000$. Thus we find that

$$\frac{dy}{dt} = 6250 \quad \text{(mi/h)}$$

at the instant in question. ◆

Example 6 illustrates the following steps in the solution of a typical related-rates problem of the sort that involves a geometric situation:

1. Draw a diagram and label as variables the various changing quantities involved in the problem.
2. Record the values of the variables and their rates of change, as given in the problem.
3. Use the diagram to determine an equation that relates the important variables in the problem.
4. Differentiate this equation implicitly with respect to time t.
5. Substitute the given numerical data in the resulting equation, and then solve for the unknown.

WARNING The most common error to be avoided is the premature substitution of the given data before differentiating implicitly. If we had substituted $z = 5$ to begin with in Example 5, our equation would have been $y^2 + 9 = 25$, and implicit differentiation would have given the absurd result $dy/dt = 0$.

We use similar triangles (rather than the Pythagorean theorem) in Example 7 to discover the needed relation between the variables.

EXAMPLE 7 A man 6 ft tall walks with a speed of 8 ft/s away from a street light that is atop an 18-ft pole. How fast is the tip of his shadow moving along the ground when he is 100 ft from the light pole?

Solution Let x be the man's distance from the pole and z the distance from the tip of his shadow to the base of the pole (Fig. 3.9.7). Although x and z are functions of time t, we do *not* attempt to find explicit formulas for either.

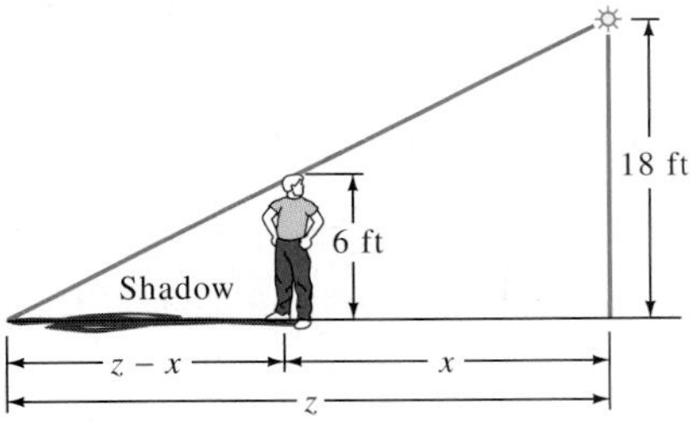

FIGURE 3.9.7 The moving shadow of Example 7.

We are given $dx/dt = 8$ (in feet per second); we want to find dz/dt when $x = 100$ (ft). We equate ratios of corresponding sides of the two similar triangles of Fig. 3.9.7 and find that

$$\frac{z}{18} = \frac{z - x}{6}.$$

It follows that

$$2z = 3x,$$

and implicit differentiation gives

$$2\frac{dz}{dt} = 3\frac{dx}{dt}.$$

We substitute $dx/dt = 8$ and find that

$$\frac{dz}{dt} = \frac{3}{2} \cdot \frac{dx}{dt} = \frac{3}{2} \cdot 8 = 12.$$

So the tip of the man's shadow is moving at 12 ft/s. ◆

Example 7 is somewhat unusual in that the answer is independent of the man's distance from the light pole—the given value $x = 100$ is superfluous because the tip of the man's shadow is moving at constant speed. Example 8 is a related-rates problem with two relationships between the variables, which is not quite so unusual.

EXAMPLE 8 Two radar stations at A and B, with B 6 km east of A, are tracking a ship. At a certain instant, the ship is 5 km from A, and this distance is increasing at the rate of 28 km/h. At the same instant, the ship is also 5 km from B, but this distance is increasing at only 4 km/h. Where is the ship, how fast is it moving, and in what direction is it moving?

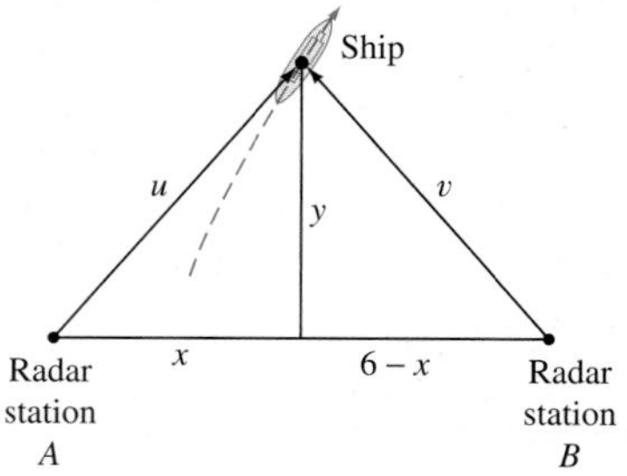

FIGURE 3.9.8 Radar stations tracking a ship (Example 8).

Solution With the distances indicated in Fig. 3.9.8, we find—again with the aid of the Pythagorean theorem—that

$$x^2 + y^2 = u^2 \quad \text{and} \quad (6 - x)^2 + y^2 = v^2. \tag{8}$$

We are given the following data: $u = v = 5$, $du/dt = 28$, and $dv/dt = 4$ at the instant in question. Because the ship is equally distant from A and B, it is clear that $x = 3$. Thus $y = 4$. Hence the ship is 3 km east and 4 km north of A.

We differentiate implicitly the two equations in (8), and we obtain

$$2x\frac{dx}{dt} + 2y\frac{dy}{dt} = 2u\frac{du}{dt}$$

and

$$-2(6 - x)\frac{dx}{dt} + 2y\frac{dy}{dt} = 2v\frac{dv}{dt}.$$

When we substitute the numerical data given and data deduced, we find that

$$3\frac{dx}{dt} + 4\frac{dy}{dt} = 140 \quad \text{and} \quad -3\frac{dx}{dt} + 4\frac{dy}{dt} = 20.$$

These equations are easy to solve: $dx/dt = dy/dt = 20$. Therefore, the ship is sailing northeast at a speed of

$$\sqrt{20^2 + 20^2} = 20\sqrt{2} \quad \text{(km/h)}$$

—*if* the figure is correct! A mirror along the line AB will reflect *another* ship, 3 km east and 4 km *south* of A, sailing *southeast* at a speed of $20\sqrt{2}$ km/h.

The lesson? Figures are important, helpful, often essential—but potentially misleading. Avoid taking anything for granted when you draw a figure. In this example there would be no real problem, for each radar station could determine whether the ship was generally to the north or to the south. ◆

3.9 TRUE/FALSE STUDY GUIDE

3.9 CONCEPTS: QUESTIONS AND DISCUSSION

1. Figure 3.9.1 shows the graphs of the two functions $f(x) = \sqrt{x}$ and $g(x) = -\sqrt{x}$ that are defined implicitly by the equation $x - y^2 = 0$. Both f and g are continuous for $x \geqq 0$. Can you think of a discontinuous function $y = h(x)$ that satisfies the same equation?
2. How many different continuous functions of x (with the same domain of definition) are implicitly defined by a given quadratic equation in x and y? A given cubic equation? A given quartic (fourth-degree) equation? How many different discontinuous functions?

3. How many different continuous functions of x are defined by the following equations?

(a) $x^2 + y^2 + 1 = 0$ (b) $x^3 + y^3 = 1$ (c) $x^4 + y^4 = 1$

4. How many different continuous functions of x are defined by the transcendental equation $\sin y = x$?

3.9 PROBLEMS

In Problems 1 through 4, first find the derivative dy/dx by implicit differentiation. Then solve the original equation for y explicitly in terms of x and differentiate to find dy/dx. Finally verify that your two results are the same by substituting the explicit expression for $y(x)$ in the implicit form of the derivative.

1. $x^2 - y^2 = 1$

2. $xy = 1$

3. $16x^2 + 25y^2 = 400$

4. $x^3 + y^3 = 1$

In Problems 5 through 14, find dy/dx by implicit differentiation.

5. $\sqrt{x} + \sqrt{y} = 1$

6. $x^4 + x^2y^2 + y^4 = 48$

7. $x^{2/3} + y^{2/3} = 1$

8. $(x - 1)y^2 = x + 1$

9. $x^2(x - y) = y^2(x + y)$

10. $x^5 + y^5 = 5x^2y^2$

11. $x \sin y + y \sin x = 1$

12. $\cos(x + y) = \sin x \sin y$

13. $2x + 3e^y = e^{x+y}$

14. $xy = e^{-xy}$

In Problems 15 through 28, use implicit differentiation to find an equation of the line tangent to the given curve at the given point.

15. $x^2 + y^2 = 25$; $(3, -4)$

16. $xy = -8$; $(4, -2)$

17. $x^2y = x + 2$; $(2, 1)$

18. $x^{1/4} + y^{1/4} = 4$; $(16, 16)$

19. $xy^2 + x^2y = 2$; $(1, -2)$

20. $\dfrac{1}{x+1} + \dfrac{1}{y+1} = 1$; $(1, 1)$

21. $12(x^2 + y^2) = 25xy$; $(3, 4)$

22. $x^2 + xy + y^2 = 7$; $(3, -2)$

23. $2e^{-x} + e^y = 3e^{x-y}$; $(0, 0)$

24. $xy = 6e^{2x-3y}$; $(3, 2)$

25. $x^{2/3} + y^{2/3} = 5$; $(8, 1)$ (Fig. 3.9.9)

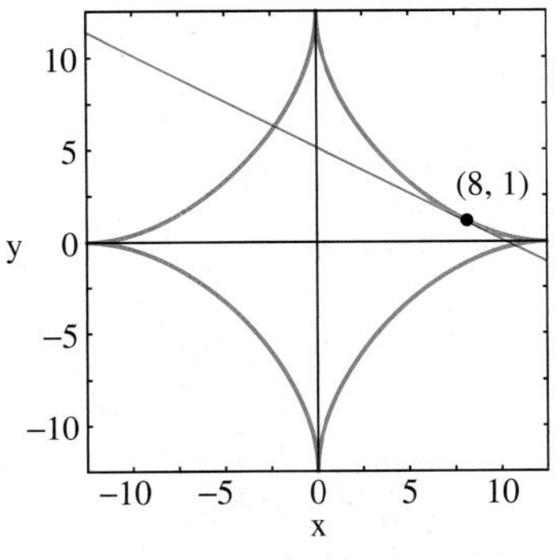

FIGURE 3.9.9 Problem 25.

26. $x^2 - xy + y^2 = 19$; $(3, -2)$ (Fig. 3.9.10)

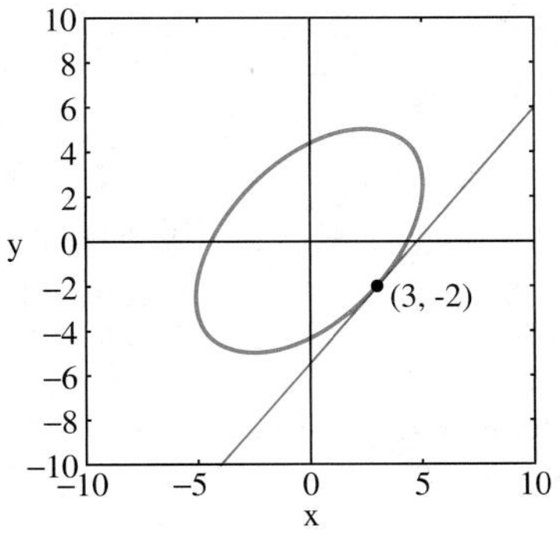

FIGURE 3.9.10 Problem 26.

27. $(x^2 + y^2)^2 = 50xy$; $(2, 4)$ (Fig. 3.9.11)

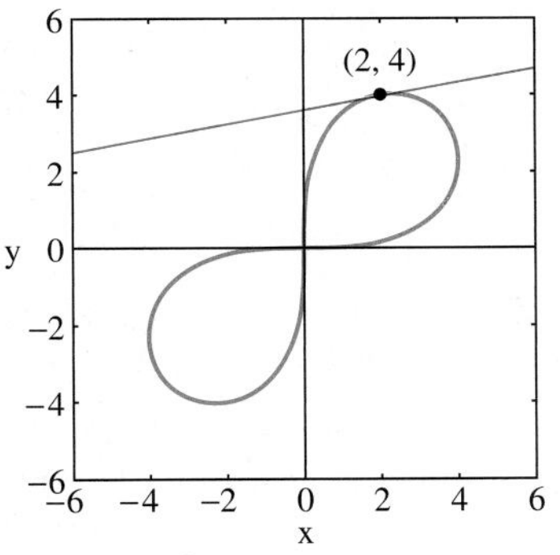

FIGURE 3.9.11 Problem 27.

28. $y^2 = x^2(x + 7)$; $(-3, 6)$ (Fig. 3.9.12)

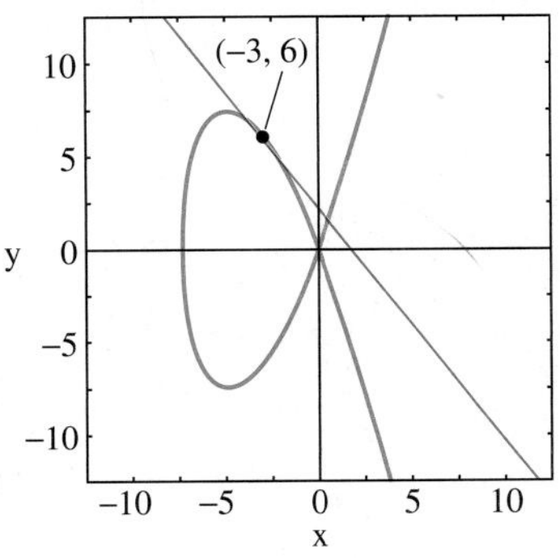

FIGURE 3.9.12 Problem 28.

29. The curve $x^3 + y^3 = 9xy$ is similar in shape and appearance to the folium of Descartes in Fig. 3.9.4. Find (a) the equation of its tangent line at the point (2, 4) and (b) the equation of its tangent line with slope -1.

30. (a) Factor the left-hand side of the equation

$$2x^2 - 5xy + 2y^2 = 0$$

to show that its graph consists of two straight lines through the origin. Hence the derivative $y'(x)$ has only two possible numerical values (the slopes of these two lines). (b) Calculate dy/dx by implicit differentiation of the equation in part (a). Verify that the expression you obtain yields the proper slope for each of the straight lines of part (a).

31. Find all points on the graph of $x^2 + y^2 = 4x + 4y$ at which the tangent line is horizontal.

32. Find the first-quadrant points of the folium of Example 3 at which the tangent line is either horizontal ($dy/dx = 0$) or vertical (where $dx/dy = 1/(dy/dx) = 0$).

33. Figure 3.9.13 shows the graph of the equation $x = ye^y$. Show first that explicit differentiation to find dx/dy and implicit differentiation to find dy/dx yield consistent results. Then find the equation of the line tangent to the graph at the point (a) $(0, 0)$; (b) $(e, 1)$.

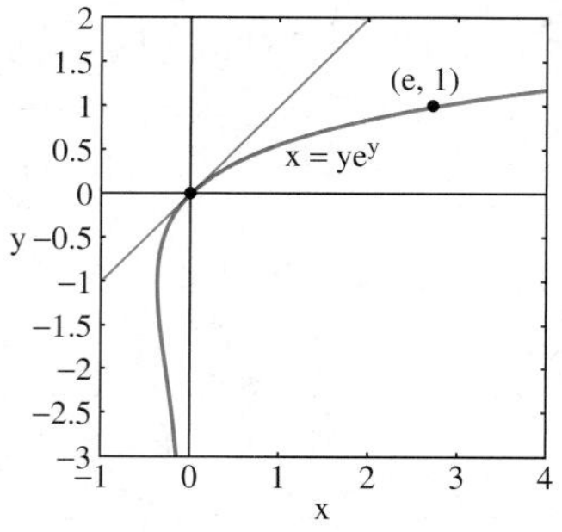

FIGURE 3.9.13 The curve $x = ye^y$ and its tangent line at the origin.

34. (a) Find the points on the curve $x = ye^y$ of Fig. 3.9.13 where the tangent line is vertical ($dx/dy = 0$). (b) Is there a point on the curve where the tangent line is horizontal? (c) Show that $x \to 0$ and $dy/dx \to -\infty$ as $y \to -\infty$. (d) Show that $y \to +\infty$ and $dy/dx \to 0$ as $x \to +\infty$.

35. The graph in Fig. 3.9.14 is a *lemniscate* with equation $(x^2 + y^2)^2 = x^2 - y^2$. Find by implicit differentiation the four points on the lemniscate where the tangent line is horizontal. Then find the two points where the tangent line is vertical—that is, where $dx/dy = 1/(dy/dx) = 0$.

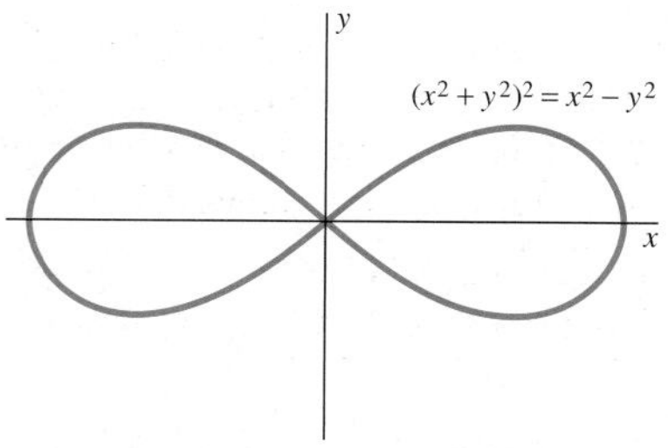

FIGURE 3.9.14 The lemniscate of Problem 35.

36. Water is being collected from a block of ice with a square base (Fig. 3.9.15). The water is produced because the ice is melting in such a way that each edge of the base of the block is decreasing at 2 in./h while the height of the block is decreasing at 3 in./h. What is the rate of flow of water into the collecting pan when the base has edge length 20 in. and the height of the block is 15 in.? Make the simplifying assumption that water and ice have the same density.

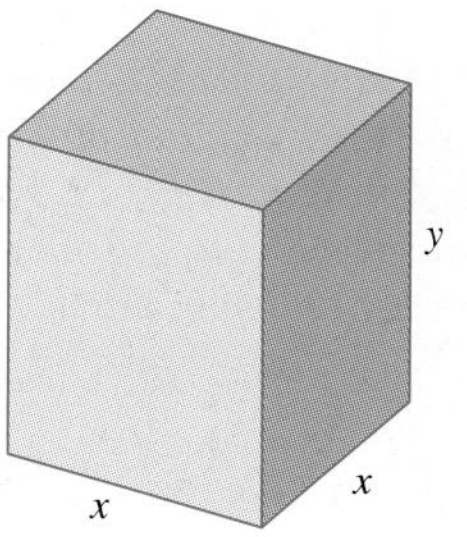

FIGURE 3.9.15 The ice block of Problem 36.

37. Sand is being emptied from a hopper at the rate of 10 ft^3/s. The sand forms a conical pile whose height is always twice its radius (Fig. 3.9.16). At what rate is the radius of the pile increasing when its height is 5 ft?

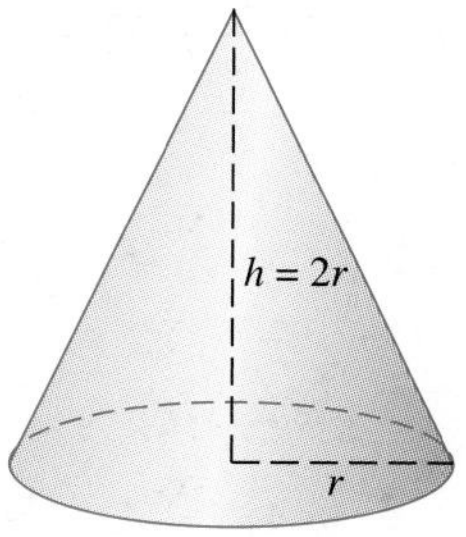

FIGURE 3.9.16 The conical sand pile of Problem 37 with volume $V = \frac{1}{3}\pi r^2 h$.

38. Suppose that water is being emptied from a spherical tank of radius 10 ft (Fig. 3.9.17). If the depth of the water in the tank is 5 ft and is decreasing at the rate of 3 ft/s, at what rate is the radius r of the top surface of the water decreasing?

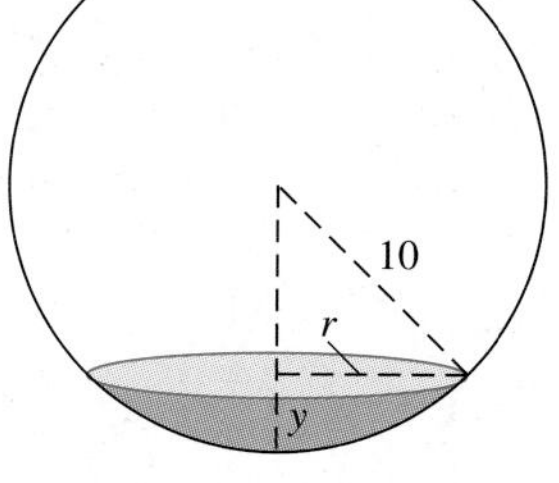

FIGURE 3.9.17 The spherical tank of Problem 38.

39. A circular oil slick of uniform thickness is caused by a spill of 1 m^3 of oil. The thickness of the oil slick is decreasing at the rate of 0.1 cm/h. At what rate is the radius of the slick increasing when the radius is 8 m?

40. Suppose that an ostrich 5 ft tall is walking at a speed of 4 ft/s directly toward a street light 10 ft high. How fast is the tip of the ostrich's shadow moving along the ground? At what rate is the ostrich's shadow decreasing in length?

41. The width of a rectangle is half its length. At what rate is its area increasing if its width is 10 cm and is increasing at 0.5 cm/s?

42. At what rate is the area of an equilateral triangle increasing if its base is 10 cm long and is increasing at 0.5 cm/s?

43. A gas balloon is being filled at the rate of 100π cm^3 of gas per second. At what rate is the radius of the balloon increasing when its radius is 10 cm?

44. The volume V (in cubic inches) and pressure p (in pounds per square inch) of a certain gas satisfy the equation $pV = 1000$. At what rate is the volume of the sample changing if the pressure is 100 lb/in.^{2} and is increasing at the rate of 2 lb/in.^{2} per second?

45. Figure 3.9.18 shows a kite in the air at an altitude of 400 ft. The kite is being blown horizontally at the rate of 10 ft/s away from the person holding the kite string at ground level. At what rate is the string being payed out when 500 ft of string is already out? (Assume that the string forms a straight line.)

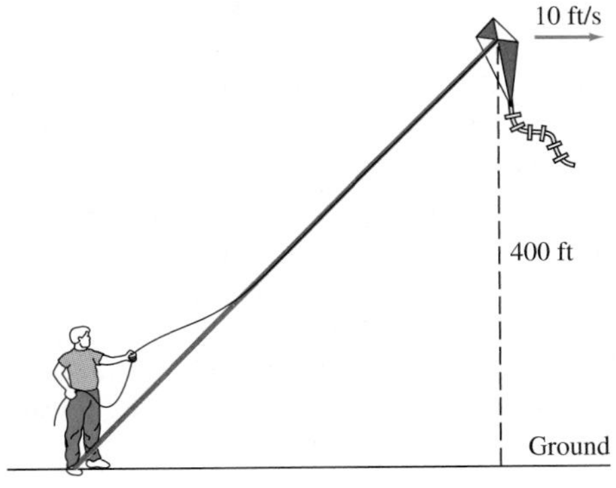

FIGURE 3.9.18 The kite of Problem 45.

46. A weather balloon that is rising vertically is being observed from a point on the ground 300 ft from the spot directly beneath the balloon. At what rate is the balloon rising when the angle between the ground and the observer's line of sight is 45° and is increasing at 1° per second?

47. An airplane flying horizontally at an altitude of 3 mi and at a speed of 480 mi/h passes directly above an observer on the ground. How fast is the distance from the observer to the airplane increasing 30 s later?

48. Figure 3.9.19 shows a spherical tank of radius a partly filled with water. The maximum depth of water in the tank is y. A formula for the volume V of water in the tank—a formula you can derive after you study Chapter 6—is $V = \frac{1}{3}\pi y^2(3a - y)$. Suppose that water is being drained from a spherical tank of radius 5 ft at the rate of 100 gal/min. Find the rate at which the depth y of water is decreasing when (a) $y = 7$ (ft); (b) $y = 3$ (ft). [*Note:* One gallon of water occupies a volume of approximately 0.1337 ft^3.]

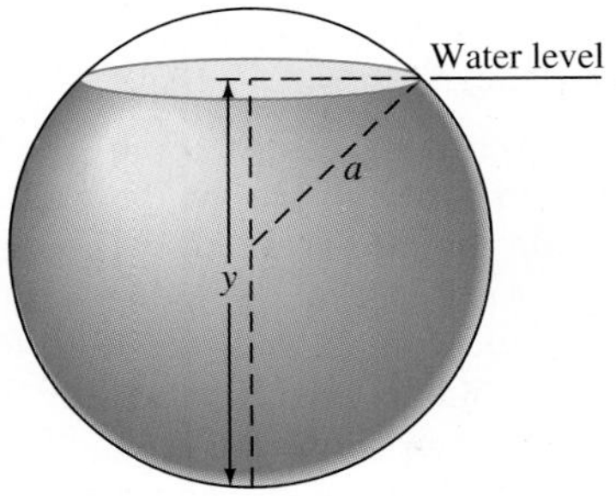

FIGURE 3.9.19 The spherical water tank of Problem 48.

49. Repeat Problem 48, but use a tank that is hemispherical, flat side on top, with radius 10 ft.

50. A swimming pool is 50 ft long and 20 ft wide. Its depth varies uniformly from 2 ft at the shallow end to 12 ft at the deep end (Fig. 3.9.20). Suppose that the pool is being filled at the rate of 1000 gal/min. At what rate is the depth of water at the deep end increasing when the depth there is 6 ft? [*Note:* One gallon of water occupies a volume of approximately 0.1337 ft^3.]

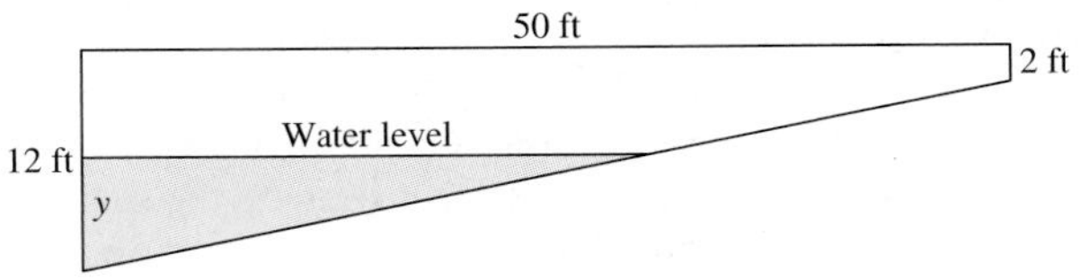

FIGURE 3.9.20 Cross section of the swimming pool of Problem 50.

51. A ladder 41 ft long that was leaning against a vertical wall begins to slip. Its top slides down the wall while its bottom moves along the level ground at a constant speed of 4 ft/s. How fast is the top of the ladder moving when it is 9 ft above the ground?

52. The base of a rectangle is increasing at 4 cm/s while its height is decreasing at 3 cm/s. At what rate is its area changing when its base is 20 cm and its height is 12 cm?

53. The height of a cone is decreasing at 3 cm/s while its radius is increasing at 2 cm/s. When the radius is 4 cm and the height is 6 cm, is the volume of the cone increasing or decreasing? At what rate is the volume changing then?

54. A square is expanding. When each edge is 10 in., its area is increasing at 120 in.^{2}/s. At what rate is the length of each edge changing then?

55. A rocket that is launched vertically is tracked by a radar station located on the ground 4 mi from the launch site. What is the vertical speed of the rocket at the instant its distance from the radar station is 5 mi and this distance is increasing at the rate of 3600 mi/h?

56. Two straight roads intersect at right angles. At 10 A.M. a car passes through the intersection headed due east at 30 mi/h. At 11 A.M. a truck heading due north at 40 mi/h passes through the intersection. Assume that the two vehicles maintain the given speeds and directions. At what rate are they separating at 1 P.M.?

57. A 10-ft ladder is leaning against a wall. The bottom of the ladder begins to slide away from the wall at a speed of 1 mi/h. (a) Find the rate at which the top of the latter is moving when it is 4 ft from the ground. If the top of the ladder maintained contact with the wall, find the speed with which it would be moving when it is (b) 1 in. above the ground; (c) 1 mm above the ground. Do you believe your answers? The key to the apparent paradox is that when the top of the ladder is about 1.65 ft high, it disengages altogether from the wall and thereafter slides away from it.

58. Two ships are sailing toward a very small island. One ship, the Pinta, is east of the island and is sailing due west at 15 mi/h. The other ship, the Niña, is north of the island and is sailing due south at 20 mi/h. At a certain time the Pinta is 30 mi from the island and the Niña is 40 mi from the island. At what rate are the two ships drawing closer together at that time?

59. At time $t = 0$, a single-engine military jet is flying due east at 12 mi/min. At the same altitude and 208 mi directly ahead of the military jet, still at time $t = 0$, a commercial jet is flying due north at 8 mi/min. When are the two planes closest to each other? What is the minimum distance between them?

60. A ship with a long anchor chain is anchored in 11 fathoms of water. The anchor chain is being wound in at the rate of 10 fathoms/min, causing the ship to move toward the spot directly above the anchor resting on the seabed. The hawsehole—the point of contact between ship and chain—is located 1 fathom above the water line. At what speed is the ship moving when there are exactly 13 fathoms of chain still out?

61. A water tank is in the shape of a cone with vertical axis and vertex downward. The tank has radius 3 ft and is 5 ft high. At first the tank is full of water, but at time $t = 0$ (in seconds), a small hole at the vertex is opened and the water begins to drain. When the height of the water in the tank has dropped to 3 ft, the water is flowing out at 2 ft^3/s. At what rate, in feet per second, is the water level dropping then?

62. A spherical tank of radius 10 ft is being filled with water at the rate of 200 gal/min. How fast is the water level rising when the maximum depth of water in the tank is 5 ft? See Problem 48 for a useful formula and a helpful note.

63. A water bucket is shaped like the frustum of a cone with height 2 ft, base radius 6 in., and top radius 12 in. Water is leaking from the bucket at 10 in.^{3}/min. At what rate is the water level falling when the depth of water in the bucket is 1 ft? [*Note:* The volume V of a conical frustum with height h and base radii a and b is

$$V = \frac{\pi h}{3}(a^2 + ab + b^2).$$

Such a frustum is shown in Fig. 3.9.21.]

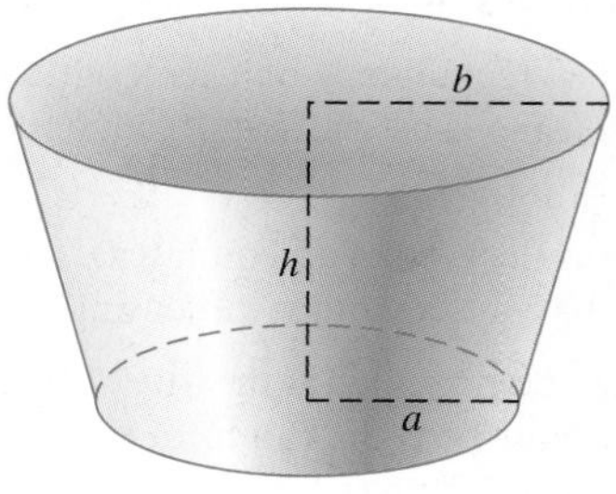

FIGURE 3.9.21 The volume of this conical frustum is given in Problem 63.

64. Suppose that the radar stations A and B of Example 8 are now 12.6 km apart. At a certain instant, a ship is 10.4 km from A and its distance from A is increasing at 19.2 km/h. At the same instant, its distance from B is 5 km and is decreasing at 0.6 km/h. Find the location, speed, and direction of motion of the ship.

65. An airplane climbing at an angle of 45° passes directly over a ground radar station at an altitude of 1 mi. A later reading shows that the distance from the radar station to the plane is 5 mi and is increasing at 7 mi/min. What is the speed of the plane then (in miles per hour)? [*Suggestion:* You may find the law of cosines useful—see Appendix C.]

66. The water tank of Problem 62 is completely full when a plug at its bottom is removed. According to *Torricelli's law*, the water drains in such a way that $dV/dt = -k\sqrt{y}$, where V is the volume of water in the tank and k is a positive empirical constant. (a) Find dy/dt as a function of the depth y. (b) Find the depth of water when the water level is falling the *least* rapidly. (You will need to compute the derivative of dy/dt with respect to y.)

67. A person 6 ft tall walks at 5 ft/s along one edge of a road 30 ft wide. On the other edge of the road is a light atop a pole 18 ft high. How fast is the length of the person's shadow (on the horizontal ground) increasing when the person is 40 ft from the point directly across the road from the pole?

68. A highway patrol officer's radar unit is parked behind a billboard 200 ft from a long straight stretch of U.S. 17. Down the highway, 200 ft from the point on the highway closest to the officer, is an emergency call box. The officer points the radar gun at the call box. A minivan passes the call box and, at that moment, the radar unit indicates that the *distance between the officer and the minivan* is increasing at 45 mi/h—that is, 66 ft/s. The posted speed limit is 55 mi/h. Does the officer have any reason to apprehend the driver of the minivan?

3.9 Project: Investigating the Folium of Descartes

Computer graphics often requires lots of mathematics, and much mathematics was used in constructing many of the figures in this book. To see one way to construct Fig. 3.9.4, use a computer algebra system to solve the equation $x^3 + y^3 = 3xy$ for y in terms of x. Verify that the three expressions you get define three different functions f, g, and h whose graphs are the three branches of the curve that are colored differently in Fig. 3.9.22. Investigate the domains of definition and the graphs of these functions to verify that they fit together precisely as shown in the figure.

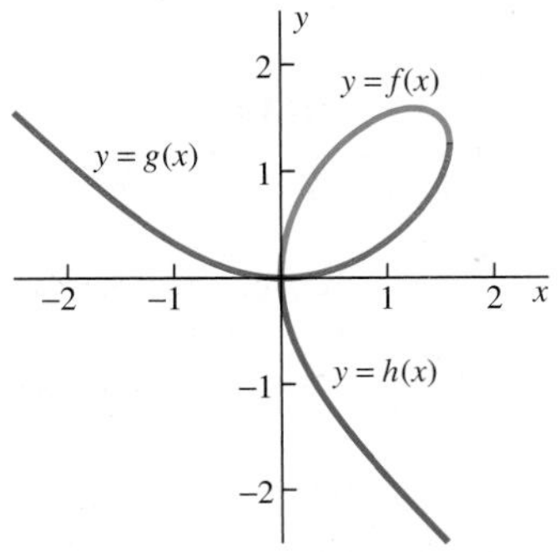

FIGURE 3.9.22 The equation $x^3 + y^3 = 3xy$ implicitly defines three functions f, g, and h.

3.10 SUCCESSIVE APPROXIMATIONS AND NEWTON'S METHOD

The solution of equations has always been a central task of mathematics. More than two millennia ago, mathematicians of ancient Babylon discovered the method of "completing the square," which leads to the *quadratic formula* for an exact solution of any second-degree equation $ax^2 + bx + c = 0$. Early in the sixteenth century, several Italian mathematicians (Cardan, del Ferro, Ferrari, and Tartaglia) discovered formulas for the exact solutions of third- and fourth-degree equations. (Because they are quite complicated, these formulas are seldom used today except in computer algebra systems.) And in 1824 a brilliant young Norwegian mathematician, Niels Henrik Abel* (1802–1829), published a proof that there is *no* general formula giving the solution of an arbitrary polynomial equation of degree 5 (or higher) in terms of algebraic combinations of its coefficients. Thus the exact solution (for all its roots) of an equation such as

$$f(x) = x^5 - 3x^3 + x^2 - 23x + 19 = 0 \quad \textbf{(1)}$$

may be quite difficult or even—as a practical matter—impossible to find. In such a case it may be necessary to resort to *approximate methods.*

For example, the graph of $y = f(x)$ in Fig. 3.10.1 indicates that Eq. (1) has three real solutions (and hence two complex ones as well). The indicated small rectangle $0.5 \leqq x \leqq 1$, $-5 \leqq y \leqq 5$ encloses one of these solutions. If we use this small rectangle as a new "viewing window" with a computer or graphics calculator, then we see that this solution is near 0.8 (Fig. 3.10.2). A few additional magnifications might yield greater accuracy, showing that the solution is approximately 0.801.

*For the complete story of Abel's remarkable achievements in his brief lifetime, see Oystein Ore's very readable biography *Niels Henrik Abel* (The University of Minnesota and Chelsea Publishing Company, 1974).

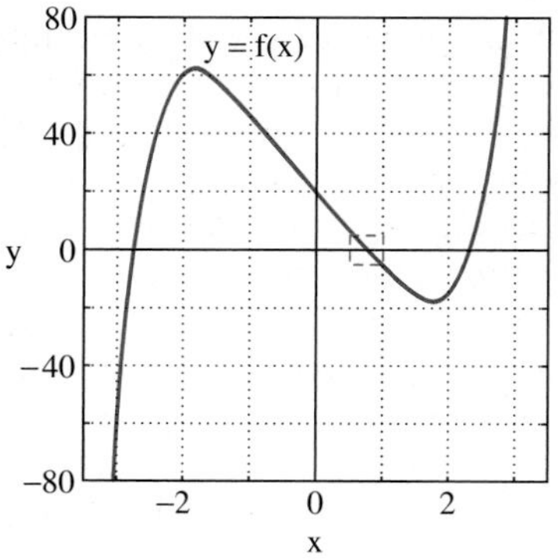

FIGURE 3.10.1 The graph $y = f(x)$ in Eq. (1).

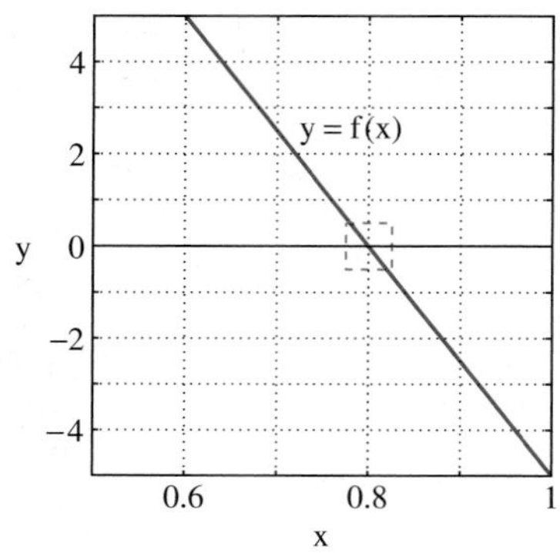

FIGURE 3.10.2 Magnification of Fig. 3.10.1 near a solution.

Graphical methods are good for three- or four-place approximations. Here we shall discuss an analytical method developed by Isaac Newton that can rapidly provide much more accurate approximations.

Iteration and the Babylonian Square Root Method

What it means to solve even so simple an equation as

$$x^2 - 2 = 0 \tag{2}$$

is open to question. The positive exact solution is $x = \sqrt{2}$. But the number $\sqrt{2}$ is irrational and hence cannot be expressed as a terminating or repeating decimal. Thus if we mean by a *solution* an exact decimal value for x, then even Eq. (2) can be solved only approximately.

The ancient Babylonians devised an effective way to generate a sequence of better and better approximations to $\sqrt{A}$, the square root of a given positive number A. Here is the *Babylonian square root method:* We begin with a first guess x_0 for the value of $\sqrt{A}$. For $\sqrt{2}$, we might guess $x_0 = 1.5$. If x_0 is too large—that is, if $x_0 > \sqrt{A}$—then

$$\frac{A}{x_0} < \frac{A}{\sqrt{A}} = \sqrt{A},$$

so A/x_0 is too small an estimate of $\sqrt{A}$. Similarly, if x_0 is too small (if $x_0 < \sqrt{A}$), then A/x_0 is too large an estimate of $\sqrt{A}$; that is, $A/x_0 > \sqrt{A}$.

Thus in each case one of the two numbers x_0 and A/x_0 is an underestimate of $\sqrt{A}$ and the other is an overestimate. The Babylonian idea was that we should get a better estimate of $\sqrt{A}$ by *averaging* x_0 and A/x_0. This yields a better approximation

$$x_1 = \frac{1}{2}\left(x_0 + \frac{A}{x_0}\right) \tag{3}$$

to $\sqrt{A}$. But why not repeat this process? We can average x_1 and A/x_1 to get a second approximation x_2, average x_2 and A/x_2 to get x_3, and so on. By repeating this process, we generate a sequence of numbers

$$x_1, \quad x_2, \quad x_3, \quad x_4, \quad \ldots$$

that we have every right to expect will consist of better and better approximations to $\sqrt{A}$.

Specifically, having calculated the nth approximation x_n, we calculate the next one by means of the *iterative formula*

$$x_{n+1} = \frac{1}{2}\left(x_n + \frac{A}{x_n}\right). \tag{4}$$

In other words, we plow each approximation to $\sqrt{A}$ back into the right-hand side in Eq. (4) to calculate the next approximation. This is an *iterative* process—the words *iteration* and *iterative* are derived from the Latin *iterare*, "to plow again."

Suppose we find that after sufficiently many steps in this iteration, $x_{n+1} \approx x_n$ accurate to the number of decimal places we are retaining in our computations. Then Eq. (4) yields

$$x_n \approx x_{n+1} = \frac{1}{2}\left(x_n + \frac{A}{x_n}\right) = \frac{1}{2x_n}(x_n^2 + A),$$

so $2x_n^2 \approx x_n^2 + A$, and hence $x_n^2 \approx A$ to some degree of accuracy.

EXAMPLE 1 With $A = 2$ we begin with the crude first guess $x_0 = 1$ to the value of $\sqrt{A}$. Then successive applications of the formula in Eq. (4) yield

$$x_1 = \frac{1}{2}\left(1 + \frac{2}{1}\right) = \frac{3}{2} = 1.5,$$

$$x_2 = \frac{1}{2}\left(\frac{3}{2} + \frac{2}{3/2}\right) = \frac{17}{12} \approx 1.416666667,$$

$$x_3 = \frac{1}{2}\left(\frac{17}{12} + \frac{2}{17/12}\right) = \frac{577}{408} \approx 1.414215686,$$

$$x_4 = \frac{1}{2}\left(\frac{577}{408} + \frac{2}{577/408}\right) = \frac{665857}{470832} \approx 1.414213562,$$

rounding results to nine decimal places. It happens that x_4 gives $\sqrt{2}$ accurate to all nine places! ◆

The Babylonian iteration defined in Eq. (4) is a method for generating a sequence of approximations to the positive root $r = \sqrt{A}$ of the particular equation $x^2 - A = 0$. We turn next to a method that gives such a sequence of approximations for more general equations.

Newton's Method

Newton's method is an iterative method for generating a sequence $x_1, x_2, x_3, \dots$ of approximations to a solution r of a given equation written in the general form

$$f(x) = 0. \tag{5}$$

We hope that this sequence of approximations will "converge" to the root r in the sense of the following definition.

DEFINITION Convergence of Approximations
We say that the sequence of approximations $x_1, x_2, x_3, \dots$ **converges** to the number r provided that we can make x_n as close to r as we please merely by choosing n sufficiently large. More precisely, for any given $\epsilon > 0$, there exists a positive integer N such that $|x_n - r| < \epsilon$ for all $n \geqq N$.

As a practical matter such convergence means, as illustrated in Example 1, that for any positive integer k, x_n and r will agree to k or more decimal places once n becomes sufficiently large.

The idea is that we begin with an *initial guess* x_0 that roughly approximates a solution r of the equation $f(x) = 0$. This initial guess may, for example, be obtained by inspection of the graph of $y = f(x)$, perhaps obtained from a computer or graphics calculator. We use x_0 to calculate an approximation x_1, use x_1 to calculate a better approximation x_2, use x_2 to calculate a still better approximation x_3, and so on.

Here is the general step in the process. Having reached the nth approximation x_n, we use the tangent line at $(x_n, f(x_n))$ to construct the next approximation x_{n+1} to the solution r as follows: Begin at the point x_n on the x-axis. Go vertically up (or down) to the point $(x_n, f(x_n))$ on the curve $y = f(x)$. Then follow the tangent line L there to the point where L meets the x-axis (Fig. 3.10.3). That point will be x_{n+1}.

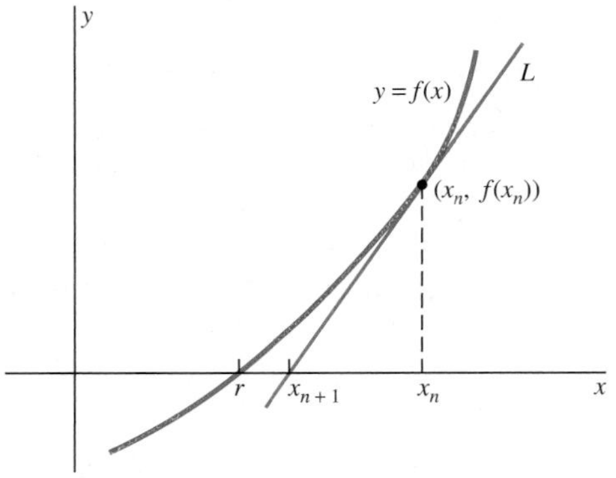

FIGURE 3.10.3 Geometry of the formula of Newton's method.

Here is a formula for x_{n+1}. We obtain it by computing the slope of the line L in two ways: from the derivative and from the two-point definition of slope. Thus

$$f'(x_n) = \frac{f(x_n) - 0}{x_n - x_{n+1}},$$

and we easily solve for

$$x_{n+1} = x_n - \frac{f(x_n)}{f'(x_n)}. \tag{6}$$

This equation is the **iterative formula** of Newton's method, so called because in about 1669, Newton introduced an algebraic procedure (rather than the geometric construction illustrated in Fig. 3.10.3) that is equivalent to the iterative use of Eq. (6). Newton's first example was the cubic equation $x^3 - 2x - 5 = 0$, for which he found the root $r \approx 2.0946$ (as we ask you to do in Problem 18).

Suppose now that we want to apply Newton's method to solve the equation

$$f(x) = 0 \tag{7}$$

to an accuracy of k decimal places (k digits to the right of the decimal correct or correctly rounded). Remember that an equation must be written precisely in the form of Eq. (7) in order to use the formula in Eq. (6). If we reach the point in our iteration at which x_n and x_{n+1} agree to k decimal places, it then follows that

$$x_n \approx x_{n+1} = x_n - \frac{f(x_n)}{f'(x_n)}; \qquad 0 \approx -\frac{f(x_n)}{f'(x_n)}; \qquad f(x_n) \approx 0.$$

Thus we have found an approximate root $x_n \approx x_{n+1}$ of Eq. (7). In practice, then, we retain k decimal places in our computations and persist until $x_n \approx x_{n+1}$ to this degree of accuracy. (We do not consider here the possibility of round-off error, an important topic in numerical analysis.)

EXAMPLE 2 Use Newton's method to find $\sqrt{2}$ accurate to nine decimal places.

Solution More generally, consider the square root of the positive number A as the positive root of the equation

$$f(x) = x^2 - A = 0.$$

Because $f'(x) = 2x$, Eq. (6) gives the iterative formula

$$x_{n+1} = x_n - \frac{x_n^2 - A}{2x_n} = \frac{1}{2}\left(x_n + \frac{A}{x_n}\right). \tag{8}$$

Thus we have derived the Babylonian iterative formula as a special case of Newton's method. The use of Eq. (8) with $A = 2$ therefore yields exactly the values of x_1, x_2, x_3, and x_4 that we computed in Example 1, and after performing another iteration we find that

$$x_5 = \frac{1}{2}\left(x_4 + \frac{2}{x_4}\right) \approx 1.414213562,$$

which agrees with x_4 to nine decimal places. The very rapid convergence here is an important characteristic of Newton's method. As a general rule (with some exceptions), each iteration doubles the number of decimal places of accuracy. ◆

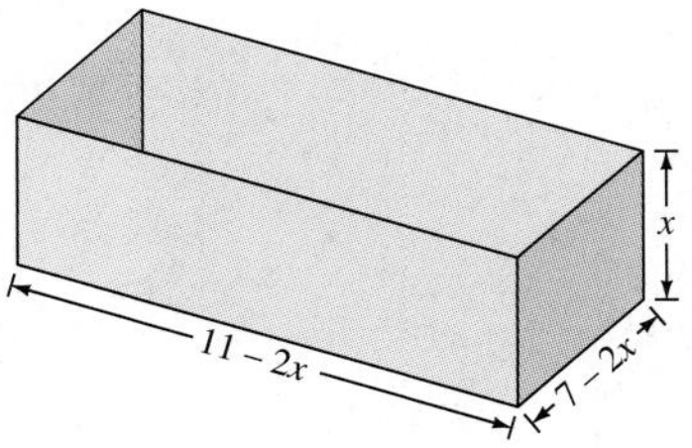

FIGURE 3.10.4 The tray of Example 3.

EXAMPLE 3 Figure 3.10.4 shows an open-topped tray constructed by the method of Example 2 in Section 3.6. We begin with a 7-by-11-in. rectangle of sheet metal. We cut a square with edge length x from each of its four corners and then fold up the resulting flaps to obtain a rectangular tray with volume

$$\begin{aligned} V(x) &= x(7 - 2x)(11 - 2x) \\ &= 4x^3 - 36x^2 + 77x, \quad 0 \leqq x \leqq 3.5. \end{aligned} \tag{9}$$

In Section 3.6 we inquired about the maximum possible volume of such a tray. Here we want to find instead the value(s) of x that will yield a tray with volume 40 in.^{3}; we will find x by solving the equation

$$V(x) = 4x^3 - 36x^2 + 77x = 40.$$

To solve this equation for x, first we write an equation of the form in Eq. (7):

$$f(x) = 4x^3 - 36x^2 + 77x - 40 = 0. \tag{10}$$

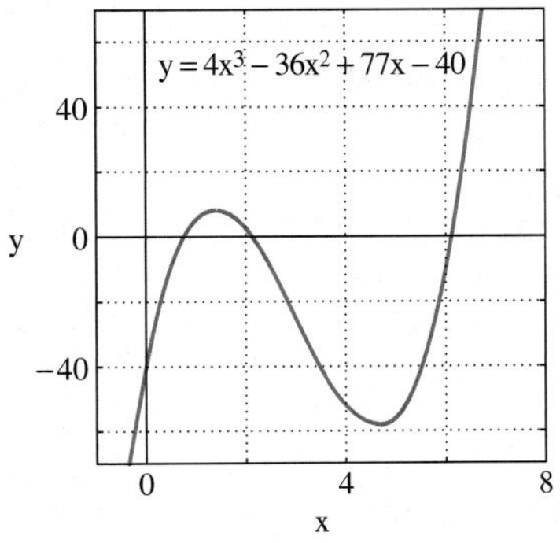

FIGURE 3.10.5 The graph of $f(x)$ in Eq. 10 of Example 3.

Figure 3.10.5 shows the graph of f. We see three solutions: a root r_1 between 0 and 1, a root r_2 slightly greater than 2, and a root r_3 slightly larger than 6. Because

$$f'(x) = 12x^2 - 72x + 77,$$

Newton's iterative formula in Eq. (6) takes the form

$$\begin{aligned} x_{n+1} &= x_n - \frac{f(x_n)}{f'(x_n)} \\ &= x_n - \frac{4x_n^3 - 36x_n^2 + 77x_n - 40}{12x_n^2 - 72x_n + 77}. \end{aligned} \tag{11}$$

Beginning with the initial guess $x_0 = 1$ (because it's reasonably close to r_1), Eq. (11) gives

$$\begin{aligned} x_1 &= 1 - \frac{4 \cdot 1^3 - 36 \cdot 1^2 + 77 \cdot 1 - 40}{12 \cdot 1^2 - 72 \cdot 1 + 77} \approx 0.7059, \\ x_2 &\approx 0.7736, \\ x_3 &\approx 0.7780, \\ x_4 &\approx 0.7780. \end{aligned}$$

Thus we obtain the root $r_1 \approx 0.7780$, retaining only four decimal places.

If we had begun with a different initial guess, the sequence of Newton iterates might well have converged to a different root of the equation $f(x) = 0$. The approximate solution obtained therefore depends on the initial guess. For example, with

$x_0 = 2$ and, later, with $x_0 = 6$, the iteration in Eq. (11) produces the two sequences

$x_0 = 2$	$x_0 = 6$
$x_1 \approx 2.1053$	$x_1 \approx 6.1299$
$x_2 \approx 2.0993$	$x_2 \approx 6.1228$
$x_3 \approx 2.0992$	$x_3 \approx 6.1227$
$x_4 \approx 2.0992$	$x_4 \approx 6.1227$

Thus the other two roots of Eq. (10) are $r_2 \approx 2.0992$ and $r_3 \approx 6.1277$.

With $x = r_1 \approx 0.7780$, the tray in Fig. 3.10.4 has the approximate dimensions 9.4440 in. by 5.4440 in. by 0.7780 in. With $x = r_2 \approx 2.0992$, its approximate dimensions are 6.8015 in. by 2.8015 in. by 2.0992 in. But the third root $r_3 \approx 6.1227$ would *not* lead to a tray that is physically possible. (Why not?) Thus the *two* values of x that yield trays with volume 40 in. are $x \approx 0.7780$ and $x \approx 2.0992$. ◆

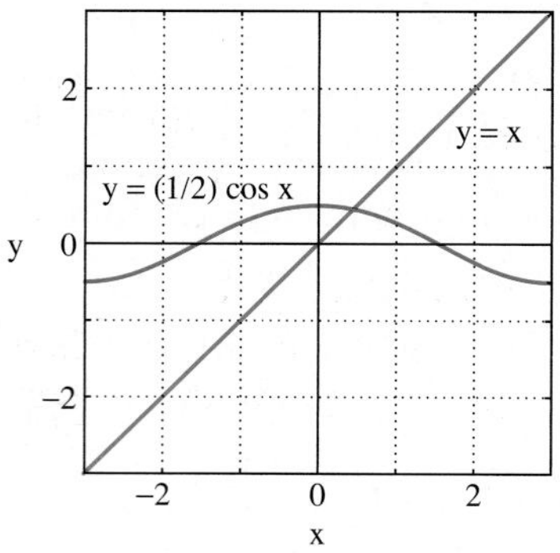

FIGURE 3.10.6 Solving the equation $x = \frac{1}{2}\cos x$ (Example 4).

EXAMPLE 4 Figure 3.10.6 indicates that the equation

$$x = \tfrac{1}{2}\cos x \tag{12}$$

has a solution r near 0.5. To apply Newton's method to approximate r, we rewrite Eq. (12) in the form

$$f(x) = 2x - \cos x = 0.$$

Because $f'(x) = 2 + \sin x$, the iterative formula of Newton's method is

$$x_{n+1} = x_n - \frac{2x_n - \cos x_n}{2 + \sin x_n}.$$

Beginning with $x_0 = 0.5$ and retaining five decimal places, this formula yields

$$x_1 \approx 0.45063, \qquad x_2 \approx 0.45018, \qquad x_3 \approx 0.45018.$$

Thus the root is 0.45018 to five decimal places. ◆

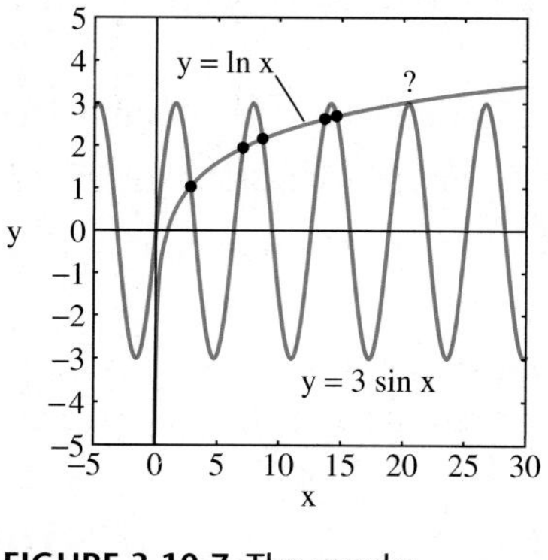

FIGURE 3.10.7 The graphs $y = 3\sin x$ and $y = \ln x$.

EXAMPLE 5 Figure 3.10.7 indicates that the equation

$$3\sin x = \ln x$$

has either five or six positive solutions. To better approximate the smallest solution $r \approx 3$, we apply Newton's method with

$$f(x) = 3\sin x - \ln x, \qquad f'(x) = 3\cos x - \frac{1}{x}.$$

Then the iterative formula of Newton's method is

$$x_{n+1} = x_n - \frac{3\sin x_n - \ln x_n}{3\cos x_n - (1/x_n)}.$$

When we begin with $x_0 = 3$ and retain five decimal places, this formula gives

$$x_1 \approx 2.79558, \qquad x_2 \approx 2.79225, \qquad x_3 \approx 2.79225.$$

Thus $r \approx 2.79225$ to five decimal places. In Problem 42 we ask you to find the remaining solutions indicated in Fig. 3.10.7. ◆

EXAMPLE 6 Newton's method is one for which "the proof is in the pudding." If it works, it's obvious that it does, and everything's fine. When Newton's method fails, it may do so spectacularly. For example, suppose that we want to solve the equation

$$x^{1/3} = 0.$$

Here $r = 0$ is the only solution. The iterative formula in Eq. (6) becomes

$$x_{n+1} = x_n - \frac{(x_n)^{1/3}}{\frac{1}{3}(x_n)^{-2/3}} = x_n - 3x_n = -2x_n.$$

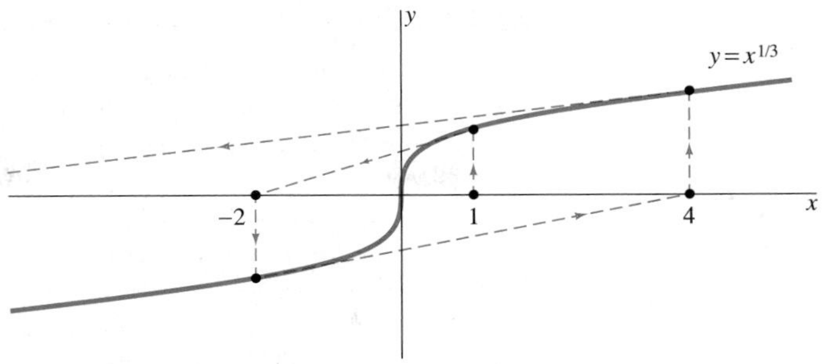

FIGURE 3.10.8 A failure of Newton's method.

If we begin with $x_0 = 1$, Newton's method yields $x_1 = -2$, $x_2 = +4$, $x_3 = -8$, and so on. Figure 3.10.8 indicates why our "approximations" are not converging. ◆

When Newton's method fails, a graph will typically indicate the reason why. Then the use of an alternative method such as repeated tabulation or successive magnification is appropriate.

Newton's Method with Calculators and Computers

With calculators and computers that permit user-defined functions, Newton's method is very easy to set up and apply repeatedly. It is helpful to interpret Newton's iteration

$$x_{n+1} = x_n - \frac{f(x_n)}{f'(x_n)}$$

as follows. Having first defined the functions f and f', we then define the "iteration function"

$$g(x) = x - \frac{f(x)}{f'(x)}.$$

Newton's method is then equivalent to the following procedure. Begin with an initial estimate x_0 of the solution of the equation

$$f(x) = 0.$$

Calculate successive approximations $x_1, x_2, x_3, \ldots$ to the exact solution by means of the iteration

$$x_{n+1} = g(x_n).$$

That is, apply the function g to each approximation to get the next.

Figure 3.10.9 shows a TI graphics calculator prepared to solve the equation

$$f(x) = x^3 - 3x^2 + 1 = 0.$$

Then we need only store the initial guess, **0.5** $\longrightarrow$ **x**, and next enter repeatedly the command **y3** $\longrightarrow$ **x**, as indicated in Fig. 3.10.10.

Figure 3.10.11 shows an HP calculator prepared to carry out the same iteration. The functions **F(X)**, **D(X)** (for $f'(x)$), and **G(X)** are each defined by pressing the

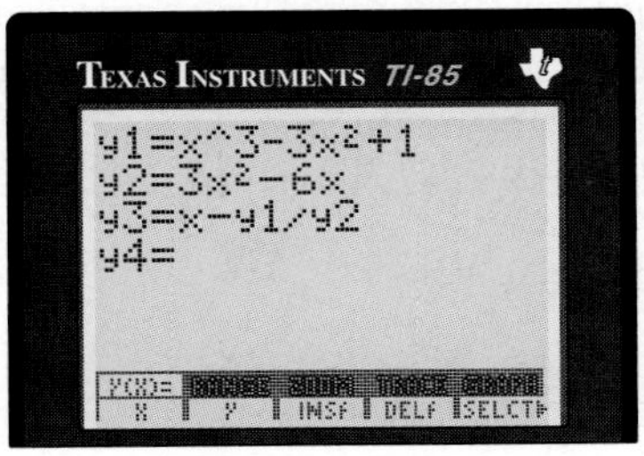

FIGURE 3.10.9 Preparing to solve the equation $x^3 - 3x^2 + 1 = 0$.

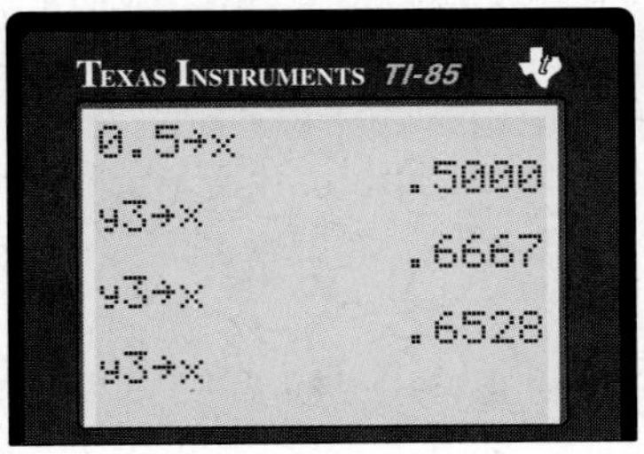

FIGURE 3.10.10 Solving the equation $x^3 - 3x^2 + 1 = 0$.

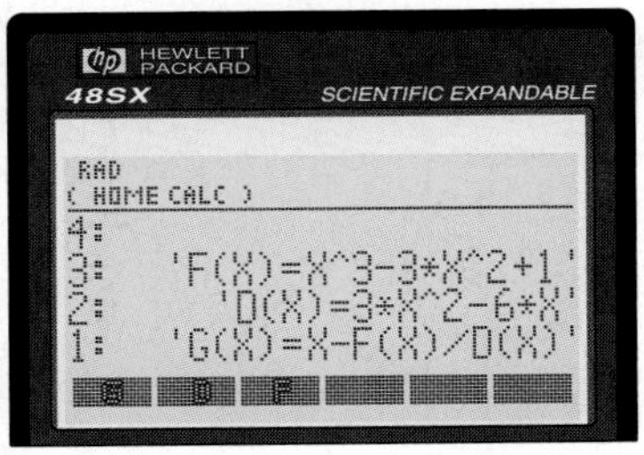

FIGURE 3.10.11 Preparing to solve the equation $x^3 - 3x^2 + 1 = 0$.

DEFINE key. Then it is necessary only to **ENTER** the initial guess x_0 and press the **G** key repeatedly to generate the desired successive appropriations.

With *Maple* or *Mathematica* you can define the functions f and g and then repeatedly enter the command $x = g(x)$, as shown in Fig. 3.10.12.

Mathematica Command	*Maple* Command	Result
f[x_] := x^3 − 3x^2 + 1	f := x −> x^3 − 3 * x^2 + 1;	
g[x_] := x − f[x]/f′[x]	g := x −> x − f(x)/D(f)(x);	
x = 0.5	x := 0.5;	0.500000
x = g[x]	x := g(x);	0.666667
x = g[x]	x := g(x);	0.652778
x = g[x]	x := g(x);	0.652704
x = g[x]	x := g(x);	0.652704

FIGURE 3.10.12 *Mathematica* and *Maple* implementations of Newton's method.

Newton's Method and Computer Graphics

Newton's method and similar iterative techniques are often used to generate vividly colored "fractal patterns," in which the same or similar structures are replicated on smaller and smaller scales at successively higher levels of magnification. To describe one way this can be done, we replace the real numbers in our Newton's method computations with *complex* numbers. We illustrate this idea with the cubic equation

$$f(x) = x^3 - 3x^2 + 1 = 0. \tag{13}$$

In the Project we ask you to approximate the three solutions

$$r_1 \approx -0.53, \qquad r_2 \approx 0.65, \qquad r_3 \approx 2.88$$

of this equation.

First, recall that a *complex number* is a number of the form $a + bi$, where $i = \sqrt{-1}$, so $i^2 = -1$. The real numbers a and b are called the *real part* and the *imaginary part*, respectively, of $a + bi$. You add, multiply, and divide complex numbers as if they were binomials, with real and imaginary parts "collected" as in the computations

$$(3 + 4i) + (5 - 7i) = (3 + 5) + (4 - 7)i = 8 - 3i,$$

$$\begin{aligned}(2 + 5i)(3 - 4i) &= 2(3 - 4i) + 5i(3 - 4i)\\ &= 6 - 8i + 15i - 20i^2 = 26 + 7i,\end{aligned}$$

and

$$\frac{2 + 5i}{3 + 4i} = \frac{2 + 5i}{3 + 4i} \cdot \frac{3 - 4i}{3 - 4i} = \frac{26 + 7i}{9 - 16i^2} = \frac{26 + 7i}{25} = 1.04 + (0.28)i.$$

The use of the *conjugate* $3 - 4i$ of the denominator $3 + 4i$ in the last computation is a very common technique for writing a complex fraction in the standard form $a + bi$. (The **conjugate** of $x + yi$ is $x - yi$; it follows that the conjugate of $x - yi$ is $x + yi$.)

Now let us substitute the complex number $z = x + iy$ into the cubic polynomial

$$f(z) = z^3 - 3z^2 + 1$$

of Eq. (13) and into its derivative $f'(z) = 3z^2 - 6z$. We find that

$$\begin{aligned}f(z) &= (x + iy)^3 - 3(x + iy)^2 + 1\\ &= (x^3 - 3xy^2 - 3x^2 + 3y^2 + 1) + (3x^2y - y^3 - 6xy)i\end{aligned} \tag{14}$$

and

$$f'(z) = 3(x + iy)^2 - 6(x + iy)$$
$$= (3x^2 - 3y^2 - 6x) + (6xy - 6y)i. \qquad \textbf{(15)}$$

Consequently, there is nothing to prevent us from applying Newton's method to Eq. (13) with complex numbers. Beginning with a *complex* initial guess $z_0 = x_0 + iy_0$, we can substitute Eqs. (14) and (15) into Newton's iterative formula

$$z_{n+1} = z_n - \frac{f(z_n)}{f'(z_n)} \qquad \textbf{(16)}$$

to generate the complex sequence $\{z_n\}$, which may yet converge to a (real) solution of Eq. (13).

With this preparation, we can now explain how Fig. 3.10.13 was generated. A computer was programmed to carry out Newton's iteration repeatedly, beginning with many thousands of initial guesses $z_0 = x_0 + iy_0$ that "fill" the rectangle $-2 \leqq x \leqq 4$, $-2.25 \leqq y \leqq 2.25$ in the complex plane. The initial point $z_0 = x_0 + iy_0$ was then color-coded according to the root (if any) to which the corresponding sequence $\{z_n\}$ converged:

Color z_0 green if $\{z_n\}$ converges to the root $r_1 \approx -0.53$;

Color z_0 red if $\{z_n\}$ converges to the root $r_2 \approx 0.65$;

Color z_0 yellow if $\{z_n\}$ converges to the root $r_3 \approx 2.88$.

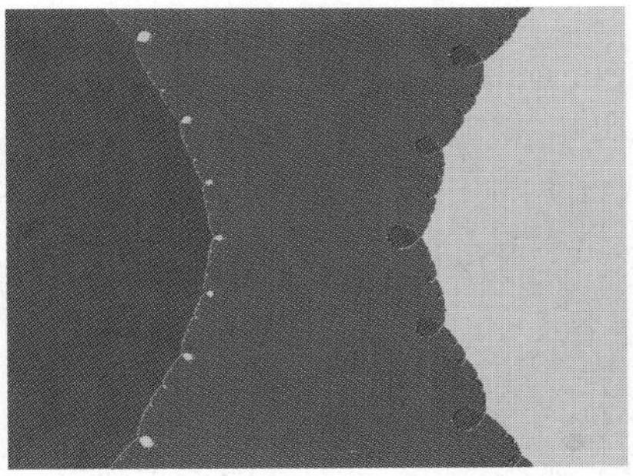

FIGURE 3.10.13 $-2 \leqq x \leqq 4$, $-2.25 \leqq y \leqq 2.25$.

Thus we use different colors to distinguish different "*Newton basins* of attraction" for the equation we are investigating. It is not surprising that a red region containing the root r_2 appears in the middle of Fig. 3.10.13, separating a green region to the left that contains r_1 and a yellow region to the right that contains r_3. But why would yellow lobes protrude from the green region into the red region and green lobes protrude from the yellow region into the red one? To see what's happening near these lobes, we generated some blowups.

Figure 3.10.14 shows a blowup of the rectangle $1.6 \leqq x \leqq 2.4$, $-0.3 \leqq y \leqq 0.3$ containing the green lobe that's visible in Fig. 3.10.13. Figure 3.10.15 ($1.64 \leqq x \leqq 1.68$, $-0.015 \leqq y \leqq 0.015$) and Fig. 3.10.16 ($1.648 \leqq x \leqq 1.650$, $-0.00075 \leqq y \leqq 0.00075$) are further magnifications. The rectangle shown in Fig. 3.10.16 corresponds to less than one millionth of a square inch of Fig. 3.10.13.

At every level of magnification, each green lobe has smaller yellow lobes protruding into the surrounding red region, and each of these yellow lobes has still smaller green lobes protruding from it, and so on ad infinitum (just like the proverbial little fleas that are bitten by still smaller fleas, and so on ad infinitum).

Figure 3.10.17 shows the Newton basins picture for the twelfth-degree polynomial equation

$$f(x) = x^{12} - 14x^{10} + 183x^8 - 612x^6 - 2209x^4 - 35374x^2 + 38025 = 0, \qquad \textbf{(17)}$$

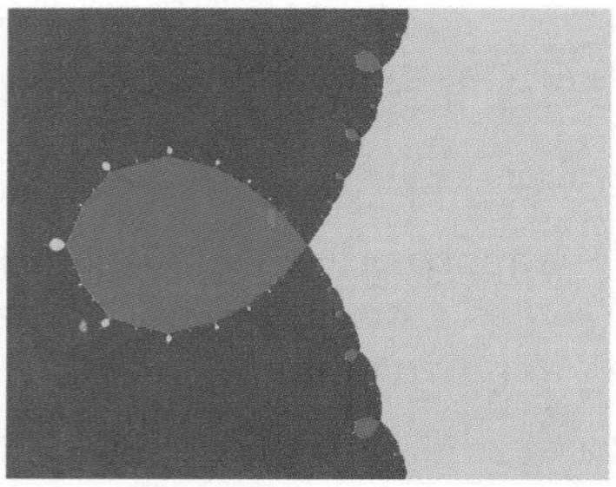

FIGURE 3.10.14 $1.6 \leqq x \leqq 2.4$, $-0.3 \leqq y \leqq 0.3$.

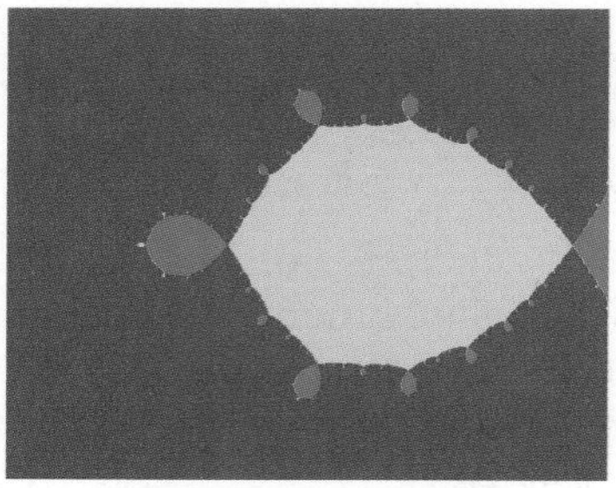

FIGURE 3.10.15 $1.64 \leqq x \leqq 1.68$, $-0.015 \leqq y \leqq 0.015$.

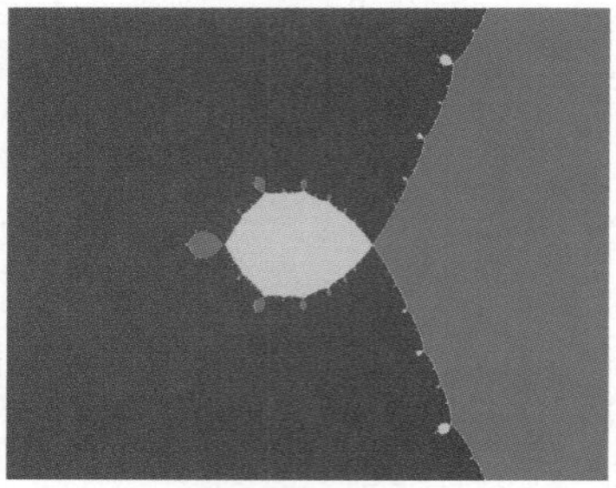

FIGURE 3.10.16 $1.648 \leqq x \leqq 1.650$, $-0.00075 \leqq y \leqq 0.00075$.

which has as its solution the twelve complex numbers

$$1, \quad 1 \pm 2i, \quad -1, \quad -1 \pm 2i,$$
$$3, \quad 3 \pm 2i, \quad -3, \quad -3 \pm 2i.$$

Twelve different colors are used to distinguish the Newton basins of these twelve solutions of Eq. (17).

Where the fractal common boundary appears to separate basins of different colors, it is studded with "flowers" like the one at the center of Fig. 3.10.17, which is magnified in Fig. 3.10.18. Each of these flowers has ten "leaves" (in the remaining ten colors). Each of these leaves has "buds" like the one shown in Fig. 3.10.19. Each of these buds is encircled with flowers that have leaves that have buds that are encircled with flowers—and so on ad infinitum.

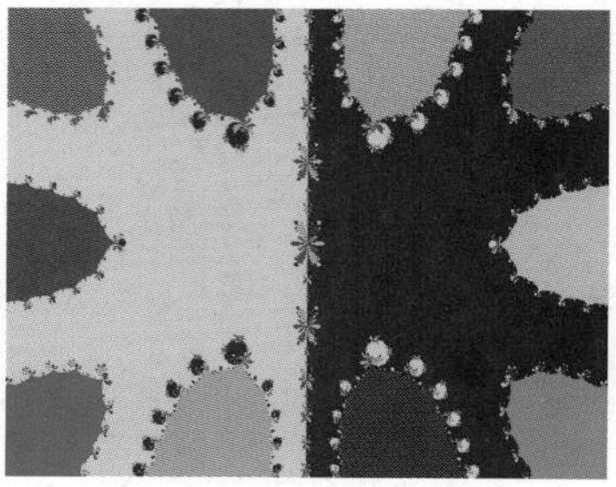

FIGURE 3.10.17 Newton basis for the twelfth-degree polynomial.

FIGURE 3.10.18 The flower at the center of Fig. 3.10.17.

FIGURE 3.10.19 A bud on a petal of the flower in Fig. 3.10.18.

3.10 TRUE/FALSE STUDY GUIDE

3.10 CONCEPTS: QUESTIONS AND DISCUSSION

1. Example 1 in this section illustrates the use of Babylonian iteration to approximate the square root of a positive number A, beginning with a positive initial guess x_0. How is the number of iterations required for six-place accuracy affected by choosing x_0 very close to zero, or very large? Does it appear that the number of decimal places of accuracy is roughly doubled with each iteration? What happens if a negative initial guess is used? What happens if A itself is negative?
2. The general rule—that each iteration of Newton's method typically doubles the number of decimal places of accuracy—does not hold when the method is used to approximate a solution r of $f(x) = 0$ if r is also a critical point of f. Investigate the "rate of convergence" to the root r if: (a) $f(x) = (x - 2)^2$, so that $r = 2$ is a double root; (b) $f(x) = (x - 1)^{2/3}$, so that the graph has a cusp and a vertical tangent at $r = 1$.

3. Consider the exotic function $f(x) = x^2 \sin(1/x)$ [with $f(0) = 0$] of Problem 88 in Section 3.7. Investigate what happens when you use a computer algebra system to attempt to approximate the root $r = 0$ by iterating $g(x) = x - f(x)/f'(x)$. Try a variety of different nonzero initial guesses and explain the results.

3.10 PROBLEMS

In Problems 1 through 20, use Newton's method to find the solution of the given equation $f(x) = 0$ in the indicated interval $[a, b]$ accurate to four decimal places. You may choose the initial guess either on the basis of a calculator graph or by interpolation between the values $f(a)$ and $f(b)$.

1. $x^2 - 5 = 0$; $[2, 3]$ (to find the positive square root of 5)

2. $x^3 - 2 = 0$; $[1, 2]$ (to find the cube root of 2)

3. $x^5 - 100 = 0$; $[2, 3]$ (to find the fifth root of 100)

4. $x^{3/2} - 10 = 0$; $[4, 5]$ (to find $10^{2/3}$)

5. $x^2 + 3x - 1 = 0$; $[0, 1]$

6. $x^3 + 4x - 1 = 0$; $[0, 1]$

7. $x^6 + 7x^2 - 4 = 0$; $[-1, 0]$

8. $x^3 + 3x^2 + 2x = 10$; $[1, 2]$

9. $x - \cos x = 0$; $[0, 2]$

10. $x^2 - \sin x = 0$; $[0.5, 1.0]$

11. $4x - \sin x = 4$; $[1, 2]$

12. $5x + \cos x = 5$; $[0, 1]$

13. $x^5 + x^4 = 100$; $[2, 3]$

14. $x^5 + 2x^4 + 4x = 5$; $[0, 1]$

15. $x + \tan x = 0$; $[2, 3]$

16. $x + \tan x = 0$; $[11, 12]$

17. $x - e^{-x} = 0$; $[0, 1]$

18. $x^3 - 2x - 5 = 0$; $[2, 3]$ (Newton's own example)

19. $e^x + x - 2 = 0$; $[0, 1]$

20. $e^{-x} - \ln x = 0$; $[1, 2]$

21. (a) Show that Newton's method applied to the equation $x^3 - a = 0$ yields the formula

$$x_{n+1} = \frac{1}{3}\left(2x_n + \frac{a}{x_n^2}\right)$$

for approximating the cube root of a. (b) Use this formula to find $\sqrt[3]{2}$ accurate to five decimal places.

22. (a) Show that Newton's method yields the formula

$$x_{n+1} = \frac{1}{k}\left[(k-1)x_n + \frac{a}{(x_n)^{k-1}}\right]$$

for approximating the kth root of the positive number a. (b) Use this formula to find $\sqrt[10]{100}$ accurate to five decimal places.

23. Equation (12) has the special form $x = G(x)$, where $G(x) = \frac{1}{2}\cos x$. For an equation of this form, the iterative formula $x_{n+1} = G(x_n)$ produces a sequence of approximations that *sometimes* converges to a root. In the case of Eq. (12), this *repeated substitution* formula is simply $x_{n+1} = \frac{1}{2}\cos x_n$. Begin with $x_0 = 0.5$ as in Example 4 and retain five decimal places in your computation of the solution of Eq. (12). [*Check:* You should find that $x_8 \approx 0.45018$.]

24. The equation $x^4 = x + 1$ has a solution between $x = 1$ and $x = 2$. Use the initial guess $x_0 = 1.5$ and the method of repeated substitution (see Problem 23) to discover that one of the solutions of this equation is approximately 1.220744. Iterate using the formula

$$x_{n+1} = (x_n + 1)^{1/4}.$$

Then compare the result with what happens when you iterate using the formula

$$x_{n+1} = (x_n)^4 - 1.$$

25. The equation $x^3 - 3x^2 + 1 = 0$ has a solution between $x = 0$ and $x = 1$. To apply the method of repeated substitution (see Problem 23) to this equation, you may write it either in the form

$$x = 3 - \frac{1}{x^2}$$

or in the form

$$x = (3x^2 - 1)^{1/3}.$$

If you begin with $x_0 = 0.5$ in the hope of finding the nearby solution (approximately 0.6527) of the original equation by using each of the preceding iterative formulas, you will observe some of the drawbacks of the method. Describe what goes wrong.

26. Show that Newton's method applied to the equation

$$\frac{1}{x} - a = 0$$

yields the iterative formula

$$x_{n+1} = 2x_n - a(x_n)^2$$

and thus provides a method for approximating the reciprocal $1/a$ without performing any divisions. Such a method is useful because, in most high-speed computers, the operation of division is more time consuming than even several additions and multiplications.

27. Prove that the equation $x^5 + x = 1$ has exactly one real solution. Then use Newton's method to find it with four places correct to the right of the decimal point.

In Problems 28 through 30, use Newton's method to find all real roots of the given equation with four digits correct to the right of the decimal point. [Suggestion: In order to determine the number of roots and their approximate locations, graph the left- and right-hand sides of each equation and observe where the graphs cross.]

28. $x^2 = \cos x$

29. $x = 2\sin x$

30. $\cos x = -\frac{1}{5}x$ (There are exactly three solutions, as indicated in Fig. 3.10.20.)

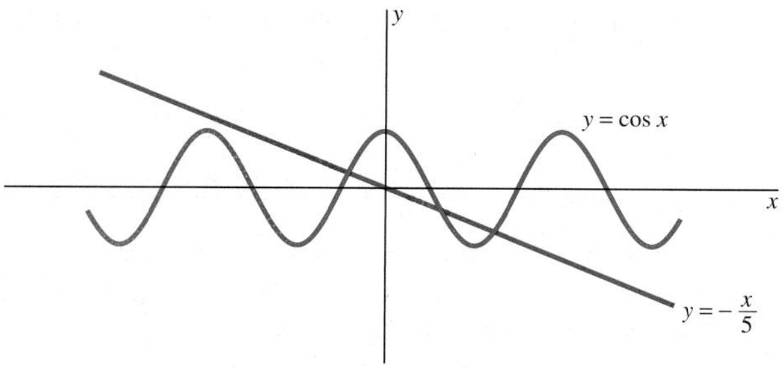

FIGURE 3.10.20 Solving the equation in Problem 30.

31. Prove that the equation $x^7 - 3x^3 + 1 = 0$ has at least one solution. Then use Newton's method to find one solution to three-place accuracy.

32. Use Newton's method to approximate $\sqrt[3]{5}$ to four-place accuracy.

33. Use Newton's method to find the value of x for which $x^3 = \cos x$.

34. Use Newton's method to find the smallest positive value of x for which $x = \tan x$.

35. In Problem 49 of Section 3.6, we dealt with the problem of minimizing the cost of building a road to two points on opposite sides of a geologic fault. This problem led to the equation

$$f(x) = 3x^4 - 24x^3 + 51x^2 - 32x + 64 = 0.$$

Use Newton's method to find, to four-place accuracy, the root of this equation that lies in the interval $[3, 4]$.

36. The moon of Planet Gzyx has an elliptical orbit with eccentricity 0.5, and its period of revolution around the planet is 10 days. If the moon is at the position $(a, 0)$ when $t = 0$, then (Fig. 3.10.21) the central angle after t days is given by *Kepler's equation*

$$\frac{2\pi t}{100} = \theta - \frac{1}{2}\sin\theta.$$

Use Newton's method to solve for θ when $t = 17$ (days). Take $\theta_0 = 1.5$ (rad) and calculate the first two approximations θ_1 and θ_2. Express θ_2 in degrees as well.

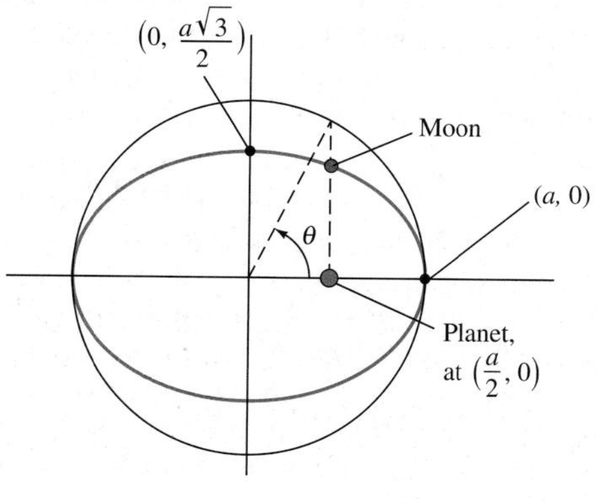

FIGURE 3.10.21 The elliptical orbit of Problem 36.

37. A great problem of Archimedes was that of using a plane to cut a sphere into two segments with volumes in a given (preassigned) ratio. Archimedes showed that the volume of a segment of height h of a sphere of radius a is $V = \frac{1}{3}\pi h^2(3a - h)$. If a plane at distance x from the center of a sphere of radius 1 cuts the sphere into two segments, one with twice the volume of the other, show that $3x^3 - 9x + 2 = 0$. Then use Newton's method to find x accurate to four decimal places.

38. The equation $f(x) = x^3 - 4x + 1 = 0$ has three distinct real roots. Approximate their locations by evaluating f at $x = -3, -2, -1, 0, 1, 2,$ and 3. Then use Newton's method to approximate each of the three roots to four-place accuracy.

39. The equation $x + \tan x = 0$ is important in a variety of applications—for example, in the study of the diffusion of heat. It has a sequence $\alpha_1, \alpha_2, \alpha_3, \dots$ of positive roots, with the nth root slightly larger than $(n - 0.5)\pi$. Use Newton's method to compute α_1 and α_2 to three-place accuracy.

40. Investigate the cubic equation

$$4x^3 - 42x^2 - 19x - 28 = 0.$$

Perhaps you can see graphically that it has only a single real solution. Find it (accurate to four decimal places). First try the initial guess $x_0 = 0$; be prepared for at least 25 iterations. Then try initial guesses $x_0 = 10$ and $x_0 = 100$.

41. A 15-ft ladder and a 20-ft ladder lean in opposite directions against the vertical walls of a hall (Fig. 3.10.22). The ladders cross at a height of 5 ft. You must find the width w of the hall. First, let x and y denote the heights of the tops of the ladders on the walls and u and v the lengths shown in the figure, so that $w = u + v$. Use similar triangles to show that

$$x = 5\left(1 + \frac{u}{v}\right), \qquad y = 5\left(1 + \frac{v}{u}\right).$$

Then apply the Pythagorean theorem to show that $t = u/v$ satisfies the equation

$$t^4 + 2t^3 - 7t^2 - 2t - 1 = 0.$$

Finally, use Newton's method to find first the possible values of t, and then those of w, accurate to four decimal places.

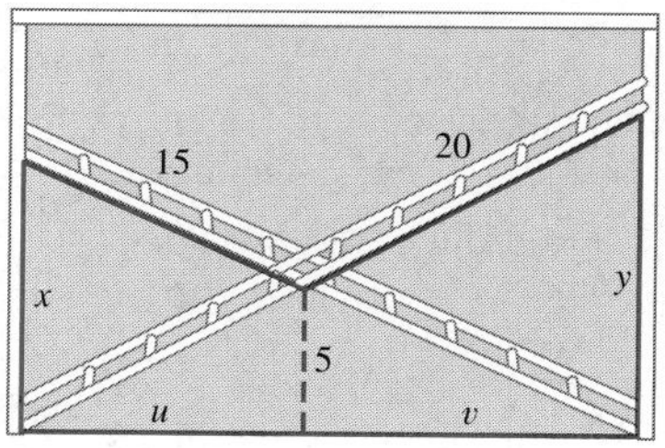

FIGURE 3.10.22 The crossing ladders of Problem 41.

42. Use Newton's method to find the remaining positive solutions of the equation $3\sin x = \ln x$ of Example 5 (Fig. 3.10.7). Do whatever is necessary to determine whether there is or is not a solution near $x = 20$.

43. The spherical asteroid problem in the Section 1.4 Project leads to the equation $(100 + \theta)\cos\theta = 100$, where $R = 1000/\theta$ is the radius of the asteroid, and it is clear from the context that $0 < \theta < \pi/2$. Use Newton's method to solve this problem.

44. This is a famous "railroad track problem." Consider a 1-mile railroad track that was constructed without leaving the usual expansion spaces between consecutive rails. Thus each rail of the track is, in effect, a single steel rail one mile long. Suppose that an increase in the temperature by 20°C increases—by thermal expansion of the steel—the length of this rail by one foot. Also assume that the ends of the track are fixed, so the rail "bows up" in the shape of a circular arc with central angle 2θ and radius R (Fig. 3.10.23). Find the resulting height x (at its midpoint) of the bowed rail above the ground.

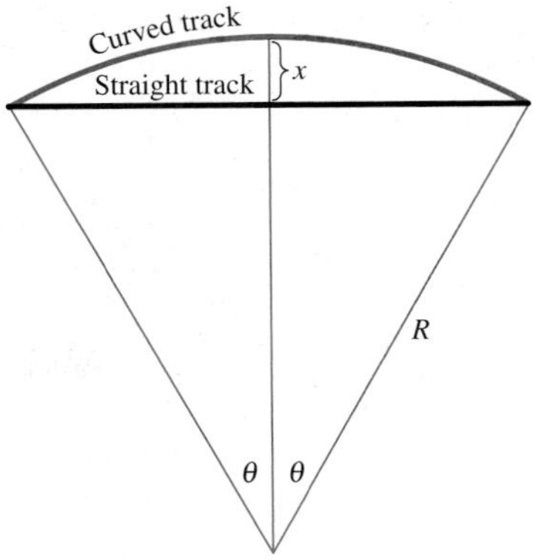

FIGURE 3.10.23 The bowed railroad track of Problem 44.

3.10 Project: How Deep Does a Floating Ball Sink?

Figure 3.10.24 shows a large cork ball of radius $a = 1$ floating in water of density 1. If the ball's density ρ is one-fourth that of water, $\rho = \frac{1}{4}$, then Archimedes' law of buoyancy implies that the ball floats in such a way that one-fourth of its total volume is submerged. Because the volume of the ball is $4\pi/3$, it follows that the volume of the part of the ball beneath the waterline is given by

$$V = \rho \cdot \frac{4\pi}{3} = \frac{1}{4} \cdot \frac{4\pi}{3} = \frac{\pi}{3}. \tag{1}$$

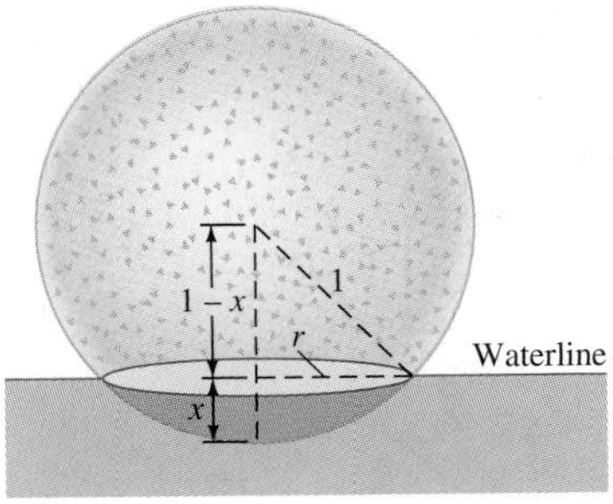

FIGURE 3.10.24 The floating cork ball.

The shape of the submerged part of the ball is that of a **spherical segment** with a circular flat top. The volume of a spherical segment of *top radius* r and *depth* $h = x$ (as in Fig. 3.10.24) is given by the formula

$$V = \frac{\pi x}{6}(3r^2 + x^2). \tag{2}$$

This formula is also due to Archimedes and holds for any depth x, whether the spherical segment is smaller or larger than a hemisphere. For instance, note that with $r = 0$ and $x = 2a$ it gives $V = \frac{4}{3}\pi a^3$, the volume of an entire sphere of radius a.

For a preliminary investigation, proceed as follows to find the depth x to which the ball sinks in the water. Equate the two expressions for V in Eqs. (1) and (2), then use the right triangle in Fig. 3.10.24 to eliminate r. You should find that x must be a solution of the cubic equation

$$f(x) = x^3 - 3x^2 + 1 = 0. \tag{3}$$

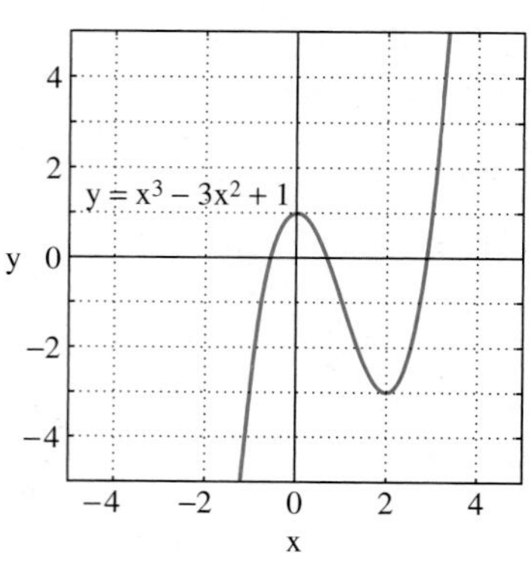

FIGURE 3.10.25 Graph for the cork-ball equation.

As the graph $y = f(x)$ in Fig. 3.10.25 indicates, this equation has three real solutions—one in $(-1, 0)$, one in $(0, 1)$, and one in $(2, 3)$. The solution between 0 and 1 gives the actual depth x to which the ball sinks (why?). You can find x using Newton's method.

Your Investigation For your very own floating ball to investigate, let its density ρ in Eq. (1) be given by

$$\rho = \frac{10 + k}{20}$$

where k denotes the last nonzero digit in the sum of the final four digits of your student I.D. number. Your objective is to find the depth to which this ball sinks in the water. Begin by deriving the cubic equation that you need to solve, explaining each step carefully. Then find all of its solutions accurate to at least four decimal places. Include in your report a sketch of a spherical ball with the waterline located accurately (to scale) in the position corresponding to your result for the desired depth.

CHAPTER 3 REVIEW: FORMULAS, CONCEPTS, DEFINITIONS

Differentiation Formulas

$$D_x(cu) = c\frac{du}{dx}$$

$$D_x(u + v) = \frac{du}{dx} + \frac{dv}{dx}$$

$$D_x(uv) = u\frac{dv}{dx} + v\frac{du}{dx}$$

$$D_x\frac{u}{v} = \frac{v\dfrac{du}{dx} - u\dfrac{dv}{dx}}{v^2}$$

$$D_x g(u) = g'(u)\frac{du}{dx}$$

$$D_x(u^r) = ru^{r-1}\frac{du}{dx}$$

$$D_x \sin u = (\cos u)\frac{du}{dx}$$

$$D_x \cos u = (-\sin u)\frac{du}{dx}$$

$$D_x \tan u = (\sec^2 u)\frac{du}{dx}$$

$$D_x \cot u = (-\csc^2 u)\frac{du}{dx}$$

$$D_x \sec u = (\sec u \tan u)\frac{du}{dx}$$

$$D_x \csc u = (-\csc u \cot u)\frac{du}{dx}$$

$$D_x \ln u = \frac{1}{u}\frac{du}{dx}$$

$$D_x e^u = e^u\frac{du}{dx}$$

Use the following list as a guide to concepts for review.

1. Definition of the derivative
2. Average rate of change of a function
3. Instantaneous rate of change of a function
4. Position function; velocity and acceleration
5. Differential, function, and operator notation for derivatives
6. The power rule
7. The binomial formula
8. Linearity of differentiation
9. The product rule
10. The reciprocal rule
11. The quotient rule
12. The chain rule
13. The generalized power rule
14. Vertical tangent lines
15. Local maxima and minima
16. $f'(c) = 0$ as a necessary condition for local extrema
17. Absolute (or global) extrema
18. Critical points
19. The closed-interval maximum-minimum method
20. Steps in the solution of applied maximum-minimum problems
21. Derivatives of the sine and cosine functions
22. Derivatives of the other four trigonometric functions
23. Derivatives of natural exponential and logarithmic functions
24. The number e
25. Inverse functions and their derivatives
26. Logarithmic differentiation
27. Implicitly defined functions
28. Implicit differentiation
29. Solving related-rates problems
30. Newton's method

CHAPTER 3 MISCELLANEOUS PROBLEMS

Find dy/dx in Problems 1 through 35.

1. $y = x^2 + \dfrac{3}{x^2}$

2. $y^2 = x^2$

3. $y = \sqrt{x} + \dfrac{1}{\sqrt[3]{x}}$

4. $y = (x^2 + 4x)^{5/2}$

5. $y = (x-1)^7(3x+2)^9$

6. $y = \dfrac{x^4 + x^2}{x^2 + x + 1}$

7. $y = \left(3x - \dfrac{1}{2x^2}\right)^4$

8. $y = x^{10}\sin 10x$

9. $xy = 9$

10. $y = \sqrt{\dfrac{1}{5x^6}}$

11. $y = \dfrac{1}{\sqrt{(x^3 - x)^3}}$

12. $y = \sqrt[3]{2x+1}\,\sqrt[5]{3x-2}$

13. $y = \dfrac{1}{1+u^2}$ where $u = \dfrac{1}{1+x^2}$

14. $x^3 = \sin^2 y$

15. $y = \left(\sqrt{x} + \sqrt[3]{2x}\right)^{7/3}$

16. $y = \sqrt{3x^5 - 4x^2}$

17. $y = \dfrac{u+1}{u-1}$, where $u = \sqrt{x+1}$

18. $y = \sin(2\cos 3x)$

19. $x^2y^2 = x + y$

20. $y = \sqrt{1 + \sin\sqrt{x}}$

21. $y = \sqrt{x + \sqrt{2x + \sqrt{3x}}}$

22. $y = \dfrac{x + \sin x}{x^2 + \cos x}$

23. $\sqrt[3]{x} + \sqrt[3]{y} = 4$

24. $x^3 + y^3 = xy$

25. $y = (1 + 2u)^3$, where $u = \dfrac{1}{(1+x)^3}$

26. $y = \cos^2(\sin^2 x)$

27. $y = \sqrt{\dfrac{\sin^2 x}{1 + \cos x}}$

28. $y = \left(1 + \sqrt{x}\right)^3\left(1 - 2\sqrt[3]{x}\right)^4$

29. $y = \dfrac{\cos 2x}{\sqrt{\sin 3x}}$

30. $x^3 - x^2y + xy^2 - y^3 = 4$

31. $y = e^x\cos x$

32. $y = e^{-2x}\sin 3x$

33. $y = [1 + (2 + 3e^x)^{-3/2}]^{2/3}$

34. $y = (e^x + e^{-x})^5$

35. $y = \cos^3\left(\sqrt[3]{1 + \ln x}\right)$

Find the derivatives of the functions defined in Problems 36 through 45.

36. $f(x) = \cos(1 - e^{-x})$

37. $f(x) = \sin^2(e^{-x})$

38. $f(x) = \ln(x + e^{-x})$

39. $f(x) = e^x\cos 2x$

40. $f(x) = e^{-2x}\sin 3x$

41. $g(t) = \ln\left(te^{t^2}\right)$

42. $g(t) = 3(e^t - \ln t)^5$

43. $g(t) = \sin(e^t)\cos(e^{-t})$

44. $f(x) = \dfrac{2+3x}{e^{4x}}$

45. $g(t) = \dfrac{1+e^t}{1-e^t}$

In Problems 46 through 51, find dy/dx by implicit differentiation.

46. $xe^y = y$

47. $\sin(xy) = x$

48. $e^x + e^y = e^{xy}$

49. $x = ye^y$

50. $e^{x-y} = xy$

51. $x\ln y = x + y$

In Problems 52 through 57, find dy/dx by logarithmic differentiation.

52. $y = \sqrt{(x^2-4)\sqrt{2x+1}}$

53. $y = \dfrac{(3-x^2)^{1/2}}{(x^4+1)^{1/4}}$

54. $y = \left[\dfrac{(x+1)(x+2)}{(x^2+1)(x^2+2)}\right]^{1/3}$

55. $y = \sqrt{x+1}\,\sqrt[3]{x+2}\,\sqrt[4]{x+3}$

56. $y = x^{(e^x)}$

57. $y = (\ln x)^{\ln x}$, $x > 1$

In Problems 58 through 61, write an equation of the line tangent to the given curve at the indicated point.

58. $y = \dfrac{x+1}{x-1}$; $(0, -1)$

59. $x = \sin 2y$; $(1, \pi/4)$

60. $x^2 - 3xy + 2y^2 = 0$; $(2, 1)$

61. $y^3 = x^2 + x$; $(0, 0)$

62. If a hemispherical bowl with radius 1 ft is filled with water to a depth of x in., then the volume of water in the bowl is

$$V = \frac{\pi}{3}(36x^2 - x^3) \quad (\text{in.}^3).$$

If the water flows out a hole at the bottom of the bowl at the rate of 36π in.^{3}/s, how fast is x decreasing when $x = 6$ in.?

63. Falling sand forms a conical sandpile. Its height h always remains twice its radius r while both are increasing. If sand is falling onto the pile at the rate of 25π ft^3/min, how fast is r increasing when $r = 5$ ft?

Find the limits in Problems 64 through 69.

64. $\displaystyle\lim_{x\to 0}\frac{x - \tan x}{\sin x}$

65. $\displaystyle\lim_{x\to 0} x\cot 3x$

66. $\displaystyle\lim_{x\to 0}\frac{\sin 2x}{\sin 5x}$

67. $\displaystyle\lim_{x\to 0} x^2\csc 2x\cot 2x$

68. $\displaystyle\lim_{x\to 0} x^2\sin\frac{1}{x^2}$

69. $\displaystyle\lim_{x\to 0^+}\sqrt{x}\sin\frac{1}{x}$

In Problems 70 through 75, identify two functions f and g such that h(x) = f(g(x)). Then apply the chain rule to find h′(x).

70. $h(x) = \sqrt[3]{x + x^4}$

71. $h(x) = \dfrac{1}{\sqrt{x^2 + 25}}$

72. $h(x) = \sqrt{\dfrac{x}{x^2+1}}$

73. $h(x) = \sqrt[3]{(x-1)^5}$

74. $h(x) = \dfrac{(x+1)^{10}}{(x-1)^{10}}$

75. $h(x) = \cos(x^2 + 1)$

76. The period T of oscillation (in seconds) of a simple pendulum of length L (in feet) is given by $T = 2\pi\sqrt{L/32}$. What is the rate of change of T with respect to L when $L = 4$ ft?

77. What is the rate of change of the volume $V = \frac{4}{3}\pi r^3$ of a sphere with respect to its surface area $A = 4\pi r^2$?

78. What is an equation for the straight line through $(1, 0)$ that is tangent to the graph of

$$h(x) = x + \frac{1}{x}$$

at a point in the first quadrant?

79. A rocket is launched vertically upward from a point 3 mi west of an observer on the ground. What is the speed of the rocket when the angle of elevation (from the horizontal) of the observer's line of sight to the rocket is 60° and is increasing at 6° per second?

80. An oil field containing 20 wells has been producing 4000 barrels of oil daily. For each new well drilled, the daily production of each well decreases by 5 barrels. How many new wells should be drilled to maximize the total daily production of the oil field?

81. A triangle is inscribed in a circle of radius R. One side of the triangle coincides with a diameter of the circle. In terms of R, what is the maximum possible area of such a triangle?

82. Five rectangular pieces of sheet metal measure 210 cm by 336 cm each. Equal squares are to be cut from all their corners, and the resulting five cross-shaped pieces of metal are to be folded and welded to form five boxes without tops. The 20 little squares that remain are to be assembled in groups of four into five larger squares, and these five larger squares are to be assembled into a cubical box with no top. What is the maximum possible total volume of the six boxes that are constructed in this way?

83. A mass of clay of volume V is formed into two spheres. For what distribution of clay is the total surface area of the two spheres a maximum? A minimum?

84. A right triangle has legs of lengths 3 m and 4 m. What is the maximum possible area of a rectangle inscribed in the triangle in the "obvious" way—with one corner at the triangle's right angle, two adjacent sides of the rectangle lying on the triangle's legs, and the opposite corner on the hypotenuse?

85. What is the maximum possible volume of a right circular cone inscribed in a sphere of radius R?

86. A farmer has 400 ft of fencing with which to build a rectangular corral. He will use some or even all of an existing straight wall 100 ft long as part of the perimeter of the corral. What is the maximum area that can be enclosed?

87. In one simple model of the spread of a contagious disease among members of a population of M people, the incidence of the disease, measured as the number of new cases per day, is given in terms of the number x of individuals already infected by

$$R(x) = kx(M - x) = kMx - kx^2,$$

where k is a positive constant. How many individuals in the population are infected when the incidence R is the greatest?

88. Three sides of a trapezoid have length L, a constant. What should be the length of the fourth side if the trapezoid is to have maximal area?

89. A box with no top must have a base twice as long as it is wide, and the total surface area of the box is to be 54 ft^2. What is the maximum possible volume of such a box?

90. A small right circular cone is inscribed in a larger one (Fig. 3.MP.1). The larger cone has fixed radius R and fixed altitude H. What is the largest fraction of the volume of the larger cone that the smaller one can occupy?

91. Two vertices of a trapezoid are at $(-2, 0)$ and $(2, 0)$, and the other two lie on the semicircle $x^2 + y^2 = 4$, $y \geq 0$. What is the maximum possible area of the trapezoid? [*Note:* The area of a trapezoid with bases b_1 and b_2 and height h is $A = h(b_1 + b_2)/2$.]

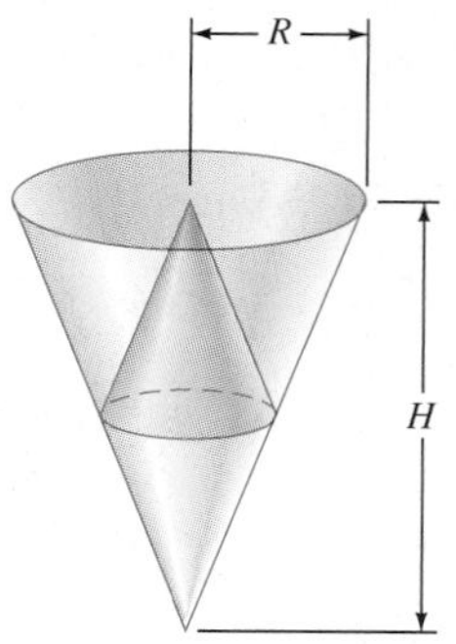

FIGURE 3.MP.1 A small cone inscribed in a larger one (Problem 90).

92. Suppose that f is a differentiable function defined on the whole real number line $\mathbf{R}$ and that the graph of f contains a point $Q(x, y)$ closest to the point $P(x_0, y_0)$ not on the graph. Show that

$$f'(x) = -\frac{x - x_0}{y - y_0}$$

at Q. Conclude that the segment PQ is perpendicular to the line tangent to the curve at Q. [*Suggestion:* Minimize the square of the distance PQ.]

93. Use the result of Problem 92 to show that the minimum distance from the point (x_0, y_0) to a point of the straight line $Ax + By + C = 0$ is

$$\frac{|Ax_0 + By_0 + C|}{\sqrt{A^2 + B^2}}.$$

94. A race track is to be built in the shape of two parallel and equal straightaways connected by semicircles on each end (Fig. 3.MP.2). The length of the track, one lap, is to be exactly 4 km. What should its design be to maximize the rectangular area within it?

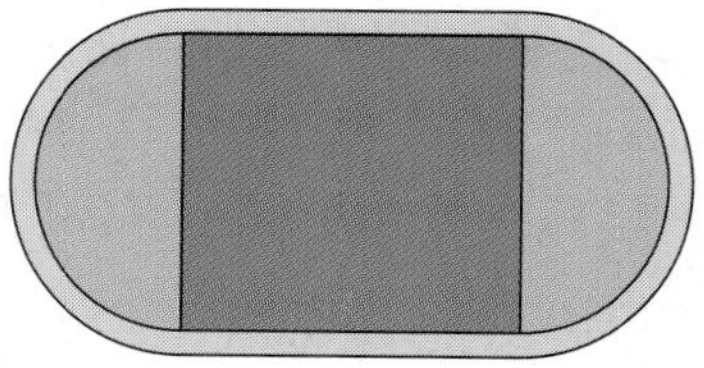

FIGURE 3.MP.2 Design the race track to maximize the rectangular area (Problem 94).

95. Two towns are located near the straight shore of a lake. Their nearest distances to points on the shore are 1 mi and 2 mi, respectively, and these points on the shore are 6 mi apart. Where should a fishing pier be located to minimize the total amount of paving necessary to build a straight road from each town to the pier?

96. A hiker finds herself in a forest 2 km from a long straight road. She wants to walk to her cabin, which is 10 km away in the forest and also 2 km from the road (Fig. 3.MP.3). She can walk at a rate of 8 km/h along the road but only 3 km/h through the forest. So she decides to walk first to the road, then along the road, and finally through the forest to the cabin. What angle θ (shown in the figure) would minimize the total time required for the hiker to reach her cabin? How much time is saved in comparison with the straight route through the forest?

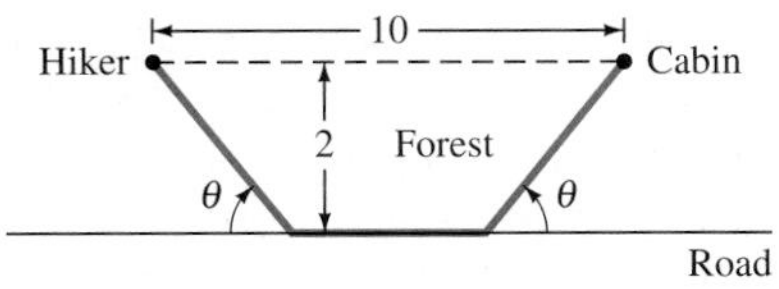

FIGURE 3.MP.3 The hiker's quickest path to the cabin (Problem 96).

97. When an arrow is shot from the origin with initial velocity v and initial angle of inclination α (from the horizontal x-axis, which represents the ground), then its trajectory is the curve

$$y = mx - \frac{16}{v^2}(1 + m^2)x^2,$$

where $m = \tan\alpha$. (a) Find the maximum height reached by the arrow in terms of m and v. (b) For what value of m (and hence, for what α) does the arrow travel the greatest horizontal distance?

98. A projectile is fired with initial velocity v and angle of elevation θ from the base of a plane inclined at 45° from the horizontal (Fig. 3.MP.4). The range of the projectile, as measured up this slope, is given by

$$R = \frac{v^2\sqrt{2}}{16}(\cos\theta\sin\theta - \cos^2\theta).$$

What value of θ maximizes R?

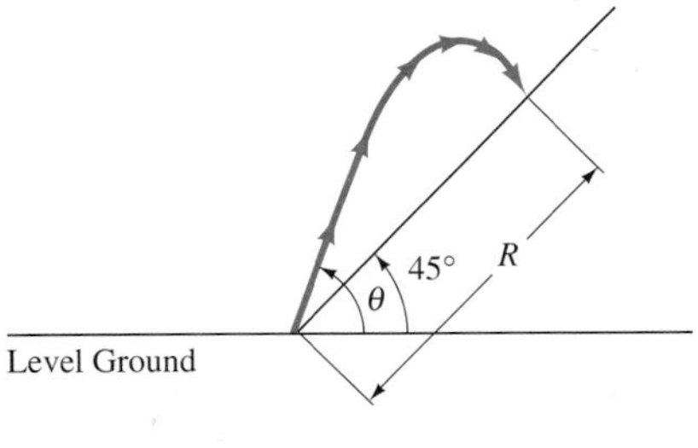

FIGURE 3.MP.4 A projectile fired uphill (Problem 98).

In Problems 99 through 110, use Newton's method to find the solution of the given equation $f(x) = 0$ in the indicated interval $[a, b]$ accurate to four decimal places.

99. $x^2 - 7 = 0$; $[2, 3]$ (to find the positive square root of 7)

100. $x^3 - 3 = 0$; $[1, 2]$ (to find the cube root of 3)

101. $x^5 - 75 = 0$; $[2, 3]$ (to find the fifth root of 75)

102. $x^{4/3} - 10 = 0$; $[5, 6]$ (to approximate $10^{3/4}$)

103. $x^3 - 3x - 1 = 0$; $[-1, 0]$

104. $x^3 - 4x - 1 = 0$; $[-1, 0]$

105. $e^{-x} - \sin x = 0$; $[0, 2]$

106. $\cos x - \ln x = 0$; $[0, 2]$

107. $x + \cos x = 0$; $[-2, 0]$

108. $x^2 + \sin x = 0$; $[-1.0, -0.5]$

109. $4x - \sin x + 4 = 0$; $[-2, -1]$

110. $5x - \cos x + 5 = 0$; $[-1, 0]$

111. Find the depth to which a wooden ball with radius 2 ft sinks in water if its density is one-third that of water. A useful formula appears in Problem 37 of Section 3.10.

112. The equation $x^2 + 1 = 0$ has no real solutions. Try finding a solution by using Newton's method and report what happens. Use the initial estimate $x_0 = 2$.

113. At the beginning of Section 3.10 we mentioned the fifth-degree equation

$$x^5 - 3x^3 + x^2 - 23x + 19 = 0;$$

its graph appears in Fig. 3.10.1. The graph makes it clear that this equation has exactly three real solutions. Find all of them, to four-place accuracy, using Newton's method.

114. The equation

$$\tan x = \frac{1}{x}$$

has a sequence $\alpha_1, \alpha_2, \alpha_3, \ldots$ of positive roots, with α_n slightly larger than $(n-1)\pi$. Use Newton's method to approximate α_1 and α_2 to three-place accuracy.

115. Criticize the following "proof" that $3 = 2$. Begin by writing

$$x^3 = x \cdot x^2 = x^2 + x^2 + \cdots + x^2 \qquad (x \text{ summands}).$$

Differentiate to obtain

$$3x^2 = 2x + 2x + \cdots + 2x \qquad (\text{still } x \text{ summands}).$$

Thus $3x^2 = 2x^2$, and "therefore" $3 = 2$.

If we substitute $z = x + h$ into the definition of the derivative, the result is

$$f'(x) = \lim_{z \to x} \frac{f(z) - f(x)}{z - x}.$$

Use this formula in Problems 116 and 117, together with the formula

$$a^3 - b^3 = (a - b)(a^2 + ab + b^2)$$

for factoring the difference of two cubes.

116. Show that

$$D_x x^{3/2} = \lim_{z \to x} \frac{z^{3/2} - x^{3/2}}{z - x} = \frac{3}{2}x^{1/2}.$$

[*Suggestion:* Factor the numerator as a difference of cubes and the denominator as a difference of squares.]

117. Prove that

$$D_x x^{2/3} = \lim_{Z \to x} \frac{z^{2/3} - x^{2/3}}{z - x} = \frac{2}{3}x^{-1/3}.$$

[*Suggestion:* Factor the numerator as a difference of squares and the denominator as a difference of cubes.]

118. A rectangular block with square base is being squeezed in such a way that its height y is decreasing at the rate of 2 cm/min while its volume remains constant. At what rate is the edge x of its base increasing when $x = 30$ cm and $y = 20$ cm?

119. Air is being pumped into a spherical balloon at the constant rate of 10 in.³/s. At what rate is the surface area of the balloon increasing when its radius is 5 in.?

120. A ladder 10 ft long is leaning against a wall. If the bottom of the ladder slides away from the wall at the constant rate of 1 mi/h, how fast (in miles per hour) is the top of the ladder moving when it is 0.01 ft above the ground?

121. A water tank in the shape of an inverted cone, axis vertical and vertex downward, has a top radius of 5 ft and height 10 ft. Water is flowing out of the tank through a hole at the vertex at the rate of 50 ft^3/min. What is the time rate of change of the water depth at the instant when the water is 6 ft deep?

122. Plane A is flying west toward an airport at an altitude of 2 mi. Plane B is flying south toward the same airport at an altitude of 3 mi. When both planes are 2 mi (ground distance) from the airport, the speed of plane A is 500 mi/h and the distance between the two planes is decreasing at 600 mi/h. What is the speed of plane B then?

123. A water tank is shaped in such a way that the volume of water in the tank is $V = 2y^{3/2}$ in.^{3} when its depth is y inches. If water flows out through a hole at the bottom of the tank at the rate of $3\sqrt{y}$ in.^{3}/min, at what rate does the water level in the tank fall? What is a practical application for such a water tank?

124. Water is being poured into the conical tank of Problem 121 at the rate of 50 ft^3/min and is draining through the hole at the bottom at the rate of $10\sqrt{y}$ ft^3/min, where y is the depth of water in the tank. (a) At what rate is the water level rising when the water is 5 ft deep? (b) Suppose that the tank is initially empty, water is poured in at 25 ft^3/min, and water continues to drain at $10\sqrt{y}$ ft^3/min. What is the maximum depth attained by the water?

125. Let L be a straight line passing through the fixed point $P(x_0, y_0)$ and tangent to the parabola $y = x^2$ at the point $Q(a, a^2)$. (a) Show that $a^2 - 2ax_0 + y_0 = 0$. (b) Apply the quadratic formula to show that if $y_0 < (x_0)^2$ (that is, if P lies below the parabola), then there are two possible values for a and thus two lines through P that are tangent to the parabola. (c) Similarly, show that if $y_0 > (x_0)^2$ (P lies above the parabola), then no line through P can be tangent to the parabola.

ADDITIONAL APPLICATIONS OF THE DERIVATIVE

4

G. W. Leibniz (1646–1716)

Gottfried Wilhelm Leibniz entered the University of Leipzig when he was 15, studied philosophy and law, graduated at 17, and received his doctorate in philosophy at 21. Upon completion of his academic work, Leibniz entered the political and governmental service of the Elector of Mainz (Germany). His serious study of mathematics did not begin until 1672 (when he was 26) when he was sent to Paris on a diplomatic mission. During the next four years there he conceived the principal features of calculus. For this work he is remembered (with Newton) as a codiscoverer of the subject. Newton's discoveries had come slightly earlier (in the late 1660s), but Leibniz's were the first to be published, beginning in 1684. Despite an unfortunate priority dispute between supporters of Newton and supporters of Leibniz that raged for more than a century, it is clear now that the discoveries were made independently.

Throughout his life, Leibniz sought a universal language incorporating notation and terminology that would provide *all* educated people with the powers of clear and correct reasoning in all subjects. But only in mathematics did he largely accomplish this goal. His differential notation for calculus is arguably the best example of a system of notation chosen so as to mirror perfectly the basic operations and processes of the subject. Indeed, it can be said that Leibniz's notation for calculus brings within the range of ordinary students problems that once required the ingenuity of an Archimedes or a Newton. For this reason, Leibniz's approach to calculus was dominant during the eighteenth century, even though Newton's somewhat different approach may have been closer to our modern understanding of the subject.

The origin of differential notation was an infinitesimal right triangle with legs dx and dy and with hypotenuse a tiny segment of the curve $y = f(x)$. Leibniz later described the moment he first visualized this "characteristic triangle" as a burst of light that was the inception of his calculus. Indeed, he sometimes referred to his calculus as "my method of the Characteristic Triangle."

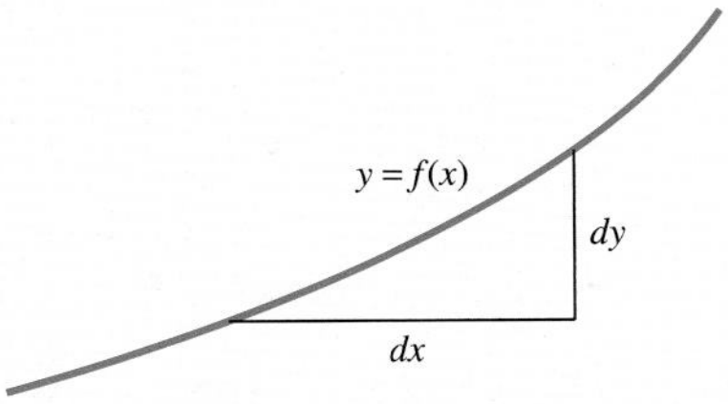

Leibniz's characteristic triangle

The following excerpt shows the opening paragraphs of Leibniz's first published article (in the 1684 *Acta Eruditorum*) in which the differential notation initially appeared. In the fifth line of the second paragraph, the product rule for differentiation is expressed as

$$d(xv) = x\,dv + v\,dx.$$

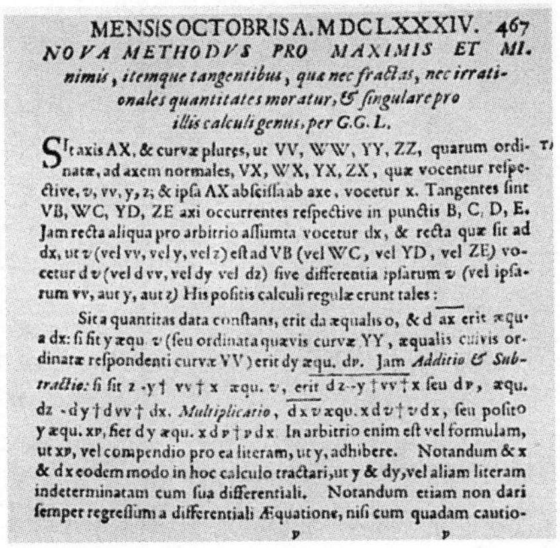

MENSIS OCTOBRIS A. MDCLXXXIV. 467

NOVA METHODUS PRO MAXIMIS ET MInimis, itemque tangentibus, quæ nec fractas, nec irrationales quantitates moratur, & singulare pro illis calculi genus, per G.G.L.

Sit axis AX, & curvæ plures, ut VV, WW, YY, ZZ, quarum ordinatæ, ad axem normales, VX, WX, YX, ZX, quæ vocentur respective, v, vv, y, z; & ipsa AX abscissa ab axe, vocetur x. Tangentes sint VB, WC, YD, ZE axi occurrentes respective in punctis B, C, D, E. Jam recta aliqua pro arbitrio assumta vocetur dx, & recta quæ sit ad dx, ut v (vel vv, vel y, vel z) est ad VB (vel WC, vel YD, vel ZE) vocetur dv (vel d vv, vel dy vel dz) sive differentia ipsarum v (vel ipsarum vv, aut y, aut z) His positis calculi regulæ erunt tales:

Sit a quantitas data constans, erit da æqualis o, & d ax erit æqu. a dx: si sit y æqu. v (seu ordinata quævis curvæ YY, æqualis cuivis ordinatæ respondenti curvæ VV) erit dy æqu. dv. Jam *Additio & Subtractio*: si sit z -y † vv † x æqu. v, erit d z--y † vv † x seu dv, æqu. dz -dy † dvv † dx. *Multiplicatio*, dx v æqu. x dv † v dx, seu posito y æqu. xv, fiet dy æqu. x dv † v dx In arbitrio enim est vel formulam, ut xv, vel compendio pro ea literam, ut y, adhibere. Notandum & x & dx eodem modo in hoc calculo tractari, ut y & dy, vel aliam literam indeterminatam cum sua differentiali. Notandum etiam non dari semper regressum a differentiali Æquatione, nisi cum quadam cautio-

4.1 INTRODUCTION

We learned in Chapter 3 how to differentiate a wide variety of algebraic and trigonometric functions. We saw that derivatives have such diverse applications as maximum-minimum problems, related-rates problems, and the solution of equations by Newton's method. The further applications of differentiation that we discuss in this chapter all depend ultimately upon a single fundamental question. Suppose that $y = f(x)$ is a differentiable function defined on the closed interval $[a, b]$ of length $\Delta x = b - a$. Then the *increment* Δy in the value of $f(x)$ as x changes from $x = a$ to $x = b = a + \Delta x$ is

$$\Delta y = f(b) - f(a). \tag{1}$$

The question is this: How is the increment Δy related to the derivative—the rate of change—of the function f at the points of the interval $[a, b]$?

An *approximate* answer is given in Section 4.2. If the function continued throughout the interval with the same rate of change $f'(a)$ that it had at $x = a$, then the change in its value would be $f'(a)(b - a) = f'(a)\,\Delta x$. This observation motivates the tentative approximation

$$\Delta y \approx f'(a)\,\Delta x. \tag{2}$$

A precise answer to the preceding question is provided by the mean value theorem of Section 4.3. This theorem implies that the exact increment is given by

$$\Delta y = f'(c)\,\Delta x \tag{3}$$

for some number c in (a, b). The mean value theorem is the central theoretical result of differential calculus, and is also the key to many of the more advanced applications of derivatives.

4.2 INCREMENTS, DIFFERENTIALS, AND LINEAR APPROXIMATION

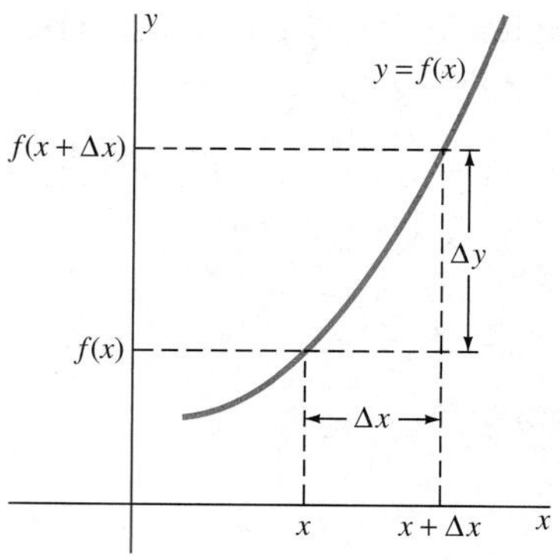

FIGURE 4.2.1 The increments Δx and Δy.

Sometimes we need a quick and simple estimate of the change in $f(x)$ that results from a given change in x. We write y for $f(x)$ and suppose first that the change in the independent variable is the *increment* Δx, so that x changes from its original value to the new value $x + \Delta x$. The change in the value of y is the **increment** Δy, computed by subtracting the old value of y from its new value:

$$\Delta y = f(x + \Delta x) - f(x). \tag{1}$$

The increments Δx and Δy are represented geometrically in Fig. 4.2.1.

Now we compare the actual increment Δy with the change that *would* occur in the value of y *if* it continued to change at the *fixed* rate $f'(x)$ while the value of the independent variable changes from x to $x + \Delta x$. This hypothetical change in y is the **differential**

$$dy = f'(x)\,\Delta x. \tag{2}$$

As Fig. 4.2.2 shows, dy is the change in height of a point that moves along the tangent line at the point $(x, f(x))$ rather than along the curve $y = f(x)$.

Think of x as fixed. Then Eq. (2) shows that the differential dy is a *linear* function of the increment Δx. For this reason, dy is called the **linear approximation** to the increment Δy. We can approximate $f(x + \Delta x)$ by substituting dy for Δy:

$$f(x + \Delta x) = y + \Delta y \approx y + dy.$$

Because $y = f(x)$ and $dy = f'(x)\Delta x$, this gives the **linear approximation formula**

$$f(x + \Delta x) \approx f(x) + f'(x)\,\Delta x. \tag{3}$$

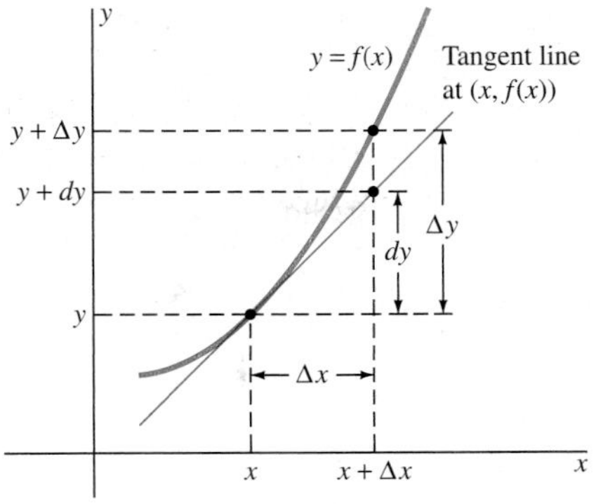

FIGURE 4.2.2 The estimate dy of the actual increment Δy.

The point is that this approximation is a "good" one, at least when Δx is relatively small. If we combine Eqs. (1), (2), and (3), we see that

$$\Delta y \approx f'(x)\,\Delta x = dy. \tag{4}$$

Thus the differential $dy = f'(x)\,\Delta x$ is a good approximation to the increment $\Delta y = f(x + \Delta x) - f(x)$.

If we replace x with a in Eq. (3), we get the approximation

$$f(a + \Delta x) \approx f(a) + f'(a)\,\Delta x. \tag{5}$$

If we now write $\Delta x = x - a$, so that $x = a + \Delta x$, the result is

$$f(x) \approx f(a) + f'(a) \cdot (x - a). \tag{6}$$

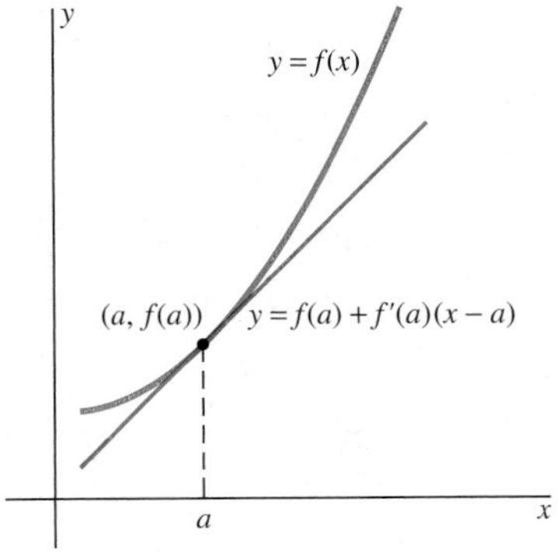

FIGURE 4.2.3 The graph of the linear approximation $L(x) = f(a) + f'(a) \cdot (x - a)$ is the line tangent to $y = f(x)$ at the point $(a, f(a))$.

Because the right-hand side

$$L(x) = f(a) + f'(a) \cdot (x - a) \tag{7}$$

in Eq. (6) is a linear function of x, we call it the **linear approximation $L(x)$ to the function $f(x)$ near the point $x = a$**. As illustrated in Fig. 4.2.3, the graph $y = L(x)$ is the straight line tangent to the graph $y = f(x)$ at the point $(a, f(a))$.

EXAMPLE 1 Find the linear approximation to the function $f(x) = \sqrt{1+x}$ near the point $a = 0$.

Solution Note that $f(0) = 1$ and that

$$f'(x) = \frac{1}{2}(1+x)^{-1/2} = \frac{1}{2\sqrt{1+x}},$$

so $f'(0) = \frac{1}{2}$. Hence Eq. (6) with $a = 0$ yields

$$f(x) \approx f(0) + f'(0) \cdot (x - 0) = 1 + \tfrac{1}{2}x = L(x).$$

Thus the desired linear approximation is

$$\sqrt{1+x} \approx 1 + \tfrac{1}{2}x. \tag{8}$$

Figure 4.2.4 illustrates the close approximation near $x = 0$ of the nonlinear function $f(x) = \sqrt{1+x}$ by its linear approximation $L(x) = 1 + \frac{1}{2}x$. ◆

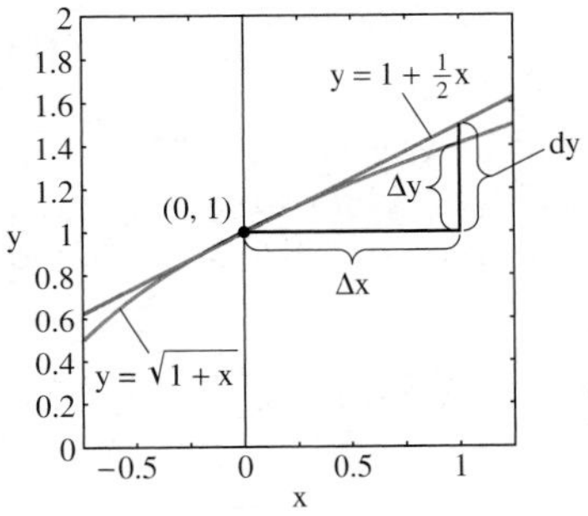

FIGURE 4.2.4 The function $f(x) = \sqrt{1+x}$ and its linear approximation $L(x) = 1 + \frac{1}{2}x$ near $a = 0$.

IMPORTANT It is evident in Fig. 4.2.4 that the value of the linear approximation $L(x) = 1 + \frac{1}{2}x$ is closer to the actual value of the function $f(x) = \sqrt{1+x}$ when x is closer to $a = 0$. For instance, the approximate values

$$\sqrt{1.1} \approx 1 + \tfrac{1}{2}(0.1) = 1.05 \qquad \text{(using } x = 0.1 \text{ in (8))}$$

and

$$\sqrt{1.03} \approx 1 + \tfrac{1}{2}(0.03) = 1.015 \qquad \text{(using } x = 0.03 \text{ in (8))}$$

are accurate to two and three decimal places (rounded), respectively. But

$$\sqrt{3} \approx 1 + \tfrac{1}{2} \cdot 2 = 2,$$

using $x = 2$, is a very poor approximation to $\sqrt{3} \approx 1.732$.

The approximation $\sqrt{1+x} \approx 1 + \frac{1}{2}x$ is a special case of the approximation

$$(1+x)^k \approx 1 + kx \qquad \textbf{(9)}$$

(k is a constant, x is near zero), an approximation with numerous applications. The derivation of (9) is similar to that in Example 1. (See Problem 39.)

EXAMPLE 2 Use the linear approximation formula to approximate $(122)^{2/3}$. Note that

$$(125)^{2/3} = \left[(125)^{1/3}\right]^2 = 5^2 = 25.$$

Solution We need to approximate a particular value of $x^{2/3}$, so our strategy is to apply Eq. (6) with $f(x) = x^{2/3}$. We first note that $f'(x) = \frac{2}{3}x^{-1/3}$. We choose $a = 125$, because we know the *exact* values

$$f(125) = (125)^{2/3} = 25 \quad \text{and} \quad f'(125) = \tfrac{2}{3}(125)^{-1/3} = \tfrac{2}{15}$$

and because 125 is relatively close to 122. Then the linear approximation in (6) to $f(x) = x^{2/3}$ near $a = 125$ takes the form

$$f(x) \approx f(125) + f'(125) \cdot (x - 125);$$

that is,

$$x^{2/3} \approx 25 + \tfrac{2}{15}(x - 125).$$

With $x = 122$ we get

$$(122)^{2/3} \approx 25 + \tfrac{2}{15}(-3) = 24.6.$$

Thus $(122)^{2/3}$ is approximately 24.6. The actual value of $(122)^{2/3}$ is about 24.5984, so the formula in (6) gives a relatively good approximation in this case. ◆

EXAMPLE 3 A hemispherical bowl of radius 10 in. is filled with water to a depth of x inches. The volume V of water in the bowl (in cubic inches) is given by the formula

$$V = \frac{\pi}{3}(30x^2 - x^3) \qquad \textbf{(10)}$$

(Fig. 4.2.5). (You will be able to derive this formula after you study Chapter 6.) Suppose that you *measure* the depth of water in the bowl to be 5 in. with a maximum possible measured error of $\frac{1}{16}$ in. Estimate the maximum error in the calculated volume of water in the bowl.

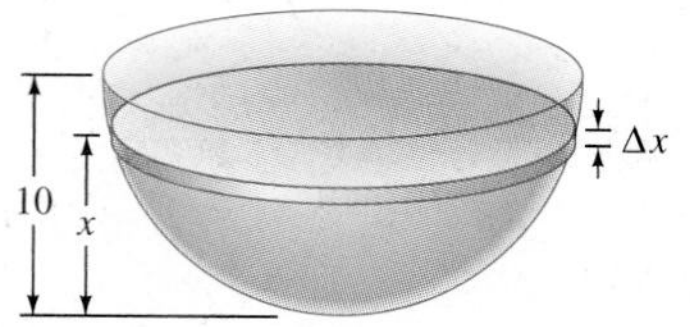

FIGURE 4.2.5 The bowl of Example 3.

Solution The error in the calculated volume $V(5)$ is the difference

$$\Delta V = V(x) - V(5)$$

between the actual volume $V(x)$ and the calculated volume. We do not know the depth x of water in the bowl. We are given only that the difference

$$\Delta x = x - 5$$

between the actual and the measured depths is numerically at most $\frac{1}{16}$ in.: $|\Delta x| \leqq \frac{1}{16}$. Because Eq. (10) yields

$$V'(x) = \frac{\pi}{3}(60x - 3x^2) = \pi(20x - x^2),$$

the linear approximation

$$\Delta V \approx dV = V'(5)\,\Delta x$$

at $x = 5$ gives

$$\Delta V \approx \pi(20 \cdot 5 - 5^2)\,\Delta x = 75\pi\,\Delta x.$$

With the common practice in science of writing $\Delta x = \pm\frac{1}{16}$ to signify that $-\frac{1}{16} \leqq \Delta x \leqq \frac{1}{16}$, this gives

$$\Delta V \approx (75\pi)\left(\pm\tfrac{1}{16}\right) \approx \pm 14.73 \quad (\text{in.}^3).$$

The formula in Eq. (10) gives the calculated volume $V(5) \approx 654.50$ in.^{3}, but we now see that this may be in error by almost 15 in.^{3} in either direction. ◆

Absolute and Relative Errors

The **(absolute) error** in a measured or approximated value is defined to be the remainder when the approximate value is subtracted from the true value. Hence

$$\text{“actual value} = \text{approximate value} + \text{error.”}$$

The **relative error** is the ratio of the (absolute) error to the true value,

$$\text{“relative error} = \frac{\text{error}}{\text{value}}\text{,”}$$

and may be given as either a numerical fraction or as a percentage of the value.

EXAMPLE 4 In Example 3, a relative error in the measured depth x of

$$\frac{\Delta x}{x} = \frac{\frac{1}{16}}{5} = 0.0125 = 1.25\%$$

leads to a relative error in the estimated volume of

$$\frac{dV}{V} \approx \frac{14.73}{654.50} \approx 0.0225 = 2.25\%.$$

The relationship between these two relative errors is of some interest. The formulas for dV and V in Example 3 give

$$\frac{dV}{V} = \frac{\pi(20x - x^2)\,\Delta x}{\frac{1}{3}\pi(30x^2 - x^3)} = \frac{3(20 - x)}{30 - x} \cdot \frac{\Delta x}{x}.$$

When $x = 5$, this gives

$$\frac{dV}{V} = (1.80)\frac{\Delta x}{x}.$$

Hence, to approximate the volume of water in the bowl with a relative error of at most 0.5%, for instance, we would need to measure the depth with a relative error of at most $(0.5\%)/1.8$, thus with a relative error of less than 0.3%. ◆

The Error in Linear Approximation

Now we consider briefly the question of the difference between the values of a function $f(x)$ and its linear approximation $L(x)$ near the point $x = a$. If we let $\Delta x = x - a$ and write

$$y = f(x), \qquad f(a + \Delta x) = f(a) + \Delta y,$$

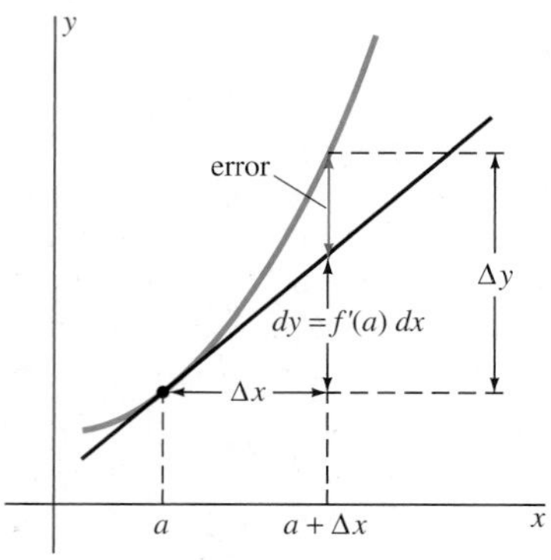

FIGURE 4.2.6 The error $\Delta y - dy$ in the linear approximation $\Delta y \approx f'(a)\,\Delta x = dy$.

and

$$L(x) = f(a) + f'(a) \cdot \Delta x = f(a) + dy,$$

it then follows that the error in the linear approximation is given by

$$f(x) - L(x) = \Delta y - dy, \tag{11}$$

as illustrated in Fig. 4.2.6. It appears in the figure that, the smaller Δx is, the closer are the corresponding points on the curve $y = f(x)$ and its tangent line $y = L(x)$. Because Eq. (11) implies that the difference in the heights of two such points is equal to $\Delta y - dy$, the figure suggests that $\Delta y - dy$ approaches zero as $\Delta x \to 0$.

But even more is true: The difference

$$\Delta y - dy = f(a + \Delta x) - f(a) - f'(a)\,\Delta x \tag{12}$$

is a function of Δx that is small *even in comparison with* Δx. To see why, let's write

$$\epsilon(\Delta x) = \frac{\Delta y - dy}{\Delta x} = \frac{f(a + \Delta x) - f(a)}{\Delta x} - f'(a)$$

and note that

$$\lim_{\Delta x \to 0} \epsilon(\Delta x) = f'(a) - f'(a) = 0.$$

Consequently, the **error**

$$\Delta y - dy = \epsilon(\Delta x) \cdot \Delta x \tag{13}$$

in the linear approximation $dy = f'(a)\,\Delta x$ to the actual increment Δy is the product of two quantities, *both* of which approach zero as $\Delta x \to 0$. If Δx is "very small"—so that $\epsilon(\Delta x)$ is also "very small"—then we might well describe their product in (13) as "*very* very small." In this case we may finally rewrite Eq. (13) in the form

$$f(a + \Delta x) - f(a) = f'(a)\,\Delta x + \epsilon(\Delta x) \cdot \Delta x, \tag{14}$$

expressing the actual increment $\Delta y = f(a + \Delta x) - f(a)$ as the sum of the very small differential $dy = f'(a)\,\Delta x$ and the very very small error $\epsilon(\Delta x) \cdot \Delta x$ in this differential.

EXAMPLE 5 If $y = f(x) = x^3$, then simple computations (with $\Delta x = x - a$) give

$$\begin{aligned} \Delta y &= f(a + \Delta x) - f(a) \\ &= (a + \Delta x)^3 - a^3 = 3a^2\,\Delta x + 3a(\Delta x)^2 + (\Delta x)^3 \end{aligned}$$

and

$$dy = f'(a)\,\Delta x = 3a^2\,\Delta x.$$

Hence

$$\Delta y - dy = 3a(\Delta x)^2 + (\Delta x)^3.$$

If $a = 1$ and $\Delta x = 0.1$, for instance, then these formulas yield

$$\Delta y = 0.331, \qquad dy = 0.3, \quad \text{and} \quad \Delta y - dy = 0.031,$$

thereby illustrating the smallness in the error $\Delta y - dy$ in the linear approximation in comparison with the values of Δy and dy. ◆

Example 6 indicates how we sometimes can use a graphing calculator or computer to specify *how accurate* a linear approximation is—in terms of its accuracy throughout an entire interval containing the point $x = a$. In concrete situations we often want to determine an interval throughout which the linear approximation provides a specified accuracy.

EXAMPLE 6 Find an interval on which the approximation

$$\sqrt{1 + x} \approx 1 + \tfrac{1}{2}x \tag{15}$$

of Example 1 is accurate to within 0.1.

Solution Accuracy to within 0.1 means that the two functions in (15) differ by less than 0.1:

$$\left|\sqrt{1+x} - \left(1 + \tfrac{1}{2}x\right)\right| < 0.1,$$

which is equivalent to

$$\sqrt{1+x} - 0.1 < 1 + \tfrac{1}{2}x < \sqrt{1+x} + 0.1.$$

Thus we want the graph of the linear approximation $y = 1 + \frac{1}{2}x$ to lie between the two curves obtained by shifting the graph $y = \sqrt{1+x}$ vertically up and down by the amount 0.1. Figure 4.2.7 shows the graphs of all these curves on the interval $-1 < x < 1.5$. The points at which the linear approximation $y = 1+\frac{1}{2}x$ emerges from the band of width 0.2 around the graph $y = \sqrt{1+x}$ are marked, and we see that a smaller interval around $x = 0$ is needed to confine the linear approximation within the desired range. Indeed, the zoom shown in Fig. 4.2.8 indicates that the approximation in (15) is accurate to within 0.1 for every x in the interval $-0.6 < x < 0.9$. ◆

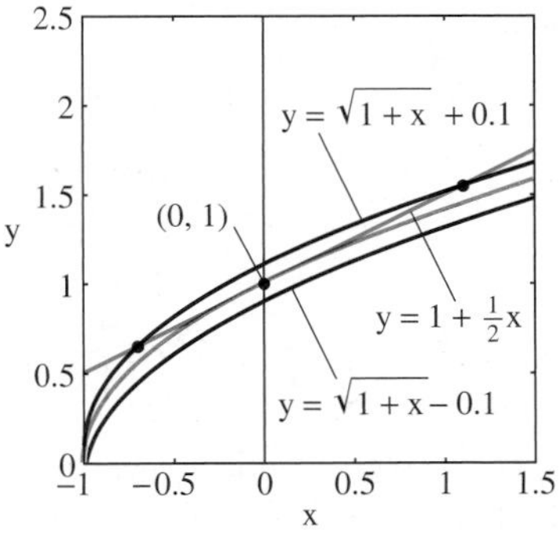

FIGURE 4.2.7 The function $f(x) = \sqrt{1+x}$ on the interval $-1 < x < 1.5$.

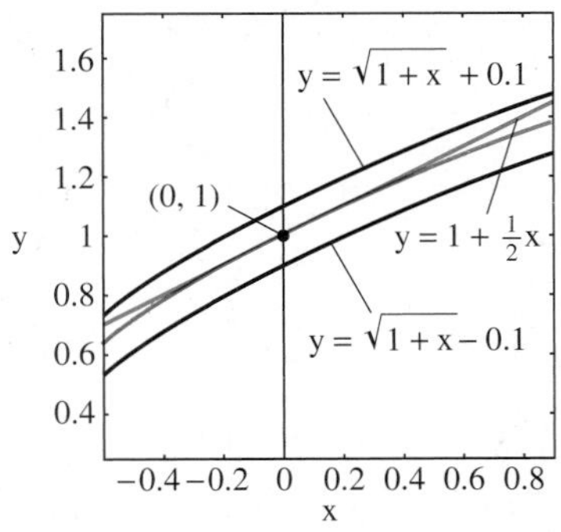

FIGURE 4.2.8 The function $f(x) = \sqrt{1+x}$ on the smaller interval $-0.6 < x < 0.9$.

Differentials

The linear approximation formula in (3) is often written with dx in place of Δx:

$$f(x + dx) \approx f(x) + f'(x)\,dx. \tag{16}$$

In this case dx is an independent variable, called the **differential** of x, and x is fixed. Thus the differentials of x and y are defined to be

$$dx = \Delta x \quad \text{and} \quad dy = f'(x)\,\Delta x = f'(x)\,dx. \tag{17}$$

From this definition it follows immediately that

$$\frac{dy}{dx} = \frac{f'(x)\,dx}{dx} = f'(x),$$

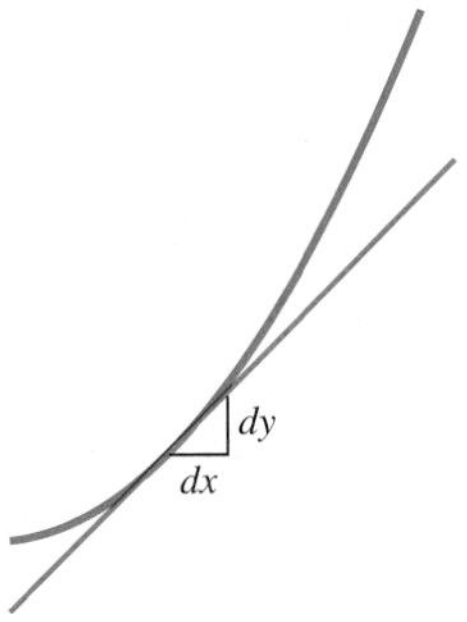

FIGURE 4.2.9 The slope of the tangent line as the ratio of the infinitesimals dy and dx.

in perfect accord with the notation we have been using. Indeed, Leibniz originated differential notation by visualizing "infinitesimal" increments dx and dy (Fig. 4.2.9), with their ratio dy/dx being the slope of the tangent line. The key to Leibniz's independent discovery of differential calculus in the 1670s was his insight that if dx and dy are sufficiently small, then the segment of the curve $y = f(x)$ and the straight line segment joining (x, y) and $(x + dx,\ y + dy)$ are virtually indistinguishable. This insight is illustrated by the successive magnifications in Figs. 4.2.10 through 4.2.12 of the curve $y = x^2$ near the point (1, 1).

Differential notation provides us with a convenient way to write derivative formulas. Suppose that $z = f(u)$, so that $dz = f'(u)\,du$. For particular choices of the function f, we get the formulas

$$d(u^n) = nu^{n-1}\,du,$$

$$d(\sin u) = (\cos u)\,du,$$

$$d(e^u) = e^u\,du,$$

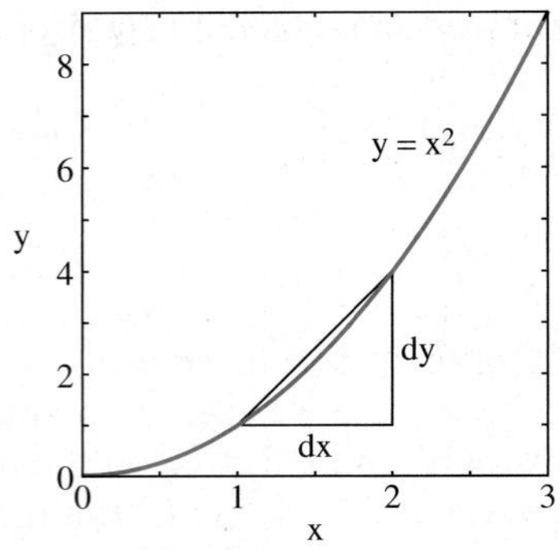

FIGURE 4.2.10 $dx = 1$.

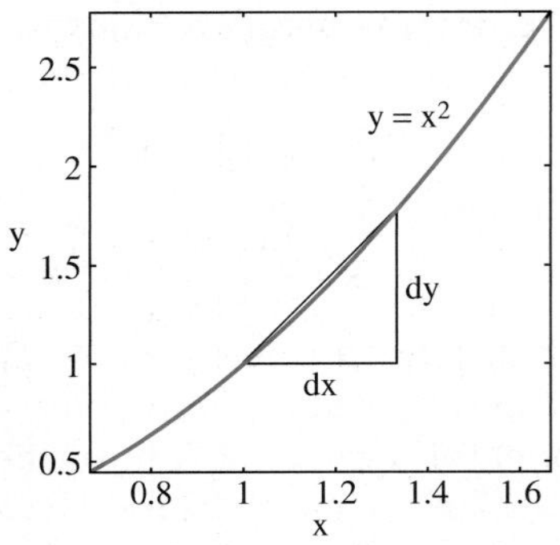

FIGURE 4.2.11 $dx = \frac{1}{3}$.

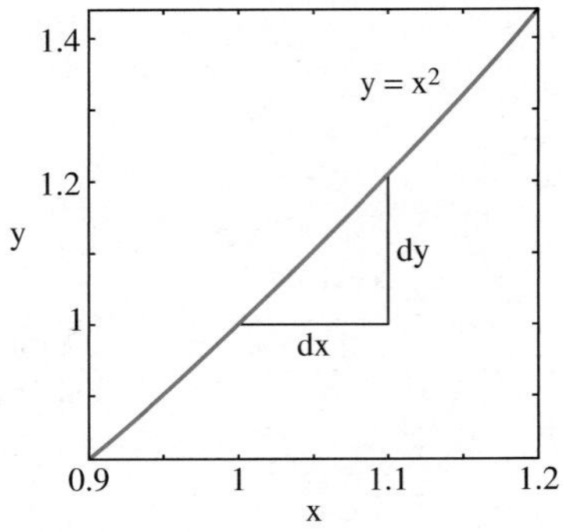

FIGURE 4.2.12 $dx = \frac{1}{10}$.

and so on. Thus we can write differentiation rules in differential form without having to identify the independent variable. The sum, product, and quotient rules take the respective forms

$$d(u+v) = du + dv,$$

$$d(uv) = u\,dv + v\,du,$$

and

$$d\left(\frac{u}{v}\right) = \frac{v\,du - u\,dv}{v^2}.$$

If $z = f(u)$ and $u = g(x)$, we may substitute $du = g'(x)\,dx$ into the formula $dz = f'(u)\,du$. This gives

$$dz = f'(g(x)) \cdot g'(x)\,dx.$$

This is the differential form of the chain rule

$$D_x f(g(x)) = f'(g(x)) \cdot g'(x).$$

Thus the chain rule appears here as though it were the result of mechanical manipulations of the differential notation. This compatibility with the chain rule is one reason for the extraordinary usefulness of differential notation in calculus.

EXAMPLE 7

(a) If $y = 3x^2 - 2x^{3/2}$, then $dy = \left(6x - 3\sqrt{x}\right) dx$.

(b) If $u = \sin^2 t - \cos 2t$, then

$$du = (2 \sin t \cos t + 2 \sin 2t)\,dt = 3 \sin 2t\,dt$$

(using the trigonometric identity $\sin 2t = 2 \sin t \cos t$).

(c) If $w = ze^z$, then

$$dw = (1 \cdot e^z + z \cdot e^z)\,dz = (1+z)e^z\,dz.$$

◆

Proof of the Chain Rule

We can now use our knowledge of the error in linear approximations to give a proof of the chain rule for the composition $f \circ g$ that does not require the assumption $g'(x) \neq 0$ that we needed in Section 3.3. Here we suppose only the existence of the derivatives $g'(a)$ and $f'(b)$ (where $b = g(a)$) of the functions $u = g(x)$ and $y = f(g(x)) = f(u)$. If we write

$$\Delta u = g(a + \Delta x) - g(a) \quad \text{and} \quad \Delta y = f(b + \Delta u) - f(b),$$

then Eq. (14) in this section—with g in place of f—gives

$$\Delta u = g'(a)\,\Delta x + \epsilon_1 \cdot \Delta x = [g'(a) + \epsilon_1]\,\Delta x \tag{18}$$

where $\epsilon_1 \to 0$ as $\Delta x \to 0$. A second application of Eq. (14)—this time with u in place of x—gives

$$\begin{aligned}\Delta y &= f'(b)\,\Delta u + \epsilon_2 \cdot \Delta u = [f'(b) + \epsilon_2]\,\Delta u \\ &= [f'(g(a)) + \epsilon_2] \cdot [g'(a) + \epsilon_1]\,\Delta x \end{aligned} \tag{19}$$

where $\epsilon_2 \to 0$ as $\Delta u \to 0$, and hence as $\Delta x \to 0$ (because Eq. (18) shows that $\Delta u \to 0$ as $\Delta x \to 0$). Finally, when we divide by Δx in Eq. (19) and then take the limit as $\Delta x \to 0$, we get

$$\frac{dy}{dx} = \lim_{\Delta x \to 0} \frac{\Delta y}{\Delta x} = \lim_{\Delta x \to 0} [f'(g(a)) + \epsilon_2] \cdot [g'(a) + \epsilon_1] = f'(g(a)) \cdot g'(a).$$

Thus we have shown that the chain rule formula $D_x [f(g(x))] = f'(g(x)) \cdot g'(x)$ holds at $x = a$.

4.2 TRUE/FALSE STUDY GUIDE

4.2 CONCEPTS: QUESTIONS AND DISCUSSION

1. Use Eqs. (11)–(13) of this section to show that the linear function $L(x) = f(a) + f'(a) \cdot (x - a)$ of x satisfies the condition
$$\lim_{x \to a} \frac{f(x) - L(x)}{x - a} = 0. \tag{20}$$
2. Any linear function $L(x) = mx + b$ satisfying Eq. (20) is called a **linearization** of the function $f(x)$ at the point $x = a$. Can a function have two different linearizations at the same point?
3. Can a function have a linearization (as in Question 2) at a point where it is not differentiable?

4.2 PROBLEMS

In Problems 1 through 16, write dy in terms of x and dx.

1. $y = 3x^2 - \dfrac{4}{x^2}$

2. $y = 2\sqrt{x} - \dfrac{3}{\sqrt[3]{x}}$

3. $y = x - \sqrt{4 - x^3}$

4. $y = \dfrac{1}{x - \sqrt{x}}$

5. $y = 3x^2(x - 3)^{3/2}$

6. $y = \dfrac{x}{x^2 - 4}$

7. $y = x(x^2 + 25)^{1/4}$

8. $y = \dfrac{1}{(x^2 - 1)^{4/3}}$

9. $y = \cos\sqrt{x}$

10. $y = x^2 \sin x$

11. $y = \sin 2x \cos 2x$

12. $y = \cos^3 3x$

13. $y = \dfrac{\sin 2x}{3x}$

14. $y = x^3 e^{-2x}$

15. $y = \dfrac{1}{1 - x\sin x}$

16. $y = \dfrac{\ln x}{x}$

In Problems 17 through 24, find—as in Example 1—the linear approximation L(x) to the given function f(x) near the point a = 0.

17. $f(x) = \dfrac{1}{1 - x}$

18. $f(x) = \dfrac{1}{\sqrt{1 + x}}$

19. $f(x) = (1 + x)^2$

20. $f(x) = (1 - x)^3$

21. $f(x) = (1 - 2x)^{3/2}$

22. $f(x) = e^{-x}$

23. $f(x) = \sin x$

24. $f(x) = \ln(1 + x)$

In Problems 25 through 34, use—as in Example 2—a linear approximation L(x) to an appropriate function f(x), with an appropriate value of a, to estimate the given number.

25. $\sqrt[3]{25}$

26. $\sqrt{102}$

27. $\sqrt[4]{15}$

28. $\sqrt{80}$

29. $65^{-2/3}$

30. $80^{3/4}$

31. $\cos 43°$

32. $\sin 32°$

33. $e^{1/10}$

34. $\ln\left(\frac{11}{10}\right)$

In Problems 35 through 38, compute the differential of each side of the given equation, regarding x and y as dependent variables (as if both were functions of some third, unspecified, variable). Then solve for dy/dx.

35. $x^2 + y^2 = 1$

36. $xe^y = 1$

37. $x^3 + y^3 = 3xy$

38. $x \ln y = 1$

39. Assuming that $D_x\, x^k = kx^{k-1}$ for any real constant k (which we shall establish in Chapter 6), derive the linear approximation formula $(1 + x)^k \approx 1 + kx$ for x near zero.

In Problems 40 through 47, use linear approximations to estimate the change in the given quantity.

40. The circumference of a circle, if its radius is increased from 10 in. to 10.5 in.

41. The area of a square, if its edge length is decreased from 10 in. to 9.8 in.

42. The surface area of a sphere, if its radius is increased from 5 in. to 5.2 in. (Fig. 4.2.13).

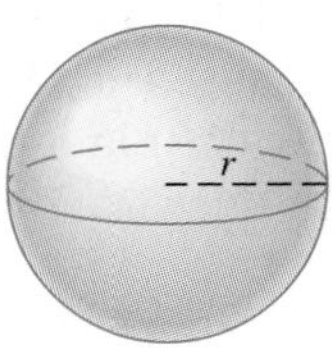

FIGURE 4.2.13 The sphere of Problem 42—area $A = 4\pi r^2$, volume $V = \frac{4}{3}\pi r^3$.

43. The volume of a cylinder, if both its height and its radius are decreased from 15 cm to 14.7 cm (Fig. 4.2.14).

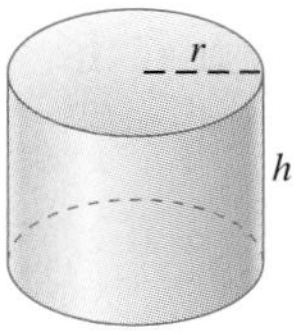

FIGURE 4.2.14 The cylinder of Problem 43—volume $V = \pi r^2 h$.

44. The volume of the conical sandpile of Fig. 4.2.15, if its radius is 14 in. and its height is increased from 7 in. to 7.1 in.

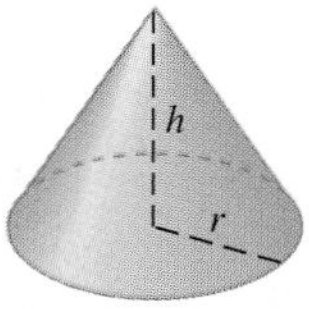

FIGURE 4.2.15 The conical sandpile of Problem 44—volume $V = \frac{1}{3}\pi r^2 h$.

45. The range $R = \frac{1}{16}v^2 \sin 2\theta$ of a shell fired at inclination angle $\theta = 45°$, if its initial velocity v is increased from 80 ft/s to 81 ft/s.

46. The range $R = \frac{1}{16}v^2 \sin 2\theta$ of a projectile fired with initial velocity $v = 80$ ft/s, if its initial inclination angle θ is increased from 45° to 46°.

47. The wattage $W = RI^2$ of a floodlight with resistance $R = 10$ ohms, if the current I is increased from 3 amperes to 3.1 amperes.

48. The equatorial radius of the earth is approximately 3960 mi. Suppose that a wire is wrapped tightly around the earth at the equator. Approximately how much must this wire be lengthened if it is to be strung all the way around the earth on poles 10 ft above the ground? Use the linear approximation formula!

49. The radius of a spherical ball is measured as 10 in., with a maximum error of $\frac{1}{16}$ in. What is the maximum resulting error in its calculated volume?

50. With what accuracy must the radius of the ball of Problem 49 be measured to ensure an error of at most 1 in.^{3} in its calculated volume?

51. The radius of a hemispherical dome is measured as 100 m with a maximum error of 1 cm (Fig. 4.2.16). What is the maximum resulting error in its calculated surface area?

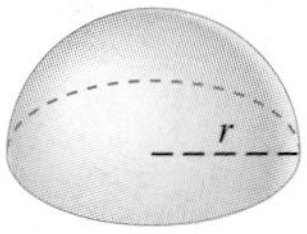

FIGURE 4.2.16 The hemisphere of Problem 51—curved surface area $A = 2\pi r^2$.

52. With what accuracy must the radius of a hemispherical dome be measured to ensure an error of at most 0.01% in its calculated surface area?

In Problems 53 through 60, a function $f(x)$ and a point $x = a$ are given. Determine graphically an open interval I centered at a so that the function $f(x)$ and its linear approximation $L(x)$ differ by less than the given value ϵ at each point of I.

53. $f(x) = x^2, \quad a = 1, \quad \epsilon = 0.2$

54. $f(x) = \sqrt{x}, \quad a = 1, \quad \epsilon = 0.1$

55. $f(x) = \dfrac{1}{x}, \quad a = 2, \quad \epsilon = 0.01$

56. $f(x) = \sqrt[3]{x}, \quad a = 8, \quad \epsilon = 0.01$

57. $f(x) = \sin x, \quad a = 0, \quad \epsilon = 0.05$

58. $f(x) = e^x, \quad a = 0, \quad \epsilon = 0.05$

59. $f(x) = \sin x, \quad a = \pi/4, \quad \epsilon = 0.02$

60. $f(x) = \tan x, \quad a = \pi/4, \quad \epsilon = 0.02$

4.3 INCREASING AND DECREASING FUNCTIONS AND THE MEAN VALUE THEOREM

The significance of the *sign* of the first derivative of a function is simple but crucial:

$f(x)$ is increasing on an interval where $f'(x) > 0$;
$f(x)$ is decreasing on an interval where $f'(x) < 0$.

Geometrically, this means that where $f'(x) > 0$, the graph of $y = f(x)$ is rising as you scan it from left to right. Where $f'(x) < 0$, the graph is falling. We can clarify the terms *increasing* and *decreasing* as follows.

DEFINITION Increasing and Decreasing Functions

The function f is **increasing** on the interval $I = (a, b)$ provided that

$$f(x_1) < f(x_2)$$

for all pairs of numbers x_1 and x_2 in I for which $x_1 < x_2$. The function f is **decreasing** on I provided that

$$f(x_1) > f(x_2)$$

for all pairs of numbers x_1 and x_2 for which $x_1 < x_2$.

Figure 4.3.1 illustrates this definition. In short, the function f is increasing on $I = (a, b)$ if the values of $f(x)$ increase as x increases [Fig. 4.3.1(a)]; f is decreasing on I if the values of $f(x)$ decrease as x increases [Fig. 4.3.1(b)].

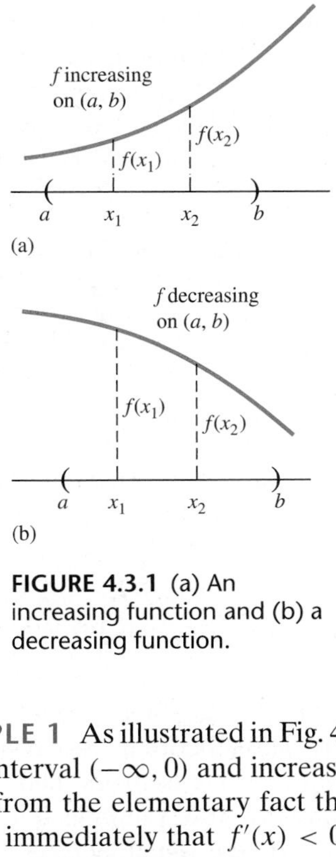

FIGURE 4.3.1 (a) An increasing function and (b) a decreasing function.

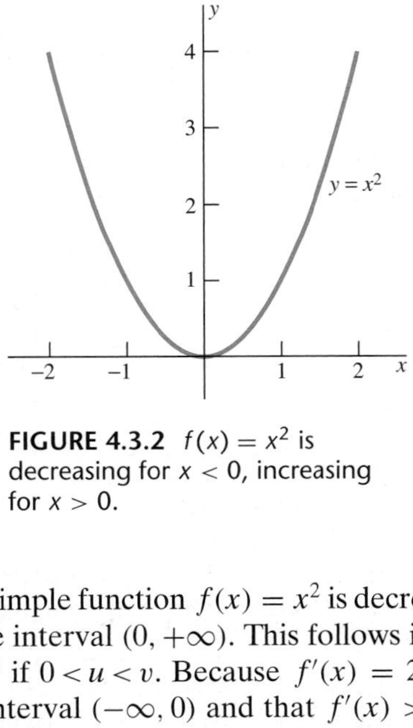

FIGURE 4.3.2 $f(x) = x^2$ is decreasing for $x < 0$, increasing for $x > 0$.

EXAMPLE 1 As illustrated in Fig. 4.3.2, the simple function $f(x) = x^2$ is decreasing on the interval $(-\infty, 0)$ and increasing on the interval $(0, +\infty)$. This follows immediately from the elementary fact that $u^2 < v^2$ if $0 < u < v$. Because $f'(x) = 2x$, we also see immediately that $f'(x) < 0$ on the interval $(-\infty, 0)$ and that $f'(x) > 0$ on the interval $(0, +\infty)$. But for more general functions, the mean value theorem of this section is needed to establish the precise relationship between the sign of the derivative of a function and its increasing-decreasing behavior. ◆

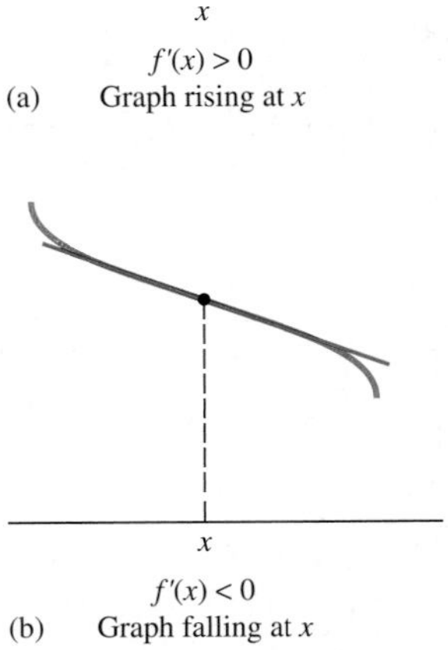

FIGURE 4.3.3 (a) A graph rising at x and (b) a graph falling at x.

REMARK We speak of a function as increasing or decreasing *on an interval,* not at a single point. Nevertheless, if we consider the sign of f', the derivative of f, at a single point, we get a useful intuitive picture of the significance of the sign of the derivative. This is because the derivative $f'(x)$ is the slope of the tangent line at the point $(x, f(x))$ on the graph of f. If $f'(x) > 0$, then the tangent line has positive slope. Therefore, it rises as you scan from left to right. Intuitively, a rising tangent would seem to correspond to a rising graph and thus to an increasing function. Similarly, we expect to see a falling graph where $f'(x)$ is negative (Fig. 4.3.3). One caution: In order to determine whether a function f is increasing or decreasing, we must examine the sign of f' on a whole interval, not merely at a single point. (See Problem 59.)

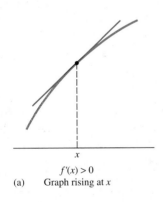

The Mean Value Theorem

Although pictures of rising and falling graphs are suggestive, they provide no actual *proof* of the significance of the sign of the derivative. To establish rigorously the connection between a graph's rising and falling and the sign of the derivative of the graphed function, we need the *mean value theorem,* stated later in this section. This theorem is the principal theoretical tool of differential calculus, and we shall see that it has many important applications.

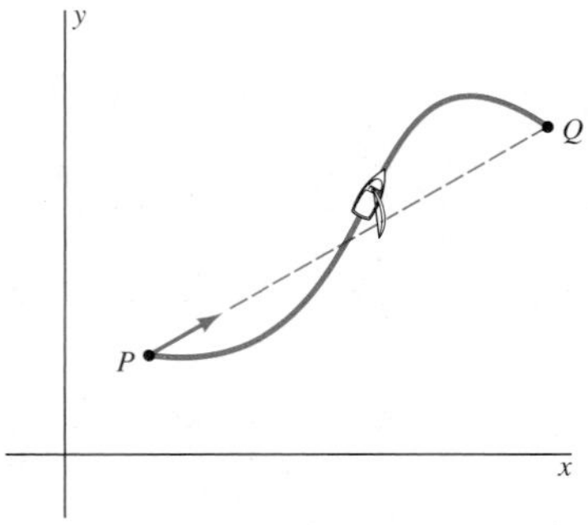

FIGURE 4.3.4 Can you sail from P to Q without ever sailing—even for an instant—in the direction PQ (the direction of the arrow)?

A Question As an introduction to the mean value theorem, we pose the following question. Suppose that P and Q are two points on the surface of the sea, with Q lying generally to the east of P (Fig. 4.3.4). Is it possible to sail a boat from P to Q, always sailing roughly east, without *ever* (even for an instant) sailing in the exact direction from P to Q? That is, can we sail from P to Q without our instantaneous line of motion ever being parallel to the line PQ?

The mean value theorem answers this question: No. There will always be at least one instant when we are sailing parallel to the line PQ, no matter which path we choose.

To paraphrase: Let the path of the sailboat be the graph of a differentiable function $y = f(x)$ with endpoints $P(a, f(a))$ and $Q(b, f(b))$. Then we say that there must be some point on this graph where the tangent line (corresponding to the instantaneous line of motion of the boat) to the curve is parallel to the line PQ that joins the curve's endpoints. This is a *geometric interpretation* of the mean value theorem.

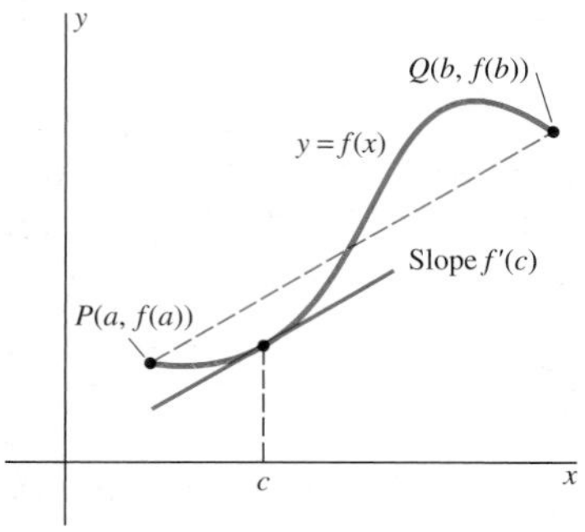

FIGURE 4.3.5 The sailboat problem in mathematical terminology.

The Geometric Formulation The slope of the line tangent at the point $(c, f(c))$ (Fig. 4.3.5) is $f'(c)$, whereas the slope of the line PQ is

$$\frac{f(b) - f(a)}{b - a}.$$

We may think of this last quotient as the average (or *mean*) value of the slope of the curve $y = f(x)$ over the interval $[a, b]$. The mean value theorem guarantees that there is a point c in (a, b) for which the line tangent to $y = f(x)$ at $(c, f(c))$ is indeed parallel to the line PQ. In the language of algebra, there's a number c in (a, b) such that

$$f'(c) = \frac{f(b) - f(a)}{b - a}. \qquad \textbf{(1)}$$

A Preliminary Result We first state a "lemma" to expedite the proof of the mean value theorem. This theorem is called *Rolle's theorem,* after Michel Rolle (1652–1719), who discovered it in 1690. In his youth Rolle studied the emerging subject of calculus but later renounced it. He argued that the subject was based on logical fallacies, and he is remembered today only for the single theorem that bears his name. It is ironic that his theorem plays an important role in the rigorous proofs of several calculus theorems.

ROLLE'S THEOREM

Suppose that the function f is continuous on the closed interval $[a, b]$ and is differentiable in its interior (a, b). If $f(a) = 0 = f(b)$, then there exists some number c in (a, b) such that $f'(c) = 0$.

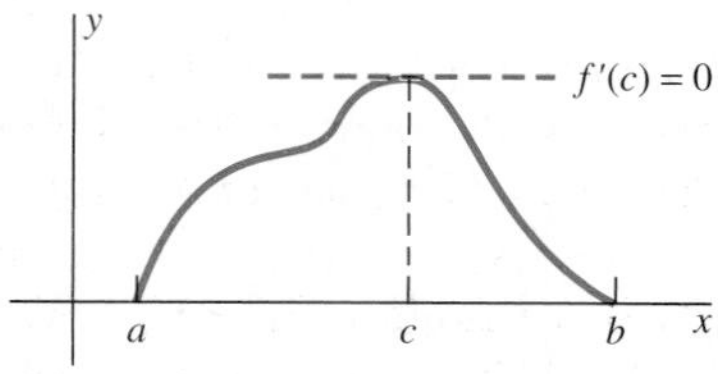

FIGURE 4.3.6 The idea of the proof of Rolle's theorem.

Figure 4.3.6 illustrates the first case in the following proof of Rolle's theorem. The idea of the proof is this: Suppose that the smooth graph $y = f(x)$ starts $(x = a)$ at height zero and ends $(x = b)$ at height zero. Then if it goes up, it must come back down. But where it stops going up and starts coming back down, its tangent line must be horizontal. Therefore the derivative is zero at that point.

PROOF OF ROLLE'S THEOREM Because f is continuous on $[a, b]$, it must attain both a maximum and a minimum value on $[a, b]$ (by the maximum value property of Section 3.5). If f has any positive values, consider its maximum value $f(c)$. Now c is not an endpoint of $[a, b]$ because $f(a) = 0$ and $f(b) = 0$. Therefore c is a point of (a, b). But we know that f is differentiable at c. So it follows from Theorem 2 of Section 3.5 that $f'(c) = 0$.

Similarly, if f has any negative values, we can consider its minimum value $f(c)$ and conclude much as before that $f'(c) = 0$.

If f has neither positive nor negative values, then f is identically zero on $[a, b]$, and it follows that $f'(c) = 0$ for *every* c in (a, b).

Thus we see that the conclusion of Rolle's theorem is justified in every case. ◄

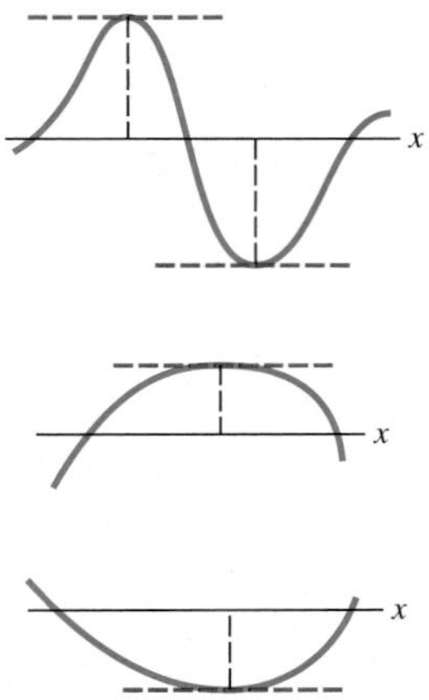

FIGURE 4.3.7 The existence of the horizontal tangent is a consequence of Rolle's theorem.

An important consequence of Rolle's theorem is that between each pair of zeros of a differentiable function, there is *at least one* point at which the tangent line is horizontal. Some possible pictures of the situation are indicated in Fig. 4.3.7.

EXAMPLE 2 Suppose that $f(x) = x^{1/2} - x^{3/2}$ on $[0, 1]$. Find a number c that satisfies the conclusion of Rolle's theorem.

Solution Note that f is continuous on $[0, 1]$ and differentiable on $(0, 1)$. Because the term $x^{1/2}$ is present, f is *not* differentiable at $x = 0$, but this is irrelevant. Also, $f(0) = 0 = f(1)$, so all of the hypotheses of Rolle's theorem are satisfied. Finally,

$$f'(x) = \tfrac{1}{2}x^{-1/2} - \tfrac{3}{2}x^{1/2} = \tfrac{1}{2}x^{-1/2}(1 - 3x),$$

so we see that $f'(c) = 0$ for $c = \frac{1}{3}$. An accurate graph of f on $[0, 1]$, including c and the horizontal tangent line, is shown in Fig. 4.3.8. ◆

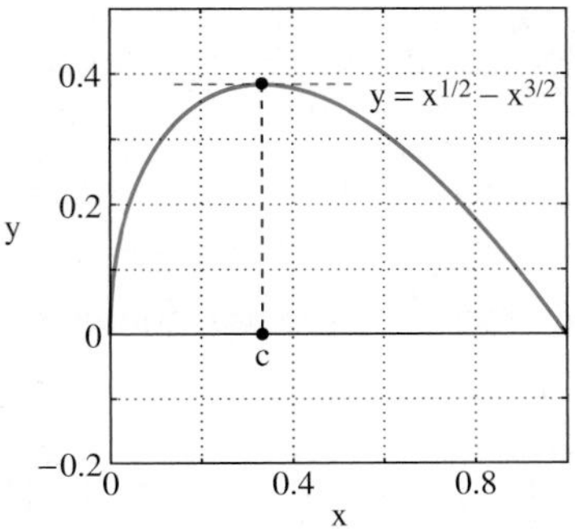

FIGURE 4.3.8 The number c of Example 2.

EXAMPLE 3 Suppose that $f(x) = 1 - x^{2/3}$ on $[-1, 1]$. Then f satisfies the hypotheses of Rolle's theorem *except* for the fact that $f'(0)$ does not exist. It is clear from the graph of f that there is *no* point where the tangent line is horizontal (Fig. 4.3.9). Indeed,

$$f'(x) = -\frac{2}{3}x^{-1/3} = -\frac{2}{3\sqrt[3]{x}},$$

so $f'(x) \neq 0$ for $x \neq 0$, and we see that $|f'(x)| \to \infty$ as $x \to 0$. Hence the graph of f has a vertical tangent line—rather than a horizontal one—at the point $(0, 1)$. Thus the conclusion of Rolle's theorem—like that of any theorem—may fail to hold if any of its hypotheses are not satisfied. ◆

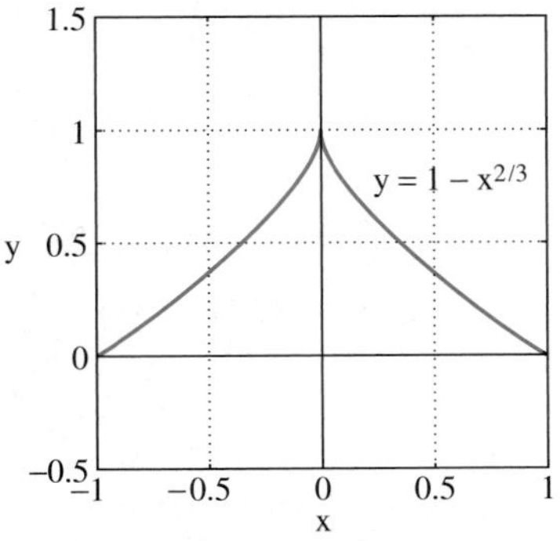

FIGURE 4.3.9 The function $f(x) = 1 - x^{2/3}$ of Example 3.

Now we are ready to state formally and prove the mean value theorem.

THE MEAN VALUE THEOREM

Suppose that the function f is continuous on the closed interval $[a, b]$ and differentiable on the open interval (a, b). Then

$$f(b) - f(a) = f'(c) \cdot (b - a) \qquad \textbf{(2)}$$

for some number c in (a, b).

COMMENT Because Eq. (2) is equivalent to Eq. (1), the conclusion of the mean value theorem is that there must be at least one point on the curve $y = f(x)$ at which the tangent line is parallel to the line joining its endpoints $P(a, f(a))$ and $Q(b, f(b))$.

MOTIVATION FOR THE PROOF OF THE MEAN VALUE THEOREM We consider the auxiliary function ϕ suggested by Fig. 4.3.10. The value of $\phi(x)$ is, by definition, the vertical height difference over x between the point $(x, f(x))$ on the curve and the corresponding point on the line PQ. It appears that a point on the

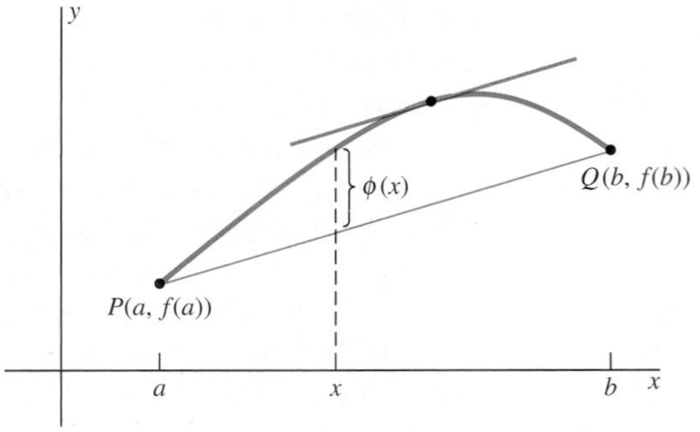

FIGURE 4.3.10 The construction of the auxiliary function ϕ.

curve $y = f(x)$ where the tangent line is parallel to PQ corresponds to a maximum or minimum of ϕ. It's also clear that $\phi(a) = 0 = \phi(b)$, so Rolle's theorem can be applied to the function ϕ on $[a, b]$. So our plan for proving the mean value theorem is this: First, we obtain a formula for the function ϕ. Second, we locate the point c such that $\phi'(c) = 0$. Finally, we show that this number c is exactly the number needed to satisfy the conclusion of the mean value theorem in Eq. (2).

PROOF OF THE MEAN VALUE THEOREM Because the line PQ passes through $P(a, f(a))$ and has slope

$$m = \frac{f(b) - f(a)}{b - a},$$

the point-slope formula for the equation of a straight line gives us the following equation for PQ:

$$y = y_{\text{line}} = f(a) + m(x - a).$$

Thus

$$\phi(x) = y_{\text{curve}} - y_{\text{line}} = f(x) - f(a) - m(x - a).$$

You may verify by direct substitution that $\phi(a) = 0 = \phi(b)$. And, because ϕ is continuous on $[a, b]$ and differentiable on (a, b), we may apply Rolle's theorem to it. Thus there is a point c somewhere in the open interval (a, b) at which $\phi'(c) = 0$. But

$$\phi'(x) = f'(x) - m = f'(x) - \frac{f(b) - f(a)}{b - a}.$$

Because $\phi'(c) = 0$, we conclude that

$$0 = f'(c) - \frac{f(b) - f(a)}{b - a}.$$

That is,

$$f(b) - f(a) = f'(c) \cdot (b - a).$$ ◄

The proof of the mean value theorem is an application of Rolle's theorem, whereas Rolle's theorem is the special case of the mean value theorem in which $f(a) = 0 = f(b)$.

EXAMPLE 4 Suppose that we drive from Kristiansand, Norway to Oslo—a road distance of almost exactly 350 km—in exactly 4 h, from time $t = 0$ to time $t = 4$. Let $f(t)$ denote the distance we have traveled at time t, and assume that f is a differentiable function. Then the mean value theorem implies that

$$350 = f(4) - f(0) = f'(c) \cdot (4 - 0) = 4f'(c)$$

and thus that

$$f'(c) = \tfrac{350}{4} = 87.5$$

at some instant c in $(0, 4)$. But $f'(c)$ is our *instantaneous* velocity at time $t = c$, and 87.5 km/h is our *average* velocity for the trip. Thus the mean value theorem implies that we must have an instantaneous velocity of exactly 87.5 km/h at least once during the trip. ◆

The argument in Example 4 is quite general—during any trip, the instantaneous velocity must at *some* instant equal the average velocity for the whole trip. For instance, it follows that if two toll stations are 70 mi apart and you drive between the two in exactly 1 h, then at some instant you must have been speeding in excess of the posted limit of 65 mi/h. Speeding tickets have been issued by the Pennsylvania State Police to speeders on the Pennsylvania Turnpike on exactly such evidence!

Consequences of the Mean Value Theorem

The first of three important consequences of the mean value theorem is the *non*trivial converse of the trivial fact that the derivative of a constant function is identically zero. That is, we prove that there can be *no* exotic function that is nonconstant but has a derivative that is identically zero. In Corollaries 1 through 3 we assume, as in Rolle's theorem and the mean value theorem, that f and g are continuous on the closed interval $[a, b]$ and differentiable on (a, b).

COROLLARY 1 Functions with Zero Derivative

If $f'(x) \equiv 0$ on (a, b) (that is, $f'(x) = 0$ for all x in (a, b)), then f is a constant function on $[a, b]$. In other words, there exists a constant C such that $f(x) \equiv C$.

PROOF Apply the mean value theorem to the function f on the interval $[a, x]$, where x is a fixed but arbitrary point of the interval $(a, b]$. We find that

$$f(x) - f(a) = f'(c) \cdot (x - a)$$

for some number c between a and x. But $f'(x)$ is always zero on the interval (a, b), so $f'(c) = 0$. Thus $f(x) - f(a) = 0$, and therefore $f(x) = f(a)$.

But this last equation holds for *all* x in $(a, b]$. Therefore, $f(x) = f(a)$ for all x in $(a, b]$ and, indeed, for all x in $[a, b]$. That is, $f(x)$ has the constant value $C = f(a)$. This establishes Corollary 1. ◂

Corollary 1 is usually applied in a different but equivalent form, which we state and prove next.

COROLLARY 2 Functions with Equal Derivatives

Suppose that $f'(x) = g'(x)$ for all x in the open interval (a, b). Then f and g differ by a constant on $[a, b]$. That is, there exists a constant K such that

$$f(x) = g(x) + K$$

for all x in $[a, b]$.

PROOF Given the hypotheses, let $h(x) = f(x) - g(x)$. Then

$$h'(x) = f'(x) - g'(x) = 0$$

for all x in (a, b). So, by Corollary 1, $h(x)$ is a constant K on $[a, b]$. That is, $f(x) - g(x) = K$ for all x in $[a, b]$; therefore,

$$f(x) = g(x) + K$$

for all x in $[a, b]$. This establishes Corollary 2. ◂

EXAMPLE 5 If $f'(x) = 6e^{2x}$ and $f(0) = 7$, what is the function $f(x)$?

Solution Because $D_x(e^{2x}) = 2e^{2x}$, we see immediately that one function with derivative $g'(x) = 6e^{2x}$ is

$$g(x) = 3e^{2x}.$$

Hence Corollary 2 implies that there exists a constant K such that

$$f(x) = g(x) + K = 3e^{2x} + K$$

on any given interval $[a, b]$ containing zero. But we can find the value of K by substituting $x = 0$:

$$f(0) = 3e^0 + K;$$
$$7 = 3 \cdot 1 + K;$$

so $K = 4$. Thus the function f is defined by

$$f(x) = 3e^{2x} + 4.$$ ◆

The following consequence of the mean value theorem verifies the remarks about increasing and decreasing functions with which we opened this section.

COROLLARY 3 Increasing and Decreasing Functions
If $f'(x) > 0$ for all x in (a, b), then f is an increasing function on $[a, b]$. If $f'(x) < 0$ for all x in (a, b), then f is a decreasing function on $[a, b]$.

PROOF Suppose, for example, that $f'(x) > 0$ for all x in (a, b). We need to show the following: If u and v are points of $[a, b]$ with $u < v$, then $f(u) < f(v)$. We apply the mean value theorem to f, but on the closed interval $[u, v]$. This is legitimate because $[u, v]$ is contained in $[a, b]$, so f satisfies the hypotheses of the mean value theorem on $[u, v]$ as well as on $[a, b]$. The result is that

$$f(v) - f(u) = f'(c) \cdot (v - u)$$

for some number c in (u, v). Because $v > u$ and because, by hypothesis, $f'(c) > 0$, it follows that

$$f(v) - f(u) > 0; \qquad \text{that is,} \quad f(u) < f(v),$$

as we wanted to show. The proof is similar in the case that $f'(x)$ is negative on (a, b). ◄

The meaning of Corollary 3 is summarized in Fig. 4.3.11. Figure 4.3.12 shows a graph $y = f(x)$ labeled in accord with this correspondence between the sign of the derivative $f'(x)$ and the increasing or decreasing behavior of the function $f(x)$.

$f'(x)$	$f(x)$
Negative	Decreasing
Positive	Increasing

FIGURE 4.3.11 Corollary 3.

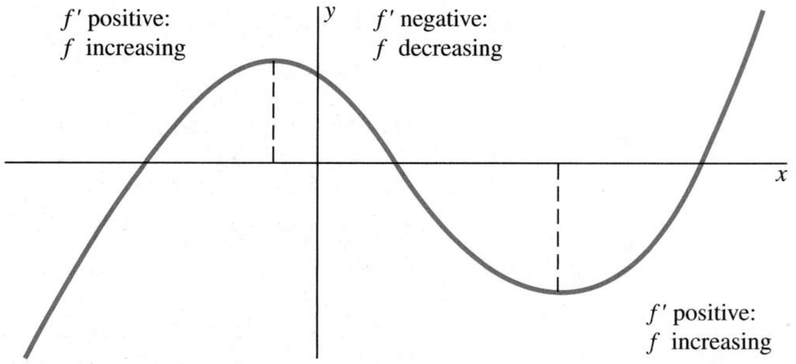

FIGURE 4.3.12 The significance of the sign of $f'(x)$.

EXAMPLE 6 Where is the function $f(x) = x^2 - 4x + 5$ increasing, and where is it decreasing?

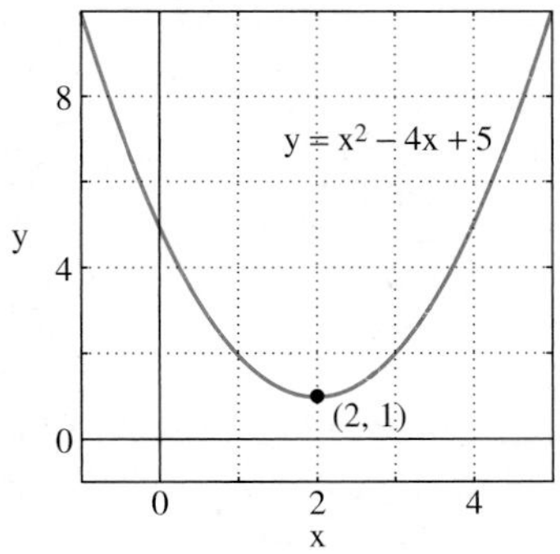

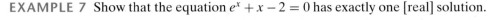

FIGURE 4.3.13 The parabola of Example 6.

Solution The derivative of f is $f'(x) = 2x - 4$. Clearly $f'(x) > 0$ if $x > 2$, whereas $f'(x) < 0$ if $x < 2$. Hence f is decreasing on $(-\infty, 2)$ and increasing on $(2, +\infty)$, as we see in Fig. 4.3.13. ◆

EXAMPLE 7 Show that the equation $e^x + x - 2 = 0$ has exactly one [real] solution.

Solution A solution of the given equation will be a zero of the function

$$f(x) = e^x + x - 2.$$

Now $f(0) = -1 < 0$ while $f(1) = e - 1 > 0$. Because f is continuous (everywhere), the intermediate value property of continuous functions therefore guarantees that $f(x)$ has *at least* one zero x_0 in the interval $(0, 1)$. We see this zero in Fig. 4.3.14, but cannot conclude from graphical evidence alone that there is no other zero somewhere (perhaps outside the viewing window of the figure).

To prove that there is no other zero, we note that f is an increasing function on the whole real line. This follows from Corollary 3 and the fact that

$$f'(x) = e^x + 1 > 1 > 0$$

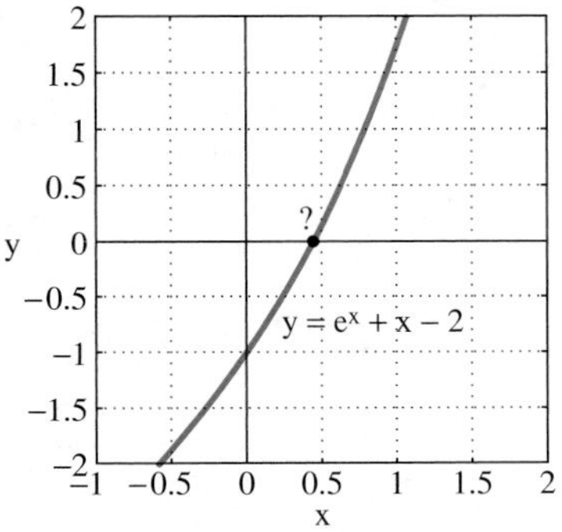

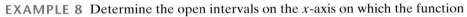

FIGURE 4.3.14 The graph $y = e^x + x - 2$.

because $e^x > 0$ for all x. Hence it follows from the definition of an increasing function that if $x < x_0$, then $f(x) < f(x_0) = 0$, while if $x > x_0$ then $f(x) > f(x_0) = 0$. Thus x_0 is the only zero of $f(x)$ and hence is the one and only real solution of the equation $e^x + x - 2 = 0$. ◆

EXAMPLE 8 Determine the open intervals on the x-axis on which the function

$$f(x) = 3x^4 - 4x^3 - 12x^2 + 5$$

is increasing and those on which it is decreasing.

Solution The derivative of f is

$$\begin{aligned} f'(x) &= 12x^3 - 12x^2 - 24x \\ &= 12x(x^2 - x - 2) = 12x(x + 1)(x - 2). \end{aligned} \quad \textbf{(3)}$$

The critical points $x = -1$, 0, and 2 separate the x-axis into the four open intervals $(-\infty, -1)$, $(-1, 0)$, $(0, 2)$, and $(2, +\infty)$ (Fig. 4.3.15). The derivative $f'(x)$ does not change sign within any of these intervals, because

- The factor $x + 1$ in Eq. (3) changes sign only at $x = -1$,
- The factor $12x$ changes sign only at $x = 0$, and
- The factor $x - 2$ changes sign only at $x = 2$.

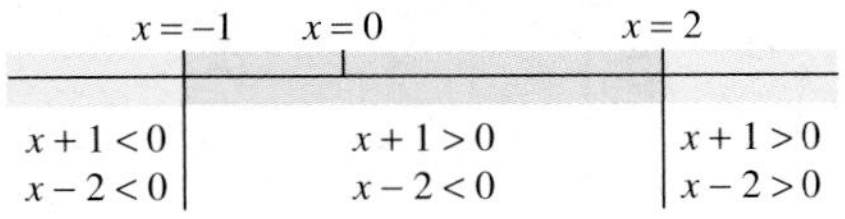

FIGURE 4.3.15 The signs of $x + 1$ and $x - 2$ (Example 8).

Figure 4.3.15 indicates the signs of $x + 1$ and $x - 2$ on each of the four intervals. We illustrate two different methods of determining the sign of $f'(x)$ on each interval.

Method 1 The second, third, and fourth columns of the next table record the signs of the factors in Eq. (3) on each of the four intervals listed in the first column. The signs of $f'(x)$ shown in the fifth column are then obtained by multiplication. The sixth column lists the resulting increasing or decreasing behavior of f on the four intervals.

Interval	$x+1$	$12x$	$x-2$	$f'(x)$	f
$(-\infty, -1)$	$-$	$-$	$-$	$-$	Decreasing
$(-1, 0)$	$+$	$-$	$-$	$+$	Increasing
$(0, 2)$	$+$	$+$	$-$	$-$	Decreasing
$(2, +\infty)$	$+$	$+$	$+$	$+$	Increasing

Method 2 Because the derivative $f'(x)$ does not change sign within any of the four intervals, we need only calculate its value at a single point in each interval. Whatever the sign at that point may be, it is the sign of $f'(x)$ throughout that interval.

In $(-\infty, -1)$: $f'(-2) = -96 < 0$; f is decreasing.
In $(-1, 0)$: $f'(-0.5) = 7.5 > 0$; f is increasing.
In $(0, 2)$: $f'(1) = -24 < 0$; f is decreasing.
In $(2, +\infty)$: $f'(3) = 144 > 0$; f is increasing.

The second method is especially convenient if the derivative is complicated but an appropriate calculator for computation of its values is available.

Finally, note that the results we have obtained in each method are consistent with the graph of $y = f(x)$ shown in Fig. 4.3.16. ◆

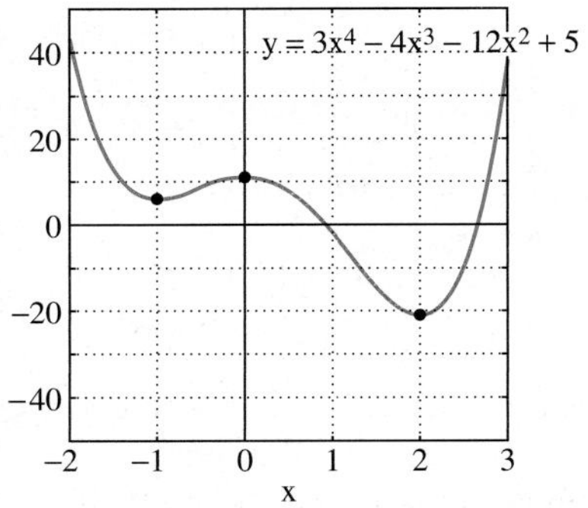

FIGURE 4.3.16 The critical points of the polynomial of Example 8.

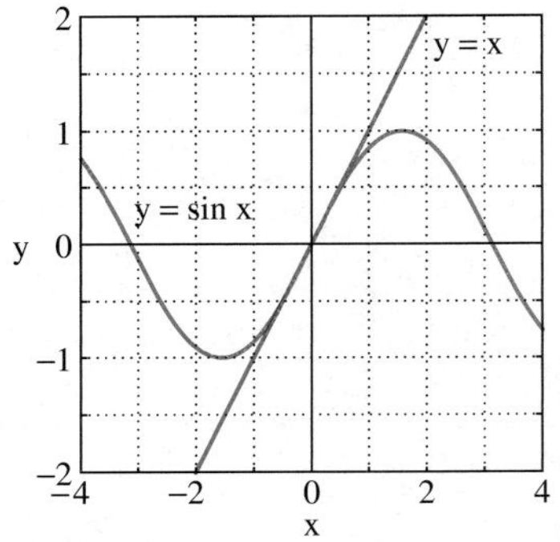

FIGURE 4.3.17 x and $\sin x$ (Example 9).

EXAMPLE 9 The graph in Fig. 4.3.17 suggests that $\sin x < x$ for all $x > 0$. To show that this is indeed so, it suffices to show that the *difference*

$$h(x) = f(x) - g(x) = x - \sin x$$

of the functions $f(x) = x$ and $g(x) = \sin x$ is positive-valued for $x > 0$. But

$$h'(x) = 1 - \cos x > 0$$

for all x in the interval $(0, 2\pi)$, where $\cos x < 1$. Hence Corollary 3 implies that h is an increasing function on the closed interval $[0, 2\pi]$. Because $h(0) = 0$, it therefore follows that $h(x) > 0$ if $0 < x \leqq 2\pi$. But if $x > 2\pi$ then certainly

$$h(x) = x - \sin x > 2\pi - \sin x > 0$$

because $|\sin x| \leqq 1$ for all x. Thus we have proved that

$$x - \sin x = h(x) > 0,$$

and hence that $x > \sin x$ for all $x > 0$. (Can you tell why it follows from this that $x < \sin x$ for all $x < 0$?) ◆

4.3 TRUE/FALSE STUDY GUIDE

4.3 CONCEPTS: QUESTIONS AND DISCUSSION

1. It's often said that "what goes up must come down." Can you translate this common saying into a mathematical statement? Does it follow from results in this section?
2. Suppose that $f'(x) > 0$ for all x in the open interval (a, b). Why does it follow that there exists an inverse function g such that $g(f(x)) = x$ for all x in (a, b)? What is the domain of definition of g?
3. Continuing Question 2, explain why it follows from results in this section that the function $f(x) = e^x$ has an inverse function ($g(x) = \ln x$) that is defined for all $x > 0$.
4. Why does it *not* follow from results in this section that the function $f(x) = \sin x$ has an inverse function g such that $g(f(x)) = x$ for all x? Determine a maximal closed interval I containing the origin such that there *does* exist a function g such that $g(f(x)) = x$ for all x in I. Does your function g agree with the function $\sin^{-1}$ on your calculator?
5. Repeat Question 4, except with (a) $f(x) = \tan x$; (b) $f(x) = \cos x$.

4.3 PROBLEMS

For the functions in Problems 1 through 6, first determine (as in Example 8) the open intervals on the x-axis on which each function is increasing and those where it is decreasing. Then use this information to match the function to its graph, one of the six shown in Fig. 4.3.18.

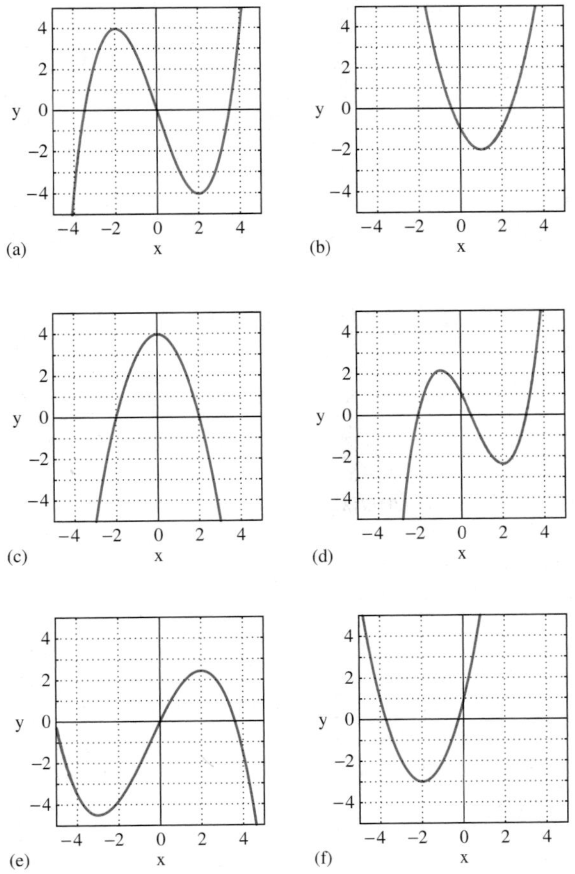

FIGURE 4.3.18 Problems 1 through 6.

1. $f(x) = 4 - x^2$
2. $f(x) = x^2 - 2x - 1$
3. $f(x) = x^2 + 4x + 1$
4. $f(x) = \frac{1}{4}x^3 - 3x$
5. $f(x) = \frac{1}{3}x^3 - \frac{1}{2}x^2 - 2x + 1$
6. $f(x) = 2x - \frac{1}{6}x^2 - \frac{1}{9}x^3$

In Problems 7 through 10, the derivative $f'(x)$ and the value $f(0)$ are given. Use the method of Example 5 to find the function $f(x)$.

7. $f'(x) = 4x$; $f(0) = 5$
8. $f'(x) = 3\sqrt{x}$; $f(0) = 4$
9. $f'(x) = \dfrac{1}{x^2}$; $f(1) = 1$
10. $f'(x) = 6e^{-3x}$; $f(0) = 3$

In Problems 11 through 24, determine (as in Example 8) the open intervals on the x-axis on which the function is increasing as well as those on which it is decreasing. If you have a graphics calculator or computer, plot the graph $y = f(x)$ to see whether it agrees with your results.

11. $f(x) = 3x + 2$
12. $f(x) = 4 - 5x$
13. $f(x) = 8 - 2x^2$
14. $f(x) = 4x^2 + 8x + 13$
15. $f(x) = 6x - 2x^2$
16. $f(x) = x^3 - 12x + 17$
17. $f(x) = x^4 - 2x^2 + 1$
18. $f(x) = \dfrac{x}{x+1}$ [*Note:* $f'(x)$ doesn't change sign at $x = -1$. Why?]
19. $f(x) = 3x^4 + 4x^3 - 12x^2$
20. $f(x) = x\sqrt{x^2 + 1}$
21. $f(x) = xe^{-x/2}$
22. $f(x) = x^2e^{-2x}$
23. $f(x) = (x - 1)^2e^{-x}$
24. $f(x) = \dfrac{\ln 2x}{x}$ for $x > 0$

In Problems 25 through 28, show that the given function satisfies the hypotheses of Rolle's theorem on the indicated interval $[a, b]$, and find all numbers x in (a, b) that satisfy the conclusion of that theorem.

25. $f(x) = x^2 - 2x$; $[0, 2]$
26. $f(x) = 9x^2 - x^4$; $[-3, 3]$

27. $f(x) = 2\sin x \cos x$; $[0, \pi]$

28. $f(x) = 5x^{2/3} - x^{5/3}$; $[0, 5]$

In Problems 29 through 31, show that the given function f does not satisfy the conclusion of Rolle's theorem on the indicated interval. Which of the hypotheses does it fail to satisfy?

29. $f(x) = 1 - |x|$; $[-1, 1]$

30. $f(x) = 1 - (2 - x)^{2/3}$; $[1, 3]$

31. $f(x) = xe^x$; $[0, 1]$

In Problems 32 through 36, show that the given function f satisfies the hypotheses of the mean value theorem on the indicated interval, and find all numbers c in that interval that satisfy the conclusion of that theorem.

32. $f(x) = x^3$; $[-1, 1]$

33. $f(x) = 3x^2 + 6x - 5$; $[-2, 1]$

34. $f(x) = \sqrt{x - 1}$; $[2, 5]$

35. $f(x) = (x - 1)^{2/3}$; $[1, 2]$

36. $f(x) = x + \dfrac{1}{x}$; $[2, 3]$

In Problems 37 through 40, show that the given function f satisfies neither the hypotheses nor the conclusion of the mean value theorem on the indicated interval.

37. $f(x) = |x - 2|$; $[1, 4]$

38. $f(x) = 1 + |x - 1|$; $[0, 3]$

39. $f(x) = [\![x]\!]$ (the greatest integer function); $[-1, 1]$

40. $f(x) = 3x^{2/3}$; $[-1, 1]$

In Problems 41 through 44, show that the given equation has exactly one solution in the indicated interval.

41. $x^5 + 2x - 3 = 0$; $[0, 1]$

42. $e^{-x} = x - 1$; $[1, 2]$

43. $x \ln x = 3$; $[2, 4]$

44. $\sin x = 3x - 1$; $[-1, 1]$

45. A car is driving along a rural road where the speed limit is 70 mi/h. At 3:00 P.M. its odometer (measuring distance traveled) reads 8075 mi. At 3:18 P.M. it reads 8100 mi. Prove that the driver violated the speed limit at some instant between 3:00 and 3:18 P.M.

46. Suppose that a car's speedometer reads 50 mi/h at 3:25 P.M. and 65 mi/h at 3:35 P.M. Prove that at some instant in this 10-minute time interval the car's acceleration was exactly 90 mi/h^2.

47. Points A and B along Interstate Highway 80 in Nebraska are 60 miles apart. Two cars both pass point A at 9:00 A.M. and both pass point B at 10:00 A.M. Show that at some instant between 9:00 and 10:00 A.M. the two cars have the same velocity. (*Suggestion:* Consider the difference $h(t) = f(t) - g(t)$ between the position functions of the two cars.)

48. Show that the function $f(x) = x^{2/3}$ does not satisfy the hypotheses of the mean value theorem on $[-1, 27]$ but that nevertheless there is a number c in $(-1, 27)$ such that

$$f'(c) = \frac{f(27) - f(-1)}{27 - (-1)}.$$

49. Prove that the function

$$f(x) = (1 + x)^{3/2} - \tfrac{3}{2}x - 1$$

is increasing on $(0, +\infty)$. Explain carefully how you could conclude that

$$(1 + x)^{3/2} > 1 + \tfrac{3}{2}x$$

for all $x > 0$.

50. Suppose that f' is a constant function on the interval $[a, b]$. Prove that f must be a linear function (a function whose graph is a straight line).

51. Suppose that $f'(x)$ is a polynomial of degree $n - 1$ on the interval $[a, b]$. Prove that $f(x)$ must be a polynomial of degree n on $[a, b]$.

52. Suppose that there are k different points of $[a, b]$ at which the differentiable function f vanishes (is zero). Prove that f' must vanish on at least $k - 1$ points of $[a, b]$.

53. (a) Apply the mean value theorem to $f(x) = \sqrt{x}$ on $[100, 101]$ to show that

$$\sqrt{101} = 10 + \frac{1}{2\sqrt{c}}$$

for some number c in $(100, 101)$. (b) Show that if $100 < c < 101$, then $10 < \sqrt{c} < 10.5$, and use this fact to conclude from part (a) that $10.0475 < \sqrt{101} < 10.0500$.

54. Prove that the equation $x^7 + x^5 + x^3 + 1 = 0$ has exactly one real solution.

55. (a) Show that $D_x \tan^2 x = D_x \sec^2 x$ on the open interval $(-\pi/2, \pi/2)$. (b) Conclude that there exists a constant C such that $\tan^2 x = \sec^2 x + C$ for all x in $(-\pi/2, \pi/2)$. Then evaluate C.

56. Explain why the mean value theorem does not apply to the function $f(x) = |x|$ on the interval $[-1, 2]$.

57. Suppose that the function f is differentiable on the interval $[-1, 2]$ and that $f(-1) = -1$ and $f(2) = 5$. Prove that there is a point on the graph of f at which the tangent line is parallel to the line with the equation $y = 2x$.

58. Let $f(x) = x^4 - x^3 + 7x^2 + 3x - 11$. Prove that the graph of f has at least one horizontal tangent line.

59. Let the function g be defined as follows:

$$g(x) = \begin{cases} \dfrac{x}{2} + x^2 \sin \dfrac{1}{x} & \text{if } x \neq 0, \\ 0 & \text{if } x = 0. \end{cases}$$

(a) Show that $g'(0) = \frac{1}{2} > 0$. (b) Sketch the graph of g near $x = 0$. Is g increasing on any open interval containing $x = 0$? [*Answer:* No.]

60. Suppose that f is increasing on every closed interval $[a, b]$ provided that $2 \leqq a < b$. Prove that f is increasing on the unbounded open interval $(2, +\infty)$. Note that the principle you discover was used implicitly in Example 6 of this section.

Approximations *Problems 61 through 64 illustrate the use of the mean value theorem to approximate numerical values of functions.*

61. Use the method of Example 9 with $f(x) = \cos x$ and $g(x) = 1 - \frac{1}{2}x^2$ to show that

$$\cos x > 1 - \tfrac{1}{2}x^2$$

for all $x > 0$ (Fig. 4.3.19).

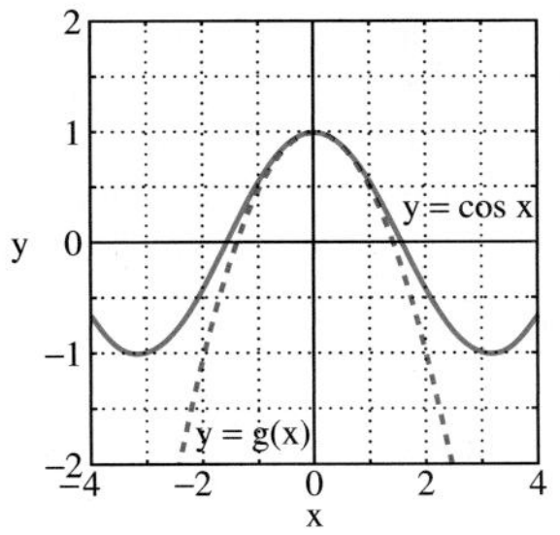

FIGURE 4.3.19 $\cos x$ and $g(x) = 1 - \frac{1}{2}x^2$ (Problem 61).

62. (a) Use the method of Example 9 and the result of Problem 61 to show that

$$\sin x > x - \tfrac{1}{6}x^3$$

for all $x > 0$ (Fig. 4.3.20). (b) Use the results of Example 9 and part (a) to calculate the sine of a 5° angle accurate to three decimal places.

63. (a) Use the results of Problem 62(a) to show that

$$\cos x < 1 - \tfrac{1}{2}x^2 + \tfrac{1}{24}x^4$$

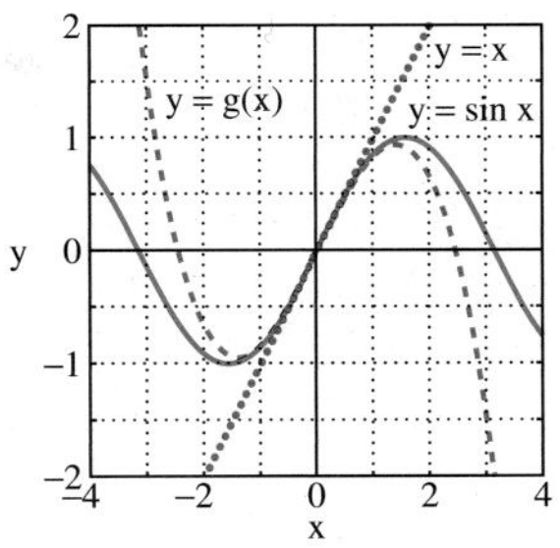

FIGURE 4.3.20 x, $\sin x$, and $g(x) = x - \frac{1}{6}x^3$ (Problem 62).

for all $x > 0$. (b) Use the results of Problem 61 and part (a) to calculate the cosine of a 10° angle accurate to three decimal places.

64. Let

$$p_n(x) = 1 - x + \frac{x^2}{2!} - \frac{x^3}{3!} + \cdots + (-1)^n \frac{x^n}{n!}$$

for each positive integer n.

(a) Use the method of Example 9 to show that $e^{-x} > p_1(x) = 1 - x$ for all $x > 0$.

(b) Use the result of part (a) to show that $e^{-x} < p_2(x) = 1 - x + \frac{1}{2}x^2$ for all $x > 0$.

(c) Use the result of part (b) to show that $e^{-x} > p_3(x) = 1 - x + \frac{1}{2}x^2 - \frac{1}{6}x^3$ for all $x > 0$.

(d) Continue one step at a time in like manner until you have shown that $p_7(x) < e^{-x} < p_8(x)$ for all $x < 0$. Finally, substitute $x = 1$ in this inequality to show that $e \approx 2.718$ accurate to three decimal places.

4.4 THE FIRST DERIVATIVE TEST AND APPLICATIONS

In Section 3.5 we discussed maximum and minimum values of a function defined on a closed and bounded interval $[a, b]$. Now we consider extreme values of functions defined on more general domains, including open or unbounded intervals as well as closed and bounded intervals.

The distinction between *absolute* and *local* extrema is important here. Let c be a point of the domain D of the function f. Then recall from Section 3.5 that $f(c)$ is the **(absolute) maximum value** of $f(x)$ on D provided that $f(c) \geqq f(x)$ for all x in D, whereas the value $f(c)$ is a **local maximum value** of $f(x)$ if it is the maximum value of $f(x)$ on some open interval containing c. Similarly, $f(c)$ is the **(absolute) minimum value** of $f(x)$ on D provided that $f(c) \leqq f(x)$ for all x in D; $f(c)$ is a **local minimum value** of $f(x)$ if it is the minimum value of $f(x)$ on some open interval containing c. Thus a local maximum value is one that is as large as or greater than any nearby value of $f(x)$, and a local minimum value is one that is as small as or less than any nearby value. Figure 4.4.1 shows a typical example of a function that has neither an absolute maximum nor an absolute minimum value. But each of the two local extrema pictured there is an (absolute) extreme value on a sufficiently small open interval.

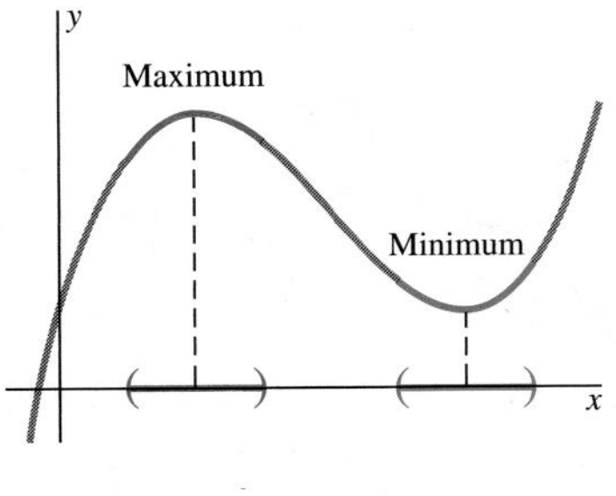

FIGURE 4.4.1 Local extrema are absolute extrema on sufficiently small intervals.

REMARK Absolute extreme values are sometimes called **global** extreme values, and local extreme values are sometimes called **relative** extreme values.

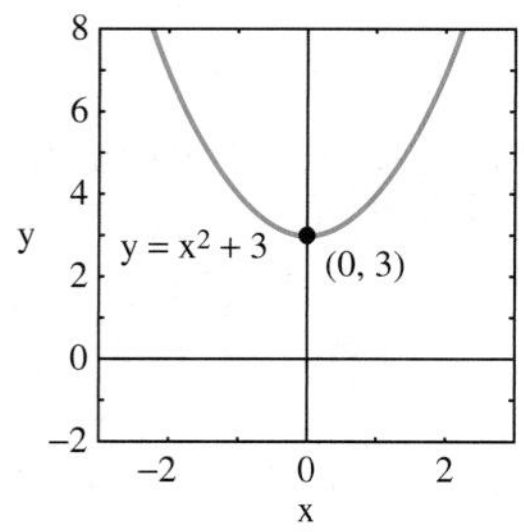

FIGURE 4.4.2 The graph of $f(x) = x^2 + 3$. The local minimum value $f(0) = 3$ is also the global minimum value of $f(x)$.

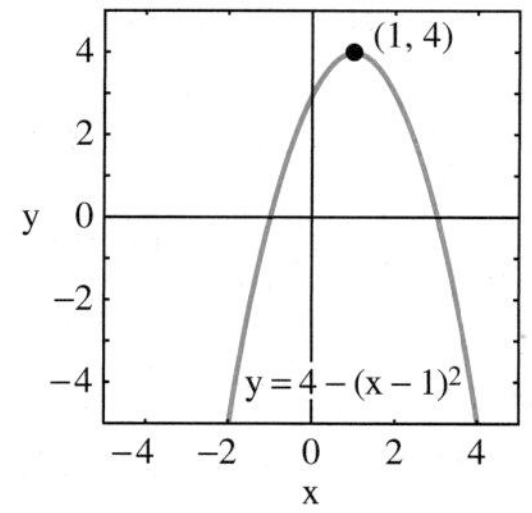

FIGURE 4.4.3 The graph of $f(x) = 4 - (x - 1)^2$. The local maximum value $f(1) = 4$ is also the global maximum value of $f(x)$.

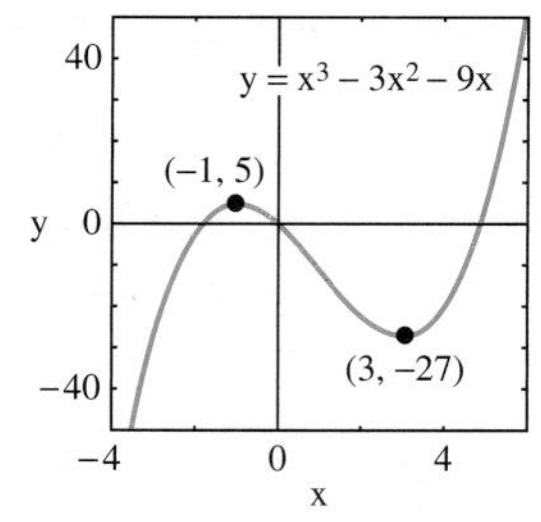

FIGURE 4.4.4 The graph of $f(x) = x^3 - 3x^2 - 9x$. The local minimum value $f(3) = -27$ clearly is not the global minimum value. Similarly, the local maximum value $f(-1) = 5$ is not the global maximum value.

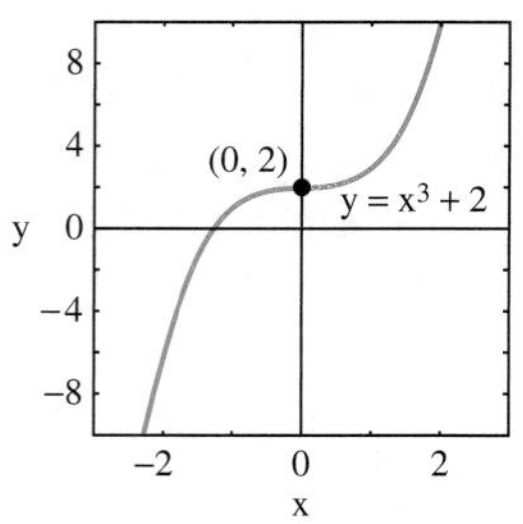

FIGURE 4.4.5 The graph of $f(x) = x^3 + 2$. The critical value $f(0) = 2$ is neither a global nor a local extreme value of $f(x)$.

Theorem 2 of Section 3.5 tells us that any extremum of the differentiable function f on an open interval I must occur at a *critical point* where the derivative vanishes:

$$f'(x) = 0.$$

But the mere fact that $f'(c) = 0$ does *not,* by itself, imply that the critical value $f(c)$ is an extreme value of f. Figures 4.4.2 through 4.4.5 illustrate different possibilities for the nature of $f(c)$: whether it is a local or global maximum or minimum value, or neither.

A Test for Local Extrema

What we need is a way to test whether, at the critical point $x = c$, the value $f(c)$ is actually a maximum or a minimum value of $f(x)$, either local or global. Figure 4.4.6 shows how such a test might be developed. Suppose that the function f is continuous at c and that c is an **interior point** of the domain of f—that is, f is defined on some open interval that contains c. If f is decreasing immediately to the left of c and increasing immediately to the right, then $f(c)$ should be a local minimum value of $f(x)$. But if f is increasing immediately to the left of c and decreasing immediately to its right, then $f(c)$ should be a local maximum. If f is increasing on both sides or decreasing on both sides, then $f(c)$ should be neither a maximum value nor a minimum value of $f(x)$.

Moreover, we know from Corollary 3 in Section 4.3 that the *sign* of the derivative $f'(x)$ determines where $f(x)$ is decreasing and where it is increasing:

- $f(x)$ is decreasing where $f'(x) < 0$;
- $f(x)$ is increasing where $f'(x) > 0$.

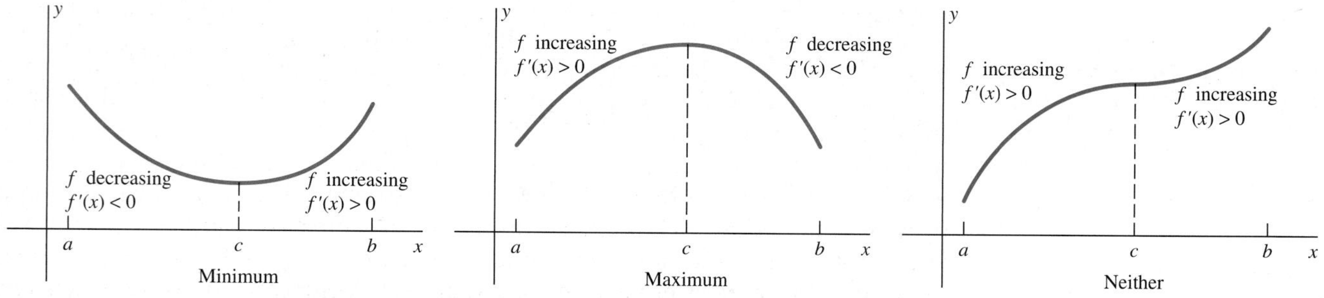

FIGURE 4.4.6 The first derivative test.

In the following test for local extrema, we say that

- $f'(x) < 0$ *to the left of* c if $f'(x) < 0$ on some interval (a, c) of numbers immediately to the left of c, and that
- $f'(x) > 0$ *to the right of* c if $f'(x) > 0$ on some interval (c, b) of numbers immediately to the right of c,

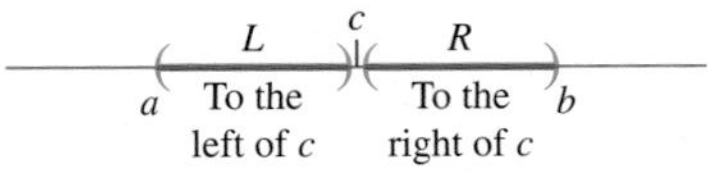

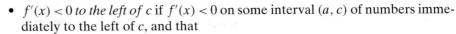

FIGURE 4.4.7 Open intervals to the left and right of the point c.

and so forth. (See Fig. 4.4.7.) Theorem 1 tells us how to use the *signs* of $f'(x)$ to the left and right of the point c to determine whether $f(x)$ has a local maximum or local minimum value at $x = c$.

THEOREM 1 The First Derivative Test for Local Extrema
Suppose that the function f is continuous on the interval I and also is differentiable there except possibly at the interior point c of I.

1. If $f'(x) < 0$ to the left of c and $f'(x) > 0$ to the right of c, then $f(c)$ is a *local minimum value* of $f(x)$ on I.
2. If $f'(x) > 0$ to the left of c and $f'(x) < 0$ to the right of c, then $f(c)$ is a *local maximum value* of $f(x)$ on I.
3. If $f'(x) > 0$ both to the left of c and to the right of c, *or* if $f'(x) < 0$ both to the left of c and to the right of c, then $f(c)$ is *neither* a maximum nor a minimum value of $f(x)$.

COMMENT Thus $f(c)$ is a local extremum if the first derivative $f'(x)$ *changes sign* as x increases through c, and the direction of this sign change determines whether $f(c)$ is a local maximum or a local minimum. A good way to remember the first derivative test for local extrema is simply to visualize Fig. 4.4.6.

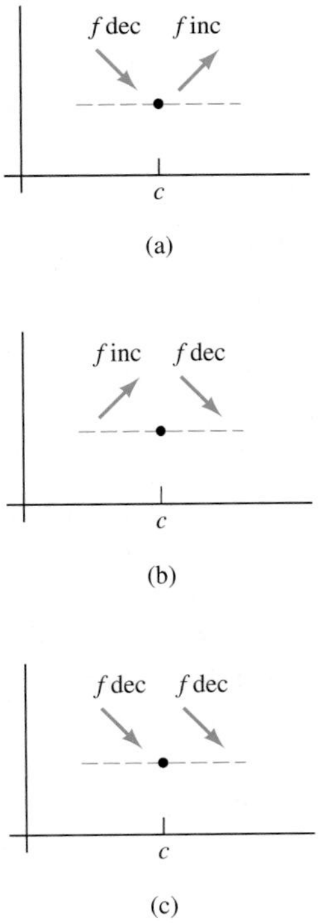

FIGURE 4.4.8 The three cases in the first derivative test.

PROOF We will prove only part 1; the other two parts have similar proofs. Suppose that the hypotheses of Theorem 1 hold: that f is continuous on the interval I, that c is an interior point of I, and that f is differentiable on I except possibly at $x = c$. Then there exist two intervals (a, c) and (c, b), each wholly contained in I, such that $f'(x) < 0$ on (a, c) and $f'(x) > 0$ on (c, b).

Suppose that x is in (a, b). Then there are three cases to consider. First, if $x < c$, then x is in (a, c) and f is decreasing on $(a, c]$, so $f(x) > f(c)$. Second, if $x > c$, then x is in (c, b) and f is increasing on $[c, b)$, so again $f(x) > f(c)$. Finally, if $x = c$, then $f(x) = f(c)$. Thus, for each x in (a, b), $f(x) \geqq f(c)$. Therefore, by definition, $f(c)$ is a local minimum value of $f(x)$. ◂

The idea of this proof is illustrated in Fig. 4.4.8. Part (a) shows f decreasing to the left of c and increasing to the right, so there must be a local minimum at $x = c$. Part (b) shows f increasing to the left of c and decreasing to the right, so $f(c)$ is a local maximum value of $f(x)$. In part (c), the derivative has the same sign on each side of c, and so there can be no extremum of any sort at $x = c$.

REMARK Figures 4.4.9 through 4.4.13 illustrate cases in which Theorem 1 applies, where the interval I is the entire real number line $\boldsymbol{R}$. In Fig. 4.4.9 through 4.4.11, the origin $c = 0$ is a critical point because $f'(0) = 0$. In Figs. 4.4.12 and 4.4.13, $c = 0$ is a critical point because $f'(0)$ does not exist.

Classification of Critical Points

Suppose that we have found the critical points of a function. Then we can attempt to classify them—as local maxima, local minima, or neither—by applying the first derivative test at each point in turn. Example 1 illustrates a procedure that can be used.

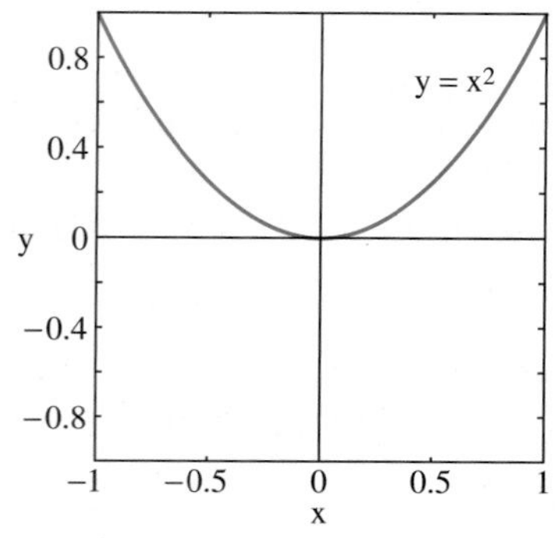

FIGURE 4.4.9 $f(x) = x^2, f'(x) = 2x$, a local minimum at $x = 0$.

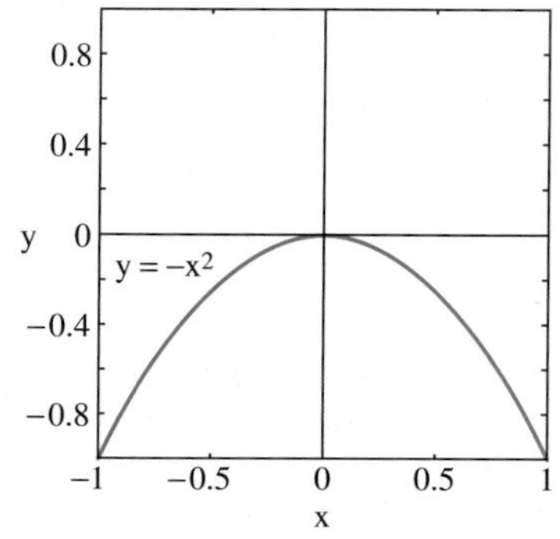

FIGURE 4.4.10 $f(x) = -x^2$, $f'(x) = -2x$, a local maximum at $x = 0$.

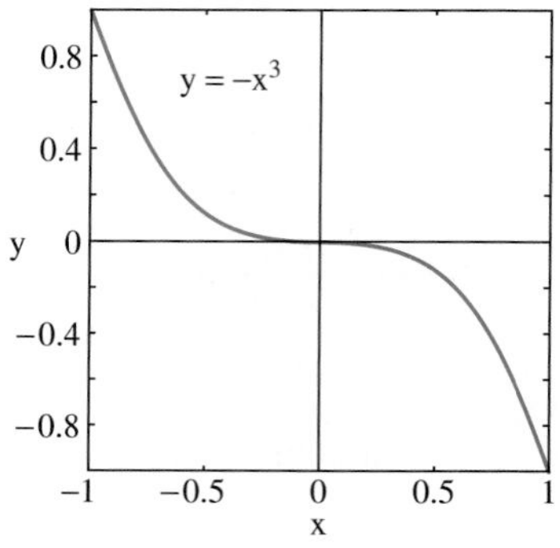

FIGURE 4.4.11 $f(x) = -x^3$, $f'(x) = -3x^2$, no extremum at $x = 0$.

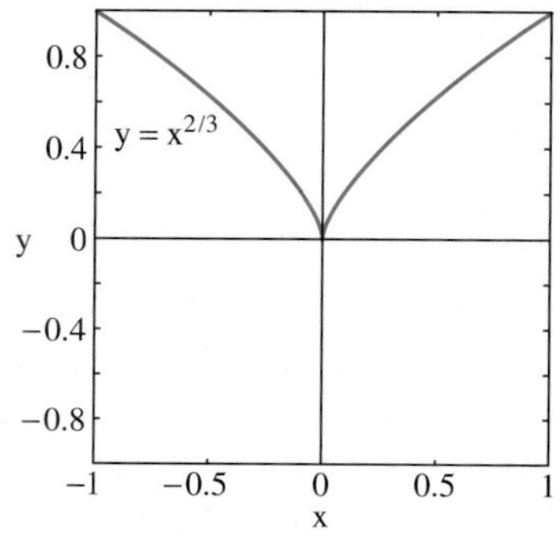

FIGURE 4.4.12 $f(x) = x^{2/3}$, $f'(x) = \frac{2}{3}x^{-1/3}$, a local minimum at $x = 0$.

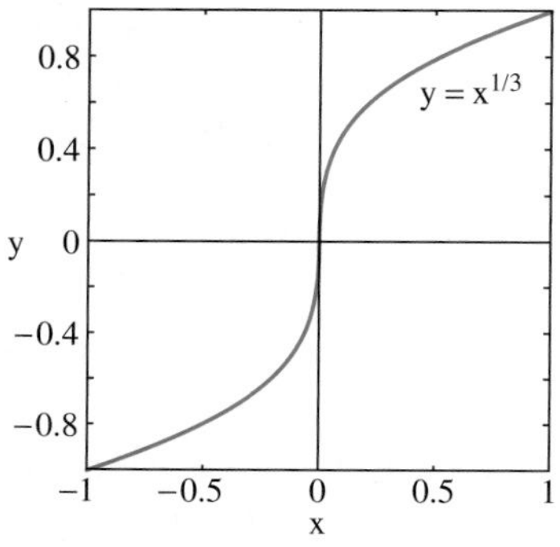

FIGURE 4.4.13 $f(x) = x^{1/3}$, $f'(x) = \frac{1}{3}x^{-2/3}$, no extremum at $x = 0$.

EXAMPLE 1 Find and classify the critical points of the function

$$f(x) = 2x^3 - 3x^2 - 36x + 7.$$

Solution The derivative is

$$f'(x) = 6x^2 - 6x - 36 = 6(x + 2)(x - 3), \qquad \textbf{(1)}$$

so the critical points [where $f'(x) = 0$] are $x = -2$ and $x = 3$. These two points separate the x-axis into the three open intervals $(-\infty, -2)$, $(-2, 3)$, and $(3, +\infty)$. The derivative $f'(x)$ cannot change sign within any of these intervals. One reason is that the factor $x + 2$ in Eq. (1) changes sign only at -2, whereas the factor $x - 3$ changes sign only at 3 (Fig. 4.4.14). As in Example 7 of Section 4.3, we illustrate here two methods of determining the signs of $f'(x)$ on the intervals $(-\infty, -2)$, $(-2, 3)$, and $(3, +\infty)$.

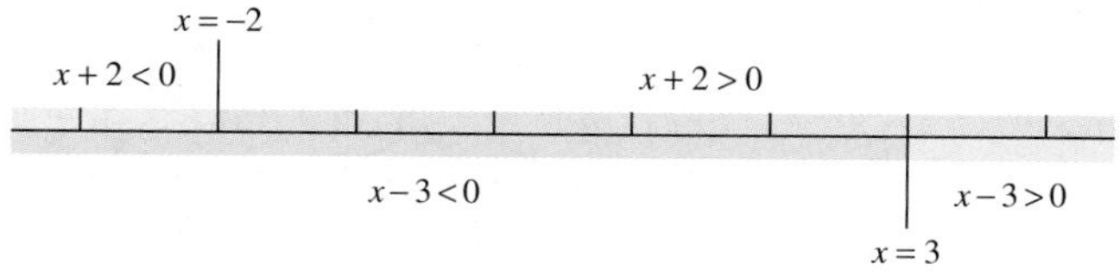

FIGURE 4.4.14 The signs of $x + 2$ and $x - 3$ (Example 1).

Method 1 The second and third columns of the following table record (from Fig. 4.4.14) the signs of the factors $x + 2$ and $x - 3$ in Eq. (1) on the three intervals listed in the first column. The signs of $f'(x)$ in the fourth column are then obtained by multiplication.

Interval	$x+2$	$x-3$	$f'(x)$
$(-\infty, -2)$	$-$	$-$	$+$
$(-2, 3)$	$+$	$-$	$-$
$(3, +\infty)$	$+$	$+$	$+$

Method 2 Because the derivative $f'(x)$ does not change sign within any of the three intervals, we need to calculate its value only at a single point in each interval:

In $(-\infty, -2)$: $f'(-3) = 36 > 0$; f' is positive;
In $(-2, 3)$: $f'(0) = -36 < 0$; f' is negative;
In $(3, +\infty)$: $f'(4) = 36 > 0$; f' is positive.

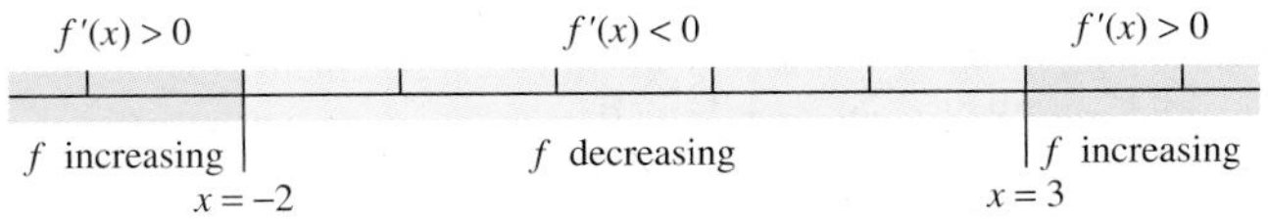

FIGURE 4.4.15 The three intervals of Example 1.

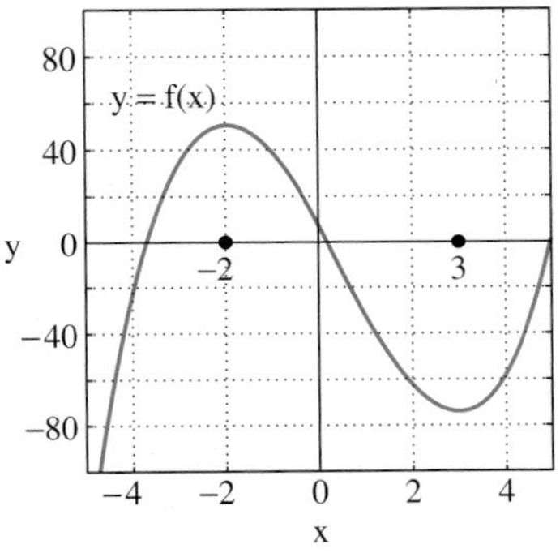

FIGURE 4.4.16 $y = f(x)$ (Example 1).

Figure 4.4.15 summarizes our information about the signs of $f'(x)$. Because $f'(x)$ is positive to the left and negative to the right of the critical point $x = -2$, the first derivative test implies that $f(-2) = 51$ is a local maximum value. Because $f'(x)$ is negative to the left and positive to the right of $x = 3$, it follows that $f(3) = -74$ is a local minimum value. The graph of $y = f(x)$ in Fig. 4.4.16 confirms this classification of the critical points $x = -2$ and $x = 3$. ◆

Open-Interval Maximum-Minimum Problems

In Section 3.6 we discussed applied maximum-minimum problems in which the values of the dependent variable are given by a function defined on a closed and bounded interval. Sometimes, though, the function f describing the variable to be maximized (or minimized) is defined on an *open* interval (a, b), possibly an *unbounded* open interval such as $(1, +\infty)$ or $(-\infty, +\infty)$, and we cannot "close" the interval by adjoining endpoints. Typically, the reason is that $|f(x)| \to +\infty$ as x approaches a or b. But if f has only a single critical point in (a, b), then the first derivative test can tell us that $f(c)$ is the desired extreme value and can even determine whether it is a maximum or a minimum value of $f(x)$.

EXAMPLE 2 Figure 4.4.17 shows the graph of the function

$$f(x) = \frac{2 \ln x}{x},$$

which is defined on the open interval $(0, +\infty)$. Because

$$f'(x) = \frac{2}{x} \cdot \frac{1}{x} - \frac{2}{x^2} \cdot \ln x = \frac{2}{x^2}(1 - \ln x),$$

there is a lone critical point at $x = e$. Note that

- If $x < e$, then $\ln x < 1$, so $f'(x) > 0$ if $x < e$;
- If $x > e$, then $\ln x > 1$, so $f'(x) < 0$ if $x > e$.

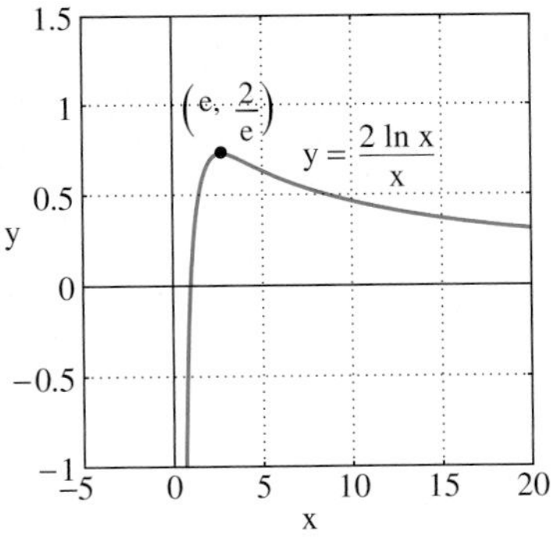

FIGURE 4.4.17 The graph $y = \dfrac{2 \ln x}{x}$.

Therefore the first derivative test implies that $f(e) = 2/e$ is a local maximum value of f. Indeed, because f is increasing if $0 < x < e$ and decreasing if $x > e$, it follows that $2/e$ is the absolute maximum value of f. ◆

EXAMPLE 3 Find the (absolute) minimum value of

$$f(x) = x + \frac{4}{x} \quad \text{for} \quad 0 < x < +\infty.$$

Solution The derivative is

$$f'(x) = 1 - \frac{4}{x^2} = \frac{x^2 - 4}{x^2}. \tag{2}$$

The roots of the equation

$$f'(x) = \frac{x^2 - 4}{x^2} = 0$$

are $x = -2$ and $x = 2$. But $x = -2$ is not in the open interval $(0, +\infty)$, so we have only the critical point $x = 2$ to consider.

We see immediately from Eq. (2) that

- $f'(x) < 0$ to the left of $x = 2$ (because $x^2 < 4$ there), and
- $f'(x) > 0$ to the right of $x = 2$ (because $x^2 > 4$ there).

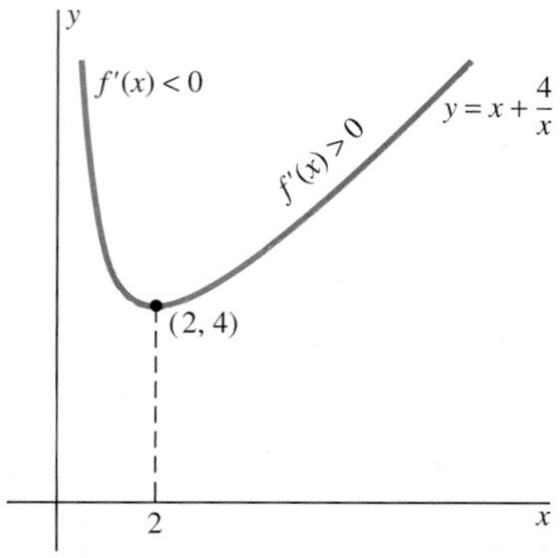

FIGURE 4.4.18 The graph of the function of Example 3.

Therefore, the first derivative test implies that $f(2) = 4$ is a local minimum value. We note also that $f(x) \to +\infty$ as either $x \to 0^+$ or as $x \to +\infty$. Hence the graph of f must resemble Fig. 4.4.18, and we see that $f(2) = 4$ is in fact the absolute minimum value of $f(x)$ on the entire interval $(0, +\infty)$. ◆

EXAMPLE 4 We must make a cylindrical can with volume 125 in.^{3} (about 2 L) by cutting its top and bottom from squares of metal and forming its curved side by bending a rectangular sheet of metal to match its ends. What radius r and height h of the can will minimize the total amount of material required for the rectangle and the two squares?

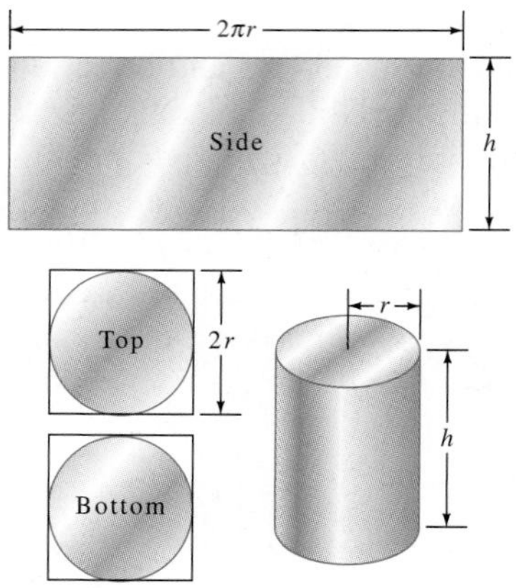

FIGURE 4.4.19 The parts to make the cylindrical can of Example 4.

Solution We assume that the corners cut from the two squares, shown in Fig. 4.4.19, are wasted but that there is no other waste. As the figure shows, the area of the total amount of sheet metal required is

$$A = 8r^2 + 2\pi rh.$$

The volume of the resulting can is then

$$V = \pi r^2 h = 125,$$

so $h = 125/(\pi r^2)$. Hence A is given as a function of r by

$$A(r) = 8r^2 + 2\pi r \cdot \frac{125}{\pi r^2} = 8r^2 + \frac{250}{r}, \quad 0 < r < +\infty.$$

The domain of A is the unbounded open interval $(0, +\infty)$ because r can have any positive value, so $A(r)$ is defined for every number r in $(0, +\infty)$. But $A(r) \to +\infty$ as $r \to 0^+$ and as $r \to +\infty$. So we cannot use the closed-interval maximum-minimum method. But we *can* use the first derivative test.

The derivative of $A(r)$ is

$$\frac{dA}{dr} = 16r - \frac{250}{r^2} = \frac{16}{r^2}\left(r^3 - \frac{125}{8}\right). \tag{3}$$

Thus the only critical point in $(0, +\infty)$ is where $r^3 = \frac{125}{8}$; that is,

$$r = \sqrt[3]{\frac{125}{8}} = \frac{5}{2} = 2.5.$$

We see immediately from Eq. (3) that

- $dA/dr < 0$ to the left of $r = \frac{5}{2}$, because $r^3 < \frac{125}{8}$ there, and
- $dA/dr > 0$ to the right, where $r^3 > \frac{125}{8}$.

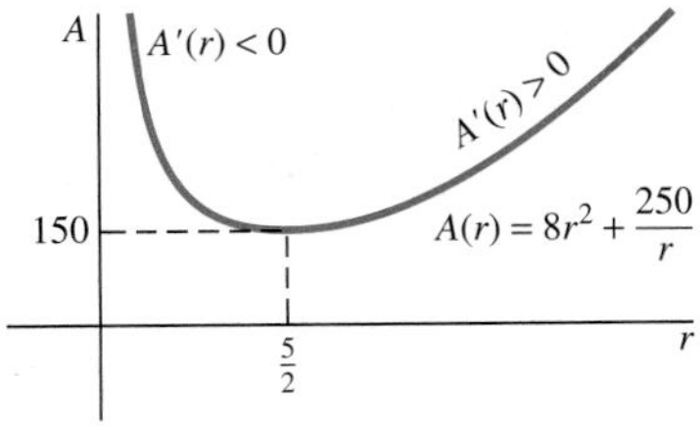

FIGURE 4.4.20 Graph of the function of Example 4.

Therefore, the first derivative test implies that a local minimum value of $A(r)$ on $(0, +\infty)$ is

$$A\left(\frac{5}{2}\right) = 8 \cdot \left(\frac{5}{2}\right)^2 + \frac{250}{\frac{5}{2}} = 150.$$

Considering that $A(r) \to +\infty$ as $r \to 0^+$ and as $r \to +\infty$, we see that the graph of $A(r)$ on $(0, +\infty)$ looks like Fig. 4.4.20. This clinches the fact that $A(\frac{5}{2}) = 150$ is the *absolute* minimum value of $A(r)$. Therefore, we minimize the amount of material required by making a can with radius $r = 2.5$ in. and height

$$h = \frac{125}{\pi(2.5)^2} = \frac{20}{\pi} \approx 6.37 \quad \text{(in.)}.$$

The total amount of material used is 150 in.^{2} ◆

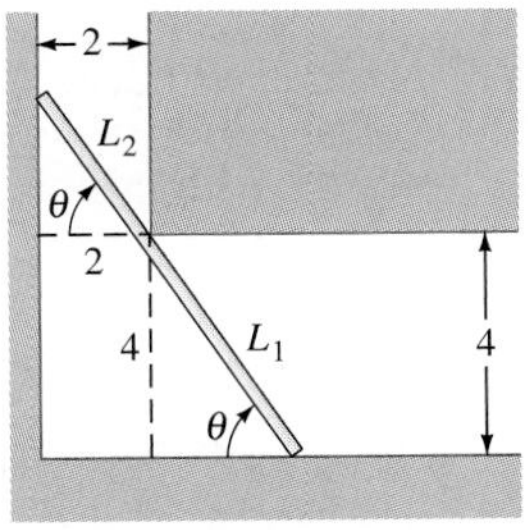

FIGURE 4.4.21 Carrying a rod around a corner (Example 5).

EXAMPLE 5 Find the length of the longest rod that can be carried horizontally around the corner from a hall 2 m wide into one that is 4 m wide.

Solution The desired length is the *minimum* length $L = L_1 + L_2$ of the rod being carried around the corner in Fig. 4.4.21. We see from the two similar triangles in the figure that

$$\frac{4}{L_1} = \sin\theta \quad \text{and} \quad \frac{2}{L_2} = \cos\theta,$$

so

$$L_1 = 4\csc\theta \quad \text{and} \quad L_2 = 2\sec\theta.$$

Therefore, the length $L = L_1 + L_2$ of the rod is given as a function of θ by

$$L(\theta) = 4\csc\theta + 2\sec\theta$$

on the open interval $(0, \pi/2)$. Note that $L(\theta) \to +\infty$ as either $\theta \to 0^+$ or as $\theta \to (\pi/2)^-$. (Why?)

The derivative of $L(\theta)$ is

$$\begin{aligned}\frac{dL}{d\theta} &= -4\csc\theta\,\cot\theta + 2\sec\theta\,\tan\theta \\ &= -\frac{4\cos\theta}{\sin^2\theta} + \frac{2\sin\theta}{\cos^2\theta} = \frac{2\sin^3\theta - 4\cos^3\theta}{\sin^2\theta\,\cos^2\theta} \\ &= \frac{(2\cos\theta)(\tan^3\theta - 2)}{\sin^2\theta}.\end{aligned} \quad \textbf{(4)}$$

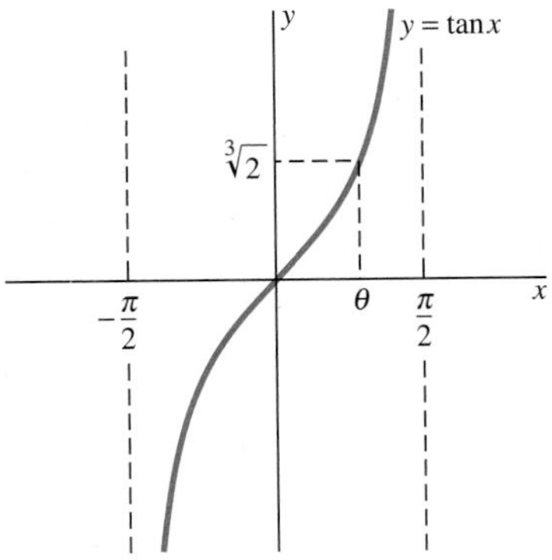

FIGURE 4.4.22 $y = \tan x$ (Example 5).

Hence $dL/d\theta = 0$ exactly when

$$\tan\theta = \sqrt[3]{2}, \quad \text{so} \quad \theta \approx 0.90 \quad \text{(rad)}.$$

We now see from Eq. (4) and from the graph of the tangent function (Fig. 4.4.22) that

- $dL/d\theta < 0$ to the left of $\theta \approx 0.90$, where $\tan\theta < \sqrt[3]{2}$, so $\tan^3\theta < 2$, and
- $dL/d\theta > 0$ to the right, where $\tan^3\theta > 2$.

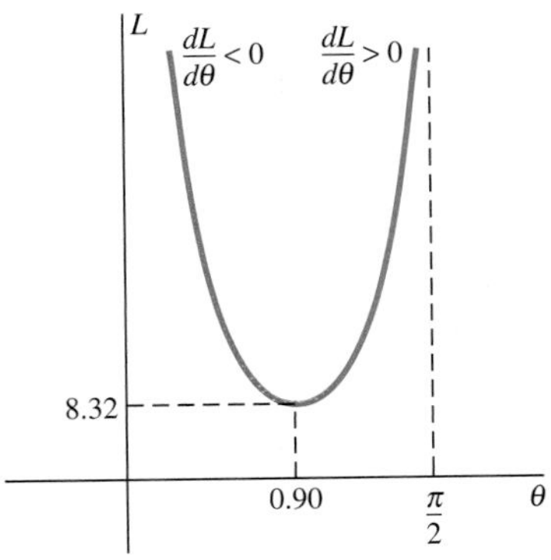

FIGURE 4.4.23 The graph of $L(\theta)$ (Example 5).

Hence the graph of L resembles Fig. 4.4.23. This means that the absolute minimum value of L—and therefore the maximum length of the rod in question—is about

$$L(0.90) = 4\csc(0.90) + 2\sec(0.90),$$

approximately 8.32 m. ◆

The method we used in Examples 3 through 5 to establish absolute extrema illustrates the following global version of the first derivative test.

THEOREM 2 The First Derivative Test for Global Extrema
Suppose that f is defined on an open interval I, either bounded or unbounded, and that f is differentiable at each point of I except possibly at the critical point c where f is continuous.

1. If $f'(x) < 0$ for all x in I with $x < c$ and $f'(x) > 0$ for all x in I with $x > c$, then $f(c)$ is the absolute minimum value of $f(x)$ on I.
2. If $f'(x) > 0$ for all x in I with $x < c$ and $f'(x) < 0$ for all x in I with $x > c$, then $f(c)$ is the absolute maximum value of $f(x)$ on I.

The proof of this theorem is essentially the same as that of Theorem 1.

REMARK When the function $f(x)$ has only one critical point c in an open interval I, Theorem 2 may apply to tell us either that $f(c)$ is the absolute minimum or that it is the absolute maximum of $f(x)$ on I. But it is good practice to verify your conclusion by sketching the graph as we did in Examples 3 through 5.

4.4 TRUE/FALSE STUDY GUIDE

4.4 CONCEPTS: QUESTIONS AND DISCUSSION

1. Suppose that the function f is continuous on the whole real line $\boldsymbol{R}$. Sketch a possible graph—if any—of f in each of the following three cases.
   (a) f has two critical points a and b but neither a local minimum nor a local maximum anywhere. Discuss separately the various possibilities: as to whether f is, or is not, differentiable at a and/or b.
   (b) f has three critical points but only a single local extremum.
   (c) f has three local minima but only a single local maximum.
2. Suppose that f is a cubic polynomial with positive leading coefficient. List the possibilities—with a typical graph of each—for the number and types of critical points of f.
3. Repeat Question 2 for a quartic (fourth-degree) polynomial.
4. Repeat Question 2 for a quintic (fifth-degree) polynomial.
5. Can you show that the function f must have an absolute minimum value if f is a polynomial of even degree with positive leading coefficient?
6. Can you show that the function f must have an absolute minimum value if
$$\lim_{x\to a^+} f(x) = \lim_{x\to b^-} f(x) = +\infty$$
and f is continuous on the open interval (a, b)?

4.4 PROBLEMS

Apply the first derivative test to classify each of the critical points of the functions in Problems 1 through 16 (local or global, maximum or minimum, or not an extremum). If you have a graphics calculator or computer, plot $y = f(x)$ to see whether the appearance of the graph corresponds to your classification of the critical points.

1. $f(x) = x^2 - 4x + 5$

2. $f(x) = 6x - x^2$

3. $f(x) = x^3 - 3x^2 + 5$

4. $f(x) = x^3 - 3x + 5$

5. $f(x) = x^3 - 3x^2 + 3x + 5$

6. $f(x) = 2x^3 + 3x^2 - 36x + 17$

7. $f(x) = 10 + 60x + 9x^2 - 2x^3$

8. $f(x) = 27 - x^3$

9. $f(x) = x^4 - 2x^2$

10. $f(x) = 3x^5 - 5x^3$

11. $f(x) = x + \dfrac{9}{x}$

12. $f(x) = x^2 + \dfrac{2}{x}$

13. $f(x) = xe^{-2x}$

14. $f(x) = x^2e^{-x/3}$

15. $f(x) = (x + 4)^2e^{-x/5}$

16. $f(x) = \dfrac{1 - \ln x}{x}$ for $x > 0$

In Problems 17 through 26, find and classify the critical points of the given function in the indicated open interval. You may find it useful to construct a table of signs as in Example 1.

17. $f(x) = \sin^2 x; \quad (0, 3)$

18. $f(x) = \cos^2 x; \quad (-1, 3)$

19. $f(x) = \sin^3 x; \quad (-3, 3)$

20. $f(x) = \cos^4 x; \quad (0, 4)$

21. $f(x) = \sin x - x \cos x; \quad (-5, 5)$

22. $f(x) = \cos x + x \sin x; \quad (-5, 5)$

23. $f(x) = \dfrac{\ln x}{x^2}; \quad (0, 5)$

24. $f(x) = \dfrac{\ln(1+x)}{1+x}; \quad (0, 5)$

25. $f(x) = e^x \sin x; \quad (-3, 3)$

26. $f(x) = x^3 e^{-x-x^2}; \quad (-3, 3)$

In Problems 27 through 50, which are applied maximum-minimum problems, use the first derivative test to verify your answer.

27. Determine two real numbers with difference 20 and minimum possible product.

28. A long rectangular sheet of metal is to be made into a rain gutter by turning up two sides at right angles to the remaining center strip (Fig. 4.4.24). The rectangular cross section of the gutter is to have area 18 in.^{2} Find the minimum possible width of the sheet.

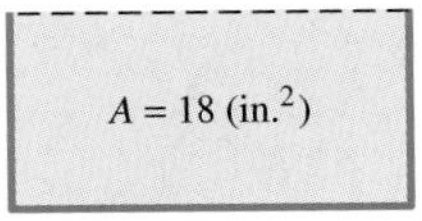

FIGURE 4.4.24 The rectangular cross section of the gutter of Problem 28.

29. Find the point (x, y) on the line $2x + y = 3$ that is closest to the point $(3, 2)$.

30. You must construct a closed rectangular box with volume 576 in.^{3} and with its bottom twice as long as it is wide (Fig. 4.4.25). Find the dimensions of the box that will minimize its total surface area.

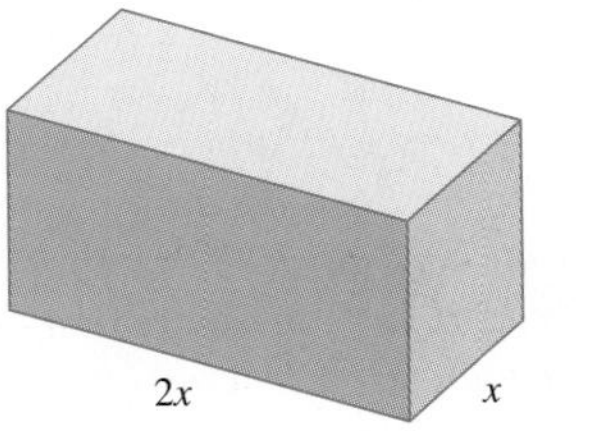

FIGURE 4.4.25 The box of Problem 30.

31. Repeat Problem 30, but use an open-topped rectangular box with volume 972 in.^{3}

32. An open-topped cylindrical pot is to have volume 125 in.^{3} What dimensions will minimize the total amount of material used in making this pot (Fig. 4.4.26)? Neglect the thickness of the material and possible wastage.

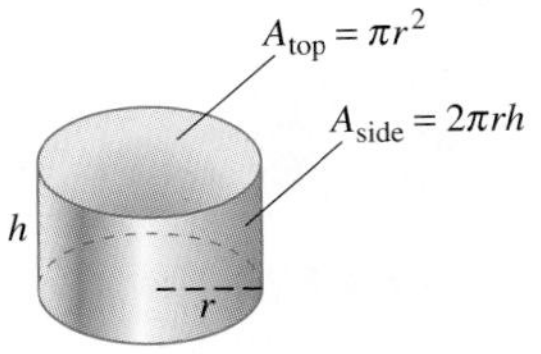

FIGURE 4.4.26 The cylinder of Problems 32, 33, 38, and 39.

33. An open-topped cylindrical pot is to have volume 250 cm^3 (Fig. 4.4.26). The material for the bottom of the pot costs 4¢/cm^2; that for its curved side costs 2¢/cm^2. What dimensions will minimize the total cost of this pot?

34. Find the point (x, y) on the parabola $y = 4 - x^2$ that is closest to the point $(3, 4)$. [*Suggestion:* The cubic equation that you should obtain has a small integer as one of its roots. *Suggestion:* Minimize the *square* of the distance.]

35. Show that the rectangle with area 100 and minimum perimeter is a square.

36. Show that the rectangular solid with a square base, volume 1000, and minimum total surface area is a cube.

37. A box with a square base and an open top is to have volume 62.5 in.^{3} Neglect the thickness of the material used to make the box, and find the dimensions that will minimize the amount of material used.

38. You need a tin can in the shape of a right circular cylinder of volume 16π cm^3 (Fig. 4.4.26). What radius r and height h would minimize its total surface area (including top and bottom)?

39. The metal used to make the top and bottom of a cylindrical can (Fig. 4.4.26) costs 4¢/in.^{2}; the metal used for the sides costs 2¢/in.^{2} The volume of the can must be exactly 100 in.^{3}. What dimensions of the can would minimize its total cost?

40. Each page of a book will contain 30 in.^{2} of print, and each page must have 2-in. margins at top and bottom and a 1-in. margin at each side. What is the minimum possible area of such a page?

41. What point or points on the curve $y = x^2$ are nearest the point $(0, 2)$? [*Suggestion:* The square of a distance is minimized exactly when the distance itself is minimized.]

42. What is the length of the shortest line segment lying wholly in the first quadrant tangent to the graph of $y = 1/x$ and with its endpoints on the coordinate axes?

43. A rectangle has area 64 cm^2. A straight line is to be drawn from one corner of the rectangle to the midpoint of one of the two more distant sides. What is the minimum possible length of such a line?

44. An oil can is to have volume 1000 in.^{3} and is to be shaped like a cylinder with a flat bottom but capped by a hemisphere (Fig. 4.4.27). Neglect the thickness of the material of the can, and find the dimensions that will minimize the total amount of material needed to construct it.

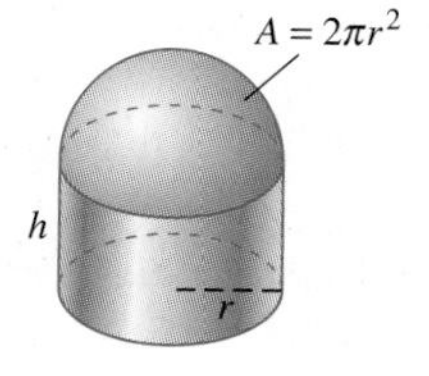

FIGURE 4.4.27 The oil can of Problem 44.

45. Find the exact length L of the longest rod that can be carried horizontally around a corner from a corridor 2 m wide into one 4 m wide. Do this by *minimizing* the length of the rod in Fig. 4.4.28 by minimizing the square of that length as a function of x.

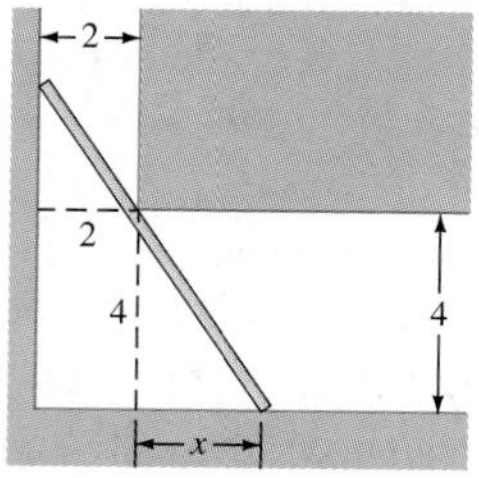

FIGURE 4.4.28 Carrying a rod around a corner (Problem 45).

46. Find the length of the shortest ladder that will reach from the ground, over a wall 8 ft high, to the side of a building 1 ft behind the wall. That is, minimize the length $L = L_1 + L_2$ shown in Fig. 4.4.29.

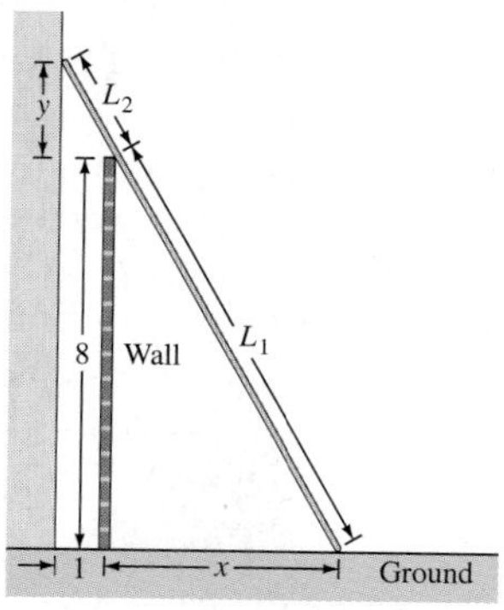

FIGURE 4.4.29 The ladder of Problem 46.

47. A sphere with fixed radius a is inscribed in a pyramid with a square base so that the sphere touches the base of the pyramid and also each of its four sides. Show that the minimum possible volume of the pyramid is $8/\pi$ times the volume of the sphere. [*Suggestion:* Use the two right triangles in Fig. 4.4.30 to show that the volume of the pyramid is

$$V = V(y) = \frac{4a^2y^2}{3(y-2a)}.$$

This can be done easily with the aid of the angle θ and *without* the formula for $\tan(\theta/2)$.] Don't forget the domain of $V(y)$.

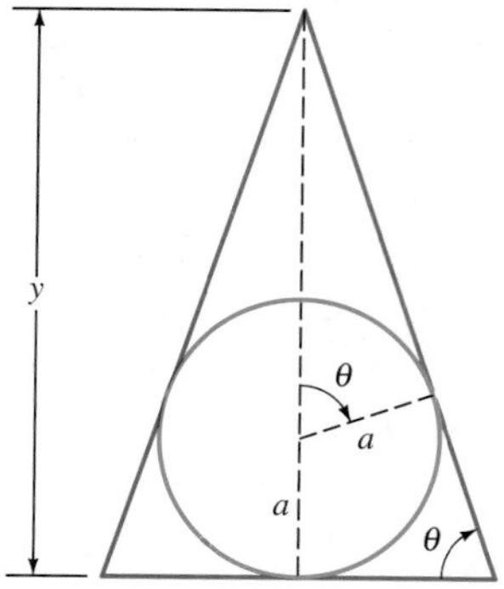

FIGURE 4.4.30 Cross section through the centers of the sphere and pyramid of Problem 47.

48. Two noisy discothèques, one four times as noisy as the other, are located on opposite ends of a block 1000 ft long. What is the quietest point on the block between the two discos? The intensity of noise at a point away from its source is proportional to the noisiness and inversely proportional to the square of the distance from the source.

49. A floored tent with fixed volume V is to be shaped like a pyramid with a square base and congruent sides (Fig. 4.4.31). What height y and base edge $2x$ would minimize its total surface area (including its floor)?

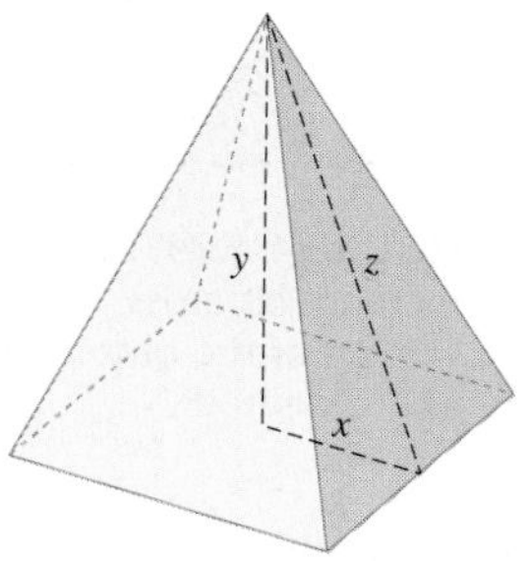

FIGURE 4.4.31 The tent of Problem 49.

50. Suppose that the distance from the building to the wall in Problem 46 is a and that the height of the wall is b. Show that the minimal length of the ladder is

$$L_{\text{min}} = \left(a^{2/3} + b^{2/3}\right)^{3/2}.$$

Problems 51 and 52 deal with square-based rectangular boxes. Such a box is said to be closed *if it has both a square base and a top (as well as four vertical sides),* open *if it has a base but no top. (Problems 51 through 55 here are in a certain sense "dual" to Problems 56 through 60 in Section 3.6. Compare corresponding problems to make sure you see the difference; one will be a closed-interval maximum-minimum problem and the other an open-interval maximum-minimum problem.)*

51. Show that, among all closed rectangular boxes with square bases and a given fixed volume, the one with minimal total surface area is a cube.

52. Show that, among all open rectangular boxes with square bases and a given fixed volume, the one with minimal total surface area has height equal to half the length of the edge of its base.

Problems 53 through 55 deal with cans in the shape of right circular cylinders. Such a can is said to be closed *if it has both a circular base and a top (as well as a curved side),* open *if it has a base but no top.*

53. Show that, among all closed cylindrical cans with a given fixed volume, the one with minimal total surface area has height equal to the diameter of its base.

54. Show that, among all open cylindrical cans with a given fixed volume, the one with minimal total surface area has height equal to the radius of its base.

55. Suppose that the base and curved side of a pop-top soft drink can have the same thickness. But the top is three times as thick as the base to prevent ripping when the can is opened. Show that, among all such cans with a given fixed volume, the one requiring the least amount (volume) of material to make—including the triply thick top—has height twice the diameter of its base. Perhaps this is why soft drink cans look somewhat taller than vegetable or soup cans.

56. Suppose that you want to construct a closed rectangular box with a square base and fixed volume V. Each of the six faces of the box—the base, top, and four vertical sides—costs a cents per square inch, and gluing each of the 12 edges costs b cents per inch of edge length. What shape should this box be in order to minimize its total cost? [*Suggestion:* Show that the critical points of the cost function are roots of a certain quartic equation that you can solve using a computer algebra system. You may even be able to solve it with pencil and paper alone; begin by grouping the two terms of highest degree.]

4.4 Project: Constructing a Candy Box with Lid

A candy maker wants to package jelly beans in boxes each having a fixed volume V. Each box is to be an open rectangular box with square base of edge length x. (See Fig. 4.4.32.) In addition, the box is to have a square lid with a two-inch rim. Thus the box-with-lid actually consists of two open rectangular boxes—the x-by-x-by-y box itself with height $y \geqq 2$ (in.) and the x-by-x-by-2 lid (which fits the box very snugly). Your job as the firm's design engineer is to determine the dimensions x and y that will minimize the total cost of the two open boxes that comprise a single candy box with lid. Assume that the box-with-lid is to be made using an attractive foil-covered cardboard that costs \$1 per square foot and that its volume is to be $V = 400 + 50n$ cubic inches. (For your personal design problem, choose an integer n between 1 and 10.)

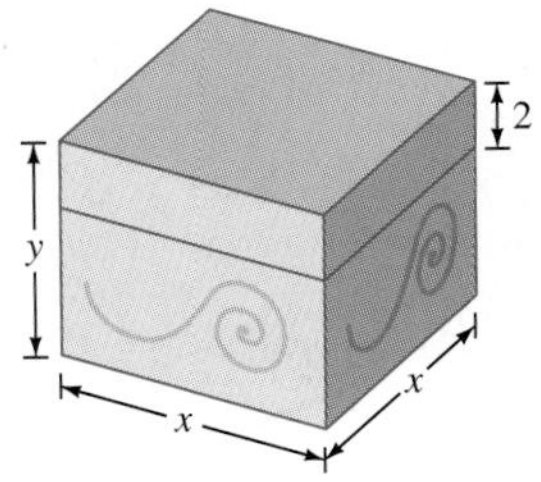

FIGURE 4.4.32 The square-based candy box with lid.

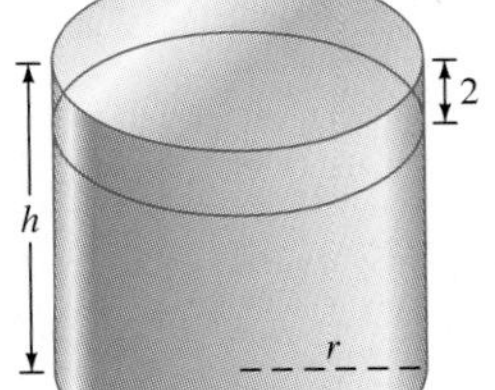

FIGURE 4.4.33 The cylindrical candy box with lid.

Your next task is to design a cylindrical box-with-lid as indicated in Fig. 4.4.33. Now the box proper and its lid are both open circular cylinders, but everything else is the same as in the previous problem—two-inch rim, \$1 per square foot foil-covered cardboard, and volume $V = 400 + 50n$. What are the dimensions of the box of minimal cost? Which is less expensive to manufacture—the optimal rectangular box-with-lid or the optimal box-with-lid in the shape of a cylinder?

4.5 SIMPLE CURVE SKETCHING

We can construct a reasonably accurate graph of the polynomial function

$$f(x) = a_n x^n + a_{n-1}x^{n-1} + \cdots + a_2 x^2 + a_1 x + a_0 \tag{1}$$

by assembling the following information.

1. **The critical points of** f—that is, the points on the graph where the tangent line is horizontal, so that $f'(x) = 0$.
2. **The increasing/decreasing behavior of** f—that is, the intervals on which f is increasing and those on which it is decreasing.
3. **The behavior of** f **"at" infinity**—that is, the behavior of f as $x \to +\infty$ and as $x \to -\infty$.

The same information often is the key to understanding the structure of a graph that has been plotted with a calculator or computer.

Behavior at Infinity

To carry out the task in item 3, we write $f(x)$ in the form

$$f(x) = x^n \left(a_n + \frac{a_{n-1}}{x} + \cdots + \frac{a_1}{x^{n-1}} + \frac{a_0}{x^n} \right).$$

Thus we conclude that the behavior of $f(x)$ as $x \to \pm\infty$ is much the same as that of its *leading term* $a_n x^n$, because all the terms that have powers of x in the denominator approach zero as $x \to \pm\infty$. In particular, if $a_n > 0$, then

$$\lim_{x \to \infty} f(x) = +\infty, \tag{2}$$

meaning that $f(x)$ increases without bound as $x \to +\infty$. Also

$$\lim_{x \to -\infty} f(x) = \begin{cases} +\infty & \text{if } n \text{ is even;} \\ -\infty & \text{if } n \text{ is odd.} \end{cases} \tag{3}$$

If $a_n < 0$, simply reverse the signs on the right-hand sides in Eqs. (2) and (3). It follows that the graph of any (nonconstant) *polynomial* function exhibits one of the four "behaviors as $x \to \pm\infty$" that are illustrated in Fig. 4.5.1.

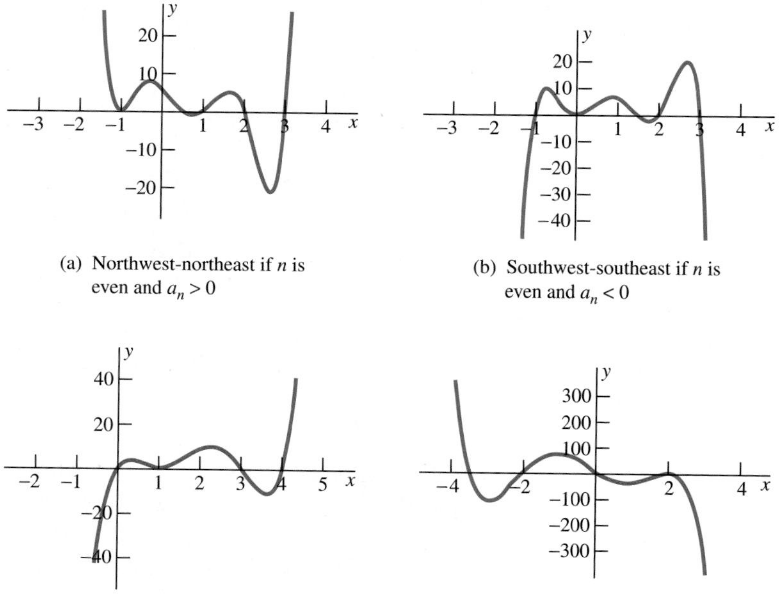

(a) Northwest-northeast if n is even and $a_n > 0$

(b) Southwest-southeast if n is even and $a_n < 0$

(c) Southwest-northeast if n is odd and $a_n > 0$

(d) Northwest-southeast if n is odd and $a_n < 0$

FIGURE 4.5.1 The behavior of polynomial graphs as $x \to \pm\infty$.

Critical Points

Every polynomial, such as $f(x)$ in Eq. (1), is differentiable everywhere. So the critical points of $f(x)$ are the roots of the polynomial equation $f'(x) = 0$—that is, solutions of

$$na_nx^{n-1} + (n-1)a_{n-1}x^{n-2} + \cdots + 2a_2x + a_1 = 0. \tag{4}$$

Sometimes we can find all (real) solutions of such an equation by factoring, but most often in practice we must resort to numerical methods aided by calculator or computer.

Increasing/Decreasing Behavior

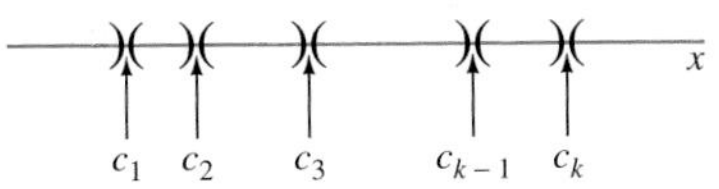

FIGURE 4.5.2 The zeros of $f'(x)$ divide the x-axis into intervals on which $f'(x)$ does not change sign.

Suppose that we have somehow found *all* the (real) solutions $c_1, c_2, \dots, c_k$ of Eq. (4). Then these solutions are the critical points of f. If they are arranged in increasing order, as in Fig. 4.5.2, then they separate the x-axis into the finite number of open intervals

$$(-\infty, c_1), \quad (c_1, c_2), \quad (c_2, c_3), \quad \dots, \quad (c_{k-1}, c_k), \quad (c_k, +\infty)$$

that also appear in the figure. The intermediate value property applied to $f'(x)$ tells us that $f'(x)$ can change sign only at the critical points of f, so $f'(x)$ has only one sign on each of these open intervals. It is typical for $f'(x)$ to be negative on some intervals and positive on others. Moreover, it's easy to find the sign of $f'(x)$ on any one such interval I: We need only substitute *any* convenient number in I into $f'(x)$.

Once we know the sign of $f'(x)$ on each of these intervals, we know where f is increasing and where it is decreasing. We then apply the first derivative test to find which of the critical values are local maxima, which are local minima, and which are neither—merely places where the tangent line is horizontal. With this information, the knowledge of the behavior of f as $x \to \pm\infty$, and the fact that f is continuous, we can sketch its graph. We plot the critical points $(c_i, f(c_i))$ and connect them with a smooth curve that is consistent with our other data.

It may also be helpful to plot the y-intercept $(0, f(0))$ and also any x-intercepts that are easy to find. But we recommend (until inflection points are introduced in Section 4.6) that you plot *only* these points—critical points and intercepts—and rely otherwise on the increasing and decreasing behavior of f.

EXAMPLE 1 Sketch the graph of $f(x) = x^3 - 27x$.

Solution Because the leading term is x^3, we see that

$$\lim_{x\to+\infty} f(x) = +\infty \quad \text{and} \quad \lim_{x\to-\infty} f(x) = -\infty.$$

Moreover, because

$$f'(x) = 3x^2 - 27 = 3(x+3)(x-3), \tag{5}$$

we see that the critical points where $f'(x) = 0$ are $x = -3$ and $x = 3$. The corresponding points on the graph of f are $(-3, 54)$ and $(3, -54)$. The critical points separate the x-axis into the three open intervals $(-\infty, -3)$, $(-3, 3)$, and $(3, +\infty)$ (Fig. 4.5.3). To determine the increasing or decreasing behavior of f on these intervals, let's

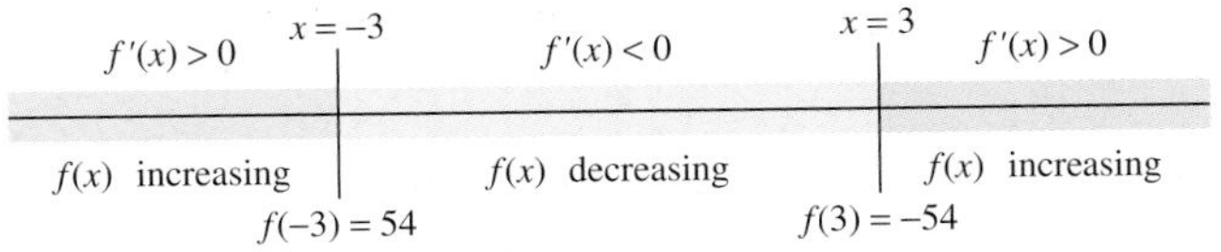

FIGURE 4.5.3 The three open intervals of Example 1.

substitute a number in each interval into the derivative in Eq. (5):

$$\begin{aligned}
&\text{On } (-\infty, -3): && f'(-4) = (3)(-1)(-7) = 21 > 0; && f \text{ is increasing;}\\
&\text{On } (-3, 3): && f'(0) = (3)(3)(-3) = -27 < 0; && f \text{ is decreasing;}\\
&\text{On } (3, +\infty): && f'(4) = (3)(7)(1) = 21 > 0; && f \text{ is increasing.}
\end{aligned}$$

We plot the critical points and the intercepts $(0, 0)$, $(3\sqrt{3}, 0)$, and $(-3\sqrt{3}, 0)$. Then we use the information about where f is increasing or decreasing to connect these points with a smooth curve. Remembering that there are horizontal tangents at the two critical points, we obtain the graph shown in Fig. 4.5.4.

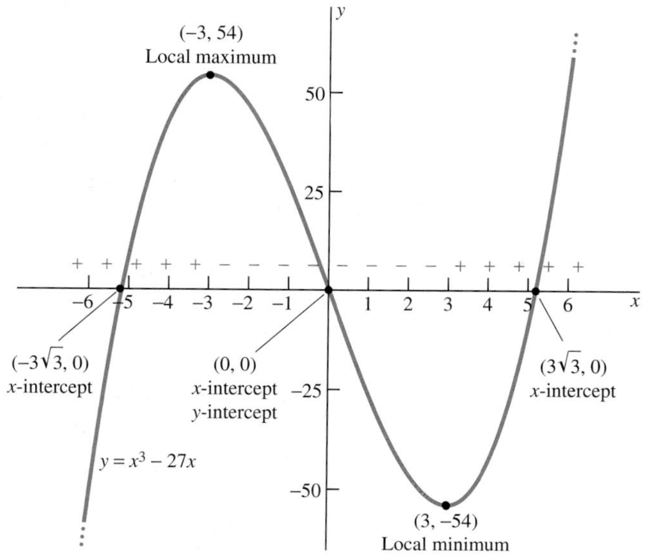

FIGURE 4.5.4 Graph of the function of Example 1.

In the figure we use plus and minus signs to mark the sign of $f'(x)$ in each interval. This makes it clear that $(-3, 54)$ is a local maximum and that $(3, -54)$ is a local minimum. The limits we found at the outset show that neither is global. ◆

EXAMPLE 2 Sketch the graph of $f(x) = 8x^5 - 5x^4 - 20x^3$.

Solution Because

$$f'(x) = 40x^4 - 20x^3 - 60x^2 = 20x^2(x + 1)(2x - 3), \tag{6}$$

the critical points where $f'(x) = 0$ are $x = -1$, $x = 0$, and $x = \frac{3}{2}$. These three critical points separate the x-axis into the four open intervals shown in Fig. 4.5.5. This time, let's determine the increasing or decreasing behavior of f by recording the signs of the factors in Eq. (6) on each of the subintervals shown in Fig. 4.5.5. In this way we

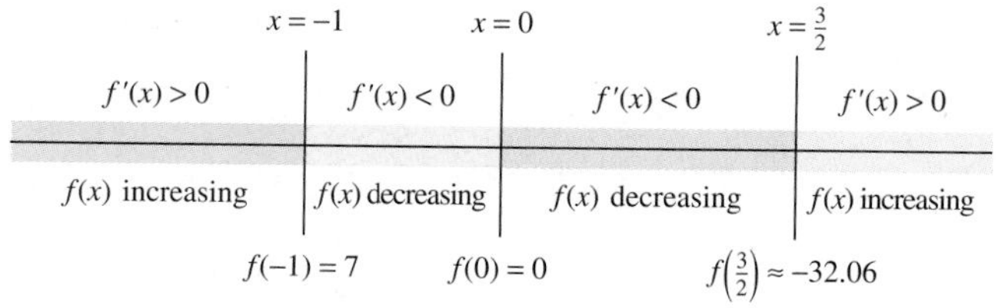

FIGURE 4.5.5 The four open intervals of Example 2.

get the following table:

Interval	$x+1$	$20x^2$	$2x-3$	$f'(x)$	f
$(-\infty, -1)$	$-$	$+$	$-$	$+$	Increasing
$(-1, 0)$	$+$	$+$	$-$	$-$	Decreasing
$\left(0, \frac{3}{2}\right)$	$+$	$+$	$-$	$-$	Decreasing
$\left(\frac{3}{2}, +\infty\right)$	$+$	$+$	$+$	$+$	Increasing

The points on the graph that correspond to the critical points are $(-1, 7)$, $(0, 0)$, and $(1.5, -32.0625)$.

We write $f(x)$ in the form

$$f(x) = x^3(8x^2 - 5x - 20)$$

in order to use the quadratic formula to find the x-intercepts. They turn out to be $(-1.30, 0)$, $(1.92, 0)$ (the abscissas are given only approximately), and the origin $(0, 0)$. The latter is also the y-intercept. We apply the first derivative test to the increasing or decreasing behavior shown in the table. It follows that $(-1, 7)$ is a local maximum, $(1.5, -32.0625)$ is a local minimum, and $(0, 0)$ is neither. The graph resembles the one shown in Fig. 4.5.6. ◆

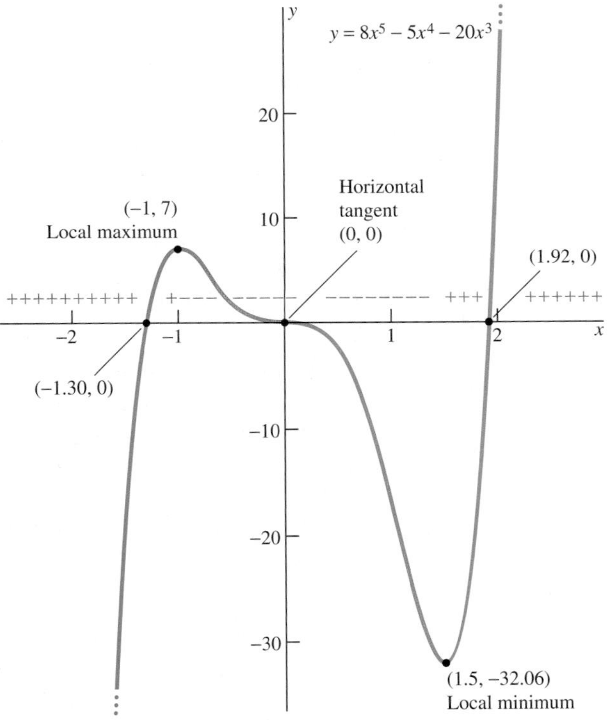

FIGURE 4.5.6 Graph of the function of Example 2.

In Example 3, the function is not a polynomial. Nevertheless, the methods of this section suffice for sketching its graph.

EXAMPLE 3 Sketch the graph of

$$f(x) = x^{2/3}(x^2 - 2x - 6) = x^{8/3} - 2x^{5/3} - 6x^{2/3}.$$

Solution The derivative of f is

$$f'(x) = \tfrac{8}{3}x^{5/3} - \tfrac{10}{3}x^{2/3} - \tfrac{12}{3}x^{-1/3}$$

$$= \tfrac{2}{3}x^{-1/3}(4x^2 - 5x - 6) = \frac{2(4x+3)(x-2)}{3x^{1/3}}. \quad \textbf{(7)}$$

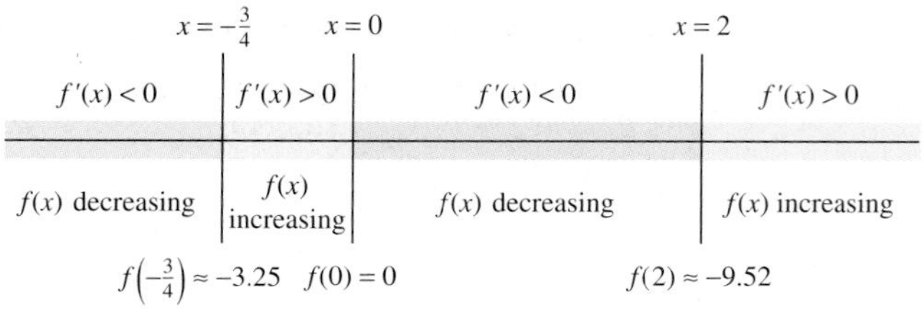

FIGURE 4.5.7 The four open intervals of Example 3.

The tangent line is horizontal at the two critical points $x = -\frac{3}{4}$ and $x = 2$, where the numerator in the last fraction of Eq. (7) is zero (and the denominator is not). Moreover, because of the presence of the factor $x^{1/3}$ in the denominator, $|f'(x)| \to +\infty$ as $x \to 0$. Thus $x = 0$ (a critical point because f is not differentiable there) is a point where the tangent line is vertical. These three critical points separate the x-axis into the four open intervals shown in Fig. 4.5.7. We determine the increasing or decreasing behavior of f by substituting a number from each interval in $f'(x)$ (Eq. (7)).

$$\text{On } \left(-\infty, -\tfrac{3}{4}\right): \quad f'(-1) = \frac{2 \cdot (-1)(-3)}{3 \cdot (-1)} < 0; \quad f \text{ is decreasing;}$$

$$\text{On } \left(-\tfrac{3}{4}, 0\right): \quad f'\left(-\tfrac{1}{2}\right) = \frac{2 \cdot (+1)\left(-\frac{5}{2}\right)}{3 \cdot \left(-\frac{1}{2}\right)^{1/3}} > 0; \quad f \text{ is increasing;}$$

$$\text{On } (0, 2): \quad f'(1) = \frac{2 \cdot (+7)(-1)}{3 \cdot (+1)} < 0; \quad f \text{ is decreasing;}$$

$$\text{On } (2, +\infty): \quad f'(3) = \frac{2 \cdot (+15)(+1)}{3 \cdot (+3)^{1/3}} > 0; \quad f \text{ is increasing.}$$

The three critical points $x = -\frac{3}{4}$, $x = 0$, and $x = 2$ give the points $(-0.75, -3.25)$, $(0, 0)$, and $(2, -9.52)$ on the graph (using approximations where appropriate).

The first derivative test now shows local minima at $(-0.75, -3.25)$ and at $(2, -9.52)$; there is a local maximum at $(0, 0)$. Although $f'(0)$ does not exist, the function f is continuous everywhere (because it involves only positive powers of x).

We use the quadratic formula to find the x-intercepts. In addition to the origin, they occur where $x^2 - 2x - 6 = 0$, and thus they are located at $(1 - \sqrt{7}, 0)$ and at $(1 + \sqrt{7}, 0)$. We then plot the approximations $(-1.65, 0)$ and $(3.65, 0)$. Finally, we note that $f(x) \to +\infty$ as $x \to \pm\infty$. So the graph has the shape shown in Fig. 4.5.8. ◆

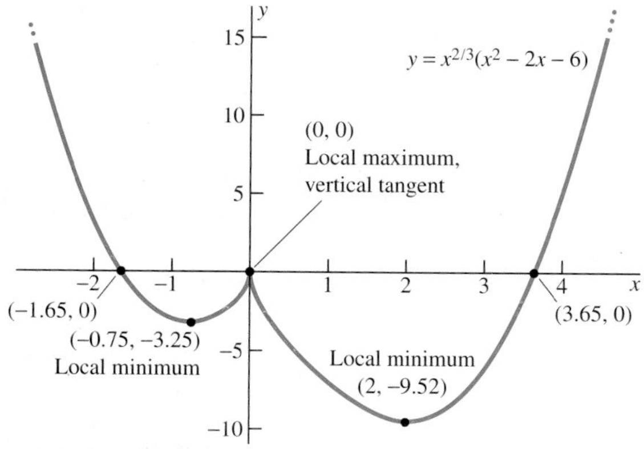

FIGURE 4.5.8 The technique is effective for nonpolynomial functions, as in Example 3.

Curve Sketching and Solution of Equations

An important application of curve-sketching techniques is the solution of an equation of the form

$$f(x) = 0. \tag{8}$$

The real (as opposed to complex) solutions of this equation are simply the x-intercepts of the graph of $y = f(x)$. Hence by sketching this graph with reasonable accuracy—either "by hand" or with a calculator or computer—we can glean information about the number of real solutions of Eq. (8) as well as their approximate locations.

For example, Figs. 4.5.9 through 4.5.11 show the graphs of the cubic polynomials on the left-hand sides of the equations

$$x^3 - 3x + 1 = 0, \tag{9}$$

$$x^3 - 3x + 2 = 0, \tag{10}$$

$$x^3 - 3x + 3 = 0. \tag{11}$$

Note that the polynomials differ only in their constant terms.

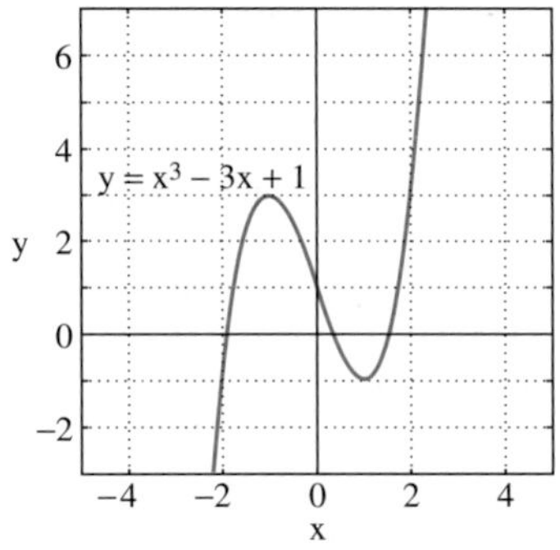

FIGURE 4.5.9 $y = x^3 - 3x + 1$.

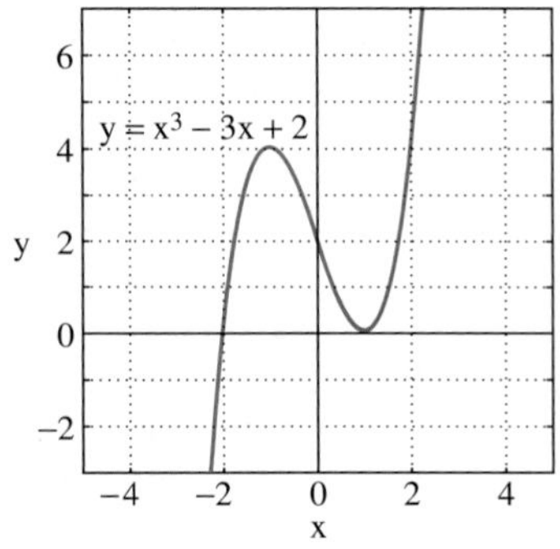

FIGURE 4.5.10 $y = x^3 - 3x + 2$.

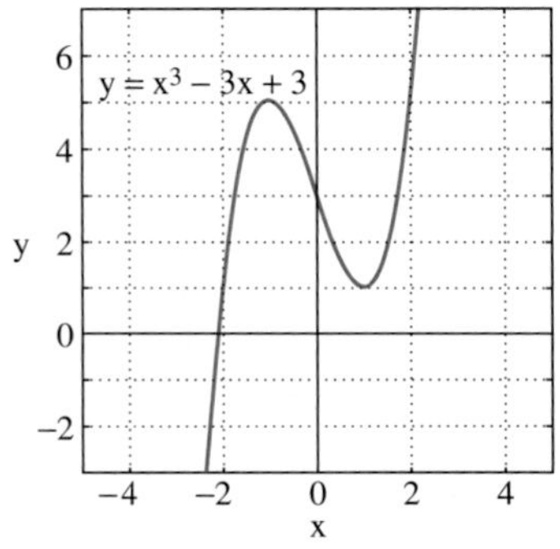

FIGURE 4.5.11 $y = x^3 - 3x + 3$.

It is clear from Fig. 4.5.9 that Eq. (9) has three real solutions, one in each of the intervals $[-2, -1]$, $[0, 1]$, and $[1, 2]$. These solutions could be approximated graphically by successive magnification or analytically by Newton's method. (As we have previously mentioned, there are even formulas—*Cardan's formulas*—for the exact solution of an arbitrary cubic equation, but they are unwieldy and are seldom used except in computer algebra programs. For example, these formulas yield (via a computer algebra system) the expressions

$$x_1 = \left(\frac{-1 + i\sqrt{3}}{2}\right)^{-1/3} + \left(\frac{-1 + i\sqrt{3}}{2}\right)^{1/3},$$

$$x_2 = \left(\frac{-1 + i\sqrt{3}}{2}\right)^{4/3} + \left(\frac{-1 + i\sqrt{3}}{2}\right)^{5/3}, \tag{12}$$

and

$$x_3 = \left(\frac{-1 + i\sqrt{3}}{2}\right)^{2/3} + \left(\frac{-1 + i\sqrt{3}}{2}\right)^{7/3}$$

for the three solutions of Eq. (9). Despite the appearance of the imaginary number $i = \sqrt{-1}$ in these three expressions, Fig. 4.5.9—with its three x-intercepts—indicates that all three solutions simplify to ordinary real numbers.)

It appears in Fig. 4.5.10 that Eq. (10) has the two real solutions $x = 1$ and $x = -2$. Once we verify that $x = 1$ is a solution, then it follows from the *factor theorem* of algebra that $x - 1$ is a factor of $x^3 - 3x + 2$. The other factor can be found by division (long or synthetic) of $x - 1$ into $x^3 - 3x + 2$; the quotient is $x^2 + x - 2$. Thus we see that

$$x^3 - 3x + 2 = (x - 1)(x^2 + x - 2) = (x - 1)^2(x + 2).$$

Hence $x = 1$ is a "double root" and $x = -2$ is a "single root" of Eq. (10), thereby accounting for the three solutions that a cubic equation "ought to have."

We see in Fig. 4.5.11 that Eq. (11) has only one real solution. It is given approximately by $x \approx -2.1038$. Problem 55 asks you to divide $x + 2.0138$ into $x^3 - 3x + 3$ to obtain a factorization of the form

$$x^3 - 3x + 3 \approx (x + 2.1038)(x^2 + bx + c). \tag{13}$$

The quadratic equation $x^2 + bx + c = 0$ has two complex conjugate solutions, which are the other two solutions of Eq. (12).

Calculator and Computer Graphing

With a graphing calculator or computer we may construct the graph of a given function with a few keystrokes. Nevertheless, the viewpoint of this section may be useful in analyzing and understanding what we see on the screen.

EXAMPLE 4 Figure 4.5.12 shows a computer-generated graph of the function

$$f(x) = x^4 - 5x^2 + x + 2. \tag{14}$$

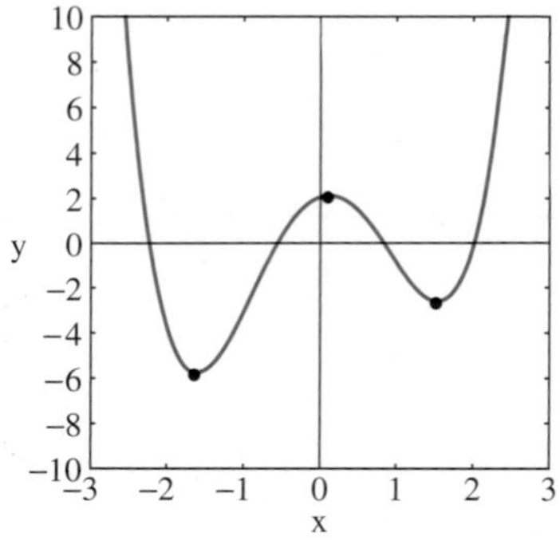

FIGURE 4.5.12 $y = x^4 - 5x^2 + x + 2$.

Three critical points are visible, separating the x-axis into two intervals on which the function f increases and two on which it decreases. In order to find these critical points, we need to solve the cubic equation

$$f'(x) = 4x^3 - 10x + 1 = 0. \tag{15}$$

For this purpose we could graph the derivative $f'(x)$ and zoom in on its solutions, or use Newton's method to approximate these solutions accurately, or simply use the "solve" command on our calculator or computer. The approximate solutions of Eq. (15) thus found are -1.6289, 0.1004, and 1.5285. The corresponding numerical values of y obtained by substitution in Eq. (14) are -5.8554, 2.0501, and -2.6947. Thus the three critical points that we see on the graph in Fig. 4.5.12 are $(-1.6289, -5.8554)$, $(0.1004, 2.0501)$, and $(1.5285, -2.6947)$. The function f is decreasing on the intervals $-\infty < x < -1.6289$ and $0.1004 < x < 1.5285$ and increasing on the intervals $-1.6289 < x < 0.1004$ and $1.55285 < x < \infty$. ◆

4.5 TRUE/FALSE STUDY GUIDE

4.5 CONCEPTS: QUESTIONS AND DISCUSSION

1. Suppose that n is a positive integer and that k is an integer such that $0 \leqq k \leqq n$. Does there always exist a polynomial of degree n having exactly k real zeros? If not, what are the exceptions?
2. Suppose that $f(x)$ is a polynomial of degree n whose graph has p local minima and q local maxima. Explain why $p + q < n$. Discuss any other necessary restrictions on p and q. For instance, if $n = 4$ is it possible that $p = 3$ and $q = 0$? If $n = 5$ is it possible that $p = 3$ and $q = 1$? Justify your answers.

3. Someone asserts that "the graphs of any two nth-degree polynomials with the same term of highest degree look essentially the same when plotted in a sufficiently large viewing window." To what extent is this a reasonable claim? Begin by testing it with two quartic polynomials both having leading term x^4. Do you need to adjust the x-scale, the y-scale, or both, to make the graphs nearly coincide?

4.5 PROBLEMS

In Problems 1 through 4, use behavior "at infinity" to match the given function with its graph in Fig. 4.5.13.

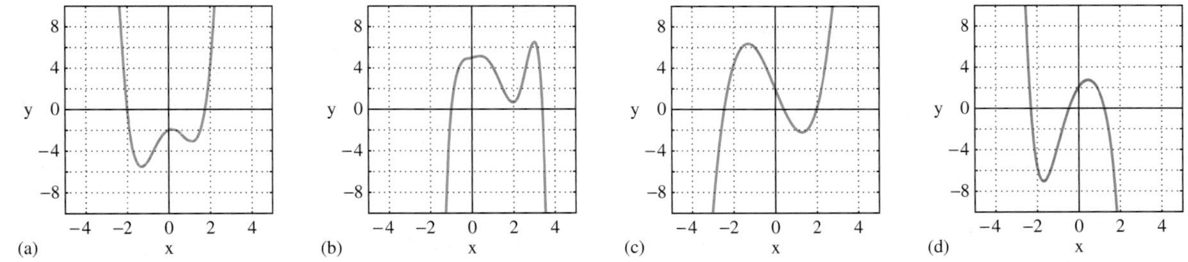

FIGURE 4.5.13 Problems 1 through 4.

1. $f(x) = x^3 - 5x + 2$

2. $f(x) = x^4 - 3x^2 + x - 2$

3. $f(x) = -\frac{1}{3}x^5 - 3x^2 + 3x + 2$

4. $f(x) = -\frac{1}{3}x^6 + 2x^5 - 3x^4 + \frac{1}{2}x + 5$

In Problems 5 through 14 a function $y = f(x)$ and its computer-generated graph are given. Find both the critical points and the increasing/decreasing intervals for $f(x)$.

5. $y = 2x^2 - 10x - 7$ (Fig. 4.5.14)

6. $y = 27 + 12x - 4x^2$ (Fig. 4.5.15)

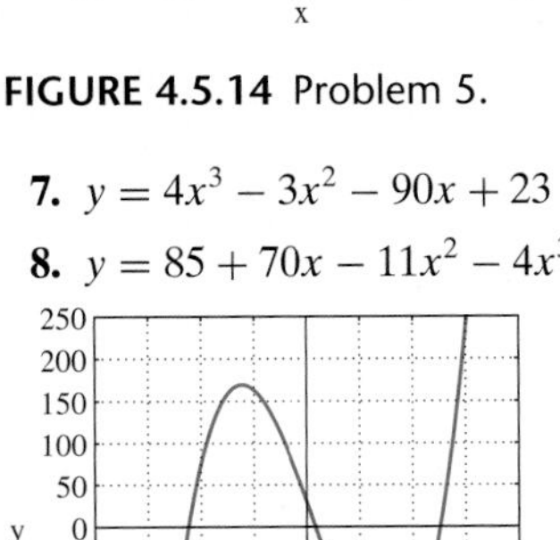

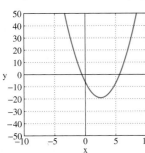

FIGURE 4.5.14 Problem 5.

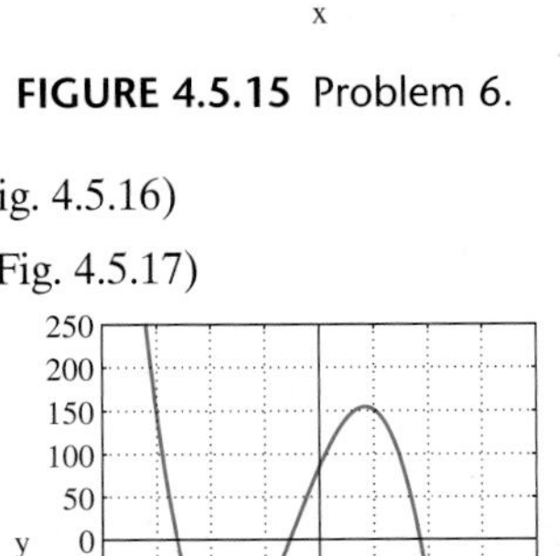

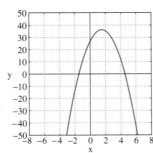

FIGURE 4.5.15 Problem 6.

7. $y = 4x^3 - 3x^2 - 90x + 23$ (Fig. 4.5.16)

8. $y = 85 + 70x - 11x^2 - 4x^3$ (Fig. 4.5.17)

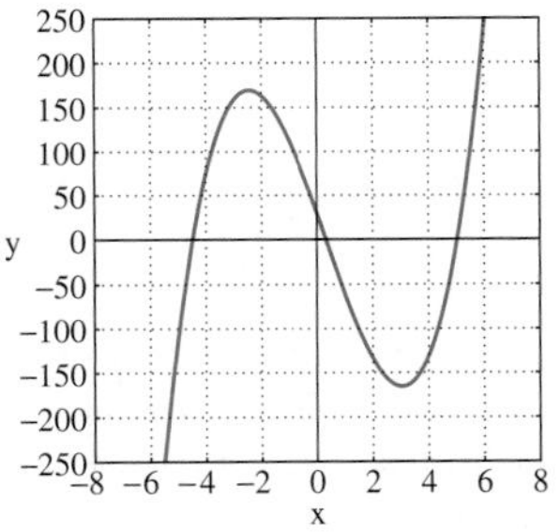

FIGURE 4.5.16 Problem 7.

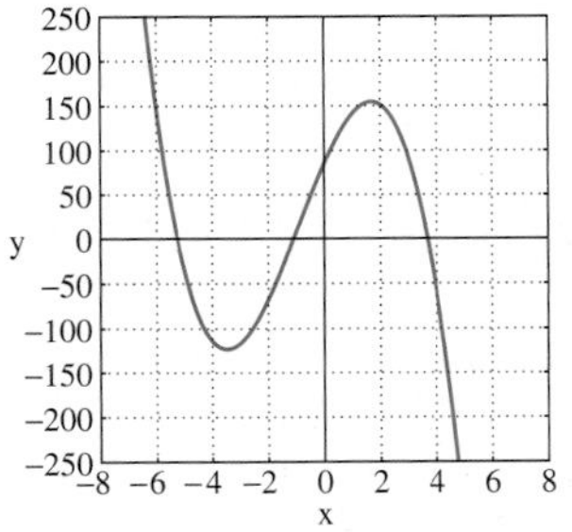

FIGURE 4.5.17 Problem 8.

9. $y = 3x^4 + 4x^3 - 36x^2 + 40$ (Fig. 4.5.18)

10. $y = 125 + 120x^2 - 2x^3 - 9x^4$ (Fig. 4.5.19)

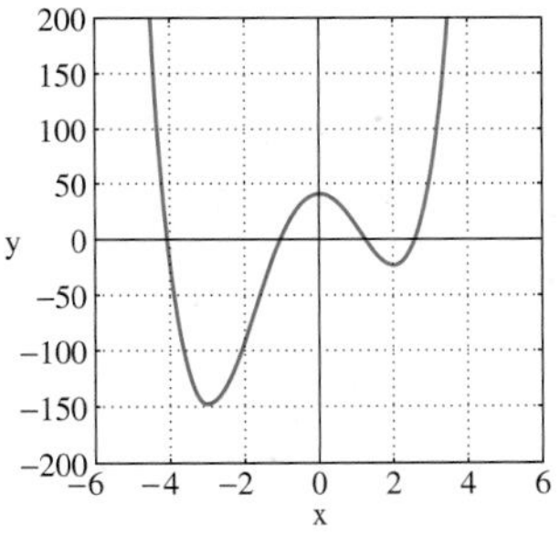

FIGURE 4.5.18 Problem 9.

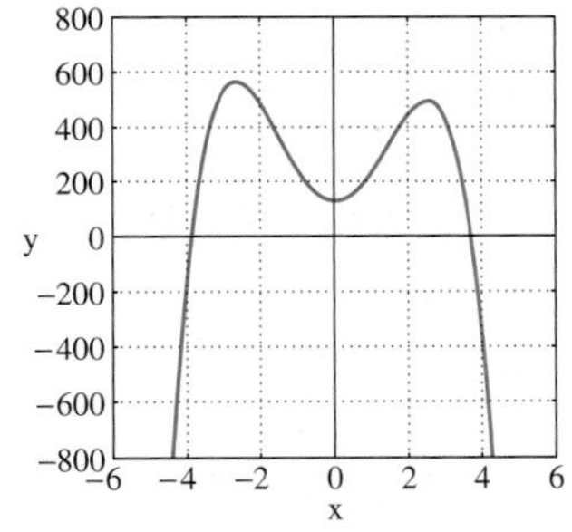

FIGURE 4.5.19 Problem 10.

11. $y = 3x^5 - 100x^3 + 960x$ (Fig. 4.5.20)

12. $y = 2x^6 - 87x^4 + 600x^2 + 3000$ (Fig. 4.5.21)

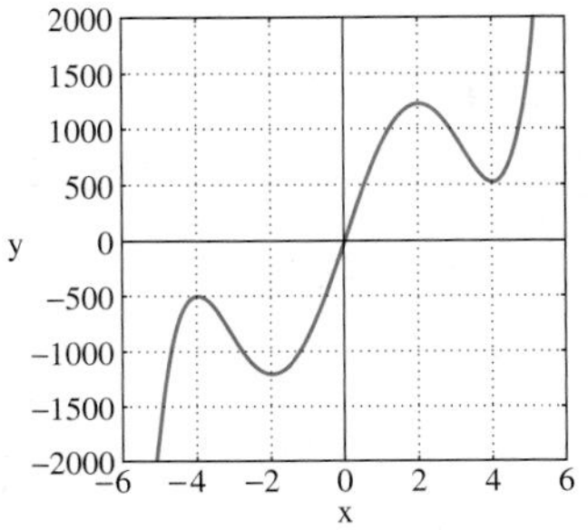

FIGURE 4.5.20 Problem 11.

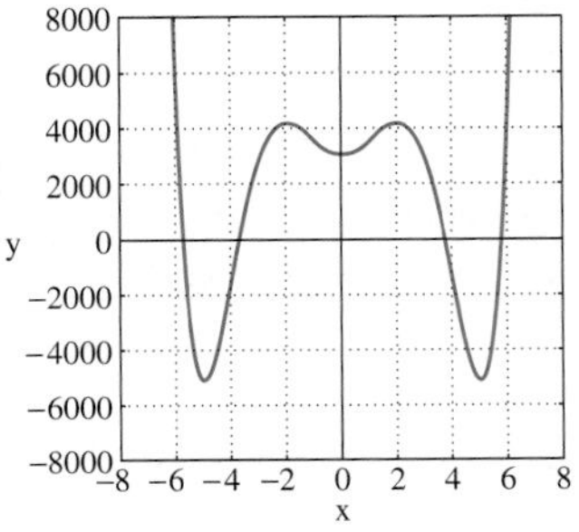

FIGURE 4.5.21 Problem 12.

13. $y = 3x^7 - 84x^5 + 448x^3$ (Fig. 4.5.22)

14. $y = 3x^8 - 52x^6 + 216x^4 - 500$ (Fig. 4.5.23)

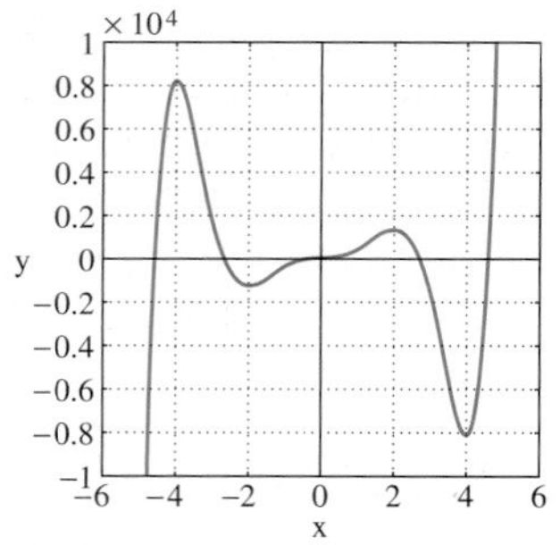

FIGURE 4.5.22 Problem 13.

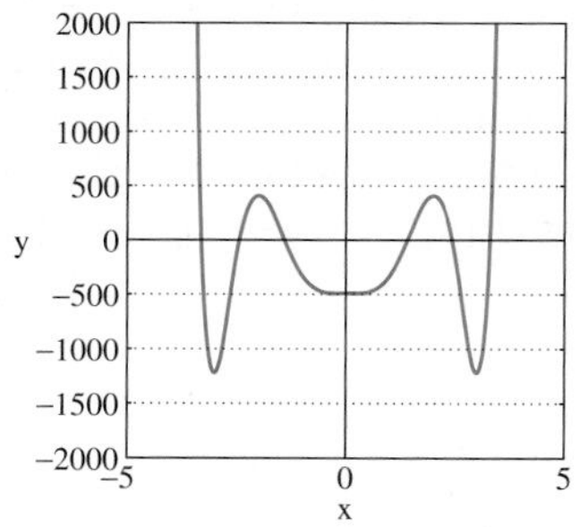

FIGURE 4.5.23 Problem 14.

In Problems 15 through 48, find the intervals on which the function f is increasing and those on which it is decreasing. Sketch the graph of $y = f(x)$ and label the local maxima and minima. Global extrema should be so identified.

15. $f(x) = 3x^2 - 6x + 5$

16. $f(x) = 5 - 8x - 2x^2$

17. $f(x) = x^3 - 12x$

18. $f(x) = x^3 + 3x$

19. $f(x) = x^3 - 6x^2 + 9x$

20. $f(x) = x^3 + 6x^2 + 9x$

21. $f(x) = x^3 + 3x^2 + 9x$

22. $f(x) = x^3 - 27x$

23. $f(x) = (x - 1)^2(x + 2)^2$

24. $f(x) = (x - 2)^2(2x + 3)^2$

25. $f(x) = 3\sqrt{x} - x\sqrt{x}$

26. $f(x) = x^{2/3}(5 - x)$

27. $f(x) = 3x^5 - 5x^3$

28. $f(x) = x^4 + 4x^3$

29. $f(x) = x^4 - 8x^2 + 7$

30. $f(x) = \dfrac{1}{x}$

31. $f(x) = 2x^2 - 3x - 9$

32. $f(x) = 6 - 5x - 6x^2$

33. $f(x) = 2x^3 + 3x^2 - 12x$

34. $f(x) = x^3 + 4x$

35. $f(x) = 50x^3 - 105x^2 + 72x$

36. $f(x) = x^3 - 3x^2 + 3x - 1$

37. $f(x) = 3x^4 - 4x^3 - 12x^2 + 8$

38. $f(x) = x^4 - 2x^2 + 1$

39. $f(x) = 3x^5 - 20x^3$

40. $f(x) = 3x^5 - 25x^3 + 60x$

41. $f(x) = 2x^3 + 3x^2 + 6x$

42. $f(x) = x^4 - 4x^3$

43. $f(x) = 8x^4 - x^8$

44. $f(x) = 1 - x^{1/3}$

45. $f(x) = x^{1/3}(4 - x)$

46. $f(x) = x^{2/3}(x^2 - 16)$

47. $f(x) = x(x - 1)^{2/3}$

48. $f(x) = x^{1/3}(2 - x)^{2/3}$

In Problems 49 through 54, the values of the function $f(x)$ at its critical points are given, together with the graph $y = f'(x)$ of its derivative. Use this information to construct a sketch of the graph $y = f(x)$ of the function.

49. $f(-3) = 78$, $f(2) = -47$; Fig. 4.5.24

50. $f(-2) = 106$, $f(4) = -110$; Fig. 4.5.25

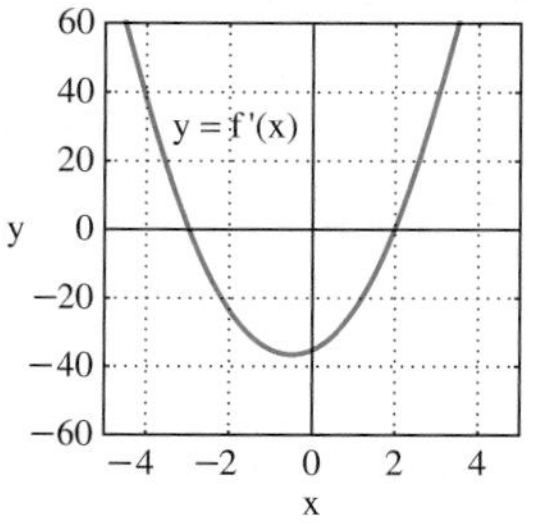

FIGURE 4.5.24 $y = f'(x)$ of Problem 49.

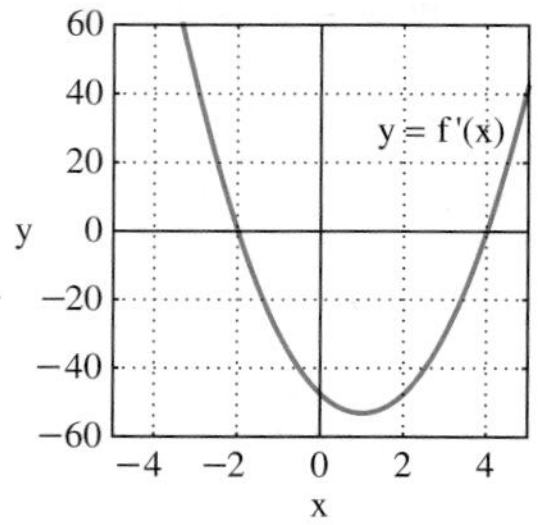

FIGURE 4.5.25 $y = f'(x)$ of Problem 50.

51. $f(-3) = -66$, $f(2) = 59$; Fig. 4.5.26

52. $f(-3) = -130$, $f(0) = 5$, $f(1) = -2$; Fig. 4.5.27

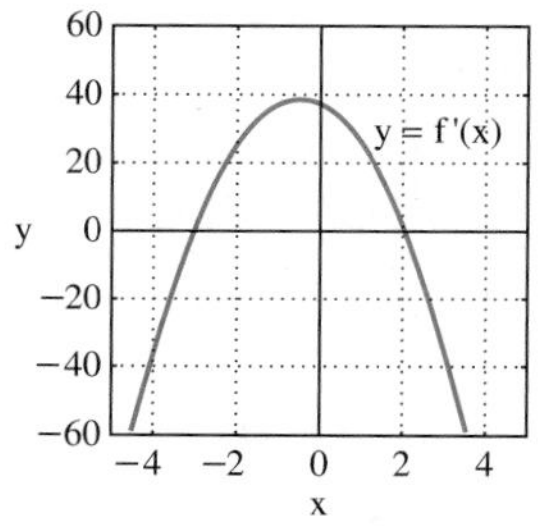

FIGURE 4.5.26 $y = f'(x)$ of Problem 51.

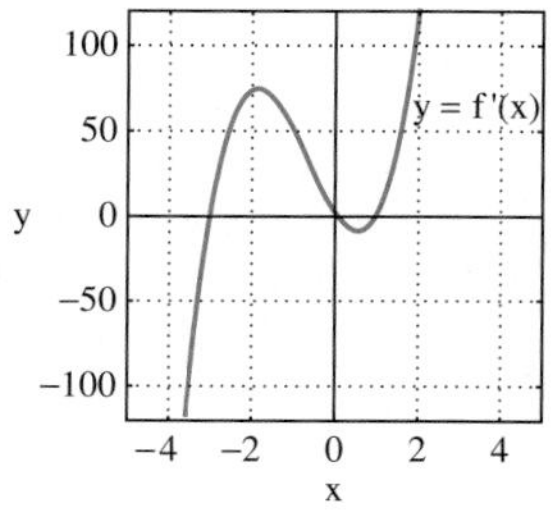

FIGURE 4.5.27 $y = f'(x)$ of Problem 52.

53. $f(-2) = -107$, $f(1) = 82$, $f(3) = 18$; Fig. 4.5.28

54. $f(-3) = 5336$, $f(0) = 17$, $f(2) = 961$, $f(4) = -495$; Fig. 4.5.29

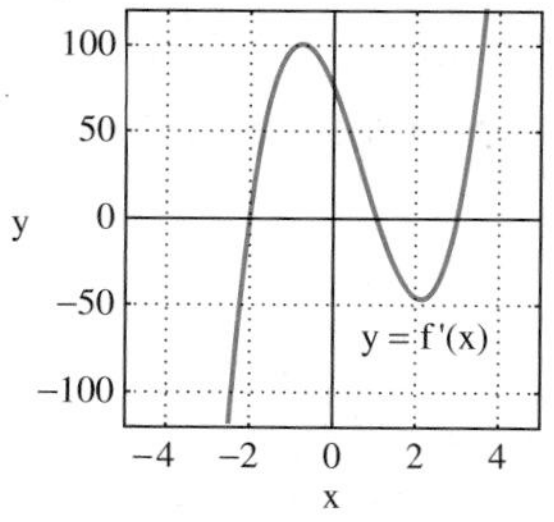

FIGURE 4.5.28 $y = f'(x)$ of Problem 53.

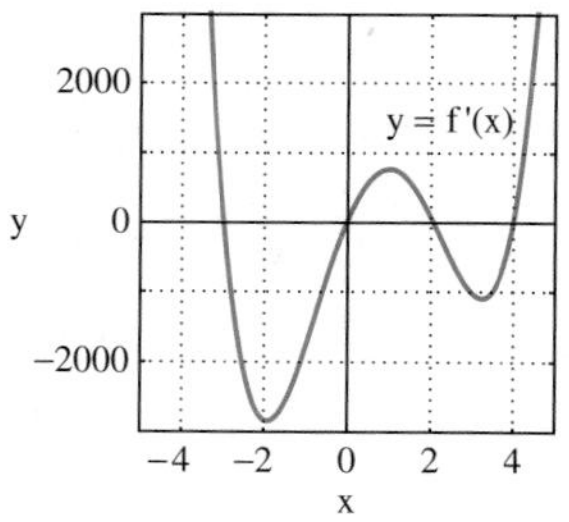

FIGURE 4.5.29 $y = f'(x)$ of Problem 54.

55. (a) Verify the approximate solution $x \approx -2.0138$ of Eq. (11). (b) Divide $x^3 - 3x + 3$ by $x + 2.1038$ to obtain the factorization in Eq. (13). (c) Use the quotient in part (b) to find (approximately) the complex conjugate pair of solutions of Eq. (11).

56. Explain why Figs. 4.5.9 and 4.5.10 imply that the cubic equation $x^3 - 3x + q = 0$ has exactly one real solution if $|q| > 2$ but has three distinct real solutions if $|q| < 2$. What is the situation if $q = -2$?

57. The computer-generated graph in Fig. 4.5.30 shows how the curve

$$y = [x(x-1)(2x-1)]^2$$

looks on any "reasonable" scale with integral units of measurement on the y-axis. Use the methods of this section to show that the graph really has the appearance shown in Fig. 4.5.31 (the values on the y-axis are in thousandths), with critical points at $0, \frac{1}{2}, \frac{1}{6}(3 \pm \sqrt{3})$, and 1.

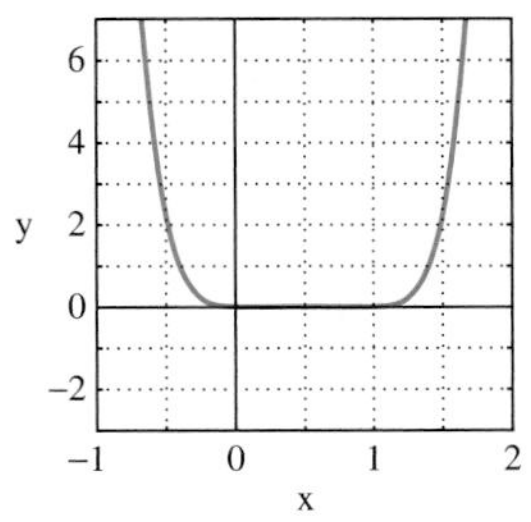

FIGURE 4.5.30 The graph $y = [x(x-1)(2x-1)]^2$ on a "reasonable scale" (Problem 57).

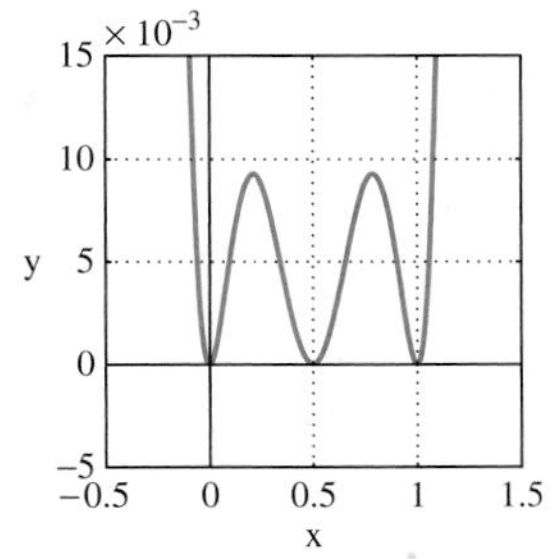

FIGURE 4.5.31 The graph $y = [x(x-1)(2x-1)]^2$ on a finer scale: $-0.005 \leqq y \leqq 0.005$ (Problem 57).

58. Use a computer algebra system to verify that the three expressions x_1, x_2, and x_3 in Eq. (12) are, indeed, distinct real solutions of Eq. (9).

Problems 59 and 60 require the use of a graphing calculator or computer algebra system. If you find it necessary to solve various equations, you may use either a graphing calculator or a "solve" command in a computer algebra system.

59. Show first that, on a "reasonable" scale with integral units of measurement on the y-axis, the graph of the polynomial

$$f(x) = \left[\tfrac{1}{6}x(9x-5)(x-1)\right]^4$$

strongly resembles the graph shown in Fig. 4.5.30, with a seemingly flat section. Then produce a plot that reveals the true structure of the graph, as in Fig. 4.5.31. Finally, find the approximate coordinates of the local maximum and minimum points on the graph.

60. This problem pertains to the plausible suggestion that two polynomials with essentially the same coefficients ought to have essentially the same roots. (a) Show, nevertheless, that the quartic equation

$$f(x) = x^4 - 55x^3 + 505x^2 + 11000x - 110000 = 0$$

has four distinct real solutions, whereas the "similar" equation

$$g(x) = x^4 - 55x^3 + 506x^2 + 11000x - 110000 = 0$$

has only two distinct real solutions (and two complex conjugate solutions). (b) Let $h(x) = f(x) + \epsilon x^2$. Note that if $\epsilon = 0$ then $h(x) = f(x)$, and if $\epsilon = 1$ then $h(x) = g(x)$. Investigate the question of *where*—between $\epsilon = 0$ and $\epsilon = 1$—the transition from four real solutions of $h(x) = 0$ to only two real solutions takes place.

4.6 HIGHER DERIVATIVES AND CONCAVITY

We saw in Section 4.3 that the sign of the first derivative f' of a differentiable function f indicates whether the graph of f is rising or falling. Here we shall see that the sign of the *second* derivative of f, the derivative of f', indicates which way the curve $y = f(x)$ is *bending,* upward or downward.

Higher Derivatives

The **second derivative** of f is the derivative of f'; it is denoted by f'', and its value at x is

$$f''(x) = D_x(f'(x)) = D_x(D_x f(x)) = D_x^2 f(x).$$

(The superscript 2 is not an exponent but only an indication that the operator D_x is to be applied twice.) The derivative of f'' is the **third derivative** f''' of f, and

$$f'''(x) = D_x(f''(x)) = D_x\left(D_x^2 f(x)\right) = D_x^3 f(x).$$

The third derivative is also denoted by $f^{(3)}$. More generally, the result of beginning with the function f and differentiating n times is succession is the **nth derivative** $f^{(n)}$ of f, with $f^{(n)}(x) = D_x^n f(x)$.

If $y = f(x)$, then the first n derivatives are written in operator notation as

$$D_x y, \quad D_x^2 y, \quad D_x^3 y, \quad \ldots, \quad D_x^n y,$$

in function notation as

$$y'(x), \quad y''(x), \quad y'''(x), \quad \dots, \quad y^{(n)}(x),$$

and in differential notation as

$$\frac{dy}{dx}, \quad \frac{d^2y}{dx^2}, \quad \frac{d^3y}{dx^3}, \quad \dots, \quad \frac{d^ny}{dx^n}.$$

The history of the curious use of superscripts in differential notation for higher derivatives involves the metamorphosis

$$\frac{d}{dx}\left(\frac{dy}{dx}\right) \to \frac{d}{dx}\frac{dy}{dx} \to \frac{(d)^2y}{(dx)^2} \to \frac{d^2y}{dx^2}.$$

EXAMPLE 1 Find the first four derivatives of

$$f(x) = 2x^3 + \frac{1}{x^2} + 16x^{7/2}.$$

Solution Write

$$f(x) = 2x^3 + x^{-2} + 16x^{7/2}.$$

Then

$$f'(x) = 6x^2 - 2x^{-3} + 56x^{5/2} = 6x^2 - \frac{2}{x^3} + 56x^{5/2},$$

$$f''(x) = 12x + 6x^{-4} + 140x^{3/2} = 12x + \frac{6}{x^4} + 140x^{3/2},$$

$$f'''(x) = 12 - 24x^{-5} + 210x^{1/2} = 12 - \frac{24}{x^5} + 210\sqrt{x},$$

and

$$f^{(4)}(x) = 120x^{-6} + 105x^{-1/2} = \frac{120}{x^6} + \frac{105}{\sqrt{x}}.$$ ◆

Example 2 shows how to find higher derivatives of implicitly defined functions.

EXAMPLE 2 Find the second derivative $y''(x)$ of a function $y(x)$ that is defined implicitly by the equation

$$x^2 - xy + y^2 = 9.$$

Solution A first implicit differentiation of the given equation *with respect to* x gives

$$2x - y - x\frac{dy}{dx} + 2y\frac{dy}{dx} = 0,$$

so

$$\frac{dy}{dx} = \frac{y - 2x}{2y - x}.$$

We obtain d^2y/dx^2 by differentiating implicitly, again with respect to x, using the quotient rule. After that, we substitute the expression we just found for dy/dx:

$$\frac{d^2y}{dx^2} = D_x\left(\frac{y-2x}{2y-x}\right) = \frac{\left(\dfrac{dy}{dx} - 2\right)(2y-x) - (y-2x)\left(2\dfrac{dy}{dx} - 1\right)}{(2y-x)^2}$$

$$= \frac{3x\dfrac{dy}{dx} - 3y}{(2y-x)^2} = \frac{3x\dfrac{y-2x}{2y-x} - 3y}{(2y-x)^2}.$$

Thus

$$\frac{d^2y}{dx^2} = -\frac{6(x^2 - xy + y^2)}{(2y - x)^3}.$$

We now substitute the original equation, $x^2 - xy + y^2 = 9$, for one final simplification:

$$\frac{d^2y}{dx^2} = -\frac{54}{(2y - x)^3}.$$

The somewhat unexpected final simplification is always available when the original equation is symmetric in x and y. ◆

The Sign of the Second Derivative

Now we shall investigate the significance of the *sign* of the second derivative. If $f''(x) > 0$ on the interval I, then the first derivative f' is an increasing function on I, because *its* derivative $f''(x)$ is positive. Thus, as we scan the graph $y = f(x)$ from left to right, we see the tangent line turning counterclockwise (Fig. 4.6.1). We describe this situation by saying that the curve $y = f(x)$ is **bending upward.** Note that a curve can bend upward without rising, as in Fig. 4.6.2.

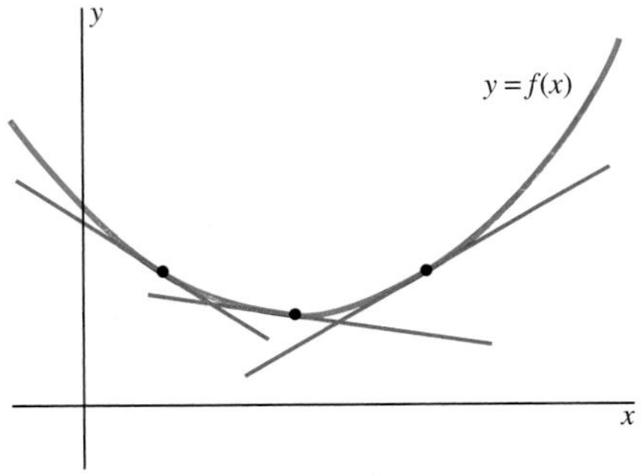

FIGURE 4.6.1 The graph is bending upward.

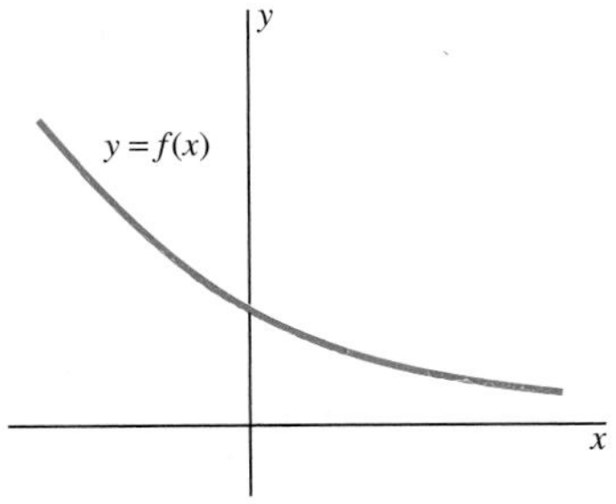

FIGURE 4.6.2 Another graph bending upward.

If $f''(x) < 0$ on the interval I, then the first derivative f' is decreasing on I, so the tangent line turns clockwise as x increases. We say in this case that the curve $y = f(x)$ is **bending downward.** Figures 4.6.3 and 4.6.4 show two ways this can happen.

The two cases are summarized in the brief table in Fig. 4.6.5.

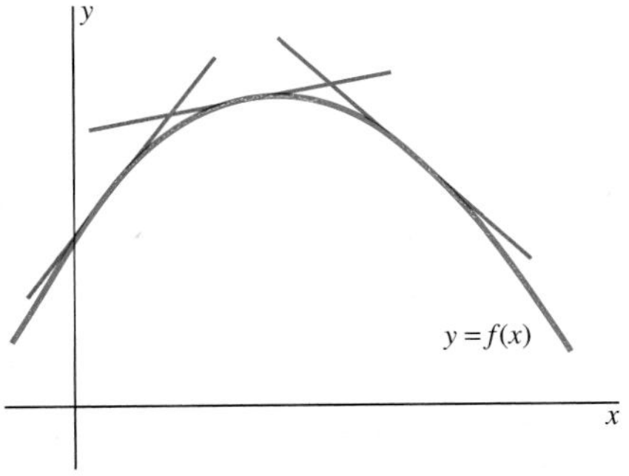

FIGURE 4.6.3 A graph bending downward.

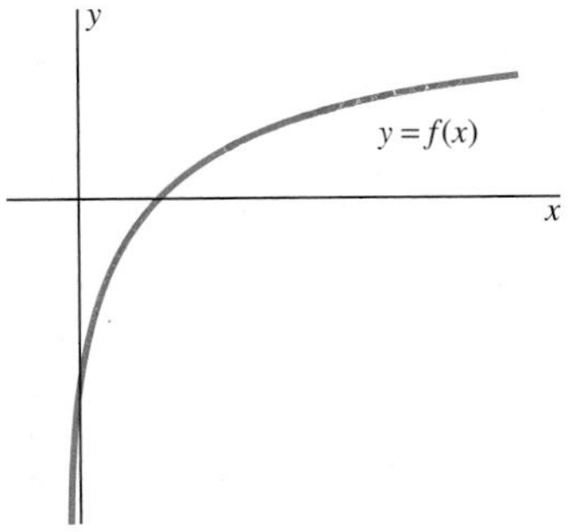

FIGURE 4.6.4 Another graph bending downward.

$f''(x)$	$y = f(x)$
Negative	Bending downward
Positive	Bending upward

FIGURE 4.6.5 Significance of the sign of $f''(x)$ on an interval.

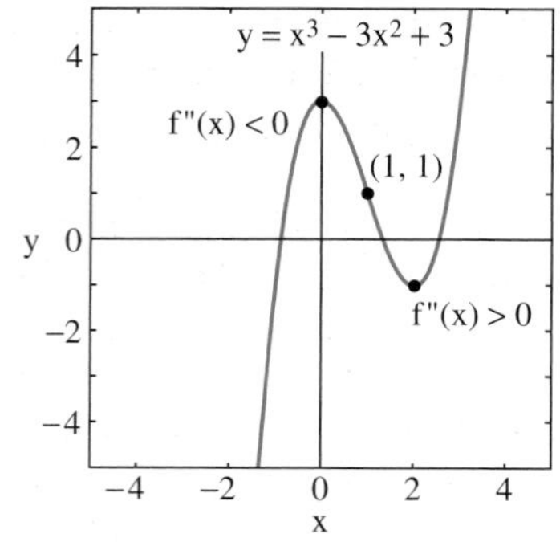

FIGURE 4.6.6 The graph of $y = x^3 - 3x^2 + 3$ (Example 3).

EXAMPLE 3 Figure 4.6.6 shows the graph of the function

$$f(x) = x^3 - 3x^2 + 3.$$

Because

$$f'(x) = 3x^2 - 6x \quad \text{and} \quad f''(x) = 6x - 6 = 6(x - 1),$$

we see that

$$f''(x) < 0 \quad \text{for} \quad x < 1,$$
$$f''(x) > 0 \quad \text{for} \quad x > 1.$$

Observe in the figure that the curve bends downward on $(-\infty, 1)$ but bends upward on $(1, +\infty)$, consistent with the correspondences in Fig. 4.6.5. ◆

The Second Derivative Test

We know from Section 3.5 that a local extremum of a differentiable function f can occur only at a critical point where $f'(c) = 0$, so the tangent line at the point $(c, f(c))$ on the curve $y = f(x)$ is horizontal. But the example $f(x) = x^3$, for which $x = 0$ is a critical point but not an extremum (Fig. 4.6.7), shows that the *necessary condition* $f'(c) = 0$ is *not* a sufficient condition from which to conclude that $f(c)$ is an extreme value of the function f.

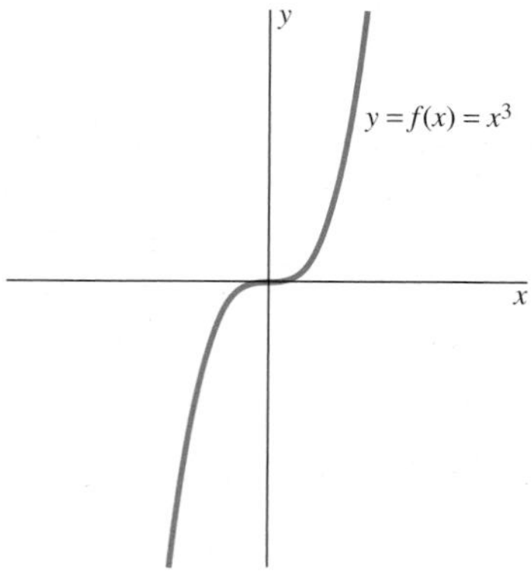

FIGURE 4.6.7 Although $f'(0) = 0$, $f(0)$ is not an extremum.

Now suppose not only that $f'(c) = 0$, but also that the curve $y = f(x)$ is bending upward on some open interval that contains the critical point $x = c$. It is apparent from Fig. 4.6.8(a) that $f(c)$ is a local minimum value. Similarly, $f(c)$ is a local maximum value if $f'(c) = 0$ while $y = f(x)$ is bending downward on some open interval containing c [Fig. 4.6.8(b)]. But the *sign* of the second derivative $f''(x)$ tells us whether $y = f(x)$ is bending upward or downward and therefore provides us with a *sufficient* condition for a local extremum.

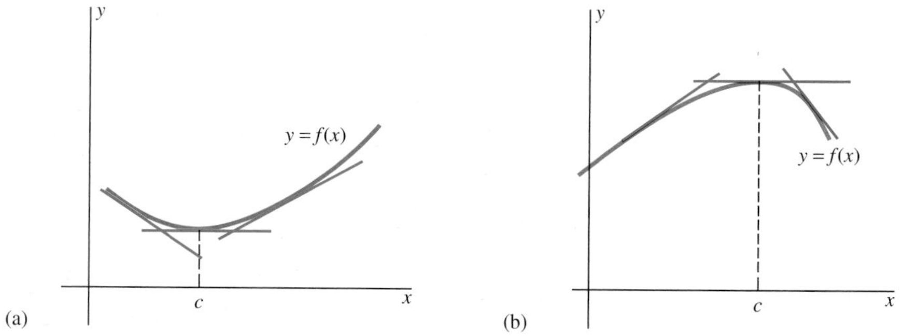

FIGURE 4.6.8 The second derivative test (Theorem 1). (a) $f''(x) > 0$: tangent turning counterclockwise; graph bending upward; local minimum at $x = c$.
(b) $f''(x) < 0$: tangent turning clockwise; graph bending downward; local maximum at $x = c$.

THEOREM 1 Second Derivative Test
Suppose that the function f is twice differentiable on the open interval I containing the critical point c at which $f'(c) = 0$. Then

1. If $f''(x) > 0$ on I, then $f(c)$ is the minimum value of $f(x)$ on I.
2. If $f''(x) < 0$ on I, then $f(c)$ is the maximum value of $f(x)$ on I.

PROOF We will prove only part 1. If $f''(x) > 0$ on I, then it follows that the first derivative f' is an increasing function on I. Because $f'(c) = 0$, we may conclude that $f'(x) < 0$ for $x < c$ in I and that $f'(x) > 0$ for $x > c$ in I. Consequently, the first derivative test of Section 4.4 implies that $f(c)$ is the minimum value of $f(x)$ on I. ◄

$f''(x)$	$f(c)$
Positive	Minimum
Negative	Maximum

FIGURE 4.6.9 Significance of the sign of $f''(x)$ on an interval containing the critical point c.

REMARK 1 Rather than memorizing verbatim the conditions in parts 1 and 2 of Theorem 1 (summarized in Fig. 4.6.9), it is easier and more reliable to remember the second derivative test by visualizing continuously turning tangent lines (Fig. 4.6.8).

REMARK 2 Theorem 1 implies that the function f has a local minimum at the critical point c if $f''(x) > 0$ on some open interval about c but a local maximum if $f''(x) < 0$ near c. But the hypothesis on $f''(x)$ in Theorem 1 is *global* in that $f''(x)$ is assumed to have the same sign at *every* point of the open interval I that contains the critical point c. There is a strictly *local* version of the second derivative test that involves only the sign of $f''(c)$ at the critical point c (rather than on a whole open interval). According to Problem 90, if $f'(c) = 0$, then $f(c)$ is a local minimum value of f if $f''(c) > 0$ but a local maximum if $f''(c) < 0$.

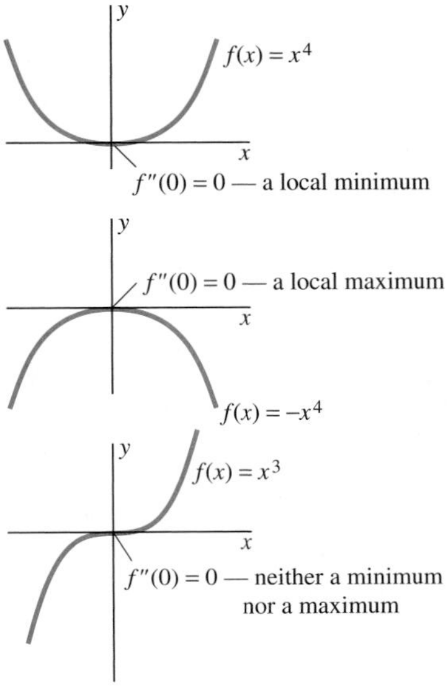

FIGURE 4.6.10 No conclusion is possible if $f'(c) = 0 = f''(c)$.

REMARK 3 The second derivative test says *nothing* about what happens if $f''(c) = 0$ at the critical point c. Consider the three functions $f(x) = x^4$, $f(x) = -x^4$, and $f(x) = x^3$. For each, $f'(0) = 0$ and $f''(0) = 0$. But their graphs, shown in Fig. 4.6.10, demonstrate that *anything* can happen at such a point—maximum, minimum, or neither.

REMARK 4 Suppose that we want to maximize or minimize the function f on the open interval I, and we find that f has only one critical point in I, a number c at which $f'(c) = 0$. If $f''(x)$ has the same sign at all points of I, then Theorem 1 implies that $f(c)$ is an absolute extremum of f on I—a minimum if $f''(x) > 0$ and a maximum if $f''(x) < 0$. This absolute interpretation of the second derivative test can be useful in applied open-interval maximum-minimum problems.

EXAMPLE 3 (continued) Consider again the function $f(x) = x^3 - 3x^2 + 3$, for which

$$f'(x) = 3x(x-2) \quad \text{and} \quad f''(x) = 6(x-1).$$

Then f has the two critical points $x = 0$ and $x = 2$, as marked in Fig. 4.6.6. Because $f''(x) < 0$ for x near zero, the second derivative test implies that $f(0) = 3$ is a local maximum value of f. And because $f''(x) > 0$ for x near 2, it follows that $f(2) = -1$ is a local minimum value. ◆

EXAMPLE 4 An open-topped rectangular box with square base has volume 500 cm^3. Find the dimensions that minimize the total area A of its base and four sides.

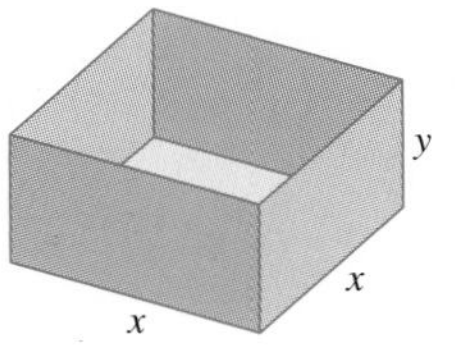

FIGURE 4.6.11 The open-topped box of Example 4.

Solution We denote by x the edge length of the square base and by y the height of the box (Fig. 4.6.11). The volume of the box is

$$V = x^2 y = 500, \tag{1}$$

and the total area of its base and four sides is

$$A = x^2 + 4xy. \tag{2}$$

When we solve Eq. (1) for $y = 500/x^2$ and substitute this into Eq. (2), we get the area function

$$A(x) = x^2 + \frac{2000}{x}, \quad 0 < x < +\infty.$$

The domain of A is the open and unbounded interval $(0, +\infty)$ because x can take on any positive value; to make the box volume 500, simply choose $y = 500/x^2$. But x cannot be zero or negative.

The first derivative of $A(x)$ is

$$A'(x) = 2x - \frac{2000}{x^2} = \frac{2(x^3 - 1000)}{x^2}. \tag{3}$$

The equation $A'(x) = 0$ yields $x^3 = 1000$, so the only critical point of A in $(0, +\infty)$ is $x = 10$. To investigate this critical point, we calculate the second derivative,

$$A''(x) = 2 + \frac{4000}{x^3}. \qquad \textbf{(4)}$$

Because it is clear that $A''(x) > 0$ on $(0, +\infty)$, it follows from the second derivative test and Remark 4 that $A(10) = 300$ is the absolute minimum value of $A(x)$ on $(0, +\infty)$. Finally, because $y = 500/x^2$, $y = 5$ when $x = 10$. Therefore, this absolute minimum corresponds to a box with base 10 cm by 10 cm and height 5 cm. ◆

Concavity and Inflection Points

A comparison of Fig. 4.6.1 with Fig. 4.6.3 suggests that the question of whether the curve $y = f(x)$ is bending upward or downward is closely related to the question of whether it lies above or below its tangent lines. The latter question refers to the important property of *concavity.*

DEFINITION Concavity

Suppose that the function f is differentiable at the point a and that L is the line tangent to the graph $y = f(x)$ at the point $(a, f(a))$. Then the function f (or its graph) is said to be

1. **Concave upward** at a if, on some open interval containing a, the graph of f lies *above* L.
2. **Concave downward** at a if, on some open interval containing a, the graph of f lies *below* L.

Figure 4.6.12(a) shows a graph that is concave upward at $(a, f(a))$. Figure 4.6.12(b) shows a graph that is concave downward at $(a, f(a))$.

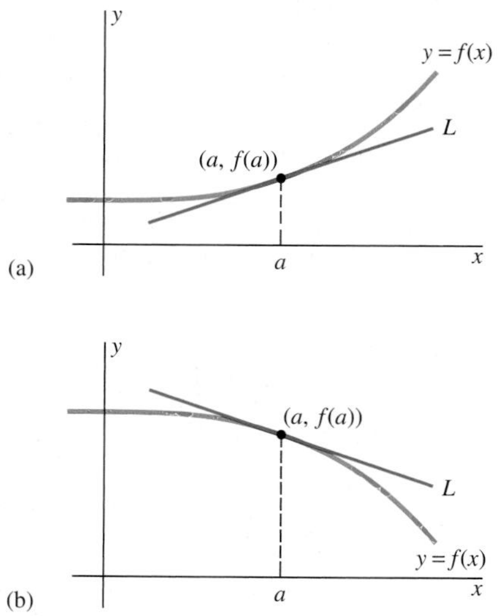

FIGURE 4.6.12 (a) At $x = a$, f is a concave upward. (b) At $x = a$, f is concave downward.

Theorem 2 establishes the connection between concavity and the sign of the second derivative. That connection is the one suggested by our discussion of bending.

THEOREM 2 Test for Concavity
Suppose that the function f is twice differentiable on the open interval I.

1. If $f''(x) > 0$ on I, then f is concave upward at each point of I.
2. If $f''(x) < 0$ on I, then f is concave downward at each point of I.

A proof of Theorem 2 based on the second derivative test is given at the end of this section.

Note The significance of the sign of the *first* derivative must not be confused with the significance of the sign of the *second* derivative. The possibilities illustrated in Figs. 4.6.13 through 4.6.16 show that the signs of f' and f'' are independent of each other.

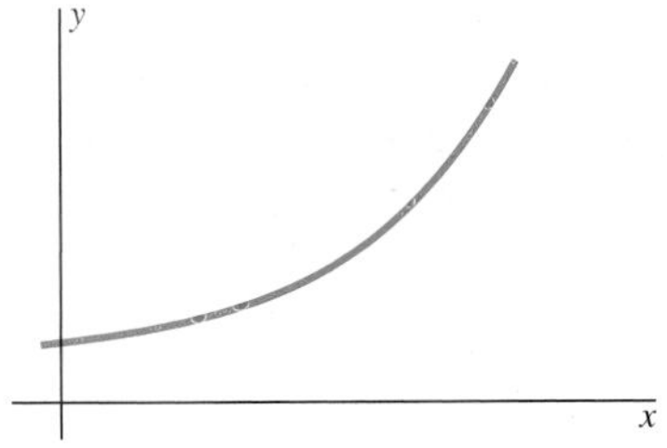

FIGURE 4.6.13 $f'(x) > 0$, f increasing; $f''(x) > 0$, f concave upward.

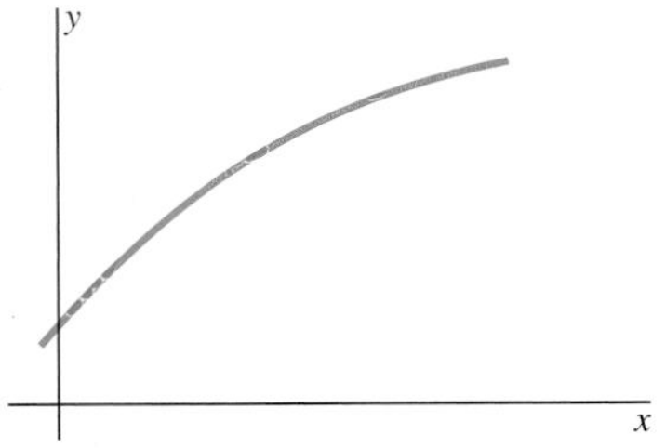

FIGURE 4.6.14 $f'(x) > 0$, f increasing; $f''(x) < 0$; f concave downward.

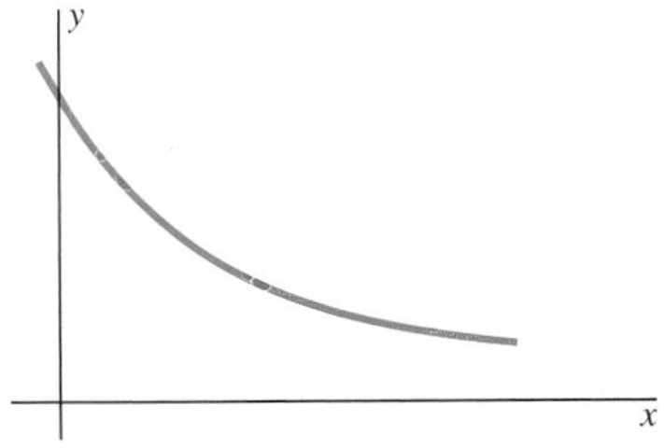

FIGURE 4.6.15 $f'(x) < 0$, f decreasing; $f''(x) > 0$, f concave upward.

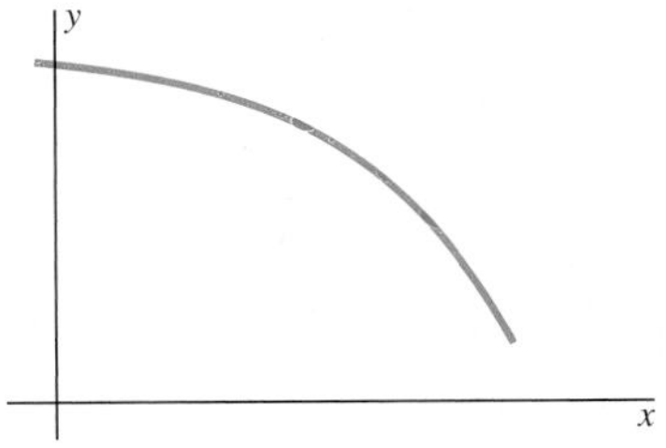

FIGURE 4.6.16 $f'(x) < 0$, f decreasing; $f''(x) < 0$, f concave downward.

EXAMPLE 3 (continued again) For the function $f(x) = x^3 - 3x + 3$, the second derivative changes sign from positive to negative at the point $x = 1$. Observe in Fig. 4.6.6 that the corresponding point $(1, 1)$ on the graph of f is where the curve changes from bending downward to bending upward. ◆

Observe that the test for concavity in Theorem 2 says nothing about the case in which $f''(x) = 0$. A point where the second derivative is zero *may or may not* be a point where the function changes from concave upward on one side to concave downward on the other. But a point like $(1, 1)$ in Fig. 4.6.6, where the concavity *does* change in this manner, is called an *inflection point* of the graph of f. More precisely, the point $x = a$ where f is continuous is an **inflection point** of the function f provided that f is concave upward on one side of $x = a$ and concave downward on the other side. We also refer to $(a, f(a))$ as an inflection point on the graph of f.

THEOREM 3 Inflection Point Test
Suppose that the function f is continuous and f'' exists on an open interval containing the point a. Then a is an inflection point of f provided that $f''(x) < 0$ on one side of a and $f''(x) > 0$ on the other side.

The fact that a point where the second derivative changes sign is an inflection point follows from Theorem 2 and the definition of an inflection point.

REMARK At the inflection point itself, either

- $f''(a) = 0$, or
- $f''(a)$ does not exist.

Thus we find *inflection points* of f by examining the *critical points* of f'. Some of the possibilities are indicated in Fig. 4.6.17. We mark the intervals of upward concavity and downward concavity by small cups opening upward and downward, respectively.

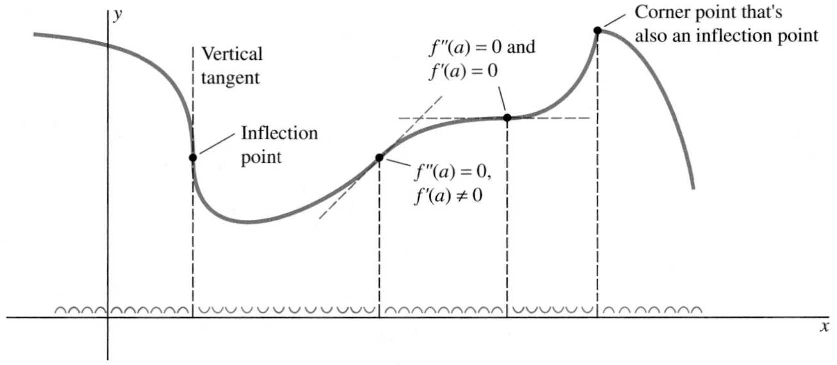

FIGURE 4.6.17 Some inflection points.

EXAMPLE 5 Figure 4.6.18 shows the graph of $f(x) = (2x^2 - 2x - 1)e^{-x}$. Two evident inflection points are marked. Find their coordinates.

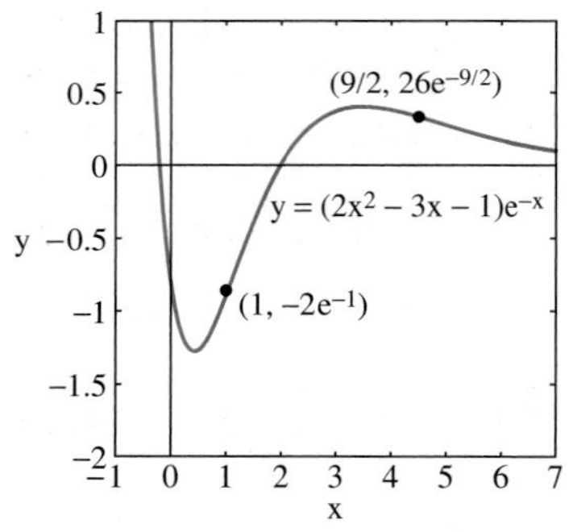

FIGURE 4.6.18 The graph $y = (2x^2 - 3x - 1)e^{-x}$ (Example 5).

Solution We calculate

$$f'(x) = (4x - 3)e^{-x} - (2x^2 - 3x - 1)e^{-x} = (-2x^2 + 7x - 2)e^{-x}$$

and

$$f''(x) = (-4x + 7)e^{-x} - (-2x^2 + 7x - 2)e^{-x} = (2x^2 - 11x + 9)e^{-x}.$$

Because e^{-x} is never zero, it follows that $f''(x) = 0$ only when

$$2x^2 - 11x + 9 = (2x - 9)(x - 1) = 0$$

—that is, when either $x = 1$ or $x = \frac{9}{2}$. Only at these two points can $f''(x)$ change sign. But

$$f''(0) = 9 > 0,$$
$$f''(2) = -5e^{-x} < 0,$$

and

$$f''(5) = 4e^{-5} > 0.$$

It therefore follows that

$$f''(x) > 0 \quad \text{if} \quad x < 1,$$
$$f''(x) < 0 \quad \text{if} \quad 1 < x < \tfrac{9}{2},$$

and

$$f''(x) > 0 \quad \text{if} \quad \tfrac{9}{2} < x.$$

Thus the graph of $f(x) = (2x^2 - 3x - 1)e^{-x}$ has inflection points where $x = 1$ and where $x = \frac{9}{2}$. These points, marked on the graph in Fig. 4.6.18, have coordinates $(1, -2e^{-1})$ and $(\frac{9}{2}, 26e^{-9/2})$. ◆

Inflection Points and Curve Sketching

Let the function f be twice differentiable for all x. Just as the critical points where $f'(x) = 0$ separate the x-axis into open intervals on which $f'(x)$ does not change sign, the *possible* inflection points where $f''(x) = 0$ separate the x-axis into open intervals on which $f''(x)$ does not change sign. On each of these intervals, the curve $y = f(x)$ either is bending downward $[f''(x) < 0]$ or is bending upward $[f''(x) > 0]$. We can determine the sign of $f''(x)$ in each of these intervals in either of two ways:

1. Evaluation of $f''(x)$ at a typical point of each interval. The sign of $f''(x)$ at that particular point is the sign of $f''(x)$ throughout the interval.
2. Construction of a table of signs of the factors of $f''(x)$. Then the sign of $f''(x)$ on each interval can be deduced from the table.

These are the same two methods we used in Sections 4.4 and 4.5 to determine the sign of $f'(x)$. We use the first method in Example 6 and the second in Example 7.

EXAMPLE 6 Sketch the graph of $f(x) = 8x^5 - 5x^4 - 20x^3$, indicating local extrema, inflection points, and concave structure.

Solution We sketched this curve in Example 2 of Section 4.5; see Fig. 4.5.6 for the graph. In that example we found the first derivative to be

$$f'(x) = 40x^4 - 20x^3 - 60x^2 = 20x^2(x+1)(2x-3),$$

so the critical points are $x = -1$, $x = 0$, and $x = \frac{3}{2}$. The second derivative is

$$f''(x) = 160x^3 - 60x^2 - 120x = 20x(8x^2 - 3x - 6).$$

When we compute $f''(x)$ at each critical point, we find that

$$f''(-1) = -100 < 0, \qquad f''(0) = 0, \qquad \text{and} \qquad f''\left(\tfrac{3}{2}\right) = 225 > 0.$$

Continuity of f'' ensures that $f''(x) < 0$ near the critical point $x = -1$ and that $f''(x) > 0$ near the critical point $x = \frac{3}{2}$. The second derivative test therefore tells us that f has a local maximum at $x = -1$ and a local minimum at $x = \frac{3}{2}$. We cannot determine from the second derivative test the behavior of f at $x = 0$.

Because $f''(x)$ exists everywhere, the possible inflection points are the solutions of the equation

$$f''(x) = 0; \qquad \text{that is,} \quad 20x(8x^2 - 3x - 6) = 0.$$

Clearly, one solution is $x = 0$. To find the other two, we use the quadratic formula to solve the equation

$$8x^2 - 3x - 6 = 0.$$

This gives

$$x = \tfrac{1}{16}\left(3 \pm \sqrt{201}\right),$$

so $x \approx 1.07$ and $x \approx -0.70$ are possible inflection points along with $x = 0$.

These three possible inflection points separate the x-axis into the intervals indicated in Fig. 4.6.19. We check the sign of $f''(x)$ on each.

$$\begin{array}{lll} \text{On } (-\infty, -0.70): & f''(-1) = -100 < 0; & f \text{ is concave downward;} \\ \text{On } (-0.70, 0): & f''\left(-\tfrac{1}{2}\right) = 25 > 0; & f \text{ is concave upward;} \\ \text{On } (0, 1.07): & f''(1) = -20 < 0; & f \text{ is concave downward;} \\ \text{On } (1.07, +\infty): & f''(2) = 800 > 0; & f \text{ is concave upward.} \end{array}$$

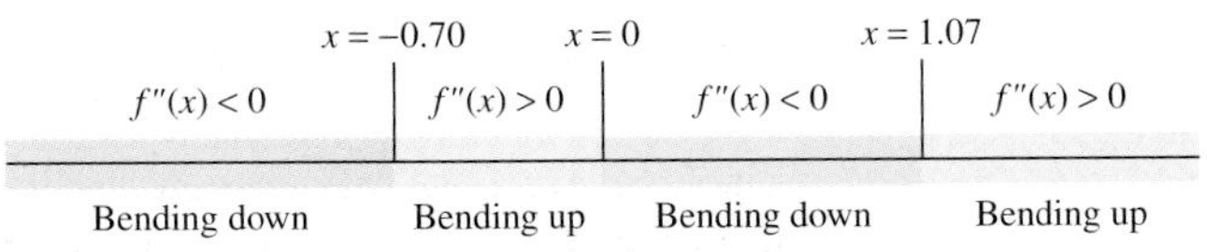

FIGURE 4.6.19 Intervals of concavity of Example 6.

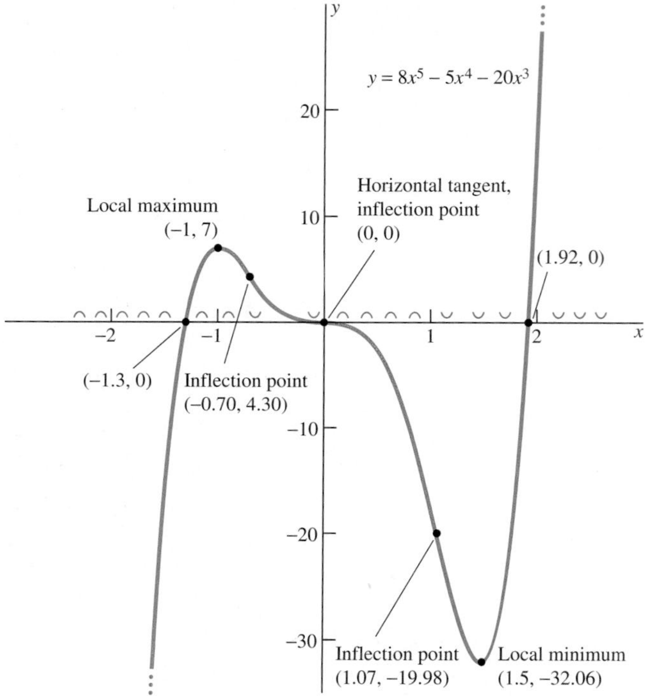

FIGURE 4.6.20 The graph of the function of Example 6.

Thus we see that the direction of concavity of f changes at each of the three points $x \approx -0.70$, $x = 0$, and $x \approx 1.07$. These three points are indeed inflection points. This information is shown in the graph of f sketched in Fig 4.6.20. ◆

EXAMPLE 7 Sketch the graph of $f(x) = 4x^{1/3} + x^{4/3}$. Indicate local extrema, inflection points, and concave structure.

Solution First,

$$f'(x) = \frac{4}{3}x^{-2/3} + \frac{4}{3}x^{1/3} = \frac{4(x+1)}{3x^{2/3}},$$

so the critical points are $x = -1$ (where the tangent line is horizontal) and $x = 0$ (where it is vertical). Next,

$$f''(x) = -\frac{8}{9}x^{-5/3} + \frac{4}{9}x^{-2/3} = \frac{4(x-2)}{9x^{5/3}},$$

so the possible inflection points are $x = 2$ (where $f''(x) = 0$) and $x = 0$ (where $f''(x)$ does not exist).

To determine where f is increasing and where it is decreasing, we construct the following table.

Interval	$x+1$	$x^{2/3}$	$f'(x)$	f
$(-\infty, -1)$	$-$	$+$	$-$	Decreasing
$(-1, 0)$	$+$	$+$	$+$	Increasing
$(0, +\infty)$	$+$	$+$	$+$	Increasing

Thus f is decreasing when $x = -1$ and increasing when $x > -1$ (Fig. 4.6.21(a)).

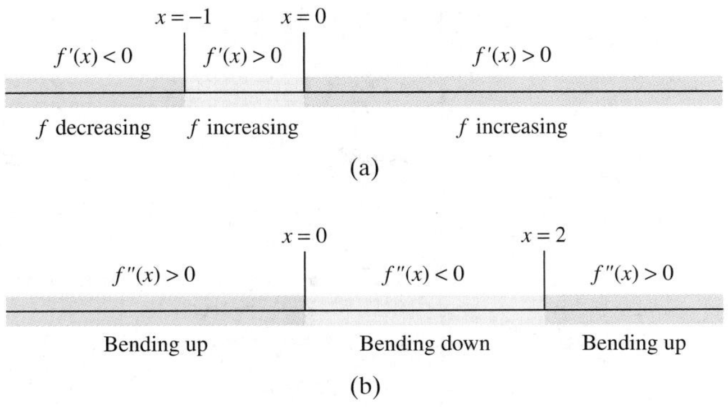

FIGURE 4.6.21 (a) Increasing and decreasing intervals of Example 7. (b) Intervals of concavity of Example 7.

To determine the concavity of f, we construct a table to find the sign of $f''(x)$ on each of the intervals separated by its zeros.

Interval	$x^{5/3}$	$x-2$	$f''(x)$	f
$(-\infty, 0)$	$-$	$-$	$+$	Concave upward
$(0, 2)$	$+$	$-$	$-$	Concave downward
$(2, +\infty)$	$+$	$+$	$+$	Concave upward

The table shows that f is concave downward on $(0, 2)$ and concave upward for $x < 0$ and for $x > 2$ (Fig. 4.6.21(b)).

We note that $f(x) \to +\infty$ as $x \to \pm\infty$, and we mark with plus signs the intervals on the x-axis where f is increasing, minus signs where it is decreasing, cups opening upward where f is concave upward, and cups opening downward where f is concave downward. We plot (at least approximately) the points on the graph of f that correspond to the zeros and discontinuities of f' and f''; these are $(-1, -3)$, $(0, 0)$, and $(2, 6\sqrt[3]{2})$. Finally, we use all this information to draw the smooth curve shown in Fig. 4.6.22. ◆

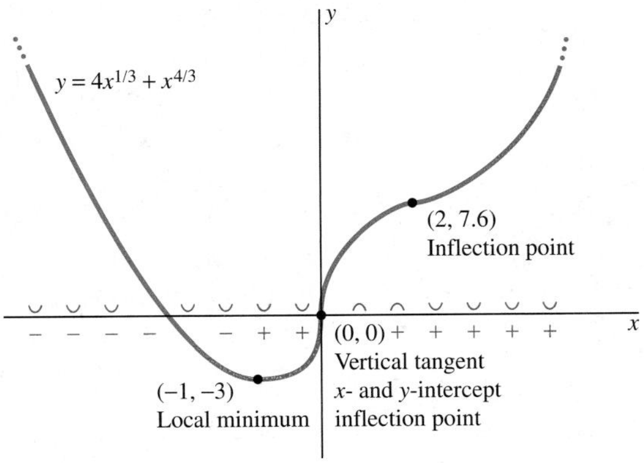

FIGURE 4.6.22 The graph of the function of Example 7.

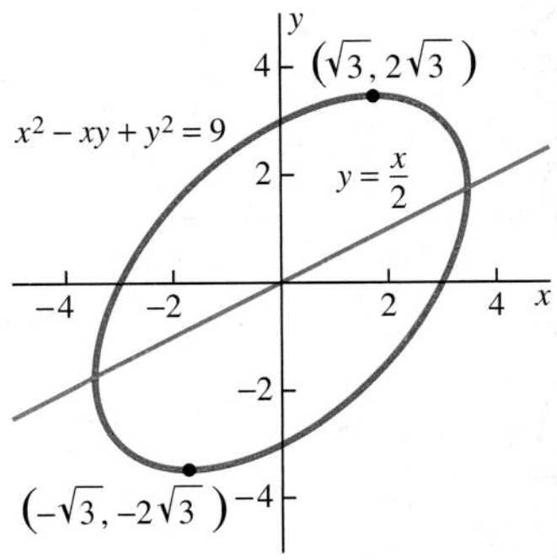

FIGURE 4.6.23 The ellipse $x^2 - xy + y^2 = 9$ is concave downward at points above the line $y = \frac{1}{2}x$, concave upward at points beneath it.

EXAMPLE 8 The graph of the equation $x^2 - xy + y^2 = 9$ is the rotated ellipse shown in Fig. 4.6.23. In Example 2 we saw that if the function $y(x)$ is implicitly defined by this equation, then

$$\frac{dy}{dx} = \frac{y - 2x}{2y - x}.$$

Hence $y = 2x$ at any critical point (x, y) at which $y'(x) = 0$. Substituting $y = 2x$ into the equation $x^2 - xy + y^2 = 9$ readily gives the two points $(\sqrt{3}, 2\sqrt{3})$ and

$(-\sqrt{3}, -2\sqrt{3})$ that are marked in Fig. 4.6.23. We also saw that

$$\frac{d^2y}{dx^2} = -\frac{54}{(2y-x)^3}.$$

It follows that $y''(x) < 0$ when $2y - x > 0$, so the graph is concave downward at any point (x, y) at which $2y > x$; that is, at points above the line $y = \frac{1}{2}x$. Similarly, $y''(x) > 0$ when $2y - x < 0$, so the graph is concave upward at any point (x, y) at which $2y < x$; that is, at points below the line $y = \frac{1}{2}x$. (See Fig. 4.6.23.) ◆

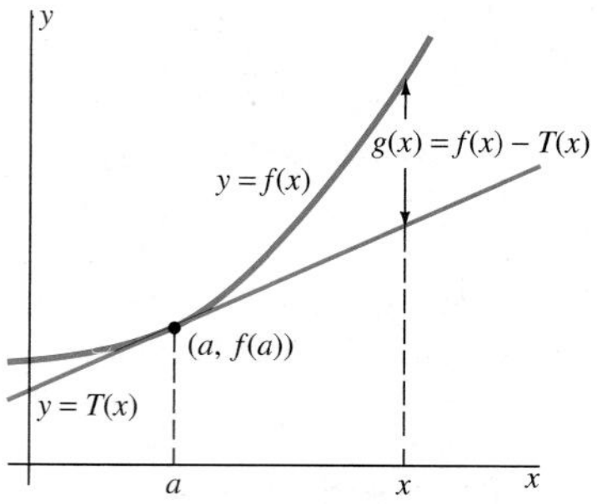

FIGURE 4.6.24 Illustrating the proof of Theorem 2.

PROOF OF THEOREM 2 We will prove only part 1—the proof of part 2 is similar. Given a fixed point a of the open interval I where $f''(x) > 0$, we want to show that the graph $y = f(x)$ lies above the tangent line at $(a, f(a))$. The tangent line in question has the equation

$$y = T(x) = f(a) + f'(a)\cdot(x-a). \qquad \textbf{(5)}$$

Consider the auxiliary function

$$g(x) = f(x) - T(x) \qquad \textbf{(6)}$$

illustrated in Fig. 4.6.24. Note first that $g(a) = g'(a) = 0$, so $x = a$ is a critical point of g. Moreover, Eq. (5) implies that $T'(x) \equiv f'(a)$ and that $T''(x) \equiv 0$, so

$$g''(x) = f''(x) - T''(x) = f''(x) > 0$$

at each point of I. Therefore, the second derivative test implies that $g(a) = 0$ is the minimum value of $g(x) = f(x) - T(x)$ on I. It follows that the curve $y = f(x)$ lies above the tangent line $y = T(x)$. ◄

4.6 TRUE/FALSE STUDY GUIDE

4.6 CONCEPTS: QUESTIONS AND DISCUSSION

1. Suppose that the function f is differentiable at the point $x = c$ where $f''(c) = 0$. Does it necessarily follow that the point $(c, f(c))$ is an inflection point of the graph of $y = f(x)$?
2. Suppose that the function f is differentiable except at the point $x = c$ where the graph of $y = f(x)$ has a vertical tangent line. Does it necessarily follow that the point $(c, f(c))$ is an inflection point of the graph of $y = f(x)$?
3. Suppose that n is a positive integer and that k is an integer such that $0 \leq k \leq n - 2$. Does there always exist a polynomial of degree n having exactly k inflection points? Justify your answer.
4. Can the graph of a function have more inflection points than critical points? Justify your answer.

4.6 PROBLEMS

Calculate the first three derivatives of the functions given in Problems 1 through 15.

1. $f(x) = 2x^4 - 3x^3 + 6x - 17$

2. $f(x) = 2x^5 + x^{3/2} - \dfrac{1}{2x}$

3. $f(x) = \dfrac{2}{(2x-1)^2}$

4. $g(t) = t^2 + \sqrt{t+1}$

5. $g(t) = (3t-2)^{4/3}$

6. $f(x) = x\sqrt{x+1}$

7. $h(y) = \dfrac{y}{y+1}$

8. $f(x) = \left(1+\sqrt{x}\right)^3$

9. $g(t) = t^2 \ln t$

10. $h(z) = \dfrac{e^z}{\sqrt{z}}$

11. $f(x) = \sin 3x$

12. $f(x) = \cos^2 2x$

13. $f(x) = \sin x \cos x$

14. $f(x) = x^2 \cos x$

15. $f(x) = \dfrac{\sin x}{x}$

In Problems 16 through 22, calculate dy/dx and d^2y/dx^2, assuming that y is defined implicitly as a function of x by the given equation.

16. $x^2 + y^2 = 4$

17. $x^2 + xy + y^2 = 3$

18. $x^{1/3} + y^{1/3} = 1$

19. $y^3 + x^2 + x = 5$

20. $\dfrac{1}{x} + \dfrac{1}{y} = 1$

21. $\sin y = xy$

22. $\sin^2 x + \cos^2 y = 1$

In Problems 23 through 30, find the exact *coordinates of the inflection points and critical points marked on the given graph.*

23. The graph of $f(x) = x^3 - 3x^2 - 45x$ (Fig. 4.6.25)

24. The graph of $f(x) = 2x^3 - 9x^2 - 108x + 200$ (Fig. 4.6.26)

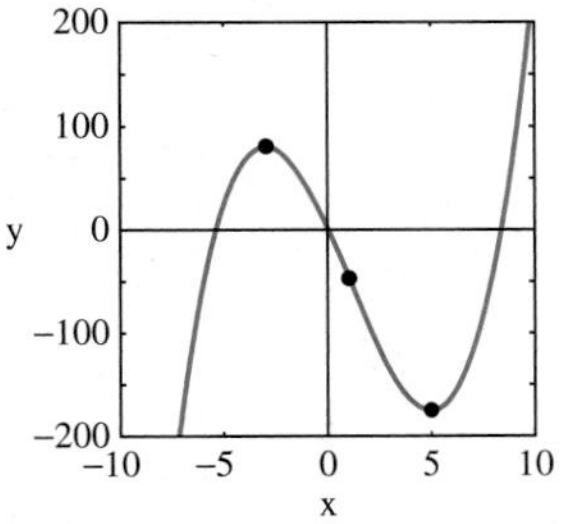

FIGURE 4.6.25 The graph of $f(x) = x^3 - 3x^2 - 45x$ (Problem 23).

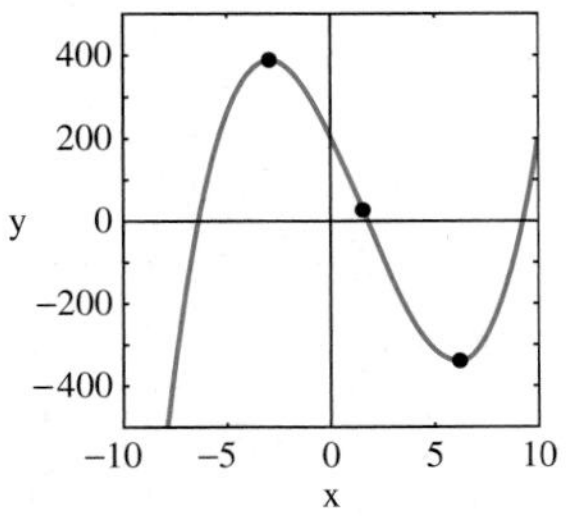

FIGURE 4.6.26 The graph of $f(x) = 2x^3 - 9x^2 - 108x + 200$ (Problem 24).

25. The graph of $f(x) = 4x^3 - 6x^2 - 189x + 137$ (Fig. 4.6.27)

26. The graph of $f(x) = -40x^3 - 171x^2 + 2550x + 4150$ (Fig. 4.6.28)

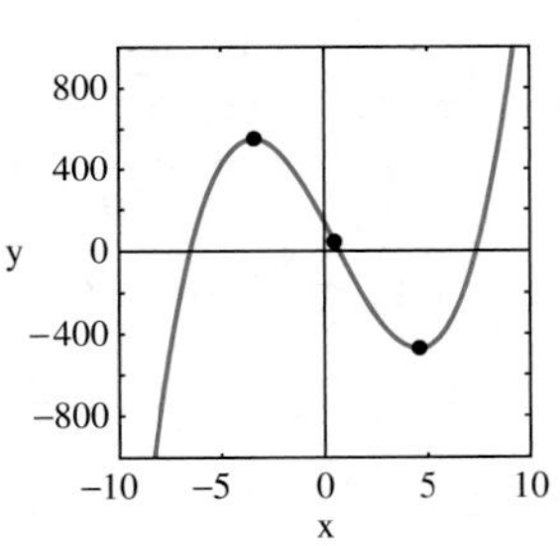

FIGURE 4.6.27 The graph of $f(x) = 4x^3 - 6x^2 - 189x + 137$ (Problem 25).

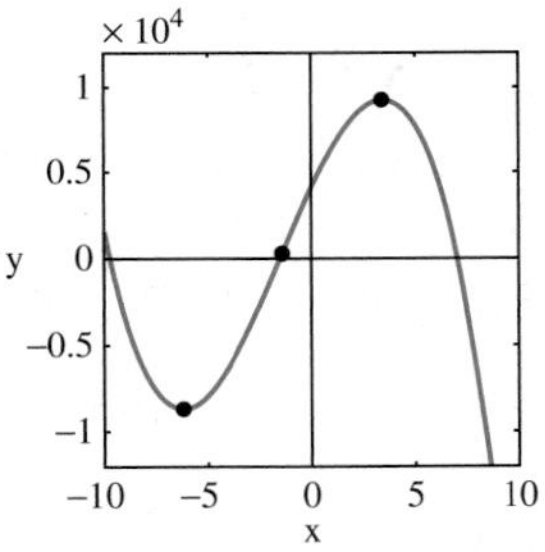

FIGURE 4.6.28 The graph of $f(x) = -40x^3 - 171x^2 + 2550x + 4150$ (Problem 26).

27. The graph of $f(x) = x^4 - 54x^2 + 237$ (Fig. 4.6.29)

28. The graph of $f(x) = x^4 - 10x^3 - 250$ (Fig. 4.6.30)

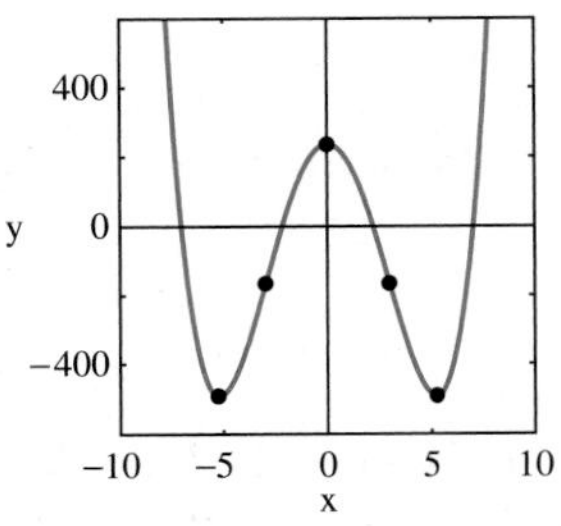

FIGURE 4.6.29 The graph of $f(x) = x^4 - 54x^2 - 237$ (Problem 27).

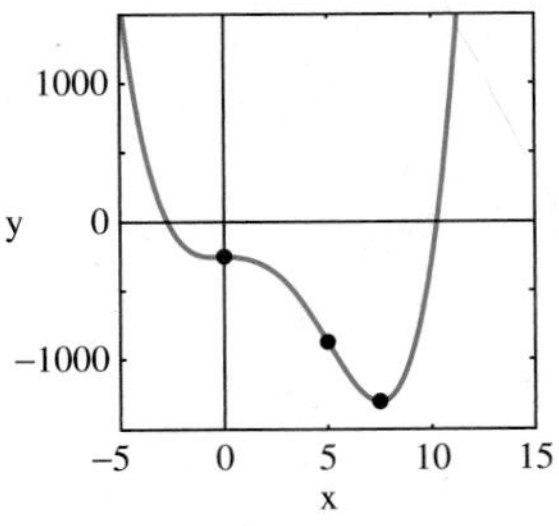

FIGURE 4.6.30 The graph of $f(x) = x^4 - 10x^3 - 250$ (Problem 28).

29. The graph of $f(x) = 3x^5 - 20x^4 + 1000$ (Fig. 4.6.31)

30. The graph of $f(x) = 3x^5 - 160x^3$ (Fig. 4.6.32)

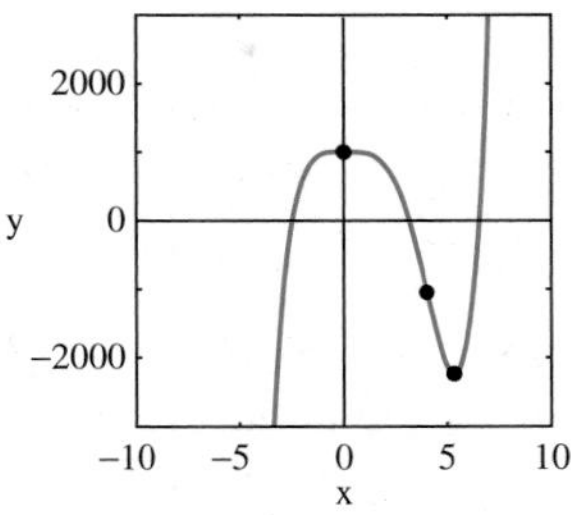

FIGURE 4.6.31 The graph of $f(x) = 3x^5 - 20x^4 + 1000$ (Problem 29).

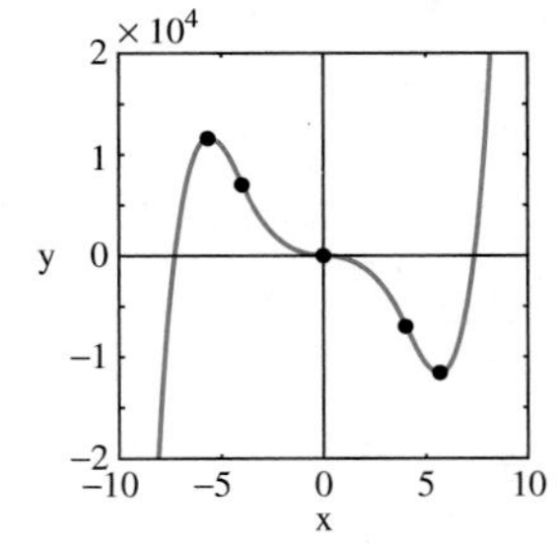

FIGURE 4.6.32 The graph of $f(x) = 3x^5 - 160x^3$ (Problem 30).

Apply the second derivative test to find the local maxima and local minima of the functions given in Problems 31 through 50, and apply the inflection point test to find all inflection points.

31. $f(x) = x^2 - 4x + 3$

32. $f(x) = 5 - 6x - x^2$

33. $f(x) = x^3 - 3x + 1$

34. $f(x) = x^3 - 3x^2$

35. $f(x) = xe^{-x}$

36. $f(x) = \dfrac{\ln x}{x}$

37. $f(x) = x^5 + 2x$

38. $f(x) = x^4 - 8x^2$

39. $f(x) = x^2(x - 1)^2$

40. $f(x) = x^3(x + 2)^2$

41. $f(x) = \sin x$ on $(0, 2\pi)$

42. $f(x) = \cos x$ on $(-\pi/2, \pi/2)$

43. $f(x) = \tan x$ on $(-\pi/2, \pi/2)$

44. $f(x) = \sec x$ on $(-\pi/2, \pi/2)$

45. $f(x) = \cos^2 x$ on $(-\pi/2, 3\pi/2)$

46. $f(x) = \sin^3 x$ on $(-\pi, \pi)$

47. $f(x) = 10(x - 1)e^{-2x}$

48. $f(x) = (x^2 - x)e^{-x}$

49. $f(x) = (x^2 - 2x - 1)e^{-x}$

50. $f(x) = xe^{-x^2}$

In Problems 51 through 62, rework the indicated problem from Section 4.4, now using the second derivative test to verify that you have found the desired absolute maximum or minimum value.

51. Problem 27

52. Problem 28

53. Problem 29

54. Problem 30

55. Problem 31

56. Problem 32

57. Problem 33

58. Problem 36

59. Problem 37

60. Problem 38

61. Problem 39

62. Problem 40

Sketch the graphs of the functions in Problems 63 through 76, indicating all critical points and inflection points. Apply the second derivative test at each critical point. Show the correct concave structure and indicate the behavior of $f(x)$ as $x \to \pm\infty$.

63. $f(x) = 2x^3 - 3x^2 - 12x + 3$

64. $f(x) = 3x^4 - 4x^3 - 5$

65. $f(x) = 6 + 8x^2 - x^4$

66. $f(x) = 3x^5 - 5x^3$

67. $f(x) = 3x^4 - 4x^3 - 12x^2 - 1$

68. $f(x) = 3x^5 - 25x^3 + 60x$

69. $f(x) = x^3(x-1)^4$

70. $f(x) = (x-1)^2(x+2)^3$

71. $f(x) = 1 + x^{1/3}$

72. $f(x) = 2 - (x-3)^{1/3}$

73. $f(x) = (x+3)\sqrt{x}$

74. $f(x) = x^{2/3}(5 - 2x)$

75. $f(x) = (4 - x)\sqrt[3]{x}$

76. $f(x) = x^{1/3}(6 - x)^{2/3}$

In Problems 77 through 82, the graph of a function $f(x)$ is shown. Match it with the graph of its second derivative $f''(x)$ in Fig. 4.6.33.

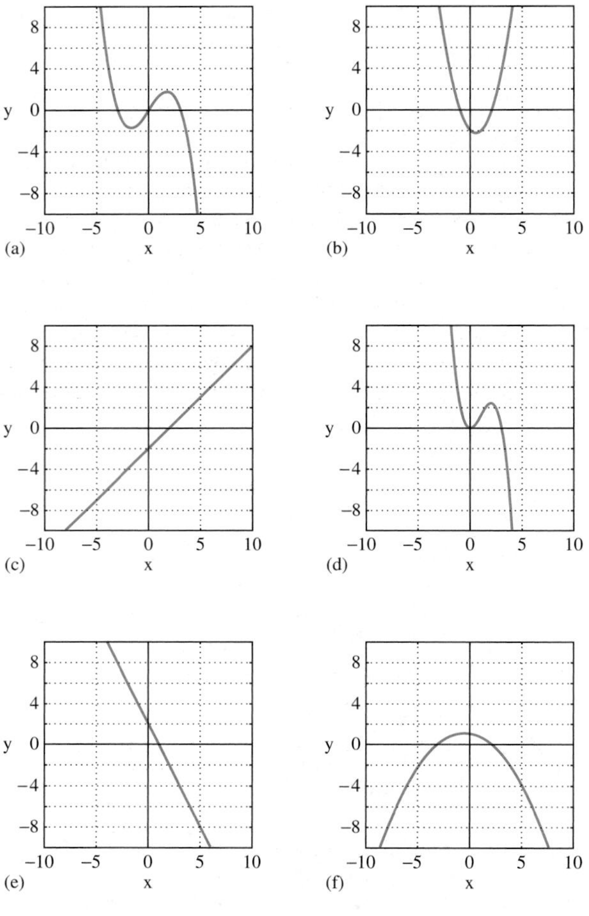

FIGURE 4.6.33

77. See Fig. 4.6.34.

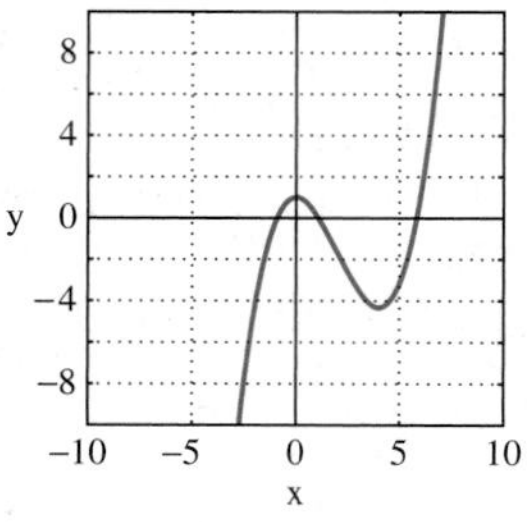

FIGURE 4.6.34

78. See Fig. 4.6.35.

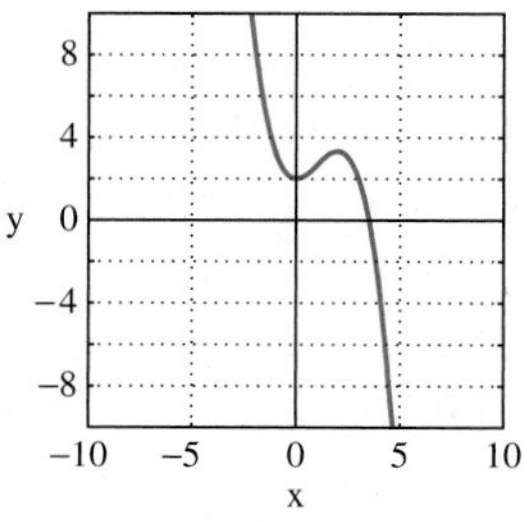

FIGURE 4.6.35

79. See Fig. 4.6.36.

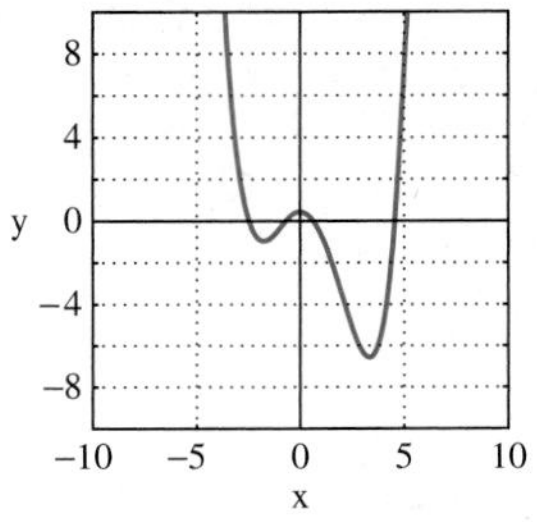

FIGURE 4.6.36

80. See Fig. 4.6.37.

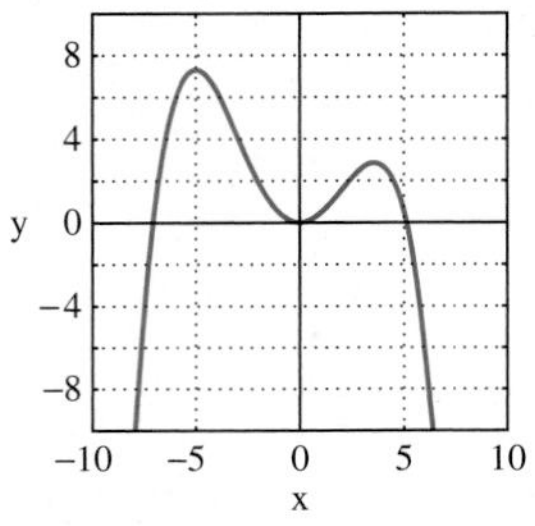

FIGURE 4.6.37

81. See Fig. 4.6.38.

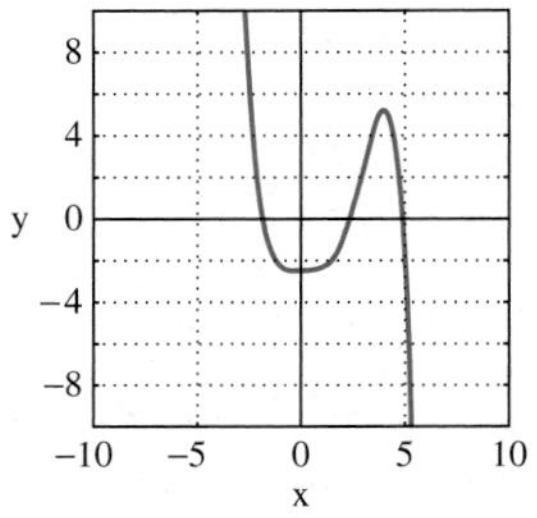

FIGURE 4.6.38

82. See Fig. 4.6.39.

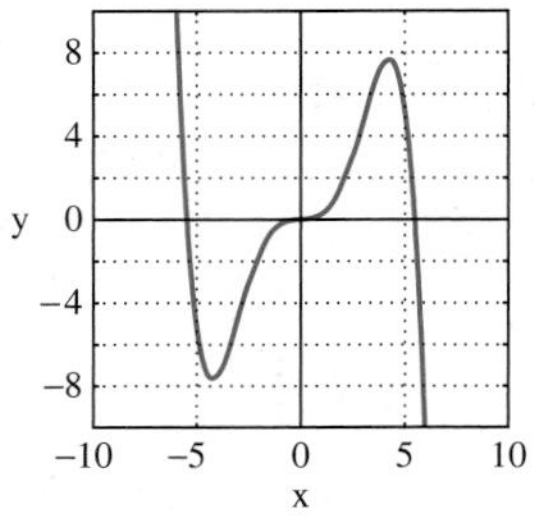

FIGURE 4.6.39

83. (a) Show first that the nth derivative of $f(x) = x^n$ is

$$f^{(n)}(x) \equiv n! = n \cdot (n-1) \cdot (n-2) \cdots 3 \cdot 2 \cdot 1.$$

(b) Conclude that if $f(x)$ is a polynomial of degree n, then $f^{(k)}(x) \equiv 0$ if $k > n$.

84. (a) Calculate the first four derivatives of $f(x) = \sin x$. (b) Explain why it follows that $D_x^{n+4} \sin x = D_x^n \sin x$ if n is a positive integer.

85. Suppose that $z = g(y)$ and that $y = f(x)$. Show that

$$\frac{d^2z}{dx^2} = \frac{d^2z}{dy^2}\left(\frac{dy}{dx}\right)^2 + \frac{dz}{dy} \cdot \frac{d^2y}{dx^2}.$$

86. Prove that the graph of a quadratic polynomial has no inflection points.

87. Prove that the graph of a cubic polynomial has exactly one inflection point.

88. Prove that the graph of a polynomial function of degree 4 has either no inflection point or exactly two inflection points.

89. Suppose that the pressure p (in atmospheres), volume V (in cubic centimeters), and temperature T (in kelvins) of n moles of carbon dioxide (CO_2) satisfy van der Waals' equation

$$\left(p + \frac{n^2a}{V^2}\right)(V - nb) = nRT,$$

where a, b, and R are empirical constants. The following experiment was carried out to find the values of these constants.

One mole of CO_2 was compressed at the constant temperature $T = 304$ K. The measured pressure-volume

(pV) data were then plotted as in Fig. 4.6.40, with the pV curve showing an inflection point coinciding with a horizontal tangent at $V = 128.1$, $p = 72.8$. Use this information to calculate a, b, and R. [*Suggestion:* Solve van der Waals' equation for p and then calculate dp/dV and d^2p/dV^2.]

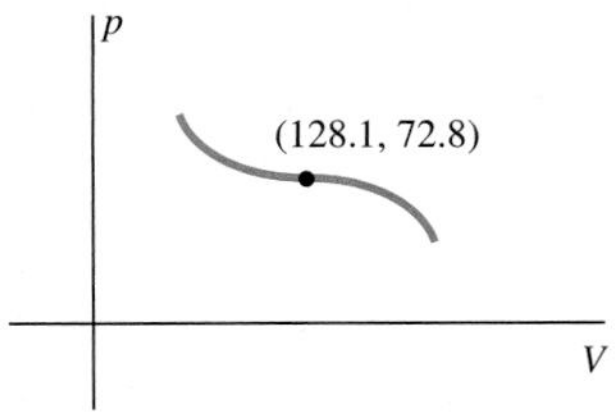

FIGURE 4.6.40 A problem involving van der Waals' equation.

90. Suppose that the function f is differentiable on an open interval containing the point c at which $f'(c) = 0$ and that the second derivative

$$f''(c) = \lim_{h\to 0} \frac{f'(c+h) - f'(c)}{h} = \lim_{h\to 0} \frac{f'(c+h)}{h}$$

exists. (a) First assume that $f''(c) > 0$. Reason that $f'(c+h)$ and h have the same sign if $h \neq 0$ is sufficiently small. Hence apply the first derivative test to show in this case that $f(c)$ is a local minimum value of f. (b) Show similarly that $f(c)$ is a local maximum value of f if $f''(c) < 0$.

Problems 91 and 92 require the use of a graphing calculator or computer algebra system. Any equations you need to solve may be solved graphically or by using a "solve" key or command; you may find it useful to look ahead to the CD-ROM discussion for the Section 4.7 project.

91. Figure 4.6.41 shows the graph of the cubic equation

$$y = 1000x^3 - 3051x^2 + 3102x + 1050$$

on a scale with x measured in units and y in tens of thousands. The graph appears to exhibit a single point near $(1, 2000)$ that is both a critical point and an inflection point. Nevertheless, this curve "has two real wiggles like a *good* cubic should." Find them! In particular, find the local extrema and inflection point (or points) on this curve. Then sketch a graph that plainly exhibits all these points—mark and label each of them.

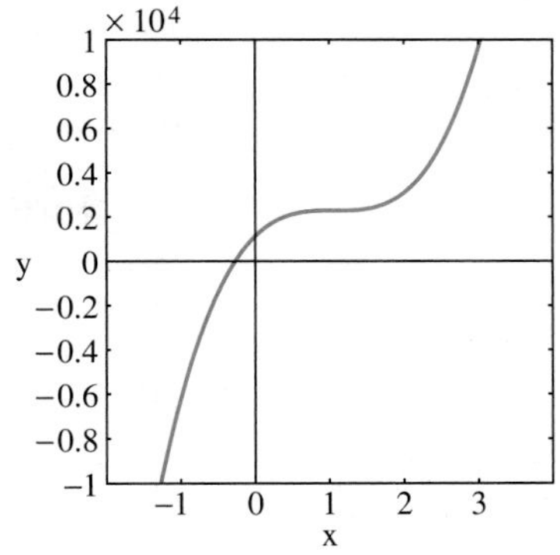

FIGURE 4.6.41 The cubic graph of Problem 91.

92. Figure 4.6.42 shows the graph of

$$y = [x(1-x)(9x-7)(4x-1)]^4$$

on a scale with x and y both measured in units. At first glance it appears that there is a local maximum near $x = \frac{1}{2}$, with "flat spots" along the x-axis to the left and to the right. But no nonconstant polynomial can have a "flat spot" where $y = 0$ on an open interval of the x-axis. (Why not?) Indeed, this graph actually has seven local extrema and six inflection points in the interval $0 \leqq x \leqq 1$. Find approximate coordinates of all thirteen of these points, then sketch a graph on a scale that makes all these points evident.

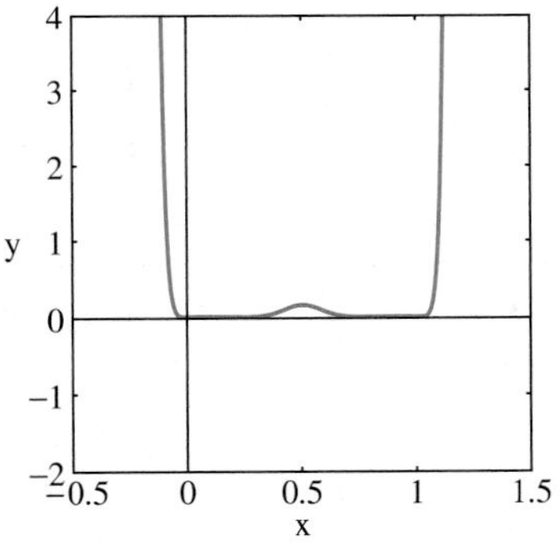

FIGURE 4.6.42 The graph of Problem 92.

4.7 CURVE SKETCHING AND ASYMPTOTES

We now extend the limit concept to include infinite limits and limits at infinity. This extension will add a powerful weapon to our arsenal of curve-sketching techniques, the notion of an *asymptote* to a curve—a straight line that the curve approaches arbitrarily close in a sense we soon make precise.

Recall from Section 2.3 that $f(x)$ is said to **increase without bound,** or **become infinite,** as x approaches a, and we write

$$\lim_{x\to a} f(x) = +\infty, \tag{1}$$

provided that $f(x)$ can be made arbitrarily large by choosing x sufficiently close (but not equal) to a. The statement that $f(x)$ **decreases without bound,** or **becomes negatively infinite,** as $x \to a$, written

$$\lim_{x \to a} f(x) = -\infty, \tag{2}$$

has an analogous definition.

EXAMPLE 1 It is apparent that

$$\lim_{x \to -2} \frac{1}{(x+2)^2} = +\infty$$

because, as $x \to -2$, $(x+2)^2$ is positive and approaches zero. (See Fig. 4.7.1.) By contrast,

$$\lim_{x \to -2} \frac{x}{(x+2)^2} = -\infty$$

because, as $x \to -2$, the denominator $(x+2)^2$ is still positive and approaches zero, but the numerator x is negative. (See Fig. 4.7.2.) Thus when x is very close to -2, we have in $x/(x+2)^2$ a negative number close to -2 divided by a very small positive number. Hence the quotient becomes a negative number of large magnitude. ◆

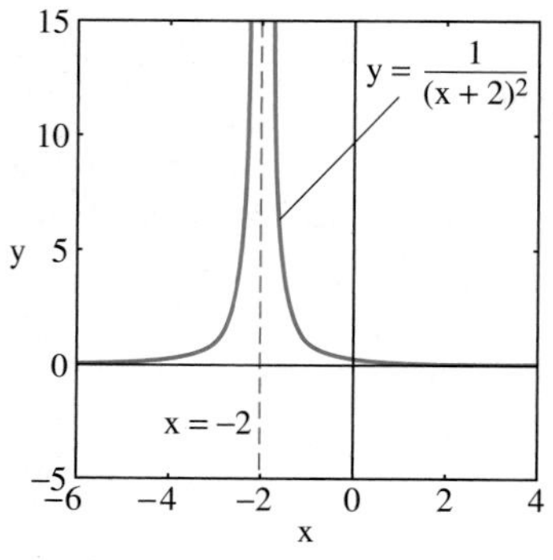

FIGURE 4.7.1 $\frac{1}{(x+2)^2} \to \infty$ as $x \to -2$, and the line $x = -2$ is a vertical asymptote.

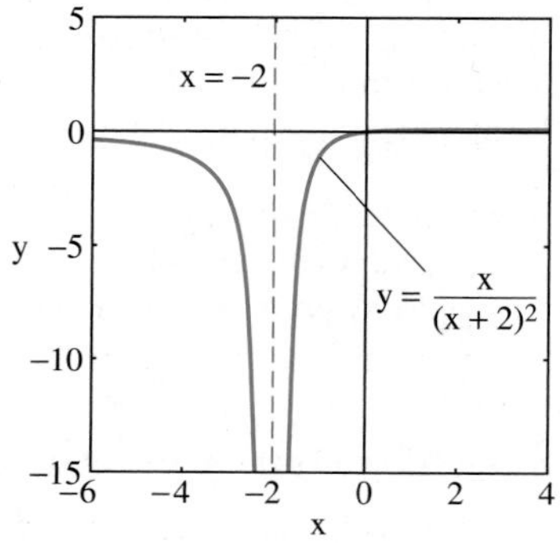

FIGURE 4.7.2 $\frac{x}{(x+2)^2} \to -\infty$ as $x \to -2$, and the line $x = -2$ is a vertical asymptote.

One-sided versions of Eqs. (1) and (2) are valid also. For instance, if n is an *odd* positive integer, then it is apparent that

$$\lim_{x \to 2^-} \frac{1}{(x-2)^n} = -\infty \quad \text{and} \quad \lim_{x \to 2^+} \frac{1}{(x-2)^n} = +\infty,$$

because $(x-2)^n$ is negative when x is to the left of 2 and positive when x is to the right of 2. The case $n = 3$ is illustrated in Fig. 4.7.3.

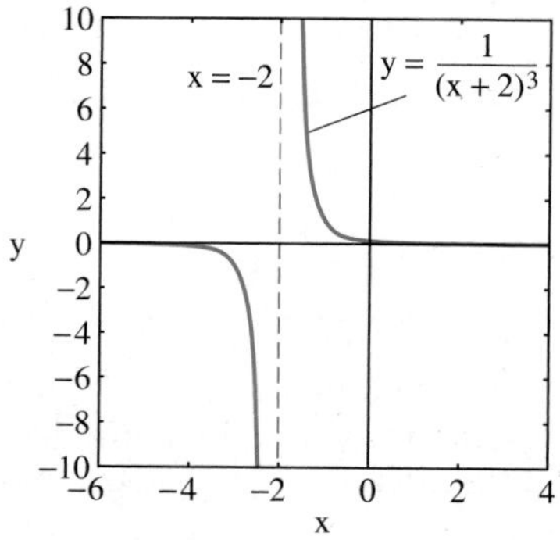

FIGURE 4.7.3 $\frac{1}{(x+2)^3}$ has infinite one-sided limits as $x \to -2$, and the line $x = -2$ is a vertical asymptote.

Vertical Asymptotes

The vertical lines at $x = -2$ in Figs. 4.7.1 through 4.7.3 are examples of *vertical asymptotes* associated with infinite limits. The line $x = a$ is a **vertical asymptote** of the curve $y = f(x)$ provided that either

$$\lim_{x \to a^-} f(x) = \pm\infty \tag{3a}$$

or

$$\lim_{x \to a^+} f(x) = \pm\infty \tag{3b}$$

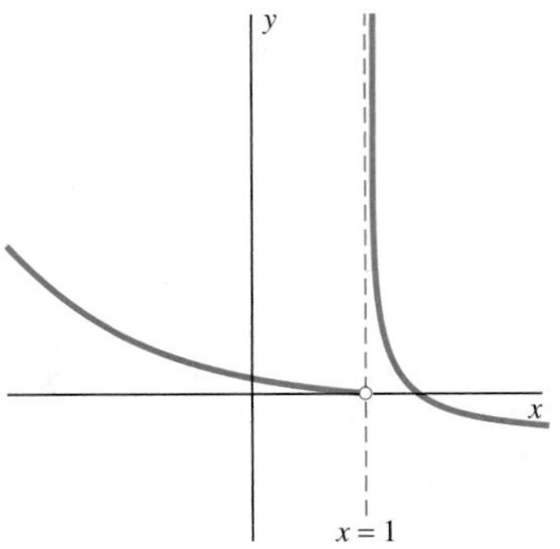

FIGURE 4.7.4 A "right-hand only" vertical asymptote.

or both. It is usually the case that both one-sided limits, rather than only one, are infinite. If so, we write

$$\lim_{x \to a} f(x) = \pm\infty. \tag{3c}$$

The geometry of a vertical asymptote is illustrated by the graphs in Figs. 4.7.1 through 4.7.3. In each case, as $x \to -2$ and $f(x) \to \pm\infty$, the point $(x, f(x))$ on the curve approaches the vertical asymptote $x = -2$ and the shape and direction of the curve are better and better approximated by the asymptote.

Figure 4.7.4 shows the graph of a function whose left-hand limit is zero at $x = 1$. But the right-hand limit there is $+\infty$, which explains why the line $x = 1$ is a vertical asymptote for this graph. The right-hand limit in Fig. 4.7.5 does not even exist, but because the left-hand limit at $x = 1$ is $-\infty$, the vertical line at $x = 1$ is again a vertical asymptote.

A vertical asymptote typically appears in the case of a rational function $f(x) = p(x)/q(x)$ at a point $x = a$ where $q(a) = 0$ but $p(a) \neq 0$. (See Examples 4 through 8 later in this section.)

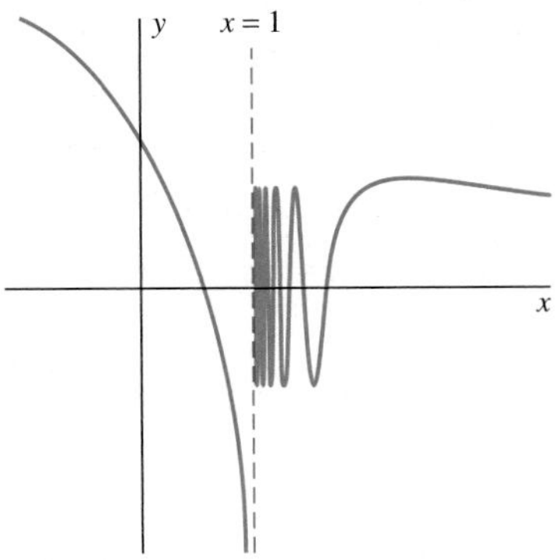

FIGURE 4.7.5 The behavior of the graph to its *left* produces the vertical asymptote.

Limits at Infinity

In Section 4.5 we discussed infinite limits at infinity in connection with the behavior of a polynomial as $x \to \pm\infty$. There is also such a thing as a *finite* limit at infinity. We say that $f(x)$ **approaches the number L as x increases without bound** and write

$$\lim_{x \to +\infty} f(x) = L \tag{4}$$

provided that $|f(x) - L|$ can be made arbitrarily small (close to zero) merely by choosing x sufficiently large. That is, given $\epsilon > 0$, there exists $M > 0$ such that

$$x > M \quad \text{implies} \quad |f(x) - L| < \epsilon. \tag{5}$$

The statement that

$$\lim_{x \to -\infty} f(x) = L$$

has a definition of similar form—merely replace the condition $x > M$ with the condition $x < -M$.

The analogues for limits at infinity of the limit laws of Section 2.2 all hold, including, in particular, the sum, product, and quotient laws. In addition, it is not difficult to show that if

$$\lim_{x \to +\infty} f(x) = L \quad \text{and} \quad \lim_{x \to +\infty} g(x) = \pm\infty,$$

then

$$\lim_{x \to +\infty} \frac{f(x)}{g(x)} = 0.$$

It follows from this result that

$$\lim_{x \to +\infty} \frac{1}{x^k} = 0 \tag{6}$$

for any choice of the positive rational number k.

Using Eq. (6) and the limit laws, we can easily evaluate limits at infinity of rational functions. The general method is this: First divide each term in both the numerator and the denominator by the highest power of x that appears in any of the terms. Then apply the limit laws.

EXAMPLE 2 Find

$$\lim_{x \to +\infty} f(x) \quad \text{if} \quad f(x) = \frac{3x^3 - x}{2x^3 + 7x^2 - 4}.$$

Solution We begin by dividing each term in the numerator and denominator by x^3:

$$\lim_{x\to+\infty} \frac{3x^3 - x}{2x^3 + 7x^2 - 4} = \lim_{x\to+\infty} \frac{3 - \dfrac{1}{x^2}}{2 + \dfrac{7}{x} - \dfrac{4}{x^3}}$$

$$= \frac{\displaystyle\lim_{x\to+\infty}\left(3 - \frac{1}{x^2}\right)}{\displaystyle\lim_{x\to+\infty}\left(2 + \frac{7}{x} - \frac{4}{x^3}\right)} = \frac{3-0}{2+0-0} = \frac{3}{2}.$$

The same computation, but with $x \to -\infty$, also gives the result

$$\lim_{x\to-\infty} f(x) = \frac{3}{2}.$$ ◆

EXAMPLE 3 Find $\displaystyle\lim_{x\to+\infty} (\sqrt{x+a} - \sqrt{x})$.

Solution We use the familiar "divide-and-multiply" technique with the conjugate of $\sqrt{x+a} - \sqrt{x}$:

$$\lim_{x\to+\infty} (\sqrt{x+a} - \sqrt{x}) = \lim_{x\to+\infty} (\sqrt{x+a} - \sqrt{x}) \cdot \frac{\sqrt{x+a} + \sqrt{x}}{\sqrt{x+a} + \sqrt{x}}$$

$$= \lim_{x\to+\infty} \frac{a}{\sqrt{x+a} + \sqrt{x}} = 0.$$ ◆

Horizontal Asymptotes

In geometric terms, the statement

$$\lim_{x\to+\infty} f(x) = L$$

means that the point $(x, f(x))$ on the curve $y = f(x)$ approaches the horizontal line $y = L$ as $x \to +\infty$. In particular, with the numbers M and ϵ of the condition in Eq. (5), the part of the curve for which $x > M$ lies between the horizontal lines $y = L - \epsilon$ and $y = L + \epsilon$ (Fig. 4.7.6). Therefore, we say that the line $y = L$ is a **horizontal asymptote** of the curve $y = f(x)$ if either

$$\lim_{x\to+\infty} f(x) = L \quad \text{or} \quad \lim_{x\to-\infty} f(x) = L.$$

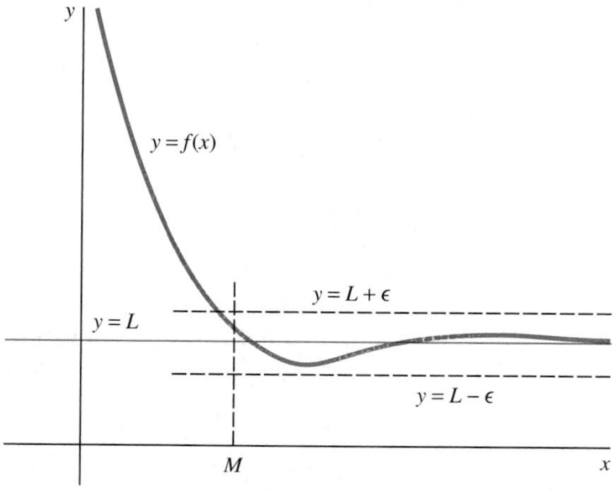

FIGURE 4.7.6 Geometry of the definition of horizontal asymptote.

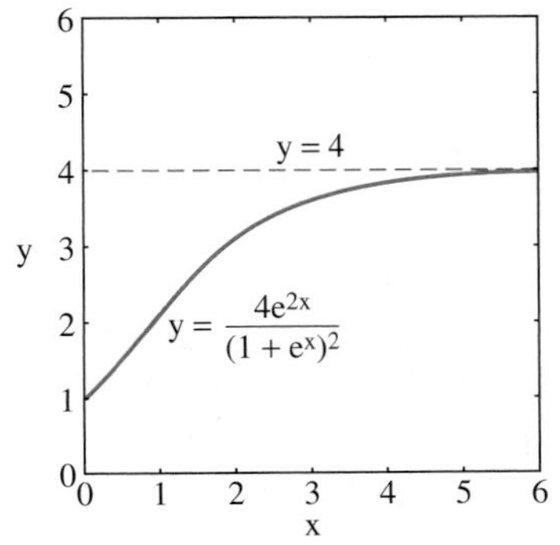

FIGURE 4.7.7 The graph $y = \dfrac{4e^{2x}}{(1+e^x)^2}$.

EXAMPLE 4 Figure 4.7.7 shows the graph of the function $f(x) = 4e^{2x}/(1+e^x)^2$. Upon dividing numerator and denominator by e^{2x}, we find that

$$f(x) = \frac{4e^{2x}}{(1+e^x)^2} = \frac{4}{(e^{-x}+1)^2} \to 4$$

as $x \to +\infty$. Thus the curve $y = 4e^{2x}/(1+e^x)^2$ has the line $y = 4$ as a horizontal asymptote. ◆

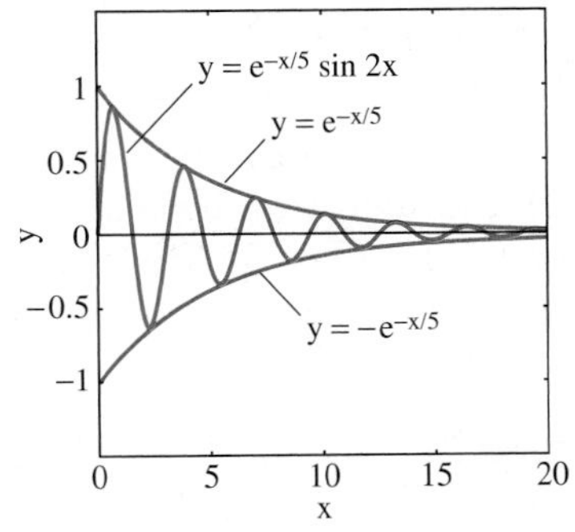

FIGURE 4.7.8 The graph $y = e^{-x/5}\sin 2x$.

EXAMPLE 5 Figure 4.7.8 shows the graph of the function $f(x) = e^{-x/5}\sin 2x$. Because $|\sin 2x| \leqq 1$ for all x and $e^{-x/5} = 1/(e^{x/5}) \to 0$ as $x \to +\infty$, the squeeze law of limits implies that $e^{-x/5}\sin 2x \to 0$ as $x \to +\infty$. Thus the curve $y = e^{-x/5}\sin 2x$ has the x-axis $y = 0$ as a horizontal asymptote. ◆

EXAMPLE 6 Sketch the graph of $f(x) = x/(x-2)$. Indicate any horizontal or vertical asymptotes.

Solution First we note that $x = 2$ is a vertical asymptote because $|f(x)| \to +\infty$ as $x \to 2$. Also,

$$\lim_{x\to\pm\infty} \frac{x}{x-2} = \lim_{x\to\pm\infty} \frac{1}{1-\dfrac{2}{x}} = \frac{1}{1-0} = 1.$$

So the line $y = 1$ is a horizontal asymptote. The first two derivatives of f are

$$f'(x) = -\frac{2}{(x-2)^2} \quad \text{and} \quad f''(x) = \frac{4}{(x-2)^3}.$$

Neither $f'(x)$ nor $f''(x)$ is zero anywhere, so the function f has no critical points and no inflection points. Because $f'(x) < 0$ for $x \neq 2$, we see that $f(x)$ is decreasing on the open intervals $(-\infty, 2)$ and $(2, +\infty)$. And because $f''(x) < 0$ for $x < 2$ and $f''(x) > 0$ for $x > 2$, the graph of f is concave downward on $(-\infty, 2)$ and concave upward on $(2, +\infty)$. The graph of f appears in Fig. 4.7.9. ◆

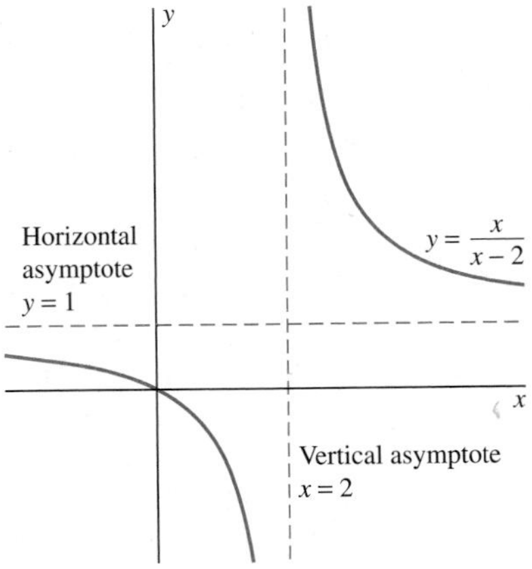

FIGURE 4.7.9 The graph for Example 6.

EXAMPLE 7 Let's reexamine the function

$$f(x) = \frac{x}{(x+2)^2}$$

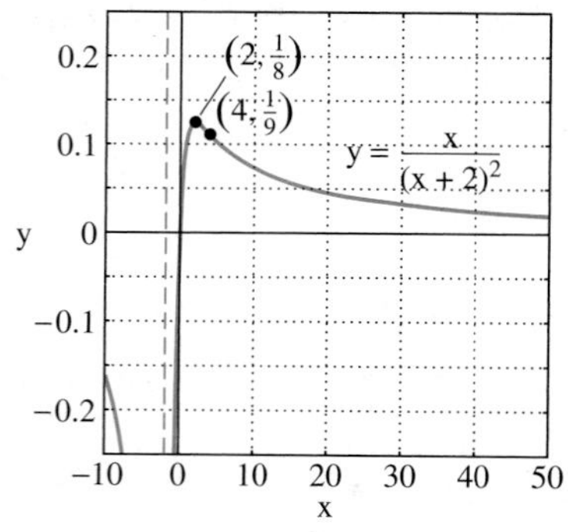

FIGURE 4.7.10 $\dfrac{x}{(x+2)^2} \to 0$ as $x \to \infty$, so the x-axis $y = 0$ is a horizontal asymptote.

whose graph was shown in Fig. 4.7.2. We note that

$$\lim_{x\to\infty} \frac{x}{(x+2)^2} = \lim_{x\to\infty} \frac{\dfrac{1}{x}}{\left(1 + \dfrac{2}{x}\right)^2} = 0,$$

so the x-axis $y = 0$ is a horizontal asymptote of the graph $y = f(x)$. We must change the viewing window to see clearly the behavior of this curve for $x > 0$. With the window $-10 < x < 40$, $-0.25 < y < 0.25$ of Fig. 4.7.10 we see that $f(x)$ appears to attain a local maximum value near the point where $x = 2$ before approaching zero as $x \to \infty$. Indeed, upon differentiating f and simplifying the result, we see that

$$f'(x) = \frac{2 - x}{(x+2)^3},$$

so the indicated maximum point on the curve is $(2, \frac{1}{8})$. The second derivative of f is

$$f''(x) = \frac{2(x-4)}{(x+2)^4},$$

and it follows that the inflection point apparent in Fig. 4.7.10 is at $(4, \frac{1}{9})$. ◆

Curve-Sketching Strategy

The curve-sketching techniques of Sections 4.5 and 4.6, together with those of this section, can be summarized as a list of steps. If you follow these steps, loosely rather than rigidly, you will obtain a qualitatively accurate sketch of the graph of a given function f:

1. Solve the equation $f'(x) = 0$ and also find where $f'(x)$ does not exist. This gives the critical points of f. Note whether the tangent line is horizontal, vertical, or nonexistent at each critical point.
2. Determine the intervals on which f is increasing and those on which it is decreasing.
3. Solve the equation $f''(x) = 0$ and also find where $f''(x)$ does not exist. These points are the *possible* inflection points of the graph.
4. Determine the intervals on which the graph of f is concave upward and those on which it is concave downward.
5. Find the y-intercept and the x-intercepts (if any) of the graph.
6. Plot and label the critical points, possible inflection points, and intercepts.
7. Determine the asymptotes (if any), discontinuities (if any), and *especially* the behavior of $f(x)$ and $f'(x)$ near discontinuities of f. Also determine the behavior of $f(x)$ as $x \to +\infty$ and as $x \to -\infty$.
8. Finally, join the plotted points with a curve that is consistent with the information you have gathered. Remember that corner points are rare and that straight sections of graph are even rarer.

You may follow these steps in any convenient order and omit any that present major computational difficulties. Many problems require fewer than all eight steps; see Example 6. But Example 8 requires them all.

EXAMPLE 8 Sketch the graph of

$$f(x) = \frac{2 + x - x^2}{(x-1)^2}.$$

Solution We notice immediately that

$$\lim_{x\to 1} f(x) = +\infty,$$

because the numerator approaches 2 as $x \to 1$ and the denominator approaches zero through *positive* values. So the line $x = 1$ is a vertical asymptote. Also,

$$\lim_{x\to\pm\infty} \frac{2+x-x^2}{(x-1)^2} = \lim_{x\to\pm\infty} \frac{\dfrac{2}{x^2}+\dfrac{1}{x}-1}{\left(1-\dfrac{1}{x}\right)^2} = -1,$$

so the line $x = -1$ is a horizontal asymptote (in both the positive and the negative directions).

Next we apply the quotient rule and simplify to find that

$$f'(x) = \frac{x-5}{(x-1)^3}.$$

Thus the only critical point in the domain of f is $x=5$, and we plot the point $(5, f(5)) = (5, -\frac{9}{8})$ on a convenient coordinate plane and mark the horizontal tangent there. To determine the increasing or decreasing behavior of f, we use both the critical point $x = 5$ and the point $x = 1$ (where f' is not defined) to separate the x-axis into open intervals. The following table shows the results:

Interval	$(x-1)^3$	$x-5$	$f'(x)$	f
$(-\infty, 1)$	$-$	$-$	$+$	Increasing
$(1, 5)$	$+$	$-$	$-$	Decreasing
$(5, +\infty)$	$+$	$+$	$+$	Increasing

After some simplifications, we find the second derivative to be

$$f''(x) = \frac{2(7-x)}{(x-1)^4}.$$

The only possible inflection point is at $x = 7$, corresponding to the point $(7, -\frac{10}{9})$ on the graph. We use both $x = 7$ and $x = 1$ (where f'' is undefined) to separate the x-axis into open intervals. The concave structure of the graph can be deduced with the aid of the next table.

Interval	$(x-1)^4$	$7-x$	$f''(x)$	f
$(-\infty, 1)$	$+$	$+$	$+$	Concave upward
$(1, 7)$	$+$	$+$	$+$	Concave upward
$(7, +\infty)$	$+$	$-$	$-$	Concave downward

The y-intercept of f is $(0, 2)$, and the equation $2 + x - x^2 = 0$ readily yields the x-intercepts $(-1, 0)$ and $(2, 0)$. We plot these intercepts, sketch the asymptotes, and finally sketch the graph with the aid of the two tables; their information now is symbolized along the x-axis in Fig. 4.7.11. ◆

Slant Asymptotes

Not all asymptotes are horizontal or vertical—some are inclined. The nonvertical line $y = mx + b$ is an **asymptote** for the curve $y = f(x)$ provided that either

$$\lim_{x\to+\infty} [f(x) - (mx + b)] = 0 \qquad \textbf{(7a)}$$

or

$$\lim_{x\to-\infty} [f(x) - (mx + b)] = 0 \qquad \textbf{(7b)}$$

(or both). These conditions mean that as $x \to +\infty$ or as $x \to -\infty$ (or both), the

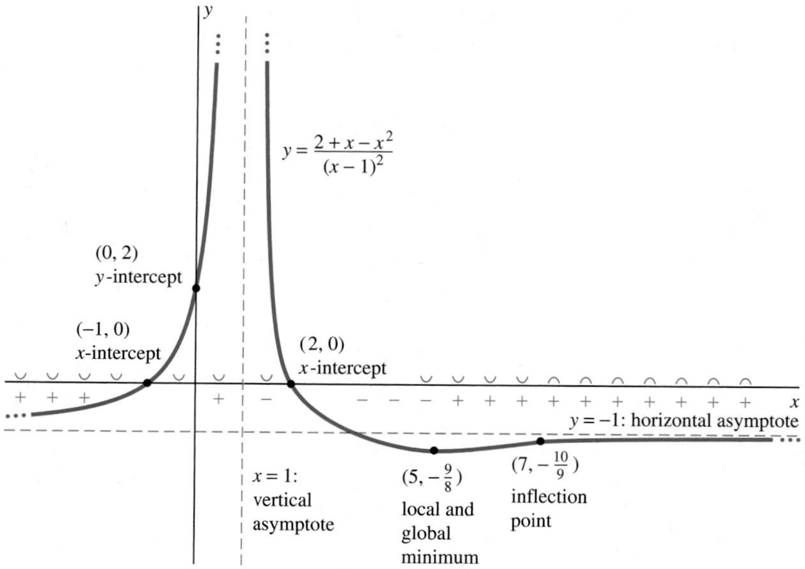

FIGURE 4.7.11 Graphing the function of Example 8.

vertical distance between the point $(x, f(x))$ on the curve and the point $(x, mx + b)$ on the line approaches zero.

Suppose that $f(x) = p(x)/q(x)$ is a rational function for which the degree of $p(x)$ is greater by 1 than the degree of $q(x)$. Then, by long division of $q(x)$ into $p(x)$, we find that $f(x)$ has the form

$$f(x) = mx + b + g(x)$$

where $m \neq 0$ and

$$\lim_{x \to \pm\infty} g(x) = 0.$$

Thus the nonvertical line $y = mx + b$ is an asymptote of the graph of $y = f(x)$. Such an asymptote is called a **slant** asymptote.

EXAMPLE 9 Sketch the graph of

$$f(x) = \frac{x^2 + x - 1}{x - 1}.$$

Solution The long division suggested previously takes the form shown next.

$$\begin{array}{r} x + 2 \\ x - 1 \overline{) \, x^2 + x - 1} \\ \underline{x^2 - x } \\ 2x - 1 \\ \underline{2x - 2} \\ 1 \end{array}$$

Thus

$$f(x) = x + 2 + \frac{1}{x - 1}.$$

So $y = x + 2$ is a slant asymptote of the curve. Also,

$$\lim_{x \to 1} |f(x)| = +\infty,$$

so $x = 1$ is a vertical asymptote. The first two derivatives of f are

$$f'(x) = 1 - \frac{1}{(x-1)^2} = \frac{x(x-2)}{(x-1)^2}$$

and

$$f''(x) = \frac{2}{(x-1)^3}.$$

It follows that f has critical points at $x = 0$ and at $x = 2$ but no inflection points. The sign of f' tells us that f is increasing on $(-\infty, 0)$ and on $(2, +\infty)$, decreasing on $(0, 1)$ and on $(1, 2)$. Examination of $f''(x)$ reveals that f is concave downward on $(-\infty, 1)$ and concave upward on $(1, +\infty)$. In particular, $f(0) = 1$ is a local maximum value and $f(2) = 5$ is a local minimum value. The graph of f looks much like the one in Fig. 4.7.12. ◆

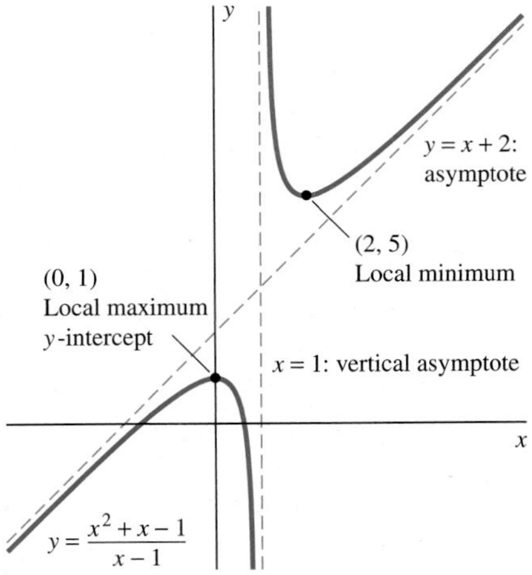

FIGURE 4.7.12 A function with slant asymptote $y = x + 2$ (Example 9).

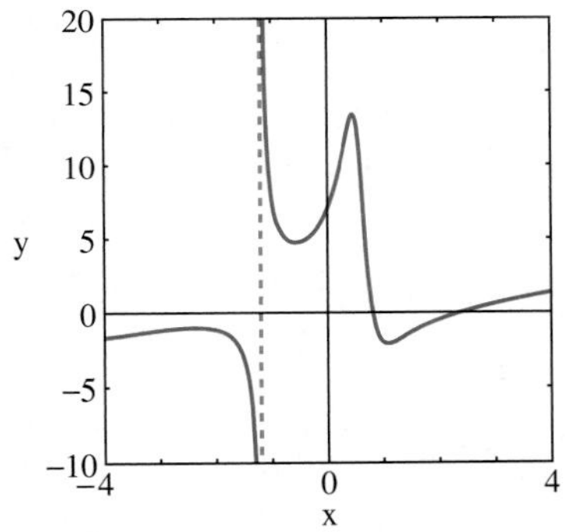

FIGURE 4.7.13 $y = \dfrac{x^4 - 5x^2 - 5x + 7}{2x^3 - 2x + 1}$.

Calculator/Computer Graphing

Instead of using concepts of calculus to construct a graph from scratch, we can go the other way. That is, we can *begin* with a graph plotted by a calculator or computer, and then use a calculator to analyze the graph and refine our understanding of it. In Sections 1.3 and 1.4 we discussed the fact that a calculator or computer graph can sometimes be misleading or incomplete. But now we can use calculus—and in particular the computation of critical points and inflection points—to make sure that the machine-generated graph exhibits all of its important features. Moreover, with graphing and automatic solution techniques we can investigate graphs of functions that would be too complicated to study without a calculator or computer.

EXAMPLE 10 Figure 4.7.13 shows a computer-generated graph of the function

$$f(x) = \frac{x^4 - 5x^2 - 5x + 7}{2x^3 - 2x + 1}. \tag{8}$$

It appears likely to have a vertical asymptote somewhere near $x = -1$. To test this hypothesis, we need to know where the denominator in (8) is zero. The graph of this denominator, shown in Fig. 4.7.14, indicates that the equation $2x^3 - 2x + 1 = 0$ has a single real solution near $x = -1.2$. We could zoom in graphically to show that the corresponding vertical asymptote is still closer to $x = -1.19$, and a calculator or computer **Solve** command yields the solution $x \approx -1.1915$ accurate to four decimal places.

Noting that the degree of the numerator in (8) exceeds that of the denominator, we find by long division that

$$f(x) = \frac{1}{2}x + \frac{-4x^2 - \frac{11}{2}x + 7}{2x^3 - 2x + 1}.$$

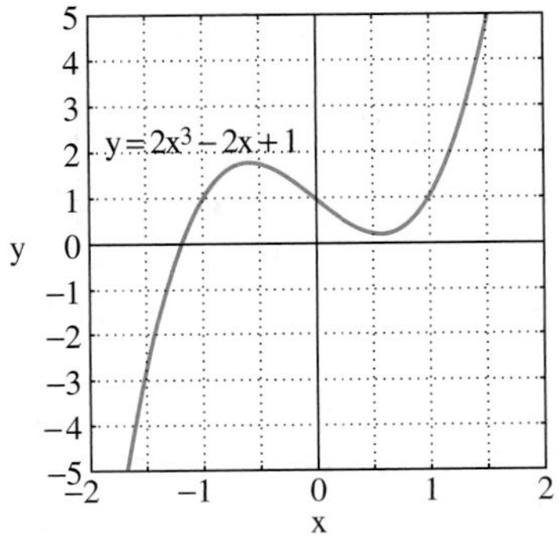

FIGURE 4.7.14 Graph of the denominator in (8).

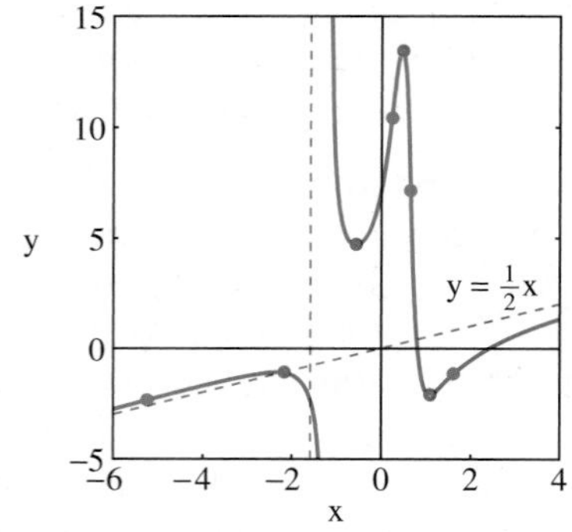

FIGURE 4.7.15 Now we see both the vertical asymptote and the slant asymptote $y = \frac{1}{2}x$.

Thus the graph $y = f(x)$ has the slant asymptote $y = \frac{1}{2}x$ (Fig. 4.7.15).

To investigate the critical points of $f(x)$, we calculate the derivative

$$f'(x) = \frac{2x^6 + 4x^4 + 24x^3 - 32x^2 - 10x + 9}{(2x^3 - 2x + 1)^2}. \tag{9}$$

The critical points of $f(x)$ are the zeros of the numerator of $f'(x)$, together with the zero of the denominator that yields the vertical asymptote. The graph of the numerator, shown in Fig. 4.7.16, indicates that the equation

$$2x^6 + 4x^4 + 24x^3 - 32x^2 - 10x + 9 = 0$$

has four real solutions, near the points $x = -2.3, -0.6, 0.5$, and 1.1. We could zoom in closer to each of these solutions, or use a calculator or computer **Solve** command to get the approximations $x \approx -2.3440, -0.5775, 0.4673$, and 1.0864 that agree with the overall structure of the graph shown in Fig. 4.7.13, where four critical points with horizontal tangent lines are apparent.

The leftmost critical point $x \approx -2.3440$ deserves closer examination. In Fig. 4.7.15 it appears to be just to the left of the point where the left branch of the graph $y = f(x)$ crosses the slant asymptote $y = \frac{1}{2}x$. The zoom shown in Fig. 4.7.17 bears out this observation.

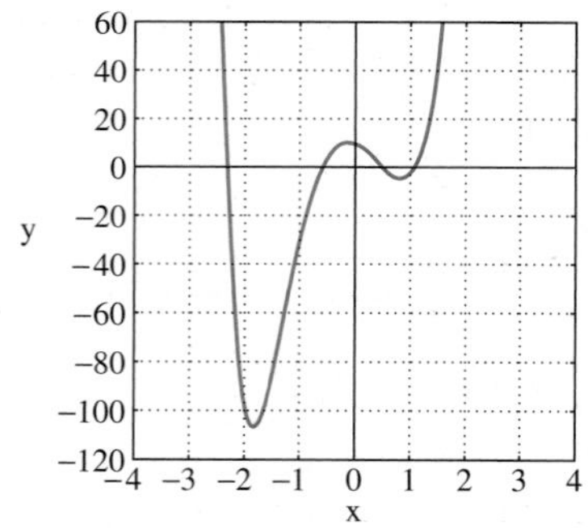

FIGURE 4.7.16 Graph of the numerator in (9).

Finally, an examination of the original graph $y = f(x)$ in Fig. 4.7.13 suggests the approximate locations of three inflection points in the first quadrant. But if the graph is to approach the slant asymptote as $x \to -\infty$, then Fig. 4.7.17 suggests the presence of a fourth inflection point somewhere to the left of the leftmost critical point. (Why?) To investigate this possibility, we calculate the second derivative

$$f''(x) = \frac{2(-16x^6 - 66x^5 + 120x^4 + 34x^3 - 18x^2 - 42x + 13)}{(2x^3 - 2x + 1)^3}. \tag{10}$$

The inflection points of $y = f(x)$ have x-coordinates given by the zeros of the numerator in (10). The graph of this numerator, shown in Fig. 4.7.18, indicates that the equation

$$2(-16x^6 - 66x^5 + 120x^4 + 34x^3 - 18x^2 - 42x + 13) = 0$$

has four real solutions—a negative one near -5.5 as well as three positive solutions between 0 and 2 that correspond to the visually apparent first-quadrant inflection points in Fig. 4.7.13. We could zoom in closer to each of these solutions, or use a calculator or computer **Solve** command to get the approximations -5.4303, 0.3152, 0.6503, and 1.3937. The larger view shown in Fig. 4.7.19 convinces us that we've found *all* the inflection points of $y = f(x)$. In particular, we see that $y = f(x)$ is concave upward to the right of the inflection point $x \approx -5.4303$, where the denominator in (10) is negative (why?), and is concave downward just to its right (consistent with what we see in Fig. 4.7.17).

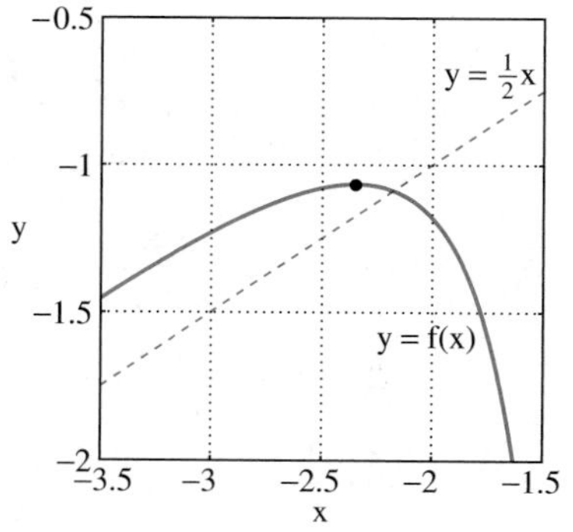

FIGURE 4.7.17 Near the leftmost critical point.

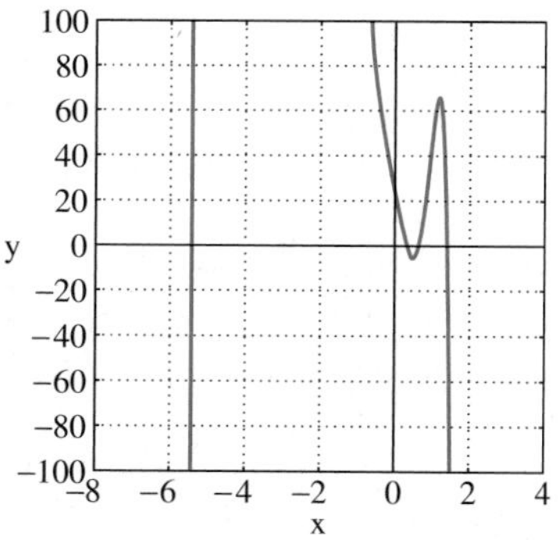

FIGURE 4.7.18 Graph of the numerator in (10).

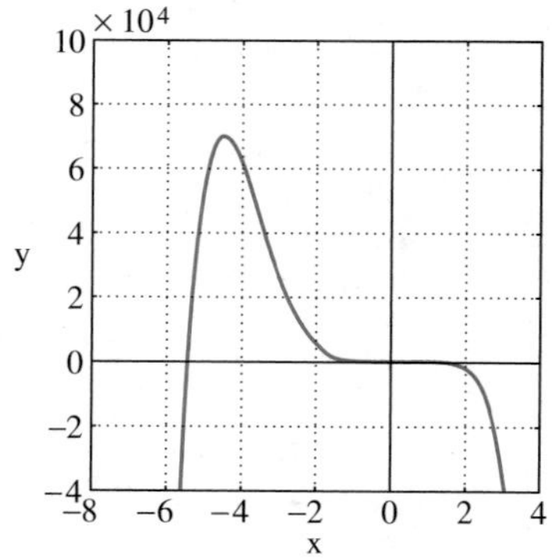

FIGURE 4.7.19 Larger view of the graph of the numerator in (10).

This thorough analysis of the graph of the function f of Eq. (8) involves a certain amount of manual labor—just to calculate and simplify the derivatives in (9) and (10) unless we use a computer algebra system for this task—but would be very challenging without the use of a graphing calculator or computer. ◆

4.7 TRUE/FALSE STUDY GUIDE

4.7 CONCEPTS: QUESTIONS AND DISCUSSION

1. Can you sketch the graph of a function that has two distinct critical points that are not separated by an inflection point? Does such a function exist?
2. Does the graph of a polynomial always have an inflection point? Or does it depend upon whether the degree n of the polynomial is even or odd? Begin by discussing separately the cases $n = 2, 3, 4$, and 5.
3. Can the graph of a polynomial have an asymptote? Does the graph of a rational function always have an asymptote? Justify your answers.
4. What can you say about the degrees of the numerator and denominator of a rational function that has a horizontal asymptote? What can you say about the degrees of the numerator and denominator of a rational function that has a slant asymptote?

4.7 PROBLEMS

Investigate the limits in Problems 1 through 16.

1. $\displaystyle\lim_{x\to+\infty} \frac{x}{x+1}$

2. $\displaystyle\lim_{x\to-\infty} \frac{x^2+1}{x^2-1}$

3. $\displaystyle\lim_{x\to 1} \frac{x^2+x-2}{x-1}$

4. $\displaystyle\lim_{x\to 1} \frac{x^2-x-2}{x-1}$

5. $\displaystyle\lim_{x\to+\infty} \frac{2x^2-1}{x^2-3x}$

6. $\displaystyle\lim_{x\to+\infty} \frac{1+e^x}{2+e^{2x}}$

7. $\displaystyle\lim_{x\to -1} \frac{x^2+2x+1}{(x+1)^2}$

8. $\displaystyle\lim_{x\to+\infty} \frac{5x^3-2x+1}{7x^3+4x^2-2}$

9. $\displaystyle\lim_{x\to 4} \frac{x-4}{\sqrt{x}-2}$

10. $\displaystyle\lim_{x\to+\infty} \frac{2x+1}{x-x\sqrt{x}}$

11. $\displaystyle\lim_{x\to-\infty} \frac{8-\sqrt[3]{x}}{2+x}$

12. $\displaystyle\lim_{x\to+\infty} \frac{4e^{6x}+5\sin 6x}{(1+2e^{2x})^3}$

13. $\displaystyle\lim_{x\to+\infty} \sqrt{\frac{4x^2-x}{x^2+9}}$

14. $\displaystyle\lim_{x\to-\infty} \frac{\sqrt[3]{x^3-8x+1}}{3x-4}$

15. $\displaystyle\lim_{x\to-\infty} \left(\sqrt{x^2+2x}-x\right)$

16. $\displaystyle\lim_{x\to-\infty} \left(2x-\sqrt{4x^2-5x}\right)$

Apply your knowledge of limits and asymptotes to match each function in Problems 17 through 28 with its graph-with-asymptotes in one of the twelve parts of Fig. 4.7.20.

17. $f(x) = \dfrac{1}{x-1}$

18. $f(x) = \dfrac{1}{1-x}$

19. $f(x) = \dfrac{1}{(x-1)^2}$

20. $f(x) = -\dfrac{1}{(1-x)^2}$

21. $f(x) = \dfrac{1}{x^2-1}$

22. $f(x) = \dfrac{1}{1-x^2}$

23. $f(x) = \dfrac{x}{x^2-1}$

24. $f(x) = \dfrac{x}{1-x^2}$

25. $f(x) = \dfrac{x}{x-1}$

26. $f(x) = \dfrac{x^2}{x^2-1}$

27. $f(x) = \dfrac{x^2}{x-1}$

28. $f(x) = \dfrac{x^3}{x^2-1}$

Sketch by hand the graph of each function in Problems 29 through 54. Identify and label all extrema, inflection points, intercepts, and asymptotes. Show the concave structure clearly as well as the behavior of the graph for $|x|$ large and for x near any discontinuities of the function.

29. $f(x) = \dfrac{2}{x-3}$

30. $f(x) = \dfrac{4}{5-x}$

31. $f(x) = \dfrac{3}{(x+2)^2}$

32. $f(x) = -\dfrac{4}{(3-x)^2}$

33. $f(x) = \dfrac{1}{(2x-3)^3}$

34. $f(x) = \dfrac{x+1}{x-1}$

35. $f(x) = \dfrac{x^2}{x^2+1}$

36. $f(x) = \dfrac{2x}{x^2+1}$

37. $f(x) = \dfrac{1}{x^2-9}$

38. $f(x) = \dfrac{x}{4-x^2}$

39. $f(x) = \dfrac{1}{x^2+x-6}$

40. $f(x) = \dfrac{2x^2+1}{x^2-2x}$

41. $f(x) = x + \dfrac{1}{x}$

42. $f(x) = 2x + e^{-x}$

43. $f(x) = \dfrac{x^2}{x-1}$

44. $f(x) = \dfrac{2x^3-5x^2+4x}{x^2-2x+1}$

45. $f(x) = \dfrac{1}{(x-1)^2}$

46. $f(x) = \dfrac{1}{(1+e^x)^2}$

47. $f(x) = \dfrac{e^x}{e^x+1}$

48. $f(x) = \dfrac{1}{e^x+e^{-x}}$

49. $f(x) = \dfrac{1}{x^2-x-2}$

50. $f(x) = \dfrac{1}{(x-1)(x+1)^2}$

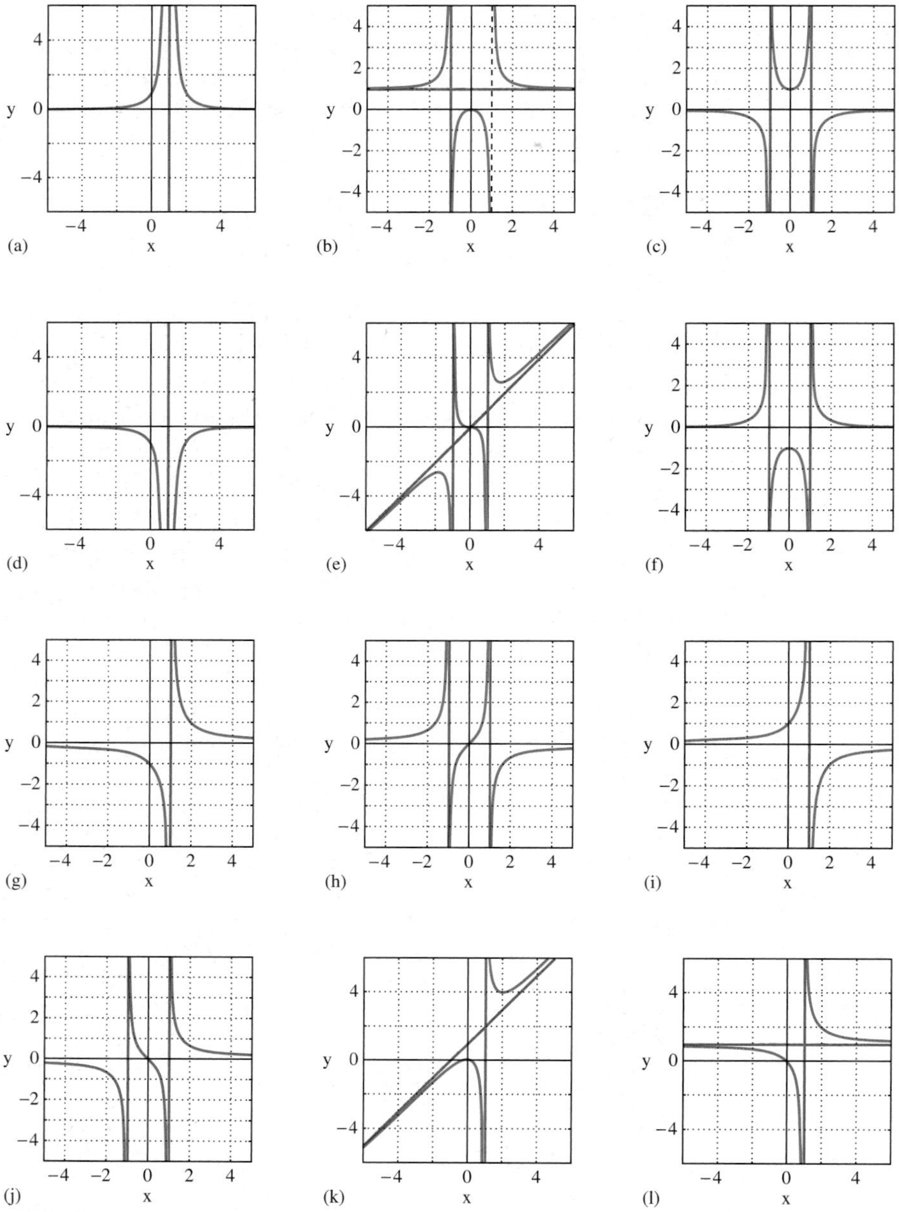

FIGURE 4.7.20 Problems 17 through 28.

51. $f(x) = \dfrac{x^2 - 4}{x}$

52. $f(x) = \dfrac{e^x - e^{-x}}{e^x + e^{-x}}$

53. $f(x) = \dfrac{x^3 - 4}{x^2}$

54. $f(x) = \dfrac{x^2 + 1}{x - 2}$

In Problems 55 through 60, you can determine by inspection the x-intercepts as well as the vertical and horizontal asymptotes of the curve y = f(x). First sketch the graph by hand, using this information, and without calculating any derivatives. Then use a calculator or computer to locate accurately the critical and inflection points of f(x). Finally, use a calculator or computer to produce graphs that display the major features of the curve.

55. $f(x) = \dfrac{(x+1)(x-3)}{x^2(x-2)}$

56. $f(x) = \dfrac{(x+1)^2(x-3)}{x^2(x-4)}$

57. $f(x) = \dfrac{(x+1)^2(x-3)}{x^3(x-2)}$

58. $f(x) = \dfrac{(x+1)^2(x-3)^2}{x^3(x-2)}$

59. $f(x) = \dfrac{(x+1)^2(x-3)^2}{x^3(x-2)^2}$

60. $f(x) = \dfrac{(x+1)(x-3)^4}{x^3(x-2)^3}$

In Problems 61 through 68, begin with a calculator- or computer-generated graph of the curve y = f(x). Then use a calculator or computer to locate accurately the vertical asymptotes and the critical and inflection points of f(x). Finally, use a calculator or computer to produce graphs that display the major features of the curve, including any vertical, horizontal, and slant asymptotes.

61. $f(x) = \dfrac{x^2}{x^3 - 3x^2 + 1}$

62. $f(x) = \dfrac{x^2}{x^3 - 3x^2 + 5}$

63. $f(x) = \dfrac{x^4 - 4x + 5}{x^3 - 3x^2 + 5}$

64. $f(x) = \dfrac{x^4 - 4x + 1}{2x^3 - 3x + 2}$

65. $f(x) = \dfrac{x^5 - 4x^2 + 1}{2x^4 - 3x + 2}$

66. $f(x) = \dfrac{x^5 - 4x^3 + 2}{2x^4 - 5x + 5}$

67. $f(x) = \dfrac{x^6 - 4x^3 + 5x}{2x^5 - 5x^3 + 5}$

68. $f(x) = \dfrac{2x^6 - 5x^4 + 6}{3x^5 - 5x^4 + 4}$

69. Suppose that

$$f(x) = x^2 + \frac{2}{x}.$$

Note that

$$\lim_{x\to\pm\infty} [f(x) - x^2] = 0,$$

so the curve $y = f(x)$ approaches the parabola $y = x^2$ as $x \to \pm\infty$. Use this observation to make an accurate sketch of the graph of f.

70. Use the method of Problem 69 to make an accurate sketch of the graph of

$$f(x) = x^3 - \frac{12}{x - 1}.$$

4.7 Project: Locating Special Points on Exotic Graphs

The investigations described here deal with fairly exotic curves having critical and inflection points that are not clearly visible on their graphs if plotted on a "natural" scale. The reason is that different scales on the x- and y-axes are required to see the unusual behavior in question. In both investigations you are to begin with a graph that you generate with calculator or computer, and then analyze the curve—locating accurately all critical and inflection points—in order to plot additional graphs that demonstrate clearly all of the major features of the curve.

Investigation A Choose in advance a single-digit integer n (perhaps the final nonzero digit of your student I.D. number). Then your task is to analyze the structure of the curve

$$y = x^7 + 5x^6 - 11x^5 - 21x^4 + 31x^3 - 57x^2 - (101 + 2n)x + (89 - 3n).$$

Find the local maximum and minimum points and the inflection point (or points) on this curve, giving their coordinates accurate to four decimal places. To display all these points, you probably will need to produce separate plots with different scales, showing different parts of this curve. In the end, use all the information accumulated to produce a careful hand sketch (*not* to scale) displaying all the maxima, minima, and inflection points with their (approximate) coordinates labeled.

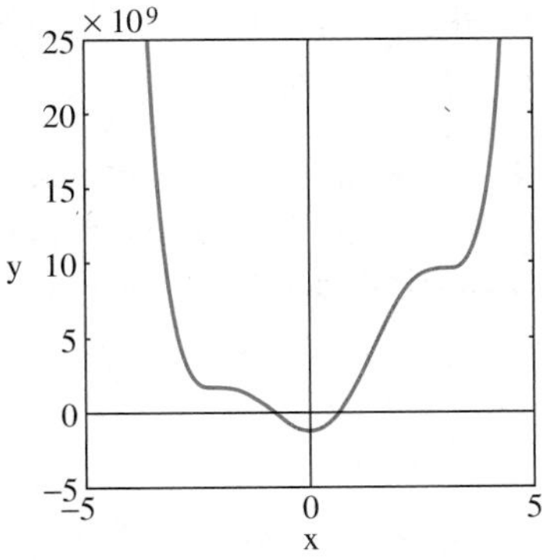

FIGURE 4.7.21 The "big picture" in Investigation B.

Investigation B Explore in the detail the structure of the graph of the function

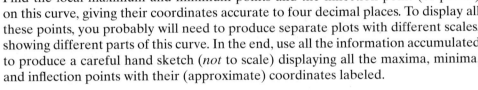

$$f(x) = -1{,}234{,}567{,}890 + 2{,}695{,}140{,}459x^2 + 605{,}435{,}400x^3 - 411{,}401{,}250x^4 - 60{,}600{,}000x^5 + 25{,}000{,}000x^6.$$

The graph $y = f(x)$ is shown in Fig. 4.7.21. At a glance, it might appear that we have only three critical points—a local minimum near the origin and two critical points that are also inflection points, as well as two more inflection points that are not critical points. Settle the matter. How many of each, in fact, are there? Find and exhibit all of them in a graph; your graph may be a neat hand sketch and need not be to scale.

4.8 INDETERMINATE FORMS AND L'HÔPITAL'S RULE

An *indeterminate form* is a certain type of expression with a limit that is not evident by inspection. There are several types of indeterminate forms. If

$$\lim_{x\to a} f(x) = 0 = \lim_{x\to a} g(x),$$

then we say that the quotient $f(x)/g(x)$ has the **indeterminate form** $0/0$ at $x = a$ (or as $x \to a$). For example, to differentiate the trigonometric functions (Section 3.7), we needed to know that

$$\lim_{x\to 0} \frac{\sin x}{x} = 1. \qquad \textbf{(1)}$$

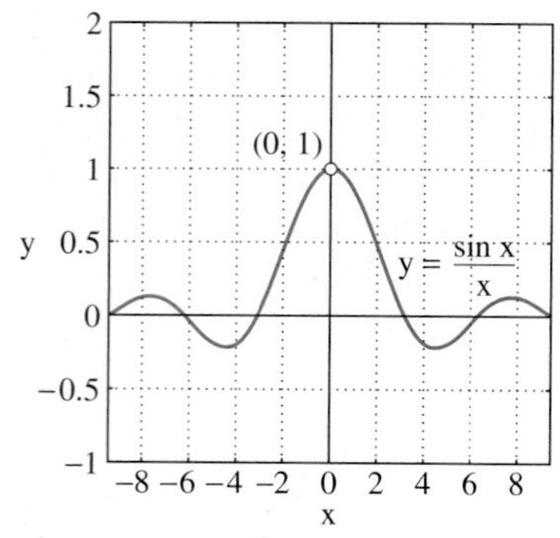

FIGURE 4.8.1 Visual evidence that the quotient $(\sin x)/x$ is near 1 when x is near zero.

Figure 4.8.1 corroborates the fact that $(\sin x)/x$ is close to 1 when x is close to zero.

The quotient $(\sin x)/x$ in Eq. (1) has the indeterminate form $0/0$ at $x = 0$ because the functions $f(x) = \sin x$ and $g(x) = x$ both approach zero as $x \to 0$. Hence the quotient law of limits cannot be used to evaluate this limit. We therefore needed a special geometric argument (see Section 2.3) to find the limit in Eq. (1). Something similar happens whenever we compute a derivative, because the quotient

$$\frac{f(x) - f(a)}{x - a},$$

whose limit as $x \to a$ is the derivative $f'(a)$, has the indeterminate form $0/0$ at $x = a$.

We can sometimes find the limit of an indeterminate form by performing a special algebraic manipulation or construction, as in our earlier computation of derivatives. Often, however, it is more convenient to apply a rule that appeared in the first calculus textbook ever published, by the Marquis de l'Hôpital, in 1696. L'Hôpital was a French nobleman who had hired the Swiss mathematician John Bernoulli as his calculus tutor, and "l'Hôpital's rule" is actually the work of Bernoulli.

THEOREM 1 L'Hôpital's Rule

Suppose that the functions f and g are differentiable and that $g'(x)$ is nonzero in some neighborhood of the point a (except possibly at a itself). Suppose also that

$$\lim_{x \to a} f(x) = 0 = \lim_{x \to a} g(x).$$

Then

$$\lim_{x \to a} \frac{f(x)}{g(x)} = \lim_{x \to a} \frac{f'(x)}{g'(x)}, \tag{2}$$

provided that the limit on the right either exists (as a finite real number) or is $+\infty$ or $-\infty$.

In essence, l'Hôpital's rule says that if $f(x)/g(x)$ has the indeterminate form $0/0$ at $x = a$, then—subject to a few mild restrictions—this quotient has the same limit at $x = a$ as does the quotient $f'(x)/g'(x)$ of *derivatives.* The proof of l'Hôpital's rule is discussed at the end of this section.

EXAMPLE 1 Find $\displaystyle\lim_{x \to 0} \frac{e^x - 1}{\sin 2x}$.

Solution The fraction whose limit we seek has the indeterminate form $0/0$ at $x = 0$. The numerator and denominator are clearly differentiable in some neighborhood of $x = 0$, and the derivative of the denominator is certainly nonzero if the neighborhood is small enough (specifically, if $|x| < \pi/4$). So l'Hôpital's rule applies, and

$$\lim_{x \to 0} \frac{e^x - 1}{\sin 2x} = \lim_{x \to 0} \frac{e^x}{2\cos 2x} = \frac{e^0}{2\cos 0} = \frac{1}{2}$$

because (by continuity) both e^x and $\cos 2x$ approach 1 as $x \to 0$. Figure 4.8.2 corroborates this limit. ◆

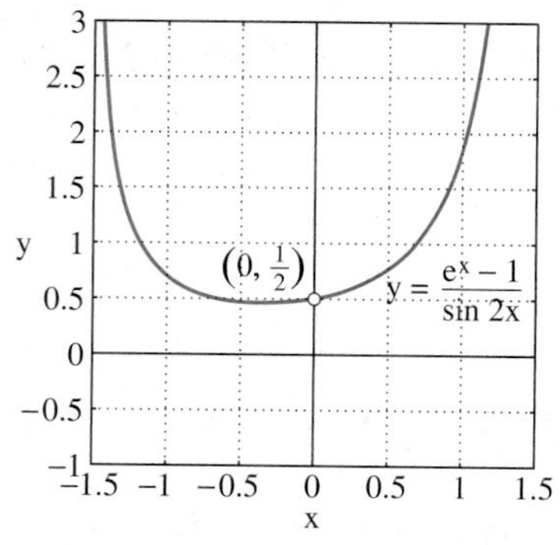

FIGURE 4.8.2 Visual evidence that the quotient $\dfrac{e^x - 1}{\sin 2x}$ is near $\frac{1}{2}$ when x is near 0.

If the quotient $f'(x)/g'(x)$ is itself indeterminate, then l'Hôpital's rule may be applied a second (or third, ...) time, as in Example 2. When the rule is applied repeatedly, however, the conditions for its applicability must be checked at each stage.

EXAMPLE 2 Find $\displaystyle\lim_{x \to 1} \frac{1 - x + \ln x}{1 + \cos \pi x}$.

Solution

$$\begin{aligned}\lim_{x\to 1}\frac{1-x+\ln x}{1+\cos\pi x} &= \lim_{x\to 1}\frac{-1+\dfrac{1}{x}}{-\pi\sin\pi x} && \text{(still of the form 0/0)}\\ &= \lim_{x\to 1}\frac{x-1}{\pi x\sin\pi x} && \text{(algebraic simplification)}\\ &= \lim_{x\to 1}\frac{1}{\pi\sin\pi x+\pi^2 x\cos\pi x} && \text{(l'Hôpital's rule again)}\\ &= -\frac{1}{\pi^2} && \text{(by inspection).}\end{aligned}$$

Because the final limit exists, so do the previous ones; the existence of the final limit in Eq. (2) implies the existence of the first. ◆

When you need to apply l'Hôpital's rule repeatedly in this way, you need only keep differentiating the numerator and denominator separately until at least one of them has a nonzero finite limit. At that point you can recognize the limit of the quotient by inspection, as in the final step in Example 2.

EXAMPLE 3 Find $\displaystyle\lim_{x\to 0}\frac{\sin x}{x+x^2}$.

Solution If we simply apply l'Hôpital's rule twice in succession, the result is the *incorrect* computation

$$\begin{aligned}\lim_{x\to 0}\frac{\sin x}{x+x^2} &= \lim_{x\to 0}\frac{\cos x}{1+2x}\\ &= \lim_{x\to 0}\frac{-\sin x}{2} = 0. && \textbf{(Wrong!)}\end{aligned}$$

The answer is wrong because $(\cos x)/(1+2x)$ is *not* an indeterminate form. Thus l'Hôpital's rule cannot be applied to it. The *correct* computation is

$$\lim_{x\to 0}\frac{\sin x}{x+x^2} = \lim_{x\to 0}\frac{\cos x}{1+2x} = \frac{\displaystyle\lim_{x\to 0}\cos x}{\displaystyle\lim_{x\to 0}(1+2x)} = \frac{1}{1} = 1.$$ ◆

The point of Example 3 is to issue a warning: Verify the hypotheses of l'Hôpital's rule *before* you apply it. It is an oversimplification to say that l'Hôpital's rule works when you need it and doesn't work when you don't, but there is still much truth in this statement.

Indeterminate Forms Involving ∞

L'Hôpital's rule has several variations. In addition to the fact that the limit in Eq. (2) is allowed to be infinite, the real number a in l'Hôpital's rule may be replaced with either $+\infty$ or $-\infty$. For example,

$$\lim_{x\to\infty}\frac{f(x)}{g(x)} = \lim_{x\to\infty}\frac{f'(x)}{g'(x)} \qquad \textbf{(3)}$$

provided that the other hypotheses are satisfied in some interval of the form $(c, +\infty)$. In particular, to use Eq. (3), we must first verify that

$$\lim_{x\to\infty} f(x) = 0 = \lim_{x\to\infty} g(x)$$

and that the right-hand limit in Eq. (3) exists. The proof of this version of l'Hôpital's rule is outlined in Problem 70.

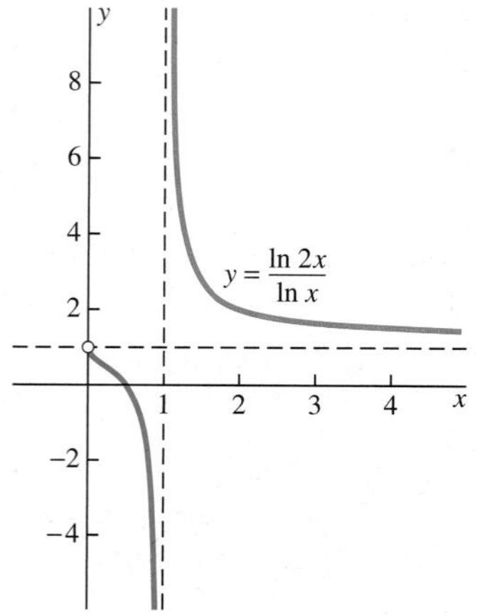

FIGURE 4.8.3 The graph $y = \dfrac{\ln 2x}{\ln x}$ has both the vertical asymptote $x = 1$ and the horizontal asymptote $y = 1$.

L'Hôpital's rule may also be used when $f(x)/g(x)$ has the **indeterminate form** ∞/∞. This means that

$$\lim_{x\to a} f(x) \quad \text{is either } +\infty \text{ or } -\infty$$

and

$$\lim_{x\to a} g(x) \quad \text{is either } +\infty \text{ or } -\infty.$$

The proof of this extension of the rule is difficult and is omitted here. [For a proof, see (for example) A. E. Taylor and W. R. Mann, *Advanced Calculus,* 3rd ed. (New York: John Wiley, 1983), p. 107.]

One-sided indeterminate forms occur, and we may speak of a $0/0$ form or an ∞/∞ form as either $x \to a^-$ or as $x \to a^+$.

EXAMPLE 4 Figure 4.8.3 shows a computer-generated graph of the function

$$f(x) = \frac{\ln 2x}{\ln x}. \qquad \textbf{(4)}$$

The vertical asymptote $x = 1$ is explained (without using l'Hôpital's rule) by the facts that

- the numerator $\ln 2x$ is positive at $x = 1$, while
- the denominator $\ln x$ approaches zero through negative values as $x \to 1^-$ and approaches zero through positive values as $x \to 1^+$.

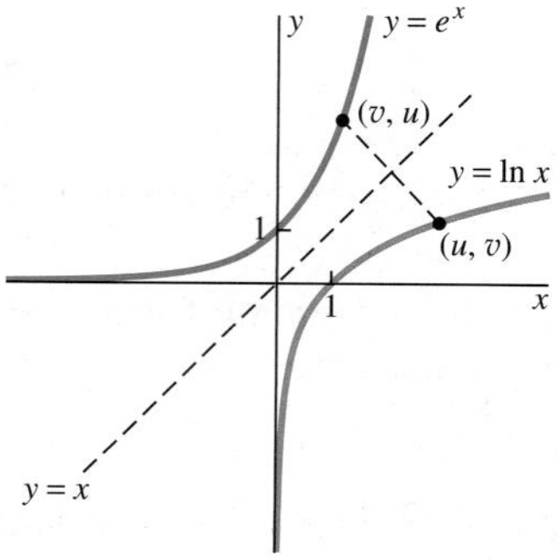

FIGURE 4.8.4 The graphs $y = e^x$ and $y = \ln x$ are reflections of one another in the line $y = x$.

The familiar graph of $y = \ln x$ (Fig. 4.8.4) reminds us that $\ln x \to -\infty$ as $x \to 0^+$ and that $\ln x \to +\infty$ as $x \to +\infty$. Consequently, we see that the function f in Eq. (4) has the indeterminate form ∞/∞, both as $x \to 0^+$ and as $x \to +\infty$. Thus l'Hôpital's rule gives

$$\lim_{x\to 0^+} \frac{\ln 2x}{\ln x} = \lim_{x\to 0^+} \frac{\frac{2}{2x}}{\frac{1}{x}} = \lim_{x\to 0^+} 1 = 1$$

and

$$\lim_{x\to +\infty} \frac{\ln 2x}{\ln x} = \lim_{x\to +\infty} \frac{\frac{2}{2x}}{\frac{1}{x}} = \lim_{x\to +\infty} 1 = 1.$$

The fact that

$$\lim_{x\to 0^+} \frac{\ln 2x}{\ln x} = 1$$

explains why the graph in Fig. 4.8.3 appears to "start" at the point (0, 1). And the fact that

$$\lim_{x\to \infty} \frac{\ln 2x}{\ln x} = 1$$

explains the horizontal asymptote $y = 1$ that we see in Fig. 4.8.3. ◆

Order of Magnitude of e^x and $\ln x$

Because $D_x e^x = e^x$ (for all x) and $D_x \ln x = 1/x$ (for all $x > 0$), the functions $\ln x$ and e^x are increasing wherever they are defined. If n is an integer and $x > n$, it follows that $e^x > e^n > 2^n$, and hence that

$$\lim_{x\to +\infty} e^x = +\infty. \qquad \textbf{(5)}$$

Similarly, if $x > 2^n$, then $\ln x > \ln 2^n = n \ln 2$, and therefore

$$\lim_{x \to +\infty} \ln x = +\infty \tag{6}$$

as well.

But the graphs in Fig. 4.8.4 suggest that as $x \to +\infty$, $e^x \to +\infty$ much more rapidly than $\ln x \to +\infty$. Indeed, l'Hôpital's rule yields

$$\lim_{x \to +\infty} \frac{e^x}{\ln x} = \lim_{x \to +\infty} \frac{e^x}{\frac{1}{x}} = \lim_{x \to +\infty} xe^x = +\infty.$$

Thus, when x is large positive, e^x is so much larger than $\ln x$ that the quotient $e^x/(\ln x)$ is large positive. This observation seems related to the facts that:

- The second derivative $D_x^2 e^x = e^x > 0$ for all x, so the curve $y = e^x$ is concave upward, and becomes steeper and steeper as x increases.
- In contrast, $D_x^2 \ln x = -1/x^2 < 0$ for all $x > 0$, so the curve $y = \ln x$ is concave downward, and becomes less and less steep as x increases.

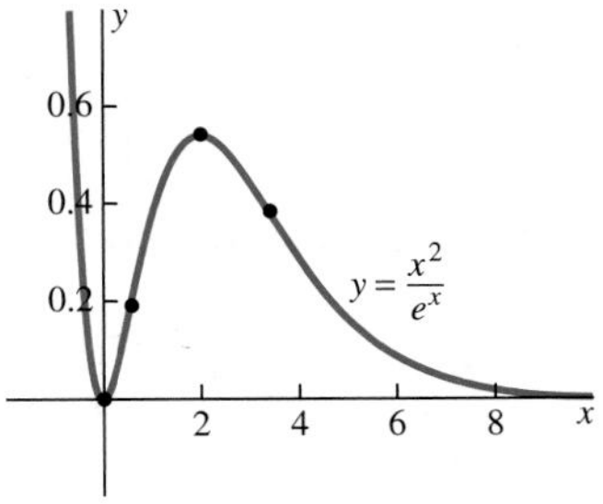

FIGURE 4.8.5 The graph $y = x^2e^{-x}$ has two local extrema, two inflection points, and the horizontal asymptote $y = 0$.

EXAMPLE 5 Explain the principal features of the graph of the function $f(x) = x^2e^{-x}$ shown in Fig. 4.8.5.

Solution The function $f(x) = x^2/e^x$ has the indeterminate form ∞/∞ as $x \to +\infty$. Hence the horizontal asymptote $y = 0$ is explained by two applications of l'Hôpital's rule:

$$\lim_{x \to \infty} \frac{x^2}{e^x} = \lim_{x \to \infty} \frac{2x}{e^x} = \lim_{x \to \infty} \frac{2}{e^x} = 0. \tag{7}$$

The two local extrema that we see in the figure result from the fact that the derivative

$$f'(x) = 2x \cdot e^{-x} - x^2 \cdot e^{-x} = (2x - x^2)e^{-x} = x(2 - x)e^{-x}$$

has the two zeros $x = 0$ and $x = 2$. The second derivative is

$$f''(x) = (2 - 2x) \cdot e^{-x} - (2x - x^2) \cdot e^{-x} = (x^2 - 4x + 2)e^{-x},$$

and the two solutions $x = 2 \pm \sqrt{2}$ of the quadratic equation $x^2 - 4x + 2 = 0$ provide the two inflection points visible in Fig. 4.8.5. ◆

The exponential function is notable for its very rapid rate of increase with increasing x. In fact, e^x increases more rapidly as $x \to +\infty$ than *any* fixed power of x. Thus the limit in (7) is a special case of the fact that

$$\lim_{x \to \infty} \frac{x^k}{e^x} = 0 \tag{8}$$

or, alternatively,

$$\lim_{x \to \infty} \frac{e^x}{x^k} = +\infty \tag{9}$$

for any fixed real number $k > 0$. For instance, if $k = n$, a positive integer, then n successive applications of l'Hôpital's rule give

$$\begin{aligned} \lim_{x \to \infty} \frac{x^n}{e^x} &= \lim_{x \to \infty} \frac{nx^{n-1}}{e^x} = \lim_{x \to \infty} \frac{n(n-1)x^{n-2}}{e^x} \\ &= \cdots = \lim_{x \to \infty} \frac{n(n-1)\cdots 2x}{e^x} = \lim_{x \to \infty} \frac{n(n-1)\cdots 2 \cdot 1}{e^x} = 0. \end{aligned}$$

In Problem 61 we ask you to consider positive nonintegral values of k.

The table in Fig. 4.8.6 illustrates the case $k = 5$ of Eq. (8). Although both $x^5 \to +\infty$ and $e^x \to +\infty$ as $x \to +\infty$, we see that e^x increases so much more rapidly than x^5 that $x^5/e^x \to 0$.

x	x^5	e^x	x^5/e^x
10	1.00×10^5	2.20×10^4	4.54×10^0
20	3.20×10^6	4.85×10^8	6.60×10^{-3}
30	2.43×10^7	1.07×10^{13}	2.27×10^{-6}
40	1.02×10^8	2.35×10^{17}	4.35×10^{-10}
50	3.13×10^8	5.18×10^{21}	6.03×10^{-14}
	$\downarrow$	$\downarrow$	$\downarrow$
	∞	∞	∞

FIGURE 4.8.6 Orders of magnitude of x^5 and e^x.

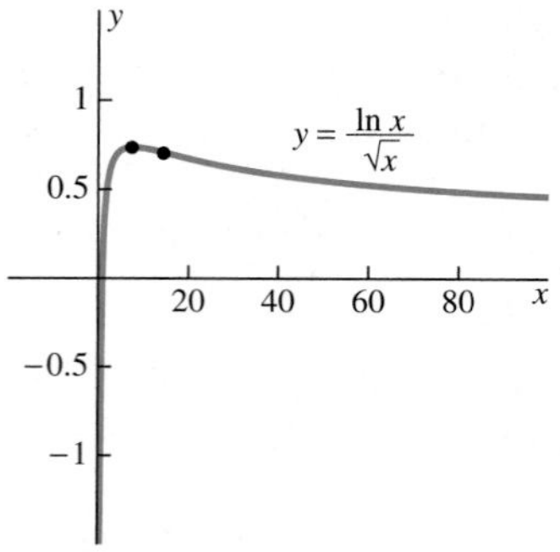

FIGURE 4.8.7 The graph $y = (\ln x)/\sqrt{x}$ has a local maximum, an inflection point, and the horizontal asymptote $y = 0$.

EXAMPLE 6 Explain the principal features of the graph of the function

$$f(x) = \frac{\ln x}{\sqrt{x}}$$

shown in Fig. 4.8.7.

Solution The function $f(x) = (\ln x)/\sqrt{x}$ has the indeterminate form ∞/∞ as $x \to +\infty$. A single application of l'Hôpital's rule yields

$$\lim_{x\to\infty} \frac{\ln x}{\sqrt{x}} = \lim_{x\to\infty} \frac{\dfrac{1}{x}}{\dfrac{1}{2\sqrt{x}}} = \lim_{x\to\infty} \frac{2}{\sqrt{x}} = 0, \qquad \textbf{(10)}$$

and thus the graph has the horizontal asymptote $y = 0$. The local maximum that we see in the figure results from the fact that the derivative

$$f'(x) = \frac{\dfrac{1}{x}\sqrt{x} - \dfrac{\ln x}{2\sqrt{x}}}{x} = \frac{2 - \ln x}{2x^{3/2}}$$

has the zero $x = e^2$. The inflection point that we see corresponds to the zero $x = e^{8/3}$ of the second derivative

$$f''(x) = \frac{-\dfrac{1}{x} \cdot 2x^{3/2} - (2 - \ln x)(3x^{1/2})}{4x^3} = \frac{-8 + 3\ln x}{4x^{5/2}}. \qquad \blacklozenge$$

In contrast with the exponential function, the natural logarithm function is notable for its very slow rate of increase with increasing x. In Problem 62 we ask you to generalize the result in Eq. (10) by showing that

$$\lim_{x\to+\infty} \frac{\ln x}{x^k} = 0 \qquad \textbf{(11)}$$

if $k > 0$. Thus

- $\ln x$ increases slower than any (positive) power of x, whereas
- e^x increases faster than any power of x.

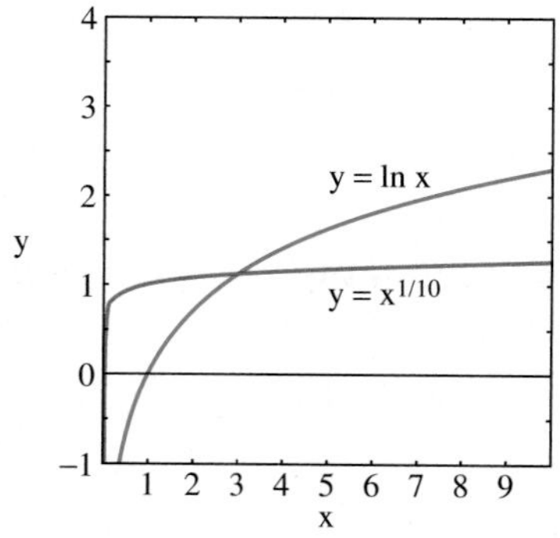

FIGURE 4.8.8 Comparing $y = \ln x$ with $y = x^{1/10}$.

REMARK Figure 4.8.8 might suggest to the unwary that $\ln x$ is greater than (rather than less than) $x^{1/10}$ when x is large positive. But Eq. (11) implies that the graph of $y = x^{1/10}$ must eventually overtake and recross the graph of $y = \ln x$. (See Problem 74.)

Proof of L'Hôpital's Rule

Suppose that the functions f and g of Theorem 1 are not merely differentiable but have continuous derivatives near $x = a$ and that $g'(a) \neq 0$. Then

$$\lim_{x \to a} \frac{f'(x)}{g'(x)} = \frac{\lim_{x \to a} f'(x)}{\lim_{x \to a} g'(x)} = \frac{f'(a)}{g'(a)} \tag{12}$$

by the quotient law for limits. In this case l'Hôpital's rule in Eq. (2) reduces to the limit

$$\lim_{x \to a} \frac{f(x)}{g(x)} = \frac{f'(a)}{g'(a)}, \tag{13}$$

which is a weak form of the rule. It actually is this weak form that is typically applied in single-step applications of l'Hôpital's rule.

EXAMPLE 7 In Example 1 we had

$$f(x) = e^x - 1, \qquad g(x) = \sin 2x$$

so

$$f'(x) = e^x, \qquad g'(x) = 2\cos 2x,$$

and $g'(0) = 2 \neq 0$. With $a = 0$, Eq. (13) therefore gives

$$\lim_{x \to 0} \frac{e^x - 1}{\sin 2x} = \lim_{x \to 0} \frac{f(x)}{g(x)} = \frac{f'(0)}{g'(0)} = \frac{1}{2}.$$ ◆

THEOREM 2 L'Hôpital's Rule (weak form)
Suppose that the functions f and g are differentiable at $x = a$, that

$$f(a) = 0 = g(a),$$

and that $g'(a) \neq 0$. Then

$$\lim_{x \to a} \frac{f(x)}{g(x)} = \frac{f'(a)}{g'(a)}. \tag{13}$$

PROOF We begin with the right-hand side of Eq. (13) and work toward the left-hand side.

$$\frac{f'(a)}{g'(a)} = \frac{\lim_{x \to a} \dfrac{f(x) - f(a)}{x - a}}{\lim_{x \to a} \dfrac{g(x) - g(a)}{x - a}} \qquad \text{(the definition of the derivative)}$$

$$= \lim_{x \to a} \frac{\dfrac{f(x) - f(a)}{x - a}}{\dfrac{g(x) - g(a)}{x - a}} \qquad \text{(the quotient law of limits)}$$

$$= \lim_{x \to a} \frac{f(x) - f(a)}{g(x) - g(a)} \qquad \text{(algebraic simplification)}$$

$$= \lim_{x \to a} \frac{f(x)}{g(x)}$$

[because $f(a) = 0 = g(a)$]. ◄

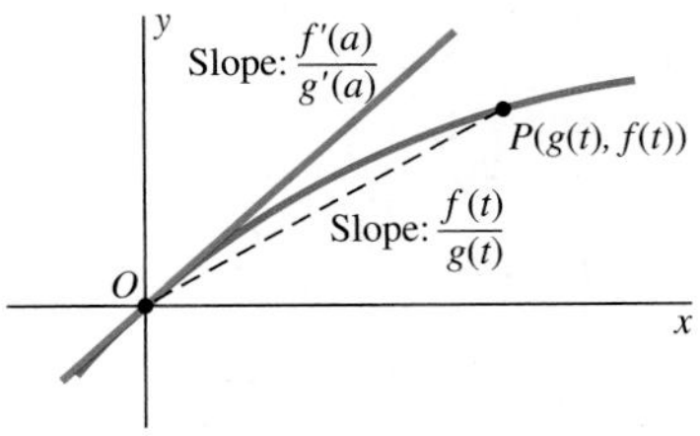

FIGURE 4.8.9 Suppose that the point $P(g(t), f(t))$ traces a continuous curve that passes through the origin O when $t = a$. Then the secant line OP approaches the tangent line at O as $t \to a$, so its slope $f(t)/g(t)$ approaches the slope $f'(a)/g'(a)$ of the tangent line at O.

Figure 4.8.9 illustrates the meaning and proof of Theorem 2. Appendix H includes a proof of the strong form of l'Hôpital's rule, the form stated in Theorem 1.

4.8 TRUE/FALSE STUDY GUIDE

4.8 CONCEPTS: QUESTIONS AND DISCUSSION

In the following questions, think of the given functions f and g as a tortoise and a hare racing toward infinity as $x \to +\infty$. Which is the tortoise and which is the hare?

1. $f(x) = x^2$ and $g(x) = x^5$
2. $f(x) = x^{1/2}$ and $g(x) = x^{1/5}$
3. $f(x) = x^{10} \ln x$ and $g(x) = \dfrac{e^x}{x^{10}}$
4. $f(x) = e^x$ and $g(x)$ is a polynomial
5. $f(x) = \ln x$ and $g(x)$ is a polynomial

4.8 PROBLEMS

Find the limits in Problems 1 through 48.

1. $\displaystyle\lim_{x\to 1} \frac{x-1}{x^2-1}$
2. $\displaystyle\lim_{x\to \infty} \frac{3x-4}{2x-5}$
3. $\displaystyle\lim_{x\to \infty} \frac{2x^2-1}{5x^2+3x}$
4. $\displaystyle\lim_{x\to 0} \frac{e^{3x}-1}{x}$
5. $\displaystyle\lim_{x\to 0} \frac{\sin x^2}{x}$
6. $\displaystyle\lim_{x\to 0^+} \frac{1-\cos\sqrt{x}}{x}$
7. $\displaystyle\lim_{x\to 1} \frac{x-1}{\sin x}$
8. $\displaystyle\lim_{x\to 0} \frac{1-\cos x}{x^3}$
9. $\displaystyle\lim_{x\to 0} \frac{e^x-x-1}{x^2}$
10. $\displaystyle\lim_{z\to \pi/2} \frac{1+\cos 2z}{1-\sin 2z}$
11. $\displaystyle\lim_{u\to 0} \frac{u\tan u}{1-\cos u}$
12. $\displaystyle\lim_{x\to 0} \frac{x-\tan x}{x^3}$
13. $\displaystyle\lim_{x\to \infty} \frac{\ln x}{\sqrt[10]{x}}$
14. $\displaystyle\lim_{r\to \infty} \frac{e^r}{(r+1)^4}$
15. $\displaystyle\lim_{x\to 10} \frac{\ln(x-9)}{x-10}$
16. $\displaystyle\lim_{t\to \infty} \frac{t^2+1}{t\ln t}$
17. $\displaystyle\lim_{x\to 0} \frac{e^x+e^{-x}-2}{x\sin x}$
18. $\displaystyle\lim_{x\to (\pi/2)^-} \frac{\tan x}{\ln(\cos x)}$
19. $\displaystyle\lim_{x\to 0} \frac{\sin 3x}{\tan 5x}$
20. $\displaystyle\lim_{x\to 0} \frac{e^x-e^{-x}}{x}$
21. $\displaystyle\lim_{x\to 1} \frac{x^3-1}{x^2-1}$
22. $\displaystyle\lim_{x\to 2} \frac{x^3-8}{x^4-16}$
23. $\displaystyle\lim_{x\to \infty} \frac{x+\sin x}{3x+\cos x}$
24. $\displaystyle\lim_{x\to \infty} \frac{\sqrt{x^2+4}}{x}$
25. $\displaystyle\lim_{x\to 0} \frac{2^x-1}{3^x-1}$
26. $\displaystyle\lim_{x\to \infty} \frac{2^x}{3^x}$
27. $\displaystyle\lim_{x\to \infty} \frac{\sqrt{x^2-1}}{\sqrt{4x^2-x}}$
28. $\displaystyle\lim_{x\to \infty} \frac{\sqrt{x^3+x}}{\sqrt{2x^3-4}}$
29. $\displaystyle\lim_{x\to 0} \frac{\ln(1+x)}{x}$
30. $\displaystyle\lim_{x\to \infty} \frac{\ln(\ln x)}{x\ln x}$
31. $\displaystyle\lim_{x\to 0} \frac{2e^x-x^2-2x-2}{x^3}$
32. $\displaystyle\lim_{x\to 0} \frac{\sin x-\tan x}{x^3}$
33. $\displaystyle\lim_{x\to 0} \frac{2-e^x-e^{-x}}{2x^2}$
34. $\displaystyle\lim_{x\to 0} \frac{e^{3x}-e^{-3x}}{2x}$
35. $\displaystyle\lim_{x\to \pi/2} \frac{2x-\pi}{\tan 2x}$
36. $\displaystyle\lim_{x\to \pi/2} \frac{\sec x}{\tan x}$
37. $\displaystyle\lim_{x\to 2} \frac{x-2\cos\pi x}{x^2-4}$
38. $\displaystyle\lim_{x\to 1/2} \frac{2x-\sin\pi x}{4x^2-1}$
39. $\displaystyle\lim_{x\to 0^+} \frac{\ln\sqrt{2x}}{\ln\sqrt[3]{3x}}$
40. $\displaystyle\lim_{x\to 0} \frac{\ln(1+x)}{\ln(1-x^2)}$
41. $\displaystyle\lim_{x\to 0} \frac{\exp(x^3)-1}{x-\sin x}$
42. $\displaystyle\lim_{x\to 0} \frac{\sqrt{1+3x}-1}{x}$
43. $\displaystyle\lim_{x\to 0} \frac{\sqrt[3]{1+4x}-1}{x}$
44. $\displaystyle\lim_{x\to 0} \frac{\sqrt{3+2x}-\sqrt{3+x}}{x}$
45. $\displaystyle\lim_{x\to 0} \frac{\sqrt[3]{1+x}-\sqrt[3]{1-x}}{x}$
46. $\displaystyle\lim_{x\to \pi/4} \frac{1-\tan x}{4x-\pi}$
47. $\displaystyle\lim_{x\to 0} \frac{\ln(1+x^2)}{e^x-\cos x}$
48. $\displaystyle\lim_{x\to 2} \frac{x^5-5x^2-12}{x^{10}-500x-24}$

Sketch the graphs of the curves in Problems 49 through 60. Even if you use a graphing calculator or computer, apply l'Hôpital's rule as necessary to verify the apparent behavior of the curve as x approaches a point where the function has an indeterminate form.

49. $y = \dfrac{\sin^2 x}{x}$
50. $y = \dfrac{\sin^2 x}{x^2}$
51. $y = \dfrac{\sin x}{x-\pi}$
52. $y = \dfrac{\cos x}{2x-\pi}$
53. $y = \dfrac{1-\cos x}{x^2}$
54. $y = \dfrac{x-\sin x}{x^3}$
55. $y = xe^{-x}$
56. $y = e^{-x}\sqrt{x}$
57. $y = xe^{-\sqrt{x}}$
58. $y = x^2e^{-2x}$
59. $y = \dfrac{\ln x}{x}$
60. $y = \dfrac{\ln x}{\sqrt{x}+\sqrt[3]{x}}$

61. Show that $\displaystyle\lim_{x\to\infty} \frac{x^k}{e^x} = 0$ if k is a positive real number.

62. Show that $\displaystyle\lim_{x\to\infty} \frac{\ln x}{x^k} = 0$ if k is a positive real number.

63. Suppose that n is a fixed positive integer larger than 1. Show that the curve $y = x^n e^{-x}$ has a single local maximum and two inflection points for $x > 0$, and also has the x-axis as an asymptote.

64. Suppose that k is an arbitrary positive real number. Show that the curve $y = x^{-k}\ln x$ has a single local maximum and a single inflection point for $x > 0$, and also has the x-axis as an asymptote.

65. Substitute $y = \dfrac{1}{x}$ in Eq. (11) to show that

$$\lim_{x\to 0^+} x^k \ln x = 0$$

if k is a positive real number.

66. Show that if n is any integer, then

$$\lim_{x\to\infty} \frac{(\ln x)^n}{x} = 0.$$

67. Suppose that $f'(x)$ is continuous. Show that

$$\lim_{h\to 0} \frac{f(x+h) - f(x-h)}{2h} = f'(x).$$

The *symmetric difference quotient* on the left can be used (with h very small) to approximate the derivative numerically, and turns out to be a better approximation than the one-sided difference quotient $[f(x+h) - f(x)]/h$.

68. Suppose that $f''(x)$ is continuous. Show that

$$\lim_{h\to 0} \frac{f(x+h) - 2f(x) + f(x-h)}{h^2} = f''(x).$$

The *second difference quotient* on the left can be used (with h very small) to approximate the second derivative numerically.

69. In his calculus textbook of 1696, l'Hôpital used a limit similar to

$$\lim_{x\to 1} \frac{\sqrt{2x - x^4} - \sqrt[3]{x}}{1 - x^{4/3}}$$

to illustrate his rule. Evaluate this limit.

70. Establish the 0/0 version of l'Hôpital's rule for the case $a = \infty$. *Suggestion:* Let $F(t) = f(1/t)$ and $G(t) = g(1/t)$. Then show that

$$\lim_{x\to\infty} \frac{f(x)}{g(x)} = \lim_{t\to 0^+} \frac{F(t)}{G(t)} = \lim_{t\to 0^+} \frac{F'(t)}{G'(t)} = \lim_{x\to\infty} \frac{f'(x)}{g'(x)},$$

using l'Hôpital's rule for the case $a = 0$.

71. Show *without* using l'Hôpital's rule that

$$\lim_{x\to\infty} \left(\frac{x}{e}\right)^x = +\infty.$$

Thus the function $f(x) = x^x$ increases even faster than the exponential function e^x as $x \to +\infty$.

72. If a chemical plant releases an amount A of a pollutant into a canal at time $t = 0$, then the resulting concentration of pollutant at time t in the water at a town on the canal a fixed distance x_0 downstream is

$$C(t) = \frac{A}{\sqrt{\pi k t}} \exp\left(-\frac{x_0^2}{4kt}\right)$$

where k is a constant. Sketch a typical graph of $C(t)$ for $t \geqq 0$. Then show that the maximum pollutant concentration that occurs at the town is

$$C_{\max} = \frac{A}{x_0}\sqrt{\frac{2}{\pi e}}.$$

73. (a) If $f(x) = x^n e^{-x}$ (where n is a fixed positive integer) is the function of Problem 63, show that the maximum value of $f(x)$ for $x \geqq 0$ is $f(n) = n^n e^{-n}$. (b) Conclude from the fact that $f(n-1)$ and $f(n+1)$ are both less than $f(n)$ that

$$\left(1 + \frac{1}{n}\right)^n < e < \left(1 - \frac{1}{n}\right)^{-n}.$$

Substitute $n = 1{,}000{,}000$ to prove that $e = 2.71828$ accurate to five decimal places.

74. (a) Approximate numerically the solution x_1 of the equation $\ln x = x^{1/10}$ that is indicated in Fig. 4.8.8. (b) Use a calculator or computer to plot the graphs of $y = \ln x$ and $y = x^{1/10}$ in a viewing window that shows a second solution x_2 of the equation $\ln x = x^{1/10}$. Then approximate x_2 numerically. *Suggestion:* Plot the graphs on successive intervals of the form $[10^n, 10^{n+1}]$ where $n = 1, 2, 3, \ldots$. By the time you locate x_2 you may well have the feeling of "going boldly where no one has gone before!"

4.9 MORE INDETERMINATE FORMS

We saw in Section 4.8 that l'Hôpital's rule can be applied to the indeterminate forms $0/0$ and ∞/∞. There are other indeterminate forms; although l'Hôpital's rule cannot be applied directly to these other forms, it may be possible to convert them into the form $0/0$ or into the form ∞/∞. If so, it may be possible to apply l'Hôpital's rule.

Suppose that

$$\lim_{x\to a} f(x) = 0 \quad \text{and} \quad \lim_{x\to a} g(x) = \infty.$$

Then we say that the product $f(x) \cdot g(x)$ has the **indeterminate form** $0 \cdot \infty$ at $x = a$ (or as $x \to a$). To find the limit of $f(x) \cdot g(x)$ at $x = a$, we can change the problem to one of the forms $0/0$ or ∞/∞ in this way:

$$f(x) \cdot g(x) = \frac{f(x)}{1/g(x)} = \frac{g(x)}{1/f(x)}.$$

Now l'Hôpital's rule may be applied if its other hypotheses are satisfied, as illustrated in Example 1.

EXAMPLE 1 Find $\lim_{x\to\infty} x \ln\left(\frac{x-1}{x+1}\right)$.

Solution We are dealing with the indeterminate form $0 \cdot \infty$, so we write

$$\lim_{x\to\infty} x \ln\left(\frac{x-1}{x+1}\right) = \lim_{x\to\infty} \frac{\ln\left(\frac{x-1}{x+1}\right)}{\frac{1}{x}}.$$

The right-hand limit has the form 0/0, so we can apply l'Hôpital's rule. First we note that

$$D_x \ln\left(\frac{x-1}{x+1}\right) = \frac{2}{x^2-1}.$$

Thus

$$\lim_{x\to\infty} x \ln\left(\frac{x-1}{x+1}\right) = \lim_{x\to\infty} \frac{\frac{2}{x^2-1}}{-\frac{1}{x^2}}$$

$$= \lim_{x\to\infty} \frac{-2x^2}{x^2-1} = \lim_{x\to\infty} \frac{-2}{1-\frac{1}{x^2}} = -2.$$

Hence the curve

$$y = x \ln \frac{x-1}{x+1}, \quad x > 1,$$

has the line $y = -2$ as a horizontal asymptote as $x \to +\infty$. (See Fig. 4.9.1.) It also has the line $x = 1$ as a vertical asymptote as $x \to 1^+$. (Why?) ◆

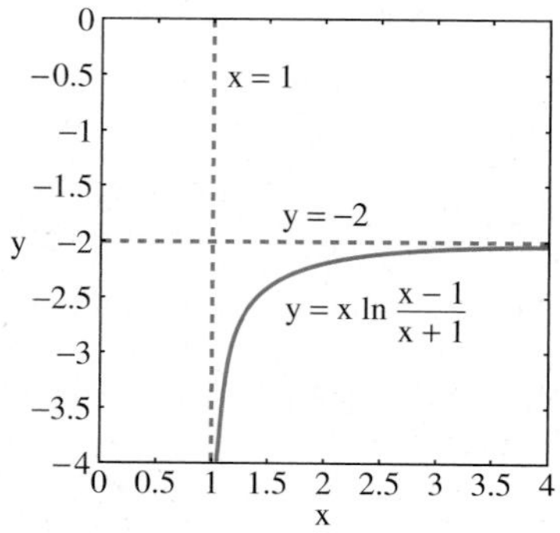

FIGURE 4.9.1 Visual corroboration of the limit in Example 1.

If

$$\lim_{x\to a} f(x) = +\infty = \lim_{x\to a} g(x),$$

then we say that $f(x) - g(x)$ has the **indeterminate form** $\infty - \infty$ as $x \to a$. To evaluate

$$\lim_{x\to a} [f(x) - g(x)],$$

we try by algebraic manipulation to convert $f(x) - g(x)$ into a form of type $0/0$ or ∞/∞ so that it may be possible to apply l'Hôpital's rule. If $f(x)$ or $g(x)$ is expressed as a fraction, we can sometimes do this by finding a common denominator. In most cases, however, subtler methods are required. Example 2 illustrates the technique of finding a common denominator. Example 3 demonstrates a factoring technique that can be effective.

EXAMPLE 2

$$\lim_{x\to 0} \left(\frac{1}{x} - \frac{1}{\sin x}\right) = \lim_{x\to 0} \frac{(\sin x) - x}{x \sin x} \qquad \text{(form 0/0)}$$

$$= \lim_{x\to 0} \frac{(\cos x) - 1}{\sin x + x\cos x} \qquad \text{(still 0/0)}$$

$$= \lim_{x\to 0} \frac{-\sin x}{2\cos x - x \sin x} = 0.$$ ◆

EXAMPLE 3

$$\lim_{x\to+\infty}\left(\sqrt{x^2+3x}-x\right)=\lim_{x\to+\infty}x\left(\sqrt{1+\frac{3}{x}}-1\right) \qquad (\text{form } \infty\cdot 0)$$

$$=\lim_{x\to+\infty}\frac{\sqrt{1+\frac{3}{x}}-1}{\frac{1}{x}} \qquad (\text{form } 0/0 \text{ now})$$

$$=\lim_{x\to+\infty}\frac{\frac{1}{2}\left(1+\frac{3}{x}\right)^{-1/2}\left(-\frac{3}{x^2}\right)}{-\frac{1}{x^2}}$$

$$=\lim_{x\to+\infty}\frac{\frac{3}{2}}{\sqrt{1+\frac{3}{x}}}=\frac{3}{2}.$$

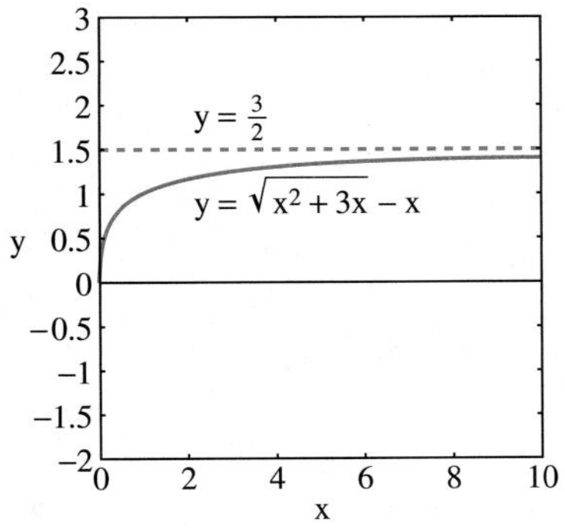

FIGURE 4.9.2 Visual corroboration of the limit in Example 3.

Thus the curve $y=\sqrt{x^2+3x}-x$, $x>0$, has the line $y=\frac{3}{2}$ as a horizontal asymptote as $x\to+\infty$. (See Fig. 4.9.2.) ◆

The Indeterminate Forms 0^0, ∞^0, and 1^∞

Suppose that we need to find the limit of a quantity

$$y=[f(x)]^{g(x)},$$

where the limits of f and g as $x\to a$ are such that one of the **indeterminate forms** 0^0, ∞^0, or 1^∞ is produced. We first compute the natural logarithm

$$\ln y=\ln\left([f(x)]^{g(x)}\right)=g(x)\ln f(x).$$

For each of the three indeterminate forms mentioned here, $g(x)\ln f(x)$ has the form $0\cdot\infty$, so we can use our earlier methods to find $L=\lim_{x\to a}\ln y$ (assuming that $f(x)>0$ near $x=a$, so that $y>0$). Then

$$\lim_{x\to a}[f(x)]^{g(x)}=\lim_{x\to a}y=\lim_{x\to a}\exp(\ln y)=\exp\left(\lim_{x\to a}\ln y\right)=e^L,$$

because the exponential function is continuous. Thus we have the following four steps for finding the limit of $[f(x)]^{g(x)}$ as $x\to a$:

1. Let $y=[f(x)]^{g(x)}$.
2. Simplify $\ln y=g(x)\ln f(x)$.
3. Evaluate $L=\lim_{x\to a}\ln y$.
4. Conclude that $\lim_{x\to a}[f(x)]^{g(x)}=e^L$.

EXAMPLE 4 Find $\lim_{x\to 0}(\cos x)^{1/x^2}$.

Solution Here we have the indeterminate form 1^∞. If we let $y=(\cos x)^{1/x^2}$, then

$$\ln y=\ln\left[(\cos x)^{1/x^2}\right]=\frac{\ln\cos x}{x^2}.$$

As $x\to 0$, $\cos x\to 1$, and so $\ln\cos x\to 0$; we are now dealing with the indeterminate form $0/0$. Hence two applications of l'Hôpital's rule yield

$$\lim_{x\to 0}\ln y=\lim_{x\to 0}\frac{\ln\cos x}{x^2}=\lim_{x\to 0}\frac{(-\sin x)/(\cos x)}{2x}=\lim_{x\to 0}\frac{-\tan x}{2x} \qquad (0/0 \text{ form})$$

$$=\lim_{x\to 0}\frac{-\sec^2 x}{2}=-\frac{1}{2}.$$

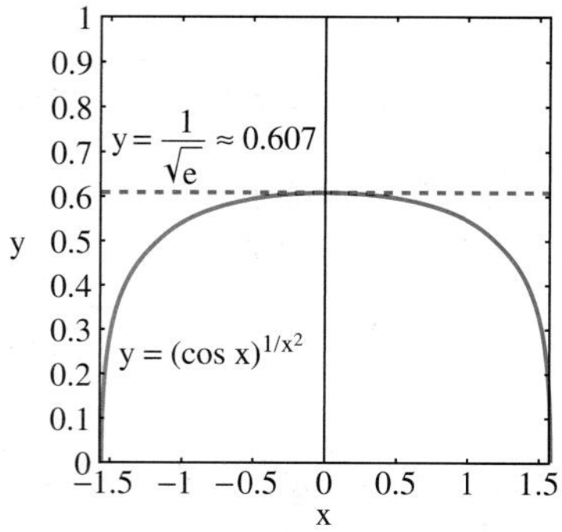

FIGURE 4.9.3 Visual corroboration of the limit in Example 4.

Consequently, as suggested by Fig. 4.9.3,

$$\lim_{x\to 0}(\cos x)^{1/x^2} = e^{-1/2} = \frac{1}{\sqrt{e}}. \quad \blacklozenge$$

EXAMPLE 5 Find $\lim_{x\to 0^+} x^{\tan x}$.

Solution This has the indeterminate form 0^0. If $y = x^{\tan x}$, then

$$\ln y = (\tan x)(\ln x) = \frac{\ln x}{\cot x}.$$

Now we have the indeterminate form ∞/∞, and l'Hôpital's rule yields

$$\lim_{x\to 0^+} \ln y = \lim_{x\to 0^+} \frac{\ln x}{\cot x} = \lim_{x\to 0^+} \frac{\frac{1}{x}}{-\csc^2 x} = -\lim_{x\to 0^+} \frac{\sin^2 x}{x}$$

$$= -\lim_{x\to 0^+}\left(\frac{\sin x}{x}\right)(\sin x) = (-1)\cdot 0 = 0.$$

Therefore, $\lim_{x\to 0^+} x^{\tan x} = e^0 = 1$. The graph of the curve $y = x^{\tan x}$, $0 < x < \pi/2$ in Fig. 4.9.4 provides corroboration of this limit. We note also a local minimum on the curve near $x = 0.4$. (See Problem 45.) ◆

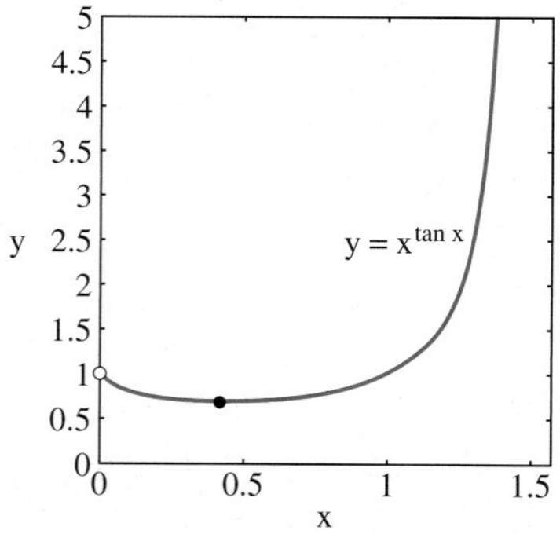

FIGURE 4.9.4 Visual corroboration of the limit in Example 5.

Although $a^0 = 1$ for any *nonzero* constant a, the form 0^0 is indeterminate—the limit is not necessarily 1 (see Problem 52). But the form 0^∞ is not indeterminate; its limit is zero. For example,

$$\lim_{x\to 0^+} x^{1/x} = 0.$$

The Number e as a Limit

Figure 4.9.5 shows the graph of the function

$$f(x) = \left(1 + \frac{1}{x}\right)^x \tag{1}$$

for $x > 0$. The graph appears to begin at the point $(0, 1)$ and to approach a horizontal asymptote as $x \to +\infty$. Note that $f(x)$ has the indeterminate form ∞^0 as $x \to 0^+$ and has the indeterminate form 0^∞ as $x \to +\infty$. In each case our strategy is to calculate the limit of $\ln f(x)$:

$$\lim\left[\ln\left(1 + \frac{1}{x}\right)^x\right] = \lim\left[x \cdot \ln\left(1 + \frac{1}{x}\right)\right] = \lim \frac{\ln\left(1 + \frac{1}{x}\right)}{\frac{1}{x}}.$$

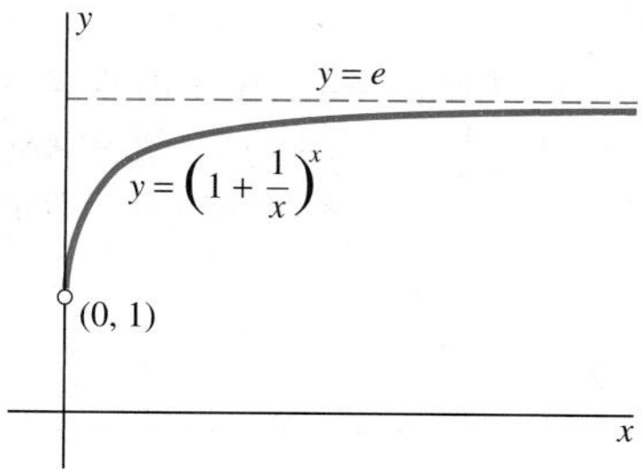

FIGURE 4.9.5 The graph $y = \left(1 + \dfrac{1}{x}\right)^x$ has the horizontal asymptote $y = e$.

The last limit here has the indeterminate form ∞/∞ as $x \to 0^+$ and has the indeterminate form $0/0$ as $x \to +\infty$. In each case we can apply l'Hôpital's rule to obtain

$$\lim\left[\ln\left(1 + \frac{1}{x}\right)^x\right] = \lim \frac{-\frac{1}{x^2}\cdot\frac{1}{1 + (1/x)}}{-\frac{1}{x^2}} = \lim \frac{x}{x+1} = \begin{cases} 0 & \text{as } x \to 0^+, \\ 1 & \text{as } x \to +\infty. \end{cases}$$

Thus we find that

$$\lim_{x\to 0^+}\left(1 + \frac{1}{x}\right)^x = e^0 = 1$$

and that

$$\lim_{x\to+\infty}\left(1+\frac{1}{x}\right)^x = e^1 = e. \tag{2}$$

The last limit shows that the horizontal asymptote in Fig. 4.9.5 is the line $y = e$.

If we write $x = n$ (a positive integer) in Eq. (2), we obtain the famous limit

$$e = \lim_{n\to\infty}\left(1+\frac{1}{n}\right)^n, \tag{3}$$

which can be used to approximate the number e. In Problem 44 we ask you to use l'Hôpital's rule similarly to derive the more general limit expression

$$e^x = \lim_{n\to\infty}\left(1+\frac{x}{n}\right)^n \tag{4}$$

for the exponential function.

The limit in (3) can be approximated with a very rudimentary calculator by substituting $n = 2^k$ (a power of 2) to get

$$e = \lim_{k\to\infty}\left(1+\frac{1}{2^k}\right)^{2^k} = \lim_{k\to\infty} v^{2^k} \tag{5}$$

where $v = 1 + (1/2^k)$. Then

$$(v^2)^2 = v^4, \quad (v^4)^2 = v^8, \quad (v^8)^2 = v^{16}, \quad \ldots, \quad \left(v^{2^{k-1}}\right)^2 = v^{2^k}.$$

Therefore we should get the value $[1 + (1/2^k)]^{2^k}$ if we enter $v = 1 + (1/2^k)$ and then press the $\boxed{x^2}$ key k times in succession. (Try this with your own calculator. Can you see how and why the process may fail if k is too large?) The entries in the table in Fig. 4.9.6 were calculated using a high-precision computer (rather than a simple calculator). They indicate that $e = 2.71828\,1828$ accurate to nine decimal places.

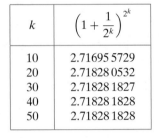

k	$\left(1+\frac{1}{2^k}\right)^{2^k}$
10	2.71695 5729
20	2.71828 0532
30	2.71828 1827
40	2.71828 1828
50	2.71828 1828

FIGURE 4.9.6 Approximating the number e.

4.9 TRUE/FALSE STUDY GUIDE

4.9 CONCEPTS: QUESTIONS AND DISCUSSION

1. List the seven indeterminate forms discussed in Sections 4.8 and 4.9. Illustrate each of these forms with your own example.
2. Explain in your own words why 0^∞ and $0^{-\infty}$ are *not* indeterminate forms.
3. Suppose that someone claims that "calculus is merely the study of indeterminate forms." Based on what you've learned so far, how would you argue for (or against) this claim?

4.9 PROBLEMS

Find the limits in Problems 1 through 34.

1. $\lim_{x\to 0} x\cot x$

2. $\lim_{x\to 0}\left(\frac{1}{x} - \cot x\right)$

3. $\lim_{x\to 0}\frac{1}{x}\ln\left(\frac{7x+8}{4x+8}\right)$

4. $\lim_{x\to 0^+}(\sin x)(\ln\sin x)$

5. $\lim_{x\to 0} x^2\csc^2 x$

6. $\lim_{x\to\infty} e^{-x}\ln x$

7. $\lim_{x\to\infty} x(e^{1/x} - 1)$

8. $\lim_{x\to 2}\left(\frac{1}{x-2} - \frac{1}{\ln(x-1)}\right)$

9. $\lim_{x\to 0^+} x\ln x$

10. $\lim_{x\to\pi/2}(\tan x)(\cos 3x)$

11. $\lim_{x\to\pi}(x-\pi)\csc x$

12. $\lim_{x\to\infty}(x - \sin x)\exp(-x^2)$

13. $\lim_{x\to 0^+}\left(\frac{1}{\sqrt{x}} - \frac{1}{\sin x}\right)$

14. $\lim_{x\to 0}\left(\frac{1}{x} - \frac{1}{e^x - 1}\right)$

15. $\lim_{x\to 1^+}\left(\frac{x}{x^2+x-2} - \frac{1}{x-1}\right)$

16. $\lim_{x\to\infty}\left(\sqrt{x+1} - \sqrt{x}\right)$

17. $\lim_{x\to 0}\left(\frac{1}{x} - \frac{1}{\ln(1+x)}\right)$

18. $\lim_{x\to\infty}\left(\sqrt{x^2+x} - \sqrt{x^2-x}\right)$

19. $\lim_{x\to\infty}\left(\sqrt[3]{x^3+2x+5}-x\right)$

20. $\lim_{x\to 0^+} x^x$

21. $\lim_{x\to 0^+} x^{\sin x}$

22. $\lim_{x\to\infty}\left(\dfrac{2x-1}{2x+1}\right)^x$

23. $\lim_{x\to\infty} (\ln x)^{1/x}$

24. $\lim_{x\to\infty}\left(1-\dfrac{1}{x^2}\right)^x$

25. $\lim_{x\to 0}\left(\dfrac{\sin x}{x}\right)^{1/x^2}$

26. $\lim_{x\to 0^+} (1+2x)^{1/(3x)}$

27. $\lim_{x\to\infty}\left(\cos\dfrac{1}{x^2}\right)^{x^4}$

28. $\lim_{x\to 0^+} (\sin x)^{\sec x}$

29. $\lim_{x\to 0^+} (x+\sin x)^x$

30. $\lim_{x\to \pi/2} (\tan x - \sec x)$

31. $\lim_{x\to 1} x^{1/(1-x)}$

32. $\lim_{x\to 1^+} (x-1)^{\ln x}$

33. $\lim_{x\to 2^+}\left(\dfrac{1}{\sqrt{x^2-4}}-\dfrac{1}{x-2}\right)$

34. $\lim_{x\to\infty}\left(\sqrt[5]{x^5-3x^4+17}-x\right)$

Figures 4.9.7 through 4.9.9 illustrate the graphs of some of the functions defined for $x > 0$ in Problems 35 through 42. In each of these problems:

(a) First use your own calculator or computer to graph the given function $f(x)$ with an x-range sufficient to suggest its behavior both as $x \to 0^+$ and as $x \to +\infty$.
(b) Then apply l'Hôpital's rule as necessary to verify this suspected behavior near zero and $+\infty$.
(c) Finally, estimate graphically and/or numerically the maximum value attained by $f(x)$ for $x \geqq 0$. If possible, find this maximum value exactly.

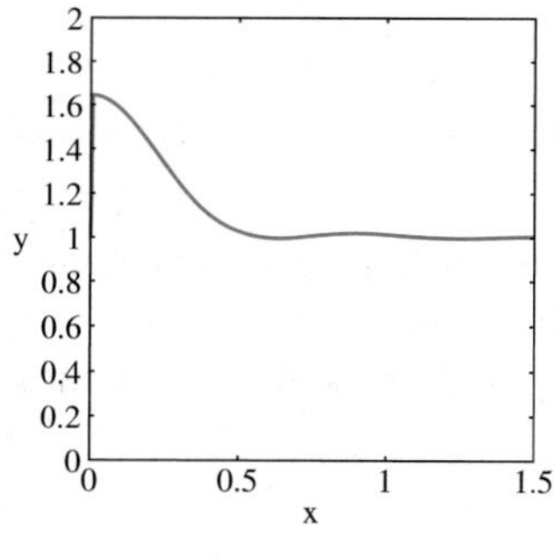

FIGURE 4.9.7

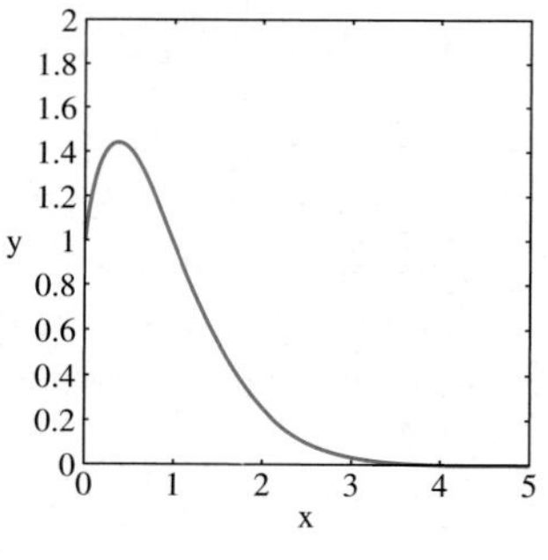

FIGURE 4.9.8

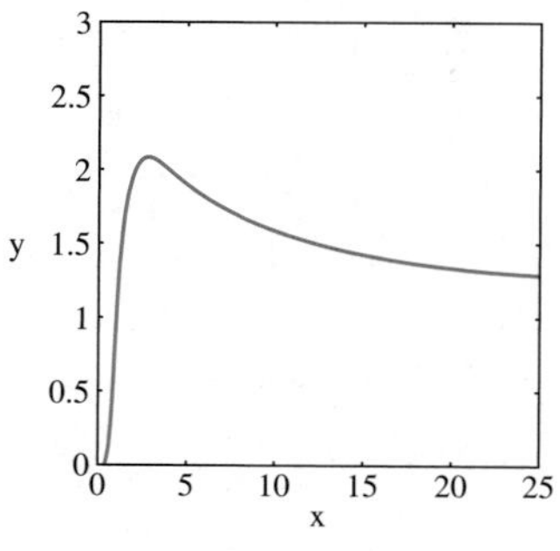

FIGURE 4.9.9

35. $f(x) = x^{1/x}$

36. $f(x) = x^{(1/x^2)}$

37. $f(x) = (x^2)^{1/x}$

38. $f(x) = x^{-x}$

39. $f(x) = (1+x^2)^{1/x}$

40. $f(x) = \left(1+\dfrac{1}{x^2}\right)^x$

41. $f(x) = (x+\sin x)^{1/x}$

42. $f(x) = \left(e^{1/x^2}\right)^{(\cos x - 1)}$

Use l'Hôpital's rule to establish the limits in Problems 43 and 44.

43. $\lim_{h\to 0}(1+hx)^{1/h} = e^x$

44. $\lim_{n\to\infty}\left(1+\dfrac{x}{n}\right)^n = e^x$

45. Estimate graphically or numerically the location of the local minimum point on the graph $y = x^{\tan x}$ shown in Fig. 4.9.4.

46. Let n be a fixed positive integer and let $p(x)$ be the polynomial

$$p(x) = x^n + a_1x^{n-1} + a_2x^{n-2} + \cdots + a_{n-1}x + a_n;$$

the numbers $a_1, a_2, \ldots, a_n$ are fixed real numbers. Prove that

$$\lim_{x\to\infty}\left([p(x)]^{1/n} - x\right) = \frac{a_1}{n}.$$

47. As we shall see in Problem 52 of Section 7.6, the surface area of the ellipsoid obtained by revolving the ellipse

$$\frac{x^2}{a^2}+\frac{y^2}{b^2}=1 \quad (a > b > 0)$$

around the x-axis is

$$A = 2\pi ab\left[\frac{b}{a}+\frac{a}{c}\sin^{-1}\left(\frac{c}{a}\right)\right],$$

where $c = \sqrt{a^2-b^2}$. Use l'Hôpital's rule to show that

$$\lim_{b\to a} A = 4\pi a^2,$$

the surface area of a sphere of radius a.

48. If the amount A_0 is invested in an account that earns interest at the annual rate $r \leqq 1$ compounded n times annually, then the amount A in the account after t years is given by

$$A = A_0\left(1+\frac{r}{n}\right)^{nt}.$$

(a) Show that A is an increasing function of n (with r and t fixed). Thus the bank that compounds more often pays more interest. (b) Use l'Hôpital's rule to show that

$$\lim_{n\to\infty} A(n) = A_0e^{rt}.$$

This is the amount after t years if the bank compounds interest "continuously." The "annual yield" is the value of this limit in the case $t = 1$. (c) If a bank advertises an annual interest rate of 8% compounded continuously, what is the annual yield?

49. Graph the function $f(x) = |\ln x|^{1/x}$ for $x > 0$ and determine its behavior as $x \to 0^+$ and as $x \to +\infty$. Estimate graphically and/or numerically the locations of any critical points or inflection points on the graph of f.

50. Graph the function $f(x) = |\ln x|^{1/|\ln x|}$ for $x > 0$ and determine its behavior as $x \to 0^+$, as $x \to +\infty$, and as x approaches 1 from either side.

51. Graph the function $f(x) = |\ln x|^{|\ln x|}$ for $x > 0$ and determine its behavior as $x \to 0^+$, as $x \to +\infty$, and as x

approaches 1 from either side. Explore both graphically (by zooming) and symbolically (by differentiating) the question of whether f is differentiable at $x = 1$.

52. Let α be a fixed real number. (a) Evaluate (in terms of α) the 0^0 indeterminate form

$$\lim_{x \to 0} \left[\exp\left(-\frac{1}{x^2}\right) \right]^{\alpha x^2}.$$

(Note that l'Hôpital's rule is not needed.) Thus the indeterminate form 0^0 may have as its limit any positive real number. Explain why. (b) Can the limit of a 0^0 indeterminate form be zero, negative, or infinite? Explain.

53. Sketch the graph of the function $f(x) = (1 + x)^{1/x}$ for $x \geqq -1, x \neq 0$. Explain why you can approximate the number e by zooming in on the *apparent* y-intercept of this graph. Do so, accurate to five decimal places.

54. This problem explores the fact that a lead ball hits the ground with greater speed than a feather when both are dropped simultaneously from the top of a tall building. Assuming that air resistance is proportional to downward velocity v, we will show in Chapter 8 that after t seconds the velocity of a dropped body of mass m is given by

$$v(t) = \frac{mg}{k}\left(1 - e^{-kt/m}\right)$$

where g is the familiar acceleration of gravity and k denotes a constant air-resistance coefficient. (a) Note that

$$\lim_{t \to \infty} v(t) = \frac{mg}{k}.$$

Thus a body's velocity tends to a finite limit after it has fallen a sufficiently long time. (b) Note that

$$\lim_{m \to 0} v(t) = 0.$$

Consequently a light "feathery" body falls very slowly through the air. (c) Show that

$$\lim_{m \to \infty} v(t) = gt = \lim_{k \to 0^+} v(t),$$

the velocity of the body after t seconds in the case of *no* air resistance. Thus a very heavy body tends to fall much as it would with no air resistance.

CHAPTER 4 REVIEW: DEFINITIONS, CONCEPTS, RESULTS

Use the following list as a guide to concepts that you may need to review.

1. Increment Δy
2. Differential dy
3. Linear approximation formula
4. Differentiation rules in differential form
5. Increasing functions and decreasing functions
6. Significance of the sign of the first derivative
7. Rolle's theorem
8. The mean value theorem
9. Consequences of the mean value theorem
10. First derivative test
11. Open-interval maximum-minimum problems
12. Graphs of polynomials
13. Calculation of higher derivatives
14. Concave-upward functions and concave-downward functions
15. Test for concavity
16. Second derivative test
17. Inflection points
18. Inflection point test
19. Infinite limits
20. Vertical asymptotes
21. Limits as $x \to \pm\infty$
22. Horizontal asymptotes
23. Slant asymptotes
24. Curve-sketching techniques
25. L'Hôpital's rule and the indeterminate forms $0/0$, ∞/∞, $0 \cdot \infty$, $\infty - \infty$, 0^0, ∞^0, and 1^∞
26. The order of magnitude of exponential and logarithmic functions
27. The number e as a limit

CHAPTER 4 MISCELLANEOUS PROBLEMS

In Problems 1 through 6, write dy in terms of x and dx.

1. $y = (4x - x^2)^{3/2}$
2. $y = 8x^3\sqrt{x^2 + 9}$
3. $y = \dfrac{x+1}{x-1}$
4. $y = \sin x^2$
5. $y = x^2 \cos\sqrt{x}$
6. $y = \dfrac{x}{\sin 2x}$

In Problems 7 through 16, estimate the indicated number by linear approximation.

7. $\sqrt{6401}$ (Note that $80^2 = 6400$.)
8. $\dfrac{1}{1.000007}$
9. $(2.0003)^{10}$ (Note that $2^{10} = 1024$.)
10. $\sqrt[3]{999}$ (Note that $10^3 = 1000$.)
11. $\sqrt[3]{1005}$
12. $\sqrt[3]{62}$
13. $26^{3/2}$
14. $\sqrt[5]{30}$
15. $\sqrt[4]{17}$
16. $\sqrt[10]{1000}$

In Problems 17 through 22, estimate by linear approximation the change in the indicated quantity.

17. The volume $V = s^3$ of a cube, if its side length s is increased from 5 in. to 5.1 in.

18. The area $A = \pi r^2$ of a circle, if its radius r is decreased from 10 cm to 9.8 cm.

19. The volume $V = \frac{4}{3}\pi r^3$ of a sphere, if its radius r is increased from 5 cm to 5.1 cm.

20. The volume $V = 1000/p$ in.^{3} of a gas, if the pressure p is decreased from 100 lb/in.^{2} to 99 lb/in.^{2}

21. The period of oscillation $T = 2\pi\sqrt{L/32}$ of a pendulum, if its length L is increased from 2 ft to 2 ft 1 in. (Time T is in seconds and L is in feet.)

22. The lifetime $L = 10^{30}/E^{13}$ of a light bulb with applied voltage E volts (V), if the voltage is increased from 110 V to 111 V. Compare your result with the exact change in the function L.

If the mean value theorem applies to the function f on the interval $[a, b]$, it ensures the existence of a solution c in the interval (a, b) of the equation

$$f'(c) = \frac{f(b) - f(a)}{b - a}.$$

In Problems 23 through 28, a function f and an interval $[a, b]$ are given. Verify that the hypotheses of the mean value theorem are satisfied for f on $[a, b]$. Then use the given equation to find the value of the number c.

23. $f(x) = x - \dfrac{1}{x}$; $[1, 3]$

24. $f(x) = x^3 + x - 4$; $[-2, 3]$

25. $f(x) = x^3$; $[-1, 2]$

26. $f(x) = x^3$; $[-2, 1]$

27. $f(x) = \frac{11}{5}x^5$; $[-1, 2]$

28. $f(x) = \sqrt{x}$; $[0, 4]$

Sketch the graphs of the functions in Problems 29 through 33. Indicate the local maxima and minima of each function and the intervals on which the function is increasing or decreasing. Show the concave structure of the graph and identify all inflection points.

29. $f(x) = x^2 - 6x + 4$

30. $f(x) = 2x^3 - 3x^2 - 36x$

31. $f(x) = 3x^5 - 5x^3 + 60x$

32. $f(x) = (3 - x)\sqrt{x}$

33. $f(x) = (1 - x)\sqrt[3]{x}$

34. Show that the equation $x^5 + x = 5$ has exactly one real solution.

Calculate the first three derivatives of the functions in Problems 35 through 44.

35. $f(x) = x^3 - 2x$

36. $f(x) = (x + 1)^{100}$

37. $g(t) = \dfrac{1}{t} - \dfrac{1}{2t + 1}$

38. $h(y) = \sqrt{3y - 1}$

39. $f(t) = 2t^{3/2} - 3t^{4/3}$

40. $g(x) = \dfrac{1}{x^2 + 9}$

41. $h(t) = \dfrac{t + 2}{t - 2}$

42. $f(z) = \sqrt[3]{z} + \dfrac{3}{\sqrt[5]{z}}$

43. $g(x) = \sqrt[3]{5 - 4x}$

44. $g(t) = \dfrac{8}{(3 - t)^{3/2}}$

In Problems 45 through 52, calculate dy/dx and d^2y/dx^2 under the assumption that y is defined implicitly as a function of x by the given equation.

45. $x^{1/3} + y^{1/3} = 1$

46. $2x^2 - 3xy + 5y^2 = 25$

47. $y^5 - 4y + 1 = \sqrt{x}$

48. $\sin xy = xy$

49. $x^2 + y^2 = 5xy + 5$

50. $x^5 + xy^4 = 1$

51. $y^3 - y = x^2 y$

52. $(x^2 - y^2)^2 = 4xy$

Sketch the graphs of the functions in Problems 53 through 72, indicating all critical points, inflection points, and asymptotes. Show the concave structure clearly.

53. $f(x) = x^4 - 32x$

54. $f(x) = 18x^2 - x^4$

55. $f(x) = x^6 - 2x^4$

56. $f(x) = x\sqrt{x - 3}$

57. $f(x) = x\sqrt[3]{4 - x}$

58. $f(x) = \dfrac{x - 1}{x + 2}$

59. $f(x) = \dfrac{x^2 + 1}{x^2 - 4}$

60. $f(x) = \dfrac{x}{x^2 - x - 2}$

61. $f(x) = \dfrac{2x^2}{x^2 - x - 2}$

62. $f(x) = \dfrac{x^3}{x^2 - 1}$

63. $f(x) = 3x^4 - 4x^3$

64. $f(x) = x^4 - 2x^2$

65. $f(x) = \dfrac{x^2}{x^2 - 1}$

66. $f(x) = x^3 - 12x$

67. $f(x) = -10 + 6x^2 - x^3$

68. $f(x) = \dfrac{x}{1 + x^2}$; note that

$$f'(x) = -\frac{(x - 1)(x + 1)}{(x^2 + 1)^2}$$

and that

$$f''(x) = \frac{2x(x^2 - 3)}{(x^2 + 1)^3}.$$

69. $f(x) = x^3 - 3x$

70. $f(x) = x^4 - 12x^2$

71. $f(x) = x^3 + x^2 - 5x + 3$

72. $f(x) = \dfrac{1}{x} + \dfrac{1}{x^2}$

73. The function

$$f(x) = \frac{1}{x^2 + 2x + 2}$$

has a maximum value, and only one. Find it.

74. You need to manufacture a cylindrical pot, without a top, with a volume of 1 ft^3. The cylindrical part of the pot is to be made of aluminum, the bottom of copper. Copper is five times as expensive as aluminum. What dimensions would minimize the total cost of the pot?

75. An open-topped rectangular box is to have a volume of 4500 cm^3. If its bottom is a rectangle whose length is twice its width, what dimensions would minimize the total area of the bottom and four sides of the box?

76. A small rectangular box must be made with a volume of 324 in.^{3} Its bottom is square and costs twice as much (per square inch) as its top and four sides. What dimensions would minimize the total cost of the material needed to make this box?

77. You must make a small rectangular box with a volume of 400 in.^{3} Its bottom is a rectangle whose length is twice its width. The bottom costs 7¢/in.^{2}; the top and four sides of the box cost 5¢/in.^{2} What dimensions would minimize the cost of the box?

78. Suppose that $f(x)$ is a cubic polynomial with exactly three distinct real zeros. Prove that the two zeros of $f'(x)$ are real and distinct.

79. Suppose that it costs $1 + (0.0003)v^{3/2}$ dollars per mile to operate a truck at v miles per hour. If there are additional costs (such as the driver's pay) of $10/hr, what speed would minimize the total cost of a 1000-mi trip?

80. The numbers $a_1, a_2, \ldots, a_n$ are fixed. Find a simple formula for the number x such that the sum of the squares of the distances of x from the n fixed numbers is as small as possible.

81. Sketch the curve $y^2 = x(x-1)(x-2)$, showing that it consists of two pieces—one bounded and the other unbounded—and has two horizontal tangent lines, three vertical tangent lines, and two inflection points. [*Suggestion:* Note that the curve is symmetric around the x-axis. Begin by determining the intervals on which the product $x(x-1)(x-2)$ is positive. Compute dy/dx and d^2y/dx^2 by implicit differentiation.]

82. Farmer Rogers wants to fence in a rectangular plot of area 2400 ft^2. She wants also to use additional fencing to build an internal divider fence parallel to two of the boundary sections (Fig. 4.MP.1). What is the minimum total length of fencing that this project will require? Verify that your answer yields the global minimum.

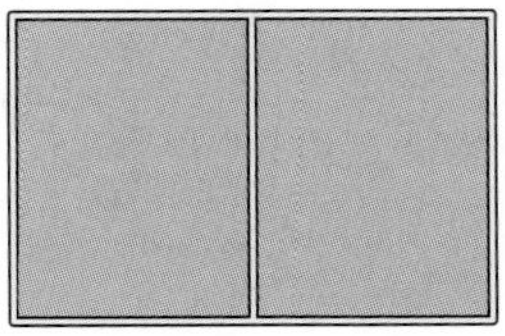

FIGURE 4.MP.1 The fencing of Problem 82.

83. Farmer Simmons wants to fence in a rectangular plot of area 1800 ft^2. He wants also to use additional fencing to build two internal divider fences, both parallel to the same two outer boundary sections (Fig. 4.MP.2). What is the minimum total length of fencing that this project will require? Verify that your answer yields the global minimum.

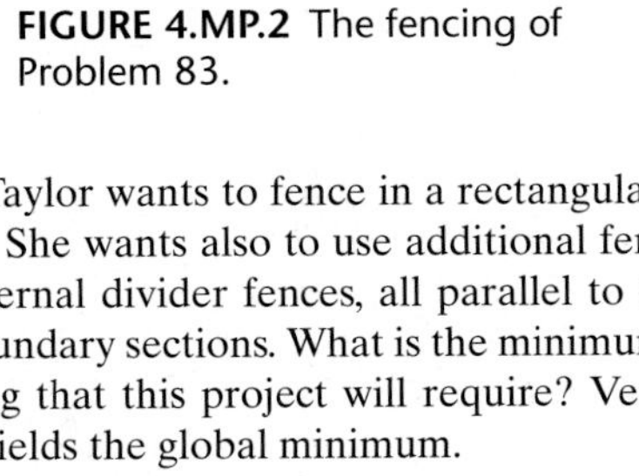

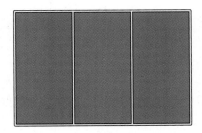

FIGURE 4.MP.2 The fencing of Problem 83.

84. Farmer Taylor wants to fence in a rectangular plot of area 2250 m^2. She wants also to use additional fencing to build three internal divider fences, all parallel to the same two outer boundary sections. What is the minimum total length of fencing that this project will require? Verify that your answer yields the global minimum.

85. Farmer Upshaw wants to fence in a rectangular plot of area A ft^2. He wants also to use additional fencing to build n (a fixed but unspecified positive integer) internal divider fences, all parallel to the same two outer boundary sections. What is the minimum total length of fencing that this project will require? Verify that your answer yields the global minimum.

86. What is the length of the shortest line segment that lies in the first quadrant with its endpoints on the coordinate axes and is also tangent to the graph of $y = 1/x^2$? Verify that your answer yields the global minimum.

87. A right triangle is formed in the first quadrant by a line segment that is tangent to the graph of $y = 1/x^2$ and whose endpoints lie on the coordinate axes. Is there a maximum possible area of such a triangle? Is there a minimum? Justify your answers.

88. A right triangle is formed in the first quadrant by a line segment that is tangent to the graph of $y = 1/x$ and whose endpoints lie on the coordinate axes. Is there a maximum possible area of such a triangle? Is there a minimum? Justify your answers.

89. A rectangular box (with a top) is to have volume 288 in.^{3}, and its base is to be exactly three times as long as it is wide. What is the minimum possible surface area of such a box? Verify that your answer yields the global minimum.

90. A rectangular box (with a top) is to have volume 800 in.^{3}, and its base is to be exactly four times as long as it is wide. What is the minimum possible surface area of such a box? Verify that your answer yields the global minimum.

91. A rectangular box (with a top) is to have volume 225 cm^3, and its base is to be exactly five times as long as it is wide. What is the minimum possible surface area of such a box? Verify that your answer yields the global minimum.

92. A rectangular box (with a top) is to have volume V, and its base is to be exactly n times as long as it is wide (n is a fixed but unspecified positive integer). What is the minimum possible surface area of such a box? Verify that your answer yields the global minimum.

93. The graph of $f(x) = x^{1/3}(1-x)^{2/3}$ is shown in Fig. 4.MP.3. Recall from Section 4.7 that this graph has a slant asymptote with equation $y = mx + b$ provided that

$$\lim_{x \to +\infty} [f(x) - (mx + b)] = 0$$

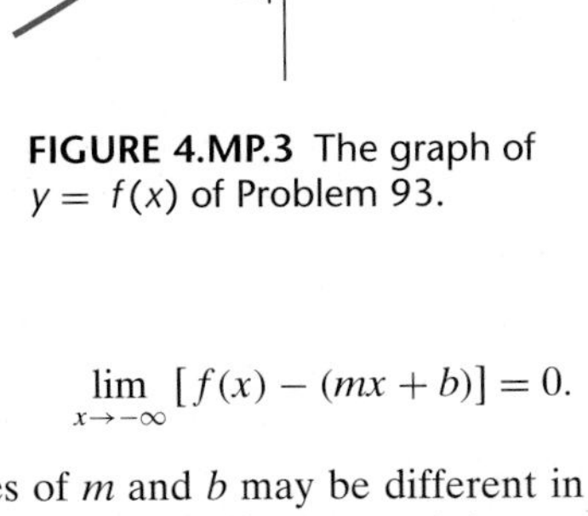

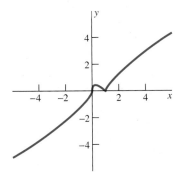

FIGURE 4.MP.3 The graph of $y = f(x)$ of Problem 93.

or that

$$\lim_{x \to -\infty} [f(x) - (mx + b)] = 0.$$

(The values of m and b may be different in the two cases $x \to +\infty$ and $x \to -\infty$.) The graph here appears to have such an asymptote as $x \to +\infty$. Find m by evaluating

$$\lim_{x \to +\infty} \frac{f(x)}{x}.$$

Then find b by evaluating

$$\lim_{x \to +\infty} [f(x) - mx].$$

Finally, find m and b for the case $x \to -\infty$.

94. You are at the southernmost point of a circular lake of radius 1 mi. Your plan is to swim a straight course to another point on the shore of the lake, then jog to the northernmost point. You can jog twice as fast as you can swim. What route gives the minimum time required for your journey?

Find the limits in Problems 95 through 109.

95. $\lim_{x \to 2} \dfrac{x-2}{x^2-4}$

96. $\lim_{x \to 0} \dfrac{\sin 2x}{x}$

97. $\lim_{x \to \pi} \dfrac{1+\cos x}{(x-\pi)^2}$

98. $\lim_{x \to 0} \dfrac{x - \sin x}{x^3}$

99. $\lim_{t \to 0} \dfrac{\tan t - \sin t}{t^3}$

100. $\lim_{x \to \infty} \dfrac{\ln(\ln x)}{\ln x}$

101. $\lim_{x \to 0} (\cot x) \ln(1+x)$

102. $\lim_{x \to 0^+} (e^{1/x} - 1) \tan x$

103. $\lim_{x \to 0} \left(\dfrac{1}{x^2} - \dfrac{1}{1-\cos x} \right)$

104. $\lim_{x \to \infty} \left(\dfrac{x^2}{x+2} - \dfrac{x^3}{x^2+3} \right)$

105. $\lim_{x \to \infty} \left(\sqrt{x^2 - x - 1} - \sqrt{x} \right)$

106. $\lim_{x \to \infty} x^{1/x}$

107. $\lim_{x \to \infty} (e^{2x} - 2x)^{1/x}$

108. $\lim_{x \to \infty} [1 - \exp(-x^2)]^{1/x^2}$

109. $\lim_{x \to \infty} x \left[\left(1 + \dfrac{1}{x}\right)^x - e \right]$ [*Suggestion:* Let $u = 1/x$, and take the limit as $u \to 0^+$.]

110. According to Problem 53 of Section 7.6, the surface area of the ellipsoid obtained by revolving around the x-axis the ellipse with equation

$$\left(\frac{x}{a}\right)^2 + \left(\frac{y}{b}\right)^2 = 1 \qquad (0 < a < b)$$

is

$$A = 2\pi ab \left[\frac{b}{a} + \frac{a}{c} \ln \left(\frac{b+c}{a} \right) \right],$$

where $c = \sqrt{b^2 - a^2}$. Use l'Hôpital's rule to show that

$$\lim_{b \to a} A = 4\pi a^2,$$

the surface area of a sphere of radius a.

THE INTEGRAL

5

Archimedes (287–212 B.C.)

Archimedes of Syracuse was the greatest mathematician of the ancient era from the fifth century B.C. to the second century A.D., when the seeds of modern mathematics sprouted in Greek communities located mainly on the shores of the Mediterranean Sea. He was famous in his own time for mechanical inventions—the so-called Archimedean screw for pumping water, lever-and-pulley devices ("give me a place to stand and I can move the earth"), a planetarium that duplicated the motions of heavenly bodies so accurately as to show eclipses of the sun and moon, and machines of war that terrified Roman soldiers in the siege of Syracuse, during which Archimedes was killed. But it is said that for Archimedes himself these inventions were merely the "diversions of geometry at play," and his writings are devoted to mathematical investigations.

Archimedes carried out many area and volume computations that now use integral calculus—ranging from areas of circles, spheres, and segments of conic sections to volumes of cones, spheres, ellipsoids, and paraboloids. It had been proved earlier in Euclid's *Elements* that the area A of a circle is proportional to the square of its radius r, so $A = \pi r^2$ for some proportionality constant π. But it was Archimedes who accurately approximated the numerical value of π, showing that it lies between the value $3\frac{1}{7}$ memorized by elementary school children and the lower bound $3\frac{10}{71}$. Euclid had also proved that the volume V of a sphere of radius r is given by $V = \mu r^3$ (μ constant), but it was Archimedes who discovered (and proved) that $\mu = 4\pi/3$. He also discovered the now-familiar volume formulas $V = \pi r^2 h$ and $V = \frac{1}{3}\pi r^2 h$ for the cylinder and the cone, respectively, of base radius r and height h.

It was long suspected that Archimedes had not originally discovered his area and volume formulas by means of the limit-based arguments he used to establish them rigorously. In 1906 an Archimedean treatise entitled *The Method* was rediscovered virtually by accident after having been lost since ancient times. In it he described a "method of discovery" based on using infinitesimals much as they were employed during the invention and exploration of calculus in the seventeenth and eighteenth centuries.

To commemorate his sphere and cylinder formulas, Archimedes requested that on his tombstone be carved a sphere inscribed in a circular cylinder. If the height of the cylinder is $h = 2r$, can you verify that the total surface areas A_C and A_S of the cylinder and sphere, and their volumes V_C and V_S, are related by Archimedes' formulas

$$A_S = \tfrac{2}{3}A_C \quad \text{and} \quad V_S = \tfrac{2}{3}V_C?$$

Thus the volumes and surface areas of the sphere and cylinder have the same 2 : 3 ratio.

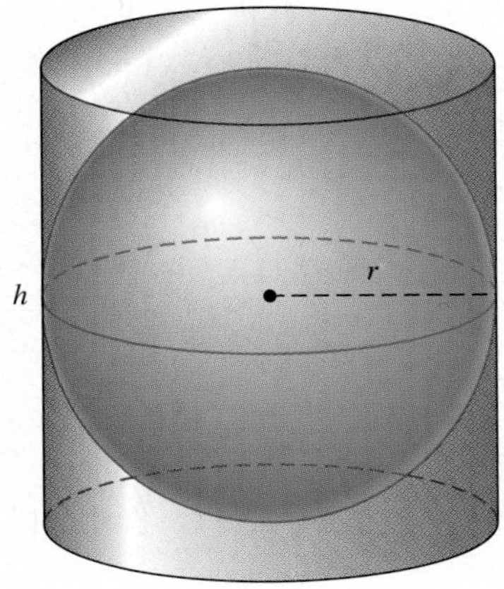

5.1 INTRODUCTION

Chapters 1 through 4 dealt with **differential calculus,** which is one of two closely related parts of *the* calculus. Differential calculus is centered on the concept of the *derivative.* Recall that the original motivation for the derivative was the problem of defining what it means for a straight line to be tangent to the graph of a function and calculating the slopes of such lines (Fig. 5.1.1). By contrast, the importance of the derivative stems from its applications to diverse problems that may, upon initial inspection, seem to have little connection with tangent lines.

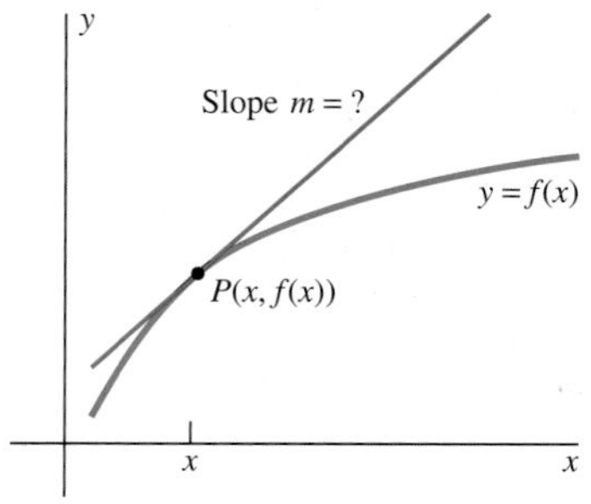

FIGURE 5.1.1 The tangent-line problem motivates differential calculus.

Integral calculus is based on the concept of the *integral.* The definition of the integral is motivated by the problem of defining and calculating the area of the region that lies between the graph of a positive-valued function f and the x-axis over a given closed interval $[a, b]$. The area of the region R of Fig. 5.1.2 is given by the *integral* of f from a to b, denoted by the symbol

$$\int_a^b f(x)\,dx. \tag{1}$$

But the integral, like the derivative, is important due to its applications in many problems that may appear unrelated to its original motivation—problems involving motion and velocity, population growth, volume, arc length, surface area, and center of gravity, among others.

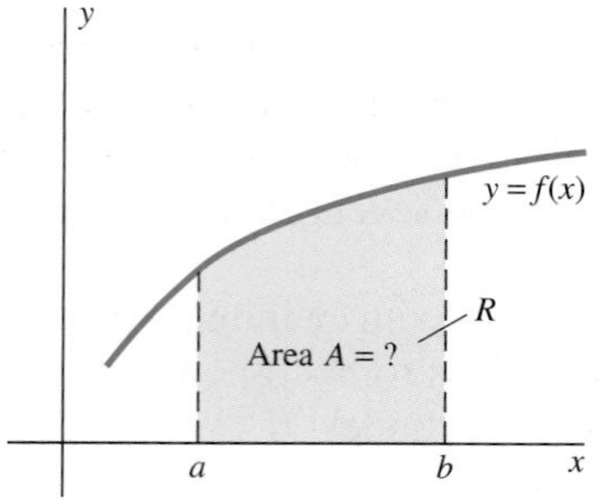

FIGURE 5.1.2 The area problem motivates integral calculus.

The principal theorem of this chapter is the *fundamental theorem of calculus* in Section 5.6. It provides a vital connection between the operations of differentiation and integration, one that provides an effective method for computing values of integrals. It turns out that, in order to apply this theorem to evaluate the integral in (1), we need to find not the derivative of the function $f(x)$ but rather a new function $F(x)$ whose derivative is $f(x)$:

$$F'(x) = f(x). \tag{2}$$

Thus we need to do "differentiation in reverse." We therefore begin in Section 5.2 with an investigation of *antidifferentiation.*

5.2 ANTIDERIVATIVES AND INITIAL VALUE PROBLEMS

The language of change is the natural language for the statement of most scientific laws and principles. For example, Newton's law of cooling says that the *rate of change* of the temperature T of a body is proportional to the difference between T and the temperature of the surrounding medium (Fig. 5.2.1). That is,

$$\frac{dT}{dt} = -k(T - A), \tag{1}$$

where k is a positive constant and A, normally assumed to be constant, is the surrounding temperature. Similarly, the *rate of change* of a population P with constant birth and death rates is proportional to the size of the population:

$$\frac{dP}{dt} = kP \qquad (k \text{ constant}). \tag{2}$$

Torricelli's law of draining (Fig. 5.2.2) implies that the *rate of change* of the depth y of water in a draining tank is proportional to the *square root* of y; that is,

$$\frac{dy}{dt} = -k\sqrt{y} \qquad (k \text{ constant}). \tag{3}$$

Mathematical models of real-world situations frequently involve equations that contain *derivatives* of unknown functions. Such equations, including Eqs. (1) through (3), are called **differential equations.**

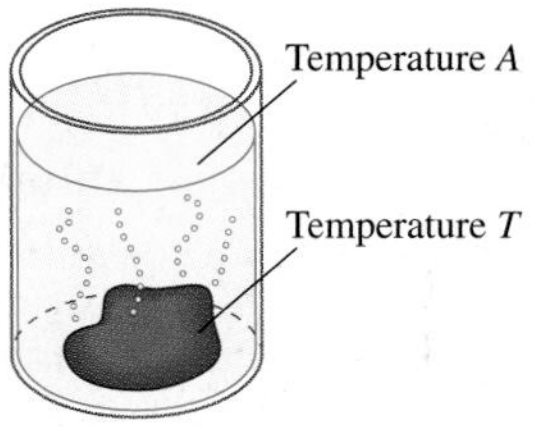

FIGURE 5.2.1 Newton's law of cooling (Eq. (1)) describes the cooling of a hot rock in cold water.

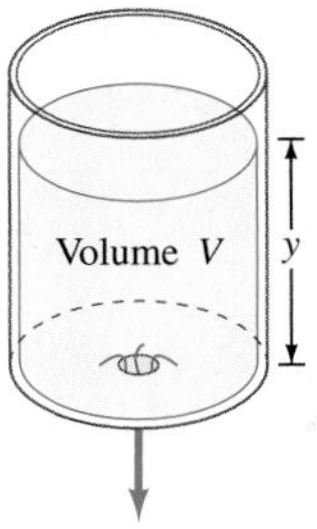

FIGURE 5.2.2 Torricelli's law of draining (Eq. (3)) describes the draining of a cylindrical water tank.

Antiderivatives

The simplest kind of differential equation has the form

$$\frac{dy}{dx} = f(x), \tag{4}$$

where f is a given (known) function and the function $y(x)$ is unknown. The process of finding a function from its derivative is the opposite of differentiation and is therefore called **antidifferentiation.** If we can find a function $y(x)$ whose derivative is $f(x)$,

$$y'(x) = f(x),$$

then we call $y(x)$ an *antiderivative* of $f(x)$.

DEFINITION Antiderivative

An **antiderivative** of the function f is a function F such that

$$F'(x) = f(x)$$

wherever $f(x)$ is defined.

The table in Fig. 5.2.3 shows some examples of functions, each paired with one of its antiderivatives. Figure 5.2.4 illustrates the operations of differentiation and antidifferentiation, beginning with the same function f and going in opposite directions. Figure 5.2.5 illustrates differentiation "undoing" the result of antidifferentiation—the derivative of the antiderivative of $f(x)$ is the original function $f(x)$.

Function $f(x)$	Antiderivative $F(x)$
1	x
$2x$	x^2
x^3	$\frac{1}{4}x^4$
$\cos x$	$\sin x$
$\sin 2x$	$-\frac{1}{2}\cos 2x$

FIGURE 5.2.3 Some antiderivatives.

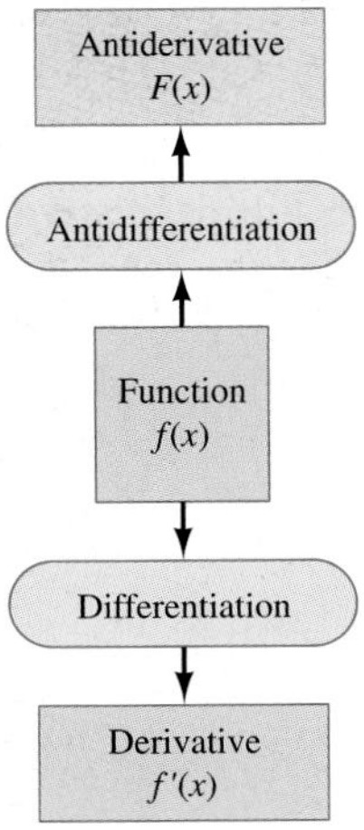

FIGURE 5.2.4 Differentiation and antidifferentiation are opposites.

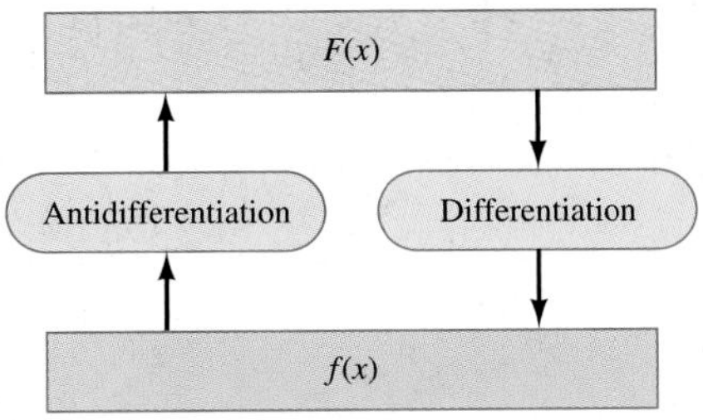

FIGURE 5.2.5 Differentiation undoes the result of antidifferentiation.

EXAMPLE 1 Given the function $f(x) = 3x^2$, $F(x) = x^3$ is an antiderivative of $f(x)$, as are the functions

$$G(x) = x^3 + 17, \qquad H(x) = x^3 + \pi, \qquad \text{and} \qquad K(x) = x^3 - \sqrt{2}.$$

Indeed, $J(x) = x^3 + C$ is an antiderivative of $f(x) = 3x^2$ for *any* choice of the *constant* C. ◆

Thus a single function has *many* antiderivatives, whereas a function can have *only one* derivative. If $F(x)$ is an antiderivative of $f(x)$, then so is $F(x) + C$ for any choice of the constant C. The converse of this statement is more subtle: If $F(x)$ is one antiderivative of $f(x)$ *on the interval* I, then *every* antiderivative of $f(x)$ on I is of the form $F(x) + C$. This follows directly from Corollary 2 of the mean value theorem in Section 4.3, according to which two functions with the same derivative on an interval differ only by a constant on that interval.

Thus the graphs of any two antiderivatives $F(x) + C_1$ and $F(x) + C_2$ of the same function $f(x)$ on the same interval I are "parallel" in the sense illustrated in Figs. 5.2.6 through 5.2.8. There we see that the constant C is the vertical distance between the curves $y = F(x)$ and $y = F(x) + C$ for each x in I. This is the geometric interpretation of Theorem 1.

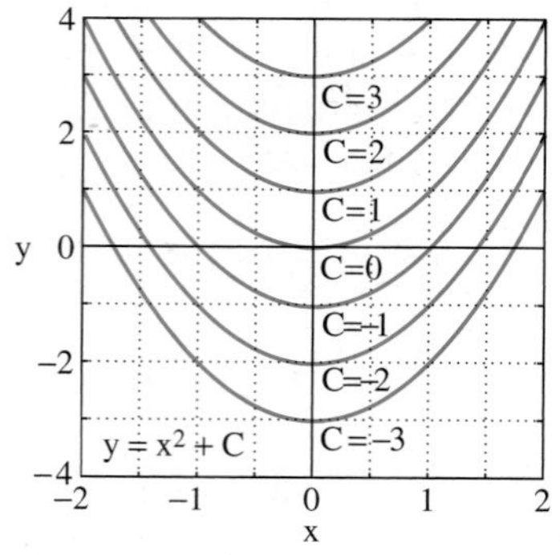

FIGURE 5.2.6 Graph of $y = x^2 + C$ for various values of C.

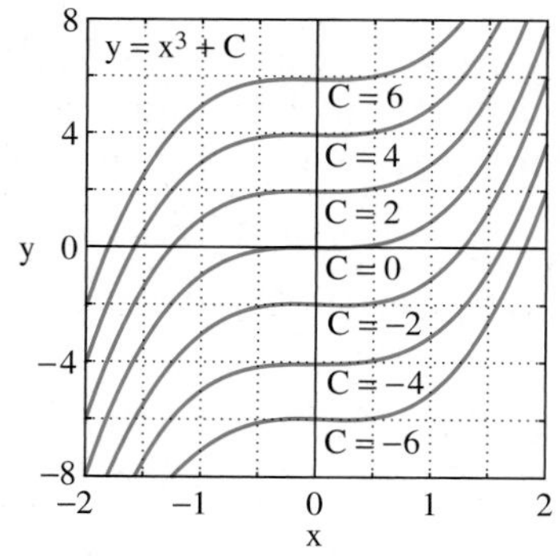

FIGURE 5.2.7 Graph of $y = x^3 + C$ for various values of C.

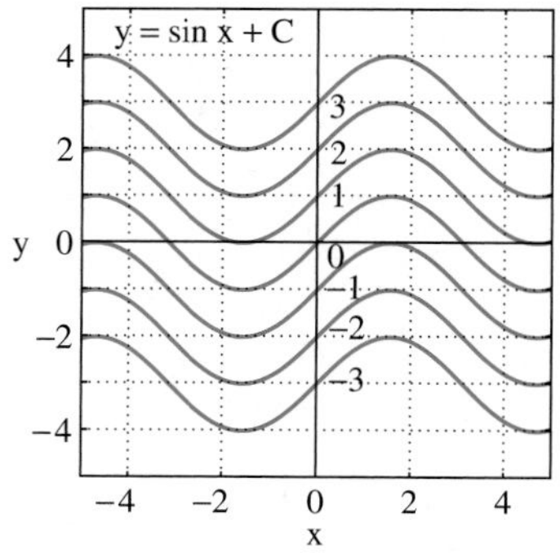

FIGURE 5.2.8 Graph of $y = \sin x + C$ for various values of C.

THEOREM 1 The Most General Antiderivative
If $F'(x) = f(x)$ at each point of the open interval I, then every antiderivative G of f on I has the form

$$G(x) = F(x) + C, \tag{5}$$

where C is a constant.

Thus if F is any single antiderivative of f on the interval I, then the *most general* antiderivative of f on I has the form $F(x) + C$, as given in Eq. (5). The collection of *all* antiderivatives of the function $f(x)$ is called the **indefinite integral** of f with respect to x and is denoted by

$$\int f(x)\,dx.$$

On the basis of Theorem 1, we write

$$\int f(x)\,dx = F(x) + C, \tag{6}$$

where $F(x)$ is any particular antiderivative of $f(x)$. Therefore,

$$\int f(x)\,dx = F(x) + C \quad \text{if and only if} \quad F'(x) = f(x).$$

The integral symbol $\int$ is made like an elongated capital S. It is, in fact, a medieval S, used by Leibniz as an abbreviation for the Latin word *summa* ("sum"). We think of the combination $\int \ \dots \ dx$ as a single symbol; we fill in the "blank" with the formula of the function whose antiderivative we seek. We may regard the differential dx as specifying the independent variable x both in the function $f(x)$ and in its antiderivatives.

EXAMPLE 2 The entries in Fig. 5.2.3 yield the indefinite integrals

$$\int 1\,dx = x + C,$$

$$\int 2x\,dx = x^2 + C,$$

$$\int x^3\,dx = \tfrac{1}{4}x^4 + C,$$

$$\int \cos x\,dx = \sin x + C,$$

and

$$\int \sin 2x\,dx = -\tfrac{1}{2}\cos 2x + C.$$

You can verify each such formula by differentiating the right-hand side. Indeed, this is the *surefire* way to check any antidifferentiation: To verify that $F(x)$ is an antiderivative of $f(x)$, compute $F'(x)$ to see whether or not you obtain $f(x)$. For instance, the differentiation

$$D_x\left(-\tfrac{1}{2}\cos 2x + C\right) = -\tfrac{1}{2}(-2\sin 2x) + 0 = \sin 2x$$

is sufficient to verify the fifth formula of this example. ◆

The differential dx in Eq. (6) specifies that the independent variable is x. But we can describe a specific antidifferentiation in terms of *any* independent variable that is convenient. For example, the indefinite integrals

$$\int 3t^2\,dt = t^3 + C, \quad \int 3y^2\,dy = y^3 + C, \quad \text{and} \quad \int 3u^2\,du = u^3 + C$$

mean exactly the same thing as

$$\int 3x^2\,dx = x^3 + C.$$

Using Integral Formulas

Every differentiation formula yields immediately—by "reversal" of the differentiation—a corresponding indefinite integral formula. The now-familiar derivatives of power functions and trigonometric and exponential functions yield the integral formulas stated in Theorem 2.

THEOREM 2 Some Integral Formulas

$$\int x^k\,dx = \frac{x^{k+1}}{k+1} + C \quad (\text{if } k \neq -1), \tag{7}$$

$$\int \cos kx\,dx = \frac{1}{k}\sin kx + C, \tag{8}$$

$$\int \sin kx\,dx = -\frac{1}{k}\cos kx + C, \tag{9}$$

$$\int \sec^2 kx\,dx = \frac{1}{k}\tan kx + C, \tag{10}$$

$$\int \csc^2 kx\,dx = -\frac{1}{k}\cot kx + C, \tag{11}$$

$$\int \sec kx\,\tan kx\,dx = \frac{1}{k}\sec kx + C, \tag{12}$$

$$\int \csc kx\,\cot kx\,dx = -\frac{1}{k}\csc kx + C, \tag{13}$$

and

$$\int e^{kx}\,dx = \frac{1}{k}e^{kx} + C. \tag{14}$$

REMARK 1 The excluded case $k = -1$ in Eq. (7) corresponds to the fact that $D_x[\ln x] = 1/x$ if $x > 0$, so

$$\int \frac{1}{x}\,dx = \ln x + C \quad (x > 0).$$

REMARK 2 Be sure you see why there is a minus sign in Eq. (9) but none in Eq. (8)!

Recall that the operation of differentiation is *linear*, meaning that

$$D_x\,[cF(x)] = cF'(x) \qquad \text{(where } c \text{ is a constant)}$$

and

$$D_x\,[F(x) \pm G(x)] = F'(x) \pm G'(x).$$

It follows in the notation of antidifferentiation that

$$\int cf(x)\,dx = c\int f(x)\,dx \quad (c \text{ is a constant}) \tag{15}$$

and

$$\int [f(x) \pm g(x)]\,dx = \int f(x)\,dx \pm \int g(x)\,dx. \tag{16}$$

We can summarize these two equations by saying that antidifferentiation is **linear.** In essence, then, we antidifferentiate a sum of functions by antidifferentiating each function individually. This is *termwise* (or *term-by-term*) antidifferentiation. Moreover, a constant coefficient in any such term is merely "carried through" the antidifferentiation.

EXAMPLE 3 Find

$$\int \left(x^3 + 3\sqrt{x} - \frac{4}{x^2}\right)dx.$$

Solution Just as in differentiation, we prepare for antidifferentiation by writing roots and reciprocals as powers with fractional or negative exponents. Thus

$$\begin{aligned}
\int \left(x^3 + 3\sqrt{x} - \frac{4}{x^2}\right)dx &= \int (x^3 + 3x^{1/2} - 4x^{-2})\,dx \\
&= \int x^3\,dx + 3\int x^{1/2}\,dx - 4\int x^{-2}\,dx && \text{[using Eqs. (15) and (16)]} \\
&= \frac{x^4}{4} + 3\cdot\frac{x^{3/2}}{\frac{3}{2}} - 4\cdot\frac{x^{-1}}{-1} + C && \text{[using Eq. (7)]} \\
&= \frac{1}{4}x^4 + 2x\sqrt{x} + \frac{4}{x} + C.
\end{aligned}$$

There's only one "$+C$" because the surefire check verifies that $\frac{1}{4}x^4 + 2x^{3/2} + 4x^{-1}$ is a particular antiderivative. Hence any other antiderivative differs from this one by only a (single) constant C. ◆

EXAMPLE 4

$$\begin{aligned}\int (2\cos 3t + 5\sin 4t + 3e^{7t})\,dt &\\ &= 2\int \cos 3t\,dt + 5\int \sin 4t\,dt + 3\int e^{7t}\,dt && \text{[using Eqs. (15) and (16)]}\\ &= 2\left(\tfrac{1}{3}\sin 3t\right) + 5\left(-\tfrac{1}{4}\cos 4t\right) + 3\left(\tfrac{1}{7}e^{7t}\right) + C && \text{[using Eqs. (8), (9), and (14)]}\\ &= \tfrac{2}{3}\sin 3t - \tfrac{5}{4}\cos 4t + \tfrac{3}{7}e^{7t} + C.\end{aligned}$$

◆

Equation (7) is the power rule "in reverse." The generalized power rule in reverse is

$$\int u^k\,du = \frac{u^{k+1}}{k+1} + C \qquad (\text{if } k \neq -1), \tag{17}$$

where

$$u = g(x) \quad \text{and} \quad du = g'(x)\,dx.$$

EXAMPLE 5 With $u = x + 5$ (so that $du = dx$), Eq. (17) yields

$$\begin{aligned}\int (x+5)^{10}\,dx &= \int u^{10}\,du\\ &= \tfrac{1}{11}u^{11} + C = \tfrac{1}{11}(x+5)^{11} + C.\end{aligned}$$

Note that, after substituting $u = x + 5$ and integrating with respect to u, our final step is to express the resulting antiderivative in terms of the original variable x. ◆

EXAMPLE 6 We want to find

$$\int \frac{20}{(4-5x)^3}\,dx.$$

We plan to use Eq. (17) with $u = 4 - 5x$. But we must get the differential $du = -5\,dx$ into the act. The "constant-multiplier rule" of Eq. (15) permits us to do this:

$$\begin{aligned}\int \frac{20}{(4-5x)^3}\,dx &= 20\int (4-5x)^{-3}\,dx\\ &= \frac{20}{-5}\int (4-5x)^{-3}(-5\,dx) \qquad (18)\\ &= -4\int u^{-3}\,du && (u = 4-5x,\ du = -5\,dx)\\ &= -4\cdot\frac{u^{-2}}{-2} + C && [\text{Eq. (7) with } k = -3].\end{aligned}$$

Thus

$$\int \frac{20}{(4-5x)^3}\,dx = \frac{2}{(4-5x)^2} + C.$$

The key step occurs in (18). There we, in effect, multiplied by the *constant* -5 inside the integral and compensated for that by dividing by -5 outside the integral. At the end it was necessary to replace u with $4 - 5x$ to express the antiderivative in terms of the original independent variable x. ◆

Very Simple Differential Equations

The technique of antidifferentiation can often be used to solve a differential equation of the special form

$$\frac{dy}{dx} = f(x) \tag{19}$$

in which the dependent variable y does not appear on the right-hand side. To *solve* the differential equation in (19) is simply to find a function $y(x)$ that satisfies Eq. (19)—a function whose derivative is the given function $f(x)$. Hence the **general solution** of Eq. (19) is the indefinite integral

$$y(x) = \int f(x)\,dx + C \tag{20}$$

of the function $f(x)$.

EXAMPLE 7 The general solution of the differential equation

$$\frac{dy}{dx} = 3x^2$$

is given by

$$y(x) = \int 3x^2\,dx = x^3 + C. \qquad \blacklozenge$$

A differential equation of the form in Eq. (19) may appear in conjunction with an **initial condition,** a condition of the form

$$y(x_0) = y_0. \tag{21}$$

This condition specifies the value $y = y_0$ that the solution function $y(x)$ must have at $x = x_0$. Once we have found the general solution in Eq. (20), we can determine the value of the constant C by substituting the information that $y = y_0$ when $x = x_0$. With this specific value of C, Eq. (20) then gives the **particular solution** of the differential equation in (19) that satisfies the initial condition in Eq. (21). The combination

$$\frac{dy}{dx} = f(x), \quad y(x_0) = y_0 \tag{22}$$

of a differential equation with an initial condition is called an **initial value problem.**

EXAMPLE 8 Solve the initial value problem

$$\frac{dy}{dx} = 2x + 3, \quad y(1) = 2. \tag{23}$$

Solution By Eq. (20) the *general solution* of the differential equation $dy/dx = 2x+3$ is given by

$$y(x) = \int (2x + 3)\,dx = x^2 + 3x + C.$$

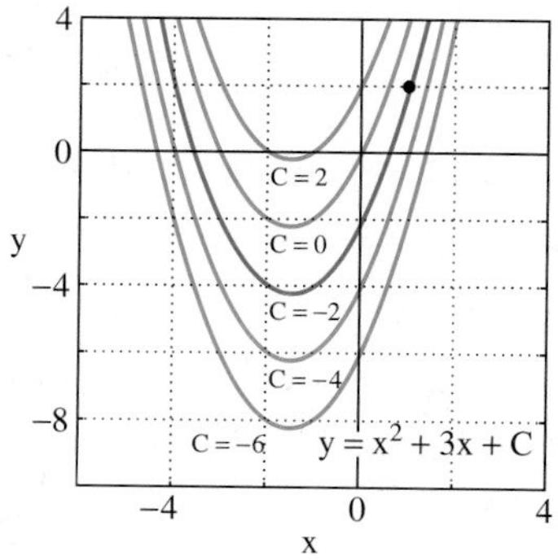

FIGURE 5.2.9 General solutions $y = x^2 + 3x + C$ of the differential equation in (22) (Example 8).

Figure 5.2.9 shows the graph $y = x^2 + 3x + C$ for various value of C. The particular solution we seek corresponds to the curve in Fig. 5.2.9 that passes through the point $(1, 2)$, thereby satisfying the initial condition

$$y(1) = (1)^2 + 3 \cdot (1) + C = 2.$$

It follows that $4 + C = 2$, and hence that $C = -2$. So the desired *particular solution* is given by

$$y(x) = x^2 + 3x - 2.$$ ◆

REMARK The method used in Example 8 may be described as "integrating both sides of a differential equation" with respect to x:

$$\int \left(\frac{dy}{dx}\right) dx = \int (2x + 3)\, dx;$$

$$y(x) = x^2 + 3x + C.$$

Rectilinear Motion

Antidifferentiation enables us, in many important cases, to analyze the motion of a particle (or "mass point") in terms of the forces acting on it. If the particle moves in *rectilinear motion* along a straight line—the x-axis, for instance—under the influence of a given (possibly variable) force, then (as in Section 3.1) the motion of the particle is described by its **position function**

$$x = x(t), \tag{24}$$

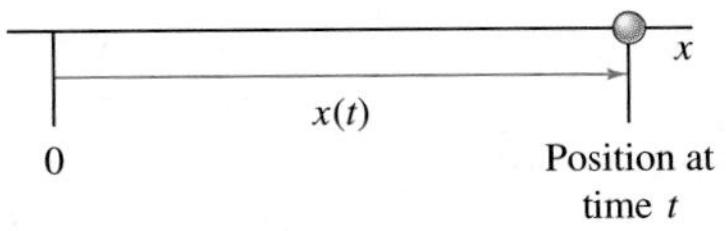

FIGURE 5.2.10 The position function $x(t)$ of a particle moving along the x-axis.

which gives its x-coordinate at time t (Fig. 5.2.10). The particle's **velocity** $v(t)$ is the time derivative of its position function,

$$v(t) = \frac{dx}{dt}, \tag{25}$$

and its **acceleration** $a(t)$ is the time derivative of its velocity:

$$a(t) = \frac{dv}{dt} = \frac{d^2x}{dt^2}. \tag{26}$$

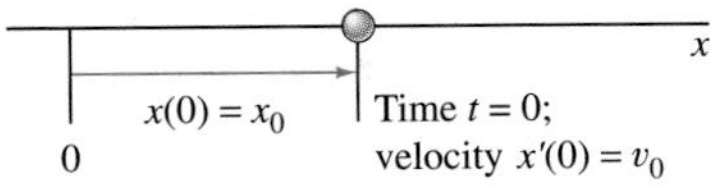

FIGURE 5.2.11 Initial data for linear motion.

In a typical situation, the following information is given (Fig. 5.2.11):

$$\begin{aligned} &a(t) && \text{the particle's acceleration;} \\ &x(0) = x_0 && \text{its } \textit{initial position;} \\ &v(0) = v_0 && \text{its } \textit{initial velocity.} \end{aligned} \tag{27}$$

In principle, we can then proceed as follows to find the particle's position function $x(t)$. First we solve the initial value problem

$$\frac{dv}{dt} = a(t), \quad v(0) = v_0 \tag{28}$$

for the velocity function $v(t)$. Knowing $v(t)$, we then solve the initial value problem

$$\frac{dx}{dt} = v(t), \quad x(0) = x_0 \tag{29}$$

for the particle's position function $x(t)$. Thus we determine $x(t)$ from the acceleration and initial data given in Eq. (28) by solving two successive initial value problems. For this purpose we can use the integral versions

$$v(t) = \int a(t)\, dt \tag{30}$$

and

$$x(t) = \int v(t)\,dt \tag{31}$$

of the derivative formulas in (25) and (26), remembering that each antidifferentiation involves an arbitrary constant.

EXAMPLE 9 A particle starts from rest (that is, with initial velocity zero) at the point $x = 10$ and moves along the x-axis with acceleration function $a(t) = 12t$. Find its resulting position function $x(t)$.

Solution First we must solve the initial value problem

$$\frac{dv}{dt} = a(t) = 12t, \quad v(0) = 0$$

to find the velocity function $v(t)$. Using Eq. (30) we get

$$v(t) = \int a(t)\,dt = \int 12t\,dt = 6t^2 + C_1.$$

(We write C_1 because we anticipate the appearance of a second constant when we integrate again to find $x(t)$.) Then substituting the initial data $t = 0$, $v = 0$ yields

$$0 = 6 \cdot 0^2 + C_1 = C_1,$$

so it follows that $v(t) = 6t^2$. Next we must solve the initial value problem

$$\frac{dx}{dt} = v(t) = 6t^2, \quad x(0) = 10$$

for $x(t)$. Using Eq. (31) we get

$$x(t) = \int v(t)\,dt = \int 6t^2\,dt = 2t^3 + C_2.$$

Then substituting the initial data $t = 0$, $x = 10$ yields

$$10 = 2 \cdot 0^3 + C_2 = C_2,$$

so it follows finally that the particle's position function is

$$x(t) = 2t^3 + 10.$$ ◆

Constant Acceleration

The solution of the initial value problems in Eqs. (28) and (29) is simplest when the given acceleration a is *constant.* We begin with

$$\frac{dv}{dt} = a \qquad (a \text{ is a constant})$$

and antidifferentiate:

$$v(t) = \int a\,dt.$$

So

$$v(t) = at + C_1. \tag{32}$$

To evaluate the constant C_1, we substitute the initial datum $v(0) = v_0$; this gives

$$v_0 = a \cdot 0 + C_1 = C_1.$$

Therefore, Eq. (32) becomes

$$v(t) = at + v_0. \tag{33}$$

Because $x'(t) = v(t)$, a second antidifferentiation yields

$$x(t) = \int v(t)\,dt$$
$$= \int (at + v_0)\,dt;$$
$$x(t) = \tfrac{1}{2}at^2 + v_0 t + C_2. \quad \textbf{(34)}$$

Now substituting the initial datum $x(0) = x_0$ gives

$$x_0 = \tfrac{1}{2}a \cdot (0)^2 + v_0 \cdot (0) + C_2 = C_2$$

in Eq. (34). Thus the position function of the particle is

$$x(t) = \tfrac{1}{2}at^2 + v_0 t + x_0. \quad \textbf{(35)}$$

WARNING Equations (33) and (35) are valid only in the case of *constant* acceleration a. They do *not* apply to problems in which the acceleration varies.

EXAMPLE 10 The skid marks made by an automobile indicate that its brakes were fully applied for a distance of 160 ft before it came to a stop. Suppose that the car in question has a constant deceleration of 20 ft/s^2 under the conditions of the skid. How fast was the car traveling when its brakes were first applied?

Solution The introduction of a convenient coordinate system is often crucial to the successful solution of a physical problem. Here we take the x-axis to be positively oriented in the direction of motion of the car. We choose the origin so that $x_0 = 0$ when $t = 0$, the time when the brakes were first applied (Fig. 5.2.12). In this coordinate system, the car's velocity $v(t)$ is a decreasing function of time t (in seconds), so its acceleration is $a = -20$ (ft/s^2) rather than $a = +20$. Hence we begin with the constant acceleration equation

$$\frac{dv}{dt} = -20.$$

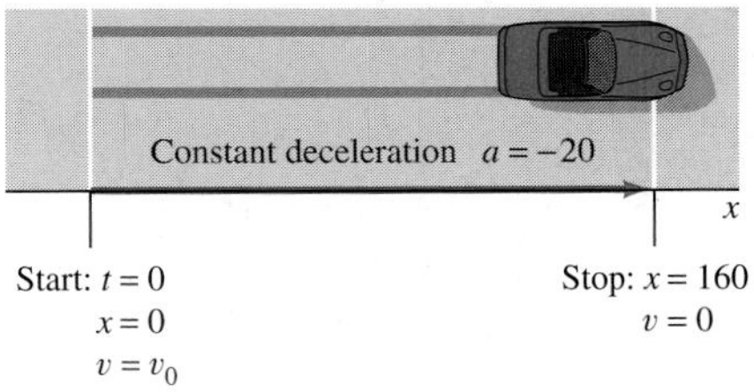

FIGURE 5.2.12 Skid marks 160 ft long (Example 10).

Antidifferentiation as in Eq. (30) gives

$$v(t) = \int (-20)\,dt = -20t + C_1.$$

Even though the initial velocity is unknown and not given, the initial data $t = 0$, $v = v_0$ still yield $C_1 = v_0$. So the car's velocity function is

$$v(t) = -20t + v_0. \quad \textbf{(36)}$$

A second antidifferentiation as in Eq. (31) gives

$$x(t) = \int (-20t + v_0)\,dt = -10t^2 + v_0 t + C_2.$$

Substituting the initial data $t = 0$, $x_0 = 0$ yields $C_2 = 0$, so the position function of the car is

$$x(t) = -10t^2 + v_0 t. \tag{37}$$

The fact that the skid marks are 160 ft long tells us that $x = 160$ when the car comes to a stop; that is,

$$x = 160 \quad \text{when} \quad v = 0.$$

Substituting these values into the velocity and position equations [Eqs. (36) and (37)] then yields the two simultaneous equations

$$-20t + v_0 = 0, \qquad -10t^2 + v_0 t = 160.$$

We now solve these for v_0 and t to find the initial velocity v_0 *and* the duration t of the car's skid. If we multiply the first equation by $-t$ and add the result to the second equation, we find that $10t^2 = 160$, so $t = 4$ when the car first comes to a stop. It follows that the velocity of the car was

$$v_0 = 20 \cdot 4 = 80 \quad \text{(ft/s)},$$

or about 55 mi/h, when the brakes were first applied. ◆

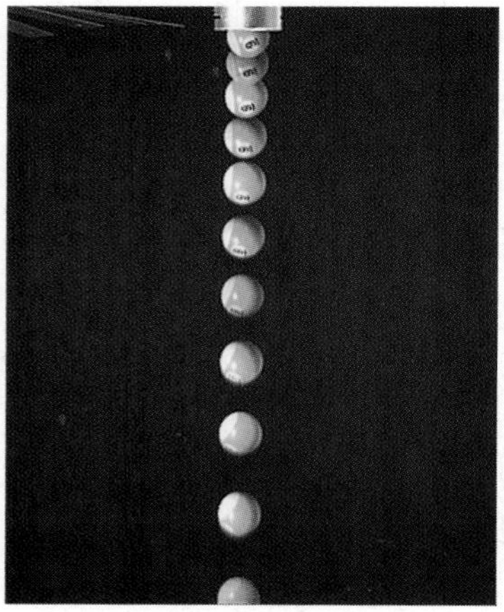

Stroboscopic photograph of a ball falling with constant acceleration due to gravity.

Vertical Motion with Constant Gravitational Acceleration

One common application of Eqs. (33) and (35) involves vertical motion near the surface of the earth. A particle in such motion is subject to a *downward* acceleration a, which is almost exactly constant if only small vertical distances are involved. The magnitude of this constant is denoted by g, approximately 32 ft/s^2 or 9.8 m/s^2. (If you need more accurate values for g, use 32.17 ft/s^2 in the fps system or 9.807 m/s^2 in the mks system.)

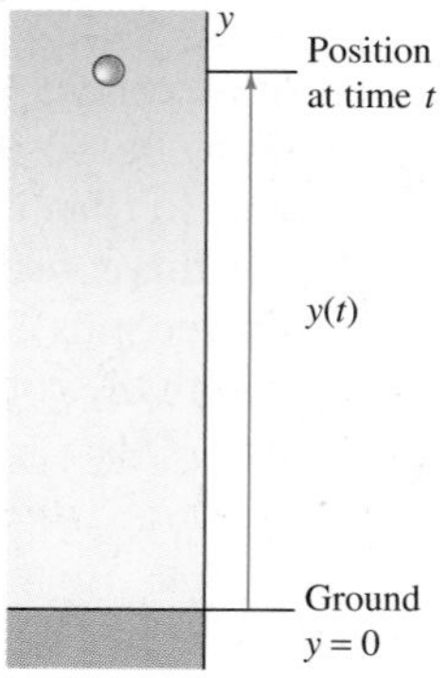

FIGURE 5.2.13 The position function $y(t)$ of a particle moving vertically.

If we neglect air resistance, we may assume that this acceleration due to gravity is the only outside influence on the moving particle. Because we deal with vertical motion here, it is natural to choose the y-axis as the coordinate system for the position of the particle and to place "ground level" where $y = 0$ (Fig. 5.2.13). If we choose the upward direction to be the positive direction, then the effect of gravity on the particle is to *decrease* its height and also to *decrease* its velocity $v = dy/dt$. Then the acceleration of the particle is

$$a = \frac{dv}{dt} = -g = -32 \quad (\text{ft/s}^2).$$

Equations (33) and (35) then become

$$v(t) = -32t + v_0 \tag{38}$$

and

$$y(t) = -16t^2 + v_0 t + y_0 \tag{39}$$

Here y_0 is the initial height of the particle in feet, v_0 is its initial velocity in feet per second, and time t is measured in seconds.

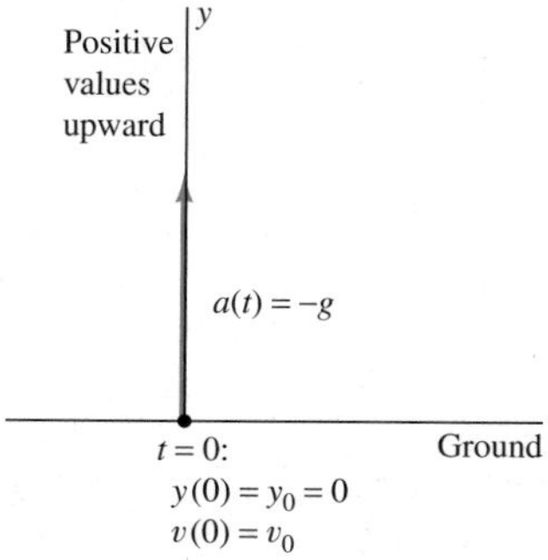

FIGURE 5.2.14 A bolt fired straight upward from a crossbow (Example 11).

EXAMPLE 11 Suppose that a bolt was fired vertically upward from a crossbow at ground level and that it struck the ground 20 s later. If air resistance may be neglected, find the initial velocity of the bolt and the maximum altitude that it reached.

Solution We set up the coordinate system illustrated in Fig. 5.2.14, with ground level corresponding to $y = 0$, with the bolt fired at time $t = 0$ (in seconds), and with the positive direction being the upward direction. Units on the y-axis are in feet.

We are given that $y = 0$ when $t = 20$. We lack any information about the initial velocity v_0. But we may use Eqs. (38) and (39) because we have set up a coordinate system in which the acceleration due to gravity acts in the negative direction. Thus

$$y(t) = -16t^2 + v_0t + y_0 = -16t^2 + v_0t$$

and

$$v(t) = -32t + v_0.$$

We use the information that $y = 0$ when $t = 20$ in the first equation:

$$0 = -16 \cdot 20^2 + 20v_0, \quad \text{and thus} \quad v_0 = 16 \cdot 20 = 320 \quad \text{(ft/s)}.$$

To find the maximum altitude of the bolt, we maximize $y(t)$ by finding the value of t for which its derivative is zero. In other words, the bolt reaches its maximum altitude when its velocity is zero:

$$\frac{dy}{dt} = -32t + v_0 = 0,$$

so at maximum altitude, $t = v_0/32 = 10$. At that time, the bolt has reached its maximum altitude of

$$y_{max} = y(10) = -16 \cdot 10^2 + 320 \cdot 10 = 1600 \quad \text{(ft)}.$$

The result seems contrary to experience. It may well suggest that air resistance cannot always be neglected, particularly not in problems involving long journeys at high velocity. ◆

5.2 TRUE/FALSE STUDY GUIDE

5.2 CONCEPTS: QUESTIONS AND DISCUSSION

List corresponding features of the graphs of a function f and its antiderivative F. Hence describe a strategy whereby—given a plot showing the graphs of f and F—you can determine which is which. Apply your strategy to the following graphs, in which h is either the derivative or the antiderivative of g.

1. Figure 5.2.15

2. Figure 5.2.16

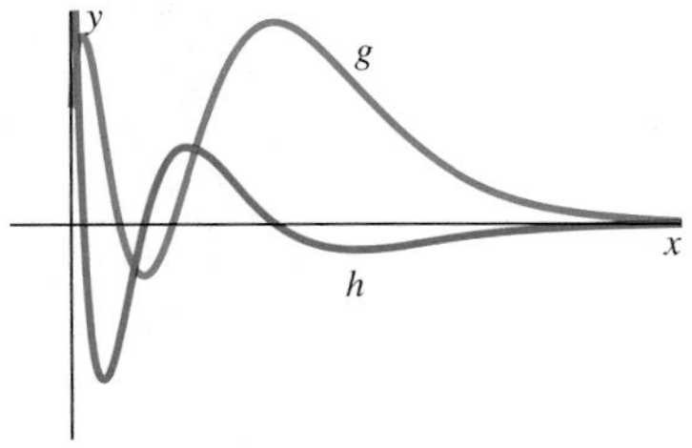

FIGURE 5.2.15

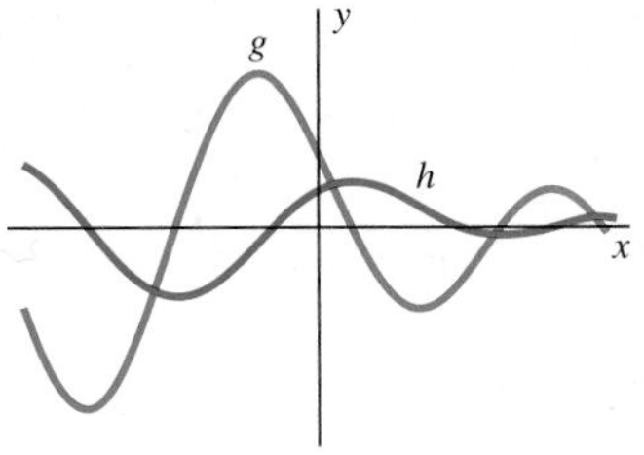

FIGURE 5.2.16

3. Figure 5.2.17

4. Figure 5.2.18

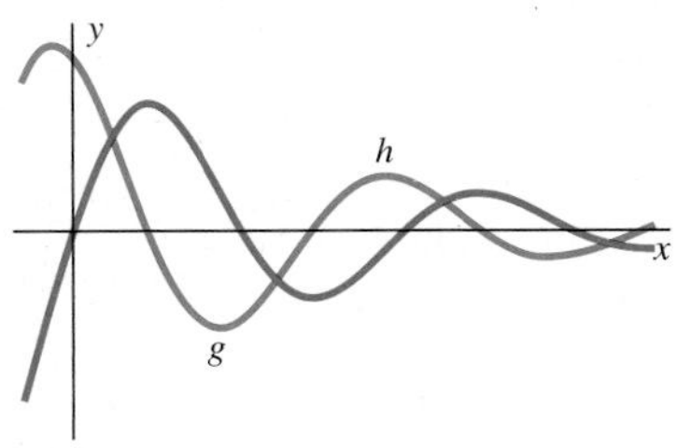

FIGURE 5.2.17

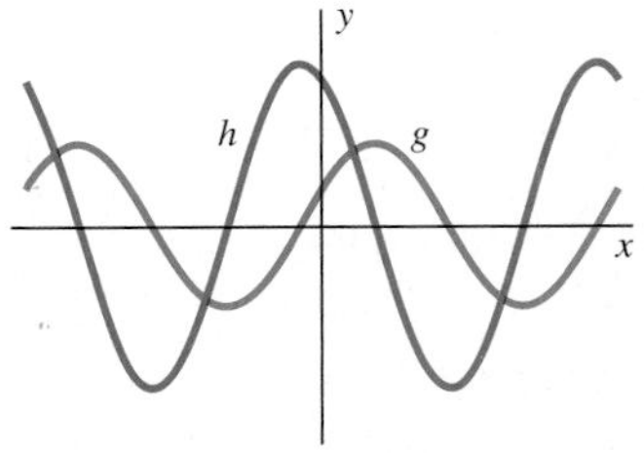

FIGURE 5.2.18

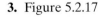

5.2 PROBLEMS

Evaluate the indefinite integrals in Problems 1 through 30.

1. $\displaystyle\int (3x^2 + 2x + 1)\,dx$

2. $\displaystyle\int (3t^4 + 5t - 6)\,dt$

3. $\displaystyle\int (1 - 2x^2 + 3x^3)\,dx$

4. $\displaystyle\int \left(-\frac{1}{t^2}\right)\,dt$

5. $\displaystyle\int \left(\frac{3}{x^3} + 2x^{3/2} - 1\right)\,dx$

6. $\displaystyle\int \left(x^{5/2} - \frac{5}{x^4} - \sqrt{x}\right)\,dx$

7. $\displaystyle\int \left(\tfrac{3}{2}t^{1/2} + 7\right)\,dt$

8. $\displaystyle\int \left(\frac{2}{x^{3/4}} - \frac{3}{x^{2/3}}\right)\,dx$

9. $\displaystyle\int \left(\sqrt[3]{x^2} + \frac{4}{\sqrt[4]{x^5}}\right)\,dx$

10. $\displaystyle\int \left(2x\sqrt{x} - \frac{1}{\sqrt{x}}\right)\,dx$

11. $\displaystyle\int (4x^3 - 4x + 6)\,dx$

12. $\displaystyle\int \left(\frac{1}{4}t^5 - \frac{5}{t^2}\right)\,dt$

13. $\displaystyle\int 7e^{x/7}\,dx$

14. $\displaystyle\int \frac{1}{7x}\,dx$

15. $\displaystyle\int (x+1)^4\,dx$

16. $\displaystyle\int (t+1)^{10}\,dt$

17. $\displaystyle\int \frac{1}{(x-10)^7}\,dx$

18. $\displaystyle\int \sqrt{z+1}\,dz$

19. $\displaystyle\int \sqrt{x}\,(1-x)^2\,dx$

20. $\displaystyle\int \sqrt[3]{x}\,(x+1)^3\,dx$

21. $\displaystyle\int \frac{2x^4 - 3x^3 + 5}{7x^2}\,dx$

22. $\displaystyle\int \frac{(3x+4)^2}{\sqrt{x}}\,dx$

23. $\displaystyle\int (9t+11)^5\,dt$

24. $\displaystyle\int \frac{1}{(3z+10)^7}\,dz$

25. $\displaystyle\int \left(e^{2x} + e^{-2x}\right)\,dx$

26. $\displaystyle\int \left(e^{x} + e^{-x}\right)^2\,dx$

27. $\displaystyle\int (5\cos 10x - 10\sin 5x)\,dx$

28. $\displaystyle\int (2\cos \pi x + 3\sin \pi x)\,dx$

29. $\displaystyle\int (3\cos \pi t + \cos 3\pi t)\,dt$

30. $\displaystyle\int (4\sin 2\pi t - 2\sin 4\pi t)\,dt$

31. Verify by differentiation that the integral formulas

$$\int \sin x\,\cos x\,dx = \tfrac{1}{2}\sin^2 x + C_1$$

and

$$\int \sin x\,\cos x\,dx = -\tfrac{1}{2}\cos^2 x + C_2$$

are both valid. Reconcile these seemingly different results. What is the relation between the constants C_1 and C_2?

32. Show that the obviously different functions

$$F_1(x) = \frac{1}{1-x} \quad \text{and} \quad F_2(x) = \frac{x}{1-x}$$

are both antiderivatives of $f(x) = 1/(1-x)^2$. What is the relation between $F_1(x)$ and $F_2(x)$?

33. Use the identities

$$\sin^2 x = \frac{1-\cos 2x}{2} \quad \text{and} \quad \cos^2 x = \frac{1+\cos 2x}{2}$$

to find the antiderivatives

$$\int \sin^2 x\,dx \quad \text{and} \quad \int \cos^2 x\,dx.$$

34. (a) First explain why $\int \sec^2 x\,dx = \tan x + C$. (b) Then use the identity $1 + \tan^2 x = \sec^2 x$ to find the antiderivative

$$\int \tan^2 x\,dx.$$

Solve the initial value problems in Problems 35 through 46.

35. $\dfrac{dy}{dx} = 2x + 1;\ y(0) = 3$

36. $\dfrac{dy}{dx} = (x-2)^3;\ y(2) = 1$

37. $\dfrac{dy}{dx} = \sqrt{x};\ y(4) = 0$

38. $\dfrac{dy}{dx} = \dfrac{1}{x^2};\ y(1) = 5$

39. $\dfrac{dy}{dx} = \dfrac{1}{\sqrt{x+2}};\ y(2) = -1$

40. $\dfrac{dy}{dx} = \sqrt{x+9};\ y(-4) = 0$

41. $\dfrac{dy}{dx} = 3x^3 + \dfrac{2}{x^2};\ y(1) = 1$

42. $\dfrac{dy}{dx} = x^4 - 3x + \dfrac{3}{x^3};\ y(1) = -1$

43. $\dfrac{dy}{dx} = (x-1)^3;\ y(0) = 2$

44. $\dfrac{dy}{dx} = \sqrt{x+5};\ y(4) = -3$

45. $\dfrac{dy}{dx} = 6e^{2x};\ y(0) = 10$

46. $\dfrac{dy}{dx} = \dfrac{3}{x};\ y(1) = 7$

In Problems 47 through 52, a particle moves along the x-axis with the given acceleration function $a(t)$, initial position $x(0)$, and initial velocity $v(0)$. Find the particle's position function $x(t)$.

47. $a(t) = 12t - 4;\ x(0) = 0,\ v(0) = -10$

48. $a(t) = 10 - 30t;\ x(0) = 5,\ v(0) = -5$

49. $a(t) = 2t^2;\ x(0) = -7,\ v(0) = 3$

50. $a(t) = 15\sqrt{t};\ x(0) = 5,\ v(0) = 7$

51. $a(t) = \sin t;\ x(0) = 0,\ v(0) = 0$

52. $a(t) = 8\cos 2t;\ x(0) = -2,\ v(0) = 4$

In Problems 53 through 56, a particle starts at the origin and travels along the x-axis with the velocity function $v(t)$ whose graph *is shown in Figs. 5.2.19 through 5.2.22. Sketch the graph of the resulting position function $x(t)$ for $0 \leqq t \leqq 10$.*

53. Figure 5.2.19

54. Figure 5.2.20

55. Figure 5.2.21

56. Figure 5.2.22

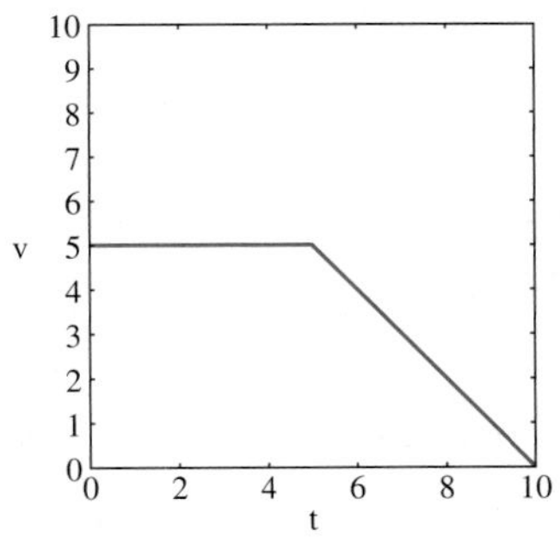

FIGURE 5.2.19 Graph of the velocity function $v(t)$ of Problem 53.

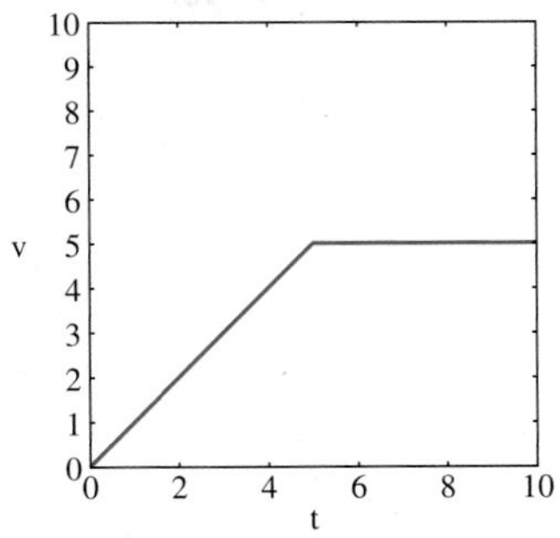

FIGURE 5.2.20 Graph of the velocity function $v(t)$ of Problem 54.

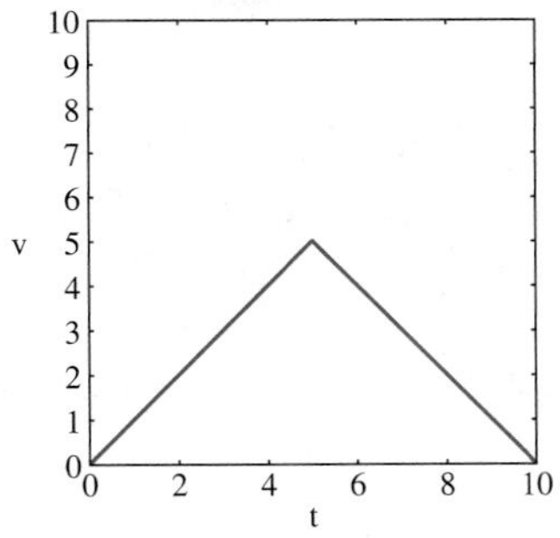

FIGURE 5.2.21 Graph of the velocity function $v(t)$ of Problem 55.

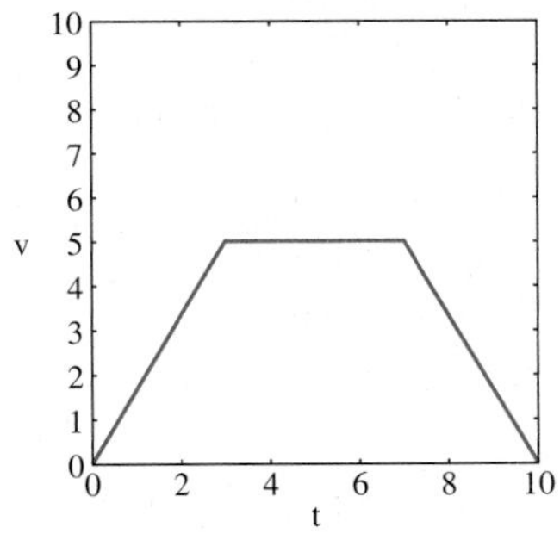

FIGURE 5.2.22 Graph of the velocity function $v(t)$ of Problem 56.

Problems 57 through 73 deal with vertical motion near the surface of the earth (with air resistance considered negligible). Use $g = 32$ ft/s^2 for the magnitude of the gravitational acceleration.

57. You throw a ball straight upward from the ground with initial velocity 96 ft/s. How high does the ball rise, and how long does it remain aloft?

58. When Alex shot a marble straight upward from ground level with his slingshot, it reached a maximum height of 400 ft. What was the marble's initial velocity?

59. Laura drops a stone into a well; it hits bottom 3 s later. How deep is the well?

60. Fran throws a rock straight upward alongside a tree (Fig. 5.2.23). The rock rises until it is even with the top of the tree and then falls back to the ground; it remains aloft for 4 s. How tall is the tree?

FIGURE 5.2.23 The tree of Problem 60.

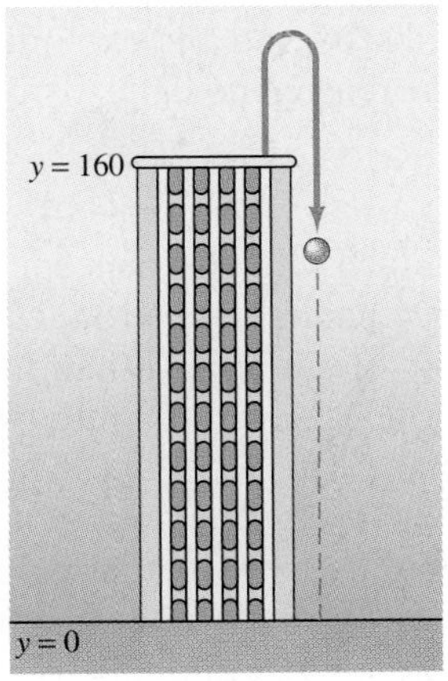

FIGURE 5.2.24 The building of Problem 61.

61. Mickey throws a ball upward with an initial velocity of 48 ft/s from the top of a building 160 ft high. The ball soon falls to the ground at the base of the building (Fig. 5.2.24). How long does the ball remain aloft, and with what speed does it strike the ground?

62. A ball is dropped from the top of a building 576 ft high. With what velocity should a second ball be thrown straight downward 3 s later so that the two balls hit the ground simultaneously?

63. A ball is dropped from the top of the Empire State Building, 960 ft above 34th Street. How long does it take for the ball to reach the street, and with what velocity does it strike the street?

64. Lynda shoots an arrow straight upward from the ground with initial velocity 320 ft/s. (a) How high is the arrow after exactly 3 s have elapsed? (b) At what time is the arrow exactly 1200 ft above the ground? (c) How many seconds after its release does the arrow strike the ground?

65. Bill throws a stone upward from the ground. The stone reaches a maximum height of 225 ft. What was its initial velocity?

66. Sydney drops a rock into a well in which the water surface is 98 m below the ground. How long does it take the rock to reach the water surface? How fast is the rock moving as it penetrates the water surface?

67. Gloria drops a tennis ball from the top of a building 400 ft high. How long does it take the ball to reach the ground? With what velocity does it strike the ground?

68. Kosmo throws a baseball straight downward from the top of a tall building. The initial speed of the ball is 25 ft/s. It hits the ground with a speed of 153 ft/s. How tall is the building?

69. A ball is thrown straight upward from ground level with an initial speed of 160 ft/s. What is the maximum height that the ball attains?

70. Carolyn drops a sandbag from the top of a tall building h feet high. At the same time Jon throws a ball upward from ground level from a point directly below the sandbag. With what (initial) velocity should the ball be thrown so that it meets the sandbag at the halfway point, where both have altitude $h/2$?

71. Kelly throws a baseball straight downward with an initial speed of 40 ft/s from the top of the Washington Monument, 555 ft high. How long does it take the baseball to reach the ground, and with what speed does it strike the ground?

72. A rock is dropped from an initial height of h feet above the surface of the earth. Show that the speed with which the rock strikes the surface is $\sqrt{2gh}$.

73. A bomb is dropped from a balloon hovering at an altitude of 800 ft. From directly below the balloon, a projectile is fired straight upward toward the bomb exactly 2 s after the bomb is released. With what initial speed should the projectile be fired in order to hit the bomb at an altitude of exactly 400 ft?

74. A car's brakes are applied when the car is moving at 60 mi/h (exactly 88 ft/s). The brakes provide a constant deceleration of 40 ft/s^2. How far does the car travel before coming to a stop?

75. A car traveling at 60 mi/h (exactly 88 ft/s) skids 176 ft after its brakes are applied. The deceleration provided by the braking system is constant. What is its value?

76. A spacecraft is in free fall toward the surface of the moon at a speed of 1000 mi/h. Its retrorockets, when fired, provide a deceleration of 20000 mi/h^2. At what height above the surface should the astronauts activate the retrorockets to ensure a "soft touchdown" ($v = 0$ at impact)? (See Fig. 5.2.25.) Ignore the effect of the moon's gravitational field.

77. (a) What initial velocity v_0 must you use to throw a ball to a maximum height of 144 ft? (b) Now suppose that you throw a ball upward with the same initial velocity v_0 on the moon, where the surface gravitational acceleration is only 5.2 ft/s^2. How high will it go, and how long will it remain aloft?

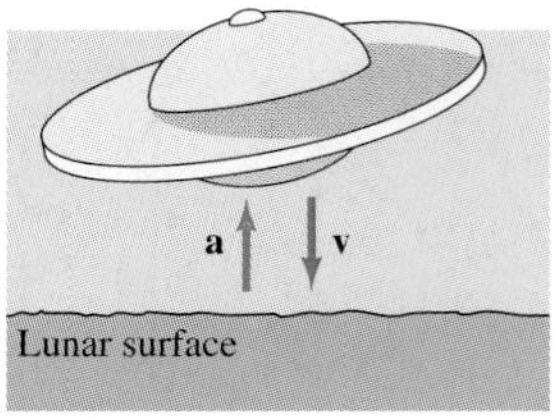

FIGURE 5.2.25 The spacecraft of Problem 76.

78. Arthur C. Clarke's *The Wind from the Sun* (1963) describes *Diana,* a spacecraft propelled by the solar wind. Its 2-mi^2 aluminized sail provides it with an acceleration of $(0.001)g = 0.032$ ft/s^2. If the *Diana* starts from rest and travels in a straight line, calculate its distance x traveled (in miles) and its velocity v (in mi/h) after 1 min, 1 h, and 1 day.

79. A driver involved in an accident claims he was going only 25 mi/h. When police tested his car, they found that when the brakes were applied at 25 mi/h, the car skidded only 45 ft before coming to a stop. The driver's skid marks at the accident scene measured 210 ft. Assuming the same (constant) deceleration, calculate the speed at which he was traveling prior to the accident.

5.3 ELEMENTARY AREA COMPUTATIONS

The indefinite integrals of Section 5.2 stem from the concept of antidifferentiation. The most fundamental type of integral is the one mentioned in Section 5.1, associated with the concept of area. It is called the *definite integral,* or simply *the* integral. Surprisingly, the quite different concepts of area and antidifferentiation have a close and deep relationship. This fact, discovered and exploited by Newton and Leibniz late in the seventeenth century, is the reason the same word, *integral,* is used in both contexts.

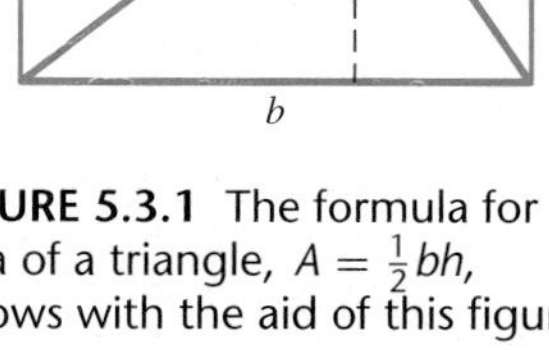

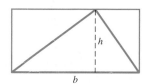

FIGURE 5.3.1 The formula for the area of a triangle, $A = \frac{1}{2}bh$, follows with the aid of this figure.

The Concept of Area

Perhaps everyone's first contact with the concept of area is the formula $A = bh$, which gives the area A of a rectangle as the product of its base length b and its height h. We next learn that the area of a triangle is half the product of its base and height. This follows because any triangle can be split into two right triangles, and every right triangle is exactly half a rectangle (Fig. 5.3.1).

Given the formula $A = \frac{1}{2}bh$ for the area of a triangle, we can—in principle—find the area of any polygonal figure (a plane region bounded by a closed "curve" consisting of a finite number of straight line segments). The reason is that any polygonal figure can be divided into nonoverlapping triangles (Fig. 5.3.2), and the area of the polygonal figure is then the sum of the areas of these triangles. This approach to area dates back several thousand years to the ancient civilizations of Egypt and Babylonia.

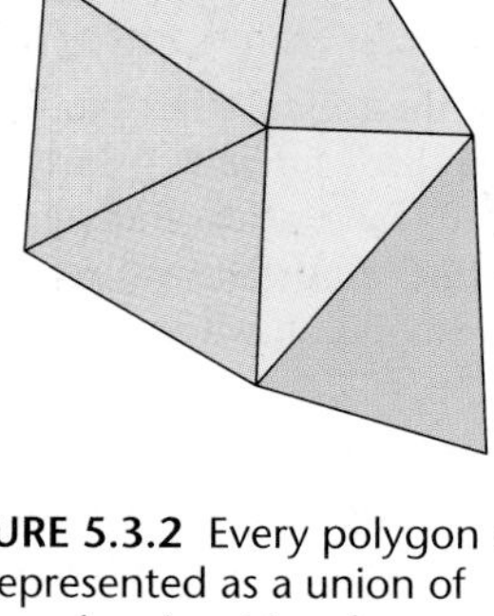

FIGURE 5.3.2 Every polygon can be represented as a union of nonoverlapping triangles.

The ancient Greeks began the investigation of areas of *curvilinear* figures in the fourth and fifth centuries B.C. Given a plane region R whose area they sought, they worked both with a polygonal P *inscribed* in R (Fig. 5.3.3) and with a polygonal Q *circumscribed* about R (Fig. 5.3.4). If the polygons P and Q have sufficiently many sides, all short, then it would appear that their areas $a(P)$ and $a(Q)$ closely approximate the area of the region R. Moreover, error control is possible: We see

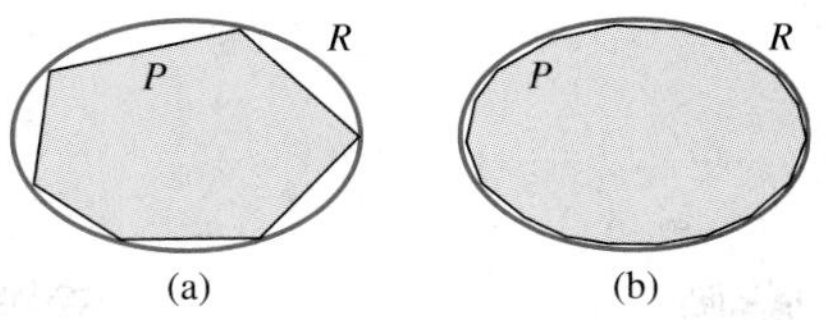

FIGURE 5.3.3 (a) A six-sided polygon P inscribed in R; (b) a many-sided inscribed polygon P more closely approximating the area of R.

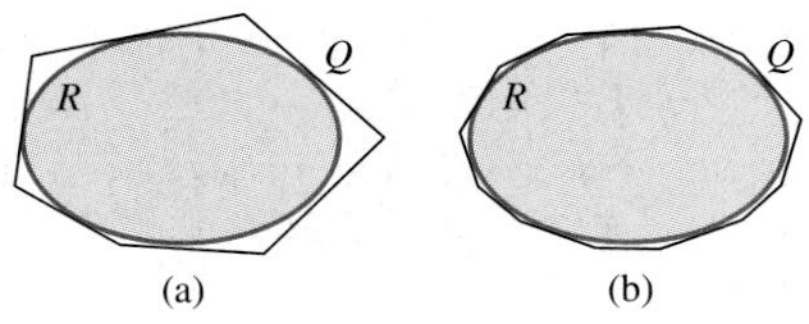

FIGURE 5.3.4 (a) A six-sided polygon Q circumscribed around R; (b) a many-sided circumscribed polygon Q more closely approximating the area of R.

that

$$a(P) < a(R) < a(Q) \tag{1}$$

because R contains the polygon P but is contained in the polygon Q.

The inequalities in (1) bracket the desired area $a(R)$. Suppose, for instance, that calculations based on triangular dissections (as in Fig. 5.3.2) yield $a(P) = 7.341$ and $a(Q) = 7.343$. Then the resulting inequality,

$$7.341 < a(R) < 7.343,$$

implies that $a(R) \approx 7.34$, accurate to two decimal places.

Our primary objective here is to describe a systematic technique by which to approximate the area of an appropriate curvilinear region using easily calculated polygonal areas.

Areas Under Graphs

We consider the type of region that is determined by a continuous positive-valued function f defined on a closed interval $[a, b]$. Suppose that we want to calculate the area A of the region R that lies *below* the curve $y = f(x)$ and *above* the interval $[a, b]$ on the x-axis (Fig. 5.3.5). The region R is bounded on the left by the vertical line $x = a$ and on the right by the vertical line $x = b$.

We divide the base interval $[a, b]$ into subintervals, all with the same length. Above each subinterval lies a vertical strip (Fig. 5.3.6), and the area of A is the sum of the areas of these strips.

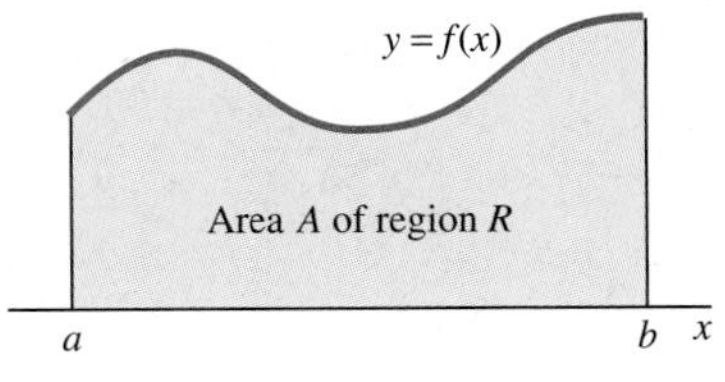

FIGURE 5.3.5 The area under the graph of $y = f(x)$ from $x = a$ to $x = b$.

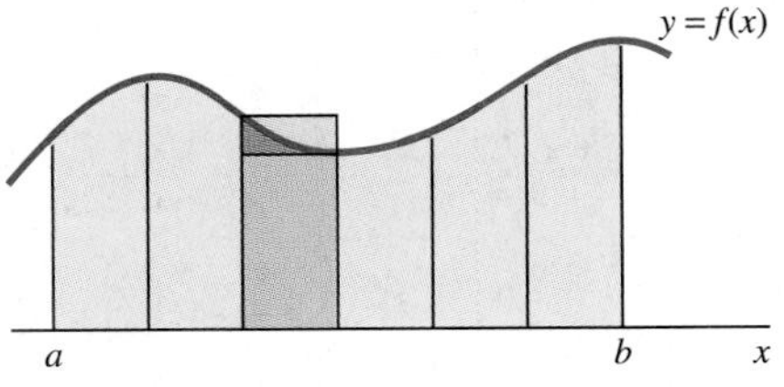

FIGURE 5.3.6 Vertical strips determined by a division of $[a, b]$ into equal-length subintervals.

On each of these base subintervals, we erect a rectangle that approximates the corresponding vertical strip. We may choose either an "inscribed" or a "circumscribed" rectangle (both possibilities are illustrated in Fig. 5.3.6), or even a rectangle that is intermediate between the two. These rectangles then make up a polygon that approximates the region R, and therefore the sum of the areas of these rectangles *approximates* the desired area A.

For example, suppose that we want to approximate the area A of the region R that lies below the parabola $y = x^2$ above the interval $[0, 3]$ on the x-axis. The computer plots in Fig. 5.3.7 show successively

- 5 inscribed and 5 circumscribed rectangles;
- 10 inscribed and 10 circumscribed rectangles;
- 20 inscribed and 20 circumscribed rectangles;
- 40 inscribed and 40 circumscribed rectangles.

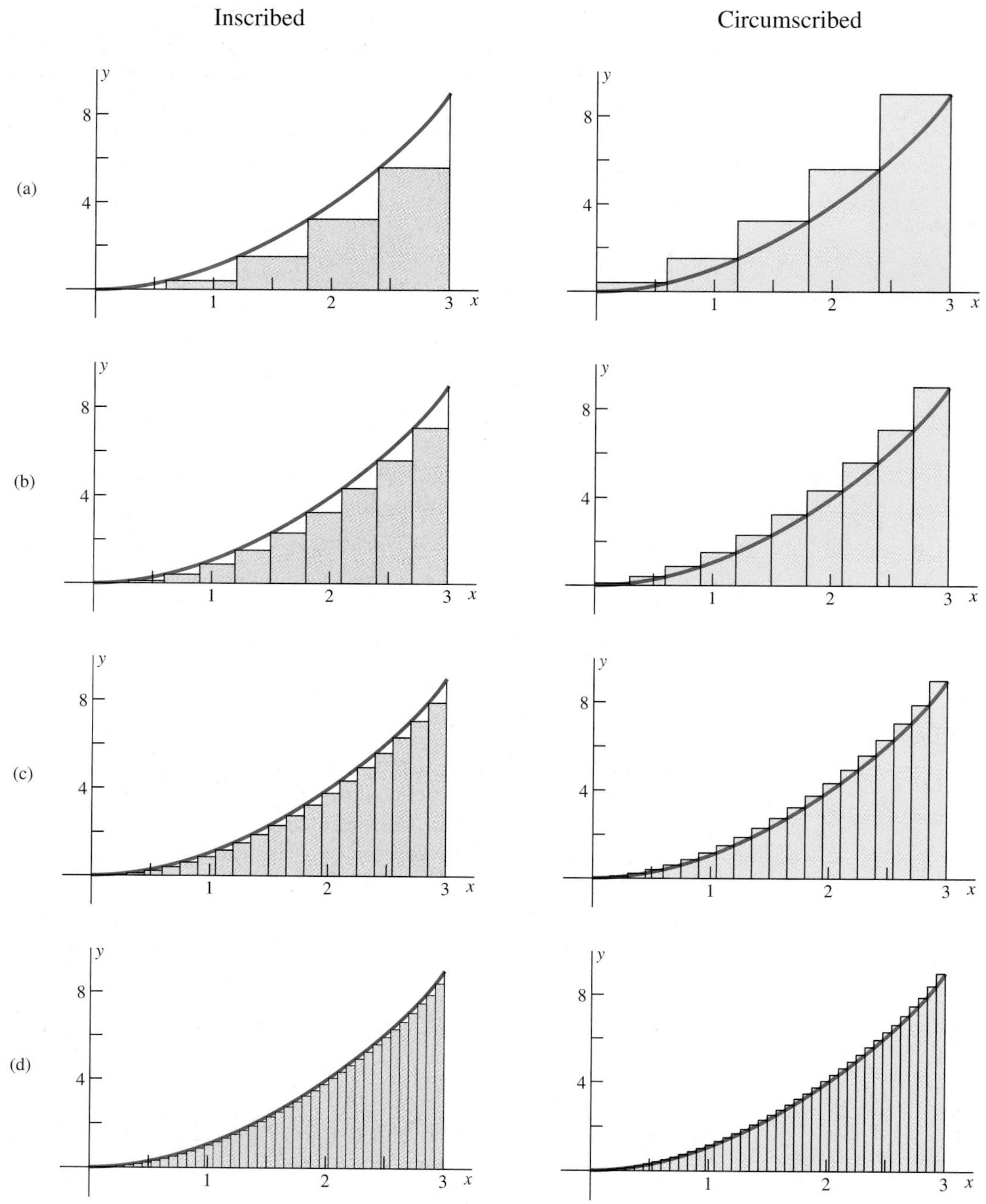

FIGURE 5.3.7 (a) Five inscribed and circumscribed polygons; (b) ten inscribed and circumscribed polygons; (c) twenty inscribed and circumscribed polygons; (d) forty inscribed and circumscribed polygons.

Each collection of inscribed rectangles gives an *underestimate* of A, whereas each collection of circumscribed rectangles gives an *overestimate* of A. The "curvilinear triangles" (by which the rectangular polygons in Fig. 5.3.7 undershoot or overshoot the region R) constitute the *errors* in these estimates. The more rectangles we use, the more accurate the approximation. Thus, to approximate accurately the area of such a region R, we need an effective way to calculate and sum the areas of collections of rectangles like those in Fig. 5.3.7.

EXAMPLE 1 As in Fig. 5.3.7, let R denote the region that lies below the graph of $f(x) = x^2$ and above the interval $[0, 3]$. Calculate the underestimate and the overestimate of the area A of R by using 5 rectangles each of width $\frac{3}{5}$. Then repeat the computations using 10 rectangles each of width $\frac{3}{10}$.

Solution First suppose that $n = 5$ rectangles are used. Let $\underline{A}_5$ denote the underestimate and $\overline{A}_5$ the overestimate obtained by using 5 rectangles based on the 5 subintervals of length $\frac{3}{5}$ (Fig. 5.3.8). From Fig. 5.3.7(a) we see that the heights of the 5 inscribed rectangles (the first of which is degenerate—its height is zero) are the values of the function $f(x) = x^2$ at the 5 *left-hand* endpoints $0, \frac{3}{5}, \frac{6}{5}, \frac{9}{5}$, and $\frac{12}{5}$. Because the base of each rectangle has length $\frac{3}{5}$, we see also that

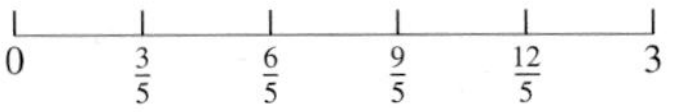

FIGURE 5.3.8 Five subintervals, each of length $\frac{3}{5}$ (Example 1).

$$\begin{aligned}\underline{A}_5 &= \tfrac{3}{5} \cdot \left[(0)^2 + \left(\tfrac{3}{5}\right)^2 + \left(\tfrac{6}{5}\right)^2 + \left(\tfrac{9}{5}\right)^2 + \left(\tfrac{12}{5}\right)^2\right] \\ &= \tfrac{3}{5} \cdot \left(0 + \tfrac{9}{25} + \tfrac{36}{25} + \tfrac{81}{25} + \tfrac{144}{25}\right) = 6.48.\end{aligned}$$

The heights of the 5 circumscribed rectangles are the values of $f(x) = x^2$ at the 5 *right-hand* endpoints $\frac{3}{5}, \frac{6}{5}, \frac{9}{5}, \frac{12}{5}$, and 3, so the corresponding overestimate is

$$\begin{aligned}\overline{A}_5 &= \tfrac{3}{5} \cdot \left[\left(\tfrac{3}{5}\right)^2 + \left(\tfrac{6}{5}\right)^2 + \left(\tfrac{9}{5}\right)^2 + \left(\tfrac{12}{5}\right)^2 + \left(\tfrac{15}{5}\right)^2\right] \\ &= \tfrac{3}{5} \cdot \left(\tfrac{9}{25} + \tfrac{36}{25} + \tfrac{81}{25} + \tfrac{144}{25} + \tfrac{225}{25}\right) = 11.88.\end{aligned}$$

These are crude approximations to the actual area A. On the basis of this information alone, our best estimate of A might well be the average of the under- and overestimates:

$$\frac{\underline{A}_5 + \overline{A}_5}{2} = \frac{6.48 + 11.88}{2} = 9.18.$$

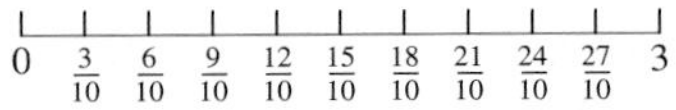

FIGURE 5.3.9 Ten subintervals, each of length $\frac{3}{10}$ (Example 1).

Let us see if doubling the number of subintervals to $n = 10$ increases the accuracy significantly. Looking at Fig. 5.3.7(b), we see that the heights of the 10 inscribed rectangles are the values of $f(x) = x^2$ at the 10 *left-hand* endpoints $0, \frac{3}{10}, \frac{6}{10}, \frac{9}{10}, \frac{12}{10}, \frac{15}{10}, \frac{18}{10}, \frac{21}{10}, \frac{24}{10}$, and $\frac{27}{10}$ of the subintervals in Fig. 5.3.9. The base of each rectangle has length $\frac{3}{10}$, so the resulting underestimate is

$$\begin{aligned}\underline{A}_{10} &= \tfrac{3}{10} \cdot \left[(0)^2 + \left(\tfrac{3}{10}\right)^2 + \left(\tfrac{6}{10}\right)^2 + \left(\tfrac{9}{10}\right)^2 + \left(\tfrac{12}{10}\right)^2 \right. \\ &\qquad \left. + \left(\tfrac{15}{10}\right)^2 + \left(\tfrac{18}{10}\right)^2 + \left(\tfrac{21}{10}\right)^2 + \left(\tfrac{24}{10}\right)^2 + \left(\tfrac{27}{10}\right)^2\right] \\ &= \tfrac{3}{10} \cdot \left(0 + \tfrac{9}{100} + \tfrac{36}{100} + \tfrac{81}{100} + \tfrac{144}{100} + \tfrac{225}{100} + \tfrac{324}{100} + \tfrac{441}{100} + \tfrac{576}{100} + \tfrac{729}{100}\right) \\ &= \tfrac{7695}{1000} = 7.695.\end{aligned}$$

Similarly, the sum of the areas of the 10 circumscribed rectangles in Fig. 5.3.7(b) is the overestimate

$$\begin{aligned}\overline{A}_{10} &= \tfrac{3}{10} \cdot \left[\left(\tfrac{3}{10}\right)^2 + \left(\tfrac{6}{10}\right)^2 + \left(\tfrac{9}{10}\right)^2 + \left(\tfrac{12}{10}\right)^2 + \left(\tfrac{15}{10}\right)^2 \right. \\ &\qquad \left. + \left(\tfrac{18}{10}\right)^2 + \left(\tfrac{21}{10}\right)^2 + \left(\tfrac{24}{10}\right)^2 + \left(\tfrac{27}{10}\right)^2 + \left(\tfrac{30}{10}\right)^2\right] \\ &= \tfrac{10395}{1000} = 10.395.\end{aligned}$$

At this point, our best estimate of the actual area A might be the average

$$\frac{\underline{A}_{10} + \overline{A}_{10}}{2} = \frac{7.695 + 10.395}{2} = 9.045. \quad \blacklozenge$$

We used a computer to calculate more refined underestimates and overestimates of the area A under the graph $y = x^2$ over $[0, 3]$, with 20, 40, 80, 160, and, finally, 320 rectangles. The results (rounded to four decimal places) are shown in Fig. 5.3.10. The average values in the final column of the table suggest that $A \approx 9$.

Number of rectangles	Underestimate	Overestimate	Average
5	6.4800	11.8800	9.1800
10	7.6950	10.3950	9.0450
20	8.3363	9.6863	9.0113
40	8.6653	9.3403	9.0028
80	8.8320	9.1695	9.0007
160	8.9158	9.0846	9.0002
320	8.9579	9.0422	9.0000

FIGURE 5.3.10 Estimate of the area under $y = x^2$ over [0,3].

Summation Notation

For more convenient computation of area estimates, as in Example 1, we need a concise notation for sums of many numbers. The symbol $\sum_{i=1}^{n} a_i$ is used to abbreviate the sum of the n numbers $a_1, a_2, a_3, \dots, a_n$:

$$\sum_{i=1}^{n} a_i = a_1 + a_2 + a_3 + \cdots + a_n. \tag{2}$$

The symbol on the left here—$\sum$ is the capital Greek letter sigma (for S, for "sum")—specifies the sum of the **terms** a_i as the **summation index** i takes on the successive integer values from 1 to n. For instance, the sum of the squares of the first 10 positive integers is

$$\begin{aligned}\sum_{i=1}^{10} i^2 &= 1^2 + 2^2 + 3^2 + 4^2 + 5^2 + 6^2 + 7^2 + 8^2 + 9^2 + 10^2 \\ &= 1 + 4 + 9 + 16 + 25 + 36 + 49 + 64 + 81 + 100 = 385.\end{aligned}$$

The particular symbol used for the summation index is immaterial:

$$\sum_{i=1}^{10} i^2 = \sum_{k=1}^{10} k^2 = \sum_{r=1}^{10} r^2 = 385.$$

EXAMPLE 2

$$\sum_{k=1}^{7} (k+1) = 2 + 3 + 4 + 5 + 6 + 7 + 8 = 35,$$

$$\sum_{n=1}^{6} 2^n = 2 + 4 + 8 + 16 + 32 + 64 = 126,$$

and

$$\sum_{j=1}^{5} \frac{(-1)^{j+1}}{j^2} = 1 - \tfrac{1}{4} + \tfrac{1}{9} - \tfrac{1}{16} + \tfrac{1}{25} = \tfrac{3019}{3600} \approx 0.8386.$$

◆

The simple rules of summation

$$\sum_{i=1}^{n} ca_i = c \sum_{i=1}^{n} a_i \tag{3}$$

and

$$\sum_{i=1}^{n} (a_i + b_i) = \left(\sum_{i=1}^{n} a_i \right) + \left(\sum_{i=1}^{n} b_i \right) \tag{4}$$

are easy to verify by writing out each sum in full.

Note that if $a_i = a$ (a constant) for $i = 1, 2, \ldots, n$, then Eq. (4) yields

$$\sum_{i=1}^{n}(a + b_i) = \sum_{i=1}^{n} a + \sum_{i=1}^{n} b_i = \Big(\underbrace{a + a + \cdots + a}_{n \text{ terms}}\Big) + \sum_{i=1}^{n} b_i,$$

and hence

$$\sum_{i=1}^{n}(a + b_i) = na + \sum_{i=1}^{n} b_i. \tag{5}$$

In particular,

$$\sum_{i=1}^{n} 1 = n. \tag{6}$$

The sum of the kth powers of the first n positive integers,

$$\sum_{i=1}^{n} i^k = 1^k + 2^k + 3^k + \cdots + n^k,$$

commonly occurs in area computations. The values of this sum for $k = 1, 2,$ and 3 are given by the following formulas (see Problems 43 and 44):

$$\sum_{i=1}^{n} i = \frac{n(n+1)}{2} = \tfrac{1}{2}n^2 + \tfrac{1}{2}n, \tag{7}$$

$$\sum_{i=1}^{n} i^2 = \frac{n(n+1)(2n+1)}{6} = \tfrac{1}{3}n^3 + \tfrac{1}{2}n^2 + \tfrac{1}{6}n, \tag{8}$$

$$\sum_{i=1}^{n} i^3 = \frac{n^2(n+1)^2}{4} = \tfrac{1}{4}n^4 + \tfrac{1}{2}n^3 + \tfrac{1}{4}n^2. \tag{9}$$

EXAMPLE 3 The sum of the first 10 positive integers is given by Eq. (7) with $n = 10$:

$$1 + 2 + 3 + \cdots + 10 = \sum_{i=1}^{10} i = \frac{10 \cdot 11}{2} = 55.$$

The sum of their squares and cubes are given by Eqs. (8) and (9):

$$1^2 + 2^2 + 3^2 + \cdots + 10^2 = \sum_{i=1}^{10} i^2 = \frac{10 \cdot 11 \cdot 21}{6} = 385$$

and

$$1^3 + 2^3 + 3^3 + \cdots + 10^3 = \sum_{i=1}^{10} i^3 = \frac{10^2 \cdot 11^2}{4} = 3025.$$ ◆

EXAMPLE 4 Consider the sum

$$\sum_{i=1}^{10}(7i^2 - 5i) = 2 + 18 + 48 + \cdots + 522 + 650.$$

Using the rules in Eqs. (3) and (4) as well as Eqs. (7) and (8), we find that

$$\sum_{i=1}^{10}(7i^2 - 5i) = 7\sum_{i=1}^{10} i^2 - 5\sum_{i=1}^{10} i$$
$$= 7 \cdot \frac{10 \cdot 11 \cdot 21}{6} - 5 \cdot \frac{10 \cdot 11}{2} = 2420.$$ ◆

EXAMPLE 5 We can use Eq. (8) to simplify the evaluation of the sum for $\underline{A}_{10}$ in Example 1, as follows:

$$\underline{A}_{10} = \tfrac{3}{10} \cdot \left[\left(\tfrac{0}{10}\right)^2 + \left(\tfrac{3}{10}\right)^2 + \left(\tfrac{6}{10}\right)^2 + \cdots + \left(\tfrac{27}{10}\right)^2\right] = \tfrac{3}{10}\sum_{i=0}^{9}\left(\tfrac{3}{10}\right)^2 i^2$$
$$= \tfrac{3}{10} \cdot \left(\tfrac{3}{10}\right)^2 \cdot \left[1^2 + 2^2 + 3^2 + \cdots + 9^2\right] = \left(\tfrac{3}{10}\right)^3 \sum_{i=1}^{9} i^2$$
$$= \frac{27}{1000} \cdot \frac{9 \cdot 10 \cdot 19}{6} = \frac{7695}{1000} = 7.695.$$ ◆

EXAMPLE 6 Evaluate the limit

$$\lim_{n\to+\infty} \frac{1 + 2 + 3 + \cdots + n}{n^2}.$$

Solution Using Eq. (7), we obtain

$$\lim_{n\to+\infty} \frac{1 + 2 + 3 + \cdots + n}{n^2} = \lim_{n\to+\infty} \frac{\frac{1}{2}n(n+1)}{n^2}$$
$$= \lim_{n\to+\infty} \frac{n+1}{2n} = \lim_{n\to+\infty}\left(\frac{1}{2} + \frac{1}{2n}\right) = \frac{1}{2},$$

because the term $1/(2n)$ has limit zero as $n \to +\infty$. ◆

Area Sums

Figure 5.3.11 shows the region R that lies below the graph of the positive-valued *increasing* function f and above the interval $[a, b]$. To approximate the area A of R, we have chosen a fixed integer n and divided the interval $[a, b]$ into n subintervals

$$[x_0, x_1], \quad [x_1, x_2], \quad [x_2, x_3], \quad \ldots, \quad [x_{n-1}, x_n],$$

all with the same length

$$\Delta x = \frac{b-a}{n}. \tag{10}$$

On each of the subintervals we have erected one inscribed rectangle and one circumscribed rectangle.

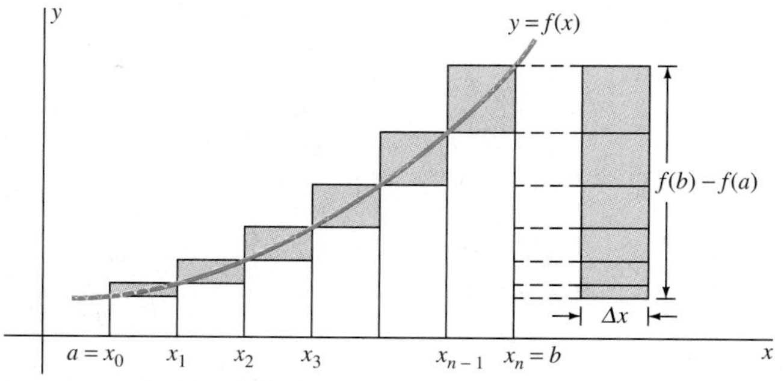

FIGURE 5.3.11 The area under $y = f(x)$ over the interval $[a, b]$.

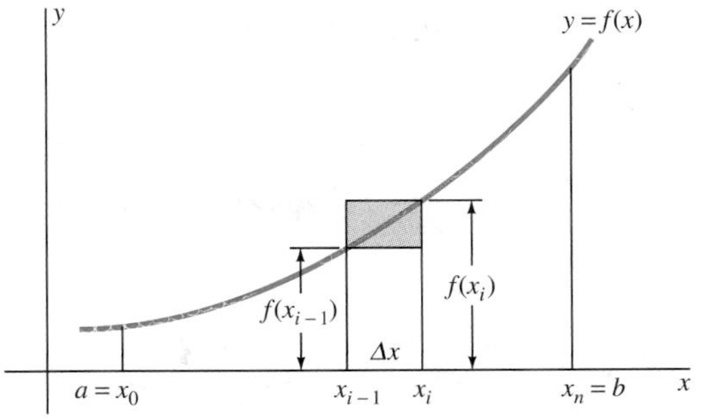

FIGURE 5.3.12 Inscribed and circumscribed rectangles on the ith subinterval $[x_{i-1}, x_i]$.

As indicated in Fig. 5.3.12, the inscribed rectangle over the ith subinterval $[x_{i-1}, x_i]$ has height $f(x_{i-1})$, whereas the ith circumscribed rectangle has height $f(x_i)$. Because the base of each rectangle has length Δx, the areas of the rectangles are

$$f(x_{i-1})\,\Delta x \quad \text{and} \quad f(x_i)\,\Delta x, \tag{11}$$

respectively. Adding the areas of the inscribed rectangles for $i = 1, 2, 3, \dots, n$, we get the underestimate

$$\underline{A}_n = \sum_{i=1}^{n} f(x_{i-1})\,\Delta x \tag{12}$$

of the actual area A. Similarly, the sum of the areas of the circumscribed rectangles is the overestimate

$$\overline{A}_n = \sum_{i=1}^{n} f(x_i)\,\Delta x. \tag{13}$$

The inequality $\underline{A}_n \leqq A \leqq \overline{A}_n$ then yields

$$\sum_{i=1}^{n} f(x_{i-1})\,\Delta x \leqq A \leqq \sum_{i=1}^{n} f(x_i)\,\Delta x. \tag{14}$$

The inequalities in (14) would be reversed if $f(x)$ were decreasing (rather than increasing) on $[a, b]$. (Why?)

Areas as Limits

An illustration such as Fig. 5.3.7 suggests that if the number n of subintervals is very large, so that Δx is small, then the areas $\underline{A}_n$ and $\overline{A}_n$ of the inscribed and circumscribed polygons will differ by very little. Hence both will be very close to the actual area A of the region R. We can also see this because, if f either is increasing or is decreasing on the whole interval $[a, b]$, then the small rectangles in Fig. 5.3.11 (representing the difference between$\overline{A}_n$ and $\underline{A}_n$) can be reassembled in a "stack," as indicated on the right in the figure. It follows that

$$|\overline{A}_n - \underline{A}_n| = |f(b) - f(a)|\,\Delta x. \tag{15}$$

But $\Delta x = (b - a)/n \to 0$ as $n \to \infty$. Thus the difference between the left-hand and right-hand sums in (14) is approaching zero as $n \to \infty$, whereas A does not change as $n \to \infty$. It follows that the area of the region R is given by

$$A = \lim_{n\to\infty} \sum_{i=1}^{n} f(x_{i-1})\,\Delta x = \lim_{n\to\infty} \sum_{i=1}^{n} f(x_i)\,\Delta x. \tag{16}$$

The meaning of these limits is simply that A can be found with any desired accuracy by calculating either sum in Eq. (16) with a sufficiently large number n of subintervals.

In applying Eq. (16), recall that

$$\Delta x = \frac{b-a}{n}. \tag{17}$$

Also note that

$$x_i = a + i\,\Delta x \tag{18}$$

for $i = 0, 1, 2, \dots, n$, because x_i is i "steps" of length Δx to the right of $x_0 = a$.

EXAMPLE 7 We can now compute exactly the area we approximated in Example 1—the area of the region under the graph of $f(x) = x^2$ over the interval $[0, 3]$. If we divide $[0, 3]$ into n subintervals all of the same length, then Eqs. (17) and (18) give

$$\Delta x = \frac{3}{n} \quad \text{and} \quad x_i = 0 + i \cdot \frac{3}{n} = \frac{3i}{n}$$

for $i = 0, 1, 2, \dots, n$. Therefore,

$$\sum_{i=1}^{n} f(x_i)\,\Delta x = \sum_{i=1}^{n} (x_i)^2\,\Delta x = \sum_{i=1}^{n} \left(\frac{3i}{n}\right)^2 \left(\frac{3}{n}\right) = \frac{27}{n^3}\sum_{i=1}^{n} i^2.$$

Then Eq. (8) for $\sum i^2$ yields

$$\sum_{i=1}^{n} f(x_i)\,\Delta x = \frac{27}{n^3}\left(\frac{1}{3}n^3 + \frac{1}{2}n^2 + \frac{1}{6}n\right) = 27\left(\frac{1}{3} + \frac{1}{2n} + \frac{1}{6n^2}\right).$$

When we take the limit as $n \to \infty$, Eq. (16) gives

$$A = \lim_{n\to\infty} 27\left(\frac{1}{3} + \frac{1}{2n} + \frac{1}{6n^2}\right) = 9,$$

because the terms $1/(2n)$ and $1/(6n^2)$ approach zero as $n \to \infty$. Thus our earlier inference from the data in Fig. 5.3.10 was correct: $A = 9$ *exactly.* ◆

EXAMPLE 8 Find the area under the graph of $f(x) = 100 - 3x^2$ from $x = 1$ to $x = 5$.

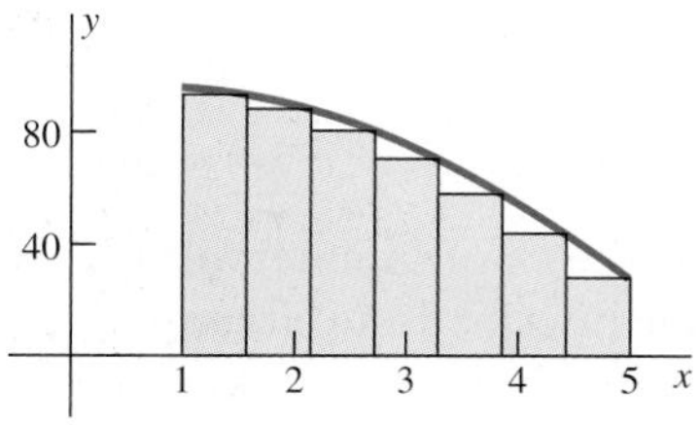

FIGURE 5.3.13 The region of Example 8.

Solution As shown in Fig. 5.3.13, the sum $\sum f(x_i)\,\Delta x$ gives the area of the inscribed rectangular polygon. With $a = 1$ and $b = 5$, Eqs. (17) and (18) give

$$\Delta x = \frac{4}{n} \quad \text{and} \quad x_i = 1 + i \cdot \frac{4}{n} = 1 + \frac{4i}{n}.$$

Therefore

$$\begin{aligned}
\sum_{i=1}^{n} f(x_i)\,\Delta x &= \sum_{i=1}^{n}\left[100 - 3\cdot\left(1 + \frac{4i}{n}\right)^2\right]\left(\frac{4}{n}\right)\\
&= \sum_{i=1}^{n}\left[97 - \frac{24i}{n} - \frac{48i^2}{n^2}\right]\left(\frac{4}{n}\right)\\
&= \frac{388}{n}\sum_{i=1}^{n} 1 - \frac{96}{n^2}\sum_{i=1}^{n} i - \frac{192}{n^3}\sum_{i=1}^{n} i^2\\
&= \frac{388}{n}\cdot n - \frac{96}{n^2}\left(\frac{1}{2}n^2 + \frac{1}{2}n\right) - \frac{192}{n^3}\left(\frac{1}{3}n^3 + \frac{1}{2}n^2 + \frac{1}{6}n\right)\\
&= 276 - \frac{144}{n} - \frac{32}{n^2}.
\end{aligned}$$

[We have applied Eqs. (6) through (8).] Consequently, the second limit in Eq. (16) yields

$$A = \lim_{n\to\infty}\left(276 - \frac{144}{n} - \frac{32}{n}\right) = 276$$

for the desired area. ◆

Historical Note—The Number π

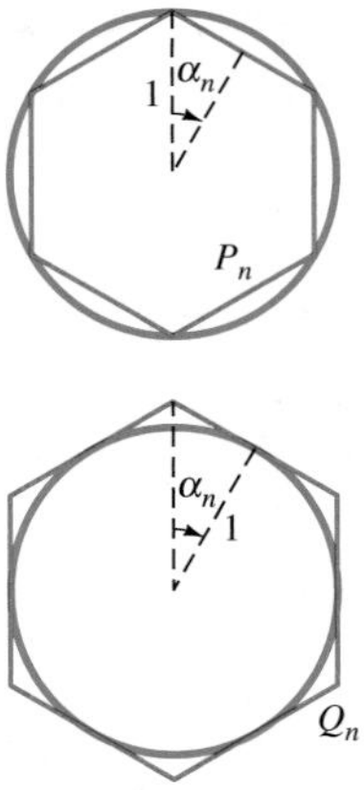

FIGURE 5.3.14 Estimating π by using inscribed and circumscribed regular polygons and the unit circle.

Mathematicians of ancient times tended to employ inscribed and circumscribed triangles rather than rectangles for area approximations. In the third century B.C., Archimedes, the greatest mathematician of antiquity, used such an approach to derive the famous estimate

$$\tfrac{223}{71} = 3\tfrac{10}{71} < \pi < 3\tfrac{1}{7} = \tfrac{22}{7}.$$

Because the area of a circle of radius r is πr^2, the number π may be *defined* to be the area of the unit circle of radius $r = 1$. We will approximate π, then, by approximating the area of the unit circle.

Let P_n and Q_n be n-sided regular polygons, with P_n inscribed in the unit circle and Q_n circumscribed around it (Fig. 5.3.14). Because both polygons are regular, all their sides and angles are equal, so we need to find the area of only *one* of the triangles that we've shown making up P_n and *one* of those making up Q_n.

Let α_n be the central angle subtended by *half* of one of the polygon's sides. The angle α_n is the same whether we work with P_n or with Q_n. In degrees,

$$\alpha_n = \frac{360^\circ}{2n} = \frac{180^\circ}{n}.$$

n	$a(P_n)$	$a(Q_n)$
6	2.598076	3.464102
12	3.000000	3.215390
24	3.105829	3.159660
48	3.132629	3.146086
96	3.139350	3.142715
180	3.140955	3.141912
360	3.141433	3.141672
720	3.141553	3.141613
1440	3.141583	3.141598
2880	3.141590	3.141594
5760	3.141592	3.141593

FIGURE 5.3.15 Data for estimating π (rounded to six-place accuracy).

We can read various dimensions and proportions from Fig. 5.3.14. For example, we see that the area $a(P_n) = \underline{A}_n$ of P_n is given by

$$\underline{A}_n = a(P_n) = n \cdot 2 \cdot \frac{1}{2}\sin\alpha_n \cos\alpha_n = \frac{n}{2}\sin 2\alpha_n = \frac{n}{2}\sin\left(\frac{360^\circ}{n}\right) \qquad \textbf{(19)}$$

and that the area of Q_n is

$$\overline{A}_n = a(Q_n) = n \cdot 2 \cdot \frac{1}{2}\tan\alpha_n = n\tan\left(\frac{180^\circ}{n}\right). \qquad \textbf{(20)}$$

We substituted selected values of n into Eqs. (19) and (20) to obtain the entries of the table in Fig. 5.3.15. Because $\underline{A}_n \leqq \pi \leqq \overline{A}_n$ for all n, we see that $\pi \approx 3.14159$ to five decimal places. Archimedes' reasoning was *not* circular—he used a direct method for computing the sines and cosines in Eqs. (19) and (20) that does not depend upon *a priori* knowledge of the value of π.*

5.3 TRUE/FALSE STUDY GUIDE

5.3 CONCEPTS: QUESTIONS AND DISCUSSION

1. When the German mathematician Carl Friedrich Gauss (1777–1855) was ten years old, his teacher asked the class to find the sum of the integers 1 through 100. Young Gauss almost immediately wrote the answer (and nothing else) on his slate. Apparently he had simply noted that the sum of the first and last of these integers is 101, as is the sum of the second and next-to-last, as is the sum of the third and second-from-last, and so forth. So he simply multiplied

*See Chapter 2 of C. H. Edwards, Jr., *The Historical Development of the Calculus* (New York: Springer-Verlag, 1979).

101 by 50—the number of such pairs—to get the sum 5050. Explain carefully how this approach can be used to show that

$$\sum_{i=1}^{n} i = \frac{n(n+1)}{2}.$$

In what way does the argument depend on whether n is even or odd?

2. The area A of the region under the curve $y = f(x)$ in Fig. 5.3.11 satisfies (for every positive integer n) the inequality $\underline{A}_n < A < \overline{A}_n$, where $\underline{A}_n$ and $\overline{A}_n$ are the sums defined in Eqs. (12) and (13). Moreover,

$$|\overline{A}_n - \underline{A}_n| = |f(b) - f(a)| \cdot \frac{b-a}{n}$$

by Eq. (15). Explain carefully why it follows—as asserted in Eq. (16)—that

$$\lim_{n\to\infty} \underline{A}_n = \lim_{n\to\infty} \overline{A}_n = A.$$

That is, explain why it follows that, given $\epsilon > 0$, there exists an integer N such that both $\underline{A}_n$ and $\overline{A}_n$ differ from A by less than ϵ if $n > N$.

5.3 PROBLEMS

Write each of the sums in Problems 1 through 8 in expanded notation.

1. $\sum_{i=1}^{5} 3^i$

2. $\sum_{i=1}^{6} \sqrt{2i}$

3. $\sum_{j=1}^{5} \frac{1}{j+1}$

4. $\sum_{j=1}^{6} (2j-1)$

5. $\sum_{k=1}^{6} \frac{1}{k^2}$

6. $\sum_{k=1}^{6} \frac{(-1)^{k+1}}{k^2}$

7. $\sum_{n=1}^{5} x^n$

8. $\sum_{n=1}^{5} (-1)^{n+1} x^{2n-1}$

Write the sums in Problems 9 through 18 in summation notation.

9. $1 + 4 + 9 + 16 + 25$

10. $1 - 2 + 3 - 4 + 5 - 6$

11. $1 + \frac{1}{2} + \frac{1}{3} + \frac{1}{4} + \frac{1}{5}$

12. $1 + \frac{1}{4} + \frac{1}{9} + \frac{1}{16} + \frac{1}{25}$

13. $\frac{1}{2} + \frac{1}{4} + \frac{1}{8} + \frac{1}{16} + \frac{1}{32} + \frac{1}{64}$

14. $\frac{1}{3} - \frac{1}{9} + \frac{1}{27} - \frac{1}{81} + \frac{1}{243}$

15. $\frac{2}{3} + \frac{4}{9} + \frac{8}{27} + \frac{16}{81} + \frac{32}{243}$

16. $1 + \sqrt{2} + \sqrt{3} + 2 + \sqrt{5} + \sqrt{6} + \sqrt{7} + 2\sqrt{2} + 3$

17. $x + \frac{x^2}{2} + \frac{x^3}{3} + \cdots + \frac{x^{10}}{10}$

18. $x - \frac{x^3}{3} + \frac{x^5}{5} - \frac{x^7}{7} + \cdots - \frac{x^{19}}{19}$

Use Eqs. (6) through (9) to find the sums in Problems 19 through 28.

19. $\sum_{i=1}^{10} (4i-3)$

20. $\sum_{j=1}^{8} (5-2j)$

21. $\sum_{i=1}^{10} (3i^2+1)$

22. $\sum_{k=1}^{6} (2k-3k^2)$

23. $\sum_{r=1}^{8} (r-1)(r+2)$

24. $\sum_{i=1}^{5} (i^3 - 3i + 2)$

25. $\sum_{i=1}^{6} (i^3 - i^2)$

26. $\sum_{k=1}^{10} (2k-1)^2$

27. $\sum_{i=1}^{100} i^2$

28. $\sum_{i=1}^{100} i^3$

Use the method of Example 6 to evaluate the limits in Problems 29 and 30.

29. $\lim_{n\to\infty} \frac{1^2 + 2^2 + 3^2 + \cdots + n^2}{n^3}$

30. $\lim_{n\to\infty} \frac{1^3 + 2^3 + 3^3 + \cdots + n^3}{n^4}$

Use Eqs. (6) through (9) to derive concise formulas in terms of n for the sums in Problems 31 and 32.

31. $\sum_{i=1}^{n} (2i-1)$

32. $\sum_{i=1}^{n} (2i-1)^2$

In Problems 33 through 42, let R denote the region that lies below the graph of $y = f(x)$ over the interval $[a, b]$ on the x-axis. Use the method of Example 1 to calculate both an underestimate $\underline{A}_n$ and an overestimate $\overline{A}_n$ for the area A of R, based on a division of $[a, b]$ into n subintervals all with the same length $\Delta x = (b-a)/n$.

33. $f(x) = x$ on $[0, 1]$; $n = 5$

34. $f(x) = x$ on $[1, 3]$; $n = 5$

35. $f(x) = 2x + 3$ on $[0, 3]$; $n = 6$

36. $f(x) = 13 - 3x$ on $[0, 3]$; $n = 6$ (Fig. 5.3.16)

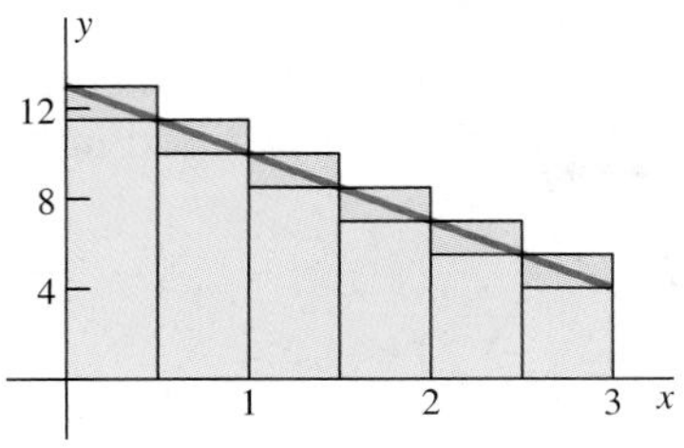

FIGURE 5.3.16 Problem 36.

37. $f(x) = x^2$ on $[0, 1]$; $n = 5$

38. $f(x) = x^2$ on $[1, 3]$; $n = 5$

39. $f(x) = 9 - x^2$ on $[0, 3]$; $n = 5$ (Fig. 5.3.17)

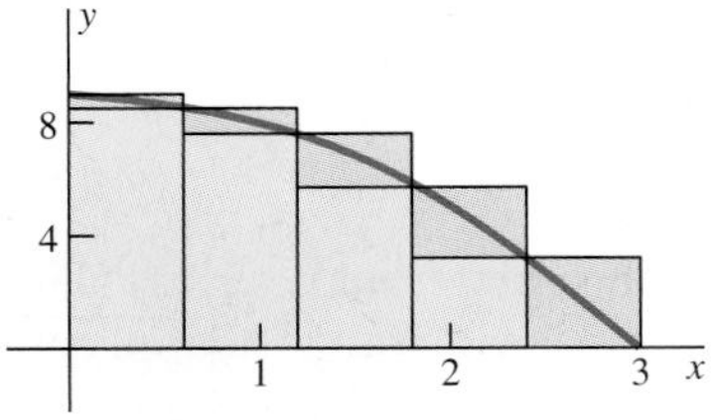

FIGURE 5.3.17 Problem 39.

40. $f(x) = 9 - x^2$ on $[1, 3]$; $n = 8$

41. $f(x) = x^3$ on $[0, 1]$; $n = 10$

42. $f(x) = \sqrt{x}$ on $[0, 1]$; $n = 10$ (Fig. 5.3.18)

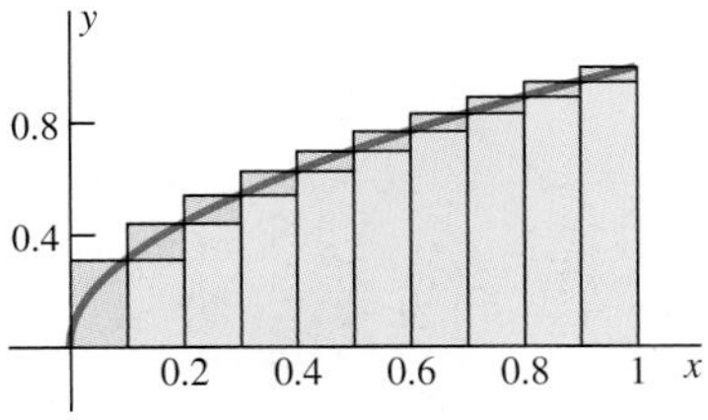

FIGURE 5.3.18 Problem 42.

43. Derive Eq. (7) by adding the equations

$$\sum_{i=1}^{n} i = 1 + 2 + 3 + \cdots + n$$

and

$$\sum_{i=1}^{n} i = n + (n-1) + (n-2) \cdots + 2 + 1.$$

44. Write the n equations obtained by substituting the values $k = 1, 2, 3, \ldots, n$ into the identity

$$(k+1)^3 - k^3 = 3k^2 + 3k + 1.$$

Add these n equations and use their sum to deduce Eq. (8) from Eq. (7).

In Problems 45 through 50, first calculate (in terms of n) the sum

$$\sum_{i=1}^{n} f(x_i)\,\Delta x$$

to approximate the area A of the region under $y = f(x)$ above the interval $[a, b]$. Then find A exactly (as in Examples 7 and 8) by taking the limit as $n \to \infty$.

45. $f(x) = x$ on $[0, 1]$

46. $f(x) = x^2$ on $[0, 2]$

47. $f(x) = x^3$ on $[0, 3]$

48. $f(x) = x + 2$ on $[0, 2]$

49. $f(x) = 5 - 3x$ on $[0, 1]$

50. $f(x) = 9 - x^2$ on $[0, 3]$

51. As in Fig. 5.3.19, the region under the graph of $f(x) = hx/b$ for $0 \leqq x \leqq b$ is a triangle with base b and height h. Use Eq. (7) to verify—with the notation of Eq. (16)—that

$$\lim_{n\to\infty} \sum_{i=1}^{n} f(x_i)\,\Delta x = \tfrac{1}{2}bh,$$

in agreement with the familiar formula for the area of a triangle.

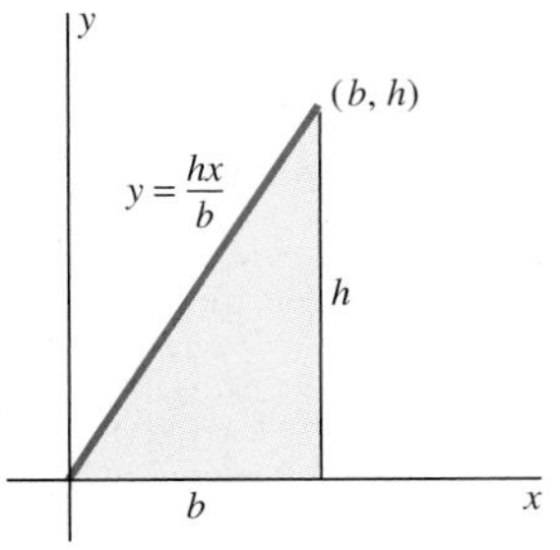

FIGURE 5.3.19 Problem 51.

In Problems 52 and 53, let A denote the area and C the circumference of a circle of radius r and let A_n and C_n denote the area and perimeter, respectively, of a regular n-sided polygon inscribed in this circle.

52. Figure 5.3.20 shows one side of the n-sided polygon subtending an angle $2\pi/n$ at the center O of the circle. Show that

$$A_n = nr^2 \sin\left(\frac{\pi}{n}\right)\cos\left(\frac{\pi}{n}\right) \quad \text{and that} \quad C_n = 2nr\sin\left(\frac{\pi}{n}\right).$$

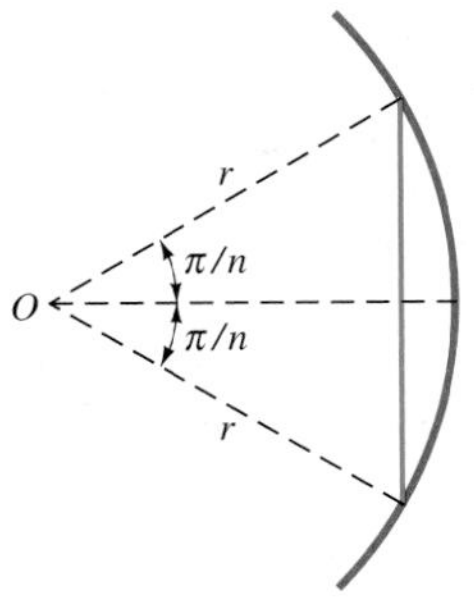

FIGURE 5.3.20 Problem 52.

53. Deduce that $A = \frac{1}{2}rC$ by taking the limit of A_n/C_n as $n \to \infty$. Then, under the assumption that $A = \pi r^2$, deduce that $C = 2\pi r$. Thus the familiar circumference formula for a circle follows from the familiar area formula for a circle.

5.4 RIEMANN SUMS AND THE INTEGRAL

Suppose that f is a positive-valued and increasing function defined on a set of real numbers that includes the interval $[a, b]$. In Section 5.3 we used inscribed and circumscribed rectangles to set up the sums

$$\sum_{i=1}^{n} f(x_{i-1})\,\Delta x \quad \text{and} \quad \sum_{i=1}^{n} f(x_i)\,\Delta x \tag{1}$$

that approximate the area A under the graph of $y = f(x)$ from $x = a$ to $x = b$. Recall that the notation in Eq. (1) is based on a division of the interval $[a, b]$ into n subintervals, all with the same length $\Delta x = (b - a)/n$, and that $[x_{i-1}, x_i]$ denotes the ith subinterval.

The approximating sums in Eq. (1) are both of the form

$$\sum_{i=1}^{n} f(x_i^\star)\,\Delta x, \tag{2}$$

where $x_i^\star$ denotes a selected point of the ith subinterval $[x_{i-1}, x_i]$ (Fig. 5.4.1). Sums of the form in (2) appear as approximations in a wide range of applications and form the basis for the definition of the integral. Motivated by our discussion of area in Section 5.3, we want to define the integral of f from a to b as some sort of limit, as $\Delta x \to 0$, of sums such as the one in (2). Our goal is to begin with a fairly general function f and define a computable real number I (the *integral* of f) that—in the special case when f is continuous and positive-valued on $[a, b]$—will equal the area under the graph of $y = f(x)$.

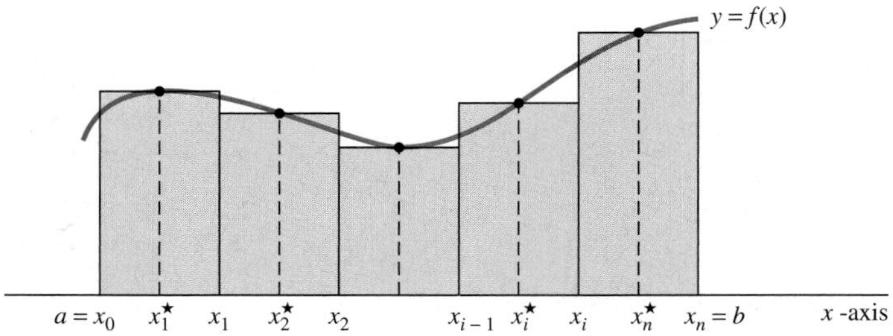

FIGURE 5.4.1 The Riemann sum in Eq. (2) as a sum of areas of rectangles.

Riemann Sums

We begin with a function f defined on $[a, b]$ that is *not* necessarily either continuous or positive valued. A **partition** P of $[a, b]$ is a collection of subintervals

$$[x_0, x_1], \quad [x_1, x_2], \quad [x_2, x_3], \quad \ldots, \quad [x_{n-1}, x_n]$$

of $[a, b]$ such that

$$a = x_0 < x_1 < x_2 < x_3 < \cdots < x_{n-1} < x_n = b,$$

as in Fig. 5.4.1. We write $\Delta x_i = x_i - x_{i-1}$ for the length of the ith subinterval $[x_{i-1}, x_i]$. To get a sum such as the one in (2), we need a point $x_i^\star$ in the ith subinterval for each i, $1 \leqq i \leqq n$. A collection of points

$$S = \{x_1^\star, x_2^\star, x_3^\star, \ldots, x_n^\star\}$$

with $x_i^\star$ in $[x_{i-1}, x_i]$ for each i (Fig. 5.4.2) is called a **selection** for the partition P.

FIGURE 5.4.2 The selected point $x_i^\star$ in the ith subinterval $[x_{i-1}, x_i]$.

DEFINITION Riemann Sum

Let f be a function defined on the interval $[a, b]$. If P is a partition of $[a, b]$ and S is a selection for P, then the **Riemann sum** for f determined by P and S is

$$R = \sum_{i=1}^{n} f(x_i^\star)\,\Delta x_i. \tag{3}$$

We also say that this Riemann sum is **associated with** the partition P.

The German mathematician G. F. B. Riemann (1826–1866) provided a rigorous definition of the integral. Various special types of "Riemann sums" had appeared in area and volume computations since the time of Archimedes, but it was Riemann who framed the preceding definition in its full generality.

The point $x_i^\star$ in Eq. (3) is simply a selected point of the ith subinterval $[x_{i-1}, x_i]$. That is, it can be *any* point of this subinterval. But when we compute Riemann sums, we usually choose the points of the selection S in some systematic manner, as illustrated in Fig. 5.4.3. There we show different Riemann sums for the function $f(x) = 2x^3 - 6x^2 + 5$ on the interval $[0, 3]$. Figure 5.4.3(a) shows rectangles associated with the *left-endpoint sum*

$$R_{\text{left}} = \sum_{i=1}^{n} f(x_{i-1})\,\Delta x, \tag{4}$$

in which each $x_i^\star$ is selected to be x_{i-1}, the *left endpoint* of the ith subinterval $[x_{i-1}, x_i]$ of length $\Delta x = (b - a)/n$. Figure 5.4.3(b) shows rectangles associated with the

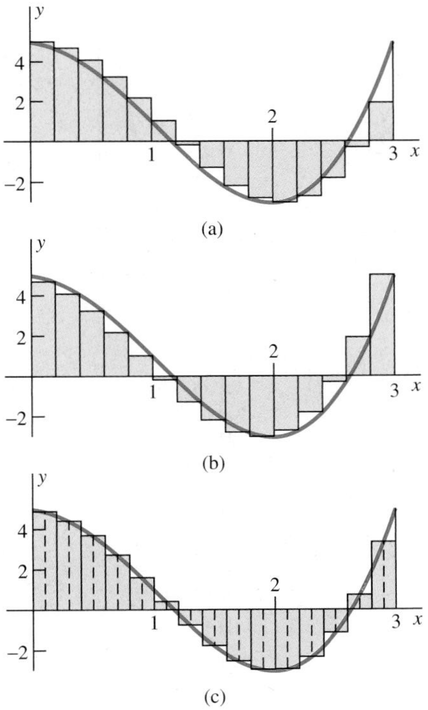

FIGURE 5.4.3 Riemann sums for $f(x) = 2x^3 - 6x^2 + 5$ on [0,3]: (a) Left-endpoint sum; (b) Right-endpoint sum; (c) Midpoint sum.

right-endpoint sum

$$R_{\text{right}} = \sum_{i=1}^{n} f(x_i)\,\Delta x, \tag{5}$$

in which each $x_i^\star$ is selected to be x_i, the *right endpoint* of $[x_{i-1}, x_i]$. In each figure, some of the rectangles are inscribed and others are circumscribed.

Figure 5.4.3(c) shows rectangles associated with the *midpoint sum*

$$R_{\text{mid}} = \sum_{i=1}^{n} f(m_i)\,\Delta x, \tag{6}$$

in which

$$x_i^\star = m_i = \frac{x_{i-1} + x_i}{2},$$

the *midpoint* of the ith subinterval $[x_{i-1}, x_i]$. The dashed lines in Fig. 5.4.3(c) represent the ordinates of f at these midpoints.

EXAMPLE 1 In Example 1 of Section 5.3 we calculated left- and right-endpoint sums for $f(x) = x^2$ on $[0, 3]$ with $n = 10$ subintervals. We now do this more concisely by using summation notation, and we also calculate the analogous midpoint sum. Figure 5.4.4 shows a typical approximating rectangle for each of these sums. With $a = 0$, $b = 3$, and $\Delta x = (b - a)/n = \frac{3}{10}$, we see that the ith subdivision point is

$$x_i = a + i \cdot \Delta x = \tfrac{3}{10}i.$$

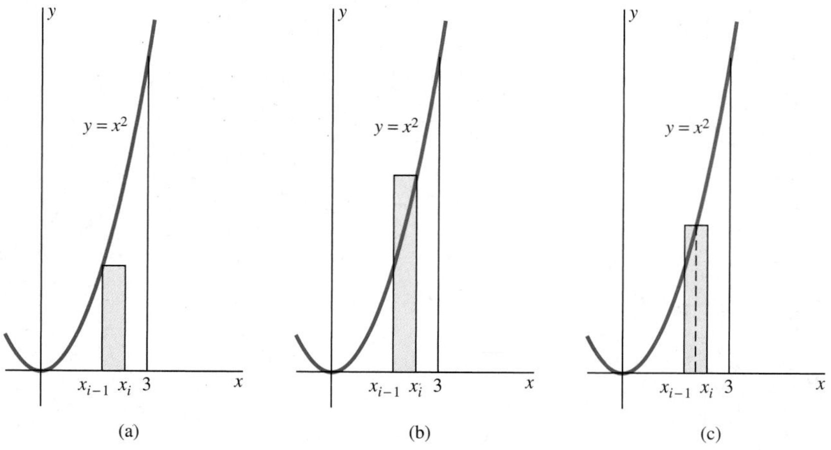

FIGURE 5.4.4 Example 1: (a) The case $x_i^\star = x_{i-1}$; (b) The case $x_i^\star = x_i$; (c) The case $x_i^\star = m_i$.

The ith subinterval, as well as its midpoint

$$m_i = \frac{x_{i-1} + x_i}{2} = \frac{1}{2}\left(\frac{3i-3}{10} + \frac{3i}{10}\right) = \frac{3}{20}(2i - 1),$$

are shown in Fig. 5.4.5. With $x_i^\star = x_{i-1} = \frac{3}{10}(i-1)$, we obtain the left-endpoint sum in Eq. (4),

$$R_{\text{left}} = \sum_{i=1}^{n} f(x_{i-1})\,\Delta x = \sum_{i=1}^{10} \left[\tfrac{3}{10}(i-1)\right]^2 \left(\tfrac{3}{10}\right)$$

$$= \tfrac{27}{1000} \cdot (0^2 + 1^2 + 2^2 + \cdots + 9^2)$$

$$= \tfrac{7695}{1000} = 7.695 \qquad \text{[using Eq. (8) of Section 5.3].}$$

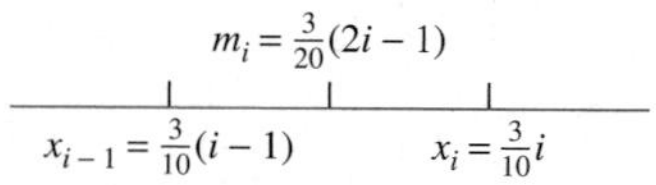

FIGURE 5.4.5 The ith subinterval of Example 1.

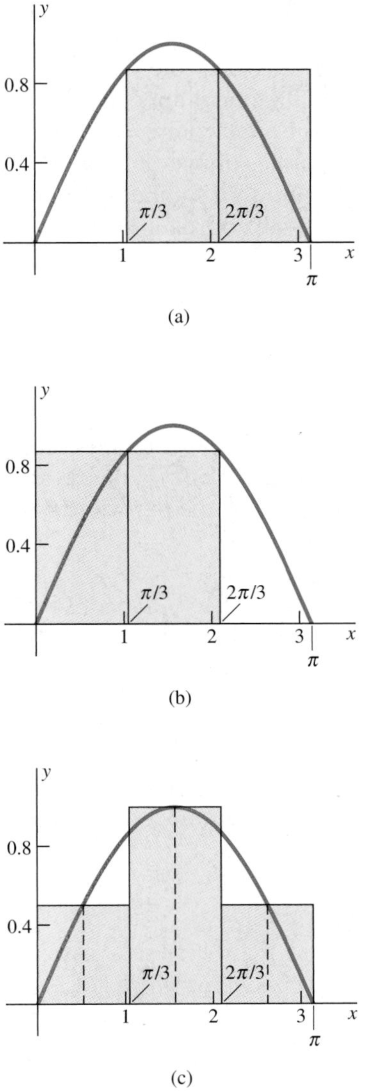

FIGURE 5.4.6 Approximating the area under $y = \sin x$ on $[0, \pi]$ (Example 2): (a) Left-endpoint sum; (b) Right-endpoint sum; (c) Midpoint sum.

With $x_i^\star = x_i = \frac{3}{10}i$, we get the right-endpoint sum in Eq. (5),

$$R_{\text{right}} = \sum_{i=1}^{n} f(x_i)\,\Delta x = \sum_{i=1}^{10} \left[\tfrac{3}{10}i\right]^2 \left(\tfrac{3}{10}\right)$$

$$= \tfrac{27}{1000} \cdot (1^2 + 2^2 + 3^2 + \cdots + 10^2)$$

$$= \tfrac{10395}{1000} = 10.395 \qquad \text{[using Eq. (8) of Section 5.3].}$$

Finally, with $x_i^\star = m_i = \frac{3}{20}(2i - 1)$, we get the midpoint sum in Eq. (6),

$$R_{\text{mid}} = \sum_{i-1}^{n} f(m_i)\,\Delta x = \sum_{i=1}^{10} \left[\tfrac{3}{20}(2i-1)\right]^2 \left(\tfrac{3}{10}\right)$$

$$= \tfrac{27}{4000} \cdot \left(1^2 + 3^2 + 5^2 + \cdots + 17^2 + 19^2\right) = \tfrac{35910}{4000} = 8.9775.$$

The midpoint sum is much closer than either endpoint sum to the actual value 9 (of the area under the graph of $y = x^2$ over $[0, 3]$) that we found in Example 7 of Section 5.3. ◆

EXAMPLE 2 Figure 5.4.6 illustrates Riemann sums for $f(x) = \sin x$ on $[0, \pi]$ based on $n = 3$ subintervals: $[0, \pi/3]$, $[\pi/3, 2\pi/3]$, and $[2\pi/3, \pi]$, of length $\Delta x = \pi/3$, and with midpoints $\pi/6$, $\pi/2$, and $5\pi/6$. The left-endpoint sum is

$$R_{\text{left}} = (\Delta x) \cdot \left(\sum_{i=1}^{n} f(x_{i-1})\right) = \frac{\pi}{3} \cdot \left(\sin 0 + \sin\frac{\pi}{3} + \sin\frac{2\pi}{3}\right)$$

$$= \frac{\pi}{3} \cdot \left(0 + \frac{\sqrt{3}}{2} + \frac{\sqrt{3}}{2}\right) = \frac{\pi\sqrt{3}}{3} \approx 1.81.$$

It is clear from the figure that the right-endpoint sum has the same value. The corresponding midpoint sum is

$$R_{\text{mid}} = \frac{\pi}{3} \cdot \left(\sin\frac{\pi}{6} + \sin\frac{\pi}{2} + \sin\frac{5\pi}{6}\right) = \frac{\pi}{3} \cdot \left(\frac{1}{2} + 1 + \frac{1}{2}\right) = \frac{2\pi}{3} \approx 2.09.$$

(We will soon be able to show that the area under one arch of the sine curve is exactly 2.) ◆

The Integral as a Limit

In the case of a function f that has both positive and negative values on $[a, b]$, it is necessary to consider the *signs* indicated in Fig. 5.4.7 when we interpret geometrically the Riemann sum in Eq. (3). On each subinterval $[x_{i-1}, x_i]$, we have a rectangle with width Δx and "height" $f(x_i^\star)$. If $f(x_i^\star) > 0$, then this rectangle stands *above* the x-axis;

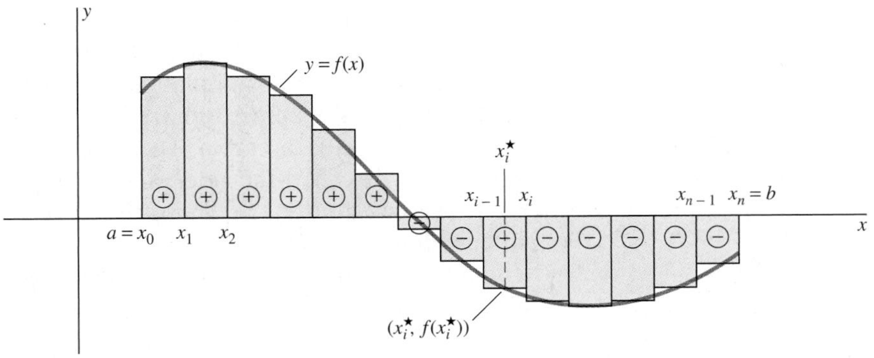

FIGURE 5.4.7 A geometric representation of the Riemann sum in Eq. (3).

if $f(x_i^\star) < 0$, it lies *below* the x-axis. The Riemann sum R is then the sum of the **signed** areas of these rectangles—that is, the sum of the areas of those rectangles that lie above the x-axis *minus* the sum of the areas of those that lie below the x-axis.

If the widths Δx_i of these rectangles are all very small, then it appears that the corresponding Riemann sum R will closely approximate the area from $x = a$ to $x = b$ under $y = f(x)$ and above the x-axis, *minus* the area that lies above the graph and below the x-axis. This suggests that the integral of f from a to b should be defined by taking the limit of the Riemann sums as the widths Δx_i all approach zero:

$$I = \lim_{\Delta x_i \to 0} \sum_{i=1}^{n} f(x_i^\star)\,\Delta x_i. \tag{7}$$

The formal definition of the integral is obtained by saying precisely what it means for this limit to exist. The **norm** of the partition P is the largest of the lengths $\Delta x_i = x_i - x_{i-1}$ of the subintervals in P and is denoted by $|P|$. Briefly, Eq. (7) means that if $|P|$ is sufficiently small, then *all* Riemann sums associated with the partition P are close to the number I.

DEFINITION The Definite Integral

The **definite integral of the function** f **from** a **to** b is the number

$$I = \lim_{|P| \to 0} \sum_{i=1}^{n} f(x_i^\star)\,\Delta x_i, \tag{8}$$

provided that this limit exists, in which case we say that f is **integrable** on $[a, b]$. Equation (8) means that, for each number $\epsilon > 0$, there exists a number $\delta > 0$ such that

$$\left| I - \sum_{i=1}^{n} f(x_i^\star)\,\Delta x_i \right| < \epsilon$$

for every Riemann sum associated with any partition P of $[a, b]$ for which $|P| < \delta$.

The customary notation for the integral of f from a to b, due to the German mathematician and philosopher G. W. Leibniz, is

$$I = \int_a^b f(x)\,dx = \lim_{|P| \to 0} \sum_{i=1}^{n} f(x_i^\star)\,\Delta x_i. \tag{9}$$

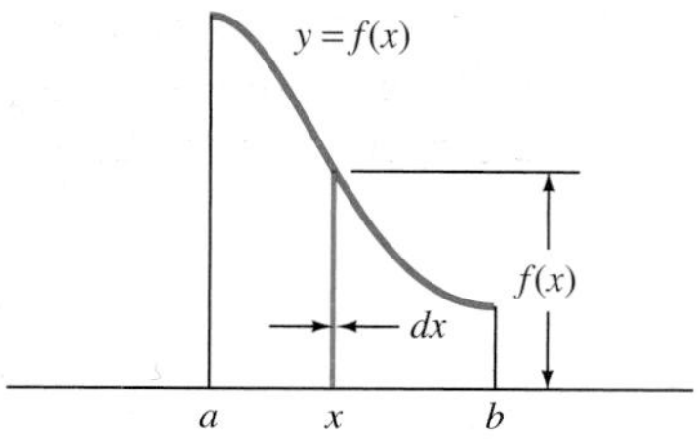

FIGURE 5.4.8 Origin of Leibniz's notation for the integral.

Considering I to be the area under $y = f(x)$ from a to b, Leibniz first thought of a narrow strip with height $f(x)$ and "infinitesimally small" width dx (as in Fig. 5.4.8), so that its area would be the product $f(x)\,dx$. He regarded the integral as a sum of areas of such strips and denoted this sum by the elongated capital S (for *summa*) that appears as the integral sign in Eq. (9).

We shall see that this integral notation is not only highly suggestive, but also is exceedingly useful in manipulations with integrals. The numbers a and b are called the **lower limit** and **upper limit,** respectively, of the integral; they are the endpoints of the interval of integration. The function $f(x)$ that appears between the integral sign and dx is called the **integrand.** The symbol dx that follows the integrand in Eq. (9) should, for the time being, be thought of as simply an indication of what the independent variable is. Like the index of summation, the independent variable x is a "dummy variable"—it may be replaced with any other variable without affecting the meaning of Eq. (9). Thus if f is integrable on $[a, b]$, we can write

$$\int_a^b f(x)\,dx = \int_a^b f(t)\,dt = \int_a^b f(u)\,du.$$

The definition given for the definite integral applies only if $a < b$, but it is convenient to include the cases $a = b$ and $a > b$ as well. The integral is defined in these cases as follows:

$$\int_a^a f(x)\,dx = 0 \tag{10}$$

and

$$\int_a^b f(x)\,dx = -\int_b^a f(x)\,dx, \tag{11}$$

provided that the right-hand integral exists. Thus *interchanging the limits of integration reverses the sign of the integral.*

Just as not all functions are differentiable, not every function is integrable. Suppose that c is a point of $[a, b]$ such that $f(x) \to +\infty$ as $x \to c$. If $[x_{k-1}, x_k]$ is the subinterval of the partition P that contains c, then the Riemann sum in Eq. (3) can be made arbitrarily large by choosing $x_k^\star$ to be sufficiently close to c. For our purposes, however, we need to know only that every continuous function is integrable. The following theorem is proved in Appendix G.

THEOREM 1 Existence of the Integral
If the function f is continuous on $[a, b]$, then f is integrable on $[a, b]$.

Although we omit the details, it is not difficult to show that the definition of the integral can be reformulated in terms of sequences of Riemann sums, as follows.

THEOREM 2 The Integral as a Limit of a Sequence
The function f is integrable on $[a, b]$ with integral I if and only if

$$\lim_{n\to\infty} R_n = I \tag{12}$$

for every sequence $\{R_n\}_1^\infty$ of Riemann sums associated with a sequence of partitions $\{P_n\}_1^\infty$ of $[a, b]$ such that $|P_n| \to 0$ as $n \to +\infty$.

Riemann Sum Computations

The reformulation in Theorem 2 of the definition of the integral is helpful because it is easier to visualize a specific sequence of Riemann sums than to visualize the vast totality of all possible Riemann sums. In the case of a continuous function f (known to be integrable by Theorem 1), the situation can be simplified even more by using only Riemann sums associated with partitions consisting of subintervals all with the same length

$$\Delta x_1 = \Delta x_2 = \cdots = \Delta x_n = \frac{b-a}{n} = \Delta x.$$

Such a partition of $[a, b]$ into equal-length subintervals is called a **regular partition** of $[a, b]$.

Any Riemann sum associated with a regular partition can be written in the form

$$\sum_{i=1}^{n} f(x_i^\star)\,\Delta x, \tag{13}$$

where the absence of a subscript in Δx signifies that the sum is associated with a regular partition. In such a case the conditions $|P| \to 0$, $\Delta x \to 0$, and $n \to +\infty$ are equivalent, so the integral of a *continuous* function can be defined quite simply:

$$\int_a^b f(x)\,dx = \lim_{n\to\infty} \sum_{i=1}^{n} f(x_i^\star)\,\Delta x = \lim_{\Delta x\to 0} \sum_{i=1}^{n} f(x_i^\star)\,\Delta x. \tag{14}$$

Consequently, we henceforth will use only regular partitions; the subintervals will thus have length and endpoints given by

$$\Delta x = \frac{b-a}{n} \quad \text{and} \quad x_i = a + i\cdot\Delta x \tag{15}$$

for $i = 0, 1, 2, 3, \dots, n$. If we select $x_i^\star = x_i$ then (14) gives

$$\int_a^b f(x)\,dx = \lim_{n\to\infty} \sum_{i=1}^{n} f(x_i)\,\Delta x. \tag{16}$$

EXAMPLE 3 Use Riemann sums to evaluate $\displaystyle\int_0^4 (x^3 - 2x)\,dx$.

Solution With $a = 0$ and $b = 4$ in (15), we have $\Delta x = 4/n$ and $x_i = 4i/n$. Hence

$$\begin{aligned}
\int_0^4 (x^3 - 2x)\,dx &= \lim_{n\to\infty} \sum_{i=1}^{n} f(x_i)\,\Delta x = \lim_{n\to\infty} \sum_{i=1}^{n} \left[\left(\frac{4i}{n}\right)^3 - 2\left(\frac{4i}{n}\right)\right]\cdot\frac{4}{n} \\
&= \lim_{n\to\infty} \frac{4}{n} \sum_{i=1}^{n} \left[\frac{64i^3}{n^3} - \frac{8i}{n}\right] \\
&= \lim_{n\to\infty} \left[\frac{256}{n^4} \sum_{i=1}^{n} i^3 - \frac{32}{n^2} \sum_{i=1}^{n} i\right].
\end{aligned}$$

We now use Eqs. (7) and (9) in Section 5.3 to convert each of the last two sums to closed form:

$$\begin{aligned}
\int_0^4 (x^3 - 2x)\,dx &= \lim_{n\to\infty} \left[\frac{256}{n^4}\cdot\frac{n^2(n+1)^2}{4} - \frac{32}{n^2}\cdot\frac{n(n+1)}{2}\right] \\
&= \lim_{n\to\infty} \left[\frac{64}{n^2}(n+1)^2 - \frac{16}{n}(n+1)\right] \\
&= \lim_{n\to\infty} \left[64\left(1 + \frac{1}{n}\right)^2 - 16\left(1 + \frac{1}{n}\right)\right];
\end{aligned}$$

$$\int_0^4 (x^3 - 2x)\,dx = 64 - 16 = 48.$$

◆

EXAMPLE 4 Use Riemann sums to evaluate $\displaystyle\int_a^b x\,dx$ (where $a < b$).

Solution With $f(x) = x$ and $x_i^\star = x_i$ (see Fig. 5.4.9), Eqs. (15) and (16) yield

$$\begin{aligned}
\int_a^b x\,dx &= \lim_{n\to\infty} \sum_{i=1}^{n} f(x_i)\,\Delta x = \lim_{n\to\infty} \sum_{i=1}^{n} (a + i\cdot\Delta x)\,\Delta x \\
&= \lim_{n\to\infty} \left[(a\,\Delta x)\sum_{i=1}^{n} 1 \;+\; (\Delta x)^2 \sum_{i=1}^{n} i\right] \\
&= \lim_{n\to\infty} \left[(a\,\Delta x)\cdot n \;+\; (\Delta x)^2\cdot\frac{n(n+1)}{2}\right]
\end{aligned}$$

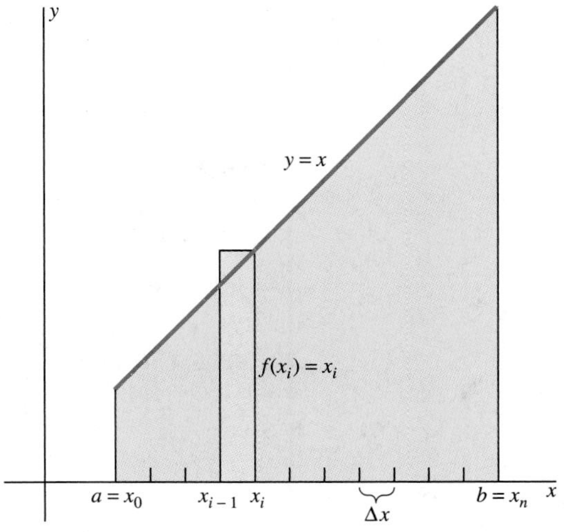

FIGURE 5.4.9 Calculating the area under $y = x$ from $x = a$ to $x = b$.

when we use Eqs. (6) and (7) in Section 5.3 to convert the sums to closed form. Then substituting $\Delta x = (b - a)/n$ gives

$$\begin{aligned}\int_a^b x\,dx &= \lim_{n\to\infty}\left[a\cdot\frac{b-a}{n}\cdot n + \left(\frac{b-a}{n}\right)^2\cdot\frac{n(n+1)}{2}\right]\\ &= \lim_{n\to\infty}\left[a(b-a) + \frac{1}{2}(b-a)^2\left(1+\frac{1}{n}\right)\right]\\ &= a(b-a) + \frac{1}{2}(b-a)^2 = (b-a)\left(a + \frac{1}{2}b - \frac{1}{2}a\right)\\ &= (b-a)\cdot\frac{1}{2}(b+a) = \frac{1}{2}(b^2 - a^2).\end{aligned}$$

Thus we see finally that

$$\int_a^b x\,dx = \frac{1}{2}b^2 - \frac{1}{2}a^2. \tag{17}$$

◆

REMARK 1 If $0 < a < b$, then $A = \int_a^b x\,dx$ is the area of the trapezoid shown in Fig. 5.4.9. Then Eq. (17) implies that

$$A = (b-a)\cdot\tfrac{1}{2}(a+b) = w\cdot\overline{h},$$

where $w = b - a$ is the width and $\overline{h} = \frac{1}{2}(a+b)$ is the average height of the trapezoid.

REMARK 2 Figures 5.4.10 and 5.4.11 illustrate two different cases in Example 4. In each case Eq. (17) agrees with the sum of the indicated *signed* areas. The minus sign in Fig. 5.4.10 represents the fact that area *beneath* the x-axis is measured with a *negative* number. The minus sign in Fig. 5.4.11 signifies that the area of the triangle over $[0, a]$ is subtracted from the area of the triangle over $[0, b]$ to get the area of the trapezoid.

The summation formulas in Eqs. (6) through (9) in Section 5.3 suffice for the integration of polynomials of low degree, but integrals of other functions may require other devices (or a computer algebra system) for the conversion of Riemann sums to closed forms whose limits can be evaluated.

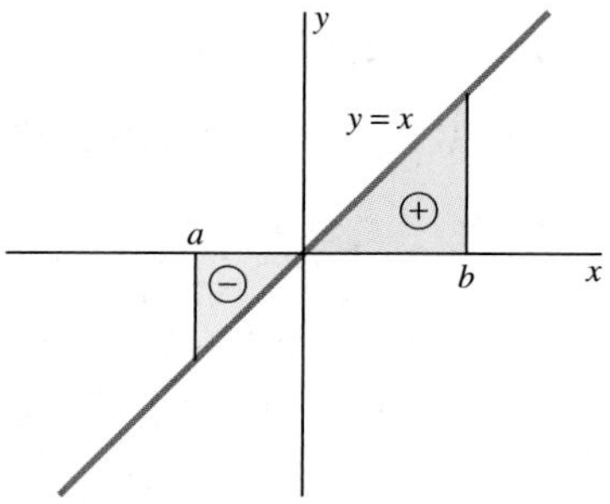

FIGURE 5.4.10 Example 4 with $a < 0 < b$.

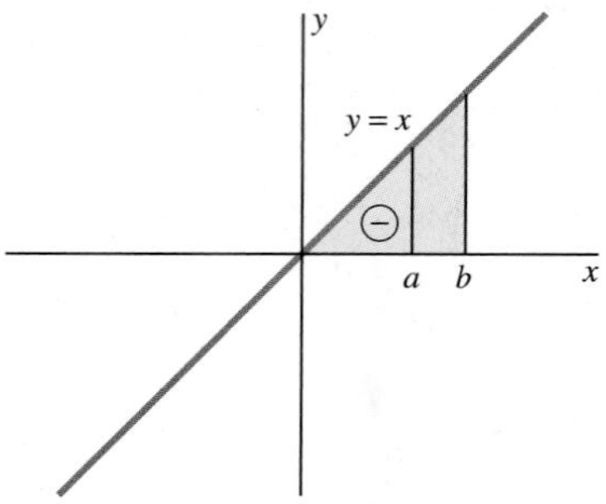

FIGURE 5.4.11 Example 4 with $0 < a < b$.

EXAMPLE 5 Use Riemann sums to evaluate $\int_0^2 e^x\,dx$.

Solution With $\Delta x = 2/n$ and $x_i = i \cdot \Delta x = 2i/n$, we have

$$\begin{aligned}\int_0^2 e^x\,dx &= \lim_{n\to\infty} \sum_{i=1}^{n} f(x_i)\,\Delta x = \lim_{n\to\infty} \sum_{i=1}^{n} e^{2i/n} \cdot \frac{2}{n} \\ &= \lim_{n\to\infty} \frac{2}{n}\left(e^{2/n} + e^{4/n} + e^{6/n} + \cdots + e^{2n/n}\right) \\ &= \lim_{n\to\infty} \frac{2e^{2/n}}{n}\left(1 + e^{2/n} + e^{4/n} + \cdots + e^{2(n-1)/n}\right) \\ &= \lim_{n\to\infty} \frac{2r}{n}\left(1 + r + r^2 + \cdots + r^{n-1}\right)\end{aligned}$$

where $r = e^{2/n}$. To convert the last sum to closed form, we use the formula

$$1 + r + r^2 + \cdots + r^{n-1} = \frac{r^n - 1}{r - 1}, \tag{18}$$

which is readily verified by multiplying the left-hand side by the denominator on the right. This gives

$$\begin{aligned}\int_0^2 e^x\,dx &= \lim_{n\to\infty} \frac{2r}{n} \cdot \frac{r^n - 1}{r - 1} \\ &= \lim_{n\to\infty} \frac{2e^{2/n}}{n} \cdot \frac{e^2 - 1}{e^{2/n} - 1} = \frac{2(e^2 - 1)}{\lim_{n\to\infty} n \cdot (1 - e^{-2/n})}.\end{aligned} \tag{19}$$

The limit in the denominator has the indeterminate form $\infty \cdot 0$ as $n \to +\infty$. We evaluate it using l'Hôpital's rule as in Section 4.9:

$$\begin{aligned}\lim_{n\to\infty} n \cdot (1 - e^{-2/n}) &= \lim_{n\to\infty} \frac{1 - e^{-2/n}}{\dfrac{1}{n}} \qquad \text{[now the indeterminate form } 0/0\text{]} \\ &= \lim_{n\to\infty} \frac{-\dfrac{2}{n^2} \cdot e^{-2/n}}{-\dfrac{1}{n^2}} = \lim_{n\to\infty} 2e^{-2/n} = 2.\end{aligned}$$

Substituting this limit in Eq. (19) finally gives

$$\int_0^2 e^x\,dx = e^2 - 1.$$

◆

5.4 TRUE/FALSE STUDY GUIDE

5.4 CONCEPTS: QUESTIONS AND DISCUSSION

1. Explain why you would generally expect the midpoint sum in Eq. (6) to be a more accurate approximation to the actual value $\int_a^b f(x)\,dx$ than either the left-endpoint sum in (4) or the right-endpoint sum in (5).
2. The result in Example 4, with $a = 0$, and Problems 49 and 50 tell us that

$$\int_0^b x\,dx = \frac{1}{2}b^2, \qquad \int_0^b x^2\,dx = \frac{1}{3}b^3, \qquad \text{and} \qquad \int_0^b x^3\,dx = \frac{1}{4}b^4$$

if $b > 0$. Assuming that the pattern holds (it does), what would you expect to be the value of $\int_a^b x^n\,dx$ with $n > 0$ and $0 < a < b$? Explain how you take into account the nonzero lower limit a.
3. Example 5 and Problem 56 imply that $\int_0^2 e^x\,dx = e^2 - 1$ and $\int_0^5 e^x\,dx = e^5 - 1$. Thinking of area under the curve $y = e^x$, what would you expect to be the value of $\int_2^5 e^x\,dx$? What would you conjecture about the value of $\int_a^b e^x\,dx$ with $0 < a < b$?

5.4 PROBLEMS

In Problems 1 through 10, express the given limit as a definite integral over the indicated interval $[a, b]$. Assume that $[x_{i-1}, x_i]$ denotes the ith subinterval of a subdivision of $[a, b]$ into n subintervals, all with the same length $\Delta x = (b - a)/n$, and that $m_i = \frac{1}{2}(x_{i-1} + x_i)$ is the midpoint of the ith subinterval.

1. $\lim_{n\to\infty} \sum_{i=1}^{n} (2x_i - 1)\Delta x$ over $[1, 3]$
2. $\lim_{n\to\infty} \sum_{i=1}^{n} (2 - 3x_{i-1})\Delta x$ over $[-3, 2]$
3. $\lim_{n\to\infty} \sum_{i=1}^{n} \left(x_i^2 + 4\right)\Delta x$ over $[0, 10]$
4. $\lim_{n\to\infty} \sum_{i=1}^{n} \left(x_i^3 - 3x_i^2 + 1\right)\Delta x$ over $[0, 3]$
5. $\lim_{n\to\infty} \sum_{i=1}^{n} \sqrt{m_i}\,\Delta x$ over $[4, 9]$
6. $\lim_{n\to\infty} \sum_{i=1}^{n} \sqrt{25 - x_i^2}\,\Delta x$ over $[0, 5]$
7. $\lim_{n\to\infty} \sum_{i=1}^{n} \frac{1}{\sqrt{1 + m_i}}\,\Delta x$ over $[3, 8]$
8. $\lim_{n\to\infty} \sum_{i=1}^{n} (\cos 2x_{i-1})\,\Delta x$ over $[0, \pi/2]$
9. $\lim_{n\to\infty} \sum_{i=1}^{n} (\sin 2\pi m_i)\,\Delta x$ over $[0, 1/2]$
10. $\lim_{n\to\infty} \sum_{i=1}^{n} e^{2x_i}\,\Delta x$ over $[0, 1]$

In Problems 11 through 20, compute the Riemann sum

$$\sum_{i=1}^{n} f(x_i^\star)\,\Delta x$$

for the indicated function and a regular partition of the given interval into n subintervals. Use $x_i^\star = x_i$, the right-hand endpoint of the ith subinterval $[x_{i-1}, x_i]$.

11. $f(x) = x^2$ on $[0, 1]$; $n = 5$
12. $f(x) = x^3$ on $[0, 1]$; $n = 5$
13. $f(x) = \frac{1}{x}$ on $[1, 6]$; $n = 5$
14. $f(x) = \sqrt{x}$ on $[0, 5]$; $n = 5$
15. $f(x) = 2x + 1$ on $[1, 4]$; $n = 6$
16. $f(x) = x^2 + 2x$ on $[1, 4]$; $n = 6$
17. $f(x) = x^3 - 3x$ on $[1, 4]$; $n = 5$
18. $f(x) = 1 + 2\sqrt{x}$ on $[2, 3]$; $n = 5$
19. $f(x) = \cos x$ on $[0, \pi]$; $n = 6$
20. $f(x) = \ln x$ on $[1, 6]$; $n = 5$
21. through **30.** Repeat Problems 11 through 20, except with $x_i^\star = x_{i-1}$, the left-hand endpoint.
31. through **40.** Repeat Problems 11 through 20, except with $x_i^\star = (x_{i-1} + x_i)/2$, the midpoint of the ith subinterval.
41. Work Problem 13 with $x_i^\star = (3x_{i-1} + 2x_i)/5$.
42. Work Problem 14 with $x_i^\star = (x_{i-1} + 2x_i)/3$.

In Problems 43 through 48, evaluate the given integral by computing

$$\lim_{n\to\infty} \sum_{i=1}^{n} f(x_i)\,\Delta x$$

for a regular partition of the given interval of integration.

43. $\int_0^2 x^2\,dx$
44. $\int_0^4 x^3\,dx$

45. $\displaystyle\int_0^3 (2x+1)\,dx$

46. $\displaystyle\int_1^5 (4-3x)\,dx$

47. $\displaystyle\int_0^3 (3x^2+1)\,dx$

48. $\displaystyle\int_0^4 (x^3-x)\,dx$

49. Show by the method of Example 4 that

$$\int_0^b x^2\,dx = \frac{1}{3}b^3$$

if $b > 0$.

50. Show by the method of Example 4 that

$$\int_0^b x^3\,dx = \frac{1}{4}b^4$$

if $b > 0$.

51. Let $f(x) = x$, and let $\{x_0, x_1, x_2, \dots, x_n\}$ be an arbitrary partition of the closed interval $[a, b]$. For each i $(1 \leqq i \leqq n)$, let $x_i^\star = (x_{i-1} + x_i)/2$. Then show that

$$\sum_{i=1}^n x_i^\star\,\Delta x_i = \frac{1}{2}b^2 - \frac{1}{2}a^2.$$

Explain why this computation proves that

$$\int_a^b x\,dx = \frac{b^2-a^2}{2}.$$

52. Suppose that f is a function continuous on $[a, b]$ and that k is a constant. Use Riemann sums to prove that

$$\int_a^b kf(x)\,dx = k\int_a^b f(x)\,dx.$$

53. Suppose that $f(x) \equiv c$, a constant. Use Riemann sums to prove that

$$\int_a^b c\,dx = c(b-a).$$

[*Suggestion:* First consider the case $a < b$.]

54. Suppose that the function f is defined on the interval $[0, 1]$ as follows:

$$f(x) = \begin{cases} \dfrac{1}{x} & \text{if } 0 < x \leqq 1, \\ 0 & \text{if } x = 0. \end{cases}$$

Show that the integral $\int_0^1 f(x)\,dx$ does not exist. [*Suggestion:* Show that, whatever n may be, the first term in the Riemann sum $\sum_{i=1}^n f(x_i^\star)\,\Delta x$ can be made arbitrarily large by the choice of the first selected point $x_i^\star$.] Why does this not contradict Theorem 1?

55. Suppose that the function f is defined as follows:

$$f(x) = \begin{cases} 0 & \text{if } x \text{ is rational,} \\ 1 & \text{if } x \text{ is irrational.} \end{cases}$$

Show that the integral $\int_0^1 f(x)\,dx$ does not exist. [*Suggestion:* Show that, whatever n may be, the Riemann sum $\sum_{i=1}^n f(x_i^\star)\,\Delta x$ has the value 0 for one possible selection of points $\{x_i^\star\}$, but the value 1 for another possible selection.] Why does this not contradict Theorem 1?

Use the method of Example 5 to verify the results in Problems 56 through 58.

56. $\displaystyle\int_0^5 e^x\,dx = e^5 - 1.$

57. $\displaystyle\int_0^3 e^{-x}\,dx = 1 - e^{-3}.$

58. $\displaystyle\int_2^5 e^x\,dx = e^5 - e^2.$

59. First show that

$$\int_0^\pi \sin x\,dx = \lim_{n\to\infty} \frac{\pi}{n}\sum_{k=1}^n \sin\frac{k\pi}{n}.$$

A computer algebra system reports that

$$\sum_{k=1}^n \sin\frac{k\pi}{n} = \cot\frac{\pi}{2n}.$$

Use this fact and l'Hôpital's rule to show finally that

$$\int_0^\pi \sin x\,dx = 2.$$

In Problems 60 through 62, verify the given result as follows: Use a computer algebra system first to set up the appropriate Riemann sum, then to simplify the sum, and finally to evaluate its limit as $n \to +\infty$.

60. $\displaystyle\int_a^b e^x\,dx = e^b - e^a.$

61. $\displaystyle\int_a^b \sin x\,dx = \cos a - \cos b.$

62. $\displaystyle\int_a^b \cos x\,dx = \sin b - \sin a.$

5.4 Project: Calculator/Computer Riemann Sums

Suppose that you want to approximate the integral

$$\int_a^b f(x)\,dx$$

numerically using *midpoint sums.* If $\Delta x = (b-a)/n$ and

$$m_i = x_i - \frac{1}{2}\Delta x = (a + i\cdot\Delta x) - \frac{1}{2}\Delta x = a + \left(i - \frac{1}{2}\right)\Delta x$$

is the midpoint of the ith subinterval $[x_{i-1}, x_i]$, then the selection $x_i^\star = m_i$ in Eq. (14) gives

$$\int_a^b f(x)\,dx = \lim_{n\to\infty} \sum_{i=1}^{n} f(m_i)\,\Delta x.$$

Many calculators and computer algebra systems include **Sum** commands that can be used to calculate easily and rapidly the midpoint sum with larger and larger values of n. A common practice is to begin with perhaps $n = 50$ subintervals, then calculate midpoint sums with successively doubled numbers of subintervals; that is, with $n = 50, 100, 200, \dots,$ until successive sums agree to the desired number of decimal places of accuracy. In the CD-ROM resource material for this project we illustrate this procedure using graphing calculators and typical computer algebra systems. You can then carry out the following investigations.

1. Approximate the integral

$$\int_0^2 e^x\,dx = e^2 - 1 \approx 6.3891$$

of Example 5 accurate to four decimal places.

2. Approximate the integral

$$\int_0^\pi \sin x\,dx = 2 = 2.0000$$

of Problem 61 accurate to four decimal places.

3. First explain why Fig. 5.4.12 and the area formula $A = \pi r^2$ for a circle of radius r imply that

$$\int_0^1 4\sqrt{1-x^2}\,dx = \pi.$$

Then use midpoint sums to approximate this integral and, thereby, the numerical value of π. Begin with $n = 50$ subintervals, then successively double n. How large must n be for you to obtain the familiar four-place approximation $\pi \approx 3.1416$?

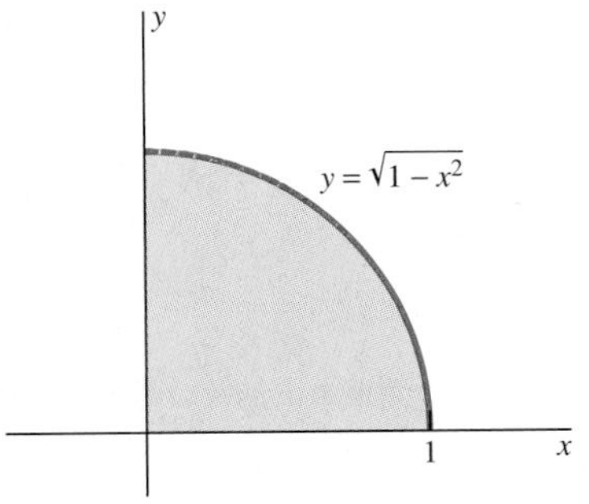

FIGURE 5.4.12 Investigation 3.

5.5 EVALUATION OF INTEGRALS

The evaluation of integrals by using Riemann sums, as in Section 5.4, is tedious and time-consuming. Fortunately, we will seldom find it necessary to evaluate an integral in this way. In 1666, Isaac Newton, while still a student at Cambridge University, discovered a much more efficient way to evaluate an integral. A few years later, Gottfried Wilhelm Leibniz, working with a different approach, discovered this method independently.

Newton's key idea was that to evaluate the *number*

$$\int_a^b f(x)\,dx,$$

we should first introduce the *function* $A(x)$ defined as follows:

$$A(x) = \int_a^x f(t)\,dt. \tag{1}$$

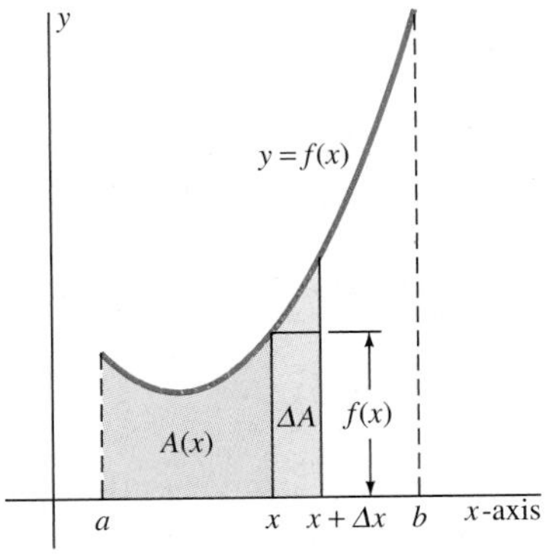

FIGURE 5.5.1 The area function $A(x)$.

The independent variable x appears as the *upper limit* of the integral in Eq. (1); the dummy variable t is used in the integrand merely to avoid confusion. If f is positive-valued, continuous, and $x > a$, then $A(x)$ is the area below the curve $y = f(x)$ above the interval $[a, x]$ (Fig. 5.5.1).

It is apparent from Fig. 5.5.1 that $A(x)$ increases as x increases. When x increases by Δx, A increases by the area ΔA of the narrow strip in Fig. 5.5.1 with base $[x, x + \Delta x]$. If Δx is very small, then the area of this strip is very close to the area $f(x)\,\Delta x$ of the rectangle with base $[x, x + \Delta x]$ and height $f(x)$. Thus

$$\Delta A \approx f(x)\,\Delta x; \qquad \frac{\Delta A}{\Delta x} \approx f(x). \tag{2}$$

Moreover, the figure makes it plausible that we get equality in the limit at $\Delta x \to 0$:

$$\frac{dA}{dx} = \lim_{\Delta x \to 0} \frac{\Delta A}{\Delta x} = f(x).$$

That is,

$$A'(x) = f(x), \tag{3}$$

so *the derivative of the area function $A(x)$ is the curve's height function $f(x)$*. In other words, Eq. (3) implies that $A(x)$ is an *antiderivative* of $f(x)$.

Figure 5.5.2 shows a physical interpretation of Eq. (3). A paint roller is laying down a 1-mm-thick coat of paint to cover the region under the curve $y = f(t)$. The paint roller is of adjustable length—as it rolls with a speed of 1 mm/s from left to right, one end traces the x-axis and the other end traces the curve $y = f(t)$. At any time t, the volume V of paint the roller has laid down equals the area of the region already painted:

$$V = A(t) \quad (\text{mm}^3).$$

Then Eq. (3) yields

$$\frac{dV}{dt} = A'(t) = f(t).$$

Thus the instantaneous rate at which the roller is depositing paint is equal to the current length of the roller.

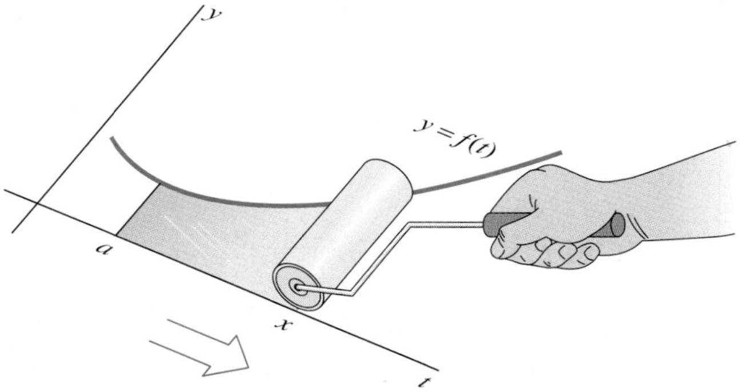

FIGURE 5.5.2 The adjustable-length paint roller.

The Evaluation Theorem

Equation (3) implies that the area function $A(x)$ defined in (1) and illustrated in Fig. 5.5.1 is *one* antiderivative of the given function $f(x)$. Now suppose that $G(x)$ is any other antiderivative of $f(x)$—perhaps one found by the methods of Section 5.2. Then

$$A(x) = G(x) + C, \tag{4}$$

because (by the second corollary to the mean value theorem) two antiderivatives of the same function (on an interval) can differ only by a constant. Also,

$$A(a) = \int_a^a f(t)\,dt = 0 \tag{5}$$

and

$$A(b) = \int_a^b f(t)\,dt = \int_a^b f(x)\,dx \tag{6}$$

by Eq. (1). So it follows that

$$\int_a^b f(x)\,dx = A(b) - A(a) = [G(b) + C] - [G(a) + C],$$

and thus

$$\int_a^b f(x)\,dx = G(b) - G(a). \tag{7}$$

Our intuitive discussion has led us to the statement of Theorem 1.

THEOREM 1 Evaluation of Integrals
If G is an antiderivative of the continuous function f on the interval $[a, b]$, then

$$\int_a^b f(x)\,dx = G(b) - G(a). \tag{7}$$

In Section 5.6 we will fill in the details of the preceding discussion, thus giving a rigorous proof of Theorem 1 (which is part of the fundamental theorem of calculus). Here we concentrate on the computational applications of this theorem. The difference $G(b) - G(a)$ is customarily abbreviated as $[G(x)]_a^b$, so Theorem 1 implies that

$$\int_a^b f(x)\,dx = \Big[G(x)\Big]_a^b = G(b) - G(a) \tag{8}$$

if G is any antiderivative of the continuous function f on the interval $[a, b]$. Thus if we can find an antiderivative G of f, we can quickly evaluate the integral *without* having to resort to the paraphernalia of limits of Riemann sums.

If $G'(x) = f(x)$, then (as in Section 5.2) we write

$$\int f(x)\,dx = G(x) + C \tag{9}$$

for the indefinite integral of f. With the indefinite integral $\int f(x)\,dx$ in place of the antiderivative $G(x)$, Eq. (8) takes the form

$$\int_a^b f(x)\,dx = \left[\int f(x)\,dx\right]_a^b. \tag{10}$$

This is the connection between the indefinite integral and the definite integral to which we have alluded in the earlier sections of Chapter 5.

EXAMPLE 1 Because

$$\int x^n\,dx = \frac{x^{n+1}}{n+1} + C \quad (\text{if } n \neq -1),$$

it follows that

$$\int_a^b x^n\,dx = \left[\frac{x^{n+1}}{n+1}\right]_a^b = \frac{b^{n+1} - a^{n+1}}{n+1}$$

if $n \neq -1$. For instance,

$$\int_0^3 x^2\,dx = \left[\tfrac{1}{3}x^3\right]_0^3 = \tfrac{1}{3} \cdot 3^3 - \tfrac{1}{3} \cdot 0^3 = 9.$$

Contrast the immediacy of this result with the complexity of the computations of Example 7 in Section 5.3. ◆

EXAMPLE 2 Because

$$\int \cos x \, dx = \sin x + C,$$

it follows that

$$\int_a^b \cos x \, dx = \Big[\sin x \Big]_a^b = \sin b - \sin a.$$

Similarly,

$$\int_a^b \sin x \, dx = \Big[-\cos x \Big]_a^b = \cos a - \cos b.$$

In particular, as we mentioned in Example 2 of Section 5.4,

$$\int_0^\pi \sin x \, dx = \Big[-\cos x \Big]_0^\pi = (-\cos \pi) - (-\cos 0) = (+1) - (-1) = 2.$$ ◆

EXAMPLE 3

$$\int_0^2 x^5 \, dx = \Big[\tfrac{1}{6}x^6 \Big]_0^2 = \tfrac{64}{6} - 0 = \tfrac{32}{3}.$$

$$\int_1^9 \left(2x - x^{-1/2} - 3\right) dx = \Big[x^2 - 2x^{1/2} - 3x \Big]_1^9 = 52.$$

$$\int_0^1 (2x+1)^3 \, dx = \Big[\tfrac{1}{8}(2x+1)^4 \Big]_0^1 = \tfrac{1}{8} \cdot (81 - 1) = 10.$$

$$\int_0^{\pi/2} \sin 2x \, dx = \Big[-\tfrac{1}{2} \cos 2x \Big]_0^{\pi/2} = -\tfrac{1}{2}(\cos \pi - \cos 0) = 1.$$

$$\int_0^1 e^{2x} \, dx = \Big[\tfrac{1}{2} e^{2x} \Big]_0^1 = \tfrac{1}{2}(e^2 - 1).$$

We have not shown the details of finding the antiderivatives, but you can (and should) check each of these results by showing that the derivative of the function within the evaluation brackets on the right is equal to the integrand on the left. In Example 4 we show the details. ◆

EXAMPLE 4 Evaluate $\displaystyle\int_1^5 \sqrt{3x+1} \, dx$.

Solution We apply the antiderivative form of the generalized power rule,

$$\int u^k \, du = \frac{u^{k+1}}{k+1} + C \quad (k \neq -1),$$

with $k = \frac{1}{2}$ and

$$u = 3x + 1, \qquad du = 3 \, dx.$$

This gives

$$\int (3x+1)^{1/2} \, dx = \tfrac{1}{3} \int (3x+1)^{1/2} \, (3 \, dx) = \tfrac{1}{3} \int u^{1/2} \, du$$

$$= \tfrac{1}{3} \cdot \frac{u^{3/2}}{\frac{3}{2}} + C = \tfrac{2}{9}(3x+1)^{3/2} + C$$

for the indefinite integral, so it follows from Eq. (10) that

$$\int_1^5 \sqrt{3x+1}\,dx = \left[\tfrac{2}{9}(3x+1)^{3/2}\right]_1^5$$
$$= \tfrac{2}{9}(16^{3/2} - 4^{3/2}) = \tfrac{2}{9}(4^3 - 2^3) = \tfrac{112}{9}.$$
◆

If the derivative $F'(x)$ of the function $F(x)$ is continuous, then the evaluation theorem, with $F'(x)$ in place of $f(x)$ and $F(x)$ in place of $G(x)$, yields

$$\int_a^b F'(x)\,dx = \Big[F(x)\Big]_a^b = F(b) - F(a). \tag{11}$$

The next example provides an immediate application.

EXAMPLE 5 Suppose that an animal population $P(t)$ initially numbers $P(0) = 100$ and that its rate of growth after t months is given by

$$P'(t) = 10 + t + (0.06)t^2.$$

What is the population after 10 months?

Solution By Eq. (11), we know that

$$P(10) - P(0) = \int_0^{10} P'(t)\,dt = \int_0^{10} [10 + t + (0.06)t^2]\,dt$$
$$= \left[10t + \tfrac{1}{2}t^2 + (0.02)t^3\right]_0^{10} = 170.$$

Thus $P(10) = 100 + 170 = 270$ individuals. ◆

EXAMPLE 6 Evaluate

$$\lim_{n\to\infty} \sum_{i=1}^{n} \frac{2i}{n^2}$$

by recognizing this limit as the value of an integral.

Solution If we write

$$\sum_{i=1}^{n} \frac{2i}{n^2} = \sum_{i=1}^{n} \left(\frac{2i}{n}\right)\left(\frac{1}{n}\right),$$

we recognize that we have a Riemann sum for the function $f(x) = 2x$ associated with a partition of the interval $[0, 1]$ into n equal-length subintervals. The ith points of subdivision is $x_i = i/n$, and $\Delta x = 1/n$. Hence it follows from the definition of the integral and from the evaluation theorem that

$$\lim_{n\to\infty} \sum_{i=1}^{n} \frac{2i}{n^2} = \lim_{n\to\infty} \sum_{i=1}^{n} 2x_i\,\Delta x = \lim_{n\to\infty} \sum_{i=1}^{n} f(x_i)\,\Delta x$$
$$= \int_0^1 f(x)\,dx = \int_0^1 2x\,dx.$$

Therefore,

$$\lim_{n\to\infty} \sum_{i=1}^{n} \frac{2i}{n^2} = \Big[x^2\Big]_0^1 = 1.$$
◆

Basic Properties of Integrals

Problems 59 through 62 outline elementary proofs of the integral properties that are stated next. We assume throughout that each function mentioned is integrable on $[a, b]$.

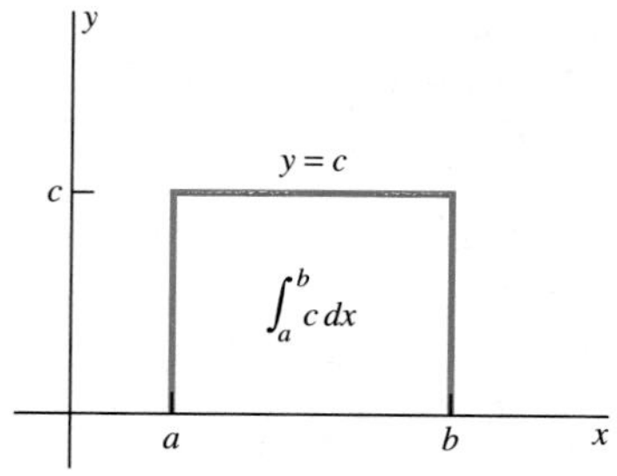

FIGURE 5.5.3 The integral of a constant is the area of a rectangle.

Integral of a Constant

$$\int_a^b c\,dx = c(b - a).$$

This property is intuitively obvious because the area represented by the integral is simply a rectangle with base $b - a$ and height c (Fig. 5.5.3).

Constant Multiple Property

$$\int_a^b cf(x)\,dx = c\int_a^b f(x)\,dx.$$

Thus a constant can be "moved across" the integral sign. For example,

$$\int_0^{\pi/2} 2\sin x\,dx = 2\int_0^{\pi/2} \sin x\,dx = 2\Big[-\cos x\Big]_0^{\pi/2} = 2.$$

Sum Property

$$\int_a^b [f(x) + g(x)]\,dx = \int_a^b f(x)\,dx + \int_a^b g(x)\,dx.$$

Thus if the functions f and g are both integrable on $[a, b]$, then *the integral of their sum is equal to the sum of their integrals*. This fact sometimes permits a "divide-and-conquer" strategy for the calculation of integrals:

$$\int_0^{\pi}\left(3\sqrt{x} + \cos\frac{x}{2}\right)dx = \int_0^{\pi} 3\sqrt{x}\,dx + \int_0^{\pi} \cos\frac{x}{2}\,dx$$

$$= \Big[2x^{3/2}\Big]_0^{\pi} + \Big[2\sin\frac{x}{2}\Big]_0^{\pi} = 2\pi^{3/2} + 2.$$

Figure 5.5.4 illustrates geometrically the sum property of integrals. The proof of the sum property illustrates a Riemann sums approach that can be adapted to all of

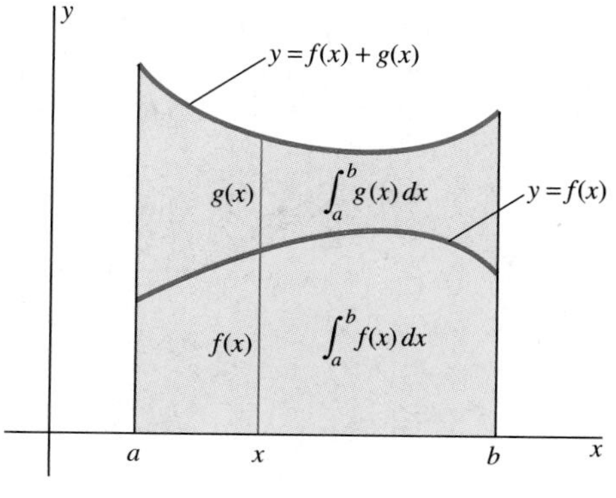

FIGURE 5.5.4 The integral of the sum of two positive-valued functions.

the properties under discussion here. Let us think of partitions of the interval $[a, b]$ into subintervals all having the same length Δx. If the functions f, g, and $f + g$ are all integrable, then Theorem 2 in Section 5.4 gives

$$\begin{aligned}\int_a^b [f(x) + g(x)]\,dx &= \lim_{\Delta x \to 0} \sum_{i=1}^{n} [f(x_i^\star) + g(x_i^\star)]\,\Delta x \\ &= \lim_{\Delta x \to 0} \left[\sum_{i=1}^{n} f(x_i^\star)\,\Delta x + \sum_{i=1}^{n} g(x_i^\star)\,\Delta x\right] \\ &= \left[\lim_{\Delta x \to 0} \sum_{i=1}^{n} f(x_i^\star)\,\Delta x\right] + \left[\lim_{\Delta x \to 0} \sum_{i=1}^{n} g(x_i^\star)\,\Delta x\right] \\ &= \int_a^b f(x)\,dx + \int_a^b g(x)\,dx.\end{aligned}$$

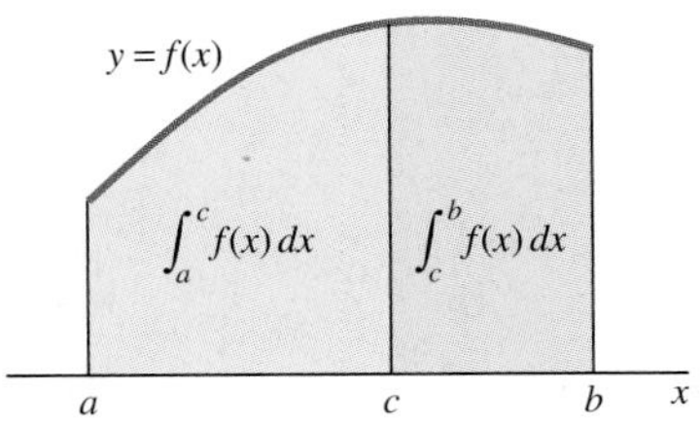

FIGURE 5.5.5 The way the interval union property works.

Interval Union Property

If $a < c < b$, then

$$\int_a^b f(x)\,dx = \int_a^c f(x)\,dx + \int_c^b f(x)\,dx.$$

Figure 5.5.5 indicates the plausibility of the interval union property.

EXAMPLE 7 If $f(x) = 2\,|x|$, then

$$f(x) = \begin{cases} -2x & \text{if } x \leqq 0, \\ 2x & \text{if } x \geqq 0. \end{cases}$$

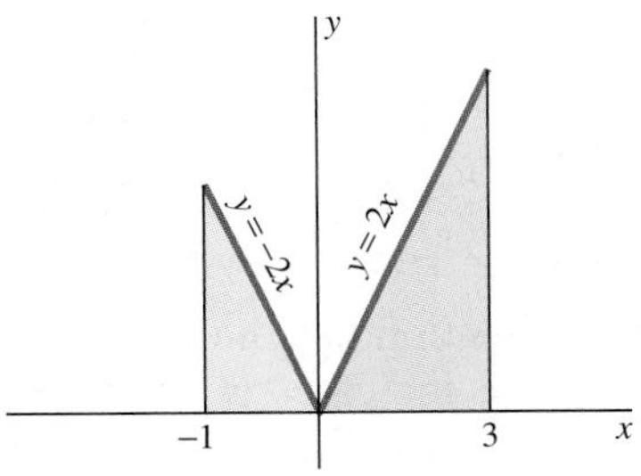

FIGURE 5.5.6 The area under the graph of $y = 2|x|$ over $[-1, 3]$.

The graph of f is shown in Fig. 5.5.6. An antiderivative of $f(x)$ is not evident, but the interval union property allows us to split the integral of f on $[-1, 3]$ into two easily calculated integrals:

$$\begin{aligned}\int_{-1}^{3} 2\,|x|\,dx &= \int_{-1}^{0} (-2x)\,dx + \int_0^3 (2x)\,dx \\ &= \Big[-x^2\Big]_{-1}^{0} + \Big[x^2\Big]_0^3 = [0 - (-1)] + [9 - 0] = 10.\end{aligned}$$

Does the result agree with Fig. 5.5.6? ◆

EXAMPLE 8 Evaluate the integral $\int_0^{2\pi} |\cos x - \sin x|\,dx$.

Solution Figure 5.5.7 shows the graph of the function $f(x) = \cos x - \sin x$ and Fig. 5.5.8 shows the graph of its absolute value $|f(x)| = |\cos x - \sin x|$ that we want

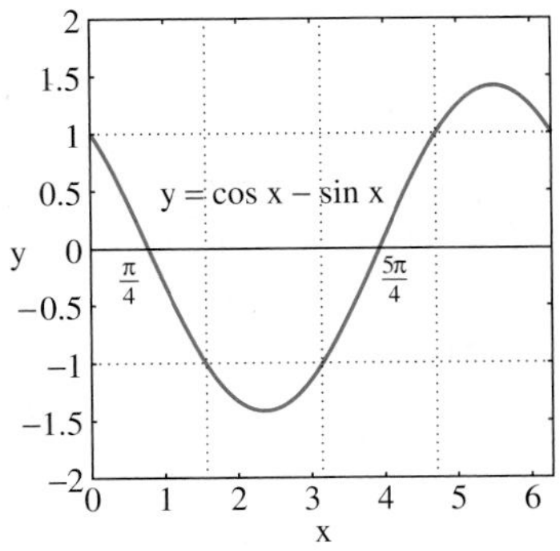

FIGURE 5.5.7 $y = \cos x - \sin x$.

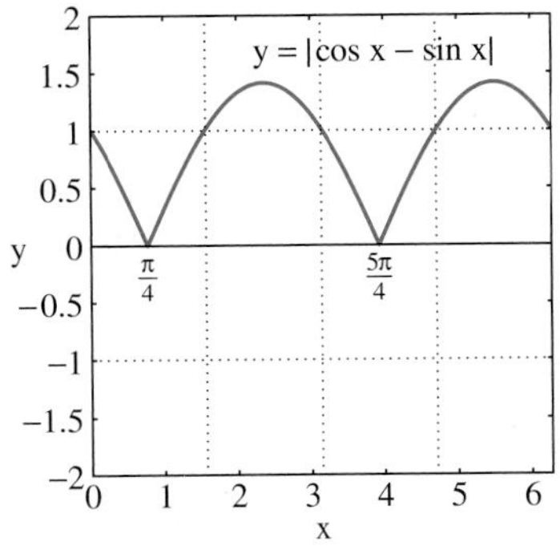

FIGURE 5.5.8 $y = |\cos x - \sin x|$.

to integrate. We readily see that $f(x) = 0$ at $x = \pi/4$ and $x = 5\pi/4$, so

$$|f(x)| = \begin{cases} \cos x - \sin x & \text{if } 0 \leqq x < \pi/4, \\ \sin x - \cos x & \text{if } \pi/4 \leqq x < 5\pi/4, \\ \cos x - \sin x & \text{if } 5\pi/4 \leqq x \leqq 2\pi. \end{cases}$$

The interval union property therefore gives

$$\begin{aligned} &\int_0^{2\pi} |\cos x - \sin x|\, dx \\ &= \int_0^{\pi/4} (\cos x - \sin x)\, dx + \int_{\pi/4}^{5\pi/4} (\sin x - \cos x)\, dx + \int_{5\pi/4}^{2\pi} (\cos x - \sin x)\, dx \\ &= \Big[\sin x + \cos x\Big]_0^{\pi/4} + \Big[-\cos x - \sin x\Big]_{\pi/4}^{5\pi/4} + \Big[\sin x + \cos x\Big]_{5\pi/4}^{2\pi} \\ &= \left(\tfrac{1}{2}\sqrt{2} + \tfrac{1}{2}\sqrt{2}\right) - (0 + 1) + \left(\tfrac{1}{2}\sqrt{2} + \tfrac{1}{2}\sqrt{2}\right) \\ &\quad - \left(-\tfrac{1}{2}\sqrt{2} - \tfrac{1}{2}\sqrt{2}\right) + (0 + 1) - \left(-\tfrac{1}{2}\sqrt{2} - \tfrac{1}{2}\sqrt{2}\right) \\ &= 4\sqrt{2}. \end{aligned}$$

◆

Comparison Properties

(1) If $f(x) \leqq g(x)$ for all x in $[a, b]$, then

$$\int_a^b f(x)\, dx \leqq \int_a^b g(x)\, dx.$$

(2) If $m \leqq f(x) \leqq M$ for all x in $[a, b]$, then

$$m(b - a) \leqq \int_a^b f(x)\, dx \leqq M(b - a).$$

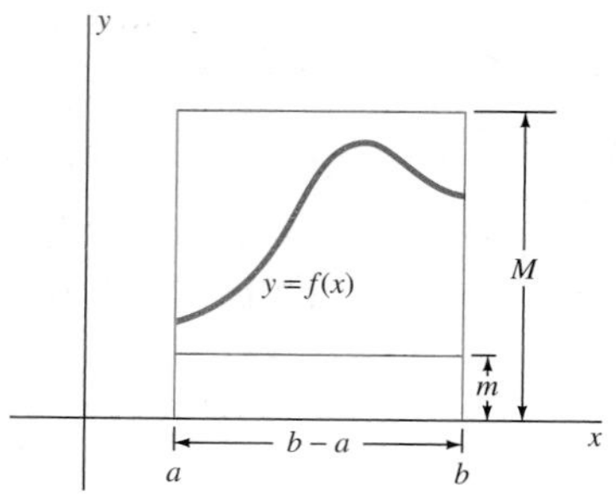

FIGURE 5.5.9 Plausibility of the comparison property.

The first comparison property says that the larger function has the larger integral. The plausibility of the second comparison property is indicated in Fig. 5.5.9. Note that m and M need not necessarily be the minimum and maximum values of $f(x)$ on $[a, b]$.

EXAMPLE 9 Figure 5.5.10 shows the graphs

$$y = \sqrt{1 + x}, \qquad y = \sqrt{1 + \sqrt{x}}, \quad \text{and} \quad y = 1.2 + (0.3)x,$$

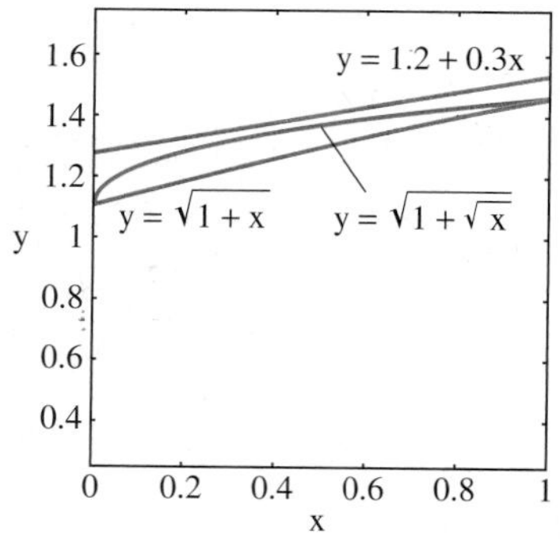

FIGURE 5.5.10 Bounding the graph of $f(x) = \sqrt{1 + \sqrt{x}}$.

and we see that

$$\sqrt{1 + x} \leqq \sqrt{1 + \sqrt{x}} \leqq 1.2 + (0.3)x \tag{12}$$

for x in $[0, 1]$. Indeed, the fact that $x \leqq \sqrt{x}$ for x in $[0, 1]$ implies that $\sqrt{1 + x} \leqq \sqrt{1 + \sqrt{x}}$ there. The graph $y = 1.2 + (0.3)x$ that lies above $y = \sqrt{1 + \sqrt{x}}$ was discovered empirically using a graphing calculator. At any rate, the inequalities in (12) and the first comparison property of integrals imply that

$$\int_0^1 \sqrt{1 + x}\, dx \leqq \int_0^1 \sqrt{1 + \sqrt{x}}\, dx \leqq \int_0^1 [1.2 + (0.3)x]\, dx;$$

thus

$$\left[\tfrac{2}{3}(1 + x)^{3/2}\right]_0^1 \leqq \int_0^1 \sqrt{1 + \sqrt{x}}\, dx \leqq \left[(1.2)x + (0.15)x^2\right]_0^1 = 1.35.$$

Now $\frac{2}{3}(2^{3/2} - 1) \approx 1.2190$, so we see finally that

$$1.21 \leqq \int_0^1 \sqrt{1 + \sqrt{x}}\, dx \leqq 1.35. \qquad \textbf{(13)}$$

It turns out (using the methods of Section 5.9) that the actual value of $\int_0^1 \sqrt{1 + \sqrt{x}}\, dx$ is 1.29 rounded to two decimal places—quite close to the average 1.28 of the upper and lower bounds in (13). ◆

The properties of integrals stated here are frequently used in computing and will be applied in the proof of the fundamental theorem of calculus in Section 5.6.

5.5 TRUE/FALSE STUDY GUIDE

5.5 CONCEPTS: QUESTIONS AND DISCUSSION

1. Let f be a continuous function defined on the closed interval $[a, b]$. Explain the difference between the functions g and h defined for t in $[a, b]$ by

$$g(t) = \int_a^b f(x)\, dx \quad \text{and} \quad h(t) = \int_a^t f(x)\, dx.$$

What is the difference between their derivatives $g'(t)$ and $h'(t)$?

2. What is the relation between the integrals

$$\int_a^b f(x)\, dx \quad \text{and} \quad \int_a^b |f(x)|\, dx?$$

What is the difference between their absolute values? Discuss separately the following cases:
(a) f is positive-valued on the interval $[a, b]$;
(b) f is negative-valued on the interval $[a, b]$;
(c) f has both positive and negative values on $[a, b]$.

5.5 PROBLEMS

Apply the evaluation theorem to evaluate the integrals in Problems 1 through 36.

1. $\displaystyle\int_0^1 \left(3x^2 + 2\sqrt{x} + 3\sqrt[3]{x}\right) dx$

2. $\displaystyle\int_1^3 \frac{6}{x^2}\, dx$

3. $\displaystyle\int_0^1 x^3(1 + x)^2\, dx$

4. $\displaystyle\int_{-2}^{-1} \frac{1}{x^4}\, dx$

5. $\displaystyle\int_0^1 (x^4 - x^3)\, dx$

6. $\displaystyle\int_1^2 (x^4 - x^3)\, dx$

7. $\displaystyle\int_{-1}^0 (x + 1)^3\, dx$

8. $\displaystyle\int_1^3 \frac{x^4 + 1}{x^2}\, dx$

9. $\displaystyle\int_0^4 \sqrt{x}\, dx$

10. $\displaystyle\int_1^4 \frac{1}{\sqrt{x}}\, dx$

11. $\displaystyle\int_{-1}^2 (3x^2 + 2x + 4)\, dx$

12. $\displaystyle\int_0^1 x^{99}\, dx$

13. $\displaystyle\int_{-1}^1 x^{99}\, dx$

14. $\displaystyle\int_0^4 \left(7x^{5/2} - 5x^{3/2}\right) dx$

15. $\displaystyle\int_1^3 (x - 1)^5\, dx$

16. $\displaystyle\int_1^2 (x^2 + 1)^3\, dx$

17. $\displaystyle\int_{-1}^0 (2x + 1)^3\, dx$

18. $\displaystyle\int_1^3 \frac{10}{(2x + 3)^2}\, dx$

19. $\displaystyle\int_1^8 x^{2/3}\, dx$

20. $\displaystyle\int_1^9 \left(1 + \sqrt{x}\right)^2 dx$

21. $\displaystyle\int_{-1}^1 (e^x - e^{-x})\, dx$

22. $\displaystyle\int_0^4 \sqrt{3t}\, dt$

23. $\displaystyle\int_0^2 \sqrt{e^{3t}}\, dt$

24. $\displaystyle\int_2^3 \frac{du}{u^2}$ $\left(\text{Note the abbreviation for } \dfrac{1}{u^2}\, du.\right)$

25. $\displaystyle\int_1^2 \frac{1}{t}\, dt$

26. $\displaystyle\int_5^{10} \frac{1}{x}\, dx$

27. $\displaystyle\int_0^1 (e^x - 1)^2\, dx$

28. $\displaystyle\int_0^{\pi/2} \cos 2x\, dx$

29. $\displaystyle\int_0^{\pi/4} \sin x \cos x\, dx$

30. $\displaystyle\int_0^{\pi} \sin^2 x \cos x\, dx$

31. $\displaystyle\int_0^{\pi} \sin 5x\,dx$

32. $\displaystyle\int_0^{2} \cos \pi t\,dt$

33. $\displaystyle\int_0^{\pi/2} \cos 3x\,dx$

34. $\displaystyle\int_0^{5} \sin \frac{\pi x}{10}\,dx$

35. $\displaystyle\int_0^{2} \cos \frac{\pi x}{4}\,dx$

36. $\displaystyle\int_0^{\pi/8} \sec^2 2t\,dt$

In Problems 37 through 42, evaluate the given limit by first recognizing the indicated sum as a Riemann sum associated with a regular partition of $[0, 1]$ *and then evaluating the corresponding integral.*

37. $\displaystyle\lim_{n\to\infty} \sum_{i=1}^{n} \left(\frac{2i}{n} - 1\right)\frac{1}{n}$

38. $\displaystyle\lim_{n\to\infty} \sum_{i=1}^{n} \frac{i^2}{n^3}$

39. $\displaystyle\lim_{n\to\infty} \frac{1+2+3+\cdots+n}{n^2}$

40. $\displaystyle\lim_{n\to\infty} \frac{1^3+2^3+3^3+\cdots+n^3}{n^4}$

41. $\displaystyle\lim_{n\to\infty} \frac{\sqrt{1}+\sqrt{2}+\sqrt{3}+\cdots+\sqrt{n}}{n\sqrt{n}}$

42. $\displaystyle\lim_{n\to\infty} \sum_{i=1}^{n} \frac{1}{n}\sin\frac{\pi i}{n}$

In Problems 43 through 48, an integral $\int_a^b f(x)\,dx$ *is given. First sketch the graph* $y = f(x)$ *on the interval* $[a, b]$. *Then, interpreting the integral as the area of a region, evaluate it using known area formulas for rectangles, triangles, and circles.*

43. $\displaystyle\int_{-2}^{2} |1 - x|\,dx$

44. $\displaystyle\int_{-3}^{3} |3x - 2|\,dx$

45. $\displaystyle\int_0^{5} (2 - |x|)\,dx$

46. $\displaystyle\int_0^{6} \big|5 - |2x|\big|\,dx$

47. $\displaystyle\int_0^{5} \sqrt{25 - x^2}\,dx$

48. $\displaystyle\int_0^{6} \sqrt{6x - x^2}\,dx$ (*Suggestion:* Complete the square.)

In Problems 49 through 54, use properties of integrals to establish each inequality without evaluating the integrals involved.

49. $\displaystyle 1 \leqq \int_0^1 \sqrt{1+x^2}\,dx \leqq \int_0^1 \sqrt{1+x}\,dx$

50. $\displaystyle\int_1^2 \sqrt{1+x}\,dx \leqq \int_1^2 \sqrt{1+x^3}\,dx \leqq \sqrt{10}$

51. $\displaystyle\int_0^1 \frac{1}{1+\sqrt{x}}\,dx \leqq \int_0^1 \frac{1}{1+x^2}\,dx$

52. $\displaystyle\int_2^5 \frac{1}{1+x^5}\,dx \leqq \int_2^5 \frac{1}{1+x^2}\,dx$

53. $\displaystyle\int_0^2 \sin\sqrt{x}\,dx \leqq 2$

54. $\displaystyle\frac{\pi}{8} \leqq \int_0^{\pi/4} \frac{1}{1+\cos^2 x}\,dx \leqq \frac{\pi}{6}$

In Problems 55 through 58, use the second comparison property of integrals to estimate—giving both a lower bound and an upper bound as in Problem 54—the value of the given integral.

55. $\displaystyle\int_0^1 \frac{1}{1+x}\,dx$

56. $\displaystyle\int_4^9 \frac{1}{1+\sqrt{x}}\,dx$

57. $\displaystyle\int_0^{\pi/6} \cos^2 x\,dx$

58. $\displaystyle\int_0^{\pi/4} \sqrt{16 + 2\sin^2 x}\,dx$

59. Use Riemann sums—as in the proof of the sum property of integrals—to establish the constant multiple property.

60. Use Riemann sums to establish the first comparison property of integrals.

61. Deduce the second comparison property of integrals from the first comparison property.

62. Use sequences of Riemann sums to establish the interval union property of the integral. Note that if R_n' and R_n'' are Riemann sums for f on the intervals $[a, c]$ and $[c, b]$, respectively, then $R_n = R_n' + R_n''$ is a Riemann sum for f on $[a, b]$.

63. Suppose that a tank initially contains 1000 gal of water and that the rate of change of its volume after the tank drains for t min is $V'(t) = (0.8)t - 40$ (in gallons per minute). How much water does the tank contain after it has been draining for a half-hour?

64. Suppose that the population of Juneau in 1970 was 125 (in thousands) and that its rate of growth t years later was $P'(t) = 8 + (0.5)t + (0.03)t^2$ (in thousands per year). What was its population in 1990?

65. Figure 5.5.11 shows the graph of $f(x) = 1/x$ on the interval $[1, 2]$, the line joining its endpoints $(1, 1)$ and $(2, \frac{1}{2})$, and its tangent line at the point $(\frac{3}{2}, \frac{2}{3})$. Use this construction to estimate the value of the integral

$$\int_1^2 \frac{1}{x}\,dx$$

(whose exact value is known to be $\ln 2 \approx 0.693$).

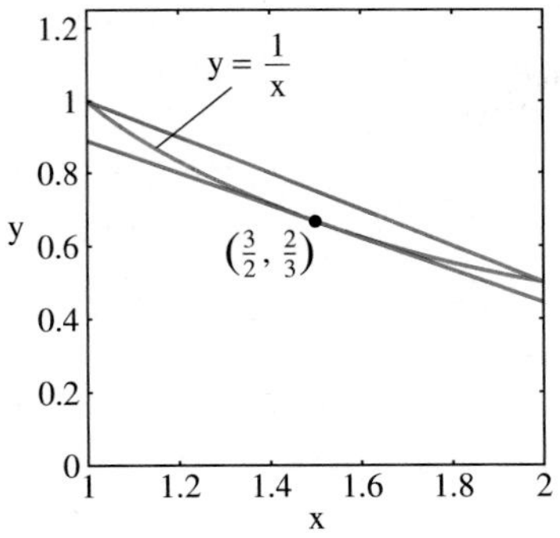

FIGURE 5.5.11 Bounding the graph of $f(x) = \dfrac{1}{x}$.

66. Figure 5.5.12 shows the graph of $f(x) = 1/(1 + x^2)$ on the interval $[0, 1]$, the line $y = L(x)$ joining its endpoints $(0, 1)$ and $(1, \frac{1}{2})$, and the line $y = L(x) + 0.07$. First graph $f(x) - L(x)$ to verify that the latter line lies above $y = f(x)$ on the interval $[0, 1]$. Then use this construction to estimate the value of the integral

$$\int_0^1 \frac{1}{1+x^2}\,dx$$

(whose exact value is known to be $\frac{1}{4}\pi \approx 0.785$).

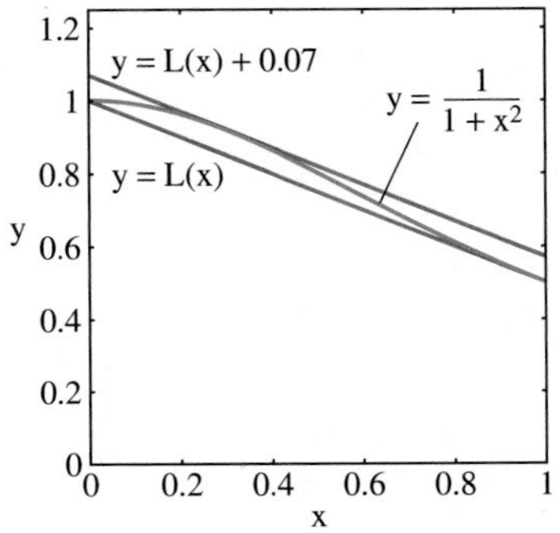

FIGURE 5.5.12 Bounding the graph of $f(x) = \dfrac{1}{1+x^2}$.

5.6 THE FUNDAMENTAL THEOREM OF CALCULUS

Newton and Leibniz are generally credited with the invention of calculus in the latter part of the seventeenth century. Actually, others had earlier calculated areas essentially equivalent to integrals and tangent line slopes essentially equivalent to derivatives. The great accomplishments of Newton and Leibniz were the discovery and computational exploitation of the inverse relationship between differentiation and integration. This relationship is embodied in the *fundamental theorem of calculus.* One part of this theorem is the evaluation theorem of Section 5.5: To evaluate

$$\int_a^b f(x)\,dx,$$

it suffices to find an antiderivative of f on $[a, b]$. The other part of the fundamental theorem tells us that doing so is usually possible, at least in theory: Every continuous function has an antiderivative.

The Average Value of a Function

The concept of the *average value* of a function is useful for the proof of the fundamental theorem and has numerous important applications in its own right. The ordinary (arithmetic) **average** of n given numbers $a_1, a_2, \ldots, a_n$ is defined to be

$$\overline{a} = \frac{a_1 + a_2 + \cdots + a_n}{n} = \frac{1}{n}\sum_{i=1}^{n} a_i. \qquad \textbf{(1)}$$

But a function f defined on an interval generally has infinitely many values $f(x)$, so we cannot simply divide the sum of all these values by their number to find the average value of $f(x)$. We introduce the proper notion with a discussion of average temperature.

EXAMPLE 1 Let the measured temperature T during a particular 24-h day at a certain location be given by the function

$$T = f(t), \quad 0 \leqq t \leqq 24$$

(with the 24-h clock running from $t = 0$ at one midnight to $t = 24$ at the following midnight). Thus, for example, the temperatures $f(1), f(2), \ldots, f(24)$ are recorded at 1-h intervals during the day. We might define the average temperature $\overline{T}$ for the day as the (ordinary arithmetic) average of the hourly temperatures:

$$\overline{T} = \frac{1}{24}\sum_{i=1}^{24} f(t_i),$$

where $t_i = i$. If we divided the day into n equal subintervals rather than into 24 1-h intervals, we would obtain the more general average

$$\overline{T} = \frac{1}{n}\sum_{i=1}^{n} f(t_i).$$

The larger n is, the closer would we expect $\overline{T}$ to be to the "true" average temperature for the entire day. It is therefore plausible to define the true average temperature by letting n increase without bound. This gives

$$\overline{T} = \lim_{n\to\infty} \frac{1}{n}\sum_{i=1}^{n} f(t_i).$$

The right-hand side resembles a Riemann sum, and we can make it into a Riemann sum by introducing the factor

$$\Delta t = \frac{b-a}{n},$$

where $a = 0$ and $b = 24$. Then

$$\begin{aligned}
\overline{T} &= \lim_{n\to\infty} \frac{1}{n}\cdot\frac{n}{b-a}\sum_{i=1}^{n} f(t_i)\cdot\frac{b-a}{n} \\
&= \lim_{n\to\infty} \frac{1}{b-a}\sum_{i=1}^{n} f(t_i)\cdot\frac{b-a}{n} \\
&= \frac{1}{b-a}\lim_{n\to\infty}\sum_{i=1}^{n} f(t_i)\,\Delta t = \frac{1}{b-a}\int_a^b f(t)\,dt.
\end{aligned}$$

Thus

$$\overline{T} = \frac{1}{24}\int_0^{24} f(t)\,dt \qquad \textbf{(2)}$$

under the assumption that f is continuous, so the Riemann sums converge to the integral as $n \to \infty$. ◆

The final result in Eq. (2) is the *integral of the function divided by the length of the interval.* Example 1 motivates the following definition:

DEFINITION Average Value of a Function

Suppose that the function f is integrable on $[a, b]$. Then the **average value** $\overline{y}$ of $y = f(x)$ for x in the interval $[a, b]$ is

$$\overline{y} = \frac{1}{b-a}\int_a^b f(x)\,dx. \qquad \textbf{(3)}$$

We can rewrite Eq. (3) in the form

$$\int_a^b f(x)\,dx = \overline{y}\cdot(b-a). \qquad \textbf{(4)}$$

If f is positive-valued on $[a, b]$, then Eq. (4) implies that the area under $y = f(x)$ over $[a, b]$ is equal to the area of a rectangle with base length $b - a$ and height $\overline{y}$ (Fig. 5.6.1).

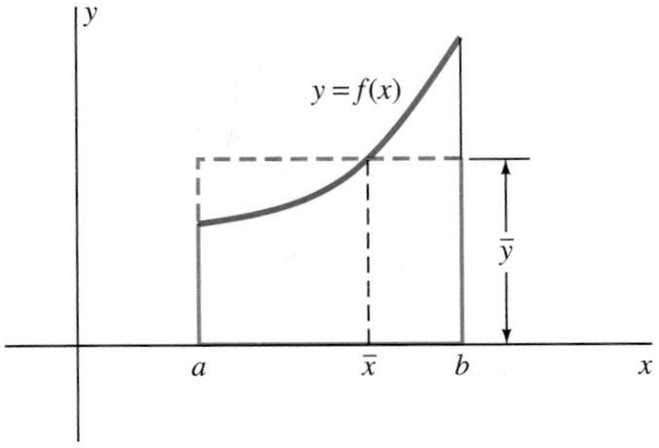

FIGURE 5.6.1 A rectangle illustrating the average value $\overline{y}$ of a function.

EXAMPLE 2 The average value of $f(x) = x^2$ for x in $[0, 2]$ is

$$\overline{y} = \frac{1}{2}\int_0^2 x^2\,dx = \frac{1}{2}\left[\frac{1}{3}x^3\right]_0^2 = \frac{4}{3}. \qquad \blacklozenge$$

EXAMPLE 3 In Athens, Georgia (USA) the mean daily temperature in degrees Fahrenheit t months after July 15 is closely approximated by

$$T = 61 + 18\cos\frac{\pi t}{6} = f(t). \tag{5}$$

Find the average temperature between September 15 ($t = 2$) and December 15 ($t = 5$).

Solution Equation (3) gives

$$\overline{T} = \frac{1}{5-2}\int_2^5 \left(61 + 18\cos\frac{\pi t}{6}\right)dt$$

$$= \frac{1}{3}\left[61t + \frac{6 \cdot 18}{\pi}\sin\frac{\pi t}{6}\right]_2^5 \approx 57^\circ\text{F}.$$

Figure 5.6.2 shows the graphs of $T = f(t)$ and $T \equiv 57$. Can you see that Eq. (4) implies that the two almost-triangular regions in the figure have equal areas? ◆

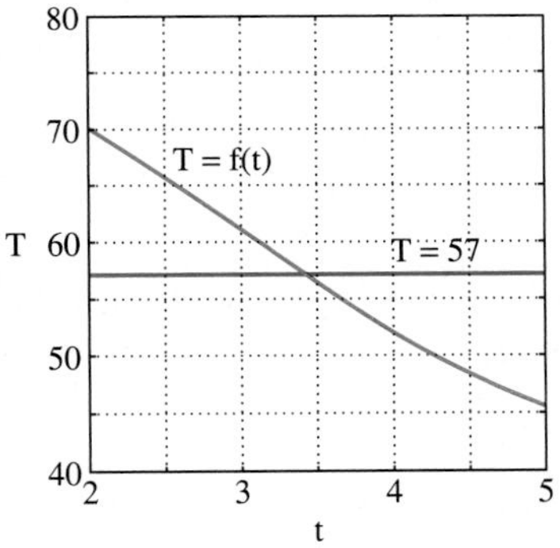

FIGURE 5.6.2 The temperature function $T = f(t)$ of Example 3.

Theorem 1 tells us that every continuous function on a closed interval *attains* its average value at some point of the interval.

THEOREM 1 Average Value Theorem

If f is continuous on $[a, b]$, then

$$f(\overline{x}) = \frac{1}{b-a}\int_a^b f(x)\,dx \tag{6}$$

for some number $\overline{x}$ in $[a, b]$.

PROOF Let $m = f(c)$ be the minimum value of $f(x)$ on $[a, b]$ and let $M = f(d)$ be its maximum value there. Then, by the comparison property of Section 5.5,

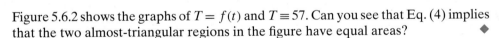

$$m = f(c) \leqq \overline{y} = \frac{1}{b-a}\int_a^b f(x)\,dx \leqq f(d) = M.$$

Because f is continuous, we can now apply the intermediate value property. The number $\overline{y}$ is between the two values m and M of f, and consequently, $\overline{y}$ itself must be a value of f. Specifically, $\overline{y} = f(\overline{x})$ for some number $\overline{x}$ between a and b. This yields Eq. (6). ◄

REMARK Whereas $\overline{y}$ denotes the average value of the function $y = f(x)$, the point $\overline{x}$ where this average value is attained is not itself an average value of x.

EXAMPLE 4 If $v(t)$ denotes the velocity function of a sports car accelerating during the time interval $a \leqq t \leqq b$, then the car's average velocity is given by

$$\overline{v} = \frac{1}{b-a} \int_a^b v(t)\, dt.$$

The average value theorem implies that $\overline{v} = v\left(\overline{t}\right)$ for some number $\overline{t}$ in $[a, b]$. Thus $\overline{t}$ is an instant at which the car's instantaneous velocity is equal to its average velocity over the entire time interval. ◆

The Fundamental Theorem

We state the fundamental theorem of calculus in two parts. The first part is the fact that every function f that is continuous on an interval I has an antiderivative on I. In particular, an antiderivative of f can be obtained by integrating f in a certain way. Intuitively, in the case $f(x) > 0$, we let $F(x)$ denote the area under the graph of f from a fixed point a of I to x, a point of I with $x > a$. We shall prove that $F'(x) = f(x)$. We show the construction of the function F in Fig. 5.6.3. More precisely, we define the function F as follows:

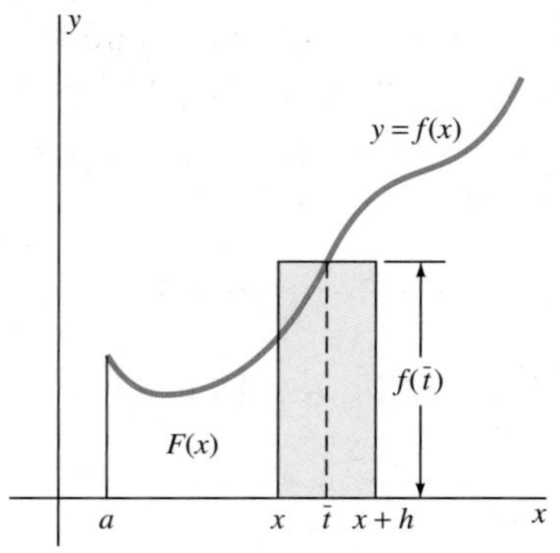

FIGURE 5.6.3 The area function F is an antiderivative of f.

$$F(x) = \int_a^x f(t)\, dt,$$

where we use the dummy variable t in the integrand to avoid confusion with the upper limit x. The proof that $F'(x) = f(x)$ will be independent of the supposition that $x > a$.

THE FUNDAMENTAL THEOREM OF CALCULUS

Suppose that f is continuous on the closed interval $[a, b]$.

Part 1: If the function F is defined on $[a, b]$ by

$$F(x) = \int_a^x f(t)\, dt, \tag{7}$$

then F is an antiderivative of f. That is, $F'(x) = f(x)$ for x in $[a, b]$.

Part 2: If G is any antiderivative of f on $[a, b]$, then

$$\int_a^b f(x)\, dx = \Big[G(x) \Big]_a^b = G(b) - G(a). \tag{8}$$

PROOF OF PART 1 By the definition of the derivative,

$$F'(x) = \lim_{h \to 0} \frac{F(x+h) - F(x)}{h} = \lim_{h \to 0} \frac{1}{h} \left(\int_a^{x+h} f(t)\, dt - \int_a^x f(t)\, dt \right).$$

But

$$\int_a^{x+h} f(t)\, dt = \int_a^x f(t)\, dt + \int_x^{x+h} f(t)\, dt$$

by the interval union property of Section 5.5. Thus

$$F'(x) = \lim_{h \to 0} \frac{1}{h} \int_x^{x+h} f(t)\,dt.$$

The average value theorem tells us that

$$\frac{1}{h} \int_x^{x+h} f(t)\,dt = f(\bar{t})$$

for some number $\bar{t}$ in $[x, x+h]$. Finally, we note that $\bar{t} \to x$ as $h \to 0$. Thus, because f is continuous, we see that

$$F'(x) = \lim_{h \to 0} \frac{1}{h} \int_x^{x+h} f(t)\,dt = \lim_{h \to 0} f(\bar{t}) = \lim_{\bar{t} \to x} f(\bar{t}) = f(x).$$

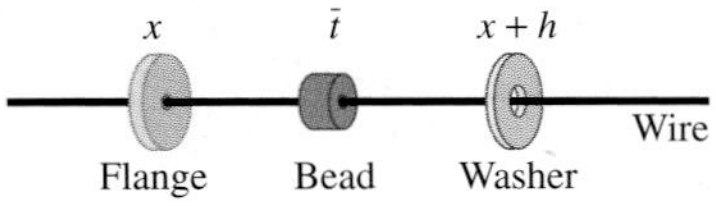

FIGURE 5.6.4 The bead at $\bar{t}$ trapped between the washer at $x+h$ and the flange at x.

Hence the function F in Eq. (7) is, indeed, an antiderivative of f. ◄

REMARK Figure 5.6.4 indicates why $\bar{t}$ must approach x as $h \to 0$. As the moving washer at $x+h$ approaches the fixed flange at x, the bead $\bar{t}$ between them has nowhere else to go.

PROOF OF PART 2 Here we apply Part 1 to give a proof of the evaluation theorem in Section 5.5. If G is *any* antiderivative of f, then—because it and the function F of Part 1 are both antiderivatives of f on the interval $[a, b]$—we know that

$$G(x) = F(x) + C$$

on $[a, b]$ for some constant C. To evaluate C, we substitute $x = a$ and obtain

$$C = G(a) - F(a) = G(a),$$

because

$$F(a) = \int_a^a f(t)\,dt = 0.$$

Hence $G(x) = F(x) + G(a)$. In other words,

$$F(x) = G(x) - G(a)$$

for all x in $[a, b]$. With $x = b$ this gives

$$G(b) - G(a) = F(b) = \int_a^b f(x)\,dx,$$

which establishes Eq. (8) ◄

Sometimes the fundamental theorem of calculus is interpreted to mean that differentiation and integration are *inverse processes.* Part 1 can be written in the form

$$\frac{d}{dx}\left(\int_a^x f(t)\,dt\right) = f(x) \qquad \textbf{(9)}$$

if f is continuous on an open interval containing a and x. That is, if we first integrate the function f (with *variable* upper limit of integration x) and then differentiate with respect to x, the result is the function f again. So differentiation "cancels" the effect of integration of continuous functions.

Moreover, Part 2 of the fundamental theorem can be written in the form

$$\int_a^x G'(t)\,dt = G(x) - G(a) \qquad \textbf{(10)}$$

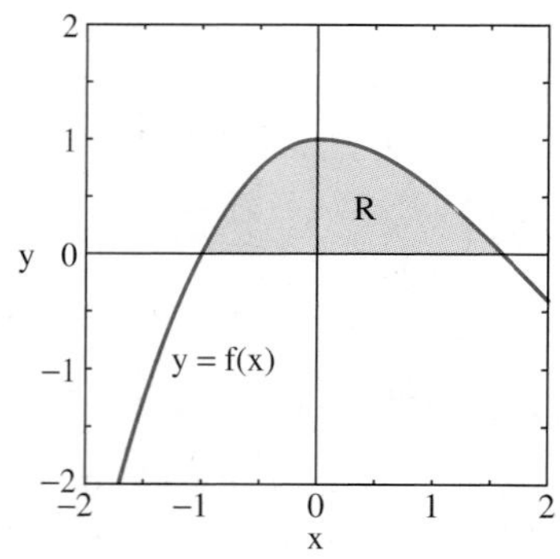

FIGURE 5.6.5 The region of Example 5.

if we assume that G' is continuous. If so, this equation means that if we first differentiate the function G and then integrate the result from a to x, the result can differ from the original function G by, at worst, the *constant* $G(a)$. This means that integration "cancels" the effect of differentiation when a is chosen so that $G(a) = 0$.

Computational Applications

Examples 1 through 4 of Section 5.5 illustrate the use of Part 2 of the fundamental theorem in the evaluation of integrals. Additional examples appear in the end-of-section problems, in this section, and in Section 5.7. Example 5 illustrates the necessity of splitting an integral into a sum of integrals when its integrand has different antiderivative formulas on different intervals.

EXAMPLE 5 Figure 5.6.5 shows the graph of the function f defined by

$$f(x) = \begin{cases} \cos x & \text{if } x \geqq 0, \\ 1 - x^2 & \text{if } x \leqq 0. \end{cases}$$

Find the area A of the region R bounded above by the graph of $y = f(x)$ and below by the x-axis.

Solution The x-intercepts shown in the figure are $x = -1$ (where $1 - x^2 = 0$ and $x < 0$) and $x = \pi/2$ (where $\cos x = 0$ and $x > 0$). Hence

$$A = \int_{-1}^{\pi/2} f(x)\,dx = \int_{-1}^{0} (1 - x^2)\,dx + \int_{0}^{\pi/2} \cos x\,dx$$

$$= \left[x - \frac{1}{3}x^3\right]_{-1}^{0} + \Big[\sin x\Big]_{0}^{\pi/2} = \frac{2}{3} + 1 = \frac{5}{3}.$$ ◆

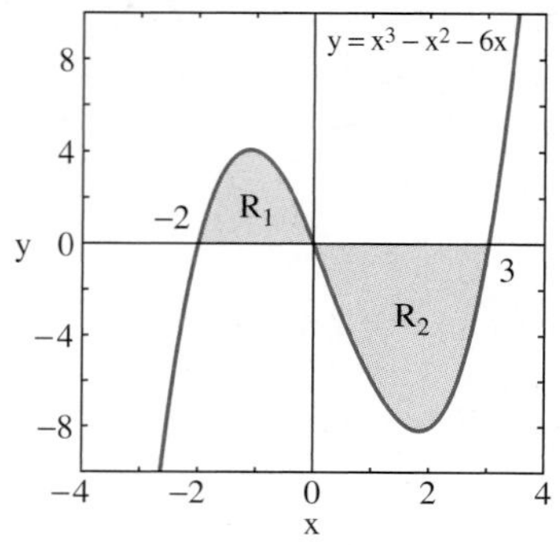

FIGURE 5.6.6 The graph $y = x^3 - x^2 - 6x$ of Example 6.

EXAMPLE 6 Figure 5.6.6 shows the graph of

$$f(x) = x^3 - x^2 - 6x.$$

Find the area A of the entire region R bounded by the graph of f and the x-axis.

Solution The region R consists of the two regions R_1 and R_2 and extends from $x = -2$ to $x = 3$. The area of R_1 is

$$A_1 = \int_{-2}^{0} (x^3 - x^2 - 6x)\,dx = \left[\tfrac{1}{4}x^4 - \tfrac{1}{3}x^3 - 3x^2\right]_{-2}^{0} = \tfrac{16}{3}.$$

But on the interval $(0, 3)$, the function $f(x)$ is negative-valued, so to get the (positive) area A_2 of R_2, we must integrate the *negative* of f:

$$A_2 = \int_{0}^{3} (-x^3 + x^2 + 6x)\,dx = \left[-\tfrac{1}{4}x^4 + \tfrac{1}{3}x^3 + 3x^2\right]_{0}^{3} = \tfrac{63}{4}.$$

Consequently the area of the entire region R is

$$A = A_1 + A_2 = \tfrac{16}{3} + \tfrac{63}{4} = \tfrac{253}{12} \approx 21.08.$$

In effect, we have integrated the *absolute value* of $f(x)$:

$$A = \int_{-2}^{3} |f(x)|\,dx$$

$$= \int_{-2}^{0} (x^3 - x^2 - 6x)\,dx + \int_{0}^{3} (-x^3 + x^2 + 6x)\,dx = \tfrac{253}{12}.$$

Compare the graph of $y = |f(x)|$ in Fig. 5.6.7 with that of $y = f(x)$ in Fig. 5.6.6. ◆

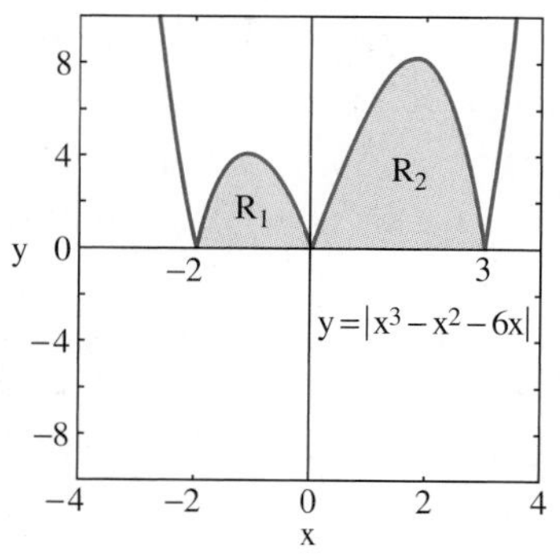

FIGURE 5.6.7 The graph $y = |x^3 - x^2 - 6x|$ of Example 6.

EXAMPLE 7 Evaluate

$$\int_{-1}^{2} |x^3 - x|\,dx.$$

Solution We note that $x^3 - x \geqq 0$ on $[-1, 0]$, that $x^3 - x \leqq 0$ on $[0, 1]$, and that $x^3 - x \geqq 0$ on $[1, 2]$. So we write

$$\begin{aligned}\int_{-1}^{2} |x^3 - x|\, dx &= \int_{-1}^{0} (x^3 - x)\, dx + \int_{0}^{1} (x - x^3)\, dx + \int_{1}^{2} (x^3 - x)\, dx \\ &= \left[\tfrac{1}{4}x^4 - \tfrac{1}{2}x^2\right]_{-1}^{0} + \left[\tfrac{1}{2}x^2 - \tfrac{1}{4}x^4\right]_{0}^{1} + \left[\tfrac{1}{4}x^4 - \tfrac{1}{2}x^2\right]_{1}^{2} \\ &= \tfrac{1}{4} + \tfrac{1}{4} + \left[2 - \left(-\tfrac{1}{4}\right)\right] = \tfrac{11}{4} = 2.75.\end{aligned}$$

◆

Part 1 of the fundamental theorem of calculus says that the derivative of an integral with respect to its upper limit is equal to the value of the integrand at the upper limit. For example, if

$$y(x) = \int_{0}^{x} t^3 \sin t\, dt,$$

then

$$\frac{dy}{dx} = x^3 \sin x.$$

Example 8 is a bit more complicated in that the upper limit of the integral is a nontrivial function of the independent variable.

EXAMPLE 8 Find $h'(x)$ given

$$h(x) = \int_{0}^{x^2} t^3 \sin t\, dt.$$

Solution Let $y = h(x)$ and $u = x^2$. Then

$$y = \int_{0}^{u} t^3 \sin t\, dt,$$

so

$$\frac{dy}{du} = u^3 \sin u$$

by the fundamental theorem of calculus. Then the chain rule yields

$$h'(x) = \frac{dy}{dx} = \frac{dy}{du} \cdot \frac{du}{dx} = (u^3 \sin u)(2x) = 2x^7 \sin x^2.$$

Initial Value Problems

Note that if

$$y(x) = \int_{a}^{x} f(t)\, dt, \tag{11}$$

then $y(a) = 0$. Hence $y(x)$ is a solution of the initial value problem

$$\frac{dy}{dx} = f(x), \quad y(a) = 0. \tag{12}$$

To get a solution of the initial value problem

$$\frac{dy}{dx} = f(x) \qquad y(a) = b, \tag{13}$$

we need only add the desired initial value:

$$y(x) = b + \int_{a}^{x} f(t)\, dt. \tag{14}$$

◆

EXAMPLE 9 Express as an integral the solution of the initial value problem

$$\frac{dy}{dx} = \sec x, \qquad y(2) = 3. \tag{15}$$

Solution With $a = 2$ and $b = 3$, Eq. (14) gives

$$y(x) = 3 + \int_2^x \sec t \, dt. \tag{16}$$

With our present knowledge, we cannot antidifferentiate $\sec t$, but for a particular value of x the integral in Eq. (16) can be approximated using Riemann sums. For instance, with $x = 4$ a calculator with an **INTEGRATE** key gives

$$\int_2^4 \sec t \, dt \approx -2.5121.$$

Hence the value of the solution in Eq. (16) at $x = 4$ is

$$y(4) \approx 3 - 2.5121 = 0.4879.$$

◆

5.6 TRUE/FALSE STUDY GUIDE

5.6 CONCEPTS: QUESTIONS AND DISCUSSION

1. Does every function defined on a closed interval have an average value there?
2. Suppose that the function f defined on the closed interval $[a, b]$ has an average value there. Does f necessarily attain this average value there? Either show that it does or give an example showing it need not.
3. Discuss the validity of the assertion that the fundamental theorem of calculus simply says that "every function is the derivative of its integral and the integral of its derivative." Does the fundamental theorem of calculus say this? Does it say more than this?
4. Suppose that a particle moving along the x-axis has position $x = f(t)$ and velocity $v = f'(t)$ at time t. Interpret the values of the integrals

$$\int_a^b f'(t)\, dt \quad \text{and} \quad \int_a^b |f'(t)|\, dt$$

in terms of the change in position of the particle and the distance it travels. What's the difference?

5.6 PROBLEMS

In Problems 1 through 12, find the average value of the given function on the specified interval.

1. $f(x) = x^4$; $[0, 2]$
2. $g(x) = \sqrt{x}$; $[1, 4]$
3. $h(x) = 3x^2\sqrt{x^3+1}$; $[0, 2]$
4. $f(x) = 8x$; $[0, 4]$
5. $g(x) = 8x$; $[-4, 4]$
6. $h(x) = x^2$; $[-4, 4]$
7. $f(x) = x^3$; $[0, 5]$
8. $g(x) = x^{-1/2}$; $[1, 4]$
9. $f(x) = \sqrt{x+1}$; $[0, 3]$
10. $g(x) = \sin 2x$; $[0, \pi/2]$
11. $f(x) = \sin 2x$; $[0, \pi]$
12. $g(t) = e^{2t}$; $[-1, 1]$

Evaluate the integrals in Problems 13 through 28.

13. $\int_{-1}^{3} dx$ (Here dx stands for $1\, dx$.)
14. $\int_1^2 (y^5 - 1)\, dy$
15. $\int_1^4 \frac{dx}{\sqrt{9x^3}}$
16. $\int_{-1}^{1} (x^3 + 2)^2\, dx$
17. $\int_1^3 \frac{3t - 5}{t^4}\, dt$
18. $\int_{-2}^{-1} \frac{x^2 - x + 3}{\sqrt[3]{x}}\, dx$
19. $\int_0^{\pi} \sin x \cos x\, dx$

20. $\displaystyle\int_{-1}^{2} |x|\, dx$

21. $\displaystyle\int_{1}^{2} \left(t - \frac{1}{2t}\right)^2 dt$

22. $\displaystyle\int_{0}^{1} e^{2x-1}\, dx$

23. $\displaystyle\int_{0}^{1} \frac{e^{2x} - 1}{e^x}\, dx$

24. $\displaystyle\int_{0}^{2} \left|x - \sqrt{x}\right|\, dx$

25. $\displaystyle\int_{-2}^{2} |x^2 - 1|\, dx$

26. $\displaystyle\int_{0}^{\pi/3} \sin 3x\, dx$

27. $\displaystyle\int_{4}^{8} \frac{1}{x}\, dx$

28. $\displaystyle\int_{6}^{11} \frac{1}{x - 1}\, dx$

In Problems 29 through 32, the graph of f and the x-axis divide the xy-plane into several regions, some of which are bounded. Find the total area of the bounded regions in each problem.

29. $f(x) = 1 - x^4$ if $x \leqq 0$; $f(x) = 1 - x^3$ if $x \geqq 0$ (Fig. 5.6.8)

30. $f(x) = (\pi/2)^2 \sin x$ on $[0, \pi/2]$; $f(x) = x(\pi - x)$ on $[\pi/2, \pi]$ (Fig. 5.6.9)

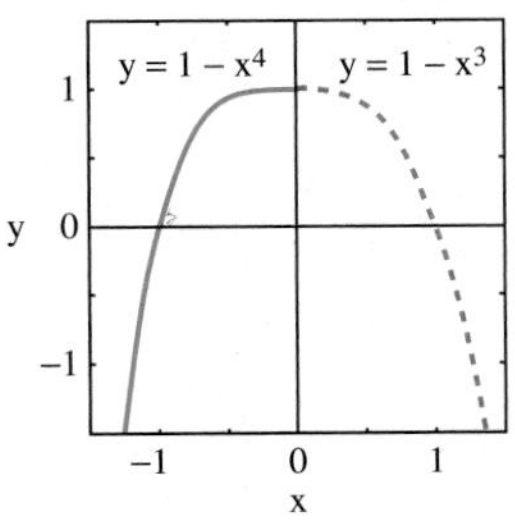

FIGURE 5.6.8 Problem 29.

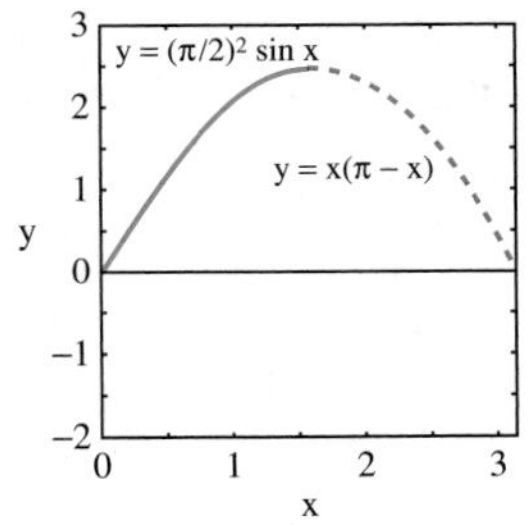

FIGURE 5.6.9 Problem 30.

31. $f(x) = x^3 - 9x$ (Fig. 5.6.10)

32. $f(x) = x^3 - 2x^2 - 15x$ (Fig. 5.6.11)

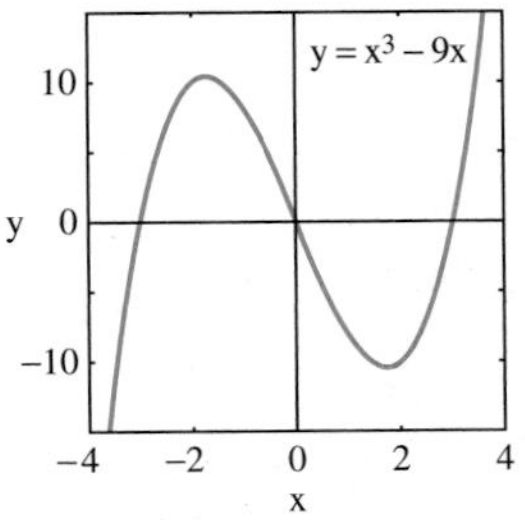

FIGURE 5.6.10 Problem 31.

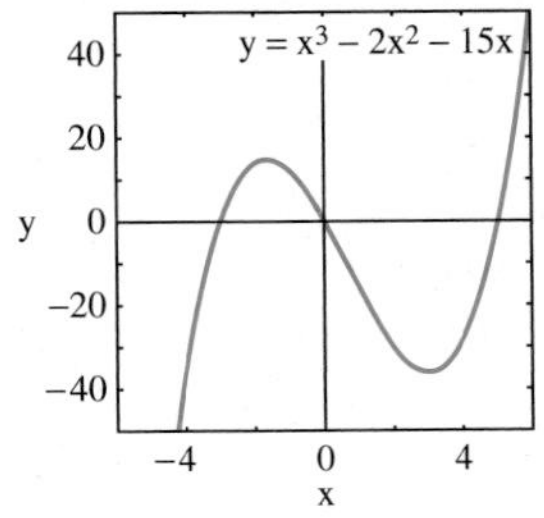

FIGURE 5.6.11 Problem 32.

33. Rosanne drops a ball from a height of 400 ft. Find the ball's average height and its average velocity between the time it is dropped and the time it strikes the ground.

34. Find the average value of the animal population $P(t) = 100 + 10t + (0.02)t^2$ over the time interval $[0, 10]$.

35. Suppose that a 5000-L water tank takes 10 min to drain and that after t minutes, the amount of water remaining in the tank is $V(t) = 50(10 - t)^2$ liters. What is the average amount of water in the tank during the time it drains?

36. On a certain day the temperature t hours past midnight was

$$T(t) = 80 + 10 \sin\left(\frac{\pi}{12}(t - 10)\right).$$

What was the average temperature between noon and 6 P.M.?

37. Suppose that a heated rod lies along the interval $0 \leqq x \leqq 10$. If the temperature at points of the rod is given by $T(x) = 4x(10 - x)$, what is the rod's average temperature?

38. Figure 5.6.12 shows a cross section at distance x from the center of a sphere of radius 1. Find the average area of the cross section for $0 \leqq x \leqq 1$.

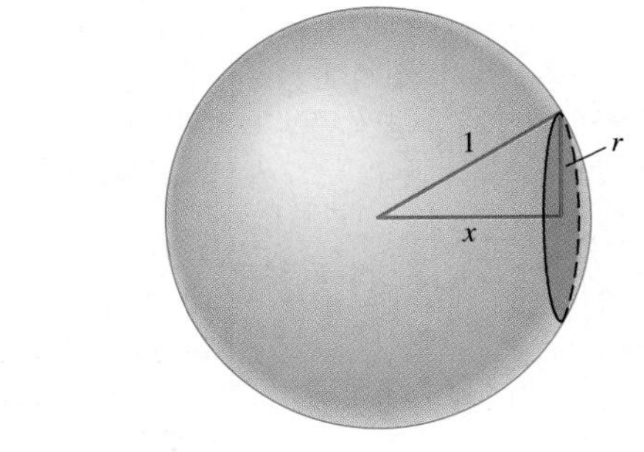

FIGURE 5.6.12 The sphere of Problem 38.

39. Figure 5.6.13 shows a cross section at distance y from the vertex of a cone with base radius 1 and height 2. Find the average area of this cross section for $0 \leqq y \leqq 2$.

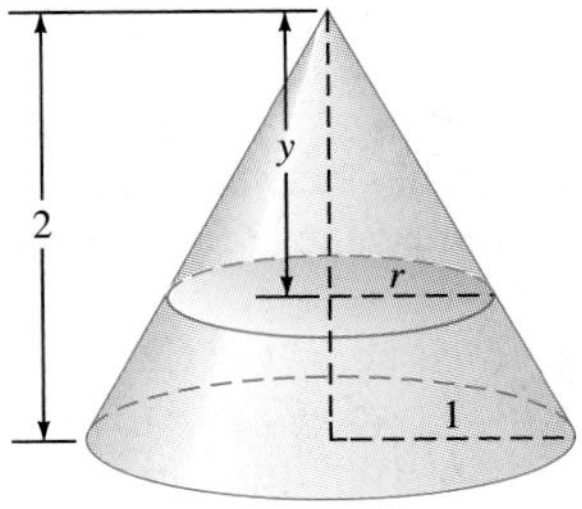

FIGURE 5.6.13 The cone of Problem 39.

40. A sports car starts from rest ($x = 0, t = 0$) and experiences constant acceleration $x''(t) = a$ for T seconds. Find, in terms of a and T, (a) its final and average velocities and (b) its final and average positions.

41. (a) Figure 5.6.14 shows a triangle inscribed in the region that lies between the x-axis and the curve $y = 9 - x^2$. Express the area of this triangle as a function $A(x)$ of the x-coordinate of its upper vertex P. (b) Find the average area $\overline{A}$ of $A(x)$ for x in the interval $[-3, 3]$. (c) Sketch a triangle as in Fig. 5.6.14 that has the area $\overline{A}$ found in part (b). How many different such triangles are there?

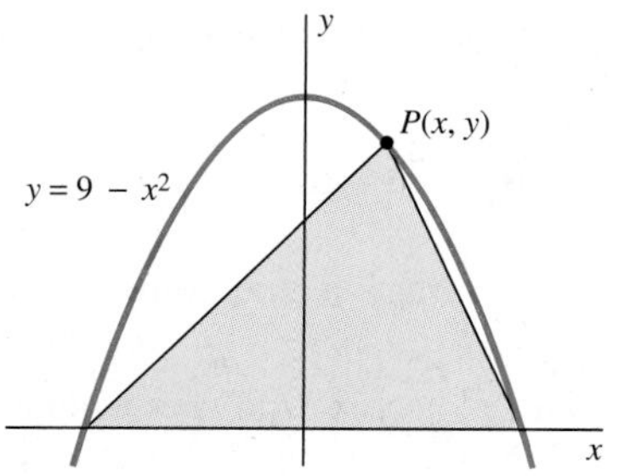

FIGURE 5.6.14 The typical triangle of Problem 41.

42. (a) Figure 5.6.15 shows a rectangle inscribed in the first-quadrant region that lies between the x-axis and the line $y = 10 - x$. Express the area of this rectangle as a function $A(x)$ of the x-coordinate of its vertex P on the line. (b) Find the average area $\overline{A}$ of $A(x)$ for x in the interval $[0, 10]$. (c) Sketch a rectangle as in Fig. 5.6.15 that has the area $\overline{A}$ found in part (b). How many different such rectangles are there?

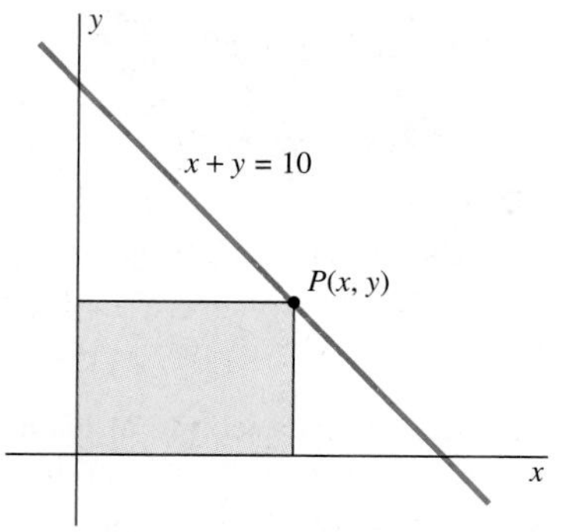

FIGURE 5.6.15 The typical rectangle of Problem 42.

43. (a) Figure 5.6.16 shows a rectangle inscribed in the semicircular region that lies between the x-axis and the graph $y = \sqrt{16 - x^2}$. Express the area of the rectangle as a function $A(x)$ of the x-coordinate of its vertex P on the line. (b) Find the average area $\overline{A}$ of $A(x)$ for x in the interval $[0, 4]$. (c) Sketch a rectangle as in Fig. 5.6.16 that has the area $\overline{A}$ found in part (b). How many different such rectangles are there?

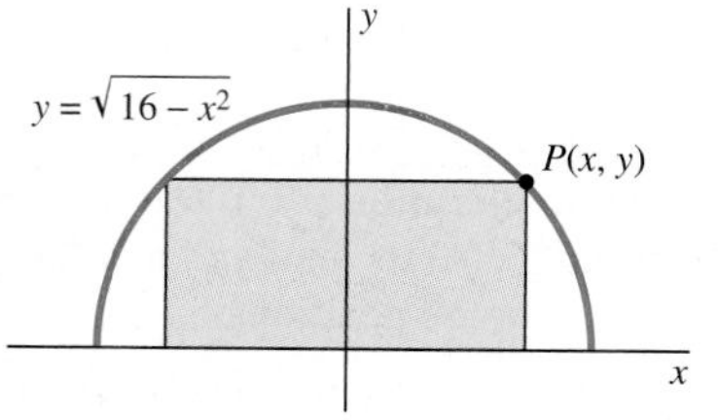

FIGURE 5.6.16 The typical rectangle of Problem 43.

44. Repeat Problem 43 in the case that the rectangle has two vertices on the x-axis and two on the parabola $y = 16 - x^2$ (rather than on the semicircle $y = \sqrt{16 - x^2}$). You may need to use a calculator or computer to find the base of a rectangle whose area is the average $\overline{A}$ of $A(x)$ for x in $[0, 4]$.

In Problems 45 through 49, apply the fundamental theorem of calculus to find the derivative of the given function.

45. $f(x) = \int_{-1}^{x} (t^2 + 1)^{17}\,dt$

46. $g(t) = \int_{0}^{t} \sqrt{x^2 + 25}\,dx$

47. $h(z) = \int_{2}^{z} \sqrt[3]{u - 1}\,du$

48. $A(x) = \int_{1}^{x} \frac{1}{t}\,dt$

49. $f(x) = \int_{x}^{10} (e^t - e^{-t})\,dt$

In Problems 50 through 53, $G(x)$ is the integral of the given function $f(t)$ over the specified interval of the form $[a, x]$, $x > a$. Apply Part 1 of the fundamental theorem of calculus to find $G'(x)$.

50. $f(t) = \dfrac{t}{t^2 + 1}$; $[2, x]$

51. $f(t) = \sqrt{t + 4}$; $[0, x]$

52. $f(t) = \sin^3 t$; $[0, x]$

53. $f(t) = \sqrt{t^3 + 1}$; $[1, x]$

In Problems 54 through 60, differentiate the function by first writing $f(x)$ in the form $g(u)$, where u denotes the upper limit of integration.

54. $f(x) = \int_{0}^{x^2} \sqrt{1 + t^3}\,dt$

55. $f(x) = \int_{2}^{3x} \sin t^2\,dt$

56. $f(x) = \int_{0}^{\sin x} \sqrt{1 - t^2}\,dt$

57. $f(x) = \int_{0}^{x^2} \sin t\,dt$

58. $f(x) = \int_{1}^{\sin x} (t^2 + 1)^3\,dt$

59. $f(x) = \int_{1}^{x^2+1} \frac{dt}{t}$

60. $f(x) = \int_{1}^{e^x} \ln(1 + t^2)\,dt$

Use integrals (as in Example 9) to solve the initial value problems in Problems 61 through 64.

61. $\dfrac{dy}{dx} = \dfrac{1}{x}$, $y(1) = 0$

62. $\dfrac{dy}{dx} = \dfrac{1}{1 + x^2}$, $y(1) = \dfrac{\pi}{4}$

63. $\dfrac{dy}{dx} = \sqrt{1 + x^2}$, $y(5) = 10$

64. $\dfrac{dy}{dx} = \tan x$, $y(1) = 2$

65. The fundamental theorem of calculus *seems* to say that

$$\int_{-1}^{1} \frac{dx}{x^2} = \left[-\frac{1}{x}\right]_{-1}^{1} = -2,$$

in apparent contradiction to the fact that $1/x^2$ is always positive. What's wrong here?

66. Prove that the average rate of change

$$\frac{f(b) - f(a)}{b - a}$$

of the differentiable function f on $[a, b]$ is equal to the average value of its derivative on $[a, b]$.

67. The graph $y = f(x), 0 \leqq x \leqq 10$ is shown in Fig. 5.6.17. Let

$$g(x) = \int_0^x f(t)\,dt.$$

(a) Find the values $g(0)$, $g(2)$, $g(4)$, $g(6)$, $g(8)$, and $g(10)$. (b) Find the intervals on which $g(x)$ is increasing and those on which it is decreasing. (c) Find the global maximum and minimum values of $g(x)$ for $0 \leqq x \leqq 10$. (d) Sketch a rough graph of $y = g(x)$.

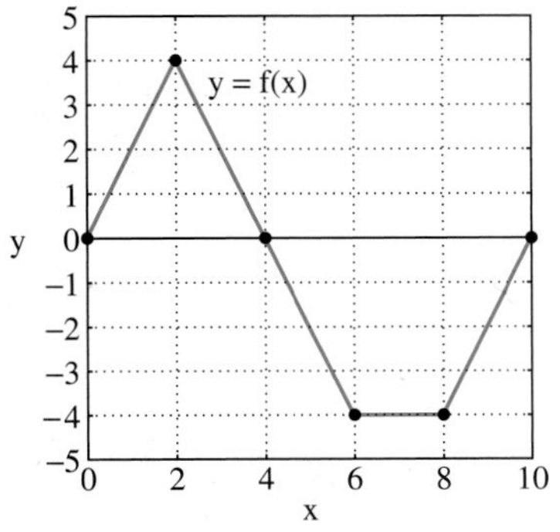

FIGURE 5.6.17 Problem 67.

68. Repeat Problem 67, except use the graph of the function f shown in Fig. 5.6.18.

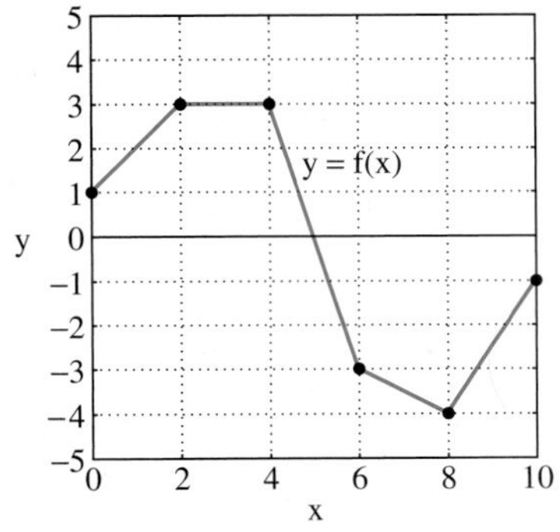

FIGURE 5.6.18 Problem 68.

69. Figure 5.6.19 shows the graph of the function $f(x) = x \sin x$ on the interval $[0, 4\pi]$. Let

$$g(x) = \int_0^x f(t)\,dt.$$

(a) Find the values of x at which $g(x)$ has local maximum and minimum values on the interval $[0, 4\pi]$. (b) Where does $g(x)$ attain its global maximum and minimum values on $[0, 4\pi]$. (c) Which points on the graph $y = f(x)$ correspond to inflection points on the graph $y = g(x)$? (d) Sketch a rough graph of $y = g(x)$.

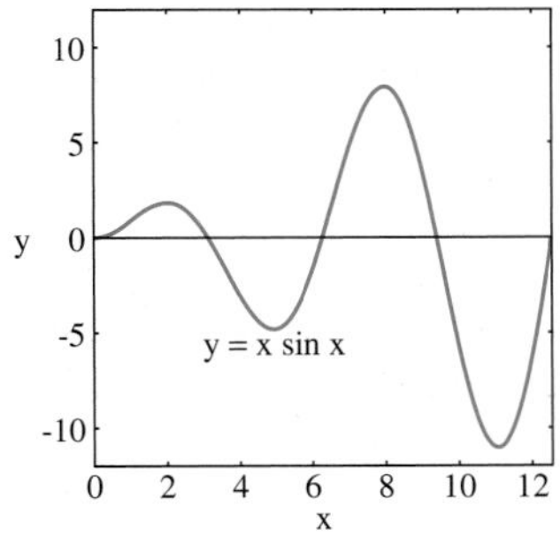

FIGURE 5.6.19 Problem 69.

70. Repeat Problem 69, except use the function

$$f(x) = \frac{\sin x}{x}$$

on the interval $[0, 4\pi]$ (as shown in Fig. 5.6.20). Take $f(0) = 1$ because $(\sin x)/x \to 1$ as $x \to 0$.

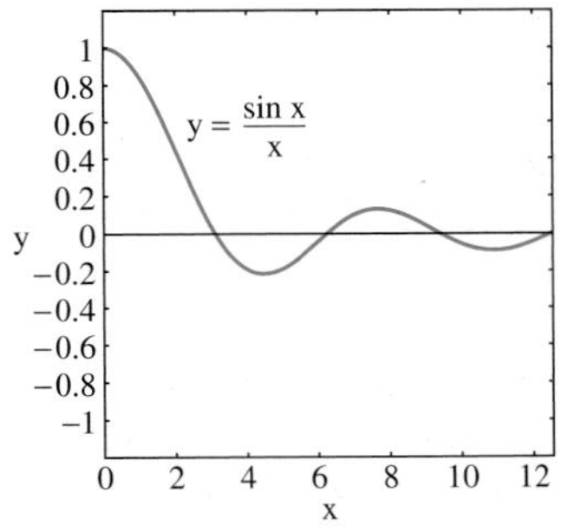

FIGURE 5.6.20 Problem 70.

5.7 INTEGRATION BY SUBSTITUTION

The fundamental theorem of calculus in the form

$$\int_a^b f(x)\,dx = \left[\int f(x)\,dx\right]_a^b \tag{1}$$

implies that we can readily evaluate the definite integral on the left if we can find the indefinite integral (that is, antiderivative) on the right. We now discuss a powerful method of antidifferentiation that amounts to "the chain rule in reverse." This method is a generalization of the "generalized power rule in reverse,"

$$\int u^n\,du = \frac{u^{n+1}}{n+1} + C \quad (n \neq -1), \tag{2}$$

which we introduced in Section 5.2.

Equation (2) is an abbreviation for the formula

$$\int [g(x)]^n g'(x)\,dx = \frac{[g(x)]^{n+1}}{n+1} + C \quad (n \neq -1) \tag{3}$$

that results when we write

$$u = g(x), \qquad du = g'(x)\,dx.$$

So to apply Eq. (2) to a given integral, we must be able to visualize the integrand as a *product* of a power of a differentiable function $g(x)$ and its derivative $g'(x)$.

EXAMPLE 1 With

$$u = 2x + 1, \qquad du = 2\,dx,$$

we see that

$$\int (2x+1)^5 \cdot 2\,dx = \int u^5\,du = \frac{u^6}{6} + C = \frac{1}{6}(2x+1)^6 + C.$$ ◆

EXAMPLE 2

(a)
$$\begin{aligned}\int 2x\sqrt{1+x^2}\,dx &= \int (1+x^2)^{1/2}\cdot 2x\,dx \\ &= \int u^{1/2}\,du \qquad [u = 1+x^2, \quad du = 2x\,dx] \\ &= \frac{u^{3/2}}{\frac{3}{2}} + C = \tfrac{2}{3}(1+x^2)^{3/2} + C.\end{aligned}$$

(b) Similarly, but with $u = 1 + e^x$ and $du = e^x\,dx$, we get

$$\begin{aligned}\int \frac{e^x}{\sqrt{1+e^x}}\,dx = \int \frac{1}{\sqrt{u}}\,du &= 2\sqrt{u} + C \\ &= 2\sqrt{1+e^x} + C.\end{aligned}$$ ◆

Equation (3) is the special case $f(u) = u^n$ of the general integral formula

$$\int f(g(x))\cdot g'(x)\,dx = \int f(u)\,du. \tag{4}$$

The right-hand side of Eq. (4) results when we make the formal substitutions

$$u = g(x), \qquad du = g'(x)\,dx$$

on the left-hand side.

One of the beauties of differential notation is that Eq. (4) is not only plausible but is, in fact, true—with the understanding that u is to be replaced with $g(x)$ after the indefinite integration on the right-hand side of Eq. (4) has been carried out. Indeed, Eq. (4) is merely an indefinite integral version of the chain rule. For if $F'(x) = f(x)$,

then

$$D_x F(g(x)) = F'(g(x)) \cdot g'(x) = f(g(x)) \cdot g'(x)$$

by the chain rule, so

$$\begin{aligned} \int f(g(x)) \cdot g'(x)\,dx &= \int F'(g(x)) \cdot g'(x)\,dx = F(g(x)) + C \\ &= F(u) + C \quad [u = g(x)] \\ &= \int f(u)\,du. \end{aligned}$$

Equation (4) is the basis for the powerful technique of indefinite **integration by substitution.** It may be used whenever the integrand function is recognized to be of the form $f(g(x)) \cdot g'(x)$.

EXAMPLE 3 Find

$$\int x^2 \sqrt{x^3 + 9}\,dx.$$

Solution Note that x^2 is, to within a ***constant*** factor, the derivative of $x^3 + 9$. We can, therefore, substitute

$$u = x^3 + 9, \qquad du = 3x^2\,dx. \tag{5}$$

The constant factor 3 can be supplied if we compensate by multiplying the integral by $\frac{1}{3}$. This gives

$$\begin{aligned} \int x^2 \sqrt{x^3 + 9}\,dx &= \frac{1}{3} \int (x^3 + 9)^{1/2} \cdot 3x^2\,dx = \frac{1}{3} \int u^{1/2}\,du \\ &= \frac{1}{3} \cdot \frac{u^{3/2}}{\frac{3}{2}} + C = \frac{2}{9} u^{3/2} + C = \frac{2}{9}(x^3 + 9)^{3/2} + C. \end{aligned}$$

An alternative way to carry out the substitution in (5) is to solve

$$du = 3x^2\,dx \quad \text{for} \quad x^2\,dx = \tfrac{1}{3}\,du,$$

and then write

$$\int (x^3 + 9)^{1/2}\,dx = \int u^{1/2} \cdot \tfrac{1}{3}\,du = \tfrac{1}{3} \int u^{1/2}\,du,$$

concluding the computation as before. ◆

The following three steps in the solution of Example 3 are worth special mention:

- The differential dx along with the rest of the integrand is "transformed," or replaced, in terms of u and du.
- Once the integration has been performed, the constant C of integration is added.
- A final resubstitution is necessary to write the answer in terms of the original variable x.

Substitution in Trigonometric Integrals

By now we know that every differentiation formula yields—upon "reversal"—a corresponding antidifferentiation formula. The familiar formulas for the derivatives of the six trigonometric functions thereby yield the following indefinite-integral formulas:

$$\int \cos u\, du = \sin u + C, \tag{6}$$

$$\int \sin u\, du = -\cos u + C, \tag{7}$$

$$\int \sec^2 u\, du = \tan u + C, \tag{8}$$

$$\int \csc^2 u\, du = -\cot u + C, \tag{9}$$

$$\int \sec u \tan u\, du = \sec u + C, \tag{10}$$

$$\int \csc u \cot u\, du = -\csc u + C. \tag{11}$$

Also, the derivatives $D_x[e^x] = e^x$ and $D_x[\ln x] = 1/x$ (for $x > 0$) yield the integral formulas

$$\int e^u\, du = e^u + C, \tag{12}$$

$$\int \frac{1}{u}\, du = \ln u + C \quad (\text{for } u > 0). \tag{13}$$

Any of these integrals can appear as the integral $\int f(u)\, du$ that results from an appropriate *u-substitution* in a given integral.

EXAMPLE 4

$$\begin{aligned}\int \sin(3x+4)\, dx &= \int (\sin u)\cdot \tfrac{1}{3}\, du \qquad (u = 3x+4, \quad du = 3\,dx)\\ &= \tfrac{1}{3}\int \sin u\, du = -\tfrac{1}{3}\cos u + C\\ &= -\tfrac{1}{3}\cos(3x+4) + C.\end{aligned}$$

◆

EXAMPLE 5

$$\begin{aligned}\int 3x\cos(x^2)\, dx &= 3\int (\cos x^2)\cdot x\, dx\\ &= 3\int (\cos u)\cdot \tfrac{1}{2}\, du \qquad (u = x^2, \quad du = 2x\,dx)\\ &= \tfrac{3}{2}\int \cos u\, du = \tfrac{3}{2}\sin u + C = \tfrac{3}{2}\sin(x^2) + C.\end{aligned}$$

◆

EXAMPLE 6

$$\int \sec^2 3x\,dx = \int (\sec^2 u)\cdot \tfrac{1}{3}\,du \qquad (u = 3x, \quad du = 3\,dx)$$

$$= \tfrac{1}{3}\tan u + C = \tfrac{1}{3}\tan 3x + C. \qquad \blacklozenge$$

EXAMPLE 7 Evaluate

$$\int 2\sin^3 x \cos x\,dx.$$

Solution None of the integrals in Eqs. (6) through (11) appears to "fit," but the substitution

$$u = \sin x, \qquad du = \cos x\,dx$$

yields

$$\int 2\sin^3 x \cos x\,dx = 2\int u^3\,du = 2\cdot\frac{u^4}{4} + C = \frac{1}{2}\sin^4 x + C. \qquad \blacklozenge$$

EXAMPLE 8 Let

$$u = 1 + \sqrt{x^3} = 1 + x^{3/2}, \quad \text{so that} \quad du = \frac{3}{2}x^{1/2}\,dx = \frac{3}{2}\sqrt{x}\,dx.$$

Then Eq. (12) yields

$$\int 3\sqrt{x}\exp\left(1+\sqrt{x^3}\right)dx = \int e^u\cdot 2\,du = 2e^u + C = 2\exp\left(1+\sqrt{x^3}\right) + C.$$

Moreover, if $x > 0$ then Eq. (13) gives

$$\int \frac{3\sqrt{x}}{1+\sqrt{x^3}}\,dx = \int \frac{2}{u}\,du = 2\ln u + C = 2\ln\left(1+\sqrt{x^3}\right) + C. \qquad \blacklozenge$$

Substitution in Definite Integrals

The method of integration by substitution can be used with definite integrals as well as with indefinite integrals. Only one additional step is required—evaluation of the final antiderivative at the original limits of integration.

EXAMPLE 9 The substitution used in Example 3 gives

$$\int_0^3 x^2\sqrt{x^3+9}\,dx = \int_{-}^{-} u^{1/2}\cdot\tfrac{1}{3}\,du \qquad (u = x^3+9, \quad du = 3x^2\,dx)$$

$$= \tfrac{1}{3}\left[\tfrac{2}{3}u^{3/2}\right]_{-}^{-} = \tfrac{2}{9}\left[(x^3+9)^{3/2}\right]_0^3 \qquad \text{(resubstitute)}$$

$$= \tfrac{2}{9}(216-27) = 42.$$

The limits on u were left "blank" above because they weren't calculated—there was no need to know them, because we planned to resubstitute for u in terms of the original variable x before using the original limits of integration.

But sometimes it is more convenient to determine the limits of integration with respect to the new variable u. With the substitution $u = x^3 + 9$, $du = 3x^2\,dx$, we see that

- $u = 9$ when $x = 0$ (lower limit);
- $u = 36$ when $x = 3$ (upper limit).

Use of these limits on u (rather than resubstitution in terms of x) gives

$$\int_0^3 x^2\sqrt{x^3+9}\,dx = \tfrac{1}{3}\int_9^{36} u^{1/2}\,du = \tfrac{1}{3}\left[\tfrac{2}{3}u^{3/2}\right]_9^{36} = 42.$$ ◆

Theorem 1 says that the "natural" way of transforming an integral's limits under a u-substitution, like the work just done, is in fact correct.

THEOREM 1 Definite Integration by Substitution

Suppose that the function g has a continuous derivative on $[a, b]$ and that f is continuous on the interval $g\left([a, b]\right)$. Let $u = g(x)$. Then

$$\int_a^b f(g(x))\cdot g'(x)\,dx = \int_{g(a)}^{g(b)} f(u)\,du. \tag{14}$$

REMARK Thus we get the new limits on u by applying the substitution function $u = g(x)$ to the old limits on x. Then:

- The new lower limit is $g(a)$, and
- The new upper limit is $g(b)$,

whether or not $g(b)$ is greater than $g(a)$.

PROOF OF THEOREM 1 Choose an antiderivative F of f, so $F' = f$. Then, by the chain rule,

$$D_x[F(g(x))] = F'(g(x))\cdot g'(x) = f(g(x))\cdot g'(x).$$

Therefore,

$$\begin{aligned}\int_a^b f(g(x))\cdot g'(x)\,dx &= \Big[F(g(x))\Big]_a^b = F(g(b)) - F(g(a))\\ &= \Big[F(u)\Big]_{u=g(a)}^{g(b)} = \int_{g(a)}^{g(b)} f(u)\,du.\end{aligned}$$

Note how we used the fundamental theorem to obtain the first and last equalities in this argument. ◄

Whether it is simpler to apply Theorem 1 and transform to new u-limits or to resubstitute $u = g(x)$ and use the old x-limits depends on the specific problem at hand. Examples 10 and 11 illustrate the technique of transforming to new limits.

EXAMPLE 10 Evaluate

$$\int_3^5 \frac{x\,dx}{(30-x^2)^2}.$$

Solution Note that $30 - x^2$ is nonzero on $[3, 5]$, so the integrand is continuous there. We substitute

$$u = 30 - x^2, \qquad du = -2x\,dx,$$

and observe that

If $x = 3$, then $u = 21$ (lower limit);

If $x = 5$, then $u = 5$ (upper limit).

Hence our substitution gives

$$\int_3^5 \frac{x\,dx}{(30-x^2)^2} = \int_{21}^5 \frac{-\frac{1}{2}\,du}{u^2} = -\frac{1}{2}\left[-\frac{1}{u}\right]_{21}^5 = -\frac{1}{2}\left(-\frac{1}{5}+\frac{1}{21}\right) = \frac{8}{105}.$$ ◆

EXAMPLE 11 Evaluate

$$\int_0^{\pi/4} \frac{\cos 2t}{1+\sin 2t}\,dt.$$

Solution We substitute

$$u = 1 + \sin 2t, \quad \text{so} \quad du = 2\cos 2t\,dt.$$

Then $u = 1$ when $t = 0$ and $u = 2$ when $t = \pi/4$. Hence Eq. (13) gives

$$\begin{aligned}\int_0^{\pi/4} \frac{\cos 2t}{1+\sin 2t}\,dt &= \tfrac{1}{2}\int_1^2 \frac{1}{u}\,du \\ &= \tfrac{1}{2}\Big[\ln u\Big]_1^2 \\ &= \tfrac{1}{2}\ln 2 \approx 0.3466.\end{aligned}$$

◆

5.7 TRUE/FALSE STUDY GUIDE

5.7 CONCEPTS: QUESTIONS AND DISCUSSION

1. In Theorem 1, the function g is continuous on the interval $[a, b]$. Give an example in which the range set $g([a, b])$ is *not* simply the closed interval with endpoints $g(a)$ and $g(b)$. Then select a nontrivial function f that is continuous on $g([a, b])$ and verify that Eq. (14) holds.
2. Suppose that the function g is continuous on the interval $[a, b]$. Use properties of continuous functions stated in Sections 2.4 and 3.5 to prove that the range set $g([a, b])$ *is* a closed interval.
3. Discuss the possible advantages and disadvantages of transforming to new u-limits when evaluating a definite integral by substitution. Perhaps you can give one example in which this makes the calculation simpler, another example in which it does not.

5.7 PROBLEMS

In Problems 1 through 10, use the indicated substitution to evaluate the given integral.

1. $\displaystyle\int (3x-5)^{17}\,dx; \quad u = 3x - 5$

2. $\displaystyle\int \frac{1}{(4x+7)^6}\,dx; \quad u = 4x + 7$

3. $\displaystyle\int x\sqrt{x^2+9}\,dx; \quad u = x^2 + 9$

4. $\displaystyle\int \frac{x^2}{\sqrt[3]{2x^3-1}}\,dx; \quad u = 2x^3 - 1$

5. $\displaystyle\int \sin 5x\,dx; \quad u = 5x$

6. $\displaystyle\int \cos kx\,dx; \quad u = kx$

7. $\displaystyle\int x\sin(2x^2)\,dx; \quad u = 2x^2$

8. $\displaystyle\int \frac{e^{\sqrt{x}}}{\sqrt{x}}\,dx; \quad u = \sqrt{x}$

9. $\displaystyle\int (1-\cos x)^5 \sin x\,dx; \quad u = 1 - \cos x$

10. $\displaystyle\int \frac{\cos 3x}{5+2\sin 3x}\,dx; \quad u = 5 + 2\sin 3x$

Evaluate the indefinite integrals in Problems 11 through 50.

11. $\displaystyle\int (x+1)^6\,dx$

12. $\displaystyle\int (2-x)^5\,dx$

13. $\displaystyle\int (4-3x)^7\,dx$

14. $\displaystyle\int \sqrt{2x+1}\,dx$

15. $\displaystyle\int \frac{dx}{\sqrt{7x+5}}$

16. $\displaystyle\int \frac{dx}{(3-5x)^2}$

17. $\displaystyle\int \sin(\pi x + 1)\,dx$

18. $\displaystyle\int \cos\frac{\pi t}{3}\,dt$

19. $\displaystyle\int \sec 2\theta \tan 2\theta\,d\theta$

20. $\displaystyle\int \csc^2 5x\,dx$

21. $\displaystyle\int e^{1-2x}\,dx$

22. $\displaystyle\int x\exp(x^2)\,dx$

23. $\displaystyle\int x^2\exp(3x^3-1)\,dx$

24. $\displaystyle\int \sqrt{x}\,e^{2x\sqrt{x}}\,dx$

25. $\displaystyle\int \frac{1}{2x-1}\,dx$

26. $\displaystyle\int \frac{1}{3x+5}\,dx$

27. $\displaystyle\int \frac{1}{x}(\ln x)^2\,dx$

28. $\displaystyle\int \frac{1}{x\ln x}\,dx$

29. $\displaystyle\int \frac{x + e^{2x}}{x^2 + e^{2x}}\,dx$

30. $\displaystyle\int (e^x + e^{-x})^2\,dx$

31. $\displaystyle\int x\sqrt{x^2 - 1}\,dx$

32. $\displaystyle\int 3t(1 - 2t^2)^{10}\,dt$

33. $\displaystyle\int x\sqrt{2 - 3x^2}\,dx$

34. $\displaystyle\int \frac{t\,dt}{\sqrt{2t^2 + 1}}$

35. $\displaystyle\int x^3\sqrt{x^4 + 1}\,dx$

36. $\displaystyle\int \frac{x^2\,dx}{\sqrt[3]{x^3 + 1}}$

37. $\displaystyle\int x^2\cos(2x^3)\,dx$

38. $\displaystyle\int t\sec^2(t^2)\,dt$

39. $\displaystyle\int xe^{-x^2}\,dx$

40. $\displaystyle\int \frac{x}{1 + x^2}\,dx$

41. $\displaystyle\int \cos^3 x\,\sin x\,dx$

42. $\displaystyle\int \sin^5 3z\,\cos 3z\,dz$

43. $\displaystyle\int \tan^3\theta\,\sec^2\theta\,d\theta$

44. $\displaystyle\int \sec^3\theta\,\tan\theta\,d\theta$

45. $\displaystyle\int \frac{\cos\sqrt{x}}{\sqrt{x}}\,dx$ [*Suggestion:* Try $u = \sqrt{x}$.]

46. $\displaystyle\int \frac{dx}{\sqrt{x}\left(1 + \sqrt{x}\right)^2}$

47. $\displaystyle\int (x^2+2x+1)^4(x+1)\,dx$

48. $\displaystyle\int \frac{(x + 2)\,dx}{(x^2 + 4x + 3)^3}$

49. $\displaystyle\int \frac{x + 2}{x^2 + 4x + 3}\,dx$

50. $\displaystyle\int \frac{2x + e^x}{(x^2 + e^x + 1)^2}\,dx$

Evaluate the definite integrals in Problems 51 through 64.

51. $\displaystyle\int_1^2 \frac{dt}{(t + 1)^3}$

52. $\displaystyle\int_0^4 \frac{dx}{\sqrt{2x + 1}}$

53. $\displaystyle\int_0^4 x\sqrt{x^2 + 9}\,dx$

54. $\displaystyle\int_1^4 \frac{\left(1 + \sqrt{x}\right)^4}{\sqrt{x}}\,dx$ [*Suggestion:* Try $u = 1 + \sqrt{x}$.]

55. $\displaystyle\int_0^8 t\sqrt{t + 1}\,dx$ [*Suggestion:* Try $u = t + 1$.]

56. $\displaystyle\int_0^{\pi/2} \sin x\,\cos x\,dx$

57. $\displaystyle\int_0^{\pi/6} \sin 2x\,\cos^3 2x\,dx$

58. $\displaystyle\int_0^{\sqrt{\pi}} t\sin\frac{t^2}{2}\,dt$

59. $\displaystyle\int_0^{\pi/2} (1 + 3\sin\theta)^{3/2}\cos\theta\,d\theta$ [*Suggestion:* Try $u = 1 + 3\sin\theta$.]

60. $\displaystyle\int_0^{\pi/2} \sec^2\frac{x}{2}\,dx$

61. $\displaystyle\int_0^{\pi/2} e^{\sin x}\cos x\,dx$ [*Suggestion:* Try $u = \sin x$.]

62. $\displaystyle\int_1^2 \frac{1 + \ln x}{x}\,dx$ [*Suggestion:* Try $u = 1 + \ln x$.]

63. $\displaystyle\int_1^2 \frac{e^{-1/x}}{x^2}\,dx$ $\left[\textit{Suggestion: } \text{Try } u = \dfrac{1}{x}.\right]$

64. $\displaystyle\int_{\pi^2/4}^{\pi^2} \frac{\sin\sqrt{x}\,\cos\sqrt{x}}{\sqrt{x}}\,dx$

Use the half-angle identities

$$\cos^2\theta = \frac{1 + \cos 2\theta}{2} \quad\text{and}\quad \sin^2\theta = \frac{1 - \cos 2\theta}{2}$$

to evaluate the integrals in Problems 65 through 68.

65. $\displaystyle\int \sin^2 x\,dx$

66. $\displaystyle\int \cos^2 x\,dx$

67. $\displaystyle\int_0^{\pi} \sin^2 3t\,dt$

68. $\displaystyle\int_0^1 \cos^2 \pi t\,dt$

Use the identity $1 + \tan^2\theta = \sec^2\theta$ *to evaluate the integrals in Problems 69 and 70.*

69. $\displaystyle\int \tan^2 x\,dx$

70. $\displaystyle\int_0^{\pi/12} \tan^2 3t\,dt$

71. Substitute $\sin^3 x = (\sin x)(1 - \cos^2 x)$ to show that

$$\int \sin^3 x\,dx = \tfrac{1}{3}\cos^3 x - \cos x + C.$$

72. Evaluate

$$\int_0^{\pi/2} \cos^3 x\,dx$$

by the method of Problem 71.

73. Substitute first $u = \sin\theta$ and then $u = \cos\theta$ to obtain

$$\int \sin\theta\,\cos\theta\,d\theta = \tfrac{1}{2}\sin^2\theta + C_1 = -\tfrac{1}{2}\cos^2\theta + C_2.$$

Reconcile these results. What is the relation between the constants C_1 and C_2? *Suggestion:* Compare the graphs (on the same screen) of

$$f(x) = \tfrac{1}{2}\sin^2\theta \quad\text{and}\quad g(x) = -\tfrac{1}{2}\cos^2\theta.$$

74. Substitute first $u = \tan\theta$ and then $u = \sec\theta$ to obtain

$$\int \sec^2\theta\,\tan\theta\,d\theta = \tfrac{1}{2}\tan^2\theta + C_1 = \tfrac{1}{2}\sec^2\theta + C_2.$$

Reconcile these results. What is the relation between the constants C_1 and C_2? *Suggestion:* Compare the graphs (on the same screen) of

$$f(x) = \tfrac{1}{2}\tan^2\theta \quad\text{and}\quad g(x) = \tfrac{1}{2}\sec^2\theta.$$

75. (a) Verify by differentiation that

$$\int \frac{dx}{(1 - x)^2} = \frac{x}{1 - x} + C_1.$$

(b) Substitute $u = 1 - x$ to show that

$$\int \frac{dx}{(1 - x)^2} = \frac{1}{1 - x} + C_2.$$

(c) Reconcile the results of parts (a) and (b). *Suggestion:* Compare the graphs (on the same screen) of

$$f(x) = \frac{x}{1 - x} \quad\text{and}\quad g(x) = \frac{1}{1 - x}.$$

76. (a) Substitute $u = x^2$ and apply part (a) of Problem 75 to show that

$$\int \frac{x\,dx}{(1-x^2)^2} = \frac{x^2}{2(1-x^2)} + C_1.$$

(b) Substitute $u = 1 - x^2$ to show that

$$\int \frac{x\,dx}{(1-x^2)^2} = \frac{1}{2(1-x^2)} + C_2.$$

(c) Reconcile the results of parts (a) and (b). *Suggestion:* Compare the graphs (on the same screen) of

$$f(x) = \frac{x^2}{2(1-x^2)} \quad \text{and} \quad g(x) = \frac{1}{2(1-x^2)}.$$

Problems 77 and 78 deal with even and odd functions. An ***even*** *function f is a function such that*

$$f(-x) = f(x)$$

for all x. This means that the graph of $y = f(x)$ is symmetric under reflection across the y-axis (Fig. 5.7.1). Examples of even functions include $f(x) = \cos x$, 1, x^2, x^4, and x^6. An ***odd*** *function f is a function such that*

$$f(-x) = -f(x)$$

for all x. This means that the graph of $y = f(x)$ is symmetric under reflections first across the y-axis, then across the x-axis (Fig. 5.7.2). Examples of odd functions are $f(x) = \sin x$, x, x^3, and x^5. Think about these reflections with the (even) cosine function (in Fig. 5.7.3) and the (odd) sine function (in Fig. 5.7.4).

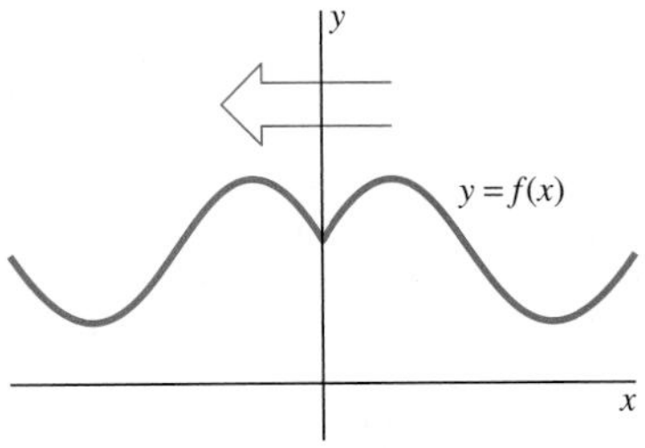

FIGURE 5.7.1 The graph of the even function $y = f(x)$ is invariant under reflection through the y-axis.

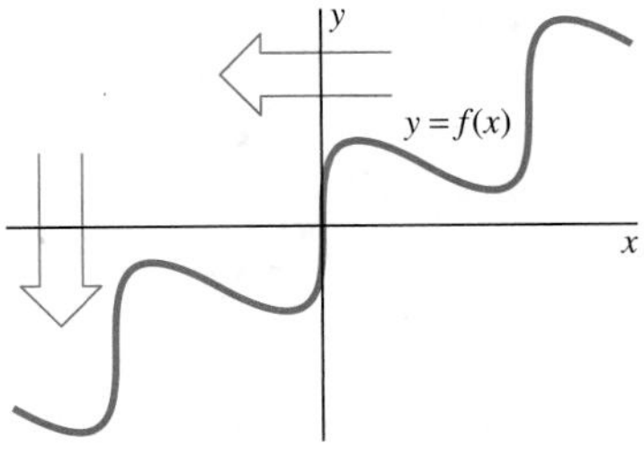

FIGURE 5.7.2 The graph of the odd function $y = f(x)$ is invariant under successive reflections through both axes.

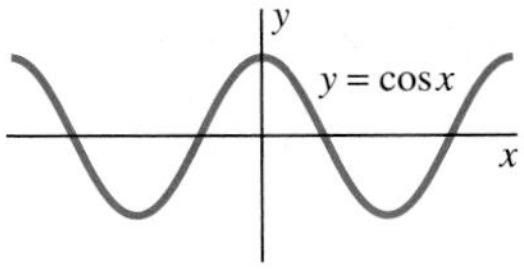

FIGURE 5.7.3 The cosine function is even.

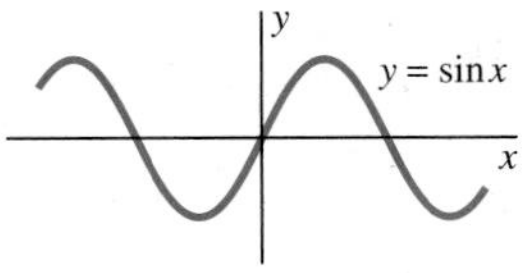

FIGURE 5.7.4 The sine function is odd.

77. See Fig. 5.7.5. If the continuous function f is odd, substitute $u = -x$ into the integral

$$\int_{-a}^{0} f(x)\,dx \quad \text{to show that} \quad \int_{-a}^{a} f(x)\,dx = 0.$$

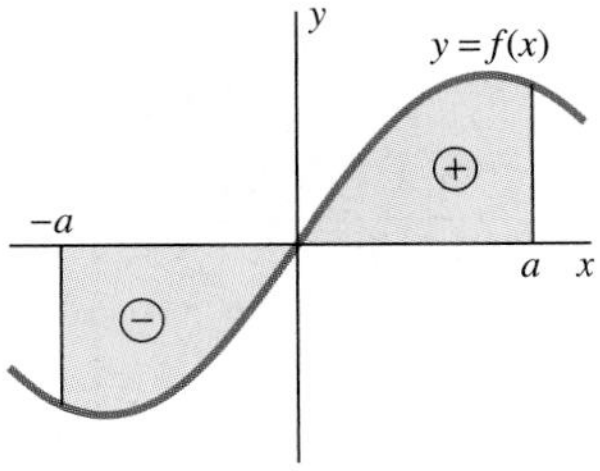

FIGURE 5.7.5 Areas cancel when f is odd (Problem 77).

78. See Fig. 5.7.6. If the continuous function f is even, use the method of Problem 77 to show that

$$\int_{-a}^{a} f(x)\,dx = 2\int_{0}^{a} f(x)\,dx.$$

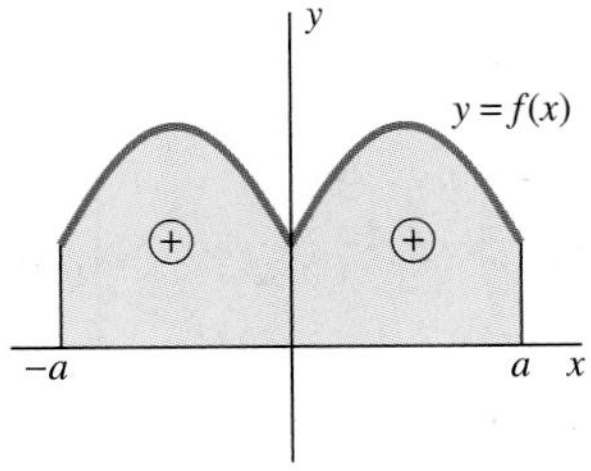

FIGURE 5.7.6 Areas add when f is even (Problem 78).

In Problems 79 and 80, use the results of Problems 77 and 78 to justify the values of the given integrals without extensive computation.

79. $\displaystyle\int_{-1}^{1} \left[\tan x + \frac{\sqrt[3]{x}}{(1+x^2)^7} - x^{17}\cos x\right] dx = 0.$

80. $\displaystyle\int_{-5}^{5} \left(3x^2 - x^{10}\sin x + x^5\sqrt{1+x^4}\,\right) dx = 2\Big[x^3\Big]_0^5 = 250.$

81. Suppose that f is continuous everywhere and that k is a constant. Show that

$$\int_{a}^{b} f(x+k)\,dx = \int_{a+k}^{b+k} f(x)\,dx.$$

In the case in which $k > 0$ and $f(x) > 0$, illustrate this formula with a sketch showing two regions with bases $[a, b]$ and $[a + k, b + k]$. Why is it plausible that these two regions have equal areas?

82. Suppose that f is continuous everywhere and that k is a constant. Show that

$$\int_{ka}^{kb} f(x)\,dx = k\int_a^b f(kx)\,dx.$$

In the case in which $k > 1$ and $f(x) > 0$, illustrate this formula with a sketch showing two regions with bases $[a, b]$ and $[ka, kb]$. Why is it plausible that the area of one of these regions is k times the area of the other?

83. (a) Verify by differentiation that

$$\int ue^u\,du = (u-1)e^u + C.$$

(b) Use part (a) to show that

$$\int_0^1 e^{\sqrt{x}}\,dx = 2.$$

84. (a) Verify by differentiation that

$$\int u\sin u\,du = \sin u - u\cos u + C.$$

(b) Use part (a) to show that

$$\int_0^{\pi^2} \sin\sqrt{x}\,dx = 2\pi.$$

5.8 AREAS OF PLANE REGIONS

In Section 5.3 we discussed the area A under the graph of a positive-valued continuous function f on the interval $[a, b]$. This discussion motivated our definition in Section 5.4 of the integral of f from a to b as the limit of Riemann sums. An important result was that

$$A = \int_a^b f(x)\,dx, \tag{1}$$

by definition.

Here we consider the problem of finding the areas of more general regions in the coordinate plane. Regions such as the ones illustrated in Fig. 5.8.1 may be bounded by the graphs of *two* (or more) different functions.

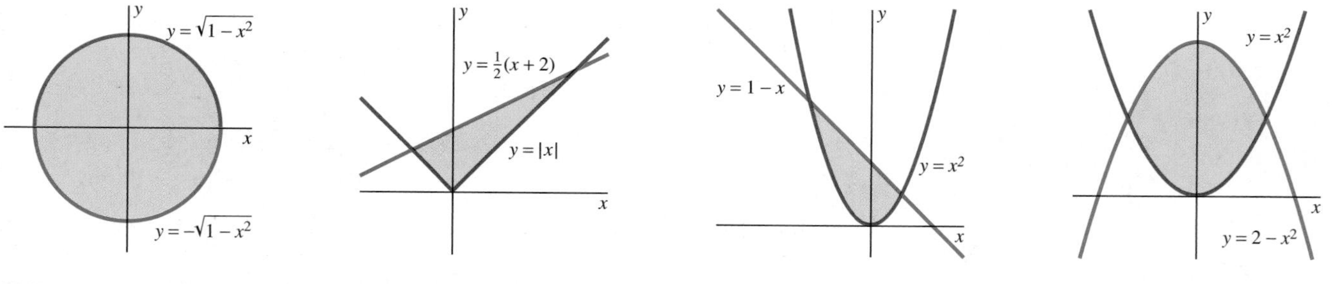

FIGURE 5.8.1 Plane regions bounded by pairs of curves.

Let f and g be continuous functions such that $f(x) \geqq g(x)$ for all x in the interval $[a, b]$. We are interested in the area A of the region R in Fig. 5.8.2, which lies *between* the graphs of $y = f(x)$ and $y = g(x)$ for x in $[a, b]$. Thus R is bounded by

- The curve $y = f(x)$, the upper boundary of R, by
- The curve $y = g(x)$, the lower boundary of R, *and* by
- The vertical lines $x = a$ and $x = b$ (if needed).

To approximate A, we consider a partition of $[a, b]$ into n subintervals, all with the same length $\Delta x = (b - a)/n$. If ΔA_i denotes the area of the region between the graphs of f and g over the ith subinterval $[x_{i-1}, x_i]$, and $x_i^\star$ is a selected number chosen in that subinterval (all this for $i = 1, 2, 3, \ldots, n$), then ΔA_i is approximately equal to the area of a rectangle with height $f(x_i^\star) - g(x_i^\star)$ and width Δx (Fig. 5.8.3).

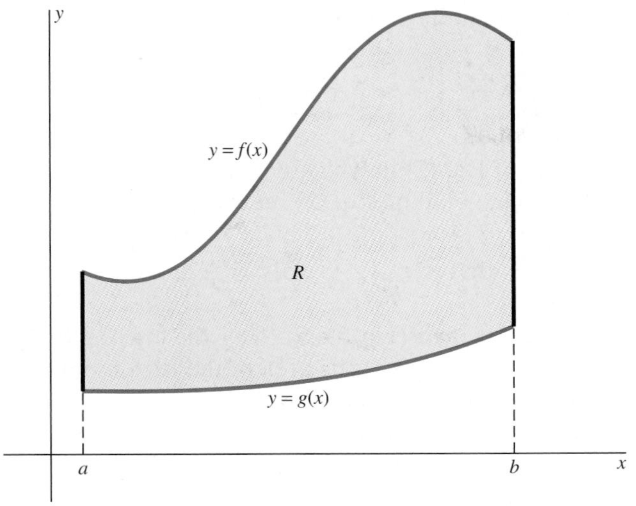

FIGURE 5.8.2 A region between two graphs.

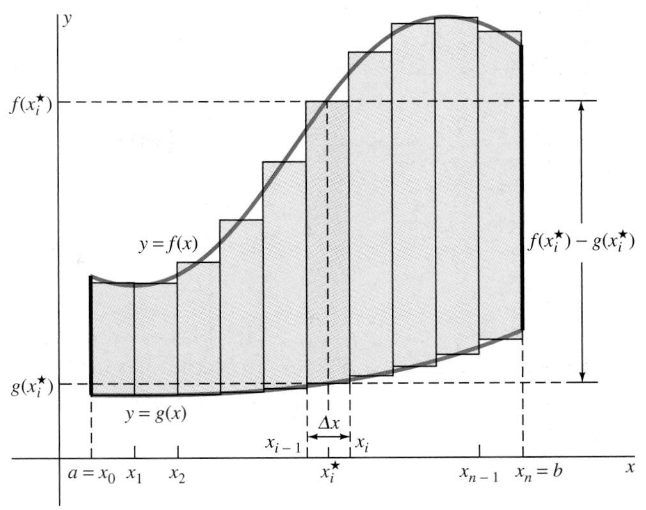

FIGURE 5.8.3 A partition of [*a*, *b*] divides *R* into vertical strips that we approximate with rectangular strips.

Hence

$$\Delta A_i \approx [f(x_i^\star) - g(x_i^\star)]\,\Delta x;$$

so

$$A = \sum_{i=1}^{n} \Delta A_i \approx \sum_{i=1}^{n} [f(x_i^\star) - g(x_i^\star)]\,\Delta x.$$

We introduce the *height* function $h(x) = f(x) - g(x)$ and observe that A is approximated by a Riemann sum for $h(x)$ associated with our partition of $[a, b]$:

$$A \approx \sum_{i=1}^{n} h(x_i^\star)\,\Delta x.$$

Both intuition and reason suggest that this approximation can be made arbitrarily accurate by choosing n to be sufficiently large (and hence $\Delta x = (b - a)/n$ to be sufficiently small). We therefore conclude that

$$A = \lim_{\Delta x \to 0} \sum_{i=1}^{n} h(x_i^\star)\,\Delta x = \int_a^b h(x)\,dx = \int_a^b [f(x) - g(x)]\,dx.$$

Because our discussion is based on an intuitive concept rather than on a precise logical definition of area, it does *not* constitute a proof of this area formula. It does, however, provide justification for the following *definition* of the area in question.

DEFINITION The Area Between Two Curves
Let f and g be continuous with $f(x) \geqq g(x)$ for x in $[a, b]$. Then the **area** A of the region bounded by the curves $y = f(x)$ and $y = g(x)$ and by the vertical lines $x = a$ and $x = b$ is

$$A = \int_a^b [f(x) - g(x)]\,dx. \qquad \textbf{(2)}$$

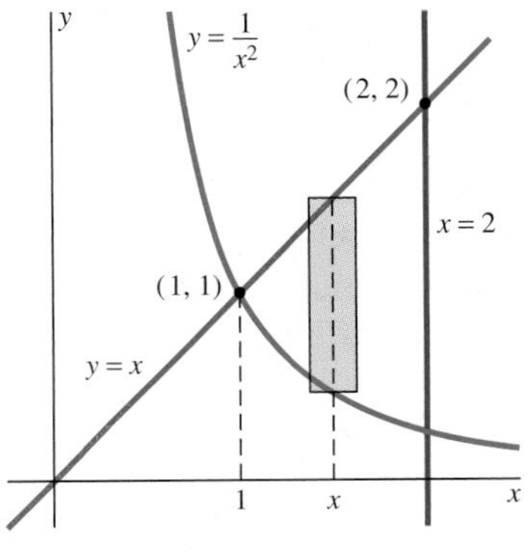

FIGURE 5.8.4 The region of Example 1.

EXAMPLE 1 Find the area of the region bounded by the lines $y = x$ and $x = 2$ and by the curve $y = 1/x^2$ (Fig. 5.8.4).

Solution Here the top curve is $y = f(x) = x$, the bottom curve is $y = g(x) = 1/x^2$, $a = 1$, and $b = 2$. The vertical line $x = 2$ is "needed" (to form the right-hand boundary of the region), whereas $x = 1$ is not. Equation (2) yields

$$A = \int_1^2 \left(x - \frac{1}{x^2}\right) dx = \left[\frac{1}{2}x^2 + \frac{1}{x}\right]_1^2 = \left(2 + \frac{1}{2}\right) - \left(\frac{1}{2} + 1\right) = 1. \quad \blacklozenge$$

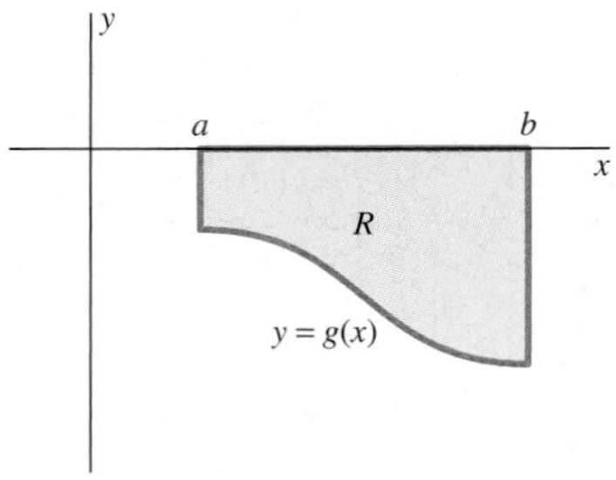

FIGURE 5.8.5 The integral $\int_a^b g(x)\,dx$ gives the negative of the geometric area for a region that lies below the x-axis.

Equation (1) is the special case of Eq. (2) in which $g(x)$ is identically zero on $[a, b]$. But if $f(x) \equiv 0$ and $g(x) \leqq 0$ on $[a, b]$, then Eq. (2) reduces to

$$A = -\int_a^b g(x)\,dx; \qquad \text{that is,} \qquad \int_a^b g(x)\,dx = -A.$$

In this case the region R lies beneath the x-axis (Fig. 5.8.5). Thus the integral from a to b of a negative-valued function is the *negative* of the area of the region bounded by its graph, the x-axis, and the vertical lines $x = a$ and $x = b$.

More generally, consider a continuous function f with a graph that crosses the x-axis at finitely many points $c_1, c_2, c_3, \ldots, c_k$ between a and b (Fig. 5.8.6). We write

$$\int_a^b f(x)\,dx = \int_a^{c_1} f(x)\,dx + \int_{c_1}^{c_2} f(x)\,dx + \cdots + \int_{c_k}^b f(x)\,dx.$$

Thus we see that

$$\int_a^b f(x)\,dx$$

is equal to the area below $y = f(x)$ and *above* the x-axis *minus* the area above $y = f(x)$ and *below* the x-axis.

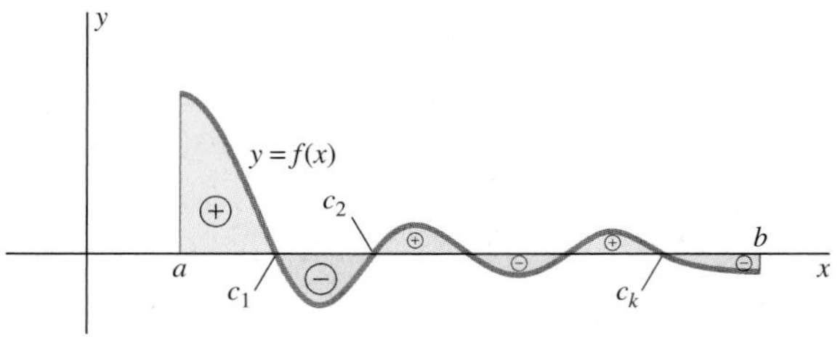

FIGURE 5.8.6 The integral $\int_a^b f(x)\,dx$ computes the area above the x-axis *minus* the area below the x-axis.

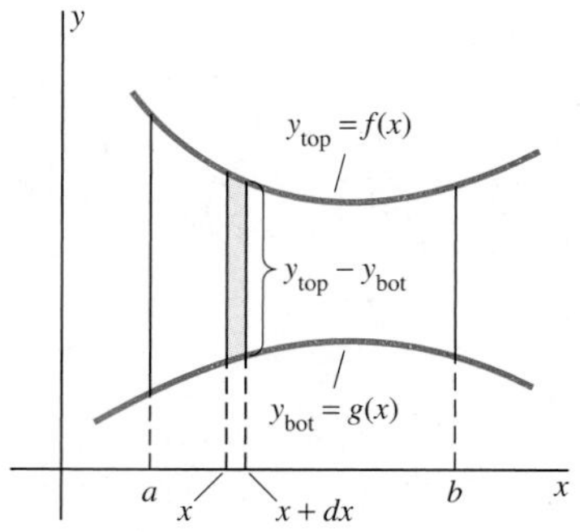

FIGURE 5.8.7 Heuristic (suggestive but nonrigorous) approach to setting up area integrals.

The following *heuristic* (suggestive, though nonrigorous) way of setting up integral formulas such as Eq. (2) can be useful. Consider the vertical strip of area that lies above the interval $[x, x + dx]$, shown shaded in Fig. 5.8.7, where we have written

$$y_{\text{top}} = f(x) \quad \text{and} \quad y_{\text{bot}} = g(x)$$

for the top and bottom boundary curves. We think of the length dx of the interval $[x, x + dx]$ as being so small that we can regard this strip as a rectangle with width dx and height $y_{\text{top}} - y_{\text{bot}}$. Its area is then

$$dA = (y_{\text{top}} - y_{\text{bot}})\,dx.$$

Think now of the region over $[a, b]$ that lies between $y_{\text{top}} = f(x)$ and $y_{\text{bot}} = g(x)$ as being made up of many such vertical strips. We can regard its area as a sum of areas of such rectangular strips. If we write $\int$ for *sum,* we get the formula

$$A = \int dA = \int_a^b (y_{\text{top}} - y_{\text{bot}})\,dx.$$

This heuristic approach bypasses the subscript notation associated with Riemann sums. Nevertheless, it *is not* and *should not* be regarded as a *complete* derivation of the last formula. It is best used only as a convenient memory device. For instance, in the figures that accompany many examples, we shall often show a strip of width dx as a visual aid in properly setting up the correct integral.

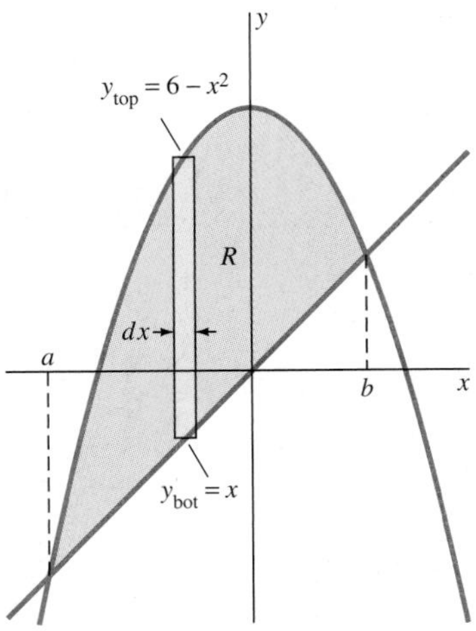

FIGURE 5.8.8 The region R of Example 2.

EXAMPLE 2 Find the area A of the region R bounded by the line $y = x$ and the parabola $y = 6 - x^2$.

Solution The region R is shown in Fig. 5.8.8. We can use Eq. (2) and take $f(x) = 6 - x^2$ and $g(x) = x$. The limits a and b will be the x-coordinates of the two points of intersection of the line and the parabola; our first order of business is to find a and b. To do so, we equate $f(x)$ and $g(x)$ and solve the resulting equation for x:

$$x = 6 - x^2; \qquad x^2 + x - 6 = 0;$$
$$(x - 2)(x + 3) = 0; \quad x = -3,\ 2.$$

Thus $a = -3$ and $b = 2$, so Eq. (2) gives

$$A = \int_{-3}^{2} (6 - x^2 - x)\,dx = \left[6x - \tfrac{1}{3}x^3 - \tfrac{1}{2}x^2\right]_{-3}^{2}$$
$$= \left[6 \cdot 2 - \tfrac{1}{3} \cdot 2^3 - \tfrac{1}{2} \cdot 2^2\right] - \left[6 \cdot (-3) - \tfrac{1}{3} \cdot (-3)^3 - \tfrac{1}{2} \cdot (-3)^2\right] = \tfrac{125}{6}. \qquad ◆$$

Subdividing Regions Before Integrating

Example 3 shows that it is sometimes necessary to subdivide a region before applying Eq. (2), typically because the formula for either the top or the bottom boundary curve (or both) changes somewhere between $x = a$ and $x = b$.

EXAMPLE 3 Find the area of the region R bounded by the line $y = \frac{1}{2}x$ and the parabola $y^2 = 8 - x$.

Solution The region R is shaded in Fig. 5.8.9. The points of intersection $(-8, -4)$ and $(4, 2)$ are found by equating $y = \frac{1}{2}x$ and $y = \pm\sqrt{8 - x}$ and then solving for x. The lower boundary of R is given by $y_{bot} = -\sqrt{8 - x}$ on $[-8, 8]$. But the upper boundary of R is given by

$$y_{top} = \tfrac{1}{2}x \quad \text{on} \quad [-8, 4], \qquad y_{top} = +\sqrt{8 - x} \quad \text{on} \quad [4, 8].$$

We must therefore divide R into the two regions R_1 and R_2, as indicated in Fig. 5.8.9. Then Eq. (2) gives

$$A = \int_{-8}^{4} \left(\tfrac{1}{2}x + \sqrt{8 - x}\right) dx + \int_{4}^{8} 2\sqrt{8 - x}\,dx$$
$$= \left[\tfrac{1}{4}x^2 - \tfrac{2}{3}(8 - x)^{3/2}\right]_{-8}^{4} + \left[-\tfrac{4}{3}(8 - x)^{3/2}\right]_{4}^{8}$$
$$= \left[\left(\tfrac{16}{4} - \tfrac{16}{3}\right) - \left(\tfrac{64}{4} - \tfrac{128}{3}\right)\right] + \left[0 + \tfrac{32}{3}\right] = 36. \qquad ◆$$

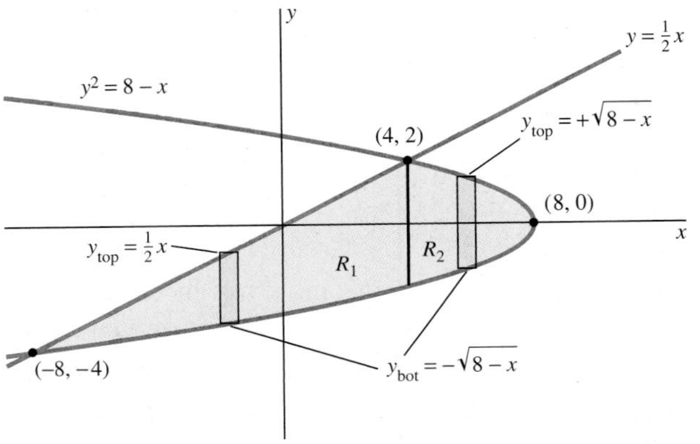

FIGURE 5.8.9 In Example 3, we split region R into two regions R_1 and R_2.

Determining Area by Integrating with Respect to y

The region in Example 3 appears to be simpler if we consider it to be bounded by graphs of functions of y rather than by graphs of functions of x. Figure 5.8.10 shows a region R bounded by the curves $x = f(y)$ and $x = g(y)$, with $f(y) \geqq g(y)$ for y in $[c, d]$, and by the horizontal lines $y = c$ and $y = d$. To approximate the

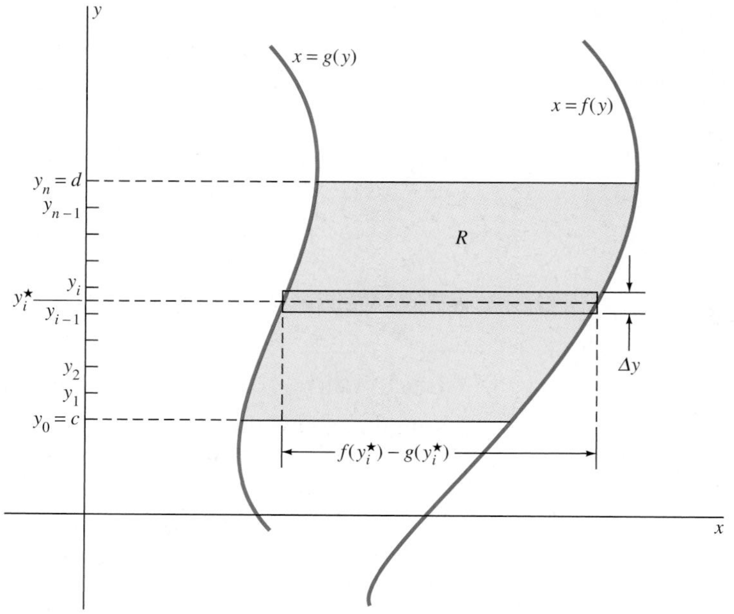

FIGURE 5.8.10 Find area by using an integral with respect to y.

area A of R, we begin with a partition of $[c, d]$ into n subintervals all with the same length $\Delta y = (d - c)/n$. We choose a point $y_i^\star$ in the ith subinterval $[y_{i-1}, y_i]$ for each i $(1 \leqq i \leqq n)$. The horizontal strip of R lying opposite $[y_{i-1}, y_i]$ is approximated by a rectangle with width Δy (measured vertically) and height $f(y_i^\star) - g(y_i^\star)$ (measured horizontally). Hence

$$A \approx \sum_{i=1}^{n} [f(y_i^\star) - g(y_i^\star)]\ \Delta y.$$

Recognizing the sum as a Riemann sum for the integral

$$\int_c^d [f(y) - g(y)]\ dy$$

motivates the following definition.

DEFINITION The Area Between Two Curves
Let f and g be continuous functions of y with $f(y) \geqq g(y)$ for y in $[c, d]$. Then the **area** A of the region bounded by the curves $x = f(y)$ and $x = g(y)$ and by the horizontal lines $y = c$ and $y = d$ is

$$A = \int_c^d [f(y) - g(y)]\ dy. \qquad \textbf{(3)}$$

In a more advanced course, we would now prove that Eqs. (2) and (3) yield the same area A for a region that can be described both in the manner shown in Fig. 5.8.2 and in the manner shown in Fig. 5.8.10.

Let us write

$$x_{\text{right}} = f(y) \quad \text{and} \quad x_{\text{left}} = g(y)$$

for the right and left boundary curves, respectively, of the region in Fig. 5.8.10. Then

Eq. (3) takes the form

$$A = \int_c^d [x_{\text{right}} - x_{\text{left}}]\,dy.$$

Comparing Example 3 with Example 4 illustrates the advantage of choosing the "right" variable of integration—the one that makes the resulting computations simpler.

EXAMPLE 4 Integrate with respect to y to find the area of the region R of Example 3.

Solution We see from Fig. 5.8.11 that Eq. (3) applies with $x_{\text{right}} = f(y) = 8 - y^2$ and $x_{\text{left}} = g(y) = 2y$ for y in $[-4, 2]$. This gives

$$A = \int_{-4}^{2} [(8 - y^2) - 2y]\,dy = \left[8y - \tfrac{1}{3}y^3 - y^2\right]_{-4}^{2} = 36.$$ ◆

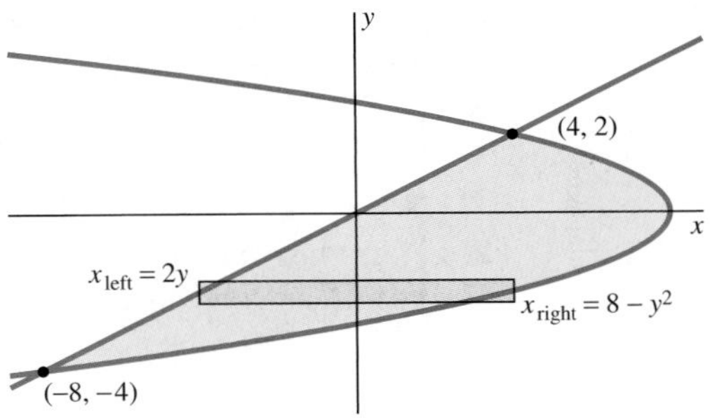

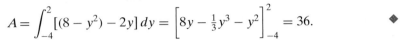

FIGURE 5.8.11 Recomputation of the area of Example 3 (Example 4).

EXAMPLE 5 Use calculus to derive the formula $A = \pi r^2$ for the area of a circle of radius r.

Solution The point is that the formula $A = \pi r^2$, however familiar it seems, requires *proof*—it certainly is not self-evident. We begin with the *definition* (as suggested in Section 5.3) of the famous number π as the area of the unit circle $x^2 + y^2 = 1$. This implies that

$$\int_0^1 \sqrt{1 - x^2}\,dx = \frac{\pi}{4}, \tag{4}$$

(see Fig. 5.8.12) despite the fact that there is no immediate or obvious way of antidifferentiating $\sqrt{1 - x^2}$ in order to explicitly evaluate the integral in (4).

Now we turn our attention to the general circle of radius r shown in Fig. 5.8.13. We apply Eq. (1) to the first quadrant and then multiply by 4. Thus we find that the

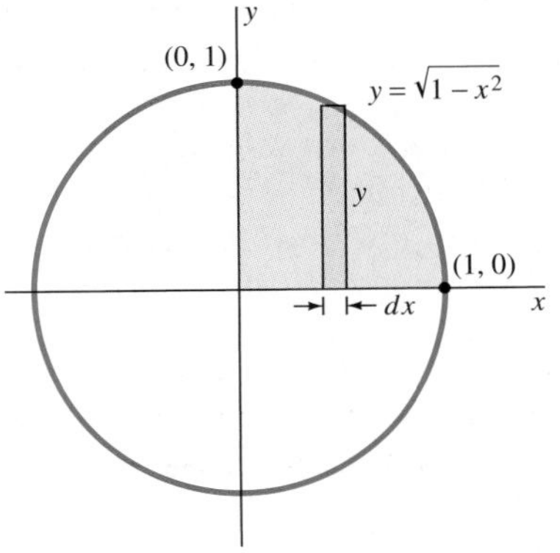

FIGURE 5.8.12 The number π is four times the shaded area.

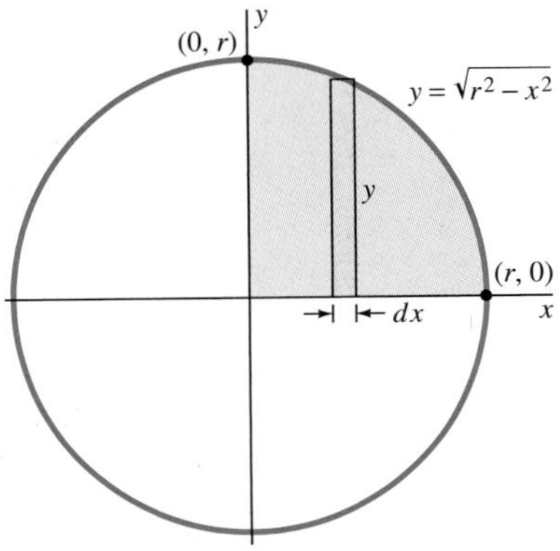

FIGURE 5.8.13 The shaded area can be expressed as an integral.

total area A of the circle is given by

$$\begin{aligned} A &= 4\int_0^r \sqrt{r^2 - x^2}\,dx = 4r\int_0^r \sqrt{1 - \frac{x^2}{r^2}}\,dx \\ &= 4r\int_0^1 r\sqrt{1 - u^2}\,du \quad \left(\textit{Substitution: } u = \frac{x}{r},\ x = ru,\ dx = r\,du\right) \\ &= 4r^2 \cdot \frac{\pi}{4}. \end{aligned}$$

Here we have applied Eq. (4)—with u in place of x—to get $A = \pi r^2$, as desired. ◆

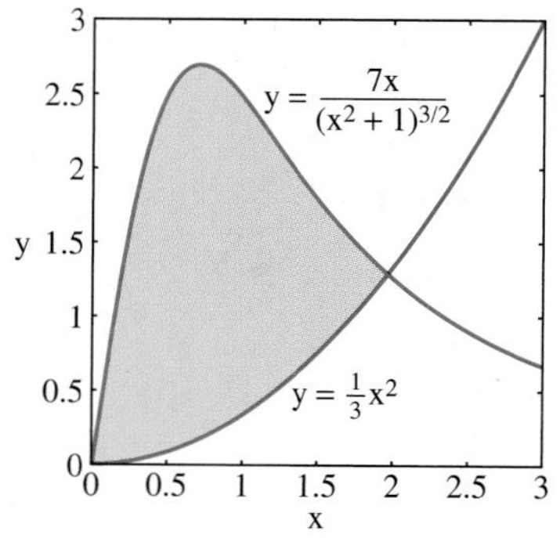

FIGURE 5.8.14 The region of Example 6.

EXAMPLE 6 Approximate the area A of the first-quadrant region shown in Fig. 5.8.14. This is the region bounded by the curves

$$y = \frac{7x}{(x^2+1)^{3/2}} \quad \text{and} \quad y = \frac{x^2}{3}.$$

Solution In order to find the exact coordinates of the first-quadrant intersection point in Fig. 5.8.14, we would need to solve the equation

$$\frac{7x}{(x^2+1)^{3/2}} = \frac{x^2}{3}.$$

We might begin by first canceling x, then cross-multiplying and squaring both sides. The result, as you can verify, simplifies to the equation

$$x^8 + 3x^6 + 3x^4 + x^2 - 441 = 0.$$

Now you see why we asked for the *approximate* area. Although it is impractical to solve this eighth-degree equation exactly, we can use a graphing calculator or computer to zoom in on the desired intersection point shown in Fig. 5.8.14. In this way we find that $x \approx 1.963$ is its approximate x-coordinate. (We could also use Newton's method or a calculator root-finder.) We can now proceed to approximate the desired area.

$$\begin{aligned} A &\approx \int_0^{1.963} \left(\frac{7x}{(x^2+1)^{3/2}} - \frac{x^2}{3}\right) dx \\ &= \int_0^{1.963} \frac{7x}{(x^2+1)^{3/2}}\,dx - \int_0^{1.963} \frac{x^2}{3}\,dx. \end{aligned}$$

We substitute $u = x^2 + 1$, $du = 2x\,dx$ to evaluate the first integral and get

$$A \approx \left[\frac{-7x}{\sqrt{x^2+1}} - \frac{x^3}{9}\right]_0^{1.963} \approx -4.018 - (-7) = 2.982.$$ ◆

5.8 TRUE/FALSE STUDY GUIDE

5.8 CONCEPTS: QUESTIONS AND DISCUSSION

The concept of area is not self-evident, and the area of a plane set must be defined *before it can be calculated. The area of a region bounded by curves can be defined in terms of areas of inscribed and circumscribed* polygons. *We say that the region R has* **area** *A—and write $a(R) = A$—provided that, given any number $\epsilon > 0$ (however small), there exist a polygon P contained in R and a polygon Q containing R such that*

$$A - \epsilon < a(P) \leqq a(Q) < A + \epsilon.$$

The polygonal areas $a(P)$ and $a(Q)$ are defined as sums of areas of nonoverlapping triangles and/or rectangles. If for some $\epsilon > 0$ there do not exist such polygons P and Q, then R does not have area. (There do exist plane sets whose area is not defined!)

1. Suppose that the function f is continuous and positive-valued on the interval $[a, b]$. Let R be the plane region bounded above by the graph of $y = f(x)$,

below by the x-axis, and on the sides by the vertical lines $x = a$ and $x = b$. Then use the fact that f is integrable and the observation that any Riemann sum is the area of a polygon to prove that the area A of R exists and is given by

$$A = \int_a^b f(x)\,dx.$$

Suggestion: Think about the maximum and minimum values of $f(x)$ on a typical subinterval of a partition of $[a, b]$.

2. Suppose that the function f is defined on the interval $[0, 1]$ as follows:

$$f(x) = \begin{cases} 0 & \text{if } x \text{ is rational,} \\ 1 & \text{if } x \text{ is irrational.} \end{cases}$$

Let R be the plane region consisting of all those points (x, y) such that $0 \leqq x \leqq 1$ and $0 \leqq y \leqq f(x)$. Show that R does not have area. Note that if the polygon P is contained in R then P is "degenerate," so that it has area $a(P) = 0$. Note also that $a(Q) \geqq 1$ if the polygon Q contains R.

5.8 PROBLEMS

Find the areas of the regions shown in Problems 1 through 10.

1. (See Fig. 5.8.15.)

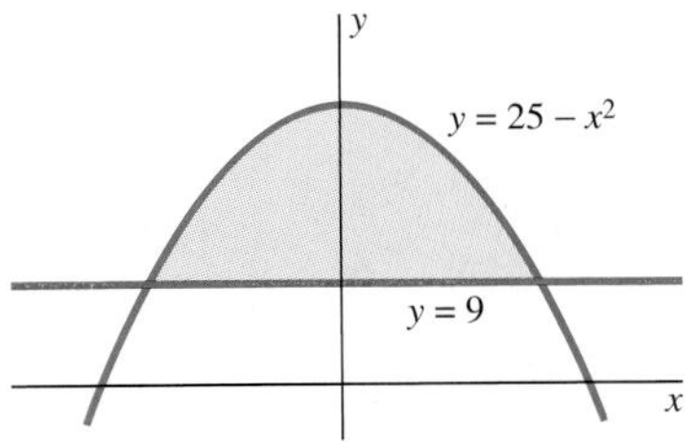

FIGURE 5.8.15 Problem 1.

2. (See Fig. 5.8.16.)

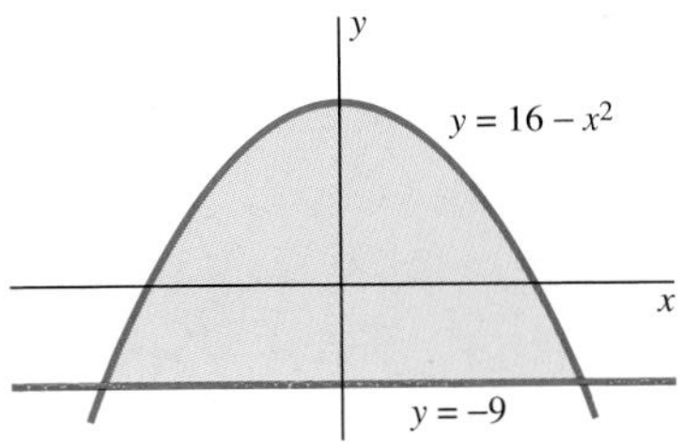

FIGURE 5.8.16 Problem 2.

3. (See Fig. 5.8.17.)

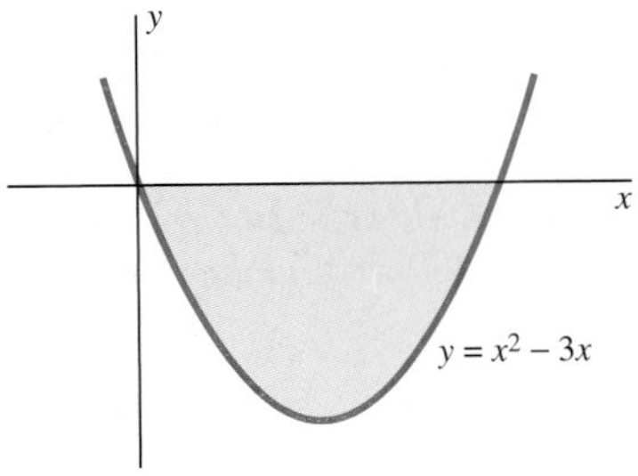

FIGURE 5.8.17 Problem 3.

4. (See Fig. 5.8.18.)

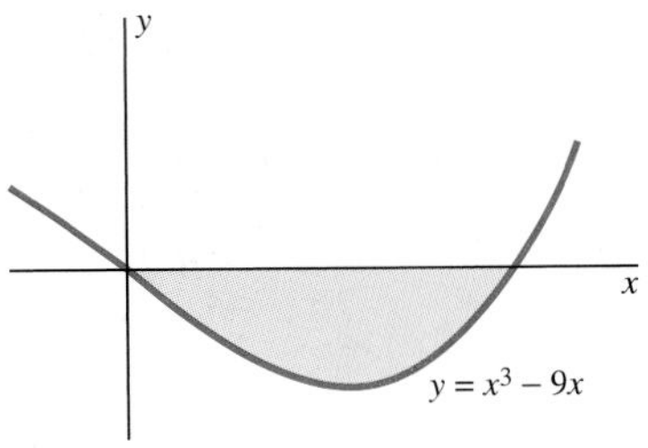

FIGURE 5.8.18 Problem 4.

5. (See Fig. 5.8.19.)

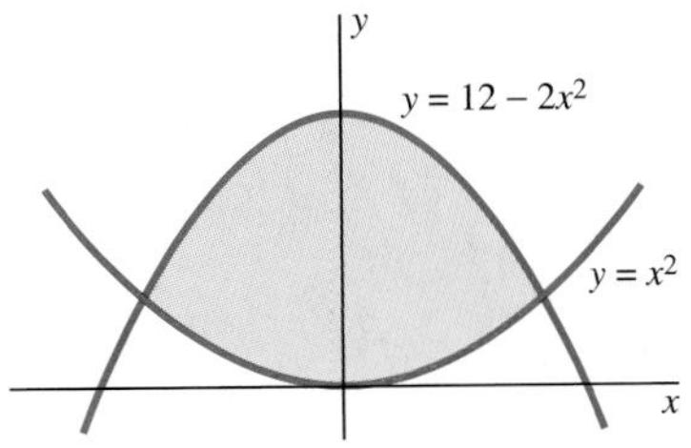

FIGURE 5.8.19 Problem 5.

6. (See Fig. 5.8.20.)

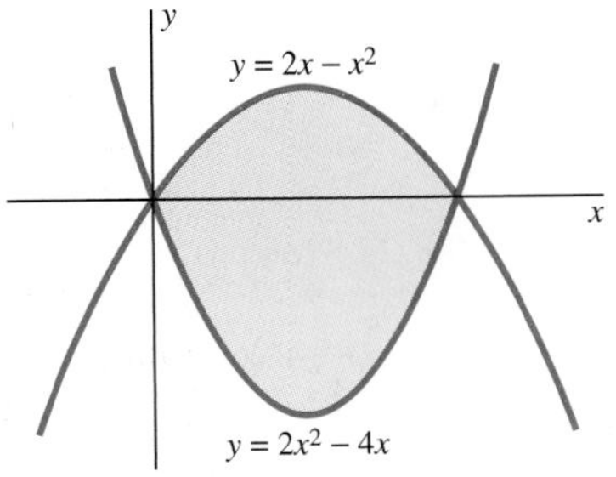

FIGURE 5.8.20 Problem 6.

7. (See Fig. 5.8.21.)

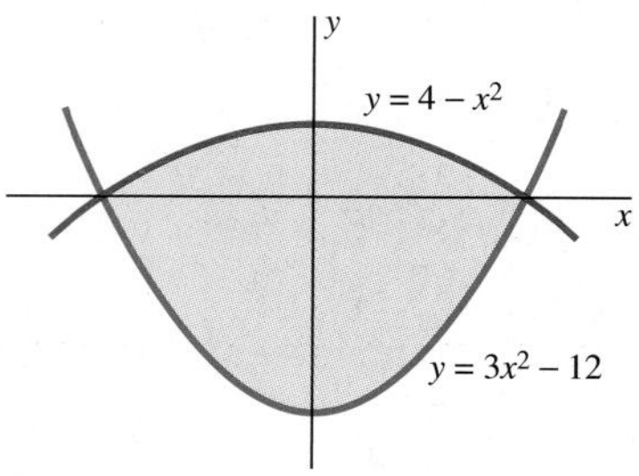

FIGURE 5.8.21 Problem 7.

8. (See Fig. 5.8.22.)

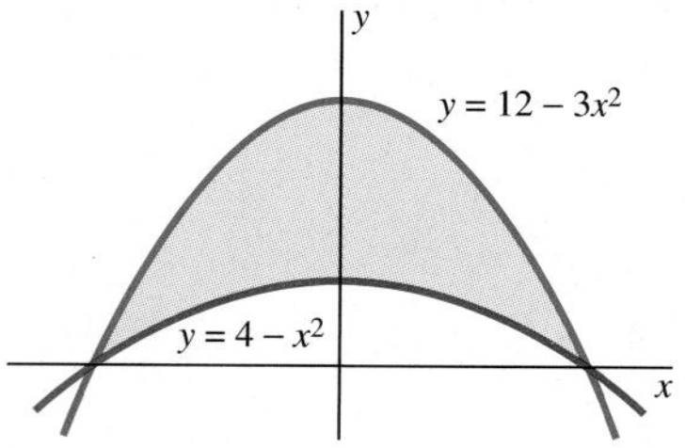

FIGURE 5.8.22 Problem 8.

9. (See Fig. 5.8.23.)

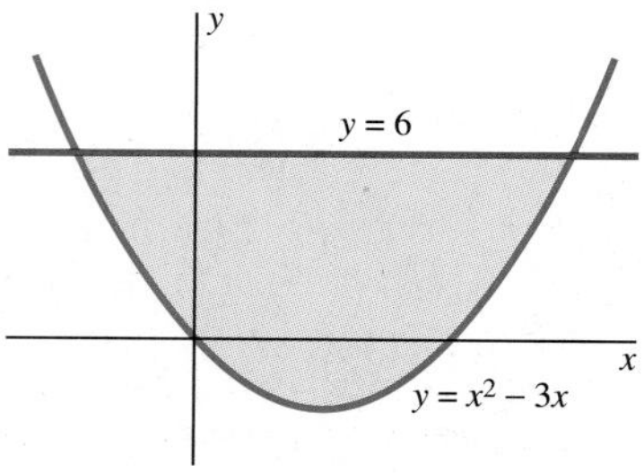

FIGURE 5.8.23 Problem 9.

10. (See Fig. 5.8.24.)

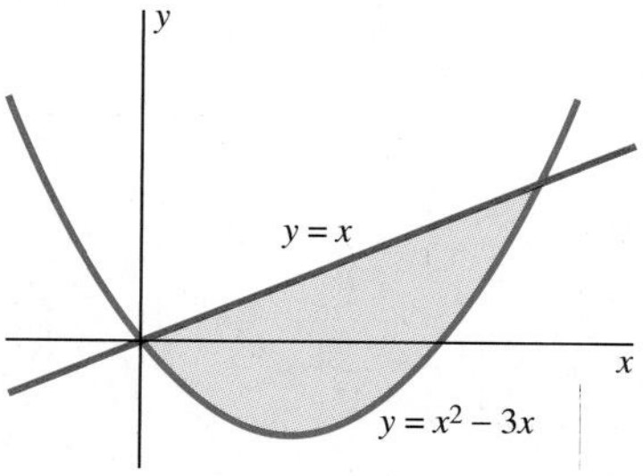

FIGURE 5.8.24 Problem 10.

Find the areas of the regions described in Problems 11 through 20.

11. The region R bounded below by the graph of $y = x^3$ and above by the graph of $y = x$ over the interval $[0, 1]$

12. The region R between the graph of $y = 1/(x+1)^2$ and the x-axis over the interval $[1, 3]$

13. The region R bounded above by the graph of $y = x^3$ and below by the graph of $y = x^4$ over the interval $[0, 1]$

14. The region R bounded above by the graph of $y = x^2$ and below by the horizontal line $y = -1$ over the interval $[-1, 2]$

15. The region R bounded above by the graph of $y = 1/(x+1)$ and below by the x-axis over the interval $[0, 2]$

16. The region R bounded above by the graph of $y = 4x - x^2$ and below by the x-axis

17. The region R bounded on the left by the graph of $x = y^2$ and on the right by the vertical line $x = 4$

18. The region R between the graphs of $y = x^4 - 4$ and $y = 3x^2$

19. The region R between the graphs of $x = 8 - y^2$ and $x = y^2 - 8$

20. The region R between the graphs of $y = x^{1/3}$ and $y = x^3$

In Problems 21 through 40, sketch the region bounded by the given curves, then find its area.

21. $y = x^2, \; y = 2x$

22. $y = x^2, \; y = 8 - x^2$

23. $x = y^2, \; x = 25$

24. $x = y^2, \; x = 32 - y^2$

25. $y = x^2, \; y = 2x + 3$

26. $y = x^2, \; y = 2x + 8$

27. $x = y^2, \; x = y + 6$

28. $x = y^2, \; x = 8 - 2y$

29. $y = \cos x, \; y = \sin x, \; 0 \leqq x \leqq \pi/4$

30. $y = \cos x, \; y = \sin x, \; -3\pi/4 \leqq x \leqq 0$

31. $x = 4y^2, \; x + 12y + 5 = 0$

32. $y = x^2, \; y = 3 + 5x - x^2$

33. $x = 3y^2, \; x = 12y - y^2 - 5$

34. $y = x^2, \; y = 4(x-1)^2$

35. $y = x + 1, \; y = 1/(x+1), \; 0 \leqq x \leqq 1$

36. $y = x + 1, \; y = e^{-x}, \; x = 1$

37. $y = e^x, \; y = e^{-x}, \; x = 1$

38. $y = 1/(x+1), \; y = 1/(10x+1), \; x = 10$

39. $y = x \exp(-x^2), \; y = 0, \; x = 1$

40. $y = 8/(x+2), \; x + y = 4$

In Problems 41 and 42, first use a calculator or computer to graph the given curves $y = f(x)$ and $y = g(x)$. You should then be able to find the coordinates of the points of intersection that will be evident in your figure. Finally, find the area of the region bounded by the two curves. Problems 43 and 44 are similar, except that the two curves bound two *regions; find the sum of the areas of these two regions.*

41. $y = x^2 - x, \quad y = 1 - x^3$

42. $y = x^3 - x, \quad y = 1 - x^4$

43. $y = x^2, \quad y = x^3 - 2x$

44. $y = x^3; \quad y = 2x^3 + x^2 - 2x$

45. Evaluate

$$\int_{-3}^{3} (4x + 5)\sqrt{9 - x^2}\, dx$$

by writing this integral as a sum of two integrals and interpreting one of them in terms of a known (circular) area.

46. Evaluate

$$\int_{0}^{3} x\sqrt{81 - x^4}\, dx$$

by making a substitution of the form $u = x^p$ (you choose p) and then interpreting the result in terms of a known area.

47. The *ellipse* $x^2/a^2 + y^2/b^2 = 1$ is shown in Fig. 5.8.25. Use the method of Example 4 to show that the area of the region it bounds is $A = \pi ab$, a pleasing generalization of the area formula for the circle.

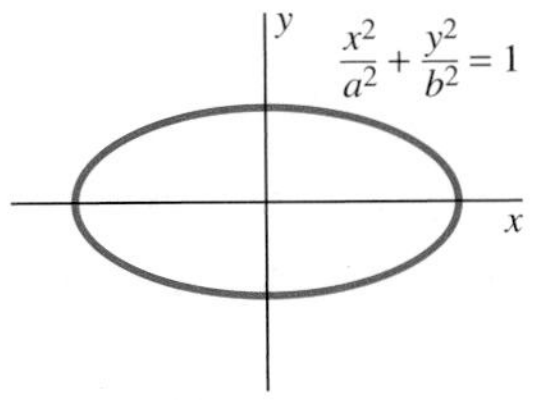

FIGURE 5.8.25 The ellipse of Problem 47.

48. Figure 5.8.26 shows a *parabolic segment* bounded by the parabola $y = x^2$ and the line $y = 1$. In the third century B.C., Archimedes showed that the area of a parabolic segment is four-thirds the area of the triangle ABC, where AB is the "base" of the parabolic segment and C is its vertex (as in Fig. 5.8.26). Verify this for the indicated parabolic segment.

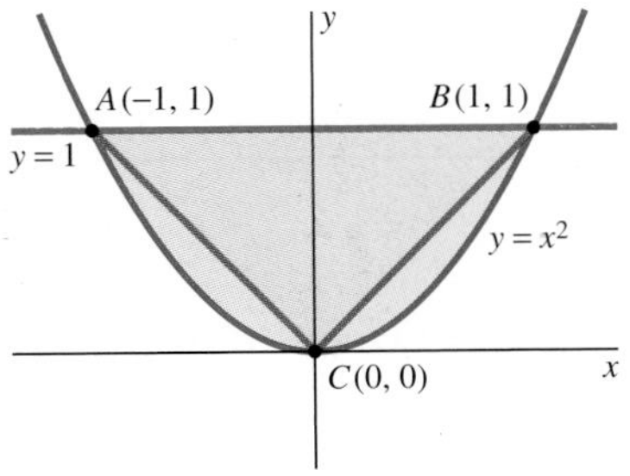

FIGURE 5.8.26 The parabolic segment of Problem 48.

49. Let A and B be the points of intersection of the parabola $y = x^2$ and the line $y = x + 2$, and let C be the point on the parabola where the tangent line is parallel to the graph of $y = x + 2$. Show that the area of the parabolic segment cut from the parabola by the line (Fig. 5.8.27) is four-thirds the area of the triangle ABC.

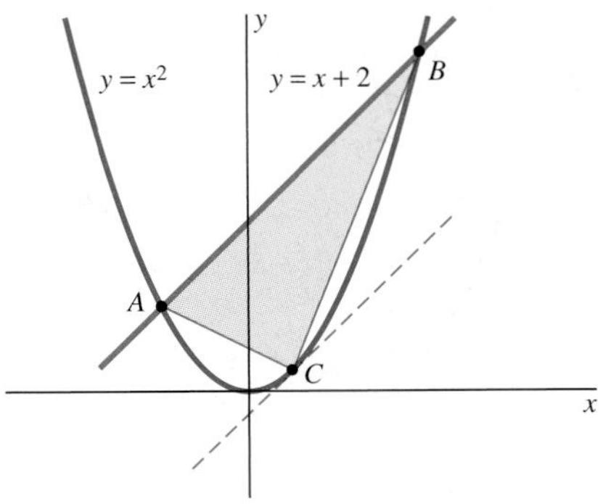

FIGURE 5.8.27 The parabolic segment of Problem 49.

50. Find the area of the unbounded region R shaded in Fig. 5.8.28—regard it as the limit as $b \to \infty$ of the region bounded by $y = 1/x^2$, $y = 0$, $x = 1$, and $x = b > 1$.

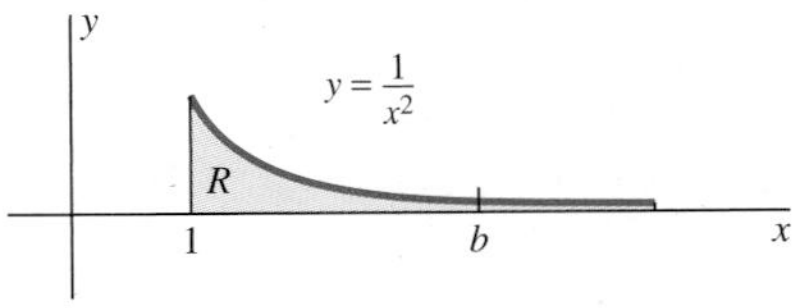

FIGURE 5.8.28 The unbounded region of Problem 50.

51. Find the total area of the bounded regions that are bounded by the x-axis and the curve $y = 2x^3 - 2x^2 - 12x$.

52. Suppose that the quadratic function

$$f(x) = px^2 + qx + r$$

is never negative on $[a, b]$. Show that the area under the graph of f from a to b is $A = \frac{1}{3}h[f(a) + 4f(m) + f(b)]$, where $h = (b - a)/2$ and $m = (a + b)/2$. [*Suggestion:* By a horizontal translation of this region, you may assume that $a = -h$, $m = 0$, and $b = h$.]

The curves defined in Problems 53 and 54 include the loops shown in Figs. 5.8.29 and 5.8.30. Find the area of the region bounded by each loop.

53. $y^2 = x(5 - x)^2$

54. $y^2 = x^2(x + 3)$

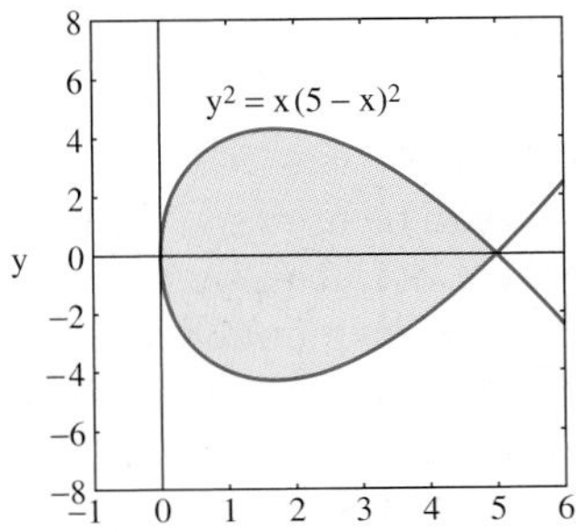

FIGURE 5.8.29 The region of Problem 53.

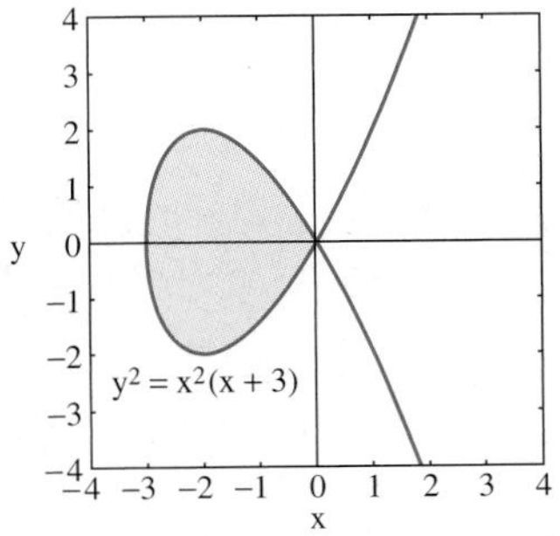

FIGURE 5.8.30 The region of Problem 54.

In Problems 55 through 58, use a calculator to approximate (graphically or otherwise) the points of intersection of the two given curves. Then integrate to find (approximately) the area of the region bounded by these curves.

55. $y = x^2, \quad y = \cos x$

56. $y = x^2 - 2x, \quad y = \sin x$

57. $y = x^2 - 1, \quad y = \dfrac{1}{1 + x^2}$

58. $y = x^4 - 16, \quad y = 2x - x^2$

59. Find a number $k > 0$ such that the area bounded by the curves $y = x^2$ and $y = k - x^2$ is 72.

60. Find a number $k > 0$ such that the line $y = k$ divides the region between the parabola $y = 100 - x^2$ and the x-axis into two regions having equal areas.

In Problems 61 through 63 the graphs $y = f(x)$ and $y = g(x)$ of the given functions f and g bound two regions R_1 and R_2 as indicated in Figs. 5.8.31 through 5.8.33. You are to find the sum A of the areas $A_1 = a(R_1)$ and $A_2 = a(R_2)$ of these two regions. If possible, obtain the exact value of A; otherwise, approximate it very accurately. You may use a calculator or a computer algebra system, both to find the points of intersection of the two graphs and to carry out the integrations required to calculate A.

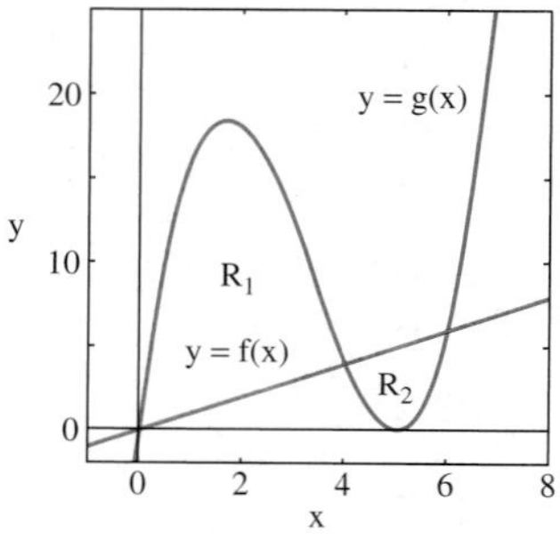

FIGURE 5.8.31 The regions of Problem 61.

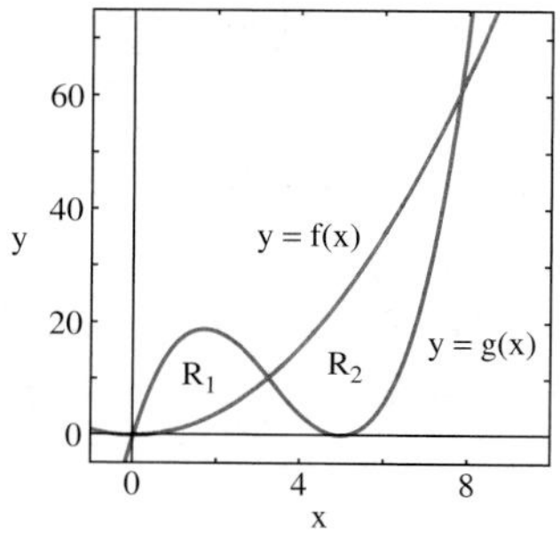

FIGURE 5.8.32 The regions of Problem 62.

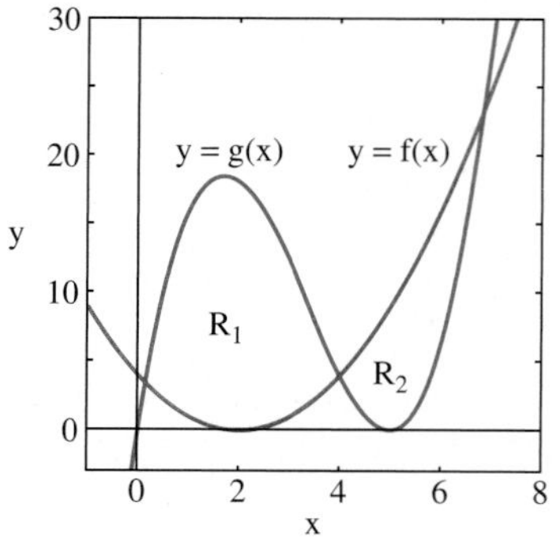

FIGURE 5.8.33 The regions of Problem 63.

61. $f(x) = x$ and $g(x) = x(x - 4)^2$. Here you should be able to find A without a calculator or a computer.

62. $f(x) = x^2$ and $g(x) = x(x - 4)^2$. Here you can solve manually for the points of intersection, but probably will want to use a calculator or computer for the integrations.

63. $f(x) = (x - 2)^2$ and $g(x) = x(x - 4)^2$. Here you will need to approximate numerically the coordinates of the points of intersection as well as the value of A.

In Problems 64 through 67 you may use a calculator or computer algebra system as in Problems 61 through 63.

64. Approximate numerically the area of the region that lies beneath the curve $y = 3 - 2x + 5\ln x$ and above the x-axis.

65. Approximate numerically the area of the region bounded by the curves $y = 10\ln x$ and $y = (x - 5)^2$.

66. Approximate numerically the area of the region bounded by the curves $y = e^x$ and $y = 10(1 + 5x - x^2)$.

67. Approximate numerically the sum of the areas of the regions bounded by the curves $y = e^{-x/2}$ and $y = x^4 - 6x^2 - 2x + 4$.

5.9 NUMERICAL INTEGRATION

The fundamental theorem of calculus,

$$\int_a^b f(x)\,dx = \Big[G(x)\Big]_a^b,$$

can be used to evaluate an integral only if a convenient formula for the antiderivative G of f can be found. But there are simple functions with antiderivatives that are not elementary functions. An **elementary function** is one that can be expressed in terms of polynomial, trigonometric, exponential, or logarithmic functions by means of finite combinations of sums, differences, products, quotients, roots, and function composition.

The problem is that elementary functions can have nonelementary antiderivatives. For example, it is known that the elementary function $f(x) = e^{-x^2}$ has no elementary antiderivative. Consequently, we cannot use the fundamental theorem of calculus to evaluate an integral such as

$$\int_0^1 e^{-x^2}\,dx.$$

Here we discuss the use of Riemann sums to approximate numerically integrals that cannot conveniently be evaluated exactly, whether or not nonelementary functions are involved. Given a continuous function f on $[a, b]$ with an integral to be approximated, consider a partition of $[a, b]$ into n subintervals, each with the same length $\Delta x = (b - a)/n$. Then the value of any Riemann sum of the form

$$S = \sum_{i=1}^{n} f(x_i^\star)\,\Delta x \tag{1}$$

may be taken to be an approximation to the value of the integral $\int_a^b f(x)\,dx$.

With $x_i^\star = x_{i-1}$ and with $x_i^\star = x_i$ in Eq. (1), we get the *left-endpoint approximation* L_n and the *right-endpoint approximation* R_n to the definite integral $\int_a^b f(x)\,dx$ associated with the partition of $[a, b]$ into n equal-length subintervals. Thus

$$L_n = \sum_{i=1}^{n} f(x_{i-1})\,\Delta x \tag{2}$$

and

$$R_n = \sum_{i=1}^{n} f(x_i)\,\Delta x. \tag{3}$$

We can simplify the notation for L_n and R_n by writing y_i for $f(x_i)$ (Fig. 5.9.1).

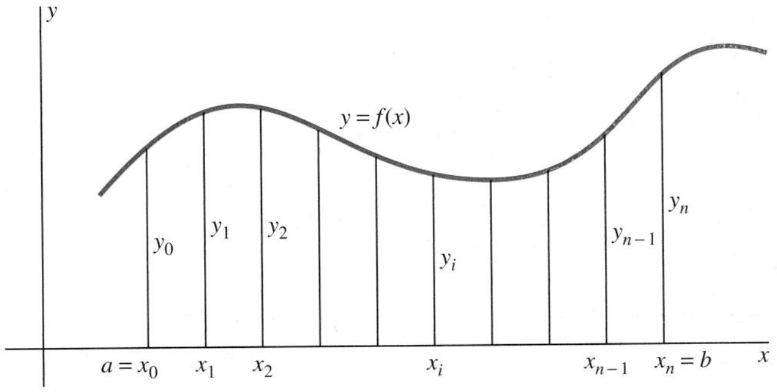

FIGURE 5.9.1 $y_i = f(x_i)$.

DEFINITION Endpoint Approximations

The **left-endpoint approximation** L_n and the **right-endpoint approximation** R_n to $\int_a^b f(x)\,dx$ with $\Delta x = (b - a)/n$ are

$$L_n = (\Delta x)(y_0 + y_1 + y_2 + \cdots + y_{n-1}) \tag{2$'$}$$

and

$$R_n = (\Delta x)(y_1 + y_2 + y_3 + \cdots + y_n). \tag{3$'$}$$

In Example 1 of Section 5.3 we calculated the left- and right-endpoint approximations to the integral

$$\int_0^3 x^2\,dx = 9 \tag{4}$$

with $n = 5$ and $n = 10$. The table in Fig. 5.9.2 shows values of L_n and R_n with larger values of n.

n	L_n	R_n	$\frac{1}{2}(L_n + R_n)$
5	6.4800	11.8800	9.1800
10	7.6950	10.3950	9.0450
20	8.3363	9.6863	9.0113
40	8.6653	9.3403	9.0028
80	8.8320	9.1695	9.0007
160	8.9158	9.0846	9.0002
320	8.9579	9.0422	9.0000

FIGURE 5.9.2 Left- and right-endpoint approximations to the integral in Eq. (4).

The final column of this table gives the *average* of the endpoint sums L_n and R_n. It is apparent that (for a given value of n) this average is a considerably more accurate approximation to the integral than is either one-sided approximation by itself.

The Trapezoidal and Midpoint Approximations

The average $T_n = (L_n + R_n)/2$ of the left-endpoint and right-endpoint approximations is called the *trapezoidal approximation* to $\int_a^b f(x)\,dx$ associated with the partition of $[a, b]$ into n equal-length subintervals. Written in full,

$$\begin{aligned} T_n &= \frac{1}{2}(L_n + R_n) \\ &= \frac{\Delta x}{2}\sum_{i=1}^{n}[f(x_{i-1}) + f(x_i)] \\ &= \frac{\Delta x}{2}\{[f(x_0) + f(x_1)] + [f(x_1) + f(x_2)] + [f(x_2) + f(x_3)] + \cdots \\ &\qquad + [f(x_{n-2}) + f(x_{n-1})] + [f(x_{n-1}) + f(x_n)]\}; \end{aligned}$$

that is,

$$T_n = \frac{\Delta x}{2}[f(x_0) + 2f(x_1) + 2f(x_2) + \cdots + 2f(x_{n-2}) + 2f(x_{n-1}) + f(x_n)]. \quad \textbf{(5)}$$

Note the 1-2-2- $\cdots$ -2-2-1 pattern of the coefficients.

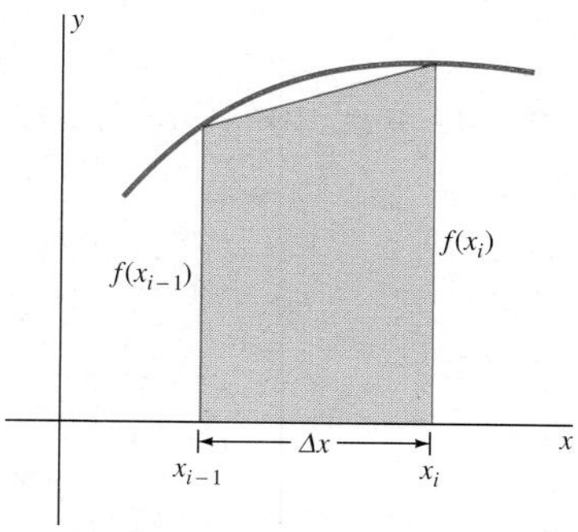

FIGURE 5.9.3 The area of the trapezoid is $\frac{1}{2}[f(x_{i-1}) + f(x_i)]\,\Delta x$.

DEFINITION **The Trapezoidal Approximation**

The **trapezoidal approximation** to

$$\int_a^b f(x)\,dx \quad \text{with} \quad \Delta x = \frac{b-a}{n}$$

is

$$T_n = \frac{\Delta x}{2}(y_0 + 2y_1 + 2y_2 + \cdots + 2y_{n-2} + 2y_{n-1} + y_n). \quad \textbf{(6)}$$

Figure 5.9.3 shows where the trapezoidal approximation gets its name. The partition points $x_0, x_1, x_2, \ldots, x_n$ are used to build trapezoids from the x-axis to the graph of the function f. The trapezoid over the ith subinterval $[x_{i-1}, x_i]$ has height Δx, and its parallel bases have widths $f(x_{i-1})$ and $f(x_i)$. So its area is

$$\frac{\Delta x}{2}[f(x_{i-1}) + f(x_i)] = \frac{\Delta x}{2}(y_{i-1} + y_i).$$

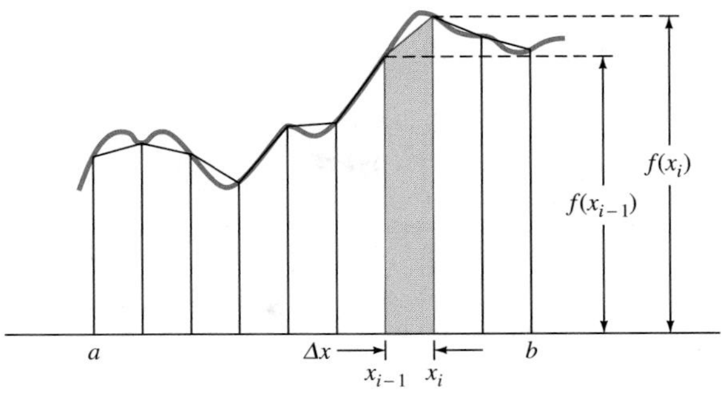

FIGURE 5.9.4 Geometry of the trapezoidal approximation.

Comparing this with Eq. (6) shows that T_n is merely the sum of the areas of the n trapezoids shown in Fig. 5.9.4.

EXAMPLE 1 Calculate the trapezoidal approximation to the integral in Eq. (4) with $n = 6$ and $\Delta x = 0.5$.

Solution The trapezoids in Fig. 5.9.5 indicate why T_6 should be a much better approximation than either of the endpoint approximations L_6 or R_6. The table in Fig. 5.9.6 shows the values of $f(x) = x^2$ that are needed to compute L_6 and R_6. The 1-2-2- $\cdots$ -2-2-1 coefficients appear in the final column. Using Eq. (6), we get

$$\begin{aligned} T_6 &= \frac{0.5}{2}[1\cdot(0) + 2\cdot(0.25) + 2\cdot(1) \\ &\quad + 2\cdot(2.25) + 2\cdot(4) + 2\cdot(6.25) + 1\cdot(9)] \\ &= 9.125 \end{aligned}$$

(as compared with the actual value 9). ◆

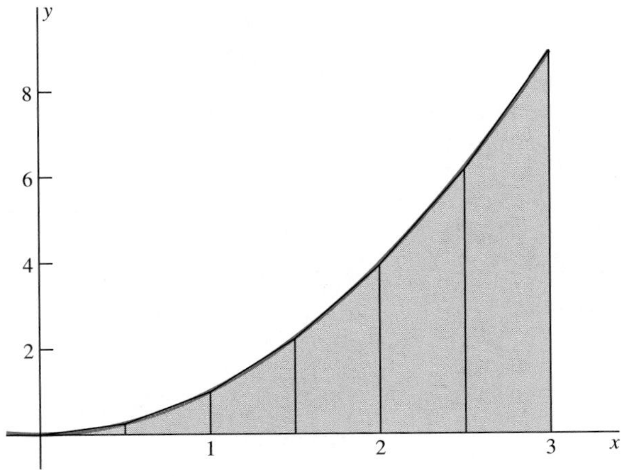

FIGURE 5.9.5 The area under $y = x^2$ (Example 1).

n	L_n	$f(x_n) = x_n^2$	Coefficients
0	0	0	1
1	0.5	0.25	2
2	1	1	2
3	1.5	2.25	2
4	2	4	2
5	2.5	6.25	2
6	3	9	1

FIGURE 5.9.6 Data for Example 1.

Another useful approximation to $\int_a^b f(x)\,dx$ is the *midpoint approximation* M_n. It is the Riemann sum obtained by choosing the point $x_i^\star$ in $[x_{i-1}, x_i]$ to be its midpoint $m_i = (x_{i-1} + x_i)/2$. Thus

$$M_n = \sum_{i=1}^{n} f(m_i)\,\Delta x = (\Delta x)\left[f(m_1) + f(m_2) + \cdots + f(m_n)\right]. \tag{7}$$

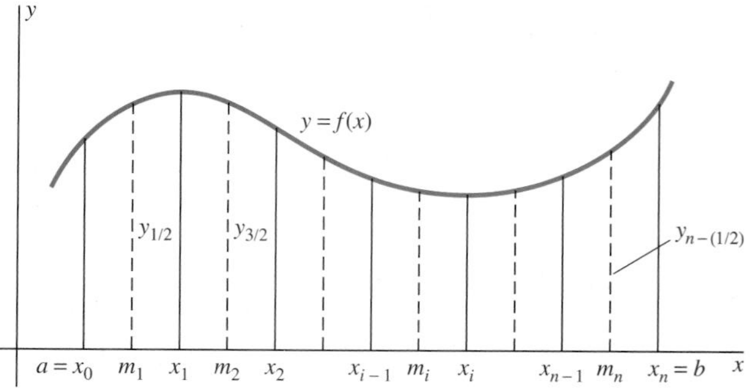

FIGURE 5.9.7 The ordinates used in the midpoint approximation.

Because m_1 is the midpoint of $[x_0, x_1]$, it is sometimes convenient to write $y_{1/2}$ for $f(m_1)$, $y_{3/2}$ for $f(m_2)$, and so on (Fig. 5.9.7).

DEFINITION The Midpoint Approximation

The **midpoint approximation** to

$$\int_a^b f(x)\,dx \quad \text{with} \quad \Delta x = \frac{b-a}{n}$$

is

$$M_n = (\Delta x)\big(y_{1/2} + y_{3/2} + y_{5/2} + \cdots + y_{n-(1/2)}\big). \qquad \textbf{(7')}$$

EXAMPLE 2 Figure 5.9.8 illustrates the midpoint approximation to the integral

$$\int_0^3 x^2\,dx = 9$$

of Example 1, with $n = 6$ and $\Delta x = 0.5$, and the table in Fig. 5.9.9 shows the values of $f(x) = x^2$ needed to compute M_6. Using Eq. (7), we obtain

$$\begin{aligned} M_6 &= (0.5)\,[1 \cdot (0.0625) + 1 \cdot (0.5625) + 1 \cdot (1.5625) \\ &\quad + 1 \cdot (3.0625) + 1 \cdot (5.0625) + 1 \cdot (7.5625)] \\ &= 8.9375. \end{aligned}$$

◆

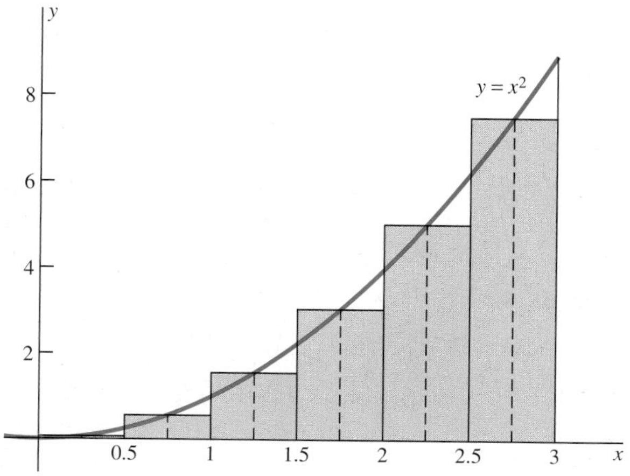

FIGURE 5.9.8 Midpoint rectangles approximating the area under $y = x^2$ (Example 2).

n	m_n	$f(m_i)$	Coefficients
1	0.25	0.0625	1
2	0.75	0.5625	1
3	1.25	1.5625	1
4	1.75	3.0625	1
5	2.25	5.0625	1
6	2.75	7.5625	1

FIGURE 5.9.9 Data for Example 2.

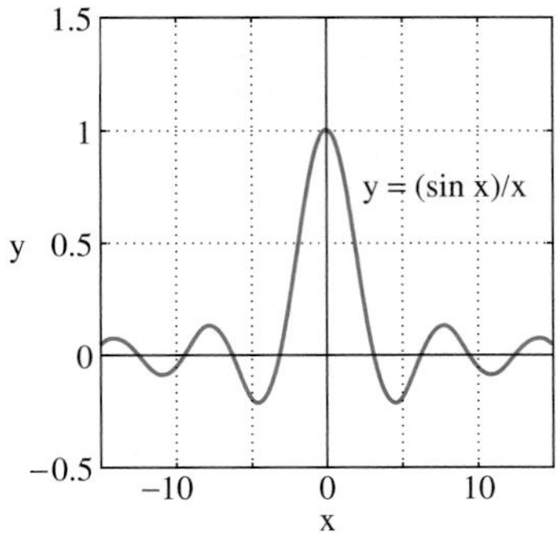

FIGURE 5.9.10 The graph of $f(x) = \dfrac{\sin x}{x}$ (Example 3).

EXAMPLE 3 Figure 5.9.10 shows the graph of the function

$$f(x) = \frac{\sin x}{x}. \tag{8}$$

Recall from Section 2.5 that

$$\lim_{x \to 0} \frac{\sin x}{x} = 1.$$

Hence we *define* $f(0) = 1$; then there will be no difficulty at $x = 0$, even though both the numerator and denominator in Eq. (8) are zero. It happens that the function $f(x)$ has *no* elementary antiderivative, so the fundamental theorem of calculus cannot be used to calculate the value of the integral

$$I = \int_0^1 \frac{\sin x}{x}\, dx. \tag{9}$$

But such integrals are important in the design of precise photographic lenses (among other applications), so we are well motivated to resort to approximating its value numerically.

With $n = 10$ and $\Delta x = 0.1$, the trapezoidal approximation is

$$\begin{aligned} T_{10} &= \frac{0.1}{2}\left[1 \cdot 1 + 2 \cdot \frac{\sin 0.1}{0.1} + 2 \cdot \frac{\sin 0.2}{0.2}\right. \\ &\qquad \left. + 2 \cdot \frac{\sin 0.3}{0.3} + \cdots + 2 \cdot \frac{\sin 0.9}{0.9} + 1 \cdot \frac{\sin 1.0}{1.0}\right]; \\ T_{10} &\approx 0.94583. \end{aligned}$$

The corresponding midpoint approximation is

$$\begin{aligned} M_{10} &= (0.1)\left[1 \cdot \frac{\sin 0.05}{0.05} + 1 \cdot \frac{\sin 0.15}{0.15} + 1 \cdot \frac{\sin 0.25}{0.25} + \cdots\right. \\ &\qquad \left. + 1 \cdot \frac{\sin 0.85}{0.85} + 1 \cdot \frac{\sin 0.95}{0.95}\right]; \\ M_{10} &\approx 0.94621. \end{aligned}$$

The actual value of the integral in Eq. (9) is $I \approx 0.94608$ (accurate to five decimal places). Thus both T_{10} and M_{10} give the correct value 0.946 when rounded to three places. But

- T_{10} underestimates I by about 0.00025, whereas
- M_{10} overestimates I by about 0.00013.

Thus in this example the midpoint approximation is somewhat more accurate than the trapezoidal approximation. ◆

Simpson's Approximation

The midpoint approximation M_n in Eq. (7) is sometimes called the **tangent-line approximation,** because the area of the rectangle with base $[x_{i-1}, x_i]$ and height $f(m_i)$ is also the area of another approximating figure. As shown in Fig. 5.9.11, we draw a segment tangent to the graph of f at the point $(m_i, f(m_i))$ and use that segment for one side of a trapezoid (somewhat like the method of trapezoidal approximation). The trapezoid and the rectangle mentioned above have the same area, and so the value of M_n is the sum of the areas of trapezoids like the one in Fig. 5.9.11.

The area of the trapezoid associated with the midpoint approximation is generally closer to the true value of

$$\int_{x_{i-1}}^{x_i} f(x)\, dx$$

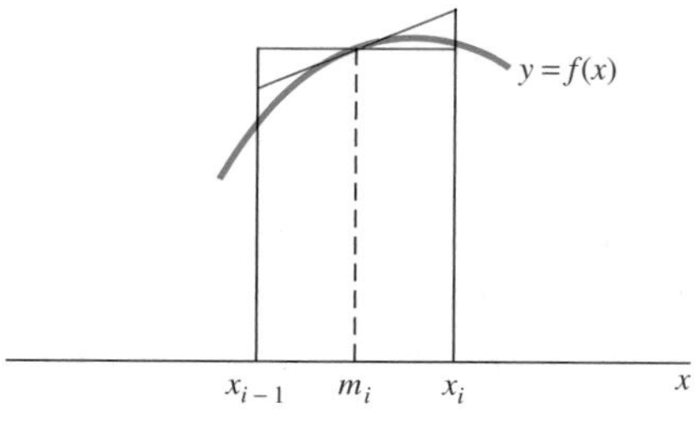

FIGURE 5.9.11 The midpoint (or tangent) approximation.

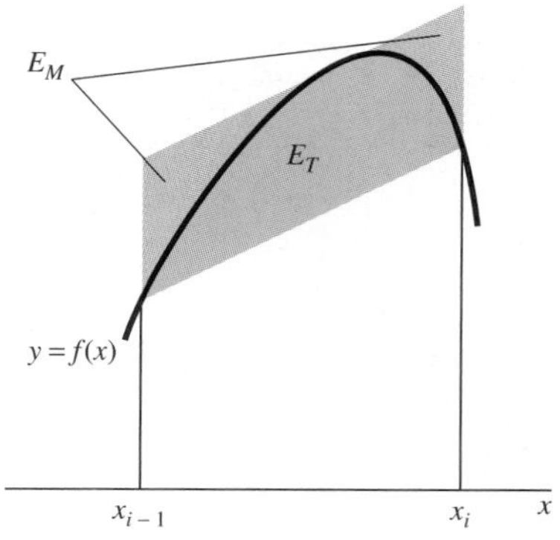

FIGURE 5.9.12 Comparison of the midpoint approximation error E_M with the trapezoidal approximation error E_T.

than is the area of the trapezoid associated with the trapezoidal approximation, as was the case in Example 3. Figure 5.9.12 shows this too, in that the midpoint error E_M (above the curve in this figure) is generally smaller than the trapezoidal error E_T (below the curve in this figure). Figure 5.9.12 also indicates that if $y = f(x)$ is concave downward, then M_n will be an overestimate and T_n will be an underestimate of $\int_a^b f(x)\,dx$. If the graph is concave upward, then the situation is reversed.

Such observations motivate the consideration of a *weighted average* of M_n and T_n, with M_n weighted more heavily than T_n, to improve our numerical estimates of the definite integral. The particular weighted average

$$S_{2n} = \tfrac{1}{3}(2M_n + T_n) = \tfrac{2}{3}M_n + \tfrac{1}{3}T_n \qquad \textbf{(10)}$$

is called *Simpson's approximation* to $\int_a^b f(x)\,dx$. The reason for the subscript $2n$ is that we associate S_{2n} with a partition of $[a, b]$ into an *even* number, $2n$, of equal-length subintervals with the endpoints

$$a = x_0 < x_1 < x_2 < \cdots < x_{2n-2} < x_{2n-1} < x_{2n} = b.$$

The midpoint and trapezoidal approximations associated with the n subintervals

$$[x_0, x_2], \quad [x_2, x_4], \quad [x_4, x_6], \quad \dots, \quad [x_{2n-4}, x_{2n-2}], \quad [x_{2n-2}, x_{2n}],$$

all with the same length $2\,\Delta x$, can then be written in the respective forms

$$M_n = (2\,\Delta x)(y_1 + y_3 + y_5 + \cdots + y_{2n-1})$$

and

$$T_n = \frac{2\,\Delta x}{2}(y_0 + 2y_2 + 2y_4 + \cdots + 2y_{2n-2} + y_{2n}).$$

We substitute these formulas for M_n and T_n into Eq. (10) and find—after a bit of algebra—that

$$S_{2n} = \frac{\Delta x}{3}(y_0 + 4y_1 + 2y_2 + 4y_3 + 2y_4 + \cdots + 2y_{2n-2} + 4y_{2n-1} + y_{2n}). \qquad \textbf{(11)}$$

To be consistent with our other approximation formulas, we next rewrite Eq. (11) with n (rather than $2n$) to denote the total number of subintervals used.

DEFINITION Simpson's Approximation

Simpson's approximation to $\int_a^b f(x)\,dx$ with $\Delta x = (b-a)/n$, associated with a partition of $[a, b]$ into an ***even*** number n of equal-length subintervals, is the sum S_n defined as

$$S_n = \frac{\Delta x}{3}(y_0 + 4y_1 + 2y_2 + 4y_3 + 2y_4 + \cdots + 2y_{n-2} + 4y_{n-1} + y_n). \qquad \textbf{(12)}$$

REMARK Note the 1-4-2-4-2- $\cdots$ -2-4-2-4-1 pattern of coefficients in Simpson's approximation. This pattern is symmetric (ending in -2-4-1), as shown, if and only if n is *even.*

EXAMPLE 4 Simpson's approximation (with $n = 6$ and $\Delta x = 0.5$) to the integral

$$\int_0^3 x^2\,dx = 9$$

of Examples 1 and 2 is

$$\begin{aligned} S_6 &= \frac{0.5}{3}[1\cdot(0)^2 + 4\cdot(0.5)^2 + 2\cdot(1)^2 + 4\cdot(1.5)^2 \\ &\qquad + 2\cdot(2)^2 + 4\cdot(2.5)^2 + 1\cdot(3)^2]; \\ S_6 &= 9 \quad \text{(exactly).} \end{aligned}$$

Problem 29 explains why Simpson's approximation to this particular integral is *exact* rather than merely a good approximation. ◆

EXAMPLE 5 Simpson's approximation (with $n = 10$ and $\Delta x = 0.1$) to the integral

$$\int_0^1 \frac{\sin x}{x}\,dx$$

of Example 3 is

$$\begin{aligned} S_{10} &= \frac{0.1}{3}\left[1\cdot 1 + 4\cdot\frac{\sin 0.1}{0.1} + 2\cdot\frac{\sin 0.2}{0.2} + 4\cdot\frac{\sin 0.3}{0.3} + \cdots \right. \\ &\qquad \left. + 2\cdot\frac{\sin 0.8}{0.8} + 4\cdot\frac{\sin 0.9}{0.9} + 1\cdot\frac{\sin 1.0}{1.0}\right]; \\ S_{10} &\approx 0.94608, \end{aligned}$$

which is accurate to *all five* decimal places shown. ◆

Examples 4 and 5 illustrate the greater accuracy of Simpson's approximation in comparison with the midpoint and trapezoidal approximations.

The numerical methods of this section are especially useful for approximating integrals of functions that are available only in graphical or in tabular form. This is often the case with functions derived from empirical data or from experimental measurements.

EXAMPLE 6 Suppose that the graph in Fig. 5.9.13 shows the velocity $v(t)$ recorded by instruments on board a submarine traveling under the polar ice cap directly toward the North Pole. Use the trapezoidal approximation and Simpson approximation to

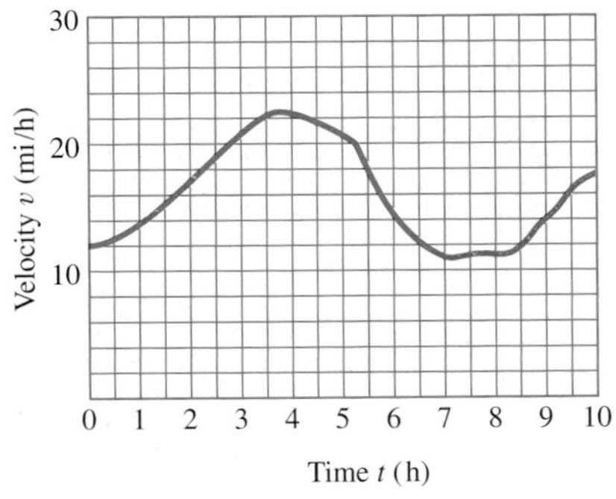

FIGURE 5.9.13 Velocity graph for the submarine of Example 6.

estimate the distance $s = \int_a^b v(t)\,dt$ traveled by the submarine during the 10-h period from $t = 0$ to $t = 10$.

Solution We read the following data from the graph:

t	0	1	2	3	4	5	6	7	8	9	10	h
v	12	14	17	21	22	21	15	11	11	14	17	mi/h

Using the trapezoidal approximation with $n = 10$ and $\Delta x = 1$, we obtain

$$\begin{aligned} s &= \int_0^{10} v(t)\,dt \\ &\approx \tfrac{1}{2}[12 + 2(14 + 17 + 21 + 22 + 21 + 15 + 11 + 11 + 14) + 17] \\ &= 160.5 \quad \text{(mi)}. \end{aligned}$$

Using Simpson's approximation with $2n = 10$ and $\Delta x = 1$, we obtain

$$\begin{aligned} s &= \int_0^{10} v(t)\,dt \\ &\approx \tfrac{1}{3}[12 + 4\cdot 14 + 2\cdot 17 + 4\cdot 21 + 2\cdot 22 + 4\cdot 21 \\ &\qquad + 2\cdot 15 + 4\cdot 11 + 2\cdot 11 + 4\cdot 14 + 17] \\ &= 161 \quad \text{(mi)} \end{aligned}$$

as an estimate of the distance traveled by the submarine during this 10-h period. ◆

Parabolic Approximations

Although we have defined Simpson's approximation S_{2n} as a weighted average of the midpoint and trapezoidal approximations, Simpson's approximation has an important interpretation in terms of **parabolic approximations** to the curve $y = f(x)$. Beginning with the partition of $[a, b]$ into $2n$ equal-length subintervals, we define the parabolic function

$$p_i(x) = A_i + B_i x + C_i x^2$$

on $[x_{2i-2}, x_{2i}]$ as follows: We choose the coefficients A_i, B_i, and C_i so that $p_i(x)$ agrees with $f(x)$ at the three points x_{2i-2}, x_{2i-1}, and x_{2i} (Fig. 5.9.14). This can be done by solving the three equations

$$\begin{aligned} A_i + B_i x_{2i-2} + C_i(x_{2i-2})^2 &= f(x_{2i-2}), \\ A_i + B_i x_{2i-1} + C_i(x_{2i-1})^2 &= f(x_{2i-1}), \\ A_i + B_i x_{2i} + C_i(x_{2i})^2 &= f(x_{2i}) \end{aligned}$$

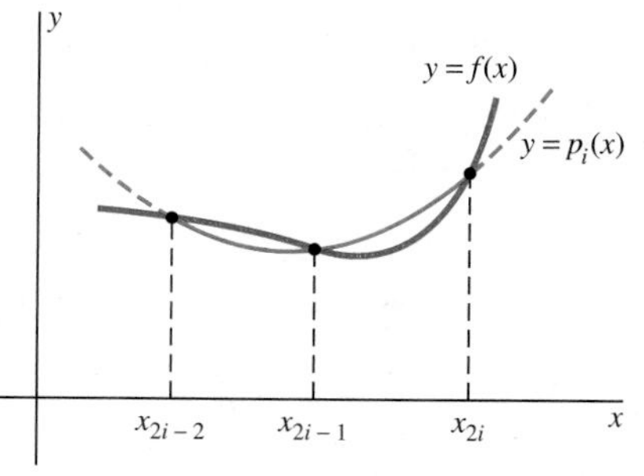

FIGURE 5.9.14 The parabolic approximation $y = p_i(x)$ to $y = f(x)$ on $[x_{2i-2}, x_{2i}]$.

for the three unknowns A_i, B_i, and C_i. A routine (but tedious) algebraic computation—see Problem 52 of Section 5.8—shows that

$$\int_{x_{2i-2}}^{x_{2i}} p_i(x)\,dx = \frac{\Delta x}{3}(y_{2i-2} + 4y_{2i-1} + y_{2i}).$$

We now approximate $\int_a^b f(x)\,dx$ by replacing $f(x)$ with $p_i(x)$ on the interval $[x_{2i-2}, x_{2i}]$ for $i = 1, 2, 3, \ldots, n$. This gives

$$\begin{aligned}\int_a^b f(x)\,dx &= \sum_{i=1}^{n} \int_{x_{2i-2}}^{x_{2i}} f(x)\,dx \approx \sum_{i=1}^{n} \int_{x_{2i-2}}^{x_{2i}} p_i(x)\,dx \\ &= \sum_{i=1}^{n} \frac{\Delta x}{3}(y_{2i-2} + 4y_{2i-1} + y_{2i}) \\ &= \frac{\Delta x}{3}(y_0 + 4y_1 + 2y_2 + 4y_3 + \cdots + 4y_{2n-3} + 2y_{2n-2} + 4y_{2n-1} + y_{2n}).\end{aligned}$$

Thus the parabolic approximation described here results in Simpson's approximation S_{2n} to $\int_a^b f(x)\,dx$.

Error Estimates

The trapezoidal approximation, the midpoint approximation, and Simpson's approximation are widely used for numerical integration, and there are *error estimates* that can be used to predict the maximum possible error in a particular approximation. The trapezoidal error ET_n, the midpoint error EM_n, and Simpson's error ES_n are defined by the equations

$$\int_a^b f(x)\,dx = T_n + ET_n, \tag{13}$$

$$\int_a^b f(x)\,dx = M_n + EM_n, \tag{14}$$

and

$$\int_a^b f(x)\,dx = S_n + ES_n \quad (n \text{ even}). \tag{15}$$

Note that each of these formulas is of the form

$$\int_a^b f(x)\,dx = \{\text{approximation}\} + \{\text{error}\}.$$

The absolute value $|ET_n|$ is the difference between the value of the integral and the trapezoidal approximation with n subintervals (and similarly for $|EM_n|$ and $|ES_n|$). Theorems 1 and 2 are proved in numerical analysis textbooks.

THEOREM 1 Trapezoidal and Midpoint Error Estimates
Suppose that the second derivative f'' is continuous on $[a, b]$ and that $|f''(x)| \leqq M_2$ for $a \leqq x \leqq b$. Then

$$|ET_n| \leqq \frac{M_2(b-a)^3}{12n^2} \tag{16}$$

and

$$|EM_n| \leqq \frac{M_2(b-a)^3}{24n^2}. \tag{17}$$

REMARK Comparing (16) and (17), we see that the maximal predicted midpoint error is *half* the predicted trapezoidal error. This is the reason for weighting M_n

twice as heavily as T_n when we calculate Simpson's approximation using the formula $S_{2n} = \frac{2}{3}M_n + \frac{1}{3}T_n$ in (10).

THEOREM 2 Simpson's Error Estimate

Suppose that the fourth derivative $f^{(4)}$ is continuous on $[a, b]$ and that $|f^{(4)}(x)| \leqq M_4$ for $a \leqq x \leqq b$. If n is even, then

$$|ES_n| \leqq \frac{M_4(b-a)^5}{180n^4}. \tag{18}$$

REMARK The factor n^4 in (18)—compared with the n^2 in (16) and (17)—explains the greater accuracy typical of Simpson's approximation. For instance, if $n = 10$, then $n^2 = 100$ but $n^4 = 10000$, so the denominator in the error formula for Simpson's approximation is much larger.

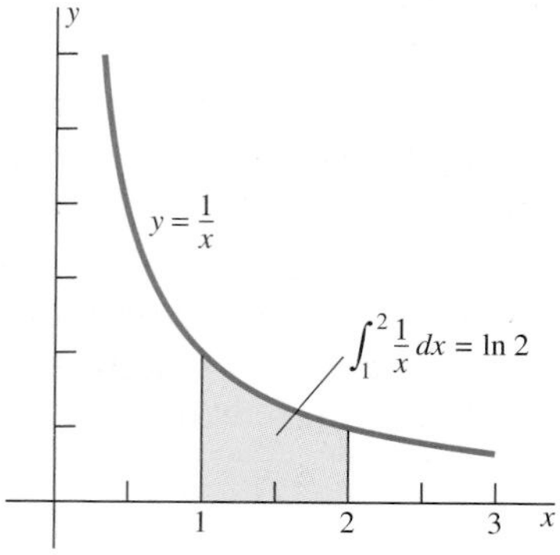

FIGURE 5.9.15 The number ln 2 as an area.

EXAMPLE 7 Because $D_x[\ln x] = 1/x$, it follows that the natural logarithm of the number 2 (given approximately by the LN key on your calculator) is the value of the integral

$$\int_1^2 \frac{1}{x}\,dx = \Big[\ln x\Big]_1^2 = \ln 2 - \ln 1 = \ln 2$$

illustrated in Fig. 5.9.15. Estimate the errors in the trapezoidal, midpoint, and Simpson's approximation to this integral by using $n = 10$ subintervals. (The actual value of ln 2 is approximately 0.693147.)

Solution With $f(x) = 1/x$ we calculate

$$f'(x) = -\frac{1}{x^2}, \qquad f''(x) = \frac{2}{x^3}$$

$$f'''(x) = -\frac{6}{x^4}, \qquad f^{(4)}(x) = \frac{24}{x^5}.$$

The maximum values of all these derivatives for $1 \leqq x \leqq 2$ occur at $x = 1$, so we may take $M_2 = 2$ and $M_4 = 24$ in Eqs. (16), (17), and (18). From Eqs. (16) and (17) we see that

$$|ET_{10}| \leqq \frac{2 \cdot 1^3}{12 \cdot 10^2} \approx 0.0016667 \quad \text{and} \quad |EM_{10}| \leqq \frac{2 \cdot 1^3}{24 \cdot 10^2} \approx 0.000833. \tag{19}$$

Hence we would expect both the trapezoidal approximation T_{10} and the midpoint approximation M_{10} to give ln 2 accurate to at least two decimal places. From Eq. (18) we see that

$$|ES_{10}| \leqq \frac{24 \cdot 1^5}{180 \cdot 10^4} \approx 0.000013, \tag{20}$$

so we would expect Simpson's approximation S_{10} to be accurate to at least four decimal places. When we calculate these approximations, we find that

$$T_{10} = \frac{0.1}{2}\left(\frac{1}{1} + \frac{2}{1.1} + \frac{2}{1.2} + \frac{2}{1.3} + \cdots + \frac{2}{1.9} + \frac{1}{2}\right) \approx 0.693771,$$

$$M_{10} = (0.1)\left(\frac{1}{1.05} + \frac{1}{1.15} + \frac{1}{1.25} + \cdots + \frac{1}{1.85} + \frac{1}{1.95}\right) \approx 0.692835,$$

and

$$S_{10} = \frac{0.1}{3}\left(\frac{1}{1} + \frac{4}{1.1} + \frac{2}{1.2} + \frac{4}{1.3} + \cdots + \frac{4}{1.9} + \frac{1}{2}\right) \approx 0.693150.$$

It follows that the values of the errors in these approximations (in comparison with the actual value $\ln 2 \approx 0.693147$) are

$$ET_{10} \approx -0.000624, \qquad EM_{10} \approx 0.000312, \quad \text{and} \quad ES_{10} \approx 0.000003.$$

Comparing these actual errors with the maximal predicted errors in (19) and (20), we see that our approximations are somewhat more accurate than predicted—M_{10} actually is accurate to three decimal places and S_{10} is accurate to five. It is fairly typical of numerical integration that the trapezoidal, midpoint, and Simpson approximations are somewhat more accurate than the "worst-case" estimates provided by Theorems 1 and 2. ◆

5.9 TRUE/FALSE STUDY GUIDE

5.9 CONCEPTS: QUESTIONS AND DISCUSSION

1. Suppose that f is an increasing function on $[a, b]$. Which is larger—the left-hand sum or the right-hand sum, each with n subintervals? What if f is a decreasing function? Draw sketches illustrating both cases.
2. Suppose that the graph of $y = f(x)$ is concave upward on $[a, b]$. Which is larger—the midpoint sum or the trapezoidal sum, each with n subintervals? What if the graph is concave downward? Draw sketches illustrating both cases.
3. Explain why you cannot set up a Simpson's sum beginning with a partition of $[a, b]$ into an odd number of subintervals. If you try to do so, what goes wrong?

5.9 PROBLEMS

In Problems 1 through 6, calculate the trapezoidal approximation T_n to the given integral, and compare T_n with the exact value of the integral. Use the indicated number n of subintervals, and round answers to two decimal places.

1. $\displaystyle\int_0^4 x\,dx \quad n=4$

2. $\displaystyle\int_1^2 x^2\,dx, \quad n=5$

3. $\displaystyle\int_0^1 \sqrt{x}\,dx, \quad n=5$

4. $\displaystyle\int_1^3 \frac{1}{x^2}\,dx, \quad n=4$

5. $\displaystyle\int_0^{\pi/2} \cos x\,dx, \quad n=3$

6. $\displaystyle\int_0^{\pi} \sin x\,dx, \quad n=4$

7. through **12.** Calculate the midpoint approximations to the integrals in Problems 1 through 6, using the indicated number of subintervals. In each case compare M_n with the exact value of the integral.

In Problems 13 through 20, calculate both the trapezoidal approximation T_n and Simpson's approximation S_n to the given integral. Use the indicated number of subintervals and round answers to four decimal places. In Problems 13 through 16, also compare these approximations with the exact value of the integral.

13. $\displaystyle\int_1^3 x^2\,dx, \quad n=4$

14. $\displaystyle\int_1^5 \frac{1}{x}\,dx, \quad n=4$

15. $\displaystyle\int_0^2 e^{-x}\,dx, \quad n=4$

16. $\displaystyle\int_0^1 \sqrt{1+x}\,dx, \quad n=4$

17. $\displaystyle\int_0^2 \sqrt{1+x^3}\,dx, \quad n=6$

18. $\displaystyle\int_0^3 \frac{1}{1+x^4}\,dx, \quad n=6$

19. $\displaystyle\int_1^5 \sqrt[3]{1+\ln x}\,dx, \quad n=8$

20. $\displaystyle\int_0^1 \frac{e^x-1}{x}\,dx, \quad n=10$

[*Note:* Make the integrand in Problem 20 continuous by assuming that its value at $x = 0$ is its limit there,

$$\lim_{x\to 0} \frac{e^x-1}{x} = 1.]$$

In Problems 21 and 22, calculate (a) the trapezoidal approximation and (b) Simpson's approximation to

$$\int_a^b f(x)\,dx,$$

where f is the given tabulated function.

21.

x	$a = 1.00$	1.25	1.50	1.75
$f(x)$	3.43	2.17	0.38	1.87

x	2.00	2.25	$2.50 = b$
$f(x)$	2.65	2.31	1.97

22.

x	$a = 0$	1	2	3	4	5	6
$f(x)$	23	8	−4	12	35	47	53

x	7	8	9	$10 = b$
$f(x)$	50	39	29	5

23. Figure 5.9.16 shows the measured rate of water flow (in liters per minute) into a tank during a 10-min period. Using 10 subintervals in each case, estimate the total amount of water that flows into the tank during this period by using (a) the trapezoidal approximation and (b) Simpson's approximation.

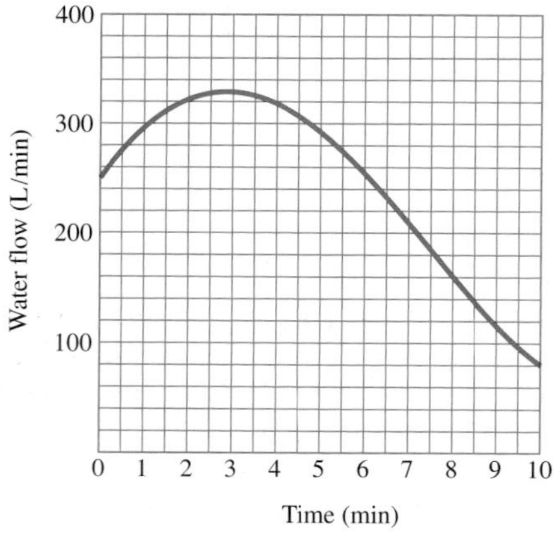

FIGURE 5.9.16 Water-flow graph for Problem 23.

24. Figure 5.9.17 shows the daily mean temperature recorded during December at Big Frog, California. Using 10 subintervals in each case, estimate the average temperature during that month by using (a) the trapezoidal approximation and (b) Simpson's approximation.

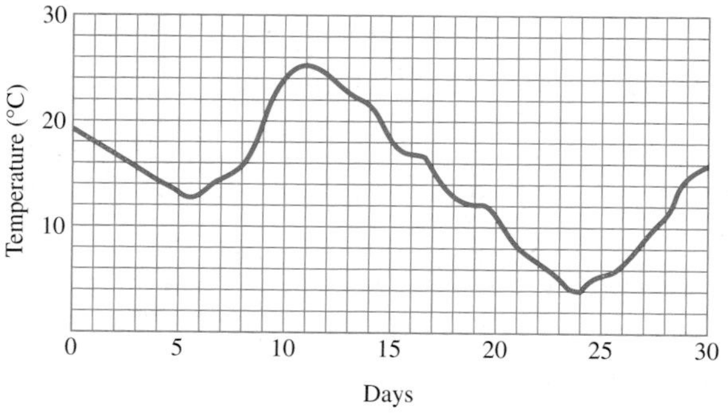

FIGURE 5.9.17 Temperature graph for Problem 24.

25. Figure 5.9.18 shows a tract of land with measurements in feet. A surveyor has measured its width w at 50-ft intervals (the values of x shown in the figure), with the following results.

x	0	50	100	150	200	250	300
w	0	165	192	146	63	42	84

x	350	400	450	500	550	600
w	155	224	270	267	215	0

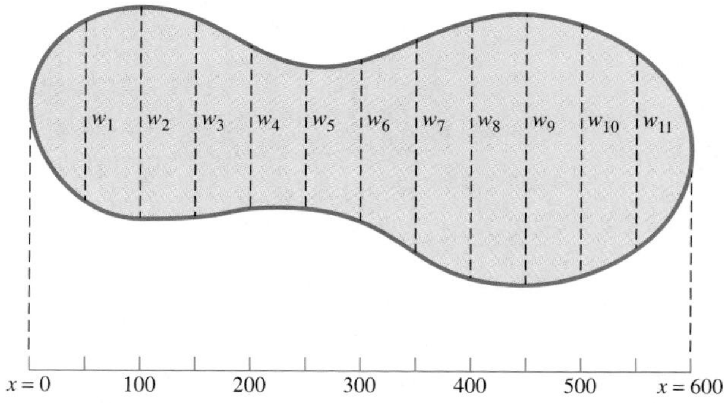

FIGURE 5.9.18 The tract of land of Problem 25.

Use (a) the trapezoidal approximation and (b) Simpson's approximation to estimate the acreage of this tract. [*Note:* An acre is 4840 yd^2.]

26. Because the number e is the base for natural logarithms, it follows that

$$\int_1^e \frac{1}{x}\,dx = 1.$$

Approximate the integrals

$$\int_1^{2.7} \frac{1}{x}\,dx \quad \text{and} \quad \int_1^{2.8} \frac{1}{x}\,dx$$

with sufficient accuracy to show that $2.7 < e < 2.8$.

Problems 27 and 28 deal with the integral

$$\ln 2 = \int_1^2 \frac{1}{x}\,dx$$

of Example 7.

27. Use the trapezoidal error estimate to determine how large n must be in order to guarantee that T_n differs from $\ln 2$ by at most 0.0005.

28. Use the Simpson's error estimate to determine how large n must be in order to guarantee that S_n differs from $\ln 2$ by at most 0.000005.

29. Deduce the following from the error estimate for Simpson's approximation: If $p(x)$ is a polynomial of degree at most 3, then Simpson's approximation with $n = 2$ subintervals gives the exact value of the integral

$$\int_a^b p(x)\,dx.$$

30. Use the result of Problem 29 to calculate (without explicit integration) the area of the region shown in Fig. 5.9.19. [*Answer:* 1331/216.]

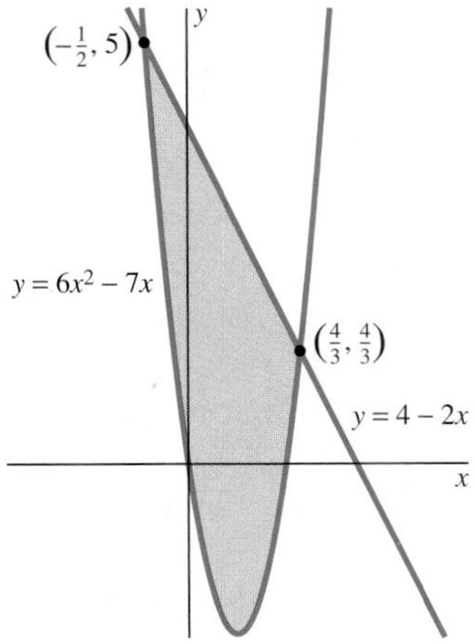

FIGURE 5.9.19 The region of Problem 30.

31. Whereas the carefully weighted average in (10) of the midpoint and trapezoidal approximations M_n and T_n gives Simpson's approximation S_{2n}, show that their equally weighted average gives the trapezoidal approximation with twice as many intervals; that is, $\frac{1}{2}(M_n + T_n) = T_{2n}$.

32. Figure 5.9.20 shows a pendulum of length L. If this pendulum is released from rest at an angle α from the vertical, then it swings back and forth with its period T (for one complete oscillation) given by

$$T = 4\sqrt{\frac{L}{g}} \int_0^{\pi/2} \frac{1}{\sqrt{1 - k^2 \sin^2 x}}\, dx,$$

where $k = \sin(\alpha/2)$. Taking $L = 1$ m and $g = 9.8$ m/s^2, use Simpson's approximation with $n = 10$ subintervals to calculate the pendulum's period of oscillation if its initial angle α is (a) 10°; (b) 50°.

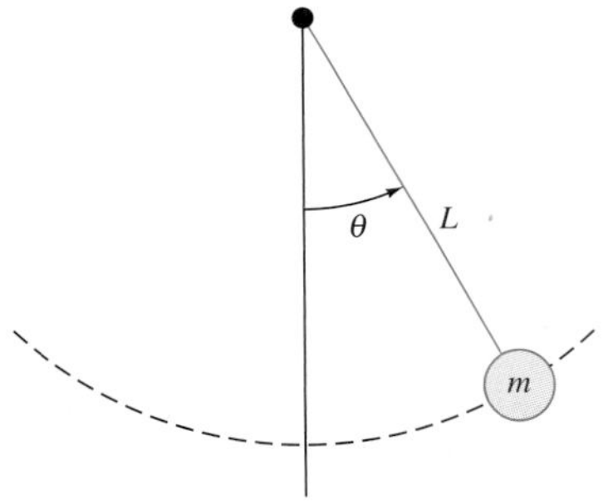

FIGURE 5.9.20 The pendulum of Problem 32.

33. Note in Example 7 that the midpoint and trapezoidal approximations gave under- and over-estimates, respectively, of the value $\int_1^2 (1/x)\, dx = \ln 2$. Show that this is a general phenomenon. That is, if $f(x)$ and $f''(x)$ are both positive for $a \leqq x \leqq b$, then

$$M_n < \int_a^b f(x)\, dx < T_n,$$

whereas if $f(x)$ is positive but $f''(x)$ is negative for $a \leqq x \leqq b$ then the inequalities are reversed.

34. Approximate the number e as follows. First apply Simpson's approximation with $n = 2$ subintervals to the integral

$$\int_0^1 e^x\, dx = e - 1$$

to obtain the approximation $5e - 4\sqrt{e} - 7 \approx 0$. Then solve for the resulting approximate value of e.

35. According to the prime number theorem, which was conjectured by the great German mathematician Carl Friedrich Gauss in 1792 (when he was 15 years old) but not proved until 1896 (independently, by Jacques Hadamard and C. J. de la Vallée Poussin), the number of prime numbers between the positive integers a and $b > a$ is given to a close approximation by the integral

$$\int_a^b \frac{1}{\ln x}\, dx.$$

The midpoint and trapezoidal approximations with $n = 1$ subinterval provide an underestimate and an overestimate of the value of this integral. (Why?) Calculate these estimates with $a = 90000$ and $b = 100000$. The actual number of prime numbers in this range is 879.

5.9 Project: Trapezoidal and Simpson Approximations

In the CD-ROM resource material for the Section 5.4 project, we illustrated calculator and computer algebra system commands that can be used to calculate the Riemann sums

L_n — the left-endpoint approximation,
R_n — the right-endpoint approximation, and
M_n — the midpoint approximation,

based on a division of $[a, b]$ into n equal-length subintervals, to approximate the integral

$$\int_a^b f(x)\, dx. \tag{1}$$

The Riemann sums L_n, R_n, and M_n suffice, in turn, to calculate the trapezoidal and Simpson sums of this section. In particular, the trapezoidal approximation is given—using Eq. (5) of this section—by

$$T_n = \tfrac{1}{2}(L_n + R_n). \tag{2}$$

Once these sums are known, Simpson's approximation based on a subdivision of $[a, b]$ into $2n$ equal-length subintervals is given—using Eq. (10) of this section—by

$$S_n = \tfrac{1}{3}\left(2M_{n/2} + T_{n/2}\right) = \tfrac{1}{6}\left(L_{n/2} + 4M_{n/2} + R_{n/2}\right). \tag{3}$$

Here, then, is a practical scheme for approximating accurately the integral in (1). Begin with a selected value of n, such as $n = 5$, and calculate the Riemann sums L_5, R_5, and M_5. Then use Eq. (3) to calculate the Simpson approximation S_{10}. Next, double the value of n and calculate similarly L_{10}, R_{10}, M_{10}, and finally S_{20}. A typical strategy is to continue in this manner, always doubling the current value of n for the next cycle of computations, until two successively calculated Simpson approximations agree to the desired number of decimal places of accuracy.

Investigation A According to Example 7, the natural logarithm (corresponding to the **LN** or, in some cases, the **LOG** key on your calculator) of the number 2 is the value of the integral

$$\ln 2 = \int_1^2 \frac{1}{x}\,dx.$$

The value of ln 2 correct to 15 decimal places is

$$\ln 2 \approx 0.69314\,71805\,59945.$$

See how many correct decimal places you can obtain in a reasonable period of time by using a Simpson's approximation procedure.

Investigation B In Section 6.8 we will study the inverse tangent function $y = \arctan x$ (y is the angle between $-\pi/2$ and $\pi/2$ such that $\tan y = x$). There we will discover that the derivative of $y = \arctan x$ is

$$\frac{dy}{dx} = \frac{1}{1+x^2}.$$

This implies that

$$\int_0^1 \frac{1}{1+x^2}\,dx = \Big[\arctan x\Big]_0^1 = \arctan 1 - \arctan 0 = \frac{\pi}{4}.$$

It follows that the number π is the value of the integral

$$\pi = \int_0^1 \frac{4}{1+x^2}\,dx.$$

The value of π to 15 decimal places is

$$\pi \approx 3.14159\,26535\,89793.$$

See how many correct decimal places you can obtain in a reasonable period of time by using a Simpson's approximation procedure.

Investigation C Taking $f(x) = 4/(1+x^2)$ as the integrand of the integral

$$\int_0^1 \frac{4}{1+x^2}\,dx = \pi$$

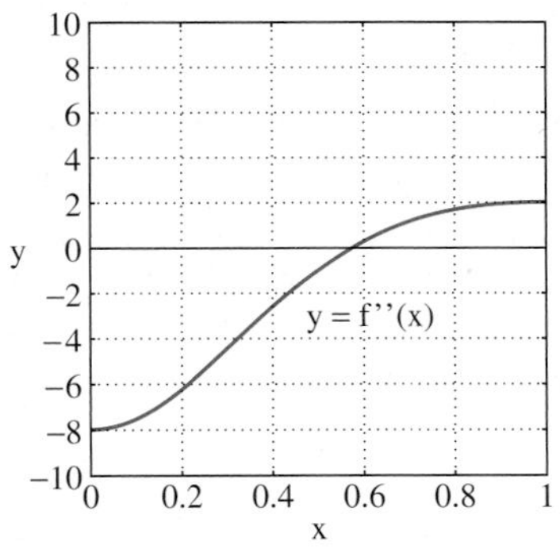

FIGURE 5.9.21 Graph of the second derivative of $f(x) = \dfrac{4}{1+x^2}$.

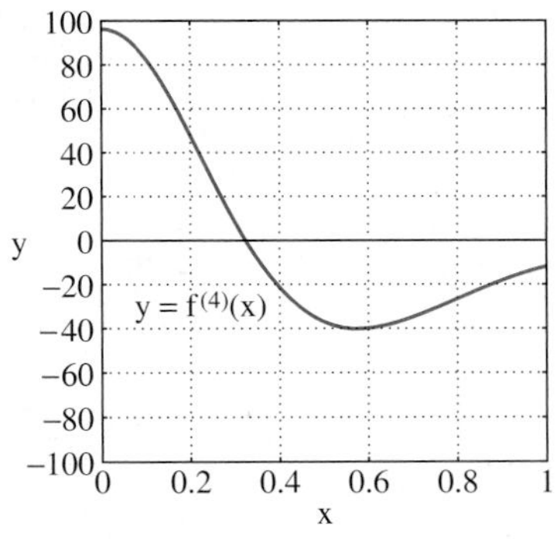

FIGURE 5.9.22 Graph of the fourth derivative of $f(x) = \dfrac{4}{1+x^2}$.

of Investigation B, it would be somewhat lengthy to first calculate and then maximize by hand the derivatives $f''(x)$ and $f^{(4)}(x)$ as needed to apply the error estimates in Theorems 1 and 2 of this section. Instead, a computer algebra system readily yields

$$f''(x) = \frac{8(3x^2-1)}{(1+x^2)^3} \quad \text{and} \quad f^{(4)}(x) = \frac{96(5x^4-10x^2+1)}{(1+x^2)^5}.$$

Figures 5.9.21 and 5.9.22 show the graphs of these two derivatives on the interval $[0, 1]$. From these graphs it is clear that each of these derivatives attains its maximum absolute value at the left endpoint $x = 0$. Thus you can take

$$M_2 = 8 \quad \text{and} \quad M_4 = 96$$

in Theorems 1 and 2. Use this information to determine how large the integer n must be so that:

1. $|EM_n| < 5 \times 10^{-6}$, so the midpoint approximation M_n will give the number π accurate to five decimal places;
2. $|ES_n| < 5 \times 10^{-11}$, so the Simpson approximation S_n will give the number π accurate to ten decimal places;
3. The Simpson approximation S_n will give the number π accurate to 15 decimal places.

CHAPTER 5 REVIEW: DEFINITIONS, CONCEPTS, RESULTS

Use the following list as a guide to concepts that you may need to review.

1. Antidifferentiation and antiderivatives
2. The most general antiderivative of a function
3. Integral formulas
4. Initial value problems
5. Velocity and acceleration
6. Solution of problems involving constant acceleration
7. Properties of area
8. Summation notation
9. The area under the graph of f from a to b
10. Inscribed and circumscribed rectangular polygons
11. A partition of $[a, b]$
12. The norm of a partition
13. A Riemann sum associated with a partition
14. The definite integral of f from a to b
15. Existence of the integral of a continuous function
16. The integral as the limit of a sequence of Riemann sums
17. Regular partitions and

$$\lim_{\Delta x \to 0} \sum_{i=1}^{n} f(x_i^\star)\,\Delta x$$

18. Riemann sums
19. The constant multiple, interval union, and comparison properties of integrals
20. Evaluation of definite integrals by using antiderivatives
21. The average value of $f(x)$ on the interval $[a, b]$
22. The average value theorem

23. The fundamental theorem of calculus
24. Indefinite integrals
25. The method of integration by substitution
26. Transforming the limits in integration by substitution
27. The area between $y = f(x)$ and $y = g(x)$ by integration with respect to x
28. The area between $x = f(y)$ and $x = g(y)$ by integration with respect to y
29. The right-endpoint and left-endpoint approximations
30. The trapezoidal approximation
31. The midpoint approximation
32. Simpson's approximation
33. Error estimates for the trapezoidal approximation and Simpson's approximation

CHAPTER 5 MISCELLANEOUS PROBLEMS

Find the indefinite integrals in Problems 1 through 24. In Problems 13 through 24, use the indicated substitution.

1. $\displaystyle\int \frac{x^5 - 2x + 5}{x^3}\,dx$

2. $\displaystyle\int \sqrt{x}\left(1 + \sqrt{x}\right)^3 dx$

3. $\displaystyle\int (1 - 3x)^9\,dx$

4. $\displaystyle\int \frac{7}{(2x + 3)^3}\,dx$

5. $\displaystyle\int \sqrt[3]{9 + 4x}\,dx$

6. $\displaystyle\int \frac{24}{\sqrt{6x + 7}}\,dx$

7. $\displaystyle\int x^3(1 + x^4)^5\,dx$

8. $\displaystyle\int 3x^2\sqrt{4 + x^3}\,dx$

9. $\displaystyle\int x\sqrt[3]{1 - x^2}\,dx$

10. $\displaystyle\int \frac{3x}{\sqrt{1 + 3x^2}}\,dx$

11. $\displaystyle\int (7\cos 5x - 5\sin 7x)\,dx$

12. $\displaystyle\int 5\sin^3 4x \cos 4x\,dx$

13. $\displaystyle\int x^3\sqrt{1 + x^4}\,dx; \quad u = x^4$

14. $\displaystyle\int \sin^2 x \cos x\,dx; \quad u = \sin x$

15. $\displaystyle\int \frac{1}{\sqrt{x}\left(1 + \sqrt{x}\right)^2}\,dx; \quad u = 1 + \sqrt{x}$

16. $\displaystyle\int \frac{1}{\sqrt{x}\left(1 + \sqrt{x}\right)^2}\,dx; \quad u = \sqrt{x}$

17. $\displaystyle\int x^2\cos 4x^3\,dx; \quad u = 4x^3$

18. $\displaystyle\int x(x + 1)^{14}\,dx; \quad u = x + 1$

19. $\displaystyle\int x(x^2 + 1)^{14}\,dx; \quad u = x^2 + 1$

20. $\displaystyle\int x^3\cos x^4\,dx; \quad u = x^4$

21. $\displaystyle\int x\sqrt{4 - x}\,dx; \quad u = 4 - x$

22. $\displaystyle\int \frac{x + 2x^3}{(x^4 + x^2)^3}\,dx; \quad u = x^4 + x^2$

23. $\displaystyle\int \frac{2x^3}{\sqrt{1 + x^4}}\,dx; \quad u = x^4$

24. $\displaystyle\int \frac{2x + 1}{\sqrt{x^2 + x}}\,dx; \quad u = x^2 + x$

Solve the initial value problems in 25 through 30.

25. $\dfrac{dy}{dx} = 3x^2 + 2x; \quad y(0) = 5$

26. $\dfrac{dy}{dx} = 3\sqrt{x}; \quad y(4) = 20$

27. $\dfrac{dy}{dx} = (2x + 1)^5; \quad y(0) = 2$

28. $\dfrac{dy}{dx} = \dfrac{2}{\sqrt{x + 5}}; \quad y(4) = 3$

29. $\dfrac{dy}{dx} = \dfrac{1}{\sqrt[3]{x}}; \quad y(1) = 1$

30. $\dfrac{dy}{dx} = 1 - \cos x; \quad y(0) = 0$

31. When its brakes are fully applied, a certain automobile has a constant deceleration of 22 ft/s^2. If its initial velocity is 90 mi/h, how long will it take to come to a stop? How many feet will it travel during that time?

32. In Hal Clement's novel *Mission of Gravity,* much of the action take place in the polar regions of the planet Mesklin, where the acceleration of gravity is 22500 ft/s^2. A stone is dropped near the north pole of Mesklin from a height of 450 ft. How long does it remain aloft? With what speed does it strike the ground?

33. An automobile is traveling along the x-axis in the positive direction. At time $t = 0$ its brakes are fully applied, and the car experiences a constant deceleration of 40 ft/s^2 while skidding. The car skids 180 ft before coming to a stop. What was its initial velocity?

34. If a car starts from rest with an acceleration of 8 ft/s^2, how far has it traveled by the time it reaches a speed of 60 mi/h?

35. On the planet Zorg, a ball dropped from a height of 20 ft hits the ground in 2 s. If the ball is dropped from the top of a 200-ft building on Zorg, how long will it take to reach the ground? With what speed will it hit?

36. Suppose that you can throw a ball from the earth's surface straight upward to a maximum height of 144 ft. (a) How high could you throw it on the planet of Problem 35? (b) How high could you throw it in the polar regions of Mesklin? (See Problem 32.)

37. Suppose that a car skids 44 ft if its velocity is 30 mi/h when the brakes are fully applied. Assuming the same constant deceleration, how far will it skid if its velocity is 60 mi/h when the brakes are fully applied?

38. The graph of the velocity of a model rocket fired at time $t = 0$ is shown in Fig. 5.MP.1. (a) At what time was the fuel exhausted? (b) At what time did the parachute open? (c) At what time did the rocket reach its maximum altitude? (d) At what time did the rocket land? (e) How high did the rocket go? (f) How high was the pole on which the rocket landed?

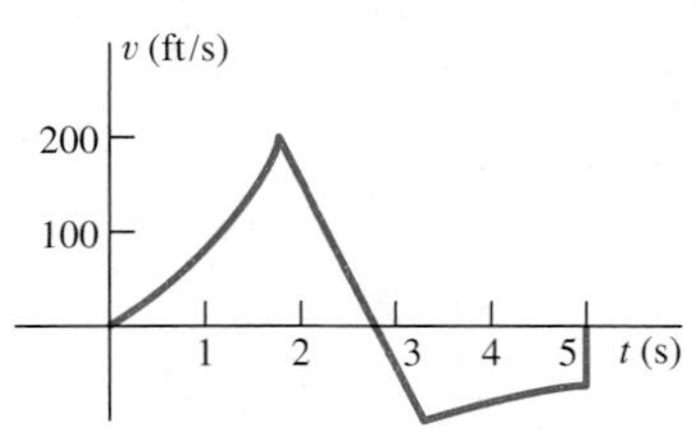

FIGURE 5.MP.1 Rocket velocity graph for Problem 38.

Find the sums in Problems 39 through 42.

39. $\displaystyle\sum_{i=1}^{100} 17$

40. $\displaystyle\sum_{k=1}^{100} \left(\frac{1}{k} - \frac{1}{k+1}\right)$

41. $\displaystyle\sum_{n=1}^{10} (3n-2)^2$

42. $\displaystyle\sum_{n=1}^{16} \sin\frac{n\pi}{2}$

In Problems 43 through 45, find the limit of the given Riemann sum associated with a regular partition of the indicated interval $[a, b]$. First express it as an integral from a to b; then evaluate that integral.

43. $\displaystyle\lim_{n\to\infty} \sum_{i=1}^{n} \frac{\Delta x}{\sqrt{x_i^\star}}; \quad [1, 2]$

44. $\displaystyle\lim_{n\to\infty} \sum_{i=1}^{n} [(x_i^\star)^2 - 3x_i^\star]\,\Delta x; \quad [0, 3]$

45. $\displaystyle\lim_{n\to\infty} \sum_{i=1}^{n} 2\pi x_i^\star \sqrt{1 + (x_i^\star)^2}\,\Delta x; \quad [0, 1]$

46. Evaluate

$$\lim_{n\to\infty} \frac{1^{10} + 2^{10} + 3^{10} + \cdots + n^{10}}{n^{11}}$$

by expressing this limit as an integral over $[0, 1]$.

47. Use Riemann sums to prove that if $f(x) \equiv c$ (a constant), then

$$\int_a^b f(x)\,dx = c(b-a).$$

48. Use Riemann sums to prove that if f is continuous on $[a, b]$ and $f(x) \geqq 0$ for all x in $[a, b]$, then

$$\int_a^b f(x)\,dx \geqq 0.$$

49. Use the comparison property of integrals (Section 5.5) to prove that

$$\int_a^b f(x)\,dx > 0$$

if f is a continuous function with $f(x) > 0$ on $[a, b]$.

Evaluate the integrals in Problems 50 through 63.

50. $\displaystyle\int_0^1 (1-x^2)^3\,dx$

51. $\displaystyle\int \left(\sqrt{2x} - \frac{1}{\sqrt{3x^3}}\right) dx$

52. $\displaystyle\int \frac{(1+\sqrt[3]{x})^2}{\sqrt{x}}\,dx$

53. $\displaystyle\int \frac{4-x^3}{2x^2}\,dx$

54. $\displaystyle\int_0^1 \frac{dt}{(3-2t)^2}$

55. $\displaystyle\int \sqrt{x}\cos x\sqrt{x}\,dx$

56. $\displaystyle\int_0^2 x^2\sqrt{9-x^3}\,dx$

57. $\displaystyle\int \frac{1}{t^2}\sin\frac{1}{t}\,dt$

58. $\displaystyle\int_1^2 \frac{2t+1}{\sqrt{t^2+t}}\,dt$

59. $\displaystyle\int \frac{\sqrt[3]{u}}{(1+u^{4/3})^3}\,du$

60. $\displaystyle\int_0^{\pi/4} \frac{\sin t}{\sqrt{\cos t}}\,dt$

61. $\displaystyle\int_1^4 \frac{(1+\sqrt{t})^2}{\sqrt{t}}\,dt$

62. $\displaystyle\int \frac{1}{u^2}\sqrt[3]{1-\frac{1}{u}}\,du$

63. $\displaystyle\int \frac{\sqrt{4x^2-1}}{x^4}\,dx$

Find the areas of the plane regions bounded by the curves given in Problems 64 through 70.

64. $y = x^3, \quad x = -1, \quad y = 1$

65. $y = x^4, \quad y = x^5$

66. $y^2 = x, \quad 3y^2 = x + 6$

67. $y = x^4, \quad y = 2 - x^2$

68. $y = x^4, \quad y = 2x^2 - 1$

69. $y = (x-2)^2, \quad y = 10 - 5x$

70. $y = x^{2/3}, \quad y = 2 - x^2$

71. Evaluate the integral

$$\int_0^2 \sqrt{2x - x^2}\,dx$$

by interpreting it as the area of a region.

72. Evaluate the integral

$$\int_1^5 \sqrt{6x - 5 - x^2}\,dx$$

by interpreting it as the area of a region.

73. Find a function f such that

$$x^2 = 1 + \int_1^x \sqrt{1 + [f(t)]^2}\,dt$$

for all $x > 1$. [*Suggestion:* Differentiate both sides of the equation with the aid of the fundamental theorem of calculus.]

74. Show that $G'(x) = \phi(h(x)) \cdot h'(x)$ if

$$G(x) = \int_a^{h(x)} \phi(t)\,dt.$$

75. Use right-endpoint and left-endpoint approximations to estimate

$$\int_0^1 \sqrt{1+x^2}\,dx$$

with error not exceeding 0.05.

76. Calculate the trapezoidal approximation and Simpson's approximation to

$$\int_0^{\pi} \sqrt{1-\cos x}\,dx$$

with six subintervals. For comparison, use an appropriate half-angle identity to calculate the exact value of this integral.

77. Calculate the midpoint approximation and trapezoidal approximation to

$$\int_1^2 \frac{1}{x+x^2}\,dx$$

with $n = 5$ subintervals. Then explain why the exact value of the integral lies between these two approximations.

In Problems 78 through 80, let $\{x_0, x_1, x_2, \dots, x_n\}$ *be a partition of* $[a, b]$, *where* $a < b$.

78. For $i = 1, 2, 3, \dots, n$, let $x_i^\star$ be given by

$$(x_i^\star)^2 = \tfrac{1}{3}[(x_{i-1})^2 + x_{i-1}x_i + (x_i)^2].$$

Show first that $x_{i-1} < x_i^\star < x_i$. Then use the algebraic identity

$$(c-d)(c^2+cd+d^2) = c^3 - d^3$$

to show that

$$\sum_{i=1}^{n} (x_i^\star)^2\,\Delta x_i = \tfrac{1}{3}(b^3 - a^3).$$

Explain why this computation proves that

$$\int_a^b x^2\,dx = \tfrac{1}{3}(b^3 - a^3).$$

79. Let $x_i^\star = \sqrt{x_{i-1}x_i}$ for $i = 1, 2, 3, \dots, n$, and assume that $0 < a < b$. Show that

$$\sum_{i=1}^{n} \frac{\Delta x_i}{(x_i^\star)^2} = \frac{1}{a} - \frac{1}{b}.$$

Then explain why this computation proves that

$$\int_a^b \frac{dx}{x^2} = \frac{1}{a} - \frac{1}{b}.$$

80. Assume that $0 < a < b$. Define

$$x_i^\star = \frac{\frac{2}{3}\left[(x_i)^{3/2} - (x_{i-1})^{3/2}\right]}{x_i - x_{i-1}}.$$

First show that $x_{i-1} < x_i^\star < x_i$. Then use this selection for the given partition to prove that

$$\int_a^b \sqrt{x}\,dx = \tfrac{2}{3}\left(b^{3/2} - a^{3/2}\right).$$

APPLICATIONS OF THE INTEGRAL 6

G.F.B. Riemann (1826–1866)

The general concept of integration traces back to the area and volume computations of ancient times, but the integrals used by Newton and Leibniz were not defined with sufficient precision for full understanding. We owe to the German mathematician G. F. Bernhard Riemann the modern definition that uses "Riemann sums." The son of a Protestant minister, Riemann studied theology and philology at Göttingen University until he finally gained his father's permission to concentrate on mathematics. He transferred to Berlin University, where he received his Ph.D. in 1851. The work he did in the next decade justifies his place on everyone's short list of the most profound and creative mathematicians of all time. But in 1862 he was stricken ill. He never fully recovered and in 1866 died prematurely at the age of 39.

Riemann's mathematical investigations were as varied as they were deep, ranging from the basic concepts of functions and integrals to such areas as non-Euclidean (differential) geometry and the distribution of prime numbers. Recall that the positive integer p is *prime* if it cannot be factored into smaller integers. In a famous paper of 1859, Riemann analyzed the approximation

$$\pi(x) \approx \int_2^x \frac{dt}{\ln t} = \text{li}(x)$$

to the number $\pi(x)$ of those primes $p \leqq x$ (with $\ln x$ denoting the natural logarithm of x). There is a remarkable correspondence between the values of $\pi(x)$ and the "logarithmic integral" approximation li(x):

x	1,000,000	10,000,000	100,000,000	1,000,000,000
li(x)	78,628	664,918	5,762,209	50,849,235
$\pi(x)$	78,498	664,579	5,761,455	50,847,543
error	0.165%	0.051%	0.013%	0.003%

Thirty years after Riemann's death, his ideas led ultimately to a proof that the percentage error in the approximation li(x) to $\pi(x)$ approaches 0 as $x \to \infty$.

In his 1851 thesis, Riemann introduced a geometric way of visualizing "multi-valued" functions such as the square root function with two values $\pm\sqrt{x}$. The following graph illustrates the *cube root* function. For each complex number $z = x + iy$ in the unit disk $x^2 + y^2 \leqq 1$, the three (complex) cube roots of z are plotted directly above z. Each root is plotted at a height equal to its real part, with color determined by its imaginary part. The result is the "Riemann surface" of the cube root function.

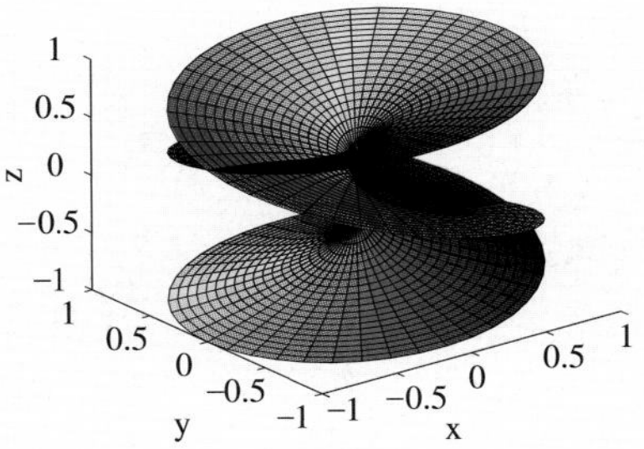

6.1 RIEMANN SUM APPROXIMATIONS

In Section 5.4 we defined the integral of the function f on the interval $[a, b]$ as a limit of Riemann sums. Specifically, let the interval $[a, b]$ be divided into n subintervals, all with the same length $\Delta x = (b - a)/n$ (Fig. 6.1.1). Then a selection of numbers $x_1^\star, x_2^\star, \dots, x_n^\star$ in these subintervals ($x_i^\star$ being a point of the ith subinterval $[x_{i-1}, x_i]$) produces a Riemann sum

$$\sum_{i=1}^{n} f(x_i^\star)\,\Delta x \tag{1}$$

whose value approximates the integral of f on $[a, b]$. The value of the integral is the limiting value (if any) of such sums as the subinterval length Δx approaches zero. That is,

$$\int_a^b f(x)\,dx = \lim_{\Delta x \to 0} \sum_{i=1}^{n} f(x_i^\star)\,\Delta x. \tag{2}$$

The wide applicability of the definite integral arises from the fact that many geometric and physical quantities can be approximated arbitrarily closely by Riemann sums. Such approximations lead to integral formulas for the computation of such quantities.

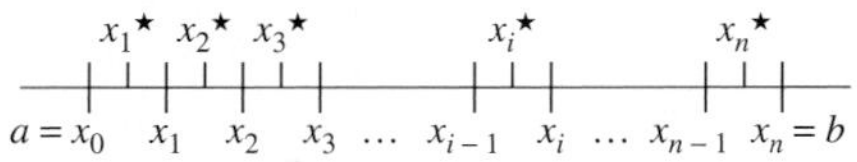

FIGURE 6.1.1 A division (or partition) of $[a, b]$ into n equal-length subintervals.

For example, suppose that $f(x)$ is positive-valued on $[a, b]$ and that our goal—as in Section 5.3—is to calculate the area A of the region that lies below the graph of $y = f(x)$ over the interval $[a, b]$. Beginning with the subdivision (or partition) of $[a, b]$ indicated in Fig. 6.1.1, let ΔA_i denote the area of the vertical "strip" that lies under $y = f(x)$ over the ith subinterval $[x_{i-1}, x_i]$. Then, as illustrated in Fig. 6.1.2, the "strip areas"

$$\Delta A_1, \quad \Delta A_2, \quad \dots, \quad \Delta A_n$$

add up to the total area A:

$$A = \sum_{i=1}^{n} \Delta A_i. \tag{3}$$

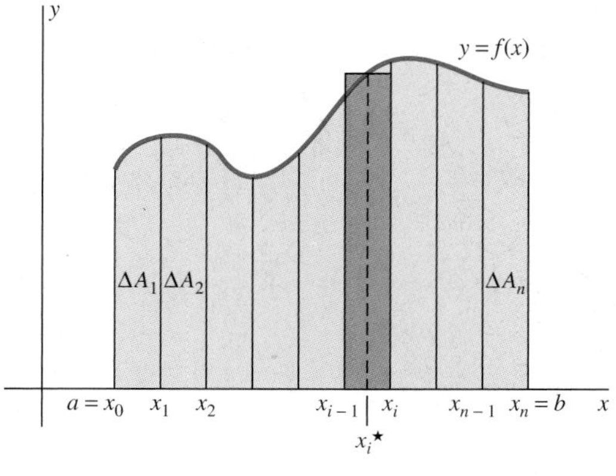

FIGURE 6.1.2 Approximating an area by means of a Riemann sum.

But the ith strip is approximated by a rectangle with base $[x_{i-1}, x_i]$ and height $f(x_i^\star)$, so its area is given approximately by

$$\Delta A_i \approx f(x_i^\star)\,\Delta x. \tag{4}$$

After we substitute Eq. (4) into Eq. (3), it becomes apparent that the total area A under the graph of f is given approximately by

$$A \approx \sum_{i=1}^{n} f(x_i^\star)\,\Delta x. \tag{5}$$

Note that the approximating sum on the right is a Riemann sum for f on $[a, b]$. Moreover,

1. It is intuitively evident that the Riemann sum in (5) approaches the actual area A as $n \to +\infty$ (which forces $\Delta x \to 0$);
2. By the definition of the integral, this Riemann sum approaches $\int_a^b f(x)\,dx$ as $n \to +\infty$.

These observations justify the *definition* of the area A by the formula

$$A = \int_a^b f(x)\,dx. \tag{6}$$

Other Quantities as Integrals

Our justification of the area formula in Eq. (6) illustrates an important general method of setting up integral formulas. Suppose that we want to calculate a certain quantity Q that is associated with an interval $[a, b]$ in such a way that subintervals of $[a, b]$ correspond to specific portions of Q (such as the portion of area lying above a particular subinterval). Then a subdivision of $[a, b]$ into n subintervals produces portions

$$\Delta Q_1, \quad \Delta Q_2, \quad \dots, \quad \Delta Q_n,$$

which add up to the quantity

$$Q = \sum_{i=1}^{n} \Delta Q_i. \tag{7}$$

Now suppose that we can find a function f such that the ith portion ΔQ_i is given approximately by

$$\Delta Q_i \approx f(x_i^\star)\,\Delta x \tag{8}$$

(for each i, $1 \leqq i \leqq n$) for a selected point $x_i^\star$ of the ith subinterval $[x_{i-1}, x_i]$ of $[a, b]$. Then substituting Eq. (8) into Eq. (7) yields the Riemann sum approximation

$$Q \approx \sum_{i=1}^{n} f(x_i^\star)\,\Delta x \tag{9}$$

analogous to the approximation in Eq. (5). The right-hand sum in Eq. (9) is a Riemann sum that approaches the integral

$$\int_a^b f(x)\,dx \quad \text{as} \quad n \to +\infty.$$

If it is also evident—for geometric or physical reasons, for example—that this Riemann sum must approach the quantity Q as $n \to +\infty$, then Eq. (9) justifies our setting up the integral formula

$$Q = \int_a^b f(x)\,dx. \tag{10}$$

Because the right-hand side in Eq. (10) can be easy to calculate (by the fundamental theorem of calculus), this gives us a practical way of finding the exact numerical value of the quantity Q.

In addition to area, the following are some of the quantities that can be calculated by using integral formulas such as Eq. (10). (The variable x is replaced with t where appropriate.)

- The *mass* of a thin rod of variable density lying along the interval $a \leqq x \leqq b$.
- The *profit* earned by a company between time $t = a$ and time $t = b$.
- The *number* of people in a city who contract a certain disease between time $t = a$ and time $t = b$.
- The *distance* traveled by a moving particle during the time interval $a \leqq t \leqq b$.
- The *volume* of water flowing into a tank during the time interval $a \leqq t \leqq b$.
- The *work* done by a variable force in moving a particle from the point $x = a$ to the point $x = b$.

In each case it is evident that a *subinterval* of $[a, b]$ determines a specific *portion* ΔQ of the whole quantity Q that corresponds to the whole interval $[a, b]$. The question is this: What function f should be integrated from a to b? Examples 1 through 3 illustrate the process of finding the needed function f by approximating the portion ΔQ_i of the quantity Q that corresponds to the subinterval $[x_{i-1}, x_i]$. An approximation of the form

$$\Delta Q_i \approx f(x_i^\star)\,\Delta x \tag{8}$$

leads to the desired integral formula

$$Q = \int_a^b f(x)\,dx. \tag{10}$$

The integral in Eq. (10) results from the summation in Eq. (9) when we make the following replacements:

$$\sum_{i=1}^{n} \text{ becomes } \int_a^b,$$

$$x_i^\star \text{ becomes } x, \quad \text{and}$$

$$\Delta x \text{ becomes } dx.$$

FIGURE 6.1.3 The tank of Example 1.

EXAMPLE 1 Suppose that water is pumped into the initially empty tank of Fig. 6.1.3. The rate of water flow into the tank at time t (in seconds) is $50 - t$ liters (L) per second. How much water flows into the tank during the first 30 s?

Solution We want to compute the amount Q of water that flows into the tank during the time interval $[0, 30]$. Think of a subdivision of $[0, 30]$ into n subintervals, all with the same length $\Delta t = 30/n$.

Next choose a point $t_i^\star$ in the ith subinterval $[t_{i-1}, t_i]$. If this subinterval is very short, then the rate of water flow between time t_{i-1} and time t_i remains approximately $50 - t_i^\star$ liters per second. So the amount ΔQ_i of water *in liters* that flows into the tank during this subinterval of time is obtained approximately by multiplying the *flow rate in liters per second* by the *duration of flow in seconds:*

$$\left[(50 - t_i^\star)\,\frac{\text{liters}}{\text{second}}\right] \cdot [\Delta t \text{ seconds}],$$

and hence

$$\Delta Q_i \approx (50 - t_i^\star)\,\Delta t \quad \text{(liters)}.$$

Therefore, the total amount Q that we seek is given approximately by

$$Q = \sum_{i=1}^{n} \Delta Q_i \approx \sum_{i=1}^{n} (50 - t_i^\star)\,\Delta t \quad \text{(liters)}.$$

We recognize that the sum on the right is a Riemann sum, and—most important—we see that it is a Riemann sum for the function $f(t) = 50 - t$. Hence we may conclude that

$$Q = \lim_{n\to\infty} \sum_{i=1}^{n} (50 - t_i^\star)\,\Delta t = \int_0^{30} (50 - t)\,dt$$

$$= \left[50t - \frac{1}{2}t^2\right]_0^{30} = 1050 \quad \text{(liters).}$$

◆

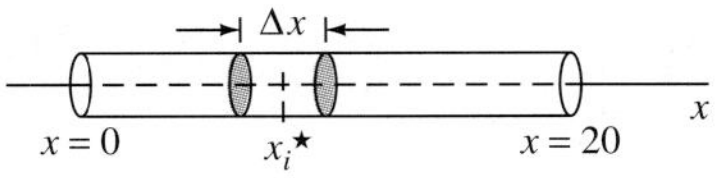

FIGURE 6.1.4 The 20-cm rod of Example 2.

EXAMPLE 2 Figure 6.1.4 shows a thin rod 20 cm long. Its (linear) *density* at the point x is $15 + 2x$ grams of mass per centimeter of the rod's length (g/cm). The rod's density thus varies from 15 g/cm at the left end $x = 0$ to 55 g/cm at the right end $x = 20$. Find the total mass M of this rod.

Solution Think of a subdivision of $[0, 20]$ into n subintervals of length $\Delta x = 20/n$ each. Figure 6.1.4 shows the short piece of the rod that corresponds to the typical ith subinterval $[x_{i-1}, x_i]$. If $x_i^\star$ is, say, the midpoint of $[x_{i-1}, x_i]$, then the mass ΔM_i of this short piece is obtained approximately by multiplying its *density in grams per centimeter* by its *length in centimeters:*

$$\left[(15 + 2x_i^\star)\,\frac{\text{grams}}{\text{centimeter}}\right] \cdot [\Delta x \text{ centimeters}].$$

That is,

$$\Delta M_i \approx (15 + 2x_i^\star)\,\Delta x \quad \text{(grams).}$$

Therefore, the total mass M of the entire rod is given approximately by

$$M = \sum_{i=1}^{n} \Delta M_i \approx \sum_{i=1}^{n} (15 + 2x_i^\star)\,\Delta x.$$

We recognize a Riemann sum on the right, as in Example 1, although this time for the function $f(x) = 15 + 2x$ on the interval $[0, 20]$. Hence we may conclude that

$$M = \lim_{n\to\infty} \sum_{i=1}^{n} (15 + 2x_i^\star)\,\Delta x = \int_0^{20} (15 + 2x)\,dx$$

$$= \left[15x + x^2\right]_0^{20} = 700 \quad \text{(g).}$$

◆

The key to setting up an integral formula as in Examples 1 and 2 is the recognition of the definite integral that corresponds to a given Riemann sum approximation to the quantity we wish to calculate.

EXAMPLE 3 Calculate Q if

$$Q = \lim_{n\to\infty} \sum_{i=1}^{n} 2x_i \exp\left(-x_i^2\right)\Delta x,$$

where $x_0, x_1, x_2, \ldots, x_n$ are the endpoints of a partition of the interval $[1, 2]$ into n subintervals, all with the same length $\Delta x = 1/n$.

Solution We recognize the given sum as a Riemann sum (with $x_i^\star = x_i$) for the integral of $f(x) = 2x\exp(-x^2) = 2xe^{-x^2}$. Hence

$$Q = \int_1^2 2xe^{-x^2}\,dx = \left[-e^{-x^2}\right]_1^2 = -e^{-4} + e^{-1} \approx 0.3496.$$

◆

Distance and Velocity

Consider a particle that travels along the x-axis with position $x(t)$ and velocity $v = x'(t)$ at time t. Suppose that it begins its motion at time $t = a$ and ends it at time $t = b$. When we integrate the velocity, we get

$$\int_a^b v(t)\,dt = \int_a^b x'(t)\,dt = \Big[x(t)\Big]_a^b = x(b) - x(a)$$

(using the fundamental theorem of calculus). Because the particle has initial position $x(a)$ and final position $x(b)$, we see that the integral

$$s = \int_a^b v(t)\,dt = x(b) - x(a) \qquad \textbf{(11)}$$

of its velocity gives the **displacement** or **net distance** s traveled by the particle. (See Fig. 6.1.5.)

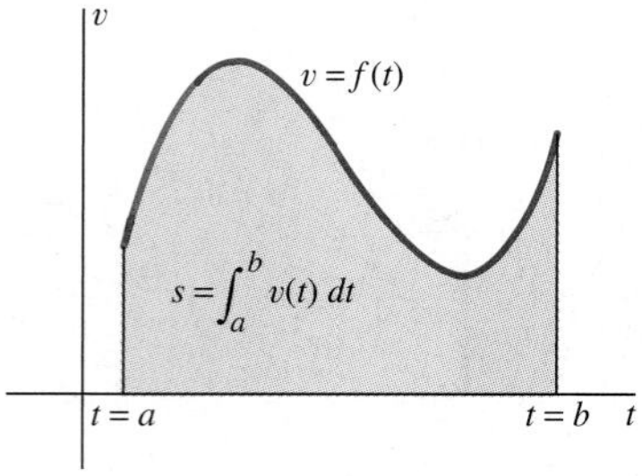

FIGURE 6.1.5 Equation (11) means that the (net) distance traveled is equal to the (signed) area under the velocity curve.

The velocity function $v(t) = x'(t)$ may have both positive and negative values, as illustrated in Fig. 6.1.6. Then forward distances ($v > 0$) and backward distances ($v < 0$) partially or even totally cancel when we compute the net distance in Eq. (11). But suppose that we want to calculate the *total distance* S traveled—irrespective of direction. We can begin with a partition of $[a, b]$ into n subintervals all having the same length $\Delta t = (b - a)/n$ and set up the approximation

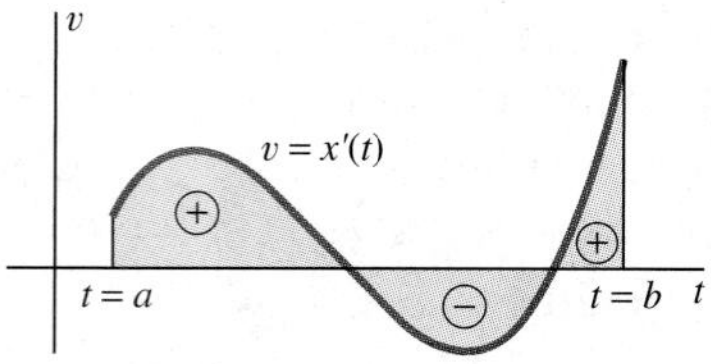

FIGURE 6.1.6 Velocity curve of a particle that first travels forward, then backward, then forward again.

$$S \approx \sum_{i=1}^{n} |v(t_i^\star)|\,\Delta t. \qquad \textbf{(12)}$$

Here $|v(t_i^\star)|$ is the *speed*—irrespective of direction—of the particle at a typical point $t_i^\star$ of the ith subinterval $[t_{i-1}, t_i]$, and thus $|v(t_i^\star)|\,\Delta t$ is the approximate distance traveled during that time interval. The approximation in (12) is a Riemann sum for the integral that gives the **total distance** traveled:

$$S = \int_a^b |v(t)|\,dt. \qquad \textbf{(13)}$$

In summary, we see that:

- The *net distance* s traveled by the particle is the integral of its (signed) *velocity* v, whereas
- The *total distance* S traveled by the particle is the integral of the (unsigned) *speed* $|v|$ of the particle.

The integral in (13) can be calculated by integrating separately over the subintervals where v is positive and those where v is negative, then adding the absolute values of the results. This is exactly the same procedure we use to find the area between the graph of a function and the x-axis when the function has both positive and negative values.

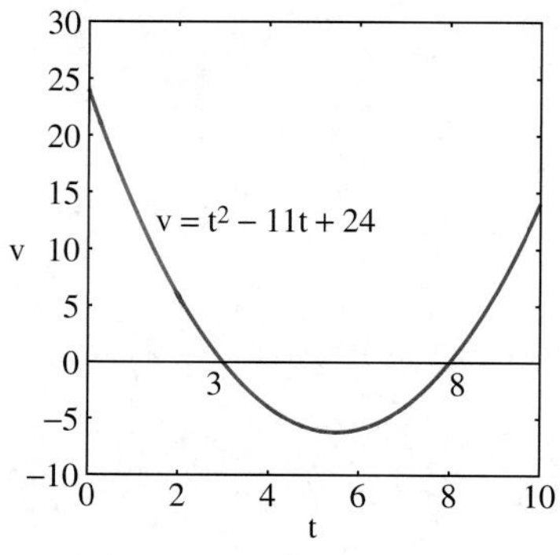

FIGURE 6.1.7 Graph of the velocity function of Example 4.

EXAMPLE 4 Suppose that the velocity of a moving particle is $v(t) = t^2 - 11t + 24$ (ft/s). Find both the net distance s and the total distance S it travels between time $t = 0$ and $t = 10$ (s).

Solution For net distance, we use Eq. (11) and find that

$$s = \int_0^{10} (t^2 - 11t + 24)\,dt = \left[\frac{1}{3}t^3 - \frac{11}{2}t^2 + 24t\right]_0^{10}$$

$$= \frac{1000}{3} - \frac{1100}{2} + 240 = \frac{70}{3} \quad \text{(ft)}.$$

To find the total distance traveled, we note from the graph of $v(t)$ in Fig. 6.1.7—or from the factorization $v(t) = (t - 3)(t - 8)$—that $v(t) > 0$ if $0 \leqq t < 3$, $v(t) < 0$ if

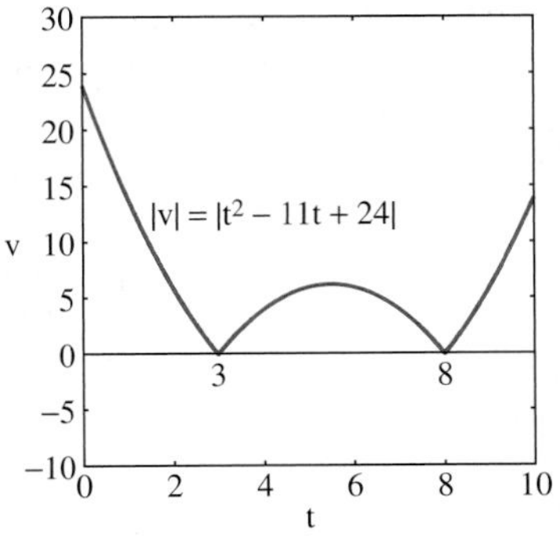

FIGURE 6.1.8 Graph of the absolute value of the velocity function of Example 4.

$3 < t < 8$, and $v(t) > 0$ if $8 < t \leqq 10$. By Eq. (13), we need to integrate the absolute value $|v(t)|$ graphed in Fig. 6.1.8. Now

$$\int_0^3 (t^2 - 11t + 24)\,dt = \left[\frac{1}{3}t^3 - \frac{11}{2}t^2 + 24\right]_0^3 = \frac{63}{2},$$

$$\int_3^8 (-t^2 + 11t - 24)\,dt = \left[-\frac{1}{3}t^3 + \frac{11}{2}t^2 - 24t\right]_3^8 = \frac{125}{6},$$

and

$$\int_8^{10} (t^2 - 11t + 24)\,dt = \left[\frac{1}{3}t^3 - \frac{11}{2}t^2 + 24\right]_8^{10} = \frac{38}{3}.$$

Thus the particle travels $\frac{63}{2}$ ft forward, then $\frac{125}{6}$ ft backward, and finally $\frac{38}{3}$ ft forward, for a total distance traveled of $S = \frac{63}{2} + \frac{125}{6} + \frac{38}{3} = 65$ ft. ◆

Fluid Flow in Circular Pipes

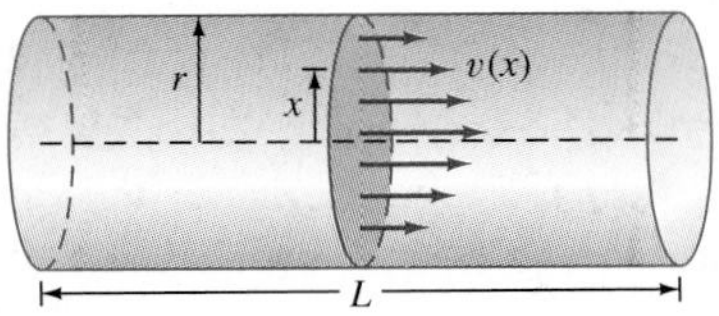

FIGURE 6.1.9 Fluid flow in a pipe of radius r and length L.

We consider the flow of fluid in a straight circular pipe of radius r. Because of the friction with the wall of the pipe, the velocity v of the fluid tends to be greatest along the centerline of the pipe, and decreases with the distance x from the center (Fig. 6.1.9). We therefore write $v(x)$ for the velocity (in units such as cm/s) at distance x. We want to calculate the total *flow rate* F (in units such as cm^3/s).

Figure 6.1.10 shows the pipe's circular cross section of radius r divided into washer-shaped *annular rings* by concentric circles whose radii are the points $x_0 = 0, x_1, x_2, \dots, x_n = r$ of a subdivision of the x-interval $[0, r]$ into n subintervals all having the same length $\Delta x = r/n$. The ith annular ring corresponds to the ith subinterval $[x_{i-1}, x_i]$ and is bounded by the circles of radii x_{i-1} and x_i. To approximate its area ΔA_i, we think of cutting this annular ring and straightening it into a long strip of width Δx, as indicated in Fig. 6.1.11). The lengths of the top and bottom edges of this strip are simply the circumferences $2\pi x_{i-1}$ and $2\pi x_i$ of the two circles bounded the original annular ring. If $\overline{x}_i = \frac{1}{2}(x_{i-1} + x_i)$ denotes the "average radius" of this annular ring—so that $2\pi\overline{x}_i$ is the average length of the straightened strip—it follows that

$$\Delta A_i \approx 2\pi\overline{x}_i\,\Delta x. \tag{14}$$

If we reason that the average velocity of the fluid flowing across this ith annular ring is accurately approximated by $v(\overline{x}_i)$, then the volume that flows across it in one unit of time is approximately a cylindrical shell with base area ΔA_i and height $v(\overline{x}_i)$. It follows that the rate of flow ΔF_i of fluid across this annular ring is given approximately by

$$\Delta F_i \approx v(\overline{x}_i)\,\Delta A_i \approx 2\pi\overline{x}_i v(\overline{x}_i)\,\Delta x.$$

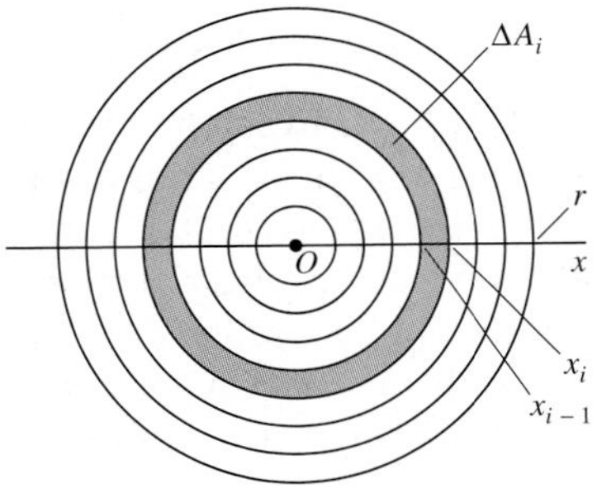

FIGURE 6.1.10 The circular cross section divided into annular rings.

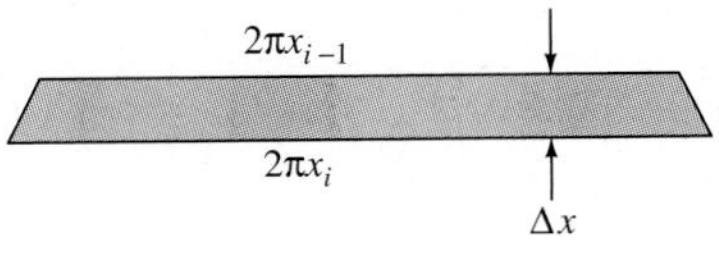

FIGURE 6.1.11 An annular ring "straightened out."

Adding the flow rates across all n annular rings that make up the entire circular cross section of the pipe, we get the approximation

$$F = \sum_{i=1}^{n} \Delta F_i \approx \sum_{i=1}^{n} 2\pi \overline{x}_i v(\overline{x}_i)\, \Delta x.$$

Finally, we see that this approximation is a Riemann sum (with $x_i^\star = \overline{x}_i$) for the integral

$$F = \int_0^r 2\pi x v(x)\, dx, \tag{15}$$

which thus gives the total flow rate F of fluid along the pipe.

EXAMPLE 5 If the velocity of the flowing fluid is exactly 1 at each point of the pipe, then the volume of fluid that flows across a circular cross section of radius r and area A in 1 second is a cylinder of volume $A \cdot 1 = A$. Thus $F = A$ in this case. Substituting $v(x) \equiv 1$ in Eq. (15) then yields

$$A = \int_0^r 2\pi x\, dx = \left[\pi x^2\right]_0^r = \pi r^2.$$

Because our derivation of Eq. (15) used only the formula $C = 2\pi r$ for the circumference of a circle—and not its area formula—this is a new and independent derivation of the formula $A = \pi r^2$ for the area of a circle of radius r. ◆

EXAMPLE 6 According to the law of laminar flow—discovered by the French physician Jean-Louis-Marie Poiseuille in 1840—the velocity function for fluid flow in a pipe of length L and radius r is

$$v(x) = \frac{P}{4\nu L}(r^2 - x^2), \tag{16}$$

where ν is the viscosity of the fluid and P is the difference in pressure at the two ends of the pipe. Note that this formula gives $v = 0$ at the wall of the pipe, where $x = r$. Substituting this velocity function in Eq. (15) gives

$$\begin{aligned} F &= \int_0^r 2\pi x \cdot \frac{P}{4\nu L}(r^2 - x^2)\, dx \\ &= \frac{\pi P}{2\nu L}\int_0^r (r^2 x - x^3)\, dx = \frac{\pi P}{2\nu L}\left[\frac{r^2x^2}{2} - \frac{x^4}{4}\right]_{x=0}^{r}; \end{aligned}$$

$$F = \frac{\pi P}{2\nu L} \cdot \frac{r^4}{4} = \frac{\pi P r^4}{8\nu L}. \tag{17}$$

The formula in (17) is called *Poiseuille's law* for laminar fluid flow in a circular pipe. With a given ratio P/L of pressure difference per length, the flow rate F is proportional to the fourth power of the radius r of the pipe. For instance, because $\sqrt[4]{2} \approx 1.19$, a 20% increase in r more than doubles the flow rate F. ◆

Flow Rates and Cardiac Output

The determination of flow rates in pipes or streams has important applications ranging from engineering and environmental studies to medical procedures. A common technique involves the injection of a known amount A of a dye or other marker into the flow at time $t = 0$, followed by measurement at periodic intervals of the concentration of the dye by a probe at a fixed point downstream.

Suppose that all the dye has passed the measurement probe by time $t = T$. Subdivide the interval $[0, T]$ into n time intervals all of the same duration $\Delta t = T/n$. If the concentration $c(t)$ of the dye in the stream is measured at times $t_1, t_2, \dots, t_n$, then we can estimate the amount of dye that passes the probe during the subinterval of time $[t_{i-1}, t_i]$. If the (unknown) constant flow rate is F, then the volume ΔV_i of fluid that passes the probe during the subinterval of time is $\Delta V_i = F\, \Delta t$. It may help to think of typical units:

$$\left(F \frac{\text{cm}^3}{\text{second}}\right) \times (\Delta t \text{ seconds}) = \Delta V_i \text{ cm}^3.$$

The amount ΔA_i (in mg, for instance) of dye in this volume of fluid is given by

$$\text{amount (mg)} = \text{concentration} \left(\frac{\text{mg}}{\text{cm}^3}\right) \times \text{volume (cm}^3\text{)}.$$

If we use the measured concentration $c(t_i)$ as the approximate concentration throughout the time interval $[t_{i-1}, t_i]$, this gives

$$\Delta A_i \approx c(t_i)\,\Delta V_i = c(t_i) \cdot F\,\Delta t = F \cdot c(t_i)\,\Delta t.$$

Because the total amount A of injected dye passes the probe by time $t = T$, we add the individual amounts ΔA_i for $i = 1, 2, \dots, n$ and get

$$A = \sum_{i=1}^{n} \Delta A_i \approx F \cdot \sum_{i=1}^{n} c(t_i)\,\Delta t$$

(using the fact that the flow rate F is constant). Finally, we recognize the Riemann sum on the right and conclude that

$$A = F \int_0^T c(t)\,dt.$$

Thus the previously unknown flow rate is given by

$$F = \frac{A}{\displaystyle\int_0^T c(t)\,dt}, \qquad \textbf{(18)}$$

in terms of the known amount A of dye injected and the downstream concentration $c(t)$ that has been measured at times $t = t_1, t_2, \dots, t_n$. We can therefore estimate F by substituting in Eq. (18) the Riemann sum approximation

$$\int_0^T c(t)\,dt \approx \sum_{i=1}^{n} c(t_i)\,\Delta t$$

—or, alternatively, a Simpson approximation to the integral.

In medicine the term *cardiac output* is used for the flow rate of blood pumped through the aorta by the heart. A typical cardiac output for a 70-kg man would be in the range of 5 to 7 liters per minute (L/min). The dye marker is injected into the heart (or into a vein entering the heart), and then concentration readings are taken by a probe inserted into the aorta leaving the heart.

t	$c(t)$	t	$c(t)$
0	0	12	1.84
2	1.93	14	0.88
4	8.17	16	0.39
6	9.00	18	0.15
8	6.34	20	0
10	3.65		

FIGURE 6.1.12 Dye concentration data in Example 7.

EXAMPLE 7 The table in Fig. 6.1.12 lists concentration readings (in mg/L) taken with an aortic probe at 2-second intervals after 6 mg of dye was injected into the heart of a patient undergoing surgery. Approximate the cardiac output of the patient.

Solution Here we have $A = 6$, $T = 20$, and $\Delta t = 2$ with t in seconds. Simpson's approximation gives

$$\begin{aligned}\int_0^{20} c(t)\,dt &\approx \frac{2}{3}[0 + 4\cdot(1.93) + 2\cdot(8.17) + 4\cdot(9.00) + 2\cdot(6.34) + 4\cdot(3.65)\\ &\quad + 2\cdot(1.84) + 4\cdot(0.88) + 2\cdot(0.39) + 4\cdot(0.15) + 0]\\ &\approx 63.95.\end{aligned}$$

The formula in (18) therefore gives

$$F = \frac{A}{\displaystyle\int_0^{20} c(t)\,dt} \approx \frac{6}{63.95} \approx 0.0938 \quad \text{(L/s)},$$

approximately 5.63 L/min, for the patient's cardiac output. ◆

6.1 TRUE/FALSE STUDY GUIDE

6.1 CONCEPTS: QUESTIONS AND DISCUSSION

1. Describe in your own words—with a minimum of calculus textbook jargon and a maximum of personal understanding—the process of using approximating sums to set up an integral formula for use in calculating a specified quantity.
2. Carry out the process described in Question 1 to derive an integral formula corresponding to an application found in a science, engineering, economics, or medical textbook. (The derivation you see there may be somewhat cursory; do it carefully, as in this section.)

6.1 PROBLEMS

In Problems 1 through 10, $x_i^\star$ denotes a selected point, and m_i the midpoint, of the ith subinterval $[x_{i-1}, x_i]$ of a partition of the indicated interval $[a, b]$ into n subintervals each of length Δx. Evaluate the given limit by computing the value of the appropriate related integral.

1. $\lim_{n\to\infty} \sum_{i=1}^{n} 2x_i^\star \,\Delta x; \quad a = 0, b = 1$

2. $\lim_{n\to\infty} \sum_{i=1}^{n} \frac{\Delta x}{(x_i^\star)^2}; \quad a = 1, b = 2$

3. $\lim_{n\to\infty} \sum_{i=1}^{n} (\sin \pi x_i^\star)\,\Delta x; \quad a = 0, b = 1$

4. $\lim_{n\to\infty} \sum_{i=1}^{n} \left[3(x_i^\star)^2 - 1\right]\Delta x; \quad a = -1, b = 3$

5. $\lim_{n\to\infty} \sum_{i=1}^{n} x_i^\star \sqrt{(x_i^\star)^2 + 9}\,\Delta x; \quad a = 0, b = 4$

6. $\lim_{n\to\infty} \sum_{i=1}^{n} \frac{1}{x_i}\,\Delta x; \quad a = 2, b = 4$

7. $\lim_{n\to\infty} \sum_{i=1}^{n} e^{-m_i}\,\Delta x; \quad a = 0, b = 1$

8. $\lim_{n\to\infty} \sum_{i=1}^{n} \sqrt{2m_i + 1}\,\Delta x; \quad a = 0, b = 4$

9. $\lim_{n\to\infty} \sum_{i=1}^{n} \frac{m_i}{m_i^2 + 9}\,\Delta x; \quad a = 0, b = 6$

10. $\lim_{n\to\infty} \sum_{i=1}^{n} 2m_i e^{-m_i^2}\,\Delta x; \quad a = 0, b = 1$

The notation in Problems 11 through 14 is the same as in Problems 1 through 10. Express the given limit as an integral involving the function f.

11. $\lim_{n\to\infty} \sum_{i=1}^{n} 2\pi x_i^\star f(x_i^\star)\,\Delta x; \quad a = 1, b = 4$

12. $\lim_{n\to\infty} \sum_{i=1}^{n} [f(x_i^\star)]^2\,\Delta x; \quad a = -1, b = 1$

13. $\lim_{n\to\infty} \sum_{i=1}^{n} \sqrt{1 + [f(x_i^\star)]^2}\,\Delta x; \quad a = 0, b = 10$

14. $\lim_{n\to\infty} \sum_{i=1}^{n} 2\pi m_i \sqrt{1 + [f(m_i)]^2}\,\Delta x; \quad a = -2, b = 3$

In Problems 15 through 18, a rod coinciding with the interval $[a, b]$ on the x-axis (units in centimeters) has the specified density function $\rho(x)$ that gives its density (in grams per centimeter) at the point x. Find the mass M of the rod.

15. $a = 0, b = 100; \quad \rho(x) = \frac{1}{5}x$

16. $a = 0, b = 25; \quad \rho(x) = 60 - 2x$

17. $a = 0, b = 10; \quad \rho(x) = x(10 - x)$

18. $a = 0, b = 10; \quad \rho(x) = 10 \sin \frac{\pi x}{10}$

In Problems 19 through 30, compute both the net distance and the total distance traveled between time $t = a$ and time $t = b$ by a particle moving with the given velocity function $v = f(t)$ along a line.

19. $v = -32; \quad a = 0, b = 10$

20. $v = 2t + 10; \quad a = 1, b = 5$

21. $v = 4t - 25; \quad a = 0, b = 10$

22. $v = |2t - 5|; \quad a = 0, b = 5$

23. $v = 4t^3; \quad a = -2, b = 3$

24. $v = t - \frac{1}{t^2}; \quad a = 0.1, b = 1$

25. $v = \sin 2t; \quad a = 0, b = \frac{\pi}{2}$

26. $v = \cos 2t; \quad a = 0, b = \frac{\pi}{2}$

27. $v = \cos \pi t; \quad a = -1, b = 1$

28. $v = \sin t + \cos t; \quad a = 0, b = \pi$

29. $v = t^2 - 9t + 14; \quad a = 0, b = 10$

30. $v = t^3 - 8t^2 + 15t; \quad a = 0, b = 6$

In Problems 31 through 34, use a calculator or computer to approximate both the net distance and the total distance traveled by a particle with the given velocity function $v(t)$ during the indicated time interval $[a, b]$. Begin by graphing $v = v(t)$ to estimate the intervals where $v(t) > 0$ and where $v(t) < 0$. You may then integrate numerically if your calculator or computer has this facility.

31. $v(t) = t^3 - 7t + 4; \quad a = 0, b = 3$

32. $v(t) = t^3 - 5t^2 + 10; \quad a = 0, b = 5$

33. $v(t) = t \sin t - \cos t; \quad a = 0, b = \pi$

34. $v(t) = \sin t + \sqrt{t} \cos t; \quad a = 0, b = 2\pi$

35. Suppose that the circular disk of Fig. 6.1.10 has mass density $\rho(x)$ (in grams per square centimeter) at distance x from the origin. Then the annular ring of Figs. 6.1.10 and 6.1.11 has density approximately $\rho(x_i^\star)$ at each point. Conclude that the mass M of this disk of radius r is given by

$$M = \int_0^r 2\pi x \rho(x)\,dx.$$

In Problems 36 and 37, use the result of Problem 35 to find the mass of a circular disk with the given radius r and density function ρ.

36. $r = 10, \quad \rho(x) = x$

37. $r = 4, \quad \rho(x) = 25 - x^2$

38. If a particle is thrown straight upward from the ground with an initial velocity of 160 ft/s, then its velocity after t seconds is $v = -32t + 160$ feet per second, and it attains its maximum height when $t = 5$ s (and $v = 0$). Use Eq. (11) to compute this maximum height. Check your answer by the methods of Section 5.2.

39. Suppose that the rate of water flow into an initially empty tank is $100 - 3t$ gallons per minute at time t (in minutes). How much water flows into the tank during the interval from $t = 10$ to $t = 20$ min?

40. Suppose that the birth rate in Calgary t years after 1970 was $13 + t$ thousands of births per year. Set up and evaluate an appropriate integral to compute the total number of births that occurred between 1970 and 1990.

41. Assume that the city of Problem 40 had a death rate of $5 + \frac{1}{2}t$ thousands per year t years after 1970. If the population of the city was 125,000 in 1970, what was its population in 1990? Consider both births and deaths.

42. The average daily rainfall in Sioux City is $r(t)$ inches per day at time t (in days), $0 \leqq t \leqq 365$. Begin with a partition of the interval [0, 365] and derive the formula

$$R = \int_0^{365} r(t)\,dt$$

for the average total annual rainfall R.

43. Take the average daily rainfall of Problem 42 to be

$$r(t) = a - b\cos\frac{2\pi t}{365},$$

where a and b are constants to be determined. If the value of $r(t)$ on January 1 ($t = 0$) is 0.1 in. and the value of $r(t)$ on July 1 ($t = 182.5$) is 0.5 in., what is the average total annual rainfall in this locale?

44. Suppose that the rate of water flow into a tank is $r(t)$ liters per minute at time t (in minutes). Use the method of Example 1 to derive the formula

$$Q = \int_a^b r(t)\,dt$$

for the amount of water that flows into the tank between times $t = a$ and $t = b$.

45. Evaluate

$$\lim_{n\to\infty} \frac{\sqrt[3]{1} + \sqrt[3]{2} + \sqrt[3]{3} + \cdots + \sqrt[3]{n}}{n^{4/3}}$$

by first finding a function f such that the limit is equal to

$$\int_0^1 f(x)\,dx.$$

46. In this problem you are to derive the volume formula $V = \frac{4}{3}\pi r^3$ for a spherical ball of radius r, assuming as *known* the formula $S = 4\pi r^2$ for the surface area of a sphere of radius r. Assume it follows that the volume of a thin spherical shell of radius r and thickness t (Fig. 6.1.13) is given approximately by $\Delta V \approx S \cdot t = 4\pi r^2 t$. Then divide the spherical ball into concentric spherical shells, analogous to the concentric annular rings of Fig. 6.1.10. Finally, interpret the sum of the volumes of these spherical shells as a Riemann sum.

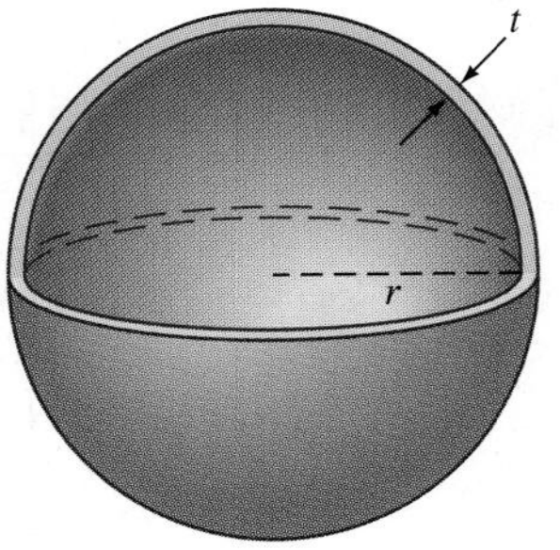

FIGURE 6.1.13 A thin spherical shell of thickness t and inner radius r (Problem 46).

47. A spherical ball has radius 1 ft and, at distance x from its center, its density is $100(1 + x)$ lb/ft^3. Use Riemann sums to find a function $f(x)$ such that the weight of the ball is

$$W = \int_0^1 f(x)\,dx$$

(in pounds). Then compute W by evaluating this integral. [*Suggestion:* Given a partition $0 = x_0 < x_1 < x_2 < \cdots < x_n = 1$ of [0, 1], estimate the weight ΔW_i of the spherical shell $x_{i-1} \leqq x \leqq x_i$ of the ball.]

48. Find the flow rate F in a circular pipe of radius r if the velocity of the fluid at distance x from the center of the pipe is given by

$$v(x) = k \cos \frac{\pi x}{2r}.$$

You may use the formula $\int u \cos u \, du = u \sin u + \cos u + C$.

49. Poiseuille discovered his law of fluid flow in the course of investigating the flow of blood in veins and arteries in the human body. With a given fixed flow rate F through a blood vessel of specified length L, Poiseuille's law in Eq. (17) shows that a decrease in the radius r requires an increase in the blood pressure P. This is why hypertension—high blood pressure—frequently results from constriction of arteries. Construct a table in which the first column shows the percentage decrease in the radius of an artery (in increments of 5% from 0% to 25%) and the second column shows the resulting percentage increase in blood pressure.

50. Find the cardiac output (in L/min) if an injection of 4 mg of dye into a patient's heart results in the aortic concentration function $c(t) = 40te^{-t}$ for $0 \leqq t \leqq 10$ (seconds). You may use the formula

$$\int ue^u \, du = (u-1)e^u + C.$$

51. The table in Fig. 6.1.14 lists concentration readings (in mg/L) taken with an aortic probe at 1-second intervals after 4.5 mg of dye was injected into the heart of a patient undergoing surgery. Approximate the patient's cardiac output.

t	$c(t)$	t	$c(t)$
0	0	6	2.21
1	2.32	7	1.06
2	9.80	8	0.47
3	10.80	9	0.18
4	7.61	10	0
5	4.38		

FIGURE 6.1.14 Dye concentration data for Problem 51.

52. Figure 6.1.15 shows the graph of the concentration function $c(t)$ recorded by an aortic probe connected to a plotter after 5.5 mg of dye was injected into the heart of a patient undergoing surgery. Approximate the cardiac output of the patient.

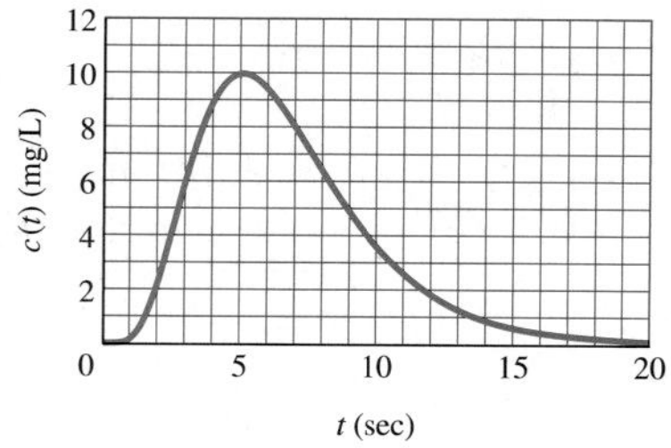

FIGURE 6.1.15 Graph of concentration $c(t)$ in Problem 52.

6.2 VOLUMES BY THE METHOD OF CROSS SECTIONS

Here we use integrals to calculate the volumes of certain solids or regions in space. We begin with an intuitive idea of volume as a measure of solids, analogous to area as a measure of plane regions. In particular, we assume that every simply expressible bounded solid region R has a volume measured by a nonnegative number $v(R)$ such that

- If R consists of two nonoverlapping pieces, then $v(R)$ is the sum of *their* volumes;
- Two different solids have the same volume if they have the same size and shape.

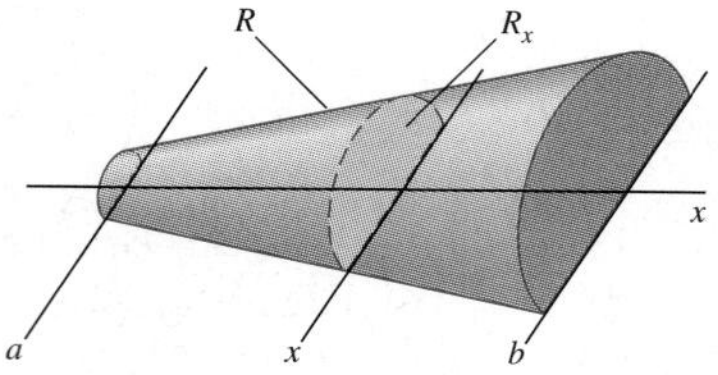

FIGURE 6.2.1 R_x is the cross section of R in the plane perpendicular to the x-axis at x.

The **method of cross sections** is a way of computing the volume of a solid that is described in terms of its cross sections (or "slices") in planes perpendicular to a fixed *reference line* (such as the x-axis or y-axis). For instance, Fig. 6.2.1 shows a solid R with volume $V = v(R)$ lying alongside the interval $[a, b]$ on the x-axis. That is, a plane perpendicular to the x-axis intersects the solid if and only if this plane meets the x-axis in a point of $[a, b]$. Let R_x denote the intersection of R with the perpendicular plane that meets the x-axis at the point x of $[a, b]$. We call R_x the (plane) **cross section** of the solid at x.

Volumes of Cylinders

This situation is especially simple if all the cross sections of R are congruent to one another and are parallel translations of each other. In this case the solid R is called a **cylinder** with **bases** R_a and R_b and **height** $h = b - a$. If R_a and R_b are circular disks, then R is the familiar **circular cylinder.** Recall that the volume formula for a circular

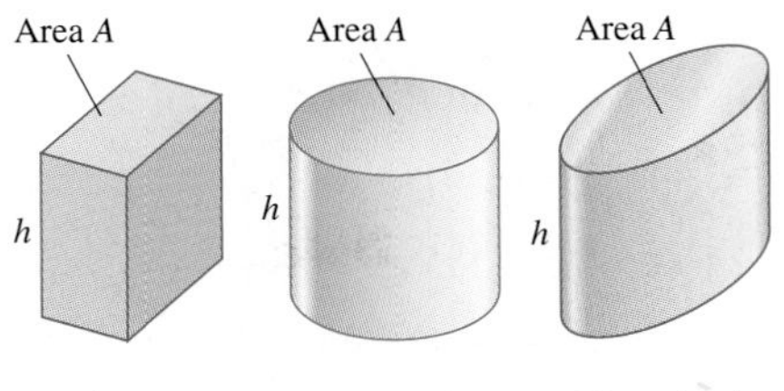

FIGURE 6.2.2 Every cylinder of height h and base area A has volume $V = Ah$.

cylinder of height h and circular base of radius r and area $A = \pi r^2$ is

$$V = \pi r^2 h = Ah.$$

Figure 6.2.2 shows several (general) cylinders with bases of various shapes. The method of cross sections is based on the fact that the volume V of any cylinder—circular or not—is equal to the product of the cylinder's height h and the area A of its base:

$$V = Ah \quad \text{(volume of a cylinder).} \tag{1}$$

More General Volumes

The volume of a more general solid, as in Fig. 6.2.1, can be approximated by using cylinders. For each x in $[a, b]$, let $A(x)$ denote the area of the cross section R_x of the solid R:

$$A(x) = \text{area}(R_x). \tag{2}$$

We shall assume that the shape of R is sufficiently simple that this **cross-sectional area function** A is continuous (and therefore integrable).

To set up an integral formula for $V = v(R)$, we begin with a partition of $[a, b]$ into n subintervals, all with the same length $\Delta x = (b - a)/n$. Let R_i denote the slab or slice of the solid R positioned alongside the ith subinterval $[x_{i-1}, x_i]$ (Fig. 6.2.3). We denote the volume of this ith slice of R by $\Delta V_i = v(R_i)$, so

$$V = \sum_{i=1}^{n} \Delta V_i.$$

To approximate ΔV_i, we select a typical point $x_i^\star$ in $[x_{i-1}, x_i]$ and consider the *cylinder* C_i whose height is Δx and whose base is the cross section $R_{x_i^\star}$ of R at $x_i^\star$. Figure 6.2.4 suggests that if Δx is small, then $v(C_i)$ is a good approximation to $\Delta V_i = v(R_i)$:

$$\Delta V_i \approx v(C_i) = \text{area}(R_{x_i^\star}) \cdot \Delta x = A(x_i^\star)\,\Delta x,$$

a consequence of Eq. (1) with $A = A(x_i^\star)$ and $h = \Delta x$.

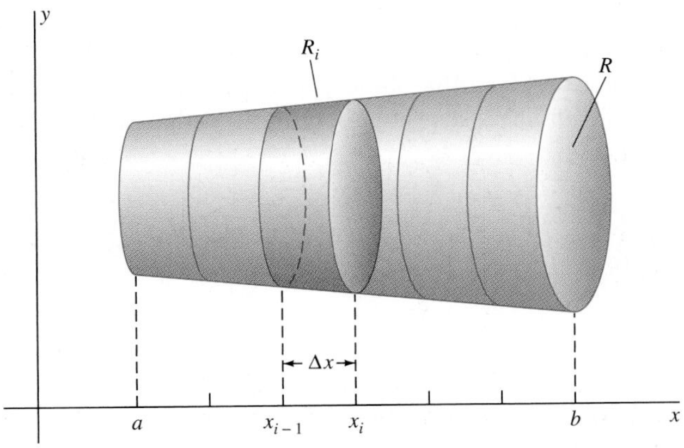

FIGURE 6.2.3 Planes through the partition points x_0, x_1, x_2, ..., x_n partition the solid R into slabs R_1, R_2, ..., R_n.

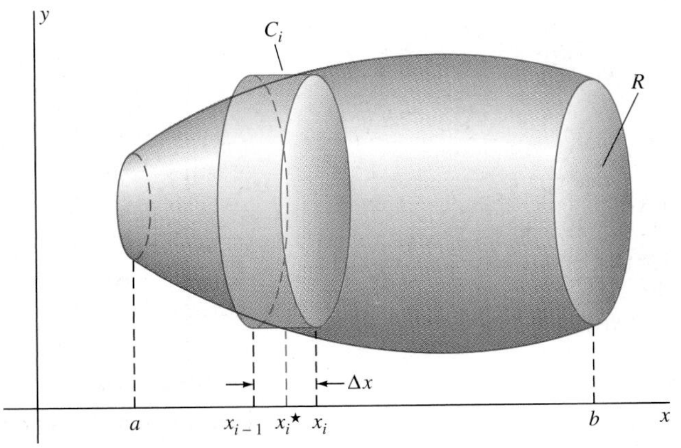

FIGURE 6.2.4 The slab R_i is approximated by the cylinder C_i of volume $A(x_i^\star)\,\Delta x$.

Then we add the volumes of these approximating cylinders for $i = 1, 2, 3, \ldots, n$. We find that

$$V = \sum_{i=1}^{n} \Delta V_i \approx \sum_{i=1}^{n} A(x_i^{\star})\,\Delta x.$$

We recognize the approximating sum on the right to be a Riemann sum that approaches $\int_a^b A(x)\,dx$ as $n \to +\infty$. This justifies the following definition of the volume of a solid R in terms of its cross-sectional area function $A(x)$.

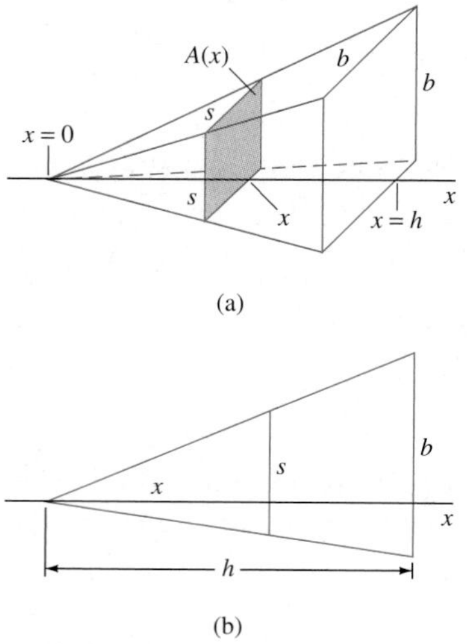

FIGURE 6.2.5 The square-based pyramid of Example 1.

DEFINITION Volumes by Cross Sections
If the solid R lies alongside the interval $[a, b]$ on the x-axis and has continuous cross-sectional area function $A(x)$, then its volume $V = v(R)$ is

$$V = \int_a^b A(x)\,dx. \tag{3}$$

Equation (3) is known as **Cavalieri's principle,** after the Italian mathematician Bonaventura Cavalieri (1598–1647), who systematically exploited the fact that the volume of a solid is determined by the areas of its cross sections perpendicular to a given reference line.

EXAMPLE 1 Figure 6.2.5(a) shows a square-based pyramid oriented so that its height h corresponds to the interval $[0, h]$ on the x-axis. Its *base* is a b-by-b square, and each cross section perpendicular to the x-axis is also a square. To find the area $A(x)$ of the s-by-s cross section at x, we equate height-to-length ratios in the similar triangles of Fig. 6.2.5(b):

$$\frac{s}{x} = \frac{b}{h}, \quad \text{so} \quad s = \frac{b}{h}x.$$

Therefore,

$$A(x) = s^2 = \frac{b^2}{h^2}x^2,$$

and Eq. (3)—with $[0, h]$ as the interval of integration—gives

$$V = \int_0^h A(x)\,dx = \int_0^h \frac{b^2}{h^2}x^2\,dx = \left[\frac{b^2}{h^2}\cdot\frac{x^3}{3}\right]_{x=0}^{x=h} = \frac{1}{3}b^2h.$$

With $A = b^2$ denoting the area of the base, our result takes the form

$$V = \tfrac{1}{3}Ah$$

for the volume of a pyramid. ◆

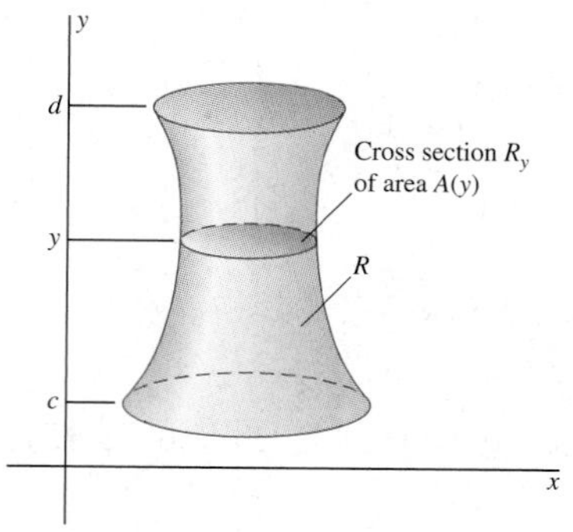

FIGURE 6.2.6 $A(y)$ is the area of the cross section R_y in the plane perpendicular to the y-axis at the point y.

Cross Sections Perpendicular to the *y*-Axis

In the case of a solid R lying alongside the interval $[c, d]$ on the y-axis, we denote by $A(y)$ the area of the solid's cross section R_y in the plane perpendicular to the y-axis at the point y of $[c, d]$ (Fig. 6.2.6). A similar discussion, beginning with a partition of $[c, d]$, leads to the volume formula

$$V = \int_c^d A(y)\,dy. \tag{4}$$

Solids of Revolution

An important special case of Eq. (3) gives the volume of a **solid of revolution.** For example, consider the solid R obtained by revolving around the x-axis the region under the graph of $y = f(x)$ over the interval $[a, b]$, where $f(x) \geqq 0$. Such a region and the resulting solid of revolution are shown in Fig. 6.2.7.

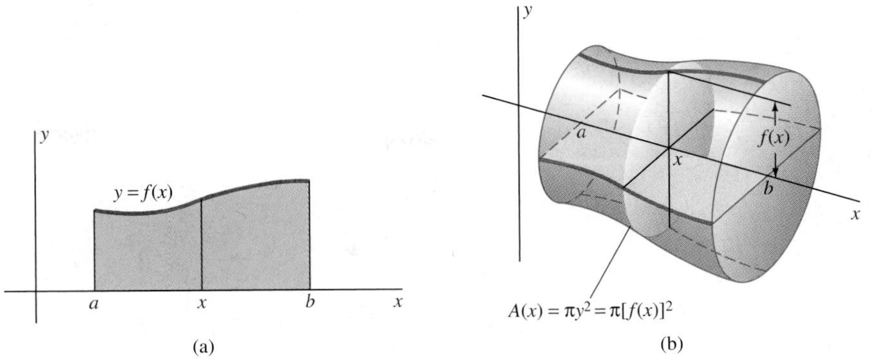

FIGURE 6.2.7 (a) A region from which we can determine the volume of a (b) solid of revolution around the x-axis.

Because the solid R is obtained by revolution, each cross section of R at x is a circular *disk* of radius $f(x)$. The cross-sectional area function is then $A(x) = \pi y^2 = \pi[f(x)]^2$, so Eq. (3) yields

$$V = \int_a^b \pi y^2\,dx = \int_a^b \pi[f(x)]^2\,dx \qquad \textbf{(5)}$$

for the **volume of a solid of revolution around the x-axis.**

Note In the expression $\pi y^2\,dx$, the differential dx tells us that the independent variable is x. We *must* express the dependent variable y in terms of x in order to perform the indicated integration.

EXAMPLE 2 Figure 6.2.8 shows the region that lies below the parabola $y^2 = x$ and above the x-axis over the interval $[0, 2]$. Find the volume V of the solid paraboloid (Fig. 6.2.9) obtained by revolving this region around the x-axis.

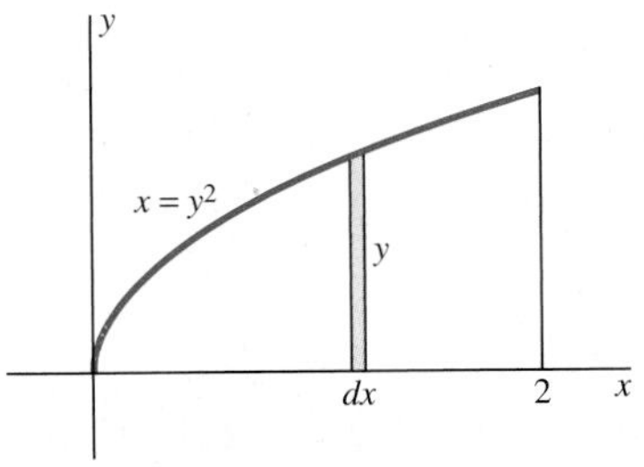

FIGURE 6.2.8 The parabolic region of Example 2.

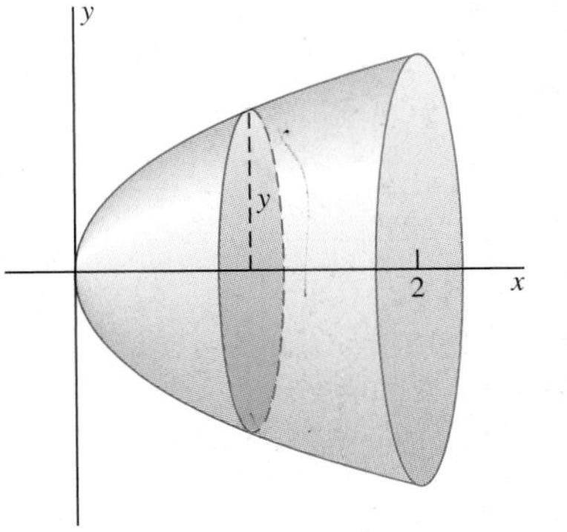

FIGURE 6.2.9 The solid paraboloid of Example 2.

Solution Because $y^2 = x$ on the parabola, the cross-sectional area function in Eq. (5) is given in terms of x by

$$A(x) = \pi y^2 = \pi x.$$

Hence integration immediately gives

$$V = \int_0^2 \pi x\,dx = \left[\tfrac{1}{2}\pi x^2\right]_0^2 = 2\pi. \qquad \blacklozenge$$

EXAMPLE 3 Use the method of cross sections to verify the familiar formula $V = \frac{4}{3}\pi R^3$ for the volume of a sphere of radius R.

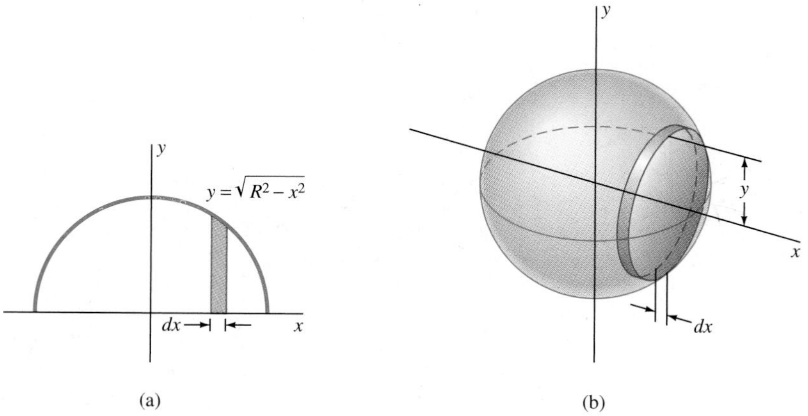

FIGURE 6.2.10 (a) A semicircular region that we rotate (b) to generate a sphere (Example 3).

Solution We think of the sphere as the solid of revolution obtained by revolving the semicircular plane region in Fig. 6.2.10 around the x-axis. This is the region bounded above by the semicircle

$$y = \sqrt{R^2 - x^2}, \quad -R \leqq x \leqq R$$

and below by the interval $[-R, R]$ on the x-axis. To use Eq. (5), we take

$$f(x) = \sqrt{R^2 - x^2}, \qquad a = -R, \quad \text{and} \quad b = R.$$

This gives

$$V = \int_{-R}^{R} \pi\left(\sqrt{R^2 - x^2}\right)^2 dx = \pi \int_{-R}^{R} (R^2 - x^2)\, dx$$

$$= \pi \left[R^2 x - \tfrac{1}{3}x^3\right]_{-R}^{R} = \tfrac{4}{3}\pi R^3. \qquad \blacklozenge$$

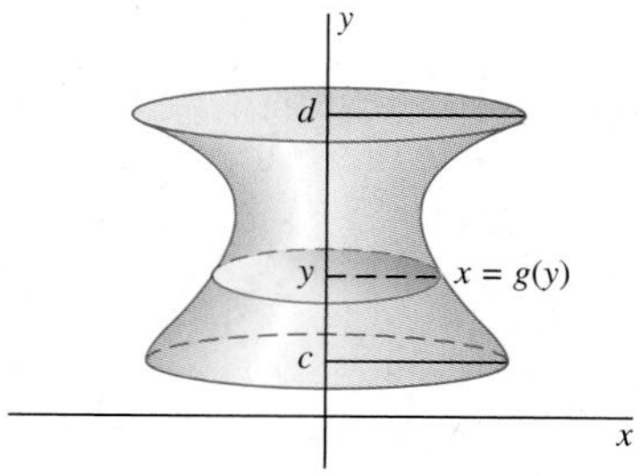

FIGURE 6.2.11 A region lying between the y-axis and the curve $x = g(y)$, $c \leqq y \leqq d$, is rotated around the y-axis.

Revolution Around the y-Axis

Figure 6.2.11 shows a solid of revolution around the y-axis. The region being revolved is bounded by the y-axis and the curve $x = g(y)$, $c \leqq y \leqq d$ (as well as the lines $y = c$ and $y = d$). In this case the *horizontal* circular cross section has radius x, and thus the cross-sectional area at y is πx^2, where $x = g(y)$. Hence the cross-sectional area function is $A(y) = \pi[g(y)]^2$. We therefore obtain the formula

$$V = \int_c^d \pi x^2\, dy = \int_c^d \pi [g(y)]^2\, dy \tag{6}$$

(contrast it with (5)) for the **volume of a solid of revolution around the y-axis.**

Note In the expression $\pi x^2\, dy$, the differential dy tells us that the independent variable is y. So here we must express the dependent variable x in terms of y before integrating.

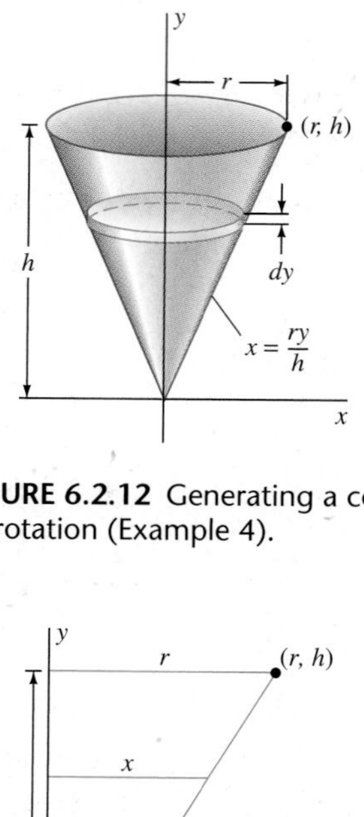

FIGURE 6.2.12 Generating a cone by rotation (Example 4).

EXAMPLE 4 Use the method of cross sections to verify the familiar formula $V = \frac{1}{3}\pi r^2 h$ for the volume of a right circular cone with base radius r and height h.

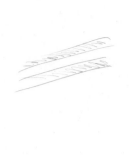

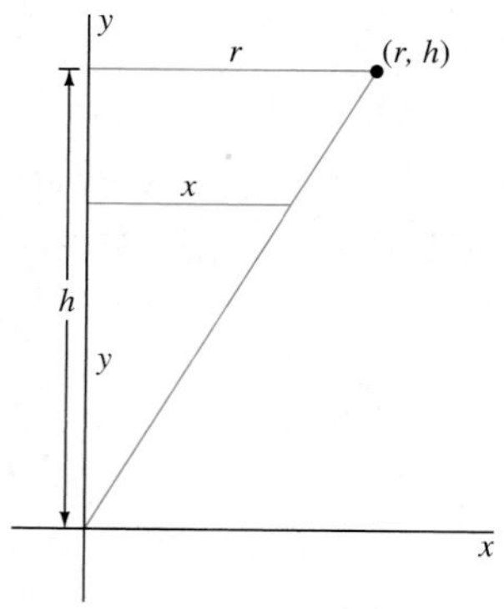

FIGURE 6.2.13 Finding the radius x of the circular cross section (Example 4).

Solution Figure 6.2.12 depicts the cone as the solid of revolution obtained by revolving around the y-axis the triangle with vertices $(0, 0)$, $(0, h)$, and (r, h). The similar triangles in Fig. 6.2.13 yield the equation $x/y = r/h$, so the radius of the circular cross section perpendicular to the y-axis at the point y is $x = ry/h$. Then Eq. (6), with $g(y) = ry/h$, gives

$$V = \int_a^b A(y)\,dy = \int_a^b \pi x^2\,dy = \int_0^h \pi \left(\frac{ry}{h}\right)^2 dy$$

$$= \frac{\pi r^2}{h^2} \int_0^h y^2\,dy = \frac{1}{3}\pi r^2 h = \frac{1}{3}Ah,$$

where $A = \pi r^2$ is the area of the base of the cone. ◆

Revolving the Region Between Two Curves

Sometimes we need to calculate the volume of a solid generated by revolving a plane region that lies between two given curves. Suppose that $f(x) \geqq g(x) \geqq 0$ for x in the interval $[a, b]$ and that the solid R is generated by revolving around the x-axis the region between $y = f(x)$ and $y = g(x)$. Then the cross section at x is an **annular ring** (or **washer**) bounded by two circles (Fig. 6.2.14). The ring has inner radius $r_{\text{in}} = g(x)$ and outer radius $r_{\text{out}} = f(x)$, so the formula for the cross-sectional area at x is

$$A(x) = \pi(r_{\text{out}})^2 - \pi(r_{\text{in}})^2 = \pi[(y_{\text{top}})^2 - (y_{\text{bot}})^2] = \pi\{[f(x)]^2 - [g(x)]^2\},$$

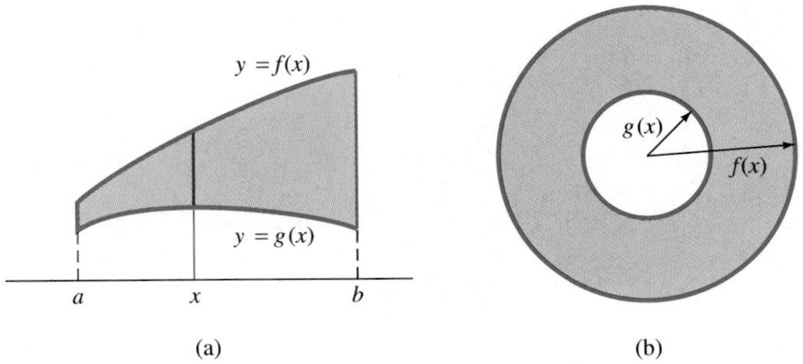

FIGURE 6.2.14 (a) The region between two positive graphs (b) is rotated around the x-axis. Cross sections are annular rings.

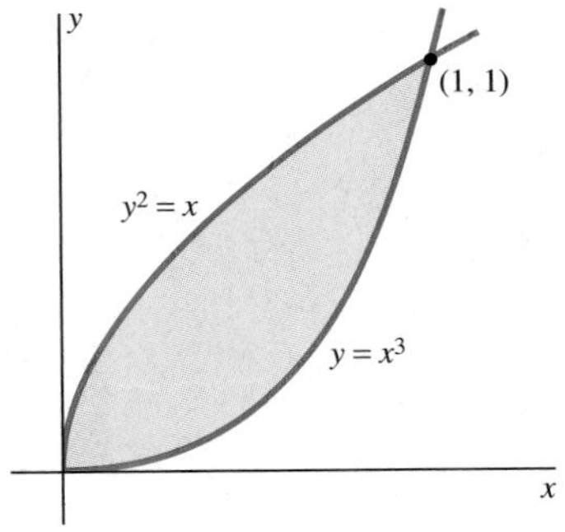

FIGURE 6.2.15 The plane region of Example 5.

where we write $y_{\text{top}} = f(x)$ and $y_{\text{bot}} = g(x)$ for the top and bottom curves of the plane region. Therefore, Eq. (3) yields

$$V = \int_a^b \pi[(y_{\text{top}})^2 - (y_{\text{bot}})^2]\,dx = \int_a^b \pi\{[f(x)]^2 - [g(x)]^2\}\,dx \qquad \textbf{(7)}$$

for the volume V of the solid.

Similarly, if $f(y) \geqq g(y) \geqq 0$ for $c \leqq y \leqq d$, then the volume of the solid obtained by revolving around the y-axis the region between $x_{\text{right}} = f(y)$ and $x_{\text{left}} = g(y)$ is

$$V = \int_c^d \pi[(x_{\text{right}})^2 - (x_{\text{left}})^2]\,dy = \int_c^d \pi\{[f(y)]^2 - [g(y)]^2\}\,dy. \qquad \textbf{(8)}$$

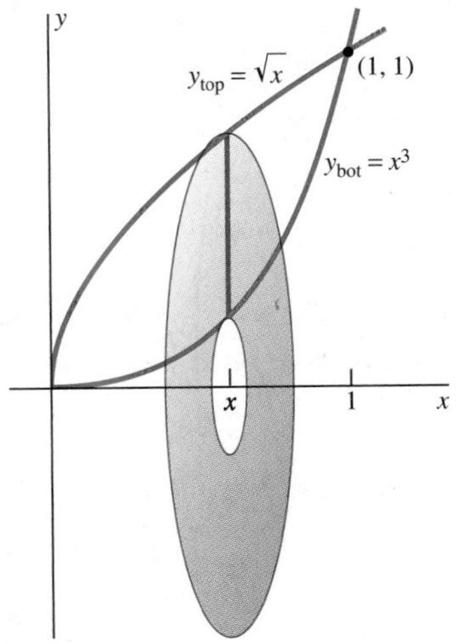

FIGURE 6.2.16 Revolution around the x-axis (Example 5).

EXAMPLE 5 Consider the plane region shown in Fig. 6.2.15, bounded by the curves $y^2 = x$ and $y = x^3$, which intersect at the points $(0, 0)$ and $(1, 1)$. If this region is revolved around the x-axis (Fig. 6.2.16), then Eq. (7) with

$$y_{\text{top}} = \sqrt{x}, \qquad y_{\text{bot}} = x^3$$

gives

$$V = \int_0^1 \pi\left[\left(\sqrt{x}\right)^2 - (x^3)^2\right] dx = \int_0^1 \pi(x - x^6)\,dx$$

$$= \pi\left[\tfrac{1}{2}x^2 - \tfrac{1}{7}x^7\right]_0^1 = \tfrac{5}{14}\pi$$

for the volume of revolution.

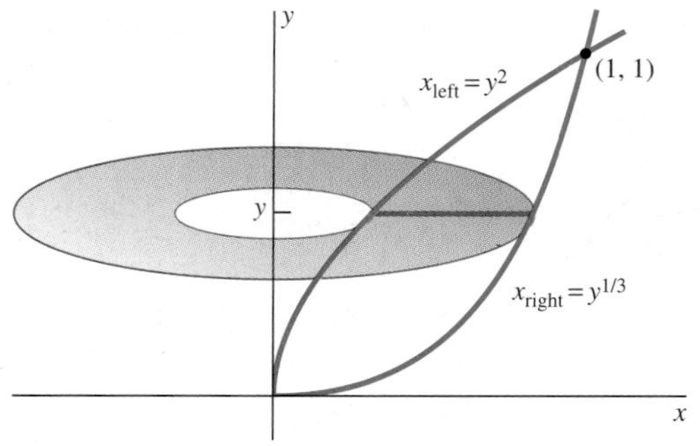

FIGURE 6.2.17 Revolution around the y-axis (Example 5).

If the same region is revolved around the y-axis (Fig. 6.2.17), then each cross section perpendicular to the y-axis is an annular ring with outer radius $x_{\text{right}} = y^{1/3}$ and inner radius $x_{\text{left}} = y^2$. Hence Eq. (8) gives the volume of revolution generated by this region as

$$V = \int_0^1 \pi \left[\left(y^{1/3}\right)^2 - (y^2)^2 \right] dy = \int_0^1 \pi \left(y^{2/3} - y^4\right) dy$$

$$= \pi \left[\tfrac{3}{5} y^{5/3} - \tfrac{1}{5} y^5 \right]_0^1 = \tfrac{2}{5}\pi.$$ ◆

EXAMPLE 6 Suppose that the plane region of Example 5 (Fig. 6.2.15) is revolved around the vertical line $x = -1$ (Fig. 6.2.18). Then each cross section of the resulting solid is an annular ring with outer radius

$$r_{\text{out}} = 1 + x_{\text{right}} = 1 + y^{1/3}$$

and inner radius

$$r_{\text{in}} = 1 + x_{\text{left}} = 1 + y^2.$$

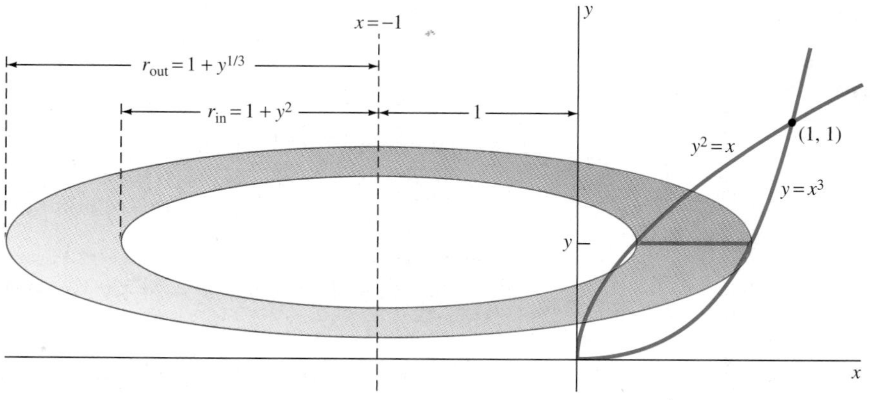

FIGURE 6.2.18 The annular ring of Example 6.

The area of such a cross section is

$$A(y) = \pi \left(1 + y^{1/3}\right)^2 - \pi (1 + y^2)^2 = \pi \left(2y^{1/3} + y^{2/3} - 2y^2 - y^4\right),$$

so the volume of the resulting solid of revolution is

$$V = \int_0^1 \pi \left(2y^{1/3} + y^{2/3} - 2y^2 - y^4\right) dy$$

$$= \pi \left[\tfrac{3}{2} y^{4/3} + \tfrac{3}{5} y^{5/3} - \tfrac{2}{3} y^3 - \tfrac{1}{5} y^5 \right]_0^1 = \tfrac{37}{30}\pi.$$ ◆

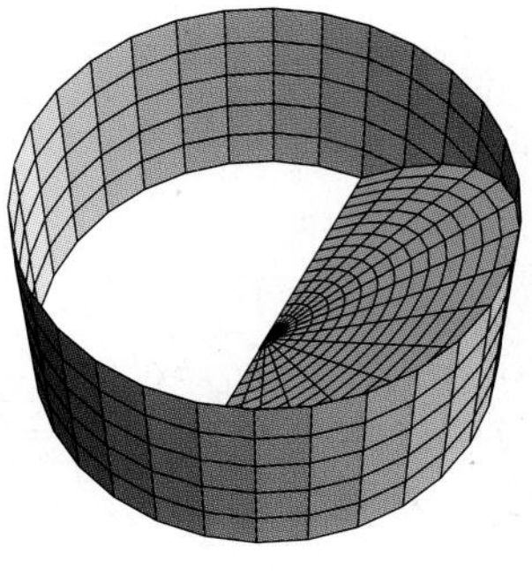

FIGURE 6.2.19 The wedge and cylinder of Example 7.

EXAMPLE 7 Find the volume of the wedge that is cut from a circular cylinder with unit radius and unit height by a plane that passes through a diameter of the base of the cylinder and through a point on the circumference of its top.

Solution The cylinder and wedge are shown in Fig. 6.2.19. To form such a wedge, fill a cylindrical glass with cider and then drink slowly, tipping the bottom up as you drink, until half the bottom of the glass is exposed; the remaining cider forms the wedge.

We choose as reference line and x-axis the line through the "edge of the wedge"—the original diameter of the base of the cylinder. We can verify with similar triangles that each cross section of the wedge perpendicular to the diameter is an isosceles right triangle. One of these triangles is shown in Fig. 6.2.20. We denote by y the equal base and height of this triangle.

To determine the cross-sectional area function $A(x)$, we must express y in terms of x. Figure 6.2.21 shows the unit circular base of the original cylinder. We apply the Pythagorean theorem to the right triangle in this figure and find that $y = \sqrt{1 - x^2}$. Hence

$$A(x) = \tfrac{1}{2}y^2 = \tfrac{1}{2}(1 - x^2),$$

so Eq. (3) gives

$$V = \int_{-1}^{1} A(x)\,dx = 2\int_{0}^{1} A(x)\,dx \qquad \text{(by symmetry)}$$

$$= 2\int_{0}^{1} \tfrac{1}{2}(1 - x^2)\,dx = \Big[x - \tfrac{1}{3}x^3\Big]_0^1 = \tfrac{2}{3}$$

for the volume of the wedge. ◆

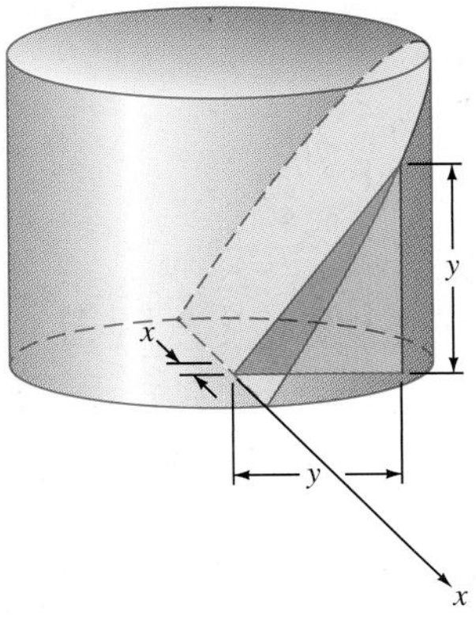

FIGURE 6.2.20 A cross section of the wedge—an isosceles triangle (Example 7).

COMMENT It is a useful habit to check answers for plausibility whenever convenient. For example, we may compare a given solid with one whose volume is known. Because the volume of the original cylinder in Example 7 is π, we have found that the wedge occupies the fraction

$$\frac{V_{\text{wedge}}}{V_{\text{cyl}}} = \frac{\frac{2}{3}}{\pi} \approx 21\%$$

of the volume of the cylinder. A glance at Fig. 6.2.19 indicates that this is plausible. An error in our computations could well have given an unbelievable answer.

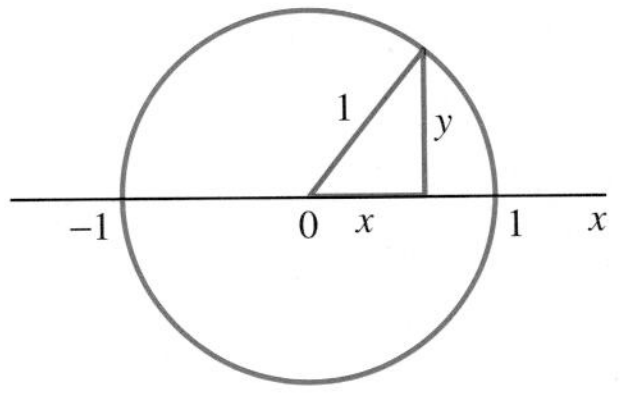

FIGURE 6.2.21 The base of the cylinder of Example 7.

HISTORICAL NOTE The wedge of Example 7 has an ancient history. Its volume was first calculated in the third century B.C. by Archimedes, who also derived the formula $V = \frac{4}{3}\pi r^3$ for the volume of a sphere of radius r. His work on the wedge is found in a manuscript that was discovered in 1906 after having been lost for centuries. Archimedes used a method of exhaustion for volume similar to that discussed for areas in Section 5.3. For more information, see pp. 73–74 of C. H. Edwards, *The Historical Development of the Calculus* (New York: Springer-Verlag, 1979).

6.2 TRUE/FALSE STUDY GUIDE

6.2 CONCEPTS: QUESTIONS AND DISCUSSION

1. Suppose that the cross sections of a solid perpendicular to two nonparallel axes are all circular disks. Is this solid necessarily a solid sphere?
2. Give your own example of a solid whose volume can be calculated by two essentially different integrals—stemming from cross sections perpendicular to nonparallel axes. Show that both integrals give the same volume for this solid.
3. Question 2 points to the need for a definition of volume that is not based on cross sections perpendicular to a particular axis. Formulate a possible definition based on collections of nonoverlapping rectangular blocks containing and contained by a solid. (Consult the definition of area given in the concepts discussion at the end of Section 5.8.)

6.2 PROBLEMS

In Problems 1 through 24, find the volume of the solid that is generated by rotating around the indicated axis the plane region bounded by the given curves.

1. $y = x^2$, $y = 0$, $x = 1$; the x-axis

2. $y = \sqrt{x}$, $y = 0$, $x = 4$; the x-axis

3. $y = x^2$, $y = 4$, $x = 0$ (first quadrant only); the y-axis (Fig. 6.2.22)

4. $y = 1/x$, $y = 0$, $x = 0.1$, $x = 1$; the x-axis (Fig. 6.2.23)

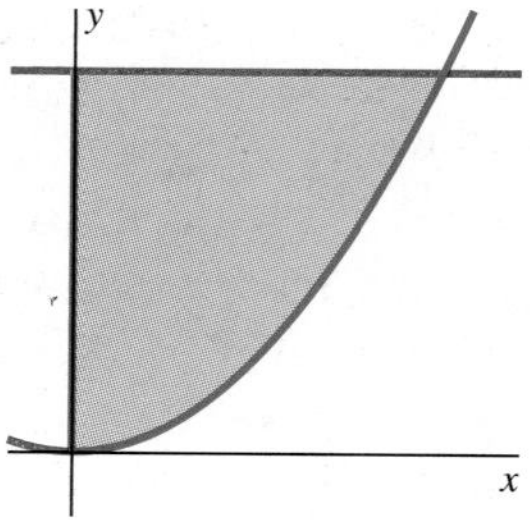

FIGURE 6.2.22 Problem 3.

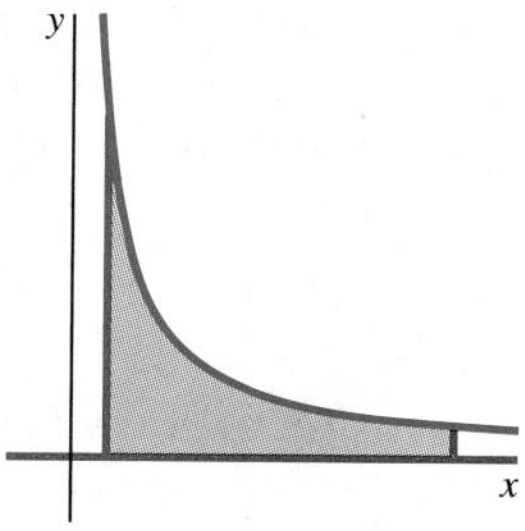

FIGURE 6.2.23 Problem 4.

5. $y = \sin x$ on $[0, \pi]$, $y = 0$; the x-axis

6. $y = 9 - x^2$, $y = 0$; the x-axis

7. $y = x^2$, $x = y^2$; the x-axis (Fig. 6.2.24)

8. $y = x^2$, $y = 4x$; the line $x = 5$ (Fig. 6.2.25)

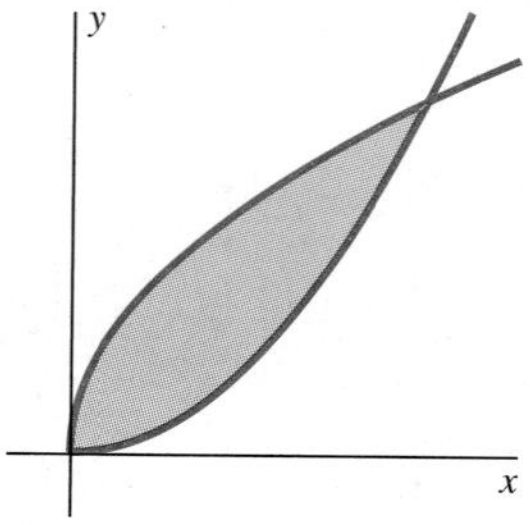

FIGURE 6.2.24 Problem 7.

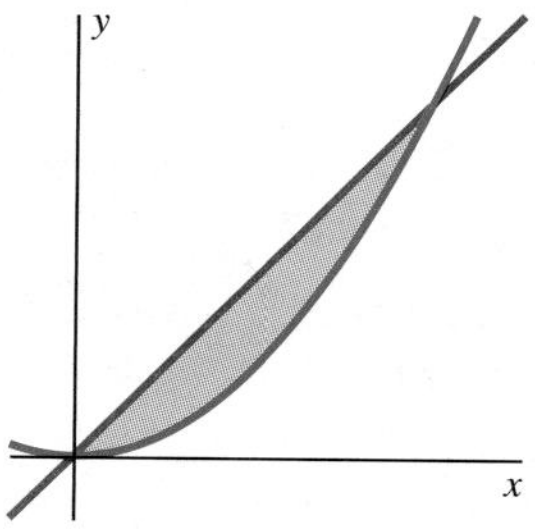

FIGURE 6.2.25 Problem 8.

9. $y = \dfrac{1}{\sqrt{x}}$, $y = 0$, $x = 1$, $x = 5$; the x-axis

10. $x = y^2$, $x = y + 6$; the y-axis

11. $y = 1 - x^2$, $y = 0$; the x-axis (Fig. 6.2.26)

12. $y = x - x^3$, $y = 0$ $(0 \leqq x \leqq 1)$; the x-axis (Fig. 6.2.27)

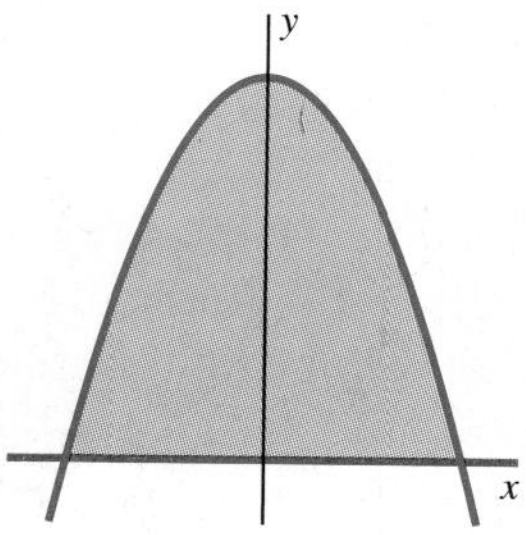

FIGURE 6.2.26 Problem 11.

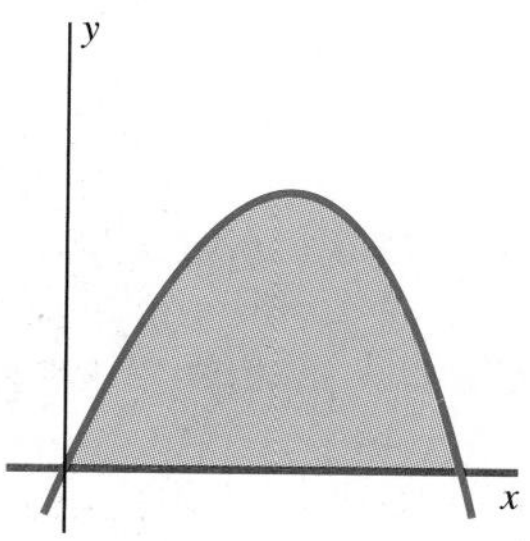

FIGURE 6.2.27 Problem 12.

13. $y = 1 - x^2$, $y = 0$; the y-axis

14. $y = e^x$, $y = 0$, $x = 0$, $x = 1$; the x-axis

15. $y = 6 - x^2$, $y = 2$; the y-axis (Fig. 6.2.28)

16. $y = 1 - x^2$, $y = 0$; the vertical line $x = 2$

17. $y = x - x^3$, $y = 0$ $(0 \leqq x \leqq 1)$; the horizontal line $y = -1$

18. $y = e^x$, $y = e^{-x}$, $x = 1$; the x-axis

19. $y = 4$, $x = 0$, $y = x^2$; the y-axis

20. $x = 16 - y^2$, $x = 0$, $y = 0$ (first quadrant only); the x-axis (Fig. 6.2.29)

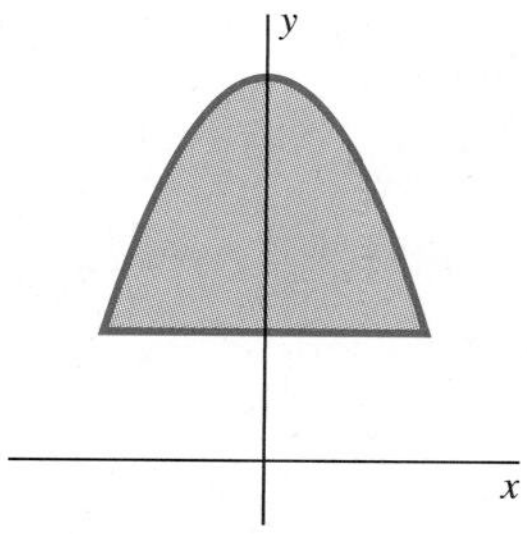

FIGURE 6.2.28 Problem 15.

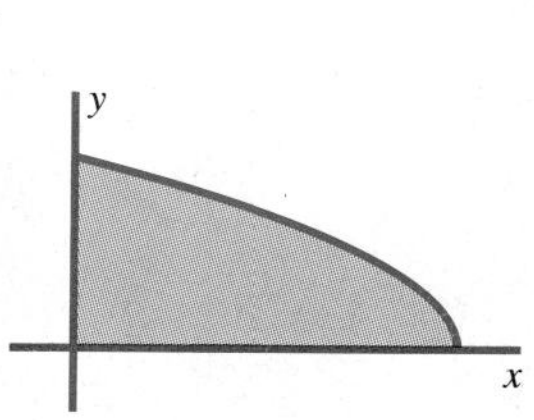

FIGURE 6.2.29 Problem 20.

21. $y = x^2$, $x = y^2$; the line $y = -2$

22. $y = x^2$, $y = 8 - x^2$; the line $y = -1$

23. $y = x^2$, $x = y^2$; the line $x = 3$

24. $y = e^{-x}$, $y = 2$, $x = 1$; the line $y = -1$

In Problems 25 through 30, find the volume obtained by revolving the region R around the x-axis. You may use the trigonometric identities

$$\cos^2 x = \frac{1 + \cos 2x}{2} \quad \text{and} \quad \sin^2 x = \frac{1 - \cos 2x}{2}$$

to help you evaluate some of the integrals.

25. R is the region between the graph $y = \sin x$ and the x-axis for $0 \leqq x \leqq \pi$.

26. R is the region between the graph $y = \cos(\frac{1}{2}\pi x)$ and the x-axis for $-1 \leqq x \leqq 1$.

27. R is the region between the curves $y = \sin x$ and $y = \cos x$ for $0 \leqq x \leqq \pi/4$.

28. R is the region between $x = -\pi/3$ and $x = \pi/3$ that is bounded by the curves $y = \cos x$ and $y = 1/2$.

29. R is bounded by the curve $y = \tan x$ and the lines $y = 0$ and $x = \pi/4$.

30. R is bounded by the curve $y = \tan x$ and the lines $x = 0$ and $y = 1$.

In Problems 31 through 34, first use a calculator or computer to approximate (graphically or otherwise) the points of intersection of the two given curves. Let R be the region bounded by these curves. Integrate to approximate the volume of the solid obtained by revolving the region R around the x-axis.

31. $y = x^3 + 1$, $y = 3x^2$

32. $y = x^4$, $y = x + 4$

33. $y = x^2, \quad y = \cos x$

34. $y = \sin x, \quad y = (x - 1)^2$

35. The region R shown in Fig. 6.2.30 is bounded by the parabolas $y^2 = 2(x - 3)$ and $y^2 = x$. Find the volume of the solid generated by rotating R around the x-axis.

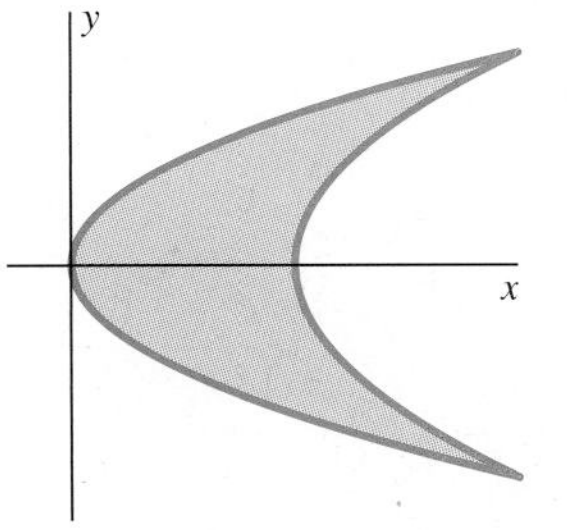

FIGURE 6.2.30 The region of Problem 35.

36. Find the volume of the ellipsoid generated by rotating around the x-axis the region bounded by the ellipse with equation

$$\left(\frac{x}{a}\right)^2 + \left(\frac{y}{b}\right)^2 = 1$$

(Fig. 6.2.31).

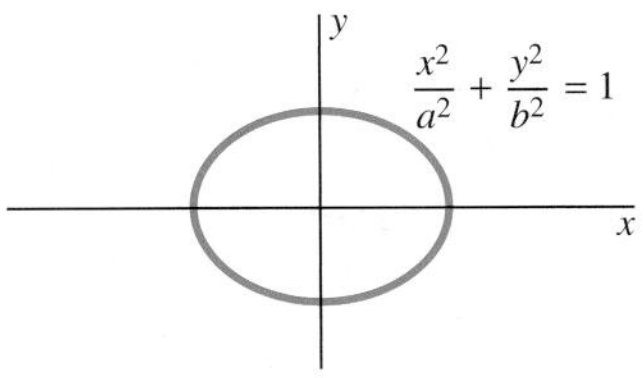

FIGURE 6.2.31 The ellipse of Problems 36 and 37.

37. Repeat Problem 36, except rotate the elliptical region around the y-axis.

38. (a) Find the volume of the unbounded solid generated by rotating the unbounded region of Fig. 6.2.32 around the x-axis. This is the region between the graph of $y = e^{-x}$ and the x-axis for $x \geqq 1$. [*Method:* Compute the volume from $x = 1$ to $x = b$, where $b > 1$. Then find the limit of this volume as $b \to +\infty$.] (b) What happens if $y = 1/\sqrt{x}$ instead?

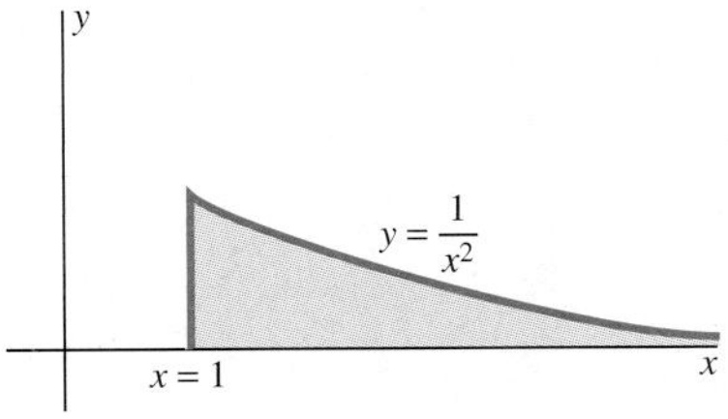

FIGURE 6.2.32 The unbounded plane region of Problem 38.

39. An observatory (Fig. 6.2.33) is shaped like a solid whose base is a circular disk with diameter AB of length $2a$ (Fig. 6.2.34). Find the volume of this solid if each cross section perpendicular to AB is a square.

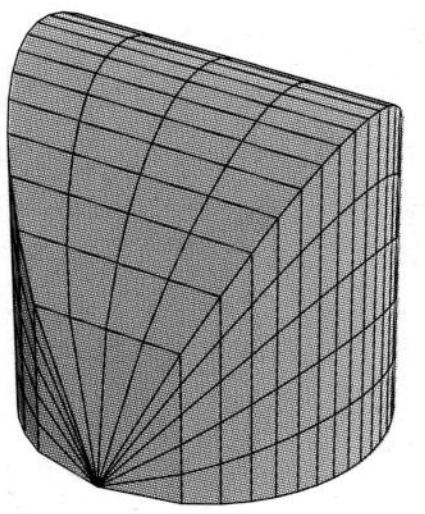

FIGURE 6.2.33 The observatory of Problem 39.

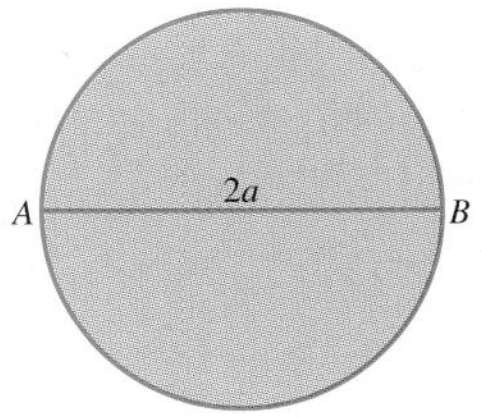

FIGURE 6.2.34 The circular base of the observatory (Problem 39).

40. The base of a certain solid is a circular disk with diameter AB of length $2a$. Find the volume of the solid if each cross section perpendicular to AB is a semicircle.

41. The base of a certain solid is a circular disk with diameter AB of length $2a$. Find the volume of the solid if each cross section perpendicular to AB is an equilateral triangle.

42. The base of a solid is the region in the xy-plane bounded by the parabolas $y = x^2$ and $x = y^2$. Find the volume of this solid if every cross section perpendicular to the x-axis is a square with *its* base in the xy-plane.

43. The *paraboloid* generated by rotating around the x-axis the region under the parabola $y^2 = 2px$, $0 \leqq x \leqq h$, is shown in Fig. 6.2.35. Show that the volume of the paraboloid is one-half that of the circumscribed cylinder also shown in the figure.

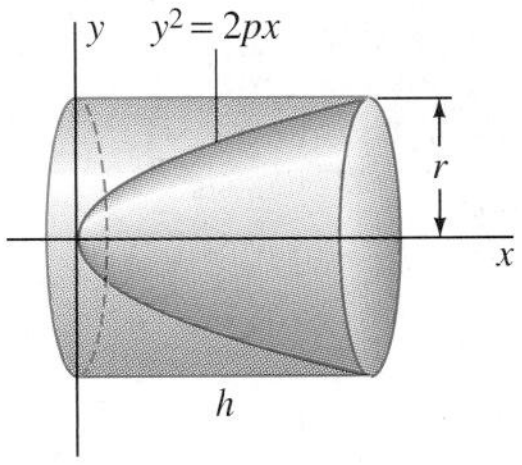

FIGURE 6.2.35 The paraboloid and cylinder of Problem 43.

44. A pyramid has height h and rectangular base with area A. Show that its volume is $V = \frac{1}{3}Ah$. [*Suggestion:* Note that each cross section parallel to the base is a rectangle.]

45. Repeat Problem 44, except make the base a triangle with area A.

46. Find the volume that remains after a hole of radius 3 is bored through the center of a solid sphere of radius 5 (Fig. 6.2.36).

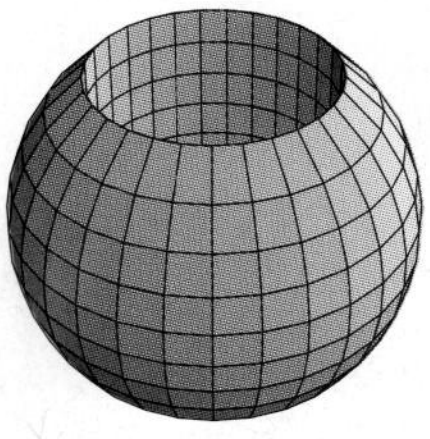

FIGURE 6.2.36 The sphere-with-hole of Problem 46.

47. Two horizontal circular cylinders both have radius a, and their axes intersect at right angles. Find the volume of their solid of intersection (Figs. 6.2.37 and 6.2.38, where $a = 1$). Is it clear to you that each horizontal cross section of the solid is a square?

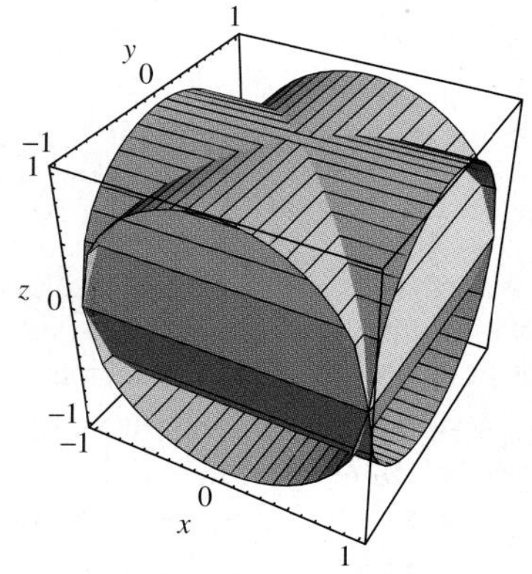

FIGURE 6.2.37 The intersecting cylinders of Problem 47.

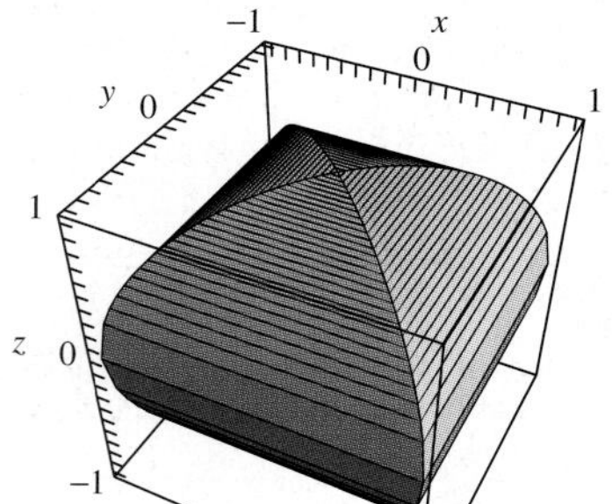

FIGURE 6.2.38 The solid of intersection (Problem 47).

48. Figure 6.2.39 shows a "spherical segment" of height h that is cut off from a sphere of radius r by a horizontal plane. Show that its volume is

$$V = \tfrac{1}{3}\pi h^2(3r - h).$$

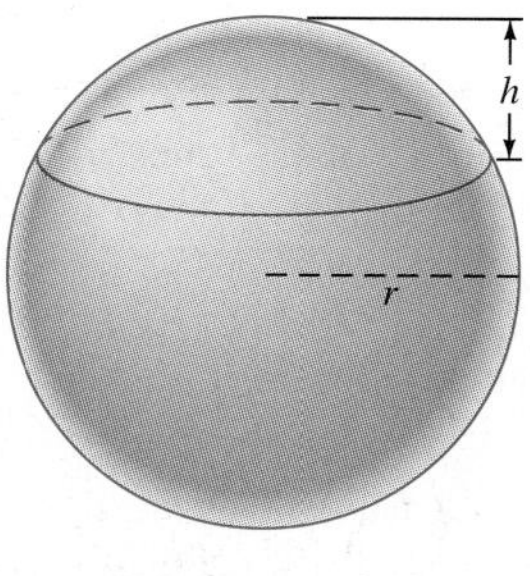

FIGURE 6.2.39 A spherical segment (Problem 48).

49. A doughnut-shaped solid, called a *torus* (Fig. 6.2.40), is generated by revolving around the y-axis the circular disk $(x - b)^2 + y^2 \leqq a^2$ centered at the point $(b, 0)$, where $0 < a < b$. Show that the volume of this torus is $V = 2\pi^2 a^2 b$. [*Suggestion:* Note that each cross section perpendicular to the y-axis is an annular ring, and recall that

$$\int_0^a \sqrt{a^2 - y^2}\,dy = \tfrac{1}{4}\pi a^2$$

because the integral represents the area of a quarter-circle of radius a.]

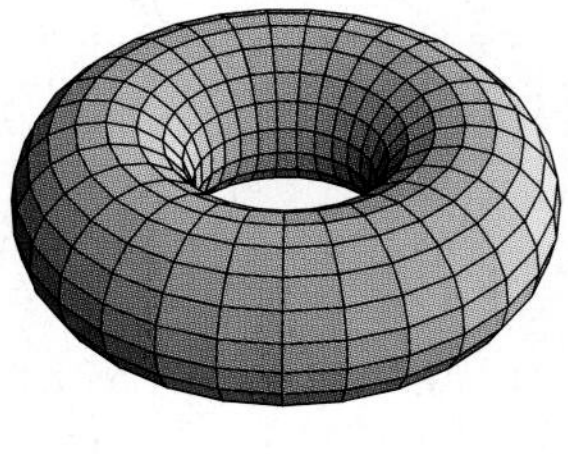

FIGURE 6.2.40 The torus of Problem 49.

50. The summit of a hill is 100 ft higher than the surrounding level terrain, and each horizontal cross section of the hill is circular. The following table gives the radius r (in feet) for selected values of the height h (in feet) above the surrounding terrain. Use Simpson's approximation to estimate the volume of the hill.

h	0	25	50	75	100
r	60	55	50	35	0

51. *Newton's Wine Barrel* Consider a barrel with the shape of the solid generated by revolving around the x-axis the region under the parabola

$$y = R - kx^2, \qquad -\tfrac{1}{2}h \leqq x \leqq \tfrac{1}{2}h$$

(Fig. 6.2.41). (a) Show that the radius of each end of the barrel is $r = R - \delta$, where $4\delta = kh^2$. (b) Then show that the volume of the barrel is

$$V = \tfrac{1}{3}\pi h\left(2R^2 + r^2 - \tfrac{2}{5}\delta^2\right).$$

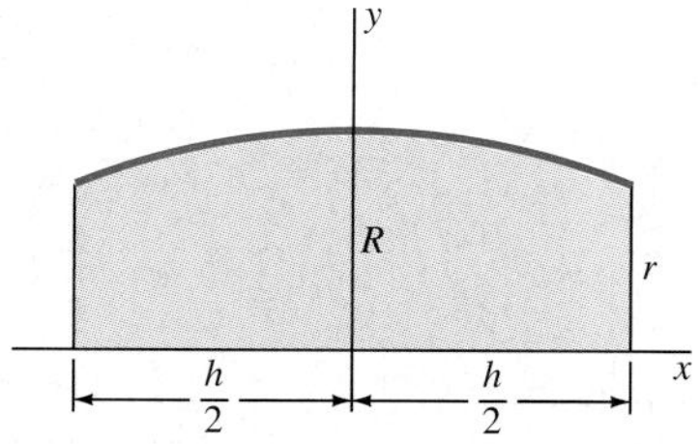

FIGURE 6.2.41 The region of Problem 51.

52. *The Clepsydra, or Water Clock* Consider a water tank whose side surface is generated by rotating the curve $y = kx^4$ around the y-axis (k is a positive constant). (a) Compute $V(y)$, the volume of water in the tank as a function of the depth y. (b) Suppose that water drains from the tank through a small hole at its bottom. Use the chain rule and Torricelli's law of draining [Eq. (3) of Section 5.2] to show that the water level in this tank falls at a *constant* rate. How could such a tank be used as a clock?

53. A contractor wants to bid on the job of leveling a 60-ft hill. It will cost $3.30/yd^3 of material in the hill to be removed. The following table, based on surveying data, shows areas of horizontal cross sections of the hill at 10-ft height intervals. Use (a) the trapezoidal approximation and (b) Simpson's approximation to estimate how much this job should cost. Round each answer to the nearest hundred dollars.

Height x (ft)	0	10	20	30	40	50	60
Area (ft^2)	1513	882	381	265	151	50	0

54. Water evaporates from an open bowl at a rate proportional to the area of the surface of the water. Show that whatever the shape of the bowl, the water level will drop at a constant rate.

55. A frustum of a right circular cone has height h and volume V. Its base is a circular disk with radius R and its top is a circular disk with radius r (Fig. 6.2.42). Apply the method of cross sections to show that

$$V = \tfrac{1}{3}\pi h(r^2 + rR + R^2).$$

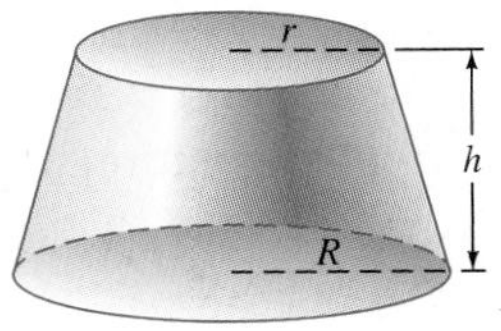

FIGURE 6.2.42 A frustum of a cone (Problem 55).

56. Find the volume of the solid of intersection of two spheres of radius a, if the center of each lies on the surface of the other.

57. Find the volume of the solid of intersection of two spheres of radii a and b (with $b < a$) if the center of the smaller sphere lies on the surface of the larger one.

58. In the third century B.C. Archimedes regarded the sphere of radius r as a solid of revolution in deriving his famous volume formula $V = \frac{4}{3}\pi r^3$. A major difference between his method and the method of this section is that his derivation used conical frusta rather than circular cylinders (Fig. 6.2.43). Figure 6.2.44 shows the approximating solid obtained by revolving around the x-axis the polygonal arc $P_0P_1P_2 \dots P_n$, where P_i denotes the point $(x_i, f(x_i))$ on the curve $y = f(x)$. The approximating slice that corresponds to the ith subinterval $[x_{i-1}, x_i]$ is the conical frustum highlighted in Fig. 6.2.44. The volume formula of Problem 55 gives the "frustum approximation"

$$V \approx \sum_{i=1}^{n} \frac{\pi}{3}\left\{[f(x_{i-1})]^2 + f(x_{i-1})f(x_i) + [f(x_i)]^2\right\}\Delta x.$$

Use continuity of f to show that this approximation leads (as $n \to +\infty$) to the same volume formula

$$V = \int_a^b \pi[f(x)]^2\,dx$$

that we derived using the method of cross sections.

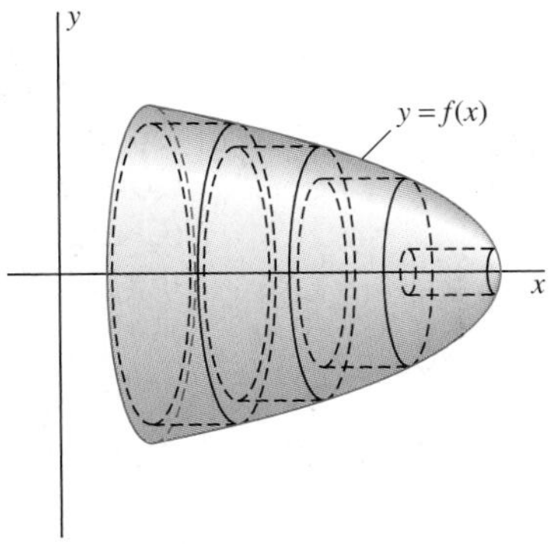

FIGURE 6.2.43 Using cylinders to approximate a solid of revolution.

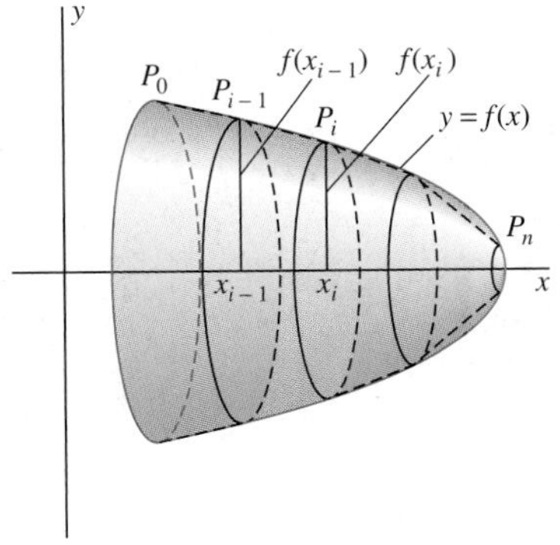

FIGURE 6.2.44 Using conical frusta to approximate a solid of revolution.

6.3 VOLUMES BY THE METHOD OF CYLINDRICAL SHELLS

The method of cross sections of Section 6.2 is a technique of approximating a solid by a stack of thin slabs or slices. In the case of a solid of revolution, these slices are circular disks or annular rings. The **method of cylindrical shells** is a second way of computing volumes of solids of revolution. It is a technique of approximating a solid of revolution by a collection of thin right cylindrical shells, and it frequently leads to simpler computations than does the method of cross sections.

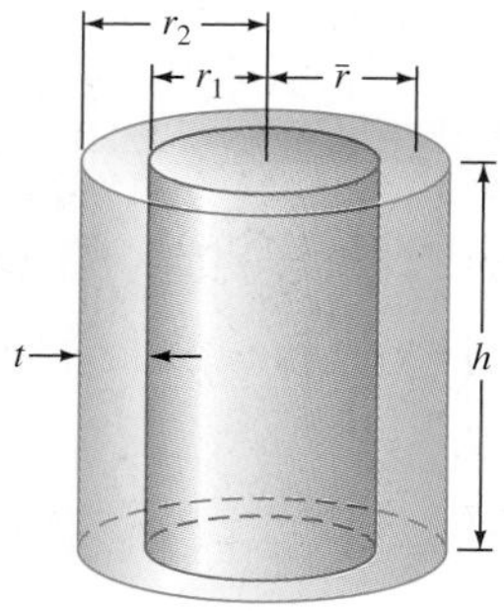

FIGURE 6.3.1 A cylindrical shell.

Volume of a Cylindrical Shell

A **cylindrical shell** is a region bounded by two concentric circular cylinders of the same height h. If, as in Fig. 6.3.1, the inner cylinder has radius r_1 and the outer one has radius r_2, then $\overline{r} = (r_1 + r_2)/2$ is the **average radius** of the cylindrical shell and $t = r_2 - r_1$ is its **thickness.** We then get the volume of the cylindrical shell by subtracting the volume of the inner cylinder from that of the outer one:

$$V = \pi r_2^2 h - \pi r_1^2 h = 2\pi \frac{r_1 + r_2}{2}(r_2 - r_1)h = 2\pi \overline{r} t h. \qquad \textbf{(1)}$$

In words, the volume of the shell is the product of 2π, its average radius, its thickness, and its height. Thus the volume of a very thin shell is closely approximated by multiplying its curved surface area by its thickness.

More General Volumes

Now suppose that we want to find the volume V of revolution generated by revolving around the y-axis the region under $y = f(x)$ from $x = a$ to $x = b$. We assume, as indicated in Fig. 6.3.2(a), that $0 \leqq a < b$ and that $f(x)$ is continuous and nonnegative on $[a, b]$. The solid will then resemble the one shown in Fig. 6.3.2(b).

To find V, we begin with a partition of $[a, b]$ into n subintervals, all with the same length $\Delta x = (b - a)/n$. Let $\overline{x}_i$ denote the midpoint of the ith subinterval $[x_{i-1}, x_i]$. Consider the rectangle in the xy-plane with base $[x_{i-1}, x_i]$ and height $f(\overline{x}_i)$. Figure 6.3.2(c) shows the cylindrical shell that is obtained by revolving this rectangle around the y-axis. This cylindrical shell approximates the solid with volume ΔV_i that is obtained by revolving the region under $y = f(x)$ and over $[x_{i-1}, x_i]$, and thus Eq. (1) gives

$$\Delta V_i \approx 2\pi \overline{x}_i f(\overline{x}_i)\, \Delta x.$$

We add the volumes of the n cylindrical shells determined by the partition of $[a, b]$. This sum should approximate V because—as Fig. 6.3.2(d) suggests—the union of these shells physically approximates the solid of revolution. Thus we obtain the

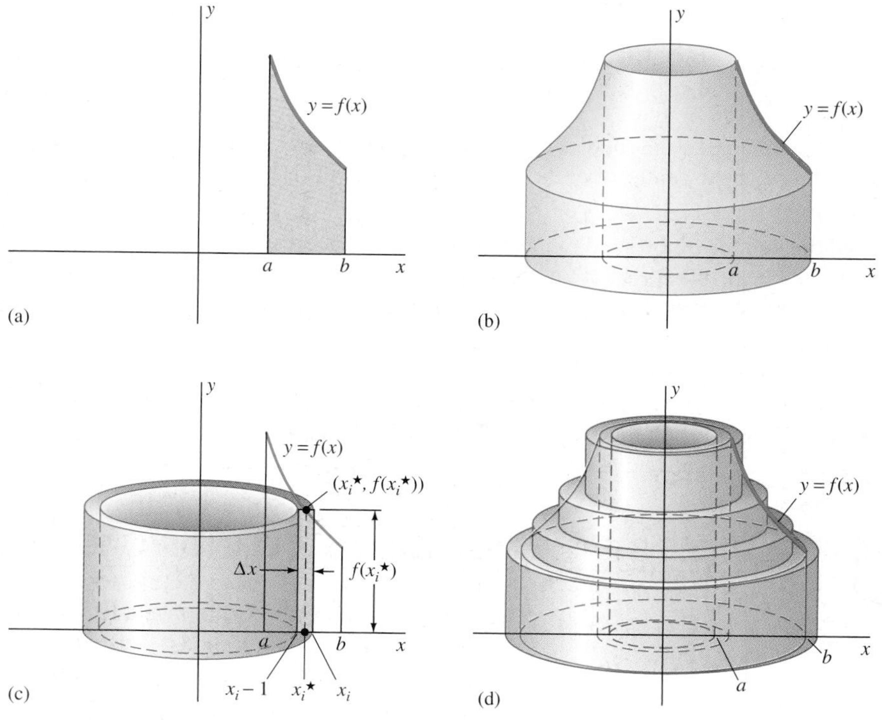

FIGURE 6.3.2 A solid of revolution—note the hole through its center—and a way to approximate it with nested cylindrical shells.

approximation

$$V = \sum_{i=1}^{n} \Delta V_i \approx \sum_{i=1}^{n} 2\pi \overline{x}_i\, f(\overline{x}_i)\, \Delta x.$$

This approximation to the volume V is a Riemann sum that approaches the integral

$$\int_a^b 2\pi x f(x)\, dx \quad \text{as} \quad \Delta x \to 0,$$

so it appears that the volume of the solid of revolution is given by

$$V = \int_a^b 2\pi x f(x)\, dx. \tag{2}$$

A complete discussion would require a proof that this formula gives the same volume as that *defined* by the method of cross sections in Section 6.2. (See Appendix G.)

It is more reliable to learn how to set up integral formulas than merely to memorize such formulas. A useful heuristic (suggestive but nonrigorous) device for setting up Eq. (2) is to picture the very narrow rectangular strip of area shown in Fig. 6.3.3. When this strip is revolved around the y-axis, it produces a thin cylindrical shell of radius x, height $y = f(x)$, and thickness dx (Fig. 6.3.4). So, if its volume is denoted by dV, we may write

$$dV = 2\pi x \cdot f(x) \cdot dx = 2\pi x f(x)\, dx.$$

This is easy to remember if you visualize Fig. 6.3.5.

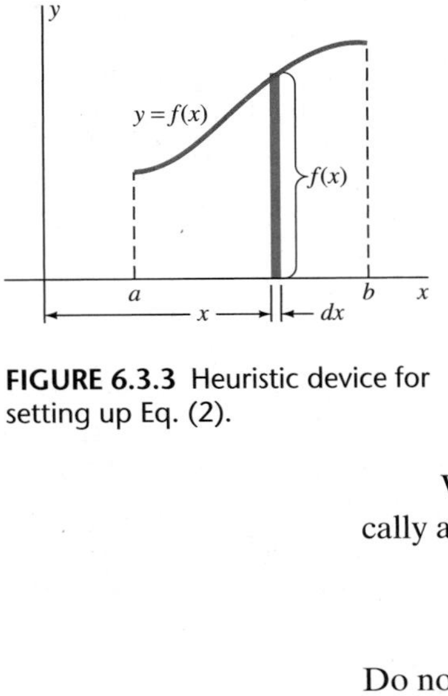

FIGURE 6.3.3 Heuristic device for setting up Eq. (2).

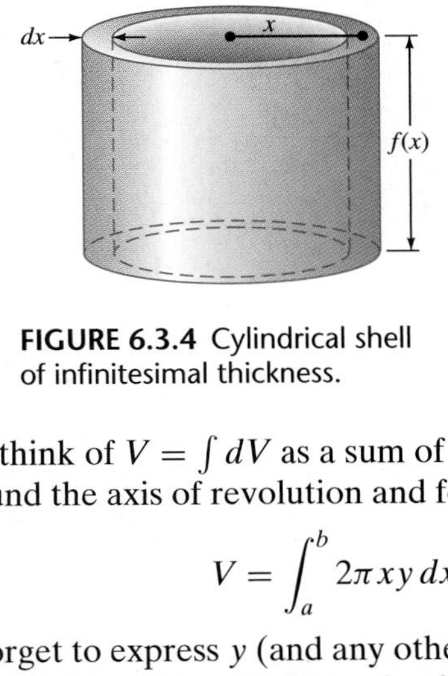

FIGURE 6.3.4 Cylindrical shell of infinitesimal thickness.

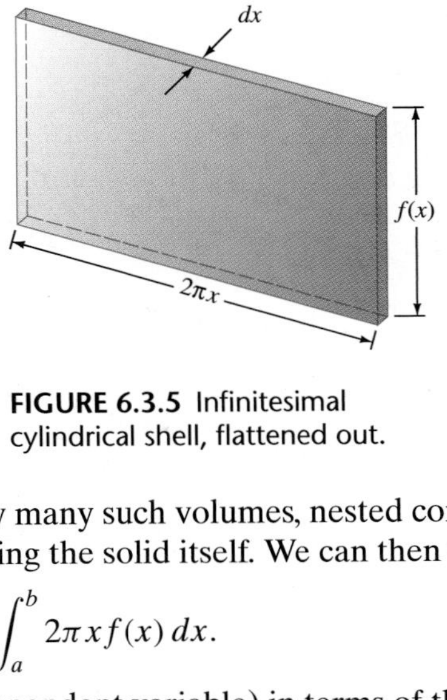

FIGURE 6.3.5 Infinitesimal cylindrical shell, flattened out.

We think of $V = \int dV$ as a sum of very many such volumes, nested concentrically around the axis of revolution and forming the solid itself. We can then write

$$V = \int_a^b 2\pi x y\, dx = \int_a^b 2\pi x f(x)\, dx.$$

Do not forget to express y (and any other dependent variable) in terms of the independent variable x (identified here by the differential dx) before you integrate.

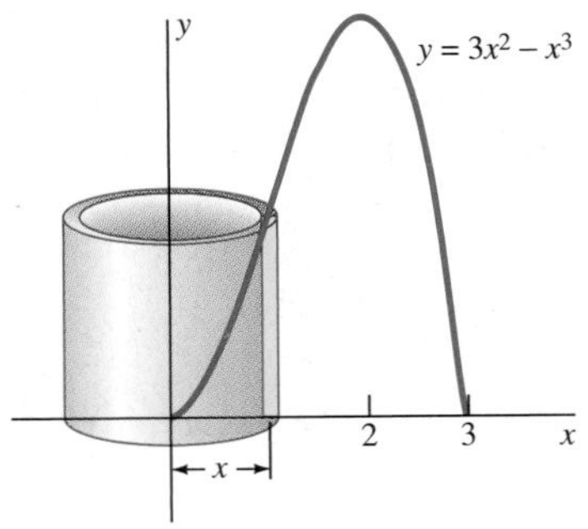

FIGURE 6.3.6 The region of Example 1: Rotate it around the y-axis.

EXAMPLE 1 Find the volume V of the solid generated by revolving around the y-axis the region under $y = 3x^2 - x^3$ from $x = 0$ to $x = 3$ (Fig. 6.3.6).

Solution Here it would be impractical to use the method of cross sections, because a cross section perpendicular to the y-axis is an annular ring, and finding its inner and outer radii would require us to solve the equation $y = 3x^2 - x^3$ for x in terms of y. We prefer to avoid this troublesome task, and Eq. (2) provides us with an alternative: We take $f(x) = 3x^2 - x^3$, $a = 0$, and $b = 3$. It immediately follows that

$$V = \int_0^3 2\pi x(3x^2 - x^3)\, dx = 2\pi \int_0^3 (3x^3 - x^4)\, dx$$

$$= 2\pi \left[\tfrac{3}{4}x^4 - \tfrac{1}{5}x^5 \right]_0^3 = \tfrac{243}{10}\pi.$$ ◆

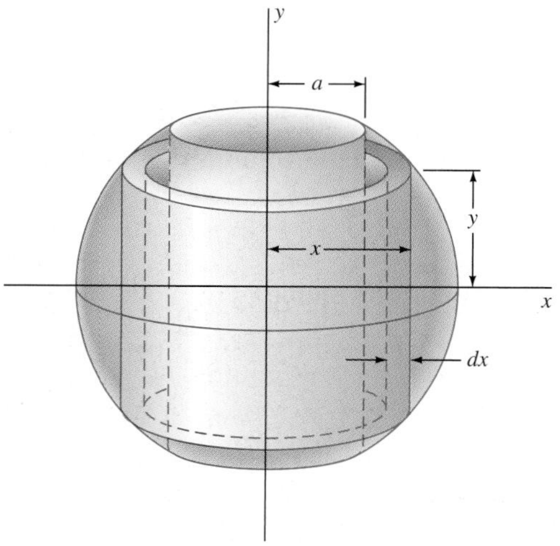

FIGURE 6.3.7 The sphere-with-hole of Example 2.

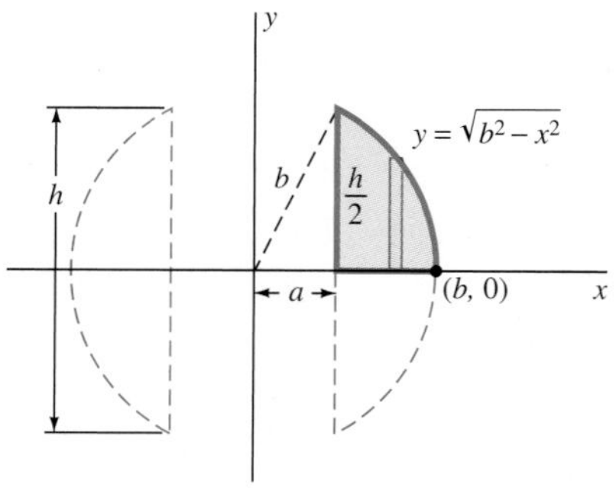

FIGURE 6.3.8 Middle cross section of the sphere-with-hole (Example 2).

EXAMPLE 2 Find the volume V of the solid that remains after you bore a circular hole of radius a through the center of a solid sphere of radius $b > a$ (Fig. 6.3.7).

Solution We think of the sphere of radius b as generated by revolving the right half of the circular disk $x^2 + y^2 = b^2$ around the y-axis, and we think of the hole as vertical and with its centerline lying on the y-axis. Then the upper *half* of the solid in question is generated by revolving around the y-axis the region shaded in Fig. 6.3.8. This is the region below the graph of $y = \sqrt{b^2 - x^2}$ (and above the x-axis) from $x = a$ to $x = b$. The volume of the entire sphere-with-hole is then double that of the upper half, and Eq. (2) gives

$$V = 2\int_a^b 2\pi x(b^2 - x^2)^{1/2}\,dx = 4\pi\left[-\tfrac{1}{3}(b^2 - x^2)^{3/2}\right]_a^b,$$

so

$$V = \tfrac{4}{3}\pi(b^2 - a^2)^{3/2}.$$ ◆

A way to check an answer such as this is to test it in some extreme cases. If $a = 0$ and $b = r$, which corresponds to drilling no hole at all through a sphere of radius r, then our result reduces to the volume $V = \frac{4}{3}\pi r^3$ of the entire sphere. If $a = b$, which corresponds to using a drill bit as large as the sphere, then $V = 0$; this, too, is correct.

Revolving the Region Between Two Curves

Now let A denote the region between the curves $y = f(x)$ and $y = g(x)$ over the interval $[a, b]$, where $0 \leqq a < b$ and $g(x) \leqq f(x)$ for x in $[a, b]$. Such a region is shown in Fig. 6.3.9. When A is rotated around the y-axis, it generates a solid of revolution. Suppose that we want to find the volume V of this solid. A development similar to that of Eq. (2) leads to the approximation

$$V \approx \sum_{i=1}^{n} 2\pi\overline{x}_i[f(\overline{x}_i) - g(\overline{x}_i)]\,\Delta x,$$

from which we may conclude that

$$V = \int_a^b 2\pi x[f(x) - g(x)]\,dx. \qquad \textbf{(3)}$$

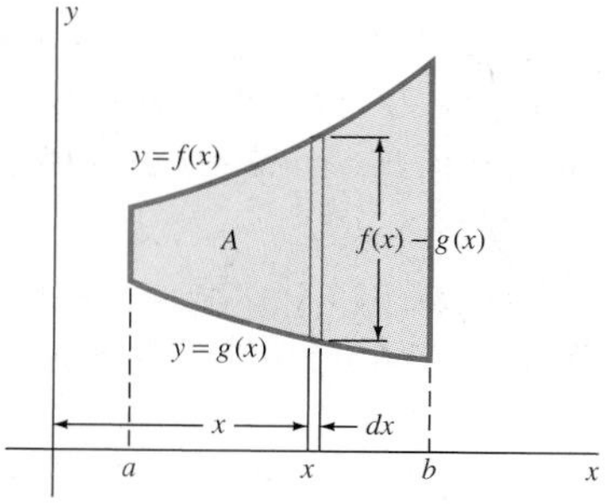

FIGURE 6.3.9 The region A between the graphs of f and g over $[a, b]$ is to be rotated around the y-axis.

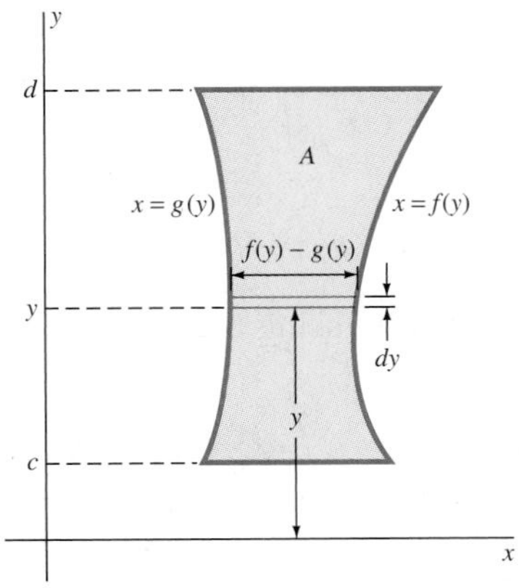

FIGURE 6.3.10 The region A is to be rotated around the x-axis.

Thus

$$V = \int_a^b 2\pi x[y_{\text{top}} - y_{\text{bot}}]\,dx, \qquad \textbf{(3')}$$

where $y_{\text{top}} = f(x)$ and $y_{\text{bot}} = g(x)$.

The method of cylindrical shells is also an effective way to compute volumes of solids of revolution around the x-axis. Figure 6.3.10 shows the region A bounded by the curves $x = f(y)$ and $x = g(y)$ for $c \leqq y \leqq d$ and by the horizontal lines $y = c$ and $y = d$. Let V be the volume obtained by revolving the region A around the x-axis. To compute V, we begin with a partition of $[c, d]$ into n subintervals, all of the same length $\Delta y = (d - c)/n$. Let $\overline{y}_i$ denote the midpoint of the ith subinterval $[y_{i-1}, y_i]$ of the partition. Then the volume of the cylindrical shell with average radius $\overline{y}_i$, height $f(\overline{y}_i) - g(\overline{y}_i)$, and thickness Δy is

$$\Delta V_i = 2\pi \overline{y}_i[f(\overline{y}_i) - g(\overline{y}_i)]\,\Delta y.$$

We add the volumes of these cylindrical shells and thus obtain the approximation

$$V \approx \sum_{i=1}^{n} 2\pi \overline{y}_i[f(\overline{y}_i) - g(\overline{y}_i)]\,\Delta y.$$

We recognize the right-hand side to be a Riemann sum for an integral with respect to y from c to d and so conclude that the volume of the solid of revolution is given by

$$V = \int_c^d 2\pi y[f(y) - g(y)]\,dy. \qquad \textbf{(4)}$$

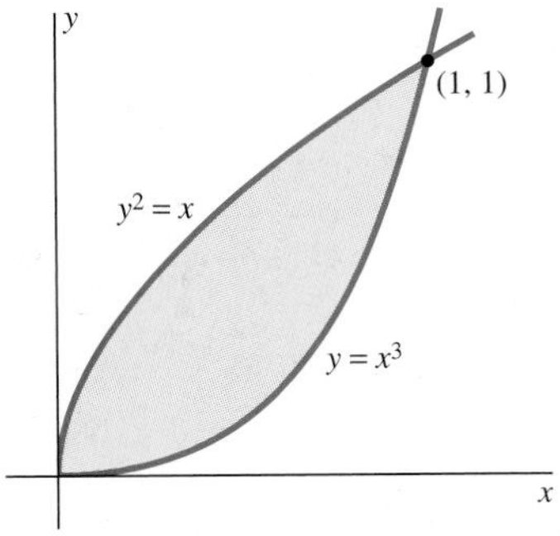

FIGURE 6.3.11 The region of Example 3.

Thus

$$V = \int_c^d 2\pi y[x_{\text{right}} - x_{\text{left}}]\,dy, \qquad \textbf{(4')}$$

where $x_{\text{right}} = f(y)$ and $x_{\text{left}} = g(y)$.

Note To use Eqs. (3′) and (4′), the integrand must be expressed in terms of the variable of integration specified by the differential.

EXAMPLE 3 Consider the region in the first quadrant bounded by the curves $y^2 = x$ and $y = x^3$ ((Fig. 6.3.11). Use the method of cylindrical shells to compute the volume of the solids obtained by revolving this region first around the y-axis and then around the x-axis.

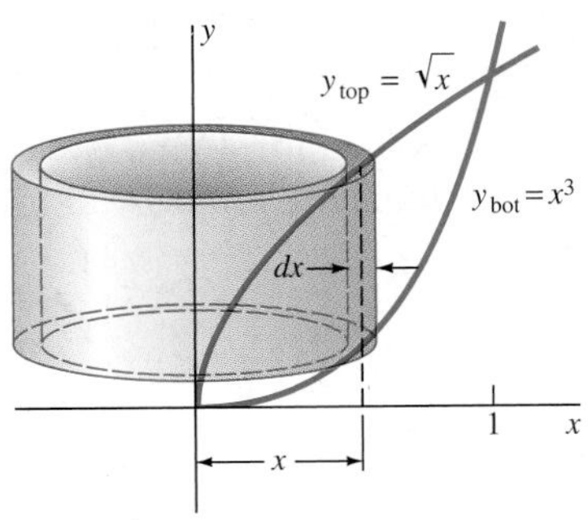

FIGURE 6.3.12 Revolution around the y-axis (Example 3).

Solution It is best to use cylindrical shells, as in Figs. 6.3.12 and 6.3.13, rather than memorized formulas, to set up the appropriate integrals. Thus the volume of revolution around the y-axis (Fig. 6.3.12) is given by

$$V = \int_0^1 2\pi x(y_{\text{top}} - y_{\text{bot}})\,dx = \int_0^1 2\pi x\left(\sqrt{x} - x^3\right)dx$$

$$= \int_0^1 2\pi\left(x^{3/2} - x^4\right)dx = 2\pi\left[\tfrac{2}{5}x^{5/2} - \tfrac{1}{5}x^5\right]_0^1 = \tfrac{2}{5}\pi.$$

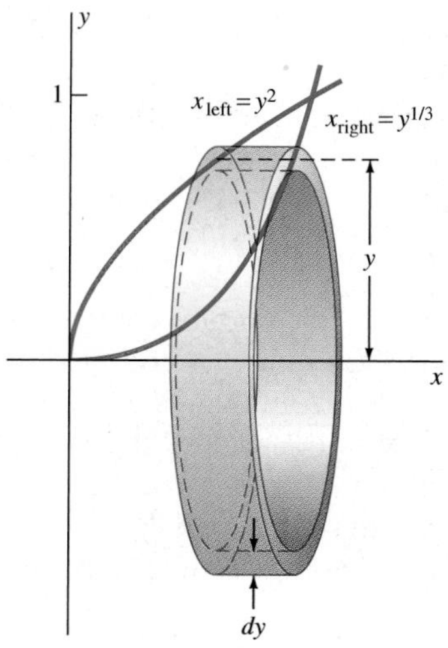

FIGURE 6.3.13 Revolution around the x-axis (Example 3).

The volume of revolution around the x-axis (Fig. 6.3.13) is given by

$$V = \int_0^1 2\pi y(x_{\text{right}} - x_{\text{left}})\,dy = \int_0^1 2\pi y\left(y^{1/3} - y^2\right)dy$$

$$= \int_0^1 2\pi\left(y^{4/3} - y^3\right)dy = 2\pi\left[\tfrac{3}{7}y^{7/3} - \tfrac{1}{4}y^4\right]_0^1 = \tfrac{5}{14}\pi.$$

The answers are the same, of course, as those we obtained by using the method of cross sections in Example 5 of Section 6.2. ◆

EXAMPLE 4 Suppose that the region of Example 3 is rotated around the vertical line $x = -1$ (Fig. 6.3.14). Then the area element

$$dA = (y_{\text{top}} - y_{\text{bot}})\,dx = \left(\sqrt{x} - x^3\right)dx$$

is revolved through a circle of radius $r = 1 + x$. Hence the volume of the resulting cylindrical shell is

$$dV = 2\pi r\,dA = 2\pi(1 + x)\left(x^{1/2} - x^3\right)dx$$

$$= 2\pi\left(x^{1/2} + x^{3/2} - x^3 - x^4\right)dx.$$

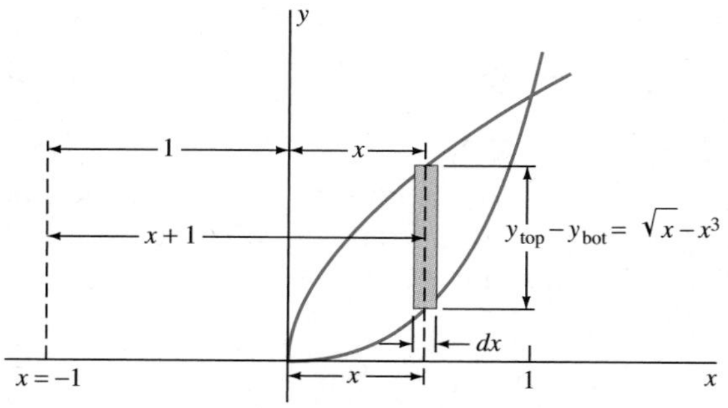

FIGURE 6.3.14 Revolution around the vertical line $x = -1$ (Example 4).

The volume of the resulting solid of revolution is then

$$V = \int_0^1 2\pi\left(x^{1/2} + x^{3/2} - x^3 - x^4\right)dx$$

$$= 2\pi\left[\tfrac{2}{3}x^{3/2} + \tfrac{2}{5}x^{5/2} - \tfrac{1}{4}x^4 - \tfrac{1}{5}x^5\right]_0^1 = \tfrac{37}{30}\pi,$$

as we found by using the method of cross sections in Example 6 of Section 6.2. ◆

We may observe finally that the method of cylindrical shells is summarized by the heuristic formula

$$V = \int_{\star}^{\star\star} 2\pi r\,dA,$$

where dA denotes the area of an infinitesimal strip that is revolved through a circle of radius r to generate a thin cylindrical shell. The asterisks indicate limits of integration that you need to find.

6.3 TRUE/FALSE STUDY GUIDE

6.3 CONCEPTS: QUESTIONS AND DISCUSSION

In the text's discussion of the solid of revolution illustrated in Fig. 6.3.2, ΔV_i denotes the part of the entire volume V that is obtained by revolving just the strip that lies under the graph of $y = f(x)$ over the ith subinterval $[x_{i-1}, x_i]$. We argued that ΔV_i is approximated by $\Delta V_i \approx 2\pi \overline{x}_i f(\overline{x}_i)\,\Delta x$.

1. Explain why continuity of the function f implies that $\Delta V_i = 2\pi \overline{x}_i f(x_i^\star)\,\Delta x$ *exactly* for some point $x_i^\star$ in the ith subinterval.
2. Then the volume V of the entire solid of revolution is given exactly by

$$V = \sum_{i=1}^{n} \Delta V_i = \sum_{i=1}^{n} 2\pi \overline{x}_i f(x_i^\star)\,\Delta x$$

$$= \sum_{i=1}^{n} 2\pi x_i^\star f(x_i^\star)\,\Delta x + \sum_{i=1}^{n} 2\pi (\overline{x}_i - x_i^\star) f(x_i^\star)\,\Delta x. \tag{5}$$

Explain why continuity of f now implies that the last sum in (5) approaches zero as $n \to +\infty$. Explain why this implies that

$$V = \int_a^b 2\pi x f(x)\,dx.$$

6.3 PROBLEMS

In Problems 1 through 28, use the method of cylindrical shells to find the volume of the solid generated by rotating around the indicated axis the region bounded by the given curves.

1. $y = x^2$, $y = 0$, $x = 2$; the y-axis

2. $x = y^2$, $x = 4$; the y-axis

3. $y = 25 - x^2$, $y = 0$; the y-axis (Fig. 6.3.15)

4. $y = 2x^2$, $y = 8$; the y-axis (Fig. 6.3.16)

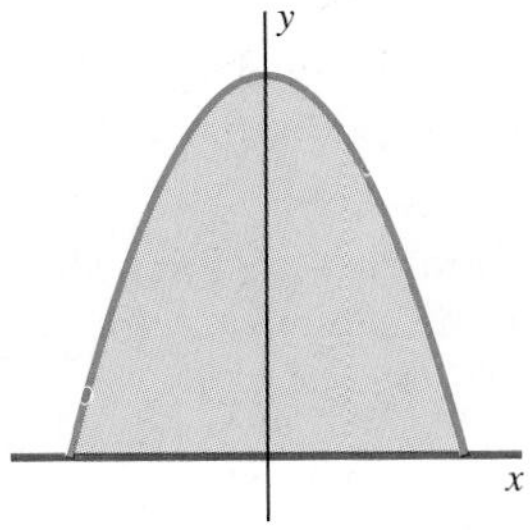

FIGURE 6.3.15 Problem 3.

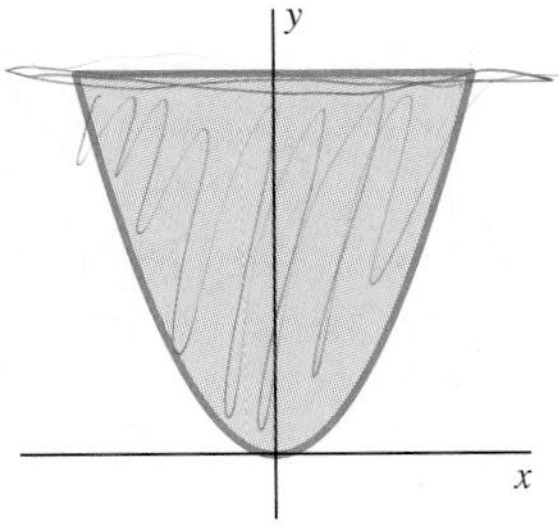

FIGURE 6.3.16 Problem 4.

5. $y = x^2$, $y = 8 - x^2$; the y-axis

6. $x = 9 - y^2$, $x = 0$; the x-axis

7. $x = y$, $x + 2y = 3$, $y = 0$; the x-axis (Fig. 6.3.17)

8. $y = x^2$, $y = 2x$; the line $y = 5$

9. $y = 2x^2$, $y^2 = 4x$; the x-axis

10. $y = 3x - x^2$, $y = 0$; the y-axis

11. $y = 4x - x^3$, $y = 0$; the y-axis (Fig. 6.3.18)

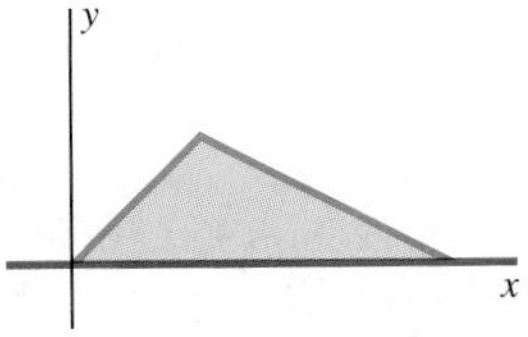

FIGURE 6.3.17 Problem 7.

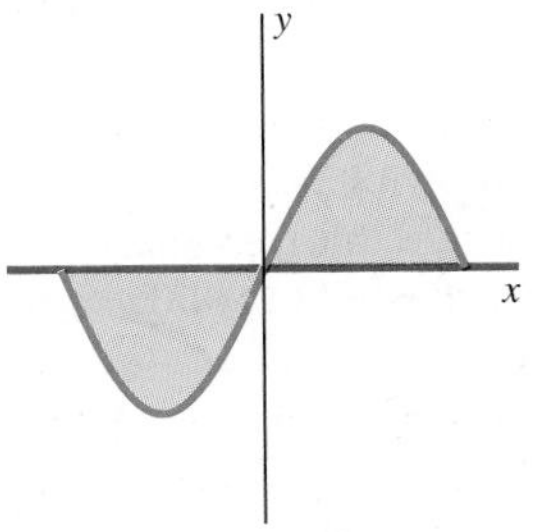

FIGURE 6.3.18 Problem 11.

12. $x = y^3 - y^4$, $x = 0$; the line $y = -2$ (Fig. 6.3.19)

13. $y = x - x^3$, $y = 0$ $(0 \leqq x \leqq 1)$; the y-axis

14. $x = 16 - y^2$, $x = 0$, $y = 0$ $(0 \leqq y \leqq 4)$; the x-axis

15. $y = x - x^3$, $y = 0$ $(0 \leqq x \leqq 1)$; the line $x = 2$ (Fig. 6.3.20)

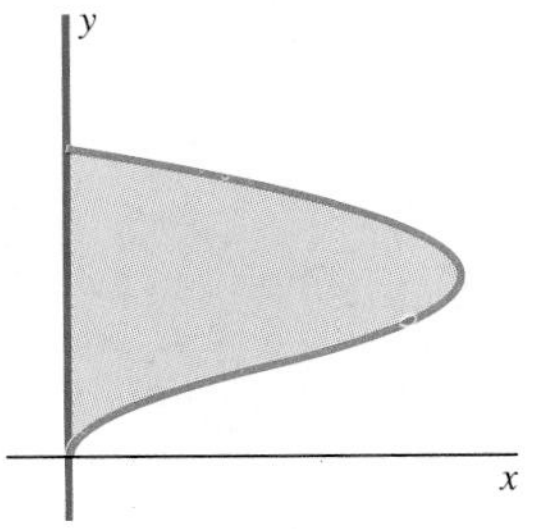

FIGURE 6.3.19 Problem 12.

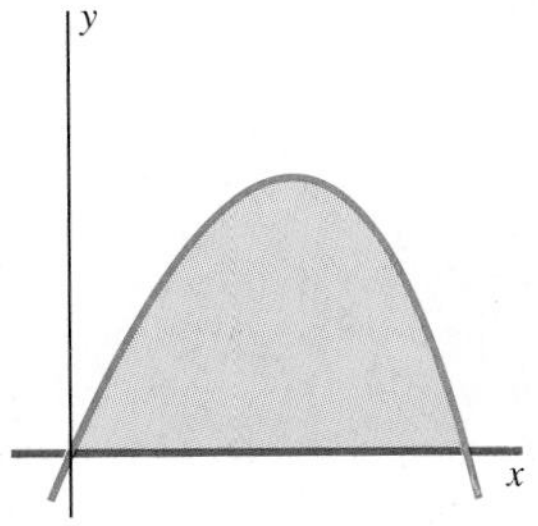

FIGURE 6.3.20 Problem 15.

16. $y = x^3$, $y = 0$, $x = 2$; the y-axis (Fig. 6.3.21)

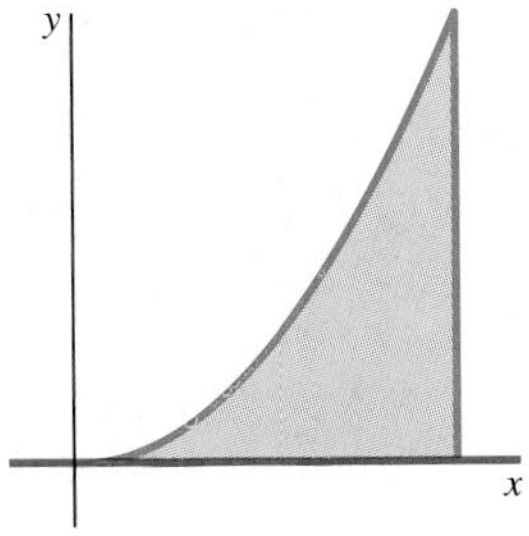

FIGURE 6.3.21 Problem 16.

17. $y = x^3$, $y = 0$, $x = 2$; the line $x = 3$

18. $y = x^3$, $y = 0$, $x = 2$; the x-axis

19. $y = x^2$, $y = 0$, $x = -1$, $x = 1$; the line $x = 2$

20. $y = x^2$, $y = x$ $(0 \leqq x \leqq 1)$; the y-axis

21. $y = x^2$, $y = x$ $(0 \leqq x \leqq 1)$; the x-axis

22. $y = x^2$, $y = x$ $(0 \leqq x \leqq 1)$; the line $y = 2$

23. $y = x^2$, $y = x$ $(0 \leqq x \leqq 1)$; the line $x = -1$

24. $y = \dfrac{1}{x^2}$, $y = 0$, $x = 1$, $x = 2$; the y-axis

25. $y = e^{-x^2}$, $y = 0$, $x = 0$, $x = 1$; the y-axis

26. $y = \dfrac{1}{1+x^2}$, $y = 0$, $x = 0$, $x = 2$; the y-axis

27. $y = \sin(x^2)$ and $y = -\sin(x^2)$ for $0 \leqq x \leqq \sqrt{\pi}$; the y-axis (Fig. 6.3.22)

28. $y = \dfrac{1}{x^2}$, $y = 0$, $x = 1$, $x = 2$; the line $x = -1$

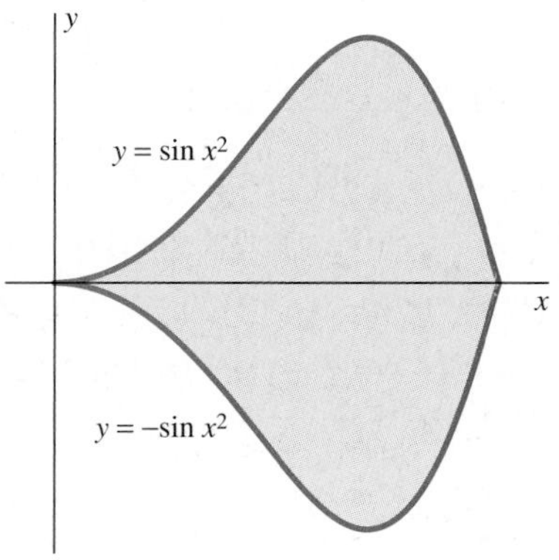

FIGURE 6.3.22 Problem 27.

In Problems 29 through 34, first use a calculator or computer to approximate (graphically or otherwise) the points of intersection of the two given curves. Let R be the region bounded by these curves. Integrate to approximate the volume of the solid obtained by revolving the region R around the y-axis. In Problems 31 through 34 you will find helpful the integral formula

$$\int u \cos u \, du = \cos u + u \sin u + C,$$

which you can verify by differentiation of the right-hand side.

29. $y = x^3 + 1$, $y = 6x - x^2$ (R lies to the right of the y-axis)

30. $y = x^4$, $y = 10x - 5$

31. $y = \cos x$, $y = x^2$

32. $y = \cos x$, $y = (x-1)^2$

33. $y = \cos x$, $y = 3x^2 - 6x + 2$

34. $y = 3\cos x$, $y = -\cos 4x$ (R lies between $x = -2$ and $x = 2$)

35. Verify the formula for the volume of a right circular cone by using the method of cylindrical shells. Apply the method to the figure generated by rotating the triangular region with vertices $(0, 0)$, $(r, 0)$, and $(0, h)$ around the y-axis.

36. Use the method of cylindrical shells to compute the volume of the paraboloid of Problem 43 in Section 6.2.

37. Use the method of cylindrical shells to find the volume of the ellipsoid obtained by revolving the elliptical region bounded by the graph of the equation

$$\left(\frac{x}{a}\right)^2 + \left(\frac{y}{b}\right)^2 = 1$$

around the y-axis.

38. Use the method of cylindrical shells to derive the formula given in Problem 48 of Section 6.2 for the volume of a spherical segment.

39. Use the method of cylindrical shells to compute the volume of the torus in Problem 49 in Section 6.2. [*Suggestion:* Substitute u for $x - b$ in the integral given by the formula in Eq. (2).]

40. (a) Find the volume of the solid generated by revolving the region bounded by the curves $y = x^2$ and $y = x + 2$ around the line $x = -2$. (b) Repeat part (a), but revolve the region around the line $x = 3$.

41. Find the volume of the solid generated by revolving the circular disk $x^2 + y^2 \leqq a^2$ around the vertical line $x = a$.

42. (a) Verify by differentiation that

$$\int xe^x \, dx = (x-1)e^x + C.$$

(b) Find the volume of the solid obtained by rotating around the y-axis the area under $y = e^x$ from $x = 0$ to $x = 1$.

43. We found in Example 2 that the volume remaining after a hole of radius a is bored through the center of a sphere of radius $b > a$ is

$$V = \tfrac{4}{3}\pi(b^2 - a^2)^{3/2}.$$

(a) Express the volume V in this formula *without* use of the hole radius a; use instead the hole height h. [*Suggestion:* Use the right triangle in Fig. 6.3.8.] (b) What is remarkable about the answer to part (a)?

44. The plane region R is bounded above and on the right by the graph of $y = 25 - x^2$, on the left by the y-axis, and below by the x-axis. A paraboloid is generated by revolving R around the y-axis. Then a vertical hole of radius 3 and centered along the y-axis is bored through the paraboloid. Find the volume of the solid that remains by using (a) the method of cross sections and (b) the method of cylindrical shells.

45. The loop of the curve $y^2 = x(5-x)^2$ bounds the region shown in Fig. 6.3.23. Find the volume of the solid obtained when this region is revolved around (a) the x-axis; (b) the y-axis; (c) the line $x = 5$.

46. The loop of the curve $y^2 = x^2(x+3)$ bounds the region shown in Fig. 6.3.24. Find the volume of the solid obtained when this region is revolved around (a) the x-axis; (b) the y-axis; (c) the line $x = -3$. *Suggestion:* If useful, substitute $u = x + 3$ before integrating.

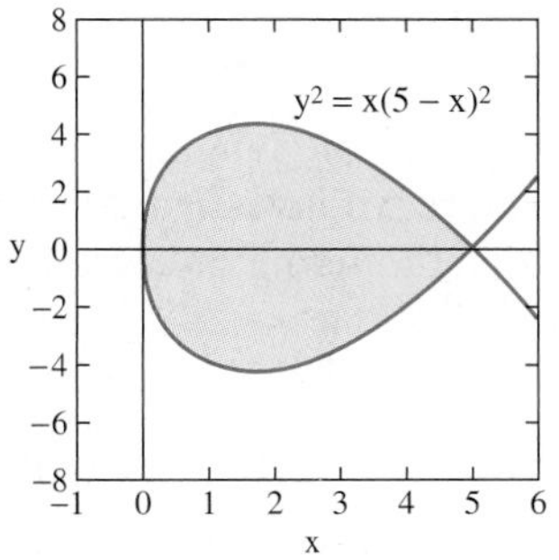

FIGURE 6.3.23 The region of Problem 45.

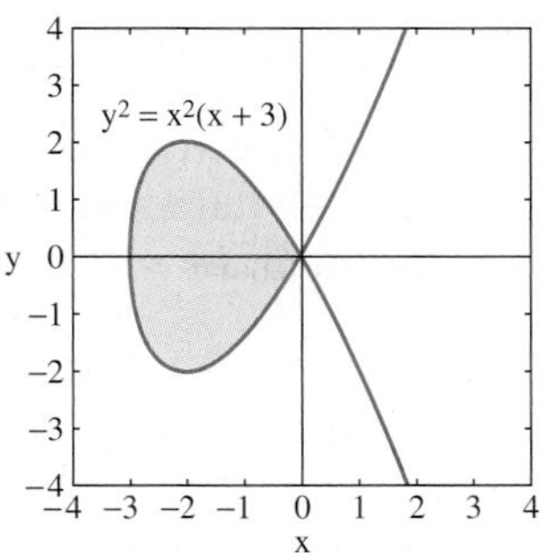

FIGURE 6.3.24 The region of Problem 46.

47. Figure 6.3.25 illustrates the solid concrete birdbath whose shape is obtained by revolving around the y-axis the region that lies between the curves

$$y = 1 + \frac{x^2}{5} - \frac{x^4}{500} \quad \text{and} \quad y = \frac{x^4}{10000}.$$

(a) Calculate the volume of concrete used in making this birdbath. (b) Calculate the volume of water it holds when full.

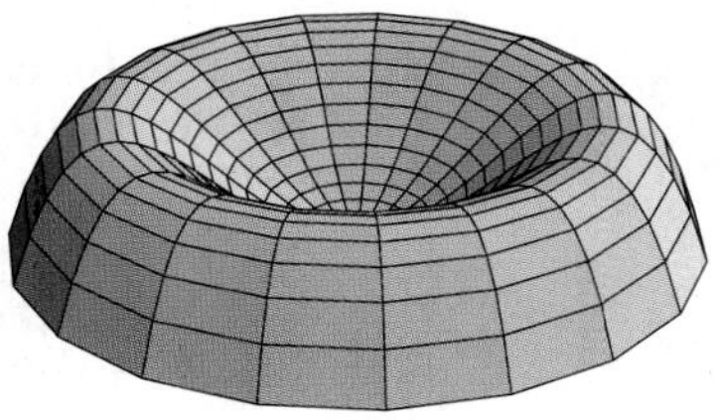

FIGURE 6.3.25 The concrete birdbath of Problem 47.

6.3 Project: Design Your Own Ring!

This project deals with the custom-made gold wedding band pictured in Fig. 6.3.26. Its shape is obtained by revolving the region A shown in Fig. 6.3.27 around the vertical axis shown there. The resulting wedding band has

- Inner radius R,
- Minimum thickness T, and
- Width W.

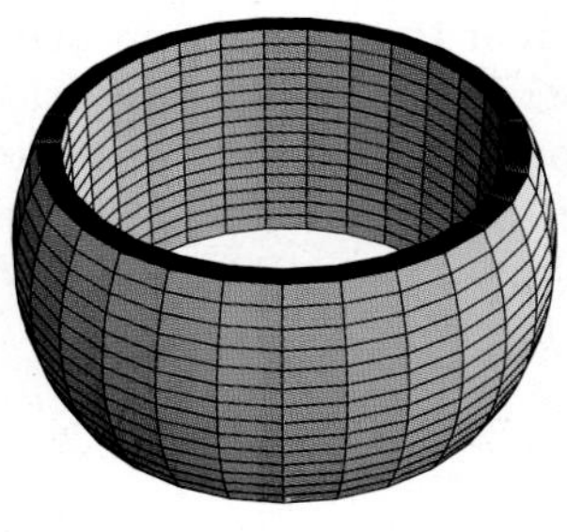

FIGURE 6.3.26 Wedding band.

The curved boundary of the region A is an arc of a circle whose center lies on the axis of revolution. For a typical wedding band, R might be anywhere from 6 to 12 mm, T might be 0.5 to 1.5 mm, and W might be 4 to 10 mm.

If a customer asks the price of a wedding band with given dimensions R, T, and W, the jeweler must first calculate the volume of the desired band to determine how much gold will be required to make it. Use the methods of this section to show that the volume V is given by the formula

$$V = \frac{\pi W}{6}(W^2 + 12RT + 6T^2). \tag{5}$$

If these dimensions are measured in millimeters, then V is given in cubic millimeters. (There are 1000 mm^3 in 1 cm^3.)

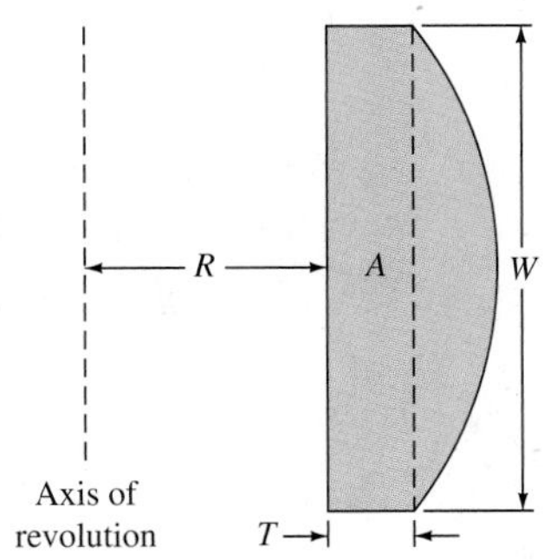

FIGURE 6.3.27 Cross section of the wedding band.

Suppose that the jeweler plans to charge the customer \$1000 per *troy ounce* of alloy (90% gold, 10% silver) used to make the ring. (The profit on the sale, covering the jeweler's time and overhead in making the ring, is fairly substantial because the price of gold is generally under \$400/oz, and that of silver under \$6/oz.) The inner radius R of the wedding band is determined by the measurement of the customer's finger (in millimeters; there are exactly 25.4 mm per inch). Suppose that the jeweler makes all wedding bands with $T = 1$ (mm). Then, for a given acceptable cost C (in dollars), the customer wants to know the maximum width W of the wedding band he or she can afford.

Investigation Measure your own ring finger to determine R (you can measure its circumference C with a piece of string and then divide by 2π). Then choose a cost figure C in the \$100 to \$500 price range. Use Eq. (5) with $T = 1$ to find the width W of a band that costs C dollars (at \$1000/oz). You will need to know that the density of the gold-silver alloy is 18.4 g/cm^3 and that 1 lb contains 12 troy ounces and 453.59 g. Use a graphics calculator or a calculator with a **SOLVE** key to solve the resulting cubic equation in W.

6.4 ARC LENGTH AND SURFACE AREA OF REVOLUTION

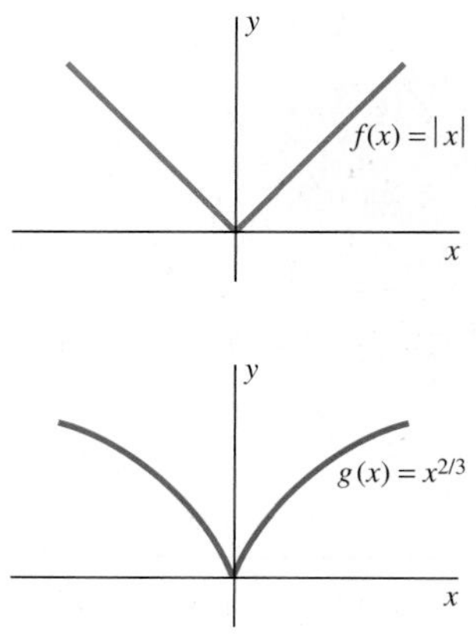

FIGURE 6.4.1 Graphs that have corner points.

If you plan to hike the Appalachian Trail, you will need to know the length of this curved path so you'll know how much equipment to take. Here we investigate how to find the length of a curved path and the closely related idea of finding the surface area of a curved surface.

A **smooth arc** is the graph of a smooth function defined on a closed interval; a **smooth function** f on $[a, b]$ is a function whose derivative f' is continuous on $[a, b]$. The continuity of f' rules out the possibility of corner points on the graph of f, points where the direction of the tangent line changes abruptly. The graphs of $f(x) = |x|$ and $g(x) = x^{2/3}$ are shown in Fig. 6.4.1; neither is smooth because each has a corner point at the origin.

The Length of a Curve

To investigate the length of a smooth arc, we begin with the length of a straight line segment, which is simply the distance between its endpoints. Then, given a smooth arc C, we pose the following question: If C were a thin wire and we straightened it without stretching it, how long would the resulting straight wire be? The answer is what we call the *length* of C.

To approximate the length s of the smooth arc C, we can inscribe in C a polygonal arc—one made up of straight line segments—and then calculate the length of this polygonal arc. We proceed in the following way, under the assumption that C is the graph of a smooth function f defined on the closed interval $[a, b]$. Consider a partition of $[a, b]$ into n subintervals, all with the same length Δx. Let P_i denote the point $(x_i, f(x_i))$ on the arc C corresponding to the ith subdivision point x_i. Our polygonal arc "inscribed in C" is then the union of the line segments P_0P_1, P_1P_2, P_2P_3, $\dots$, $P_{n-1}P_n$. So an approximation to the length s of C is

$$s \approx \sum_{i=1}^{n} |P_{i-1}P_i|, \tag{1}$$

the sum of the lengths of these line segments (Fig. 6.4.2). Our plan is to take the limit of this sum as $n \to \infty$: We want to evaluate

$$s = \lim_{n\to\infty} \sum_{i=1}^{n} |P_{i-1}P_i|.$$

The length of the typical line segment $P_{i-1}P_i$ is

$$|P_{i-1}P_i| = [(x_i - x_{i-1})^2 + (f(x_i) - f(x_{i-1}))^2]^{1/2}.$$

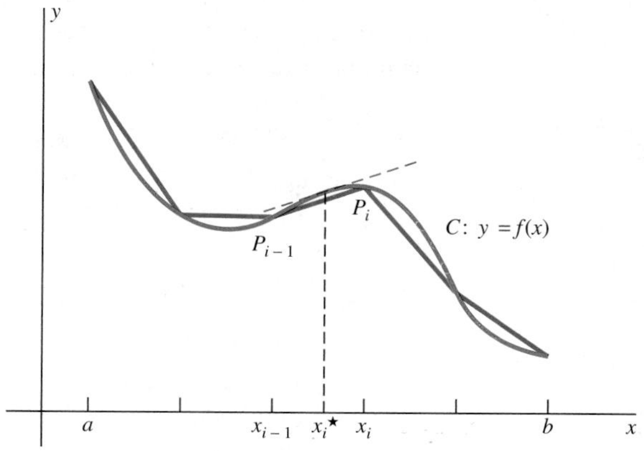

FIGURE 6.4.2 A polygonal arc inscribed in the smooth curve C.

We apply the mean value theorem to the function f on the interval $[x_{i-1}, x_i]$ and thereby conclude the existence of a point $x_i^\star$ in this interval such that

$$f(x_i) - f(x_{i-1}) = f'(x_i^\star) \cdot (x_i - x_{i-1}).$$

Hence

$$|P_{i-1}P_i| = \left[1 + \left(\frac{f(x_i) - f(x_{i-1})}{x_i - x_{i-1}}\right)^2\right]^{1/2} \cdot (x_i - x_{i-1})$$
$$= \sqrt{1 + [f'(x_i^\star)]^2}\,\Delta x,$$

where $\Delta x = x_i - x_{i-1}$.

We next substitute this expression for $|P_{i-1}P_i|$ into Eq. (1) and get the approximation

$$s \approx \sum_{i=1}^{n} \sqrt{1 + [f'(x_i^\star)]^2}\,\Delta x.$$

This sum is a Riemann sum for the function $\sqrt{1 + [f'(x)]^2}$ on $[a, b]$, and therefore—because f' is continuous—such sums approach the integral

$$\int_a^b \sqrt{1 + [f'(x)]^2}\,dx$$

as $\Delta x \to 0$. But our approximation ought to approach, as well, the actual length s as $\Delta x \to 0$. On this basis we *define* the **length** s of the smooth arc C to be

$$s = \int_a^b \sqrt{1 + [f'(x)]^2}\,dx = \int_a^b \sqrt{1 + \left(\frac{dy}{dx}\right)^2}\,dx. \qquad \textbf{(2)}$$

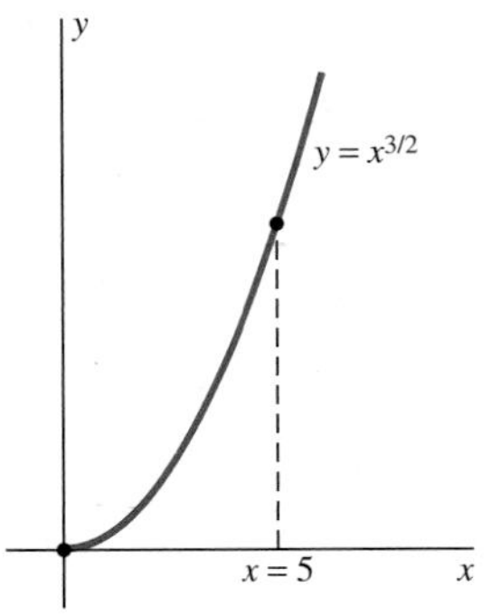

FIGURE 6.4.3 The semicubical parabola of Example 1.

EXAMPLE 1 Find the length of the so-called semicubical parabola (it's not really a parabola) $y = x^{3/2}$ on $[0, 5]$ (Fig. 6.4.3).

Solution We first compute the integrand in Eq. (2):

$$\sqrt{1 + \left(\frac{dy}{dx}\right)^2} = \sqrt{1 + \left(\tfrac{3}{2}x^{1/2}\right)^2} = \sqrt{1 + \tfrac{9}{4}x} = \tfrac{1}{2}(4 + 9x)^{1/2}.$$

Hence the length of the arc $y = x^{3/2}$ over the interval $[0, 5]$ is

$$s = \int_0^5 \tfrac{1}{2}(4 + 9x)^{1/2}\,dx = \left[\tfrac{1}{27}(4 + 9x)^{3/2}\right]_0^5 = \tfrac{335}{27} \approx 12.41.$$

As a plausibility check, the endpoints of the arc are $(0, 0)$ and $(5, 5\sqrt{5})$, so the straight line segment connecting these points has length $5\sqrt{6} \approx 12.25$. This is, as it should be, somewhat less than the calculated length of the arc. ◆

EXAMPLE 2 A manufacturer needs to make corrugated metal sheets 36 in. wide with cross sections in the shape of the curve

$$y = \tfrac{1}{2}\sin \pi x, \quad 0 \leqq x \leqq 36$$

(Fig. 6.4.4). How wide must the original flat sheets be for the manufacturer to produce these corrugated sheets?

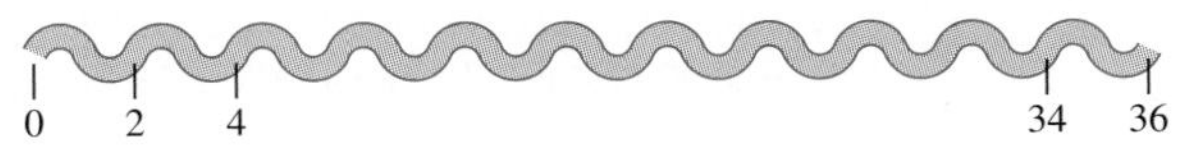

FIGURE 6.4.4 The corrugated metal sheet in the shape of $y = \frac{1}{2}\sin \pi x$ (Example 2).

Solution If

$$f(x) = \tfrac{1}{2}\sin \pi x, \qquad \text{then} \quad f'(x) = \tfrac{1}{2}\pi \cos \pi x.$$

Hence Eq. (2) yields the arc length of the graph of f over $[0, 36]$:

$$s = \int_0^{36} \sqrt{1 + \left(\tfrac{1}{2}\pi\right)^2 \cos^2 \pi x}\, dx = 36 \int_0^1 \sqrt{1 + \left(\tfrac{1}{2}\pi\right)^2 \cos^2 \pi x}\, dx.$$

These integrals cannot be evaluated in terms of elementary functions. Because of this, we cannot apply the fundamental theorem of calculus. So we estimate their values with the aid of Simpson's approximation (Section 5.9). Both with $n = 6$ and with $n = 12$ subintervals we find that

$$\int_0^1 \sqrt{1 + \left(\tfrac{1}{2}\pi\right)^2 \cos^2 \pi x}\, dx \approx 1.46$$

inches. Therefore the manufacturer should use flat sheets of approximate width $36 \cdot 1.46 \approx 52.6$ in. ◆

Arc Length by Integration with Respect to *y*

In the case of a smooth arc given as a graph $x = g(y)$ for y in $[c, d]$, a similar discussion beginning with a subdivision of $[c, d]$ leads to the formula

$$s = \int_c^d \sqrt{1 + [g'(y)]^2}\, dy = \int_c^d \sqrt{1 + \left(\frac{dx}{dy}\right)^2}\, dy \qquad \textbf{(3)}$$

for its length. We can compute the length of a more general curve, such as a circle, by dividing it into a finite number of smooth arcs and then applying to each of these arcs whichever of Eqs. (2) and (3) is required.

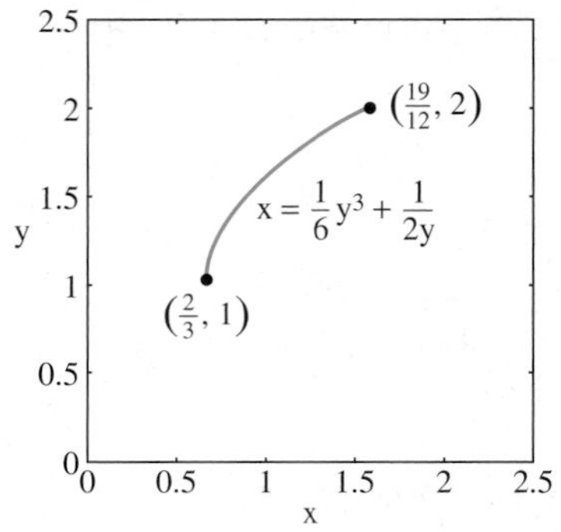

FIGURE 6.4.5 The curve of Example 3 (although we can calculate its arc length without even visualizing it).

EXAMPLE 3 Find the length s of the curve (Fig. 6.4.5)

$$x = \frac{1}{6}y^3 + \frac{1}{2y}, \qquad 1 \leqq y \leqq 2.$$

Solution Here y is the natural independent variable, so we use the arc-length formula in Eq. (3). First we calculate

$$1 + \left(\frac{dx}{dy}\right)^2 = 1 + \left(\frac{1}{2}y^2 - \frac{1}{2y^2}\right)^2 = 1 + \frac{1}{4}y^4 - \frac{1}{2} + \frac{1}{4y^4}$$

$$= \frac{1}{4}y^4 + \frac{1}{2} + \frac{1}{4y^4} = \left(\frac{1}{2}y^2 + \frac{1}{2y^2}\right)^2.$$

Thus we can "get out from under the radical" in Eq. (3):

$$s = \int_c^d \sqrt{1 + \left(\frac{dx}{dy}\right)^2}\, dy = \int_1^2 \left(\frac{1}{2}y^2 + \frac{1}{2y^2}\right) dy$$

$$= \left[\frac{1}{6}y^3 - \frac{1}{2y}\right]_1^2 = \frac{17}{12}.$$ ◆

A Symbolic Device

There is a convenient symbolic device that we can employ to remember both Eqs. (2) and (3) simultaneously. We think of two nearby points $P(x, y)$ and $Q(x+dx, y+dy)$ on the smooth arc C and denote by ds the length of the arc that joins P and Q. Imagine that P and Q are so close together that ds is, for all practical purposes, equal to the length of the straight line segment PQ. Then the Pythagorean theorem applied

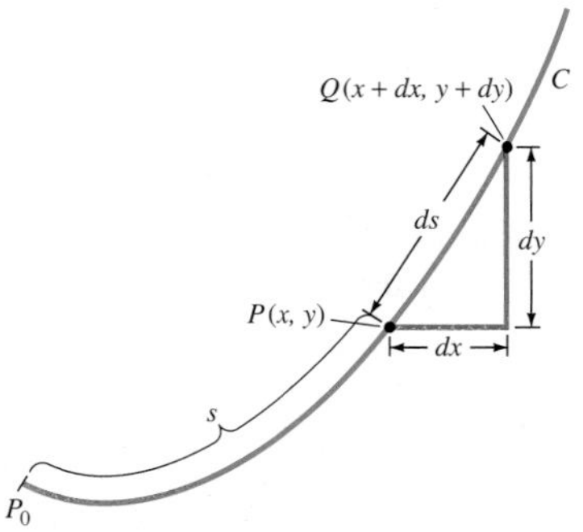

FIGURE 6.4.6 Heuristic development of the arc-length formula.

to the small right triangle in Fig. 6.4.6 gives

$$ds = \sqrt{(dx)^2 + (dy)^2} \tag{4}$$

$$= \sqrt{1 + \left(\frac{dy}{dx}\right)^2}\, dx \tag{4'}$$

$$= \sqrt{1 + \left(\frac{dx}{dy}\right)^2}\, dy. \tag{4''}$$

Thinking of the entire length s of C as the sum of small pieces such as ds, we write

$$s = \int_{\star}^{\star\star} ds. \tag{5}$$

Then formal (symbolic) substitution of the expressions in Eqs. (4′) and (4″) for ds in Eq. (5) yields Eqs. (2) and (3); only the limits of integration remain to be determined.

Cones and Conical Frusta

A **surface of revolution** is a surface obtained by revolving an arc or curve around an axis that lies in the same plane as the arc. The surface of a cylinder or of a sphere and the curved surface of a cone are important as examples of surfaces of revolution.

Our basic approach to finding the area of such a surface is this: First we inscribe a polygonal arc in the curve to be revolved. We then regard the area of the surface generated by revolving the polygonal arc to be an approximation to the surface generated by revolving the original curve. Because a surface generated by revolving a polygonal arc around an axis consists of frusta (sections) of cones, we can calculate its area in a reasonably simple way.

This approach to surface area originated with Archimedes. For example, he used this method to establish the formula $A = 4\pi r^2$ for the surface area of a sphere of radius r.

We will need the formula

$$A = 2\pi \bar{r} L \tag{6}$$

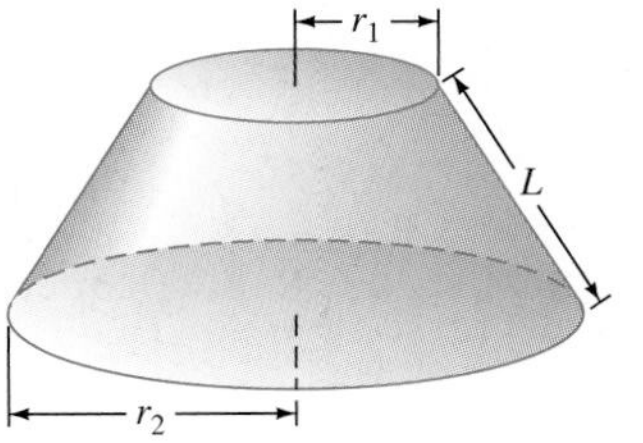

FIGURE 6.4.7 A frustum of a cone; the slant height is L.

for the curved surface area of a conical frustum with average radius $\bar{r} = \frac{1}{2}(r_1 + r_2)$ and *slant height* L (Fig. 6.4.7). Equation (6) follows from the formula

$$A = \pi r L \tag{7}$$

for the area of a conical surface with base radius r and slant height L (Fig. 6.4.8). It is easy to derive Eq. (7) by "unrolling" the conical surface onto a sector of a circle

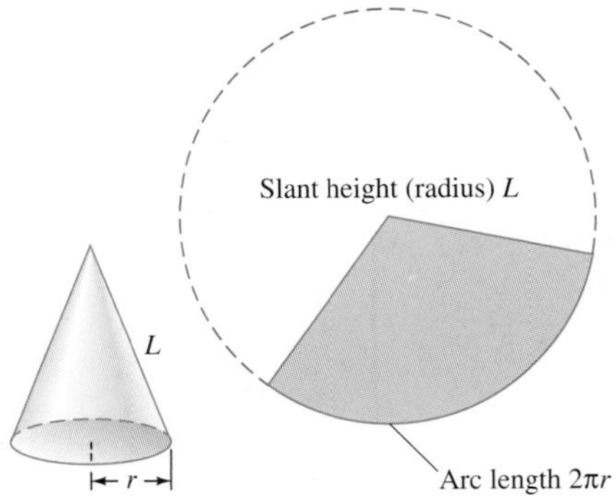

FIGURE 6.4.8 Surface area of a cone: Cut along L, then unroll the cone onto the circular sector.

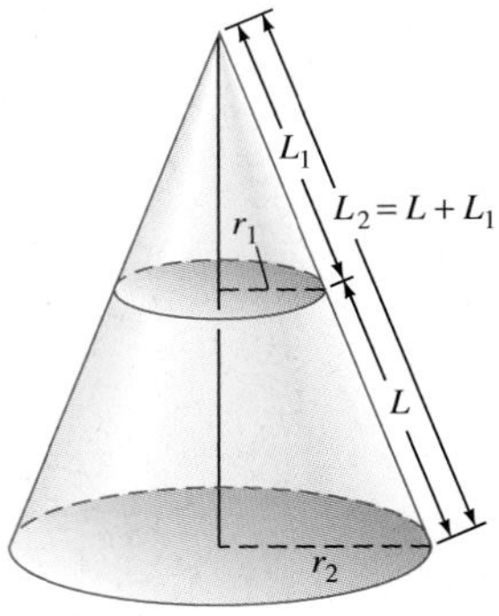

FIGURE 6.4.9 Derivation of Eq. (6).

of radius L, because the area of this sector is

$$A = \frac{2\pi r}{2\pi L} \cdot \pi L^2 = \pi r L.$$

To derive Eq. (6) from Eq. (7), we think of the frustum as the lower section of a cone with slant height $L_2 = L + L_1$ (Fig. 6.4.9). Then subtracting the area of the upper conical section from that of the entire cone gives

$$A = \pi r_2 L_2 - \pi r_1 L_1 = \pi r_2(L + L_1) - \pi r_1 L_1 = \pi(r_2 - r_1)L_1 + \pi r_2 L$$

for the area of the frustum. But the similar right triangles in Fig. 6.4.9 yield the proportion

$$\frac{r_1}{L_1} = \frac{r_2}{L_2} = \frac{r_2}{L + L_1},$$

from which we find that $(r_2 - r_1)L_1 = r_1 L$. Hence the area of the frustum is

$$A = \pi r_1 L + \pi r_2 L = 2\pi \bar{r} L,$$

where $\bar{r} = \frac{1}{2}(r_1 + r_2)$. So we have verified Eq. (6).

Surface Area of Revolution

Suppose that the surface S has area A and is generated by revolving around the x-axis the smooth arc $y = f(x)$, $a \leqq x \leqq b$; suppose also that $f(x)$ is never negative on $[a, b]$. To approximate A we begin with a division of $[a, b]$ into n subintervals, each of length Δx. As in our discussion of arc length leading to Eq. (2), let P_i denote the point $(x_i, f(x_i))$ on the arc. Then, as before, the line segment $P_{i-1}P_i$ has length

$$L_i = |P_{i-1}P_i| = \sqrt{1 + [f'(x_i^\star)]^2}\,\Delta x$$

for some point $x_i^\star$ in the ith subinterval $[x_{i-1}, x_i]$.

The conical frustum obtained by revolving the segment $P_{i-1}P_i$ around the x-axis has slant height L_i and, as shown in Fig. 6.4.10, average radius

$$\bar{r}_i = \tfrac{1}{2}\left[f(x_{i-1}) + f(x_i)\right].$$

Because $\bar{r}_i$ lies between the values $f(x_{i-1})$ and $f(x_i)$, the intermediate value property of continuous functions (Section 2.4) yields a point $x_i^{\star\star}$ in $[x_{i-1}, x_i]$ such that $\bar{r}_i = f(x_i^{\star\star})$. By Eq. (6), the area of this conical frustum is, therefore,

$$2\pi \bar{r}_i L_i = 2\pi f(x_i^{\star\star})\sqrt{1 + [f'(x_i^\star)]^2}\,\Delta x.$$

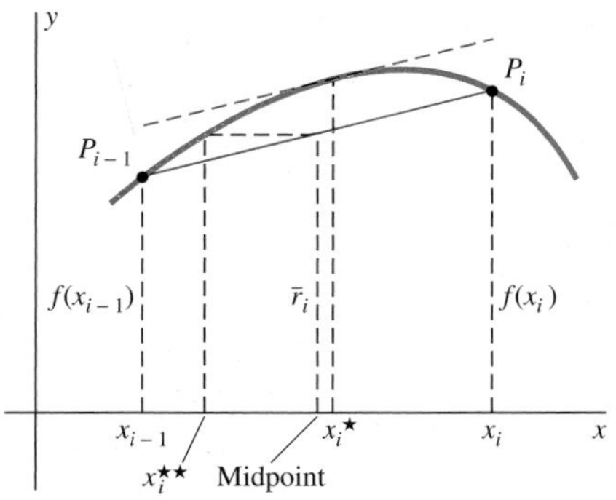

FIGURE 6.4.10 Approximating a surface area of revolution by the surface of a frustum of a cone.

We add the areas of these conical frusta for $i = 1, 2, 3, \ldots, n$. This gives the approximation

$$A \approx \sum_{i=1}^{n} 2\pi f(x_i^{\star\star})\sqrt{1 + [f'(x_i^{\star})]^2}\,\Delta x.$$

If $x_i^\star$ and $x_i^{\star\star}$ were the *same* point of the ith subinterval $[x_{i-1}, x_i]$, then this approximation would be a Riemann sum for the integral

$$\int_a^b 2\pi f(x)\sqrt{1 + [f'(x)]^2}\,dx.$$

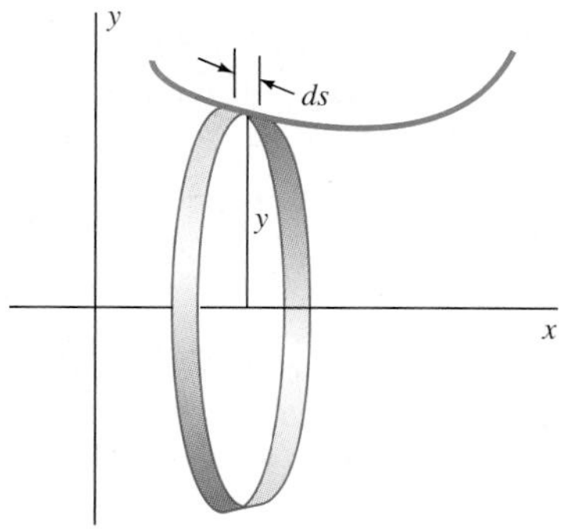

FIGURE 6.4.11 The tiny arc ds generates a ribbon with circumference $2\pi y$ when it is revolved around the x-axis.

Even though the numbers $x_i^\star$ and $x_i^{\star\star}$ are generally not equal, it still follows (from a result in Appendix G) that our approximation approaches the integral above as $\Delta x \to 0$. Intuitively, this is easy to believe; after all, as $\Delta x \to 0$, the difference between $x_i^\star$ and $x_i^{\star\star}$ also approaches zero.

We therefore *define* the **area** A of the surface generated by revolving around the x-axis the smooth arc $y = f(x)$, $a \leqq x \leqq b$, by the formula

$$A = \int_a^b 2\pi f(x)\sqrt{1 + [f'(x)]^2}\,dx. \tag{8}$$

If we write y for $f(x)$ and ds for $\sqrt{1 + (dy/dx)^2}\,dx$, as in Eq. (4′), then we can abbreviate Eq. (8) as

$$A = \int_a^b 2\pi y\,ds \quad (x\text{-axis}). \tag{9}$$

This abbreviated formula is conveniently remembered by thinking of $dA = 2\pi y\,ds$ as the area of the narrow frustum obtained by revolving the tiny arc ds around the x-axis in a circle of radius y (Fig. 6.4.11).

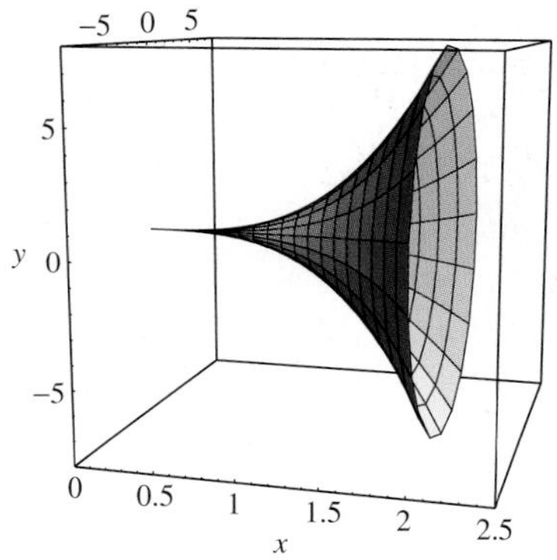

FIGURE 6.4.12 The "horn" generated by revolving the curve $y = x^3$, $0 \leqq x \leqq 2$, around the x-axis.

EXAMPLE 4 Figure 6.4.12 shows the horn-shaped surface generated by revolving the curve $y = x^3$, $0 \leqq x \leqq 2$, around the x-axis. Find its surface area of revolution.

Solution Substituting $y = x^3$ and

$$ds = \sqrt{1 + [y'(x)]^2}\,dx = \sqrt{1 + 9x^4}\,dx$$

in Eq. (9), we get

$$A = \int_0^2 2\pi x^3(1 + 9x^4)^{1/2}\,dx \qquad (\text{let } u = 1 + 9x^4)$$

$$= \left[\frac{\pi}{27}(1 + 9x^4)^{3/2}\right]_0^2 = \frac{\pi}{27}\left(145^{3/2} - 1\right) \approx 203.04.$$ ◆

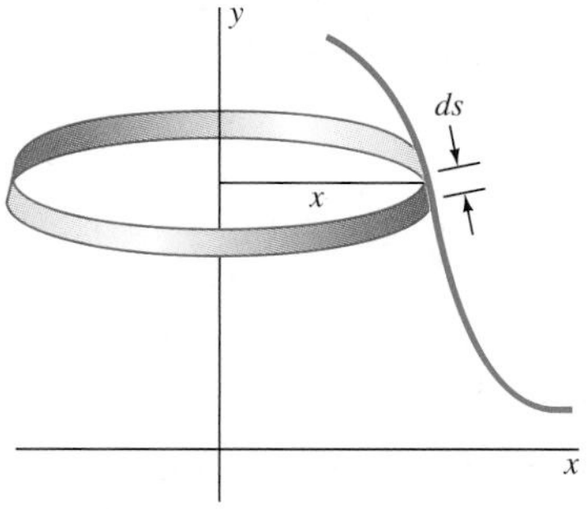

FIGURE 6.4.13 The tiny arc ds generates a ribbon with circumference $2\pi x$ when it is revolved around the y-axis.

If the smooth arc being revolved around the x-axis is given instead by $x = g(y)$, $c \leqq y \leqq d$, then an approximation based on a subdivision of $[c, d]$ leads to the area formula

$$A = \int_c^d 2\pi y\sqrt{1 + [g'(y)]^2}\,dy. \tag{10}$$

We can obtain Eq. (10) by making the formula substitution $ds = \sqrt{1 + (dx/dy)^2}\,dy$ of Eq. (4″) into the abbreviated formula in Eq. (9) for surface area of revolution and then replacing a and b with the correct limits of integration.

Revolution Around the y-Axis

Now let us consider the surface generated by revolving a smooth arc around the y-axis rather than around the x-axis In Fig. 6.4.13 we see that the average radius of the narrow frustum obtained by revolving the tiny arc ds is now x instead of y. This suggests the abbreviated formula

$$A = \int_a^b 2\pi x\, ds \quad (y\text{-axis}) \tag{11}$$

for a surface area of revolution around the y-axis. If the smooth arc is given by $y = f(x)$, $a \leqq x \leqq b$, then the symbolic substitution $ds = \sqrt{1 + (dy/dx)^2}\, dx$ gives

$$A = \int_a^b 2\pi x \sqrt{1 + [f'(x)]^2}\, dx. \tag{12}$$

But if the smooth arc is presented in the form $x = g(y)$, $c \leqq y \leqq d$, then the symbolic substitution of $ds = \sqrt{1 + (dx/dy)^2}\, dy$ into Eq. (11) gives

$$A = \int_c^d 2\pi g(y) \sqrt{1 + [g'(y)]^2}\, dy. \tag{13}$$

Equations (12) and (13) may be verified by using approximations similar to the one leading to Eq. (8).

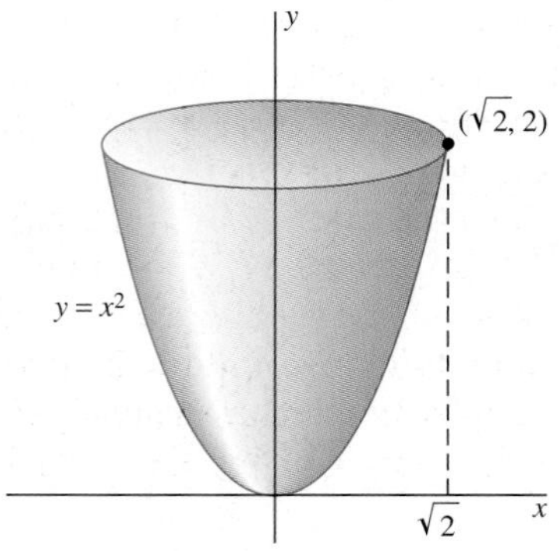

FIGURE 6.4.14 The paraboloid of Example 5.

EXAMPLE 5 Find the area of the paraboloid shown in Fig. 6.4.14, which is obtained by revolving the parabolic arc $y = x^2$, $0 \leqq x \leqq \sqrt{2}$, around the y-axis.

Solution Following the suggestion that precedes the example, we get

$$\begin{aligned} A &= \int_{\star}^{\star\star} 2\pi x\, ds = \int_a^b 2\pi x \sqrt{1 + \left(\frac{dy}{dx}\right)^2}\, dx \\ &= \int_0^{\sqrt{2}} 2\pi x \sqrt{1 + (2x)^2}\, dx \\ &= \int_0^{\sqrt{2}} \frac{\pi}{4}(1 + 4x^2)^{1/2} \cdot 8x\, dx = \left[\frac{\pi}{6}(1 + 4x^2)^{3/2}\right]_0^{\sqrt{2}} = \frac{13}{3}\pi. \end{aligned}$$

◆

SURFACE AREA SUMMARY

In conclusion, we have *four* formulas for areas of surfaces of revolution, summarized in the table in Fig. 6.4.15. Which of these formulas is appropriate for computing the area of a given surface depends on two factors:

1. Whether the smooth arc that generates the surface is presented in the form $y = f(x)$ or in the form $x = g(y)$, and
2. Whether this arc is to be revolved around the x-axis or around the y-axis.

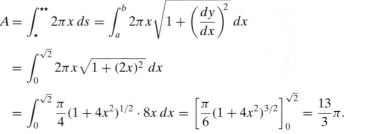

Description of curve C	Axis of revolution: x-axis		Axis of revolution: y-axis	
$y = f(x)$, $a \leqq x \leqq b$	$\int_a^b 2\pi f(x)\sqrt{1 + [f'(x)]^2}\, dx$	**(8)**	$\int_a^b 2\pi x\sqrt{1 + [f'(x)]^2}\, dx$	**(12)**
$x = g(y)$, $c \leqq y \leqq d$	$\int_c^d 2\pi y\sqrt{1 + [g'(y)]^2}\, dy$	**(10)**	$\int_c^d 2\pi g(y)\sqrt{1 + [g'(y)]^2}\, dy$	**(13)**

FIGURE 6.4.15 Area formulas for surfaces of revolution.

Memorizing the four formulas in the table is unnecessary. We suggest that you instead remember the abbreviated formulas in Eqs. (9) and (11) in conjunction with Figs. 6.4.11 and 6.4.13 and make either the substitution

$$y = f(x), \qquad ds = \sqrt{1 + \left(\frac{dy}{dx}\right)^2}\, dx$$

or the substitution

$$x = g(y), \qquad ds = \sqrt{1 + \left(\frac{dx}{dy}\right)^2}\, dy,$$

depending on whether the smooth arc is presented as a function of x or as a function of y. It may also be helpful to note that all four of these surface-area formulas have the form

$$A = \int_{\star}^{\star\star} 2\pi r\, ds, \tag{14}$$

where r denotes the radius of the circle around which the arc length element ds is revolved.

As in earlier sections, we again caution you to identify the independent variable by examining the differential and to express every dependent variable in terms of the independent variable before you antidifferentiate. That is, either express everything, including ds, in terms of x (and dx) or everything in terms of y (and dy).

The decision of which abbreviated formula—Eq. (9) or Eq. (11)—to use is determined by the axis of revolution. In contrast, the decision of whether the variable of integration should be x or y is made by the way in which the smooth arc is given: as a function of x or as a function of y. In some problems, either x or y may be used as the variable of integration, but the integral is usually much simpler to evaluate if you make the correct choice. Experience is very helpful here. Right now, try Example 5 with independent variable y.

6.4 TRUE/FALSE STUDY GUIDE

6.4 CONCEPTS: QUESTIONS AND DISCUSSION

Frequently a concept is enriched when it is viewed from different perspectives.

1. Show that the arc length method of this section gives

$$C = 8\int_0^{1/\sqrt{2}} \frac{1}{\sqrt{1-x^2}}\, dx$$

for the circumference of the unit circle $x^2 + y^2 = 1$. In Section 6.8 we will see that

$$D_x \sin^{-1} x = \frac{1}{\sqrt{1-x^2}},$$

where $y = \sin^{-1} x$ denotes the angle in $[-\pi/2, \pi/2]$ such that $\sin y = x$. Conclude that

$$C = 8\Big[\sin^{-1} x\Big]_0^{1/\sqrt{2}} = 8\left(\frac{\pi}{4} - 0\right) = 2\pi.$$

2. Suppose that P_n denotes the perimeter of a regular polygon with n sides inscribed in the unit circle. Assume that the limit $P = \lim_{n\to\infty} P_n$ exists (it does). Can you conclude from the definition of the integral as a limit of Riemann sums that $P = C$ (the circumference of the circle of Question 1)?
3. Can you *prove* that the limit $P = \lim_{n\to\infty} P_n$ in Question 2 exists? Can you show that if Q_m is the perimeter of a regular polygon with m sides

circumscribed about the unit circle, then $P_n < Q_m$? Can you see that this implies that there is a limit as to how large P_n can be?

4. Problem 52 in Section 5.3 implies that the perimeter P_n of Question 2 is given by

$$P_n = 2n \sin \frac{\pi}{n}.$$

Use l'Hôpital's rule to show that $P = \lim_{n\to\infty} P_n = 2\pi$.

6.4 PROBLEMS

In Problems 1 through 10, set up and simplify the integral that gives the length of the given smooth arc. Do not evaluate the integral.

1. $y = x^2, \quad 0 \leqq x \leqq 1$

2. $y = x^{5/2}, \quad 1 \leqq x \leqq 3$

3. $y = 2x^3 - 3x^2, \quad 0 \leqq x \leqq 2$

4. $y = x^{4/3}, \quad -1 \leqq x \leqq 1$

5. $y = 1 - x^2, \quad 0 \leqq x \leqq 100$

6. $x = 4y - y^2, \quad 0 \leqq y \leqq 1$

7. $x = y^4, \quad -1 \leqq y \leqq 2$

8. $y = e^x, \quad 0 \leqq x \leqq 1$

9. $y = \ln x, \quad 1 \leqq x \leqq 2$

10. $y = \ln(\cos x), \quad 0 \leqq x \leqq \pi/4$

In Problems 11 through 20, set up and simplify the integral that gives the surface area of revolution generated by rotation of the given smooth arc around the given axis. Do not evaluate the integral.

11. $y = x^2, \quad 0 \leqq x \leqq 4;$ the x-axis

12. $y = x^2, \quad 0 \leqq x \leqq 4;$ the y-axis

13. $y = x - x^2, \quad 0 \leqq x \leqq 1;$ the x-axis

14. $y = x^2, \quad 0 \leqq x \leqq 1;$ the line $y = 4$

15. $y = x^2, \quad 0 \leqq x \leqq 1;$ the line $x = 2$

16. $y = x - x^3, \quad 0 \leqq x \leqq 1;$ the x-axis

17. $y = \ln(x^2 - 1), \quad 2 \leqq x \leqq 3;$ the x-axis

18. $y = \sqrt{x}, \quad 1 \leqq x \leqq 4;$ the y-axis

19. $y = \ln(x + 1), \quad 0 \leqq x \leqq 1;$ the line $x = -1$

20. $y = x^{5/2}, \quad 1 \leqq x \leqq 4;$ the line $y = -2$

Find the lengths of the smooth arcs in Problems 21 through 28.

21. $y = \frac{2}{3}(x^2 + 1)^{3/2}$ from $x = 0$ to $x = 2$

22. $x = \frac{2}{3}(y - 1)^{3/2}$ from $y = 1$ to $y = 5$

23. $y = \dfrac{1}{6}x^3 + \dfrac{1}{2x}$ from $x = 1$ to $x = 3$

24. $x = \dfrac{1}{8}y^4 + \dfrac{1}{4y^2}$ from $y = 1$ to $y = 2$

25. $8x^2y - 2x^6 = 1$ from $\left(1, \frac{3}{8}\right)$ to $\left(2, \frac{129}{32}\right)$

26. $12xy - 4y^4 = 3$ from $\left(\frac{7}{12}, 1\right)$ to $\left(\frac{67}{24}, 2\right)$

27. $y = \frac{1}{2}(e^x + e^{-x})$ from $x = 0$ to $x = 1$

28. $y = \frac{1}{8}x^2 - \ln x$ from $x = 1$ to $x = 2$

In Problems 29 through 35, find the area of the surface of revolution generated by revolving the given curve around the indicated axis.

29. $y = \sqrt{x}, 0 \leqq x \leqq 1;$ the x-axis

30. $y = x^3, 1 \leqq x \leqq 2;$ the x-axis

31. $y = \dfrac{1}{5}x^5 + \dfrac{1}{12x^3}, 1 \leqq x \leqq 2;$ the y-axis

32. $x = \dfrac{1}{8}y^4 + \dfrac{1}{4y^2}, 1 \leqq y \leqq 2;$ the x-axis

33. $y^3 = 3x, 0 \leqq x \leqq 9;$ the y-axis

34. $y = \frac{1}{2}(e^x + e^{-x}), 0 \leqq x \leqq 1;$ the x-axis

35. $y = x^2 - \frac{1}{8}\ln x, 1 \leqq x \leqq 2;$ the y-axis

36. Prove that the length of one arch of the sine curve $y = \sin x$ is equal to half the circumference of the ellipse $2x^2 + y^2 = 2$. [*Suggestion:* Substitute $x = \cos\theta$ into the arc length integral for the ellipse.]

37. Use Simpson's approximation with $n = 6$ subintervals to estimate the length of the sine arch of Problem 36.

38. Use Simpson's approximation with $n = 10$ subintervals to estimate the length of the parabola $y = x^2$ from $x = 0$ to $x = 1$.

39. Verify Eq. (6) for the curved surface area of a conical frustum. Think of the frustum as being generated by revolving around the y-axis the line segment from $(r_1, 0)$ to (r_2, h).

40. By considering a sphere of radius r to be a surface of revolution, derive the formula $A = 4\pi r^2$ for its surface area.

41. Find the total length of the *astroid* shown in Fig. 6.4.16. The equation of its graph is $x^{2/3} + y^{2/3} = 1$.

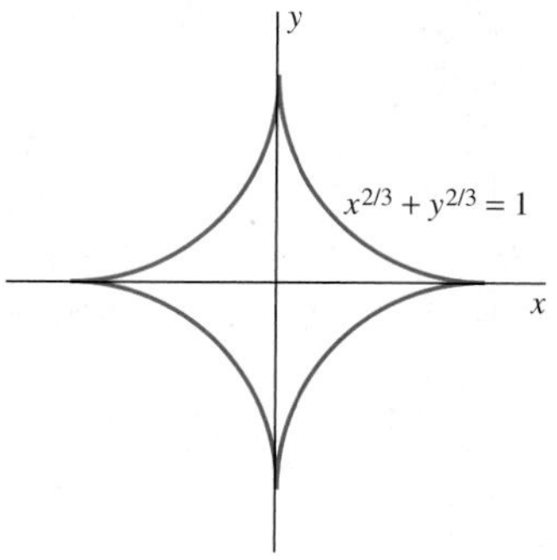

FIGURE 6.4.16 The astroid of Problem 41.

42. Find the area of the surface generated by revolving the astroid of Problem 41 around the y-axis (Fig. 6.4.17).

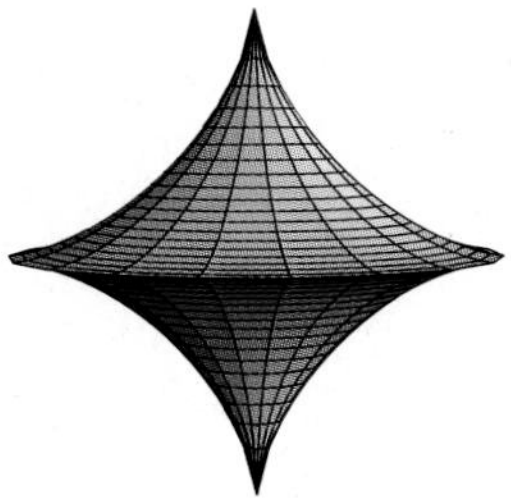

FIGURE 6.4.17 The surface of Problem 42.

43. Figure 6.4.18 shows a *spherical zone* of height h—it is cut out of the sphere by two parallel planes that intersect the sphere. Show that the surface area of this zone is $A = 2\pi rh$, where r is the radius of the sphere and h (the *height* of the zone) is the distance between the two planes. Note that A depends only on the height of the zone, and not (otherwise) on the specific location of the two planes relative to the sphere.

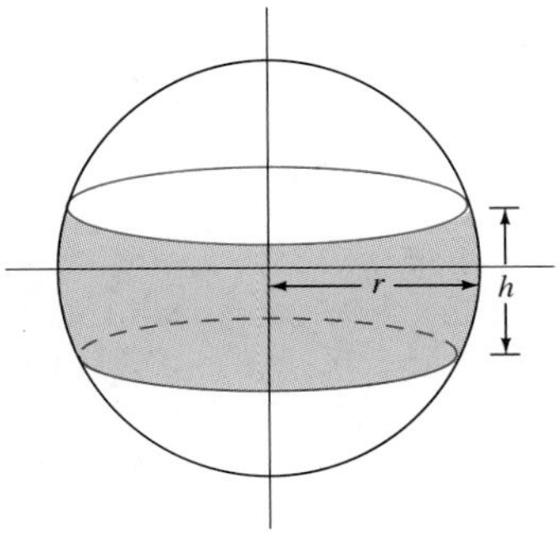

FIGURE 6.4.18 The spherical zone of Problem 43.

44. Figure 6.4.19 shows a loop of the curve $32y^2 = x^2(4 - x^2)$. Find the surface area generated by revolving this loop around the x-axis.

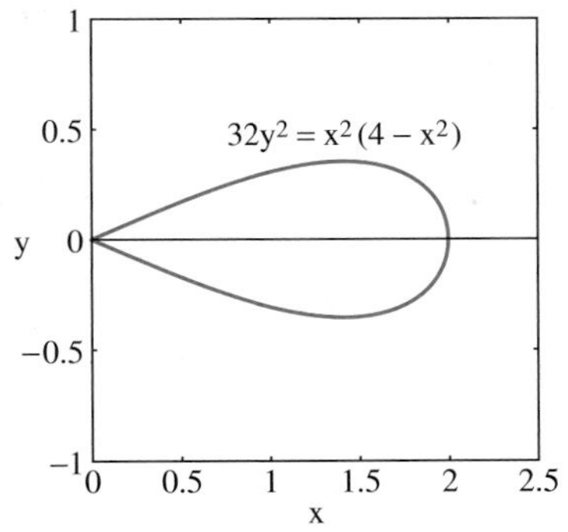

FIGURE 6.4.19 The loop of Problem 44.

45. Figure 6.4.20 shows a cable for a suspension bridge. The cable has the shape of a parabola with equation $y = kx^2$. The suspension bridge has total span $2S$ and the height of the cable (relative to its lowest point) is H at each end. Show that the total length of the cable is given by

$$L = 2\int_0^S \sqrt{1 + \frac{4H^2}{S^4}x^2}\,dx.$$

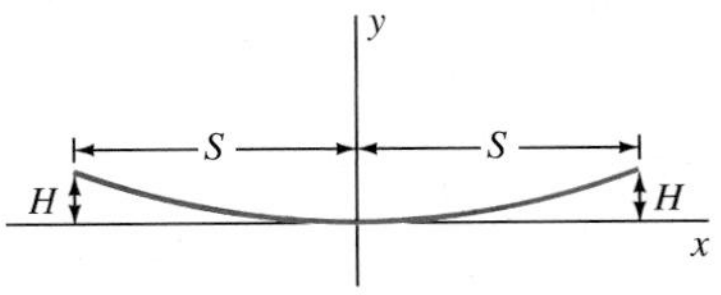

FIGURE 6.4.20 The parabolic supporting cable of a suspension bridge.

46. Italian engineers have proposed a single-span suspension bridge across the Strait of Messina (8 km wide) between Italy and Sicily. The plans include suspension towers 380 meters high at each end. Use the integral in Problem 45 to approximate the length L of the parabolic suspension cables for this proposed bridge. Assuming that the given dimensions are exact, use Simpson's approximation to estimate the integral with sufficient accuracy to determine L to the nearest meter.

6.5 FORCE AND WORK

The concept of *work* is introduced to measure the cumulative effect of a force in moving a body from one position to another. In the simplest case, a particle is moved along a straight line by the action of a *constant* force. The work done by such a force is defined to be the product of the force and the distance through which it acts. Thus if the constant force has magnitude F and the particle is moved through the distance d, then the work done by the force is given by

$$W = F \cdot d. \tag{1}$$

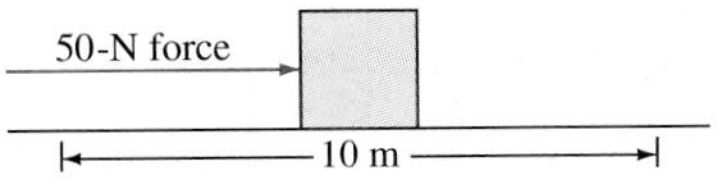

FIGURE 6.5.1 A 50-N force does 500 N·m of work in pushing a box 10 m.

EXAMPLE 1 If a constant horizontal force of 50 newtons (N) is applied to a heavy box to push it a distance of 10 m along a rough floor (Fig. 6.5.1), then the work done by the force is

$$W = 50 \cdot 10 = 500$$

newton-meters (N · m). Note the units; because of the definition of work, units of work are always products of force units and distance units. For another example, to lift a weight of 75 lb a vertical distance of 5 ft, a constant force of 75 lb must be applied. The work done by this force is

$$W = 75 \cdot 5 = 375$$

foot-pounds (ft · lb). ◆

Work Done by a Variable Force

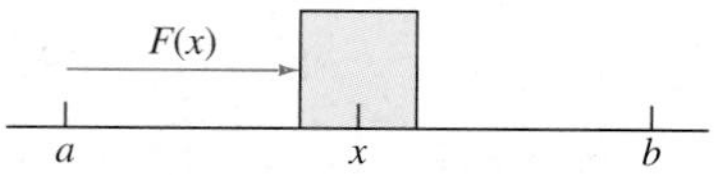

FIGURE 6.5.2 A variable force pushing a particle from a to b.

Here we use the integral to generalize the definition of work to the case in which a particle is moved along a straight line by a *variable* force. Given a **force function** $F(x)$ defined at each point x of the straight line segment $[a, b]$, we want to define the work W done by this variable force in pushing the particle from the point $x = a$ to the point $x = b$ (Fig. 6.5.2).

We begin with the usual partition of the interval $[a, b]$ into n subintervals, all with the same length $\Delta x = (b - a)/n$. For each i $(1 \leqq i \leqq n)$, let $x_i^\star$ be an arbitrary point of the ith subinterval $[x_{i-1}, x_i]$. The key idea is to approximate the actual work ΔW_i done by the *variable* force $F(x)$ in moving the particle from x_{i-1} to x_i by the work $F(x_i^\star)\,\Delta x$ (force × distance) done in moving a particle the distance Δx from x_{i-1} to x_i (Fig. 6.5.3). Thus

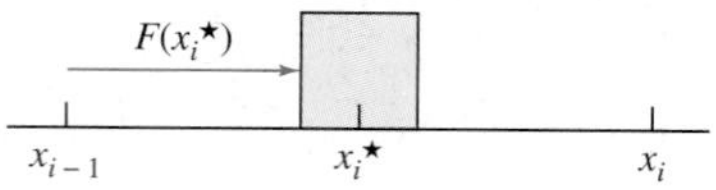

FIGURE 6.5.3 The *constant* force $F(x_i^\star)$ acting through the ith subinterval.

$$\Delta W_i \approx F(x_i^\star)\,\Delta x. \tag{2}$$

We approximate the total work W by summing from $i = 1$ to $i = n$:

$$W = \sum_{i=1}^{n} \Delta W_i \approx \sum_{i=1}^{n} F(x_i^\star)\,\Delta x. \tag{3}$$

But the final sum in (3) is a Riemann sum for $F(x)$ on the interval $[a, b]$, and as $n \to +\infty$ (and $\Delta x \to 0$), such sums approach the *integral* of $F(x)$ from $x = a$ to $x = b$. We therefore are motivated to *define* the **work** W done by the force $F(x)$ in moving the particle from $x = a$ to $x = b$ to be

$$W = \int_a^b F(x)\,dx. \tag{4}$$

The following heuristic way of setting up Eq. (4) is useful in obtaining integrals for work problems. Imagine that dx is so small a number that the value of $F(x)$ does not change appreciably on the tiny interval from x to $x + dx$. Then the work done by the force F in moving a particle from x to $x + dx$ should be very close to

$$dW = F(x)\,dx.$$

The natural additive property of work then implies that we could obtain the total work W by adding these tiny elements of work:

$$W = \int_{\star}^{\star\star} dW = \int_a^b F(x)\,dx.$$

Elastic Springs

Consider a spring whose left end is held fixed and whose right end is free to move along the x-axis. We assume that the right end is at the origin $x = 0$ when the spring has its **natural length**—that is, when the spring is in its rest position, neither compressed nor stretched by outside forces.

According to **Hooke's law** for elastic springs, the force $F(x)$ that must be exerted on the spring to hold its right end at the point x is proportional to the displacement x of the right end from its rest position. That is,

$$F(x) = kx, \tag{5}$$

where k is a positive constant. The constant k, called the **spring constant,** is a characteristic of the particular spring under study.

Figure 6.5.4 shows the arrangement of such a spring along the x-axis. The right end of the spring is held at position x on the x-axis by a force $F(x)$. The figure shows the situation for $x > 0$, so the spring is stretched. The force that the spring exerts on its right-hand end is directed to the left, so—as the figure indicates—the external force $F(x)$ must act to the right. The right is the positive direction here, so $F(x)$ must be a positive number. Because x and $F(x)$ have the same sign, k must also be positive. You can check that k is positive as well in the case $x < 0$.

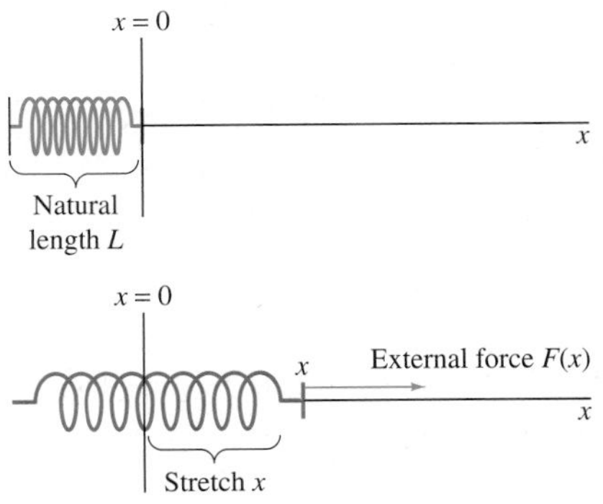

FIGURE 6.5.4 The stretch x is proportional to the impressed force F.

Work in the sense of physics is different than work in the sense of physiology. At this moment the weightlifter is doing no work in the physics sense because he is holding the weight still.

EXAMPLE 2 Suppose that a spring has a natural length of 1 ft and that a force of 10 lb is required to hold it compressed to a length of 6 in. How much work is done in stretching the spring from its natural length to a total length of 2 ft?

Solution To move the free end from $x = 0$ (the natural-length position) to $x = 1$ (stretched by 1 ft), we must exert a variable force $F(x)$ determined by Hooke's law. We are given that $F = -10$ (lb) when $x = -0.5$ (ft), so Eq. (5), $F = kx$, implies that the spring constant for this spring is $k = 20$ (lb/ft). Thus $F(x) = 20x$, and so—using Eq. (4)—we find that the work done in stretching this spring in the manner given is

$$W = \int_0^1 20x\,dx = \Big[10x^2\Big]_0^1 = 10 \quad (\text{ft} \cdot \text{lb}). \qquad \blacklozenge$$

Work Done Against Gravity

According to Newton's law of gravitation, the force that must be exerted on a body to hold it at a distance r from the center of the earth is inversely proportional to r^2 (if $r \geqq R$, the radius of the earth). In other words, if $F(r)$ denotes the holding force, then

$$F(r) = \frac{k}{r^2} \tag{6}$$

for some positive constant k. The value of this force at the surface of the earth, where $r = R \approx 4000$ mi (about 6370 km), is called the **weight** of the body.

Given the weight $F(R)$ of a particular body, we can find the corresponding value of k by using Eq. (6):

$$k = R^2 \cdot F(R).$$

The work that must be done to lift the body vertically from the surface to a distance $R_1 > R$ from the center of the earth is then

$$W = \int_R^{R_1} \frac{k}{r^2}\,dr. \tag{7}$$

If distance is measured in miles and force in pounds, then this integral gives the work in mile-pounds. This is a very unconventional unit of work. We shall multiply by 5280 (ft/mi) to convert any such result into foot-pounds.

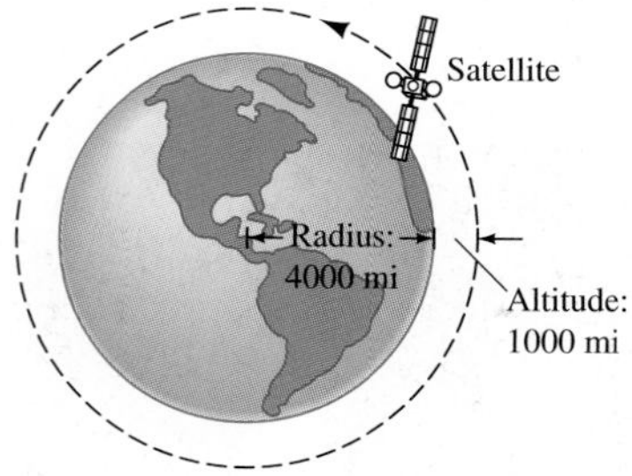

FIGURE 6.5.5 A satellite in orbit 1000 mi above the surface of the earth (Example 3).

EXAMPLE 3 (Satellite Launch) How much work must be done to lift a 1000-lb satellite vertically from the earth's surface to an orbit 1000 mi above the surface? See Fig. 6.5.5, and take $R = 4000$ (mi) to be the radius of the earth.

Solution Because $F = 1000$ (lb) when $r = R = 4000$ (mi), we find from Eq. (6) that

$$k = 4000^2 \cdot 1000 = 16 \times 10^9 \quad (\text{mi}^2\text{·lb}).$$

Then by Eq. (7), the work done is

$$W = \int_{4000}^{5000} \frac{k}{r^2}\,dr = \left[-\frac{k}{r}\right]_{4000}^{5000}$$
$$= (16 \times 10^9) \cdot \left(\tfrac{1}{4000} - \tfrac{1}{5000}\right) = 8 \times 10^5 \quad (\text{mi·lb}).$$

We multiply by 5280 (ft/mi) and write the answer as

$$4.224 \times 10^9 = 4{,}224{,}000{,}000 \quad (\text{ft·lb}).$$ ◆

We can instead express the answer to Example 3 in terms of the power that the launch rocket must provide. **Power** is the rate at which work is done. For instance, 1 **horsepower** (hp) is defined to be 33,000 ft·lb/min. If the ascent to orbit takes 15 min and if only 2% of the power generated by the rocket is effective in lifting the satellite (the rest is used to lift the rocket and its fuel), we can convert the answer in Example 3 to horsepower. The *average* power that the rocket engine must produce during the 15-min ascent is

$$P = \frac{50 \cdot (4.224 \times 10^9)}{15 \cdot 33{,}000} \approx 427{,}000 \quad (\text{hp}).$$

The factor of 50 in the numerator comes from the 2% "efficiency" of the rocket: The total power must be multiplied by $1/(0.02) = 50$.

Work Done in Filling a Tank

Examples 2 and 3 are applications of Eq. (4) for calculating the work done by a variable force in moving a particle a certain distance. Another common type of force-work problem involves the summation of work done by constant forces that act through different distances. For example, consider the problem of pumping a fluid from ground level—where we take $y = 0$—up into an aboveground tank (Fig. 6.5.6).

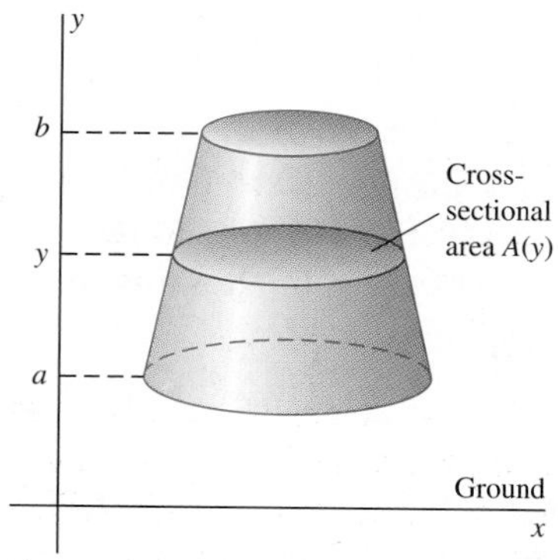

FIGURE 6.5.6 An aboveground tank.

It is convenient to think of the tank as being filled in thin, horizontal layers of fluid, each lifted from its ground to its final position in the tank. No matter how the fluid actually behaves as the tank is filled, this simple way of thinking about the process gives us a way to compute the work done in the filling process. But when we think of filling the tank in this way, we must allow for the fact that different layers of fluid are lifted different distances to reach their final positions in the tank.

Suppose that the bottom of the tank is at height $y = a$ and that its top is at height $y = b > a$. Let $A(y)$ be the cross-sectional area of the tank at height y. Consider a partition of $[a, b]$ into n subintervals, all with the same length Δy. Then the volume

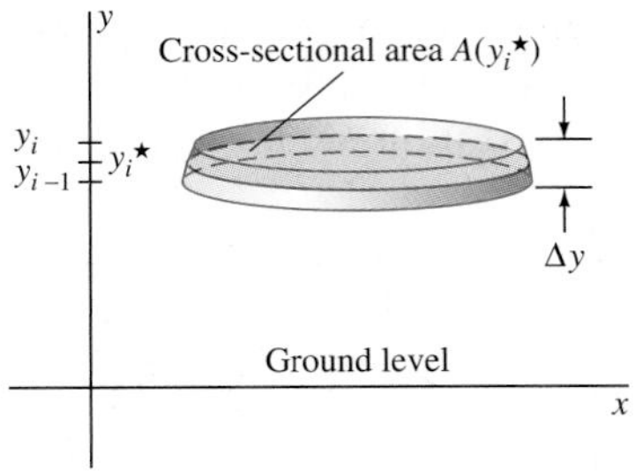

FIGURE 6.5.7 A thin horizontal slice of fluid with volume $\Delta V = A(y_i^\star)\,\Delta y$. Each particle of this slice must be lifted (from the ground at $y = 0$) a distance between y_{i-1} and y_i.

of the horizontal slice (Fig. 6.5.7) of the tank that corresponds to the ith subinterval $[y_{i-1}, y_i]$ is

$$\Delta V_i = \int_{y_{i-1}}^{y_i} A(y)\,dy = A(y_i^\star)\,\Delta y$$

for some number $y_i^\star$ in $[y_{i-1}, y_i]$; this is a consequence of the average value theorem for integrals (Section 5.6). If ρ is the density of the fluid (in pounds per cubic foot, for example), then the force required to lift this slice from the ground to its final position in the tank is simply the (constant) weight of the slice:

$$F_i = \rho\,\Delta V_i = \rho A(y_i^\star)\,\Delta y.$$

But what about the distance through which this force must act? The fluid in question is lifted from ground level to the level of the subinterval $[y_{i-1}, y_i]$, so every particle of the fluid is lifted at least the distance y_{i-1} and at most the distance y_i (remember, the fluid begins its journey at ground level, where $y = 0$). Hence the work ΔW_i needed to lift this ith slice of fluid satisfies the inequalities

$$F_i y_{i-1} \leqq \Delta W_i \leqq F_i y_i;$$

that is,

$$\rho y_{i-1} A(y_i^\star)\,\Delta y \leqq \Delta W_i \leqq \rho y_i A(y_i^\star)\,\Delta y.$$

Now we add these inequalities for $i = 1, 2, 3, \dots, n$ and find thereby that the total work $W = \sum \Delta W_i$ satisfies the inequalities

$$\sum_{i=1}^{n} \rho y_{i-1} A(y_i^\star)\,\Delta y \leqq W \leqq \sum_{i-1}^{n} \rho y_i A(y_i^\star)\,\Delta y.$$

If the three points y_{i-1}, y_i, and $y_i^\star$ of $[y_{i-1}, y_i]$ were the same, then both the last two sums would be Riemann sums for the function $f(y) = \rho y A(y)$ on $[a, b]$. Although the three points are not the same, it still follows—from a result stated in Appendix G—that both sums approach

$$\int_a^b \rho y A(y)\,dy \quad \text{as} \quad \Delta y \to 0.$$

The squeeze law of limits therefore gives the formula

$$W = \int_a^b \rho y A(y)\,dy. \qquad \textbf{(8)}$$

This is the work W done in pumping fluid of density ρ from the ground into a tank that has horizontal cross-sectional area $A(y)$ and is located between heights $y = a$ and $y = b$ above the ground.

A quick heuristic way to set up Eq. (8), and many variants of it, is to think of a thin, horizontal slice of fluid with volume $dV = A(y)\,dy$ and weight $\rho\,dV = \rho A(y)\,dy$. The work required to lift this slice a distance y is

$$dW = y \cdot \rho\,dV = \rho y A(y)\,dy,$$

so the total work required to fill the tank is

$$W = \int_\star^{\star\star} dW = \int_a^b \rho y A(y)\,dy,$$

because the horizontal slices lie between $y = a$ and $y = b$.

EXAMPLE 4 Suppose that it took 20 yr to construct the great pyramid of Khufu at Gizeh, Egypt. This pyramid is 500 ft high and has a square base with edge length 750 ft. Suppose also that the pyramid is made of rock with density $\rho = 120$ lb/ft^3. Finally, suppose that each laborer did 160 ft·lb/h of work in lifting rocks from ground level to their final position in the pyramid and worked 12 h daily for 330 days/yr. How many laborers would have been required to construct the pyramid?

Solution We assume a constant labor force throughout the 20-yr construction period. We think of the pyramid as being made up of thin, horizontal slabs of rock, each slab lifted (just like a slice of liquid) from ground level to its ultimate height. Hence we can use Eq. (8) to compute the work W required.

The great pyramid of Khufu.

Figure 6.5.8 shows a vertical cross section of the pyramid. The horizontal cross section at height y is a square with edge length s. We see from the similar triangles in Fig. 6.5.8 that

$$\frac{s}{750} = \frac{500 - y}{500}, \qquad \text{so} \quad s = \frac{3}{2}(500 - y).$$

Hence the cross-sectional area at height y is

$$A(y) = \tfrac{9}{4}(500 - y)^2.$$

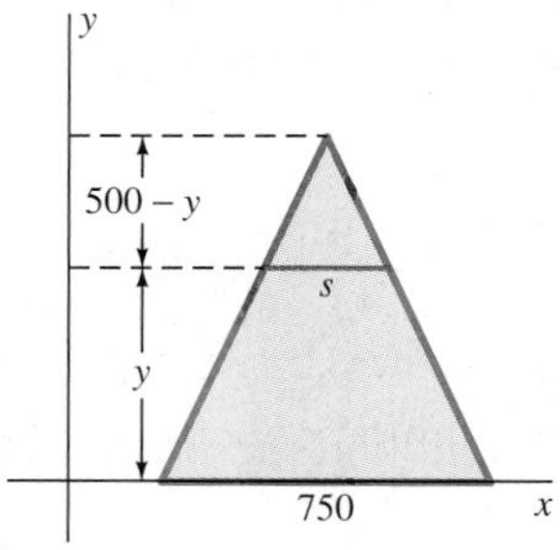

FIGURE 6.5.8 Vertical cross section of Khufu's pyramid.

Equation (8) therefore gives

$$W = \int_0^{500} 120 \cdot y \cdot \tfrac{9}{4}(500 - y)^2\,dy$$

$$= 270 \int_0^{500} (250{,}000y - 1000y^2 + y^3)\,dy$$

$$= 270\left[125{,}000y^2 - \tfrac{1000}{3}y^3 + \tfrac{1}{4}y^4\right]_0^{500},$$

so $W \approx 1.406 \times 10^{12}$ ft·lb.

Because each laborer does

$$160 \cdot 12 \cdot 330 \cdot 20 \approx 1.267 \times 10^7 \quad \text{ft·lb}$$

of work, the construction of the pyramid would—under our assumptions—have required

$$\frac{1.406 \times 10^{12}}{1.267 \times 10^7} \approx 111{,}000$$

laborers. ◆

Emptying a Tank

Suppose now that the tank shown in Fig. 6.5.9 is already filled with a liquid of density ρ lb/ft^3, and we want to pump all this liquid from the tank up to the level $y = h$ above the top of the tank. We imagine a thin, horizontal slice of liquid at height y. If its

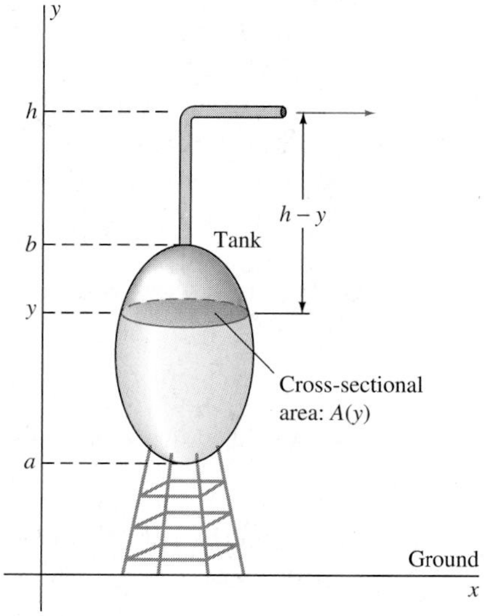

FIGURE 6.5.9 Pumping liquid from a tank to a higher level.

thickness is dy, then its volume is $dV = A(y)\,dy$, so its weight is $\rho\,dV = \rho A(y)\,dy$. This slice must be lifted the distance $h - y$, so the work done to lift the slice is

$$dW = (h - y)\rho\,dV = \rho(h - y)A(y)\,dy.$$

Hence the total amount of work done on all the liquid originally in the tank is

$$W = \int_a^b \rho(h - y)A(y)\,dy. \tag{9}$$

Problem 14 asks you to use Riemann sums to set up this integral.

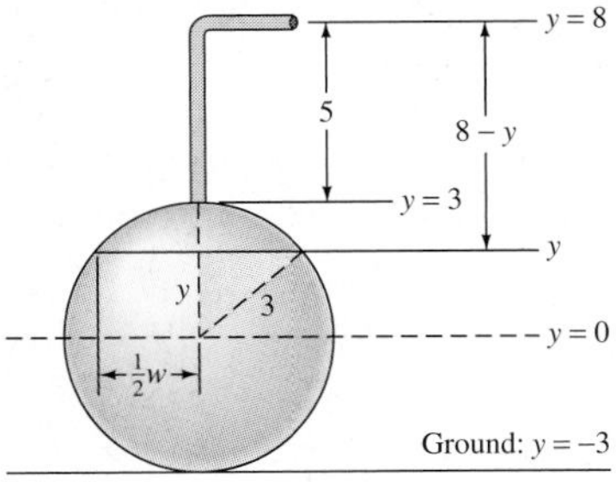

FIGURE 6.5.10 End view of the cylindrical tank of Example 5.

EXAMPLE 5 A cylindrical tank of radius 3 ft and length 10 ft is lying on its side on horizontal ground. If this tank initially is full of gasoline weighing 40 lb/ft^3, how much work is done in pumping all this gasoline to a point 5 ft above the top of the tank?

Solution Figure 6.5.10 shows an end view of the tank. To exploit its circular symmetry, we choose $y = 0$ at the *center* of the circular vertical section, so the tank lies between $y = -3$ and $y = 3$. A horizontal cross section of the tank that meets the y-axis is a rectangle of length 10 ft and width w. From the right triangle in Fig. 6.5.10, we see that

$$\tfrac{1}{2}w = \sqrt{9 - y^2},$$

so the area of this cross section is

$$A(y) = 10w = 20\sqrt{9 - y^2}.$$

This cross section must be lifted from its initial position y to the final position $5+3 = 8$, so it is to be lifted the distance $8 - y$. Thus Eq. (9) with $\rho = 40$, $a = -3$, and $b = 3$ yields

$$\begin{aligned} W &= \int_{-3}^{3} 40 \cdot (8 - y) \cdot 20\sqrt{9 - y^2}\,dy \\ &= 6400\int_{-3}^{3} \sqrt{9 - y^2}\,dy - 800\int_{-3}^{3} y\sqrt{9 - y^2}\,dy. \end{aligned}$$

We attack the two integrals separately. First,

$$\int_{-3}^{3} y\sqrt{9 - y^2}\,dy = \left[-\tfrac{1}{3}(9 - y^2)^{3/2}\right]_{-3}^{3} = 0.$$

Second,

$$\int_{-3}^{3} \sqrt{9 - y^2}\,dy = \tfrac{1}{2}\pi \cdot 3^2 = \tfrac{9}{2}\pi,$$

because the integral is simply the area of a semicircle of radius 3. Hence

$$W = 6400 \cdot \tfrac{9}{2}\pi = 28800\pi,$$

approximately 90,478 ft·lb. ◆

REMARK As in Example 5, you may use as needed in the problems the integral

$$\int_0^a \sqrt{a^2 - x^2}\,dx = \tfrac{1}{4}\pi a^2, \tag{10}$$

which corresponds to the area of a quarter-circle of radius a.

Force Exerted by a Liquid

The **pressure** p at depth h in a liquid is the force per unit area exerted by the liquid at that depth. Pressure is given by

$$p = \rho h, \tag{11}$$

where ρ is the (weight) density of the liquid. For example, at a depth of 10 ft in water, for which $\rho = 62.4$ lb/ft^3, the pressure is $62.4 \cdot 10 = 624$ lb/ft^2. Hence if a thin, flat plate of area 5 ft^2 is suspended in a horizontal position at a depth of 10 ft in water, then the water exerts a downward force of $624 \cdot 5 = 3120$ lb on the top face of the plate and an equal upward force on its bottom face.

It is an important fact that at a given depth in a liquid, the pressure is the same in all directions. But if a flat plate is submerged in a vertical position in the liquid, then the pressure on the face of the plate is *not* constant, because by Eq. (11) the pressure increases with increasing depth. Consequently, the total force exerted on a vertical plate must be computed by integration.

Consider a thin, vertical, flat plate submerged in a liquid of density ρ (Fig. 6.5.11). The surface of the liquid is at the line $y = c$, and the plate lies alongside the interval $a \leqq y \leqq b$. The width of the plate at depth $c - y$ is some function of y, which we denote by $w(y)$.

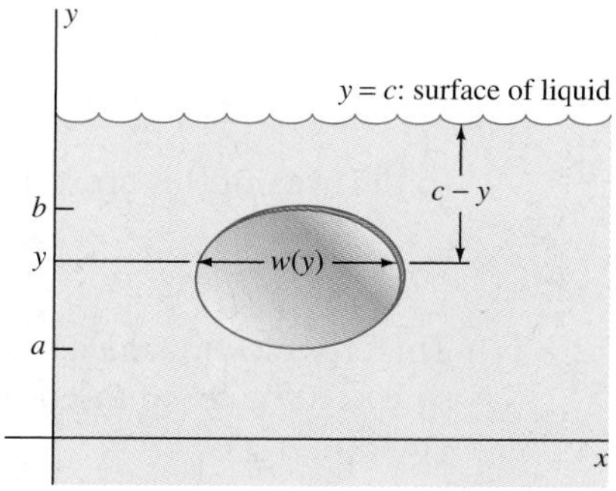

FIGURE 6.5.11 A thin plate suspended vertically in a liquid.

To compute the total force F exerted by the liquid on either face of this plate, we begin with a partition of $[a, b]$ into n subintervals, all with the same length Δy, and denote by $y_i^\star$ the midpoint of the subinterval $[y_{i-1}, y_i]$. The horizontal strip of the plate opposite this ith subinterval is approximated by a rectangle of width $w(y_i^\star)$ and height Δy, and its average depth in the liquid is $c - y_i^\star$. Hence the force ΔF_i exerted by the liquid on this horizontal strip is given approximately by

$$\Delta F_i \approx \rho(c - y_i^\star)w(y_i^\star)\,\Delta y. \tag{12}$$

The total force on the entire plate is given approximately by

$$F = \sum_{i=1}^{n} \Delta F_i \approx \sum_{i=1}^{n} \rho(c - y_i^\star)w(y_i^\star)\,\Delta y.$$

We obtain the exact value of F by taking the limit of such Riemann sums as $\Delta y \to 0$:

$$F = \int_a^b \rho(c - y)w(y)\,dy. \tag{13}$$

EXAMPLE 6 A cylindrical tank 8 ft in diameter is lying on its side and is half full of oil of density $\rho = 75$ lb/ft^3. Find the total force F exerted by the oil on one end of the tank.

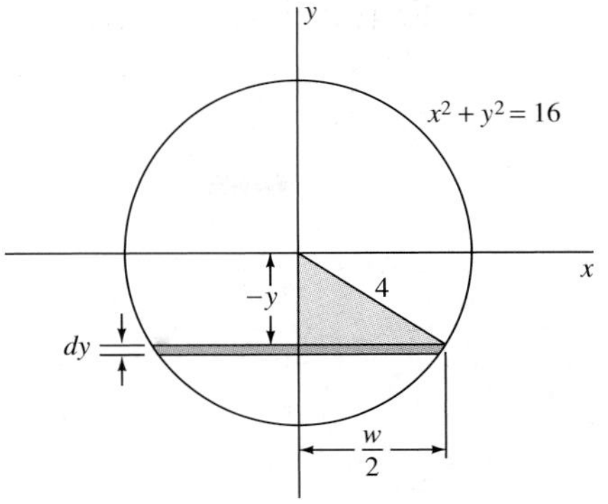

FIGURE 6.5.12 View of one end of the tank of Example 6.

Solution We locate the y-axis as indicated in Fig. 6.5.12, so that the surface of the oil is at the level $y = 0$. The oil lies alongside the interval $-4 \leqq y \leqq 0$. We see from the right triangle in the figure that the width of the oil at depth $-y$ (and thus at location y) is

$$w(y) = 2\sqrt{16 - y^2}.$$

Hence Eq. (13) gives

$$F = \int_{-4}^{0} 75(-y)\left(2\sqrt{16 - y^2}\right) dy = 75\left[\tfrac{2}{3}(16 - y^2)^{3/2}\right]_{-4}^{0} = 3200 \quad \text{(lb)}. \qquad \blacklozenge$$

6.5 TRUE/FALSE STUDY GUIDE

6.5 CONCEPTS: QUESTIONS AND DISCUSSION

1. Suppose that you lift a heavy weight from the floor to high over your head and then let it fall back to the floor. An observer says you did no work because the weight wound up in its initial position. You disagree. Who's right?
2. Consider a water tank in the shape of a solid of revolution around a vertical axis. The work formula in Eq. (8) was derived using horizontal slices. Is it possible to use cylindrical shells to derive an integral formula to use in calculating the work required to fill the tank?

6.5 PROBLEMS

In Problems 1 through 5, find the work done by the given force $F(x)$ in moving a particle along the x-axis from $x = a$ to $x = b$.

1. $F(x) = 10; \quad a = -2, b = 1$

2. $F(x) = 3x - 1; \quad a = 1, b = 5$

3. $F(x) = \dfrac{10}{x^2}; \quad a = 1, b = 10$

4. $F(x) = -3\sqrt{x}; \quad a = 0, b = 4$

5. $F(x) = \sin \pi x; \quad a = -1, b = 1$

6. A spring has a natural length of 1 m, and a force of 10 N is required to hold it stretched to a total length of 2 m. How much work is done in compressing this spring from its natural length to a length of 60 cm?

7. A spring has a natural length of 2 ft, and a force of 15 lb is required to hold it compressed at a length of 18 in. How much work is done in stretching this spring from its natural length to a length of 3 ft?

8. Apply Eq. (4) to compute the amount of work done in lifting a 100-lb weight a height of 10 ft, assuming that this work is done against the constant force of gravity.

9. Compute the amount of work (in foot-pounds) done in lifting a 1000-lb weight from an orbit 1000 mi above the earth's surface to one 2000 mi above the earth's surface. Use the value of k given in Example 3.

10. A cylindrical tank of radius 5 ft and height 10 ft is resting on the ground with its axis vertical. Use Eq. (8) to compute the amount of work done in filling this tank with water pumped in from ground level. (Use $\rho = 62.4$ lb/ft^3 for the weight density of water.)

11. A conical tank is resting on its base, which is at ground level, and its axis is vertical. The tank has radius 5 ft and height

10 ft (Fig. 6.5.13). Compute the work done in filling this tank with water ($\rho = 62.4$ lb/ft^3) pumped in from ground level.

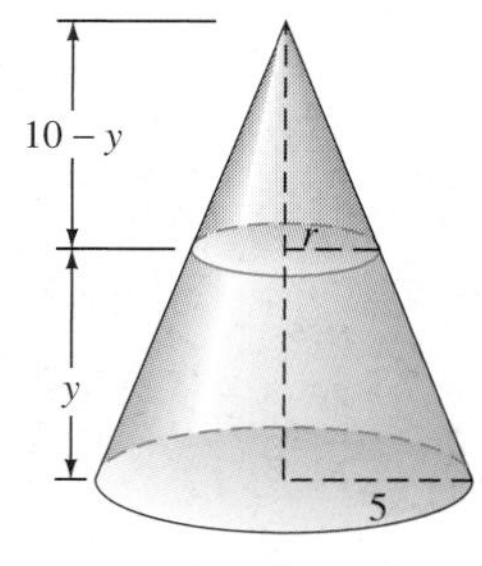

FIGURE 6.5.13 The conical tank of Problem 11.

12. Repeat Problem 11, except that now the tank is upended: Its vertex is at ground level and its base is 10 ft above the ground.

13. A tank whose lowest point is 10 ft above the ground has the shape of a cup obtained by rotating the parabola $x^2 = 5y$, $-5 \leqq y \leqq 5$, around the y-axis (Fig. 6.5.14). The units on the coordinate axes are in feet. How much work is done in filling this tank with oil of density 50 lb/ft^3 if the oil is pumped in from ground level?

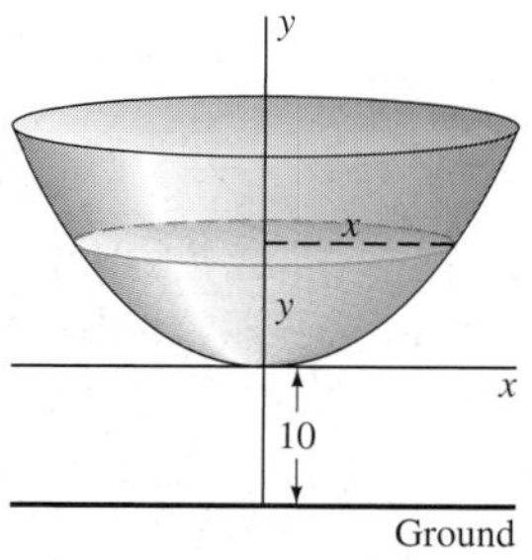

FIGURE 6.5.14 The cup-shaped tank of Problem 13.

14. Suppose that the tank of Fig. 6.5.9 is filled with fluid of density ρ and that all this fluid must be pumped from the tank to the level $y = h$ above the top of the tank. Use Riemann sums, as in the derivation of Eq. (8), to obtain the formula

$$W = \int_a^b \rho(h - y)A(y)\,dy$$

for the work required to do so.

15. Use the formula in Problem 14 to find the amount of work done in pumping the water in the tank of Problem 10 to a height of 5 ft above the top of the tank.

16. Gasoline at a service station is stored in a cylindrical tank buried on its side, with the highest part of the tank 5 ft below the surface. The tank is 6 ft in diameter and 10 ft long. The density of gasoline is 45 lb/ft^3. Assume that the filler cap of each automobile gas tank is 2 ft above the ground (Fig. 6.5.15). (a) How much work is done in emptying all the gasoline from this tank, initially full, into automobiles?

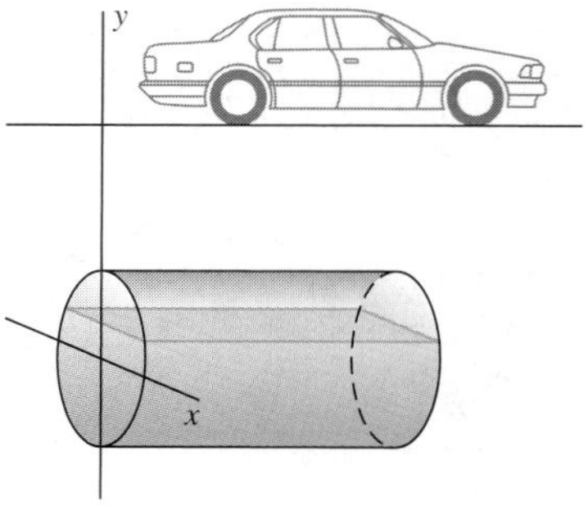

FIGURE 6.5.15 The gasoline tank of Problem 16.

(b) Recall that 1 hp is equivalent to 33,000 ft·lb/min. For electrical conversions, 1 kW (1000 W) is the same as 1.341 hp. The charge for use of electricity generated by a power company is typically about 7.2¢/kWh. Assume that the electrical motor in the gas pump at this station is 30% efficient. How much does it cost to pump all the gasoline from this tank into automobiles?

17. Consider a spherical water tank whose radius is 10 ft and whose center is 50 ft above the ground. How much work is required to fill this tank by pumping water up from ground level? [*Suggestion:* It may simplify your computations to take $y = 0$ at the center of the tank and to think of the distance each horizontal slice of water must be lifted.]

18. A hemispherical tank of radius 10 ft is located with its flat side down atop a tower 60 ft high (Fig. 6.5.16). How much work is required to fill this tank with oil of density 50 lb/ft^3 if the oil is to be pumped into the tank from ground level?

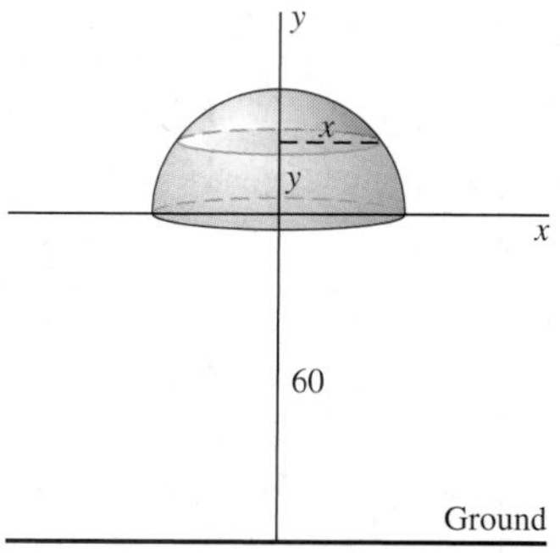

FIGURE 6.5.16 The hemispherical tank of Problem 18.

19. Water is being drawn from a well 100 ft deep, using a bucket that scoops up 100 lb of water. The bucket is pulled up at the rate of 2 ft/s, but it has a hole in the bottom through which water leaks out at the rate of 0.5 lb/s. How much work is done in pulling the bucket to the top of the well? Neglect the weight of the bucket, the weight of the rope, and the work done in overcoming friction. [*Suggestion:* Take $y = 0$ at the level of the water surface in the well, so that $y = 100$ at ground level. Let $\{y_0, y_1, y_2, \dots, y_n\}$ be a partition of $[0, 100]$ into n equal-length subintervals. Estimate the amount of work ΔW_i required to raise the bucket from y_{i-1} to y_i. Then set up the sum $W = \sum \Delta W_i$ and proceed to the appropriate integral by letting $n \to +\infty$.]

20. A rope that is 100 ft long and weighs 0.25 lb per linear foot hangs from the edge of a very tall building. How much work is required to pull this rope to the top of the building?

21. Suppose that we plug the hole in the leaky bucket of Problem 19. How much work do we do in lifting the mended bucket, full of water, to the surface, using the rope of Problem 20? Ignore friction and the weight of the bucket, but allow for the weight of the rope.

22. Consider a volume V of gas in a cylinder fitted with a piston at one end, where the pressure p of the gas is a function $p(V)$ of its volume (Fig. 6.5.17). Let A be the area of the face of the piston. Then the force exerted on the piston by gas in the cylinder is $F = pA$. Assume that the gas expands from volume V_1 to volume V_2. Show that the work done by the force F is then given by

$$W = \int_{V_1}^{V_2} p(V)\,dV.$$

[*Suggestion:* If x is the length of the cylinder (from its fixed end to the face of the piston), then $F = A \cdot p(Ax)$. Apply Eq. (4) and substitute $V = Ax$ into the resulting integral.]

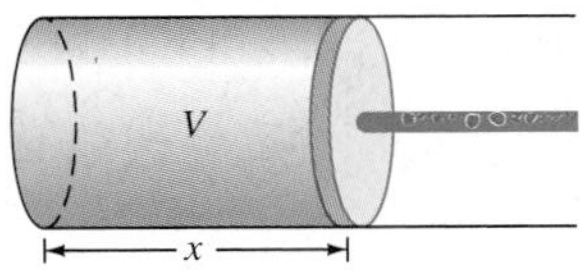

FIGURE 6.5.17 A cylinder fitted with a piston (Problem 22).

23. The pressure p and volume V of the steam in a small steam engine satisfy the condition $pV^{1.4} = c$ (where c is a constant). In one cycle, the steam expands from a volume $V_1 = 50$ in.^{3} to $V_2 = 500$ in.^{3} with an initial pressure of 200 lb/in.^{2} Use the formula in Problem 22 to compute the work, in foot-pounds, done by this engine in each such cycle.

24. A tank in the shape of a hemisphere of radius 60 is resting on its flat base with the curved surface on top. It is filled with alcohol of density 40 lb/ft^3. How much work is done in pumping all the alcohol to the level of the top of the tank?

25. A tank has the shape of the surface generated by rotating around the y-axis the graph of $y = x^4$, $0 \leqq x \leqq 1$. The tank is initially full of oil of density 60 lb/ft^3. The units on the coordinate axes are in feet. How much work is done in pumping all the oil to the level of the top of the tank?

26. A cylindrical tank of radius 3 ft and length 20 ft is lying on its side on horizontal ground. Gasoline weighing 40 lb/ft^3 is at ground level and is to be pumped into the tank. Find the work required to fill the tank.

27. The base of a spherical storage tank of radius 12 ft is at ground level. Find the amount of work done in filling the tank with oil of density 50 lb/ft^3 if all the oil is initially at ground level.

28. A 20-lb monkey is attached to a 50-ft chain that weighs 0.5 lb per (linear) foot. The other end of the chain is attached to the 40-ft-high ceiling of the monkey's cage (Fig. 6.5.18). Find the amount of work the monkey does in climbing up her chain to the ceiling.

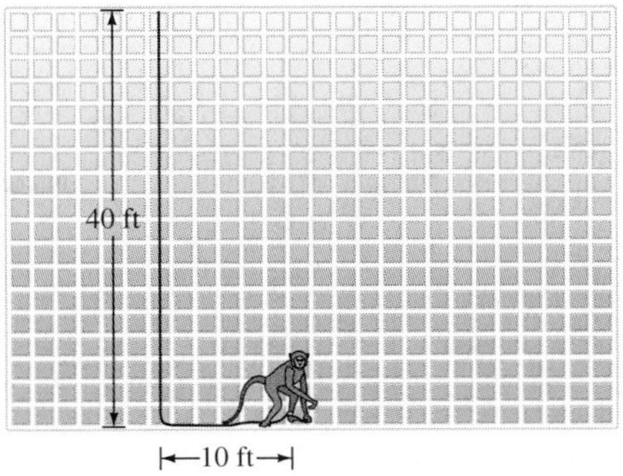

FIGURE 6.5.18 The monkey of Problem 28.

29. Tom is flying his kite at a height of 500 ft above the ground. Suppose that the kite string weighs $\frac{1}{16}$ oz per (linear) foot and is stretched in a straight line at a 45° angle to the ground. How much work was done by the wind in lifting the string from ground level up to its flying position?

30. The center of a spherical tank of radius R is at a distance $H > R$ above the ground. A liquid of weight density ρ is at ground level. Show that the work required to pump the initially empty tank full of this liquid is the same as that to lift the full tank the distance H (ignoring the weight of the tank itself).

31. A water trough 10 ft long has a square cross section that is 2 ft wide. If the trough is full of water (density $\rho = 62.4$ lb/ft^3), find the force exerted by the water on one end of the trough.

32. Repeat Problem 31 for a trough whose cross section is an equilateral triangle with edges 2 ft long.

33. Repeat Problem 31 for a trough whose cross section is a trapezoid 3 ft high, 2 ft wide at the bottom, and 4 ft wide at the top.

34. Find the force on one end of the cylindrical tank of Example 5 if the tank is filled with oil of density 50 lb/ft^3. Remember that

$$\int_0^a \sqrt{a^2 - y^2}\,dy = \tfrac{1}{4}\pi a^2,$$

because the integral represents the area of a quarter-circle of radius a.

In Problems 35 through 38, a gate in the vertical face of a dam is described. Find the total force of water on this gate if its top is 10 *ft beneath the surface of the water.*

35. A square of edge length 5 ft whose top is parallel to the water surface.

36. A circle of radius 3 ft.

37. An isosceles triangle 5 ft high and 8 ft wide at the top.

38. A semicircle of radius 4 ft whose top edge is its diameter (also parallel to the water surface).

39. Suppose that the dam of Fig. 6.5.19 is $L = 200$ ft long and $T = 30$ ft thick at its base. Find the force of water on the dam if the water is 100 ft deep and the *slanted* end of the dam faces the water.

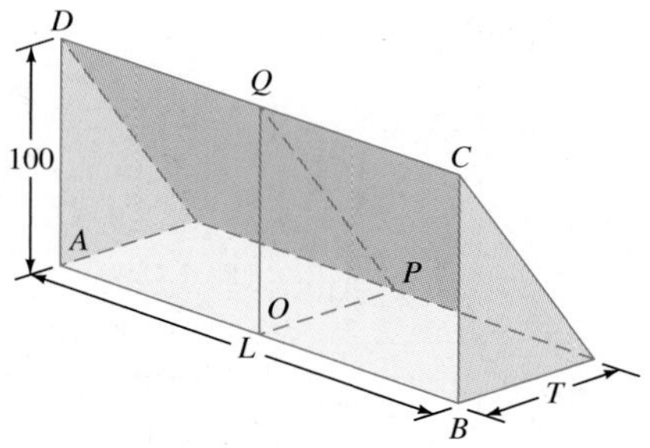

FIGURE 6.5.19 View of a model of a dam (Problem 39).

40. The concrete birdbath of Problem 47 in Section 6.3 is obtained by revolving around the y-axis the region that lies between the curves

$$y = 1 + \frac{x^2}{5} - \frac{x^4}{500} \quad \text{and} \quad y = \frac{x^4}{10000}$$

(with x and y in inches). When placed on its stand, the bottom of this birdbath is 40 inches above the ground. How much work (in ft·lb) is done in filling this birdbath with water lifted from ground level? If necessary, use a computer algebra system or table of integrals to help solve this problem.

6.6 CENTROIDS OF PLANE REGIONS AND CURVES

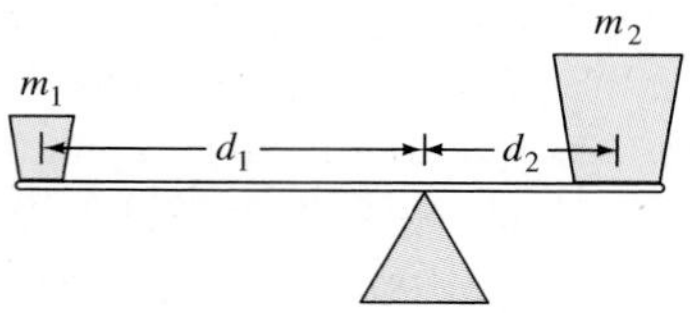

FIGURE 6.6.1 Law of the lever: The weights balance when $m_1 d_1 = m_2 d_2$.

According to the **law of the lever,** two masses m_1 and m_2 on opposite sides and at respective distances d_1 and d_2 from the fulcrum of a lever will balance provided that $m_1 d_1 = m_2 d_2$. (See Fig. 6.6.1.) Think of the x-axis as the location of a (weightless) lever arm supporting various point masses and of the origin as a fulcrum. Then a more general form of the law of the lever states that (particle) masses $m_0, m_1, m_2, \dots, m_n$ with respective coordinates $x_0, x_1, x_2, \dots, x_n$ will balance provided that

$$\sum_{i=0}^{n} m_i x_i = m_0 x_0 + m_1 x_1 + m_2 x_2 + \cdots + m_n x_n = 0. \tag{1}$$

Now consider arbitrary masses $m_1, m_2, \dots, m_n$ at the points $x_1, x_2, \dots, x_n$. Then a single particle with mass

$$m = m_1 + m_2 + \cdots + m_n = \sum_{i=1}^{n} m_i$$

at position $-\overline{x}$ will balance these n masses provided that

$$-m\overline{x} + \sum_{i=1}^{n} m_i x_i = 0;$$

that is, provided that

$$\overline{x} = \frac{1}{m} \sum_{i=1}^{n} m_i x_i. \tag{2}$$

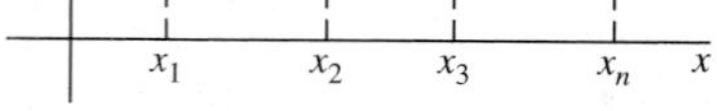

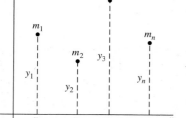

FIGURE 6.6.2 Point masses in the plane and their moment arms about the x-axis.

Thus the original n masses act on the lever like a single particle of mass m located at the point $\overline{x}$. The point $\overline{x}$ is called the *center of mass,* and the sum $\sum_{i=1}^{n} m_i x_i$ that appears in Eq. (2) is called the *moment* of the system of masses about the origin.

Now consider a system of n particles with masses $m_1, m_2, \dots, m_n$ located in the plane (see Fig. 6.6.2) at the points with respective coordinates $(x_1, y_1), (x_2, y_2), \dots, (x_n, y_n)$. In analogy with the one-dimensional case just discussed, we define the **moment** M_y of this system of masses **about the** y**-axis** and its **moment** M_x **about the** x**-axis** by means of the equations

$$M_y = \sum_{i=1}^{n} m_i x_i \quad \text{and} \quad M_x = \sum_{i=1}^{n} m_i y_i. \tag{3}$$

The **center of mass** of this system of n particles is the point $(\overline{x}, \overline{y})$ with coordinates defined to be

$$\overline{x} = \frac{M_y}{m} \quad \text{and} \quad \overline{y} = \frac{M_x}{m} \tag{4}$$

where $m = m_1 + m_2 + \cdots + m_n$ is the sum of the masses. Thus $(\overline{x}, \overline{y})$ is the point where a single particle of mass m would have the same moments as the original whole system about the two coordinate axes. In elementary physics it is shown that, if the xy-plane were a rigid and weightless plastic sheet with our n particles imbedded in it, then it would balance (horizontally) on the point of an icepick at the point $(\overline{x}, \overline{y})$.

Laminas and Thin Plates

As the number of particles we consider increases while their masses decrease in proportion, their aggregate more and more closely resembles a plane region of varying density. Let us first define the center of mass $(\overline{x}, \overline{y})$ and the moments about the coordinate axes of a thin plate, or **lamina,** of constant density ρ, one that occupies a bounded plane region R. Because ρ is constant, the numbers $\overline{x}$ and $\overline{y}$ should be independent of the value of ρ, so we will assume that $\rho = 1$ (for convenience) in our definitions and computations. In this case $(\overline{x}, \overline{y})$ is called the **centroid** of the plane region R. We shall also define $M_y(R)$ and $M_x(R)$, the moments of the plane region R about the coordinate axes, with $\rho = 1$. The corresponding moments of a lamina of constant density $\rho \neq 1$ would then be $\rho M_y(R)$ and $\rho M_x(R)$.

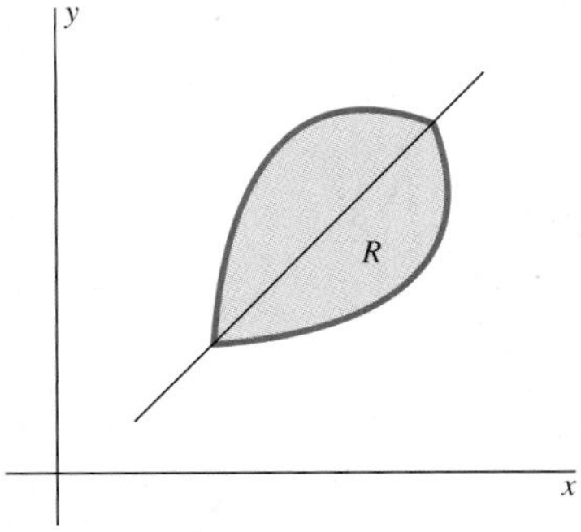

FIGURE 6.6.3 A line of symmetry.

Our definitions will be based on the following two physical principles. The first is quite natural: If a region has a line of symmetry, as in Fig. 6.6.3, then its centroid lies on this line.

Symmetry Principle

If the plane region R is symmetric with respect to the line L—that is, if R is carried onto itself when the plane is rotated through an angle of 180° about the line L—then the centroid of R (considered as a lamina of constant density) lies on L.

We will find the second principle very useful in locating centroids of regions that are unions of simple regions.

Additivity of Moments

If R is the union of the two nonoverlapping regions S and T, then

$$M_y(R) = M_y(S) + M_y(T) \quad \text{and} \quad M_x(R) = M_x(S) + M_x(T). \tag{5}$$

For example, the symmetry principle implies that the centroid of a rectangle is its geometric center: the intersection of the perpendicular bisectors of its sides. We assume also that the moments of a rectangle R with area A and centroid $(\overline{x}, \overline{y})$ are $M_y(R) = A\overline{x}$ and $M_x(R) = A\overline{y}$. Knowing the centroid of a rectangle, our strategy will be to calculate the moments of a more general region by using additivity of moments and the integral, and finally to define the centroid of this more general region by analogy with the equations in (4).

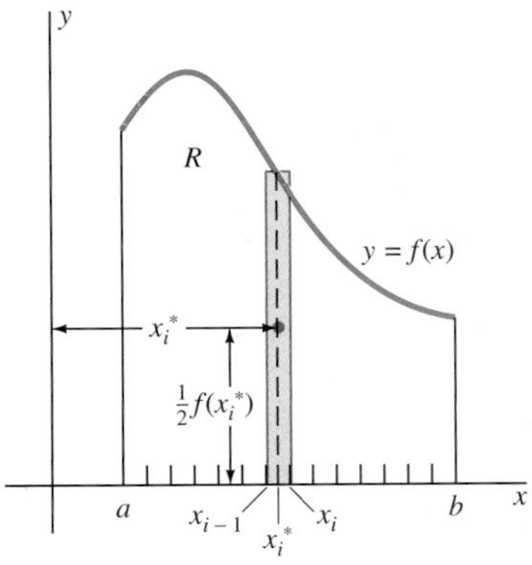

FIGURE 6.6.4 Locating the centroid of R by approximating R with rectangles.

Integral Formulas for Centroids

Suppose that the function f is continuous and nonnegative on $[a, b]$, and suppose also that the region R is the region between the graph of f and the x-axis for $a \leqq x \leqq b$. We begin with a regular partition of $[a, b]$ into n subintervals all having the same length $\Delta x = (b-a)/n$. Denote by $x_i^\star$ the midpoint of the ith subinterval $[x_{i-1}, x_i]$. As shown in Fig. 6.6.4, the rectangle with base $[x_{i-1}, x_i]$ and height $f(x_i^\star)$ has area $f(x_i^\star)\,\Delta x$ and centroid $(x_i^\star, \frac{1}{2}f(x_i^\star))$. If P_n denotes the union of these rectangles for $i = 1, 2, 3, \ldots, n$, then—because moments are additive—the moments of the rectangular polygon P_n

about the y-axis and the x-axis are

$$M_y(P_n) = \sum_{i=1}^{n} x_i^\star \cdot f(x_i^\star)\,\Delta x$$

and

$$M_x(P_n) = \sum_{i=1}^{n} \frac{1}{2} f(x_i^\star) \cdot f(x_i^\star)\,\Delta x,$$

respectively.

We define the moments $M_y(R)$ and $M_x(R)$ of the region R itself by taking the limits of $M_y(P_n)$ and $M_x(P_n)$ as $\Delta x \to 0$. Because the preceding sums are Riemann sums, their limits are the definite integrals

$$M_y(R) = \int_a^b x f(x)\,dx \tag{6}$$

and

$$M_x(R) = \int_a^b \frac{1}{2}[f(x)]^2\,dx. \tag{7}$$

Finally, we define the **centroid** $(\overline{x}, \overline{y})$ of R by

$$\overline{x} = \frac{M_y(R)}{A} \quad \text{and} \quad \overline{y} = \frac{M_x(R)}{A}, \tag{8}$$

where $A = \int_a^b f(x)\,dx$ is the area of R. Thus

$$\overline{x} = \frac{1}{A}\int_a^b x f(x)\,dx \tag{9}$$

and

$$\overline{y} = \frac{1}{A}\int_a^b \frac{1}{2}[f(x)]^2\,dx \tag{10}$$

are the coordinates of the centroid of the region under $y = f(x)$ from $x = a$ to $x = b$.

By the symmetry principle, the centroid of a circular disk is its center. But the centroid of a semicircle is more interesting.

EXAMPLE 1 Find the centroid of the upper half D of the circular disk with center $(0, 0)$ and radius r.

Solution By symmetry, the centroid of D lies on the y-axis, so $\overline{x} = 0$. The semicircular disk lies under $y = \sqrt{r^2 - x^2}$ from $x = -r$ to $x = r$, as shown in Fig. 6.6.5. Hence Eq. (7) yields

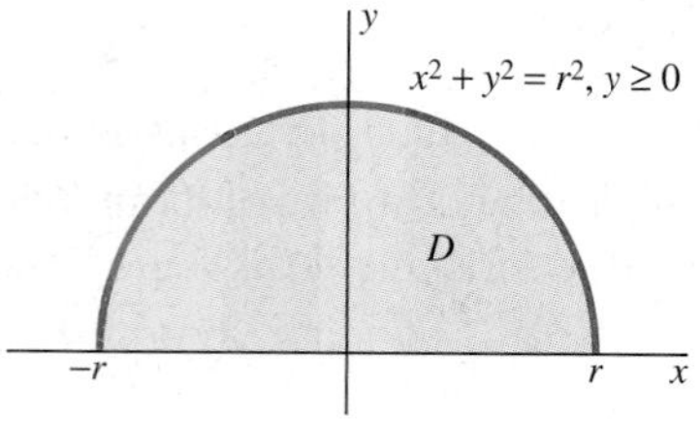

FIGURE 6.6.5 The half-disk D of Example 1.

$$M_x(D) = \int_{-r}^{r} \frac{1}{2}\left(\sqrt{r^2 - x^2}\right)^2 dx$$

$$= \frac{1}{2}\int_{-r}^{r} (r^2 - x^2)\,dx = \left[r^2 x - \frac{x^3}{3}\right]_{-r}^{r} = \frac{2}{3}r^3.$$

Because $A = \frac{1}{2}\pi r^2$, the second formula in (8) gives

$$\overline{y} = \frac{M_x(D)}{A} = \frac{\frac{2}{3}r^3}{\frac{1}{2}\pi r^2} = \frac{4r}{3\pi} \approx (0.4244)r.$$

Thus the centroid of D is the point $(0, 4r/3\pi)$. Note that our computed value for $\overline{y}$ has the dimensions of length (because r is a length), as it should. Any answer having other dimensions would be suspect. Do you feel that these coordinates are plausible as well? ◆

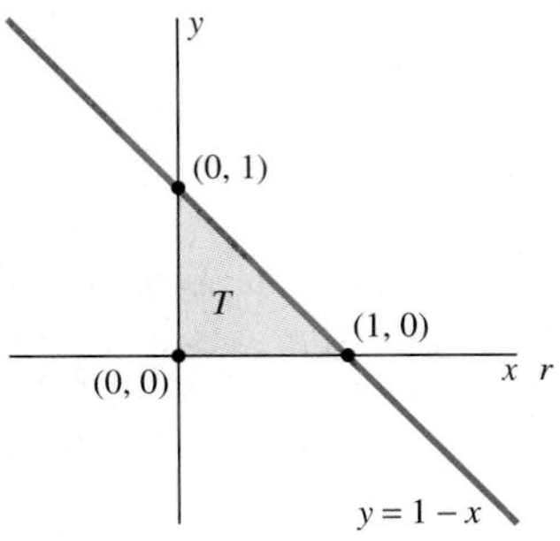

FIGURE 6.6.6 The triangle T of Example 2.

EXAMPLE 2 Find the centroid of the triangle T with vertices (0, 0), (1, 0), and (0, 1).

Solution Figure 6.6.6 shows the triangle T. Note first that $\overline{x} = \overline{y}$ by symmetry. Equation (6), with $y = f(x) = 1 - x$, gives

$$M_y(T) = \int_0^1 x(1-x)\,dx = \left[\frac{1}{2}x^2 - \frac{1}{3}x^3\right]_0^1 = \frac{1}{6}.$$

So, by the first equation in (8),

$$\overline{x} = \frac{M_y(T)}{A} = \frac{\frac{1}{6}}{\frac{1}{2}} = \frac{1}{3}.$$

Thus the centroid of T is $(\frac{1}{3}, \frac{1}{3})$. ◆

Additivity of moments can be used to define the moments and centroid of any plane region that is the union of a finite number of nonoverlapping regions shaped like the one in Fig. 6.6.7. For example, suppose that $f(x) \geqq g(x) \geqq 0$ on $[a, b]$ and that R is the region between the graphs of $y = f(x)$ and $y = g(x)$ for $a \leqq x \leqq b$. If R_f and R_g denote the regions under the graphs of f and g, respectively, then $M_y(R) + M_y(R_g) = M_y(R_f)$ by additivity of moments. Therefore

$$M_y(R) = M_y(R_f) - M_y(R_g) = \int_a^b xf(x)\,dx - \int_a^b xg(x)\,dx$$

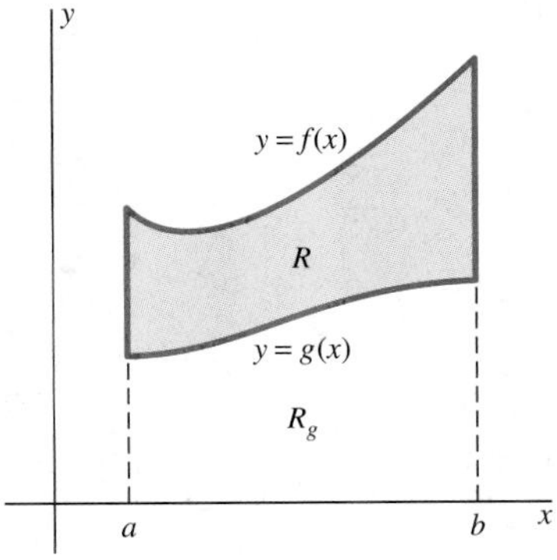

FIGURE 6.6.7 The region R between $y = f(x)$ and $y = g(x)$.

by Eq. (6). Thus

$$M_y(R) = \int_a^b x \cdot [f(x) - g(x)]\,dx. \tag{11}$$

Similarly,

$$M_x(R) = M_x(R_f) - M_x(R_g) = \int_a^b \frac{1}{2}[f(x)]^2\,dx - \int_a^b \frac{1}{2}[g(x)]^2\,dx$$

by Eq. (7), and therefore

$$M_x(R) = \int_a^b \frac{1}{2}\{[f(x)]^2 - [g(x)]^2\}\,dx. \tag{12}$$

We then define the centroid of R by means of the equations in (8), with

$$A = \int_a^b [f(x) - g(x)]\,dx. \tag{13}$$

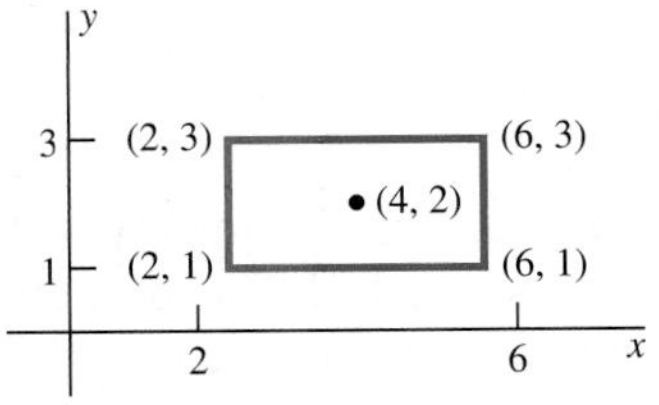

FIGURE 6.6.8 The rectangle of Example 3.

EXAMPLE 3 Let R be the rectangle of Fig. 6.6.8. Following two applications of the symmetry principle, we conclude that its centroid is at the point (4, 2). Let us test Eqs. (11) and (12) to see if they yield the same result. With $a = 2$, $b = 6$, $f(x) \equiv 3$, and $g(x) \equiv 1$, we get

$$M_y(R) = \int_2^6 x(3-1)\,dx = \left[x^2\right]_2^6 = 32,$$

$$M_x(R) = \int_2^6 \frac{1}{2}[(3)^2 - (1)^2]\,dx = \left[4x\right]_2^6 = 16.$$

Because the region has area $A = 4 \cdot 2 = 8$, we find using (8) that

$$\overline{x} = \frac{M_y}{A} = \frac{32}{8} = 4 \quad \text{and} \quad \overline{y} = \frac{M_x}{A} = \frac{16}{8} = 2,$$

exactly as expected. ◆

An interesting theorem relating centroids and volumes of revolution is named for the Greek mathematician who stated it during the third century A.D.

FIRST THEOREM OF PAPPUS Volume of Revolution

Suppose that a plane region R is rotated around an axis in its plane, thereby generating a solid of revolution with volume V. Assume that the axis does not intersect the interior of R. Then V is the product of the area A of R and the distance traveled by the centroid of R during one complete rotation.

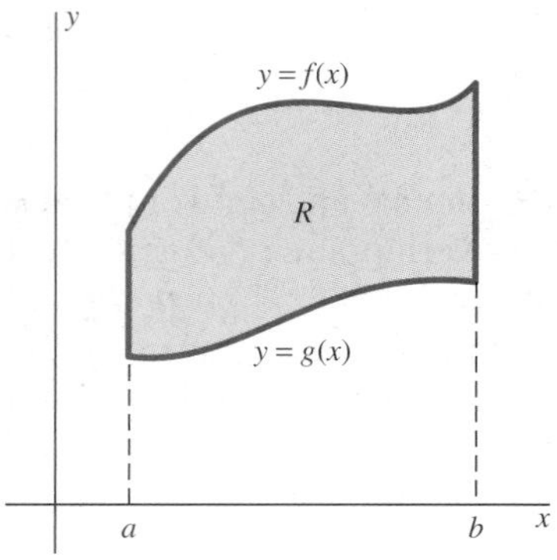

FIGURE 6.6.9 A region R between the graphs of two functions.

PROOF (For the special case of a region shaped like the one shown in Fig. 6.6.9.) This is the region between the two graphs $y = f(x)$ and $y = g(x)$ for $a \leqq x \leqq b$, and we shall take the axis of rotation to be the y-axis. Then, in one complete rotation around the y-axis, the distance traveled by the centroid of R is $d = 2\pi\overline{x}$. By the method of cylindrical shells of Section 6.3, the volume of the solid generated is

$$\begin{aligned} V &= \int_a^b 2\pi x\,[f(x) - g(x)]\,dx \\ &= 2\pi\, M_y(R) \qquad \text{(by Eq. (11))} \\ &= 2\pi\overline{x}\cdot A \qquad \text{(by Eq. (8))}, \end{aligned}$$

and thus $V = d \cdot A$, as desired. ◄

EXAMPLE 4 Find the volume V of the sphere of radius r generated by rotation of the semicircle D of Example 1 around the x-axis.

Solution The area of D is $A = \frac{1}{2}\pi r^2$, and we found in Example 1 that $\overline{y} = 4r/3\pi$. Hence Pappus's theorem gives

$$V = 2\pi\,\overline{y}\,A = 2\pi \cdot \frac{4r}{3\pi} \cdot \frac{\pi r^2}{2} = \frac{4}{3}\pi r^3. \quad ◆$$

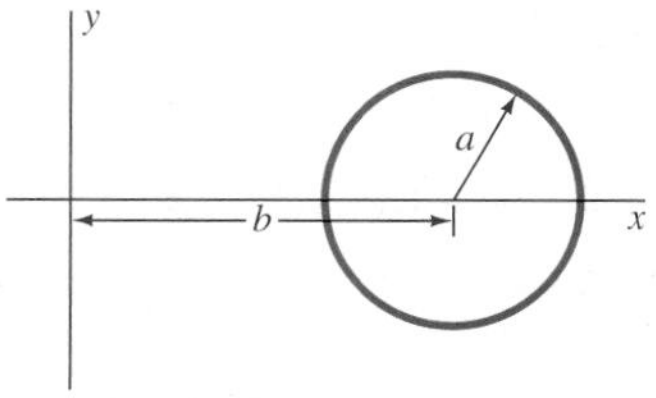

FIGURE 6.6.10 Rotate the circular disk about the y-axis to produce a torus (Example 5).

EXAMPLE 5 Consider the circular disk of Fig. 6.6.10, with radius a and center at the point $(0, b)$ where $0 < a < b$. Find the volume V of the solid torus generated by rotating this disk around the y-axis.

Solution The centroid of the circle is its center $(b, 0)$, so $\overline{x} = b$. Hence the centroid of the disk is moved once around a circle of radius b, thus through the distance $d = 2\pi b$. Consequently

$$V = d \cdot A = 2\pi b \cdot \pi a^2 = 2\pi^2 a^2 b.$$

Note that this result is dimensionally correct. ◆

Moments and Centroids of Curves

Moments and centroids of plane *curves* are defined in a way that is quite analogous to the method for plane regions, so we shall present this topic in less detail. The moments $M_y(C)$ and $M_x(C)$ of the plane curve C about the y- and x-axes, respectively, are defined to be

$$M_y(C) = \int x\,ds \quad \text{and} \quad M_x(C) = \int y\,ds \qquad \textbf{(14)}$$

(with appropriate limits to be inserted when the integrals are evaluated). The **centroid** of C is then defined to be the point with coordinates

$$\overline{x} = \frac{1}{s}M_y(C) = \frac{1}{s}\int_C x\,ds \quad \text{and} \quad \overline{y} = \frac{1}{s}M_x(C) = \frac{1}{s}\int_C y\,ds \qquad \textbf{(15)}$$

where s is the arc length of C.

The meaning of the integrals in (14) and (15) is that of the notation of Section 6.4. That is, ds is a symbol to be replaced (before evaluation of the integral) by either

$$ds = \sqrt{1+\left(\frac{dy}{dx}\right)^2}\,dx \quad \text{or} \quad ds = \sqrt{1+\left(\frac{dx}{dy}\right)^2}\,dy,$$

depending on whether C is a smooth arc of the form $y = f(x)$ or one of the form $x = g(y)$. For example, if C is described by $y = f(x)$, $a \leqq x \leqq b$, then

$$M_y(C) = \int_a^b x\sqrt{1+[f'(x)]^2}\,dx \tag{16}$$

and

$$M_x(C) = \int_a^b f(x)\sqrt{1+[f'(x)]^2}\,dx. \tag{17}$$

EXAMPLE 6 Let J denote the upper half of the circle (not the *disk*) of radius r. Thus the arc J is the graph of

$$y = \sqrt{r^2 - x^2}, \quad -r \leqq x \leqq r.$$

Find the centroid of J.

Solution Note first that $M_y(J) = 0$ by symmetry. Now

$$\frac{dy}{dx} = -\frac{x}{\sqrt{r^2 - x^2}},$$

so

$$ds = \sqrt{1 + \frac{x^2}{r^2 - x^2}}\,dx = \frac{r}{\sqrt{r^2 - x^2}}\,dx.$$

Hence the second formula in (14) yields

$$M_x(J) = \int_{-r}^{r} \sqrt{r^2 - x^2}\,\frac{r}{\sqrt{r^2 - x^2}}\,dx = \int_{-r}^{r} r\,dx = 2r^2.$$

Because $s = \pi r$, the coordinates of the centroid of J are

$$\overline{x} = \frac{M_y(J)}{s} = 0 \quad \text{and} \quad \overline{y} = \frac{M_x(J)}{s} = \frac{2r^2}{\pi r} = \frac{2r}{\pi}.$$

The answer is dimensionally correct. Is it plausible? ◆

The first theorem of Pappus has an analogue for surface area of revolution.

SECOND THEOREM OF PAPPUS Surface Area of Revolution
Suppose that the plane curve C is rotated around an axis in its plane that does not intersect C. Then the area A of the surface of revolution generated is equal to the product of the length of C and the distance traveled by the centroid of C.

PROOF (For the special case in which C is a smooth arc described by $y = f(x)$, $a \leq x \leq b$, and the axis of revolution is the y-axis.) The distance traveled by the centroid of C is $d = 2\pi\overline{x}$. By Eq. (12) in Section 6.4, the area of the surface of revolution is

$$\begin{aligned} A &= \int_a^b 2\pi x\sqrt{1+[f'(x)]^2}\,dx \\ &= 2\pi\,M_y(C) \quad \text{(Eq. (16))} \\ &= 2\pi\overline{x}s = d \cdot s \quad \text{(Eq. (15))}, \end{aligned}$$

as desired. ◄

EXAMPLE 7 Find the surface area A of the sphere of radius r generated by revolving around the x-axis the semicircular arc of Example 6.

Solution Because we found that $\overline{y} = 2r/\pi$ and we know that $S = \pi r$, the second theorem of Pappus gives

$$A = 2\pi\,\overline{y}s = 2\pi \cdot \frac{2r}{\pi} \cdot \pi r = 4\pi r^2.$$ ◆

EXAMPLE 8 Find the surface area A of the torus of Example 5.

Solution Now we think of rotating the circle (*not* the disk) of radius a centered at the point $(b, 0)$. Naturally, the centroid of the circle is located at its center $(b, 0)$; this follows from the symmetry principle and can be verified independently by computations like those in Example 6. Hence the distance traveled by the centroid is $d = 2\pi b$. Because the circumference of the circle is $S = 2\pi a$, the second theorem of Pappus gives

$$A = 2\pi b \cdot 2\pi a = 4\pi^2 ab.$$

Again, note that the answer is dimensionally correct. ◆

6.6 TRUE/FALSE STUDY GUIDE

6.6 CONCEPTS: QUESTIONS AND DISCUSSION

1. Example 6 shows that the centroid of a curve need not be a point of the curve. Must the centroid of a plane region be a point of that region?
2. How could you check the results in Examples 4 and 8 for plausibility?

6.6 PROBLEMS

In Problems 1 through 18, find the centroid of the plane region bounded by the given curves.

1. $x = 0, x = 4, y = 0, y = 6$
2. $x = 1, x = 3, y = 2, y = 4$
3. $x = -1, x = 3, y = -2, y = 4$
4. $x = 0, y = 0, x + y = 3$
5. $x = 0, y = 0, x + 2y = 4$
6. $y = 0, y = x, x + y = 2$
7. $y = 0, y = x^2, x = 2$
8. $y = x^2, y = 9$
9. $y = 0, y = x^2 - 4$
10. $x = -2, x = 2, y = 0, y = x^2 + 1$
11. $y = 4 - x^2, y = 0$
12. $y = x^2, y = 18 - x^2$
13. $y = 3x^2,\ y = 0, x = 1$
14. $x = y^2, y = 0, x = 4$ (*above* the x-axis)
15. $y = x, y = 6 - x^2$
16. $y = x^2, y^2 = x$
17. $y = x^2, y = x^3$
18. $y = \sin x \ \ (0 \leqq x \leqq \pi), y = 0$
19. Find the centroid of the first quadrant of the circular disk $x^2 + y^2 \leqq r^2$ by direct computation, as in Example 1.
20. Apply the first theorem of Pappus to find the centroid of the first quadrant of the circular disk $x^2 + y^2 \leqq r^2$. Use the facts that $\overline{x} = \overline{y}$ by symmetry and that rotation of this quarter disk about either coordinate axis gives a solid hemisphere with volume $V = \frac{2}{3}\pi r^3$.
21. Find the centroid of the arc consisting of the first-quadrant portion of the circle $x^2 + y^2 = r^2$ by direct computation, as in Example 6.
22. Apply the second theorem of Pappus to find the centroid of the quarter-circular arc of Problem 21. Note that $\overline{x} = \overline{y}$ by symmetry, and that rotation of this arc around either coordinate axis gives a hemisphere with surface area $A = 2\pi r^2$.
23. Show by direct computation that the centroid of the triangle with vertices $(0, 0)$, $(r, 0)$, and $(0, h)$ is the point $(r/3, h/3)$. Verify that this point lies on the line from the vertex $(0, 0)$ to the midpoint of the opposite side of the triangle and two-thirds of the way from the vertex to that midpoint.
24. Apply the first theorem of Pappus and the result of Problem 23 to verify the formula $V = \frac{1}{3}\pi r^2 h$ for the volume of the cone generated by rotating the triangle around the y-axis.
25. Apply the second theorem of Pappus to show that the lateral surface area of the cone of Problem 24 is $A = \pi r L$, where $L = \sqrt{r^2 + h^2}$ is the *slant height* of the cone.

26. (a) Use additivity of moments to find the centroid of the trapezoid shown in Fig. 6.6.11. (b) Apply the first theorem of Pappus and the result of part (a) to show that the volume of the conical frustum generated by rotating the trapezoid around the y-axis is

$$V = \frac{\pi h}{3}\left(r_1^2 + r_1 r_2 + r_2^2\right).$$

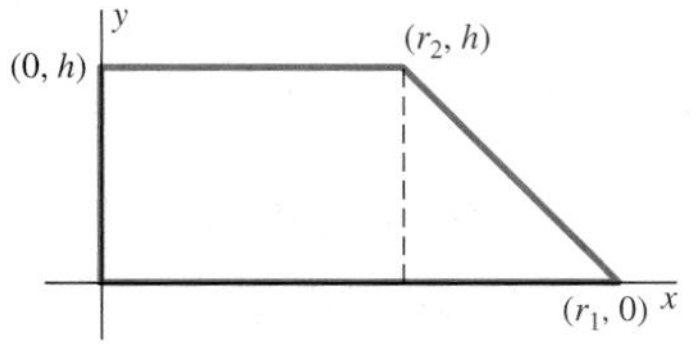

FIGURE 6.6.11 The trapezoid of Problem 26.

27. Apply the second theorem of Pappus to show that the lateral surface area of the conical frustum in Problem 26 is $A = \pi(r_1 + r_2)L$, where

$$L = \sqrt{(r_1 - r_2)^2 + h^2}$$

is the slant height of the frustum.

28. (a) Apply the second theorem of Pappus to verify that the curved surface area of a right circular cylinder with height h and base radius r is $A = 2\pi rh$. (b) Explain how this formula also follows from the result of Problem 27.

29. (a) Use additivity of moments to find the centroid of the plane region shown in Fig. 6.6.12, which consists of a semicircular region of radius a resting atop a rectangular region of width $2a$ and height h. (b) Then apply the first theorem of Pappus to find the volume generated by rotating this region around the x-axis.

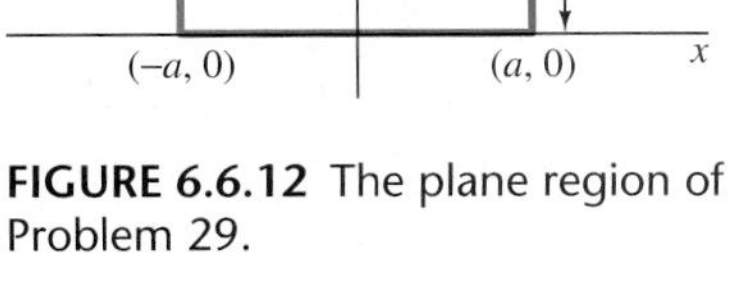

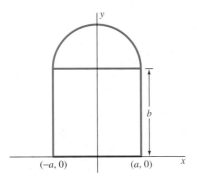

FIGURE 6.6.12 The plane region of Problem 29.

30. (a) Consider the plane region shown in Fig. 6.6.13, bounded by $x^2 = 2py$, $x = 0$, and $y = h = r^2/2p$ $(p > 0)$. Show that its area is $A = 2rh/3$ and that the x-coordinate of its centroid is $\overline{x} = 3r/8$. (b) Use Pappus's theorem and the result of part (a) to show that the volume of a paraboloid of revolution with radius r and height h is $V = \frac{1}{2}\pi r^2 h$.

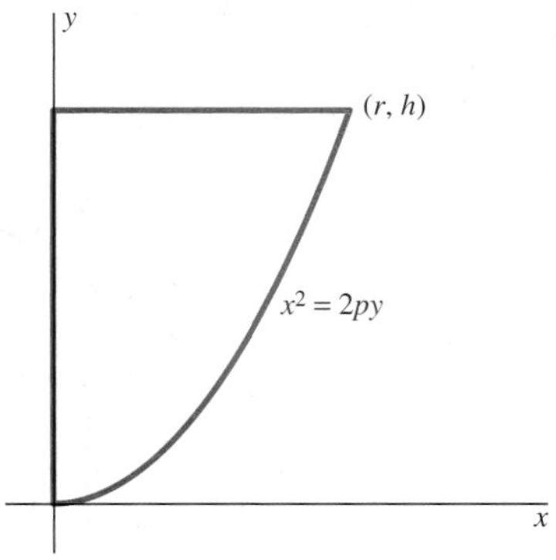

FIGURE 6.6.13 The region of Problem 30.

31. Figure 6.6.14 shows an arc of the unit circle $x^2 + y^2 = 1$ with central angle 2α and height $h = 1 - \cos\alpha$. Show that the centroid of this arc is the point

$$\left(0, \frac{\sin\alpha}{\alpha}\right).$$

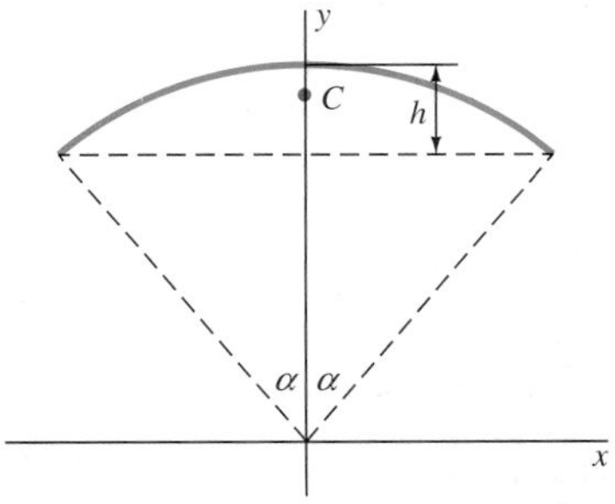

FIGURE 6.6.14 The circular arc of Problem 31.

32. The centroid C shown in Fig. 6.6.14 appears to lie about one-third of the way from the arc to the horizontal chord joining its endpoints. Prove that this is indeed so for small values of α by showing that $\lim_{\alpha \to 0}(d/h) = 1/3$, where d denotes the distance from the centroid C to the highest point (0, 1) of the circular arc.

33. The region in the first quadrant bounded by the graphs of $y = x$ and $y = x^2$ is rotated around the line $y = x$. Find first the centroid of the region, then find the volume of revolution thereby generated.

34. Let R be the region in the xy-plane that is bounded by the curves $y = x^m$ and $y = x^n$, where m and n are positive integers such that $m < n$. Use a computer algebra system to show that the centroid $(\overline{x}, \overline{y})$ of R has coordinates

$$\overline{x} = \frac{(m+1)(n+1)}{(m+2)(n+2)} \quad \text{and} \quad \overline{y} = \frac{(m+1)(n+1)}{(2m+1)(2n+1)}.$$

Can you see by visualizing (or plotting) the figure why it is natural to conjecture that, if m is sufficiently large and $n = m + 1$, then the centroid $(\overline{x}, \overline{y})$ does *not* lie within R? Find specific values of m and n such that the centroid of R does not lie within R.

6.7 THE NATURAL LOGARITHM AS AN INTEGRAL

Our introduction of the functions e^x and $\ln x$ in Section 3.8 was informal and based on an intuitive conception rather than a precise definition of exponentials. Here we provide a solid foundation and careful development of the natural exponential and logarithm functions and their properties.

It is simplest to make the definition of the natural logarithm function our starting point. Guided by the properties of logarithms outlined in Section 3.8, we want to define $\ln x$ for $x > 0$ in such a way that

$$\ln 1 = 0 \quad \text{and} \quad D_x \ln x = \frac{1}{x}. \tag{1}$$

To do so, we recall part 1 of the fundamental theorem of calculus (Section 5.6), from which it follows that

$$D_x\left(\int_a^x f(t)\,dt\right) = f(x) \tag{2}$$

if f is continuous on an interval containing a and x. In order that $\ln x$ satisfy the equations in (1), we take $a = 1$ and $f(t) = 1/t$.

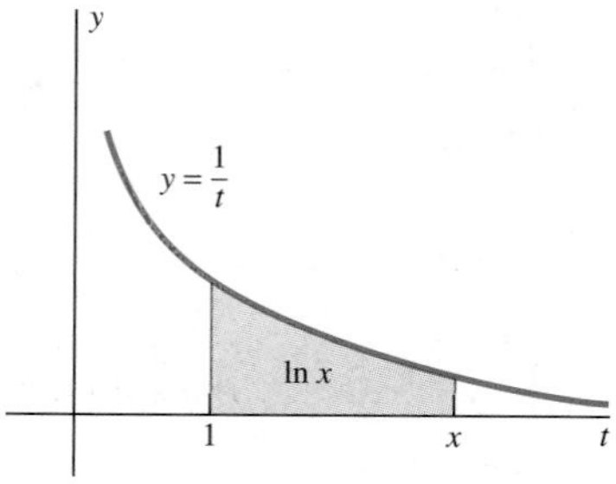

FIGURE 6.7.1 The natural logarithm function defined by means of an integral.

DEFINITION The Natural Logarithm

The **natural logarithm** $\ln x$ of the positive number x is defined to be

$$\ln x = \int_1^x \frac{1}{t}\,dt. \tag{3}$$

Note that $\ln x$ is *not* defined for $x \leqq 0$. Geometrically the value $\ln x$ of the natural logarithm of x is equal to:

- The area under the graph of $y = 1/t$ from $t = 1$ to $t = x$ if $x > 1$ (Fig. 6.7.1);
- The negative of this area if $0 < x < 1$;
- Zero if $x = 1$.

EXAMPLE 1 The number

$$\ln 2 = \int_1^2 \frac{1}{t}\,dt = \int_1^2 \frac{1}{x}\,dx$$

is equal to the area under the graph of $y = 1/x$ from $x = 1$ to $x = 2$. Examining the inscribed and circumscribed rectangles in Fig. 6.7.2, we see immediately that

$$\frac{1}{2} < \ln 2 < 1.$$

We can use Simpson's approximation to estimate $\ln 2$ more closely. The regular partition of $[1, 2]$ into $n = 10$ subintervals, each of length $\Delta x = \frac{1}{10}$, and with

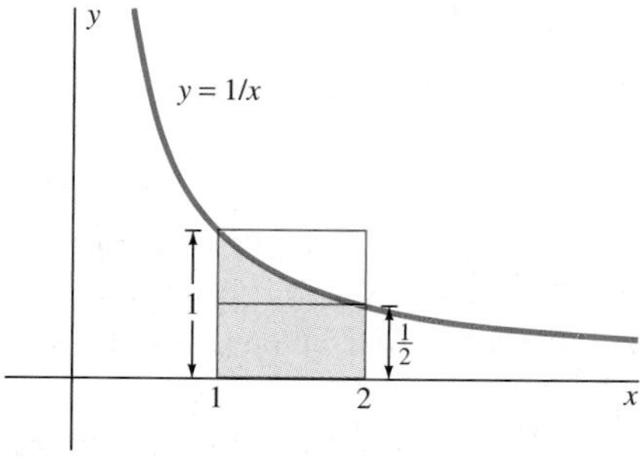

FIGURE 6.7.2 Using rectangles to estimate ln 2.

endpoints $1, \frac{11}{10}, \frac{12}{10}, \dots, \frac{19}{10}, 2$ yields

$$\ln 2 = \frac{\frac{1}{10}}{3} \cdot \left(1 + 4 \cdot \frac{10}{11} + 2 \cdot \frac{10}{12} + 4 \cdot \frac{10}{13} + 2 \cdot \frac{10}{14} + 4 \cdot \frac{10}{15} + 2 \cdot \frac{10}{16} + 4 \cdot \frac{10}{17} + 2 \cdot \frac{10}{18} + 4 \cdot \frac{10}{19} + \frac{1}{2}\right) \approx 0.69315.$$

This approximation is accurate to five decimal places because the actual value of ln 2 to six places is 0.693147. ◆

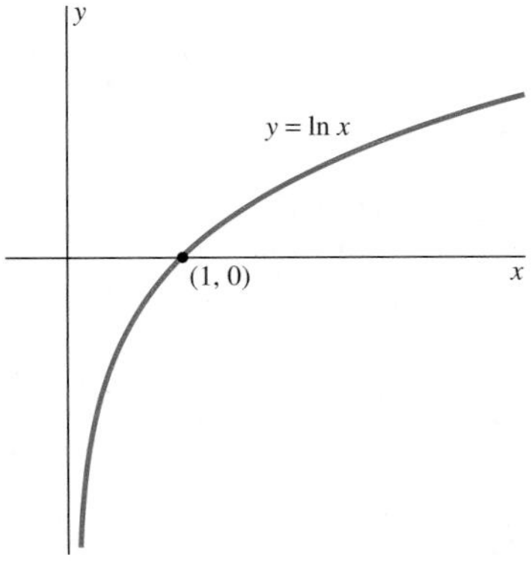

FIGURE 6.7.3 The graph of the natural logarithm function.

The Graph of $y = \ln x$

The fact that $D_x \ln x = 1/x$ follows immediately from the fundamental theorem of calculus in (2). And by Theorem 2 in Section 3.4, the fact that the function $\ln x$ is differentiable for $x > 0$ implies that it is continuous for $x > 0$. Because

$$D_x \ln x = \frac{1}{x} > 0 \quad \text{and} \quad D_x^2 \ln x = -\frac{1}{x^2} < 0$$

for $x > 0$, we see that $\ln x$ is an increasing function whose graph is concave downward everywhere (by Theorem 2 in Section 4.6). Using the laws of logarithms (in Theorem 1) we see, from the facts that

$$\ln 2^n = n \ln 2 \to +\infty \quad \text{and} \quad \ln 2^{-n} = -n \ln 2 \to -\infty$$

as $n \to +\infty$, that

$$\lim_{x\to\infty} \ln x = +\infty \quad \text{and} \quad \lim_{x\to 0^+} \ln x = -\infty. \tag{4}$$

When we assemble these facts, we see that the graph of $y = \ln x$ has the familiar shape shown in Fig. 6.7.3.

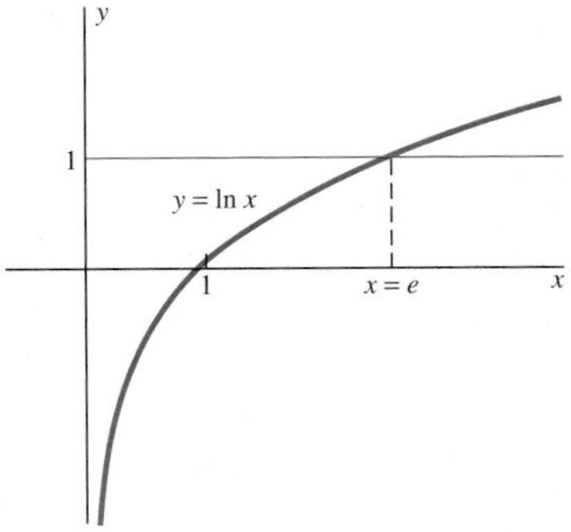

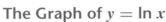

FIGURE 6.7.4 The fact that $\ln e = 1$ is expressed graphically here.

The Number e

Because $\ln x$ is an increasing function, the intermediate value property of continuous functions implies that the curve $y = \ln x$ crosses the horizontal line $y = 1$ precisely once. (See Fig. 6.7.4.) The x-coordinate of the point of intersection is the famous number $e \approx 2.71828$ introduced differently in Section 3.8.

DEFINITION The Number e
The number e is the (unique) real number such that

$$\ln e = 1. \tag{5}$$

The number e has been used to denote the number whose natural logarithm is 1 ever since this number was introduced by the great Swiss mathematician Leonhard Euler (1707–1783), who used e for "exponential." [Euler also popularized the use of π for the area (approximately 3.14159) of the unit circle as well as the symbol i for the imaginary number $\sqrt{-1}$.]

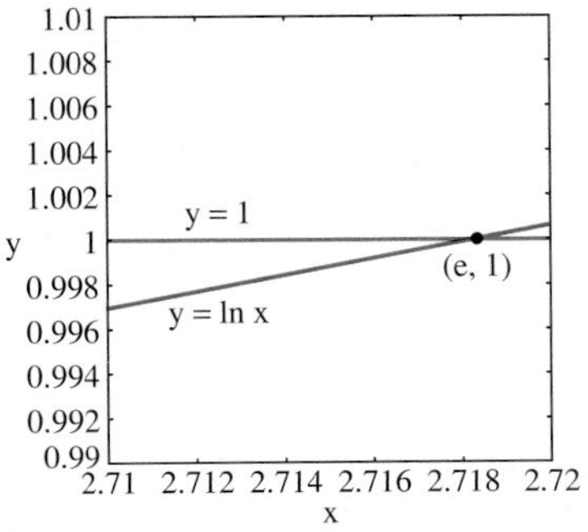

FIGURE 6.7.5 Zooming in on the number e.

EXAMPLE 2 With a graphing calculator or computer you can zoom in on the intersection of the graphs of $y = \ln x$ and $y = 1$ to verify the first few decimal places of e. For instance, the viewing window $2.71 \leqq x \leqq 2.72$, $0.99 \leqq y \leqq 1.01$ of Fig. 6.7.5 suffices to verify that $e \approx 2.718$ to three decimal places. ◆

The Laws of Logarithms

We now use our ability to differentiate logarithms to establish rigorously the laws of logarithms.

THEOREM 1 Laws of Logarithms
If x and y are positive numbers and r is a rational number, then

$$\ln xy = \ln x + \ln y; \qquad (6)$$

$$\ln\left(\frac{1}{x}\right) = -\ln x; \qquad (7)$$

$$\ln\left(\frac{x}{y}\right) = \ln x - \ln y; \qquad (8)$$

$$\ln(x^r) = r\ln x. \qquad (9)$$

The restriction that r is rational is removed later in this section, when we examine general exponential functions (those with bases other than e).

PROOF OF EQUATION (6) We temporarily fix y, so that we may regard x as the independent variable and y as a constant in what follows. Then

$$D_x \ln xy = \frac{D_x(xy)}{xy} = \frac{y}{xy} = \frac{1}{x} = D_x \ln x.$$

Thus $\ln xy$ and $\ln x$ have the same derivative with respect to x. We antidifferentiate both and conclude that

$$\ln xy = \ln x + C$$

for some constant C. To evaluate C, we substitute $x = 1$ into both sides of the last equation. The fact that $\ln 1 = 0$ then implies that $C = \ln y$, and this is enough to establish Eq. (6). ◄

PROOF OF EQUATION (7) We differentiate $\ln(1/x)$:

$$D_x\left(\ln\frac{1}{x}\right) = \frac{-\dfrac{1}{x^2}}{\dfrac{1}{x}} = -\frac{1}{x} = D_x(-\ln x).$$

Thus $\ln(1/x)$ and $-\ln x$ have the same derivative. Hence antidifferentiation gives

$$\ln\left(\frac{1}{x}\right) = -\ln x + C,$$

where C is a constant. We substitute $x = 1$ into this last equation. Because $\ln 1 = 0$, it follows that $C = 0$, and this proves Eq. (7). ◄

PROOF OF EQUATION (8) Because $x/y = x \cdot (1/y)$, Eq. (8) follows immediately from Eqs. (6) and (7). ◄

PROOF OF EQUATION (9) We know that $D_x x^r = rx^{r-1}$ if r is rational. So

$$D_x(\ln x^r) = \frac{rx^{r-1}}{x^r} = \frac{r}{x} = D_x(r\ln x).$$

Antidifferentiation then gives

$$\ln(x^r) = r\ln x + C$$

for some constant C. As before, substituting $x = 1$ then gives $C = 0$, which proves Eq. (9). We show later in this section that Eq. (9) holds whether or not r is rational. ◄

The proofs of Eqs. (6), (7), and (9) are all quite similar—we differentiate the left-hand side, apply the fact that two functions with the same derivative (on an interval) differ by a constant C (on that interval), and evaluate C by using the fact that $\ln 1 = 0$.

Logarithms and Experimental Data

Certain empirical data can be explained by assuming that the observed dependent variable is a **power function** of the independent variable x. In other words, y is described by a mathematical model of the form

$$y = kx^m,$$

where k and m are constants. If so, the laws of logarithms imply that

$$\ln y = \ln k + m \ln x.$$

An experimenter can then plot values of $\ln y$ against values of $\ln x$. If the power-function model is valid, the resulting data points will lie on a straight line of slope m and with y-intercept $\ln k$ (Fig. 6.7.6). This technique makes it easy to see whether or not the data lie on a straight line. And if they do, this technique makes it easy also to measure the slope and y-intercept of the line and thereby find the values of k and m.

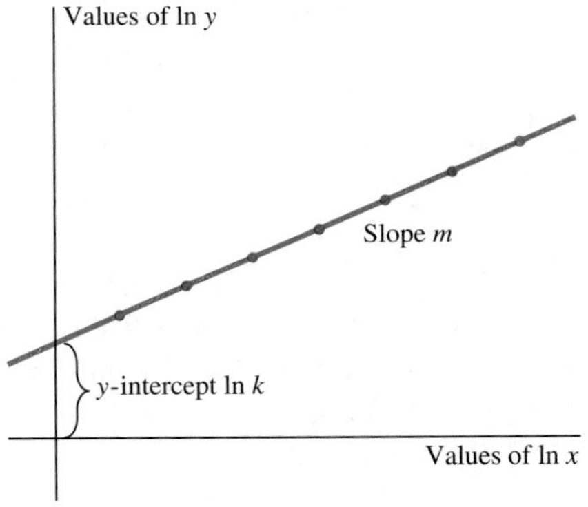

FIGURE 6.7.6 Plotting the logarithms of data may reveal a hidden relationship.

EXAMPLE 3 (Planetary Motion) The table in Fig. 6.7.7 gives the period of revolution T and the major semiaxis a of the elliptical orbit of each of the first six planets around the sun, together with the logarithms of these numbers. If we plot $\ln T$ against $\ln a$, it is immediately apparent that the resulting points lie on a straight line of slope $m = \frac{3}{2}$. Hence T and a satisfy an equation of the form $T = ka^{3/2}$, so

$$T^2 = Ca^3.$$

This means that the square of the period T is proportional to the cube of the major semiaxis a. This is Kepler's third law of planetary motion, which Johannes Kepler (1571–1630) discovered empirically in 1619. ◆

Planet	T (in days)	a (in 10^6 km)	$\ln T$	$\ln a$
Mercury	87.97	58	4.48	4.06
Venus	224.70	108	5.41	4.68
Earth	365.26	149	5.90	5.00
Mars	686.98	228	6.53	5.43
Jupiter	4332.59	778	8.37	6.66
Saturn	10,759.20	1426	9.28	7.26

FIGURE 6.7.7 Data for Example 3.

The Natural Exponential Function

We know that the natural logarithm function $\ln x$ is continuous and increasing for $x > 0$ and that it attains arbitrarily large positive and negative values (because of the limits in (4)). It follows that $\ln x$ has an inverse function that is defined for all x. To see this, let y be any (fixed) real number. If a and b are positive numbers such that $\ln a < y < \ln b$, then the intermediate value property gives a number $x > 0$, with x between a and b, such that $\ln x = y$. Because $\ln x$ is an increasing function, there is only *one* such number x such that $\ln x = y$ (Fig. 6.7.8). Because y determines precisely one such value x, we see that x is a *function* of y.

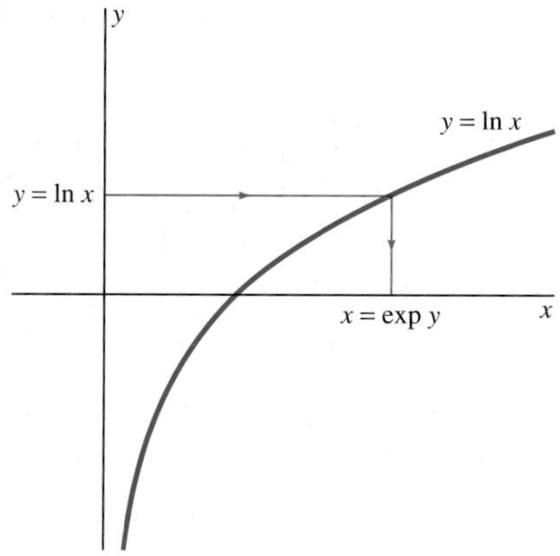

FIGURE 6.7.8 To get $x = \exp y$, move straight over from y to the graph $y = \ln x$, then move straight down (or up) to x.

This function x of y is the inverse function of the natural logarithm function, and it is called the *natural exponential function*. It is commonly denoted by exp (for "exponential"), so

$$x = \exp y \quad \text{provided that} \quad y = \ln x.$$

Interchanging x and y yields the following definition.

DEFINITION The Natural Exponential Function

The **natural exponential function** exp is defined for all x as follows:

$$\exp x = y \quad \text{if and only if} \quad \ln y = x. \tag{10}$$

Thus $\exp x$ is simply that (positive) number whose natural logarithm is x. It is an immediate consequence of Eq. (10) that

$$\ln(\exp x) = x \quad \text{for all } x \tag{11}$$

and that

$$\exp(\ln y) = y \quad \text{for all } y > 0. \tag{12}$$

As in the case of the graphs of $y = a^x$ and $y = \log_a x$ discussed informally in Section 3.8, the fact that $\exp x$ and $\ln x$ are inverse functions implies that the graphs of $y = \exp x$ and $y = \ln x$ are reflections of each other across the line $y = x$. (See Fig. 6.7.9.) Therefore, the graph of the exponential function looks like the one shown in Fig. 6.7.10. In particular, $\exp x$ is positive-valued for all x, and

$$\exp 0 = 1, \tag{13}$$

$$\lim_{x\to\infty} \exp x = +\infty, \quad \text{and} \tag{14}$$

$$\lim_{x\to-\infty} \exp x = 0. \tag{15}$$

These facts follow from the equation $\ln 1 = 0$ and the limits in (4).

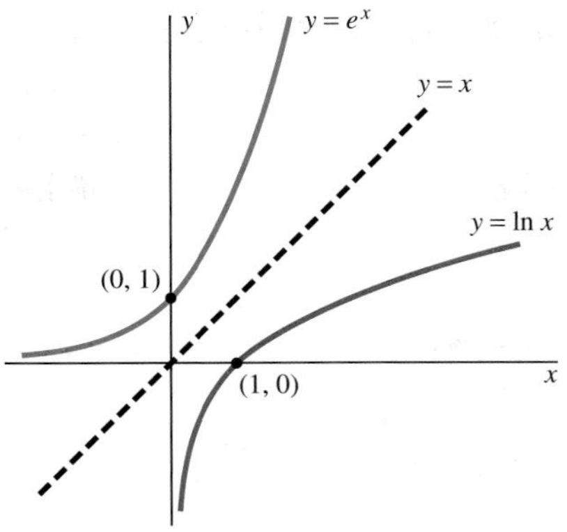

FIGURE 6.7.9 The graphs $y = e^x$ and $y = \ln x$ are reflections of each other across the 45° line $y = x$.

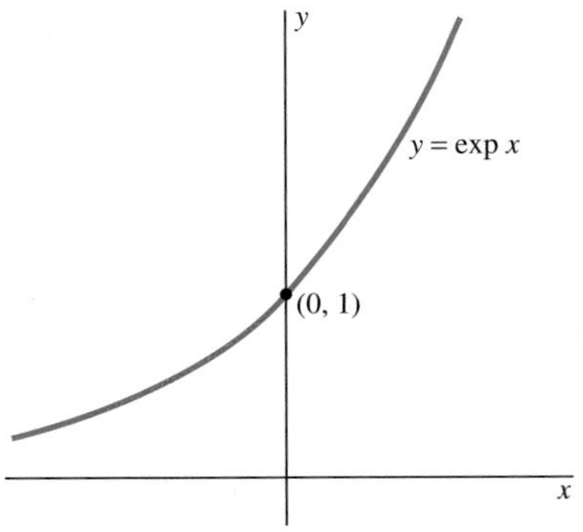

FIGURE 6.7.10 The graph of the natural exponential function, exp.

Exponentials and Powers of *e*

Recall that we have now defined the number $e \approx 2.71828$ as the number whose natural logarithm is 1. If r is any rational number, it follows that

$$\ln(e^r) = r \ln e = r.$$

But Eq. (10) implies that $\ln(e^r) = r$ if and only if

$$\exp r = e^r.$$

Thus $\exp x$ is equal to e^x (e raised to the power x) if x is a rational number. We therefore *define* e^x for irrational as well as rational values of x by

$$e^x = \exp x. \qquad \textbf{(16)}$$

This is our first legitimate instance of powers with irrational exponents.

Equation (16) is the reason for calling exp the natural exponential function. With this notation, Eqs. (10) through (12) become

$$e^x = y \quad \text{if and only if} \quad \ln y = x, \qquad \textbf{(17)}$$

$$\ln(e^x) = x \quad \text{for all } x, \qquad \textbf{(18)}$$

and

$$e^{\ln x} = x \quad \text{for all } x > 0. \qquad \textbf{(19)}$$

To justify Eq. (16), we should show rigorously that powers of e satisfy the laws of exponents. We can do this immediately.

THEOREM 2 Laws of Exponents

If x and y are real numbers and r is rational, then

$$e^x e^y = e^{x+y}, \qquad \textbf{(20)}$$

$$e^{-x} = \frac{1}{e^x}, \qquad \textbf{(21)}$$

and

$$(e^x)^r = e^{rx}. \qquad \textbf{(22)}$$

PROOF The laws of logarithms and Eq. (18) give

$$\ln(e^x e^y) = \ln(e^x) + \ln(e^y) = x + y = \ln(e^{x+y}).$$

Then Eq. (20) follows from the fact that ln is an increasing function and therefore is one-to-one—if $x_1 \neq x_2$, then $\ln x_1 \neq \ln x_2$. Similarly,

$$\ln\left([e^x]^r\right) = r\ln(e^x) = rx = \ln(e^{rx}).$$

So Eq. (22) follows in the same way. The proof of Eq. (21) is almost identical. We will see presently that the restriction that r is rational in Eq. (22) is unnecessary; that is, the equation $(e^x)^y = e^{xy}$ holds for *all* real numbers x and y. ◀

General Exponential Functions

The natural exponential function e^x and the natural logarithm function $\ln x$ are often called the exponential and logarithm functions with *base e*. We now define general exponential and logarithm functions, with the forms a^x and $\log_a x$, whose base is a positive number $a \neq 1$. But it is convenient here to reverse the order of treatment, so we first consider the general exponential function.

If r is a rational number, then the law of exponents in Eq. (22) gives

$$a^r = (e^{\ln a})^r = e^{r\ln a}.$$

We therefore *define* arbitrary powers (rational *and* irrational) of the positive number a in this way:

$$a^x = e^{x\ln a} \qquad \textbf{(23)}$$

for all x. Thus

$$3^{\sqrt{2}} = e^{\sqrt{2}\ln 3} \approx e^{1.5537} \approx 4.7289$$

and

$$(0.5)^{-\pi} = e^{-\pi\ln(0.5)} \approx e^{2.1776} \approx 8.8251.$$

Then $f(x) = a^x$ is called the **exponential function with base** a. Note that $a^x > 0$ for all x and that $a^0 = e^0 = 1$ for all $a > 0$.

The *laws of exponents* for general exponentials follow almost immediately from the definition in Eq. (23) and from the laws of exponents for the natural exponential function:

$$a^x a^y = a^{x+y}, \qquad \textbf{(24)}$$

$$a^{-x} = \frac{1}{a^x}, \qquad \textbf{(25)}$$

and

$$(a^x)^y = a^{xy} \qquad \textbf{(26)}$$

for all x and y. To prove Eq. (24), we write

$$a^x a^y = e^{x\ln a}e^{y\ln a} = e^{(x\ln a)+(y\ln a)} = e^{(x+y)\ln a} = a^{x+y}.$$

To derive Eq. (26), note first from Eq. (23) that $\ln a^x = x\ln a$. Then

$$(a^x)^y = e^{y\ln(a^x)} = e^{xy\ln a} = a^{xy}.$$

This follows for all real numbers x and y, so the restriction that r is rational in the formula $(e^x)^r = e^{rx}$ (Eq. (22)) has now been removed.

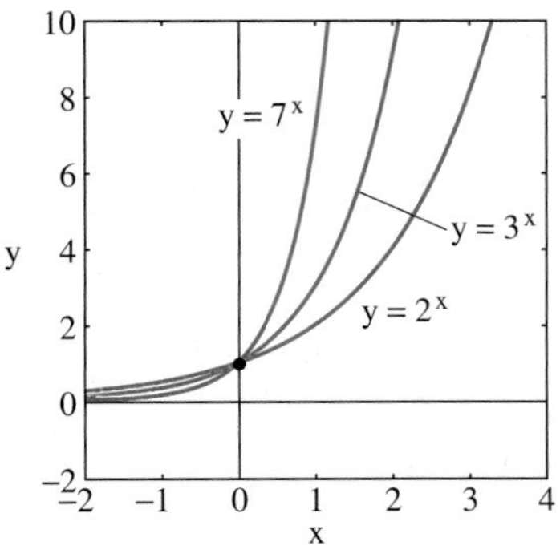

FIGURE 6.7.11 If $a > 1$ then $\lim_{x\to-\infty} a^x = 0$, $\lim_{x\to\infty} a^x = +\infty$.

If $a > 1$, so that $\ln a > 0$, then Eqs. (14) and (15) immediately give us the results

$$\lim_{x\to\infty} a^x = +\infty \quad \text{and} \quad \lim_{x\to-\infty} a^x = 0. \tag{27}$$

Because

$$D_x a^x = D_x(e^{x \ln a}) = (\ln a)e^{x \ln a} = a^x \ln a \tag{28}$$

is positive for all x if $a > 1$, we see that—in this case—$f(x) = a^x$ is an *increasing* function of x. The graph $y = a^x$ then resembles that of the natural exponential function $y = e^x$, but its relative steepness depends on the magnitude of a (Fig. 6.7.11).

If $0 < a < 1$, then $\ln a < 0$. In this case it therefore follows from (28) that $f(x) = a^x$ is a *decreasing* function, and the values of the two limits in (27) are interchanged (Fig. 6.7.12). Whether $a > 1$ or $0 < a < 1$, it follows from (28) that $f''(x) = a^x(\ln a)^2 > 0$ for all x, so the graphs in both Figs. 6.7.11 and 6.7.12 are concave upward for all x.

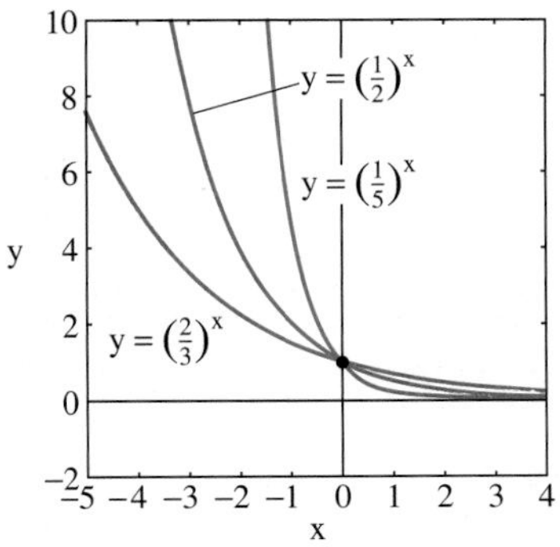

FIGURE 6.7.12 If $0 < a < 1$ then $\lim_{x\to-\infty} a^x = +\infty$, $\lim_{x\to\infty} a^x = 0$.

Derivatives and Integrals

If $u = u(x)$ is a differentiable function of x, then Eq. (28) combined with the chain rule gives

$$D_x a^u = (a^u \ln a)\frac{du}{dx}. \tag{29}$$

The corresponding integral formula is

$$\int a^u \, du = \frac{a^u}{\ln a} + C. \tag{30}$$

But rather than using these general formulas, it usually is simpler to rely solely on the definition in Eq. (23), as in Examples 3, 4, and 5.

EXAMPLE 4 To differentiate $f(x) = 3^{x^2}$, we may first write

$$3^{x^2} = (e^{\ln 3})^{x^2} = e^{x^2 \ln 3}.$$

Then

$$D_x 3^{x^2} = D_x e^{x^2 \ln 3} = e^{x^2 \ln 3} D_x(x^2 \ln 3) = 3^{x^2}(\ln 3)(2x).$$ ◆

EXAMPLE 5 Find $\displaystyle\int \frac{10^{\sqrt{x}}}{\sqrt{x}}\,dx$.

Solution We first write $10^{\sqrt{x}} = (e^{\ln 10})^{\sqrt{x}} = e^{\sqrt{x}\ln 10}$. Then

$$\begin{aligned}\int \frac{10^{\sqrt{x}}}{\sqrt{x}}\,dx &= \int \frac{e^{\sqrt{x}\ln 10}}{\sqrt{x}}\,dx \\ &= \int \frac{2e^u}{\ln 10}\,du \qquad \left(u = \sqrt{x}\ln 10, \quad du = \frac{\ln 10}{2\sqrt{x}}\,dx\right) \\ &= \frac{2e^u}{\ln 10} + C = \frac{2}{\ln 10}10^{\sqrt{x}} + C.\end{aligned}$$ ◆

EXAMPLE 6 The function

$$P(t) = 3 \cdot (1.07)^t$$

describes a population that starts with $P(0) = 3$ (million) bacteria at time $t = 0$ (h) and increases in number by 7% every hour. After 10 hours the population is

$$P(10) = 3 \cdot (1.07)^{10} \approx 5.90 \quad \text{(million)},$$

so the population has almost doubled. The derivative of $P(t)$ is

$$P'(t) = D_t[3 \cdot (1.07)^t] = 3 \cdot D_t\left(e^{t \ln(1.07)}\right)$$
$$= 3 \cdot [\ln(1.07)]e^{t \ln(1.07)} = 3(\ln 1.07)(1.07)^t,$$

so at time $t = 10$ the rate of growth of this bacteria population is

$$P'(10) = 3(\ln 1.07)(1.07)^{10} \approx 0.40 \quad \text{(million per hour).}$$

◆

Whether or not the exponent r is rational, the **general power function** $f(x) = x^r$ is now defined for $x > 0$ by

$$x^r = e^{r \ln x}.$$

We may now prove the power rule of differentiation for an *arbitrary* (constant) exponent as follows:

$$D_x x^r = D_x(e^{r \ln x}) = e^{r \ln x} D_x(r \ln x) = x^r \cdot \frac{r}{x} = r x^{r-1}.$$

For example, we now know that

$$D_x x^\pi = \pi x^{\pi - 1} \approx (3.14159)x^{2.14159}.$$

General Logarithm Functions

If $a > 1$, then the general exponential function a^x is continuous and increasing for all x and attains all positive values. It therefore has an inverse function that is defined for all $x > 0$. This inverse function of a^x is called the **logarithm function with base** a and is denoted by $\log_a x$. Thus

$$y = \log_a x \quad \text{if and only if} \quad x = a^y. \tag{31}$$

The logarithm function with base e is the natural logarithm function: $\log_e x = \ln x$.

The following *laws of logarithms* are easy to derive from the laws of exponents in Eqs. (24) through (26).

$$\log_a xy = \log_a x + \log_a y, \tag{32}$$

$$\log_a \left(\frac{1}{x}\right) = -\log_a x, \tag{33}$$

$$\log_a x^y = y \log_a x. \tag{34}$$

These formulas hold for any positive base $a \neq 1$ and for all positive values of x and y; in Eq. (34), y may be negative or zero as well.

Logarithms with one base are related to logarithms with another base, and the relationship is most easily expressed by the formula

$$(\log_a b)(\log_b c) = \log_a c. \tag{35}$$

This formula holds for all values of a, b, and c for which it makes sense—the bases a and b are positive numbers other than 1 and c is positive. The proof of this formula is outlined in Problem 39. Equation (35) should be easy to remember—it is as if some arcane cancellation law applies.

If we take $c = a$ in Eq. (35), this gives

$$(\log_a b)(\log_b a) = 1, \tag{36}$$

which in turn, with $b = e$, gives

$$\ln a = \frac{1}{\log_a e}. \tag{37}$$

If we replace a with e, b with a, and c with x in Eq. (35), we obtain

$$(\log_e a)(\log_a x) = \log_e x,$$

so

$$\log_a x = \frac{\log_e x}{\log_e a} = \frac{\ln x}{\ln a}. \tag{38}$$

On most calculators, the **LOG** key denotes common (base 10) logarithms: $\log x = \log_{10} x$. In contrast, in many programming languages, such as BASIC, and some symbolic algebra programs, such as *Mathematica,* only the natural logarithm appears explicitly—as `LOG(X)` (in BASIC) and as `Log[x]` (in *Mathematica*). To get $\log_{10} x$ we write `LOG(X)/LOG(10)` and `Log[10,x]`, respectively.

Differentiating both sides of Eq. (38) yields

$$D_x \log_a x = \frac{1}{x \ln a} = \frac{\log_a e}{x}. \tag{39}$$

For example,

$$D_x \log_{10} x = \frac{\log_{10} e}{x} \approx \frac{0.4343}{x}.$$

If we combine the formula $D_x \ln |x| = 1/x$ (see Eq. (22) in Section 3.8) and Eq. (38) with u in place of x, the chain rule gives

$$D_x \log_a |u| = \frac{D_x \ln |u|}{\ln a} = \frac{1}{u \ln a} \cdot \frac{du}{dx} = \frac{\log_a e}{u} \cdot \frac{du}{dx} \tag{40}$$

if u is a differentiable function of x.

EXAMPLE 7

$$D_x \log_2 \sqrt{x^2+1} = \frac{1}{2} D_x \log_2(x^2+1) = \frac{1}{2} \cdot \frac{\log_2 e}{x^2+1} \cdot 2x \approx \frac{(1.4427)x}{x^2+1}.$$

In the last step we substituted $\log_2 e = 1/(\ln 2) \approx 1/0.68315 \approx 1.4427$. ◆

6.7 TRUE/FALSE STUDY GUIDE

6.7 CONCEPTS: QUESTIONS AND DISCUSSION

1. Contrast the "e^x first" approach to exponentials and logarithms in Section 3.8 with the "$\ln x$ first" approach in this section. Outline each approach and point out key differences. What, if anything, is not fully defined in each approach? How is the number e introduced in each approach? How do the derivations of the differentiation formulas $D_x e^x = e^x$ and $D_x \ln x = 1/x$ differ in the two approaches?
2. Outline the way in which the precise definitions of the exponential and logarithm functions in this section are used to define and differentiate the power function $f(x) = x^r$ (for both positive and negative values of x and both rational and irrational values of the exponent r).

6.7 PROBLEMS

In Problems 1 through 24, find the derivative of the given function $f(x)$.

1. $f(x) = 10^x$

2. $f(x) = 2^{1/x^2}$

3. $f(x) = \dfrac{3^x}{4^x}$

4. $f(x) = \log_{10} \cos x$

5. $f(x) = 7^{\cos x}$

6. $f(x) = 2^x 3^{x^2}$

7. $f(x) = 2^{x\sqrt{x}}$

8. $f(x) = \log_{100} 10^x$

9. $f(x) = 2^{\ln x}$

10. $f(x) = 7^{8^x}$

11. $f(x) = 17^x$

12. $f(x) = 2^{\sqrt{x}}$

13. $f(x) = 10^{1/x}$

14. $f(x) = 3^{\sqrt{1-x^2}}$

15. $f(x) = 2^{2^x}$

16. $f(x) = \log_2 x$

17. $f(x) = \log_3 \sqrt{x^2+4}$

18. $f(x) = \log_{10}(e^x)$

19. $f(x) = \log_3(2^x)$

20. $f(x) = \log_{10}(\log_{10} x)$

21. $f(x) = \log_2(\log_3 x)$

22. $f(x) = \pi^x + x^\pi + \pi^\pi$

23. $f(x) = \exp(\log_{10} x)$

24. $f(x) = \pi^{x^3}$

Evaluate the integrals given in Problems 25 through 32.

25. $\displaystyle\int 3^{2x}\,dx$

26. $\displaystyle\int x \cdot 10^{-x^2}\,dx$

27. $\displaystyle\int \frac{2^{\sqrt{x}}}{\sqrt{x}}\,dx$

28. $\displaystyle\int \frac{10^{1/x}}{x^2}\,dx$

29. $\displaystyle\int x^2 7^{x^3+1}\,dx$

30. $\displaystyle\int \frac{1}{x\log_{10} x}\,dx$

31. $\displaystyle\int \frac{\log_2 x}{x}\,dx$

32. $\displaystyle\int (2^x)3^{(2^x)}\,dx$

33. The heart rate R (in beats per minute) and weight W (in pounds) of various mammals were measured, with the results shown in Fig. 6.7.13. Use the method of Example 3 to find a relation between the two of the form $R = kW^m$.

W	25	67	127	175	240	975
R	131	103	88	81	75	53

FIGURE 6.7.13 Data for Problem 33.

34. During the adiabatic expansion of a certain diatomic gas, its volume V (in liters) and pressure p (in atmospheres) were measured, with the results shown in Fig. 6.7.14. Use the method of Example 3 to find a relation between V and p of the form $p = kV^m$.

V	1.46	2.50	3.51	5.73	7.26
p	28.3	13.3	8.3	4.2	3.0

FIGURE 6.7.14 Data for Problem 34.

35. Find the highest point on the curve $f(x) = x \cdot 2^{-x}$ for $x > 0$.

36. Approximate the area of the first-quadrant region bounded by the curves $y = 2^{-x}$ and $y = (x-1)^2$. One of the points of intersection of these two curves should be evident, but you will need to approximate the other point.

37. Approximate the volume of the solid generated by rotation of the region of Problem 36 around the x-axis.

38. Approximate the area of the first-quadrant region bounded by the curves $y = 3^{2-x}$ and $y = (3x-4)^2$. You will need to approximate the two points of intersection of these two curves.

39. Prove Eq. (35). [*Suggestion:* Let $x = \log_a b$, $y = \log_b c$, and $z = \log_a c$. Then show that $a^z = a^{xy}$, and conclude that $z = xy$.]

40. Consider the function

$$f(x) = \frac{1}{1 + 2^{1/x}} \quad \text{for} \quad x \neq 0.$$

Show that both the left-hand and right-hand limits of $f(x)$ at $x = 0$ exist but are unequal.

41. Find dy/dx if $y = \log_x 2$.

6.7 Project: Natural Functional Equations

Provide complete details in the proofs of Fact 1 and Fact 2 that are outlined here.

Fact 1 If f is a continuous function such that

$$f(x+y) = f(x) + f(y) \tag{1}$$

for all real numbers x and y, then $f(x) = kx$ for some constant k.

OUTLINE OF PROOF Let us substitute t for y in Eq. (1) and then integrate from $t = 0$ to $t = y$ with x held constant:

$$\int_{t=0}^{y} f(x+t)\,dt = \int_{t=0}^{y} f(x)\,dt + \int_{t=0}^{y} f(t)\,dt.$$

Then substituting $u = u(t) = x + t$, $du = dt$ on the left-hand side yields the equation

$$\int_{u=x}^{x+y} f(u)\,du = y \cdot f(x) + \int_{t=0}^{y} f(t)\,dt,$$

from which we find that

$$y \cdot f(x) = \int_{t=0}^{x+y} f(t)\,dt - \int_{t=0}^{x} f(t)\,dt - \int_{t=0}^{y} f(t)\,dt.$$

The right-hand side is symmetric in the variables x and y, so interchanging them gives

$$y \cdot f(x) = x \cdot f(y), \quad \text{so that} \quad \frac{f(x)}{x} = \frac{f(y)}{y}$$

for all x and y. Because x and y are independent, it follows that the function $f(x)/x$ must be constant-valued, and therefore

$$f(x) = kx$$

for some constant k. ◄

This subtle proof is due to H. N. Shapiro, "A micronote on a functional equation," *Amer. Math. Monthly* **80** (1973), p. 1041. There exist discontinuous functions that satisfy Eq. (1). But it is known that any such function must be truly bizarre—its graph must intersect every circular disk in the xy-plane. Can you see that this implies not only that such a function not only is discontinuous at every point, but also is unbounded near every point?

Fact 2 If f is a positive-valued continuous function such that

$$f(x + y) = f(x) \cdot f(y) \qquad \textbf{(2)}$$

for all real numbers x and y, then

$$f(x) = e^{kx}$$

for some constant k.

OUTLINE OF PROOF Let $g(x) = \ln(f(x))$. Taking the natural logarithm of both sides in Eq. (2) gives

$$g(x + y) = \ln(f(x + y)) = \ln(f(x) \cdot f(y)) = \ln(f(x)) + \ln(f(y)) = g(x) + g(y).$$

Application of Fact 1 to g now yields

$$\ln(f(x)) = kx, \quad \text{so that} \quad f(x) = e^{kx}.$$ ◄

Challenge Use similar methods to establish the following result:

Fact 3 If f is a continuous function such that

$$f(xy) = f(x) \cdot f(y) \qquad \textbf{(3)}$$

for all positive real numbers x and y, then

$$f(x) = x^k$$

for some constant k.

6.8 INVERSE TRIGONOMETRIC FUNCTIONS

Recall that the function f is said to be **one-to-one** on its domain of definition D if, given x_1 and x_2 in D, $x_1 \neq x_2$ implies that $f(x_1) \neq f(x_2)$: "Different inputs give different outputs." (To prove that f is one-to-one, it is usually easier to prove the contrapositive—that if $f(x_1) = f(x_2)$, then $x_1 = x_2$.) What is important here is that if f is one-to-one on its domain of definition, then it has an inverse function f^{-1}. This inverse function is *defined* by

$$f^{-1}(x) = y \quad \text{if and only if} \quad f(y) = x. \qquad \textbf{(1)}$$

For example, from Section 3.8 we are familiar with the pair of inverse functions

$$f(x) = e^x \quad \text{and} \quad f^{-1}(x) = \ln x.$$

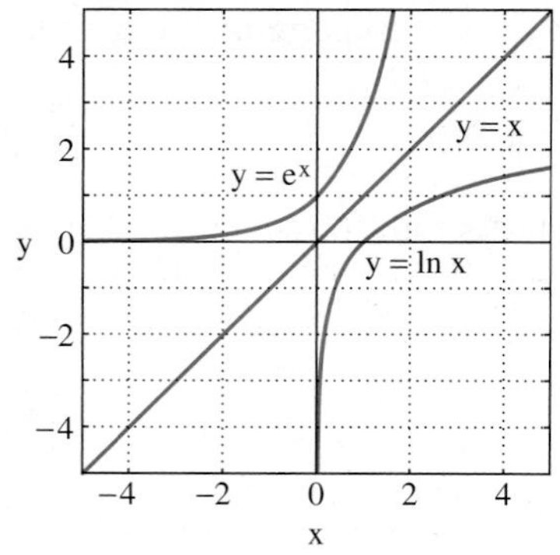

FIGURE 6.8.1 The graphs $y = e^x$ and $y = \ln x$ are reflections across the line $y = x$.

From a geometric viewpoint, Eq. (1) implies that the graphs $y = f(x)$ and $y = f^{-1}(x)$ are reflections across the 45° line $y = x$, like the familiar graphs $y = e^x$ and $y = \ln x$ in Fig. 6.8.1.

The Inverse Tangent Function

Here we want to define the inverses of the trigonometric functions, beginning with the inverse tangent function. We must, however, confront the fact that the trigonometric functions fail to be one-to-one because the period of each of the six is π or 2π. For example, $\tan x = 1$ if x is $\pi/4$ *or* $\pi/4$ plus any integral multiple of π. These many values of x, all with tangent equal to 1, correspond to the multiple points of intersection of the graph $y = \tan x$ and the horizontal line $y = 1$ in Fig. 6.8.2.

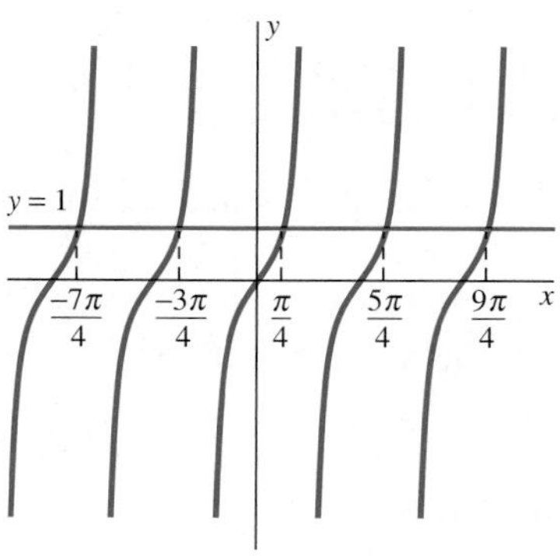

FIGURE 6.8.2 The tangent function takes on every real number value infinitely often.

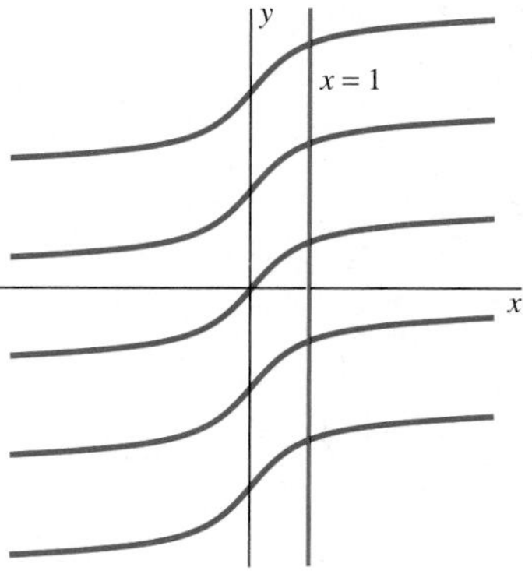

FIGURE 6.8.3 Simply reflecting the graph of $y = \tan x$ across the 45°-line $y = x$ does not produce the graph of a function.

Figure 6.8.3 is the reflection of Fig. 6.8.2 across the 45° line $y = x$. The multiple intersections of $x = \tan y$ and the vertical line $x = 1$ indicate that we must make a choice in order to define $\tan^{-1} 1$. That is, we *cannot* define $y = \tan^{-1} x$, the inverse of the tangent function, by saying simply that y is the number such that $\tan y = x$. There are *many* such values of y, and we must specify just which particular one of these is to be used. (Note that the symbol -1 in the notation $\tan^{-1} x$ is not an exponent—it does *not* mean $(\tan x)^{-1}$.)

We do this by suitably restricting the domain of the tangent function. Because the function $\tan x$ is increasing on $(-\pi/2, \pi/2)$ and its range of values is $(-\infty, +\infty)$, for each x in $(-\infty, +\infty)$ there is *one* number y in $(-\pi/2, \pi/2)$ such that $\tan y = x$. This observation leads to the following definition of the *inverse tangent* (or *arctangent*) function, denoted by $\tan^{-1} x$ or $\arctan x$.

DEFINITION The Inverse Tangent Function

The **inverse tangent** (or **arctangent**) **function** is defined as follows:

$$y = \tan^{-1} x \quad \text{if and only if} \quad \tan y = x \quad \text{and} \quad -\pi/2 < y < \pi/2 \tag{2}$$

where x is an arbitrary real number.

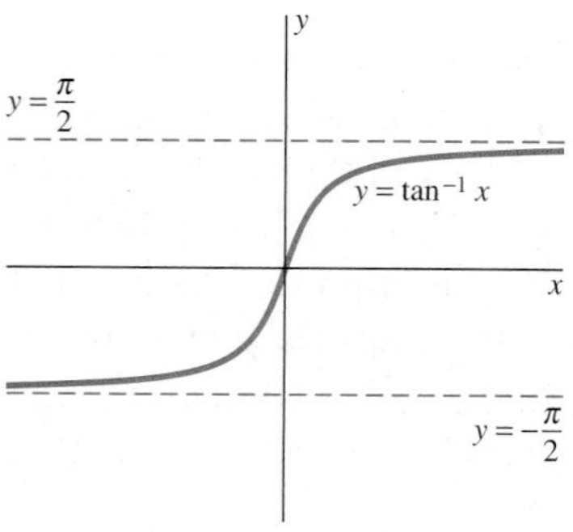

FIGURE 6.8.4 The inverse tangent function $\tan^{-1} x$ is defined for all x.

Because the tangent function attains all real values, $\tan^{-1} x$ is defined for all real numbers x; $\tan^{-1} x$ is that number y in the interval $(-\pi/2, \pi/2)$ whose tangent is x. The graph of $y = \tan^{-1} x$ is the reflection of the graph of $y = \tan x$, $-\pi/2 < x < \pi/2$, across the line $y = x$ (Fig. 6.8.4).

It follows from Eq. (2) that

$$\tan(\tan^{-1} x) = x \quad \text{for all } x \tag{3a}$$

and

$$\tan^{-1}(\tan x) = x \quad \text{if } -\pi/2 < x < \pi/2. \tag{3b}$$

Because the derivative of $\tan x$ is positive for all x in the interval $(-\pi/2, \pi/2)$, it follows from Theorem 1 in Section 3.8 that $\tan^{-1} x$ is differentiable for all x. We may, therefore, differentiate both sides of the identity in Eq. (3a). First we write that identity in the form

$$\tan y = x,$$

where $y = \tan^{-1} x$. Then differentiation with respect to x yields

$$(\sec^2 y)\frac{dy}{dx} = 1;$$

$$\frac{dy}{dx} = \frac{1}{\sec^2 y} = \frac{1}{1+\tan^2 y} = \frac{1}{1+x^2}.$$

Thus

$$D_x \tan^{-1} x = \frac{1}{1+x^2}, \tag{4}$$

and if u is any differentiable function of x, then

$$D_x \tan^{-1} u = \frac{1}{1+u^2} \cdot \frac{du}{dx}. \tag{5}$$

The definition of the inverse cotangent function is similar to that of the inverse tangent function, except that we begin by restricting the cotangent function to the interval $(0, \pi)$, where it is a decreasing function attaining all real values. Thus the **inverse cotangent** (or **arccotangent**) **function** is defined as

$$y = \cot^{-1} x \quad \text{if and only if} \quad \cot y = x \quad \text{and} \quad 0 < y < \pi \tag{6}$$

where x is any real number. Then differentiation of both sides of the identity $\cot(\cot^{-1} x) = x$ leads, as in the derivation of Eq. (4), to

$$D_x \cot^{-1} x = -\frac{1}{1+x^2}.$$

If u is a differentiable function of x, then the chain rule gives

$$D_x \cot^{-1} u = -\frac{1}{1+u^2} \cdot \frac{du}{dx}. \tag{7}$$

Upon comparing Eqs. (5) and (7) we see that $D_x \cot^{-1} x = -D_x \tan^{-1} x$, so the two inverse functions are closely related. Indeed, perhaps you can show—using either calculus or just the definitions of the two functions—that $\tan^{-1} x + \cot^{-1} x \equiv \pi/2$.

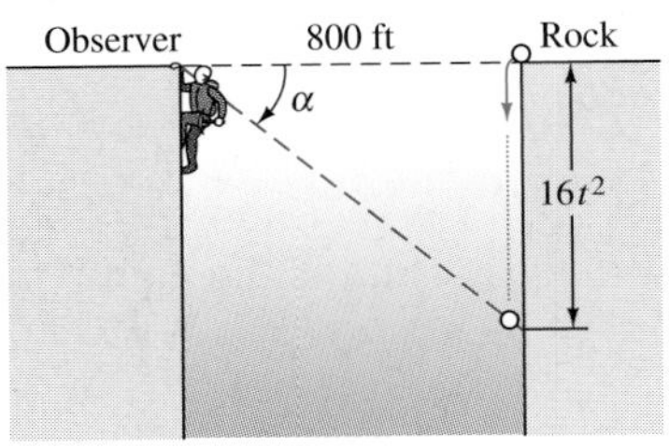

FIGURE 6.8.5 The falling rock of Example 1.

EXAMPLE 1 A mountain climber on one edge of a deep canyon 800 ft wide sees a large rock fall from the opposite edge at time $t = 0$. As he watches the rock plummet downward, his eyes first move slowly, then faster, then more slowly again. Let α be the angle of depression of his line of sight below the horizontal. At what angle α would the rock *seem* to be moving the most rapidly? That is, when would $d\alpha/dt$ be maximal?

Solution From our study of constant acceleration in Section 5.2, we know that the rock will fall $16t^2$ feet in the first t seconds. We refer to Fig. 6.8.5 and see that the value of α at time t will be

$$\alpha = \alpha(t) = \tan^{-1}\left(\frac{16t^2}{800}\right) = \tan^{-1}\left(\frac{t^2}{50}\right).$$

Hence

$$\frac{d\alpha}{dt} = \frac{1}{1+\left(\dfrac{t^2}{50}\right)^2}\cdot\frac{2t}{50} = \frac{100t}{t^4+2500}.$$

To find when $d\alpha/dt$ is maximal, we find when *its* derivative is zero:

$$\frac{d}{dt}\left(\frac{d\alpha}{dt}\right) = \frac{100(t^4+2500)-(100t)(4t^3)}{(t^4+2500)^2} = \frac{100(2500-3t^4)}{(t^4+2500)^2}.$$

So $d^2\alpha/dt^2$ is zero when $3t^4 = 2500$—that is, when

$$t = \sqrt[4]{\tfrac{2500}{3}} \approx 5.37 \quad \text{(s)}.$$

This is the value of t when $d\alpha/dt$ is maximal, and at this time we have $t^2 = 50/\sqrt{3}$. So the angle at this time is

$$\alpha = \arctan\left(\frac{1}{50}\cdot\frac{50}{\sqrt{3}}\right) = \arctan\left(\frac{1}{\sqrt{3}}\right) = \frac{\pi}{6}.$$

The *apparent* speed of the falling rock is greatest when the climber's line of sight is 30° below the horizontal. The actual speed of the rock is then $32t$ with $t \approx 5.37$ and thus is about 172 ft/s. ◆

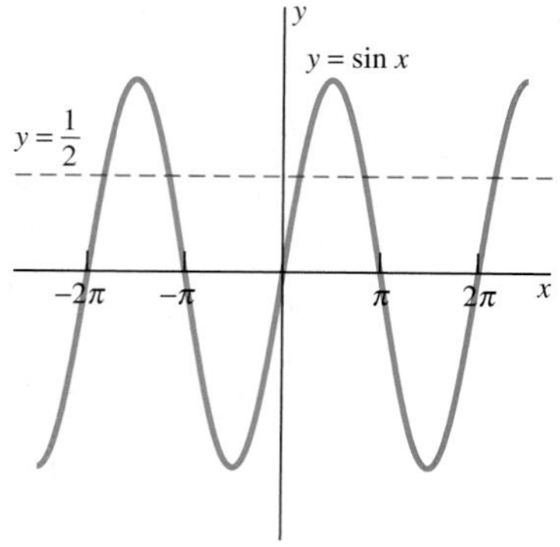

FIGURE 6.8.6 Multiple values of x such that $\sin x = \frac{1}{2}$.

The Inverse Sine Function

Figure 6.8.6 shows the graph of the sine function and the horizontal line $y = \frac{1}{2}$. Because this line meets the graph of the sine function in more than one point (indeed, at infinitely many points), the sine function takes on the value $\frac{1}{2}$ for many different values of x. For example, $\sin x = \frac{1}{2}$ if x is *either* $\pi/6$ plus any integral multiple of 2π *or* $5\pi/6$ plus any integral multiple of 2π.

Figure 6.8.7 is the reflection of Fig. 6.8.6 across the 45° line $y = x$. The multiple intersections of $x = \sin y$ and the vertical line $x = \frac{1}{2}$ indicate that we must make a choice in order to define $\sin^{-1}(\frac{1}{2})$. That is, we *cannot* define $y = \sin^{-1} x$, the inverse of the sine function, by saying merely that y is the number such that $\sin y = x$. There are *many* such values of y, and we must specify just which particular one of these is to be used.

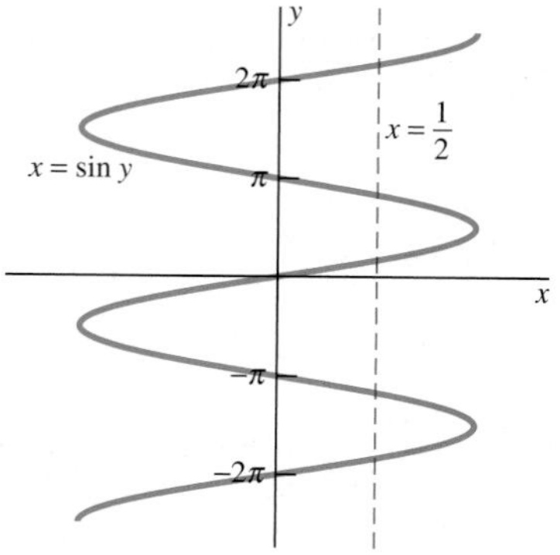

FIGURE 6.8.7 There are many possible choices for $y = \sin^{-1}\frac{1}{2}$.

We do this by suitably restricting the domain of the sine function. Because the function $\sin x$ is increasing on $[-\pi/2, \pi/2]$ and its range of values is $[-1, 1]$, for each x in $[-1, 1]$ there is *one* number y in $[-\pi/2, \pi/2]$ such that $\sin y = x$. This observation leads to the following definition of the *inverse sine* (or *arcsine*) function, denoted by $\sin^{-1} x$ or by $\arcsin x$.

DEFINITION The Inverse Sine Function

The **inverse sine** (or **arcsine**) **function** is defined as follows:

$$y = \sin^{-1} x \quad \text{if and only if} \quad \sin y = x \quad \text{and} \quad -\pi/2 \leqq y \leqq \pi/2 \qquad \textbf{(8)}$$

where $-1 \leqq x \leqq 1$.

Thus, if x is between -1 and $+1$ (inclusive), then $\sin^{-1} x$ is that number y between $-\pi/2$ and $\pi/2$ such that $\sin y = x$. Even more briefly, $\arcsin x$ is the angle (in radians) nearest zero whose sine is x. For instance,

$$\sin^{-1} 1 = \frac{\pi}{2}, \qquad \sin^{-1} 0 = 0, \qquad \sin^{-1}(-1) = -\frac{\pi}{2},$$

and $\sin^{-1} 2$ does not exist.

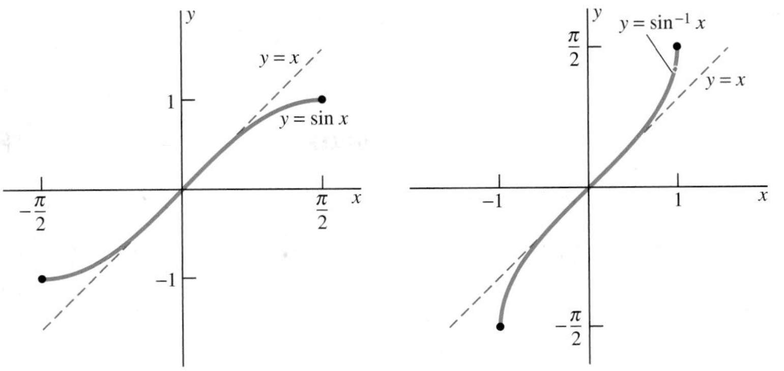

(a) Graph of the sine function

(b) Graph of the arcsine function

FIGURE 6.8.8 The graphs $y = \sin x$ and $y = \sin^{-1} x$ are reflections of each other across the line $y = x$.

Because interchanging x and y in the equation $\sin y = x$ yields $y = \sin x$, it follows from Eq. (8) that the graph of $y = \sin^{-1} x$ is the reflection of the graph of $y = \sin x$, $-\pi/2 \leqq x \leqq \pi/2$, across the line $y = x$ (Fig. 6.8.8).

It also follows from Eq. (8) that

$$\sin(\sin^{-1} x) = x \quad \text{if } -1 \leqq x \leqq 1 \tag{9a}$$

and

$$\sin^{-1}(\sin x) = x \quad \text{if } -\pi/2 \leqq x \leqq \pi/2. \tag{9b}$$

Because the derivative of $\sin x$ is positive for $-\pi/2 < x < \pi/2$, it follows from Theorem 1 of Section 3.8 that $\sin^{-1} x$ is differentiable on $(-1, 1)$. We may, therefore, differentiate both sides of the identity in (9a), but we begin by writing it in the form

$$\sin y = x,$$

where $y = \sin^{-1} x$. Then differentiation with respect to x gives

$$(\cos y)\frac{dy}{dx} = 1,$$

so

$$\frac{dy}{dx} = \frac{1}{\cos y} = \frac{1}{\sqrt{1 - \sin^2 y}} = \frac{1}{\sqrt{1 - x^2}}.$$

We are correct in taking the positive square root in this computation because $\cos y > 0$ for $-\pi/2 < y < \pi/2$. Thus

$$D_x \sin^{-1} x = \frac{1}{\sqrt{1 - x^2}} \tag{10}$$

provided that $-1 < x < 1$. When we combine this result with the chain rule, we get

$$D_x \sin^{-1} u = \frac{1}{\sqrt{1 - u^2}} \cdot \frac{du}{dx} \tag{11}$$

if u is a differentiable function with values in the interval $(-1, 1)$.

EXAMPLE 2 If $y = \sin^{-1} x^2$, then Eq. (11) with $u = x^2$ yields

$$\frac{dy}{dx} = \frac{1}{\sqrt{1 - (x^2)^2}} \cdot 2x = \frac{2x}{\sqrt{1 - x^4}}.$$

◆

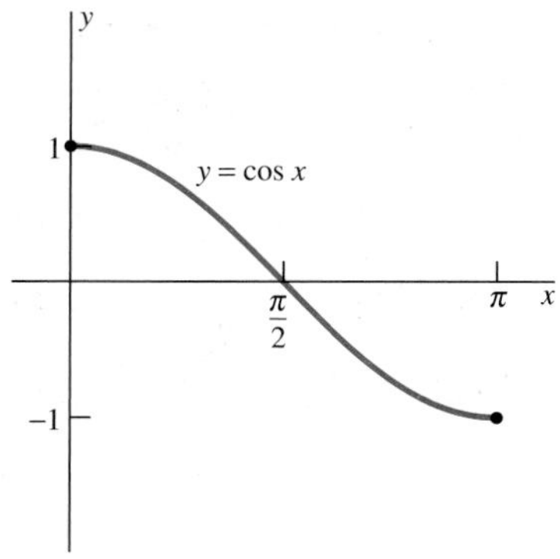

FIGURE 6.8.9 The cosine function is decreasing on the interval $0 \leqq x \leqq \pi$.

The definition of the inverse cosine function is similar to that of the inverse sine function, except that we begin by restricting the cosine function to the interval $[0, \pi]$, where it is a decreasing function (Fig. 6.8.9). Thus the **inverse cosine** (or **arccosine**) **function** is defined by means of the rule

$$y = \cos^{-1} x \quad \text{if and only if} \quad \cos y = x \quad \text{and} \quad 0 \leqq y \leqq \pi \tag{12}$$

where $-1 \leqq x \leqq 1$. Thus $\cos^{-1} x$ is the angle in $[0, \pi]$ whose cosine is x. For instance,

$$\cos^{-1} 1 = 0, \qquad \cos^{-1} 0 = \frac{\pi}{2}, \qquad \cos^{-1}(-1) = \pi.$$

We may compute the derivative of $\cos^{-1} x$, also written $\arccos x$, by differentiating both sides of the identity

$$\cos(\cos^{-1} x) = x \quad (-1 < x < 1).$$

The computations are similar to those for $D_x \sin^{-1} x$ and lead to the result

$$D_x \cos^{-1} x = -\frac{1}{\sqrt{1 - x^2}}.$$

And if u denotes a differentiable function of x, the chain rule then gives

$$D_x \cos^{-1} u = -\frac{1}{\sqrt{1 - u^2}} \cdot \frac{du}{dx}. \tag{13}$$

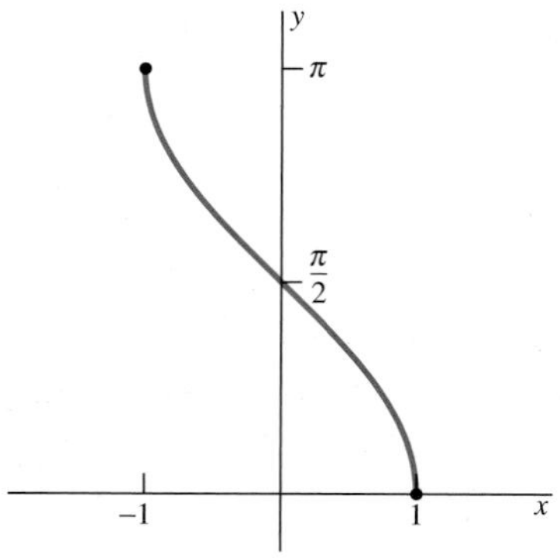

FIGURE 6.8.10 The graph $y = \cos^{-1} x$ of the arccosine function.

Figure 6.8.10 shows the graph of $y = \cos^{-1} x$ as the reflection of the graph of $y = \cos x$, $0 \leqq x \leqq \pi$, across the line $y = x$.

The Inverse Secant Function

Figure 6.8.11 shows that the secant function is increasing on each of the intervals $[0, \pi/2)$ and $(\pi/2, \pi]$. On the union of these two intervals, the secant function attains all real values y such that $|y| \geqq 1$. We may, therefore, define the inverse secant function, denoted by $\sec^{-1} x$ or by $\operatorname{arcsec} x$, by restricting the secant function to the union of the two intervals $[0, \pi/2)$ and $(\pi/2, \pi]$.

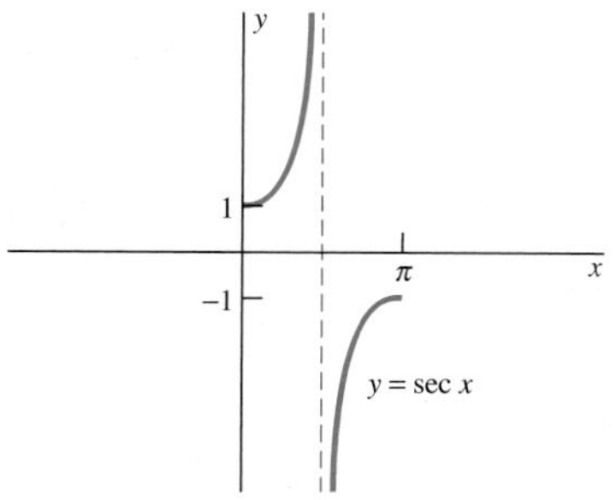

FIGURE 6.8.11 Restriction of the secant function to the union of the two intervals $[0, \pi/2)$ and $(\pi/2, \pi]$.

DEFINITION The Inverse Secant Function

The **inverse secant** (or **arcsecant**) **function** is defined as follows:

$$y = \sec^{-1} x \quad \text{if and only if} \quad \sec y = x \quad \text{and} \quad 0 \leqq y \leqq \pi \tag{14}$$

where $|x| \geqq 1$.

REMARK 1 Because $\sec(\pi/2)$ is not defined, the restriction $0 \leqq y \leqq \pi$ in (14) implies that the range of the inverse secant function is the union $[0, \pi/2) \cup (\pi/2, \pi]$.

REMARK 2 Some textbooks offer alternative definitions of the inverse secant based on different intervals of definition of $\sec x$. The definition given here, however, satisfies the condition that

$$\sec^{-1} x = \cos^{-1} \frac{1}{x} \quad (\text{if } |x| > 1),$$

which is convenient for calculator-computer calculations. (See Problem 62.) Moreover, our definition of $\sec^{-1} x$ is the same as that used in computer algebra systems such as *Maple* and *Mathematica*. (See Problem 61.)

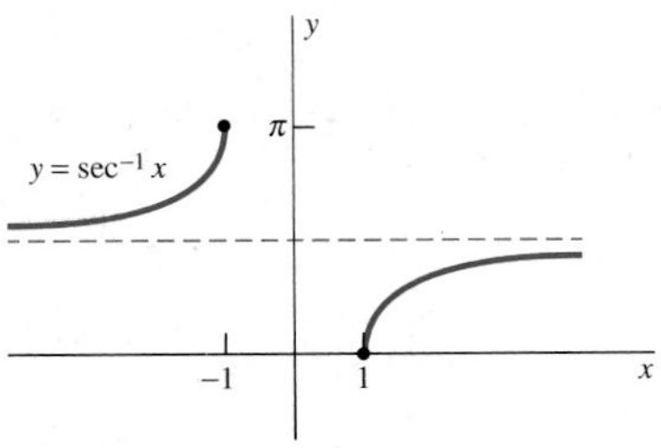

FIGURE 6.8.12 The graph of $y = \operatorname{arcsec} x = \sec^{-1} x$.

The graph of $y = \sec^{-1} x$ is the reflection of the graph of $y = \sec x$, suitably restricted to the intervals $0 \leqq x < \pi/2$ and $\pi/2 < x \leqq \pi$, across the line $y = x$ (Fig. 6.8.12). It follows from the definition of the inverse secant that

$$\sec(\sec^{-1} x) = x \quad \text{if } |x| \geqq 1, \tag{15a}$$

$$\sec^{-1}(\sec x) = x \quad \text{for } x \text{ in } [0, \pi/2) \cup (\pi/2, \pi]. \tag{15b}$$

Following the now-familiar pattern, we find $D_x \sec^{-1} x$ by differentiating both sides of Eq. (15a) in the form

$$\sec y = x,$$

where $y = \sec^{-1} x$. This yields

$$(\sec y \tan y)\,\frac{dy}{dx} = 1,$$

so

$$\frac{dy}{dx} = \frac{1}{\sec y \tan y} = \frac{1}{\pm x\sqrt{x^2 - 1}},$$

because $\tan y = \pm\sqrt{\sec^2 y - 1} = \pm\sqrt{x^2 - 1}$.

To obtain the correct choice of sign here, note what happens in the two cases $x > 1$ and $x < -1$. In the first case, $0 < y < \pi/2$ and $\tan y > 0$, so we choose the plus sign. If $x < -1$, then $\pi/2 < y < \pi$ and $\tan y < 0$, so we take the minus sign. Thus

$$\frac{dy}{dx} = \frac{1}{|x|\sqrt{x^2 - 1}} \quad (|x| > 1). \tag{16}$$

If u is a differentiable function of x with values that exceed 1 in magnitude, then by the chain rule we have

$$D_x \sec^{-1} u = \frac{1}{|u|\sqrt{u^2 - 1}} \cdot \frac{du}{dx}. \tag{17}$$

EXAMPLE 3 The function $f(x) = \sec^{-1} e^x$ is defined if $x > 0$, because then $e^x > 1$. Then by Eq. (17),

$$D_x \sec^{-1} e^x = \frac{e^x}{|e^x|\sqrt{e^{2x} - 1}} = \frac{1}{\sqrt{e^{2x} - 1}}$$

because $|e^x| = e^x$ for all x. ◆

The **inverse cosecant** (or **arccosecant**) **function** is the inverse of the function $y = \csc x$, where x is restricted to the union of the intervals $[-\pi/2, 0)$ and $(0, \pi/2]$. Thus

$$y = \csc^{-1} x \quad \text{if and only if} \quad \csc y = x \quad \text{and} \quad -\pi/2 < y < \pi/2 \tag{18}$$

where $|x| \geqq 1$. Its derivative formula, which has a derivation similar to that of the inverse secant function, is

$$D_x \csc^{-1} u = -\frac{1}{|u|\sqrt{u^2 - 1}} \cdot \frac{du}{dx}. \tag{19}$$

SUMMARY

The following table summarizes the domains, ranges, and derivatives of the six inverse trigonometric functions.

Function	Domain of Definition	Range of Values	Derivative
$\sin^{-1} x$	$-1 \leqq x \leqq 1$	$-\pi/2 \leqq y \leqq \pi/2$	$\dfrac{1}{\sqrt{1-x^2}}$
$\cos^{-1} x$	$-1 \leqq x \leqq 1$	$0 \leqq y \leqq \pi$	$-\dfrac{1}{\sqrt{1-x^2}}$
$\tan^{-1} x$	$-\infty < x < +\infty$	$-\pi/2 < y < \pi/2$	$\dfrac{1}{1+x^2}$
$\cot^{-1} x$	$-\infty < x < +\infty$	$0 < y < \pi$	$-\dfrac{1}{1+x^2}$
$\sec^{-1} x$	$\lvert x\rvert \geqq 1$	$0 \leqq y < \pi/2,\ \pi/2 < y \leqq \pi$	$\dfrac{1}{\lvert x\rvert\sqrt{x^2-1}}$
$\csc^{-1} x$	$\lvert x\rvert \geqq 1$	$-\pi/2 < y < 0,\ 0 < y < \pi/2$	$-\dfrac{1}{\lvert x\rvert\sqrt{x^2-1}}$

It is worth noting that

- $\sin^{-1} x$ has the range $[-\pi/2, \pi/2]$ and $\tan^{-1} x$ has the range $(-\pi/2, \pi/2)$, whereas
- $\cos^{-1} x$ has the range $[0, \pi]$ and $\sec^{-1} x$ has the range $[0, \frac{1}{2}\pi) \cup (\frac{1}{2}\pi, \pi]$.

Observe also the "difference only in sign" of the derivatives of function/cofunction pairs of inverse functions. ◆

Integrals Involving Inverse Trigonometric Functions

The derivatives of the six inverse trigonometric functions are all simple *algebraic* functions. As a consequence, inverse trigonometric functions typically occur when we integrate algebraic functions. Moreover, as mentioned earlier, the derivatives of $\cos^{-1} x$, $\cot^{-1} x$, and $\csc^{-1} x$ differ only in sign from the derivatives of their respective cofunctions. For this reason only the arcsine, arctangent, and arcsecant functions are necessary for integration, and only these three are in common use. That is, you need commit to memory the integral formulas only for the latter three functions. They follow immediately from Eqs. (5), (11), and (17) and may be written in the forms shown next:

$$\int \frac{du}{\sqrt{1-u^2}} = \sin^{-1} u + C, \tag{20}$$

$$\int \frac{du}{1+u^2} = \tan^{-1} u + C, \tag{21}$$

$$\int \frac{du}{u\sqrt{u^2-1}} = \sec^{-1} |u| + C. \tag{22}$$

It is easy to verify that the absolute value on the right-hand side in Eq. (22) follows from the one in Eq. (17). (See Problem 57.) And remember that because $\sec^{-1} |u|$ is undefined unless $|u| \geqq 1$, the definite integral

$$\int_a^b \frac{du}{u\sqrt{u^2-1}}$$

is meaningful only when the limits a and b are both at least 1 or both at most -1.

EXAMPLE 4 It follows immediately from Eq. (21) that

$$\int_0^1 \frac{dx}{1+x^2} = \left[\tan^{-1} x\right]_0^1 = \tan^{-1} 1 - \tan^{-1} 0 = \frac{\pi}{4}. \quad \blacklozenge$$

EXAMPLE 5 The substitution $u = 3x$, $du = 3\,dx$ gives

$$\int \frac{1}{1+9x^2}\,dx = \frac{1}{3}\int \frac{3}{1+(3x)^2}\,dx$$
$$= \frac{1}{3}\int \frac{du}{1+u^2} = \frac{1}{3}\tan^{-1} u + C = \frac{1}{3}\tan^{-1} 3x + C. \quad \blacklozenge$$

EXAMPLE 6 The substitution $u = \frac{1}{2}x$, $du = \frac{1}{2}\,dx$ gives

$$\int \frac{1}{\sqrt{4-x^2}}\,dx = \int \frac{1}{2\sqrt{1-(x/2)^2}}\,dx$$
$$= \int \frac{1}{\sqrt{1-u^2}}\,du = \arcsin u + C = \arcsin\left(\frac{x}{2}\right) + C. \quad \blacklozenge$$

EXAMPLE 7 The substitution $u = x\sqrt{2}$, $du = \sqrt{2}\,dx$ gives

$$\int_1^{\sqrt{2}} \frac{1}{x\sqrt{2x^2-1}}\,dx = \int_{\sqrt{2}}^{2} \frac{1}{u\sqrt{u^2-1}}\,du$$
$$= \left[\sec^{-1} |u|\right]_{\sqrt{2}}^{2} = \sec^{-1} 2 - \sec^{-1}\sqrt{2}$$
$$= \frac{\pi}{3} - \frac{\pi}{4} = \frac{\pi}{12}. \quad \blacklozenge$$

6.8 TRUE/FALSE STUDY GUIDE

6.8 CONCEPTS: QUESTIONS AND DISCUSSION

1. Suppose that f and g are inverse functions. Outline the way the derivative of g can be found if the derivative of f is known.
2. Discuss the differences and similarities in the domains and ranges of the six inverse trigonometric functions.

6.8 PROBLEMS

Find the values indicated in Problems 1 through 4.

1. (a) $\sin^{-1}\left(\frac{1}{2}\right)$; (b) $\sin^{-1}\left(-\frac{1}{2}\right)$; (c) $\sin^{-1}\left(\frac{1}{2}\sqrt{2}\right)$; (d) $\sin^{-1}\left(-\frac{1}{2}\sqrt{3}\right)$

2. (a) $\cos^{-1}\left(\frac{1}{2}\right)$; (b) $\cos^{-1}\left(-\frac{1}{2}\right)$; (c) $\cos^{-1}\left(\frac{1}{2}\sqrt{2}\right)$; (d) $\cos^{-1}\left(-\frac{1}{2}\sqrt{3}\right)$

3. (a) $\tan^{-1} 0$; (b) $\tan^{-1} 1$; (c) $\tan^{-1}(-1)$; (d) $\tan^{-1}\sqrt{3}$

4. (a) $\sec^{-1} 1$; (b) $\sec^{-1}(-1)$; (c) $\sec^{-1} 2$; (d) $\sec^{-1}\left(-\sqrt{2}\right)$

Differentiate the functions in Problems 5 through 26.

5. $f(x) = \sin^{-1}(x^{100})$

6. $f(x) = \arctan(e^x)$

7. $f(x) = \sec^{-1}(\ln x)$

8. $f(x) = \ln(\tan^{-1} x)$

9. $f(x) = \arcsin(\tan x)$

10. $f(x) = x \arctan x$

11. $f(x) = \sin^{-1} e^x$

12. $f(x) = \arctan\sqrt{x}$

13. $f(x) = \cos^{-1} x + \sec^{-1}\left(\frac{1}{x}\right)$

14. $f(x) = \cot^{-1}\left(\frac{1}{x^2}\right)$

15. $f(x) = \csc^{-1} x^2$

16. $f(x) = \arccos\left(\frac{1}{\sqrt{x}}\right)$

17. $f(x) = \dfrac{1}{\arctan x}$

18. $f(x) = (\arcsin x)^2$

19. $f(x) = \tan^{-1}(\ln x)$

20. $f(x) = \operatorname{arcsec}\sqrt{x^2+1}$

21. $f(x) = \tan^{-1} e^x + \cos^{-1} e^{-x}$

22. $f(x) = \exp(\arcsin x)$

23. $f(x) = \sin(\arctan x)$

24. $f(x) = \sec(\sec^{-1} e^x)$

25. $f(x) = \dfrac{\arctan x}{(1+x^2)^2}$

26. $f(x) = (\sin^{-1} 2x^2)^{-2}$

In Problems 27 through 30, find dy/dx by implicit differentiation. Then find the line tangent to the graph of the equation at the indicated point P.

27. $\tan^{-1} x + \tan^{-1} y = \dfrac{\pi}{2}$; $P(1, 1)$

28. $\sin^{-1} x + \sin^{-1} y = \dfrac{\pi}{2}$; $P\left(\frac{1}{2}, \frac{1}{2}\sqrt{3}\right)$

29. $(\sin^{-1} x)(\sin^{-1} y) = \dfrac{\pi^2}{16}$; $P\left(\frac{1}{2}\sqrt{2}, \frac{1}{2}\sqrt{2}\right)$

30. $(\sin^{-1} x)^2 + (\sin^{-1} y)^2 = \dfrac{5\pi^2}{36}$; $P\left(\frac{1}{2}, \frac{1}{2}\sqrt{3}\right)$

Evaluate or antidifferentiate, as appropriate, in Problems 31 through 55.

31. $\displaystyle\int_0^1 \frac{dx}{1+x^2}$

32. $\displaystyle\int_0^{1/2} \frac{dx}{\sqrt{1-x^2}}$

33. $\displaystyle\int_{\sqrt{2}}^2 \frac{dx}{x\sqrt{x^2-1}}$

34. $\displaystyle\int_{-2}^{-2/\sqrt{3}} \frac{dx}{x\sqrt{x^2-1}}$

35. $\displaystyle\int_0^3 \frac{dx}{9+x^2}$

36. $\displaystyle\int_0^{\sqrt{12}} \frac{dx}{\sqrt{16-x^2}}$

37. $\displaystyle\int \frac{dx}{\sqrt{1-4x^2}}$

38. $\displaystyle\int \frac{dx}{9x^2+4}$

39. $\displaystyle\int \frac{dx}{x\sqrt{x^2-25}}$

40. $\displaystyle\int \frac{dx}{x\sqrt{4x^2-9}}$

41. $\displaystyle\int \frac{e^x}{1+e^{2x}}\,dx$

42. $\displaystyle\int \frac{x^2}{x^6+25}\,dx$

43. $\displaystyle\int \frac{dx}{x\sqrt{x^6-25}}$

44. $\displaystyle\int \frac{\sqrt{x}}{1+x^3}\,dx$

45. $\displaystyle\int \frac{dx}{\sqrt{x(1-x)}}$

46. $\displaystyle\int \frac{\sec x \tan x}{1+\sec^2 x}\,dx$

47. $\displaystyle\int \frac{x^{49}}{1+x^{100}}\,dx$

48. $\displaystyle\int \frac{x^4}{\sqrt{1-x^{10}}}\,dx$

49. $\displaystyle\int \frac{1}{x[1+(\ln x)^2]}\,dx$

50. $\displaystyle\int \frac{\arctan x}{1+x^2}\,dx$

51. $\displaystyle\int_0^1 \frac{1}{1+(2x-1)^2}\,dx$

52. $\displaystyle\int_0^1 \frac{x^3}{1+x^4}\,dx$

53. $\displaystyle\int_1^e \frac{dx}{x\sqrt{1-(\ln x)^2}}$

54. $\displaystyle\int_1^2 \frac{dx}{x\sqrt{x^2-1}}$

55. $\displaystyle\int_1^3 \frac{dx}{2\sqrt{x}\,(1+x)}$ [*Suggestion:* Let $u = x^{1/2}$.]

56. Conclude from the formula $D_x \cos^{-1} x = -D_x \sin^{-1} x$ that $\sin^{-1} x + \cos^{-1} x = \pi/2$ if $0 \leqq x \leqq 1$.

57. The integral formula in (22) is equivalent to

$$D_u \sec^{-1} |u| = \frac{1}{u\sqrt{u^2-1}} \quad \text{if } |u| > 1. \qquad \mathbf{(22')}$$

This is the same as (16) if $u > 1$. Assuming that $u < -1$, substitute $x = -u$ in $y = \sec^{-1} |u|$ and use the chain rule $dy/du = (dy/dx)(dx/du)$ to verify (22′).

In Problems 58 through 60, substitute $u = ax$ (assuming that $a > 0$) to derive the given integral formula.

58. $\displaystyle\int \frac{1}{\sqrt{a^2-u^2}}\,du = \sin^{-1}\left(\frac{u}{a}\right) + C \quad (u < a).$

59. $\displaystyle\int \frac{1}{a^2+u^2}\,du = \frac{1}{a}\tan^{-1}\left(\frac{u}{a}\right) + C.$

60. $\displaystyle\int \frac{1}{u\sqrt{u^2-a^2}}\,du = \frac{1}{a}\sec^{-1}\left|\frac{u}{a}\right| + C \quad (u > a).$

61. If $f(x) = \sec^{-1} x$ then the computer algebra systems *Mathematica* and *Maple* both give

$$f'(x) = \frac{1}{x^2\sqrt{1-\dfrac{1}{x^2}}}$$

for the derivative of f. Verify carefully that (if either $x < -1$ or $x > 1$) this result is equivalent to the derivative formula for $\sec^{-1} x$ given in this section.

62. Show that

$$D_x \sec^{-1} x = D_x \cos^{-1} \frac{1}{x} \quad \text{if } |x| > 1,$$

and conclude that

$$\sec^{-1} x = \cos^{-1} \frac{1}{x} \quad \text{if } |x| > 1.$$

This fact can be used to find arcsecant values on a calculator that has a key for the arccosine function, usually written **INV** **COS** or **COS⁻¹**, but no arcsecant key.

63. Some calculus textbooks define the inverse secant function as that function g such that $y = g(x)$ if and only if $\sec y = x$ with y in either $[0, \pi/2)$ or $[\pi, 3\pi/2)$ (the latter instead of the interval $(\pi/2, \pi]$ used in this text). In contrast with Fig. 6.8.12, show that the graph of this "alternative arcsecant function" is as shown in Fig. 6.8.13. Then show that its derivative is given by

$$g'(x) = \frac{1}{x\sqrt{x^2-1}}$$

(with no absolute value on the right).

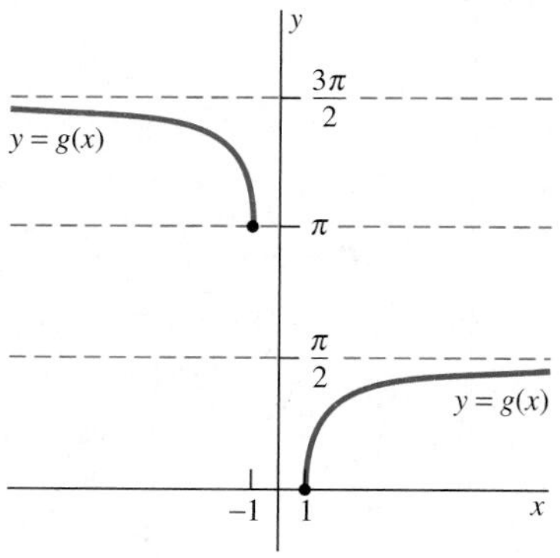

FIGURE 6.8.13 Graph of the alternative inverse secant function of Problem 63.

64. (a) Deduce from the addition formula for tangents (Problem 28 in Appendix C) that

$$\arctan x + \arctan y = \arctan \frac{x+y}{1-xy}$$

provided that $xy < 1$. (b) Apply part (a) to show that each of the following numbers is equal to $\pi/4$: (i) $\arctan(\frac{1}{2}) + \arctan(\frac{1}{3})$; (ii) $2\arctan(\frac{1}{3}) + \arctan(\frac{1}{7})$; (iii) $\arctan(\frac{120}{119}) - \arctan(\frac{1}{239})$; (iv) $4\arctan(\frac{1}{5}) - \arctan(\frac{1}{239})$.

65. A billboard to be built *parallel* to a highway will be 12 m high and its bottom will be 4 m above the eye level of the average passing motorist. How far from the highway should the billboard be placed in order to maximize the vertical angle it subtends at the motorist's eyes?

66. Use inverse trigonometric functions to prove that the vertical angle subtended by a rectangular painting on a wall is greatest when the painting is hung with its center at the level of the observer's eyes.

67. Show that the circumference of a circle of radius a is $2\pi a$ by finding the length of the circular arc

$$y = \sqrt{a^2 - x^2}$$

from $x = 0$ to $x = a/\sqrt{2}$ and then multiplying by 8.

68. Find the volume generated by revolving around the y-axis the area under $y = 1/(1 + x^4)$ from $x = 0$ to $x = 1$.

69. The unbounded region R is bounded on the left by the y-axis, below by the x-axis, and above by the graph of $y = 1/(1+x^2)$. Show that the area of R is finite by evaluating

$$\lim_{a\to\infty} \int_0^a \frac{dx}{1+x^2}.$$

70. A building 250 ft high is equipped with an external elevator. The elevator starts at the top at time $t = 0$ and descends at the constant rate of 25 ft/s. You are watching the elevator from a window that is 100 ft above the ground and in a building 50 ft from the elevator. At what height does the elevator appear to you to be moving the fastest?

71. Suppose that the function f is defined for all x such that $|x| > 1$ and has the property that

$$f'(x) = \frac{1}{x\sqrt{x^2 - 1}}$$

for all such x. (a) Explain why there exist two constants A and B such that

$$f(x) = \operatorname{arcsec} x + A \quad \text{if } x > 1;$$
$$f(x) = -\operatorname{arcsec} x + B \quad \text{if } x < -1.$$

(b) Determine the values of A and B so that $f(2) = 1 = f(-2)$. Then sketch the graph of $y = f(x)$.

In some computing languages the arctangent is the only inverse trigonometric function that is programmed directly, so it is necessary to express $\sin^{-1} x$ *and* $\sec^{-1} x$ *in terms of* $\tan^{-1} x$. *In Problems 72 and 73 verify each given identity by differentiating both sides. What else must be done?*

72. $\sin^{-1} x = \tan^{-1}\left(\dfrac{x}{\sqrt{1-x^2}}\right)$.

73. (a) $\sec^{-1} x = \tan^{-1}\sqrt{x^2 - 1}$ if $x > 1$;

(b) $\sec^{-1} x = \pi - \tan^{-1}\sqrt{x^2 - 1}$ if $x < -1$.

In Problems 74 through 76, estimate the absolute maximum value of $f(x)$ *for* $x > 0$. *You may want to begin by locating the pertinent critical point graphically.*

74. $f(x) = x^{-1/2}\tan^{-1} x$

75. $f(x) = e^{-x/10}\tan^{-1} x$

76. $f(x) = e^{-x/100}\sec^{-1} x$

6.9 HYPERBOLIC FUNCTIONS

The **hyperbolic cosine** and the **hyperbolic sine** of the real number x are denoted by $\cosh x$ and $\sinh x$ and are defined to be

$$\cosh x = \frac{e^x + e^{-x}}{2} \quad \text{and} \quad \sinh x = \frac{e^x - e^{-x}}{2}. \tag{1}$$

These particular combinations of familiar exponentials are useful in certain applications of calculus and are also helpful in evaluating certain integrals. The other four hyperbolic functions—the hyperbolic tangent, cotangent, secant, and cosecant—are defined in terms of $\cosh x$ and $\sinh x$ by analogy with trigonometry:

$$\tanh x = \frac{\sinh x}{\cosh x} = \frac{e^x - e^{-x}}{e^x + e^{-x}},$$
$$\coth x = \frac{\cosh x}{\sinh x} = \frac{e^x + e^{-x}}{e^x - e^{-x}} \quad (x \neq 0); \tag{2}$$

$$\operatorname{sech} x = \frac{1}{\cosh x} = \frac{2}{e^x + e^{-x}},$$
$$\operatorname{csch} x = \frac{1}{\sinh x} = \frac{2}{e^x - e^{-x}} \quad (x \neq 0). \tag{3}$$

The trigonometric terminology and notation for these hyperbolic functions stems from the fact that these functions satisfy a list of identities that, apart from an occasional difference of sign, much resemble familiar trigonometric identities:

$$\cosh^2 x - \sinh^2 x = 1; \tag{4}$$

$$1 - \tanh^2 x = \text{sech}^2\, x; \tag{5}$$

$$\coth^2 x - 1 = \text{csch}^2\, x; \tag{6}$$

$$\sinh(x + y) = \sinh x \cosh y + \cosh x \sinh y; \tag{7}$$

$$\cosh(x + y) = \cosh x \cosh y + \sinh x \sinh y; \tag{8}$$

$$\sinh 2x = 2\sinh x \cosh x; \tag{9}$$

$$\cosh 2x = \cosh^2 x + \sinh^2 x; \tag{10}$$

$$\cosh^2 x = \tfrac{1}{2}(\cosh 2x + 1); \tag{11}$$

$$\sinh^2 x = \tfrac{1}{2}(\cosh 2x - 1). \tag{12}$$

The identities in Eqs. (4), (7), and (8) follow directly from the definitions of $\cosh x$ and $\sinh x$, as in Example 1.

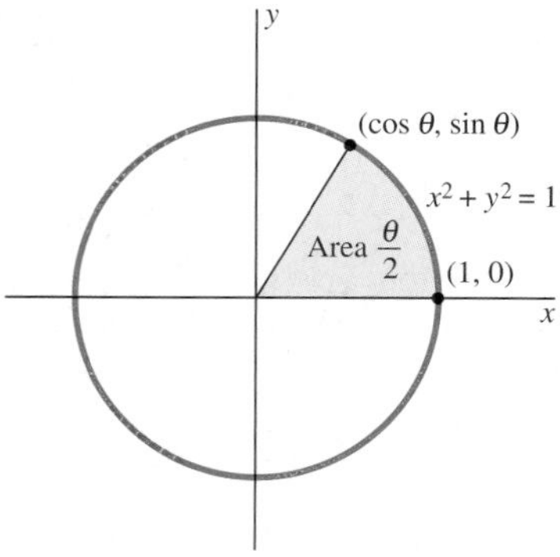

FIGURE 6.9.1 Relation of the ordinary cosine and sine functions to the circle $x^2 + y^2 = 1$.

EXAMPLE 1 To establish the "fundamental identity" in Eq. (4), we simply substitute the definitions of $\cosh x$ and $\sinh x$ on the left-hand side and write

$$\begin{aligned}\cosh^2 x - \sinh^2 x &= \tfrac{1}{4}(e^x + e^{-x})^2 - \tfrac{1}{4}(e^x - e^{-x})^2 \\ &= \tfrac{1}{4}(e^{2x} + 2 + e^{-2x}) - \tfrac{1}{4}(e^{2x} - 2 + e^{-2x}) = 1.\end{aligned}$$

The other identities listed previously may be derived from Eqs. (4), (7), and (8) in ways that parallel the derivations of the corresponding trigonometric identities. ◆

The trigonometric functions are sometimes called the *circular* functions because the point $(\cos\theta, \sin\theta)$ lies on the circle $x^2 + y^2 = 1$ for all θ (Fig. 6.9.1). Similarly, the identity in Eq. (4) tells us that the point $(\cosh\theta, \sinh\theta)$ lies on the hyperbola $x^2 - y^2 = 1$, and this is how the name *hyperbolic* function originated (Fig. 6.9.2).

The graphs of $y = \cosh x$ and $y = \sinh x$ are easy to construct. Add (for cosh) or subtract (for sinh) the ordinates of the graphs of $y = \frac{1}{2}e^x$ and $y = \frac{1}{2}e^{-x}$. The graphs of the other four hyperbolic functions can then be constructed by dividing ordinates. The graphs of all six are shown in Fig. 6.9.3.

These graphs show a striking difference between the hyperbolic functions and the ordinary trigonometric functions: None of the hyperbolic functions is periodic. They do, however, have even-odd properties, as the circular functions do. Like cosine and secant, the two functions cosh and sech are even, because

$$\cosh(-x) = \cosh x \quad \text{and} \quad \text{sech}(-x) = \text{sech}\, x$$

for all x. The other four hyperbolic functions, like the sine and tangent functions, are odd:

$$\sinh(-x) = -\sinh x, \qquad \tanh(-x) = -\tanh x,$$

and so on.

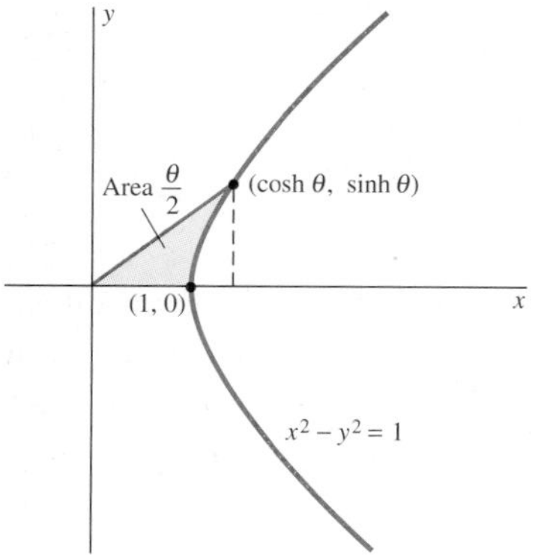

FIGURE 6.9.2 Relation of the hyperbolic cosine and hyperbolic sine functions to the hyperbola $x^2 - y^2 = 1$.

Derivatives and Integrals of Hyperbolic Functions

The formulas for the derivatives of the hyperbolic functions parallel those for the trigonometric functions, with occasional sign differences. For example,

$$D_x \cosh x = D_x\left(\tfrac{1}{2}e^x + \tfrac{1}{2}e^{-x}\right) = \tfrac{1}{2}e^x - \tfrac{1}{2}e^{-x} = \sinh x.$$

The chain rule then gives

$$D_x \cosh u = (\sinh u)\,\frac{du}{dx} \tag{13}$$

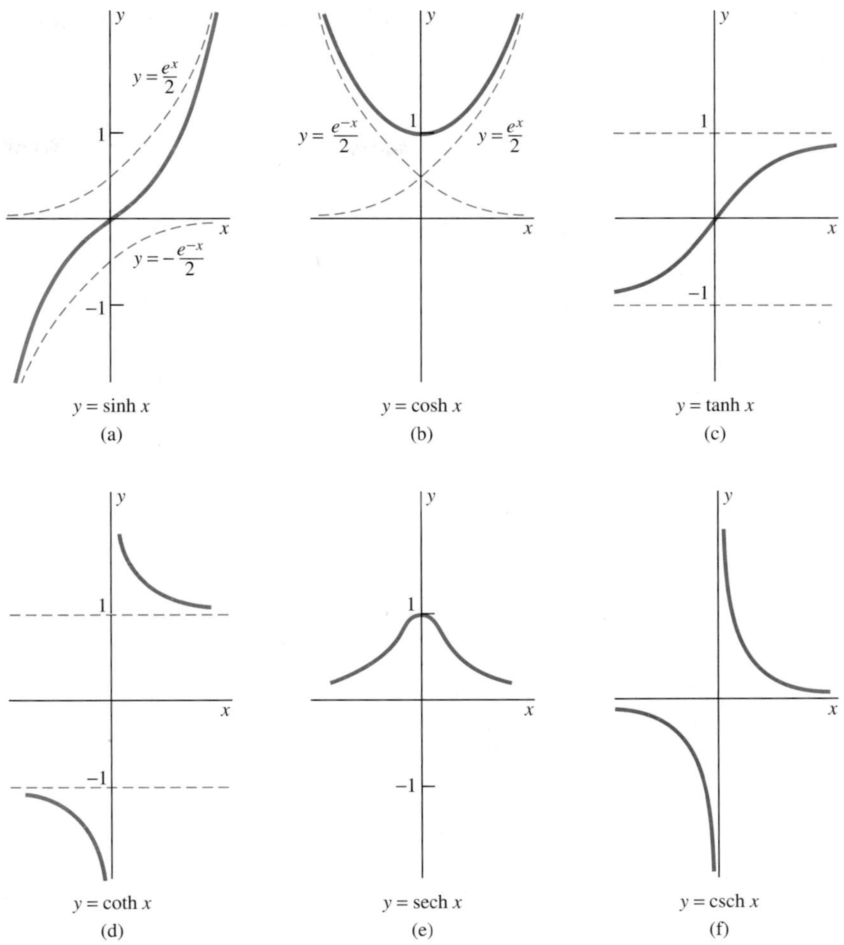

FIGURE 6.9.3 Graphs of the six hyperbolic functions.

if u is a differentiable function of x. The other five differentiation formulas are

$$D_x \sinh u = (\cosh u)\,\frac{du}{dx}, \tag{14}$$

$$D_x \tanh u = (\operatorname{sech}^2 u)\,\frac{du}{dx}, \tag{15}$$

$$D_x \coth u = (-\operatorname{csch}^2 u)\,\frac{du}{dx}, \tag{16}$$

$$D_x \operatorname{sech} u = (-\operatorname{sech} u \tanh u)\,\frac{du}{dx}, \tag{17}$$

$$D_x \operatorname{csch} u = (-\operatorname{csch} u \coth u)\,\frac{du}{dx}. \tag{18}$$

Equation (14) is derived exactly as Eq. (13) is. Then Eqs. (15) through (18) follow from Eqs. (13) and (14) with the aid of the quotient rule and the identities in Eqs. (5) and (6).

As indicated in Example 2, the differentiation of hyperbolic functions using Eqs. (13) through (18) is very similar to the differentiation of trigonometric functions.

EXAMPLE 2

(a) $D_x \cosh 2x = 2 \sinh 2x$.

(b) $D_x \sinh^2 x = 2 \sinh x \cosh x$.

(c) $D_x(x \tanh x) = \tanh x + x \operatorname{sech}^2 x$.

(d) $D_x \operatorname{sech}(x^2) = -2x \operatorname{sech}(x^2) \tanh(x^2)$. ◆

The antiderivative versions of the differentiation formulas in Eqs. (13) through (18) are the following integral formulas:

$$\int \sinh u\,du = \cosh u + C, \tag{19}$$

$$\int \cosh u\,du = \sinh u + C, \tag{20}$$

$$\int \operatorname{sech}^2 u\,du = \tanh u + C, \tag{21}$$

$$\int \operatorname{csch}^2 u\,du = -\coth u + C, \tag{22}$$

$$\int \operatorname{sech} u \tanh u\,du = -\operatorname{sech} u + C, \tag{23}$$

$$\int \operatorname{csch} u \coth u\,du = -\operatorname{csch} u + C. \tag{24}$$

The integrals in Example 3 illustrate the fact that simple hyperbolic integrals may be treated in much the same way as simple trigonometric integrals.

EXAMPLE 3

(a) With $u = 3x$, we have

$$\int \cosh 3x\,dx = \int (\cosh u)\left(\tfrac{1}{3}du\right) = \tfrac{1}{3}\sinh u + C = \tfrac{1}{3}\sinh 3x + C.$$

(b) With $u = \sinh x$, we have

$$\int \sinh x \cosh x\,dx = \int u\,du = \tfrac{1}{2}u^2 + C = \tfrac{1}{2}\sinh^2 x + C.$$

(c) Using Eq. (12), we find that

$$\int \sinh^2 x\,dx = \int \tfrac{1}{2}(\cosh 2x - 1)\,dx = \tfrac{1}{4}\sinh 2x - \tfrac{1}{2}x + C.$$

(d) Finally, using Eq. (5), we see that

$$\begin{aligned}\int_0^1 \tanh^2 x\,dx &= \int_0^1 (1 - \operatorname{sech}^2 x)\,dx = \Big[x - \tanh x\Big]_0^1 \\ &= 1 - \tanh 1 = 1 - \frac{e - e^{-1}}{e + e^{-1}} = \frac{2}{e^2 + 1} \\ &\approx 0.238406.\end{aligned}$$

◆

Inverse Hyperbolic Functions

Figure 6.9.3 shows that

- The functions $\sinh x$ and $\tanh x$ are increasing for all x;
- The functions $\coth x$ and $\operatorname{csch} x$ are decreasing and defined for all $x \neq 0$;
- The function $\cosh x$ is increasing on the half-line $x \geqq 0$; and
- The function $\operatorname{sech} x$ is decreasing on the half-line $x \geqq 0$.

It follows that each of the six hyperbolic functions can be "inverted" on the indicated domain where it is either increasing or decreasing. The resulting inverse hyperbolic functions and their domains of definition are listed in the following table:

Inverse Hyperbolic Function	Defined for
$\sinh^{-1} x$	All x
$\cosh^{-1} x$	$x \geqq 1$
$\tanh^{-1} x$	$\lvert x\rvert < 1$
$\coth^{-1} x$	$\lvert x\rvert > 1$
$\operatorname{sech}^{-1} x$	$0 < x \leqq 1$
$\operatorname{csch}^{-1} x$	$x \neq 0$

EXAMPLE 4 Find the numerical value of $\tanh^{-1}\left(\frac{1}{2}\right)$.

Solution If $y = \tanh^{-1} x$, then

$$\begin{aligned}
\tanh y &= x; \\
\frac{e^y - e^{-y}}{e^y + e^{-y}} &= x \qquad \text{[by Eq. (2)];} \\
e^y - e^{-y} &= xe^y + xe^{-y}; \\
(1 - x)e^y &= (1 + x)e^{-y}; \\
e^{2y} &= \frac{1 + x}{1 - x}; \\
y &= \tfrac{1}{2} \ln \frac{1 + x}{1 - x}.
\end{aligned}$$

Hence, with $x = \frac{1}{2}$, we find that

$$\tanh^{-1}\left(\tfrac{1}{2}\right) = \tfrac{1}{2} \ln 3 \approx 0.549306.$$

◆

Scientific calculators ordinarily are used to find values of hyperbolic and inverse hyperbolic functions. Many calculators give values only of $\sinh^{-1}$, $\cosh^{-1}$, and $\tanh^{-1}$. Values of the other three inverse hyperbolic functions can then be found by using the identities

$$\operatorname{sech}^{-1} x = \cosh^{-1}\left(\frac{1}{x}\right), \tag{25}$$

$$\operatorname{csch}^{-1} x = \sinh^{-1}\left(\frac{1}{x}\right), \tag{26}$$

and

$$\coth^{-1} x = \tanh^{-1}\left(\frac{1}{x}\right). \tag{27}$$

For example,

$$\coth^{-1} 2 = \tanh^{-1}\left(\tfrac{1}{2}\right) \approx 0.549306.$$

Derivatives of Inverse Hyperbolic Functions

Here are the derivatives of the six inverse hyperbolic functions:

$$D_x \sinh^{-1} x = \frac{1}{\sqrt{1 + x^2}}, \tag{28}$$

$$D_x \cosh^{-1} x = \frac{1}{\sqrt{x^2 - 1}}, \tag{29}$$

$$D_x \tanh^{-1} x = \frac{1}{1 - x^2}, \tag{30}$$

$$D_x \coth^{-1} x = \frac{1}{1 - x^2}, \tag{31}$$

$$D_x \operatorname{sech}^{-1} x = -\frac{1}{x\sqrt{1 - x^2}}, \tag{32}$$

$$D_x \operatorname{csch}^{-1} x = -\frac{1}{|x|\sqrt{1 + x^2}}. \tag{33}$$

We can derive these formulas by the standard method of finding the derivative of the inverse of a function when the derivative of the function itself is known. The only requirement is that the inverse function is known in advance to be differentiable.

EXAMPLE 5 To differentiate $\tanh^{-1} x$, we begin with the inverse function relation

$$\tanh(\tanh^{-1} x) = x$$

and substitute $u = \tanh^{-1} x$. Then, because this equation is actually an identity,

$$D_x \tanh u = D_x x = 1,$$

so

$$(\operatorname{sech}^2 u)\frac{du}{dx} = 1.$$

Thus

$$\begin{aligned} D_x \tanh^{-1} x &= \frac{du}{dx} = \frac{1}{\operatorname{sech}^2 u} = \frac{1}{1 - \tanh^2 u} \\ &= \frac{1}{1 - \tanh^2(\tanh^{-1} x)} = \frac{1}{1 - x^2}. \end{aligned}$$

This establishes Eq. (30). We can use similar methods to verify the formulas for the derivatives of the other five hyperbolic functions. ◆

The hyperbolic functions are defined in terms of the natural exponential e^x, so it's no surprise that their inverses may be expressed in terms of $\ln x$. (See Example 4.) In fact,

$$\sinh^{-1} x = \ln\left(x + \sqrt{x^2 + 1}\right) \quad \text{for all } x; \tag{34}$$

$$\cosh^{-1} x = \ln\left(x + \sqrt{x^2 - 1}\right) \quad \text{for all } x \geqq 1; \tag{35}$$

$$\tanh^{-1} x = \tfrac{1}{2} \ln\left(\frac{1 + x}{1 - x}\right) \quad \text{for } |x| < 1; \tag{36}$$

$$\coth^{-1} x = \tfrac{1}{2} \ln\left(\frac{x + 1}{x - 1}\right) \quad \text{for } |x| > 1; \tag{37}$$

$$\operatorname{sech}^{-1} x = \ln\left(\frac{1 + \sqrt{1 - x^2}}{x}\right) \quad \text{if } 0 < x \leqq 1; \tag{38}$$

$$\operatorname{csch}^{-1} x = \ln\left(\frac{1}{x} + \frac{\sqrt{1 + x^2}}{|x|}\right) \quad \text{if } x \neq 0. \tag{39}$$

Each of these identities may be established by showing that each side has the same derivative and also that the two sides agree for at least one value of x in every interval of their respective domains.

EXAMPLE 6 To establish the identity in (34), we begin by differentiating each side:

$$D_x \ln\left(x + \sqrt{x^2+1}\right) = \frac{1 + \dfrac{x}{\sqrt{x^2+1}}}{x + \sqrt{x^2+1}} = \frac{1}{\sqrt{x^2+1}} = D_x \sinh^{-1} x.$$

Thus

$$\sinh^{-1} x = \ln\left(x + \sqrt{x^2+1}\right) + C.$$

But $\sinh^{-1}(0) = 0 = \ln(0 + \sqrt{0+1})$. This implies that $C = 0$ and thus establishes Eq. (34). It is not quite so easy to show that $C = 0$ in the proofs of Eqs. (37) and (39); see Problems 64 and 65. ◆

Equations (34) through (39) may be used to calculate the values of inverse hyperbolic functions. This is convenient if you own a calculator whose repertoire does not include the inverse hyperbolic functions or if you are programming in a language such as BASIC, most forms of which do not include these functions.

Integrals Involving Inverse Hyperbolic Functions

The principal applications of inverse hyperbolic functions are to the evaluation of algebraic integrals. The differentiation formulas in Eqs. (28) through (33) may, in the usual way, be written as the following integral formulas:

$$\int \frac{du}{\sqrt{u^2+1}} = \sinh^{-1} u + C, \tag{40}$$

$$\int \frac{du}{\sqrt{u^2-1}} = \cosh^{-1} u + C, \tag{41}$$

$$\int \frac{du}{1-u^2} = \tanh^{-1} u + C \quad \text{if } |u| < 1, \tag{42a}$$

$$\int \frac{du}{1-u^2} = \coth^{-1} u + C \quad \text{if } |u| > 1, \tag{42b}$$

$$\int \frac{du}{1-u^2} = \tfrac{1}{2} \ln\left|\frac{1+u}{1-u}\right| + C, \tag{42c}$$

$$\int \frac{du}{u\sqrt{1-u^2}} = -\operatorname{sech}^{-1} |u| + C, \tag{43}$$

$$\int \frac{du}{u\sqrt{1+u^2}} = -\operatorname{csch}^{-1} |u| + C. \tag{44}$$

The distinction between the two cases $|u| < 1$ and $|u| > 1$ in Eqs. (42a) and (42b) results from the fact that the inverse hyperbolic tangent is defined for $|x| < 1$, whereas the inverse hyperbolic cotangent is defined for $|x| > 1$.

EXAMPLE 7 The substitution $u = 2x$, $dx = \frac{1}{2}\,du$ yields

$$\int \frac{dx}{\sqrt{4x^2-1}} = \frac{1}{2}\int \frac{du}{\sqrt{u^2+1}} = \frac{1}{2}\sinh^{-1} 2x + C.$$ ◆

EXAMPLE 8

$$\int_0^{1/2} \frac{dx}{1-x^2} = \Big[\tanh^{-1} x\Big]_0^{1/2}$$
$$= \frac{1}{2}\left[\ln\left|\frac{1+x}{1-x}\right|\right]_0^{1/2} = \frac{1}{2}\ln 3 \approx 0.549306.$$ ◆

EXAMPLE 9

$$\int_2^5 \frac{dx}{1-x^2} = \left[\coth^{-1} x\right]_2^5 = \frac{1}{2}\left[\ln\left|\frac{1+x}{1-x}\right|\right]_2^5$$
$$= \frac{1}{2}\left[\ln\left(\frac{6}{4}\right) - \ln 3\right] = -\frac{1}{2}\ln 2 \approx -0.346574.$$ ◆

6.9 TRUE/FALSE STUDY GUIDE

6.9 CONCEPTS: QUESTIONS AND DISCUSSION

1. Discuss the differences and similarities in the domains and ranges of the six inverse hyperbolic functions.
2. Beginning with the graphs of the hyperbolic functions in Fig. 6.9.3, sketch the graphs of the six inverse hyperbolic functions. Then identify the domain and range of each.
3. Discuss the analogy between trigonometric identities (such as $\cos^2 x + \sin^2 x = 1$) and hyperbolic identities (such as $\cosh^2 x - \sinh^2 x = 1$).

6.9 PROBLEMS

Find the derivatives of the functions in Problems 1 through 14.

1. $f(x) = \cosh(3x - 2)$

2. $f(x) = \sinh \sqrt{x}$

3. $f(x) = x^2 \tanh\left(\frac{1}{x}\right)$

4. $f(x) = \operatorname{sech} e^{2x}$

5. $f(x) = \coth^3 4x$

6. $f(x) = \ln \sinh 3x$

7. $f(x) = e^{\operatorname{csch} x}$

8. $f(x) = \cosh \ln x$

9. $f(x) = \sin(\sinh x)$

10. $f(x) = \tan^{-1}(\tanh x)$

11. $f(x) = \sinh x^4$

12. $f(x) = \sinh^4 x$

13. $f(x) = \dfrac{1}{x + \tanh x}$

14. $f(x) = \cosh^2 x - \sinh^2 x$

Evaluate the integrals in Problems 15 through 28.

15. $\int x \sinh x^2\, dx$

16. $\int \cosh^2 3u\, du$

17. $\int \tanh^2 3x\, dx$

18. $\int \dfrac{\operatorname{sech} \sqrt{x} \tanh \sqrt{x}}{\sqrt{x}}\, dx$

19. $\int \sinh^2 2x \cosh 2x\, dx$

20. $\int \tanh 3x\, dx$

21. $\int \dfrac{\sinh x}{\cosh^3 x}\, dx$

22. $\int \sinh^4 x\, dx$

23. $\int \coth x \operatorname{csch}^2 x\, dx$

24. $\int \operatorname{sech} x\, dx$

25. $\int \dfrac{\sinh x}{1 + \cosh x}\, dx$

26. $\int \dfrac{\sinh \ln x}{x}\, dx$

27. $\int \dfrac{1}{(e^x + e^{-x})^2}\, dx$

28. $\int \dfrac{e^x + e^{-x}}{e^x - e^{-x}}\, dx$

Find the derivatives of the functions in Problems 29 through 38.

29. $f(x) = \sinh^{-1} 2x$

30. $f(x) = \cosh^{-1}(x^2 + 1)$

31. $f(x) = \tanh^{-1} \sqrt{x}$

32. $f(x) = \coth^{-1} \sqrt{x^2 + 1}$

33. $f(x) = \operatorname{sech}^{-1}\left(\frac{1}{x}\right)$

34. $f(x) = \operatorname{csch}^{-1} e^x$

35. $f(x) = (\sinh^{-1} x)^{3/2}$

36. $f(x) = \sinh^{-1}(\ln x)$

37. $f(x) = \ln(\tanh^{-1} x)$

38. $f(x) = \dfrac{1}{\tanh^{-1} 3x}$

Use inverse hyperbolic functions to evaluate the integrals in Problems 39 through 48.

39. $\int \dfrac{dx}{\sqrt{x^2 + 9}}$

40. $\int \dfrac{dy}{\sqrt{4y^2 - 9}}$

41. $\int_{1/2}^{1} \dfrac{dx}{4 - x^2}$

42. $\int_{5}^{10} \dfrac{dx}{4 - x^2}$

43. $\int \dfrac{dx}{x\sqrt{4 - 9x^2}}$

44. $\int \dfrac{dx}{x\sqrt{x^2 + 25}}$

45. $\int \dfrac{e^x}{\sqrt{e^{2x} + 1}}\, dx$

46. $\int \dfrac{x}{\sqrt{x^4 - 1}}\, dx$

47. $\int \dfrac{1}{\sqrt{1 - e^{2x}}}\, dx$

48. $\int \dfrac{\cos x}{\sqrt{1 + \sin^2 x}}\, dx$

49. Apply the definitions in Eq. (1) to prove the identity in Eq. (7).

50. Derive the identities in Eqs. (5) and (6) from the identity in Eq. (4).

51. Deduce the identities in Eqs. (10) and (11) from the identity in Eq. (8).

52. Suppose that A and B are constants. Show that the function $x(t) = A \cosh kt + B \sinh kt$ is a solution of the differential equation

$$\frac{d^2x}{dt^2} = k^2 x(t).$$

53. Find the length of the curve $y = \cosh x$ over the interval $[0, a]$.

54. Find the volume of the solid obtained by revolving around the x-axis the area under $y = \sinh x$ from $x = 0$ to $x = \pi$.

55. Show that the area $A(\theta)$ of the shaded sector in Fig. 6.9.2 is $\theta/2$. This corresponds to the fact that the area of the sector of the unit circle between the positive x-axis and the radius to the point $(\cos\theta, \sin\theta)$ is $\theta/2$. [*Suggestion:* Note first that

$$A(\theta) = \tfrac{1}{2}\cosh\theta\sinh\theta - \int_1^{\cosh\theta} \sqrt{x^2-1}\,dx.$$

Then use the fundamental theorem of calculus to show that $A'(\theta) = \frac{1}{2}$ for all θ.]

56. Evaluate the following limits:

(a) $\displaystyle\lim_{x\to 0} \frac{\sinh x}{x}$; (b) $\displaystyle\lim_{x\to\infty} \tanh x$; (c) $\displaystyle\lim_{x\to\infty} \frac{\cosh x}{e^x}$.

57. Use the method of Example 4 to find the numerical value of $\sinh^{-1} 1$.

58. Apply Eqs. (34) and (39) to verify the identity

$$\operatorname{csch}^{-1} x = \sinh^{-1}\left(\frac{1}{x}\right) \quad \text{if } x \neq 0.$$

59. Establish the formula for $D_x \sinh^{-1} x$ in Eq. (28).

60. Establish the formula for $D_x \operatorname{sech}^{-1} x$ in Eq. (32).

61. Prove Eq. (36) by differentiating both sides.

62. Establish Eq. (34) by solving the equation

$$x = \sinh y = \frac{e^y - e^{-y}}{2}$$

for y in terms of x.

63. Establish Eq. (37) by solving the equation

$$x = \coth y = \frac{e^y + e^{-y}}{e^y - e^{-y}}$$

for y in terms of x.

64. (a) Differentiate both sides of Eq. (37) to show that they differ by a constant C. (b) Then prove that $C = 0$ by using the definition of $\coth x$ to show that $\coth^{-1} 2 = \frac{1}{2}\ln 3$.

65. (a) Differentiate both sides of Eq. (39) to show that they differ by a constant C. (b) Then prove that $C = 0$ by using the definition of csch$x$ to show that $\operatorname{csch}^{-1} 1 = \ln(1+\sqrt{2})$.

66. Estimate (graphically or numerically) the points of intersection of the curves $y = x + 2$ and $y = \cosh x$. Then approximate the area of the region bounded by these two curves.

In Problems 67 and 68, show first that $f(x) \to 0$ as $x \to +\infty$. Then estimate (graphically or numerically) the absolute maximum value of $f(x)$ for $x > 0$. Differentiate $f(x)$ to verify that you have an approximate critical point.

67. $f(x) = e^{-2x}\tanh x$ **68.** $f(x) = e^{-x}\sinh^{-1} x$

Problems 69 and 70 deal with the hanging cable illustrated in Fig. 6.9.4. If the cable is flexible and has uniform density, then elementary principles of physics can be used to show that its shape function $y = y(x)$ satisfies the differential equation

$$\frac{d^2y}{dx^2} = k\sqrt{1 + \left(\frac{dy}{dx}\right)^2} \qquad \textbf{(45)}$$

where k is a constant determined by the density and tension of the cable.

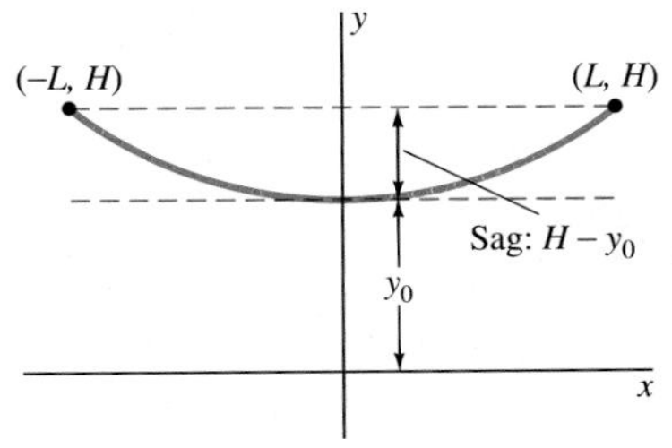

FIGURE 6.9.4 A flexible uniform cable suspended between two points at equal heights.

69. Verify that the function

$$y(x) = y_0 + \frac{1}{k}(-1 + \cosh kx) \qquad \textbf{(46)}$$

satisfies the differential equation in (45) and also satisfies the initial conditions $y(0) = y_0$, $y'(0) = 0$. A curve with this shape is called a *catenary,* from the Latin word *catena* (chain).

70. A high-voltage line is to be strung between two 50-ft towers 200 ft apart. (a) If the line sags 20 ft at its middle (where $x = 0$)—so $y_0 = 30$ (ft) in Eq. (46)—estimate graphically or numerically the value of the parameter k. (b) Then approximate the total length of the high-voltage line.

CHAPTER 6 REVIEW: DEFINITIONS, CONCEPTS, RESULTS

Use the following list as a guide to concepts that you may need to review.

1. The general method of setting up an integral formula for a quantity by approximating it and then recognizing the approximation to be a Riemann sum that corresponds to the desired integral: If the interval $[a, b]$ is partitioned into n subintervals of equal length $\Delta x = (b-a)/n$ and if $x_i^\star$ denotes a point of the ith subinterval, then

$$\lim_{n\to\infty} \sum_{i=1}^{n} f(x_i^\star)\,\Delta x = \int_a^b f(x)\,dx.$$

2. Net distance traveled as the integral of velocity:

$$s = \int_a^b v(t)\,dt.$$

3. The method of cross sections for computing volumes:

$$V = \int_a^b A(x)\,dx,$$

where $A(x)$ denotes the area of a slice with infinitesimal thickness dx.

4. Determining the volume of a solid of revolution by the method of cross sections, with the cross sections being either disks or annular rings.
5. Determining the volume of a solid of revolution by the method of cylindrical shells:
$$V = \int_{\star}^{\star\star} 2\pi r\, dA,$$
where r denotes the radius of the circle through which the area element dA is revolved.
6. The arc length of a smooth arc described either in the form $y = f(x)$, $a \leqq x \leqq b$, or in the form $x = g(y)$, $c \leqq y \leqq d$:
$$s = \int_{\star}^{\star\star} ds \qquad \left(ds = \sqrt{(dx)^2 + (dy)^2}\,\right),$$
where
$$ds = \sqrt{1 + [f'(x)]^2}\, dx \quad \text{for } y = f(x);$$
$$ds = \sqrt{1 + [g'(y)]^2}\, dy \quad \text{for } x = g(y).$$
7. Determining the area of the surface of revolution generated by revolving a smooth arc, given in the form $y = f(x)$ or in the form $x = g(y)$, around either the x-axis or the y-axis:
$$A = \int_{\star}^{\star\star} 2\pi r\, ds,$$
where r denotes the radius of the circle through which the arc length element ds is revolved.
8. The work done by a force function in moving a particle along a straight line segment:
$$W = \int_a^b F(x)\, dx$$
if the force $F(x)$ acts from $x = a$ to $x = b$.
9. Hooke's law and the work done in stretching or compressing an elastic spring.
10. Work done against the varying force of gravity.
11. Work done in filling a tank or in pumping the liquid in a tank to another level:
$$W = \int_a^b \rho h(y) A(y)\, dy,$$
where $h(y)$ denotes the vertical distance that the horizontal fluid slice of volume $dV = A(y)\, dy$, at the height y, must be lifted.
12. The force exerted by a liquid on the face of a submerged vertical plate:
$$F = \int_{\star}^{\star\star} \rho h\, dA.$$
where h denotes the depth of the horizontal area element dA beneath the surface of the fluid of weight density ρ.
13. The laws of exponents.
14. The laws of logarithms.
15. The definition of the natural logarithm function.
16. The definition of the number e.
17. The definition of the natural exponential function $\exp(x)$.
18. The inverse function relationship between $\ln x$ and $\exp(x)$.
19. Differentiation of $\ln u$ and e^u, where u is a differentiable function of x.
20. The definition of general exponential and logarithm functions.
21. Differentiation of a^u and $\log_a u$.
22. The definitions of the six inverse trigonometric functions.
23. The derivatives of the inverse trigonometric functions.
24. The integral formulas corresponding to the derivatives of the inverse sine, inverse tangent, and inverse secant functions.
25. The definitions and derivatives of the hyperbolic functions.
26. The definition and derivatives of the inverse hyperbolic functions.
27. The use of inverse hyperbolic functions to evaluate certain algebraic integrals.

CHAPTER 6 MISCELLANEOUS PROBLEMS

In Problems 1 through 3, find both the net distance and the total distance traveled between times $t = a$ and $t = b$ by a particle moving along a line with the given velocity function $v = f(t)$.

1. $v = t^2 - t - 2$; $a = 0, b = 3$
2. $v = \left|t^2 - 4\right|$; $a = 1, b = 4$
3. $v = \pi \sin \frac{1}{2}\pi(2t - 1)$; $a = 0, b = \frac{3}{2}$

In Problems 4 through 8, a solid extends along the x-axis from $x = a$ to $x = b$, and its cross-section area at x is $A(x)$. Find its volume.

4. $A(x) = x^3$; $a = 0, b = 1$
5. $A(x) = \sqrt{x}$; $a = 1, b = 4$
6. $A(x) = x^3$; $a = 1, b = 2$
7. $A(x) = \pi(x^2 - x^4)$; $a = 0, b = 1$
8. $A(x) = x^{100}$; $a = -1, b = 1$

9. Suppose that rainfall begins at time $t = 0$ and that the rate after t hours is $(t + 6)/12$ inches per hour. How many inches of rain fall during the first 12 h?
10. The base of a certain solid is the region in the first quadrant bounded by the curves $y = x^3$ and $y = 2x - x^2$. Find the solid's volume if each cross section perpendicular to the x-axis is a square with one edge in the base of the solid.
11. Find the volume of the solid generated by revolving around the x-axis the first-quadrant region of Problem 10.
12. Find the volume of the solid generated by revolving the region bounded by $y = 2x^4$ and $y = x^2 + 1$ around (a) the x-axis; (b) the y-axis.
13. A wire made of copper (density 8.5 g/cm^3) is shaped like a helix that spirals around the x-axis from $x = 0$ to $x = 20$. Each cross section of this wire perpendicular to the x-axis is a circular disk of radius 0.25 cm. What is the total mass of the wire?

14. Derive the formula $V = \frac{1}{3}\pi h(r_1^2 + r_1 r_2 + r_2^2)$ for the volume of a frustum of a cone with height h and base radii r_1 and r_2.

15. Suppose that the point P lies on a line perpendicular to the xy-plane at the origin O, with $|OP| = h$. Consider the "elliptical cone" that consists of all points on line segments from P to points on and within the ellipse with equation

$$\left(\frac{x}{a}\right)^2 + \left(\frac{y}{b}\right)^2 = 1.$$

Show that the volume of this elliptical cone is $V = \frac{1}{3}\pi abh$.

16. Figure 6.MP.1 shows the region R bounded by the ellipse $(x/a)^2 + (y/b)^2 = 1$ and by the line $x = a - h$, where $0 < h < a$. Revolution of R around the x-axis generates a "segment of an ellipsoid" of radius r, height h, and volume V. Show that

$$r^2 = \frac{b^2(2ah - h^2)}{a^2} \quad \text{and that} \quad V = \frac{1}{3}\pi r^2 h \frac{3a - h}{2a - h}.$$

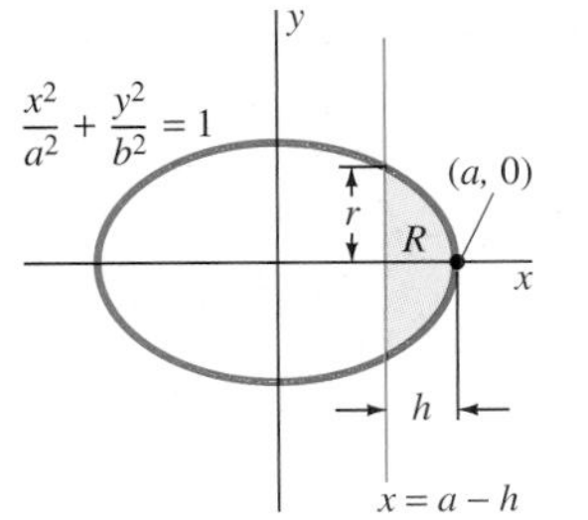

FIGURE 6.MP.1 A segment of an ellipse (Problem 16).

17. Figure 6.MP.2 shows the region R bounded by the hyperbola $(x/a)^2 - (y/b)^2 = 1$ and the line $x = a + h$, where $h > 0$. Revolving R around the x-axis generates a "segment of a hyperboloid" of radius r, height h, and volume V. Show that

$$r^2 = \frac{b^2(2ah + h^2)}{a^2} \quad \text{and that} \quad V = \frac{1}{3}\pi r^2 h \frac{3a + h}{2a + h}.$$

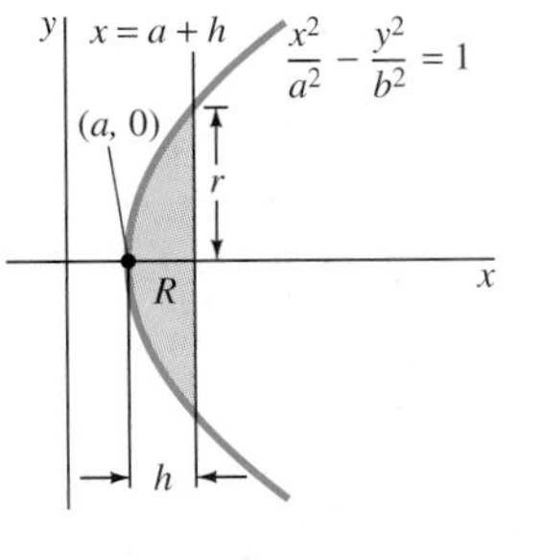

FIGURE 6.MP.2 The region R of Problem 17.

In Problems 18 through 20, the function $f(x)$ is nonnegative and continuous for $x \geqq 1$. When the region lying under $y = f(x)$ from $x = 1$ to $x = t$ is revolved around the indicated axis, the volume of the resulting solid is $V(t)$. Find the function $f(x)$.

18. $V(t) = \pi\left(1 - \frac{1}{t}\right)$; the x-axis

19. $V(t) = \frac{1}{6}\pi[(1 + 3t)^2 - 16]$; the x-axis

20. $V(t) = \frac{2}{9}\pi[(1 + 3t^2)^{3/2} - 8]$; the y-axis

21. Use the integral formula

$$\int u \sin u \, du = \sin u - u \cos u + C$$

to find the volume of the solid generated by revolving around the y-axis the first-quadrant region bounded by $y = x$ and $y = \sin(\frac{1}{2}\pi x)$.

22. Use the method of cylindrical shells to find the volume of the solid generated by revolving around the line $x = -2$ the region bounded by $y = x^2$ and $y = x + 2$.

23. Find the length of the curve $y = \frac{1}{3}x^{3/2} - x^{1/2}$ from $x = 1$ to $x = 4$.

24. Find the area of the surface generated by revolving the curve of Problem 23 around (a) the x-axis; (b) the y-axis.

25. Find the length of the curve $x = \frac{3}{8}(y^{4/3} - 2y^{2/3})$ from $y = 1$ to $y = 8$.

26. Find the area of the surface generated by revolving the curve of Problem 25 around (a) the x-axis; (b) the y-axis.

27. Find the area of the surface generated by revolving the curve of Problem 23 around the line $x = 1$.

28. If $-r \leqq a < b \leqq r$, then a "spherical zone" of "height" $h = b - a$ is generated by revolving around the x-axis the circular arc $y = \sqrt{r^2 - x^2}$, $a \leqq x \leqq b$. Show that the area of this spherical zone is $A = 2\pi rh$, the same as that of a cylinder of radius r and height h.

29. Apply the result of Problem 28 to show that the surface area of a sphere of radius r is $A = 4\pi r^2$.

30. Let R denote the region bounded by the curves $y = 2x^3$ and $y^2 = 4x$. Find the volumes of the solids obtained by revolving the region R around (a) the x-axis; (b) the y-axis; (c) the line $y = -1$; (d) the line $x = 2$. In each case use both the method of cross sections and the method of cylindrical shells.

31. Find the natural length L of a spring if five times as much work is required to stretch it from a length of 2 ft to a length of 5 ft as is required to stretch it from a length of 2 ft to a length of 3 ft.

32. A steel beam weighing 1000 lb hangs from a 50-ft cable that weighs 5 lb per linear foot. How much work is done in winding in 25 ft of the cable with a windlass?

33. A spherical tank of radius R (in feet) is initially full of oil of density ρ lb/ft^3. Find the total work done in pumping all the oil from the sphere to a height of $2R$ above the top of the tank.

34. How much work is done by a colony of ants in building a conical anthill of height and diameter 1 ft, using sand initially at ground level and with a density of 150 lb/ft^3?

35. The gravitational attraction below the earth's surface is directly proportional to the distance from the center of the earth. Suppose that a straight cylindrical hole of radius 1 ft is dug from the earth's surface to its center. Assume that the earth has radius 3960 mi and uniform density 350 lb/ft^3. How much work, in foot-pounds, is done in lifting a 1-lb weight from the bottom of this hole to its top?

36. How much work is done in digging the hole of Problem 35—that is, in lifting all the material it initially contained to the earth's surface?

37. Suppose that a dam is shaped like a trapezoid of height 100 ft, 300 ft long at the top and 200 ft long at the bottom. When the water level behind the dam is even with its top, what is the total force that the water exerts on the dam?

38. Suppose that a dam has the same top and bottom lengths as the dam of Problem 37 and the same vertical height of 100 ft, but that its face toward the water is slanted at an angle of 30° from the vertical. What is the total force of water pressure on this dam?

39. For $c > 0$, the graphs of $y = c^2x^2$ and $y = c$ bound a plane region. Revolve this region around the horizontal line $y = -1/c$ to form a solid. For what value of c is the volume of this solid maximal? Minimal?

Find the centroids of the curves in Problems 40 through 43.

40. $y = \dfrac{x^5}{5} + \dfrac{1}{12x^3}, \quad 1 \leqq x \leqq 2$

41. $x = \dfrac{y^4}{8} + \dfrac{1}{4y^2}, \quad 1 \leqq y \leqq 2$

42. $y = \dfrac{x^{3/2}}{3} - x^{1/2}, \quad 1 \leqq x \leqq 4$

43. $x = \dfrac{3}{8}(y^{4/3} - 2y^{2/3}), \quad 1 \leqq y \leqq 8$

44. Find the centroid of the plane region in the first quadrant that is bounded by the curves $y = x^3$ and $y = 2x - x^2$.

45. Find the centroid of the plane region bounded by the curves $x = 2y^4$ and $x = y^2 + 1$.

46. Let T be the plane triangle with vertices $(0, 0)$, (a, b), and $(c, 0)$. Show that the centroid of T is the point of intersection of its medians.

47. Use the first theorem of Pappus to find the y-coordinate of the centroid of the upper half of the ellipse

$$\frac{x^2}{a^2} + \frac{y^2}{b^2} = 1.$$

Use the facts that the area of this semiellipse is $A = \frac{1}{2}\pi ab$ and the volume of the ellipsoid it generates when rotated around the x-axis is $V = \frac{4}{3}\pi ab^2$.

48. (a) Use the first theorem of Pappus to find the centroid of the first-quadrant part of the annular ring with boundary circles $x^2 + y^2 = a^2$ and $x^2 + y^2 = b^2$, with $0 < a < b$. (b) Show that the limiting position of this centroid as $b \to a$ is the centroid of a quarter-circular arc of radius a.

49. Let T be the triangle in the first quadrant with one vertex at $(0, 0)$ and the opposite side L, of length w, joining the other two vertices (a, b) and (c, d), where $a > c > 0$ and $d > b > 0$. Let A denote the area of T, $\bar{y}$ the y-coordinate of the centroid of T, p the perpendicular distance from $(0, 0)$ to L, V the volume generated by revolving T around the x-axis, and S the surface area generated by revolving L around the x-axis. Derive the following formulas in the order listed.

(a) $A = \frac{1}{2}(ad - bc)$

(b) $\bar{y} = \frac{1}{3}(b + d)$

(c) $V = \frac{1}{3}\pi(b + d)(ad - bc)$

(d) $p = \dfrac{ad - bc}{w}$

(e) $S = \pi(b + d)w$

(f) $V = \frac{1}{3}pS$

50. Suppose that n is an even positive integer. Let J be an n-sided regular polygon inscribed in the circle of radius r centered at the origin. Let S be the surface area generated by rotating J around a diameter of the circle through two opposite vertices of J; let V be the volume of the solid enclosed by that surface. Conclude from part (f) of Problem 49 that

$$V = \frac{1}{3}\left(r \cos \frac{\pi}{n}\right) \cdot S.$$

Archimedes deduced from this result that, if the surface area of a sphere of radius r is $4\pi r^2$, then its volume is $\frac{4}{3}\pi r^3$. Supply the details of his reasoning.

51. Suppose that n is a positive integer. Let R_n denote the region bounded by the curves $y = x$ and $y = x^n$ for $0 \leqq x \leqq 1$. Show that the limiting position as $n \to +\infty$ of the centroid of R_n is the centroid of the triangle with vertices $(0, 0)$, $(1, 0)$, and $(1, 1)$. Why does this seem plausible?

52. Let the region R be the union of the semicircular disk $x^2 + y^2 \leqq 9$, $x \geqq 0$, and the square with vertices $(1, 0)$, $(-1, 0)$, $(1, -2)$, and $(-1, -2)$. (a) Find the centroid of R. (b) Then find the volume of the solid obtained by rotating R around the line $y = -4$.

Evaluate the indefinite integrals in Problems 53 through 64.

53. $\displaystyle\int \frac{dx}{1 - 2x}$

54. $\displaystyle\int \frac{\sqrt{x}}{1 + x^{3/2}}\,dx$

55. $\displaystyle\int \frac{3 - x}{1 + 6x - x^2}\,dx$

56. $\displaystyle\int \frac{e^x - e^{-x}}{e^x + e^{-x}}\,dx$

57. $\displaystyle\int \frac{\sin x}{2 + \cos x}\,dx$

58. $\displaystyle\int \frac{e^{-1/x^2}}{x^3}\,dx$

59. $\displaystyle\int \frac{10^{\sqrt{x}}}{\sqrt{x}}\,dx$

60. $\displaystyle\int \frac{1}{x(\ln x)^2}\,dx$

61. $\displaystyle\int e^x\sqrt{1 + e^x}\,dx$

62. $\displaystyle\int \frac{1}{x}\sqrt{1 + \ln x}\,dx$

63. $\displaystyle\int 2^x 3^x\,dx$

64. $\displaystyle\int \frac{dx}{x^{1/3}(1 + x^{2/3})}$

65. A grain warehouse holds B bushels of grain, which is deteriorating in such a way that only $B \cdot 2^{-t/12}$ bushels will be salable after t months. Meanwhile, the grain's market price is increasing linearly: After t months it will be $2 + \frac{1}{12}t$ dollars per bushel. After how many months should the grain to sold to maximize the revenue obtained?

66. You have borrowed \$1000 at 10% annual interest, compounded continuously, to plant timber on a tract of land. Your agreement is to repay the loan, plus interest, when the timber is cut and sold. If the cut timber can be sold after t years for $800\exp(\frac{1}{2}\sqrt{t})$ dollars, when should you cut and sell to maximize the profit?

67. Blood samples from 1000 students are to be tested for a certain disease known to occur in 1% of the population. Each test costs \$5, so it would cost \$5000 to test the samples individually. Suppose, however, that "lots" made up of x samples each are formed by pooling halves of individual samples, and that these lots are tested first (for \$5 each). Only in case a lot tests positive—the probability of this is $1 - (0.99)^x$—will the x samples used to make up this lot be tested individually. (a) Show that the total expected number of tests is

$$f(x) = \frac{1000}{x}[(1)(0.99)^x + (x+1)(1-(0.99)^x)]$$

$$= 1000 + \frac{1000}{x} - 1000 \cdot (0.99)^x$$

if $x \geqq 2$. (b) Show that the value of x that minimizes $f(x)$ is a root of the equation

$$x = \frac{(0.99)^{-x/2}}{\sqrt{\ln(100/99)}}.$$

Because the denominator is approximately 0.1, it may be convenient to solve instead the simpler equation $x = 10 \cdot (0.99)^{-x/2}$. (c) From the results in parts (a) and (b), compute the minimum (expected) cost of using this batch method to test the original 1000 samples.

68. Find the length of the curve $y = \frac{1}{2}x^2 - \frac{1}{4}\ln x$ from $x = 1$ to $x = e$.

Differentiate the functions in Problems 69 through 88.

69. $f(x) = \sin^{-1} 3x$

70. $f(x) = \tan^{-1} 7x$

71. $g(t) = \sec^{-1} t^2$

72. $g(t) = \tan^{-1} e^t$

73. $f(x) = \sin^{-1}(\cos x)$

74. $f(x) = \sinh^{-1} 2x$

75. $g(t) = \cosh^{-1} 10t$

76. $h(u) = \tanh^{-1}\left(\frac{1}{u}\right)$

77. $f(x) = \sin^{-1}\left(\frac{1}{x^2}\right)$

78. $f(x) = \tan^{-1}\left(\frac{1}{x}\right)$

79. $f(x) = \arcsin \sqrt{x}$

80. $f(x) = x \sec^{-1} x^2$

81. $f(x) = \tan^{-1}(1 + x^2)$

82. $f(x) = \sin^{-1} \sqrt{1 - x^2}$

83. $f(x) = e^x \sinh e^x$

84. $f(x) = \ln \cosh x$

85. $f(x) = \tanh^2 3x + \operatorname{sech}^2 3x$

86. $f(x) = \sinh^{-1} \sqrt{x^2 - 1}$

87. $f(x) = \cosh^{-1} \sqrt{x^2 - 1}$

88. $f(x) = \tanh^{-1}(1 - x^2)$

Evaluate the integrals in Problems 89 through 108.

89. $\displaystyle\int \frac{dx}{\sqrt{1 - 4x^2}}$

90. $\displaystyle\int \frac{dx}{1 + 4x^2}$

91. $\displaystyle\int \frac{dx}{\sqrt{4 - x^2}}$

92. $\displaystyle\int \frac{dx}{4 + x^2}$

93. $\displaystyle\int \frac{e^x}{\sqrt{1 - e^{2x}}}\,dx$

94. $\displaystyle\int \frac{x}{1 + x^4}\,dx$

95. $\displaystyle\int \frac{1}{\sqrt{9 - 4x^2}}\,dx$

96. $\displaystyle\int \frac{1}{9 + 4x^2}\,dx$

97. $\displaystyle\int \frac{x^2}{1 + x^6}\,dx$

98. $\displaystyle\int \frac{\cos x}{1 + \sin^2 x}\,dx$

99. $\displaystyle\int \frac{1}{x\sqrt{4x^2 - 1}}\,dx$

100. $\displaystyle\int \frac{1}{x\sqrt{x^4 - 1}}\,dx$

101. $\displaystyle\int \frac{1}{\sqrt{e^{2x} - 1}}\,dx$

102. $\displaystyle\int x^2 \cosh x^3\,dx$

103. $\displaystyle\int \frac{\sinh \sqrt{x}}{\sqrt{x}}\,dx$

104. $\displaystyle\int \operatorname{sech}^2(3x - 2)\,dx$

105. $\displaystyle\int \frac{\arctan x}{1 + x^2}\,dx$

106. $\displaystyle\int \frac{1}{\sqrt{4x^2 - 1}}\,dx$

107. $\displaystyle\int \frac{1}{\sqrt{4x^2 + 9}}\,dx$

108. $\displaystyle\int \frac{x}{\sqrt{x^4 + 1}}\,dx$

109. Find the volume generated by revolving around the y-axis the region under $y = 1/\sqrt{1 - x^4}$ from $x = 0$ to $x = 1/\sqrt{2}$.

110. Find the volume generated by revolving around the y-axis the region under $y = 1/\sqrt{x^4 + 1}$ from $x = 0$ to $x = 1$.

111. Use Eqs. (35) through (38) of Section 6.9 to show that

(a) $\coth^{-1} x = \tanh^{-1}\left(\frac{1}{x}\right)$;

(b) $\operatorname{sech}^{-1} x = \cosh^{-1}\left(\frac{1}{x}\right)$.

112. Show that $x''(t) = k^2 x(t)$ if

$$x(t) = A \cosh kt + B \sinh kt,$$

where A and B are constants. Determine A and B if (a) $x(0) = 1$, $x'(0) = 0$; (b) $x(0) = 0$, $x'(0) = 1$.

113. Use Newton's method to find the least positive solution of the equation $\cos x \cosh x = 1$. Begin by sketching the graphs of $y = \cos x$ and $y = \operatorname{sech} x$.

114. (a) Verify by differentiation that

$$\int \sec x\,dx = \sinh^{-1}(\tan x) + C.$$

(b) Show similarly that

$$\int \operatorname{sech} x\,dx = \tan^{-1}(\sinh x) + C.$$

115. Figure 6.MP.3 shows the graphs of $f(x) = x^{1/2}$, $g(x) = \ln x$, and $h(x) = x^{1/3}$ plotted on the interval $[0.2, 10]$. You can see that the graph of f remains above the graph of $\ln x$, whereas the graph of h dips below the graph of $\ln x$. But because $\ln x$ increases *less* rapidly than any positive power of x, the graph of h must eventually cross the graph of $\ln x$ and rise above it. Finally, it is easy to believe that, for a suitable choice of p between 2 and 3, the graph of $j(x) = x^{1/p}$ never dips below the graph of $\ln x$ but does drop down just far enough to be tangent to the graph of $\ln x$ at a certain point. (a) Show that $f(x) > \ln x$ for all $x > 0$ by finding the global minimum value of $f(x) - \ln x$ on the interval $(0, +\infty)$. (b) Use Newton's method to find the value at which $h(x)$ crosses the graph of $\ln x$ and rises above it—the value of x *not* shown in Fig. 6.MP.3. (c) Find the value of p for which the graph of $j(x)$ is tangent to the graph of $\ln x$ at the point $(q, \ln q)$.

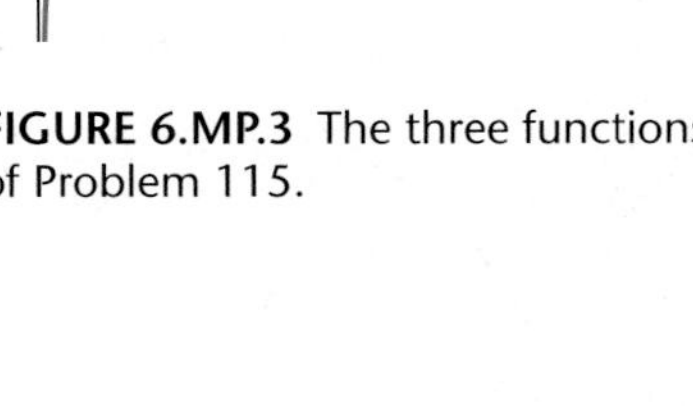

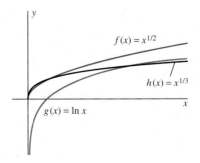

FIGURE 6.MP.3 The three functions of Problem 115.

TECHNIQUES OF INTEGRATION 7

Leonhard Euler (1707–1783)

The most prolific mathematician in all history was Leonhard Euler, who was born in 1707 in Basel, Switzerland, the home of the Bernoulli family of mathematicians. His father preferred a theological career for his son, but young Euler learned mathematics from John Bernoulli and thereby found his true vocation. During his lifetime Euler published more than 500 books and papers. His work continued unabated even after he lost his sight in 1766. Upon his death in 1783, he left behind more than 300 additional manuscripts whose publication continued in a steady flow for another half century. His collected works fill approximately 75 substantial volumes.

No other mathematician of the past more directly affects the modern student of mathematics, because it was largely Euler who shaped the notation and terminology still used today in teaching high school algebra and trigonometry as well as calculus. His *Introductio in Analysin Infinitorium* (Introduction to Infinitesimal Analysis) is the earliest mathematics textbook whose exposition would (in translation from the original Latin) be accessible to a modern student. Here are some now-familiar notations whose use was popularized and standardized by Euler:

e	for the base of natural logarithms;
a, b, c	for the sides of the triangle ABC;
i	for the square root of -1;
$\sum$	for the summation symbol;
$f(x)$	for function notation;
π	for the area of the unit circle;

and the trigonometric abbreviations sin, cos, tang, cot, sec, and cosec, which are close to their current forms. It was Euler's *Introductio* that once and for all based calculus squarely on the function concept. His 1755 and 1768 calculus treatises provide the original source for much of the content and methods of modern calculus courses and texts.

Euler originally discovered so many of the standard formulas and identities of mathematics that it is customary to attribute a formula to the first mathematician *after* Euler to *re*discover it. But the identity $e^{ix} = \cos x + i \sin x$ relating the exponential and trigonometric functions is still known as Euler's formula. Substituting $x = \pi$ yields the relation $e^{i\pi} + 1 = 0$, which links five of the most important constants in mathematics.

The photograph—part of a page from Chapter VII of the *Introductio*—shows the first appearance in public print of the number $e \approx 2.71828$. Immediately following its definition as the sum of the infinite series

$$\sum_{n=0}^{\infty} \frac{1}{n!} = 1 + \frac{1}{1!} + \frac{1}{2!} + \cdots + \frac{1}{n!} + \cdots,$$

Euler gives the numerical value of e accurate to 23 decimal places.

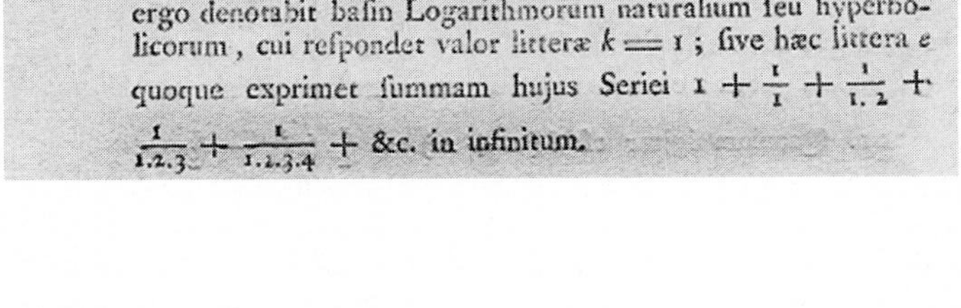

90 DE QUANTITATUM EXPONENTIALIUM

LIB. I. (116) inventam, $a = 1 + \frac{1}{1} + \frac{1}{1.2} + \frac{1}{1.2.3} + \frac{1}{1.2.3.4} +$ &c., qui termini, si in fractiones decimales convertantur atque actu addantur, præbebunt hunc valorem pro $a =$ 2,71828182845904523536028, cujus ultima adhuc nota veritati est consentanea. Quod si jam ex hac basi Logarithmi construantur, ii vocari solent Logarithmi *naturales* seu *hyperbolici*, quoniam quadratura hyperbolæ per istiusmodi Logarithmos exprimi potest. Ponamus autem brevitatis gratia pro numero hoc 2,718281828459 &c. constanter litteram e, quæ ergo denotabit basin Logarithmorum naturalium seu hyperbolicorum, cui respondet valor litteræ $k = 1$; sive hæc littera e quoque exprimet summam hujus Seriei $1 + \frac{1}{1} + \frac{1}{1.2} + \frac{1}{1.2.3} + \frac{1}{1.2.3.4} +$ &c. in infinitum.

7.1 INTRODUCTION

We saw in Chapter 6 that many geometric and physical quantities can be expressed as definite integrals. The fundamental theorem of calculus reduces the problem of calculating the definite integral

$$\int_a^b f(x)\,dx$$

to that of finding an antiderivative $G(x)$ of $f(x)$. Once this is accomplished, then

$$\int_a^b f(x)\,dx = \Big[G(x)\Big]_a^b = G(b) - G(a).$$

But as yet we have relied largely on trial-and-error methods for finding the required antiderivative $G(x)$. In some cases a knowledge of elementary derivative formulas, perhaps in combination with a simple substitution, allows us to integrate a given function. This approach can, however, be inefficient and time-consuming, especially in view of the following surprising fact: Some simple-looking integrals, such as

$$\int e^{-x^2}\,dx, \qquad \int \frac{\sin x}{x}\,dx, \qquad \text{and} \qquad \int \sqrt{1+x^4}\,dx,$$

cannot be evaluated in terms of finite combinations of the familiar algebraic and elementary transcendental functions. For example, the antiderivative

$$H(x) = \int_0^x e^{-t^2}\,dt$$

of $\exp(-x^2)$ has no finite expression in terms of elementary functions. Any attempt to find such an expression will, therefore, inevitably be unsuccessful.

The presence of such integrals indicates that we cannot hope to reduce integration to a routine process such as differentiation. In fact, finding antiderivatives is an art, the mastery of which depends on experience and practice. Nevertheless, there are a number of techniques whose systematic use can substantially reduce our dependence on chance and intuition alone. This chapter deals with some of these systematic techniques of integration.

7.2 INTEGRAL TABLES AND SIMPLE SUBSTITUTIONS

Integration would be a simple matter if we had a list of integral formulas, an *integral table,* in which we could locate any integral that we needed to evaluate. But the diversity of integrals that we encounter is too great for such an all-inclusive integral table to be practical. It is more sensible to print or memorize a short table of integrals of the sort seen frequently and to learn techniques by which the range of applicability of this short table can be extended. We begin with the list of integrals in Fig. 7.2.1, which are familiar from earlier chapters. Each formula is equivalent to one of the basic derivative formulas.

A table of 113 integral formulas appears on the inside back cover of this book. Even more extensive integral tables are readily available. For example, the volume of *Standard Mathematical Tables and Formulas,* edited by William H. Beyer and published by the CRC Press, Inc. (Boca Raton, Florida), contains over 700 integral formulas. But even such a lengthy table can be expected to include only a small fraction of the integrals we may need to evaluate. Thus it is necessary to learn techniques for deriving new formulas and for transforming a given integral either into one that's already familiar or into one that appears in an accessible table.

$$\int u^n\,du = \frac{u^{n+1}}{n+1} + C \quad [n \neq -1] \qquad \textbf{(1)}$$

$$\int \frac{du}{u} = \ln|u| + C \qquad \textbf{(2)}$$

$$\int e^u\,du = e^u + C \qquad \textbf{(3)}$$

$$\int \cos u\,du = \sin u + C \qquad \textbf{(4)}$$

$$\int \sin u\,du = -\cos u + C \qquad \textbf{(5)}$$

$$\int \sec^2 u\,du = \tan u + C \qquad \textbf{(6)}$$

$$\int \csc^2 u\,du = -\cot u + C \qquad \textbf{(7)}$$

$$\int \sec u \tan u\,du = \sec u + C \qquad \textbf{(8)}$$

$$\int \csc u \cot u\,du = -\csc u + C \qquad \textbf{(9)}$$

$$\int \frac{du}{\sqrt{1-u^2}} = \sin^{-1} u + C \qquad \textbf{(10)}$$

$$\int \frac{du}{1+u^2} = \tan^{-1} u + C \qquad \textbf{(11)}$$

$$\int \frac{du}{u\sqrt{u^2-1}} = \sec^{-1}|u| + C \qquad \textbf{(12)}$$

FIGURE 7.2.1 A short table of integrals.

The principal such technique is the *method of substitution,* which we first considered in Section 5.7. Recall that if

$$\int f(u)\,du = F(u) + C,$$

then

$$\int f(g(x)) \cdot g'(x)\,dx = F(g(x)) + C.$$

Thus the substitution

$$u = g(x), \qquad du = g'(x)\,dx$$

transforms the integral

$$\int f(g(x)) \cdot g'(x)\,dx \quad \text{into the simpler integral} \quad \int f(u)\,du.$$

The key to making this simplification lies in spotting the composition $f(g(x))$ in the given integrand. For this integrand to be converted into a function of u alone, the remaining factor must be a *constant* multiple of the derivative $g'(x)$ of the "inside function" $g(x)$. In this case we replace $f(g(x))$ with the simpler $f(u)$ and $g'(x)\,dx$ with the simpler du. Chapter 6 contains numerous illustrations of this method of substitution, and the problems at the end of this section provide an opportunity to review it.

EXAMPLE 1 Find $\displaystyle\int \frac{1}{x}(1+\ln x)^5\,dx.$

Solution We need to spot *both* the inner function $g(x)$ and its derivative $g'(x)$. If we choose $g(x) = 1 + \ln x$, then $g'(x) = 1/x$. Hence the given integral is of the form discussed above with $f(u) = u^5$, $u = 1 + \ln x$, and $du = dx/x$. Therefore,

$$\int \frac{1}{x}(1+\ln x)^5\,dx = \int u^5\,du = \tfrac{1}{6}u^6 + C = \tfrac{1}{6}(1+\ln x)^6 + C.$$ ◆

EXAMPLE 2 Find $\displaystyle\int \frac{x}{1+x^4}\,dx.$

Solution Here it is not so clear what the inside function is. But, looking at the integral formula in Eq. (11) (Fig. 7.2.1), we try the substitution $u = x^2$, $du = 2x\,dx$. We take advantage of the factor $x\,dx = \frac{1}{2}\,du$ that is available in the integrand and compute as follows:

$$\int \frac{x}{1+x^4}\,dx = \frac{1}{2}\int \frac{du}{1+u^2} = \tfrac{1}{2}\tan^{-1} u + C = \tfrac{1}{2}\tan^{-1} x^2 + C.$$

Note that the substitution $u = x^2$ would have been of no use had the integrand been either $1/(1+x^4)$ or $x^2/(1+x^4)$. ◆

Example 2 illustrates how to make a substitution that converts a given integral into a familiar one. This is a kind of *pattern-matching.* Often an integral that does not appear in any integral table can be transformed into one that does by using the techniques of this chapter. In Example 3 we employ an appropriate substitution to match the given integral with the standard integral formula

$$\int \frac{u^2}{\sqrt{a^2-u^2}}\,du = \frac{a^2}{2}\sin^{-1}\left(\frac{u}{a}\right) - \frac{u}{2}\sqrt{a^2-u^2} + C, \qquad \textbf{(13)}$$

which is Formula (56) (inside the back cover).

EXAMPLE 3 Find $\displaystyle\int \frac{x^2}{\sqrt{25-16x^2}}\,dx.$

Solution So that $25-16x^2$ will be equal to a^2-u^2 in Eq. (13), we take $a=5$ and $u=4x$. Then $du=4\,dx$, and so $dx=\frac{1}{4}\,du$. This gives

$$\begin{aligned}\int \frac{x^2}{\sqrt{25-16x^2}}\,dx &= \int \frac{\left(\frac{1}{4}u\right)^2}{\sqrt{25-u^2}}\cdot\frac{1}{4}\,du = \frac{1}{64}\int \frac{u^2}{\sqrt{25-u^2}}\,du\\ &= \frac{1}{64}\left[\frac{25}{2}\sin^{-1}\left(\frac{u}{5}\right)-\frac{u}{2}\sqrt{25-u^2}\right]+C\\ &= \frac{25}{128}\sin^{-1}\left(\frac{4x}{5}\right)-\frac{x}{32}\sqrt{25-16x^2}+C.\end{aligned}$$ ◆

In Section 7.6 we will see how to derive integral formulas such as that in Eq. (13).

Computer Algebra Systems

Systems such as *Derive, Maple,* and *Mathematica* have integral formulas stored internally and can perform pattern matching substitutions like the one used in Example 3. For instance, the *Mathematica* command

```
Integrate[ x^2 / Sqrt[ 25 - 16 x^2 ], x ]
```

and the *Maple* command

```
int( x^2 / sqrt(25 - 16*x^2), x );
```

as well as the corresponding *Derive* command, all produce precisely the same result as that found in Example 3 (except without adding the arbitrary constant of integration, which computer algebra systems generally omit).

Sometimes different methods (whether manual, table, or computer methods) produce integrals that appear to differ. For instance, *Derive* and *Maple* yield the same antiderivative $\frac{1}{2}\tan^{-1}x^2$ of $x/(1+x^4)$ found in Example 2, whereas *Mathematica* returns the function $-\frac{1}{2}\tan^{-1}x^{-2}$ as the result. Naturally we wonder whether $\frac{1}{2}\tan^{-1}x^2 \equiv -\frac{1}{2}\tan^{-1}x^{-2}$. Figure 7.2.2 shows that the answer is No! In Problem 55 we ask you to reconcile these apparently different antiderivatives of the same function.

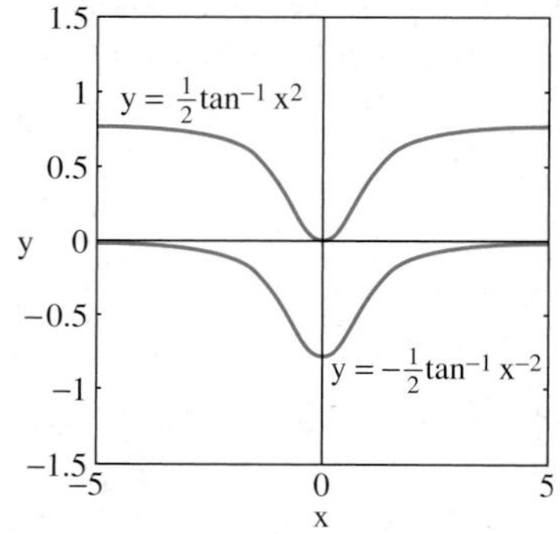

FIGURE 7.2.2 The graphs $y=\frac{1}{2}\tan^{-1}x^2$ and $y=-\frac{1}{2}\tan^{-1}x^{-2}$ do not agree. What *is* the relationship between them?

If computer algebra systems can match patterns with tables of integrals stored in computer memory, you may wonder why manual integration techniques should still be learned. One answer is that a hand computation may yield an integral in a simpler or more convenient form than a computer result. For instance, a computer algebra system may yield a result of the form

$$\int \frac{1}{x}(1+\ln x)^5\,dx = \ln x + \tfrac{5}{2}(\ln x)^2 + \tfrac{10}{3}(\ln x)^3 + \tfrac{5}{2}(\ln x)^4 + (\ln x)^5 + \tfrac{1}{6}(\ln x)^6$$

that looks considerably less appealing than the hand result $\frac{1}{6}(1+\ln x)^6$ of Example 1. Is the relationship between the two obvious? See Problem 54.

7.2 TRUE/FALSE STUDY GUIDE

7.2 CONCEPTS: QUESTIONS AND DISCUSSION

As illustrated at the end of this section (and in Problems 54 through 57), two different methods may yield antiderivatives $G(x)$ and $H(x)$ quite different in appearance, even though they are supposed to be antiderivatives of the same function $f(x)$. Discuss different means for reconciling these antiderivatives, such as:

1. Calculating numerical values of $G(x)$ and $H(x)$ for selected values of x. What can you conclude from such results?

2. Using a calculator or computer to graph $G(x)$ and $H(x)$ simultaneously. What would it mean if the two graphs do not coincide, but intersect at one or more isolated points?
3. Graphing the derivatives $G'(x)$ and $H'(x)$ to see if the graphs coincide. If so, does it follow that $G(x)$ and $H(x)$ are both antiderivatives of $f(x)$?
4. Using a calculator or computer to calculate the numerical values of

$$\int_a^b f(x)\,dx, \qquad G(b) - G(a), \qquad \text{and} \qquad H(b) - H(a)$$

for selected values of a and b. What can you conclude from such results?
5. Simply differentiating $G(x)$ and $H(x)$. Will a glance at the results always settle the matter immediately?

7.2 PROBLEMS

Evaluate the integrals in Problems 1 through 30.

1. $\int (2-3x)^4\,dx$
2. $\int \frac{1}{(1+2x)^2}\,dx$
3. $\int x^2\sqrt{2x^3-4}\,dx$
4. $\int \frac{5t}{5+2t^2}\,dt$
5. $\int \frac{2x}{\sqrt[3]{2x^2+3}}\,dx$
6. $\int x\sec^2 x^2\,dx$
7. $\int \frac{\cot\sqrt{y}\csc\sqrt{y}}{\sqrt{y}}\,dy$
8. $\int \sin\pi(2x+1)\,dx$
9. $\int (1+\sin\theta)^5\cos\theta\,d\theta$
10. $\int \frac{\sin 2x}{4+\cos 2x}\,dx$
11. $\int e^{-\cot x}\csc^2 x\,dx$
12. $\int \frac{e^{\sqrt{x+4}}}{\sqrt{x+4}}\,dx$
13. $\int \frac{(\ln t)^{10}}{t}\,dt$
14. $\int \frac{t}{\sqrt{1-9t^2}}\,dt$
15. $\int \frac{1}{\sqrt{1-9t^2}}\,dt$
16. $\int \frac{e^{2x}}{1+e^{2x}}\,dx$
17. $\int \frac{e^{2x}}{1+e^{4x}}\,dx$
18. $\int \frac{e^{\arctan x}}{1+x^2}\,dx$
19. $\int \frac{3x}{\sqrt{1-x^4}}\,dx$
20. $\int \sin^3 2x\cos 2x\,dx$
21. $\int \tan^4 3x\sec^2 3x\,dx$
22. $\int \frac{1}{1+4t^2}\,dt$
23. $\int \frac{\cos\theta}{1+\sin^2\theta}\,d\theta$
24. $\int \frac{\sec^2\theta}{1+\tan\theta}\,d\theta$
25. $\int \frac{(1+\sqrt{x})^4}{\sqrt{x}}\,dx$
26. $\int t^{-1/3}\sqrt{t^{2/3}-1}\,dt$
27. $\int \frac{1}{(1+t^2)\arctan t}\,dt$
28. $\int \frac{\sec 2x\tan 2x}{(1+\sec 2x)^{3/2}}\,dx$
29. $\int \frac{1}{\sqrt{e^{2x}-1}}\,dx$
30. $\int \frac{x}{\sqrt{\exp(2x^2)-1}}\,dx$

In Problems 31 through 35, evaluate the given integral by making the indicated substitution.

31. $\int x^2\sqrt{x-2}\,dx;\quad u = x-2$
32. $\int \frac{x^2}{\sqrt{x+3}}\,dx;\quad u = x+3$
33. $\int \frac{x}{\sqrt{2x+3}}\,dx;\quad u = 2x+3$
34. $\int x\sqrt[3]{x-1}\,dx;\quad u = x-1$
35. $\int \frac{x}{\sqrt[3]{x+1}}\,dx;\quad u = x+1$

In Problems 36 through 50, evaluate the given integral. First make a substitution that transforms it into a standard form. The standard forms with the given formula numbers are inside the back cover of this book. If a computer algebra system is available, compare and reconcile (if necessary) the result found using the integral table formula with a machine result.

36. $\int \frac{1}{100+9x^2}\,dx;$ Formula (17)
37. $\int \frac{1}{100-9x^2}\,dx;$ Formula (18)
38. $\int \sqrt{9-4x^2}\,dx;$ Formula (54)
39. $\int \sqrt{4+9x^2}\,dx;$ Formula (44)
40. $\int \frac{1}{\sqrt{16x^2+9}}\,dx;$ Formula (45)
41. $\int \frac{x^2}{\sqrt{16x^2+9}}\,dx;$ Formula (49)
42. $\int \frac{x^2}{\sqrt{25+16x^2}}\,dx;$ Formula (49)
43. $\int x^2\sqrt{25-16x^2}\,dx;$ Formula (57)
44. $\int x\sqrt{4-x^4}\,dx;$ Formula (54)
45. $\int e^x\sqrt{9+e^{2x}}\,dx;$ Formula (44)
46. $\int \frac{\cos x}{(\sin^2 x)\sqrt{1+\sin^2 x}}\,dx;$ Formula (50)

47. $\displaystyle\int \frac{\sqrt{x^4-1}}{x}\,dx;$ Formula (47)

48. $\displaystyle\int \frac{e^{3x}}{\sqrt{25+16e^{2x}}}\,dx;$ Formula (49)

49. $\displaystyle\int \frac{(\ln x)^2}{x}\sqrt{1+(\ln x)^2}\,dx;$ Formula (48)

50. $\displaystyle\int x^8\sqrt{4x^6-1}\,dx;$ Formula (48)

51. The substitution $u = x^2$, $x = \sqrt{u}$, $dx = du/(2\sqrt{u})$ appears to lead to this result:

$$\int_{-1}^{1} x^2\,dx = \tfrac{1}{2}\int_{1}^{1} \sqrt{u}\,du = 0.$$

Do you believe this result? If not, why not?

52. Use the fact that $x^2 + 4x + 5 = (x+2)^2 + 1$ to evaluate

$$\int \frac{1}{x^2+4x+5}\,dx.$$

53. Use the fact that $1 - (x-1)^2 = 2x - x^2$ to evaluate

$$\int \frac{1}{\sqrt{2x-x^2}}\,dx.$$

54. Use the binomial expansion

$$(1+t)^6 = 1 + 6t + 15t^2 + 20t^3 + 15t^4 + 6t^5 + t^6$$

to reconcile the result of Example 1 with the machine result listed at the end of this section. Are the two results precisely equal?

55. Establish the precise relationship between the two functions

$$\tfrac{1}{2}\tan^{-1} x^2 \quad \text{and} \quad -\tfrac{1}{2}\tan^{-1} x^{-2}$$

graphed in Fig. 7.2.2. Are both actually antiderivatives of $x/(1+x^4)$?

56. With $u = x$ and $a = 1$, Formula (54) inside the back cover yields

$$\int \sqrt{x^2+1}\,dx = \tfrac{1}{2}x\sqrt{x^2+1} + \tfrac{1}{2}\ln\left|x+\sqrt{x^2+1}\right|,$$

whereas *Maple* and *Mathematica* both give

$$\int \sqrt{x^2+1}\,dx = \tfrac{1}{2}x\sqrt{x^2+1} + \tfrac{1}{2}\sinh^{-1} x.$$

Consult Section 6.9 to reconcile these two results.

57. According to Formula (44) inside the back covers of this book,

$$\int \sqrt{x^2+1}\,dx = G(x) + C$$

where

$$G(x) = \tfrac{1}{2}x\sqrt{x^2+1} + \tfrac{1}{2}\ln\left(x+\sqrt{x^2+1}\right).$$

But another calculus book states that $\int \sqrt{x^2+1}\,dx = H(x) + C$ where

$$H(x) = \tfrac{1}{8}\Big[\left(x+\sqrt{x^2+1}\right)^2 + 4\ln\left(x+\sqrt{x^2+1}\right) - \left(x+\sqrt{x^2+1}\right)^{-2}\Big].$$

Which book is correct? Use a computer algebra system if you wish.

7.3 INTEGRATION BY PARTS

One reason for transforming a given integral into another is to make its evaluation easier. There are two general ways to accomplish this. We have seen the first, integration by substitution. The second is *integration by parts.*

The formula for integration by parts is a simple consequence of the product rule for derivatives,

$$D_x(uv) = v\frac{du}{dx} + u\frac{dv}{dx}.$$

If we write this formula in the form

$$u(x)v'(x) = D_x[u(x)v(x)] - v(x)u'(x), \tag{1}$$

then antidifferentiation gives

$$\int u(x)v'(x)\,dx = u(x)v(x) - \int v(x)u'(x)\,dx. \tag{2}$$

This is the formula for **integration by parts.** With $du = u'(x)\,dx$ and $dv = v'(x)\,dx$, Eq. (2) becomes

$$\int u\,dv = uv - \int v\,du. \tag{3}$$

To apply the integration by parts formula to a given integral, we must first factor its integrand into two "parts," u and dv, the latter including the differential dx. We try to choose the parts in accordance with two principles:

1. The antiderivative $v = \int dv$ is easy to find.
2. The new integral $\int v\,du$ is easier to compute than the original integral $\int u\,dv$.

An effective strategy is to choose for dv the most complicated factor that can readily be integrated. Then we differentiate the other part, u, to find du.

We begin with two examples in which we have little flexibility in choosing the parts u and dv.

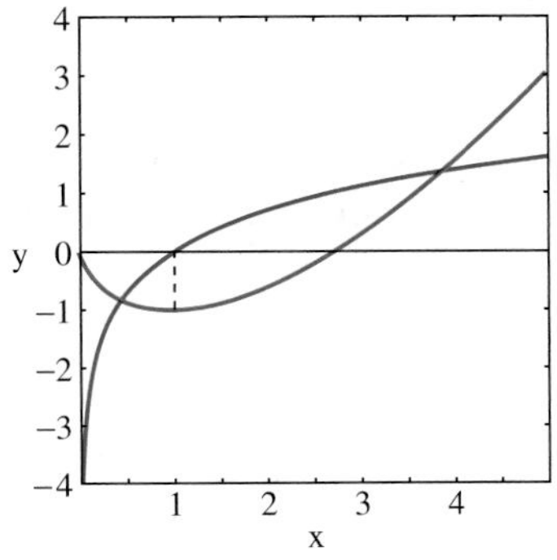

FIGURE 7.3.1 Graphs of the functions $\ln x$ and $x \ln x - x$ of Example 1. If you did not recognize either graph, but noted that the zero of one corresponds to a critical point of the other, which would you conclude is the antiderivative?

EXAMPLE 1 Find $\displaystyle\int \ln x\,dx$. (See Fig. 7.3.1.)

Solution Here there is little alternative to the natural choice $u = \ln x$ and $dv = dx$. It is helpful to systematize the procedure of integration by parts by writing u, dv, du, and v in a rectangular array like this:

$$\begin{array}{llll} \text{Let} & u = \ln x & \text{and} & dv = dx. \\ \text{Then} & du = \dfrac{1}{x}\,dx & \text{and} & v = x. \end{array}$$

The first line in the array specifies the choice of u and dv; the second line is computed from the first. Then Eq. (3) gives

$$\int \ln x\,dx = x \ln x - \int dx = x \ln x - x + C. \quad \blacklozenge$$

COMMENT 1 The constant of integration appears only at the last step. We know that once we have found one antiderivative, any other may be obtained by adding a constant C to the one we have found.

COMMENT 2 In computing $v = \int dv$, we ordinarily take the constant of integration to be zero. Had we written $v = x + C_1$ in Example 1, the answer would have been

$$\begin{aligned} \int \ln x\,dx &= (x + C_1) \ln x - \int \left(1 + \frac{C_1}{x}\right) dx \\ &= x \ln x + C_1 \ln x - (x + C_1 \ln x) + C = x \ln x - x + C \end{aligned}$$

as before, so introducing the extra constant C_1 has no effect.

EXAMPLE 2 Find $\displaystyle\int \arcsin x\,dx$.

Solution Again, there is only one plausible choice for u and dv:

$$\begin{array}{llll} \text{Let} & u = \arcsin x & \text{and} & dv = dx. \\ \text{Then} & du = \dfrac{dx}{\sqrt{1 - x^2}} & \text{and} & v = x. \end{array}$$

Then Eq. (3) gives

$$\begin{aligned} \int \arcsin x\,dx &= x \arcsin x - \int \frac{x}{\sqrt{1 - x^2}}\,dx \\ &= x \arcsin x + \sqrt{1 - x^2} + C. \quad \blacklozenge \end{aligned}$$

EXAMPLE 3 Find $\displaystyle\int xe^{-x}\,dx$.

Solution Here we appear to have some flexibility. Suppose that we try

$$u = e^{-x}, \qquad dv = x\,dx$$

so that

$$du = -e^{-x}\,dx, \qquad v = \tfrac{1}{2}x^2.$$

Then integration by parts gives

$$\int xe^{-x}\,dx = \tfrac{1}{2}x^2e^{-x} + \tfrac{1}{2}\int x^2e^{-x}\,dx.$$

The new integral on the right looks more troublesome than the original problem on the left! Let us begin anew.

$$\text{Let} \quad u = x \quad \text{and} \quad dv = e^{-x}\,dx.$$
$$\text{Then} \quad du = dx \quad \text{and} \quad v = -e^{-x}.$$

Now integration by parts gives

$$\int xe^{-x}\,dx = -xe^{-x} + \int e^{-x}\,dx = -xe^{-x} - e^{-x} + C.$$

◆

Integration by parts can be applied to definite integrals as well as to indefinite integrals. We integrate Eq. (1) from $x = a$ to $x = b$ and apply the fundamental theorem of calculus. This gives

$$\int_a^b u(x)v'(x)\,dx = \int_a^b D_x[u(x)v(x)]\,dx - \int_a^b v(x)u'(x)\,dx$$
$$= \Big[u(x)v(x)\Big]_a^b - \int_a^b v(x)u'(x)\,dx.$$

In the notation of Eq. (3), this equation would be written

$$\int_{x=a}^{x=b} u\,dv = \Big[uv\Big]_a^b - \int_{x=a}^{x=b} v\,du, \qquad \textbf{(4)}$$

although we must not forget that u and v are functions of x. For example, with $u = x$ and $dv = e^{-x}\,dx$, as in Example 3, we obtain

$$\int_0^1 xe^{-x}\,dx = \Big[-xe^{-x}\Big]_0^1 + \int_0^1 e^{-x}\,dx = -e^{-1} + \Big[-e^{-x}\Big]_0^1 = 1 - \frac{2}{e}.$$

EXAMPLE 4 Find $\int x^2e^{-x}\,dx.$

Solution If we choose $u = x^2$, then $du = 2x\,dx$, so we will reduce the exponent of x by this choice.

$$\text{Let} \quad u = x^2 \quad \text{and} \quad dv = e^{-x}\,dx.$$
$$\text{Then} \quad du = 2x\,dx \quad \text{and} \quad v = -e^{-x}.$$

Then integration by parts gives

$$\int x^2e^{-x}\,dx = -x^2e^{-x} + 2\int xe^{-x}\,dx.$$

We apply integration by parts a second time to the right-hand integral and obtain the result

$$\int xe^{-x}\,dx = -xe^{-x} - e^{-x}$$

of Example 3. Substitution then yields

$$\int x^2e^{-x}\,dx = -x^2e^{-x} - 2xe^{-x} - 2e^{-x} + C$$
$$= -(x^2 + 2x + 2)e^{-x} + C.$$

In effect, we have annihilated the original factor x^2 by integrating by parts twice in succession. See Fig. 7.3.2. ◆

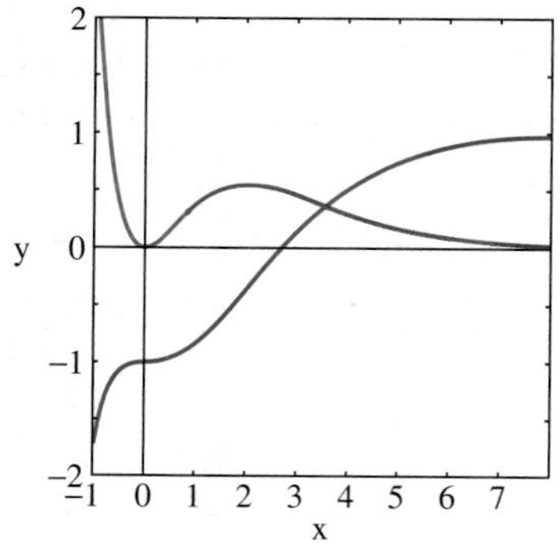

FIGURE 7.3.2 Graphs of the functions x^2e^{-x} and $-(x^2 + 2x + 2)e^{-x} + 1$. Noting only that the zero of one corresponds to a critical point of the other, which do you conclude is the antiderivative?

EXAMPLE 5 Find $\int e^{2x} \sin 3x\,dx$.

Solution This is another example in which repeated integration by parts succeeds, but with a twist:

$$\text{Let} \quad u = \sin 3x, \qquad dv = e^{2x}\,dx.$$
$$\text{Then} \quad du = 3\cos 3x\,dx, \qquad v = \tfrac{1}{2}e^{2x}.$$

Therefore,

$$\int e^{2x}\sin 3x\,dx = \tfrac{1}{2}e^{2x}\sin 3x - \tfrac{3}{2}\int e^{2x}\cos 3x\,dx.$$

At first it might appear that little progress has been made, for the integral on the right is as difficult to integrate as the one on the left. We ignore this objection and try again, applying integration by parts to the new integral:

$$\text{Let} \quad u = \cos 3x, \qquad dv = e^{2x}\,dx.$$
$$\text{Then} \quad du = -3\sin 3x\,dx, \qquad v = \tfrac{1}{2}e^{2x}.$$

Now we find that

$$\int e^{2x}\cos 3x\,dx = \tfrac{1}{2}e^{2x}\cos 3x + \tfrac{3}{2}\int e^{2x}\sin 3x\,dx.$$

When we substitute this result into the previous equation, we discover that

$$\int e^{2x}\sin 3x\,dx = \tfrac{1}{2}e^{2x}\sin 3x - \tfrac{3}{4}e^{2x}\cos 3x - \tfrac{9}{4}\int e^{2x}\sin 3x\,dx.$$

So we are back where we started. Or *are* we? In fact we are *not*, because we can *solve* this last equation for the desired integral. We add the right-hand integral here to both sides of the last equation. This gives

$$\tfrac{13}{4}\int e^{2x}\sin 3x\,dx = \tfrac{1}{4}e^{2x}(2\sin 3x - 3\cos 3x) + C_1,$$

so

$$\int e^{2x}\sin 3x\,dx = \tfrac{1}{13}e^{2x}(2\sin 3x - 3\cos 3x) + C.$$

(See Fig. 7.3.3.) ◆

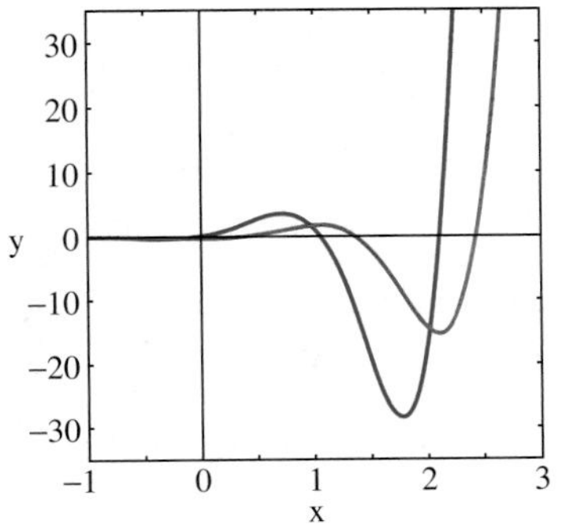

FIGURE 7.3.3 Graphs of the functions $e^{2x}\sin 3x$ and $\frac{1}{13}e^{2x}(2\sin 3x - 3\cos 3x)$. Noting only that the zeros of one correspond to the critical points of the other, which do you conclude is the antiderivative?

EXAMPLE 6 Find a reduction formula for $\int \sec^n x\,dx$.

Solution The idea is that n is a (large) positive integer and that we want to express the given integral in terms of the integral of a lower power of $\sec x$. The easiest power of $\sec x$ to integrate is $\sec^2 x$, so we proceed as follows:

$$\text{Let} \quad u = \sec^{n-2} x, \qquad dv = \sec^2 x\,dx.$$
$$\text{Then} \quad du = (n-2)\sec^{n-2} x \tan x\,dx, \qquad v = \tan x.$$

This gives

$$\int \sec^n x\,dx = \sec^{n-2} x \tan x - (n-2)\int \sec^{n-2} x \tan^2 x\,dx$$
$$= \sec^{n-2} x \tan x - (n-2)\int (\sec^{n-2} x)(\sec^2 x - 1)\,dx.$$

Hence

$$\int \sec^n x\,dx = \sec^{n-2} x \tan x - (n-2)\int \sec^n x\,dx + (n-2)\int \sec^{n-2} x\,dx.$$

We solve this equation for the original integral and find that

$$\int \sec^n x\,dx = \frac{\sec^{n-2} x \tan x}{n-1} + \frac{n-2}{n-1}\int \sec^{n-2} x\,dx. \tag{5}$$

This is the desired reduction formula. For example, if we take $n = 3$ in this formula, we find that

$$\int \sec^3 x\,dx = \tfrac{1}{2}\sec x \tan x + \tfrac{1}{2}\int \sec x\,dx$$

and thus

$$\int \sec^3 x\,dx = \tfrac{1}{2}\sec x \tan x + \tfrac{1}{2}\ln|\sec x + \tan x| + C. \tag{6}$$

In the last step we used the integral formula

$$\int \sec x\,dx = \ln|\sec x + \tan x| + C, \tag{7}$$

which is tricky to derive systematically (see Section 7.4) but is easy to verify by differentiation:

$$\begin{aligned} D_x(\ln|\sec x + \tan x|) &= \frac{D_x(\sec x + \tan x)}{\sec x + \tan x} \\ &= \frac{\sec x \tan x + \sec^2 x}{\sec x + \tan x} = \frac{(\sec x)(\tan x + \sec x)}{\sec x + \tan x} = \sec x. \end{aligned}$$

The reason for using the reduction formula in Eq. (5) to integrate $\sec^n x$ is that—if n is a positive integer—repeated application of the formula must yield either Eq. (7) or the elementary integral

$$\int \sec^2 x\,dx = \tan x + C.$$

◆

EXAMPLE 7 With $n = 4$ in Eq. (5) we get

$$\begin{aligned} \int \sec^4 x\,dx &= \tfrac{1}{3}\sec^2 x \tan x + \tfrac{2}{3}\int \sec^2 x\,dx \\ &= \tfrac{1}{3}\sec^2 x \tan x + \tfrac{2}{3}\tan x + C, \end{aligned} \tag{8}$$

and with $n = 5$ we get

$$\begin{aligned} \int \sec^5 x\,dx &= \tfrac{1}{4}\sec^3 x \tan x + \tfrac{3}{4}\int \sec^3 x\,dx \\ &= \tfrac{1}{4}\sec^3 x \tan x + \tfrac{3}{8}\sec x \tan x + \tfrac{3}{8}\ln|\sec x + \tan x| + C, \end{aligned} \tag{9}$$

using Eq. (6) in the last step. ◆

7.3 TRUE/FALSE STUDY GUIDE

7.3 CONCEPTS: QUESTIONS AND DISCUSSION

1. Given an integral $\int f(x)\,dx$, state in your own words, with a minimum of jargon, your strategy for factoring the integrand into a product $f(x) = g(x)h(x)$ so that the separation into parts $u = g(x)$, $dv = h(x)\,dx$ is effective in evaluating the integral.

2. Give an example of an integral $\int f(x)\,dx$ for which the integrand can be factored in at least three different ways—some that lead to a successful integration by parts and some that do not.

7.3 PROBLEMS

Use integration by parts to compute the integrals in Problems 1 through 34.

1. $\int xe^{2x}\,dx$
2. $\int x^2e^{2x}\,dx$
3. $\int t\sin t\,dt$
4. $\int t^2\sin t\,dt$
5. $\int x\cos 3x\,dx$
6. $\int x\ln x\,dx$
7. $\int x^3\ln x\,dx$
8. $\int e^{3z}\cos 3z\,dz$
9. $\int \arctan x\,dx$
10. $\int \frac{\ln x}{x^2}\,dx$
11. $\int \sqrt{y}\ln y\,dy$
12. $\int x\sec^2 x\,dx$
13. $\int (\ln t)^2\,dt$
14. $\int t(\ln t)^2\,dt$
15. $\int x\sqrt{x+3}\,dx$
16. $\int x^3\sqrt{1-x^2}\,dx$
17. $\int x^5\sqrt{x^3+1}\,dx$
18. $\int \sin^2\theta\,d\theta$
19. $\int \csc^3\theta\,d\theta$
20. $\int \sin(\ln t)\,dt$
21. $\int x^2\arctan x\,dx$
22. $\int \ln(1+x^2)\,dx$
23. $\int \sec^{-1}\sqrt{x}\,dx$
24. $\int x\tan^{-1}\sqrt{x}\,dx$
25. $\int \tan^{-1}\sqrt{x}\,dx$
26. $\int x^2\cos 4x\,dx$
27. $\int x\csc^2 x\,dx$
28. $\int x\arctan x\,dx$
29. $\int x^3\cos x^2\,dx$
30. $\int e^{-3x}\sin 4x\,dx$
31. $\int \frac{\ln x}{x\sqrt{x}}\,dx$
32. $\int \frac{x^7}{(1+x^4)^{3/2}}\,dx$
33. $\int x\cosh x\,dx$
34. $\int e^x\cosh x\,dx$

In Problems 35 through 38, first make a substitution of the form $t = x^k$ and then integrate by parts.

35. $\int x^3\sin x^2\,dx$
36. $\int x^7\cos x^4\,dx$
37. $\int \exp(-\sqrt{x})\,dx$
38. $\int x^2\sin x^{3/2}\,dx$

In Problems 39 through 42, use the method of cylindrical shells to calculate the volume of the solid obtained by revolving the region R around the y-axis.

39. R is bounded below by the x-axis and above by the curve $y = \cos x$, $-\pi/2 \leqq x \leqq \pi/2$.

40. R is bounded below by the x-axis and above by the curve $y = \sin x$, $0 \leqq x \leqq \pi$.

41. R is bounded below by the x-axis, on the right by the line $x = e$, and above by the curve $y = \ln x$.

42. R is bounded below by the x-axis, on the left by the y-axis, on the right by the line $x = 1$, and above by the curve $y = e^{-x}$.

In Problems 43 through 45, first estimate graphically or numerically the points of intersection of the two given curves, then approximate the volume of the solid that is generated when the region bounded by these two curves is revolved around the y-axis.

43. $y = x^2$ and $y = \cos x$

44. $y = 10x - x^2$ and $y = e^x - 1$

45. $y = x^2 - 2x$ and $y = \ln(x+1)$

46. Use integration by parts to evaluate

$$\int 2x\arctan x\,dx,$$

with $dv = 2x\,dx$, but let $v = x^2 + 1$ rather than $v = x^2$. Is there a reason why v should not be chosen in this way?

47. Use integration by parts to evaluate $\int xe^x\cos x\,dx$.

48. Use integration by parts to evaluate $\int \sin 3x\cos x\,dx$.

Derive the reduction formulas given in Problems 49 through 54. Throughout, n denotes a positive integer with an appropriate side condition (such as $n \geqq 1$ or $n \geqq 2$).

49. $\int x^n e^x\,dx = x^n e^x - n\int x^{n-1}e^x\,dx$

50. $\int x^n e^{-x^2}\,dx = -\frac{1}{2}x^{n-1}e^{-x^2} + \frac{n-1}{2}\int x^{n-2}e^{-x^2}\,dx$

51. $\int (\ln x)^n\,dx = x(\ln x)^n - n\int (\ln x)^{n-1}\,dx$

52. $\int x^n\cos x\,dx = x^n\sin x - n\int x^{n-1}\sin x\,dx$

53. $\int \sin^n x\,dx = -\frac{\sin^{n-1}x\cos x}{n} + \frac{n-1}{n}\int \sin^{n-2}x\,dx$

54. $\int \cos^n x\,dx = \frac{\cos^{n-1}x\sin x}{n} + \frac{n-1}{n}\int \cos^{n-2}x\,dx$

Use appropriate reduction formulas from the preceding list to evaluate the integrals in Problems 55 through 57.

55. $\displaystyle\int_0^1 x^3 e^x\,dx$

56. $\displaystyle\int_0^1 x^5 e^{-x^2}\,dx$

57. $\displaystyle\int_1^e (\ln x)^3\,dx$

58. Apply the reduction formula in Problem 53 to show that for each positive integer n,

$$\int_0^{\pi/2} \sin^{2n} x\,dx = \frac{\pi}{2}\cdot\frac{1}{2}\cdot\frac{3}{4}\cdot\frac{5}{6}\cdots\frac{2n-1}{2n}$$

and

$$\int_0^{\pi/2} \sin^{2n+1} x\,dx = \frac{2}{3}\cdot\frac{4}{5}\cdot\frac{6}{7}\cdot\frac{8}{9}\cdots\frac{2n}{2n+1}.$$

59. Derive the formula

$$\int \ln(x+10)\,dx = (x+10)\ln(x+10) - x + C$$

in three different ways: (a) by substituting $u = x + 10$ and applying the result of Example 1; (b) by integrating by parts with $u = \ln(x+10)$ and $dv = dx$, noting that

$$\frac{x}{x+10} = 1 - \frac{10}{x+10};$$

and (c) by integrating by parts with $u = \ln(x+10)$ and $dv = dx$, but with $v = x + 10$.

60. Derive the formula

$$\int x^3 \tan^{-1} x\,dx = \tfrac{1}{4}(x^4 - 1)\tan^{-1} x - \tfrac{1}{12}x^3 + \tfrac{1}{4}x + C$$

by integrating by parts with $u = \tan^{-1} x$ and $v = \frac{1}{4}(x^4 - 1)$.

61. Let $J_n = \int_0^1 x^n e^{-x}\,dx$ for each integer $n \geqq 0$. (a) Show that

$$J_0 = 1 - \frac{1}{e} \quad\text{and that}\quad J_n = nJ_{n-1} - \frac{1}{e}$$

for $n \geqq 1$. (b) Deduce by mathematical induction that

$$J_n = n! - \frac{n!}{e}\sum_{k=0}^{n}\frac{1}{k!}$$

for each integer $n \geqq 0$. (c) Explain why $J_n \to 0$ as $n \to +\infty$. (d) Conclude that

$$e = \lim_{n\to\infty}\sum_{k=0}^{n}\frac{1}{k!}.$$

62. Let m and n be positive integers. Derive the reduction formula

$$\int x^m(\ln x)^n\,dx = \frac{x^{m+1}}{m+1}(\ln x)^n - \frac{n}{m+1}\int x^m(\ln x)^{n-1}\,dx.$$

63. An advertisement for a symbolic algebra program claims that an engineer worked for three weeks on the integral

$$\int (k\ln x - 2x^3 + 3x^2 + b)^4\,dx,$$

which deals with turbulence in an aerospace application. The advertisement said that the engineer never got the same answer twice in the three weeks. Explain how you could use the reduction formula of Problem 62 to find the engineer's integral (but don't actually do it). Can you see any reason why it should have taken three weeks?

64. Figure 7.3.4 shows the region bounded by the x-axis and the graph of $y = \frac{1}{2}x^2 \sin x$, $0 \leqq x \leqq \pi$. Use Formulas (42) and (43) (inside the back cover)—which are derived by integration by parts—to find (a) the area of this region; (b) the volume obtained by revolving this region around the y-axis.

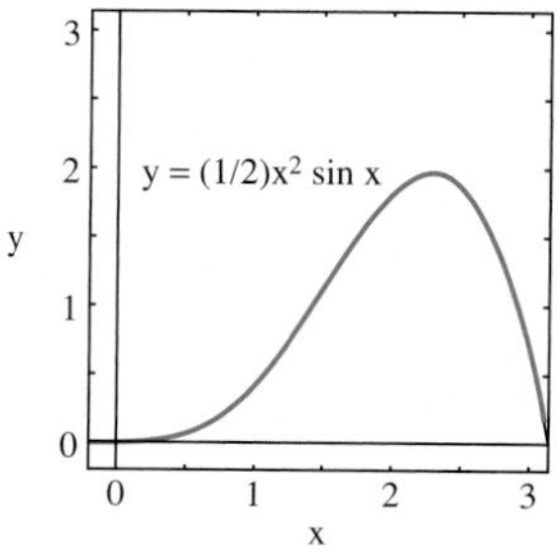

FIGURE 7.3.4 The region of Problem 64.

65. The top shown in Fig. 7.3.5 has the shape of the solid obtained by revolving the region of Problem 64 around the x-axis. Find the volume of this top.

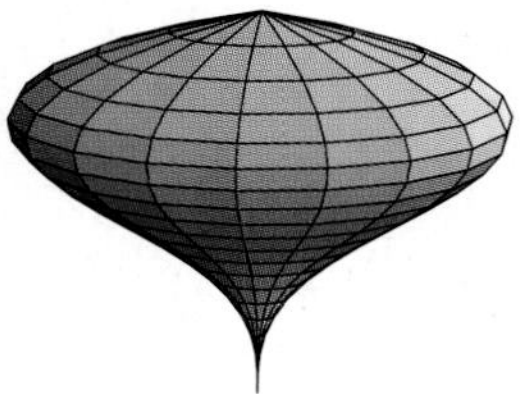

FIGURE 7.3.5 The top of Problem 65.

66. A particle is set in motion at time $t = 0$ and moves to the right along the x-axis. (a) Suppose that its acceleration at time t is $a = 100e^{-t}$. Show that the particle moves infinitely far to the right along the x-axis. (b) Suppose that its acceleration at time t is $a = 100(1-t)e^{-t}$. Show that the particle never moves beyond a certain point to the right of its initial position and find that point. Explain why the particle "effectively" comes to a stop at that point.

67. Find the area and centroid of the region that is bounded by the curves $y = x^2$ and $y = 2^x$ for $2 \leqq x \leqq 4$.

68. For each positive integer k, let

$$I_k = \int_0^{\pi/2} \sin^k x\,dx.$$

(a) Show that $I_{2n} \geqq I_{2n+1} \geqq I_{2n+2}$ for each positive integer n.
(b) Use Problem 58 to show that

$$\lim_{n\to\infty}\frac{I_{2n+2}}{I_{2n}} = 1.$$

(c) Conclude from parts (a) and (b) that

$$\lim_{n\to\infty} \frac{I_{2n+1}}{I_{2n}} = 1.$$

(d) Conclude from part (c) and Problem 58 that

$$\lim_{n\to\infty} \frac{2}{1}\cdot\frac{2}{3}\cdot\frac{4}{3}\cdot\frac{4}{5}\cdot\frac{6}{5}\cdot\frac{6}{7}\cdots\frac{2n}{2n-1}\cdot\frac{2n}{2n+1} = \frac{\pi}{2}.$$

This result is usually written as the *infinite product*

$$\frac{\pi}{2} = \frac{2}{1}\cdot\frac{2}{3}\cdot\frac{4}{3}\cdot\frac{4}{5}\cdot\frac{6}{5}\cdot\frac{6}{7}\cdots,$$

which was discovered by the English mathematician John Wallis in 1655.

7.4 TRIGONOMETRIC INTEGRALS

Here we discuss the evaluation of certain integrals in which the integrand is either a power of a trigonometric function or the product of two such powers. Such integrals are among the most common trigonometric integrals in applications of calculus.

To evaluate the integrals

$$\int \sin^2 u\,du \quad \text{and} \quad \int \cos^2 u\,du$$

that appear in numerous applications, we use the **half-angle identities**

$$\sin^2\theta = \tfrac{1}{2}(1 - \cos 2\theta), \qquad \textbf{(1)}$$

$$\cos^2\theta = \tfrac{1}{2}(1 + \cos 2\theta) \qquad \textbf{(2)}$$

of Eqs. (11) and (10) in Appendix C.

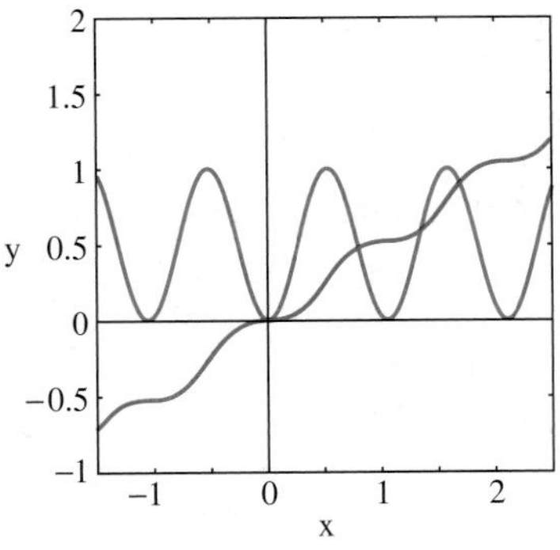

FIGURE 7.4.1 Graphs of the functions $\sin^2 3x$ and $\frac{1}{12}(6x - \sin 6x)$ of Example 1. The zeros of which correspond to the critical points of the other? So which is the antiderivative?

EXAMPLE 1 Find $\int \sin^2 3x\,dx$. (See Fig. 7.4.1.)

Solution The identity in Eq. (1)—with $3x$ in place of θ—yields

$$\int \sin^2 3x\,dx = \int \tfrac{1}{2}(1 - \cos 6x)\,dx$$

$$= \tfrac{1}{2}\left(x - \tfrac{1}{6}\sin 6x\right) + C = \tfrac{1}{12}(6x - \sin 6x) + C. \qquad \blacklozenge$$

To integrate $\tan^2 x$ and $\cot^2 x$, we use the identities

$$1 + \tan^2 x = \sec^2 x \quad \text{and} \quad 1 + \cot^2 x = \csc^2 x. \qquad \textbf{(3)}$$

The first of these follows from the fundamental identity $\sin^2 x + \cos^2 x = 1$ upon division of both sides by $\cos^2 x$. To obtain the second formula in (3), we divide both sides of the fundamental identity by $\sin^2 x$.

EXAMPLE 2 Compute the antiderivative $\int \cot^2 3x\,dx$.

Solution By using the second identity in (3) with $3x$ in place of x, we obtain

$$\int \cot^2 3x\,dx = \int (\csc^2 3x - 1)\,dx$$

$$= \int (\csc^2 u - 1)\left(\tfrac{1}{3}du\right) \qquad (u = 3x)$$

$$= \tfrac{1}{3}(-\cot u - u) + C = -\tfrac{1}{3}\cot 3x - x + C. \qquad \blacklozenge$$

Integrals of Products of Sines and Cosines

The substitution $u = \sin x$, $du = \cos x\,dx$ gives

$$\int \sin^3 x \cos x\,dx = \int u^3 du = \tfrac{1}{4}u^4 + C = \tfrac{1}{4}\sin^4 x + C.$$

This substitution, or the similar substitution $u = \cos x$, $du = -\sin x\,dx$, can be used to evaluate an integral of the form

$$\int \sin^m x \cos^n x\,dx \qquad \textbf{(4)}$$

in the first of the following two cases:

- *Case 1:* At least one of the two numbers m and n is an *odd positive integer.* If so, then the other may be any real number.
- *Case 2:* Both m and n are *nonnegative even integers.*

Suppose, for example, that $m = 2k + 1$ is an odd positive integer. Then we isolate one $\sin x$ factor and use the identity $\sin^2 x = 1 - \cos^2 x$ to express the remaining $\sin^{m-1} x$ factor in terms of $\cos x$, as follows:

$$\int \sin^m x \cos^n x\,dx = \int \sin^{m-1} x \cos^n x \sin x\,dx = \int (\sin^2 x)^k \cos^n x \sin x\,dx$$
$$= \int (1 - \cos^2 x)^k \cos^n x \sin x\,dx.$$

Now the substitution $u = \cos x$, $du = -\sin x\,dx$ yields

$$\int \sin^m x \cos^n x\,dx = -\int (1 - u^2)^k u^n du.$$

The exponent $k = (m - 1)/2$ is a nonnegative integer because m is an odd positive integer. Thus the factor $(1 - u^2)^k$ of the integrand is a polynomial in the variable u, and so its product with u^n is easy to integrate.

In essence, this method consists of peeling off one copy of $\sin x$ (if m is odd) and then converting the remaining sines into cosines. If n is odd, then we can split off one copy of $\cos x$ and convert the remaining cosines into sines.

EXAMPLE 3

(a)
$$\int \sin^3 x \cos^2 x\,dx = \int (1 - \cos^2 x)\cos^2 x \sin x\,dx$$
$$= \int (u^4 - u^2)\,du \qquad (u = \cos x)$$
$$= \tfrac{1}{5}u^5 - \tfrac{1}{3}u^3 + C = \tfrac{1}{5}\cos^5 x - \tfrac{1}{3}\cos^3 x + C.$$

(b)
$$\int \cos^5 x\,dx = \int (1 - \sin^2 x)^2 \cos x\,dx$$
$$= \int (1 - u^2)^2\,du \qquad (u = \sin x)$$
$$= \int (1 - 2u^2 + u^4)\,du = u - \tfrac{2}{3}u^3 + \tfrac{1}{5}u^5 + C$$
$$= \sin x - \tfrac{2}{3}\sin^3 x + \tfrac{1}{5}\sin^5 x + C.$$

◆

In case 2 of the sine-cosine integral in Eq. (4), with both m and n nonnegative even integers, we use the half-angle formulas in Eqs. (1) and (2) to halve the even powers of $\sin x$ and $\cos x$. If we repeat this process with the resulting powers of $\cos 2x$ (if necessary), we get integrals involving odd powers, and we have seen how to handle these in Case 1.

EXAMPLE 4 Use of Eqs. (1) and (2) gives

$$\begin{aligned}\int \sin^2 x \cos^2 x\, dx &= \int \tfrac{1}{2}(1-\cos 2x)\tfrac{1}{2}(1+\cos 2x)\, dx\\ &= \tfrac{1}{4}\int (1-\cos^2 2x)\, dx = \tfrac{1}{4}\int \left[1-\tfrac{1}{2}(1+\cos 4x)\right] dx\\ &= \tfrac{1}{8}\int (1-\cos 4x)\, dx = \tfrac{1}{8}x - \tfrac{1}{32}\sin 4x + C.\end{aligned}$$

In the third step we used Eq. (2) with $\theta = 2x$. ◆

EXAMPLE 5 Here we apply Eq. (2), first with $\theta = 3x$ and then with $\theta = 6x$.

$$\begin{aligned}\int \cos^4 3x\, dx &= \int \left[\tfrac{1}{2}(1+\cos 6x)\right]^2 dx\\ &= \tfrac{1}{4}\int (1 + 2\cos 6x + \cos^2 6x)\, dx\\ &= \tfrac{1}{4}\int \left(\tfrac{3}{2} + 2\cos 6x + \tfrac{1}{2}\cos 12x\right) dx\\ &= \tfrac{3}{8}x + \tfrac{1}{12}\sin 6x + \tfrac{1}{96}\sin 12x + C.\end{aligned}$$ ◆

Integrals of Products of Secants and Tangents

To integrate $\tan x$, we use the substitution

$$u = \cos x, \qquad du = -\sin x\, dx$$

and get

$$\int \tan x\, dx = \int \frac{\sin x}{\cos x}\, dx = -\int \frac{1}{u}\, du = -\ln|u| + C.$$

Thus

$$\int \tan x\, dx = -\ln|\cos x| + C = \ln|\sec x| + C. \tag{5}$$

In Eq. (5) we used the fact that $|\sec x| = 1/|\cos x|$.

Similarly,

$$\int \cot x\, dx = \ln|\sin x| + C = -\ln|\csc x| + C. \tag{6}$$

The first person to integrate $\sec x$ may well have spent much time doing so. Here is one of several methods. First we "prepare" the function for integration:

$$\sec x = \frac{1}{\cos x} = \frac{\cos x}{\cos^2 x} = \frac{\cos x}{1-\sin^2 x}.$$

Next we use the algebraic identity

$$\frac{1}{1+z} + \frac{1}{1-z} = \frac{2}{1-z^2},$$

which you can verify by finding a common denominator on the right. Similarly, we see that

$$\frac{2\cos x}{1-\sin^2 x} = \frac{\cos x}{1+\sin x} + \frac{\cos x}{1-\sin x}.$$

Therefore,

$$\begin{aligned}\int \sec x\, dx &= \tfrac{1}{2}\int \left(\frac{\cos x}{1+\sin x} + \frac{\cos x}{1-\sin x}\right) dx\\ &= \tfrac{1}{2}(\ln|1+\sin x| - \ln|1-\sin x|) + C.\end{aligned}$$

It's customary to simplify this result:

$$\begin{aligned}\int \sec x\, dx &= \tfrac{1}{2}\ln\left|\frac{1+\sin x}{1-\sin x}\right| + C = \tfrac{1}{2}\ln\left|\frac{(1+\sin x)^2}{1-\sin^2 x}\right| + C\\ &= \ln\left|\frac{(1+\sin x)^2}{\cos^2 x}\right|^{1/2} + C = \ln\left|\frac{1+\sin x}{\cos x}\right| + C\\ &= \ln|\sec x + \tan x| + C.\end{aligned}$$

After we verify by differentiation that

$$\int \sec x\, dx = \ln|\sec x + \tan x| + C, \tag{7}$$

we could always "derive" this result by using an unmotivated trick:

$$\begin{aligned}\int \sec x\, dx &= \int (\sec x)\frac{\tan x + \sec x}{\sec x + \tan x}\, dx\\ &= \int \frac{\sec x \tan x + \sec^2 x}{\sec x + \tan x}\, dx = \ln|\sec x + \tan x| + C.\end{aligned}$$

A similar technique yields

$$\int \csc x\, dx = -\ln|\csc x + \cot x| + C. \tag{8}$$

EXAMPLE 6 The substitution $u = \frac{1}{2}x$, $du = \frac{1}{2}\, dx$ gives

$$\begin{aligned}\int_0^{\pi/2} \sec\frac{x}{2}\, dx &= 2\int_0^{\pi/4} \sec u\, du\\ &= 2\Big[\ln|\sec u + \tan u|\Big]_0^{\pi/4} = 2\ln\left(1+\sqrt{2}\right) \approx 1.76275.\end{aligned}$$ ◆

An integral of the form

$$\int \tan^m x \sec^n x\, dx \tag{9}$$

can be routinely evaluated in either of the following two cases:

- *Case 1: m is an odd positive integer.*
- *Case 2: n is an even positive integer.*

In Case 1, we split off the factor $\sec x \tan x$ to form, along with dx, the differential $\sec x \tan x\, dx$ of $\sec x$. We then use the identity $\tan^2 x = \sec^2 x - 1$ to convert the remaining even power of $\tan x$ into powers of $\sec x$. This prepares the integrand for the substitution $u = \sec x$.

EXAMPLE 7

$$\int \tan^3 x \sec^3 x\,dx = \int (\sec^2 x - 1)\sec^2 x \sec x \tan x\,dx$$
$$= \int (u^4 - u^2)\,du \qquad (u = \sec x)$$
$$= \tfrac{1}{5}u^5 - \tfrac{1}{3}u^3 + C = \tfrac{1}{5}\sec^5 x - \tfrac{1}{3}\sec^3 x + C. \quad ◆$$

To evaluate the integral in Eq. (9) in Case 2, we split off $\sec^2 x$ to form, along with dx, the differential of $\tan x$. We then use the identity $\sec^2 x = 1 + \tan^2 x$ to convert the remaining even power of $\sec x$ into powers of $\tan x$. This prepares the integrand for the substitution $u = \tan x$.

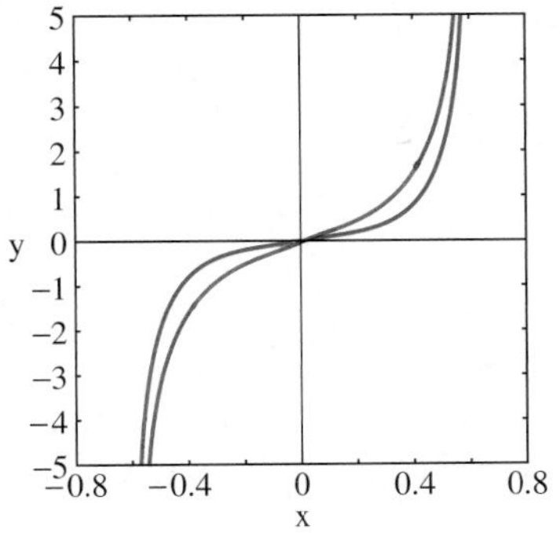

FIGURE 7.4.2 After an initial attempt to evaluate $\int \sec^6 2x\,dx$ using the two methods of Example 8, a computer was used to plot the two alleged antiderivatives. Why does this figure indicate the presence of an error in the calculations? What is the relationship between any two antiderivatives of a given function?

EXAMPLE 8 **Method 1.** Use of the secant-tangent form $\sec^2 u = 1 + \tan^2 u$ of the fundamental identity of trigonometry gives

$$\int \sec^6 2x\,dx = \int (1 + \tan^2 2x)^2 \sec^2 2x\,dx$$
$$= \tfrac{1}{2}\int (1 + \tan^2 2x)^2 (2\sec^2 2x)\,dx$$
$$= \tfrac{1}{2}\int (1 + u^2)^2\,du \qquad (u = \tan 2x, \quad du = 2\sec^2 2x\,dx)$$
$$= \tfrac{1}{2}\int (1 + 2u^2 + u^4)\,du = \tfrac{1}{2}u + \tfrac{1}{3}u^3 + \tfrac{1}{10}u^5 + C$$
$$= \tfrac{1}{2}\tan 2x + \tfrac{1}{3}\tan^3 2x + \tfrac{1}{10}\tan^5 2x + C. \qquad \textbf{(10)}$$

Method 2. Alternatively, we could apply the reduction formula

$$\int \sec^n x\,dx = \frac{\sec^{n-2} x \tan x}{n-1} + \frac{n-2}{n-1}\int \sec^{n-2} x\,dx$$

of Section 7.3, first with $n = 6$ and then with $n = 4$. This gives

$$\int \sec^6 2x\,dx = \tfrac{1}{2}\int \sec^6 u\,du \qquad (u = 2x)$$
$$= \tfrac{1}{2}\left(\tfrac{1}{5}\sec^4 u \tan u + \tfrac{4}{5}\int \sec^4 u\,du\right)$$
$$= \tfrac{1}{10}\sec^4 u \tan u + \tfrac{2}{5}\left(\tfrac{1}{3}\sec^2 u \tan u + \tfrac{2}{3}\int \sec^2 u\,du\right)$$
$$= \tfrac{1}{10}\sec^4 2x \tan 2x + \tfrac{2}{15}\sec^2 2x \tan 2x + \tfrac{4}{15}\tan 2x + C. \qquad \textbf{(11)}$$

(See Figs. 7.4.2 and 7.4.3.) ◆

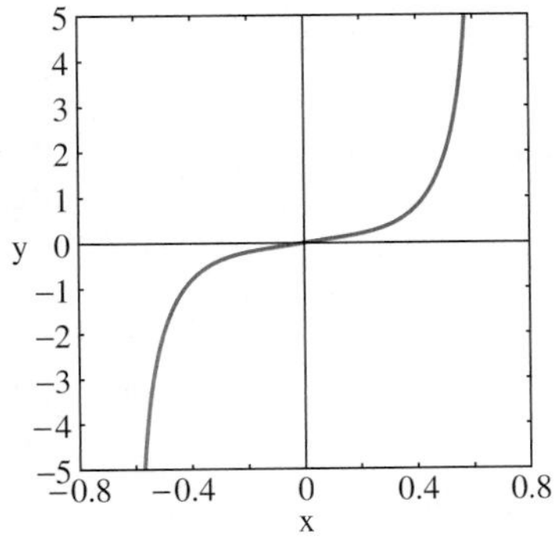

FIGURE 7.4.3 After the error made initially was found and corrected, this figure resulted when the two antiderivatives in (10) and (11) were plotted simultaneously (each with $C = 0$). It indicates that the two antiderivative formulas are, in fact, equivalent.

Similar methods are effective with integrals of the form

$$\int \csc^m x \cot^n x\,dx,$$

because the cotangent and cosecant functions satisfy analogous differentiation formulas and trigonometric identities:

$$1 + \tan^2 x = \sec^2 x, \qquad D_x \tan x = \sec^2 x, \qquad D_x \sec x = \sec x \tan x$$

and

$$1 + \cot^2 x = \csc^2 x, \qquad D_x \cot x = -\csc^2 x, \qquad D_x \csc x = -\csc x \cot x.$$

The method of Case 1 succeeds with the integral

$$\int \tan^n x \, dx$$

only when n is an odd positive integer, but there is another approach that works equally well whether n is even or odd. We split off the factor $\tan^2 x$ and replace it with $\sec^2 x - 1$:

$$\begin{aligned}\int \tan^n x \, dx &= \int (\tan^{n-2} x)(\sec^2 x - 1)\, dx \\ &= \int \tan^{n-2} x \sec^2 x \, dx - \int \tan^{n-2} x \, dx.\end{aligned}$$

We integrate what we can and find that

$$\int \tan^n x \, dx = \frac{\tan^{n-1} x}{n-1} - \int \tan^{n-2} x \, dx. \qquad \textbf{(12)}$$

Equation (12) is another example of a reduction formula. Its use effectively reduces the original exponent from n to $n - 2$. If we apply Eq. (12) repeatedly, we eventually get either

$$\int \tan^2 x \, dx = \int (\sec^2 x - 1)\, dx = \tan x - x + C$$

or

$$\int \tan x \, dx = \ln|\sec x| + C.$$

EXAMPLE 9 Two applications of Eq. (12) give

$$\begin{aligned}\int \tan^6 x \, dx &= \tfrac{1}{5}\tan^5 x - \int \tan^4 x \, dx \\ &= \tfrac{1}{5}\tan^5 x - \left(\tfrac{1}{3}\tan^3 x - \int \tan^2 x \, dx\right) \\ &= \tfrac{1}{5}\tan^5 x - \tfrac{1}{3}\tan^3 x + \tan x - x + C.\end{aligned}$$ ◆

Finally, in the case of an integral involving an unusual mixture of trigonometric functions—tangents and cosecants, for example—expressing the integrand entirely in terms of sines and cosines may yield an expression that's easy to integrate.

7.4 TRUE/FALSE STUDY GUIDE

7.4 CONCEPTS: QUESTIONS AND DISCUSSION

In the following questions, the term "trigonometric integral" means an integral of a positive whole-number power of a trigonometric function or an integral of the product of two such powers.

1. Describe the types of trigonometric integrals that can be evaluated using the methods of this section.
2. Give examples of several trigonometric integrals that cannot be evaluated using the methods of this section.

7.4 PROBLEMS

Evaluate the integrals in Problems 1 through 44.

1. $\int \sin^2 2x\,dx$
2. $\int \cos^2 5x\,dx$
3. $\int \sec^2 \frac{x}{2}\,dx$
4. $\int \tan^2 \frac{x}{2}\,dx$
5. $\int \tan 3x\,dx$
6. $\int \cot 4x\,dx$
7. $\int \sec 3x\,dx$
8. $\int \csc 2x\,dx$
9. $\int \frac{1}{\csc^2 x}\,dx$
10. $\int \sin^2 x \cot^2 x\,dx$
11. $\int \sin^3 x\,dx$
12. $\int \sin^4 x\,dx$
13. $\int \sin^2\theta \cos^3\theta\,d\theta$
14. $\int \sin^3 t \cos^3 t\,dt$
15. $\int \cos^5 x\,dx$
16. $\int \frac{\sin t}{\cos^3 t}\,dt$
17. $\int \frac{\sin^3 x}{\sqrt{\cos x}}\,dx$
18. $\int \sin^3\phi \cos^4\phi\,d\phi$
19. $\int \sin^5 2z \cos^2 2z\,dz$
20. $\int \sin^{3/2} x \cos^3 x\,dx$
21. $\int \frac{\sin^3 4x}{\cos^2 4x}\,dx$
22. $\int \cos^6 4\theta\,d\theta$
23. $\int \sec^4 t\,dt$
24. $\int \tan^3 x\,dx$
25. $\int \cot^3 2x\,dx$
26. $\int \tan\theta \sec^4\theta\,d\theta$
27. $\int \tan^5 2x \sec^2 2x\,dx$
28. $\int \cot^3 x \csc^2 x\,dx$
29. $\int \csc^6 2t\,dt$
30. $\int \frac{\sec^4 t}{\tan^2 t}\,dt$
31. $\int \frac{\tan^3\theta}{\sec^4\theta}\,d\theta$
32. $\int \frac{\cot^3 x}{\csc^2 x}\,dx$
33. $\int \frac{\tan^3 t}{\sqrt{\sec t}}\,dt$
34. $\int \frac{1}{\cos^4 2x}\,dx$
35. $\int \frac{\cot\theta}{\csc^3\theta}\,d\theta$
36. $\int \sin^2 3\alpha \cos^2 3\alpha\,d\alpha$
37. $\int \cos^3 5t\,dt$
38. $\int \tan^4 x\,dx$
39. $\int \cot^4 3t\,dt$
40. $\int \tan^2 2t \sec^4 2t\,dt$
41. $\int \sin^5 2t \cos^{3/2} 2t\,dt$
42. $\int \cot^3\xi \csc^{3/2}\xi\,d\xi$
43. $\int \frac{\tan x + \sin x}{\sec x}\,dx$
44. $\int \frac{\cot x + \csc x}{\sin x}\,dx$

In Problems 45 through 48, find the area of the region bounded by the two given curves.

45. The x-axis and the curve $y = \sin^3 x$, from $x = 0$ to $x = \pi$
46. $y = \cos^2 x$ and $y = \sin^2 x$, from $x = -\pi/4$ to $x = \pi/4$
47. $y = \sin x \cos x$ and $y = \sin^2 x$, from $x = \pi/4$ to $x = \pi$
48. $y = \cos^3 x$ and $y = \sin^3 x$, from $x = \pi/4$ to $x = 5\pi/4$

In Problems 49 and 50, first graph the integrand function and guess the value of the integral. Then verify your guess by actually evaluating the integral.

49. $\int_0^{2\pi} \sin^3 x \cos^2 x\,dx$
50. $\int_0^{\pi} \sin^5 2x\,dx$

In Problems 51 through 54, find the volume of the solid generated by revolving the given region R around the x-axis.

51. R is bounded by the x-axis and the curve $y = \sin^2 x$, $0 \leqq x \leqq \pi$.
52. R is the region of Problem 46.
53. R is bounded by $y = 2$ and $y = \sec x$ for $-\pi/3 \leqq x \leqq \pi/3$.
54. R is bounded by $y = 4\cos x$ and $y = \sec x$ for $-\pi/3 \leqq x \leqq \pi/3$.
55. Let R denote the region that lies between the curves $y = \tan^2 x$ and $y = \sec^2 x$ for $0 \leqq x \leqq \pi/4$. Find: (a) the area of R; (b) the volume of the solid obtained by revolving R around the x-axis.
56. Find the length of the graph of $y = \ln(\cos x)$ from $x = 0$ to $x = \pi/4$.
57. Find
$$\int \tan x \sec^4 x\,dx$$
in two different ways. Then show that your two results are equivalent.
58. Find
$$\int \cot^3 x\,dx$$
in two different ways. Then show that your two results are equivalent.

Problems 59 through 62 are applications of the trigonometric identities

$$\sin A \sin B = \tfrac{1}{2}[\cos(A - B) - \cos(A + B)],$$
$$\sin A \cos B = \tfrac{1}{2}[\sin(A - B) + \sin(A + B)],$$
$$\cos A \cos B = \tfrac{1}{2}[\cos(A - B) + \cos(A + B)].$$

59. Find $\int \sin 3x \cos 5x\,dx$.
60. Find $\int \sin 2x \sin 4x\,dx$.
61. Find $\int \cos x \cos 4x\,dx$.

62. Suppose that m and n are positive integers with $m \neq n$. Show that

(a) $\displaystyle\int_0^{2\pi} \sin mx \sin nx \, dx = 0;$

(b) $\displaystyle\int_0^{2\pi} \cos mx \sin nx \, dx = 0;$

(c) $\displaystyle\int_0^{2\pi} \cos mx \cos nx \, dx = 0.$

63. Substitute $\sec x \csc x = (\sec^2 x)/(\tan x)$ to derive the formula

$$\int \sec x \csc x \, dx = \ln|\tan x| + C.$$

64. Show that

$$\csc x = \frac{1}{2\sin\left(\frac{1}{2}x\right)\cos\left(\frac{1}{2}x\right)},$$

then apply the result of Problem 63 to derive the formula

$$\int \csc x \, dx = \ln\left|\tan\frac{x}{2}\right| + C.$$

65. Substitute $x = \frac{1}{2}\pi - u$ into the integral formula of Problem 64 to show that

$$\int \sec x \, dx = \ln\left|\cot\left(\frac{\pi}{4} - \frac{x}{2}\right)\right| + C.$$

66. Use appropriate trigonometric identities to deduce from the result of Problem 65 that

$$\int \sec x \, dx = \ln|\sec x + \tan x| + C.$$

67. Show first that the reduction formula in Eq. (12) gives

$$\int \tan^4 x \, dx = \tfrac{1}{3}\tan^3 x - \tan x + x + C.$$

Then compare this result with the alleged antiderivative

$$\int \tan^4 x \, dx = \tfrac{1}{12}(\sec^3 x)(9x\cos x + 3x\cos 3x - 4\sin 3x)$$

given by some versions of *Mathematica.*

68. Compare the result given in Example 9 with the integral

$$\int \tan^6 x \, dx$$

as given by your favorite computer algebra system.

7.5 RATIONAL FUNCTIONS AND PARTIAL FRACTIONS

We now discuss methods with which every rational function can be integrated in terms of elementary functions. Recall that a rational function $R(x)$ is a function that can be expressed as the quotient of two polynomials. That is,

$$R(x) = \frac{P(x)}{Q(x)}, \tag{1}$$

where $P(x)$ and $Q(x)$ are polynomials. The **method of partial fractions** is an *algebraic* technique that decomposes $R(x)$ into a sum of terms:

$$R(x) = \frac{P(x)}{Q(x)} = p(x) + F_1(x) + F_2(x) + \cdots + F_k(x), \tag{2}$$

where $p(x)$ is a polynomial and each expression $F_i(x)$ is a fraction that can be integrated with little difficulty.

EXAMPLE 1 We can verify (by finding a common denominator on the right) that

$$\frac{x^3-1}{x^3+x} = 1 - \frac{1}{x} + \frac{x-1}{x^2+1}. \tag{3}$$

It follows that

$$\int \frac{x^3-1}{x^3+x}\,dx = \int\left(1 - \frac{1}{x} + \frac{x}{x^2+1} - \frac{1}{x^2+1}\right)dx$$
$$= x - \ln|x| + \tfrac{1}{2}\ln(x^2+1) - \tan^{-1} x + C.$$

The key to this simple integration lies in finding the decomposition given in Eq. (3). The existence of such a decomposition and the technique of finding it are what the method of partial fractions is about. (See Fig. 7.5.1.) ◆

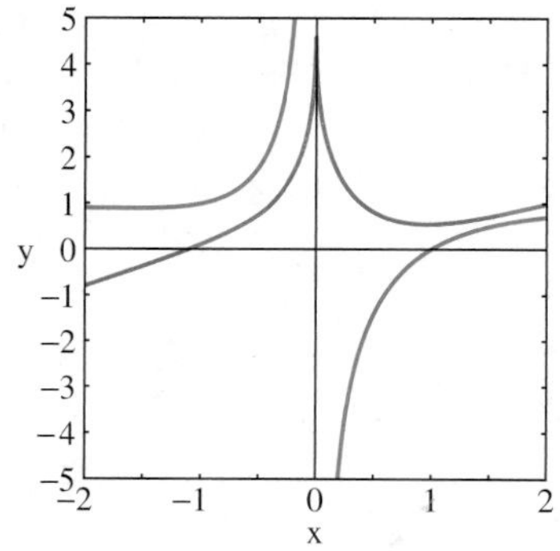

FIGURE 7.5.1 Graphs of the function $f(x) = (x^3 - 1)/(x^3 + x)$ of Example 1 and its indefinite integral with $C = 0$. Which is which?

According to a theorem proved in advanced algebra, every rational function can be written in the form in Eq. (2) with each $F_i(x)$ being a fraction either of the form

$$\frac{A}{(ax+b)^n} \tag{4}$$

or of the form

$$\frac{Bx+C}{(ax^2+bx+c)^n} \tag{5}$$

(where A, B, C, a, b, and c are constants). Here the quadratic polynomial ax^2+bx+c is **irreducible:** It is not a product of linear factors with real coefficients. This is the same as saying that the equation $ax^2+bx+c=0$ has no real roots, and the quadratic formula tells us that this is the case exactly when its discriminant is negative: $b^2-4ac<0$.

Fractions of the forms in Eqs. (4) and (5) are called **partial fractions,** and the sum in Eq. (2) is called the **partial-fraction decomposition** of $R(x)$. Thus Eq. (3) gives the partial-fraction decomposition of $(x^3-1)/(x^3+x)$. A partial fraction of the form in Eq. (4) may be integrated immediately, and we will see in Section 7.7 how to integrate one of the form in Eq. (5).

The first step in finding the partial-fraction decomposition of $R(x)$ is to find the polynomial $p(x)$ in Eq. (2). It turns out that $p(x)\equiv 0$ provided that the degree of the numerator $P(x)$ is *less than* that of the denominator $Q(x)$; such a rational function $R(x)=P(x)/Q(x)$ is said to be **proper.** If $R(x)$ is not proper, then $p(x)$ may be found by dividing $Q(x)$ into $P(x)$, as in Example 2.

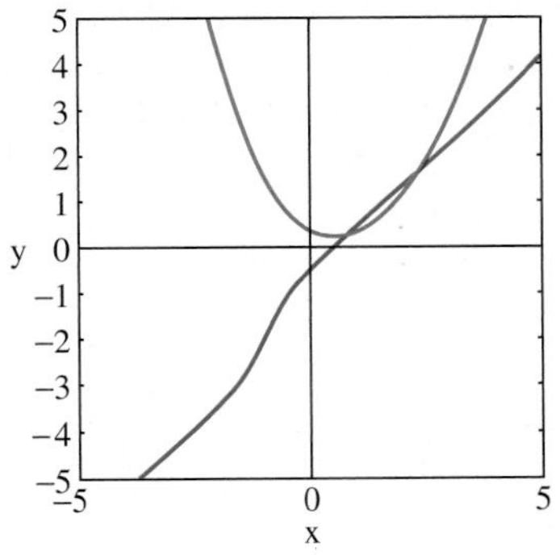

FIGURE 7.5.2 Graphs of the function $f(x)=(x^3+x^2+x-1)/(x^2+2x+2)$ of Example 2 and its indefinite integral with $C=0$. Which is which?

EXAMPLE 2 Find $\displaystyle\int \frac{x^3+x^2+x-1}{x^2+2x+2}\,dx$. (See Fig. 7.5.2.)

Solution Long division of denominator into numerator may be carried out as follows:

$$\begin{array}{rl}
 & x - 1 \longleftarrow\quad p(x) \quad \text{(quotient)}\\
x^2+2x+2\,\big) & x^3 + x^2 + x - 1\\
 & x^3 + 2x^2 + 2x\\
\hline
 & -x^2 - x - 1\\
 & -x^2 - 2x - 2\\
\hline
 & x + 1 \longleftarrow\quad r(x) \quad \text{(remainder)}
\end{array}$$

As in simple arithmetic,

$$\text{“fraction} = \text{quotient} + \frac{\text{remainder}}{\text{divisor}}\text{.”}$$

Thus

$$\frac{x^3+x^2+x-1}{x^2+2x+2} = (x-1) + \frac{x+1}{x^2+2x+2},$$

and hence

$$\begin{aligned}\int \frac{x^3+x^2+x-1}{x^2+2x+2}\,dx &= \int \left(x-1+\frac{x+1}{x^2+2x+2}\right)dx\\ &= \tfrac{1}{2}x^2 - x + \tfrac{1}{2}\ln(x^2+2x+2) + C.\end{aligned}$$

◆

By using long division as in Example 2, any rational function $R(x)$ can be written as a sum of a polynomial $p(x)$ and a *proper* rational fraction,

$$R(x) = p(x) + \frac{r(x)}{Q(x)}.$$

To see how to integrate an arbitrary rational function, we therefore need only see how to find the partial-fraction decomposition of a proper rational fraction.

To obtain such a decomposition, the first step is to factor the denominator $Q(x)$ into a product of linear factors (those of the form $ax + b$) and irreducible quadratic factors (those of the form $ax^2 + bx + c$ with $b^2 - 4ac < 0$). This is always possible in principle but may be difficult in practice. But once we have found the factorization of $Q(x)$, we can obtain the partial-fraction decomposition by routine algebraic methods (described next). Each linear or irreducible quadratic factor of $Q(x)$ leads to one or more partial fractions of the forms in Eqs. (4) and (5).

Linear Factors

Let $R(x) = P(x)/Q(x)$ be a *proper* rational fraction, and suppose that the linear factor $ax + b$ occurs n times in the complete factorization of $Q(x)$. That is, $(ax + b)^n$ is the highest power of $ax + b$ that divides "evenly" into $Q(x)$. In this case we call n the **multiplicity** of the factor $ax + b$.

RULE 1 Linear Factor Partial Fractions

The part of the partial-fraction decomposition of $R(x)$ that corresponds to the linear factor $ax + b$ of multiplicity n is a sum of n partial fractions, specifically

$$\frac{A_1}{ax+b} + \frac{A_2}{(ax+b)^2} + \cdots + \frac{A_n}{(ax+b)^n}, \tag{6}$$

where $A_1, A_2, \ldots, A_n$ are constants.

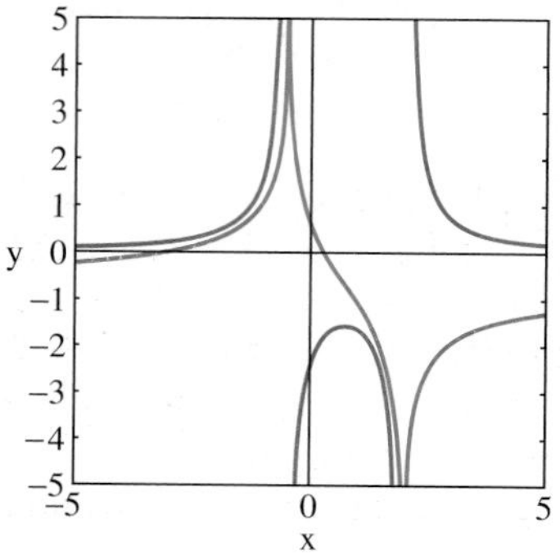

FIGURE 7.5.3 Graphs of the function $f(x) = 5/[(2x+1)(x-2)]$ of Example 3 and its indefinite integral with $C = 0$. Which is which?

If *all* the factors of $Q(x)$ are linear, then the partial-fraction decomposition of $R(x)$ is a sum of expressions like the one in Eq. (6). The situation is especially simple if each of these linear factors is *nonrepeated*—that is, if each has multiplicity $n = 1$. In this case, the expression in (6) reduces to its first term, and the partial-fraction decomposition of $R(x)$ is a sum of such terms. The solutions in Examples 3 and 4 illustrate how the constant numerators can be determined.

EXAMPLE 3 Find $\displaystyle\int \frac{5}{(2x+1)(x-2)}\,dx$. (See Fig. 7.5.3.)

Solution The linear factors in the denominator are distinct, so we seek a partial-fraction decomposition of the form

$$\frac{5}{(2x+1)(x-2)} = \frac{A}{2x+1} + \frac{B}{x-2}.$$

To find the constants A and B, we multiply both sides of this *identity* by the left-hand (common) denominator $(2x+1)(x-2)$. The result is

$$5 = A(x-2) + B(2x+1) = (A + 2B)x + (-2A + B).$$

Next we equate coefficients of x and coefficients of 1 on the left-hand and right-hand sides of this equation. This yields the equations

$$A + 2B = 0,$$
$$-2A + B = 5,$$

which we readily solve for $A = -2$, $B = 1$. Hence

$$\frac{5}{(2x+1)(x-2)} = \frac{-2}{2x+1} + \frac{1}{x-2},$$

and therefore

$$\int \frac{5}{(2x+1)(x-2)}\,dx = -\ln|2x+1| + \ln|x-2| + C = \ln\left|\frac{x-2}{2x+1}\right| + C. \quad ◆$$

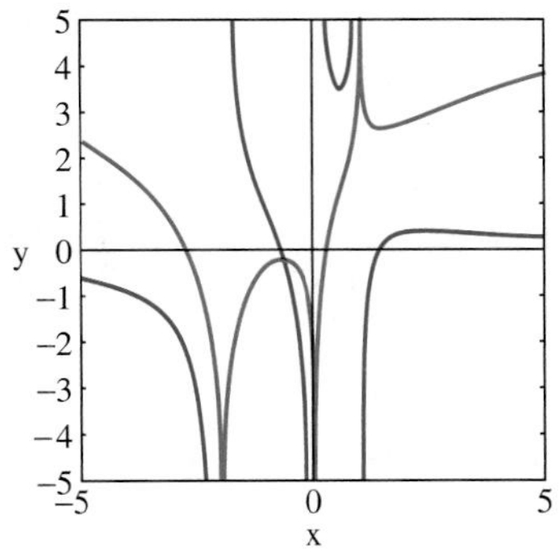

FIGURE 7.5.4 Graphs of the function $f(x) = (4x^2 - 3x - 4)/(x^3 + x^2 - 2x)$ of Example 4 and its indefinite integral with $C = 0$. Which is which?

EXAMPLE 4 Find $\displaystyle\int \frac{4x^2 - 3x - 4}{x^3 + x^2 - 2x}\,dx$. (See Fig. 7.5.4.)

Solution The rational function to be integrated is proper, so we immediately factor its denominator:

$$x^3 + x^2 - 2x = x(x^2 + x - 2) = x(x - 1)(x + 2).$$

We are dealing with three nonrepeated linear factors, so the partial-fraction decomposition has the form

$$\frac{4x^2 - 3x - 4}{x^3 + x^2 - 2x} = \frac{A}{x} + \frac{B}{x - 1} + \frac{C}{x + 2}.$$

To find the constants A, B, and C, we multiply both sides of this equation by the common denominator $x(x - 1)(x + 2)$ and find thereby that

$$4x^2 - 3x - 4 = A(x - 1)(x + 2) + Bx(x + 2) + Cx(x - 1). \qquad \textbf{(7)}$$

Then we collect coefficients of like powers of x on the right:

$$4x^2 - 3x - 4 = (A + B + C)x^2 + (A + 2B - C)x + (-2A).$$

Because two polynomials are (identically) equal only if the coefficients of corresponding powers of x are the same, we conclude that

$$A + B + C = 4,$$
$$A + 2B - C = -3,$$
$$-2A = -4.$$

We solve these simultaneous equations and thus find that $A = 2$, $B = -1$, and $C = 3$.

There is an alternative way to find A, B, and C that is especially effective in the case of nonrepeated linear factors. Substitute the values of $x = 0$, $x = 1$, and $x = -2$ (the zeros of the linear factors of the denominator) into Eq. (7). Substituting $x = 0$ into Eq. (7) immediately gives $-4 = -2A$, so $A = 2$. Substituting $x = 1$ into Eq. (7) gives $-3 = 3B$, so $B = -1$. Substituting $x = -2$ gives $18 = 6C$, so $C = 3$.

With these values of $A = 2$, $B = -1$, and $C = 3$, however obtained, we find that

$$\int \frac{4x^2 - 3x - 4}{x^3 + x^2 - 2x}\,dx = \int \left(\frac{2}{x} - \frac{1}{x - 1} + \frac{3}{x + 2}\right) dx$$
$$= 2\ln|x| - \ln|x - 1| + 3\ln|x + 2| + C.$$

Laws of logarithms allow us to write this antiderivative in the more compact form

$$\int \frac{4x^2 - 3x - 4}{x^3 + x^2 - 2x}\,dx = \ln\left|\frac{x^2(x + 2)^3}{x - 1}\right| + C. \qquad \blacklozenge$$

EXAMPLE 5 Find $\displaystyle\int \frac{x^3 - 4x - 1}{x(x - 1)^3}\,dx$.

Solution Here we have the linear factor x of multiplicity 1 but also the linear factor $x - 1$ of multiplicity 3. According to Rule 1, the partial-fraction decomposition of the integrand has the form

$$\frac{x^3 - 4x - 1}{x(x - 1)^3} = \frac{A}{x} + \frac{B}{x - 1} + \frac{C}{(x - 1)^2} + \frac{D}{(x - 1)^3}.$$

To find the constants A, B, C, and D, we multiply both sides of this equation by the least common denominator $x(x - 1)^3$. We find that

$$x^3 - 4x - 1 = A(x - 1)^3 + Bx(x - 1)^2 + Cx(x - 1) + Dx.$$

We expand and then collect coefficients of like powers of x on the right-hand side. This yields

$$x^3 - 4x + 1 = (A + B)x^3 + (-3A - 2B + C)x^2 + (3A + B - C + D)x - A.$$

Then we equate coefficients of like powers of x on each side of this equation. We get the four simultaneous equations

$$\begin{aligned} A + B &= 1, \\ -3A - 2B + C &= 0, \\ 3A + B - C + D &= -4, \\ -A &= -1. \end{aligned}$$

The last equation gives $A = 1$, and then the first equation gives $B = 0$. Next, the second equation gives $C = 3$. When we substitute these values into the third equation, we finally get $D = -4$. Hence

$$\begin{aligned} \int \frac{x^3 - 4x - 1}{x(x-1)^3}\,dx &= \int \left(\frac{1}{x} + \frac{3}{(x-1)^2} - \frac{4}{(x-1)^3} \right) dx \\ &= \ln|x| - \frac{3}{x-1} + \frac{2}{(x-1)^2} + C. \end{aligned}$$

◆

Quadratic Factors

Suppose that $R(x) = P(x)/Q(x)$ is a proper rational fraction and that the irreducible quadratic factor $ax^2 + bx + c$ occurs n times in the factorization. That is, $(ax^2 + bx + c)^n$ is the highest power of $ax^2 + bx + c$ that divides evenly into $Q(x)$. As before, we call n the **multiplicity** of the quadratic factor $ax^2 + bx + c$.

RULE 2 Quadratic Factor Partial Fractions

The part of the partial-fraction decomposition of $R(x)$ that corresponds to the irreducible quadratic factor $ax^2 + bx + c$ of multiplicity n is a sum of n partial fractions. It has the form

$$\frac{B_1 x + C_1}{ax^2 + bx + c} + \frac{B_2 x + C_2}{(ax^2 + bx + c)^2} + \cdots + \frac{B_n x + C_n}{(ax^2 + bx + c)^n}, \tag{8}$$

where $B_1, B_2, \ldots, B_n, C_1, C_2, \ldots,$ and C_n are constants.

If $Q(x)$ has both linear and irreducible quadratic factors, then the partial-fraction decomposition of $R(x)$ is simply the sum of the expressions of the form in (6) that correspond to the linear factors plus the sum of the expressions of the form in (8) that correspond to the quadratic factors. In the case of an irreducible quadratic factor of multiplicity $n = 1$, the expression in (8) reduces to its first term alone.

The most important case is that of a nonrepeated quadratic factor of the sum of squares form $x^2 + k^2$ (where k is a positive constant). The corresponding partial fraction $(Bx + C)/(x^2 + k^2)$ is readily integrated by using the familiar integrals

$$\int \frac{x}{x^2 + k^2}\,dx = \frac{1}{2}\ln(x^2 + k^2) + C,$$

$$\int \frac{1}{x^2 + k^2}\,dx = \frac{1}{k}\arctan\frac{x}{k} + C.$$

We will discuss in Section 7.7 the integration of more general partial fractions involving irreducible quadratic factors.

EXAMPLE 6 Find $\displaystyle\int \frac{5x^3 - 3x^2 + 2x - 1}{x^4 + x^2}\,dx.$

Solution The denominator $x^4 + x^2 = x^2(x^2 + 1)$ is the product of an irreducible quadratic factor and a repeated linear factor. The partial-fraction decomposition of the integrand takes the form

$$\frac{5x^3 - 3x^2 + 2x - 1}{x^4 + x^2} = \frac{A}{x} + \frac{B}{x^2} + \frac{Cx + D}{x^2 + 1}.$$

We multiply both sides by $x^4 + x^2$ and obtain

$$\begin{aligned} 5x^3 - 3x^2 + 2x - 1 &= Ax(x^2 + 1) + B(x^2 + 1) + (Cx + D)x^2 \\ &= (A + C)x^3 + (B + D)x^2 + Ax + B. \end{aligned}$$

As before, we equate coefficients of like powers of x. This yields the four simultaneous equations

$$\begin{aligned} A + C &= 5, \\ B + D &= -3, \\ A &= 2, \\ B &= -1. \end{aligned}$$

These equations are easily solved for $A = 2$, $B = -1$, $C = 3$, and $D = -2$. Thus

$$\begin{aligned} \int \frac{5x^3 - 3x^2 + 2x - 1}{x^4 + x^2}\,dx &= \int \left(\frac{2}{x} - \frac{1}{x^2} + \frac{3x - 2}{x^2 + 1}\right) dx \\ &= 2\ln|x| + \frac{1}{x} + \frac{3}{2}\int \frac{2x\,dx}{x^2 + 1} - 2\int \frac{dx}{x^2 + 1} \\ &= 2\ln|x| + \frac{1}{x} + \frac{3}{2}\ln(x^2 + 1) - 2\tan^{-1} x + C. \end{aligned}$$ ◆

REMARK Numerical solution for the coefficients in a partial-fraction decomposition frequently is more tedious than in Examples 3 through 6. But computer algebra systems can do this work automatically. For instance, if we write

```
f := (5*x^3 - 3*x^2 + 2*x -1)/(x^4 + x^2)
```

then either the *Mathematica* command `Apart[ f ]` or the *Maple* command

```
convert( f, parfrac, x )
```

quickly produces the partial-fraction decomposition found in Example 6. Figure 7.5.5 shows this decomposition as generated by a graphing calculator.

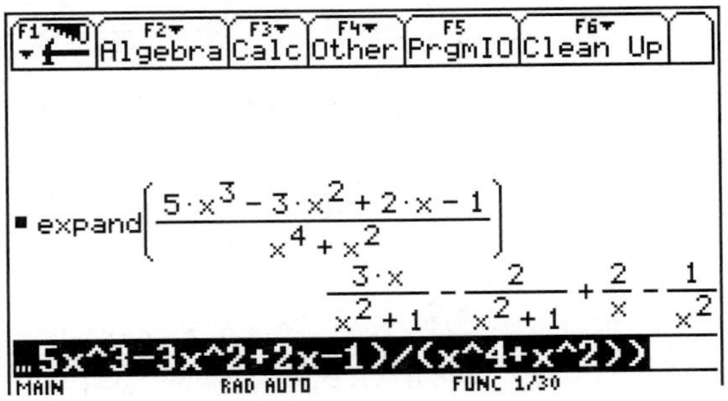

FIGURE 7.5.5 Using the TI-92 **expand** command to produce the partial-fraction decomposition in Example 6.

7.5 TRUE/FALSE STUDY GUIDE

7.5 CONCEPTS: QUESTIONS AND DISCUSSION

1. Suppose that a factorization of the polynomial $Q(x)$ into two or more linear factors of the form $x - a$ is known. Explain why the methods of this section suffice to integrate any rational function of the form $f(x) = P(x)/Q(x)$. Will the integral

$$\int f(x)\,dx$$

ever be a rational function? If so, give an example illustrating this possibility.

2. Suppose that a factorization of the polynomial $Q(x)$ into two or more quadratic factors of the form $x^2 + a^2$ is known. Explain why the methods of this section suffice to integrate any rational function of the form $f(x) = P(x)/Q(x)$. Will the integral

$$\int f(x)\,dx$$

ever be a rational function? If so, give an example illustrating this possibility.

7.5 PROBLEMS

Find the integrals in Problems 1 through 36.

1. $\displaystyle\int \frac{x^2}{x+1}\,dx$
2. $\displaystyle\int \frac{x^3}{2x-1}\,dx$
3. $\displaystyle\int \frac{1}{x^2-3x}\,dx$
4. $\displaystyle\int \frac{x}{x^2+4x}\,dx$
5. $\displaystyle\int \frac{1}{x^2+x-6}\,dx$
6. $\displaystyle\int \frac{x^3}{x^2+x-6}\,dx$
7. $\displaystyle\int \frac{1}{x^3+4x}\,dx$
8. $\displaystyle\int \frac{1}{(x+1)(x^2+1)}\,dx$
9. $\displaystyle\int \frac{x^4}{x^2+4}\,dx$
10. $\displaystyle\int \frac{1}{(x^2+1)(x^2+4)}\,dx$
11. $\displaystyle\int \frac{x-1}{x+1}\,dx$
12. $\displaystyle\int \frac{2x^3-1}{x^2+1}\,dx$
13. $\displaystyle\int \frac{x^2+2x}{(x+1)^2}\,dx$
14. $\displaystyle\int \frac{2x-4}{x^2-x}\,dx$
15. $\displaystyle\int \frac{1}{x^2-4}\,dx$
16. $\displaystyle\int \frac{x^4}{x^2+4x+4}\,dx$
17. $\displaystyle\int \frac{x+10}{2x^2+5x-3}\,dx$
18. $\displaystyle\int \frac{x+1}{x^3-x^2}\,dx$
19. $\displaystyle\int \frac{x^2+1}{x^3+2x^2+x}\,dx$
20. $\displaystyle\int \frac{x^2+x}{x^3-x^2-2x}\,dx$
21. $\displaystyle\int \frac{4x^3-7x}{x^4-5x^2+4}\,dx$
22. $\displaystyle\int \frac{2x^2+3}{x^4-2x^2+1}\,dx$
23. $\displaystyle\int \frac{x^2}{(x+2)^3}\,dx$
24. $\displaystyle\int \frac{x^2+x}{(x^2-4)(x+4)}\,dx$
25. $\displaystyle\int \frac{1}{x^3+x}\,dx$
26. $\displaystyle\int \frac{6x^3-18x}{(x^2-1)(x^2-4)}\,dx$
27. $\displaystyle\int \frac{x+4}{x^3+4x}\,dx$
28. $\displaystyle\int \frac{4x^4+x+1}{x^5+x^4}\,dx$
29. $\displaystyle\int \frac{x}{(x+1)(x^2+1)}\,dx$
30. $\displaystyle\int \left(\frac{x+2}{x^2+4}\right)^2 dx$
31. $\displaystyle\int \frac{x^2-10}{2x^4+9x^2+4}\,dx$
32. $\displaystyle\int \frac{x^2}{x^4-1}\,dx$
33. $\displaystyle\int \frac{x^3+x^2+2x+3}{x^4+5x^2+6}\,dx$
34. $\displaystyle\int \frac{x^2+4}{(x^2+1)^2(x^2+2)}\,dx$
35. $\displaystyle\int \frac{x^4+3x^2-4x+5}{(x-1)^2(x^2+1)}\,dx$
36. $\displaystyle\int \frac{2x^3+5x^2-x+3}{(x^2+x-2)^2}\,dx$

In Problems 37 through 40, make a preliminary substitution before using the method of partial fractions to find the integrals.

37. $\displaystyle\int \frac{e^{4t}}{(e^{2t}-1)^3}\,dt$
38. $\displaystyle\int \frac{\cos\theta}{\sin^2\theta-\sin\theta-6}\,d\theta$
39. $\displaystyle\int \frac{1+\ln t}{t(3+2\ln t)^2}\,dt$
40. $\displaystyle\int \frac{\sec^2 t}{\tan^3 t+\tan^2 t}\,dt$

In Problems 41 through 44, find the area of the region R between the curve and the x-axis over the given interval.

41. $y = \dfrac{x-9}{x^2-3x}$, $1 \leqq x \leqq 2$
42. $y = \dfrac{x+5}{3+2x-x^2}$, $0 \leqq x \leqq 2$
43. $y = \dfrac{3x-15-2x^2}{x^3-9x}$, $1 \leqq x \leqq 2$
44. $y = \dfrac{x^2+10x+16}{x^3+8x^2+16x}$, $2 \leqq x \leqq 5$

In Problems 45 through 48, find the volume of the solid obtained by revolving the region R around the y-axis.

45. The region R of Problem 41
46. The region R of Problem 42
47. The region R of Problem 43
48. The region R of Problem 44

In Problems 49 and 50, find the volume of the solid obtained by revolving the region R around the x-axis.

49. The region R of Problem 41

50. The region R of Problem 42

51. The plane region R shown in Fig. 7.5.6 is bounded by the curve

$$y^2 = \frac{1-x}{1+x}x^2, \quad 0 \leqq x \leqq 1.$$

Find the volume generated by revolving R around the x-axis.

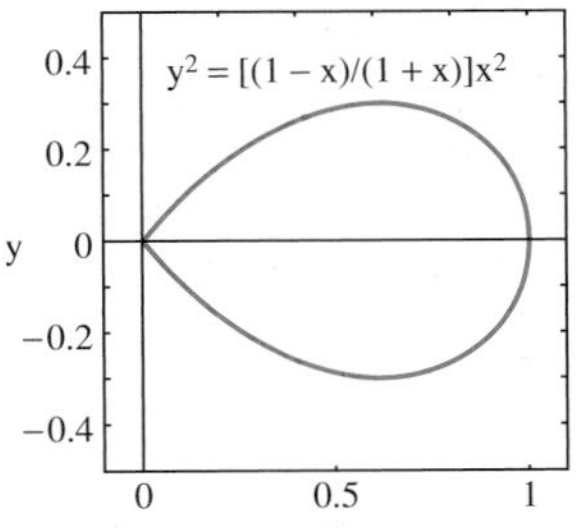

FIGURE 7.5.6 The region of Problem 51.

52. Figure 7.5.7 shows the region bounded by the curve

$$y^2 = \frac{(1-x)^2}{(1+x)^2}x^4, \quad 0 \leqq x \leqq 1.$$

Find the volume generated by revolving this region around: (a) the x-axis; (b) the y-axis.

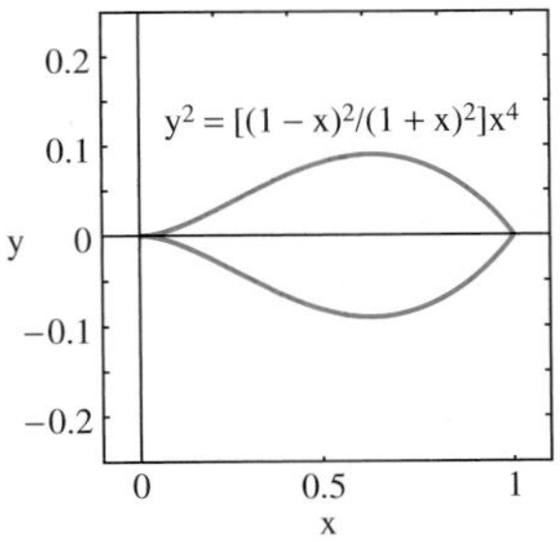

FIGURE 7.5.7 The region of Problem 52.

In Problems 53 through 56, write the general form of a partial-fraction decomposition of the given rational function $f(x)$ (with coefficients A, B, C, ... remaining to be determined). Then use a computer algebra system (as in the remark following Example 6) to find the numerical values of the coefficients in the decomposition. Finally, find the indefinite integral $\int f(x)\,dx$ both by hand and using the computer algebra system, and resolve any apparent discrepancy between the two results.

53. $f(x) = \dfrac{98(x^3 - 50x + 100)}{x^2(x^2 - 12x + 35)}$

54. $f(x) = \dfrac{16(2x^3 + 77x - 99)}{(x^2 + 10x + 21)^2}$

55. $f(x) = \dfrac{324(x^3 + 8)}{(x^2 - x - 6)(x^2 + x - 20)^2}$

56. $f(x) = \dfrac{500(4x^4 - 23x^2 + 16)}{(x^2 - 4)^2(x - 3)^2}$

In Problems 57 and 58, first use a computer algebra system to find a partial-fraction decomposition of the given integrand. Then proceed as in Problems 53 through 56 to evaluate the given integral, both manually and automatically.

57. $\displaystyle\int \frac{2(54x^4 + 859x^2 - 581x + 85)}{18x^5 - 21x^4 + 458x^3 - 526x^2 + 200x - 25}\,dx$

58. $\displaystyle\int \frac{3750x^5 + 125x^4 - 9900x^3 - 495x^2 + 2x - 20}{625x^6 - 2450x^4 - 199x^2 - 4}\,dx$

In Problems 59 through 61, find values of the coefficients a, b, and c (not all zero) such that the given indefinite integral involves no logarithms, and is therefore a rational function.

59. $\displaystyle\int \frac{ax^2 + bx + c}{x^2(x-1)}\,dx$

60. $\displaystyle\int \frac{ax^2 + bx + c}{x^3(x-1)^2}\,dx$

61. $\displaystyle\int \frac{ax^2 + bx + c}{x^3(x-4)^4}\,dx$

7.6 TRIGONOMETRIC SUBSTITUTION

The method of *trigonometric substitution* can be very effective in dealing with integrals when the integrands contain algebraic expressions such as $(a^2 - u^2)^{1/2}$, $(u^2 - a^2)^{3/2}$, and $1/(a^2 + u^2)^2$. There are three basic trigonometric substitutions:

If the integral involves	then substitute	and use the identity
$a^2 - u^2$	$u = a\sin\theta$	$1 - \sin^2\theta = \cos^2\theta$
$a^2 + u^2$	$u = a\tan\theta$	$1 + \tan^2\theta = \sec^2\theta$
$u^2 - a^2$	$u = a\sec\theta$	$\sec^2\theta - 1 = \tan^2\theta$

What we mean by the substitution $u = a\sin\theta$ is, more precisely, the *inverse* trigonometric substitution

$$\theta = \sin^{-1}\frac{u}{a}, \qquad -\frac{\pi}{2} \leqq \theta \leqq \frac{\pi}{2},$$

where $|u| \leqq a$. Suppose, for example, that an integral contains the expression $(a^2 - u^2)^{1/2}$. Then this substitution yields

$$(a^2 - u^2)^{1/2} = (a^2 - a^2\sin^2\theta)^{1/2} = (a^2\cos^2\theta)^{1/2} = a\cos\theta.$$

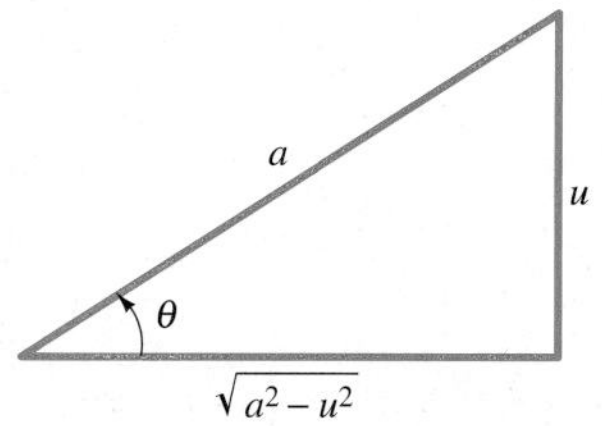

FIGURE 7.6.1 The reference triangle for the substitution $u = a\sin\theta$.

We chose the nonnegative square root in the last step because $\cos\theta \geqq 0$ for $-\pi/2 \leqq \theta \leqq \pi/2$. Thus the troublesome factor $(a^2 - u^2)^{1/2}$ becomes $a\cos\theta$ and, meanwhile, $du = a\cos\theta\,d\theta$. If the trigonometric integral that results from the substitution can be evaluated by earlier methods of this chapter, the result will ordinarily involve $\theta = \sin^{-1}(u/a)$ and trigonometric functions of θ. The final step will be to express the answer in terms of the original variable. For this purpose the values of the various trigonometric functions can be read from the right triangle in Fig. 7.6.1, which contains an angle θ such that $\sin\theta = u/a$ (if u is negative, then θ is negative).

EXAMPLE 1 Evaluate $\displaystyle\int \frac{x^3}{\sqrt{1-x^2}}\,dx$, where $|x| < 1$.

Solution Here $a = 1$ and $u = x$, so we substitute

$$x = \sin\theta, \qquad dx = \cos\theta\,d\theta.$$

This gives

$$\begin{aligned}\int \frac{x^3}{\sqrt{1-x^2}}\,dx &= \int \frac{\sin^3\theta\cos\theta}{\sqrt{1-\sin^2\theta}}\,d\theta\\ &= \int \sin^3\theta\,d\theta = \int (\sin\theta)(1-\cos^2\theta)\,d\theta\\ &= \tfrac{1}{3}\cos^3\theta - \cos\theta + C.\end{aligned}$$

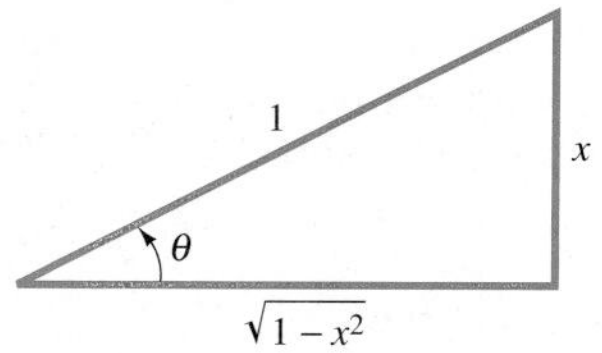

FIGURE 7.6.2 Reference triangle for the substitution $x = \sin\theta$.

Figure 7.6.2, in which $\sin\theta = x$, reminds us that

$$\cos\theta = (1-\sin^2\theta)^{1/2} = \sqrt{1-x^2},$$

so our final result in terms of x is

$$\int \frac{x^3}{\sqrt{1-x^2}}\,dx = \frac{1}{3}(1-x^2)^{3/2} - \sqrt{1-x^2} + C. \qquad \blacklozenge$$

Example 2 illustrates the use of trigonometric substitution to find integrals like those in Formulas (44) through (62) inside the back cover.

EXAMPLE 2 Find $\int \sqrt{a^2 - u^2}\,du$, where $|u| \leqq a$.

Solution The substitution $u = a\sin\theta$, $du = a\cos\theta\,d\theta$ gives

$$\begin{aligned}\int \sqrt{a^2-u^2}\,du &= \int \sqrt{a^2 - a^2\sin^2\theta}\,(a\cos\theta)\,d\theta\\ &= \int a^2\cos^2\theta\,d\theta = \tfrac{1}{2}a^2\int (1+\cos 2\theta)\,d\theta\\ &= \tfrac{1}{2}a^2\left(\theta + \tfrac{1}{2}\sin 2\theta\right) + C = \tfrac{1}{2}a^2(\theta + \sin\theta\cos\theta) + C.\end{aligned}$$

(We used the identity $\sin 2\theta = 2\sin\theta\cos\theta$ in the last step.) Now from Fig. 7.6.1 we see that

$$\sin\theta = \frac{u}{a} \quad \text{and} \quad \cos\theta = \frac{\sqrt{a^2-u^2}}{a}.$$

Hence

$$\begin{aligned}\int \sqrt{a^2-u^2}\,du &= \frac{1}{2}a^2\left(\sin^{-1}\frac{u}{a} + \frac{u}{a}\cdot\frac{\sqrt{a^2-u^2}}{a}\right) + C\\ &= \frac{u}{2}\sqrt{a^2-u^2} + \frac{a^2}{2}\sin^{-1}\frac{u}{a} + C.\end{aligned}$$

Thus we have obtained Formula (54) inside the back cover. ◆

What we mean by the substitution $u = a\tan\theta$ in an integral that contains a^2+u^2 is the substitution

$$\theta = \tan^{-1}\frac{u}{a}, \qquad -\frac{\pi}{2} < \theta < \frac{\pi}{2}.$$

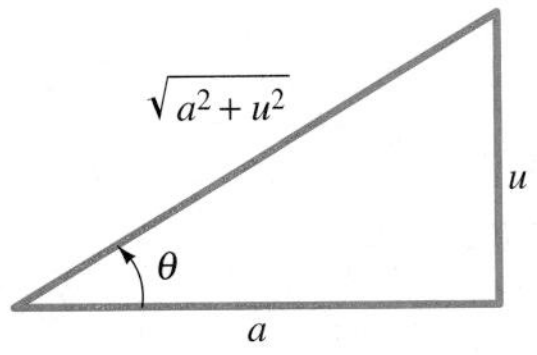

FIGURE 7.6.3 The reference triangle for the substitution $u = a\tan\theta$.

In this case

$$\sqrt{a^2+u^2} = \sqrt{a^2+a^2\tan^2\theta} = \sqrt{a^2\sec^2\theta} = a\sec\theta,$$

under the assumption that $a > 0$. We take the positive square root in the last step here because $\sec\theta > 0$ for $-\pi/2 < \theta < \pi/2$. The values of the various trigonometric functions of θ under this substitution can be read from the right triangle of Fig. 7.6.3, which shows a [positive or negative] acute angle θ such that $\tan\theta = u/a$.

EXAMPLE 3 Find $\displaystyle\int \frac{1}{(4x^2+9)^2}\,dx.$

Solution The factor $4x^2+9$ corresponds to u^2+a^2 with $u = 2x$ and $a = 3$. Hence the substitution $u = a\tan\theta$ amounts to

$$2x = 3\tan\theta, \qquad x = \tfrac{3}{2}\tan\theta, \qquad dx = \tfrac{3}{2}\sec^2\theta\,d\theta.$$

This gives

$$\begin{aligned}\int \frac{1}{(4x^2+9)^2}\,dx &= \int \frac{\frac{3}{2}\sec^2\theta}{(9\tan^2\theta+9)^2}\,d\theta\\ &= \tfrac{3}{2}\int \frac{\sec^2\theta}{(9\sec^2\theta)^2}\,d\theta = \tfrac{1}{54}\int \frac{1}{\sec^2\theta}\,d\theta\\ &= \tfrac{1}{54}\int \cos^2\theta\,d\theta = \tfrac{1}{108}(\theta + \sin\theta\cos\theta) + C.\end{aligned}$$

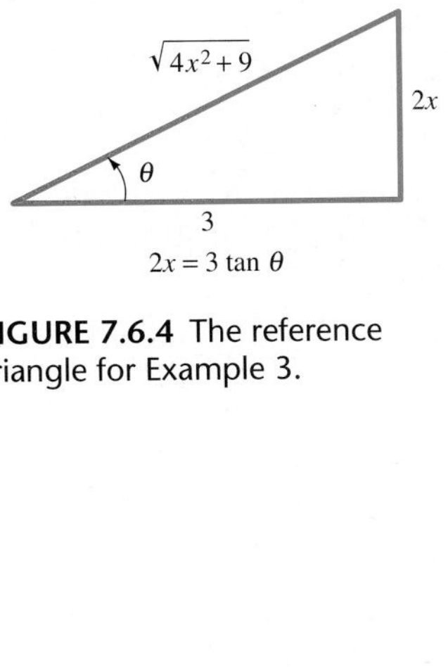

FIGURE 7.6.4 The reference triangle for Example 3.

(The integration in the last step is the same as in Example 2.) Now $\theta = \tan^{-1}(2x/3)$, and the triangle of Fig. 7.6.4 gives

$$\sin\theta = \frac{2x}{\sqrt{4x^2+9}}, \qquad \cos\theta = \frac{3}{\sqrt{4x^2+9}}.$$

Hence

$$\begin{aligned}\int \frac{1}{(4x^2+9)^2}\,dx &= \frac{1}{108}\left[\tan^{-1}\left(\frac{2x}{3}\right) + \frac{2x}{\sqrt{4x^2+9}}\cdot\frac{3}{\sqrt{4x^2+9}}\right] + C\\ &= \frac{1}{108}\tan^{-1}\left(\frac{2x}{3}\right) + \frac{x}{18(4x^2+9)} + C.\end{aligned}$$

◆

What we mean by the substitution $u = a\sec\theta$ in an integral that contains $u^2 - a^2$ is the substitution

$$\theta = \sec^{-1}\frac{u}{a}, \quad 0 \leqq \theta \leqq \pi,$$

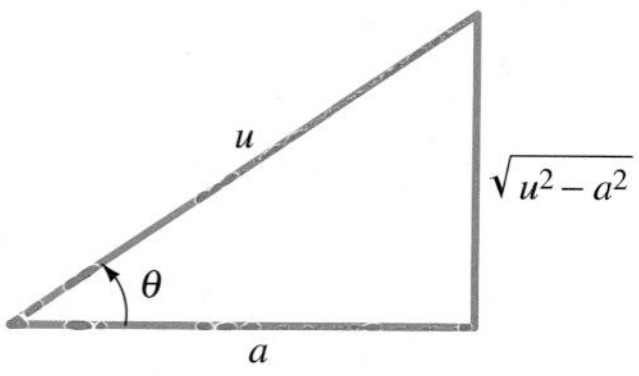

FIGURE 7.6.5 The reference triangle for the substitution $u = a\sec\theta$.

where $|u| \geqq a > 0$ (because of the domain and range of the inverse secant function). Then

$$\sqrt{u^2 - a^2} = \sqrt{a^2\sec^2\theta - a^2} = \sqrt{a^2\tan^2\theta} = \pm a\tan\theta.$$

Here we must take the plus sign if $u > a$, so that $0 < \theta < \pi/2$ and $\tan\theta > 0$. If $u < -a$, so $\pi/2 < \theta < \pi$ and $\tan\theta < 0$, we take the minus sign. In either case the values of the various trigonometric functions of θ can be read from the right triangle in Fig. 7.6.5.

EXAMPLE 4 Find $\displaystyle\int \frac{\sqrt{x^2 - 25}}{x}\,dx$, where $x > 5$.

Solution We substitute $x = 5\sec\theta$, $dx = 5\sec\theta\tan\theta\,d\theta$. Then

$$\sqrt{x^2 - 25} = \sqrt{25(\sec^2\theta - 1)} = 5\tan\theta,$$

because $x > 5$ implies that $0 < \theta < \pi/2$, so $\tan\theta > 0$. Hence this substitution gives

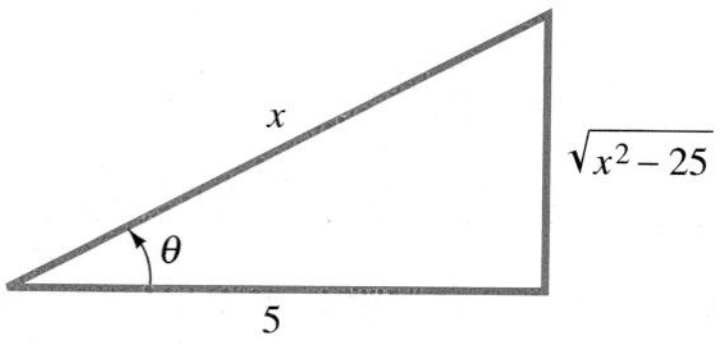

FIGURE 7.6.6 The reference triangle for the substitution $x = 5\sec\theta$.

$$\begin{aligned}\int \frac{\sqrt{x^2 - 25}}{x}\,dx &= \int \frac{5\tan\theta}{5\sec\theta}(5\sec\theta\tan\theta)\,d\theta \\ &= 5\int \tan^2\theta\,d\theta = 5\int(\sec^2\theta - 1)\,d\theta \\ &= 5\tan\theta - 5\theta + C = \sqrt{x^2 - 25} - 5\sec^{-1}\left(\frac{x}{5}\right) + C.\end{aligned}$$

The substitutions in the last step may be read from the reference triangle in Fig. 7.6.6. ◆

Hyperbolic substitutions may be used in a similar way—and with the same effect—as trigonometric substitutions. The three basic hyperbolic substitutions, which are not ordinarily memorized, are listed in the following table for reference.

If the integral involves	then substitute	and use the identity
$a^2 - u^2$	$u = a\tanh\theta$	$1 - \tanh^2\theta = \operatorname{sech}^2\theta$
$a^2 + u^2$	$u = a\sinh\theta$	$1 + \sinh^2\theta = \cosh^2\theta$
$u^2 - a^2$	$u = a\cosh\theta$	$\cosh^2\theta - 1 = \sinh^2\theta$

EXAMPLE 5 Find $\displaystyle\int \frac{1}{\sqrt{x^2 - 1}}\,dx$, where $x > 1$.

Solution For purposes of comparison, we evaluate this integral both by trigonometric substitution and by hyperbolic substitution. The trigonometric substitution

$$x = \sec\theta, \qquad dx = \sec\theta\tan\theta\,d\theta, \qquad \tan\theta = \sqrt{x^2 - 1}$$

gives

$$\begin{aligned}\int \frac{1}{\sqrt{x^2 - 1}}\,dx &= \int \frac{\sec\theta\tan\theta}{\tan\theta}\,d\theta = \int \sec\theta\,d\theta \\ &= \ln|\sec\theta + \tan\theta| + C \qquad \text{[Eq. (7), Section 7.4]} \\ &= \ln\left|x + \sqrt{x^2 - 1}\right| + C.\end{aligned}$$

Using instead the hyperbolic substitution $x = \cosh\theta$, $dx = \sinh\theta\,d\theta$, we have

$$\sqrt{x^2 - 1} = \sqrt{\cosh^2\theta - 1} = \sinh\theta.$$

We take the positive square root here, because $x > 1$ implies that $\theta = \cosh^{-1} x > 0$ and thus that $\sinh\theta > 0$. Hence

$$\int \frac{1}{\sqrt{x^2-1}}\,dx = \int \frac{\sinh\theta}{\sinh\theta}\,d\theta = \int 1\,d\theta = \theta + C = \cosh^{-1} x + C.$$

The two results appear to differ, but Eq. (35) in Section 6.9 shows that they are equivalent. ◆

7.6 TRUE/FALSE STUDY GUIDE

7.6 CONCEPTS: QUESTIONS AND DISCUSSION

1. For each of the three trigonometric substitutions discussed in this section, describe—perhaps by means of examples—the types of integrals for which substitution permits evaluation.
2. Explain why these trigonometric substitutions do not appear to suffice for the evaluation of integrals such as

$$\int \sqrt{1+x^3}\,dx \quad \text{and} \quad \int \sqrt{x^4-1}\,dx.$$

7.6 PROBLEMS

Use trigonometric substitutions to evaluate the integrals in Problems 1 through 36.

1. $\displaystyle\int \frac{1}{\sqrt{16-x^2}}\,dx$

2. $\displaystyle\int \frac{1}{\sqrt{4-9x^2}}\,dx$

3. $\displaystyle\int \frac{1}{x^2\sqrt{4-x^2}}\,dx$

4. $\displaystyle\int \frac{1}{x^2\sqrt{x^2-25}}\,dx$

5. $\displaystyle\int \frac{x^2}{\sqrt{16-x^2}}\,dx$

6. $\displaystyle\int \frac{x^2}{\sqrt{9-4x^2}}\,dx$

7. $\displaystyle\int \frac{1}{(9-16x^2)^{3/2}}\,dx$

8. $\displaystyle\int \frac{1}{(25+16x^2)^{3/2}}\,dx$

9. $\displaystyle\int \frac{\sqrt{x^2-1}}{x^2}\,dx$

10. $\displaystyle\int x^3\sqrt{4-x^2}\,dx$

11. $\displaystyle\int x^3\sqrt{9+4x^2}\,dx$

12. $\displaystyle\int \frac{x^3}{\sqrt{x^2+25}}\,dx$

13. $\displaystyle\int \frac{\sqrt{1-4x^2}}{x}\,dx$

14. $\displaystyle\int \frac{1}{\sqrt{1+x^2}}\,dx$

15. $\displaystyle\int \frac{1}{\sqrt{9+4x^2}}\,dx$

16. $\displaystyle\int \sqrt{1+4x^2}\,dx$

17. $\displaystyle\int \frac{x^2}{\sqrt{25-x^2}}\,dx$

18. $\displaystyle\int \frac{x^3}{\sqrt{25-x^2}}\,dx$

19. $\displaystyle\int \frac{x^2}{\sqrt{1+x^2}}\,dx$

20. $\displaystyle\int \frac{x^3}{\sqrt{1+x^2}}\,dx$

21. $\displaystyle\int \frac{x^2}{\sqrt{4+9x^2}}\,dx$

22. $\displaystyle\int (1-x^2)^{3/2}\,dx$

23. $\displaystyle\int \frac{1}{(1+x^2)^{3/2}}\,dx$

24. $\displaystyle\int \frac{1}{(4-x^2)^2}\,dx$

25. $\displaystyle\int \frac{1}{(4-x^2)^3}\,dx$

26. $\displaystyle\int \frac{1}{(4x^2+9)^3}\,dx$

27. $\displaystyle\int \sqrt{9+16x^2}\,dx$

28. $\displaystyle\int (9+16x^2)^{3/2}\,dx$

29. $\displaystyle\int \frac{\sqrt{x^2-25}}{x}\,dx$

30. $\displaystyle\int \frac{\sqrt{9x^2-16}}{x}\,dx$

31. $\displaystyle\int x^2\sqrt{x^2-1}\,dx$

32. $\displaystyle\int \frac{x^2}{\sqrt{4x^2-9}}\,dx$

33. $\displaystyle\int \frac{1}{(4x^2-1)^{3/2}}\,dx$

34. $\displaystyle\int \frac{1}{x^2\sqrt{4x^2-9}}\,dx$

35. $\displaystyle\int \frac{\sqrt{x^2-5}}{x^2}\,dx$

36. $\displaystyle\int (4x^2-5)^{3/2}\,dx$

Use hyperbolic substitutions to evaluate the integrals in Problems 37 through 41.

37. $\displaystyle\int \frac{1}{\sqrt{25+x^2}}\,dx$

38. $\displaystyle\int \sqrt{1+x^2}\,dx$

39. $\displaystyle\int \frac{\sqrt{x^2-4}}{x^2}\,dx$

40. $\displaystyle\int \frac{1}{\sqrt{1+9x^2}}\,dx$

41. $\displaystyle\int x^2\sqrt{1+x^2}\,dx$

42. Use the result of Example 2 to show that the area bounded by the ellipse

$$\frac{x^2}{a^2}+\frac{y^2}{b^2}=1$$

of Fig. 7.6.7 is given by $A = \pi ab$. (The special case $b = a$ is the familiar circular area formula $A = \pi a^2$.)

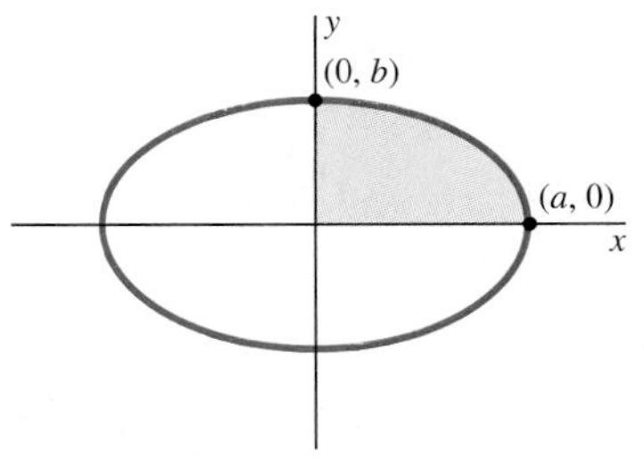

FIGURE 7.6.7 The ellipse of Problem 42.

43. Derive the formula $A = \frac{1}{2}a^2\theta$ for the area of a circular sector with radius a and central angle θ by calculating and adding the areas of the right triangle OAC and the region ABC of Fig. 7.6.8.

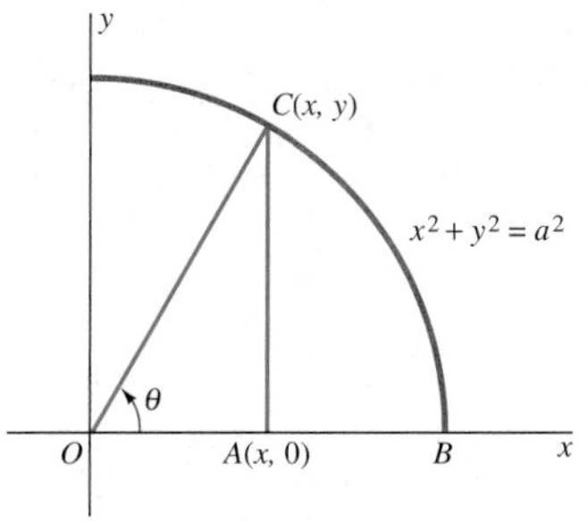

FIGURE 7.6.8 The circular sector of Problem 43.

44. Compute the arc length of the parabola $y = x^2$ over the interval $[0, 1]$.

45. Compute the area of the surface obtained by revolving around the x-axis the parabolic arc of Problem 44.

46. Show that the length of one arch of the sine curve $y = \sin x$ is equal to half the circumference of the ellipse $x^2 + \frac{1}{2}y^2 = 1$. [*Suggestion:* Substitute $x = \cos\theta$ into the arc-length integral for the ellipse.] (See Fig. 7.6.9.)

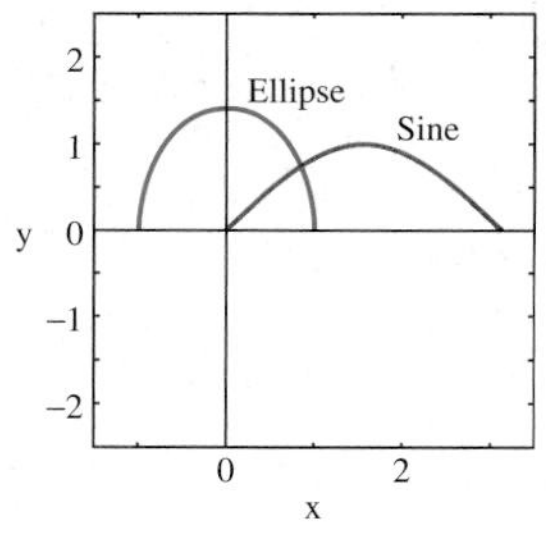

FIGURE 7.6.9 Two arcs with the same length (Problem 46).

47. Compute the arc length of the curve $y = \ln x$ over the interval $[1, 2]$.

48. Compute the area of the surface obtained by revolving around the y-axis the curve of Problem 47.

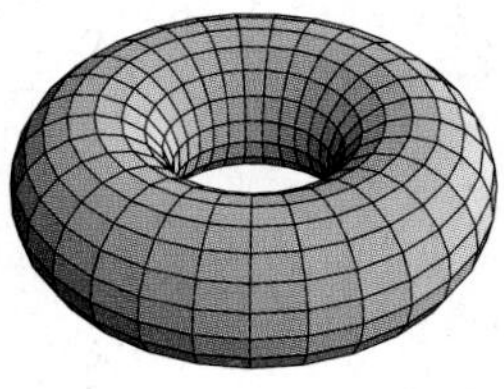

FIGURE 7.6.10 The torus of Problem 49.

49. A torus (see Fig. 7.6.10) is obtained by revolving around the y-axis the circle

$$(x - b)^2 + y^2 = a^2 \quad (0 < a \leqq b).$$

Show that the surface area of the torus is $4\pi^2 ab$.

50. Find the area under the curve $y = \sqrt{9 + x^2}$ over the interval $[0, 4]$.

51. Find the area of the surface obtained by revolving around the x-axis the curve $y = \sin x$, $0 \leqq x \leqq \pi$. (See Fig. 7.6.11.)

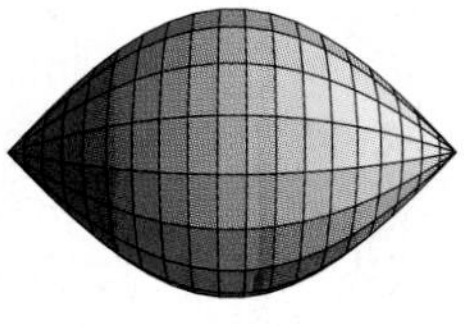

FIGURE 7.6.11 The pointed football of Problem 51.

52. An ellipsoid of revolution is obtained by revolving the ellipse $x^2/a^2 + y^2/b^2 = 1$ around the x-axis. Suppose that $a > b$. Show that the ellipsoid has surface area

$$A = 2\pi ab\left[\frac{b}{a} + \frac{a}{c}\sin^{-1}\left(\frac{c}{a}\right)\right],$$

where $c = \sqrt{a^2 - b^2}$. Assume that $a \approx b$, so that $c \approx 0$ and $\sin^{-1}(c/a) \approx c/a$. Conclude that $A \approx 4\pi a^2$.

53. Suppose that $b > a$ for the ellipsoid of revolution of Problem 52. Show that its surface area is then

$$A = 2\pi ab\left[\frac{b}{a} + \frac{a}{c}\ln\left(\frac{b + c}{a}\right)\right],$$

where $c = \sqrt{b^2 - a^2}$. Use the fact that $\ln(1+x) \approx x$ if $x \approx 0$, and thereby conclude that $A \approx 4\pi a^2$ if $a \approx b$.

54. A road is to be built from the point $(2, 1)$ to the point $(5, 3)$, following the path of the parabola

$$y = -1 + 2\sqrt{x - 1}.$$

Calculate the length of this road (the units on the coordinate axes are in miles). [*Suggestion:* Substitute $x = \sec^2\theta$ into the arc-length integral.]

55. Suppose that the cost of the road in Problem 54 is $\sqrt{x}$ million dollars per mile. Calculate the total cost of the road.

56. A kite is flying at a height of 500 ft and at a horizontal distance of 100 ft from the stringholder on the ground. The kite string weighs 1/16 oz/ft and is hanging in the shape of the parabola $y = x^2/20$ that joins the stringholder at $(0, 0)$ to the kite at $(100, 500)$ (Fig. 7.6.12). Calculate the work (in foot-pounds) done in lifting the kite string from the ground to its present position.

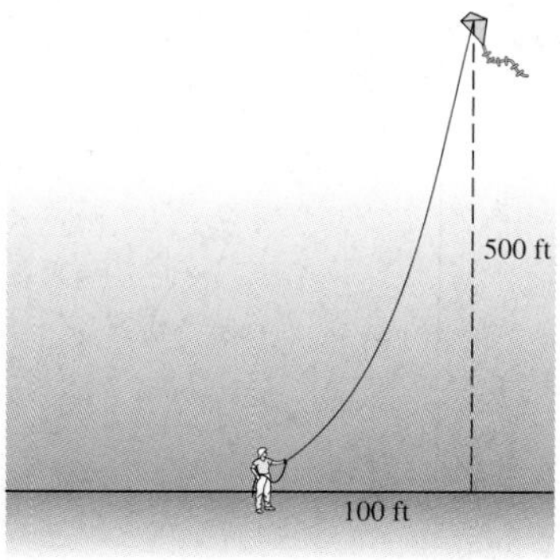

FIGURE 7.6.12 The kite string of Problem 56.

7.7 INTEGRALS INVOLVING QUADRATIC POLYNOMIALS

Many integrals containing a square root or negative power of a quadratic polynomial $ax^2 + bx + c$ can be simplified by the process of *completing the square.* For example,

$$x^2 + 2x + 2 = (x+1)^2 + 1,$$

and hence the substitution $u = x + 1$, $du = dx$ yields

$$\int \frac{1}{x^2 + 2x + 2}\,dx = \int \frac{1}{u^2 + 1}\,du = \tan^{-1} u + C = \tan^{-1}(x+1) + C.$$

In general, the objective is to convert $ax^2 + bx + c$ into either a sum or difference of two squares—either $u^2 \pm a^2$ or $a^2 - u^2$—so that the method of trigonometric substitution can then be used. To see how this works in practice, suppose first that $a = 1$, so that the quadratic in question has the form $x^2 + bx + c$. The sum $x^2 + bx$ of the first two terms can be completed to a perfect square by adding $b^2/4$, the square of half the coefficient of x, and in turn subtracting $b^2/4$ from the constant term c. This gives

$$\begin{aligned} x^2 + bx + c &= \left(x^2 + bx + \frac{b^2}{4}\right) + \left(c - \frac{b^2}{4}\right) \\ &= \left(x + \frac{b}{2}\right)^2 + \left(c - \frac{b^2}{4}\right). \end{aligned}$$

With $u = x + \frac{1}{2}b$, this result is of the form $u^2 + A^2$ or $u^2 - A^2$, depending on the sign of $c - \frac{1}{4}b^2$. If the coefficient a of x^2 is not 1, we first factor it out and proceed as before:

$$ax^2 + bx + c = a\left(x^2 + \frac{b}{a}x + \frac{c}{a}\right).$$

EXAMPLE 1 Find $\displaystyle\int \frac{1}{9x^2 + 6x + 5}\,dx$.

Solution The first step is to complete the square:

$$\begin{aligned} 9x^2 + 6x + 5 &= 9\left(x^2 + \tfrac{2}{3}x\right) + 5 = 9\left(x^2 + \tfrac{2}{3}x + \tfrac{1}{9}\right) - 1 + 5 \\ &= 9\left(x + \tfrac{1}{3}\right)^2 + 4 = (3x+1)^2 + 2^2. \end{aligned}$$

Hence

$$\begin{aligned} \int \frac{1}{9x^2 + 6x + 5}\,dx &= \int \frac{1}{(3x+1)^2 + 4}\,dx \\ &= \frac{1}{3}\int \frac{1}{u^2 + 4}\,du \qquad (u = 3x + 1) \\ &= \frac{1}{6}\int \frac{\frac{1}{2}}{\left(\frac{1}{2}u\right)^2 + 1}\,du \\ &= \frac{1}{6}\int \frac{1}{v^2 + 1}\,dv \qquad \left(v = \tfrac{1}{2}u\right) \\ &= \frac{1}{6}\tan^{-1} v + C = \frac{1}{6}\tan^{-1}\left(\frac{u}{2}\right) + C \\ &= \frac{1}{6}\tan^{-1}\left(\frac{3x+1}{2}\right) + C. \end{aligned}$$

◆

EXAMPLE 2 Find $\displaystyle\int \frac{1}{\sqrt{9+16x-4x^2}}\,dx.$

Solution First we complete the square:

$$\begin{aligned} 9+16x-4x^2 &= 9-4(x^2-4x) \\ &= 9-4(x^2-4x+4)+16 = 25-4(x-2)^2. \end{aligned}$$

Hence

$$\begin{aligned} \int \frac{1}{\sqrt{9+16x-4x^2}}\,dx &= \int \frac{1}{\sqrt{25-4(x-2)^2}}\,dx \\ &= \frac{1}{5}\int \frac{1}{\sqrt{1-\frac{4}{25}(x-2)^2}}\,dx \\ &= \frac{1}{2}\int \frac{1}{\sqrt{1-u^2}}\,du \qquad \left(u=\tfrac{2}{5}(x-2)\right) \\ &= \tfrac{1}{2}\sin^{-1}u + C = \tfrac{1}{2}\sin^{-1}\left(\tfrac{2}{5}(x-2)\right)+C. \end{aligned}$$

An alternative approach is to make the trigonometric substitution

$$2(x-2) = 5\sin\theta, \qquad 2\,dx = 5\cos\theta\,d\theta$$

immediately after completing the square. This yields

$$\begin{aligned} \int \frac{1}{\sqrt{9+16x-4x^2}}\,dx &= \int \frac{1}{\sqrt{25-4(x-2)^2}}\,dx \\ &= \int \frac{\frac{5}{2}\cos\theta}{\sqrt{25-25\sin^2\theta}}\,d\theta \\ &= \frac{1}{2}\int 1\,d\theta = \frac{1}{2}\theta + C \\ &= \frac{1}{2}\arcsin\frac{2(x-2)}{5} + C. \end{aligned}$$

◆

Some integrals that contain a quadratic expression can be split into two simpler integrals. Examples 3 and 4 illustrate this technique.

EXAMPLE 3 Find $\displaystyle\int \frac{2x+3}{9x^2+6x+5}\,dx.$

Solution Because $D_x(9x^2+6x+5) = 18x+6$, this would be a simpler integral if the numerator $2x+3$ were a constant multiple of $18x+6$. Our strategy is to write

$$2x+3 = A\cdot(18x+6)+B$$

so that we can split the given integral into a sum of two integrals, one of which has numerator $18x+6$ in its integrand. By matching coefficients in

$$2x+3 = 18Ax + (6A+B),$$

we find that $A = \frac{1}{9}$ and $B = \frac{7}{3}$. Hence

$$\int \frac{2x+3}{9x^2+6x+5}\,dx = \frac{1}{9}\int \frac{18x+6}{9x^2+6x+5}\,dx + \frac{7}{3}\int \frac{1}{9x^2+6+5}\,dx.$$

The first integral on the right is a logarithm, and the second is given in Example 1. Thus

$$\int \frac{2x+3}{9x^2+6x+5}\,dx = \frac{1}{9}\ln(9x^2+6x+5) + \frac{7}{18}\tan^{-1}\left(\frac{3x+1}{2}\right) + C.$$

Alternatively, we could first complete the square in the denominator. The substitution $u = 3x + 1$, $x = \frac{1}{3}(u - 1)$, $dx = \frac{1}{3}\,du$ then gives

$$\begin{aligned}\int \frac{2x+3}{(3x+1)^2+4}\,dx &= \int \frac{\frac{2}{3}(u-1)+3}{u^2+4}\cdot\frac{1}{3}\,du\\ &= \frac{1}{9}\int\frac{2u}{u^2+4}\,du + \frac{7}{9}\int\frac{1}{u^2+4}\,du\\ &= \frac{1}{9}\ln(u^2+4) + \frac{7}{18}\tan^{-1}\left(\frac{u}{2}\right) + C\\ &= \frac{1}{9}\ln(9x^2+6x+5) + \frac{7}{18}\tan^{-1}\left(\frac{3x+1}{2}\right) + C.\end{aligned}$$

◆

EXAMPLE 4 Find $\displaystyle\int \frac{2+6x}{(3+2x-x^2)^2}\,dx$ given $|x-1| < 2$.

Solution Because $D_x(3+2x-x^2) = 2-2x$, we first write

$$\int \frac{2+6x}{(3+2x-x^2)^2}\,dx = -3\int\frac{2-2x}{(3+2x-x^2)^2}\,dx + 8\int\frac{1}{(3+2x-x^2)^2}\,dx.$$

Then let $u = 3+2x-x^2$, $du = (2-2x)\,dx$ in the first integral to obtain

$$-3\int\frac{2-2x}{(3+2x-x^2)^2}\,dx = -3\int\frac{du}{u^2} = \frac{3}{u} + C_1 = \frac{3}{3+2x-x^2} + C_1.$$

Therefore,

$$\int \frac{2+6x}{(3+2x-x^2)^2}\,dx = \frac{3}{3+2x-x^2} + 8\int\frac{1}{(3+2x-x^2)^2}\,dx. \quad \textbf{(1)}$$

(We can drop the constant C_1 because it can be absorbed by the constant C, which we will obtain when we evaluate the remaining integral.) To evaluate the remaining integral, we complete the square:

$$3+2x-x^2 = 4-(x^2-2x+1) = 4-(x-1)^2.$$

Because $|x-1| < 2$, this suggests the substitution

$$x - 1 = 2\sin\theta, \qquad dx = 2\cos\theta\,d\theta,$$

with which

$$3+2x-x^2 = 4-4\sin^2\theta = 4\cos^2\theta.$$

This substitution yields

$$\begin{aligned}8\int\frac{1}{(3+2x-x^2)^2}\,dx &= 8\int\frac{2\cos\theta}{(4\cos^2\theta)^2}\,d\theta = \int\sec^3\theta\,d\theta\\ &= \tfrac{1}{2}\sec\theta\tan\theta + \tfrac{1}{2}\int\sec\theta\,d\theta \qquad \text{[Section 7.3, Eq. (5)]}\\ &= \tfrac{1}{2}\sec\theta\tan\theta + \tfrac{1}{2}\ln|\sec\theta+\tan\theta| + C\\ &= \frac{x-1}{3+2x-x^2} + \tfrac{1}{2}\ln\left|\frac{x+1}{\sqrt{3+2x-x^2}}\right| + C. \quad \textbf{(2)}\end{aligned}$$

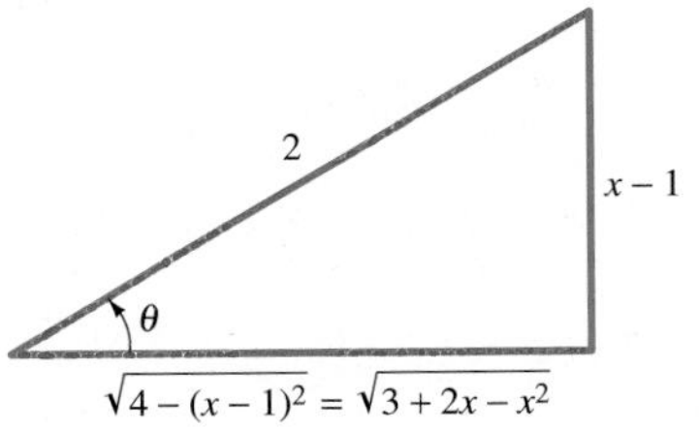

FIGURE 7.7.1 The reference triangle for Example 4.

In the last step we read the values of $\sec\theta$ and $\tan\theta$ from the right triangle in Fig. 7.7.1. When we substitute Eq. (2) into Eq. (1), we finally obtain the result

$$\int\frac{2+6x}{(3+2x-x^2)^2}\,dx = \frac{x+2}{3+2x-x^2} + \frac{1}{2}\ln\left|\frac{x+1}{\sqrt{3+2x-x^2}}\right| + C.$$

◆

The method of Example 4 can be used to evaluate a general integral of the form

$$\int \frac{Ax+B}{(ax^2+bx+c)^n}\,dx, \tag{3}$$

where n is a positive integer. By splitting such an integral into two simpler ones and then completing the square in the quadratic expression in the denominator, the problem of evaluating the integral in Eq. (3) can be reduced to that of computing

$$\int \frac{1}{(a^2 \pm u^2)^n}\,du. \tag{4}$$

If the sign in the denominator of Eq. (4) is the plus sign, then the substitution $u = a\tan\theta$ transforms the integral into the form

$$\int \cos^m \theta\, d\theta.$$

(See Problem 35.) This integral can be handled by the methods of Section 7.4 or by using the reduction formula

$$\int \cos^k \theta\, d\theta = \frac{1}{k}\cos^{k-1}\theta \sin\theta + \frac{k-1}{k}\int \cos^{k-2}\theta\, d\theta$$

of Problem 54 in Section 7.3.

If the sign of the denominator in Eq. (4) is the minus sign, then the substitution $u = a\sin\theta$ transforms the integral into the form

$$\int \sec^m \theta\, d\theta.$$

(See Problem 36.) This integral may be evaluated with the aid of the reduction formula

$$\int \sec^k \theta\, d\theta = \frac{1}{k-1}\sec^{k-2}\theta \tan\theta + \frac{k-2}{k-1}\int \sec^{k-2}\theta\, d\theta$$

[Eq. (5) of Section 7.3].

7.7 TRUE/FALSE STUDY GUIDE

7.7 CONCEPTS: QUESTIONS AND DISCUSSION

1. Suppose that the factorization of the polynomial $Q(x)$ yields two or more quadratic factors. Explain why the methods of this section suffice to integrate any rational function of the form $f(x) = P(x)/Q(x)$.
2. Will the integral $\int f(x)\,dx$ in Question 1 ever be a rational function? If so, give an example illustrating this possibility.

7.7 PROBLEMS

Evaluate the antiderivatives in Problems 1 through 34.

1. $\displaystyle\int \frac{1}{x^2+4x+5}\,dx$

2. $\displaystyle\int \frac{2x+5}{x^2+4x+5}\,dx$

3. $\displaystyle\int \frac{5-3x}{x^2+4x+5}\,dx$

4. $\displaystyle\int \frac{x+1}{(x^2+4x+5)^2}\,dx$

5. $\displaystyle\int \frac{1}{\sqrt{3-2x-x^2}}\,dx$

6. $\displaystyle\int \frac{x+3}{\sqrt{3-2x-x^2}}\,dx$

7. $\displaystyle\int x\sqrt{3-2x-x^2}\,dx$

8. $\displaystyle\int \frac{1}{4x^2+4x-3}\,dx$

9. $\displaystyle\int \frac{3x+2}{4x^2+4x-3}\,dx$

10. $\displaystyle\int \sqrt{4x^2+4x-3}\,dx$

11. $\displaystyle\int \frac{1}{x^2+4x+13}\,dx$

12. $\displaystyle\int \frac{1}{\sqrt{2x-x^2}}\,dx$

13. $\displaystyle\int \frac{1}{3+2x-x^2}\,dx$

14. $\displaystyle\int x\sqrt{8+2x-x^2}\,dx$

15. $\displaystyle\int \frac{2x-5}{x^2+2x+2}\,dx$

16. $\displaystyle\int \frac{2x-1}{4x^2+4x-15}\,dx$

17. $\displaystyle\int \frac{x}{\sqrt{5+12x-9x^2}}\,dx$

18. $\displaystyle\int (3x-2)\sqrt{9x^2+12x+8}\,dx$

19. $\displaystyle\int (7-2x)\sqrt{9+16x-4x^2}\,dx$

20. $\displaystyle\int \frac{2x+3}{\sqrt{x^2+2x+5}}\,dx$

21. $\displaystyle\int \frac{x+4}{(6x-x^2)^{3/2}}\,dx$

22. $\displaystyle\int \frac{x-1}{(x^2+1)^2}\,dx$

23. $\displaystyle\int \frac{2x+3}{(4x^2+12x+13)^2}\,dx$

24. $\displaystyle\int \frac{x^3}{(1-x^2)^4}\,dx$

25. $\displaystyle\int \frac{3x-1}{x^2+x+1}\,dx$

26. $\displaystyle\int \frac{3x-1}{(x^2+x+1)^2}\,dx$

27. $\displaystyle\int \frac{1}{(x^2-4)^2}\,dx$

28. $\displaystyle\int (x-x^2)^{3/2}\,dx$

29. $\displaystyle\int \frac{x^2+1}{x^3+x^2+x}\,dx$

30. $\displaystyle\int \frac{x^2+2}{(x^2+1)^2}\,dx$

31. $\displaystyle\int \frac{2x^2+3}{x^4-2x^2+1}\,dx$

32. $\displaystyle\int \frac{x^2+4}{(x^2+1)^2(x^2+2)}\,dx$

33. $\displaystyle\int \frac{3x+1}{(x^2+2x+5)^2}\,dx$

34. $\displaystyle\int \frac{x^3-2x}{x^2+2x+2}\,dx$

35. Show that the substitution $u = a\tan\theta$ gives

$$\int \frac{1}{(a^2+u^2)^n}\,du = \frac{1}{a^{2n-1}}\int \cos^{2n-2}\theta\,d\theta.$$

36. Show that the substitution $u = a\sin\theta$ gives

$$\int \frac{1}{(a^2-u^2)^n}\,du = \frac{1}{a^{2n-1}}\int \sec^{2n-2}\theta\,d\theta.$$

In Problems 37 through 39 the region R lies between the curve $y = 1/(x^2-2x+5)$ and the x-axis from $x=0$ to $x=5$.

37. Find the area of the region R.

38. Find the volume of the solid generated by revolving R around the y-axis.

39. Find the volume of the solid generated by revolving R around the x-axis.

In Problems 40 through 42 the region R lies between the curve $y = 1/(4x^2-20x+29)$ and the x-axis from $x=1$ to $x=4$.

40. Find the area of the region R.

41. Find the volume of the solid generated by revolving R around the y-axis.

42. Find the volume of the solid generated by revolving R around the x-axis.

43. Your task is to build a road that joins the points $(0,0)$ and $(3,2)$ and follows the path of the circle with equation $(4x+4)^2+(4y-19)^2 = 377$. Find the length of this road. (Units on the coordinate axes are measured in miles.)

44. Suppose that the road of Problem 43 (Fig. 7.7.2) costs $10/(1+x)$ million dollars per mile. (a) Calculate its total cost. (b) With the same cost per mile, calculate the total cost of a straight line road from $(0,0)$ to $(3,2)$. You should find that it is *more* expensive than the *longer* circular road!

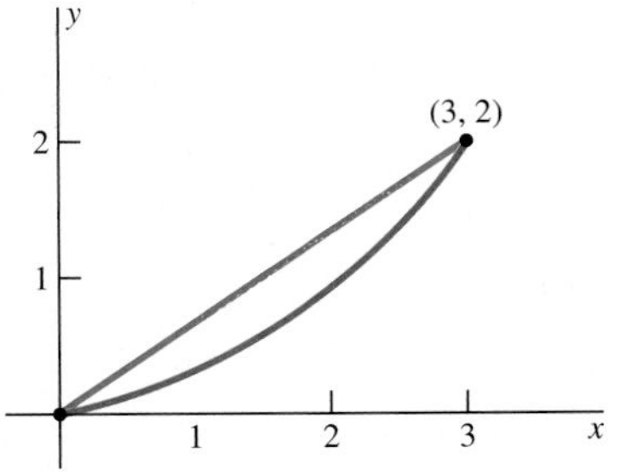

FIGURE 7.7.2 According to Problem 44, the circular-arc road is less expensive to construct than the shorter straight road!

In Problems 45 through 47, factor the denominator by first noting by inspection a root r of the denominator and then employing long division by x − r. Finally, use the method of partial fractions to aid in finding the indicated antiderivative.

45. $\displaystyle\int \frac{3x+2}{x^3+x^2-2}\,dx$

46. $\displaystyle\int \frac{1}{x^3+8}\,dx$

47. $\displaystyle\int \frac{x^4+2x^2}{x^3-1}\,dx$

48. (a) Find constants a and b such that

$$x^4+1 = (x^2+ax+1)(x^2+bx+1).$$

(b) Prove that

$$\int_0^1 \frac{x^2+1}{x^4+1}\,dx = \frac{\pi}{2\sqrt{2}}.$$

[*Suggestion:* If u and v are positive numbers and $uv = 1$, then

$$\arctan u + \arctan v = \frac{1}{2}\pi.]$$

49. Factor x^4+x^2+1 with the aid of ideas suggested in Problem 48. Then evaluate

$$\int \frac{2x^3+3x}{x^4+x^2+1}\,dx.$$

50. Evaluate the integral to show that

$$\int_0^1 \frac{16(x-1)}{x^4-2x^3+4x-4}\,dx = \pi.$$

This integral was (in effect) used by D. Bailey, P. Borwein, and S. Plouffe as a starting point in their recent determination of the 5 billionth hexagesimal digit of the number π (it's a 9). [*Suggestion:* Long divide to verify that x^2-2 is a factor of the denominator and to find the other factor.]

In Problems 51 through 54, write the general form of a partial-fraction decomposition of the given rational function $f(x)$ (with coefficients A, B, C, … remaining to be determined). Then use a computer algebra system (as in the remark following Example 6 in Section 7.5) to find the numerical values of the coefficients in the decomposition. Finally, find the indefinite integral $\int f(x)\,dx$

both by hand and by using the computer algebra system, and resolve any apparent discrepancy between the two results.

51. $\displaystyle\int \frac{7x^4 + 28x^3 + 50x^2 + 67x + 23}{(x-1)(x^2+2x+2)^2}\,dx$

52. $\displaystyle\int \frac{35 + 84x + 55x^2 - x^3 + 5x^4 - 4x^5}{(x^2+1)^2(x^2+6x+10)}\,dx$

53. $\displaystyle\int \frac{32x^5 + 16x^4 + 19x^3 - 98x^2 - 107x - 15}{(x^2-2x-15)(4x^2+4x+5)^2}\,dx$

54. $\displaystyle\int \frac{63x^5 + 302x^4 + 480x^3 + 376x^2 - 240x - 300}{(x^2+6x+10)^2(4x^2+4x+5)^2}\,dx$

In Problems 55 through 58, find values of the coefficients a, b, c, and d (not all zero) such that the given indefinite integral involves no logarithms or inverse tangents, and is therefore a rational function.

55. $\displaystyle\int \frac{ax+b}{(x^2+4x+5)^2}\,dx$

56. $\displaystyle\int \frac{ax^2+bx+c}{(x^2+4x+5)^2}\,dx$

57. $\displaystyle\int \frac{ax^2+bx+c}{(x^2+2x+2)(x^2+4x+5)}\,dx$

58. $\displaystyle\int \frac{ax^3+bx^2+cx+d}{(x^2+4x+5)^3}\,dx$

7.8 IMPROPER INTEGRALS

To show the existence of the definite integral, we have relied until now on the existence theorem stated in Section 5.4. This is the theorem that guarantees the existence of the definite integral

$$\int_a^b f(x)\,dx$$

provided that the function f is *continuous* on the closed and bounded interval $[a, b]$. Certain applications of calculus, however, lead naturally to the formulation of integrals in which either

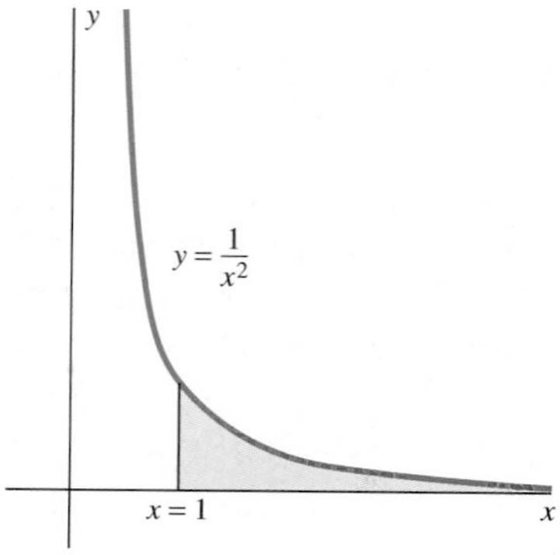

FIGURE 7.8.1 The shaded area cannot be measured by using our earlier techniques.

1. The interval of integration is not bounded; it has one of the forms

$$[a, +\infty), \quad (-\infty, a], \quad \text{or} \quad (-\infty, +\infty); \quad \text{or}$$

2. The integrand has an infinite discontinuity at some point c in the interval $[a, b]$:

$$\lim_{x \to c} f(x) = \pm\infty.$$

An example of Case 1 is the integral

$$\int_1^\infty \frac{1}{x^2}\,dx.$$

A geometric interpretation of this integral is the area of the unbounded region (shaded in Fig. 7.8.1) that lies between the curve $y = 1/x^2$ and the x-axis and to the right of the vertical line $x = 1$. An example of Case 2 is the integral

$$\int_0^1 \frac{1}{\sqrt{x}}\,dx.$$

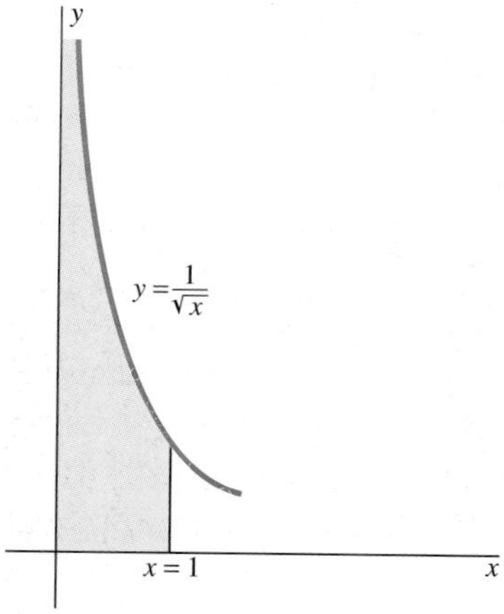

FIGURE 7.8.2 Another area that must be measured with an improper integral.

This integral may be interpreted as the area of the unbounded region (shaded in Fig. 7.8.2) that lies under the curve $y = 1/\sqrt{x}$ from $x = 0$ to $x = 1$.

Such integrals are called **improper integrals.** The natural interpretation of an improper integral is the area of an unbounded region. It is perhaps surprising that such an area can nevertheless be finite, and here we shall show how to find such areas—that is, how to evaluate improper integrals.

To see why improper integrals require special care, let us consider the integral

$$\int_{-1}^1 \frac{1}{x^2}\,dx.$$

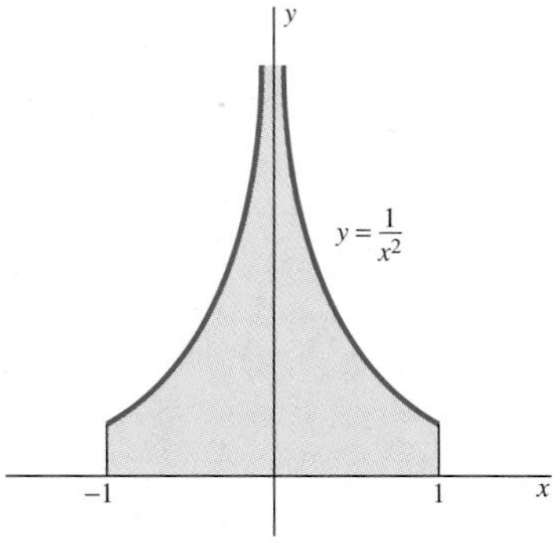

FIGURE 7.8.3 The area under $y = 1/x^2$, $-1 \leqq x \leqq 1$.

This integral is improper because its integrand $f(x) = 1/x^2$ is unbounded as $x \to 0$, and thus f is not continuous at $x = 0$. If we thoughtlessly applied the fundamental theorem of calculus, we would obtain

$$\int_{-1}^{1} \frac{1}{x^2}\,dx = \left[-\frac{1}{x}\right]_{-1}^{1} = (-1) - (+1) = -2. \qquad \textbf{(Wrong!)}$$

The negative answer is obviously incorrect, because the area shown in Fig. 7.8.3 lies above the x-axis and hence the integral cannot be negative. This simple example emphasizes that we cannot ignore the hypotheses—*continuous* function and *bounded closed* interval—of the fundamental theorem of calculus.

Infinite Limits of Integration

Suppose that the function f is continuous and nonnegative on the unbounded interval $[a, +\infty)$. Then, for any fixed $t > a$, the area $A(t)$ of the region under $y = f(x)$ from $x = a$ to $x = t$ (shaded in Fig. 7.8.4) is given by the (ordinary) definite integral

$$A(t) = \int_a^t f(x)\,dx.$$

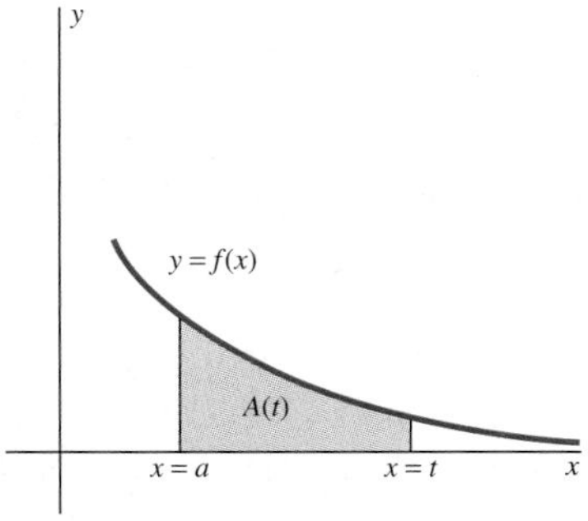

FIGURE 7.8.4 The shaded area $A(t)$ exists provided that f is continuous.

Suppose now that we let $t \to +\infty$ and find that the limit of $A(t)$ exists. Then we may regard this limit as the area of the unbounded region that lies under $y = f(x)$ and over $[a, +\infty)$. For f continuous on $[a, +\infty)$, we therefore *define*

$$\int_a^{\infty} f(x)\,dx = \lim_{t\to\infty} \int_a^t f(x)\,dx \qquad \textbf{(1)}$$

provided that this limit exists (as a finite real number). If this limit exists, we say that the improper integral on the left **converges;** if the limit does not exist, we say that the improper integral **diverges.** If $f(x)$ is nonnegative on $[a, +\infty)$, then the limit in Eq. (1) either exists or is infinite, and in the latter case we write

$$\int_a^{\infty} f(x)\,dx = +\infty$$

and say that the improper integral **diverges to infinity.**

If the function f has both positive and negative values on $[a, +\infty)$, then the improper integral can diverge *by oscillation*—that is, without diverging to infinity. This occurs with $\int_0^{\infty} \sin x\,dx$, because it is easy to verify that $\int_0^t \sin x\,dx$ is zero if t is an even integral multiple of π but is 2 if t is an odd integral multiple of π. Thus $\int_0^t \sin x\,dx$ oscillates between 0 and 2 as $t \to +\infty$, and so the limit in Eq. (1) does not exist.

We handle an infinite lower limit of integration similarly: We define

$$\int_{-\infty}^{b} f(x)\,dx = \lim_{t\to-\infty} \int_t^b f(x)\,dx \qquad \textbf{(2)}$$

provided that the limit exists. If the function f is continuous on the whole real line, we define

$$\int_{-\infty}^{\infty} f(x)\,dx = \int_{-\infty}^{c} f(x)\,dx + \int_c^{\infty} f(x)\,dx \qquad \textbf{(3)}$$

for any convenient choice of c, provided that both improper integrals on the right-hand side converge. Note that the leftmost integral in Eq. (3) is *not* necessarily equal to

$$\lim_{t\to\infty} \int_{-t}^{t} f(x)\,dx.$$

(See Problem 52.)

It makes no difference what value of c is used in Eq. (3), because if $c < d$, then

$$\int_{-\infty}^{c} f(x)\,dx + \int_{c}^{\infty} f(x)\,dx = \int_{-\infty}^{c} f(x)\,dx + \int_{c}^{d} f(x)\,dx + \int_{d}^{\infty} f(x)\,dx$$
$$= \int_{-\infty}^{d} f(x)\,dx + \int_{d}^{\infty} f(x)\,dx,$$

under the assumption that the limits involved all exist.

EXAMPLE 1 Investigate the improper integrals

(a) $\displaystyle\int_{1}^{\infty} \frac{1}{x^2}\,dx$ and

(b) $\displaystyle\int_{-\infty}^{0} \frac{1}{\sqrt{1-x}}\,dx.$

Solution

(a) $$\int_{1}^{\infty} \frac{1}{x^2}\,dx = \lim_{t\to\infty} \int_{1}^{t} \frac{1}{x^2}\,dx = \lim_{t\to\infty} \left[-\frac{1}{x}\right]_1^t = \lim_{t\to\infty}\left(-\frac{1}{t}+1\right) = 1.$$

Thus this improper integral converges, and this is the area of the region shaded in Fig. 7.8.1.

(b) $$\int_{-\infty}^{0} \frac{1}{\sqrt{1-x}}\,dx = \lim_{t\to-\infty} \int_{t}^{0} \frac{1}{\sqrt{1-x}}\,dx$$
$$= \lim_{t\to-\infty} \left[-2\sqrt{1-x}\right]_t^0 = \lim_{t\to-\infty} (2\sqrt{1-t}-2) = +\infty.$$

Thus the second improper integral of this example diverges to $+\infty$ (Fig. 7.8.5). ◆

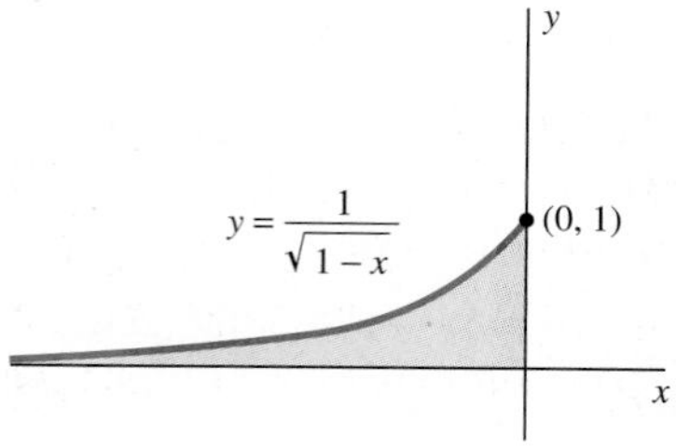

FIGURE 7.8.5 The unbounded region represented by the improper integral in Example 1(b).

EXAMPLE 2 Investigate the improper integral $\displaystyle\int_{-\infty}^{\infty} \frac{1}{1+x^2}\,dx.$

Solution The choice $c = 0$ in Eq. (3) gives

$$\int_{-\infty}^{\infty} \frac{1}{1+x^2}\,dx = \int_{-\infty}^{0} \frac{1}{1+x^2}\,dx + \int_{0}^{\infty} \frac{1}{1+x^2}\,dx$$
$$= \lim_{s\to-\infty} \int_{s}^{0} \frac{1}{1+x^2}\,dx + \lim_{t\to\infty} \int_{0}^{t} \frac{1}{1+x^2}\,dx$$
$$= \lim_{s\to-\infty} \left[\tan^{-1} x\right]_s^0 + \lim_{t\to\infty} \left[\tan^{-1} x\right]_0^t$$
$$= \lim_{s\to-\infty} (-\tan^{-1} s) + \lim_{t\to\infty} (\tan^{-1} t) = \frac{\pi}{2} + \frac{\pi}{2} = \pi.$$

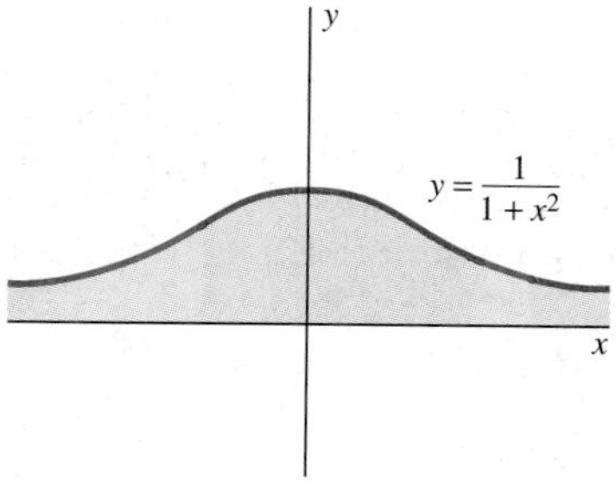

FIGURE 7.8.6 The area measured by the integral in Example 2.

The shaded region in Fig. 7.8.6 is a geometric interpretation of the integral of Example 2. ◆

Infinite Integrands

Suppose that the function f is continuous and nonnegative on $[a, b)$ but that $f(x) \to +\infty$ as $x \to b^-$. The graph of such a function appears in Fig. 7.8.7. The area $A(t)$ of the region lying under $y = f(x)$ from $x = a$ to $x = t < b$ is the value of the (ordinary) definite integral

$$A(t) = \int_{a}^{t} f(x)\,dx.$$

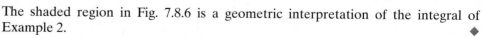

If the limit of $A(t)$ exists as $t \to b^-$, then this limit may be regarded as the area of the (unbounded) region under $y = f(x)$ from $x = a$ to $x = b$. For f continuous on

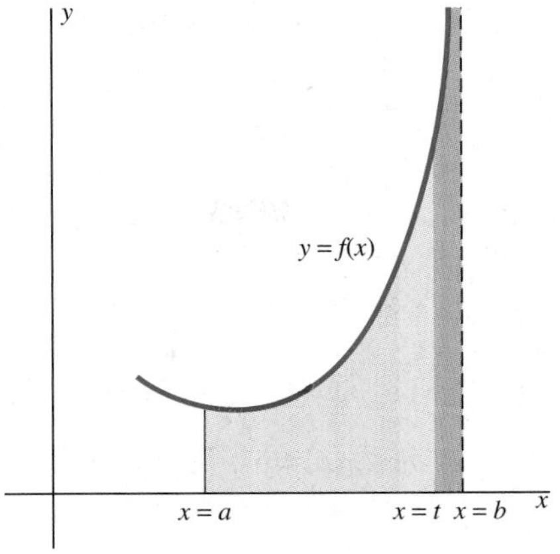

FIGURE 7.8.7 An improper integral of the second type: $f(x) \to \infty$ as $x \to b^-$.

$[a, b)$, we therefore *define*

$$\int_a^b f(x)\,dx = \lim_{t \to b^-} \int_a^t f(x)\,dx, \tag{4}$$

provided that this limit exists (as a finite number), in which case we say that the improper integral on the left **converges;** if the limit does not exist, we say that the integral **diverges.** If

$$\int_a^b f(x)\,dx = \lim_{t \to b^-} \int_a^t f(x)\,dx = \infty,$$

then we say that the improper integral **diverges to infinity.**

If f is continuous on $(a, b]$ but the limit of $f(x)$ as $x \to a^+$ is infinite, then we *define*

$$\int_a^b f(x)\,dx = \lim_{t \to a^+} \int_t^b f(x)\,dx, \tag{5}$$

provided that the limit exists. If f is continuous at every point of $[a, b]$ except for the point c in (a, b) and one or both one-sided limits of f at c are infinite, then we *define*

$$\int_a^b f(x)\,dx = \int_a^c f(x)\,dx + \int_c^b f(x)\,dx \tag{6}$$

provided that both improper integrals on the right converge.

EXAMPLE 3 Investigate the improper integrals

(a) $\displaystyle\int_0^1 \frac{1}{\sqrt{x}}\,dx$ and

(b) $\displaystyle\int_1^2 \frac{1}{(x-2)^2}\,dx.$

Solution (a) The integrand $1/\sqrt{x}$ becomes infinite as $x \to 0^+$, so

$$\int_0^1 \frac{1}{\sqrt{x}}\,dx = \lim_{t \to 0^+} \int_t^1 \frac{1}{\sqrt{x}}\,dx$$

$$= \lim_{t \to 0^+} \Big[2\sqrt{x}\Big]_t^1 = \lim_{t \to 0^+} 2\big(1 - \sqrt{t}\,\big) = 2.$$

Thus the area of the unbounded region shown in Fig. 7.8.2 is 2.

(b) Here the integrand becomes infinite as x approaches the right-hand endpoint, so

$$\begin{aligned}\int_1^2 \frac{1}{(x-2)^2}\,dx &= \lim_{t\to 2^-}\int_1^t \frac{1}{(x-2)^2}\,dx\\ &= \lim_{t\to 2^-}\left[-\frac{1}{x-2}\right]_1^t\\ &= \lim_{t\to 2^-}\left(-1-\frac{1}{t-2}\right) = +\infty.\end{aligned}$$

Hence this improper integral diverges to infinity (Fig. 7.8.8). It follows that the improper integral

$$\int_1^3 \frac{1}{(x-2)^2}\,dx = \int_1^2 \frac{1}{(x-2)^2}\,dx + \int_2^3 \frac{1}{(x-2)^2}\,dx$$

also diverges, because not both of the right-hand improper integrals converge. (You can verify that the second one also diverges to $+\infty$.) ◆

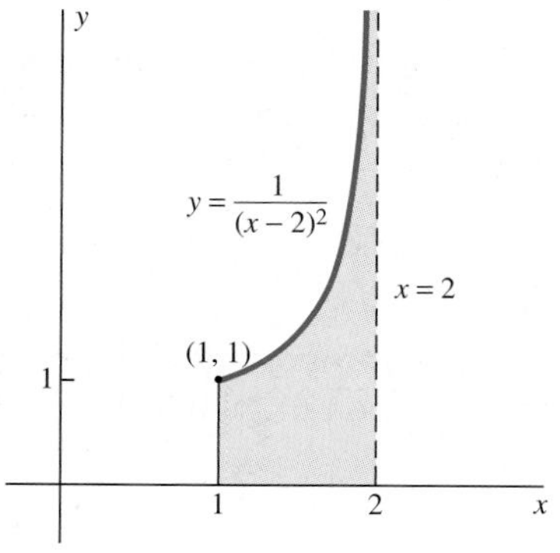

FIGURE 7.8.8 The unbounded region represented by the improper integral in Example 3(b).

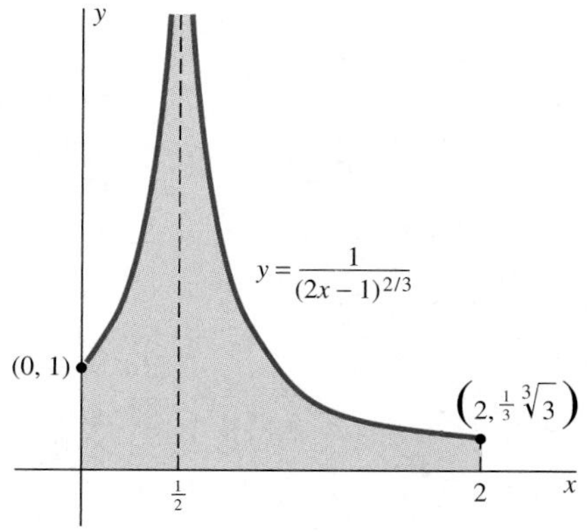

FIGURE 7.8.9 The region of Example 4.

EXAMPLE 4 Investigate the improper integral

$$\int_0^2 \frac{1}{(2x-1)^{2/3}}\,dx.$$

Solution This improper integral corresponds to the region shaded in Fig. 7.8.9. The integrand has an infinite discontinuity at the point $c=\frac{1}{2}$ within the interval of integration, so we write

$$\int_0^2 \frac{1}{(2x-1)^{2/3}}\,dx = \int_0^{1/2} \frac{1}{(2x-1)^{2/3}}\,dx + \int_{1/2}^2 \frac{1}{(2x-1)^{2/3}}\,dx$$

and investigate separately the two improper integrals on the right. We find that

$$\begin{aligned}\int_0^{1/2} \frac{1}{(2x-1)^{2/3}}\,dx &= \lim_{t\to(1/2)^-}\int_0^t \frac{1}{(2x-1)^{2/3}}\,dx\\ &= \lim_{t\to(1/2)^-}\left[\tfrac{3}{2}(2x-1)^{1/3}\right]_0^t\\ &= \lim_{t\to(1/2)^-}\tfrac{3}{2}\left[(2t-1)^{1/3}-(-1)^{1/3}\right] = \tfrac{3}{2},\end{aligned}$$

and

$$\int_{1/2}^{2} \frac{1}{(2x-1)^{2/3}}\,dx = \lim_{t\to(1/2)^+} \int_{t}^{2} \frac{1}{(2x-1)^{2/3}}\,dx$$

$$= \lim_{t\to(1/2)^+} \left[\tfrac{3}{2}(2x-1)^{1/3}\right]_t^2$$

$$= \lim_{t\to(1/2)^+} \tfrac{3}{2}\left[3^{1/3} - (2t-1)^{1/3}\right] = \tfrac{3}{2}\sqrt[3]{3}.$$

Therefore,

$$\int_{0}^{2} \frac{1}{(2x-1)^{2/3}}\,dx = \tfrac{3}{2}\left(1+\sqrt[3]{3}\right).$$ ◆

Special Functions and Improper Integrals

Special functions in advanced mathematics frequently are defined by means of improper integrals. An important example is the **gamma function** $\Gamma(t)$ that the prolific Swiss mathematician Leonhard Euler (1707–1783) introduced to "interpolate" values of the factorial function

$$n! = 1\cdot 2\cdot 3\cdots(n-1)\cdot n.$$

The gamma function is defined for all real numbers $t > 0$ by the improper integral

$$\Gamma(t) = \int_0^\infty x^{t-1}e^{-x}\,dx. \qquad \textbf{(7)}$$

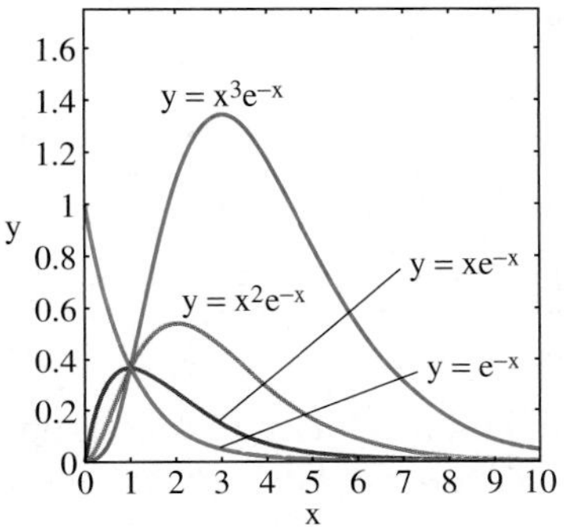

FIGURE 7.8.10 The graphs $y = x^{t-1}e^{-x}$ for $t = 1, 2, 3, 4$.

Thus for a fixed positive number t, the value $\Gamma(t)$ is the area under the curve $y = x^{t-1}e^{-x}$ from $x = 0$ to ∞ (Fig. 7.8.10). It turns out that the improper integral in (7) converges for all $t > 0$. The following example treats the illustrative cases $t = 1$ and $t = 2$.

EXAMPLE 5 If $t = 1$, then

$$\Gamma(1) = \int_0^\infty e^{-x}\,dx = \lim_{b\to\infty}\int_0^b e^{-x}\,dx$$

$$= \lim_{b\to\infty}\left[e^{-x}\right]_0^b = \lim_{b\to\infty}(1-e^{-b}) = 1.$$

If $t = 2$, then integration by parts with $u = x$, $dv = e^{-x}\,dx$ yields

$$\Gamma(2) = \int_0^\infty xe^{-x}\,dx = \lim_{b\to\infty}\int_0^b xe^{-x}\,dx$$

$$= \lim_{b\to\infty}\left(\left[-xe^{-x}\right]_0^b + \int_0^b e^{-x}\,dx\right)$$

$$= \lim_{b\to\infty}(0 - be^{-b}) + \int_0^\infty e^{-x}\,dx = 0 + 1 = 1$$

as well. ◆

Because $0! = 1$ by definition and $1! = 1$, Example 5 can be interpreted to say that $\Gamma(1) = 0!$ and $\Gamma(2) = 1!$. More generally, it turns out—see Problems 47 and 48—that

$$\Gamma(n+1) = n! \qquad \textbf{(8)}$$

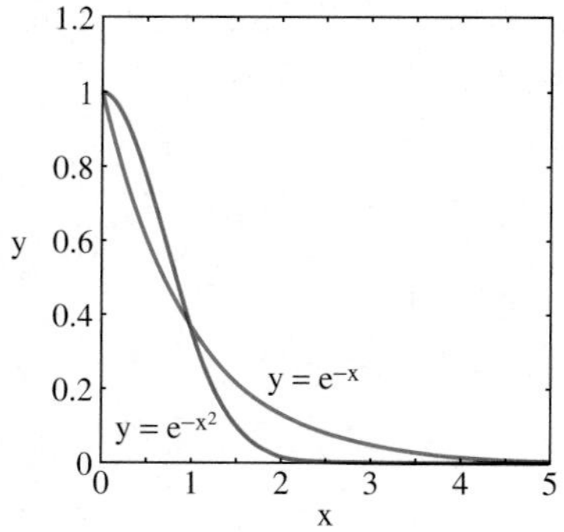

FIGURE 7.8.11 The graphs $y = e^{-x}$ and $y = e^{-x^2}$.

for every nonnegative integer n. But remember, the gamma function is defined for all positive *real numbers*. In Problem 57 we ask you to show that

$$\Gamma\left(\tfrac{1}{2}\right) = 2\int_0^\infty e^{-x^2}\,dx. \tag{9}$$

Figure 7.8.11 shows the graph $y = e^{-x^2}$ in comparison with $y = e^{-x}$ for $x \geqq 0$. Because $e^{-x^2} < e^{-x}$ for $x > 1$ and the improper integral

$$\int_0^\infty e^{-x}\,dx$$

converges by Example 5, it is plausible that the area under $y = e^{-x^2}$ for $x \geqq 0$ is finite, and hence that the improper integral in (9) converges as well. Indeed, we will see in Section 14.4 that $\Gamma(\frac{1}{2}) = \sqrt{\pi}$.

Escape Velocity

We saw in Section 6.5 how to compute the work W_r required to lift a body of mass m from the surface of a planet of mass M and radius R to a distance $r > R$ from the center of the planet. According to Eq. (7) there, the answer is

$$W_r = \int_R^r \frac{GMm}{x^2}\,dx.$$

So the work required to move the mass m "infinitely far" from the planet is

$$W = \lim_{r\to\infty} W_r = \int_R^\infty \frac{GMm}{x^2}\,dx = \lim_{r\to\infty}\left[-\frac{GMm}{x}\right]_R^r = \frac{GMm}{R}.$$

EXAMPLE 6 Suppose that the mass is projected with initial velocity v straight upward from the planet's surface, as in Jules Verne's novel *From the Earth to the Moon* (1865), in which a spacecraft was fired from an immense cannon. Then the initial kinetic energy $\frac{1}{2}mv^2$ of the mass is available to supply this work—by conversion into potential energy. From the equation

$$\frac{1}{2}mv^2 = \frac{GMm}{R},$$

we find that

$$v = \sqrt{\frac{2GM}{R}}.$$

Substituting appropriate numerical values for the constants G, M, and R yields the value $v \approx 11{,}175$ m/s (about 25,000 mi/h) for the *escape velocity* from the earth. ◆

Present Value of a Perpetuity

Consider a perpetual annuity, under which you and your heirs (and theirs, ad infinitum) will be paid A dollars annually. The question we pose is this: What is the fair market value of such an annuity? What should you pay to purchase it?

EXAMPLE 7 If the interest is compounded continuously at the annual rate r, then a dollar deposited in a savings account would grow to e^{rt} dollars in t years. Hence e^{-rt} dollars deposited now would yield \$1 after t years. Consequently, the **present value** of the amount you (and your heirs) will receive between time $t = 0$ (the present) and time $t = T > 0$ is defined to be

$$P_T = \int_0^T Ae^{-rt}\,dt.$$

Hence the total present value of the perpetual annuity is

$$P = \lim_{T\to\infty} P_T = \int_0^\infty Ae^{-rt}\,dt = \lim_{T\to\infty}\left[-\frac{A}{r}e^{-rt}\right]_0^T = \frac{A}{r}.$$

Thus $A = rP$. For instance, at an annual interest rate of 8% ($r = 0.08$), you should be able to purchase for $P = (\$50000)/(0.08) = \$625{,}000$ a perpetuity that pays you (and your heirs) an annual sum of \$50000 forever. ◆

Statistics and Probability Integrals

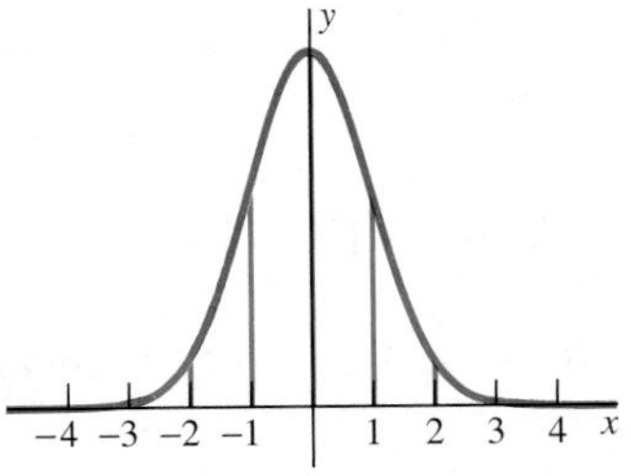

FIGURE 7.8.12 The bell-shaped curve $y = \dfrac{1}{\sqrt{2\pi}} e^{-\frac{1}{2}x^2}$.

Figure 7.8.12 shows the famous *bell-shaped curve* with equation

$$y = \frac{1}{\sqrt{2\pi}} \exp\left(-\tfrac{1}{2}x^2\right).$$

Problem 63 gives the value

$$\frac{1}{\sqrt{2\pi}} \int_{-\infty}^{\infty} \exp\left(-\tfrac{1}{2}x^2\right)dx = 1. \tag{10}$$

Thus the area of the region under the bell-shaped curve—with "infinite tails" extending in both directions—is exactly 1. By numerical integration (using Simpson's approximation, for instance) it can be verified that

$$\frac{1}{\sqrt{2\pi}} \int_{-1}^{1} \exp\left(-\tfrac{1}{2}x^2\right)dx \approx 0.6827, \qquad \frac{1}{\sqrt{2\pi}} \int_{-2}^{2} \exp\left(-\tfrac{1}{2}x^2\right)dx \approx 0.9545,$$

$$\text{and} \quad \frac{1}{\sqrt{2\pi}} \int_{-3}^{3} \exp\left(-\tfrac{1}{2}x^2\right)dx \approx 0.9973. \tag{11}$$

It follows that:

- Just over two-thirds of the area under the bell-shaped curve lies between $x = -1$ and $x = 1$;
- Almost 95.5% of the area lies between $x = -2$ and $x = 2$; and
- Almost 99.75% of the total area lies between $x = -3$ and $x = 3$.

The bell-shaped curve is used in describing the distributions of a wide range of typical attributes of individual members of large populations. Many such attributes—such as the height, weight, IQ, or SAT score of a college student—are distributed among members of the population in a manner that has both random and systematic aspects. For instance, let $X(s)$ denote the height of the college-age male s. Then X is a real-valued function defined on the set S of all college-age males. In statistics, such a real-valued function defined on a population set is called a **random variable** on S. Given a random variable X defined on a set S and two numbers a and b, we may ask what is the fraction or proportion $P\{a \leqq X(s) \leqq b\}$ of elements s of the set S such that the value $X(s)$ lies between a and b (inclusive). This fraction may be regarded as the **probability** $P\{a \leqq X(s) \leqq b\}$ that $X(s)$ lies in $[a, b]$.

A **normal** random variable X with **mean value** μ and **standard deviation** σ is one such that

$$P\left\{a \leqq \frac{X(s) - \mu}{\sigma} \leqq b\right\} = \frac{1}{\sqrt{2\pi}} \int_a^b \exp\left(-\tfrac{1}{2}x^2\right)dx \tag{12}$$

if $a < b$. You can easily verify that

$$a \leqq \frac{X(s) - \mu}{\sigma} \leqq b \quad \text{if and only if} \quad \mu + a\sigma \leqq X(s) \leqq \mu + b\sigma, \tag{13}$$

so Eq. (12) is equivalent to

$$P\{\mu + a\sigma \leqq X(s) \leqq \mu + b\sigma\} = \frac{1}{\sqrt{2\pi}} \int_a^b \exp\left(-\tfrac{1}{2}x^2\right)dx. \tag{14}$$

The next example reveals the meanings of the parameters μ and σ.

EXAMPLE 8 If we take $a = 0$ and $b = +\infty$ in Eqs. (13) and (14), we see that

$$P\{\mu \leqq X(s) < +\infty\} = \frac{1}{\sqrt{2\pi}} \int_0^\infty \exp\left(-\tfrac{1}{2}x^2\right) dx = \tfrac{1}{2}$$

(using (10) and the symmetry of the graph in Fig. 7.8.12). Thus the value $X(s)$ is greater than or equal to the mean value μ for *half* the elements of the set S. If we take $a = -1$ and $b = 1$, we see that

$$P\{\mu - \sigma \leqq X(s) \leqq \mu + \sigma\} = \frac{1}{\sqrt{2\pi}} \int_{-1}^{1} \exp\left(-\tfrac{1}{2}x^2\right) dx \approx 0.6827$$

(using (11)). Thus the value $X(s)$ lies within one standard deviation σ of the mean value μ for a bit over two-thirds of the elements of S. Similarly,

$$P\{\mu - 2\sigma \leqq X(s) \leqq \mu + 2\sigma\} = \frac{1}{\sqrt{2\pi}} \int_{-2}^{2} \exp\left(-\tfrac{1}{2}x^2\right) dx \approx 0.9545,$$

so the value $X(s)$ lies within two standard deviations of the mean value for over 95% of the elements of S. ◆

Numerical values of the integral in (12) and (14) are frequently calculated using values of the special function

$$\operatorname{erf}(x) = \frac{2}{\sqrt{\pi}} \int_0^x \exp\left(-t^2\right) dx \qquad \textbf{(15)}$$

(*erf* is for "error function"), which is tabulated in mathematical handbooks and included in most computer algebra systems. In Problem 64 we ask you to show that

$$\frac{1}{\sqrt{2\pi}} \int_0^u \exp\left(-\tfrac{1}{2}x^2\right) dx = \tfrac{1}{2}\operatorname{erf}\left(\frac{u}{\sqrt{2}}\right). \qquad \textbf{(16)}$$

EXAMPLE 9 Scoring for the Scholastic Aptitude Test (SAT) has been changed in recent years, so that the same performance now receives a higher score than before. This test was originally designed so that the SAT score $X(s)$ of a college-bound student s would (theoretically) be a random variable with mean value $\mu = 1000$ and standard deviation $\sigma = 100$. (a) In this case, what percentage of students tested should score at least 1250? (b) What percentage should score no more than 850?

Solution (a) If we write $\mu + a\sigma = 1000 + 100a = 1250$, we find that $a = 2.5$. Then the proportion of students with scores of at least 1250—that is, $1250 \leqq X(s) < +\infty$—is given (with $b = +\infty$) by

$$\begin{aligned}
P\{1250 \leqq X(s) < +\infty\} &= \frac{1}{\sqrt{2\pi}} \int_{2.5}^{\infty} \exp\left(-\tfrac{1}{2}x^2\right) dx \qquad \text{[using (14)]} \\
&= \frac{1}{\sqrt{2\pi}} \int_0^\infty \exp\left(-\tfrac{1}{2}x^2\right) dx - \frac{1}{\sqrt{2\pi}} \int_0^{2.5} \exp\left(-\tfrac{1}{2}x^2\right) dx \\
&= \frac{1}{2} - \frac{1}{2}\operatorname{erf}\left(\frac{2.5}{\sqrt{2}}\right) \qquad \text{[using (16)]} \\
&\approx \frac{1}{2}(1 - 0.98758) = 0.00621
\end{aligned}$$

(using the approximation $\operatorname{erf}(2.5/\sqrt{2}) \approx 0.98758$ given by a computer algebra system). Thus about 0.62% of all students tested—fewer than one in a hundred—should score at least 1250 (under the original grading system).

(b) If we write $\mu + b\sigma = 1000 + 100b = 850$, we find that $b = -1.5$. Then the proportion of students with scores at most 850—that is, $-\infty < X(s) \leqq 850$—is given (with $a = -\infty$) by

$$\begin{aligned}
P\{-\infty < X(s) \leqq 850\} &= \frac{1}{\sqrt{2\pi}} \int_{-\infty}^{-1.5} \exp\left(-\tfrac{1}{2}x^2\right) dx && \text{[using (14)]} \\
&= \frac{1}{\sqrt{2\pi}} \int_{1.5}^{\infty} \exp\left(-\tfrac{1}{2}x^2\right) dx && \text{[by symmetry]} \\
&= \frac{1}{\sqrt{2\pi}} \int_{0}^{\infty} \exp\left(-\tfrac{1}{2}x^2\right) dx - \frac{1}{\sqrt{2\pi}} \int_{0}^{1.5} \exp\left(-\tfrac{1}{2}x^2\right) dx \\
&= \frac{1}{2} - \frac{1}{2}\operatorname{erf}\left(\frac{1.5}{\sqrt{2}}\right) && \text{[using (16)]} \\
&\approx \frac{1}{2}(1 - 0.86639) \approx 0.06681
\end{aligned}$$

(using the value $\operatorname{erf}(1.5/\sqrt{2}) \approx 0.86639$ given by a computer algebra system). Thus about 6.68% of all students tested—about one in fifteen—should score 850 or less. ◆

Random Sampling

A **binary event** is one with two possible outcomes that occur with probabilities p and $q = 1 - p$, respectively. Examples include:

- The toss of a coin with possible outcomes "heads" and "tails."
- Polling the preference of a randomly selected voter who may be either a Democrat or a Republican.
- Quality testing an electronic component chosen at random from a shipment; it may be either defective or not.

Given a particular binary event, the result of carrying out the event N times—for instance, tossing a coin N times—is called an N-**event sample** (or a sample of **size** N). Given an N-event sample s, let $X(s)$ denote the number of "successful" outcomes with probability p. For instance, if s consists of N tosses of a coin, then $X(s)$ might denote the number of heads obtained. Then X is a random variable defined on the set S of all N-event samples.

If the sample size N is large, then the *central limit theorem* of advanced probability theory implies that X is closely approximated by a normal random variable whose mean value and standard deviation are given by

$$\mu = Np \quad \text{and} \quad \sigma = \sqrt{Npq}. \tag{17}$$

The probability that the number of successes lies between $\mu + a\sigma$ and $\mu + b\sigma$ is then given approximately by

$$P\{\mu + a\sigma \leqq X(s) \leqq \mu + b\sigma\} \approx \frac{1}{\sqrt{2\pi}} \int_a^b \exp\left(-\tfrac{1}{2}x^2\right) dx. \tag{18}$$

It is common statistical practice to assume equality in (18) if $N > 30$.

EXAMPLE 10 Suppose that a fair coin (one with $p = q = \frac{1}{2}$) is tossed $N = 400$ times. (a) Calculate the approximate probability that the number of heads obtained is between 185 and 215 inclusive. (b) Calculate the approximate probability that the number is more than 230.

Solution First note that by the equations in (17), the mean and standard deviation of $X(s)$ are

$$\mu = Np = 400 \cdot \tfrac{1}{2} = 200 \quad \text{and} \quad \sigma = \sqrt{Npq} = \sqrt{400 \cdot \tfrac{1}{2} \cdot \tfrac{1}{2}} = 10.$$

(a) If we write $\mu + a\sigma = 200 + 10a = 185$ and $\mu + b\sigma = 200 + 10b = 215$, we find that $a = -1.5$ and $b = 1.5$. Hence the probability that the number of heads lies in the interval [185, 215] is

$$\begin{aligned} P\{185 \leqq X(s) \leqq 215\} &\approx \frac{1}{\sqrt{2\pi}} \int_{-1.5}^{1.5} \exp\left(-\tfrac{1}{2}x^2\right) dx && \text{[using (18)]} \\ &= 2 \cdot \frac{1}{\sqrt{2\pi}} \int_{0}^{1.5} \exp\left(-\tfrac{1}{2}x^2\right) dx && \text{[by symmetry]} \\ &= 2 \cdot \frac{1}{2} \operatorname{erf}\left(\frac{1.5}{\sqrt{2}}\right) && \text{[using (16)]} \\ &\approx 0.86639 && \text{[computer algebra system].} \end{aligned}$$

In common language, the probability of getting 185 to 215 heads in 400 tosses of a fair coin is thus about 87%, which is about 7 chances out of 8.

(b) If we write $\mu + a\sigma = 200 + 10a = 230$, we get $a = 3$. Then the probability that the number of heads is at least 230—that is, $230 \leqq X(s) < +\infty$—is given (with $b = +\infty$) by

$$\begin{aligned} &P\{230 \leqq X(s) < +\infty\} \\ &\quad\approx \frac{1}{\sqrt{2\pi}} \int_{3}^{\infty} \exp\left(-\tfrac{1}{2}x^2\right) dx && \text{[using (18)]} \\ &\quad= \frac{1}{\sqrt{2\pi}} \int_{0}^{\infty} \exp\left(-\tfrac{1}{2}x^2\right) dx - \frac{1}{\sqrt{2\pi}} \int_{0}^{3} \exp\left(-\tfrac{1}{2}x^2\right) dx \\ &\quad= \frac{1}{2} - \frac{1}{2} \operatorname{erf}\left(\frac{3}{\sqrt{2}}\right) && \text{[using (16)]} \\ &\quad\approx \frac{1}{2}(1 - 0.99730) = 0.00135 \end{aligned}$$

(using the value $\operatorname{erf}(3/\sqrt{2}) \approx 0.99730$ given by a computer algebra system). Thus there is less than two chances in a thousand of obtaining at least 230 heads in 400 tosses of a fair coin. ◆

REMARK The decimal-place answers calculated in Example 10 are only rough approximations, although the final conclusions—"about 7 chances in 8" in part (a) and "less than two chances in a thousand" in part (b)—are correct. One reason is that (18) is, after all, only an approximation whose accuracy we have not discussed. Moreover, because a count of coin tosses is discrete rather than continuous, one might well argue that in part (a) we should calculate $P\{184.5 \leqq X(s) \leqq 215.5\}$ rather than $P\{185 \leqq X(s) \leqq 215\}$, and "extend by half integers" similarly in part (b). Such issues as these are discussed in statistics courses.

7.8 TRUE/FALSE STUDY GUIDE

7.8 CONCEPTS: QUESTIONS AND DISCUSSION

1. List the different types of improper integrals that are discussed in this section. For each type, give examples of both convergent and divergent improper integrals of that type.
2. Discuss and illustrate with examples the difference between "divergence to infinity" and "divergence by oscillation."
3. Discuss the relation between improper integrals and areas of unbounded regions in the plane.

7.8 PROBLEMS

Determine whether or not the improper integrals in Problems 1 through 38 converge. Evaluate those that do converge.

1. $\displaystyle\int_2^{\infty} \frac{1}{x\sqrt{x}}\,dx$

2. $\displaystyle\int_1^{\infty} \frac{1}{x^{2/3}}\,dx$

3. $\displaystyle\int_0^{4} \frac{1}{x\sqrt{x}}\,dx$

4. $\displaystyle\int_0^{8} \frac{1}{x^{2/3}}\,dx$

5. $\displaystyle\int_1^{\infty} \frac{1}{x+1}\,dx$

6. $\displaystyle\int_3^{\infty} \frac{1}{\sqrt{x+1}}\,dx$

7. $\displaystyle\int_5^{\infty} \frac{1}{(x-1)^{3/2}}\,dx$

8. $\displaystyle\int_0^{4} \frac{1}{\sqrt{4-x}}\,dx$

9. $\displaystyle\int_0^{9} \frac{1}{(9-x)^{3/2}}\,dx$

10. $\displaystyle\int_0^{3} \frac{1}{(x-3)^2}\,dx$

11. $\displaystyle\int_{-\infty}^{-2} \frac{1}{(x+1)^3}\,dx$

12. $\displaystyle\int_{-\infty}^{0} \frac{1}{\sqrt{4-x}}\,dx$

13. $\displaystyle\int_{-1}^{8} \frac{1}{\sqrt[3]{x}}\,dx$

14. $\displaystyle\int_{-4}^{4} \frac{1}{(x+4)^{2/3}}\,dx$

15. $\displaystyle\int_2^{\infty} \frac{1}{\sqrt[3]{x-1}}\,dx$

16. $\displaystyle\int_{-\infty}^{\infty} \frac{x}{(x^2+4)^{3/2}}\,dx$

17. $\displaystyle\int_{-\infty}^{\infty} \frac{x}{x^2+4}\,dx$

18. $\displaystyle\int_0^{\infty} e^{-(x+1)}\,dx$

19. $\displaystyle\int_0^{1} \frac{e^{\sqrt{x}}}{\sqrt{x}}\,dx$

20. $\displaystyle\int_0^{2} \frac{x}{x^2-1}\,dx$

21. $\displaystyle\int_0^{\infty} xe^{-3x}\,dx$

22. $\displaystyle\int_{-\infty}^{2} e^{2x}\,dx$

23. $\displaystyle\int_0^{\infty} xe^{-x^2}\,dx$

24. $\displaystyle\int_{-\infty}^{\infty} |x|e^{-x^2}\,dx$

25. $\displaystyle\int_0^{\infty} \frac{1}{1+x^2}\,dx$

26. $\displaystyle\int_0^{\infty} \frac{x}{1+x^2}\,dx$

27. $\displaystyle\int_0^{\infty} \cos x\,dx$

28. $\displaystyle\int_0^{\infty} \sin^2 x\,dx$

29. $\displaystyle\int_1^{\infty} \frac{\ln x}{x}\,dx$

30. $\displaystyle\int_2^{\infty} \frac{1}{x\ln x}\,dx$

31. $\displaystyle\int_2^{\infty} \frac{1}{x(\ln x)^2}\,dx$

32. $\displaystyle\int_1^{\infty} \frac{\ln x}{x^2}\,dx$

33. $\displaystyle\int_0^{\pi/2} \frac{\cos x}{\sqrt{\sin x}}\,dx$

34. $\displaystyle\int_0^{\pi/2} \frac{\sin x}{(\cos x)^{4/3}}\,dx$

35. $\displaystyle\int_0^{1} \ln x\,dx$

36. $\displaystyle\int_0^{1} \frac{\ln x}{x}\,dx$

37. $\displaystyle\int_0^{1} \frac{\ln x}{x^2}\,dx$

38. $\displaystyle\int_0^{\infty} e^{-x}\cos x\,dx$

In Problems 39 through 42, the given integral is improper both because the interval of integration is unbounded and because the integrand is unbounded near zero. Investigate its convergence by expressing it as a sum of two integrals—one from 0 to 1, the other from 1 to ∞. Evaluate those integrals that converge.

39. $\displaystyle\int_0^{\infty} \frac{1}{x+x^2}\,dx$

40. $\displaystyle\int_0^{\infty} \frac{1}{x^2+x^4}\,dx$

41. $\displaystyle\int_0^{\infty} \frac{1}{x^{1/2}+x^{3/2}}\,dx$

42. $\displaystyle\int_0^{\infty} \frac{1}{x^{2/3}+x^{4/3}}\,dx$

In Problems 43 through 46, find all real number values of k for which the given improper integral converges. Evaluate the integral for those values of k.

43. $\displaystyle\int_0^{1} \frac{1}{x^k}\,dx$

44. $\displaystyle\int_1^{\infty} \frac{1}{x^k}\,dx$

45. $\displaystyle\int_0^{1} x^k \ln x\,dx$

46. $\displaystyle\int_1^{\infty} \frac{1}{x(\ln x)^k}\,dx$

47. Beginning with the definition of the gamma function in Eq. (7), integrate by parts to show that

$$\Gamma(x+1) = x\Gamma(x)$$

for every positive real number x.

48. Explain how to apply the result of Problem 47 n times in succession to show that if n is a positive integer, then $\Gamma(n+1) = n!\Gamma(1) = n!$.

Problems 49 through 51 deal with Gabriel's horn, the surface obtained by revolving the curve $y = 1/x$, $x \geqq 1$, around the x-axis (Fig. 7.8.13).

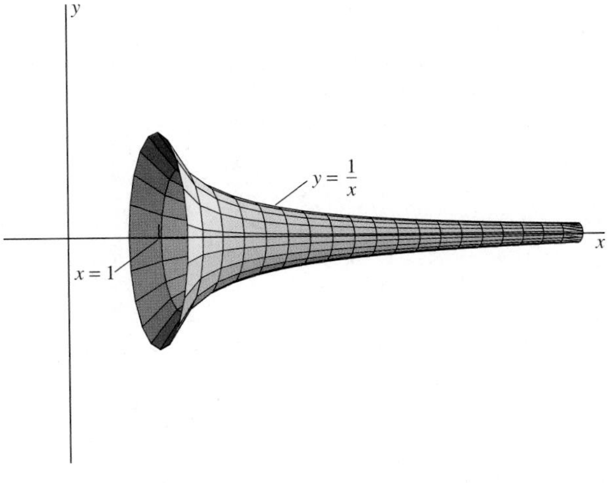

FIGURE 7.8.13 Gabriel's horn (Problems 49 through 51).

49. Show that the area under the curve $y = 1/x$, $x \geqq 1$, is infinite.

50. Show that the volume of revolution enclosed by Gabriel's horn is finite, and compute it.

51. Show that the surface area of Gabriel's horn is infinite. [*Suggestion:* Let A_t denote the surface area from $x = 1$ to $x = t > 1$. Prove that $A_t > 2\pi \ln t$.] In any case, the implication is that we could fill Gabriel's horn with a finite amount of paint (Problem 50), but no finite amount suffices to paint its surface.

52. Show that

$$\int_{-\infty}^{\infty} \frac{1+x}{1+x^2}\,dx$$

diverges, but that

$$\lim_{t\to\infty} \int_{-t}^{t} \frac{1+x}{1+x^2}\,dx = \pi.$$

53. Use the substitution $x = e^{-u}$ and the fact that $\Gamma(n+1) = n!$ (Problem 48) to prove that if m and n are fixed but arbitrary positive integers, then

$$\int_0^1 x^m (\ln x)^n\,dx = \frac{n!(-1)^n}{(m+1)^{n+1}}.$$

54. Consider a perpetual annuity under which you and your heirs will be paid at the rate of $10 + t$ thousand dollars per year t years hence. Thus you will receive $20,000 10 years from now, your heir will receive $110,000 100 years from now, and so on. Assuming a constant annual interest rate of 10%, show that the present value of this perpetuity is

$$P = \int_0^\infty (10+t)e^{-t/10}\,dt,$$

and then evaluate this improper integral.

55. A "semi-infinite" uniform rod occupies the nonnegative x-axis ($x \geqq 0$) and has linear density δ; that is, a segment of length dx has mass $\delta\,dx$. Show that the force of gravitational attraction that the rod exerts on a point mass m at $(-a, 0)$ is

$$F = \int_0^\infty \frac{Gm\delta}{(a+x)^2}\,dx = \frac{Gm\delta}{a}.$$

56. A rod of linear density δ (see Problem 55) occupies the entire y-axis. A point mass m is located at $(a, 0)$ on the x-axis, as indicated in Fig. 7.8.14. Show that the total (horizontal) gravitational attraction that the rod exerts on m is

$$F = \int_{-\infty}^{\infty} \frac{Gm\delta\cos\theta}{r^2}\,dy = \frac{2Gm\delta}{a},$$

where $r^2 = a^2 + y^2$ and $\cos\theta = a/r$.

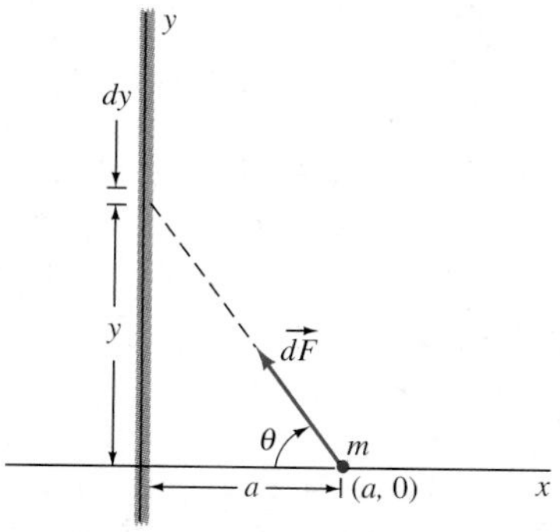

FIGURE 7.8.14 Gravitational attraction exerted on a point mass by an infinite rod (Problem 56).

57. Verify Eq. (9) by substituting $x = u^2$ in the integral that defines the value $\Gamma(\frac{1}{2})$.

58. Given the fact that

$$\int_0^\infty e^{-x^2}\,dx = \tfrac{1}{2}\sqrt{\pi},$$

find the volume of the unbounded solid that is obtained by revolving around the x-axis the unbounded region R between the x-axis and the curve $y = e^{-x^2}$ for $x \geqq 0$.

59. Find the volume of the unbounded solid that is obtained by revolving around the y-axis the region R of Problem 58.

60. Recall from Problem 47 that $\Gamma(x+1) = x\Gamma(x)$ if $x > 0$. Suppose that n is a positive integer. Use Eq. (9) to establish that

$$\Gamma\left(n+\tfrac{1}{2}\right) = \frac{1\cdot 3\cdot 5\cdots(2n-1)}{2^n}\sqrt{\pi}.$$

61. (a) Suppose that $k > 1$. Use integration by parts to show that

$$\int_0^\infty x^k \exp(-x^2)\,dx = \frac{k-1}{2}\int_0^\infty x^{k-2}\exp(-x^2)\,dx.$$

(b) Suppose that n is a positive integer. Prove that

$$\int_0^\infty x^{n-1}\exp(-x^2)\,dx = \frac{1}{2}\Gamma\left(\frac{n}{2}\right).$$

62. Suppose that you win the Florida lottery and decide to use part of your winnings to purchase a perpetual annuity that will pay you and your heirs $10,000 per year (forever). Assuming an annual interest rate of 6%, what is a fair price for an insurance company to charge you for such an annuity?

63. Deduce Eq. (10) in this section from the value

$$\int_{-\infty}^{\infty} \exp(-t^2)\,dt = \sqrt{\pi}$$

established in Section 14.4.

64. Deduce Eq. (16) in this section from Eq. (15).

In Problems 65 through 70, determine the value of k by using a calculator or computer to evaluate

$$\int_0^b f(x)\,dx$$

for successively larger values of b. Continue until it seems certain that the four-place value of k is an integer.

65. $\displaystyle\int_0^\infty x^5 e^{-x}\,dx = 60k$

66. $\displaystyle\int_0^\infty \frac{\sin x}{x}\,dx = \frac{\pi}{k}$

67. $\displaystyle\int_0^\infty \frac{1}{x^2+2}\,dx = \frac{\pi}{k\sqrt{2}}$

68. $\displaystyle\int_0^\infty \frac{1-e^{-3x}}{x}\,dx = \frac{\ln 10}{k}$

69. $\displaystyle\int_0^\infty \exp(-x^2)\cos 2x\,dx = \frac{\pi}{ke}$

70. $\displaystyle\int_0^\infty \sin(x^2)\,dx = \frac{1}{k}\sqrt{\frac{\pi}{2}}$

71. Suppose that the intelligence quotient (IQ) scores of middle-school students constitute a normal random variable with a mean of 100 and a standard deviation of 15. (a) Calculate the percentage of students that have an IQ score between 90 and 110, inclusive. (b) Calculate the percentage that have an IQ score of 125 or higher.

72. Suppose that the heights of adult U.S. males constitute a normal random variable with a mean of 69 inches and a standard deviation of 3 inches. (a) Calculate the percentage of adult U.S. males who have heights between 67 inches and 72 inches, inclusive. (b) Calculate the percentage having height 76 inches or more.

73. Suppose that the fair coin of Example 10 is tossed $N = 1000$ times. (a) Calculate the probability that the number of heads obtained is between 425 and 475, inclusive. (b) Calculate the probability that 500 or more heads are obtained.

74. Suppose that a biased coin has probability $p = 0.6$ of heads and $q = 0.4$ of tails is tossed $N = 600$ times. (a) Calculate the probability that the number of heads obtained is between 345 and 375, inclusive. (b) Calculate the probability that fewer than 350 heads are obtained.

75. A student randomly guesses the answers on a 50-question true-false test. (a) What is the probability that he passes (60% or more correct)? (b) What is the probability that he makes a C or better (70% or more correct)?

76. A machine prints circuit boards of which 1% are defective. What is the probability that 10 or more circuit boards out of 500 tested are defective?

77. Suppose that 55% of registered voters in a metropolitan population actually favor the Democratic candidate for mayor. If 750 registered voters are randomly polled, what is the probability that between 41% and 49% (inclusive) will favor the Republican candidate?

78. Let

$$I_n = \int_1^\infty \frac{(\ln x)^n}{x^2}\,dx.$$

Show that $(n+1)I_n = I_{n+1}$ for each integer $n \geqq 0$. Conclude that $I_n = n!$ if n is a positive integer.

CHAPTER 7 SUMMARY

When you confront the problem of evaluating a particular integral, you must first decide which of the several methods of this chapter to try. There are only two *general* methods of integration:

- Integration by substitution (Section 7.2), and
- Integration by parts (Section 7.3).

These are the analogues for integration of the chain rule and product rule, respectively, for differentiation.

Look first at the given integral to see if you can spot a substitution that would transform it into an elementary or familiar integral or one likely to be found in an integral table. In the case of an integral

$$\int f(x)g(x)\,dx,$$

whose integral is an unfamiliar product of two functions, one of which is easily differentiated and the other easily integrated, then an attempt to integrate by parts is indicated.

Beyond these two general methods, the chapter deals with a number of *special* methods. In the case of an integral that is obviously trigonometric, $\int \text{trig}(x)\,dx$, the simple "spin-off" methods of Section 7.4 may succeed. Remember that reduction formulas [such as Eq. (5) and Problems 53 and 54 of Section 7.3] are available for integrating an integral power of a single trigonometric function.

Any integral of a rational function—that is, an integral of the form

$$\int \frac{p(x)}{q(x)}\,dx,$$

where the integrand is a quotient of polynomials—can be evaluated by the method of partial fractions (Section 7.5). If the degree of the numerator is not less than that of the denominator—that is, if the rational function is not proper—first use long division to express it as the sum of a polynomial (easily integrated) and a proper rational fraction. Then decompose the latter into partial fractions. Partial fractions corresponding to linear factors are easily integrated, and those corresponding to irreducible quadratic factors can be integrated by completing the square and making (if necessary) a trigonometric substitution. As we explained in Section 7.7, the trigonometric integrals that result can always be evaluated.

In the case of an integral involving $\sqrt{ax^2 + bx + c}$, first complete the square (Section 7.7) and then rationalize the integral by making an appropriate trigonometric substitution (Section 7.6). This will leave you with a trigonometric integral.

Some additional special substitutions are introduced in the Miscellaneous Problems that follow. Notable among these is the substitution

$$u = \tan \frac{\theta}{2},$$

which transforms any integral $\int R(\sin\theta, \cos\theta)\, d\theta$ of a rational function of $\sin\theta$ and $\cos\theta$ into an integral of a rational function of u. The latter integral can then be evaluated by the method of partial fractions.

A final comment: Computer algebra systems are increasingly used for the evaluation of integrals such as those studied in this chapter. Nevertheless, the availability of these systems is no panacea. For instance, such computer systems are likely to be stumped by the integral

$$\int (1 + \ln x)\sqrt{1 + (x \ln x)^2}\, dx.$$

But you probably notice that the substitution

$$u = x \ln x, \qquad du = (1 + \ln x)\, dx$$

transforms this integral into the integral

$$\int \sqrt{1 + u^2}\, du,$$

which is amenable to trigonometric substitution (and can be found in almost any integral table). Thus the human factor remains—thankfully—essential. ◆

CHAPTER 7 MISCELLANEOUS PROBLEMS

Evaluate the integrals in Problems 1 through 100.

1. $\displaystyle\int \frac{1}{(1+x)\sqrt{x}}\, dx$ [*Suggestion:* Let $x = u^2$.]

2. $\displaystyle\int \frac{\sec^2 t}{1 + \tan t}\, dt$

3. $\displaystyle\int \sin x \sec x\, dx$

4. $\displaystyle\int \frac{\csc x \cot x}{1 + \csc^2 x}\, dx$

5. $\displaystyle\int \frac{\tan\theta}{\cos^2\theta}\, d\theta$

6. $\displaystyle\int \csc^4 x\, dx$

7. $\displaystyle\int x \tan^2 x\, dx$

8. $\displaystyle\int x^2 \cos^2 x\, dx$

9. $\displaystyle\int x^5 \sqrt{2 - x^3}\, dx$

10. $\displaystyle\int \frac{1}{\sqrt{x^2+4}}\, dx$

11. $\displaystyle\int \frac{x^2}{\sqrt{25+x^2}}\, dx$

12. $\displaystyle\int (\cos x)\sqrt{4 - \sin^2 x}\, dx$

13. $\displaystyle\int \frac{1}{x^2 - x + 1}\, dx$

14. $\displaystyle\int \sqrt{x^2 + x + 1}\, dx$

15. $\displaystyle\int \frac{5x + 31}{3x^2 - 4x + 11}\, dx$

16. $\displaystyle\int \frac{x^4+1}{x^2+1}\, dx$

17. $\displaystyle\int \sqrt{x^4 + x^7}\, dx$

18. $\displaystyle\int \frac{\sqrt{x}}{1+x}\, dx$ [*Suggestion:* Let $x = u^2$.]

19. $\displaystyle\int \frac{\cos x}{\sqrt{4 - \sin^2 x}}\, dx$

20. $\displaystyle\int \frac{\cos 2x}{\cos x}\, dx$

21. $\displaystyle\int \frac{\tan x}{\ln(\cos x)}\, dx$

22. $\displaystyle\int \frac{x^7}{\sqrt{1 - x^4}}\, dx$

23. $\displaystyle\int \ln(1+x)\, dx$

24. $\displaystyle\int x \sec^{-1} x\, dx$

25. $\displaystyle\int \sqrt{x^2+9}\, dx$

26. $\displaystyle\int \frac{x^2}{\sqrt{4 - x^2}}\, dx$

27. $\displaystyle\int \sqrt{2x - x^2}\, dx$

28. $\displaystyle\int \frac{4x - 2}{x^3 - x}\, dx$

29. $\displaystyle\int \frac{x^4}{x^2 - 2}\, dx$

30. $\displaystyle\int \frac{\sec x \tan x}{\sec x + \sec^2 x}\, dx$

31. $\displaystyle\int \frac{x}{(x^2 + 2x + 2)^2}\, dx$

32. $\displaystyle\int \frac{x^{1/3}}{x^{1/2} + x^{1/4}}\, dx$ [*Suggestion:* Let $x = u^{12}$.]

33. $\displaystyle\int \frac{1}{1 + \cos 2\theta}\, d\theta$

34. $\displaystyle\int \frac{\sec x}{\tan x}\, dx$

35. $\displaystyle\int \sec^3 x \tan^3 x\, dx$

36. $\displaystyle\int x^2 \tan^{-1} x\, dx$

37. $\int x(\ln x)^3\,dx$

38. $\int \dfrac{1}{x\sqrt{1+x^2}}\,dx$

39. $\int e^x\sqrt{1+e^{2x}}\,dx$

40. $\int \dfrac{x}{\sqrt{4x-x^2}}\,dx$

41. $\int \dfrac{1}{x^3\sqrt{x^2-9}}\,dx$

42. $\int \dfrac{x}{(7x+1)^{17}}\,dx$

43. $\int \dfrac{4x^2+x+1}{4x^3+x}\,dx$

44. $\int \dfrac{4x^3-x+1}{x^3+1}\,dx$

45. $\int \tan^2 x\sec x\,dx$

46. $\int \dfrac{x^2+2x+2}{(x+1)^3}\,dx$

47. $\int \dfrac{x^4+2x+2}{x^5+x^4}\,dx$

48. $\int \dfrac{8x^2-4x+7}{(x^2+1)(4x+1)}\,dx$

49. $\int \dfrac{3x^5-x^4+2x^3-12x^2-2x+1}{(x^3-1)^2}\,dx$

50. $\int \dfrac{x}{x^4+4x^2+8}\,dx$

51. $\int (\ln x)^6\,dx$

52. $\int \dfrac{\left(1+x^{2/3}\right)^{3/2}}{x^{1/3}}\,dx$ [*Suggestion:* Let $x=u^3$.]

53. $\int \dfrac{(\arcsin x)^2}{\sqrt{1-x^2}}\,dx$

54. $\int \dfrac{1}{x^{3/2}\left(1+x^{1/3}\right)}\,dx$ [*Suggestion:* Let $x=u^6$.]

55. $\int \tan^3 z\,dz$

56. $\int \sin^2\omega\cos^4\omega\,d\omega$

57. $\int \dfrac{xe^{x^2}}{1+e^{2x^2}}\,dx$

58. $\int \dfrac{\cos^3 x}{\sqrt{\sin x}}\,dx$

59. $\int x^3e^{-x^2}\,dx$

60. $\int \sin\sqrt{x}\,dx$

61. $\int \dfrac{\arcsin x}{x^2}\,dx$

62. $\int \sqrt{x^2-9}\,dx$

63. $\int x^2\sqrt{1-x^2}\,dx$

64. $\int x\sqrt{2x-x^2}\,dx$

65. $\int \dfrac{x-2}{4x^2+4x+1}\,dx$

66. $\int \dfrac{2x^2-5x-1}{x^3-2x^2-x+2}\,dx$

67. $\int \dfrac{e^{2x}}{e^{2x}-1}\,dx$

68. $\int \dfrac{\cos x}{\sin^2 x-3\sin x+2}\,dx$

69. $\int \dfrac{2x^3+3x^2+4}{(x+1)^4}\,dx$

70. $\int \dfrac{\sec^2 x}{\tan^2 x+2\tan x+2}\,dx$

71. $\int \dfrac{x^3+x^2+2x+1}{x^4+2x^2+1}\,dx$

72. $\int \sin x\cos 3x\,dx$

73. $\int x^5\sqrt{x^3-1}\,dx$

74. $\int \ln(x^2+2x)\,dx$

75. $\int \dfrac{\sqrt{1+\sin x}}{\sec x}\,dx$

76. $\int \dfrac{1}{x^{2/3}\left(1+x^{2/3}\right)}\,dx$

77. $\int \dfrac{\sin x}{\sin 2x}\,dx$

78. $\int \sqrt{1+\cos t}\,dt$

79. $\int \sqrt{1+\sin t}\,dt$

80. $\int \dfrac{\sec^2 t}{1-\tan^2 t}\,dt$

81. $\int \ln(x^2+x+1)\,dx$

82. $\int e^x\sin^{-1}(e^x)\,dx$

83. $\int \dfrac{\arctan x}{x^2}\,dx$

84. $\int \dfrac{x^2}{\sqrt{x^2-25}}\,dx$

85. $\int \dfrac{x^3}{(x^2+1)^2}\,dx$

86. $\int \dfrac{1}{x\sqrt{6x-x^2}}\,dx$

87. $\int \dfrac{3x+2}{(x^2+4)^{3/2}}\,dx$

88. $\int x^{3/2}\ln x\,dx$

89. $\int \dfrac{\sqrt{1+\sin^2 x}}{\sec x\csc x}\,dx$

90. $\int \dfrac{\exp\left(\sqrt{\sin x}\right)}{(\sec x)\sqrt{\sin x}}\,dx$

91. $\int xe^x\sin x\,dx$

92. $\int x^2e^{x^{3/2}}\,dx$

93. $\int \dfrac{\arctan x}{(x-1)^3}\,dx$

94. $\int \ln\left(1+\sqrt{x}\right)dx$

95. $\int \dfrac{2x+3}{\sqrt{3+6x-9x^2}}\,dx$

96. $\int \dfrac{1}{\sqrt{e^{2x}-1}}\,dx$

97. $\int \dfrac{x^4}{(x-1)^2}\,dx$ [*Suggestion:* Let $u=x-1$.]

98. $\int x^{3/2}\tan^{-1}\left(\sqrt{x}\right)dx$

99. $\int \operatorname{arcsec}\left(\sqrt{x}\right)dx$

100. $\int x\sqrt{\dfrac{1-x^2}{1+x^2}}\,dx$

101. Find the area of the surface generated by revolving the curve $y=\cosh x$, $0\leqq x\leqq 1$, around the x-axis.

102. Find the length of the curve $y=e^{-x}$, $0\leqq x\leqq 1$.

103. (a) Find the area A_t of the surface generated by revolving the curve $y=e^{-x}$, $0\leqq x\leqq t$, around the x-axis. (b) Find $\lim_{t\to\infty} A_t$.

104. (a) Find the area A_t of the surface generated by revolving the curve $y=1/x$, $1\leqq x\leqq t$, around the x-axis. (b) Find $\lim_{t\to\infty} A_t$.

105. Find the area of the surface generated by revolving the curve $y=\sqrt{x^2-1}$, $1\leqq x\leqq 2$, around the x-axis.

106. (a) Derive the reduction formula

$$\int x^m(\ln x)^n\,dx=\frac{1}{m+1}x^{m+1}(\ln x)^n-\frac{n}{m+1}\int x^m(\ln x)^{n-1}\,dx.$$

(b) Evaluate $\int_1^e x^3(\ln x)^3\,dx$.

107. Derive the reduction formula

$$\int \sin^m x\cos^n x\,dx=-\frac{1}{m+n}\sin^{m-1}x\cos^{n+1}x+\frac{m-1}{m+n}\int \sin^{m-2}x\cos^n x\,dx.$$

108. Use the reduction formulas of Problem 107 here and Problem 54 of Section 7.3 to evaluate

$$\int_0^{\pi/2}\sin^6 x\cos^5 x\,dx.$$

109. Find the area bounded by the curve $y^2=x^5(2-x)$, $0\leqq x\leqq 2$. [*Suggestion:* Substitute $x=2\sin^2\theta$, then use the result of Problem 58 of Section 7.3.]

110. Show that

$$0 < \int_0^1 \frac{t^4(1-t)^4}{1+t^2}\,dt$$

and that

$$\int_0^1 \frac{t^4(1-t)^4}{1+t^2}\,dt = \frac{22}{7} - \pi.$$

111. Evaluate

$$\int_0^1 t^4(1-t)^4\,dt;$$

then apply the results of Problem 110 to conclude that

$$\frac{22}{7} - \frac{1}{630} < \pi < \frac{22}{7} - \frac{1}{1260}.$$

Thus $3.1412 < \pi < 3.1421$.

112. Find the length of the curve $y = \frac{4}{5}x^{5/4}$, $0 \leqq x \leqq 1$.

113. Find the length of the curve $y = \frac{4}{3}x^{3/4}$, $1 \leqq x \leqq 4$.

114. An initially empty water tank is shaped like a cone whose axis is vertical. Its vertex is at the bottom; the cone is 9 ft deep and has a top radius of 4.5 ft. Beginning at time $t = 0$, water is poured into this tank at 50 ft^3/min. Meanwhile, water leaks from a hole at the tank's bottom at the rate of $10\sqrt{y}$ cubic feet per minute, where y is the depth of water in the tank. (This is consistent with Torricelli's law of draining.) How long does it take to fill the tank?

115. (a) Evaluate

$$\int \frac{1}{1+e^x+e^{-x}}\,dx.$$

(b) Explain why your substitution in part (a) suffices to integrate any rational function of e^x.

116. (a) The equation $x^3 + x + 1 = 0$ has at least one real root r. Use Newton's method to find it, accurate to at least two places. (b) Use long division to find (approximately) the irreducible quadratic factor of $x^3 + x + 1$. (c) Use the factorization obtained in part (b) to evaluate (approximately)

$$\int_0^1 \frac{1}{x^3+x+1}\,dx.$$

117. Evaluate $\displaystyle\int \frac{1}{1+e^x}\,dx$.

118. The integral

$$\int \frac{1+2x^2}{x^5(1+x^2)^3}\,dx = \int \frac{x+2x^3}{(x^4+x^2)^3}\,dx$$

would require you to solve 11 equations in 11 unknowns if you were to use the method of partial fractions to evaluate it. Use the substitution $u = x^4 + x^2$ to evaluate it much more simply.

119. Evaluate

$$\int \sqrt{\tan\theta}\,d\theta.$$

[*Suggestion:* First substitute $u = \tan\theta$. Then substitute $u = x^2$. Finally, use the method of partial fractions; see Problem 48 of Section 7.7.]

120. Prove that if $p(x)$ is a polynomial, then the substitution $u^n = (ax+b)/(cx+d)$ transforms the integral

$$\int p(x)\left(\frac{ax+b}{cx+d}\right)^{1/n} dx$$

into the integral of a rational function of u. (The substitution indicated here is called a *rationalizing substitution;* its name comes from the fact that it converts the integrand into a *rational* function of u.)

In Problems 121 through 129, use the rationalizing substitution indicated in Problem 120 to evaluate the integral.

121. $\displaystyle\int x^3\sqrt{3x-2}\,dx$

122. $\displaystyle\int x^3\sqrt[3]{x^2+1}\,dx$

123. $\displaystyle\int \frac{x^3}{(x^2-1)^{4/3}}\,dx$

124. $\displaystyle\int x^2(x-1)^{3/2}\,dx$

125. $\displaystyle\int \frac{x^5}{\sqrt{x^3+1}}\,dx$

126. $\displaystyle\int x^7\sqrt[3]{x^4+1}\,dx$

127. $\displaystyle\int \sqrt{\frac{1+x}{1-x}}\,dx$

128. $\displaystyle\int \frac{x}{\sqrt{x+1}}\,dx$

129. $\displaystyle\int \frac{\sqrt[3]{x+1}}{x}\,dx$

130. Substitute $x = u^2$ to find

$$\int \sqrt{1+\sqrt{x}}\,dx.$$

131. Substitute $u^2 = 1 + e^{2x}$ to find

$$\int \sqrt{1+e^{2x}}\,dx.$$

132. Find the area A of the surface obtained by revolving the curve $y = \frac{2}{3}x^{3/2}$, $3 \leqq x \leqq 8$, around the x-axis. [*Suggestion:* Substitute $x = u^2$ into the surface area integral. *Note:* $A \approx 732.39$.]

133. Find the area bounded by one loop of the curve

$$y^2 = x^2(1-x), \quad 0 \leqq x \leqq 1.$$

134. Find the area bounded by the loop of the curve

$$y^2 = x^2\left(\frac{1-x}{1+x}\right), \quad 0 \leqq x \leqq 1.$$

More General Trigonometric Integrals

As a last resort, any trigonometric integral can be transformed into an integral

$$\int R(\sin\theta, \cos\theta)\,d\theta \tag{1}$$

of sines and cosines. If the integrand in Eq. (1) is a quotient of polynomials in the variables $\sin\theta$ *and* $\cos\theta$*, then the special substitution*

$$u = \tan\frac{\theta}{2} \tag{2}$$

suffices for its evaluation.

To carry out the substitution indicated in Eq. (2), we must express $\sin\theta$*,* $\cos\theta$*, and* $d\theta$ *in terms of* u *and* du*. Note first that*

$$\theta = 2\tan^{-1}u, \quad \text{so} \quad d\theta = \frac{2\,du}{1+u^2}. \tag{3}$$

We see from the triangle in Fig. 7.MP.1 that

$$\sin\frac{\theta}{2} = \frac{u}{\sqrt{1+u^2}}, \qquad \cos\frac{\theta}{2} = \frac{1}{\sqrt{1+u^2}}.$$

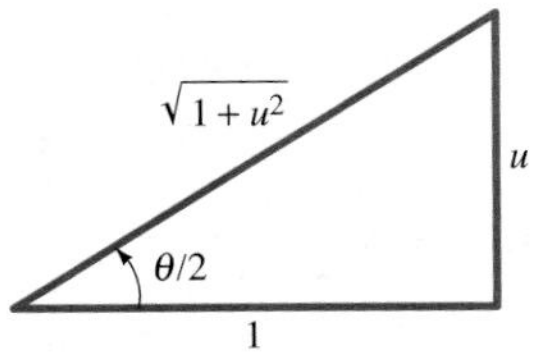

FIGURE 7.MP.1 The special rationalizing substitution $u = \tan\dfrac{\theta}{2}$.

Hence

$$\sin\theta = 2\sin\frac{\theta}{2}\cos\frac{\theta}{2} = \frac{2u}{1+u^2}, \tag{4}$$

$$\cos\theta = \cos^2\frac{\theta}{2} - \sin^2\frac{\theta}{2} = \frac{1-u^2}{1+u^2}. \tag{5}$$

These substitutions will convert the integral in Eq. (1) into an integral of a rational function of u. The latter can then be evaluated by the methods of Section 7.5.

EXAMPLE With the substitutions in Eqs. (3) and (5),

$$\int \frac{1}{5+3\cos\theta}\,d\theta = \int \frac{1}{5+3\cdot\dfrac{1-u^2}{1+u^2}}\cdot\frac{2}{1+u^2}\,du$$

$$= \int \frac{2}{8+2u^2}\,du = \int \frac{1}{4+u^2}\,du$$

$$= \frac{1}{2}\tan^{-1}\frac{u}{2} + C = \frac{1}{2}\tan^{-1}\left(\frac{1}{2}\tan\frac{\theta}{2}\right) + C.$$

◆

In Problems 135 through 142, use the rationalizing substitution given in Eqs. (2) through (5).

135. $\displaystyle\int \frac{1}{1+\cos\theta}\,d\theta$

136. $\displaystyle\int \frac{1}{5+4\cos\theta}\,d\theta$

137. $\displaystyle\int \frac{1}{1+\sin\theta}\,d\theta$

138. $\displaystyle\int \frac{1}{(1-\cos\theta)^2}\,d\theta$

139. $\displaystyle\int \frac{1}{\sin\theta+\cos\theta}\,d\theta$

140. $\displaystyle\int \frac{1}{2+\sin\phi+\cos\phi}\,d\phi$

141. $\displaystyle\int \frac{\sin\theta}{2+\cos\theta}\,d\theta$

142. $\displaystyle\int \frac{\sin\theta-\cos\theta}{\sin\theta+\cos\theta}\,d\theta$

143. (a) Substitute $u = \tan(\theta/2)$ to show that

$$\int \sec\theta\,d\theta = \ln\left|\frac{1+\tan\dfrac{\theta}{2}}{1-\tan\dfrac{\theta}{2}}\right| + C.$$

(b) Use the trigonometric identity

$$\tan\frac{\theta}{2} = \sqrt{\frac{1-\cos\theta}{1+\cos\theta}}$$

to derive our earlier formula

$$\int \sec\theta\,d\theta = \ln|\sec\theta+\tan\theta| + C$$

from the solution in part (a).

144. (a) Use the method of Problem 143 to show that

$$\int \csc\theta\,d\theta = \ln\left|\tan\frac{\theta}{2}\right| + C.$$

(b) Use trigonometric identities to derive the formula

$$\int \csc\theta\,d\theta = \ln|\csc\theta-\cot\theta| + C$$

from part (a).

DIFFERENTIAL EQUATIONS

8

John Bernoulli (1667–1748)

In the eighteenth century the remarkable Swiss family Bernoulli was to mathematics what the Bach family was to music. Eight different Bernoullis were sufficiently prominent that more than two centuries later, they rate entries in the *Dictionary of Scientific Biography.* The brothers James (Jakob, 1654–1705) and John (Johann, 1667–1748) Bernoulli played crucial rules in the early development of Leibniz's version of the calculus based on infinitely small differentials, which in continental European science predominated over Newton's version based more explicitly on limits of ratios. It was James Bernoulli who introduced the word "integral" in suggesting the name *calculus integralis* (instead of Leibniz's original *calculus summatorius*) for the subject inverse to the *calculus differentialis.*

John Bernoulli first studied mathematics under his older brother James at the university in Basel, Switzerland, but soon they were on an equal footing in mathematical understanding. In 1691, John Bernoulli visited Paris and there met the young Marquis de l'Hôpital (1661–1704), who was anxious to learn the secrets of the new infinitesimal calculus. In return for a generous monthly stipend, Bernoulli agreed to tutor the wealthy Marquis and continued the lessons (and the financial arrangement) by mail after his return to Basel. The result of this correspondence was the first differential calculus textbook, published by l'Hôpital in 1696. This text is remembered mainly for its inclusion of the theorem concerning indeterminate forms now known as *l'Hôpital's rule* (Section 4.8), although it was actually discovered by John Bernoulli.

The Bernoulli brothers were pioneers in the use of differential equations to model physical phenomena, and they introduced now-familiar techniques such as "separation of variables" to solve these differential equations. For instance, both James and John worked on the *catenary problem,* which asks for the shape of a hanging cable suspended between two fixed points, assuming that it is inelastic (unstretchable) but perfectly flexible. They showed that the shape function $y(x)$ of the cable satisfies the differential equation

$$a\frac{d^2y}{dx^2} = \sqrt{1 + \left(\frac{dy}{dx}\right)^2},$$

(where the constant $a = T/\rho$ is the ratio of the cable's tension at its lowest point and its density [assumed constant]). They succeeded in solving this differential equation to show that the shape of the hanging cable is described by

$$y = a\cosh\left(\frac{x}{a}\right).$$

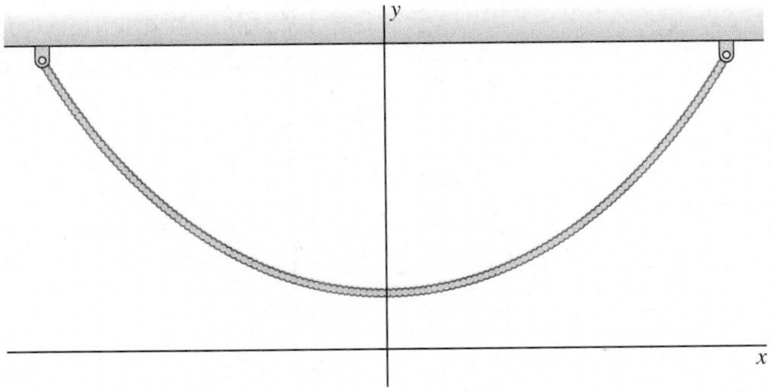

What is the shape of a hanging cable?

8.1 SIMPLE EQUATIONS AND MODELS

Mathematical models of changing real-world phenomena frequently involve differential equations—that is, equations containing *derivatives* of unknown functions. For instance, in Section 5.2 we mentioned the simple differential equations

$$\frac{dP}{dt} = kP \quad \text{(natural population growth)} \tag{1}$$

and

$$\frac{dy}{dt} = -k\sqrt{y} \quad \text{(Torricelli's law).} \tag{2}$$

Each equation involves the independent variable t, a proportionality constant k, and an unknown function $P(t)$ or $y(t)$ of t. The dependent variable P in Eq. (1) denotes a population for which the time rate of change is proportional to the size of the population. The dependent variable y in Eq. (2) denotes the depth of water in a tank draining slowly through a small hole in its bottom.

A (first-order) **differential equation** is an equation that can be written in the form

$$\frac{dy}{dx} = F(x, y) \tag{3}$$

where $F(x, y)$ is a given expression involving the independent variable x and the dependent variable y, the latter representing an unknown function $y(x)$ of x. A **solution** of Eq. (3) is a specific function $y(x)$ such that $y'(x) = F(x, y(x))$ for all x in some appropriate interval I. That is, Eq. (3) reduces to an identity in x when the dependent variable y is replaced with the solution function $y(x)$.

A differential equation frequently appears together with an **initial condition** $y(a) = b$ that specifies a desired value b of the solution at the point $x = a$. The two together constitute an **initial value problem**

$$\frac{dy}{dx} = F(x, y), \qquad y(a) = b, \tag{4}$$

which asks for a particular solution $y(x)$ of the differential equation that also satisfies the given initial condition.

EXAMPLE 1 To verify that a given function $y(x)$ is a solution of Eq. (3), it suffices to calculate the derivative dy/dx and then to verify that it is equal to $F(x, y(x))$. For instance, if $y = x^7$ then

$$\frac{dy}{dx} = 7x^6 = 7 \cdot \frac{x^7}{x} = \frac{7y}{x}$$

if $x > 0$. This calculation verifies that the function $y(x) = x^7$ is a solution (on the interval $x > 0$) of the differential equation

$$\frac{dy}{dx} = \frac{7y}{x}. \tag{5}$$

You can verify similarly that $y(x) = Cx^7$ satisfies Eq. (5) for any value of the constant C. ◆

When a solution of a differential equation contains an arbitrary constant C, we call the solution a **general solution** of the equation. A general solution actually describes an infinite collection of **particular solutions** of the differential equation, because different choices of C yield different solutions of the equation. Thus the choices $C = 11$ and $C = 23$ in Example 1 yield the two particular solutions $y_1(x) = 11x^7$

and $y_2(x) = 23x^7$. The former satisfies the initial value problem

$$\frac{dy}{dx} = \frac{7y}{x}, \qquad y(1) = 11,$$

while $y_2(x)$ satisfies the initial condition $y(1) = 23$.

Equations with One Variable Missing

If the dependent variable y does not appear explicitly on the right-hand side in (3), then the differential equation reduces to the form

$$\frac{dy}{dx} = f(x) \tag{6}$$

where f is a given function of x. As we noted in Section 5.2, the solution of a differential equation of this simple form reduces to integration:

$$y(x) = \int y'(x)\,dx = \int f(x)\,dx + C.$$

If the indefinite integral $F(x) = \int f(x)\,dx$ can be evaluated, then $y(x) = F(x) + C$ is a general solution of Eq. (6). For instance, a general solution of the differential equation $dy/dx = 21x^6$ is given by

$$y(x) = \int 21x^6\,dx = 3x^7 + C.$$

If, on the other hand, the independent variable x does not appear explicitly on the right-hand side in (3), then the differential equation reduces to the form

$$\frac{dy}{dx} = g(y), \tag{7}$$

which implies that $y'(x) = g(y(x))$ if $y(x)$ is a solution. To solve the equation in (7), we can first divide both sides by $g(y(x))$ and then attempt to integrate with respect to x:

$$\int \frac{y'(x)}{g(y(x))}\,dx = \int 1\,dx; \qquad [\text{substitute } y = y(x),\ dy = y'(x)\,dx]$$

$$\int \frac{1}{g(y)}\,dy = x + C. \tag{8}$$

If the indefinite integral

$$G(y) = \int \frac{1}{g(y)}\,dy$$

in (8) can be evaluated, then we may call the resulting equation

$$G(y) = x + C \tag{9}$$

an **implicit solution** of Eq. (7)—whether or not we can solve explicitly for y as a function of x.

REMARK Note that Eq. (8) results formally from the differential equation in (7) if we first divide both sides by $g(y)$ and multiply by dx to "separate the variables,"

$$\frac{1}{g(y)}\,dy = dx,$$

then integrate each side with respect to its "own" variable—y on the left and x on the right.

EXAMPLE 2 Solve the initial value problem

$$\frac{dy}{dx} = y^2, \qquad y(0) = 2. \tag{10}$$

Solution Separating the variables as in Eqs. (8) and (9), we write

$$\int \frac{1}{y^2}\,dy = \int dx; \qquad -\frac{1}{y} = x + C.$$

This is an implicit general solution from which we readily obtain the explicit general solution

$$y(x) = -\frac{1}{x + C}. \tag{11}$$

We find C by substituting the initial values $x = 0$, $y = 2$. This gives $C = -\frac{1}{2}$, so the desired particular solution of the initial value problem in (10) is

$$y(x) = -\frac{1}{x - \frac{1}{2}} = \frac{2}{1 - 2x}.$$

The graph of this solution, passing through the point (0, 2), is highlighted in Fig. 8.1.1. ◆

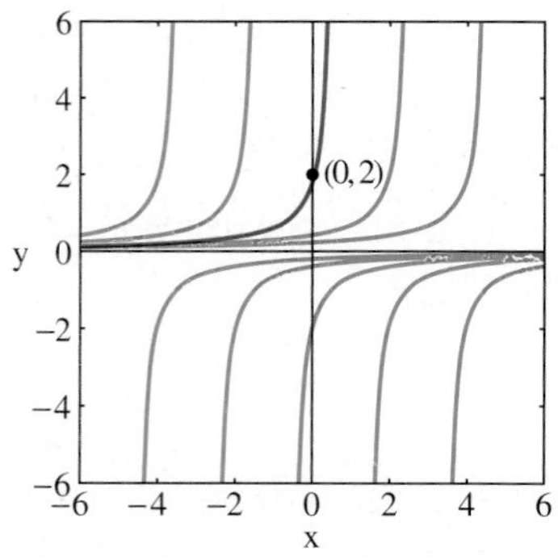

FIGURE 8.1.1 Solution curves of the differential equation $y' = y^2$.

The graph of a solution of a differential equation is called a **solution curve** of the equation. Figure 8.1.1 shows a variety of different solution curves of the differential equation $dy/dx = y^2$ of Example 2. It appears that these solution curves fill the xy-plane. Indeed, given any fixed point (a, b) of the plane with $b \neq 0$, we can substitute $x = a$ and $y = b$ in Eq. (11) and solve for C to obtain the solution curve that passes through this point. The x-axis is also a solution curve of the differential equation. (Why?)

The Natural Growth Equation

With $x(t)$ in place of $P(t)$ in Eq. (1), we have the differential equation

$$\frac{dx}{dt} = kx, \tag{12}$$

which serves as the mathematical model for an extraordinarily wide range of natural phenomena. It is easily solved if we first "separate the variables" and then integrate:

$$\frac{dx}{x} = k\,dt;$$

$$\int \frac{dx}{x} = \int k\,dt;$$

$$\ln x = kt + C.$$

We apply the exponential function to both sides of the last equation to solve for x:

$$x = e^{\ln x} = e^{kt+C} = e^{kt}e^{C} = Ae^{kt}.$$

Here, $A = e^C$ is a constant that remains to be determined. But we see that A is simply the value $x_0 = x(0)$ of $x(t)$ when $t = 0$, and thus $A = x_0$.

THEOREM 1 The Natural Growth Equation
The solution of the initial value problem

$$\frac{dx}{dt} = kx, \qquad x(0) = x_0 \tag{13}$$

is

$$x(t) = x_0 e^{kt}. \tag{14}$$

As a consequence, Eq. (12) is often called the **exponential growth equation,** or the **natural growth equation.** We see from Eq. (14) that, with $x_0 > 0$, the solution $x(t)$ is an increasing function if $k > 0$ and a decreasing function if $k < 0$. (The situation $k < 0$ is sometimes called *exponential decay.*) These two cases are illustrated in Figs. 8.1.2 and 8.1.3, respectively. The remainder of this section concerns examples of natural phenomena for which this differential equation serves as a mathematical model.

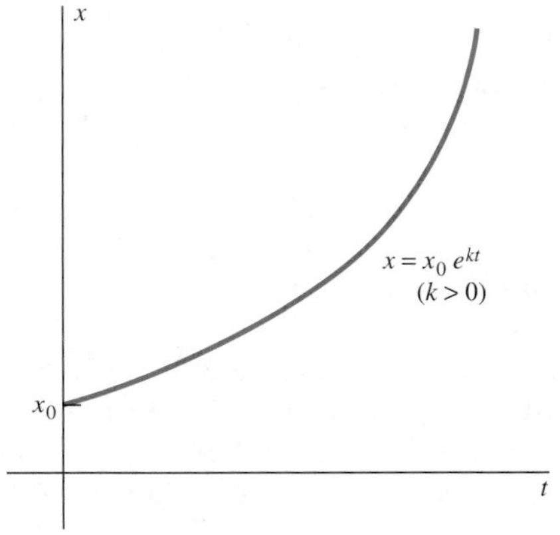

FIGURE 8.1.2 Solution of the exponential growth equation for $k > 0$.

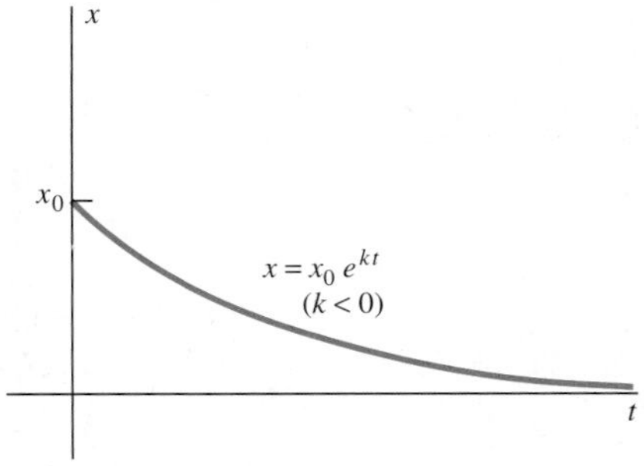

FIGURE 8.1.3 Solution of the exponential growth equation—now actually a *decay* equation—for the case $k < 0$.

Population Growth

When we compare Eqs. (1), (12), and (14), we see that a population $P(t)$ with growth rate proportional to its size is given by

$$P(t) = P_0 e^{kt}, \tag{15}$$

where $P_0 = P(0)$. If t is measured in years, then the proportionality constant k in (15) is called the **annual growth rate,** which can be positive, negative, or zero. Its value is often given as a percentage (its decimal value multiplied by 100). If k is close to zero, then this value is fairly close to the annual percentage increase (or decrease) of the population each year.

EXAMPLE 3 According to data posted at `www.census.gov`, the world's total population reached 6 billion persons in mid-1999, and was then increasing at the rate of about 212 thousand persons each day. Assuming that natural population growth at this rate continues, we want to answer these questions:

(a) What is the annual growth rate k?
(b) What will the world population be at the middle of the 21st century?
(c) How long will it take for the world population to increase tenfold—thereby reaching the 60 billion that some demographers believe to be the maximum for which the planet can provide adequate food supplies?

Solution **(a)** We measure the world population $P(t)$ in billions and measure time in years. We take $t = 0$ to correspond to mid-1999, so that $P_0 = 6$. The fact that $P(t)$ is increasing by 212,000, or 0.000212 billion, persons per day at time $t = 0$ means that

$$P'(0) = (0.000212)(365.25) \approx 0.07743$$

billion per year. From the natural growth equation $dP/dt = kP$ with $t = 0$ we now obtain

$$k = \frac{P'(0)}{P(0)} \approx \frac{0.07743}{6} \approx 0.0129.$$

Thus the world population was growing at the rate of about 1.29% annually in mid-1999. This value of k gives the world population function

$$P(t) = P_0 e^{kt} = 6e^{(0.0129)t}.$$

(b) With $t = 51$ we obtain the prediction

$$P(51) = 6e^{(0.0129)(51)} \approx 11.58$$

(billion) for the world population in mid-2050, so the population will almost double in the 51 years following 1999.

(c) The world population should reach 60 billion when

$$60 = 6e^{(0.0129)t}; \quad \text{that is, when} \quad t = \frac{\ln 10}{0.0129} \approx 178,$$

thus in the year 2177. ◆

Note Actually, the rate of growth of the world population is expected to slow somewhat during the next half-century, and the best current prediction for the population in the year 2050 is "only" 9.1 billion. A simple mathematical model cannot be expected to mirror precisely the complexity of the real world.

Radioactive Decay and Radiocarbon Dating

Consider a sample of material that contains $N(t)$ atoms of a certain radioactive isotope at time t. Many experiments have confirmed that a constant fraction of these radioactive atoms will spontaneously decay (into atoms of another element or another isotope of the same element) during each given unit of time. Consequently, the sample behaves exactly like a population with a constant death rate but with no births occurring. To write a model for $N(t)$, we use Eq. (1) with N in place of P and with $-k$ in place of k (so that $k > 0$ corresponds to a decreasing number of atoms). We thus obtain the differential equation

$$\frac{dN}{dt} = -kN. \tag{16}$$

From the solution (14) of Eq. (12), with k replaced with $-k$, we conclude that

$$N(t) = N_0 e^{-kt}, \tag{17}$$

where $N_0 = N(0)$, the number of radioactive atoms of the original isotope present in the sample at time $t = 0$.

The value of the *decay constant* k depends on the particular isotope with which we are dealing. If k is large, then the isotope decays rapidly. If k is near zero, the isotope decays quite slowly and thus may be a relatively persistent factor in its environment. The decay constant k is often specified in terms of another empirical parameter that is more convenient, the *half-life* of the isotope. The **half-life** τ of a sample of a radioactive isotope is the time required for *half* of that sample to decay. To find the relationship between k and τ, we set

$$t = \tau \quad \text{and} \quad N = \tfrac{1}{2}N_0$$

in Eq. (17), so that

$$\tfrac{1}{2}N_0 = N_0 e^{-k\tau}. \tag{18}$$

When we solve for τ, we find that

$$\tau = \frac{\ln 2}{k}. \tag{19}$$

Note that the concept of half-life is meaningful—the value of τ depends *only* on k and thus depends only on the particular isotope involved. It does *not* depend on the amount of that isotope present.

The method of *radiocarbon dating* is based on the fact that the radioactive carbon isotope ^{14}C has a known half-life of about 5700 yr. Living organic matter maintains a constant level of ^{14}C by "breathing" air (or by consuming organic matter that does so). But air contains ^{14}C along with the much more common, stable isotope ^{12}C of carbon, mostly in the gas CO_2. Thus all living organisms maintain the same percentage of ^{14}C as in air, because organic processes seem to make no distinction between the two isotopes. But when an organism dies, it ceases to metabolize carbon, and the process of radioactive decay begins to deplete its ^{14}C content. The fraction of ^{14}C in the air remains roughly constant because new ^{14}C is continuously generated by the bombardment of nitrogen atoms in the upper atmosphere by cosmic rays, and this generation has long been in steady-state equilibrium with the loss of ^{14}C through radioactive decay.

EXAMPLE 4 A specimen of charcoal found at Stonehenge contains 63% as much ^{14}C as a sample of present-day charcoal. What is the age of the sample?

Solution We take $t = 0$ (in years) as the time of death of the tree from which the Stonehenge charcoal was made. From Eq. (18), we know that

$$\tfrac{1}{2}N_0 = N_0 e^{-5700k},$$

so

$$k = \frac{\ln 2}{\tau} = \frac{\ln 2}{5700} \approx 0.0001216.$$

We are given that $N = (0.63)N_0$ at present, so we solve the equation

$$(0.63)N_0 = N_0 e^{-kt}$$

with this value of k. We thus find that

$$t = -\frac{\ln(0.63)}{0.0001216} \approx 3800 \quad \text{(yr)}.$$

Therefore, the sample is about 3800 yr old. If it is connected in any way with the builders of Stonehenge, our computations suggest that this observatory, monument, or temple—whichever it may be—dates from almost 1800 B.C. ◆

EXAMPLE 5 According to one cosmological theory, there were equal amounts of the uranium isotopes ^{235}U and ^{238}U at the creation of the universe in the "big bang." At present there are 137.7 ^{238}U atoms for each ^{235}U atom. Using the known half-lives

$$4.51 \text{ billion yr for } ^{238}U,$$

$$0.71 \text{ billion yr for } ^{235}U,$$

calculate the age of the universe.

Solution Let $N_8(t)$ and $N_5(t)$ be the numbers of ^{238}U and ^{235}U atoms, respectively, at time t, in billions of years after the creation of the universe. Then

$$N_8(t) = N_0 e^{-kt} \quad \text{and} \quad N_5(t) = N_0 e^{-ct},$$

where N_0 is the initial number of atoms of each isotope. Also,

$$k = \frac{\ln 2}{4.51} \quad \text{and} \quad c = \frac{\ln 2}{0.71},$$

a consequence of Eq. (19). We divide the equation for N_8 by the equation for N_5 and find that when t has the value corresponding to "now,"

$$137.7 = \frac{N_8}{N_5} = e^{(c-k)t}.$$

Finally, we solve this equation for t:

$$t = \frac{\ln(137.7)}{\left(\frac{1}{0.71} - \frac{1}{4.51}\right)\ln 2} \approx 5.99.$$

Thus we estimate the age of the universe to be about 6 billion years, which is roughly on the same order of magnitude as recent estimates of 10 to 15 billion years (based on astronomical observations of the rate of expansion of the universe). ◆

More Natural Growth and Decay Models

Continuously Compounded Interest Consider a savings account that is opened with an initial deposit of A_0 dollars and earns interest at the annual rate r. If there are $A(t)$ dollars in the account at time t and the interest is compounded at time $t + \Delta t$, this means that $rA(t)\,\Delta t$ dollars in interest are added to the account then. So

$$A(t + \Delta t) = A(t) + rA(t)\,\Delta t,$$

and thus

$$\frac{\Delta A}{\Delta t} = \frac{A(t + \Delta t) - A(t)}{\Delta t} = rA(t).$$

Continuous compounding of interest results from taking the limit as $\Delta t \to 0$, so

$$\frac{dA}{dt} = rA. \tag{20}$$

This is an exponential growth equation with solution

$$A(t) = A_0 e^{rt}. \tag{21}$$

EXAMPLE 6 If $A_0 = \$1000$ is invested at an annual interest rate of 6% compounded continuously, then $r = 0.06$, and Eq. (21) gives

$$A(1) = 1000e^{(0.06)(1)} = \$1061.84$$

for the value of the investment after one year. Hence the *effective annual interest rate* is 6.184%. Thus the more often interest is compounded, the more rapidly savings grow, but bank advertisements sometimes overemphasize this advantage. For instance, 6% compounded *monthly* multiplies your investment by

$$1 + \frac{0.06}{12} = 1.005$$

at the end of each month, so an initial investment of \$1000 would grow in one year to

$$(1000)(1.005)^{12} = \$1061.68,$$

only 16¢ less than would be yielded by continuous compounding. ◆

Drug Elimination The amount $A(t)$ of a certain drug in the human bloodstream, as measured by the excess above the natural level of the drug in the bloodstream, typically declines at a rate proportional to that excess amount. That is,

$$\frac{dA}{dt} = -\lambda A, \quad \text{so} \quad A(t) = A_0 e^{-\lambda t}. \tag{22}$$

The parameter λ is called the *elimination constant* of the drug, and $T = 1/\lambda$ is called the *elimination time.*

EXAMPLE 7 The elimination time for alcohol varies from one person to another. If a person's "sobering time" $T = 1/\lambda$ is 2.5 h, how long will it take the excess bloodstream alcohol concentration to be reduced from 0.10% to 0.02%?

Solution We assume that the normal concentration of alcohol in the blood is zero, so any amount is an excess amount. In this problem, we have $\lambda = 1/2.5 = 0.4$, so Eq. (22) yields

$$0.02 = (0.10)e^{-(0.4)t}.$$

Thus

$$t = -\frac{\ln(0.2)}{0.4} \approx 4.02 \quad \text{(h)}.$$

◆

Sales Decline According to marketing studies, if advertising for a particular product is halted and other market conditions—such things as number and promotion of competing products, their prices, and so on—remain unchanged, then the sales of the unadvertised product will decline at a rate that is proportional at any time t to the current sales S. That is,

$$\frac{dS}{dt} = -\lambda S, \quad \text{so} \quad S(t) = S_0 e^{-\lambda t}.$$

Here S_0 denotes the initial value of the sales, which we take to be sales in the final month of advertising. If we take months as the units for time t, then $S(t)$ gives the number of sales t months after advertising is halted, and λ might be called the *sales decay constant.*

Linguistics Consider a basic list of N_0 words in use in a given language at time $t = 0$. Let $N(t)$ denote the number of these words that are still in use at time t—those that have neither disappeared from the language nor been replaced. According to one theory in linguistics, the rate of decrease of N is proportional to N. That is,

$$\frac{dN}{dt} = -\lambda N, \quad \text{so} \quad N(t) = N_0 e^{-\lambda t}.$$

If t is measured in millennia (as is standard in linguistics), then $k = e^{-\lambda}$ is the fraction of the words in the original list that survive for 1000 yr.

Torricelli's Law

Suppose that a water tank has a hole with area a at its bottom and that water is draining from the hole. Denote by $y(t)$ the depth (in feet) of water in the tank at time t (in seconds) and by $V(t)$ the volume of water (in cubic feet) in the tank then. It is plausible—and true under ideal conditions—that the velocity of the stream of water exiting through the hole is

$$v = \sqrt{2gy} \quad (g \approx 32 \text{ ft/s}^2), \tag{23}$$

which is the velocity that a drop of water would acquire in falling freely from the water surface to the hole. This is *Torricelli's law of draining.*

As indicated in Fig. 8.1.4, the amount of water that leaves through the bottom hole during a short time interval dt amounts to a cylinder with base area a and height $v\,dt$. Hence the resulting change dV in the volume of water in the tank is given by

$$dV = -av\,dt = -a\sqrt{2gy}\,dt. \tag{24}$$

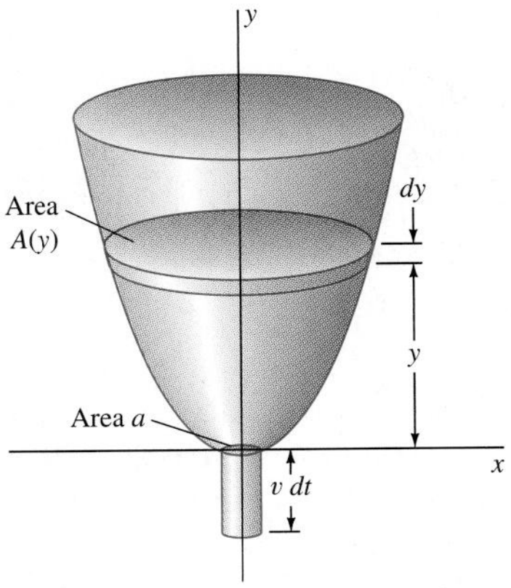

FIGURE 8.1.4 Derivation of Torricelli's law.

But if $A(y)$ denotes the horizontal cross-sectional area of the tank at height y above the hole, then

$$dV = A(y)\,dy, \tag{25}$$

as usual. Comparing Eqs. (24) and (25), we see that $y(t)$ satisfies the differential equation

$$A(y)\frac{dy}{dt} = -a\sqrt{2gy}. \tag{26}$$

In some applications this is a very convenient form of Torricelli's law. In other situations you may prefer to work with the differential equation in (24) in the form

$$\frac{dV}{dt} = -a\sqrt{2gy} \tag{27}$$

or, if the area of the bottom hole is unknown, the form

$$\frac{dV}{dt} = -c\sqrt{y}, \tag{28}$$

where $c = a\sqrt{2g}$ is a positive constant.

Cylindrical water tanks are common (Fig. 8.1.5). In this case the cross-sectional area function in Eq. (26) is constant: $A(y) \equiv A$. Consequently Eq. (26) reduces to the simple differential equation

$$\frac{dy}{dt} = -k\sqrt{y} \tag{29}$$

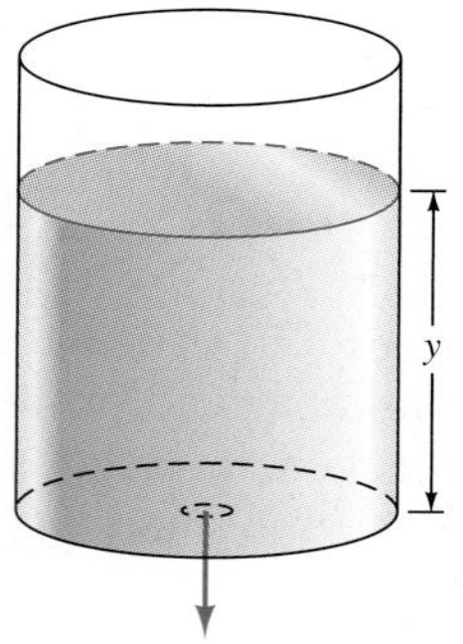

FIGURE 8.1.5 A cylindrical water tank has constant cross-sectional area.

where $k = (a/A)\sqrt{2g}$ is a positive constant that frequently is determined from given tank-draining data (rather than from a knowledge of the areas a and A).

EXAMPLE 8 The water in a draining cylindrical tank is 10 ft deep at noon. At 1:00 P.M. it is 5 ft deep. When will the tank be empty?

Solution If we write Eq. (29) in the form

$$\frac{1}{\sqrt{y}} \cdot \frac{dy}{dt} = -k,$$

then integration yields

$$2\sqrt{y} = -kt + C.$$

Substituting the initial data $y = 10$ when $t = 0$ (noon) gives $C = 2\sqrt{10}$, so

$$2\sqrt{y} = -kt + 2\sqrt{10}. \tag{30}$$

Then substituting the additional data $y = 5$ when $t = 1$ (1:00 P.M.) gives $k = 2\sqrt{10} - 2\sqrt{5}$. Substituting this value in Eq. (30) and dividing by 2 yields

$$\sqrt{y} = (\sqrt{5} - \sqrt{10})t + \sqrt{10}. \tag{31}$$

Finally, the tank is empty when $y = 0$ in Eq. (31), and thus when

$$t = \frac{\sqrt{10}}{\sqrt{10} - \sqrt{5}} \approx 3.414,$$

about 3 h 25 min. So we see that—whereas a natural (but naive and **wrong**) guess might have been 2:00 P.M. (one more hour for the remaining 5 ft of water to drain)—the tank actually is not empty until about 3:25 P.M. You should use Eq. (31) to show that the actual water depth in the tank at 2:00 P.M. is about 1.72 ft, and the depth at 3:00 P.M. is about 2 in., so it takes about 25 min for the last 2 in. of water in the tank to drain! ◆

EXAMPLE 9 A hemispherical tank has top radius 4 ft and, at time $t = 0$, is full of water. At that moment a circular hole of diameter 1 in. is opened in the bottom of the tank. How long will it take for all the water to drain from the tank?

Solution From the right triangle in Fig. 8.1.6, we see that

$$A(y) = \pi r^2 = \pi[16 - (4 - y)^2] = \pi(8y - y^2).$$

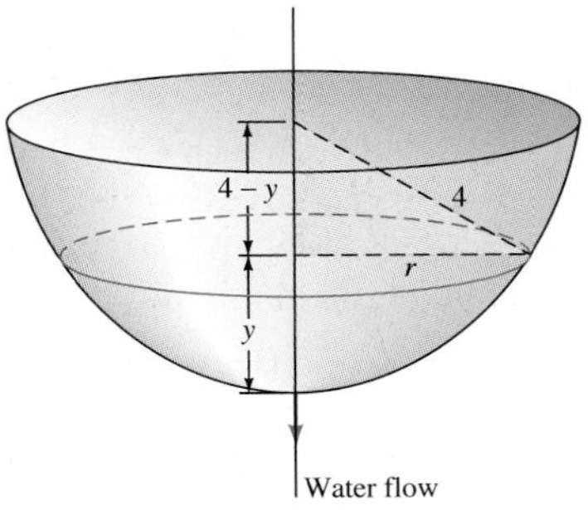

FIGURE 8.1.6 Draining a hemispherical tank.

With $g = 32$ ft/s^2, Eq. (26) takes the form

$$\pi(8y - y^2)\frac{dy}{dt} = -\pi\left(\frac{1}{24}\right)^2\sqrt{64y};$$

$$\int (8y^{1/2} - y^{3/2})\,dy = -\int \frac{1}{72}\,dt + C;$$

$$\frac{16}{3}y^{3/2} - \frac{2}{5}y^{5/2} = -\frac{1}{72}t + C.$$

Now $y(0) = 4$, so

$$C = \tfrac{16}{3} \cdot 4^{3/2} - \tfrac{2}{5} \cdot 4^{5/2} = \tfrac{448}{15}.$$

The tank is empty when $y = 0$—that is, when

$$t = 72 \cdot \tfrac{448}{15} \approx 2150 \quad \text{(s)},$$

about 35 min 50 s. So it takes slightly less than 36 min for the tank to drain. ◆

8.1 TRUE/FALSE STUDY GUIDE

8.1 CONCEPTS: QUESTIONS AND DISCUSSION

1. Do the solution curves of the natural growth equation $dy/dx = ky$ (with k a constant) fill the entire xy-plane?
2. Do the solution curves of the differential equation $dy/dx = k\sqrt{y}$ (with k a constant) fill the entire xy-plane?
3. Can one solve the differential equation $dy/dx = xy$ simply by integrating both sides with respect to x? Justify your answer.

8.1 PROBLEMS

In each of Problems 1 through 10, first find a general solution of the given differential equation. Then find a particular solution that satisfies the given initial condition.

1. $\dfrac{dy}{dx} = 2y, \quad y(1) = 3$

2. $\dfrac{dy}{dx} = -3y, \quad y(5) = -10$

3. $\dfrac{dy}{dx} = 2y^2, \quad y(7) = 3$

4. $\dfrac{dy}{dx} = \dfrac{7}{y}, \quad y(0) = 6$

5. $\dfrac{dy}{dx} = 2\sqrt{y}, \quad y(0) = 9$

6. $\dfrac{dy}{dx} = 6y^{2/3}, \quad y(1) = 8$

7. $\dfrac{dy}{dx} = 1 + y, \quad y(0) = 5$

8. $\dfrac{dy}{dx} = (2 + y)^2, \quad y(5) = 3$

9. $\dfrac{dy}{dx} = e^{-y}, \quad y(0) = 2$

10. $\dfrac{dy}{dx} = 2\sec y, \quad y(0) = 0$

In Problems 11 through 15, a function $y = g(x)$ is described by a geometric property of its graph. Write a differential equation of the form $dy/dx = F(x, y)$ having the function g as a solution.

11. The slope of the graph of g at the point (x, y) is the sum of x and y.

12. The line tangent to the graph of g at the point (x, y) intersects the x-axis at the point $(x/2, 0)$.

13. Every straight line normal to the graph of g passes through the point $(0, 1)$.

14. The graph of g is normal to every curve of the form $y = kx^2$ (k is a constant) where they meet.

15. The line tangent to the graph of g at the point (x, y) passes through the point $(-y, x)$.

In Problems 16 through 20, write—in the manner of Eqs. (1) and (2) of this section—a differential equation that is a mathematical model of the situation described.

16. The time rate of change of a population $P = P(t)$ is proportional to the square root of P.

17. The time rate of change of the velocity $v = v(t)$ of a coasting motorboat is proportional to the square of v.

18. The acceleration dv/dt of a Lamborghini is proportional to the difference between 250 km/h and the velocity of the car.

19. In a city having a fixed population P of persons, the time rate of change of the number N of those persons who have heard a certain rumor is proportional to the number of those who have not yet heard the rumor.

20. In a city with a fixed population P of persons, the time rate of change of the number N of those persons infected with a certain contagious disease is proportional to the product of the number who have the disease and the number who do not.

21. *Continuously Compounded Interest* Suppose that \$1000 is deposited in a savings account that pays 8% annual interest compounded continuously. At what rate (in dollars per year) is it earning interest after 5 yr? After 20 yr?

22. *Population Growth* Coopersville had a population of 25,000 in 1970 and a population of 30,000 in 1980. Assume that its population will continue to grow exponentially at a constant rate. What population can the Coopersville city planners expect in the year 2010?

23. *Population Growth* In a certain culture of bacteria, the number of bacteria increased sixfold in 10 h. Assuming natural growth, how long did it take for their number to double?

24. *Radiocarbon Dating* Carbon extracted from an ancient skull recently unearthed contained only one-sixth as much radioactive ^{14}C as carbon extracted from present-day bone. How old is the skull?

25. *Radiocarbon Dating* Carbon taken from a relic purported to date from A.D. 30 contained 4.6×10^{10} atoms of ^{14}C per gram. Carbon extracted from a present-day specimen of the same substance contained 5.0×10^{10} atoms of ^{14}C per gram. Compute the approximate age of the relic. What is your opinion as to its authenticity?

26. *Continuously Compounded Interest* Upon the birth of their first child, a couple deposited \$5000 in a savings account that pays 6% annual interest compounded continuously. The interest payments are allowed to accumulate. How much will the account contain when the child is ready to go to college at age 18?

27. *Continuously Compounded Interest* You discover in your attic an overdue library book on which your great-great-great-grandfather owed a fine of 30¢ exactly 100 years ago. If an overdue fine grows exponentially at a 5% annual interest rate compounded continuously, how much would you have to pay if you returned the book today?

28. *Drug Elimination* Suppose that sodium pentobarbital will anesthetize a dog when its bloodstream contains at least 45 mg of sodium pentobarbital per kilogram of body weight of the dog. Suppose also that sodium pentobarbital is eliminated exponentially from a dog's bloodstream, with a half-life of 5 h. What single dose should be administered to anesthetize a 50-kg dog for 1 h?

29. *Sales Decline* Moonbeam Motors has discontinued advertising of their sports-utility vehicle. The company plans to resume advertising when sales have declined to 75% of their initial rate. If after 1 week without advertising, sales have declined to 95% of their initial rate, when should the company expect to resume advertising?

30. *Linguistics* The English language evolves in such a way that 77% of all words disappear (or are replaced) every 1000 yr. Of a basic list of words used by Chaucer in A.D. 1400, what percentage should we expect to find still in use today?

31. *Radioactive Decay* The half-life of radioactive cobalt is 5.27 yr. Suppose that a nuclear accident has left the level of cobalt radiation in a certain region at 100 times the level acceptable for human habitation. How long will it be before the region is again habitable? (Ignore the likely presence of other radioactive substances.)

32. *Radioactive Decay* Suppose that a rare mineral deposit formed in an ancient cataclysm—such as the collision of a meteorite with the earth—originally contained the uranium isotope ^{238}U (which has a half-life of 4.51×10^9 yr) but none of the lead isotope ^{207}Pb, the end product of the radioactive decay of ^{238}U. If the ratio of ^{238}U atoms to ^{207}Pb atoms in the mineral deposit today is 0.9, when did the cataclysm occur?

33. A bacteria population $P(t)$ undergoing natural growth numbers 49 at 12 noon. (a) Suppose that there are 294 bacteria at 1:00 P.M. Write a formula giving $P(t)$ after t hours. (b) How many bacteria are there at 1:40 P.M.? (c) At what time—to the nearest minute—will there be 20 thousand bacteria?

34. The amount $A(t)$ of atmospheric pollutants in a certain mountain valley satisfies the natural growth equation and triples every 7.5 years. (a) If the initial amount is 10 pu ("pollution units"), write a formula for $A(t)$ giving the amount (in pu) present after t years. (b) What will be the amount of pollutants present in the valley atmosphere after 5 years? (c) If it will be dangerous to stay in the valley when the amount of pollutants reaches 100 pu, how long will this take?

35. An accident at a nuclear power plant has left the surrounding area polluted with radioactive material that undergoes natural decay. The initial amount of radioactive material present is 15 su ("safe units") and 5 months later it is 10 su. (a) Write a formula giving the amount $A(t)$ of radioactive material (in su) remaining after t months. (b) What amount of radioactive material will remain after 8 months? (c) How long—total number of months or fraction thereof—will it be until $A(t) = 1$ (su), so it is safe for people to return to the area?

36. There are now about 3300 different human *language families* in the whole world. Assume that all of these are derived from a single original language and that a language family develops into 1.5 language families every 6 thousand years. About how long ago was the single original human language spoken?

37. Thousand of years ago ancestors of the Native Americans crossed the Bering Strait from Asia and entered the western hemisphere. Since then, they have fanned out across North and South America. The single language that the original Native Americans spoke has since split into many language families. Assume (as in Problem 36) that the number of these language families has been multiplied by 1.5 every 6000 years. There are now 150 Native American language families in the western hemisphere. About when did the ancestors of today's Native Americans arrive?

38. In 1998 there were 40 million Internet users in the world and this number was then doubling every 100 days. Assuming that this rate of growth continued, how long would it be until all the world's 6 billion human beings were using the Internet?

39. A tank shaped like a vertical cylinder initially contains water to a depth of 9 ft (Fig. 8.1.7). A bottom plug is pulled at time $t = 0$ (t in hours). After 1 h the depth has dropped to 4 ft. How long will it take all the water to drain from this tank?

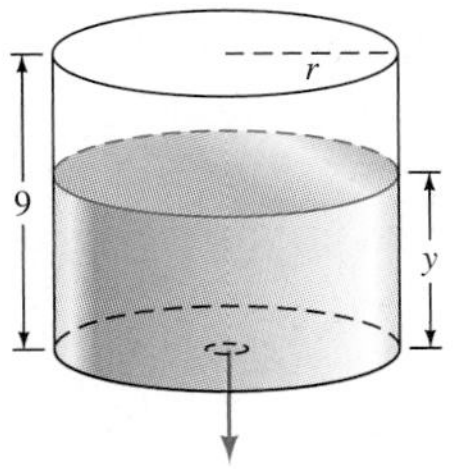

FIGURE 8.1.7 The cylindrical tank of Problem 39.

40. Suppose that the tank of Problem 39 has a radius of 3 ft and that its bottom hole is circular with radius 1 in. How long will it take for the water, initially 9 ft deep, to drain completely?

41. A water tank is in the shape of a right circular cone with its axis vertical and its vertex at the bottom. The tank is 16 ft high and the radius of its top is 5 ft. At time $t = 0$, a plug at its vertex is removed and the tank, initially full of water, begins to drain. After 1 h the water in the tank is 9 ft deep. When will the tank be empty (Fig. 8.1.8)?

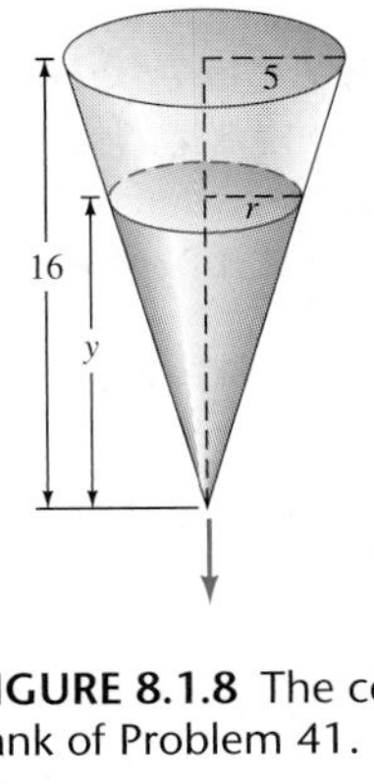

FIGURE 8.1.8 The conical tank of Problem 41.

42. Suppose that a cylindrical tank (axis vertical) initially containing V_0 liters of water drains through a bottom hole in T minutes. Use Torricelli's law to show that the volume of water in the tank after $t \leqq T$ minutes is $V(t) = V_0[1-(t/T)]^2$.

43. The shape of a water tank is obtained by revolving the curve $y = x^{4/3}$ around the y-axis (units on the coordinate axes are in feet). A plug at the bottom is removed at 12 noon, when the water depth in the tank is 12 ft. At 1 P.M. the water depth is 6 ft. When will the tank be empty?

44. The shape of a water tank is obtained by revolving the parabola $y = x^2$ around the y-axis (units on the coordinate axes are in feet; see Fig. 8.1.9). The water depth is 4 ft at 12 noon; at that time, a plug in a circular hole at the bottom of the tank is removed. At 1 P.M. the water level is 1 ft. (a) Find the water depth $y(t)$ after t hours. (b) When will the tank be empty? (c) What is the radius of the circular hole at the bottom?

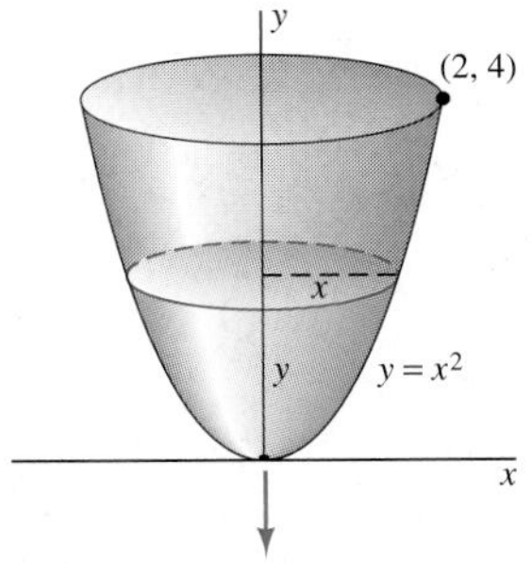

FIGURE 8.1.9 The tank of Problem 44.

45. A cylindrical tank of length 5 ft and radius 3 ft is situated with its axis horizontal. If a circular bottom hole of radius 1 in. is opened and the tank is initially half full of xylene, how long will it take the liquid to drain completely?

46. A spherical tank of radius 25 cm is full of mercury when a circular bottom hole of radius 5 mm is opened. How long will it be before all of the mercury drains from the tank? (Use $g = 9.8$ m/s^2.)

47. *The Clepsydra, or Water Clock* A 12-h water clock is to be designed with the dimensions shown in Fig. 8.1.10, shaped like the surface obtained by revolving the curve $y = f(x)$ around the y-axis. What equation should this curve have, *and* what radius should the bottom hole have, so that the water level will fall at the *constant* rate of 4 in./h?

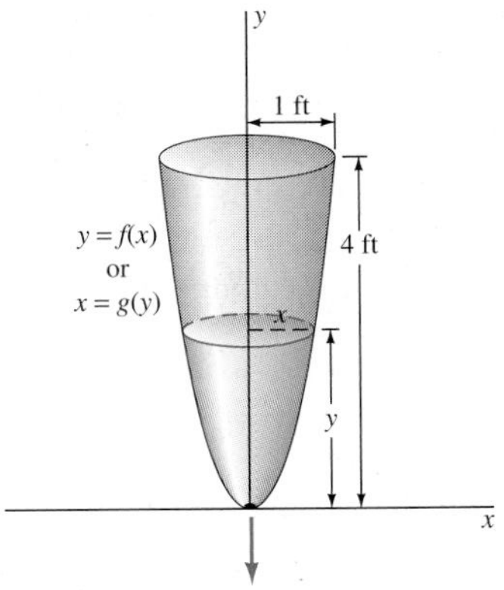

FIGURE 8.1.10 The clepsydra of Problem 47.

8.2 SLOPE FIELDS AND EULER'S METHOD

Consider a differential equation of the form $dy/dx = F(x, y)$, where $F(x, y)$ contains both the variables x and y. To solve it, we might think of integrating both sides with respect to x, and hence write $y(x) = \int F(x, y(x))\,dx + C$. Unfortunately, this does not provide a solution of the differential equation because the integral involves the unknown function $y(x)$ itself. In fact, there exists *no* straightforward procedure by which a general differential equation can be solved explicitly. Indeed, the solutions of such a simple-looking differential equation as $dy/dx = x^2 + y^2$ cannot be expressed in terms of the ordinary elementary functions studied in calculus. Nevertheless, the graphical and numerical methods of this section can be used to construct *approximate* solutions of differential equations that suffice for many practical purposes.

Slope Fields and Graphical Solutions

There is a simple geometric way to think about solutions of a given differential equation $dy/dx = F(x, y)$. At each point (x, y) of the xy-plane at which F is defined, the value of $F(x, y)$ determines a slope $m = y'(x) = F(x, y)$. A solution of the differential equation is simply a function whose graph has this "correct slope" at each point through which it passes. Thus a **solution curve** of the differential equation $dy/dx = F(x, y)$—the graph of a solution of this equation—is simply a curve in the xy-plane whose tangent line at each point (x, y) has slope $m = F(x, y)$. For instance, Fig. 8.2.1 shows a solution curve of the differential equation $dy/dx = x - y$ together with lines tangent at three typical points.

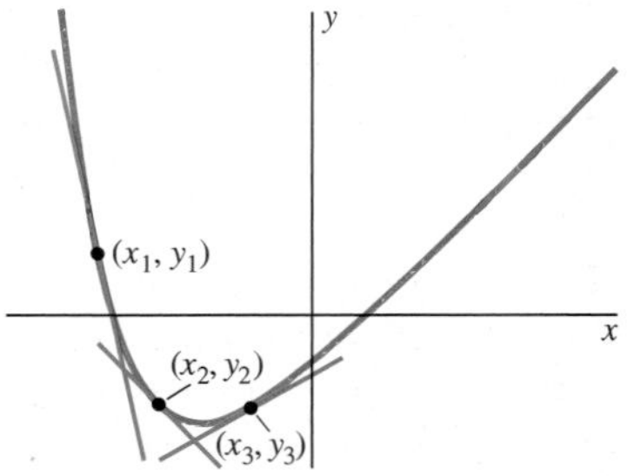

FIGURE 8.2.1 A solution curve for the differential equation $y' = x - y$ together with tangent lines having

- slope $m_1 = x_1 - y_1$ at the point (x_1, y_1);
- slope $m_2 = x_2 - y_2$ at the point (x_2, y_2); and
- slope $m_3 = x_3 - y_3$ at the point (x_3, y_3).

This geometric viewpoint suggests a *graphical method* for constructing approximate solutions of the differential equation $dy/dx = F(x, y)$. Through each of a representative collection of points (x, y) in the plane we draw a short line segment having the proper slope $m = F(x, y)$. All these line segments constitute a **slope field** (or **direction field**) for the equation $dy/dx = F(x, y)$.

This slope field suggests visually the general shapes of solution curves of the differential equation. Through each point a solution curve should proceed in such a direction that its tangent line is nearly parallel to the nearby line segments of the slope field. Beginning at any initial point (a, b), we can attempt to sketch freehand a solution curve that threads its way through the slope field, following the visible line segments as closely as possible.

EXAMPLE 1 Construct a slope field for the differential equation $dy/dx = x - y$ and use it to sketch a solution curve that passes through the point $(-4, 4)$.

Solution Figure 8.2.2 shows a table of slopes for the given equation. The numerical slope $m = x - y$ appears at the intersection of the horizontal x-row and the vertical y-column of the table. If you inspect the pattern of upper-left to lower-right diagonals in this table, you can see that it was easily and quickly constructed. (Of course, a more complicated function $F(x, y)$ on the right-hand side of the differential equation would

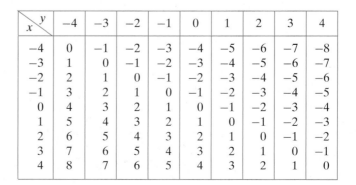

x \ y	−4	−3	−2	−1	0	1	2	3	4
−4	0	−1	−2	−3	−4	−5	−6	−7	−8
−3	1	0	−1	−2	−3	−4	−5	−6	−7
−2	2	1	0	−1	−2	−3	−4	−5	−6
−1	3	2	1	0	−1	−2	−3	−4	−5
0	4	3	2	1	0	−1	−2	−3	−4
1	5	4	3	2	1	0	−1	−2	−3
2	6	5	4	3	2	1	0	−1	−2
3	7	6	5	4	3	2	1	0	−1
4	8	7	6	5	4	3	2	1	0

FIGURE 8.2.2 Values of the slope $y' = x - y$ for $-4 \le x, y \le 4$.

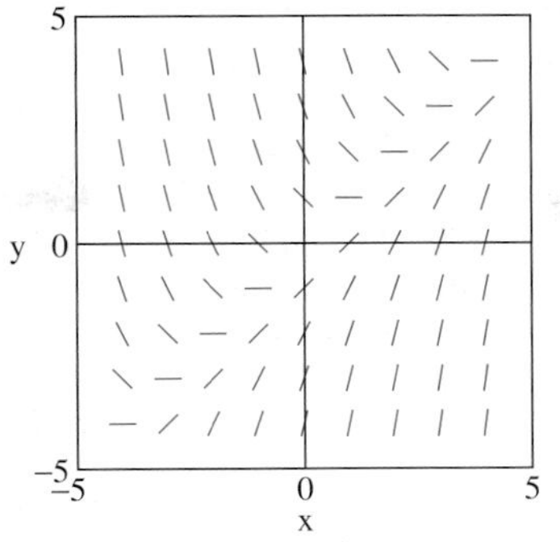

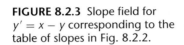

FIGURE 8.2.3 Slope field for $y' = x - y$ corresponding to the table of slopes in Fig. 8.2.2.

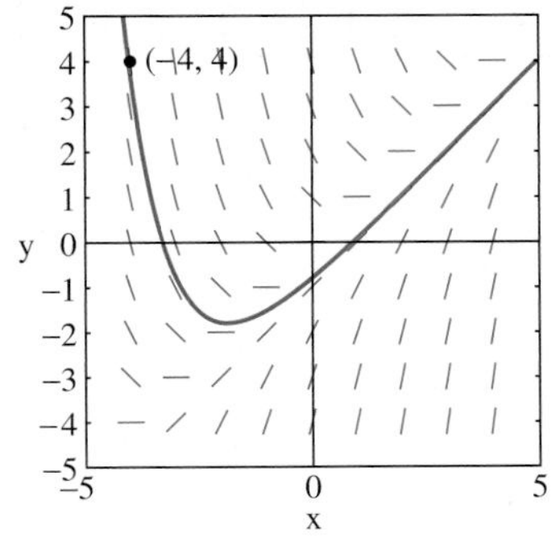

FIGURE 8.2.4 The solution curve through (−4, 4).

necessitate more complicated calculations.) Figure 8.2.3 shows the corresponding slope field; Fig. 8.2.4 shows a solution curve sketched through the point $(-4, 4)$ in such a way as to follow this slope field as closely as possible. At each point the solution curve appears to proceed in the direction indicated by the nearby line segments of the slope field. ◆

The 81 slope segments in Fig. 8.2.3 are tedious to construct by hand. Fortunately, most computer algebra systems include commands for quick and ready construction of slope fields with as many line segments as desired; such commands are illustrated in the project material for this section. The more line segments are constructed, the more accurately solution curves can be visualized and sketched. Figure 8.2.4 shows a "finer" slope field for the differential equation $dy/dx = x - y$ of Example 1, together with typical solution curves threading through this slope field.

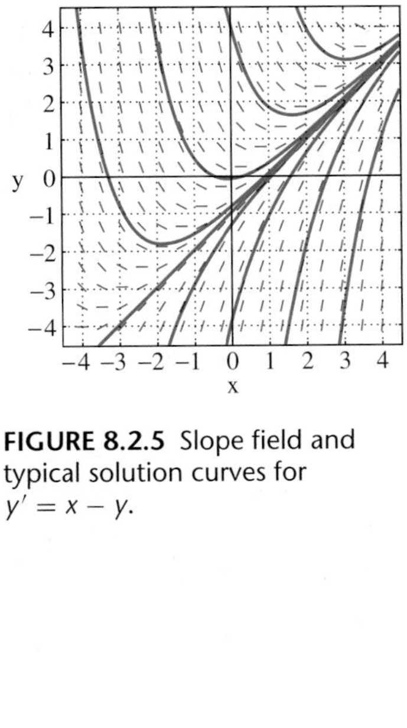

FIGURE 8.2.5 Slope field and typical solution curves for $y' = x - y$.

If you look closely at Fig. 8.2.5, you may detect a solution curve that appears to be a straight line! Indeed, you can verify that the linear function $y = x - 1$ is a solution of the equation $dy/dx = x - y$, and it appears likely that the other solution curves approach this straight line as an asymptote as $x \to +\infty$. This inference illustrates the fact that a slope field can suggest tangible information about solutions that is not at all evident from the differential equation itself. Can you, by tracing the appropriate solution curve in this figure, infer that $y(3) \approx 2$ if $y(x)$ is the solution of the initial value problem $dy/dx = x - y$, $y(-4) = 4$?

The next two examples illustrate the use of slope fields to glean useful information in physical situations that are modeled by differential equations. Example 2 is based on the fact that a baseball moving through the air at moderate velocity v (less than about 300 ft/s) encounters air resistance that is approximately proportional to the magnitude of v. If the baseball is thrown straight downward from the top of a tall building or from a hovering helicopter, then it experiences both the downward acceleration of gravity and an upward acceleration due to air resistance. If the y-axis is directed *downward*, then the ball's velocity $v = dy/dx$ and its gravitational acceleration $g = 32$ ft/s^2 are both positive, whereas its acceleration due to air resistance is negative. Hence its total acceleration has the form

$$\frac{dv}{dt} = g - kv. \tag{1}$$

A typical value of the air resistance proportionality constant might be $k = 0.16$.

EXAMPLE 2 Suppose that you throw a baseball straight downward from a helicopter hovering at an altitude of 3000 ft. You wonder whether someone standing on the ground directly below could conceivably catch the ball. To estimate the speed with which the ball will land, you use your laptop's computer algebra system to construct a slope field for the differential equation

$$\frac{dv}{dt} = 32 - (0.16)v. \tag{2}$$

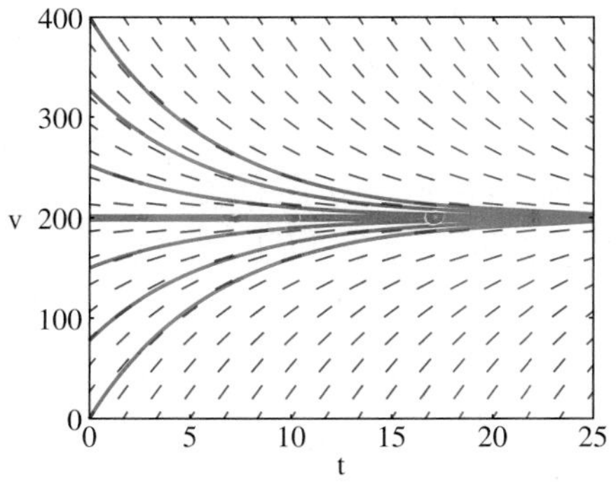

FIGURE 8.2.6 Slope field and typical solution curves for $v' = 32 - 0.16v$.

The result is shown in Fig. 8.2.6, together with a number of solution curves corresponding to different values of the initial velocity $v(0)$ with which you might throw the baseball downward. Note that all these solution curves appear to approach the horizontal line $v = 200$ as an asymptote. This means that—however you throw it—the baseball should approach the *limiting velocity* $v = 200$ ft/s instead of accelerating indefinitely (as it would in the absence of air resistance). The useful fact that 60 mi/h is the same speed as 88 ft/s yields

$$v = 200\,\frac{\text{ft}}{\text{s}} \times \frac{60 \text{ mi/h}}{88 \text{ ft/s}} \approx 136.36\,\frac{\text{mi}}{\text{h}}.$$

Perhaps a catcher accustomed to 100 mi/h fastballs would have some chance of fielding this speeding ball. ◆

COMMENT If the initial velocity of the ball is $v(0) = 200$, then Eq. (2) gives $v'(0) = 32 - (0.16) \cdot (200) = 0$, so the ball experiences *no* initial acceleration. Its velocity therefore remains unchanged, and hence $v(t) \equiv 200$ is a constant "equilibrium solution" of the differential equation. If the initial velocity is greater than 200, then the initial acceleration given in Eq. (2) is negative, so the ball slows as it falls. But if the initial velocity is less than 200, then the initial acceleration given in (2) is positive, so the ball gains speed as it falls. It therefore seems quite reasonable that, because of air resistance, the baseball will approach a limiting velocity of 200 ft/s, whatever its initial velocity might be. You might like to verify that—in the absence of air resistance—this ball would hit the ground at over 300 mi/h.

In Section 8.5 we will discuss in detail the logistic differential equation

$$\frac{dP}{dt} = kP(M - P), \qquad \textbf{(3)}$$

which often is used to model a population $P(t)$ that inhabits an environment with *carrying capacity M*. This means that M is the maximum population that this environment can sustain on a long-term basis (in terms of available food and space, for instance).

EXAMPLE 3 If we take $k = 0.0004$ and $M = 150$, then the logistic equation in (3) takes the form

$$\frac{dP}{dt} = (0.0004)P \cdot (150 - P) = (0.06)P - (0.0004)P^2. \qquad \textbf{(4)}$$

The positive term $(0.06)P$ on the right in (4) corresponds to natural growth at a 6% annual rate (with time t measured in years). The negative term $-(0.0004)P^2$ represents the inhibition of growth due to limited resources in the environment.

Figure 8.2.7 shows a slope field for Eq. (4) together with a number of solution curves corresponding to possible different values of the initial population $P(0)$. Note that all these solution curves appear to approach the horizontal line $P = 150$ as an asymptote. This means that—whatever the initial population—the population $P(t)$ approaches the *limiting population* $P = 150$ as $t \to +\infty$. ◆

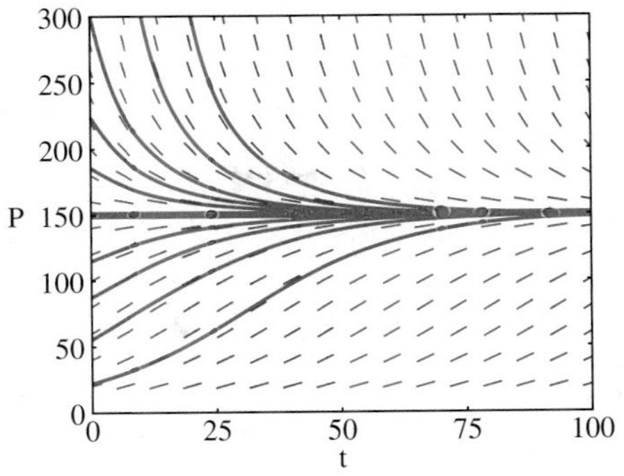

FIGURE 8.2.7 Slope field and typical solution curves for $P' = 0.06P - 0.0004P^2$.

COMMENT If the initial population is $P(0) = 150$, then Eq. (4) gives

$$P'(0) = (0.0004)(150) \cdot (150 - 150) = 0,$$

so the population experiences *no* initial (instantaneous) change. It therefore remains unchanged, and hence $P(t) \equiv 150$ is a constant "equilibrium solution" of the differential equation in (4). If the initial population is greater than 150, then the initial rate of change given by (4) is negative, so the population immediately begins to decrease. But if the initial population is less than 150 (but positive), then the initial rate of change given by (4) is positive, so the population immediately begins to increase. It therefore seems quite reasonable that the population will approach a limiting value of 150, whatever the (positive) initial population.

Euler's Method and Numerical Solutions

A computer plotter can be programmed to draw a solution curve that begins at the point (x_0, y_0) and threads its way through the slope field of a given differential equation $dy/dx = F(x, y)$. The procedure that the plotter carries out can be described as follows.

- The plotter pen begins at the initial point (x_0, y_0) and moves a tiny distance along the slope segment through (x_0, y_0). This takes it to the point (x_1, y_1).
- At (x_1, y_1) the pen changes direction and now moves a tiny distance along the slope segment through this new starting point (x_1, y_1). This takes it to the next starting point (x_2, y_2).
- At (x_2, y_2) the pen direction changes again and now moves a tiny distance along the slope segment through (x_2, y_2). This takes it to the next starting point (x_3, y_3).

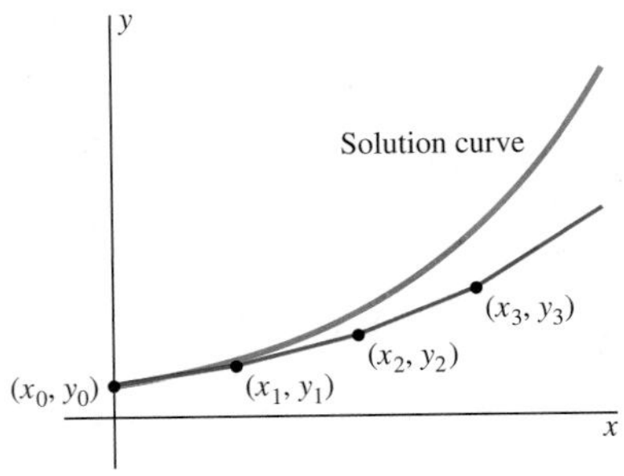

FIGURE 8.2.8 The first few steps in approximating a solution curve.

Figure 8.2.8 illustrates the result of continuing in this fashion—by a sequence of discrete straight-line steps from one starting point to the next. In this figure we see a polygonal curve consisting of line segments that connect the successive points (x_0, y_0), (x_1, y_1), (x_2, y_2), (x_3, y_3), ... But suppose that each "tiny distance" that the pen travels along a slope segment—before the course correction that sends it along a fresh new slope segment—is so small that the naked eye cannot distinguish the individual line segments constituting the polygonal curve. Then the resulting polygonal curve looks like a smooth, continuously turning solution curve of the differential equation. Indeed, this is (in essence) how the solution curves shown in the figures of this chapter were computer-generated.

Euler did not have a computer plotter (it was the 18th century), and his idea was to do all this numerically rather than graphically. To approximate the solution of the initial value problem

$$\frac{dy}{dx} = F(x, y), \qquad y(x_0) = y_0, \tag{5}$$

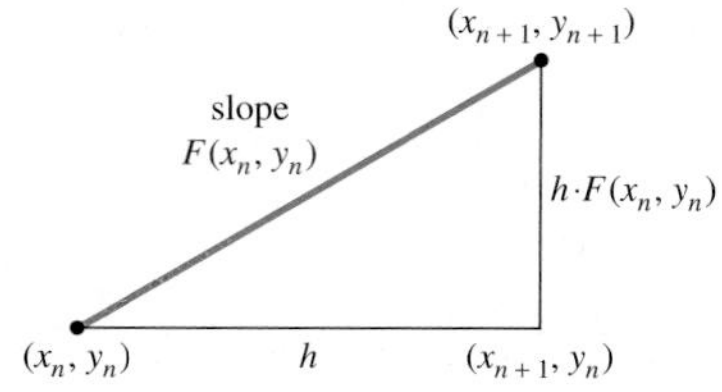

FIGURE 8.2.9 The step from (x_n, y_n) to (x_{n+1}, y_{n+1}).

we first choose a fixed (horizontal) **step size** h to use in making each step from one point to the next. Suppose that we have begun at the initial point (x_0, y_0) and, after n steps, have reached the point (x_n, y_n). Then the step from (x_n, y_n) to the next point (x_{n+1}, y_{n+1}) is illustrated in Fig. 8.2.9. The slope of the direction segment through (x_n, y_n) is $m = F(x_n, y_n)$. Hence a horizontal change of size h from x_n to x_{n+1} corresponds to a vertical change of size $m \cdot h = h \cdot F(x_n, y_n)$ from y_n to y_{n+1}. Therefore the coordinates of the new point (x_{n+1}, y_{n+1}) are given in terms of the old coordinates by

$$x_{n+1} = x_n + h, \qquad y_{n+1} = y_n + h \cdot F(x_n, y_n). \tag{6}$$

Given the initial value problem in (5), **Euler's method** with step size h consists of beginning with the initial point (x_0, y_0) and applying the formulas

$$\begin{aligned} x_1 &= x_0 + h, & y_1 &= y_0 + h \cdot F(x_0, y_0); \\ x_2 &= x_1 + h, & y_2 &= y_1 + h \cdot F(x_1, y_1); \\ x_3 &= x_2 + h, & y_3 &= y_2 + h \cdot F(x_2, y_2); \\ &\vdots & &\vdots \end{aligned} \tag{7}$$

to calculate successive points (x_1, y_1), (x_2, y_2), (x_3, y_3), ... on an approximate solution curve.

But we do not ordinarily sketch the corresponding polygonal approximation. Instead, the numerical result of applying Euler's method is the sequence of *approximations*

$$y_1, \quad y_2, \quad y_3, \quad \dots, \quad y_n, \quad \dots$$

to the *true values*

$$y(x_1), \quad y(x_2), \quad y(x_3), \quad \dots, \quad y(x_n), \quad \dots$$

at the points $x_1, x_2, x_3, \dots, x_n, \dots$ of the *exact* (although unknown) solution of the initial value problem. These results typically are presented in the form of a table of approximate values of the desired solution.

EXAMPLE 4 (a) Apply Euler's method to approximate the solution of the initial value problem

$$\frac{dy}{dx} = x + \frac{1}{5}y, \qquad y(0) = -3, \tag{8}$$

with step size $h = 1$ on the interval $[0, 5]$. (b) Repeat part (a), but use step size $h = 0.2$ and the interval $[0, 1]$.

Solution (a) With $x_0 = 0$, $y_0 = -3$, $F(x, y) = x + \frac{1}{5}y$, and $h = 1$, the equations in (7) yield the approximate values

$$y_1 = y_0 + h \cdot \left[x_0 + \tfrac{1}{5}y_0\right] = (-3) + 1 \cdot \left[0 + \tfrac{1}{5}(-3)\right] = -3.6,$$

$$y_2 = y_1 + h \cdot \left[x_1 + \tfrac{1}{5}y_1\right] = (-3.6) + 1 \cdot \left[1 + \tfrac{1}{5}(-3.6)\right] = -3.32,$$

$$y_3 = y_2 + h \cdot \left[x_2 + \tfrac{1}{5}y_2\right] = (-3.32) + 1 \cdot \left[2 + \tfrac{1}{5}(-3.32)\right] = -1.984,$$

$$y_4 = y_3 + h \cdot \left[x_3 + \tfrac{1}{5}y_3\right] = (-1.984) + 1 \cdot \left[3 + \tfrac{1}{5}(-1.984)\right] = 0.6192, \quad \text{and}$$

$$y_5 = y_4 + h \cdot \left[x_4 + \tfrac{1}{5}y_4\right] = (0.6912) + 1 \cdot \left[4 + \tfrac{1}{5}(0.6912)\right] \approx 4.74304$$

at the points $x_1 = 1$, $x_2 = 2$, $x_3 = 3$, $x_4 = 4$, and $x_5 = 5$. Note how the result of each calculation feeds into the next. The resulting table of approximate values is shown next.

x	0	1	2	3	4	5
Approx. y	-3	-3.6	-3.32	-1.984	0.6912	4.74304

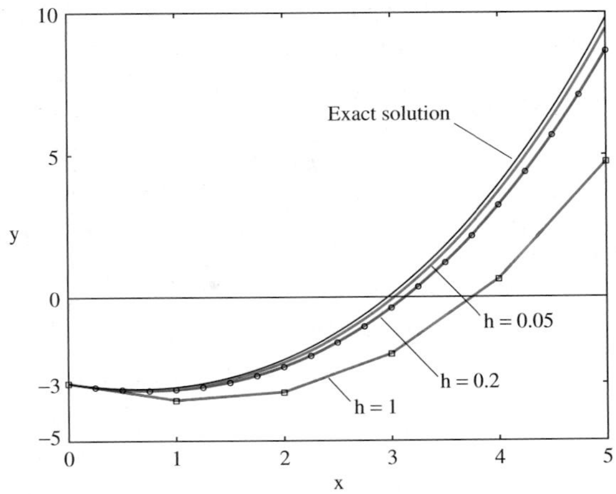

FIGURE 8.2.10 Graphs of Euler approximations with step sizes $h = 1$, $h = 0.2$, and $h = 0.05$.

Figure 8.2.10 shows the graph of this approximation, together with the graphs of the Euler approximations obtained with step sizes $h = 0.2$ and $h = 0.05$. The exact solution is the highest curve in the figure. We see that decreasing the step size increases the accuracy, but with any given step size, the accuracy decreases with distance from the initial point.

(b) We begin anew with $x_0 = 0$, $y_0 = 3$, $F(x, y) = x + \frac{1}{5}y$, and $h = 0.2$. The equations in (7) yield the approximate values

$$y_1 = y_0 + h \cdot \left[x_0 + \tfrac{1}{5}y_0\right] = (-3) + (0.2) \cdot \left[0 + \tfrac{1}{5}(-3)\right] = -3.12,$$

$$y_2 = y_1 + h \cdot \left[x_1 + \tfrac{1}{5}y_1\right] = (-3.12) + (0.2) \cdot \left[0.2 + \tfrac{1}{5}(-3.12)\right] = -3.205,$$

$$y_3 = y_2 + h \cdot \left[x_2 + \tfrac{1}{5}y_2\right] = (-3.205) + (0.2) \cdot \left[0.4 + \tfrac{1}{5}(-3.205)\right] = -3.253,$$

$$y_4 = y_3 + h \cdot \left[x_3 + \tfrac{1}{5}y_3\right] = (-3.253) + (0.2) \cdot \left[0.6 + \tfrac{1}{5}(-3.253)\right] = -3.263, \quad \text{and}$$

$$y_5 = y_4 + h \cdot \left[x_4 + \tfrac{1}{5}y_4\right] = (-3.263) + (0.2) \cdot \left[0.8 + \tfrac{1}{5}(-3.263)\right] \approx -3.234$$

at the points $x_1 = 0.2$, $x_2 = 0.4$, $x_3 = 0.6$, $x_4 = 0.8$, and $x_5 = 1$. The resulting table of values is next.

x	0	0.2	0.4	0.6	0.8	1
Approx. y	−3	−3.12	−3.205	−3.253	−3.263	−3.234

High accuracy with Euler's method usually requires a very small step size, and hence a larger number of steps than can reasonably be carried out by hand. The project material for this section contains calculator and computer programs for automating Euler's method. One of these programs was used to calculate the entries in the table in Fig. 8.2.11. We see that 500 Euler steps (with step size $h = 0.002$) from $x = 0$ to $x = 1$ yield values with errors not exceeding 0.001. ◆

x	Approx. y with $h = 0.2$	Approx. y with $h = 0.02$	Approx. y with $h = 0.002$	Actual Value of y
0	−3.000	−3.000	−3.000	−3.000
0.2	−3.120	−3.104	−3.102	−3.102
0.4	−3.205	−3.172	−3.168	−3.168
0.6	−3.253	−3.201	−3.196	−3.195
0.8	−3.263	−3.191	−3.184	−3.183
1	−3.234	−3.140	−3.130	−3.129

FIGURE 8.2.11 Euler approximations with step sizes $h = 0.2$, $h = 0.02$, and $h = 0.002$.

EXAMPLE 5 Suppose that the baseball of Example 2 is simply dropped (instead of being thrown downward) from the helicopter. Then its velocity $v(t)$ after t seconds satisfies the initial value problem

$$\frac{dv}{dt} = 32 - (0.16)v, \qquad v(0) = 0. \tag{9}$$

We use Euler's method with $h = 1$ to track the ball's increasing velocity at 1-second intervals for the first 10 seconds of fall. With $t_0 = 0$, $v_0 = 0$, $F(t, v) = 32 - (0.16)v$, and $h = 1$, the equations in (7) yield the approximate values

$$v_1 = v_0 + h \cdot [32 - (0.16)v_0] = (0) + 1 \cdot [32 - (0.16)(0)] = 32,$$

$$v_2 = v_1 + h \cdot [32 - (0.16)v_1] = (32) + 1 \cdot [32 - (0.16)(32)] = 58.88,$$

$$v_3 = v_2 + h \cdot [32 - (0.16)v_2] = (58.88) + 1 \cdot [32 - (0.16)(58.88)] \approx 81.46,$$

$$v_4 = v_3 + h \cdot [32 - (0.16)v_3] = (81.46) + 1 \cdot [32 - (0.16)(81.46)] \approx 100.43, \quad \text{and}$$

$$v_5 = v_4 + h \cdot [32 - (0.16)v_4] = (100.43) + 1 \cdot [32 - (0.16)(100.43)] \approx 116.36.$$

Continuing in this fashion, we complete the column of values of v corresponding to $h = 1$ in the table of Fig. 8.2.12. (We have rounded velocity to the nearest foot per second). The values corresponding to $h = 0.1$ were calculated using a computer, and we see that they are accurate to within about 1 ft/s. Note also that after 10 seconds the falling ball has attained about 80% of its limiting velocity of 200 ft/s. ◆

t	Approx. v with $h = 1$	Approx. v with $h = 0.1$	Actual Value of v
1	32	30	30
2	59	55	55
3	81	77	76
4	100	95	95
5	116	111	110
6	130	124	123
7	141	135	135
8	150	145	144
9	158	153	153
10	165	160	160

FIGURE 8.2.12 Euler approximations in Example 5 with step sizes $h = 1$ and $h = 0.1$.

Existence and Uniqueness

The initial value problem $dy/dx = 1/x$, $y(0) = 0$ has *no* solution, because no solution $y(x) = \ln|x| + C$ of the differential equation is defined at $x = 0$. On the other hand, you can readily verify that the initial value problem $dy/dx = 2\sqrt{y}$, $y(0) = 0$ has the *two* different solutions $y_1(x) = x^2$ and $y_2(x) \equiv 0$. Consequently, before we speak of "the" solution of a given initial value problem, it is necessary to know that it has *one and only one* solution.

A general existence-uniqueness theorem of differential equations implies that the initial value problem $dy/dx = F(x, y)$, $y(a) = b$ has one and only one solution on some open interval containing the point $x = a$ provided that both the function F and its partial derivative $\partial F/\partial y$ are continuous near the point (a, b) in the xy-plane. (Continuity and partial derivatives of functions of two variables are defined in Chapter 13. Briefly, continuity of F means that $F(x, y)$ is close to $F(a, b)$ if (x, y) is close to (a, b). The partial derivative $\partial F/\partial y$ denotes the derivative of F with respect to y, with x regarded as a constant.) For many subsequent applications in this chapter, it will suffice to know that the hypotheses of this general theorem are satisfied if $F(x, y)$ is a polynomial in the two variables x and y. A precise statement and further discussion of the theorem can be found in Section 1.3 of Edwards and Penney, *Differential Equations: Computing and Modeling,* 2nd ed. (Upper Saddle River, N. J.: Prentice Hall, 2000).

8.2 TRUE/FALSE STUDY GUIDE

8.2 CONCEPTS: QUESTIONS AND DISCUSSION

1. Compare the advantages of graphical and numerical approximation of solutions of differential equations.
2. For what purposes does a graph of one or more solutions give more useful information than a table of values, and for what purposes is a table preferable?
3. Simple examples of the failure of existence and uniqueness of solutions of initial value problem are given near the end of this section. Provide two or three similar examples of your own.

8.2 PROBLEMS

In Problems 1 through 10, we have provided the direction field of the indicated differential equation, together with one or more solution curves. Sketch likely solution curves through the additional points marked in each direction field. (One method: Photocopy the direction field and draw your solution curves in a second color. Another method: Use a computer algebra program to construct and print the given direction field.)

1. $\dfrac{dy}{dx} = -y - \sin x$ (Fig. 8.2.13)

2. $\dfrac{dy}{dx} = x + y$ (Fig. 8.2.14)

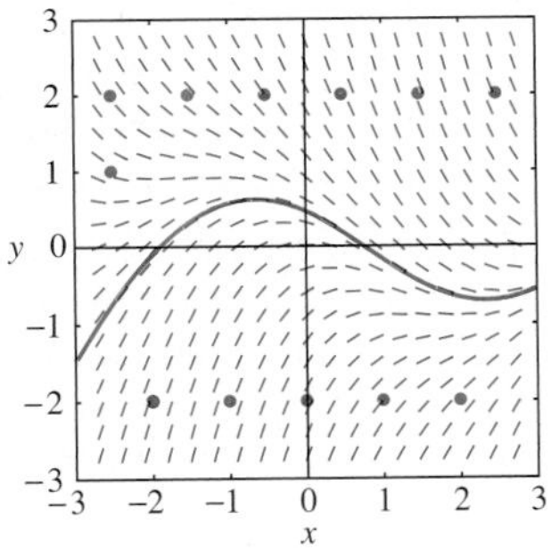

FIGURE 8.2.13

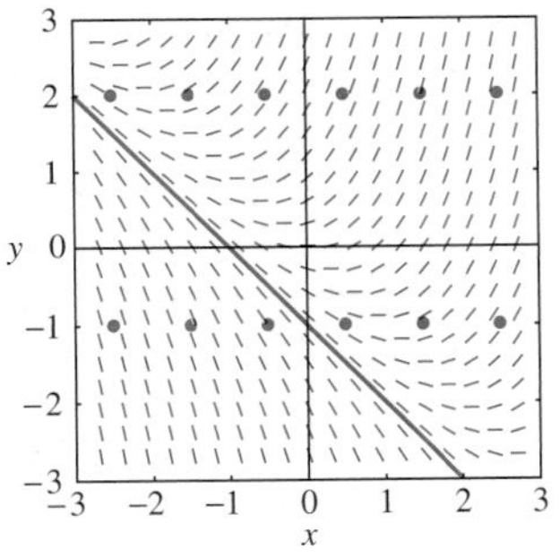

FIGURE 8.2.14

3. $\dfrac{dy}{dx} = y - \sin x$ (Fig. 8.2.15)

4. $\dfrac{dy}{dx} = x - y$ (Fig. 8.2.16)

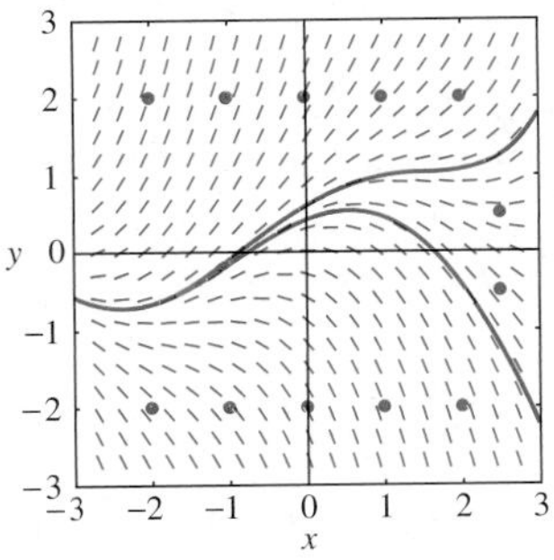

FIGURE 8.2.15

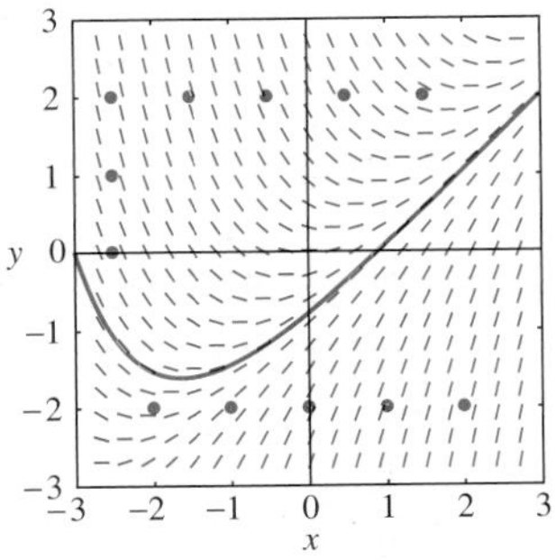

FIGURE 8.2.16

5. $\dfrac{dy}{dx} = y - x + 1$ (Fig. 8.2.17)

6. $\dfrac{dy}{dx} = x - y + 1$ (Fig. 8.2.18)

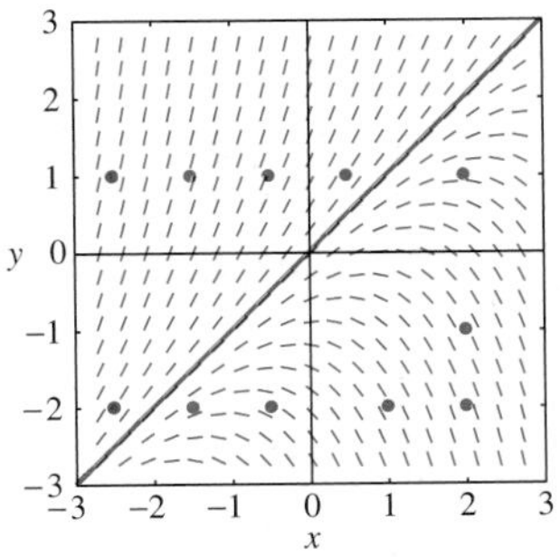

FIGURE 8.2.17

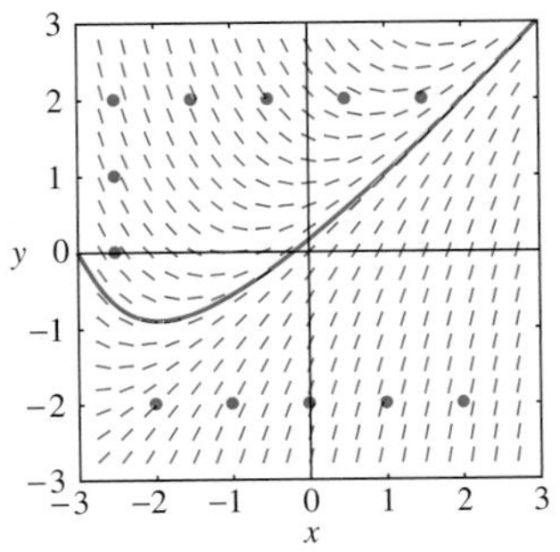

FIGURE 8.2.18

7. $\dfrac{dy}{dx} = \sin x + \sin y$ (Fig. 8.2.19)

8. $\dfrac{dy}{dx} = x^2 - y$ (Fig. 8.2.20)

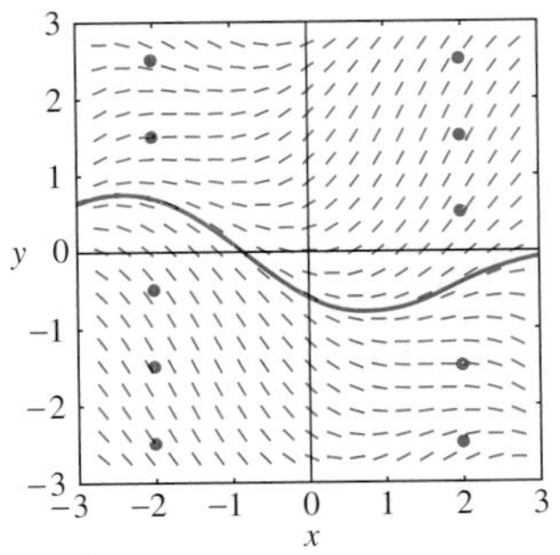

FIGURE 8.2.19

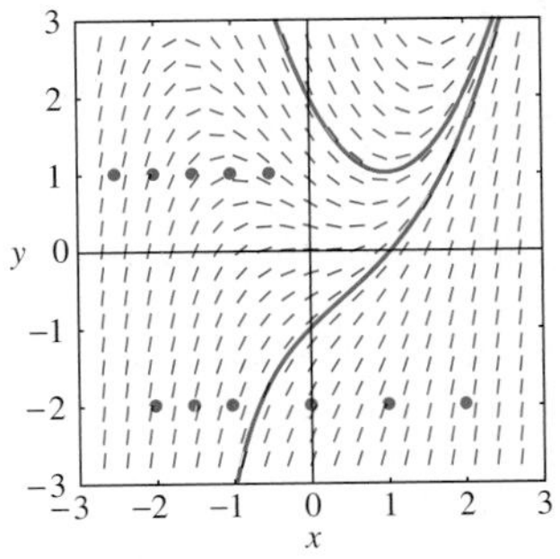

FIGURE 8.2.20

9. $\dfrac{dy}{dx} = x^2 - y - 2$ (Fig. 8.2.21)

10. $\dfrac{dy}{dx} = -x^2 + \sin y$ (Fig. 8.2.22)

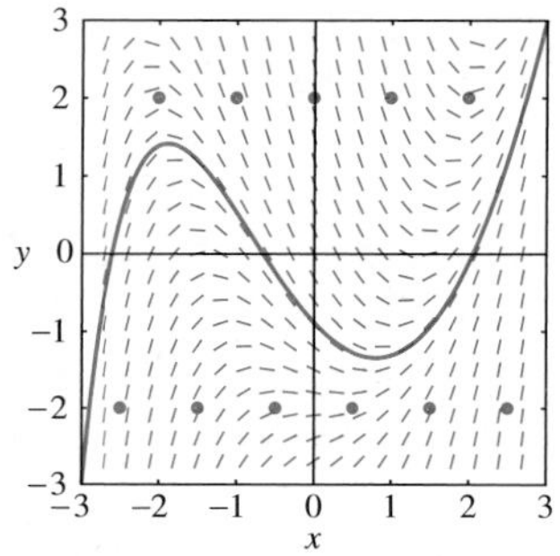

FIGURE 8.2.21

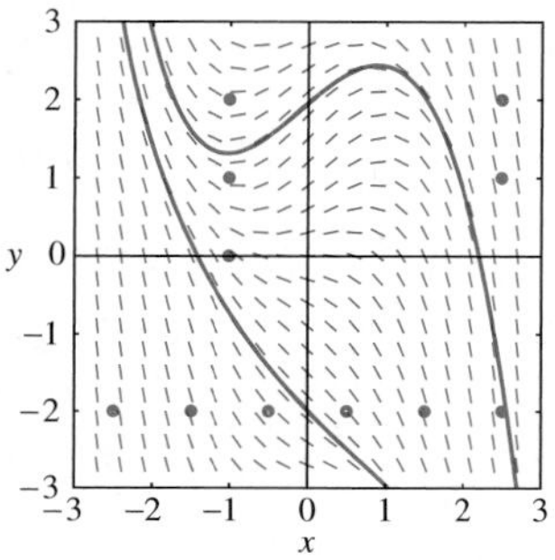

FIGURE 8.2.22

In Problems 11 through 20, an initial value problem and its exact solution $y(x)$ are given. Apply Euler's method twice to approximate this solution on the interval $[0, \frac{1}{2}]$, first with step size $h = 0.25$, then with step size $h = 0.1$. Compare the three-place values of the two approximations at $x = \frac{1}{2}$ with the value $y(\frac{1}{2})$ of the exact solution.

11. $\frac{dy}{dx} = -y, \quad y(0) = 2; \quad y(x) = 2e^{-x}$

12. $\frac{dy}{dx} = 2y, \quad y(0) = \frac{1}{2}; \quad y(x) = \frac{1}{2}e^{2x}$

13. $\frac{dy}{dx} = y + 1, \quad y(0) = 1; \quad y(x) = 2e^x - 1$

14. $\frac{dy}{dx} = x - y, \quad y(0) = 1; \quad y(x) = 2e^{-x} + x - 1$

15. $\frac{dy}{dx} = y - x - 1, \quad y(0) = 1; \quad y(x) = 2 + x - e^x$

16. $\frac{dy}{dx} = -2xy, \quad y(0) = 2; \quad y(x) = 2\exp(-x^2)$

17. $\frac{dy}{dx} = -3x^2y, \quad y(0) = 3; \quad y(x) = 3\exp(-x^3)$

18. $\frac{dy}{dx} = e^{-y}, \quad y(0) = 0; \quad y(x) = \ln(x + 1)$

19. $\frac{dy}{dx} = \frac{1}{4}(1 + y^2), \quad y(0) = 1; \quad y(x) = \tan\left(\frac{x + \pi}{4}\right)$

20. $\frac{dy}{dx} = 2xy^2, \quad y(0) = 1; \quad y(x) = \frac{1}{1 - x^2}$

In Problems 21 and 22, first use the method of Example 1 to construct a slope field for the given differential equation. Then sketch the solution curve corresponding to the given initial condition. Finally, use this solution curve to estimate the desired value of the solution $y(x)$.

21. $\frac{dy}{dx} = x + y, \quad y(0) = 0; \quad y(-4) = ?$

22. $\frac{dy}{dx} = y - x, \quad y(4) = 0; \quad y(-4) = ?$

Problems 23 and 24 are like Problems 21 and 22, but now use a computer algebra system to plot and print out a slope field for the given differential equation. If you wish (and know how), you can check your manually sketched solution curve by plotting it with the computer.

23. $\frac{dy}{dx} = x^2 + y^2 - 1, \quad y(0) = 0; \quad y(2) = ?$

24. $\frac{dy}{dx} = x + \frac{1}{2}y^2, \quad y(-2) = 0; \quad y(2) = ?$

25. You bail out of the helicopter of Example 2 and your parachute opens immediately. Now $k = 1.6$ in Eq. (1), so your downward velocity satisfies the initial value problem

$$\frac{dv}{dt} = 32 - (1.6)v, \qquad v(0) = 0.$$

To investigate your chances of survival, construct a slope field for this differential equation and sketch the appropriate solution curve. What will your limiting velocity be? Will a strategically located haystack do any good? How long will it take you to reach 95% of your limiting velocity?

26. Suppose that the deer population $P(t)$ in a small forest satisfies the logistic equation

$$\frac{dP}{dt} = (0.0225)P - (0.0003)P^2.$$

Construct a slope field and appropriate solution curve to answer the following questions: If there are 25 deer at time $t = 0$ and t is measured in months, how long will it take for the number of deer to double? What will be the limiting deer population?

27. Use Euler's method with a programmable calculator or computer system to find the desired solutions in Problem 23. Begin with step size $h = 0.1$, and then use successively smaller step sizes $h = 0.01, h = 0.001, \ldots,$ until successive approximate solutions at $x = 2$ agree to two decimal places.

28. Use Euler's method with a programmable calculator or computer system to find the desired solutions in Problem 24. Begin with step size $h = 0.1$, and then use successively smaller step sizes $h = 0.01, h = 0.001, \ldots,$ until successive approximate solutions at $x = 2$ agree to two decimal places.

Problems 29 through 32 illustrate the fact that, if the hypotheses of the theorem cited near the end of this section are not satisfied, then the initial value problem $dy/dx = F(x, y)$, $y(a) = b$ may have either no solutions, finitely many solutions, or infinitely many solutions.

29. Show that on the interval $[0, \pi]$, the functions $y_1(x) \equiv 1$ and $y_2(x) = \cos x$ both satisfy the initial value problem

$$\frac{dy}{dx} + \sqrt{1 - y^2} = 0, \qquad y(0) = 1.$$

Why does this fact not contradict the existence-uniqueness theorem cited in this section? Explain your answer carefully.

30. Find by inspection two different solutions of the initial value problem

$$\frac{dy}{dx} = 3y^{2/3}, \qquad y(0) = 0.$$

Why does the existence of different solutions not contradict the existence-uniqueness theorem of this section?

31. Use Fig. 8.2.23 as a suggestion for showing that the initial value problem

$$\frac{dy}{dx} = 3y^{2/3}, \qquad y(-1) = -1$$

has infinitely many solutions. Why does this not contradict the existence-uniqueness theorem of this section?

32. Verify that if k is a constant, then the function $y(x) = kx$ satisfies the differential equation $x(dy/dx) = y$. Hence conclude that the initial value problem

$$x\frac{dy}{dx} = y, \qquad y(0) = 0$$

has infinitely many solutions on any open interval containing $x = 0$.

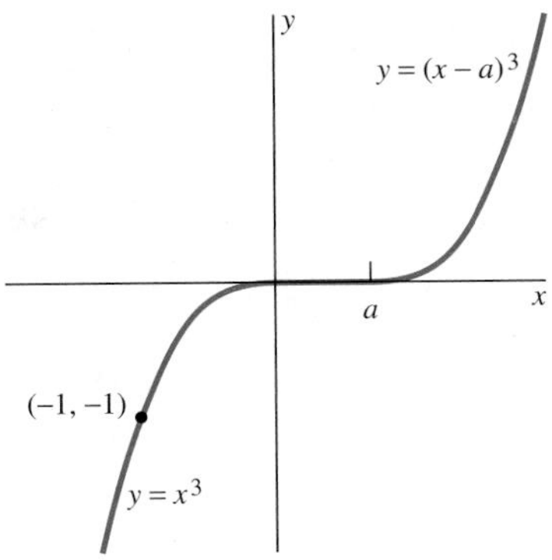

FIGURE 8.2.23 A suggestion for Problem 31.

8.2 Project: Computer-Assisted Slope Fields and Euler's Method

The CD-ROM project material illustrates the use of *Maple, Mathematica,* and MATLAB to construct slope fields and solution curves for a given differential equation. Use your system to carry out the following investigation.

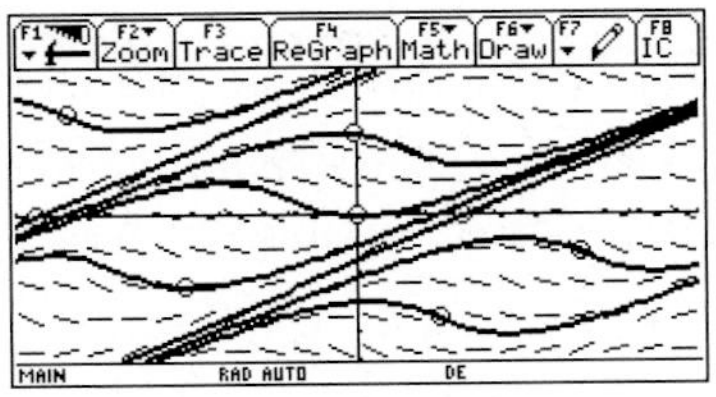

FIGURE 8.2.24 Direction field and solution curves for $y' = \sin(x - y)$ generated by a TI-89/92+ graphing calculator.

Investigation Plot a direction field and typical solution curves for the differential equation $dy/dx = \sin(x-y)$, but with a bigger viewing window than that of Fig. 8.2.24. With $-10 \leq x \leq 10$, $-10 \leq y \leq 10$, for instance, a number of apparent straight-line solution curves should be visible. (a) Substitute $y = ax + b$ in the differential equation to determine what the coefficients a and b must be in order to get a solution. (b) Perhaps your computer algebra system will give the general solution

$$y(x) = x - 2\tan^{-1}\left(\frac{x - 2 + C}{x - C}\right).$$

Can you determine a value of the arbitrary constant C that yields the linear solution $y(x) = x - \frac{1}{2}\pi$ determined by the initial condition $y(\frac{1}{2}\pi) = 0$? (Verify this.)

Famous Numbers Investigation

The CD-ROM project material illustrates *Maple, Mathematica,* and MATLAB routines for the implementation of Euler's method. The problems below describe the numbers $e \approx 2.71828$, $\ln 2 \approx 0.69315$, and $\pi \approx 3.14159$ as specific values of solutions of certain initial value problems. In each case, apply Euler's method with $n = 50, 100, 200, \ldots$ subintervals (doubling n each time). How many subintervals are needed to obtain—twice in succession—the correct value of the target number rounded to three decimal places?

1. The number $e = y(1)$, where $y(x)$ is the solution of the initial value problem $dy/dx = y$, $y(0) = 1$.
2. The number $\ln 2 = y(2)$, where $y(x)$ is the solution of the initial value problem $dy/dx = 1/x$, $y(1) = 0$.
3. The number $\pi = y(1)$, where $y(x)$ is the solution of the initial value problem

$$\frac{dy}{dx} = \frac{4}{1 + x^2}, \qquad y(0) = 0.$$

Also, explain in each of these problems what the point is—why the indicated famous number is, indeed, the expected result.

8.3 SEPARABLE EQUATIONS AND APPLICATIONS

In Section 8.1 we saw that a general solution of a differential equation $dy/dx = R(x, y)$ can be written in terms of integrals if either the independent variable x or the dependent variable y is missing from the expression $R(x, y)$ on the right-hand side. The method we used there also can be used if $R(x, y)$ can be expressed as the *product* of a function $g(x)$ of x and a function $h(y)$ of y. In this case the differential equation takes the form

$$\frac{dy}{dx} = g(x)h(y). \tag{1}$$

Such a differential equation is said to be **separable** because—upon formal multiplication of both sides by dx and by $f(y) = 1/h(y)$—it takes the symbolic form

$$f(y)\,dy = g(x)\,dx, \tag{2}$$

in which the variables x and y (and their respective differentials dx and dy) are *separated* on opposite sides of the equation. The equation in (2) is literally an equation relating differentials, a "differential equation," but we take Eq. (2) to be concise notation for the more familiar "derivative" equation

$$f(y)\frac{dy}{dx} = g(x), \tag{3}$$

and abuse the terminology slightly by referring to Eq. (3) as a *differential equation* as well.

The process of rewriting Eq. (1) in the form in (2) is called *separating the variables*. It is then tempting to find a solution of (1) simply by integrating each side in (2) with respect to its "own" variable:

$$\int f(y)\,dy = \int g(x)\,dx. \tag{4}$$

Indeed, if the antiderivatives $F(y) = \int f(y)\,dy$ and $G(x) = \int g(x)\,dx$ can be found, then the resulting equation

$$F(y) = G(x) + C \tag{5}$$

provides an *implicit solution* of Eq. (3). That is, any differentiable function $y(x)$ defined implicitly by (5) is an actual (explicit) solution of Eq. (3).

To verify this claim, suppose that $y = y(x)$ satisfies Eq. (5) for some fixed value of the arbitrary constant C. Then differentiating each side yields

$$\begin{aligned} D_x[F(y(x))] &= D_x[G(x) + C]; \\ F'(y(x)) \cdot y'(x) &= G'(x) && \text{[using the chain rule];} \\ f(y(x))\frac{dy}{dx} &= g(x) && \text{[because } F' = f \text{ and } G' = g\text{].} \end{aligned}$$

Thus $y(x)$ does, indeed, satisfy the differential equation in (3).

EXAMPLE 1 (a) Solve the differential equation $dy/dx = -6xy$ with the initial condition $y(0) = 7$. (b) Repeat with the initial condition $y(0) = -4$.

Solution (a) Separation of variables gives

$$\frac{1}{y}\,dy = -6x\,dx.$$

Integration then yields

$$\int \frac{1}{y}\,dy = \int (-6x)\,dx;$$

$$\ln|y| = -3x^2 + C. \qquad \textbf{(6)}$$

The initial condition $y(0) = 7$ indicates a positive-valued solution, so we replace $|y|$ with y in the implicit solution in (6). This gives

$$\ln y = -3x^2 + C, \quad \text{so that} \quad y(x) = e^{-3x^2+C} = e^C e^{-3x^2}.$$

Then substituting $x = 0$ and $y = 7$ gives $C = \ln 7$, so the desired particular solution is

$$y(x) = e^{\ln 7} e^{-3x^2} = 7e^{-3x^2}.$$

(b) Beginning with Eq. (6), we note that the initial condition $y(0) = -4$ indicates a negative-valued solution, so we replace $|y|$ with $-y$ in the implicit solution there. This gives

$$\ln(-y) = -3x^2 + C, \quad \text{so that} \quad y(x) = -e^{-3x^2+C} = -e^C e^{-3x^2}.$$

Then substituting $x = 0$, $y = -4$ gives $C = \ln 4$, so the desired particular solution is

$$y(x) = -e^{\ln 4} e^{-3x^2} = -4e^{-3x^2}.$$ ◆

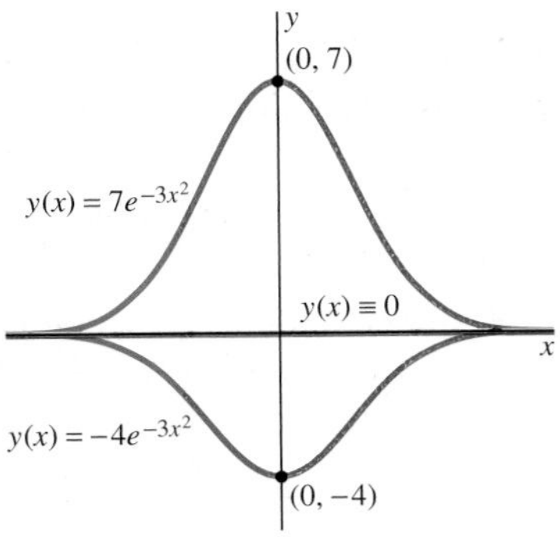

FIGURE 8.3.1 Three different solutions of the differential equation $\frac{dy}{dx} = -6xy$.

REMARK 1 An alternative approach in Example 1 would be to solve Eq. (6) for

$$|y| = e^C e^{-3x^2} = Be^{-3x^2}, \quad \text{so that} \quad y(x) = \pm Be^{-3x^2}.$$

The constant $B = e^C$ is necessarily positive. (Why?) In spite of this, we can accommodate both positive and negative values of y by writing $A = \pm B$ (thereby absorbing the sign in the coefficient A). This gives the general solution

$$y(x) = Ae^{-3x^2} \qquad \textbf{(7)}$$

of the differential equation $dy/dx = -6xy$. The values $A = 7$ and $A = -4$ give the two particular solutions found in parts (a) and (b) of Example 1. The graphs of both of these solutions are shown in color in Fig. 8.3.1.

REMARK 2 The value $A = 0$ in Eq. (7) gives the trivial solution $y(x) \equiv 0$ of the differential equation $dy/dx = -6xy$. Examine the solution

$$\ln|y| = -3x^2 + C \qquad \textbf{(6)}$$

obtained by the method of separation of variables. Can you see that *no* value of C in Eq. (6) yields the trivial solution $y(x) \equiv 0$? This observation illustrates the fact that constant-valued solutions of a differential equation can be "lost" when we separate the variables. That is, if $y = y_0$ is a root of the equation $h(y) = 0$, then the constant-valued solution $y(x) \equiv y_0$ of the differential equation $dy/dx = g(x)h(y)$ may not satisfy the new differential equation obtained upon division by $h(y)$. Therefore, it is important to note in advance any such constant-valued solutions of a given separable differential equation.

EXAMPLE 2 Find all solutions of the differential equation $dy/dx = 6x(y-1)^{2/3}$.

Solution Obviously we intend to divide both sides by the factor $(y-1)^{2/3}$ to separate the variables. But first we note that $(y-1)^{2/3} = 0$ when $y(x) \equiv 1$, and that the latter is a solution of the given differential equation. Putting aside this trivial solution for the time being, we proceed to the separation of variables and integration. This leads to

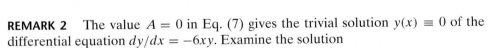

$$\int \frac{1}{3(y-1)^{2/3}}\,dy = \int 2x\,dx;$$

$$(y-1)^{1/3} = x^2 + C.$$

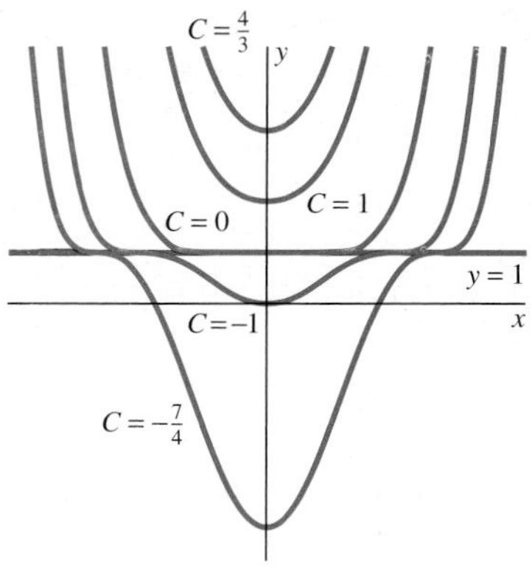

FIGURE 8.3.2 Solution curves of the differential equation $\frac{dy}{dx} = 6x(y-1)^{2/3}$ corresponding to different values of C in Eq. (8).

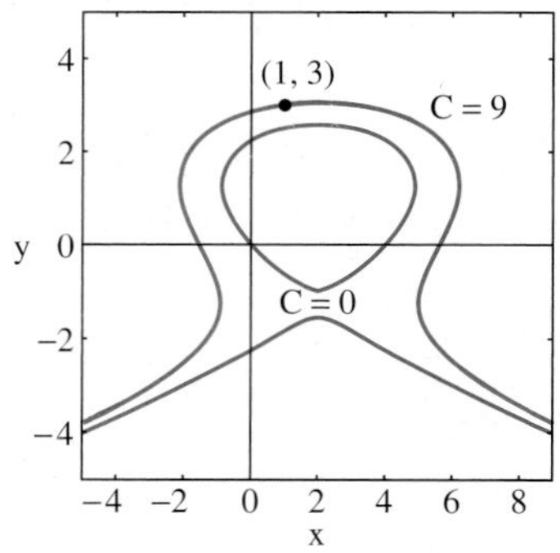

FIGURE 8.3.3 Solution curves of the differential equation $\frac{dy}{dx} = \frac{4-2x}{3y^2-5}$ corresponding to different values of C in Eq. (10).

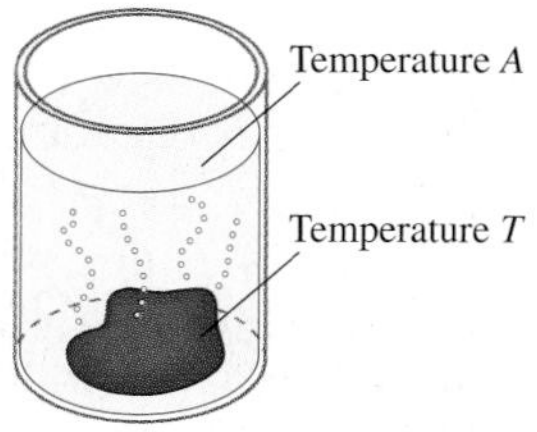

FIGURE 8.3.4 Newton's law of cooling, Eq. (11), describes the cooling of a hot rock in water.

We can solve this implicit solution for the general solution

$$y(x) = 1 + (x^2 + C)^3. \tag{8}$$

Positive values of the arbitrary constant C give the solution curves that lie above the line $y = 1$ (see Fig. 8.3.2) and negative values of C yield those that dip below this line. The value $C = 0$ gives the solution $y(x) = 1 + x^6$, rather than the so-called "singular" solution $y(x) \equiv 1$. Apparently the latter solution was lost when the variables were separated. (See Questions 1 through 3 at the end of this section.) ◆

EXAMPLE 3 Solve the initial value problem

$$\frac{dy}{dx} = \frac{4-2x}{3y^2-5}, \qquad y(1) = 3. \tag{9}$$

Solution We proceed to separate the variables and integrate. This gives

$$\int (3y^2 - 5)\,dy = \int (4 - 2x)\,dx;$$

$$y^3 - 5y = 4x - x^2 + C. \tag{10}$$

If we substitute $x = 1$ and $y = 3$ in the implicit solution in Eq. (10), we find that $C = 9$. Hence the desired particular solution $y(x)$ is defined implicitly by the equation

$$y^3 - 5y = 4x - x^2 + 9.$$

The graph of this equation—generated by the contour plotting command of a computer algebra system—is the upper solution curve shown in Fig. 8.3.3. (See Question 4 at the end of this section.) ◆

Cooling and Heating

According to **Newton's law of cooling** (or heating), the time rate of change of the temperature $u(t)$ of a body immersed in a medium of constant temperature A (Fig. 8.3.4) is proportional to the temperature difference $u - A$. It follows that

$$\frac{du}{dt} = -k(u - A), \tag{11}$$

where the proportionality constant k is positive. Note that if $u > A$ then $du/dt < 0$, so that the body is cooling. But if $u < A$ then $du/dt > 0$, so the temperature u is increasing.

EXAMPLE 4 A 4-lb roast, initially at 50°F, is placed in a 375° oven at 5:00 P.M. At 6:15 P.M. the temperature of the roast is 125°F. When will it be ready to serve medium rare (at 150°F)?

Solution With $A = 375$, Eq. (11) gives

$$\frac{du}{dt} = -k(u - 375) = k(375 - u).$$

While the roast is cooking, its temperature satisfies the inequality $u < 375$. That noted, we separate the variables and integrate:

$$\int \frac{1}{375 - u}\,du = \int k\,dt;$$

$$-\ln(375 - u) = kt + C;$$

$$375 - u = e^{-(kt+C)} = Be^{-kt}$$

where $B = e^{-C}$. Next, $u(0) = 50$ implies that $B = 375 - 50 = 325$, so

$$u(t) = 375 - 325e^{-kt}.$$

We also know that $u = 125$ when $t = 75$. Substituting these values then yields $k = -\frac{1}{75}\ln\left(\frac{250}{325}\right) \approx 0.003498$. We finally solve the equation $150 = 375 - 325e^{-kt}$ for

$$t = -\frac{\ln\left(\frac{225}{325}\right)}{k} \approx -\frac{\ln\left(\frac{225}{325}\right)}{0.003498} \approx 105.12$$

minutes. This is the total cooking time required for the roast. Because it was placed in the oven at 5:00 P.M., it should be removed at about 6:45 P.M. ◆

Linear Differential Equations

The differential equation

$$\frac{dx}{dt} = ax + b \tag{12}$$

is said to be **linear** in the dependent variable x (and its derivative) if the coefficients a and b do not involve x. The general case in which a and b may involve the independent variable t is discussed in Section 8.4.

Equation (12) is separable if a and b are *constants* (with $a \neq 0$). Separating the variables and integrating, we get

$$\int \frac{a}{ax+b}\,dx = \int a\,dt;$$

$$\ln|ax+b| = at + C.$$

It follows that

$$ax + b = \pm e^C e^{at} = Ke^{at}.$$

If an initial condition $x(0) = x_0$ is given, we see immediately that $K = ax_0 + b$, so

$$ax + b = (ax_0 + b)e^{at}.$$

Finally, we readily solve for the solution

$$x(t) = x_0 e^{at} + \frac{b}{a}(e^{at} - 1) \tag{13}$$

of the initial value problem $dx/dt = ax + b$, $x(0) = x_0$.

The remaining examples of this section illustrate the numerous applications of the solution in (13) of a linear differential equation with constant coefficients.

Population Growth with Immigration Consider a national population $P(t)$ with constant birth and death rates β and δ (in births or deaths per year per unit of population). If also there is a constant net immigration rate of I persons entering the country annually, then P satisfies the linear differential equation

$$\frac{dP}{dt} = kP + I \tag{14}$$

where $k = \beta - \delta$. According to Eq. (13) with P, k, and I in place of x, a, and b, the solution of Eq. (14) for which $P(0) = P_0$ is

$$P(t) = P_0 e^{kt} + \frac{I}{k}(e^{kt} - 1). \tag{15}$$

The first term on the right-hand side is the net effect after t years of natural population growth via births and deaths; the second term represents the effect of immigration.

EXAMPLE 5 In the year 2000 the U.S. population was approximately $P_0 = 280$ million and about 14.6 births and 8.6 deaths per thousand population were occurring annually. In addition, a net immigration into the country at the rate of about 960 thousand people per year was occurring. Let's examine the effects of these birth, death, and immigration rates, assuming that they hold constant for the next 20 years. With

$$\beta = \frac{14.6}{1000} = 0.0146, \qquad \delta = \frac{8.6}{1000} = 0.0086, \quad k = \beta - \delta = 0.006,$$

and $I = 0.96$ (counting people by the million in each case), the differential equation in (14) takes the form

$$\frac{dP}{dt} = (0.006)P + 0.96.$$

Its solution, given by (15) with $P_0 = 280$, is

$$P(t) = 280e^{(0.006)t} + 160\left(e^{(0.006)t} - 1\right).$$

The predicted population for the year 2020 is $P(20) \approx 336.1$ million. Of the 20-year predicted U.S. population increase of $336.1 - 280 = 56.1$ million, the amount due to the 0.6% natural growth rate (as if there were no immigration) is

$$280e^{(0.006)(20)} - 280 \approx 35.7 \quad \text{(million)},$$

and the remaining amount due ultimately to immigration is

$$160\left(e^{(0.006)(20)} - 1\right) \approx 20.4 \quad \text{(million)}.$$

◆

Savings Account with Continuous Deposits Consider a savings account that contains A_0 dollars initially and earns interest at the annual rate r compounded continuously (as in Section 8.1). Now suppose that deposits are added to this account at the rate of Q dollars per year. To simplify the mathematical model, we assume that these deposits are made continuously rather than (for instance) monthly. We may then regard the amount $A(t)$ in the account at time t as a "population" of dollars, with a natural (annual) growth rate r and "immigration" (deposits) at the rate of Q dollars annually. Then by merely changing the notation in Eqs. (14) and (15), we get the differential equation

$$\frac{dA}{dt} = rA + Q, \tag{16}$$

which has the solution

$$A(t) = A_0e^{rt} + \frac{Q}{r}(e^{rt} - 1). \tag{17}$$

EXAMPLE 6 Suppose that you wish to arrange, at the time of her birth, for your daughter to have \$100,000 available for her college expenses when she is 18 years old. You plan to do so by making frequent, small—essentially continuous—deposits in a mutual fund at the rate of Q dollars per year. This fund will accumulate 9% annual interest compounded continuously. What should Q be so that you achieve your goal?

Solution With $A_0 = 0$ and $r = 0.09$, we want the value of Q so that Eq. (17) yields the result

$$A(18) = 100000.$$

That is, we must find Q so that

$$100000 = \frac{Q}{0.09}\left(e^{(0.09)(18)} - 1\right).$$

When we solve this equation we find that $Q \approx 2220.53$. Thus you should deposit \$2220.53 per year, or about \$185.04 per month, in order to have \$100,000 in the fund after 18 years. You may wish to verify that your total deposits will be \$39,939.50 and that the total interest accumulated will be \$60,030.50. (You should also remember that you will have to pay taxes on an interest income averaging about \$3335 per year.) ◆

Diffusion of Information and Spread of Disease Let $N(t)$ denote the number of people (in a fixed population P) who by time t have heard a certain news item spread by the mass media. Under certain common conditions, the time rate of increase of N will be proportional to the number of people who have not yet heard the news. Thus

$$\frac{dN}{dt} = k(P - N). \qquad \textbf{(18)}$$

If $N(0) = 0$, the solution of Eq. (18) is

$$N(t) = P \cdot (1 - e^{-kt}). \qquad \textbf{(19)}$$

If P and some later value $N(t_1)$ are known, we can then solve for k and thereby determine $N(t)$ for all t. Problem 35 illustrates this situation.

Different infectious diseases spread in different ways. A simple model may be built on the assumption that some infectious diseases spread like information—in a fixed population P, the rate of increase of the number $N(t)$ of people infected with the disease is proportional to the number $P - N$ who are not yet infected. Then N satisfies the differential equation in (18). See Problems 39 and 40 for applications.

8.3 TRUE/FALSE STUDY GUIDE

8.3 CONCEPTS: QUESTIONS AND DISCUSSION

Questions 1 through 3 pertain to the general solution $y(x) = 1 + (x^2 + C)^3$ of the differential equation $dy/dx = 6x(y - 1)^{2/3}$ discussed in Example 2.

1. Is there a value of the arbitrary constant C that yields the constant-valued solution $y(x) \equiv 1$?
2. Does every point of the xy-plane lie on precisely one solution curve of the form $y(x) = 1 + (x^2 + C)^3$? (See Fig. 8.3.2.)
3. Find two different solutions of the differential equation, both of which satisfy the initial condition $y(1) = 1$. Can you show that the entire singular solution curve $y \equiv 1$ consists of points where the differential equation has two different solutions? If so, why does this fact not contradict the existence-uniqueness theorem discussed at the end of Section 8.1?
4. Examine the solution curves of the differential equation

$$\frac{dy}{dx} = \frac{4 - 2x}{3y^2 - 5}$$

shown in Fig. 8.3.3. (a) Explain why the solution satisfying the initial condition $y(0) = 0$ is defined on the interval $0 \leqq x \leqq 4$ but not on the interval $-2 \leqq x \leqq 6$. (b) Explain why the solution satisfying the initial condition $y(1) = 3$ is defined on the interval $-1 \leqq x \leqq 5$ but not on the interval $-3 \leqq x \leqq 7$.

8.3 PROBLEMS

Find general solutions (implicit if necessary, explicit if possible) of the differential equations in Problems 1 through 10.

1. $\frac{dy}{dx} = 2x\sqrt{y}$

2. $\frac{dy}{dx} = 2xy^2$

3. $\frac{dy}{dx} = x^2y^2$

4. $\frac{dy}{dx} = (xy)^{3/2}$

5. $\frac{dy}{dx} = 2x\sqrt{y-1}$

6. $\frac{dy}{dx} = 4x^3(y-4)^2$

7. $\frac{dy}{dx} = \frac{1+\sqrt{x}}{1+\sqrt{y}}$

8. $\frac{dy}{dx} = \frac{x+x^3}{y+y^3}$

9. $\frac{dy}{dx} = \frac{x^2+1}{x^2(3y^2+1)}$

10. $\frac{dy}{dx} = \frac{(x^3-1)y^3}{x^2(2y^3-3)}$

Solve the initial value problems in Problems 11 through 20.

11. $\frac{dy}{dx} = y^2, \quad y(0) = 1$

12. $\frac{dy}{dx} = \sqrt{y}, \quad y(0) = 4$

13. $\frac{dy}{dx} = \frac{1}{4y^3}, \quad y(0) = 1$

14. $\frac{dy}{dx} = \frac{1}{x^2y}, \quad y(1) = 2$

15. $\frac{dy}{dx} = \sqrt{xy^3}, \quad y(0) = 4$

16. $\frac{dy}{dx} = \frac{x}{y}, \quad y(3) = 5$

17. $\frac{dy}{dx} = -\frac{x}{y}, \quad y(12) = -5$

18. $y^2\frac{dy}{dx} = x^2+2x+1, \quad y(1) = 2$

19. $\frac{dy}{dx} = 3x^2y^2 - y^2, \quad y(0) = 1$

20. $\frac{dy}{dx} = 2xy^3(2x^2+1), \quad y(1) = 1$

In Problems 21 through 30, use the method of derivation of Eq. (13), rather than the equation itself, to find the solution of the given initial value problem.

21. $\frac{dy}{dx} = y+1; \quad y(0) = 1$

22. $\frac{dy}{dx} = 2-y; \quad y(0) = 3$

23. $\frac{dy}{dx} = 2y-3; \quad y(0) = 2$

24. $\frac{dy}{dx} = \frac{1}{4} - \frac{y}{16}; \quad y(0) = 20$

25. $\frac{dx}{dt} = 2(x-1); \quad x(0) = 0$

26. $\frac{dx}{dt} = 2-3x; \quad x(0) = 4$

27. $\frac{dx}{dt} = 5(x+2); \quad x(0) = 25$

28. $\frac{dx}{dt} = -3-4x; \quad x(0) = -5$

29. $\frac{dv}{dt} = 10(10-v); \quad v(0) = 0$

30. $\frac{dv}{dt} = -5(10-v); \quad v(0) = -10$

31. Zembla had a population of 1.5 million in 1990. Assume that this country's population is growing continuously at a 4% annual rate and that Zembla absorbs 50,000 newcomers per year. What will its population be in the year 2010?

32. When a cake is removed from an oven, the temperature of the cake is 210°F. The cake is left to cool at room temperature, which is 70°F. After 30 min the temperature of the cake is 140°F. When will it be 100°F?

33. Payments are made continuously on a mortgage of original amount P_0 dollars at the constant rate of c dollars per month. Let $P(t)$ denote the balance (amount still owed) after t months and let r denote the monthly interest rate paid by the mortgage holder. (For example, $r = 0.06/12 = 0.005$ if the annual interest rate is 6%.) Derive the differential equation

$$\frac{dP}{dt} = rP - c, \qquad P(0) = P_0.$$

34. Your cousin must pay off an auto loan of \$3600 continuously over a period of 36 months. Apply the result of Problem 33 to determine the monthly payment required if the annual interest rate is (a) 12%; (b) 18%.

35. A rumor about phenylethylamine in the drinking water began to spread one day in a city with a population of 100,000. Within a week, 10,000 people had heard this rumor. Assuming that the rate of increase of the number of people who have heard the rumor is proportional to the number who have not yet heard it, how long will it be until half the population of the city has heard the rumor?

36. Rework Example 5, assuming that new conditions, effective in the year 2000, result in 17 births and 7 deaths annually per thousand population, as well as an immigration rate of 1.5 million persons per year.

37. You would like to be a multimillionaire but cannot rely on winning the lottery. How much would you need to invest per month—in effect, continuously—in an investment account that pays an annual interest rate of 10%, compounded continuously, in order for the account to be worth \$5 million after 30 years?

38. Just before midday the body of an apparent homicide victim is found in a room that is kept at a constant temperature of 70°F. At 12 noon the temperature of the body is 80°F and at 1 P.M. it is 75°F. Assume that the temperature of the body at the time of death was 98.6°F and that is has cooled in accord with Newton's law of cooling. What was the time of death?

39. Pottstown has a fixed population of 10,000 people. On January 1, 1000 people have the flu; on April 1, 2000 people have it. Assume that the rate of increase of the number $N(t)$ who have the flu is proportional to the number who don't have it. How many will have the disease on October 1?

40. Let $x(t)$ denote the number of people in Athens, Georgia, of population 100,000, who have the Tokyo flu. The rate of change of $x(t)$ is proportional to the number of those in Athens who do not yet have the disease. Suppose that 20,000 have the flu on March 1 and that 60,000 have it on March 16. (a) Set up and solve a differential equation to find $x(t)$. (b) On what date will the number of people infected with the disease reach 80,000? (c) What happens in the long run?

41. Early one morning it began to snow at a constant rate. At 7 A.M. a snowplow set off to clear a road. By 8 A.M. it had traveled 2 miles, but it took two more hours (until 10 A.M.) for the snowplow to clear an addition 2 miles of road. (a) Let $t = 0$ when it began to snow and let $x(t)$ denote the distance traveled by the snowplow at time t. Assuming that the snowplow clears snow from the road at a constant rate (in cubic feet per hour, say), show that

$$k\frac{dx}{dt} = \frac{1}{t}$$

where k is a constant. (b) What time did it begin to snow? (*Answer:* 6 A.M.)

42. A snowplow sets off at 7 A.M. as in Problem 41. Suppose now that by 8 A.M. it had traveled 4 miles and that by 9 A.M. it had moved an additional 3 miles. What time did it begin to snow? This is a more difficult snowplow problem because now a transcendental equation must be solved numerically to find the value of k. (*Answer:* 4:27 A.M.)

43. (*The catenary*) Suppose that a uniform flexible cable is suspended between two points $(\pm L, H)$ at equal height located symmetrically on either side of the y-axis (Fig. 8.3.5). Principles of physics can be used to show that the shape $y = y(x)$ of the hanging cable satisfies the differential equation

$$a\frac{d^2y}{dx^2} = \sqrt{1 + \left(\frac{dy}{dx}\right)^2},$$

where the constant $a = T/\rho$ is the ratio of the tension T of the cable at its lowest point to its linear (constant) density ρ. Note that the lowest point of the cable occurs where $x = 0$ and that $y'(0) = 0$. Substitute $v = dy/dx$ in this second-order differential equation to show that it becomes the first-order equation

$$a\frac{dv}{dt} = \sqrt{1 + v^2}.$$

Solve this differential equation for $y'(x) = v(x) = \sinh(x/a)$. Then integrate to get the shape function

$$y(x) = a\cosh\left(\frac{x}{a}\right) + C$$

of the hanging cable. This curve is called a *catenary*, from the Latin word for *chain*.

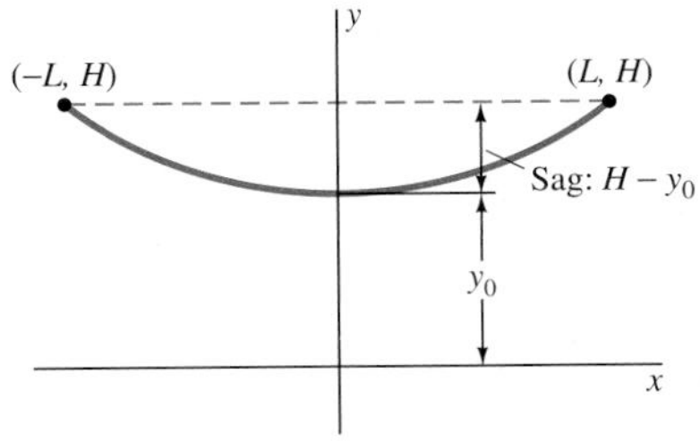

FIGURE 8.3.5 A flexible uniform cable suspended between two points at equal heights.

8.4 LINEAR EQUATIONS AND APPLICATIONS

A **linear** first-order differential equation is one that can be written in the form

$$\frac{dy}{dx} + P(x) \cdot y = Q(x) \qquad (1)$$

where $P(x)$ and $Q(x)$ are given functions of x. If the "coefficients" P and Q are actually constants, then the differential equation is separable and can be solved as in Section 8.3.

In this section we discuss the general case of variable coefficients. We will show how to multiply each side in (1) by an appropriately selected function $\rho(x)$ so that each side of the resulting equation can be integrated, thereby eliminating the unknown derivative term dy/dx. Such a function $\rho(x)$, multiplication by which makes it possible to integrate the equation, is called an **integrating factor** for the differential equation. The following example indicates how this can happen.

EXAMPLE 1 The differential equation

$$x^3\frac{dy}{dx} + x^2y = 2x^3 + 1$$

is not of the form in (1). But division of both sides by x^3 gives the linear equation

$$\frac{dy}{dx} + \frac{1}{x}y = 2 + \frac{1}{x^3}, \tag{2}$$

which *is* of the form in (1), with $P(x) = 1/x$ and $Q(x) = 2 + (1/x^3)$. For whatever reason, let us multiply both sides in (2) by the factor $\rho(x) = x$. The result is

$$x\frac{dy}{dx} + y = 2x + \frac{1}{x^2}.$$

We now recognize the left-hand side as the derivative $D_x(x \cdot y) = xy' + y$ of the product $x \cdot y$. Thus the last equation can be written as follows:

$$D_x(x \cdot y) = 2x + \frac{1}{x^2}.$$

We can now integrate both sides with respect to x. An indefinite integral of the derivative on the left is simply the product $x \cdot y$. Because the integrals of the two sides differ by a constant, integration yields

$$x \cdot y = \int \left(2x + \frac{1}{x^2}\right) dx = x^2 - \frac{1}{x} + C.$$

Finally, we divide by x to solve for y and thereby obtain the general solution

$$y(x) = x - \frac{1}{x^2} + \frac{C}{x}$$

of the linear differential equation in (2). ◆

The crucial integrating factor $\rho(x) = x$ in Example 1 was simply "pulled out of a hat." Next we will show that the integrating factor given by

$$\rho(x) = \exp\left(\int P(x)\,dx\right) \tag{3}$$

always effects the solution of the linear equation $y' + P(x)y = Q(x)$. Note first that, in Example 1 where $P(x) = 1/x$, Eq. (3) gives

$$\rho(x) = \exp\left(\int \frac{1}{x}\,dx\right) = e^{\ln x} = x,$$

and this is how we actually obtained the integrating factor $\rho(x) = x$ of Example 1—it was not "pulled out of a hat."

In general, we begin with the linear equation $y' + Py = Q$ and multiply both sides by the integrating factor $\rho(x)$ in Eq. (3). The result is

$$e^{\int P(x)\,dx}\frac{dy}{dx} + P(x)e^{\int P(x)\,dx}y = Q(x)e^{\int P(x)\,dx}. \tag{4}$$

Because

$$D_x\left[\int P(x)\,dx\right] = P(x),$$

the left-hand side in Eq. (4) is the derivative of the *product* $y(x) \cdot e^{\int P(x)\,dx}$, so Eq. (4) is equivalent to

$$D_x\left[y(x) \cdot e^{\int P(x)\,dx}\right] = Q(x)e^{\int P(x)\,dx}.$$

Integrating both sides of this equation yields

$$y(x)e^{\int P(x)\,dx} = \int \left(Q(x)e^{\int P(x)\,dx}\right) dx + C.$$

Finally solving for y, we obtain the general solution of the linear first-order differential equation in (1):

$$y(x) = e^{-\int P(x)\,dx}\left[\int \left(Q(x)e^{\int P(x)\,dx}\right) dx + C\right]. \tag{5}$$

There is no need to memorize such a formula. In a specific problem it generally is simpler to use the *method* by which we developed the formula. That is, in order to solve an equation written in the form in (1) with the coefficient functions $P(x)$ and $Q(x)$ displayed explicitly, you should carry out the following steps.

Method: Solution of First-Order Linear Equations

1. Begin by calculating the integrating factor $\rho(x) = e^{\int P(x)\,dx}$.
2. Then multiply both sides of the differential equation by $\rho(x)$.
3. Next, recognize the left-hand side of the resulting equation as the derivative of a product:
$$D_x[\rho(x)y(x)] = \rho(x)Q(x).$$
4. Finally, integrate this equation to obtain
$$\rho(x)y(x) = \int \rho(x)Q(x)\,dx + C,$$
then solve for y to obtain the general solution of the original differential equation.

REMARK 1 Given an initial condition $y(x_0) = y_0$, you can (as usual) substitute $x = x_0$ and $y = y_0$ in the general solution and solve for the value of C that yields the particular solution satisfying this initial condition.

REMARK 2 The integrating factor $\rho(x)$ is determined only to within a multiplicative constant. If we replace

$$\int P(x)\,dx \quad \text{with} \quad \int P(x)\,dx + K$$

in Eq. (3), the result is

$$\rho(x) = e^{K+\int P(x)\,dx} = e^K e^{\int P(x)\,dx}.$$

But the constant factor e^K does not affect the result of multiplying both sides of the differential equation in (1) by $\rho(x)$. Hence we may choose for $\int P(x)\,dx$ any convenient antiderivative of $P(x)$.

EXAMPLE 2 Solve the initial value problem

$$\frac{dy}{dx} - y = \frac{11}{8}e^{-x/3}, \qquad y(0) = 0.$$

Solution Here we have $P(x) \equiv -1$ and $Q(x) = \frac{11}{8}e^{-x/3}$, so the integrating factor is

$$\rho(x) = e^{\int(-1)\,dx} = e^{-x}.$$

Multiplying both sides of the given equation by e^{-x} yields

$$e^{-x}\frac{dy}{dx} - e^{-x}y = \frac{11}{8}e^{-4x/3}, \tag{6}$$

which we recognize as

$$\frac{d}{dx}(e^{-x}y) = \frac{11}{8}e^{-4x/3}.$$

Hence integration with respect to x gives

$$e^{-x}y = \int \frac{11}{8}e^{-4x/3}\,dx = -\frac{33}{32}e^{-4x/3} + C,$$

and then multiplication by e^x gives the general solution

$$y(x) = Ce^x - \frac{33}{32}e^{-x/3}. \tag{7}$$

Substituting $x = y = 0$ now gives $C = \frac{33}{32}$, so the desired particular solution is

$$y(x) = \frac{33}{32}e^x - \frac{33}{32}e^{-x/3} = \frac{33}{32}\left(e^x - e^{-x/3}\right). \quad ◆$$

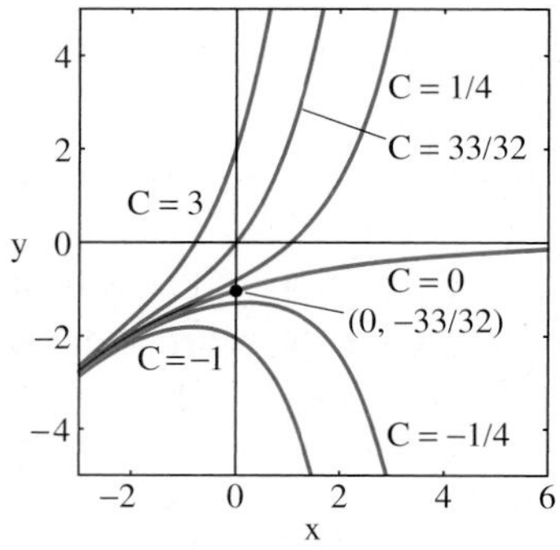

FIGURE 8.4.1 Solution curves of the differential equation $\dfrac{dy}{dx} - y = \dfrac{11}{8}e^{-x/3}$ corresponding to different values of C in Eq. (7).

REMARK Figure 8.4.1 shows some typical solution curves for Eq. (6), including the one passing through the origin. Note that some solutions grow rapidly in the positive direction as x increases, while others grow rapidly in the negative direction. Such behavior of a given solution curve is determined by its initial condition $y(0) = y_0$. The two types of behavior in Example 2 are separated by the particular solution

$$y(x) = -\frac{33}{32}e^{-x/3}$$

for which $C = 0$ in Eq. (7), so $y_0 = -\frac{33}{32}$. If $y_0 > -\frac{33}{32}$ then $C > 0$ in Eq. (7), so the term e^x eventually dominates the behavior of $y(x)$, and hence $y(x) \to +\infty$ as $x \to +\infty$. But if $y_0 < -\frac{33}{32}$ then $C < 0$, so both terms in $y(x)$ are negative and therefore $y(x) \to -\infty$ as $x \to +\infty$. Thus the initial condition $y_0 = -\frac{33}{32}$ is *critical* in the sense that solutions that start above $-\frac{33}{32}$ on the y-axis grow in the positive direction, whereas solutions that start lower than $-\frac{33}{32}$ grow in the negative direction as $x \to +\infty$. The interpretation of a mathematical model often hinges on finding such a critical condition that separates one kind of behavior of a solution from a different kind of behavior.

EXAMPLE 3 Find a general solution of

$$(x^2+1)\frac{dy}{dx} + 3xy = 6x. \tag{8}$$

Solution After dividing both sides of the equation by x^2+1, we recognize the result

$$\frac{dy}{dx} + \frac{3x}{x^2+1}y = \frac{6x}{x^2+1}$$

as a first-order linear equation with $P(x) = 3x/(x^2+1)$ and $Q(x) = 6x/(x^2+1)$. Multiplication by the integrating factor

$$\rho(x) = \exp\left(\int \frac{3x}{x^2+1}\,dx\right) = \exp\left(\tfrac{3}{2}\ln(x^2+1)\right) = (x^2+1)^{3/2}$$

yields

$$(x^2+1)^{3/2}\frac{dy}{dx} + 3x(x^2+1)^{1/2}y = 6x(x^2+1)^{1/2},$$

and thus

$$D_x\left[(x^2+1)^{3/2}y\right] = 6x(x^2+1)^{1/2}.$$

Integration then yields

$$(x^2+1)^{3/2}y = \int 6x(x^2+1)^{1/2}\,dx = 2(x^2+1)^{3/2} + C.$$

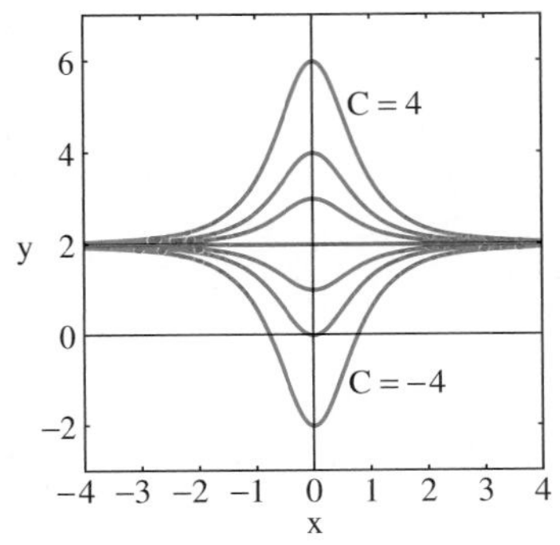

FIGURE 8.4.2 Solution curves of the differential equation $(x^2+1)\dfrac{dy}{dx}+3xy=6x$ corresponding to different values of C in Eq. (9).

Multiplying both sides by $(x^2+1)^{-3/2}$ then gives the general solution

$$y(x) = 2 + C(x^2+1)^{-3/2}. \tag{9}$$

◆

REMARK Figure 8.4.2 shows some typical solution curves for Eq. (8). Note that as $x \to +\infty$, all other solution curves approach the constant solution curve $y(x) \equiv 2$ that corresponds to $C = 0$ in Eq. (9). This constant solution can be described as an *equilibrium solution* of the differential equation, because $y(0) = 2$ implies that $y(x) = 2$ for all x (and thus the value of the solution remains forever where it begins).

A Closer Look at the Method

The preceding derivation of the solution in Eq. (5) of the linear first-order equation $y' + Py = Q$ bears closer examination. Suppose that the coefficient functions $P(x)$ and $Q(x)$ are continuous on the (possibly unbounded) open interval I. Then the antiderivatives

$$\int P(x)\,dx \quad\text{and}\quad \int \left(Q(x)e^{\int P(x)\,dx}\right)dx$$

exist on I. Our derivation of Eq. (5) shows that *if* $y = y(x)$ is a solution of Eq. (1) on I, *then* $y(x)$ is given by the formula in (5) for some choice of the constant C. Conversely, you may verify by direct substitution that the function $y(x)$ given in Eq. (5) satisfies Eq. (1). Finally, given a point x_0 of I and any number y_0, there is—as previously noted—a unique value of C such that $y(x_0) = y_0$. Consequently, we have proved the following existence-uniqueness theorem.

THEOREM 1 The Linear First-Order Equation

If the functions $P(x)$ and $Q(x)$ are continuous on the open interval I containing the point x_0, then the initial value problem

$$\frac{dy}{dx} + P(x)y = Q(x), \qquad y(x_0) = y_0 \tag{10}$$

has a unique solution $y(x)$ on I, given by the formula in (5) with an appropriate value of C.

REMARK 1 Theorem 1 gives a solution on the *entire* interval I for a *linear* differential equation, in contrast with the existence-uniqueness theorem mentioned in the last paragraphs of Section 8.2. That theorem guarantees only a solution on a possibly smaller interval J.

REMARK 2 Theorem 1 tells us that every solution of Eq. (1) is included in the general solution given in Eq. (5). Thus a linear first-order differential equation has no "singular solution" that is not included in its general solution.

REMARK 3 The appropriate value of the constant C in Eq. (5)—as needed to solve the initial value problem in (10)—can be selected "automatically" by writing

$$\rho(x) = \exp\left(\int_{x_0}^{x} P(t)\,dt\right),$$
$$y(x) = \frac{1}{\rho(x)}\left[y_0 + \int_{x_0}^{x} \rho(t)Q(t)\,dt\right]. \tag{11}$$

The indicated limits x_0 and x effect a choice of indefinite integrals in Eq. (5) that guarantees in advance that $\rho(x_0) = 1$ and that $y(x_0) = y_0$ (as you can verify directly by substituting $x = x_0$ in the equations in (11)).

EXAMPLE 4 Solve the initial value problem

$$x^2\frac{dy}{dx} + xy = \sin x, \qquad y(1) = y_0. \tag{12}$$

Solution Division by x^2 gives the linear first-order equation

$$\frac{dy}{dx} + \frac{1}{x}y = \frac{\sin x}{x^2}$$

in which $P(x) = 1/x$ and $Q(x) = (\sin x)/x^2$. With $x_0 = 1$ the integrating factor in (11) is

$$\rho(x) = \exp\left(\int_1^x \frac{1}{t}\,dt\right) = \exp(\ln x) = x,$$

so the desired particular solution is given by

$$y(x) = \frac{1}{x}\left[y_0 + \int_1^x \frac{\sin t}{t}\,dt\right]. \tag{13}$$

In accord with Theorem 1, this solution is defined on the whole positive x-axis. ◆

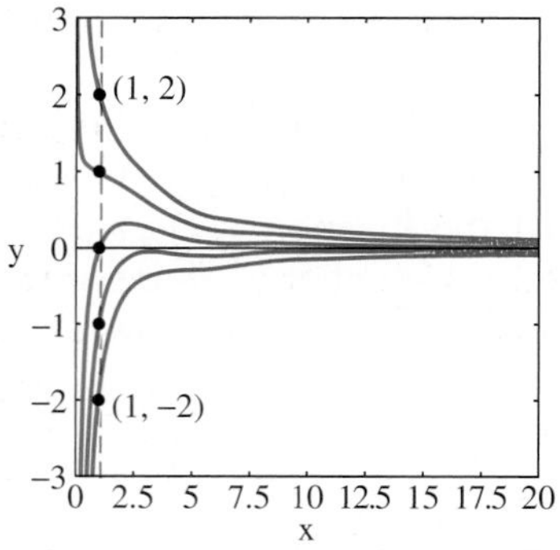

FIGURE 8.4.3 Solution curves of the differential equation $x^2\dfrac{dy}{dx} + xy = \sin x$ corresponding to different values of y_0 in Eq. (13).

COMMENT Ordinarily an integral such as the one in Eq. (13) might (for a given value of x) need to be approximated numerically—using Simpson's approximation, for instance—to find the value $y(x)$ of the solution at x. But the integral serves perfectly well to define the function $y(x)$. For instance, using a computer algebra system, one can use the *Maple* command

```
y := (1/x)*(1 + int(sin(t)/t, t=1..x));
```

or the *Mathematica* command

```
y = (1/x)*(1 + Integrate[Sin[t]/t, { t, 1, x } ]);
```

to define the particular solution with $y(1) = 1$, then plot this solution in the usual manner. Figure 8.4.3 shows solution curves plotted in this way with initial values $y(1) = y_0$ ranging from $y_0 = -2$ to $y_0 = 2$. It appears that on each solution curve, $y(x) \to 0$ as $x \to +\infty$.

Mixture Problems

As a first application of linear first-order equations, we consider a tank containing a solution—a mixture of solute and solvent—such as salt dissolved in water. There is both inflow and outflow, and we want to compute the *amount* $x(t)$ of solute in the tank at time t, given the amount $x(0) = x_0$ at time $t = 0$. Suppose that solution with a concentration of c_i grams of solute per liter of solution flows into the tank at the constant rate of r_i liters per second, and that the solution in the tank—kept thoroughly mixed by stirring—flows out at the constant rate of r_o liters per second.

To set up a differential equation for $x(t)$, we estimate the change Δx in x during the brief time interval $[t, t + \Delta t]$. The amount of solute that flows into the tank during Δt seconds is $r_i c_i\,\Delta t$ grams. To check this, observe how the cancellation of units in

$$\left(r_i\ \frac{\text{liters}}{\text{second}}\right) \times \left(c_i\ \frac{\text{grams}}{\text{liter}}\right) \times (\Delta t\ \ \text{seconds})$$

yields a quantity measured in grams.

The amount of solute that flows out of the tank during the same time interval depends on the concentration $c_o(t)$ of solute in the solution at time t. But as noted

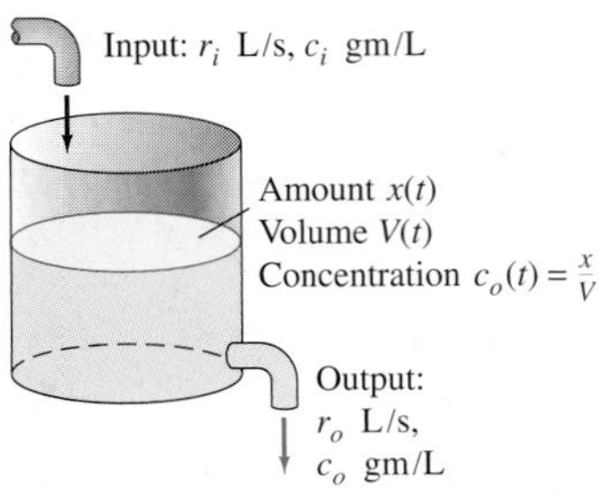

FIGURE 8.4.4 The single-tank mixture problem.

in Fig. 8.4.4, $c_o(t) = x(t)/V(t)$, where $V(t)$ denotes the volume (not constant unless $r_i = r_o$) of solution in the tank at time t. Then

$$\Delta x = \{\text{grams input}\} - \{\text{grams output}\} \approx r_i c_i \, \Delta t - r_o c_o \, \Delta t. \tag{14}$$

We now divide by Δt:

$$\frac{\Delta x}{\Delta t} \approx r_i c_i - r_o c_o.$$

Finally, we take the limit as $\Delta t \to 0$; if all the functions involved are continuous and $x(t)$ is differentiable, then the error in this approximation also approaches zero, and we obtain the differential equation

$$\frac{dx}{dt} = r_i c_i - r_o c_o, \tag{15}$$

in which r_i, c_i, and r_o are constants, but c_o denotes the variable concentration

$$c_o(t) = \frac{x(t)}{V(t)} \tag{16}$$

of solute in the tank at time t. Thus the amount $x(t)$ of solution in the tank satisfies the differential equation

$$\frac{dx}{dt} = r_i c_i - \frac{r_o}{V} x. \tag{17}$$

If $V_0 = V(0)$, then $V(t) = V_0 + (r_i - r_o)t$, so Eq. (17) is a linear first-order differential equation for the amount $x(t)$ of solute in the tank at time t.

IMPORTANT Equation (17) is *not* one you should commit to memory. It is the process we used to obtain that equation—examination of the behavior of the system over the short time interval $[t, t+\Delta t]$—that you should strive to understand, because it is a very useful tool for obtaining all sorts of differential equations.

REMARK In deriving Eq. (17) we used g/L mass/volume units for convenience. Any other consistent system of units can be used to measure amounts of solute and volumes of solution. In the following example we measure both in cubic kilometers.

EXAMPLE 5 Assume that Lake Erie has a volume of 480 km^3 and that its rates of inflow (from Lake Huron) and outflow (to Lake Ontario) are both 350 km^3 per year. Suppose that at the time $t = 0$ (in years), the pollutant concentration of Lake Erie—caused by past industrial pollution that has now been ordered to cease—is five times that of Lake Huron. If the outflow from Lake Erie henceforth is perfectly mixed lake water, how long will it take to reduce the pollution concentration in Lake Erie to twice that of Lake Huron?

Solution Here our "mixing tank" is Lake Erie and $x(t)$ denotes the volume of pollutants in the lake after t years. If c denotes the constant (although unknown) pollutant concentration in Lake Huron, then the initial concentration of pollutants in Lake Erie is $5c$. Summarizing the given information, we therefore have

$$V = 480 \quad (\text{km}^3),$$

$$r_i = r_o = r = 350 \quad (\text{km}^3/\text{yr}),$$

$$c_i = c \quad \text{(the pollutant concentration of Lake Huron),} \quad \text{and}$$

$$x_0 = x(0) = 5cV.$$

The question is this: When is $x(t) = 2cV$? With this notation, Eq. (17) is the separable equation

$$\frac{dx}{dt} = rc - \frac{r}{V}x, \tag{18}$$

which we rewrite in the linear first-order form

$$\frac{dx}{dt} + px = q \tag{19}$$

with constant coefficients $p = r/V$, $q = rc$, and integrating factor $\rho = e^{pt}$. You can either solve this equation directly or apply the formula in (11). The latter gives

$$\begin{aligned} x(t) &= e^{-pt}\left[x_0 + \int_0^t qe^{pt}\,dt\right] = e^{-pt}\left[x_0 + \frac{q}{p}(e^{pt} - 1)\right] \\ &= e^{-rt/V}\left[5cV + \frac{rc}{r/V}\left(e^{rt/V} - 1\right)\right]; \\ x(t) &= cV + 4cVe^{-rt/V}. \end{aligned} \tag{20}$$

To find when $x(t) = 2cV$, we therefore need only solve the equation

$$2cV = cV + 4cVe^{-rt/V},$$

which simplifies upon cancellation of cV to $4e^{-rt/V} = 1$:

$$t = \frac{V}{r}\ln 4 = \frac{480}{350}\ln 4 \approx 1.091 \quad \text{(years)}.$$ ◆

EXAMPLE 6 A 120-gal tank initially contains 90 lb of salt dissolved in 90 gal of water. Brine containing 2 lb/gal of salt flows into the tank at the rate of 4 gal/min, and the well-stirred mixture flows out of the tank at the rate of 3 gal/min. How much salt does the tank contain when it is full?

Solution The interesting feature of this example is that, due to the differing rates of inflow and outflow, the volume of brine in the tank increases steadily with $V(t) = 90 + t$ gallons. The change Δx in the amount x of salt in the tank from time t to time $t + \Delta t$ is given by

$$\Delta x \approx 4 \cdot 2 \cdot \Delta t \;-\; 3 \cdot \left(\frac{x}{90 + t}\right)\Delta t,$$

so the differential equation that models this example is

$$\frac{dx}{dt} + \frac{3}{90 + t}x = 8.$$

An integrating factor is

$$\rho(x) = \exp\left(\int \frac{3}{90 + t}\,dt\right) = e^{3\ln(90+t)} = (90 + t)^3,$$

which gives

$$\begin{aligned} D_t[(90 + t)^3 x] &= 8 \cdot (90 + t)^3; \\ (90 + t)^3 x &= 2(90 + t)^4 + C. \end{aligned}$$

Substituting $x(0) = 90$ gives $C = -(90)^4$, so the amount of salt in the tank at time t is

$$x(t) = 2 \cdot (90 + t) - \frac{90^4}{(90 + t)^3}.$$

The tank is full after 30 min, and when $t = 30$, we have

$$x(30) = 2 \cdot (90 + 30) - \frac{90^4}{120^3} \approx 202 \quad \text{(lb)}$$

of salt in the tank. ◆

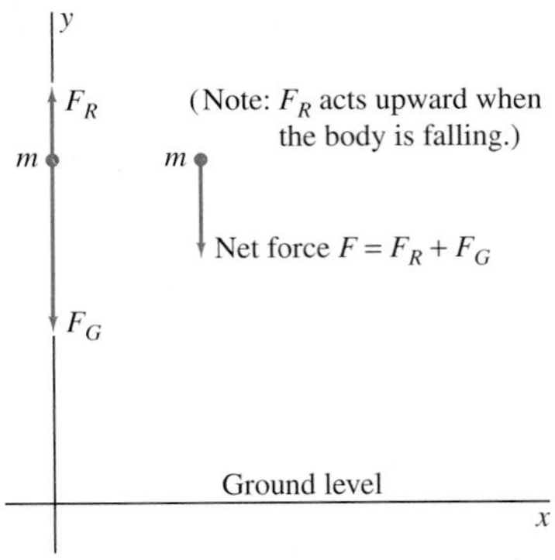

FIGURE 8.4.5 Vertical motion with air resistance.

Motion with Resistance

In Section 8.2 we explored graphical and numerical solutions of the differential equation

$$\frac{dv}{dt} = g - kv \tag{21}$$

satisfied by the *downward* velocity v of a ball that is dropped from a hovering helicopter and is thereafter affected both by gravity and by air resistance proportional to v. Here we explore symbolic solutions of similar problems.

Suppose that a ball is thrown straight up in the air. We set up a coordinate system with the y-axis directed *upward* and with $y = 0$ at ground level. As illustrated in Fig. 8.4.5, the ball of mass m is acted on by the force $F_G = -mg$ of gravity—negative because it is directed downward—and by the force F_R of air resistance. If the air resistance is proportional to the velocity of the ball, then

$$F_R = -Rv \tag{22}$$

where $R > 0$ is a proportionality constant. Observe that the minus sign here correctly indicates that the force acts downward ($F_R < 0$) when the motion is upward ($v > 0$) and that the force acts upward ($F_R > 0$) when the motion is downward ($v < 0$).

By Newton's second law of motion, the total force $F = F_G + F_R$ is equal to the acceleration $ma = m(dv/dt)$ of the ball, and hence

$$m\frac{dv}{dt} = -mg - Rv. \tag{23}$$

Division by the mass m of the ball yields the equation

$$\frac{dv}{dt} = -g - kv \tag{24}$$

where $k = R/m > 0$ and $g \approx 32$ ft/s^2 (or 9.8 m/s^2). (You should verify that if the positive y-axis were directed downward instead of upward, then this derivation would yield Eq. (21) with a single sign changed on the right-hand side.)

EXAMPLE 7 Suppose that the ball of Section 8.2 (with $k = 0.16$ in fps units) is projected straight upward from the ground—perhaps by a pitching machine—with initial velocity 160 ft/s. Find

(a) the maximum height it attains;
(b) its time of ascent to that height and the time of its descent back to the ground;
(c) the velocity with which it strikes the ground.

Before proceeding with the solution, let's first consider (for the purpose of later comparison) the case of *no* air resistance. With $k = 0$ in Eq. (24), we readily calculate

$$\frac{dv}{dt} = -32, \quad v = -32t + 160, \quad \text{and} \quad y = \int v\,dt = -16t^2 + 160t$$

(with $v_0 = 160$ and $y_0 = 0$) as in Section 5.2. The ball reaches its maximum height when $v = -32t + 160 = 0$, thus when $t = 5$ (s). Hence its maximum height is $y(5) = 400$ (ft). It strikes the ground after 10 seconds, with velocity $v(10) = -160$ ft/s (and thus with the same speed as its launch from the ground).

Solution If we rewrite Eq. (24) in the form

$$\frac{dv}{dt} + (0.16)v = -32,$$

we see that we have a linear differential equation with integrating factor

$$\rho(t) = e^{\int (0.16)\,dt} = e^{(0.16)t}.$$

After we multiply both sides of the differential equation by $\rho(t)$ and integrate, we get

$$e^{(0.16)t}\frac{dv}{dt} + (0.16)e^{(0.16)t}v = -32e^{(0.16)t};$$

$$D_t\left(e^{(0.16)t}v\right) = -32e^{(0.16)t};$$

$$e^{(0.16)t}v = \int \left[-32e^{(0.16)t}\right]dt = -200e^{(0.16)t} + C_1.$$

Substituting $t = 0$, $v = 160$ gives $C_1 = 360$, so after t seconds the velocity of the ball is

$$v(t) = e^{-(0.16)t}\left(-200e^{(0.16)t} + 360\right) = -200 + 360e^{-(0.16)t}. \qquad \textbf{(25)}$$

Now an easy antidifferentiation yields

$$y = \int v\,dt = \int \left(-200 + 360e^{-(0.16)t}\right)dt = -200t - 2250e^{-(0.16)t} + C_2.$$

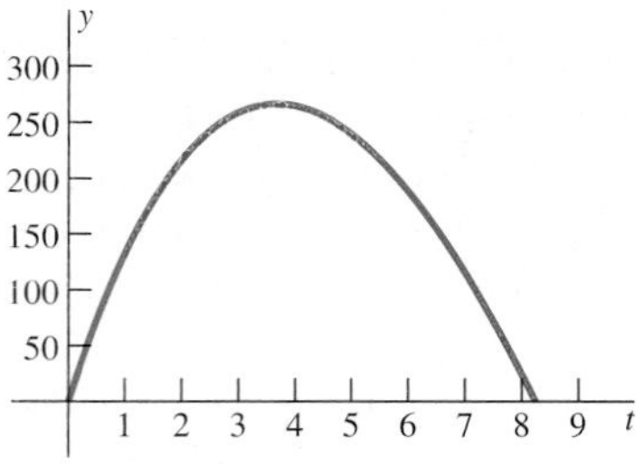

FIGURE 8.4.6 Graph of the ball's height function $y(t)$.

Finally, substituting $t = 0$, $y = 0$ gives $C_2 = 2250$, so the height of the ball after t seconds in the air is

$$y(t) = -200t + 2250\left(1 - e^{-(0.16)t}\right). \qquad \textbf{(26)}$$

Figure 8.4.6 shows the graph of $y(t)$. We see at a glance that air resistance has reduced the maximum height of the ball from 400 ft to under 300 ft, and its total time in the air from 10 s to a bit over 8 s:

(a) The ball reaches its apex when $y'(t) = v(t) = -200 + 360e^{-(0.16)t} = 0$, so its time of ascent is

$$t = \frac{\ln(360/200)}{0.16} \approx 3.67 \quad \text{(s)}.$$

Thus its maximum height in the air is given by Eq. (26) as $y(3.67) \approx 265.27$ ft.

(b) The ball returns to the ground when

$$y(t) = 0; \quad \text{that is, when} \quad -200t + 2250\left(1 - e^{-(0.16)t}\right) = 0.$$

Using Newton's method, or the `Solve` command on a calculator or computer algebra system, we find that $t \approx 8.24$ s. Thus the time of descent of the ball (from its apex back to the ground) is approximately $8.24 - 3.76 = 4.57$ s. Note that this time of descent is somewhat greater than its time of ascent, as every major-league catcher knows.

(c) Finally, the velocity with which the ball strikes the ground is given by Eq. (25) as $v(8.24) \approx -95.44$ ft/s. Note that the ball's impact speed is considerably less than its initial launch speed of 160 ft/s. ◆

REMARK Suppose that instead of striking the ground, the ball of Example 7 falls into a virtually bottomless pit (such as a deep mine shaft). Note that Eq. (25) implies that it does not fall with uniformly increasing speed—as it would in a vacuum—but instead approaches a *limiting speed* of 200 ft/s (as we observed graphically and numerically in Section 8.2). This phenomenon is analyzed more generally in Problem 43. It explains why one occasionally reads of someone whose parachute failed to open completely but nevertheless survived (perhaps with the aid of a conveniently located haystack).

8.4 TRUE/FALSE STUDY GUIDE

8.4 CONCEPTS: QUESTIONS AND DISCUSSION

1. Write several differential equations that are linear but not separable.
2. Write several differential equations that are both linear and separable. Compare their solutions by separation of variables with their solutions using the method of this section. Are the two general solutions identical in each case? If not, reconcile the two results. Does one method seem generally easier than the other? Does this depend on the particular differential equation?
3. Consider the up-and-down motion of a ball subject both to gravity and to a force $F_R = -Rv^p$ of air resistance that is proportional to some *power* $p \neq 1$ of the velocity. Is the resulting differential equation—analogous to (24)—still linear? If not, can you think of a value of p for which you might still be able to solve for the velocity function $v(t)$?

8.4 PROBLEMS

Find general solutions of the differential equations in Problems 1 through 20. If an initial condition is given, find the corresponding particular solution.

1. $\dfrac{dy}{dx} + y = 2, \quad y(0) = 0$

2. $\dfrac{dy}{dx} - 2y = 3e^{2x}, \quad y(0) = 0$

3. $\dfrac{dy}{dx} + 3y = 2xe^{-3x}$

4. $\dfrac{dy}{dx} - 2xy = e^{x^2}$

5. $x\dfrac{dy}{dx} + 2y = 3x, \quad y(1) = 5$

6. $x\dfrac{dy}{dx} + 5y = 7x^2, \quad y(2) = 5$

7. $2x\dfrac{dy}{dx} + y = 10\sqrt{x}$

8. $3x\dfrac{dy}{dx} + y = 12x$

9. $x\dfrac{dy}{dx} - y = x, \quad y(1) = 7$

10. $2x\dfrac{dy}{dx} - 3y = 9x^3$

11. $x\dfrac{dy}{dx} + y = 3xy, \quad y(1) = 0$

12. $x\dfrac{dy}{dx} + 3y = 2x^5, \quad y(2) = 1$

13. $\dfrac{dy}{dx} + y = e^x, \quad y(0) = 1$

14. $x\dfrac{dy}{dx} - 3y = x^3, \quad y(1) = 10$

15. $\dfrac{dy}{dx} + 2xy = x, \quad y(0) = -2$

16. $\dfrac{dy}{dx} = (1 - y)\cos x, \quad y(\pi) = 2$

17. $(1 + x)\dfrac{dy}{dx} + y = \cos x, \quad y(0) = 1$

18. $x\dfrac{dy}{dx} = 2y + x^3 \cos x$

19. $\dfrac{dy}{dx} + y \cot x = \cos x$

20. $\dfrac{dy}{dx} = 1 + x + y + xy, \quad y(0) = 0$

21. Express the general solution of $dy/dx = 1 + 2xy$ in terms of the **error function**

$$\operatorname{erf}(x) = \frac{2}{\sqrt{\pi}} \int_0^x e^{-t^2}\, dt.$$

22. Express the solution of the initial value problem

$$2x\frac{dy}{dx} = y + 2x\cos x, \qquad y(1) = 0$$

as an integral as in Example 4.

23. A tank contains 1000 L of a solution consisting of 100 kg of salt dissolved in water. Pure water is pumped into the tank at the rate of 5 L/s, and the mixture—kept uniform by stirring—is pumped out at the same rate. How long will it be until only 10 kg of salt remains in the tank?

24. Consider a reservoir with a volume of 8 billion cubic feet (ft^3) and an initial pollutant concentration of 0.25%. There is a daily inflow of 500 million ft^3 of water with a pollutant concentration of 0.05% and an equal daily outflow of the well-mixed water in the reservoir. How long will it take to reduce the pollutant concentration in the reservoir to 0.10%?

25. Rework Example 5 for the case of Lake Ontario. The only differences are that this lake has a volume of 1640 km^3 and an inflow-outflow rate of 410 km^3/year.

26. A tank initially contains 60 gal of pure water. Brine containing 1 lb of salt per gallon enters the tank at 2 gal/min, and the (perfectly mixed) solution leaves the tank at 3 gal/min; thus the tank is empty after exactly 1 h. (a) Find the amount of salt in the tank after t minutes. (b) What is the maximum amount of salt ever in the tank?

27. A 400-gal tank initially contains 100 gal of brine containing 50 lb of salt. Brine containing 1 lb of salt per gallon enters the tank at the rate of 5 gal/s, and the well-mixed brine in the tank flows out at the rate of 3 gal/s. How much salt will the tank contain when it is full of brine?

28. Consider the *cascade* of two tanks shown in Fig. 8.4.7, with $V_1 = 100$ (gal) and $V_2 = 200$ (gal) the volumes of brine in the two tanks. Each tank also initially contains 50 lb of salt. The three flow rates indicated in the figure are each 5 gal/min, with pure water flowing into tank 1. (a) Find the amount $x(t)$ of salt in tank 1 at time t. (b) Suppose that $y(t)$ is the amount of salt in tank 2 at time t. Show first that

$$\frac{dy}{dt} = \frac{5x}{100} - \frac{5y}{200},$$

then solve for $y(t)$, using the function $x(t)$ found in part (a). (c) Finally, find the maximum amount of salt ever in tank 2.

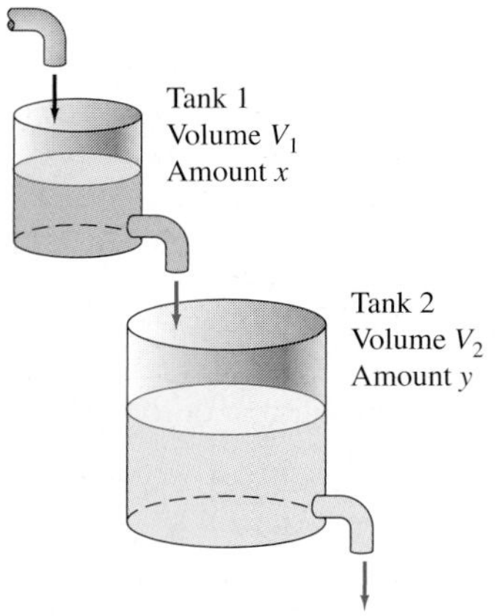

FIGURE 8.4.7 A cascade of two tanks.

29. A 30-year-old woman accepts an engineering position with a starting salary of \$30,000 per year. Her salary $S(t)$ increases exponentially, with $S(t) = 30e^{t/20}$ thousand dollars after t years. Meanwhile, 12% of her salary is deposited continuously in a retirement account, which accumulates interest at a continuous annual rate of 6%. (a) Estimate ΔA in terms of Δt to derive the differential equation satisfied by the amount $A(t)$ in her retirement account after t years. (b) Compute $A(40)$, the amount available for her retirement at age 70.

30. Suppose that a falling hailstone with density $\delta = 1$ starts from rest with negligible radius $r = 0$. Thereafter its radius is $r = kt$ (k is a constant) as it grows by accretion during its fall. Set up and solve the initial value problem

$$\frac{d}{dt}(mv) = mg, \qquad v(0) = 0,$$

where m is the variable mass of the hailstone, $v = dy/dt$ is its velocity, and the positive y-axis points downward. Then show that $dv/dt = g/4$. Thus the hailstone falls as though it were under *one-fourth* the influence of gravity.

31. The acceleration of a Maserati is proportional to the difference between 250 km/h and the velocity of this sports car. If this machine can accelerate from rest to 100 km/h in 10 s, how long will it take for the car to accelerate from rest to 200 km/h?

32. Suppose that a body moves through a resisting medium with resistance proportional to its velocity v, so that $dv/dt = -kv$. (a) Show that its velocity and position at time t are given by

$$v(t) = v_0 e^{-kt}$$

and

$$x(t) = x_0 + \left(\frac{v_0}{k}\right)(1 - e^{-kt}).$$

(b) Conclude that the body travels only a finite distance, and find that distance.

33. Suppose that a motorboat is moving at 40 ft/s when its motor suddenly quits, and that 10 s later the boat has slowed to 20 ft/s. Assume, as in Problem 32, that the resistance it encounters while coasting is proportional to its velocity. How far will the boat coast in all?

34. Consider a body that moves horizontally through a medium whose resistance is proportional to the *square* of the velocity v, so that $dv/dt = -kv^2$. Show that

$$v(t) = \frac{v_0}{1 + v_0 kt}$$

and that

$$x(t) = x_0 + \frac{1}{k}\ln(1 + v_0 kt).$$

Note that, in contrast with the result of Problem 32, $x(t) \to +\infty$ as $t \to +\infty$.

35. Assuming resistance proportional to the square of the velocity (as in Problem 34), how far does the motorboat of Problem 33 coast in the first minute after its motor quits?

36. Assume that a body moving with velocity v encounters resistance of the form $dv/dt = -kv^{3/2}$. Show that

$$v(t) = \frac{4v_0}{\left(kt\sqrt{v_0} + 2\right)^2}$$

and that

$$x(t) = x_0 + \frac{2}{k}\sqrt{v_0}\left(1 - \frac{2}{kt\sqrt{v_0} + 2}\right).$$

Conclude that under a $\frac{3}{2}$-power resistance a body coasts only a finite distance before coming to a stop.

37. Suppose that a car starts from rest, its engine providing an acceleration of 10 ft/s^2, while air resistance provides 0.1 ft/s^2 of deceleration for each foot per second of the car's velocity. (a) Find the car's maximum possible (limiting) velocity. (b) Find how long it takes the car to attain 90% of its limiting velocity, and how far it travels while doing so.

38. Rework both parts of Problem 37, with the sole difference that the deceleration due to air resistance now is $(0.001)v^2$ ft/s^2 when the car's velocity is v feet per second.

39. A motorboat weighs 32,000 lb and its motor provides a thrust of 5000 lb. Assume that the water resistance is

100 pounds for each foot per second of the speed v of the boat. Then

$$1000\frac{dv}{dt} = 5000 - 100v.$$

If the boat starts from rest, what is the maximum velocity that it can attain?

40. It is proposed to dispose of nuclear wastes—in drums with weight $W = 640$ lb and volume 8 ft^3—by dropping them into the ocean ($v_0 = 0$). The force equation for a drum falling through water is

$$m\frac{dv}{dt} = -W + B + F_R,$$

where the buoyant force B is equal to the weight (at 62.5 lb/ft^3) of the volume of water displaced by the drum (Archimedes' principle) and F_R is the force of water resistance, found empirically to be 1 lb for each foot per second of the velocity of a drum. If the drums are likely to burst upon an impact of more than 75 ft/s, what is the maximum depth to which they can be dropped in the ocean without likelihood of bursting?

41. Rework Example 7, but for a crossbow bolt launched straight upward from the ground with an initial velocity of 49 m/s.

42. Consider anew the ball of Examples 2 and 5 in Section 8.2, which is dropped from a helicopter hovering at a height of 3000 ft. (a) Solve the initial value problem

$$\frac{dv}{dt} = 32 - (0.16)v, \qquad v(0) = 0$$

to find its velocity $v(t)$ after t seconds. (b) Show that the limiting velocity of the ball is exactly 200 ft/s. (c) Find the ball's time of descent to the ground.

43. Solve the general initial velocity problem $dv/dt = -g - kv$, $v(0) = v_0$ for a projectile moving in a vertical line subject to gravity and air resistance. Show that

$$v(t) = v_0 e^{-kt} + \frac{g}{k}(e^{-kt} - 1).$$

Conclude that the projectile approaches a limiting *terminal velocity* given by

$$v_\tau = \lim_{t\to\infty} v(t) = -\frac{g}{k}.$$

44. A woman bails out of an airplane at an altitude of 10,000 ft, falls freely for 20 s, then opens her parachute. How long will it take her to reach the ground? Assume linear air resistance of k feet per second per second, taking $k = 0.15$ without the parachute and $k = 1.5$ with the parachute. (*Suggestion:* First determine her height and velocity when the parachute opens.)

8.5 POPULATION MODELS

In Section 8.1 we introduced the exponential differential equation $dP/dt = kP$, with solution $P(t) = P_0 e^{kt}$, as a mathematical model for natural population growth that occurs as a result of constant birth and death rates. Here we present a more general population model that accommodates birth and death rates that are not necessarily constant. As before, however, our population function $P(t)$ will be a *continuous* approximation to the actual population, which of course grows by integral increments.

Suppose that the population changes only by the occurrence of births and deaths—there is no immigration to or emigration from outside the country or environment under consideration. It is customary to track the growth or decline of a population in terms of its *birth rate* and *death rate* functions defined as follows:

- $\beta(t)$ is the number of births per unit of population per unit of time at time t;
- $\delta(t)$ is the number of deaths per unit of population per unit of time at time t.

Then the numbers of births and deaths that occur during the [short] time interval $[t, t + \Delta t]$ is given (approximately) by

$$\text{births: } \beta(t)\cdot P(t)\cdot \Delta t; \qquad \text{deaths: } \delta(t)\cdot P(t)\cdot \Delta t.$$

Hence the change ΔP in the population during the time interval $[t, t + \Delta t]$ of length Δt is

$$\Delta P = \{\text{births}\} - \{\text{deaths}\} \approx \beta(t)\cdot P(t)\cdot \Delta t - \delta(t)\cdot P(t)\cdot \Delta t,$$

so

$$\frac{\Delta P}{\Delta t} \approx [\beta(t) - \delta(t)]\cdot P(t).$$

The error in this approximation should approach zero as $\Delta t \to 0$, so—taking the limit—we get the differential equation

$$\frac{dP}{dt} = (\beta - \delta) \cdot P, \tag{1}$$

in which we write $\beta = \beta(t)$, $\delta = \delta(t)$, and $P = P(t)$ for brevity. Equation (1) is the **general population equation.** If β and δ are constant, Eq. (1) reduces to the natural growth equation with $k = \beta - \delta$. But it also includes the possibility that β and δ are variable functions of t. The birth and death rates need not be known in advance; they may well depend on the unknown function $P(t)$.

EXAMPLE 1 Suppose that an alligator population numbers 100 initially, and that its death rate is $\delta = 0$ (so none of the alligators is dying). If the birth rate is $\beta = (0.0005)P$—and thus the rate increases as the population does—then Eq. (1) gives the initial value problem

$$\frac{dP}{dt} = (0.0005)P^2, \qquad P(0) = 100$$

(with t in years). Upon separating the variables we get

$$\int \frac{1}{P^2}\,dP = \int (0.0005)\,dt;$$

$$-\frac{1}{P} = (0.0005)t + C.$$

Substituting $t = 0$, $P = 100$ gives $C = -1/100$, and then we readily solve for

$$P(t) = \frac{2000}{20 - t}.$$

For instance, $P(10) = 2000/10 = 200$, so after 10 years the alligator population has doubled. But we see that $P(t) \to +\infty$ as $t \to 20$, so a real "population explosion" occurs in 20 years. Indeed, the direction field and solution curves shown in Fig. 8.5.1 indicate that a population explosion always occurs, whatever the size of the (positive) initial population $P(0) = P_0$. In particular, it appears that the population always becomes unbounded in a *finite* period of time. ◆

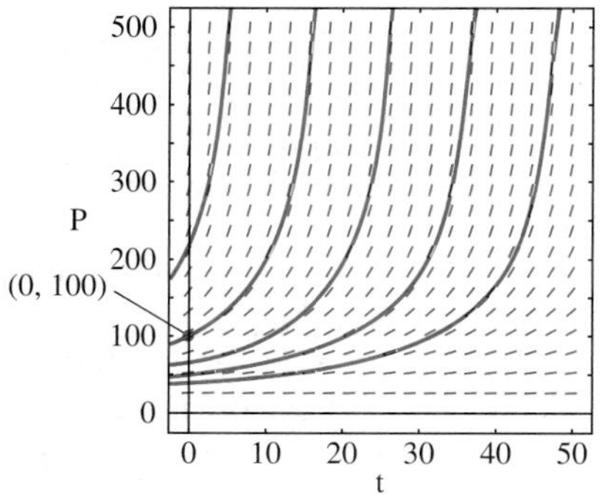

FIGURE 8.5.1 Direction field and solution curves for the equation $dP/dt = (0.0005)P^2$ in Example 1.

Bounded Populations and the Logistic Equation

In situations as diverse as the human population of a nation and a fruit fly population in a closed container, it is often observed that the birth rate decreases as the population itself increases. The reasons may range from increased scientific or cultural sophistication to limitations on available food or space. Suppose, for example, that the birth rate β is a *linear* decreasing function of the population size P, so that $\beta = \beta_0 - \beta_1 P$ where β_0 and β_1 are positive constants. If the death rate $\delta = \delta_0$ remains constant, then Eq. (1) takes the form

$$\frac{dP}{dt} = (\beta_0 - \beta_1 P - \delta_0)P;$$

that is,

$$\frac{dP}{dt} = aP - bP^2 \tag{2}$$

where $a = \beta_0 - \delta_0$ and $b = \beta_1$.

If the coefficients a and b are both positive, then Eq. (2) is called the **logistic equation.** For the purpose of relating the behavior of the population $P(t)$ to the values of the parameters in the equation, it is useful to rewrite the logistic equation in the form

$$\frac{dP}{dt} = kP(M - P) \tag{3}$$

where $k = b$ and $M = a/b$ are constants.

EXAMPLE 2 In Example 3 of Section 8.2 we explored graphically a population modeled by the logistic equation

$$\frac{dP}{dt} = (0.0004)P \cdot (150 - P) = (0.06)P - (0.0004)P^2. \tag{4}$$

To solve this differential equation symbolically, we separate the variables and integrate. We get

$$\int \frac{1}{P(150 - P)}\,dP = \int (0.0004)\,dt;$$

$$\frac{1}{150}\int \left(\frac{1}{P} + \frac{1}{150 - P}\right) dP = \int (0.0004)\,dt \qquad \text{[partial fractions];}$$

$$\ln|P| - \ln|150 - P| = (0.06)t + C;$$

$$\frac{P}{150 - P} = \pm e^C e^{(0.06)t} = Be^{(0.06)t} \qquad [\text{where } B = \pm e^C].$$

If we substitute $t = 0$ and $P = P_0$ into this last equation, we find that $B = P_0/(150 - P_0)$. Hence

$$\frac{P}{150 - P} = \frac{P_0 e^{(0.06)t}}{150 - P_0}.$$

Finally, this last equation is easy to solve for the population

$$P(t) = \frac{150P_0}{P_0 + (150 - P_0)e^{-(0.06)t}} \tag{5}$$

at time t in terms of the initial population $P_0 = P(0)$. Figure 8.5.2 shows a number of solution curves corresponding to different values of the initial population ranging from $P_0 = 20$ to $P_0 = 300$. Note that all these solution curves appear to approach the horizontal line $P = 150$ as an asymptote. Indeed, you should be able to see directly from Eq. (5) that $P(t) \to 150$ as $t \to +\infty$, whatever the initial value $P_0 > 0$. ◆

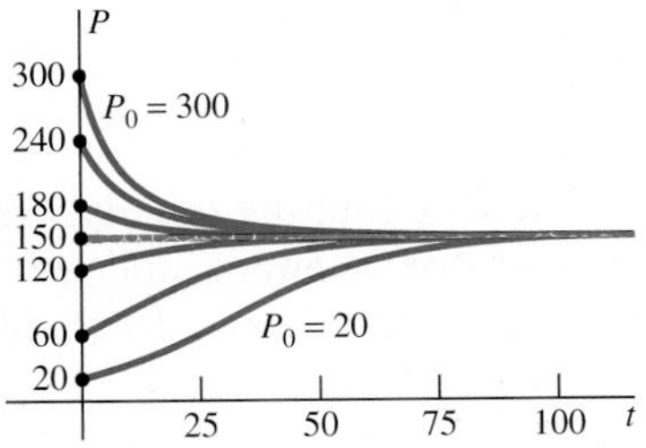

FIGURE 8.5.2 Typical solution curves for the logistic equation $P' = 0.06P - 0.0004P^2$.

Limited Populations and Carrying Capacity

The finite limiting population discovered in Example 2 is characteristic of logistic populations. In Problem 31 we ask you to use the method of solution of Example 2 to show that the solution of the logistic initial value problem

$$\frac{dP}{dt} = kP(M - P), \quad P(0) = P_0 \tag{6}$$

is

$$P(t) = \frac{MP_0}{P_0 + (M - P_0)e^{-kMt}}. \tag{7}$$

Actual populations are positive-valued. If $P_0 = M$ then (7) reduces to the unchanging (constant-valued) *equilibrium population* $P(T) \equiv M$. Otherwise, the behavior of a logistic population depends upon whether $0 < P_0 < M$ or $P_0 > M$. If $0 < P_0 < M$ then we see from Eqs. (6) and (7) that $P'(t) > 0$ and

$$P(t) = \frac{MP_0}{P_0 + (M - P_0)e^{-kMt}} = \frac{MP_0}{P_0 + \{\text{positive number}\}} < \frac{MP_0}{P_0} = M.$$

But if $P_0 > M$ then we see from Eqs. (6) and (7) that $P'(t) < 0$ and

$$P(t) = \frac{MP_0}{P_0 + (M - P_0)e^{-kMt}} = \frac{MP_0}{P_0 + \{\text{negative number}\}} > \frac{MP_0}{P_0} = M.$$

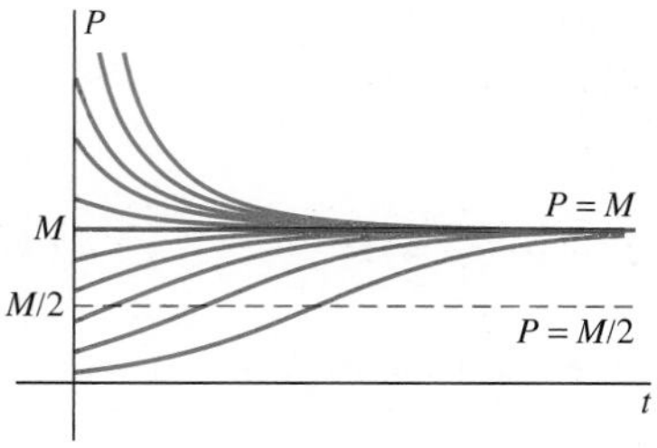

FIGURE 8.5.3 Typical solution curves for the logistic equation $P' = kP(M - P)$. Each solution curve that starts below the line $P = M/2$ has an inflection point on this line. (See Problem 33.)

In each case, the "positive number" or "negative number" in the denominator has absolute value less than P_0 and—because of the exponential factor—approaches zero as $t \to +\infty$. It follows that

$$\lim_{t \to \infty} P(t) = \frac{MP_0}{P_0 + 0} = M. \tag{8}$$

Thus a population that satisfies the logistic equation does *not* grow without bound like a naturally growing population modeled by the exponential equation $dP/dt = kP$. Instead, it approaches the finite **limiting population** M as $t \to +\infty$. As illustrated by the typical solution curves of the logistic equation shown in Fig. 8.5.3, the population $P(t)$ steadily increases and approaches M from below if $0 < P_0 < M$, but steadily decreases and approaches M from above if $P_0 > M$. Sometimes M is called the **carrying capacity** of the environment, considering it to be the maximum population that the environment can support on a long-term basis.

EXAMPLE 3 Suppose that in 1885 the population of a certain country was 50 million and was growing at the rate of 750,000 people per year at that time. Suppose also that in 1940 its population was 100 million and was then growing at the rate of 1 million per year. Assume that this population satisfies the logistic equation. Determine both the limiting population M and the predicted population for the year 2000.

Solution We substitute the two given pairs of data in Eq. (6) and find that

$$0.75 = 50k(M - 50) \quad \text{and} \quad 1.00 = 100k(M - 100).$$

We solve simultaneously for $M = 200$ and $k = 0.0001$. Thus the limiting population of the country in question is 200 million. With these values of M and k, and with $t = 0$ corresponding to the year 1940 (in which $P_0 = 100$), we find that—according to Eq. (7)—the population in the year 2000 will be

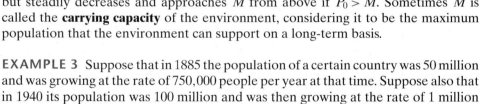

$$P(60) = \frac{200 \cdot 100}{100 + (200 - 100)e^{-(0.0001)(200)(60)}},$$

about 153.7 million people. ◆

HISTORICAL NOTE The logistic equation was introduced (around 1840) as a possible model for human population growth by the Belgian mathematician and demographer P. F. Verhulst (1804–1849). In Examples 4 and 5 we compare natural growth and logistic model fits to census data for the U.S. population in the 19th century.

EXAMPLE 4 The U.S. population in 1800 was 5.308 million and in 1900 was 76.212 million. If we take $P_0 = 5.308$ (with $t = 0$ in 1800) in the natural growth model $P(t) = P_0 e^{rt}$ and substitute $t = 100$, $P = 76.212$, we find that

$$76.212 = 5.308e^{100r}, \quad \text{so} \quad r = \frac{1}{100} \ln \frac{76.212}{5.308} \approx 0.026643.$$

Thus our natural growth model for the U.S. population during the 19th century is

$$P(t) = (5.308)e^{(0.026643)t} \tag{9}$$

(with t in years and P in millions). Because $e^{0.026643} \approx 1.02700$, the average population growth between 1800 and 1900 was about 2.7% per year. This rate of growth was not sustained during the 20th century. Whereas Eq. (9) predicts $P(150) = 288.780$ and $P(200) = 1094.240$ (over a billion), the actual population of the United States in the year 2000 was "only" about 280 million. ◆

EXAMPLE 5 The U.S. population in the year 1850 was 23.192 million. If we take $P_0 = 5.308$ and substitute the data pairs $t = 50$, $P = 23.192$ (for 1850) and $t = 100$, $P = 76.212$ (for 1900) in the logistic model formula in Eq. (7), we get the two equations

$$\begin{aligned} \frac{(5.308)M}{5.308 + (M - 5.308)e^{-50kM}} &= 23.192, \\ \frac{(5.308)M}{5.308 + (M - 5.308)e^{-100kM}} &= 76.212 \end{aligned} \tag{10}$$

in the two unknowns k and M. A computer algebra system yields the approximate solution

$$k = 0.000167716, \qquad M = 188.121$$

of the simultaneous equations in (10). Substituting these values in Eq. (7) yields the logistic model

$$P(t) = \frac{998.546}{5.308 + (182.813)e^{-(0.031551)t}}, \tag{11}$$

which agrees with the U.S. population data for 1850 and 1900. It predicts $P(150) = 144.354$, a fairly good approximation to the actual 1950 U.S. population of 151.326 million. Nevertheless, by the year 2000, the U.S. population had already greatly exceeded the limiting population of 188.121 million predicted by Eq. (11). ◆

The moral of Examples 4 and 5 is simply that one cannot expect too much of models that are based on severely limited information (such as a mere pair of data points). Much of the science of statistics is devoted to the analysis of large "data sets" to formulate useful (and perhaps reliable) mathematical models.

More Applications of the Logistic Equation

We next describe some situations that illustrate the varied circumstances in which the logistic equation is a satisfactory mathematical model.

1. *Limited environment situation.* A certain environment can support a population of at most M individuals. It is then reasonable to expect the growth rate $\beta - \delta$ (the combined birth and death rates) to be proportional to $M - P$, because we may think of $M - P$ as the potential for further expansion. Then $\beta - \delta = k(M - P)$, so that

$$\frac{dP}{dt} = (\beta - \delta)P = kP(M - P).$$

The classic example of a limited environment situation is a fruit fly population in a closed container.

2. *Competition situation.* If the birth rate β is constant but the death rate δ is proportional to P, so that $\delta = \alpha P$, then

$$\frac{dP}{dt} = (\beta - \alpha P)P = kP(M - P).$$

This might be a reasonable working hypothesis in a study of a cannibalistic population, in which all deaths result from chance encounters between individuals. Of course, competition between individuals is not usually so deadly, nor its effects so immediate and decisive.

3. *Joint proportion situation.* Let $P(t)$ denote the number of individuals in a constant-size susceptible population M who are infected with a certain contagious and incurable disease. The disease is spread by chance encounters. Then $P'(t)$ should be proportional to the product of the number P of individuals having the disease and the number $M - P$ of those not having it, and therefore $dP/dt = kP(M - P)$. Again we discover that the mathematical model is the logistic equation. The mathematical description of the spread of a rumor in a population of M individuals is identical.

EXAMPLE 6 Suppose that at time $t = 0$ (weeks), 10 thousand people in a city with population $M = 100$ thousand have heard a certain rumor. After 1 week the number $P(t)$ of those who have heard it has increased to $P(1) = 20$ thousand. Assuming that $P(t)$ satisfies a logistic equation, when will 80% of the city's population have heard the rumor?

Solution Substituting $P_0 = 10$ and $M = 100$ (thousand) in Eq. (7), we get

$$P(t) = \frac{1000}{10 + 90e^{-100kt}}. \qquad \textbf{(12)}$$

Then substituting $t = 1$, $P = 20$ gives the equation

$$20 = \frac{1000}{10 + 90e^{-100k}}$$

that is readily solved for

$$e^{-100k} = \frac{4}{9}, \quad \text{so} \quad k = \frac{1}{100}\ln\frac{9}{4} \approx 0.008109.$$

With $P(t) = 80$, Eq. (12) takes the form

$$80 = \frac{1000}{10 + 90e^{-100kt}},$$

which we solve for $e^{-100kt} = \frac{1}{36}$. It follows that 80% of the population has heard the rumor when

$$t = \frac{\ln 36}{100k} = \frac{\ln 36}{\ln(9/4)} \approx 4.42,$$

thus after about four weeks and three days. ◆

Doomsday versus Extinction

Consider a population $P(t)$ of unsophisticated animals that rely solely on chance encounters between females and males for reproductive purposes. It is reasonable to expect such encounters to occur at a rate that is proportional to the product of the number $P/2$ of males and the number $P/2$ of females, hence at a rate proportional to P^2. We therefore assume that births occur at the rate kP^2 (per unit time, with k constant). The birth rate (in births per unit time per unit of population) is then given by $\beta = kP$. If the death rate δ is constant, then the general population equation in (1) yields the differential equation

$$\frac{dP}{dt} = kP^2 - \delta P = kP(P - M) \tag{13}$$

(where $M = \delta/k > 0$) as a mathematical model of the population.

Note that the right-hand side in Eq. (13) is the *negative* of the right-hand side in the logistic equation in (7). We will see that the constant M is now a **threshold population,** with the way the population behaves in the future depending critically on whether the initial population P_0 is less than or greater than M.

EXAMPLE 7 Consider an animal population $P(t)$ that is modeled by the equation

$$\frac{dP}{dt} = (0.0004)P \cdot (P - 150) = (0.0004)P^2 - (0.06)P. \tag{14}$$

We want to find $P(t)$ in the following two cases: (a) $P(0) = 200$; (b) $P(0) = 100$.

Solution To solve the equation in (14), we separate the variables and integrate. We get

$$\int \frac{1}{P(P-150)}\,dP = \int (0.0004)\,dt;$$

$$-\frac{1}{150}\int \left(\frac{1}{P} - \frac{1}{P-150}\right) dP = \int (0.0004)\,dt \qquad \text{[partial fractions]};$$

$$\ln|P| - \ln|P - 150| = -(0.06)t + C;$$

$$\frac{P}{P-150} = \pm e^C e^{-(0.06)t} = Be^{-(0.06)t} \qquad [\text{where } B = \pm e^C]. \tag{15}$$

(a) Substituting $t = 0$ and $P = 200$ in (15) gives $B = 4$. With this value of B we solve Eq. (15) for

$$P(t) = \frac{600e^{-(0.06)t}}{4e^{-(0.06)t} - 1}. \tag{16}$$

Note that as t increases and approaches $T = (\ln 4)/0.06 \approx 23.105$, the positive denominator on the right in (16) decreases and approaches zero. Consequently $P(t) \to +\infty$ as $t \to T^-$. This is a *doomsday* situation—a real population *explosion*.

(b) Substituting $t = 0$ and $P = 100$ in (15) gives $B = -2$. With this value of B we solve Eq. (15) for

$$P(t) = \frac{300e^{-(0.06)t}}{2e^{-(0.06)t} + 1} = \frac{300}{2 + e^{(0.06)t}}. \tag{17}$$

Note that as t increases without bound, the positive denominator in the right in Eq. (17) approaches $+\infty$. Consequently $P(t) \to 0$ as $t \to +\infty$. This is an (eventual) *extinction* situation. ◆

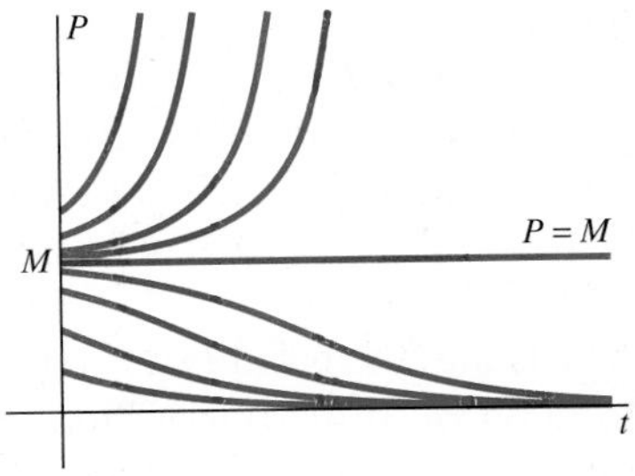

FIGURE 8.5.4 Typical solution curves for the explosion-extinction equation $P' = kP(P - M)$.

Thus the population in Example 7 either explodes or is an endangered species threatened with extinction, depending on whether or not its initial size exceeds the threshold population $M = 150$. An approximation to this phenomenon is sometimes observed with animal populations, such as the alligator population in certain areas of the southern United States.

Figure 8.5.4 shows typical solution curves that illustrate the two possibilities for a population $P(t)$ satisfying Eq. (13). If $P_0 = M$ (exactly!), then the population remains constant. But this equilibrium situation is highly unstable. If P_0 exceeds M, no matter how slightly, then $P(t)$ increases rapidly without bound, whereas if the initial (positive) population is less than M (however slightly), then it decreases—more gradually—toward zero as $t \to +\infty$.

Predator-Prey Populations

Consider the two differential equations

$$\frac{dx}{dt} = px - axy, \tag{18}$$

$$\frac{dy}{dt} = -qy + bxy \tag{19}$$

with positive coefficients a, b, p, and q. In ecology this system of simultaneous equations is often used to model a *prey population* (such as rabbits) and a *predator population* (such as foxes) that occupy the same environment. Typically, there is a food supply ample to feed the prey, whereas the predators feed on the prey—thereby impeding the growth of the prey population and promoting the growth of the predator population.

Note that if a were zero then Eq. (18) would reduce to the natural growth equation $dx/dt = px$, and thereby would imply that the prey population $x(t)$ would increase without bound as $t \to +\infty$. On the other hand, if b were zero, then Eq. (19) would reduce to the natural decay equation $dy/dx = -qy$, and would therefore imply that the predator population $y(t)$ would decline and die out as $t \to +\infty$. If a and b are positive, then the negative term $-axy$ in (18) represents a decrease in the rate of growth of the prey due to life-threatening "interaction" with the predators, while the positive term bxy in (19) represents a concomitant decrease in the rate of decline of the predators (who prosper at the expense of the prey).

If the coefficients in Eqs. (18) and (19) are all positive, then it generally is impossible to solve for x and y explicitly as elementary functions of the time variable t. But if we think of y as a function of x, then the chain rule in the form

$$\frac{dy}{dt} = \frac{dy}{dx}\frac{dx}{dt}$$

yields

$$\frac{dy}{dx} = \frac{dy/dt}{dx/dt} = \frac{-qy + bxy}{px - axy}. \tag{20}$$

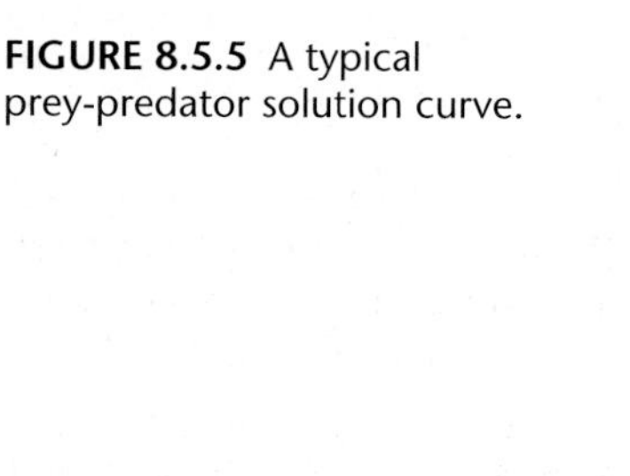

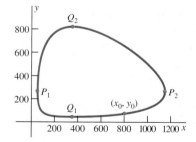

FIGURE 8.5.5 A typical prey-predator solution curve.

In Problem 34 we ask you to separate the variables in Eq. (20) to derive the implicit general solution

$$x^q y^p = Ce^{bx}e^{ay} \tag{21}$$

of Eq. (20). The value of the arbitrary constant C can be determined by substitution of specific numerical values of the coefficients a, b, p, and q together with initial values $x_0 = x(0)$ and $y_0 = y(0)$ of the two populations.

The graph of Eq. (21) is typically a closed curve in the shape of an oval in the first quadrant of the xy-plane. For instance, with the values $a = 0.0003$, $b = 0.0002$, $p = 0.08$, and $q = 0.07$ of the coefficients and with initial populations of $x_0 = 800$ rabbits and $y_0 = 75$ foxes, the implicit plotting command of a computer algebra system generates the solution curve shown in Fig. 8.5.5.

We can "zoom in" on the leftmost and rightmost points P_1 and P_2 on the curve by plotting it in appropriately selected smaller viewing windows. Of course we may also zoom in on the lowest and highest points Q_1 and Q_2. Doing so, we determine the approximate coordinates $P_1(50, 267)$, $P_2(1146, 267)$, $Q_1(350, 45)$, and $Q_2(350, 820)$ of these extreme points on the solution curve.

A more detailed analysis of Eqs. (18) and (19) reveals that the point $(x(t), y(t))$ traverses the solution curve repeatedly in a counterclockwise direction as time t advances. Consequently, if we begin with $x_0 = 800$ rabbits and $y_0 = 75$ foxes, then:

- The numbers of rabbits and foxes both increase initially until there are 1146 rabbits and 267 foxes (at P_2);

- Next the number of rabbits decreases and the number of foxes continues to increase until there are 350 rabbits and 820 foxes (at Q_2);
- Next the numbers of rabbits and foxes both decrease until there are 50 rabbits and 267 foxes (at P_1);
- Next the number of rabbits increases and the number of foxes continues to decrease until there are 350 rabbits and 45 foxes (at Q_1);
- Finally the numbers of rabbits and foxes both increase until there are again 800 rabbits and 75 foxes.

This process of cyclical variation of the rabbit and fox populations continues indefinitely. In particular, we see that the number of rabbits oscillates between 50 and 1146, while the number of foxes oscillates between 45 and 820. This classical model of a predator-prey situation was developed in the 1920s by the Italian mathematician Vito Volterra (1860–1940) in order to analyze the cyclic variations observed in the shark and food fish populations in the Adriatic Sea. The *time lag* of the predator population behind that of the prey population is worthy of note. A similar (but more complicated) analysis might help explain—and even predict—the time lag of economic prosperity of a nation subsequent to certain fiscal policies of its government.

8.5 TRUE/FALSE STUDY GUIDE

8.5 CONCEPTS: QUESTIONS AND DISCUSSION

1. Describe the general features of populations that are modeled by the differential equation $dP/dt = kP(M - P)$ with $k > 0$ and $M > 0$. In what way do solutions depend on the value of the initial population $P(0) = P_0$?
2. Describe the general features of populations that are modeled by the differential equation $dP/dt = kP(P - M)$ with $k > 0$ and $M > 0$. In what way do solutions depend on the value of the initial population $P(0) = P_0$?
3. What are the similarities and differences between the two situations in Questions 1 and 2? Distinguish the roles played by the critical population M in the two cases.
4. Explain the behaviors of the rabbit and fox populations starting at the points P_1, P_2, Q_1, and Q_2 in Fig. 8.5.5; do so by calculating the signs of the derivatives dx/dt and dy/dt at these points.

8.5 PROBLEMS

Use partial fractions to solve the initial value problems in Problems 1 through 8.

1. $\dfrac{dx}{dt} = x - x^2, \quad x(0) = 2$

2. $\dfrac{dx}{dt} = 10x - x^2, \quad x(0) = 1$

3. $\dfrac{dx}{dt} = 1 - x^2, \quad x(0) = 3$

4. $\dfrac{dx}{dt} = 9 - 4x^2, \quad x(0) = 0$

5. $\dfrac{dx}{dt} = 3x(5 - x), \quad x(0) = 8$

6. $\dfrac{dx}{dt} = 3x(x - 5), \quad x(0) = 2$

7. $\dfrac{dx}{dt} = 4x(7 - x), \quad x(0) = 11$

8. $\dfrac{dx}{dt} = 7x(x - 13), \quad x(0) = 17$

9. The time rate of change of a rabbit population P is proportional to the square root of P. At time $t = 0$ (in months) the population numbers 100 rabbits and is increasing at the rate of 20 rabbits per month. How many rabbits will there be one year later?

10. Suppose that the fish population $P(t)$ in a lake was attacked by a disease at time $t = 0$, with the result that the fish ceased to reproduce (so that the birth rate was $\beta = 0$) and the death rate δ (deaths per week per fish) was thereafter proportional to $1/\sqrt{P}$. If there were initially 900 fish in the lake and 441 were left after 6 weeks, how long did it take all the fish in the lake to die?

11. Suppose that when a certain lake is stocked with fish, the birth and death rates β and δ are both inversely proportional to $\sqrt{P}$. (a) Show that

$$P(t) = \left(\tfrac{1}{2}kt - \sqrt{P_0}\,\right)^2$$

where k is a constant. (b) If $P_0 = 100$ and after 6 months there are 169 fish in the lake, how many will there be after one year?

12. The time rate of change of an alligator population P in a swamp is proportional to the square of P. The swamp contained a dozen alligators in 1988, two dozen in 1998. When will there be four dozen alligators in the swamp? What happens thereafter?

13. Consider a prolific breed of rabbits whose birth and death rates, β and δ, are each proportional to the rabbit population $P = P(t)$, with $\beta > \delta$. (a) Show that

$$P(t) = \frac{P_0}{1 - kP_0 t}$$

where k is a constant. Note that $P(t) \to +\infty$ as $t \to 1/(kP_0)$. This is doomsday. (b) Suppose that $P_0 = 6$ and that there are nine rabbits after ten months. When does doomsday occur?

14. Repeat part (a) of Problem 13 in the case $\beta < \delta$. What now happens to the rabbit population in the long run?

15. Suppose that the population $P(t)$ (in millions) of Ruritania satisfies the differential equation

$$\frac{dP}{dt} = k \cdot P \cdot (200 - P) \quad (k \text{ constant}).$$

Its population in 1940 was 100 million and was then growing at the rate of 1 million per year. Predict this country's population for the year 2000.

16. Suppose that a community contains 15000 people who are susceptible to Michaud's syndrome, a contagious disease. At time $t = 0$ the number $N(t)$ of people who have developed Michaud's syndrome is 5000 and is increasing at the rate of 500 per day. Assume that $N'(t)$ is proportional to the product of the numbers of those who have caught the disease and those who have not. How long will it take for another 5000 people to develop Michaud's syndrome?

17. As the salt KNO_3 dissolves in methanol, the number $x(t)$ of grams of the salt in solution after t seconds satisfies the differential equation

$$\frac{dx}{dt} = (0.8)x - (0.004)x^2.$$

(a) If $x = 50$ when $t = 0$, how long will it take an additional 50 g of the salt to dissolve? (b) What is the maximum amount of the salt that will ever dissolve in the methanol?

18. A population $P(t)$ (t in months) of squirrels satisfies the differential equation

$$\frac{dP}{dt} = (0.001)P^2 - kP \quad (k \text{ constant}).$$

If $P(0) = 100$ and $P'(0) = 8$, how long will it take for this population to double to 200 squirrels?

19. Consider an animal population $P(t)$ (t in years) that satisfies the differential equation

$$\frac{dP}{dt} = kP^2 - (0.01)P \quad (k \text{ constant}).$$

Suppose also that $P(0) = 200$ and that $P'(0) = 2$. (a) When is $P = 1000$? (b) When will doomsday occur for this population?

20. Suppose that the number $x(t)$ (t in months) of alligators in a swamp satisfies the differential equation

$$\frac{dx}{dt} = (0.0001)x^2 - (0.01)x.$$

(a) If initially there are 25 alligators, solve this equation to determine what happens to this alligator population in the long run. (b) Repeat part (a), but use 150 alligators initially.

21. Consider a population $P(t)$ satisfying the logistic equation $dP/dt = aP - bP^2$, where $B = aP$ is the time rate at which births occur and $D = bP^2$ is the rate at which deaths occur. If the initial population is $P(0) = P_0$ and B_0 births per month and D_0 deaths per month are occurring at time $t = 0$, show that the limiting population is $M = B_0 P_0 / D_0$.

22. Consider a rabbit population $P(t)$ satisfying the logistic equation as in Problem 21. If the initial population is 120 rabbits and there are 8 births per month and 6 deaths per month occurring at time $t = 0$, how many months does it take for $P(t)$ to reach 95% of the limiting population M?

23. Consider a rabbit population $P(t)$ satisfying the logistic equation as in Problem 21. If the initial population is 240 rabbits and there are 9 births per month and 12 deaths per month occurring at time $t = 0$, how many months does it take for $P(t)$ to reach 105% of the limiting population M?

24. Consider a population $P(t)$ satisfying the extinction/explosion equation

$$\frac{dP}{dt} = aP^2 - bP,$$

where $B = aP^2$ is the time rate at which births occur and $D = bP$ is the rate at which deaths occur. If the initial population is $P(0) = P_0$ and B_0 births per month and D_0 deaths per month are occurring at time $t = 0$, show that the threshold population is $M = D_0 P_0 / B_0$.

25. Consider an alligator population $P(t)$ satisfying the extinction/explosion equation as in Problem 24. If the initial population is 100 alligators and there are 10 births per month and 9 deaths per month occurring at time $t = 0$, how many months does it take for $P(t)$ to reach 10 times the threshold population M?

26. Consider an alligator population $P(t)$ satisfying the extinction/explosion equation as in Problem 24. If the initial population is 110 alligators and there are 11 births per month and 12 deaths per month occurring at time $t = 0$, how many months does it take for $P(t)$ to reach 10% of the threshold population M?

27. Suppose that at time $t = 0$, half of a "logistic" population of 100,000 persons has heard a certain rumor, and that the number of those who have heard it is then increasing at the rate of 1000 persons per day. How long will it take for this rumor to spread to 80% of the population? (*Suggestion:* Find the value of k by substituting $P(0)$ and $P'(0)$ in the logistic equation in (3).)

28. The data in the table in Fig. 8.5.6 are given for a certain population $P(t)$ that satisfies the logistic equation in (3). (a) What is the limiting population M? (*Suggestion:* Use the approximation

$$P'(t) \approx \frac{P(t+h) - P(t-h)}{2h}$$

Year	P (millions)
1924	24.63
1925	25.00
1926	25.38
⋮	⋮
1974	47.04
1975	47.54
1976	48.04

FIGURE 8.5.6 Population data for Problem 28.

with $h=1$ to estimate the values of $P'(t)$ when $P=25.00$ and when $P=47.54$. Then substitute these values in the logistic equation and solve for k and M.) (b) Use the values of k and M found in part (a) to determine when $P = 75$. (*Suggestion:* Take $t = 0$ to correspond to the year 1925.)

29. During the period from 1790 to 1930, the U.S. population $P(t)$ (t in years) grew from 3.9 million to 123.2 million. Throughout this period, $P(t)$ remained close to the solution of the initial value problem

$$\frac{dP}{dt} = (0.03135)P - (0.0001489)P^2, \qquad P(0) = 3.9.$$

(a) What population does this logistic equation predict for the year 1930? (b) What limiting population does this equation predict? (c) Has this logistic equation continued since 1930 to accurately model the U.S. population? [This problem is based on a computation by Verhulst, who in 1845 used the 1790–1840 U.S. population data to predict accurately the U.S. population through the year 1930 (long after his own death, of course).]

30. Consider two populations $P_1(t)$ and $P_2(t)$, both of which satisfy the logistic equation with the same limiting population M, but with different values k_1 and k_2 of the constant k in Eq. (3). Assume that $k_1 < k_2$. Which population approaches M more rapidly? You can reason *geometrically* by examining slope fields (especially if appropriate software is available), *symbolically* by analyzing the solution given in Eq. (7), or *numerically* by substituting successive values of t.

31. (a) Derive the solution given in Eq. (7) for the logistic initial value problem in (6). (b) How does this solution behavior (as t increases) if the initial value P_0 is negative?

32. (a) Derive the solution

$$P(t) = \frac{MP_0}{P_0 + (M - P_0)e^{kMt}}$$

satisfying the initial condition $P(0) = P_0$ for the explosion/extinction equation in (13). (b) How does this solution behave (as t increases) if the initial value P_0 is negative?

33. If $P(t)$ satisfies the logistic equation in (3), use the chain rule to show that

$$P''(t) = 2k^2 P \cdot \left(P - \tfrac{1}{2}M\right) \cdot (P - M).$$

Conclude that

$$P''(t) > 0 \quad \text{if } 0 < P < \tfrac{1}{2}M; \qquad P''(t) = 0 \quad \text{if } P = \tfrac{1}{2}M;$$

$$P''(t) < 0 \quad \text{if } \tfrac{1}{2}M < P < M; \quad \text{and} \quad P''(t) > 0 \quad \text{if } P > M.$$

In particular, it follows that any solution curve that crosses the horizontal line $P = \frac{1}{2}M$ has an inflection point where it crosses that line, and therefore resembles one of the lower S-shaped curves in Fig. 8.5.3.

34. Derive the solution in Eq. (21) of the separable differential equation in (20).

8.5 Project: Predator-Prey Equations and Your Own Game Preserve

You own a large forested game preserve that you originally stocked with F_0 foxes and R_0 rabbits. The following differential equations model the numbers $R(t)$ of rabbits and $F(t)$ of foxes t months later.

$$\frac{dR}{dt} = (0.01)pR - (0.0001)aRF,$$

$$\frac{dF}{dt} = -(0.01)qF + (0.0001)bRF$$

where p and q are the two largest digits (with $p < q$), and a and b the two smallest nonzero digits (with $a < b$), in your student I.D. number.

The numbers of foxes and rabbits will oscillate periodically, as in the situation illustrated by Fig. 8.5.5. Choose your initial numbers F_0 of foxes and R_0 of rabbits—perhaps several hundred of each—so that the resulting solution curve in the RF-plane is a fairly eccentric closed curve. (The eccentricity may be increased if you begin with a relatively large number of rabbits and a relatively small number of foxes, as any game preserve manager would naturally do.)

Your task is then to determine the maximum and minimum number of rabbits and foxes that will ever be observed in your game preserve. Use the implicit curve-plotting facility of a graphing calculator or computer algebra system to zoom in on the rightmost, leftmost, lowest, and highest points of the solution curve with sufficient precision to determine their coordinates accurate to the nearest integer.

8.6 LINEAR SECOND-ORDER EQUATIONS

A **second-order** differential equation is one that involves the second derivative y'' (and perhaps also the first derivative y') of the dependent variable y. It is called **linear** provided that it is linear in the unknown function $y(x)$ and its derivatives. Thus a **linear second-order** differential equation is one of the form

$$A(x)\frac{d^2y}{dx^2} + B(x)\frac{dy}{dx} + C(x)y = F(x), \tag{1}$$

where the *coefficients* $A(x)$, $B(x)$, $C(x)$, and $F(x)$ are given continuous functions of the independent variable on an appropriate open interval I where we may hope to determine a solution. A **solution** of (1) is simply a function $y = y(x)$ that satisfies the differential equation at every point of the interval I.

The coefficient functions in (1) need *not* be linear in x. Thus the second-order equation

$$e^x y'' + (\cos x)y' + \left(1 + \sqrt{x}\right)y = \tan^{-1} x$$

is linear even though the coefficients are quite nonlinear functions of x. By contrast, the equations

$$y'' = yy' \quad \text{and} \quad y'' + 3(y')^2 + 4y^3 = 0$$

are not linear, because products or powers of y or its derivatives appear in each.

In this section we restrict our attention to the case in which $F(x) \equiv 0$ in Eq. (1). Such a linear equation is said to be **homogeneous.** Thus a homogeneous linear second-order differential equation is one of the form

$$A(x)\frac{d^2y}{dx^2} + B(x)\frac{dy}{dx} + C(x)y = 0. \tag{2}$$

Homogeneous linear equations have the useful feature that we can construct new solutions by forming *linear combinations* of known solutions. In particular, any **linear combination** $y = c_1y_1 + c_2y_2$ of two known solutions y_1 and y_2 of Eq. (2) is another solution of the differential equation.

THEOREM 1 Linear Combinations of Solutions
Suppose that the two functions y_1 and y_2 are both solutions of the homogeneous linear equation in (2) and that c_1 and c_2 are constants. Then the new function y defined by

$$y(x) = c_1y_1(x) + c_2y_2(x)$$

also satisfies Eq. (2).

PROOF The fact that y_1 and y_2 are both solutions of Eq. (2) means that

$$Ay_1'' + By_1' + Cy_1 = 0 \quad \text{and} \quad Ay_2'' + By_2' + Cy_2 = 0.$$

It follows (using linearity of differentiation) that

$$\begin{aligned} Ay'' + By' + Cy &= A(c_1y_1 + c_2y_2)'' + B(c_1y_1 + c_2y_2)' + C(c_1y_1 + c_2y_2) \\ &= A(c_1y_1'' + c_2y_2'') + B(c_1y_1' + c_2y_2') + C(c_1y_1 + c_2y_2) \\ &= c_1(Ay_1'' + By_1' + Cy_1) + c_2(Ay_2'' + By_2' + Cy_2) \\ &= c_1 \cdot 0 + c_2 \cdot 0 = 0. \end{aligned}$$

Thus we have verified that the linear combination $y = c_1y_1 + c_2y_2$ is also a solution of Eq. (2). ◀

Theorem 1 implies that any constant multiple cy of the solution y is also a solution of the differential equation—choose $c_2 = 0$. But we do not regard y and cy as "really different" solutions. The two solutions y_1 and y_2 are called **independent** provided that neither is a constant multiple of the other. We can always determine whether two given solutions y_1 and y_2 are independent by noting whether either of the two quotients y_1/y_2 or y_2/y_1 is a constant. For instance, the following pairs of functions are obviously independent:

$$x \quad \text{and} \quad x^2; \qquad e^x \quad \text{and} \quad e^{-2x}; \qquad \cos x \quad \text{and} \quad \sin x.$$

Theorem 1 provides us with the means to solve initial value problems by forming linear combinations of independent solutions.

EXAMPLE 1 (a) Verify that $y_1(x) = x^2$ and $y_2(x) = x^{-3}$ are independent solutions (for $x > 0$) of the homogeneous linear equation

$$x^2 y'' + 2xy' - 6y = 0.$$

(b) Find a solution satisfying the two initial conditions $y(1) = 10$ and $y'(1) = 5$.

Solutions For part (a), we readily calculate

$$x^2 y_1'' + 2xy_1' - 6y_1 = x^2 \cdot 2 + 2x \cdot 2x - 6 \cdot x^2 = (2 + 4 - 6)x^2 = 0$$

and

$$x^2 y_2'' + 2xy_2' - 6y_2 = x^2 \cdot 12x^{-5} + 2x \cdot (-3x^{-4}) - 6 \cdot x^{-3} = (12 - 6 - 6)x^{-3} = 0.$$

Thus both y_1 and y_2 are solutions of the given differential equation. Because neither $y_1/y_2 = x^5$ nor $y_2/y_1 = x^{-5}$ is a constant, it follows that these two solutions are independent.

To solve part (b), we see (by Theorem 1) that the linear combination

$$y(x) = c_1 y_1(x) + c_2 y_2(x) = c_1 x^2 + c_2 x^{-3}$$

is also a solution (for any choice of the constants c_1 and c_2). When we impose the given initial conditions on this new solution and its derivative $y'(x) = 2c_1 x - 3c_2 x^{-4}$, we get the equations

$$c_1 + c_2 = 10 \quad \text{and} \quad 2c_1 - 3c_2 = 5.$$

We readily solve these equations—for instance, by substituting $c_1 = 10 - c_2$ from the first equation into the second—for $c_1 = 7$ and $c_2 = 3$. Consequently, a solution of the initial value problem

$$x^2 y'' + 2xy' - 6y = 0, \qquad y(1) = 10, \quad y'(1) = 5$$

is $y(x) = 7x^2 + 3x^{-3}$. ◆

The leading coefficient function $A(x) = x^2$ in the differential equation of Example 1 vanishes at $x = 0$. On any interval where the function $A(x)$ is nonzero, we can divide by $A(x)$ to write the homogeneous equation $Ay'' + By' + Cy = 0$ in the form

$$\frac{d^2y}{dx^2} + P(x)\frac{dy}{dx} + Q(x)y = 0. \tag{3}$$

One key to solving homogeneous second-order equations is the fact that *every* solution of such an equation is a linear combination of any two given independent solutions. Unlike Theorem 1, the theorem that follows is not elementary, and its proof is therefore omitted.

THEOREM 2 General Solutions

Suppose that y_1 and y_2 are independent solutions of the homogeneous linear equation in (3) on an interval I where the coefficient functions $P(x)$ and $Q(x)$ are continuous. If y is any given solution of the equation on this interval, then there exist constants c_1 and c_2 such that

$$y(x) = c_1 y_1(x) + c_2 y_2(x)$$

for all x in I.

As a result of Theorem 2, we know *every* solution of a homogeneous linear second-order differential equation once we know just *two* independent solutions y_1 and y_2. For this reason, we may call $y = c_1 y_1 + c_2 y_2$ *the* **general solution** of the differential equation—there is no other solution of it. For instance, $y(x) = c_1 x^2 + c_2 x^{-3}$ is the general solution (for $x > 0$) of the differential equation in Example 1, and no function not of this form can satisfy the equation.

Because the general solution of a second-order differential equation involves *two* arbitrary constants, we can hope to satisfy *two* initial conditions by selecting the values of c_1 and c_2 appropriately. This is why a second-order linear initial value problem typically takes the form

$$Ay'' + By' + Cy = F, \qquad y(a) = b_0, \quad y'(a) = b_1 \tag{4}$$

with two initial conditions specifying given values of the solution y and its derivative y' at the same point $x = a$.

A general existence-uniqueness theorem asserts that an initial value problem of the form in (4) has exactly one solution on an interval where the coefficient functions are continuous and $A(x)$ is nonzero. For instance, the coefficient functions in the differential equation of Example 1 are all continuous and $A(x) = x^2$ is nonzero for $x > 0$. Therefore the only solution for $x > 0$ of the initial value problem

$$x^2 y'' + 6xy' - 2y = 0, \qquad y(1) = 10, \quad y'(1) = 5$$

is the solution $y(x) = 7x^2 + 3x^{-3}$ that we found in part (b) of Example 1.

Constant-Coefficient Equations

We saw in Section 8.4 that the general solution of a linear first-order equation is given by an explicit integral formula involving the coefficients in the equation. In contrast, this is far from true for second-order linear equations. It can be a formidable task to find the two independent solutions that are needed to construct a general solution of a given homogeneous linear second-order equation. For instance, the general solution of the differential equation $6y'' + 2xy' - x^2 y = 0$—which may appear to resemble superficially the equation of Example 1—cannot be expressed simply in terms of familiar elementary functions. This difficulty somehow arises from the variable coefficients in the last equation.

Fortunately, it *is* possible to solve explicitly a homogeneous linear second-order differential equation of the form

$$a\frac{d^2y}{dx^2} + b\frac{dy}{dx} + cy = 0 \tag{5}$$

with *constant coefficients* a, b, and c. And we will see in Section 8.7 that such differential equations have important physical applications.

To construct the general solution $y = c_1 y_1 + c_2 y_2$ of the differential equation

$$ay'' + by' + cy = 0 \tag{6}$$

with $a \neq 0$, we need to find two independent particular solutions y_1 and y_2. The key idea is to find a plausible *form* of a possible solution of Eq. (6). Because—if r

is a constant—$D_x(e^{rx})$ is a constant multiple of $y(x) = e^{rx}$, it follows that y'' and y' would have that form as well. Indeed, substitution of $y(x) = e^{rx}$, $y'(x) = re^{rx}$, and $y'' = r^2 e^{rx}$ into Eq. (6) yields the equation

$$ar^2 e^{rx} + bre^{rx} + ce^{rx} = (ar^2 + br + c)e^{rx} = 0.$$

Because e^{rx} is never zero, the product $(ar^2 + br + c)e^{rx}$ is zero exactly when

$$ar^2 + br + c = 0. \tag{7}$$

This is a simple quadratic equation that we can solve to find the value (or values) of r such that $y = e^{rx}$ is, indeed, a solution of Eq. (6).

Equation (7) is called the **characteristic equation** (or **auxiliary equation**) of the homogeneous linear second-order differential equation $ay'' + by' + cy = 0$. We can solve it by any of various techniques; the familiar quadratic formula yields the solutions

$$r_1, r_2 = \frac{-b \pm \sqrt{b^2 - 4ac}}{2a},$$

in which the two possible choices of sign yield the two roots r_1 and r_2. If $b^2 - 4ac > 0$ then both are real, and the two distinct possibilities r_1 and r_2 for the value of r yield the two independent solutions $y_1 = e^{r_1 x}$ and $y_2 = e^{r_2 x}$ of the differential equation in (6). (Do you see why neither is a constant multiple of the other?) Remember that once we have found two independent solutions, we have found them all. Examples 2 through 4 illustrate how to proceed if r_1 and r_2 are distinct and real. It remains to be seen how we should proceed should $b^2 - 4ac$ be zero or negative.

CASE 1 Distinct Real Roots

Suppose that the characteristic equation of the linear homogeneous differential equation $ay'' + by' + cy = 0$ has unequal real roots r_1 and r_2. Then a general solution of this differential equation is

$$y(x) = c_1 e^{r_1 x} + c_2 e^{r_2 x}. \tag{8}$$

Thus, in this case, the solution of the differential equation reduces to the simple matter of solving a quadratic equation.

EXAMPLE 2 Solve the differential equation $3y'' + 7y' + 2y = 0$.

Solution We can solve the characteristic equation $3r^2 + 7r + 2 = 0$ by factoring:

$$3r^2 + 7r + 2 = (3r + 1)(r + 2) = 0.$$

The roots $r_1 = -\frac{1}{3}$ and $r_2 = -2$ are real and distinct. Therefore the general solution of the differential equation—given in (8)—is

$$y(x) = c_1 e^{-x/3} + c_2 e^{-2x}.$$

◆

REMARK A first-order differential equation generally has only a single solution curve passing through a given initial point (a, b) and, therefore, only a single tangent line through the initial point is tangent to a solution curve of the equation. By contrast, Fig. 8.6.1 illustrates the fact that a second-order differential equation generally has infinitely many solution curves passing through a given point (a, b_0)—one for each (real) value of the initial slope $y'(a) = b_1$. That is, every nonvertical straight line through (a, b_0) is tangent to some solution curve. Figure 8.6.1 shows several solution curves of the differential equation $3y'' + 7y' + 2y = 0$ all having the same initial value $y(0) = 1$; Fig. 8.6.2 shows several solution curves all having the same initial slope $y'(0) = 1$.

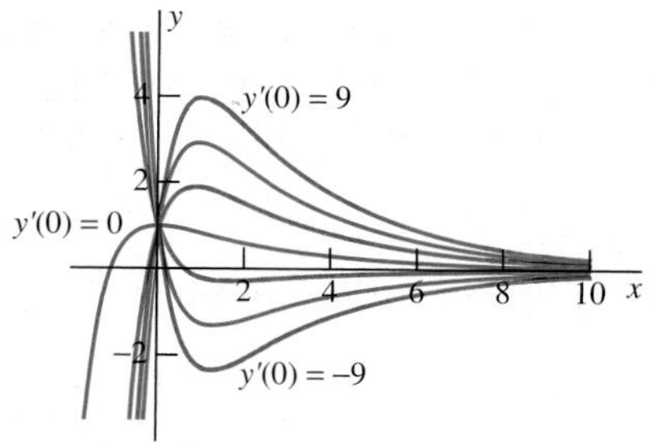

FIGURE 8.6.1 Solutions of $3y'' + 7y' + 2y = 0$ with the same initial value $y(0) = 1$ but with different values of the initial slope ranging from $y'(0) = -9$ to $y'(0) = 9$.

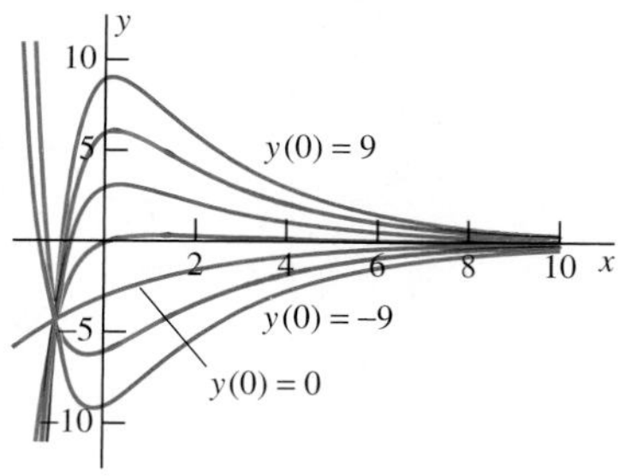

FIGURE 8.6.2 Solutions of $3y'' + 7y' + 2y = 0$ with the same initial slope $y'(0) = 1$ but with different initial values ranging from $y(0) = -9$ to $y(0) = 9$.

EXAMPLE 3 Solve the differential equation $5y'' - 2y' = 0$.

Solution Here the coefficient of y in the differential equation is zero, corresponding to the constant term zero in its characteristic equation: $5r^2 - 2r = 0$. The factorization

$$5r^2 - 2r = r(5r - 2) = 5r\left(r - \tfrac{2}{5}\right) = 0$$

reveals the distinct real roots $r_1 = 0$ and $r_2 = \frac{2}{5}$. Because $e^{r_1 x} = e^{0 \cdot x} = e^0 = 1$, the general solution of the given equation is

$$y(x) = c_1 + c_2 e^{2x/5}.$$

As illustrated in Fig. 8.6.3, the solution curves with a given value of c_1 and different values of c_2 all have the line $y = c_1$ as an asymptote as $x \to -\infty$. ◆

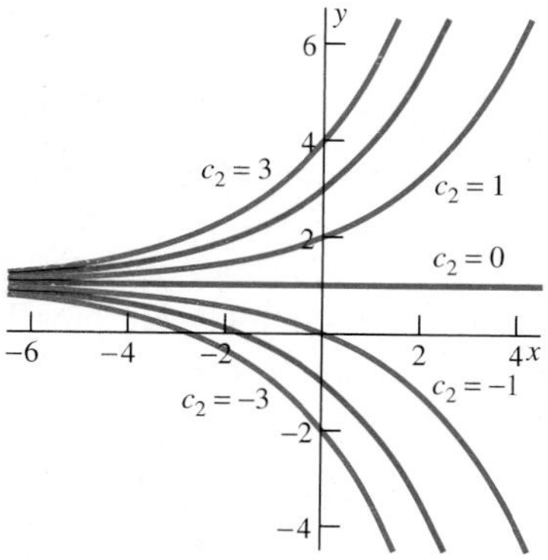

FIGURE 8.6.3 Solutions $y(x) = 1 + c_2 e^{2x/5}$ of $5y'' - 2y' = 0$ with different values of c_2 ranging from $c_2 = -3$ to $c_2 = 3$.

EXAMPLE 4 Solve the differential equation $y'' - 4y = 0$.

Solution Now the coefficient of y' in the differential equation is zero, and hence the coefficient of r is zero in its characteristic equation: $r^2 - 4 = 0$. The factorization $r^2 - 4 = (r + 2)(r - 2)$ reveals the roots $r_1 = -2$ and $r_2 = 2$, and thus we obtain the general solution

$$y(x) = c_1 e^{-2x} + c_2 e^{2x}.$$

But you can readily verify (by differentiating y_3 and y_4 twice each) that $y_3(x) = \cosh 2x$ and $y_4(x) = \sinh 2x$ are also independent solutions of the differential equation $y'' - 4y = 0$. Therefore

$$y(x) = c_1 \cosh 2x + c_2 \sinh 2x$$

is another general solution of the same differential equation. Thus there is nothing unique about the *form* of a given general solution—any two different pairs of independent solutions of the same linear homogeneous second-order differential equation will yield two different expressions for the general solution. Moreover, the graphs of the two functions in one such pair can appear quite different from those in the other pair. Figure 8.6.4 shows the graphs of the four solutions e^{-2x}, e^{2x}, $\cosh 2x$, and $\sinh 2x$ of $y'' - 4y = 0$. ◆

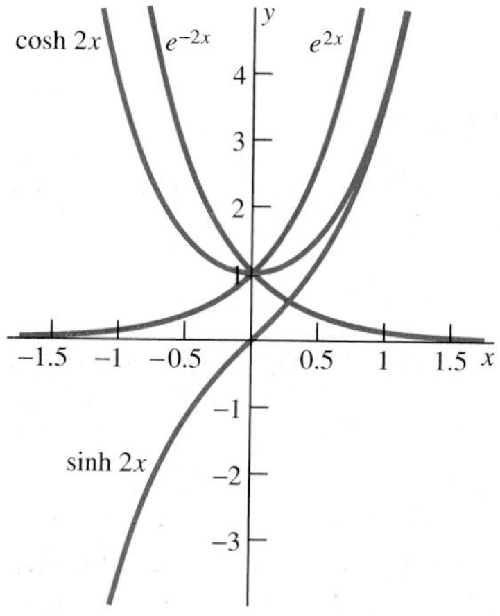

FIGURE 8.6.4 Four different solutions of the equation $y'' - 4y = 0$.

REMARK Because $\cosh 2x$ and $\sinh 2x$ are solutions of the linear differential equation $y'' - 4y = 0$ with general solution $y(x) = c_1 e^{-2x} + c_2 e^{2x}$, it follows from Theorem 2 that each of the functions $\cosh 2x$ and $\sinh 2x$ can be expressed as a linear combination of the functions e^{-2x} and e^{2x}. Of course this is no surprise, because

$$\cosh 2x = \tfrac{1}{2}e^{2x} + \tfrac{1}{2}e^{-2x} \quad \text{and} \quad \sinh 2x = \tfrac{1}{2}e^{2x} - \tfrac{1}{2}e^{-2x}$$

by the definitions of the hyperbolic sine and cosine functions (Section 6.9).

If $b^2 - 4ac = 0$, then the characteristic equation $ar^2 + br + c = 0$ has equal (and necessarily real) roots $r_1 = r_2 = -b/(2a)$. Hence, at first, we obtain only the single solution $y_1(x) = e^{r_1 x}$ of the corresponding differential equation. In this case the problem that remains is to produce the "missing" second independent solution of the differential equation.

A double root $r = r_1$ can occur only if the characteristic equation factors as

$$ar^2 + br + c = a(r - r_1)^2 = a\left(r^2 - 2r_1 r + r_1^2\right) = 0.$$

The corresponding differential equation is then a constant multiple of the equation

$$y'' - 2r_1 y' + r_1^2 y = 0.$$

But it is easy to verify (and you should do so) that $y_2 = xe^{r_1 x}$ is a second solution of this differential equation. Obviously the solutions $y_1(x) = e^{r_1 x}$ and $y_2(x) = xe^{r_1 x}$ are independent, and therefore enable us to write a general solution of the given differential equation.

CASE 2 Equal Real Roots

Suppose that the characteristic equation of the linear homogeneous differential equation $ay'' + by' + cy = 0$ has equal real roots $r_1 = r_2$. Then a general solution of this differential equation is

$$y(x) = c_1 e^{r_1 x} + c_2 x e^{r_1 x} = (c_1 + c_2 x)e^{r_1 x}. \tag{9}$$

EXAMPLE 5 Solve the initial value problem

$$4y'' + 12y' + 9y = 0, \qquad y(0) = 4, \quad y'(0) = -3.$$

Solution The characteristic equation

$$4r^2 + 12r + 9 = (2r + 3)^2 = 0$$

of the given differential equation has equal real roots $r_1 = r_2 = -\frac{3}{2}$. Hence the general solution—as given in Eq. (9)—is

$$y(x) = c_1 e^{-3x/2} + c_2 x e^{-3x/2}.$$

Differentiation gives

$$y'(x) = -\tfrac{3}{2}c_1 e^{-3x/2} + c_2 e^{-3x/2} - \tfrac{3}{2}c_2 x e^{-3x/2}.$$

Therefore the given initial conditions yield the simultaneous equations

$$\begin{aligned} y(0) &= \quad c_1 \qquad = 4, \\ y'(0) &= -\tfrac{3}{2}c_1 + c_2 = -3, \end{aligned}$$

which imply that $c_1 = 4$ and $c_2 = 3$. Thus the solution of the given initial value problem is

$$y(x) = 4e^{-3x/2} + 3xe^{-3x/2} = (4 + 3x)e^{-3x/2}.$$

This particular solution of the differential equation is illustrated in Fig. 8.6.5, together with several others of the form $y(x) = c_1 e^{-3x/2} + 3xe^{-3x/2}$. ◆

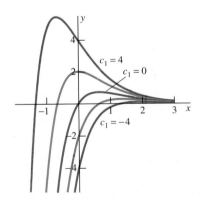

FIGURE 8.6.5 Solutions $y(x) = (c_1 + 3x)e^{-3x/2}$ of $4y'' + 12y' + 9y = 0$ with different values of c_1 ranging from $c_1 = -4$ to $c_1 = 4$.

Complex Roots

If $b^2 - 4ac < 0$, then solution of the characteristic equation $ar^2 + br + c = 0$ using the quadratic formula gives

$$r = \frac{-b \pm \sqrt{b^2 - 4ac}}{2a} = \frac{-b \pm \sqrt{(-1)\cdot(4ac - b^2)}}{2a} = p \pm iq$$

where

$$p = -\frac{b}{2a}, \qquad q = \frac{\sqrt{4ac - b^2}}{2a}, \qquad \text{and} \qquad i = \sqrt{-1}.$$

Thus we obtain a complex conjugate pair of roots.

But what does $y = e^{rx}$ mean when r is a complex number? The answer stems from **Euler's formula**

$$e^{i\theta} = \cos\theta + i\sin\theta, \tag{10}$$

which we discuss in detail in Section 10.4. If $r = p + iq$, we therefore define the complex exponential e^{rx} (for x real) by writing

$$e^{rx} = e^{(p+iq)x} = e^{px+iqx} = e^{px}e^{iqx},$$

so

$$e^{rx} = e^{px}(\cos qx + i\sin qx), \tag{11}$$

using Euler's formula with $\theta = qx$.

As defined by Eq. (11), the exponential e^{rx} is a complex-valued function of the real variable x. Such a function is differentiated by differentiating separately its real and imaginary parts. That is,

$$\begin{aligned} D_x(e^{rx}) &= D_x(e^{px}\cos qx) + i\,D_x(e^{px}\sin qx) \\ &= (pe^{px}\cos qx - qe^{px}\sin qx) + i(pe^{px}\sin qx + qe^{px}\cos qx) \\ &= (p + iq)(e^{px}\cos qx + ie^{px}\sin qx) = re^{rx}. \end{aligned}$$

Thus $D_x e^{rx} = re^{rx}$ when r is complex, exactly as when r is real.

This familiar differentiation formula is the basis for the fact that $y = e^{rx}$ is a solution of the homogeneous linear differential equation $ay'' + by' + cy = 0$ precisely when r is a root of the associated characteristic equation $ar^2 + br + c = 0$. In the case of complex conjugate roots $r_1, r_2 = p \pm qi$, we get the general complex-valued solution

$$y(x) = C_1e^{r_1x} + C_2e^{r_2x},$$

which we can rewrite as

$$\begin{aligned} y(x) &= C_1e^{(p+iq)x} + C_2e^{(p-iq)x} \\ &= C_1e^{px}(\cos qx + i\sin qx) + C_2e^{px}(\cos qx - i\sin qx) \\ &= (C_1 + C_2)e^{px}\cos qx + i(C_1 - C_2)e^{px}\sin qx; \end{aligned}$$

thus

$$y(x) = c_1e^{px}\cos qx + c_2e^{px}\sin qx$$

where $c_1 = C_1 + C_2$ and $c_2 = i(C_1 - C_2)$. In the last line here, we have expressed the solution $y(x)$ as a linear combination of the real-valued functions $e^{px}\cos qx$ and $e^{px}\sin qx$. Thus the complex conjugate roots $p \pm iq$ of the characteristic equation lead to the independent *real-valued* solutions $y_1 = e^{px}\cos qx$ and $y_2 = e^{px}\sin qx$ of the differential equation.

CASE 3 Complex Conjugate Roots

Suppose that the characteristic equation of the linear homogeneous differential equation $ay'' + by' + cy = 0$ has complex conjugate roots $p \pm iq$ (with $q \neq 0$). Then a general solution of the differential equation is

$$y(x) = e^{px}(c_1\cos qx + c_2\sin qx). \tag{12}$$

EXAMPLE 6 The characteristic equation $r^2 + 4 = 0$ of the differential equation

$$y'' + 4y = 0$$

has complex conjugate roots $\pm 2i$. With $p = 0$ and $q = 2$ in Eq. (12) we get the general solution

$$y(x) = c_1 \cos 2x + c_2 \sin 2x.$$

Figures 8.6.6 and 8.6.7 show some typical solution curves. Each is shaped like the graph of a constant multiple of the sine or cosine of $2x$. Figure 8.6.6 illustrates the effect of varying the *joint amplitude* $c = \sqrt{c_1^2 + c_2^2}$ of the constants c_1 and c_2. Figure 8.6.7 illustrates the effect of varying their ratio. ◆

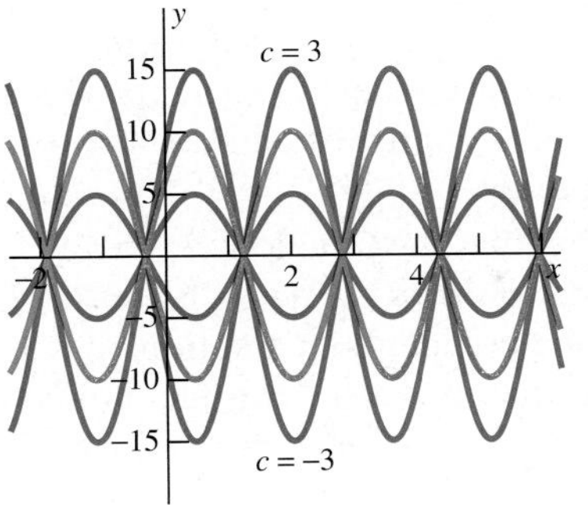

FIGURE 8.6.6 Solutions $y(x) = c(3\cos 2x + 4\sin 2x)$ of $y'' + 4y = 0$ with different values of c ranging from $c = -3$ to $c = 3$.

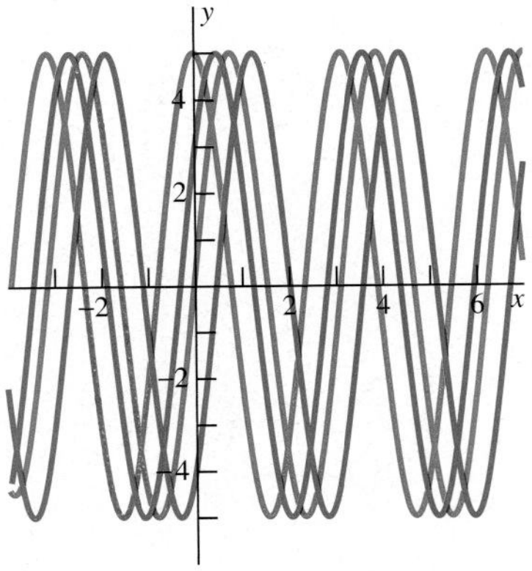

FIGURE 8.6.7 Solutions $y_1 = 5\cos 2x$, $y_2 = 5\sin 2x$, $y_3 = 3\cos 2x + 4\sin 2x$, and $y_4 = -4\cos 2x + 3\sin 2x$ of $y'' + 4y = 0$. Can you determine which is which?

EXAMPLE 7 Solve the initial value problem

$$9y'' + 6y' + 325y = 0, \qquad y(0) = 12, \quad y'(0) = 50.$$

Solution The roots of the characteristic equation $9r^2 + 6r + 325 = 0$ are given by

$$r = \frac{-6 \pm \sqrt{(-6)^2 - 4 \cdot 9 \cdot 325}}{2 \cdot 9} = \frac{-6 \pm 6\sqrt{-324}}{18} = -\frac{1}{3} \pm 6i.$$

Therefore the general solution is

$$y(x) = e^{-x/3}(c_1 \cos 6x + c_2 \sin 6x).$$

Its derivative is

$$y'(x) = -\tfrac{1}{3}e^{-x/3}(c_1 \cos 6x + c_2 \sin 6x) + 6e^{-x/3}(c_2 \cos 6x - c_1 \sin 6x).$$

When we impose the given initial conditions we obtain the simultaneous equations

$$y(0) = c_1 = 12,$$
$$y'(0) = -\tfrac{1}{3}c_1 + 6c_2 = 50$$

with solution $c_1 = 12$, $c_2 = 9$. Thus the solution of the given initial value problem is

$$y(x) = e^{-x/3}(12\cos 6x + 9\sin 6x).$$

The graph of this solution is shown in Fig. 8.6.8. It is worth nothing that $y(x) \to 0$ as $x \to +\infty$. ◆

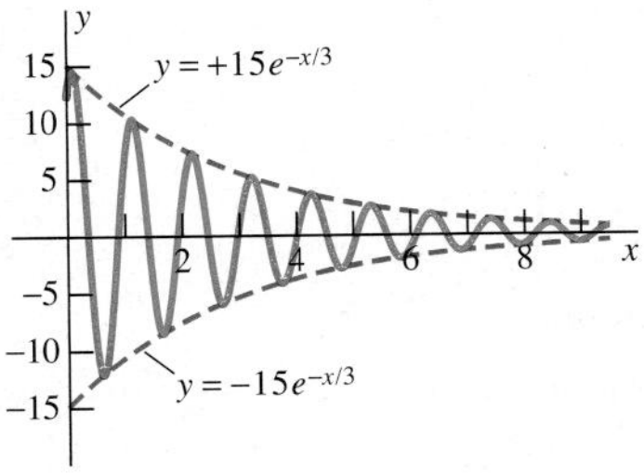

FIGURE 8.6.8 The solution $y(x) = e^{-x/3}(12\cos 6x + 9\sin 6x)$ of the initial value problem in Example 7 oscillates between the dashed "envelope curves" $y = +15e^{-x/3}$ and $y = -15e^{-x/3}$.

8.6 TRUE/FALSE STUDY GUIDE

8.6 CONCEPTS: QUESTIONS AND DISCUSSION

1. Complete the entries in the following table summarizing the various cases for the solutions of the constant-coefficient differential equation $ay'' + by' + cy = 0$.

Roots of $ar^2 + br + c = 0$	General Solution

2. For each case in Question 1, give your own example of such a differential equation and its general solution—preferably one with rather large and interesting integer coefficients. (*Suggestion:* Begin with the numerical roots r_1 and r_2 and work backward to construct the corresponding differential equation.)
3. Show that the function $y(x) = |x^{-3}|$ is a solution for $x \neq 0$ of the differential equation $x^2y'' + 2xy' - 6y = 0$ of Example 1. Is $y(x)$ a linear combination of the independent solutions $y_1(x) = x^2$ and $y_2(x) = x^{-3}$ of this equation? If not, then why does this fact not contradict Theorem 2?

8.6 PROBLEMS

Find general solutions of the differential equations in Problems 1 through 14.

1. $y'' - 7y' + 10y = 0$
2. $y'' + 2y' - 15y = 0$
3. $4y'' - 4y' - 3y = 0$
4. $12y'' + 13y' + 3y = 0$
5. $y'' + 4y' + y = 0$
6. $4y'' - 4y' - 19y = 0$
7. $4y'' + 12y' + 9y = 0$
8. $9y'' - 30y' + 25y = 0$
9. $25y'' - 20y' + 4y = 0$
10. $49y'' + 126y' + 81y = 0$
11. $y'' + 6y' + 13y = 0$
12. $y'' - 10y' + 74y = 0$
13. $9y'' + 6y' + 226y = 0$
14. $9y'' + 90y' + 226y = 0$

Solve the initial value problems in Problems 15 through 26.

15. $2y'' - 11y' + 12y = 0$; $y(0) = 5$, $y'(0) = 15$
16. $y'' - 2y' - 35y = 0$; $y(0) = 12$, $y'(0) = 0$
17. $y'' - 18y' + 77y = 0$; $y(0) = 4$, $y'(0) = 8$
18. $12y'' - y' - 6y = 0$; $y(0) = 2$, $y'(0) = 10$
19. $y'' + 22y' + 121y = 0$; $y(0) = 2$, $y'(0) = -25$
20. $9y'' + 42y' + 49y = 0$; $y(0) = 6$, $y'(0) = -11$
21. $y'' + 25y = 0$; $y(0) = 7$, $y'(0) = 10$
22. $9y'' + 100y = 0$; $y(0) = 99$, $y'(0) = 100$
23. $y'' + 4y' + 20y = 0$; $y(0) = 9$, $y'(0) = 10$
24. $y'' + 10y' + 106y = 0$; $y(0) = 11$, $y'(0) = -10$
25. $4y'' + 4y' + 101y = 0$; $y(0) = 10$, $y'(0) = 25$
26. $100y'' + 20y' + 10001y = 0$; $y(0) = 30$, $y'(0) = -33$

Each of Problems 27 through 34 gives the general solution of a homogeneous linear second-order differential equation with constant coefficients. Find that equation.

27. $y(x) = c_1 + c_2e^{-10x}$
28. $y(x) = c_1e^{10x} + c_2e^{-10x}$
29. $y(x) = c_1e^{-10x} + c_2xe^{-10x}$
30. $y(x) = c_1e^{10x} + c_2e^{100x}$
31. $y(x) = c_1 + c_2x$
32. $y(x) = e^x\left[c_1\exp\left(x\sqrt{2}\right) + c_2\exp\left(-x\sqrt{2}\right)\right]$
33. $y(x) = e^{-5x}\left(c_1\cos\frac{x}{5} + c_2\sin\frac{x}{5}\right)$
34. $y(x) = e^{-x/5}\,(c_1\cos 5x + c_2\sin 5x)$
35. Given: The differential equation $y''+25y=0$. (a) Show that this equation has infinitely many different solutions $y(x)$ such that $y(0) = y(\pi) = 0$. (b) Show that this equation has no nontrivial solution $y(x)$ such that $y(0) = y(3) = 0$.
36. Suppose that $y(x)$ is a solution of the equation $ay'' + by' + cy=0$ and that a, b, and c are all positive. Show that $y(x) \to 0$ as $x \to \pm\infty$.

8.7 MECHANICAL VIBRATIONS

Many natural phenomena exhibit either steady growth (or decline) or periodic oscillations (ebb and flow). Phenomena of steady growth are typically modeled by first-order differential equations (as in Section 8.5), whereas periodic phenomena and vibrations are typically modeled by second-order differential equations.

The motion of a mass attached to a spring serves as a relatively simple example of the vibrations that occur in more complex mechanical systems. For many such systems, the analysis of these vibrations is a problem in the solution of linear differential equations with constant coefficients (Section 8.6).

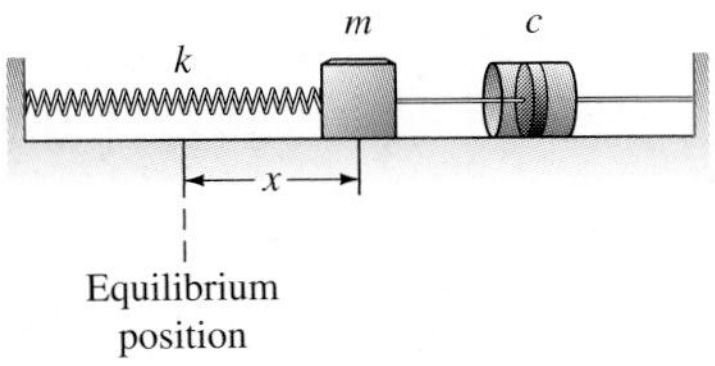

FIGURE 8.7.1 A mass-spring-dashpot system.

We consider a body of mass m attached to one end of an ordinary spring that resists compression as well as stretching; the other end of the spring is attached to a fixed structure, as shown in Fig. 8.7.1. Assume that the body rests on a frictionless horizontal plane so that it can move only back and forth as the spring compresses and stretches. Denote by x the distance of the body from its **equilibrium position**—its position when the spring is unstretched. We take $x > 0$ when the spring is stretched, and thus $x < 0$ when it is compressed.

According to Hooke's law, the restorative force F_S that the spring exerts on the mass is proportional to the distance x that the spring has been stretched or compressed. Because this is the same as the displacement x of the mass m from its equilibrium position, it follows that

$$F_S = -kx. \tag{1}$$

The positive constant of proportionality k is called the **spring constant.** Note that F_S and x have opposite signs: $F_S < 0$ when $x > 0$, $F_S > 0$ when $x < 0$.

Figure 8.7.1 shows the mass attached to a dashpot—a device that, like a shock absorber, provides a force directed opposite to the instantaneous direction of motion of the mass m. We assume that the dashpot is so designed that this force F_R is proportional to the velocity $v = dx/dt$ of the mass; that is, that

$$F_R = -cv = -c\frac{dx}{dt}. \tag{2}$$

The positive constant c is the **damping constant** of the dashpot. More generally, we may regard Eq. (2) as specifying frictional forces in our system (including air resistance to the motion of the mass).

If, in addition to the forces F_S and F_R, the mass is subjected to a given **external force** $F_E = F(t)$, then the total force acting on the mass is $F_T = F_S + F_R + F_E$. Using Newton's second law in the form

$$F_T = ma = m\frac{d^2x}{dt^2} = mx'',$$

we obtain the linear second-order differential equation

$$mx'' + cx' + kx = F(t) \tag{3}$$

that governs the motion of the mass.

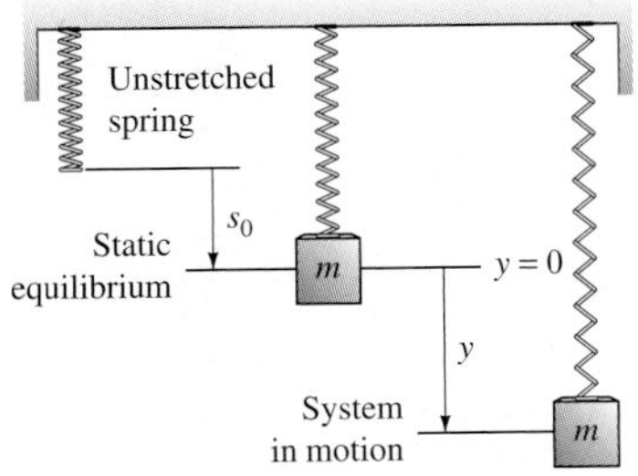

FIGURE 8.7.2 A mass suspended vertically from a spring.

For an alternative example, we might attach the mass to the lower end of a spring that is suspended vertically from a fixed support, as in Fig. 8.7.2. In this case the weight $W = mg$ of the mass would stretch the spring a distance s_0 determined by Eq. (1) with $F_S = -W$ and $x = s_0$. That is, $mg = ks_0$, so that $s_0 = mg/k$. This gives the **static** equilibrium position of the mass. If y denotes the displacement of the mass in motion, measured downward from its static equilibrium position, then we ask you to show (in Problem 23) that $y = y(t)$ satisfies Eq. (3); specifically, that

$$my'' + cy' + ky = F(t) \tag{4}$$

if we include damping and external forces.

Free Undamped Motion

If there is no external force acting on the spring, then $F(t) \equiv 0$ in Eq. (3), and we call the resulting motion **free.** Thus the homogeneous equation

$$mx'' + cx' + kx = 0 \tag{5}$$

describes free motion of a mass on a spring with dashpot but with no external force applied.

If we have only a mass on a spring, with neither damping nor external force, then $c = 0$ as well, so Eq. (5) reduces to the equation

$$mx'' + kx = 0 \tag{6}$$

that models **free undamped motion.** It is convenient to rewrite Eq. (6) in the form

$$x'' + \omega_0^2 x = 0 \tag{7}$$

where

$$\omega_0 = \sqrt{\frac{k}{m}} \tag{8}$$

is the **natural frequency** of vibration of the mass on the spring. [The Greek letter ω (omega) is often used to denote frequency.]

EXAMPLE 1 A body with mass $m = \frac{1}{2}$ kilograms (kg) is attached to the end of a spring that is stretched 2 meters (m) by a force of 100 Newtons (N). This body is displaced one-half meter to the right (from its equilibrium position when the spring is unstretched) and then released from rest. Describe the motion that results.

Solution The spring constant is $k = 100/2 = 50$ (N/m), so the position function $x(t)$ of the body satisfies the initial value problem

$$\tfrac{1}{2}x'' + 50x = 0; \qquad x(0) = \tfrac{1}{2}, \quad x'(0) = 0. \tag{9}$$

The equivalent different equation $x'' + 100x = 0$ has characteristic equation $r^2 + 100 = 0$ with roots $r = \pm 10i$. Therefore the general solution of the differential equation in (9) and its derivative are

$$x(t) = A\cos 10t + B\sin 10t \quad \text{and} \quad x'(t) = -10A\sin 10t + 10B\cos 10t.$$

(We write A and B for the coefficients merely to avoid subscripts.) The initial conditions immediately give $A = \frac{1}{2}$ and $B = 0$. Thus the position function of the body is

$$x(t) = \tfrac{1}{2}\cos 10t.$$

This function describes a back-and-forth oscillation between the rightmost position $x = \frac{1}{2}$ (when $t = 0, \pi/5, 2\pi/5, \ldots$) and its leftmost position $x = -\frac{1}{2}$ (when $t = \pi/10, 3\pi/10, 5\pi/10, \ldots$). ◆

The general solution of Eq. (7) is

$$x(t) = A\cos\omega_0 t + B\sin\omega_0 t. \tag{10}$$

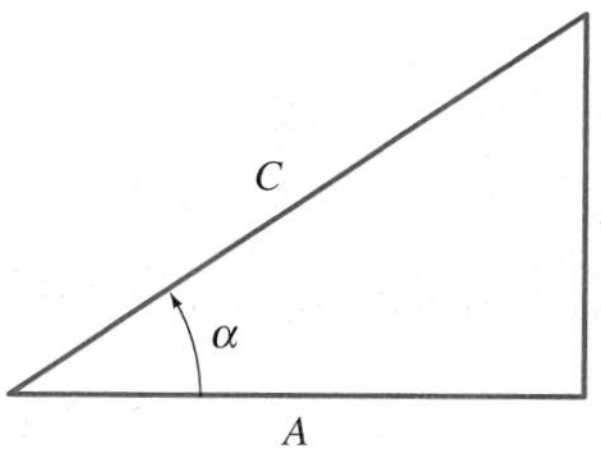

FIGURE 8.7.3 The angle α.

To analyze the motion described by this solution, we choose constants C and α so that

$$C = \sqrt{A^2 + B^2}, \qquad \cos\alpha = \frac{A}{C}, \quad \text{and} \quad \sin\alpha = \frac{B}{C}, \tag{11}$$

as indicated in Fig. 8.7.3. Note that, although $\tan\alpha = B/A$, the angle α is *not* given by the principal branch of the inverse tangent function (which gives values only in the interval $-\pi/2 < x < \pi/2$). Instead, α is the angle between 0 and 2π whose cosine and sine have the signs given in (11), where either A or B or both may be negative. Thus

$$\alpha = \begin{cases} \tan^{-1}(B/A) & \text{if} \quad A > 0,\ B > 0 \quad \text{(first quadrant)}, \\ \pi + \tan^{-1}(B/A) & \text{if} \quad A < 0 \quad \text{(second or third quadrant)}, \\ 2\pi + \tan^{-1}(B/A) & \text{if} \quad A > 0,\ B < 0 \quad \text{(fourth quadrant)}, \end{cases}$$

where $\tan^{-1}(B/A)$ is the angle in $(-\pi/2, \pi/2)$ given by a calculator or computer.

In any event, from (10) and (11) we get

$$x(t) = C\left(\frac{A}{C}\cos\omega_0 t + \frac{B}{C}\sin\omega_0 t\right) = C(\cos\alpha\cos\omega_0 t + \sin\alpha\sin\omega_0 t).$$

With the aid of the cosine addition formula, we find that

$$x(t) = C\cos(\omega_0 t - \alpha). \tag{12}$$

Thus the mass oscillates to and fro about its equilibrium position with

Amplitude	C,
Circular frequency	ω_0, and
Phase angle	α.

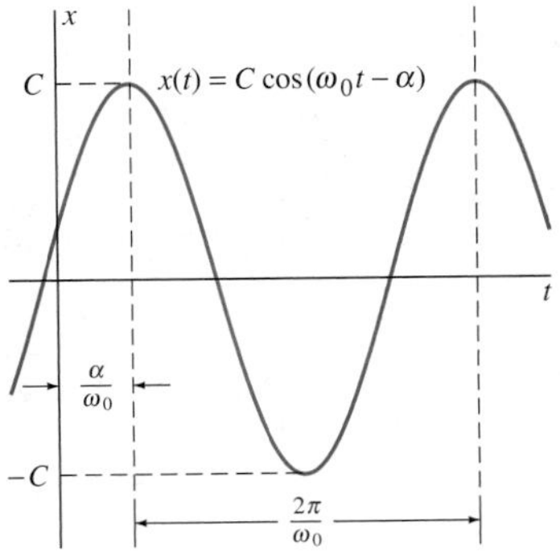

FIGURE 8.7.4 Simple harmonic motion.

Such motion is called **simple harmonic motion.** A typical graph of $x(t)$ is shown in Fig. 8.7.4. If time t is measured in seconds, the circular frequency ω_0 has dimensions of radians per second (rad/s). The **period** of the motion is the time required for the system to complete one full oscillation, so it is given by

$$T = \frac{2\pi}{\omega_0} \tag{13}$$

seconds; its **frequency** is

$$\nu = \frac{1}{T} = \frac{\omega_0}{2\pi} \tag{14}$$

in hertz (Hz), which measures the number of complete cycles per second. Note that frequency is measured in cycles per second, whereas circular frequency has the dimensions of radians per second.

If the initial position $x(0) = x_0$ and initial velocity $x'(0) = v_0$ of the mass are given, we first determine the values of the coefficients A and B in Eq. (10), then find the amplitude C and phase angle α by carrying out the transformation of $x(t)$ to the form in Eq. (12), as indicated previously.

EXAMPLE 2 Suppose that the mass $m = \frac{1}{2}$ (kg) of Example 1 is attached to the same spring with Hooke's constant $k = 50$ (N/m). But now it is set in motion with initial position $x(0) = \frac{1}{2}$ (m) and initial velocity $x'(0) = -10$ (m/s). (Thus the mass is displaced to the right and moving to the left at time $t = 0$.) Find the position function of the body as well as the amplitude, frequency, period of oscillation, and phase angle of its motion.

Solution With $m = \frac{1}{2}$ and $k = 50$, Eq. (6) yields $\frac{1}{2}x'' + 50x = 0$; that is,

$$x'' + 100x = 0.$$

Consequently, we see from Eq. (8) that the circular frequency will be $\omega_0 = \sqrt{100} = 10$ (rad/s). Hence the body will oscillate with

$$\text{frequency:} \quad \frac{10}{2\pi} \approx 1.59 \quad \text{Hz}$$

and

$$\text{period: } \frac{2\pi}{10} \approx 0.63 \quad \text{s.}$$

We now impose the initial conditions $x(0) = 0.5$ and $x'(0) = -10$ on the general solution $x(t) = A\cos 10t + B\sin 10t$, and it follows that $A = 0.5$ and $B = -1$. So the position function of the body is

$$x(t) = \tfrac{1}{2}\cos 10t - \sin 10t.$$

Hence its amplitude of motion is

$$C = \sqrt{\left(\tfrac{1}{2}\right)^2 + (1)^2} = \tfrac{1}{2}\sqrt{5} \approx 1.12 \quad \text{(m)}.$$

To find the phase angle, we use the cosine addition formula to write

$$x(t) = \frac{\sqrt{5}}{2}\left(\frac{1}{\sqrt{5}}\cos 10t - \frac{2}{\sqrt{5}}\sin 10t\right) = \frac{\sqrt{5}}{2}\cos(10t - \alpha).$$

Thus we require

$$\cos\alpha = \frac{1}{\sqrt{5}} > 0 \quad \text{and} \quad \sin\alpha = -\frac{2}{\sqrt{5}} < 0.$$

Hence α is the fourth-quadrant angle

$$\alpha = 2\pi - \tan^{-1}\left(\frac{\frac{2}{5}\sqrt{5}}{\frac{1}{5}\sqrt{5}}\right) \approx 5.1760 \quad \text{(rad)}.$$

In the form in which the amplitude and phase angle are made explicit, the position function is

$$x(t) \approx \frac{\sqrt{5}}{2}\cos(10t - 5.1760).$$

◆

Free Damped Motion

We assume now that $c > 0$ in Eq. (5) and consider **damped motion** of a mass on a spring (still with no external force). The characteristic equation of the differential equation $mx'' + cx' + kx = 0$ has roots

$$r = \frac{-c \pm \sqrt{c^2 - 4km}}{2m}. \tag{15}$$

Therefore the type of motion that occurs depends on whether $c^2 > 4km$ (distinct real roots), $c^2 = 4km$ (equal real roots), or $c^2 < 4km$ (complex conjugate roots).

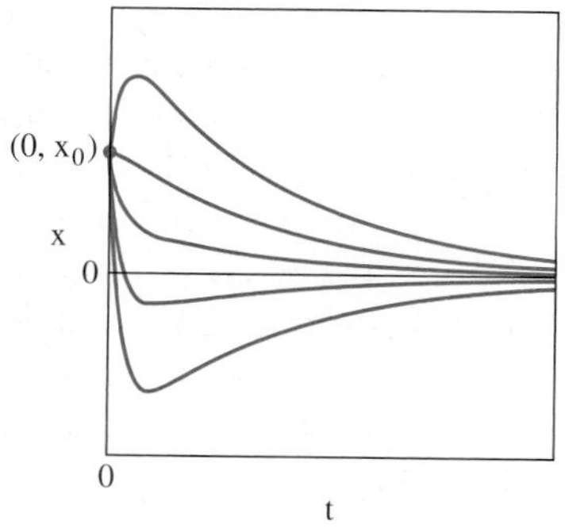

FIGURE 8.7.5 Overdamped motion. $x(t) = c_1e^{r_1t} + c_2e^{r_2t}$ with $r_1 < 0$ and $r_2 < 0$. Solution curves are graphed with the same initial position x_0 and different initial velocities.

Overdamped Case $c^2 > 4km$. Because c is relatively large in this case, we are dealing with a strong resistance in comparison with a relatively weak spring (or a small mass). In this case Eq. (15) gives negative distinct real roots $r_1 = -p_1$ and $r_2 = -p_2$ (where $p_1, p_2 > 0$). Hence the position function has the form

$$x(t) = c_1e^{-p_1t} + c_2e^{-p_2t}. \tag{16}$$

Obviously $x(t) \to 0$ as $t \to +\infty$, and hence the mass m settles to its equilibrium position without any oscillations. That is, any would-be oscillations are damped out. (See Problem 36.) Figure 8.7.5 shows some typical graphs of the position function in this overdamped case.

Critically Damped Case $c^2 = 4km$. In this case Eq. (15) gives equal real roots $r = -p = -c/(2m) < 0$. The position function therefore has the form

$$x(t) = (c_1 + c_2t)e^{-pt}. \tag{17}$$

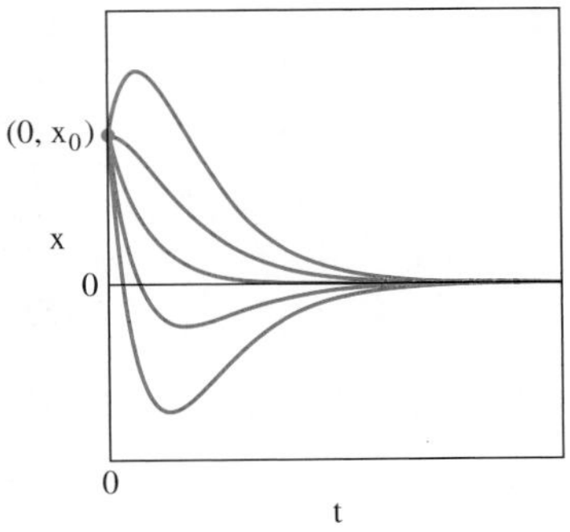

FIGURE 8.7.6 Critically damped motion. $x(t) = (c_1 + c_2 t)e^{-pt}$ with $p > 0$. Solution curves are graphed with the same initial position x_0 and different initial velocities.

Figure 8.7.6 shows some typical graphs of the position function in this critically damped case. The damping is just large enough to damp out any would-be oscillations, but even a slight decrease in the damping brings us to the remaining case, the one that shows the most dramatic behavior.

Underdamped Case $c^2 < 4km$. Now Eq. (15) gives the complex conjugate roots

$$r = \frac{-c \pm \sqrt{-(4km - c^2)}}{2m} = -\frac{c}{2m} \pm i\sqrt{\frac{k}{m} - \left(\frac{c}{2m}\right)^2}.$$

Let us write

$$p = \frac{c}{2m} \quad \text{and} \quad \omega_1 = \sqrt{\omega_0^2 - p^2} \tag{18}$$

(recalling from (8) the undamped circular frequency $\omega_0 = \sqrt{k/m}$). Then the complex conjugate roots of the characteristic equation are $r = -p \pm i\omega_1$, so the general solution of $mx'' + cx' + kx = 0$ is

$$x(t) = e^{-pt}(A\cos\omega_1 t + B\sin\omega_1 t) = Ce^{-pt}\left(\frac{A}{C}\cos\omega_1 t + \frac{B}{C}\sin\omega_1 t\right)$$

where $C = \sqrt{A^2 + B^2}$. Using the cosine addition formula, it follows that $x(t)$ can be written in the form

$$x(t) = Ce^{-pt}\cos(\omega_1 t - \alpha) \tag{19}$$

similar to Eq. (12), with

$$C = \sqrt{A^2 + B^2}, \quad \cos\alpha = \frac{A}{C}, \quad \text{and} \quad \sin\alpha = \frac{B}{C}.$$

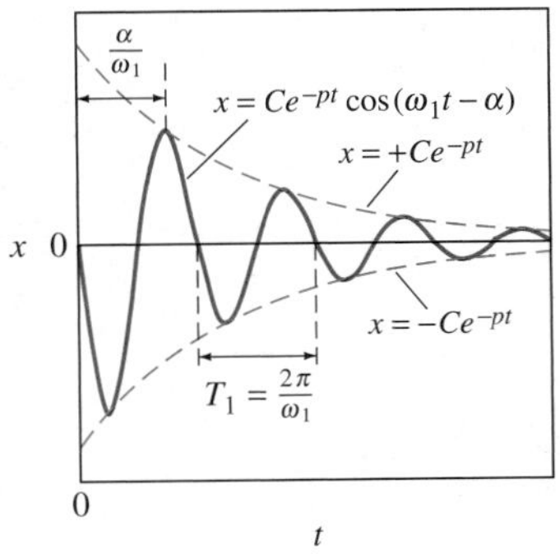

FIGURE 8.7.7 Underdamped oscillations: $x(t) = Ce^{-pt}\cos(\omega_1 t - \alpha)$.

The solution in (19) represents exponentially damped oscillations of the body around its equilibrium position. The graph of $x(t)$ lies between the curves $x = -Ce^{-pt}$ and $x = Ce^{-pt}$ and touches them when $\omega_1 t - \alpha$ is an integral multiple of π. The motion is not actually periodic, but it is nevertheless useful to call ω_1 its **circular frequency,** $T_1 = 2\pi/\omega_1$ its **pseudoperiod** of oscillation, and Ce^{-pt} its **time-varying amplitude.** Most of these quantities are shown in the typical graph of underdamped motion shown in Fig. 8.7.7. Note from Eq. (18) that in this case ω_1 is less than the undamped circular frequency ω_0, so T_1 is larger than the period T of oscillation of the same mass without damping on the same spring. Thus the action of the dashpot has at least three effects:

1. It exponentially damps the oscillations, in accord with the time-varying amplitude.
2. It slows the motion; that is, the dashpot decreases the frequency of the motion.
3. It delays the motion—this is the effect of the phase angle in Eq. (19).

EXAMPLE 3 The mass and spring of Example 2 are now attached also to a dashpot that provides 6 N of resistance for each meter per second of velocity. The mass is set in motion with the same initial position $x(0) = \frac{1}{2}$ (m) and the same initial velocity $x'(0) = -10$ (m/s). Find the position function of the mass as well as its new frequency of oscillation, its pseudoperiod, and the phase angle of its motion.

Solution Rather than memorizing the various formulas given in the preceding discussion, it is better practice in a particular case to set up the differential equation and then solve it directly. Recall that $m = \frac{1}{2}$ and $k = 50$; we are now given $c = 6$ in mks units. Hence Eq. (3) is $\frac{1}{2}x'' + 6x' + 50x = 0$; that is,

$$x'' + 12x' + 100x = 0.$$

The roots of the characteristic equation $r^2 + 12r + 100 = 0$ are

$$r_1,\ r_2 = \frac{-12 \pm \sqrt{144 - 400}}{2} = -6 \pm 8i,$$

so the general solution is

$$x(t) = e^{-6t}(A\cos 8t + B\sin 8t). \tag{20}$$

The new circular frequency is $\omega_1 = 8$ (rad/s), and the pseudoperiod and new frequency are

$$T_1 = \frac{2\pi}{8} \approx 0.79 \text{ (s)}$$

and

$$\frac{1}{T_1} = \frac{8}{2\pi} \approx 1.27 \text{ (Hz)}$$

(in contrast with 0.63 s and 1.59 Hz, respectively, in the undamped case).

From Eq. (20) we compute

$$x'(t) = e^{-6t}(-8A\sin 8t + 8B\cos 8t) - 6e^{-6t}(A\cos 8t + B\sin 8t).$$

The initial conditions therefore produce the equations

$$x(0) = A = \tfrac{1}{2} \quad \text{and} \quad x'(0) = -6A + 8B = -10,$$

so $A = \frac{1}{2}$ and $B = -\frac{7}{8}$. Thus

$$x(t) = e^{-6t}\left(\tfrac{1}{2}\cos 8t - \tfrac{7}{8}\sin 8t\right),$$

and so with

$$C = \sqrt{\left(\tfrac{1}{2}\right)^2 + \left(\tfrac{7}{8}\right)^2} = \tfrac{1}{8}\sqrt{65}$$

we have

$$x(t) = \frac{\sqrt{65}}{8}e^{-6t}\left(\frac{4}{\sqrt{65}}\cos 8t - \frac{7}{\sqrt{65}}\sin 8t\right).$$

We require

$$\cos\alpha = \frac{4}{\sqrt{65}} > 0 \quad \text{and} \quad \sin\alpha = -\frac{7}{\sqrt{65}} < 0,$$

so α is the fourth-quadrant angle

$$\alpha = 2\pi - \tan^{-1}\left(\tfrac{7}{4}\right) \approx 5.2315 \text{ (rad)}.$$

In summary, the position function of the oscillating mass is given approximately by

$$x(t) = \frac{\sqrt{65}}{8}e^{-6t}\cos(8t - 5.2315). \tag{21}$$

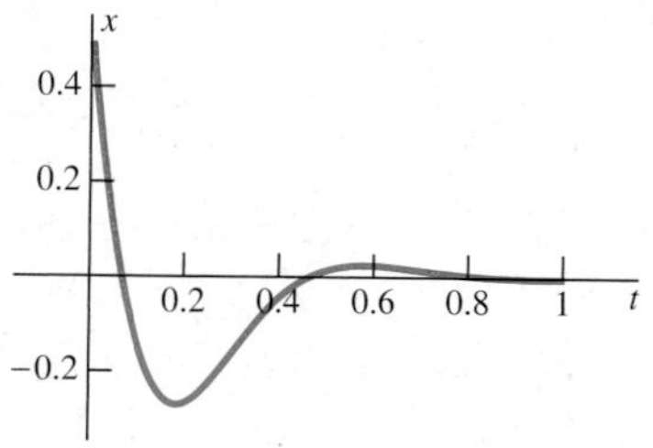

FIGURE 8.7.8 Graph of the position function in Eq. (21).

Figure 8.7.8 shows the graph of this position function. Although the oscillations theoretically occur indefinitely, we see that—from a practical point of view—they are effectively damped out after a second or so. ◆

Forced Oscillations

Masses on springs in mechanical systems often are subject to periodic external forces. A typical example would be a car being driven down a road with periodic pavement oscillations. The mass is the car itself; the dashpot and spring consist of its suspension (shock absorbers and leaf or coil springs). The road provides the external force and the driver feels the periodic motion of the car.

If we include in Eq. (3) a periodic external force $F(t) = F_0 \cos \omega t$ with amplitude F_0 and circular frequency ω, we get the nonhomogeneous differential equation

$$mx'' + cx' + kx = F_0 \cos \omega t. \tag{22}$$

In Section 8.6 we solved only homogeneous differential equations, but a nonhomogeneous equation of the special form in (22) frequently can be solved by a method of "shrewd guessing." Because derivatives of sines and cosines of ωt are again sines and cosines of ωt, it is reasonable to guess that Eq. (22) might have a *particular solution* of the form

$$x_p(t) = A \cos \omega t + B \sin \omega t. \tag{23}$$

If so, we can hope to discover the values of A and B by substituting (23) for x in Eq. (22) and then collecting coefficients of $\cos \omega t$ and $\sin \omega t$.

Now let $x_c(t)$ denote the *general solution*—involving arbitrary constants c_1 and c_2—of the associated (force-free) equation $mx'' + c'x + kx = 0$. Then the sum

$$x(t) = x_c(t) + x_p(t) \tag{24}$$

will be a general solution of Eq. (22), because we find that

$$\begin{aligned} mx'' + cx' + kx &= m(x_c + x_p)'' + c(x_c + x_p)' + k(x_c + x_p) \\ &= (mx_c'' + cx_c' + kx_c) + (mx_p'' + cx_p' + kx_p) \\ &= 0 + F_0 \cos \omega t = F_0 \cos \omega t. \end{aligned}$$

In summary, the general solution of the nonhomogeneous equation in (22) is the sum of the particular solution x_p and the general solution x_c of the associated homogeneous equation. Finally, we can impose given initial conditions on $x(t)$ to determine the numerical values of the constants c_1 and c_2 that appear in x_c. Examples 4 and 5 illustrate this procedure.

EXAMPLE 4 Suppose that $m = 1$, $c = 0$, $k = 9$, $F_0 = 80$, and $\omega = 5$, so that the nonhomogeneous differential equation in (22) is

$$x'' + 9x = 80 \cos 5t. \tag{25}$$

Find $x(t)$ if $x(0) = x'(0) = 0$.

Solution The associated homogeneous differential equation $x'' + 9x = 0$ has general solution

$$x_c(t) = c_1 \cos 3t + c_2 \sin 3t.$$

The particular solution given by (23) with $\omega = 5$ takes the form $x_p(t) = A \cos 5t + B \sin 5t$. Substituting x_p for x in the nonhomogeneous equation in (25) yields

$$(-25A \cos 5t - 25B \sin 5t) + 9(A \cos 5t + B \sin 5t) = 80 \cos 5t.$$

When we group and compare coefficients of $\cos 5t$ and $\sin 5t$ on the two sides of this equation, we see that $-16A = 80$ and $-16B = 0$. Consequently $A = -5$ and $B = 0$, so the particular solution is $x_p(t) = -5 \cos 5t$. The general solution $x(t) = x_c(t) + x_p(t)$ in (24) is therefore

$$x(t) = c_1 \cos 3t + c_2 \sin 3t - 5 \cos 5t.$$

Finally, we apply the given initial conditions $x(0) = x'(0) = 0$ to $x(t)$ and its derivative

$$x'(t) = -3c_1 \sin 3t + 3c_2 \cos 3t + 25 \sin 5t.$$

We thereby obtain $x(0) = c_1 - 5 = 0$ and $x'(0) = 3c_2 = 0$, so $c_1 = 5$ and $c_2 = 0$. This gives the desired solution

$$x(t) = 5 \cos 3t - 5 \cos 5t$$

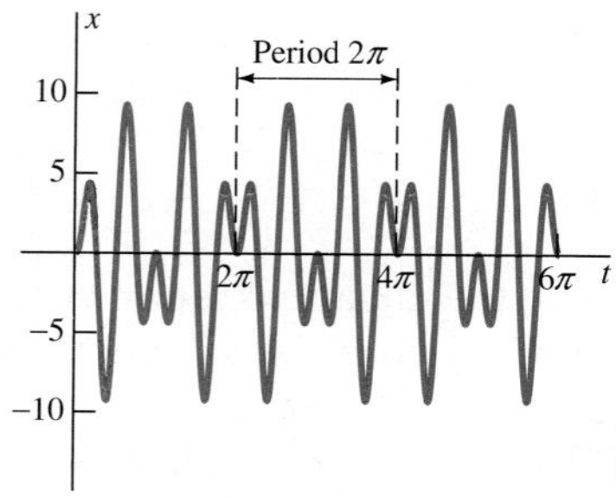

FIGURE 8.7.9 The response function $x(t) = 5\cos 3t - 5\cos 5t$ in Example 4.

of the original initial value problem. As indicated in Fig. 8.7.9, the period of $x(t)$ is the least common integral multiple 2π of the periods $2\pi/3$ and $2\pi/5$ of the two cosine terms. ◆

REMARK Suppose that the frequency ω of the external force $F(t) = 80\cos\omega t$ in Example 4 had been equal to the natural frequency $\omega_0 = 3$ of the mass-and-spring system. Then substituting the "trial solution" $x_p(t) = A\cos 3t + B\sin 3t$ in the nonhomogeneous differential equation would have led to the contradictory equation $0 = 80\cos 3t$. (Verify this for yourself.) Thus we would not have been able to determine A and B in this way. The case in which the external and natural frequencies are equal leads to the phenomenon of **resonance,** with oscillations of larger and larger amplitudes occurring. (See Problem 37.) This phenomenon does not occur when nonzero damping is present, as in Example 5.

EXAMPLE 5 Suppose that $m = 1$, $c = 2$, $k = 26$, $F_0 = 82$, and $\omega = 4$, so the nonhomogeneous differential equation in (22) is

$$x'' + 2x' + 26x = 82\cos 4t. \tag{26}$$

Find $x(t)$ if $x(0) = 6$ and $x'(0) = 0$.

Solution The associated homogeneous differential equation $x'' + 2x' + 26x = 0$ has characteristic equation

$$r^2 + 2r + 26 = (r + 1)^2 + 25 = 0$$

with complex conjugate roots $r = -1 \pm 5i$. Hence its general solution is

$$x_c(t) = e^{-t}(c_1\cos 5t + c_2\sin 5t).$$

The particular solution given in (23) with $\omega = 4$ is $x_p(t) = A\cos 4t + B\sin 4t$. When we substitute this trial solution in the nonhomogeneous equation in (26), collect like terms, and equate coefficients of $\cos 4t$ and $\sin 4t$, we get the equations

$$10A + 8B = 82,$$
$$-8A + 10B = 0$$

with solution $A = 5$, $B = 4$. This gives the particular solution

$$x_p(t) = 5\cos 4t + 4\sin 4t$$

of Eq. (26). The general solution $x(t) = x_c(t) + x_p(t)$ in (24) is therefore

$$x(t) = e^{-t}(c_1\cos 5t + c_2\sin 5t) + 5\cos 4t + 4\sin 4t.$$

Finally, we apply the given initial conditions $x(0) = 6$ and $x'(0) = 0$ to $x(t)$ and its derivative

$$x'(t) = -e^{-t}(c_1\cos 5t + c_2\sin 5t) + e^{-t}(-5c_1\sin 5t + 5c_2\cos 5t) - 20\sin 4t + 16\cos 4t.$$

Thereby we get the simultaneous equations

$$x(0) = c_1 + 5 = 6,$$
$$x'(0) = -c_1 + 5c_2 + 16 = 0$$

with solution $c_1 = 1$, $c_2 = -3$. These coefficients yield the desired solution

$$x(t) = e^{-t}(\cos 5t - 3\sin 5t) + 5\cos 4t + 4\sin 4t \tag{27}$$

of the original initial value problem. ◆

REMARK The solution in (27) is the sum of a **transient solution**

$$x_{tr}(t) = e^{-t}(\cos 5t - 3\sin 5t)$$

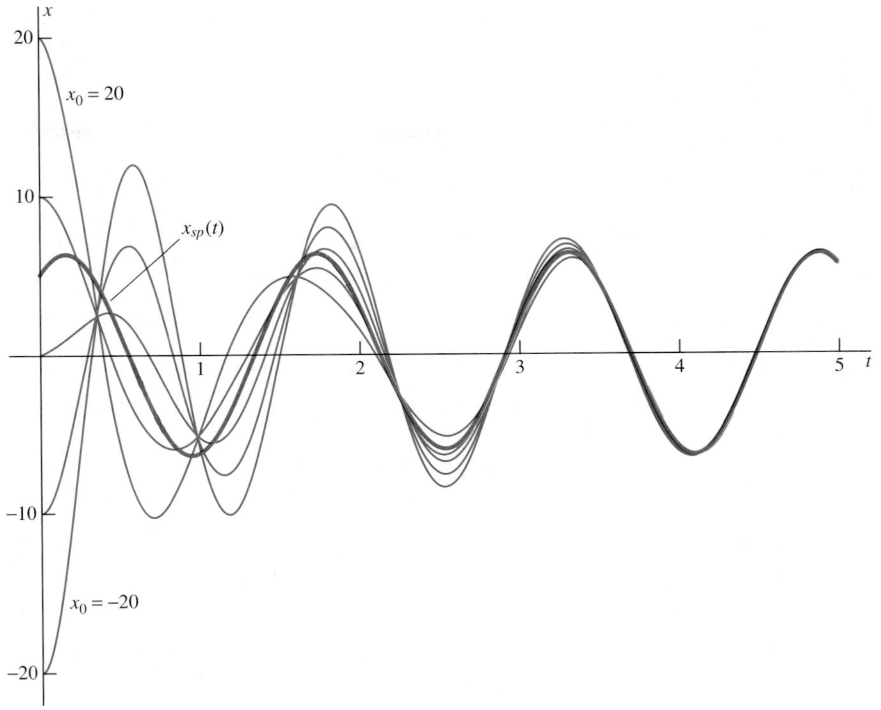

FIGURE 8.7.10 Solutions of the initial value problem in (28) with $x_0 = -20, -10, 0, 10, 20$, and also the steady periodic solution $x_{sp}(t)$.

—thus named because it dies out as $t \to +\infty$—and the **steady periodic solution**

$$\begin{aligned} x_{sp}(t) &= 5\cos 4t + 4\sin 4t \\ &= \sqrt{41}\left(\frac{5}{\sqrt{41}}\cos 4t + \frac{4}{\sqrt{41}}\sin 4t\right) = \sqrt{41}\cos\left(4t - \tan^{-1}\frac{4}{5}\right) \end{aligned}$$

that represents a motion in which the mass perpetually continues to oscillate with constant amplitude $\sqrt{41}$ and circular frequency $\omega = 4$.

Figure 8.7.10 shows graphs of the solution $x(t) = x_{tr}(t) + x_{sp}(t)$ of the initial value problem

$$x'' + 2x' + 26x = 82\cos 4t; \qquad x(0) = x_0, \quad x'(0) = 0 \tag{28}$$

for the different initial positions $x_0 = -20, -10, 0, 10$, and 20. Here we see clearly what it means for the transient solution $x_{tr}(t)$ to "die out with the passage of time," leaving only the steady periodic solution $x_{sp}(t)$. Indeed, because $x_{tr}(t) \to 0$ exponentially, within a very few cycles the graphs of the full solution $x(t)$ and the steady periodic solution $x_{sp}(t)$ are virtually indistinguishable.

8.7 TRUE/FALSE STUDY GUIDE

8.7 CONCEPTS: QUESTIONS AND DISCUSSION

In each of Questions 1 through 4, describe differences between the indicated varieties of motion of a mass attached to a spring.

1. Damped and undamped free motions
2. Overdamped and underdamped free motions
3. Damped and undamped forced motions
4. Transient and steady periodic forced motions

8.7 PROBLEMS

Problems 1 through 4 concern undamped free motion of a mass m on a spring with Hooke's (spring) constant k. Suppose that the mass is set in motion with initial position $x(0) = x_0$ and initial velocity $x'(0) = v_0$. Write the position function of the mass in the form $x(t) = C\cos(\omega_0 t - \alpha)$.

1. $m = 2, k = 50; \quad x_0 = 4, v_0 = 15$

2. $m = 3, k = 48; \quad x_0 = -6, v_0 = 32$

3. $m = 4, k = 36; \quad x_0 = -5, v_0 = -36$

4. $m = 5, k = 80; \quad x_0 = 15, v_0 = -32$

Problems 5 through 10 deal with damped free motion of a mass m that is attached both to a spring with Hooke's constant k and to a dashpot with damping constant c. Suppose that the mass is set in motion with initial position $x(0) = x_0$ and initial velocity $x'(0) = v_0$. Find the position function $x(t)$ of the mass. Determine whether the resulting motion is overdamped, critically damped, or underdamped; in the latter case, write the position function in the form $x(t) = Ce^{-pt}\cos(\omega_1 t - \alpha)$.

5. $m = \frac{1}{2},\ c = 3,\ k = 4,\ x_0 = 2,\ v_0 = 0$

6. $m = 3,\ c = 30,\ k = 63,\ x_0 = 2,\ v_0 = 2$

7. $m = 1,\ c = 8,\ k = 16,\ x_0 = 5,\ v_0 = -10$

8. $m = 2,\ c = 12,\ k = 50,\ x_0 = 0,\ v_0 = -8$

9. $m = 2,\ c = 16,\ k = 40,\ x_0 = 5,\ v_0 = 4$

10. $m = 1,\ c = 10,\ k = 125,\ x_0 = 6,\ v_0 = 50$

The initial value problems in Problems 11 through 14 describe forced undamped motion of a mass on a spring. Express the position function $x(t)$ as the sum of two oscillations (as in Example 4). Throughout, primes denote derivatives with respect to t.

11. $x'' + 9x = 10\cos 2t; \quad x(0) = x'(0) = 0$

12. $x'' + 4x = 5\sin 3t; \quad x(0) = x'(0) = 0$

13. $x'' + 100x = 300\sin 5t; \quad x(0) = 0, x'(0) = 0$

14. $x'' + 25x = 90\cos 4t; \quad x(0) = 25, x'(0) = 10$

In Problems 15 through 18, find the steady periodic solution of the given differential equation. If initial conditions are given, also find the transient solution.

15. $x'' + 4x' + 4x = 130\cos 3t$

16. $x'' + 3x' + 5x = -500\cos 5t$

17. $x'' + 4x' + 5x = 40\cos 3t; \quad x(0) = x'(0) = 0$

18. $x'' + 8x' + 25x = 200\cos t + 520\sin t; \quad x(0) = 5, x'(0) = 0$

19. Determine the period and frequency of the simple harmonic motion of a 4-kg mass on the end of a spring with spring constant 16 N/m.

20. Determine the period and frequency of the simple harmonic motion of a body of mass 0.75 kg on the end of a spring with spring constant 48 N/m.

21. A mass of 3 kg is attached to the end of a spring that is stretched 20 cm by a force of 15 N. It is set in motion with initial position $x_0 = 0$ and initial velocity $v_0 = -10$ m/s. Find the amplitude, period, and frequency of the resulting motion.

22. A body with mass 250 g is attached to the end of a spring that is stretched 25 cm by a force of 9 N. At time $t = 0$ the body is pulled 1 m to the right, stretching the spring, and set in motion with an initial velocity of 5 m/s to the left. (a) Find $x(t)$ in the form $C\cos(\omega_0 t + \alpha)$. (b) Find the amplitude and period of motion of the body.

23. Derive Eq. (4) describing the motion of a mass attached to the bottom of a vertically suspended spring. (*Suggestion:* First denote by $x(t)$ the displacement of the mass below the unstretched position of the spring; set up the differential equation for x. Then substitute $y = x - x_0$ in this differential equation.)

24. Consider a floating cylindrical buoy with radius r, height h, and uniform density $\rho \leqq 0.5$ (recall that the density of water is 1 g/cm^3). The buoy is initially suspended at rest with its bottom at the top surface of the water and is released at time $t = 0$. Thereafter it is acted on by two forces: a downward gravitational force equal to its weight $mg = \rho\pi r^2 hg$ and an upward force of buoyancy equal to the weight $\pi r^2 xg$ of water displaced, where $x = x(t)$ is the depth of the bottom of the buoy beneath the surface at time t (Fig. 8.7.11). Conclude that the buoy undergoes simple harmonic motion around its equilibrium position $x_e = \rho h$ with period $p = 2\pi\sqrt{\rho h/g}$. Compute p and the amplitude of the motion if $\rho = 0.5$ g/cm^3, $h = 200$ cm, and $g = 980$ cm/s^2.

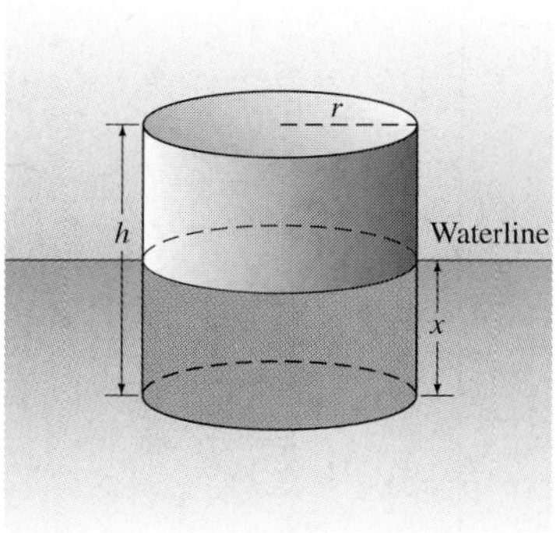

FIGURE 8.7.11 The buoy of Problem 24.

25. A cylindrical buoy weighing 100 lb (thus of mass $m = 3.125$ slugs in ft-lb-s (fps) units) floats in water with its axis vertical (as in Problem 24). When depressed slightly and released, it oscillates up and down four times every 10 s. Assume that friction is negligible. Find the radius of the buoy.

26. Assume that the earth is a solid sphere of uniform density, with mass M and radius $R = 3960$ (mi). For a particle of mass m *within* the earth at distance r from the center of the earth, the gravitational force attracting m toward the center is $F_r = -GM_r m/r^2$, where M_r is the mass of the part of the earth within a sphere of radius r. (a) Show that $F_r = -GMmr/R^3$. (b) Now suppose that a small hole is drilled straight through the center of the earth, thus connecting two antipodal points on its surface. Let a particle of mass m be dropped at time $t = 0$ into this hole with initial speed zero, and let $r(t)$ be its distance from the center of the earth at time t (Fig. 8.7.12). Conclude from Newton's second law and part (a) that $r''(t) = -k^2 r(t)$, where $k^2 = GM/R^3 = g/R$. (c) Take $g = 32.2$ ft/s^2, and conclude from part (b) that the particle undergoes simple harmonic motion back and forth between the ends of the hole, with a period of about 84 min. (d) Look up (or derive) the period of a satellite that just skims the surface of the earth; compare it with the result in part (c). How do you explain the coincidence? Or *is* it a coincidence? (e) With what speed (in miles per hour) does the particle pass through the center of the earth? (f) Look up (or derive) the orbital velocity of a satellite that just skims the surface of the earth; compare it with the result in part (e). How do you explain the coincidence? Or *is* it a coincidence?

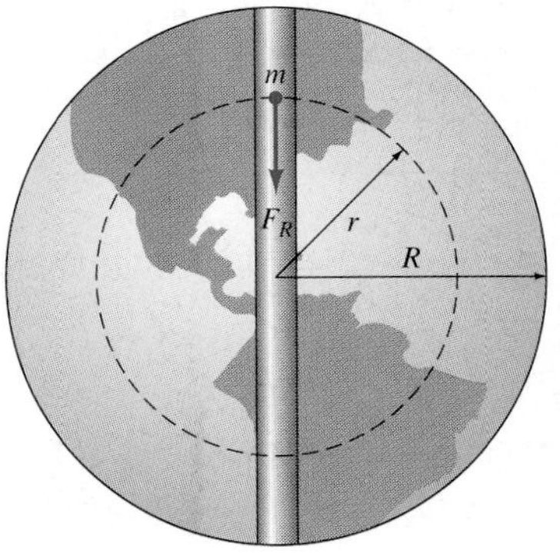

FIGURE 8.7.12 A mass m falling down a hole through the center of the earth (Problem 26).

27. Suppose that the mass in a free mass-spring-dashpot system with $m = 10$, $c = 9$, and $k = 2$ is set in motion with $x(0) = 0$ and $x'(0) = 5$. (a) Find the position function $x(t)$ and show that its graph looks as indicated in Fig. 8.7.13. (b) Find how far the mass moves to the right before starting back toward the origin.

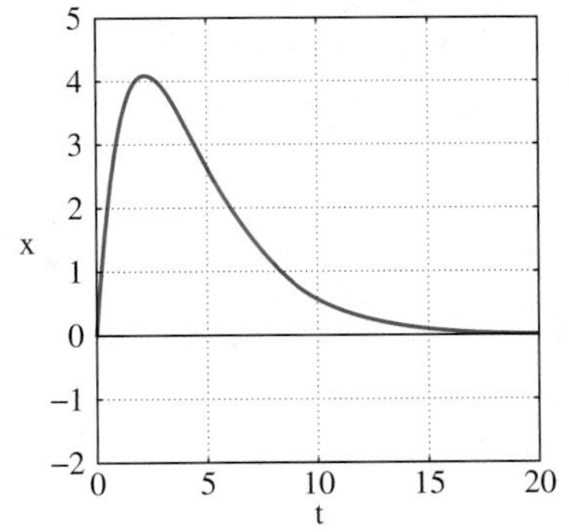

FIGURE 8.7.13 The position function $x(t)$ of Problem 27.

28. Suppose that the mass in a free mass-spring-dashpot system with $m = 25$, $c = 10$, and $k = 226$ is set in motion with $x(0) = 20$ and $x'(0) = 41$. (a) Find the position function $x(t)$ and show that its graph looks as indicated in Fig. 8.7.14. (b) Find the pseudoperiod of the oscillations and the equations of the "envelope curves" that are dashed in the figure.

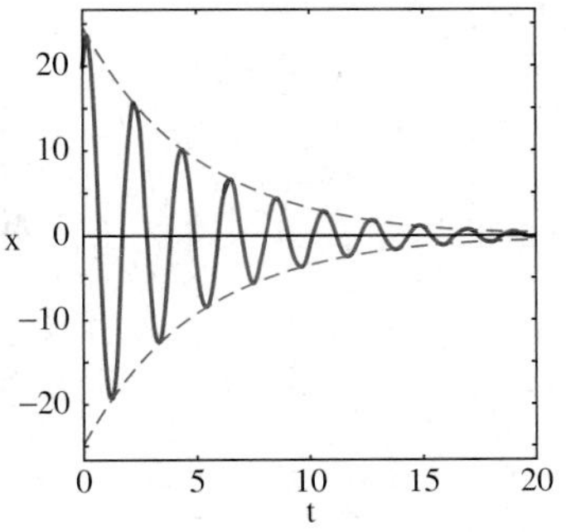

FIGURE 8.7.14 The position function $x(t)$ of Problem 28.

29. A 12-lb weight (mass $m = 0.375$ slugs in fps units) is attached both to a vertically suspended spring that it stretches 6 in. and to a dashpot that provides 3 lb of resistance for every foot per second of velocity. (a) If the weight is pulled down 1 ft below its static equilibrium position and then released from rest at time $t = 0$, find its position function $x(t)$. (b) Find the frequency, time-varying amplitude, and phase angle of the motion.

30. This problem deals with a highly simplified model of a car of weight 3200 lb (mass $m = 100$ slugs in fps units). Assume that the suspension system acts like a single spring and its shock absorbers like a single dashpot, so that its vertical vibrations satisfy Eq. (4) with appropriate values of the coefficients. (a) Find the stiffness coefficient k of the spring if the car undergoes free vibrations at 80 cycles per minute (cycles/min) when its shock absorbers are disconnected. (b) With the shock absorbers connected the car is set into vibration by driving it over a bump, and the resulting damped vibrations have a frequency of 78 cycles/min. After how long will the time-varying amplitude be 1% of its initial value?

Problems 31 through 36 deal with damped free vibrations of a mass-spring-dashpot system whose position function satisfies the equation $mx'' + cx' + kx = 0$. The mass is set in motion with initial position $x(0) = x_0$ and initial velocity $x'(0) = v_0$. Recall the notation $p = c/(2m)$ and $\omega_0 = \sqrt{k/m}$ in Eqs. (18) and (8), respectively. The system is critically damped or overdamped as specified in each problem.

31. (Critically damped) Show in this case that

$$x(t) = (x_0 + v_0 t + p x_0 t)e^{-pt}.$$

32. (Critically damped) Deduce from Problem 31 that the mass passes through $x = 0$ at some instant $t > 0$ if and only if x_0 and $v_0 + p x_0$ have opposite signs.

33. (Critically damped) Deduce from Problem 31 that $x(t)$ has a local maximum or minimum at some instant $t > 0$ if and only if v_0 and $v_0 + p x_0$ have the same sign.

34. (Overdamped) Show in this case that

$$x(t) = \frac{1}{2\gamma}\left[(v_0 - r_2 x_0)e^{r_1 t} - (v_0 - r_1 x_0)e^{r_2 t}\right],$$

where $r_1, r_2 = -p \pm \sqrt{p^2 - \omega_0^2}$ and $\gamma = (r_1 - r_2)/2 > 0$.

35. (Overdamped) If $x_0 = 0$, deduce from Problem 34 that

$$x(t) = \frac{v_0}{\gamma} e^{-pt} \sinh \gamma t.$$

36. (Overdamped) Prove that in this case the mass can pass through its equilibrium position $x = 0$ at most once.

37. Consider the mass-and-spring system of Example 4, except with the external force $F(t) = 60\cos 3t$ having frequency ω equal to the natural frequency $\omega_0 = 3$ of the system. Then the position function of the mass satisfies the differential equation $x'' + 9x = 60\cos 3t$. (a) Show that this nonhomogeneous differential equation has *no* solution of the form $x(t) = A\cos 3t + B\sin 3t$. (As suggested in the text, try to find one and observe what happens.) (b) Verify that $x_p(t) = 10t\sin 3t$ is a particular solution of $x'' + 9x = 60\cos 3t$. The graph of $x_p(t)$, which is shown in Fig. 8.7.15, indicates that any solution

$$x(t) = c_1\cos 3t + c_2\sin 3t + 10t\sin 3t$$

of this equation exhibits oscillations of unbounded magnitude as $t \to +\infty$.

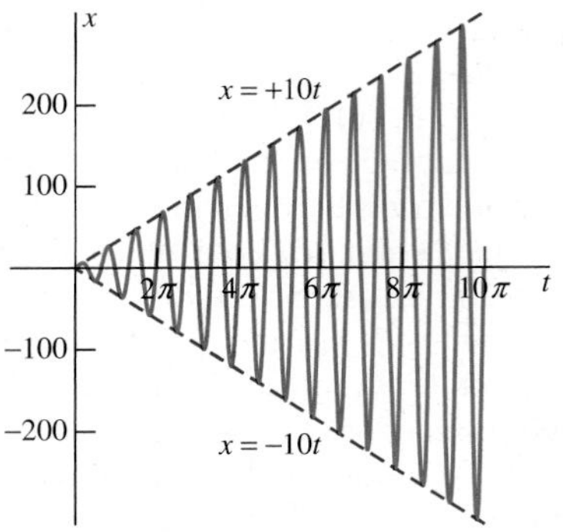

FIGURE 8.7.15 The resonance solution $x_p(t) = 10t\sin 3t$ of the differential equation $x'' + 9x = 60\cos 3t$ oscillates between the lines $x = -10t$ and $x = +10t$.

CHAPTER 8 REVIEW: DEFINITIONS, CONCEPTS, RESULTS

Use the following list as a guide to concepts you may need to review.

1. General and particular solutions of a differential equation
2. Solution curves of a differential equation
3. Solution of the equation $dy/dx = f(x, y)$ if either x or y is missing
4. The exponential equation $dx/dt = kx$ and applications to natural growth and decay
5. Torricelli's law and draining tanks
6. Slope fields and approximate solution curves
7. Euler's method and approximate numerical solutions
8. Separable differential equations
9. Separation of variables
10. Newton's law of cooling
11. The constant-coefficient linear equation $dx/dt = ax + b$ and applications
12. The linear first-order equation $dy/dx + P(x)y = Q(x)$
13. The integrating factor $\rho(x) = \exp\left(\int P(x)\,dx\right)$ for a linear equation
14. Mixture problems
15. Motion with resistance proportional to velocity
16. The general population equation $dP/dt = (\beta - \delta)P$
17. Bounded populations and the logistic equation $dP/dt = kP(M - P)$
18. The explosion-extinction equation $dP/dt = kP(P - M)$
19. The characteristic equation of a homogeneous linear second-order differential equation
20. The general solution of a differential equation of the form $ay'' + by' + cy = 0$
21. Oscillation of a mass-spring-dashpot system—free and forced, damped and undamped

CHAPTER 8 MISCELLANEOUS PROBLEMS

In Problems 1 through 26, find the general solution of the given differential equation. If an initial condition is given, find the corresponding particular solution.

1. $\dfrac{dy}{dx} = 2x + \cos x; \quad y(0) = 0$

2. $\dfrac{dy}{dx} = 3\sqrt{x} + \dfrac{1}{\sqrt{x}}; \quad y(1) = 10$

3. $\dfrac{dy}{dx} = (y + 1)^2$

4. $\dfrac{dy}{dx} = \sqrt{y + 1}$

5. $\dfrac{dy}{dx} = 3x^2y^2; \quad y(0) = 1$

6. $\dfrac{dy}{dx} = \sqrt[3]{xy}; \quad y(1) = 1$

7. $x^2y^2\dfrac{dy}{dx} = 1$

8. $\sqrt{xy}\dfrac{dy}{dx} = 1$

9. $\dfrac{dy}{dx} = y^2\cos x; \quad y(0) = 1$

10. $\dfrac{dy}{dx} = \sqrt{y}\sin x; \quad y(0) = 4$

11. $\dfrac{dy}{dx} = \dfrac{y^2\left(1 - \sqrt{x}\right)}{x^2\left(1 - \sqrt{y}\right)}$

12. $\dfrac{dy}{dx} = \dfrac{\sqrt{y}(x+1)^3}{\sqrt{x}(y+1)^3}$

13. $x^3 + 3y - x\dfrac{dy}{dx} = 0$

14. $xy^2 + 3y^2 - x^2\dfrac{dy}{dx} = 0$

15. $3y + x^4\dfrac{dy}{dx} = 2xy$

16. $2xy^2 + x^2\dfrac{dy}{dx} = y^2$

17. $2x^2y + x^3\dfrac{dy}{dx} = 1$

18. $\dfrac{dy}{dx} = 1 + x^2 + y^2 + x^2y^2$

19. $\dfrac{dy}{dx} + 3y = 3x^2e^{-3x}$

20. $3x^5y^2 + x^3\dfrac{dy}{dx} = 2y^2$

21. $x\dfrac{dy}{dx} + 3y = 3x^{-3/2}$

22. $(x^2 - 1)\dfrac{dy}{dx} + (x - 1)y = 1$

23. $9x^2y^2 + x^{3/2}\dfrac{dy}{dx} = y^2$

24. $2y + (x + 1)\dfrac{dy}{dx} = 3x + 3$

25. $\dfrac{dy}{dx} = e^x + y$

26. $y + x\dfrac{dy}{dx} = 2e^{2x}$

Each of the differential equations in Problems 27 and 28 is both separable and linear. Derive and reconcile the two general solutions you find by the two indicated methods.

27. $\dfrac{dy}{dx} = 3(y + 7)x^2$

28. $\dfrac{dy}{dx} = \dfrac{2xy + 2x}{x^2 + 1}$

Solve the initial value problems in 29 and 30.

29. $\dfrac{dx}{dt} = x^2 + 5x + 6;\quad x(0) = 5$

30. $\dfrac{dx}{dt} = 2x^2 + x - 15;\quad x(0) = 10$

31. *Radioactive Decay* A certain moon rock contains equal numbers of potassium atoms and argon atoms. Assume that all the argon is present because of radioactive decay of potassium (its half-life is about 1.28×10^9 yr) and that 1 out of every 9 potassium atom disintegrations yields an atom of argon. What is the age of the rock, measured from the time it contained only potassium?

32. *Newton's Law of Cooling* If a body is cooling in a medium with constant temperature A, then—according to Newton's law of cooling (Section 8.3)—the rate of change of temperature T of the body is proportional to $T - A$. We plan to cool a pitcher of buttermilk initially at 25°C by setting it on the front porch, where the temperature is 0°C. If the temperature of the buttermilk drops to 15°C after 20 min, when will it be at 5°C?

33. When sugar is dissolved in water, the amount A of sugar that remains undissolved after t minutes satisfies the differential equation $dA/dt = -kA$ $(k > 0)$. If 25% of the sugar dissolves in 1 min, how long does it take for half the sugar to dissolve?

34. The intensity I of light at a depth x meters below the surface of a lake satisfies the differential equation $dI/dx = -(1.4)I$. (a) At what depth is the intensity half the intensity I_0 at the surface (where $x = 0$)? (b) What is the intensity at a depth of 10 m (as a fraction of I_0)? (c) At what depth will the intensity be 1% of its value at the surface?

35. The barometric pressure p (in inches of mercury) at an altitude x miles above sea level satisfies the differential equation $dp/dx = -(0.2)p$ with initial condition $p(0) = 29.92$. (a) Calculate the barometric pressure at 10,000 ft and again at 30,000 ft. (b) Without prior conditioning, few people can survive when the pressure drops to less than 15 in. of mercury. How high is that? (c) The highest mountain in North America is Mt. McKinley, in Denali National Park, Alaska, U.S.A. What is the atmospheric pressure at its summit, approximately 20,320 ft above sea level?

36. An accident at a nuclear power plant has left the surrounding area polluted with a radioactive element that decays at a rate proportional to its current amount $A(t)$. The initial level of radiation is 10 times the maximum amount S that is safe, and 100 days later it is still 7 times that amount. (a) Set up and solve a differential equation to find $A(t)$. (b) How long (to the nearest day after the original accident) will it be before it is safe for people to return to the area?

37. Suppose that a nuclear accident was confined to a single room of a nuclear research laboratory but has left that room contaminated with polonium-210, which has a half-life of 140 days. If the initial contamination of the room is five times the amount safe for long-term human exposure, how long should laboratory workers wait before entering the room to decontaminate it?

38. Suppose that the national government's current annual budget is \$2 trillion, but only \$1.85 trillion in taxes is being collected annually (so the current deficit is \$150 billion per year). Suppose also that both the annual budget and the annual tax revenues increase exponentially. If revenues increase at 3% annually, what annual percentage increase in the national budget will yield a balanced budget seven years in the future? You may choose either a symbolic approach or a graphical approach (in which case you need to determine the budget's rate of increase so that the budget and revenue graphs intersect seven years from now).

Solve the initial value problems in 39 through 44. Primes denote derivatives with respect to x.

39. $6y'' - 19y' + 15y = 0;\quad y(0) = 13,\ y'(0) = 21$

40. $50y'' - 5y' - 28y = 0;\quad y(0) = 25,\ y'(0) = -10$

41. $121y'' + 154y' + 49y = 0;\quad y(0) = 11,\ y'(0) = 10$

42. $169y'' - 130y' + 25y = 0;\quad y(0) = 26,\ y'(0) = 39$

43. $100y'' + 20y' + 10001y = 0;\quad y(0) = 10,\ y'(0) = 9$

44. $100y'' + 2000y' + 10001y = 0;\quad y(0) = 1,\ y'(0) = -9$

45. (a) In 1979 the typical microcomputer contained 29 thousand transistors. Assuming natural growth at the annual rate r, write a formula giving the number $N(t)$ of transistors

in a typical microcomputer t years later. (b) In 1993 the typical microcomputer CPU contained 3.1 million transistors. Find the annual growth rate r of part (a), expressed as a percentage. (c) At the rate you found in part (b), how many months are required to double the number of transistors in a typical microcomputer? (d) Assuming the rate of parts (b) and (c) remains constant, how many transistors (rounded to the nearest million) did the typical microcomputer contain in the year 2001?

46. Dinosaurs became extinct late in the Cretaceous Era, about 70 million years ago. Suppose that you find a dinosaur bone containing exactly one atom of ^{14}C. Obtain a sound *underestimate* of the weight of the dinosaur. You will need to look up Avogadro's number and perhaps read about what it tells you.

47. Suppose that the fish population $P(t)$ in a lake is attacked by disease at time $t = 0$, with the result that

$$\frac{dP}{dt} = -k\sqrt{P} \quad (k > 0)$$

thereafter. If there were initially 900 fish in the lake and only 441 remained after 6 weeks, how long would it take for all of the fish in the lake to die?

48. Prove that the solution of the initial value problem

$$\frac{dP}{dt} = k\sqrt{P}, \qquad P(0) = P_0 \quad (P_0 > 0)$$

is given by

$$P(t) = \left(\tfrac{1}{2}kt + \sqrt{P_0}\right)^2.$$

49. Suppose that the population of Fremont satisfies the differential equation of Problem 48. (a) If $P = 100{,}000$ in 1970 and $P = 121{,}000$ in 1980, what should the population be in 2000? (b) When will the population reach 200,000?

50. The population $P(t)$ of a certain group of rabbits satisfies the initial value problem

$$\frac{dP}{dt} = kP^2, \quad P(0) = P_0,$$

where k is a positive constant. Derive the solution

$$P(t) = \frac{P_0}{1 - kP_0t}.$$

51. In Problem 50, suppose that $P_0 = 2$ and that there are 4 rabbits after 3 months. What happens in the next 3 months?

52. Suppose that a motorboat is traveling at $v_0 = 40$ ft/s when its motor is cut off at time $t = 0$. Thereafter its deceleration (due to water resistance) is given by $dv/dt = -kv^2$, where k is a positive constant. (a) Solve this differential equation to show that the speed of the boat at time $t > 0$ is

$$v(t) = \frac{40}{1 + 40kt}$$

feet per second. (b) If the speed of the boat after 10 s is 20 ft/s, how long does it take (since the motor was cut off) for the boat to slow to 5 ft/s?

53. Suppose that the fish population $P(t)$ in a lake is attacked by disease at time $t = 0$, with the result that

$$\frac{dP}{dt} = -3\sqrt{P}$$

thereafter. Time t is measured in weeks. Initially there are $P_0 = 900$ fish in the lake. How long will it take for all of the fish to die?

54. A race car sliding along a level surface is decelerated by frictional forces proportional to its speed. Suppose that it decelerates initially at 2 m/s^2 and travels a total distance of 1800 m. What was its initial velocity? (See Problem 32 of Section 8.4.)

55. A home mortgage of \$120,000 is to be paid off continuously over a period of 25 yr. Apply the result of Problem 33 in Section 8.3 to determine the monthly payment if the annual interest rate, compounded continuously, is (a) 8%; (b) 12%.

56. A powerboat weighs 32000 lb and its motor provides a thrust of 5000 lb. Assume that the water resistance is 100 lb for each foot per second of the boat's speed. Then the velocity $v(t)$ (in ft/s) of the boat at time t (in seconds) satisfies the differential equation

$$1000\frac{dv}{dt} = 5000 - 100v.$$

Find the maximum velocity that the boat can attain if it starts from rest.

57. The temperature in my freezer is $-16°$C and the room temperature is a constant $20°$C. At 11 P.M. one evening the power goes off because of an ice storm. At 6 A.M. the next morning I see that the temperature in the freezer has risen to $-10°$C. At what time will the temperature in the freezer reach the critical value of $0°$C if the power is not restored?

58. Suppose that the action of fluorocarbons depletes the ozone in the upper atmosphere by 0.25% annually, so that the amount A of ozone in the upper atmosphere satisfies the differential equation

$$\frac{dA}{dt} = -\frac{1}{400}A \quad (t \text{ in years}).$$

(a) What percentage of the original amount A_0 of upper-atmospheric ozone will remain 25 yr from now? (b) How long will it take for the amount of upper-atmospheric ozone to be reduced to half its initial amount?

59. A car starts from rest and travels along a straight and level road. Its engine provides a constant acceleration of a feet per second per second. Air resistance and road friction cause a deceleration of ρ feet per second per second for every foot per second of the car's velocity v. (a) Show that the velocity of the car after t seconds is

$$v(t) = \frac{a}{\rho}(1 - e^{-\rho t}).$$

(b) If $a = 17.6$ ft/s^2 and $\rho = 0.1$, find v when $t = 10$ s and find the limiting velocity of the car as $t \to +\infty$. (Give each answer in miles per hour as well as in feet per second.)

60. Immediately after an accident in a nuclear power plant, the level of radiation there was 10 times the safe limit. After 6 months it dropped to 9 times the safe limit. Assuming exponential decay, how long (in years) after the accident will the radiation level drop to the safe limit?

61. A 22-yr-old engineer accepts a position with a starting salary of \$30,000/yr. Her annual salary S increases exponentially, with

$$S(t) = 30e^{(0.05)t}$$

POLAR COORDINATES AND PARAMETRIC CURVES

9

Pierre de Fermat (1601–1665)

Pierre de Fermat exemplifies the distinguished tradition of great amateurs in mathematics. Like his contemporary René Descartes, he was educated as a lawyer. But unlike Descartes, Fermat actually practiced law as his profession and served in the regional parliament. His ample leisure time was, however, devoted to mathematics and to other intellectual pursuits, such as the study of ancient Greek manuscripts.

In a margin of one such manuscript (by the Greek mathematician Diophantus) was found a handwritten note that has remained an enigma ever since. Fermat asserts that for *no* integer $n > 2$ do positive integers x, y, and z exist such that $x^n + y^n = z^n$. For instance, although $15^2 + 8^2 = 17^2$, the sum of two (positive integer) cubes cannot be a cube. "I have found an admirable proof of this," Fermat wrote, "but this margin is too narrow to contain it." Despite the publication of many incorrect proofs, "Fermat's last theorem" remained unproved for three and one-half centuries. But in a June 1993 lecture, the British mathematician Andrew Wiles of Princeton University announced a long and complex proof of Fermat's last theorem. Although the proof as originally proposed contained some gaps, these have been repaired, and experts in the field agree that Fermat's last *conjecture* is, finally, a *theorem*.

Descartes and Fermat shared in the discovery of analytic geometry. But whereas Descartes typically used geometrical methods to solve algebraic equations (see the Chapter 1 opening), Fermat concentrated on the investigation of geometric curves defined by algebraic equations. For instance, he introduced the translation and rotation methods of this chapter (and Chapter 11) to show that the graph of an equation of the form $Ax^2 + Bxy + Cy^2 + Dx + Ey + F = 0$ is generally a conic section. Most of his mathematical work remained unpublished during his lifetime, but it contains numerous tangent line (derivative) and area (integral) computations.

The brilliantly colored left-hand photograph is a twentieth-century example of a geometric object defined by means of algebraic operations. Starting with the point $P(a, b)$ in the xy-plane, we interpret P as the complex number $c = a + bi$ and define the sequence $\{z_n\}$ of points of the complex plane iteratively (as in Section 3.10) by the equations

$$z_0 = c, \qquad z_{n+1} = z_n^2 + c \quad (\text{for } n \geqq 0).$$

If this sequence of points remains inside the circle $x^2 + y^2 = 4$ for all n, then the original point $P(a, b)$ is colored black. Otherwise, the color assigned to P is determined by the speed with which this sequence "escapes" that circular disk. The set of all black points is the famous *Mandelbrot set,* discovered in 1980 by the French mathematician Benoit Mandelbrot.

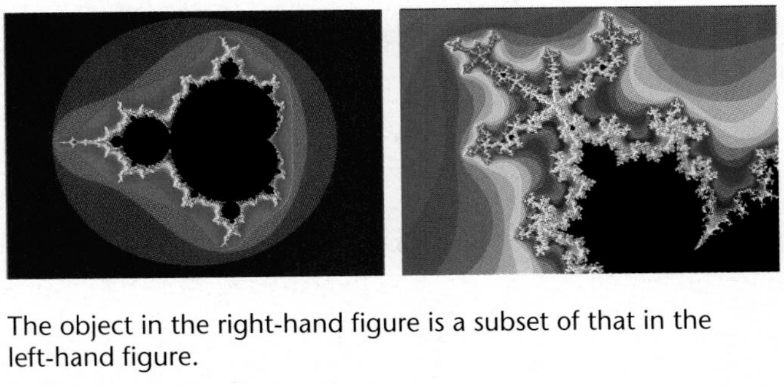

The object in the right-hand figure is a subset of that in the left-hand figure.

9.1 ANALYTIC GEOMETRY AND THE CONIC SECTIONS

Plane analytic geometry, a central topic of this chapter, is the use of algebra and calculus to study the properties of curves in the xy-plane. The ancient Greeks used deductive reasoning and the methods of axiomatic Euclidean geometry to study lines, circles, and the **conic sections** (parabolas, ellipses, and hyperbolas). The properties of conic sections have played an important role in diverse scientific applications since the seventeenth century, when Kepler discovered—and Newton explained—the fact that the orbits of planets and other bodies in the Solar System are conic sections.

The French mathematicians Descartes and Fermat, working almost independently of one another, initiated analytic geometry in 1637. The central idea of analytic geometry is the correspondence between an equation $F(x, y) = 0$ and its **locus** (typically, a curve), the set of all those points (x, y) in the plane with coordinates that satisfy this equation.

A central idea of analytic geometry is this: Given a geometric locus or curve, its properties can be derived algebraically or analytically from its defining equation $F(x, y) = 0$. For example, suppose that the equation of a given curve turns out to be the linear equation

$$Ax + By = C, \tag{1}$$

where A, B, and C are constants with $B \neq 0$. This equation may be written in the form

$$y = mx + b, \tag{2}$$

where $m = -A/B$ and $b = C/B$. But Eq. (2) is the slope-intercept equation of the straight line with slope m and y-intercept b. Hence the given curve is this straight line. We use this approach in Example 1 to show that a specific geometrically described locus is a particular straight line.

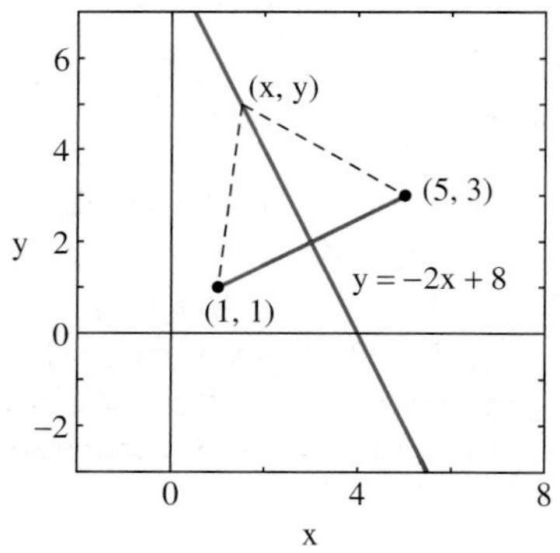

FIGURE 9.1.1 The perpendicular bisector of Example 1.

EXAMPLE 1 Prove that the set of all points equidistant from the points (1, 1) and (5, 3) is the perpendicular bisector of the line segment that joins these two points.

Solution The typical point $P(x, y)$ in Fig. 9.1.1 is equally distant from (1, 1) and (5, 3) if and only if

$$\begin{aligned}(x-1)^2 + (y-1)^2 &= (x-5)^2 + (y-3)^2;\\ x^2 - 2x + 1 + y^2 - 2y + 1 &= x^2 - 10x + 25 + y^2 - 6y + 9;\\ 2x + y &= 8;\\ y &= -2x + 8.\end{aligned} \tag{3}$$

Thus the given locus is the straight line in Eq. (3) whose slope is −2. The straight line through (1, 1) and (5, 3) has equation

$$y - 1 = \tfrac{1}{2}(x - 1) \tag{4}$$

and thus has slope $\frac{1}{2}$. Because the product of the slopes of these two lines is −1, it follows (from Theorem 2 in Appendix B) that these lines are perpendicular. If we solve Eqs. (3) and (4) simultaneously, we find that the intersection of these lines is, indeed, the midpoint (3, 2) of the given line segment. Thus the locus described is the perpendicular bisector of this line segment. ◆

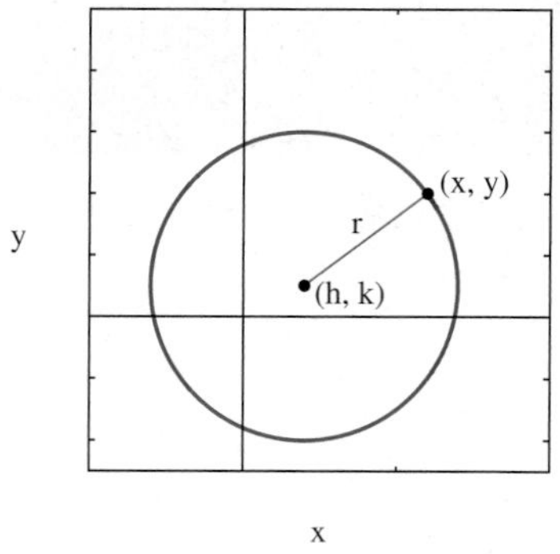

FIGURE 9.1.2 The circle with center (h, k) and radius r.

The circle shown in Fig. 9.1.2 has center (h, k) and radius r. It may be described geometrically as the set or locus of all points $P(x, y)$ whose distance from (h, k) is r.

The distance formula then gives

$$(x-h)^2+(y-k)^2=r^2 \tag{5}$$

as the equation of this circle. In particular, if $h = k = 0$, then Eq. (5) takes the simple form

$$x^2+y^2=r^2. \tag{6}$$

We can see directly from this equation, without further reference to the definition of *circle,* that a circle centered at the origin has the following symmetry properties:

- *Symmetry around the x-axis:* The equation of the curve is unchanged when y is replaced with $-y$.
- *Symmetry around the y-axis:* The equation of the curve is unchanged when x is replaced with $-x$.
- *Symmetry with respect to the origin:* The equation of the curve is unchanged when x is replaced with $-x$ and y is replaced with $-y$.
- *Symmetry around the* 45° *line $y = x$:* The equation is unchanged when x and y are interchanged.

The relationship between Eqs. (5) and (6) is an illustration of the *translation principle* stated informally in Section 1.2. Imagine a translation (or "slide") of the plane that moves each point (x, y) to the new position $(x + h, y + k)$. Under such a translation, a curve C is moved to a new position. The equation of the new translated curve is easy to obtain from the old equation—we simply replace x with $x - h$ and y with $y - k$. Conversely, we can recognize a translated circle from its equation: Any equation of the form

$$x^2+y^2+Ax+By+C=0 \tag{7}$$

can be rewritten in the form

$$(x-h)^2+(y-k)^2=p$$

by completing squares, as in Example 2 of Section 1.2. Thus the graph of Eq. (7) is either a circle (if $p > 0$), a single point (if $p = 0$), or no points at all (if $p < 0$). We use this approach in Example 2 to discover that the locus described is a particular circle.

EXAMPLE 2 Determine the locus of a point $P(x, y)$ if its distance $|AP|$ from $A(7, 1)$ is twice its distance $|BP|$ from $B(1, 4)$.

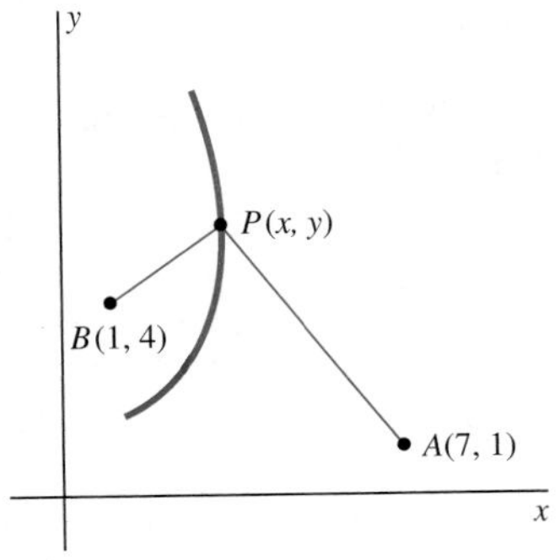

FIGURE 9.1.3 The locus of Example 2.

Solution The points A, B, and P appear in Fig. 9.1.3, along with a curve through P that represents the given locus. From

$$|AP|^2 = 4|BP|^2 \quad (\text{because } |AP| = 2|BP|),$$

we get the equation

$$(x-7)^2+(y-1)^2=4[(x-1)^2+(y-4)^2].$$

Hence

$$3x^2+3y^2+6x-30y+18=0;$$
$$x^2+y^2+2x-10y=-6;$$
$$(x+1)^2+(y-5)^2=20.$$

Thus the locus is a circle with center $(-1, 5)$ and radius $r = \sqrt{20} = 2\sqrt{5}$. ◆

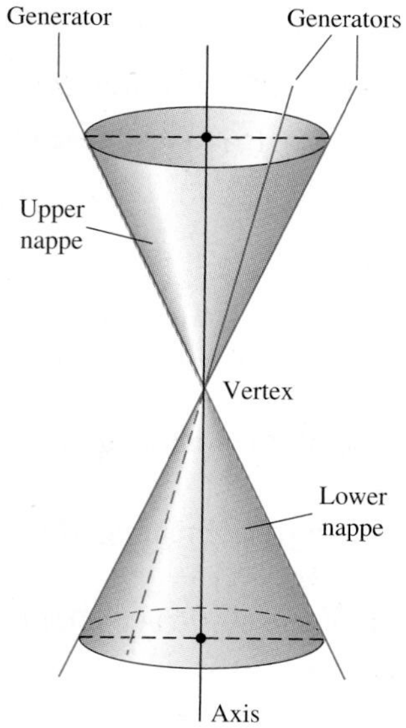

FIGURE 9.1.4 A cone with two nappes.

Conic Sections

Conic sections are so named because they are the curves formed by a plane intersecting a cone. The cone used is a right circular cone with two *nappes* extending infinitely far in both directions (Fig. 9.1.4). There are three types of conic sections, as illustrated in Fig. 9.1.5. If the cutting plane is parallel to some generator of the cone (a line that, when revolved around an axis, forms the cone), then the curve of intersection is a *parabola.* If the plane is not parallel to a generator, then the curve of intersection is either a single closed curve—an *ellipse*—or a *hyperbola* with two *branches.*

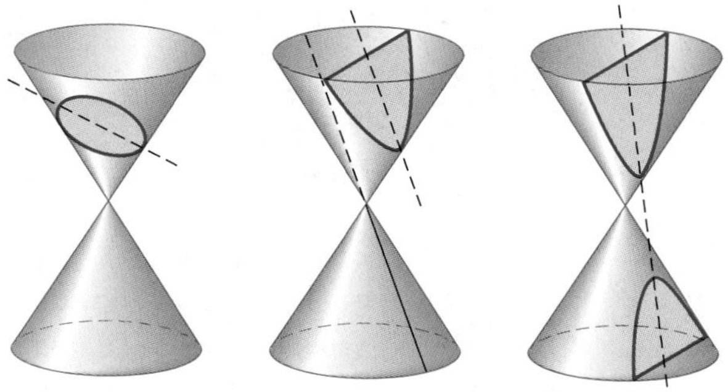

FIGURE 9.1.5 The conic sections.

In Appendix J we use the methods of three-dimensional analytic geometry to show that if an appropriate xy-coordinate system is set up in the intersecting plane, then the equations of the three conic sections take the following forms:

$$\text{Parabola:} \quad y^2 = kx; \tag{8}$$

$$\text{Ellipse:} \quad \frac{x^2}{a^2} + \frac{y^2}{b^2} = 1; \tag{9}$$

$$\text{Hyperbola:} \quad \frac{x^2}{a^2} - \frac{y^2}{b^2} = 1. \tag{10}$$

In Section 9.6 we discuss these conic sections on the basis of definitions that are two-dimensional—they do not require the three-dimensional setting of a cone and an intersecting plane. Example 3 illustrates one such approach to the conic sections.

EXAMPLE 3 Let e be a given positive number (*not* to be confused with the natural logarithm base; in the context of conic sections, e stands for *eccentricity*). Determine the locus of a point $P(x, y)$ if its distance from the fixed point $F(p, 0)$ is e times its distance from the vertical line L whose equation is $x = -p$ (Fig. 9.1.6).

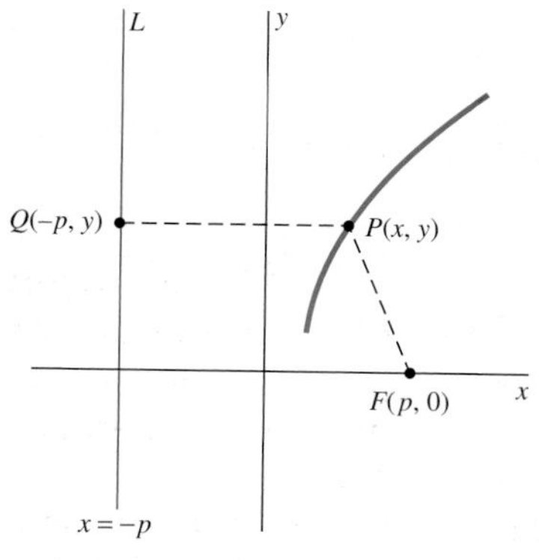

FIGURE 9.1.6 The locus of Example 3.

Solution Let PQ be the perpendicular from P to the line L. Then the condition

$$|PF| = e|PQ|$$

takes the analytic form

$$\sqrt{(x-p)^2 + y^2} = e|x - (-p)|.$$

That is,

$$(x^2 - 2px + p^2) + y^2 = e^2(x^2 + 2px + p^2),$$

so

$$x^2(1 - e^2) - 2p(1 + e^2)x + y^2 = -p^2(1 - e^2). \tag{11}$$

- *Case 1:* $e = 1$. Then Eq. (11) reduces to

$$y^2 = 4px. \tag{12}$$

We see upon comparison with Eq. (8) that the locus of P is a *parabola* if $e = 1$.

- *Case 2:* $e < 1$. Dividing both sides of Eq. (11) by $1 - e^2$, we get

$$x^2 - 2p \cdot \frac{1+e^2}{1-e^2}x + \frac{y^2}{1-e^2} = -p^2.$$

We now complete the square in x. The result is

$$\left(x - p \cdot \frac{1+e^2}{1-e^2}\right)^2 + \frac{y^2}{1-e^2} = p^2\left[\left(\frac{1+e^2}{1-e^2}\right)^2 - 1\right] = a^2.$$

This equation has the form

$$\frac{(x-h)^2}{a^2} + \frac{y^2}{b^2} = 1, \tag{13}$$

where

$$h = +p \cdot \frac{1+e^2}{1-e^2} \quad \text{and} \quad b^2 = a^2(1-e^2). \tag{14}$$

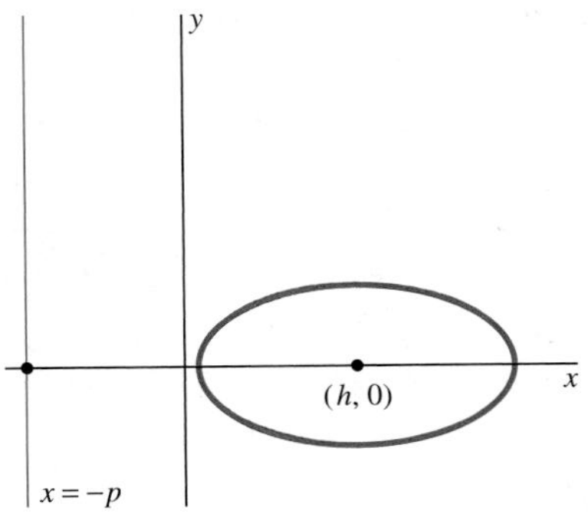

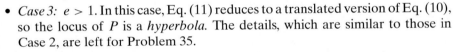

FIGURE 9.1.7 An ellipse: $e < 1$ (Example 3).

When we compare Eqs. (9) and (13), we see that if $e < 1$, then the locus of P is an *ellipse* with (0, 0) translated to $(h, 0)$, as illustrated in Fig. 9.1.7.

- *Case 3:* $e > 1$. In this case, Eq. (11) reduces to a translated version of Eq. (10), so the locus of P is a *hyperbola*. The details, which are similar to those in Case 2, are left for Problem 35.

Thus the locus in Example 3 is a *parabola* if $e = 1$, an *ellipse* if $e < 1$, and a *hyperbola* if $e > 1$. The number e is called the **eccentricity** of the conic section. The point $F(p, 0)$ is commonly called its **focus** in the parabolic case. Figure 9.1.8 shows the parabola of Case 1; Fig. 9.1.9 illustrates the hyperbola of Case 3. ◆

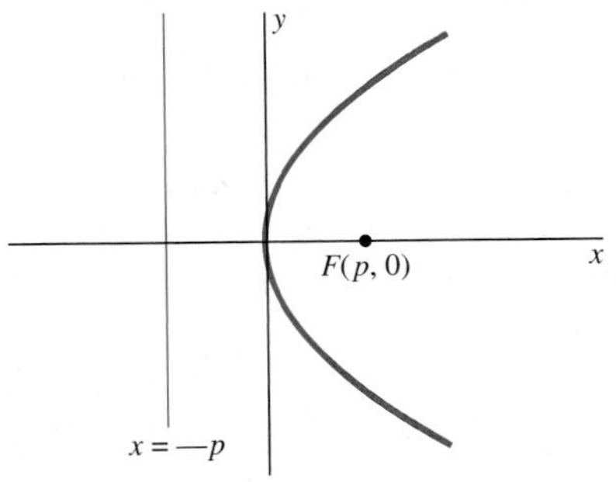

FIGURE 9.1.8 A parabola: $e = 1$ (Example 3).

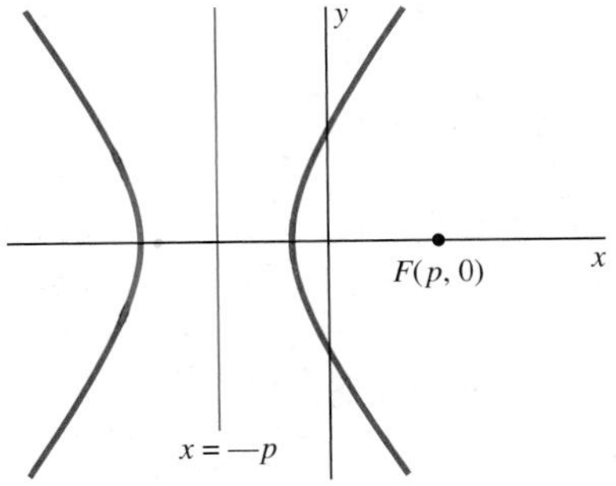

FIGURE 9.1.9 A hyperbola: $e > 1$ (Example 3).

If we begin with Eqs. (8) through (10), we can derive the general characteristics of the three conic sections shown in Figs. 9.1.7 through 9.1.9. For example, in the case of the parabola of Eq. (8) with $k < 0$, the curve passes through the origin, $x \geqq 0$ at each of the curve's points, $y \to \pm\infty$ as $x \to \infty$, and the graph is symmetric around the x-axis (because the curve is unchanged when y is replaced with $-y$).

In the case of the ellipse of Eq. (9), the graph must be symmetric around both coordinate axes. At each point (x, y) of the graph, we must have $|x| \leqq a$ and $|y| \leqq b$. The graph intersects the axes at the four points $(\pm a, 0)$ and $(0, \pm b)$.

Finally, the hyperbola of Eq. (10)—or its alternative form

$$y = \pm\frac{b}{a}\sqrt{x^2 - a^2}$$

—is symmetric around both coordinate axes. Its meets the x-axis at the two points $(\pm a, 0)$ and has one branch consisting of points with $x \geqq a$ and has another branch where $x \leqq -a$. Also, $|y| \to \infty$ as $|x| \to \infty$.

9.1 TRUE/FALSE STUDY GUIDE

9.1 CONCEPTS: QUESTIONS AND DISCUSSION

You may want to use the implicit plotting facility of a computer algebra system to investigate the following questions.

1. The graph of the equation $x^2 - y^2 = 0$ consists of the two lines $x - y = 0$ and $x + y = 0$ through the origin. What is the graph of the equation $x^n - y^n = 0$? Does it depend on whether the positive integer n is even or odd? Explain your answers.
2. How do the graphs of the equations $x^3 + y^3 = 1$ and $x^4 + y^4 = 1$ differ from the unit circle $x^2 + y^2 = 1$ (and from each other)? How does the graph of the equation $x^n + y^n = 1$ change as the positive integer n gets larger and larger? Discuss the possibility of a "limiting set" as $n \to +\infty$. Do these questions depend on whether n is even or odd?
3. The graph of the equation $x^2 - y^2 = 1$ is a hyperbola. Discuss (as in Question 2) the graph of the equation $x^n - y^n = 1$.

9.1 PROBLEMS

In Problems 1 through 6, write an equation of the specified straight line.

1. The line through the point $(1, -2)$ that is parallel to the line with equation $x + 2y = 5$

2. The line through the point $(-3, 2)$ that is perpendicular to the line with equation $3x - 4y = 7$

3. The line that is tangent to the circle $x^2 + y^2 = 25$ at the point $(3, -4)$

4. The line that is tangent to the curve $y^2 = x + 3$ at the point $(6, -3)$

5. The line that is perpendicular to the curve $x^2 + 2y^2 = 6$ at the point $(2, -1)$

6. The perpendicular bisector of the line segment with endpoints $(-3, 2)$ and $(5, -4)$

In Problems 7 through 16, find the center and radius of the circle described in the given equation.

7. $x^2 + 2x + y^2 = 4$

8. $x^2 + y^2 - 4y = 5$

9. $x^2 + y^2 - 4x + 6y = 3$

10. $x^2 + y^2 + 8x - 6y = 0$

11. $4x^2 + 4y^2 - 4x = 3$

12. $4x^2 + 4y^2 + 12y = 7$

13. $2x^2 + 2y^2 - 2x + 6y = 13$

14. $9x^2 + 9y^2 - 12x = 5$

15. $9x^2 + 9y^2 + 6x - 24y = 19$

16. $36x^2 + 36y^2 - 48x - 108y = 47$

In Problems 17 through 20, show that the graph of the given equation consists either of a single point or of no points.

17. $x^2 + y^2 - 6x - 4y + 13 = 0$

18. $2x^2 + 2y^2 + 6x + 2y + 5 = 0$

19. $x^2 + y^2 - 6x - 10y + 84 = 0$

20. $9x^2 + 9y^2 - 6x - 6y + 11 = 0$

In Problems 21 through 24, write the equation of the specified circle.

21. The circle with center $(-1, -2)$ that passes through the point $(2, 3)$

22. The circle with center $(2, -2)$ that is tangent to the line $y = x + 4$

23. The circle with center $(6, 6)$ that is tangent to the line $y = 2x - 4$

24. The circle that passes through the points $(4, 6)$, $(-2, -2)$, and $(5, -1)$

In Problems 25 through 30, derive the equation of the set of all points $P(x, y)$ that satisfy the given condition. Then sketch the graph of the equation.

25. The point $P(x, y)$ is equally distant from the two points $(3, 2)$ and $(7, 4)$.

26. The distance from P to the point $(-2, 1)$ is half the distance from P to the point $(4, -2)$.

27. The point P is three times as far from the point $(-3, 2)$ as it is from the point $(5, 10)$.

28. The distance from P to the line $x = -3$ is equal to its distance from the point $(3, 0)$.

29. The sum of the distances from P to the points $(4, 0)$ and $(-4, 0)$ is 10.

30. The sum of the distances from P to the points $(0, 3)$ and $(0, -3)$ is 10.

31. Find all the lines through the point $(2, 1)$ that are tangent to the parabola $y = x^2$.

32. Find all lines through the point $(-1, 2)$ that are normal to the parabola $y = x^2$.

33. Find all lines that are normal to the curve $xy = 4$ and simultaneously are parallel to the line $y = 4x$.

34. Find all lines that are tangent to the curve $y = x^3$ and are also parallel to the line $3x - y = 5$.

35. Suppose that $e > 1$. Show that Eq. (11) of this section can be written in the form

$$\frac{(x-h)^2}{a^2} - \frac{y^2}{b^2} = 1,$$

thus showing that its graph is a hyperbola. Find a, b, and h in terms of p and e.

9.2 POLAR COORDINATES

A familiar way to locate a point in the coordinate plane is by specifying its rectangular coordinates (x, y)—that is, by giving its abscissa x and ordinate y relative to given perpendicular axes. In some problems it is more convenient to locate a point by means of its *polar coordinates.* The polar coordinates give its position relative to a fixed reference point O (the **pole**) and to a given ray (the **polar axis**) beginning at O.

For convenience, we begin with a given xy-coordinate system and then take the origin as the pole and the nonnegative x-axis as the polar axis. Given the pole O and the polar axis, the point P with **polar coordinates** r and θ, written as the ordered pair (r, θ), is located as follows. First find the terminal side of the angle θ, given in radians, where θ is measured counterclockwise (if $\theta > 0$) from the x-axis (the polar axis) as its initial side. If $r \geqq 0$, then P is on the terminal side of this angle at the distance r from the origin. If $r < 0$, then P lies on the ray opposite the terminal side at the distance $|r| = -r > 0$ from the pole (Fig. 9.2.1). The **radial coordinate** r can be described as the *directed* distance of P from the pole along the terminal side of the angle θ. Thus, if r is positive, the point P lies in the same quadrant as θ, whereas if r is negative, then P lies in the opposite quadrant. If $r = 0$, the angle θ does not matter; the polar coordinates $(0, \theta)$ represent the origin whatever the **angular coordinate** θ might be. The origin, or pole, is the only point for which $r = 0$.

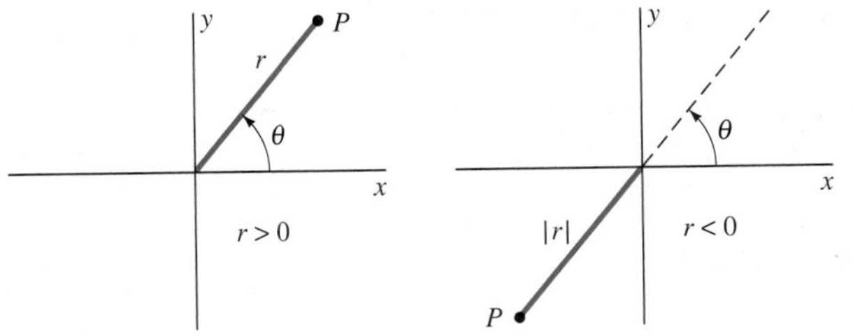

FIGURE 9.2.1 The difference between the two cases $r > 0$ and $r < 0$.

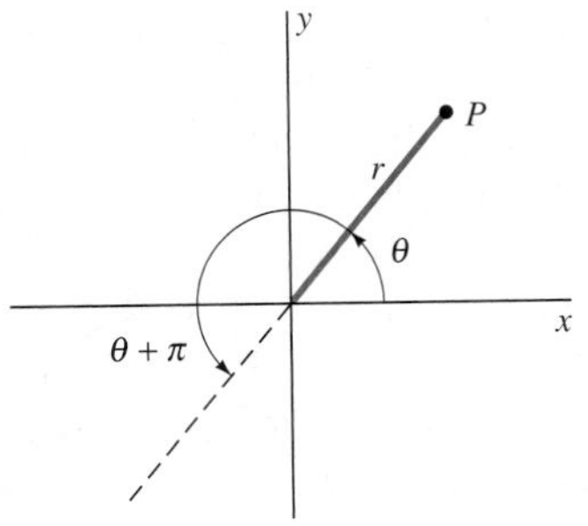

FIGURE 9.2.2 The polar coordinates (r, θ) and $(-r, \theta + \pi)$ represent the same point P (Example 1).

EXAMPLE 1 Polar coordinates differ from rectangular coordinates in that any point has more than one representation in polar coordinates. For example, the polar coordinates (r, θ) and $(-r, \theta + \pi)$ represent the same point P, as shown in Fig. 9.2.2. More generally, this point P has the polar coordinates $(r, \theta + n\pi)$ for any even integer n *and* the coordinates $(-r, \theta + n\pi)$ for any odd integer n. Thus the polar-coordinate pairs

$$\left(2, \frac{\pi}{3}\right), \quad \left(-2, \frac{4\pi}{3}\right), \quad \left(2, \frac{7\pi}{3}\right), \quad \text{and} \quad \left(-2, -\frac{2\pi}{3}\right)$$

all represent the same point P in Fig. 9.2.3. (The rectangular coordinates of P are $(1, \sqrt{3})$.) ◆

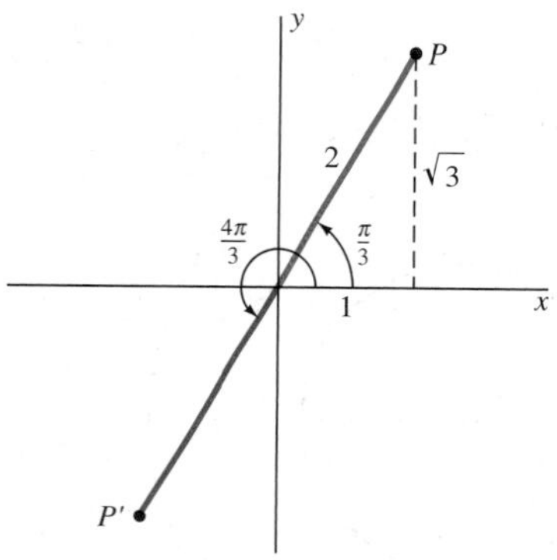

FIGURE 9.2.3 The point P of Example 1 can be described in many different ways using polar coordinates.

To convert polar coordinates into rectangular coordinates, we use the basic relations

$$x = r\cos\theta, \quad y = r\sin\theta \tag{1}$$

that we read from the right triangle in Fig. 9.2.4. Converting in the opposite direction, we have

$$r^2 = x^2 + y^2, \qquad \tan\theta = \frac{y}{x} \quad \text{if } x \neq 0. \tag{2}$$

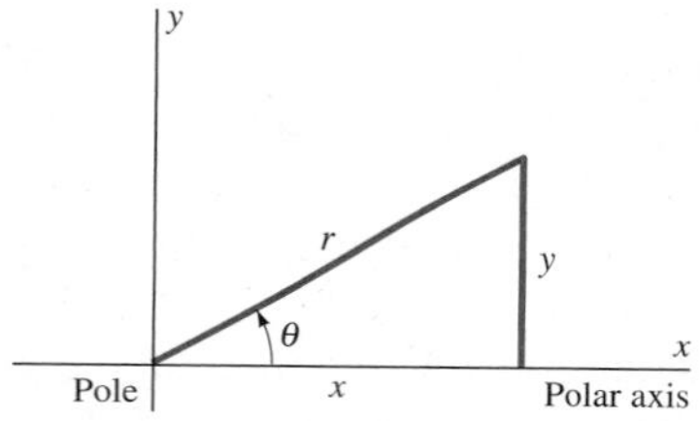

FIGURE 9.2.4 Read Eqs. (1) and (2)—conversions between polar and rectangular coordinates—from this figure.

Some care is required in making the correct choice of θ in the formula $\tan\theta = y/x$. If $x > 0$, then (x, y) lies in either the first or fourth quadrant, so $-\pi/2 < \theta < \pi/2$, which is the range of the inverse tangent function. Hence if $x > 0$, then $\theta = \arctan(y/x)$. But if $x < 0$, then (x, y) lies in the second or third quadrant. In this case a proper choice for the angle is $\theta = \pi + \arctan(y/x)$. In any event, the signs of x and y in Eq. (1) with $r > 0$ indicate the quadrant in which θ lies.

Polar Coordinate Equations

Some curves have simpler equations in polar coordinates than in rectangular coordinates; this is an important reason for the usefulness of polar coordinates. The **graph** of an equation in the polar-coordinate variables r and θ is the set of all those points P such that P has some pair of polar coordinates (r, θ) that satisfy the given equation. The graph of a polar equation $r = f(\theta)$ can be constructed by computing a table of values of r against θ and then plotting the corresponding points (r, θ) on polar-coordinate graph paper.

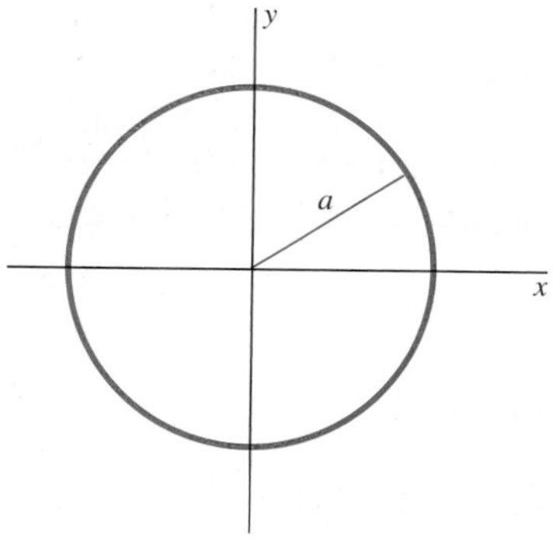

FIGURE 9.2.5 The circle $r = a$ centered at the origin (Example 2).

EXAMPLE 2 One reason for the importance of polar coordinates is that many real-world problems involve circles, and the polar-coordinate equation (or *polar equation*) of the circle with center $(0, 0)$ and radius $a > 0$ (Fig. 9.2.5) is very simple:

$$r = a. \tag{3}$$

Note that if we begin with the rectangular-coordinates equation $x^2 + y^2 = a^2$ of this circle and transform it using the first relation in (2), we get the polar-coordinate equation $r^2 = a^2$. Then Eq. (3) results upon taking positive square roots. ◆

EXAMPLE 3 Construct the polar-coordinate graph of the equation $r = 2\sin\theta$.

Solution Figure 9.2.6 shows a table of values of r as a function of θ. The corresponding points (r, θ) are plotted in Fig. 9.2.7, using the rays at multiples of $\pi/6$ and the circles (centered at the pole) of radii 1 and 2 to locate these points. A visual inspection of the smooth curve connecting these points suggests that it is a circle of radius 1. Let us assume for the moment that this is so. Note then that the point $P(r, \theta)$ moves *once around this circle counterclockwise* as θ increases from 0 to π and then moves around this circle a *second time* as θ increases from π to 2π. This is because the negative values of r for θ between π and 2π give—in this example—the same geometric points as do the positive values of r for θ between 0 and π. (Why?) ◆

The verification that the graph of $r = 2\sin\theta$ is the indicated circle illustrates the general procedure for transferring back and forth between polar and rectangular coordinates, using the relations in (1) and (2).

θ	r
0	0.00
$\pi/6$	1.00
$\pi/3$	1.73
$\pi/2$	2.00
$2\pi/3$	1.73
$5\pi/6$	1.00
π	0.00
$7\pi/6$	−1.00
$4\pi/3$	−1.73
$3\pi/2$	−2.00
$5\pi/3$	−1.73
$11\pi/6$	−1.00
2π	0.00
	(data rounded)

FIGURE 9.2.6 Values of $r = 2\sin\theta$ (Example 3).

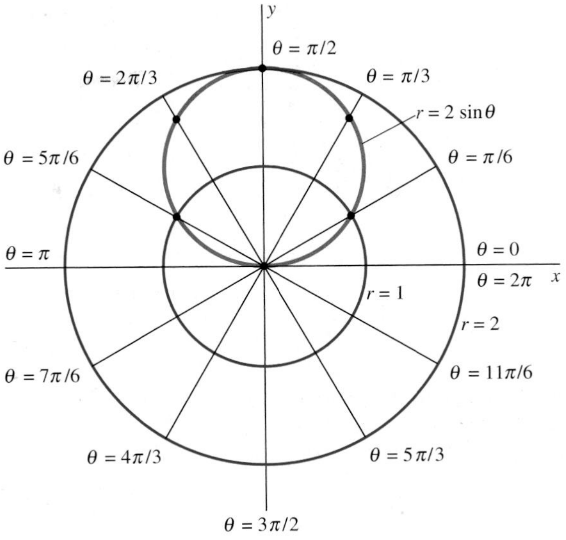

FIGURE 9.2.7 The graph of the polar equation $r = 2\sin\theta$ (Example 3).

EXAMPLE 4 To transform the equation $r = 2\sin\theta$ of Example 3 into rectangular coordinates, we first multiply both sides by r to get

$$r^2 = 2r\sin\theta.$$

Equations (1) and (2) now give

$$x^2 + y^2 = 2y.$$

Finally, after we complete the square in y, we have

$$x^2 + (y-1)^2 = 1,$$

the rectangular-coordinate equation (or *rectangular equation*) of a circle whose center is (0, 1) and whose radius is 1. ◆

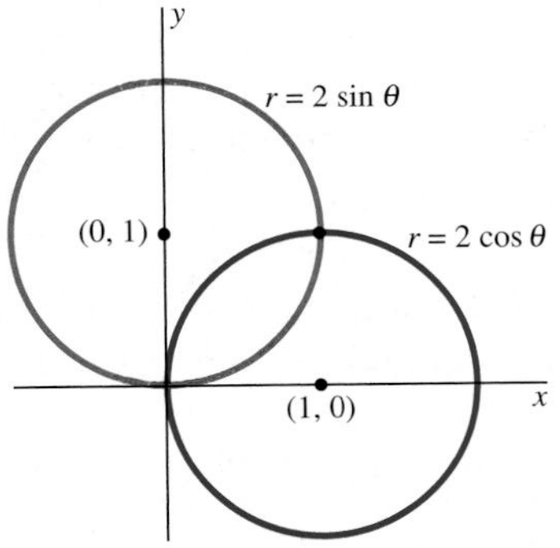

FIGURE 9.2.8 The graphs of the circles whose equations appear in Eq. (4) with $a = 1$.

More generally, the graphs of the equations

$$r = 2a\sin\theta \quad \text{and} \quad r = 2a\cos\theta \tag{4}$$

are circles of radius a centered, respectively, at the points $(0, a)$ and $(a, 0)$. This is illustrated (with $a = 1$) in Fig. 9.2.8.

By substituting the equations given in (1), we can transform the rectangular equation $ax + by = c$ of a straight line into

$$ar\cos\theta + br\sin\theta = c.$$

Let us take $a = 1$ and $b = 0$. Then we see that the polar equation of the vertical line $x = c$ is $r = c\sec\theta$, as we can deduce directly from Fig. 9.2.9.

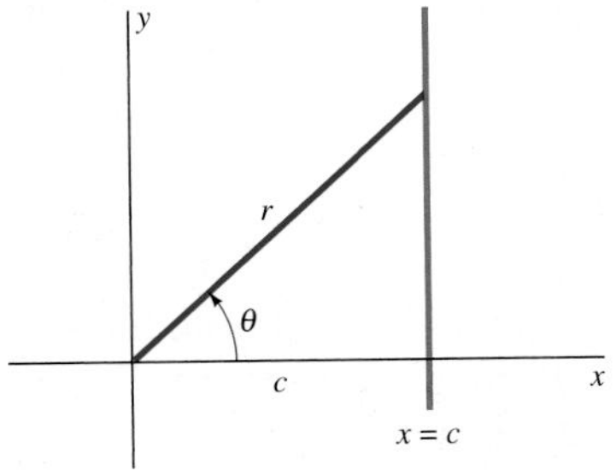

FIGURE 9.2.9 Finding the polar equation of the vertical line $x = c$.

EXAMPLE 5 Sketch the graph of the polar equation $r = 2 + 2\sin\theta$.

Solution If we scan the second column of the table in Fig. 9.2.6, mentally adding 2 to each entry for r, we see that

- r increases from 2 to 4 as θ increases from 0 to $\pi/2$;
- r decreases from 4 to 2 as θ increases from $\pi/2$ to π;
- r decreases from 2 to 0 as θ increases from π to $3\pi/2$;
- r increases from 0 to 2 as θ increases from $3\pi/2$ to 2π.

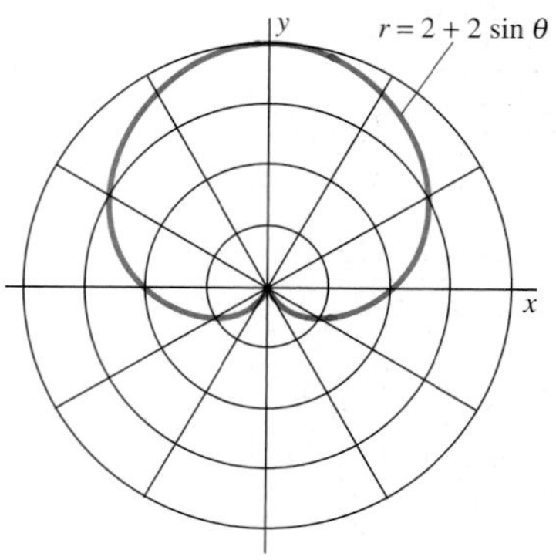

FIGURE 9.2.10 A cardioid (Example 5).

This information tells us that the graph resembles the curve shown in Fig. 9.2.10. This heart-shaped graph is called a **cardioid.** The graphs of the equations

$$r = a(1 \pm \sin\theta) \quad \text{and} \quad r = a(1 \pm \cos\theta)$$

are all cardioids, differing only in size (determined by a), axis of symmetry (horizontal or vertical), and the direction in which the cusp at the pole points. ◆

EXAMPLE 6 Sketch the graph of the equation $r = 2\cos 2\theta$.

Solution Rather than constructing a table of values of r as a function of θ and then plotting individual points, let us begin with a *rectangular-coordinate graph* of r as a function of θ. In Fig. 9.2.11, we see that $r = 0$ if θ is an odd integral multiple of $\pi/4$, and that r is alternately positive and negative on successive intervals of length $\pi/2$ from one odd integral multiple of $\pi/4$ to the next.

Now let's think about how r changes as θ increases, beginning at $\theta = 0$. As θ increases from 0 to $\pi/4$, r decreases in value from 2 to 0, and so we draw the first portion (labeled "1") of the polar curve in Fig. 9.2.12. As θ increases from $\pi/4$ to $3\pi/4$, r first decreases from 0 to -2 and then increases from -2 to 0. Because r is now negative, we draw the second and third portions (labeled "2" and "3") of the polar curve in the third and fourth quadrants (rather than in the first and second quadrants) in Fig. 9.2.12. Continuing in this fashion, we draw the fourth through eighth portions of the polar curve, with those portions where r is negative in the quadrants opposite those in which θ lies. The arrows on the resulting polar curve in Fig. 9.2.12 indicate the direction of motion of the point $P(r, \theta)$ along the curve as θ increases. The whole graph consists of four loops, each of which begins and ends at the pole. ◆

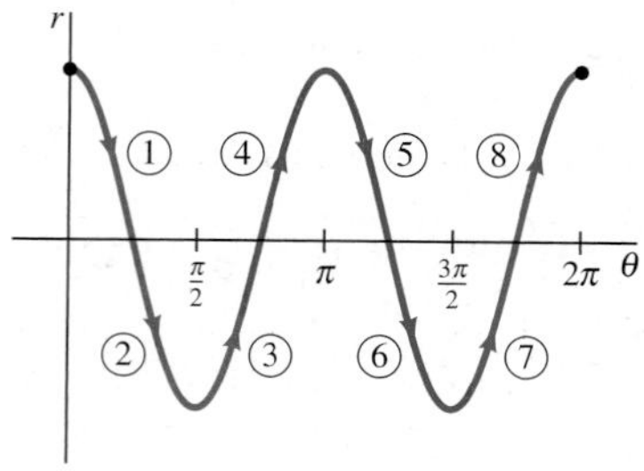

FIGURE 9.2.11 The rectangular coordinates graph of $r = 2\cos 2\theta$ as a function of θ. Numbered portions of the graph correspond to numbered portions of the polar-coordinates graph in Fig. 9.2.12.

The curve in Example 6 is called a *four-leaved rose.* The equations $r = a\cos n\theta$ and $r = a\sin n\theta$ represent "roses" with $2n$ "leaves," or loops, if n is even and $n \geqq 2$ but with n loops if n is odd and $n \geqq 3$.

The four-leaved rose exhibits several types of symmetry. The following are some *sufficient* conditions for symmetry in polar coordinates:

- *For symmetry around the x-axis:* The equation is unchanged when θ is replaced with $-\theta$.
- *For symmetry around the y-axis:* The equation is unchanged when θ is replaced with $\pi - \theta$.
- *For symmetry with respect to the origin:* The equation is unchanged when r is replaced with $-r$.

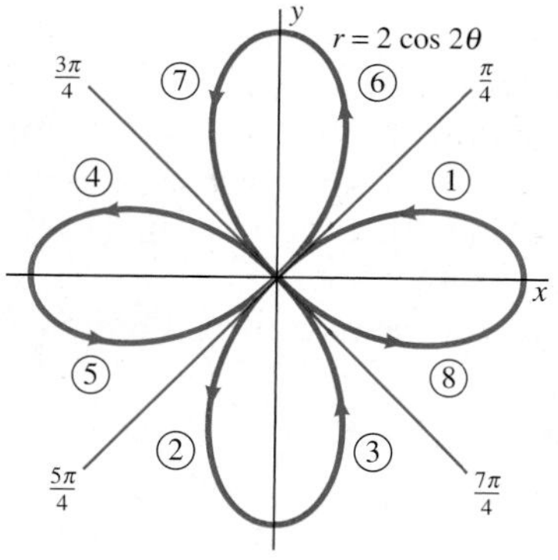

FIGURE 9.2.12 A four-leaved rose (Example 6).

Because $\cos 2\theta = \cos(-2\theta) = \cos 2(\pi - \theta)$, the equation $r = 2\cos 2\theta$ of the four-leaved rose satisfies the first two symmetry conditions, and therefore its graph is symmetric around both the x-axis and the y-axis. Thus it is also symmetric around the origin. Nevertheless, this equation does *not* satisfy the third condition, the one for symmetry around the origin. This illustrates that although the symmetry conditions given are *sufficient* for the symmetries described, they are not *necessary* conditions.

EXAMPLE 7 Figure 9.2.13 shows the lemniscate with equation

$$r^2 = -4\sin 2\theta.$$

To see why it has loops only in the second and fourth quadrants, we examine a table of signs of values of $-4\sin 2\theta$.

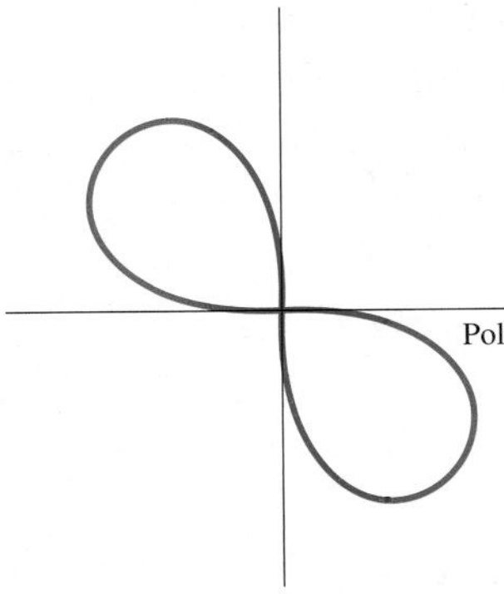

FIGURE 9.2.13 The lemniscate $r^2 = -4\sin 2\theta$ (Example 7).

θ	2θ	$-4\sin 2\theta$
$0 < \theta < \frac{1}{2}\pi$	$0 < 2\theta < \pi$	Negative
$\frac{1}{2}\pi < \theta < \pi$	$\pi < 2\theta < 2\pi$	Positive
$\pi < \theta < \frac{3}{2}\pi$	$2\pi < 2\theta < 3\pi$	Negative
$\frac{3}{2}\pi < \theta < 2\pi$	$3\pi < 2\theta < 4\pi$	Positive

When θ lies in the first or the third quadrant, the quantity $-4\sin 2\theta$ is negative, so the equation $r^2 = -4\sin 2\theta$ cannot be satisfied for any real values of r. ◆

Example 6 illustrates a peculiarity of graphs of polar equations, caused by the fact that a single point has multiple representations in polar coordinates. The point with polar coordinates $(2, \pi/2)$ clearly lies on the four-leaved rose, but these coordinates do *not* satisfy the equation $r = 2\cos 2\theta$. This means that a point may have one pair of polar coordinates that satisfy a given equation and others that do not. Hence we must be careful to understand this: The graph of a polar equation consists of all those points with *at least one* polar-coordinate representation that satisfies the given equation.

Another result of the multiplicity of polar coordinates is that the simultaneous solution of two polar equations does not always give all the points of intersection of their graphs. For instance, consider the circles $r = 2\sin\theta$ and $r = 2\cos\theta$ shown in Fig. 9.2.8. The origin is clearly a point of intersection of these two circles. Its polar representation $(0, \pi)$ satisfies the equation $r = 2\sin\theta$, and its representation $(0, \pi/2)$ satisfies the other equation, $r = 2\cos\theta$. But the origin has no *single* polar representation that satisfies both equations simultaneously! If we think of θ as increasing uniformly with time, then the corresponding moving points on the two circles pass through the origin at different times. Hence the origin cannot be discovered as a point of intersection of the two circles merely by solving their equations $r = 2\sin\theta$ and $r = 2\cos\theta$ simultaneously in a straightforward manner. But one fail-safe way to find *all* points of intersection of two polar-coordinate curves is to graph both curves.

EXAMPLE 8 Find all points of intersection of the graphs of the equations $r = 1 + \sin\theta$ and $r^2 = 4\sin\theta$.

Solution The graph of $r = 1 + \sin\theta$ is a scaled-down version of the cardioid of Example 5. In Problem 52 we ask you to show that the graph of $r^2 = 4\sin\theta$ is the figure-eight curve shown with the cardioid in Fig. 9.2.14. The figure shows four points of intersection: A, B, C, and O. Can we find all four using algebra?

Given the two equations, we begin by eliminating r. Because

$$(1 + \sin\theta)^2 = r^2 = 4\sin\theta,$$

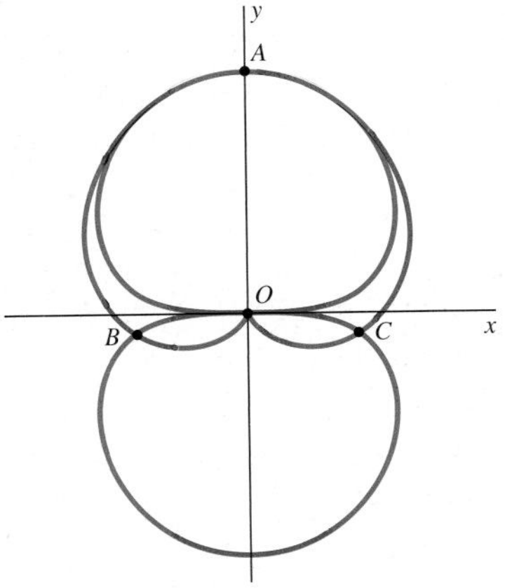

FIGURE 9.2.14 The cardioid $r = 1 + \sin\theta$ and the figure eight $r^2 = 4\sin\theta$ meet in four points (Example 8).

it follows that

$$\sin^2\theta - 2\sin\theta + 1 = 0;$$
$$(\sin\theta - 1)^2 = 0;$$

and thus that $\sin\theta = 1$. So θ must be an angle of the form $\frac{1}{2}\pi + 2n\pi$ where n is an integer. All points on the cardioid and all points on the figure-eight curve are produced by letting θ range from 0 to 2π, so $\theta = \pi/2$ will produce all the solutions that we can obtain by simple algebraic elimination. The only such point is $A(2, \pi/2)$, and the other three points of intersection are detected only when the two equations are graphed. ◆

Calculator/Computer-Generated Polar Curves

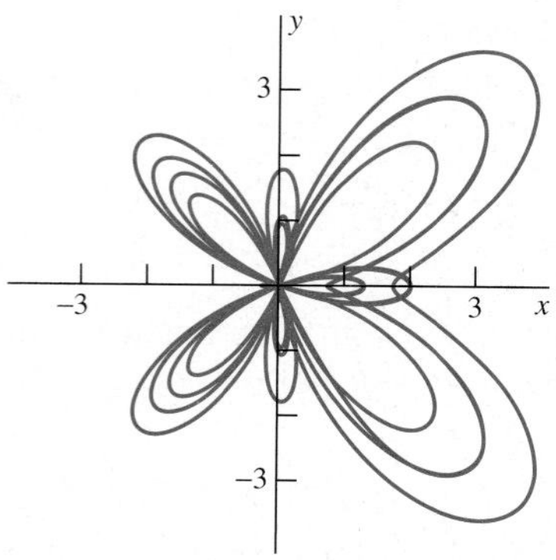

FIGURE 9.2.15 $r = e^{\cos\theta} - 2\cos 4\theta + \sin^3(\theta/4)$.

It might take you quite a while to construct by hand the "butterfly curve" shown in Fig. 9.2.15. But most graphing calculators and computer algebra systems have facilities for plotting polar curves. For instance, with a TI calculator set in "polar graph mode," one need only enter and graph the equation

```
r = e∧(cos(θ)) - 2*cos(4θ) + sin(θ/4)∧3
```

on the interval $0 \leqq \theta \leqq 8\pi$. With *Maple* and *Mathematica* the graphics package commands

```
polarplot(exp(cos(t)) - 2*cos(4*t) + sin(t/4)∧3, t=0..8*Pi);
```

and

```
PolarPlot[ Exp[Cos[t]] - 2*Cos[4*t] + Sin[t/4]∧3, { t, 0, 8*Pi } ];
```

(respectively) give the same result (with t in place of θ).

Because of the presence of the term $\sin^3(\theta/4)$, the more usual interval $0 \leqq \theta \leqq 2\pi$ gives only a part of the curve shown in Fig. 9.2.15. (Try it to see for yourself.) But

$$\sin^3\left(\frac{\theta + 8\pi}{4}\right) = \sin^3\left(\frac{\theta}{4} + 2\pi\right) = \sin^3\left(\frac{\theta}{4}\right),$$

so values of $\sin^3(\theta/4)$ repeat themselves when θ exceeds 8π. Therefore the interval $0 \leqq \theta \leqq 8\pi$ suffices to give the entire butterfly curve. You might try plotting a butterfly curve with the term $\sin^3(\theta/4)$ replaced with $\sin^5(\theta/12)$—as originally recommended by Temple H. Fay in his article "The Butterfly Curve" (*American Mathematical Monthly,* May 1989, p. 442). What range of values of θ will now be required to obtain the whole butterfly?

9.2 TRUE/FALSE STUDY GUIDE

9.2 CONCEPTS: QUESTIONS AND DISCUSSION

1. Figures 9.2.16 through 9.2.18 illustrate the polar curve $r = a + b\cos\theta$ for various values of a and b. What determines whether the curve exhibits a cusp (Fig. 9.2.16), a loop (Fig. 9.2.17), or neither (Fig. 9.2.18)? Does your answer apply also to polar curves of the form $r = a + b\sin\theta$? Given a and b, what is the difference between the curves $r = a + b\cos\theta$ and $r = a + b\sin\theta$?

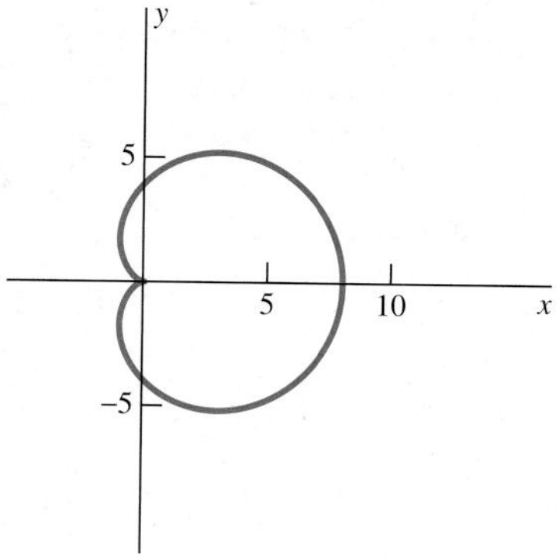

FIGURE 9.2.16 $r = 4 + 4\cos\theta$.

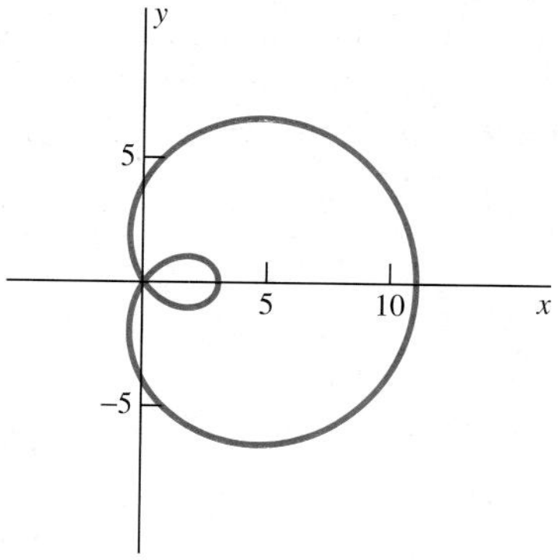

FIGURE 9.2.17 $r = 4 + 7\cos\theta$.

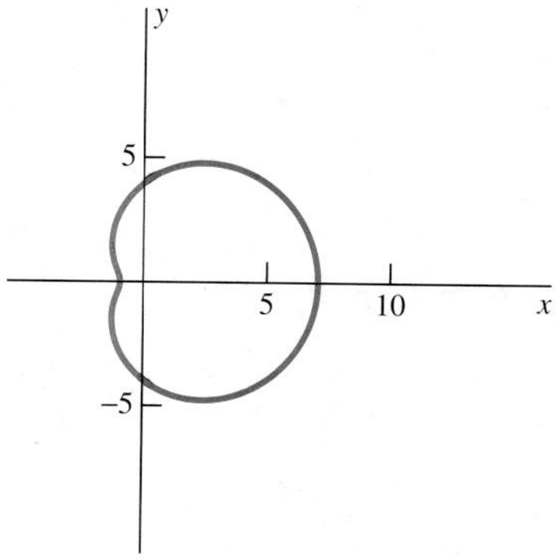

FIGURE 9.2.18 $r = 4 + 3\cos\theta$.

2. Figures 9.2.19 and 9.2.20 show the graphs of the equations $r = \cos 3\theta$ and $r = \sin 4\theta$. Given a positive integer n, what is the difference between the "rose graphs" $r = \cos n\theta$ and $r = \sin n\theta$? Explain precisely how the number of leaves in the complete graph depends on n. What determines whether $0 \leqq \theta \leqq \pi$ or $0 \leqq \theta \leqq 2\pi$ gives all the leaves?

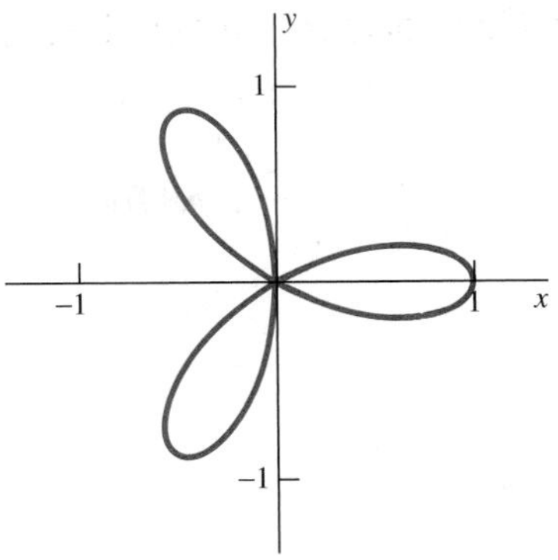

FIGURE 9.2.19 $r = \cos 3\theta$.

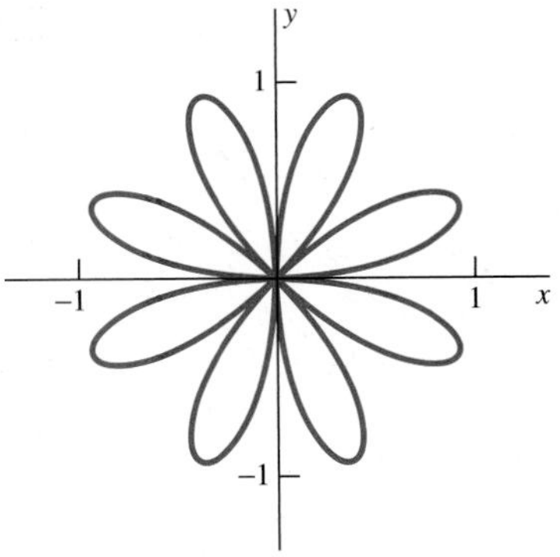

FIGURE 9.2.20 $r = \sin 4\theta$.

9.2 PROBLEMS

1. Plot the points with the given polar coordinates, and then find the rectangular coordinates of each.

(a) $(1, \pi/4)$ (b) $(-2, 2\pi/3)$ (c) $(1, -\pi/3)$
(d) $(3, 3\pi/2)$ (e) $(2, -\pi/4)$ (f) $(-2, -7\pi/6)$
(g) $(2, 5\pi/6)$

2. Find two polar-coordinate representations, one with $r > 0$ and the other with $r < 0$, for the points with the given rectangular coordinates.

(a) $(-1, -1)$ (b) $(\sqrt{3}, -1)$ (c) $(2, 2)$
(d) $(-1, \sqrt{3})$ (e) $(\sqrt{2}, -\sqrt{2})$ (f) $(-3, \sqrt{3})$

In Problems 3 through 10, express the given rectangular equation in polar form.

3. $x = 4$ **4.** $y = 6$
5. $x = 3y$ **6.** $x^2 + y^2 = 25$
7. $xy = 1$ **8.** $x^2 - y^2 = 1$
9. $y = x^2$ **10.** $x + y = 4$

In Problems 11 through 18, express the given polar equation in rectangular form.

11. $r = 3$ **12.** $\theta = 3\pi/4$
13. $r = -5\cos\theta$ **14.** $r = \sin 2\theta$
15. $r = 1 - \cos 2\theta$ **16.** $r = 2 + \sin\theta$
17. $r = 3\sec\theta$ **18.** $r^2 = \cos 2\theta$

For the curves described in Problems 19 through 28, write equations in both rectangular and polar form.

19. The vertical line through $(2, 0)$

20. The horizontal line through $(1, 3)$

21. The line with slope -1 through $(2, -1)$

22. The line with slope 1 through $(4, 2)$

23. The line through the points $(1, 3)$ and $(3, 5)$

24. The circle with center $(3, 0)$ that passes through the origin

25. The circle with center $(0, -4)$ that passes through the origin

26. The circle with center $(3, 4)$ and radius 5

27. The circle with center $(1, 1)$ that passes through the origin

28. The circle with center $(5, -2)$ that passes through the point $(1, 1)$

In Problems 29 through 32, transform the given polar-coordinate equation into a rectangular-coordinate equation, then match the equation with its graph among those in Figs. 9.2.21 through 9.2.24.

29. $r = -4\cos\theta$ **30.** $r = 5\cos\theta + 5\sin\theta$
31. $r = -4\cos\theta + 3\sin\theta$ **32.** $r = 8\cos\theta - 15\sin\theta$

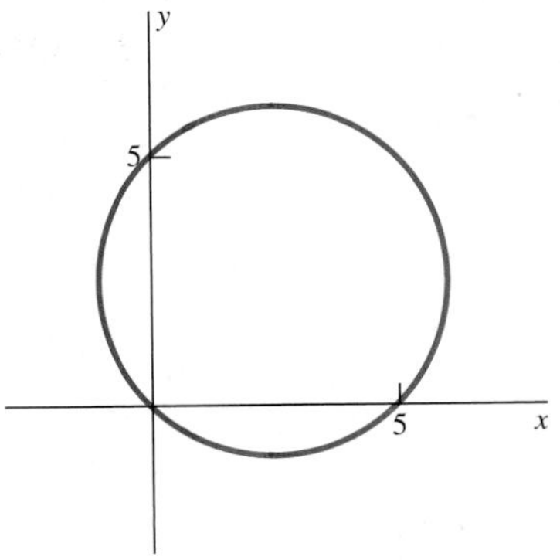

FIGURE 9.2.21

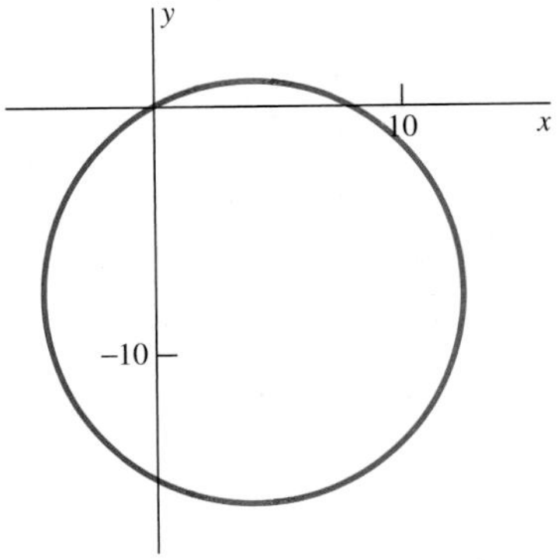

FIGURE 9.2.22

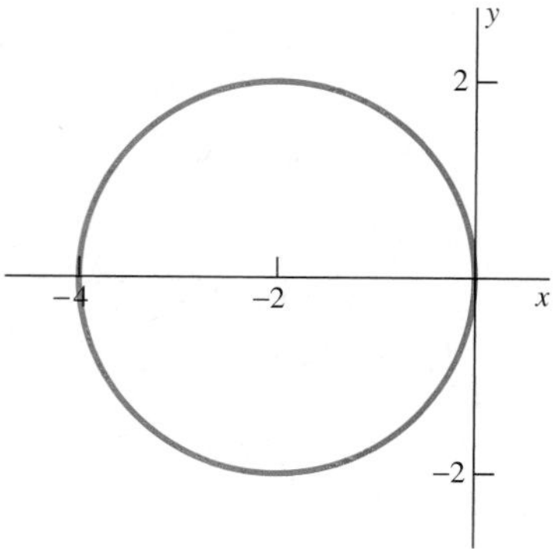

FIGURE 9.2.23

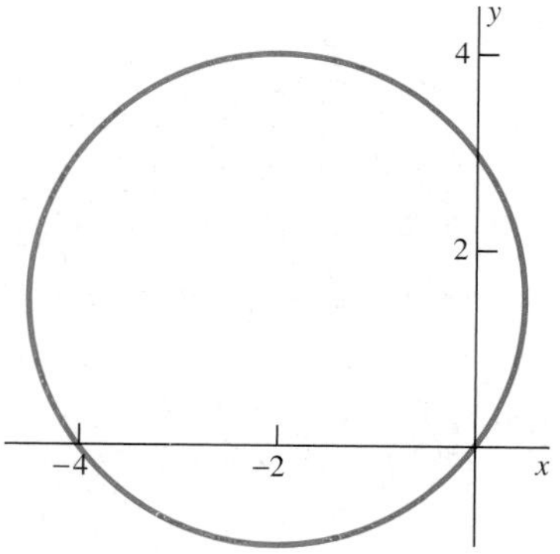

FIGURE 9.2.24

The graph of a polar equation of the form $r = a + b\cos\theta$ *(or* $r = a + b\sin\theta$*) is called a* limaçon *(from the French word for* snail*). In Problems 33 through 36, match the given polar-coordinate equation with its graph among the limaçons in Figs. 9.2.25 through 9.2.28.*

33. $r = 8 + 6\cos\theta$

34. $r = 7 + 7\cos\theta$

35. $r = 5 + 9\cos\theta$

36. $r = 3 + 11\cos\theta$

37. Show that the graph of the polar equation $r = a\cos\theta + b\sin\theta$ is a circle if $a^2 + b^2 \neq 0$. Express the center (h, k) and radius r of this circle in terms of a and b.

38. Show that if $0 < a < b$, then the limaçon with polar equation $r = a + b\cos\theta$ has an inner loop (as in Figs. 9.2.25 and 9.2.27). In this case, find (in terms of a and b) the range of values of θ that correspond to points of the inner loop.

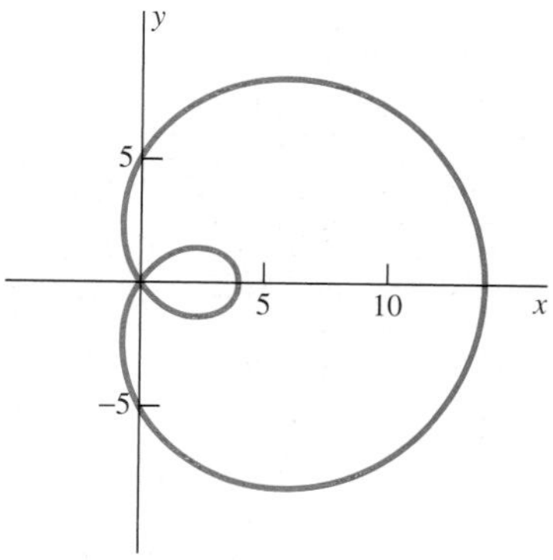

FIGURE 9.2.25

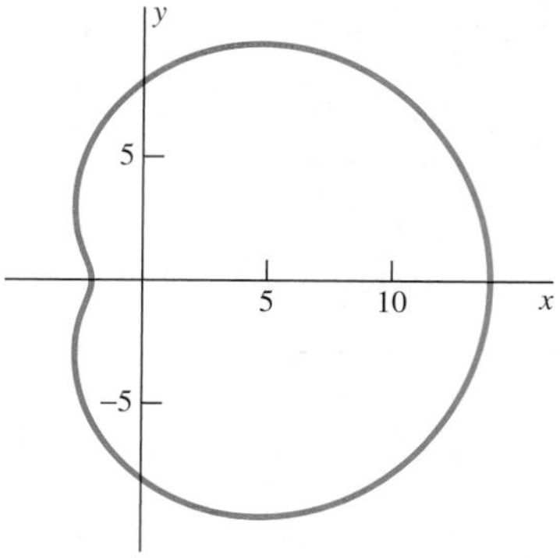

FIGURE 9.2.26

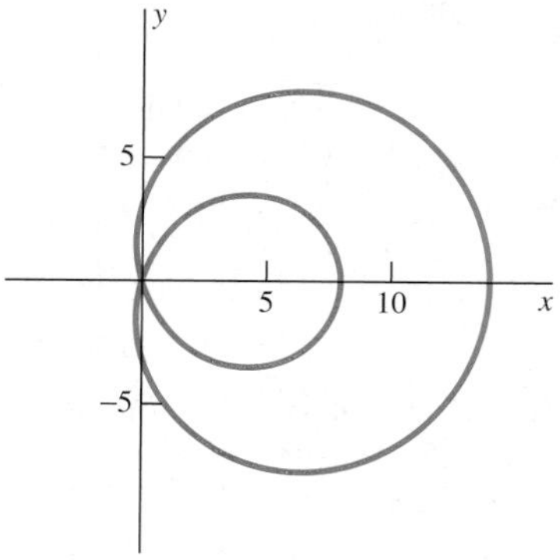

FIGURE 9.2.27

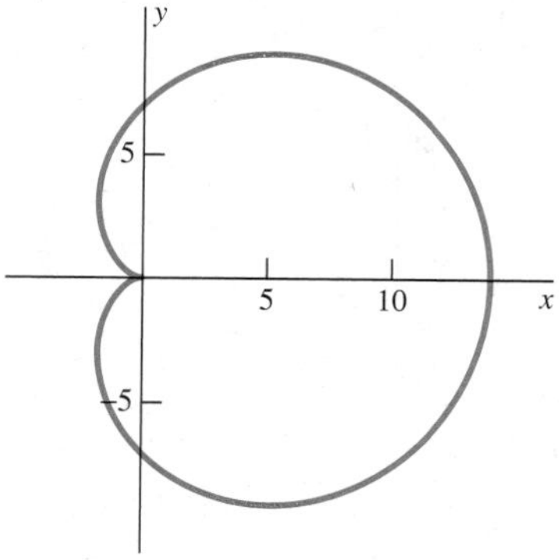

FIGURE 9.2.28

Sketch the graphs of the polar equations in Problems 39 through 52. Indicate any symmetries around either coordinate axis or the origin.

39. $r = 2\cos\theta$ (circle)

40. $r = 2\sin\theta + 2\cos\theta$ (circle)

41. $r = 1 + \cos\theta$ (cardioid)

42. $r = 1 - \sin\theta$ (cardioid)

43. $r = 2 + 4\sin\theta$ (limaçon)

44. $r = 4 + 2\cos\theta$ (limaçon)

45. $r^2 = 4\sin 2\theta$ (lemniscate)

46. $r^2 = 4\cos 2\theta$ (lemniscate)

47. $r = 2\sin 2\theta$ (four-leaved rose)

48. $r = 3\sin 3\theta$ (three-leaved rose)

49. $r = 3\cos 3\theta$ (three-leaved rose)

50. $r = 3\theta$ (spiral of Archimedes)

51. $r = 2\sin 5\theta$ (five-leaved rose)

52. $r^2 = 4\sin\theta$ (figure eight)

In Problems 53 through 58, find all points of intersection of the curves with the given polar equations.

53. $r = 1, \quad r = \cos\theta$

54. $r = \sin\theta, \quad r^2 = 3\cos^2\theta$

55. $r = \sin\theta, \quad r = \cos 2\theta$

56. $r = 1 + \cos\theta, \quad r = 1 - \sin\theta$

57. $r = 1 - \cos\theta, \quad r^2 = 4\cos\theta$

58. $r^2 = 4\sin\theta, \quad r^2 = 4\cos\theta$

59. (a) The straight line L passes through the point with polar coordinates (p, α) and is perpendicular to the line segment joining the pole and the point (p, α). Write the polar-coordinate equation of L. (b) Show that the rectangular-coordinate equation of L is

$$x\cos\alpha + y\sin\alpha = p.$$

60. Find a rectangular-coordinate equation of the cardioid with polar equation $r = 1 - \cos\theta$.

61. Use polar coordinates to identify the graph of the rectangular-coordinate equation $a^2(x^2 + y^2) = (x^2 + y^2 - by)^2$.

62. Plot the polar equations

$$r = 1 + \cos\theta \quad \text{and} \quad r = -1 + \cos\theta$$

on the same coordinate plane. Comment on the results.

63. Figures 9.2.29 and 9.2.30 show the graphs of the equations $r = \cos(5\theta/3)$ and $r = \cos(5\theta/2)$. Why does one have five

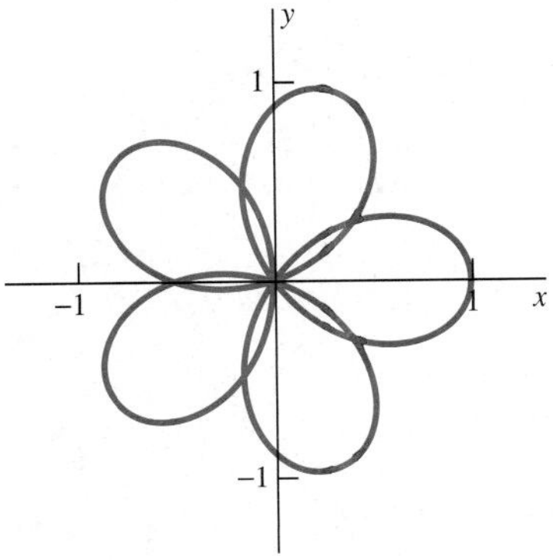

FIGURE 9.2.29 $r = \cos\left(\dfrac{5\theta}{3}\right)$.

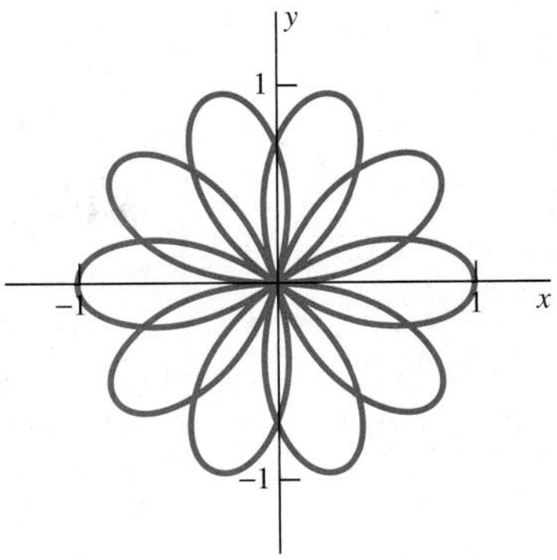

FIGURE 9.2.30 $r = \cos\left(\dfrac{5\theta}{2}\right)$.

(overlapping) loops while the other has ten loops? In each case, what range of values of θ is required to obtain all the loops? In the more general case $r = (\cos p\theta/q)$ where p and q are positive integers, is it p or q (or both) that determine the number of loops and the range of values of θ required to show all the loops in the complete graph?

64. Figures 9.2.31 and 9.2.32 show the graphs of the equations $r = 1 + 4\sin 3\theta$ and $r = 1 + 4\cos 4\theta$. What determines whether a polar curve of the form $r = a + b\sin(n\theta)$—with a and b positive constants and n a positive integer—has both larger and smaller loops? What determines whether the smaller loops are within or outside of the larger ones?

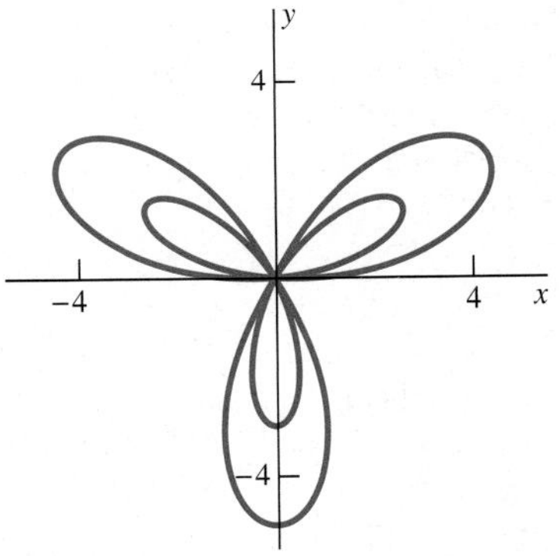

FIGURE 9.2.31 $r = 1 + 4\sin 3\theta$.

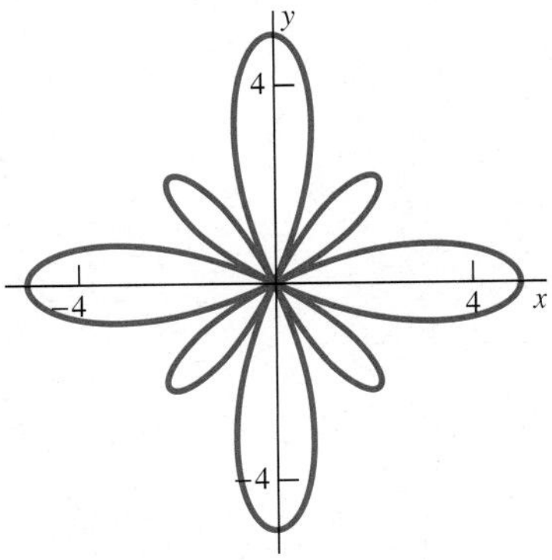

FIGURE 9.2.32 $r = 1 + 4\cos 4\theta$.

9.3 AREA COMPUTATIONS IN POLAR COORDINATES

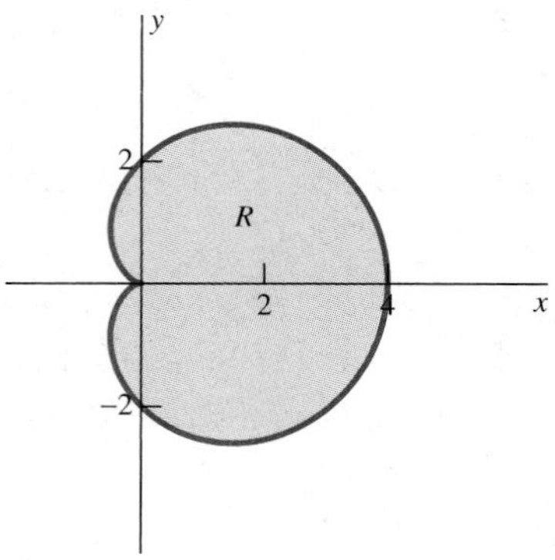

FIGURE 9.3.1 What is the area of the region R bounded by the cardioid $r = 2(1 + \cos\theta)$?

The graph of the polar-coordinate equation $r = f(\theta)$ may bound an area, as does the cardioid $r = 2(1 + \cos\theta)$—see Fig. 9.3.1. To calculate the area of this region, we may find it convenient to work directly with polar coordinates rather than to change to rectangular coordinates.

To see how to set up an area integral using polar coordinates, we consider the region R of Fig. 9.3.2. This region is bounded by the two radial lines $\theta = \alpha$ and $\theta = \beta$ and by the curve $r = f(\theta)$, $\alpha \leqq \theta \leqq \beta$. To approximate the area A of R, we begin with a partition

$$\alpha = \theta_0 < \theta_1 < \theta_2 < \cdots < \theta_n = \beta$$

of the interval $[\alpha,\ \beta]$ into n subintervals, all with the same length $\Delta\theta = (\beta - \alpha)/n$. We select a point $\theta_i^\star$ in the ith subinterval $[\theta_{i-1},\ \theta_i]$ for $i = 1, 2, \dots, n$.

Let ΔA_i denote the area of the sector bounded by the lines $\theta = \theta_{i-1}$ and $\theta = \theta_i$ and by the curve $r = f(\theta)$. We see from Fig. 9.3.2 that for small values of $\Delta\theta$, ΔA_i is approximately equal to the area of the *circular* sector that has radius $r_i^\star = f(\theta_i^\star)$ and is bounded by the same lines. That is,

$$\Delta A_i \approx \tfrac{1}{2}(r_i^\star)^2\,\Delta\theta = \tfrac{1}{2}\,[f(\theta_i^\star)]^2\,\Delta\theta.$$

We add the areas of these sectors for $i = 1, 2, \dots, n$ and thereby find that

$$A = \sum_{i=1}^{n} \Delta A_i \approx \sum_{i=1}^{n} \tfrac{1}{2}\,[f(\theta_i^\star)]^2\,\Delta\theta.$$

The right-hand sum is a Riemann sum for the integral

$$\int_\alpha^\beta \tfrac{1}{2}[f(\theta)]^2\,d\theta.$$

Hence, if f is continuous, the value of this integral is the limit, as $\Delta\theta \to 0$, of the preceding sum. We therefore conclude that the *area A of the region R bounded by the lines $\theta = \alpha$ and $\theta = \beta$ and the curve $r = f(\theta)$* is

$$A = \int_\alpha^\beta \tfrac{1}{2}[f(\theta)]^2\,d\theta. \qquad \textbf{(1)}$$

The infinitesimal sector shown in Fig. 9.3.3, with radius r, central angle $d\theta$, and area $dA = \frac{1}{2}r^2\,d\theta$, serves as a useful device for remembering Eq. (1) in the abbreviated

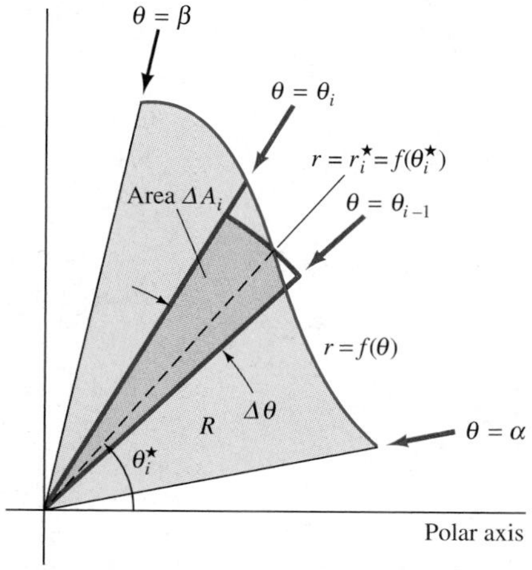

FIGURE 9.3.2 We obtain the area formula from Riemann sums.

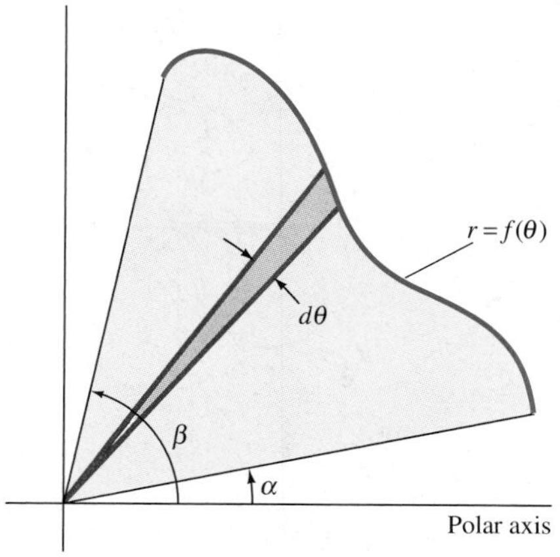

FIGURE 9.3.3 Nonrigorous derivation of the area formula in polar coordinates.

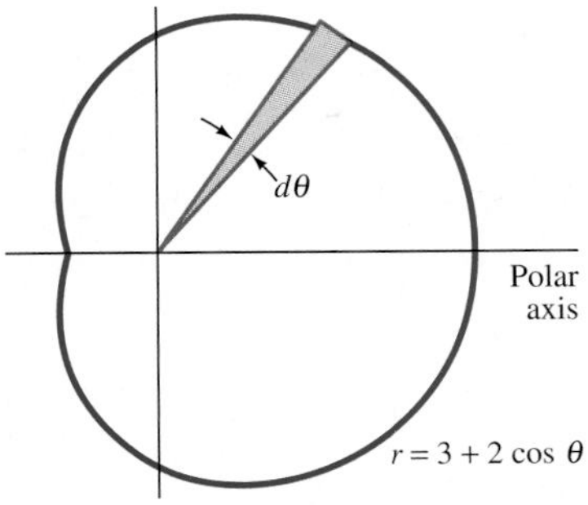

FIGURE 9.3.4 The limaçon of Example 1.

form

$$A = \int_{\alpha}^{\beta} \tfrac{1}{2}r^2\,d\theta. \tag{2}$$

EXAMPLE 1 Find the area of the region bounded by the limaçon with equation $r = 3 + 2\cos\theta$, $0 \leqq \theta \leqq 2\pi$ (Fig. 9.3.4).

Solution We could apply Eq. (2) with $\alpha = 0$ and $\beta = 2\pi$. Here, instead, we will make use of symmetry. We will calculate the area of the upper half of the region and then double the result. Note that the infinitesimal sector shown in Fig. 9.3.4 sweeps out the upper half of the limaçon as θ increases from 0 to π (Fig. 9.3.5). Hence

$$\begin{aligned} A &= 2\int_{\alpha}^{\beta} \tfrac{1}{2}r^2\,d\theta = \int_0^{\pi} (3 + 2\cos\theta)^2\,d\theta \\ &= \int_0^{\pi} (9 + 12\cos\theta + 4\cos^2\theta)\,d\theta. \end{aligned}$$

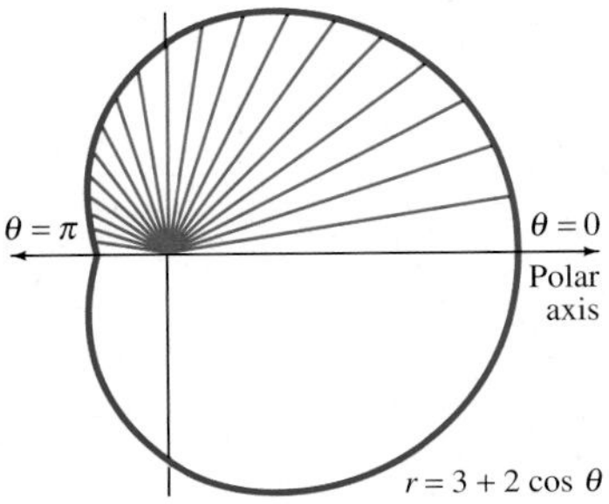

FIGURE 9.3.5 Infinitesimal sectors from $\theta = 0$ to $\theta = \pi$ (Example 1).

Because

$$4\cos^2\theta = 4 \cdot \frac{1 + \cos 2\theta}{2} = 2 + 2\cos 2\theta,$$

we now get

$$\begin{aligned} A &= \int_0^{\pi} (11 + 12\cos\theta + 2\cos 2\theta)\,d\theta \\ &= \Big[11\theta + 12\sin\theta + \sin 2\theta\Big]_0^{\pi} = 11\pi. \end{aligned}$$

◆

EXAMPLE 2 Find the area bounded by each loop of the limaçon with equation $r = 1 + 2\cos\theta$ (Fig. 9.3.6).

Solution The equation $1 + 2\cos\theta = 0$ has two solutions for θ in the interval $[0, 2\pi]$: $\theta = 2\pi/3$ and $\theta = 4\pi/3$. The upper half of the outer loop of the limaçon corresponds to values of θ between 0 and $2\pi/3$, where r is positive. Because the curve is symmetric around the x-axis, we can find the total area A_1 bounded by the outer loop by integrating from 0 to $2\pi/3$ and then doubling. Thus

$$\begin{aligned} A_1 &= 2\int_0^{2\pi/3} \tfrac{1}{2}(1 + 2\cos\theta)^2\,d\theta = \int_0^{2\pi/3} (1 + 4\cos\theta + 4\cos^2\theta)\,d\theta \\ &= \int_0^{2\pi/3} (3 + 4\cos\theta + 2\cos 2\theta)\,d\theta \\ &= \Big[3\theta + 4\sin\theta + \sin 2\theta\Big]_0^{2\pi/3} = 2\pi + \tfrac{3}{2}\sqrt{3}. \end{aligned}$$

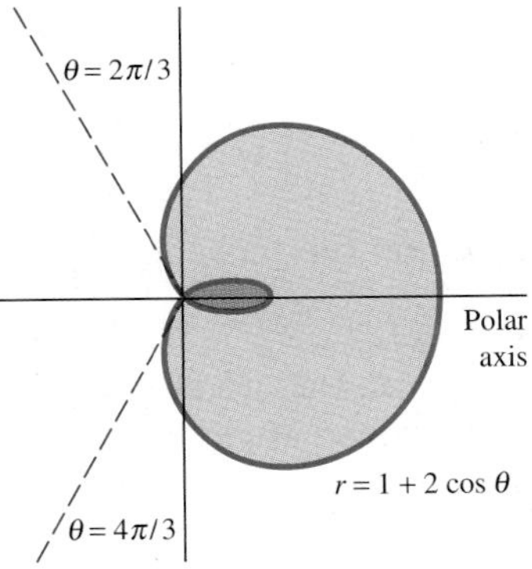

FIGURE 9.3.6 The limaçon of Example 2.

The inner loop of the limaçon corresponds to values of θ between $2\pi/3$ and $4\pi/3$, where r is negative. Hence the area bounded by the inner loop is

$$\begin{aligned} A_2 &= \int_{2\pi/3}^{4\pi/3} \tfrac{1}{2}(1 + 2\cos\theta)^2\,d\theta \\ &= \tfrac{1}{2}\Big[3\theta + 4\sin\theta + \sin 2\theta\Big]_{2\pi/3}^{4\pi/3} = \pi - \tfrac{3}{2}\sqrt{3}. \end{aligned}$$

The area of the region lying *between* the two loops of the limaçon is then

$$A = A_1 - A_2 = 2\pi + \tfrac{3}{2}\sqrt{3} - \left(\pi - \tfrac{3}{2}\sqrt{3}\right) = \pi + 3\sqrt{3}.$$

◆

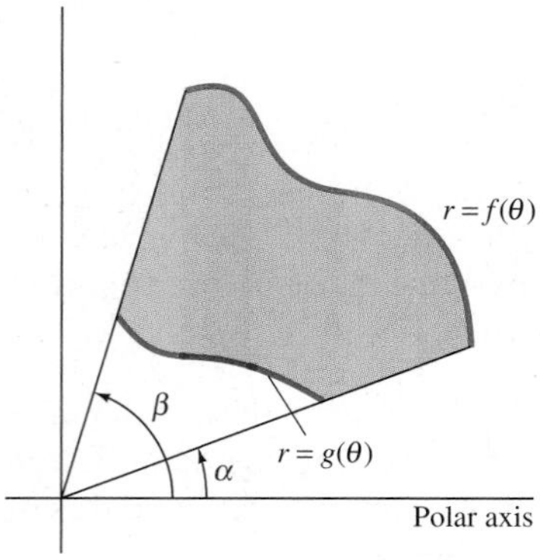

FIGURE 9.3.7 The area between the graphs of f and g.

The Area Between Two Polar Curves

Now consider two curves $r = f(\theta)$ and $r = g(\theta)$, with $f(\theta) \geqq g(\theta) \geqq 0$ for $\alpha \leqq \theta \leqq \beta$. Then we can find the area of the region bounded by these curves and the rays (radial lines) $\theta = \alpha$ and $\theta = \beta$ (Fig. 9.3.7) by subtracting the area bounded by the inner curve from that bounded by the outer curve. That is, the area A between the two curves is given by

$$A = \int_\alpha^\beta \tfrac{1}{2}[f(\theta)]^2\,d\theta - \int_\alpha^\beta \tfrac{1}{2}[g(\theta)]^2\,d\theta,$$

so that

$$A = \frac{1}{2}\int_\alpha^\beta \{[f(\theta)]^2 - [g(\theta)]^2\}\,d\theta. \tag{3}$$

With r_{outer} for the outer curve and r_{inner} for the inner curve, we get the abbreviated formula

$$A = \frac{1}{2}\int_\alpha^\beta [(r_{\text{outer}})^2 - (r_{\text{inner}})^2]\,d\theta \tag{4}$$

for the area of the region shown in Fig. 9.3.8.

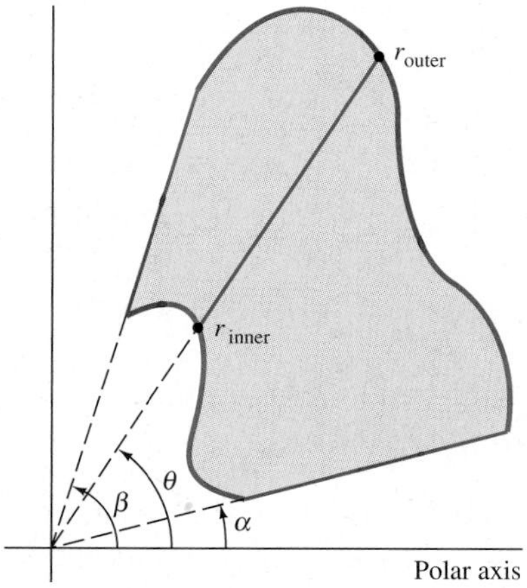

FIGURE 9.3.8 The radial line segment illustrates the radii r_{inner} and r_{outer} of Eq. (4).

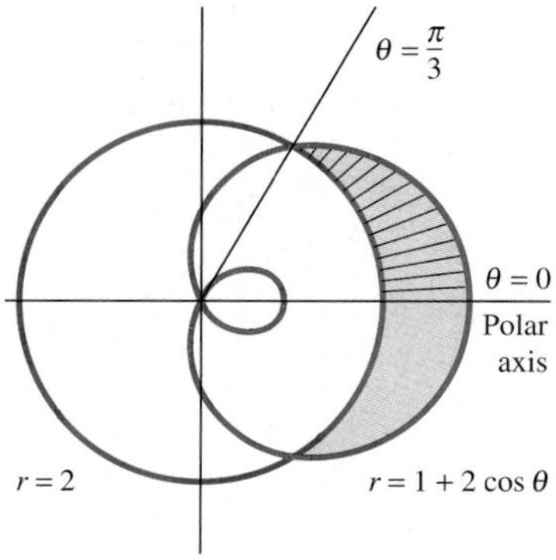

FIGURE 9.3.9 The region of Example 3.

EXAMPLE 3 Find the area A of the region that lies within the limaçon $r = 1 + 2\cos\theta$ and outside the circle $r = 2$.

Solution The circle and limaçon are shown in Fig. 9.3.9, with the area A between them shaded. The points of intersection of the circle and limaçon are given by

$$1 + 2\cos\theta = 2, \quad \text{so} \quad \cos\theta = \tfrac{1}{2},$$

and the figure shows that we should choose the solutions $\theta = \pm\pi/3$. These two values of θ are the needed limits of integration. When we use Eq. (3), we find that

$$\begin{aligned} A &= \frac{1}{2}\int_{-\pi/3}^{\pi/3} [(1 + 2\cos\theta)^2 - 2^2]\,d\theta \\ &= \int_0^{\pi/3} (4\cos\theta + 4\cos^2\theta - 3)\,d\theta \qquad \text{(by symmetry)} \\ &= \int_0^{\pi/3} (4\cos\theta + 2\cos 2\theta - 1)\,d\theta \\ &= \Big[4\sin\theta + \sin 2\theta - \theta\Big]_0^{\pi/3} = \frac{15\sqrt{3} - 2\pi}{6}. \end{aligned}$$

◆

9.3 TRUE/FALSE STUDY GUIDE

9.3 CONCEPTS: QUESTIONS AND DISCUSSION

1. Give an example of a plane region whose area can be calculated both by a rectangular-coordinate integral and a polar-coordinate integral, but the latter is easier to evaluate.
2. Give an example of a plane region whose area can be calculated both by a rectangular-coordinate integral and a polar-coordinate integral, but the former is easier to evaluate.
3. Give an example of an unbounded plane region such that its polar-coordinate area integral is improper but convergent.

9.3 PROBLEMS

In Problems 1 through 6, sketch the plane region bounded by the given polar curve $r = f(\theta)$, $\alpha \leqq \theta \leqq \beta$, and the rays $\theta = \alpha$, $\theta = \beta$.

1. $r = \theta, \quad 0 \leqq \theta \leqq \pi$

2. $r = \theta, \quad 0 \leqq \theta \leqq 2\pi$

3. $r = 1/\theta, \quad \pi \leqq \theta \leqq 3\pi$

4. $r = 1/\theta, \quad 3\pi \leqq \theta \leqq 5\pi$

5. $r = e^{-\theta}, \quad 0 \leqq \theta \leqq \pi$

6. $r = e^{-\theta}, \quad \pi/2 \leqq \theta \leqq 3\pi/2$

In Problems 7 through 16, find the area bounded by the given curve.

7. $r = 2\cos\theta$

8. $r = 4\sin\theta$

9. $r = 1 + \cos\theta$

10. $r = 2 - 2\sin\theta$ (Fig. 9.3.10)

11. $r = 2 - \cos\theta$

12. $r = 3 + 2\sin\theta$ (Fig. 9.3.11)

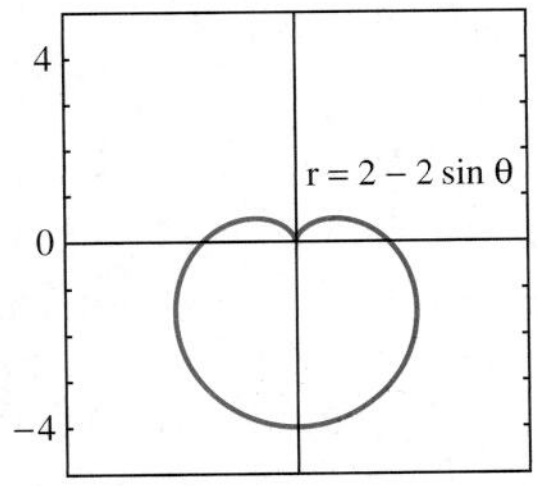

FIGURE 9.3.10 The cardioid of Problem 10.

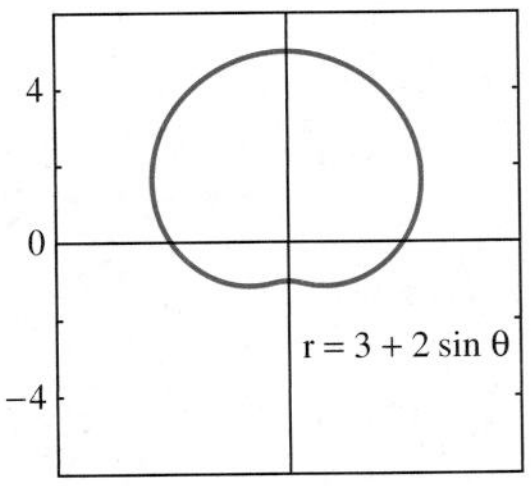

FIGURE 9.3.11 The limaçon of Problem 12.

13. $r = -4\cos\theta$

14. $r = 5(1 + \sin\theta)$

15. $r = 3 - \cos\theta$

16. $r = 2 + \sin\theta + \cos\theta$

In Problems 17 through 24, find the area bounded by one loop of the given curve.

17. $r = 2\cos 2\theta$

18. $r = 3\sin 3\theta$ (Fig. 9.3.12)

19. $r = 2\cos 4\theta$ (Fig. 9.3.13)

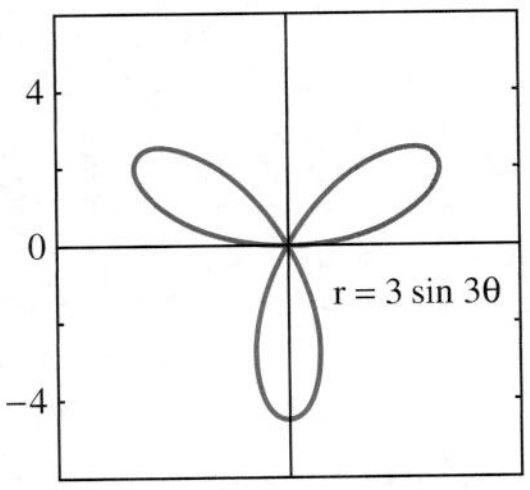

FIGURE 9.3.12 The three-leaved rose of Problem 18.

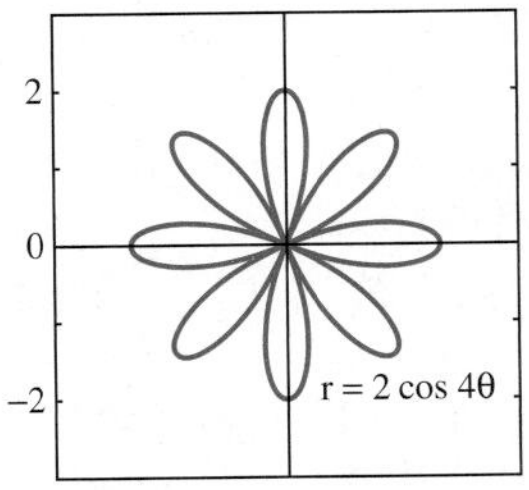

FIGURE 9.3.13 The eight-leaved rose of Problem 19.

20. $r = \sin 5\theta$ (Fig. 9.3.14)

21. $r^2 = 4\sin 2\theta$

22. $r^2 = 4\cos 2\theta$ (Fig. 9.3.15)

23. $r^2 = 4\sin\theta$

24. $r = 6\cos 6\theta$

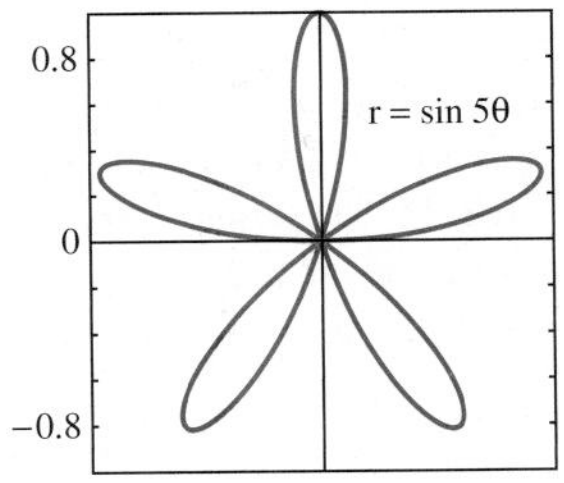

FIGURE 9.3.14 The five-leaved rose of Problem 20.

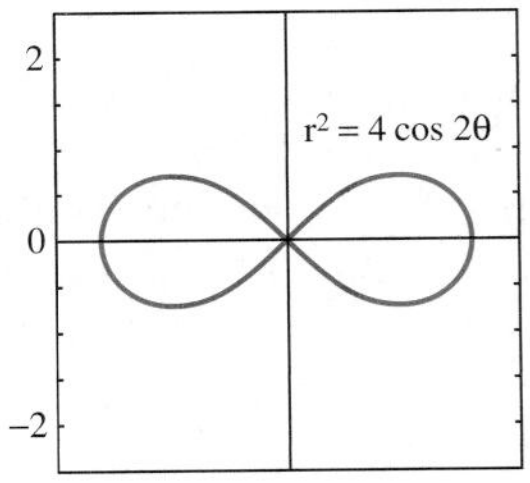

FIGURE 9.3.15 The lemniscate of Problem 22.

In Problems 25 through 36, find the area of the region described.

25. Inside $r = 2\sin\theta$ and outside $r = 1$

26. Inside both $r = 4\cos\theta$ and $r = 2$

27. Inside both $r = \cos\theta$ and $r = \sqrt{3}\sin\theta$

28. Inside $r = 2 + \cos\theta$ and outside $r = 2$

29. Inside $r = 3 + 2\cos\theta$ and outside $r = 4$

30. Inside $r^2 = 2\cos 2\theta$ and outside $r = 1$

31. Inside $r^2 = \cos 2\theta$ and $r^2 = \sin 2\theta$ (Fig. 9.3.16)

32. Inside the large loop and outside the small loop of $r = 1 - 2\sin\theta$ (Fig. 9.3.17)

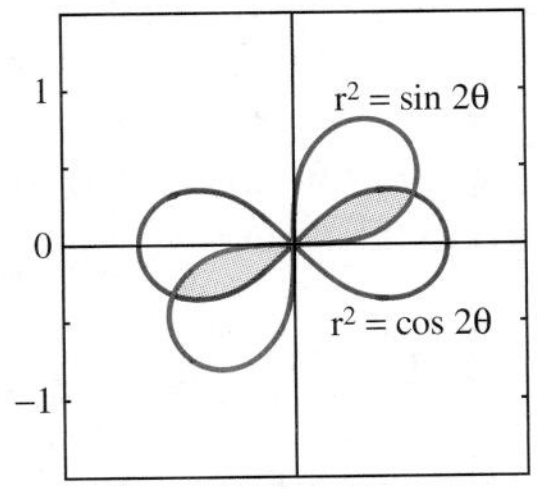

FIGURE 9.3.16 Problem 31.

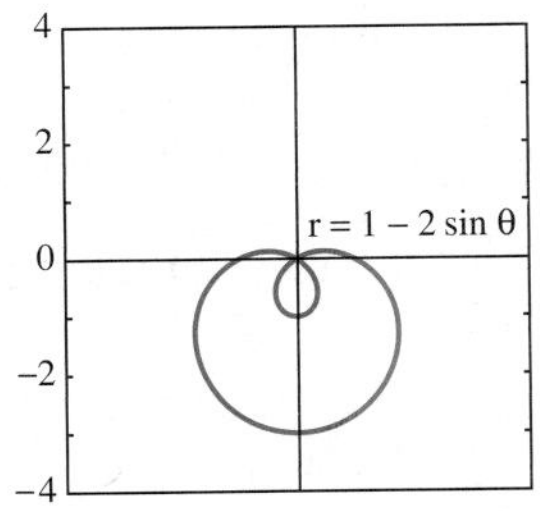

FIGURE 9.3.17 Problem 32.

33. Inside $r = 2(1 + \cos\theta)$ and outside $r = 1$

34. Inside the figure-eight curve $r^2 = 4\cos\theta$ and outside $r = 1 - \cos\theta$

35. Inside both $r = 2\cos\theta$ and $r = 2\sin\theta$

36. Inside $r = 2 + 2\sin\theta$ and outside $r = 2$

37. Find the area of the circle $r = \sin\theta + \cos\theta$ by integration in polar coordinates (Fig. 9.3.18). Check your answer by writing the equation of the circle in rectangular coordinates, finding its radius, and then using the familiar formula for the area of a circle.

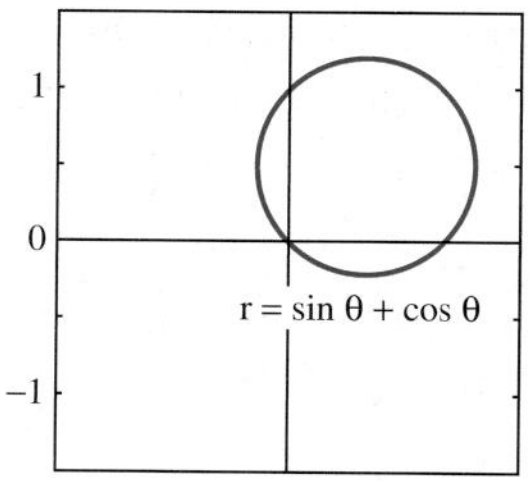

FIGURE 9.3.18 The circle $r = \sin\theta + \cos\theta$ (Problem 37).

38. Find the area of the region that lies interior to all three circles $r = 1$, $r = 2\cos\theta$, and $r = 2\sin\theta$.

39. The *spiral of Archimedes*, shown in Fig. 9.3.19, has the simple equation $r = a\theta$ (a is a constant). Let A_n denote the area bounded by the nth turn of the spiral, where $2(n-1)\pi \leqq \theta \leqq 2n\pi$, and by the portion of the polar axis joining its endpoints. For each $n \geqq 2$, let $R_n = A_n - A_{n-1}$ denote the area between the $(n-1)$th and the nth turns. Then derive the following results of Archimedes:

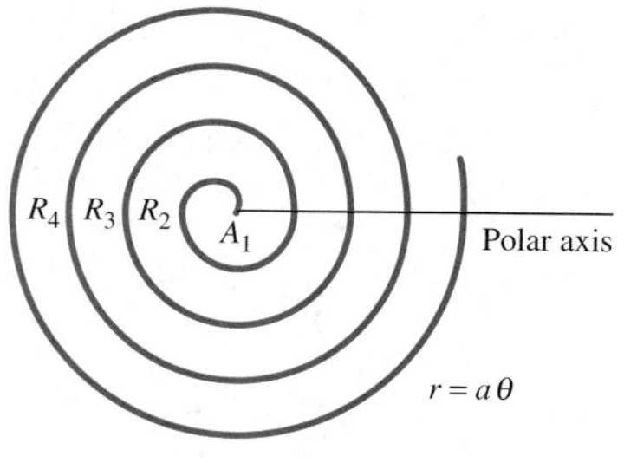

FIGURE 9.3.19 The spiral of Archimedes (Problem 39).

(a) $A_1 = \frac{1}{3}\pi(2\pi a)^2$; (b) $A_2 = \frac{7}{12}\pi(4\pi a)^2$;

(c) $R_2 = 6A_1$; (d) $R_{n+1} = nR_2$ for $n \geqq 2$.

40. Two circles both have radius a, and each circle passes through the center of the other. Find the area of the region that lies within both circles.

41. A polar curve of the form $r = ae^{-k\theta}$ is called a *logarithmic spiral,* and the portion given by $2(n-1)\pi \leqq \theta \leqq 2n\pi$ is called the nth *turn* of this spiral. Figure 9.3.20 shows the first five turns of the logarithmic spiral $r = e^{-\theta/10}$, and the area of the region lying between the second and third turns is shaded. Find:

(a) The area of the region that lies between the first and second turns.

(b) The area of the region that lies between the $(n-1)$th and nth turns for $n > 1$.

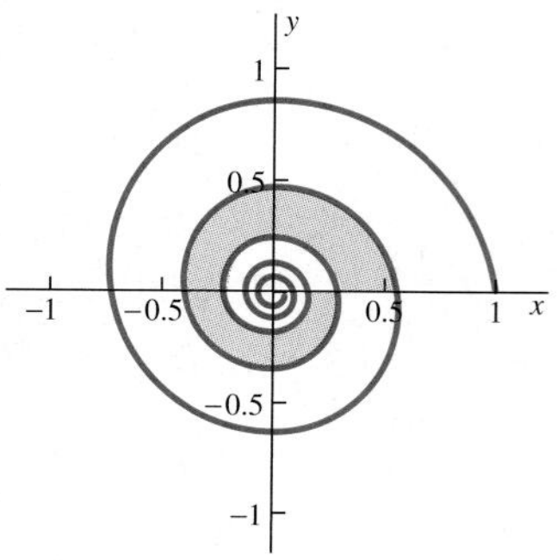

FIGURE 9.3.20 The logarithmic spiral of Problem 41.

42. Figure 9.3.21 shows the first turn of the logarithmic spiral $r = 2e^{-\theta/10}$ together with the two circles, both centered at (0, 0), through the endpoints of the spiral. Find the areas of the two shaded regions and verify that their sum is the area of the annular region between the two circles.

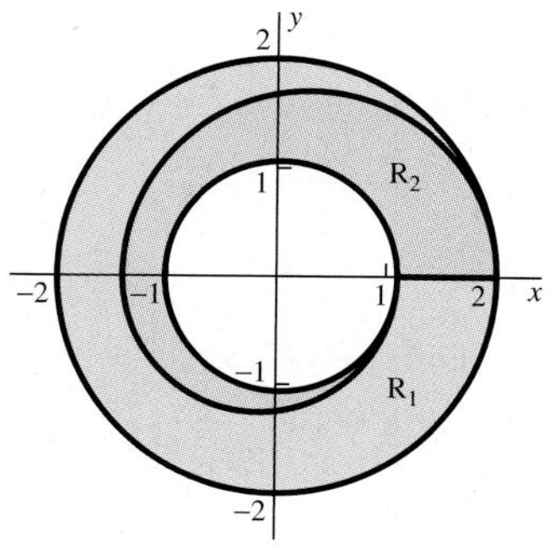

FIGURE 9.3.21 The two regions of Problem 42.

43. The shaded region R in Fig. 9.3.22 is bounded by the cardioid $r = 1 + \cos\theta$, the spiral $r = e^{-\theta/5}$, $0 \leqq \theta \leqq \pi$, and the spiral $r = e^{\theta/5}$, $-\pi \leqq \theta \leqq 0$. Graphically estimate the points of intersection of the cardioid and the spirals, then approximate the area of the region R.

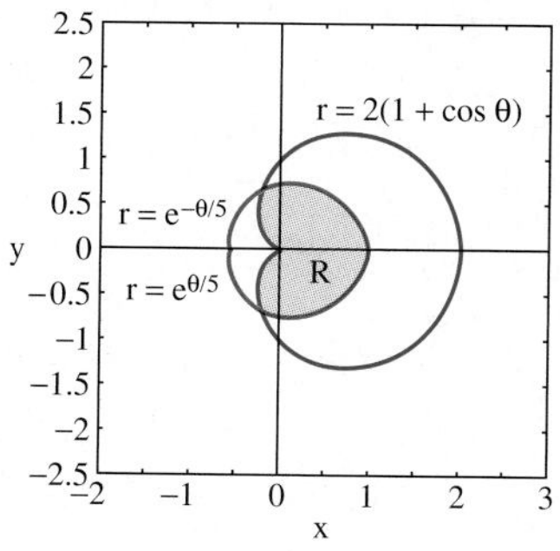

FIGURE 9.3.22 The region of Problem 43.

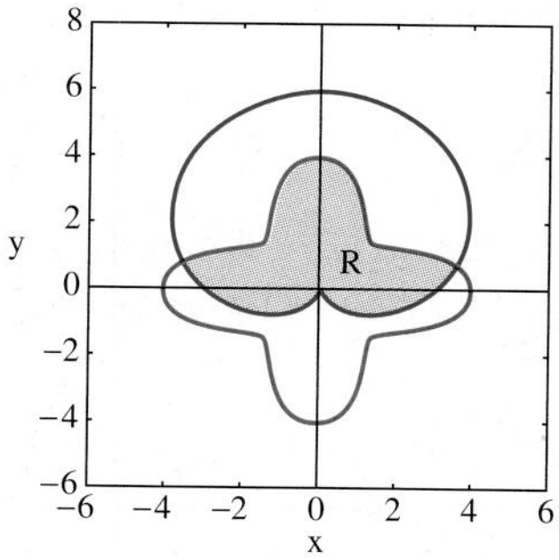

FIGURE 9.3.23 The region of Problem 44.

44. The shaded region R in Fig. 9.3.23 lies inside both the cardioid $r = 3 + 3\sin\theta$ and the polar curve $r = 3 + \cos 4\theta$. Graphically estimate the points of intersection of the two curves; then approximate the area of the region R.

9.4 PARAMETRIC CURVES

Until now we have encountered *curves* mainly as graphs of equations. An equation of the form $y = f(x)$ or of the form $x = g(y)$ determines a curve by giving one of the coordinate variables explicitly as a function of the other. An equation of the form $F(x, y) = 0$ may also determine a curve, but then each variable is given implicitly as a function of the other.

Another important type of curve is the trajectory of a point moving in the coordinate plane. The motion of the point can be described by giving its position $(x(t), y(t))$ at time t. Such a description involves expressing both the rectangular-coordinate variables x and y as functions of a third variable, or *parameter,* t rather than as functions of one another. In this context a **parameter** is an independent variable (not a constant, as is sometimes meant in popular usage). This approach motivates the following definition.

DEFINITION Parametric Curve

A **parametric curve** C in the plane is a pair of functions

$$x = f(t), \quad y = g(t), \tag{1}$$

that give x and y as continuous functions of the real number t (the parameter) in some interval I.

Each value of the parameter t determines a point $(f(t), g(t))$, and the set of all such points is the **graph** of the curve C. Often the distinction between the curve—the pair of **coordinate functions** f and g—and the graph is not made. Therefore, we may refer interchangeably to the curve and to its graph when the context makes clear the intended meaning. The two equations in (1) are called the **parametric equations** of the curve.

The graph of a parametric curve may be sketched by plotting enough points to indicate its likely shape. In some cases we can eliminate the parameter t and thus obtain an equation in x and y. This equation may give us more information about the shape of the curve.

t	x	y
0	1	0
$\pi/4$	$1/\sqrt{2}$	$1/\sqrt{2}$
$\pi/2$	0	1
$3\pi/4$	$-1/\sqrt{2}$	$1/\sqrt{2}$
π	-1	0
$5\pi/4$	$-1/\sqrt{2}$	$-1/\sqrt{2}$
$3\pi/2$	0	-1
$7\pi/4$	$1/\sqrt{2}$	$-1/\sqrt{2}$
2π	1	0

FIGURE 9.4.1 A table of values for Example 1.

EXAMPLE 1 Determine the graph of the curve

$$x = \cos t, \quad y = \sin t, \quad 0 \leqq t \leqq 2\pi. \tag{2}$$

Solution Figure 9.4.1 shows a table of values of x and y that correspond to multiples of $\pi/4$ for the parameter t. These values give the eight points highlighted in Fig. 9.4.2, all of which lie on the unit circle. This suggests that the graph is, in fact, the unit circle. To verify this, we note that the fundamental identity of trigonometry gives

$$x^2 + y^2 = \cos^2 t + \sin^2 t \equiv 1,$$

so every point of the graph lies on the circle with equation $x^2 + y^2 = 1$. Conversely, the point of the circle with angular (polar) coordinate t is the point $(\cos t, \sin t)$ of the graph. Thus the graph is precisely the unit circle. ◆

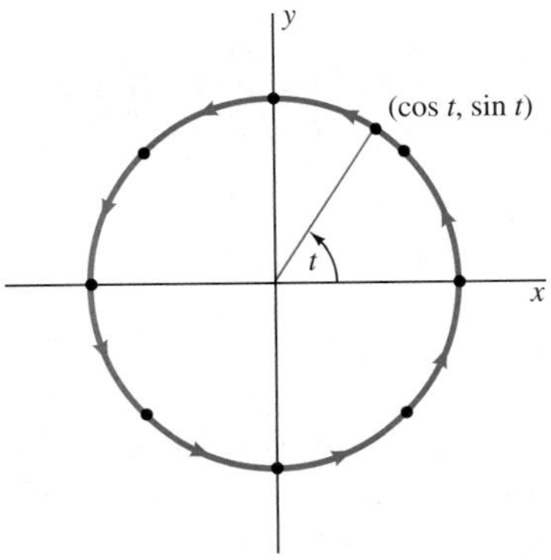

FIGURE 9.4.2 The graph of the parametric functions of Example 1.

What is lost in the process in Example 1 is the information about how the graph is produced as t goes from 0 to 2π. But this is easy to determine by inspection. As t travels from 0 to 2π, the point $(\cos t, \sin t)$ begins at $(1, 0)$ and travels counterclockwise around the circle, ending at $(1, 0)$ when $t = 2\pi$.

A given figure in the plane may be the graph of different curves. To speak more loosely, a given curve may have different **parametrizations.**

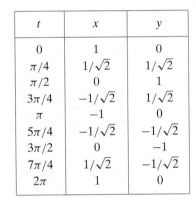

EXAMPLE 2 The graph of the parametric curve

$$x = \frac{1 - t^2}{1 + t^2}, \quad y = \frac{2t}{1 + t^2}, \quad -\infty < t < +\infty$$

also lies on the unit circle, because we find that $x^2 + y^2 = 1$ here as well. If t begins at 0 and increases, then the point $P(x(t), y(t))$ begins at $(1, 0)$ and travels along the upper half of the circle. If t begins at 0 and decreases, then the point $P(x(t), y(t))$ travels along the lower half of the circle. As t approaches either $+\infty$ or $-\infty$, the point P approaches the point $(-1, 0)$. Thus the graph consists of the unit circle with the single point $(-1, 0)$ deleted. A slight modification of the curve of Example 1,

$$x = \cos t, \quad y = \sin t, \quad -\pi < t < \pi,$$

is a different parametrization of this same graph. ◆

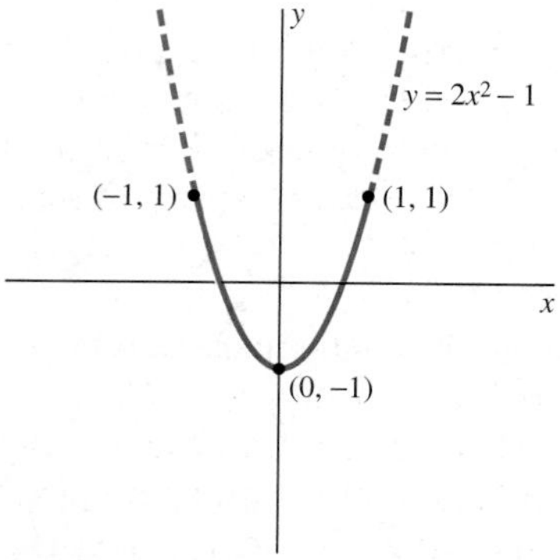

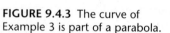

FIGURE 9.4.3 The curve of Example 3 is part of a parabola.

EXAMPLE 3 Eliminate the parameter to determine the graph of the parametric curve

$$x = t - 1, \quad y = 2t^2 - 4t + 1, \quad 0 \leqq t \leqq 2.$$

Solution We substitute $t = x + 1$ (from the equation for x) into the equation for y. This yields

$$y = 2(x + 1)^2 - 4(x + 1) + 1 = 2x^2 - 1$$

for $-1 \leqq x \leqq 1$. Thus the graph of the given curve is a portion of the parabola $y = 2x^2 - 1$ (Fig. 9.4.3). As t increases from 0 to 2, the point $(t - 1, 2t^2 - 4t + 1)$ travels along the parabola from $(-1, 1)$ to $(1, 1)$. ◆

REMARK The parabolic arc of Example 3 can be reparametrized with

$$x = \sin t, \quad y = 2\sin^2 t - 1.$$

Now, as t increases, the point $(\sin t, 2\sin^2 t - 1)$ travels back and forth along the parabola between the two points $(-1, 1)$ and $(1, 1)$, rather like the bob of a pendulum.

The parametric curve of Example 3 is one in which we can eliminate the parameter and thus obtain an explicit equation $y = f(x)$. Moreover, any explicitly presented curve $y = f(x)$ can be viewed as a parametric curve by writing

$$x = t, \quad y = f(t),$$

with the parameter t taking on values in the original domain of f. By contrast, the circle of Example 1 illustrates a parametric curve whose graph is not the graph of any single function. (Why not?) Example 4 exhibits another way in which parametric curves can differ from graphs of functions—they can have self-intersections.

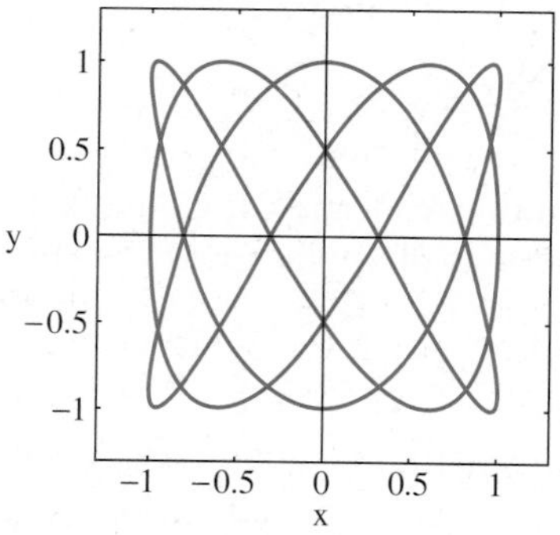

FIGURE 9.4.4 The Lissajous curve with $a = 3$, $b = 5$.

EXAMPLE 4 The parametric equations

$$x = \cos at, \quad y = \sin bt$$

(with a and b constant) define the *Lissajous curves* that typically appear on oscilloscopes in physics and electronics laboratories. The Lissajous curve with $a = 3$ and $b = 5$ is shown in Fig. 9.4.4. You probably would not want to calculate and plot by hand enough points to produce a Lissajous curve. Figure 9.4.4 was plotted with a computer program that generated it almost immediately. But it is perhaps more instructive to watch a slower graphing calculator plot a parametric curve like this, because the curve is traced by a point that moves on the screen as the parameter t increases (from 0 to 2π in this case). For instance, with a TI calculator set in "parametric graph mode," one need only enter and graph the equations

```
XT = cos(3T)      YT = sin(5T)
```

on the interval $0 \leqq t \leqq 2\pi$. With *Maple* and *Mathematica* the commands

```
plot([cos(3*t),sin(5*t),t=0..2*pi]);
```

and

```
ParametricPlot[{Cos[3*t],Sin[5*t]},{t,0,2*Pi}];
```

(respectively) give the same figure. ◆

The use of parametric equations $x = x(t)$, $y = y(t)$ is most advantageous when elimination of the parameter is either impossible or would lead to an equation $y = f(x)$ that is considerably more complicated than the original parametric equations. This often happens when the curve is a geometric locus or the path of a point moving under specified conditions.

EXAMPLE 5 The curve traced by a point P on the edge of a rolling circle is called a **cycloid.** The circle rolls along a straight line without slipping or stopping. (You will see a cycloid if you watch a patch of bright paint on the tire of a bicycle that crosses your path.) Find parametric equations for the cycloid if the line along which the circle rolls is the x-axis, the circle is above the x-axis but always tangent to it, and the point P begins at the origin.

Solution Evidently the cycloid consists of a series of arches. We take as parameter t the angle (in radians) through which the circle has turned since it began with P at the origin. This is the angle TCP in Fig. 9.4.5.

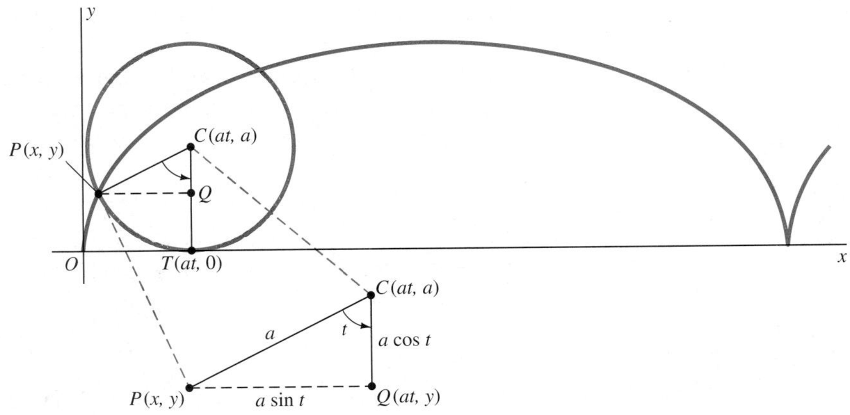

FIGURE 9.4.5 The cycloid and the right triangle CPQ (Example 5).

The distance the circle has rolled is $|OT|$, so this is also the length of the circumference subtended by the angle TCP. Thus $|OT| = at$ if a is the radius of the circle, so the center C of the rolling circle has coordinates (at, a) when the angle TCP is t. The right triangle CPQ in Fig. 9.4.5 provides us with the relations

$$at - x = a \sin t \quad \text{and} \quad a - y = a \cos t.$$

Therefore the cycloid—the path of the moving point P—has parametric equations

$$x = a(t - \sin t), \quad y = a(1 - \cos t). \tag{3}$$

◆

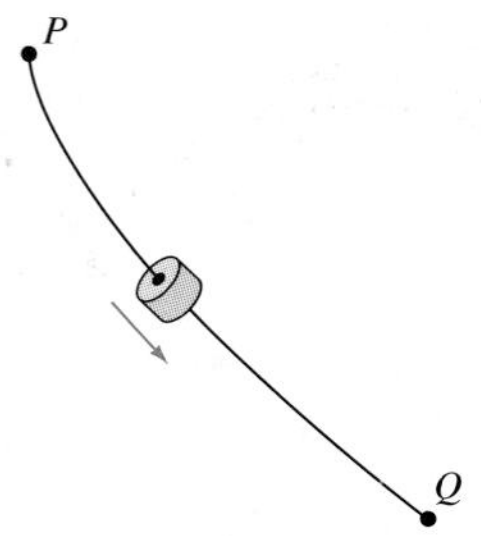

FIGURE 9.4.6 A bead sliding down a wire—the brachistochrone problem.

HISTORICAL NOTE Figure 9.4.6 shows a bead sliding down a frictionless wire from point P to point Q. The *brachistochrone problem* asks what shape the wire should be to minimize the bead's time of descent from P to Q. In June of 1696, John Bernoulli proposed the brachistochrone problem as a public challenge, with a 6-month deadline (later extended to Easter 1697 at Leibniz's request). Isaac Newton, then retired from academic life and serving as Warden of the Mint in London, received Bernoulli's challenge on January 29, 1697. The very next day he communicated his own solution—the curve of minimal descent time is an arc of an inverted cycloid—to the Royal Society of London.

Lines Tangent to Parametric Curves

The parametric curve $x = f(t)$, $y = g(t)$ is called **smooth** if the derivatives $f'(t)$ and $g'(t)$ are continuous and never simultaneously zero. In some neighborhood of each point of its graph, a smooth parametric curve can be described in one or possibly both of the forms $y = F(x)$ and $x = G(y)$. To see why this is so, suppose (for example) that $f'(t) > 0$ on the interval I. Then $f(t)$ is an increasing function on I and therefore has an inverse function $t = \phi(x)$ there. If we substitute $t = \phi(x)$ into the equation $y = g(t)$, then we get

$$y = g(\phi(x)) = F(x).$$

We can use the chain rule to compute the slope dy/dx of the line tangent to a smooth parametric curve at a given point. Differentiating $y = F(x)$ with respect to t yields

$$\frac{dy}{dt} = \frac{dy}{dx} \cdot \frac{dx}{dt},$$

so

$$\frac{dy}{dx} = \frac{dy/dt}{dx/dt} = \frac{g'(t)}{f'(t)} \tag{4}$$

at any point where $f'(t) \neq 0$. The tangent line is vertical at any point where $f'(t) = 0$ but $g'(t) \neq 0$.

Equation (4) gives $y' = dy/dx$ as a function of t. Another differentiation with respect to t, again with the aid of the chain rule, results in the formula

$$\frac{dy'}{dt} = \frac{dy'}{dx} \cdot \frac{dx}{dt},$$

so

$$\frac{d^2y}{dx^2} = \frac{dy'}{dx} = \frac{dy'/dt}{dx/dt}. \tag{5}$$

EXAMPLE 6 Calculate dy/dx and d^2y/dx^2 for the cycloid with the parametric equations in (3).

Solution We begin with

$$x = a(t - \sin t), \quad y = a(1 - \cos t). \tag{3}$$

Then Eq. (4) gives

$$\frac{dy}{dx} = \frac{dy/dt}{dx/dt} = \frac{a \sin t}{a(1 - \cos t)} = \frac{\sin t}{1 - \cos t}. \tag{6}$$

This derivative is zero when y is an odd integral multiple of π, so the tangent line is horizontal at the midpoint of each arch of the cycloid. The endpoints of the arches correspond to even integral multiples of π, where both the numerator and the denominator in Eq. (6) are zero. These are isolated points (called *cusps*) at which the cycloid fails to be a smooth curve. (See Fig. 9.4.7.)

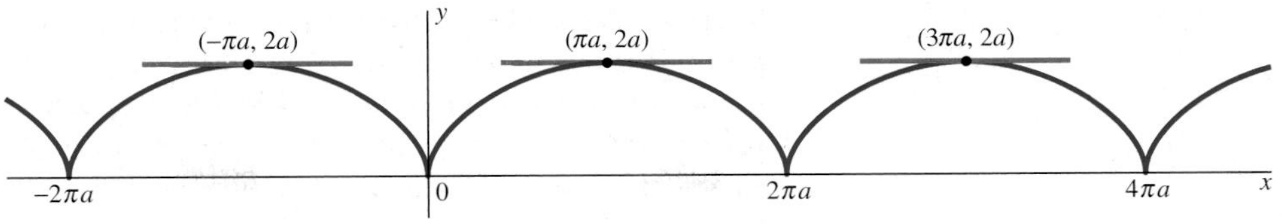

FIGURE 9.4.7 Horizontal tangents and cusps of the cycloid.

Next, Eq. (5) yields

$$\frac{d^2y}{dx^2} = \frac{(\cos t)(1-\cos t) - (\sin t)(\sin t)}{(1-\cos t)^2 \cdot a(1-\cos t)} = -\frac{1}{a(1-\cos t)^2}.$$

Because $d^2y/dx^2 < 0$ for all t (except for the isolated even integral multiples of π), this shows that each arch of the cycloid is concave downward (Fig. 9.4.5). ◆

REMARK In Fig. 9.4.7 it appears that the cycloid has a vertical tangent line at each cusp point $(2n\pi a, 0)$. We can verify this observation by calculating the limit as $t \to 2n\pi$ of the derivative in (6). Using l'Hôpital's rule, we get

$$\lim_{t\to 2n\pi} \frac{dy}{dx} = \lim_{t\to 2n\pi} \frac{\sin t}{1-\cos t} = \lim_{t\to 2n\pi} \frac{\cos t}{\sin t} = \pm\infty,$$

because $\cos t \to 1$ and $\sin t \to 0$ as $t \to 2n\pi$. The limit is $+\infty$ or $-\infty$ according as t approaches $2n\pi$ from the right or the left. In either event, we conclude that the tangent line is, indeed, vertical at the cusp point.

EXAMPLE 7 It would be impractical to attempt to graph the curve

$$x^3 = 2y^6 - 5y^4 + 9y \qquad \textbf{(7)}$$

by solving for y as a function of x. However, we can parametrize this curve by defining

$$y = t, \quad x = (2t^6 - 5t^4 + 9t)^{1/3}. \qquad \textbf{(8)}$$

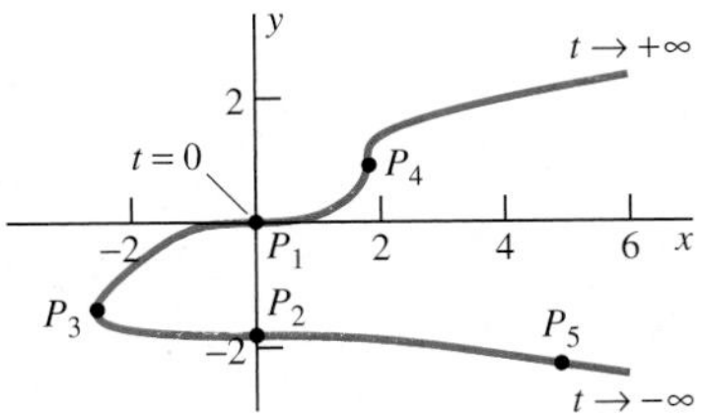

FIGURE 9.4.8 The parametric curve of Example 7.

Figure 9.4.8 shows a computer plot of this parametric curve for $-2.5 \leqq t \leqq 2.5$. We see at least four likely critical and inflection points. It appears that there are horizontal tangent lines at the points P_1 and P_2 on the y-axis, and vertical tangent lines at P_3 and P_4. Let's investigate the character of these points by calculating the pertinent derivatives.

To investigate the possibility of horizontal and vertical tangent lines, we use Eq. (4) to calculate the first derivative

$$\frac{dy}{dx} = \frac{dy/dt}{dx/dt} = \frac{3(2t^6 - 5t^4 + 9t)^{2/3}}{12t^5 - 20t^3 + 9}. \qquad \textbf{(9)}$$

Using a computer algebra system, we find that the only real zeros of the polynomial $2t^6 - 5t^4 + 9t$ in the numerator are $t = 0$ and $t \approx -1.8065$. These values of t yield the points $P_1(0, 0)$ and $P_2(-0.00002422, -1.86065)$, respectively, that are shown in the figure. Thus P_2 does not lie precisely on the y-axis, after all.

The denominator polynomial $12t^5 - 20t^3 + 9$ in (9) has only the single real zero $t \approx -2.5587$, which yields the single point $P_3(-2.5587, -1.3941)$ on the curve where the tangent line is vertical. In particular, there is *no* vertical tangent line near the point P_4 indicated in the figure.

To investigate the possibility of possible inflection points, we use Eq. (5) and a computer algebra system to calculate the second derivative

$$\frac{d^2y}{dx^2} = \frac{d}{dt}\left(\frac{dy}{dx}\right) \div \frac{dx}{dt}$$

$$= -\frac{6(2t^6 - 5t^4 + 9t)^{1/3}(36t^{10} - 150t^8 + 50t^6 + 594t^5 - 450t^3 - 81)}{(12t^5 - 20t^3 + 9)^3}. \qquad \textbf{(10)}$$

The two trinomials that appear in the numerator and denominator here are the same as those in (9), and correspond to the three critical points already found. Our computer algebra system reports that the tenth-degree numerator polynomial in (10) has only two real zeros: $t \approx 1.0009$ and $t \approx -2.2614$. These two zeros of the second derivative yield the two points $P_4(1.8172, 1.0009)$ and $P_5(4.8820, -2.2614)$ that are shown in the figure. It is visually clear that the concavity of the curve changes at P_4—where $dy/dx \approx 0.9063$ so the tangent line is steep but not vertical—but the character of the remaining point is not so obvious. However, you can graph the second derivative in (10) to verify that it is positive to the right and negative to the left of P_5—so this final candidate is, indeed, also an inflection point.

Finally, because our viewing window in Fig. 9.4.8 is large enough to include all the critical points and inflection points on the curve in (7)—and since it is clear from the equations in (8) that $|x|$ and $|y| \to \infty$ as $|t| \to \infty$—we are assured that the figure shows all of the principal features of the curve. ◆

Polar Curves as Parametric Curves

A curve given in polar coordinates by the equation $r = f(\theta)$ can be regarded as a parametric curve with parameter θ. To see this, we recall that the equations $x = r\cos\theta$ and $y = r\sin\theta$ allow us to change from polar to rectangular coordinates. We replace r with $f(\theta)$, and this gives the parametric equations

$$x = f(\theta)\cos\theta, \quad y = f(\theta)\sin\theta, \tag{11}$$

which express x and y in terms of the parameter θ.

EXAMPLE 8 The *spiral of Archimedes* has the polar-coordinate equation $r = a\theta$ (Fig. 9.4.9). The equations in (11) give the spiral the parametrization

$$x = a\theta\cos\theta, \quad y = a\theta\sin\theta.$$ ◆

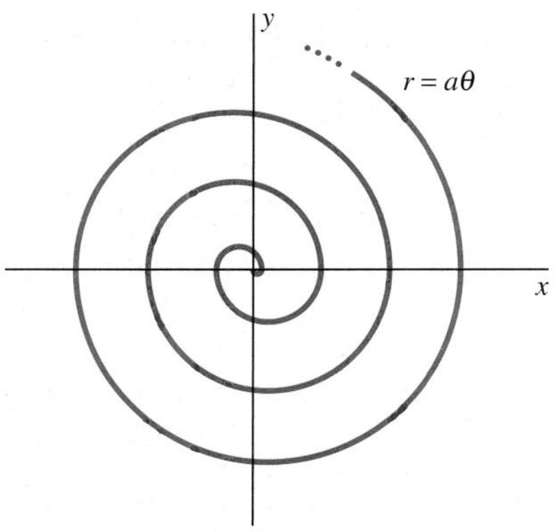

FIGURE 9.4.9 The spiral of Archimedes (Example 8).

The slope dy/dx of a tangent line can be computed in terms of polar coordinates as well as rectangular coordinates. Given a polar-coordinate curve $r = f(\theta)$, we use the parametrization shown in (11). Then Eq. (4), with θ in place of t, gives

$$\frac{dy}{dx} = \frac{dy/d\theta}{dx/d\theta} = \frac{f'(\theta)\sin\theta + f(\theta)\cos\theta}{f'(\theta)\cos\theta - f(\theta)\sin\theta}, \tag{12}$$

or, alternatively, denoting $f'(\theta)$ by r',

$$\frac{dy}{dx} = \frac{r' \sin\theta + r\cos\theta}{r'\cos\theta - r\sin\theta}. \tag{13}$$

Equation (13) has the following useful consequence. Let ψ denote the angle between the tangent line at P and the radius OP (extended) from the origin (Fig. 9.4.10). Then

$$\cot\psi = \frac{1}{r}\cdot\frac{dr}{d\theta} \quad (0 \leqq \psi \leqq \pi). \tag{14}$$

In Problem 32 we indicate how Eq. (14) can be derived from Eq. (13).

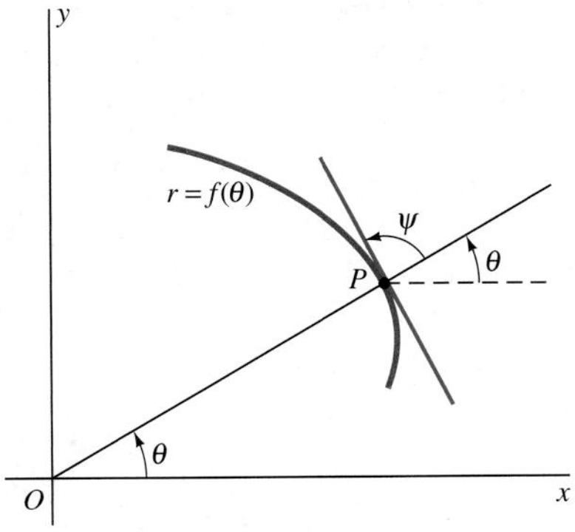

FIGURE 9.4.10 The interpretation of the angle ψ. [See Eq. (14).]

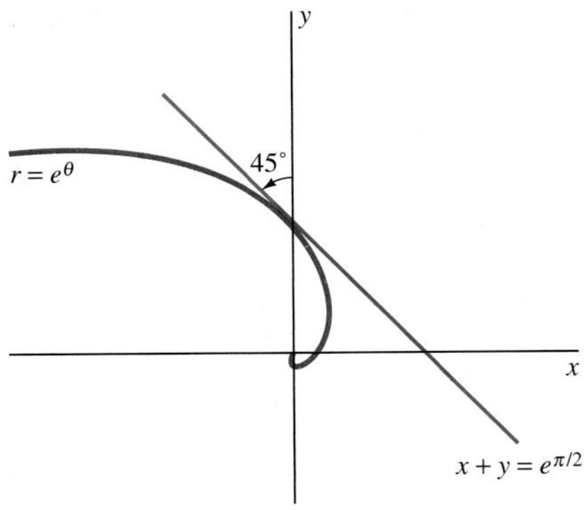

FIGURE 9.4.11 The angle ψ is always 45° for the logarithmic spiral (Example 9).

EXAMPLE 9 Consider the *logarithmic spiral* with polar equation $r = e^{\theta}$. Show that $\psi = \pi/4$ at every point of the spiral, and write an equation of its tangent line at the point $(e^{\pi/2}, \pi/2)$.

Solution Because $dr/d\theta = e^{\theta}$, Eq. (14) tells us that $\cot\psi = e^{\theta}/e^{\theta} = 1$. Thus $\psi = \pi/4$. When $\theta = \pi/2$, Eq. (13) gives

$$\frac{dy}{dx} = \frac{e^{\pi/2}\sin(\pi/2) + e^{\pi/2}\cos(\pi/2)}{e^{\pi/2}\cos(\pi/2) - e^{\pi/2}\sin(\pi/2)} = -1.$$

But when $\theta = \pi/2$, we have $x = 0$ and $y = e^{\pi/2}$. It follows that an equation of the desired tangent line is

$$y - e^{\pi/2} = -x; \quad \text{that is,} \quad x + y = e^{\pi/2}.$$

The line and the spiral appear in Fig. 9.4.11. ◆

9.4 TRUE/FALSE STUDY GUIDE

9.4 CONCEPTS: QUESTIONS AND DISCUSSION

1. Pick two points A and B in the plane. Then define a parametrization $P(t) = (x(t), y(t))$ of the line segment $\overline{AB}$ such that $P(0) = A$ and $P(1) = B$.
2. Pick two points A and B equidistant from the origin. Then define a parametrization of a circular arc AB such that $P(0) = A$ and $P(1) = B$.
3. Pick two points A and B on the parabola $y = x^2$. Then define a parametrization of the parabola such that $P(0) = A$ and $P(1) = B$.
4. Let A and B be two points on a given parametric curve. Is it always possible to define a parametrization of the curve such that $P(0) = A$ and $P(1) = B$?

9.4 PROBLEMS

In Problems 1 through 12, eliminate the parameter and then sketch the curve.

1. $x = t + 1, \quad y = 2t - 1$

2. $x = t^2 + 1, \quad y = 2t^2 - 1$

3. $x = t^2, \quad y = t^3$

4. $x = \sqrt{t}, \quad y = 3t - 2$

5. $x = t + 1, \quad y = 2t^2 - t - 1$

6. $x = t^2 + 3t, \quad y = t - 2$

7. $x = e^t, \quad y = 4e^{2t}$

8. $x = 2e^t, \quad y = 2e^{-t}$

9. $x = 5\cos t, \quad y = 3\sin t$

10. $x = \sinh t, \quad y = \cosh t$

11. $x = 2\cosh t, \quad y = 3\sinh t$

12. $x = \sec t, \quad y = \tan t$

In Problems 13 through 16, first eliminate the parameter and sketch the curve. Then describe the motion of the point $(x(t), y(t))$ as t varies in the given interval.

13. $x = \sin 2\pi t, \quad y = \cos 2\pi t; \quad 0 \leqq t \leqq 1$

14. $x = 3 + 2\cos t, \quad y = 5 - 2\sin t; \quad 0 \leqq t \leqq 2\pi$

15. $x = \sin^2 \pi t, \quad y = \cos^2 \pi t; \quad 0 \leqq t \leqq 2$

16. $x = \cos t, \quad y = \sin^2 t; \quad -\pi \leqq t \leqq \pi$

In Problems 17 through 20, (a) first write the equation of the line tangent to the given parametric curve at the point that corresponds to the given value of t, and (b) then calculate d^2y/dx^2 to determine whether the curve is concave upward or concave downward at this point.

17. $x = 2t^2 + 1, \quad y = 3t^3 + 2; \quad t = 1$

18. $x = \cos^3 t, \quad y = \sin^3 t; \quad t = \pi/4$

19. $x = t\sin t, \quad y = t\cos t; \quad t = \pi/2$

20. $x = e^t, \quad y = e^{-t}; \quad t = 0$

In Problems 21 through 24, find the angle ψ between the radius OP and the tangent line at the point P that corresponds to the given value of θ.

21. $r = \exp(\theta\sqrt{3}), \quad \theta = \pi/2$

22. $r = 1/\theta, \quad \theta = 1$

23. $r = \sin 3\theta, \quad \theta = \pi/6$

24. $r = 1 - \cos\theta, \quad \theta = \pi/3$

In Problems 25 through 28, find

(a) The points on the curve where the tangent line is horizontal.

(b) The slope of each tangent line at any point where the curve intersects the x-axis.

25. $x = t^2, \quad y = t^3 - 3t$ (Fig. 9.4.12)

26. $x = \sin t, \quad y = \sin 2t$ (Fig. 9.4.13)

27. $r = 1 + \cos\theta$

28. $r^2 = 4\cos 2\theta$ (See Fig. 9.3.15.)

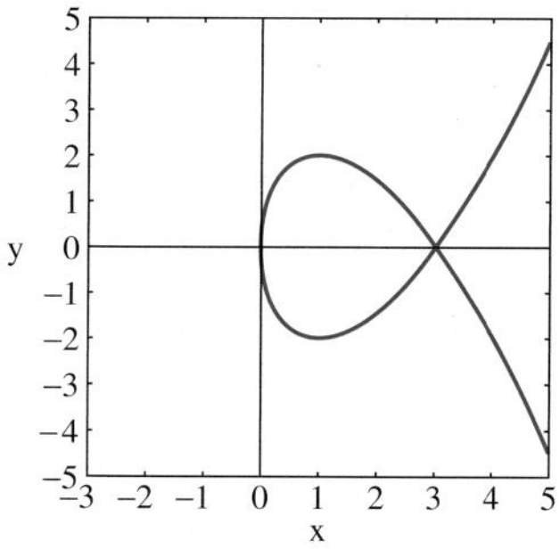

FIGURE 9.4.12 The curve of Problem 25.

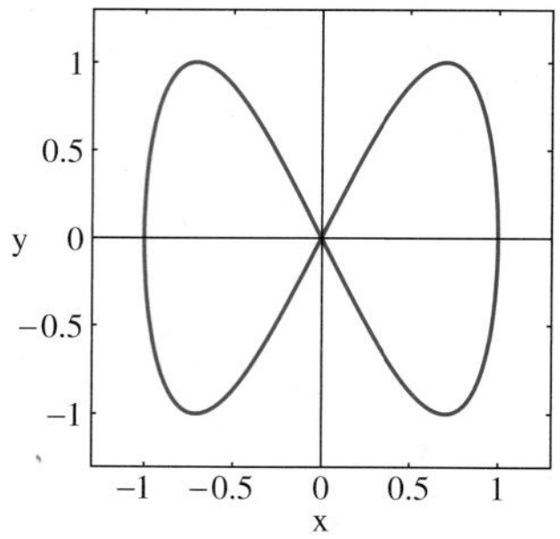

FIGURE 9.4.13 The curve of Problem 26.

29. The curve C is determined by the parametric equations $x = e^{-t}$, $y = e^{2t}$. Calculate dy/dx and d^2y/dx^2 directly from these parametric equations. Conclude that C is concave upward at every point. Then sketch C.

30. The graph of the folium of Descartes with rectangular equation $x^3 + y^3 = 3xy$ appears in Fig. 9.4.14. Parametrize its loop as follows: Let P be the point of intersection of the line $y = tx$ with the loop; then solve for the coordinates x and y of P in terms of t.

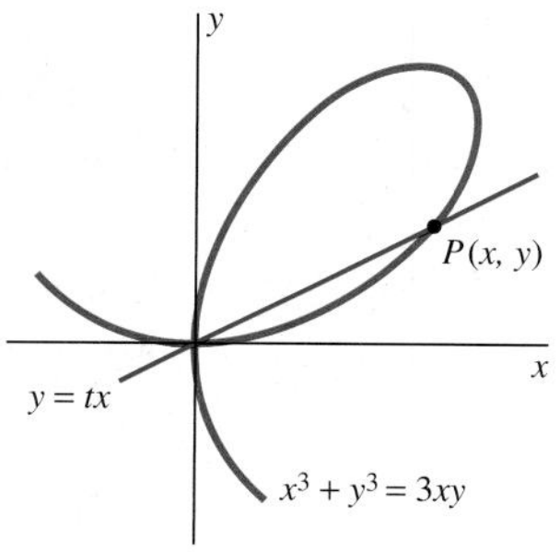

FIGURE 9.4.14 The loop of the folium of Descartes (Problem 30).

31. Parametrize the parabola $y^2 = 4px$ by expressing x and y as functions of the slope m of the tangent line at the point $P(x, y)$ of the parabola.

32. Let P be a point of the curve with polar equation $r = f(\theta)$, and let ψ be the angle between the extended radius OP and the tangent line at P. Let α be the angle of inclination of this tangent line, measured counterclockwise from the horizontal. Then $\psi = \alpha - \theta$. Verify Eq. (14) by substituting $\tan\alpha = dy/dx$ from Eq. (13) and $\tan\theta = y/x = (\sin\theta)/(\cos\theta)$ into the identity

$$\cot\psi = \frac{1}{\tan(\alpha - \theta)} = \frac{1 + \tan\alpha\tan\theta}{\tan\alpha - \tan\theta}.$$

33. Let P_0 be the highest point of the circle of Fig. 9.4.5—the circle that generates the cycloid of Example 5. Show that the line through P_0 and the point P of the cycloid (the point P is shown in Fig. 9.4.5) is tangent to the cycloid at P. This fact gives a geometric construction of the line tangent to the cycloid.

34. A circle of radius b rolls without slipping inside a circle of radius $a > b$. The path of a point fixed on the circumference of the rolling circle is called a *hypocycloid* (Fig. 9.4.15). Let P begin its journey at $A(a, 0)$ and let t be the angle AOC, where O is the center of the large circle and C is the center of the rolling circle. Show that the coordinates of P are given by the parametric equations

$$x = (a-b)\cos t + b\cos\left(\frac{a-b}{b}t\right),$$

$$y = (a-b)\sin t - b\sin\left(\frac{a-b}{b}t\right).$$

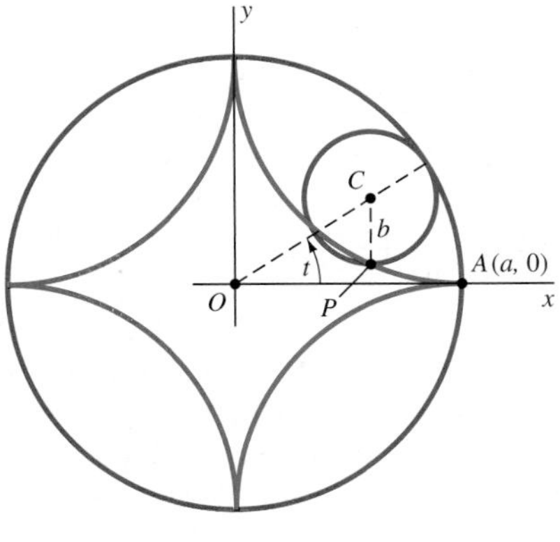

FIGURE 9.4.15 The hypocycloid of Problem 34.

35. If $b = a/4$ in Problem 34, show that the parametric equations of the hypocycloid reduce to

$$x = a\cos^3 t, \qquad y = a\sin^3 t.$$

36. (a) Prove that the hypocycloid of Problem 35 is the graph of the equation

$$x^{2/3} + y^{2/3} = a^{2/3}.$$

(b) Find all points of this hypocycloid where its tangent line is either horizontal or vertical, and find the intervals on which it is concave upward and those on which it is concave downward. (c) Sketch this hypocycloid.

37. Consider a point P on the spiral of Archimedes, the curve shown in Fig. 9.4.16 with polar equation $r = a\theta$.

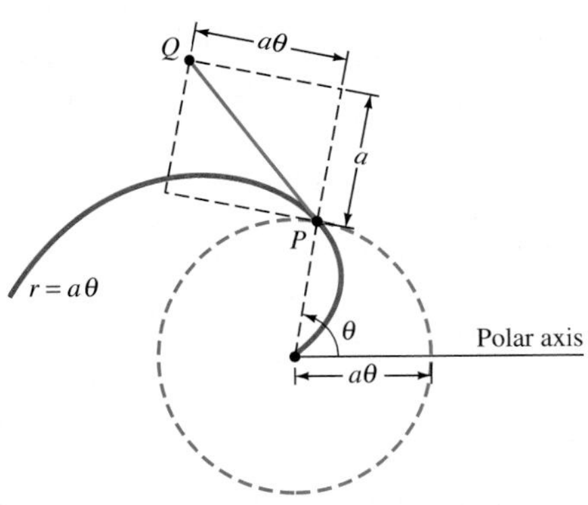

FIGURE 9.4.16 The segment PQ is tangent to the spiral (a result of Archimedes; see Problem 37).

Archimedes viewed the path of P as compounded of two motions, one with speed a directly away from the origin O and another a circular motion with unit angular speed around O. This suggests Archimedes' result that the line PQ in the figure is tangent to the spiral at P. Prove that this is indeed true.

38. (a) Deduce from Eq. (6) that if t is not an integral multiple of 2π, then the slope of the tangent line at the corresponding point of the cycloid is $\cot(t/2)$. (b) Conclude that at the cusp of the cycloid where t is an integral multiple of 2π, the cycloid has a vertical tangent line.

39. A *loxodrome* is a curve $r = f(\theta)$ such that the tangent line at P and the radius OP in Fig. 9.4.10 make a constant angle. Use Eq. (14) to prove that every loxodrome is of the form $r = Ae^{k\theta}$, where A and k are constants. Thus every loxodrome is a logarithmic spiral similar to the one considered in Example 9.

40. Let a curve be described in polar coordinates by $r = f(\theta)$ where f is continuous. If $f(\alpha) = 0$, then the origin is the point of the curve corresponding to $\theta = \alpha$. Deduce from the parametrization $x = f(\theta)\cos\theta$, $y = f(\theta)\sin\theta$ that the line tangent to the curve at this point makes the angle α with the positive x-axis. For example, the cardioid $r = f(\theta) = 1 - \sin\theta$ shown in Fig. 9.4.17 is tangent to the y-axis at the origin. And, indeed, $f(\pi/2) = 0$. The y-axis is the line $\theta = \alpha = \pi/2$.

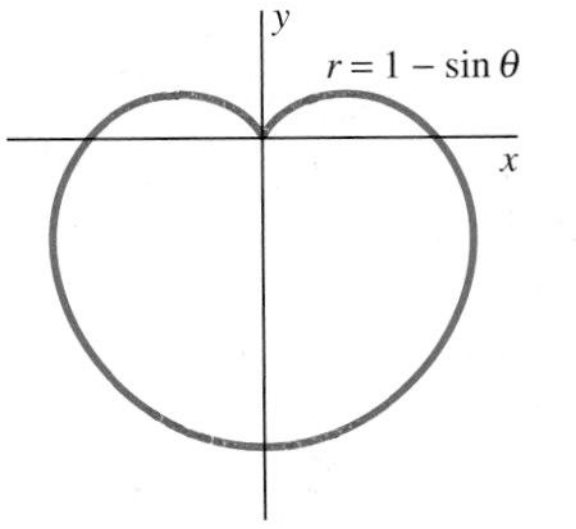

FIGURE 9.4.17 The cardioid of Problem 40.

41. Use the technique of Problem 30 to parametrize the first-quadrant loop of the folium-like curve $x^5 + y^5 = 5x^2y^2$.

42. A line segment of length $2a$ has one endpoint constrained to lie on the x-axis and the other endpoint constrained to lie on the y-axis, but its endpoints are free to move along those axes. As they do so, its midpoint sweeps out a locus in the xy-plane. Obtain a rectangular-coordinate equation of this locus and thereby identify this curve.

In Problems 43–46, investigate (as in Example 7) the given curve and construct a sketch that shows all the critical points and inflection points on it.

43. $x = y^3 - 3y^2 + 1$

44. $x = y^4 - 3y^3 + 5y$

45. $x^3 = y^5 - 5y^3 + 4$

46. $x^5 = 5y^6 - 17y^3 + 13y$

9.4 Project: Trochoid Investigations

A *trochoid* is traced by a point P on a spoke of a wheel of radius a as it rolls along the x-axis. If the distance of P from the center of the rolling wheel is $b > 0$, show that the trochoid is described by the parametric equations

$$x = at - b\sin t, \qquad y = a - b\cos t.$$

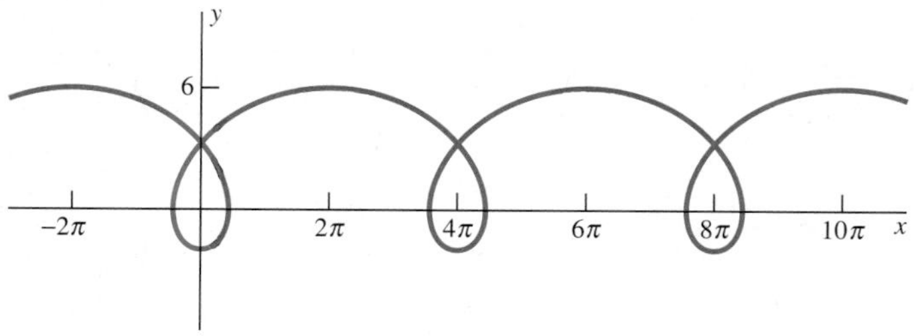

FIGURE 9.4.18 The trochoid with $a = 2$ and $b = 4$.

Note that the trochoid is a familiar cycloid if $b = a$. We allow the possibility that $b > a$. Figure 9.4.18 shows the trochoid with $a = 2$ and $b = 4$. Experiment with different values of a and b. What determines whether the trochoid has loops, cusps, or neither?

Hypotrochoids

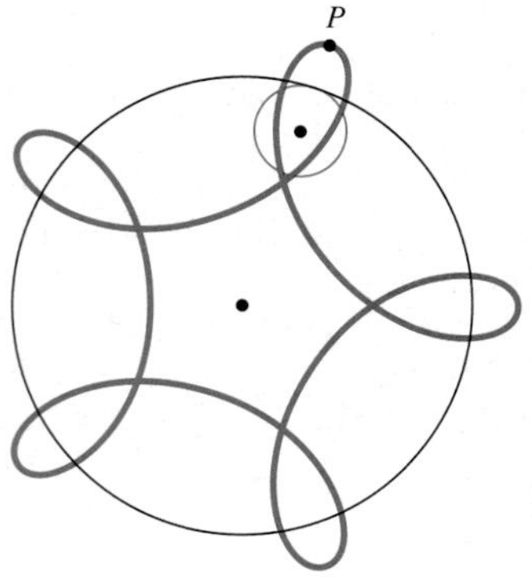

FIGURE 9.4.19 The hypotrochoid with $a = 10$, $b = 2$, $c = 4$.

A hypotrochoid is to a hypocycloid (Problem 34) as a trochoid is to a cycloid. Thus a hypotrochoid is traced by a point P on a spoke of a wheel of radius b as it rolls around inside a circle of radius a. If the distance of P from the center of the rolling wheel is $c > 0$, show that the hypotrochoid is described by the parametric equations

$$x = (a - b)\cos t + c\cos\left(\frac{a-b}{b}t\right), \qquad y = (a - b)\sin t - c\sin\left(\frac{a-b}{b}t\right).$$

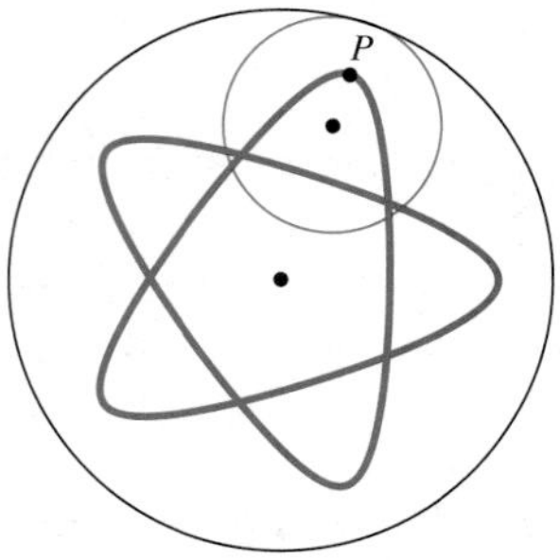

FIGURE 9.4.20 The hypotrochoid with $a = 10$, $b = 4$, $c = 2$.

Note that the hypotrochoid is a hypocycloid if $c = b$. There are a number of different ways a hypotrochoid can look. Figures 9.4.19 and 9.4.20 illustrate two possibilities. Experiment with different values of a, b, and c. What determines whether the trochoid has loops, cusps, or neither? If there are loops, what determines how many there are? Does a hypotrochoid always repeat itself after a finite number of turns around the origin? What happens if a is an integer but b is an irrational number?

Epitrochoids

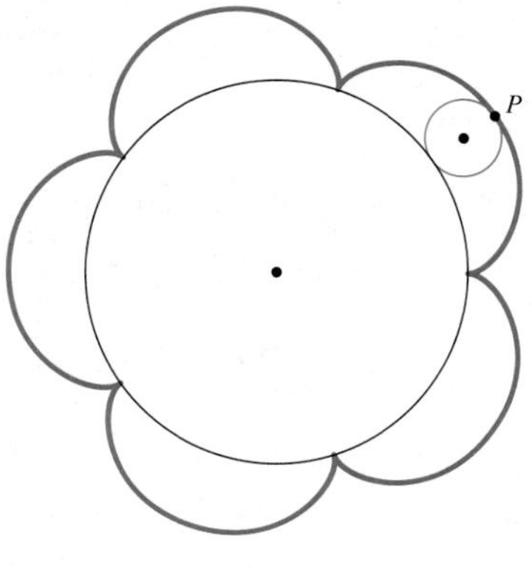

FIGURE 9.4.21 The epitrochoid with $a = 10$, $b = 2$, $c = 2$.

An *epitrochoid* is generated in the same way as a hypotrochoid, except now the small circle rolls around on the outside of the large circle. With the same notation otherwise, show that the epitrochoid is described by the parametric equations

$$x = (a + b)\cos t - c\cos\left(\frac{a+b}{b}t\right), \qquad y = (a + b)\sin t - c\sin\left(\frac{a+b}{b}t\right).$$

If $b = c$—so the point P lies on the rim of the rolling circle, then the epitrochoid is an *epicycloid* (illustrated in Fig. 9.4.21). Experiment with different values of a, b, and c, and investigate for epitrochoids the same questions posed previously for hypotrochoids.

9.5 INTEGRAL COMPUTATIONS WITH PARAMETRIC CURVES

In Chapter 6 we discussed the computation of a variety of geometric quantities associated with the graph $y = f(x)$ of a nonnegative function on the interval $[a, b]$. These included the following.

- The area under the curve:

$$A = \int_a^b y\,dx. \tag{1}$$

- The volume of revolution around the x-axis:

$$V_x = \int_a^b \pi y^2\,dx. \tag{2a}$$

- The volume of revolution around the y-axis:

$$V_y = \int_a^b 2\pi xy\,dx. \tag{2b}$$

- The arc length of the curve:

$$s = \int_0^s ds = \int_a^b \sqrt{1 + (dy/dx)^2}\,dx. \tag{3}$$

- The area of the surface of revolution around the x-axis:

$$S_x = \int_{x=a}^b 2\pi y\,ds. \tag{4a}$$

- The area of the surface of revolution around the y-axis:

$$S_y = \int_{x=a}^b 2\pi x\,ds. \tag{4b}$$

We substitute $y = f(x)$ into each of these integrals before we integrate from $x = a$ to $x = b$.

We now want to compute these same quantities for a smooth parametric curve

$$x = f(t), \quad y = g(t), \quad \alpha \leqq t \leqq \beta. \tag{5}$$

The area, volume, arc length, and surface integrals in Eqs. (1) through (4) can then be evaluated by making the formal substitutions

$$\begin{aligned} x &= f(t), & y &= g(t), \\ dx &= f'(t)\,dt, & dy &= g'(t)\,dt, \quad \text{and} \\ ds &= \sqrt{[f'(t)]^2 + [g'(t)]^2}\,dt. \end{aligned} \tag{6}$$

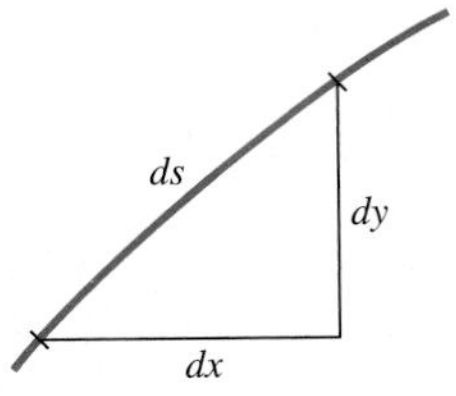

FIGURE 9.5.1 Nearly a right triangle for dx and dy close to zero.

The infinitesimal "right triangle" in Fig. 9.5.1 serves as a convenient device for remembering the latter substitution for ds. The Pythagorean theorem then leads to the symbolic manipulation

$$ds = \sqrt{dx^2 + dy^2} = \sqrt{\left(\frac{dx}{dt}\right)^2 + \left(\frac{dy}{dt}\right)^2}\,dt = \sqrt{[f'(t)]^2 + [g'(t)]^2}\,dt. \tag{7}$$

It simplifies the discussion to assume that the graph of the parametric curve in (5) resembles Fig. 9.5.2, in which $y = g(t) \geqq 0$ and $x = f(t)$ is either increasing on

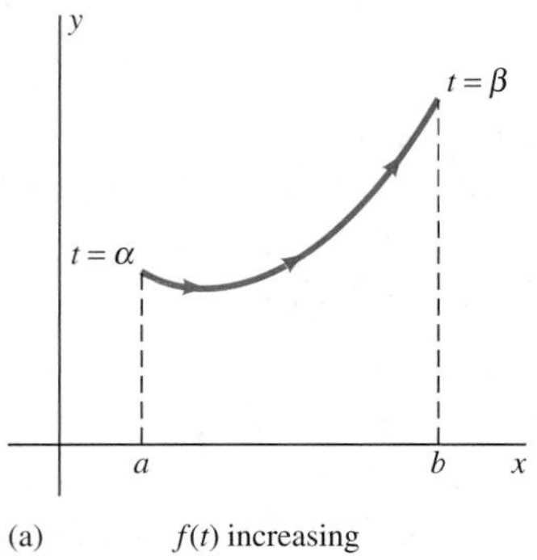

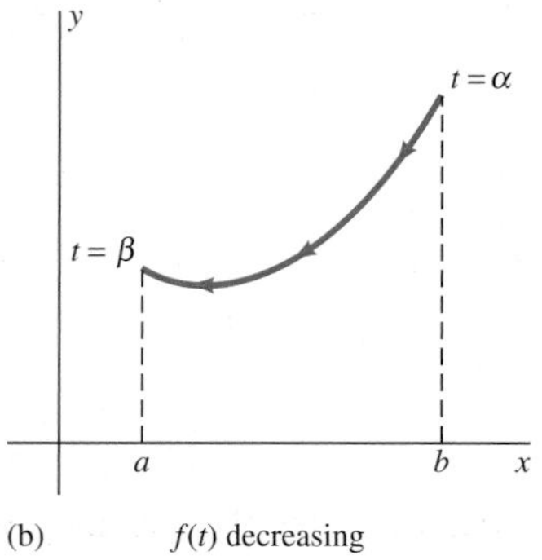

FIGURE 9.5.2 Tracing a parametrized curve: (a) $f(t)$ increasing; (b) $f(t)$ decreasing.

the entire interval $\alpha \leqq t \leqq \beta$ or is decreasing there. The two parts of Fig. 9.5.2 illustrate the two possibilities—whether as t increases the curve is traced in the positive x-direction from left to right, or in the negative x-direction from right to left. How and whether to take this direction of motion into account depends on which integral we are computing.

CASE 1 *Area and Volume of Revolution* To evaluate the integrals in (1) and (2), which involve dx, we integrate *either* from $t = \alpha$ to $t = \beta$ *or* from $t = \beta$ to $t = \alpha$—the proper choice of limits on t being the one that corresponds to traversing the curve in the positive x-direction *from left to right.* Specifically,

$$A = \int_\alpha^\beta g(t)\,f'(t)\,dt \quad \text{if } f(\alpha) < f(\beta),$$

whereas

$$A = \int_\beta^\alpha g(t)\,f'(t)\,dt \quad \text{if } f(\beta) < f(\alpha).$$

The validity of this method of evaluating the integrals in Eqs. (1) and (2) follows from Theorem 1 of Section 5.7, on integration by substitution.

CASE 2 *Arc Length and Surface Area* To evaluate the integrals in (3) and (4), which involve ds rather than dx, we integrate from $t = \alpha$ to $t = \beta$ irrespective of the direction of motion along the curve. To see why this is so, recall from Eq. (4) of Section 9.4 that $dy/dx = g'(t)/f'(t)$ if $f'(t) \neq 0$ on $[\alpha, \beta]$. Hence

$$s = \int_a^b \sqrt{1 + \left(\frac{dy}{dx}\right)^2}\,dx = \int_{f^{-1}(a)}^{f^{-1}(b)} \sqrt{1 + \left[\frac{g'(t)}{f'(t)}\right]^2}\,f'(t)\,dt.$$

Assuming that $f'(t) > 0$ if $f(\alpha) = a$ and $f(\beta) = b$, whereas $f'(t) < 0$ if $f(\alpha) = b$ and $f(\beta) = a$, it follows in either event that

$$s = \int_\alpha^\beta \sqrt{1 + \left[\frac{g'(t)}{f'(t)}\right]^2}\,|f'(t)|\,dt,$$

and so

$$s = \int_\alpha^\beta \sqrt{[f'(t)]^2 + [g'(t)]^2}\,dt = \int_\alpha^\beta \sqrt{\left(\frac{dx}{dt}\right)^2 + \left(\frac{dy}{dt}\right)^2}\,dt. \qquad \textbf{(8)}$$

This formula, derived under the assumption that $f'(t) \neq 0$ on $[\alpha, \beta]$, may be taken to be the *definition* of arc length for an arbitrary smooth parametric curve. Similarly, the area of a surface of revolution is defined for smooth parametric curves as the result of first substituting Eq. (6) into Eq. (4a) or (4b) and then integrating from $t = \alpha$ to $t = \beta$.

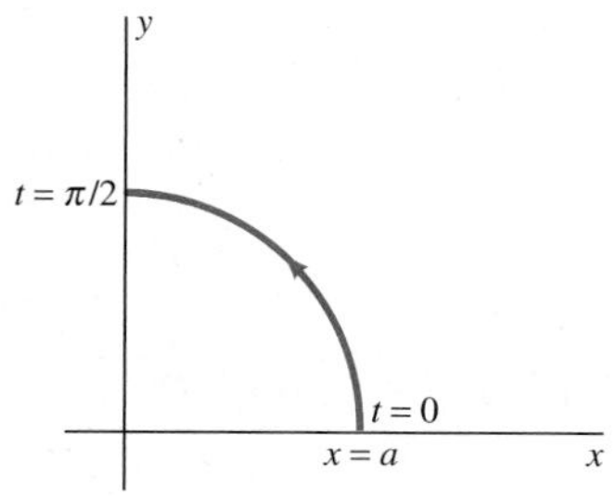

FIGURE 9.5.3 The quarter-circle of Example 1.

EXAMPLE 1 Use the parametrization $x = a\cos t$, $y = a\sin t$ $(0 \leqq t \leqq 2\pi)$ of the circle with center $(0, 0)$ and radius a to find (a) the area A of this circle; (b) the volume V of the sphere obtained by revolving the circle around the x-axis; and (c) the surface area S of this sphere.

Solution (a) The left-to-right direction along the quarter circle shown in Fig. 9.5.3 is from $t = \pi/2$ to $t = 0$, and $dx = -a\sin t\,dt$. Therefore Eq. (1) and multiplication by 4 give

$$A = 4\int_{t=\pi/2}^{0} y\,dx = 4\int_{\pi/2}^{0} (a\sin t)(-a\sin t)\,dt$$
$$= 4a^2\int_0^{\pi/2} \sin^2 t\,dt = 2a^2\int_0^{\pi/2}(1-\cos 2t)\,dt$$
$$= 2a^2\left[t - \frac{1}{2}\sin 2t\right]_0^{\pi/2} = 2a^2\cdot\frac{\pi}{2} = \pi a^2$$

for yet another derivation of the familiar formula $A = \pi a^2$ for the area of a circle of radius a.

(b) To calculate the volume of the sphere, we apply Eq. (2a) and double to get

$$V = 2\int_{t=\pi/2}^{0} \pi y^2\,dx$$
$$= 2\int_{\pi/2}^{0} \pi(a\sin t)^2(-a\sin t\,dt) = 2\pi a^3\int_0^{\pi/2}(1-\cos^2 t)\sin t\,dt$$
$$= 2\pi a^3\left[-\cos t + \frac{1}{3}\cos^3 t\right]_0^{\pi/2} = \frac{4}{3}\pi a^3.$$

(c) To find the surface area of the sphere, we calculate first the arc-length differential

$$ds = \sqrt{(-a\sin t)^2 + (a\cos t)^2}\,dt = a\,dt$$

of the parametrized curve. Then Eq. (4a) gives

$$S = 2\int_{t=0}^{\pi/2} 2\pi y\,ds = 2\int_0^{\pi/2} 2\pi(a\sin t)\cdot a\,dt$$
$$= 4\pi a^2\int_0^{\pi/2}\sin t\,dt = 4\pi a^2\Big[-\cos t\Big]_0^{\pi/2} = 4\pi a^2.$$ ◆

Of course, the results of Example 1 are familiar. In contrast, Example 2 requires the methods of this section.

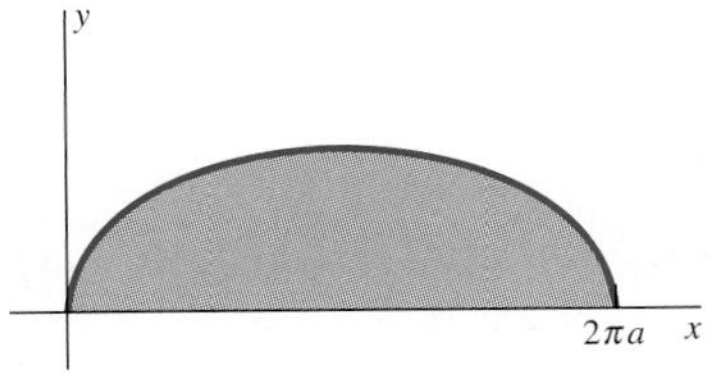

FIGURE 9.5.4 The cycloidal arch of Example 2.

EXAMPLE 2 Find the area under, and the arc length of, the cycloidal arch of Fig. 9.5.4. Its parametric equations are

$$x = a(t - \sin t), \quad y = a(1-\cos t), \quad 0 \leqq t \leqq 2\pi.$$

Solution Because $dx = a(1-\cos t)\,dt$ and the left-to-right direction along the curve is from $t = 0$ to $t = 2\pi$, Eq. (1) gives

$$A = \int_{t=0}^{2\pi} y\,dx$$
$$= \int_0^{2\pi} a(1-\cos t)\cdot a(1-\cos t)\,dt = a^2\int_0^{2\pi}(1-\cos t)^2\,dt$$

for the area. Now we use the half-angle identity

$$1 - \cos t = 2\sin^2\left(\frac{t}{2}\right)$$

and a consequence of Problem 58 in Section 7.3:

$$\int_0^{\pi} \sin^{2n} u\,du = \pi\cdot\frac{1}{2}\cdot\frac{3}{4}\cdot\frac{5}{6}\cdots\frac{2n-1}{2n}.$$

We thereby get

$$A = 4a^2 \int_0^{2\pi} \sin^4\left(\frac{t}{2}\right) dt = 8a^2 \int_0^{\pi} \sin^4 u \, du \qquad \left(u = \frac{t}{2}\right)$$
$$= 8a^2 \cdot \pi \cdot \frac{1}{2} \cdot \frac{3}{4} = 3\pi a^2$$

for the area under one arch of the cycloid. The arc-length differential is

$$ds = \sqrt{a^2(1 - \cos t)^2 + (a \sin t)^2}\, dt$$
$$= a\sqrt{2(1 - \cos t)}\, dt = 2a \sin\left(\frac{t}{2}\right) dt,$$

so Eq. (3) gives

$$s = \int_0^{2\pi} 2a \sin\frac{t}{2}\, dt = \left[-4a \cos\frac{t}{2}\right]_0^{2\pi} = 8a$$

for the length of one arch of the cycloid. ◆

Parametric Polar Coordinates

Suppose that a parametric curve is determined by giving its polar coordinates

$$r = r(t), \qquad \theta = \theta(t), \qquad \alpha \leqq t \leqq \beta$$

as functions of the parameter t. Then this curve is described in rectangular coordinates by the parametric equations

$$x(t) = r(t)\cos\theta(t), \qquad y(t) = r(t)\sin\theta(t), \qquad \alpha \leqq t \leqq \beta,$$

giving x and y as functions of t. The latter parametric equations may then be used in the integral formulas in Eqs. (1) through (4).

To compute ds, we first calculate the derivatives

$$\frac{dx}{dt} = (\cos\theta)\frac{dr}{dt} - (r\sin\theta)\frac{d\theta}{dt}, \qquad \frac{dy}{dt} = (\sin\theta)\frac{dr}{dt} + (r\cos\theta)\frac{d\theta}{dt}.$$

Upon substituting these expressions for dx/dt and dy/dt in Eq. (8) and making algebraic simplifications, we find that the arc-length differential in parametric polar coordinates is

$$ds = \sqrt{\left(\frac{dr}{dt}\right)^2 + \left(r\frac{d\theta}{dt}\right)^2}\, dt. \tag{9}$$

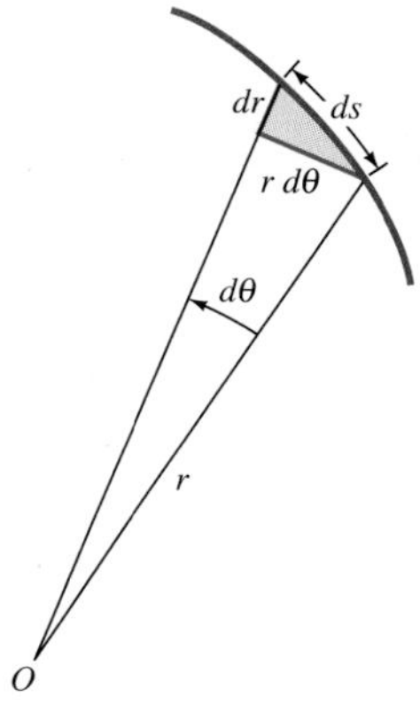

FIGURE 9.5.5 The differential triangle in polar coordinates.

In the case of a curve with the explicit polar-coordinate equation $r = f(\theta)$, we may use θ itself as the parameter. Then Eq. (9) takes the simpler form

$$ds = \sqrt{\left(\frac{dr}{d\theta}\right)^2 + r^2}\, d\theta. \tag{10}$$

The formula $ds = \sqrt{(dr)^2 + (r\,d\theta)^2}$, equivalent to Eq. (9), is easy to remember with the aid of the tiny "almost-triangle" shown in Fig. 9.5.5.

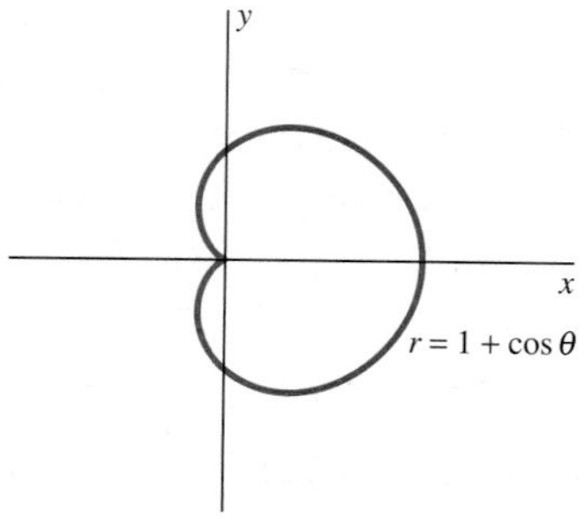

FIGURE 9.5.6 The cardioid of Example 3.

EXAMPLE 3 Find the perimeter (arc length) s of the cardioid with polar equation $r = 1 + \cos\theta$ (Fig. 9.5.6.). Find also the surface area S generated by revolving the cardioid around the x-axis.

Solution Because $dr/d\theta = -\sin\theta$, Eq. (10) and the identity

$$1 + \cos\theta = 2\cos^2\left(\frac{\theta}{2}\right) \tag{11}$$

give

$$ds = \sqrt{(-\sin\theta)^2 + (1+\cos\theta)^2}\,d\theta = \sqrt{2(1+\cos\theta)}\,d\theta$$
$$= \sqrt{4\cos^2\left(\frac{\theta}{2}\right)}\,d\theta = \left|2\cos\left(\frac{\theta}{2}\right)\right|\,d\theta.$$

Hence $ds = 2\cos(\theta/2)\,d\theta$ on the upper half of the cardioid, where $0 \leqq \theta \leqq \pi$, and thus $\cos(\theta/2) \geqq 0$. Therefore

$$s = 2\int_0^{\pi} 2\cos\frac{\theta}{2}\,d\theta = 8\left[\sin\frac{\theta}{2}\right]_0^{\pi} = 8.$$

The surface area of revolution around the x-axis (Fig. 9.5.7) is given by

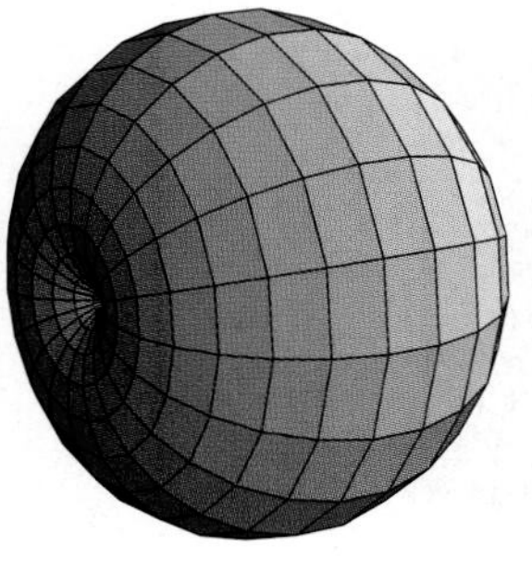

FIGURE 9.5.7 The surface generated by rotating the cardioid around the x-axis.

$$S = \int_{\theta=0}^{\pi} 2\pi y\,ds$$
$$= \int_{\theta=0}^{\pi} 2\pi(r\sin\theta)\,ds = \int_0^{\pi} 2\pi(1+\cos\theta)(\sin\theta)\cdot 2\cos\left(\frac{\theta}{2}\right)d\theta$$
$$= 16\pi\int_0^{\pi}\cos^4\frac{\theta}{2}\sin\frac{\theta}{2}\,d\theta = 16\pi\left[-\frac{2}{5}\cos^5\frac{\theta}{2}\right]_0^{\pi} = \frac{32\pi}{5},$$

using the identity

$$\sin\theta = 2\sin\left(\frac{\theta}{2}\right)\cos\left(\frac{\theta}{2}\right)$$

as well as the identity in Eq. (11). ◆

9.5 TRUE/FALSE STUDY GUIDE

9.5 CONCEPTS: QUESTIONS AND DISCUSSION

1. If the circle of radius a is parametrized by $x = a\cos t$, $y = a\sin t$ as in Example 1, explain carefully why the integral

$$\int_{t=0}^{2\pi} y\,dx = \int_{t=0}^{\pi} y\,dx + \int_{t=\pi}^{2\pi} y\,dx$$

does *not* give the correct area of the circle. Relate the two integrals on the right to the upper and lower halves of the circle.

2. If the circle of radius a is parametrized by $x = a\sin\pi t$, $y = a\cos\pi t$, explain carefully why the integral

$$\int_{t=0}^{2} y\,dx = \int_{t=0}^{1/2} y\,dx + \int_{t=1/2}^{3/2} y\,dx + \int_{t=3/2}^{2} y\,dx$$

does give the correct area. Relate the three integrals on the right to appropriate parts of the circular area.

9.5 PROBLEMS

In Problems 1 through 6, find the area of the region that lies between the given parametric curve and the x-axis.

1. $x = t^3,\ y = 2t^2 + 1;\quad -1 \leqq t \leqq 1$
2. $x = e^{3t},\ y = e^{-t};\quad 0 \leqq t \leqq \ln 2$
3. $x = \cos t,\ y = \sin^2 t;\quad 0 \leqq t \leqq \pi$
4. $x = 2 - 3t,\ y = e^{2t};\quad 0 \leqq t \leqq 1$
5. $x = \cos t,\ y = e^t;\quad 0 \leqq t \leqq \pi$
6. $x = 1 - e^t,\ y = 2t + 1;\quad 0 \leqq t \leqq 1$

In Problems 7 through 10, find the volume obtained by revolving around the x-axis the region described in the given problem.

7. Problem 1
8. Problem 2
9. Problem 3
10. Problem 5

In Problems 11 through 16, find the arc length of the given curve.

11. $x = 2t,\ y = \frac{2}{3}t^{3/2};\quad 5 \leqq t \leqq 12$
12. $x = \frac{1}{2}t^2,\ y = \frac{1}{3}t^3;\quad 0 \leqq t \leqq 1$

13. $x = \sin t - \cos t$, $y = \sin t + \cos t$; $\frac{1}{4}\pi \leqq t \leqq \frac{1}{2}\pi$

14. $x = e^t \sin t$, $y = e^t \cos t$; $0 \leqq t \leqq \pi$

15. $r = e^{\theta/2}$; $0 \leqq \theta \leqq 4\pi$

16. $r = \theta$; $2\pi \leqq \theta \leqq 4\pi$

In Problems 17 through 22, find the area of the surface of revolution generated by revolving the given curve around the indicated axis.

17. $x = 1 - t$, $y = 2\sqrt{t}$, $1 \leqq t \leqq 4$; the x-axis

18. $x = 2t^2 + t^{-1}$, $y = 8\sqrt{t}$, $1 \leqq t \leqq 2$; the x-axis

19. $x = t^3$, $y = 2t + 3$, $-1 \leqq t \leqq 1$; the y-axis

20. $x = 2t + 1$, $y = t^2 + t$, $0 \leqq t \leqq 3$; the y-axis

21. $r = 4 \sin \theta$, $0 \leqq \theta \leqq \pi$; the x-axis

22. $r = e^\theta$, $0 \leqq \theta \leqq \frac{1}{2}\pi$; the y-axis

23. Find the volume generated by revolving around the x-axis the region under the cycloidal arch of Example 2.

24. Find the area of the surface generated by revolving around the x-axis the cycloidal arch of Example 2.

25. Use the parametrization $x = a \cos t$, $y = b \sin t$ to find: (a) the area bounded by the ellipse $x^2/a^2 + y^2/b^2 = 1$; (b) the volume of the ellipsoid generated by revolving this ellipse around the x-axis.

26. Find the area bounded by the loop of the parametric curve $x = t^2$, $y = t^3 - 3t$ of Problem 25 in Section 9.4.

27. Use the parametrization $x = t \cos t$, $y = t \sin t$ of the Archimedean spiral to find the arc length of the first full turn of this spiral (corresponding to $0 \leqq t \leqq 2\pi$).

28. The circle $(x - b)^2 + y^2 = a^2$ with radius $a < b$ and center $(b, 0)$ can be parametrized by

$$x = b + a \cos t, \quad y = a \sin t, \quad 0 \leqq t \leqq 2\pi.$$

Find the surface area of the torus obtained by revolving this circle around the y-axis (Fig. 9.5.8).

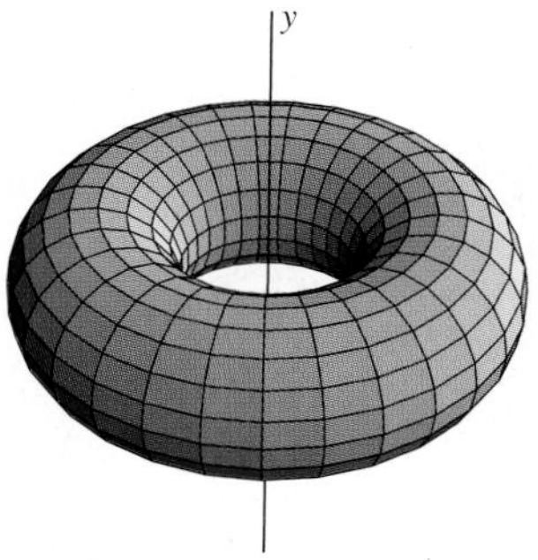

FIGURE 9.5.8 The torus of Problem 28.

29. The *astroid* (four-cusped hypocycloid) has equation $x^{2/3} + y^{2/3} = a^{2/3}$ (Fig. 9.4.15) and the parametrization

$$x = a \cos^3 t, \quad y = a \sin^3 t, \quad 0 \leqq t \leqq 2\pi.$$

Find the area of the region bounded by the astroid.

30. Find the total length of the astroid of Problem 29.

31. Find the area of the surface obtained by revolving the astroid of Problem 29 around the x-axis.

32. Find the area of the surface generated by revolving the lemniscate $r^2 = 2a^2 \cos 2\theta$ around the y-axis (Fig. 9.5.9). [*Suggestion:* Use Eq. (10); note that $r\,dr = -2a^2 \sin 2\theta\, d\theta$.]

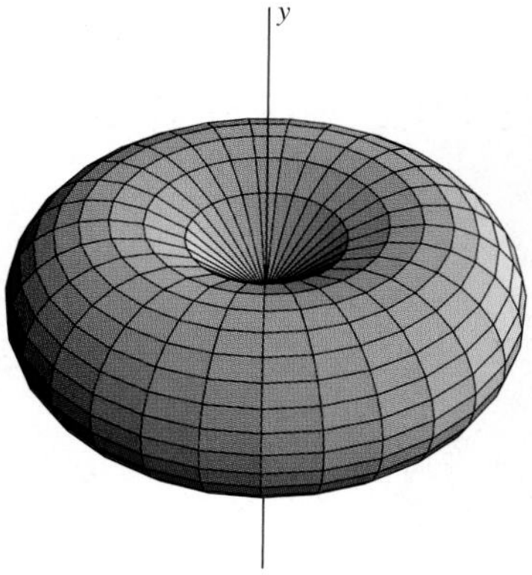

FIGURE 9.5.9 The surface generated by rotating the lemniscate of Problem 32 around the y-axis.

33. Figure 9.5.10 shows the graph of the parametric curve

$$x = t^2\sqrt{3}, \quad y = 3t - \tfrac{1}{3}t^3.$$

The shaded region is bounded by the part of the curve for which $-3 \leqq t \leqq 3$. Find its area.

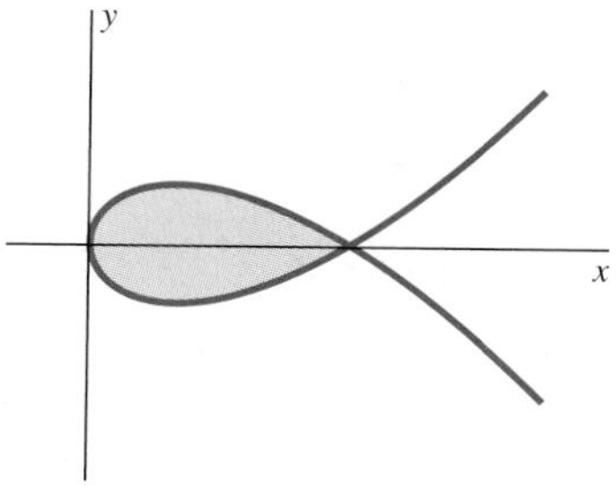

FIGURE 9.5.10 The parametric curve of Problems 33 through 36.

34. Find the arc length of the loop of the curve of Problem 33.

35. Find the volume of the solid obtained by revolving around the x-axis the shaded region in Fig. 9.5.10.

36. Find the surface area of revolution generated by revolving around the x-axis the loop of Fig. 9.5.10.

37. (a) With reference to Problem 30 and Fig. 9.4.14 in Section 9.4, show that the arc length of the first-quadrant loop of the folium of Descartes is

$$s = 6\int_0^1 \frac{\sqrt{1 + 4t^2 - 4t^3 - 4t^5 + 4t^6 + t^8}}{(1 + t^3)^2}\, dt.$$

(b) Use a programmable calculator or a computer to approximate this length.

38. Find the surface area generated by rotating around the y-axis the cycloidal arch of Example 2. [*Suggestion:* $\sqrt{x^2} = x$ only if $x \geqq 0$.]

39. Find the volume generated by rotating around the y-axis the region under the cycloidal arch of Example 2.

40. Suppose that after a string is wound clockwise around a circle of radius a, its free end is at the point $A(a, 0)$. (See Fig. 9.5.11.) Now the string is unwound, always stretched tight so the unwound portion TP is tangent to the circle at T. The locus of the string's free endpoint P is called the **involute** of the circle.

(a) Show that the parametric equations of the involute (in terms of the angle t of Fig. 9.5.11) are

$$x = a(\cos t + t \sin t), \quad y = a(\sin t - t \cos t).$$

(b) Find the length of the involute from $t = 0$ to $t = \pi$.

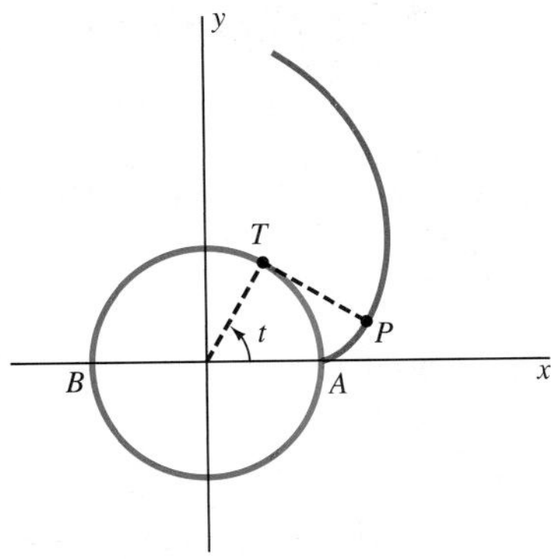

FIGURE 9.5.11 The involute of a circle.

41. Suppose that the circle of Problem 40 is a water tank and the "string" is a rope of length πa. It is anchored at the point B opposite A. Figure 9.5.12 depicts the total area that can be grazed by a cow tied to the free end of the rope. Find this total area. (The three labeled arcs of the curve in the figure represent, respectively, an involute APQ generated as the cow unwinds the rope in the counterclockwise direction, a semicircle QR of radius πa centered at B, and an involute RSA generated as the cow winds the rope around the tank proceeding in the counterclockwise direction from B to A. These three arcs form a closed curve that resembles a cardioid, and the cow can reach every point that lies inside this curve and outside the original circle.)

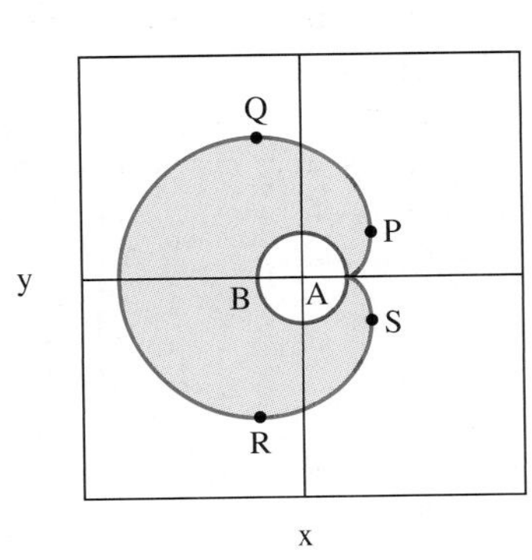

FIGURE 9.5.12 The area that the cow of Problem 41 can graze.

42. Now suppose that the rope of the previous problem has length $2\pi a$ and is anchored at the point A before being wound completely around the tank. Now find the total area that the cow can graze. Figure 9.5.13 shows an involute APQ, a semicircle QR of radius $2\pi a$ centered at A, and an involute RSA. The cow can reach every point that lies inside the outer curve and outside the original circle.

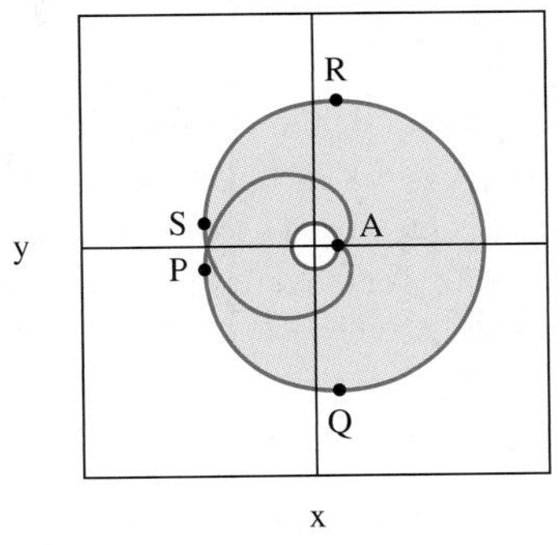

FIGURE 9.5.13 The area that the cow of Problem 42 can graze.

In Problems 43 through 54, use a graphing calculator or computer algebra system as appropriate. Approximate (by integrating numerically) the desired quantity if it cannot be calculated exactly.

43. Find the total arc length of the 3-leaved rose $r = 3 \sin 3\theta$ of Fig. 9.3.12.

44. Find the total surface area generated by rotating around the y-axis the 3-leaved rose of Problem 43.

45. Find the total length of the 4-leaved rose $r = 2 \cos 2\theta$ of Fig. 9.2.12.

46. Find the total surface area generated by revolving around the x-axis the 4-leaved rose of Problem 45.

47. Find the total arc length of the limaçon (both loops) $r = 5 + 9 \cos \theta$ of Fig. 9.2.25.

48. Find the total surface area generated by revolving around the x-axis the limaçon of Problem 47.

49. Find the total arc length (all seven loops) of the polar curve $r = \cos(\frac{7}{3}\theta)$ of Fig. 9.5.14.

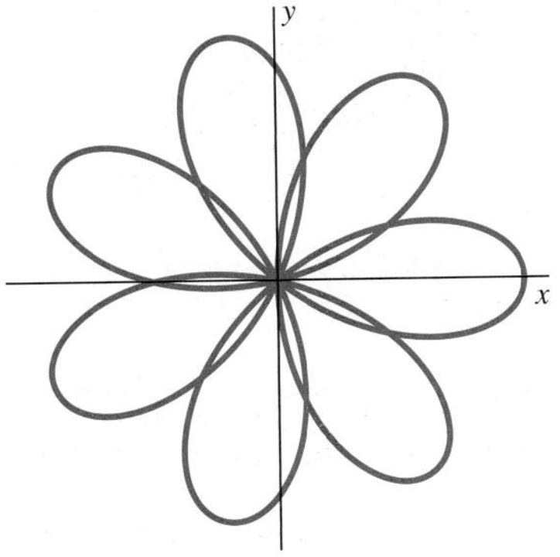

FIGURE 9.5.14 The curve $r = \cos(\frac{7}{3}\theta)$ of Problem 49.

50. Find the total arc length of the figure-8 curve $x = \sin t$, $y = \sin 2t$ of Fig. 9.4.13.

51. Find the total surface area and volume generated by revolving around the x-axis the figure-8 curve of Problem 50.

52. Find the total surface area and volume generated by revolving around the y-axis the figure-8 curve of Problem 50.

53. Find the total arc length of the Lissajous curve $x = \cos 3t$, $y = \sin 5t$ of Fig. 9.4.4.

54. Find the total arc length of the epitrochoid $x = 8\cos t - 5\cos 4t$, $y = 8\sin t - 5\sin 4t$ of Fig. 9.5.15.

55. Frank A. Farris of Santa Clara University, while designing a computer laboratory exercise for his calculus students, discovered an extremely lovely curve with the parametrization

$$x(t) = \cos t + \tfrac{1}{2}\cos 7t + \tfrac{1}{3}\sin 17t,$$

$$y(t) = \sin t + \tfrac{1}{2}\sin 7t + \tfrac{1}{3}\cos 17t.$$

For information on what these equations represent, see his article "Wheels on Wheels on Wheels—Surprising Symmetry" in the June 1996 issue of *Mathematics Magazine*. Plot these equations so you can enjoy this extraordinary figure, then numerically integrate to approximate the length of its graph. What kind of symmetry does the graph have? Is this predictable from the coefficients of t in the parametric equations?

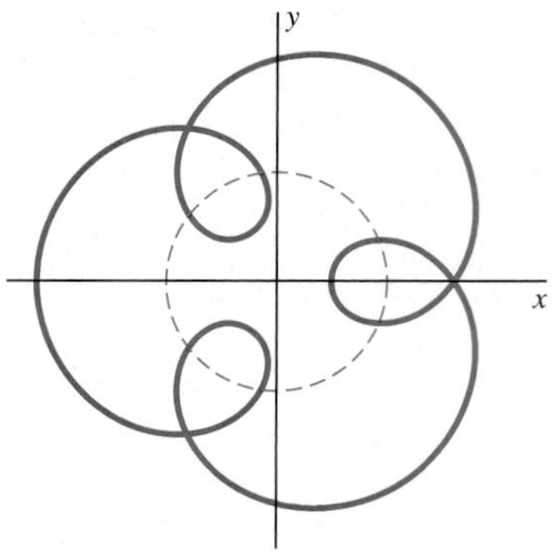

FIGURE 9.5.15 The epitrochoid of Problem 54.

9.5 Project: Moon Orbits and Race Tracks

The investigations in this project call for the use of numerical integration techniques (using a calculator or computer) to approximate the parametric arc-length integral

$$s = \int_a^b \sqrt{[x'(t)]^2 + [y'(t)]^2}\, dt. \tag{1}$$

Consider the ellipse with equation

$$\frac{x^2}{a^2} + \frac{y^2}{b^2} = 1 \quad (a > b) \tag{2}$$

and *eccentricity* $\epsilon = \sqrt{1 - (b/a)^2}$. Substitute the parametrization

$$x = a\cos t, \quad y = b\sin t \tag{3}$$

into Eq. (1) to show that the perimeter of the ellipse is given by the *elliptic integral*

$$p = 4a \int_0^{\pi/2} \sqrt{1 - \epsilon^2 \cos^2 t}\, dt. \tag{4}$$

This integral is known to be nonelementary if $0 < \epsilon < 1$. A common simple approximation to it is

$$p \approx \pi(A + R), \tag{5}$$

where

$$A = \frac{1}{2}(a + b) \quad \text{and} \quad R = \sqrt{\frac{a^2 + b^2}{2}}$$

denote the arithmetic mean and root-square mean, respectively, of the semiaxes a and b of the ellipse.

Investigation A As a warm-up, consider the ellipse whose major and minor semiaxes a and b are, respectively, the largest and smallest nonzero digits of your student I.D. number. For this ellipse, compare the arc-length estimate given by (5) and by numerical evaluation of the integral in (4).

Investigation B If we ignore the perturbing effects of the sun and the planets other than the earth, the orbit of the moon is an almost perfect ellipse with the earth at one focus. Assume that this ellipse has major semiaxis $a = 384{,}403$ km (exactly) and eccentricity $\epsilon = 0.0549$ (exactly). Approximate the perimeter p of this ellipse [using Eq. (4)] to the nearest meter.

Investigation C Suppose that you are designing an elliptical auto racetrack. Choose semiaxes for *your* racetrack so that its perimeter will be somewhere between a half mile and two miles. Your task is to construct a table with *time* and *speed* columns that an observer can use to determine the average speed of a particular car as it circles the track. The times listed in the first column should correspond to speeds up to perhaps 150 mi/h. The observer clocks a car's circuit of the track and locates its time for the lap in the first column of the table. The corresponding figure in the second column then gives the car's average speed (in miles per hour) for that circuit of the track. Your report should include a convenient table to use in this way—so you can successfully sell it to racetrack patrons attending the auto races.

9.6 CONIC SECTIONS AND APPLICATIONS

Here we discuss in more detail the three types of conic sections—parabolas, ellipses, and hyperbolas—that were introduced in Section 9.1.

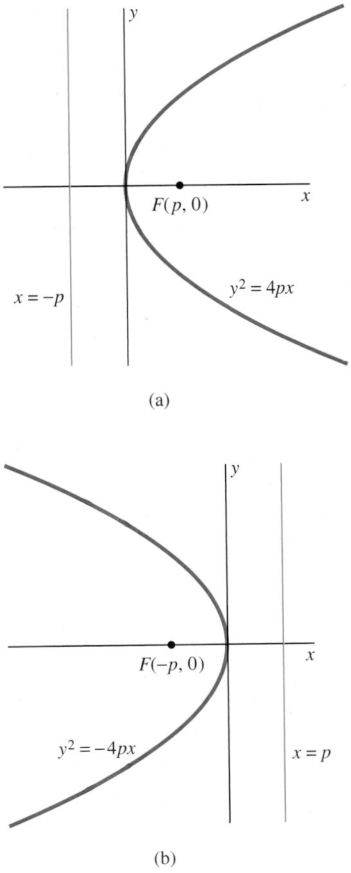

FIGURE 9.6.1 Two parabolas with vertical directrices.

The Parabola

The case $e = 1$ of Example 3 in Section 9.1 is motivation for this formal definition.

DEFINITION The Parabola
A **parabola** is the set of all points P in the plane that are equidistant from a fixed point F (called the **focus** of the parabola) and a fixed line L (called the parabola's **directrix**) not containing F.

If the focus of the parabola is $F(p, 0)$ and its directrix is the vertical line $x = -p$, $p > 0$, then it follows from Eq. (12) of Section 9.1 that the equation of this parabola is

$$y^2 = 4px. \quad \textbf{(1)}$$

When we replace x with $-x$ both in the equation and in the discussion that precedes it, we get the equation of the parabola whose focus is $(-p, 0)$ and whose directrix is the vertical line $x = p$. The new parabola has equation

$$y^2 = -4px. \quad \textbf{(2)}$$

The old and new parabolas appear in Fig. 9.6.1.

We could also interchange x and y in Eq. (1). This would give the equation of a parabola whose focus is $(0, p)$ and whose directrix is the horizontal line $y = -p$.

This parabola opens upward, as in Fig. 9.6.2(a); its equation is

$$x^2 = 4py. \tag{3}$$

Finally, we replace y with $-y$ in Eq. (3). This gives the equation

$$x^2 = -4py \tag{4}$$

of a parabola opening downward, with focus $(0, -p)$ and with directrix $y = p$, as in Fig. 9.6.2(b).

Each of the parabolas discussed so far is symmetric around one of the coordinate axes. The line around which a parabola is symmetric is called the **axis** of the parabola. The point of a parabola midway between its focus and its directrix is called the **vertex** of the parabola. The vertex of each parabola that we discussed in connection with Eqs. (1) through (4) is the origin $(0, 0)$.

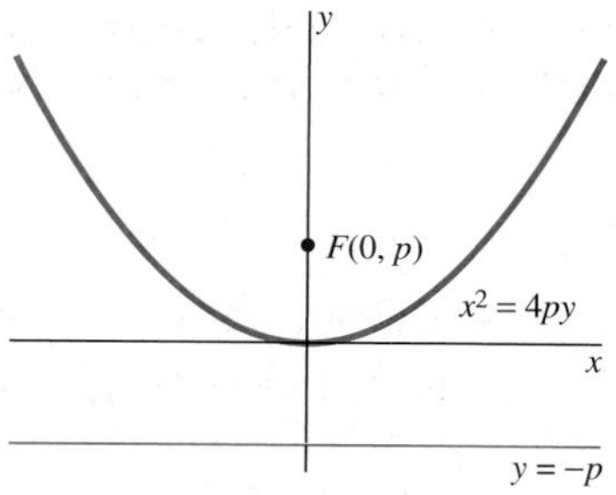

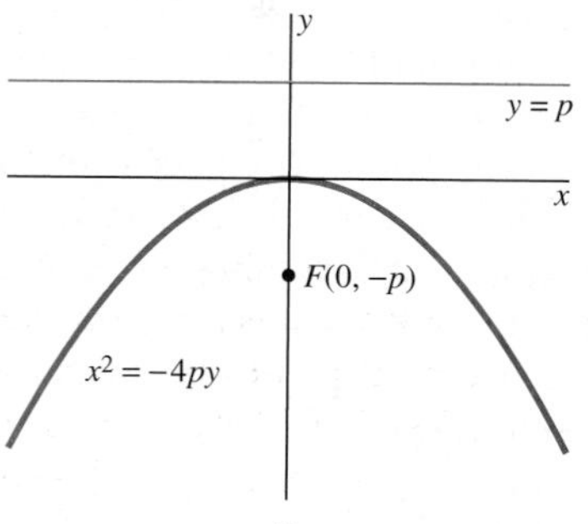

FIGURE 9.6.2 Two parabolas with horizontal directrices: (a) opening upward; (b) opening downward.

EXAMPLE 1 Determine the focus, directrix, axis, and vertex of the parabola $x^2 = 12y$.

Solution We write the given equation as $x^2 = 4 \cdot (3y)$. In this form it matches Eq. (3) with $p = 3$. Hence the focus of the given parabola is $(0, 3)$ and its directrix is the horizontal line $y = -3$. The y-axis is its axis of symmetry, and the parabola opens upward from its vertex at the origin. ◆

Suppose that we begin with the parabola of Eq. (1) and translate it in such a way that its vertex moves to the point (h, k). Then the translated parabola has equation

$$(y - k)^2 = 4p(x - h). \tag{1a}$$

The new parabola has focus $F(p + h, k)$ and its directrix is the vertical line $x = -p + h$ (Fig. 9.6.3). Its axis is the horizontal line $y = k$.

We can obtain the translates of the other three parabolas in Eqs. (2) through (4) in the same way. If the vertex is moved from the origin to the point (h, k), then the three equations take these forms:

$$(y - k)^2 = -4p(x - h), \tag{2a}$$

$$(x - h)^2 = 4p(y - k), \quad \text{and} \tag{3a}$$

$$(x - h)^2 = -4p(y - k). \tag{4a}$$

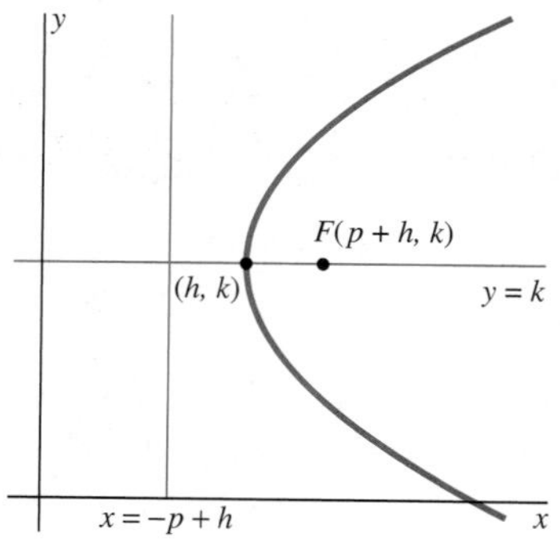

FIGURE 9.6.3 A translation of the parabola $y^2 = 4px$.

Equations (1a) and (2a) both take the general form

$$y^2 + Ax + By + C = 0 \quad (A \neq 0), \tag{5}$$

whereas Eqs. (3a) and (4a) both take the general form

$$x^2 + Ax + By + C = 0 \quad (B \neq 0). \tag{6}$$

What is significant about Eqs. (5) and (6) is what they have in common: Both are linear in one of the coordinate variables and quadratic in the other. In fact, we can reduce *any* such equation to one of the standard forms in Eqs. (1a) through (4a) by completing the square in the coordinate variable that appears quadratically. This means that the graph of any equation of the form of either Eqs. (5) or (6) is a parabola. The features of the parabola can be read from the standard form of its equation, as in Example 2.

EXAMPLE 2 Determine the graph of the equation

$$4y^2 - 8x - 12y + 1 = 0.$$

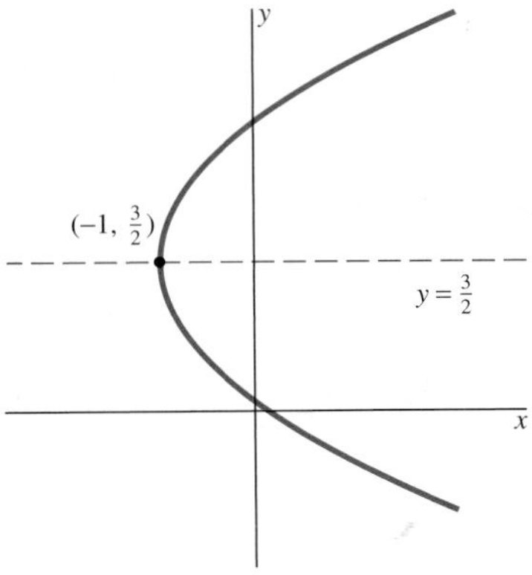

FIGURE 9.6.4 The parabola of Example 2.

Solution This equation is linear in x and quadratic in y. We divide through by the coefficient of y^2 and then collect on one side of the equation all terms that include y:

$$y^2 - 3y = 2x - \tfrac{1}{4}.$$

Then we complete the square in the variable y and thus find that

$$y^2 - 3y + \tfrac{9}{4} = 2x - \tfrac{1}{4} + \tfrac{9}{4} = 2x + 2 = 2(x + 1).$$

The final step is to write in the form $4p(x - h)$ the terms on the right-hand side that include x:

$$\left(y - \tfrac{3}{2}\right)^2 = 4 \cdot \tfrac{1}{2} \cdot (x + 1).$$

This equation has the form of Eq. (1a) with $p = \frac{1}{2}$, $h = -1$, and $k = \frac{3}{2}$. Thus the graph is a parabola that opens to the right from its vertex at $(-1, \frac{3}{2})$. Its focus is at $(-\frac{1}{2}, \frac{3}{2})$, its directrix is the vertical line $x = -\frac{3}{2}$, and its axis is the horizontal line $y = \frac{3}{2}$. It appears in Fig. 9.6.4. ◆

Applications of Parabolas

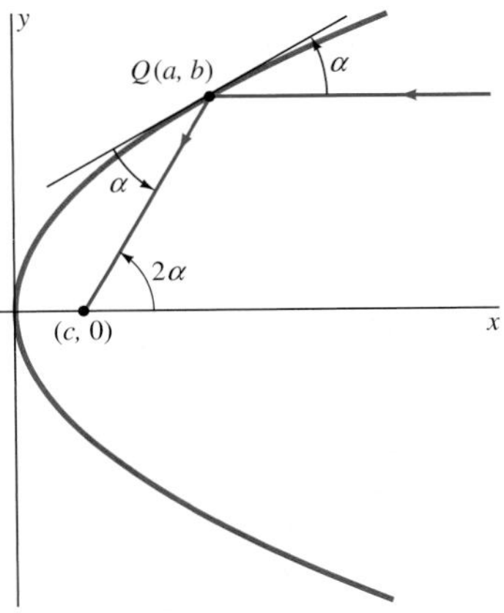

FIGURE 9.6.5 The reflection property of the parabola: $\alpha = \beta$.

The parabola $y^2 = 4px (p > 0)$ is shown in Fig. 9.6.5 along with an incoming ray of light traveling to the left and parallel to the x-axis. This light ray strikes the parabola at the point $Q(a, b)$ and is reflected toward the x-axis, which it meets at the point $(c, 0)$. The light ray's angle of reflection must equal its angle of incidence, which is why both of these angles—measured with respect to the tangent line L at Q—are labeled α in the figure. The angle vertical to the angle of incidence is also equal to α. Hence, because the incoming ray is parallel to the x-axis, the angle the reflected ray makes with the x-axis at $(c, 0)$ is 2α.

Using the points Q and $(c, 0)$ to compute the slope of the reflected light ray, we find that

$$\frac{b}{a - c} = \tan 2\alpha = \frac{2 \tan \alpha}{1 - \tan^2 \alpha}.$$

(The second equality follows from a trigonometric identity in Problem 64 of Section 6.8.) But the angle α is related to the slope of the tangent line L at Q. To find that slope, we begin with

$$y = 2\sqrt{px} = 2(px)^{1/2}$$

and compute

$$\frac{dy}{dx} = \left(\frac{p}{x}\right)^{1/2}.$$

Hence the slope of L is both $\tan \alpha$ and dy/dx evaluated at (a, b); that is,

$$\tan \alpha = \left(\frac{p}{a}\right)^{1/2}.$$

Therefore,

$$\frac{b}{a - c} = \frac{2 \tan \alpha}{1 - \tan^2 \alpha} = \frac{2\sqrt{\dfrac{p}{a}}}{1 - \dfrac{p}{a}} = \frac{2\sqrt{pa}}{a - p} = \frac{b}{a - p},$$

because $b = 2\sqrt{pa}$. Hence $c = p$. The surprise is that c is independent of a and b and depends only on the equation $y^2 = 4px$ of the parabola. Therefore *all* incoming light rays parallel to the x-axis will be reflected to the single point $F(p, 0)$. This is why F is called the *focus* of the parabola.

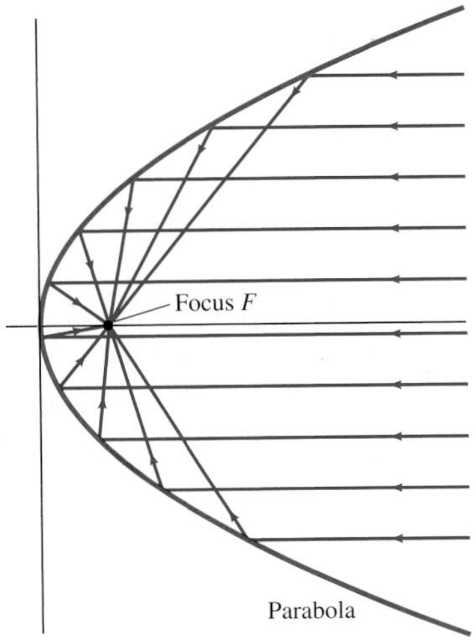

FIGURE 9.6.6 Incident rays parallel to the axis reflect through the focus.

This **reflection property** of the parabola is exploited in the design of parabolic mirrors. Such a mirror has the shape of the surface obtained by revolving a parabola around its axis of symmetry. Then a beam of incoming light rays parallel to the axis will be focused at F, as shown in Fig. 9.6.6. The reflection property can also be used in reverse—rays emitted at the focus are reflected in a beam parallel to the axis, thus keeping the light beam intense. Moreover, applications are not limited to light rays alone; parabolic mirrors are used in visual and radio telescopes, radar antennas, searchlights, automobile headlights, microphone systems, satellite ground stations, and solar heating devices.

Galileo discovered early in the seventeenth century that the trajectory of a projectile fired from a gun is a parabola (under the assumptions that air resistance can be ignored and that the gravitational acceleration remains constant). Suppose that a projectile is fired with initial velocity v_0 at time $t = 0$ from the origin and at an angle α of inclination from the horizontal x-axis. Then the initial velocity of the projectile splits into the components

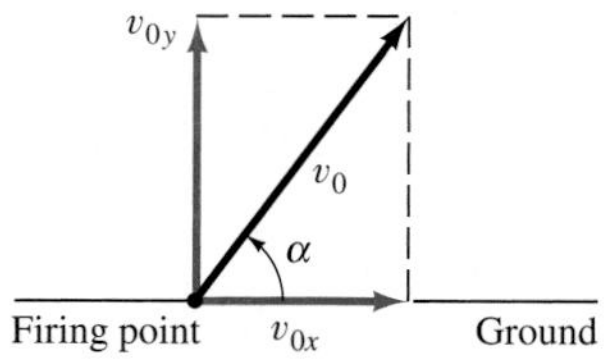

FIGURE 9.6.7 Resolution of the initial velocity v_0 into its horizontal and vertical components.

$$v_{0x} = v_0 \cos\alpha \quad \text{and} \quad v_{0y} = v_0 \sin\alpha,$$

as indicated in Fig. 9.6.7. The fact that the projectile continues to move horizontally with *constant* speed v_{0x}, together with Eq. (34) of Section 5.2, implies that its x- and y-coordinates after t seconds are

$$x = (v_0 \cos\alpha)t, \tag{7}$$

$$y = -\tfrac{1}{2}gt^2 + (v_0 \sin\alpha)t. \tag{8}$$

By substituting $t = x/(v_0 \cos\alpha)$ from Eq. (7) into Eq. (8) and then completing the square, we can derive (as in Problem 70) an equation of the form

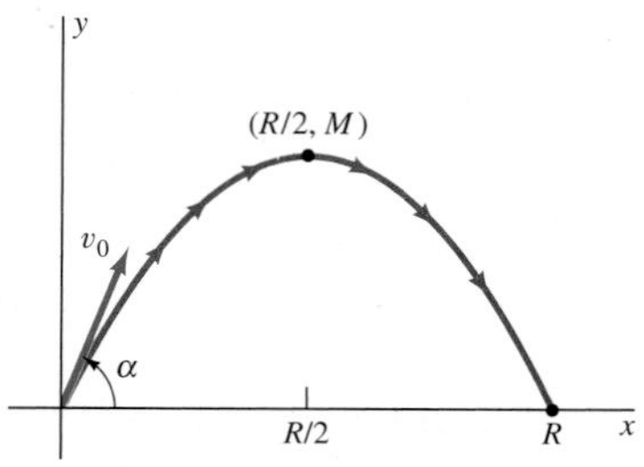

FIGURE 9.6.8 The trajectory of the projectile, showing its maximum altitude M and its range R.

$$y - M = -4p\left(x - \tfrac{1}{2}R\right)^2. \tag{9}$$

Here,

$$M = \frac{v_0^2 \sin^2\alpha}{2g} \tag{10}$$

is the maximum height attained by the projectile, and

$$R = \frac{v_0^2 \sin 2\alpha}{g} \tag{11}$$

is its **range,** the horizontal distance the projectile will travel before it returns to the ground. Thus its trajectory is the parabola shown in Fig. 9.6.8.

The Ellipse

An ellipse is a conic section with eccentricity e less than 1, as in Example 3 of Section 9.1.

DEFINITION The Ellipse

Suppose that $e < 1$, and let F be a fixed point and L a fixed line not containing F. The **ellipse** with **eccentricity** e, **focus** F, and **directrix** L is the set of all points P such that the distance $|PF|$ is e times the (perpendicular) distance from P to the line L.

The equation of the ellipse is especially simple if F is the point $(c, 0)$ on the x-axis and L is the vertical line $x = c/e^2$. The case $c > 0$ is shown in Fig. 9.6.9. If Q is the point $(c/e^2, y)$, then PQ is the perpendicular from $P(x, y)$ to L. The condition $|PF| = e|PQ|$ then gives

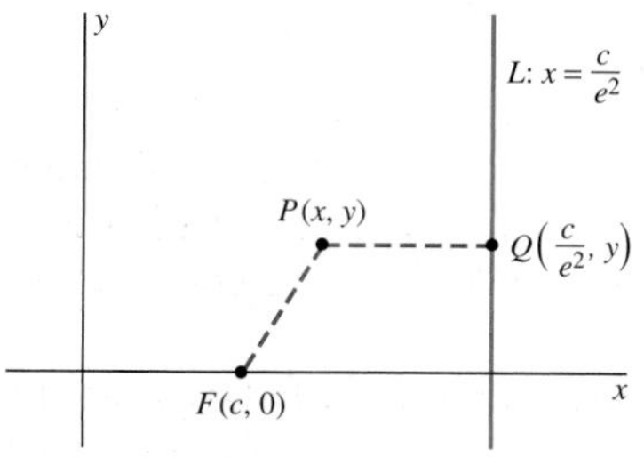

FIGURE 9.6.9 Ellipse: focus F, directrix L, eccentricity e.

$$(x - c)^2 + y^2 = e^2\left(x - \frac{c}{e^2}\right)^2;$$

$$x^2 - 2cx + c^2 + y^2 = e^2x^2 - 2cx + \frac{c^2}{e^2};$$

$$x^2(1 - e^2) + y^2 = c^2\left(\frac{1}{e^2} - 1\right) = \frac{c^2}{e^2}(1 - e^2).$$

Thus

$$x^2(1-e^2)+y^2=a^2(1-e^2),$$

where

$$a=\frac{c}{e}. \tag{12}$$

We divide both sides of the next-to-last equation by $a^2(1-e^2)$ and get

$$\frac{x^2}{a^2}+\frac{y^2}{a^2(1-e^2)}=1.$$

Finally, with the aid of the fact that $e<1$, we may let

$$b^2=a^2(1-e^2)=a^2-c^2. \tag{13}$$

Then the equation of the ellipse with focus $(c, 0)$ and directrix $x=c/e^2=a/e$ takes the simple form

$$\frac{x^2}{a^2}+\frac{y^2}{b^2}=1. \tag{14}$$

We see from Eq. (14) that this ellipse is symmetric around both coordinate axes. Its x-intercepts are $(\pm a, 0)$ and its y-intercepts are $(0, \pm b)$. The points $(\pm a, 0)$ are called the **vertices** of the ellipse, and the line segment joining them is called its **major axis.** The line segment joining $(0, b)$ and $(0, -b)$ is called the **minor axis** [note from Eq. (13) that $b<a$]. The alternative form

$$a^2=b^2+c^2 \tag{15}$$

of Eq. (13) is the Pythagorean relation for the right triangle of Fig. 9.6.10. Indeed, visualization of this triangle is an excellent way to remember Eq. (15). The numbers a and b are the lengths of the major and minor **semiaxes,** respectively.

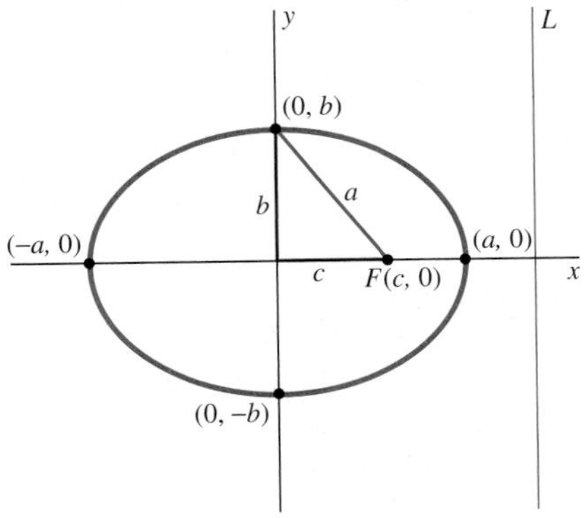

FIGURE 9.6.10 The parts of an ellipse.

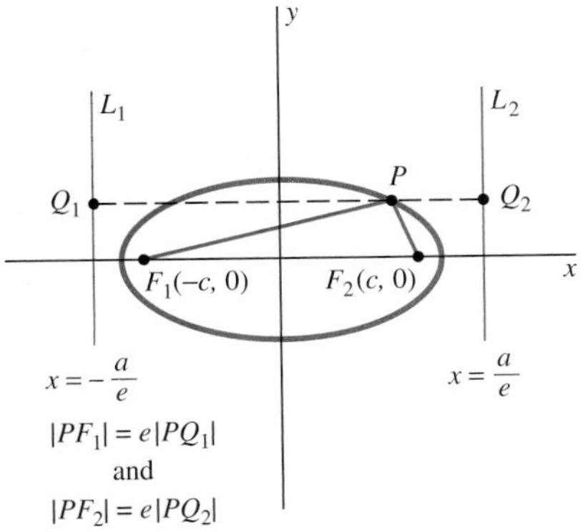

FIGURE 9.6.11 The ellipse as a conic section: two foci, two directrices.

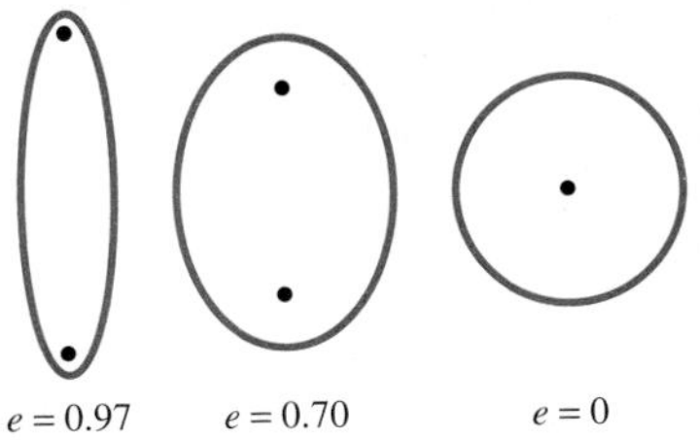

FIGURE 9.6.12 The relation between the eccentricity of an ellipse and its shape.

Because $a=c/e$, the directrix of the ellipse in Eq. (14) is $x=a/e$. If we had begun instead with the focus $(-c, 0)$ and directrix $x=-a/e$, we would still have obtained Eq. (14), because only the squares of a and c are involved in its derivation. Thus the ellipse in Eq. (14) has *two* foci, $(c, 0)$ and $(-c, 0)$, and *two* directrices, $x=a/e$ and $x=-a/e$ (Fig. 9.6.11).

The larger the eccentricity $e<1$, the more elongated the ellipse. (Remember that $e=1$ is the eccentricity of every parabola). But if $e=0$, then Eq. (13) gives $b=a$, so Eq. (14) reduces to the equation of a circle of radius a. Thus a circle is an ellipse of eccentricity zero. Compare the three cases shown in Fig. 9.6.12.

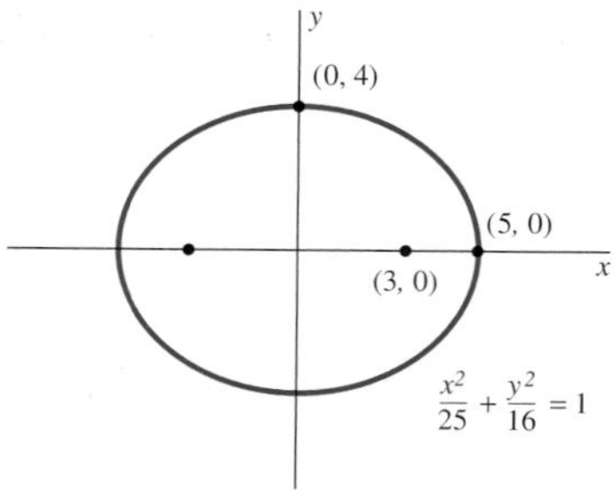

FIGURE 9.6.13 The ellipse of Example 3.

EXAMPLE 3 Find an equation of the ellipse with foci $(\pm 3, 0)$ and vertices $(\pm 5, 0)$.

Solution We are given $c = 3$ and $a = 5$, so Eq. (13) gives $b = 4$. Thus Eq. (14) gives

$$\frac{x^2}{25} + \frac{y^2}{16} = 1$$

for the desired equation. This ellipse is shown in Fig. 9.6.13. ◆

If the two foci of an ellipse are on the y-axis, such as $F_1(0, c)$ and $F_2(0, -c)$, then the equation of the ellipse is

$$\frac{x^2}{b^2} + \frac{y^2}{a^2} = 1, \tag{16}$$

and it is still true that $a^2 = b^2 + c^2$, as in Eq. (15). But now the major axis of length $2a$ is vertical and the minor axis of length $2b$ is horizontal. The derivation of Eq. (16) is similar to that of Eq. (14); see Problem 79. Figure 9.6.14 shows the case of an ellipse whose major axis is vertical. The vertices of such an ellipse are at $(0, \pm a)$; they are always the endpoints of the major axis.

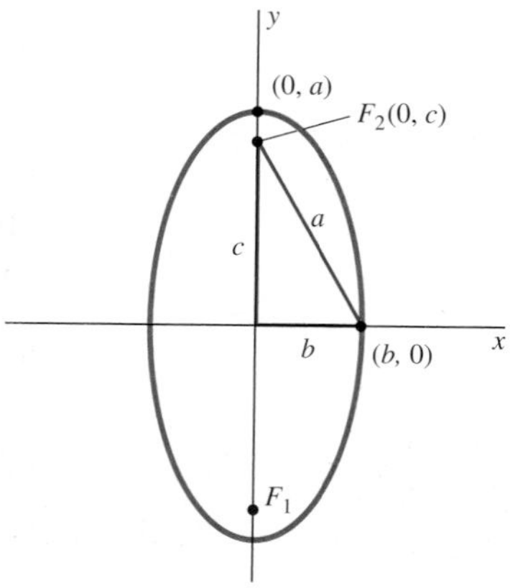

FIGURE 9.6.14 An ellipse with vertical major axis.

In practice there is little chance of confusing Eqs. (14) and (16). The equation or the given data will make it clear whether the major axis of the ellipse is horizontal or vertical. Just use the equation to read the ellipse's intercepts. The two intercepts that are farthest from the origin are the endpoints of the major axis; the other two are the endpoints of the minor axis. The two foci lie on the major axis, each at distance c from the center of the ellipse—which will be the origin if the equation of the ellipse has the form of either Eq. (14) or Eq. (16).

EXAMPLE 4 Sketch the graph of the equation

$$\frac{x^2}{16} + \frac{y^2}{25} = 1.$$

Solution The x-intercepts are $(\pm 4, 0)$; the y-intercepts are $(0, \pm 5)$. So the major axis is vertical. We take $a = 5$ and $b = 4$ in Eq. (15) and find that $c = 3$. The foci are thus at $(0, \pm 3)$. Hence this ellipse has the appearance of the one shown in Fig. 9.6.15 ◆

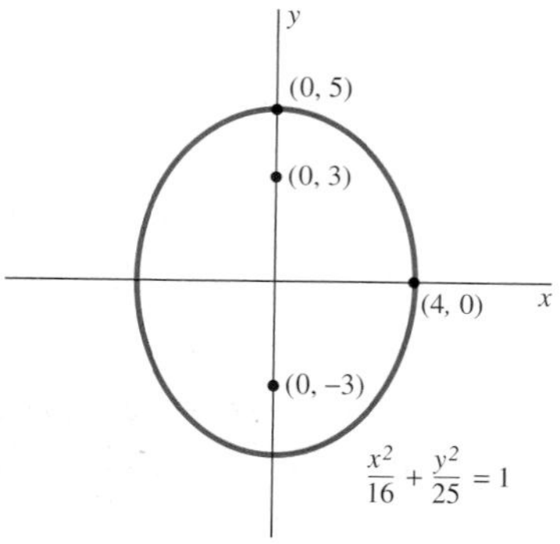

FIGURE 9.6.15 The ellipse of Example 4.

Any equation of the form

$$Ax^2 + Cy^2 + Dx + Ey + F = 0, \tag{17}$$

in which the coefficients A and C of the squared variables are *both nonzero* and *have the same sign*, may be reduced to the form

$$A(x - h)^2 + C(y - k)^2 = G$$

by completing the square in x and y. We may assume that A and C are both positive. Then if $G < 0$, there are no points that satisfy Eq. (17), and the graph is the empty set. If $G = 0$, then there is exactly one point on the locus—the single point (h, k). And if $G > 0$, we can divide both sides of the last equation by G and get an equation that resembles one of these two:

$$\frac{(x-h)^2}{a^2} + \frac{(y-k)^2}{b^2} = 1, \tag{18a}$$

$$\frac{(x-h)^2}{b^2} + \frac{(y-k)^2}{a^2} = 1. \tag{18b}$$

Which equation should you choose? Select the one that is consistent with the condition $a \geqq b > 0$. Finally, note that either of the equations in (18) is the equation of a translated ellipse. Thus, apart from the exceptional cases already noted, the graph of Eq. (17) is an ellipse if $AC > 0$.

EXAMPLE 5 Determine the graph of the equation

$$3x^2 + 5y^2 - 12x + 30y + 42 = 0.$$

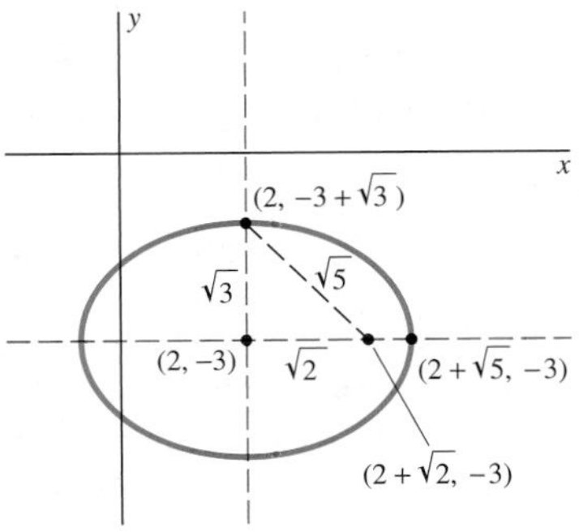

FIGURE 9.6.16 The ellipse of Example 5.

Solution We collect terms containing x, terms containing y, and complete the square in each variable. This gives

$$3(x^2 - 4x) + 5(y^2 + 6y) = -42;$$
$$3(x^2 - 4x + 4) + 5(y^2 + 69 + 9) = 15;$$
$$\frac{(x-2)^2}{5} + \frac{(y+3)^2}{3} = 1.$$

Thus the given equation is that of a translated ellipse with center at $(2, -3)$. Its horizontal major semiaxis has length $a = \sqrt{5}$ and its minor semiaxis has length $b = \sqrt{3}$ (Fig. 9.6.16). The distance from the center to each focus is $c = \sqrt{2}$ and the eccentricity is $e = c/a = \sqrt{2/5}$. ◆

Applications of Ellipses

EXAMPLE 6 The orbit of the earth is an ellipse with the sun at one focus. The planet's maximum distance from the center of the sun is 94.56 million miles and its minimum distance is 91.44 million miles. What are the major and minor semiaxes of the earth's orbit, and what is its eccentricity?

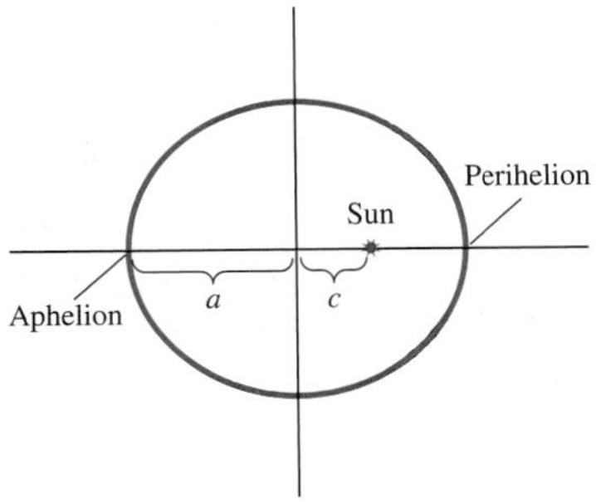

FIGURE 9.6.17 The orbit of the earth with its eccentricity exaggerated (Example 6).

Solution As Fig. 9.6.17 shows, we have

$$a + c = 94.56 \quad \text{and} \quad a - c = 91.44,$$

with units in millions of miles. We conclude from these equations that $a = 93.00$, that $c = 1.56$, and then that

$$b = \sqrt{(93.00)^2 - (1.56)^2} \approx 92.99$$

million miles. Finally,

$$e = \frac{c}{a} = \frac{1.56}{93.00} \approx 0.017,$$

a number relatively close to zero. This means that the earth's orbit is nearly circular. Indeed, the major and minor semiaxes are so nearly equal that, on any usual scale, the earth's orbit would appear to be a perfect circle. But the difference between uniform circular motion and the earth's actual motion has some important aspects, including the facts that the sun is 1.56 million miles off center and that the orbital speed of the earth is not constant. ◆

EXAMPLE 7 One of the most famous comets is Halley's comet, named for Edmund Halley (1656–1742), a disciple of Newton. By studying the records of the paths of earlier comets, Halley deduced that the comet of 1682 was the same one that had been sighted in 1607, in 1531, in 1456, and in 1066 (an omen at the Battle of Hastings). In 1682 Halley predicted that this comet would return in 1759, in 1835, and in 1910; he was correct each time. The period of Halley's comet is about 76 years—it can vary a couple of years in either direction because of perturbations of its orbit by the planet Jupiter. The orbit of Halley's comet is an ellipse with the sun at one focus. In terms of astronomical units (1 AU is the mean distance from the earth to the sun), the major and minor semiaxes of this elliptical orbit are 18.09 AU and 4.56 AU, respectively. What are the maximum and minimum distances from the sun of Halley's comet?

Solution We are given that $a = 18.09$ (all distance measurements are in astronomical units) and that $b = 4.56$, so

$$c = \sqrt{(18.09)^2 - (4.56)^2} \approx 17.51.$$

Hence the maximum distance of the comet from the sun is $a + c \approx 35.60$ AU, and its minimum distance is $a - c \approx 0.58$ AU. The eccentricity of its orbit is

$$e = \frac{c}{a} = \frac{17.51}{18.09} \approx 0.97,$$

a very eccentric orbit (but see Problem 77). ◆

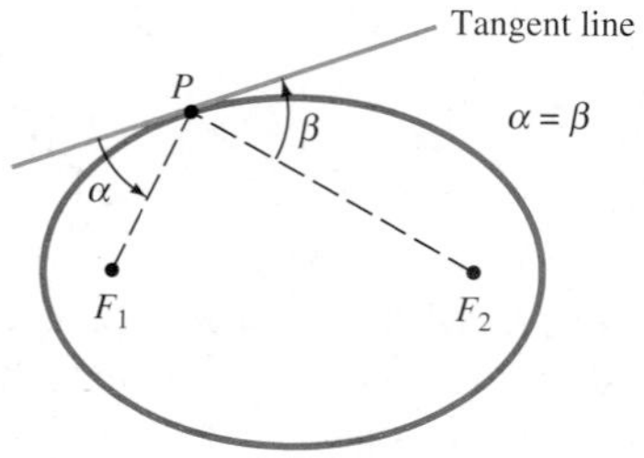

FIGURE 9.6.18 The reflection property: $\alpha = \beta$.

The *reflection property* of the ellipse states that the tangent line at a point P of an ellipse makes equal angles with the two lines PF_1 and PF_2 from P to the two foci of the ellipse (Fig. 9.6.18). This property is the basis of the "whispering gallery" phenomenon, which has been observed in the so-called whispering gallery of the U.S. Senate. Suppose that the ceiling of a large room is shaped like half an ellipsoid obtained by revolving an ellipse around its major axis. Sound waves, like light waves, are reflected with equal angles of incidence and reflection. Thus if two diplomats are holding a quiet conversation near one focus of the ellipsoidal surface, a reporter standing near the other focus—perhaps 50 feet away—might be able to eavesdrop on their conversation even if the conversation were inaudible to others in the same room.

Some billiard tables are manufactured in the shape of an ellipse. The foci of such tables are plainly marked for the convenience of enthusiasts of this unusual game.

A more serious application of the reflection property of ellipses is the nonsurgical kidney-stone treatment called *shockwave lithotripsy.* An ellipsoidal reflector with a transducer (an energy transmitter) at one focus is positioned outside the patient's body so that the offending kidney stone is located at the other focus. The stone then is pulverized by reflected shockwaves emanating from the transducer. (For further details, see the COMAP *Newsletter* **20**, November, 1986.)

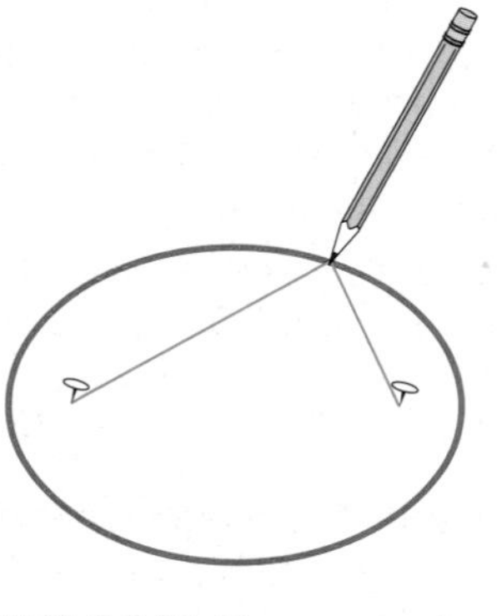

FIGURE 9.6.19 One way to draw an ellipse.

An alternative definition of the ellipse with foci F_1 and F_2 and major axis of length $2a$ is this: It is the locus of a point P such that the sum of the distances $|PF_1|$ and $|PF_2|$ is the constant $2a$. (See Problem 82.) This fact gives us a convenient way to draw the ellipse by using two tacks placed at F_1 and F_2, a string of length $2a$, and a pencil (Fig. 9.6.19).

The Hyperbola

A hyperbola is a conic section defined in the same way as is an ellipse, except that the eccentricity e of a hyperbola is greater than 1.

DEFINITION The Hyperbola

Suppose that $e > 1$, and let F be a fixed point and L a fixed line not containing F. Then the **hyperbola** with **eccentricity** e, **focus** F, and **directrix** L is the set of all points P such that the distance $|PF|$ is e times the (perpendicular) distance from P to the line L.

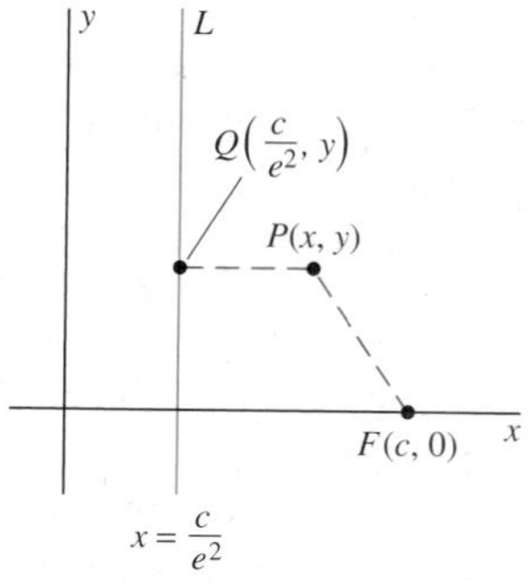

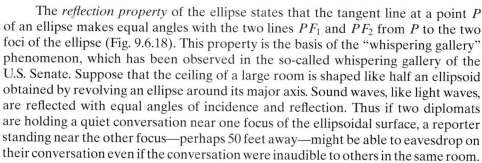

FIGURE 9.6.20 The definition of the hyperbola.

As with the ellipse, the equation of a hyperbola is simplest if F is the point $(c, 0)$ on the x-axis and L is the vertical line $x = c/e^2$. The case $c > 0$ is shown in Fig. 9.6.20. If Q is the point $(c/e^2, y)$, then PQ is the perpendicular from $P(x, y)$ to L. The condition $|PF| = e|PQ|$ gives

$$(x - c)^2 + y^2 = e^2\left(x - \frac{c}{e^2}\right)^2;$$

$$x^2 - 2cx + c^2 + y^2 = e^2x^2 - 2cx + \frac{c^2}{e^2};$$

$$(e^2 - 1)x^2 - y^2 = c^2\left(1 - \frac{1}{e^2}\right) = \frac{c^2}{e^2}(e^2 - 1).$$

Thus

$$(e^2 - 1)x^2 - y^2 = a^2(e^2 - 1),$$

where

$$a = \frac{c}{e}. \tag{19}$$

If we divide both sides of the next-to-last equation by $a^2(e^2 - 1)$, we get

$$\frac{x^2}{a^2} - \frac{y^2}{a^2(e^2 - 1)} = 1.$$

To simplify this equation, we let

$$b^2 = a^2(e^2 - 1) = c^2 - a^2. \tag{20}$$

This is permissible because $e > 1$. So the equation of the hyperbola with focus $(c, 0)$ and directrix $x = c/e^2 = a/e$ takes the form

$$\frac{x^2}{a^2} - \frac{y^2}{b^2} = 1. \tag{21}$$

The minus sign on the left-hand side is the only difference between the equation of a hyperbola and that of an ellipse. Of course, Eq. (20) also differs from the relation

$$b^2 = a^2(1 - e^2) = a^2 - c^2$$

for the case of the ellipse.

The hyperbola of Eq. (21) is clearly symmetric around both coordinate axes and has x-intercepts $(\pm a, 0)$. But it has no y-intercept. If we rewrite Eq. (21) in the form

$$y = \pm\frac{b}{a}\sqrt{x^2 - a^2}, \tag{22}$$

then we see that there are points on the graph only if $|x| \geqq a$. Hence the hyperbola has two **branches,** as shown in Fig. 9.6.21. We also see from Eq. (22) that $|y| \to \infty$ as $|x| \to \infty$.

The x-intercepts $V_1(-a, 0)$ and $V_2(a, 0)$ are the **vertices** of the hyperbola, and the line segment joining them is its **transverse axis** (Fig. 9.6.22). The line segment

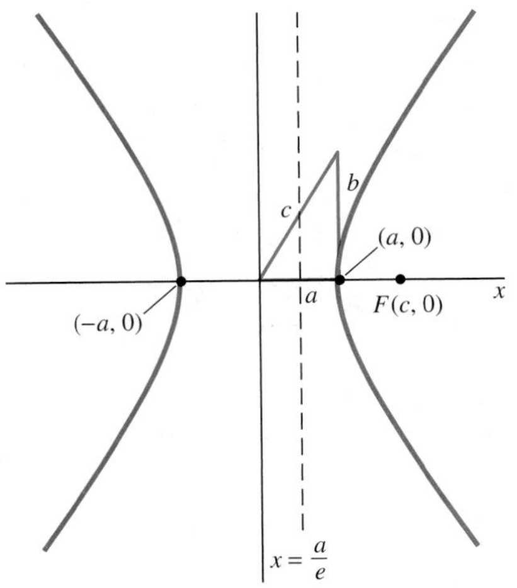

FIGURE 9.6.21 A hyperbola has two *branches.*

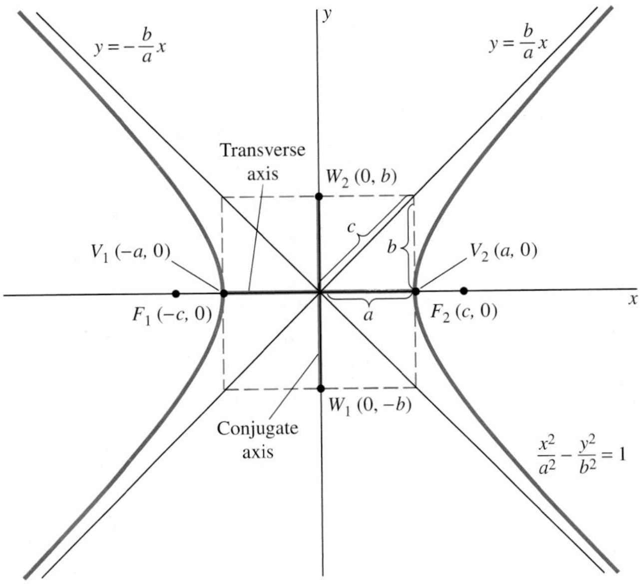

FIGURE 9.6.22 The parts of a hyperbola.

joining $W_1(0, -b)$ and $W_2(0, b)$ is its **conjugate axis.** The alternative form

$$c^2 = a^2 + b^2 \tag{23}$$

of Eq. (20) is the Pythagorean relation for the right triangle shown in Fig. 9.6.22.

The lines $y = \pm bx/a$ that pass through the **center** (0, 0) and the opposite vertices of the rectangle in Fig. 9.6.22 are **asymptotes** of the two branches of the hyperbola in both directions. That is, if

$$y_1 = \frac{bx}{a} \quad \text{and} \quad y_2 = \frac{b}{a}\sqrt{x^2 - a^2},$$

then

$$\lim_{x\to\infty}(y_1 - y_2) = 0 = \lim_{x\to-\infty}(y_1 - (-y_2)). \tag{24}$$

To verify the first limit (for instance), note that

$$\lim_{x\to\infty}\frac{b}{a}\left(x - \sqrt{x^2 - a^2}\right) = \lim_{x\to\infty}\frac{b}{a}\cdot\frac{\left(x - \sqrt{x^2 - a^2}\right)\left(x + \sqrt{x^2 - a^2}\right)}{x + \sqrt{x^2 - a^2}}$$

$$= \lim_{x\to\infty}\frac{b}{a}\cdot\frac{a^2}{x + \sqrt{x^2 - a^2}} = 0.$$

Just as in the case of the ellipse, the hyperbola with focus $(c, 0)$ and directrix $x = a/e$ also has focus $(-c, 0)$ and directrix $x = -a/e$ (Fig. 9.6.23). Because $c = ae$ by Eq. (19), the foci $(\pm ae, 0)$ and the directrices $x = \pm a/e$ take the same forms in terms of a and e for both the hyperbola ($e > 1$) and the ellipse ($e < 1$).

If we interchange x and y in Eq. (21), we obtain

$$\frac{y^2}{a^2} - \frac{x^2}{b^2} = 1. \tag{25}$$

This hyperbola has foci at $(0, \pm c)$. The foci as well as this hyperbola's transverse axis lie on the y-axis. Its asymptotes are $y = \pm ax/b$, and its graph generally resembles the one in Fig. 9.6.23.

When we studied the ellipse, we saw that its orientation—whether the major axis is horizontal or vertical—is determined by the relative sizes of a and b. In the

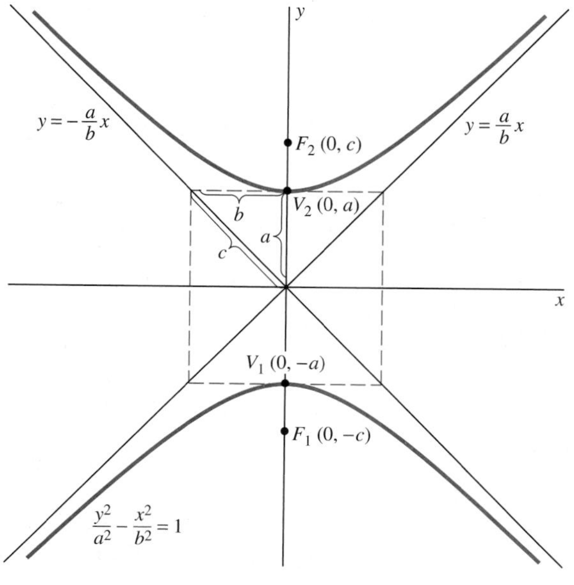

FIGURE 9.6.23 The hyperbola of Eq. (25) has horizontal directrices.

case of the hyperbola, the situation is quite different, because the relative sizes of a and b make no such difference: They affect only the slopes of the asymptotes. The direction in which the hyperbola opens—horizontal as in Fig. 9.6.22 or vertical as in Fig. 9.6.23—is determined by the signs of the terms that contain x^2 and y^2.

EXAMPLE 8 Sketch the graph of the hyperbola with equation

$$\frac{y^2}{9} - \frac{x^2}{16} = 1.$$

Solution This is an equation of the form in Eq. (25), so the hyperbola opens vertically. Because $a = 3$ and $b = 4$, we find that $c = 5$ by using Eq. (23): $c^2 = a^2 + b^2$. Thus the vertices are $(0, \pm 3)$, the foci are the two points $(0, \pm 5)$, and the asymptotes are the two lines $y = \pm 3x/4$. This hyperbola appears in Fig. 9.6.24. ◆

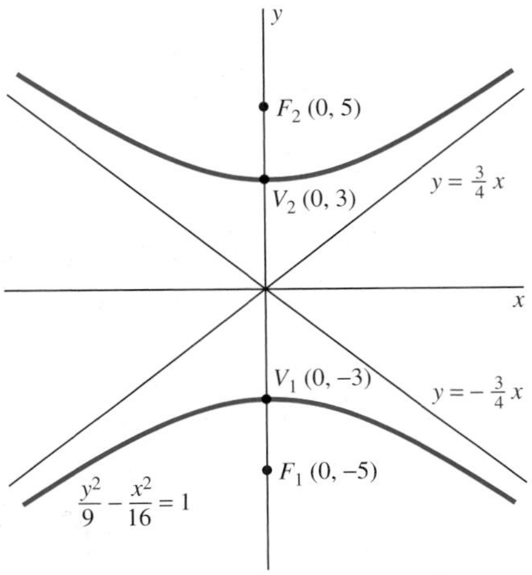

FIGURE 9.6.24 The hyperbola of Example 8.

EXAMPLE 9 Find an equation of the hyperbola with foci $(\pm 10, 0)$ and asymptotes $y = \pm 4x/3$.

Solution Because $c = 10$, we have

$$a^2 + b^2 = 100 \quad \text{and} \quad \frac{b}{a} = \frac{4}{3}.$$

Thus $b = 8$ and $a = 6$, and the standard equation of the hyperbola is

$$\frac{x^2}{36} - \frac{y^2}{64} = 1.$$ ◆

As we noted earlier, any equation of the form

$$Ax^2 + Cy^2 + Dx + Ey + F = 0 \tag{26}$$

with both A and C nonzero can be reduced to the form

$$A(x - h)^2 + B(y - k)^2 = G$$

by completing the square in x and y. Now suppose that the coefficients A and C of the quadratic terms have *opposite signs.* For example, suppose that $A = p^2$ and $B = -q^2$. The last equation then becomes

$$p^2(x - h)^2 - q^2(y - k)^2 = G. \tag{27}$$

If $G = 0$, then factorization of the difference of squares on the left-hand side yields the equations

$$p(x-h)+q(y-k)=0 \quad \text{and} \quad p(x-h)-q(y-k)=0$$

of two straight lines through (h, k) with slopes $m = \pm p/q$. If $G \neq 0$, then division of Eq. (27) by G gives an equation that looks either like

$$\frac{(x-h)^2}{a^2} - \frac{(y-k)^2}{b^2} = 1 \quad (\text{if } G > 0)$$

or like

$$\frac{(y-k)^2}{a^2} - \frac{(x-h)^2}{b^2} = 1 \quad (\text{if } G < 0).$$

Thus if $AC < 0$ in Eq. (26), the graph is either a pair of intersecting straight lines or a hyperbola.

EXAMPLE 10 Determine the graph of the equation

$$9x^2 - 4y^2 - 36x + 8y = 4.$$

Solution We collect the terms that contain x and those that contain y, and we then complete the square in each variable. We find that

$$9(x-2)^2 - 4(y-1)^2 = 36,$$

so

$$\frac{(x-2)^2}{4} - \frac{(y-1)^2}{9} = 1.$$

Hence the graph is a hyperbola with a horizontal transverse axis and center $(2, 1)$. Because $a = 2$ and $b = 3$, we find that $c = \sqrt{13}$. The vertices of the hyperbola are $(0, 1)$ and $(4, 1)$, and its foci are the two points $(2 \pm \sqrt{13}, 1)$. Its asymptotes are the two lines

$$y - 1 = \pm\tfrac{3}{2}(x-2),$$

translates of the asymptotes $y = \pm 3x/2$ of the hyperbola $\frac{1}{4}x^2 - \frac{1}{9}y^2 = 1$. Figure 9.6.25 shows the graph of the translated hyperbola. ◆

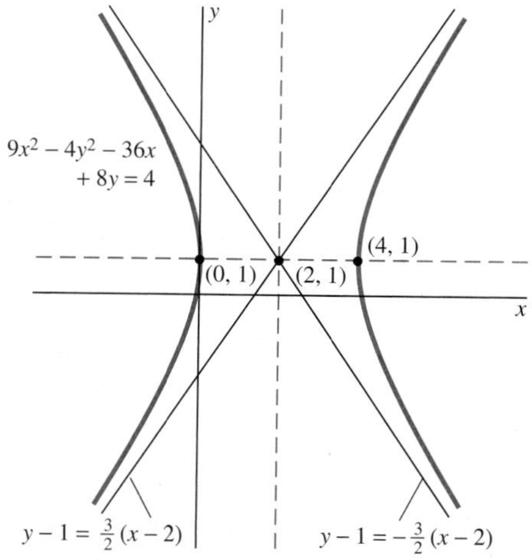

FIGURE 9.6.25 The hyperbola of Example 10, a translate of the hyperbola $x^2/4 - y^2/9 = 1$.

Applications of Hyperbolas

The *reflection property* of the hyperbola takes the same form as that for the ellipse. If P is a point on a hyperbola, then the two lines PF_1 and PF_2 from P to the two foci make equal angles with the tangent line at P. In Fig. 9.6.26 this means that $\alpha = \beta$.

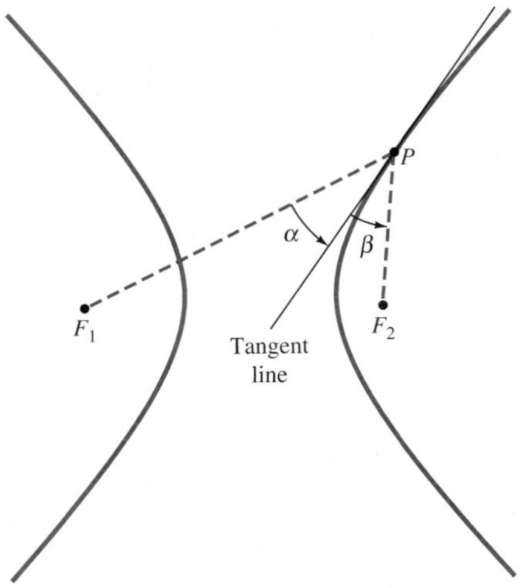

FIGURE 9.6.26 The reflection property of the hyperbola.

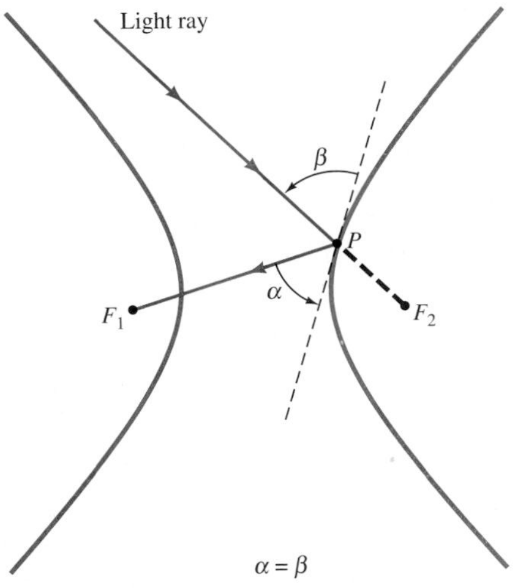

FIGURE 9.6.27 How a hyperbolic mirror reflects a ray aimed at one focus: $\alpha = \beta$ again.

For an important application of this reflection property, consider a mirror that is shaped like one branch of a hyperbola and is reflective on its outer (convex) surface. An incoming light ray aimed toward one focus will be reflected toward the other focus (Fig. 9.6.27). Figure 9.6.28 indicates the design of a reflecting telescope that makes use of the reflection properties of the parabola and the hyperbola. The parallel incoming light rays first are reflected by the parabola toward its focus at F. Then they are intercepted by an auxiliary hyperbolic mirror with foci at E and F and reflected into the eyepiece located at E.

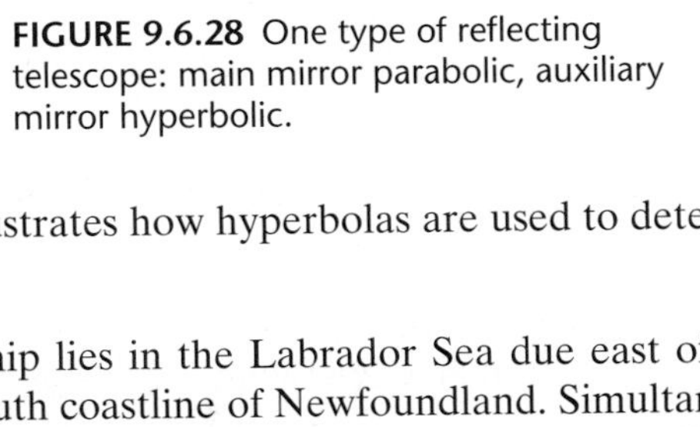

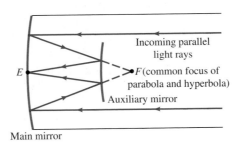

FIGURE 9.6.28 One type of reflecting telescope: main mirror parabolic, auxiliary mirror hyperbolic.

Example 11 illustrates how hyperbolas are used to determine the positions of ships at sea.

EXAMPLE 11 A ship lies in the Labrador Sea due east of Wesleyville, point A, on the long north-south coastline of Newfoundland. Simultaneous radio signals are transmitted by radio stations at A and at St. John's, point B, which is on the coast 200 km due south of A. The ship receives the signal from A 500 microseconds (μs) before it receives the signal from B. Assume that the speed of radio signals is 300 m/μs. How far out at sea is the ship?

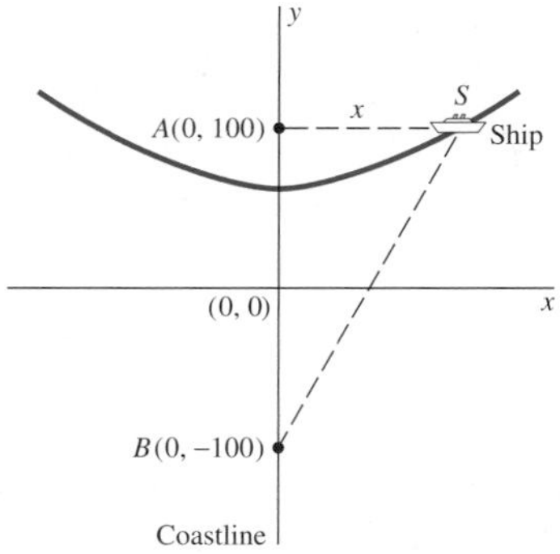

FIGURE 9.6.29 A navigation problem (Example 11).

Solution The situation is diagrammed in Fig. 9.6.29. The difference between the distances of the ship at S from A and B is

$$|SB| - |SA| = 500 \cdot 300 = 150{,}000$$

meters; that is, 150 km. Thus (by Problem 88) the ship lies on a hyperbola with foci A and B. From Fig. 9.6.29 we see that $c = 100$, so $a = \frac{1}{2} \cdot 150 = 75$, and thus

$$b = \sqrt{c^2 - a^2} = \sqrt{100^2 - 75^2} = 25\sqrt{7}.$$

In the coordinate system of Fig. 9.6.29, the hyperbola has equation

$$\frac{y^2}{75^2} - \frac{x^2}{7 \cdot 25^2} = 1.$$

We substitute $y = 100$ because the ship is due east of A. Thus we find that the ship's distance from the coastline is $x = \frac{175}{3} \approx 58.3$ km. ◆

Conics in Polar Coordinates

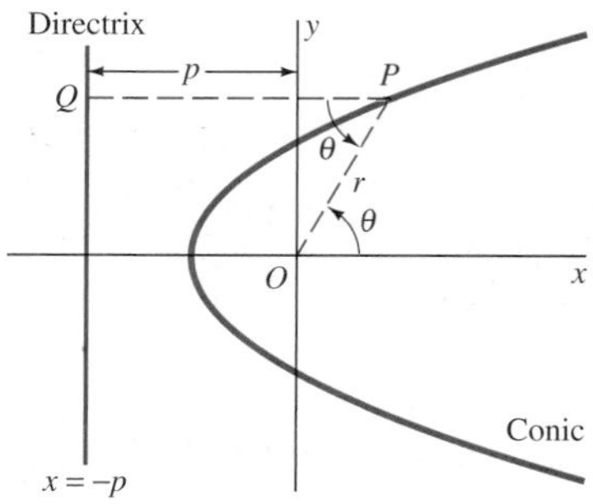

FIGURE 9.6.30 A conic section: $|OP| = e\,|PQ|$.

In order to investigate orbits of satellites—such as planets or comets orbiting the sun or natural or artificial moons orbiting a planet—we need equations of the conic sections in polar coordinates. As a bonus, we find that all three conic sections have the same general equation in polar coordinates.

To derive the polar equation of a conic section, suppose its focus is the origin O and that its directrix is the vertical line $x = -p$ (with $p > 0$). In the notation of Fig. 9.6.30, the fact that $|OP| = e|PQ|$ then tells us that $r = e(p + r\cos\theta)$. Solution of this equation for r yields

$$r = \frac{pe}{1 - e\cos\theta}.$$

If the directrix is the vertical line $x = +p > 0$ to the right of the origin, then a similar calculation gives the same result, except with a change of sign in the denominator.

Polar Coordinate Equation of a Conic Section

The polar equation of a conic section with eccentricity e, focus O, and directrix $x = \pm p$ is

$$r = \frac{pe}{1 \pm e\cos\theta}. \tag{28}$$

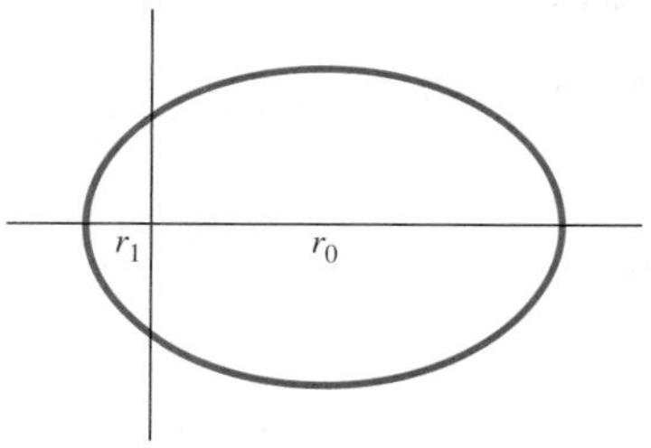

FIGURE 9.6.31 The maximal radius $r_0 = \dfrac{pe}{1-e}$ and the minimal radius $r_1 = \dfrac{pe}{1+e}$ of the ellipse.

Figure 9.6.31 shows an ellipse with eccentricity $e < 1$ and directrix $x = -p$. Its vertices correspond to $\theta = 0$ and $\theta = \pi$, where maximal and minimal radii r_0 and r_1 occur. It follows that the length $2a$ of its major axis is

$$2a = r_0 + r_1 = \frac{pe}{1-e} + \frac{pe}{1+e} = \frac{2pe}{1-e^2}.$$

Cross multiplication gives the relation

$$pe = a(1 - e^2), \tag{29}$$

and substituting in (28) then yields the equation

$$r = \frac{a(1 - e^2)}{1 \pm e\cos\theta} \tag{30}$$

of an ellipse with eccentricity e and major semiaxis a.

EXAMPLE 12 Sketch the graph of the equation

$$r = \frac{16}{5 - 3\cos\theta}.$$

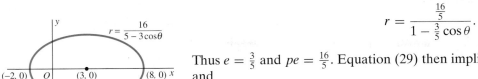

Solution First we divide numerator and denominator by 5 and find that

$$r = \frac{\frac{16}{5}}{1 - \frac{3}{5}\cos\theta}.$$

Thus $e = \frac{3}{5}$ and $pe = \frac{16}{5}$. Equation (29) then implies that $a = 5$. Finally, $c = ae = 3$ and

$$b = \sqrt{a^2 - c^2} = 4.$$

So we have here an ellipse with major semiaxis $a = 5$, minor semiaxis $b = 4$, and center at (3, 0) in Cartesian coordinates. The ellipse is shown in Fig. 9.6.32. ◆

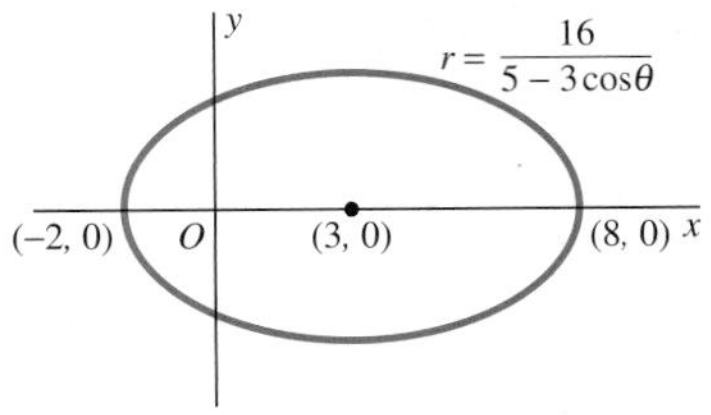

FIGURE 9.6.32 The ellipse of Example 12.

REMARK 1 The limiting form of Eq. (30) as $e \to 0$ is the equation $r = a$ of a circle. Because $p \to \infty$ as $e \to 0$ with a fixed in Eq. (29), we may therefore regard any circle as an ellipse with eccentricity zero and with directrix at infinity.

REMARK 2 If we begin with an ellipse with eccentricity $e < 1$ and directrix $x = -p$, then the limiting form of Eq. (30) as $e \to 1^-$ is the equation

$$r = \frac{p}{1 - \cos\theta} \tag{31}$$

of a parabola. For instance, Fig. 9.6.33 shows a parabola and an ellipse of eccentricity $e = 0.99$, both with directrix $p = -1$. Observe that the two curves appear to almost coincide near the origin where $30° < \theta < 330°$. This sort of approximation of an ellipse by a parabola is useful in studying comets with highly eccentric elliptical orbits.

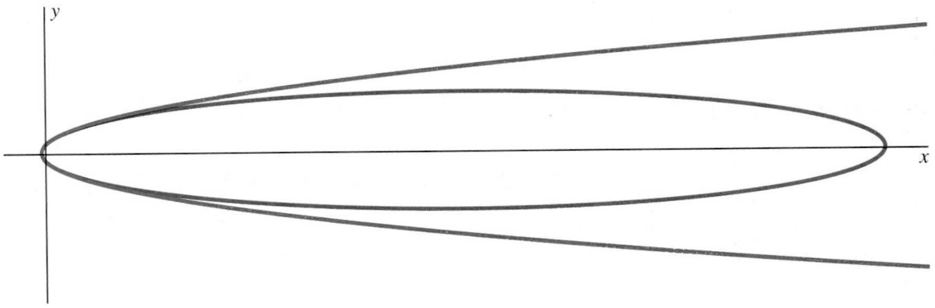

FIGURE 9.6.33 The parabola $r = \dfrac{1}{1 - \cos\theta}$ and the ellipse $r = \dfrac{0.99}{1 - 0.99\cos\theta}$.

EXAMPLE 13 A certain comet is known to have a highly eccentric elliptical orbit with the sun at one focus. Two successive observations as this comet approached the sun gave the measurements $r = 6$ AU when $\theta = 60°$, and $r = 2$ AU when $\theta = 90°$ (relative to a fixed polar-coordinate system). Estimate the position of the comet at its point of closest approach to the sun.

Solution Because the elliptical orbit is highly eccentric, we assume that near the sun it can be approximated closely by a parabola. The angle $\theta = \alpha$ of the axis is unknown, but a preliminary sketch indicates that α will be less than the initial angle of observation; thus $0 < \alpha < \pi/3$. Using the polar coordinate system with this unknown polar axis and counterclockwise angular variable $\phi = \theta - \alpha$ (Fig. 9.6.34), the equation in (31) of the parabola takes the form

$$r = \frac{p}{1 - \cos\phi} = \frac{p}{1 - \cos(\theta - \alpha)}. \tag{32}$$

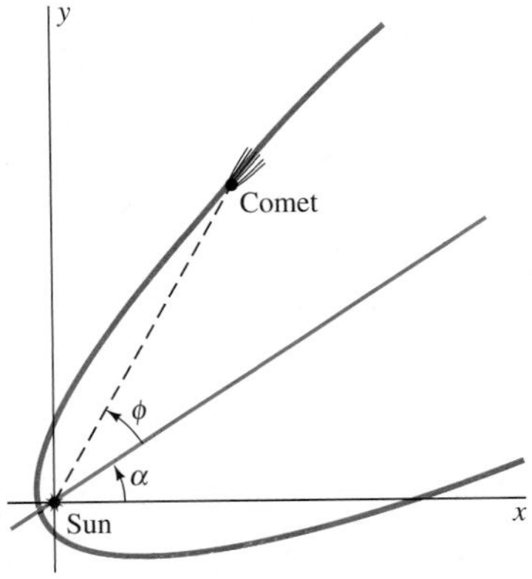

FIGURE 9.6.34 The comet of Example 13.

The vertex of a parabola is its point closest to its focus (Problem 65), so the minimum distance of the comet from the sun will be $r = p/2$ when $\theta = \pi + \alpha$. Our problem, then, is to determine the values of p and α.

Substituting the given observational data into Eq. (32) yields the two equations

$$6 = \frac{p}{1 - \cos(\pi/3 - \alpha)} \quad \text{and} \quad 2 = \frac{p}{1 - \cos(\pi/2 - \alpha)}. \tag{33}$$

Elimination of p yields

$$6 - 6\cos\left(\frac{\pi}{3} - \alpha\right) = 2 - 2\cos\left(\frac{\pi}{2} - \alpha\right);$$

$$6 - 6\left(\frac{1}{2}\cos\alpha + \frac{\sqrt{3}}{2}\sin\alpha\right) = 2 - 2\sin\alpha.$$

We therefore need to solve the single equation

$$3\cos\alpha + \left(3\sqrt{3} - 2\right)\sin\alpha - 4 = 0.$$

A calculator or computer yields the approximate root $\alpha = 0.3956 \approx 22.67°$. Then the second equation in (33) gives $p = 2(1 - \sin\alpha) \approx 1.2293$(AU). Since 1 astronomical unit is about 93 million miles, the comet's closest approach to the sun is about $\frac{1}{2}p = (0.5)(1.2293)(93) \approx 57.16$ million miles. ◆

9.6 TRUE/FALSE STUDY GUIDE

9.6 CONCEPTS: QUESTIONS AND DISCUSSION

1. Summarize the definitions and alternative constructions of the parabola, ellipse, and hyperbola.
2. Compare the reflection properties of the three types of conic sections.
3. Summarize the applications of the conic sections. You might like to consult an encyclopedia or do a web search.

9.6 PROBLEMS

In Problems 1 through 5, find the equation and sketch the graph of the parabola with vertex V and focus F.

1. $V(0, 0)$, $F(3, 0)$

2. $V(0, 0)$, $F(0, -2)$

3. $V(2, 3)$, $F(2, 1)$

4. $V(-1, -1)$, $F(-3, -1)$

5. $V(2, 3)$, $F(0, 3)$

In Problems 6 through 10, find the equation and sketch the graph of the parabola with the given focus and directrix.

6. $F(1, 2)$, $x = -1$

7. $F(0, -3)$, $y = 0$

8. $F(1, -1)$, $x = 3$

9. $F(0, 0)$, $y = -2$

10. $F(-2, 1)$, $x = -4$

In Problems 11 through 18, sketch the parabola with the given equation. Show and label its vertex, focus, axis, and directrix.

11. $y^2 = 12x$

12. $x^2 = -8y$

13. $y^2 = -6x$

14. $x^2 = 7y$

15. $x^2 - 4x - 4y = 0$

16. $y^2 - 2x + 6y + 15 = 0$

17. $4x^2 + 4x + 4y + 13 = 0$

18. $4y^2 - 12y + 9x = 0$

In Problems 19 through 33, find an equation of the ellipse specified.

19. Vertices $(\pm 4, 0)$ and $(0, \pm 5)$

20. Foci $(\pm 5, 0)$, major semiaxis 13

21. Foci $(0, \pm 8)$, major semiaxis 17

22. Center $(0, 0)$, vertical major axis 12, minor axis 8

23. Foci $(\pm 3, 0)$, eccentricity $\frac{3}{4}$

24. Foci $(0, \pm 4)$, eccentricity $\frac{2}{3}$

25. Center $(0, 0)$, horizontal major axis 20, eccentricity $\frac{1}{2}$

26. Center $(0, 0)$, horizontal minor axis 10, eccentricity $\frac{1}{2}$

27. Foci $(\pm 2, 0)$, directrices $x = \pm 8$

28. Foci $(0, \pm 4)$, directrices $y = \pm 9$

29. Center $(2, 3)$, horizontal axis 8, vertical axis 4

30. Center $(1, -2)$, horizontal major axis 8, eccentricity $\frac{3}{4}$

31. Foci $(-2, 1)$ and $(4, 1)$, major axis 10

32. Foci $(-3, 0)$ and $(-3, 4)$, minor axis 6

33. Foci $(-2, 2)$ and $(4, 2)$, eccentricity $\frac{1}{3}$

Sketch the graphs of the equations in Problems 34 through 38. Indicate centers, foci, and lengths of axes.

34. $4x^2 + y^2 = 16$

35. $4x^2 + 9y^2 = 144$

36. $4x^2 + 9x^2 = 24x$

37. $9x^2 + 4y^2 - 32y + 28 = 0$

38. $2x^2 + 3y^2 + 12x - 24y + 60 = 0$

In Problems 39 through 52, find an equation of the hyperbola described.

39. Foci $(\pm 4, 0)$, vertices $(\pm 1, 0)$

40. Foci $(0, \pm 3)$, vertices $(0, \pm 2)$

41. Foci $(\pm 5, 0)$, asymptotes $y = \pm 3x/4$

42. Vertices $(\pm 3, 0)$, asymptotes $y = \pm 3x/4$

43. Vertices $(0, \pm 5)$, asymptotes $y = \pm x$

44. Vertices $(\pm 3, 0)$, eccentricity $e = \frac{5}{3}$

45. Foci $(0, \pm 6)$, eccentricity $e = 2$

46. Vertices $(\pm 4, 0)$ and passing through $(8, 3)$

47. Foci $(\pm 4, 0)$, directrices $x = \pm 1$

48. Foci $(0, \pm 9)$, directrices $y = \pm 4$

49. Center $(2, 2)$, horizontal transverse axis of length 6, eccentricity $e = 2$

50. Center $(-1, 3)$, vertices $(-4, 3)$ and $(2, 3)$, foci $(-6, 3)$ and $(4, 3)$

51. Center $(1, -2)$, vertices $(1, 1)$ and $(1, -5)$, asymptotes $3x - 2y = 7$ and $3x + 2y = -1$

52. Focus $(8, -1)$, asymptotes $3x - 4y = 13$ and $3x + 4y = 5$

Sketch the graphs of the equations given in Problems 53 through 58; indicates centers, foci, and asymptotes.

53. $x^2 - y^2 - 2x + 4y = 4$

54. $x^2 - 2y^2 + 4x = 0$

55. $y^2 - 3x^2 - 6y = 0$

56. $x^2 - y^2 - 2x + 6y = 9$

57. $9x^2 - 4y^2 + 18x + 8y = 31$

58. $4y^2 - 9x^2 - 18x - 8y = 41$

In each of Problems 59 through 64, identify and sketch the conic section with the given polar equation.

59. $r = \dfrac{6}{1 + \cos\theta}$

60. $r = \dfrac{6}{1 + 2\cos\theta}$

61. $r = \dfrac{3}{1 - \cos\theta}$

62. $r = \dfrac{8}{8 - 2\cos\theta}$

63. $r = \dfrac{6}{2 - \sin\theta}$

64. $r = \dfrac{12}{3 + 2\cos\theta}$

65. Prove that the point of the parabola $y^2 = 4px$ closest to its focus is its vertex.

66. Find an equation of the parabola that has a vertical axis and passes through the points $(2, 3)$, $(4, 3)$, and $(6, -5)$.

67. Show that an equation of the line tangent to the parabola $y^2 = 4px$ at the point (x_0, y_0) is

$$2px - y_0 y + 2px_0 = 0.$$

Conclude that the tangent line intersects the x-axis at the point $(-x_0, 0)$. This fact provides a quick method for constructing a line tangent to a parabola at a given point.

68. A comet's orbit is a parabola with the sun at its focus. When the comet is $100\sqrt{2}$ million miles from the sun, the line from the sun to the comet makes an angle of 45° with the axis of the parabola (Fig. 9.6.35). What will be the minimum distance between the comet and the sun? [*Suggestion:* Write the equation of the parabola with the origin at the focus, then use the result of Problem 65.]

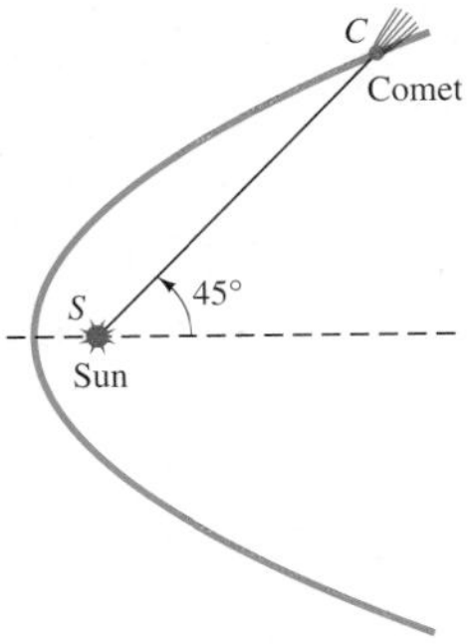

FIGURE 9.6.35 The comet of Problem 68 in parabolic orbit around the sun.

69. Suppose that the angle of Problem 68 increases from 45° to 90° in 3 days. How much longer will be required for the comet to reach its point of closest approach to the sun? Assume that the line segment from the sun to the comet sweeps out area at a constant rate (Kepler's second law).

70. Use Eqs. (7) and (8) to derive Eq. (9) with the values of M and R given in Eqs. (10) and (11).

71. Deduce from Eq. (11) that, given a fixed initial velocity v_0, the maximum range of the projectile is $R_{max} = v_0^2/g$ and is attained when $\alpha = 45°$.

In Problems 72 through 74, assume that a projectile is fired with initial velocity $v_0 = 50$ m/s from the origin and at an angle of inclination α. Use $g = 9.8$ m/s².

72. If $\alpha = 45°$, find the range of the projectile and the maximum height it attains.

73. For what value or values of α is the range $R = 125$ m?

74. Find the range of the projectile and the length of time it remains above the ground if (a) $\alpha = 30°$; (b) $\alpha = 60°$.

75. The book *Elements of Differential and Integral Calculus* by William Granville, Percey Smith, and William Longley (Ginn and Company: Boston, 1929) lists a number of "curves for reference"; the curve with equation $\sqrt{x} + \sqrt{y} = \sqrt{a}$ is called a parabola. Verify that the curve in question actually is a parabola, or show that it is not.

76. The 1992 edition of the study guide for the national actuarial examinations has a problem quite similar to this one: Every point on the plane curve K is equally distant from the point $(-1, -1)$ and the line $x + y = 1$, and K has equation

$$x^2 + Bxy + Cy^2 + Dx + Ey + F = 0.$$

Which is the value of D: -2, 2, 4, 6, or 8?

77. (a) The orbit of the comet Kahoutek is an ellipse of extreme eccentricity $e = 0.999925$; the sun is at one focus of this ellipse. The minimum distance between the sun and Kahoutek is 0.13 AU. What is the maximum distance between Kahoutek and the sun? (b) The orbit of the comet Hyakutake is an ellipse of extreme eccentricity $e = 0.999643856$; the sun is at one focus of this ellipse. The minimum distance between the sun and Hyakutake is 0.2300232 AU. What is the maximum distance between Hyakutake and the sun?

78. The orbit of the planet Mercury is an ellipse of eccentricity $e = 0.206$. Its maximum and minimum distances from the sun are 0.467 and 0.307 AU, respectively. What are the major and minor semiaxes of the orbit of Mercury? Does "nearly circular" accurately describe the orbit of Mercury?

79. Derive Eq. (16) for an ellipse whose foci lie on the y-axis.

80. Show that the line tangent to the ellipse

$$\frac{x^2}{a^2} + \frac{y^2}{b^2} = 1$$

at the point $P(x_0, y_0)$ of that ellipse has equation

$$\frac{x_0 x}{a^2} + \frac{y_0 y}{b^2} = 1.$$

81. Use the result of Problem 80 to establish the reflection property of the ellipse. [*Suggestion:* Let m be the slope of the line normal to the ellipse at $P(x_0, y_0)$ and let m_1 and m_2 be the slopes of the lines PF_1 and PF_2, respectively, from P to the two foci F_1 and F_2 of the ellipse. Show that

$$\frac{m - m_1}{1 + m_1 m} = \frac{m_2 - m}{1 + m_2 m};$$

then use the identity for $\tan(A - B)$.]

82. Given $F_1(-c, 0)$ and $F_2(c, 0)$ with $a > c > 0$, prove that the ellipse

$$\frac{x^2}{a^2} + \frac{y^2}{b^2} = 1$$

(with $b^2 = a^2 - c^2$) is the locus of those points P such that $|PF_1| + |PF_2| = 2a$.

83. Find an equation of the ellipse with horizontal and vertical axes that passes through the points $(-1, 0)$, $(3, 0)$, $(0, 2)$, and $(0, -2)$.

84. Derive an equation for the ellipse with foci $(3, -3)$ and $(-3, 3)$ and major axis of length 10. Note that the foci of this ellipse lie on neither a vertical line nor a horizontal line.

85. Show that the graph of the equation

$$\frac{x^2}{15 - c} - \frac{y^2}{c - 6} = 1$$

is (a) a hyperbola with foci $(\pm 3, 0)$ if $6 < c < 15$ and (b) an ellipse if $c < 6$. (c) Identify the graph in the case $c > 15$.

86. Establish that the line tangent to the hyperbola

$$\frac{x^2}{a^2} - \frac{y^2}{b^2} = 1$$

at the point $P(x_0, y_0)$ has equation

$$\frac{x_0 x}{a^2} - \frac{y_0 y}{b^2} = 1.$$

87. Use the result of Problem 86 to establish the reflection property of the hyperbola. (See the suggestion for Problem 81.)

88. Suppose that $0 < a < c$, and let $b = \sqrt{c^2 - a^2}$. Show that the hyperbola $x^2/a^2 - y^2/b^2 = 1$ is the locus of a point P such that the *difference* between the distances $|PF_1|$ and $|PF_2|$ is equal to $2a$ (F_1 and F_2 are the foci of the hyperbola).

89. Derive an equation for the hyperbola with vertices $(\pm 3/\sqrt{2}, \pm 3/\sqrt{2})$ and foci $(\pm 5, \pm 5)$. Use the difference definition of a hyperbola implied by Problem 88.

90. Two radio signaling stations at A and B lie on an east-west line, with A 100 mi west of B. A plane is flying west on a line 50 mi north of the line AB. Radio signals are sent (traveling at 980 ft/μs) simultaneously from A and B, and the one sent from B arrives at the plane 400 μs before the one sent from A. Where is the plane?

91. Two radio signaling stations are located as in Problem 90 and transmit radio signals that travel at the same speed. But now we know only that the plane is generally somewhere north of the line AB, that the signal from B arrives 400 μs before the one sent from A, and that the signal sent from A and reflected by the plane takes a total of 600 μs to reach B. Where is the plane?

92. A comet has a parabolic orbit with the sun at one focus. When the comet is 150 million miles from the sun, the sun-comet line makes an angle of 45° with the axis of the parabola. What will be the minimum distance between the comet and the sun?

93. A satellite has an elliptical orbit with the center of the earth (take its radius to be 4000 mi) at one focus. The lowest point of its orbit is 500 mi above the North Pole and the highest point is 5000 mi above the South Pole. What is the height of the satellite above the surface of the earth when the satellite crosses the equatorial plane?

94. Find the closest approach to the sun of a comet as in Example 13 of this section; assume that $r = 2.5$ AU when $\theta = 45°$ and that $r = 1$ AU when $\theta = 90°$.

95. An ellipse has semimajor axis a and semiminor axis b. Use the polar-coordinate equation of an ellipse to derive the formula $A = \pi ab$ for its area.

96. The orbit of a certain comet approaching the sun is the parabola

$$r = \frac{1}{1 - \cos\theta}.$$

The units for r are in astronomical units. Suppose that it takes 15 days for the comet to travel from the position $\theta = 60°$ to the position $\theta = 90°$. How much longer will it require for the comet to reach its point of closest approach to the sun? Assume that the radius from the sun to the comet sweeps out area at a constant rate as the comet moves (Kepler's second law of planetary motion).

CHAPTER 9 REVIEW: CONCEPTS AND DEFINITIONS

Use the following list as a guide to additional concepts that you may need to review.

1. Conic sections
2. The relationship between rectangular and polar coordinates
3. The graph of an equation in polar coordinates
4. The area formula in polar coordinates
5. Definition of a parametric curve and of a smooth parametric curve
6. The slope of the line tangent to a smooth parametric curve (both in rectangular and in polar coordinates)
7. Integral computations with parametric curves [Eqs. (1) through (4) of Section 9.5]
8. Arc length of a parametric curve

REVIEW OF CONIC SECTIONS

The parabola with focus $(p, 0)$ and directrix $x = -p$ has eccentricity $e = 1$ and equation $y^2 = 4px$. The accompanying table compares the properties of an ellipse and a hyperbola, each with foci $(\pm c, 0)$ and major axis of length $2a$.

	Ellipse	Hyperbola
Eccentricity	$e = \dfrac{c}{a} < 1$	$e = \dfrac{c}{a} > 1$
a, b, c relation	$a^2 = b^2 + c^2$	$c^2 = a^2 + b^2$
Equation	$\dfrac{x^2}{a^2} + \dfrac{y^2}{b^2} = 1$	$\dfrac{x^2}{a^2} - \dfrac{y^2}{b^2} = 1$
Vertices	$(\pm a, 0)$	$(\pm a, 0)$
y-intercepts	$(0, \pm b)$	None
Directrices	$x = \pm\dfrac{a}{e}$	$x = \pm\dfrac{a}{e}$
Asymptotes	None	$y = \pm\dfrac{bx}{a}$

CHAPTER 9 MISCELLANEOUS PROBLEMS

Sketch the graphs of the equations in Problems 1 through 30. In Problems 1 through 18, if the graph is a conic section, label its center, foci, and vertices.

1. $x^2 + y^2 - 2x - 2y = 2$
2. $x^2 + y^2 = x + y$
3. $x^2 + y^2 - 6x + 2y + 9 = 0$
4. $y^2 = 4(x + y)$
5. $x^2 = 8x - 2y - 20$
6. $x^2 + 2y^2 - 2x + 8y + 8 = 0$
7. $9x^2 + 4y^2 = 36x$
8. $x^2 - y^2 = 2x - 2y - 1$
9. $y^2 - 2x^2 = 4x + 2y + 3$
10. $9y^2 - 4x^2 = 8x + 18y + 31$
11. $x^2 + 2y^2 = 4x + 4y - 12$
12. $y^2 - 6y + 4x + 5 = 0$
13. $9(x^2 - 2x + 1) = 4(y^2 + 9)$
14. $(x^2 - 4)(y^2 - 1) = 0$
15. $x^2 - 8x + y^2 - 2y + 16 = 0$
16. $(x - 1)^2 + 4(y - 2)^2 = 1$
17. $(x^2 - 4x + y^2 - 4y + 8)(x + y)^2 = 0$
18. $x = y^2 + 4y + 5$
19. $r = -2\cos\theta$
20. $\cos\theta + \sin\theta = 0$
21. $r = \dfrac{1}{\sin\theta + \cos\theta}$
22. $r\sin^2\theta = \cos\theta$
23. $r = 3\csc\theta$
24. $r = 2(\cos\theta - 1)$
25. $r^2 = 4\cos\theta$
26. $r\theta = 1$
27. $r = 3 - 2\sin\theta$
28. $r = \dfrac{1}{1 + \cos\theta}$
29. $r = \dfrac{4}{2 + \cos\theta}$
30. $r = \dfrac{4}{1 - 2\cos\theta}$

In Problems 31 through 38, find the area of the region described.

31. Inside both $r = 2\sin\theta$ and $r = 2\cos\theta$
32. Inside $r^2 = 4\cos\theta$
33. Inside $r = 3 - 2\sin\theta$ and outside $r = 4$
34. Inside $r^2 = 2\sin 2\theta$ and outside $r = 2\sin\theta$
35. Inside $r = 2\sin 2\theta$ and outside $r = \sqrt{2}$
36. Inside $r = 3\cos\theta$ and outside $r = 1 + \cos\theta$
37. Inside $r = 1 + \cos\theta$ and outside $r = \cos\theta$
38. Between the loops of $r = 1 - 2\sin\theta$

In Problems 39 through 43, eliminate the parameter and sketch the curve.

39. $x = 2t^3 - 1, \quad y = 2t^3 + 1$
40. $x = \cosh t, \quad y = \sinh t$
41. $x = 2 + \cos t, \quad y = 1 - \sin t$
42. $x = \cos^4 t, \quad y = \sin^4 t$
43. $x = 1 + t^2, \quad y = t^3$

In Problems 44 through 48, write an equation of the line tangent to the given curve at the indicated point.

44. $x = t^2, \quad y = t^3; \quad t = 1$
45. $x = 3\sin t, \quad y = 4\cos t; \quad t = \pi/4$
46. $x = e^t, \quad y = e^{-t}; \quad t = 0$
47. $r = \theta; \quad \theta = \pi/2$
48. $r = 1 + \sin\theta; \quad \theta = \pi/3$

In Problems 49 through 52, find the area of the region between the given curve and the x-axis.

49. $x = 2t + 1, \quad y = t^2 + 3; \quad -1 \leqq t \leqq 2$
50. $x = e^t, \quad y = e^{-t}; \quad 0 \leqq t \leqq 10$
51. $x = 3\sin t, \quad y = 4\cos t; \quad 0 \leqq t \leqq \pi/2$
52. $x = \cosh t, \quad y = \sinh t; \quad 0 \leqq t \leqq 1$

In Problems 53 through 57, find the arc length of the given curve.

53. $x = t^2, \quad y = t^3; \quad 0 \leqq t \leqq 1$
54. $x = \ln(\cos t), \quad y = t; \quad 0 \leqq t \leqq \pi/4$
55. $x = 2t, \quad y = t^3 + \dfrac{1}{3t}; \quad 1 \leqq t \leqq 2$

56. $r = \sin\theta; \quad 0 \leqq \theta \leqq \pi$

57. $r = \sin^2(\theta/3); \quad 0 \leqq \theta \leqq \pi$

In Problems 58 through 62, find the area of the surface generated by revolving the given curve around the x-axis.

58. $x = t^2 + 1, \quad y = 3t; \quad 0 \leqq t \leqq 2$

59. $x = 4\sqrt{t}, \quad y = \dfrac{t^3}{3} + \dfrac{1}{2t^2}; \quad 1 \leqq t \leqq 4$

60. $r = \cos\theta$

61. $r = e^{\theta/2}; \quad 0 \leqq \theta \leqq \pi$

62. $x = e^t \cos t, \quad y = e^t \sin t; \quad 0 \leqq t \leqq \pi/2$

63. Consider the rolling circle of radius a that was used to generate the cycloid in Example 5 of Section 9.4. Suppose that this circle is the rim of a disk, and let Q be a point of this disk at distance $b < a$ from its center. Find parametric equations for the curve traced by Q as the circle rolls along the x-axis. Assume that Q begins at the point $(0, a - b)$. Sketch this curve, which is called a **trochoid.**

64. If the smaller circle of Problem 34 in Section 9.4 rolls around the *outside* of the larger circle, the path of the point P is called an **epicycloid.** Show that it has parametric equations

$$x = (a + b)\cos t - b\cos\left(\frac{a+b}{b}t\right),$$

$$y = (a + b)\sin t - b\sin\left(\frac{a+b}{b}t\right).$$

65. Suppose that $b = a$ in Problem 64. Show that the epicycloid is then the cardioid $r = 2a(1 - \cos\theta)$ translated a units to the right.

66. Find the area of the surface generated by revolving the lemniscate $r^2 = 2a^2\cos 2\theta$ around the x-axis.

67. Find the volume generated by revolving around the y-axis the area under the cycloid

$$x = a(t - \sin t), \quad y = a(1 - \cos t), \quad 0 \leqq t \leqq 2\pi.$$

68. Show that the length of one arch of the hypocycloid of Problem 34 in Section 9.4 is $s = 8b(a - b)/a$.

69. Find a polar-coordinate equation of the circle that passes through the origin and is centered at the point with polar coordinates (p, α).

70. Find a simple equation of the parabola whose focus is the origin and whose directrix is the line $y = x + 4$. Recall from Miscellaneous Problem 93 of Chapter 3 that the distance from the point (x_0, y_0) to the line with equation $Ax + By + C = 0$ is

$$\frac{|Ax_0 + By_0 + C|}{\sqrt{A^2 + B^2}}.$$

71. A **diameter** of an ellipse is a chord through its center. Find the maximum and minimum lengths of diameters of the ellipse with equation

$$\frac{x^2}{a^2} + \frac{y^2}{b^2} = 1.$$

72. Use calculus to prove that the ellipse of Problem 71 is normal to the coordinate axes at each of its four vertices.

73. The parabolic arch of a bridge has base width b and height h at its center. Write its equation, choosing the origin on the ground at the left end of the arch.

74. Use methods of calculus to find the points of the ellipse

$$\frac{x^2}{a^2} + \frac{y^2}{b^2} = 1$$

that are nearest to and farthest from (a) the center $(0, 0)$; (b) the focus $(c, 0)$.

75. Consider a line segment QR that contains a point P such that $|QP| = a$ and $|PR| = b$. Suppose that Q is constrained to move on the y-axis, whereas R must remain on the x-axis. Prove that the locus of P is an ellipse.

76. Suppose that $a > 0$ and that F_1 and F_2 are two fixed points in the plane with $|F_1F_2| > 2a$. Imagine a point P that moves in such a way that $|PF_2| = 2a + |PF_1|$. Prove that the locus of P is one branch of a hyperbola with foci F_1 and F_2. Then—as a consequence—explain how to construct points on a hyperbola by drawing appropriate circles centered at its foci.

77. Let Q_1 and Q_2 be two points on the parabola $y^2 = 4px$. Let P be the point of the parabola at which the tangent line is parallel to Q_1Q_2. Prove that the horizontal line through P bisects the segment Q_1Q_2.

78. Determine the locus of a point P such that the product of its distances from the two fixed points $F_1(-a, 0)$ and $F_2(a, 0)$ is a^2.

79. Find the eccentricity of the conic section with equation $3x^2 - y^2 + 12x + 9 = 0$.

80. Find the area bounded by the loop of the *strophoid*

$$r = \sec\theta - 2\cos\theta$$

shown in Fig. 9.MP.1.

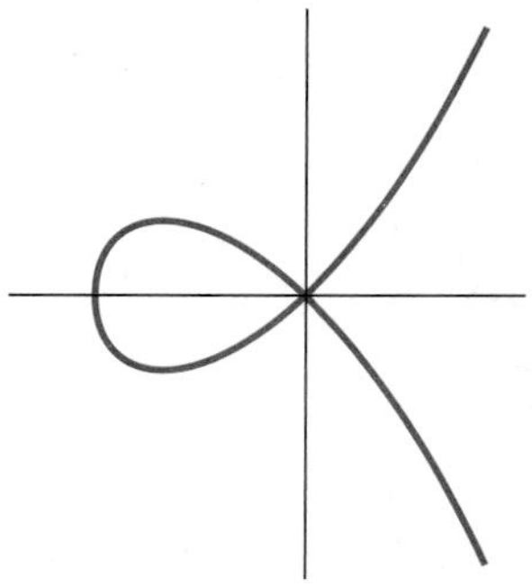

FIGURE 9.MP.1 The strophoid of Problem 80.

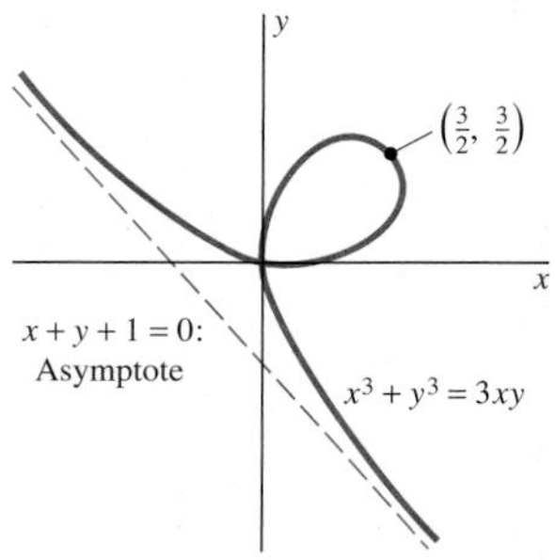

FIGURE 9.MP.2 The folium of Descartes $x^3 + y^3 = 3xy$ (Problem 81).

81. Find the area bounded by the loop of the *folium of Descartes* with equation $x^3 + y^3 = 3xy$ shown in Fig. 9.MP.2. (*Suggestion:* Change to polar coordinates and then substitute $u = \tan\theta$ to evaluate the area integral.)

82. Use the method of Problem 81 to find the area bounded by the first-quadrant loop of the curve $x^5 + y^5 = 5x^2y^2$ (similar to the folium of Problem 81).

83. The graph of a conic section in the xy-plane has intercepts at $(5, 0)$, $(-5, 0)$, $(0, 4)$, and $(0, -4)$. Deduce all the information you can about this conic. Can you determine whether it is a parabola, a hyperbola, or an ellipse? What if you also know that the graph of this conic is normal to the y-axis at the point $(0, 4)$?

INFINITE SERIES

10

Srinivasa Ramanujan
(1887–1920)

On a cold January day in 1913, the eminent Cambridge mathematics professor G. H. Hardy received a letter from an unknown 25-year-old clerk in the accounting department of a government office in Madras, India. Its author, Srinivasa Ramanujan, had no university education, he admitted—he had flunked out—but "after leaving school I have employed the spare time at my disposal to work at Mathematics . . . I have not trodden through the conventional regular course . . . but am striking out a new path for myself." The ten pages that followed listed in neat handwritten script approximately 50 formulas, most dealing with integrals and infinite series that Ramanujan had discovered, and asked Hardy's advice whether they contained anything of value. The formulas were of such exotic and unlikely appearance that Hardy at first suspected a hoax, but he and his colleague J. E. Littlewood soon realized that they were looking at the work of an extraordinary mathematical genius.

Thus began one of the most romantic episodes in the history of mathematics. In April 1914 Ramanujan arrived in England a poor, self-taught Indian mathematical amateur called to collaborate as an equal with the most sophisticated professional mathematicians of the day. For the next three years a steady stream of remarkable discoveries poured forth from his pen. But in 1917 he fell seriously ill, apparently with tuberculosis. The following year he returned to India to attempt to regain his health but never recovered, and he died in 1920 at the age of 32. Up to the very end he worked feverishly to record his final discoveries. He left behind notebooks outlining work whose completion occupied prominent mathematicians throughout the twentieth century.

With the possible exception of Euler, no one before or since has exhibited Ramanujan's virtuosity with infinite series. An example of his discoveries is the infinite series

$$\frac{1}{\pi} = \frac{\sqrt{8}}{9801} \sum_{n=0}^{\infty} \frac{(4n)!}{(n!)^4} \cdot \frac{(1103 + 26390n)}{396^{4n}},$$

whose first term yields the familiar approximation $\pi \approx 3.14159$, and with each additional term giving π to roughly eight more decimal places of accuracy. For instance, just four terms of Ramanujan's series are needed to calculate the 30-place approximation

$$\pi \approx 3.14159\,26535\,89793\,23846\,26433\,83279$$

that suffices for virtually any imaginable "practical" application—if the universe were a sphere with a radius of 10 billion light years, then this value of π would give its circumference accurate to the nearest hundredth of an inch. But in recent years Ramanujan's ideas have been used to calculate the value of π accurate to a *billion* decimal places. Indeed, such gargantuan computations of π are commonly used to check the accuracy of new supercomputers.

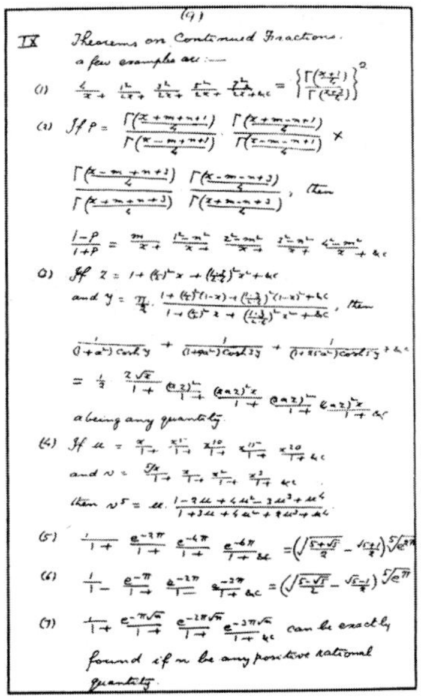
(9)
IX Theorems on Continued Fractions.
a few examples are:—

A typical page of Ramanujan's letter to Hardy, listing formulas Ramanujan had discovered, but with no hint of proof or derivation.

10.1 INTRODUCTION

In the fifth century B.C., the Greek philosopher Zeno proposed the following paradox: In order for a runner to travel a given distance, the runner must first travel halfway, then half the remaining distance, then half the distance that yet remains, and so on *ad infinitum.* But, Zeno argued, it is clearly impossible for a runner to accomplish infinitely many such tasks in a finite period of time, so motion from one point to another is impossible.

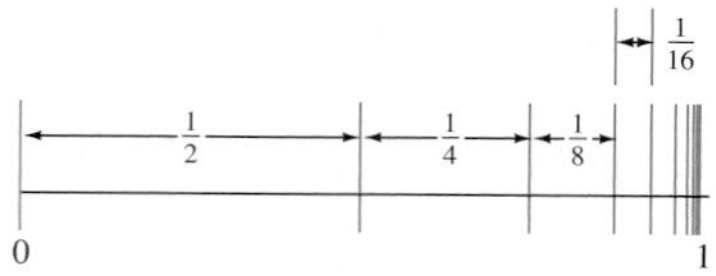

FIGURE 10.1.1 Subdivision of an interval to illustrate Zeno's paradox.

Zeno's paradox suggests the infinite subdivision of [0, 1] indicated in Fig. 10.1.1. There is one subinterval of length $1/2^n$ for each integer $n = 1, 2, 3, \dots$. If the length of the interval is the sum of the lengths of the subintervals into which it is divided, then it would appear that

$$1 = \frac{1}{2} + \frac{1}{4} + \frac{1}{8} + \frac{1}{16} + \cdots + \frac{1}{2^n} + \cdots,$$

with infinitely many terms somehow adding up to 1. But the formal infinite sum

$$1 + 2 + 3 + \cdots + n + \cdots$$

of all the positive integers seems meaningless—it does not appear to add up to *any* (finite) value.

The question is this: What, if anything, do we mean by the sum of an *infinite* collection of numbers? This chapter explores conditions under which an *infinite* sum

$$a_1 + a_2 + a_3 + \cdots + a_n + \cdots,$$

known as an *infinite series,* is meaningful. We discuss methods for computing the sum of an infinite series and applications of the algebra and calculus of infinite series. Infinite series are important in science and mathematics because many functions either arise most naturally in the form of infinite series or have infinite series representations (such as the Taylor series of Section 10.4) that are useful for numerical computations.

10.2 INFINITE SEQUENCES

An **infinite sequence** of real numbers is an ordered, unending list

$$a_1,\ a_2,\ a_3,\ a_4,\ \dots,\ a_n,\ a_{n+1},\ \dots \qquad \textbf{(1)}$$

of numbers. That this list is *ordered* implies that it has a first term a_1, a second term a_2, a third term a_3, and so forth. That the sequence is unending, or *infinite,* implies that (for every n) the ***n*th term** a_n has a successor a_{n+1}. Thus, as indicated by the final ellipsis in (1), an infinite sequence never ends and—despite the fact that we write explicitly only a finite number of terms—it actually has an infinite number of terms. Concise notation for the infinite sequence in (1) is

$$\{a_n\}_{n=1}^{\infty}, \qquad \{a_n\}_1^{\infty}, \qquad \text{or simply} \qquad \{a_n\}. \qquad \textbf{(2)}$$

Frequently an infinite sequence $\{a_n\}$ of numbers can be described "all at once" by a single function f that gives the successive terms of the sequence as successive values of the function:

$$a_n = f(n) \quad \text{for} \quad n = 1, 2, 3, \dots. \qquad \textbf{(3)}$$

Here $a_n = f(n)$ is simply a *formula for the nth term* of the sequence. Conversely, if the sequence $\{a_n\}$ is given in advance, we can regard (3) as the definition of the function f having the set of positive integers as its domain of definition. Ordinarily we will use the subscript notation a_n in preference to the function notation $f(n)$.

EXAMPLE 1 The following table exhibits several particular infinite sequences. Each is described in three ways: in the concise sequential notation $\{a_n\}$ of (2), by writing the formula as in (3) for its nth term, and in extended list notation as in (1). Note that n need not begin with the initial value 1.

$\left\{\frac{1}{n}\right\}_1^\infty$	$a_n = \frac{1}{n}$	$1, \frac{1}{2}, \frac{1}{3}, \frac{1}{4}, \dots, \frac{1}{n}, \dots$
$\left\{\frac{1}{10^n}\right\}_0^\infty$	$a_n = \frac{1}{10^n}$	$1, \frac{1}{10}, \frac{1}{100}, \frac{1}{1000}, \dots, \frac{1}{10^n}, \dots$
$\{\sqrt{3n-7}\}_3^\infty$	$a_n = \sqrt{3n-7}$	$\sqrt{2}, \sqrt{5}, \sqrt{8}, \sqrt{11}, \dots, \sqrt{3n-7}, \dots$
$\left\{\sin\frac{n\pi}{2}\right\}_1^\infty$	$a_n = \sin\frac{n\pi}{2}$	$1, 0, -1, 0, \dots, \sin\frac{n\pi}{2}, \dots$
$\{3+(-1)^n\}_1^\infty$	$a_n = 3+(-1)^n$	$2, 4, 2, 4, \dots, 3+(-1)^n, \dots$

Sometimes it is inconvenient or impossible to give an explicit formula for the nth term of a particular sequence. The following example illustrates how sequences may be defined in other ways. ◆

EXAMPLE 2 Here we give the first ten terms of each sequence.

(a) The sequence of prime integers (those positive integers n having precisely two divisors, 1 and n with $n > 1$),

$$2, 3, 5, 7, 11, 13, 17, 19, 23, 29, \dots$$

(b) The sequence whose nth term is the nth decimal digit of the number

$$\pi = 3.14159265358979323846\dots,$$

$$1, 4, 1, 5, 9, 2, 6, 5, 3, 5, \dots$$

(c) The **Fibonacci sequence** $\{F_n\}$, which may be defined by

$$F_1 = 1, \quad F_2 = 1, \quad \text{and} \quad F_{n+1} = F_n + F_{n-1} \quad \text{for} \quad n \geqq 2.$$

Thus each term after the second is the sum of the preceding two terms:

$$1, 1, 2, 3, 5, 8, 13, 21, 34, 55, \dots$$

This is an example of a *recursively defined* sequence in which each term (after the first few) is given by a formula involving its predecessors. The 13th-century Italian mathematician Fibonacci asked the following question: If we start with a single pair of rabbits that gives birth to a new pair after two months, and each such new pair does the same, how many pairs of rabbits will we have after n months? See Problems 55 and 56.

(d) If the amount $A_0 = 100$ dollars is invested in a savings account that draws 10% interest compounded annually, then the amount A_n in the account at the end of n years is defined (for $n \geqq 1$) by the *iterative formula* $A_n = (1.10)A_{n-1}$ (rounded to the nearest number of cents) in terms of the preceding amount:

$$110.00, 121.00, 133.10, 146.41, 161.05, 177.16,$$
$$194.87, 214.36, 235.79, 259.37, \dots$$ ◆

Limits of Sequences

The limit of a sequence is defined in much the same way as the limit of an ordinary function (Section 2.2).

DEFINITION Limit of a Sequence

We say that the sequence $\{a_n\}$ **converges** to the real number L, or has the **limit** L, and we write

$$\lim_{n\to\infty} a_n = L, \tag{4}$$

provided that a_n can be made as close to L as we please merely by choosing n to be sufficiently large. That is, given any number $\epsilon > 0$, there exists an integer N such that

$$|a_n - L| < \epsilon \quad \text{for all } n \geqq N. \tag{5}$$

If the sequence $\{a_n\}$ does *not* converge, then we say that $\{a_n\}$ **diverges.**

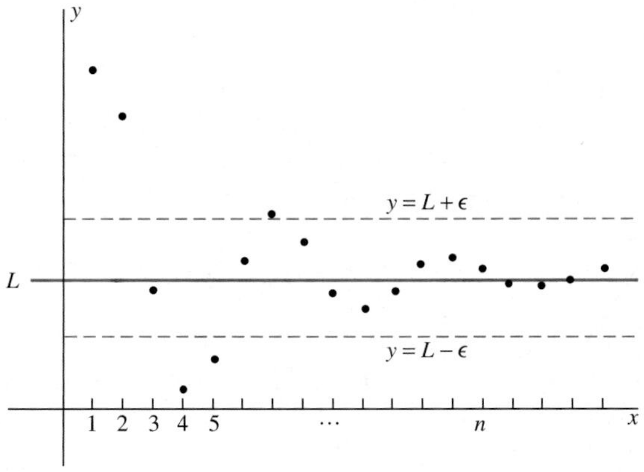

FIGURE 10.2.1 The point (n, a_n) approaches the line $y = L$ as $n \to +\infty$.

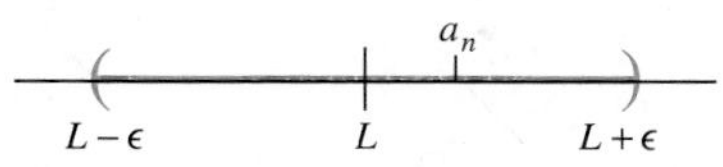

FIGURE 10.2.2 The inequality $|a_n - L| < \epsilon$ means that a_n lies somewhere between $L - \epsilon$ and $L + \epsilon$.

Figure 10.2.1 illustrates geometrically the definition of the limit of a sequence. Because

$$|a_n - L| < \epsilon \quad \text{means that} \quad L - \epsilon < a_n < L + \epsilon,$$

the condition in (5) means that if $n \geqq N$, then the point (n, a_n) lies between the horizontal lines $y = L - \epsilon$ and $y = L + \epsilon$. Alternatively, if $n \geqq N$, then the number a_n lies between the points $L - \epsilon$ and $L + \epsilon$ on the real line (Fig. 10.2.2).

EXAMPLE 3 Suppose that we want to establish rigorously the intuitively evident fact that the sequence $\{1/n\}_1^\infty$ converges to zero,

$$\lim_{n\to\infty} \frac{1}{n} = 0. \tag{6}$$

Because $L = 0$ here, we need only convince ourselves that to each positive number ϵ there corresponds an integer N such that

$$\left|\frac{1}{n}\right| = \frac{1}{n} < \epsilon \quad \text{if} \quad n \geqq N.$$

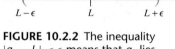

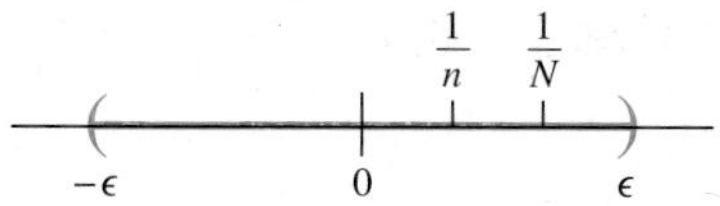

FIGURE 10.2.3 If $N > \dfrac{1}{\epsilon}$ and $n \geqq N$ then $0 < \dfrac{1}{n} \leqq \dfrac{1}{N} < \epsilon$.

But evidently it suffices to choose any fixed integer $N > 1/\epsilon$. Then $n \geqq N$ implies immediately that

$$\frac{1}{n} \leqq \frac{1}{N} < \epsilon,$$

as desired (Fig. 10.2.3). ◆

EXAMPLE 4 (a) The sequence $\{(-1)^n\}$ diverges because its successive terms "oscillate" between the two values $+1$ and -1. Hence $(-1)^n$ cannot approach any single value as $n \to \infty$. (b) The terms of the sequence $\{n^2\}$ increase without bound as $n \to \infty$. Thus the sequence $\{n^2\}$ diverges. In this case, we might also say that $\{n^2\}$ diverges *to infinity*. ◆

Using Limit Laws

The limit laws in Section 2.2 for limits of functions have natural analogues for limits of sequences. Their proofs are based on techniques similar to those used in Appendix D.

THEOREM 1 Limit Laws for Sequences

If the limits

$$\lim_{n\to\infty} a_n = A \quad \text{and} \quad \lim_{n\to\infty} b_n = B$$

exist (so A and B are real numbers), then

1. $\lim_{n\to\infty} ca_n = cA$ (c any real number);
2. $\lim_{n\to\infty} (a_n + b_n) = A + B$;
3. $\lim_{n\to\infty} a_n b_n = AB$;
4. $\lim_{n\to\infty} \dfrac{a_n}{b_n} = \dfrac{A}{B}$.

In part 4 we must assume that $B \neq 0$ (so that $b_n \neq 0$ for all sufficiently large values of n).

THEOREM 2 Substitution Law for Sequences

If $\lim_{n\to\infty} a_n = A$ and the function f is continuous at $x = A$, then

$$\lim_{n\to\infty} f(a_n) = f(A).$$

THEOREM 3 Squeeze Law for Sequences

If $a_n \leqq b_n \leqq c_n$ for all n and

$$\lim_{n\to\infty} a_n = L = \lim_{n\to\infty} c_n,$$

then $\lim_{n\to\infty} b_n = L$ as well.

These theorems can be used to compute limits of many sequences formally, without recourse to the definition. For example, Eq. (6) and the product law of limits yield

$$\lim_{n\to\infty} \frac{1}{n^k} = 0 \tag{7}$$

for every positive integer k.

EXAMPLE 5 Eq. (7) and the limit laws give (after dividing numerator and denominator by the highest power of n that is present)

$$\lim_{n\to\infty} \frac{7n^2}{5n^2 - 3} = \lim_{n\to\infty} \frac{7}{5 - \dfrac{3}{n^2}}$$

$$= \frac{\lim_{n\to\infty} 7}{\left(\lim_{n\to\infty} 5\right) - 3 \cdot \left(\lim_{n\to\infty} \dfrac{1}{n^2}\right)} = \frac{7}{5 - 3 \cdot 0} = \frac{7}{5}.$$ ◆

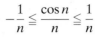

EXAMPLE 6 Show that $\lim_{n\to\infty} \frac{\cos n}{n} = 0.$

Solution This follows from the squeeze law and the fact that $1/n \to 0$ as $n \to \infty$, because

$$-\frac{1}{n} \leqq \frac{\cos n}{n} \leqq \frac{1}{n}$$

for every positive integer n. ◆

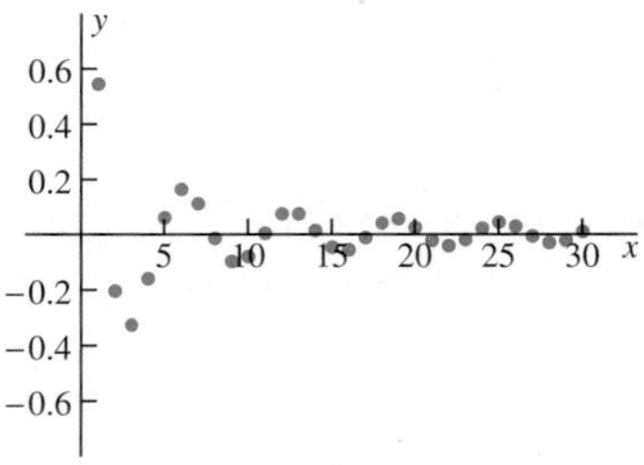

FIGURE 10.2.4 The points $(n, (\cos n)/n)$ for $n = 1, 2, \ldots, 30$.

REMARK With a typical graphing calculator (in "dot plot mode") or computer algebra system (using its "list plot" facility), one can plot the points (n, a_n) in the xy-plane corresponding to a given sequence $\{a_n\}$. Figure 10.2.4 shows such a plot for the sequence of Example 6 and provides visual evidence of its convergence to zero.

EXAMPLE 7 Show that if $a > 0$, then $\lim_{n\to\infty} \sqrt[n]{a} = 1$.

Solution We apply the substitution law with $f(x) = a^x$, $a_n = 1/n$, and $A = 0$. Because $1/n \to 0$ as $n \to \infty$ and f is continuous at $x = 0$, this gives

$$\lim_{n\to\infty} a^{1/n} = \lim_{n\to\infty} f(1/n) = f(0) = a^0 = 1.$$ ◆

EXAMPLE 8 The limit laws and the continuity of $f(x) = \sqrt{x}$ at $x = 4$ yield

$$\lim_{n\to\infty} \sqrt{\frac{4n-1}{n+1}} = \left(\lim_{n\to\infty} \frac{4 - \frac{1}{n}}{1 + \frac{1}{n}} \right)^{1/2} = \sqrt{4} = 2.$$ ◆

EXAMPLE 9 Show that if $|r| < 1$, then $\lim_{n\to\infty} r^n = 0$.

Solution Because $|r^n| = |(-r)^n|$, we may assume that $0 < r < 1$. Then $1/r = 1 + a$ for some number $a > 0$, so the binomial formula yields

$$\frac{1}{r^n} = (1+a)^n = 1 + na + \{\text{positive terms}\} > 1 + na;$$

$$0 < r^n < \frac{1}{1+na}.$$

Now $1/(1 + na) \to 0$ as $n \to \infty$. Therefore, the squeeze law implies that $r^n \to 0$ as $n \to \infty$. ◆

Figure 10.2.5 shows the graph of a function f such that $\lim_{x\to\infty} f(x) = L$. If the sequence $\{a_n\}$ is defined by the formula $a_n = f(n)$ for each positive integer n,

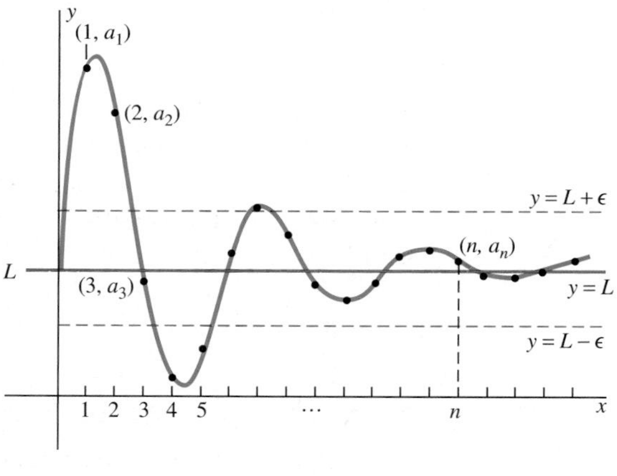

FIGURE 10.2.5 If $\lim_{x\to\infty} f(x) = L$ and $a_n = f(n)$, then $\lim_{n\to\infty} a_n = L$.

then all the points $(n, f(n))$ lie on the graph of $y = f(x)$. It therefore follows from the definition of the limit of a function that $\lim_{n\to\infty} a_n = L$ as well.

THEOREM 4 Limits of Functions and Sequences
If $a_n = f(n)$ for each positive integer n, then

$$\lim_{x\to\infty} f(x) = L \quad \text{implies that} \quad \lim_{n\to\infty} a_n = L. \tag{8}$$

The converse of the statement in (8) is generally false. For example, take $f(x) = \sin \pi x$ and, for each positive integer n, let $a_n = f(n) = \sin n\pi$. Then $\sin n\pi \equiv 0$, but $\sin nx$ oscillates between 1 and -1, so

$$\lim_{n\to\infty} a_n = \lim_{n\to\infty} \sin n\pi = 0, \quad \text{but}$$

$$\lim_{x\to\infty} f(x) = \lim_{x\to\infty} \sin \pi x \quad \text{does not exist.}$$

Because of (8) we can use **l'Hôpital's rule for sequences:** If $a_n = f(n), b_n = g(n)$, and $f(x)/g(x)$ has the indeterminate form ∞/∞ as $x \to \infty$, then

$$\lim_{n\to\infty} \frac{a_n}{b_n} = \lim_{x\to\infty} \frac{f(x)}{g(x)} = \lim_{x\to\infty} \frac{f'(x)}{g'(x)}, \tag{9}$$

provided that f and g satisfy the other hypotheses of l'Hôpital's rule, including the important assumption that the right-hand limit exists.

EXAMPLE 10 Show that $\lim_{n\to\infty} \dfrac{\ln n}{n} = 0.$

Solution The function $(\ln x)/x$ is defined for all $x \geqq 1$ and agrees with the given sequence $\{(\ln n)/n\}$ when $x = n$, a positive integer. Because $(\ln x)/x$ has the indeterminate form ∞/∞ as $x \to \infty$, l'Hôpital's rule gives

$$\lim_{n\to\infty} \frac{\ln n}{n} = \lim_{x\to\infty} \frac{\ln x}{x} = \lim_{x\to\infty} \frac{\frac{1}{x}}{1} = 0.$$ ◆

EXAMPLE 11 Show that $\lim_{n\to\infty} \sqrt[n]{n} = 1.$

Solution First we note that

$$\ln \sqrt[n]{n} = \ln n^{1/n} = \frac{\ln n}{n} \to 0 \quad \text{as } n \to \infty,$$

by Example 10. By the substitution law with $f(x) = e^x$, this gives

$$\lim_{n\to\infty} n^{1/n} = \lim_{n\to\infty} \exp\left(\ln n^{1/n}\right) = e^0 = 1.$$ ◆

EXAMPLE 12 Find $\lim_{n\to\infty} \dfrac{3n^3}{e^{2n}}.$

Solution We apply l'Hôpital's rule repeatedly, although we must be careful at each intermediate step to verify that we still have an indeterminate form. Thus we find that

$$\lim_{n\to\infty} \frac{3n^3}{e^{2n}} = \lim_{x\to\infty} \frac{3x^3}{e^{2x}} = \lim_{x\to\infty} \frac{9x^2}{2e^{2x}} = \lim_{x\to\infty} \frac{18x}{4e^{2x}} = \lim_{x\to\infty} \frac{18}{8e^{2x}} = 0.$$ ◆

Bounded Monotonic Sequences

The set of all *rational* numbers has by itself all of the most familiar elementary algebraic properties of the entire real number system. To guarantee the existence of irrational numbers, we must assume in addition a "completeness property" of the real numbers. Otherwise, the real line might have "holes" where the irrational numbers ought to be. One way of stating this completeness property is in terms of the convergence of an important type of sequence, a bounded monotonic sequence.

The sequence $\{a_n\}_1^\infty$ is said to be **increasing** if

$$a_1 \leqq a_2 \leqq a_3 \leqq \cdots \leqq a_n \leqq \cdots$$

and **decreasing** if

$$a_1 \geqq a_2 \geqq a_3 \geqq \cdots \geqq a_n \geqq \cdots.$$

The sequence $\{a_n\}$ is **monotonic** if it is either increasing or decreasing. The sequence $\{a_n\}$ is **bounded** if there is a number M such that $|a_n| \leqq M$ for all n. The following assertion may be taken to be an axiom for the real number system.

Bounded Monotonic Sequence Property

Every bounded monotonic infinite sequence converges—that is, has a finite limit.

Suppose, for example, that the increasing sequence $\{a_n\}_1^\infty$ is bounded above by a number M, meaning that $a_n \leqq M$ for all $n \geqq 1$. Because the sequence is also bounded below (by a_1, for instance), the bounded monotonic sequence property implies that

$$\lim_{n\to\infty} a_n = A \quad \text{for some real number } A \leqq M,$$

as in Fig. 10.2.6(a). If the increasing sequence $\{a_n\}$ is not bounded above, then it follows that

$$\lim_{n\to\infty} a_n = +\infty$$

as in Fig. 10.2.6(b). (See Problem 52.) Figure 10.2.7 illustrates the graph of a typical bounded increasing sequence, with the heights of the points (n, a_n) steadily rising toward A.

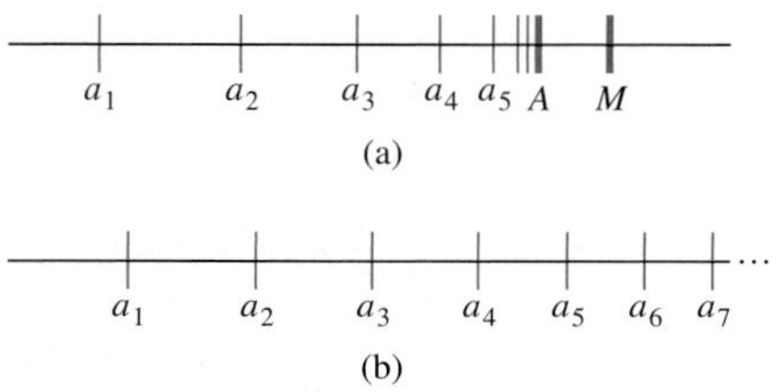

FIGURE 10.2.6 (a) If the increasing sequence $\{a_n\}$ is bounded above by M, then its terms "pile up" at some point $A \leqq M$. (b) If the sequence is unbounded, then its terms "keep going" and diverge to infinity.

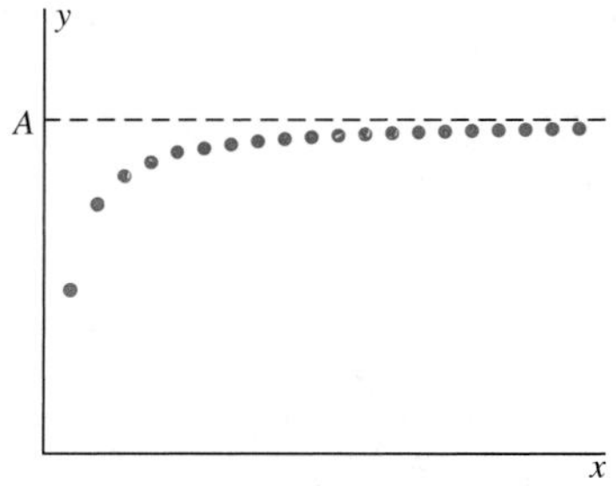

FIGURE 10.2.7 Graph of a bounded increasing sequence with limit A.

EXAMPLE 13 Investigate the sequence $\{a_n\}$ that is defined recursively by

$$a_1 = \sqrt{6}, \qquad a_{n+1} = \sqrt{6 + a_n} \quad \text{for} \quad n \geqq 1. \tag{10}$$

Solution The first four terms of this sequence are

$$\sqrt{6}, \quad \sqrt{6+\sqrt{6}}, \quad \sqrt{6+\sqrt{6+\sqrt{6}}}, \quad \sqrt{6+\sqrt{6+\sqrt{6+\sqrt{6}}}}. \tag{11}$$

If the sequence $\{a_n\}$ converges, then its limit A would seem to be the natural interpretation of the infinite expression

$$\sqrt{6+\sqrt{6+\sqrt{6+\sqrt{6+\cdots}}}}.$$

A calculator gives 2.449, 2.907, 2.984, and 2.997 for the approximate values of the terms in (11). This suggests that the sequence may be bounded above by $M = 3$. Indeed, if we assume that a particular term a_n satisfies the inequality $a_n < 3$, then it follows that

$$a_{n+1} = \sqrt{6 + a_n} < \sqrt{6 + 3} = 3;$$

that is, $a_{n+1} < 3$ as well. Can you see that this implies that *all* terms of the sequence are less than 3? (If there were a first term not less than 3, then its predecessor would be less than 3, and we would have a contradiction. This is a "proof by mathematical induction.")

In order to apply the bounded monotonic sequence property to conclude that the sequence $\{a_n\}$ converges, it remains to show that it is an increasing sequence. But

$$(a_{n+1})^2 - (a_n)^2 = (6 + a_n) - (a_n)^2 = (2 + a_n)(3 - a_n) > 0$$

because $a_n < 3$. Because all terms of the sequence are positive (why?), it therefore follows that $a_{n+1} > a_n$ for all $n \geqq 1$, as desired.

Now that we know that the limit A of the sequence $\{a_n\}$ exists, we can write

$$A = \lim_{n\to\infty} a_{n+1} = \lim_{n\to\infty} \sqrt{6 + a_n} = \sqrt{6 + A},$$

and thus $A^2 = 6 + A$. The roots of this quadratic equation are -2 and 3. Because $A > 0$ (why?), we conclude that $A = \lim_{n\to\infty} a_n = 3$, and so

$$\sqrt{6 + \sqrt{6 + \sqrt{6 + \sqrt{6 + \cdots}}}} = 3. \qquad \textbf{(12)}$$

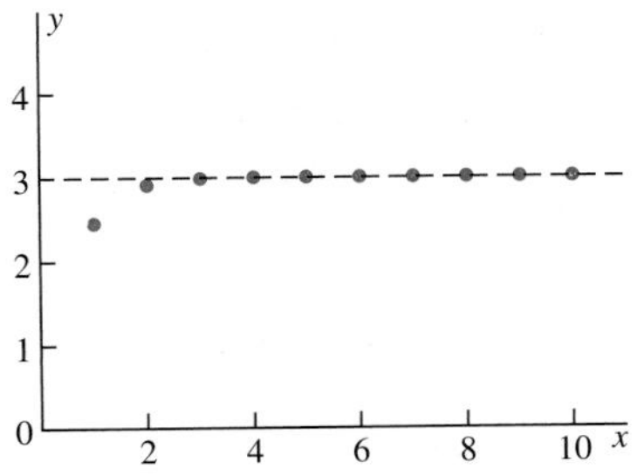

FIGURE 10.2.8 Graph of the sequence of Example 13.

The graph in Fig. 10.2.8 of the first ten terms of the sequence $\{a_n\}$ shows that the convergence to its limit 3 is quite rapid. ◆

To indicate what the bounded monotonic sequence property has to do with the "completeness property" of the real numbers, in Problem 63 we outline a proof, using this property, of the existence of the number $\sqrt{2}$. In Problems 61 and 62, we outline a proof of the equivalence of the bounded monotonic sequence property and another common statement of the completeness of the real number system—the *least upper bound property.*

10.2 TRUE/FALSE STUDY GUIDE

10.2 CONCEPTS: QUESTIONS AND DISCUSSION

1. Can a sequence $\{a_n\}_1^\infty$ converge to two different numbers?
2. Suppose it is known that every open interval containing the point L contains all but finitely many members of the sequence $\{a_n\}_1^\infty$. Does this imply that $\lim_{n\to\infty} a_n = L$?
3. Suppose that the sequence $\{a_n\}_1^\infty$ is obtained by interspersing the members of the two convergent infinite sequences $\{p_n\}_1^\infty$ and $\{q_n\}_1^\infty$. Does it follow that the sequence $\{a_n\}_1^\infty$ also converges?

10.2 PROBLEMS

In Problems 1 through 8, find a pattern in the sequence with given terms a_1, a_2, a_3, a_4 and (assuming that it continues as indicated) write a formula for the general term a_n of the sequence.

1. $1, 4, 9, 16, \dots$

2. $2, 7, 12, 17, \dots$

3. $\frac{1}{3}, \frac{1}{9}, \frac{1}{27}, \frac{1}{81}, \dots$

4. $1, -\frac{1}{2}, \frac{1}{4}, -\frac{1}{8}, \dots$

5. $\frac{1}{2}, \frac{1}{5}, \frac{1}{8}, \frac{1}{11}, \dots$

6. $\frac{1}{2}, \frac{1}{5}, \frac{1}{10}, \frac{1}{17}, \dots$

7. $0, 2, 0, 2, \dots$

8. $10, 5, 10, 5, \dots$

In Problems 9 through 42, determine whether or not the sequence $\{a_n\}$ converges, and find its limit if it does converge.

9. $a_n = \dfrac{2n}{5n - 3}$

10. $a_n = \dfrac{1 - n^2}{2 + 3n^2}$

11. $a_n = \dfrac{n^2 - n + 7}{2n^3 + n^2}$

12. $a_n = \dfrac{n^3}{10n^2 + 1}$

13. $a_n = 1 + \left(\frac{9}{10}\right)^n$

14. $a_n = 2 - \left(-\frac{1}{2}\right)^n$

15. $a_n = 1 + (-1)^n$

16. $a_n = \dfrac{1 + (-1)^n}{\sqrt{n}}$

17. $a_n = \dfrac{1 + (-1)^n \sqrt{n}}{\left(\frac{3}{2}\right)^n}$

18. $a_n = \dfrac{\sin n}{3^n}$

19. $a_n = \dfrac{\sin^2 n}{\sqrt{n}}$

20. $a_n = \sqrt{\dfrac{2 + \cos n}{n}}$

21. $a_n = n \sin \pi n$

22. $a_n = n \cos \pi n$

23. $a_n = \pi^{-(\sin n)/n}$

24. $a_n = 2^{\cos \pi n}$

25. $a_n = \dfrac{\ln n}{\sqrt{n}}$

26. $a_n = \dfrac{\ln 2n}{\ln 3n}$

27. $a_n = \dfrac{(\ln n)^2}{n}$

28. $a_n = n \sin\left(\dfrac{1}{n}\right)$

29. $a_n = \dfrac{\tan^{-1} n}{n}$

30. $a_n = \dfrac{n^3}{e^{n/10}}$

31. $a_n = \dfrac{2^n + 1}{e^n}$

32. $a_n = \dfrac{\sinh n}{\cosh n}$

33. $a_n = \left(1 + \dfrac{1}{n}\right)^n$

34. $a_n = (2n + 5)^{1/n}$

35. $a_n = \left(\dfrac{n - 1}{n + 1}\right)^n$

36. $a_n = (0.001)^{-1/n}$

37. $a_n = \sqrt[n]{2^{n+1}}$

38. $a_n = \left(1 - \dfrac{2}{n^2}\right)^n$

39. $a_n = \left(\dfrac{2}{n}\right)^{3/n}$

40. $a_n = (-1)^n (n^2 + 1)^{1/n}$

41. $a_n = \left(\dfrac{2 - n^2}{3 + n^2}\right)^n$

42. $a_n = \dfrac{\left(\frac{2}{3}\right)^n}{1 - \sqrt[n]{n}}$

In Problems 43 through 50, investigate the given sequence $\{a_n\}$ numerically or graphically. Formulate a reasonable guess for the value of its limit. Then apply limit laws to verify that your guess is correct.

43. $a_n = \dfrac{n - 2}{n + 13}$

44. $a_n = \dfrac{2n + 3}{5n - 17}$

45. $a_n = \sqrt{\dfrac{4n^2 + 7}{n^2 + 3n}}$

46. $a_n = \left(\dfrac{n^3 - 5}{8n^3 + 7n}\right)^{1/3}$

47. $a_n = e^{-1/\sqrt{n}}$

48. $a_n = n \sin \dfrac{2}{n}$

49. $a_n = 4 \tan^{-1} \dfrac{n - 1}{n + 1}$

50. $a_n = 3 \sin^{-1} \sqrt{\dfrac{3n - 1}{4n + 1}}$

51. Prove that if $\lim_{n\to\infty} a_n = A \neq 0$, then the sequence $\{(-1)^n a_n\}$ diverges.

52. Prove that if the increasing sequence $\{a_n\}$ is not bounded, then $\lim_{n\to\infty} a_n = +\infty$. (It's largely a matter of saying precisely what this means.)

53. Suppose that $A > 0$. Given $x_1 \neq 0$ but otherwise arbitrary, define the sequence $\{x_n\}$ recursively by

$$x_{n+1} = \frac{1}{2} \cdot \left(x_n + \frac{A}{x_n}\right) \quad \text{if} \quad n \geqq 1.$$

Prove that if $L = \lim_{n\to\infty} x_n$ exists, then $L = \pm\sqrt{A}$.

54. Suppose that A is a fixed real number. Given $x_1 \neq 0$ but otherwise arbitrary, define the sequence $\{x_n\}$ recursively by

$$x_{n+1} = \frac{1}{3} \cdot \left(2x_n + \frac{A}{(x_n)^2}\right) \quad \text{if} \quad n \geqq 1.$$

Prove that if $L = \lim_{n\to\infty} x_n$ exists, then $L = \sqrt[3]{A}$.

55. (a) Suppose that every newborn pair of rabbits becomes productive after two months, and thereafter gives birth to a new pair of rabbits every month. If we begin with a single newborn pair of rabbits, denote by F_n the total number of pairs of rabbits we have after n months. Explain carefully why $\{F_n\}$ is the Fibonacci sequence of Example 2. (b) If, instead, every newborn pair of rabbits becomes productive after three months, denote by $\{G_n\}$ the number of pairs of rabbits we have after n months. Give a recursive definition of the sequence $\{G_n\}$ and calculate its first ten terms.

56. Let $\{F_n\}$ be the Fibonacci sequence of Example 2, and assume that

$$\tau = \lim_{n\to\infty} \frac{F_{n+1}}{F_n}$$

exists. (It does.) Show that $\tau = \frac{1}{2}(1+\sqrt{5})$. (*Suggestion:* Write $a_n = F_n/F_{n-1}$ and show that $a_{n+1} = 1 + (1/a_n)$.)

57. Let the sequence $\{a_n\}$ be defined recursively as follows:

$$a_1 = 2; \qquad a_{n+1} = \tfrac{1}{2}(a_n + 4) \quad \text{for } n \geqq 1.$$

(a) Prove by induction on n that $a_n < 4$ for each n and that $\{a_n\}$ is an increasing sequence. (b) Find the limit of this sequence.

58. Investigate as in Example 13 the sequence $\{a_n\}$ that is defined recursively by

$$a_1 = \sqrt{2}, \qquad a_{n+1} = \sqrt{2 + a_n} \quad \text{for} \quad n \geqq 1.$$

In particular, show that

$$\sqrt{2 + \sqrt{2 + \sqrt{2 + \sqrt{2 + \cdots}}}} = 2.$$

Verify the results stated in Problems 59 and 60.

59. $\sqrt{20 + \sqrt{20 + \sqrt{20 + \sqrt{20 + \cdots}}}} = 5.$

60. $\sqrt{90 + \sqrt{90 + \sqrt{90 + \sqrt{90 + \cdots}}}} = 10.$

Problems 61 and 62 deal with the least upper bound *property of the real numbers: If the nonempty set S of real numbers has an upper bound, then S has a least upper bound. The number M is an* ***upper bound*** *for the set S if $x \leqq M$ for all x in S. The upper bound L of S is a* ***least upper bound*** *for S if no number smaller than L is an upper bound for S. You can easily show that if the set S has least upper bounds L_1 and L_2, then $L_1 = L_2$; in other words, if a least upper bound for a set exists, then it is unique.*

61. Prove that the least upper bound property implies the bounded monotonic sequence property. (*Suggestion:* If $\{a_n\}$ is a bounded increasing sequence and A is the least upper bound of the set $\{a_n : n \geqq 1\}$ of terms of the sequence, you can prove that $A = \lim_{n\to\infty} a_n$.)

62. Prove that the bounded monotonic sequence property implies the least upper bound property. (*Suggestion:* For each positive integer n, let a_n be the least integral multiple of $1/10^n$ that is an upper bound of the set S. Prove that $\{a_n\}$ is a bounded decreasing sequence and then that $A = \lim_{n\to\infty} a_n$ is a least upper bound for S.)

63. For each positive integer n, let a_n be the largest integral multiple of $1/10^n$ such that $a_n^2 \leqq 2$. (a) Prove that $\{a_n\}$ is a bounded increasing sequence, so $A = \lim_{n\to\infty} a_n$ exists. (b) Prove that if $A^2 > 2$, then $a_n^2 > 2$ for n sufficiently large. (c) Prove that if $A^2 < 2$, then $a_n^2 < B$ for some number $B < 2$ and all sufficiently large n. (d) Conclude that $A^2 = 2$.

64. Investigate the sequence $\{a_n\}$, where

$$a_n = \left[\!\left[n + \tfrac{1}{2} + \sqrt{n} \right]\!\right].$$

You may need a computer or programmable calculator to discover what is remarkable about this sequence.

10.2 Project: Nested Radicals and Continued Fractions

This project is an investigation of the relation

$$\sqrt{q + p\sqrt{q + p\sqrt{q + p\sqrt{q + \cdots}}}} = p + \cfrac{q}{p + \cfrac{q}{p + \cfrac{q}{p + \cdots}}} \tag{1}$$

where p and q are positive. We ask not only whether this equation could possibly be true, but also what it means. In the following two numerical explorations, you can (for instance) take p and q to be the last two nonzero digits in your student I.D. number.

Exploration 1 Define the infinite sequence $\{a_n\}$ recursively by

$$a_1 = \sqrt{q} \quad \text{and} \quad a_{n+1} = \sqrt{q + pa_n} \quad \text{for} \quad n \geqq 1. \tag{2}$$

Use a calculator or computer to approximate enough terms of this sequence numerically to determine whether it appears to converge. Assuming that it does, write the first several terms symbolically and conclude that $A = \lim_{n\to\infty} a_n$ is a natural interpretation of the *nested radical* on the left-hand side in (1). Finally, take the limit in the recursion in (2) to show that A is the positive solution of the quadratic equation $x^2 - px - q = 0$. Does the quadratic formula then yield a result consistent with your numerical evidence?

Exploration 2 Define the infinite sequence $\{b_n\}$ recursively by

$$b_1 = p \quad \text{and} \quad b_{n+1} = p + \frac{q}{b_n} \quad \text{for} \quad n \geqq 1. \tag{3}$$

Use a calculator or computer to approximate enough terms of this sequence numerically to determine whether or not it appears to converge. Assuming that it does, write the first several terms symbolically and conclude that $B = \lim_{n\to\infty} b_n$ is a natural interpretation of the *continued fraction* on the right-hand side in (1). Finally, take the limit in the recursion in (3) to show that B is also the positive solution of the quadratic equation $x^2 - px - q = 0$. Conclude thereby that Eq. (1) is indeed true.

10.3 INFINITE SERIES AND CONVERGENCE

An **infinite series** is an expression of the form

$$\sum_{n=1}^{\infty} a_n = a_1 + a_2 + a_3 + \cdots + a_n + \cdots, \tag{1}$$

where $\{a_n\}$ is an infinite sequence of real numbers. The number a_n is called the ***n*th term** of the series. The symbol $\sum_{n=1}^{\infty} a_n$ is simply an abbreviation for the right-hand side of Eq. (1). In this section we discover what is meant by the **sum** of an infinite series.

EXAMPLE 1 Consider the infinite series

$$\sum_{n=1}^{\infty} \frac{1}{2^n} = \frac{1}{2} + \frac{1}{4} + \frac{1}{8} + \frac{1}{16} + \cdots + \frac{1}{2^n} + \cdots \tag{2}$$

that was mentioned in Section 10.1; its nth term is $a_n = 1/2^n$. Although we cannot literally add an infinite number of terms, we can add any finite number of the terms in (2). For instance, the sum of the first five terms is

$$\frac{1}{2} + \frac{1}{4} + \frac{1}{8} + \frac{1}{16} + \frac{1}{32} = \frac{31}{32} = 0.96875.$$

n	Sum of First n Terms
5	0.96875000
10	0.99902344
15	0.99996948
20	0.99999905
25	0.99999997

FIGURE 10.3.1 Sums of terms in the infinite series of Example 1.

We could add five more terms, then five more, and so forth. The table in Fig. 10.3.1 shows what happens. It appears that the sums get closer and closer to 1 as we add more and more terms. If indeed this is so, then it is natural to say that the sum of the (whole) infinite series in (2) is 1, and hence to write

$$\sum_{n=1}^{\infty} \frac{1}{2^n} = \frac{1}{2} + \frac{1}{4} + \frac{1}{8} + \frac{1}{16} + \cdots + \frac{1}{2^n} + \cdots = 1. \quad \blacklozenge$$

Motivated by Example 1, we introduce the *partial sums* of the general infinite series in (1). The **nth partial sum** S_n of the series is the sum of its first n terms:

$$S_n = a_1 + a_2 + a_3 + \cdots + a_n = \sum_{k=1}^{n} a_k. \tag{3}$$

Thus each infinite series has not only an infinite sequence of terms, but also an **infinite sequence of partial sums** $S_1, S_2, S_3, \ldots, S_n, \ldots,$ where

$$S_1 = a_1,$$
$$S_2 = a_1 + a_2,$$
$$S_3 = a_1 + a_2 + a_3,$$
$$\vdots$$
$$S_{10} = a_1 + a_2 + a_3 + a_4 + a_5 + a_6 + a_7 + a_8 + a_9 + a_{10},$$

and so forth. We define the sum of the infinite series to be the limit of its sequence of partial sums, provided that this limit exists.

DEFINITION The Sum of an Infinite Series

We say that the infinite series

$$\sum_{n=1}^{\infty} a_n \quad \textbf{converges} \text{ (or is } \textbf{convergent}\text{)}$$

with **sum** S provided that the limit of its sequence of partial sums,

$$S = \lim_{n\to\infty} S_n, \tag{4}$$

exists (and is finite). Otherwise we say that the series **diverges** (or is **divergent**). If a series diverges, then it has no sum.

Thus the sum of an infinite series is a limit of finite sums,

$$S = \sum_{n=1}^{\infty} a_n = \lim_{N\to\infty} \sum_{n=1}^{N} a_n,$$

provided that this limit exists.

EXAMPLE 1 (continued) Show that the series

$$\sum_{n=1}^{\infty}\left(\frac{1}{2}\right)^n = \frac{1}{2}+\frac{1}{4}+\frac{1}{8}+\frac{1}{16}+\cdots$$

converges. Then find its sum.

Solution The first four partial sums are

$$S_1 = \frac{1}{2}, \qquad S_2 = \frac{3}{4}, \qquad S_3 = \frac{7}{8}, \qquad \text{and} \qquad S_4 = \frac{15}{16}.$$

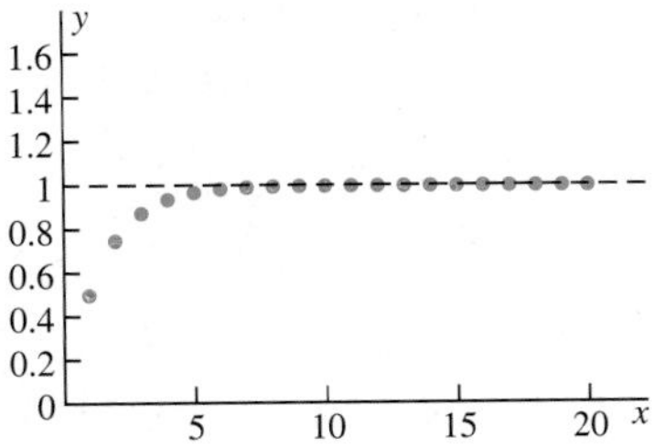

FIGURE 10.3.2 Graph of the first 20 partial sums of the infinite series in Example 1.

It seems likely that $S_n = (2^n - 1)/2^n$, and indeed this follows easily by induction, because

$$S_{n+1} = S_n + \frac{1}{2^{n+1}} = \frac{2^n - 1}{2^n} + \frac{1}{2^{n+1}} = \frac{2^{n+1} - 2 + 1}{2^{n+1}} = \frac{2^{n+1} - 1}{2^{n+1}}.$$

Hence the sum of the given series is

$$S = \lim_{n\to\infty} S_n = \lim_{n\to\infty} \frac{2^n - 1}{2^n} = \lim_{n\to\infty}\left(1 - \frac{1}{2^n}\right) = 1.$$

The graph in Fig. 10.3.2 illustrates the convergence of the partial sums to the number 1. ◆

EXAMPLE 2 Show that the series

$$\sum_{n=1}^{\infty}(-1)^{n+1} = 1 - 1 + 1 - 1 + \cdots$$

diverges.

Solution The sequence of partial sums of this series is

$$1, 0, 1, 0, 1, \ldots,$$

which has no limit. Therefore the series diverges. ◆

EXAMPLE 3 Show that the infinite series

$$\sum_{n=1}^{\infty}\frac{1}{n(n+1)}$$

converges. Then find its sum.

Solution We need a formula for the nth partial sum S_n so that we can evaluate its limit as $n \to \infty$. To find such a formula, we begin with the observation that the nth term of the series is

$$a_n = \frac{1}{n(n+1)} = \frac{1}{n} - \frac{1}{n+1}.$$

(In more complicated cases, such as those in Problems 50 through 55, such a decomposition can be obtained by the method of partial fractions.) It follows that the sum of the first n terms of the given series is

$$\begin{aligned} S_n &= \left(1 - \frac{1}{2}\right) + \left(\frac{1}{2} - \frac{1}{3}\right) + \left(\frac{1}{3} - \frac{1}{4}\right) \\ &\quad + \left(\frac{1}{4} - \frac{1}{5}\right) + \cdots + \left(\frac{1}{n} - \frac{1}{n+1}\right) \\ &= 1 - \frac{1}{n+1} = \frac{n}{n+1}. \end{aligned}$$

Hence

$$\sum_{n=1}^{\infty} \frac{1}{n(n+1)} = \lim_{n\to\infty} \frac{n}{n+1} = 1.$$ ◆

The sum for S_n in Example 3, called a *telescoping* sum, provides us with a way to find the sums of certain series. The series in Examples 1 and 2 are examples of a more common and more important type of series, the *geometric series.*

DEFINITION Geometric Series

The series $\sum_{n=0}^{\infty} a_n$ is said to be a **geometric series** if each term after the first is a fixed multiple of the term immediately before it. That is, there is a number r, called the **ratio** of the series, such that

$$a_{n+1} = ra_n \quad \text{for all } n \geqq 0.$$

If we write $a = a_0$ for the initial constant term, then $a_1 = ar, a_2 = ar^2, a_3 = ar^3$, and so forth. Thus every geometric series takes the form

$$a + ar + ar^2 + ar^3 + \cdots = \sum_{n=0}^{\infty} ar^n. \tag{5}$$

Note that the summation begins at $n = 0$ (rather than at $n = 1$). It is therefore convenient to regard the sum

$$S_n = a(1 + r + r^2 + r^3 + \cdots + r^n)$$

of the first $n + 1$ terms as the nth partial sum of the series.

EXAMPLE 4 The infinite series

$$\sum_{n=0}^{\infty} \frac{2}{3^n} = 2 + \frac{2}{3} + \frac{2}{9} + \cdots + \frac{2}{3^n} + \cdots$$

is a geometric series whose first term is $a = 2$ and whose ratio is $r = \frac{1}{3}$. ◆

THEOREM 1 The Sum of a Geometric Series

If $|r| < 1$, then the geometric series in Eq. (5) converges, and its sum is

$$S = \sum_{n=0}^{\infty} ar^n = \frac{a}{1-r}. \tag{6}$$

If $|r| \geqq 1$ and $a \neq 0$, then the geometric series diverges.

PROOF If $r = 1$, then $S_n = (n+1)a$, so the series certainly diverges if $a \neq 0$. If $r = -1$ and $a \neq 0$, then the series diverges by an argument like the one in Example 2. So we may suppose that $|r| \neq 1$. Then the elementary identity

$$1 + r + r^2 + r^3 + \cdots + r^n = \frac{1 - r^{n+1}}{1 - r}$$

follows if we multiply each side by $1 - r$. Hence the nth partial sum of the geometric series is

$$S_n = a(1 + r + r^2 + r^3 + \cdots + r^n) = a\left(\frac{1}{1-r} - \frac{r^{n+1}}{1-r}\right).$$

If $|r| < 1$, then $r^{n+1} \to 0$ as $n \to \infty$, by Example 9 in Section 10.2. So in this case the geometric series converges to

$$S = \lim_{n\to\infty} a \cdot \left(\frac{1}{1-r} - \frac{r^{n+1}}{1-r}\right) = \frac{a}{1-r}.$$

But if $|r| > 1$, then $\lim_{n\to\infty} r^{n+1}$ does not exist, so $\lim_{n\to\infty} S_n$ does not exist. This establishes the theorem. ◄

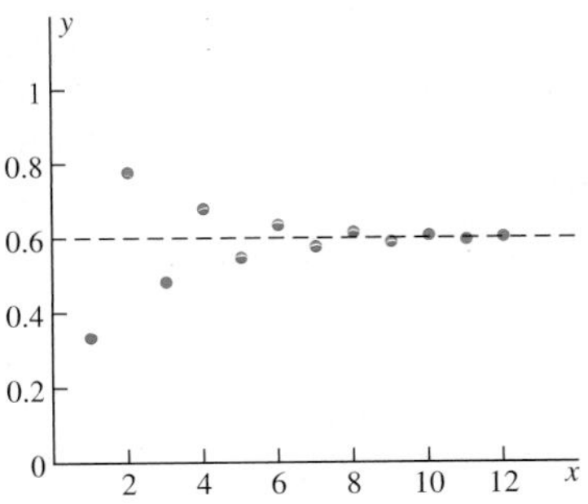

FIGURE 10.3.3 Graph of the first dozen partial sums of the infinite series in Example 5.

EXAMPLE 5 With $a = 1$ and $r = -\frac{2}{3}$, we find that

$$1 - \frac{2}{3} + \frac{4}{9} - \frac{8}{27} + \cdots = \sum_{n=0}^{\infty}\left(-\frac{2}{3}\right)^n = \frac{1}{1-\left(-\frac{2}{3}\right)} = \frac{3}{5}.$$

The graph in Fig. 10.3.3 shows the partial sums of this series approaching its sum $\frac{3}{5}$ alternately from above and below. ◆

EXAMPLE 6 Determine whether or not the infinite series $\displaystyle\sum_{n=1}^{\infty}\frac{2^{2n-1}}{3^n}$ converges.

Solution If we write this series in the form

$$\sum_{n=1}^{\infty}\frac{2^{2n-1}}{3^n} = \frac{2}{3} + \frac{8}{9} + \frac{32}{27} + \frac{128}{81} + \cdots = \frac{2}{3}\left(1 + \frac{4}{3} + \frac{16}{9} + \frac{64}{27} + \cdots\right),$$

then we recognize it as a geometric series with first term $a = \frac{2}{3}$ and ratio $r = \frac{4}{3}$. Because $r > 1$, the second part of Theorem 1 implies that this series diverges. ◆

Theorem 2 implies that the operations of addition and of multiplication by a constant can be carried out term by term in the case of *convergent* series. Because the sum of an infinite series is the limit of its sequence of partial sums, this theorem follows immediately from the limit laws for sequences (Theorem 1 of Section 10.2).

THEOREM 2 Termwise Addition and Multiplication
If the series $A = \sum a_n$ and $B = \sum b_n$ converge to the indicated sums and c is a constant, then the series $\sum(a_n + b_n)$ and $\sum ca_n$ also converge, with sums

1. $\sum(a_n + b_n) = A + B$;

2. $\sum ca_n = cA$.

The geometric series in Eq. (6) may be used to find the rational number represented by a given infinite repeating decimal.

EXAMPLE 7

$$0.55555\cdots = \frac{5}{10} + \frac{5}{100} + \frac{5}{1000} + \cdots = \frac{5}{10}\left(1 + \frac{1}{10} + \frac{1}{100} + \cdots\right)$$

$$= \sum_{n=0}^{\infty}\frac{5}{10}\left(\frac{1}{10}\right)^n = \frac{\frac{5}{10}}{1-\frac{1}{10}} = \frac{5}{10}\cdot\frac{10}{9} = \frac{5}{9}.$$

In a more complicated situation, we may need to use the termwise algebra of Theorem 2:

$$0.7282828\cdots = \frac{7}{10} + \frac{28}{10^3} + \frac{28}{10^5} + \frac{28}{10^7} + \cdots$$

$$= \frac{7}{10} + \frac{28}{10^3}\left(1 + \frac{1}{10^2} + \frac{1}{10^4} + \cdots\right)$$

$$= \frac{7}{10} + \frac{28}{1000}\sum_{n=0}^{\infty}\left(\frac{1}{100}\right)^n = \frac{7}{10} + \frac{28}{1000}\left(\frac{1}{1-\frac{1}{100}}\right)$$

$$= \frac{7}{10} + \frac{28}{1000}\cdot\frac{100}{99} = \frac{7}{10} + \frac{28}{990} = \frac{721}{990}.$$

This technique can be used to show that every repeating infinite decimal represents a rational number. Consequently, the decimal expansions of irrational numbers such as π, e, and $\sqrt{2}$ must be nonrepeating as well as infinite. Conversely, if p and q are integers with $q \neq 0$, then long division of q into p yields a repeating decimal expansion for the rational number p/q, because such a division can yield at each stage only q possible different remainders. ◆

EXAMPLE 8 Suppose that Paul and Mary toss a fair six-sided die in turn until one of them wins by getting the first "six." If Paul tosses first, calculate the probability that he will win the game.

Solution Because the die is fair, the probability that Paul gets a "six" on the first round is $\frac{1}{6}$. The probability that he gets the game's first "six" on the second round is $\left(\frac{5}{6}\right)^2\left(\frac{1}{6}\right)$—the product of the probability $\left(\frac{5}{6}\right)^2$ that neither Paul nor Mary rolls a "six" in the first round and the probability $\frac{1}{6}$ that Paul rolls a "six" in the second round. Paul's probability p of getting the first "six" in the game is the *sum* of his probabilities of getting it in the first round, in the second round, in the third round, and so on. Hence

$$\begin{aligned} p &= \frac{1}{6} + \left(\frac{5}{6}\right)^2\left(\frac{1}{6}\right) + \left(\frac{5}{6}\right)^2\left(\frac{5}{6}\right)^2\left(\frac{1}{6}\right) + \cdots \\ &= \frac{1}{6}\left[1 + \left(\frac{5}{6}\right)^2 + \left(\frac{5}{6}\right)^4 + \cdots\right] \\ &= \frac{1}{6} \cdot \frac{1}{1 - \left(\frac{5}{6}\right)^2} = \frac{1}{6} \cdot \frac{36}{11} = \frac{6}{11}. \end{aligned}$$

Because he has the advantage of tossing first, Paul has more than the fair probability $\frac{1}{2}$ of getting the first "six" and thus winning the game. ◆

Theorem 3 is often useful in showing that a given series does *not* converge.

THEOREM 3 The *n*th-Term Test for Divergence

If either

$$\lim_{n \to \infty} a_n \neq 0$$

or this limit does not exist, then the infinite series $\sum a_n$ diverges.

PROOF We want to show under the stated hypothesis that the series $\sum a_n$ diverges. It suffices to show that *if* the series $\sum a_n$ does converge, then $\lim_{n\to\infty} a_n = 0$. So suppose that $\sum a_n$ converges with sum $S = \lim_{n\to\infty} S_n$, where

$$S_n = a_1 + a_2 + a_3 + \cdots + a_n$$

is the nth partial sum of the series. Because $a_n = S_n - S_{n-1}$,

$$\lim_{n\to\infty} a_n = \lim_{n\to\infty} (S_n - S_{n-1}) = \left(\lim_{n\to\infty} S_n\right) - \left(\lim_{n\to\infty} S_{n-1}\right) = S - S = 0.$$

Consequently, if $\lim_{n\to\infty} a_n \neq 0$, then the series $\sum a_n$ diverges. ◄

REMARK It is important to remember also the *contrapositive* of the nth-term divergence test: *If the infinite series* $\sum a_n$ *converges with sum* S, *then its sequence* $\{a_n\}$ *of terms converges to* 0. Thus we have *two* sequences associated with the single infinite series $\sum a_n$: its sequence $\{a_n\}$ of *terms* and its sequence $\{S_n\}$ of *partial sums.* And (assuming that the series converges to S) these two sequences have generally different limits:

$$\lim_{n\to\infty} a_n = 0 \quad \text{and} \quad \lim_{n\to\infty} S_n = S.$$

EXAMPLE 9 The series

$$\sum_{n=1}^{\infty} (-1)^{n-1} n^2 = 1 - 4 + 9 - 16 + 25 - \cdots$$

diverges because $\lim_{n\to\infty} a_n$ does not exist, whereas the series

$$\sum_{n=1}^{\infty} \frac{n}{3n+1} = \frac{1}{4} + \frac{2}{7} + \frac{3}{10} + \frac{4}{13} + \cdots$$

diverges because

$$\lim_{n\to\infty} \frac{n}{3n+1} = \frac{1}{3} \neq 0.$$

◆

WARNING The converse of Theorem 3 is *false*! The condition

$$\lim_{n\to\infty} a_n = 0$$

is necessary *but not sufficient* to guarantee convergence of the series

$$\sum_{n=1}^{\infty} a_n.$$

That is, a series may satisfy the condition $a_n \to 0$ as $n \to \infty$ and yet diverge. An important example of a divergent series with terms that approach zero is the **harmonic series**

$$\sum_{n=1}^{\infty} \frac{1}{n} = 1 + \frac{1}{2} + \frac{1}{3} + \frac{1}{4} + \frac{1}{5} + \cdots. \tag{7}$$

THEOREM 4
The harmonic series diverges.

PROOF The nth term of the harmonic series in (7) is $a_n = 1/n$, and Fig. 10.3.4 shows the graph of the related function $f(x) = 1/x$ on the interval $1 \leqq x \leqq n+1$. For each integer k, $1 \leqq k \leqq n$, we have erected on the subinterval $[k, k+1]$ a rectangle with height $f(k) = 1/k$. All of these n rectangles have base length 1, and their respective heights are the successive terms $1, 1/2, 1/3, \ldots, 1/n$ of the harmonic series. Hence the sum of their areas is the nth partial sum

$$S_n = 1 + \frac{1}{2} + \frac{1}{3} + \frac{1}{4} + \cdots + \frac{1}{n}.$$

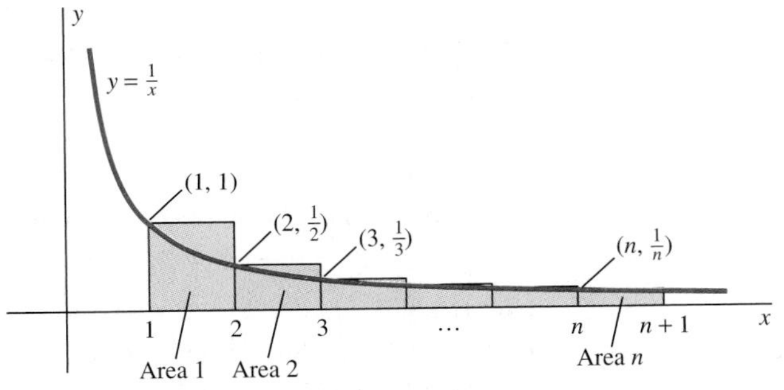

FIGURE 10.3.4 Idea of the proof of Theorem 4.

of the series. Because these rectangles circumscribe the area under the curve $y = 1/x$ from $x = 1$ to $x = n + 1$, we therefore see that S_n must exceed this area. That is,

$$S_n > \int_1^{n+1} \frac{1}{x}\,dx = \Big[\ln x\Big]_1^{n+1} = \ln(n+1).$$

But $\ln(n + 1)$ takes on arbitrarily large positive values with increasing n. Because $S_n > \ln(n + 1)$, it follows that the partial sums of the harmonic series also take on arbitrarily large positive values. Now the terms of the harmonic series are positive, so its sequence of partial sums is increasing. We may therefore conclude that $S_n \to +\infty$ as $n \to +\infty$, and hence that the harmonic series diverges. ◀

If the sequence of partial sums of the series $\sum a_n$ diverges to infinity, then we say that the series **diverges to infinity,** and we write

$$\sum_{n=1}^{\infty} a_n = \infty.$$

The series $\sum(-1)^{n+1}$ of Example 2 is a series that diverges but does not diverge to infinity. In the nineteenth century it was common to say that such a series was divergent by *oscillation;* today we say merely that it diverges.

Our proof of Theorem 4 shows that

$$\sum_{n=1}^{\infty} \frac{1}{n} = \infty.$$

But the partial sums of the harmonic series diverge to infinity very slowly. If N_A denotes the smallest integer such that

$$\sum_{n=1}^{N_A} \frac{1}{n} \geqq A,$$

then with the aid of a programmable calculator you can verify that $N_5 = 83$. With the aid of a computer and refinements of estimates like those in the proof of Theorem 4, one can show that

$$\begin{aligned} N_{10} &= 12367, \\ N_{20} &= 272{,}400{,}600, \\ N_{100} &\approx 1.5 \times 10^{43}, \qquad \text{and} \\ N_{1000} &\approx 1.1 \times 10^{434}. \end{aligned}$$

Thus you would need to add more than a quarter of a billion terms of the harmonic series to get a partial sum that exceeds 20. At this point each of the next few terms would be approximately $0.00000004 = 4 \times 10^{-9}$. The number of terms you'd have to add to reach 1000 is far greater than the estimated number of elementary particles in the entire universe (10^{80}). If you enjoy such large numbers, see the article "Partial sums of infinite series, and how they grow," by R. P. Boas, Jr., in *American Mathematical Monthly* **84** (1977): 237–248.

Theorem 5 says that if two infinite series have the same terms from some point on, then either both series converge or both series diverge. The proof is left for Problem 63.

THEOREM 5 Series that Are Eventually the Same
If there exists a positive integer k such that $a_n = b_n$ for all $n > k$, then the series $\sum a_n$ and $\sum b_n$ either both converge or both diverge.

It follows that a *finite* number of terms can be changed, deleted from, or adjoined to an infinite series without altering its convergence or divergence (although the *sum* of a convergent series will generally be changed by such alterations). In particular, taking $b_n = 0$ for $n \leqq k$ and $b_n = a_n$ for $n > k$, we see that the series

$$\sum_{n=1}^{\infty} a_n = a_1 + a_2 + a_3 + \cdots + a_k + a_{k+1} + \cdots$$

and the series

$$\sum_{n=k+1}^{\infty} a_n = a_{k+1} + a_{k+2} + a_{k+3} + a_{k+4} + \cdots$$

that is obtained by deleting its first k terms either both converge or both diverge.

10.3 TRUE/FALSE STUDY GUIDE

10.3 CONCEPTS: QUESTIONS AND DISCUSSION

1. Can one ever obtain a convergent infinite series by interspersing the terms of two divergent series?
2. Suppose that an infinite series has the property that, given any positive number, all but finitely many terms of the series are positive and less than this number. Does it follows that this series converges? What if it's true that, given any positive number, all but finitely many partial sums of the series are greater than this number? Does it then follow that this series diverges?
3. Can one determine whether a given infinite series converges or diverges merely by computing a sufficiently large number of partial sums?
4. Can one determine the sum—accurate to a given fixed number of decimal places—of a convergent geometric series merely by computing a sufficiently large number of partial sums?

10.3 PROBLEMS

In Problems 1 through 37, determine whether the given infinite series converges or diverges. If it converges, find its sum.

1. $1 + \frac{1}{3} + \frac{1}{9} + \cdots + \frac{1}{3^n} + \cdots$

2. $1 + e^{-1} + e^{-2} + e^{-3} + \cdots + e^{-n} + \cdots$

3. $1 + 3 + 5 + 7 + \cdots + (2n-1) + \cdots$

4. $\frac{1}{2} + \frac{1}{\sqrt{2}} + \frac{1}{\sqrt[3]{2}} + \cdots + \frac{1}{\sqrt[n]{2}} + \cdots$

5. $1 - 2 + 4 - 8 + 16 - \cdots + (-2)^n + \cdots$

6. $1 - \frac{1}{4} + \frac{1}{16} - \cdots + \left(-\frac{1}{4}\right)^n + \cdots$

7. $4 + \frac{4}{3} + \frac{4}{9} + \frac{4}{27} + \cdots + \frac{4}{3^n} + \cdots$

8. $\frac{1}{3} + \frac{2}{9} + \frac{4}{27} + \frac{8}{81} + \cdots + \frac{2^{n-1}}{3^n} + \cdots$

9. $1 + (1.01) + (1.01)^2 + (1.01)^3 + \cdots + (1.01)^n + \cdots$

10. $1 + \frac{1}{\sqrt{2}} + \frac{1}{\sqrt[3]{3}} + \cdots + \frac{1}{\sqrt[n]{n}} + \cdots$

11. $\sum_{n=0}^{\infty} \frac{(-1)^n n}{n+1}$

12. $\sum_{n=1}^{\infty} \left(\frac{e}{10}\right)^n$

13. $\sum_{n=0}^{\infty} (-1)^n \left(\frac{3}{e}\right)^n$

14. $\sum_{n=0}^{\infty} \frac{3^n - 2^n}{4^n}$

15. $\sum_{n=1}^{\infty} \left(\sqrt{2}\right)^{1-n}$

16. $\sum_{n=1}^{\infty} \left(\frac{2}{n} - \frac{1}{2^n}\right)$

17. $\sum_{n=1}^{\infty} \frac{n}{10n+17}$

18. $\sum_{n=1}^{\infty} \frac{\sqrt{n}}{\ln(n+1)}$

19. $\sum_{n=1}^{\infty} (5^{-n} - 7^{-n})$

20. $\sum_{n=0}^{\infty} \frac{1}{1 + \left(\frac{9}{10}\right)^n}$

21. $\sum_{n=1}^{\infty} \left(\frac{e}{\pi}\right)^n$

22. $\sum_{n=1}^{\infty} \left(\frac{\pi}{e}\right)^n$

23. $\sum_{n=0}^{\infty} \left(\frac{100}{99}\right)^n$

24. $\sum_{n=0}^{\infty} \left(\frac{99}{100}\right)^n$

25. $\sum_{n=0}^{\infty} \frac{1 + 2^n + 3^n}{5^n}$

26. $\sum_{n=0}^{\infty} \frac{1 + 2^n + 5^n}{3^n}$

27. $\sum_{n=0}^{\infty} \frac{7 \cdot 5^n + 3 \cdot 11^n}{13^n}$

28. $\sum_{n=1}^{\infty} \sqrt[n]{2}$

29. $\sum_{n=1}^{\infty} \left[\left(\frac{7}{11}\right)^n - \left(\frac{3}{5}\right)^n\right]$

30. $\sum_{n=1}^{\infty} \frac{2n}{\sqrt{4n^2+3}}$

31. $\sum_{n=1}^{\infty} \frac{n^2 - 1}{3n^2 + 1}$

32. $\sum_{n=1}^{\infty} \sin^n 1$

33. $\sum_{n=1}^{\infty} \tan^n 1$

34. $\sum_{n=1}^{\infty} (\arcsin 1)^n$

35. $\sum_{n=1}^{\infty} (\arctan 1)^n$

36. $\sum_{n=1}^{\infty} \arctan n$

37. $\sum_{n=2}^{\infty} \frac{1}{n \ln n}$ (*Suggestion:* Mimic the proof of Theorem 4 to show divergence.)

38. Use the method of Example 6 to verify that

(a) $0.666666666\cdots = \frac{2}{3}$; (b) $0.111111111\cdots = \frac{1}{9}$;

(c) $0.249999999\cdots = \frac{1}{4}$; (d) $0.999999999\cdots = 1$.

In Problems 39 through 43, find the rational number represented by the given repeating decimal.

39. $0.4747\,4747\ldots$ **40.** $0.2525\,2525\ldots$

41. $0.123\,123\,123\ldots$ **42.** $0.3377\,3377\,3377\ldots$

43. $3.14159\,14159\,14159\ldots$

In Problems 44 through 49, find the set of all those values of x for which the given series is a convergent geometric series, then express the sum of the series as a function of x.

44. $\sum_{n=1}^{\infty} (2x)^n$ **45.** $\sum_{n=1}^{\infty} \left(\frac{x}{3}\right)^n$

46. $\sum_{n=1}^{\infty} (x-1)^n$ **47.** $\sum_{n=1}^{\infty} \left(\frac{x-2}{3}\right)^n$

48. $\sum_{n=1}^{\infty} \left(\frac{x^2}{x^2+1}\right)^n$ **49.** $\sum_{n=1}^{\infty} \left(\frac{5x^2}{x^2+16}\right)^n$

In Problems 50 through 55, express the nth partial sum of the infinite series as a telescoping sum (as in Example 3) and thereby find the sum of the series if it converges.

50. $\sum_{n=1}^{\infty} \frac{1}{4n^2-1}$ **51.** $\sum_{n=1}^{\infty} \frac{1}{9n^2+3n-2}$

52. $\sum_{n=1}^{\infty} \ln \frac{n+1}{n}$ **53.** $\sum_{n=1}^{\infty} \frac{1}{16n^2-8n-3}$

54. $\sum_{n=1}^{\infty} \frac{1}{n(n+2)}$ **55.** $\sum_{n=2}^{\infty} \frac{1}{n^2-1}$

In Problems 56 through 60, use a computer algebra system to find the partial fraction decomposition of the general term, then apply the method of Problems 50 through 55 to sum the series.

56. $\sum_{n=1}^{\infty} \frac{2n+1}{n^2(n+1)^2}$ **57.** $\sum_{n=1}^{\infty} \frac{6n^2+2n-1}{n(n+1)(4n^2-1)}$

58. $\sum_{n=1}^{\infty} \frac{2}{n(n+1)(n+2)}$ **59.** $\sum_{n=1}^{\infty} \frac{6}{n(n+1)(n+2)(n+3)}$

60. $\sum_{n=3}^{\infty} \frac{6n}{n^4-5n^2+4}$

61. Prove: If $\sum a_n$ diverges and c is a nonzero constant, then $\sum ca_n$ diverges.

62. Suppose that $\sum a_n$ converges and that $\sum b_n$ diverges. Prove that $\sum (a_n + b_n)$ diverges.

63. Let S_n and T_n denote the nth partial sums of $\sum a_n$ and $\sum b_n$, respectively. Suppose that k is a fixed positive integer and that $a_n = b_n$ for all $n \geqq k$. Show that $S_n - T_n = S_k - T_k$ for all $n > k$. Hence prove Theorem 5.

64. A ball has *bounce coefficient* $r < 1$ if, when it is dropped from a height h, it bounces back to a height of rh (Fig. 10.3.5). Suppose that such a ball is dropped from the initial height a and subsequently bounces infinitely many times. Use a geometric series to show that the total up-and-down distance it travels in all its bouncing is

$$D = a \cdot \frac{1+r}{1-r}.$$

Note that D is *finite.*

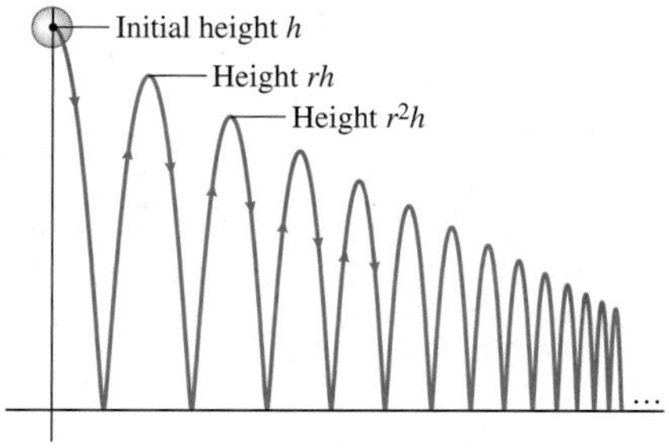

FIGURE 10.3.5 Successive bounces of the ball of Problems 64 and 65.

65. A ball with bounce coefficient $r = 0.64$ (see Problem 64) is dropped from an initial height of $a = 4$ ft. Use a geometric series to compute the total time required for it to complete its infinitely many bounces. The time required for a ball to drop h feet (from rest) is $\sqrt{2h/g}$ seconds, where $g = 32$ ft/s^2.

66. Suppose that the government spends \$1 billion and that each recipient of a fraction of this wealth spends 90% of the dollars that he or she receives. In turn, the secondary recipients spend 90% of the dollars they receive, and so on. How much total spending thereby results from the original injection of \$1 billion into the economy?

67. A tank initially contains a mass M_0 of air. Each stroke of a vacuum pump removes 5% of the air in the container. Compute: (a) The mass M_n of air remaining in the tank after n strokes of the pump; (b) $\lim_{n\to\infty} M_n$.

68. Paul and Mary toss a fair coin in turn until one of them wins the game by getting the first "head." Calculate for each the probability that he or she wins the game.

69. Peter, Paul, and Mary toss a fair coin in turn until one of them wins by getting the first "head." Calculate for each the probability that he or she wins the game. Check your answer by verifying that the sum of the three probabilities is 1.

70. Peter, Paul, and Mary roll a fair die in turn until one of them wins by getting the first "six." Calculate for each the probability that he or she wins the game. Check your answer by verifying that the sum of the three probabilities is 1.

71. A pane of a certain type of glass reflects half the incident light, absorbs one-fourth, and transmits one-fourth. A window is made of two panes of this glass separated by a small space (Fig. 10.3.6). What fraction of the incident light I is transmitted by the double window?

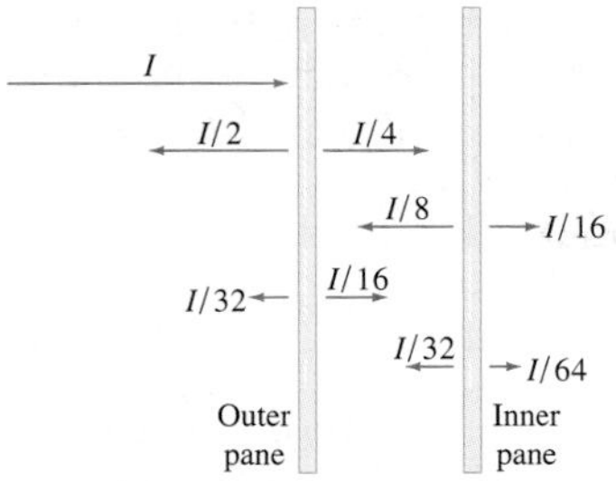

FIGURE 10.3.6 The double-pane window of Problem 71.

72. Criticize the following evaluation of the sum of an infinite series:

Let $x = 1 - 2 + 4 - 8 + 16 - 32 + 64 - \cdots$.

Then $2x = 2 - 4 + 8 - 16 + 32 - 64 + \cdots$.

Add the equations to obtain $3x = 1$. Thus $x = \frac{1}{3}$, and "therefore"

$$1 - 2 + 4 - 8 + 16 - 32 + 64 - \cdots = \tfrac{1}{3}.$$

10.3 Project: Numerical Summation and Geometric Series

With a modern calculator or computer, the computation of partial sums of infinite series—historically a tedious and time-consuming task—is now (ordinarily) a simple matter. Graphing calculators and computer algebra systems typically include one-line command such as

`sum(seq(a,k), k,1,n))`	TI calculator
`sum( a(k), k = 1..n )`	*Maple*
`Sum[ a[k], { k, 1, n } ]`	*Mathematica*

for the calculation of the nth partial sum of the infinite series $\sum_{k=1}^{\infty} a_k$ whose kth term is denoted by $a(k)$. For instance, we can check numerically the fact that

$$\sum_{k=0}^{\infty} \left(\frac{1}{5}\right)^k = \frac{5}{4}$$

by very quickly calculating the first seven partial sums 1.0000, 1.2000, 1.2400, 1.2480, 1.2496, 1.2499, and 1.2500. While not conclusive, this numerical evidence is nevertheless reassuring.

Investigation A Calculate partial sums of the geometric series

$$\sum_{n=0}^{\infty} r^n$$

with $r = 0.2, 0.5, 0.75, 0.9$, and 0.99. For each value of r, calculate the partial sums S_n with $n = 10, 20, 30, \ldots,$ continuing until two successive results agree to four or five decimal places. (For $r = 0.9$ and 0.99, you may decide to use $n = 100, 200, 300, \ldots$.) How does the apparent rate of convergence—as measured by the number of terms required for the desired accuracy—depend on the value of r?

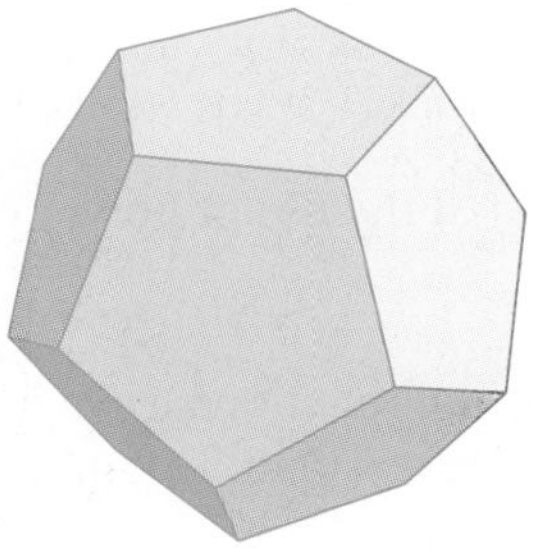

FIGURE 10.3.7 The 12-sided dodecahedron.

Investigation B Archaeological evidence indicates that the ancient (pre-Roman) Etruscans played dice using a dodecahedral die having 12 pentagonal faces numbered 1 through 12 (Fig. 10.3.7). One could simulate such a die by drawing a random card from a deck of 12 cards numbered 1 through 12. Here let's think of a deck having k cards numbered 1 through k. For your own personal value of k, begin with the largest digit in the sum of the digits in your student ID number. This is your value of k unless this digit is less than 5, in which case subtract it from 10 to get your value of k.

(a) John and Mary draw alternately from a shuffled deck of k cards. The first one to draw an ace—the card numbered 1—wins. Assume that John draws first. Use the formula for the sum of a geometric series to calculate (both as a rational number and as a four-place decimal) the probability J that John wins, and similarly the probability M that Mary wins. Check that $J + M = 1$.

(b) Now John, Mary, and Paul draw alternately from the deck of k cards. Calculate separately their respective probabilities of winning, given that John draws first and Mary draws second. Check that $J + M + P = 1$.

10.4 TAYLOR SERIES AND TAYLOR POLYNOMIALS

The infinite series we studied in Section 10.3 have *constant* terms, and the sum of such a series (assuming it converges) is a *number.* In contrast, much of the practical importance of infinite series derives from the fact that many functions have useful representations as infinite series with *variable* terms.

EXAMPLE 1 If we write $r = x$ for the ratio in a geometric series, then Theorem 1 in Section 10.3 gives the infinite series representation

$$\frac{1}{1-x} = \sum_{n=0}^{\infty} x^n = 1 + x + x^2 + x^3 + \cdots \tag{1}$$

of the function $f(x) = 1/(1-x)$. That is, for each fixed number x with $|x| < 1$, the infinite series in (1) converges to the number $1/(1-x)$. The nth partial sum

$$S_n(x) = 1 + x + x^2 + x^3 + \cdots + x^n \tag{2}$$

of the geometric series in (1) is now an nth-degree *polynomial* that approximates the function $f(x) = 1/(1-x)$. The convergence of the infinite series for $|x| < 1$ suggests that the approximation

$$\frac{1}{1-x} \approx 1 + x + x^2 + x^3 + \cdots + x^n \tag{3}$$

should then be accurate if n is sufficiently large. Figure 10.4.1 shows the graphs of $1/(1-x)$ and the three approximations $S_1(x)$, $S_2(x)$, and $S_3(x)$. It appears that the approximations are more accurate when n is larger and when x is closer to zero. ◆

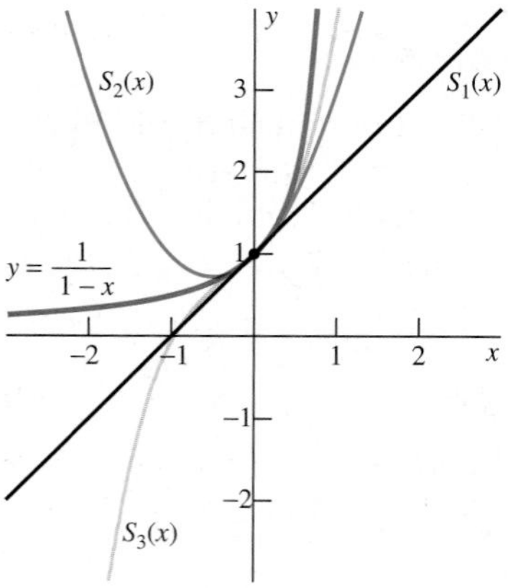

FIGURE 10.4.1 Graphs of the partial sums $S_1(x)$, $S_2(x)$, and $S_3(x)$ of the power series $\sum_{n=0}^{\infty} x^n = \frac{1}{1-x}$ of Example 1.

REMARK The approximation in (3) could be used to calculate numerical quotients with a calculator that has only $+$, $-$, $\times$ keys (but no $\div$ key). For instance,

$$\begin{aligned}\frac{329}{73} &= \frac{3.29}{0.73} = 3.29 \times \frac{1}{1-0.27}\\ &\approx (3.29)[1 + (0.27) + (0.27)^2 + \cdots + (0.27)^{10}]\\ &\approx (3.29)(1.36986); \qquad \text{thus}\\ \frac{329}{73} &\approx 4.5068,\end{aligned}$$

accurate to four decimal places. This is a simple illustration of the use of polynomial approximation for numerical computation.

The definitions of the various elementary transcendental functions leave it unclear how to compute their values precisely, except at a few isolated points. For example,

$$\ln x = \int_1^x \frac{1}{t}\,dt \quad (x > 0)$$

by definition, so obviously $\ln 1 = 0$, but no other value of $\ln x$ is obvious. The natural exponential function is the inverse of $\ln x$, so it is clear that $e^0 = 1$, but it is not at all clear how to compute e^x for $x \neq 0$. Indeed, even such an innocent-looking expression as $\sqrt{x}$ is not computable (precisely and in a finite number of steps) unless x happens to be the square of a rational number.

But *any* value of a polynomial

$$P(x) = c_0 + c_1 x + c_2 x^2 + \cdots + c_n x^n$$

with known coefficients $c_0, c_1, c_2, \ldots, c_n$ is easy to calculate—as in the preceding remark, only addition and multiplication are required. One goal of this section is to use the fact that polynomial values are so readily computable to help us calculate approximate values of functions such as $\ln x$ and e^x.

Polynomial Approximations

Suppose that we want to calculate (or, at least, closely approximate) a specific value $f(x_0)$ of a given function f. It would suffice to find a polynomial $P(x)$ with a graph that is very close to that of f on some interval containing x_0. For then we could use the value $P(x_0)$ as an approximation to the actual value of $f(x_0)$. Once we know how to find such an approximating polynomial $P(x)$, the next question would be how accurately $P(x_0)$ approximates the desired value $f(x_0)$.

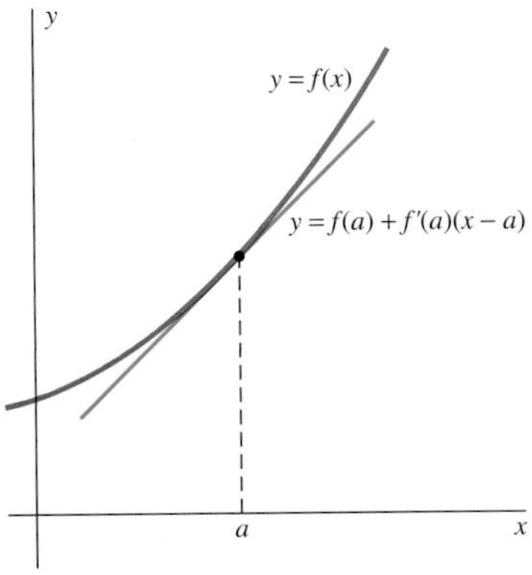

FIGURE 10.4.2 The tangent line at $(a, f(a))$ is the best linear approximation to $y = f(x)$ near a.

The simplest example of polynomial approximation is the linear approximation

$$f(x) \approx f(a) + f'(a)(x - a)$$

obtained by writing $\Delta x = x - a$ in the linear approximation formula, Eq. (3) of Section 4.2. The graph of the first-degree polynomial

$$P_1(x) = f(a) + f'(a)(x - a) \qquad \textbf{(4)}$$

is the line tangent to the curve $y = f(x)$ at the point $(a, f(a))$; see Fig. 10.4.2. This first-degree polynomial agrees with f and with its first derivative at $x = a$. That is,

$$P_1(a) = f(a) \quad \text{and} \quad P_1'(a) = f'(a).$$

EXAMPLE 2 Suppose that $f(x) = \ln x$ and that $a = 1$. Then $f(1) = 0$ and $f'(1) = 1$, so $P_1(x) = x - 1$. Hence we expect that $\ln x \approx x - 1$ for x near 1. With $x = 1.1$, we find that

$$P_1(1.1) = 0.1000, \quad \text{whereas} \quad \ln(1.1) \approx 0.0953.$$

The error in this approximation is about 5%.

To better approximate $\ln x$ near $x = 1$, let us find a second-degree polynomial

$$P_2(x) = c_0 + c_1 x + c_2 x^2$$

that not only has the same value and the same first derivative as does f at $x = 1$, but also has the same second derivative there: $P_2''(1) = f''(1) = -1$. To satisfy these conditions, we must have

$$P_2(1) = c_2 + c_1 + c_0 = 0,$$
$$P_2'(1) = 2c_2 + c_1 = 1, \qquad \text{and}$$
$$P_2''(1) = 2c_2 = -1.$$

When we solve these equations, we find that $c_0 = -\frac{3}{2}$, $c_1 = 2$, and $c_2 = -\frac{1}{2}$, so

$$P_2(x) = -\tfrac{3}{2} + 2x - \tfrac{1}{2}x^2.$$

With $x = 1.1$ we find that $P_2(1.1) = 0.0950$, which is accurate to three decimal places because $\ln(1.1) \approx 0.0953$. The graph of $y = P_2(x)$ is a parabola through $(1, 0)$ with the same value, slope, *and curvature* there as $y = \ln x$ (Fig. 10.4.3). ◆

The tangent line and the parabola used in the computations of Example 2 illustrate one general approach to polynomial approximation. To approximate the function $f(x)$ near $x = a$, we look for an nth-degree polynomial

$$P_n(x) = c_0 + c_1 x + c_2 x^2 + \cdots + c_n x^n$$

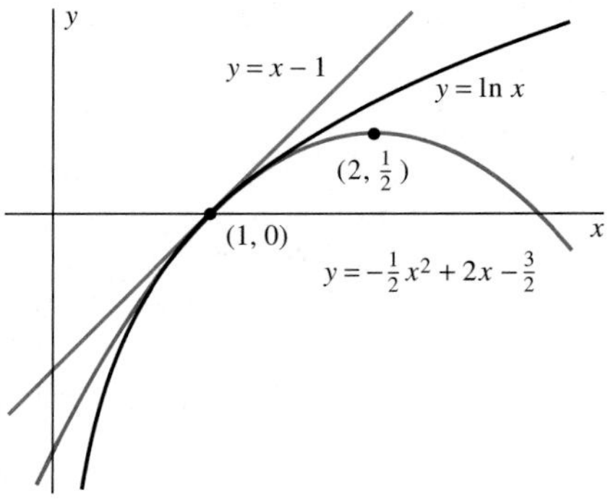

FIGURE 10.4.3 The linear and parabolic approximations to $y = \ln x$ near the point (1, 0) (Example 2).

such that its value at a and the values of its first n derivatives at a agree with the corresponding values of f. That is, we require that

$$\begin{aligned} P_n(a) &= f(a), \\ P_n'(a) &= f'(a), \\ P_n''(a) &= f''(a), \\ &\vdots \\ P_n^{(n)}(a) &= f^{(n)}(a). \end{aligned} \tag{5}$$

We can use these $n + 1$ conditions to evaluate the values of the $n + 1$ coefficients $c_0, c_1, c_2, \dots, c_n$.

The algebra involved is much simpler, however, if we begin with $P_n(x)$ expressed as an nth-degree polynomial in powers of $x - a$ rather than in powers of x:

$$P_n(x) = c_0 + c_1(x - a) + c_2(x - a)^2 + \cdots + c_n(x - a)^n. \tag{6}$$

Then substituting $x = a$ in Eq. (6) yields

$$c_0 = P_n(a) = f(a)$$

by the first condition in Eq. (5). Substituting $x = a$ into

$$P_n'(x) = c_1 + 2c_2(x - a) + 3c_3(x - a)^2 + \cdots + nc_n(x - a)^{n-1}$$

yields

$$c_1 = P_n'(a) = f'(a)$$

by the second condition in Eq. (5). Next, substituting $x = a$ into

$$P_n''(x) = 2c_2 + 3 \cdot 2c_3(x - a) + \cdots + n(n - 1)c_n(x - a)^{n-2}$$

yields $2c_2 = P_n''(a) = f''(a)$, so

$$c_2 = \tfrac{1}{2} f''(a).$$

We continue this process to find $c_3, c_4, \dots, c_n$. In general, the constant term in the kth derivative $P_n^{(k)}(x)$ is $k!c_k$, because it is the kth derivative of the kth-degree term $b_k(x - a)^k$ in $P_n(x)$:

$$P_n^{(k)}(x) = k!c_k + \{\text{powers of } x - a\}.$$

(Recall that $k! = 1 \cdot 2 \cdot 3 \cdots (k - 1) \cdot k$ denotes the *factorial* of the positive integer k, read "k factorial.") So when we substitute $x = a$ into $P_n^{(k)}(x)$, we find that

$$k!c_k = P_n^{(k)}(a) = f^{(k)}(a)$$

and thus that

$$c_k = \frac{f^{(k)}(a)}{k!} \tag{7}$$

for $k = 1, 2, 3, \dots, n$.

Indeed, Eq. (7) holds also for $k = 0$ if we use the universal convention that $0! = 1$ and agree that the zeroth derivative $g^{(0)}$ of the function g is just g itself. With such conventions, our computations establish the following theorem.

THEOREM 1 The *n*th-Degree Taylor Polynomial

Suppose that the first n derivatives of the function $f(x)$ exist at $x = a$. Let $P_n(x)$ be the nth-degree polynomial

$$\begin{aligned} P_n(x) &= \sum_{k=0}^{n} \frac{f^{(k)}(a)}{k!}(x-a)^k \\ &= f(a) + f'(a)(x-a) + \frac{f''(a)}{2!}(x-a)^2 + \cdots + \frac{f^{(n)}(a)}{n!}(x-a)^n. \end{aligned} \tag{8}$$

Then the values of $P_n(x)$ and its first n derivatives agree, at $x = a$, with the values of f and its first n derivatives there. That is, the equations in (5) all hold.

The polynomial in Eq. (8) is called the ***n*th-degree Taylor polynomial of the function f at the point $x = a$**. Note that $P_n(x)$ is a polynomial in powers of $x - a$ rather than in powers of x. To use $P_n(x)$ effectively for the approximation of $f(x)$ near a, we must be able to compute the value $f(a)$ and the values of its derivatives $f'(a)$, $f''(a)$, and so on, all the way to $f^{(n)}(a)$.

The line $y = P_1(x)$ is simply the line tangent to the curve $y = f(x)$ at the point $(a, f(a))$. Thus $y = f(x)$ and $y = P_1(x)$ have the same slope at this point. Now recall from Section 4.6 that the second derivative measures the way the curve $y = f(x)$ is bending as it passes through $(a, f(a))$. Therefore, let us call $f''(a)$ the "concavity" of $y = f(x)$ at $(a, f(a))$. Then, because $P_2''(a) = f''(a)$, it follows that $y = P_2(x)$ has the same value, the same slope, *and* the same concavity at $(a, f(a))$ as does $y = f(x)$. Moreover, $P_3(x)$ and $f(x)$ will also have the same rate of change of concavity at $(a, f(a))$. Such observations suggest that the larger n is, the more closely the nth-degree Taylor polynomial will approximate $f(x)$ for x near a.

EXAMPLE 3 Find the nth-degree Taylor polynomial of $f(x) = \ln x$ at $a = 1$.

Solution The first few derivatives of $f(x) = \ln x$ are

$$f'(x) = \frac{1}{x}, \quad f''(x) = -\frac{1}{x^2}, \quad f^{(3)}(x) = \frac{2}{x^3}, \quad f^{(4)}(x) = -\frac{3!}{x^4}, \quad f^{(5)}(x) = \frac{4!}{x^5}.$$

The pattern is clear:

$$f^{(k)}(x) = (-1)^{k-1}\frac{(k-1)!}{x^k} \quad \text{for } k \geqq 1.$$

Hence $f^{(k)}(1) = (-1)^{k-1}(k-1)!$, so Eq. (8) gives

$$P_n(x) = (x-1) - \frac{1}{2}(x-1)^2 + \frac{1}{3}(x-1)^3 - \frac{1}{4}(x-1)^4 + \cdots + \frac{(-1)^{n-1}}{n}(x-1)^n.$$

With $n = 2$ we obtain the quadratic polynomial

$$P_2(x) = (x-1) - \tfrac{1}{2}(x-1)^2 = -\tfrac{1}{2}x^2 + 2x - \tfrac{3}{2},$$

the same as in Example 2. With the third-degree Taylor polynomial

$$P_3(x) = (x-1) - \tfrac{1}{2}(x-1)^2 + \tfrac{1}{3}(x-1)^3$$

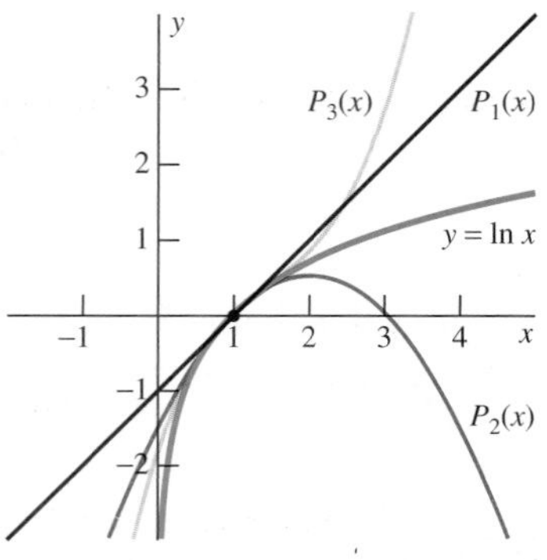

FIGURE 10.4.4 The first three Taylor polynomials approximating $f(x) = \ln x$ near $x = 1$.

we can go a step further in approximating $\ln(1.1) = 0.095310179\ldots \approx 0.0953$. The value

$$P_3(1.1) = (0.1) - \tfrac{1}{2}(0.1)^2 + \tfrac{1}{3}(0.1)^3 \approx 0.095333 \approx 0.0953$$

is accurate to four decimal places (rounded). In Fig. 10.4.4 we see that, the higher the degree and the closer x is to 1, the more accurate the approximation $\ln x \approx P_n(x)$ appears to be. ◆

In the common case $a = 0$, the nth-degree Taylor polynomial in Eq. (8) reduces to

$$P_n(x) = f(0) + f'(0)x + \frac{f''(0)}{2!}x^2 + \cdots + \frac{f^{(n)}(0)}{n!}x^n. \tag{9}$$

EXAMPLE 4 Find the nth-degree Taylor polynomial for $f(x) = e^x$ at $a = 0$.

Solution This is the easiest of all Taylor polynomials to compute, because $f^{(k)}(x) = e^x$ for all $k \geqq 0$. Hence $f^{(k)}(0) = 1$ for all $k \geqq 0$, so Eq. (9) yields

$$P_n(x) = 1 + x + \frac{x^2}{2!} + \frac{x^3}{3!} + \cdots + \frac{x^n}{n!}.$$ ◆

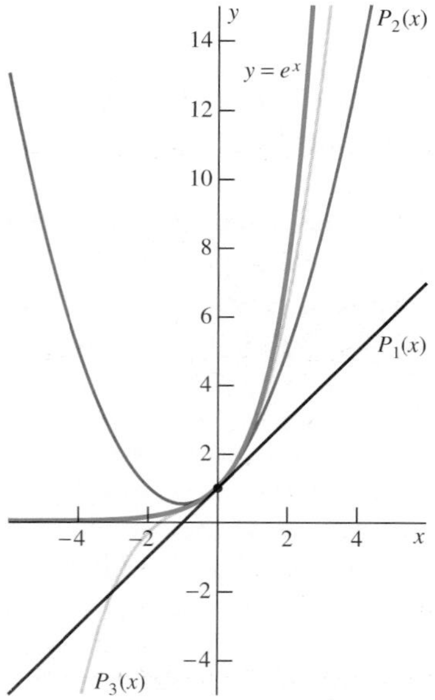

FIGURE 10.4.5 The first three Taylor polynomials approximating $f(x) = e^x$ near $x = 0$.

The first few Taylor polynomials of the natural exponential function at $a = 0$ are, therefore,

$$\begin{aligned}
P_0(x) &= 1,\\
P_1(x) &= 1 + x,\\
P_2(x) &= 1 + x + \tfrac{1}{2}x^2,\\
P_3(x) &= 1 + x + \tfrac{1}{2}x^2 + \tfrac{1}{6}x^3,\\
P_4(x) &= 1 + x + \tfrac{1}{2}x^2 + \tfrac{1}{6}x^3 + \tfrac{1}{24}x^4,\\
P_5(x) &= 1 + x + \tfrac{1}{2}x^2 + \tfrac{1}{6}x^3 + \tfrac{1}{24}x^4 + \tfrac{1}{120}x^5.
\end{aligned}$$

Figure 10.4.5 shows the graphs of $P_1(x)$, $P_2(x)$, and $P_3(x)$. The table in Fig. 10.4.6 shows how these polynomials approximate $f(x) = e^x$ for $x = 0.1$ and for $x = 0.5$. At least for these two values of x, the closer x is to $a = 0$, the more rapidly $P_n(x)$ appears to approach $f(x)$ as n increases.

$x = 0.1$

n	$P_n(x)$	e^x	$e^x - P_n(x)$
0	1.00000	1.10517	0.10517
1	1.10000	1.10517	0.00517
2	1.10500	1.10517	0.00017
3	1.10517	1.10517	0.00000
4	1.10517	1.10517	0.00000

$x = 0.5$

n	$P_n(x)$	e^x	$e^x - P_n(x)$
0	1.00000	1.64872	0.64872
1	1.50000	1.64872	0.14872
2	1.62500	1.64872	0.02372
3	1.64583	1.64872	0.00289
4	1.64844	1.64872	0.00028
5	1.64879	1.64872	0.00002

FIGURE 10.4.6 Approximating $y = e^x$ with Taylor polynomials at $a = 0$.

Taylor's Formula

The closeness with which the polynomial $P_n(x)$ approximates the function $f(x)$ is measured by the difference

$$R_n(x) = f(x) - P_n(x),$$

for which

$$f(x) = P_n(x) + R_n(x). \tag{10}$$

This difference $R_n(x)$ is called the **nth-degree remainder for** $f(x)$ **at** $x = a$. It is the *error* made if the value $f(x)$ is replaced with the approximation $P_n(x)$.

The theorem that lets us estimate the error, or remainder, $R_n(x)$ is called **Taylor's formula,** after Brook Taylor (1685–1731), a follower of Newton who introduced Taylor polynomials in an article published in 1715. The particular expression for $R_n(x)$ that we give next is called the *Lagrange form* for the remainder because it first appeared in 1797 in a book written by the French mathematician Joseph Louis Lagrange (1736–1813).

THEOREM 2 Taylor's Formula

Suppose that the $(n+1)$th derivative of the function f exists on an interval containing the points a and b. Then

$$\begin{aligned} f(b) = f(a) + f'(a)(b-a) + \frac{f''(a)}{2!}(b-a)^2 \\ + \frac{f^{(3)}(a)}{3!}(b-a)^3 + \cdots + \frac{f^{(n)}(a)}{n!}(b-a)^n + \frac{f^{(n+1)}(z)}{(n+1)!}(b-a)^{n+1} \end{aligned} \tag{11}$$

for some number z between a and b.

REMARK With $n = 0$, Eq. (11) reduces to the equation

$$f(b) = f(a) + f'(z)(b-a),$$

the conclusion of the mean value theorem (Section 4.3). Thus Taylor's formula is a far-reaching generalization of the mean value theorem of differential calculus.

A proof of Taylor's formula is given in Appendix I. If we replace b with x in Eq. (11), we get the **nth-degree Taylor formula with remainder at** $x = a$,

$$\begin{aligned} f(x) = f(a) + f'(a)(x-a) + \frac{f''(a)}{2!}(x-a)^2 + \frac{f^{(3)}(a)}{3!}(x-a)^3 \\ + \cdots + \frac{f^{(n)}(a)}{n!}(x-a)^n + \frac{f^{(n+1)}(z)}{(n+1)!}(x-a)^{n+1}, \end{aligned} \tag{12}$$

where z is some number between a and x. Thus the nth-degree remainder term is

$$R_n(x) = \frac{f^{(n+1)}(z)}{(n+1)!}(x-a)^{n+1}, \tag{13}$$

which is easy to remember—it's the same as the *last* term of $P_{n+1}(x)$, except that $f^{(n+1)}(a)$ is replaced with $f^{(n+1)}(z)$.

EXAMPLE 3 (continued) To estimate the accuracy of the approximation

$$\ln 1.1 \approx 0.095333,$$

we substitute $x = 1$ into the formula

$$f^{(k)}(x) = (-1)^{k-1}\frac{(k-1)!}{x^k}$$

for the kth derivative of $f(x) = \ln x$ and get

$$f^{(k)}(1) = (-1)^{k-1}(k-1)!.$$

Hence the third-degree Taylor formula *with remainder* at $x = 1$ is

$$\ln x = (x-1) - \frac{1}{2}(x-1)^2 + \frac{1}{3}(x-1)^3 - \frac{3!}{4!z^4}(x-1)^4$$

with z between $a = 1$ and x. With $x = 1.1$ this gives

$$\ln(1.1) \approx 0.095333 - \frac{(0.1)^4}{4z^4},$$

where $1 < z < 1.1$. The value $z = 1$ gives the largest possible magnitude $(0.1)^4/4 = 0.000025$ of the remainder term. It follows that

$$0.095308 < \ln(1.1) < 0.095334,$$

so we can conclude that $\ln(1.1) = 0.0953$ to four-place accuracy. ◆

Taylor Series

If the function f has derivatives of all orders, then we can write Taylor's formula (Eq. (11)) with any degree n that we please. Ordinarily, the exact value of z in the Taylor remainder term in Eq. (13) is unknown. Nevertheless, we can sometimes use Eq. (13) to show that the remainder approaches zero as $n \to +\infty$:

$$\lim_{n\to\infty} R_n(x) = 0 \tag{14}$$

for some particular *fixed* value of x. Then Eq. (10) gives

$$f(x) = \lim_{n\to\infty}[P_n(x) + R_n(x)] = \lim_{n\to\infty} P_n(x) = \lim_{n\to\infty}\sum_{k=0}^{n}\frac{f^{(k)}(a)}{k!}(x-a)^k;$$

that is,

$$f(x) = \sum_{k=0}^{\infty}\frac{f^{(k)}(a)}{k!}(x-a)^k. \tag{15}$$

The infinite series

$$\sum_{n=0}^{\infty}\frac{f^{(n)}(a)}{n!}(x-a)^n = f(a) + f'(a)(x-a) + \frac{f''(a)}{2!}(x-a)^2 + \cdots + \frac{f^{(n)}(a)}{n!}(x-a)^n + \cdots \tag{16}$$

is called the **Taylor series** of the function f at $x = a$. Its partial sums are the successive Taylor polynomials of f at $x = a$.

We can write the Taylor series of a function f without knowing that it converges. But if the limit in Eq. (14) can be established, then it follows as in Eq. (15) that the Taylor series in Eq. (16) actually converges to $f(x)$. If so, then we can approximate the value of $f(x)$ sufficiently accurately by calculating the value of a Taylor polynomial of f of sufficiently high degree.

EXAMPLE 5 In Example 4 we noted that if $f(x) = e^x$, then $f^{(k)}(x) = e^x$ for all integers $k \geqq 0$. Hence the Taylor formula

$$f(x) = f(0) + f'(0)x + \frac{f''(0)}{2!}x^2 + \cdots + \frac{f^{(n)}(0)}{n!}x^n + \frac{f^{(n+1)}(z)}{(n+1)!}x^{n+1}$$

at $a = 0$ gives

$$e^x = 1 + x + \frac{x^2}{2!} + \frac{x^3}{3!} + \cdots + \frac{x^n}{n!} + \frac{e^z x^{n+1}}{(n+1)!} \tag{17}$$

for some z between 0 and x. If x and hence z are negative then $e^z < 1$, whereas $e^z < e^x$ if both are positive. Thus the remainder term $R_n(x)$ satisfies the inequalities

$$0 < |R_n(x)| < \frac{|x|^{n+1}}{(n+1)!} \quad \text{if } x < 0,$$

$$0 < |R_n(x)| < \frac{e^x x^{n+1}}{(n+1)!} \quad \text{if } x > 0.$$

Therefore, the fact that

$$\lim_{n\to\infty} \frac{x^n}{n!} = 0 \tag{18}$$

for all x (see Problem 55) implies that $\lim_{n\to\infty} R_n(x) = 0$ for all x. This means that the Taylor series for e^x converges to e^x for all x, and we may write

$$e^x = \sum_{n=0}^{\infty} \frac{x^n}{n!} = 1 + x + \frac{x^2}{2!} + \frac{x^3}{3!} + \frac{x^4}{4!} + \cdots. \tag{19}$$

The series in Eq. (19) is the most famous and most important of all Taylor series.

With $x = 1$, Eq. (19) yields a numerical series

$$e = \sum_{n=0}^{\infty} \frac{1}{n!} = 1 + \frac{1}{1!} + \frac{1}{2!} + \frac{1}{3!} + \frac{1}{4!} + \cdots \tag{20}$$

for the number e itself. The 10th and 20th partial sums of this series give the approximations

$$e \approx 1 + \frac{1}{1!} + \frac{1}{2!} + \cdots + \frac{1}{10!} \approx 2.7182818$$

and

$$e \approx 1 + \frac{1}{1!} + \frac{1}{2!} + \cdots + \frac{1}{20!} \approx 2.71828\,18284\,59045\,235,$$

both of which are accurate to the number of decimal places shown. ◆

EXAMPLE 6 To find the Taylor series at $a = 0$ for $f(x) = \cos x$, we first calculate the derivatives

$$\begin{aligned} f(x) &= \cos x, & f'(x) &= -\sin x, \\ f''(x) &= -\cos x, & f^{(3)}(x) &= \sin x, \\ f^{(4)}(x) &= \cos x, & f^{(5)}(x) &= -\sin x, \\ &\vdots & &\vdots \\ f^{(2n)}(x) &= (-1)^n \cos x, & f^{(2n+1)}(x) &= (-1)^{n+1} \sin x. \end{aligned}$$

It follows that

$$f^{(2n)}(0) = (-1)^n \quad \text{but} \quad f^{(2n+1)}(0) = 0,$$

so the Taylor polynomials and Taylor series of $f(x) = \cos x$ include only terms of *even* degree. The Taylor formula of degree $2n$ for $\cos x$ at $a = 0$ is

$$\cos x = 1 - \frac{x^2}{2!} + \frac{x^4}{4!} - \cdots + (-1)^n \frac{x^{2n}}{(2n)!} + (-1)^{n+1} \frac{\cos z}{(2n+2)!} x^{2n+2},$$

where z is between 0 and x. Because $|\cos x| \leqq 1$ for all z, it follows from Eq. (18) that the remainder term approaches zero as $n \to \infty$ *for all* x. Hence the desired Taylor series of $f(x) = \cos x$ at $a = 0$ converges to $\cos x$ for all x, so we may write

$$\cos x = \sum_{n=0}^{\infty} \frac{(-1)^n x^{2n}}{(2n)!} = 1 - \frac{x^2}{2!} + \frac{x^4}{4!} - \frac{x^6}{6!} + \cdots. \quad \textbf{(21)}$$

◆

In Problem 41 we ask you to show similarly that the Taylor series at $a = 0$ of $f(x) = \sin x$ is

$$\sin x = \sum_{n=0}^{\infty} \frac{(-1)^n x^{2n+1}}{(2n+1)!} = x - \frac{x^3}{3!} + \frac{x^5}{5!} - \frac{x^7}{7!} + \cdots. \quad \textbf{(22)}$$

Figures 10.4.7 and 10.4.8 illustrate the increasingly better approximations to $\cos x$ and $\sin x$ that we get by using more and more terms of the series in Eqs. (21) and (22).

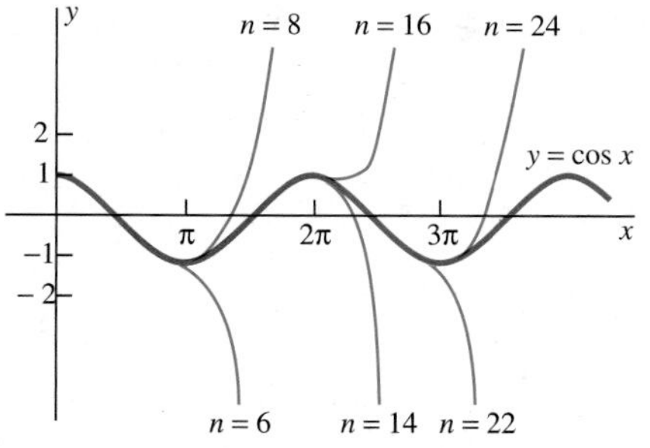

FIGURE 10.4.7 Approximating cos *x* with *n*th-degree Taylor polynomials.

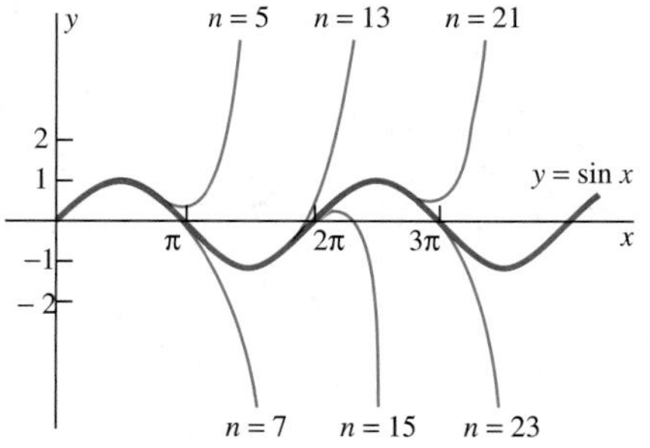

FIGURE 10.4.8 Approximating sin *x* with *n*th-degree Taylor polynomials.

The case $a = 0$ of Taylor's series in (16) is called the **Maclaurin series** of the function $f(x)$,

$$\sum_{n=0}^{\infty} \frac{f^{(n)}(0)}{n!} x^n = f(0) + f'(0)x + \frac{f''(0)}{2!}x^2 + \frac{f^{(3)}(0)}{3!}x^3 + \cdots. \quad \textbf{(23)}$$

Colin Maclaurin (1698–1746) was a Scottish mathematician who used this series as a basic tool in a calculus book he published in 1742. The three Maclaurin series

$$e^x = \sum_{n=0}^{\infty} \frac{x^n}{n!} = 1 + x + \frac{x^2}{2!} + \frac{x^3}{3!} + \frac{x^4}{4!} + \cdots, \quad \textbf{(19)}$$

$$\cos x = \sum_{n=0}^{\infty} \frac{(-1)^n x^{2n}}{(2n)!} = 1 - \frac{x^2}{2!} + \frac{x^4}{4!} - \frac{x^6}{6!} + \cdots, \quad \text{and} \quad \textbf{(21)}$$

$$\sin x = \sum_{n=0}^{\infty} \frac{(-1)^n x^{2n+1}}{(2n+1)!} = x - \frac{x^3}{3!} + \frac{x^5}{5!} - \frac{x^7}{7!} + \cdots \quad \textbf{(22)}$$

(which actually were discovered by Newton) bear careful examination and comparison. Observe that:

- The terms in the *even* cosine series are the *even*-degree terms in the exponential series but with alternating signs.
- The terms in the *odd* sine series are the *odd*-degree terms in the exponential series but with alternating signs.

Equations (19), (21), and (22) are *identities* that hold for all values of x. Consequently, new series can be derived by substitution, as in Examples 7 and 8.

EXAMPLE 7 Substituting $x = -t^2$ into Eq. (19) yields

$$e^{-t^2} = 1 - t^2 + \frac{t^4}{2!} - \frac{t^6}{3!} + \cdots + (-1)^n \frac{t^{2n}}{n!} + \cdots. \quad \blacklozenge$$

EXAMPLE 8 Substituting $x = 2t$ into Eq. (22) gives

$$\sin 2t = 2t - \frac{4}{3}t^3 + \frac{4}{15}t^5 - \frac{8}{315}t^7 + \cdots. \quad \blacklozenge$$

Euler's Formula

The sum of an infinite series $\sum c_n$ with complex terms $c_n = a_n + ib_n$ is defined by

$$\sum_{n=1}^{\infty} c_n = \sum_{n=1}^{\infty} a_n + i \sum_{n=1}^{\infty} b_n$$

provided that the two infinite series of real terms on the right-hand side converge, in which case we say that the series of complex terms on the left-hand side converges.

It can be shown that the exponential series in (19) converges whenever the number x is replaced with a complex number $z = x + iy$. Consequently, the exponential function e^z can be *defined* (for complex as well as for real arguments) by means of the series

$$e^z = \sum_{n=0}^{\infty} \frac{z^n}{n!} = 1 + z + \frac{z^2}{2!} + \frac{z^3}{3!} + \frac{z^4}{4!} + \cdots.$$

If we substitute the *pure imaginary* number $z = i\theta$ (with θ real), we get

$$\begin{aligned} e^{i\theta} &= \sum_{n=0}^{\infty} \frac{(i\theta)^n}{n!} = 1 + i\theta + \frac{(i\theta)^2}{2!} + \frac{(i\theta)^3}{3!} + \frac{(i\theta)^4}{4!} + \cdots \\ &= 1 + i\theta - \frac{\theta^2}{2!} - \frac{i\theta^3}{3!} + \frac{\theta^4}{4!} + \frac{i\theta^5}{5!} - \cdots \\ &= \left(1 - \frac{\theta^2}{2!} + \frac{\theta^4}{4!} - \cdots\right) + i\left(\theta - \frac{\theta^3}{3!} + \frac{\theta^5}{5!} - \cdots\right), \end{aligned}$$

using the facts that $i^2 = -1$, $i^3 = -i$, $i^4 = 1$, and so on. We recognize the Maclaurin series for $\cos\theta$ and $\sin\theta$ on the right-hand side and conclude that

$$e^{i\theta} = \cos\theta + i\sin\theta$$

for every real number θ. This is the famous **Euler's Formula.** For instance, with $\theta = \pi$ it gives $e^{i\pi} = \cos\pi + i\sin\pi = -1$, and hence the extraordinary relation

$$e^{i\pi} + 1 = 0$$

relating the five most important special numbers in mathematics: 0, 1, i, π, and e.

The Number π

In Section 5.3 we described how Archimedes used polygons inscribed in and circumscribed about the unit circle to show that $3\frac{10}{71} < \pi < 3\frac{1}{7}$. With the aid of electronic computers, π has been calculated to well over a *billion* decimal places. We describe now some of the methods that have been used for such computations. [For a chronicle of humanity's perennial fascination with the number π, see Peter Beckmann, *A History of* π, New York: St. Martin's Press, 1971.]

We begin with the elementary algebraic identity

$$\frac{1}{1+x} = 1 - x + x^2 - x^3 + \cdots + (-1)^{k-1}x^{k-1} + \frac{(-1)^k x^k}{1+x}, \tag{24}$$

which can be verified by multiplying both sides by $1 + x$. We substitute t^2 for x and

$n + 1$ for k and thus find that

$$\frac{1}{1+t^2} = 1 - t^2 + t^4 - t^6 + \cdots + (-1)^n t^{2n} + \frac{(-1)^{n+1}t^{2n+2}}{1+t^2}.$$

Because $D_t \tan^{-1} t = 1/(1+t^2)$, integrating both sides of this last equation from $t = 0$ to $t = x$ gives

$$\tan^{-1} x = x - \frac{x^3}{3} + \frac{x^5}{5} - \frac{x^7}{7} + \cdots + (-1)^n \frac{x^{2n+1}}{2n+1} + R_{2n+1}, \tag{25}$$

where

$$|R_{2n+1}| = \left| \int_0^x \frac{t^{2n+2}}{1+t^2}\,dx \right| \leqq \left| \int_0^x t^{2n+2}\,dx \right| = \frac{|x|^{2n+3}}{2n+3}. \tag{26}$$

This estimate of the error makes it clear that

$$\lim_{n\to\infty} R_n = 0$$

if $|x| \leqq 1$. Hence we obtain the Taylor series for the inverse tangent function:

$$\tan^{-1} x = \sum_{n=0}^{\infty} (-1)^n \frac{x^{2n+1}}{2n+1} = x - \frac{x^3}{3} + \frac{x^5}{5} - \frac{x^7}{7} + \cdots, \tag{27}$$

valid for $-1 \leqq x \leqq 1$.

If we substitute $x = 1$ into Eq. (27), we obtain *Leibniz's series*

$$\frac{\pi}{4} = 1 - \frac{1}{3} + \frac{1}{5} - \frac{1}{7} + \cdots.$$

Although this is a beautiful series, it is not an effective way to compute π. But the error estimate in Eq. (26) shows that we can use Eq. (25) to calculate $\tan^{-1} x$ if $|x|$ is small. For example, if $x = \frac{1}{5}$, then the fact that

$$\frac{1}{9 \cdot 5^9} \approx 0.000000057 < 0.0000001$$

implies that the approximation

$$\tan^{-1}\left(\tfrac{1}{5}\right) \approx \tfrac{1}{5} - \tfrac{1}{3}\left(\tfrac{1}{5}\right)^3 + \tfrac{1}{5}\left(\tfrac{1}{5}\right)^5 - \tfrac{1}{7}\left(\tfrac{1}{5}\right)^7$$

is accurate to six decimal places.

Accurate inverse tangent calculations lead to accurate computations of the number π. For example, we can use the addition formula for the tangent function to show (Problem 52) that

$$\frac{\pi}{4} = 4 \tan^{-1}\left(\frac{1}{5}\right) - \tan^{-1}\left(\frac{1}{239}\right). \tag{28}$$

HISTORICAL NOTE In 1706, John Machin used Eq. (28) to calculate the first 100 decimal places of π. (In Problem 54 we ask you to use it to show that $\pi = 3.14159$ to five decimal places.) In 1844 the lightning-fast mental calculator Zacharias Dase of Germany computed the first 200 decimal places of π, using the related formula

$$\frac{\pi}{4} = \tan^{-1}\left(\frac{1}{2}\right) + \tan^{-1}\left(\frac{1}{5}\right) + \tan^{-1}\left(\frac{1}{8}\right). \tag{29}$$

You might enjoy verifying this formula. (See Problem 53.) A recent computation of 1 million decimal places of π used the formula

$$\frac{\pi}{4} = 12 \tan^{-1}\left(\frac{1}{18}\right) + 8 \tan^{-1}\left(\frac{1}{57}\right) - 5 \tan^{-1}\left(\frac{1}{239}\right).$$

For derivations of this formula and others like it, with further discussion of the computations of the number π, see the article "An algorithm for the calculation of π" by George Miel in the *American Mathematical Monthly* **86** (1979), pp. 694–697. Although few practical applications require more than ten or twelve decimal places of π, these computations provide dramatic evidence of the power of Taylor's formula. Moreover, the number π continues to serve as a challenge both to human ingenuity and to the accuracy and efficiency of modern electronic computers. For an account of how investigations of the Indian mathematical genius Srinivasa Ramanujan (1887–1920) have led recently to the computation of over a billion decimal places of π, see the article "Ramanujan and pi," Jonathan M. Borwein and Peter B. Borwein, *Scientific American* (Feb. 1988), pp. 112–117.

10.4 TRUE/FALSE STUDY GUIDE

10.4 CONCEPTS: QUESTIONS AND DISCUSSION

1. Suppose that we take Euler's formula $e^{i\theta} = \cos\theta + i \sin\theta$ as a starting point and *define* the exponential $e^z = e^{x+iy}$ by writing
$$e^z = e^x e^{iy} = e^x(\cos y + i \sin y).$$
Can you then *prove* on this basis that $e^{z+w} = e^z e^w$ if $z = x+iy$ and $w = u+iv$ are complex numbers?
2. Can you use the definition of e^z in Question 1 to prove that $D_x e^{kx} = ke^{kx}$ if $k = a + bi$ is a complex constant and x is a real variable?

10.4 PROBLEMS

In Problems 1 through 10, find Taylor's formula for the given function f at $a = 0$. Find both the Taylor polynomial $P_n(x)$ of the indicated degree n and the remainder term $R_n(x)$.

1. $f(x) = e^{-x}, \quad n = 5$

2. $f(x) = \sin x, \quad n = 4$

3. $f(x) = \cos x, \quad n = 4$

4. $f(x) = \dfrac{1}{1-x}, \quad n = 4$

5. $f(x) = \sqrt{1+x}, \quad n = 3$

6. $f(x) = \ln(1+x), \quad n = 4$

7. $f(x) = \tan x, \quad n = 3$

8. $f(x) = \arctan x, \quad n = 2$

9. $f(x) = \sin^{-1} x, \quad n = 2$

10. $f(x) = x^3 - 3x^2 + 5x - 7, \quad n = 4$

In Problems 11 through 20, find the Taylor polynomial with remainder by using the given values of a and n.

11. $f(x) = e^x; \quad a = 1, n = 4$

12. $f(x) = \cos x; \quad a = \pi/4, n = 3$

13. $f(x) = \sin x; \quad a = \pi/6, n = 3$

14. $f(x) = \sqrt{x}; \quad a = 100, n = 3$

15. $f(x) = \dfrac{1}{(x-4)^2}; \quad a = 5, n = 5$

16. $f(x) = \tan x; \quad a = \pi/4, n = 4$

17. $f(x) = \cos x; \quad a = \pi, n = 4$

18. $f(x) = \sin x; \quad a = \pi/2, n = 4$

19. $f(x) = x^{3/2}; \quad a = 1, n = 4$

20. $f(x) = \dfrac{1}{\sqrt{1-x}}; \quad a = 0, n = 4$

In Problems 21 through 28, find the Maclaurin series of the given function f by substituting in one of the known series in Eqs. (19), (21), and (22).

21. $f(x) = e^{-x}$

22. $f(x) = e^{2x}$

23. $f(x) = e^{-3x}$

24. $f(x) = \exp(x^3)$

25. $f(x) = \sin 2x$

26. $f(x) = \sin \dfrac{x}{2}$

27. $f(x) = \sin x^2$

28. $f(x) = \sin^2 x = \frac{1}{2}(1 - \cos 2x)$

In Problems 29 through 40, find the Taylor series [Eq. (16)] of the given function at the indicated point a.

29. $f(x) = \ln(1+x), \quad a = 0$

30. $f(x) = \dfrac{1}{1-x}, \quad a = 0$

31. $f(x) = e^{-x}, \quad a = 0$

32. $f(x) = \sin x, \quad a = \pi/2$

33. $f(x) = \ln x, \quad a = 1$

34. $f(x) = e^{2x}, \quad a = 0$

35. $f(x) = \cos x, \quad a = \pi/4$

36. $f(x) = \dfrac{1}{(1-x)^2}, \quad a = 0$

37. $f(x) = \dfrac{1}{x}, \quad a = 1$

38. $f(x) = \cos x, \quad a = \pi/2$

39. $f(x) = \sin x, \quad a = \pi/4$

40. $f(x) = \sqrt{1+x}, \quad a = 0$

41. Derive, as in Example 5, the Taylor series in Eq. (22) of $f(x) = \sin x$ at $a = 0$.

42. Granted that it is valid to differentiate the sine and cosine Taylor series in a term-by-term manner, use these series to verify that $D_x \cos x = -\sin x$ and $D_x \sin x = \cos x$.

43. Use the differentiation formulas $D_x \sinh x = \cosh x$ and $D_x \cosh x = \sinh x$ to derive the Maclaurin series

$$\cosh x = \sum_{n=0}^{\infty} \frac{x^{2n}}{(2n)!} \quad \text{and} \quad \sinh x = \sum_{n=0}^{\infty} \frac{x^{2n+1}}{(2n+1)!}$$

for the hyperbolic sine and cosine functions. What is their relationship to the Maclaurin series of the ordinary sine and cosine functions?

44. Derive the Maclaurin series stated in Problem 43 by substituting the known Maclaurin series for the exponential function in the definitions

$$\cosh x = \frac{e^x + e^{-x}}{2} \quad \text{and} \quad \sinh x = \frac{e^x - e^{-x}}{2}$$

of the hyperbolic functions.

The sum commands listed for several computer algebra systems in the Section 10.3 Project can be used to calculate Taylor polynomials efficiently. For instance, when the TI graphing calculator definitions

```
Y1 = sin(x)

Y2 = sum(seq((-1)∧(N-1)*X∧(2N-1)/
        (2N-1)!,N,1,7))
```

are graphed, the result is Fig. 10.4.9, showing that the 13th-degree Taylor polynomial $P_{13}(x)$ approximates $\sin x$ rather closely if $-3\pi/2 < x < 3\pi/2$ but not outside this range. By plotting several successive Taylor polynomials of a function $f(x)$ simultaneously, we can get a visual sense of the way in which they approximate the function. Do this for each function given in Problems 45 through 50.

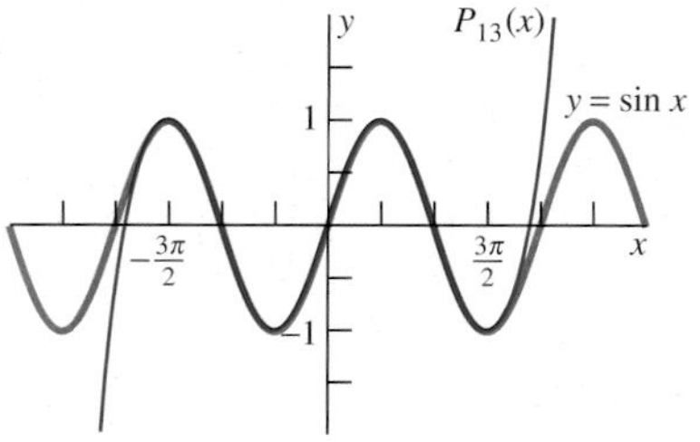

FIGURE 10.4.9 Graphs of $y = \sin x$ and its 13th-degree Taylor polynomial $P_{13}(x)$.

45. $f(x) = e^{-x}$

46. $f(x) = \sin x$

47. $f(x) = \cos x$

48. $f(x) = \ln(1 + x)$

49. $f(x) = \dfrac{1}{1+x}$

50. $f(x) = \dfrac{1}{1-x^2}$

51. Let the function

$$f(x) = \sum_{n=0}^{\infty} \frac{(-1)^n x^n}{(2n)!} = 1 - \frac{x}{2!} + \frac{x^2}{4!} - \frac{x^3}{6!} + \cdots$$

be defined by replacing x with $\sqrt{x}$ in the Maclaurin series for $\cos x$. Plot partial sums of this series to verify graphically that $f(x)$ agrees with the function $g(x)$ defined by

$$g(x) = \begin{cases} \cos\sqrt{x} & \text{if } x \geqq 0, \\ \cosh\sqrt{|x|} & \text{if } x < 0. \end{cases}$$

52. Beginning with $\alpha = \tan^{-1}(\frac{1}{5})$, use the addition formula

$$\tan(A + B) = \frac{\tan A + \tan B}{1 - \tan A \tan B}$$

to show in turn that (a) $\tan 2\alpha = \frac{5}{12}$; (b) $\tan 4\alpha = \frac{120}{119}$; (c) $\tan(\pi/4 - 4\alpha) = -\frac{1}{239}$. Finally, show that part (c) implies Eq. (28).

53. Apply the addition formula for the tangent function to verify Eq. (29).

54. Every young person deserves the thrill, just once, of calculating personally the first several decimal places of the number π. The seemingly random nature of this decimal expansion demands an explanation; how, indeed, are the digits 3.14159 26535 89793 . . . determined? For a partial answer, set your calculator to display nine decimal places. Then add enough terms of the arctangent series in (27) with $x = \frac{1}{5}$ to calculate $\arctan(\frac{1}{5})$ accurate to nine places. Next, calculate the value of $\arctan(\frac{1}{239})$ similarly. Finally, substitute these numerical results in Eq. (28) and solve for π. How many accurate decimal places do you get?

55. Prove that

$$\lim_{n \to \infty} \frac{x^n}{n!} = 0$$

if x is a real number. [*Suggestion:* Choose an integer k such that $k > |2x|$, and let $L = |x|^k/k!$. Then show that

$$\frac{|x|^n}{n!} < \frac{L}{2^{n-k}}$$

if $n > k$.]

56. Suppose that $0 < x \leqq 1$. Integrate both sides of the identity

$$\frac{1}{1+t} = 1 - t + t^2 - t^3 + \cdots + (-1)^n t^n + \frac{(-1)^{n+1} t^{n+1}}{1+t}$$

from $t = 0$ to $t = x$ to show that

$$\ln(1 + x) = x - \frac{x^2}{2} + \frac{x^3}{3} - \cdots + (-1)^n \frac{x^{n+1}}{n+1} + R_n,$$

where $\lim_{n \to \infty} R_n = 0$. Hence conclude that

$$\ln(1 + x) = \sum_{n=1}^{\infty} (-1)^{n+1} \frac{x^n}{n}$$

if $0 < x \leqq 1$.

57. Criticize the following "proof" that $2 = 1$. Substituting $x = 1$ into the result in Problem 56 yields the fact that

$$\ln 2 = 1 - \tfrac{1}{2} + \tfrac{1}{3} - \tfrac{1}{4} + \cdots.$$

If

$$S = 1 + \tfrac{1}{2} + \tfrac{1}{3} + \tfrac{1}{4} + \cdots,$$

then

$$\ln 2 = S - 2 \cdot \left(\tfrac{1}{2} + \tfrac{1}{4} + \tfrac{1}{6} + \tfrac{1}{8} + \cdots\right) = S - S = 0.$$

Hence $2 = e^{\ln 2} = e^0 = 1$.

58. Deduce from the result of Problem 56 first that

$$\ln(1 - x) = -\sum_{n=1}^{\infty} \frac{x^n}{n} = -x - \frac{x^2}{3} - \frac{x^3}{3} - \cdots$$

and then that

$$\ln \frac{1+x}{1-x} = \sum_{n \text{ odd}} \frac{2x^n}{n} = 2\left(x + \frac{x^3}{3} + \frac{x^5}{5} + \cdots\right)$$

if $0 \leqq x \leqq 1$.

59. Approximate the number $\ln 2 \approx 0.69315$ first by substituting $x = 1$ in the Maclaurin series of Problem 56, and then by substituting $x = \frac{1}{3}$ (why?) in the second series of Problem 58. Which approach appears to require the fewest terms to yield the value of $\ln 2$ accurate to a given number of decimal places?

10.4 Project: Calculating Logarithms on a Deserted Island

The problem is that you're stranded for life on a desert island with only a very basic calculator that does not calculate natural logarithms. So to get modern science going on this miserable island, you need to use the infinite series for $\ln[(1 + x)/(1 - x)]$ in Problem 58 to produce a simple table of logarithms (with five-place accuracy, say), giving $\ln x$ at least for the integers $x = 1, 2, 3, \ldots, 9$, and 10.

The most direct way might be to use the series for $\ln[(1+x)/(1-x)]$ to calculate first $\ln 2$, $\ln 3$, $\ln 5$, and $\ln 7$. Then use the law of logarithms $\ln xy = \ln x + \ln y$ to fill in the other entries in the table by simple addition of logarithms already computed. Unfortunately, larger values of x result in series that are more slowly convergent. So you could save yourself time and work by exercising some ingenuity: Calculate from scratch some four *other* logarithms from which you can build up the rest. For example, if you know $\ln 2$ and $\ln 1.25$, then $\ln 10 = \ln 1.25 + 3 \ln 2$. (Why?) Be as ingenious as you wish. Can you complete your table of ten logarithms by calculating directly (using the series) *fewer* than four logarithms to begin with?

For a finale, calculate somehow (from scratch, and accurate to five rounded decimal places) the natural logarithm $\ln(pq.rs)$, where p, q, r, and s denote the last four *nonzero* digits in your student I.D. number.

10.5 THE INTEGRAL TEST

A Taylor series (as in Section 10.4) is a special type of infinite series with *variable* terms. We saw that Taylor's formula can sometimes be used—as in the case of the exponential, sine, and cosine series—to establish the convergence of such a series.

But given an infinite series $\sum a_n$ with *constant* terms, it is the exception rather than the rule when a simple formula for the nth partial sum of that series can be found and used directly to determine whether the series converges or diverges. There are, however, several *convergence tests* that use the *terms* of an infinite series rather than its partial sums. Such a test, when successful, will tell us whether or not the series converges. Once we know that the series $\sum a_n$ does converge, it is then a separate matter to find its sum S. It may be necessary to approximate S by adding sufficiently many terms; in this case we shall need to know how many terms are required for the desired accuracy.

Here and in Section 10.6, we concentrate our attention on **positive-term series**—that is, series with terms that are all positive. If $a_n > 0$ for all n, then

$$S_1 < S_2 < S_3 < \cdots < S_n < \cdots,$$

so the sequence $\{S_n\}$ of partial sums of the series is increasing. Hence there are just two possibilities. If the sequence $\{S_n\}$ is *bounded*—there exists a number M such that $S_n < M$ for all n—then the bounded monotonic sequence property (Section 10.2) implies that $S = \lim_{n\to\infty} S_n$ exists, so the series $\sum a_n$ *converges.* Otherwise, it diverges to infinity (by Problem 52 in Section 10.2).

A similar alternative holds for improper integrals. Suppose that the function f is continuous and positive-valued for $x \geqq 1$. Then it follows (from Problem 51) that the improper integral

$$\int_1^{\infty} f(x)\,dx = \lim_{b\to\infty}\int_1^b f(x)\,dx \tag{1}$$

either converges (the limit is a real number) or diverges to infinity (the limit is $+\infty$). This analogy between positive-term series and improper integrals of positive functions is the key to the **integral test.** We compare the behavior of the series $\sum a_n$ with that of the improper integral in Eq. (1), where f is an appropriately chosen function. [Among other things, we require that $f(n) = a_n$ for all n.]

THEOREM 1 The Integral Test

Suppose that $\sum a_n$ is a positive-term series and that f is a positive-valued, decreasing, continuous function for $x \geqq 1$. If $f(n) = a_n$ for all integers $n \geqq 1$, then the series and the improper integral

$$\sum_{n=1}^{\infty} a_n \quad \text{and} \quad \int_1^{\infty} f(x)\,dx$$

either both converge or both diverge.

PROOF Because f is a decreasing function, the rectangular polygon with area

$$S_n = a_1 + a_2 + a_3 + \cdots + a_n$$

shown in Fig. 10.5.1 contains the region under $y = f(x)$ from $x = 1$ to $x = n + 1$. Hence

$$\int_1^{n+1} f(x)\,dx \leqq S_n. \tag{2}$$

Similarly, the rectangular polygon with area

$$S_n - a_1 = a_2 + a_3 + a_4 + \cdots + a_n$$

shown in Fig. 10.5.2 is contained in the region under $y = f(x)$ from $x = 1$ to $x = n$. Hence

$$S_n - a_1 \leqq \int_1^n f(x)\,dx. \tag{3}$$

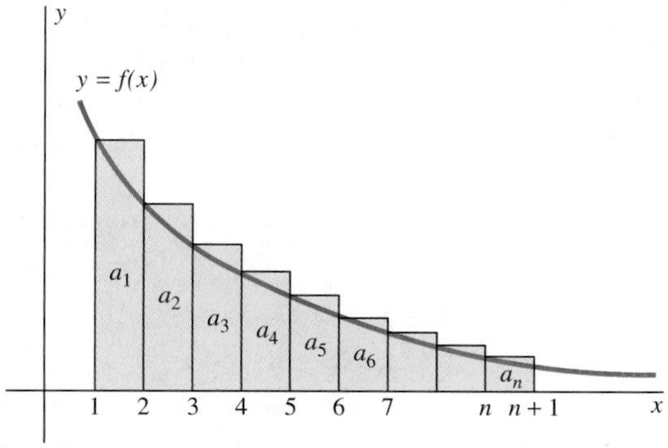

FIGURE 10.5.1 Underestimating the partial sums with an integral.

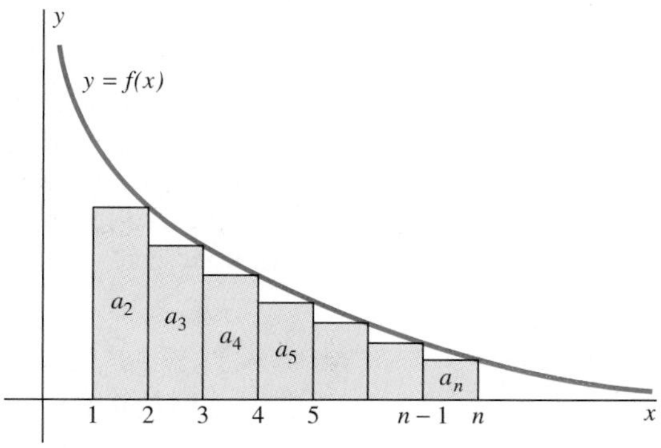

FIGURE 10.5.2 Overestimating the partial sums with an integral.

Suppose first that the improper integral $\int_1^\infty f(x)\,dx$ diverges (necessarily to $+\infty$). Then

$$\lim_{n\to\infty}\int_1^{n+1} f(x)\,dx = +\infty,$$

so it follows from (2) that $\lim_{n\to\infty} S_n = +\infty$ as well, and hence the infinite series $\sum a_n$ likewise diverges.

Now suppose instead that the improper integral $\int_1^\infty f(x)\,dx$ converges and has the (finite) value I. Then (3) implies that

$$S_n \leqq a_1 + \int_1^n f(x)\,dx \leqq a_1 + I,$$

so the increasing sequence $\{S_n\}$ is bounded. Thus the infinite series

$$\sum_{n=1}^{\infty} a_n = \lim_{n\to\infty} S_n$$

converges as well. Hence we have shown that the infinite series and the improper integral either both converge or both diverge. ◂

EXAMPLE 1 We used a version of the integral test to prove in Section 10.3 that the harmonic series

$$\sum_{n=1}^{\infty}\frac{1}{n} = 1 + \frac{1}{2} + \frac{1}{3} + \frac{1}{4} + \cdots$$

diverges. Using the test as stated in Theorem 1 is a little simpler: We note that $f(x) = 1/x$ is positive, continuous, and decreasing for $x \geqq 1$ and that $f(n) = 1/n$ for each positive integer n. Now

$$\int_1^\infty \frac{1}{x}\,dx = \lim_{b\to\infty}\int_1^b \frac{1}{x}\,dx = \lim_{b\to\infty}\Big[\ln x\Big]_1^b = \lim_{b\to\infty}(\ln b - \ln 1) = +\infty.$$

Thus the improper integral diverges and, therefore, so does the harmonic series. ◆

The harmonic series is the case $p = 1$ of the ***p*-series**

$$\sum_{n=1}^{\infty}\frac{1}{n^p} = 1 + \frac{1}{2^p} + \frac{1}{3^p} + \cdots + \frac{1}{n^p} + \cdots. \tag{4}$$

Whether the p-series converges or diverges depends on the value of p.

EXAMPLE 2 Show that the p-series converges if $p > 1$ but diverges if $0 < p \leqq 1$.

Solution The case $p = 1$ has already been settled in Example 1. If $p > 0$ but $p \neq 1$, then the function $f(x) = 1/x^p$ satisfies the conditions of the integral test, and

$$\int_1^\infty \frac{1}{x^p}\,dx = \lim_{b\to\infty}\int_1^b \frac{1}{x^p}\,dx = \lim_{b\to\infty}\left[-\frac{1}{(p-1)x^{p-1}}\right]_1^b$$

$$= \lim_{b\to\infty}\frac{1}{p-1}\left(1 - \frac{1}{b^{p-1}}\right).$$

If $p > 1$, then

$$\int_1^\infty \frac{1}{x^p}\,dx = \frac{1}{p-1} < \infty,$$

so the integral and the series both converge. But if $0 < p < 1$, then

$$\int_1^\infty \frac{1}{x^p}\,dx = \lim_{b\to\infty}\frac{1}{1-p}(b^{1-p} - 1) = \infty,$$

and in this case the integral and the series both diverge. ◆

As specific examples, the series

$$\sum_{n=1}^{\infty} \frac{1}{n^2} = 1 + \frac{1}{2^2} + \frac{1}{3^2} + \cdots + \frac{1}{n^2} + \cdots$$

converges ($p = 2 > 1$), whereas the series

$$\sum_{n=1}^{\infty} \frac{1}{\sqrt{n}} = 1 + \frac{1}{\sqrt{2}} + \frac{1}{\sqrt{3}} + \cdots + \frac{1}{\sqrt{n}} + \cdots$$

diverges ($p = \frac{1}{2} \leqq 1$).

Now suppose that the positive-term series $\sum a_n$ converges by the integral test and that we wish to approximate its sum by adding sufficiently many of its initial terms. The difference between the sum S of the series and its nth partial sum S_n is the **remainder**

$$R_n = S - S_n = a_{n+1} + a_{n+2} + a_{n+3} + \cdots. \tag{5}$$

This remainder is the error made when the sum is estimated by using in its place the partial sum S_n.

THEOREM 2 The Integral Test Remainder Estimate

Suppose that the infinite series and improper integral

$$\sum_{n=1}^{\infty} a_n \quad \text{and} \quad \int_1^{\infty} f(x)\,dx$$

satisfy the hypotheses of the integral test, and suppose in addition that both converge. Then

$$\int_{n+1}^{\infty} f(x)\,dx \leqq R_n \leqq \int_n^{\infty} f(x)\,dx, \tag{6}$$

where R_n is the remainder given in Eq. (5).

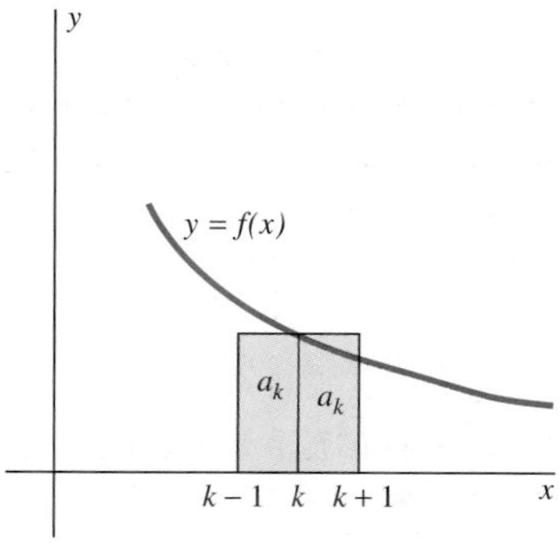

FIGURE 10.5.3 Establishing the integral test remainder estimate.

PROOF We see from Fig. 10.5.3 that

$$\int_k^{k+1} f(x)\,dx \leqq a_k \leqq \int_{k-1}^{k} f(x)\,dx$$

for $k = n+1, n+2, \ldots$. We add these inequalities for all such values of k, and the result is the inequality in (6), because

$$R_n = \sum_{k=n+1}^{\infty} a_k,$$

$$\sum_{k=n+1}^{\infty} \int_k^{k+1} f(x)\,dx = \int_{n+1}^{\infty} f(x)\,dx,$$

and

$$\sum_{k=n+1}^{\infty} \int_{k-1}^{k} f(x)\,dx = \int_n^{\infty} f(x)\,dx. \qquad \blacktriangleleft$$

If we substitute $R_n = S - S_n$, then it follows from (6) that the sum S of the series satisfies the inequality

$$S_n + \int_{n+1}^{\infty} f(x)\,dx \leqq S \leqq S_n + \int_n^{\infty} f(x)\,dx. \tag{7}$$

If the nth partial sum S_n is known and the difference

$$\int_n^{n+1} f(x)\,dx$$

between the two integrals is small, then (7) provides an accurate estimate of the sum S of the infinite series.

EXAMPLE 3 We will see in Section 10.8 that the exact sum of the p-series with $p=2$ is $\pi^2/6$, thus giving the beautiful formula

$$\frac{\pi^2}{6} = 1 + \frac{1}{2^2} + \frac{1}{3^2} + \frac{1}{4^2} + \cdots. \qquad \textbf{(8)}$$

Use this series to approximate the number π by applying the integral test remainder estimate, first with $n=50$, then with $n=200$.

Solution Obviously we take $f(x)=1/x^2$ in the remainder estimate. Because

$$\int_n^\infty \frac{1}{x^2}\,dx = \lim_{b\to\infty}\left[-\frac{1}{x}\right]_n^b = \lim_{b\to\infty}\left(\frac{1}{n}-\frac{1}{b}\right) = \frac{1}{n},$$

Eq. (7) gives

$$S_n + \frac{1}{n+1} \leqq \frac{\pi^2}{6} \leqq S_n + \frac{1}{n}, \qquad \textbf{(9)}$$

where

$$S_n = 1 + \frac{1}{2^2} + \frac{1}{3^2} + \cdots + \frac{1}{n^2}$$

is the nth partial sum of the series in (8). Upon multiplying by 6 and taking square roots, (9) gives the inequality

$$\sqrt{6\left(S_n + \frac{1}{n+1}\right)} \leqq \pi \leqq \sqrt{6\left(S_n + \frac{1}{n}\right)}. \qquad \textbf{(10)}$$

You could add the first 50 terms in (8) one by one in a few minutes using a simple four-function calculator, but this kind of arithmetic is precisely the task for which a modern calculator or computer algebra system is designed. A one-line instruction such as the calculator command `sum(seq(1/n^2,n,1,50))` yields

$$S_{50} = \sum_{n=1}^{50} \frac{1}{n^2} \approx 1.625132734.$$

Then, using (9) for illustration rather than (10), we calculate

$$1.62513273 + \frac{1}{51} < \frac{\pi^2}{6} < 1.62513274 + \frac{1}{50};$$

$$1.64474057 < \frac{\pi^2}{6} < 1.64513274;$$

$$3.14140787 < \pi < 3.14178237.$$

Finally, rounding down on the left and up on the right (why?), we conclude that $3.1414 < \pi < 3.1418$. The average of these two bounds is the traditional four-place approximation $\pi \approx 3.1416$.

The 200th partial sum of the series in (8) is

$$S_{200} = \sum_{n=1}^{200} \frac{1}{n^2} \approx 1.639946546.$$

Substituting this sum and $n = 200$ in (10), we get

$$3.14158081 < \pi < 3.14160457.$$

This proves that $\pi \approx 3.1416$ rounded accurate to four decimal places. ◆

EXAMPLE 4 Show that the series

$$\sum_{n=2}^{\infty} \frac{1}{n(\ln n)^2} \tag{11}$$

converges, and determine how many terms you would need to add to find its sum accurate to within 0.01. That is, how large must n be in order that the remainder satisfy the inequality $R_n < 0.01$?

Solution We begin the sum at $n = 2$ because $\ln 1 = 0$. Let $f(x) = 1/[x(\ln x)^2]$. Then

$$\int_n^{\infty} \frac{1}{x(\ln x)^2}\,dx = \lim_{b\to\infty}\left[-\frac{1}{\ln x}\right]_n^b = \lim_{b\to\infty}\left(\frac{1}{\ln n} - \frac{1}{\ln b}\right) = \frac{1}{\ln n}.$$

Substituting $n = 2$ shows that the series in (11) converges (by the integral test). Our calculations and the right-hand inequality in (6) now give $R_n < 1/(\ln n)$, so we need

$$\frac{1}{\ln n} \leqq 0.01; \quad \ln n \geqq 100; \quad n \geqq e^{100} \approx 2.7 \times 10^{43}.$$

A computer that could calculate a billion (10^9) terms per second would require about 8.5×10^{26} years—far longer than the expected lifetime of the universe—to sum this many terms. But you can check that accuracy to only one decimal place—that is, $R_n < 0.05$—would require only about $n = 4.85 \times 10^8$ (fewer than a half billion) terms, well within the range of a powerful desktop computer. ◆

10.5 TRUE/FALSE STUDY GUIDE

10.5 CONCEPTS: QUESTIONS AND DISCUSSION

1. What might it mean to say that one infinite series converges more slowly than another? Perhaps that more terms of one than the other must be added to make the remainder R_n less than a preassigned error? If so, compare the rates at which the series $\sum n^{-2}, \sum n^{-3/2}, \sum n^{-4/3}, \dots, \sum n^{-101/100}, \dots,$ converge.
2. Can you use infinite series such as those listed in Question 1 to illustrate the claim that, however slowly one infinite series converges, there's another one that converges even more slowly?
3. Can you think of a way in which the convergent infinite series in Question 1 seem to resemble more and more closely the divergent harmonic series $\sum n^{-1}$? Discuss the possibility that two infinite series can resemble each other arbitrarily closely, yet one converges and the other diverges.

10.5 PROBLEMS

In Problems 1 through 30, use the integral test to test the given series for convergence.

1. $\sum_{n=1}^{\infty} \frac{n}{n^2+1}$
2. $\sum_{n=1}^{\infty} \frac{n}{e^{n^2}}$
3. $\sum_{n=1}^{\infty} \frac{1}{\sqrt{n+1}}$
4. $\sum_{n=1}^{\infty} \frac{1}{(n+1)^{4/3}}$
5. $\sum_{n=1}^{\infty} \frac{1}{n^2+1}$
6. $\sum_{n=1}^{\infty} \frac{1}{n(n+1)}$
7. $\sum_{n=2}^{\infty} \frac{1}{n \ln n}$
8. $\sum_{n=1}^{\infty} \frac{\ln n}{n}$
9. $\sum_{n=1}^{\infty} \frac{1}{2^n}$
10. $\sum_{n=1}^{\infty} \frac{n}{e^n}$

11. $\sum_{n=1}^{\infty} \frac{n^2}{e^n}$

12. $\sum_{n=1}^{\infty} \frac{1}{17n - 13}$

13. $\sum_{n=1}^{\infty} \frac{\ln n}{n^2}$

14. $\sum_{n=1}^{\infty} \frac{n+1}{n^2}$

15. $\sum_{n=1}^{\infty} \frac{n}{n^4 + 1}$

16. $\sum_{n=1}^{\infty} \frac{1}{n^3 + n}$

17. $\sum_{n=1}^{\infty} \frac{2n+5}{n^2 + 5n + 17}$

18. $\sum_{n=1}^{\infty} \ln\left(\frac{n+1}{n}\right)$

19. $\sum_{n=1}^{\infty} \ln\left(1 + \frac{1}{n^2}\right)$

20. $\sum_{n=1}^{\infty} \frac{2^{1/n}}{n^2}$

21. $\sum_{n=1}^{\infty} \frac{n}{4n^2 + 5}$

22. $\sum_{n=1}^{\infty} \frac{n}{(4n^2 + 5)^{3/2}}$

23. $\sum_{n=2}^{\infty} \frac{1}{n\sqrt{\ln n}}$

24. $\sum_{n=2}^{\infty} \frac{1}{n(\ln n)^3}$

25. $\sum_{n=1}^{\infty} \frac{1}{4n^2 + 9}$

26. $\sum_{n=1}^{\infty} \frac{n+1}{n+100}$

27. $\sum_{n=1}^{\infty} \frac{n}{n^4 + 2n^2 + 1}$

28. $\sum_{n=1}^{\infty} \frac{1}{(n+1)^3}$

29. $\sum_{n=1}^{\infty} \frac{\arctan n}{n^2 + 1}$

30. $\sum_{n=3}^{\infty} \frac{1}{n(\ln n)[\ln(\ln n)]}$

In Problems 31 through 34, tell why the integral test does not apply to the given series.

31. $\sum_{n=1}^{\infty} \frac{(-1)^n}{n}$

32. $\sum_{n=1}^{\infty} e^{-n} \sin n$

33. $\sum_{n=1}^{\infty} \frac{2 + \sin n}{n^2}$

34. $\sum_{n=1}^{\infty} \left(\frac{\sin n}{n}\right)^4$

In Problems 35 through 38, determine the values of p for which the given series converges.

35. $\sum_{n=1}^{\infty} \frac{1}{p^n}$

36. $\sum_{n=1}^{\infty} \frac{n}{(n^2 + 1)^p}$

37. $\sum_{n=2}^{\infty} \frac{1}{n(\ln n)^p}$

38. $\sum_{n=3}^{\infty} \frac{1}{n(\ln n)\,[\ln(\ln n)]^p}$

In Problems 39 through 42, find the least positive integer n such that the remainder R_n in Theorem 2 is less than E.

39. $\sum_{n=1}^{\infty} \frac{1}{n^2}$; $E = 0.0001$

40. $\sum_{n=1}^{\infty} \frac{1}{n^2}$; $E = 0.00005$

41. $\sum_{n=1}^{\infty} \frac{1}{n^3}$; $E = 0.00005$

42. $\sum_{n=1}^{\infty} \frac{1}{n^6}$; $E = 2 \times 10^{-11}$

In Problems 43 through 46, find the sum of the given series accurate to the indicated number k of decimal places. Begin by finding the smallest value of n such that the remainder satisfies the inequality $R_n < 5 \times 10^{-(k+1)}$. Then use a calculator to compute the partial sum S_n and round off appropriately.

43. $\sum_{n=1}^{\infty} \frac{1}{n^{3/2}}$; $k = 2$

44. $\sum_{n=1}^{\infty} \frac{1}{n^3}$; $k = 3$

45. $\sum_{n=1}^{\infty} \frac{1}{n^5}$; $k = 5$

46. $\sum_{n=1}^{\infty} \frac{1}{n^7}$; $k = 7$

In Problems 47 and 48, use a computer algebra system (if necessary) to determine the values of p for which the given infinite series converges.

47. $\sum_{n=1}^{\infty} \frac{\ln n}{n^p}$

48. $\sum_{n=1}^{\infty} \frac{1}{p^{\ln n}}$

49. Deduce from the inequalities in (2) and (3) with the function $f(x) = 1/x$ that

$$\ln n \leqq 1 + \frac{1}{2} + \frac{1}{3} + \cdots + \frac{1}{n} \leqq 1 + \ln n$$

for $n = 1, 2, 3, \ldots$. If a computer adds 1 millions terms of the harmonic series per second, how long will it take for the partial sum to reach 50?

50. (a) Let

$$c_n = 1 + \frac{1}{2} + \frac{1}{3} + \cdots + \frac{1}{n} - \ln n$$

for $n = 1, 2, 3, \ldots$. Deduce from Problem 49 that $0 \leqq c_n \leqq 1$ for all n. (b) Note that

$$\int_n^{n+1} \frac{1}{x}\,dx \geqq \frac{1}{n+1}.$$

Conclude that the sequence $\{c_n\}$ is decreasing. Therefore the sequence $\{c_n\}$ converges. The number

$$\gamma = \lim_{n\to\infty} c_n = \lim_{n\to\infty} \left(1 + \frac{1}{2} + \frac{1}{3} + \cdots + \frac{1}{n} - \ln n\right) \approx 0.57722$$

is known as **Euler's constant.**

51. Suppose that the function f is continuous and positive-valued for $x \geqq 1$. Let

$$b_n = \int_1^n f(x)\,dx$$

for $n = 1, 2, 3, \ldots$. (a) Suppose that the increasing sequence $\{b_n\}$ is bounded, so that $B = \lim_{n\to\infty} b_n$ exists. Prove that

$$\int_1^{\infty} f(x)\,dx = B.$$

(b) Prove that if the sequence $\{b_n\}$ is not bounded, then

$$\int_1^{\infty} f(x)\,dx = +\infty.$$

10.5 Project: The Number π, Once and for All

When we replace the parameter p in the p-series $\sum 1/n^p$ with the variable x, we get one of the most important transcendental functions in higher mathematics, the **Riemann zeta function**

$$\zeta(x) = \sum_{n=1}^{\infty} \frac{1}{n^x} = 1 + \frac{1}{2^x} + \frac{1}{3^x} + \frac{1}{4^x} + \cdots.$$

REMARK One can substitute a complex number $x = a + bi$ in the zeta function. Now that Fermat's last theorem has been proved, the most famous unsolved conjecture in mathematics is the **Riemann hypothesis**—that $\zeta(a + bi) = 0$ implies that $a = \frac{1}{2}$; that is, that the only complex zeros of the Riemann zeta function have real part $\frac{1}{2}$. (The smallest such example is approximately $\frac{1}{2} + 14.13475i$.) The truth of the Riemann hypothesis would have profound implications in number theory, including information about the distribution of the prime numbers.

In Problems 1 through 4, use the given value of the zeta function and the integral-test remainder estimate (as in Example 3 of this section) with the given value of n to determine how accurately the value of the number π is thereby determined. Knowing that

$$\pi \approx 3.14159\,26535\,89793\,23846,$$

write each final answer in the form $\pi \approx 3.abcde\ldots$, giving precisely those digits that are correct or correctly rounded.

1. $\zeta(2) = \dfrac{\pi^2}{6}$ with $n = 25$.

2. $\zeta(4) = \dfrac{\pi^4}{90}$ with $n = 20$.

3. $\zeta(6) = \dfrac{\pi^6}{945}$ with $n = 15$.

4. $\zeta(8) = \dfrac{\pi^8}{9450}$ with $n = 10$.

5. Finally, use one of the preceding four problems and your own careful choice of n to show that $\pi \approx 3.141592654$ with all digits correct or correctly rounded.

Euler showed that if n is even then $\zeta(n)$ is a rational multiple of π^n (as in the cases $n = 2, 4, 6, 8$ cited above). Because any integral power of π is irrational, it follows that the number $\zeta(n)$ is irrational if n is even. But little was known about $\zeta(n)$ for n odd until 1978, when Roger Apéry proved that $\zeta(3)$ is irrational. In Section 7.7 of Andrews, Askey, and Roy, *Special Functions* (Cambridge Univ. Press: 1999), the authors show that there exist infinite sequences $\{A_n\}$ and $\{B_n\}$ of integers such that

$$0 < |A_n + B_n\zeta(3)| < 3 \cdot \left(\frac{9}{10}\right)^n$$

for each integer $n \geqq 1$. Can you explain why this implies that $\zeta(3)$ is irrational? (Assume, to the contrary, that $\zeta(3) = p/q$ is rational.)

10.6 COMPARISON TESTS FOR POSITIVE-TERM SERIES

With the integral test we attempt to determine whether or not an infinite series converges by comparing it with an improper integral. The methods of this section involves comparing the terms of the *positive-term* series $\sum a_n$ with those of another positive-term series $\sum b_n$ whose convergence or divergence is known. We have already developed two families of *reference series* for the role of the known series $\sum b_n$; these are the geometric series of Section 10.3 and the p-series of Section 10.5. They are well adapted for our new purposes because their convergence or divergence is quite easy to determine. Recall that the geometric series $\sum r^n$ converges if $|r| < 1$ and diverges if $|r| \geqq 1$, and that the p-series $\sum 1/n^p$ converges if $p > 1$ and diverges if $0 < p \leqq 1$.

Let $\sum a_n$ and $\sum b_n$ be positive-term series. Then we say that the series $\sum b_n$ **dominates** the series $\sum a_n$ provided that $a_n \leqq b_n$ for all n. Theorem 1 says that the positive-term series $\sum a_n$ converges if it is dominated by a convergent series and diverges if it dominates a positive-term divergent series.

THEOREM 1 Comparison Test

Suppose that $\sum a_n$ and $\sum b_n$ are positive-term series. Then

1. $\sum a_n$ converges if $\sum b_n$ converges and $a_n \leqq b_n$ for all n;
2. $\sum a_n$ diverges if $\sum b_n$ diverges and $a_n \geqq b_n$ for all n.

PROOF Denote the nth partial sums of the series $\sum a_n$ and $\sum b_n$ by S_n and T_n, respectively. Then $\{S_n\}$ and $\{T_n\}$ are increasing sequences. To prove part (1), suppose that $\sum b_n$ converges, so $T = \lim_{n\to\infty} T_n$ exists (so that T is a real number). Then the fact that $a_n \leqq b_n$ for all n implies that $S_n \leqq T_n \leqq T$ for all n. Thus the sequence $\{S_n\}$ of partial sums of $\sum a_n$ is bounded and increasing and therefore converges. Thus $\sum a_n$ converges.

Part (2) is merely a restatement of part (1). If the series $\sum a_n$ converged, then the fact that $\sum a_n$ dominates $\sum b_n$ would imply—by part (1), with a_n and b_n interchanged—that $\sum b_n$ converged. But $\sum b_n$ diverges, so it follows that $\sum a_n$ must also diverge. ◀

We know by Theorem 5 of Section 10.3 that the convergence or divergence of an infinite series is not affected by the insertion or deletion of a finite number of terms. Consequently, the conditions $a_n \leqq b_n$ and $a_n \geqq b_n$ in the two parts of the comparison test really need to hold only for all $n \geqq k$, where k is some fixed positive integer. Thus we can say that the positive-term series $\sum a_n$ converges if it is "eventually dominated" by the convergent positive-term series $\sum b_n$.

EXAMPLE 1 Because

$$\frac{1}{n(n+1)(n+2)} < \frac{1}{n^3}$$

for all $n \geqq 1$, the series

$$\sum_{n=1}^{\infty} \frac{1}{n(n+1)(n+2)} = \frac{1}{1\cdot 2\cdot 3} + \frac{1}{2\cdot 3\cdot 4} + \frac{1}{3\cdot 4\cdot 5} + \cdots$$

is dominated by the series $\sum 1/n^3$, which is a convergent p-series with $p = 3$. Both are positive-term series, and hence the series $\sum 1/[n(n+1)(n+2)]$ converges by part (1) of the comparison test. ◆

EXAMPLE 2 Because

$$\frac{1}{\sqrt{2n-1}} > \frac{1}{\sqrt{2n}}$$

for all $n \geqq 1$, the positive-term series

$$\sum_{n=1}^{\infty} \frac{1}{\sqrt{2n-1}} = 1 + \frac{1}{\sqrt{3}} + \frac{1}{\sqrt{5}} + \frac{1}{\sqrt{7}} + \cdots$$

dominates the series

$$\sum_{n=1}^{\infty} \frac{1}{\sqrt{2n}} = \frac{1}{\sqrt{2}} \sum_{n=1}^{\infty} \frac{1}{n^{1/2}}.$$

But $\sum 1/n^{1/2}$ is a divergent p-series with $p=\frac{1}{2}$, and a constant nonzero multiple of a divergent series diverges. So part (2) of the comparison test implies that the series $\sum 1/\sqrt{2n-1}$ also diverges. ◆

EXAMPLE 3 Test the series

$$\sum_{n=0}^{\infty} \frac{1}{n!} = 1 + \frac{1}{1!} + \frac{1}{2!} + \frac{1}{3!} + \cdots$$

for convergence.

Solution We note first that if $n \geqq 1$, then

$$\begin{aligned} n! &= n(n-1)(n-2)\cdots 3\cdot 2\cdot 1 \\ &\geqq 2\cdot 2\cdot 2\cdots 2\cdot 2\cdot 1 \qquad \text{(the same number of factors);} \end{aligned}$$

that is, $n! \geqq 2^{n-1}$ for $n \geqq 1$. Thus

$$\frac{1}{n!} \leqq \frac{1}{2^{n-1}} \quad \text{for } n \geqq 1,$$

so the series

$$\sum_{n=0}^{\infty} \frac{1}{n!} \quad \text{is dominated by the series} \quad 1 + \sum_{n=1}^{\infty} \frac{1}{2^{n-1}} = 1 + \sum_{n=0}^{\infty} \frac{1}{2^n},$$

which is a convergent geometric series (after the first term). Both are positive-term series, so by the comparison test the given series converges. We saw in Section 10.4 that the sum of the series is the number e, so

$$e = 1 + \frac{1}{1!} + \frac{1}{2!} + \frac{1}{3!} + \cdots + \frac{1}{n!} + \cdots.$$

Indeed, this series provides perhaps the simplest way of showing that

$$e \approx 2.71828\,1828\,459045\,23536. \quad ◆$$

Limit Comparison of Terms

Suppose that $\sum a_n$ is a positive-term series such that $a_n \to 0$ as $n \to +\infty$. Then, in connection with the nth-term divergence test of Section 10.3, the series $\sum a_n$ has at least a *chance* of converging. How do we choose an appropriate positive-term series $\sum b_n$ with which to compare it? A good idea is to express b_n as a *simple* function of n, simpler than a_n but such that a_n and b_n approach zero at the same rate as $n \to +\infty$. If the formula for a_n is a fraction, we can try discarding all but the terms of largest magnitude in its numerator and denominator to form b_n. For example, if

$$a_n = \frac{3n^2 + n}{n^4 + \sqrt{n}},$$

then we reason that n is small in comparison with $3n^2$, and that $\sqrt{n}$ is small in comparison with n^4, when n is quite large. This suggests that we choose $b_n = 3n^2/n^4 = 3/n^2$. The series $\sum 3/n^2$ converges ($p=2$), but when we attempt to compare $\sum a_n$ and $\sum b_n$, we find that $a_n \geqq b_n$ (rather than $a_n \leqq b_n$). Consequently, the comparison test does not apply immediately—the fact that $\sum a_n$ dominates a convergent series does *not* imply that $\sum a_n$ itself converges. Theorem 2 provides a convenient way of handling such a situation.

THEOREM 2 Limit Comparison Test
Suppose that $\sum a_n$ and $\sum b_n$ are positive-term series. If the limit

$$L = \lim_{n\to\infty} \frac{a_n}{b_n}$$

exists and $0 < L < +\infty$, then either both series converge or both series diverge.

PROOF Choose two fixed positive numbers P and Q such that $P < L < Q$. Then $P < a_n/b_n < Q$ for n sufficiently large, and so

$$Pb_n < a_n < Qb_n$$

for all sufficiently large values of n. If $\sum b_n$ converges, then $\sum a_n$ is eventually dominated by the convergent series $\sum Qb_n = Q\sum b_n$, so part (1) of the comparison test implies that $\sum a_n$ also converges. If $\sum b_n$ diverges, then $\sum a_n$ eventually dominates the divergent series $\sum Pb_n = P\sum b_n$, so part (2) of the comparison test implies that $\sum a_n$ also diverges. Thus the convergence of either series implies the convergence of the other. ◀

EXAMPLE 4 With

$$a_n = \frac{3n^2 + n}{n^4 + \sqrt{n}} \quad \text{and} \quad b_n = \frac{1}{n^2}$$

(motivated by the discussion preceding Theorem 2), we find that

$$\lim_{n\to\infty} \frac{a_n}{b_n} = \lim_{n\to\infty} \frac{3n^4 + n^3}{n^4 + \sqrt{n}} = \lim_{n\to\infty} \frac{3 + \dfrac{1}{n}}{1 + \dfrac{1}{n^{7/2}}} = 3.$$

Because $\sum 1/n^2$ is a convergent p-series ($p = 2$), the limit comparison test tells us that the series

$$\sum_{n=1}^{\infty} \frac{3n^2 + n}{n^4 + \sqrt{n}}$$

also converges. ◆

EXAMPLE 5 Test for convergence: $\displaystyle\sum_{n=1}^{\infty} \frac{1}{2n + \ln n}$.

Solution Because $\lim_{n\to\infty} (\ln n)/n = 0$ (by l'Hôpital's rule), $\ln n$ is very small in comparison with $2n$ when n is large. We therefore take $a_n = 1/(2n + \ln n)$ and, ignoring the constant coefficient 2, we take $b_n = 1/n$. Then we find that

$$\lim_{n\to\infty} \frac{a_n}{b_n} = \lim_{n\to\infty} \frac{n}{2n + \ln n} = \lim_{n\to\infty} \frac{1}{2 + \dfrac{\ln n}{n}} = \frac{1}{2}.$$

Because the harmonic series $\sum 1/n = \sum b_n$ diverges, it follows that the given series $\sum a_n$ also diverges. ◆

It is important to remember that if $L = \lim_{n\to\infty} (a_n/b_n)$ is either zero or infinite, then the limit comparison test does not apply. (See Problem 52 for a discussion of what conclusions may sometimes be drawn in these cases.) Note, for example, that if $a_n = 1/n^2$ and $b_n = 1/n$, then $\lim_{n\to\infty} (a_n/b_n) = 0$. But in this case $\sum a_n$ converges, whereas $\sum b_n$ diverges.

Estimating Remainders

Suppose that $0 \leqq a_n \leqq b_n$ for all n and we know that $\sum b_n$ converges, so the comparison test implies that $\sum a_n$ converges as well. Let us write $s = \sum a_n$ and $S = \sum b_n$. If a numerical estimate is available for the remainder

$$R_n = S - S_n = b_{n+1} + b_{n+2} + \cdots$$

in the dominating series $\sum b_n$, then we can use it to estimate the remainder

$$r_n = s - s_n = a_{n+1} + a_{n+2} + \cdots$$

in the series $\sum a_n$. The reason is that $0 \leqq a_n \leqq b_n$ (for all n) implies that $0 \leqq r_n \leqq R_n$. We can apply this fact if, for instance, we have used the integral test remainder estimate to calculate an upper bound for R_n—which is, then, an upper bound for r_n as well.

EXAMPLE 6 The series

$$\sum_{n=1}^{\infty} a_n = \sum_{n=1}^{\infty} \frac{1}{n^3 + \sqrt{n}}$$

converges because it is dominated by the convergent p-series

$$\sum_{n=1}^{\infty} b_n = \sum_{n=1}^{\infty} \frac{1}{n^3}.$$

It therefore follows by the integral test remainder estimate (Section 10.5) that

$$0 < r_n \leqq R_n \leqq \int_n^{\infty} \frac{1}{x^3}\,dx = \lim_{b\to\infty} \left[-\frac{1}{2x^2}\right]_n^b = \frac{1}{2n^2}.$$

Now a calculator gives

$$s_{100} = \sum_{n=1}^{100} \frac{1}{n^3 + \sqrt{n}} \approx 0.680284 \quad \text{and} \quad R_{100} \leqq \frac{1}{2 \cdot 100^2} = 0.00005.$$

It follows that $0.680284 \leqq s \leqq 0.680334$. In particular,

$$\sum_{n=1}^{\infty} \frac{1}{n^3 + \sqrt{n}} \approx 0.6803$$

rounded accurate to four decimal places. ◆

Rearrangement and Grouping

We close our discussion of positive-term series with the observation that the sum of a convergent *positive*-term series is not altered by grouping or rearranging its terms. For example, let $\sum a_n$ be a convergent positive-term series and consider

$$\sum_{n=1}^{\infty} b_n = (a_1 + a_2 + a_3) + a_4 + (a_5 + a_6) + \cdots.$$

That is, the new series has terms

$$b_1 = a_1 + a_2 + a_3,$$

$$b_2 = a_4,$$

$$b_3 = a_5 + a_6,$$

and so on. Then every partial sum T_n of $\sum b_n$ is equal to some partial sum $S_{n'}$ of $\sum a_n$. Because $\{S_n\}$ is an increasing sequence with limit $S = \sum a_n$, it follows easily that $\{T_n\}$ is an increasing sequence with the same limit. Thus $\sum b_n = S$ as well. The argument is more subtle if terms of $\sum a_n$ are moved "out of place," as in

$$\sum_{n=1}^{\infty} b_n = a_1 + a_2 + a_4 + a_3 + a_6 + a_8 + a_5 + a_{10} + a_{12} + \cdots,$$

but the same conclusion holds: Any rearrangement of a convergent *positive*-term series also converges, and it converges to the same sum.

Similarly, it is easy to prove that any grouping or rearrangement of a divergent positive-term series also diverges. But these observations all fail in the case of an infinite series with both positive and negative terms. For example, the series $\sum(-1)^n$ diverges, but it has the convergent grouping

$$(-1+1)+(-1+1)+(-1+1)+\cdots = 0+0+0+\cdots = 0.$$

It follows from Problem 56 of Section 10.4 that

$$\ln 2 = 1 - \frac{1}{2} + \frac{1}{3} - \frac{1}{4} + \frac{1}{5} - \cdots,$$

but the rearrangement

$$1 + \frac{1}{3} - \frac{1}{2} + \frac{1}{5} + \frac{1}{7} - \frac{1}{4} + \frac{1}{9} + \frac{1}{11} - \frac{1}{6} + \cdots$$

converges instead to $\frac{3}{2}\ln 2$. This series for ln 2 even has rearrangements that converge to zero and others that diverge to $+\infty$. (See Problem 64 of Section 10.7.)

10.6 TRUE/FALSE STUDY GUIDE

10.6 CONCEPTS: QUESTIONS AND DISCUSSION

1. Can you give examples of a pair of positive-term infinite series $\sum a_n$ and $\sum b_n$ such that $\lim_{n\to\infty}(a_n/b_n) = 0$ and (a) both series converge; (b) both diverge; (c) one converges and the other diverges?
2. Can you give an example of two convergent positive-term infinite series $\sum a_n$ and $\sum b_n$ such that $\lim_{n\to\infty}(a_n/b_n) = 1$ but neither series dominates the other?
3. Can you give an example of two positive-term infinite series $\sum a_n$ and $\sum b_n$ that either both converge or both diverge, but the limit $\lim_{n\to\infty}(a_n/b_n)$ does not exist?

10.6 PROBLEMS

Use comparison tests to determine whether the infinite series in Problems 1 through 36 converge or diverge.

1. $\displaystyle\sum_{n=1}^{\infty} \frac{1}{n^2+n+1}$

2. $\displaystyle\sum_{n=1}^{\infty} \frac{n^3+1}{n^4+2}$

3. $\displaystyle\sum_{n=1}^{\infty} \frac{1}{n+\sqrt{n}}$

4. $\displaystyle\sum_{n=1}^{\infty} \frac{1}{n+n^{3/2}}$

5. $\displaystyle\sum_{n=1}^{\infty} \frac{1}{1+3^n}$

6. $\displaystyle\sum_{n=1}^{\infty} \frac{10n^2}{n^4+1}$

7. $\displaystyle\sum_{n=2}^{\infty} \frac{10n^2}{n^3-1}$

8. $\displaystyle\sum_{n=1}^{\infty} \frac{n^2-n}{n^4+2}$

9. $\displaystyle\sum_{n=1}^{\infty} \frac{1}{\sqrt{37n^3+3}}$

10. $\displaystyle\sum_{n=1}^{\infty} \frac{1}{\sqrt{n^2+1}}$

11. $\displaystyle\sum_{n=1}^{\infty} \frac{\sqrt{n}}{n^2+n}$

12. $\displaystyle\sum_{n=1}^{\infty} \frac{1}{3+5^n}$

13. $\displaystyle\sum_{n=2}^{\infty} \frac{1}{\ln n}$

14. $\displaystyle\sum_{n=1}^{\infty} \frac{1}{n-\ln n}$

15. $\displaystyle\sum_{n=1}^{\infty} \frac{\sin^2 n}{n^2+1}$

16. $\displaystyle\sum_{n=1}^{\infty} \frac{\cos^2 n}{3^n}$

17. $\displaystyle\sum_{n=1}^{\infty} \frac{n+2^n}{n+3^n}$

18. $\displaystyle\sum_{n=1}^{\infty} \frac{1}{2^n+3^n}$

19. $\displaystyle\sum_{n=2}^{\infty} \frac{1}{n^2 \ln n}$

20. $\displaystyle\sum_{n=1}^{\infty} \frac{1}{n^{1+\sqrt{n}}}$

21. $\displaystyle\sum_{n=1}^{\infty} \frac{\ln n}{n^2}$

22. $\displaystyle\sum_{n=1}^{\infty} \frac{\arctan n}{n}$

23. $\displaystyle\sum_{n=1}^{\infty} \frac{\sin^2(1/n)}{n^2}$

24. $\displaystyle\sum_{n=1}^{\infty} \frac{e^{1/n}}{n}$

25. $\displaystyle\sum_{n=1}^{\infty} \frac{\ln n}{e^n}$

26. $\displaystyle\sum_{n=1}^{\infty} \frac{n^2+2}{n^3+3n}$

27. $\sum_{n=1}^{\infty} \frac{n^{3/2}}{n^2+4}$

28. $\sum_{n=1}^{\infty} \frac{1}{n \cdot 2^n}$

29. $\sum_{n=1}^{\infty} \frac{3}{4+\sqrt{n}}$

30. $\sum_{n=1}^{\infty} \frac{n^2+1}{e^n(n+1)^2}$

31. $\sum_{n=1}^{\infty} \frac{2n^2-1}{n^2 \cdot 3^n}$

32. $\sum_{n=1}^{\infty} \frac{1}{\sqrt[3]{2n^4+1}}$

33. $\sum_{n=1}^{\infty} \frac{2+\sin n}{n^2}$

34. $\sum_{n=1}^{\infty} \frac{\ln n}{n^3}$

35. $\sum_{n=1}^{\infty} \frac{(n+1)^n}{n^{n+1}}$ $\left[\textit{Suggestion: } \lim_{n\to\infty}\left(1+\frac{1}{n}\right)^n = e.\right]$

36. $\sum_{n=1}^{\infty} \left(\frac{\sin n}{n}\right)^4$

In Problems 37 through 40, calculate the sum of the first ten terms of the series, then estimate the error made in using this partial sum to approximate the sum of the series.

37. $\sum_{n=1}^{\infty} \frac{1}{n^2+1}$

38. $\sum_{n=1}^{\infty} \frac{1}{3^n+1}$

39. $\sum_{n=1}^{\infty} \frac{\cos^2 n}{n^2}$

40. $\sum_{n=2}^{\infty} \frac{1}{(n+1)(\ln n)^2}$

In Problems 41 through 44, first determine the smallest positive integer n such that the remainder satisfies the inequality $R_n < 0.005$. Then use a calculator or computer to approximate the sum of the series accurate to two decimal places.

41. $\sum_{n=1}^{\infty} \frac{1}{n^3+1}$

42. $\sum_{n=1}^{\infty} \frac{n}{(n+1)2^n}$

43. $\sum_{n=1}^{\infty} \frac{\cos^4 n}{n^4}$

44. $\sum_{n=1}^{\infty} \frac{1}{n^{2+(1/n)}}$

45. Show that if $\sum a_n$ is a convergent positive-term series, then the series $\sum \sin(a_n)$ also converges.

46. (a) Prove that $\ln n < n^{1/8}$ for all sufficiently large values of n. (b) Explain why part (a) shows that the series $\sum 1/(\ln n)^8$ diverges.

47. Prove that if $\sum a_n$ is a convergent positive-term series, then $\sum (a_n/n)$ converges.

48. Suppose that $\sum a_n$ is a convergent positive-term series and that $\{c_n\}$ is a sequence of positive numbers with limit zero. Prove that $\sum a_n c_n$ converges.

49. Use the result of Problem 48 to prove that if $\sum a_n$ and $\sum b_n$ are convergent positive-term series, then $\sum a_n b_n$ converges.

50. Prove that the series

$$\sum_{n=1}^{\infty} \frac{1}{1+2+3+\cdots+n}$$

converges.

51. Use the result of Problem 50 in Section 10.5 to prove that the series

$$\sum_{n=1}^{\infty} \frac{1}{1+\frac{1}{2}+\frac{1}{3}+\cdots+\frac{1}{n}}$$

diverges.

52. Adapt the proof of the limit-comparison test to prove the following two results. (a) Suppose that $\sum a_n$ and $\sum b_n$ are positive-term series and that $\sum b_n$ converges. If

$$L = \lim_{n\to\infty} \frac{a_n}{b_n} = 0,$$

then $\sum a_n$ converges. (b) Suppose that $\sum a_n$ and $\sum b_n$ are positive-term series and that $\sum b_n$ diverges. If

$$L = \lim_{n\to\infty} \frac{a_n}{b_n} = +\infty,$$

then $\sum a_n$ diverges.

10.7 ALTERNATING SERIES AND ABSOLUTE CONVERGENCE

In Sections 10.5 and 10.6 we considered only positive-term series. Now we discuss infinite series that have both positive terms and negative terms. An important example is a series with terms that are alternatively positive and negative. An **alternating series** is an infinite series of the form

$$\sum_{n=1}^{\infty} (-1)^{n+1} a_n = a_1 - a_2 + a_3 - a_4 + a_5 - \cdots \tag{1}$$

or of the form $\sum_{n=1}^{\infty} (-1)^n a_n$, where $a_n > 0$ for all n. For example, the *alternating harmonic series*

$$\sum_{n=1}^{\infty} \frac{(-1)^{n+1}}{n} = 1 - \frac{1}{2} + \frac{1}{3} - \frac{1}{4} + \frac{1}{5} - \cdots$$

and the geometric series

$$\sum_{n=0}^{\infty}\left(-\frac{1}{2}\right)^n = 1 - \frac{1}{2} + \frac{1}{4} - \frac{1}{8} + \frac{1}{16} - \cdots$$

are both alternating series. Theorem 1 shows that both these series converge because the sequence of absolute values of their terms is decreasing and has limit zero.

THEOREM 1 Alternating Series Test
If the alternating series in Eq. (1) satisfies the two conditions

1. $a_n \geqq a_{n+1} > 0$ for all n and
2. $\lim_{n\to\infty} a_n = 0$,

then the infinite series converges.

PROOF We first consider the even-numbered partial sums $S_2, S_4, S_6, \dots, S_{2n}, \dots$. We may write

$$S_{2n} = (a_1 - a_2) + (a_3 - a_4) + \cdots + (a_{2n-1} - a_{2n}).$$

Because $a_k - a_{k+1} \geqq 0$ for all k, the sequence $\{S_{2n}\}$ is increasing. Also, because

$$S_{2n} = a_1 - (a_2 - a_3) - \cdots - (a_{2n-2} - a_{2n-1}) - a_{2n},$$

$S_{2n} \leqq a_1$ for all n. So the increasing sequence $\{S_{2n}\}$ is bounded above. Hence the limit

$$S = \lim_{n\to\infty} S_{2n}$$

exists by the bounded monotonic sequence property of Section 10.2. It remains only for us to verify that the odd-numbered partial sums $S_1, S_3, S_5, \dots$ also converge to S. But $S_{2n+1} = S_{2n} + a_{2n+1}$ and $\lim_{n\to\infty} a_{2n+1} = 0$, so

$$\lim_{n\to\infty} S_{2n+1} = \left(\lim_{n\to\infty} S_{2n}\right) + \left(\lim_{n\to\infty} a_{2n+1}\right) = S.$$

Thus $\lim_{n\to\infty} S_n = S$, and therefore the series in Eq. (1) converges. ◀

FIGURE 10.7.1 The even partial sums $\{S_{2n}\}$ increase and the odd partial sums $\{S_{2n+1}\}$ decrease.

Figure 10.7.1 illustrates the way in which the partial sums of a convergent alternating series (with positive first term) approximate its sum S, with the even partial sums $\{S_{2n}\}$ approaching S from below and the odd partial sums $\{S_{2n+1}\}$ approaching S from above.

EXAMPLE 1 The series

$$\sum_{n=1}^{\infty}\frac{(-1)^{n+1}}{2n-1} = 1 - \frac{1}{3} + \frac{1}{5} - \frac{1}{7} + \frac{1}{9} - \cdots$$

satisfies the conditions of Theorem 1 and therefore converges. The alternating series test does not tell us the sum of this series, but we saw in Section 10.4 that its sum is $\pi/4$. The graph in Fig. 10.7.2 of the partial sums of this series illustrates the typical convergence of an alternating series, with its partial sums approaching its sum alternately from above and below. ◆

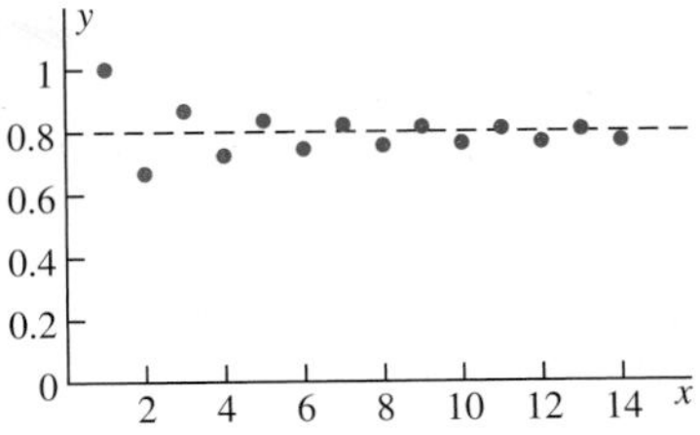

FIGURE 10.7.2 Graph of the first 14 partial sums of the alternating series in Example 1.

EXAMPLE 2 The series

$$\sum_{n=1}^{\infty}\frac{(-1)^{n+1}n}{2n-1} = 1 - \frac{2}{3} + \frac{3}{5} - \frac{4}{7} + \frac{5}{9} - \cdots$$

is an alternating series, and by expanding we verify that $n(2n+1) > (n+1)(2n-1)$, so it follows that

$$a_n = \frac{n}{2n-1} > \frac{n+1}{2n+1} = a_{n+1}$$

for all $n \geqq 1$. But

$$\lim_{n\to\infty} a_n = \frac{1}{2} \neq 0,$$

so the alternating series test *does not apply.* (This fact alone does not imply that the series in question diverges—many series in Sections 10.5 and 10.6 converge even though the alternating series test does not apply. But the series of this example diverges by the nth-term divergence test.) ◆

If a series converges by the alternating series test, then Theorem 2 shows how to approximate its sum with any desired degree of accuracy—*if* you have a computer fast enough to add a large number of its terms.

THEOREM 2 Alternating Series Remainder Estimate
Suppose that the series $\sum(-1)^{n+1}a_n$ satisfies the conditions of the alternating series test and therefore converges. Let S denote the sum of the series. Denote by $R_n = S - S_n$ the error made in replacing S with the nth partial sum S_n of the series. Then this **remainder** R_n has the same sign as the next term $(-1)^{n+2}a_{n+1}$ of the series, and

$$0 \leqq |R_n| < a_{n+1}. \qquad \textbf{(2)}$$

In particular, the *sum S of a convergent alternating series lies between any two consecutive partial sums.* This follows from the proof of Theorem 1, where we saw that $\{S_{2n}\}$ is an increasing sequence and that $\{S_{2n+1}\}$ is a decreasing sequence, both converging to S. The resulting inequalities

$$S_{2n-1} > S > S_{2n} = S_{2n-1} - a_{2n}$$

and

$$S_{2n} < S < S_{2n+1} = S_{2n} + a_{2n+1}$$

(see Fig. 10.7.3) imply the inequality in (2).

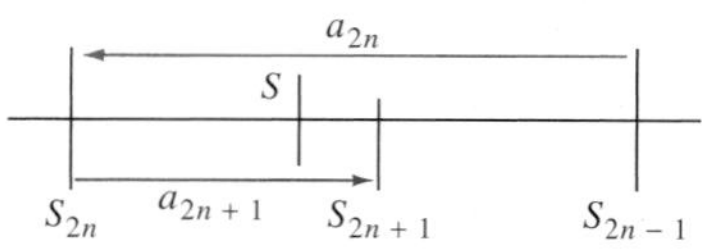

FIGURE 10.7.3 Illustrating the proof of the alternating series remainder estimate.

REMARK The inequality in (2) means the following. Suppose that you are given an *alternating series* that satisfies the conditions of Theorem 2 and has sum S. Then, if S is replaced with a partial sum S_n, the error made is numerically less than the first term a_{n+1} not retained and has the same sign as this first neglected term. **Important:** This error estimate *does not* apply to other types of series.

EXAMPLE 3 We saw in Section 10.4 that

$$e^x = \sum_{n=0}^{\infty} \frac{x^n}{n!}$$

for all x and thus (with $x = -1$) that

$$\frac{1}{e} = e^{-1} = 1 - 1 + \frac{1}{2!} - \frac{1}{3!} + \frac{1}{4!} - \cdots.$$

Use this alternating series to compute e^{-1} accurate to four decimal places.

Solution To attain four-place accuracy, we want the error to be less than a half unit in the fourth place. Thus we want

$$|R_n| < \frac{1}{(n+1)!} \leqq 0.00005.$$

If we use a calculator to compute the reciprocals of the factorials of the first several integers, we find that the least value of n for which this inequality holds is $n = 7$. Then

$$e^{-1} = 1 - \frac{1}{1!} + \frac{1}{2!} - \frac{1}{3!} + \frac{1}{4!} - \frac{1}{5!} + \frac{1}{6!} - \frac{1}{7!} + R_7 \approx 0.367857 + R_7.$$

(Relying on a common "+2 rule of thumb," we are carrying six decimal places because we want four-place accuracy in the final answer.) Now the first neglected term $1/8!$ is positive, so the inequality in (2) gives

$$0 < R_7 < \frac{1}{8!} < 0.000025.$$

Therefore

$$S_7 \approx 0.367857 < e^{-1} < S_7 + 0.000025 \approx 0.367882.$$

The two bounds here both round to $e^{-1} \approx 0.3679$. Although this approximation is accurate to four decimal places, its reciprocal

$$e = 1/e^{-1} \approx 1/(0.3679) \approx 2.7181 \approx 2.718$$

gives the number e accurate to only three decimal places. ◆

Absolute Convergence

The series

$$\sum_{n=1}^{\infty} \frac{(-1)^{n+1}}{n} = 1 - \frac{1}{2} + \frac{1}{3} - \frac{1}{4} + \frac{1}{5} - \cdots$$

converges, but if we simply replace each term with its absolute value, we get the *divergent* series

$$1 + \frac{1}{2} + \frac{1}{3} + \frac{1}{4} + \frac{1}{5} + \cdots.$$

In contrast, the *convergent* series

$$\sum_{n=0}^{\infty} \frac{(-1)^n}{2^n} = 1 - \frac{1}{2} + \frac{1}{4} - \frac{1}{8} + \cdots = \frac{2}{3}$$

has the property that the associated positive-term series

$$1 + \frac{1}{2} + \frac{1}{4} + \frac{1}{8} + \cdots = 2$$

also converges. Theorem 3 tells us that if a series of *positive* terms converges, then we may insert minus signs in front of any of the terms—every other one, for instance—and the resulting series will also converge.

THEOREM 3 Absolute Convergence Implies Convergence
If the series $\sum |a_n|$ converges, then so does the series $\sum a_n$.

PROOF Suppose that the series $\sum |a_n|$ converges. Note that

$$0 \leqq a_n + |a_n| \leqq 2|a_n|$$

for all n. Let $b_n = a_n + |a_n|$. It then follows from the comparison test that the positive-term series $\sum b_n$ converges, because it is dominated by the convergent series $\sum 2|a_n|$. It is easy to verify, too, that the termwise difference of two convergent series also converges. Hence we now see that the series

$$\sum a_n = \sum (b_n - |a_n|) = \sum b_n - \sum |a_n|$$

converges. ◄

Thus we have another convergence test, one not limited to positive-term series nor limited to alternating series: Given the series $\sum a_n$, test the series $\sum |a_n|$ for convergence. If the latter converges, then so does the former. (But the converse is *not* true!) This phenomenon motivates us to make the following definition.

DEFINITION Absolute Convergence

The series $\sum a_n$ is said to **converge absolutely** (and is called **absolutely convergent**) provided that the series

$$\sum |a_n| = |a_1| + |a_2| + |a_3| + \cdots + |a_n| + \cdots$$

converges.

Thus we have explained the title of Theorem 3, and we can rephrase the theorem as follows: *If a series converges absolutely, then it converges.* The two examples preceding Theorem 3 show that a convergent series may either converge absolutely or fail to do so:

$$1 - \frac{1}{2} + \frac{1}{4} - \frac{1}{8} + \frac{1}{16} - \cdots$$

is an absolutely convergent series because

$$1 + \frac{1}{2} + \frac{1}{4} + \frac{1}{8} + \frac{1}{16} + \cdots$$

converges, whereas

$$1 - \frac{1}{2} + \frac{1}{3} - \frac{1}{4} + \frac{1}{5} - \cdots$$

is a series that, though convergent, is *not* absolutely convergent. A series that converges but does not converge absolutely is said to be **conditionally convergent.** Consequently, the terms *absolutely convergent, conditionally convergent,* and *divergent* are simultaneously all inclusive and mutually exclusive: Any given numerical series belongs to exactly one of those three classes.

There is some advantage in the application of Theorem 3, because to apply it we test the *positive*-term series $\sum |a_n|$ for convergence—and we have a variety of tests, such as comparison tests or the integral test, designed for use on positive-term series.

Note also that absolute convergence of the series $\sum a_n$ means that a *different* series $\sum |a_n|$ converges, and the two sums will generally differ. For example, with $a_n = \left(-\frac{1}{3}\right)^n$, the formula for the sum of a geometric series gives

$$\sum_{n=0}^{\infty} a_n = \sum_{n=0}^{\infty} \left(-\frac{1}{3}\right)^n = \frac{1}{1 - \left(-\frac{1}{3}\right)} = \frac{3}{4},$$

whereas

$$\sum_{n=0}^{\infty} |a_n| = \sum_{n=0}^{\infty} \left(\frac{1}{3}\right)^n = \frac{1}{1 - \frac{1}{3}} = \frac{3}{2}.$$

EXAMPLE 4 Discuss the convergence of the series

$$\sum_{n=1}^{\infty} \frac{\cos n}{n^2} = \cos 1 + \frac{\cos 2}{4} + \frac{\cos 3}{9} + \cdots.$$

Solution Let $a_n = (\cos n)/n^2$. Then

$$|a_n| = \frac{|\cos n|}{n^2} \leqq \frac{1}{n^2}$$

for all $n \geqq 1$. Hence the positive-term series $\sum |a_n|$ converges by the comparison test, because it is dominated by the convergent p-series $\sum(1/n^2)$. Thus the given series is absolutely convergent, and it therefore converges by Theorem 3. ◆

One reason for the importance of absolute convergence is the fact (proved in advanced calculus) that the terms of an absolutely convergent series may be regrouped or rearranged without changing the sum of the series. As we suggested at the end of Section 10.6, this is *not* true of conditionally convergent series.

The Ratio Test and the Root Test

Our next two convergence tests involve a way of measuring the rate of growth or decrease of the sequence $\{a_n\}$ of terms of a series to determine whether $\sum a_n$ converges absolutely or diverges.

THEOREM 4 The Ratio Test

Suppose that the limit

$$\rho = \lim_{n \to \infty} \left| \frac{a_{n+1}}{a_n} \right| \tag{3}$$

either exists or is infinite. Then the infinite series $\sum a_n$ of nonzero terms

1. Converges absolutely if $\rho < 1$;
2. Diverges if $\rho > 1$.

If $\rho = 1$, the ratio test is inconclusive.

PROOF If $\rho < 1$, choose a (fixed) number r with $\rho < r < 1$. Then Eq. (3) implies that there exists an integer N such that $|a_{n+1}| \leqq r|a_n|$ for all $n \geqq N$. It follows that

$$|a_{N+1}| \leqq r|a_N|,$$

$$|a_{N+2}| \leqq r|a_{N+1}| \leqq r^2|a_N|,$$

$$|a_{N+3}| \leqq r|a_{N+2}| \leqq r^3|a_N|,$$

and in general that

$$|a_{N+k}| \leqq r^k|a_N| \quad \text{for } k \geqq 0.$$

Hence the series

$$|a_N| + |a_{N+1}| + |a_{N+2}| + \cdots$$

is dominated by the geometric series

$$|a_N|(1 + r + r^2 + r^3 + \cdots),$$

and the latter converges because $|r| < 1$. Thus the series $\sum |a_n|$ converges, so the series $\sum a_n$ converges absolutely.

If $\rho > 1$, then Eq. (3) implies that there exists a positive integer N such that $|a_{n+1}| > |a_n|$ for all $n \geqq N$. It follows that $|a_n| > |a_N| > 0$ for all $n > N$. Thus the sequence $\{a_n\}$ cannot approach zero as $n \to +\infty$, and consequently, by the nth-term divergence test, the series $\sum a_n$ diverges. ◄

To see that $\sum a_n$ may either converge or diverge if $\rho = 1$, consider the divergent series $\sum(1/n)$ and the convergent series $\sum(1/n^2)$. You should verify that, for both series, the value of the ratio ρ is 1.

EXAMPLE 5 Consider the series

$$\sum_{n=1}^{\infty} \frac{(-1)^n 2^n}{n!} = -2 + \frac{4}{2!} - \frac{8}{3!} + \frac{16}{4!} - \cdots.$$

Then

$$\rho = \lim_{n\to\infty}\left|\frac{a_{n+1}}{a_n}\right| = \lim_{n\to\infty}\left|\frac{\dfrac{(-1)^{n+1}2^{n+1}}{(n+1)!}}{\dfrac{(-1)^n 2^n}{n!}}\right| = \lim_{n\to\infty}\frac{2}{n+1} = 0.$$

Because $\rho < 1$, the series converges absolutely. ◆

EXAMPLE 6 Test for convergence: $\displaystyle\sum_{n=1}^{\infty}\frac{n}{2^n}$.

Solution We have

$$\rho = \lim_{n\to\infty}\left|\frac{a_{n+1}}{a_n}\right| = \lim_{n\to\infty}\frac{\dfrac{n+1}{2^{n+1}}}{\dfrac{n}{2^n}} = \lim_{n\to\infty}\frac{n+1}{2n} = \frac{1}{2}.$$

Because $\rho < 1$, this series converges (absolutely). ◆

EXAMPLE 7 Test for convergence: $\displaystyle\sum_{n=1}^{\infty}\frac{3^n}{n^2}$.

Solution Here we have

$$\rho = \lim_{n\to\infty}\left|\frac{a_{n+1}}{a_n}\right| = \lim_{n\to\infty}\frac{\dfrac{3^{n+1}}{(n+1)^2}}{\dfrac{3^n}{n^2}} = \lim_{n\to\infty}\frac{3n^2}{(n+1)^2} = 3.$$

In this case $\rho > 1$, so the given series diverges. ◆

THEOREM 5 The Root Test

Suppose that the limit

$$\rho = \lim_{n\to\infty}\sqrt[n]{|a_n|} \tag{4}$$

exists or is infinite. Then the infinite series $\sum a_n$

1. Converges absolutely if $\rho < 1$;
2. Diverges if $\rho > 1$.

If $\rho = 1$, the root test is inconclusive.

PROOF If $\rho < 1$, choose a (fixed) number r such that $\rho < r < 1$. Then $|a_n|^{1/n} < r$, and hence $|a_n| < r^n$, for n sufficiently large. Thus the series $\sum |a_n|$ is eventually dominated by the convergent geometric series $\sum r^n$. Therefore $\sum |a_n|$ converges, and so the series $\sum a_n$ converges absolutely.

If $\rho > 1$, then $|a_n|^{1/n} > 1$, and hence $|a_n| > 1$, for n sufficiently large. Therefore the nth-term test for divergence implies that the series $\sum a_n$ diverges. ◄

The ratio test is generally simpler to apply than the root test, and therefore it is ordinarily the one to try first. But there are certain series for which the root test succeeds and the ratio test fails, as in Example 8.

EXAMPLE 8 Consider the series

$$\sum_{n=0}^{\infty} \frac{1}{2^{n+(-1)^n}} = \frac{1}{2} + \frac{1}{1} + \frac{1}{8} + \frac{1}{4} + \frac{1}{32} + \frac{1}{16} + \cdots.$$

Then $a_{n+1}/a_n = 2$ if n is even, whereas $a_{n+1}/a_n = \frac{1}{8}$ if n is odd. So the limit required for the ratio test does not exist. But

$$\lim_{n\to\infty} |a_n|^{1/n} = \lim_{n\to\infty} \left| \frac{1}{2^{n+(-1)^n}} \right|^{1/n} = \lim_{n\to\infty} \frac{1}{2} \left| \frac{1}{2^{(-1)^n/n}} \right| = \frac{1}{2},$$

so the given series converges by the root test. (Its convergence also follows from the fact that it is a rearrangement of the positive-term convergent geometric series $\sum 1/2^n$.) ◆

10.7 TRUE/FALSE STUDY GUIDE

10.7 CONCEPTS: QUESTIONS AND DISCUSSION

1. Can you give an example of a divergent alternating series $\sum(-1)^{n+1}a_n$ such that $\lim_{n\to\infty} a_n = 0$? In view of the alternating series test stated in Theorem 1 of this section, how is such an example possible?
2. Give your own example of an infinite series such that both $\sum a_n$ and $\sum |a_n|$ converge but have different sums, both of which you can calculate. What can you conclude about the series if $\sum a_n$ and $\sum |a_n|$ have the same sum?
3. Can you give an example of a conditionally convergent positive-term series? Why or why not?

10.7 PROBLEMS

Determine whether or not the alternating series in Problems 1 through 20 converge or diverge.

1. $\displaystyle\sum_{n=1}^{\infty} \frac{(-1)^{n+1}}{n^2}$

2. $\displaystyle\sum_{n=1}^{\infty} \frac{(-1)^{n+1}}{\sqrt{n^2+1}}$

3. $\displaystyle\sum_{n=1}^{\infty} \frac{(-1)^n n}{3n+2}$

4. $\displaystyle\sum_{n=1}^{\infty} \frac{(-1)^n n}{3n^2+2}$

5. $\displaystyle\sum_{n=1}^{\infty} \frac{(-1)^{n+1} n}{\sqrt{n^2+2}}$

6. $\displaystyle\sum_{n=1}^{\infty} \frac{(-1)^{n+1} n^2}{\sqrt{n^5+5}}$

7. $\displaystyle\sum_{n=2}^{\infty} \frac{(-1)^{n+1} n}{\ln n}$

8. $\displaystyle\sum_{n=1}^{\infty} \frac{(-1)^n \ln n}{\sqrt{n}}$

9. $\displaystyle\sum_{n=1}^{\infty} \frac{(-1)^n n}{2^n}$

10. $\displaystyle\sum_{n=1}^{\infty} n \cdot \left(-\tfrac{2}{3}\right)^{n+1}$

11. $\displaystyle\sum_{n=1}^{\infty} \frac{(-1)^n n}{\sqrt{2^n+1}}$

12. $\displaystyle\sum_{n=1}^{\infty} \left(-\frac{n\pi}{10}\right)^{n+1}$

13. $\displaystyle\sum_{n=1}^{\infty} \frac{1}{n^{2/3}} \sin\left(\frac{n\pi}{2}\right)$

14. $\displaystyle\sum_{n=1}^{\infty} \frac{\cos n\pi}{n^{3/2}}$

15. $\displaystyle\sum_{n=1}^{\infty} (-1)^n \sin\left(\frac{1}{n}\right)$

16. $\displaystyle\sum_{n=1}^{\infty} (-1)^n n \sin\left(\frac{\pi}{n}\right)$

17. $\displaystyle\sum_{n=1}^{\infty} \frac{(-1)^{n+1}}{\sqrt[n]{2}}$

18. $\displaystyle\sum_{n=1}^{\infty} \frac{(-1.01)^{n+1}}{n^4}$

19. $\displaystyle\sum_{n=1}^{\infty} \frac{(-1)^{n+1}}{\sqrt[n]{n}}$

20. $\displaystyle\sum_{n=1}^{\infty} \frac{(-1)^{n+1} n!}{(2n)!}$

Determine whether the series in Problems 21 through 42 converge absolutely, converge conditionally, or diverge.

21. $\displaystyle\sum_{n=1}^{\infty} \frac{(-1)^{n+1}}{2^n}$

22. $\displaystyle\sum_{n=1}^{\infty} \frac{1}{n^2+1}$

23. $\displaystyle\sum_{n=1}^{\infty} \frac{(-1)^n \ln n}{n}$

24. $\displaystyle\sum_{n=1}^{\infty} \frac{1}{n^n}$

25. $\displaystyle\sum_{n=1}^{\infty} \left(\frac{10}{n}\right)^n$

26. $\displaystyle\sum_{n=1}^{\infty} \frac{3^n}{n!n}$

27. $\displaystyle\sum_{n=0}^{\infty} \frac{(-10)^n}{n!}$

28. $\displaystyle\sum_{n=1}^{\infty} \frac{(-1)^{n+1} n!}{n^n}$

29. $\displaystyle\sum_{n=1}^{\infty} (-1)^{n+1}\left(\frac{n}{n+1}\right)^n$

30. $\displaystyle\sum_{n=1}^{\infty} \frac{n!n^2}{(2n)!}$

31. $\displaystyle\sum_{n=1}^{\infty} \left(\frac{\ln n}{n}\right)^n$

32. $\displaystyle\sum_{n=0}^{\infty} \frac{(-1)^n 2^{3n}}{7^n}$

33. $\displaystyle\sum_{n=0}^{\infty} (-1)^n\left(\sqrt{n+1} - \sqrt{n}\right)$

34. $\displaystyle\sum_{n=1}^{\infty} n \cdot \left(\tfrac{3}{4}\right)^n$

35. $\displaystyle\sum_{n=1}^{\infty} \left[\ln\left(\frac{1}{n}\right)\right]^n$

36. $\displaystyle\sum_{n=0}^{\infty} \frac{(n!)^2}{(2n)!}$

37. $\sum_{n=1}^{\infty} \frac{(-1)^{n+1}3^n}{n(2^n+1)}$

38. $\sum_{n=1}^{\infty} \frac{(-1)^{n+1}\arctan n}{n}$

39. $\sum_{n=1}^{\infty} \frac{(-1)^{n+1}n!}{1\cdot 3\cdot 5\cdots(2n-1)}$

40. $\sum_{n=1}^{\infty} (-1)^{n+1}\frac{1\cdot 3\cdot 5\cdots(2n-1)}{1\cdot 4\cdot 7\cdots(3n-2)}$

41. $\sum_{n=1}^{\infty} \frac{(n+2)!}{3^n(n!)^2}$

42. $\sum_{n=1}^{\infty} \frac{(-1)^{n+1}n^n}{3^{n^2}}$

In Problems 43 through 48, sum the indicated number of initial terms of the given alternating series. Then apply the alternating series remainder estimate to estimate the error in approximating the sum of the series with this partial sum. Finally, approximate the sum of the series, writing precisely the number of decimal places that thereby are guaranteed to be correct (after rounding).

43. $\sum_{n=1}^{\infty} \frac{(-1)^{n+1}}{n^3}$, 5 terms

44. $\sum_{n=1}^{\infty} \frac{(-1)^{n+1}}{3^n}$, 8 terms

45. $\sum_{n=1}^{\infty} \frac{(-1)^{n+1}}{n!}$, 6 terms

46. $\sum_{n=1}^{\infty} \frac{(-1)^{n+1}}{n^n}$, 7 terms

47. $\sum_{n=1}^{\infty} \frac{(-1)^{n+1}}{n}$, 12 terms

48. $\sum_{n=1}^{\infty} \frac{(-1)^{n+1}}{n^2}$, 15 terms

In Problems 49 through 54, sum enough terms (tell how many) to approximate the sum of the series, writing the sum rounded to the indicated number of correct decimal places.

49. $\sum_{n=1}^{\infty} \frac{(-1)^{n+1}}{n^4}$, 3 decimal places

50. $\sum_{n=1}^{\infty} \frac{(-1)^{n+1}}{n^5}$, 4 decimal places

51. $\frac{1}{\sqrt{e}} = \sum_{n=0}^{\infty} \frac{(-1)^n}{n!2^n}$, 4 decimal places

52. $\cos 1 = \sum_{n=0}^{\infty} \frac{(-1)^n}{(2n)!}$, 5 decimal places

53. $\sin 60^\circ = \sum_{n=0}^{\infty} \frac{(-1)^n}{(2n+1)!}\left(\frac{\pi}{3}\right)^{2n+1}$, 5 decimal places

54. $\ln(1.1) = \sum_{n=1}^{\infty} \frac{(-1)^{n+1}}{n\cdot 10^n}$, 7 decimal places

In Problems 55 and 56, show that the indicated alternating series $\sum(-1)^{n+1}a_n$ satisfies the condition that $a_n \to 0$ as $n \to +\infty$, but nevertheless diverges. Tell why the alternating series test does not apply. It may be informative to graph the first 10 or 20 partial sums.

55. $a_n = \begin{cases} \dfrac{1}{n} & \text{if } n \text{ is odd,} \\ \dfrac{1}{n^2} & \text{if } n \text{ is even.} \end{cases}$

56. $a_n = \begin{cases} \dfrac{1}{\sqrt{n}} & \text{if } n \text{ is odd,} \\ \dfrac{1}{n^3} & \text{if } n \text{ is even.} \end{cases}$

57. Give an example of a pair of convergent series $\sum a_n$ and $\sum b_n$ such that $\sum a_n b_n$ diverges.

58. Prove that $\sum |a_n|$ diverges if the series $\sum a_n$ diverges.

59. Prove that

$$\lim_{n\to\infty} \frac{a^n}{n!} = 0$$

(for any real number a) by applying the ratio test to show that the infinite series $\sum a^n/n!$ converges.

60. (a) Suppose that r is a (fixed) number such that $|r| < 1$. Use the ratio test to prove that the series $\sum_{n=0}^{\infty} nr^n$ converges. Let S denote its sum. (b) Show that

$$(1-r)S = \sum_{n=1}^{\infty} r^n.$$

Show how to conclude that

$$\sum_{n=0}^{\infty} nr^n = \frac{r}{(1-r)^2}.$$

61. Let

$$H_n = \sum_{k=1}^{n} \frac{1}{k} \quad \text{and} \quad S_n = \sum_{k=1}^{n} \frac{(-1)^{k+1}}{k}$$

denote the nth partial sums of the harmonic and alternating harmonic series, respectively. (a) Show that $S_{2n} = H_{2n} - H_n$ for all $n \geqq 1$. (b) Problem 50 in Section 10.5 says that

$$\lim_{n\to\infty} (H_n - \ln n) = \gamma$$

(where $\gamma \approx 0.57722$ denotes Euler's constant). Explain why it follows that

$$\lim_{n\to\infty} (H_{2n} - \ln 2n) = \gamma.$$

(c) Conclude from parts (a) and (b) that $\lim_{n\to\infty} S_{2n} = \ln 2$. Thus

$$\ln 2 = 1 - \frac{1}{2} + \frac{1}{3} - \frac{1}{4} + \frac{1}{5} - \frac{1}{6} + \cdots.$$

62. Suppose that $\sum a_n$ is a conditionally convergent infinite series. For each n, let

$$a_n^+ = \frac{a_n + |a_n|}{2} \quad \text{and} \quad a_n^- = \frac{a_n - |a_n|}{2}.$$

(a) Explain why $\sum a_n^+$ consists of the positive terms of $\sum a_n$ and why $\sum a_n^-$ consists of the negative terms of $\sum a_n$. (b) Given a real number r, show that some rearrangement of the conditionally convergent series $\sum a_n$ converges to r. *Suggestion:* If r is positive, for instance, begin with the first partial sum of the positive series $\sum a_n^+$ that exceeds r. Then add just enough terms of the negative series $\sum a_n^-$ so that the cumulative sum is less than r. Next add just enough terms of the positive series that the cumulative sum is greater than r, and continue in this way to define the desired rearrangement. Why does it follow that this rearranged infinite series converges to r?

63. Use the method of Problem 62 to write the first dozen terms of a rearrangement of the alternating harmonic series (Problem 61) that converges to 1 rather than to ln 2.

64. Describe a way to rearrange the terms of the alternating harmonic series to obtain (a) A rearranged series that converges to -2; (b) A rearranged series that diverges to $+\infty$.

65. Here is another rearrangement of the alternating harmonic series of Problem 61:

$$1 - \frac{1}{2} - \frac{1}{4} - \frac{1}{6} - \frac{1}{8}$$
$$+ \frac{1}{3} - \frac{1}{10} - \frac{1}{12} - \frac{1}{14} - \frac{1}{16}$$
$$+ \frac{1}{5} - \frac{1}{18} - \frac{1}{20} - \frac{1}{22} - \frac{1}{24}$$
$$+ \frac{1}{7} - \frac{1}{26} - \frac{1}{28} - \frac{1}{30} - \frac{1}{32} + \cdots.$$

Use a computer to collect evidence about the value of its sum.

10.8 POWER SERIES

The most important infinite series representations of functions are those whose terms are constant multiples of (successive) integral powers of the independent variable x—that is, series that resemble "infinite polynomials." For example, we discussed in Section 10.4 the geometric series

$$\frac{1}{1-x} = 1 + x + x^2 + x^3 + \cdots \quad (|x| < 1) \tag{1}$$

and the Taylor series

$$e^x = \sum_{n=0}^{\infty} \frac{x^n}{n!} = 1 + x + \frac{x^2}{2!} + \frac{x^3}{3!} + \frac{x^4}{4!} + \cdots, \tag{2}$$

$$\cos x = \sum_{n=0}^{\infty} \frac{(-1)^n x^{2n}}{(2n)!} = 1 - \frac{x^2}{2!} + \frac{x^4}{4!} - \frac{x^6}{6!} + \cdots, \quad \text{and} \tag{3}$$

$$\sin x = \sum_{n=0}^{\infty} \frac{(-1)^n x^{2n+1}}{(2n+1)!} = x - \frac{x^3}{3!} + \frac{x^5}{5!} - \frac{x^7}{7!} + \cdots. \tag{4}$$

There we used Taylor's formula to show that the series in Eqs. (2) through (4) converge, for all x, to the functions e^x, $\cos x$, and $\sin x$, respectively. Here we investigate the convergence of a "power series" without knowing in advance the function (if any) to which it converges.

All the infinite series in Eqs. (1) through (4) have the form

$$\sum_{n=0}^{\infty} a_n x^n = a_0 + a_1 x + a_2 x^2 + \cdots + a_n x^n + \cdots \tag{5}$$

with the constant *coefficients* $a_0, a_1, a_2, \ldots$. An infinite series of this form is called a **power series** in (powers of) x. In order that the initial terms of the two sides of Eq. (5) agree, we adopt here the convention that $x^0 = 1$ even if $x = 0$.

Convergence of Power Series

The partial sums of the power series in (5) are the *polynomials*

$$s_1(x) = a_0 + a_1 x, \qquad s_2(x) = a_0 + a_1 x + a_2 x^2, \qquad s_3(x) = a_0 + a_1 x + a_2 x^2 + a_3 x^3,$$

and so forth. The nth partial sum is an nth-degree polynomial. When we ask *where* the power series converges, we seek those values of x for which the limit

$$s(x) = \lim_{n\to\infty} s_n(x)$$

exists. The sum $s(x)$ of a power series is then a function of x that is defined wherever the series converges.

The power series in (5) obviously converges when $x = 0$. In general, it will converge for some nonzero values of x and diverge for others. Because of the way in which powers of x are involved, the ratio test of Section 10.7 is particularly effective in determining the values of x for which a given power series converges.

Assume that the limit

$$\rho = \lim_{n\to\infty} \left| \frac{a_{n+1}}{a_n} \right| \tag{6}$$

exists. This is the limit that we need if we want to apply the ratio test to the series $\sum a_n$ of constants. To apply the ratio test to the power series in Eq. (5), we write $u_n = a_n x^n$ and compute the limit

$$\lim_{n\to\infty} \left| \frac{u_{n+1}}{u_n} \right| = \lim_{n\to\infty} \left| \frac{a_{n+1}x^{n+1}}{a_n x^n} \right| = \rho\,|x|. \tag{7}$$

If $\rho = 0$, then $\sum a_n x^n$ converges absolutely for all x. If $\rho = +\infty$, then $\sum a_n x^n$ diverges for all $x \neq 0$. If ρ is a positive real number, we see from Eq. (7) that $\sum a_n x^n$ converges absolutely for all x such that $\rho \cdot |x| < 1$—that is, when

$$|x| < R = \frac{1}{\rho} = \lim_{n\to\infty} \left| \frac{a_n}{a_{n+1}} \right|. \tag{8}$$

In this case the ratio test also implies that $\sum a_n x^n$ diverges if $|x| > R$ but is inconclusive when $x = \pm R$. We have therefore proved Theorem 1, under the additional hypothesis that the limit in Eq. (6) exists. In Problems 69 and 70 we outline a proof that does not require this additional hypothesis.

THEOREM 1 Convergence of Power Series

If $\sum a_n x^n$ is a power series, then either

1. The series converges absolutely for all x, or
2. The series converges only when $x = 0$, or
3. There exists a number $R > 0$ such that $\sum a_n x^n$ converges absolutely if $|x| < R$ and diverges if $|x| > R$.

The number R of Case 3 is called the **radius of convergence** of the power series $\sum a_n x^n$. We write $R = \infty$ in Case 1 and $R = 0$ in Case 2. The set of all real numbers x for which the series converges is called its **interval of convergence** (Fig. 10.8.1); note that this set *is* an interval. If $0 < R < \infty$, then the interval of convergence is one of the intervals

$$(-R, R), \quad (-R, R], \quad [-R, R), \quad \text{or} \quad [-R, R].$$

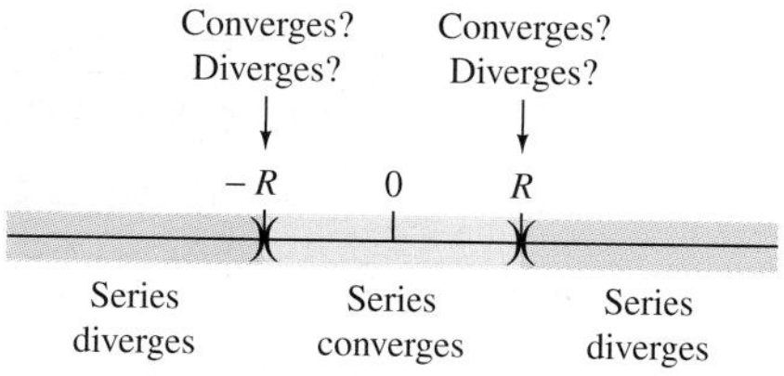

FIGURE 10.8.1 The interval of convergence if $0 < R = \lim_{n\to\infty} \left| \dfrac{a_n}{a_{n+1}} \right| < \infty$.

When we substitute either of the endpoints $x = \pm R$ into the series $\sum a_n x^n$, we obtain an infinite series with constant terms whose convergence must be determined separately. Because these will be numerical series, the earlier tests of this chapter are appropriate.

EXAMPLE 1 Find the interval of convergence of the series

$$\sum_{n=1}^{\infty} \frac{x^n}{n \cdot 3^n}.$$

Solution With $u_n = x^n/(n \cdot 3^n)$ we find that

$$\lim_{n\to\infty} \left| \frac{u_{n+1}}{u_n} \right| = \lim_{n\to\infty} \left| \frac{\dfrac{x^{n+1}}{(n+1)\cdot 3^{n+1}}}{\dfrac{x^n}{n \cdot 3^n}} \right| = \lim_{n\to\infty} \frac{n\,|x|}{3(n+1)} = \frac{|x|}{3}.$$

Now $|x|/3 < 1$ provided that $|x| < 3$, so the ratio test implies that the given series converges absolutely if $|x| < 3$ and diverges if $|x| > 3$. When $x = 3$, we have the divergent harmonic series $\sum(1/n)$, and when $x = -3$ we have the convergent alternating series $\sum(-1)^n/n$. Thus the interval of convergence of the given power series is $[-3, 3)$. We see dramatically in Fig. 10.8.2 the difference between convergence at $x = -3$ and divergence at $x = +3$. ◆

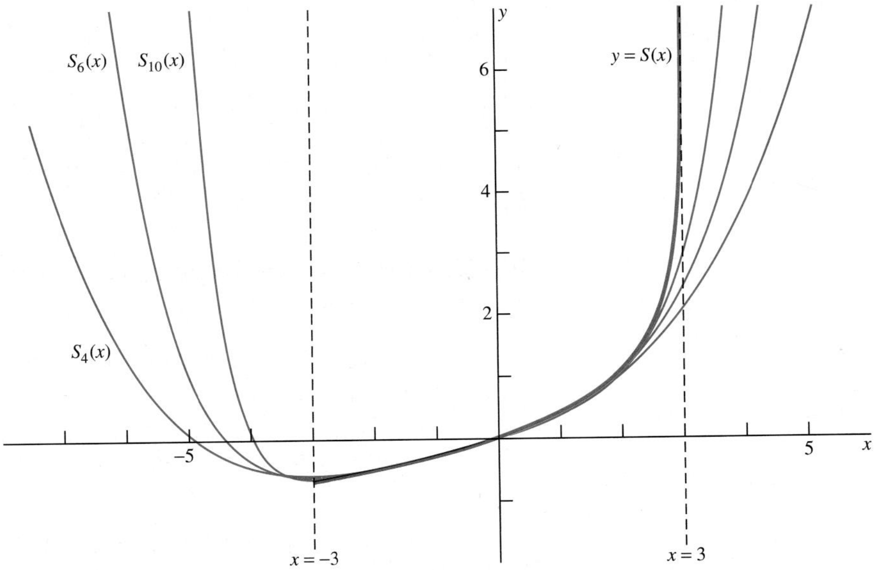

FIGURE 10.8.2 Graphs of the partial sums $S_4(x)$, $S_6(x)$, and $S_{10}(x)$ of the power series $S(x) = \sum_{n=1}^{\infty} \frac{x^n}{n \cdot 3^n}$ of Example 1. We see convergence at $x = -3$, but apparently $S(x) \to \infty$ as x approaches $+3$, where the series diverges harmonically.

EXAMPLE 2 Find the interval of convergence of the power series

$$\sum_{n=0}^{\infty} \frac{(-2)^n x^n}{(2n)!} = 1 - \frac{2x}{2!} + \frac{4x^2}{4!} - \frac{8x^3}{6!} + \frac{16x^4}{8!} - \cdots.$$

Solution With $u_n = (-2)^n x^n/(2n)!$ we find that

$$\lim_{n\to\infty} \left| \frac{u_{n+1}}{u_n} \right| = \lim_{n\to\infty} \left| \frac{\dfrac{(-2)^{n+1} x^{n+1}}{(2n+2)!}}{\dfrac{(-2)^n x^n}{(2n)!}} \right| = \lim_{n\to\infty} \frac{2|x|}{(2n+1)(2n+2)} = 0$$

for all x [using the fact that $(2n+2)! = (2n)!(2n+1)(2n+2)$]. Hence the ratio test implies that the given power series converges for all x, and its interval of convergence is therefore $(-\infty, +\infty)$, the entire real line. ◆

REMARK The power series of Example 2 results upon substituting $\sqrt{2x}$ for x in the Taylor series for $\cos x$ [Eq. (3)]. But only for $x > 0$ does the sum $S(x)$ of the series exhibit the oscillatory character of the function $\cos\sqrt{2x}$ (Fig. 10.8.3). For $x < 0$ the power series converges to the quite different (and nonoscillatory) function $\cosh\sqrt{|2x|}$.

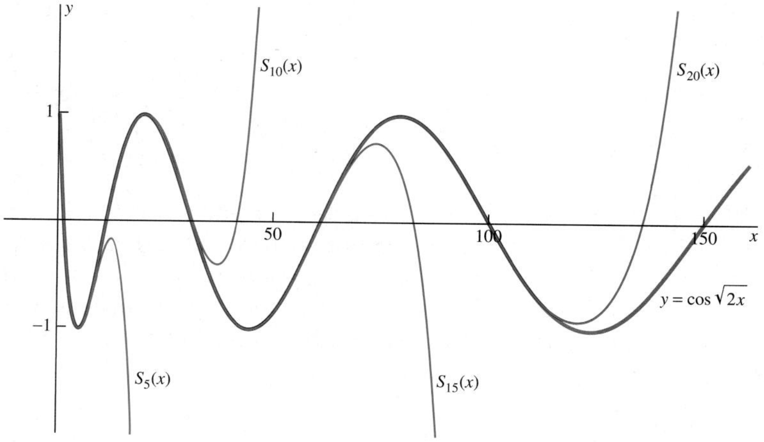

FIGURE 10.8.3 Graphs of the partial sums $S_5(x)$, $S_{10}(x)$, $S_{15}(x)$, and $S_{20}(x)$ of the power series $S(x) = \sum_{n=0}^{\infty} \frac{(-2x)^n}{(2n)!}$ of Example 2, which converges to $\cos\sqrt{2x}$ for $x > 0$.

EXAMPLE 3 Find the interval of convergence of the series $\sum_{n=1}^{\infty} n^n x^n$.

Solution With $u_n = n^n x^n$ we find that

$$\lim_{n\to\infty}\left|\frac{u_{n+1}}{u_n}\right| = \lim_{n\to\infty}\left|\frac{(n+1)^{n+1}x^{n+1}}{n^n x^n}\right| = \lim_{n\to\infty}(n+1)\left(1+\frac{1}{n}\right)^n |x| = +\infty$$

for all $x \neq 0$, because

$$\lim_{n\to\infty}\left(1+\frac{1}{n}\right)^n = e.$$

Thus the given series diverges for all $x \neq 0$, and its interval of convergence consists of the single point $x = 0$. ◆

EXAMPLE 4 Use the ratio test to verify that the Taylor series for $\cos x$ in Eq. (3) converges for all x.

Solution With $u_n = (-1)^n x^{2n}/(2n)!$ we find that

$$\lim_{n\to\infty}\left|\frac{u_{n+1}}{u_n}\right| = \lim_{n\to\infty}\left|\frac{\dfrac{(-1)^{n+1}x^{2n+2}}{(2n+2)!}}{\dfrac{(-1)^n x^{2n}}{(2n)!}}\right| = \lim_{n\to\infty}\frac{x^2}{(2n+1)(2n+2)} = 0$$

for all x, so the series converges for all x. ◆

IMPORTANT In Example 4, the ratio test tells us only that the series for $\cos x$ converges to *some* number, *not* necessarily the particular number $\cos x$. The argument of Section 10.4, using Taylor's formula with remainder, is required to establish that the sum of the series is actually $\cos x$.

Power Series in Powers of $x - c$

An infinite series of the form

$$\sum_{n=0}^{\infty} a_n(x-c)^n = a_0 + a_1(x-c) + a_2(x-c)^2 + \cdots, \qquad \textbf{(9)}$$

where c is a constant, is called a **power series in** (powers of) $x - c$. By the same reasoning that led us to Theorem 1, with x^n replaced with $(x-c)^n$ throughout, we conclude that either

1. The series in Eq. (9) converges absolutely for all x, or
2. The series converges only when $x - c = 0$—that is, when $x = c$—or
3. There exists a number $R > 0$ such that the series in Eq. (9) converges absolutely if $|x - c| < R$ and diverges if $|x - c| > R$.

As in the case of a power series with $c = 0$, the number R is called the **radius of convergence** of the series, and the **interval of convergence** of the series $\sum a_n(x-c)^n$ is the set of all numbers x for which it converges (Fig. 10.8.4). As before, when $0 < R < \infty$, the convergence of the series at the endpoints $x = c - R$ and $x = c + R$ of its interval of convergence must be checked separately.

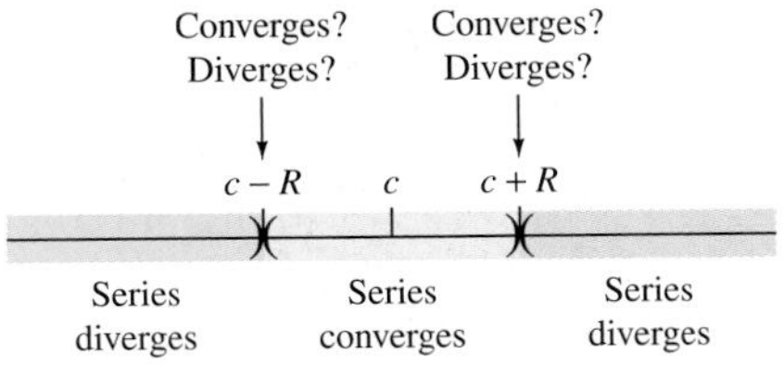

FIGURE 10.8.4 The interval of convergence of $\sum_{n=0}^{\infty} a_n(x-c)^n$.

EXAMPLE 5 Determine the interval of convergence of the series

$$\sum_{n=1}^{\infty} \frac{(-1)^n(x-2)^n}{n \cdot 4^n}.$$

Solution We let $u_n = (-1)^n(x-2)^n/(n \cdot 4^n)$. Then

$$\lim_{n\to\infty}\left|\frac{u_{n+1}}{u_n}\right| = \lim_{n\to\infty}\left|\frac{\dfrac{(-1)^{n+1}(x-2)^{n+1}}{(n+1)\cdot 4^{n+1}}}{\dfrac{(-1)^n(x-2)^n}{n\cdot 4^n}}\right|$$

$$= \lim_{n\to\infty} \frac{|x-2|}{4}\cdot\frac{n}{n+1} = \frac{|x-2|}{4}.$$

Hence the given series converges when $|x - 2| < 4$, so the radius of convergence is $R = 4$. Because $c = 2$, the series converges when $-2 < x < 6$ and diverges if either $x < -2$ or $x > 6$. When $x = -2$, the series reduces to the divergent harmonic series, and when $x = 6$ it reduces to the convergent alternating series $\sum(-1)^n/n$. Thus the interval of convergence of the given power series is $(-2, 6]$. ◆

Power Series Representations of Functions

Power series are important tools for computing (or approximating) values of functions. Suppose that the series $\sum a_n x^n$ converges to the value $f(x)$; that is,

$$f(x) = a_0 + a_1 x + a_2 x^2 + \cdots + a_n x^n + \cdots$$

for each x in the interval of convergence of the power series. Then we call $\sum a_n x^n$ a **power series representation** of $f(x)$. For example, the geometric series $\sum x^n$ in Eq. (1) is a power series representation of the function $f(x) = 1/(1-x)$ on the interval $(-1, 1)$.

We saw in Section 10.4 how Taylor's formula with remainder can often be used to find a power series representation of a given function. Recall that the nth-degree Taylor's formula for $f(x)$ at $x = a$ is

$$f(x) = f(a) + f'(a)(x-a) + \frac{f''(a)}{2!}(x-a)^2 + \frac{f^{(3)}(a)}{3!}(x-a)^3 + \cdots + \frac{f^{(n)}(a)}{n!}(x-a)^n + R_n(x). \quad \textbf{(10)}$$

The remainder $R_n(x)$ is given by

$$R_n(x) = \frac{f^{(n+1)}(z)}{(n+1)!}(x-a)^{n+1},$$

where z is some number between a and x. If we let $n \to +\infty$ in Eq. (10), we obtain Theorem 2.

THEOREM 2 Taylor Series Representations

Suppose that the function f has derivatives of all orders on some interval containing a and also that

$$\lim_{n\to\infty} R_n(x) = 0 \quad \textbf{(11)}$$

for each x in that interval. Then

$$f(x) = \sum_{n=0}^{\infty} \frac{f^{(n)}(a)}{n!}(x-a)^n \quad \textbf{(12)}$$

for each x in the interval.

The power series in Eq. (12) is the **Taylor series** of the function f **at** $x = a$ (or *in powers of* $x - a$, or *with center* a). If $a = 0$, we obtain the power series

$$f(x) = \sum_{n=0}^{\infty} \frac{f^{(n)}(0)}{n!}x^n = f(0) + f'(0)x + \frac{f''(0)}{2!}x^2 + \cdots, \quad \textbf{(13)}$$

commonly called the **Maclaurin series** of f. Thus the power series in Eqs. (2) through (4) are the Maclaurin series of the functions e^x, $\cos x$, and $\sin x$, respectively.

EXAMPLE 6 New power series can be constructed from old ones. For instance, upon replacing x with $-x$ in the Maclaurin series for e^x, we obtain

$$e^{-x} = 1 - x + \frac{x^2}{2!} - \frac{x^3}{3!} + \cdots + (-1)^n\frac{x^n}{n!} + \cdots.$$

Let us now add the series for e^x and e^{-x} and divide by 2. This gives

$$\cosh x = \frac{e^x + e^{-x}}{2} = \frac{1}{2}\left(1 + x + \frac{x^2}{2!} + \frac{x^3}{3!} + \frac{x^4}{4!} + \cdots\right) + \frac{1}{2}\left(1 - x + \frac{x^2}{2!} - \frac{x^3}{3!} + \frac{x^4}{4!} - \cdots\right),$$

so

$$\cosh x = 1 + \frac{x^2}{2!} + \frac{x^4}{4!} + \frac{x^6}{6!} + \cdots.$$

Similarly,

$$\sinh x = x + \frac{x^3}{3!} + \frac{x^5}{5!} + \frac{x^7}{7!} + \cdots.$$

Note the strong resemblance to Eqs. (3) and (4), the series for $\cos x$ and $\sin x$, respectively.

Upon replacing x with $-x^2$ in the series for e^x, we obtain

$$e^{-x^2} = \sum_{n=0}^{\infty} (-1)^n \frac{x^{2n}}{n!} = 1 - x^2 + \frac{x^4}{2!} - \frac{x^6}{3!} + \cdots.$$

Because this power series converges to $\exp(-x^2)$ for all x, it must be the Maclaurin series for $\exp(-x^2)$. (See Problem 66.) Think how tedious it would be to compute the derivatives of $\exp(-x^2)$ needed to write its Maclaurin series directly from Eq. (13). ◆

EXAMPLE 7 Sometimes a function is originally defined by means of a power series. One of the most important "higher transcendental functions" of applied mathematics is the Bessel function $J_0(x)$ of order zero defined by

$$J_0(x) = \sum_{n=0}^{\infty} \frac{(-1)^n x^{2n}}{2^{2n}(n!)^2} = 1 - \frac{x^2}{4} + \frac{x^4}{64} - \frac{x^6}{2304} + \cdots.$$

Only terms of even degree appear, so let us write $u_n = (-1)^n x^{2n}/[2^{2n}(n!)^2]$ for the nth term in this series (not counting its constant term). Then

$$\lim_{n\to\infty} \left| \frac{u_{n+1}}{u_n} \right| = \lim_{n\to\infty} \left| \frac{\dfrac{(-1)^{n+1} x^{2n+2}}{2^{2n+2}[(n+1)!]^2}}{\dfrac{(-1)^n x^{2n}}{2^{2n}(n!)^2}} \right| = \lim_{n\to\infty} \frac{x^2}{4(n+1)^2} = 0$$

for all x, so the ratio test implies that $J_0(x)$ is defined on the whole real line. The series for $J_0(x)$ resembles somewhat the cosine series, but the graph of $J_0(x)$ exhibits *damped* oscillations (Fig. 10.8.5). Bessel functions are important in such applications as the distribution of temperature in a cylindrical steam pipe and distribution of thermal neutrons in a cylindrical reactor. ◆

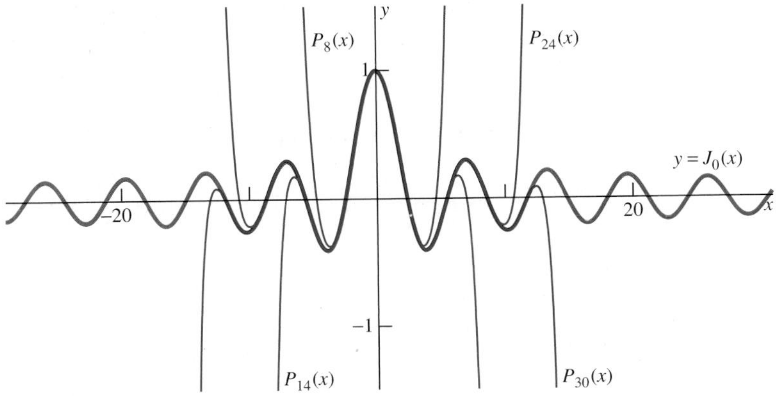

FIGURE 10.8.5 Graphs of the Bessel function $J_0(x)$ and its Taylor polynomials $P_8(x)$, $P_{14}(x)$, $P_{24}(x)$, and $P_{30}(x)$.

The Binomial Series

Example 8 gives one of the most famous and useful of all series, the *binomial series,* which was discovered by Newton in the 1660s. It is the infinite series generalization of the (finite) binomial theorem of elementary algebra.

EXAMPLE 8 Suppose that α is a nonzero real number. Show that the Maclaurin series of $f(x) = (1+x)^\alpha$ is

$$\begin{aligned}(1+x)^\alpha &= 1 + \sum_{n=1}^{\infty} \frac{\alpha(\alpha-1)(\alpha-2)\cdots(\alpha-n+1)}{n!}x^n \\ &= 1 + \alpha x + \frac{\alpha(\alpha-1)}{2!}x^2 + \frac{\alpha(\alpha-1)(\alpha-2)}{3!}x^3 + \cdots. \end{aligned} \tag{14}$$

Also determine the interval of convergence of this **binomial series.**

Solution To derive the series itself, we simply list all the derivatives of $f(x) = (1+x)^\alpha$, including its "zeroth" derivative:

$$\begin{aligned} f(x) &= (1+x)^\alpha \\ f'(x) &= \alpha(1+x)^{\alpha-1} \\ f''(x) &= \alpha(\alpha-1)(1+x)^{\alpha-2} \\ f^{(3)}(x) &= \alpha(\alpha-1)(\alpha-2)(1+x)^{\alpha-3}, \\ &\vdots \\ f^{(n)}(x) &= \alpha(\alpha-1)(\alpha-2)\cdots(\alpha-n+1)(1+x)^{\alpha-n}. \end{aligned}$$

Thus

$$f^{(n)}(0) = \alpha(\alpha-1)(\alpha-2)\cdots(\alpha-n+1).$$

If we substitute this value of $f^{(n)}(0)$ into the Maclaurin series formula in Eq. (13), we get the binomial series in Eq. (14).

To determine the interval of convergence of the binomial series, we let

$$u_n = \frac{\alpha(\alpha-1)(\alpha-2)\cdots(\alpha-n+1)}{n!}x^n.$$

We find that

$$\begin{aligned} \lim_{n\to\infty}\left|\frac{u_{n+1}}{u_n}\right| &= \lim_{n\to\infty}\left|\frac{\dfrac{\alpha(\alpha-1)(\alpha-2)\cdots(\alpha-n)x^{n+1}}{(n+1)!}}{\dfrac{\alpha(\alpha-1)(\alpha-2)\cdots(\alpha-n+1)x^n}{n!}}\right| \\ &= \lim_{n\to\infty}\left|\frac{(\alpha-n)x}{n+1}\right| = |x|. \end{aligned}$$

Hence the ratio test shows that the binomial series converges absolutely if $|x| < 1$ and diverges if $|x| > 1$. Its convergence at the endpoints $x = \pm 1$ depends on the value of α; we shall not pursue this problem. Problem 67 outlines a proof that the sum of the binomial series actually is $(1+x)^\alpha$ if $|x| < 1$. ◆

If $\alpha = k$, a positive integer, then the coefficient of x^n is zero for $n > k$, and the binomial series reduces to the binomial formula

$$(1+x)^k = \sum_{n=0}^{k} \frac{k!}{n!(k-n)!}x^n.$$

Otherwise Eq. (14) is an infinite series. For example, with $\alpha = \frac{1}{2}$, we obtain

$$\begin{aligned}\sqrt{1+x} &= 1 + \frac{\frac{1}{2}}{1!}x + \frac{\left(\frac{1}{2}\right)\left(-\frac{1}{2}\right)}{2!}x^2 + \frac{\left(\frac{1}{2}\right)\left(-\frac{1}{2}\right)\left(-\frac{3}{2}\right)}{3!}x^3 \\ &\quad + \frac{\left(\frac{1}{2}\right)\left(-\frac{1}{2}\right)\left(-\frac{3}{2}\right)\left(-\frac{5}{2}\right)}{4!}x^4 + \cdots \\ &= 1 + \tfrac{1}{2}x - \tfrac{1}{8}x^2 + \tfrac{1}{16}x^3 - \tfrac{5}{128}x^4 + \cdots. \end{aligned} \tag{15}$$

If we replace x with $-x$ and take $\alpha = -\frac{1}{2}$, we get the series

$$\frac{1}{\sqrt{1-x}} = 1 + \frac{-\frac{1}{2}}{1!}(-x) + \frac{\left(-\frac{1}{2}\right)\left(-\frac{3}{2}\right)}{2!}(-x)^2 + \cdots + \frac{1 \cdot 3 \cdot 5 \cdots (2n-1)}{n! \cdot 2^n}x^n + \cdots,$$

which in summation notation takes the form

$$\frac{1}{\sqrt{1-x}} = 1 + \sum_{n=1}^{\infty} \frac{1 \cdot 3 \cdot 5 \cdots (2n-1)}{2 \cdot 4 \cdot 6 \cdots (2n)}x^n. \tag{16}$$

We will find this series quite useful in Example 12 and in Problem 68.

Differentiation and Integration of Power Series

Sometimes it is inconvenient to compute the repeated derivatives of a function in order to find its Taylor series. An alternative method of finding new power series is by the differentiation and integration of known power series.

Suppose that a power series representation of the function $f(x)$ is known. Then Theorem 3 (we leave its proof to advanced calculus) implies that the function $f(x)$ may be differentiated by separately differentiating the individual terms in its power series. That is, the power series obtained by termwise differentiation converges to the derivative $f'(x)$. Similarly, a function can be integrated by termwise integration of its power series.

THEOREM 3 Termwise Differentiation and Integration

Suppose that the function f has a power series representation

$$f(x) = \sum_{n=0}^{\infty} a_n x^n = a_0 + a_1 x + a_2 x^2 + a_3 x^3 + \cdots$$

with nonzero radius of convergence R. Then f is differentiable on $(-R, R)$ and

$$f'(x) = \sum_{n=1}^{\infty} n a_n x^{n-1} = a_1 + 2a_2 x + 3a_3 x^2 + 4a_4 x^3 + \cdots. \tag{17}$$

Also,

$$\int_0^x f(t)\,dt = \sum_{n=0}^{\infty} \frac{a_n x^{n+1}}{n+1} = a_0 x + \tfrac{1}{2}a_1 x^2 + \tfrac{1}{3}a_2 x^3 + \cdots \tag{18}$$

for each x in $(-R, R)$. Moreover, the power series in Eqs. (17) and (18) have the same radius of convergence R.

REMARK 1 Although we omit the proof of Theorem 3, we observe that the radius of convergence of the series in Eq. (17) is

$$R = \lim_{n\to\infty} \left| \frac{n a_n}{(n+1)a_{n+1}} \right| = \left(\lim_{n\to\infty} \frac{n}{n+1} \right) \cdot \left(\lim_{n\to\infty} \left| \frac{a_n}{a_{n+1}} \right| \right) = \lim_{n\to\infty} \left| \frac{a_n}{a_{n+1}} \right|.$$

Thus, by Eq. (8), the power series for $f(x)$ and the power series for $f'(x)$ have the same radius of convergence (under the assumption that the preceding limit exists).

REMARK 2 Theorem 3 has this important consequence: If both the power series $\sum a_n x^n$ and $\sum b_n x^n$ converge and, for all x with $|x| < R$ $(R > 0)$, $\sum a_n x^n = \sum b_n x^n$, then $a_n = b_n$ for all n. In particular, the Taylor series of a function is its unique power series representation (if any). (See Problem 66.)

EXAMPLE 9 Termwise differentiation of the geometric series for

$$f(x) = \frac{1}{1-x}$$

yields

$$\frac{1}{(1-x)^2} = D_x\left(\frac{1}{1-x}\right) = D_x\,(1 + x + x^2 + x^3 + \cdots)$$
$$= 1 + 2x + 3x^2 + 4x^3 + \cdots.$$

Thus

$$\frac{1}{(1-x)^2} = \sum_{n=1}^{\infty} nx^{n-1} = \sum_{n=0}^{\infty} (n+1)x^n.$$

The series converges to $1/(1-x)^2$ if $-1 < x < 1$. ◆

EXAMPLE 10 Replacing x with $-t$ in the geometric series of Example 9 gives

$$\frac{1}{1+t} = 1 - t + t^2 - t^3 + \cdots + (-1)^n t^n + \cdots.$$

Because $D_t \ln(1+t) = 1/(1+t)$, termwise integration from $t = 0$ to $t = x$ now gives

$$\ln(1+x) = \int_0^x \frac{1}{1+t}\,dt$$
$$= \int_0^x (1 - t + t^2 - \cdots + (-1)^n t^n + \cdots)\,dt;$$

$$\ln(1+x) = x - \frac{1}{2}x^2 + \frac{1}{3}x^3 - \frac{1}{4}x^4 + \cdots + \frac{(-1)^{n+1}}{n}x^n + \cdots \qquad \textbf{(19)}$$

if $|x| < 1$. ◆

EXAMPLE 11 Find a power series representation for the arctangent function.

Solution Because $D_t \tan^{-1} t = 1/(1+t^2)$, termwise integration of the geometric series

$$\frac{1}{1+t^2} = 1 - t^2 + t^4 - t^6 + t^8 - \cdots$$

gives

$$\tan^{-1} x = \int_0^x \frac{1}{1+t^2}\,dt = \int_0^x (1 - t^2 + t^4 - t^6 + t^8 - \cdots)\,dt$$

if x is in the interval $(-1, 1)$ where the geometric series converges. Therefore

$$\tan^{-1} x = \sum_{n=1}^{\infty} \frac{(-1)^{n+1} x^{2n-1}}{2n-1} = x - \tfrac{1}{3}x^3 + \tfrac{1}{5}x^5 - \tfrac{1}{7}x^7 + \tfrac{1}{9}x^9 - \cdots \qquad \textbf{(20)}$$

if $-1 < x < 1$. Figure 10.8.6 illustrates both the convergence of the power series within this interval and the divergence outside it. ◆

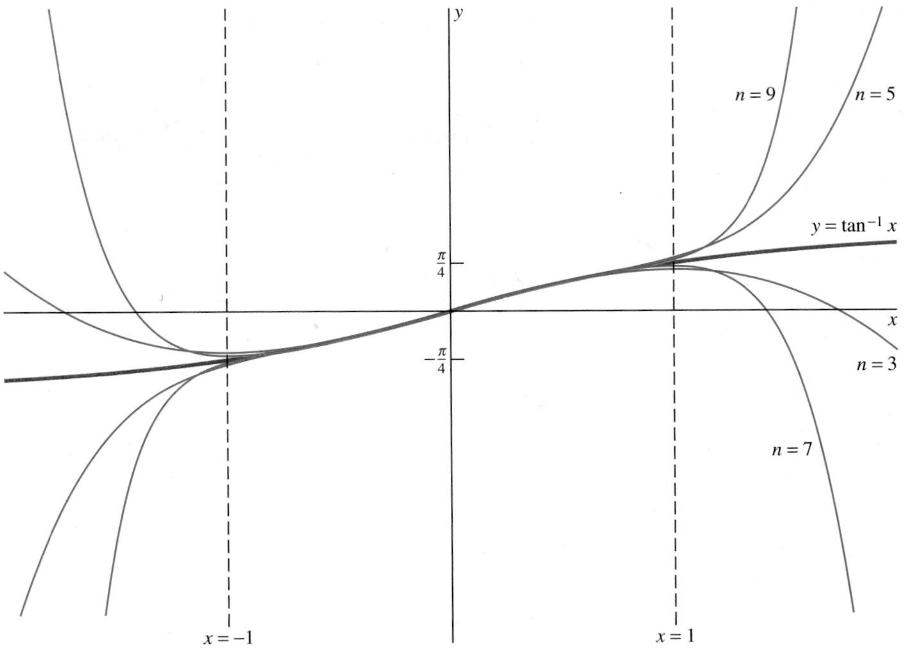

FIGURE 10.8.6 The graphs of the Taylor polynomials of degrees $n = 3, 5, 7$, and 9 illustrate the convergence within the interval $-1 < x < 1$ and divergence outside this interval.

EXAMPLE 12 Find a power series representation for the arcsine function.

Solution First we substitute t^2 for x in Eq. (16). This yields

$$\frac{1}{\sqrt{1-t^2}} = 1 + \sum_{n=1}^{\infty} \frac{1 \cdot 3 \cdot 5 \cdots (2n-1)}{2 \cdot 4 \cdot 6 \cdots (2n)} t^{2n}$$

if $|t| < 1$. Because $D_t \sin^{-1} t = 1/\sqrt{1-t^2}$, termwise integration of this series from $t = 0$ to $t = x$ gives

$$\sin^{-1} x = \int_0^x \frac{1}{\sqrt{1-t^2}}\,dt = x + \sum_{n=1}^{\infty} \frac{1 \cdot 3 \cdot 5 \cdots (2n-1)}{2 \cdot 4 \cdot 6 \cdots (2n)} \cdot \frac{x^{2n+1}}{2n+1} \tag{21}$$

if $|x| < 1$. Problem 68 shows how to use this series to derive the series

$$\frac{\pi^2}{6} = 1 + \frac{1}{2^2} + \frac{1}{3^2} + \frac{1}{4^2} + \cdots + \frac{1}{n^2} + \cdots,$$

which we used in Example 3 of Section 10.5 to approximate the number π. ◆

10.8 TRUE/FALSE STUDY GUIDE

10.8 CONCEPTS: QUESTIONS AND DISCUSSION

1. Suppose that you started with the Maclaurin series of the sine and cosine functions as their definitions. How many of the familiar properties of $\cos x$ and $\sin x$—such as their derivatives and addition formulas—could you establish using only these series?
2. Use the Maclaurin series for the sine and cosine functions and the corresponding hyperbolic series in Example 6 to explore relations between function pairs trig ix and trigh x, where trig denotes one of the cos/sin/tan trigonometric functions, and trigh denotes the corresponding hyperbolic function.

10.8 PROBLEMS

Find the interval of convergence of each power series in Problems 1 through 30.

1. $\sum_{n=1}^{\infty} nx^n$

2. $\sum_{n=1}^{\infty} \frac{x^n}{\sqrt{n}}$

3. $\sum_{n=1}^{\infty} \frac{nx^n}{2^n}$

4. $\sum_{n=1}^{\infty} \frac{(-1)^n x^n}{n^{1/2}5^n}$

5. $\sum_{n=1}^{\infty} n!x^n$

6. $\sum_{n=1}^{\infty} \frac{(-1)^n x^n}{n^n}$

7. $\sum_{n=1}^{\infty} \frac{3^n x^n}{n^3}$

8. $\sum_{n=1}^{\infty} \frac{(-4)^n x^n}{\sqrt{2n+1}}$

9. $\sum_{n=1}^{\infty} (-1)^n n^{1/2}(2x)^n$

10. $\sum_{n=1}^{\infty} \frac{n^2 x^n}{3n-1}$

11. $\sum_{n=1}^{\infty} \frac{(-1)^n nx^n}{2^n(n+1)^3}$

12. $\sum_{n=1}^{\infty} \frac{n^{10}x^n}{10^n}$

13. $\sum_{n=1}^{\infty} \frac{(\ln n)x^n}{3^n}$

14. $\sum_{n=2}^{\infty} \frac{(-1)^n 4^n x^n}{n \ln n}$

15. $\sum_{n=0}^{\infty} (5x-3)^n$

16. $\sum_{n=1}^{\infty} \frac{(2x-1)^n}{n^4+16}$

17. $\sum_{n=1}^{\infty} \frac{2^n(x-3)^n}{n^2}$

18. $\sum_{n=1}^{\infty} \frac{n!}{n^n}x^n$ (Do not test the endpoints; the series diverges at each.)

19. $\sum_{n=1}^{\infty} \frac{(2n)!}{n!}x^n$

20. $\sum_{n=1}^{\infty} \frac{1 \cdot 3 \cdot 5 \cdots (2n+1)}{n!}x^n$ (Do not test the endpoints; the series diverges at each.)

21. $\sum_{n=1}^{\infty} \frac{n^3(x+1)^n}{3^n}$

22. $\sum_{n=1}^{\infty} \frac{(-1)^{n+1}(x-2)^n}{n^2}$

23. $\sum_{n=1}^{\infty} \frac{(3-x)^n}{n^3}$

24. $\sum_{n=1}^{\infty} \frac{(-1)^{n+1}10^n}{n!}(x-10)^n$

25. $\sum_{n=1}^{\infty} \frac{n!}{2^n}(x-5)^n$

26. $\sum_{n=1}^{\infty} \frac{(-1)^{n+1}}{n \cdot 10^n}(x-2)^n$

27. $\sum_{n=0}^{\infty} x^{(2^n)}$

28. $\sum_{n=0}^{\infty} \left(\frac{x^2+1}{5}\right)^n$

29. $\sum_{n=1}^{\infty} \frac{(-1)^n x^n}{1 \cdot 3 \cdot 5 \cdots (2n-1)}$

30. $\sum_{n=1}^{\infty} \frac{1 \cdot 3 \cdot 5 \cdots (2n-1)}{2 \cdot 5 \cdot 8 \cdots (3n-1)}x^n$

In Problems 31 through 42, use power series established in this section to find a power series representation of the given function. Then determine the radius of convergence of the resulting series.

31. $f(x) = \frac{x}{1-x}$

32. $f(x) = \frac{1}{10+x}$

33. $f(x) = x^2 e^{-3x}$

34. $f(x) = \frac{x}{9-x^2}$

35. $f(x) = \sin(x^2)$

36. $f(x) = \cos^2 2x = \frac{1}{2}(1+\cos 4x)$

37. $f(x) = \sqrt[3]{1-x}$

38. $f(x) = (1+x^2)^{3/2}$

39. $f(x) = (1+x)^{-3}$

40. $f(x) = \frac{1}{\sqrt{9+x^3}}$

41. $f(x) = \frac{\ln(1+x)}{x}$

42. $f(x) = \frac{x - \arctan x}{x^3}$

In Problems 43 through 48, find a power series representation for the given function $f(x)$ by using termwise integration.

43. $f(x) = \int_0^x \sin t^3 \, dt$

44. $f(x) = \int_0^x \frac{\sin t}{t} \, dt$

45. $f(x) = \int_0^x \exp(-t^3) \, dt$

46. $f(x) = \int_0^x \frac{\arctan t}{t} \, dt$

47. $f(x) = \int_0^x \frac{1-\exp(-t^2)}{t^2} \, dt$

48. $\tanh^{-1} x = \int_0^x \frac{1}{1-t^2} \, dt$

Beginning with the geometric series $\sum_{n=0}^{\infty} x^n$ as in Example 9, differentiate termwise to find the sums (for $|x| < 1$) of the power series in Problems 49 through 51.

49. $\sum_{n=1}^{\infty} nx^n$

50. $\sum_{n=1}^{\infty} n(n-1)x^n$

51. $\sum_{n=1}^{\infty} n^2 x^n$

52. Use the power series of the preceding problems to sum the numerical series

$$\sum_{n=1}^{\infty} \frac{n}{2^n} \quad \text{and} \quad \sum_{n=1}^{\infty} \frac{n^2}{3^n}.$$

53. Verify by termwise differentiation of its Maclaurin series that the exponential function $y = e^x$ satisfies the differential equation $dy/dx = y$. (Thus the exponential series arises naturally as a power series that is its own termwise derivative.)

54. Verify by termwise differentiation of their Maclaurin series that the sine function $y = \sin x$ and the cosine function $y = \cos x$ both satisfy the differential equation

$$\frac{d^2y}{dx^2} + y = 0.$$

55. Verify by termwise differentiation of the hyperbolic sine and cosine series in Example 6 that each of the functions $\cosh x$ and $\sinh x$ is the derivative of the other, and that each satisfies the differential equation $y'' - y = 0$.

56. In elementary mathematics one sees various definitions (some circular!) of the trigonometric functions. One approach to a rigorous foundation for these functions is to begin by defining $\cos x$ and $\sin x$ by means of their Maclaurin series. For instance, never having heard of sine, cosine, or the number π, we might define the function

$$S(x) = \sum_{n=1}^{\infty} \frac{(-1)^{n-1} x^{2n-1}}{(2n-1)!}$$

and verify using the ratio test that this series converges for all x. Use a computer algebra system to plot graphs of high-degree partial sums $s_n(x)$ of this series. Does it appear that the function $S(x)$ appears to have a zero somewhere near the number 3? Solve the equation $s_n(x) = 0$ numerically (for some large values of n) to verify that this least positive zero of the sine function is approximately 3.14159 (and thus the famous number π makes a fresh new appearance).

57. The Bessel function of order 1 is defined by

$$J_1(x) = \sum_{n=0}^{\infty} \frac{(-1)^n x^{2n+1}}{2^{2n+1} n!(n+1)!} = \frac{x}{2} - \frac{x^3}{16} + \frac{x^5}{384} - \cdots.$$

Verify that this series converges for all x and that the derivative of the Bessel function of order zero is given by $J_0'(x) = -J_1(x)$. Are the graphs in Fig. 10.8.7 consistent with this latter fact?

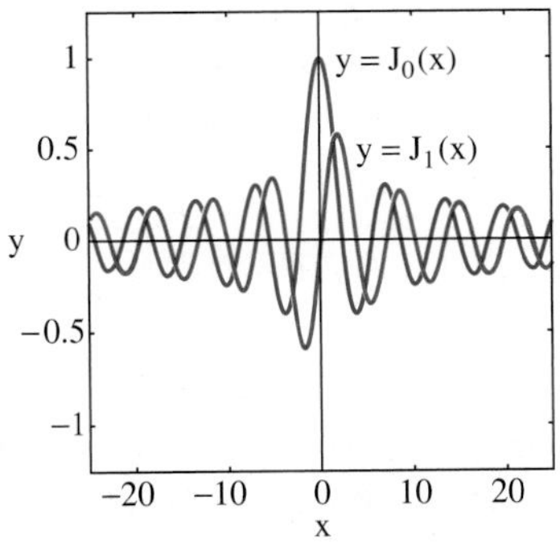

FIGURE 10.8.7 Graphs of the Bessel functions $J_0(x)$ and $J_1(x)$. Note that their zeros are interlaced, like the zeros of the cosine and sine functions.

58. Verify by termwise integration that

$$\int x J_0(x)\,dx = x J_1(x) + C.$$

59. Bessel's equation of order n is the second-order differential equation

$$x^2 y'' + x y' + (x^2 - n^2) y = 0.$$

Verify by termwise differentiation that $y = J_0(x)$ satisfies Bessel's equation of order zero.

60. Verify that $y = J_1(x)$ satisfies Bessel's equation of order 1 (Problem 59).

61. First use the sine series to find the Taylor series of $f(x) = (\sin x)/x$. Then use a graphing calculator or computer to illustrate the approximation of $f(x)$ by its Taylor polynomials with center $a = 0$.

62. First find the Taylor series of the function

$$g(x) = \int_0^x \frac{\sin t}{t}\,dt.$$

Then determine where this power series converges. Finally, use a graphing calculator or computer to illustrate the approximation of $g(x)$ by its Taylor polynomials with center $a = 0$.

63. Deduce from the arctangent series (Example 11) that

$$\pi = \frac{6}{\sqrt{3}} \sum_{n=0}^{\infty} \frac{(-1)^n}{2n+1} \left(\frac{1}{3}\right)^n.$$

Then use this alternating series to show that $\pi = 3.14$ accurate to two decimal places.

64. Substitute the Maclaurin series for $\sin x$, and then assume the validity of termwise integration of the resulting series, to derive the formula

$$\int_0^{\infty} e^{-t} \sin xt\,dt = \frac{x}{1+x^2} \qquad (|x| < 1).$$

Use the fact from Section 7.8 that

$$\int_0^{\infty} t^n e^{-t}\,dt = \Gamma(n+1) = n!.$$

65. (a) Deduce from the Maclaurin series for e^t that

$$\frac{1}{x^x} = \sum_{n=0}^{\infty} \frac{(-1)^n}{n!} (x \ln x)^n.$$

(b) Assuming the validity of termwise integration of the series in part (a), use the integral formula of Problem 53 in Section 7.8 to conclude that

$$\int_0^1 \frac{1}{x^x}\,dx = \sum_{n=1}^{\infty} \frac{1}{n^n}.$$

66. Suppose that $f(x)$ is represented by the power series

$$\sum_{n=0}^{\infty} a_n x^n$$

for all x in some open interval centered at $x = 0$. Show by repeated differentiation of the series, substituting $x = 0$ after each differentiation, that $a_n = f^{(n)}(0)/n!$ for all $n \geqq 0$. Thus the only power series in x that represents a function at and near $x = 0$ is its Maclaurin series.

67. (a) Consider the binomial series

$$f(x) = \sum_{n=0}^{\infty} \frac{\alpha(\alpha-1)(\alpha-2)\cdots(\alpha-n+1)}{n!} x^n,$$

which converges (to *something*) if $|x| < 1$. Compute the derivative $f'(x)$ by termwise differentiation, and show that it satisfies the differential equation $(1+x)f'(x) = \alpha f(x)$. (b) Solve the differential equation in part (a) to obtain $f(x) = C(1+x)^\alpha$ for some constant C. Finally, show that $C = 1$. Thus the binomial series converges to $(1+x)^\alpha$ if $|x| < 1$.

68. (a) Show by direct integration that

$$\int_0^1 \frac{\arcsin x}{\sqrt{1-x^2}}\,dx = \frac{\pi^2}{8}.$$

(b) Use the result of Problem 58 in Section 7.3 to show that

$$\int_0^1 \frac{x^{2n+1}}{\sqrt{1-x^2}}\,dx = \frac{2\cdot 4\cdot 6\cdots(2n)}{1\cdot 3\cdot 5\cdots(2n+1)}.$$

(c) Substitute the series of Example 10 for $\arcsin x$ into the integral of part (a); then use the integral of part (b) to integrate termwise. Conclude that

$$\int_0^1 \frac{\arcsin x}{\sqrt{1-x^2}}\,dx = 1 + \frac{1}{3^2} + \frac{1}{5^2} + \frac{1}{7^2} + \cdots.$$

(d) Note that

$$\sum_{n=1}^{\infty} \frac{1}{n^2} = \sum_{n=1}^{\infty} \frac{1}{(2n-1)^2} + \sum_{n=1}^{\infty} \frac{1}{(2n)^2}.$$

Use this information and parts (a) and (c) to show that

$$\sum_{n=1}^{\infty} \frac{1}{n^2} = \frac{\pi^2}{6}.$$

69. Prove that if the power series $\sum a_n x^n$ converges for some $x = x_0 \neq 0$, then it converges absolutely for all x such that $|x| < |x_0|$. [*Suggestion:* Conclude from the fact that $\lim_{n\to\infty} a_n x_0^n = 0$ that $|a_n x^n| \leqq |x/x_0|^n$ for all n sufficiently large. Thus the series $\sum |a_n x^n|$ is eventually dominated by the geometric series $\sum |x/x_0|^n$, which converges if $|x| < |x_0|$.]

70. Suppose that the power series $\sum a_n x^n$ converges for some but not all nonzero values of x. Let S be the set of real numbers for which the series converges absolutely. (a) Conclude from Problem 69 that the set S is bounded above. (b) Let λ be the least upper bound of the set S. (See Problem 61 of Section 10.2.) Then show that $\sum a_n x^n$ converges absolutely if $|x| < \lambda$ and diverges if $|x| > \lambda$. Explain why this proves Theorem 1 without the additional hypothesis that $\lim_{n\to\infty} |a_{n+1}/a_n|$ exists.

10.9 POWER SERIES COMPUTATIONS

Power series often are used to approximate numerical values of functions and integrals. *Alternating* power series (such as the sine and cosine series) are especially common and useful. Recall the alternating series remainder (or "error") estimate of Theorem 2 in Section 10.7. It applies to a convergent alternating series $\sum(-1)^{n+1}a_n$ whose terms are decreasing (so $a_n > a_{n+1}$ for every n). If we write

$$\sum_{k=1}^{\infty} (-1)^{k+1} a_k = (a_1 - a_2 + a_3 - \cdots \pm a_n) + E, \tag{1}$$

then $E = \mp a_{n+1} \pm a_{n+2} \mp a_{n+3} \pm \cdots$ is the error made when the series is *truncated*—the terms following $(-1)^{n+1}a_n$ are simply chopped off and discarded, and the n-term partial sum is used in place of the actual sum of the whole series. The remainder estimate then says that the error E has the same sign as the first term not retained, and is less in magnitude than this first neglected term; that is, $|E| < a_{n+1}$.

EXAMPLE 1 Use the first four terms of the binomial series

$$\sqrt{1+x} = 1 + \tfrac{1}{2}x - \tfrac{1}{8}x^2 + \tfrac{1}{16}x^3 - \tfrac{5}{128}x^4 + \cdots \tag{2}$$

to estimate the number $\sqrt{105}$ and to estimate the accuracy in the approximation.

Solution If $x > 0$ then the binomial series is, after the first term, an alternating series. In order to match the pattern on the left-hand side in Eq. (2), we first write

$$\sqrt{105} = \sqrt{100+5} = 10\sqrt{1+\tfrac{5}{100}} = 10\sqrt{1+0.05}.$$

Then with $x = 0.05$ the series in (2) gives

$$\begin{aligned}\sqrt{105} &= 10\left[1 + \tfrac{1}{2}(0.05) - \tfrac{1}{8}(0.05)^2 + \tfrac{1}{16}(0.05)^3 + E\right]\\ &= 10\,[1.02469531 + E] = 10.2469531 + 10E.\end{aligned}$$

Note that the error $10E$ in our approximation $\sqrt{105} \approx 10.2469531$ is 10 times the error E in the truncated series itself. It follows from the remainder estimate that E

is negative and that

$$|10E| < 10 \cdot \tfrac{5}{128}(0.05)^4 \approx 0.0000024.$$

Therefore,

$$10.2469531 - 0.0000024 = 10.2469507 < \sqrt{105} < 10.2469531,$$

so it follows that $\sqrt{105} \approx 10.24695$ rounded accurate to five decimal places. ◆

REMARK Suppose that we had been asked in advance to approximate $\sqrt{105}$ accurate to five decimal places. A convenient way to do this is to continue writing terms of the series until it is clear that they have become too small in magnitude to affect the fifth decimal place. A good rule of thumb is to use two more decimal places in the computations than are required in the final answer. Thus we use seven decimal places in this case and get

$$\begin{aligned}\sqrt{105} &= 10 \cdot (1 + 0.05)^{1/2}\\ &\approx 10 \cdot (1 + 0.025 - 0.0003125 + 0.0000078 - 0.0000002 + \cdots)\\ &\approx 10.246951 \approx 10.24695.\end{aligned}$$

EXAMPLE 2 Figure 10.9.1 shows the graph of the function $f(x) = (\sin x)/x$. Approximate (accurate to three decimal places) the area

$$A = \int_{-\pi}^{\pi} \frac{\sin x}{x}\,dx = 2\int_0^{\pi} \frac{\sin x}{x}\,dx \tag{3}$$

of the shaded region lying under the "principal arch" from $x = -\pi$ to π.

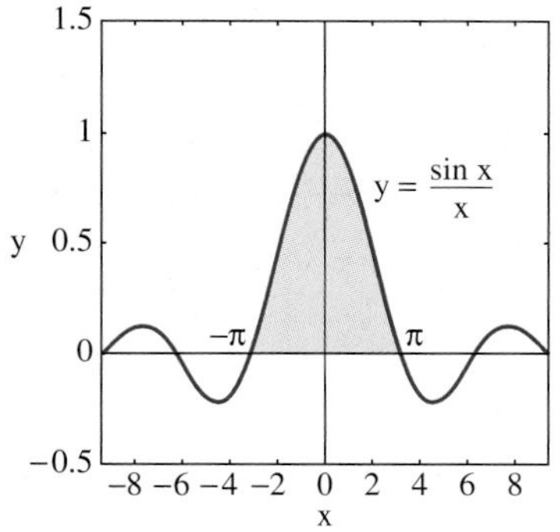

FIGURE 10.9.1 The graph $y = \dfrac{\sin x}{x}$ of Example 2.

Solution When we substitute the Taylor series for $\sin x$ in Eq. (3) and integrate termwise, we get

$$\begin{aligned}A &= 2\int_0^{\pi} \frac{1}{x}\left(x - \frac{x^3}{3!} + \frac{x^5}{5!} - \frac{x^7}{7!} + \cdots\right)dx\\ &= 2\int_0^{\pi}\left(1 - \frac{x^2}{3!} + \frac{x^4}{5!} - \frac{x^6}{7!} + \cdots\right)dx\\ &= 2\left[x - \frac{x^3}{3!3} + \frac{x^5}{5!5} - \frac{x^7}{7!7} + \cdots\right]_0^{\pi},\end{aligned}$$

and thus

$$A = 2\pi - \frac{2\pi^3}{3!3} + \frac{2\pi^5}{5!5} - \frac{2\pi^7}{7!7} + \frac{2\pi^9}{9!9} - \frac{2\pi^{11}}{11!11} + \cdots.$$

Following the "+2 rule of thumb" and retaining five decimal places, we calculate

$$A = 6.28319 - 3.44514 + 1.02007 - 0.17122 + 0.01825 - 0.00134 + 0.00007 - \cdots.$$

The sum of the first six terms gives $A \approx 3.70381$. Because we are summing an alternating series, the error in this approximation is positive and less than the next term 0.00007. Neglecting possible roundoff in the last place, we would conclude that $3.70381 < A < 3.70388$. Thus $A \approx 3.704$ rounded accurate to three decimal places. ◆

The Algebra of Power Series

Theorem 1, which we state without proof, implies that power series may be added and multiplied much like polynomials. The guiding principle is that of collecting coefficients of like powers of x.

THEOREM 1 Adding and Multiplying Power Series
Let $\sum a_n x^n$ and $\sum b_n x^n$ be power series with nonzero radii of convergence. Then

$$\sum_{n=0}^{\infty} a_n x^n + \sum_{n=0}^{\infty} b_n x^n = \sum_{n=0}^{\infty} (a_n + b_n) x^n \tag{4}$$

and

$$\left(\sum_{n=0}^{\infty} a_n x^n\right)\left(\sum_{n=0}^{\infty} b_n x^n\right) = \sum_{n=0}^{\infty} c_n x^n$$

$$= a_0 b_0 + (a_0 b_1 + a_1 b_0)x + (a_0 b_2 + a_1 b_1 + a_2 b_0)x^2 + \cdots, \tag{5}$$

where

$$c_n = a_0 b_n + a_1 b_{n-1} + a_2 b_{n-2} + \cdots + a_{n-1} b_1 + a_n b_0. \tag{6}$$

The series in Eqs. (4) and (5) converge for any x that lies interior to the intervals of convergence of both $\sum a_n x^n$ and $\sum b_n x^n$.

Thus if $\sum a_n x^n$ and $\sum b_n x^n$ are power series representations of the functions $f(x)$ and $g(x)$, respectively, then the product power series $\sum c_n x^n$ found by "ordinary multiplication" and collection of terms is a power series representation of the product function $f(x)g(x)$. This fact can also be used to divide one power series by another, *provided* that the quotient is known to have a power series representation.

EXAMPLE 3 Assume that the tangent function has a power series representation $\tan x = \sum a_n x^n$ (it does). Use the Maclaurin series for $\sin x$ and $\cos x$ to find a_0, a_1, a_2, and a_3.

Solution We multiply series to obtain

$$\sin x = \tan x \cos x$$
$$= (a_0 + a_1 x + a_2 x^2 + a_3 x^3 + \cdots)\left(1 - \frac{x^2}{2} + \frac{x^4}{24} - \cdots\right).$$

If we multiply each term in the first factor by each term in the second, then collect coefficients of like powers, the result is

$$\sin x = a_0 + a_1 x + \left(a_2 - \tfrac{1}{2}a_0\right)x^2 + \left(a_3 - \tfrac{1}{2}a_1\right)x^3 + \cdots.$$

But because

$$\sin x = x - \tfrac{1}{6}x^3 + \tfrac{1}{120}x^5 - \cdots,$$

comparison of coefficients gives the equations

$$\begin{aligned} a_0 \qquad\qquad\qquad &= 0, \\ a_1 \qquad\qquad &= 1, \\ -\tfrac{1}{2}a_0 \qquad + a_2 \qquad &= 0, \\ -\tfrac{1}{2}a_1 \qquad + a_3 &= -\tfrac{1}{6}. \end{aligned}$$

Thus we find that $a_0 = 0$, $a_1 = 1$, $a_2 = 0$, and $a_3 = \frac{1}{3}$. So

$$\tan x = x + \tfrac{1}{3}x^3 + \cdots.$$

Things are not always as they first appear. A computer algebra system gives the continuation

$$\tan x = x + \tfrac{1}{3}x^3 + \tfrac{2}{15}x^5 + \tfrac{17}{315}x^7 + \tfrac{62}{2835}x^9 + \tfrac{1382}{155{,}925}x^{11} + \cdots \tag{7}$$

of the tangent series. For the general form of the nth coefficient, see K. Knopp's *Theory and Application of Infinite Series* (New York: Hafner Press, 1971), p. 204.

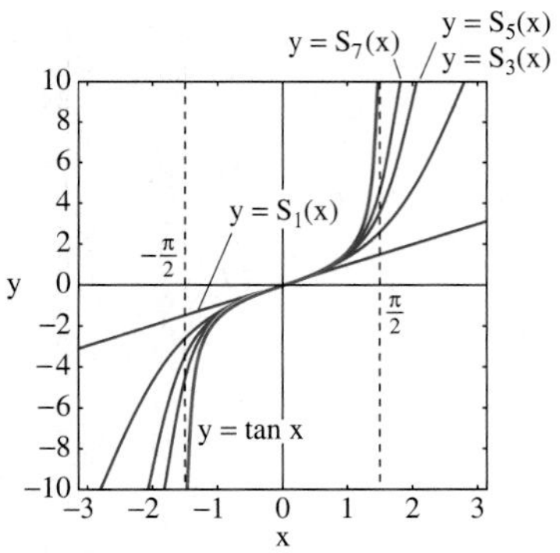

FIGURE 10.9.2 The graphs of $y = \tan x$ and the first four partial sums of the power series in (7).

You may also check that the first few terms agree with the result of ordinary division of the Maclaurin series for $\cos x$ into the Maclaurin series for $\sin x$:

$$\begin{array}{r|l} & x + \frac{1}{3}x^3 + \frac{2}{15}x^5 + \cdots \\ \hline 1 - \frac{1}{2}x^2 + \frac{1}{24}x^4 - \cdots & x - \frac{1}{6}x^3 + \frac{1}{120}x^5 - \cdots \end{array}$$

Figure 10.9.2 shows the approximation of the tangent function (on $-\pi/2 < x < \pi/2$) by the first four odd-degree polynomial partial sums corresponding to the terms exhibited in Eq. (7). Evidently these polynomial approximations have difficulty "keeping up" with $\tan x$ as it approaches $\pm\infty$ as $x \to \pm\pi/2$. ◆

Power Series and Indeterminate Forms

According to Theorem 3 of Section 10.8, a power series is differentiable and therefore continuous within its interval of convergence. It follows that

$$\lim_{x \to c} \sum_{n=0}^{\infty} a_n (x - c)^n = a_0. \quad \textbf{(8)}$$

Examples 4 and 5 illustrate the use of this simple observation to find the limit of the indeterminate form $f(x)/g(x)$. The technique is to first substitute power series representations for $f(x)$ and $g(x)$.

EXAMPLE 4 Find $\displaystyle\lim_{x \to 0} \frac{\sin x - \arctan x}{x^2 \ln(1 + x)}$.

Solution The power series of Eqs. (4), (19), and (20) in Section 10.8 give

$$\begin{aligned} \sin x - \arctan x &= \left(x - \tfrac{1}{6}x^3 + \tfrac{1}{120}x^5 - \cdots\right) - \left(x - \tfrac{1}{3}x^3 + \tfrac{1}{5}x^5 - \cdots\right) \\ &= \tfrac{1}{6}x^3 - \tfrac{23}{120}x^5 + \cdots \end{aligned}$$

and

$$x^2 \ln(1 + x) = x^2 \cdot \left(x - \tfrac{1}{2}x^2 + \tfrac{1}{3}x^3 + \cdots\right) = x^3 - \tfrac{1}{2}x^4 + \tfrac{1}{3}x^5 - \cdots.$$

Hence

$$\begin{aligned} \lim_{x \to 0} \frac{\sin x - \arctan x}{x^2 \ln(1 + x)} &= \lim_{x \to 0} \frac{\frac{1}{6}x^3 - \frac{23}{120}x^5 + \cdots}{x^3 - \frac{1}{2}x^4 + \cdots} \\ &= \lim_{x \to 0} \frac{\frac{1}{6} - \frac{23}{120}x^2 + \cdots}{1 - \frac{1}{2}x + \cdots} = \frac{1}{6}. \end{aligned}$$ ◆

EXAMPLE 5 Find $\displaystyle\lim_{x \to 1} \frac{\ln x}{x - 1}$.

Solution We first replace x with $x - 1$ in the power series for $\ln(1 + x)$ used in Example 4. [Equation (8) makes it clear that this method requires all series to have center c if we are taking limits as $x \to c$.] This gives us

$$\ln x = (x - 1) - \tfrac{1}{2}(x - 1)^2 + \tfrac{1}{3}(x - 1)^3 - \cdots.$$

Hence

$$\begin{aligned} \lim_{x \to 1} \frac{\ln x}{x - 1} &= \lim_{x \to 1} \frac{(x - 1) - \frac{1}{2}(x - 1)^2 + \frac{1}{3}(x - 1)^3 - \cdots}{x - 1} \\ &= \lim_{x \to 1} \left[1 - \tfrac{1}{2}(x - 1) + \tfrac{1}{3}(x - 1)^2 - \cdots\right] = 1. \end{aligned}$$ ◆

The method of Examples 4 and 5 provides a useful alternative to l'Hôpital's rule, especially when repeated differentiation of numerator and denominator is inconvenient or too time-consuming. (See Problems 59 and 60.)

Numerical and Graphical Error Estimation

The following examples show how to investigate the accuracy in a power-series partial-sum approximation for a specified interval of values of x. We will take the statement that a given approximation is "accurate to p decimal places" to mean that its error E is numerically less than half a unit in the pth decimal place; that is, that $|E| < 0.5 \times 10^{-p}$. For instance, four-place accuracy means that $|E| < 0.00005$. (Note that $p = 4$ is the number of zeros here.) Nevertheless, we should remember that in some cases a result accurate to within a half unit in the pth place may round "the wrong way," so that the result rounded to p places may still be in error by a unit in the pth decimal place (as in Problem 12).

EXAMPLE 6 Consider the polynomial approximation

$$\sin x \approx x - \frac{x^3}{3!} + \frac{x^5}{5!} - \cdots + (-1)^{n+1}\frac{x^{2n-1}}{(2n-1)!} \tag{9}$$

obtained by truncating the alternating Taylor series of the sine function.

(a) How accurate is the cubic approximation $P_3(x) \approx x - x^3/3!$ for angles from $0°$ to $10°$? Use this approximation to estimate $\sin 10°$.

(b) How many terms in (9) are needed to guarantee six-place accuracy in calculating $\sin x$ for angles from $0°$ to $45°$? Use the corresponding polynomial to approximate $\sin 30°$ and $\sin 40°$.

(c) For what values of x does the fifth-degree approximation yield five-place accuracy?

Solution (a) Of course we must substitute x in radians in (9), so we deal here with values of x in the interval $0 \leqq x \leqq \pi/18$. For any such x, the error E is positive (Why?) and is bounded by the magnitude of the next term:

$$|E| < \frac{x^5}{5!} \leqq \frac{(\pi/18)^5}{5!} \approx 0.00000135 < 0.000005.$$

We count five zeros on the right, and thus we have five-place accuracy. For instance, substituting $x = \pi/18$ in the cubic polynomial $P_3(x)$ gives

$$\sin 10° = \sin\left(\frac{\pi}{18}\right) \approx \frac{\pi}{18} - \frac{1}{3!}\cdot\left(\frac{\pi}{18}\right)^3$$
$$\approx 0.1736468 \approx 0.17365.$$

This five-place approximation $\sin 10° \approx 0.17365$ is correct, because the actual seven-place value of $\sin 10°$ is $0.1736482 \approx 0.17365$.

Solution (b) For any x in the interval $0 \leqq x \leqq \pi/4$, the error E made if we use the polynomial value in (9) in place of the actual value $\sin x$ is bounded by the first neglected term,

$$|E| < \frac{x^{2n+1}}{(2n+1)!} \leqq \frac{(\pi/4)^{2n+1}}{(2n+1)!}.$$

The table in Fig. 10.9.3 shows calculator values for $n = 1, 2, 3, \ldots$ of this maximal error (rounded to eight decimal places). For six-place accuracy we want $|E| < 0.0000005$, so we see that $n = 4$ will suffice. We therefore use the seventh-degree Taylor polynomial

$$P_7(x) = x - \frac{x^3}{3!} + \frac{x^5}{5!} - \frac{x^7}{7!} \tag{10}$$

to approximate $\sin x$ for $0 \leqq x \leqq \pi/4$. With $x = \pi/6$ we get

$$\sin 30° \approx \frac{\pi}{6} - \frac{(\pi/6)^3}{3!} + \frac{(\pi/6)^5}{5!} - \frac{(\pi/6)^7}{7!} \approx 0.49999999 \approx \frac{1}{2},$$

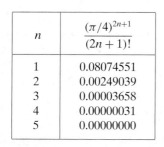

n	$\dfrac{(\pi/4)^{2n+1}}{(2n+1)!}$
1	0.08074551
2	0.00249039
3	0.00003658
4	0.00000031
5	0.00000000

FIGURE 10.9.3 Estimating the error in Example 6(b).

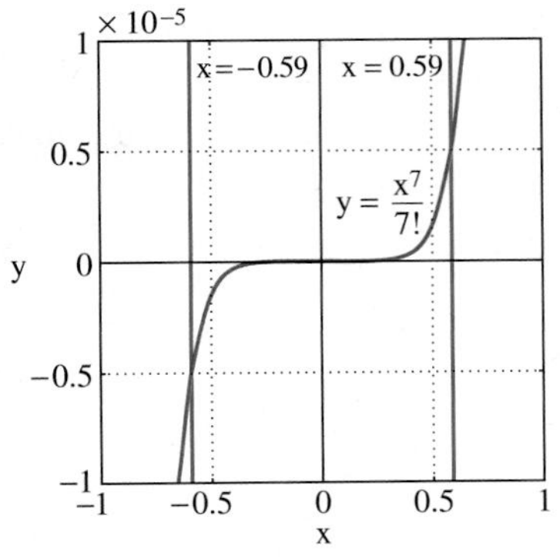

FIGURE 10.9.4 The graph of the maximal error $y = \frac{x^7}{7!}$ in Example 6(c).

as expected. Substituting $x = 2\pi/9$ in (10) similarly gives $\sin 40° \approx 0.64278750$, whereas the actual eight-place value of $\sin 40°$ is $0.64278761 \approx 0.642788$.

Solution (c) The fifth-degree approximation

$$\sin x \approx P_5(x) = x - \frac{x^3}{3!} + \frac{x^5}{5!} \qquad \textbf{(11)}$$

gives five-place accuracy when x is such that the error E satisfies the inequality

$$|E| < \frac{|x|^7}{7!} = \frac{|x|^7}{5040} \leqq 0.000005;$$

that is, when $|x| \leqq [(5040) \cdot (0.000005)]^{1/7} \approx 0.5910$ (radians). In degrees, this corresponds to angles between $-33.86°$ and $+33.86°$. In Fig. 10.9.4 the graph of $y = x^7/7!$ in the viewing window $-1 \leqq x \leqq 1$, $-0.00001 \leqq y \leqq 0.00001$ provides visual corroboration of this analysis—we see clearly that $x^7/7!$ remains between -0.000005 and 0.000005 when x is between -0.59 and 0.59. ◆

EXAMPLE 7 Suppose now that we want to approximate $f(x) = \sin x$ with three-place accuracy on the whole interval from 0° to 90°. Now it makes sense to begin with a Taylor series centered at the midpoint $x = \pi/4$ of the interval. Because the function $f(x)$ and its successive derivatives are $\sin x$, $\cos x$, $-\sin x$, $-\cos x$, and so forth, their values at $x = \pi/4$ are $\frac{1}{2}\sqrt{2}$, $\frac{1}{2}\sqrt{2}$, $-\frac{1}{2}\sqrt{2}$, $-\frac{1}{2}\sqrt{2}$, and so forth. Consequently Taylor's formula with remainder (Section 10.4) for $f(x) = \sin x$ centered at $x = \pi/4$ takes the form

$$\sin x = \frac{\sqrt{2}}{2} \cdot \left[1 + \left(x - \frac{\pi}{4}\right) - \frac{1}{2!}\left(x - \frac{\pi}{4}\right)^2 - \frac{1}{3!}\left(x - \frac{\pi}{4}\right)^3 + \cdots \pm \frac{1}{n!}\left(x - \frac{\pi}{4}\right)^n\right] + E(x) \qquad \textbf{(12)}$$

where

$$|E(x)| = \left|\frac{f^{(n+1)}(z)}{(n+1)!}\left(x - \frac{\pi}{4}\right)^{n+1}\right| \leqq \frac{1}{(n+1)!}\left|x - \frac{\pi}{4}\right|^{n+1} \qquad \textbf{(13)}$$

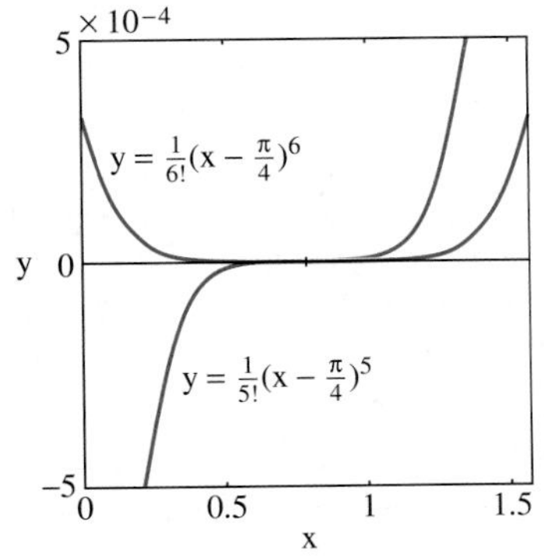

FIGURE 10.9.5 Comparing errors in Example 7.

for some z in the interval $0 \leqq x \leqq \pi/2$. Observe that the corresponding Taylor series is not alternating—if $x > \pi/4$ it has instead a "$++--++--$" pattern of signs—but we can still use the remainder estimate in (13). For three-place accuracy we need to choose n so that $y = E(x)$ remains within the viewing window $-0.0005 \leqq y \leqq 0.0005$ on the whole interval $0 \leqq x \leqq \pi/2$. Looking at the graphs plotted in Fig. 10.9.5, we see that this is so if $n = 5$ but not if $n = 4$. The desired approximation is therefore

$$\sin x \approx \frac{\sqrt{2}}{2} \cdot \left[1 + \left(x - \frac{\pi}{4}\right) - \frac{1}{2!}\left(x - \frac{\pi}{4}\right)^2 - \frac{1}{3!}\left(x - \frac{\pi}{4}\right)^3 + \frac{1}{4!}\left(x - \frac{\pi}{4}\right)^4 + \frac{1}{5!}\left(x - \frac{\pi}{4}\right)^5\right].$$

For instance, substituting $x = 0$ we get $\sin 0° \approx 0.00020 \approx 0.000$ as desired, and $x = \pi/2$ gives $\sin 90° \approx 1.00025 \approx 1.000$. ◆

10.9 TRUE/FALSE STUDY GUIDE

10.9 CONCEPTS: QUESTIONS AND DISCUSSION

1. Outline how you might use the binomial series (as in Example 1) to construct a *table of roots*—perhaps the square roots, cube roots, and fourth roots of the first 100 positive integers.

2. Give your own examples of several integrals for which numerical approximation using series (as in Example 2) would be useful.
3. Give your own examples of several indeterminate forms for which numerical evaluation using series (as in Examples 4 and 5) would be useful.

10.9 PROBLEMS

In Problems 1 through 10, use an infinite series to approximate the indicated number accurate to three decimal places.

1. $\sqrt[3]{65}$
2. $\sqrt[4]{630}$
3. $\sin(0.5)$
4. $e^{-0.2}$
5. $\tan^{-1}(0.5)$
6. $\ln(1.1)$
7. $\sin\left(\dfrac{\pi}{10}\right)$
8. $\cos\left(\dfrac{\pi}{20}\right)$
9. $\sin 10^\circ$
10. $\cos 5^\circ$

In Problems 11 through 22, use power series to approximate the value of the given integrals accurate to four decimal places.

11. $\displaystyle\int_0^1 \frac{\sin x}{x}\,dx$
12. $\displaystyle\int_0^1 \frac{\sin x}{\sqrt{x}}\,dx$
13. $\displaystyle\int_0^{1/2} \frac{\arctan x}{x}\,dx$
14. $\displaystyle\int_0^1 \sin x^2\,dx$
15. $\displaystyle\int_0^{1/10} \frac{\ln(1+x)}{x}\,dx$
16. $\displaystyle\int_0^{1/2} \frac{1}{\sqrt{1+x^4}}\,dx$
17. $\displaystyle\int_0^{1/2} \frac{1-e^{-x}}{x}\,dx$
18. $\displaystyle\int_0^{1/2} \sqrt{1+x^3}\,dx$
19. $\displaystyle\int_0^1 e^{-x^2}\,dx$
20. $\displaystyle\int_0^1 \frac{1-\cos x}{x^2}\,dx$
21. $\displaystyle\int_0^{1/2} \sqrt[3]{1+x^2}\,dx$
22. $\displaystyle\int_0^{1/2} \frac{x}{\sqrt{1+x^3}}\,dx$

In Problems 23 through 28, use power series rather than l'Hôpital's rule to evaluate the given limit.

23. $\displaystyle\lim_{x\to 0} \frac{1+x-e^x}{x^2}$
24. $\displaystyle\lim_{x\to 0} \frac{x-\sin x}{x^3\cos x}$
25. $\displaystyle\lim_{x\to 0} \frac{1-\cos x}{x(e^x-1)}$
26. $\displaystyle\lim_{x\to 0} \frac{e^x-e^{-x}-2x}{x-\arctan x}$
27. $\displaystyle\lim_{x\to 0} \left(\frac{1}{x}-\frac{1}{\sin x}\right)$
28. $\displaystyle\lim_{x\to 1} \frac{\ln(x^2)}{x-1}$

In Problems 29 through 32, calculate the indicated number with the required accuracy using Taylor's formula for an appropriate function centered at the given point $x = a$.

29. $\sin 80^\circ$; $a = \pi/4$, four decimal places

30. $\cos 35^\circ$; $a = \pi/4$, four decimal places

31. $\cos 47^\circ$; $a = \pi/4$, six decimal places

32. $\sin 58^\circ$; $a = \pi/3$, six decimal places

In Problems 33 through 36, determine the number of decimal places of accuracy the given appropriate formula yields for $|x| \leqq 0.1$.

33. $e^x \approx 1 + x + \frac{1}{2}x^2 + \frac{1}{6}x^3 + \frac{1}{24}x^4$

34. $\sin x \approx x - \frac{1}{6}x^3 + \frac{1}{120}x^5$

35. $\ln(1+x) \approx x - \frac{1}{2}x^2 + \frac{1}{3}x^3 - \frac{1}{4}x^4$

36. $\sqrt{1+x} \approx 1 + \frac{1}{2}x - \frac{1}{8}x^2$

37. Show that the approximation in Problem 33 gives the value of e^x accurate to within 0.001 if $|x| \leqq 0.5$. Then calculate $\sqrt[3]{e}$ accurate to two decimal places.

38. For what values of x is the approximation $\sin x \approx x - \frac{1}{6}x^3$ accurate to five decimal places?

39. (a) Show that the values of the cosine function for angles between 40° and 50° can be calculated with five-place accuracy using the approximation

$$\cos x \approx \frac{\sqrt{2}}{2}\left[1 - \left(x - \frac{\pi}{4}\right) - \frac{1}{2}\left(x - \frac{\pi}{4}\right)^2 + \frac{1}{6}\left(x - \frac{\pi}{4}\right)^3\right].$$

(b) Show that this approximation yields eight-place accuracy for angles between 44° and 46°.

40. Extend the approximation in Problem 39 to one that yields the values of $\cos x$ accurate to five decimal places for angles between 30° and 60°.

In Problems 41 through 44, use termwise integration of an appropriate power series to approximate the indicated area or volume accurate to two decimal places.

41. Figure 10.9.1 shows the region that lies between the graph of $y = (\sin x)/x$ and the x-axis from $x = -\pi$ to $x = \pi$. Substitute $\sin^2 x = \frac{1}{2}(1 - \cos 2x)$ to approximate the volume of the solid that is generated by revolving this region around the x-axis.

42. Approximate the area of the region that lies between the graph of $y = (1 - \cos x)/x^2$ and the x-axis from $x = -2\pi$ to $x = 2\pi$ (Fig. 10.9.6).

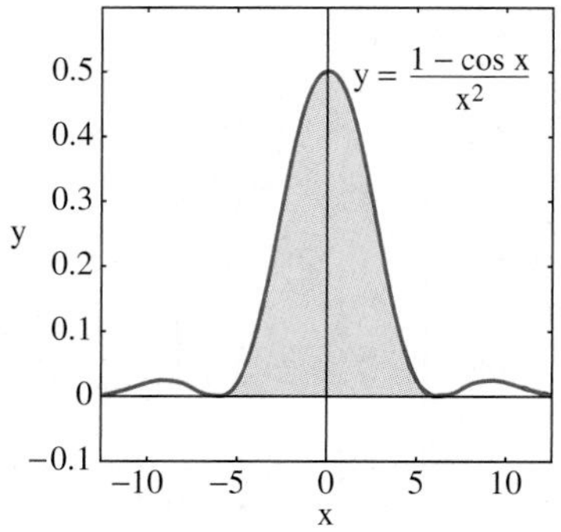

FIGURE 10.9.6 The region of Problem 42.

43. Approximate the volume of the solid generated by rotating the region of Problem 42 around the y-axis.

44. Approximate the volume of the solid generated by rotating the region of Problem 42 around the x-axis.

45. Derive the geometric series by long division of $1-x$ into 1.

46. Derive the series for $\tan x$ listed in Example 3 by long division of the Maclaurin series of $\cos x$ into the Maclaurin series of $\sin x$.

47. Derive the geometric series representation of $1/(1-x)$ by finding $a_0, a_1, a_2, \ldots$ such that

$$(1-x)(a_0 + a_1 x + a_2 x^2 + a_3 x^3 + \cdots) = 1.$$

48. Derive the first five coefficients in the binomial series for $\sqrt{1+x}$ by finding a_0, a_1, a_2, a_3, and a_4 such that

$$(a_0 + a_1 x + a_2 x^2 + a_3 x^3 + a_4 x^4 + \cdots)^2 = 1 + x.$$

49. Use the method of Example 3 to find the coefficients a_0, a_1, a_2, a_3, and a_4 in the series

$$\sec x = \frac{1}{\cos x} = \sum_{n=0}^{\infty} a_n x^n.$$

50. Multiply the geometric series for $1/(1-x)$ and the series for $\ln(1-x)$ to show that if $|x| < 1$, then

$$\frac{\ln(1-x)}{1-x} = -x - \left(1 + \tfrac{1}{2}\right)x^2 - \left(1 + \tfrac{1}{2} + \tfrac{1}{3}\right)x^3 - \left(1 + \tfrac{1}{2} + \tfrac{1}{3} + \tfrac{1}{4}\right)x^4 - \cdots.$$

51. Take as known the logarithmic series

$$\ln(1+x) = x - \tfrac{1}{2}x^2 + \tfrac{1}{3}x^3 - \tfrac{1}{4}x^4 + \cdots.$$

Find the first four coefficients in the series for e^x by finding a_0, a_1, a_2, and a_3 such that

$$1 + x = e^{\ln(1+x)} = \sum_{n=0}^{\infty} a_n \left(x - \tfrac{1}{2}x^2 + \tfrac{1}{3}x^3 - \tfrac{1}{4}x^4 + \cdots\right)^n.$$

This is exactly how the power series for e^x was first discovered (by Newton)!

52. Use the method of Example 3 to show that

$$\frac{x}{\sin x} = 1 + \frac{1}{6}x^2 + \frac{7}{360}x^4 + \cdots.$$

53. Show that long division of power series gives

$$\frac{2+x}{1+x+x^2} = 2 - x - x^2 + 2x^3 - x^4 - x^5 + 2x^6 - x^7 - x^8 + 2x^9 - x^{10} - x^{11} + \cdots.$$

Show also that the radius of convergence of this series is $R = 1$.

54. Use the series in Problem 53 to approximate with two-place accuracy the value of the integral

$$\int_0^{1/2} \frac{x+2}{x^2+x+1}\,dx.$$

Compare your estimate with the exact result given by a computer algebra system.

Use the power series in Problem 53 to approximate with two-place accuracy the rather formidable integrals in Problems 55 and 56. Compare your estimates with the exact values given by a computer algebra system.

55. $\displaystyle\int_0^{1/2} \frac{1}{1+x^2+x^4}\,dx$

56. $\displaystyle\int_0^{1/2} \frac{1}{1+x^4+x^8}\,dx$

In Problems 57 and 58, graph the given function and several of its Taylor polynomials of the indicated degrees.

57. $f(x) = \dfrac{\sin x}{x}$; degrees $n = 2, 4, 6, \ldots$.

58. $f(x) = \displaystyle\int_0^x \frac{\sin t}{t}\,dt$; degrees $n = 3, 5, 7, \ldots$.

59. Use known power series to evaluate $\displaystyle\lim_{x\to 0} \frac{\sin x - \tan x}{\sin^{-1} x - \tan^{-1} x}$.

60. Substitute series such as

$$\sin(\tan x) = x + \frac{x^3}{6} - \frac{x^5}{40} - \frac{55x^7}{1008} + \cdots$$

provided by a computer algebra system to evaluate

$$\lim_{x\to 0} \frac{\sin(\tan x) - \tan(\sin x)}{\sin^{-1}(\tan^{-1} x) - \tan^{-1}(\sin^{-1} x)}.$$

61. (a) First use the parametrization $x(t) = a\cos t$, $y(t) = b\sin t$, $0 \leqq t \leqq 2\pi$ of the ellipse $(x/a)^2 + (y/b)^2 = 1$ to show that its perimeter (arc length) p is given by

$$p = 4a \int_0^{\pi/2} \sqrt{1 - \epsilon^2 \cos^2 t}\,dt$$

where $\epsilon = \sqrt{1-(b/a)^2}$ is the *eccentricity* of the ellipse. This so-called *elliptic integral* is nonelementary, and so must be approximated numerically. (b) Use the binomial series to expand the integrand in the perimeter formula in part (a). Then integrate termwise—using Formula 113 from the table of integrals inside the back cover—to show that the perimeter of the ellipse is given in terms of its major semiaxis and eccentricity by the power series

$$p = 2\pi a\left(1 - \frac{1}{4}\epsilon^2 - \frac{3}{64}\epsilon^4 - \frac{5}{256}\epsilon^6 - \frac{175}{16384}\epsilon^8 - \cdots\right).$$

62. The *arithmetic mean* of the major and minor semiaxes of the ellipse of Problem 61 is $A = \frac{1}{2}(a+b)$; their *root-square mean* is $R = \sqrt{\frac{1}{2}(a^2+b^2)}$. Substitute $b = a\sqrt{1-\epsilon^2}$ and use the binomial series to derive the expansions

$$A = a\left(1 - \frac{1}{4}\epsilon^2 - \frac{1}{16}\epsilon^4 - \frac{1}{32}\epsilon^6 - \frac{5}{256}\epsilon^8 - \cdots\right)$$

and

$$R = a\left(1 - \frac{1}{4}\epsilon^2 - \frac{1}{32}\epsilon^4 - \frac{1}{128}\epsilon^6 - \frac{5}{2048}\epsilon^8 - \cdots\right).$$

Something wonderful happens when you average these two series; show that

$$\frac{1}{2}(A+R)=a\left(1-\frac{1}{4}\epsilon^2-\frac{3}{64}\epsilon^6-5\frac{5}{256}\epsilon^6-\frac{180}{16384}\epsilon^8-\cdots\right),$$

and then note that the first four terms of the series within the parentheses here are the same as in the ellipse perimeter series of Problem 61(b). Conclude that the perimeter p of the ellipse is given by

$$p=\pi(A+R)+\frac{5\pi a}{8192}\epsilon^8+\cdots. \qquad \textbf{(14)}$$

If ϵ is quite small—as in a nearly circular ellipse—then the difference between the exact value of p and the simple approximation

$$p \approx \pi(A+R)=\pi\left(\tfrac{1}{2}(a+b)+\sqrt{\tfrac{1}{2}(a^2+b^2)}\right)$$

is extremely small. For instance, suppose that the orbit of the moon around the earth is an ellipse with major semi-axis a exactly 238,857 miles long and eccentricity ϵ exactly 0.0549. Then use Eq. (14) and a computer algebra system with extended-precision arithmetic to find the perimeter of the moon's orbit accurate *to the nearest inch;* give your answer in miles-feet-inches format.

10.9 Project: Calculating Trigonometric Functions on a Deserted Island

Again (as in the 10.4 Project) you're stranded for life on a desert island with only a very basic calculator that doesn't know about transcendental functions. Now your task is to use the (alternating) sine and cosine series to construct a table presenting (with five-place accuracy) the sines, cosines, and tangents of angles from 0° to 90° in increments of 5°.

To begin with, you can find the sine, cosine, and tangent of an angle of 45° from the familiar 1-1-$\sqrt{2}$ right triangle. Then you can find the values of these functions at an angle of 60° from an equilateral triangle. Once you know all about 45° and 60° angles, you can use the sine and cosine addition formulas

$$\sin(\alpha \pm \beta) = \sin\alpha\cos\beta \pm \cos\alpha\sin\beta$$

and

$$\cos(\alpha \pm \beta) = \cos\alpha\cos\beta \mp \sin\alpha\sin\beta$$

and/or equivalent forms to find the sine, cosine, and tangent of such angles as 15°, 30°, 75°, and 90°.

But algebra and simple trigonometric identities will probably never give you the sine or cosine or an angle of 5°. For this you will need to use the power series for sine and cosine. Sum enough terms (and then some) so you know your result is accurate to nine decimal places. Then fill in all the entries in your table, rounding them to five places. Tell—honestly—whether your entries agree with those your *real* calculator gives.

Finally, explain what strategy you would use to complete a similar table of values of trigonometric functions with angles in increments of 1° rather than 5°.

10.10 SERIES SOLUTIONS OF DIFFERENTIAL EQUATIONS

In Section 8.6 we saw that solving a homogeneous linear differential equation with constant coefficients can be reduced to the algebraic problem of finding the roots of its characteristic equation. There is no simple or similarly routine procedure for solving linear differential equations with *variable* coefficients. Even such a simple-looking equation as $y'' - xy = 0$ has no solution that can be expressed in terms of the standard elementary functions of calculus. One of the most important applications of power series is their use to solve such differential equations.

The Power Series Method

The *power series method* for solving a differential equation consists of substituting the power series

$$y=\sum_{n=0}^{\infty} c_n x^n = c_0 + c_1x + c_2x^2 + c_3x^3 + \cdots \qquad \textbf{(1)}$$

in the differential equation, and then attempting to determine what the values of the coefficients $c_0, c_1, c_2, c_3, \ldots$ must be in order that the series in (1) will actually satisfy the given differential equation. At first glance this might seem to be a formidable problem, because we have infinitely many unknowns $c_0, c_1, c_2, c_3, \ldots$ to find. Nevertheless, we will see that the method frequently succeeds. When it does, we obtain a power series representation of a solution, in contrast to the closed form solutions that result from the solution techniques we saw in Chapter 8.

Before we can substitute the series in (1) in a differential equation, we must first know what to substitute for the derivatives $y', y'', \ldots$ of the unknown function $y(x)$. But recall from Theorem 3 in Section 10.8 that the derivative of a power series can be calculated by termwise differentiation. Hence the first and second derivatives of the series in (1) are given by

$$y' = \sum_{n=1}^{\infty} nc_n x^{n-1} = c_1 + 2c_2x + 3c_3x^2 + \cdots \tag{2}$$

and

$$y'' = \sum_{n=2}^{\infty} n(n-1)c_n x^{n-2} = 2c_2 + 6c_3x + 12c_4x^2 + \cdots. \tag{3}$$

Also, these two series have the same radius of convergence as the original series in (1).

The process of determining the coefficients $c_0, c_1, c_2, c_3, \ldots$ in the series so that it will satisfy a given differential equation depends also on the following consequence of termwise differentiation: If two power series represent the same function on an open interval, then they are identical series. That is, they are one and the same power series. (See Problem 66 in Section 10.8.) In particular, *if* $\sum a_n x^n \equiv 0$ *on an open interval, then it follows that* $a_n = 0$ *for all* n. This fact is sometimes called the **identity principle** for power series.

EXAMPLE 1 Solve the equation $y' + 2y = 0$.

Solution We substitute the series

$$y = \sum_{n=0}^{\infty} c_n x^n \quad \text{and} \quad y' = \sum_{n=1}^{\infty} nc_n x^{n-1},$$

and obtain

$$\sum_{n=1}^{\infty} nc_n x^{n-1} + 2\sum_{n=0}^{\infty} c_n x^n = 0. \tag{4}$$

To compare coefficients here, we need the general term in each sum to be the term containing x^n. To accomplish this, we shift the index of summation in the first sum. To see how to do this, note that

$$\sum_{n=1}^{\infty} nc_n x^{n-1} = c_1 + 2c_2x + 3c_3x^2 + \cdots = \sum_{n=0}^{\infty} (n+1)c_{n+1}x^n.$$

Thus we can replace n with $n+1$ if, at the same time, we start counting one step lower; that is, at $n = 0$ rather than at $n = 1$. This is a shift of $+1$ in the index of summation. The result of making this shift in Eq. (4) is the identity

$$\sum_{n=0}^{\infty} (n+1)c_{n+1}x^n + 2\sum_{n=0}^{\infty} c_n x^n = 0;$$

that is,

$$\sum_{n=0}^{\infty} [(n+1)c_{n+1} + 2c_n]x^n = 0.$$

If this equation holds on some interval, then it follows from the identity principle that $(n+1)c_{n+1} + 2c_n = 0$ for all $n \geqq 0$; consequently,

$$c_{n+1} = -\frac{2c_n}{n+1} \tag{5}$$

for all $n \geqq 0$. Equation (5) is a **recurrence relation** from which we can successively compute $c_1, c_2, c_3, \ldots$ in terms of c_0; the latter will turn out to be the arbitrary constant that we expect to find in a general solution of a first-order differential equation.

With $n = 0$, Eq. (5) gives

$$c_1 = -\frac{2c_0}{1}.$$

With $n = 1$, Eq. (5) gives

$$c_2 = -\frac{2c_1}{2} = +\frac{2^2 c_0}{1 \cdot 2} = \frac{2^2 c_0}{2!}.$$

With $n = 2$, Eq. (5) gives

$$c_3 = -\frac{2c_2}{3} = -\frac{2^3 c_0}{1 \cdot 2 \cdot 3} = -\frac{2^3 c_0}{3!}.$$

By now it should be clear that after n such steps, we will have

$$c_n = (-1)^n \frac{2^n c_0}{n!}, \quad n \geqq 1.$$

(This is easy to prove by induction on n.) Consequently, our solution takes the form

$$y(x) = \sum_{n=0}^{\infty} c_n x^n = \sum_{n=0}^{\infty} (-1)^n \frac{2^n c_0}{n!} x^n = c_0 \sum_{n=0}^{\infty} \frac{(-2x)^n}{n!} = c_0 e^{-2x}.$$

In the final step we have used the familiar exponential series to identify our power series solution as the same solution $y(x) = c_0 e^{-2x}$ we could have obtained by the method of separation of variables. ◆

Shift of Index of Summation

In the solution of Example 1 we wrote

$$\sum_{n=1}^{\infty} n c_n x^{n-1} = \sum_{n=0}^{\infty} (n+1) c_{n+1} x^n \tag{6}$$

by shifting the index of summation by $+1$ in the series on the left. That is, we simultaneously *increased* the index of summation by 1 (replacing n with $n+1$, $n \to n+1$) and *decreased* the starting point by 1, from $n = 1$ to $n = 0$, thereby obtaining the series on the right. This procedure is valid because each infinite series in (6) is simply a compact notation for the single series

$$c_1 + 2c_2 x + 3c_3 x^2 + 4c_4 x^3 + \cdots. \tag{7}$$

More generally, we can shift the index of summation by k in an infinite series by simultaneously *increasing* the summation index by k ($n \to n + k$) and *decreasing* the starting point by k. For instance, a shift by $+2$ ($n \to n+2$) yields

$$\sum_{n=3}^{\infty} a_n x^{n-1} = \sum_{n=1}^{\infty} a_{n+2} x^{n+1}.$$

If k is negative we interpret a "decrease by k" as an increase by $-k = |k|$. Thus a shift by -2 ($n \to n-2$) in the index of summation yields

$$\sum_{n=1}^{\infty} nc_n x^{n-1} = \sum_{n=3}^{\infty} (n-2)c_{n-2}x^{n-3};$$

we have *decreased* the index of summation by 2, but *increased* the starting point by 2, from $n = 1$ to $n = 3$. You should check that the summation on the right is merely another representation of the series in (7).

EXAMPLE 2 Solve the equation $(x-3)y' + 2y = 0$.

Solution As before, we substitute

$$y = \sum_{n=0}^{\infty} c_n x^n \quad \text{and} \quad y' = \sum_{n=1}^{\infty} nc_n x^{n-1}$$

to obtain

$$(x-3)\sum_{n=1}^{\infty} nc_n x^{n-1} + 2\sum_{n=0}^{\infty} c_n x^n = 0,$$

so that

$$\sum_{n=1}^{\infty} nc_n x^n - 3\sum_{n=1}^{\infty} nc_n x^{n-1} + 2\sum_{n=0}^{\infty} c_n x^n = 0.$$

In the first sum we can replace $n = 1$ with $n = 0$ with no effect on the sum. In the second sum we shift the index of summation by $+1$. This yields

$$\sum_{n=0}^{\infty} nc_n x^n - 3\sum_{n=0}^{\infty} (n+1)c_{n+1}x^n + 2\sum_{n=0}^{\infty} c_n x^n = 0;$$

that is,

$$\sum_{n=0}^{\infty} [nc_n - 3(n+1)c_{n+1} + 2c_n]\, x^n = 0.$$

The identity principle then gives

$$nc_n - 3(n+1)c_{n+1} + 2c_n = 0,$$

from which we obtain the recurrence relation

$$c_{n+1} = \frac{n+2}{3(n+1)}c_n \quad \text{for} \quad n \geqq 0.$$

We apply this formula with $n = 0$, $n = 1$, and $n = 2$ in turn, and find that

$$c_1 = \frac{2}{3}c_0, \quad c_2 = \frac{3}{3 \cdot 2}c_1 = \frac{3}{3^2}c_0, \quad \text{and} \quad c_3 = \frac{4}{3 \cdot 3}c_2 = \frac{4}{3^3}c_0.$$

This is almost enough to make the pattern evident; it is not difficult to show by induction on n that

$$c_n = \frac{n+1}{3^n}c_0 \quad \text{if} \quad n \geqq 1.$$

Hence our proposed power series solution is

$$y(x) = c_0 \sum_{n=0}^{\infty} \frac{n+1}{3^n}x^n. \tag{8}$$

Its radius of convergence is

$$\rho = \lim_{n\to\infty} \left| \frac{c_n}{c_{n+1}} \right| = \lim_{n\to\infty} \frac{3n+3}{n+2} = 3.$$

Thus the series in (8) converges if $-3 < x < 3$ and diverges if $|x| > 3$. In this particular example we can explain why. An elementary solution (obtained by separation of variables) of our differential equation is $y = 1/(3-x)^2$. If we differentiate termwise the geometric series

$$\frac{1}{3-x} = \frac{\frac{1}{3}}{1-\frac{x}{3}} = \frac{1}{3}\sum_{n=0}^{\infty}\frac{x^n}{3^n},$$

we get a constant multiple of the series in (8). Thus this series (with the arbitrary constant c_0 appropriately chosen) represents the solution

$$y(x) = \frac{1}{(3-x)^2}$$

on the interval $-3 < x < 3$, and the singularity at $x = 3$ is the reason why the radius of convergence of the power series solution turned out to be $\rho = 3$. ◆

EXAMPLE 3 Solve the equation $x^2 y' = y - x - 1$.

Solution We make the usual substitutions $y = \sum c_n x^n$ and $y' = \sum n c_n x^{n-1}$, which yield

$$x^2 \sum_{n=1}^{\infty} n c_n x^{n-1} = -1 - x + \sum_{n=0}^{\infty} c_n x^n,$$

so that

$$\sum_{n=1}^{\infty} n c_n x^{n+1} = -1 - x + \sum_{n=0}^{\infty} c_n x^n.$$

Because of the presence of the two terms -1 and $-x$ on the right-hand side, we need to split off the first two terms, $c_0 + c_1 x$, of the series on the right for comparison. If we also shift the index of summation on the left by -1 (replace $n = 1$ with $n = 2$ and n with $n - 1$), we get

$$\sum_{n=2}^{\infty}(n-1)c_{n-1}x^n = -1 - x + c_0 + c_1 x + \sum_{n=2}^{\infty} c_n x^n.$$

Because the left-hand side contains neither a constant term nor a term containing x to the first power, the identity principle now yields $c_0 = 1$, $c_1 = 1$, and $c_n = (n-1)c_{n-1}$ for $n \geqq 2$. It follows that

$$c_2 = 1 \cdot c_1 = 1!, \qquad c_3 = 2 \cdot c_2 = 2!, \qquad c_4 = 3 \cdot c_3 = 3!,$$

and, in general, that

$$c_n = (n-1)! \quad \text{for} \quad n \geqq 2.$$

Thus we obtain the power series

$$y(x) = 1 + x + \sum_{n=2}^{\infty}(n-1)!\,x^n.$$

But the radius of convergence of this series is

$$\rho = \lim_{n\to\infty}\frac{(n-1)!}{n!} = \lim_{n\to\infty}\frac{1}{n} = 0,$$

so the series converges only for $x = 0$. What does this mean? Simply that the given differential equation does not have a [convergent] power series solution of the assumed form $y = \sum c_n x^n$. This example serves as a warning that the simple act of writing $y = \sum c_n x^n$ involves an assumption that may be false. ◆

EXAMPLE 4 Solve the equation $y'' + y = 0$.

Solution If we assume a solution of the form

$$y = \sum_{n=0}^{\infty} c_n x^n,$$

we find that

$$y' = \sum_{n=1}^{\infty} n c_n x^{n-1} \quad \text{and} \quad y'' = \sum_{n=2}^{\infty} n(n-1) c_n x^{n-2}.$$

Substituting for y and y'' in the differential equation then yields

$$\sum_{n=2}^{\infty} n(n-1) c_n x^{n-2} + \sum_{n=0}^{\infty} c_n x^n = 0.$$

We shift the index of summation in the first sum by +2 (replace $n = 2$ with $n = 0$ and n with $n + 2$). This gives

$$\sum_{n=0}^{\infty} (n+2)(n+1) c_{n+2} x^n + \sum_{n=0}^{\infty} c_n x^n = 0.$$

The identity $(n+2)(n+1)c_{n+2} + c_n = 0$ now follows from the identity principle, and thus we obtain the recurrence relation

$$c_{n+2} = -\frac{c_n}{(n+1)(n+2)} \tag{9}$$

for $n \geqq 0$. It is evident that this formula will determine the coefficients c_n with even subscript in terms of c_0 and those of odd subscript in terms of c_1; c_0 and c_1 are not predetermined, and thus will be the two arbitrary constants we expect to find in a general solution of a second-order equation.

When we apply the recurrence relation in (9) with $n = 0, 2,$ and 4 in turn, we get

$$c_2 = -\frac{c_0}{2!}, \quad c_4 = \frac{c_0}{4!}, \quad \text{and} \quad c_6 = -\frac{c_0}{6!}.$$

Taking $n = 1, 3,$ and 5 in turn, we find that

$$c_3 = -\frac{c_1}{3!}, \quad c_5 = \frac{c_1}{5!}, \quad \text{and} \quad c_7 = -\frac{c_1}{7!}.$$

Again, the pattern is clear; we leave it for you to show (by induction) that for $k \geqq 1$,

$$c_{2k} = \frac{(-1)^k c_0}{(2k)!} \quad \text{and} \quad c_{2k+1} = \frac{(-1)^k c_1}{(2k+1)!}.$$

Thus we get the power series solution

$$y(x) = c_0\left(1 - \frac{x^2}{2!} + \frac{x^4}{4!} - \frac{x^6}{6!} + \cdots\right) + c_1\left(x - \frac{x^3}{3!} + \frac{x^5}{5!} - \frac{x^7}{7!} + \cdots\right);$$

that is, $y(x) = c_0 \cos x + c_1 \sin x$. Note that we have no problem with the radius of convergence here; the Taylor series for the sine and cosine functions converge for all x. ◆

Power Series Definitions of Functions

The solution of Example 4 can bear further comment. Suppose that we had never heard of the sine and cosine functions, let alone their Taylor series. We would then have discovered the two power series solutions

$$C(x) = \sum_{n=0}^{\infty} \frac{(-1)^n x^{2n}}{(2n)!} = 1 - \frac{x^2}{2!} + \frac{x^4}{4!} - \cdots \tag{10}$$

and

$$S(x) = \sum_{n=0}^{\infty} \frac{(-1)^n x^{2n+1}}{(2n+1)!} = x - \frac{x^3}{3!} + \frac{x^5}{5!} - \cdots \tag{11}$$

of the differential equation $y'' + y = 0$. It is clear that $C(0) = 1$ and that $S(0) = 0$. After verifying that the two series in (10) and (11) converge for all x, we can differentiate them term by term to find that

$$C'(x) = -S(x) \quad \text{and} \quad S'(x) = C(x). \tag{12}$$

Consequently $C'(0) = 0$ and $S'(0) = 1$. Thus with the aid of the power series method (all the while knowing nothing about the sine and cosine functions), we have discovered that $y = C(x)$ is the unique solution of

$$y'' + y = 0$$

that satisfies the initial conditions $y(0) = 1$ and $y'(0) = 0$, and that $y = S(x)$ is the unique solution that satisfies the initial conditions $y(0) = 0$ and $y'(0) = 1$. It follows that $C(x)$ and $S(x)$ are linearly independent, and—recognizing the importance of the differential equation $y'' + y = 0$—we can agree to call C the *cosine* function and S the *sine* function. Indeed, all the usual properties of these two functions can be established, using only their initial values (at $x = 0$) and the derivatives in (12); there is no need to refer to triangles or even to angles. (Can you use the series in (10) and (11) to show that $[C(x)]^2 + [S(x)]^2 = 1$ for all x?) This demonstrates that the cosine and sine functions are fully determined by the differential equation $y'' + y = 0$ of which they are the natural linearly independent solutions. Figures 10.10.1 and 10.10.2 show how the geometric character of the graphs of $\cos x$ and $\sin x$ is revealed by the graphs of the Taylor polynomial approximations that we get by truncating the infinite series in (10) and (11).

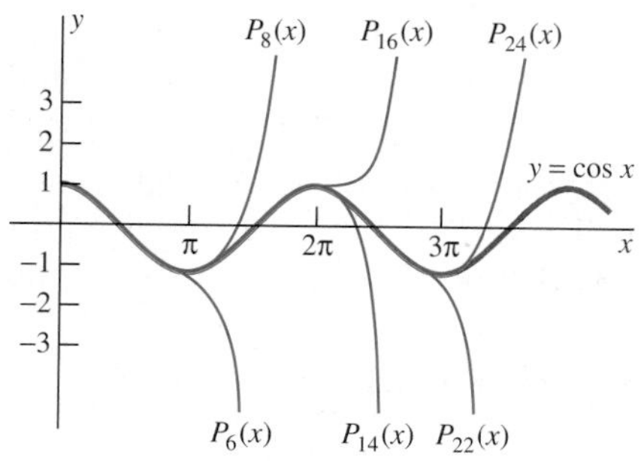

FIGURE 10.10.1 Graphs of $\cos x$ and its Taylor polynomial approximations $P_6(x)$, $P_8(x)$, $P_{14}(x)$, $P_{16}(x)$, $P_{22}(x)$, and $P_{24}(x)$.

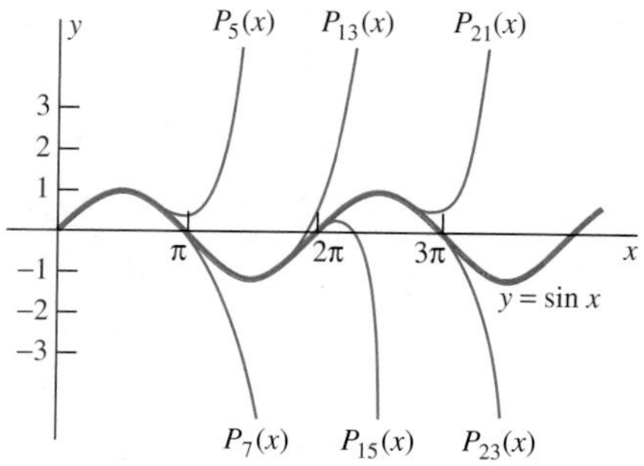

FIGURE 10.10.2 Graphs of $\sin x$ and its Taylor polynomial approximations $P_5(x)$, $P_7(x)$, $P_{13}(x)$, $P_{15}(x)$, $P_{21}(x)$, and $P_{23}(x)$.

This is by no means an uncommon situation. Many important special functions of mathematics occur in the first instance as power series solutions of differential equations, and thus are in practice *defined* by means of these power series. Example 5 introduces in this manner the *Airy functions* that appear in applications ranging from the propagation of radio waves to vibrations in atoms and molecules.

EXAMPLE 5 Solve the Airy equation $y'' - xy = 0$.

Solution Substituting $y = \sum c_n x^n$ and $y'' = \sum n(n-1)c_n x^{n-2}$ as usual yields

$$\sum_{n=2}^{\infty} n(n-1)c_n x^{n-2} - \sum_{n=0}^{\infty} c_n x^{n+1} = 0.$$

A shift of indices—replacing n with $n+2$ in the first sum and with $n-1$ in the second—yields

$$\sum_{n=0}^{\infty}(n+2)(n+1)c_{n+2}x^n - \sum_{n=1}^{\infty} c_{n-1}x^n = 0.$$

Splitting off the term corresponding to $n = 0$ in the first sum and combining the remaining terms, we get

$$2c_2 + \sum_{n=1}^{\infty}\left[(n+2)(n+1)c_{n+2} - c_{n-1}\right]x^n = 0.$$

The identity principle now gives $c_2 = 0$—because there is no other constant term on the left-hand side—and the recurrence relation $(n+2)(n+1)c_{n+2} - c_{n-1} = 0$ for $n \geqq 1$. Replacement of n with $n+1$ gives the recurrence relation

$$c_{n+3} = \frac{c_n}{(n+2)(n+3)} \tag{13}$$

for $n \geqq 0$. Thus each coefficient (after the first three) depends on the third previous one. Hence the fact that $c_2 = 0$ implies that

$$c_2 = c_5 = c_8 = \cdots = 0.$$

Beginning with c_0 as an arbitrary constant, we apply (13) with $n = 0$, $n = 3$, and $n = 6$ in turn and calculate

$$c_3 = \frac{c_0}{2\cdot 3} = \frac{c_0}{6}, \quad c_6 = \frac{c_3}{5\cdot 6} = \frac{c_0}{180}, \quad \text{and} \quad c_9 = \frac{c_6}{8\cdot 9} = \frac{c_0}{12960}.$$

Beginning with c_1 as a second arbitrary constant, we calculate similarly

$$c_4 = \frac{c_1}{3\cdot 4} = \frac{c_1}{12}, \quad c_7 = \frac{c_4}{6\cdot 7} = \frac{c_1}{504}, \quad \text{and} \quad c_{10} = \frac{c_7}{9\cdot 10} = \frac{c_1}{45360}.$$

When we collect the terms that involve c_0 and those that involve c_1, we get the general solution

$$y(x) = c_0\left(1 + \frac{x^3}{6} + \frac{x^6}{180} + \frac{x^9}{12960} + \cdots\right) + c_1\left(x + \frac{x^4}{12} + \frac{x^7}{504} + \frac{x^{10}}{45360} + \cdots\right)$$

of the Airy equation, with arbitrary constants c_0 and c_1. We see here the independent (why?) particular solutions

$$y_1(x) = 1 + \frac{x^3}{6} + \frac{x^6}{180} + \frac{x^9}{12960} + \cdots \quad \text{and} \quad y_2(x) = x + \frac{x^4}{12} + \frac{x^7}{504} + \frac{x^{10}}{45360} + \cdots.$$

Recognizing the pattern of coefficients is not so easy as in Example 4, but you can verify that the terms shown agree with the formulas

$$y_1(x) = 1 + \sum_{k=1}^{\infty}\frac{1\cdot 4\cdots(3k-2)}{(3k)!}x^{3k} \quad \text{and} \quad y_2(x) = x + \sum_{k=1}^{\infty}\frac{2\cdot 5\cdots(3k-1)}{(3k+1)!}x^{3k+1}.$$

The special combinations

$$\mathrm{Ai}(x) = \frac{y_1(x)}{3^{2/3}\Gamma\left(\frac{2}{3}\right)} - \frac{y_2(x)}{3^{1/3}\Gamma\left(\frac{1}{3}\right)} \quad \text{and} \quad \mathrm{Bi}(x) = \frac{y_1(x)}{3^{1/6}\Gamma\left(\frac{2}{3}\right)} + \frac{y_2(x)}{3^{-1/6}\Gamma\left(\frac{1}{3}\right)}$$

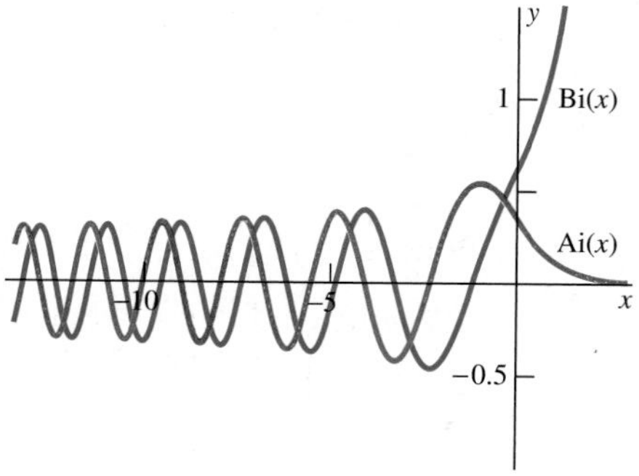

FIGURE 10.10.3 The graphs $y = \text{Ai}(x)$ and $y = \text{Bi}(x)$ of the Airy functions.

—with $\Gamma(x)$ denoting the gamma function defined in Section 7.8—are the standard *Airy functions,* which appear in mathematical tables and computer algebra systems. Their graphs in Fig. 10.10.3 exhibit oscillatory behavior for $x < 0$, but $\text{Ai}(x)$ decreases exponentially and $\text{Bi}(x)$ increases exponentially as $x \to \infty$. ◆

10.10 TRUE/FALSE STUDY GUIDE

10.10 CONCEPTS: QUESTIONS AND DISCUSSION

1. Suppose that the exponential function $E(x) = e^x$ is *defined* as the solution of the initial value problem $y' = y$, $y(0) = 1$. Beginning with this definition, what properties of the function $E(x)$ can be established?
2. Suppose that the hyperbolic functions $\text{Ch}(x) = \cosh x$ and $\text{Sh}(x) = \sinh x$ are *defined* as the solutions of the differential equation $y'' = y$ that satisfy the initial conditions $y(0) = 1$, $y'(0) = 0$ and $y(0) = 0$, $y'(0) = 1$, respectively. Beginning with these definitions, what properties of the functions $\text{Ch}(x)$ and $\text{Sh}(x)$ can be established? Can you discover a connection between these functions and the function $E(x)$ of Question 1?

10.10 PROBLEMS

In Problems 1 through 10, find a power series solution of the given differential equation. Determine the radius of convergence of the resulting series, and use your knowledge of familiar Maclaurin series and the binomial series to identify the series solution in terms of familiar elementary functions. (Of course, no one can prevent you from checking your work by also solving the equations by the methods of Chapter 8!)

1. $y' = y$

2. $y' = 4y$

3. $2y' + 3y = 0$

4. $y' + 2xy = 0$

5. $y' = x^2 y$

6. $(x - 2)y' + y = 0$

7. $(2x - 1)y' + 2y = 0$

8. $2(x + 1)y' = y$

9. $(x - 1)y' + 2y = 0$

10. $2(x - 1)y' = 3y$

In Problems 11 through 14, use the method of Example 4 to find two linearly independent power series solutions of the given differential equation. Determine the radius of convergence of each series, and identify the general solution in terms of familiar elementary functions.

11. $y'' = y$

12. $y'' = 4y$

13. $y'' + 9y = 0$

14. $y'' + y = x$

Show (as in Example 3) that the power series method fails to yield a power series solution of the form $y = \sum c_n x^n$ for the differential equations in Problems 15 through 18.

15. $xy' + y = 0$

16. $2xy' = y$

17. $x^2 y' + y = 0$

18. $x^3 y' = 2y$

In Problems 19 through 22, first derive a recurrence relation giving c_n for $n \geqq 2$ in terms of c_0 or c_1 (or both). Then apply the given initial conditions to find the values of c_0 and c_1. Next determine c_n (in terms of n, as in the text) and, finally, identify the particular solution in terms of familiar elementary functions.

19. $y'' + 4y = 0$; $y(0) = 0$, $y'(0) = 3$

20. $y'' - 4y = 0$; $y(0) = 2$, $y'(0) = 0$

21. $y'' - 2y' + y = 0$; $y(0) = 0$, $y'(0) = 1$

22. $y'' + y' - 2y = 0$; $y(0) = 1$, $y'(0) = -2$

23. Show that the equation

$$x^2 y'' + x^2 y' + y = 0$$

has no power series solution of the form $y = \sum c_n x^n$.

24. Use the power series method to discover the solution

$$J_0(x) = \sum_{k=0}^{\infty} \frac{(-1)^k x^{2k}}{2^{2k}(k!)^2} = 1 - \frac{x^2}{4} + \frac{x^4}{64} - \frac{x^6}{2304} + \cdots$$

of the Bessel equation $xy'' + y' + xy = 0$. Explain why the series method does not yield an independent second solution.

25. (a) Show that the solution of the initial value problem

$$y' = 1 + y^2, \qquad y(0) = 0$$

is $y(x) = \tan x$. (b) Because $y(x) = \tan x$ is an odd function with $y'(0) = 1$, its Taylor series is of the form

$$y = x + c_3x^3 + c_5x^5 + c_7x^7 + \cdots.$$

Substitute this series in $y' = 1 + y^2$ and equate like powers of x to derive the following relations:

$$3c_3 = 1, \qquad 5c_5 = 2c_3,$$
$$7c_7 = 2c_5 + (c_3)^2, \qquad 9c_9 = 2c_7 + 2c_3c_5,$$
$$11c_{11} = 2c_9 + 2c_3c_7 + (c_5)^2.$$

(c) Conclude that

$$\tan x = x + \frac{1}{3}x^3 + \frac{2}{15}x^5 + \frac{17}{315}x^7 + \frac{62}{2835}x^9 + \frac{1382}{155925}x^{11} + \cdots.$$

CHAPTER 10 REVIEW: DEFINITIONS, CONCEPTS, RESULTS

Use the following list as a guide to concepts that you may need to review.

1. Definition of the limit of a sequence
2. The limit laws for sequences
3. The bounded monotonic sequence property
4. Definition of the sum of an infinite series
5. Formula for the sum of a geometric series
6. The nth-term test for divergence
7. Divergence of the harmonic series
8. The nth-degree Taylor polynomial of the function f at the point $x = a$
9. Taylor's formula with remainder
10. The Taylor series of the elementary transcendental functions
11. The integral test
12. Convergence of p-series
13. The comparison and limit comparison tests
14. The alternating series test
15. Absolute convergence: definition *and* the fact that it implies convergence
16. The ratio test
17. The root test
18. Power series; radius of convergence and interval of convergence
19. The binomial series
20. Termwise differentiation and integration of power series
21. The use of power series to approximate values of functions and integrals
22. The sum and product of two power series
23. The use of power series to evaluate indeterminate forms
24. Power series solution of differential equations

CHAPTER 10 MISCELLANEOUS PROBLEMS

In Problems 1 through 15, determine whether or not the sequence $\{a_n\}$ converges, and find its limit if it does converge.

1. $a_n = \dfrac{n^2+1}{n^2+4}$

2. $a_n = \dfrac{8n-7}{7n-8}$

3. $a_n = 10 - (0.99)^n$

4. $a_n = n \sin \pi n$

5. $a_n = \dfrac{1+(-1)^n\sqrt{n}}{n+1}$

6. $a_n = \sqrt{\dfrac{1+(-0.5)^n}{n+1}}$

7. $a_n = \dfrac{\sin 2n}{n}$

8. $a_n = 2^{-(\ln n)/n}$

9. $a_n = (-1)^{\sin(n\pi/2)}$

10. $a_n = \dfrac{(\ln n)^3}{n^2}$

11. $a_n = \dfrac{1}{n}\sin\dfrac{1}{n}$

12. $a_n = \dfrac{n-e^n}{n+e^n}$

13. $a_n = \dfrac{\sinh n}{n}$

14. $a_n = \left(1+\dfrac{2}{n}\right)^{2n}$

15. $a_n = (2n^2+1)^{1/n}$

Determine whether each infinite series in Problems 16 through 30 converges or diverges.

16. $\displaystyle\sum_{n=1}^{\infty} \frac{(n^2)!}{n^n}$

17. $\displaystyle\sum_{n=1}^{\infty} \frac{(-1)^{n+1}\ln n}{n}$

18. $\displaystyle\sum_{n=0}^{\infty} \frac{3^n}{2^n+4^n}$

19. $\displaystyle\sum_{n=0}^{\infty} \frac{n!}{e^{n^2}}$

20. $\displaystyle\sum_{n=1}^{\infty} \frac{1}{n^{3/2}}\sin\frac{1}{n}$

21. $\displaystyle\sum_{n=0}^{\infty} \frac{(-2)^n}{3^n+1}$

22. $\sum_{n=1}^{\infty} 2^{-(2/n^2)}$

23. $\sum_{n=2}^{\infty} \frac{(-1)^n n}{(\ln n)^3}$

24. $\sum_{n=1}^{\infty} \frac{(-1)^n}{10^{1/n}}$

25. $\sum_{n=1}^{\infty} \frac{\sqrt{n}+\sqrt[3]{n}}{n^2+n^3}$

26. $\sum_{n=1}^{\infty} \frac{(-1)^{n+1}}{n^{[1+(1/n)]}}$

27. $\sum_{n=1}^{\infty} \frac{(-1)^{n+1}\arctan n}{\sqrt{n}}$

28. $\sum_{n=1}^{\infty} n \sin \frac{1}{n}$

29. $\sum_{n=3}^{\infty} \frac{1}{n(\ln n)(\ln \ln n)}$

30. $\sum_{n=3}^{\infty} \frac{1}{n(\ln n)(\ln \ln n)^2}$

Find the interval of convergence of the power series in Problems 31 through 40.

31. $\sum_{n=0}^{\infty} \frac{2^n x^n}{n!}$

32. $\sum_{n=0}^{\infty} \frac{(3x)^n}{2^{n+1}}$

33. $\sum_{n=1}^{\infty} \frac{(x-1)^n}{n \cdot 3^n}$

34. $\sum_{n=0}^{\infty} \frac{(2x-3)^n}{4^n}$

35. $\sum_{n=1}^{\infty} \frac{(-1)^n x^n}{4n^2-1}$

36. $\sum_{n=0}^{\infty} \frac{(2x-1)^n}{n^2+1}$

37. $\sum_{n=0}^{\infty} \frac{n! x^{2n}}{10^n}$

38. $\sum_{n=2}^{\infty} \frac{x^n}{\ln n}$

39. $\sum_{n=0}^{\infty} \frac{1+(-1)^n}{2(n!)} x^n$

40. $\sum_{n=1}^{\infty} \left(1+\frac{1}{n}\right)^n (x-1)^n$

Find the set of all values of x for which the series in Problems 41 through 43 converge.

41. $\sum_{n=1}^{\infty} (x-n)^n$

42. $\sum_{n=1}^{\infty} (\ln x)^n$

43. $\sum_{n=0}^{\infty} \frac{e^{nx}}{n!}$

44. Find the rational number that has repeated decimal expansion 2.7 1828 1828 1828 . . . .

45. Give an example of two convergent numerical series $\sum a_n$ and $\sum b_n$ such that the series $\sum a_n b_n$ diverges.

46. Prove that if $\sum a_n$ is a convergent positive-term series, then $\sum a_n^2$ converges.

47. Let the sequence $\{a_n\}$ be defined recursively as follows:

$$a_1 = 1; \qquad a_{n+1} = 1 + \frac{1}{1+a_n} \quad \text{if } n \geqq 1.$$

The limit of the sequence $\{a_n\}$ is the value of the *continued fraction*

$$1 + \cfrac{1}{2 + \cfrac{1}{2 + \cfrac{1}{2 + \cfrac{1}{2 + \cdots}}}}.$$

Assuming that $A = \lim_{n\to\infty} a_n$ exists, prove that $A = \sqrt{2}$.

48. Let $\{F_n\}_1^\infty$ be the Fibonacci sequence of Example 2 in Section 10.2. (a) Prove that $0 < F_n \leqq 2^n$ for all $n \geqq 1$, and hence conclude that the power series

$$F(x) = \sum_{n=1}^{\infty} F_n x^n$$

converges if $|x| < \frac{1}{2}$. (b) Show that $(1-x-x^2)F(x) = x$, so

$$F(x) = \frac{x}{1-x-x^2}.$$

49. We say that the *infinite product* indicated by

$$\prod_{n=1}^{\infty} (1+a_n) = (1+a_1)(1+a_2)(1+a_3)\cdots$$

converges provided that the infinite series

$$S = \sum_{n=1}^{\infty} \ln(1+a_n)$$

converges, in which case the value of the infinite product is, by definition, e^S. Use the integral test to prove that

$$\prod_{n=1}^{\infty} \left(1+\frac{1}{n}\right)$$

diverges.

50. Prove that the infinite product (see Problem 49)

$$\prod_{n=1}^{\infty} \left(1+\frac{1}{n^2}\right)$$

converges, and use the integral test remainder estimate to approximate its value. The actual value of this infinite product is known to be

$$\frac{\sinh \pi}{\pi} \approx 3.67607\,79103\,74977\,72069\,56975.$$

In Problems 51 through 55, use infinite series to approximate the indicated number accurate to three decimal places.

51. $\sqrt[5]{1.5}$

52. $\ln(1.2)$

53. $\int_0^{0.5} e^{-x^2}\,dx$

54. $\int_0^{0.5} \sqrt[3]{1+x^4}\,dx$

55. $\int_0^1 \frac{1-e^{-x}}{x}\,dx$

56. Substitute the Maclaurin series for $\sin x$ into that for e^x to obtain

$$e^{\sin x} = 1 + x + \tfrac{1}{2}x^2 - \tfrac{1}{8}x^4 + \cdots.$$

57. Substitute the Maclaurin series for the cosine and then integrate termwise to derive the formula

$$\int_0^\infty e^{-t^2} \cos 2xt\,dt = \frac{\sqrt{\pi}}{2} e^{-x^2}.$$

Use the reduction formula

$$\int_0^\infty t^{2n} e^{-t^2}\,dt = \frac{2n-1}{2} \int_0^\infty t^{2n-2} e^{-t^2}\,dt$$

that follows from the one derived in Problem 50 of Section 7.3. The validity of this improper termwise integration is subject to verification.

58. Prove that

$$\tanh^{-1} x = \int_0^x \frac{1}{1-t^2}\,dt = \sum_{n=0}^{\infty} \frac{x^{2n+1}}{2n+1}$$

if $|x| < 1$.

59. Prove that

$$\sinh^{-1} x = \int_0^x \frac{1}{\sqrt{1+t^2}}\,dt = \sum_{n=0}^{\infty} (-1)^n \frac{1\cdot 3\cdot 5\cdots(2n-1)}{2\cdot 4\cdot 6\cdots(2n)} \cdot \frac{x^{2n+1}}{2n+1}$$

if $|x| < 1$.

60. Suppose that $\tan y = \sum a_n y^n$. Determine a_0, a_1, a_2, and a_3 by substituting the inverse tangent series [Eq. (27) of Section 10.4] into the equation

$$x = \tan(\tan^{-1} x) = \sum_{n=0}^{\infty} a_n (\tan^{-1} x)^n.$$

61. According to *Stirling's series*, the value of $n!$ for large n is given to a close approximation by

$$n! \approx \sqrt{2\pi n}\left(\frac{n}{e}\right)^n e^{\mu(n)},$$

where

$$\mu(n) = \frac{1}{12n} - \frac{1}{360n^3} + \frac{1}{1260n^5}.$$

Substitute $\mu(n)$ into Maclaurin's series for e^x to show that

$$e^{\mu(n)} = 1 + \frac{1}{12n} + \frac{1}{288n^2} - \frac{139}{51840n^3} + \cdots.$$

Can you show that the next term in the last series is $-571/(2{,}488{,}320n^4)$?

62. Define

$$T(n) = \int_0^{\pi/4} \tan^n x\,dx$$

for $n \geqq 0$. (a) Show by "reduction" of the integral that

$$T(n+2) = \frac{1}{n+1} - T(n)$$

for $n \geqq 0$. (b) Conclude that $T(n) \to 0$ as $n \to \infty$. (c) Show that $T(0) = \pi/4$ and that $T(1) = \frac{1}{2}\ln 2$. (d) Prove by induction on n that

$$T(2n) = (-1)^{n+1}\left(1 - \frac{1}{3} + \frac{1}{5} - \cdots \pm \frac{1}{2n-1} - \frac{\pi}{4}\right).$$

(e) Conclude from parts (b) and (d) that

$$1 - \frac{1}{3} + \frac{1}{5} - \frac{1}{7} + \cdots = \frac{\pi}{4}.$$

(f) Prove by induction on n that

$$T(2n+1) = \frac{1}{2}(-1)^n\left(1 - \frac{1}{2} + \frac{1}{3} - \cdots \pm \frac{1}{n} - \ln 2\right).$$

(g) Conclude from parts (b) and (f) that

$$1 - \tfrac{1}{2} + \tfrac{1}{3} - \tfrac{1}{4} + \cdots = \ln 2.$$

63. Prove as follows that the number e is irrational. First suppose to the contrary that $e = p/q$, where p and q are positive integers. Note that $q > 1$. Write

$$\frac{p}{q} = e = 1 + \frac{1}{1!} + \frac{1}{2!} + \frac{1}{3!} + \cdots + \frac{1}{q!} + R_q,$$

where $0 < R_q < 3/(q+1)!$. (Why?) Then show that multiplying of both sides of this equation by $q!$ would lead to the contradiction that one side of the result is an integer but the other side is not.

64. Evaluate the infinite product (see Problem 49)

$$\prod_{n=2}^{\infty} \frac{n^2}{n^2-1}$$

by finding an explicit formula for

$$\prod_{n=2}^{k} \frac{n^2}{n^2-1} \quad (k \geqq 2)$$

and then taking the limit as $k \to \infty$.

65. Find a continued fraction representation (see Problem 47)

$$a_0 + \cfrac{1}{a_1 + \cfrac{1}{a_2 + \cfrac{1}{a_3 + \cfrac{1}{a_4 + \cdots}}}}$$

of $\sqrt{5}$.

66. Evaluate

$$1 + \frac{1}{2} - \frac{2}{3} + \frac{1}{4} + \frac{1}{5} - \frac{2}{6} + \frac{1}{7} + \frac{1}{8} - \frac{2}{9} + \frac{1}{10} + \cdots.$$

APPENDICES

APPENDIX A: REAL NUMBERS AND INEQUALITIES

The **real numbers** are already familiar to you. They are just those numbers ordinarily used in most measurements. The mass, velocity, temperature, and charge of a body are measured with real numbers. Real numbers can be represented by **terminating** or **nonterminating** decimal expansions; in fact, every real number has a nonterminating decimal expansion because a terminating expansion can be padded with infinitely many zeros:

$$\frac{3}{8} = 0.375 = 0.375000000\ldots.$$

Any **repeating** decimal, such as

$$\frac{7}{22} = 0.31818181818\ldots,$$

represents a **rational** number, one that is the ratio of two integers. Conversely, every rational number is represented by a repeating decimal like the two displayed above. But the decimal expansion of an **irrational** number (a real number that is not rational), such as

$$\sqrt{2} = 1.414213562\ldots \quad \text{or} \quad \pi = 3.14159265358979\ldots,$$

is both nonterminating and nonrepeating.

The geometric interpretation of real numbers as points on the **real line** (or *real number line*) $\boldsymbol{R}$ should also be familiar to you. Each real number is represented by precisely one point of $\boldsymbol{R}$, and each point of $\boldsymbol{R}$ represents precisely one real number. By convention, the positive numbers lie to the right of zero and the negative numbers to the left, as in Fig. A.1.

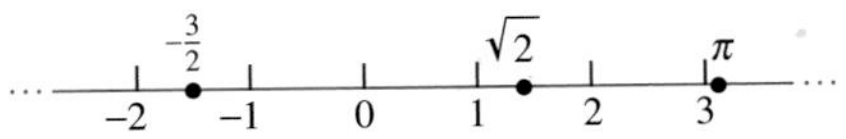

FIGURE A.1 The real line $\boldsymbol{R}$.

The following properties of inequalities of real numbers are fundamental and often used:

$$\begin{aligned} &\text{If } a < b \text{ and } b < c, \text{ then } a < c. \\ &\text{If } a < b, \text{ then } a + c < b + c. \\ &\text{If } a < b \text{ and } c > 0, \text{ then } ac < bc. \\ &\text{If } a < b \text{ and } c < 0, \text{ then } ac > bc. \end{aligned} \tag{1}$$

The last two statements mean that an inequality is preserved when its members are multiplied by a *positive* number but is *reversed* when they are multiplied by a *negative* number.

ABSOLUTE VALUE

The (nonnegative) distance along the real line between zero and the real number a is the **absolute value** of a, written $|a|$. Equivalently,

$$|a| = \begin{cases} a & \text{if } a \geqq 0; \\ -a & \text{if } a < 0. \end{cases} \tag{2}$$

The notation $a \geqq 0$ means that a is *either* greater than zero *or* equal to zero. Equation (2) implies that $|a| \geqq 0$ for every real number a and that $|a| = 0$ if and only if $a = 0$.

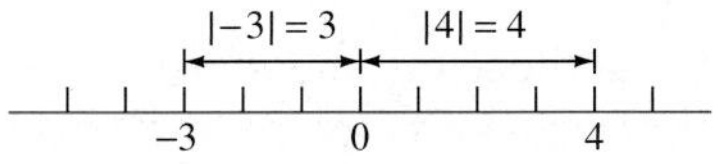

FIGURE A.2 The absolute value of a real number is simply its distance from zero (Example 1).

EXAMPLE 1 As Fig. A.2 shows,

$$|4| = 4 \quad \text{and} \quad |-3| = 3.$$

Moreover, $|0| = 0$ and $|\sqrt{2} - 2| = 2 - \sqrt{2}$, the latter being true because $2 > \sqrt{2}$. Thus $\sqrt{2} - 2 < 0$, and hence

$$\left|\sqrt{2} - 2\right| = -\left(\sqrt{2} - 2\right) = 2 - \sqrt{2}.$$

◆

The following properties of absolute values are frequently used:

$$\begin{aligned} |a| &= |-a| = \sqrt{a^2} \geqq 0, \\ |ab| &= |a|\,|b|, \\ -|a| &\leqq a \leqq |a|, \\ \text{and} \quad |a| < b \quad &\text{if and only if} \quad -b < a < b. \end{aligned} \tag{3}$$

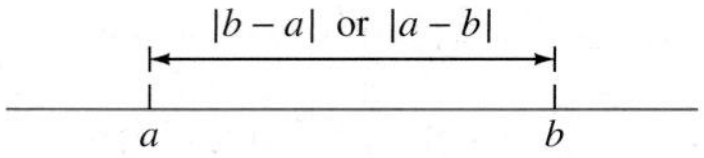

FIGURE A.3 The distance between a and b.

The **distance** between the real numbers a and b is defined to be $|a - b|$ (or $|b - a|$; there's no difference). This distance is simply the length of the line segment of the real line $\boldsymbol{R}$ with endpoints a and b (Fig. A.3).

The properties of inequalities and of absolute values in Eqs. (1) through (3) imply the following important theorem.

THEOREM 1 Triangle Inequality

For all real numbers a and b,

$$|a + b| \leqq |a| + |b|. \tag{4}$$

PROOF There are several cases to consider, depending upon whether the two numbers a and b are positive or negative and which has the larger absolute value. If both are positive, then so is $a + b$; in this case,

$$|a + b| = a + b = |a| + |b|. \tag{5}$$

If $a > 0$ but $b < 0$ and $|b| < |a|$, then

$$0 < a + b < a,$$

so

$$|a + b| = a + b < a = |a| < |a| + |b|, \tag{6}$$

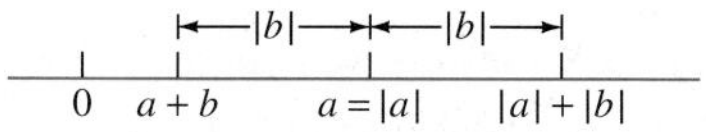

FIGURE A.4 The triangle inequality with $a > 0$, $b < 0$, and $|b| < |a|$.

as illustrated in Fig. A.4. The other cases are similar. In particular, we see that the triangle inequality is actually an equality [as in Eq. (5)] unless a and b have different signs, in which case it is a strict inequality [as in Eq. (6)]. ◄

INTERVALS

Suppose that S is a set (collection) of real numbers. It is common to describe S by the notation

$$S = \{x : \text{condition}\},$$

where the "condition" is true for those numbers x in S and false for those numbers x not in S. The most important sets of real numbers in calculus are *intervals.* If $a < b$, then the **open interval** (a, b) is defined to be the set

$$(a, b) = \{x : a < x < b\}$$

of real numbers, and the **closed interval** $[a, b]$ is

$$[a, b] = \{x : a \leqq x \leqq b\}.$$

Thus a closed interval contains its endpoints, whereas an open interval does not. We also use the **half-open intervals**

$$[a, b) = \{x : a \leqq x < b\} \quad \text{and} \quad (a, b] = \{x : a < x \leqq b\}.$$

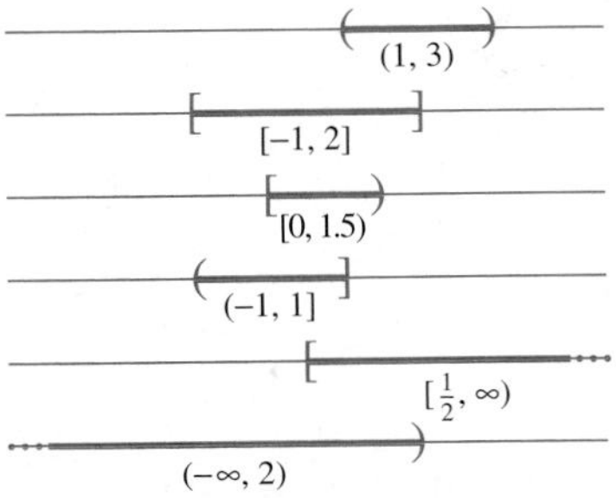

FIGURE A.5 Some examples of intervals of real numbers.

Thus the open interval $(1, 3)$ is the set of those real numbers x such that $1 < x < 3$, the closed interval $[-1, 2]$ is the set of those real numbers x such that $-1 \leqq x \leqq 2$, and the half-open interval $(-1, 2]$ is the set of those real numbers x such that $-1 < x \leqq 2$. In Fig. A.5 we show examples of such intervals as well as some **unbounded** intervals, which have forms such as

$$[a, +\infty) = \{x : x \geqq a\},$$

$$(-\infty, a] = \{x : x \leqq a\},$$

$$(a, +\infty) = \{x : x > a\},$$

$$\text{and} \quad (-\infty, a) = \{x : x < a\}.$$

The symbols $+\infty$ and $-\infty$, denoting "plus infinity" and "minus infinity," are merely notational conveniences and do *not* represent real numbers—the real line $\boldsymbol{R}$ does *not* have "endpoints at infinity." The use of these symbols is motivated by the brief and natural descriptions $[\pi, +\infty)$ and $(-\infty, 2)$ for the sets

$$\{x : x \geqq \pi\} \quad \text{and} \quad \{x : x < 2\}$$

of all real numbers x such that $x \geqq \pi$ and $x < 2$, respectively.

INEQUALITIES

The set of solutions of an inequality involving a variable x is often an interval or a union of intervals, as in the next examples. The **solution set** of such an inequality is simply the set of all those real numbers x that satisfy the inequality.

EXAMPLE 2 Solve the inequality $2x - 1 < 4x + 5$.

Solution Using the properties of inequalities listed in (1), we proceed much as if we were solving an equation for x: We isolate x on one side of the inequality. Here we begin with

$$2x - 1 < 4x + 5$$

and it follows that

$$-1 < 2x + 5;$$

$$-6 < 2x;$$

$$-3 < x.$$

Hence the solution set is the unbounded interval $(-3, +\infty)$. ◆

EXAMPLE 3 Solve the inequality $-13 < 1 - 4x \leqq 7$.

Solution We simplify the given inequality as follows:

$$-13 < 1 - 4x \leqq 7;$$
$$-7 \leqq 4x - 1 < 13;$$
$$-6 \leqq 4x < 14;$$
$$-\tfrac{3}{2} \leqq x < \tfrac{7}{2}.$$

Thus the solution set of the given inequality is the half-open interval $[-\frac{3}{2}, \frac{7}{2})$. ◆

EXAMPLE 4 Solve the inequality $|3 - 5x| < 2$.

Solution From the fourth property of absolute values in (3), we see that the given inequality is equivalent to

$$-2 < 3 - 5x < 2.$$

We now simplify as in the previous two examples:

$$-5 < -5x < -1;$$
$$\tfrac{1}{5} < x < 1.$$

Thus the solution set is the open interval $(\frac{1}{5}, 1)$. ◆

EXAMPLE 5 Solve the inequality

$$\frac{5}{|2x - 3|} < 1.$$

Solution It is usually best to begin by eliminating a denominator containing the unknown. Here we multiply each term by the *positive* quantity $|2x - 3|$ to obtain the equivalent inequality

$$|2x - 3| > 5.$$

It follows from the last property in (3) that this is so if and only if either

$$2x - 3 < -5 \quad \text{or} \quad 2x - 3 > 5.$$

The solutions of these *two* inequalities are the open intervals $(-\infty, -1)$ and $(4, +\infty)$, respectively. Hence the solution set of the original inequality consists of all those numbers x that lie in *either* of these two open intervals. ◆

The **union** of the two sets S and T is the set $S \cup T$ given by

$$S \cup T = \{x : \text{either } x \in S \text{ or } x \in T \text{ or both}\}.$$

Thus the solution set in Example 5 can be written in the form $(-\infty, -1) \cup (4, +\infty)$.

EXAMPLE 6 In accord with Boyle's law, the pressure p (in pounds per square inch) and volume V (in cubic inches) of a certain gas satisfy the condition $pV = 100$. Suppose that $50 \leqq V \leqq 150$. What is the range of possible values of the pressure p?

Solution If we substitute $V = 100/p$ in the given inequality $50 \leqq V \leqq 150$, we get

$$50 \leqq \frac{100}{p} \leqq 150.$$

It follows that *both*

$$50 \leqq \frac{100}{p} \quad \textit{and} \quad \frac{100}{p} \leqq 150;$$

that is, that both

$$p \leqq 2 \quad \text{and} \quad p \geqq \tfrac{2}{3}.$$

Thus the pressure p must lie in the closed interval $[\frac{2}{3}, 2]$. ◆

The **intersection** of the two sets S and T is the set $S \cap T$ defined as follows:

$$S \cap T = \{x : \text{both } x \in S \text{ and } x \in T\}.$$

Thus the solution set in Example 6 is the set $(-\infty, 2] \cap [\frac{2}{3}, +\infty) = [\frac{2}{3}, 2]$.

APPENDIX A PROBLEMS

Simplify the expressions in Problems 1 through 12 by writing each without using absolute value symbols.

1. $|3 - 17|$

2. $|-3| + |17|$

3. $\left|-0.25 - \frac{1}{4}\right|$

4. $|5| - |-7|$

5. $|(-5)(4 - 9)|$

6. $\dfrac{|-6|}{|4| + |-2|}$

7. $|(-3)^3|$

8. $\left|3 - \sqrt{3}\right|$

9. $\left|\pi - \frac{22}{7}\right|$

10. $-|7 - 4|$

11. $|x - 3|$, given $x < 3$

12. $|x - 5| + |x - 10|$, given $|x - 7| < 1$

Solve the inequalities in Problems 13 through 31. Write each solution set in interval notation.

13. $2x - 7 < -3$

14. $1 - 4x > 2$

15. $3x - 4 \geqq 17$

16. $2x + 5 \leqq 9$

17. $2 - 3x < 7$

18. $6 - 5x > -9$

19. $-3 < 2x + 5 < 7$

20. $4 \leqq 3x - 5 \leqq 10$

21. $-6 \leqq 5 - 2x < 2$

22. $3 < 1 - 5x < 7$

23. $|3 - 2x| < 5$

24. $|5x + 3| \leqq 4$

25. $|1 - 3x| > 2$

26. $1 < |7x - 1| < 3$

27. $2 \leqq |4 - 5x| \leqq 4$

28. $\dfrac{1}{2x + 1} > 3$

29. $\dfrac{2}{7 - 3x} \leqq -5$

30. $\dfrac{2}{|3x - 4|} < 1$

31. $\dfrac{1}{|1 - 5x|} \geqq -\dfrac{1}{3}$

32. Solve the inequality $x^2 - x - 6 > 0$. [*Suggestion:* Conclude from the factorization $x^2 - x - 6 = (x - 3)(x + 2)$ that the quantities $x - 3$ and $x + 2$ are either both positive or both negative. Consider the two cases separately to deduce that the solution set is $(-\infty, -2) \cup (3, \infty)$.]

Use the method of Problem 32 to solve the inequalities in Problems 33 through 36.

33. $x^2 - 2x - 8 > 0$

34. $x^2 - 3x + 2 < 0$

35. $4x^2 - 8x + 3 \geqq 0$

36. $2x \geqq 15 - x^2$

37. In accord with Boyle's law, the pressure p (in pounds per square inch) and volume V (in cubic inches) of a certain gas satisfy the condition $pV = 800$. What is the range of possible values of the pressure, given $100 \leqq V \leqq 200$?

38. The relationship between the Fahrenheit temperature F and the Celsius temperature C is given by $F = 32 + \frac{9}{5}C$. If the temperature on a certain day ranged from a low of 70°F to a high of 90°F, what was the range of the temperature in degrees Celsius?

39. An electrical circuit contains a battery supplying E volts in series with a resistance of R ohms, as shown in Fig. A.6. Then the current of I amperes that flows in the circuit satisfies Ohm's law, $E = IR$. If $E = 100$ and $25 < R < 50$, what is the range of possible values of I?

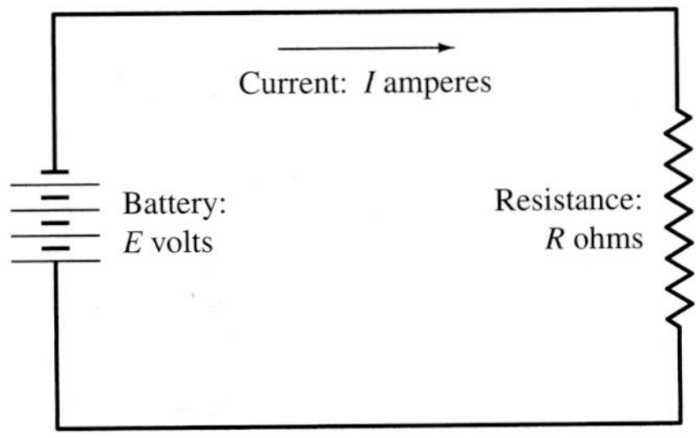

FIGURE A.6 A simple electric circuit.

40. The period T (in seconds) of a simple pendulum of length L (in feet) is given by $T = 2\pi\sqrt{L/32}$. If $3 < L < 4$, when is the range of possible values of T?

41. Use the properties of inequalities in (1) to show that the sum of two positive numbers is positive.

42. Use the properties of inequalities in (1) to show that the product of two positive numbers is positive.

43. Prove that the product of two negative numbers is positive and that the product of a positive number and a negative number is negative.

44. Suppose that $a < b$ and that a and b are either both positive or both negative. Prove that $1/a > 1/b$.

45. Apply the triangle inequality twice to show that

$$|a + b + c| \leqq |a| + |b| + |c|$$

for arbitrary real numbers a, b, and c.

46. Write $a = (a - b) + b$ to deduce from the triangle inequality that

$$|a| - |b| \leqq |a - b|$$

for arbitrary real numbers a and b.

47. Deduce from the definition in (2) that $|a| < b$ if and only if $-b < a < b$.

APPENDIX B: THE COORDINATE PLANE AND STRAIGHT LINES

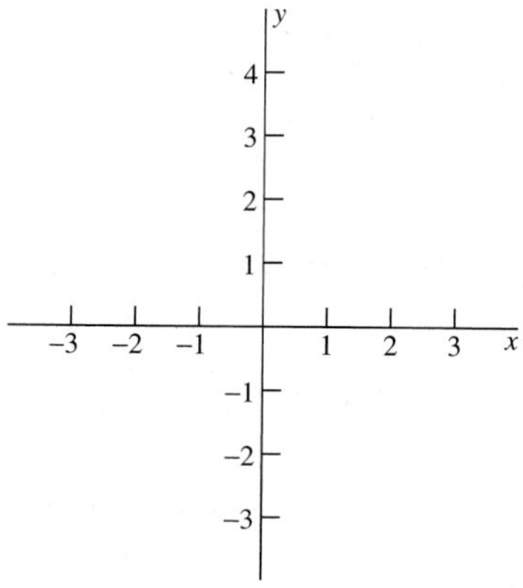

FIGURE B.1 The coordinate plane.

Imagine the flat, featureless, two-dimensional plane of Euclid's geometry. Install a copy of the real number line $\boldsymbol{R}$, with the line horizontal and the positive numbers to the right. Add another copy of $\boldsymbol{R}$ perpendicular to the first, with the two lines crossing where the number zero is located on each. The vertical line should have the positive numbers above the horizontal line, as in Fig. B.1; the negative numbers thus will be below it. The horizontal line is called the x-**axis** and the vertical line is called the y-**axis.**

With these added features, we call the plane the **coordinate plane,** because it's now possible to locate any point there by a pair of numbers, called the *coordinates of the point*. Here's how: If P is a point in the plane, draw perpendiculars from P to the coordinate axes, as shown in Fig. B.2. One perpendicular meets the x-axis at the x-**coordinate** (or **abscissa**) of P, labeled x_1 in Fig. B.2. The other meets the y-axis in the y-**coordinate** (or **ordinate**) y_1 of P. The pair of numbers (x_1, y_1), in that order, is called the **coordinate pair** for P, or simply the **coordinates** of P. To be concise, we speak of "the point $P(x_1, y_1)$."

This coordinate system is called the **rectangular coordinate system,** or the **Cartesian coordinate system** (because its use was popularized, beginning in the 1630s, by the French mathematician and philosopher René Descartes [1596–1650]). The plane, thus coordinatized, is denoted by $\boldsymbol{R}^2$ because two copies of $\boldsymbol{R}$ are used; it is known also as the **Cartesian plane.**

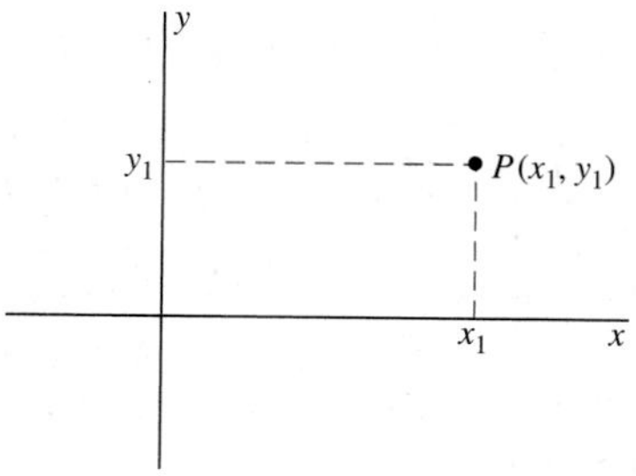

FIGURE B.2 The point P has rectangular coordinates (x_1, y_1).

Rectangular coordinates are easy to use, because $P(x_1, y_1)$ and $Q(x_2, y_2)$ denote the same point if and only if $x_1 = x_2$ and $y_1 = y_2$. Thus when you know that P and Q are two different points, you may conclude that P and Q have different abscissas, different ordinates, or both.

The point of symmetry $(0, 0)$ where the coordinate axes meet is called the **origin.** All points on the x-axis have coordinates of the form $(x, 0)$. Although the *real number* x is not the same as the geometric point $(x, 0)$, there are situations in which it is useful to think of the two as the same. Similar remarks apply to points $(0, y)$ on the y-axis.

The concept of distance in the coordinate plane is based on the **Pythagorean theorem:** If ABC is a right triangle with its right angle at the point C, with hypotenuse of length c and the other two sides of lengths a and b (as in Fig. B.3), then

$$c^2 = a^2 + b^2. \tag{1}$$

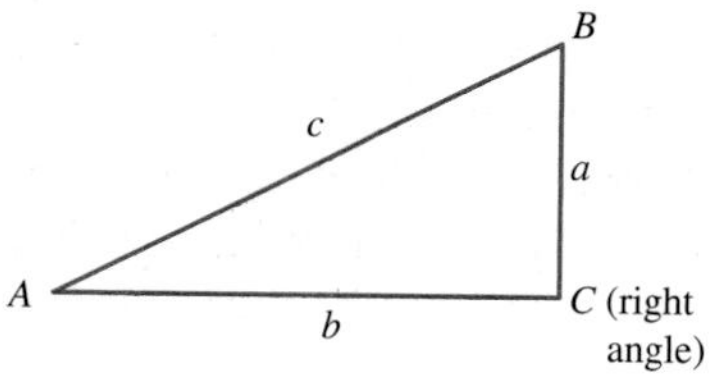

FIGURE B.3 The Pythagorean theorem.

The converse of the Pythagorean theorem is also true: If the three sides of a given triangle satisfy the Pythagorean relation in Eq. (1), then the angle opposite side c must be a right angle.

The *distance* $d(P_1, P_2)$ between the points P_1 and P_2 is, by definition, the length of the straight-line segment joining P_1 and P_2. The following formula gives $d(P_1, P_2)$ in terms of the coordinates of the two points.

Distance Formula

The **distance** between the two points $P_1(x_1, y_1)$ and $P_2(x_2, y_2)$ is

$$d(P_1, P_2) = \sqrt{(x_2 - x_1)^2 + (y_2 - y_1)^2}. \tag{2}$$

PROOF If $x_1 \neq x_2$ and $y_1 \neq y_2$, then Eq. (2) follows from the Pythagorean theorem. Use the right triangle with vertices P_1, P_2, and $P_3(x_2, y_1)$ shown in Fig. B.4.

If $x_1 = x_2$, then P_1 and P_2 lie in a vertical line. In this case

$$d(P_1, P_2) = |y_1 - y_2| = \sqrt{(y_1 - y_2)^2}.$$

This agrees with Eq. (2) because $x_1 = x_2$. The remaining case $(y_1 = y_2)$ is similar. ◄

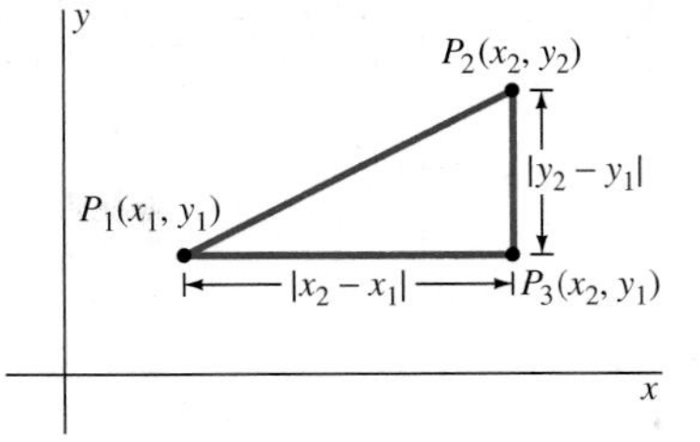

FIGURE B.4 Use this triangle to deduce the distance formula.

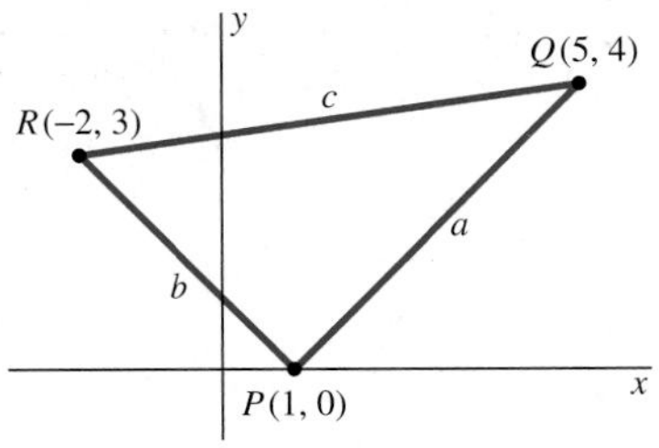

FIGURE B.5 Is this a right triangle (Example 1)?

EXAMPLE 1 Show that the triangle PQR with vertices $P(1, 0)$, $Q(5, 4)$, and $R(-2, 3)$ is a right triangle (Fig. B.5).

Solution The distance formula gives

$$a^2 = [d(P, Q)]^2 = (5-1)^2 + (4-0)^2 = 32,$$
$$b^2 = [d(P, R)]^2 = (-2-1)^2 + (3-0)^2 = 18, \quad \text{and}$$
$$c^2 = [d(Q, R)]^2 = (-2-5)^2 + (3-4)^2 = 50.$$

Because $a^2 + b^2 = c^2$, the *converse* of the Pythagorean theorem implies that RPQ is a right angle. (The right angle is at P because P is the vertex opposite the longest side, QR.) ◆

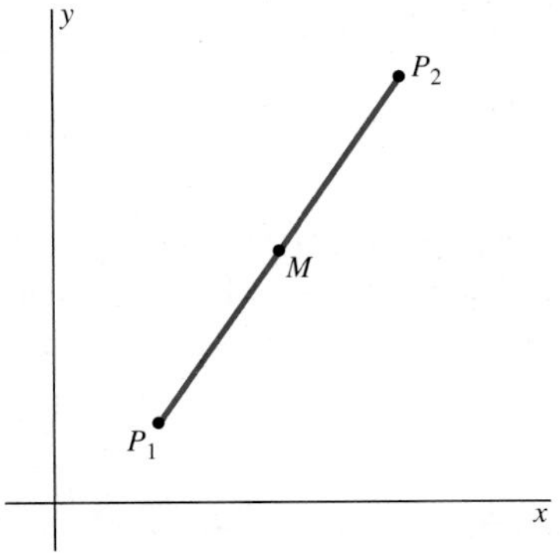

FIGURE B.6 The midpoint M.

Another application of the distance formula is an expression for the coordinates of the midpoint M of the line segment P_1P_2 with endpoints P_1 and P_2 (Fig. B.6). Recall from geometry that M is the one (and only) point of the line segment P_1P_2 that is equally distant from P_1 and P_2. The following formula tells us that the coordinates of M are the *averages* of the corresponding coordinates of P_1 and P_2.

Midpoint Formula

The **midpoint** of the line segment with endpoints $P_1(x_1, y_1)$ and $P_2(x_2, y_2)$ is the point $M(\overline{x}, \overline{y})$ with coordinates

$$\overline{x} = \tfrac{1}{2}(x_1 + x_2) \quad \text{and} \quad \overline{y} = \tfrac{1}{2}(y_1 + y_2). \tag{3}$$

PROOF If you substitute the coordinates of P_1, M, and P_2 in the distance formula, you find that $d(P_1, M) = d(P_2, M)$. All that remains is to show that M lies on the line segment P_1P_2. We ask you to do this, and thus complete the proof, in Problem 31. ◄

STRAIGHT LINES AND SLOPE

We want to define the *slope* of a straight line, a measure of its rate of rise or fall from left to right. Given a nonvertical straight line L in the coordinate plane, choose two points $P_1(x_1, y_1)$ and $P_2(x_2, y_2)$ on L. Consider the **increments** Δx and Δy (read "delta x" and "delta y") in the x- and y-coordinates from P_1 to P_2. These are defined as follows:

$$\Delta x = x_2 - x_1 \quad \text{and} \quad \Delta y = y_2 - y_1. \tag{4}$$

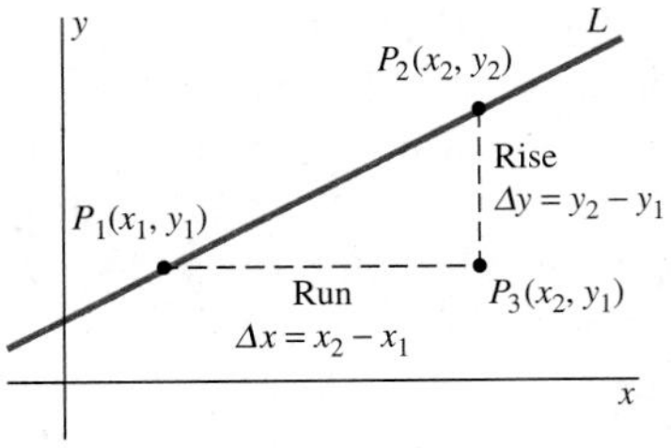

FIGURE B.7 The slope of a straight line.

Engineers (and others) call Δx the **run** from P_1 to P_2 and Δy the **rise** from P_1 to P_2, as in Fig. B.7. The **slope** m of the nonvertical line L is then defined to be the ratio of the rise to the run:

$$m = \frac{\Delta y}{\Delta x} = \frac{y_2 - y_1}{x_2 - x_1}. \tag{5}$$

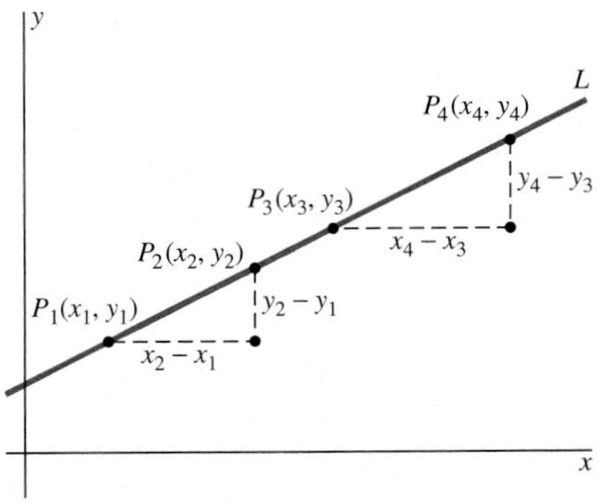

FIGURE B.8 The result of the slope computation does not depend on which two points of L are used.

This is also the definition of a line's slope in civil engineering (and elsewhere). In a surveying text you are likely to find the memory aid

$$\text{“slope} = \frac{\text{rise}}{\text{run}}.\text{”}$$

Recall that corresponding sides of similar (that is, equal-angled) triangles have equal ratios. Hence, if $P_3(x_3, y_3)$ and $P_4(x_4, y_4)$ are two other points of L, then the similarity of the triangles in Fig. B.8 implies that

$$\frac{y_4 - y_3}{x_4 - x_3} = \frac{y_2 - y_1}{x_2 - x_1}.$$

Therefore, the slope m as defined in Eq. (5) does *not* depend on the particular choice of P_1 and P_2.

If the line L is horizontal, then $\Delta y = 0$. In this case Eq. (5) gives $m = 0$. If L is vertical, then $\Delta x = 0$, so the slope of L is *not defined*. Thus we have the following statements:

- Horizontal lines have slope zero.
- Vertical lines have no defined slope.

EXAMPLE 2

(a) The slope of the line through the points $(3, -2)$ and $(-1, 4)$ is

$$m = \frac{4 - (-2)}{(-1) - 3} = \frac{6}{-4} = -\frac{3}{2}.$$

(b) The points $(3, -2)$ and $(7, -2)$ have the same y-coordinate. Therefore, the line through them is horizontal and thus has slope zero.

(c) The points $(3, -2)$ and $(3, 4)$ have the same x-coordinate. Thus the line through them is vertical, and so its slope is undefined. ◆

EQUATIONS OF STRAIGHT LINES

Our immediate goal is to be able to write equations of given straight lines. That is, if L is a straight line in the coordinate plane, we wish to construct a mathematical sentence—an equation—about points (x, y) in the plane. We want this equation to be *true* when (x, y) is a point on L and *false* when (x, y) is not a point on L. Clearly this equation will involve x and y and some numerical constants determined by L itself. For us to write this equation, the concept of the slope of L is essential.

Suppose, then, that $P(x_0, y_0)$ is a fixed point on the nonvertical line L of slope m. Let $P(x, y)$ be any *other* point on L. We apply Eq. (5) with P and P_0 in place of P_1 and P_2 to find that

$$m = \frac{y - y_0}{x - x_0};$$

that is,

$$y - y_0 = m(x - x_0). \quad \textbf{(6)}$$

Because the point (x_0, y_0) satisfies Eq. (6), as does every other point of L, and because no other points of the plane can do so, Eq. (6) is indeed an equation for the given line L. In summary, we have the following result.

The Point-Slope Equation

The point $P(x, y)$ lies on the line with slope m through the fixed point (x_0, y_0) if and only if its coordinates satisfy the equation

$$y - y_0 = m(x - x_0). \quad \textbf{(6)}$$

Equation (6) is called the **point-slope** equation of L, partly because the coordinates of the point (x_0, y_0) and the slope m of L may be read directly from this equation.

EXAMPLE 3 Write an equation for the straight line L through the points $P_1(1, -1)$ and $P_2(3, 5)$.

Solution The slope m of L may be obtained from the two given points:

$$m = \frac{5 - (-1)}{3 - 1} = 3.$$

Either P_1 or P_2 will do for the fixed point. We use $P_1(1, -1)$. Then, with the aid of Eq. (6), the point-slope equation of L is

$$y + 1 = 3(x - 1).$$

If simplification is appropriate, we may write $3x - y = 4$ or $y = 3x - 4$. ◆

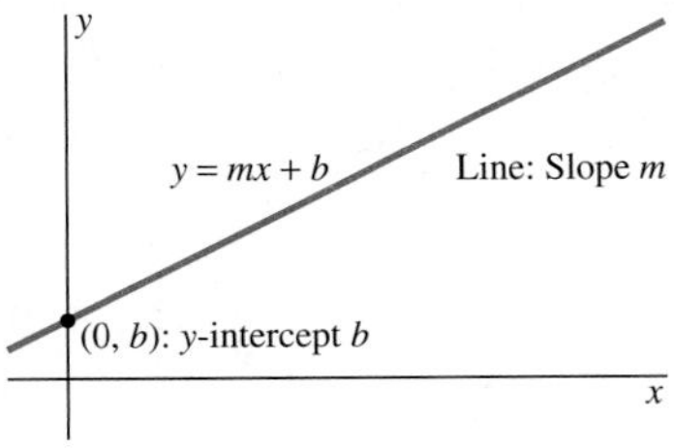

FIGURE B.9 The straight line with equation $y = mx + b$ has a slope m and y-intercept b.

Equation (6) can be written in the form

$$y = mx + b \quad \textbf{(7)}$$

where $b = y_0 - mx_0$ is a constant. Because $y = b$ when $x = 0$, the y-**intercept** of L is the point $(0, b)$ shown in Fig. B.9. Equations (6) and (7) are different forms of the equation of a straight line.

The Slope-Intercept Equation

The point $P(x, y)$ lies on the line with slope m and y-intercept b if and only if the coordinates of P satisfy the equation

$$y = mx + b. \quad \textbf{(7)}$$

Perhaps you noticed that both Eq. (6) and Eq. (7) can be written in the form of the general linear equation

$$Ax + By = C, \quad \textbf{(8)}$$

where A, B, and C are constants. Conversely, if $B \neq 0$, then Eq. (8) can be written in the form of Eq. (7) if we divide each term by B. Therefore Eq. (8) represents a straight line with its slope being the coefficient of x *after* solution of the equation for y. If $B = 0$, then Eq. (8) reduces to the equation of a vertical line: $x = K$ (where K is a constant). If $A = 0$ and $B \neq 0$, then Eq. (8) reduces to the equation of a horizontal line: $y = H$ (where H is a constant). Thus we see that Eq. (8) is always an equation of a straight line unless $A = B = 0$. Conversely, every straight line in the coordinate plane—even a vertical one—has an equation of the form in (8).

PARALLEL LINES AND PERPENDICULAR LINES

If the line L is not horizontal, then it must cross the x-axis. Then its **angle of inclination** is the angle ϕ measured counterclockwise from the positive x-axis to L. It follows that $0° < \phi < 180°$ if ϕ is measured in degrees. Figure B.10 makes it clear that this angle ϕ and the slope m of a nonvertical line are related by the equation

$$m = \frac{\Delta y}{\Delta x} = \tan \phi. \tag{9}$$

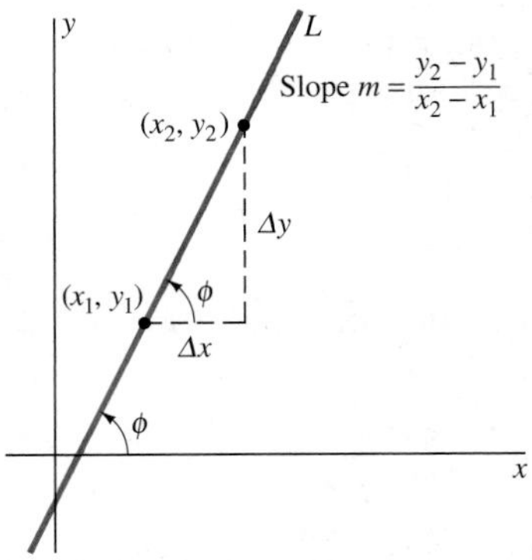

FIGURE B.10 How is the angle of inclination ϕ related to the slope m?

This is true because if ϕ is an acute angle in a right triangle, then $\tan \phi$ is the ratio of the leg opposite ϕ to the leg adjacent to ϕ.

Your intuition correctly assures you that two lines are parallel if and only if they have the same angle of inclination. So it follows from Eq. (9) that two parallel nonvertical lines have the same slope and that two lines with the same slope must be parallel. This completes the proof of Theorem 1.

THEOREM 1 Slopes of Parallel Lines
Two nonvertical lines are parallel if and only if they have the same slope.

Theorem 1 can also be proved without the use of the tangent function. The two lines shown in Fig. B.11 are parallel if and only if the two right triangles are similar, which is equivalent to the slopes of the lines being equal.

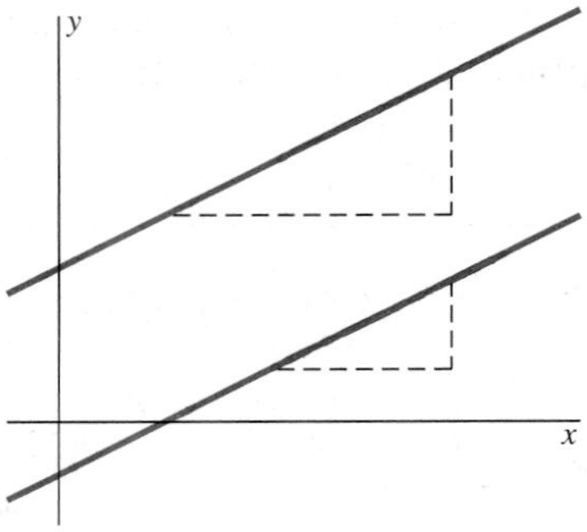

FIGURE B.11 Two parallel lines.

EXAMPLE 4 Write an equation of the line L that passes through the point $P(3, -2)$ and is parallel to the line L' with the equation $x + 2y = 6$.

Solution When we solve the equation of L' for y, we get $y = -\frac{1}{2}x + 3$. So L' has slope $m = -\frac{1}{2}$. Because L has the same slope, its point-slope equation is then

$$y + 2 = -\tfrac{1}{2}(x - 3);$$

if you prefer, $x + 2y = -1$. ◆

THEOREM 2 Slopes of Perpendicular Lines
Two lines L_1 and L_2 with slopes m_1 and m_2, respectively, are perpendicular if and only if

$$m_1 m_2 = -1. \tag{10}$$

That is, the slope of each is the *negative reciprocal* of the slope of the other.

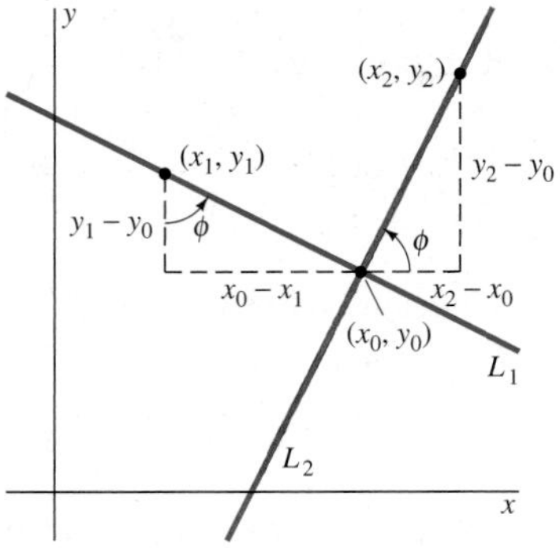

FIGURE B.12 Illustration of the proof of Theorem 2.

PROOF If the two lines L_1 and L_2 are perpendicular and the slope of each exists, then neither is horizontal or vertical. Thus the situation resembles the one shown in Fig. B.12, in which the two lines meet at the point (x_0, y_0). It is easy to see that the two right triangles of the figure are similar, so equality of ratios of corresponding sides yields

$$m_2 = \frac{y_2 - y_0}{x_2 - x_0} = \frac{x_0 - x_1}{y_1 - y_0} = -\frac{x_1 - x_0}{y_1 - y_0} = -\frac{1}{m_1}.$$

Thus Eq. (10) holds if the two lines are perpendicular. This argument can be reversed to prove the converse—that the lines are perpendicular if $m_1 m_2 = -1$. ◀

EXAMPLE 5 Write an equation of the line L through the point $P(3, -2)$ that is perpendicular to the line L' with equation $x + 2y = 6$.

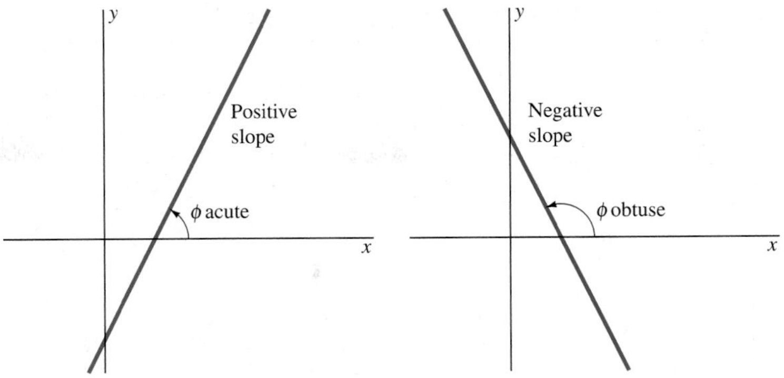

FIGURE B.13 Positive and negative slope; effect on ϕ.

Solution As we saw in Example 4, the slope of L' is $m' = -\frac{1}{2}$. By Theorem 2, the slope of L is $m = -1/m' = 2$. Thus L has the point-slope equation

$$y + 2 = 2(x - 3);$$

equivalently, $2x - y = 8$. ◆

You will find it helpful to remember that the *sign* of the slope m of the line L indicates whether L runs upward or downward as your eyes move from left to right. If $m > 0$, then the angle of inclination ϕ of L must be an acute angle, because $m = \tan\phi$. In this case, L "runs upward" to the right. If $m < 0$, then ϕ is obtuse, so L "runs downward." Figure B.13 shows the geometry behind these observations.

GRAPHICAL INVESTIGATION

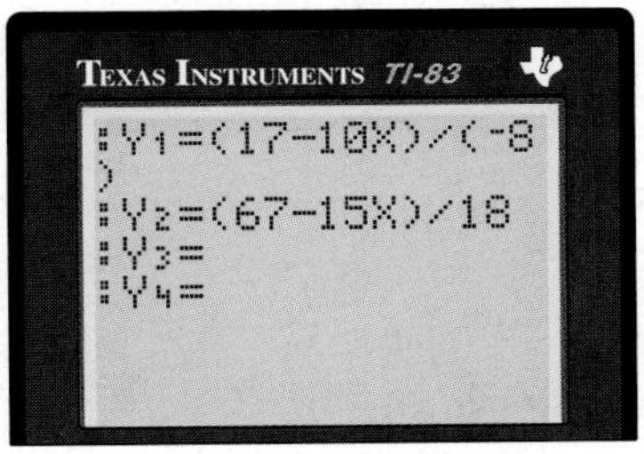

FIGURE B.14 A calculator prepared to graph the lines in Eq. (12) (Example 6).

Many mathematical problems require the simultaneous solution of a pair of linear equations of the form

$$\begin{aligned} a_1x + b_1y &= c_1, \\ a_2x + b_2y &= c_2. \end{aligned} \tag{11}$$

The graph of these two equations consists of a pair of straight lines in the xy-plane. If these two lines are not parallel, then they must intersect at a single point whose coordinates (x_0, y_0) constitute the solution of (11). That is, $x = x_0$ and $y = y_0$ are the (only) values of x and y for which both equations in (11) are true.

In elementary algebra you studied various elimination and substitution methods for solving linear systems such as the one in (11). Example 6 illustrates an alternative *graphical method* that is sometimes useful when a graphing utility—a graphics calculator or a computer with a graphing program—is available.

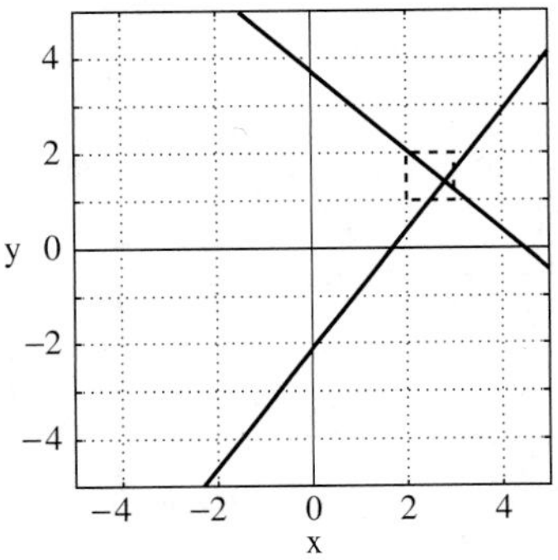

FIGURE B.15 $-5 \leqq x \leqq 5$, $-5 \leqq y \leqq 5$ (Example 6).

EXAMPLE 6 We want to investigate the simultaneous solution of the linear equations

$$\begin{aligned} 10x - 8y &= 17, \\ 15x + 18y &= 67. \end{aligned} \tag{12}$$

With many graphics calculators, it is necessary first to solve each equation for y:

$$\begin{aligned} y &= (17 - 10x)/(-8), \\ y &= (67 - 15x)/18. \end{aligned} \tag{13}$$

Figure B.14 shows a calculator prepared to graph the two lines represented by the equations in (12), and Fig. B.15 shows the result in the *viewing window* $-5 \leqq x \leqq 5$, $-5 \leqq y \leqq 5$.

Before proceeding, note that in Fig. B.15 the two lines *appear* to be perpendicular. But their slopes, $(-10)/(-8) = \frac{5}{4}$ and $(-15)/18 = -\frac{5}{6}$, are *not* negative

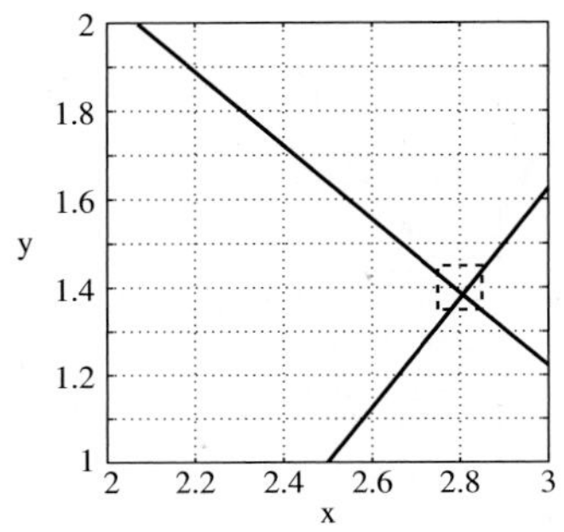

FIGURE B.16 $2 \leqq x \leqq 3$, $1 \leqq y \leqq 2$ (Example 6).

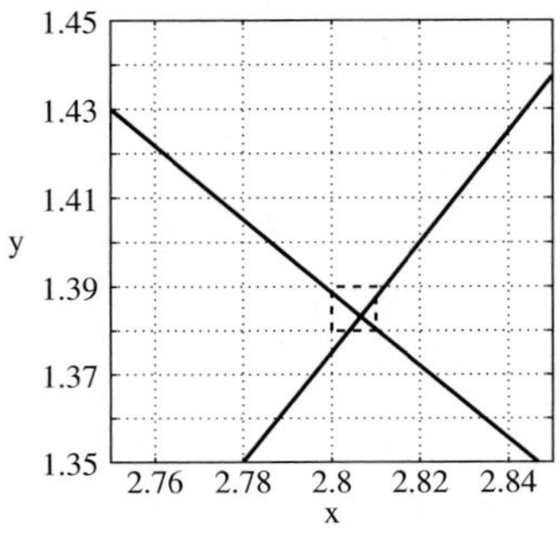

FIGURE B.17 $2.75 \leqq x \leqq 2.85$, $1.35 \leqq y \leqq 1.45$ (Example 6).

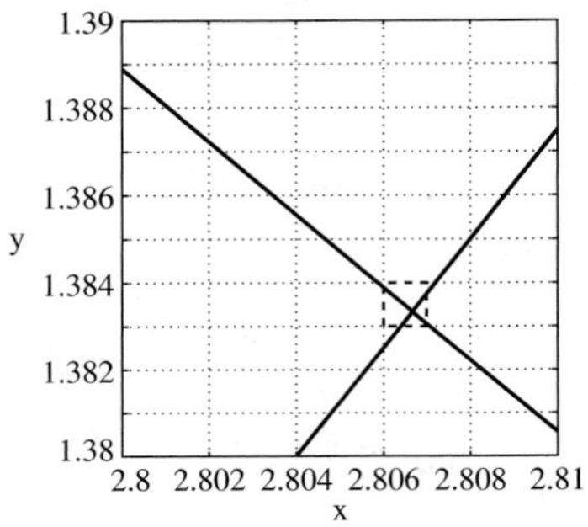

FIGURE B.18 $2.80 \leqq x \leqq 2.81$, $1.38 \leqq y \leqq 1.39$ (Example 6).

reciprocals of one another. It follows from Theorem 2 that the two linear are *not* perpendicular.

Figures B.16, B.17, and B.18 show successive magnifications produced by "zooming in" on the point of intersection of the two lines. The dashed-line box in each figure is the viewing window for the next figure. Looking at Fig. B.18, we see that the intersection point is given by the approximations

$$x \approx 2.807, \qquad y \approx 1.383, \tag{14}$$

rounded to three decimal places.

The result in (14) can be checked by equating the right-hand sides in (13) and solving for x. This gives $x = 421/150 \approx 2.8067$. Substituting the exact value of x into either equation in (13) then yields $y = 83/60 \approx 1.3833$.

The graphical method illustrated by Example 6 typically produces approximate solutions that are sufficiently accurate for practical purposes. But the method is especially useful for *nonlinear* equations, for which exact algebraic techniques of solution may not be available. ◆

APPENDIX B PROBLEMS

Three points A, B, and C lie on a single straight line if and only if the slope of AB is equal to the slope of BC. In Problems 1 through 4, plot the three given points and then determine whether or not they lie on a single line.

1. $A(-1, -2)$, $B(2, 1)$, $C(4, 3)$

2. $A(-2, 5)$, $B(2, 3)$, $C(8, 0)$

3. $A(-1, 6)$, $B(1, 2)$, $C(4, -2)$

4. $A(-3, 2)$, $B(1, 6)$, $C(8, 14)$

In Problems 5 and 6, use the concept of slope to show that the four points given are the vertices of a parallelogram.

5. $A(-1, 3)$, $B(5, 0)$, $C(7, 4)$, $D(1, 7)$

6. $A(7, -1)$, $B(-2, 2)$, $C(1, 4)$, $D(10, 1)$

In Problems 7 and 8, show that the three given points are the vertices of a right triangle.

7. $A(-2, -1)$, $B(2, 7)$, $C(4, -4)$

8. $A(6, -1)$, $B(2, 3)$, $C(-3, -2)$

In Problems 9 through 13, find the slope m and y-intercept b of the line with the given equation. Then sketch the line.

9. $2x = 3y$

10. $x + y = 1$

11. $2x - y + 3 = 0$

12. $3x + 4y = 6$

13. $2x = 3 - 5y$

In Problems 14 through 23, write an equation of the straight line L described.

14. L is vertical and has x-intercept 7.

15. L is horizontal and passes through $(3, -5)$.

16. L has x-intercept 2 and y-intercept -3.

17. L passes through $(2, -3)$ and $(5, 3)$.

18. L passes through $(-1, -4)$ and has slope $\frac{1}{2}$.

19. L passes through $(4, 2)$ and has angle of inclination $135°$.

20. L has slope 6 and y-intercept 7.

21. L passes through $(1, 5)$ and is parallel to the line with equation $2x + y = 10$.

22. L passes through $(-2, 4)$ and is perpendicular to the line with equation $x + 2y = 17$.

23. L is the perpendicular bisector of the line segment that has endpoints $(-1, 2)$ and $(3, 10)$.

24. Find the perpendicular distance from the point $(2, 1)$ to the line with equation $y = x + 1$.

25. Find the perpendicular distance between the parallel lines $y = 5x + 1$ and $y = 5x + 9$.

26. The points $A(-1, 6)$, $B(0, 0)$, and $C(3, 1)$ are three consecutive vertices of a parallelogram. What are the coordinates of the fourth vertex? (What happens if the word *consecutive* is omitted?)

27. Prove that the diagonals of the parallelogram of Problem 26 bisect each other.

28. Show that the points $A(-1, 2)$, $B(3, -1)$, $C(6, 3)$, and $D(2, 6)$ are the vertices of a *rhombus*—a parallelogram with all four sides having the same length. Then prove that the diagonals of this rhombus are perpendicular to each other.

29. The points $A(2, 1)$, $B(3, 5)$, and $C(7, 3)$ are the vertices of a triangle. Prove that the line joining the midpoints of AB and BC is parallel to AC.

30. A **median** of a triangle is a line joining a vertex to the midpoint of the opposite side. Prove that the medians of the triangle of Problem 29 intersect in a single point.

31. Complete the proof of the midpoint formula in Eq. (3). It is necessary to show that the point M lies on the segment P_1P_2. One way to do this is to show that the slope of MP_1 is equal to the slope of MP_2.

32. Let $P(x_0, y_0)$ be a point of the circle with center $C(0, 0)$ and radius r. Recall that the line tangent to the circle at the point P is perpendicular to the radius CP. Prove that the equation of this tangent line is $x_0x + y_0y = r^2$.

33. The Fahrenheit temperature F and the absolute temperature K satisfy a linear equation. Moreover, $K = 273.16$ when $F = 32$, and $K = 373.16$ when $F = 212$. Express K in terms of F. What is the value of F when $K = 0$?

34. The length L (in centimeters) of a copper rod is a linear function of its Celsius temperature C. If $L = 124.942$ when $C = 20$ and $L = 125.134$ when $C = 110$, express L in terms of C.

35. The owner of a grocery store finds that she can sell 980 gal of milk each week at \$1.69/gal and 1220 gal of milk each week at \$1.49/gal. Assume a linear relationship between price and sales. How many gallons would she then expect to sell each week at \$1.56/gal?

36. Figure B.19 shows the graphs of the equations

$$17x - 10y = 57,$$

$$25x - 15y = 17.$$

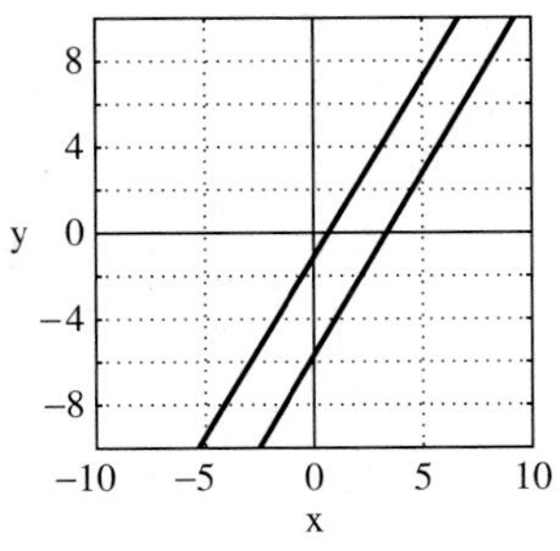

FIGURE B.19 The lines of Problem 36.

Are these two lines parallel? If not, find their point of intersection. If you have a graphing utility, find the solution by graphical approximation as well as by exact algebraic methods.

In Problems 37 through 46, use a graphics calculator or computer to approximate graphically (with three digits to the right of the decimal correct or correctly rounded) the solution of the given linear equation. Then check your approximate solution by solving the system by an exact algebraic method.

37. $2x + 3y = 5$
$2x + 5y = 12$

38. $6x + 4y = 5$
$8x - 6y = 13$

39. $3x + 3y = 17$
$3x + 5y = 16$

40. $2x + 3y = 17$
$2x + 5y = 20$

41. $4x + 3y = 17$
$5x + 5y = 21$

42. $4x + 3y = 15$
$5x + 5y = 29$

43. $5x + 6y = 16$
$7x + 10y = 29$

44. $5x + 11y = 21$
$4x + 10y = 19$

45. $6x + 6y = 31$
$9x + 11y = 37$

46. $7x + 6y = 31$
$11x + 11y = 47$

47. Justify the phrase "no other point of the plane can do so" that follows the first appearance of Eq. (6).

48. The discussion of the linear equation $Ax + By = C$ in Eq. (8) does not include a description of the graph of this equation if $A = B = 0$. What is the graph in this case?

APPENDIX C: REVIEW OF TRIGONOMETRY

In elementary trigonometry, the six basic trigonometric functions of an acute angle θ in a right triangle are defined as ratios between pairs of sides of the triangle. As in Fig. C.1, where "adj" stands for "adjacent," "opp" for "opposite," and "hyp" for "hypotenuse,"

$$\cos\theta = \frac{\text{adj}}{\text{hyp}}, \quad \sin\theta = \frac{\text{opp}}{\text{hyp}}, \quad \tan\theta = \frac{\text{opp}}{\text{adj}},$$
$$\sec\theta = \frac{\text{hyp}}{\text{adj}}, \quad \csc\theta = \frac{\text{hyp}}{\text{opp}}, \quad \cot\theta = \frac{\text{adj}}{\text{opp}}. \tag{1}$$

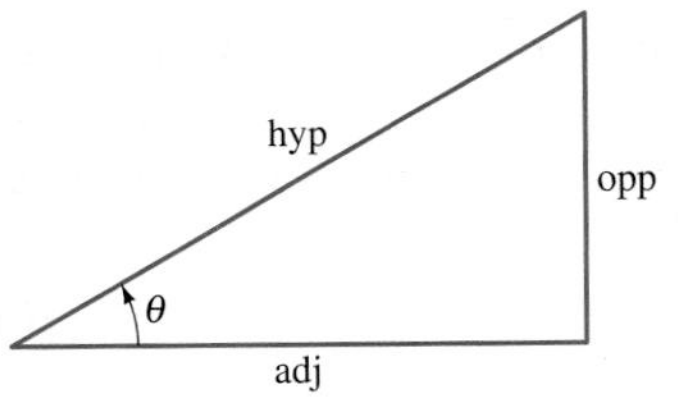

FIGURE C.1 The sides and angle θ of a right triangle.

We generalize these definitions to *directed* angles of arbitrary size in the following way. Suppose that the initial side of the angle θ is the positive x-axis, so its vertex is

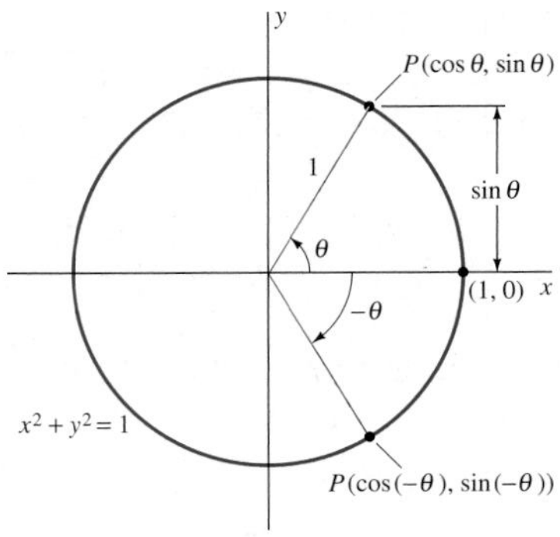

FIGURE C.2 Using the unit circle to define the trigonometric functions.

at the origin. The angle is **directed** if a direction of rotation from its initial side to its terminal side is specified. We call θ a **positive angle** if this rotation is counterclockwise and a **negative angle** if it is clockwise.

Let $P(x, y)$ be the point at which the terminal side of θ intersects the *unit* circle $x^2 + y^2 = 1$. Then we define

$$\cos\theta = x, \qquad \sin\theta = y, \qquad \tan\theta = \frac{y}{x},$$
$$\sec\theta = \frac{1}{x}, \qquad \csc\theta = \frac{1}{y}, \qquad \cot\theta = \frac{x}{y}. \tag{2}$$

We assume that $x \neq 0$ in the case of $\tan\theta$ and $\sec\theta$ and that $y \neq 0$ in the case of $\cot\theta$ and $\csc\theta$. If the angle θ is positive and acute, then it is clear from Fig. C.2 that the definitions in (2) agree with the right triangle definitions in (1) in terms of the coordinates of P. A glance at the figure also shows which of the functions are positive for angles in each of the four quadrants. Figure C.3 summarizes this information.

Here we discuss primarily the two most basic trigonometric functions, the sine and the cosine. From (2) we see immediately that the other four trigonometric functions are defined in terms of $\sin\theta$ and $\cos\theta$ by

$$\tan\theta = \frac{\sin\theta}{\cos\theta}, \qquad \sec\theta = \frac{1}{\cos\theta},$$
$$\cot\theta = \frac{\cos\theta}{\sin\theta}, \qquad \csc\theta = \frac{1}{\sin\theta}. \tag{3}$$

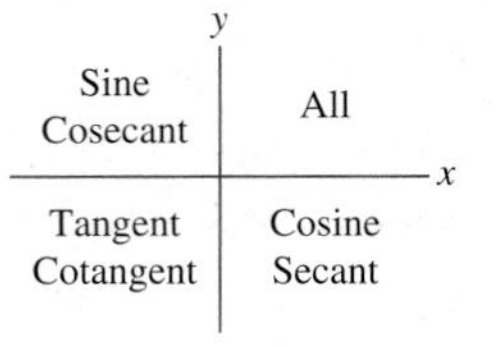

Positive in quadrants shown

FIGURE C.3 The signs of the trigonometric functions.

Next, we compare the angles θ and $-\theta$ in Fig. C.4. We see that

$$\cos(-\theta) = \cos\theta \quad \text{and} \quad \sin(-\theta) = -\sin\theta. \tag{4}$$

Because $x = \cos\theta$ and $y = \sin\theta$ in (2), the equation $x^2 + y^2 = 1$ of the unit circle translates immediately into the **fundamental identity of trigonometry,**

$$\cos^2\theta + \sin^2\theta = 1. \tag{5}$$

Dividing each term of this fundamental identity by $\cos^2\theta$ gives the identity

$$1 + \tan^2\theta = \sec^2\theta. \tag{5′}$$

Similarly, dividing each term in Eq. (5) by $\sin^2\theta$ yields the identity

$$1 + \cot^2\theta = \csc^2\theta. \tag{5″}$$

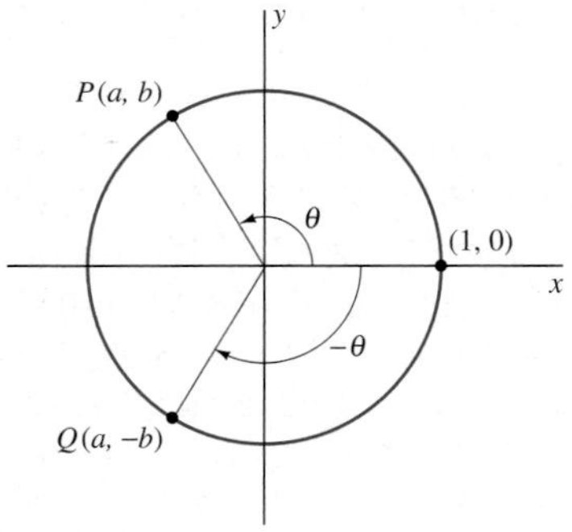

FIGURE C.4 The effect of replacing θ with $-\theta$ in the sine and cosine functions.

(See Problem 9 of this appendix.)

In Problems 41 and 42 we outline derivations of the **addition formulas**

$$\sin(\alpha + \beta) = \sin\alpha\cos\beta + \cos\alpha\sin\beta, \tag{6}$$
$$\cos(\alpha + \beta) = \cos\alpha\cos\beta - \sin\alpha\sin\beta. \tag{7}$$

With $\alpha = \theta = \beta$ in Eqs. (6) and (7), we get the **double-angle formulas**

$$\sin 2\theta = 2\sin\theta\cos\theta, \tag{8}$$
$$\cos 2\theta = \cos^2\theta - \sin^2\theta \tag{9}$$
$$= 2\cos^2\theta - 1 \tag{9a}$$
$$= 1 - 2\sin^2\theta, \tag{9b}$$

where Eqs. (9a) and (9b) are obtained from Eq. (9) by use of the fundamental identity in Eq. (5).

If we solve Eq. (9a) for $\cos^2\theta$ and Eq. (9b) for $\sin^2\theta$, we get the **half-angle formulas**

$$\cos^2\theta = \tfrac{1}{2}(1 + \cos 2\theta), \tag{10}$$
$$\sin^2\theta = \tfrac{1}{2}(1 - \cos 2\theta). \tag{11}$$

Equations (10) and (11) are especially important in integral calculus.

RADIAN MEASURE

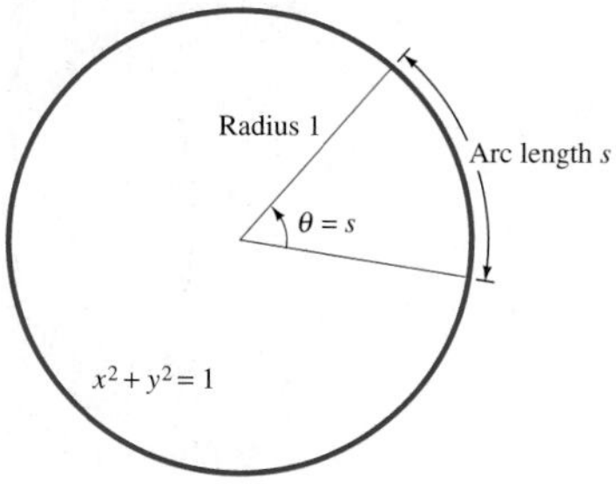

FIGURE C.5 The radian measure of an angle.

In elementary mathematics, angles frequently are measured in *degrees*, with 360° in one complete revolution. In calculus it is more convenient—and often essential—to measure angles in *radians*. The **radian measure** of an angle is the length of the arc it subtends in (that is, the arc it cuts out of) the unit circle when the vertex of the angle is at the center of the circle (Fig. C.5).

Recall that the area A and circumference C of a circle of radius r are given by the formulas

$$A = \pi r^2 \quad \text{and} \quad C = 2\pi r,$$

where the irrational number π is approximately 3.14159. Because the circumference of the unit circle is 2π and its central angle is 360°, it follows that

$$2\pi \text{ rad} = 360^\circ; \qquad 180^\circ = \pi \text{ rad} \approx 3.14159 \text{ rad}. \tag{12}$$

Using Eq. (12) we can easily convert back and forth between radians and degrees:

$$1 \text{ rad} = \frac{180^\circ}{\pi} \approx 57^\circ\, 17'\, 44.8'', \tag{12a}$$

$$1^\circ = \frac{\pi}{180} \text{ rad} \approx 0.01745 \text{ rad}. \tag{12b}$$

Radians	Degrees
0	0
$\pi/6$	30
$\pi/4$	45
$\pi/3$	60
$\pi/2$	90
$2\pi/3$	120
$3\pi/4$	135
$5\pi/6$	150
π	180
$3\pi/2$	270
2π	360
4π	720

FIGURE C.6 Some radian-degree conversions.

Figure C.6 shows radian-degree conversions for some common angles.

Now consider an angle of θ radians at the center of a circle of radius r (Fig. C.7). Denote by s the length of the arc subtended by θ; denote by A the area of the sector of the circle bounded by this angle. Then the proportions

$$\frac{s}{2\pi r} = \frac{A}{\pi r^2} = \frac{\theta}{2\pi}$$

give the formulas

$$s = r\theta \qquad (\theta \text{ in radians}) \tag{13}$$

and

$$A = \tfrac{1}{2} r^2 \theta \qquad (\theta \text{ in radians}). \tag{14}$$

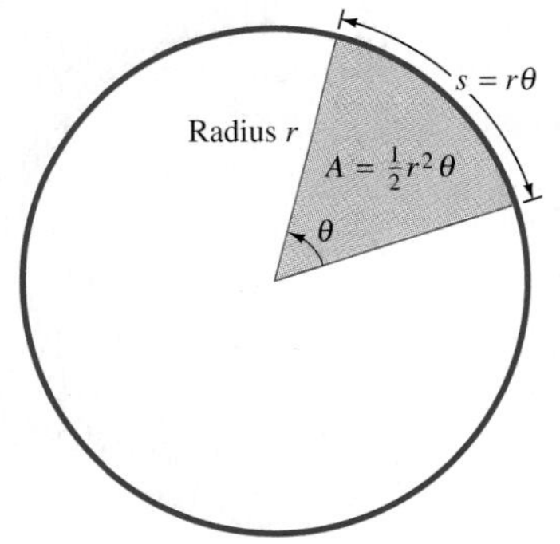

FIGURE C.7 The area of a sector and arc length of a circular arc.

The definitions in (2) refer to trigonometric functions of *angles* rather than trigonometric functions of *numbers*. Suppose that t is a real number. Then the number $\sin t$ is, *by definition*, the sine of an angle of t radians—recall that a positive angle is directed counterclockwise from the positive x-axis, whereas a negative angle is directed clockwise. Briefly, $\sin t$ is the sine of an angle of t *radians*. The other trigonometric functions of the number t have similar definitions. Hence, when we write $\sin t$, $\cos t$, and so on, with t a real number, it is *always* in reference to an angle to t *radians*.

When we need to refer to the sine of an angle of t *degrees*, we will henceforth write $\sin t^\circ$. The point is that $\sin t$ and $\sin t^\circ$ are quite different functions of the variable t. For example, you would get

$$\sin 1^\circ \approx 0.0175 \quad \text{and} \quad \sin 30^\circ = 0.5$$

on a calculator set in degree mode. But in radian mode, a calculator would give

$$\sin 1 \approx 0.8415 \quad \text{and} \quad \sin 30 \approx -0.9880.$$

The relationship between the functions $\sin t$ and $\sin t^\circ$ is

$$\sin t^\circ = \sin\left(\frac{\pi t}{180}\right). \tag{15}$$

The distinction extends even to programming languages. In FORTRAN, the function `SIN` is the radian sine function, and you must write $\sin t^\circ$ in the form `SIND(T)`. In BASIC you must write `SIN(Pi*T/180)` to get the correct value of the sine of an angle of t degrees.

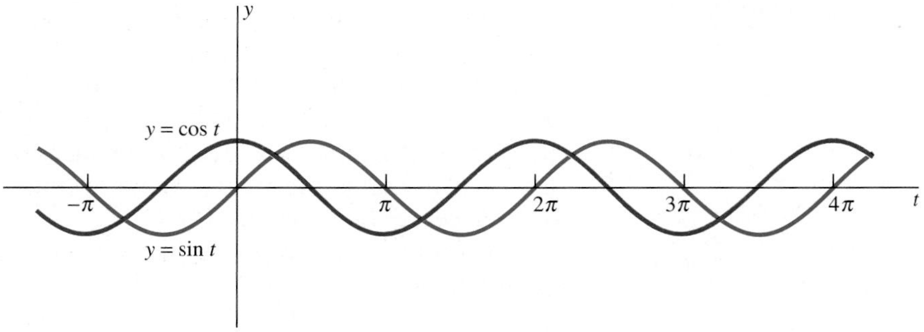

FIGURE C.8 Periodicity of the sine and cosine functions.

An angle of 2π rad corresponds to one revolution around the unit circle. This implies that the sine and cosine functions have **period** 2π, meaning that

$$\sin(t + 2\pi) = \sin t, \qquad \cos(t + 2\pi) = \cos t. \tag{16}$$

It follows from the equations in (16) that

$$\sin(t + 2n\pi) = \sin t \quad \text{and} \quad \cos(t + 2n\pi) = \cos t \tag{17}$$

for every integer n. This periodicity of the sine and cosine functions is evident in their graphs (Fig. C.8). From the equations in (3), the other four trigonometric functions also must be periodic, as their graphs in Figs. C.9 and C.10 show.

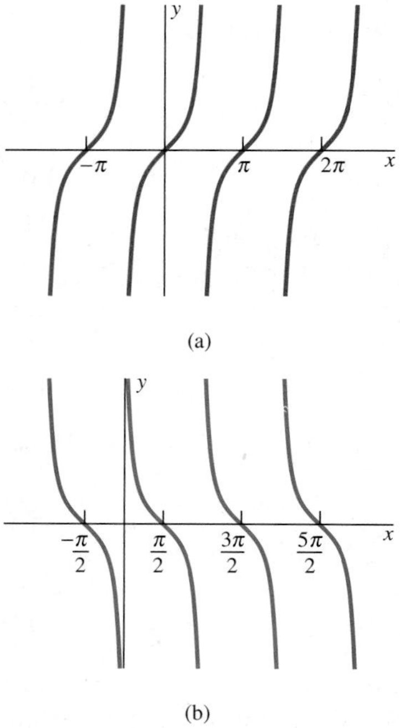

FIGURE C.9 The graphs of (a) the tangent function and (b) the cotangent function.

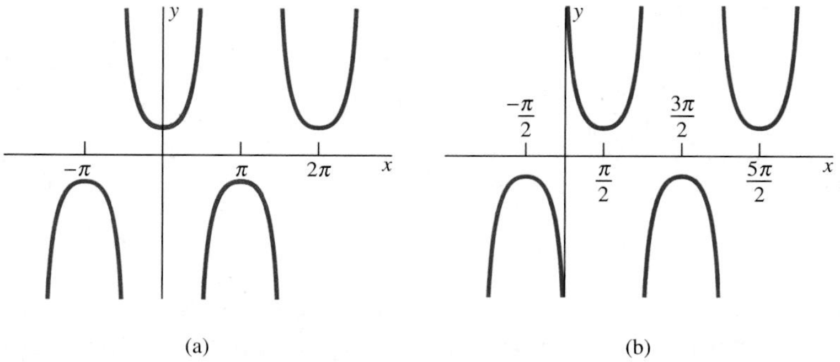

FIGURE C.10 The graphs of (a) the secant function and (b) the cosecant function.

We see from the equations in (2) that

$$\begin{aligned} &\sin 0 = 0, && \sin\frac{\pi}{2} = 1, && \sin\pi = 0, \\ &\cos 0 = 1, && \cos\frac{\pi}{2} = 0, && \cos\pi = -1. \end{aligned} \tag{18}$$

The trigonometric functions of $\pi/6$, $\pi/4$, and $\pi/3$ (the radian equivalents of 30°, 45°, and 60°, respectively) are easy to read from the well-known triangles of Fig. C.11. For instance,

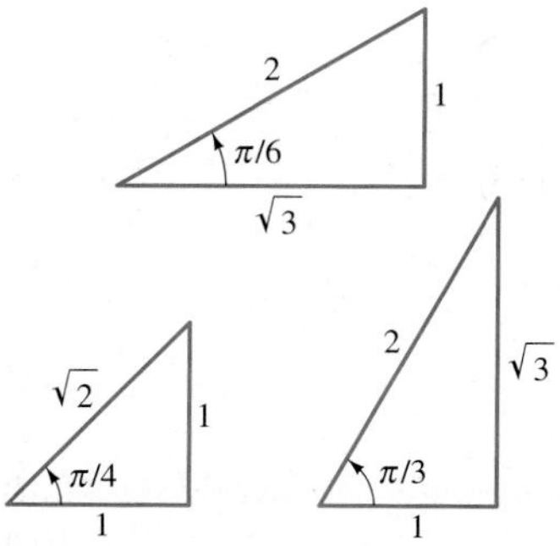

FIGURE C.11 Familiar right triangles.

$$\begin{aligned} &\sin\frac{\pi}{6} = \cos\frac{\pi}{3} = \frac{1}{2} = \frac{\sqrt{1}}{2}, \\ &\sin\frac{\pi}{4} = \cos\frac{\pi}{4} = \frac{1}{\sqrt{2}} = \frac{\sqrt{2}}{2}, \quad \text{and} \\ &\sin\frac{\pi}{3} = \cos\frac{\pi}{6} = \frac{\sqrt{3}}{2}. \end{aligned} \tag{19}$$

To find the values of trigonometric functions of angles larger than $\pi/2$, we can use their periodicity and the identities

$$\begin{aligned} \sin(\pi \pm \theta) &= \mp \sin\theta, \\ \cos(\pi \pm \theta) &= -\cos\theta \quad \text{and} \\ \tan(\pi \pm \theta) &= \pm\tan\theta \end{aligned} \tag{20}$$

(Problems 38, 39, and 40) as well as similar identities for the cosecant, secant, and cotangent functions.

EXAMPLE 1

$$\begin{aligned}
\sin\frac{5\pi}{4} &= \sin\left(\pi + \frac{\pi}{4}\right) = -\sin\frac{\pi}{4} = -\frac{\sqrt{2}}{2}; \\
\cos\frac{2\pi}{3} &= \cos\left(\pi - \frac{\pi}{3}\right) = -\cos\frac{\pi}{3} = -\frac{1}{2}; \\
\tan\frac{2\pi}{4} &= \tan\left(\pi - \frac{\pi}{4}\right) = -\tan\frac{\pi}{4} = -1; \\
\sin\frac{7\pi}{6} &= \sin\left(\pi + \frac{\pi}{6}\right) - \sin\frac{\pi}{6} = -\frac{1}{2}; \\
\cos\frac{5\pi}{3} &= \cos\left(2\pi - \frac{\pi}{3}\right) = \cos\left(-\frac{\pi}{3}\right) = \cos\frac{\pi}{3} = \frac{1}{2}; \\
\sin\frac{17\pi}{6} &= \sin\left(2\pi + \frac{5\pi}{6}\right) = \sin\frac{5\pi}{6} \\
&= \sin\left(\pi - \frac{\pi}{6}\right) = \sin\frac{\pi}{6} = \frac{1}{2}.
\end{aligned}$$

◆

EXAMPLE 2 Find the solutions (if any) of the equation

$$\sin^2 x - 3\cos^2 x + 2 = 0$$

that lie in the interval $[0, \pi]$.

Solution Using the fundamental identity in Eq. (5), we substitute $\cos^2 x = 1 - \sin^2 x$ into the given equation to obtain

$$\begin{aligned} \sin^2 x - 3(1 - \sin^2 x) + 2 &= 0; \\ 4\sin^2 x - 1 &= 0; \\ \sin x &= \pm\tfrac{1}{2}. \end{aligned}$$

Because $\sin x \geqq 0$ for x in $[0, \pi]$, $\sin x = -\frac{1}{2}$ is impossible. But $\sin x = \frac{1}{2}$ for $x = \pi/6$ and for $x = \pi - \pi/6 = 5\pi/6$. These are the solutions of the given equation that lie in $[0, \pi]$. ◆

APPENDIX C PROBLEMS

Express in radian measure the angles in Problems 1 through 5.

1. $40°$

2. $-270°$

3. $315°$

4. $210°$

5. $-150°$

In Problems 6 through 10, express in degrees the angles given in radian measure.

6. $\dfrac{\pi}{10}$

7. $\dfrac{2\pi}{5}$

8. 3π

9. $\dfrac{15\pi}{4}$

10. $\dfrac{23\pi}{60}$

In Problems 11 through 14, evaluate the six trigonometric functions of x at the given values.

11. $x = -\dfrac{\pi}{3}$

12. $x = \dfrac{3\pi}{4}$

13. $x = \dfrac{7\pi}{6}$

14. $x = \dfrac{5\pi}{3}$

Find all solutions x of each equation in Problems 15 through 23.

15. $\sin x = 0$

16. $\sin x = 1$

17. $\sin x = -1$

18. $\cos x = 0$

19. $\cos x = 1$

20. $\cos x = -1$

21. $\tan x = 0$

22. $\tan x = 1$

23. $\tan x = -1$

24. Suppose that $\tan x = \frac{3}{4}$ and that $\sin x < 0$. Find the values of the other five trigonometric functions of x.

25. Suppose that $\csc x = -\frac{5}{3}$ and that $\cos x > 0$. Find the values of the other five trigonometric functions of x.

Deduce the identities in Problems 26 and 27 from the fundamental identity

$$\cos^2\theta + \sin^2\theta = 1$$

and from the definitions of the other four trigonometric functions.

26. $1 + \tan^2\theta = \sec^2\theta$

27. $1 + \cot^2\theta = \csc^2\theta$

28. Deduce from the addition formulas for the sine and cosine the addition formula for the tangent:

$$\tan(x + y) = \frac{\tan x + \tan y}{1 - \tan x \,\tan y}.$$

In Problems 29 through 36, use the method of Example 1 to find the indicated values.

29. $\sin \dfrac{5\pi}{6}$

30. $\cos \dfrac{7\pi}{6}$

31. $\sin \dfrac{11\pi}{6}$

32. $\cos \dfrac{19\pi}{6}$

33. $\sin \dfrac{2\pi}{3}$

34. $\cos \dfrac{4\pi}{3}$

35. $\sin \dfrac{5\pi}{3}$

36. $\cos \dfrac{10\pi}{3}$

37. Apply the addition formula for the sine, cosine, and tangent functions (the latter from Problem 28) to show that if $0 < \theta < \pi/2$, then

(a) $\cos\left(\dfrac{\pi}{2} - \theta\right) = \sin\theta$;

(b) $\sin\left(\dfrac{\pi}{2} - \theta\right) = \cos\theta$;

(c) $\cot\left(\dfrac{\pi}{2} - \theta\right) = \tan\theta$.

The prefix *co-* is an abbreviation for the adjective *complementary,* which describes two angles whose sum is $\pi/2$. For example, $\pi/6$ and $\pi/3$ are complementary angles, so (a) implies that $\cos\pi/6 = \sin\pi/3$.

Suppose that $0 < \theta < \pi/2$. Derive the identities in Problems 38 through 40.

38. $\sin(\pi \pm \theta) = \mp\sin\theta$

39. $\cos(\pi \pm \theta) = -\cos\theta$

40. $\tan(\pi \pm \theta) = \pm\tan\theta$

41. The points $A(\cos\theta, -\sin\theta)$, $B(1, 0)$, $C(\cos\phi, \sin\phi)$, and $D(\cos(\theta + \phi), \sin(\theta + \phi))$ are shown in Fig. C.12; all are points on the unit circle. Deduce from the fact that the line segments AC and BD have the same length (because they are subtended by the same central angle $\theta + \phi$) that

$$\cos(\theta + \phi) = \cos\theta\,\cos\phi - \sin\theta\,\sin\phi.$$

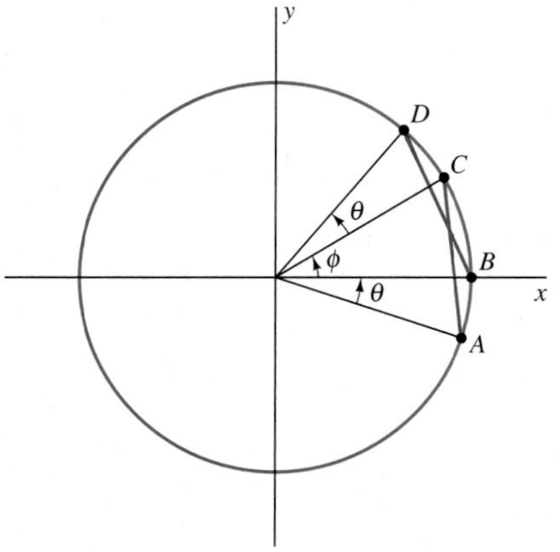

FIGURE C.12 Deriving the cosine addition formula (Problem 41).

42. (a) Use the triangles shown in Fig. C.13 to deduce that

$$\sin\left(\theta + \frac{\pi}{2}\right) = \cos\theta \quad\text{and}\quad \cos\left(\theta + \frac{\pi}{2}\right) = -\sin\theta.$$

(b) Use the results of Problem 41 and part (a) to derive the addition formula for the sine function.

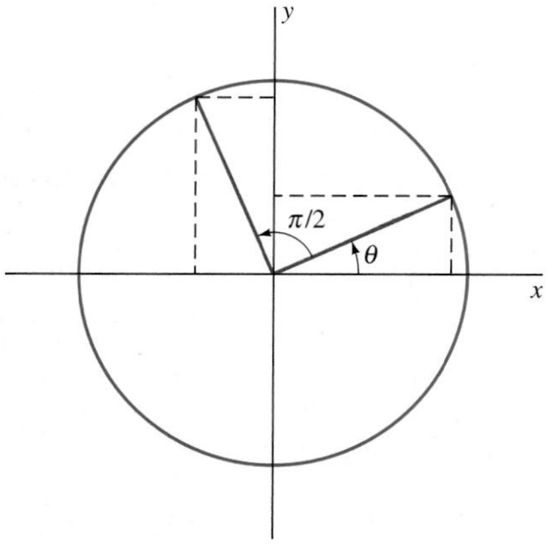

FIGURE C.13 Deriving the identities of Problem 42.

In Problems 43 through 48, find all solutions of the given equation that lie in the interval $[0, \pi]$.

43. $3\sin^2 x - \cos^2 x = 2$

44. $\sin^2 x = \cos^2 x$

45. $2\cos^2 x + 3\sin^2 x = 3$

46. $2\sin^2 x + \cos x = 2$

47. $8\sin^2 x\cos^2 x = 1$

48. $\cos 2\theta - 3\cos\theta = -2$

APPENDIX D: PROOFS OF THE LIMIT LAWS

Recall the definition of the limit:

$$\lim_{x\to a} F(x) = L$$

provided that, given $\epsilon > 0$, there exists a number $\delta > 0$ such that

$$0 < |x - a| < \delta \quad \text{implies that} \quad |F(x) - L| < \epsilon. \qquad \textbf{(1)}$$

Note that the number ϵ comes *first*. *Then* a value of $\delta > 0$ must be found so that the implication in (1) holds. To prove that $F(x) \to L$ as $x \to a$, you must, in effect, be able to stop the next person you see and ask him or her to pick a positive number ϵ at random. Then you must *always* be ready to respond with a positive number δ. This number δ must have the property that the implication in (1) holds for your number δ and the given number ϵ. The **only** restriction on x is that

$$0 < |x - a| < \delta,$$

as given in (1).

To do all this, you will ordinarily need to give an explicit method—a recipe or formula—for producing a value of δ that works for each value of ϵ. As Examples 1 through 3 show, the method will depend on the particular function F under study as well as the values of a and L.

EXAMPLE 1 Prove that $\lim_{x\to 3} (2x - 1) = 5$.

Solution Given $\epsilon > 0$, we must find $\delta > 0$ such that

$$|(2x - 1) - 5)| < \epsilon \quad \text{if} \quad 0 < |x - 3| < \delta.$$

Now

$$|(2x - 1) - 5| = |2x - 6| = 2|x - 3|,$$

so

$$0 < |x - 3| < \frac{\epsilon}{2} \quad \text{implies that} \quad |(2x - 1) - 5| < 2 \cdot \frac{\epsilon}{2} = \epsilon.$$

Hence, given $\epsilon > 0$, it suffices to choose $\delta = \epsilon/2$. This illustrates the observation that the required number δ is generally a function of the given number ϵ. ◆

EXAMPLE 2 Prove that $\lim_{x\to 2} (3x^2 + 5) = 17$.

Solution Given $\epsilon > 0$, we must find $\delta > 0$ such that

$$0 < |x - 2| < \delta \quad \text{implies that} \quad |(3x^2 + 5) - 17| < \epsilon.$$

Now

$$|(3x^2 + 5) - 17| = |3x^2 - 12| = 3 \cdot |x + 2| \cdot |x - 2|.$$

Our problem, therefore, is to show that $|x + 2| \cdot |x - 2|$ can be made as small as we please by choosing $x - 2$ sufficiently small. The idea is that $|x + 2|$ cannot be too large if $|x - 2|$ is fairly small. For example, if $|x - 2| < 1$, then

$$|x + 2| = |(x - 2) + 4| \leqq |x - 2| + 4 < 5.$$

Therefore,

$$0 < |x - 2| < 1 \quad \text{implies that} \quad |(3x^2 + 5) - 17| < 15 \cdot |x - 2|.$$

Consequently, let us choose δ to be the minimum of the two numbers 1 and $\epsilon/15$. Then

$$0 < |x-2| < \delta \quad \text{implies that} \quad |(3x^2+5)-17| < 15 \cdot \frac{\epsilon}{15} = \epsilon,$$

as desired. ◆

EXAMPLE 3 Prove that

$$\lim_{x\to a} \frac{1}{x} = \frac{1}{a} \quad \text{if} \quad a \neq 0.$$

Solution For simplicity, we will consider only the case in which $a > 0$ (the case $a < 0$ is similar).

Suppose that $\epsilon > 0$ is given. We must find a number δ such that

$$0 < |x-a| < \delta \quad \text{implies that} \quad \left|\frac{1}{x} - \frac{1}{a}\right| < \epsilon.$$

Now

$$\left|\frac{1}{x} - \frac{1}{a}\right| = \left|\frac{a-x}{ax}\right| = \frac{|x-a|}{a|x|}.$$

The idea is that $1/|x|$ cannot be too large if $|x-a|$ is fairly small. For example, if $|x-a| < a/2$, then $a/2 < x < 3a/2$. Therefore,

$$|x| > \frac{a}{2}, \quad \text{so} \quad \frac{1}{|x|} < \frac{2}{a}.$$

In this case it would follow that

$$\left|\frac{1}{x} - \frac{1}{a}\right| < \frac{2}{a^2} \cdot |x-a|$$

if $|x-a| < a/2$. Thus, if we choose δ to be the minimum of the two numbers $a/2$ and $a^2\epsilon/2$, then

$$0 < |x-a| < \delta \quad \text{implies that} \quad \left|\frac{1}{x} - \frac{1}{a}\right| < \frac{2}{a^2} \cdot \frac{a^2\epsilon}{2} = \epsilon.$$

Therefore

$$\lim_{x\to a} \frac{1}{x} = \frac{1}{a} \quad \text{if} \quad a \neq 0,$$

as desired. ◆

We are now ready to give proofs of the limit laws stated in Section 2.2.

Constant Law

If $f(x) \equiv C$, a constant, then

$$\lim_{x\to a} f(x) = \lim_{x\to a} C = C.$$

PROOF Because $|C - C| = 0$, we merely choose $\delta = 1$, regardless of the previously given value of $\epsilon > 0$. Then, if $0 < |x-a| < \delta$, it is automatic that $|C - C| < \epsilon$. ◄

Addition Law

If $\lim_{x\to a} F(x) = L$ and $\lim_{x\to a} G(x) = M$, then

$$\lim_{x\to a} [F(x) + G(x)] = L + M.$$

PROOF Let $\epsilon > 0$ be given. Because L is the limit of $F(x)$ as $x \to a$, there exists a number $\delta_1 > 0$ such that

$$0 < |x - a| < \delta_1 \quad \text{implies that} \quad |F(x) - L| < \frac{\epsilon}{2}.$$

Because M is the limit of $G(x)$ as $x \to a$, there exists a number $\delta_2 > 0$ such that

$$0 < |x - a| < \delta_2 \quad \text{implies that} \quad |G(x) - M| < \frac{\epsilon}{2}.$$

Let $\delta = \min\{\delta_1, \delta_2\}$. Then $0 < |x - a| < \delta$ implies that

$$|(F(x) + G(x)) - (L + M)| \leqq |F(x) - L| + |G(x) - M| < \frac{\epsilon}{2} + \frac{\epsilon}{2} = \epsilon.$$

Therefore

$$\lim_{x \to a} [F(x) + G(x)] = L + M,$$

as desired.

Product Law

If $\lim_{x \to a} F(x) = L$ and $\lim_{x \to a} G(x) = M$, then

$$\lim_{x \to a} [F(x) \cdot G(x)] = L \cdot M.$$

PROOF Given $\epsilon > 0$, we must find a number $\delta > 0$ such that

$$0 < |x - a| < \delta \quad \text{implies that} \quad |F(x) \cdot G(x) - L \cdot M| < \epsilon.$$

But first, the triangle inequality gives the result

$$\begin{aligned} |F(x) \cdot G(x) - L \cdot M| &= |F(x) \cdot G(x) - L \cdot G(x) + L \cdot G(x) - L \cdot M| \\ &\leqq |G(x)| \cdot |F(x) - L| + |L| \cdot |G(x) - M|. \end{aligned} \tag{2}$$

Because $\lim_{x \to a} F(x) = L$, there exists $\delta_1 > 0$ such that

$$0 < |x - a| < \delta_1 \quad \text{implies that} \quad |F(x) - L| < \frac{\epsilon}{2(|M| + 1)}. \tag{3}$$

And because $\lim_{x \to a} G(x) = M$, there exists $\delta_2 > 0$ such that

$$0 < |x - a| < \delta_2 \quad \text{implies that} \quad |G(x) - M| < \frac{\epsilon}{2(|L| + 1)}. \tag{4}$$

Moreover, there is a *third* number $\delta_3 > 0$ such that

$$0 < |x - a| < \delta_3 \quad \text{implies that} \quad |G(x) - M| < 1,$$

which in turn implies that

$$|G(x)| < |M| + 1, \tag{5}$$

We now choose $\delta = \min\{\delta_1, \delta_2, \delta_3\}$. Then we substitute (3), (4), and (5) into (2) and, finally, see that $0 < |x - a| < \delta$ implies that

$$\begin{aligned} |F(x) \cdot G(x) - L \cdot M| &< (|M| + 1) \cdot \frac{\epsilon}{2(|M| + 1)} + |L| \cdot \frac{\epsilon}{2(|L| + 1)} \\ &< \frac{\epsilon}{2} + \frac{\epsilon}{2} = \epsilon, \end{aligned}$$

as desired. The use of $|M| + 1$ and $|L| + 1$ in the denominators avoids the technical difficulty that arises should either L or M be zero.

Substitution Law

If $\lim_{x\to a} g(x) = L$ and $\lim_{x\to L} f(x) = f(L)$, then

$$\lim_{x\to a} f(g(x)) = f(L).$$

PROOF Let $\epsilon > 0$ be given. We must find a number $\delta > 0$ such that

$$0 < |x - a| < \delta \quad \text{implies that} \quad |f(g(x)) - f(L)| < \epsilon.$$

Because $f(y) \to f(L)$ as $y \to L$, there exists $\delta_1 > 0$ such that

$$0 < |y - L| < \delta_1 \quad \text{implies that} \quad |f(y) - f(L)| < \epsilon. \tag{6}$$

Also, because $g(x) \to L$ as $x \to a$, we can find $\delta > 0$ such that

$$0 < |x - a| < \delta \quad \text{implies that} \quad |g(x) - L| < \delta_1;$$

that is, such that

$$|y - L| < \delta_1,$$

where $y = g(x)$. From (6) we see that $0 < |x - a| < \delta$ implies that

$$|f(g(x)) - f(L)| = |f(y) - f(L)| < \epsilon,$$

as desired. ◄

Reciprocal Law

If $\lim_{x\to a} g(x) = L$ and $L \neq 0$, then

$$\lim_{x\to a} \frac{1}{g(x)} = \frac{1}{L}.$$

PROOF Let $f(x) = 1/x$. Then, as we saw in Example 3,

$$\lim_{x\to a} f(x) = \lim_{x\to a} \frac{1}{x} = \frac{1}{L} = f(L).$$

Hence the substitution law gives the result

$$\lim_{x\to a} \frac{1}{g(x)} = \lim_{x\to a} f(g(x)) = f(L) = \frac{1}{L},$$

as desired. ◄

Quotient Law

If $\lim_{x\to a} F(x) = L$ and $\lim_{x\to a} G(x) = M \neq 0$, then

$$\lim_{x\to a} \frac{F(x)}{G(x)} = \frac{L}{M}.$$

PROOF It follows immediately from the product and reciprocal laws that

$$\lim_{x\to a} \frac{F(x)}{G(x)} = \lim_{x\to a} F(x) \cdot \frac{1}{G(x)} = \left(\lim_{x\to a} F(x)\right)\left(\lim_{x\to a} \frac{1}{G(x)}\right) = L \cdot \frac{1}{M} = \frac{L}{M},$$

as desired. ◄

Squeeze Law

Suppose that $f(x) \leqq g(x) \leqq h(x)$ in some deleted neighborhood of a and that

$$\lim_{x\to a} f(x) = L = \lim_{x\to a} h(x).$$

Then

$$\lim_{x\to a} g(x) = L.$$

PROOF Given $\epsilon > 0$, we choose $\delta_1 > 0$ and $\delta_2 > 0$ such that

$$0 < |x - a| < \delta_1 \quad \text{implies that} \quad |f(x) - L| < \epsilon$$

and

$$0 < |x - a| < \delta_2 \quad \text{implies that} \quad |h(x) - L| < \epsilon.$$

Let $\delta = \min\{\delta_1, \delta_2\}$. Then $\delta > 0$. Moreover, if $0 < |x - a| < \delta$, then both $f(x)$ and $h(x)$ are points of the open interval $(L - \epsilon, L + \epsilon)$. So

$$L - \epsilon < f(x) \leqq g(x) \leqq h(x) < L + \epsilon.$$

Thus

$$0 < |x - a| < \delta \quad \text{implies that} \quad |g(x) - L| < \epsilon,$$

as desired. ◄

APPENDIX D PROBLEMS

In Problems 1 through 10, apply the definition of the limit to establish the given equality.

1. $\lim_{x \to a} x = a$

2. $\lim_{x \to 2} 3x = 6$

3. $\lim_{x \to 2} (x + 3) = 5$

4. $\lim_{x \to -3} (2x + 1) = -5$

5. $\lim_{x \to 1} x^2 = 1$

6. $\lim_{x \to a} x^2 = a^2$

7. $\lim_{x \to -1} (2x^2 - 1) = 1$

8. $\lim_{x \to a} \dfrac{1}{x^2} = \dfrac{1}{a^2}$

9. $\lim_{x \to a} \dfrac{1}{x^2 + 1} = \dfrac{1}{a^2 + 1}$

10. $\lim_{x \to a} \dfrac{1}{\sqrt{x}} = \dfrac{1}{\sqrt{a}}$ if $a > 0$

11. Suppose that

$$\lim_{x \to a} f(x) = L \quad \text{and} \quad \lim_{x \to a} f(x) = M.$$

Apply the definition of the limit to prove that $L = M$. Thus the limit of the function f at $x = a$ is unique if it exists.

12. Suppose that C is a constant and that $f(x) \to L$ as $x \to a$. Apply the definition of the limit to prove that

$$\lim_{x \to a} C \cdot f(x) = C \cdot L.$$

13. Suppose that $L \neq 0$ and that $f(x) \to L$ as $x \to a$. Use the method of Example 3 and the definition of the limit to show directly that

$$\lim_{x \to a} \frac{1}{f(x)} = \frac{1}{L}.$$

14. Use the algebraic identity

$$x^n - a^n = (x - a)(x^{n-1} + x^{n-2}a + x^{n-3}a^2 + \cdots + xa^{n-2} + a^{n-1})$$

to show directly from the definition of the limit that $\lim_{x \to a} x^n = a^n$ if n is a positive integer.

15. Apply the identity

$$\left|\sqrt{x} - \sqrt{a}\right| = \frac{|x - a|}{\sqrt{x} + \sqrt{a}}$$

to show directly from the definition of the limit that $\lim_{x \to a} \sqrt{x} = \sqrt{a}$ if $a > 0$.

16. Suppose that $f(x) \to f(a) > 0$ as $x \to a$. Prove that there exists a neighborhood of a on which $f(x) > 0$; that is, prove that there exists $\delta > 0$ such that

$$|x - a| < \delta \quad \text{implies that} \quad f(x) > 0.$$

APPENDIX E: THE COMPLETENESS OF THE REAL NUMBER SYSTEM

Here we present a self-contained treatment of those consequences of the completeness of the real number system that are relevant to this text. Our principal objective is to prove the intermediate value theorem and the maximum value theorem. We begin with the least upper bound property of the real numbers, which we take to be an axiom.

DEFINITION Upper Bound and Lower Bound

The set S of real numbers is said to be **bounded above** if there is a number b such that $x \leqq b$ for every number x in S, and the number b is then called an **upper bound** for S. Similarly, if there is a number a such that $x \geqq a$ for every number x in S, then S is said to be **bounded below,** and a is called a **lower bound** for S.

DEFINITION Least Upper Bound and Greatest Lower Bound
The number λ is said to be a **least upper bound** for the set S of real numbers provided that

1. λ is an upper bound for S, and
2. If b is an upper bound for S, then $\lambda \leqq b$.

Similarly, the number γ is said to be a **greatest lower bound** for S if γ is a lower bound for S and $\gamma \geqq a$ for every lower bound a of S.

EXERCISE Prove that if the set S has a least upper bound λ, then it is unique. That is, prove that if λ and μ are least upper bounds for S, then $\lambda = \mu$.

It is easy to show that the greatest lower bound γ of a set S, if any, is also unique. At this point you should construct examples to illustrate that a set with a least upper bound λ may or may not contain λ and that a similar statement is true of the set's greatest lower bound.

We now state the *completeness axiom* of the real number system.

Least Upper Bound Axiom

If the nonempty set S of real numbers has an upper bound, then it has a least upper bound.

By working with the set T consisting of the numbers $-x$, where x is in S, it is not difficult to show the following consequence of the least upper bound axiom: If the nonempty set S of real numbers is bounded below, then S has a greatest lower bound. Because of this symmetry, we need only one axiom, not two; results for least upper bounds also hold for greatest lower bounds, provided that some attention is paid to the directions of the inequalities.

The restriction that S be nonempty is annoying but necessary. If S is the "empty" set of real numbers, then 15 is an upper bound for S, but S has no least upper bound because $14, 13, 12, \ldots, 0, -1, -2, \ldots$ are also upper bounds for S.

DEFINITION Increasing, Decreasing, and Monotonic Sequences
The infinite sequence $x_1, x_2, x_3, \ldots, x_k, \ldots$ is said to be **nondecreasing** if $x_n \leqq x_{n+1}$ for every $n \geqq 1$. This sequence is said to be **nonincreasing** if $x_n \geqq x_{n+1}$ for every $n \geqq 1$. If the sequence $\{x_n\}$ is either nonincreasing or nondecreasing, then it is said to be **monotonic.**

Theorem 1 gives the **bounded monotonic sequence property** of the set of real numbers. (Recall that a set S of real numbers is said to be **bounded** if it is contained in an interval of the form $[a, b]$.)

THEOREM 1 Bounded Monotonic Sequences
Every bounded monotonic sequence of real numbers converges.

PROOF Suppose that the sequence

$$S = \{x_n\} = \{x_1, x_2, x_3, \ldots, x_k, \ldots\}$$

is bounded and nondecreasing. By the least upper bound axiom, S has a least upper bound λ. We claim that λ is the limit of the sequence $\{x_n\}$. Consider an open interval centered at λ—that is, an interval of the form $I = (\lambda - \epsilon, \lambda + \epsilon)$, where $\epsilon > 0$. Some terms of the sequence must lie within I, else $\lambda - \epsilon$ would be an upper bound for

S that is less than its least upper bound λ. But if x_N is in I, then—because we are dealing with a nondecreasing sequence—$x_N \leqq x_k \leqq \lambda$ for all $k \geqq N$. That is, x_k is in I for all $k \geqq N$. Because ϵ is an arbitrary positive number, λ is—almost by definition—the limit of the sequence $\{x_n\}$. Thus we have shown that a bounded nonincreasing sequence converges. A similar proof can be constructed for nonincreasing sequences by working with the greatest lower bound. ◄

Therefore, the least upper bound axiom implies the bounded monotonic sequence property of the real numbers. With just a little effort, you can prove that the two are logically equivalent. That is, if you take the bounded monotonic sequence property as an axiom, then the least upper bound property follows as a theorem. The *nested interval property* of Theorem 2 is also equivalent to the least upper bound property, but we shall prove only that it follows from the least upper bound property, because we have chosen the latter as the fundamental completeness axiom for the real number system.

THEOREM 2 Nested Interval Property of the Real Numbers

Suppose that $I_1, I_2, I_3, \dots, I_n, \dots$ is a sequence of closed intervals (so I_n is of the form $[a_n, b_n]$ for each positive integer n) such that

1. I_n contains I_{n+1} for each $n \geqq 1$, and
2. $\lim_{n\to\infty} (b_n - a_n) = 0.$

Then there exists exactly one real number c such that c belongs to I_n for all n. Thus

$$\{c\} = I_1 \cap I_2 \cap I_3 \cap \cdots.$$

PROOF It is clear from hypothesis (2) of Theorem 2 that there is at most one such number c. The sequence $\{a_n\}$ of the left-hand endpoints of the intervals is a bounded (by b_1) nondecreasing sequence and thus has a limit a by the bounded monotonic sequence property. Similarly, the sequence $\{b_n\}$ has a limit b. Because $a_n \leqq b_n$ for all n, it follows easily that $a \leqq b$. It is clear that $a_n \leqq a \leqq b \leqq b_n$ for all $n \geqq 1$, so a and b belong to every interval I_n. But then hypothesis (2) of Theorem 2 implies that $a = b$, and clearly this common value—call it c—is the number satisfying the conclusion of Theorem 2. ◄

We can now use these results to prove several important theorems used in the text.

THEOREM 3 Intermediate Value Property of Continuous Functions

If the function f is continuous on the interval $[a, b]$ and $f(a) < K < f(b)$, then $K = f(c)$ for some number c in (a, b).

PROOF Let $I_1 = [a, b]$. Suppose that I_n has been defined for $n \geqq 1$. We describe (inductively) how to define I_{n+1}, and this shows in particular how to define I_2, I_3, and so forth. Let a_n be the left-hand endpoint of I_n, b_n be its right-hand endpoint, and m_n be its midpoint. There are now three cases to consider: $f(m_n) > K$, $f(m_n) < K$, and $f(m_n) = K$.

If $f(m_n) > K$, then $f(a_n) < K < f(m_n)$; in this case, let $a_{n+1} = a_n$, $b_{n+1} = m_n$, and $I_{n+1} = [a_{n+1}, b_{n+1}]$.

If $f(m_n) < K$, then let $a_{n+1} = m_n$, $b_{n+1} = b_n$, and $I_{n+1} = [a_{n+1}, b_{n+1}]$.

If $f(m_n) = K$, then we simply let $c = m_n$ and the proof is complete. Otherwise, at each stage we bisect I_n and let I_{n+1} be the half of I_n on which f takes on values both above and below K.

It is easy to show that the sequence $\{I_n\}$ of intervals satisfies the hypotheses of Theorem 2. Let c be the (unique) real number common to all the intervals I_n. We will show that $f(c) = K$, and this will conclude the proof.

The sequence $\{b_n\}$ has limit c, so by the continuity of f, the sequence $\{f(b_n)\}$ has limit $f(c)$. But $f(b_n) > K$ for all n, so the limit of $\{f(b_n)\}$ can be no less than K; that is, $f(c) \geqq K$. By considering the sequence $\{a_n\}$, it follows that $f(c) \leqq K$ as well. Therefore, $f(c) = K$. ◄

LEMMA 1

If f is continuous on the closed interval $[a, b]$, then f is bounded there.

PROOF Suppose by way of contradiction that f is not bounded on $I_1 = [a, b]$. Bisect I_1 and let I_2 be either half of I_1 on which f is unbounded. (If f is unbounded on both halves, let $I_2 = I_1$.) In general, let I_{n+1} be a half of I_n on which f is unbounded.

Again it is easy to show that the sequence $\{I_n\}$ of closed intervals satisfies the hypotheses of Theorem 2. Let c be the number common to them all. Because f is continuous, there exists a number $\epsilon > 0$ such that f is bounded on the interval $(c - \epsilon, c + \epsilon)$. But for sufficiently large values of n, I_n is a subset of $(c - \epsilon, c + \epsilon)$. This contradiction shows that f must be bounded on $[a, b]$. ◄

THEOREM 4 Maximum Value Property of Continuous Functions

If the function f is continuous on the closed and bounded interval $[a, b]$, then there exists a number c in $[a, b]$ such that $f(x) \leqq f(c)$ for all x in $[a, b]$.

PROOF Consider the set $S = \{f(x) \mid a \leqq x \leqq b\}$. By Lemma 1, this set is bounded, and it is certainly nonempty. Let λ be the least upper bound of S. Our goal is to show that λ is a value $f(x)$ of f.

With $I_1 = [a, b]$, bisect I_1 as before. Note that λ is the least upper bound of the values of f on at least one of the two halves of I_1; let I_2 be that half. Having defined I_n, let I_{n+1} be the half of I_n on which λ is the least upper bound of the values of f. Let c be the number common to all these intervals. It then follows from the continuity of f, much as in the proof of Theorem 3, that $f(c) = \lambda$. And it is clear that $f(x) \leqq \lambda$ for all x in $[a, b]$. ◄

The technique we are using in these proof is called the *method of bisection*. We now use it once again to establish the *Bolzano–Weierstrass property* of the real number system.

DEFINITION Limit Point

Let S be a set of real numbers. The number p is said to be a **limit point** of S if every open interval containing p also contains points of S other than p.

BOLZANO–WEIERSTRASS THEOREM

Every bounded infinite set of real numbers has a limit point.

PROOF Let I_0 be a closed interval containing the bounded infinite set S of real numbers. Bisect I_0. Let I_1 be one of the resulting closed half-intervals of I_0 that contains infinitely many points of S. If I_n has been chosen, let I_{n+1} be one of the closed half-intervals of I_n containing infinitely many points of S. An application of Theorem 2 yields a number p common to all the intervals I_n. If J is an open interval containing p, then J contains I_n for some sufficiently large value of n and thus contains infinitely many points of S. Therefore p is a limit point of S. ◄

Our final goal is in sight: We can now prove that a sequence of real numbers converges if and only if it is a *Cauchy sequence*.

DEFINITION Cauchy Sequence
The sequence $\{a_n\}_1^\infty$ is said to be a **Cauchy sequence** if, for every $\epsilon > 0$, there exists an integer N such that

$$|a_m - a_n| < \epsilon$$

for all $m, n \geqq N$.

LEMMA 2 Convergent Subsequences
Every bounded sequence of real numbers has a convergent subsequence.

PROOF If $\{a_n\}$ has only a finite number of values, then the conclusion of Lemma 2 follows easily. We therefore focus our attention on the case in which $\{a_n\}$ is an infinite set. It is easy to show that this set is also bounded, and thus we may apply the Bolzano–Weierstrass theorem to obtain a limit point p of $\{a_n\}$.

For each integer $k \geqq 1$, let $a_{n(k)}$ be a term of the sequence $\{a_n\}$ such that

1. $n(k+1) > n(k)$ for all $k \geqq 1$, and
2. $|a_{n(k)} - p| < \dfrac{1}{k}$.

It is then easy to show that $\{a_{n(k)}\}$ is a convergent (to p) subsequence of $\{a_n\}$. ◄

THEOREM 6 Convergence of Cauchy Sequences
A sequence of real numbers converges if and only if it is a Cauchy sequence.

PROOF It follows immediately from the triangle inequality that every convergent sequence is a Cauchy sequence. Thus suppose that the sequence $\{a_n\}$ is a Cauchy sequence.

Choose N such that

$$|a_m - a_n| < 1$$

if $m, n \geqq N$. It follows that if $n \geqq N$, then a_n lies in the closed interval $[a_N - 1,\ a_N + 1]$. This implies that the sequence $\{a_n\}$ is bounded, and thus by Lemma 2 it has a convergent subsequence $\{a_{n(k)}\}$. Let p be the limit of this subsequence.

We claim that $\{a_n\}$ itself converges to p. Given $\epsilon > 0$, choose M such that

$$|a_m - a_n| < \frac{\epsilon}{2}$$

if $m, n \geqq M$. Next choose K such that $n(K) \geqq M$ and

$$|a_{n(K)} - p| < \frac{\epsilon}{2}.$$

Then if $n \geqq M$,

$$|a_n - p| \leqq |a_n - a_{n(K)}| + |a_{n(K)} - p| < \epsilon.$$

Therefore, $\{a_n\}$ converges to p by definition. ◄

APPENDIX F: EXISTENCE OF THE INTEGRAL

When the basic computational algorithms of the calculus were discovered by Newton and Leibniz in the latter half of the seventeenth century, the logical rigor that had been a feature of the Greek method of exhaustion was largely abandoned. When computing the area A under the curve $y = f(x)$, for example, Newton took it as intuitively obvious that the area function existed, and he proceeded to compute it as the antiderivative of the height function $f(x)$. Leibniz regarded A as an infinite sum of infinitesimal area elements, each of the form $dA = f(x)\,dx$, but in practice computed the area

$$A = \int_a^b f(x)\,dx$$

by antidifferentiation just as Newton did—that is, by computing

$$A = \Big[D^{-1} f(x)\Big]_a^b.$$

The question of the *existence* of the area function—one of the conditions that a function f must satisfy in order for its integral to exist—did not at first seem to be of much importance. Eighteenth-century mathematicians were mainly occupied (and satisfied) with the impressive applications of calculus to the solution of real-world problems and did not concentrate on the logical foundations of the subject.

The first attempt at a precise definition of the integral and a proof of its existence for continuous functions was that of the French mathematician Augustin Louis Cauchy (1789–1857). Curiously enough, Cauchy was trained as an engineer, and much of his research in mathematics was in fields that we today regard as applications-oriented: hydrodynamics, waves in elastic media, vibrations of elastic membranes, polarization of light, and the like. But he was a prolific researcher, and his writings cover the entire spectrum of mathematics, with occasional essays into almost unrelated fields.

Around 1824, Cauchy defined the integral of a continuous function in a way that is familiar to us, as a limit of left-endpoint approximations:

$$\int_a^b f(x)\,dx = \lim_{\Delta x \to 0} \sum_{i=1}^{n} f(x_{i-1})\,\Delta x.$$

This is a much more complicated sort of limit than the ones we discussed in Chapter 2. Cauchy was not entirely clear about the nature of the limit process involved in this equation, nor was he clear about the precise role that the hypothesis of the continuity of f played in proving that the limit exists.

A complete definition of the integral, as we gave in Section 5.4, was finally produced in the 1850s by the German mathematician Georg Bernhard Riemann. Riemann was a student of Gauss; he met Gauss upon his arrival at Göttingen, Germany, for the purpose of studying theology, when he was about 20 years old and Gauss was about 70. Riemann soon decided to study mathematics and became known as one of the truly great mathematicians of the nineteenth century. Like Cauchy, he was particularly interested in applications of mathematics to the real world; his research emphasized electricity, heat, light, acoustics, fluid dynamics, and—as you might infer from the fact that Wilhelm Weber was a major influence on Riemann's education—magnetism. Riemann also made significant contributions to mathematics itself, particularly in the field of complex analysis. A major conjecture of his, involving the zeta function

$$\zeta(s) = \sum_{n=1}^{\infty} \frac{1}{n^s}, \tag{1}$$

remains unsolved to this day. This conjecture has important consequences in the theory of the distribution of prime numbers because

$$\zeta(k) = \prod \left(1 - \frac{1}{p^k}\right)^{-1},$$

where the product $\prod$ is taken over all primes p. [The zeta function is defined in Eq. (1) for complex numbers s to the right of the vertical line at $x = 1$ and is extended to other complex numbers by the requirement that it be differentiable.] Riemann died of tuberculosis shortly before his fortieth birthday.

Here we give a proof of the existence of the integral of a continuous function. We will follow Riemann's approach. Specifically, suppose that the function f is continuous on the closed and bounded interval $[a, b]$. We will prove that the definite integral

$$\int_a^b f(x)\,dx$$

exists. That is, we will demonstrate the existence of a number I that satisfies the following condition: For every $\epsilon > 0$ there exists $\delta > 0$ such that, for *every* Riemann sum R associated with *any* partition P with $|P| < \delta$,

$$|I - R| < \epsilon.$$

(Recall that the norm $|P|$ of the partition P is the length of the longest subinterval in the partition.) In other words, every Riemann sum associated with every sufficiently "fine" partition is close to the number I. If this happens, then the definite integral

$$\int_a^b f(x)\,dx$$

is said to **exist,** and I is its **value.**

Now we begin the proof. Suppose throughout that f is a function continuous on the closed interval $[a, b]$. Given $\epsilon > 0$, we need to show the existence of a number $\delta > 0$ such that

$$\left| I - \sum_{i=1}^{n} f(x_i^\star)\,\Delta x_i \right| < \epsilon \qquad \textbf{(2)}$$

for every Riemann sum associated with any partition P of $[a, b]$ with $|P| < \delta$.

Given a partition P of $[a, b]$ into n subintervals that are *not necessarily of equal length,* let p_i be a point in the subinterval $[x_{i-1}, x_i]$ at which f attains its minimum value $f(p_i)$. Similarly, let $f(q_i)$ be its maximum value there. These numbers exist for $i = 1, 2, 3, \ldots, n$ because of the maximum value property of continuous functions (Theorem 4 of Appendix E.)

In what follows we will denote the resulting lower and upper Riemann sums associated with P by

$$L(P) = \sum_{i=1}^{n} f(p_i)\,\Delta x_i \qquad \textbf{(3a)}$$

and

$$U(P) = \sum_{i=1}^{n} f(q_i)\,\Delta x_i, \qquad \textbf{(3b)}$$

respectively. Then Lemma 1 is obvious.

LEMMA 1

For any partition P of $[a, b]$, $L(P) \leqq U(P)$.

Now we need a definition. The partition P' is called a **refinement** of the partition P if each subinterval of P' is contained in some subinterval of P. That is, P' is obtained from P by adding more points of subdivision to P.

LEMMA 2

Suppose that P' is a refinement of P. Then

$$L(P) \leqq L(P') \leqq U(P') \leqq U(P). \tag{4}$$

PROOF The inequality $L(P') \leqq U(P')$ is a consequence of Lemma 1. We will show that $L(P) \leqq L(P')$; the proof that $U(P') \leqq U(P)$ is similar.

The refinement P' is obtained from P by adding one or more points of subdivision to P. So all we need show is that the Riemann sum $L(P)$ cannot be decreased by adding a single point of subdivision. Thus we will suppose that the partition P' is obtained from P by dividing the kth subinterval $[x_{k-1}, x_k]$ of P into two subintervals $[x_{k-1}, z]$ and $[z, x_k]$ by means of the new subdivision point z.

The only resulting effect on the corresponding Riemann sum is to replace the term

$$f(p_k) \cdot (x_k - x_{k-1})$$

in $L(P)$ with the two-term sum

$$f(u) \cdot (z - x_{k-1}) + f(v) \cdot (x_k - z),$$

where $f(u)$ is the minimum of f on $[x_{k-1}, z]$ and $f(v)$ is the minimum of f on $[z, x_k]$. But

$$f(p_k) \leqq f(u) \quad \text{and} \quad f(p_k) \leqq f(v).$$

Hence

$$\begin{aligned} f(u) \cdot (z - x_{k-1}) + f(v) \cdot (x_k - z) &\geqq f(p_k) \cdot (z - x_{k-1}) + f(p_k) \cdot (x_k - z) \\ &= f(p_k) \cdot (z - x_{k-1} + x_k - z) \\ &= f(p_k) \cdot (x_k - x_{k-1}). \end{aligned}$$

So the replacement of $f(p_k) \cdot (x_k - x_{k-1})$ cannot decrease the sum $L(P)$ in question, and therefore $L(P) \leqq L(P')$. Because this is all we needed to show, we have completed the proof of Lemma 2. ◀

To prove that all the Riemann sums for sufficiently fine partitions are close to some number I, we must first give a construction of I. This is accomplished through Lemma 3.

LEMMA 3

Let P_n denote the regular partition of $[a, b]$ into 2^n subintervals of equal length. Then the (sequential) limit

$$I = \lim_{n \to \infty} L(P_n) \tag{5}$$

exists.

PROOF We begin with the observation that each partition P_{n+1} is a refinement of P_n, so (by Lemma 2)

$$L(P_1) \leqq L(P_2) \leqq \cdots \leqq L(P_n) \leqq \cdots.$$

Therefore $\{L(P_n)\}$ is a nondecreasing sequence of real numbers. Moreover,

$$L(P_n) = \sum_{i=1}^{2^n} f(p_i)\Delta x_i \leqq M \sum_{i-1}^{2^n} \Delta x_i = M(b - a),$$

where M is the maximum value of f on $[a, b]$.

Theorem 1 of Appendix E guarantees that a bounded monotonic sequence of real numbers must converge. Thus the number

$$I = \lim_{n\to\infty} L(P_n)$$

exists. This establishes Eq. (5), and the proof of Lemma 3 is complete. ◀

It is proved in advanced calculus that if f is continuous on $[a, b]$, then—for every number $\epsilon > 0$—there exists a number $\delta > 0$ such that

$$|f(u) - f(v)| < \epsilon$$

for every two points u and v of $[a, b]$ such that

$$|u - v| < \delta.$$

This property of a function is called **uniform continuity** of f on the interval $[a, b]$. Thus the theorem from advanced calculus that we need to use states that every continuous function on a closed and bounded interval is uniformly continuous there.

Note The fact that f is continuous on $[a, b]$ means that for each number u in the interval and each $\epsilon > 0$, there exists $\delta > 0$ such that if v is a number in the interval with $|u - v| < \delta$, then $|f(u) - f(v)| < \epsilon$. But *uniform* continuity is a more stringent condition. It means that given $\epsilon > 0$, you can find not only a value δ_1 that "works" for u_1, a value δ_2 that works for u_2, and so on, but more: You can find a universal value of $\delta > 0$ that works for *all* values of u in the interval. This should not be obvious when you notice the possibility that $\delta_1 = 1$, $\delta_2 = \frac{1}{2}$, $\delta_3 = \frac{1}{3}$, and so on. In any case, it is clear that uniform continuity of f on an interval implies its continuity there.

Remember that throughout we have a continuous function f defined on the closed interval $[a, b]$.

LEMMA 4

Suppose that $\epsilon > 0$ is given. Then there exists a number $\delta > 0$ such that if P is a partition of $[a, b]$ with $|P| < \delta$ and P' is a refinement of P, then

$$|R(P) - R(P')| < \frac{\epsilon}{3} \qquad \textbf{(6)}$$

for any two Riemann sums $R(P)$ associated with P and $R(P')$ associated with P'.

PROOF Because f must be uniformly continuous on $[a, b]$, there exists a number $\delta > 0$ such that if

$$|u - v| < \delta, \quad \text{then} \quad |f(u) - f(v)| < \frac{\epsilon}{3(b-a)}.$$

Suppose now that P is a partition of $[a, b]$ with $|P| < \delta$. Then

$$|U(P) - L(P)| = \sum_{i=1}^{n} |f(q_i) - f(p_i)|\Delta x_i < \frac{\epsilon}{3(b-a)} \sum_{i=1}^{n} \Delta x_i = \frac{\epsilon}{3}.$$

This is valid because $|p_i - q_i| < \delta$, for both p_i and q_i belong to the same subinterval $[x_{i-1}, x_i]$ of P, and $|P| < \delta$.

Now, as shown in Fig. F.1, we know that $L(P)$ and $U(P)$ differ by less than $\epsilon/3$. We know also that

$$L(P) \leqq R(P) \leqq U(P)$$

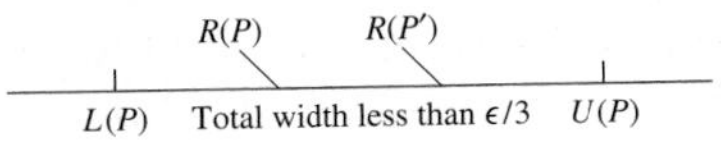

FIGURE F.1 Part of the proof of Lemma 4.

for every Riemann sum $R(P)$ associated with P. But

$$L(P) \leqq L(P') \leqq U(P') \leqq U(P)$$

by Lemma 2, because P' is a refinement of P; moreover,

$$L(P') \leqq R(P') \leqq U(P')$$

for every Riemann sum $R(P')$ associated with P'.

As Fig. F.1 shows, both the numbers $R(P)$ and $R(P')$ lie in the interval $[L(P), U(P)]$ of length less than $\epsilon/3$, so Eq. (6) follows. This concludes the proof of Lemma 4. ◄

THEOREM 1 Existence of the Integral
If f is continuous on the closed and bounded interval $[a, b]$, then the integral

$$\int_a^b f(x)\,dx$$

exists.

PROOF Suppose that $\epsilon > 0$ is given. We must show the existence of a number $\delta > 0$ such that, for every partition P of $[a, b]$ with $|P| < \delta$, we have

$$|I - R(P)| < \epsilon.$$

where I is the number given in Lemma 3 and $R(P)$ is an arbitrary Riemann sum for f associated with P.

We choose the number δ provided by Lemma 4 such that

$$|R(P) - R(P')| < \frac{\epsilon}{3}$$

if $|P| < \delta$ and P' is a refinement of P.

By Lemma 3, we can choose an integer N so large that

$$|P_N| < \delta \quad \text{and} \quad |L(P_N) - I| < \frac{\epsilon}{3}. \tag{7}$$

Given an arbitrary partition P such that $|P| < \delta$, let P' be a common refinement of both P and P_N. You can obtain such a partition P', for example, by using all the points of subdivision of both P and P_N to form the subintervals of $[a, b]$ that constitute P'.

Because P' is a refinement of both P and P_N and both the latter partitions have mesh less than δ, Lemma 4 implies that

$$|R(P) - R(P')| < \frac{\epsilon}{3} \quad \text{and} \quad |L(P_N) - R(P')| < \frac{\epsilon}{3}. \tag{8}$$

Here $R(P)$ and $R(P')$ are (arbitrary) Riemann sums associated with P and P', respectively.

Given an arbitrary Riemann sum $R(P)$ associated with the partition P with mesh less than δ, we see that

$$\begin{aligned}|I - R(P)| &= |I - L(P_N) + L(P_N) - R(P') + R(P') - R(P)|\\ &\leqq |I - L(P_N)| + |L(P_N) - R(P')| + |R(P') - R(P)|.\end{aligned}$$

In the last sum, both of the last two terms are less than $\epsilon/3$ by virtue of the inequalities in (8). We also know, by (7), that the first term is less than $\epsilon/3$. Consequently,

$$|I - R(P)| < \epsilon.$$

This establishes Theorem 1. ◄

We close with an example that shows that some hypothesis of continuity is required for integrability.

EXAMPLE 1 Suppose that f is defined for $0 \leqq x \leqq 1$ as follows:

$$f(x) = \begin{cases} 1 & \text{if } x \text{ is irrational;} \\ 0 & \text{if } x \text{ is rational.} \end{cases}$$

Then f is not continuous anywhere. (Why?) Given a partition P of $[0, 1]$, let p_i be a rational point and q_i an irrational point of the ith subinterval of P for each i, $1 \leqq i \leqq n$. As before, f attains its minimum value 0 at each p_i and its maximum value 1 at each q_i. Also

$$L(P) = \sum_{i=1}^{n} f(p_i)\Delta x_i = 0, \quad \text{whereas} \quad U(P) = \sum_{i=1}^{n} f(q_i)\Delta x_i = 1.$$

Thus if we choose $\epsilon = \frac{1}{2}$, then there is *no* number I that can lie within ϵ of both $L(P)$ and $U(P)$, no matter how small the mesh of P. It follows that f is *not* integrable on $[0, 1]$. ◆

REMARK This is not the end of the story of the integral. Integrals of highly discontinuous functions are important in many applications of physics, and near the beginning of the twentieth century a number of mathematicians, most notably Henri Lebesgue (1875–1941), developed more powerful integrals. The Lebesgue integral, in particular, always exists when the Riemann integral does, and gives the same value; but the Lebesgue integral is sufficiently powerful to integrate even functions that are continuous nowhere. It reports that

$$\int_0^1 f(x)\,dx = 1$$

for the function f of Example 1. Other mathematicians have developed integrals with domains far more general than sets of real numbers or subsets of the plane or space.

APPENDIX G: APPROXIMATIONS AND RIEMANN SUMS

Several times in Chapter 6 our attempt to compute some quantity Q led to the following situation. Beginning with a regular partition of an appropriate interval $[a, b]$ into n subintervals, each of length Δx, we found an approximation A_n to Q of the form

$$A_n = \sum_{i=1}^{n} g(u_i)h(v_i)\,\Delta x, \tag{1}$$

where u_i and v_i are two (generally different) points of the ith subinterval $[x_{i-1}, x_i]$. For example, in our discussion of surface area of revolution that precedes Eq. (8) of Section 6.4, we found the approximation

$$\sum_{i=1}^{n} 2\pi f(u_i)\sqrt{1 + [f'(v_i)]^2}\,\Delta x \tag{2}$$

to the area of the surface generated by revolving the curve $y = f(x)$, $a \leqq x \leqq b$, around the x-axis. (In Section 6.4 we wrote $x_i^{\star\star}$ for u_i and $x_i^\star$ for v_i.) Note that the expression in (2) is the same as the right-hand side in Eq. (1); take $g(x) = 2\pi f(x)$ and $h(x) = \sqrt{1 + [f'(x)]^2}$.

In such a situation we observe that if u_i and v_i were the *same* point $x_i^\star$ of $[x_{i-1}, x_i]$ for each i ($i = 1, 2, 3, \ldots, n$), then the approximation in Eq. (1) would be a Riemann sum for the function $g(x)h(x)$ on $[a, b]$. This leads us to suspect that

$$\lim_{\Delta x \to 0} \sum_{i=1}^{n} g(u_i)h(v_i)\,\Delta x = \int_a^b g(x)h(x)\,dx. \tag{3}$$

In Section 6.4, we assumed the validity of Eq. (3) and concluded from the approximation in (2) that the surface area of revolution ought to be defined to be

$$A = \lim_{\Delta x \to 0} \sum_{i=1}^{n} 2\pi f(u_i)\sqrt{1 + [f'(v_i)]^2}\,\Delta x = \int_a^b 2\pi f(x)\sqrt{1 + [f'(x)]^2}\,dx.$$

Theorem 1 guarantees that Eq. (3) holds under mild restrictions on the functions g and h.

THEOREM 1 A Generalization of Riemann Sums
Suppose that h and g' are continuous on $[a, b]$. Then

$$\lim_{\Delta x \to 0} \sum_{i=1}^{n} g(u_i)h(v_i)\,\Delta x = \int_a^b g(x)h(x)\,dx, \tag{3}$$

where u_i and v_i are arbitrary points of the ith subinterval of a regular partition of $[a, b]$ into n subintervals, each of length Δx.

PROOF Let M_1 and M_2 denote the maximum values on $[a, b]$ of $|g'(x)|$ and $|h(x)|$, respectively. Note that

$$\sum_{i=1}^{n} g(u_i)h(v_i)\,\Delta x = R_n + S_n, \quad \text{where} \quad R_n = \sum_{i=1}^{n} g(v_i)h(v_i)\,\Delta x$$

is a Riemann sum approaching $\int_a^b g(x)h(x)dx$ as $\Delta x \to 0$, and

$$S_n = \sum_{i=1}^{n} [g(u_i) - g(v_i)]\,h(v_i)\,\Delta x.$$

To prove Eq. (3) it is sufficient to show that $S_n \to 0$ as $\Delta x \to 0$. The mean value theorem gives

$$\begin{aligned} |g(u_i) - g(v_i)| &= |g'(\overline{x}_i)| \cdot |u_i - v_i| \qquad [\overline{x}_i \text{ in } (u_i, v_i)] \\ &\leqq M_1\,\Delta x, \end{aligned}$$

because both u_i and v_i are points of the interval $[x_{i-1}, x_i]$ of length Δx. Then

$$\begin{aligned} |S_n| &\leqq \sum_{i=1}^{n} |g(u_i) - g(v_i)| \cdot |h(v_i)|\,\Delta x \leqq \sum_{i=1}^{n} (M_1\,\Delta x) \cdot (M_2\,\Delta x) \\ &= (M_1 M_2\,\Delta x) \sum_{i=1}^{n} \Delta x = M_1 M_2 (b - a)\,\Delta x, \end{aligned}$$

from which it follows that $S_n \to 0$ as $\Delta x \to 0$, as desired. ◂

As an application of Theorem 1, let us give a rigorous derivation of Eq. (2) of Section 6.3,

$$V = \int_a^b 2\pi x f(x)\,dx, \tag{4}$$

for the volume of the solid generated by revolving around the y-axis the region between the graph of $y = f(x)$ and the x-axis for $a \leqq x \leqq b$. Beginning with the usual regular partition of $[a, b]$, let $f(x_i^\flat)$ and $f(x_i^\sharp)$ denote the minimum and maximum values of f on the ith subinterval $[x_{i-1}, x_i]$. Denote by $x_i^\star$ the midpoint of this subinterval. From Fig. G.1, we see that the part of the solid generated by revolving the region below $y = f(x)$, $x_{i-1} \leqq x \leqq x_i$, contains a cylindrical shell with average radius $x_i^\star$, thickness Δx, and height $f(x_i^\flat)$ and is contained in another cylindrical shell with the same average radius and thickness but with height $f(x_i^\sharp)$. Hence the volume ΔV_i of this part of the solid satisfies the inequalities

$$2\pi x_i^\star f(x_i^\flat)\,\Delta x \leqq \Delta V_i \leqq 2\pi x_i^\star f(x_i^\sharp)\,\Delta x.$$

We add these inequalities for $i = 1, 2, 3, \dots, n$ and find that

$$\sum_{i=1}^{n} 2\pi x_i^\star f(x_i^\flat)\,\Delta x \leqq V \leqq \sum_{i=1}^{n} 2\pi x_i^\star f(x_i^\sharp)\,\Delta x.$$

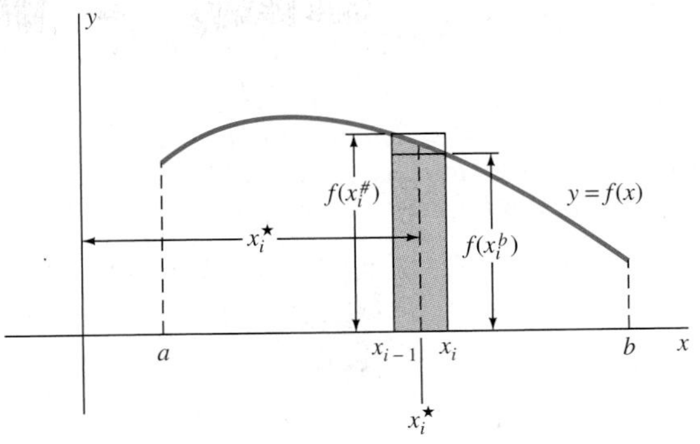

FIGURE G.1 A careful estimate of the volume of a solid of revolution around the y-axis.

Because Theorem 1 implies that both of the last two sums approach $\int_a^b 2\pi f(x)dx$, the squeeze law of limits now implies Eq. (4).

We will occasionally need a generalization of Theorem 1 that involves the notion of a continuous function $F(x, y)$ of two variables. We say that F is *continuous* at the point (x_0, y_0) provided that the value $F(x, y)$ can be made arbitrarily close to $F(x_0, y_0)$ merely by choosing the point (x, y) sufficiently close to (x_0, y_0). We discuss continuity of functions of two variables in Chapter 13. Here it will suffice to accept the following facts: If $g(x)$ and $h(y)$ are continuous functions of the single variables x and y, respectively, then simple combinations such as

$$g(x) \pm h(y), \quad g(x)h(y), \quad \text{and} \quad \sqrt{[g(x)]^2 + [h(y)]^2}$$

are continuous functions of the two variables x and y.

Now consider a regular partition of $[a, b]$ into n subintervals, each of length Δx, and let u_i and v_i denote arbitrary points of the ith subinterval $[x_{i-1}, x_i]$. Theorem 2—we omit the proof—tells us how to find the limit as $\Delta x \to 0$ of a sum such as

$$\sum_{i=1}^{n} F(u_i, v_i)\,\Delta x.$$

THEOREM 2 A Further Generalization

Let $F(x, y)$ be continuous for x and y both in the interval $[a, b]$. Then, in the notation of the preceding paragraph,

$$\lim_{\Delta x \to 0} \sum_{i=1}^{n} F(u_i, v_i)\,\Delta x = \int_a^b F(x, x)\,dx. \qquad \textbf{(5)}$$

Theorem 1 is the special case $F(x, y) = g(x)h(y)$ of Theorem 2. Moreover, the integrand $F(x, x)$ on the right in Eq. (5) is merely an ordinary function of the single variable x. As a formal matter, the integral corresponding to the sum in Eq. (5) is obtained by replacing the summation symbol with an integral sign, changing both u_i and v_i to x, replacing Δx with dx, and inserting the correct limits of integration. For example, if the interval $[a, b]$ is $[0, 4]$, then

$$\begin{aligned}\lim_{\Delta x \to 0} \sum_{i=1}^{n} \sqrt{9u_i^2 + v_i^4}\,\Delta x &= \int_0^4 \sqrt{9x^2 + x^4}\,dx \\ &= \int_0^4 x(9 + x^2)^{1/2}\,dx = \left[\frac{1}{3}(9 + x^2)^{3/2}\right]_0^4 \\ &= \frac{1}{3}\left[(25)^{3/2} - (9)^{3/2}\right] = \frac{98}{3}.\end{aligned}$$

APPENDIX G PROBLEMS

In Problems 1 through 7, u_i and v_i are arbitrary points of the ith subinterval of a regular partition of $[a, b]$ into n subintervals, each of length Δx. Express the given limit as an integral from a to b, then compute the value of this integral.

1. $\displaystyle\lim_{\Delta x \to 0} \sum_{i=1}^{n} u_i v_i \, \Delta x; \quad a = 0, \; b = 1$

2. $\displaystyle\lim_{\Delta x \to 0} \sum_{j=1}^{n} (3u_j + 5v_j) \, \Delta x; \quad a = -1, \; b = 3$

3. $\displaystyle\lim_{\Delta x \to 0} \sum_{i=1}^{n} u_i \sqrt{4 - v_i^2} \, \Delta x; \quad a = 0, \; b = 2$

4. $\displaystyle\lim_{\Delta x \to 0} \sum_{i=1}^{n} \frac{u_i}{\sqrt{16 + v_i^2}} \, \Delta x; \quad a = 0, \; b = 3$

5. $\displaystyle\lim_{\Delta x \to 0} \sum_{i=1}^{n} \sin u_i \cos v_i \, \Delta x; \quad a = 0, \; b = \pi/2$

6. $\displaystyle\lim_{\Delta x \to 0} \sum_{i=1}^{n} \sqrt{\sin^2 u_i + \cos^2 v_i} \, \Delta x; \quad a = 0, \; b = \pi$

7. $\displaystyle\lim_{\Delta x \to 0} \sum_{k=1}^{n} \sqrt{u_k^4 + v_k^7} \, \Delta x; \quad a = 0, \; b = 2$

8. Explain how Theorem 1 applies to show that Eq. (8) of Section 6.4 follows from the discussion that precedes it in that section.

9. Use Theorem 1 to derive Eq. (10) of Section 6.4.

APPENDIX H: L'HÔPITAL'S RULE AND CAUCHY'S MEAN VALUE THEOREM

Here we give a proof of l'Hôpital's rule,

$$\lim_{x \to a} \frac{f(x)}{g(x)} = \lim_{x \to a} \frac{f'(x)}{g'(x)}, \tag{1}$$

under the hypotheses of Theorem 1 in Section 4.8. The proof is based on a generalization of the mean value theorem due to the French mathematician Augustin Louis Cauchy. Cauchy used this generalization in the early nineteenth century to give rigorous proofs of several calculus results not previously established firmly.

CAUCHY'S MEAN VALUE THEOREM

Suppose that the functions f and g are continuous on the closed and bounded interval $[a, b]$ and differentiable on (a, b). Then there exists a number c in (a, b) such that

$$[f(b) - f(a)]\, g'(c) = [g(b) - g(a)]\, f'(c). \tag{2}$$

REMARK 1 To see that this theorem is indeed a generalization of the (ordinary) mean value theorem, we take $g(x) = x$. Then $g'(x) \equiv 1$, and the conclusion in Eq. (2) reduces to the fact that

$$f(b) - f(a) = (b - a)\, f'(c)$$

for some number c in (a, b).

REMARK 2 Equation (2) has a geometric interpretation like that of the ordinary mean value theorem. Let us think of the equations $x = g(t)$, $y = f(t)$ as describing the motion of a point $P(x, y)$ moving along a curve C in the xy-plane as t increases from a to b (Fig. H.1). That is, $P(x, y) = P(g(t), f(t))$ is the location of the point P at time t. Under the assumption that $g(b) \neq g(a)$, the slope of the line L connecting the endpoints of the curve C is

$$m = \frac{f(b) - f(a)}{g(b) - g(a)}. \tag{3}$$

But if $g'(c) \neq 0$, then the chain rule gives

$$\frac{dy}{dx} = \frac{dy/dt}{dx/dt} = \frac{f'(c)}{g'(c)} \tag{4}$$

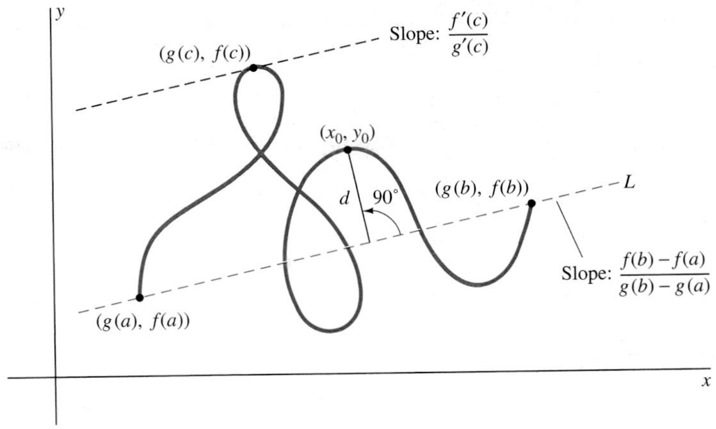

FIGURE H.1 The idea of Cauchy's mean value theorem.

for the slope of the line tangent to the curve C at the point $(g(c), f(c))$. But if $g(b) \neq g(a)$ and $g'(c) \neq 0$, then Eq. (2) may be written in the form

$$\frac{f(b) - f(a)}{g(b) - g(a)} = \frac{f'(c)}{g'(c)}, \tag{5}$$

so the two slopes in Eqs. (3) and (4) are equal. Thus Cauchy's mean value theorem implies that (under our assumptions) there is a point on the curve C where the tangent line is *parallel* to the line joining the endpoints of C. This is exactly what the (ordinary) mean value theorem says for an explicitly defined curve $y = f(x)$. This geometric interpretation motivates the following proof of Cauchy's mean value theorem.

PROOF The line L through the endpoints in Fig. H.1 has point-slope equation

$$y - f(a) = \frac{f(b) - f(a)}{g(b) - g(a)}\,[x - g(a)],$$

which can be rewritten in the form $Ax + By + C = 0$ with

$$A = g(b) - f(a), \quad B = -[g(b) - g(a)], \quad \text{and}$$
$$C = f(a)\,[g(b) - g(a)] - g(a)\,[f(b) - f(a)]. \tag{6}$$

According to Miscellaneous Problem 93 of Chapter 3, the (perpendicular) distance from the point (x_0, y_0) to the line L is

$$d = \frac{|Ax_0 + By_0 + C|}{\sqrt{A^2 + B^2}}.$$

Figure H.1 suggests that the point $(g(c), f(c))$ will maximize this distance d for points on the curve C.

We are motivated, therefore, to define the auxiliary function

$$\phi(t) = Ag(t) + Bf(t) + C, \tag{7}$$

with the constants A, B, and C as defined in (6). Thus $\phi(t)$ is essentially a constant multiple of the distance from $(g(t), f(t))$ to the line L in Fig. H.1.

Now $\phi(a) = 0 = \phi(b)$ (why?), so Rolle's theorem (Section 4.3) implies the existence of a number c in (a, b) such that

$$\phi'(c) = Ag'(c) + Bf'(c) = 0. \tag{8}$$

We substitute the values of A and B from Eq. (6) into (8) and obtain the equation

$$[f(b) - f(a)]\,g'(c) - [g(b) - g(a)]\,f'(c) = 0.$$

This is the same as Eq. (2) in the conclusion of Cauchy's mean value theorem, and the proof is complete. ◄

Note Although the assumptions that $g(b) \neq g(a)$ and $g'(c) \neq 0$ were needed for our geometric interpretation of the theorem, they were not used in its proof—only in the motivation for the method of proof.

PROOF OF L'HÔPITAL'S RULE Suppose that $f(x)/g(x)$ has the indeterminate form 0/0 at $x = a$. We may invoke continuity of f and g to allow the assumption that $f(a) = 0 = f(b)$. That is, we simply define $f(a)$ and $g(a)$ to be zero in case their values at $x = a$ are not originally given.

Now we restrict our attention to values of $x \neq a$ in a fixed neighborhood of a on which both f and g are differentiable. Choose one such value of x and hold it temporarily constant. Then apply Cauchy's mean value theorem on the interval $[a, x]$. (If $x < a$, use the interval $[x, a]$.) We find that there is a number z between a and x that behaves as c does in Eq. (2). Hence, by virtue of Eq. (2), we obtain the equation

$$\frac{f(x)}{g(x)} = \frac{f(x) - f(a)}{g(x) - g(a)} = \frac{f'(z)}{g'(z)}.$$

Now z depends on x, but z is trapped between x and a, so z is forced to approach a as $x \to a$. We conclude that

$$\lim_{x \to a} \frac{f(x)}{g(x)} = \lim_{z \to a} \frac{f'(z)}{g'(z)} = \lim_{x \to a} \frac{f'(x)}{g'(x)},$$

under the assumption that the right-hand limit exists. Thus we have verified l'Hôpital's rule in the form of Eq. (1). ◀

APPENDIX I: PROOF OF TAYLOR'S FORMULA

Several different proofs of Taylor's formula (Theorem 2 of Section 10.4) are known, but none of them seems very well motivated—each requires some "trick" to begin the proof. The trick we employ here (suggested by C. R. MacCluer) is to begin by introducing an auxiliary function $F(x)$, defined as follows:

$$F(x) = f(b) - f(x) - f'(x)(b - x) - \frac{f''(x)}{2!}(b - x)^2$$

$$- \cdots - \frac{f^{(n)}(x)}{n!}(b - x)^n - K(b - x)^{n+1}, \qquad \textbf{(1)}$$

where the *constant* K is chosen so that $F(a) = 0$. To see that there *is* such a value of k, we could substitute $x = a$ on the right and $F(x) = F(a) = 0$ on the left in Eq. (1) and then solve routinely for K, but we have no need to do this explicitly.

Equation (1) makes it quite obvious that $F(b) = 0$ as well. Therefore, Rolle's theorem (Section 4.3) implies that

$$F'(z) = 0 \qquad \textbf{(2)}$$

for some point z of the open interval (a, b) (under the assumption that $a < b$). To see what Eq. (2) means, we differentiate both sides of Eq. (1) and find that

$$F'(x) = -f'(x) + [f'(x) - f''(x)(b - x)]$$

$$+ \left[f''(x)(b - x) - \frac{1}{2!}f^{(3)}(x)(b - x)^2\right]$$

$$+ \left[\frac{1}{2!}f^{(3)}(x)(b - x)^2 - \frac{1}{3!}f^{(4)}(x)(b - x)^3\right]$$

$$+ \cdots + \left[\frac{1}{(n-1)!}f^{(n)}(x)(b - x)^{n-1} - \frac{1}{n!}f^{(n+1)}(x)(b - x)^n\right]$$

$$+ (n + 1)K(b - x)^n.$$

Upon careful inspection of this result, we see that all terms except the final two cancel in pairs. Thus the sum "telescopes" to give

$$F'(x) = (n+1)K(b-x)^n - \frac{f^{(n+1)}(x)}{n!}(b-x)^n. \tag{3}$$

Hence Eq. (2) means that

$$(n+1)K(b-z)^n - \frac{f^{(n+1)}(z)}{n!}(b-z)^n = 0.$$

Consequently we can cancel $(b-z)^n$ and solve for

$$K = \frac{f^{(n+1)}(z)}{(n+1)!}. \tag{4}$$

Finally, we return to Eq. (1) and substitute $x = a$, $f(x) = 0$, and the value of K given in Eq. (4). The result is the equation

$$0 = f(b) - f(a) - f'(a)(b-a) - \frac{f''(a)}{2!}(b-a)^2$$
$$- \cdots - \frac{f^{(n)}(a)}{n!}(b-a)^n - \frac{f^{(n+1)}(z)}{(n+1)!}(b-a)^{n+1},$$

which is equivalent to the desired Taylor's formula, Eq. (11) of Section 10.4. ◄

APPENDIX J: CONIC SECTIONS AS SECTIONS OF A CONE

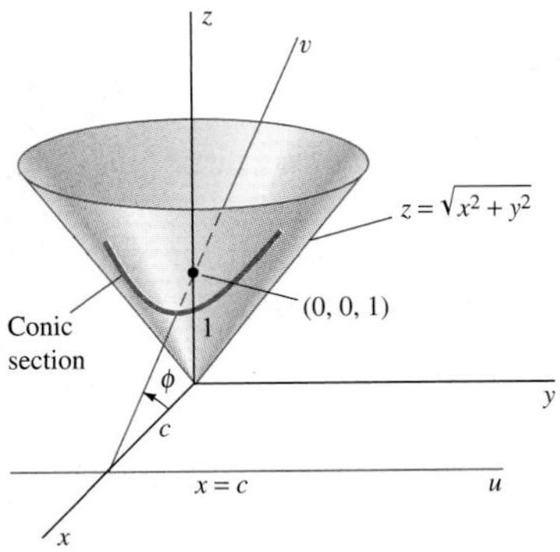

FIGURE J.1 Finding an equation for a conic section.

The parabola, hyperbola, and ellipse that we studied in Chapter 9 were originally introduced by the ancient Greek mathematicians as plane sections (traces) of a right circular cone. Here we show that the intersection of a plane and a cone is, indeed, one of the three conic sections as defined in Chapter 9.

Figure J.1 shows the cone with equation $z = \sqrt{x^2 + y^2}$ and its intersection with a plane $\mathcal{P}$ that passes through the point (0, 0, 1) and the line $x = c > 0$ in the xy-plane. An equation of $\mathcal{P}$ is

$$z = 1 - \frac{x}{c}. \tag{1}$$

The angle between $\mathcal{P}$ and the xy-plane is $\phi = \tan^{-1}(1/c)$. We want to show that the conic section obtained by intersecting the cone and the plane is

A parabola if $\phi = 45^\circ$ $(c = 1)$,

An ellipse if $\phi < 45^\circ$ $(c > 1)$,

A hyperbola if $\phi > 45^\circ$ $(c < 1)$.

We begin by introducing uv-coordinates in the plane $\mathcal{P}$ as follows. The u-coordinate of the point (x, y, z) of $\mathcal{P}$ is $u = y$. The v-coordinate of the same point is its perpendicular distance from the line $x = c$. This explains the u- and v-axes indicated in Fig. J.1. Figure J.2 shows the cross section in the plane $y = 0$ exhibiting the relation between v, x, and z. We see that

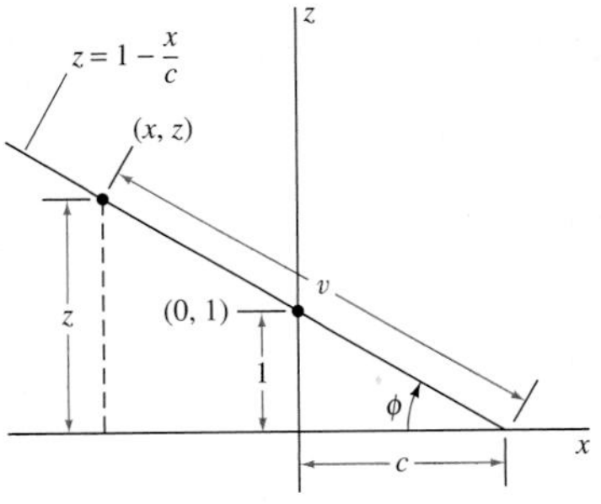

FIGURE J.2 Computing coordinates in the uv-plane.

$$z = v \sin\phi = \frac{v}{\sqrt{1+c^2}}. \tag{2}$$

Equations (1) and (2) give

$$x = c(1-z) = c\left(1 - \frac{v}{\sqrt{1+c^2}}\right). \tag{3}$$

We had $z^2 = x^2 + y^2$ for the equation of the cone. We make the following substitutions in this equation: Replace y with u, and replace z and x with the expressions on the right-hand sides of Eqs. (2) and (3), respectively. These replacements yield

$$\frac{v^2}{1+c^2} = c^2\left(1 - \frac{v}{\sqrt{1+c^2}}\right)^2 + u^2.$$

After we simplify, this last equation takes the form

$$u^2 + \frac{c^2-1}{c^2+1}v^2 - \frac{2c^2}{\sqrt{1+c^2}}v + c^2 = 0. \qquad \textbf{(4)}$$

This is the equation of the curve in the uv-plane. We examine the three cases for the angle ϕ.

Suppose first that $\phi = 45°$. Then $c = 1$, so Eq. (4) contains a term that includes u^2, another term that includes v, and a constant term. So the curve is a parabola; see Eq. (6) of Section 9.6.

Suppose next that $\phi < 45°$. Then $c > 1$, and both the coefficients of u^2 and v^2 in Eq.(4) are positive. Thus the curve is an ellipse; see Eq. (17) of Section 9.6.

Finally, if $\phi > 45°$, then $c < 1$, and the coefficients of u^2 and v^2 in Eq. (4) have opposite signs. So the curve is a hyperbola; see Eq. (26) of Section 9.6.

APPENDIX K: PROOF OF THE LINEAR APPROXIMATION THEOREM

Under the hypothesis of continuous differentiability of the linear approximation theorem stated in Section 13.6, we want to prove that the increment

$$\Delta f = f(\mathbf{a}+\mathbf{h}) - f(\mathbf{a})$$

is given by

$$\Delta f = f'(\mathbf{a})\mathbf{h} + \epsilon(\mathbf{h})\mathbf{h} \qquad \textbf{(1)}$$

where $\epsilon(\mathbf{h}) = [\epsilon_1(\mathbf{h}) \;\; \epsilon_2(\mathbf{h}) \;\; \ldots \;\; \epsilon_n(\mathbf{h})]$ is a row matrix such that each element $\epsilon_i(\mathbf{h})$ approaches zero as $\mathbf{h} \to \mathbf{0}$.

To analyze the increment Δf, we split the jump from $\mathbf{a}$ to $\mathbf{a}+\mathbf{h}$ into n separate steps, in each of which only a single coordinate changes. Let $\mathbf{e}_i$ denote the unit n-vector with 1 in the ith position, and write

$$\mathbf{a}_0 = \mathbf{a} \quad \text{and} \quad \mathbf{a}_i = \mathbf{a}_{i-1} + h_i\mathbf{e}_i \qquad \textbf{(2)}$$

for $i = 1, 2, \ldots, n$, so that $\mathbf{a}_n = \mathbf{a}+\mathbf{h}$. Then

$$\begin{aligned}\Delta f &= f(\mathbf{a}_n) - f(\mathbf{a}_0)\\ &= [f(\mathbf{a}_n) - f(\mathbf{a}_{n-1})] + [f(\mathbf{a}_{n-1}) - f(\mathbf{a}_{n-2})] + \cdots\\ &\quad + [f(\mathbf{a}_2) - f(\mathbf{a}_1)] + [f(\mathbf{a}_1) - f(\mathbf{a}_0)];\end{aligned}$$

that is,

$$\Delta f = \sum_{i=1}^{n} [f(\mathbf{a}_i) - f(\mathbf{a}_{i-1})]. \qquad \textbf{(3)}$$

The ith term in this sum is given by

$$\begin{aligned}f(\mathbf{a}_i) - f(\mathbf{a}_{i-1}) &= f(a_1+h_1, \ldots, a_{i-1}+h_{i-1}, a_i+h_i, a_{i+1}, \ldots, a_n)\\ &\quad - f(a_1+h_1, \ldots, a_{i-1}+h_{i-1}, a_i, a_{i+1}, \ldots, a_n)\\ &= g_i(1) - g_i(0),\end{aligned}$$

where the differentiable function g_i is defined by

$$g_i(t) = f(a_1+h_1, \ldots, a_{i-1}+h_{i-1}, a_i+th_i, a_{i+1}, \ldots, a_n).$$

The mean value theorem then yields

$$\begin{aligned} f(\mathbf{a}_i) - f(\mathbf{a}_{i-1}) &= g_i(1) - g_i(0) = g_i'(\overline{t_i})(1-0) \\ &= D_i f(a_1 + h_1, \ldots, a_{i-1} + h_{i-1}, a_i + \overline{t_i} h_i, a_{i+1}, \ldots, a_n) \cdot h_i \\ &= D_i f(\mathbf{a}_{i-1} + \overline{t_i} h_i \mathbf{e}_i) \cdot h_i \end{aligned}$$

for some $\overline{t_i}$ between 0 and 1. Substitution in (2) then gives

$$\begin{aligned} \Delta f &= \sum_{i=1}^{n} D_i f(\mathbf{a}_{i-1} + \overline{t_i} h_i \mathbf{e}_i) \cdot h_i \\ &= \sum_{i=1}^{n} [D_i f(\mathbf{a}) + D_i f(\mathbf{a}_{i-1} + \overline{t_i} h_i \mathbf{e}_i) - D_i f(\mathbf{a})] \cdot h_i. \end{aligned}$$

Thus

$$\begin{aligned} \Delta f &= \sum_{i=1}^{n} [D_i f(\mathbf{a} + \epsilon_i(\mathbf{h})] \cdot h_i \\ &= f'(\mathbf{a})\mathbf{h} + [\epsilon_i(\mathbf{h}) \quad \epsilon_i(\mathbf{h}) \quad \ldots \quad \epsilon_i(\mathbf{h})]\,\mathbf{h} \end{aligned}$$

where

$$\epsilon_i(\mathbf{h}) = D_i f(\mathbf{a}_{i-1} + \overline{t_i} h_i \mathbf{e}_i) - D_i f(\mathbf{a}) \to 0$$

(by continuity of $D_i f$ at $\mathbf{a}$) as $\mathbf{h} \to \mathbf{0}$ (and hence $\mathbf{a}_{i-1} \to \mathbf{a}$ by (2)). We have therefore established (1) and hence completed the proof. ◄

APPENDIX L: UNITS OF MEASUREMENT AND CONVERSION FACTORS

MKS SCIENTIFIC UNITS

- *Length* in meters (m); *mass* in kilograms (kg), *time* in seconds (s)
- *Force* in newtons (N); a force of 1 N imparts an acceleration of 1 m/s^2 to a mass of 1 kg.
- *Work* in joules (J); 1 J is the work done by a force of 1 N acting through a distance of 1 m.
- *Power* in watts (W); 1 W is 1 J/s.

BRITISH ENGINEERING UNITS (FPS)

- *Length* in feet (ft), *force* in pounds (lb), *time* in seconds (s)
- *Mass* in slugs; 1 lb of force imparts an acceleration of 1 ft/s^2 to a mass of 1 slug. A mass of m slugs at the surface of the earth has a *weight* of $w = mg$ pounds (lb), where $g \approx 32.17$ ft/s^2.
- *Work* in ft · lb, *power* in ft · lb/s.

CONVERSION FACTORS

1 in. = 2.54 cm = 0.0254m, 1 m ≈ 3.2808 ft

1 mi = 5280 ft; 60 mi/h = 88 ft/s

1 lb ≈ 4.4482 N; 1 slug ≈ 14.594 kg

1 hp = 550 ft · lb/s ≈ 745.7 W

- *Gravitational acceleration:* $g \approx 32.17\ \text{ft/s}^2 \approx 9.807\ \text{m/s}^2$
- *Atmospheric pressure:* 1 atm is the pressure exerted by a column of mercury 76 cm high; 1 atm $\approx 14.70\ \text{lb/in.}^2 \approx 1.013 \times 10^5\ \text{N/m}^2$
- *Heat energy:* 1 Btu $\approx$ 778 ft · lb $\approx$ 252 cal, 1 cal $\approx$ 4.184 J

APPENDIX M: FORMULAS FROM ALGEBRA, GEOMETRY, AND TRIGONOMETRY

LAWS OF EXPONENTS

$$a^m a^n = a^{m+n}, \qquad (a^m)^n = a^{mn}, \qquad (ab)^n = a^n b^n, \qquad a^{m/n} = \sqrt[n]{a^m};$$

in particular,

$$a^{1/2} = \sqrt{a}.$$

If $a \neq 0$, then

$$a^{m-n} = \frac{a^m}{a^n}, \qquad a^{-n} = \frac{1}{a^n}, \qquad \text{and} \quad a^0 = 1.$$

QUADRATIC FORMULA

The quadratic equation

$$ax^2 + bx + c = 0 \quad (a \neq 0)$$

has solutions

$$x = \frac{-b \pm \sqrt{b^2 - 4ac}}{2a}.$$

FACTORING

$$\begin{aligned}
a^2 - b^2 &= (a - b)(a + b) \\
a^3 - b^3 &= (a - b)(a^2 + ab + b^2) \\
a^4 - b^4 &= (a - b)(a^3 + a^2 b + ab^2 + b^3) \\
&= (a - b)(a + b)(a^2 + b^2) \\
a^5 - b^5 &= (a - b)(a^4 + a^3 b + a^2 b^2 + ab^3 + b^4)
\end{aligned}$$

(The pattern continues.)

$$\begin{aligned}
a^3 + b^3 &= (a + b)(a^2 - ab + b^2) \\
a^5 + b^5 &= (a + b)(a^4 - a^3 b + a^2 b^2 - ab^3 + b^4) \\
a^7 + b^7 &= (a + b)(a^6 - a^5 b + a^4 b^2 - a^3 b^3 + a^2 b^4 - ab^5 + b^6)
\end{aligned}$$

(The pattern continues for odd exponents.)

BINOMIAL FORMULA

$$\begin{aligned}
(a + b)^n = a^n + na^{n-1}b &+ \frac{n(n-1)}{1 \cdot 2} a^{n-2} b^2 \\
&+ \frac{n(n-1)(n-2)}{1 \cdot 2 \cdot 3} a^{n-3} b^3 + \cdots + nab^{n-1} + b^n
\end{aligned}$$

if n is a positive integer.

AREA AND VOLUME

In Fig. M.1, the symbols have the following meanings.

A: area	b: length of base	r: radius
B: area of base	C: circumference	V: volume
h: height	ℓ: length	w: width

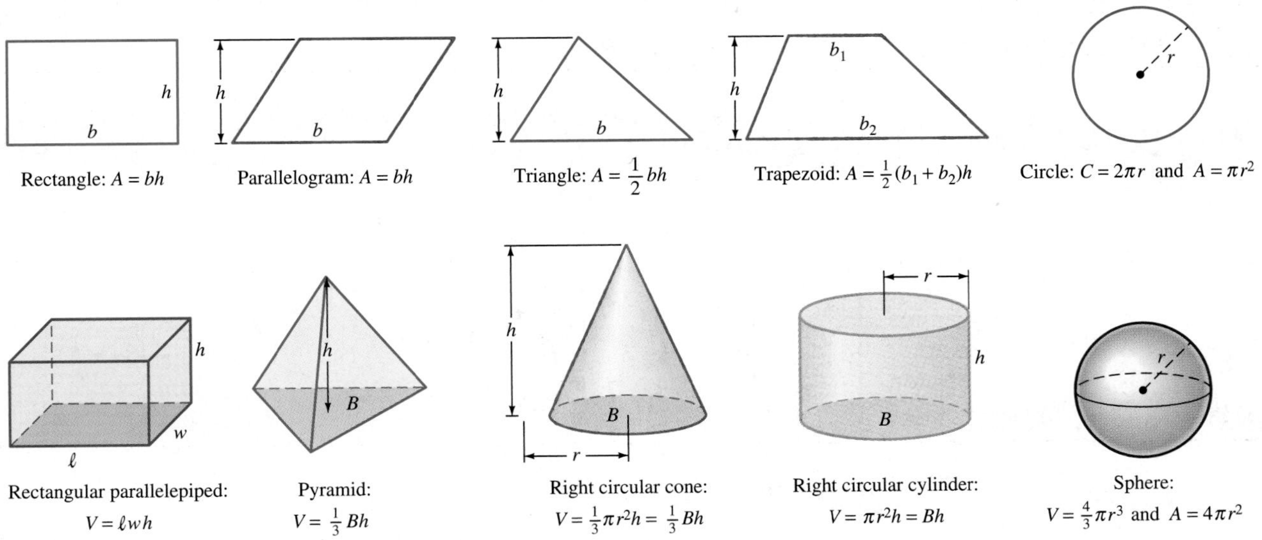

FIGURE M.1 The basic geometric shapes.

PYTHAGOREAN THEOREM

In a right triangle with legs a and b and hypotenuse c,

$$a^2 + b^2 = c^2.$$

FORMULAS FROM TRIGONOMETRY

$$\sin(-\theta) = -\sin\theta$$

$$\cos(-\theta) = \cos\theta$$

$$\sin^2\theta + \cos^2\theta = 1$$

$$\sin 2\theta = 2\sin\theta\cos\theta$$

$$\cos 2\theta = \cos^2\theta - \sin^2\theta$$

$$\sin(\alpha + \beta) = \sin\alpha\cos\beta + \cos\alpha\sin\beta$$

$$\cos(\alpha + \beta) = \cos\alpha\cos\beta - \sin\alpha\sin\beta$$

$$\tan(\alpha + \beta) = \frac{\tan\alpha + \tan\beta}{1 - \tan\alpha\tan\beta}$$

$$\sin^2\frac{\theta}{2} = \frac{1 - \cos\theta}{2}$$

$$\cos^2\frac{\theta}{2} = \frac{1 + \cos\theta}{2}$$

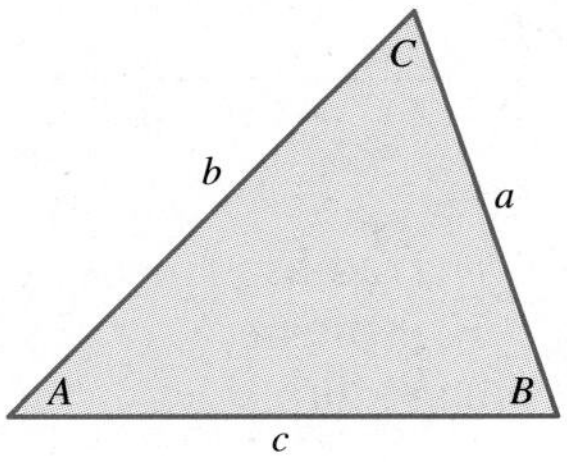

FIGURE M.2 An arbitrary triangle.

For an arbitrary triangle (Fig. M.2):

Law of cosines: $$c^2 = a^2 + b^2 - 2ab\cos C.$$

Law of sines: $$\frac{\sin A}{a} = \frac{\sin B}{b} = \frac{\sin C}{c}.$$

APPENDIX N: THE GREEK ALPHABET

A	α	alpha	I	ι	iota	P	ρ	rho
B	β	beta	K	κ	kappa	Σ	σ	sigma
Γ	γ	gamma	Λ	λ	lambda	T	τ	tau
Δ	δ	delta	M	μ	mu	Υ	υ	upsilon
E	ϵ	epsilon	N	ν	nu	Φ	ϕ	phi
Z	ζ	zeta	Ξ	ξ	xi	X	χ	chi
H	η	eta	O	o	omicron	Ψ	ψ	psi
Θ	θ	theta	Π	π	pi	Ω	ω	omega

ANSWERS TO ODD-NUMBERED PROBLEMS

SECTION 1.1 (PAGE 9)

1. a. $-\dfrac{1}{a}$; **b.** a; **c.** $a^{-1/2}$; **d.** a^{-2}

3. a. $\dfrac{1}{a^2+5}$; **b.** $\dfrac{a^2}{1+5a^2}$; **c.** $\dfrac{1}{a+5}$; **d.** $\dfrac{1}{a^4+5}$

5. $a = \dfrac{1}{3}$

7. $a = 3$ or $a = -3$

9. $a = 100$ **11.** $3h$ **13.** $2ah + h^2$

15. $-\dfrac{h}{a(a+h)}$ **17.** $\{-1, 0, 1\}$ **19.** $\{-1, 1\}$

21. The set $\boldsymbol{R}$ of all real numbers

23. The set $\boldsymbol{R}$ of all real numbers

25. $\left[\frac{5}{3}, +\infty\right)$ **27.** $\left(-\infty, \frac{1}{2}\right]$

29. $(-\infty, 3) \cup (3, +\infty)$

31. The set $\boldsymbol{R}$ of all real numbers

33. $[0, 16]$ **35.** $(-\infty, 0) \cup (0, +\infty)$

37. $C(A) = 2\sqrt{\pi A},\quad 0 \leqq A < +\infty$

39. $C(F) = \frac{5}{9}(F - 32),\quad F > -459.67$

41. $A(x) = x\sqrt{16 - x^2},\quad 0 \leqq x \leqq 4$

43. $C(x) = 3x^2 + \dfrac{1296}{x},\quad 0 < x < +\infty$

45. $A(r) = 2\pi r^2 + \dfrac{2000}{r},\quad 0 < r < +\infty$

47. $V(x) = x(50 - 2x)^2,\quad 0 \leqq x \leqq 25$

49. Drill ten new wells.

51. CEILING$(x) = -$FLOOR$(-x)$

53. The set of all integral multiples of $\frac{1}{10}$

55. ROUND4$(x) = \frac{1}{10000}$ROUND$(10000x)$

57. 0.38 **59.** 1.24 **61.** 0.72

63. 3.21 **65.** 1.62

SECTION 1.2 (PAGE 21)

1. $2y = 3x$ **3.** $y = -5$

5. $y = 2x - 7$ **7.** $x + y = 6$

9. $y - 5 = -2(x - 1)$ **11.** Center $(2, 0)$, radius 2

13. Center $(-1, -1)$, radius 2

15. Center $\left(-\frac{1}{2}, \frac{1}{2}\right)$, radius 1

17. Opens upward, vertex at $(3, 0)$

19. Opens upward, vertex at $(-1, 3)$

21. Opens upward, vertex at $(-2, 3)$

23. Circle, center $(3, -4)$, radius 5

25. There are no points on the graph.

27. The graph is the straight line segment connecting the two points $(-1, 7)$ and $(1, -3)$ (including those two points).

29. The graph is the parabola that opens downward, symmetric around the y-axis, with vertex at $(0, 10)$ and x-intercepts $\pm\sqrt{10}$.

31.

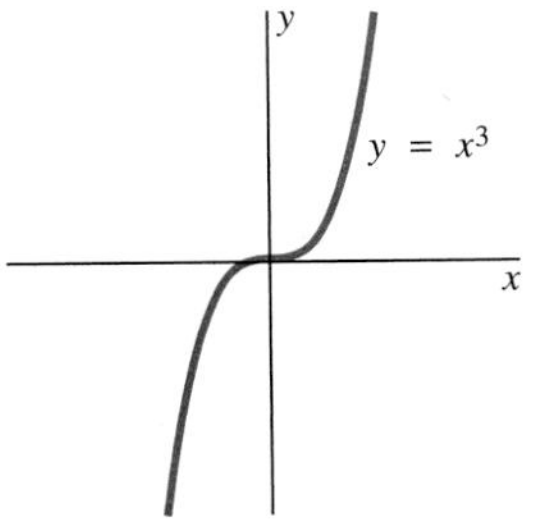

33.

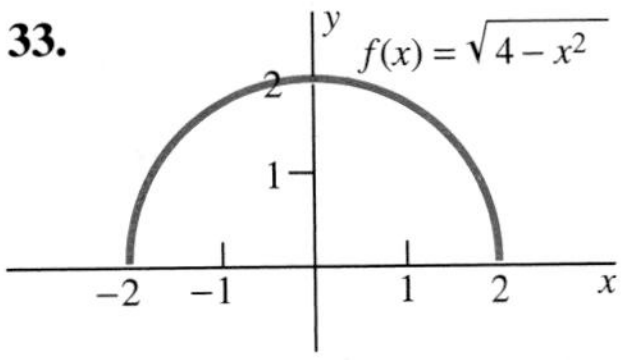

35.

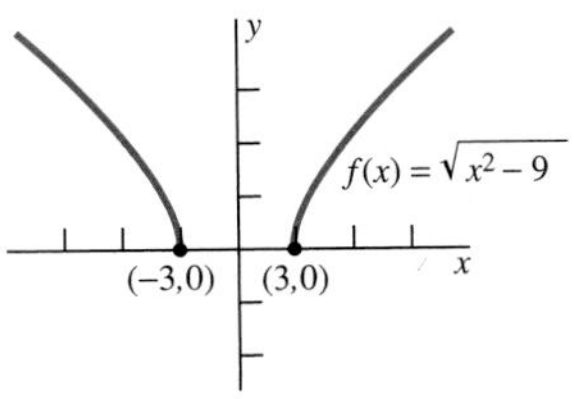

37.

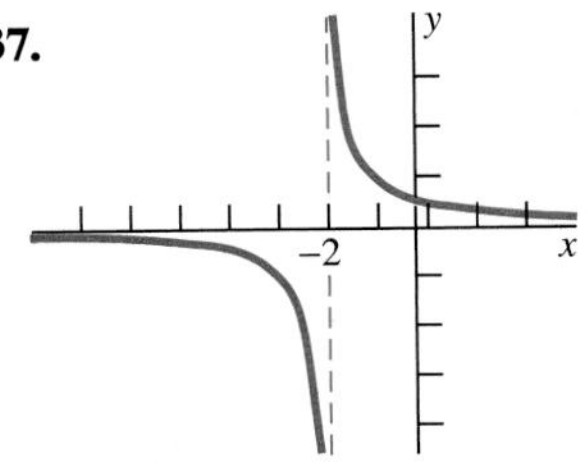

39.

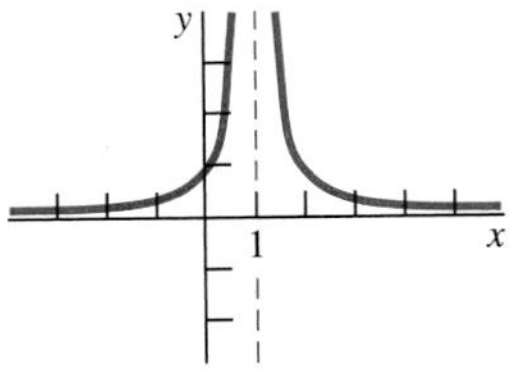

41.

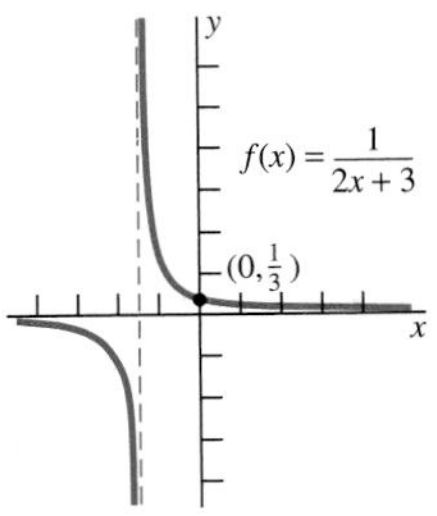

43.

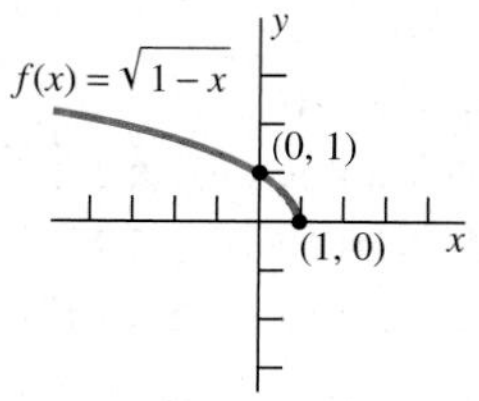

45.

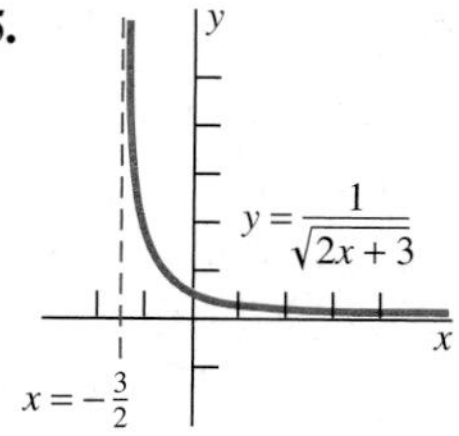

47.

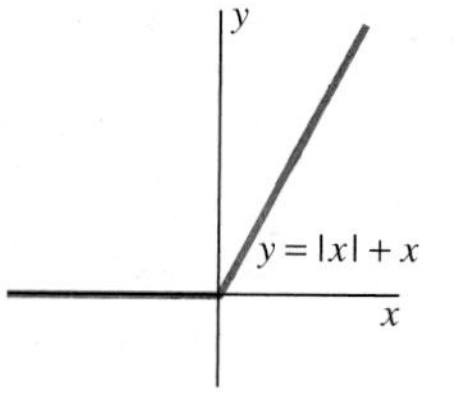

49.

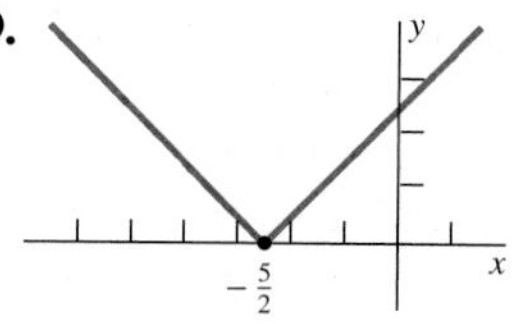

51.

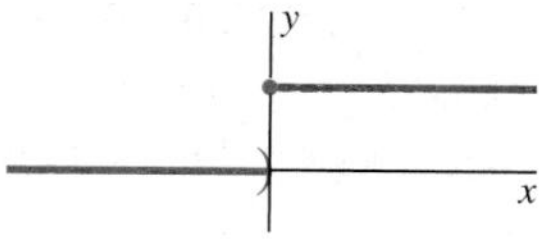

53.

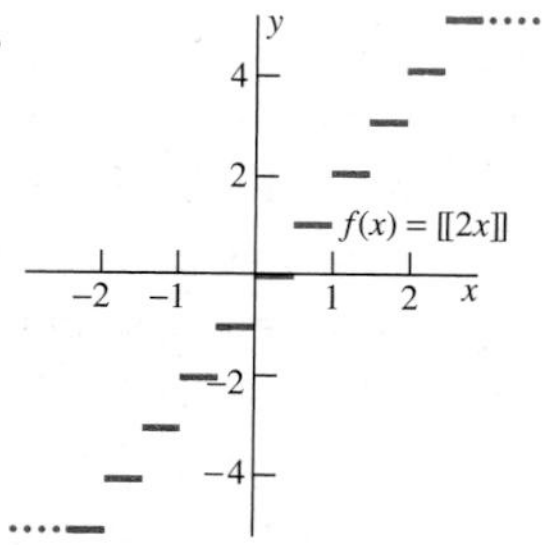

55.

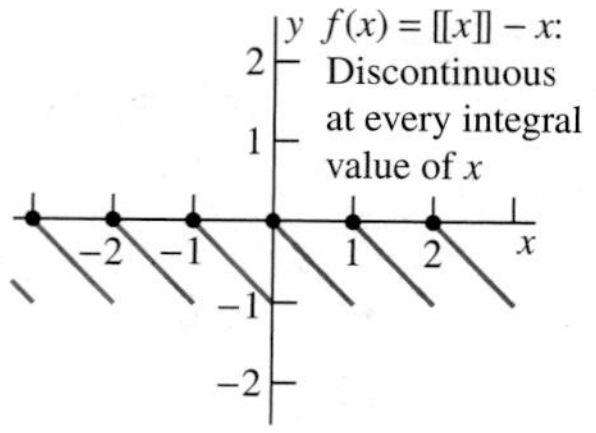

57. (1.5, 2.5) **59.** (2.25, 1.75) **61.** (2.25, 8.5)

63. $\left(-\frac{4}{3}, \frac{25}{3}\right)$ **65.** 144 ft **67.** 625

69. $f(x) = |x+1|, \quad -2 \leqq x \leqq 2$

71. $f(x) = [\![2x]\!], \quad -1 \leqq x < 2$

73. $x(t) = \begin{cases} 45t & \text{if } 0 \leqq t \leqq 1, \\ 75t - 30 & \text{if } 1 < t \leqq 2. \end{cases}$

75. $x(t) = \begin{cases} 60t & \text{if } 0 \leqq t \leqq 1, \\ 90 - 30t & \text{if } 1 < t \leqq 3. \end{cases}$

77. In dollars, $C(p) = (0.03)p + 0.68$, $1 \leqq p \leqq 100$. When $p = 50$, $C = \$2.18$. Fixed cost: \$0.68. Marginal cost: 3¢ per page.

79. $C(x) = \begin{cases} 8 & \text{if } 0 < x \leqq 8, \\ 8 - (0.8) \cdot [\![8 - x]\!] & \text{if } 8 < x \leqq 16. \end{cases}$

81. $V(0.5) \approx 3.36$, $V(5) \approx 0.336$ (L)

SECTION 1.3 (PAGE 32)

1. $(f+g)(x) = x^2 + 3x - 2$, x in $\mathbf{R}$. $(f \cdot g)(x) = x^3 + 3x^2 - x - 3$, x in $\mathbf{R}$. $(f/g)(x) = (x+1)/(x^2+2x-3)$, $x \neq -3, 1$.

3. $(f+g)(x) = \sqrt{x} + \sqrt{x-2}$, $2 \leqq x < +\infty$. $(f \cdot g)(x) = \sqrt{x^2 - 2x}$, $2 \leqq x < +\infty$. $(f/g)(x) = \sqrt{x/(x-2)}$, $2 < x < +\infty$.

5. All three functions have domain $(-2, 2)$. And

$$(f+g)(x) = \sqrt{x^2+1} + \frac{1}{\sqrt{4-x^2}},$$

$$(f \cdot g)(x) = \frac{\sqrt{x^2+1}}{\sqrt{4-x^2}},$$

$$\left(\frac{f}{g}\right)(x) = \sqrt{4 + 3x^2 - x^4}.$$

7. Matches Fig. 1.3.30

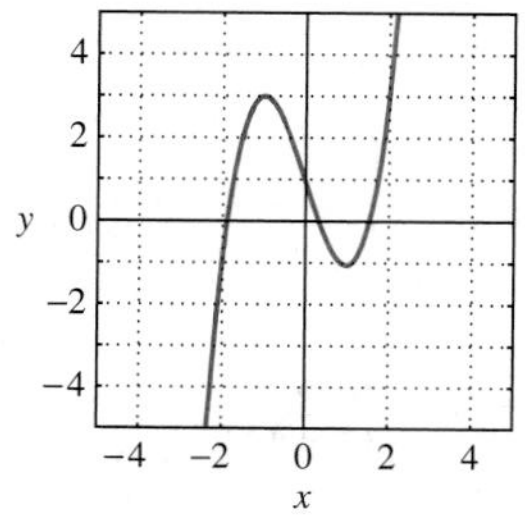

9. Matches Fig. 1.3.31

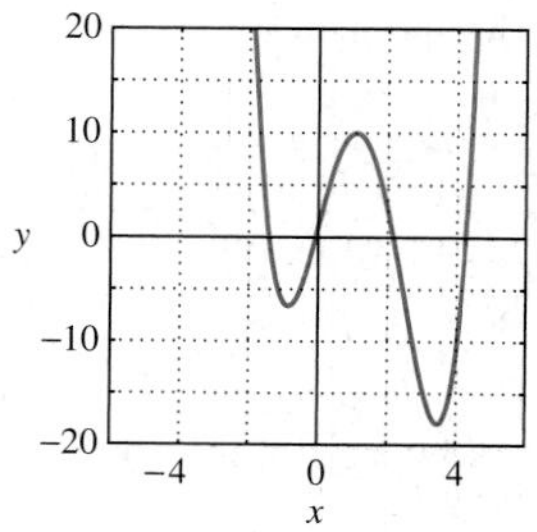

11. Matches Fig. 1.3.27

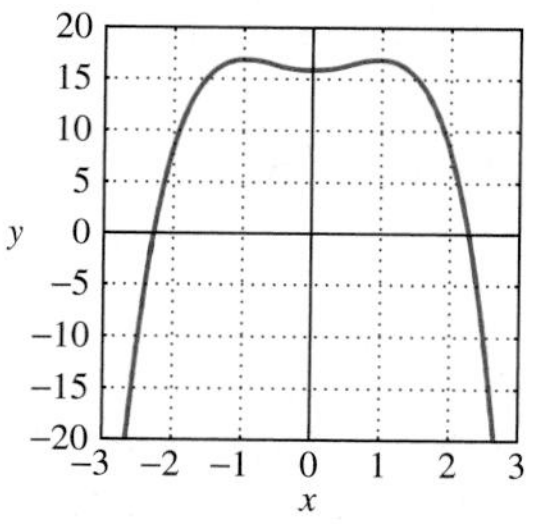

13. Matches Fig. 1.3.34

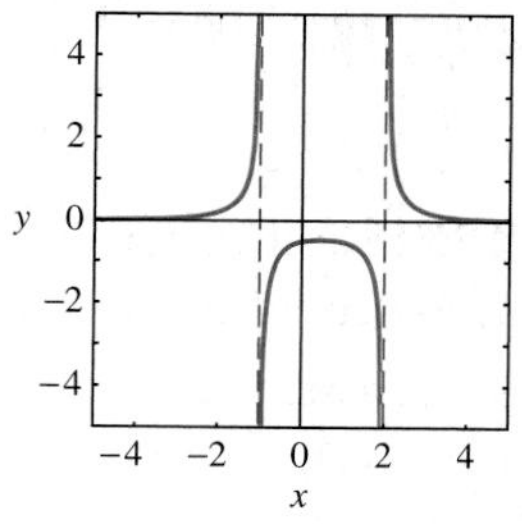

15. Matches Fig. 1.3.33

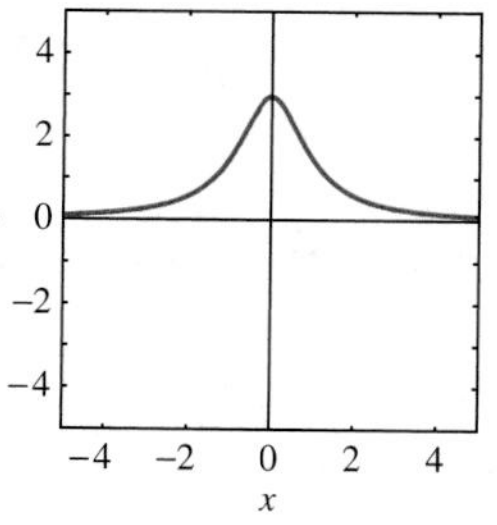

17. Matches Fig. 1.3.38

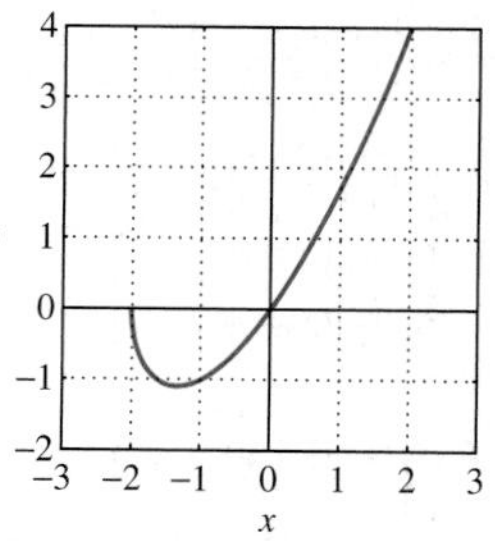

19. Matches Fig. 1.3.39

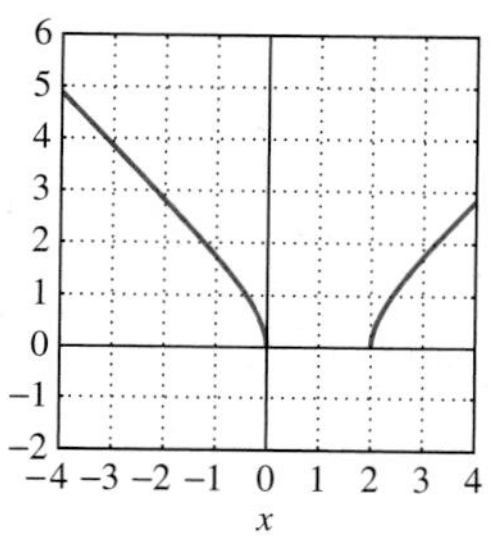

21. $-1.88, 0.35, 1.53$ **23.** -2.10
25. -1.30, 1 (exactly), 2.30 **27.** $-5.70, -2.22, 7.91$
29. $-10.20, -7.31, 1.98, 3.25$ 7.28

31.

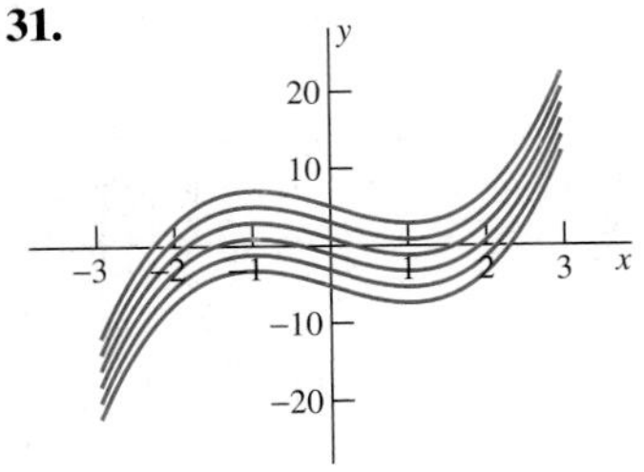

33.

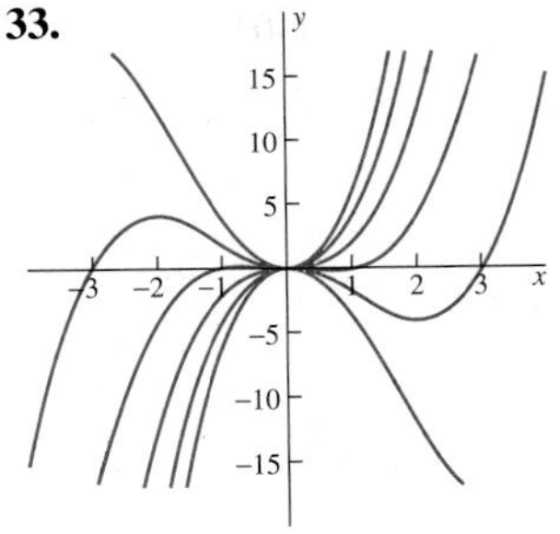

35.

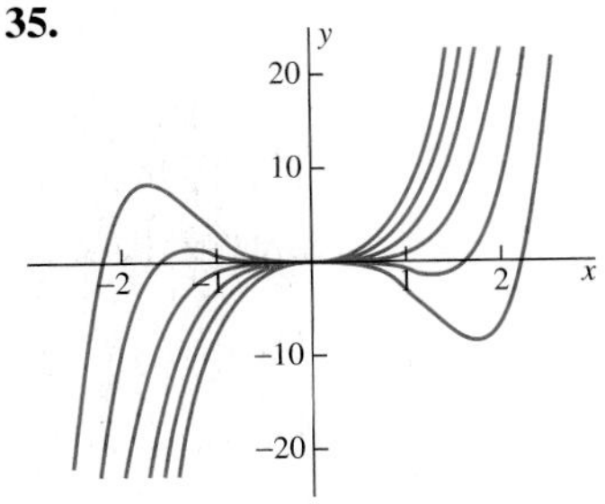

37.

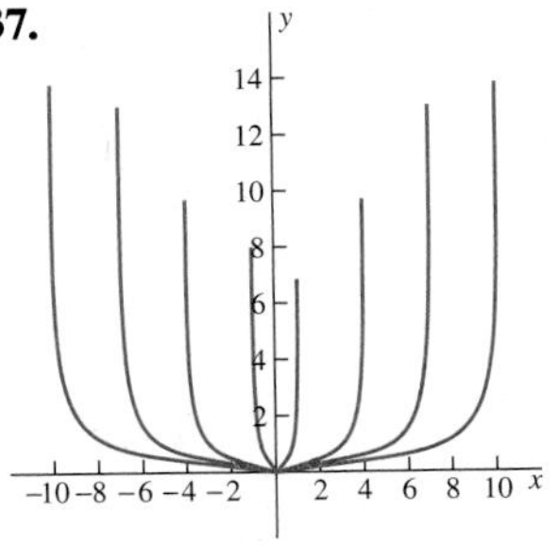

SECTION 1.4 (PAGE 43)

1. Matches Fig. 1.4.29
3. Matches Fig. 1.4.27
5. Matches Fig. 1.4.35
7. Matches Fig. 1.4.31
9. Matches Fig. 1.4.34
11. $f(g(x)) = -4x^2 - 12x - 8$; $g(f(x)) = -2x^2 + 5$
13. $f(g(x)) = \sqrt{x^4 + 6x^2 + 6}$; $g(f(x)) = x^2$, $|x| \geqq \sqrt{3}$
15. $f(g(x)) = g(f(x)) = x$
17. $f(g(x)) = \sin x^3 = \sin(x^3)$; $g(f(x)) = (\sin x)^3 = \sin^3 x$
19. $f(g(x)) = 1 + \tan^2 x$ if x is not an odd integral multiple of $\pi/2$; $g(f(x)) = \tan(1 + x^2)$ if $x^2 + 1$ is not an odd integral multiple of $\pi/2$

Problems 21 through 30 have many correct answers; we give only the most natural.

21. $k = 2$, $g(x) = 2 + 3x$ **23.** $k = \frac{1}{2}$, $g(x) = 2x - x^2$
25. $k = \frac{3}{2}$, $g(x) = 5 - x^2$ **27.** $k = -1$, $g(x) = x + 1$
29. $k = -\frac{1}{2}$, $g(x) = x + 10$ **31.** Exactly one solution
33. Exactly one solution **35.** Exactly five solutions
37. Exactly three solutions **39.** Exactly six solutions
41. 3.322 months **43.** 27.0046 years
45. 98.149 years **47.** $x \approx 4.84891$

CHAPTER 1 MISCELLANEOUS PROBLEMS (PAGE 49)

1. $x \geqq 4$ **3.** $x \neq \pm 3$ **5.** $x \geqq 0$ **7.** $x \leqq \frac{2}{3}$
9. $\boldsymbol{R}$ **11.** $4 \leqq p \leqq 8$ **13.** $2 < I < 4$
15. $V(S) = (S/6)^{3/2}$, $0 < S < +\infty$
17. $A(P) = \left(P^2\sqrt{3}\right)/36$, $0 < P < +\infty$
19. $y = 2x + 11$ **21.** $2y + 10 = x$
23. $x + 2y = 11$ **25.** Matches Fig. 1.MP.6
27. Matches Fig. 1.MP.4 **29.** Matches Fig. 1.MP.3
31. Matches Fig. 1.MP.7 **33.** Matches Fig. 1.MP.8
35. The straight line with x-intercept $\frac{7}{2}$ and y-intercept $-\frac{7}{5}$
37. Circle, center $(1, 0)$, radius 1
39. Parabola, opening upward, vertex at $(1, -3)$
41. The graph is that of $y = 1/x$ (Fig. 1.2.12) translated 5 units to the left.
43. The graph is that of $y = |x|$ (Fig. 1.2.11) translated 3 units to the right.
45. Key step: $|a + b + c| = |(a + b) + c| \leqq |a + b| + |c|$.
47. $(-\infty, -2) \cup (3, +\infty)$ **49.** $(-\infty, -2) \cup (4, +\infty)$
51. -1.140 and 6.140 **53.** 1.191 and 2.309
55. -5.021 and 0.896 **57.** $(2.50, 0.75)$
59. $(1.75, -1.25)$ **61.** $(-2.0625, 0.96875)$
63. $x \approx 0.4505$ **65.** Exactly three solutions
67. Exactly three solutions **69.** Exactly three solutions

SECTION 2.1 (PAGE 62)

1. $m(a) \equiv 0$; $y \equiv 5$ **3.** $m(a) = 2a$; $y = 4x - 4$
5. $m(a) \equiv 4$; $y = 4x - 5$ **7.** $m(a) = 4a - 3$; $y = 5x - 4$
9. $m(a) = 4a + 6$; $y = 14x - 8$
11. $m(a) = -\frac{2}{100}a + 2$; $25y = 49x + 1$
13. $m(a) = 8a$; $y = 16x - 15$
15. $(0, 10)$ **17.** $(1, 0)$ **19.** $(50, 25)$
21. $(1, -16)$ **23.** $(35, 1225)$
25. $m(x) = 2x$; $y = -4x - 4$, $4y = x + 18$
27. $m(x) = 4x + 3$; $y = 11x - 13$, $x + 11y = 101$
29. $y(3) = 144$ (ft) **31.** 625
33. $y = 12x - 36$ **35.** $(1, 1)$
37. 12; $y = 12x - 16$ **39.** $\dfrac{1}{2}$; $y = \frac{1}{2}(x + 1)$
41. -1; $y = -x + 2$ **43.** 0; $y \equiv 1$
45. $-\dfrac{1}{\sqrt{2}}$; $y - \dfrac{\sqrt{2}}{2} = -\dfrac{\sqrt{2}}{2}\left(x - \dfrac{\pi}{4}\right)$
47. 0; $y \equiv 5$

SECTION 2.2 (PAGE 73)

1. 36 **3.** 0 **5.** $\dfrac{2}{3}$
7. 125 **9.** 3 **11.** $16\sqrt{2}$
13. 1 **15.** 4 **17.** 0
19. $-\dfrac{1}{3}$ **21.** $-\dfrac{3}{2}$ **23.** 0
25. 0 **27.** $\dfrac{3}{4}$ **29.** $-\dfrac{1}{9}$
31. 4 **33.** $\dfrac{1}{4}$ **35.** -32
37. $m(x) = 3x^2$; $y = 12x - 16$

39. $m(x) = -\dfrac{2}{x^3}$; $x + 4y = 3$

41. $m(x) = -\dfrac{2}{(x-1)^2}$; $y = 6 - 2x$

43. $m(x) = -\dfrac{1}{2(x+2)^{3/2}}$; $x + 16y = 10$

45. $m(x) = \dfrac{1}{\sqrt{2x+5}}$; $3y = x + 7$

47. 2 **49.** $\dfrac{1}{6}$ **51.** $-\dfrac{3}{8}$

53. 1 **55.** $\dfrac{1}{6}$

57. The limit seems to be the number $e \approx 2.71828$.

59. -0.3333

61. This limit does not exist.

63. For instance, $f(10^{-20}) \approx 0.9106$.

SECTION 2.3 (PAGE 85)

1. 0 **3.** $\dfrac{1}{2}$

5. $-\infty$ (or "does not exist")

7. 5 **9.** 0 **11.** $\dfrac{1}{3}$

13. 0 **15.** 1 **17.** $\dfrac{1}{2}$

19. $\dfrac{1}{2}$ **21.** $\dfrac{1}{3}$ **23.** $\dfrac{1}{4}$

25. 0 **27.** 0 **29.** 3

31. This limit does not exist.

33. 0 **35.** 0

37. $+\infty$ (or "does not exist")

39. -1 **41.** 1 **43.** -1

45. 2 **47.** -1

49. $f(x) \to +\infty$ as $x \to 1^+$ and $f(x) \to -\infty$ as $x \to 1^-$.

51. $f(x) \to -\infty$ as $x \to -1^+$ and $f(x) \to +\infty$ as $x \to -1^-$.

53. $f(x) \to -\infty$ as $x \to -2^+$ and $f(x) \to +\infty$ as $x \to -2^-$.

55. $f(x) \to +\infty$ as $x \to 1$.

57. $f(x) \to -\frac{1}{4}$ as $x \to 2$, but f has no limit at -2 because

$$\lim_{x \to -2^+} f(x) = -\infty \quad \text{and} \quad \lim_{x \to -2^-} f(x) = +\infty.$$

59. $f(x) \to 4$ as $x \to 2^+$, $f(x) \to -4$ as $x \to 2^-$.

61. For every real number a, $\lim_{x \to a} f(x) = 2$.

63. If n is an integer, then $f(x) \to 10n - 1$ as $x \to n^-$ and $f(x) \to 10n$ as $x \to n^+$. The limit of $f(x)$ at $x = a$ exists if and only if $10a$ is not an integer.

65. If n is an integer, then $f(x) \to -\frac{1}{2}$ as $x \to n^+$, but $f(x) \to \frac{1}{2}$ as $x \to n^-$. The limit of $f(x)$ at $x = a$ exists if and only if a is not an integer.

67. For every real number a, $f(x) \to -1$ as $x \to a$.

69. The limit of $g(x)$ at $x = a$ exists if and only if a is not an integral multiple of $\frac{1}{10}$. If b is an integral multiple of $\frac{1}{10}$, then $g(x) \to b - \frac{1}{10}$ as $x \to b^-$ and $g(x) \to b$ as $x \to b^+$.

71. $f(x) \to 0$ as $x \to 0$. **73.** $f(x) \to 1$ as $x \to 0$.

75. Given $\epsilon > 0$, let $\delta = \epsilon/7$. **77.** Given $\epsilon > 0$, let $\delta = \epsilon^2$.

79. Given $\epsilon > 0$, let δ be the minimum of the two numbers 1 and $\epsilon/5$.

81. Given $\epsilon > 0$, let δ be the minimum of the two numbers 1 and $\epsilon/29$.

83. Case 1: $a = 0$. See Problem 78. Case 2: $a > 0$. Given $\epsilon > 0$, let δ be the minimum of the two numbers 1 and $\epsilon/(2a+1)$. Case 3: $a < 0$. Given $\epsilon > 0$, let δ be the minimum of the two numbers 1 and $\epsilon/|2a - 1|$.

SECTION 2.4 (PAGE 97)

1. If a is a real number, then

$$\begin{aligned}\lim_{x \to a} f(x) &= \lim_{x \to a} (2x^5 - 7x^2 + 13) \\ &= \left(\lim_{x \to a} 2x^5\right) - \left(\lim_{x \to a} 7x^2\right) + \left(\lim_{x \to a} 13\right) \\ &= \left(\lim_{x \to a} 2\right)\left(\lim_{x \to a} x\right)^5 - \left(\lim_{x \to a} 7\right)\left(\lim_{x \to a} x\right)^2 + 13 \\ &= 2a^5 - 7a^2 + 13 = f(a).\end{aligned}$$

3. Begin with the quotient law for limits.

5. Begin with the substitution law for limits.

7. Use the quotient law for limits and Theorem 1.

9. Use the quotient law for limits.

11. Use the substitution law for limits.

13. Use the quotient law for limits and Theorem 1.

15. Continuous on $\boldsymbol{R}$, the set of all real numbers.

17. Continuous on its domain, $(-\infty, -3) \cup (-3, +\infty)$.

19. Continuous on $\boldsymbol{R}$.

21. Continuous on its domain, $(-\infty, 5) \cup (5, +\infty)$.

23. Continuous on its domain, $(-\infty, 2) \cup (2, +\infty)$.

25. Continuous on $(-\infty, 1) \cup (1, +\infty)$.

27. Continuous on its domain, $(-\infty, 0) \cup (0, 1) \cup (1, +\infty)$.

29. Continuous on its domain, $(-2, 2)$.

31. Continuous on its domain, $(-\infty, 0) \cup (0, +\infty)$.

33. Continuous on its domain, the set of all real numbers other than the integral multiples of $\pi/2$.

35. Continuous on $\boldsymbol{R}$.

37. Nonremovable discontinuity at $x = -3$.

39. Nonremovable discontinuity at $x = -2$, removable discontinuity at $x = 2$.

41. Nonremovable discontinuities at $x = \pm 1$.

43. Nonremovable discontinuity at $x = 17$.

45. Removable discontinuity at $x = 0$.

47. Removable discontinuity at $x = 0$.

49. $c = 4$

51. $c = 0$

53. Apply the intermediate value property of continuous functions to $f(x) = x^2 - 5$ on $[2, 3]$.

55. Apply the intermediate value property of continuous functions to $f(x) = x^3 - 3x^2 + 1$ on $[0, 1]$.

57. Apply the intermediate value property of continuous functions to $f(x) = x^4 + 2x - 1$ on $[0, 1]$.

59. Examine $f(x)$ at the endpoints of the interval $[-3, -2]$, $[0, 1]$, and $[1, 2]$.

61. Your salary after t years have elapsed will be $S(t) = 25 \cdot (1.06)^{[\![t]\!]}$, with a discontinuity at each positive integer in the domain of S.

63. It is easy to see that $h(a)$ and $h(b)$ have opposite signs and that h is continuous on $[a, b]$. Apply the intermediate value property of continuous functions to h.

65. Given $a > 0$, let $f(x) = x^2 - a$. Apply the intermediate value theorem to f on the interval $[0, a+1]$.

67. To show that

$$\lim_{x \to a} \cos x = \cos a,$$

let $h = x - a$, so that $x = a + h$. Then use the cosine addition formula (from the endpapers).

69. Suppose by way of contradiction that f has limit L at $x = a$, an arbitrary real number. Choose $\epsilon = \frac{1}{4}$ and use the fact that every neighborhood of a contains both rational and irrational numbers and the formal definition of limit to obtain a contradiction.

71. 0.74

73. Left continuous but not right continuous at $x = 0$.

75. Left continuous but not right continuous at $x = 0$.

77. Right continuous but not left continuous at each odd integral multiple of $\pi/2$.

CHAPTER 2 MISCELLANEOUS PROBLEMS (PAGE 99)

1. 4 **3.** 0 **5.** $-\dfrac{5}{3}$

7. -2 **9.** 0 **11.** 4

13. 8 **15.** $\dfrac{1}{6}$ **17.** $-\dfrac{1}{54}$

19. -1 **21.** 1

23. This limit does not exist.

25. $+\infty$ (or "does not exist")

27. $+\infty$ (or "does not exist")

29. $-\infty$ (or "does not exist")

31. 3 **33.** $\dfrac{3}{2}$ **35.** 0

37. $\dfrac{9}{4}$ **39.** 2

41. $m(x) = 4x$; $y = 4x + 1$

43. $m(x) = 6x + 4$; $y = 10x - 8$

45. $m(x) = 4x - 3$; $y = x - 1$

47. $m(x) = 4x + 3$

49. $m(x) = \dfrac{1}{(3-x)^2}$ **51.** $m(x) = 1 + \dfrac{1}{x^2}$

53. $m(x) = -\dfrac{2}{(x-1)^2}$ **55.** $a = 6 \pm 2\sqrt{5}$

57. Nonremovable discontinuity at $x = -1$, removable discontinuity at $x = 1$

59. Nonremovable discontinuity at $x = -3$, removable discontinuity at $x = 1$

61. Apply the intermediate value theorem to $f(x) = x^5 + x - 1$ on $[0, 1]$.

63. Apply the intermediate value theorem to $g(x) = x - \cos x$ on $[0, \pi/2]$.

65. There are three such lines; their slopes are $\frac{1}{4}$, $\frac{1}{2}$, and $-\frac{1}{6}$.

SECTION 3.1 (PAGE 112)

1. $f'(x) \equiv 4$ **3.** $h'(z) = -2z + 25$

5. $\dfrac{dy}{dx} = 4x + 3$ **7.** $\dfrac{dz}{du} = 10u - 3$

9. $\dfrac{dx}{dy} = -10y + 17$ **11.** $f'(x) \equiv 2$

13. $f'(x) = 2x$ **15.** $f'(x) = -\dfrac{2}{(2x+1)^2}$

17. $f'(x) = \dfrac{1}{\sqrt{2x+1}}$ **19.** $f'(x) = \dfrac{1}{(1-2x)^2}$

21. $x(0) = 100$ **23.** $x(2.5) = 99$

25. $x(-2) = 120$ **27.** $y(2) = 64$ (ft)

29. $y(3) = 194$ (ft) **31.** Matches Fig. 3.1.28(e)

33. Matches Fig. 3.1.28(f) **35.** Matches Fig. 3.1.28(d)

37. $A(C) = \dfrac{C}{2\pi}$ **39.** 500 ft; 10 s

41. **a.** 2.5 months; **b.** 50 chipmunks per month

43. $v(20) \approx 51$ (mi/h); $v(40) \approx 74$ (mi/h)

45. $\dfrac{dV}{dr} = 4\pi r^2$

47. $V'(30) = -\dfrac{25\pi}{12}$ (cubic inches per second), so that air is leaking out at approximately 6.545 cubic inches per second.

49. **a.** $V'(6) = -144\pi$ (cm³/h); **b.** -156π (cm³/h)

51. At $t = 2$ (s); $y'(2) = 0$ (m/s)

53. Average rate of change: 0.6 (thousands per year). The instantaneous rate of change is 0.6 when $t = (50 \pm 10\sqrt{7})/9$.

55. **a.** $f'_-(0) = 1$, $f'_+(0) = 2$; **b.** $f'_-(0) = 0 = f'_+(0)$

57. $f'_-(3) = 0 = f'_+(3)$

59. $f'_-(0) = 0$, $f'_+(0) = 2$; $f'(x) = 1 + \dfrac{|x|}{x}$

SECTION 3.2 (PAGE 123)

1. $f'(x) = 6x - 1$ **3.** $f'(x) = 12x + 5$

5. $h'(x) = 3(x+1)^2$ **7.** $f'(y) = 12y^2 - 1$

9. $g'(x) = \dfrac{1}{(x-1)^2} - \dfrac{1}{(x+1)^2}$ **11.** $h'(x) = -\dfrac{6x+3}{(x^2+x+1)^2}$

13. $g'(t) = 5t^4 + 4t^3 + 3t^2 + 4t$ **15.** $g'(z) = \dfrac{4-3z}{6z^3}$

17. $g'(y) = 30y^4 + 48y^3 + 48y^2 - 8y - 6$

19. $g'(t) = \dfrac{3-t}{(1+t)^3}$ **21.** $v'(t) = -\dfrac{3}{(t-1)^4}$

23. $g'(x) = -\dfrac{6x^3+15}{(x^3+7x-5)^2}$ **25.** $g'(x) = \dfrac{4x^3 - 13x^2 + 12x}{(2x-3)^2}$

27. $h'(x) = 3x^2 - 30x^4 - 6x^{-5}$ **29.** $y'(x) = \dfrac{2x^5 + 4x^2 - 15}{x^4}$

31. $y'(x) = 3 + \dfrac{1}{2x^3}$ **33.** $y'(x) = \dfrac{-12x^2 + 6x - 3}{(3x^2 - 3x)^2}$

35. $y'(x) = \dfrac{x^4 + 31x^2 - 10x - 36}{(x^2+9)^2}$

37. $y'(x) = \dfrac{30x^5(5x^5 - 8)}{(15x^5 - 4)^2}$

39. $y'(x) = \dfrac{x(x+2)}{(x+1)^2}$ **41.** $12x - y = 16$

43. $x + y = 3$ **45.** $5x - y = 10$

47. $18x - y = -25$ **49.** $3x + y = 0$

51. **a.** It contracts; **b.** -0.06427 cm^3 per °C

53. $14400\pi \approx 45239$ cm^3 per cm

55. $y = 3x + 2$

57. Suppose that the line L is tangent at the two points (a, a^2) and (b, b^2). Use the derivative to show that $b = a$.

59. The x-intercept is $\dfrac{n-1}{n}x_0$.

61. $3[f(x)]^2 \cdot f'(x)$

63. Let $u_i(x) = f(x)$ for $1 \leqq i \leqq n$. Then the left-hand side in Eq. (16) is $D_x[(f(x))^n]$. Now compute and simplify the right-hand side.

65. $g'(x) = 17(x^3 - 17x + 35)^{16} \cdot (3x^2 - 17)$

67. If $f(x) = \dfrac{1}{1+x^2}$ then

$$f'(x) = -\frac{2x}{(1+x^2)^2},$$

and if $f(x) = \dfrac{x^2}{1+x^2}$ then

$$f'(x) = \frac{2x}{(1+x^2)^2}.$$

Thus there can be only one horizontal tangent. If $n = 0$ it is tangent at $(0, 1)$; if $n = 2$ then it is tangent at the origin.

69. If n is an integer and $n \geqq 3$, then

$$D_x\left[\frac{x^n}{1+x^2}\right] = \frac{x^{n-1}[n + (n-2)x^2]}{(1+x^2)^2}.$$

It is now easy to show that the only horizontal tangent to the graph is tangent at the origin.

71. $(0, 0)$, $\left(-\sqrt{3}, \frac{9}{8}\right)$, and $\left(\sqrt{3}, \frac{9}{8}\right)$

73. $f'(0) = 0$ **75.** $f'(1) = 6$

77. $f'(1) = 1$

SECTION 3.3 (PAGE 132)

1. $\dfrac{dy}{dx} = 15(3x+4)^4$ **3.** $\dfrac{dy}{dx} = -\dfrac{3}{(3x-2)^2}$

5. $\dfrac{dy}{dx} = 3(x^2 + 3x + 4)^2(2x + 3)$

7. $\dfrac{dy}{dx} = 7(2-x)^4(3+x)^6 - 4(2-x)^3(3+x)^7$

9. $\dfrac{dy}{dx} = -\dfrac{6x+22}{(3x-4)^4}$

11. $\dfrac{dy}{dx} = 12[1 + (1+x)^3]^3(1+x)^2$

13. $\dfrac{dy}{dx} = -\dfrac{6(x^2+1)^2}{x^7}$

15. $\dfrac{dy}{dx} = 48(4x-1)^3[1 + (4x-1)^4]^2$

17. $\dfrac{dy}{dx} = [(1 - x^{-4})^3 - 3x^{-4}(1 - x^{-4})^2] \cdot (-4x^{-5})$

$= \cdots = \dfrac{16 - 36x^4 + 24x^8 - 4x^{12}}{x^{17}}$

19.
$\dfrac{dy}{dx} = [2x^{-2}(x^{-2} - x^{-8})^3 + 3x^{-4}(x^{-2} - x^{-8})^2(1 - 4x^{-6})] \cdot (-2x^{-3})$

$= \cdots = \dfrac{28 - 66x^6 + 48x^{12} - 10x^{18}}{x^{29}}$

21. $u(x) = 2x - x^2$, $n = 3$, $f'(x) = 3(2x - x^2)^2(2 - 2x)$

23. $u(x) = 1 - x^2$, $n = -4$, $f'(x) = \dfrac{8x}{(1-x^2)^5}$

25. $u(x) = \dfrac{x+1}{x-1}$, $n = 7$, $f'(x) = -14 \cdot \dfrac{(x+1)^6}{(x-1)^8}$

27. $g'(y) = 1 + 10(2y - 3)^4$

29. $F'(s) = \dfrac{3(s^9 - 3s^3 + 2)}{s^7}$

31. $f'(u) = (1+u)^2(1+u^2)^3(11u^2 + 8u + 3)$

33. $h'(v) = \dfrac{2(v-1)(v^2 - 2v + 2)}{v^3(2-v)^3}$

35. $F'(z) = \dfrac{40 - 250z^4}{(5z^5 - 4z + 3)^{11}}$

37. $\dfrac{dy}{dx} = 4(x^3)^3 \cdot 3x^2 = 12x^{11}$

39. $\dfrac{dy}{dx} = 2(x^2 - 1) \cdot 2x = 4x^3 - 4x$

41. $\dfrac{dy}{dx} = 4(x+1)^3 = 4x^3 + 12x^2 + 12x + 4$

43. $\dfrac{dy}{dx} = -2x(x^2+1)^{-2} = -\dfrac{2x}{(x^2+1)^2}$

45. $f'(x) = 3x^2\cos(x^3) = 3x^2\cos x^3$

47. $g'(z) = 6(\sin 2z)^2(\cos 2z) = 6\sin^2 2z \cos 2z$

49. 40π (in.^{2}/s) **51.** 40 (in.^{2}/s)

53. Decreasing at 600 in.^{3}/h **55.** $G'(1) = -18$

57. $400\pi \approx 1256.64$ (cm^3/s)

59. $r = 5$ (cm)

61. Total melting time: $T = \dfrac{\sqrt[3]{2}}{\sqrt[3]{2} - 1} \approx 4.8473$ (h); all melted by about 2:50:50 P.M. on the same day.

63. $\dfrac{du}{dx} = \dfrac{du}{dv} \cdot \dfrac{dv}{dx} = \dfrac{du}{dv} \cdot \dfrac{dv}{dw} \cdot \dfrac{dw}{dx}$

65. $h'(x) = \dfrac{1}{2\sqrt{x+4}}$

67. $h'(x) = D_x\left[(x^2+4)\sqrt{x^2+4}\right] = \cdots = 3x\sqrt{x^2+4}$

SECTION 3.4 (PAGE 138)

1. $f'(x) = 10x^{3/2} - x^{-3/2}$

3. $f'(x) = \dfrac{1}{\sqrt{2x+1}}$

5. $f'(x) = -3x^{-3/2} - \dfrac{3}{2}x^{1/2}$

7. $D_x(2x+3)^{3/2} = 3\cdot(2x+3)^{1/2}$

9. $D_x(3-2x^2)^{-3/2} = \dfrac{6x}{(3-2x^2)^{5/2}}$

11. $D_x(x^3+1)^{1/2} = \dfrac{1}{2}(x^3+1)^{-1/2}\cdot 3x^2$

13. $D_x(2x^2+1)^{1/2} = \dfrac{2x}{\sqrt{2x^2+1}}$

15. $D_t\left(t^{3/2}\sqrt{2}\right) = \dfrac{3\sqrt{t}}{\sqrt{2}}$

17. $D_x(2x^2-x+7)^{3/2} = \dfrac{3}{2}(4x-1)\sqrt{2x^2-x+7}$

19. $D_x(x-2x^3)^{-4/3} = \dfrac{4(6x^2-1)}{3(x-2x^3)^{7/3}}$

21. $f'(x) = \dfrac{1-2x^2}{\sqrt{1-x^2}}$

23. $f'(t) = \dfrac{1}{2}\left(\dfrac{t^2+1}{t^2-1}\right)^{-1/2}\cdot\dfrac{(t^2-1)(2t)-(t^2+1)(2t)}{(t^2-1)^2}$

$= -\dfrac{2t}{(t^2-1)^{3/2}\sqrt{t^2+1}}$

25. $f'(x) = 3\left(x-\dfrac{1}{x}\right)^2\left(1+\dfrac{1}{x^2}\right)$

27. $f'(v) = -\dfrac{v+2}{2v^2\sqrt{v+1}}$

29. $D_x(1-x^2)^{1/3} = -\dfrac{2x}{3(1-x^2)^{2/3}}$

31. $f'(x) = \dfrac{3(1-2x)}{\sqrt{3-4x}}$

33. $f'(x) = \dfrac{2-24x-14x^2}{3(2x+4)^{2/3}}$

35. $g'(t) = \dfrac{3t^4-3t^2-2t-2}{t^3\sqrt{3t^2+1}}$

37. $f'(x) = \dfrac{23-24x}{(3x+4)^6}$

39. $f'(x) = \dfrac{x+3}{(3x+4)^{4/3}(2x+1)^{1/2}}$

41. $h'(y)$

$= \dfrac{y^{5/3}\left[\frac{1}{2}(1+y)^{-1/2}-\frac{1}{2}(1-y)^{-1/2}\right]-\frac{5}{3}y^{2/3}[(1+y)^{1/2}+(1-y)^{1/2}]}{y^{10/3}}$

$= \cdots = \dfrac{(7y-10)\sqrt{1+y}-(7y+10)\sqrt{1-y}}{6y^{8/3}\sqrt{1-y^2}}$

43. $g'(t) = \dfrac{1}{2}\left[t+(t+t^{1/2})^{1/2}\right]^{-1/2}\cdot\left[1+\dfrac{1}{2}(t+t^{1/2})^{-1/2}\right.$
$\left.\times\left(1+\dfrac{1}{2}t^{-1/2}\right)\right]$

45. There are no horizontal tangents, but there is a vertical line tangent to the graph at $(0,0)$.

47. There is a horizontal line tangent to the graph at $\left(\frac{1}{3}, \frac{2}{9}\sqrt{3}\right)$ and a vertical line tangent to the graph at $(0,0)$.

49. There are no horizontal or vertical tangent lines.

51. $2y = x+4$

53. $2x+y=1$

55. $y=2x$

57. Matches Fig. 3.4.13(d)

59. Matches Fig. 3.4.13(b)

61. Matches Fig. 3.4.13(e)

63. $\dfrac{\pi^2}{32} \approx 0.308$ (seconds per foot)

65. $\left(-\frac{2}{5}\sqrt{5}, -\frac{1}{5}\sqrt{5}\right)$ and $\left(\frac{2}{5}\sqrt{5}, \frac{1}{5}\sqrt{5}\right)$

67. $x+4y=18$

69. $3x+2y=5$ and $3x-2y=-5$

71. Equation (2) is an *identity*, and if two functions have identical graphs on an interval, then their derivatives will also be identically equal to each other on that interval.

73. In $f'(a) = \lim_{x\to a}\dfrac{x^{1/3}-a^{1/3}}{x-a}$, replacement of $x-a$ with

$$(x^{1/3}-a^{1/3})(x^{2/3}+x^{1/3}a^{1/3}+a^{2/3})$$

yields a useful cancellation.

75. The preamble to Problems 72 through 75 implies that $x-a$ can be written as the product of $x^{1/q}-a^{1/q}$ and

$$x^{(q-1)/q}+x^{(q-2)/q}a^{1/q}+x^{(q-3)/q}a^{2/q}+\cdots$$
$$+x^{1/q}a^{(q-2)/q}+a^{(q-1)/q}.$$

SECTION 3.5 (PAGE 148)

In the answers to Problems 1 through 39, we first give the maximum (if any), then the minimum (if any).

1. $f(-1)=2$; none

3. None; $f(0)=0$

5. $f(4)=2$; $f(2)=0$

7. $f(1)=2$; $f(-1)=0$

9. $f(3)=-\frac{1}{6}$; $f(2)=-\frac{1}{2}$

11. $f(3)=7$; $f(-2)=-8$

13. $h(1)=3$; $h(3)=-5$

15. $g(4)=9$; $g(1)=0$

17. $f(4)=52$; $f(-2)=f(1)=-2$

19. $h(1)=h(4)=5$; $h(2)=4$

21. $f(-1)=5$; $f(1)=1$

23. $f\left(-\frac{2}{3}\right)=9$; $f(1)=-16$

25. $f(-1)=10$; $f(3)=-22$

27. $f(2)=56$; $f(-2)=-56$

29. $f(5)=13$; $f\left(\frac{7}{3}\right)=5$

31. $f(1)=17$; $f(0)=0$

33. $f(3)=\frac{3}{4}$; $f(0)=0$

35. $f(-1)=\frac{1}{2}$; $f(3)=-\frac{1}{6}$

37. $f\left(\frac{1}{2}\sqrt{2}\right)=\frac{1}{2}$; $f\left(-\frac{1}{2}\sqrt{2}\right)=-\frac{1}{2}$

39. $f\left(\frac{3}{2}\right)=3\cdot 2^{-4/3}\approx 1.190551$; $f(3)=-3$

41. Consider the cases $A=0$ and $A\neq 0$.

43. $f'(x)=0$ if x is not an integer; $f'(x)$ does not exist if x is an integer.

45. Apply the test of the discriminant to the quadratic equation $f'(x)=0$.

47. Matches Fig. 3.5.15(c).

49. Matches Fig. 3.5.15(d).

51. Matches Fig. 3.5.15(a).

We ignore the character and values of the endpoint extrema in Problems 53 through 59.

53. Global minimum value approximately 6.828387610996 at $x=-1+\frac{1}{3}\sqrt{30}\approx 0.825741858351$.

55. Global minimum value approximately -8.669500829438 at $x\approx -0.762212740507$.

57. Global maximum value approximately 8.976226903748 at $x\approx 1.323417756580$.

59. Global maximum value approximately 30.643243080334 at $x \approx -1.911336401963$, local minimum value approximately -5.767229705222 at $x \approx -0.460141424682$, local maximum value approximately 21.047667292488 at $x \approx 0.967947424014$.

SECTION 3.6 (PAGE 159)

1. 25 and 25
3. 1250
5. 500 (in.^{3})
7. 1152
9. 250
11. 11250 (yd^2)
13. 128
15. Approximately 3.967°C
17. 1000 (cm^3)
19. 0.25 (m^3) (all cubes, no open-topped boxes)
21. Two equal squares yield minimum total area 200; a single square yields maximum area 400.
23. 30000 (m^2)
25. Approximately 9259.259 in.^{3}
27. Five presses
29. The minimizing value of x is $-2 + \frac{10}{3}\sqrt{6}$. Result: Install 6 inches of insulation for an annual savings of about \$285.
31. Either \$1.10 or \$1.15
33. Radius $\frac{2}{3}R$, height $\frac{1}{3}H$
35. Let R denote the [constant] radius of the circle.
37. $\frac{2000}{27}\pi\sqrt{3}$
39. Maximum 4, minimum $\sqrt[3]{16}$
41. $\frac{1}{2}\sqrt{3}$
43. Each plank has length $\frac{1}{2}\sqrt{7-\sqrt{17}} \approx 0.848071$, width $\frac{1}{8}(\sqrt{34} - 3\sqrt{2}) \approx 0.198539$, and area $\frac{1}{2}\sqrt{142+34\sqrt{17}} \approx 0.673500$.
45. The boater should make landfall $\frac{2}{3}\sqrt{3} \approx 1.155$ km from the point on the shore closest to the island.
47. At $P\left(\frac{1}{3}\sqrt{3}, 0\right)$
49. Approximately 3.45246
51. To minimize the sum, make the sphere of radius $5\sqrt{10/(\pi+6)}$ and the edge length of the cube twice that amount. To maximize the sum, make the edge length of the cube zero.
53. The maximum volume is approximately 95.406 (ft^3).
55. In Problem 53, $x = 4$ maximizes the volume V_1, and $V_1(4) = \frac{128}{3}\sqrt{5}$. In Problem 54, $x = 8$ maximizes the volume V_2, and $V_2(8) = \frac{256}{3}\sqrt{10}$.
57. The volume is maximized when the length of the base is $\sqrt{A/3}$ and the height is half that.
59. The volume is maximized when the radius of the cylinder is $\sqrt{A/(3\pi)}$ and its height is the same.
61. The global maximum value of the area is 1 and occurs when the point of tangency is $(1, \frac{1}{2})$.

SECTION 3.7 (PAGE 171)

1. $f'(x) = 6\sin x\cos x$
3. $f'(x) = \cos x - x\sin x$
5. $f'(x) = \dfrac{x\cos x - \sin x}{x^2}$
7. $f'(x) = \cos^3 x - 2\sin^2 x\cos x$
9. $g'(t) = 4(1+\sin t)^3 \cdot \cos t$
11. $g'(t) = \dfrac{\sin t - \cos t}{(\sin t + \cos t)^2}$
13. $f'(x) = 3x^2\sin x - 4x\cos x + 2\sin x$
15. $f'(x) = -2\sin 2x\sin 3x + 3\cos 2x\cos 3x$
17. $g'(t) = 3t^2\sin^2 2t + 4t^3\sin 2t\cos 2t$
19. $g'(t) = \dfrac{5}{2}(\cos 3t + \cos 5t)^{3/2} \cdot (-3\sin 3t - 5\sin 5t)$
21. $\dfrac{dy}{dx} = \dfrac{\sin\sqrt{x}\,\cos\sqrt{x}}{\sqrt{x}}$
23. $\dfrac{dy}{dx} = 2x\cos(3x^2-1) - 6x^3\sin(3x^2-1)$
25. $\dfrac{dy}{dx} = -3\sin 2x\sin 3x + 2\cos 3x\cos 2x$
27. $\dfrac{dy}{dx} = -\dfrac{3\sin 3x\sin 5x + 5\cos 3x\cos 5x}{\sin^2 5x}$
29. $\dfrac{dy}{dx} = 4x\sin x^2\cos x^2$
31. $\dfrac{dy}{dx} = \dfrac{\cos 2\sqrt{x}}{\sqrt{x}}$
33. $\dfrac{dy}{dx} = \sin x^2 + 2x^2\cos x^2$
35. $\dfrac{dy}{dx} = \dfrac{\sin\sqrt{x} + \sqrt{x}\,\cos\sqrt{x}}{2\sqrt{x}}$
37. $\dfrac{dy}{dx} = \dfrac{(x-\cos x)^3 + 6x(x-\cos x)^2(1+\sin x)}{2\sqrt{x}}$
39. $\dfrac{dy}{dx} = -2x[\sin(\sin x^2)]\cos x^2$
41. $\dfrac{dx}{dt} = 7t^6\sec^2 t^7$
43. $\dfrac{dx}{dt} = 7\tan^6 t\sec^2 t$
45. $\dfrac{dx}{dt} = 7t^6\tan 5t + 5t^7\sec^2 5t$
47. $\dfrac{dx}{dt} = \dfrac{\sec\sqrt{t} + \sqrt{t}\,\sec\sqrt{t}\,\tan\sqrt{t}}{2\sqrt{t}}$
49. $\dfrac{dx}{dt} = \dfrac{2}{t^3}\csc\left(\dfrac{1}{t^2}\right)\cot\left(\dfrac{1}{t^2}\right)$
51. $\dfrac{dx}{dt} = 5\cot 3t\sec 5t\tan 5t - 3\csc^2 3t\sec 5t$
53. $\dfrac{dx}{dt} = t\sec^2 t + \sec t\csc t - t\csc^2 t$
55. $\dfrac{dx}{dt} = [\sec(\sin t)\tan(\sin t)] \cdot \cos t$
57. $\dfrac{dx}{dt} = \cos^2 t - \sin^2 t = \cos 2t$
59. $\dfrac{dx}{dt} = -\dfrac{5\csc^2 5t}{2\sqrt{1+\cot 5t}}$
61. $y = -x$
63. $y = 2x - 2 + \dfrac{4}{\pi}$

65. At every integral multiple of $\pi/2$

67. The tangent line is horizontal at all points of the form $(n\pi + \frac{1}{4}\pi, \frac{1}{2})$ and at all points of the form $(n\pi + \frac{3}{4}\pi, -\frac{1}{2})$ where n is an integer.

69. $y = x \pm 2$

71. See Appendix C for various trigonometric identities.

$$D_x \cot x = D_x \frac{\cos x}{\sin x} = \frac{-\sin^2 x - \cos^2 x}{\sin^2 x} = -\frac{1}{\sin^2 x} = -\csc^2 x,$$

$$D_x \sec x = D_x \frac{1}{\cos x} = -\frac{-\sin x}{\cos^2 x} = \frac{1}{\cos x} \cdot \frac{\sin x}{\cos x} = \sec x \tan x,$$

and

$$D_x \csc x = D_x \frac{1}{\sin x} = -\frac{\cos x}{\sin^2 x} = -\frac{1}{\sin x} \cdot \frac{\cos x}{\sin x} = -\csc x \cot x.$$

73. $\alpha = \pi/4$

75. Approximately 0.4224 mi/s; that is, about 1521 mi/h

77. $\dfrac{2000\pi}{27} \approx 232.71$ ft/s; that is, about 158.67 mi/h

79. $\theta = \pi/3$

81. $\frac{8}{3}\pi R^3$, twice the volume of the sphere!

83. $\frac{3}{4}\sqrt{3}$

85. $A(\theta) = \dfrac{s^2(\theta - \sin\theta)}{2\theta^2}$

87. Show that if n is a positive integer,

$$h = \frac{2}{(4n+1)\pi}, \quad \text{and} \quad k = \frac{2}{(4n-1)\pi},$$

then

$$\frac{f(h) - f(0)}{h} = 1 \quad \text{and} \quad \frac{f(k) - f(0)}{k} = -1.$$

39. $f'(x) = \dfrac{22x^2 + 8x - 24}{(2x+1)(x^2-4)}$

41. $f'(x) = \dfrac{13x}{(x^2-4)(x^2+9)}$

43. $f'(x) = -\dfrac{2}{(x-1)(x+1)}$

45. $g'(t) = \dfrac{2}{t(t^2+1)}$

47. $\dfrac{dy}{dx} = 2^x \ln 2$

49. $\dfrac{dy}{dx} = \dfrac{2x^{\ln x} \ln x}{x}$

51. $\dfrac{dy}{dx} = \dfrac{2 + (\ln x)\ln(\ln x)}{2x^{1/2}\ln x} \cdot (\ln x)^{\sqrt{x}}$

53. $\dfrac{dy}{dx} = \dfrac{(3x - 4x^2 - x^4)(1+x^2)^{1/2}}{(1+x^3)^{7/3}}$

55. $\dfrac{dy}{dx} = \left[\dfrac{2x^3}{x^2+1} + 2x\ln(x^2+1)\right] \cdot (x^2+1)^{x^2}$

57. $\dfrac{dy}{dx} = \dfrac{(2 + \ln x)\left(\sqrt{x}\right)^{\sqrt{x}}}{4\sqrt{x}}$

59. $y = 3e^2 x - 2e^2$

61. $y = x - 1$

63. It appears that $f^{(n)}(x) = 2^n e^{2x}$.

65. The first maximum point occurs where $x = \arctan 6$ and the first minimum point occurs where $x = \pi + \arctan 6$.

67. 1.118 (to three places)

69. The (rounded) values for $k=7$ and $k=8$ are 2.718281692545 and 2.718281814868.

71. Because $\ln y = \ln u + \ln v + \ln w - \ln p - \ln q - \ln r$,

$$\frac{1}{y}\cdot\frac{dy}{dx} = \frac{1}{u}\cdot\frac{du}{dx} + \frac{1}{v}\cdot\frac{dv}{dx} + \frac{1}{w}\cdot\frac{dw}{dx} - \frac{1}{p}\cdot\frac{dp}{dx} - \frac{1}{q}\cdot\frac{dq}{dx} - \frac{1}{r}\cdot\frac{dr}{dx},$$

and the generalization is obvious.

73. (a) If $f(x) = \log_{10} x$, then it follows from the definition of the derivative that

$$f'(1) = \lim_{h\to 0} \log_{10}(1+h)^{1/h}.$$

(b) With $h = \pm 0.0001$ the value of $\log_{10}(1+h)^{1/h}$ is approximately 0.4343.

SECTION 3.8 (PAGE 187)

1. $f'(x) = 2e^{2x}$

3. $f'(x) = 2x\exp(x^2)$

5. $f'(x) = -\dfrac{2}{x^3}\exp\left(\dfrac{1}{x^2}\right)$

7. $g'(t) = \dfrac{2 + t^{1/2}}{2}\exp(t^{1/2})$

9. $g'(t) = (1 + 2t - t^2)e^{-t}$

11. $g'(t) = (-\sin t)\exp(\cos t)$

13. $g'(t) = \dfrac{te^{-t} + e^{-t} - 1}{t^2}$

15. $f'(x) = \dfrac{(-1)e^x - (1-x)e^x}{(e^x)^2} = \dfrac{x-2}{e^x}$

17. $f'(x) = e^x\exp(e^x)$

19. $f'(x) = 2e^x\cos(2e^x)$

21. $f'(x) = \dfrac{3}{3x-1}$

23. $f'(x) = \dfrac{1}{1+2x}$

25. $f'(x) = \dfrac{3x^2-1}{3(x^3-x)}$

27. $f'(x) = -\dfrac{\sin(\ln x)}{x}$

29. $f'(x) = -\dfrac{1}{x(\ln x)^2}$

31. $f'(x) = \dfrac{2x^2+1}{x(x^2+1)}$

33. $f'(x) = -\tan x$

35. $f'(t) = 2t\ln(\cos t) - t^2\tan t$

37. $g'(t) = (2 + \ln t)\ln t$

SECTION 3.9 (PAGE 195)

1. $\dfrac{dy}{dx} = \pm\dfrac{x}{\sqrt{x^2-1}} = \dfrac{x}{y}$

3. $\dfrac{dy}{dx} = \mp\dfrac{16x}{5\sqrt{400-16x^2}}$

5. $\dfrac{dy}{dx} = -\sqrt{\dfrac{y}{x}}$

7. $\dfrac{dy}{dx} = -\left(\dfrac{y}{x}\right)^{1/3}$

9. $\dfrac{dy}{dx} = \dfrac{3x^2 - 2xy - y^2}{3y^2 + 2xy + x^2}$

11. $\dfrac{dy}{dx} = -\dfrac{\sin y + y\cos x}{x\cos y + \sin x}$

13. $\dfrac{dy}{dx} = \dfrac{3e^y + 2x - 2}{(e^x - 3)e^y}$

15. $y + 4 = \dfrac{3}{4}(x - 3)$

17. $3x + 4y = 10$

19. $y \equiv -2$

21. $4x = 3y$

23. $4y = 5x$

25. $x + 2y = 10$

27. $11y = 2x + 40$

29. **a.** $5y = 4x + 12$; **b.** $y - \dfrac{9}{2} = -\left(x - \dfrac{9}{2}\right)$

31. $\left(2, 2 \pm \sqrt{8}\right)$

33. **a.** $y = x$; **b.** $x + e = 2ey$

35. All four points where $|x| = \frac{1}{4}\sqrt{6}$ and $|y| = \frac{1}{4}\sqrt{2}$

37. $\dfrac{4}{5\pi} \approx 0.25645$ (ft/s)

39. $\dfrac{32\pi}{125} \approx 0.80425$ (m/h)

41. 20 (cm^2/s)

43. 0.25 (cm/s)

45. 6 (ft/s)

47. 384 (mi/h)

49. (a) Decreasing at $\dfrac{191}{1300\pi} \approx 0.047$ (ft/min); (b) decreasing at $\dfrac{1337}{5100\pi} \approx 0.083$ (ft/min)

51. Moving downward at $\dfrac{160}{9} \approx 17.78$ (ft/s)

53. Increasing at 16π (cm^3/s)

55. 6000 (mi/h)

57. (a) Moving downward at $\frac{11}{15}\sqrt{21} \approx 3.36$ (ft/s), about 2.29 mi/h; (b) moving downward at $\frac{88}{5} \cdot (9.999965) \approx 176$ (ft/s), about 120 mi/h; (c) moving downward at $\frac{22}{15} \cdot (3048) \approx 4470$ (ft/s), about 3048 mi/h. The results in parts (b) and (c) are not plausible.

59. $32\sqrt{13} \approx 115.38$ (mi)

61. $\dfrac{50}{81\pi} \approx 0.1965$ (ft/s)

63. Falling at $\dfrac{10}{81\pi} \approx 0.0393$ (in./min)

65. $5\sqrt{2}$ mi/min; that is, about 424.26 mi/h

67. Lengthening at 2 ft/s

SECTION 3.10 (PAGE 209)

Note: When we used Newton's method, we used Mathematica *3.0 and carried at least 40 decimal digits in all calculations; answers shown here are correct or correctly rounded to the number of digits shown. Your answers may show differences in the last, or even the last few, decimal places depending on the hardware and software you use.*

1. 2.2361 **3.** 2.5119 **5.** 0.3028

7. 0.7402 **9.** 0.7391 **11.** 1.2361

13. 2.3393 **15.** 2.0288 **17.** 0.5671

19. 0.4429

21. $x_{n+1} = \dfrac{1}{3}\left(2x_n + \dfrac{a}{(x_n)^2}\right)$; 1.25992

23. $x_{14} = 0.4501836113 = x_{15}$ (to ten places)

25. The first formula yields the wrong root 2.879385, as does the second. Use $x = \dfrac{1}{\sqrt{3 - x}}$.

27. Let $f(x) = x^5 + x - 1$. Use the intermediate value theorem to show that there is at *least* one solution of $f(x) = 0$; use the fact that f is increasing on $\boldsymbol{R}$ to show that there is at *most* one real solution; to four places, $x = 0.7549$.

29. 0 and ± 1.8955

31. -1.3578, 0.7147, and 1.2570

33. $x \approx 0.865474033102$

35. $x \approx 3.452462314058$

37. 0.2261

39. 2.028758 and 4.913180

41. $t \approx 0.4909$, $w \approx 13.0164$

43. $\theta \approx 0.0199966678$; $R \approx 50008.3319$ (ft), about 9.47 mi

CHAPTER 3 MISCELLANEOUS PROBLEMS (PAGE 213)

1. $\dfrac{dy}{dx} = 2x - \dfrac{6}{x^3}$

3. $\dfrac{dy}{dx} = \dfrac{3x^{5/6} - 2}{6x^{4/3}}$

5. $\dfrac{dy}{dx} = (x-1)^6(3x+2)^8(48x - 13)$

7. $\dfrac{dy}{dx} = 4\left(3x - \dfrac{1}{2x^2}\right)^3 \cdot \left(3 + \dfrac{1}{x^3}\right)$

9. $\dfrac{dy}{dx} = -\dfrac{y}{x} = -\dfrac{9}{x^2}$

11. $\dfrac{dy}{dx} = -\dfrac{3(3x^2 - 1)}{2(x^3 - x)^{5/2}}$

13. $\dfrac{dy}{dx} = \dfrac{4x(x^2+1)}{(x^4 + 2x^2 + 2)^2}$

15. $\dfrac{dy}{dx} = \dfrac{7}{3}\left(x^{1/2} + (2x)^{1/3}\right)^{4/3} \cdot \left(\dfrac{1}{2}x^{-1/2} + \dfrac{2^{1/3}}{3x^{2/3}}\right)$

17. $\dfrac{dy}{dx} = -\dfrac{1}{\left(\sqrt{x+1} - 1\right)^2 \sqrt{x+1}}$

19. $\dfrac{dy}{dx} = \dfrac{1 - 2xy^2}{2x^2y - 1} = -\dfrac{(x+2y)y}{(2x+y)x}$

21. $$\frac{dy}{dx} = \frac{1}{2}\left(x + \left[2x + (3x)^{1/2}\right]^{1/2}\right)^{-1/2} \cdot \left(1 + \frac{1}{2}\left[2x + (3x)^{1/2}\right]^{-1/2} \cdot \left[2 + \frac{3}{2}(3x)^{-1/2}\right]\right).$$

The symbolic algebra program *Mathematica* writes this answer without fractional exponents as follows:

$$\frac{dy}{dx} = \frac{1 + \dfrac{2 + \dfrac{\sqrt{3}}{2\sqrt{x}}}{2\sqrt{2x + \sqrt{3x}}}}{2\sqrt{x + \sqrt{2x + \sqrt{3x}}}}.$$

23. $\dfrac{dy}{dx} = -\left(\dfrac{y}{x}\right)^{2/3}$

25. $\dfrac{dy}{dx} = -18 \cdot \dfrac{(x^3 + 3x^2 + 3x + 3)^2}{(x+1)^{10}}$

27. $\dfrac{dy}{dx} = \left(\dfrac{1 + \cos x}{\sin^2 x}\right)^{1/2} \cdot \dfrac{2\sin x \cos x + 2\sin x \cos^2 x + \sin^3 x}{2(1 + \cos x)^2}$

29. $\dfrac{dy}{dx} = -\dfrac{4\sin 2x \sin 3x + 3\cos 2x \cos 3x}{2(\sin 3x)^{3/2}}$

31. $\dfrac{dy}{dx} = e^x(\cos x - \sin x)$

33. $\dfrac{dy}{dx} = -\dfrac{3e^x}{(2 + 3e^x)^{5/2}\left[1 + (2 + 3e^x)^{-3/2}\right]^{1/3}}$

35. $\dfrac{dy}{dx} = -\dfrac{\cos^2\left([1 + \ln x]^{1/3}\right)\sin\left([1 + \ln x]^{1/3}\right)}{x[1 + \ln x]^{2/3}}$

37. $f'(x) = -2e^{-x}\sin(e^{-x})\cos(e^{-x})$

39. $f'(x) = e^x(\cos 2x - 2\sin 2x)$

41. $g'(t) = \dfrac{1 + 2t^2}{t}$

43. $g'(t) = e^t\cos(e^t)\cos(e^{-t}) + e^{-t}\sin(e^t)\sin(e^{-t})$

45. $g'(t) = \dfrac{2e^t}{(1 - e^t)^2}$

47. $\dfrac{dy}{dx} = \dfrac{1 - ye^{yx}\cos(e^{xy})}{xe^{xy}\cos(e^{xy})}$

49. $\dfrac{dx}{dy}=e^y+ye^y$, and so $\dfrac{dy}{dx}=\dfrac{1}{e^y+ye^y}=\dfrac{y}{ye^y+yx}=\dfrac{y}{x+xy}$.

51. $\dfrac{dy}{dx}=\dfrac{(1-\ln y)y}{x-y}$

53. $\dfrac{dy}{dx}=\dfrac{xy(3x^2+1)}{(x^2-3)(x^4+1)}=-\dfrac{3x^3+x}{(3-x^2)^{1/2}(x^4+1)^{5/4}}$

55.
$$\frac{dy}{dx}=\frac{(13x^2+55x+54)y}{12(x+1)(x+2)(x+3)}=\frac{13x^2+55x+54}{12(x+1)^{1/2}(x+2)^{2/3}(x+3)^{3/4}}$$

57. $\dfrac{dy}{dx}=\dfrac{1+\ln(\ln x)}{x}\cdot(\ln x)^{\ln x}$

59. $x\equiv 1$ **61.** $x\equiv 0$ **63.** $\dfrac{1}{2}$ (ft/min)

65. $\dfrac{1}{3}$ **67.** $\dfrac{1}{4}$

69. Use $-1\leq \sin u\leq 1$ for all u and the squeeze law to show that the limit is zero.

71. $h'(x)=-\dfrac{x}{(x^2+25)^{3/2}}$ **73.** $h'(x)=\dfrac{5}{3}(x-1)^{2/3}$

75. $h'(x)=-2x\sin(x^2+1)$ **77.** $\dfrac{dV}{dA}=\dfrac{1}{4}\sqrt{\dfrac{A}{\pi}}$

79. $\dfrac{2}{5}\pi\approx 1.2566$ (mi/s), about 4524 mi/h

81. R^2

83. For maximum surface area $(72\pi V^2)^{1/3}$, make two equal spheres. For minimum surface area $(36\pi V^2)^{1/3}$, make only one sphere.

85. $\dfrac{32}{81}\pi R^3$ **87.** $\dfrac{M}{2}$ **89.** 36 (ft^3)

91. $3\sqrt{3}\approx 5.196$

93. Case 1: $A=0$ and $B\neq 0$. Case 2: $A\neq 0$ and $B=0$. Case 3: $A\neq 0$ and $B\neq 0$ (this is by far the longest and hardest case).

95. The pier should be built two miles from the point on the shore nearest the first town.

97. **a.** $y_{\max}=\dfrac{m^2v^2}{64(m^2+1)}$; **b.** $m=1$, $\alpha=\pi/4$

Note: When we used Newton's method, we used Mathematica *3.0 and carried at least 40 decimal digits in all calculations; answers shown here are correct or correctly rounded to the number of digits shown. Your answers may show differences in the last, or even the last few, decimal places depending on the hardware and software you use.*

99. 2.6458 **101.** 2.3714

103. −0.3473 **105.** 0.588532744

107. −0.7391 **109.** −1.2361

111. Approximately 1.547852572 ft

113. There are exactly three real solutions; approximations thereto are −2.722493355, 0.8012614801, and 2.309976541.

115. We have no formula for finding the derivative of the sum of a *variable* number of terms.

117. Begin this way: $z^{2/3}-x^{2/3}=(z^{1/3}-x^{1/3})(z^{1/3}+x^{1/3})$ and $z-x=(z^{1/3}-z^{1/3})(z^{2/3}+z^{1/3}x^{1/3}+x^{2/3})$.

119. 4 in.^{2}/s

121. $-\dfrac{50}{9\pi}\approx -1.7684$ (ft/min)

123. 1 in./min—a constant rate

125. Think of $a^2-2ax_0+y_0=0$ as a quadratic equation in the unknown a. Use the discriminant to determine the number of solutions of this equation.

SECTION 4.2 (PAGE 225)

1. $dy=\left(6x+\dfrac{8}{x^3}\right)dx$ **3.** $dy=\dfrac{3x^2+2\sqrt{4-x^3}}{2\sqrt{4-x^3}}\,dx$

5. $dy=\dfrac{3}{2}(7x^2-12x)(x-3)^{1/2}\,dx$

7. $dy=\dfrac{3x^2+50}{2(x^2+25)^{3/4}}\,dx$ **9.** $dy=-\dfrac{\sin\sqrt{x}}{2\sqrt{x}}\,dx$

11. $dy=(2\cos^2 2x-2\sin^2 2x)\,dx$

13. $dy=\dfrac{2x\cos 2x-\sin 2x}{3x^2}\,dx$

15. $dy=\dfrac{x\cos x+\sin x}{(1-x\sin x)^2}\,dx$

17. $f(x)\approx 1+x$ **19.** $f(x)\approx 1+2x$

21. $f(x)\approx 1-3x$ **23.** $f(x)\approx x$

25. $L(x)=2+\dfrac{1}{27}x$, so $\sqrt[3]{25}\approx\dfrac{79}{27}\approx 2.9259$.

27. $L(x)=\dfrac{3}{2}+\dfrac{1}{32}x$, so $\sqrt[4]{15}\approx\dfrac{63}{32}=1.96875$.

29. $L(x)=\dfrac{5}{48}-\dfrac{1}{1536}x$, so $65^{-2/3}\approx\dfrac{95}{1536}\approx 0.06185$.

31. $L(x)=\dfrac{\sqrt{2}}{2}\left(\dfrac{\pi}{4}+1\right)-\dfrac{\sqrt{2}}{2}x$, so $\cos 43^\circ\approx\dfrac{\pi+90}{90\sqrt{2}}\approx 0.7318$.

33. $L(x)=x+1$, so $e^{0.1}\approx 1.1$.

35. $2x\,dx+2y\,dy=0$, so $\dfrac{dy}{dx}=-\dfrac{x}{y}$.

37. $3x^2\,dx+3y^2\,dy=3y\,dx+3x\,dy$, so $\dfrac{dy}{dx}=\dfrac{y-x^2}{y^2-x}$.

39. $L(x)=1+kx$

41. The area decreases by approximately 4 square inches.

43. The volume decreases by approximately $\frac{405}{2}\pi\approx 636.17\ \text{cm}^3$.

45. The range is increased by approximately 10 ft.

47. The wattage increases by approximately 6 watts.

49. $25\pi\approx 78.5398$ in.^{3} **51.** $4\pi\approx 12.57\ \text{m}^2$

53. (0.55, 1.45) **55.** (1.74, 2.30)

57. (−0.67, 0.67) **59.** (0.54, 1.01)

SECTION 4.3 (PAGE 235)

1. Increasing if $x<0$, decreasing if $x>0$; matching graph: (c).

3. Decreasing if $x<-2$, increasing if $x>-2$; matching graph: (f).

5. Increasing if $x<-1$, decreasing on $(-1,2)$, increasing if $x>2$; matching graph: (d).

7. $f(x)=2x^2+5$ **9.** $f(x)=-\dfrac{1}{x}+2$

11. Increasing on $\boldsymbol{R}$

13. Increasing if $x < 0$, decreasing if $x > 0$

15. Increasing if $x < \frac{3}{2}$, decreasing if $x > \frac{3}{2}$

17. Decreasing if $x < -1$, increasing on $(-1, 0)$, decreasing on $(0, 1)$, increasing if $x > 1$

19. Decreasing if $x < -2$, increasing on $(-2, 0)$, decreasing on $(0, 1)$, increasing if $x > 1$

21. Increasing if $x < 2$, decreasing if $x > 2$

23. Decreasing on $(0, 1)$, increasing on $(1, 3)$, decreasing if $x > 3$

25. $f(0) = 0 = f(2)$, $f'(x) = 2x - 2$; $c = 1$

27. $c = \pi/4$ and $c = 3\pi/4$ **29.** $f'(0)$ does not exist.

31. $f(1) = e \neq 0$ **33.** $c = -\dfrac{1}{2}$ **35.** $c = \dfrac{35}{27}$

37. The average slope is $\frac{1}{3}$, but $|f'(x)| = 1$ where it is defined.

39. The average slope is 1, but $f'(x) = 0$ wherever it is defined.

41. If $f(x) = x^5 + 2x - 3$, then $f'(x) > 0$ for all x, so the equation $f(x) = 0$ can have at most one solution in any interval. Because $f(1) = 0$, the equation $f(x) = 0$ has exactly one solution in $[0, 1]$.

43. Let $f(x) = -3 + x \ln x$. Show that $f(2) < 0 < f(4)$ (so that $f(x) = 0$ has at least one solution in $[2, 4]$) and that $f'(x) > 0$ on $[2, 4]$ (so that $f(x) = 0$ has at most one solution there).

45. Compute the average speed of the car between 3:00 P.M. and 3:18 P.M., then apply the mean value theorem to the position function of the car.

47. Let $f(t)$ be the distance the first car has traveled (starting at point A at time $t = 0$) and let $g(t)$ be the corresponding function for the second car. Apply Rolle's theorem to $h(t) = f(t) - g(t)$.

49. Note first that $f'(x) = \frac{3}{2}[(1 + x)^{1/2} - 1]$.

51. You may assume that $f'(x) = a_{n-1}x^{n-1} + a_{n-2}x^{n-2} + \cdots + a_1x + a_0$ where $a_{n-1} \neq 0$. Construct a polynomial $p(x)$ such that $p'(x) = f'(x)$. Conclude that $f(x) = p(x) + C$ on $[a, b]$.

53. Apply the mean value theorem to $f(x)$ on $[100, 101]$.

55. $C = -1$

57. First show that the average slope of the graph of f on $[-1, 2]$ is 2.

59. Use the definition of the derivative to show that $g'(0) = \frac{1}{2}$. Then show that $g'(x)$ takes on values close to -1 and close to 1 in every subinterval containing $x = 0$.

61. Let $h(x) = 1 - \frac{1}{2}x^2 - \cos x$. Compute $h'(x)$ and use the result in Example 8.

63. Let $K(x) = 1 - \frac{1}{2}x^2 + \frac{1}{24}x^4 - \cos x$. Compute $K'(x)$ and apply the result in Problem 61.

SECTION 4.4 (PAGE 244)

1. Global minimum at $x = 2$

3. Local maximum at $x = 0$, local minimum at $x = 2$

5. No extrema

7. Local minimum at -2, local maximum at $x = 5$

9. Global minimum at $x = -1$ and at $x = 1$, local maximum at $x = 0$

11. Local maximum at $x = -3$, local minimum at $x = 3$

13. Global maximum at $x = \frac{1}{2}$

15. Global minimum at $x = -4$, local maximum at $x = 6$

17. Global maximum at $x = \pi/2$

19. Global minimum at $x = -\pi/2$, global maximum at $x = \pi/2$

21. Global minimum at $x = -\pi$, global maximum at $x = \pi$

23. Global maximum at $x = \sqrt{e}$

25. Global minimum at $x = -\pi/4$, global maximum at $x = 3\pi/4$

27. -10 and 10 **29.** $(1, 1)$

31. Base 9 in. by 18 in., height 6 in.

33. Height and diameter both $10\pi^{-1/3} \approx 6.828$ cm

35. The perimeter is minimized when all four sides have length 10.

37. Square base of edge length 5 in., height 2.5 in.

39. Base radius $(25/\pi)^{1/3} \approx 1.9965$ in., height $(1600/\pi)^{1/3} \approx 7.9859$ in. (four times the radius of the base)

41. $\left(\pm\sqrt{3/2}, 3/2\right)$ **43.** 8 cm

45. $\left(20 + 12\sqrt[3]{4} + 12\sqrt[3]{16}\right)^{1/2} \approx 8.323876$ m

47. Minimum volume of pyramid: $\frac{32}{3}a^3$; ratio of volume of smallest pyramid to volume of sphere: $8/\pi$

49. Height $(6V)^{1/3}$, base edge $(\frac{9}{2}V^2)^{1/6}$

51. Base edge and height both $V^{1/3}$

53. Radius of base $(V/2\pi)^{1/3}$, height double that

55. Simplifying assumption: The volume of material used to make the can is accurately approximated by multiplying the area of the top by its thickness, the area of the bottom by its thickness, and the area of the curved side by its thickness, then adding these products.

SECTION 4.5 (PAGE 255)

1. Matches (c) **3.** Matches (d)

5. Critical point: $a = \frac{5}{2}$; decreasing on $(-\infty, a)$, increasing on $(a, +\infty)$

7. Critical points: $a = -\frac{5}{2}$, $b = 3$; increasing on $(-\infty, a)$ and on $(b, +\infty)$, decreasing on (a, b)

9. Critical points at $x = -3$, $x = 0$, and $x = 2$; decreasing on $(-\infty, -3)$ and on $(0, 2)$, increasing on $(-3, 0)$ and on $(2, +\infty)$

11. Critical points at $x = -4$, $x = -2$, $x = 2$, and $x = 4$; increasing on $(-\infty, -4)$, on $(-2, 2)$, and on $(4, +\infty)$, decreasing on $(-4, -2)$ and on $(2, 4)$

13. Critical points at $x = -4$, $x = -2$, $x = 0$, $x = 2$, and $x = 4$; increasing on $(-\infty, -4)$, on $(-2, 2)$, and on $(4, +\infty)$, decreasing on $(-4, -2)$ and on $(2, 4)$

15. Decreasing on $(-\infty, 1)$, increasing on $(1, +\infty)$; global minimum at $(1, 2)$

17. Increasing on $(-\infty, -2)$ and on $(2, +\infty)$, decreasing on $(-2, 2)$; local maximum at $(-2, 16)$, local minimum at $(2, -16)$

19. Increasing on $(-\infty, 1)$ and on $(3, +\infty)$, decreasing on $(0, 3)$; local maximum at $(1, 4)$, local minimum at $(3, 0)$

21. Increasing on $\boldsymbol{R} = (-\infty, +\infty)$; no extrema

23. Decreasing on $(-\infty, -2)$ and on $(-\frac{1}{2}, 1)$, increasing on $(-2, -\frac{1}{2})$ and on $(1, +\infty)$; global minimum at $(-2, 0)$ and at $(1, 0)$, local maximum at $(-\frac{1}{2}, \frac{81}{16})$

25. Increasing on $(0, 1)$, decreasing on $(1, +\infty)$; local minimum at $(0, 0)$, global maximum at $(1, 2)$

27. Increasing on $(-\infty, -1)$ and on $(1, +\infty)$, decreasing on $(-1, 1)$; local maximum at $(-1, 2)$, local minimum at $(1, -2)$

29. Decreasing on $(-\infty, -2)$ and on $(0, 2)$, increasing on $(-2, 0)$ and on $(2, +\infty)$; global minimum at $(-2, -9)$ and at $(2, -9)$, local maximum at $(0, 7)$

31. Decreasing on $(-\infty, \frac{3}{4})$, increasing on $(\frac{3}{4}, +\infty)$; global minimum at $(\frac{3}{4}, -\frac{81}{8})$

33. Increasing on $(-\infty, -2)$ and on $(1, +\infty)$, decreasing on $(-2, 1)$; local maximum at $(-2, 20)$, local minimum at $(2, -7)$

35. Increasing on $(-\infty, \frac{3}{5})$ and on $(\frac{4}{5}, +\infty)$, decreasing on $(\frac{3}{5}, \frac{4}{5})$; local maximum at $(\frac{3}{4}, \frac{81}{5})$, local minimum at $(\frac{4}{5}, 16)$

37. Decreasing on $(-\infty, -1)$ and on $(0, 2)$, increasing on $(-1, 0)$ and on $(2, +\infty)$; local minimum at $(-1, 3)$, local maximum at $(0, 8)$, global minimum at $(2, -24)$

39. Increasing on $(-\infty, -2)$ and on $(2, +\infty)$, decreasing on $(-2, 2)$; local maximum at $(-2, 64)$, local minimum at $(2, -64)$

41. Increasing on $\boldsymbol{R} = (-\infty, +\infty)$; no extrema

43. Increasing on $(-\infty, -\sqrt{2}\,)$ and on $(0, \sqrt{2}\,)$, decreasing on $(-\sqrt{2}, 0)$ and also on $(\sqrt{2}, +\infty)$; global maximum at $(-\sqrt{2}, 16)$ and at $(\sqrt{2}, 16)$, local minimum at $(0, 0)$

45. Increasing on $(-\infty, 0)$ and on $(0, 1)$ (it is also correct to say that f is increasing on $(-\infty, 1)$), decreasing on $(1, +\infty)$; global maximum at $(1, 3)$; vertical tangent at $(0, 0)$

47. Increasing on $(-\infty, \frac{3}{5})$ and on $(1, +\infty)$, decreasing on $(\frac{3}{5}, 1)$; local maximum where $x = \frac{3}{5}$ (the ordinate is approximately 0.3257), local minimum at $(1, 0)$

49. Plot $y = 2x^3 + 3x^2 - 36x - 3$ to see the graph.

51. Plot $y = -2x^3 - 3x^2 + 36x + 15$ to see the graph.

53. Plot $y = 3x^4 - 8x^3 - 30x^2 + 72x + 45$ to see the graph.

55. (a) $f(-2.1038034027) \approx 7.58 \times 10^{-9}$; (b) Approximate factorization:

$$f(x) \approx [x + 2.1038034027] \cdot [x^2 - (2.1023803403)x + 1.4244923418].$$

(c) $1.0519017014 \pm 0.5652358517i$

57. Decreasing if $x < 0$, if $\frac{1}{6}(3 - \sqrt{3}) < x < \frac{1}{2}$, and if $\frac{1}{6}(3 + \sqrt{3}) < x < 1$, increasing if $0 < x < \frac{1}{6}(3 - \sqrt{3})$, if $\frac{1}{2} < x < \frac{1}{6}(3 + \sqrt{3})$, and if $1 < x$; [equal] global minima where $x = 0, x = \frac{1}{2}, x = 1$, [equal] local maxima at $x = \frac{1}{6}(3 \pm \sqrt{3})$

59. [Equal] global minima at $(0, 0)$, $(\frac{5}{9}, 0)$, and $(1, 0)$; local maxima *very near* the two points $(0.22925, 0.0000559441)$ and $(0.807787, 0.0000119091)$

SECTION 4.6 (PAGE 268)

1. $f'(x) = 8x^3 - 9x^2 + 6$, $f''(x) = 24x^2 - 18x$, $f'''(x) = 48x - 18$

3. $f'(x) = -8(2x - 1)^{-3}$, $f''(x) = 48(2x - 1)^{-4}$, $f'''(x) = -384(2x - 1)^{-5}$

5. $g'(t) = 4(3t - 2)^{1/3}$, $g''(t) = 4(3t - 2)^{-2/3}$, $g'''(t) = -8(3t - 2)^{-5/3}$

7. $h'(y) = (y + 1)^{-2}$, $h''(y) = -2(y + 1)^{-3}$, $h'''(y) = 6(y + 1)^{-4}$

9. $g'(t) = t(1 + 2\ln t)$, $g''(t) = 3 + 2\ln t$, $g'''(t) = \dfrac{2}{t}$

11. $f'(x) = 3\cos 3x$, $f''(x) = -9\sin 3x$, $f'''(x) = -27\cos 3x$

13. $f'(x) = \cos^2 x - \sin^2 x$, $f''(x) = -4\sin x\cos x$, $f'''(x) = 4\sin^2 x - 4\cos^2 x$

15. $f'(x) = \dfrac{x\cos x - \sin x}{x^2}$, $f''(x) = \dfrac{(2 - x^2)\sin x - 2x\cos x}{x^3}$,

$f'''(x) = \dfrac{(6 - x^2)x\cos x + (3x^2 - 6)\sin x}{x^4}$

17. $y'(x) = -\dfrac{2x + y}{x + 2y}$, $y''(x) = -\dfrac{18}{(x + 2y)^3}$

19. $y'(x) = -\dfrac{2x + 1}{3y^2}$, $y''(x) = -\dfrac{2[(2x + 1)^2 + 3y^3]}{9y^5}$

21. $y'(x) = -\dfrac{y}{x - \cos y}$, $y''(x) = -\dfrac{y^2\sin y + 2y\cos y - 2xy}{(x - \cos y)^3}$

23. Critical points: $(-3, 81)$ and $(5, -175)$; inflection point: $(1, -47)$

25. Critical points: $(-3.5, 553.5)$ and $(4.5, -470.5)$; inflection point: $(0.5, 41.5)$

27. Critical points: $(0, 237)$, $(-3\sqrt{3}, -492)$, and $(3\sqrt{3}, -492)$; inflection points: $(-3, -168)$ and $(3, -168)$

29. Critical points: $(0, 1000)$ and $(\frac{16}{3}, -\frac{181144}{81})$ (approximately $(5.333, -2236.345679)$); inflection points: $(0, 1000)$ and $(4, -1048)$

31. Global minimum at $(2, 1)$; no inflection points

33. Local maximum at $(-1, 3)$, local minimum at $(1, -1)$; inflection point: $(0, 1)$

35. Global maximum at $(1, e^{-1})$, inflection point at $(2, 2e^{-2})$

37. No critical points; inflection point: $(0, 0)$

39. Global minimum value 0 at $x = 0$ and at $x = 1$; local maximum at $(\frac{1}{2}, \frac{1}{16})$; inflection points: $(\frac{1}{6}(3 \pm \sqrt{3}), \frac{1}{36})$

41. Global maximum at $(\pi/2, 1)$, global minimum at $(3\pi/2, -1)$; inflection point: $(\pi, 0)$

43. No critical points, no extrema; inflection point: $(0, 0)$

45. Global maximum value 1 at $x = 0$ and at $x = \pi$, global minimum at $(\pi/2, 0)$; inflection points: $(-\pi/4, 1/2)$, $(\pi/4, 1/2)$, $(3\pi/4, 1/2)$, and $(5\pi/4, 1/2)$

47. Global maximum at $(\frac{3}{2}, 5e^{-3})$, inflection point at $(2, 10e^{-4})$

49. Global minimum at $(2 - \sqrt{3}, -1.119960)$, local maximum at $(2 + \sqrt{3}, 0.130831)$ (ordinates approximate); inflection points: $(1, -2e^{-1})$ and $(5, 14e^{-5})$

51. -10 and 10

53. $(1, 1)$

55. 9 in. wide, 18 in. long, 6 in. high

57. Radius $5/\sqrt[3]{\pi}$ cm, height double that

59. Square base of edge length 5 in., height 2.5 in.

61. Radius $(25/\pi)^{1/3}$ in., height four times that

63. Increasing for $x < -1$ and for $x > 2$, decreasing on $(-1, 2)$, local maximum at $(-1, 10)$, local minimum at $(2, -17)$; inflection point: $(0.5, -3.5)$

65. Increasing for $x < -2$ and on $(0, 2)$, decreasing otherwise, global maximum at $(-2, 22)$ and at $(2, 22)$, local minimum at $(0, 6)$; inflection points: $(\pm\frac{2}{3}\sqrt{3}, \frac{134}{9})$

67. Decreasing if $x < -1$ and on $(0, 2)$, increasing otherwise; local minimum at $(-1, -6)$, local maximum at $(0, -1)$, global minimum at $(2, -33)$; there are two inflection points: $(\frac{1}{3}(1 \pm \sqrt{7}), \frac{1}{27}(-311 \mp 80\sqrt{7}))$

69. Increasing if $x < \frac{3}{7}$ and if $x > 1$, decreasing otherwise; local maximum at $(\frac{3}{7}, \frac{6912}{823543})$, local minimum at $(1, 0)$; inflection points at $(0, 0)$ and also at the two points for which $x = \frac{1}{7}(3 \pm \sqrt{2})$

71. Increasing for all x, no extrema; critical point, vertical tangent, and inflection point at $(0, 1)$

73. Increasing on $[0, +\infty)$, global minimum at $(0, 0)$; inflection point: $(1, 4)$

75. Increasing if $x < 1$, decreasing if $x > 1$, critical point, inflection point, and vertical tangent at $(0, 0)$, global maximum at $(1, 3)$, inflection point at $(-2, -6\sqrt[3]{2})$

77. Matches (c) **79.** Matches (b) **81.** Matches (d)

83. Part (a): Key step in a proof by induction: Assume that for some integer $k \geqq 1$, $f^{(k)}(x) = k!$ if $f(x) = x^k$; let $g(x) = x^{k+1} = x \cdot f(x)$, apply the product rule to compute $g'(x)$, then apply the inductive assumption to $g'(x)$.

85. Apply the product rule to $\dfrac{dz}{dx} = \dfrac{dz}{dy} \cdot \dfrac{dy}{dx}$.

89. $b = \dfrac{1}{3}V = 42.7$, $a = 3V^2p \approx 3{,}583{,}859$, $R = \dfrac{8Vp}{3T} \approx 81.8$

91. Local maximum at $(1, 2101)$, local minimum at $(1.034, 2100.980348)$, inflection point at $(1.017, 2100.990174)$ (coordinates exact). To see these points clearly, plot $y = f(x)$ on the interval $[0.96, 1.07]$.

SECTION 4.7 (PAGE 281)

1. 1 **3.** 3 **5.** 2 **7.** 1

9. 4 **11.** 0 **13.** 2

15. $+\infty$ (or "does not exist")

17. Matches (g) **19.** Matches (a)

21. Matches (f) **23.** Matches (j)

25. Matches (l) **27.** Matches (k)

29. No extrema or inflection points, sole intercept $(0, -\frac{2}{3})$, vertical asymptote $x = 3$, (two-way) horizontal asymptote $y = 0$

31. No extrema or inflection points, sole intercept $(0, \frac{3}{4})$, vertical asymptote $x = -2$, (two-way) horizontal asymptote $y = 0$

33. No extrema or inflection points, sole intercept $(0, -\frac{1}{27})$, vertical asymptote $x = \frac{3}{2}$, (two-way) horizontal asymptote $y = 0$

35. Global minimum and sole intercept $(0, 0)$, inflection points at $(\pm\frac{1}{3}\sqrt{3}, \frac{1}{4})$, horizontal asymptote $y = 1$

37. Local maximum and sole intercept $(0, -\frac{1}{9})$, vertical asymptotes $x = \pm 3$, (two-way) horizontal asymptote $y = 0$

39. Local maximum at $(-\frac{1}{2}, -\frac{4}{25})$, sole intercept $(0, -\frac{1}{6})$, vertical asymptotes $x = -3$ and $x = 2$, (two-way) horizontal asymptote $y = 0$

41. Local maximum at $(-1, -2)$, local minimum at $(1, 2)$, no inflection points or intercepts, vertical asymptote $x = 0$, slant asymptote $y = x$

43. Local maximum and sole intercept $(0, 0)$, local minimum at $(2, 4)$, vertical asymptote $x = 1$, slant asymptote $y = x + 1$

45. No extrema or inflection points, sole intercept $(0, 1)$, vertical asymptote $x = 1$, (two-way) horizontal asymptote $y = 0$

47. The graph is increasing everywhere, concave upward for $x < 0$, concave downward for $x > 0$, no extrema, and inflection point and sole intercept $(0, \frac{1}{2})$, right-hand horizontal asymptote $y = 1$, left-hand horizontal asymptote $y = 0$

49. Local maximum and sole extremum $(\frac{1}{2}, -\frac{4}{9})$, sole intercept $(0, -\frac{1}{2})$, vertical asymptotes $x = -1$ and $x = 2$, (two-way) horizontal asymptote $y = 0$

51. Only intercepts $(\pm 2, 0)$, no inflection points or extrema, vertical asymptote $x = 0$, slant asymptote $y = x$

53. Sole intercept $(\sqrt[3]{4}, 0)$, local maximum $(-2, -3)$, no other extrema, no inflection points, vertical asymptote $x = 0$, slant asymptote $y = x$

55. With all coordinates approximate, there are local minima at $(-1.9095, -0.3132)$ and $(1.3907, 3.2649)$ and a local maximum at $(4.5188, 0.1630)$. There are inflection points at $(-2.8119, -0.2768)$ and $(6.0623, 0.1449)$, a horizontal asymptote $y = 0$, and vertical asymptotes $x = 0$ and $x = 2$.

57. Horizontal asymptote $y = 0$, vertical asymptotes $x = 0$ and $x = 2$; local minima at $(-2.8173, -0.1783)$ and $(1.4695, 5.5444)$, local maxima at $(-1, 0)$ and $(4.3478, 0.1998)$, and inflection points at $(-4.3611, -0.1576)$, $(-1.2569, -0.0434)$, and $(5.7008, 0.1769)$ (numbers with decimal points are approximations)

59. Horizontal asymptote $y = 0$, vertical asymptotes $x = 0$ and $x = 2$, local minima at $(-2.6643, -0.2160)$, $(1.2471, 14.1117)$, and $(3, 0)$; local maxima at $(-1, 0)$ and $(5.4172, 0.1296)$; there are inflection points at $(-4.0562, -0.1900)$, $(-1.2469, -0.0538)$, $(3.3264, 0.0308)$, and $(7.4969, 0.1147)$ (numbers with decimal points are approximations)

61. The x-axis is a horizontal asymptote; there are vertical asymptotes at $x = -0.5321$, $x = 0.6527$, and $x = 2.8794$. There is a local minimum at $(0, 0)$ and a local maximum at $(\sqrt[3]{2}, -0.9008)$. There are no inflection points. (Numbers with decimal points are approximations.)

63. The line $y = x + 3$ is a slant asymptote in both the positive and negative directions; thus there is no horizontal asymptote. There is a vertical asymptote at $x = -1.1038$. There are local maxima at $(-2.3562, -1.8292)$ and $(2.3761, 18.5247)$, local minima at $(0.8212, 0.6146)$ and $(5.0827, 11.0886)$. There are inflection points at $(1.9433, 11.3790)$ and $(2.7040, 16.8013)$. (Numbers with decimal points are approximations.)

65. The line $2y = x$ is a slant asymptote in both the positive and negative directions; thus there is no horizontal asymptote. There also are no vertical asymptotes. There is a local maximum at $(0.2201, 0.6001)$, a local minimum at $(0.8222, -2.9690)$, and inflection points at $(-2.2417, -1.2782)$, $(-0.5946, -0.1211)$, $(0.6701, -1.6820)$, and $(0.9649, -2.2501)$. (Numbers with decimal points are approximations.)

67. The line $2y = x$ is a slant asymptote in both the positive and negative directions; thus there is no horizontal asymptote. There is a vertical asymptote at $x = -1.7277$. There are local maxima at $(-3.1594, -2.3665)$ and $(1.3381, 1.7792)$, local minima at $(-0.5379, -0.3591)$ and $(1.8786, 1.4388)$. There are inflection points at $(0, 0)$, $(0.5324, 0.4805)$, $(1.1607, 1.4294)$, and $(1.4627, 1.6727)$. (Numbers with decimal points are approximations.)

69. The graph of f is decreasing for $0 < x < 1$ and for $x < 0$, increasing for $x > 1$. It is concave upward for $x < -\sqrt[3]{2}$ and also for $x > 0$, concave downward for $-\sqrt[3]{2} < x < 0$. The only intercept is at $(-\sqrt[3]{2}, 0)$; this is also the only inflection point. There is a local minimum at $(1, 3)$. The y-axis is a vertical asymptote.

SECTION 4.8 (PAGE 290)

1. $\frac{1}{2}$ **3.** $\frac{2}{5}$ **5.** 0 **7.** 0

9. $\frac{1}{2}$ **11.** 2 **13.** 0 **15.** 1

17. 1 **19.** $\frac{3}{5}$ **21.** $\frac{3}{2}$ **23.** $\frac{1}{3}$

25. $\frac{\ln 2}{\ln 3}$ **27.** $\frac{1}{2}$ **29.** 1 **31.** $\frac{1}{3}$

33. $-\frac{1}{2}$ **35.** 1 **37.** $\frac{1}{4}$ **39.** $\frac{3}{2}$

41. 6 **43.** $\frac{4}{3}$ **45.** $\frac{2}{3}$ **47.** 0

49. $\lim_{x\to 0} f(x) = 0$ **51.** $\lim_{x\to\pi} \frac{\sin x}{x-\pi} = -1$

53. $\lim_{x\to 0} \frac{1-\cos x}{x^2} = \frac{1}{2}$

55. $\lim_{x\to\infty} f(x) = 0$, $\lim_{x\to-\infty} f(x) = -\infty$

57. $\lim_{x\to\infty} f(x) = 0$

59. $\lim_{x\to 0+} f(x) = -\infty$ and $\lim_{x\to\infty} f(x) = 0$

61. Assume that the result holds for $n = k$ where k is some fixed positive integer, then apply l'Hôpital's rule to

$$\lim_{x\to\infty} \frac{x^{k+1}}{e^x}.$$

63. Global maximum at $(n, n^n e^{-n})$, inflection points at the two points where $x^2 - 2nx + n^2 - n = 0$.

65. With $y = 1/x$ we have

$$\lim_{x\to 0+} x^k \ln x = \lim_{y\to\infty} \frac{-\ln y}{y^k} = -\left(\lim_{y\to\infty} \frac{1}{ky^k}\right) = 0.$$

67. Holding x fixed, apply l'Hôpital's rule to

$$\lim_{h\to 0} \frac{f(x+h) - f(x-h)}{2h}.$$

69. 1

71. If x is large then $\frac{x}{e} > \frac{e^2}{e}$.

73. $f(n-1) < n^n e^{-n}$ leads to $e < \left(1 - \frac{1}{n}\right)^{-n}$, etc.

SECTION 4.9 (PAGE 295)

1. 1 **3.** $\frac{3}{8}$ **5.** 1

7. 1 **9.** 0 **11.** -1

13. $-\infty$ (or "does not exist")

15. $-\infty$ (or "does not exist")

17. $-\frac{1}{2}$ **19.** 0 **21.** 1 **23.** 1

25. $e^{-1/6} \approx 0.8464817249$

27. $e^{-1/2} \approx 0.6065306597$

29. 1

31. $e^{-1} \approx 0.3678795512$

33. $-\infty$ (or "does not exist")

35. $f(x) \to 1$ as $x \to +\infty$, $f(x) \to 0$ as $x \to 0^+$; the global maximum is $f(e) = e^{1/e} \approx 1.4446678610$.

37. $f(x) \to 1$ as $x \to +\infty$, $f(x) \to 0$ as $x \to 0^+$; the global maximum is $f(e) = e^{2/e} \approx 2.0870652286$.

39. $f(x) \to 1$ as $x \to +\infty$ and as $x \to 0^+$; the global maximum value of $f(x)$ occurs at the solution of $2x^2 = (1+x^2)\ln(1+x^2)$ near $x = 2$; it is approximately $f(1.9802913004) \approx 2.2361202715$.

41. $f(x) \to 1$ as $x \to +\infty$, $f(x) \to 0$ as $x \to 0^+$; the global maximum is approximately $f(1.2095994645) \approx 1.8793598343$.

43. $\lim_{h\to 0} (1 + hx)^{1/h} = e^x$

45. Approximately $(0.4099776300, 0.6787405265)$

49. $f(x) \to 1$ as $x \to +\infty$. $f(x) \to +\infty$ as $x \to 0^+$; there is a global minimum at $(1, 0)$, an inflection point with approximate coordinates $(0.8358706352, 0.1279267691)$, another near $(1.1163905964, 0.1385765415)$, yet another near $(8.9280076968, 1.0917274397)$, and a local maximum near $(5.8312001357, 1.1021470392)$.

51. $f(x) \to +\infty$ as $x \to 0^+$ and as $x \to +\infty$; the global minimum value of $f(x)$ is $e^{-1/e}$ and occurs at both $x = e^{-1/e}$ and $x = e^{1/e}$. There is a cusp at $(1, 1)$ ($|f'(x)| \to +\infty$ as $x \to 1$) and there is also a local maximum at $(1, 1)$

53. The graph of $y = f(x)$ on the interval $[-0.00001, 0.00001]$ shows clearly that $e \approx 2.71828$ to five places. The graphical method succeeds because

$$\lim_{x\to 0} \left(1 + \frac{1}{x}\right)^x = e.$$

CHAPTER 4 MISCELLANEOUS PROBLEMS (PAGE 297)

1. $dy = \frac{3}{2}(4x - x^2)^{1/2}(4 - 2x)\,dx$

3. $dy = -\frac{2}{(x-1)^2}\,dx$

5. $dy = \left(2x\cos\sqrt{x} - \frac{1}{2}x^{3/2}\sin\sqrt{x}\right)dx$

7. $\frac{12801}{160} = 80.00625$ **9.** 1025.536

11. $\frac{601}{60} \approx 10.0167$ **13.** 132.5

15. $\frac{65}{32} = 2.03125$ **17.** $\Delta V \approx 7.5$ (in.^{3})

19. $\Delta V \approx 10\pi$ (cm^3) **21.** $\Delta T \approx \frac{\pi}{96} \approx 0.0327$ (s)

23. $c = \sqrt{3}$ **25.** $c = 1$

27. $c = \left(\frac{11}{4}\right)^{1/4}$.

29. Decreasing for $x < 3$, increasing for $x > 3$, concave upward everywhere, global minimum at $(3, -5)$

31. Increasing for all x, inflection points at

$\left(-\frac{1}{2}\sqrt{2}, -\frac{233}{8}\sqrt{2}\right)$, $(0, 0)$, and $\left(\frac{1}{2}\sqrt{2}, \frac{233}{8}\sqrt{2}\right)$

33. Increasing for $x < \frac{1}{4}$, decreasing for $x > \frac{1}{4}$, vertical tangent at $(0, 0)$, global maximum at $x = \frac{1}{4}$, inflection points where $x = 0$ and where $x = -\frac{1}{2}$

35. $f'(x) = 3x^2 - 2$, $f''(x) = 6x$, and $f'''(x) \equiv 6$

37. $g'(t) = \dfrac{2}{(2t+1)^2} - \dfrac{1}{t^2}$, $g''(t) = \dfrac{2}{t^3} - \dfrac{8}{(2t+1)^3}$, $g'''(t) = \dfrac{48}{(2t+1)^4} - \dfrac{6}{t^4}$

39. $f'(t) = 3t^{1/2} - 4t^{1/3}$, $f''(t) = \dfrac{3}{2}t^{-1/2} - \dfrac{4}{3}t^{-2/3}$, $f'''(t) = \dfrac{8}{9}t^{-5/3} - \dfrac{3}{4}t^{-3/2}$

41. $h'(t) = -\dfrac{4}{(t-2)^2}$, $h''(t) = \dfrac{8}{(t-2)^3}$, $h'''(t) = -\dfrac{24}{(t-2)^4}$

43. $g'(x) = -\dfrac{4}{3(5-4x)^{2/3}}$, $g''(x) = -\dfrac{32}{9(5-4x)^{5/3}}$, $g'''(x) = -\dfrac{640}{27(5-4x)^{8/3}}$

45. $\dfrac{dy}{dx} = -\left(\dfrac{y}{x}\right)^{2/3}$, $\dfrac{d^2y}{dx^2} = \dfrac{2}{3}\left(\dfrac{y}{x^5}\right)^{1/3}$

47. $\dfrac{dy}{dx} = \dfrac{1}{2(5y^4-4)\sqrt{x}}$, $\dfrac{d^2y}{dx^2} = \dfrac{40y^4 - 25y^8 - 20x^{1/3}y^3 - 16}{4x^{3/2}(5y^4-4)^3}$

49. $\dfrac{dy}{dx} = \dfrac{2x-5y}{5x-2y}$, $\dfrac{d^2y}{dx^2} = -\dfrac{210}{(5x-2y)^3}$

51. $\dfrac{dy}{dx} = -\dfrac{2xy}{x^2+1-3y^2}$, $y''(x) = \dfrac{2y[3x^4 - 9y^4 + 6(x^2+1)y^2 + 2x^2 - 1]}{(x^2+1-3y^2)^3}$

53. Global minimum at $(2, -48)$, concave upward everywhere, intercepts $(0, 0)$ and $(\sqrt[3]{32}, 0)$

55. Decreasing for $x < a = -\frac{2}{3}\sqrt{3}$, increasing for $a < x < 0$, decreasing for $0 < x < b = \frac{2}{3}\sqrt{3}$, increasing for $x > b$. Global minima at $x = a$ and $x = b$, local maximum at $x = 0$, inflection points where $x = \pm\frac{2}{5}\sqrt{5}$

57. Increasing if $x < 3$, decreasing if $x > 3$; global maximum at $(3, 3)$ intercepts at $(0, 0)$ and $(4, 0)$, a vertical tangent and inflection point at the latter, and an inflection point at $(6, -6\sqrt[3]{2})$

59. Increasing if $x < -2$ and if $-2 < x < 0$, decreasing if $0 < x < 2$ and if $x > 2$; local maximum at $(0, -\frac{1}{4})$, no other extrema, no inflection points, no x-intercepts; vertical asymptotes $x = \pm 2$ and horizontal asymptote $y = 1$

61. Increasing if $-4 < x < -1$ and if $-1 < x < 0$, decreasing if $x < -4$, if $0 < x < 2$ and if $x > 2$. Local maximum and sole intercept $(0, 0)$, local minimum at $(-4, \frac{16}{9})$, vertical asymptotes $x = -1$ and $x = 2$, horizontal asymptote $y = 2$, inflection point with approximate coordinates $(-6.107243, 1.801610)$

63. Decreasing for $x < 1$, increasing for $x > 1$; concave upward for $x < 0$ and for $x > \frac{2}{3}$, concave downward on $(0, \frac{2}{3})$, global minimum at $(1, -1)$, inflection points at $(0, 0)$ and at $(\frac{2}{3}, -\frac{16}{27})$, no asymptotes, and $f(x) \to +\infty$ as $x \to +\infty$ and as $x \to -\infty$

65. Increasing if $x < -2$ and if $-2 < x < 0$, decreasing if $0 < x < 2$ and if $x > 2$; local maximum and sole intercept at $(0, 0)$, no inflection points, vertical asymptotes $x = \pm 1$, horizontal asymptote $y = 1$

67. Decreasing if $x < 0$ and if $x > 4$, increasing on $(0, 4)$, local minimum at $(0, -10)$, inflection point at $(2, 6)$, local maximum at $(4, 22)$; four intercepts, at $(-1.180140, 0)$, $(1.488872, 0)$, $(0, -10)$, and $(5.691268, 0)$ (numbers with decimals are approximations)

69. Increasing if $x < -1$, local maximum at $(-1, 2)$, decreasing on $(-1, 1)$, inflection point and intercept at $(0, 0)$, local minimum at $(1, -2)$, increasing if $x > 1$, intercepts $(\pm\sqrt{3}, 0)$

71. Increasing if $x < -\frac{5}{3}$, local maximum at $(-\frac{5}{3}, \frac{256}{27})$, decreasing on $(-\frac{5}{3}, 1)$, inflection point at $(-\frac{1}{3}, \frac{128}{27})$, local minimum and intercept at $(1, 0)$, increasing if $x > 1$, another intercept at $(-3, 0)$

73. Maximum value $1 = f(-1)$

75. 15 cm wide, 30 cm long, 10 cm high

77. 5 in. wide, 10 in. long, 8 in. high

79. $100 \cdot \left(\frac{2}{9}\right)^{2/5} \approx 54.79$ mi/h

81. Two horizontal tangents, where $x = 1 - \frac{1}{3}\sqrt{3}$ and $y \approx \pm 0.6204$; vertical tangent lines at the x-intercepts 0, 1, and 2; inflection points where $x \approx 2.4679$ and $y \approx \pm 1.3019$

83. 240 ft

85. $2\sqrt{2A(n+2)}$ ft

87. If x is the abscissa of the point of tangency, then the area of the triangle is $A(x) = 9/(4x)$, $0 < x < +\infty$, so there is neither a maximum nor a minimum area.

89. 288 in.^{2}

91. 270 cm^2

93. In both cases $m = 1$ and $b = -\frac{2}{3}$.

95. $\dfrac{1}{4}$ **97.** $\dfrac{1}{2}$ **99.** $\dfrac{1}{2}$ **101.** 1

103. $-\infty$ (or "does not exist")

105. $+\infty$ (or "does not exist")

107. e^2 **109.** $-\dfrac{e}{2}$

SECTION 5.2 (PAGE 314)

1. $x^3 + x^2 + x + C$ **3.** $\dfrac{3}{4}x^4 - \dfrac{2}{3}x^3 + x + C$

5. $-\dfrac{3}{2}x^{-2} + \dfrac{4}{5}x^{5/2} - x + C$ **7.** $t^{3/2} + 7t + C$

9. $\dfrac{3}{5}x^{5/3} - 16x^{-1/4} + C$ **11.** $x^4 - 2x^2 + 6x + C$

13. $49e^{x/7} + C$ **15.** $\dfrac{1}{5}(x+1)^5 + C$

17. $-\dfrac{1}{6}(x-10)^{-6} + C$ **19.** $\dfrac{2}{3}x^{3/2} - \dfrac{4}{5}x^{5/2} + \dfrac{2}{7}x^{7/2} + C$

21. $\dfrac{2}{21}x^3 - \dfrac{3}{14}x^2 - \dfrac{5}{7}x^{-1} + C$ **23.** $\dfrac{1}{54}(9t+11)^6 + C$

25. $\dfrac{1}{2}e^{2x} - \dfrac{1}{2}e^{-2x} + C$ **27.** $\dfrac{1}{2}\sin 10x + 2\cos 5x + C$

29. $\dfrac{3}{\pi}\sin \pi t + \dfrac{1}{3\pi}\sin 3\pi t + C$

31. $D_x\left(\frac{1}{2}\sin^2 x + C_1\right) = 2\sin x \cos x = D_x\left(-\frac{1}{2}\cos^2 x + C_2\right)$; $C_2 - C_1 = \frac{1}{2}$

33. $\frac{1}{2}x - \frac{1}{4}\sin 2x + C$; $\frac{1}{2}x + \frac{1}{4}\sin 2x + C$

35. $y(x) = x^2 + x + 3$ **37.** $y(x) = \frac{2}{3}x^{3/2} - \frac{16}{3}$

39. $y(x) = 2\sqrt{x+2} - 5$ **41.** $y(x) = \frac{3}{4}x^4 - 2x^{-1} + \frac{9}{4}$

43. $y(x) = \frac{1}{4}(x-1)^4 + \frac{7}{4}$ **45.** $y(x) = 3e^{2x} + 7$

47. $v(t) = 6t^2 - 4t - 10$, $x(t) = 2t^3 - 2t^2 - 10t$

49. $v(t) = \frac{2}{3}t^3 + 3$, $x(t) = \frac{1}{6}t^4 + 3t - 7$

51. $v(t) = 1 - \cos t$, $x(t) = t - \sin t$

53. $x(t) = 5t$ if $0 \leqq t \leqq 5$; $x(t) = 10t - \frac{1}{2}t^2 - \frac{25}{2}$ if $5 \leqq t \leqq 10$

55. $x(t) = \frac{1}{2}t^2$ if $0 \leqq t \leqq 5$, $x(t) = 10t - \frac{1}{2}t^2 - 25$ if $5 \leqq t \leqq 10$

57. 144 ft; 6 s **59.** 144 ft **61.** 5 s; 112 ft/s

63. $2\sqrt{15} \approx 7.746$ s; $64\sqrt{15} \approx 247.87$ ft/s

65. 120 ft/s **67.** 5 s; 160 ft/s **69.** 400 ft

71. $\frac{1}{4}\left(-5 + 2\sqrt{145}\right) \approx 4.77$ s; $16\sqrt{145} \approx 192.6655$ ft/s

73. $\frac{544}{3} \approx 181.33$ ft/s **75.** 22 ft/s^2

77. 96 ft/s; $\frac{11520}{13} \approx 886$ ft **79.** $\frac{25}{3}\sqrt{42} \approx 54$ mi/h

SECTION 5.3 (PAGE 326)

1. $3 + 9 + 27 + 81 + 243$ **3.** $\frac{1}{2} + \frac{1}{3} + \frac{1}{4} + \frac{1}{5} + \frac{1}{6}$

5. $1 + \frac{1}{4} + \frac{1}{9} + \frac{1}{16} + \frac{1}{25} + \frac{1}{36}$

7. $x + x^2 + x^3 + x^4 + x^5$

9. $\sum_{n=1}^{5} n^2$ **11.** $\sum_{k=1}^{5} \frac{1}{k}$ **13.** $\sum_{m=1}^{6} \frac{1}{2^m}$

15. $\sum_{n=1}^{5} \left(\frac{2}{3}\right)^n$ **17.** $\sum_{n=1}^{10} \frac{1}{n}x^n$ **19.** 190

21. 1165 **23.** 224 **25.** 350

27. 338350 **29.** $\frac{1}{3}$ **31.** n^2

33. $\frac{2}{5}, \frac{3}{5}$ **35.** $\frac{33}{2}, \frac{39}{2}$ **37.** $\frac{6}{25}, \frac{11}{25}$

39. $\frac{378}{25}, \frac{513}{25}$ **41.** $\frac{81}{400}, \frac{121}{400}$

43. $2 \cdot \sum_{i=1}^{n} i = (n+1) + (n+1) + \cdots + (n+1)$ (n terms)

45. $\frac{n(n+1)}{2n^2} \to \frac{1}{2}$ as $n \to +\infty$

47. $\frac{81n^2(n+1)^2}{4n^4} \to \frac{81}{4}$ as $n \to +\infty$

49. $5n \cdot \frac{1}{n} - \frac{3n(n+1)}{2n^2} \to \frac{7}{2}$ as $n \to +\infty$

51. $\frac{bh}{n^2} \cdot \frac{n(n+1)}{2} \to \frac{1}{2}bh$ as $n \to +\infty$

53. $\lim_{n\to\infty} \frac{A_n}{C_n} = \frac{r}{2}$

SECTION 5.4 (PAGE 337)

1. $\int_1^3 (2x-1)\,dx$ **3.** $\int_0^{10} (x^2+4)\,dx$

5. $\int_4^9 \sqrt{x}\,dx$ **7.** $\int_3^8 \frac{1}{\sqrt{1+x}}\,dx$

9. $\int_0^{1/2} \sin 2\pi x\,dx$ **11.** $\frac{11}{25} = 0.44$

13. $\frac{29}{20} = 1.45$ **15.** $\frac{39}{2} = 19.5$

17. $\frac{294}{5} = 58.8$ **19.** $-\frac{\pi}{6} \approx -0.523598776$

21. $\frac{6}{25} = 0.24$ **23.** $\frac{137}{60} \approx 2.283333333$

25. $\frac{33}{2} = 16.5$ **27.** $\frac{132}{5} = 26.4$

29. $\frac{\pi}{6} \approx 0.523598776$ **31.** $\frac{33}{100} = 0.33$

33. $\frac{6086}{3465} \approx 1.756421356$ **35.** 18

37. $\frac{1623}{40} = 40.575$ **39.** 0

41. $\frac{259775}{141372} \approx 1.837527940$ **43.** $\frac{8}{3} \approx 2.666666667$

45. 12 **47.** 30

49. Choose $x_i^\star = x_i = \frac{bi}{n}$ and $\Delta x = \frac{b}{n}$.

51. Choose $x_i^\star = \frac{x_{i-1} + x_i}{2}$; note that $\Delta x = x_i - x_{i-1}$ for each meaningful value of i.

53. Case 1: $a < b$. Let $\mathcal{P} = \{x_0, x_1, x_2, \dots, x_n\}$ be a partition of $[a, b]$ and let $\{x_i^\star\}$ be a selection for $\mathcal{P}$. Note that $\Delta x_i = x_i - x_{i-1}$ for $1 \leqq i \leqq n$. Begin with the equation

$$\int_a^b c\,dx = \lim_{n\to\infty} \sum_{i=1}^{n} c\,\Delta x_i$$

and continue by expanding and simplifying the right-hand side. Don't forget Case 2.

55. Whatever partition $\mathcal{P}$ is given, a selection $\{x_i^\star\}$ with all $x_i^\star$ irrational can be made. (An explanation follows the solution of Problem 69 of Section 2.4 in the Student Solutions Manual.) Show that for every such selection, we have

$$\sum_{i=1}^{n} f(x_i^\star)\,\Delta x_i = 1.$$

Then show that there are also selections for which every such Riemann sum has the value zero.

57. Follow Example 5, but choose $x_i^\star = x_i = \dfrac{3i}{n}$ and $\Delta x = \dfrac{3}{n}$.

59. Choose $x_i^\star = x_i = \dfrac{k\pi}{n}$ and $\Delta x = \dfrac{\pi}{n}$.

61. Let
$$h = b - a, \quad x_i^\star = x_i = a + \frac{ih}{n}, \quad \text{and} \quad \Delta x = \frac{h}{n}.$$
Your computer algebra system should report a result similar to this:
$$\sum_{i=1}^{n} \sin\left(a + \frac{ih}{n}\right) = \csc\frac{h}{2n}\ \sin\frac{h}{2}\ \sin\left(\frac{1}{2}\left[2a + h + \frac{h}{n}\right]\right).$$
If so, you will probably need to use one of the trigonometric identities that immediately precede Problems 59 through 62 in Section 7.4.

SECTION 5.5 (PAGE 347)

1. $\dfrac{55}{12} \approx 4.583333333$ **3.** $\dfrac{49}{60} \approx 0.816666667$

5. $-\dfrac{1}{20} = -0.05$ **7.** $\dfrac{1}{4}$

9. $\dfrac{16}{3} \approx 5.333333333$ **11.** 24

13. 0 **15.** $\dfrac{32}{3} \approx 10.666666667$

17. 0 **19.** $\dfrac{93}{5} = 18.6$ **21.** 0

23. $\dfrac{2}{3}(e^3 - 1) \approx 12.723691282$

25. $\ln 2 \approx 0.693147181$

27. $\dfrac{1}{2}e^2 - 2e + \dfrac{5}{2} \approx 0.757964393$

29. $\dfrac{1}{4}$ **31.** $\dfrac{2}{5}$ **33.** $-\dfrac{1}{3}$

35. $\dfrac{4}{\pi} \approx 1.273239545$

37. 0 **39.** $\dfrac{1}{2}$ **41.** $\dfrac{2}{3}$ **43.** 5

45. $-\dfrac{5}{2}$ **47.** $\dfrac{25\pi}{4} \approx 19.63495409$

49. First show that $1 \leqq \sqrt{1 + x^2} \leqq \sqrt{1 + x}$ if $0 \leqq x \leqq 1$.

51. First show that $\dfrac{1}{1 + \sqrt{x}} \leqq \dfrac{1}{1 + x^2}$ if $0 \leqq x \leqq 1$.

53. Note first that $\sin t \leqq 1$ for all t.

55. $\dfrac{1}{2} \leqq \displaystyle\int_0^1 \frac{1}{1 + x}\,dx \leqq 1$ **57.** $\dfrac{\pi}{8} \leqq \displaystyle\int_0^{\pi/6} \cos^2 x\,dx \leqq \frac{\pi}{6}$

59. Key step: $\displaystyle\lim_{\Delta x \to 0} \sum_{i=1}^{n} cf(x_i^\star)\,\Delta x = \lim_{\Delta x \to 0} c \cdot \sum_{i=1}^{m} f(x_i^\star)\,\Delta x$

61. If $f(x) \leqq M$ for all x in $[a, b]$, let $g(x) \equiv M$ and use the first comparison property.

63. $1000 + \displaystyle\int_0^{30} V'(t)\,dt = 160$ (gal)

65. First deduce from Fig. 5.5.11 that $\dfrac{12 - 4x}{9} \leqq \dfrac{1}{x} \leqq \dfrac{3 - x}{2}$.

SECTION 5.6 (PAGE 356)

1. $\dfrac{16}{5}$ **3.** $\dfrac{26}{3}$ **5.** 0

7. $\dfrac{125}{4} = 31.25$ **9.** $\dfrac{14}{9} \approx 1.555555556$

11. 0 **13.** 4 **15.** $\dfrac{1}{3}$

17. $-\dfrac{22}{81} \approx -0.271604938$ **19.** 0

21. $\dfrac{35}{24} \approx 1.458333333$ **23.** $\dfrac{(e - 1)^2}{e} \approx 1.086161$

25. 4 **27.** $\ln 2 \approx 0.693147181$

29. $\dfrac{31}{20} = 1.55$ **31.** $\dfrac{81}{2} = 40.5$

33. Average height $\frac{800}{3} \approx 266.666666667$ (ft), average velocity -80 ft/s

35. $\dfrac{1}{10}\displaystyle\int_0^{10} V(t)\,dt = \frac{5000}{3} \approx 1666.666667$ (L)

37. $\dfrac{1}{10}\displaystyle\int_0^{10} T(x)\,dx = \frac{200}{3} \approx 66.666667$

39. $\dfrac{1}{2}\displaystyle\int_0^{2} A(y)\,dy = \frac{\pi}{3} \approx 1.047197551$

41. **a.** $A(x) = 27 - 3x^2$, $-3 \leqq x \leqq 3$; **b.** 18; **c.** two

43. **a.** $A(x) = 2x\sqrt{16 - x^2}$, $0 \leqq x \leqq 4$; **b.** $\dfrac{32}{3}$; **c.** two

45. $f'(x) = (x^2 + 1)^{17}$ **47.** $h'(z) = (z - 1)^{1/3}$

49. $f'(x) = e^{-x} - e^x$ **51.** $G'(x) = f(x) = \sqrt{x + 4}$

53. $G'(x) = f(x) = \sqrt{x^3 + 1}$

55. $f'(x) = 3\sin 9x^2$ **57.** $f'(x) = 2x\sin x^2$

59. $f'(x) = \dfrac{2x}{x^2 + 1}$ **61.** $y(x) = \displaystyle\int_1^x \frac{1}{t}\,dt$

63. $y(x) = 10 + \displaystyle\int_5^x \sqrt{1 + t^2}\,dt$

65. The integrand is not continuous on $[-1, 1]$.

67. (a) $g(x) = x^2$ if $0 \leqq x \leqq 2$; $g(x) = 8x - x^2 - 8$ if $2 \leqq x \leqq 6$; $g(x) = 28 - 4x$ if $6 \leqq x \leqq 8$; $g(x) = x^2 - 20x + 92$ if $8 \leqq x \leqq 10$; (b) Increasing on $(0, 4)$ and decreasing on $(4, 10)$; global maximum at $(4, 8)$, global minimum at $(10, -8)$

69. (a) $x = 0$, π, 2π, 3π, and 4π. (b) Global maximum at $x = 3\pi$, global minimum at $x = 4\pi$. (c) Inflection points where $x \approx 2.028757838$, 4.913180439, 7.978665712, and 11.08553841.

SECTION 5.7 (PAGE 365)

1. $\dfrac{1}{54}(3x - 5)^{18} + C$ **3.** $\dfrac{1}{3}(x^2 + 9)^{3/2} + C$

5. $-\dfrac{1}{5}\cos 5x + C$ **7.** $-\dfrac{1}{4}\cos(2x^2) + C$

9. $\dfrac{1}{6}(1 - \cos x)^6 + C$ **11.** $\dfrac{1}{7}(x + 1)^7 + C$

13. $-\frac{1}{24}(4-3x)^8 + C$

15. $\frac{2}{7}(7x+5)^{1/2} + C$

17. $-\frac{1}{\pi}\cos(\pi x + 1) + C$

19. $\frac{1}{2}\sec 2\theta + C$

21. $-\frac{1}{2}e^{1-2x} + C$

23. $\frac{1}{9}\exp(3x^3 - 1) + C$

25. $\frac{1}{2}\ln|2x-1| + C$

27. $\frac{1}{3}(\ln x)^3 + C$

29. $\frac{1}{2}\ln(x^2 + e^{2x}) + C$

31. $\frac{1}{3}(x^2-1)^{3/2} + C$

33. $-\frac{1}{9}(2-3x^2)^{3/2} + C$

35. $\frac{1}{6}(x^4+1)^{3/2} + C$

37. $\frac{1}{6}\sin(2x^3) + C$

39. $-\frac{1}{2}\exp(-x^2) + C$

41. $-\frac{1}{4}\cos^4 x + C$

43. $\frac{1}{4}\tan^4\theta + C$

45. $2\sin(x^{1/2}) + C$

47. $\frac{1}{10}(x+1)^{10} + C$

49. $\frac{1}{2}\ln|x^2+4x+3| + C$

51. $\frac{5}{72} \approx 0.069444444$

53. $\frac{98}{3} \approx 32.666666667$

55. $\frac{1192}{15} \approx 79.466666667$

57. $\frac{15}{128} = 0.1171875$

59. $\frac{62}{15} \approx 4.133333333$

61. $e - 1 \approx 1.718281828$

63. $\frac{\sqrt{e}-1}{e} \approx 0.238651219$

65. $\frac{1}{2}x - \frac{1}{2}\sin x\cos x + C$

67. $\frac{\pi}{2} \approx 1.570796327$

69. $-x + \tan x + C$

71. $-\cos x + \frac{1}{3}\cos^3 x + C$

73. If $\frac{1}{2}\sin^2\theta + C_1 = -\frac{1}{2}\cos^2\theta + C_2$, then $C_2 - C_1 = \frac{1}{2}$.

75. $\frac{1}{1-x} - \frac{x}{1-x} = \frac{1-x}{1-x} \equiv 1$ if $x \neq 1$.

77. Note that $\int_{-a}^{0} f(x)\,dx = -\int_{0}^{a} f(x)\,dx$.

79. The tangent function is odd; the product (or quotient) of an odd function and an even function is odd.

81. Substitute $u = x + k$ in the first integral and simplify.

83. Substitute $u = x^{1/2}$, $x = u^2$, $dx = 2u\,du$.

SECTION 5.8 (PAGE 375)

1. $\frac{256}{3}$

3. $\frac{9}{2}$

5. 32

7. $\frac{128}{3}$

9. $\frac{11\sqrt{33}}{2} \approx 31.595095$

11. $\frac{1}{4}$

13. $\frac{1}{20}$

15. $\ln 3 \approx 1.098612$

17. $\frac{32}{3}$

19. $\frac{128\sqrt{2}}{3} \approx 60.339779$

21. $\frac{4}{3}$

23. $\frac{500}{3}$

25. $\frac{32}{3}$

27. $\frac{125}{6} \approx 20.833333$

29. $\frac{1}{3}$

31. $\frac{16}{3}$

33. $\frac{16}{3}$

35. $\frac{27}{2}$

37. $\frac{(e-1)^2}{e} \approx 1.086161$

39. $\frac{e-1}{2e} \approx 0.316060$

41. $\frac{4}{3}$

43. $\frac{5}{12} + \frac{8}{3} = \frac{37}{12} \approx 3.08333$

45. $\frac{45\pi}{2} \approx 70.685835$

47. $A = 4\int_0^a \frac{b}{a}(a^2 - x^2)^{1/2}\,dx = \frac{4b}{a}\int_0^a \sqrt{a^2 - x^2}\,dx$

49. The area of the parabolic segment is $\frac{9}{2}$.

51. $\frac{32}{3} + \frac{63}{2} = \frac{253}{6} \approx 42.166667$

53. $\frac{40\sqrt{5}}{3} \approx 29.814240$

55. Approximately 1.09475

57. Approximately 3.00044

59. $k = 18$

61. $\frac{253}{12} \approx 21.083333$

63. Approximately 25.3622616057

65. Approximately 86.1489054767

67. Approximately 16.8330174093

SECTION 5.9 (PAGE 389)

1. $T_4 = 8$; true value: 8

3. $T_5 \approx 0.65$; true value: $\frac{2}{3}$

5. $T_3 \approx 0.98$; true value: 1

7. $M_4 = 8$; true value: 8

9. $M_5 \approx 0.67$; true value: $\frac{2}{3}$

11. $M_3 \approx 1.01$; true value 1

13. $T_4 = 8.75$, $S_4 = \frac{26}{3} \approx 8.67$; true value: $\frac{26}{3}$

15. $T_4 \approx 0.882604$, $S_4 \approx 0.864956$; true value: $1 - e^{-2} \approx 0.864665$

17. $T_6 \approx 3.26$, $S_6 \approx 3.24$; true value: approximately 3.24131

19. $T_8 \approx 5.013970$, $S_8 \approx 5.019676$; true value: approximately 5.02005

21. $T_6 = 3.02$, $S_6 \approx 3.07167$

23. $T_{10} = 2441$, $S_{10} = \frac{7342}{3} \approx 2447.33$

25. $T_{12} = 91150$ (square feet, about 2.093 acres), $S_{12} = \frac{281600}{3}$ (square feet, about 2.155 acres)

27. $n = 19$

29. If $p(x)$ is a polynomial of degree at most three, then $p^{(4)}(x) \equiv 0$.

31. Expand the sum $M_n + T_n$ using the definitions.

33. If $f''(x) > 0$ on $[a, b]$, then the graph of f is concave upward there; now examine Fig. 5.9.11. Don't forget Case 2: $f''(x) < 0$ on $[a, b]$.

35. The midpoint rule yields approximately 872.476; the trapezoidal rule yields approximately 872.600. Note that the graph of $y = 1/(\ln x)$ is concave upward for $x > 0$.

CHAPTER 5 MISCELLANEOUS PROBLEMS (PAGE 394)

1. $-\frac{5}{2}x^{-2}+2x^{-1}+\frac{1}{3}x^3+C$ **3.** $-\frac{1}{30}(1-3x)^{10}+C$

5. $\frac{3}{16}(9+4x)^{4/3}+C$ **7.** $\frac{1}{24}(1+x^4)^6+C$

9. $-\frac{3}{8}(1-x^2)^{4/3}+C$

11. $\frac{1}{35}(25\cos 7x + 49\sin 5x)+C$

13. $\frac{1}{6}(1+x^4)^{3/2}+C$ **15.** $-\frac{2}{1+\sqrt{x}}+C$

17. $\frac{1}{12}\sin 4x^3+C$ **19.** $\frac{1}{30}(x^2+1)^{15}+C$

21. $\frac{2}{5}(4-x)^{5/2}-\frac{8}{3}(4-x)^{3/2}+C$

23. $\sqrt{1+x^4}+C$ **25.** $y(x)=x^3+x^2+5$

27. $y(x)=\frac{1}{12}(2x+1)^6+\frac{23}{12}$ **29.** $y(x)=\frac{3x^{2/3}-1}{2}$

31. 6 (s); 396 (ft) **33.** 120 (ft/s)

35. Impact time: $2\sqrt{10}$ (s); impact speed: $20\sqrt{10}$ (ft/s)

37. 176 (ft) **39.** 1700 **41.** 2845 **43.** $2\sqrt{2}-2$

45. $\int_0^1 2\pi x\sqrt{1+x^2}\,dx=\frac{2\pi}{3}(2\sqrt{2}-1)$

47. Show that every Riemann sum is equal to $c(b-a)$.

49. If m is the global minimum value of $f(x)$ on $[a,b]$, then $m>0$. Now apply the second comparison property.

51. $\frac{2\sqrt{2}}{3}x^{3/2}+\frac{2\sqrt{3}}{3x^{1/2}}+C$ **53.** $-\frac{2}{x}-\frac{1}{4}x^2+C$

55. $\frac{2}{3}\sin x^{3/2}+C$ **57.** $\cos\frac{1}{t}+C$

59. $-\frac{3}{8}(1+u^{4/3})^{-2}+C$ **61.** $\frac{38}{3}$

63. Let $u=1/x$; result: $\frac{(4x^2-1)^{3/2}}{3x^3}+C$

65. $\frac{1}{30}\approx 0.033333333$ **67.** $\frac{44}{15}\approx 2.933333333$

69. $\frac{125}{6}\approx 20.833333333$ **71.** $\frac{\pi}{2}\approx 1.570796327$

73. One solution is $f(x)=\sqrt{4x^2-1}$.

75. Use $n \geqq 5$. $L_5\approx 1.10873$, $R_5\approx 1.19154$. The integrand is increasing on $[0,1]$, so the value of the integral is 1.15015 ± 0.04143. Its true value is

$$\frac{1}{2}\left[\sqrt{2}+\ln\left(1+\sqrt{2}\right)\right]\approx 1.147793574696319.$$

77. $M_5\approx 0.28667$, $T_5\approx 0.28971$. They bound the true value of the integral because the second derivative of the integrand is positive on $[1,2]$.

79. Show that $x_{i-1}<\sqrt{x_{i-1}x_i}<x_i$ and that

$$\sum_{i=1}^{n}\frac{1}{(x_i^\star)^2}\Delta x_i=\sum_{i=1}^{n}\left(\frac{1}{x_{i-1}}-\frac{1}{x_i}\right),$$

then expand the right-hand side.

SECTION 6.1 (PAGE 406)

1. 1 **3.** $\frac{2}{\pi}$ **5.** $\frac{98}{3}$

7. $1-\frac{1}{e}\approx 0.632120559$ **9.** $\frac{1}{2}\ln 5\approx 0.814718956$

11. $\int_1^4 2\pi f(x)\,dx$ **13.** $\int_0^{10}\sqrt{1+[f(x)]^2}\,dx$

15. $M=1000$ (grams) **17.** $M=\frac{500}{3}$ (grams)

19. Net distance: −320; total distance: 320

21. Net distance: −50; total distance: 106.25

23. Net distance: 65; total distance: 97

25. Net distance: 1; total distance: 1

27. Net distance: 0; total distance: $\frac{4}{\pi}$

29. Net distance: $\frac{70}{3}$; total distance: 65

31. Net distance: $\frac{3}{4}$; total distance approximately 7.73330

33. Net distance: π; total distance approximately 4.26379

35. Total mass: approximately $\sum_{i=1}^{n} 2\pi x_i^\star \rho(x_i^\star)\,\Delta x$

37. $M=\frac{625\pi}{2}\approx 981.747704$

39. 550 (gal) **41.** 695000

43. $a=0.2$, $b=0.1$; average rainfall: 73 inches per year

45. $\frac{3}{4}$

47. $W=\frac{700\pi}{3}\approx 733.038286$ (pounds)

49.

100	0.000
95	22.774
90	52.416
85	91.569
80	144.141
75	216.049

51. Approximately 7.035 liters per minute

SECTION 6.2 (PAGE 416)

1. $\frac{\pi}{5}$ **3.** 8π **5.** $\frac{1}{2}\pi^2$ **7.** $\frac{3}{10}\pi$

9. $\pi\ln 5\approx 5.05619832$ **11.** $\frac{16}{15}\pi\approx 3.351032164$

13. $\frac{\pi}{2}$ **15.** 8π

17. $\frac{121}{210}\pi\approx 1.810155767$ **19.** 8π

21. $\frac{49}{30}\pi\approx 5.131268001$ **23.** $\frac{17}{10}\pi\approx 5.340707511$

25. $\frac{1}{2}\pi^2\approx 4.934802201$ **27.** $\frac{\pi}{2}$

29. $\frac{1}{4}\pi(4-\pi)\approx 0.674191553$

31. Approximately $2.998 + 267.442 = 270.440$

33. Approximately 3.67743

35. 9π **37.** $\frac{4}{3}\pi a^2 b$ **39.** $\frac{16}{3}a^3$ **41.** $\frac{4\sqrt{3}}{3}a^3$

43. Paraboloid volume: $\pi p h^2$

47. $\frac{16}{3}a^3$ **49.** Torus volume: $2\pi^2 a^2 b$

51. Barrel volume: $\frac{\pi h}{3}\left(2R^2 + r^2 - \frac{2}{5}\delta^2\right)$

53. $T_6 \approx 3037.83$ and $S_6 \approx 3000.56$; to the nearest hundred, \$3000

57. $\frac{\pi b^3}{12a}(8a - 3b)$

SECTION 6.3 (PAGE 425)

1. 8π **3.** $\frac{625\pi}{2} \approx 981.747704$

5. 16π **7.** π

9. $\frac{6\pi}{5} \approx 3.769911$ **11.** $\frac{256\pi}{15} \approx 53.616515$

13. $\frac{4\pi}{15} \approx 0.837758$ **15.** $\frac{11\pi}{15} \approx 2.303835$

17. $\frac{56\pi}{5} \approx 35.185838$ **19.** $\frac{8\pi}{3} \approx 8.377580$

21. $\frac{2\pi}{15} \approx 0.418879$ **23.** $\frac{\pi}{2}$

25. $\frac{\pi(e-1)}{e} \approx 1.985865$ **27.** 4π

29. Approximately 23.2990983139

31. Approximately 1.0602688478

33. Approximately 8.1334538068

37. $\frac{4}{3}\pi a^2 b$ **39.** $V = 2\pi^2 a^2 b$

41. $V = 2\pi^2 a^3$ **43.** (a) $V = \frac{\pi}{6}h^3$

45. **a.** $\frac{625\pi}{12} \approx 163.624617$; **b.** $\frac{400\pi\sqrt{5}}{7} \approx 401.417985$;
c. $\frac{1600\pi\sqrt{5}}{21} \approx 535.223980$

47. **a.** $400\pi \approx 1256.637061$; **b.** $\frac{250\pi}{3} \approx 261.799388$

SECTION 6.4 (PAGE 436)

1. $\int_0^1 \sqrt{1+4x^2}\,dx$ **3.** $\int_0^2 \sqrt{1+36x^2(x-1)^2}\,dx$

5. $\int_0^{100} \sqrt{1+4x^2}\,dx$ **7.** $\int_{-1}^2 \sqrt{1+16y^6}\,dy$

9. $\int_1^2 \left(1+\frac{1}{x^2}\right)^{1/2} dx$ **11.** $\int_0^4 2\pi x^2\sqrt{1+4x^2}\,dx$

13. $\int_0^1 2\pi(x-x^2)\sqrt{4x^2-4x+2}\,dx$

15. $\int_0^1 2\pi(2-x)\sqrt{1+4x^2}\,dx$

17. $\int_2^3 2\pi x \cdot \frac{x^2+1}{x^2-1}\,dx$ **19.** $\int_0^1 2\pi\sqrt{x^2+2x+2}\,dx$

21. $\frac{22}{3}$ **23.** $\frac{14}{3}$

25. $\frac{123}{32} = 3.84375$ **27.** $\frac{e^2-1}{2e} \approx 1.175201$

29. $\frac{5\sqrt{5}-1}{6}\pi \approx 5.330414$

31. $\frac{339\pi}{16} \approx 66.562494$

33. $\frac{82\sqrt{82}-1}{9}\pi \approx 258.846843$

35. $\frac{115\pi}{12} \approx 30.106930$

37. $S_6 = \frac{\pi}{18}\left(4+2\sqrt{2}+2\sqrt{5}+4\sqrt{7}\right) \approx 3.819403$

41. Avoid the problem when $x = 0$ as follows: Let $\alpha = \frac{1}{4}\sqrt{2}$. Then

$$L = 8\int_\alpha^1 \frac{1}{x^{1/3}}\,dx = 6.$$

45. First establish that $y(x) = \frac{H}{S^2}x^2$.

SECTION 6.5 (PAGE 445)

1. 30 **3.** 9 **5.** 0 **7.** 15 ft·lb

9. 2.816×10^9 ft·lb

11. $13000\pi \approx 40840.7045$ ft·lb

13. $\frac{125000\pi}{3} \approx 130899.694$ ft·lb

15. $156000\pi \approx 490088.454$ ft·lb

17. With water density $\rho = 62.4$ lb/ft^3: $4160000\pi \approx 13069025$ ft·lb

19. 8750 ft·lb **21.** 11250 ft·lb

23. $\frac{2500}{12}(10 - 10^{3/5}) \approx 1253.943395$ ft·lb

25. $16\pi \approx 50.265482$ ft·lb

27. $1382400\pi \approx 4.342938 \times 10^6$ ft·lb

29. $\frac{15625\sqrt{2}}{32} \approx 690.533966$ ft·lb

31. 249.6 lb **33.** 748.8 lb **35.** 19500 lb

37. 14560 lb

39. Approximately 6.24×10^7 lb

SECTION 6.6 (PAGE 454)

1. $(2, 3)$ **3.** $(1, 1)$ **5.** $\left(\frac{4}{3}, \frac{2}{3}\right)$

7. $\left(\frac{3}{2}, \frac{6}{5}\right)$ **9.** $\left(0, -\frac{8}{5}\right)$ **11.** $\left(0, \frac{8}{5}\right)$

13. $\left(\frac{3}{4}, \frac{9}{10}\right)$ **15.** $\left(-\frac{1}{2}, 2\right)$ **17.** $\left(\frac{3}{5}, \frac{12}{35}\right)$

19. $\left(\frac{4r}{3\pi}, \frac{4r}{3\pi}\right)$ **21.** $\left(\frac{2r}{\pi}, \frac{2r}{\pi}\right)$

25. $A = \left(2 \cdot \frac{\pi r}{2}\right)\sqrt{r^2 + h^2}$

27. $A = 2\pi \cdot \left(\frac{r_1 + r_2}{2}\right)[(r_2 - r_1)^2 + h^2]^{1/2}$

29. $\bar{y} = \frac{4a^2 + 3\pi ab + 6b^2}{12b + 3\pi a}$; $V = \frac{1}{3}\pi a(4a^2 + 3\pi ab + 6b^2)$

33. The centroid is at $\left(\frac{1}{2}, \frac{2}{5}\right)$, at distance $\frac{\sqrt{2}}{20}$ from the axis of rotation.

SECTION 6.7 (PAGE 465)

1. $f'(x) = 10^x \ln 10$ **3.** $f'(x) = \left(\frac{3}{4}\right)^x \ln\left(\frac{3}{4}\right)$

5. $f'(x) = -(7^{\cos x} \ln 7) \cdot \sin x$

7. $f'(x) = \frac{3}{2}x^{1/2}\left(2^{x^{3/2}} \ln 2\right)$ **9.** $f'(x) = \frac{1}{x}(2^{\ln x} \ln 2)$

11. $f'(x) = 17^x \ln 17$ **13.** $f'(x) = -\frac{1}{x^2}(10^{1/x} \ln 10)$

15. $f'(x) = 2^{2^x} \cdot 2^x \cdot (\ln 2)^2$ **17.** $f'(x) = \frac{x}{(x^2 + 4)\ln 3}$

19. $f'(x) = \frac{\ln 2}{\ln 3}$ **21.** $f'(x) = \frac{1}{(x \ln x)\ln 2}$

23. $f'(x) = \frac{\exp(\log_{10} x)}{x \ln 10}$ **25.** $\frac{3^{2x}}{2 \ln 3} + C$

27. $\frac{2 \cdot 2^{\sqrt{x}}}{\ln 2} + C$ **29.** $\frac{7^{(x^3+1)}}{3 \ln 7} + C$

31. $\frac{(\ln x)^2}{2 \ln 2} + C$ **33.** $R \approx (290.903) \cdot W^{-0.2473}$

35. $\left(\frac{1}{\ln 2}, \frac{1}{2^{1/(\ln 2)} \ln 2}\right)$ **37.** $V \approx 1.343088216395$

39. $a^z = c$, $a^x = b$, and $b^y = c$, so $a^{xy} = b^y = c = a^z$.

41. $\frac{dy}{dx} = -\frac{\ln 2}{x(\ln x)^2}$

SECTION 6.8 (PAGE 475)

1. $\frac{\pi}{6}, -\frac{\pi}{6}, \frac{\pi}{4}, -\frac{\pi}{3}$ **3.** $0, \frac{\pi}{4}, -\frac{\pi}{4}, \frac{\pi}{3}$

5. $f'(x) = \frac{100x^{99}}{\sqrt{1 - x^{200}}}$

7. $f'(x) = \frac{1}{x|\ln x|\sqrt{(\ln x)^2 - 1}}$

9. $f'(x) = \frac{\sec^2 x}{\sqrt{1 - \tan^2 x}}$ **11.** $f'(x) = \frac{e^x}{\sqrt{1 - e^{2x}}}$

13. $f'(x) = -\frac{2}{\sqrt{1 - x^2}}$ **15.** $f'(x) = -\frac{2}{x\sqrt{x^4 - 1}}$

17. $f'(x) = -\frac{1}{(1 + x^2)(\arctan x)^2}$

19. $f'(x) = \frac{1}{x[1 + (\ln x)^2]}$

21. $f'(x) = \frac{2e^x}{1 + e^{2x}}$

23. $f'(x) = \frac{\cos(\arctan x)}{1 + x^2} = \frac{1}{(1 + x^2)^{3/2}}$

25. $f'(x) = \frac{1 - 4x \arctan x}{(1 + x^2)^3}$

27. $y = 2 - x$ **29.** $x + y = \sqrt{2}$

31. $\frac{\pi}{4}$ **33.** $\frac{\pi}{12}$

35. $\frac{\pi}{12}$ **37.** $\frac{1}{2}\arcsin 2x + C$

39. $\frac{1}{5}\operatorname{arcsec}\frac{|x|}{5} + C$ **41.** $\arctan(e^x) + C$

43. $\frac{1}{15}\operatorname{arcsec}\frac{|x^3|}{5} + C$

45. $\arcsin(2x - 1) + C_1 = 2\arcsin\sqrt{x} + C_2$

47. $\frac{1}{50}\arctan(x^{50}) + C$

49. $\arctan(\ln x) + C$ **51.** $\frac{\pi}{4}$

53. $\frac{\pi}{2}$ **55.** $\frac{\pi}{12}$

57. Suppose that $u < -1$ and let $x = -u$; apply the chain rule.

59. The substitution yields $\frac{1}{a}\int \frac{1}{1 + x^2}\,dx$.

61. In the case $x < -1$, note that $-x > 0$, so that $\sqrt{(-x)^2} = -x$.

65. 8 m

67. The circumference is $8\int_0^{a/\sqrt{2}} \frac{a}{\sqrt{a^2 - x^2}}\,dx$.

69. The area is $\frac{\pi}{2}$.

71. $A = 1 - \frac{\pi}{3}$ and $B = 1 + \frac{2\pi}{3}$

75. Approximately (2.689220, 0.928343)

SECTION 6.9 (PAGE 484)

1. $f'(x) = 3\sinh(3x - 2)$

3. $f'(x) = 2x \tanh\left(\frac{1}{x}\right) - \operatorname{sech}^2\left(\frac{1}{x}\right)$

5. $f'(x) = -12\coth^2 4x \operatorname{csch}^2 4x$

7. $f'(x) = -(\operatorname{csch} x \coth x)\exp(\operatorname{csch} x)$

9. $f'(x) = (\cosh x)\cos(\sinh x)$

11. $f'(x) = 4x^3 \cosh x^4$

13. $f'(x) = -\frac{1 + \operatorname{sech}^2 x}{(x + \tanh x)^2}$

15. $\frac{1}{2}\cosh x^2 + C$ **17.** $x - \frac{1}{3}\tanh 3x + C$

19. $\frac{1}{6}\sinh^3 2x + C$ **21.** $-\frac{1}{2}\operatorname{sech}^2 x + C$

23. $-\frac{1}{2}\coth^2 x + C_1 = -\frac{1}{2}\operatorname{csch}^2 x + C_2$

25. $\ln(1+\cosh x) + C$

27. $\frac{1}{4}\tanh x + C_1 = -\frac{1}{4}e^{-x}\operatorname{sech} x + C_2$

29. $f'(x) = \dfrac{2}{\sqrt{1+4x^2}}$ **31.** $f'(x) = \dfrac{1}{2(1-x)\sqrt{x}}$

33. $f'(x) = \dfrac{x}{|x|\sqrt{x^2-1}}$ **35.** $f'(x) = \dfrac{3\sqrt{\sinh^{-1} x}}{2\sqrt{1+x^2}}$

37. $f'(x) = \dfrac{1}{(1-x^2)\tanh^{-1} x}$

39. $\operatorname{arcsinh}\left(\dfrac{x}{3}\right) + C$

41. $\dfrac{1}{2}\left[\tanh^{-1}\left(\dfrac{1}{2}\right) - \tanh^{-1}\left(\dfrac{1}{4}\right)\right] = \dfrac{1}{4}\ln\left(\dfrac{9}{5}\right) \approx 0.146947$

43. $-\dfrac{1}{2}\operatorname{sech}^{-1}\left|\dfrac{3x}{2}\right| + C$

45. $\sinh^{-1}(e^x) + C$

47. $-\operatorname{sech}^{-1}(e^x) + C$

49. Expand and simplify $\sinh x \cosh y + \cosh x \sinh y - \sinh(x+y)$.

51. First substitute y for x in Eq. (8) of the section.

53. $\sinh a$

55. Begin with the equation for $A(\theta)$, differentiate both sides with respect to θ, and then simplify.

57. Let $y = \sinh^{-1}(1)$; use Eq. (1) and the quadratic formula to solve for e^y.

67. Here,

$$f'(x) = -\frac{e^{4x} - 2e^{2x} - 1}{2e^{2x}(e^{2x}+1)^2},$$

and therefore $f'(x) = 0$ when $x = \frac{1}{2}\ln(1+\sqrt{2}) \approx 0.440687$.

CHAPTER 6 MISCELLANEOUS PROBLEMS (PAGE 486)

1. Net: $-\dfrac{3}{2}$; total: $\dfrac{31}{6}$ **3.** Net: 1; total : 3

5. $\dfrac{14}{3}$ **7.** $\dfrac{2\pi}{15}$ **9.** 12 in. **11.** $\dfrac{41\pi}{105}$

13. $\dfrac{85\pi}{8}$ **15.** $V = \displaystyle\int_0^h \frac{\pi ab}{h^2} z^2\, dz$

17. $V = \displaystyle\int_a^{a+h} \pi y^2\, dx = \frac{\pi b^2}{a^2}\int_a^{a+h}(x^2 - a^2)\, dx$

19. $f(x) = \sqrt{1+3x}$ **21.** $\dfrac{8}{\pi} - \dfrac{2\pi}{3}$

23. $\dfrac{10}{3}$ **25.** $\dfrac{63}{8}$ **27.** $\dfrac{52\pi}{5}$

29. Substitute $2r$ for h in the formula derived in Problem 28.

31. 1 (ft) **33.** $4\pi\rho R^4$

35. 10454400 (ft·lb) **37.** 72800000 (lb)

39. No maximum; minimum when $c = \frac{1}{3}\sqrt{5}$

41. $\left(\dfrac{393}{352}, \dfrac{268}{165}\right)$ **43.** $\left(\dfrac{111}{112}, \dfrac{1136}{245}\right)$

45. $\left(\dfrac{16}{21}, 0\right)$ **47.** $\dfrac{4b}{3\pi}$

49. (a) A is the sum of the area of a triangle and the area of a parallelogram, minus the area of another triangle. (b) Use the result in Problem 46. (c) $V = 2\pi\bar{y}A$. (d) $A = pw/2$. (e) $S = 2\pi(b+d)/2$. (f) $V = 2\pi\bar{y}A$ again.

51. Centroid: $\left(\dfrac{n+1}{3(n+2)}, \dfrac{2(n+1)}{3(2n+1)}\right)$

53. $-\dfrac{1}{2}\ln|1-2x| + C$ **55.** $\dfrac{1}{2}\ln|1+6x-x^2| + C$

57. $-\ln(2+\cos x) + C$ **59.** $\dfrac{2\cdot 10^{\sqrt{x}}}{\ln 10} + C$

61. $\dfrac{2}{3}(1+e^x)^{3/2} + C$ **63.** $\dfrac{6^x}{\ln 6} + C$

65. Sell immediately!

67. 90 lots of 11 samples and one lot of 10 samples; total cost about \$978

69. $f'(x) = \dfrac{3}{\sqrt{1-9x^2}}$ **71.** $g'(t) = \dfrac{2}{t\sqrt{t^4-1}}$

73. $f'(x) = -\dfrac{\sin x}{|\sin x|}$

75. $g'(t) = \dfrac{10}{\sqrt{100t^2-1}}$, $t > \dfrac{1}{10}$

77. $f'(x) = -\dfrac{2}{x\sqrt{x^4-1}}$

79. $f'(x) = \dfrac{1}{2\sqrt{1-x}\sqrt{x}}$

81. $f'(x) = \dfrac{2x}{x^4+2x^2+2}$

83. $f'(x) = e^{2x}\cosh e^x + e^x \sinh e^x$

85. $f'(x) \equiv 0$ **87.** $f'(x) = \dfrac{x}{|x|\sqrt{x^2+1}}$

89. $\dfrac{1}{2}\sin^{-1} 2x + C$ **91.** $\arcsin\left(\dfrac{x}{2}\right) + C$

93. $\arcsin(e^x) + C$ **95.** $\dfrac{1}{2}\arcsin\left(\dfrac{2x}{3}\right) + C$

97. $\dfrac{1}{3}\arctan(x^3) + C$ **99.** $\operatorname{arcsec}|2x| + C$

101. $\operatorname{arcsec}(e^x) + C$ **103.** $2\cosh\sqrt{x} + C$

105. $\dfrac{1}{2}(\arctan x)^2 + C$ **107.** $\dfrac{1}{2}\sinh^{-1}\left(\dfrac{2x}{3}\right) + C$

109. $\dfrac{\pi^2}{6}$

111. Use (in order) Eqs. (36), (37), (35), and (38) of Section 6.9.

113. Approximately 4.7300407449; the difficulty is in showing that there is no smaller positive solution.

115. (c) $p = e$

SECTION 7.2 (PAGE 495)

1. $-\frac{1}{15}(2-3x)^5+C$

3. $\frac{1}{9}(2x^3-4)^{3/2}+C$

5. $\frac{3}{4}(2x^2+3)^{2/3}+C$

7. $-2\csc\sqrt{y}+C$

9. $\frac{1}{6}(1+\sin\theta)^6+C$

11. $\exp(-\cos x)+C$

13. $\frac{1}{11}(\ln t)^{11}+C$

15. $\frac{1}{3}\arcsin(3t)+C$

17. $\frac{1}{2}\arctan(e^{2x})+C$

19. $\frac{3}{2}\arcsin(x^2)+C$

21. $\frac{1}{15}\tan^5 3x+C$

23. $\arctan(\sin\theta)+C$

25. $\frac{2}{5}\left(1+\sqrt{x}\right)^5+C$

27. $\ln|\arctan t|+C$

29. $\operatorname{arcsec}(e^x)+C=\arctan\left(\sqrt{e^{2x}-1}\right)+C$

31. $\frac{2}{7}(x-2)^{7/2}+\frac{8}{5}(x-2)^{5/2}+\frac{8}{3}(x-2)^{3/2}+C$

33. $\frac{1}{6}(2x+3)^{3/2}-\frac{3}{2}(2x+3)^{1/2}+C$

35. $\frac{3}{5}(x+1)^{5/3}-\frac{3}{2}(x+1)^{2/3}+C$

37. $\frac{1}{60}\ln\left|\frac{3x+10}{3x-10}\right|+C$

39. $\frac{1}{2}x(4+9x^2)^{1/2}+\frac{2}{3}\ln(3x+(4+9x^2)^{1/2})+C$

41. $\frac{1}{32}x\sqrt{16x^2+9}-\frac{9}{128}\ln\left(4x+\sqrt{16x^2+9}\right)+C$

43. $\frac{1}{128}x(32x^2-25)\sqrt{25-16x^2}+\frac{625}{512}\arcsin\left(\frac{4x}{5}\right)+C$

45. $\frac{1}{2}e^x\sqrt{9+e^{2x}}+\frac{9}{2}\ln\left(e^x+\sqrt{9+e^{2x}}\right)+C$

47. $\frac{1}{2}\sqrt{x^4-1}-\frac{1}{2}\operatorname{arcsec}(x^2)+C$

49. $\frac{1}{8}(\ln x)[2(\ln x)^2+1]\sqrt{1+(\ln x)^2}$
$-\frac{1}{8}\ln\left(\ln x+\sqrt{1+(\ln x)^2}\right)+C$

51. Illegal substitution: $x=\sqrt{u}\geqq 0$, but $x<0$ for many x in $[-1,1]$.

53. $\arcsin(x-1)+C$

55. $\frac{1}{2}\tan^{-1}x^2=\frac{\pi}{2}-\frac{1}{2}\tan^{-1}x^{-2}$ if $x\neq 0$; both *are* antiderivatives of $x/(1+x^4)$.

57. $G'(x)=H'(x)=\sqrt{1+x^2}$ and $G(0)=H(0)$, so $G(x)=H(x)$ for all x.

SECTION 7.3 (PAGE 501)

1. $\frac{1}{2}xe^{2x}-\frac{1}{4}e^{2x}+C$

3. $-t\cos t+\sin t+C$

5. $\frac{1}{3}x\sin 3x+\frac{1}{9}\cos 3x+C$

7. $\frac{1}{4}x^4\ln x-\frac{1}{16}x^4+C$

9. $x\arctan x-\frac{1}{2}\ln(1+x^2)+C$

11. $\frac{2}{3}y^{3/2}\ln y-\frac{4}{9}y^{3/2}+C$

13. $t(\ln t)^2-2t\ln t+2t+C$

15. $\frac{2}{5}(x-2)(x+3)^{3/2}+C$

17. $\frac{2}{45}(x^3+1)^{3/2}(3x^2-2)+C$

19. $-\frac{1}{2}\csc\theta\cot\theta+\frac{1}{2}\ln|\csc\theta-\cot\theta|+C$

21. $\frac{1}{3}x^2\arctan x-\frac{1}{6}x^2+\frac{1}{6}\ln(x^2+1)+C$

23. $x\operatorname{arcsec}(x^{1/2})-(x-1)^{1/2}+C$

25. $(x+1)\arctan(x^{1/2})-x^{1/2}+C$

27. $-x\cot x+\ln|\sin x|+C$

29. $\frac{1}{2}x^2\sin x^2+\frac{1}{2}\cos x^2+C$

31. $-\frac{2\ln x}{x^{1/2}}-\frac{4}{x^{1/2}}+C$

33. $x\sinh x-\cosh x+C$

35. $\frac{1}{2}(-x^2\cos x^2+\sin x^2)+C$

37. $-2\sqrt{x}\exp\left(-\sqrt{x}\right)-2\exp\left(-\sqrt{x}\right)+C$

39. $\pi(\pi-2)\approx 3.5864190939$

41. $\frac{\pi(e^2+1)}{2}\approx 13.1774985055$

43. $V\approx 1.06027$

45. $V\approx 22.7894$

47. $\frac{1}{2}xe^x\cos x+\frac{1}{2}(x-1)e^x\sin x+C$

49. Let $u=x^n$ and $dv=e^x\,dx$.

51. Let $u=(\ln x)^n$ and $dv=dx$.

53. Let $u=(\sin x)^{n-1}$ and $dv=\sin x\,dx$.

55. $6-2e\approx 0.5634363431$

57. $6-2e\approx 0.5634363431$ [*sic*]

59. (a), (b), and (c): $(x+10)\ln(x+10)-x+C$

61. (a) $J_0=1-\frac{1}{e}$; $J_n=nJ_{n-1}-\frac{1}{e}$

65. $\frac{\pi^2}{80}\left(2\pi^4-10\pi^2+15\right)$

67. $\bar{x}=\frac{12[3-14\ln 2+15(\ln 2)^2]}{(-36+56\ln 2)(\ln 2)}\approx 3.090471,$

$\bar{y}=\frac{225-372\ln 2}{45-70\ln 2}\approx 9.331797$

SECTION 7.4 (PAGE 509)

1. $\frac{1}{2}(x-\sin x\cos x)+C$

3. $2\tan\frac{x}{2}+C$

5. $\frac{1}{3}\ln|\sec x|+C$

7. $\frac{1}{3}\ln|\sec 3x+\tan 3x|+C$

9. $\frac{1}{2}(x-\sin x\cos x)+C$

11. $\frac{1}{3}\cos^3 x-\cos x+C$

13. $\frac{1}{3}\sin^3\theta-\frac{1}{5}\sin^5\theta+C$

15. $\frac{1}{5}\sin^5 x - \frac{2}{3}\sin^3 x + \sin x + C$

17. $\frac{2}{5}(\cos x)^{5/2} - 2(\cos x)^{1/2} + C$

19. $-\frac{1}{14}\cos^7 2z + \frac{1}{5}\cos^5 2z - \frac{1}{6}\cos^3 2z + C$

21. $\frac{1}{4}(\sec 4x + \cos 4x) + C$

23. $\tan t + \frac{1}{3}\tan^3 t + C$

25. $-\frac{1}{4}\csc^2 3x - \frac{1}{2}\ln|\sin 2x| + C$

27. $\frac{1}{12}\sec^6 2x - \frac{1}{4}\sec^4 2x + \frac{1}{4}\sec^2 2x + C$

29. $-\frac{1}{10}\cot^5 2t - \frac{1}{3}\cot^3 2t - \frac{1}{2}\cot 2t + C$

31. $\frac{1}{4}\sin^4\theta + C$

33. $\frac{2}{3}(\sec t)^{3/2} + 2(\cos t)^{1/2} + C$

35. $\frac{1}{3}\sin^3\theta + C$

37. $\frac{1}{5}\sin 5t - \frac{1}{15}\sin^3 5t + C$

39. $-\frac{1}{9}\cot^3 3t + \frac{1}{3}\cot 3t + t + C$

41. $-\frac{1}{5}(\cos 2t)^{5/2} + \frac{2}{9}(\cos 2t)^{9/2} - \frac{1}{13}(\cos 2t)^{13/2} + C$

43. $\frac{1}{2}\sin^2 x - \cos x + C$

45. $\frac{4}{3}$ **47.** $\frac{3\pi + 4}{8}$ **49.** 0 **51.** $\frac{3\pi^2}{8}$

53. $\frac{2\pi}{3}\left(4\pi - 3\sqrt{3}\right) \approx 15.436149$

55. **a.** $\frac{\pi}{4}$; **b.** $\frac{\pi}{4}(8 - \pi) \approx 3.815784$

57. $\frac{1}{4}\sec^4 x + C$ **59.** $\frac{1}{4}\cos 2x - \frac{1}{16}\cos 8x + C$

61. $\frac{1}{6}\sin 3x + \frac{1}{10}\sin 5x + C$ **63.** $\ln|\tan x| + C$

SECTION 7.5 (PAGE 516)

1. $\frac{1}{2}x^2 - x + \ln|x+1| + C$ **3.** $\frac{1}{3}\ln\left|\frac{x-3}{x}\right| + C$

5. $\frac{1}{5}(\ln|x-2| - \ln|x-3|) + C$

7. $\frac{1}{4}\ln|x| - \frac{1}{8}\ln(x^2+4) + C$

9. $\frac{1}{3}x^3 - 4x + 8\arctan\left(\frac{x}{2}\right) + C$

11. $x - 2\ln|x+1| + C$ **13.** $x + \frac{1}{x+1} + C$

15. $\frac{1}{4}\ln|x-2| - \frac{1}{4}\ln|x+2| + C$

17. $\frac{3}{2}\ln|2x-1| - \ln|x+3| + C$

19. $\frac{2}{x+1} + \ln|x| + C$

21. $\frac{1}{2}(3\ln|x-2| + \ln|x-1| + \ln|x+1| + 3\ln|x+2|) + C$

23. $\frac{4}{x+2} - \frac{2}{(x+2)^2} + \ln|x+2| + C$

25. $\frac{1}{2}\ln\left(\frac{x^2}{x^2+1}\right) + C$

27. $\ln|x| - \frac{1}{2}\ln(x^2+4) + \frac{1}{2}\arctan\left(\frac{x}{2}\right) + C$

29. $-\frac{1}{2}\ln|x+1| + \frac{1}{4}\ln(x^2+1) + \frac{1}{2}\arctan x + C$

31. $\arctan\left(\frac{x}{2}\right) - \frac{3\sqrt{2}}{2}\arctan\left(x\sqrt{2}\right) + C$

33. $\frac{1}{2}\ln(x^2+3) + \frac{\sqrt{2}}{2}\arctan\left(\frac{x\sqrt{2}}{2}\right) + C$

35. $x + \frac{1}{2}\ln|x-1| - \frac{5}{2(x-1)} + \frac{3}{4}\ln(x^2+1) + 2\arctan x + C$

37. $-\frac{1}{2(e^{2t}-1)} - \frac{1}{4(e^{2t}-1)^2} + C$

39. $\frac{1}{4}\ln|3 + 2\ln t| + \frac{1}{4(3 + 2\ln t)} + C$

41. $5\ln 2 \approx 3.465736$

43. $\frac{1}{3}(23\ln 2 - 7\ln 5) \approx 1.558773$

45. $V = 2\pi(1 + 6\ln 2) \approx 32.414218$

47. $V = 2\pi(7\ln 5 - 2 - 10\ln 2) \approx 14.668684$

49. $V = \frac{\pi}{2}(13 + 16\ln 2) \approx 37.841041$

51. $V = \frac{\pi}{3}(-4 + 6\ln 2) \approx 0.166382$

53. $93\ln|x-7| + 49\ln|x-5| - 44\ln|x| - \frac{280}{x} + C$

55. $-\frac{104}{3}\ln|x-4| - \frac{48}{x-4} + \frac{567}{16}\ln|x-3| - \frac{37}{48}\ln|x+5|$
$+ \frac{39}{2(x+5)} + C$

57. $-\frac{1}{3x-1} + \arctan\frac{x}{5} + 2\ln|3x-1| + 2\ln|2x-1|$
$+ \ln(x^2+25) + C$

59. Choose $a = 0, b \neq 0$ (but otherwise arbitrary), and $c = -b$.

61. Choose a and b not both zero (but otherwise arbitrary) and let $c = -(8a + 8b)/5$.

SECTION 7.6 (PAGE 521)

1. $\arcsin\left(\frac{x}{4}\right) + C$ **3.** $-\frac{\sqrt{4-x^2}}{4x} + C$

5. $8\arcsin\left(\frac{x}{4}\right) - \frac{x\sqrt{16-x^2}}{2} + C$

7. $\dfrac{x}{9\sqrt{9-16x^2}} + C$

9. $\ln\left|x+\sqrt{x^2-1}\right| - \dfrac{\sqrt{x^2-1}}{x} + C$

11. $\dfrac{243}{16}\left[\dfrac{1}{5}\cdot\dfrac{(9+4x^2)^{5/2}}{243} - \dfrac{1}{3}\cdot\dfrac{(9+4x^2)^{3/2}}{27}\right] + C$
$= \dfrac{8x^4+6x^2-27}{40}\sqrt{9+4x^2} + C$

13. $\ln(4x^2) - \ln|2x| - \ln\left(1+\sqrt{1-4x^2}\right) + \sqrt{1-4x^2} + C$

15. $\dfrac{1}{2}\ln\left|\dfrac{\sqrt{9+4x^2}}{3} + \dfrac{2x}{3}\right| + C$

17. $\dfrac{25}{2}\arcsin\left(\dfrac{x}{5}\right) - \dfrac{x}{2}\sqrt{25-x^2} + C$

19. $\dfrac{1}{2}\left[x\sqrt{1+x^2} - \ln\left(x+\sqrt{1+x^2}\right)\right] + C$

21. $\dfrac{2}{27}\left[\dfrac{3x\sqrt{4+9x^2}}{4} - \ln\left(\dfrac{3x+\sqrt{4+9x^2}}{2}\right)\right] + C$

23. $\dfrac{x}{\sqrt{1+x^2}} + C$

25. $\dfrac{1}{512}\left[\dfrac{32x}{(4-x^2)^2} + \dfrac{12x}{4-x^2} + 3\ln\left|\dfrac{2+x}{2-x}\right|\right] + C$

27. $\dfrac{9}{8}\left[\dfrac{4x\sqrt{9+16x^2}}{9} + \ln\left(\dfrac{4x\sqrt{9+16x^2}}{3}\right)\right] + C$

29. $\sqrt{x^2-25} - 5\arctan\left(\dfrac{\sqrt{x^2-25}}{5}\right) + C$

31. $\dfrac{1}{4}x^3\sqrt{x^2-1} - \dfrac{1}{8}x\sqrt{x^2-1} - \dfrac{1}{8}\ln\left|x+\sqrt{x^2-1}\right| + C$

33. $-\dfrac{x}{\sqrt{4x^2-1}} + C$

35. $\ln\left|\dfrac{x+\sqrt{x^2-5}}{\sqrt{5}}\right| - \dfrac{\sqrt{x^2-5}}{x} + C$

37. $\sinh^{-1}\left(\dfrac{x}{5}\right) + C$

39. $\cosh^{-1}\left(\dfrac{x}{2}\right) - \dfrac{\sqrt{x^2-4}}{x} + C$

41. $\dfrac{1}{8}\left[x(2x^2+1)\sqrt{1+x^2} - \sinh^{-1}x\right] + C$

43. The area A of sector OBC is

$$A = \frac{\pi a^2}{4} - \frac{a^2}{2}\arcsin\left(\frac{x}{a}\right).$$

Now use the fact that $x = a\cos\theta$.

45. $\dfrac{\pi}{32}\left[18\sqrt{5} - \ln\left(2+\sqrt{5}\right)\right] \approx 3.809730$

47. $\ln\left(\sqrt{2}+1\right) - \ln\left(\sqrt{5}+1\right) + \ln 2 + \sqrt{5} - \sqrt{2} \approx 1.222016$

49. The surface area is

$$S = 4\pi a\int_{b-a}^{b+a}\frac{x}{\sqrt{a^2-(x-b)^2}}\,dx.$$

Now substitute $x = b + a\sin\theta$.

51. $A = 2\pi\left[\sqrt{2} + \ln(1+\sqrt{2})\right] \approx 14.4236$

53. The surface area is

$$A = \frac{4\pi b}{a^2}\int_0^a\sqrt{a^4+(b^2-a^2)x^2}\,dx.$$

Note that as $b \to a^+$, $\dfrac{b+c}{a} \approx 1 + \dfrac{c}{a}$.

55. $\dfrac{20}{3}$ (million dollars)

SECTION 7.7 (PAGE 526)

1. $\arctan(x+2) + C$

3. $-\dfrac{3}{2}\ln(x^2+4x+5) + 11\arctan(x+2) + C$

5. $\arcsin\left(\dfrac{x+1}{2}\right) + C$

7. $-\dfrac{1}{3}(3-2x-x^2)^{3/2} - 2\arcsin\left(\dfrac{x+1}{2}\right)$
$-\dfrac{x+1}{2}\sqrt{3-2x-x^2} + C$

9. $\dfrac{3}{4}\ln\left|\dfrac{\sqrt{4x^2+4x-3}}{2}\right| + \dfrac{1}{8}\ln\left|\dfrac{2x-1}{\sqrt{4x^2+4x-3}}\right| + C$

11. $\dfrac{1}{3}\arctan\left(\dfrac{x+2}{3}\right) + C$

13. $\dfrac{1}{4}\ln\left|\dfrac{x+1}{x-3}\right| + C$

15. $\ln(x^2+2x+2) - 7\arctan(x+1) + C$

17. $\dfrac{2}{9}\arcsin\left(\dfrac{3x-2}{3}\right) - \dfrac{1}{9}\sqrt{5+12x-9x^2} + C$

19. $\dfrac{25}{4}\left[3\arcsin\left(\dfrac{2x-4}{5}\right) + \dfrac{3}{25}(2x-4)\sqrt{9+16x-4x^2}\right.$
$\left. + \dfrac{2}{75}(9+16x-4x^2)^{3/2}\right] + C$

21. $\dfrac{7x-12}{\sqrt{6x-x^2}} + C$

23. $-\dfrac{1}{4(4x^2+12x+13)} + C$

25. $\dfrac{3}{2}\ln(x^2+x+1) - \dfrac{5\sqrt{3}}{3}\arctan\left(\dfrac{\sqrt{3}}{3}[2x+1]\right) + C$

27. $-\dfrac{1}{16}\left(\dfrac{2x}{x^2-4} + \ln\left|\dfrac{x-2}{\sqrt{x^2-4}}\right|\right) + C$ by trigonometric substitution;
$\dfrac{1}{32}\ln\left|\dfrac{x+2}{x-2}\right| - \dfrac{x}{8(x^2-4)} + C$ by partial fractions (the answers are the same).

29. $\ln|x| - \dfrac{2\sqrt{3}}{3}\arctan\left(\dfrac{\sqrt{3}}{3}[2x+1]\right) + C$

31. $\dfrac{1}{4}\ln\left|\dfrac{x+1}{x-1}\right| - \dfrac{5x}{2(x^2-1)} + C$

33. $\dfrac{3x^2+4x+1}{8(x^2+2x+5)} - \dfrac{1}{8}\arctan\left(\dfrac{x+1}{2}\right) + C$

37. The substitution $x = 1 + 2\tan u$ yields $A = \dfrac{\pi}{4}$.

39. The substitution $x = 1 + 2\tan u$ yields $V = \dfrac{5\pi^2 + 8\pi}{160} \approx 0.465505$.

41. The substitution $x = \dfrac{5}{2} + \tan u$ yields

$$V = \frac{5\pi}{2}\arctan\left(\frac{3}{2}\right) \approx 7.718844.$$

43. The length is $\dfrac{\sqrt{377}}{4}\arcsin\left(\dfrac{260}{377}\right) \approx 3.694049$ mi.

45. $\ln|x+1| - \dfrac{1}{2}(x^2+2x+2) + \arctan(x+1) + C$

47. $\dfrac{1}{2}x^2 + \ln|x-1| + \dfrac{1}{2}\ln(x^2+x+1) + \dfrac{\sqrt{3}}{3}\arctan\left(\dfrac{2x+1}{\sqrt{3}}\right) + C$

49. $\dfrac{1}{2}\ln(x^4+x^2+1) - \dfrac{2\sqrt{3}}{3}\arctan\left(\dfrac{\sqrt{3}}{2x^2+1}\right) + C$

51. $7\ln|x-1| + \dfrac{11}{2}\arctan(x+1) + \dfrac{11(x+1) - 6(x+1)^2}{2(x^2+2x+2)} + C$

53. $\dfrac{7}{8}\ln|x-5| + \dfrac{9}{8}\ln|x+3| - \dfrac{11}{64}\arctan\left(\dfrac{2x+1}{2}\right)$

$$-\frac{11}{32}\cdot\frac{2x+1}{4x^2+4x+5} + \frac{3}{32}\cdot\frac{(2x+1)^2}{4x^2+4x+5} + C$$

55. Choose $a \neq 0$ and $b = 2a$ to obtain $-\dfrac{a}{2(x^2+4x+5)} + C$.

57. The only solution is $a = b = c = 0$.

59. π

63. Substitute $t = \dfrac{x}{\sqrt{2}}$.

65. $\displaystyle\int_0^{60} x^5 e^{-x}\,dx \approx 119.9999999999999999258$

67. If $b = 10240$ then $k \approx 2.000176$.

69. If $b = 10$ and if $b = 100$ then $k \approx 3.5449077018110321$.

71. **a.** About 49.50%; **b.** about 4.78%

73. **a.** About 90.44%; **b.** about 0.04%

75. **a.** About 7.86%; **b.** about 0.23%

77. About 97.23%

SECTION 7.8 (PAGE 539)

1. $\sqrt{2}$ **3.** $+\infty$ **5.** $+\infty$

7. 1 **9.** $+\infty$ **11.** $-\dfrac{1}{2}$

13. $\dfrac{9}{2}$ **15.** $+\infty$

17. Does not converge

19. $2(e-1)$ **21.** $\dfrac{1}{9}$ **23.** $\dfrac{1}{2}$ **25.** $\dfrac{\pi}{2}$

27. Does not converge **29.** $+\infty$

31. $\dfrac{1}{\ln 2}$ **33.** 2 **35.** -1 **37.** $-\infty$

39. The first integral diverges; the second (from 1 to $+\infty$) converges to $\ln 2$.

41. Both converge to $\dfrac{\pi}{2}$.

43. Converges to $\dfrac{1}{1-k}$ if $k < 1$, diverges if $k \geqq 1$.

45. Converges to $-\dfrac{1}{(k+1)^2}$ if $k > -1$, diverges if $k \leqq -1$.

47. Write the definition of $\Gamma(t+1)$, integrate by parts with $u = x^t$ and $dv = e^{-x}\,dx$, and use Problem 61 of Section 4.8.

49. The area is $A = \displaystyle\int_1^{\infty} \frac{1}{x}\,dx$.

51. The surface area is $S = 2\pi\displaystyle\int_1^{\infty} \frac{\sqrt{x^4+1}}{x^3}\,dx$.

CHAPTER 7 MISCELLANEOUS PROBLEMS (PAGE 542)

1. $2\arctan\sqrt{x} + C$

3. $\ln|\sec x| + C$

5. $\dfrac{1}{2}\sec^2\theta + C$

7. $x\tan x + \ln|\cos x| - \dfrac{1}{2}x^2 + C$

9. $-\dfrac{2}{9}x^3(2-x^3)^{3/2} - \dfrac{4}{45}(2-x^3)^{5/2} + C$

11. $\dfrac{1}{2}x\sqrt{25+x^2} - \dfrac{25}{2}\ln\left(x + \sqrt{25+x^2}\right) + C$

13. $\dfrac{2\sqrt{3}}{3}\arctan\left(\dfrac{\sqrt{3}}{3}[2x-1]\right) + C$

15. $\dfrac{103\sqrt{29}}{87}\arctan\left(\dfrac{3x-2}{\sqrt{29}}\right) + \dfrac{5}{6}\ln(3x^2-4x+11) + C$

17. $\dfrac{2}{9}(1+x^3)^{3/2} + C$

19. $\arcsin\left(\dfrac{\sin x}{2}\right) + C$

21. $-\ln|\ln(\cos x)| + C$

23. $(1+x)\ln(1+x) - x + C$

25. $\dfrac{1}{2}x\sqrt{x^2+9} + \dfrac{9}{2}\ln\left(x + \sqrt{x^2+9}\right) + C$

27. $\dfrac{1}{2}\arcsin(x-1) + \dfrac{1}{2}(x-1)\sqrt{2x-x^2} + C$

29. $\dfrac{1}{3}x^3 + 2x + \sqrt{2}\ln\left|\dfrac{x-\sqrt{2}}{x+\sqrt{2}}\right| + C$

31. $\dfrac{x^2+x}{2(x^2+2x+2)} - \dfrac{1}{2}\arcsin(x+1) + C$

33. $\frac{1}{2}\tan\theta + C$

35. $\frac{1}{5}\sec^5 x - \frac{1}{3}\sec^3 x + C$

37. $\dfrac{x^2}{8}[4(\ln x)^3 - 6(\ln x)^2 + 6(\ln x) - 3] + C$

39. $\dfrac{1}{2}e^x\sqrt{1+e^{2x}} + \dfrac{1}{2}\ln\left(e^x + \sqrt{1+e^{2x}}\right) + C$

41. $\dfrac{1}{54}\operatorname{arcsec}\left(\dfrac{x}{3}\right) + \dfrac{\sqrt{x^2-9}}{18x^2} + C$

43. $\ln|x| + \frac{1}{2}\arctan(2x) + C$

45. $\frac{1}{2}(\sec x \tan x - \ln|\sec x + \tan x|) + C$

47. $\ln|x+1| - \dfrac{2}{3x^3} + C$

49. $\ln|x-1| + \dfrac{1}{x-1} + \ln(x^2+x+1) - \dfrac{2}{x^2+x+1} + C$

51. $x(\ln x)^2 - 6x(\ln x)^5 + 30x(\ln x)^4 - 120x(\ln x)^3 + 360x(\ln x)^2 - 720x\ln x + 720x + C$

53. $\frac{1}{3}(\arcsin x)^3 + C$

55. $\frac{1}{2}\sec^2 z + \ln|\cos z| + C$

57. $\frac{1}{2}\arctan(\exp(x^2)) + C$

59. $-\dfrac{x^2+1}{2}\exp(-x^2) + C$

61. $-\dfrac{1}{x}\arcsin x + \ln\left(1 - \sqrt{1-x^2}\right) - \ln|x| + C$

63. $\frac{1}{8}x(2x^2-1)\sqrt{1-x^2} + \frac{1}{8}\arcsin x + C$

65. $\dfrac{1}{4}\ln|2x+1| + \dfrac{5}{4(2x+1)} + C$

67. $\frac{1}{2}\ln|e^{2x}-1| + C$

69. $2\ln|x+1| + \dfrac{3}{x+1} - \dfrac{5}{3(x+1)^3} + C$

71. $\dfrac{1}{2}\ln(x^2+1) + \arctan x - \dfrac{1}{2(x^2+1)} + C$

73. $\frac{2}{45}(3x^6 - x^3 - 2)\sqrt{x^3-1} + C$

75. $\frac{2}{3}(1+\sin x)^{3/2} + C$

77. $\frac{1}{2}\ln|\sec x + \tan x| + C$

79. $-2\sqrt{1-\sin t} + C$ if $\cos t \geqq 0$, $2\sqrt{1-\sin t} + C$ if $\cos t \leqq 0$

81. $x\ln(x^2+x+1) - 2x + \sqrt{3}\arctan\left(\dfrac{2x+1}{\sqrt{3}}\right) + \dfrac{1}{2}\ln(x^2+x+1) + C$

83. $-\dfrac{1}{x}\arctan x + \ln|x| - \dfrac{1}{2}\ln(x^2+1) + C$

85. $\dfrac{1}{2}\ln(x^2+1) + \dfrac{1}{2(x^2+1)} + C$

87. $\dfrac{x-6}{2\sqrt{x^2+4}} + C$

89. $\frac{1}{3}(1+\sin^2 x)^{3/2} + C$

91. $\frac{1}{2}e^x(x\sin x - x\cos x + \cos x) + C$

93. $-\dfrac{\arctan x}{2(x-1)^2} + \dfrac{1}{4}\left(\dfrac{1}{2}\ln(x^2+1) - \dfrac{1}{x-1} - \ln|x-1|\right) + C$

95. $\dfrac{11}{9}\arcsin\left(\dfrac{3x-1}{2}\right) - \dfrac{2}{9}\sqrt{3+6x-9x^2} + C$

97. $\dfrac{1}{3}x^3 + x^2 + 3x - \dfrac{1}{x-1} + 4\ln|x-1| + C$

99. $x\,\mathrm{arcsec}(\sqrt{x}) - \sqrt{x-1} + C$

101. $\dfrac{\pi}{4}(4 + e^2 + e^{-2}) \approx 8.838652$

103. $A_t = \pi\left[\sqrt{2} + \ln\left(1+\sqrt{2}\right) - e^{-t}\sqrt{1+e^{-2t}} - \ln\left(e^{-t} + \sqrt{1+e^{-2t}}\right)\right]$;
$A = \pi\left[\sqrt{2} + \ln\left(1+\sqrt{2}\right)\right] \approx 7.211800$

105. $2\pi\sqrt{2}\left[\sqrt{\dfrac{7}{2}} - \dfrac{1}{4}\ln\left(2 + \sqrt{\dfrac{7}{2}}\right) - \dfrac{1}{2}\sqrt{\dfrac{1}{2}} + \dfrac{1}{4}\ln\left(1 + \sqrt{\dfrac{1}{2}}\right)\right] \approx 11.663529$

107. Assume that m and n are integers with $n \geqq 0$ and $m \geqq 2$. Let $u = (\sin x)^{m-1}$ and $dv = (\cos x)^n \sin x\,dx$.

109. $\dfrac{5\pi}{4} \approx 3.926991$

111. The value of the integral is $\frac{1}{630}$.

113. $\frac{1}{2}\left[5\sqrt{6} - 3\sqrt{2} + \ln\left(\sqrt{6} + \sqrt{3} - 2 - \sqrt{2}\right)\right] \approx 3.869983$

115. $\dfrac{2\sqrt{3}}{3}\arctan\left(\dfrac{2e^x+1}{\sqrt{3}}\right) + C$

117. $-\ln(1+e^{-x}) + C$

119. $\dfrac{\sqrt{2}}{4}\left[2\arctan\left(-1+\sqrt{2\tan\theta}\right) + 2\arctan\left(1+\sqrt{2\tan\theta}\right) + \ln\left(\tan\theta - \sqrt{2\tan\theta} + 1\right) - \ln\left(\tan\theta + \sqrt{2\tan\theta} + 1\right)\right] + C$

121. $\dfrac{(3x-2)^{3/2}}{3780}(945x^3 + 540x^2 + 288x + 128) + C$

123. $\dfrac{3(x^2-3)}{4(x^2-1)^{1/3}} + C$

125. $\dfrac{2}{9}(x^2-2)\sqrt{x^3+1} + C$

127. $2\arctan\left(\dfrac{1+x}{1-x}\right)^{1/2} - \sqrt{1-x^2} + C$

129. $3(x+1)^{1/3} + \ln\left|(x+1)^{1/3} - 1\right| - \sqrt{3}\arctan\left(\dfrac{2(x+1)^{1/3}+1}{\sqrt{3}}\right) - \dfrac{1}{2}\ln\left((x+1)^{2/3} + (x+1)^{1/3} + 1\right) + C$

131. $\sqrt{1+e^{2x}} + \dfrac{1}{2}\ln\left|\dfrac{-1+\sqrt{1+e^{2x}}}{1+\sqrt{1+e^{2x}}}\right| + C$

133. $\dfrac{8}{15}$

135. $\dfrac{\sin\theta}{1+\cos\theta} + C$

137. $-\dfrac{2+2\cos\theta}{1+\sin\theta+\cos\theta} + C$

139. $\sqrt{2}\left[\ln\left(-1+\sqrt{2}+\dfrac{1-\cos\theta}{\sin\theta}\right) - \dfrac{1}{2}\ln\left(1 + \dfrac{2-2\cos\theta}{\sin\theta} - \dfrac{(1-\cos\theta)^2}{\sin^2\theta}\right)\right] + C$

141. $-\ln(2+\cos\theta) + C$

SECTION 8.1 (PAGE 557)

1. $y(x) = 3e^{2x-2}$

3. $y(x) = \dfrac{3}{43 - 6x}$

5. $y(x) = (x+3)^2$

7. $y(x) = 6e^x - 1$

9. $y(x) = \ln(x + e^2)$

11. $\dfrac{dy}{dx} = x + y$

13. $\dfrac{dy}{dx} = \dfrac{x}{1-y}$

15. $\dfrac{dy}{dx} = \dfrac{y-x}{y+x}$

17. $\dfrac{dv}{dt} = -kv^2 \quad (k > 0 \text{ constant})$

19. $\dfrac{dN}{dt} = k(P - N) \quad (k > 0 \text{ constant})$

21. \$119.35; \$396.24

23. Approximately 3 h 52 min

25. Between 650 and 700 years old

27. \$44.52

29. About 39 days

31. About 35 years

33. **a.** $P(t) = 49 \cdot 6^t$; **b.** 971; **c.** 3:21 P.M.

35. **a.** $A(t) = 15 \cdot \left(\frac{2}{3}\right)^{t/5}$; **b.** approximately 7.84; **c.** After about 33.4 months

37. About 74000 years ago

39. Three hours

41. About 1 h 19 min

43. 1:20 P.M.

45. About 6 min 3 s

47. Approximately 0.02887 in.

SECTION 8.2 (PAGE 567)

1.

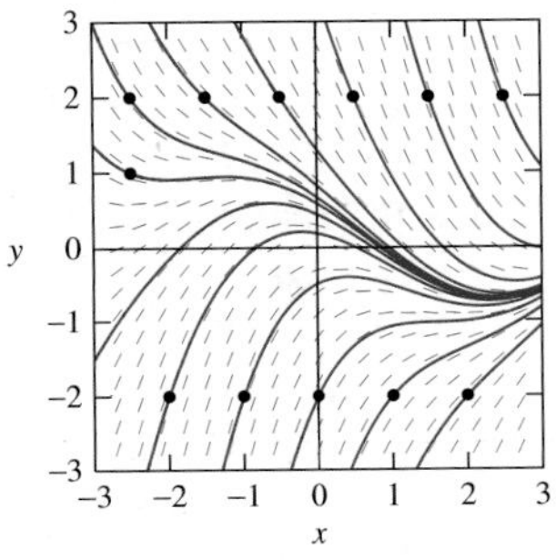

3.

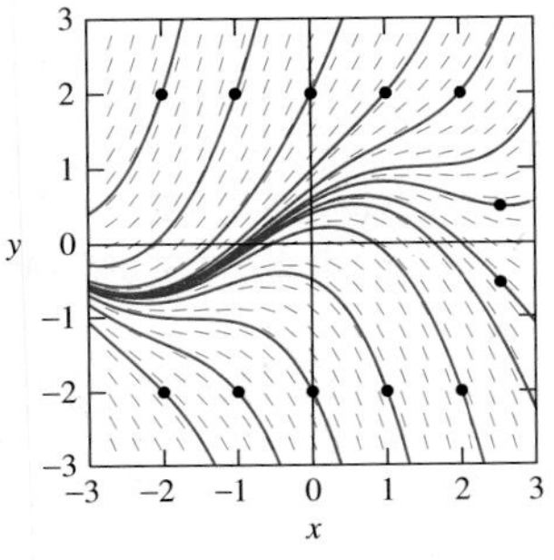

5.

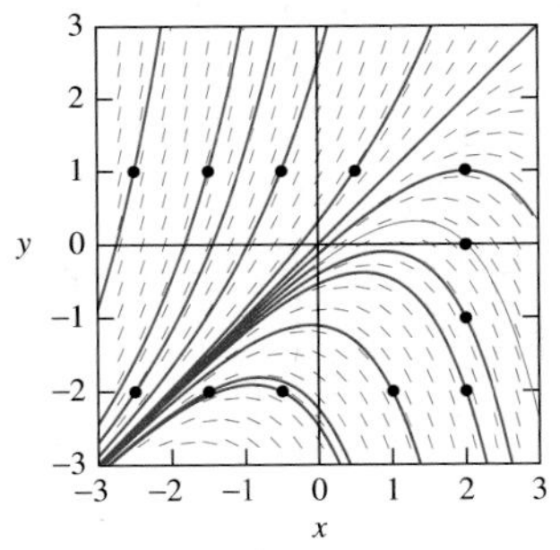

7.

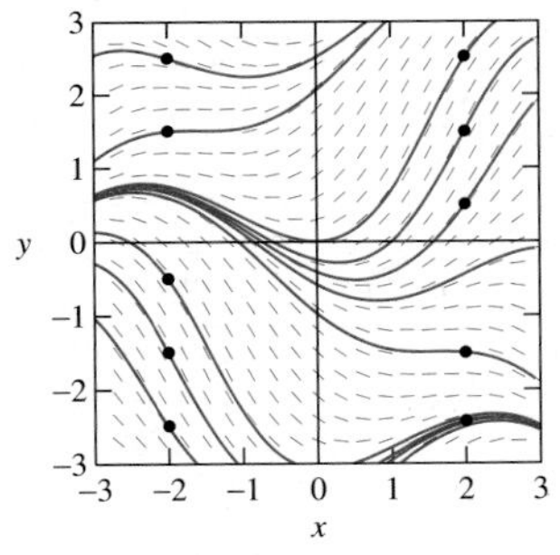

9.

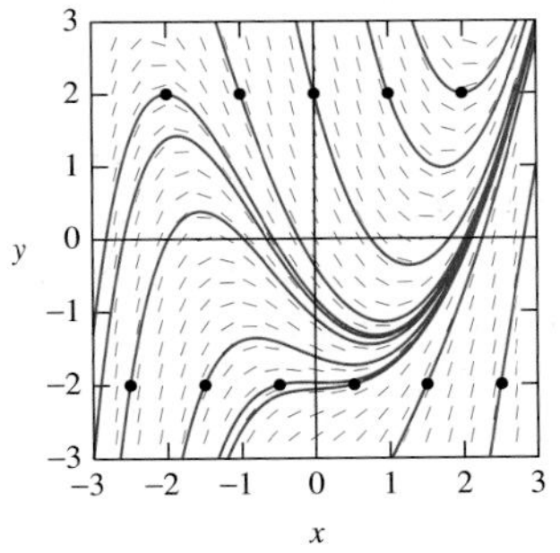

11. To three places: approximations: 1.125 and 1.181; true value: 1.213.

13. To three places: approximations: 2.125 and 2.221; true value: 2.297.

15. To three places: approximations: 0.938 and 0.889; true value: 0.851.

17. To three places: approximations: 2.859 and 2.737; true value: 2.647.

19. To three places: approximations: 1.267 and 1.278; true value: 1.287.

21. Your figure should indicate that $y(-4) \approx 3$; the exact solution is $y(-4) = 3 + e^{-4} \approx 3.018316$.

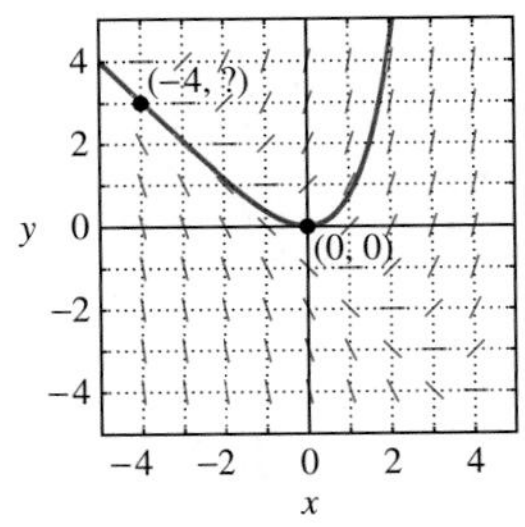

23. Your figure should indicate that $y(2) \approx 1$. The exact value is closer to 1.004. See the *Solutions Manual* for additional details.

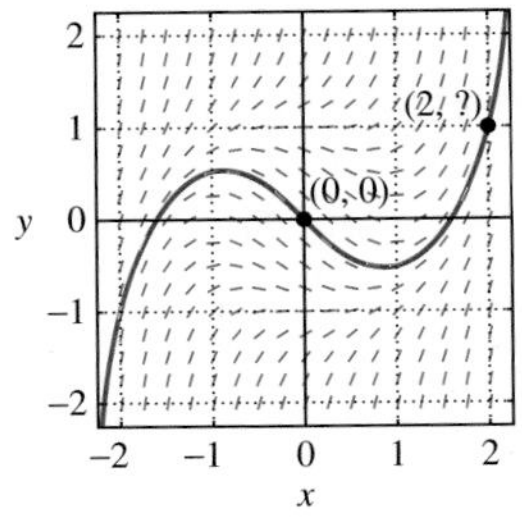

25. Your figure should indicate that the limiting velocity is about 20 ft/s (quite survivable) and that the time to reach 19 ft/s is a little less than two seconds. An exact solution gives $v(t) = 19$ when $t = \frac{5}{8} \ln 20 \approx 1.872333$.

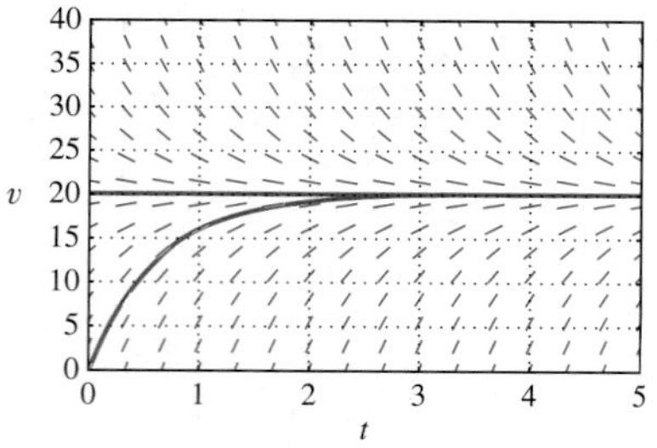

27. With $h = 10^{-4}$ and $h = 10^{-5}$ we find that $y(2) \approx 1.004$ to three places.

29. No guarantee of uniqueness because $D_y(-\sqrt{1-y^2})$ is not continuous at $(0, 1)$.

31. For fixed $a \geqq 0$, let $y(x) = x^3$ if $x \leqq 0$, $y(x) = 0$ if $0 \leqq x \leqq a$, and $y(x) = (x-a)^3$ if $a \leqq x$. Then $y(x)$ is (for each a) a solution of the given initial value problem. No contradiction because uniqueness is guaranteed only "near" the point $(-1, 1)$.

SECTION 8.3 (PAGE 576)

1. $y(x) = \frac{1}{4}(x^2 + C)^2$

3. $y(x) = \dfrac{3}{C - x^3}$

5. $y(x) = 1 + \frac{1}{4}(x^2 + C)^2$

7. $3y + 2y^{3/2} = 3x + 2x^{3/2} + C$

9. $y^3 + y = x - \dfrac{1}{x} + C$

11. $y(x) = \dfrac{1}{1 - x}$

13. $y(x) = (x+1)^{1/4}$

15. $y(x) = \dfrac{36}{(3 - 2x^{3/2})^2}$

17. $y(x) = -\sqrt{169 - x^2}$

19. $y(x) = \dfrac{1}{1 + x - x^3}$

21. $y(x) = 2e^x - 1$

23. $y(x) = \dfrac{1}{2}(e^{2x} + 3)$

25. $x(t) = 1 - e^{2t}$

27. $x(t) = 27e^{5t} - 2$

29. $v(t) = 10(1 - e^{-10t})$

31. Approximately 4.87 million

33. Justify $P(t + \Delta t) - P(t) \approx r\,P(t)\,\Delta t - c\,\Delta t$, divide both sides by Δt, and finally let $\Delta t \to 0$.

35. After about 46 days

37. About \$2183.15 per month

39. About 3679

41. At 6:00 A.M.

SECTION 8.4 (PAGE 587)

1. $\rho(x) = e^x$; $y(x) = 2(1 - e^{-x})$

3. $\rho(x) = e^{3x}$; $y(x) = e^{-3x}(x^2 + C)$

5. $\rho(x) = x^2$; $y(x) = x + 4x^{-2}$

7. $\rho(x) = x^{1/2}$; $y(x) = 5x^{1/2} + Cx^{-1/2}$

9. $\rho(x) = x^{-1}$; $y(x) = 7x + x \ln x$

11. $\rho(x) = xe^{-3x}$; $y(x) \equiv 0$

13. $\rho(x) = e^x$; $y(x) = \cosh x$

15. $\rho(x) = \exp(x^2)$; $y(x) = \dfrac{1 - 5\exp(-x^2)}{2}$

17. $\rho(x) = 1 + x$; $y(x) = \dfrac{1 + \sin x}{1 + x}$

19. $\rho(x) = \sin x$; $y(x) = C \csc x + \dfrac{1}{2} \sin x = C_1 \csc x - \dfrac{1}{2} \cos x \cot x$

21. $\rho(x) = \exp(-x^2)$; $y(x) = [\exp(x^2)]\left[\dfrac{\sqrt{\pi}}{2} \operatorname{erf}(x) + C\right]$

23. $t = 200 \ln 10$ seconds, about 7 min 40.517 s

25. $4 \ln 4 \approx 5.5452$ years

27. 393.75 lb

29. (a) $A(t) = 360[e^{(0.06)t} - e^{(0.05)t}]$; (b) \$1,308,283.30 (minus taxes)

31. $t = \dfrac{10 \ln 5}{\ln(5/3)} \approx 31.507$ s

33. $\dfrac{400}{\ln 2} \approx 577$ ft

35. $v(t) = \dfrac{400}{t + 10}$; $x(60) = 400 \ln 7 \approx 778.4$ ft

37. (a) 100 ft/s; (b) $t = 10 \ln 10 \approx 23.03$ s; about 1402.6 ft

39. 50

41. Maximum height: about 108.28 m; impact speed: about 43.23 m/s

43. Limiting velocity: $v_\tau = -\dfrac{g}{k}$

SECTION 8.5 (PAGE 597)

1. $x(t) = \dfrac{2}{2 - e^{-t}}$

3. $x(t) = \dfrac{2e^{2t} + 1}{2e^{2t} - 1}$

5. $x(t) = \dfrac{40}{8 - 3e^{-15t}}$

7. $x(t) = \dfrac{77}{11 - 4e^{-28t}}$

9. 484 rabbits

11. a. $P(t) = \left(\dfrac{1}{2}kt + \sqrt{P_0}\right)^2$; **b.** 256 fish

13. a. $P(t) = \dfrac{P_0}{1 - kP_0 t}$; **b.** 30 months

15. $\dfrac{200}{1 + e^{-6/5}} \approx 153.7$ million

17. a. $\dfrac{5}{4} \ln 3 \approx 1.373$ (seconds); **b.** $\lim_{t \to \infty} x(t) = 200$

19. a. $100 \ln \dfrac{9}{5} \approx 58.779$ years; **b.** $100 \ln 2 \approx 69.315$ years

23. About 44.22 months

25. About 24.41 months

27. A little less than 35 days

29. (a) $P(140) \approx 127.008$; (b) About 210.544 million; (c) In 2000, we have $P \approx 196.169$, whereas the actual population was about 281.422 million.

SECTION 8.6 (PAGE 608)

1. $y(x) = c_1 e^{2x} + c_2 e^{5x}$

3. $y(x) = c_1 e^{-x/2} + c_2 e^{3x/2}$

5. $y(x) = c_1 \exp\left(\left[-2 + \sqrt{3}\right]x\right) + c_2 \exp\left(\left[-2 - \sqrt{3}\right]x\right)$

7. $y(x) = (c_1 + c_2 x)e^{-3x/2}$
9. $y(x) = (c_1 + c_2 x)e^{2x/5}$
11. $y(x) = e^{-3x}(c_1 \cos 2x + c_2 \sin 2x)$
13. $y(x) = e^{-x/3}(c_1 \cos 5x + c_2 \sin 5x)$
15. $y(x) = 2e^{3x/2} + 3e^{4x}$
17. $y(x) = 9e^{7x} - 5e^{11x}$
19. $y(x) = (2 - 3x)e^{-11x}$
21. $y(x) = 7\cos 5x + 2\sin 5x$
23. $y(x) = e^{-2x}(9\cos 4x + 7\sin 4x)$
25. $y(x) = e^{-x/2}(10\cos 5x + 6\sin 5x)$
27. $y'' + 10y' = 0$
29. $y'' + 20y' + 100y = 0$
31. $y'' = 0$
33. $25y'' + 250y' + 626y = 0$
35. General solution: $y(x) = c_1 \cos 5x + c_2 \sin 5x$. (a) $y(x) = c_2 \sin x$ is a solution for every real value of the constant c_2; (b) the two "initial conditions" (they are actually *boundary conditions*) imply that $c_1 = c_2 = 0$.

SECTION 8.7 (PAGE 618)

1. $x(t) = 5\cos\left(5t - \tan^{-1}\frac{3}{4}\right)$
3. $x(t) = 13\cos\left(3t - \pi - \tan^{-1}\frac{12}{5}\right)$
5. $x(t) = 4e^{-2t} - 2e^{-4t}$; overdamped
7. $x(t) = (5 + 10t)e^{-4t}$; critically damped
9. $x(t) = e^{-4t}(5\cos 2t + 12\sin 2t) = 13e^{-4t}\cos\left(2t - \tan^{-1}\frac{12}{5}\right)$; underdamped
11. $x(t) = 2\cos 2t - 2\cos 3t$
13. $x(t) = 4\sin 5t - 2\sin 10t$
15. $x_{sp}(t) = -\frac{50}{13}\cos 3t + \frac{120}{13}\sin 3t = 10\cos\left(3t - \pi + \tan^{-1}\frac{12}{5}\right)$
17. $x_{sp}(t) = 3\sin 3t - \cos 3t = \sqrt{10}\cos(3t - \pi + \tan^{-1} 3)$; $x_{tr}(t) = e^{-2t}(\cos t - 7\sin t) = 5e^{-2t}\sqrt{2}\cos(t + \tan^{-1} 7)$
19. π s
21. Amplitude 2 m, frequency 5 rad/s, period $2\pi/5$ s
23. Begin with

$$mx'' + cx' + kx = F(t) + mg; \quad x(0) = x_0, \quad x'(0) = v_0.$$

25. The radius is approximately 3.8078 in.
27. **a.** $x(t) = 50(e^{-2t/5} - e^{-t/2})$; **b.** 4.096
29. (a) The position function is

$$x(t) = \frac{2}{\sqrt{3}}e^{-4t}\cos\left(4t\sqrt{3} - \frac{\pi}{6}\right).$$

(b) Frequency $4\sqrt{3}$ rad/s, time-varying amplitude $2/\sqrt{3}$ ft, phase angle $\pi/6$.
37. (a) If $x(t) = A\cos 3t + B\sin 3t$, then $x'' + 9x \equiv 0$.

CHAPTER 8 MISCELLANEOUS PROBLEMS (PAGE 620)

1. $y(x) = x^2 + \sin x$
3. $y(x) = -1 - \dfrac{1}{x + C}$
5. $y(x) = \dfrac{1}{1 - x^3}$
7. $y(x) = \left(C - \dfrac{3}{x}\right)^{1/3}$
9. $y(x) = \dfrac{1}{1 - \sin x}$
11. $y^{-1} - 2y^{-1/2} = x^{-1} - 2x^{-1/2} + C$
13. Linear: $y(x) = Cx^3 + x^3 \ln x$
15. Separable: $y(x) = C\exp\left(\dfrac{1 - x}{x^3}\right)$
17. Linear: $y(x) = \dfrac{C + \ln x}{x^2}$
19. Linear: $y(x) = (x^3 + C)e^{-3x}$
21. Linear: $y(x) = 2x^{-3/2} + Cx^{-3}$
23. Separable: $y(x) = \dfrac{x^{1/2}}{6x^2 + Cx^{1/2} + 2}$
25. Linear: $y(x) = (x + C)e^x$
27. Both methods yield $y(x) = -7 + C\exp(x^3)$.
29. $x(t) = \dfrac{21e^t - 16}{8 - 7e^t}$
31. Approximately 4.2521×10^9 years old
33. About 2 min 25 s
35. (a) Approximately 9.604 in.; (b) approximately 18230 ft; (c) about 13.86 in. of mercury (the summit of Mt. McKinley is 20320 ft above sea level)
37. A little over 325 days
39. $y(x) = 4e^{3x/2} + 9e^{5x/3}$
41. $y(x) = (17x + 11)e^{-7x/11}$
43. $y(x) = e^{-x/10}(10\cos 10x + \sin 10x)$
45. **a.** $N(t) = 29e^{rt}$; **b.** about 49.8%; **c.** about 16.7 months; **d.** about 1668 million
47. 20 weeks
49. 169 thousand in the year 2000; 200 thousand about May 26, 2011
51. $P(t) \to +\infty$ as $t \to 6^-$
53. 20 weeks
55. At 8%, the monthly payment is \$925.21; at 12% it is \$1262.87.
57. At about 9:34 P.M. on the following day
59. $v(10) \approx 111.253$ ft/s, about 75.854 mi/h; the limiting velocity is 176 ft/s, exactly 120 mi/h.
61. (b) Approximately \$1,308,283.30
63. $\alpha \approx 0.39148754$; the limiting population is about 2.152×10^6 (cells).

SECTION 9.1 (PAGE 630)

1. $x + 2y + 3 = 0$
3. $3x - 4y = 25$
5. $x + y = 1$
7. Center $(-1, 0)$, radius $\sqrt{5}$
9. Center $(2, -3)$, radius 4
11. Center $(\frac{1}{2}, 0)$, radius 1
13. Center $(\frac{1}{2}, -\frac{3}{2})$, radius 3
15. Center $(-\frac{1}{3}, \frac{4}{3})$, radius 2
17. The single point $(3, 2)$
19. There are no points on the graph.
21. $(x + 1)^2 + (y + 2)^2 = 34$

23. $(x-6)^2+(y-6)^2=\frac{4}{5}$

25. The locus is the perpendicular bisector of the segment joining the two given points; it has equation $2x+y=13$.

27. The circle with center $(6, 11)$ and radius $3\sqrt{2}$

29. The locus has equation $9x^2+25y^2=225$; it is an ellipse with center $(0, 0)$, horizontal major axis of length 10, vertical minor axis of length 6, and intercepts $(\pm 5, 0)$ and $(0, \pm 3)$.

31. There are two such lines, with equations $y-1=\left(4\pm 2\sqrt{3}\right)\cdot(x-2)$.

33. There are two such lines, with equations $y-1=4(x-4)$ and $y+1=4(x+4)$.

SECTION 9.2 (PAGE 637)

1. **a.** $\left(\frac{1}{2}\sqrt{2}, \frac{1}{2}\sqrt{2}\right)$; **b.** $\left(1, -\sqrt{3}\right)$; **c.** $\left(\frac{1}{2}, -\frac{1}{2}\sqrt{3}\right)$; **d.** $(0, -3)$; **e.** $\left(\sqrt{2}, -\sqrt{2}\right)$; **f.** $\left(\sqrt{3}, -1\right)$; **g.** $\left(-\sqrt{3}, 1\right)$

3. $r=4\sec\theta$

5. $\theta=\tan^{-1}\left(\dfrac{1}{3}\right)$

7. $r^2=\sec\theta\,\csc\theta$

9. $r=\sec\theta\,\tan\theta$

11. $x^2+y^2=9$

13. $x^2+y^2+5x=0$

15. $(x^2+y^2)^3=4y^4$

17. $x=3$

19. $r=2\sec\theta$; $x=2$

21. $r=\dfrac{1}{\cos\theta+\sin\theta}$; $x+y=1$

23. $r=\dfrac{2}{\sin\theta-\cos\theta}$; $y=x+2$

25. $r+8\sin\theta=0$; $x^2+y^2+8y=0$

27. $r=2(\cos\theta+\sin\theta)$; $x^2+y^2=2x+2y$

29. Matches Fig. 9.2.23.

31. Matches Fig. 9.2.24.

33. Matches Fig. 9.2.26.

35. Matches Fig. 9.2.25.

37. Circle, center $(\frac{1}{2}a, \frac{1}{2}b)$, radius $\frac{1}{2}\sqrt{a^2+b^2}$

39. Circle, center $(1, 0)$, radius 1, symmetric around the x-axis

41. Cardioid, cusp at the origin (where $\theta=\pi$), symmetric around the x-axis

43. Limaçon, symmetric around the y-axis

45. Lemniscate lying in the first and third quadrants, symmetric around the lines $y=x$ and $y=-x$ and with respect to the pole

47. Four-leaved rose, symmetric around both coordinate axes, around both lines $y=\pm x$, and with respect to the pole

49. Three-leaved rose, symmetric around the x-axis, unchanged through any rotation around the origin of an integral multiple of $2\pi/3$

51. Five-leaved rose, symmetric around the y-axis, unchanged through any rotation around the origin of an integral multiple of $2\pi/5$

53. The only point of intersection has coordinates $(1, 0)$.

55. The points of intersection are $(\frac{1}{2}, \frac{1}{6}\pi)$, $(\frac{1}{2}, \frac{5}{6}\pi)$, $(-1, \frac{3}{2}\pi)$, and $(0, 0)$.

57. The points of intersection are $(0, 0)$, $(2, \pi)$,

$$\left(2\sqrt{2}-2, \cos^{-1}\left(3-2\sqrt{2}\right)\right) \text{ and}$$

$$\left(2\sqrt{2}-2, -\cos^{-1}\left(3-2\sqrt{2}\right)\right).$$

61. The polar equation can be written in the form $r=\pm a+b\sin\theta$. If $|a|=|b|$ and neither is zero, then the graph is a cardioid. If $|a|\neq|b|$ and neither a nor b is zero, then the graph is a limaçon. If either a or b is zero and the other is not, then the graph is a circle. If $a=b=0$ then the graph consists of the pole alone.

SECTION 9.3 (PAGE 643)

1.

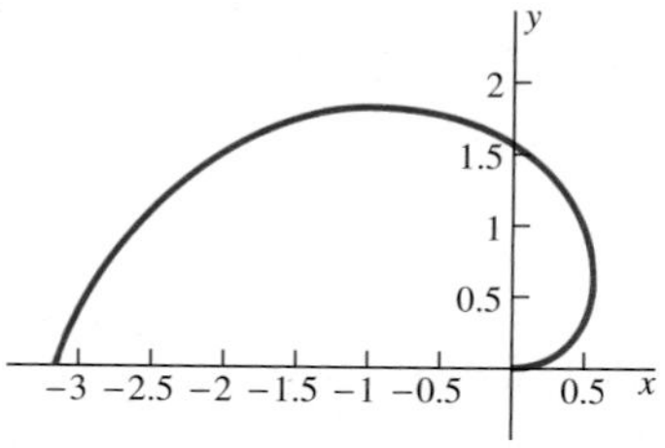

3.

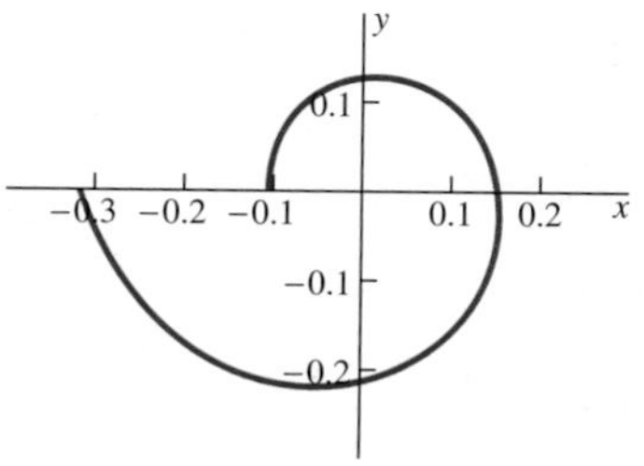

5.

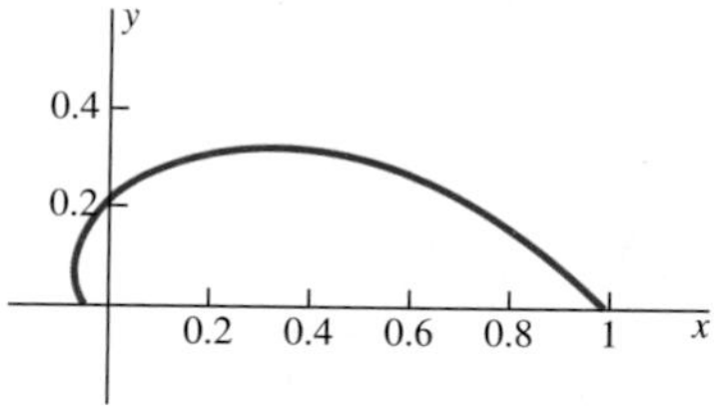

7. π

9. $\dfrac{3}{2}\pi$

11. $\dfrac{9}{2}\pi$

13. 4π

15. $\dfrac{19}{2}\pi$

17. $\dfrac{1}{2}\pi$

19. $\dfrac{1}{4}\pi$

21. 2

23. 4

25. $\dfrac{2\pi+3\sqrt{3}}{6}$

27. $\dfrac{5\pi-6\sqrt{3}}{24}$

29. $\dfrac{39\sqrt{3}-10\pi}{6}$

31. $\dfrac{2-\sqrt{2}}{2}$

33. $\dfrac{20\pi+21\sqrt{3}}{6}$

35. $\dfrac{\pi-2}{2}$

37. $\left(x-\frac{1}{2}\right)^2+\left(y-\frac{1}{2}\right)^2=\frac{1}{2}$; area $\frac{1}{2}\pi$

39. **a.** $A_1=\dfrac{1}{2}\displaystyle\int_0^{2\pi} a^2\theta^2\,d\theta$;

b. $A_2=\dfrac{1}{2}\displaystyle\int_{2\pi}^{4\pi} a^2\theta^2\,d\theta$;

c. $R_2=A_2-A_1$;

d. If $n\geqq 2$, then $A_n=\dfrac{1}{2}\displaystyle\int_{2(n-1)\pi}^{2n\pi} a^2\theta^2\,d\theta$.

41. **a.** $\frac{5}{2}\left(1-e^{-2\pi/5}\right)^2$; **b.** $\frac{5}{2}e^{-2(n-1)\pi/5}\left(1-e^{-2\pi/5}\right)^2$

43. Approximately 1.58069

SECTION 9.4 (PAGE 652)

1. $y=2x-3$

3. $y^2=x^3$

5. $y=2x^2-5x+2$

7. $y=4x^2, x>0$

9. $9x^2+25y^2=225$

11. $9x^2-4y^2=16$

13. $x^2+y^2=1$

15. $y=1-x, 0\leqq x\leqq 1$

17. $9x=4y+7$; concave upward

19. $2\pi x+4y=\pi^2$; concave downward

21. $\psi=\dfrac{\pi}{6}$

23. $\psi=\dfrac{\pi}{2}$

25. Horizontal tangents at (1, −2) and (1, 2); vertical tangent at (0, 0) and *no* tangent line at (3, 0).

27. Horizontal tangents at $\left(\frac{3}{4}, \pm\frac{3}{4}\sqrt{3}\right)$ and at (0, 0); vertical tangent at (2, 0).

29. $\dfrac{dy}{dx}=-2e^{3t}$ and $\dfrac{d^2y}{dx^2}=6e^{4t}$.

31. $x=\dfrac{p}{m^2}, y=\dfrac{2p}{m}, -\infty<m<+\infty$

33. The slope of the line containing P_0 and P is

$$\frac{1+\cos\theta}{\sin\theta},$$

and this is also the value of dy/dx at the point P.

35. The identities $\cos 3t=\cos^3 t-3\sin^2 t\cos t$ and $\sin 3t=3\sin t\cos^2 t-\sin^3 t$ will be very helpful.

41. $x=\dfrac{5t^2}{1+t^5}, y=\dfrac{5t^3}{1+t^5}, 0\leqq t<+\infty$

43. No horizontal tangents; vertical tangents at (−3, 2) and (1, 0); inflection point at (−1, 1).

45. Horizontal tangents at (0, −2.3077), (0, 1), and (0, 2.1433) (numbers with decimal points are approximations); vertical tangents at (−1.8559, 1.7321), (2.4324, 1.7321), and (1.5874, 0); inflection points at (−5.1505, −3.1103), (0, −2.3077), (2.0370, −1.0443), (1.5874, 0), (0, 1), (0, 2.1433), and (4.2661, 2.8565). To see the graph, use a computer algebra system to plot the parametric equations $x=(t^5-5t^3+4)^{1/3}$, $y=t$ with the [suggested] range $-2.7\leqq t\leqq 2.7$.

SECTION 9.5 (PAGE 659)

1. $\dfrac{22}{5}$

3. $\dfrac{4}{3}$

5. $\dfrac{1}{2}(e^{\pi}+1)$

7. $\dfrac{358\pi}{35}$

9. $\dfrac{16\pi}{15}$

11. $\dfrac{74}{3}$

13. $\dfrac{\pi\sqrt{2}}{4}$

15. $(e^{2\pi}-1)\sqrt{5}$

17. $\dfrac{8\pi}{3}\left(5^{3/2}-2^{3/2}\right)$

19. $\dfrac{2\pi}{27}\left(13^{3/2}-8\right)$

21. $16\pi^2$

23. $5\pi^2a^3$

25. **a.** πab; **b.** $\dfrac{4}{3}\pi ab^2$

27. $\dfrac{1}{2}\left[2\pi\sqrt{1+4\pi^2}+\ln\left(2\pi+\sqrt{1+4\pi^2}\right)\right]$

29. $\dfrac{3}{8}\pi a^2$

31. $\dfrac{12}{5}\pi a^2$

33. $\dfrac{216\sqrt{3}}{5}$

35. $\dfrac{243\pi\sqrt{3}}{4}$

37. The length is $2\displaystyle\int_0^1 \frac{3\sqrt{t^8+4t^6-4t^5-4t^3+4t^2+1}}{(t^3+1)^2}\,dt$ ≈ 4.9174887217.

39. $6\pi^3a^3$

41. $\dfrac{5}{6}\pi^3a^2$

43. $\displaystyle\int_0^{\pi}\sqrt{45+36\cos 6\theta}\,d\theta\approx 20.0473398308$

45. $\displaystyle\int_0^{2\pi}\sqrt{10-6\cos 4\theta}\,d\theta\approx 19.3768964411$

47. $\displaystyle\int_0^{2\pi}\sqrt{106+90\cos\theta}\,d\theta\approx 61.0035813739$

49. $\displaystyle\int_0^{3\pi}\frac{1}{3}\sqrt{29-20\cos\left(\tfrac{14}{3}\theta\right)}\,d\theta\approx 16.3428333739$

51. (a) Approximately 16.0570275666; (b) $\dfrac{16\pi}{15}$

53. $\displaystyle\int_0^{2\pi}\sqrt{25\cos^2 5t+9\sin^2 3t}\,dt\approx 24.6029616185$

55. Length: $\displaystyle\int_0^{2\pi}\sqrt{[x'(t)]^2+[y'(t)]^2}\,dt\approx 39.4035787129$

SECTION 9.6 (PAGE 678)

1. Opens to the right; equation $y^2=12x$

3. Opens downward; equation $(x-2)^2=-8(y-3)$

5. Opens to the left; equation $(y-3)^2=-8(x-3)$

7. Opens downward; equation $x^2=-6(y+\frac{3}{2})$

9. Opens upward; equation $x^2=4(y+1)$

11. Opens to the right, vertex at (0, 0), axis the x-axis, focus at (3, 0), directrix $x=-3$

13. Opens to the left, vertex at (0, 0), axis the x-axis, focus at $(-\frac{3}{2}, 0)$, directrix $x=\frac{3}{2}$

15. Opens upward, vertex at (2, −1), axis $x=2$, focus at (2, 0), directrix $y=-2$

17. Opens downward, vertex at $(-\frac{1}{2}, -3)$, axis $x=-\frac{1}{2}$, focus at $(-\frac{1}{2}, \frac{13}{4})$, directrix $y=-\frac{11}{4}$

19. $\left(\dfrac{x}{4}\right)^2+\left(\dfrac{y}{5}\right)^2=1$

21. $\left(\dfrac{x}{13}\right)^2+\left(\dfrac{y}{17}\right)^2=1$

23. $\left(\dfrac{x}{4}\right)^2+\left(\dfrac{y}{\sqrt{7}}\right)^2=1$

25. $\dfrac{x^2}{100}+\dfrac{y^2}{75}=1$

27. $\dfrac{x^2}{16}+\dfrac{y^2}{12}=1$

29. $\left(\dfrac{x-2}{4}\right)^2+\left(\dfrac{y-3}{2}\right)^2=1$

31. $\left(\dfrac{x-1}{5}\right)^2+\left(\dfrac{y-1}{4}\right)^2=1$

33. $\dfrac{(x-1)^2}{81}+\dfrac{(y-2)^2}{72}=1$

35. Center (0, 0), foci $(\pm 2\sqrt{5}, 0)$, axes 12 and 8

37. Center (0, 4), foci $(0, 4 \pm \sqrt{5})$, axes 6 and 4

39. $\dfrac{x^2}{1} - \dfrac{y^2}{15} = 1$

41. $\left(\dfrac{x}{4}\right)^2 - \left(\dfrac{y}{3}\right)^2 = 1$

43. $\left(\dfrac{y}{5}\right)^2 - \left(\dfrac{x}{5}\right)^2 = 1$

45. $\dfrac{y^2}{9} - \dfrac{x^2}{27} = 1$

47. $\dfrac{x^2}{4} - \dfrac{y^2}{12} = 1$

49. $\dfrac{(x-2)^2}{9} - \dfrac{(y-2)^2}{27} = 1$

51. $\left(\dfrac{y+2}{3}\right)^2 - \left(\dfrac{x-1}{2}\right)^2 = 1$

53. Center (1, 2), foci $(1 \pm \sqrt{2}, 2)$, asymptotes $y - 2 = \pm(x - 1)$

55. Center (0, 3), foci $(0, 3 \pm 2\sqrt{3})$, asymptotes $y = 3 \pm x\sqrt{3}$

57. Center $(-1, 1)$, foci $(-1 \pm \sqrt{13}, 1)$, asymptotes $y - 1 = \pm\frac{3}{2}(x + 1)$

59. Parabola, opening to the left, vertex (3, 0), axis the x-axis

61. Parabola, opening to the right, vertex $(-\frac{3}{2}, 0)$, axis the x-axis

63. Ellipse, center (0, 2), vertices at (0, 6) and $(0, -2)$

65. Minimize $(x - p)^2 + y^2$ where $x = y^2/(4p)$.

69. About 16 h 38 min

71. Maximize $R(\alpha) = (v_0^2 \sin 2\alpha)/g$.

73. Approximately 14° 40′ 13″ and 75° 19′ 47″

75. Square the given equation twice to eliminate radicals, convert to polar form, rotate 45° by replacing θ with $\theta + (\pi/4)$, and finally return to Cartesian coordinates. You will recognize the equation as that of a parabola.

77. **a.** About 322 billion miles; **b.** about 120 billion miles

79. With focus $F(0, c)$ and directrix the line L: $y = c/e^2$ $(0 < e < 1)$, begin with $|PF| = e \cdot |PL|$, eliminate radicals, simplify, replace $a^2(1 - e^2)$ with b^2, and convert the resulting Cartesian equation to "standard form."

81. Go to **www.augsburg.edu/depts/math/MATtours/ellipses.1.09.0.html**.

83. The only solution is $\dfrac{(x-1)^2}{4} + \dfrac{3y^2}{16} = 1$.

85. (c) In this case there are no points on the graph.

89. $16x^2 + 50xy + 16y^2 = 369$

91. If A is at $(-50, 0)$ and B is at (50, 0), then the x-coordinate of the plane is approximately 41.3395 (in mi).

93. 2000 mi

95. Begin with $r = pe/(1 - e\cos\theta)$ and first show that the area of the ellipse is

$$A = 2\int_0^{\pi} \frac{1}{2} r^2 \, d\theta \quad \text{where } I = \int_0^{\pi} \frac{1}{(1 - e\cos\theta)^2} \, d\theta.$$

Then use the substitution discussed after Miscellaneous Problem 134 of Chapter 7.

CHAPTER 9 MISCELLANEOUS PROBLEMS (PAGE 681)

1. Circle, center (1, 1), radius 2

3. Circle, center $(3, -1)$, radius 1

5. Parabola, vertex $(4, -2)$, focus $(4, -25)$, opening downward

7. Ellipse, center (2, 0), vertices at (0, 0), (4, 0), (2, 3), and $(2, -3)$, foci $(2, \pm\sqrt{5})$

9. Hyperbola, center $(-1, 1)$, foci $(-1, 1 \pm \sqrt{3})$, vertices $(-1, 1 \pm \sqrt{2})$

11. There are no points on the graph.

13. Hyperbola, center (1, 0), vertices (3, 0) and $(-1, 0)$, foci $(1 \pm \sqrt{3}, 0)$

15. Circle, center (4, 1), radius 1

17. The graph consists of the straight line $y = -x$ together with the isolated point (2, 2); it is not a conic section.

19. Circle, center $(-1, 0)$, radius 1

21. The straight line with Cartesian equation $y = x + 1$

23. The horizontal line $y = 3$

25. A pair of tangent ovals through the origin; the figure is symmetric around the y-axis

27. A limaçon symmetric around the y-axis

29. Ellipse, center $(-\frac{4}{3}, 0)$, horizontal semimajor axis of length $\frac{8}{3}$, semiminor axis of length $\frac{4}{3}\sqrt{3}$, vertices $(-4, 0)$, $(0, \pm\frac{4}{3}\sqrt{3})$, and $(\frac{4}{3}, 0)$, foci $(-\frac{8}{3}, 0)$ and (0, 0)

31. $\dfrac{\pi - 2}{2}$

33. $\dfrac{39\sqrt{3} - 10\pi}{6}$

35. 2

37. $\dfrac{5\pi}{4}$

39. The straight line $y = x + 2$

41. The circle with center (2, 1) and radius 1

43. The "semicubical parabola" with Cartesian equation $y^2 = (x - 1)^3$

45. $y = -\frac{4}{3}(x - 3\sqrt{2})$

47. $4x + 2\pi y = \pi^2$

49. 24

51. 3π

53. $\dfrac{13\sqrt{13} - 8}{27}$

55. $\dfrac{43}{6}$

57. $1 + \dfrac{9\sqrt{5}}{10} \arcsin \dfrac{\sqrt{5}}{3} - \dfrac{\sqrt{31}}{8} - \dfrac{9\sqrt{5}}{10} \arcsin \dfrac{\sqrt{5}}{6}$

59. $\dfrac{471295\pi}{1024}$

61. $\dfrac{\pi(e^{\pi} + 1)\sqrt{5}}{2}$

63. Suppose that the circle rolls to the right through a central angle θ. Then $x = a\theta - b\sin\theta$, $y = a - b\cos\theta$.

65. If the epicycloid is shifted a units to the left, its equations will be

$$x = 2a\cos\theta - a\cos 2\theta - a, \quad y = 2a\sin\theta - a\sin 2\theta.$$

Now compute and simplify $r^2 = x^2 + y^2$.

67. $6\pi^3 a^3$

69. $r = 2p\cos(\theta - \alpha)$

71. Maximum $2a$, minimum $2b$

73. $y = \dfrac{4hx(b - x)}{b^2}$

75. The ellipse has equation $\left(\dfrac{x}{a}\right)^2 + \left(\dfrac{y}{b}\right)^2 = 1$.

79. $e = 2$

81. $A = 9\displaystyle\int_0^{\pi/4} \frac{\sec^2\theta \tan^2\theta}{(1 + \tan^3\theta)^2} \, d\theta = \frac{3}{2}$.

83. If $B < \frac{5}{2}$, then the conic is an ellipse; if $B > \frac{5}{2}$, it is a hyperbola. If $B = \frac{5}{2}$ the graph is a degenerate parabola: two parallel lines. If the graph is normal to the y-axis at the point $(0, 4)$, then the graph is the ellipse with equation

$$\left(\frac{x}{5}\right)^2 + \left(\frac{y}{4}\right)^2 = 1.$$

SECTION 10.2 (PAGE 691)

1. $a_n = n^2$ for $n \geqq 1$.
3. $a_n = 3^{-n}$ for $n \geqq 1$.
5. $a_n = (3n-1)^{-1}$ for $n \geqq 1$.
7. $a_n = 1 + (-1)^n$ for $n \geqq 1$.

9. $\frac{2}{5}$	**11.** 0	**13.** 1
15. Diverges	**17.** 0	**19.** 0
21. 0	**23.** 1	**25.** 0
27. 0	**29.** 0	**31.** 0
33. e	**35.** $\frac{1}{e^2}$	**37.** 2
39. 1	**41.** Diverges	**43.** 1
45. 2	**47.** 1	**49.** π

51. To begin, suppose (without loss of generality) that $A > 0$.
53. Let $L = \lim_{n\to\infty} x_n$. Then $L = \lim_{n\to\infty} x_{n+1}$.
55. (b) $G_1 = G_2 = G_3 = 1$; $G_{n+1} = G_n + G_{n-2}$ for $n \geqq 3$. Check: $G_{25} = 5896$.
57. (b) 4

SECTION 10.3 (PAGE 701)

1. $\frac{3}{2}$
3. Diverges (the kth partial sum is k^2).
5. Diverges (geometric with ratio -2).
7. 6
9. Diverges (geometric with ratio 1.01).
11. Diverges by the nth-term test.
13. Diverges (geometric with ratio $-3/e$).
15. $2 + \sqrt{2}$
17. Diverges by the nth-term test.
19. $\frac{1}{12}$
21. $\frac{e}{\pi - e}$
23. Diverges (geometric with ratio $\frac{100}{99}$).
25. $\frac{65}{12}$ **27.** $\frac{247}{8}$ **29.** $\frac{1}{4}$
31. Diverges by the nth-term test.
33. Diverges (geometric with ratio $\tan 1 > 1$).
35. $\frac{\pi}{4 - \pi}$
37. Diverges: Show that $S_k \geqq \int_2^{k+1} \frac{1}{x \ln x}\,dx > \ln(\ln(k+1))$.

39. $\frac{47}{99}$ **41.** $\frac{41}{333}$ **43.** $\frac{314156}{99999}$
45. Converges to $\frac{x}{3-x}$ if $-3 < x < 3$.
47. Converges to $\frac{x-2}{5-x}$ if $-1 < x < 5$.
49. Converges to $\frac{5x^2}{16 - 4x^2}$ if $-2 < x < 2$.
51. $\frac{1}{6}$
53. $\frac{1}{4}$ (Beaverbock's constant)
55. $\frac{3}{4}$ **57.** 2 **59.** $\frac{1}{3}$
61. Use the converse of part 2 of Theorem 2.
65. 4.5 s
67. $M_n \to 0$ as $n \to +\infty$.
69. Peter: $\frac{4}{7}$; Paul: $\frac{2}{7}$; Mary: $\frac{1}{7}$
71. $\frac{1}{12}$ of the incident light

SECTION 10.4 (PAGE 715)

1. $e^{-x} = 1 - x + \frac{x^2}{2!} - \frac{x^3}{3!} + \frac{x^4}{4!} - \frac{x^5}{5!} + \frac{x^6}{6!}e^{-z}$ for some number z between 0 and x.
3. $\cos x = 1 - \frac{x^2}{2!} + \frac{x^4}{4!} - \frac{x^5}{5!}\sin z$ for some number z between 0 and x.
5. $\sqrt{1+x} = 1 + \frac{x}{2} - \frac{x^2}{8} + \frac{x^3}{16} - \frac{5x^4}{128(1+z)^{7/2}}$ for some number z between 0 and x.
7. $\tan x = x + \frac{x^3}{3} + \frac{x^4}{4!}(16\sec^4 z \tan z + 8\sec^2 z \tan^3 z)$ for some number z between 0 and x.
9. $\arcsin x = x + \frac{x^3(1+2z^2)}{3!(1-z^2)^{5/2}}$ for some number z between 0 and x.
11. $e^x = e + e(x-1) + \frac{e}{2}(x-1)^2 + \frac{e}{6}(x-1)^3 + \frac{e}{24}(x-1)^4 + \frac{e^z}{120}(x-1)^5$ for some number z between 1 and x.
13. $\sin x = \frac{1}{2} + \frac{\sqrt{3}}{2}\left(x - \frac{\pi}{6}\right) - \frac{1}{4}\left(x - \frac{\pi}{6}\right)^2 - \frac{\sqrt{3}}{12}\left(x - \frac{\pi}{6}\right)^3 + \frac{\sin z}{24}\left(x - \frac{\pi}{6}\right)^4$ for some number z between $\pi/6$ and x.
15. $\frac{1}{(x-4)^2} = 1 - 2(x-5) + 3(x-5)^2 - 4(x-5)^3 + 5(x-5)^4 - 6(x-5)^5 + \frac{(x-5)^6}{720} \cdot \frac{5040}{(z-4)^8}$ for some number z between 5 and x.
17. $\cos x = -1 + \frac{(x-\pi)^2}{2} - \frac{(x-\pi)^4}{24} - \frac{\sin z}{120}(x-\pi)^5$ for some number z between π and x.

19. $x^{3/2} = 1+\frac{3}{2}(x-1)+\frac{3}{8}(x-1)^2-\frac{1}{16}(x-1)^3+\frac{3}{128}(x-1)^4 - \frac{(x-1)^5}{120}\cdot\frac{45}{32z^{7/2}}$ for some number z between 1 and x.

21. $e^{-x} = \sum_{n=0}^{\infty}\frac{(-1)^n x^n}{n!}$. This representation is valid for all x.

23. $e^{-3x} = \sum_{n=0}^{\infty}\frac{(-1)^n 3^n x^n}{n!}$. This representation is valid for all x.

25. $\sin 2x = \sum_{n=0}^{\infty}\frac{(-1)^n (2x)^{2n+1}}{(2n+1)!}$. This representation is valid for all x.

27. $\sin(x^2) = \sum_{n=0}^{\infty}\frac{(-1)^n x^{4n+2}}{(2n+1)!}$. This representation is valid for all x.

29. $\ln(1+x) = \sum_{n=1}^{\infty}\frac{(-1)^{n+1} x^n}{n}$. This representation is valid if $-1 < x \leq 1$.

31. $e^{-x} = \sum_{n=0}^{\infty}\frac{(-1)^n x^n}{n!}$. This representation is valid for all x.

33. $\ln x = \sum_{n=1}^{\infty}\frac{(-1)^{n+1}(x-1)^n}{n}$. This representation is valid if $0 < x \leq 2$.

35. $\cos x = \frac{\sqrt{2}}{2} - \frac{\sqrt{2}}{2}\left(x-\frac{\pi}{4}\right) - \frac{\sqrt{2}}{2!\cdot 2}\left(x-\frac{\pi}{4}\right)^2 + \frac{\sqrt{2}}{3!\cdot 2}\left(x-\frac{\pi}{4}\right)^3 + \frac{\sqrt{2}}{4!\cdot 2}\left(x-\frac{\pi}{4}\right)^4 - \cdots$.
This representation is valid for all x.

37. $\frac{1}{x} = \sum_{n=0}^{\infty}(-1)^n(x-1)^n$. This representation is valid for $0 < x < 2$.

39. $\sin x = \frac{\sqrt{2}}{2} + \frac{\sqrt{2}}{2}\left(x-\frac{\pi}{4}\right) - \frac{\sqrt{2}}{2!\cdot 2}\left(x-\frac{\pi}{4}\right)^2 - \frac{\sqrt{2}}{3!\cdot 2} \times \left(x-\frac{\pi}{4}\right)^3 + \frac{\sqrt{2}}{4!\cdot 2}\left(x-\frac{\pi}{4}\right)^4 + \frac{\sqrt{2}}{5!\cdot 2}\left(x-\frac{\pi}{4}\right)^5 - \cdots$.
This representation is valid for all x.

45. Given $f(x) = e^{-x}$, its plot together with that of

$$P_3(x) = 1 - x + \frac{x^2}{2!} - \frac{x^3}{3!}$$

are shown next.

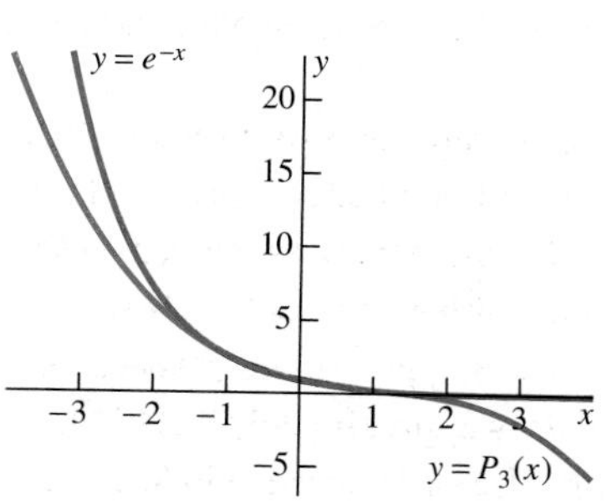

The graphs of $f(x) = e^{-x}$ and

$$P_6(x) = 1 - x + \frac{x^2}{2!} - \frac{x^3}{3!} + \frac{x^4}{4!} - \frac{x^5}{5!} + \frac{x^6}{6!}$$

are shown together next.

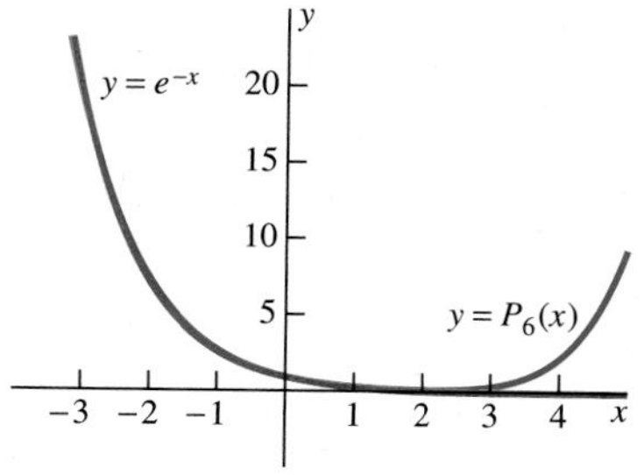

47. One of the Taylor polynomials for $f(x) = \cos x$ is $P_4(x) = 1 - \frac{x^2}{2!} + \frac{x^4}{4!}$. The graphs of f and P_4 are shown together, next.

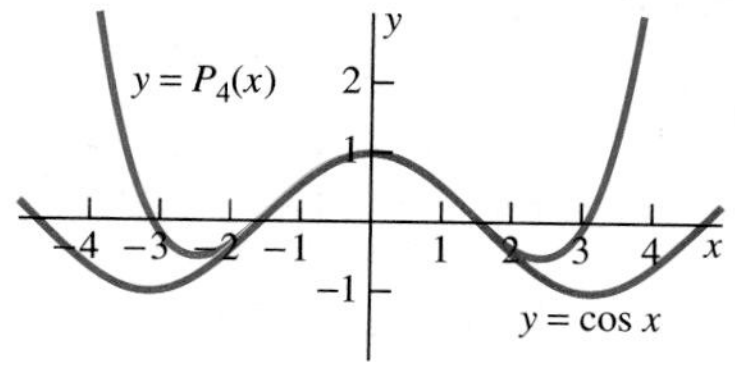

49. One of the Taylor polynomials for $f(x) = \frac{1}{1+x}$ is $P_4(x) = 1 - x + x^2 - x^3 + x^4$. The graphs of f and P_4 are shown together, next.

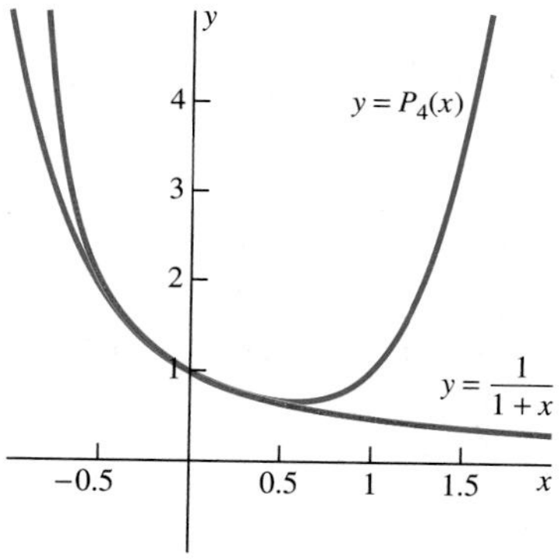

51. The graph of the Taylor polynomial

$$P_4(x) = 1 - \frac{x}{2!} + \frac{x^2}{4!} - \frac{x^3}{6!} + \frac{x^4}{8!}$$

of $f(x)$ and the graph of $g(x)$ are shown together, next.

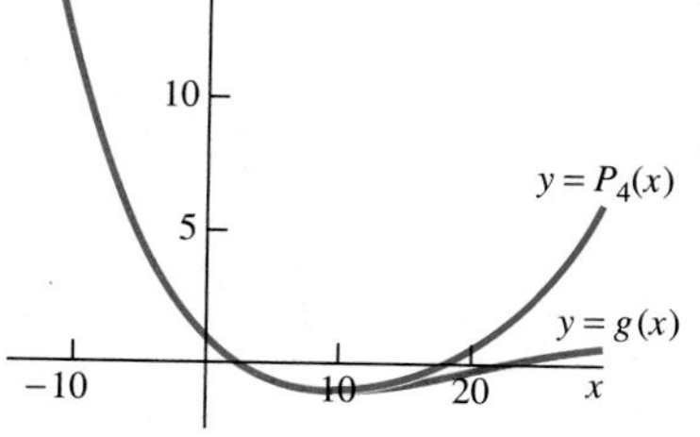

57. By Theorem 4 of Section 10.3, S is not a number. Hence attempts to do arithmetic with S will generally lead to false or meaningless results.

59. Results: With $x = 1$ in the Maclaurin series in Problem 56, we find that

$$a = \sum_{n=1}^{50} \frac{(-1)^{n+1}}{n} \approx 0.68324716057591818842565811649.$$

With $x = \frac{1}{3}$ in the second series in Problem 58, we find that

$$b = \sum_{\substack{n=1\\ n \text{ odd}}}^{49} \frac{2}{n \cdot 3^n} \approx 0.69314718055994530941723210107.$$

Because $|a - \ln 2| \approx 0.009900019984$, whereas $|b - \ln 2| \approx 2.039 \times 10^{-26}$, it is clear that the second series of Problem 58 is *far* superior to the series of Problem 56 for the accurate approximation of ln 2.

SECTION 10.5 (PAGE 722)

1. Diverges: $\displaystyle\int_0^\infty \frac{x}{x^2+1}\,dx = \left[\frac{1}{2}\ln(x^2+1)\right]_0^\infty = +\infty.$

3. Diverges: $\displaystyle\int_0^\infty (x+1)^{-1/2}\,dx = \left[2(x+1)^{1/2}\right]_0^\infty = +\infty.$

5. Converges: $\displaystyle\int_0^\infty \frac{1}{x^2+1}\,dx = \left[\arctan x\right]_0^\infty = \frac{\pi}{2} < +\infty.$

7. Diverges: $\displaystyle\int_2^\infty \frac{1}{x\ln x}\,dx = \left[\ln(\ln x)\right]_2^\infty = +\infty.$

9. Converges (to 1): $\displaystyle\int_0^\infty 2^{-x}\,dx = \left[-\frac{1}{2^x \ln 2}\right]_0^\infty = \frac{1}{\ln 2} < +\infty.$

11. Converges: $\displaystyle\int_0^\infty x^2 e^{-x}\,dx = -\left[(x^2 + 2x + 2)e^{-x}\right]_0^\infty = 2 < +\infty.$

13. Converges: $\displaystyle\int_1^\infty \frac{\ln x}{x^2}\,dx = \left[-\frac{1+\ln x}{x}\right]_1^\infty = \frac{1+0}{1} - \lim_{x\to\infty}\frac{1+\ln x}{x} = 1 < +\infty.$

15. Converges: $\displaystyle\int_0^\infty \frac{x}{x^4+1}\,dx = \left[\frac{1}{2}\arctan(x^2)\right]_0^\infty = \frac{\pi}{4} < +\infty.$

17. Diverges: $\displaystyle\int_1^\infty \frac{2x+5}{x^2+5x+17}\,dx = \left[\ln(x^2 + 5x + 17)\right]_1^\infty = +\infty.$

19. Converges: $\displaystyle\int_1^\infty \ln\left(1+\frac{1}{x^2}\right)dx = \frac{\pi}{2} - \ln 2 < +\infty.$

21. Diverges: $\displaystyle\int_1^\infty \frac{x}{4x^2+5}\,dx = \left[\frac{1}{8}\ln(4x^2+5)\right]_1^\infty = +\infty.$

23. Diverges: $\displaystyle\int_2^\infty \frac{1}{x\sqrt{\ln x}}\,dx = \int_2^\infty \frac{(\ln x)^{-1/2}}{x}\,dx = \left[2(\ln x)^{1/2}\right]_2^\infty = +\infty.$

25. Converges: $\displaystyle\int_1^\infty \frac{1}{4x^2+9}\,dx = \left[\frac{1}{6}\arctan\left(\frac{2x}{3}\right)\right]_1^\infty = \frac{\pi}{12} - \frac{1}{6}\arctan\left(\frac{2}{3}\right) < +\infty.$

27. Converges: $\displaystyle\int_1^\infty \frac{x}{(x^2+1)^2}\,dx = \left[-\frac{1}{2(x^2+1)}\right]_1^\infty = \frac{1}{4} < +\infty.$

29. Converges: $\displaystyle\int_1^\infty \frac{\arctan x}{x^2+1}\,dx = \left[\frac{1}{2}(\arctan x)^2\right]_1^\infty = \frac{3\pi^2}{32} < +\infty.$

31. This is not a positive-term series.

33. The terms of this series are not monotonically decreasing.

35. Diverges if $0 < p \leqq 1$, converges if $p > 1$.

37. Diverges if $p \leqq 1$, converges otherwise.

39. $n > 10000$

41. $n > 100$

43. $n > 160{,}000$

45. $n \geqq 15$

47. $p > 1$

49. Sloppy answer: Over 604,414 centuries. A more precise answer: A little over 922,460 centuries.

51. Apply Theorem 4 and Problem 52 of Section 10.2.

SECTION 10.6 (PAGE 729)

1. Converges: Dominated by the p-series with $p = 2$.

3. Diverges by limit-comparison with the harmonic series.

5. Converges: Dominated by the geometric series with ratio $\frac{1}{3}$.

7. Diverges by limit-comparison with the harmonic series.

9. Converges: Dominated by the p-series with $p = \frac{3}{2}$.

11. Converges: Dominated by the p-series with $p = \frac{3}{2}$.

13. Diverges by comparison with the harmonic series.

15. Converges: Dominated by the p-series with $p = 2$.

17. Converges: Dominated by a geometric series with ratio $\frac{2}{3}$.

19. Converges by comparison with the p-series with $p = 2$.

21. Converges: Dominated by the p-series with $p = \frac{3}{2}$ (among others).

23. Converges: Dominated by the p-series with $p = 2$.

25. Converges: Dominated by the geometric series with ratio $\dfrac{2}{e}$.

27. Diverges by limit-comparison with the p-series with $p = \frac{1}{2}$.

29. Diverges by limit-comparison with the p-series with $p = \frac{1}{2}$.

31. Converges by comparison with a geometric series with ratio $\frac{2}{3}$ and by limit comparison with a geometric series with ratio $\frac{1}{3}$.

33. Converges by comparison with the p-series with $p = 2$.

35. Diverges by limit-comparison with the harmonic series.

37. $S_{10} \approx 0.981793$ with error less than 0.094882.

39. $S_{10} \approx 0.528870$ with error less than 0.1.

41. $n = 10$; the sum is approximately 0.686503.

43. $n = 3$; the sum is approximately 0.100714.

45. Use the converse of Theorem 3 in Section 10.3.

47. Use the comparison test.

49. Apply the converse of Theorem 3 in Section 10.3 and the result in Problem 48.

51. Use the result in Problem 50 in Section 10.5.

SECTION 10.7 (PAGE 737)

1. Converges (to $\frac{1}{12}\pi^2$) by the alternating series test.

3. Diverges by the nth-term test for divergence.

5. Diverges by the nth-term test for divergence.

7. Diverges by the nth-term test for divergence.

9. Converges (to $-\frac{2}{9}$) by the alternating series test.

11. Converges by the alternating series test. (The sum is approximately -0.1782434556.)

13. Converges by the alternating series test. (The sum is approximately 0.711944418056.)

15. Converges by the alternating series test. (The sum is roughly -0.550796848134.)

17. Diverges by the nth-term test for divergence.

19. Diverges by the nth-term test for divergence.

21. Converges absolutely by the ratio test. (The sum is $\frac{1}{3}$.)

23. Converges by the alternating series test, but only conditionally by the integral test. (The sum is approximately 0.159868903742.)

25. Converges absolutely by the root test. (The sum is approximately 186.724948614024.)

27. Converges absolutely by the ratio test. (The sum is $e^{-10} \approx 0.00004539992976$.)

29. Diverges by the nth-term test for divergence.

31. Converges absolutely by the root test. (The sum is approximately 0.187967875056.)

33. Converges by the alternating series test, but only conditionally by the comparison test. (The sum is approximately 0.760209625219.)

35. Diverges by the nth-term test for divergence.

37. Diverges by the nth-term test for divergence.

39. Converges absolutely by the ratio test. (The sum is approximately 0.586781998767.)

41. Converges absolutely by the ratio test. (The sum is approximately 2.807109464185.)

43. 0.9044; 0.005; 0.90

45. 0.6319; 0.0002; 0.632

47. 0.6532; 0.08; 0.7

49. $n = 6$; 0.947 (the sum is $\frac{7}{720}\pi^4$)

51. $n = 5$; 0.6065

53. $n = 4$; 0.86603

55. The sequence of terms is not monotonically decreasing; the series diverges by comparison with the harmonic series.

57. Let $a_n = b_n = \dfrac{(-1)^n}{\sqrt{n}}$.

63. $1 + \frac{1}{3} - \frac{1}{2} + \frac{1}{5} - \frac{1}{4} + \frac{1}{7} + \frac{1}{9} - \frac{1}{6} + \frac{1}{11} + \frac{1}{13} - \frac{1}{8} + \frac{1}{15}$

65. It converges to zero.

SECTION 10.8 (PAGE 750)

1. $(-1, 1)$

3. $(-2, 2)$

5. $[0, 0]$

7. $\left[-\frac{1}{3}, \frac{1}{3}\right]$

9. $\left(-\frac{1}{2}, \frac{1}{2}\right)$

11. $[-2, 2]$

13. $(-3, 3)$

15. $\left(\frac{2}{5}, \frac{4}{5}\right)$

17. $\left[\frac{5}{2}, \frac{7}{2}\right]$

19. $[0, 0]$

21. $(-4, 2)$

23. $[2, 4]$

25. $[5, 5]$

27. $(-1, 1)$

29. $(-\infty, +\infty)$

31. $f(x) = x + x^2 + x^3 + x^4 + x^5 + \cdots;\ R = 1$

33. $f(x) = \sum_{n=0}^{\infty} \dfrac{(-1)^n 3^n x^{n+2}}{n!};\ R = +\infty$

35. $f(x) = \sum_{n=0}^{\infty} \dfrac{(-1)^n x^{4n+2}}{(2n+1)!};\ R = +\infty$

37. $f(x) = 1 - \frac{1}{3}x - \frac{2}{3^2}\cdot\frac{x^2}{2!} - \frac{2\cdot 5}{3^3}\cdot\frac{x^3}{3!} - \frac{2\cdot 5\cdot 8}{3^4}\cdot\frac{x^4}{4!} - \frac{2\cdot 5\cdot 8\cdot 11}{3^5}\cdot\frac{x^5}{5!} - \cdots;\ R = 1$

39. $f(x) = (1+x)^{-3} = 1 - 3x + 3\cdot 4\cdot\frac{x^2}{2!} - 3\cdot 4\cdot 5\cdot\frac{x^3}{3!} + 3\cdot 4\cdot 5\cdot 6\cdot\frac{x^4}{4!} - \cdots;\ R = 1$

41. $f(x) = \sum_{n=0}^{\infty} \dfrac{(-1)^n x^n}{n+1};\ R = 1$

43. $f(x) = \sum_{n=0}^{\infty} \dfrac{(-1)^n x^{6n+4}}{(2n+1)!\cdot(6n+4)};\ R = +\infty$

45. $f(x) = \sum_{n=0}^{\infty} \dfrac{(-1)^n x^{3n+1}}{n!\cdot(3n+1)};\ R = +\infty$

47. $f(x) = \sum_{n=1}^{\infty} \dfrac{(-1)^{n+1} x^{2n-1}}{n!\cdot(2n-1)};\ R = +\infty$

49. $\dfrac{x}{(1-x)^2},\ -1 < x < 1$

51. $\dfrac{x(1+x)}{(1-x)^3},\ -1 < x < 1$

61. $f(x) = \sum_{n=0}^{\infty} \dfrac{(-1)^n x^{2n}}{(2n+1)!},\ -\infty < x < +\infty$

SECTION 10.9 (PAGE 758)

1. $65^{1/3} = 4 \cdot \left(1 + \frac{1}{64}\right)^{1/3} \approx 4 + \frac{4}{3} \cdot \frac{1}{64} \approx 4.021.$

3. $\sin(0.5) \approx \frac{1}{2} - \frac{1}{3! \cdot 2^3} \approx 0.479.$

5. 0.464

7. $\sin\left(\frac{\pi}{10}\right) \approx \frac{\pi}{10} - \frac{\pi^3}{3! \cdot 10^3} \approx 0.309.$

9. $\sin\left(\frac{\pi}{18}\right) \approx \frac{\pi}{18} - \frac{\pi^3}{3! \cdot 18^3} \approx 0.174.$

11. $\int_0^1 \frac{\sin x}{x}\,dx \approx 1 - \frac{1}{3!3} + \frac{1}{5!5} - \frac{1}{7!7} \approx 0.9641.$

13. $\int_0^{1/2} \frac{\arctan x}{x}\,dx \approx \frac{1}{2} - \frac{1}{2^3 \cdot 3^2} + \frac{1}{2^5 \cdot 5^2} - \frac{1}{2^7 \cdot 7^2} \approx 0.4872.$

15. $\int_0^{0.1} \frac{\ln(1+x)}{x}\,dx \approx \frac{1}{10} - \frac{1}{4 \cdot 10^2} + \frac{1}{9 \cdot 10^3} \approx 0.0976.$

17. $\int_0^{1/2} \frac{1 - e^{-x}}{x}\,dx \approx \frac{1}{2} - \frac{1}{2! \cdot 2 \cdot 2^2} + \frac{1}{3! \cdot 3 \cdot 2^3} - \frac{1}{4! \cdot 4 \cdot 2^4}$
$+ \frac{1}{5! \cdot 5 \cdot 2^5} \approx 0.4438.$

19. $0.7468241328124270 \approx 0.7468$

21. $0.5132555590033423 \approx 0.5133$

23. $-\frac{1}{2} - \frac{x}{6} - \frac{x^2}{24} - \cdots \to -\frac{1}{2}$ as $x \to +\infty$.

25. $\lim_{x\to 0} \dfrac{\frac{1}{2!} - \frac{x^2}{4!} + \frac{x^4}{6!} - \cdots}{1 + \frac{x}{2!} + \frac{x^2}{3!} + \cdots} = \frac{1}{2}.$

27. $\lim_{x\to 0} \dfrac{-\frac{x}{3!} + \frac{x^3}{5!} - \frac{x^5}{7!} + \cdots}{1 - \frac{x^2}{3!} + \frac{x^4}{5!} - \cdots} = \frac{0}{1} = 0.$

29. $\sin 80^\circ \approx 1 - \frac{1}{2!} \cdot \left(\frac{\pi}{18}\right)^2 \approx 0.9848.$

31. 0.681998

33. Six-place accuracy

35. Five-place accuracy

37. $e^{1/3} \approx 1.39$

39. **a.** $|R_3(x)| < 0.000002$; **b.** $|R_3(x)| < 0.000000003$

41. $V = 2\pi \int_0^{\pi} \frac{\sin^2 x}{x^2}\,dx = \frac{(2\pi)^2}{2!} - \frac{(2\pi)^4}{4! \cdot 3} + \frac{(2\pi)^6}{6! \cdot 5} - \cdots$
$\approx 8.9105091465101038.$

43. $V = 2\pi \int_0^{2\pi} \frac{1 - \cos x}{x}\,dx = \frac{(2\pi)^3}{2! \cdot 2} - \frac{(2\pi)^5}{4! \cdot 4} + \frac{(2\pi)^7}{6! \cdot 6} - \cdots$
$\approx 15.316227983254.$

47. $a_0 + (a_1 - a_0)x + (a_2 - a_1)x^2 + (a_3 - a_2)x^3 + (a_4 - a_3)x^4 + \cdots = 1$

49. $a_0 + a_1 x + \left(a_2 - \frac{1}{2}a_0\right)x^2 + \left(a_3 - \frac{1}{2}a_1\right)x^3$
$+ \left(a_4 - \frac{1}{2}a_2 + \frac{1}{24}a_0\right)x^4 + \cdots = 1;$
$\sec x = 1 + \frac{1}{2}x^2 + \frac{5}{24}x^4 + \frac{61}{720}x^6 + \cdots.$

51. $1 + x = a_0 + a_1 x + \left(a_2 - \frac{1}{2}a_1\right)x^2 + \left(a_3 - a_2 + \frac{1}{3}a_1\right)x^3 + \cdots.$

53. Apply Theorem 1 to determine R.

55. $\int_0^{1/2} \frac{1}{1 + x^2 + x^4}\,dx = \frac{1}{2} - \frac{1}{2^3 \cdot 3} + \frac{1}{2^7 \cdot 7} - \frac{1}{2^9 \cdot 9} + \frac{1}{2^{13} \cdot 13}$
$- \cdots \approx 0.4592398250.$

59. -1

SECTION 10.10 (PAGE 768)

1. $y(x) = a_0 \sum_{n=0}^{\infty} \frac{x^n}{n!} = a_0 e^x; \quad R = +\infty$

3. $y(x) = a_0 \sum_{n=0}^{\infty} \frac{(-1)^n}{n!}\left(\frac{3x}{2}\right)^n = a_0 e^{-3x/2}; \quad R = +\infty$

5. $y(x) = a_0\left[1 + \frac{1}{1!} \cdot \frac{x^3}{3} + \frac{1}{2!}\left(\frac{x^3}{3}\right)^2 + \frac{1}{3!}\left(\frac{x^3}{3}\right)^3 + \cdots\right]$
$= a_0 \exp\left(\frac{x^3}{3}\right); \quad R = +\infty$

7. $y(x) = a_0 \sum_{n=0}^{\infty} 2^n x^n = a_0 \sum_{n=0}^{\infty} (2x)^n = \frac{a_0}{1 - 2x}; \quad R = \frac{1}{2}$

9. $y(x) = a_0 \sum_{n=0}^{\infty} (n+1)x^n = \frac{a_0}{(1-x)^2}; \quad R = 1$

11. $y(x) = a_0\left(1 + \frac{x^2}{2!} + \frac{x^4}{4!} + \frac{x^6}{6!} + \cdots\right)$
$+ a_1\left(x + \frac{x^3}{3!} + \frac{x^5}{5!} + \frac{x^7}{7!} + \cdots\right)$
$= a_0 \cosh x + a_1 \sinh x; \quad R = +\infty$

13. $y(x) = a_0\left(1 - \frac{9x^2}{2!} + \frac{9^2 x^4}{4!} - \frac{9^3 x^6}{6!} + \cdots\right)$
$+ a_1\left(x - \frac{9x^3}{3!} + \frac{9^2 x^5}{5!} - \frac{9^3 x^7}{7!} + \cdots\right)$
$= a_0 \cos 3x + \frac{a_1}{3}\sin 3x = c_1 \cos 3x + c_2 \sin 3x;$
$R = +\infty$

15. $a_0 = 0$ and $(n+1)a_n = 0$ if $n \geqq 1$, so $y(x) \equiv 0$.

17. $a_0 = a_1 = 0$ and $(n-1)a_{n-1} + a_n = 0$ for $n \geqq 2$, so $y(x) \equiv 0$.

19. $y(x) = \frac{3}{2}\sin 2x$

21. $y(x) = xe^x$

23. $c_1 = c_2 = 0$ and

$$c_n = -\frac{n-1}{n^2 - n + 1} c_{n-1}$$

if $n \geqq 2$, so $y(x) \equiv 0$.

25. $x + \frac{1}{3}x^3 + \frac{2}{15}x^5 + \frac{17}{315}x^7 + \frac{62}{2835}x^9 + \frac{1382}{155925}x^{11}$
$+ \frac{21844}{6081075}x^{13} + \frac{929569}{638512875}x^{15} + \cdots$

CHAPTER 10 MISCELLANEOUS PROBLEMS (PAGE 769)

1. 1 **3.** 10 **5.** 0 **7.** 0

9. The limit does not exist.

11. 0

13. $+\infty$ (or "Does not exist.")

15. 1

17. Converges by the alternating series test. (The sum is approximately 0.080357603217.)

19. Converges by the ratio test. (The sum is approximately 1.405253880284.)

21. Converges by the comparison test and Theorem 3 of Section 10.7. (The sum is approximately 0.230836643803.)

23. Diverges by the nth-term test for divergence.

25. Converges by the comparison test. (The sum is approximately 1.459973884376.)

27. Converges by the alternating series test. (The sum is approximately 0.378868816198.)

29. Diverges by the integral test.

31. Converges by the ratio test; the sum is e^{2x} and the radius of convergence is $+\infty$.

33. The interval of convergence is $[-2, 4)$.

35. The interval of convergence is $[-1, 1]$.

37. The series converges only if $x = 0$.

39. The series converges to $\cosh x$ on $(-\infty, +\infty)$.

41. Diverges for all x by the nth-term test for divergence.

43. Converges for all x to $\exp(e^x)$.

45. Let $a_n = b_n = (-1)^n \cdot n^{-1/2}$.

51. 1.084

53. 0.461

55. 0.797

65. $a_0 = 2$ and $a_n = 4$ for all $n \geqq 1$.

REFERENCES FOR FURTHER STUDY

References 2, 3, 7, and 10 may be consulted for historical topics pertinent to calculus. Reference 14 provides a more theoretical treatment of single-variable calculus topics than ours. References 4, 5, 8, and 15 include advanced topics in multivariable calculus. Reference 11 is a standard work on infinite series. References 1, 9, and 13 are differential equations textbooks. Reference 6 discusses topics in calculus together with computing and programming in BASIC. Those who would like to pursue the topic of fractals should look at reference 12.

1. Boyce, William E. and Richard C. DiPrima, *Elementary Differential Equations* (5th ed.). New York: John Wiley, 1991.
2. Boyer, Carl B., *A History of Mathematics* (2nd ed.). New York: John Wiley, 1991.
3. Boyer, Carl B., *The History of the Calculus and Its Conceptual Development.* New York: Dover Publications, 1959.
4. Buck, R. Creighton, *Advanced Calculus* (3rd ed.). New York: McGraw-Hill, 1977.
5. Courant, Richard and Fritz John, *Introduction to Calculus and Analysis.* Vols. I and II. New York: Springer-Verlag, 1982.
6. Edwards, C. H., Jr., *Calculus and the Personal Computer.* Englewood Cliffs, NJ: Prentice-Hall, 1986.
7. Edwards, C. H., Jr., *The Historical Development of the Calculus.* New York: Springer-Verlag, 1979.
8. Edwards, C. H., Jr., *Advanced Calculus of Several Variables.* New York: Academic Press, 1973.
9. Edwards, C. H., Jr. and David E. Penney, *Differential Equations with Boundary Value Problems: Computing and Modeling* (2nd ed.). Upper Saddle River, NJ: Prentice Hall, 2000.
10. Kline, Morris, *Mathematical Thought from Ancient to Modern Times.* Vols. I, II, and III. New York: Oxford University Press, 1972.
11. Knopp, Konrad, *Theory and Application of Infinite Series* (2nd ed.). New York: Hafner Press, 1990.
12. Peitgen, H.-O. and P. H. Richter, *The Beauty of Fractals.* New York: Springer-Verlag, 1986.
13. Simmons, George E., *Differential Equations with Applications and Historical Notes.* New York: McGraw-Hill, 1972.
14. Spivak, Michael E., *Calculus* (2nd ed.). Berkeley: Publish or Perish, 1980.
15. Taylor, Angus E. and W. Robert Mann, *Advanced Calculus* (3rd ed.). New York: John Wiley, 1983.

INDEX

Entries in bold type indicate the page (or pages) where a term is formally defined.

D

E

T

U

V

TABLE OF INTEGRALS

ELEMENTARY FORMS

1 $\int u\,dv = uv - \int v\,du$

2 $\int u^n\,du = \frac{1}{n+1}u^{n+1} + C \quad \text{if } n \neq -1$

3 $\int \frac{du}{u} = \ln|u| + C$

4 $\int e^u\,du = e^u + C$

5 $\int a^u\,du = \frac{a^u}{\ln a} + C$

6 $\int \sin u\,du = -\cos u + C$

7 $\int \cos u\,du = \sin u + C$

8 $\int \sec^2 u\,du = \tan u + C$

9 $\int \csc^2 u\,du = -\cot u + C$

10 $\int \sec u \tan u\,du = \sec u + C$

11 $\int \csc u \cot u\,du = -\csc u + C$

12 $\int \tan u\,du = \ln|\sec u| + C$

13 $\int \cot u\,du = -\ln|\csc u| + C$

14 $\int \sec u\,du = \ln|\sec u + \tan u| + C$

15 $\int \csc u\,du = -\ln|\csc u + \cot u| + C$

16 $\int \frac{du}{\sqrt{a^2 - u^2}} = \sin^{-1}\frac{u}{a} + C$

17 $\int \frac{du}{a^2 + u^2} = \frac{1}{a}\tan^{-1}\frac{u}{a} + C$

18 $\int \frac{du}{a^2 - u^2} = \frac{1}{2a}\ln\left|\frac{u+a}{u-a}\right| + C$

19 $\int \frac{du}{u\sqrt{u^2 - a^2}} = \frac{1}{a}\sec^{-1}\left|\frac{u}{a}\right| + C$

TRIGONOMETRIC FORMS

20 $\int \sin^2 u\,du = \frac{1}{2}u - \frac{1}{4}\sin 2u + C$

21 $\int \cos^2 u\,du = \frac{1}{2}u + \frac{1}{4}\sin 2u + C$

22 $\int \tan^2 u\,du = \tan u - u + C$

23 $\int \cot^2 u\,du = -\cot u - u + C$

24 $\int \sin^3 u\,du = -\frac{1}{3}(2 + \sin^2 u)\cos u + C$

25 $\int \cos^3 u\,du = \frac{1}{3}(2 + \cos^2 u)\sin u + C$

26 $\int \tan^3 u\,du = \frac{1}{2}\tan^2 u + \ln|\cos u| + C$

27 $\int \cot^3 u\,du = -\frac{1}{2}\cot^2 u - \ln|\sin u| + C$

28 $\int \sec^3 u\,du = \frac{1}{2}\sec u \tan u + \frac{1}{2}\ln|\sec u + \tan u| + C$

29 $\int \csc^3 u\,du = -\frac{1}{2}\csc u \cot u + \frac{1}{2}\ln|\csc u - \cot u| + C$

30 $\int \sin au \sin bu\,du = \frac{\sin(a-b)u}{2(a-b)} - \frac{\sin(a+b)u}{2(a+b)} + C \quad \text{if } a^2 \neq b^2$

31 $\int \cos au \cos bu\,du = \frac{\sin(a-b)u}{2(a-b)} + \frac{\sin(a+b)u}{2(a+b)} + C \quad \text{if } a^2 \neq b^2$

32 $\int \sin au \cos bu\,du = -\frac{\cos(a-b)u}{2(a-b)} - \frac{\cos(a+b)u}{2(a+b)} + C \quad \text{if } a^2 \neq b^2$

33 $\int \sin^n u\,du = -\frac{1}{n}\sin^{n-1} u \cos u + \frac{n-1}{n}\int \sin^{n-2} u\,du$

34 $\int \cos^n u\,du = \frac{1}{n}\cos^{n-1} u \sin u + \frac{n-1}{n}\int \cos^{n-2} u\,du$

35 $\int \tan^n u\,du = \frac{1}{n-1}\tan^{n-1} u - \int \tan^{n-2} u\,du \quad \text{if } n \neq 1$

36 $\int \cot^n u\,du = -\frac{1}{n-1}\cot^{n-1} u - \int \cot^{n-2} u\,du \quad \text{if } n \neq 1$

37 $\int \sec^n u\,du = \frac{1}{n-1}\sec^{n-2} u \tan u + \frac{n-2}{n-1}\int \sec^{n-2} u\,du \quad \text{if } n \neq 1$

38 $\int \csc^n u\,du = -\frac{1}{n-1}\csc^{n-2} u \cot u + \frac{n-2}{n-1}\int \csc^{n-2} u\,du \quad \text{if } n \neq 1$

39a $\int \sin^n u \cos^m u\,du = -\frac{\sin^{n-1} u \cos^{m+1} u}{n+m} + \frac{n-1}{n+m}\int \sin^{n-2} u \cos^m u\,du \quad \text{if } n \neq -m$

39b $\int \sin^n u \cos^m u\,du = -\frac{\sin^{n+1} u \cos^{m-1} u}{n+m} + \frac{m-1}{n+m}\int \sin^n u \cos^{m-2} u\,du \quad \text{if } m \neq -n$

40 $\int u \sin u\,du = \sin u - u\cos u + C$

41 $\int u \cos u\,du = \cos u + u \sin u + C$

42 $\int u^n \sin u\,du = -u^n \cos u + n\int u^{n-1}\cos u\,du$

43 $\int u^n \cos u\,du = u^n \sin u - n\int u^{n-1}\sin u\,du$

(Table of Integrals continues from previous page)

FORMS INVOLVING $\sqrt{u^2 \pm a^2}$

44 $\displaystyle\int \sqrt{u^2 \pm a^2}\,du = \frac{u}{2}\sqrt{u^2 \pm a^2} \pm \frac{a^2}{2}\ln\left|u + \sqrt{u^2 \pm a^2}\right| + C$

45 $\displaystyle\int \frac{du}{\sqrt{u^2 \pm a^2}} = \ln\left|u + \sqrt{u^2 \pm a^2}\right| + C$

46 $\displaystyle\int \frac{\sqrt{u^2 + a^2}}{u}\,du = \sqrt{u^2 + a^2} - a\ln\left(\frac{a + \sqrt{u^2 + a^2}}{u}\right) + C$

47 $\displaystyle\int \frac{\sqrt{u^2 - a^2}}{u}\,du = \sqrt{u^2 - a^2} - a\sec^{-1}\frac{u}{a} + C$

48 $\displaystyle\int u^2\sqrt{u^2 \pm a^2}\,du = \frac{u}{8}(2u^2 \pm a^2)\sqrt{u^2 \pm a^2} - \frac{a^4}{8}\ln\left|u + \sqrt{u^2 \pm a^2}\right| + C$

49 $\displaystyle\int \frac{u^2\,du}{\sqrt{u^2 \pm a^2}} = \frac{u}{2}\sqrt{u^2 \pm a^2} \mp \frac{a^2}{2}\ln\left|u + \sqrt{u^2 \pm a^2}\right| + C$

50 $\displaystyle\int \frac{du}{u^2\sqrt{u^2 \pm a^2}} = \mp\frac{\sqrt{u^2 \pm a^2}}{a^2 u} + C$

51 $\displaystyle\int \frac{\sqrt{u^2 \pm a^2}}{u^2}\,du = -\frac{\sqrt{u^2 \pm a^2}}{u} + \ln\left|u + \sqrt{u^2 \pm a^2}\right| + C$

52 $\displaystyle\int \frac{du}{(u^2 \pm a^2)^{3/2}} = \pm\frac{u}{a^2\sqrt{u^2 \pm a^2}} + C$

53 $\displaystyle\int (u^2 \pm a^2)^{3/2}\,du = \frac{u}{8}(2u^2 \pm 5a^2)\sqrt{u^2 \pm a^2} + \frac{3a^4}{8}\ln\left|u + \sqrt{u^2 \pm a^2}\right| + C$

FORMS INVOLVING $\sqrt{a^2 - u^2}$

54 $\displaystyle\int \sqrt{a^2 - u^2}\,du = \frac{u}{2}\sqrt{a^2 - u^2} + \frac{a^2}{2}\sin^{-1}\frac{u}{a} + C$

55 $\displaystyle\int \frac{\sqrt{a^2 - u^2}}{u}\,du = \sqrt{a^2 - u^2} - a\ln\left|\frac{a + \sqrt{a^2 - u^2}}{u}\right| + C$

56 $\displaystyle\int \frac{u^2\,du}{\sqrt{a^2 - u^2}} = -\frac{u}{2}\sqrt{a^2 - u^2} + \frac{a^2}{2}\sin^{-1}\frac{u}{a} + C$

57 $\displaystyle\int u^2\sqrt{a^2 - u^2}\,du = \frac{u}{8}(2u^2 - a^2)\sqrt{a^2 - u^2} + \frac{a^4}{8}\sin^{-1}\frac{u}{a} + C$

58 $\displaystyle\int \frac{du}{u^2\sqrt{a^2 - u^2}} = -\frac{\sqrt{a^2 - u^2}}{a^2 u} + C$

59 $\displaystyle\int \frac{\sqrt{a^2 - u^2}}{u^2}\,du = -\frac{\sqrt{a^2 - u^2}}{u} - \sin^{-1}\frac{u}{a} + C$

60 $\displaystyle\int \frac{du}{u\sqrt{a^2 - u^2}} = -\frac{1}{a}\ln\left|\frac{a + \sqrt{a^2 - u^2}}{u}\right| + C$

61 $\displaystyle\int \frac{du}{(a^2 - u^2)^{3/2}} = \frac{u}{a^2\sqrt{a^2 - u^2}} + C$

62 $\displaystyle\int (a^2 - u^2)^{3/2}\,du = \frac{u}{8}(5a^2 - 2u^2)\sqrt{a^2 - u^2} + \frac{3a^4}{8}\sin^{-1}\frac{u}{a} + C$

EXPONENTIAL AND LOGARITHMIC FORMS

63 $\displaystyle\int ue^u\,du = (u - 1)e^u + C$

64 $\displaystyle\int u^n e^u\,du = u^n e^u - n\int u^{n-1}e^u\,du$

65 $\displaystyle\int \ln u\,du = u\ln u - u + C$

66 $\displaystyle\int u^n\ln u\,du = \frac{u^{n+1}}{n+1}\ln u - \frac{u^{n+1}}{(n+1)^2} + C$

67 $\displaystyle\int e^{au}\sin bu\,du = \frac{e^{au}}{a^2 + b^2}(a\sin bu - b\cos bu) + C$

68 $\displaystyle\int e^{au}\cos bu\,du = \frac{e^{au}}{a^2 + b^2}(a\cos bu + b\sin bu) + C$

INVERSE TRIGONOMETRIC FORMS

69 $\displaystyle\int \sin^{-1}u\,du = u\sin^{-1}u + \sqrt{1 - u^2} + C$

70 $\displaystyle\int \tan^{-1}u\,du = u\tan^{-1}u - \frac{1}{2}\ln(1 + u^2) + C$

71 $\displaystyle\int \sec^{-1}u\,du = u\sec^{-1}u - \ln\left|u + \sqrt{u^2 - 1}\right| + C$

72 $\displaystyle\int u\sin^{-1}u\,du = \frac{1}{4}(2u^2 - 1)\sin^{-1}u + \frac{u}{4}\sqrt{1 - u^2} + C$

73 $\displaystyle\int u\tan^{-1}u\,du = \frac{1}{2}(u^2 + 1)\tan^{-1}u - \frac{u}{2} + C$

74 $\displaystyle\int u\sec^{-1}u\,du = \frac{u^2}{2}\sec^{-1}u - \frac{1}{2}\sqrt{u^2 - 1} + C$

75 $\displaystyle\int u^n\sin^{-1}u\,du = \frac{u^{n+1}}{n+1}\sin^{-1}u - \frac{1}{n+1}\int \frac{u^{n+1}}{\sqrt{1 - u^2}}\,du \quad \text{if } n \neq -1$

76 $\displaystyle\int u^n\tan^{-1}u\,du = \frac{u^{n+1}}{n+1}\tan^{-1}u - \frac{1}{n+1}\int \frac{u^{n+1}}{1 + u^2}\,du \quad \text{if } n \neq -1$

77 $\displaystyle\int u^n\sec^{-1}u\,du = \frac{u^{n+1}}{n+1}\sec^{-1}u - \frac{1}{n+1}\int \frac{u^n}{\sqrt{u^2 - 1}}\,du \quad \text{if } n \neq -1$

(Table of Integrals continues from previous page)

HYPERBOLIC FORMS

78 $\displaystyle\int \sinh u\,du = \cosh u + C$

79 $\displaystyle\int \cosh u\,du = \sinh u + C$

80 $\displaystyle\int \tanh u\,du = \ln(\cosh u) + C$

81 $\displaystyle\int \coth u\,du = \ln|\sinh u| + C$

82 $\displaystyle\int \operatorname{sech} u\,du = \tan^{-1}|\sinh u| + C$

83 $\displaystyle\int \operatorname{csch} u\,du = \ln\left|\tanh\frac{u}{2}\right| + C$

84 $\displaystyle\int \sinh^2 u\,du = \frac{1}{4}\sinh 2u - \frac{u}{2} + C$

85 $\displaystyle\int \cosh^2 u\,du = \frac{1}{4}\sinh 2u + \frac{u}{2} + C$

86 $\displaystyle\int \tanh^2 u\,du = u - \tanh u + C$

87 $\displaystyle\int \coth^2 u\,du = u - \coth u + C$

88 $\displaystyle\int \operatorname{sech}^2 u\,du = \tanh u + C$

89 $\displaystyle\int \operatorname{csch}^2 u\,du = -\coth u + C$

90 $\displaystyle\int \operatorname{sech} u \tanh u\,du = -\operatorname{sech} u + C$

91 $\displaystyle\int \operatorname{csch} u \coth u\,du = -\operatorname{csch} u + C$

MISCELLANEOUS ALGEBRAIC FORMS

92 $\displaystyle\int u(au+b)^{-1}\,du = \frac{u}{a} - \frac{b}{a^2}\ln|au+b| + C$

93 $\displaystyle\int u(au+b)^{-2}\,du = \frac{1}{a^2}\left(\ln|au+b| + \frac{b}{au+b}\right) + C$

94 $\displaystyle\int u(au+b)^n\,du = \frac{(au+b)^{n+1}}{a^2}\left(\frac{au+b}{n+2} - \frac{b}{n+1}\right) + C \quad \text{if } n \neq -1, -2$

95 $\displaystyle\int \frac{du}{(a^2 \pm u^2)^n} = \frac{1}{2a^2(n-1)}\left(\frac{u}{(a^2 \pm u^2)^{n-1}} + (2n-3)\int \frac{du}{(a^2 \pm u^2)^{n-1}}\right) \quad \text{if } n \neq 1$

96 $\displaystyle\int u\sqrt{au+b}\,du = \frac{2}{15a^2}(3au-2b)(au+b)^{3/2} + C$

97 $\displaystyle\int u^n\sqrt{au+b}\,du = \frac{2}{a(2n+3)}\left(u^n(au+b)^{3/2} - nb\int u^{n-1}\sqrt{au+b}\,du\right)$

98 $\displaystyle\int \frac{u\,du}{\sqrt{au+b}} = \frac{2}{3a^2}(au-2b)\sqrt{au+b} + C$

99 $\displaystyle\int \frac{u^n\,du}{\sqrt{au+b}} = \frac{2}{a(2n+1)}\left(u^n\sqrt{au+b} - nb\int \frac{u^{n-1}\,du}{\sqrt{au+b}}\right)$

100a $\displaystyle\int \frac{du}{u\sqrt{au+b}} = \frac{1}{\sqrt{b}}\ln\left|\frac{\sqrt{au+b}-\sqrt{b}}{\sqrt{au+b}+\sqrt{b}}\right| + C \quad \text{if } b > 0$

100b $\displaystyle\int \frac{du}{u\sqrt{au+b}} = \frac{2}{\sqrt{-b}}\tan^{-1}\sqrt{\frac{au+b}{-b}} + C \quad \text{if } b < 0$

101 $\displaystyle\int \frac{du}{u^n\sqrt{au+b}} = -\frac{\sqrt{au+b}}{b(n-1)u^{n-1}} - \frac{(2n-3)a}{(2n-2)b}\int \frac{du}{u^{n-1}\sqrt{au+b}} \quad \text{if } n \neq 1$

102 $\displaystyle\int \sqrt{2au-u^2}\,du = \frac{u-a}{2}\sqrt{2au-u^2} + \frac{a^2}{2}\sin^{-1}\frac{u-a}{a} + C$

103 $\displaystyle\int \frac{du}{\sqrt{2au-u^2}} = \sin^{-1}\frac{u-a}{a} + C$

104 $\displaystyle\int u^n\sqrt{2au-u^2}\,du = -\frac{u^{n-1}(2au-u^2)^{3/2}}{n+2} + \frac{(2n+1)a}{n+2}\int u^{n-1}\sqrt{2au-u^2}\,du$

105 $\displaystyle\int \frac{u^n\,du}{\sqrt{2au-u^2}} = -\frac{u^{n-1}}{n}\sqrt{2au-u^2} + \frac{(2n-1)a}{n}\int \frac{u^{n-1}\,du}{\sqrt{2au-u^2}}$

106 $\displaystyle\int \frac{\sqrt{2au-u^2}}{u}\,du = \sqrt{2au-u^2} + a\sin^{-1}\frac{u-a}{a} + C$

107 $\displaystyle\int \frac{\sqrt{2au-u^2}}{u^n}\,du = \frac{(2au-u^2)^{3/2}}{(3-2n)au^n} + \frac{n-3}{(2n-3)a}\int \frac{\sqrt{2au-u^2}}{u^{n-1}}\,du$

108 $\displaystyle\int \frac{du}{u^n\sqrt{2au-u^2}} = \frac{\sqrt{2au-u^2}}{a(1-2n)u^n} + \frac{n-1}{(2n-1)a}\int \frac{du}{u^{n-1}\sqrt{2au-u^2}}$

109 $\displaystyle\int \left(\sqrt{2au-u^2}\right)^n du = \frac{u-a}{n+1}(2au-u^2)^{n/2} + \frac{na^2}{n+1}\int \left(\sqrt{2au-u^2}\right)^{n-2} du$

110 $\displaystyle\int \frac{du}{\left(\sqrt{2au-u^2}\right)^n} = \frac{u-a}{(n-2)a^2}\left(\sqrt{2au-u^2}\right)^{2-n} + \frac{n-3}{(n-2)a^2}\int \frac{du}{\left(\sqrt{2au-u^2}\right)^{n-2}}$

DEFINITE INTEGRALS

111 $\displaystyle\int_0^\infty u^n e^{-u}\,du = \Gamma(n+1) = n! \quad (n \geqq 0)$

112 $\displaystyle\int_0^\infty e^{-au^2}\,du = \frac{1}{2}\sqrt{\frac{\pi}{a}} \quad (a > 0)$

113 $$\int_0^{\pi/2} \sin^n u\,du = \int_0^{\pi/2} \cos^n u\,du = \begin{cases} \dfrac{1\cdot 3\cdot 5\cdot \cdots \cdot (n-1)}{2\cdot 4\cdot 6\cdot \cdots \cdot n}\cdot\dfrac{\pi}{2} & \text{if } n \text{ is an even integer and } n \geqq 2 \\[2ex] \dfrac{2\cdot 4\cdot 6\cdot \cdots \cdot (n-1)}{3\cdot 5\cdot 7\cdot \cdots \cdot n} & \text{if } n \text{ is an odd integer and } n \geqq 3 \end{cases}$$